MEANINGS OF PREFIXES AND SUFFIXES

Prefix or Suffix	Meaning	Example
a-, an-	not, without	avirulent (lacking virulence) anaerobic (without air)
aer-	air	aerobic
anti-	against	antiseptic
-ase	enzyme	penicillinase
chlor-	green	chlorophyll
-chrom-	color	metachromatic (staining differently with the same dye)
-cide	causing death	germicide
co-, com-, con-	together	coenzyme
cyan-	blue	pyocyanin (a blue bacterial pigment)
de-	down, from	dehydrate (remove water)
-dem-	people, district	epidemic
end-	within	endospore (spore within a cell)
-enter-	intestine	enteritis (inflammation of the intestine)
epi-	upon	epidermis
erythro-	red	erythrocyte (red blood cell)
eu-	well, normal	Eubacteriales (the "true" bacteria) eukaryotic (true nucleus)
exo-	outside	exotoxin (toxin found outside a cell)
extra-	outside of	extracellular
flav-	yellow	flavoprotein (a protein containing a yellow enxyme
-gen	produce, originate	antigen (a substance that induces a production of antibodies)
glyc-	sweet	glycemia (the presence of sugar in the blood
hetero-	other	heterotroph (organism that obtains enery from organic compounds)
homo-	common, same	homologous (similar in structure or origin
hydr-	water	dehydrate
hyper-	excessive, above	hypersensitive
hypo-	under	hypotonic (having low osmotic pressure)
iso-	same, equal	isotonic (having the same osmotic pressure)
-itis	inflammation	appendicitis, meningitis
leuko-	white	leukocyte (white blood cell)
ly-, -lys, -lyt-	loosen; dissolve	bacteriolysis (dissolution of bacteria)
meso-	middle	mesophilic (preferring moderate temperatures)
meta-	changed	metachromatic (showing a change of color)
micro-	small; one-millionth part	microscopic
milli-	one-thousandth part	millimeter (10^{-3} meter)
mito-	thread	mitochondrion (small, rod-shaped or granular body in cytoplasm)
mono-	single	monotrichous (having a single flagellum)
multi-	many	multinuclear (having many nuclei)
myc-	fungus	mycotic (caused by fungus)
myx-	mucus	myxomycete (slime mold)
-oid	resembling	lymphoid (resembling lymphocytes)
-ose	a sugar	lactose (milk sugar)

MICROBIOLOGY

A Human Perspective

Second Edition

MICROBIOLOGY

A Human Perspective

Eugene W. Nester, C. Evans Roberts

University of Washington

Nancy N. Pearsall

Denise G. Anderson

University of Washington

Martha T. Nester

Boston, Massachusetts Burr Ridge, Illinois Dubuque, Iowa
Madison, Wisconsin New York, New York San Francisco, California St. Louis, Missouri

WCB/McGraw-Hill

A Division of The ***McGraw·Hill*** *Companies*

MICROBIOLOGY: A HUMAN PERSPECTIVE

This book is printed on recycled, acid-free paper containing 10% postconsumer waste.

3 4 5 6 7 8 9 0 VNHVNH 0

ISBN 0-697-28602-9

Vice President, Editorial director: Keven T. Kane
Publisher: James M. Smith
Sponsoring editor: Ronald E. Worthington, Ph.D.
Developmental editor: Terrance Stanton
Marketing manager: Thomas C. Lyon
Project manager: Marilyn M. Sulzer
Production supervisor: Laura Fuller
Cover designer: Kristyn Kalnes
Cover photo: Lab Technician © Tony Stone Images
Inset: © Pasteur/Phototake
Interior designer: Jeff Storm
Photo research coordinator: Lori Hancock
Art editor: Jodi K. Banowetz
Compositor: PC&F, Inc.
Typeface: 10/12 Palatino
Printer: Von Hoffman Press, Inc.

Library of Congress Cataloging-in-Publication Data

Microbiology : a human perspective --2nd ed. / Eugene W. Nester . . .
[et al.]
p. cm.
Rev. ed. of: Microbiology / Eugene W. Nester. c1995.
Includes bibliographical references and index.
ISBN 0-697-28602-9
1. Microbiology. I. Nester, Eugene W.
QR41.2.N47 1998
579--DC21 97-15264
CIP

www.mhhe.com

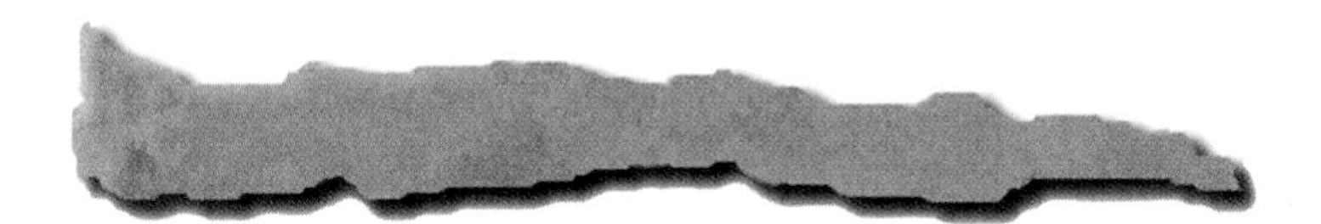

We dedicate this book to our students; we hope it helps to enrich their lives and to make them better informed citizens,

to our families whose patience and endurance made completion of this project a reality,

to Anne Nongthanat Panarak Roberts in recognition of her invaluable help, patience, and understanding,

to our colleagues for continuing encouragement and advice.

BRIEF CONTENTS

part one Life and Death of Microorganisms

part two The Microbial World

part three Microorganisms and Humans

part four Applied Microbiology

CONTENTS

chapter three
Functional Anatomy of Prokaryotes and Eukaryotes 45

chapter four
Dynamics of Bacterial Growth 85

chapter six
Informational Macromolecules: Function, Synthesis, and Regulation of Activity 131

chapter seven
Bacterial Genetics 153

chapter eight
Microbiology and Biotechnology 182

chapter nine
Control of Microbial Growth 201

part two The Microbial World

chapter ten
Classification and Identification of Prokaryotes 222

chapter eleven
Prokaryotic Microorganisms: Eubacteria and Archaea 242

chapter twelve
Eukaryotic Microorganisms: Algae, Fungi, and Protozoa 275

chapter thirteen
The Nature and Study of Viruses 297

chapter fourteen
Viruses of Animals and Plants 319

part three Microorganisms and Humans

chapter fifteen
Nonspecific Immunity 344

chapter sixteen
Specific Immunity 359

chapter seventeen
Functions and Applications of Immune Responses 378

chapter eighteen
Immunologic Disorders 395

chapter nineteen
Interactions Between Microorganisms and Humans 409

chapter twenty
Epidemiology and Public Health 428

chapter twenty-one
Antimicrobial Medicine 447

chapter twenty-two
Skin Infections 471

chapter twenty-seven
Genitourinary Infections 599

chapter twenty-eight
Nervous System Infections 626

chapter twenty-nine
Wound Infections 651

chapter thirty
Blood and Lymphatic Infections 672

chapter thirty-one
HIV Disease and Complications of Immunodeficiency 695

chapter thirty-four
Microbiology of Food and Beverages 760

PREFACE

As everyone who tries to teach or learn microbiology knows, the amount of new information to be sifted and sorted has grown enormously in recent years. The number of new microbiological terms is enough to constitute a new language. Donald P. Hayes writes about the "Growing Inaccessibility of Science" (*Nature* 356:739, April 30, 1992) and the flight of readers from journals that fail to make their articles intelligible to their readership.

A colleague of ours who had recently published a textbook was told by one of his students, "Great book. Too bad I have no time to read it." There is good reason to believe that students generally have less time to cover more material than they did a few years ago. We have tried to avoid the chatty, superficial style and lavish illustrations used by some of the more recent texts, in favor of clarity and conciseness.

Organization and Approach

A textbook is not a novel. Few students will read a text from cover to cover, and instructors generally like to pick and choose among chapters. We have therefore used judicious redundancy to help the chapters stand alone. The text is organized into 34 chapters grouped into four parts. The first nine chapters (Part One) focus on presenting the basic principles of microbiology, with an emphasis on indicating the relevance of these principles to human health. Because basic chemistry concepts are important to the study of microbiology, we have included a chemistry chapter in Part One for those students with little or no background in this area. Part One also contains a chapter on microbiology and biotechnology in which we present up-to-date, relevant information on genetic engineering and technological advances, such as the polymerase chain reaction, which has an important impact on human society today.

Part Two (chapters 10–14) provides a survey of the major types of microorganisms and viruses, and a discussion on how they are classified and identified.

In Parts Three and Four, previous users of the book will notice some reorganization of chapter content and sequence. We have added an additional chapter on immunology. This rapidly expanding topic is now covered in chapters 15 through 18. Host-parasite interactions are now covered in chapter 19, and the coverage of antimicrobial and antiviral medicines has been moved forward, to chapter 21. Part Three (chapters 15–31) covers the many important aspects of the interactions between microorganisms and humans, including immunology, epidemiology, a survey of infectious diseases, and the action of antimicrobial and antiviral medicines. In the chapters on infectious diseases, we employ an "organ-system" approach to organizing the material presented. The significant change in this section is a chapter devoted to blood and lymphatic infections (chapter 30), and a full chapter (31) devoted to AIDS and complications of immunodeficiency disorders.

Also new in the disease-focused chapters are case studies that relate factual information to actual clinical situations. Finally, Part Four (chapters 32–34) discusses environmental microbiology and the microbiology of food and beverages.

Experience has shown that few instructors attempt to cover multicellular parasites, but those who do feel strongly that they should be included. We, therefore, have placed a section on multicellular parasitology in an appendix (appendix VII). Reference sections on diseases and microorganisms are found in appendices IV and V. There are also appendices covering microbial mathematics (appendix I), pH (appendix II), glucose degradation pathways (appendix III), and immunizing agents (appendix VI).

Learning Aids for the Student

In order for students to succeed in their study of microbiology, they must be able to understand the material presented, utilize the text as a tool for learning, and enjoy reading the text. Therefore, we have included many aids to make the study of microbiology efficient and enjoyable. Each chapter contains the following:

1. ***Chapter Opening Key Concepts*** The chapter opening key concepts were written to help the student preview the major concepts to be covered and understood in the chapter.
2. ***Chapter Opening "Glimpse of History"*** The chapter opening "Glimpse of History" sections are designed to help the student "see" the relevance of historical events in microbiology to what they study today and

understand the importance of the scientific method. The "Glimpse of History" from page 3 is reproduced here.

3. *Chapter Opening Introduction* A brief preview of the chapter's contents, which relates it to other chapters within the text. This introduction is designed to help the student put the chapter into perspective at the start of their study of a specific topic.
4. *Declarative Statement Heads* Clear statements to help introduce students to each important topic.
5. *Boldfaced Key Terms* Important terms are emphasized and clearly defined when they are first presented.
6. *Cross-reference Notes* These notes refer the student to major topics that are difficult and which may need to be reviewed in order to understand the current material. An example of the usage of a cross-reference note from page 7 has been reproduced here.

> In addition to well-accepted diseases, maladies that were attributed to other causes have now been shown to be caused by microorganisms. One well-known example is peptic ulcers. This common affliction has recently been shown to be caused by a bacterium and to be treatable by antibacterial agents. It seems likely that other maladies whose causes are not understood will be shown to be caused by microorganisms or other agents of infectious disease.
>
> **• Ulcers, p. 564**

7. *End-of-page Footnotes* These footnotes provide students with on-page explanations and definitions for particularly difficult concepts and terms.
8. *Review Questions After Major Chapter Sections* These review questions help students test their knowledge and understanding of the section's factual material and major concepts before continuing on to the next section of the chapter.
9. *Chapter Review Questions* At the end of each chapter, a set of 10 questions tests students' knowledge of the entire chapter.
10. *Summary Tables* The summary tables are designed to help students organize their thoughts on important facts or characteristics presented for a specific topic.
11. *New Case Presentation* In the chapters (22–31) that focus on human diseases, we now include *case presentations:* actual clinical reports that relate the chapter's information to the experiences of patients and physicians, nurses, and medical personnel.
12. *"Perspective" Box Readings* Most chapters contain at least one "Perspective" box reading. These box readings cover a wide variety of interesting topics. Topics include recent developments in biotechnology, items of medical significance, applications of microbial activities, and descriptions of unusual microorganisms. Perspective 2.2 from page 27 is shown here.

A Glimpse of History

Microbiology was born in 1674 when Antony van Leeuwenhoek (1632–1723), an inquisitive Dutch drapery merchant, peered at a drop of lake water through a carefully ground glass lens. What he observed through this simple magnifying glass was undoubtedly one of the most startling and amazing sights that humans have ever beheld—the first glimpse of the world of microbes. As recorded in a letter to the Royal Society of London, van Leeuwenhoek saw

"... Very many little animalcules, whereof some were roundish, while others a bit bigger consisted of an oval. On these last, I saw two little legs near the head, and two little fins at the hind most end of the body. Others were somewhat longer than an oval, and these were very slow a-moving, and few in number. These animalcules had diverse colours, some being whitish and transparent; others with green and very glittering little scales, others again were green in the middle, and before and behind white; others yet were ashed grey. And the motion of most of these animalcules in the water was so swift, and so various, upwards, downwards, and round about, that 'twas wonderful to see. . . ."

From descriptions, it is clear that van Leeuwenhoek saw representatives of all the groups of microorganisms.

Perspective 2.2

Water–The Universal Requirement for Life

The properties of water depend on the properties of its constituent atoms, hydrogen (H) and oxygen (O), as well as on the bonds that join the atoms together. Since the O, covalently bonded to H, attracts electrons more strongly than the H, water is a highly polar molecule, with the O having a negative charge and the H a positive charge (see figure 2.3). This polar nature of water is very important in biology. First, it enables each water molecule to form hydrogen bonds with four other water molecules, leading to a strong attraction between water molecules (figure 1).

The strong attraction requires that a large amount of heat be applied to break the bonds between water molecules. This explains why water is a liquid at room temperature and that a great deal of heat is required to convert it into a gas. This fact is significant because water vapor (gas) is not important in biological systems, whereas the liquid form is very important.

The polar nature of water also accounts for its ability to dissolve a large number of compounds, another feature important for life as we know it. Water-soluble compounds contain atoms with positive or negative charges; in water they ionize, or split, into charged atoms. A positive charge means that the ion has one proton more than it has electrons. A negative charge means that the ion has one more electron than it has protons. For example, NaCl dissolves in water to form Na^+ ions, termed **cations,** because they move towards the cathode in an electrical field, and Cl^- ions, termed **anions,** because they move toward the positively charged anode. In solution, ions such as Na^+ and Cl^- tend to be surrounded by water molecules in such a way that the O^- of HOH forms weak bonds with Na^+, and the H^+ forms weak bonds with Cl^- (figure 2).

This prevents the Na+ and Cl^- from coming together, and it accounts for the solubility of NaCl in water. Nonpolar molecules such as fats and hydrocarbons are insoluble in water, but thus are soluble in uncharged (nonpolar) solvents such as benzene. Indeed, when a nonpolar liquid such as oil is mixed with a polar liquid such as water, the two do not mix. This phenomenon explains why oil and vinegar, which is dilute acetic acid, separate in salad dressing.

Another important feature of aqueous solutions is their level of acidity, or pH. This aspect is discussed in appendix II.

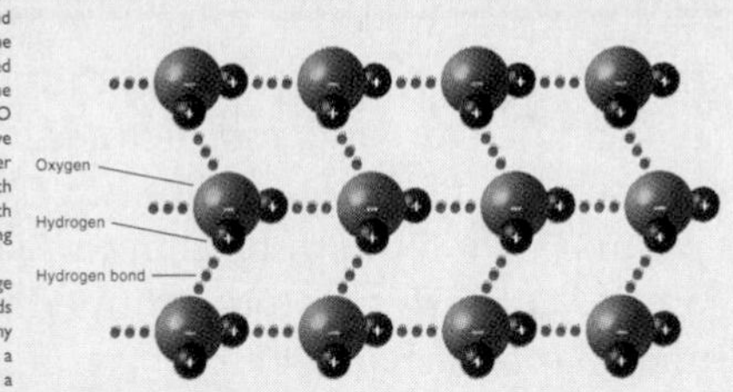

Figure 1
Water molecules held in a latticework, illustrating the hydrogen bonding among molecules.

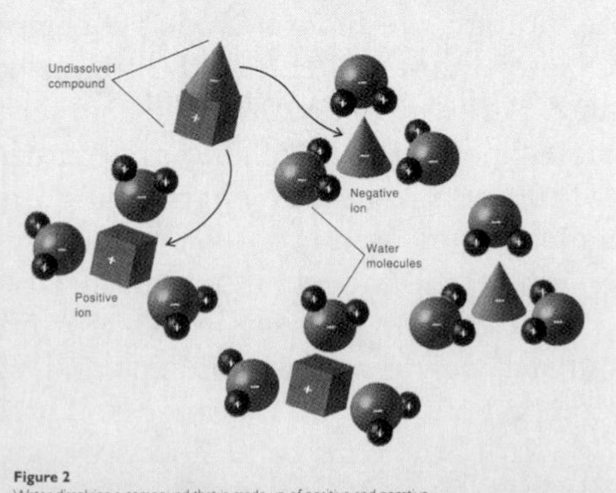

Figure 2
Water dissolving a compound that is made up of positive and negative charges. The charged regions of the water molecules surround the ions and keep them from associating with each other.

13. *Chapter Summary in Outline Form* This outline of important chapter concepts is designed to serve as a guide for chapter study.

14. ***End-of-chapter Critical Thinking Questions*** The "Critical Thinking Questions" are meant to aid the student in reviewing, integrating, and applying the concepts and principles learned from the chapter.
15. ***End-of-chapter Further Readings*** References are provided for further investigation or study of important topics.

In addition to these chapter aids, the text also features the following:

1. ***Chapter Specific Study Cards*** Chapter specific study cards for chapters 22–31 are bound within Part Three of the text. These cards can easily be torn out of the text to aid the student in studying key diseases, their causative microorganisms and significant features; even when they don't have their book in hand.
2. ***Multimedia-supported Microbiological Concepts and Processes Illustrations*** Throughout the text, the reader will find *illustrations* of microbiological concepts and processes that can be supplemented with full color video and/or animations from an interactive Multimedia program, *Microbes in Motion,* available from WCB/McGraw-Hill. The reader will be able to easily recognize these figures, as the figure legends are preceded by a CD icon. Figure 3.27, page 64, is one example of such an illustration.
3. Preview Links to *Microbes in Motion* CD-Rom. Each chapter-opening page contains a "Preview Link" that refers to specific parts of *Microbes in Motion* that will help a student comprehend the material or explore concepts further.

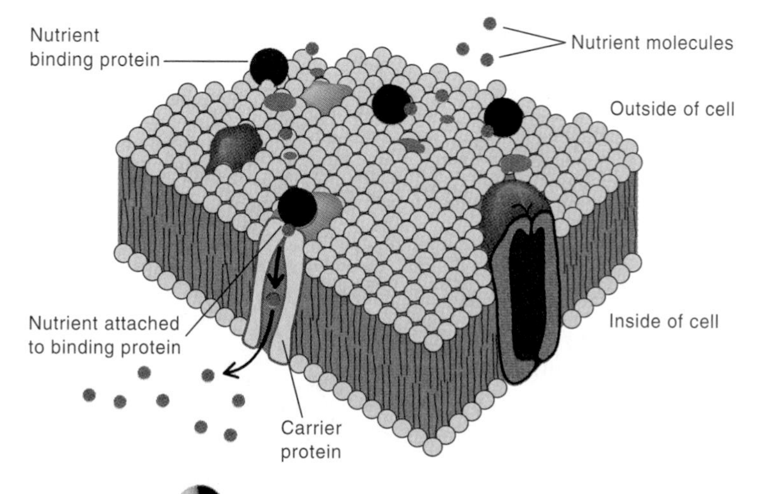

Figure 3.27
Active transport by means of specific binding proteins. These proteins are in the periplasm in gram-negative cells and are attached to the cytoplasmic membrane in gram-positive cells. The binding proteins interact with carrier proteins in the membrane. This process is rapid and results in the accumulation of the small molecules inside the cell.

Microbes in Motion
The following books and chapters in the *Microbes in Motion* CD-ROM may serve as a useful preview or supplement to your reading:
Bacterial Structure and Function: Internal Structures; Cell Walls; Bacterial Groups. *Viral Structure and Function:* Viral Structure.

4. ***Glossary*** Definitions of boldface terms are assembled in the glossary at the end of the text. A pronunciation guide for the names of microorganisms is also included.
5. ***Appendices*** The appendices aid the student with extra review information on microbial mathematics, pH, metabolic pathways, bacteria of medical importance, parasitic diseases, and immunizing agents.
6. ***Index*** An easy to use, comprehensive index is provided.
7. ***Endsheets with Reference Material*** The endsheets of the text contain useful reference information and a summary table of "Diseases and Their Causative Microorganisms."

Supplementary Materials

For the Instructor

1. An *Instructor's Manual with Test Item File,* written by Clementine A. deAngelis, Tarrant County Junior College, features a chapter introduction, chapter objectives, important terms, and answers for in-text review questions and end-of-chapter critical thinking questions for each text chapter. Suggestions for appropriate audiovisual materials are also included. In addition, the manual offers an average of 50 objective test questions per chapter that can be used to generate exams. (ISBN 28604)
2. A *Computerized Testing and Classroom Management Service* is offered free upon request to adopters of this text. The service provides a database of objective questions for preparing exams, and a grade-recording program. The software requires no programming experience and is available in Windows and MacIntosh formats: Windows 3.5 (ISBN 14021), and Mac 3.5 (ISBN 14023)
3. A set of 200 full-color acetate *transparencies* is available free to adopters and may be used to supplement classroom lectures. (ISBN 28607)
4. A set of 50 color *projection slides* derived from in-text photos is available. (ISBN 28606)

5. The second edition of *Experiments in Microbiology: A Health Science Perspective* by John Kleyn, Marie Gilstrap, and Mary Bicknell, University of Washington, has been prepared to supplement this text. This manual contains references to this text, but can also be easily used to supplement other microbiology textbooks. Like this text, the laboratory manual provides a human health perspective in its presentation of laboratory techniques and microbiological principles. Each exercise in this manual is also available as a customized, one-color separate. Specific lab exercises can be combined with one's own materials. Contact the local Wm. C. Brown Publishers representative for more details on the custom publishing service. (ISBN 28598)

For the Student

1. A *Student Study Guide* by William O'Dell, University of Nebraska–Omaha, contains chapter introductions, key concepts, summaries, terms and definitions tests, reading material tests, and review questions to aid the student's comprehension of each chapter within the text. (ISBN 28605)
2. A set of 300, 3″ × 5″ *Microbiology Study Cards* prepared by Kent M. Van de Graaff, F. Brent Johnson, Brigham Young University, and Christopher H. Creek feature complete descriptions of terms, clearly labeled drawings, clinical information on diseases, and much more. (ISBN 10979)

Acknowledgments

We thank our colleagues in the Department of Microbiology at the University of Washington who have lent their support of this project over many years. Our special thanks goes to: Neal Groman for his contribution to the material on pathogenesis, James T. Staley for advice on classification of the prokaryotes, Mary Bicknell for reviewing the manuscript, and Tom Fritsche, Sam Eng, and the late Fritz Schoenknecht for their help with illustrating clinical material.

In addition, we thank the many others who helped at various times in the preparation of the manuscript. They include Sharon Bradley, Virginia Dillard, Ramona Finical, Lynn Gottlieb, Jane Halsey, Kirsten Nester, Catherine Pousson, Katie Rhodes, Jean Roberts, and Wendy Staley.

We wish to extend a special thank you to our many colleagues across the country for reviewing the manuscript and making many helpful suggestions.

Steve K. Alexander
University of Mary Hardin-Baylor

Peter Amborse
Columbia Greene Community College

D. Andy Anderson
East Utah State University

Rodney P. Anderson
Ohio Northern University

Len M. Archer
Florida Hospital College of Health Sciences

Gail Baker
LaGuardia Community College

Russell G. Barnekow, Jr.
Southwest Missouri State University

Robert Bauman, Jr.
Amarillo College

David Berryhill
North Dakota State University

Peter Biesemeyer
North Country Community College

Edward Botan
Nashua Scientific Consultants

Thomas Byrne
Roane State Community College

Alfred E. Brown
Auburn University

Wade Collier
Manatee Community College

John R. Crooks
Iowa Wesleyan College

Clementine A. deAngelis
Tarrant County Junior College

Joseph DePierro
Nassau Community College

Ellen Digan
Manchester Community College

Elizabeth Doller
Parker College of Chiropractic

Bruce Evans
Huntington College

Edward R. Fliss
Missouri Baptist College

Steven Forst
University of Wisconsin–Milwaukee

Pamela B. Fouche
Walters State Community College

Marc Franco
South Seattle Community College

Rebecca Halyard
Clayton State College

Anne Heise
Washtenaw Community College

Dr. Donald Hicks
Los Angeles City College

Sonja L. Hyduke
Polk Community College

William J. Lembeck
Louisiana State University at Eunice

Lisa Lemke
Sussex County Community College

William Marshall
New York Medical College

John Natalini
Quincy University

William D. O'Dell
University of Nebraska at Omaha

Leba Sarkis
Aims Community College

Thomas Secrest
Austin Community College

Brian Shmaefsky
Kingwood College

C. V. Sommer
University of Wisconsin–Milwaukee

Philip Stukus
Denison University

Donald L. Terpening
Ulster County Community College

Jun Tsuji
Siena Heights College

Ed Ward
Alabama A&M University

Vernon L. Wranosky
Colby Community College

Once again, Iris Nichols has completed, reviewed, and updated the line art widely acclaimed in our past editions, and created many new figures. We are indebted to her and to Will Agranoff, who assisted with the computer-generated line art pieces. We are also grateful for the skillful assistance of the Wm. C. Brown/McGraw-Hill staff, especially Liz Sievers, our editor; Terry Stanton, developmental editor; Lori Hancock, photo editor; Jodi Banowetz, art editor; Jeff Storm, designer; Sarah Lane, copy editor; and Marla Irion, the original production editor on the project. Our special appreciation goes to Marilyn Sulzer, who as project manager was instrumental in coordinating the many complexities of the publishing process and in meeting the highest standards for accuracy and quality.

We hope very much that this text will be interesting and educational for students and a help to their instructors. We would appreciate any comments and suggestions from our readers.

E. W. N.
C. E. R.
N. N. P.
D. G. A.
M. T. N.

Department of Microbiology
University of Washington
Seattle, WA 98195
gnester@u.washington.edu

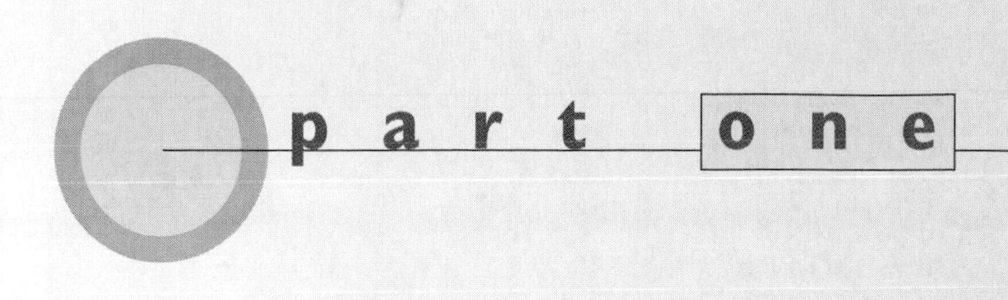

Life and Death of Microorganisms

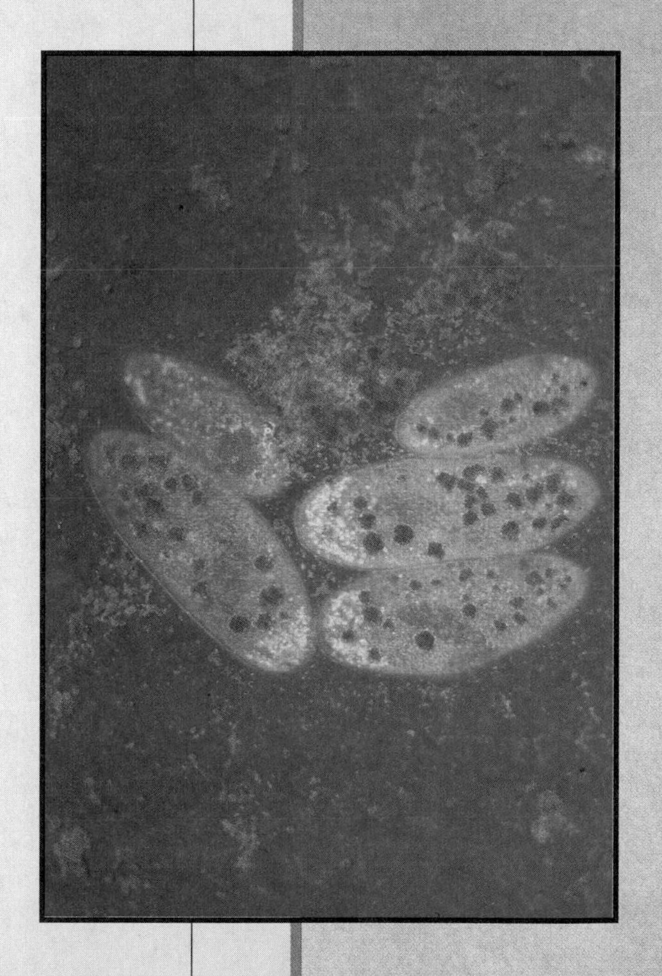

Paramecium caudatum, a protozoan, ingesting carmine red particles, which *Paramecium* cannot distinguish from bacteria. The red clumps in the body are food vacuoles containing indigestible particles of carmine red (X150).

Microbiology in the Biological World

KEY CONCEPTS

1. Microorganisms have determined the course of history because of the diseases they cause.
2. New infectious diseases appear as lifestyles change, people travel to exotic places, and techniques for growing and identifying organisms and viruses improve.
3. Cells are the basic units of life, and all cells must carry out the same critical functions in order to survive.
4. Two major types of cells exist: the prokaryotes, which do not contain a "true" nucleus or other membrane-bound internal structures, and the eukaryotes, which do contain a true nucleus.
5. The microbial world consists of prokaryotes and eukaryotes as well as nonliving agents, the viruses and viroids.
6. All prokaryotes can be divided into two very distinct groups, the Eubacteria and the Archaea, based on differences in chemical composition.

A laboratory technician examining bacteria under a light microscope. The bacteria have been cultured by inoculating agar plates (seen here) with sample material such as urine, blood, or feces. The organisms present multiply and may be distinguished by the appearance of the colonies and by microscopic examination of differentially stained smears.

Microbes in Motion

The following books and chapters in the *Microbes in Motion* CD-ROM may serve as a useful preview or supplement to your reading:

Bacterial Structure and Function: Internal Structures; Cell Walls; Bacterial Groups. *Viral Structure and Function:* Viral Structure.

A Glimpse of History

Microbiology was born in 1674 when Antony van Leeuwenhoek (1632–1723), an inquisitive Dutch drapery merchant, peered at a drop of lake water through a carefully ground glass lens. What he observed through this simple magnifying glass was undoubtedly one of the most startling and amazing sights that humans have ever beheld—the first glimpse of the world of microbes. As recorded in a letter to the Royal Society of London, van Leeuwenhoek saw

"... Very many little animalcules, whereof some were roundish, while others a bit bigger consisted of an oval. On these last, I saw two little legs near the head, and two little fins at the hind most end of the body. Others were somewhat longer than an oval, and these were very slow a-moving, and few in number. These animalcules had diverse colours, some being whitish and transparent; others with green and very glittering little scales, others again were green in the middle, and before and behind white; others yet were ashed grey. And the motion of most of these animalcules in the water was so swift, and so various, upwards, downwards, and round about, that 'twas wonderful to see...."

From descriptions, it is clear that van Leeuwenhoek saw representatives of all the groups of microorganisms.

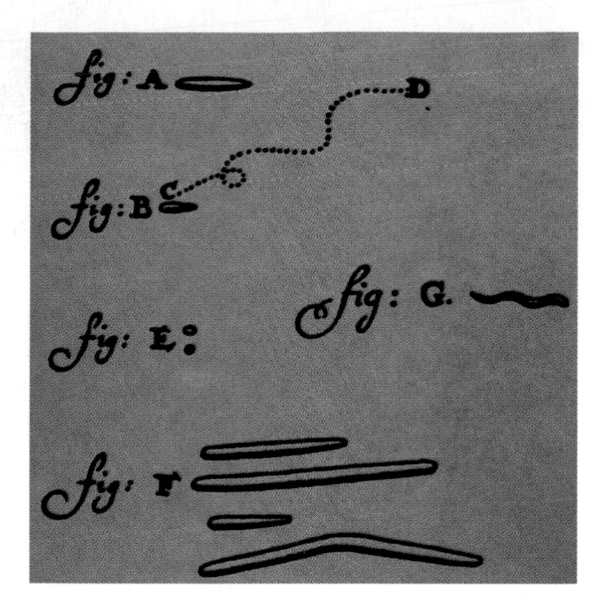

Figure 1.1
Van Leeuwenhoek's figures of bacteria from the human mouth enlarged one and one-half times from the engravings published in *Arc. Nat. Det.*, 1695.

Microbiology is the study of **microorganisms** and other agents, the **viruses,** that can be seen well only with the aid of a microscope. Microorganisms are very diverse. Indeed, the only feature they have in common is their small size. Even though they seem to be relatively simple, you will soon learn that they can be extraordinarily complex. Microorganisms and viruses are studied for many reasons, but one important reason for this course is that they cause deadly diseases. Microorganisms and viruses have killed far more people than have ever been killed in war. The outcomes of many battles have been determined not by the number of soldiers killed by swords or bullets, but by the number who survived infectious diseases. Modern sanitation, vaccination, and effective antibiotic treatments have reduced the incidence of some of the worst diseases, such as smallpox, bubonic plague, and influenza, to a small fraction of their former numbers. However, another disease, acquired immunodeficiency syndrome (AIDS), has risen as a modern-day plague. Frequent newspaper accounts serve notice that infectious diseases continue to be a problem.

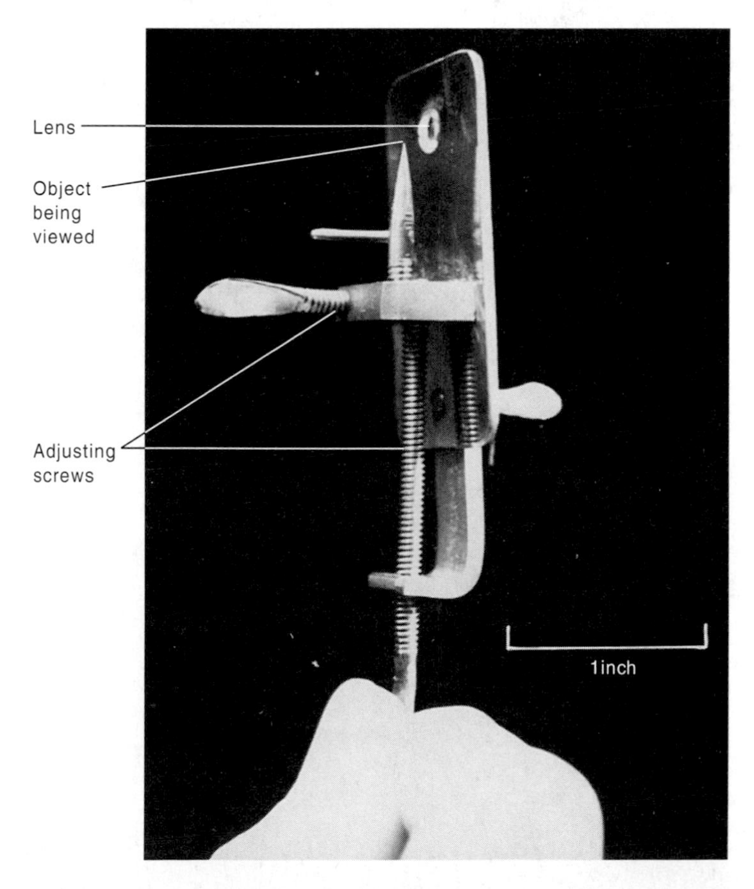

Figure 1.2
Model of Van Leeuwenhoek microscope. The original, made in 1673, could magnify the object being viewed almost 300 times. The object being viewed is brought into focus with the adjusting screws. This replica was made according to the directions given in the *American Biology Teacher* 30:537, 1958. Note its small size.

Microorganisms Discovered

To see the microbial world, van Leeuwenhoek used magnifying glasses. It is quite surprising that not only was he able to see some of the bacteria, but he even described their numerous shapes (figure 1.1).

His simple microscope (figure 1.2) increased the size of the object he was viewing only about 300 times. Microscopes commonly used in the laboratory today magnify objects over 1,000 times.

Perspective 1.1

Microbial Sleuthing

One important method used to understand the experiments of any scientist is to review the original literature. This literature, published in a scientific journal and written by the scientist who performed the experiments, describes how the experiments were performed and their outcome. The founder of microbiology, Antony van Leeuwenhoek, did not publish his observations in a journal but rather described his observations on "animalcules" in over 150 letters sent to the Royal Society of London. In 1981, Brian Ford, an English biologist and author, examined these original letters. As he leafed through them, he made a surprising and totally unexpected discovery. An envelope was pasted to the back of the last page of a letter that was dated 1674. Inside it were four packets of folded notepaper, and the contents of each were described in van Leeuwenhoek's own handwriting. One contained thinly sliced cork, another the inner part of a tree, and still another contained pieces cut from the optic nerve of a cow. The fourth packet was empty. Ford further analyzed the section of cork. It was clear that van Leeuwenhoek had done an excellent job of cutting the cork, as well as could be done today. Apparently, he used a sharp shaving razor, because the specimen also contained his own red blood cells that were probably deposited from the blade. The appearance of this section of cork as viewed with a microscope that van Leeuwenhoek made is shown in figure 1.

This is typical of the kind of specimens seen by van Leeuwenhoek through his microscope. It is quite remarkable that the image of this specimen of cork is similar to what is seen when it is viewed through a modern microscope. Although van Leeuwenhoek's microscopes were simple, they were very effective and their use had far-reaching consequences for all of science.

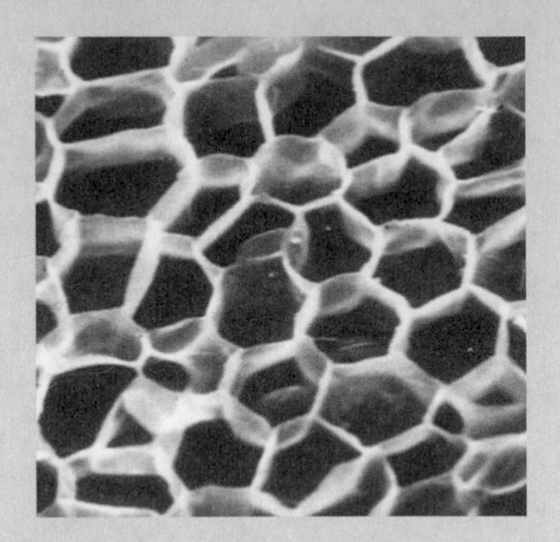

Figure 1
Fine cork section prepared for the microscope in May 1674. This specimen from Antony van Leeuwenhoek, the "father of the microscope," beautifully reveals the way cells were discovered.

Seeing small objects through the lens of a microscope depends not only on magnifying the size of the specimen being viewed, but also on illuminating it. Therefore, van Leeuwenhoek must have developed an unusually good method for lighting the specimens that he reported seeing. He inadvertently discovered **dark-field illumination,** a technique in which faint objects appear brightly lit against a dark background (figure 1.3).

• **Dark-field microscopy, p. 48**

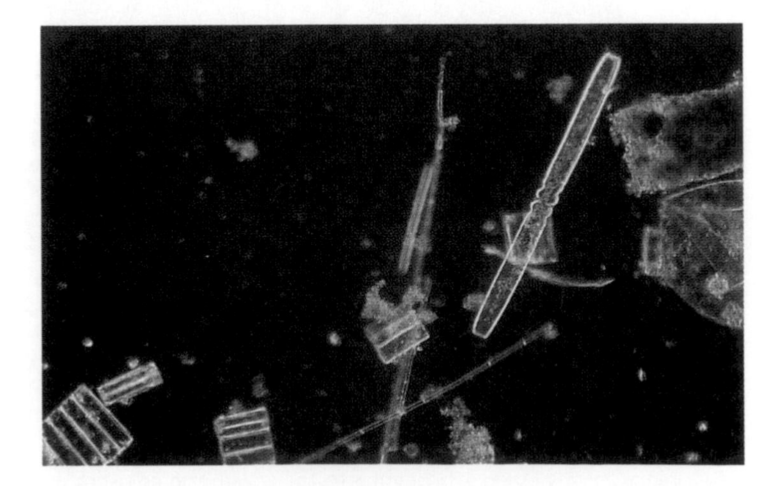

Figure 1.3
Dark-field photomicrograph of diatoms from a drop of pond water (X100).

The Theory of Spontaneous Generation Was Revived with the Discovery of the Microbial World

One of the most intriguing questions raised by the discovery of microorganisms was, "Where did these microscopic forms originate?" At the end of the seventeenth century, the theory of spontaneous generation—that organisms visible to the naked eye can arise spontaneously from nonliving matter—had just been completely debunked by Francesco Redi, an Italian biologist and physician. By a simple experiment, he demonstrated conclusively that worms found on rotting meat originated from the eggs of flies and not from the decaying meat. To prove this, he covered the meat with gauze fine enough to prevent flies from depositing their eggs. No worms appeared.

Despite these findings, the possibility that organic matter could spontaneously give rise to microscopic organisms was quickly suggested. It took about 200 years to disprove this hypothesis. Because gauze could not prevent the development of microscopic forms, new experiments were needed. The experiment designed to determine whether microbes could arise from nonliving material consisted first of boiling organic material in a vessel to kill all forms of life (sterilize) and then sealing the vessel to prevent the air from entering. If the solution became turbid after standing, then one could conclude that microorganisms had arisen from the organic material in the vessel, thus supporting the theory of spontaneous generation. Unfortunately, this experiment did not consider several alternative possibilities: that the flask might be improperly sealed, that microorganisms might be present in the air, or that boiling might not kill all forms of life. When this experiment was carried out in different laboratories, different investigators obtained different results.

Experiments of Pasteur

The two giants in science who did the most to disprove the theory of spontaneous generation were the French chemist Louis Pasteur, considered by many to be the father of modern microbiology, and the English physicist John Tyndall. The refutation published by Pasteur in 1861 was a masterpiece of logic. First, he demonstrated that air was filled with microorganisms. Pasteur filtered air through a cotton plug, trapping organisms that he then examined with a microscope. Many of these trapped organisms could not be distinguished microscopically from those that had previously been observed by others in many infusions.[1] Pasteur further showed that if the cotton plug was dropped into a sterilized infusion, it became turbid because the organisms quickly multiplied. Perhaps Pasteur's most important experiment was that which demonstrated that sterile infusions would remain sterile in specially constructed flasks even when they were left open to the air. Organisms from the air would settle in the bends of these swan-necked flasks (figure 1.4) and never reach the fluid. Only when the flasks were tipped would bacteria be able to enter the broth and grow. These simple and elegant experiments finally ended even the claims put forth by some individuals that unheated air or the infusions themselves contained a "vital force" necessary for spontaneous generation.

[1]infusion—broth that contains nutrients that allow microorganisms to grow

Experiments of Tyndall

Although most people in the scientific community, especially in France, were convinced by Pasteur's experiments, other respected scientists refused to give up the concept of spontaneous generation. This persistence undoubtedly stemmed in part from the fact that some scientists were still unable to verify Pasteur's results. John Tyndall finally provided a logical explanation for the differences in experimental results obtained in different laboratories, since he also obtained different results on different days. Tyndall realized that different infusions required different boiling times in order to be sterilized. Thus, boiling for 5 minutes would sterilize some materials, while others, most notably hay infusions, could be boiled for 5 hours and still contain living organisms! Furthermore, if hay was present in the laboratory, it became almost impossible to sterilize even the infusions that had previously been sterilized by boiling for 5 minutes. What was in hay that caused these effects? Tyndall finally realized that heat-resistant forms of life were being brought into his laboratory on the hay. These heat-resistant life forms were transferred to all other infusions in his laboratory, thereby making everything difficult to sterilize. Tyndall concluded that some microorganisms must be able to exist in two forms: a cell that is readily killed by boiling and one that is heat resistant. In the same year (1876), a German botanist, Ferdinand Cohn, also discovered the heat-resistant forms of bacteria, now termed **endospores.**

The extreme heat resistance of endospores explained the differences between Pasteur's results and those of other investigators. Organisms that produce endospores are commonly found in the soil and most likely were present in hay infusions. Because Pasteur used only infusions prepared from sugar or yeast extract, his broth most likely did not contain

• **Endospore, p. 74**

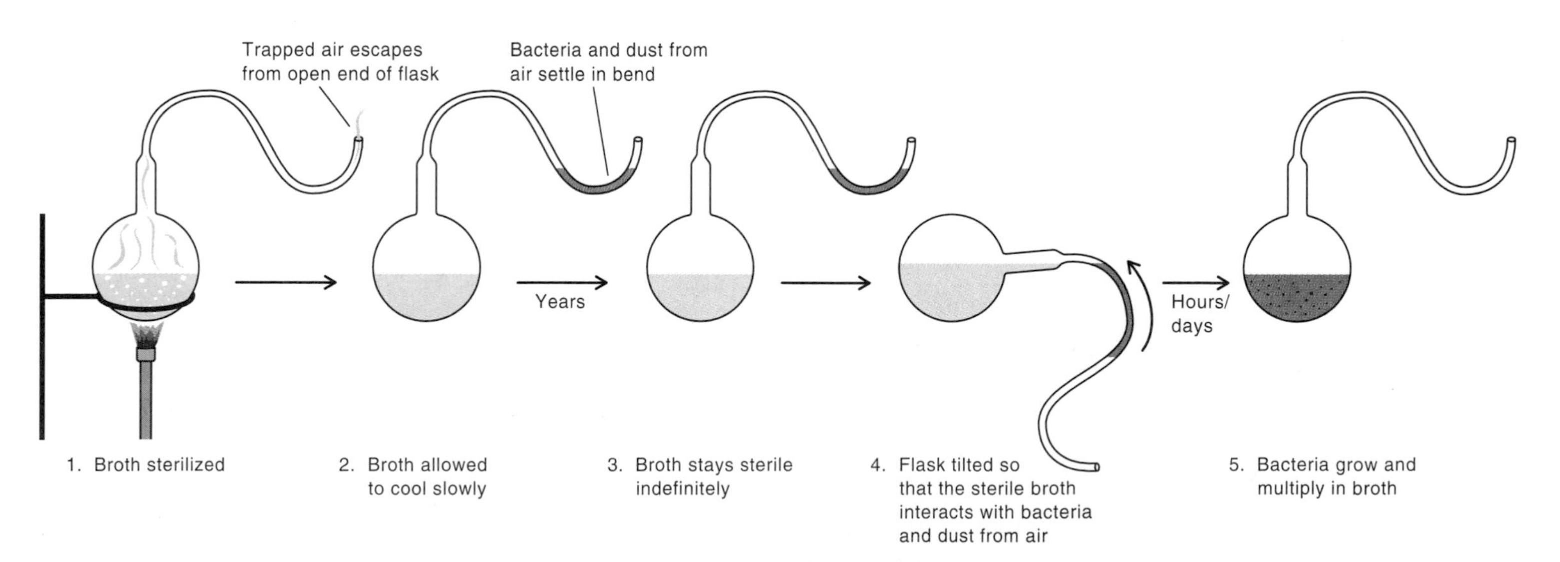

Figure 1.4
Pasteur's experiment with the swan-necked flask. (1–3) If the flask remains upright, no microbial growth occurs. (4 and 5) If microorganisms trapped in the neck reach the sterile liquid, they grow.

endospores. At the time these experiments on spontaneous generation were performed, the importance of the source of the infusion was not appreciated. In hindsight, the knowledge of the infusion source turned out to be critical to the results that were observed and the conclusions that were drawn.

These experiments point out an important lesson for all scientists. It is absolutely essential to reproduce all conditions of an experiment as closely as possible if one is trying to repeat an experiment and compare results with those of another investigator. It may seem surprising that the concept of spontaneous generation, as envisioned by Needham and others, was disproved only a little over a century ago. How far the science of microbiology and all biological sciences have advanced over the last 100 years!

Table 1.1 lists some of the more important advances in microbiology that have been made in the course of history. Many of these advances will be covered in more detail in *A Glimpse of History* in succeeding chapters.

Table 1.1 Some Major Milestones in Microbiology

Date	Event
1500 B.C.	Ancient Egyptians ferment cereal grains to make beer.
1546 A.D.	Italian physician Girolamo Fracastoro suggests that invisible organisms may cause disease.
1676	Antony van Leeuwenhoek observes bacteria and protozoa using his homemade microscope.
1796	Edward Jenner introduces a vaccination procedure for smallpox.
1838–1839	Mathias Schleiden and Theodor Schwann independently propose that all organisms are composed of cells,the basic unit of life.
1847–1850	Physician Ignaz Semmelweis demonstrates that childhood fever is a contagious disease transmitted by physicians to their patients during childbirth.
1853–1854	John Snow demonstrates the epidemic spread of cholera through a water supply contaminated with human sewage.
1857	Louis Pasteur demonstrates that yeast can degrade sugar to ethanol and carbon dioxide and multiply in the process.
1867	Joseph Lister publishes the first work on antiseptic surgery, beginning the trend toward modern aseptic techniques in medicine.
1881	Robert Koch introduces the use of pure culture techniques for handling bacteria in the laboratory.
	Walter and Fanny Hesse introduce agar-agar as a solidifying gel for culture media.
1882	Koch identifies the causative agent of tuberculosis.
1884	Koch states Koch's postulates.
	Elie Metchnikoff discovers phagocytic cells and thus begins the study of immunology.
	Hans Christian Gram devises the gram stain technique, which differentiates bacteria.
1887	Julius Petri, a German bacteriologist, adapts two plates to form a container for holding media and culturing microbes—the petri dish.
1908	Paul Ehrlich develops the drug salvarsan to treat syphilis, thereby starting the use of chemotherapy to treat diseases.
1911	F. Peyton Rous discovers that a virus can cause cancer in chickens.
1928	Frederick Griffith discovers genetic transformation in bacteria, thereby raising a key question in genetics: What caused the transformation?
1929	Alexander Fleming discovers and describes the properties of the first antibiotic, penicillin.
1944	Oswald Avery, Colin MacLeod, and Maclyn McCarty demonstrate that Griffith's transforming principle is DNA.
	Joshua Lederberg and Edward Tatum demonstrate that DNA can be transferred from one bacterium to another.
1953	James Watson, Francis Crick, Rosalind Franklin, and Maurice Wilkins determine the structure of DNA.
1957	D. Carlton Gajdusek demonstrates the slow infectious nature of the disease kuru, which is later shown to be caused by a prion.
1971	Theodor Diener demonstrates the fundamental differences between viroids and viruses.
1973	Herbert Boyer and Stanley Cohen, using plasmids, are the first to clone DNA.
1975	Cesar Milstein, Georges Kohler, and Niels Kai Jerne develop the technique for making monoclonal antibodies.
1976	Michael Bishop and Harold Varmus discover the cancer-causing genes, called *oncogenes* in viruses, and find that such genes are present in normal tissues.
1982	Stanley Prusiner isolates a protein from a slow disease infection and suggests that it might direct its own replication. He suggests the agent be termed *prion*.
1983	Luc Montagnier of France and Robert Gallo of the United States independently isolate and characterize the human immunodeficiency virus (HIV) that causes AIDS.
	Kerry Mullis invents the polymerase chain reaction.
1993	Three "new" diseases appear in the United States, emphasizing that emerging diseases are a continued threat to people's health.
1994	The Food and Drug Administration approves the first genetically engineered food for human consumption, a slower-ripening tomato.
1995	The Food and Drug Administration approves the first protease inhibitor, a major weapon against the progression of AIDS.
	The first complete nucleotide sequence of the chromosome of a bacterium is reported (*Haemophilus influenzae*).
1997	The first complete nucleotide sequence of all of the chromosomes of a eukaryote is reported (yeast).

Medical Microbiology—Past Triumphs

About the time that spontaneous generation was finally disproven to everyone's satisfaction, the Golden Age of Medical Microbiology was born. Between the years 1875 and 1918, most bacteria that cause diseases were identified, and early work on viruses was begun. Once people realized that these invisible agents could cause disease, efforts were made to prevent their spread from sick to healthy people. The great successes that have been made in the area of human health in the last 100 years have resulted largely from advances in the prevention and treatment of infectious diseases.

Consider two diseases: smallpox, a viral disease, and bubonic plague, a bacterial disease. The agents that cause these diseases are two of the greatest killers the world has ever known. Approximately 10 million people have died of smallpox in the course of history. However, as a result of an active worldwide vaccination program, no cases have been reported since 1977, and the disease will probably never reappear.

Between 1346 and 1350, one-third of the entire population of Europe, approximately 25 million people, died from bubonic plague. Now, generally less than 100 people in the entire world die each year from this disease. The discovery of antibiotics in the early twentieth century provided an increasingly important weapon against bacterial diseases. However, although progress has been impressive, a great deal still remains to be done, especially in the treatment of viral diseases and of diseases that are prevalent in developing countries.

Medical Microbiology—Future Challenges

Some people may believe that medical microbiology is no longer an active field of research since, in their opinion, all important disease-causing organisms have been identified and infectious diseases are well under control. This perception is far from the truth. For example, about 750 million cases of infectious diseases of all types occur each year in the United States. These diseases lead to 200,000 deaths annually and result in tens of billions of dollars in health care costs alone. Respiratory infections and diarrheal diseases are the leading causes of illness and deaths.

In addition to well-accepted diseases, maladies that were attributed to other causes have now been shown to be caused by microorganisms. One well-known example is peptic ulcers. This common affliction has recently been shown to be caused by a bacterium and to be treatable by antibacterial agents. It seems likely that other maladies whose causes are not understood will be shown to be caused by microorganisms or other agents of infectious disease.

• Ulcers, p. 564

Seemingly "new" diseases also continue to arise. In fact, in the past twenty years, there has been much media coverage in the United States of several "new" diseases. These include Legionnaires' disease, toxic shock syndrome, Lyme disease, acquired immunodeficiency syndrome (AIDS), and three that emerged in 1993. In the summer of that year, a mysterious flulike disease struck the Southwest, resulting in 33 deaths. The causative agent was identified as a virus, hantavirus, carried by deer mice and spread in their droppings. Also in that same year, more than 500 residents of the state of Washington became ill with a strain of *E. coli* present in undercooked beef prepared at a fast-food restaurant. The organism synthesized a potent toxin and caused hemolytic-uremic syndrome. Three children died. Further, in 1993, 400,000 people in Milwaukee became ill with a diarrheal disease, crytosporidiosis, that resulted from improper chlorination of the water supply.

It is a great credit to the biomedical research community that the causative agents for all of these diseases have been identified. In fact, none of these diseases is new, but changes in their occurrence and distribution have brought them to the attention of the scientific community. The bacteria causing Legionnaires' disease and Lyme disease have only been isolated and identified in the past decade or two, as have the viruses that cause AIDS. However, much information must still be learned about the mechanisms by which these agents cause disease, which, in turn, will lead to better ways of treating and preventing them.

A number of factors account for the fact that seemingly "new" diseases arise almost spontaneously, even in industrially advanced countries. As people live longer, their ability to ward off infectious agents is impaired and, as a result, organisms that usually are unable to cause disease become potentially deadly agents. Also, as lifestyles change, new opportunities arise for infectious agents to cause disease. For example, the use of vaginal tampons by women has resulted in an environment in which the *Staphylococcus* bacterium can grow and produce a toxin that causes toxic shock syndrome. The hantavirus will likely cause more deaths as cities spread into rural areas, bringing people into close contact with animals that previously were isolated from human beings. New diseases may emerge because some agents have the ability to change abruptly and thereby gain the opportunity to infect new hosts. It is possible that one of the agents that causes AIDS arose from a virus that at one time could only infect monkeys. Are there other agents in existence that may be ready to cause new human diseases? The answer is undoubtedly yes.

Not only are new diseases appearing, but many infectious diseases that were on the wane in the United States have started to increase again. One reason for this resurgence is that thousands of U.S. citizens and foreign visitors enter this country daily. About one in five visitors comes from a country where such diseases as malaria, cholera, plague, and yellow fever still exist. In developed countries, these diseases have been largely eliminated through sanitation, vaccination, and quarantine. However, an international traveler incubating a disease in his or her body could theoretically circle the globe, touch down in many countries, and expose many people before he or she developed recognizable signs of the disease. As a result, these diseases threaten to recur in economically developed countries.

Another reason that certain diseases are on the rise is that in both developed and developing countries many childhood diseases such as measles, mumps, whooping cough, and diphtheria have been so effectively controlled by childhood vaccinations that many parents have become lax about having their children vaccinated. A dramatic increase in the number of those infected has resulted.

A third reason for the rise in infectious diseases is that the increasing use of medications that prolong the life of the elderly and of treatments that lower the disease resistance of patients weaken the ability of the immune system to fight diseases. People infected with the human immunodeficiency virus (HIV), the virus responsible for AIDS, are a high-risk group for infections that their immune systems would normally resist. For these reasons, tuberculosis (TB) has increased both in the United States and worldwide. Nearly half the world's population is infected with the bacterium that causes TB, although for most infected people the infection is inactive. However, many thousands of new cases of TB are reported annually in the United States alone, involving primarily the elderly, minority groups, and people infected with HIV. Furthermore, the organism causing these new cases of TB is resistant to the drugs that once were effective in curing the disease.

Until a few years ago, it seemed unlikely that the terrible loss of life associated with the plagues of the Middle Ages or with the pandemic influenza outbreak of 1918 and 1919 would ever recur. However, the emergence of the disease AIDS, with over 1 million people infected with HIV, over 500,000 cases of AIDS, and more than 300,000 deaths in the United States alone as of 1996, dramatizes the fact that microorganisms and viruses can still cause serious, incurable, life-threatening disease.

Beneficial Applications of Microbiology—Past and Present

Although it is often presumed that microorganisms are only important because they cause disease, this is far from the truth. Human life would not exist on this planet without the activities of bacteria. Some bacteria degrade dead plants and animals to small molecules, which serve as the nutrients necessary to sustain the growth of all living organisms. Other bacteria can take nitrogen gas (N_2) from the air and convert (fix) it into the form of nitrogen (NH_4^+) fertilizer required for plant growth. Without these nitrogen-fixing bacteria, life as we know it could not exist. In addition to the crucial roles that microorganisms play in maintaining life, they also have made life more comfortable for people over the centuries. For example, microorganisms have been used for centuries in food production. As early as 2100 B.C., Egyptian bakers used yeasts to make bread. The excavation of early tombs in Egypt revealed that by 1500 B.C. the ancient Egyptians employed a highly complex and sophisticated procedure for fermenting cereal grains to produce beer. Virtually every human culture that has domesticated milk-producing animals also has developed fermented milk products, such as yogurt, cheeses, and buttermilk.

Biotechnology—New Applications for Microorganisms

In addition to infectious diseases, microorganisms have been written up in the popular press in recent years because of their potential for solving many problems associated with an industrial society. Biotechnology is a multibillion dollar industry that is revolutionizing all aspects of life. Microbiology and immunology are its cornerstones.

Consider the following biotechnological applications of microorganisms:

1. **Interferon,** insulin, human growth hormone, blood-clotting factors, and enzymes that dissolve blood clots and thick mucus in cystic fibrosis patients can be produced by microorganisms.
2. **Microorganisms** are being developed to produce vaccines against rabies, gonorrhea, herpes, leprosy, malaria, and hepatitis.
3. **Bacteria can destroy** such chemical pollutants as polychlorinated biphenyls (PCBs) and DDT, which contaminate soil and waters. They are also being used to assist the cleanup of oil spills. (See perspective 1.2.)
4. **A bacterium is** being used to make plants resistant to insect attacks, viral diseases, and frost.
5. **Bacteria can synthesize** a variety of different compounds that can be used in various ways, such as cellulose used in stereo headsets; hydroxybutyric acid, used to make disposable diapers biodegradable; and ethanol, used in gasoline to reduce smog.
6. **Bananas genetically engineered** by a bacterium may be used as vaccines to prevent certain diarrheal diseases in children.

These examples represent only a few of the ways that microorganisms are being used to promote human welfare. In the past, microorganisms were considered to be dangerous because they caused disease, and the importance of medical microbiology cannot be minimized. However, the current and future use of microorganisms in various ways that increase the quality of human life will receive increasing attention.

Perspective 1.2

Cleaning Up the Environment with Bacterial Pac-Men

One of the beneficial ways in which bacteria are currently being used is in degrading toxic pollutants in the environment. One such harmful pollutant in the United States is the compound PCB (polychlorinated biphenyl). PCBs are carbon-based, nonflammable liquids that are excellent electrical insulators. As such, they were widely used in electrical devices and power transformers. In 1976, their use was discontinued because it was found that exposure of humans and animals to PCBs resulted in skin sores, liver damage, and cancer from the accumulation of PCBs in tissues, where they remain for years.

However, by the time the use of PCBs was banned, large amounts had already been released into the environment. In 1987, the General Electric Company reported that almost 150 tons of PCBs had been released into the Hudson River from its plants. The General Electric Company recently reported encouraging test results from **bioremediation,** the process of using bacteria to chew up and detoxify the PCB pollutants.

Bacteria, which require oxygen to grow and live in river and pond waters, can remove chlorine atoms from PCBs, thereby detoxifying them. However, much of the PCBs are buried in the mud of lake and river bottoms where oxygen is not available. The bacteria in these environments detoxify PCBs very slowly and only remove some of the chlorine atoms.

To overcome these problems, researchers have taken the sediments contaminated with PCBs and added nutrients to encourage the bacteria living in them to grow and attack the PCBs. One investigator said, "We don't know exactly how the bacteria do it, but we have a concept of a Pac-Man that goes around and bites off chlorines." Once the PCBs are partially dechlorinated, the bacteria growing in air remove the rest of the chlorine atoms. The bacteria gain energy from the process of removing the chlorine atoms and degrading the organic molecules that result. Once the chlorines are removed, the PCBs are no longer toxic. Thus, both the bacteria and people benefit from the actions of the microorganisms.

The strategy of feeding bacteria that occur in a contaminated environment so that they will flourish and degrade the pollutant was also tried on the oil-soaked beaches of southeast Alaska where almost 11 million gallons of oil were spilled in the spring of 1989. The effectiveness of the bacteria is still being debated.

These are just two examples of the ways that bacteria may be used to help humanity restore a polluted environment with a minimum amount of environmental disruption. The area of bioremediation appears to have a very bright future, and numerous companies are being formed to take advantage of the ability of microorganisms to feed on a very large and diverse group of chemicals.

To understand these "applied" aspects of microbiology, you must have some knowledge of the properties of microorganisms, their structure, their growth properties, and how they can change their properties. This information will help you make rational decisions on such hotly debated issues as whether genetically engineered organisms should be released into the environment, what the likelihood is of receiving AIDS from an infected individual, and whether the widespread use of antibiotics in animal feeds should be stopped.

REVIEW QUESTIONS

1. What two contributions to microscopy did van Leeuwenhoek make?
2. Why was the source of infusions used in studies on spontaneous generation so important?
3. What two viral diseases have been especially destructive in the past? What viral disease is especially deadly today and in the foreseeable future?
4. List three applications of bacteria that enhance the quality of life.

Cell Theory

The studies of Pasteur and others who helped disprove the theory of spontaneous generation demonstrated that microscopic and macroscopic (visible) organisms arise only from preexisting organisms. To gain some understanding of the place microorganisms occupy in the biological world, it is helpful to consider their similarities with and differences from other organisms. Microorganisms are generally composed of single **cells** (*cella* means "small room"), and therefore microbiology emphasizes the study of single-celled organisms. A cell is the fundamental unit of all living matter. It consists of chemicals and structures separated from other cells and the environment by a membrane barrier and, in bacteria, usually by a tough cell wall.

Before the 1830s, when most people were only aware of the organisms they could see, the entire organism was thought to be the fundamental unit of structure. However, in 1838, a German botanist, Matthias Schleiden, and in 1839, a German zoologist, Theodor Schwann, published statements that all organisms were composed of cells, which were the fundamental units of life. They both recognized that cells carried out all of the basic functions that were attributed to living organisms. Schleiden and Schwann's hypothesis, called the **cell theory,** was one of the most important conceptual advances in biological thinking. Previously, scientists had focused their attention on the great differences in size, color, shape, and movement of whole living organisms. With the proposal of the cell theory, they transferred their attention to the cell and the similarities of all cells even in organisms having very different appearances. In the century and a half since the publication of the cell theory, the work of thousands of investigators in the various biological sciences has determined just how basic these similarities are and how much variety nature allows. The study of bacteria and other unicellular organisms has been especially important to an understanding of cell structures and their functions.

Similarity in Function and Composition of All Cells

All cells including bacteria must solve similar problems in order to survive and live. Many cells, especially bacteria, have one major function in life—to reproduce exact copies of themselves. In order to duplicate itself, a cell must contain genetic instructions for the synthesis of all of its parts. In addition, cell duplication requires that cells be able to manufacture (synthesize) the components of living matter from the simple foodstuffs available to them. Because these synthetic reactions require energy, all cells must also have the means to both obtain and use energy, which they accomplish by degrading foodstuffs.

The chemicals that carry out these functions are similar in all cells, from single-celled microorganisms to the cells that are part of a much larger organism. For example, all cells have protein molecules that are **enzymes,** the biological catalysts that speed up chemical reactions. These reactions are involved in the formation and use of energy as well as in the synthesis of the structures of which the cells are made. All cells contain the very large molecules termed **macromolecules,** such as **deoxyribonucleic acid (DNA),** which stores genetic information, and **ribonucleic acid (RNA),** which is involved in protein synthesis.

A few smaller molecules also found in all cells have similar functions. For example, **adenosine triphosphate (ATP)** serves as a storage form of energy, while **coenzymes,** which are synthesized from vitamins in the diet, are necessary for the functioning of certain enzymes. The role of all of these small molecules will be covered in greater detail in several later chapters of the text.

In addition to being similar in function, each of the molecules has the same chemical nature in all cells. For example, nucleic acids from all organisms are made up of the same component parts. Proteins are mostly composed of the same twenty **amino acids.** As you will learn later, some bacteria do not have to be provided certain nutrients in their diets in order to multiply. For example, *E. coli* does not require any amino acids in its diet. This does not mean that the proteins of such bacteria do not contain these amino acids, but rather that they can synthesize them. The details of their chemical composition are covered in chapter 2. Some of the important molecules found in all cells are given in table 1.2.

Table 1.2 Important Molecules in Cells

Molecule	Function
Macromolecules	
Deoxyribonucleic acid (DNA)	Store genetic information
Ribonucleic acid (RNA)	Are involved in protein synthesis
Proteins	As enzymes speed up chemical reactions
	Form part of structure of cell
Small Molecules	
Adenosine triphosphate (ATP)	Are a storage form of energy
Coenzymes	Are required for enzyme function; are synthesized from vitamins
Amino acids	Make up proteins

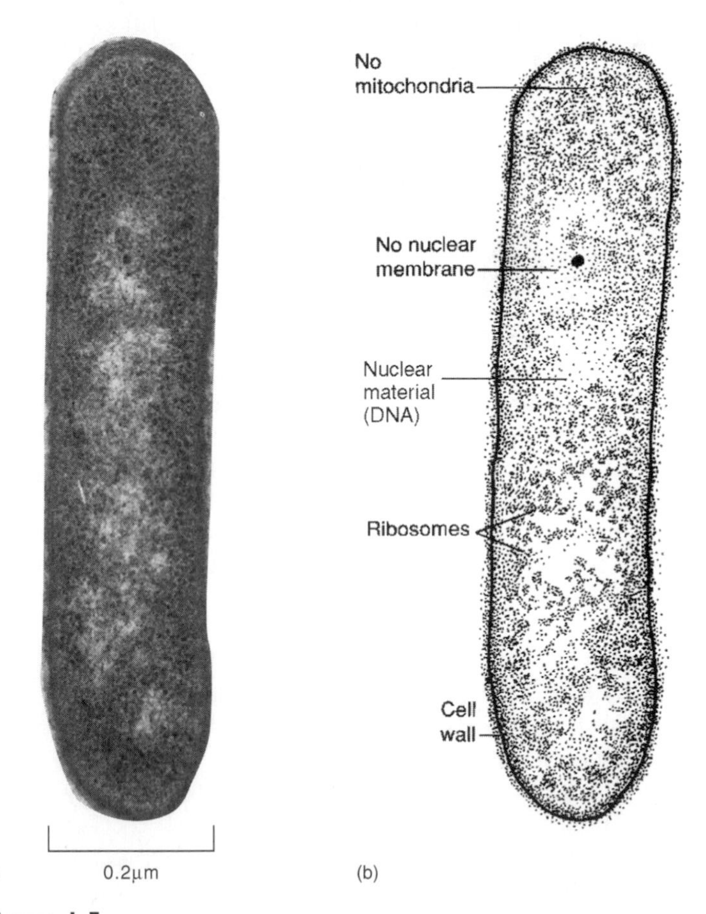

Figure 1.5
Prokaryotic cell (*Eikenella corrodens*). Note that there are no internal structures that are enclosed by a membrane. (*a*) Electron photomicrograph (X70,000). (*b*) Schematic drawing of (*a*).

Basic Cell Types

Although all cells carry out the same basic functions using the same chemicals, major differences exist in the structures that cells use to carry out these functions. Two different types of cells can be readily recognized from their appearance when viewed with a high-powered microscope. One that appears to be much smaller and simpler than the other is termed **prokaryotic,** which means "prenucleus." Prokaryotic cells do not have a true nucleus, because their nuclear material (DNA) is not surrounded by a membrane (figure 1.5 *a* and *b*). Additionally, they do not contain other membrane-bound internal structures, often referred to as **organelles,** such as **mitochondria,** the structures concerned with the production of energy from foodstuffs, and **chloroplasts,** the structures that convert the sun's energy into chemical energy. Organisms that are prokaryotic in their structure are often referred to as **prokaryotes.**

The larger, more complex cell type is termed **eukaryotic,** which means "true nucleus" (figure 1.6 *a* and *b*). In these cells, the nuclear material is surrounded by a double membrane that forms the true nucleus. In addition, these cells usually contain several different intracellular membrane-bound structures, such as mitochondria and chloroplasts.

All algae, fungi, protozoa, and multicellular organisms are eukaryotic and therefore are frequently called **eukaryotes.** Other important differences in these two cell types are given in table 1.3. These differences are discussed in greater detail in chapter 3.

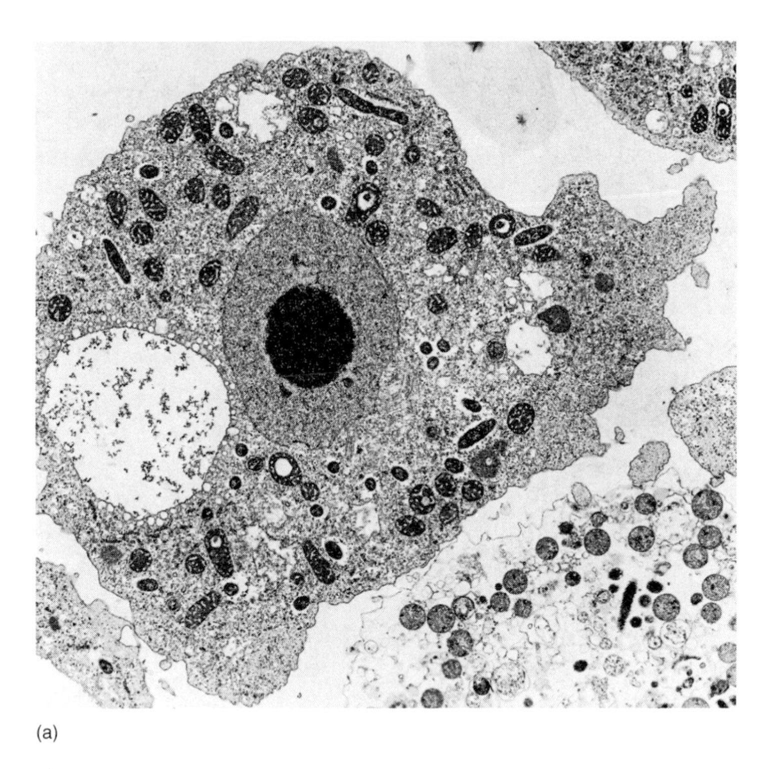

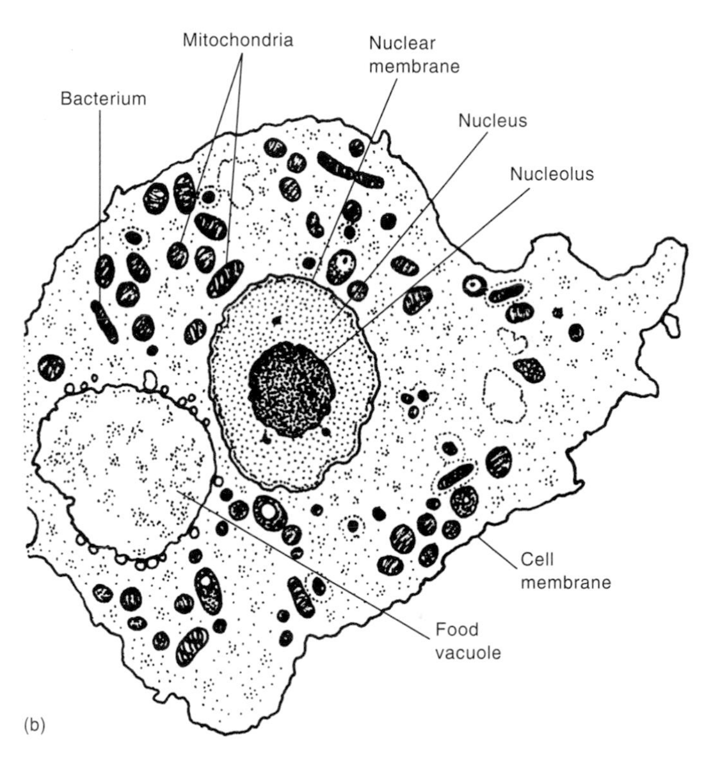

Figure 1.6
Eukaryotic cell. (*a*) Electron photomicrograph of a yeast cell. (*b*) Diagram of (*a*). Note that this cell contains several internal structures, the mitochondria and nucleus, that are surrounded by membranes. The cell is considerably larger than the prokaryotic cell.

Table 1.3 Comparison of Prokaryotic and Eukaryotic Cells

Features of Cells	Prokaryotic	Eukaryotic
Size (diameter)	0.3–2 μm	2–20 μm
Genetic structures		
Chromosome number	1–4 (all identical)	>1 (all different)
Nuclear membrane	No	Yes
Cell wall	Unique chemical components	Not always present; usually many different components
Cytoplasmic structures		
Mitochondria	No	Yes
Chloroplasts	No	In some cells
Ribosomes	70S*	80S

*The size of the ribosome is indicated by the speed at which it sediments upon centrifugation (S value). The larger the S value, the heavier and larger the object.

Perspective 1.3

Prokaryotes and Eukaryotes in the *Guinness Book of Records*

We might assume that because prokaryotes have been so extensively studied over the past hundred years no major surprises are left to be discovered. However, this is far from the truth. Large, peculiar-looking organisms can be seen when the intestinal tracts from sturgeon fish from both the Red Sea and from the Great Barrier Reef in Australia are examined (figure 1). However, the organisms cannot be cultured in the laboratory. Because of their large size—600 µm long and 80 µm wide, which is about the size of a comma and is clearly visible without any magnification— it was generally believed that the organisms in question had to be eukaryotic cells, even though a nuclear membrane could not be seen. From other studies it has become absolutely certain that they are prokaryotes.

It has always been believed that a significant size difference exists between eukaryotic and prokaryotic cells and that the appearance of large cells in the fossil record must have occurred at the time eukaryotes appeared. The discovery of a prokaryote larger than most eukaryotes now questions both of these assumptions. With the unexpected finding of such a large prokaryote, it might also be expected that very small eukaryotes would be discovered. This, in fact, was recently reported. The picoplankton* of estuaries and oceans consist largely of prokaryotes. However, eukaryotic forms are also important. In a Mediterranean lagoon the main component of the phytoplankton is a eukaryote classified as a green alga. Its length is about 1 µm and its width is about 7 µm. The organism does not have a cell wall but does have a nucleus, one mitochondrion, and a reduced compartment of cytoplasm.

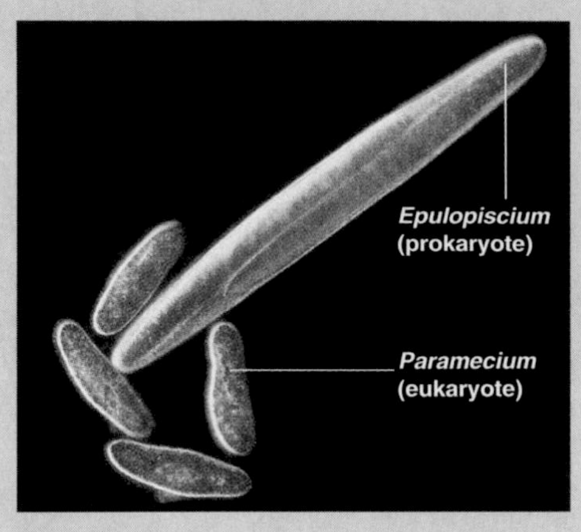

Figure 1
Photomicrograph of the largest known bacterium, *Epulopiscium* (0.4 mm), mixed with paramecia. Note how large this prokaryote is compared with the four eukaryotic paramecia.

*picoplankton—microorganisms floating freely in water

Some Cell Types Appear to Be Intermediate Between Prokaryotes and Eukaryotes

Some cell types have been found that seem to have properties of both eukaryotic and prokaryotic cells. Such cells resemble eukaryotes in that they have a true nucleus, but unlike eukaryotes they do not have any mitochondria or chloroplasts. Since they seem to be an ancestor of the eukaryotes, some scientists have given them the name *Archezoa* (from *arches*, meaning "first" or "primitive"). However, other scientists believe the organisms should be considered primitive eukaryotes and should not be given a special name. A member of these organisms is *Giardia intestinalis,* a water-borne agent that causes a diarrheal disease, shown in figure 1.7. Another member of this group is *Trichomonas vaginalis,* which causes trichomoniasis, a sexually transmitted disease that is widespread. Both organisms are usually classified as protozoa. Another cell that has always been classified as a bacterium is a member of the genus *Gemmate.* This organism is unusual in that it has a membrane-bound nuclear body but no other membrane-bound organelles (figure 1.8). As greater numbers of organisms are carefully studied, it is likely that additional organisms will be discovered that have properties of both major groups.

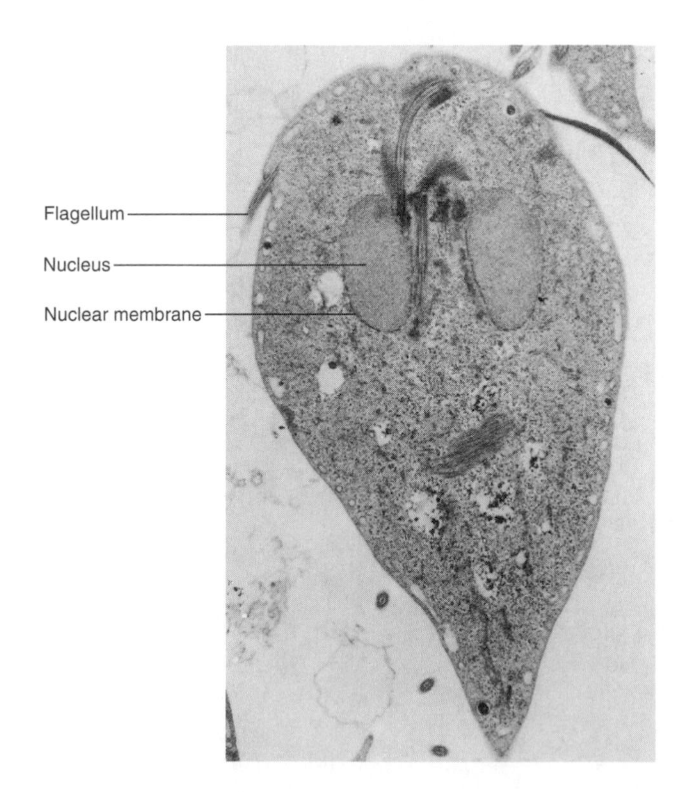

Figure 1.7
Giardia intestinalis. Scanning electron photomicrograph (X9,000). Note that the nucleus is obvious but there is no evidence of mitochondria.

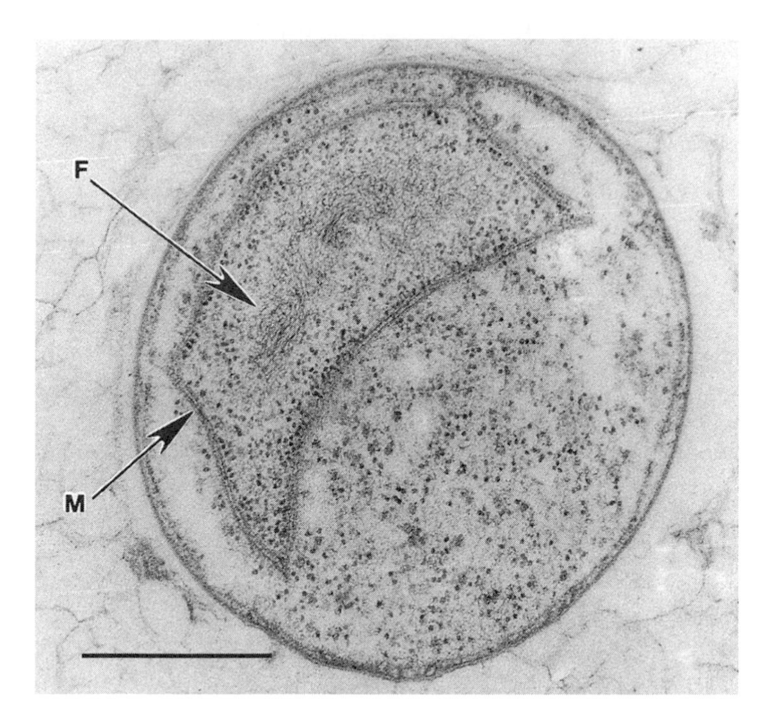

Figure 1.8
The prokaryote *Gemmata obscuriglobus* as viewed by transmission electron microscope of a thin section of cells. The DNA fibrils of the nuclear body are labeled (F) and these are surrounded by a distinct membrane (labeled M). The bar marker is 0.5μm. Courtesy of Dr. John Fuerst.

Prokaryotes Can Be Divided into Two Major Groups Termed *Domains*—the Eubacteria and the Archaea

The bacteria are a very heterogeneous group. The only property they have in common is that they are all prokaryotes. A cursory microscopic examination reveals that they have a number of different shapes. Further, if certain structures found in all prokaryotes are analyzed using recently developed techniques, it becomes even more obvious that prokaryotes can be quite different. When the RNA in the ribosomes, the protein-synthesizing structures, of a large number of prokaryotes was carefully analyzed, it became clear that these seemingly similar organisms could be divided into two groups, or domains. Within each domain, the RNA structure is very similar; however, it differs greatly from one domain to the other. One domain is called the **Eubacteria** (true bacteria), and the other is the **Archaea** (ancient organisms). When the chemical nature of other bacterial structures was analyzed, it became clear that eubacteria and archaea differ from one another in many other ways. These differences will be discussed in chapter 10, after you have learned more about the chemistry of and structures in the cell.

These differences in the chemical composition of certain cell structures have led to the general conclusion that, from an evolutionary point of view, the Eubacteria and Archaea belong in two unrelated groups. The eukaryotes of the domain Eukarya make up the third major group of organisms. It is important to emphasize that the differences between the two types of prokaryotic cells relate to fundamental differences in the chemical make-up of the individual cells and not to their physical appearance or to what the cells are capable of doing. Indeed, these two cell types appear virtually identical when viewed with a high-powered microscope (figure 1.9 *a* and *b*).

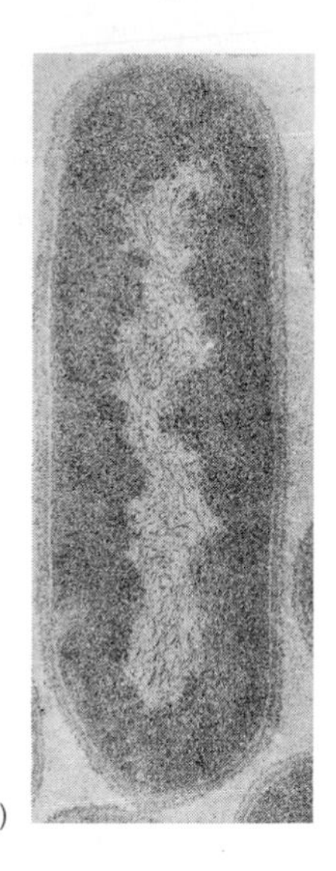
(a)

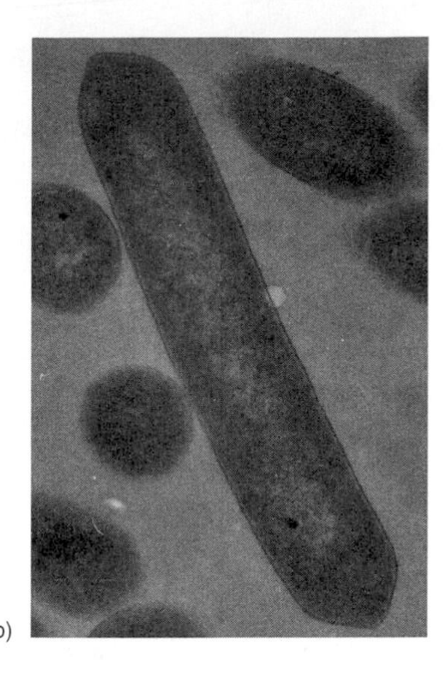
(b)

Figure 1.9
Electron photomicrograph of (*a*) a eubacterium, *E. coli* (X38,000), and (*b*) an archaebacterium, *Methanobacterium foricum* (X111,000). Note that there are no major differences between the two organisms that can be seen using an electron microscope. The differences are at the chemical level.

These studies on cells illustrate an important point in microbiology as well as in all of biology. The microscope is a very powerful tool for observing the structures that make up cells, and it can readily distinguish between eukaryotic and prokaryotic cells. However, its use is limited to studying what is visible with the microscope. Chemical analysis of cell structures provides information far beyond what can be seen with the microscope. This explains why so many current studies in biology are at the chemical or molecular level. Indeed, some of the most exciting advances being made in biology today use the tools of **molecular biology.**

Eubacteria Comprise Most Prokaryotes Familiar to Microbiologists

Of the two groups, the Eubacteria will be covered more extensively in this course, since they are the more important in causing disease and have been more intensively studied.

We will use the term *bacteria* throughout this text to refer to the Eubacteria. Members of the domain Archaea have little importance in microbiology related to health and disease. However, the Archaea are important in many biological processes in the environment. Their relationship to eubacteria and eukaryotic cells is shown in the following figure. If the words *the* or *domain* precede the terms *Eubacteria* or *Archaea,* then the first letter of Eubacteria and Archaea generally will be capitalized: for example, "The Eubacteria differ from the Archaea in many ways." Otherwise, the first letter is not capitalized: for example "Scientists have isolated archaea from the ocean." At this time, only an overview of the major properties of the Eubacteria will be given. These prokaryotes form a very heterogeneous group, as can be seen in table 1.4.

The features that are used to separate these organisms into groups include (1) the appearance of the cells, (2) the means by which they move, (3) their nutritional characteristics, (4) their ability to multiply in the absence of living cells, and (5) their possession of a rigid cell wall. You will read much more about these individual groups throughout the course so it is not necessary to memorize this table. As you peruse this table, it should be clear that enormous diversity exists between the groups. However, some features of bacteria deserve special mention.

Most but not all bacteria have a rigid cell wall that has an unusual chemical structure. Some bacteria are motile (capable of movement) and others are not. If they are motile, bacteria move by several different mechanisms. Their nutritional requirements vary markedly. A few bacteria can convert the nitrogen in the air into the nitrogen of nucleic acids and proteins (nitrogen fixation).

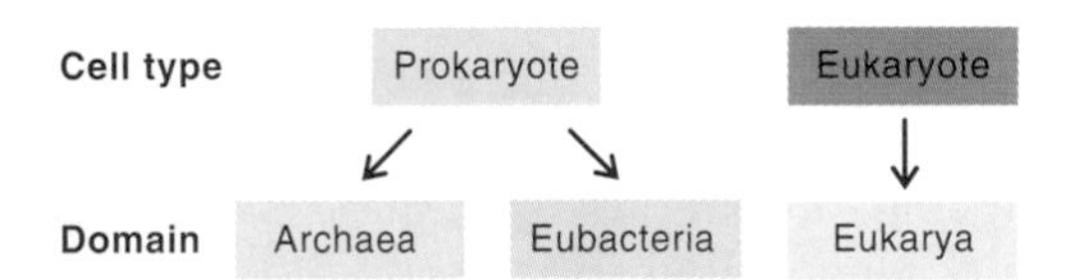

Many of these same groups can convert the energy in sunlight into chemical energy. Most eubacteria can grow in a broth medium, but others can grow only inside other living cells.

Archaea Often Grow Under Extreme Conditions of Temperature and Salinity

The Archaea are also a diverse group of prokaryotes but probably not as diverse as the Eubacteria. The distinguishing feature of the Archaea as a group is their ability to grow in environments in which most organisms cannot survive. For example, some archaea can grow only in salt concentrations ten times higher than that found in seawater. These organisms grow in such habitats as the Great Salt Lake and the Dead Sea. Other archaea grow best at extremely high temperatures; for example, one member grows best at temperatures above 105°C (100°C is the temperature at which water boils at sea level). Thermophilic archaea can be found in the boiling hot springs at Yellowstone National Park. However, the Archaea are spread far beyond extreme environments. They are widely distributed in the oceans and they constitute over 30% of the picoplankton in the cold surface waters of Antarctica and Alaska. Many of the Archaea produce methane (CH_4) from carbon dioxide (CO_2), and some require certain sulfur compounds for their growth.

Since the Archaea are commonly associated with extreme environments, it is possible that they were the earliest forms of life on earth. It is likely that the surface of the earth in early times was much hotter than it is today. This implies that the early living organisms were able to withstand high temperatures, a feature that is associated with many of the

Table 1.4 Major Groups of Eubacteria*

Bacteria	Distinguishing Features
Typical bacteria	Rigid cell walls; unicellular; multiply by binary fission; if motile, by flagella
Prosthecate and budding bacteria	May have unusual shapes, appendages (prosthecae); most are prosthecate bacteria; divide by budding; most budding bacteria are prosthecate
Mycelial bacteria	Often moldlike in appearance
Filamentous sheathed bacteria	Individual cells enclosed in a common sheath
Mycoplasma sp. or pleuropneumonialike organisms (PPLOs)	Very heterogeneous group in shape; properties distinct from other bacteria; no cell wall and may not divide by binary fission; sterols in cytoplasmic membrane
Gliding bacteria	Organisms have flexible cell wall; may have a developmental cycle
Spirochetes	Helical cells with flexible cell walls; movement by axial filament
Rickettsia sp.	Obligate intracellular parasites; "leaky" cytoplasmic membrane
Chlamydia sp.	Obligate intracellular parasites that lack some enzymes required for energy production
Cyanobacteria	Flexible cell wall; gliding motility; gain energy by photosynthesis; reproduction primarily by binary fission

*These groupings are for convenience and do not follow the accepted classification scheme.

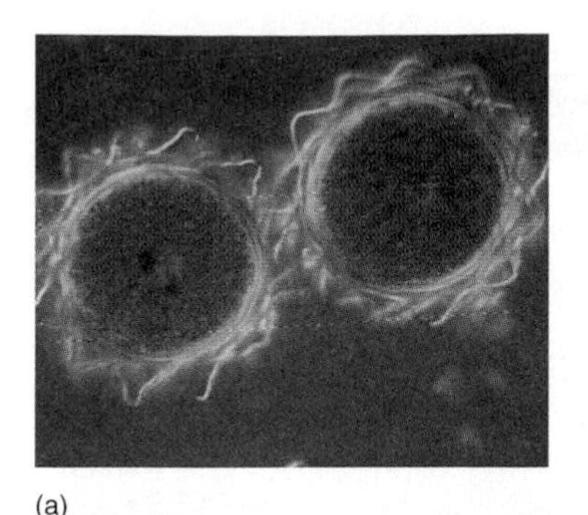
(a)

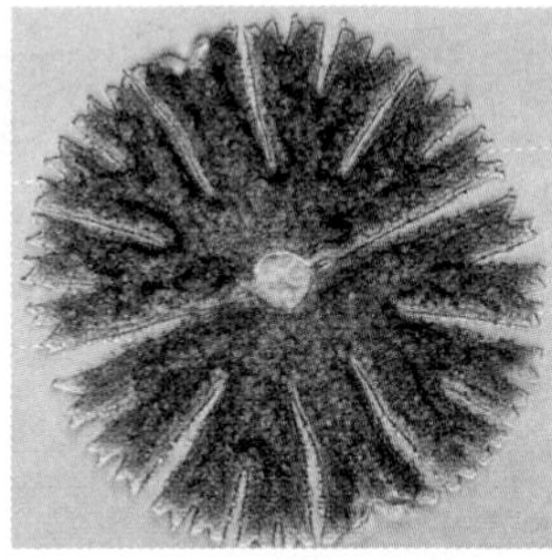
(b)

(c)

Figure 1.10
Algae. (*a*) *Volvox* (X400), a motile alga that grows in colonies. Each colony contains up to 50,000 cells. (*b*) *Micrasterias*, a green alga composed of two symmetrical halves (X100). (*c*) Diatom organism that serves as food for aquatic animals, as viewed through a scanning electron microscope (X2,400).

Archaea. For these reasons, they were considered to be *primitive* bacteria. Determining whether in fact the Archaea were the earliest life forms must await further analysis of their ribosomes. The evolutionary implications of the two bacterial domains, Eubacteria and Archaea, are considered in chapter 10.

REVIEW QUESTIONS

1. State the cell theory.
2. List three basic functions all cells perform.
3. List the two basic cell types and how they can be distinguished from one another.
4. List four important characteristics that distinguish different groups of bacteria.

Members of the Microbial World

No general agreement exists about what organisms and agents, other than the Eubacteria, Archaea, and viruses, should be included in the microbial world. No clear dividing line can be drawn between the microscopic and macroscopic world. In this text, all single-celled organisms, both prokaryotes and eukaryotes as well as viruses, will be considered members. These organisms include bacteria as well as three eukaryotic groups: **algae, fungi,** and **protozoa.** Multicellular organisms that are closely related to the single-celled eukaryotes will be discussed briefly.

Algae, Fungi, and Protozoa Are Eukaryotic Members of the Microbial World

Algae and fungi can be unicellular or multicellular, but only the single-celled members will be considered in detail in this text. All protozoa are unicellular. These three groups of organisms can be distinguished from one another by the criteria indicated in table 1.5. They are considered in more detail in chapter 12. Some photographs of algae, fungi, and protozoa are shown in figures 1.10 *a–c*, 1.11 *a* and *b*, and 1.12 *a* and *b*, respectively.

Table 1.5 Distinguishing Features of Eukaryotic Members of the Microbial World

Algae
Gain energy from sunlight (photosynthetic)
Fungi
Not photosynthetic
Most are multicellular, consisting of long filaments.
Protozoa
Not photosynthetic
Unicellular
Cells may be extremely complex.

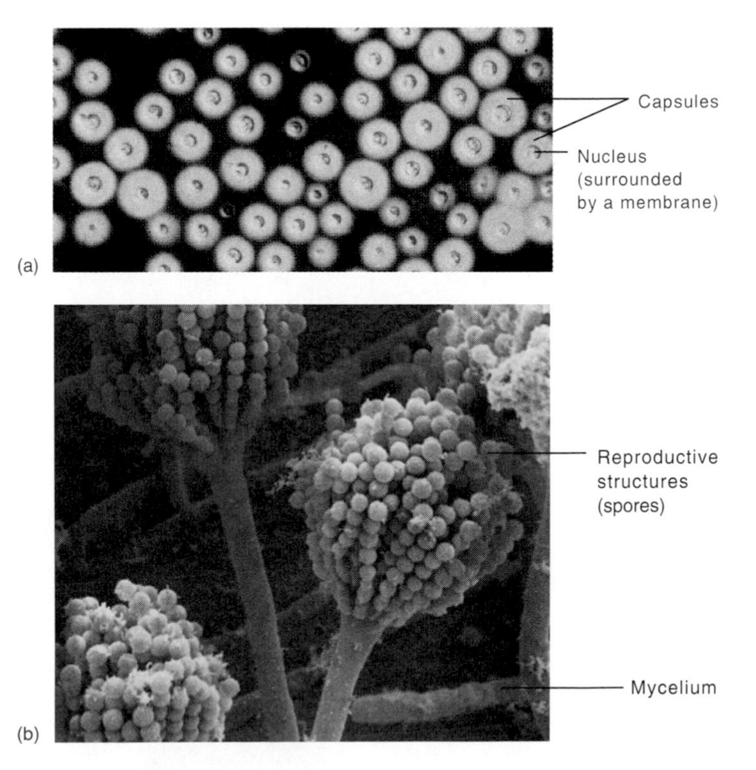

Figure 1.11
The two forms of fungi: (*a*) yeast form and (*b*) mold form. (*a*) Living cells of *Cryptococcus neoformans* stained with India ink to reveal the large capsules that surround the cell (X150). (*b*) *Aspergillus*, a typical mold whose dark reproductive structures rise above the mycelium (X1,375).

Viruses, Viroids, and Prions Are Nonliving Members of the Microbial World

Other members of the microbial world that we will discuss are the **viruses, viroids,** and **prions** (table 1.6).

Neither viruses, viroids, nor prions are cells and, therefore, they are not organisms. They are often referred to as **infectious agents.** A **virus** is actually a piece of genetic material, either DNA or RNA (not both), surrounded by a protective protein coat (figure 1.13 *a–c*). A **viroid** is a small piece of RNA that does not have a protein coat (figure 1.14). The RNA does not code for any protein.

These agents can multiply only inside living cells, since they do not have the machinery necessary to carry out the functions required for cell growth. Viruses are responsible for a large number of diseases in both animals and plants. Viroids have been shown to cause a number of plant diseases, and some scientists speculate that they may also cause diseases in humans.

A group of very unusual agents responsible for at least six diseases in humans and animals are **prions.** They appear to be proteins without nucleic acid, either DNA or RNA, although some scientists dispute this (table 1.6). Thus, prions differ from viruses and viroids, although some scientists categorize them as unconventional viruses. Their mode of replication is a mystery. Prions are discussed more fully in several subsequent chapters.

The members of the microbial world are summarized in figure 1.15.

• **Nucleotide, p. 37**
• **Prion, p. 301**

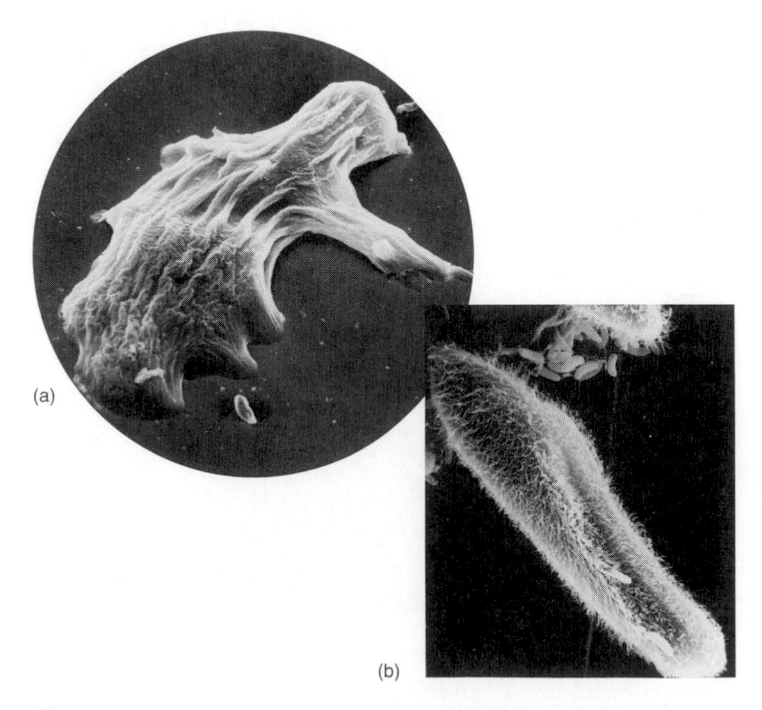

Figure 1.12
Protozoa. (*a*) *Amoeba*, whose cells continually change shape by extending pseudopodia, their means of locomotion (X180). (*b*) *Paramecium*, a ciliated protozoan (X1,200).

Table 1.6 Distinguishing Characteristics of Viruses, Viroids, and Prions

Viruses	Viroids	Prions
Are obligately intracellular agents	Are obligately intracellular agents	Are obligately intracellular agents
Contain either DNA or RNA, surrounded by a protein coat	Contain only RNA; no protein coat	Contain only protein, no DNA or RNA
Contain very few, if any, enzymes	Contain no enzymes	Contain no enzymes

Sizes of Members of the Microbial World

The relative sizes of the members of the microbial world are given in figure 1.16. As you can readily see, the sizes have a tremendous range. The smallest viruses are about 1 million times smaller than the largest eukaryotic cells. Even within a single group wide variations exist. The variation in size of bacteria was recently expanded by the discovery of a bacterium that is longer than 0.5 μm. In fact, it is so big that it is visible to the naked eye. Likewise, a eukaryotic cell in the picoplankton was recently discovered that is no larger than a typical bacterium. (See perspective 1.3.)

As you might expect, the small size and broad size range of members of the microbial world have required the use of measurements not commonly used in everyday life. The use of logarithms has proven to be enormously helpful in the study of microorganisms. A brief discussion of measurements and logarithms is given in appendix I.

Nomenclature of Organisms

In biology, **nomenclature** refers to the system by which organisms are named. Almost all organisms are named according to the **binomial system** of genus and species, as devised by the Swedish botanist Carl von Linné, or Carolus Linnaeus (1707–1778), as he is more commonly called. (His name, like the names of the organisms he classified, became Latinized.)

In the binomial system the first word in the name is the **genus,** with the first letter always capitalized; the second word is the **species** name, which is not capitalized. Both words are always italicized. For example, *Escherichia coli* is a member of the genus *Escherichia*. Like many bacteria, it was named after the person who first isolated and described it (the German physician Theodor

• **Large bacterium, p. 12**

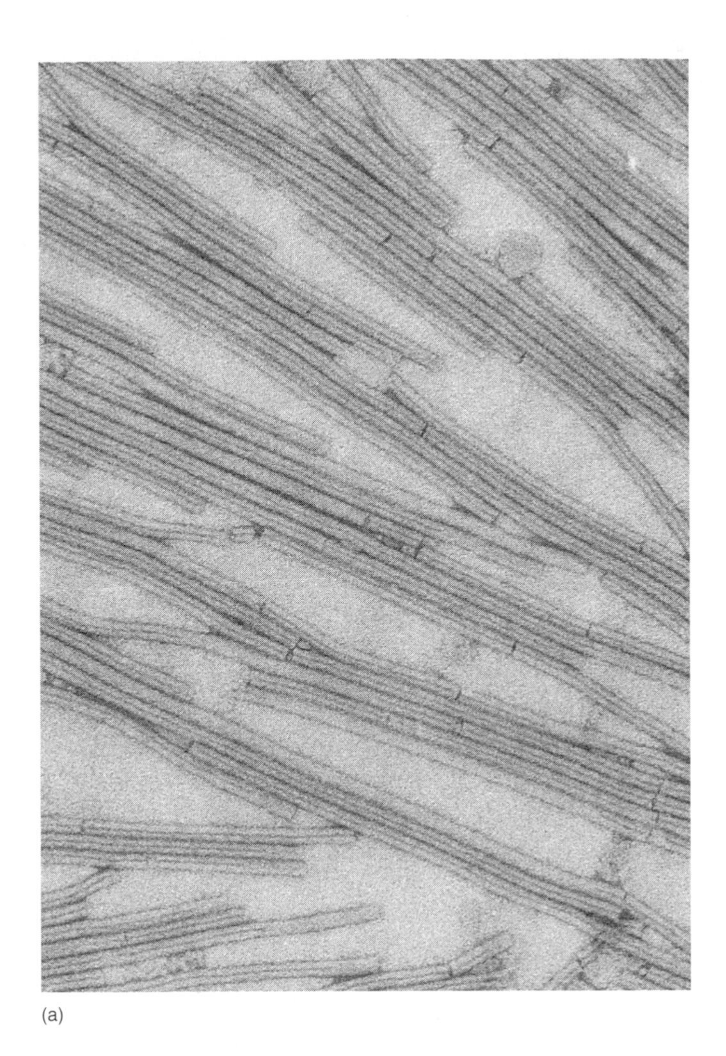

(a)

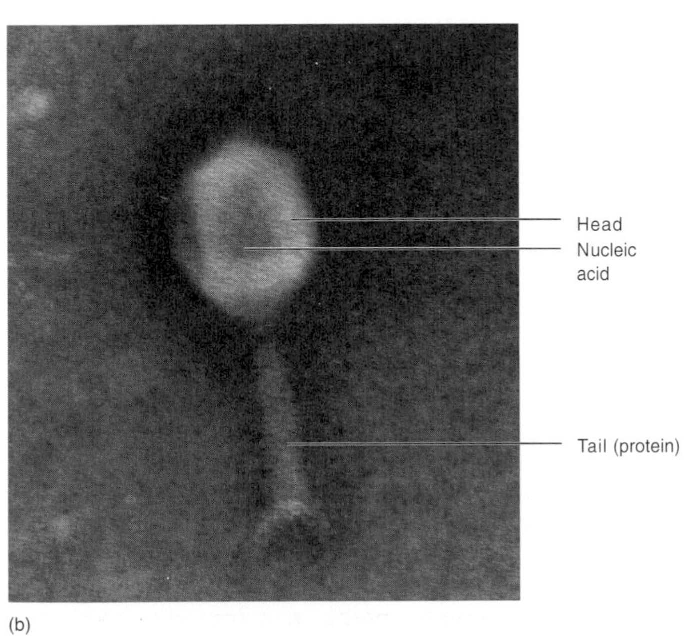

(b)

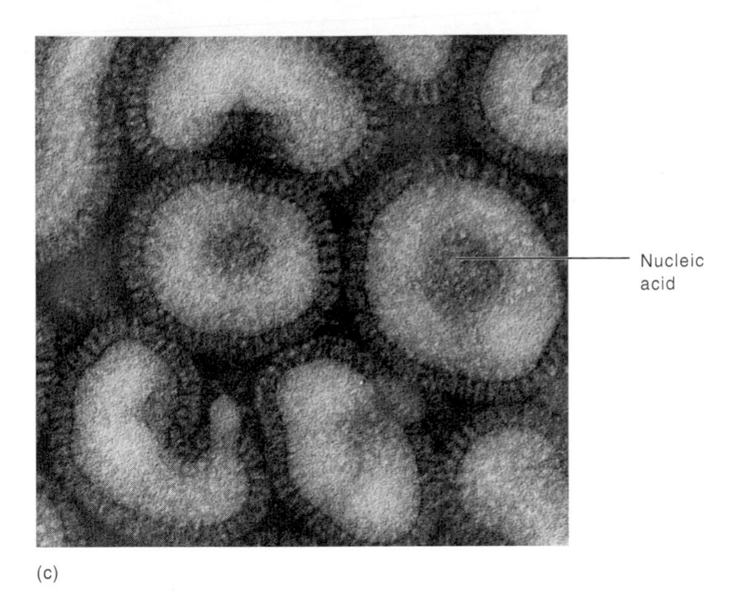

(c)

Figure 1.13
Examples of viruses that infect three kinds of organisms.
(*a*) Tobacco mosaic virus that infects tobacco plants. A long, hollow protein coat surrounds a molecule of RNA (X144,000).
(*b*) A bacterial virus (bacteriophage) that invades bacteria.
(*c*) Influenza virus, thin section (X283,000).

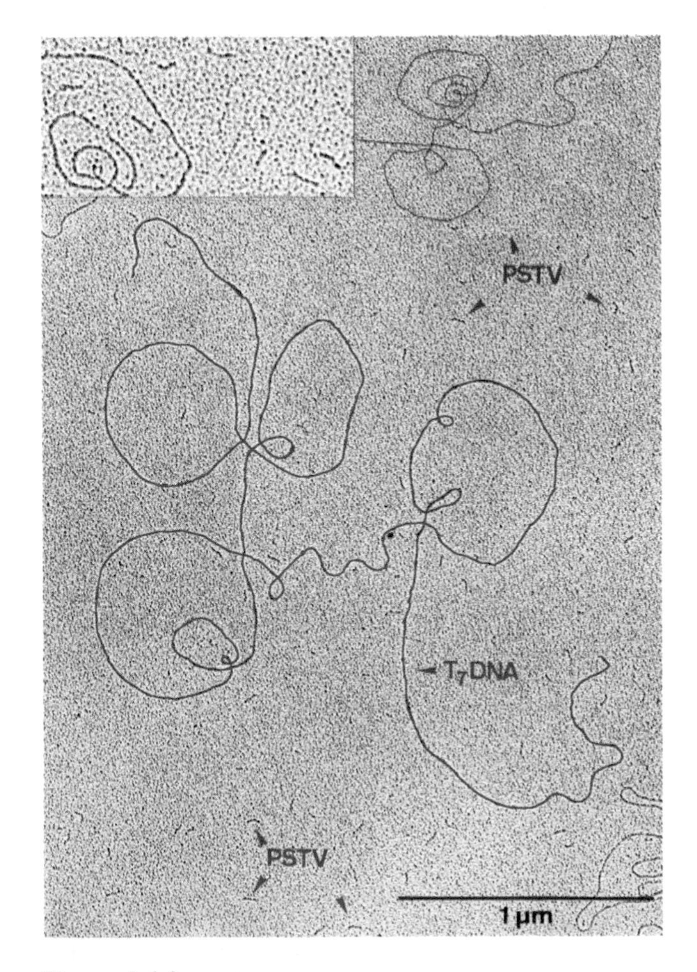

Figure 1.14
The size of a viroid compared to a molecule of DNA from the bacteriophage T_7. The arrows from PSTV point to the potato spindle tuber viroids (PSTV); the other arrow points to T_7 DNA. The PSTV consists of about 350 nucleotides. Courtesy of Dr. T. Diener, United States Department of Agriculture.

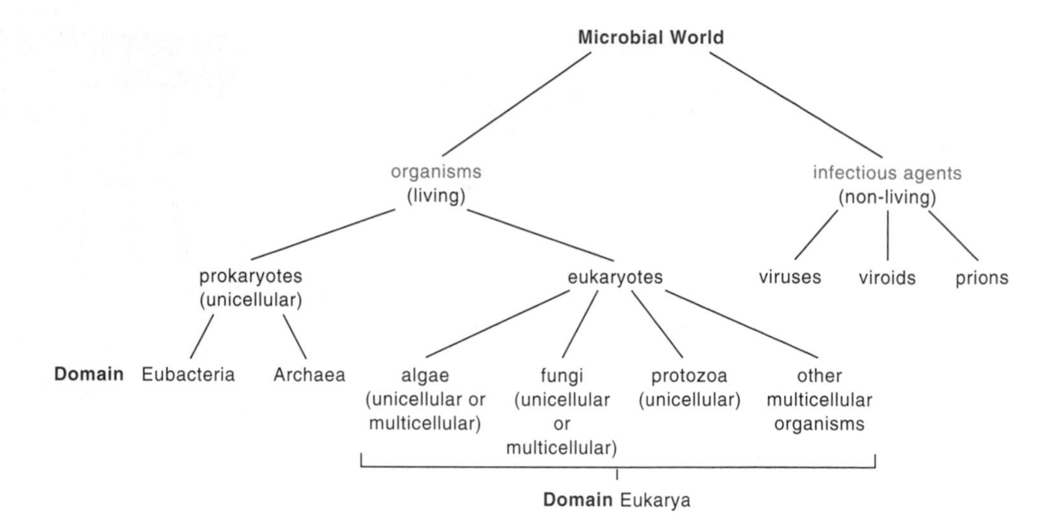

Figure 1.15
The microbial world.

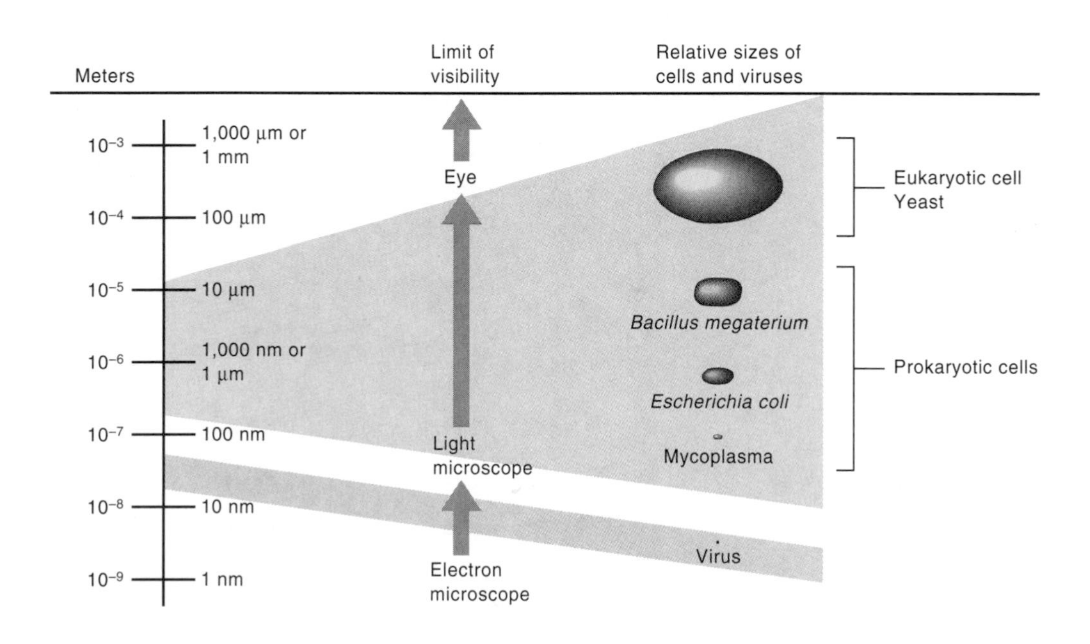

The basic unit of length is the meter (m), and all other units are fractions of a meter.

nanometer (nm) = 10^{-9} meter = .000000001 meter
micrometer (µm) = 10^{-6} meter = .000001 meter
millimeter (mm) = 10^{-3} meter = .001 meter
1 meter = 39.4 inches

These units of measurement correspond to units in an older but still widely used convention.

1 angstrom (Å) = 10^{-10} meter
1 micron (µ) = 10^{-6} meter

Figure 1.16
Measurements of organisms and viruses.

Escherich, 1857–1911). In this case, the species name comes from the location (colon) where the organism is most commonly found.

The genus name is commonly abbreviated, with the first letter capitalized: that is, *E. coli.* A number of different species are included in the same genus. For example, in the genus *Legionella,* many species have been identified, including *L. pneumophila* and *L. micdadei.* Members of the same species may vary from one another in minor ways. For example, one member may not be able to break down a foodstuff that another can. This slight distinction generally does not justify giving organisms different species names, and so they are designated as **strains** or **varieties** of the same species. For example, two common strains of *E. coli* are K-12, isolated from a patient at Stanford University Hospital, and *E. coli* ML, which is reputed to have been isolated from the bowel of a famous French scientist whose initials are ML. Both strains are identical in their growth characteristics, but they differ in many minor aspects. The names and a description of most bacteria that are currently recognized are given in a reference text, *Bergey's Manual of Systematic Bacteriology,* which is published in four volumes.

• **Bergey's Manual, p. 224**

To be properly identified, an organism must have both a genus and a species name. However, many organisms are commonly referred to by less precise terms that serve to place them only in a certain group. Since terms are especially common when describing organisms of medical importance, you will frequently encounter them especially in later chapters of this text. For example, bacillus refers to rod-shaped bacteria; thus, all bacilli (plural) are rod-shaped (cylindrical) organisms. These include members of the genus *Bacillus*, as well as of the genera *Clostridium*, *Lactobacillus*, and *Escherichia*. Bacillus is neither capitalized nor italicized and thus can be differentiated from the genus *Bacillus*. Other common names used to describe the shapes of bacteria are cocci for spherical-shaped organisms and spirilla for helical-shaped organisms. Thus, a genus name that ends in -coccus, such as *Streptococcus* or *Staphylococcus*, must contain organisms that are round.

CHAPTER REVIEW QUESTIONS

1. Name the prokaryotic members of the microbial world.
2. Name three nonliving members of the microbial world and what are their major properties.
3. What are three eukaryotic members of the microbial world and what are their major properties?
4. In the designation *Escherichia coli* K-12, what is the genus? What is the species? What is the strain?
5. Name the three major molecules in the cell and give their functions.
6. What are the distinguishing features of typical bacteria?
7. How can you distinguish between the mycelial bacteria and fungi microscopically?
8. Where would you go to isolate members of the Archaea?
9. How did endospores complicate studies on spontaneous generation?
10. What are two functions of proteins?

Summary

Microorganisms are, in large part, responsible for determining the course of human history. The use of modern sanitation facilities, vaccinations, and antibiotics has dramatically reduced the incidence of infectious disease.

I. Microorganisms Discovered
   A. Antony van Leeuwenhoek discovered microorganisms over 300 years ago by viewing water samples through lenses that magnified objects 300 times.
   B. The theory of spontaneous generation was revived with the discovery of the microbial world.
      1. Differing results from different investigators led to the controversy about whether living organisms could arise from dead organic matter. This controversy was not resolved until the 1860s.
      2. Pasteur demonstrated that the air is filled with microorganisms and showed that swan-necked flasks containing sterile infusions could remain sterile indefinitely.
      3. Tyndall and, independently, Cohn discovered that heat-resistant forms of bacteria, or endospores, are present in certain infusions.

II. Medical Microbiology—Past Triumphs
   A. Between 1875 and 1918, most disease-causing bacteria were identified.

III. Medical Microbiology—Future Challenges
   A. "New" diseases are appearing. These include Legionnaires' disease, toxic shock syndrome, Lyme disease, AIDS, hantavirus disease, cryptosporidiosis, and hemolytic-uremic syndrome.
   B. Many diseases that were on the wane are now increasing in frequency. These include mumps, whooping cough, diphtheria, and tuberculosis.
   C. Organisms are becoming increasingly resistant to antibiotics.

IV. Beneficial Applications of Microbiology—Past and Present
   A. Human life could not exist without the activity of microorganisms.
   B. Microorganisms have been used for centuries for food production.

V. Biotechnology—New Applications for Microorganisms
   A. Microorganisms are now being developed to produce vaccines, clean up the environment, and carry out many other processes designed to make life more comfortable.

VI. Cell Theory
   A. Schleiden and Schwann in the mid-1800s proposed the cell theory—that cells are the basic units of life.

VII. Similarity in Function and Composition of All Cells
   A. All cells growing independently of other cells, such as bacteria, have one basic function—to reproduce. To do this, they must generate energy and synthesize the components of living matter.
   B. All cells are composed of the same macromolecules, such as nucleic acids (DNA and RNA) and proteins, which, in turn, are composed of the same subunits.

VIII. Basic Cell Types
   A. There are two cell types: prokaryotic and eukaryotic. Prokaryotic cells are simple, without membrane-bound internal structures. Eukaryotic cells are larger, are more complex, and have several internal membrane-bound structures. All bacteria are prokaryotic; algae, fungi, and protozoa are eukaryotic.
   B. Some cell types have been found that appear to be intermediate between prokaryotes and eukaryotes; they have a true nucleus but no mitochondria.
   C. Prokaryotes can be divided into two major domains: the Eubacteria and the Archaea.

Summary (continued)

1. Both groups are similar microscopically but differ in the chemical composition of several structures.
2. The two groups are not closely related to each other or to the eukaryotes.

D. Eubacteria comprise most prokaryotes familiar to microbiologists.
1. This group is very diverse.
2. Typical bacteria are heterogeneous, but they do share some obvious properties.

E. Archaea often grow under extreme conditions of temperature and salinity.

IX. Members of the Microbial World
A. The members include all unicellular organisms, which include all prokaryotes.
B. Algae, fungi, and protozoa are the eukaryotic members of the microbial world.
C. Viruses, viroids, and prions are nonliving members of the microbial world.

X. Sizes of Members of the Microbial World

XI. Nomenclature of Organisms
A. All organisms are named according to the binomial system of genus and species.
B. Names and descriptions of most bacteria are published in *Bergey's Manual of Systematic Bacteriology.*

Critical Thinking Questions

1. What single experiment do you think clinched the argument against spontaneous generation?
2. If you were asked to nominate one of the individuals mentioned in this chapter for the Nobel Prize, who would it be? Make a statement supporting your choice.
3. Chlamydias and rickettsias were once classified as viruses because of their small size and obligate intracellular growth requirement. What techniques can be used to show that they are really bacteria?

Further Reading

Brock, Thomas (ed.). 1961. *Milestones in microbiology.* American Society for Microbiology, Washington, D.C.: Prentice Hall, reprinted 1975. A collection of historically significant papers in microbiology that cover the past 300 years. The editor discusses briefly the significance of each paper to the development of microbiology.

Brock, T. D. (ed.). 1989. *Microbes and infectious diseases.* Scientific American Books, New York. A series of articles that provide historical and modern ideas on the role of microorganisms and viruses as disease-causing agents.

DeKruif, P. 1966. *Microbe hunters.* Harcourt Brace Jovanovich, New York. The stories of the early microbiologists told by a microbiologist in a very interesting fashion.

Ford, B. J. 1985. *Single lens: the story of the simple microscope.* Harper & Row Publishers, Inc., New York. A fascinating and highly readable account of the development of simple microscopes and the people who made them.

Garrett, L. 1994. *The coming plague: newly emerging diseases in a world out of balance.* Farrar, Strauss, & Giroux, New York. A very lengthy book warning about the numerous "new" diseases that have arisen in the last few years.

2 Biochemistry of the Molecules of Life

KEY CONCEPTS

1. Four elements—carbon, oxygen, hydrogen, and nitrogen—make up over 98% of all living matter. Two other elements—phosphorus and sulfur—are also very important.
2. The bonds that hold atoms together to form molecules result from electrons interacting with each other. Bonds vary in strength, which gives molecules characteristic properties.
3. Weak bonds are important in biological systems, since they often determine the most important properties of the molecules and are responsible for their proper functioning.
4. All life is based on the bonding properties of water, which comprises over 90% of a cell's weight.
5. Macromolecules consist of many repeating subunits, each subunit consisting of a small, simple molecule. The subunits are first synthesized and then bonded together to form the macromolecule.

Water molecules represented by ball and stick models. The white balls represent hydrogen atoms; the red balls represent oxygen atoms.

PREVIEW LINK

Microbes in Motion

The following books and chapters in the *Microbes in Motion* CD-ROM may serve as a useful preview or supplement to your reading:

Microbial Metabolism and Growth: Growth; Metabolism. *Fungal Structure and Function*: Metabolism and Growth.

A Glimpse of History

Louis Pasteur (1822–1895) is often considered the Father of Bacteriology. His contributions to this science, especially in its early formative years, were enormous and are discussed in many of the succeeding chapters. Pasteur started his scientific career as a chemist, initially working in the science of crystallography. It was not until he was on the Faculty of Science at Lille, France, a town with important brewing industries, that he became interested in the biological aspect of chemical problems. All of his later studies—including his work on spontaneous generation, infectious disease, and protection against infectious diseases through vaccination—employed the analytical experimental methods and thinking of a trained chemist.

His first chemical studies were performed on two compounds, tartaric and paratartaric acids, which formed the thick crusts found within wine barrels. These two substances form crystals that have the same number and arrangement of atoms, yet they differ in their ability to twist (rotate) a plane of polarized light* that passes through the crystal: Tartaric acid twists the light, but paratartaric acid does not. Therefore, the two molecules must differ in some way, even though they appear to be identical. Pasteur was intrigued by these observations and set about to understand how the crystals differ from one another. Looking at the crystals under a microscope, he saw that paratartaric acid consists of two different kinds of crystals. Using tweezers, he separated the two and dissolved each kind in a separate flask of water. When he shone polarized light through each solution of paratartaric acid, he found that one solution twisted the light to the left and the other twisted it to the right. Thus, the two components of the mixture neutralized each other and, as a result, the mixture did not rotate the light. Pasteur concluded that paratartaric acid is a mixture of two compounds, one being the **mirror image,** or **isomer,** of the other. This mixture can be conceived of as a mixture of right- and left-handed molecules, represented as a right and left hand facing each other (figure 2.1).

*Polarized light—light that moves only in one plane

Pasteur thus solved a long-standing mystery in chemistry. He followed up these studies with another remarkable discovery four years later. One day he observed that mold was growing in a solution that previously did not twist a plane of polarized light. Instead of throwing out the moldy solution, Pasteur decided to check its effect on polarized light. Much to his surprise, the solution now rotated the light. On the basis of his previous studies, he reasoned that the original solution must have been a mixture of two components that rotated light in opposite directions and thus neutralized each other. He concluded that the mold had destroyed one of the components of the mixture, thereby revealing the light rotation by a single component.

These observations had great biological significance. Pasteur recognized that molecules synthesized by living organisms have a preferred handedness, whereas molecules synthesized by chemical means are always a mixture of right- and left-handed molecules. Further, living organisms can only use one type, as noted by Pasteur when he found that an inactive solution was turned into an active solution by a mold growing in it. Indeed, an organism would starve to death if it consumed only the wrong-handed sugars, because such sugars are indigestible.

The mirror images of the same molecule have greatly different properties. For example, the amino acid phenylalanine, one of the key ingredients in the artificial sweetener aspartame, makes aspartame sweet when it is in one form but makes it bitter when it is in the other form.

Thus, what Pasteur studied as a straightforward problem in chemistry has implications far beyond what he ever imagined. It is often difficult to predict where research will lead or what the significance of seemingly unimportant observations will be.

Mirror

L-Amino acid

D-Amino acid

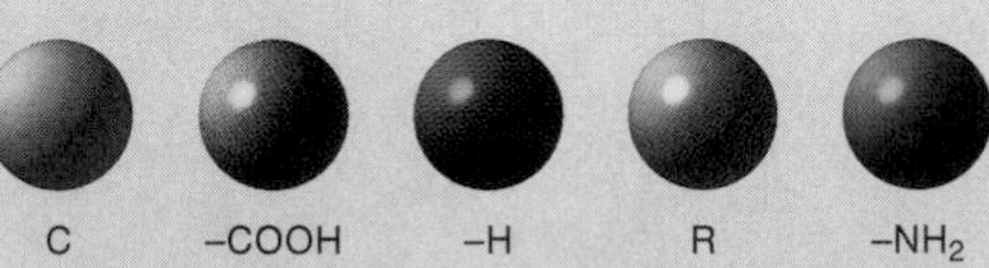

Figure 2.1
Mirror image forms of an amino acid. The joining of a carbon atom to four different groups leads to asymmetry in the molecule. Thus, the molecule can exist in either the L- or the D-form, each being the mirror image of the other. There is no way that they can be rotated in space to give two identical molecules.

Understanding how microorganisms live and die in nature, cause disease in humans, and play a role in genetic engineering depends on knowledge of certain basic facts of physics and chemistry as well as microbiology. This chapter covers some of these fundamental facts. It will serve as background and reference material for various aspects of microbiology covered throughout the text. For some students this may be a review of material from other courses. For others, it may be a first encounter with the chemistry of biological molecules. The discussion proceeds from the lowest level of organization, the atom, to a higher level, the macromolecule.

Elements and Atoms

An **element** is a pure substance that consists of a single type of **atom.** Although 92 naturally occurring elements exist, four elements comprise over 98% of all living material by weight. These elements are carbon (abbreviated C), hydrogen (H), oxygen (O), and nitrogen (N). Two other elements, phosphorus (P) and sulfur (S), together make up an additional 1% of existing elements. The relative abundance of these atoms in a cell follows this same pattern (table 2.1).

But for rare exceptions, all of the remaining elements together account for less than 0.5% of living material. In general, the basic chemical composition of all living cells is remarkably similar.

Elements are composed of atoms, the basic unit of all matter. Atoms, in turn, are composed of three major components: **electrons,** which are negatively charged; **protons,** which are positively charged; and **neutrons,** which are not charged. Since the number of protons equals the number of electrons, the atom as a whole does not have a charge. The protons and neutrons are found in the nucleus of the atom, which is the heaviest part. The electrons, which are very light, orbit the nucleus. The relative sizes of the parts of an atom can be illustrated by the following analogy. If a single atom were enlarged to the size of a football stadium, the nucleus would be the size of a marble and it would be positioned somewhere above the 50-yard line. The electrons would resemble fruit flies zipping around the nucleus. Their orbits would be mostly inside the stadium, but on occasion they would travel outside it.

The atom of each different element has different numbers of electrons, which are arranged in shells orbiting the nucleus. The electron structure of the element is very important because the electrons in the atoms participate in chemical reactions and form bonds with other atoms. Bonding with other atoms forms molecules.

Each element is identified by two numbers: its **atomic number** and its **atomic weight.** The atomic number is the number of protons, which is equal to the number of electrons. For example, hydrogen has one proton, and thus its atomic number is 1; oxygen has eight protons, and its atomic number is 8. The atomic weight is the sum of the number of protons and neutrons, since electrons are too light to contribute to the weight. The atomic weight of the element hydrogen is 1, abbreviated ^{1}H. It is the lightest element known. The atomic weight of oxygen is approximately 16 and is abbreviated ^{16}O.

Formation of Molecules: Chemical Bonds

Most atoms do not occur free in nature but associate (bond) with other atoms to form molecules. A **molecule** is two or more atoms held together by chemical bonds. The atoms that comprise a molecule may be of the same or different elements. For example, H_2 is a molecule of hydrogen gas formed from two atoms of hydrogen; water is an association of hydrogen and oxygen atoms (H_2O). The **molecular weight** of a molecule is the sum of the number of protons and neutrons that comprise the molecule. Thus, the molecular weight of water is 18 (16 + 1 + 1).

Table 2.1 Atomic Structure of Elements Commonly Found in the Living World

Element	Symbol	Atomic Number (Total Number of Protons)	Number of Possible Covalent Bonds*	% of Atoms in Cells
Hydrogen	H	1	1	49
Carbon	C	6	4	25
Nitrogen	N	7	3	0.5
Oxygen	O	8	2	25
Phosphorus	P	15	3	0.1
Sulfur	S	16	2	0.4

*The number of electrons required to fill the outer shell equals the number of possible covalent bonds. The number of electrons in a completed outer shell varies depending on the distance of the shell from the nucleus.

Perspective 2.1

Isotopes–Valuable Tools for the Study of Biological Systems

The analysis of living cells depends on a variety of techniques, many of which will be discussed later in the text. One important tool in the analysis of living cells involves the use of **isotopes,** different forms of the same element that have different atomic weights. The nuclei of certain elements can gain or lose neutrons and thereby gain or lose weight. For example, the most common form of the hydrogen atom contains one proton and no neutrons and has an atomic weight of 1 (^{1}H). However, another form also exists in nature in very low amounts. This isotope, ^{2}H (deuterium), contains a neutron. A third even heavier isotope, ^{3}H (tritium), is not found in nature but can be made by a nuclear reaction in which stable atoms are bombarded with high-energy particles. This latter isotope is unstable and gives off radiation (decays) in the form of rays or electrons, which can be very sensitively measured by a radioactivity counter. Once the atom has finished disintegrating, it no longer gives off radiation but is stable.

An important feature of radioactive isotopes is that their other properties are very similar to their nonradioactive counterparts. For example, tritium combines with oxygen to form water and with carbon to form hydrocarbons, and both molecules have biological properties very similar to their nonradioactive counterparts. The only difference is that the molecules containing tritium can be detected by the radiation they emit.

Isotopes are used in numerous ways in biological research and some of these will be covered in later chapters. They are frequently added to growing cells used to label particular molecules in cells. For example, when tritiated thymidine (a component of DNA) is added to growing bacteria, the thymidine will become a part of the DNA of the cells. The DNA molecules can then be detected easily by their radioactivity. The shape and size of the radioactive DNA molecule can actually be seen by placing the DNA on photographic film and storing the material in the dark for several days to allow the radioactivity to make its marks on the photographic emulsion. The position of the radioactivity in the cell can be determined from the position of the developed silver grains produced by the radioactivity, making it possible to locate the position of the molecules in the cell. For example, DNA is in the nucleus only (figure 1 *a*), whereas RNA, synthesized from ^{3}H uridine (tritiated uridine, a component of RNA), is found throughout the cytoplasm (figure 1 *b*). This technique involving radioisotopes and photography is termed **autoradiography.**

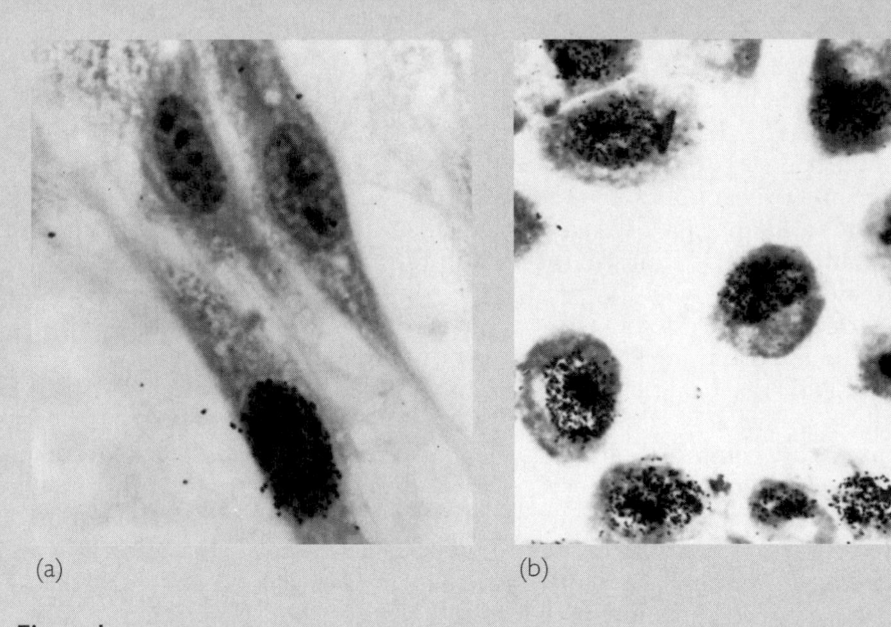

Figure 1
Autoradiography of (*a*) cells labeled with tritiated thymine, which is incorporated into DNA only, and (*b*) cells labeled with tritiated uridine, which is incorporated into RNA.

- **Thymidine, p. 39**
- **Uridine, p. 39**

The chemical bonds that hold atoms together to form molecules are of various types, each characterized by its strength. Some bonds are much stronger than other bonds (see table 2.3). An important strong bond is the **covalent bond,** commonly indicated by a line joining the two atoms (for example, H—O). Especially important weak bonds are **ionic** and **hydrogen bonds.** Atoms and molecules will always try to achieve maximum stability. An atom is most stable when its outermost shell is filled with electrons. If the outermost shell is unfilled, atoms tend to acquire electrons by bonding to nearby atoms to fill this shell and achieve maximum stability.

Covalent Bonds Are Formed When Atoms Share Electrons

One common way for atoms to fill their outer shell is to share electrons and the outer shells of both atoms at the same time. This sharing of electrons creates strong bonds between the atoms called **covalent bonds.** An example of covalent bonding involves carbon atoms, the most important single atom in all of biology. Since carbon has four electrons but requires a total of eight to fill its outer shell, it must borrow four electrons from other atoms to achieve maximum stability. It can do this by combining with four hydrogen atoms to form CH_4, which is the gas methane

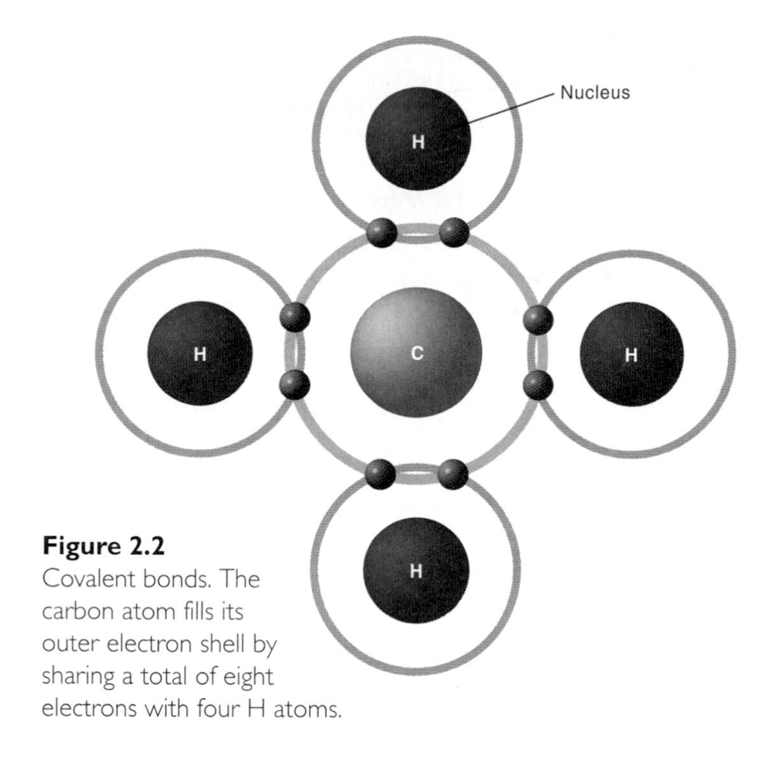

Figure 2.2
Covalent bonds. The carbon atom fills its outer electron shell by sharing a total of eight electrons with four H atoms.

(figure 2.2). Each H atom requires one additional electron to fill its outer shell, which it gains by sharing the electron of the carbon atom. Once the atoms fill their outer shell with electrons, the molecule is highly stable and cannot form additional covalent bonds.

Other elements require different numbers of electrons to fill their outer shells, numbers that determine the number of covalent bonds that the elements can make. This information is given in table 2.1 for the elements most important in the living world. The dash or colon between two atoms represents two shared electrons (C—H or C:H), a single bond. Sometimes electrons are shared in groups of two pairs, called *double bonds* (for example, C=O).

All covalent bonds are strong. An important rule in biological systems is that the stronger the bond, the more difficult it is to break. Therefore, covalent bonds do not break unless they are exposed to strong chemicals or are supplied with large amounts of energy, generally as heat. Molecules formed by covalent bonds never fall apart at physiological temperatures; only at temperatures much higher will they decompose. Since biological systems function only within a narrow temperature range and cannot tolerate high temperatures, they utilize the protein catalysts called **enzymes,** which can break these covalent bonds at temperatures at which life can exist. The way in which enzymes function will be considered in chapter 5.

Covalent bonds also vary in the ways in which electrons are shared between the atoms. In covalent bonds between identical atoms such as H—H, the electrons are shared equally, and the bond is termed a **nonpolar covalent bond.**

• **Enzymes, p. 108**

Nonpolar covalent bonds also exist between different atoms, such as C—H, if both atoms have a similar attraction for electrons. However, if one atom has a much greater attraction for electrons than the other, the electrons are shared unequally, and **polar covalent bonds** result so that one part of the molecule has a positive charge and another part has a negative charge (table 2.2). An example is the water molecule (see perspective 2.2 and figure 2.3).

Table 2.2 Nonpolar and Polar Covalent Bonds

Type of Covalent Bond	Atoms Involved and Charge Distribution	
Nonpolar	C—C C—H H—H	C and H have equal attractions for electrons, so there is an equivalent charge on each atom.
Polar	O—H N—H O—C N—C	The O and N atoms have a stronger attraction for electrons than do C and H, so the O and N have a negative charge.

Figure 2.3
Formation of covalent bonds in a water molecule. Oxygen has a greater attraction for the shared electrons than do the hydrogen atoms. This results in the electrons being closer to the oxygen atom and conferring on it a negative charge. Each of the hydrogen atoms, in turn, has a positive charge.

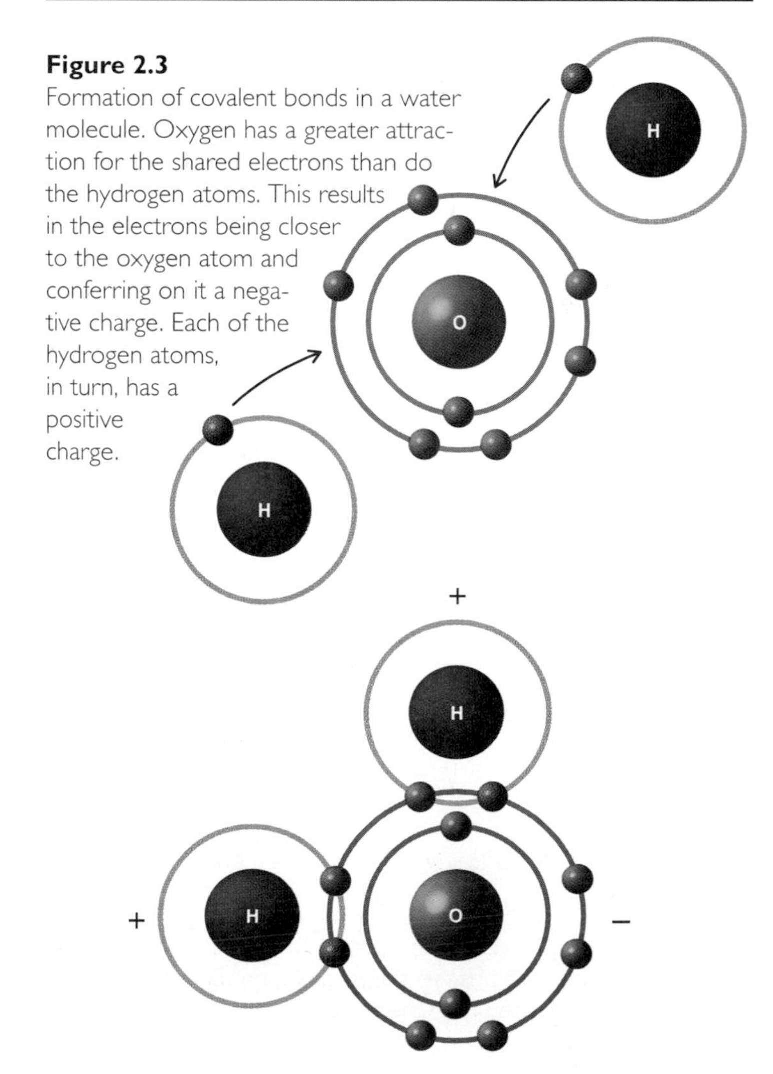

Ionic Bonds Are Formed by the Loss and Gain of Electrons Between Atoms

If electrons from one atom are attracted very strongly by another atom that is nearby, the electrons completely leave the first atom and become a part of the outer electron shell of the second, without any sharing. This bond is termed an **ionic bond** and is an example of extreme polarity. The atom that gains the electrons becomes negatively charged, while the atom that gives up the electrons becomes positively charged. Such charged atoms are termed **ions,** hence the name of the bond. Note that an atom can fill its outer shell by either gaining electrons or by losing electrons (figure 2.4), and in this way becomes negatively or positively charged, respectively. The charged ions, one positive and the other negative, are attracted to one another and this attraction tends to hold the atoms together.

Ionic bonds are the most common type of bonds in inorganic molecules, but they are not very important in organic molecules. Nevertheless, they are important in biology because they are common among the weak forces holding ions, atoms, and molecules together. In water (aqueous) solutions, ionic bonds are about 100 times weaker than covalent bonds because water molecules tend to move between the ions, thereby greatly reducing their attraction for one another (see perspective 2.2). Thus, in aqueous solutions, these bonds are readily broken and so are considered weak bonds. However, in the absence of water, which rarely occurs in biological systems, they are very strong.

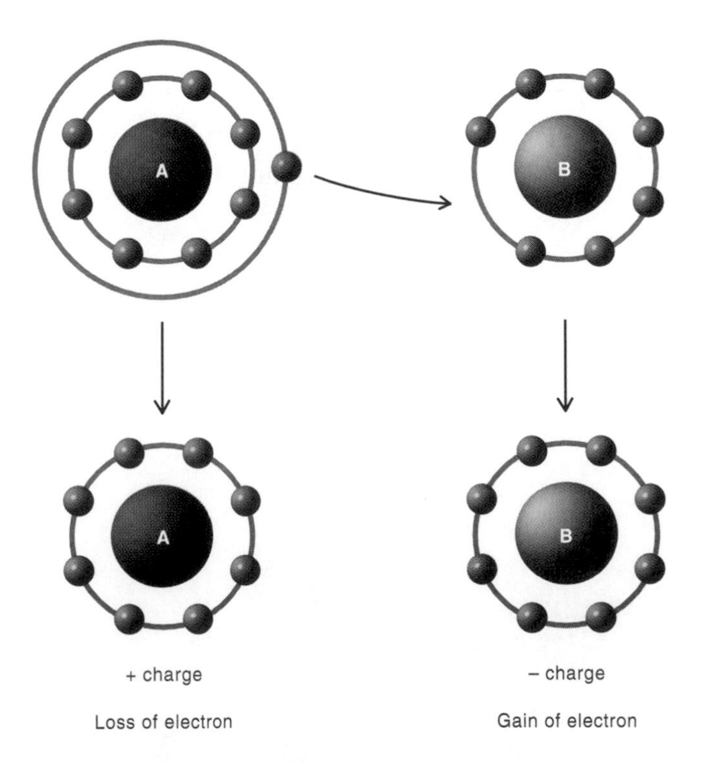

Figure 2.4
Ionic bond. Atom A gives up an electron to atom B. The result is that atom A acquires a positive charge and atom B a negative charge. Both atoms then have their outer electron shells filled, leading to stability.

Hydrogen Bonds Are Weak but Biologically Very Important

As already mentioned, when a molecule contains two atoms that have different degrees of attraction for electrons, the atom that has the greatest attraction will tend to be negatively charged and the other atom will tend to be positively charged. For example, in a water molecule, the O is negatively charged and the H is positively charged (see figure 2.3). Therefore, a hydrogen nucleus in a water molecule is attracted to the other atoms with a negative charge, such as the oxygen atom of an adjacent water molecule (see perspective 2.2). This attraction of a positively charged hydrogen atom to different negatively charged atoms is called a **hydrogen bond.**

Another important atom in biology, the nitrogen atom (N), has a strong attraction for electrons and tends to be negatively charged when covalently bonded to hydrogen. Hydrogen bonding is responsible for some important properties of many molecules of biological importance, such as DNA and proteins. Hydrogen bonds are responsible for holding the two strands of DNA together, which will be considered in chapter 6; their role in protein structure will be discussed later in this chapter. This bonding plays a very important role in the properties of water, the most abundant molecule on earth. Hydrogen bonds differ from covalent bonds in several ways, the most important of which is that hydrogen bonds are much weaker (table 2.3).

If atoms have the opportunity to form either covalent or hydrogen bonds, they will always form the stronger, more stable covalent bonds, thereby following the general rule that molecules will always assume their most stable form. Hydrogen bonds are constantly made and broken at room temperature, indicating that enough energy is produced by the movement of electrons, even at room temperature, to

Table 2.3 Relative Strength of Bonds

Name of Bond	Relative Strength
Covalent Bonds ([—] Strong)	
H—H	1.0
C—C	0.8
C—H	1.0
C—O	0.8
C—N	0.7
Hydrogen Bonds ([· · ·] Weak)	
H· · ·O—O	
H· · ·N—O	0.02–0.1
H· · ·N—N	

Perspective 2.2

Water–The Universal Requirement for Life

The properties of water depend on the properties of its constituent atoms, hydrogen (H) and oxygen (O), as well as on the bonds that join the atoms together. Since the O, covalently bonded to H, attracts electrons more strongly than the H, water is a highly polar molecule, with the O having a negative charge and the H a positive charge (see figure 2.3). This polar nature of water is very important in biology. First, it enables each water molecule to form hydrogen bonds with four other water molecules, leading to a strong attraction between water molecules (figure 1).

The strong attraction requires that a large amount of heat be applied to break the bonds between water molecules. This explains why water is a liquid at room temperature and that a great deal of heat is required to convert it into a gas. This fact is significant because water vapor (gas) is not important in biological systems, whereas the liquid form is very important.

The polar nature of water also accounts for its ability to dissolve a large number of compounds, another feature important for life as we know it. Water-soluble compounds contain atoms with positive or negative charges; in water they ionize, or split, into charged atoms. A positive charge means that the ion has one proton more than it has electrons. A negative charge means that the ion has one more electron than it has protons. For example, NaCl dissolves in water to form Na^+ ions, termed **cations,** because they move towards the cathode in an electrical field, and Cl^- ions, termed **anions,** because they move toward the positively charged anode. In solution, ions such as Na^+ and Cl^- tend to be surrounded by water molecules in such a way that the O^- of HOH forms weak bonds with Na^+, and the H^+ forms weak bonds with Cl^- (figure 2).

This prevents the Na+ and Cl^- from coming together, and it accounts for the solubility of NaCl in water. Nonpolar molecules such as fats and hydrocarbons are insoluble in water, but thus are soluble in uncharged (nonpolar) solvents such as benzene. Indeed, when a nonpolar liquid such as oil is mixed with a polar liquid such as water, the two do not mix. This phenomenon explains why oil and vinegar, which is dilute acetic acid, separate in salad dressing.

Another important feature of aqueous solutions is their level of acidity, or pH. This aspect is discussed in appendix II.

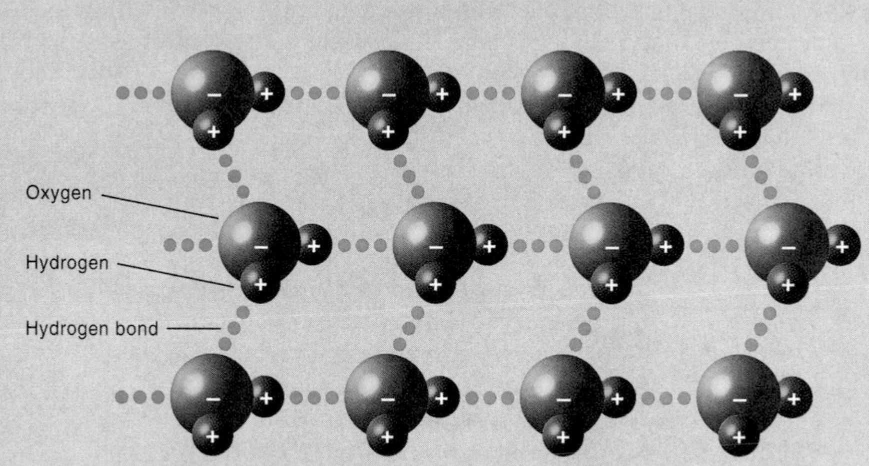

Figure 1
Water molecules held in a latticework, illustrating the hydrogen bonding among molecules.

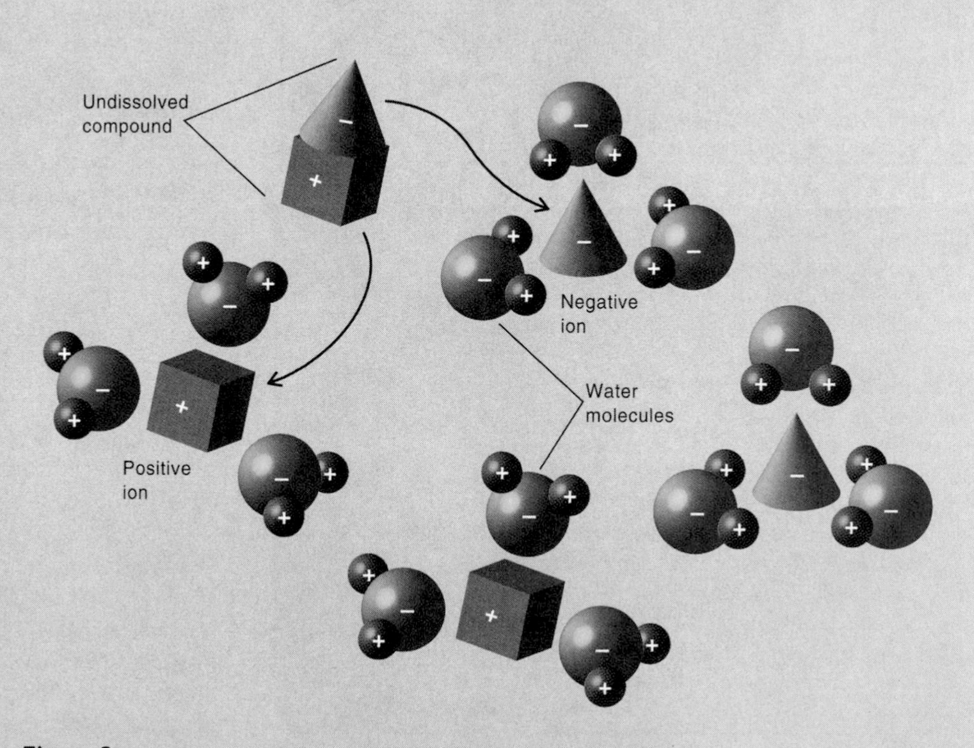

Figure 2
Water dissolving a compound that is made up of positive and negative charges. The charged regions of the water molecules surround the ions and keep them from associating with each other.

break these weak bonds. Since the average lifetime of a single hydrogen bond is only a fraction of a second at room temperature, cells do not need catalysts (enzymes) to speed up the rate at which these bonds are formed and broken.

Although a single hydrogen bond is too weak to bind molecules together, a large number can hold molecules together firmly. Thus, weak bonds are of biological significance only when they occur in large numbers in a given molecule, such as in holding the two strands of the DNA molecule together.

Another important feature of hydrogen bonds is that they are effective only when the surfaces that are being held together are close to each other. Such closeness is possible only when the atoms of two molecules form **complementary structures** such that a positive charge on the surface of one molecule is matched by a negative charge on the other. One way that molecules can get close to one another is through a lock-and-key relationship; that is, a protruding group of one molecule (key) may fit into a cavity (lock) of the other (figure 2.5). If the structures are not complementary, the key will not fit the lock and the two molecules will not get close and be held together. This strict requirement for complementarity between molecules for weak bonding determines which molecules are next to each other in the cell and is the reason that most molecules can bond weakly with only a small number of other molecules.

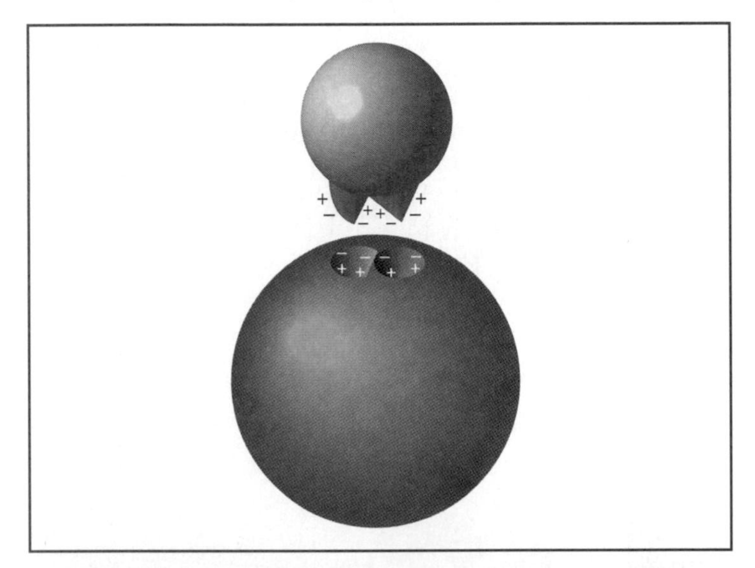

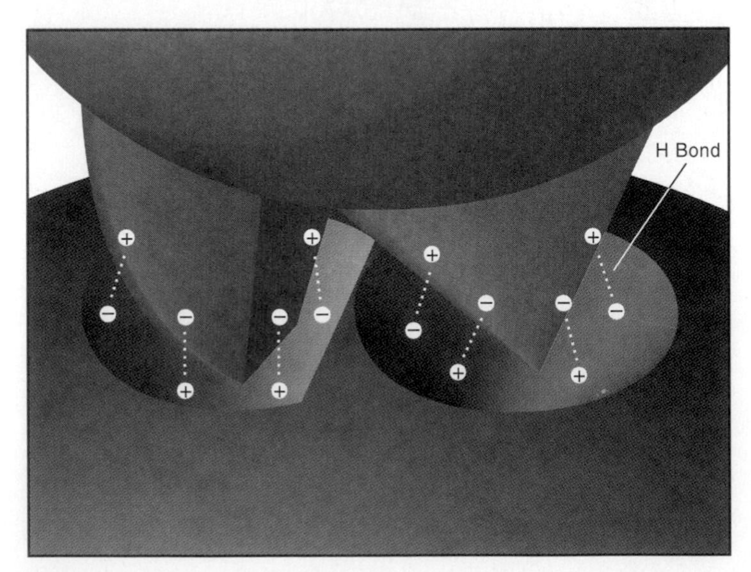

Figure 2.5
Lock-and-key fit of molecules that have complementary structures. The two molecules are complementary because the atoms on the "key" have the shape and charge to interact with atoms in the "lock." The interaction results in the formation of many weak bonds that hold the complementary molecules loosely together.

REVIEW QUESTIONS

1. List the four major elements that make up most of all living matter.
2. Give the major components of atoms and their properties.
3. Compare the relative strengths of covalent, hydrogen, and ionic bonds. Which of these bonds requires enzymes for breakage?
4. Give two examples of hydrogen bonds' importance in macromolecule structure.

Important Molecules of Life

Unquestionably, water is the most important molecule in the world. Over 90% of the weight of any cell is composed of water, and all of the chemical reactions associated with life are based on this compound. Therefore, it is important to understand its most important properties and how it functions in the life of a cell.

Small Molecules in the Cell Include a Variety of Inorganic and Organic Compounds

All cells contain a variety of small organic and inorganic molecules, many of which occur in the form of ions. About 1% of the total dry weight of a bacterial cell is composed of inorganic ions, principally Na^{+1} (sodium),[1] K^{+} (potassium), Mg^{+2} (magnesium), Ca^{+2} (calcium), Fe^{+2} (iron), Cl^{-} (chloride), PO_4^{-3} (phosphate), and SO_4^{-2} (sulfate). Many of the positively charged ions are required in very minute amounts for the functioning of certain enzymes. The negatively charged phosphate ion plays an important role in energy metabolism, discussed in chapter 5.

In biology, the most important molecules are **organic molecules,** defined as those that contain at least one carbon atom covalently bonded to another carbon or hydrogen atom. The biologically important small organic molecules are mainly compounds that are being metabolized[2] or have accumulated as a result of metabolism and those that serve

[1]Na^{+1} (sodium). The superscript number is the difference in the number of electrons and protons in the atom. If the number is positive, there are more protons; if the number is negative, there are more electrons.

[2]Metabolize—to break down or degrade through a series of enzymatic reactions

as the building blocks of various macromolecules. These organic molecules can be grouped into categories based upon distinctive groups of atoms they contain (table 2.4).

As table 2.4 indicates, the most important covalent bonds in these molecules are between atoms of carbon and carbon (C—C), carbon and hydrogen (C—H), carbon and oxygen (C—O), carbon and nitrogen (C—N), oxygen and hydrogen (O—H), and nitrogen and hydrogen (N—H). Some of the distinctive groups are the hydroxyl, amino, and carboxyl groups.

Table 2.4 Classification of Small Molecules by Functional Groups

Group	Name of Group	Example of Group in Biologically Important Molecule	Name of Compound	Class of Molecule Found in
H | —C—H | H	Methyl	H | H—C—H | H	Methane	
—O—H	Hydroxyl	H H | | H—C—C—O—H | | H H	Ethanol (ethyl alcohol)	Alcohols
O ‖ —C—O—H	Carboxyl	H O | ‖ H—C—C—O—H | H	Acetic acid	Acids
H / —N \\ H	Amino	H H \\ / N O | ‖ H—C—C—O—H	Glycine	Amines and amino acids
O ‖ —C—	Keto	H O H | ‖ | H—C—C—C—H | | H H	Acetone	Ketones
O ‖ —C—H	Aldehyde	H O | ‖ H—C—C—H | H	Acetaldehyde	Aldehydes
—S—H	Sulfhydryl	H H \\ / H N O | | ‖ H—S—C—C—C—O—H | H	Cysteine	A few amino acids
O ‖ —C— and O ‖ —C—O—H	Ketone and Carboxyl	H O O | ‖ ‖ H—C—C—C—OH | H	Pyruvic acid	Keto acids
—O—H and O ‖ —C—O—H	Hydroxyl and Carboxyl	H | H O O | | ‖ H—C—C—C—OH | | H H	Lactic acid	Hydroxy acids

Key functional groups are shaded.

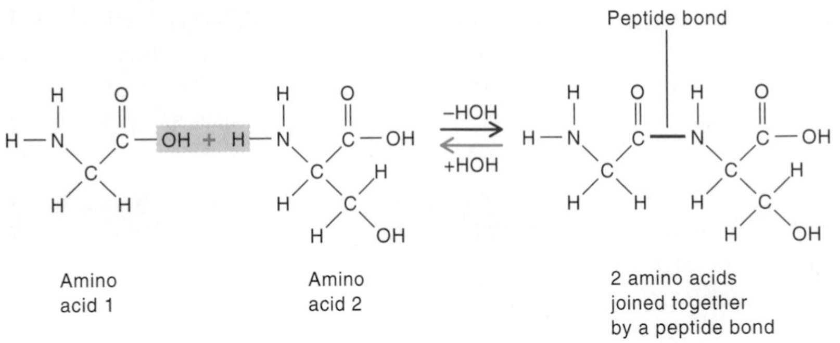

Peptide bond

Amino acid 1 — Amino acid 2 — 2 amino acids joined together by a peptide bond

Figure 2.6
Formation of a covalent bond (peptide) by dehydration synthesis between two amino acids.

Some molecules of biological importance contain several distinctive groups. For example, every amino acid has both an amino (NH_2) and a carboxyl (COOH) or acid group, hence the name *amino acid.* If it only had a carboxyl group, it would be an acid. An acid that contains a keto group (C=O) is often termed a *keto acid.* An example of a keto acid is pyruvic acid, a compound important in energy metabolism (see table 2.4). Unfortunately, many other compounds are commonly designated by trivial names that provide no clue as to what functional groups are in the molecule. For example, there is no hint that lactic acid is a hydroxyl acid. Moreover, some of these molecules may be referred to by more than one name. As an example, inside the cell, organic acids exist as ions, rather than as acids themselves. Accordingly, *pyruvate,* the ionized form of pyruvic acid, is often used interchangeably with *pyruvic acid,* and *acetate* is commonly used in place of *acetic acid.*

All Very Large Molecules in the Cell (Macromolecules) Consist of Repeating Subunits Called *Monomers*

The three major classes of biologically important **macromolecules** are **proteins, polysaccharides,** and **nucleic acids.** (The lipids are also large but not large enough to qualify as macromolecules. Nevertheless, because of their biological importance, they are also briefly discussed in this chapter.) We will discuss first the general features of macromolecules and then specific macromolecules.

Macromolecules are characterized by their large size. Although they consist of at least several thousand atoms each, their structure and synthesis have several features in common. Understanding these common features makes the structure of these large molecules surprisingly simple to learn and understand.

Structural Similarities

All macromolecules are **polymers,** large molecules formed by joining together the same small molecules, the **subunits.** Each class of macromolecule is composed of different subunits, which have similar structures. For example, amino acids are the subunits of proteins. Although 20 different amino acids exist, each one has a **carboxyl group** (—COOH) and an **amino group** (—NH_2) bonded to carbon atom 2.

Biosynthetic Similarities

Macromolecule synthesis involves two steps: first the synthesis of subunits and then the joining of them together, one by one. The synthesis of the various subunits involves almost 100 different chemical reactions and is beyond the scope of this book. However, an overview of their synthesis is given in chapter 5.

The overall process of joining two subunits, such as two amino acids in a protein molecule, involves a chemical reaction in which H_2O (HOH) is removed, termed a **dehydration synthesis** (figure 2.6). When the macromolecule is broken down into its subunits, the reverse reaction occurs, and a water molecule (H^+ and OH^-) is added back, a reaction termed a **hydrolytic reaction,** or **hydrolysis** (figure 2.6).

This type of reversible reaction involving the removal and addition of HOH molecules is common to the synthesis and degradation of a large number of molecules and requires the action of specific enzymes. We will now discuss each of the three major macromolecules: proteins, polysaccharides, and nucleic acids.

Amino Acids Joined Together Make Up Proteins

Twenty amino acids, all of which share certain features, are found in proteins. These features are illustrated in figure 2.7, which shows a generalized amino acid.

These shared features are (1) a carboxyl (acid [—COOH]) group at one end and (2) an amino (basic [—NH_2]) group bonded to the same carbon atom. This carbon atom is also bonded to a side chain or backbone (which is shaded in figure 2.8), which is characteristic for each amino acid. The side chain labeled *R* may be a very simple group, such as —H or —CH_3; a longer chain of carbon atoms; or one of various kinds of ring structures, such as the benzene ring found in tyrosine and phenylalanine. The amino acids can be subdivided into several different groups (see figure 2.8). Those grouped together have similar side chains that give similar properties to the amino acids.

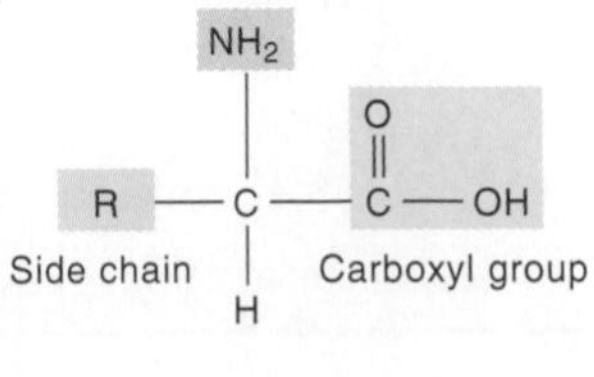

Amino group — Side chain — Carboxyl group

Figure 2.7
Generalized amino acid, illustrating the three groups that all amino acids possess. The R side chain differs with each amino acid and determines its properties.

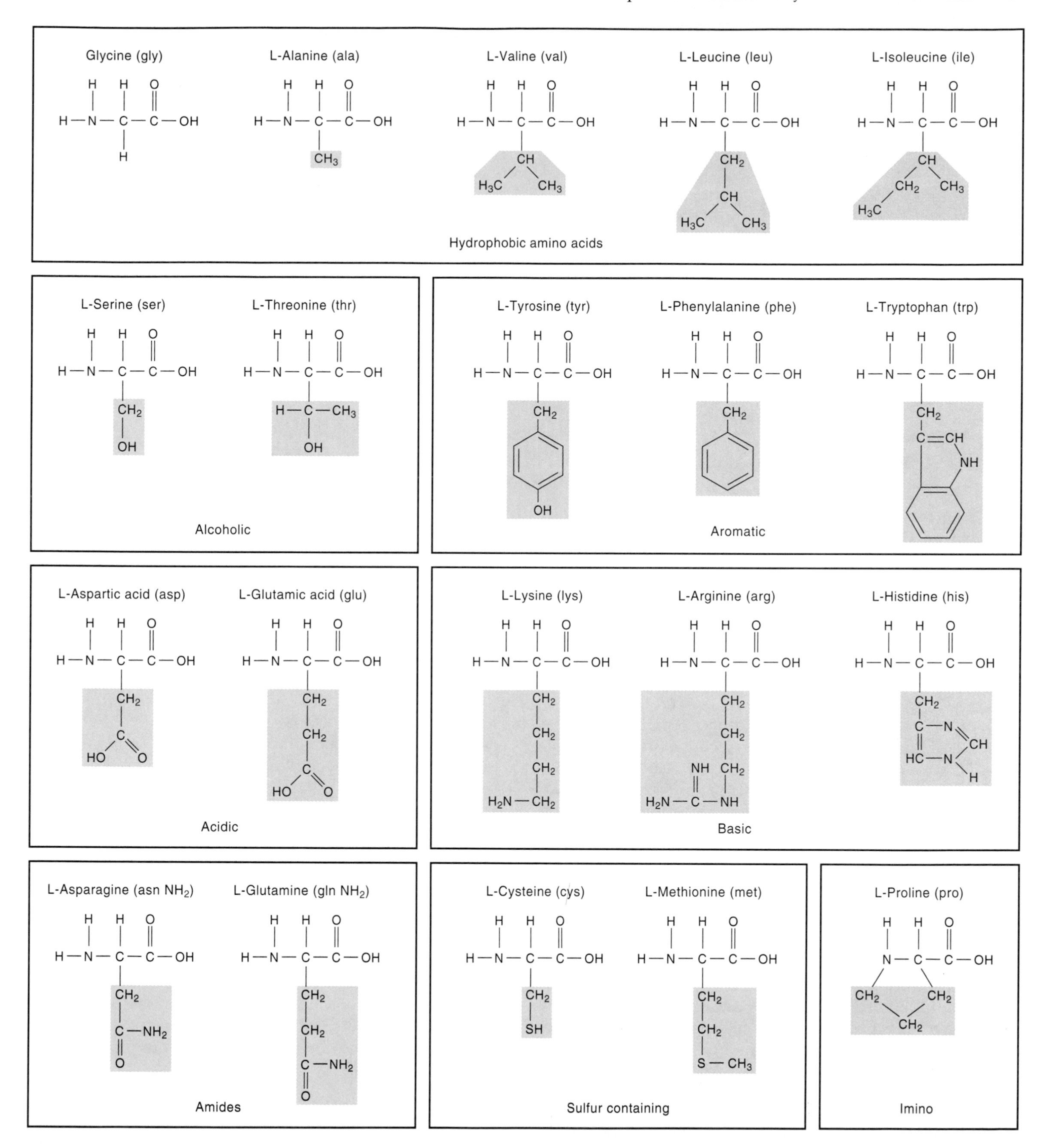

Figure 2.8
Amino acids. All amino acids have one feature in common: a carboxyl group and an amino group on the carbon atom next to the carboxyl group. Although the remainder of the molecule differs for each amino acid, the acids can be grouped because of certain features they have in common. The R group that gives the amino acid its distinctive feature is shaded.

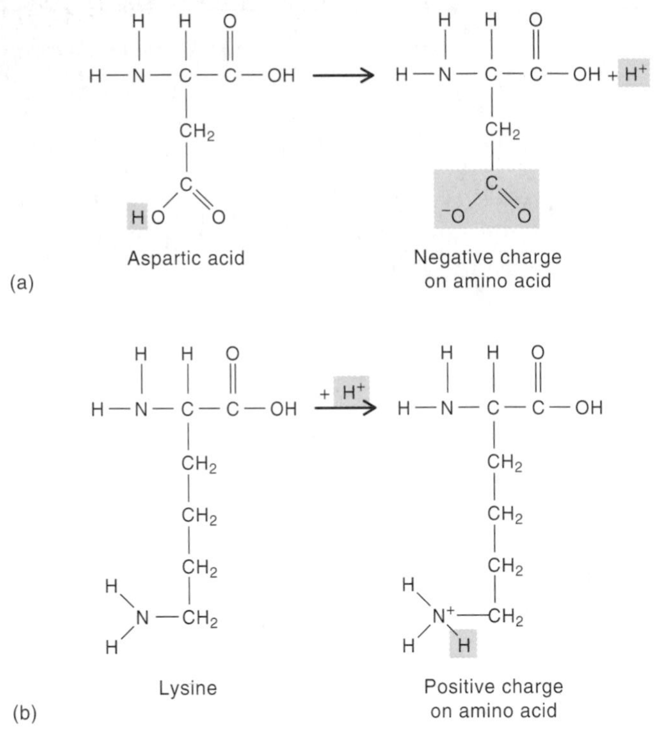

Figure 2.9
(*a*) Acidic (aspartic acid) and (*b*) basic (lysine) amino acids. At a neutral pH (7), aspartic acid is negatively charged and lysine is positively charged because it has one less proton or one additional proton, respectively.

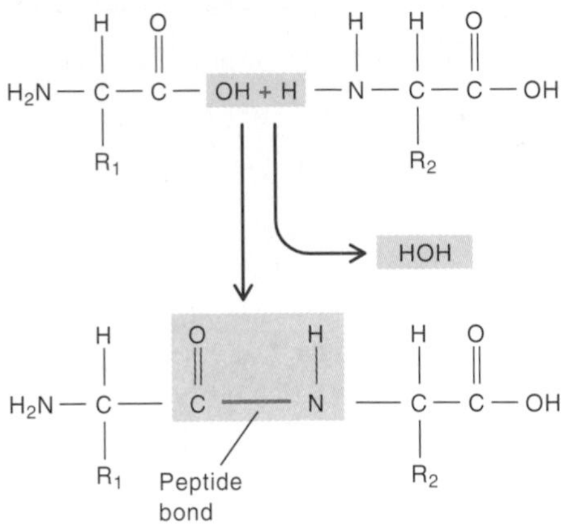

Figure 2.10
Peptide bond formation by dehydration synthesis.

If these side chains contain carboxyl or amino groups, they readily form ions, which give negative (COO^-) or positive (NH_3^+) electrical charges to the amino acid (figure 2.9). Such amino acids are termed *acidic* or *basic amino acids,* respectively. They are readily soluble in water because the charged water molecules can hydrogen bond to the positive or negative charges of the amino acids. Thus, they are termed *hydrophilic* (water loving). Those that contain many methyl (CH_3) groups are uncharged and therefore do not interact with water molecules. They are thus less soluble in water and termed *hydrophobic* (water fearing). Other amino acids have distinctive chemical structures and can be grouped according to these structures.

All amino acids, except glycine, can exist in either a left-handed or a right-handed isomer known as the *L-form* or *D-form,* respectively (see figure 2.1). Advancing our understanding of these different forms was the first of Pasteur's many great contributions to science. Only L-amino acids occur in proteins, and they are designated the **natural amino acids.** D-amino acids, the **unnatural amino acids,** are rare in nature and are found in only a few materials mostly associated with bacteria—primarily cell walls and antibiotics.

Peptide Bonds and Their Synthesis

The 20 amino acids in proteins are held together by **peptide bonds,** a unique type of covalent linkage formed when the carboxyl group of one amino acid reacts with the amino group of another amino acid, with the release of water (dehydration synthesis [figure 2.10]).

The chain of amino acids formed when a large number of amino acids are joined by peptide bonds is called a **polypeptide chain** (figure 2.11). One end of the chain has a free amino (—NH_2) group, which is termed the *N terminal,* or *amino terminal,* end. The other end has a free carboxyl (—COOH) group, which is termed the *C terminal,* or *carboxy terminal,* end. Some proteins consist of a single polypeptide chain, but some consist of one or more chains joined together by weak bonds. The total molecule comprising one or more polypeptide chains is the protein. The individual polypeptide chains generally do not have biological activity by themselves. Proteins vary in size, but an average-size protein consists of about 400 amino acids.

Levels of Protein Structure

Three features characterize each polypeptide chain. The first and most important is its composition—the number and sequence of the amino acids that comprise it. The second characteristic feature is its final shape, and the third is its interactions with the same or different polypeptide chains. Not all polypeptide chains interact with other polypeptide chains.

Composition The sequence of amino acids in a protein is termed its **primary structure** (figure 2.12 *a*). The substitution of even one amino acid for another in a protein of a thousand amino acids may completely destroy the protein's function, as will be discussed in chapter 7.

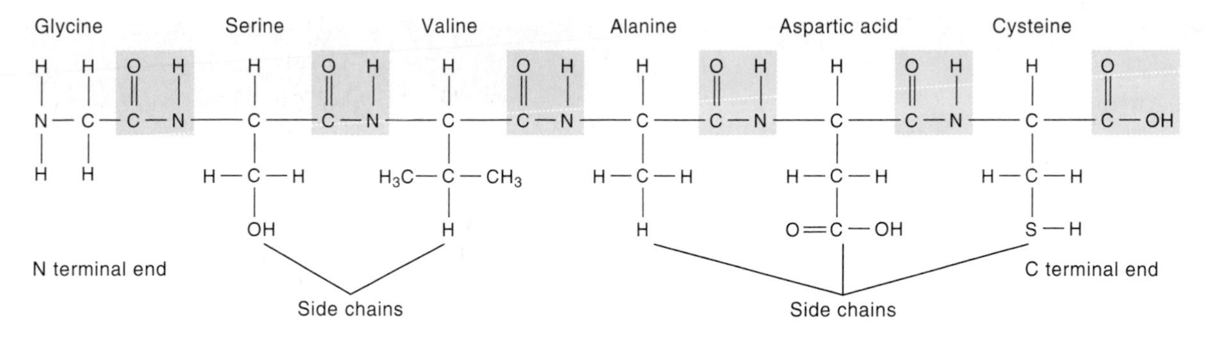

Figure 2.11
Polypeptide chain. The shaded areas represent the atoms involved in forming the peptide bonds. This chain is much shorter than would be found in enzymes.

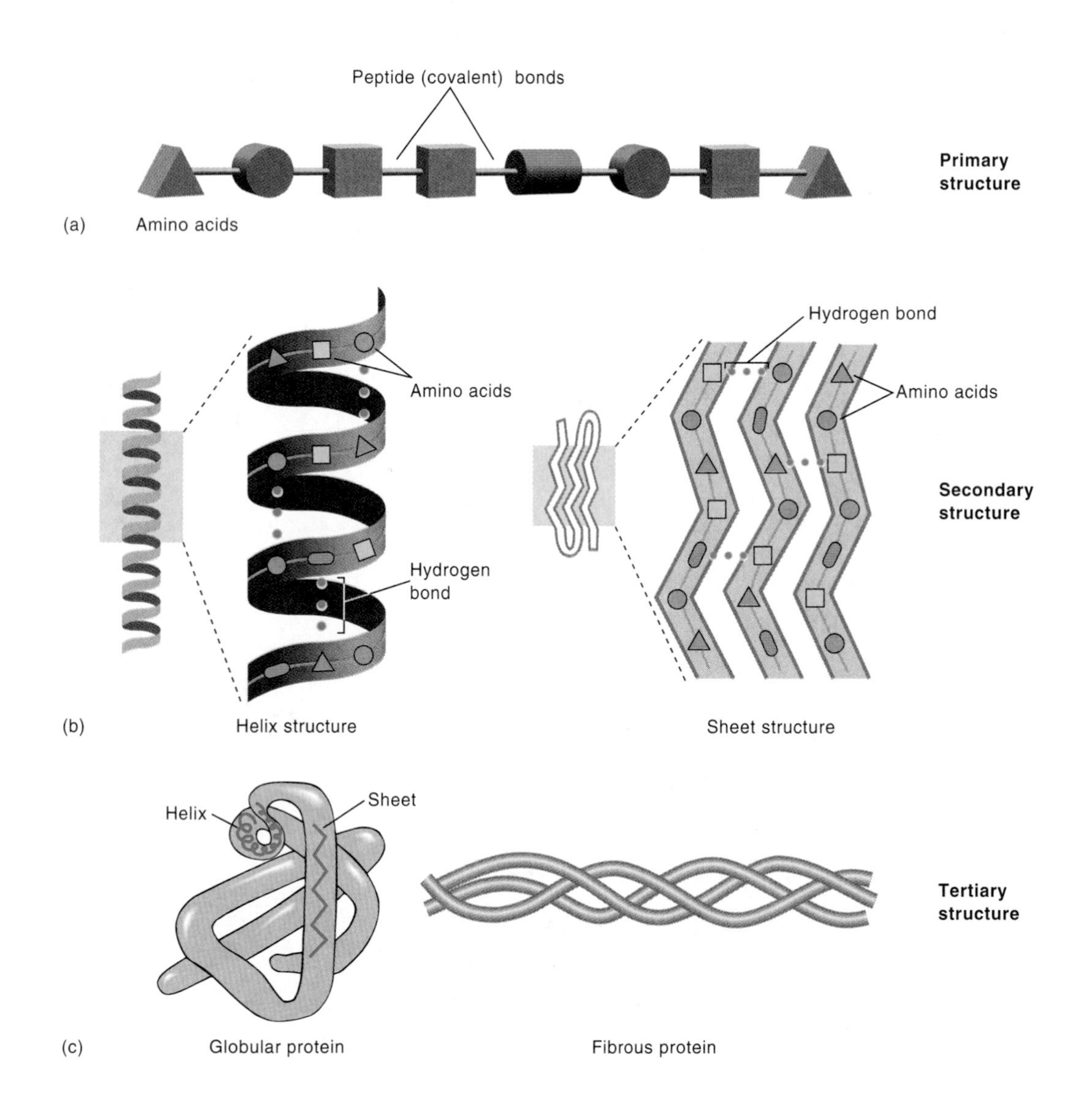

Figure 2.12
Protein structures. (*a*) The primary structure is determined by the amino acid composition. (*b*) The secondary structure results from folding of the various parts of the protein into two major patterns, helices and pleats. Various bonds, both weak and covalent, are important in determining the folding patterns. (*c*)The tertiary structure is defined by the final overall shape of the molecule. Most proteins are globular.

Perspective 2.3

Alzheimer's Disease—a Result of Protein Misfolding?

The brains of people with Alzheimer's disease are filled with abnormal structures called *plaques,* which consist of deposits of fibrils of a protein termed ß *amyloid.* ß amyloid is cut out of a larger protein, the amyloid precursor protein. The formation of ß amyloid from precursor protein occurs in all people, not only those with Alzheimer's disease. However, the fibril protein deposits resulting in plaque formation do not form in normal people. Why do fibrils form in those destined to suffer from Alzheimer's disease? There is evidence that the amount of ß amyloid produced in some people is much higher than that produced in other people, perhaps because of changes in the amyloid precursor protein. The increased production of the protein may result in fibril formation because the folding of the protein into its tertiary structure does not occur properly as a result of its higher concentration. Rather, the high concentration of the protein results in the proteins grabbing onto each other and aggregating into the fibrils rather than folding into their proper configuration (figure 3).

Other proteins besides ß amyloids are also apparently involved in deposition of ß amyloid. Mutations in the genes of two proteins called *presenilins* are found in most people suffering from Alzheimer's disease. These proteins may play a role in how the precursor protein is cut to yield the ß amyloid protein. Proteins of a certain length appear to be especially dangerous in forming plaques since they are the most commonly occurring proteins in the plaques of Alzheimer's patients. People with mutations in the genes which code for presenilins contain the ß amyloid fibrils of this length, suggesting that the mutations results in the formation of the altered size ß amyloid protein.

There exist about a dozen different diseases that are characterized by the formation of protein deposits in the brain, including the infectious neurodegenerative diseases caused by prions. Can these diseases also have their basis in abnormal protein folding? The understanding of how proteins fold is an important area of study not only for protein chemists but also for medical doctors interested in the prevention of a number of very puzzling diseases whose causes remain a mystery.

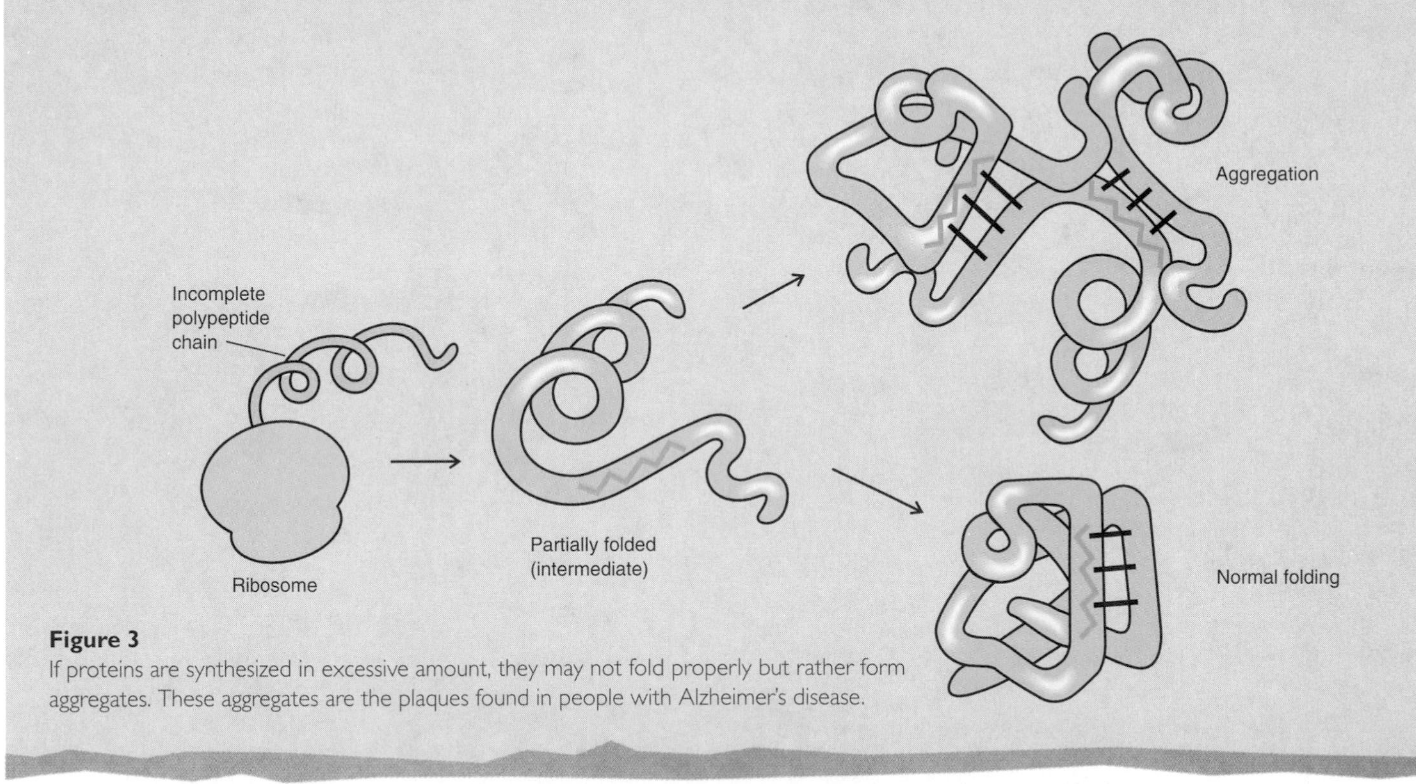

Figure 3
If proteins are synthesized in excessive amount, they may not fold properly but rather form aggregates. These aggregates are the plaques found in people with Alzheimer's disease.

Protein Shapes Once the protein is made, it assumes a very specific three-dimensional shape as a result of a variety of bonds that form between many different amino acids. Within any single protein, a string of certain amino acids will often form a helix, and another group of amino acids will form sheets (figure 2.12 *b*). These two major structures are called the **secondary structure** of the protein. They vary in number and position within the molecule, and these variations determine the final three-dimensional shape, the tertiary structure.The conformation of the protein depends on a variety of different bonds between the various amino acids. These include hydrogen bonds between H and O atoms, as well as other weak and strong covalent bonds.

The combinations of strong covalent bonds and weak bonds between the amino acids that compose the protein

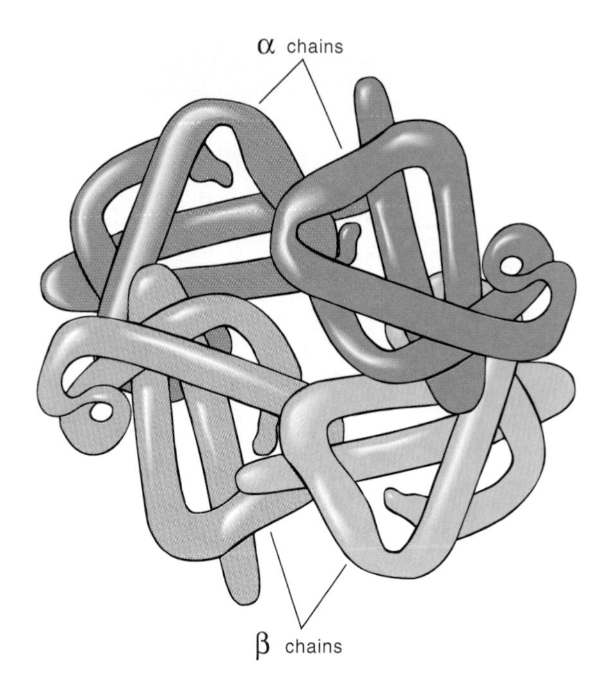

Figure 2.13
Quaternary structure of hemoglobin, a protein containing four polypeptide chains of two identical subunits: α chains (top) and β chains (bottom). Different colors are used to distinguish the four chains.

(primary structure) result in the final shape of the protein. Most proteins are essentially spherical, or **globular,** but some are long fibers, **fibrous** (figure 2.12 *c*). This is the **tertiary structure** of the protein.

Interaction with Other Polypeptide Chains Some proteins consist of a single polypeptide chain. However, others consist of more than one chain held together by weak bonds between amino acids of the different chains. These proteins may be composed of two or more of the same chains or different chains. Each polypeptide chain has its own specific shape, its tertiary structure. When it interacts with another polypeptide chain, the two chains also assume a specific shape. This resulting shape is termed the **quaternary structure** of the protein (figure 2.13). Of course, only proteins that consist of more than one polypeptide chain have a quaternary structure.

Sometimes different proteins, each having different functions, associate with one another to make even larger structures termed **multiprotein complexes,** the next level of organization above the quaternary level.

Protein Denaturation The three-dimensional shape of the molecule is flexible, changing if the temperature, acidity of the solution, or the binding of ions to the protein changes. The protein must have a specific shape in order to carry out its function. This explains why enzymes, the protein catalysts, have an optimum pH and temperature for their functioning and may require inorganic ions to

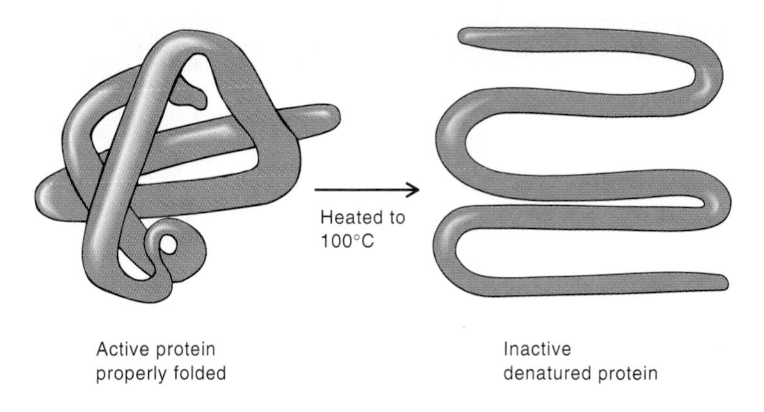

Figure 2.14
Denaturation of a protein.

function. If the temperature is raised too high, or the solution becomes very acidic or basic, numerous weak bonds will be broken and the molecule will lose its proper shape, in a process termed **denaturation.** Boiling an egg denatures the egg white protein (albumen), which results in a readily observable change (figure 2.14). This change is not reversible since cooling a boiled egg does not restore the protein to its original shape. Very high temperatures (fever) in the human body can also prevent proteins from functioning properly, which has serious consequences.

Substituted Proteins

Many proteins of biological importance contain molecules that are covalently bound to the amino acids. These proteins are called **substituted proteins.** They include the **glycoproteins,** in which sugar molecules (see next section) are joined to the protein, and **lipoproteins,** in which lipids are covalently joined to the protein. Both kinds of proteins are found in the membranes of cells.

Polysaccharides Are Polymers of Monosaccharide Subunits

Carbohydrates are compounds containing principally carbon, hydrogen, and oxygen atoms in a ratio of approximately 1:2:1. The chief distinguishing feature of all carbohydrates is that they contain a large number of alcohol groups (—OH) in which the C is also bonded to an H atom to form H—C—OH, giving them their distinctive ratio of carbon, hydrogen, and oxygen atoms. **Polysaccharides** are the macromolecular form of carbohydrates. They are linear or branched polymers of their subunits, the **monosaccharides.** The term **sugar** is often applied to monosaccharides and **disaccharides,** which are two monosaccharides joined together by covalent bonds. Carbohydrates also usually have an aldehyde group and, less commonly, a keto group (see table 2.4). Some carbohydrates, the amino sugars, also contain an amino group.

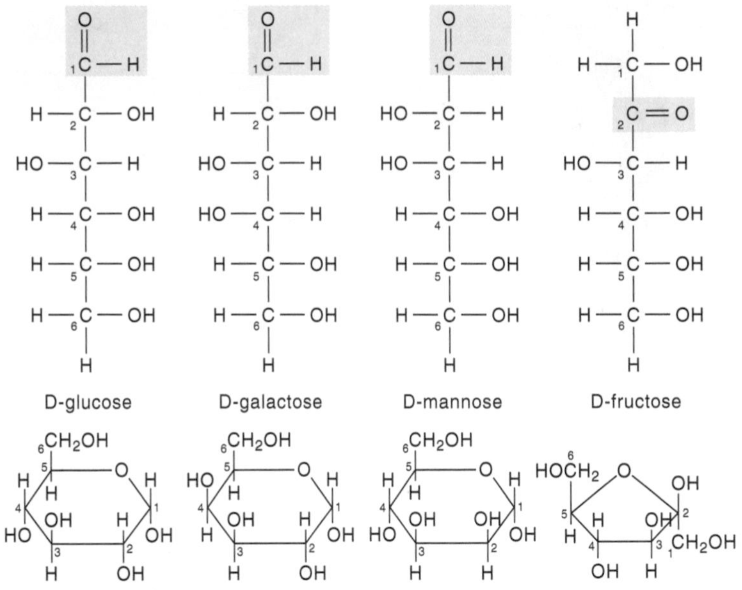

Figure 2.15
Formulas of some common sugars represented in two different ways: in a linear structure and in a ring structure. The ring structures result from the reaction of the —OH group on carbon atom 5 with the aldehyde group on carbon atom 1, resulting in an O bridge between carbon atoms 1 and 5, except in the case of fructose. Note that fructose is unusual in that is has a keto group (C═O) on carbon atom 2 of the linear structure. The aldehyde and keto groups are shaded.

Monosaccharides

Monosaccharides are classified by the number of carbon atoms they contain. In nature, the most common are the 5-carbon (pentose) and the 6-carbon (hexose) sugars. The 5-carbon sugars, ribose and deoxyribose, are the sugars in nucleic acids. Note that these sugars are identical except that deoxyribose has one less molecule of oxygen than does ribose (the prefix *de* is Latin for "away from"). Thus, deoxyribose is ribose "away from" oxygen.

Common hexoses include glucose, galactose, and mannose. They all contain the same atoms but differ in the arrangements of the —H and —OH groups relative to the carbon atoms (figure 2.15). For example, glucose and galactose are identical structures except for the arrangement of the —H and —OH groups attached to carbon 4. Mannose and glucose differ with respect to the arrangement of the —H and —OH groups joined to carbon 2. These small changes result in three distinct sugars with different properties. For example, glucose has a sweet taste, but it is not as sweet as the same weight of fructose, the major sugar in many carbonated soft drinks. Mannose also is sweet, but has a bitter aftertaste. Glucose and mannose contain aldehyde groups; fructose has a keto group. It is interesting to speculate how these small chemical differences give such different tastes.

Sugars, like amino acids, have carbon atoms bonded to four different groups and so can also exist in two forms, D and L, which are mirror images of one another (see figure 2.1). By convention, if the —OH group on carbon 5 in hexoses (carbon 4 in pentoses) is written on the left, the sugar is L; if on the right, it is a member of the D series. Most monosaccharides in living organisms are of the D-configuration, which is the opposite situation observed with amino acids.

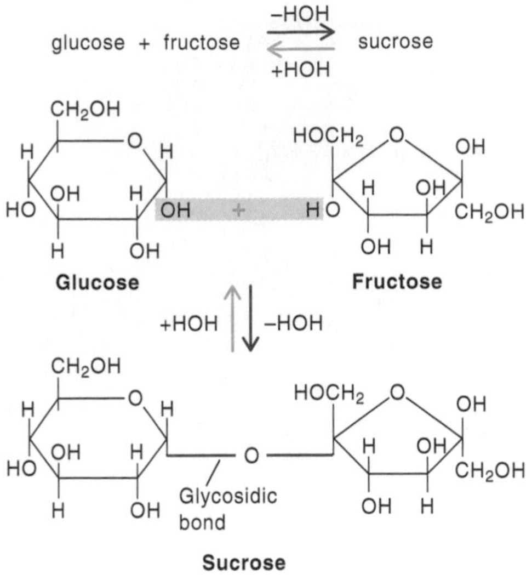

Figure 2.16
Formation of sucrose from glucose and fructose by dehydration synthesis.

Disaccharides

The two most common disaccharides are the milk sugar, **lactose,** and the common table sugar, **sucrose.**

Lactose consists of glucose and galactose, while sucrose, which comes from sugar cane or sugar beets, is composed of glucose and fructose. The linkage joining them is termed a **glycosidic linkage.** It results from the reaction between hydroxyl groups of two sugar units, with the loss of a molecule of water (figure 2.16). Note that this is another example of dehydration synthesis, similar to the joining of two amino acids.

Polysaccharides

Polysaccharides, which are found in many different places in nature, serve different functions. Some serve as the storage forms for carbon and energy; others are parts of the cells' structures. **Cellulose,** a polymer of glucose subunits, is the principal constituent of plant cell walls, and it is also synthesized by some bacteria. It is the most abundant organic compound in the world. **Glycogen,** a carbohydrate storage product of animals and some bacteria, and **dextran,** which is also synthesized by bacteria, resemble cellulose in some ways (figure 2.17).

Although all three are composed of glucose subunits, they differ from one another in many important ways, which include (1) the size of the polymer, (2) the degree of chain branching, (3) the particular carbon atoms involved

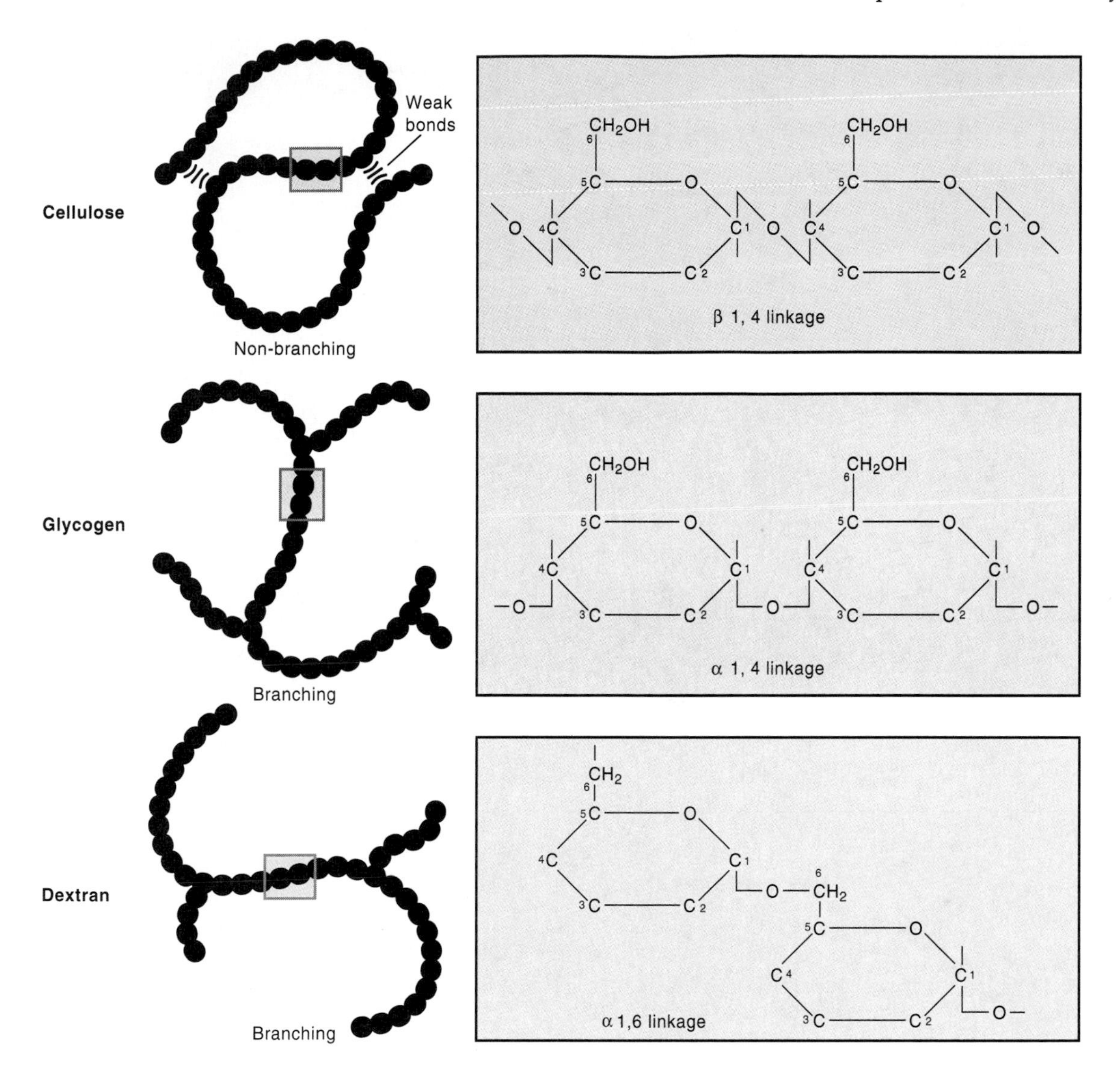

Figure 2.17
Structure of important polysaccharides. The three molecules shown consist of the same subunit, glucose, but they are distinctly different molecules because of differences in the atoms that join the molecules together, the degree of branching, and the bonds involved in branching. Hydrogen bonds are responsible for some of the three-dimensional shapes of some of the polymers.

in the covalent bond between two sugar molecules, and (4) the orientation of the bond between the sugar molecules—one orientation is termed *alpha* (α) and the other *beta* (β). Note that, unlike proteins and nucleic acids, polysaccharides can have side chains of monosaccharides branching from the main linear chain. Thus, it is possible to have a large variety of polysaccharides made up of the same subunits. Polysaccharides can also contain a variety of monosaccharide subunits in the same molecule.

Nucleic Acids Are Polymers of Nucleotide Subunits

Nucleic acids include deoxyribonucleic acid (DNA) and ribonucleic acid (RNA). These two nucleic acids have some features in common, but they differ in other respects. The major similarity is that both are composed of subunits called **nucleotides.**

Deoxyribonucleic Acid (DNA)

The nucleotides of DNA are composed of three units: a nitrogen-containing ring compound (nitrogen base), which is covalently bonded to a 5-carbon sugar molecule (deoxyribose), which is bonded to a phosphate molecule (figures 2.18 and 2.19).

Because of the sugar, the subunit of DNA is termed **deoxyribonucleotide,** although it is more frequently referred to simply as *nucleotide*. The four different nitrogen-containing bases can be divided into two groups according to their structures: two **purines,** adenine and guanine, and two **pyrimidines,** cytosine and thymine (see figures 2.18 and 2.19).

The nucleotide subunits are joined by a covalent bond that links the phosphate of one nucleotide of one subunit to the sugar of the adjacent nucleotide (figure 2.20). Thus, the phosphate is a bridge that joins the 3-carbon atom of one

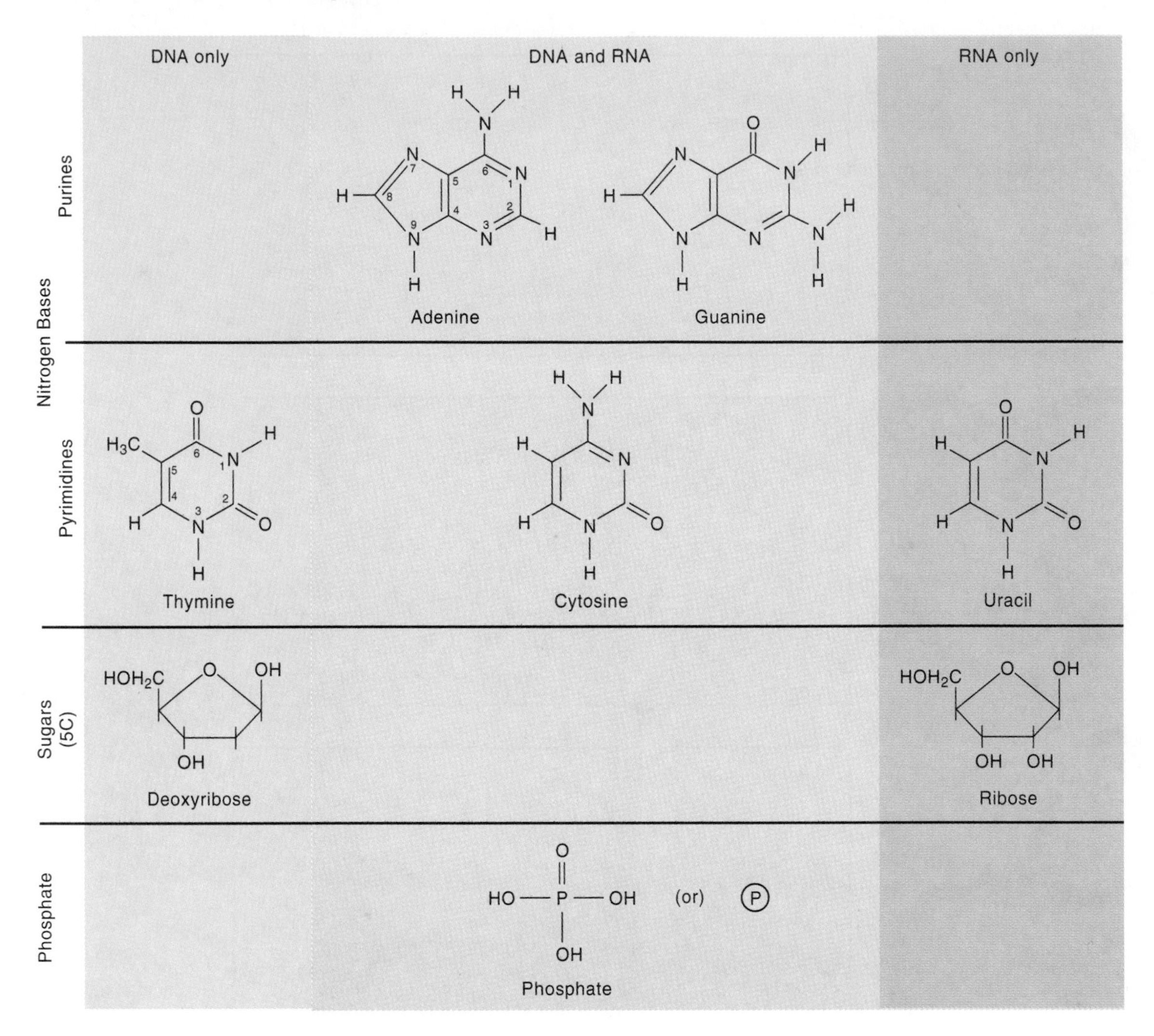

Figure 2.18
Components of RNA and DNA. Carbon atoms in the ring structures are not shown but are represented by numbers in a purine and a pyrimidine according to a convention for numbering the rings.

sugar to the 5-carbon atom of the other, resulting in a backbone composed of alternating sugar and phosphate molecules. Water is removed in the reaction that joins the sugar and phosphate molecules in an **ester linkage,** which is the name of the covalent bond formed between an alcohol and an acid (see figure 2.20).

The DNA of a typical bacterium is a single molecule composed of 30 million nucleotides arranged in the form of a **double-stranded helix** (figure 2.21). The cell carries its genetic information in its DNA, which is coded in the order of sequence of its purines and pyrimidines. This code is then converted into a specific arrangement of amino acids in the protein molecules of the cell. The details of this process and of DNA structure are covered in chapter 6.

Ribonucleic Acid (RNA)

The structure of RNA is similar to but differs from DNA in several ways. First, RNA contains the pyrimidine uracil in place of thymine and the sugar ribose in place of deoxyribose (see figure 2.18). Thus, the subunit of RNA is a ribonucleotide, although this term, like deoxyribonucleotide, is also frequently shortened to *nucleotide.* Also, whereas DNA is a long, double-stranded helix, RNA is considerably shorter and occurs as a chain of ribonucleotides that may be straight or in loops. Three types of RNA are found in the cell, all of which function in converting the information coded in DNA into cellular components (see chapter 6).

The most important properties of all four groups of macromolecules, proteins, polysaccharides, and DNA and RNA are summarized in table 2.5.

Lipids Are Heterogeneous in Structure but All Are Insoluble in Water

Lipids represent a very heterogeneous group of biologically important molecules. The one feature they have in common is that they are only slightly soluble in water but are very soluble in most organic solvents such as ether, benzene, and chloroform. The difference in solubility is due to

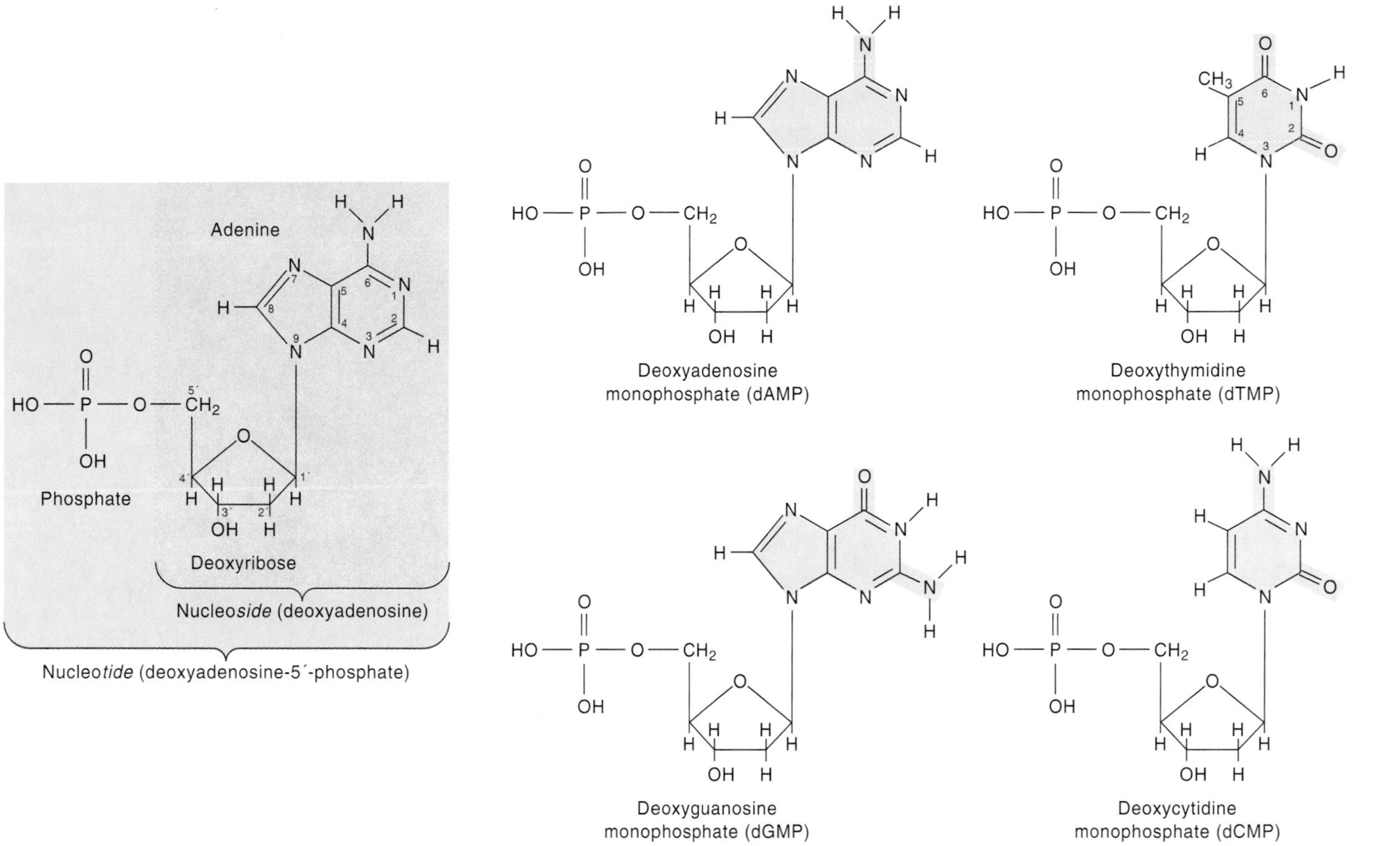

Adenine

Phosphate

Deoxyribose

Nucleo*side* (deoxyadenosine)

Nucleo*tide* (deoxyadenosine-5′-phosphate)

Deoxyadenosine monophosphate (dAMP)

Deoxythymidine monophosphate (dTMP)

Deoxyguanosine monophosphate (dGMP)

Deoxycytidine monophosphate (dCMP)

Figure 2.19
Components of DNA, the mononucleotides. Four mononucleotides are present in DNA and four in RNA. The nucleotides of RNA have the same structure as those in DNA except that, in RNA, ribose replaces deoxyribose and uracil replaces thymine. The mononucleotides in DNA are more properly termed *deoxyribonucleotides*.

Table 2.5 Structure and Function of Macromolecules

Name	Subunit	Bond Joining Subunits	Atoms in Bond	Some Functions of Macromolecules
Protein	Amino acid	Peptide	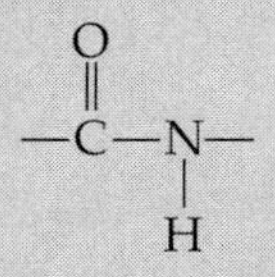 $-\mathrm{C}(=\mathrm{O})-\mathrm{N}(\mathrm{H})-$	Catalysts; structural portion of cell organelles
Ribonucleic acid	Nucleotide	Diester	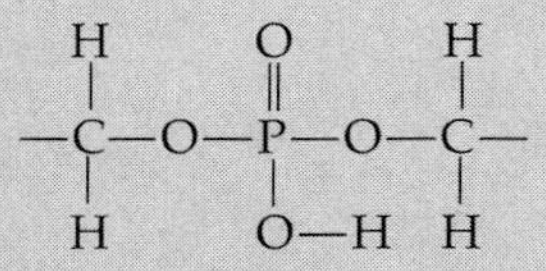 $-\mathrm{CH_2}-\mathrm{O}-\mathrm{P}(=\mathrm{O})(\mathrm{O{-}H})-\mathrm{O}-\mathrm{CH_2}-$	Various roles in protein synthesis
Deoxyribonucleic acid	Deoxynucleotide	Diester	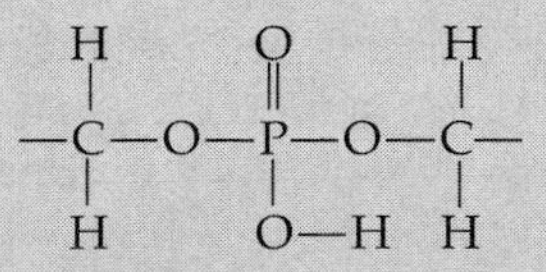 $-\mathrm{CH_2}-\mathrm{O}-\mathrm{P}(=\mathrm{O})(\mathrm{O{-}H})-\mathrm{O}-\mathrm{CH_2}-$	Carrier of genetic information
Polysaccharide	Monosaccharide	Glycosidic	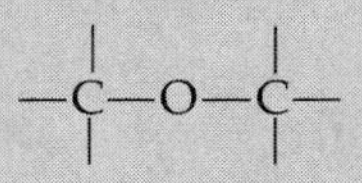 $-\mathrm{C}-\mathrm{O}-\mathrm{C}-$	Structural component of plant cell wall; storage products

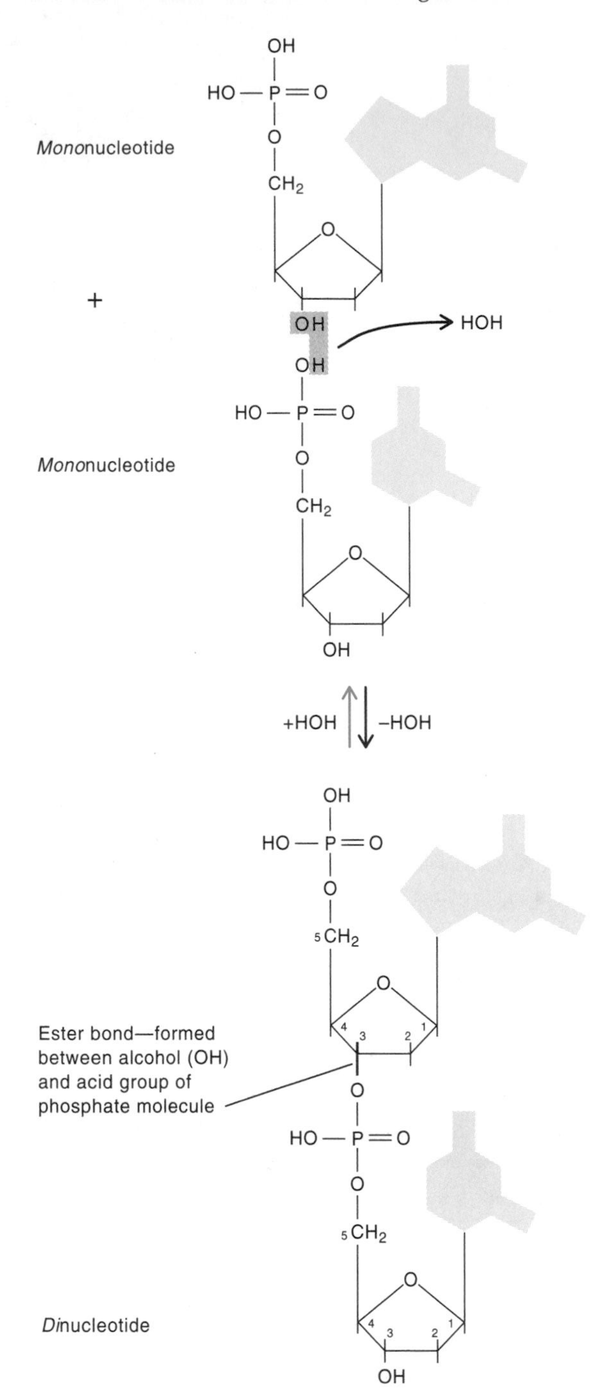

Figure 2.20
Formation of an ester bond in joining two nucleotides.

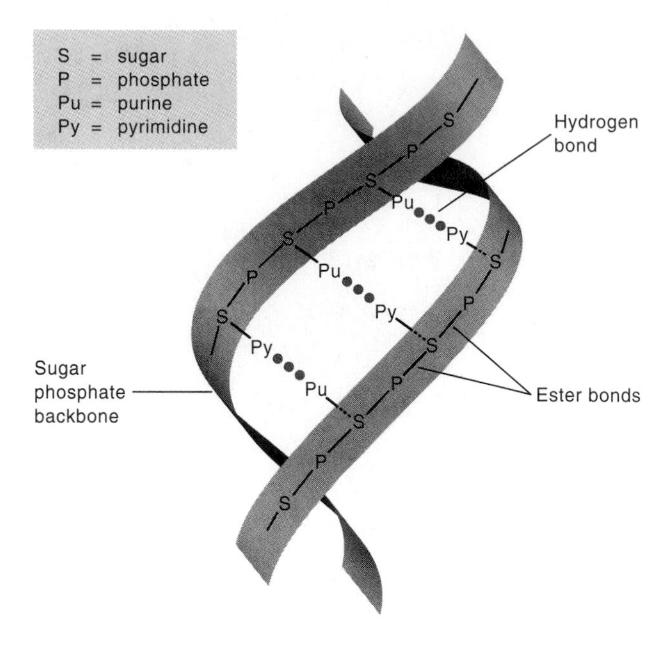

Figure 2.21
DNA double-stranded helix. The two strands are held together by hydrogen bonds between specific purines and pyrimidines on the inside of the helix. The backbone (ribbon) consists of sugar-phosphate molecules. Only a very short segment of a DNA molecule is illustrated.

their nonpolar, hydrophobic nature. Lipids are not macromolecules in the sense of the term used so far, because they have molecular weights of no more than a few thousand. Furthermore, they are not composed of similar subunits but of a wide variety of substances that differ in their chemical structure. Lipids can be divided into two general classes: the **simple** and the **compound lipids,** which differ in their chemical composition.

Simple Lipids

Simple lipids contain only carbon, hydrogen, and oxygen. The most common are the **fats,** a combination of **fatty acids** and **glycerol** that are solid at room temperature (figure 2.22).

Fatty acids are molecules consisting of long chains of C atoms bonded to H atoms (which are therefore termed **hydrocarbons**) with an acidic group (COOH) on the end. Hundreds of different fatty acids exist. If the hydrocarbon portion contains no double bonds, the fatty acid is termed **saturated.** If it contains one or more double bonds, it is **unsaturated.** Three of the most common in nature are palmitic acid (16 carbon atoms), stearic acid (18 carbon atoms), and oleic acid (18 carbon atoms and one double bond in the molecule). Because oleic acid has only one double bond, it is a **monounsaturated** fatty acid (figure 2.22). Other fatty acids contain numerous double bonds that make them **polyunsaturated.** Different fats and oils are therefore often referred to as highly *saturated* or *unsaturated.* Unsaturated fats tend to be liquid and are then called **oils.** These unsaturated fatty acids develop kinks in their long tails that prevent tight packing. Since glycerol has three hydroxyl groups, a maximum of three fatty acid molecules, either the same or different, can be bonded to it through an ester linkage. Note that water is eliminated in the formation of this ester bond (through dehydration synthesis). If only one fatty acid is bound to glycerol the fat is called a *monoglyceride;* when two fatty acids are joined, it is called a *diglyceride;* when three fatty acids are bound, the fat is

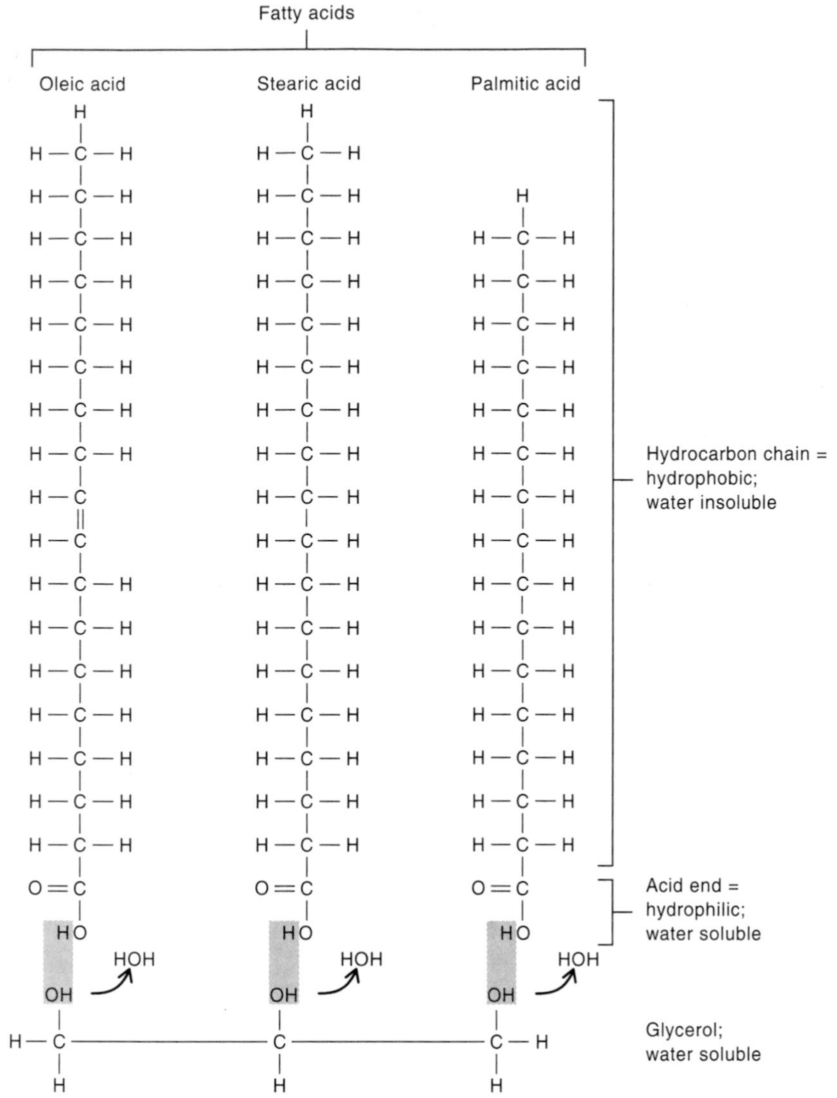

Fatty acids

Oleic acid
Stearic acid
Palmitic acid

Hydrocarbon chain = hydrophobic; water insoluble

Acid end = hydrophilic; water soluble

Glycerol; water soluble

Figure 2.22
Formation and general formula of a fat. Ester bonds are formed between the three alcohol groups of the glycerol and the acid groups of the fatty acids, to yield one glycerol molecule bonded to three fatty acid molecules. The gycerol end of the molecule is water soluble; the fatty acid portion (hydrocarbon chain) is insoluble in water.

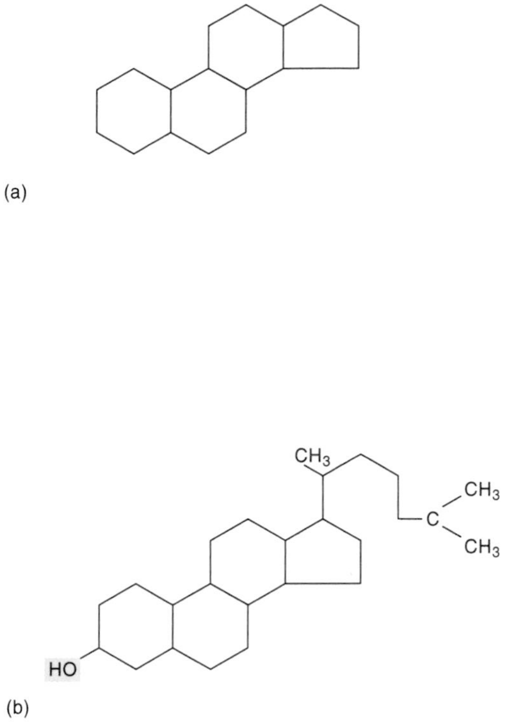

(a)

(b)

Figure 2.23
Steroid. (*a*) General formula, showing the four-membered ring and (*b*) the sterol, cholesterol. The carbon atoms in the ring structures are not shown.

called a *triglyceride.* Fatty acids are stored in the body as an energy reserve by forming triglycerides.

Another very important group of simple lipids is the **steroids.** All members of this group have the four-membered ring structure shown in figure 2.23 *a*. These compounds differ from the fats in chemical structure, but both are classified as lipids because they are both insoluble in water. If a hydroxyl group is attached to one of the rings, the steroid is called a **sterol;** in this particular case, it is cholesterol (figure 2.23 b).

Sterols are commonly found in the cytoplasmic membrane of eukaryotic cells but rarely in prokaryotic cells (see chapter 3). Other important compounds in this general group of lipids are certain hormones and vitamins.

Compound Lipids

Compound lipids contain fatty acids and glycerol and elements other than carbon, hydrogen, and oxygen. Some of the most important members of this group in biology are the phospholipids, so named because they contain a phosphate molecule in addition to the fatty acids and glycerol (figure 2.24). The phosphate is further linked to a variety of other polar molecules, such as alcohol, sugar, or certain amino acids. This latter group is often referred to as a **polar head group.**

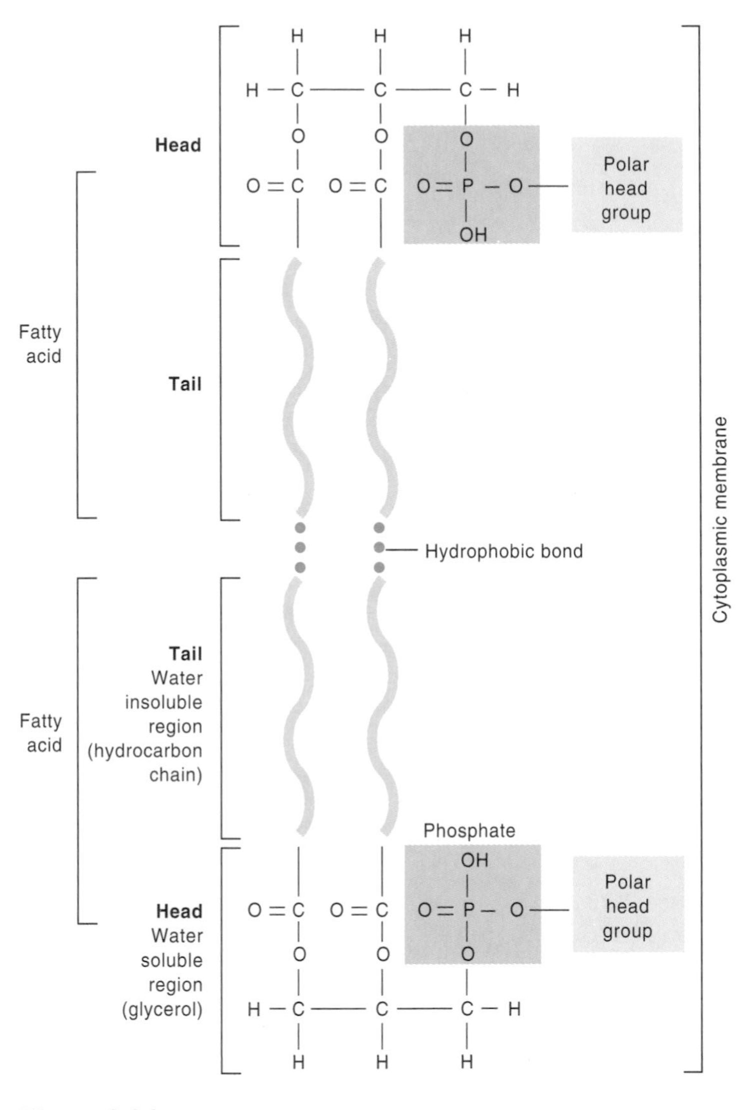

Figure 2.24
Phospholipid bilayer. The two molecules consist of long fatty acid chains of hydrocarbons, which are insoluble in water, and tails that are soluble. The tails are composed of glycerol, phosphate, and the polar head group. The two molecules in the bilayer are held together by weak hydrophobic bonds.

In bacteria, phospholipids occur as a double layer, a **bilayer,** in the cytoplasmic membrane, a structure that separates the outside of the cell from its internal contents (figure 2.24). The chemical structure of the bilayer gives it the properties required of the cytoplasmic membrane. This membrane acts as the major barrier to the entrance to and exit from the cell of substances. This barrier results from the phospholipids consisting of two parts, each with different properties. The end with the phosphate bonded to the polar head group is hydrophilic and therefore soluble in water. The long fatty acid chains consisting of only C and H atoms are hydrophobic and therefore water insoluble. The hydrophilic regions orient themselves toward the external or internal (cytoplasmic) environment, while the long chain fatty acids orient themselves away from the aqueous areas. In this phospholipid bilayer, the water soluble substances cannot pass through the hydrophobic portion. Hydrophobic regions of molecules always orient themselves away from water, and these hydrophobic regions often form weak bonds, termed **hydrophobic bonds,** between portions of the molecules.

Other compound lipids are found in the outer covering of bacterial cells and will be discussed in chapter 3. These include the **lipoproteins,** covalent associations of proteins and lipids, and the **lipopolysaccharides,** molecules of lipid linked with polysaccharides through covalent bonds.

REVIEW QUESTIONS

1. List the three major classes of macromolecules in the cell. What three properties do they share?
2. What are the subunits of (1) proteins, (2) nucleic acids, and (3) polysaccharides? What are the names of the covalent bonds that join each of their subunits?
3. Name four different ways in which polysaccharides can differ from one another.
4. What are the two major subdivisions of nucleic acids and how do they differ from one another?
5. How are lipids subdivided into groups and what feature is common to all lipids?
6. How do an element and its isotopic form differ? What are the two kinds of isotopes?
7. Differentiate an element, an atom, and a molecule.
8. Distinguish between a nucleoside and a nucleotide.
9. Define the primary structure of a protein.
10. Water dissolves many substances. Explain how its structure makes it a good solvent.

Summary

I. Elements and Atoms
   - A. An element is a pure substance that consists of a single type of atom. Atoms are the basic units of all matter. They consist of three major components: electrons, protons, and neutrons.

II. Formation of Molecules: Chemical Bonds
   - A. Chemical bonds are of two types: strong and weak. The stronger the bond, the more energy is required to break it.
   - B. Strong bonds are usually covalent bonds formed when atoms share electrons to fill their outer shell and thereby achieve maximum stability. Covalent bonds vary in their distribution of shared electrons, which results in the molecule having a positive and negative charge at different sites.
   - C. Ionic bonds are formed by the loss and gain of electrons between atoms. In aqueous solutions, they are weak.
   - D. Hydrogen bonds are weak but biologically very important. They hold the two strands of DNA together and are important in determining the shape of proteins. They result from the attraction of positively charged H atoms to negatively charged N or O atoms.

III. Important Molecules of Life
   - A. Small molecules in the cell include organic and inorganic compounds.
     1. The inorganic molecules include many that are required for enzyme function.
     2. Organic molecules are mainly compounds that are being metabolized or molecules that are the subunits of macromolecules.
   - B. All very large molecules in the cell (macromolecules) consist of repeating subunits called *monomers*. There are three important macromolecules. Proteins are chains of amino acids that form a polypeptide. Polysaccharides are chains of monosaccharides that form branching structures. The nucleic acids are DNA and RNA, which are chains of nucleotides.
   - C. Proteins are polymers of amino acids.
     1. Amino acids consist of a molecule with a carboxyl group and an amino group bonded to the same carbon atom. The carbon atom is bonded to a side chain. Twenty different amino acids, each differing in their side chain, are present in proteins.
     2. In peptide bond synthesis, a covalent bond is formed between the amino group of one amino acid and the carboxyl group of the adjacent amino acid with the removal of HOH (through dehydration synthesis).
     3. Three features characterize a protein: (1) its composition, or primary structure, (2) its shape, whether globular or fibrous, and (3) its interaction with one or several polypeptide chains. A variety of weak bonds are involved in maintaining the three-dimensional shape of proteins.
     4. Substitute proteins contain covalently bonded molecules other than amino acids. These include glycoproteins (sugars) and lipoproteins (lipids).
   - D. Polysaccharides are polymers of monosaccharide (carbohydrate) subunits. Carbohydrates (sugars) contain a large number of alcohol groups (—OH) in which the C atom is also bonded to an H atom to form H—C—OH.
     1. Monosaccharides are the subunits of polysaccharides. The most common are pentoses (5 carbon) and hexoses (6 carbon).
     2. Disaccharides are two monosaccharides joined with a loss of water (through dehydration synthesis).
     3. Polysaccharides vary in size, their degree of branching, the bonding of their monosaccharides to one another, and the monosaccharides involved.
   - E. Nucleic acids, which include deoxyribonucleic acid (DNA) and ribonucleic acid (RNA), are polymers of nucleotide subunits and are unbranched.
     1. The nucleotides of DNA are composed of three units: a (1) nitrogen base (purine [adenine or guanine] or pyrimidine [thymine or cytosine]) covalently bonded to (2) deoxyribose, which, in turn, is bonded to (3) a phosphate molecule. A phosphate bonded to the sugar molecule joins the nucleotides together. DNA occurs in the cell as a double-stranded helix in which the two strands are held together by hydrogen bonding between adenine and thymine and between guanine and cytosine.
     2. The nucleotides of RNA are the same as those in DNA except that ribose replaces deoxyribose, and uracil replaces thymine. RNA is shorter in length and does not occur as a double helix. Three different types of RNA exist in the cell.
   - F. Lipids are heterogeneous in structure but all are insoluble in water.
     1. All lipids have one property in common—they are insoluble in water but soluble in organic solvents. This difference in solubility is due to their nonpolar, hydrophobic nature.
     2. They are not composed of similar subunits but rather of a variety of substances that differ in chemical structure.
     3. Simple lipids, which contain only C, H, and O, include fats and steroids. Fats are composed of glycerol covalently bonded to fatty acids. Steroids have a four-membered ring structure.
     4. Compound lipids contain fatty acids and glycerol and often elements other than C, H, and O. They include phospholipids, lipoproteins, and lipopolysaccharides. They all play important roles in the cell envelope of bacteria.
     5. Phospholipids consist of two parts, each with different properties. One end is hydrophilic and therefore is soluble in water. The other end, containing only C and H, is hydrophobic and therefore insoluble in water but soluble in organic solvents.

Critical Thinking Questions

1. What are the properties of water that make it critical for life on this planet?
2. What are the properties of carbon that make it so important in molecules of biological importance? Could another element be substituted for it? Why or why not?

Further Reading

Brown, T. L., H. E. LeMay, and B. E. Bursten. 1991. *Chemistry: Central Science.* Englewood Cliffs, N.J.: Prentice Hall.

Chang, R. 1990. *Chemistry,* 4th ed. McGraw-Hill, New York. An introductory textbook that gives examples of biological significance in chemistry.

Doolittle, R. F. 1985. Proteins. *Scientific American* **253**(4):88.

Felsenfeld, G. 1985. DNA. *Scientific American* **253**(4):58.

Garrett, R. H., and C. M. Grisham. 1995. *Biochemistry.* Philadelphia: Saunders College Publishing.

Zubay, G. 1993. *Biochemistry,* 3d ed. Dubuque, Iowa: W. C. Brown.

3 Functional Anatomy of Prokaryotes and Eukaryotes

KEY CONCEPTS

1. The effectiveness of a microscope is based on its ability to visually separate two objects that are close together.
2. The many different microscopes that have been developed differ primarily in their lenses and their methods of illuminating the specimens being studied.
3. All bacteria must be able to carry out the functions required for life, and their small cells must therefore contain the structures to carry out these functions.
4. The composition of the rigid bacterial cell wall determines various properties of the organism, including its susceptibility to penicillin and its staining characteristics.
5. The cytoplasmic membrane largely determines what material gets into and out of a cell.
6. In bacteria, DNA occurs as a covalently closed, circular molecule without a surrounding membrane.
7. The structure of ribosomes differs in eukaryotic and prokaryotic cells; for this reason, some antibacterial medicines, such as streptomycin, kill bacteria but are harmless to human cells.
8. Some bacteria can develop into endospores, a type of cell that can survive adverse conditions such as high temperatures.

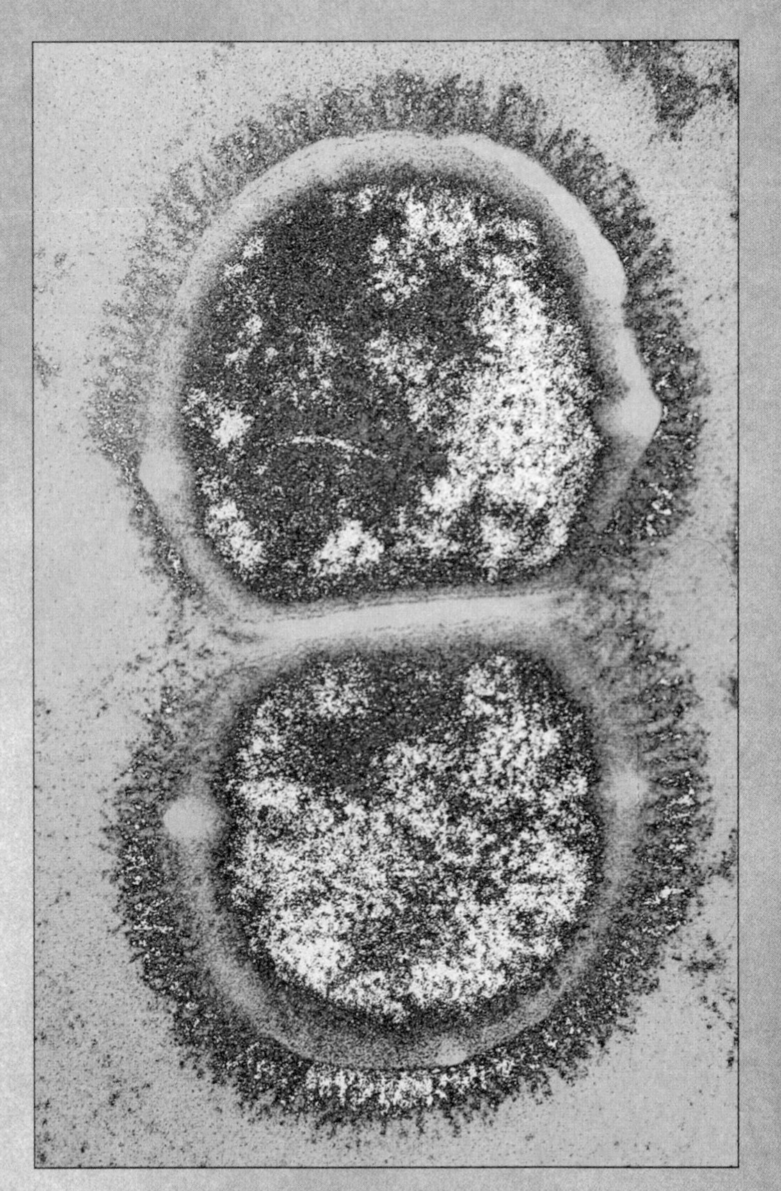

TEM of dividing cells of *Neisseria gonorrhoeae*. The image has been color enhanced (×33,500).

PREVIEW LINK

Microbes in Motion

The following books and chapters in the *Microbes in Motion* CD-ROM may serve as a useful preview or supplement to your reading:

Bacterial Structure and Function: Cell Membrane, Cell Wall; Gram-Positive Cell; Gram-Negative Cell, External Structures; Internal Structures. *Fungal Structure and Function:* General Eucaryotic Structures; Metabolism and Growth.

A Glimpse of History

Antibiotics are antimicrobial medicines that play an enormous role in treating a wide variety of infectious diseases. Their success in killing or inhibiting the growth of microorganisms without affecting the host depends on anatomic and physiological differences between host and parasite.

One of the first antibiotics that proved effective in treating infections was penicillin, which was discovered by Alexander Fleming in 1929 while he was carrying out experiments on bacteria of the genus *Staphylococcus*. As Fleming was examining bacterial colonies in a petri dish, he happened to remove the dish cover to get a better view. Shortly thereafter, Fleming went on vacation and left the petri dish on his lab bench. Upon returning, he examined the dish and noticed that the agar medium was contaminated with a mold. Apparently, a spore of the mold had landed on the agar while Fleming was examining the plates several weeks before his vacation. While he was gone, the spore developed into a mycelium.* What caught Fleming's attention was the appearance of the colonies of staphylococci. They had dissolved as they neared the area where the mold was growing. Obviously, something quite extraordinary and unexpected had happened. Fleming correctly reasoned that the mold must have secreted something that caused the *Staphylococcus* colonies to dissolve. He proceeded to isolate the mold, which proved to be a member of the genus *Penicillium;* thus, he named the bacteria-destroying substance *penicillin.* Even though Fleming was unable to purify penicillin, he found that even impure preparations were remarkably effective in killing many different kinds of bacteria in the laboratory. Fleming made some additional important observations. He noted that penicillin was not toxic to the white blood cells of the body, nor did it harm rabbits and white mice when he injected it into them. Thus, Fleming believed that someday penicillin would be an important therapeutic agent. However, he became so discouraged with his inability to purify it, that he abandoned further studies on the material.

Ten years later, in 1940, a group of Oxford University scientists in England headed by Doctors Howard Florey and Ernst Chain purified enough penicillin to begin testing its effectiveness in treating humans. Penicillin proved to be so effective in treating many infections, that, within three years, it was being produced on an industrial scale in the United States and was enormously useful in the latter part of World War II. Fifty years later, penicillin remains one of the most important antibiotics. Fleming, Chain, and Florey shared the Nobel Prize in Medicine in 1945 for their pioneering studies of this antibiotic.

The usefulness of penicillin and most other effective antimicrobial agents relies on their **selective toxicity,** their ability to inhibit or kill the infectious agent without affecting the host. Penicillin is highly selective in its toxicity because it inhibits the synthesis of a compound that is found only in prokaryotic cells. The information in this chapter describes the major features of bacteria and some structures that differ between prokaryotic and eukaryotic cells; these differences are responsible for the success of antibiotics.

*Mycelium—mold growth that resembles thin filaments

Microbiology as a science became possible as techniques for working with invisible organisms were developed. Thus, this area of science was born as a result of the microscopic observations of Antony van Leeuwenhoek on his "animalcules." He used only a single lens microscope whose magnification was limited to approximately 300-fold to view these "animalcules." The details of cell structure cannot be observed at this magnification, so further microscopic studies of the internal structures of cells had to await the invention of a microscope that could enlarge objects about 1,000 times. However, even at a 1,000-fold magnification, many structures of prokaryotic cells cannot be seen clearly, and it was not until the development of the electron microscope in the 1930s that definitive studies on prokaryotic cell structure became possible.

In addition to the development of more powerful microscopes, techniques of staining were devised that made bacteria and their internal structures more readily visible. Thus, preparing microorganisms for viewing along with developing highly specialized microscopes made all structures of the cell visible.

Microscopic Techniques: Instruments

A large number of instruments have been developed for viewing organisms. They differ primarily in their lenses and the source of illumination.

The Compound Microscope Has Two Sets of Magnifying Lenses

The most commonly used instrument for observing any cell is the **light microscope,** so named because visible light illuminates the object being studied. Many variations of this instrument exist. The light microscope used by van Leeuwenhoek, consisting of only a single magnifying lens, is called a **simple microscope.** The most commonly used microscope today has two sets of magnifying lenses—an **objective lens** and an **ocular lens**—and is called a **compound microscope.** A modern compound microscope, with its major components labeled, is shown in figure 3.1.

The magnification achieved with such a microscope is the product of the magnification of each of the individual lenses. Most compound microscopes have several objective lenses, with varying magnifying capacities, thereby making several different magnifications possible with the same instrument. A high-powered compound microscope can magnify objects up to 2,000 times.

The usefulness of a microscope depends not so much on its degree of magnification, but rather on its ability to separate clearly, or **resolve,** two objects that are very close together. The **resolving power** of a microscope is defined as the minimum distance that can exist between two points such that the points are observed as separate entities. The maximum resolving power of the best light microscope functioning under optimal conditions is 0.2 μm. The resolving power determines how much detail actually can be seen (figure 3.2). It depends on the quality and type of lens, the magnification, and how the specimen under observation has been prepared. (The renowned English microscopist Robert Hooke was less successful than van Leeuwenhoek in

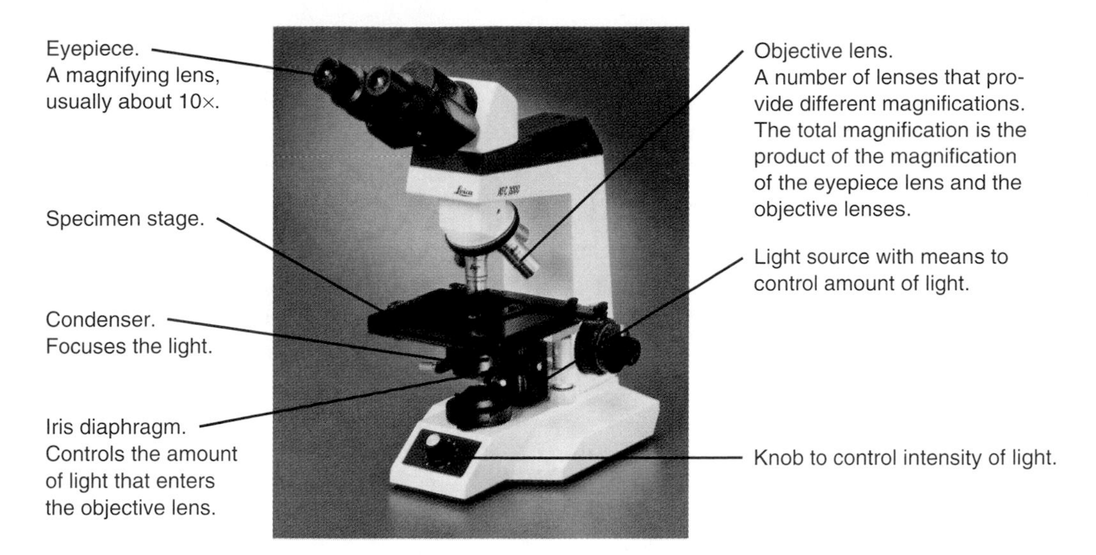

Figure 3.1
Compound microscope.

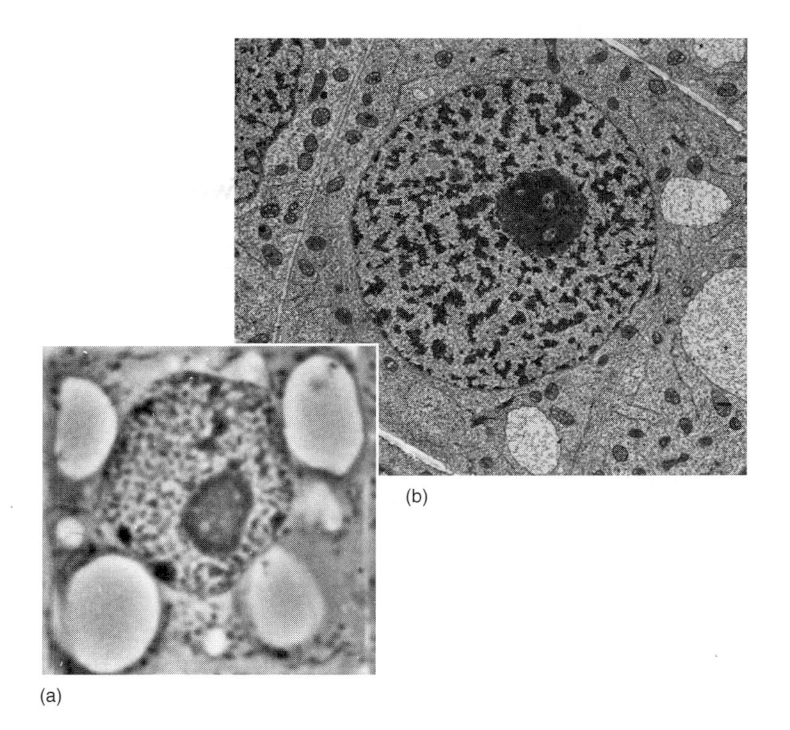

Figure 3.2
Comparison of the resolving power of a light microscope (*a*) and an electron microscope (*b*). The same preparation (onion root tip) was magnified 450 times. Note the difference in the degree of detail that can be seen at the same magnification.

distinguishing the details of microorganisms, even though he invented the compound microscope, which had greater magnification. The reason for Hooke's lack of success was that the lenses he employed had serious optical defects that resulted in blurred images.) The resolving power of a microscope is limited by the wavelength of the light used for illumination: The shorter the wavelength of this light, the greater the resolution that can be attained. Thus, to increase resolving power, microscopes have been developed that employ illumination of a wavelength shorter than visible light (compare *a* and *b* in figure 3.2 to see this effect). These microscopes, which include the electron microscope, are discussed in the next section.

The resolving power of the microscope can also be increased in another way—by using oil, rather than air, as the material between the specimen being viewed and the objective lens. The oil allows a wider cone of light to pass from the light source through the specimen into the objective lens. The objective lens that is designed to be used with oil is called an **oil immersion lens.**

The Contrast Between Cells and Their Surrounding Media Must Be Increased for Good Observations

Although bacteria suspended in a drop of liquid are large enough to be seen with a light microscope, they are very difficult to see because they are transparent and usually colorless. This problem can be overcome in two ways: (1) by using special kinds of light microscopes and (2) by staining the cells.

Phase Contrast Microscope

The **phase contrast microscope,** commonly used in research laboratories for observing living microorganisms, is a special kind of compound microscope. It has optical devices that increase the contrast between the microbes and the surrounding medium. Cellular components are denser than the surrounding medium, so light passing through a cell's components is slowed down more than it is by the surrounding medium. Since different cell structures slow down the light in varying degrees, some of the structures of the cell can be discerned by phase contrast microscopy. Even though the cells are not magnified with this form of microscopy to any greater extent than they are with the ordinary light microscope, the cells and some structures stand out from the background and are clearly visible.

Thus, it is possible with a phase contrast microscope to observe living organisms clearly and study their movements in the medium in which they are growing. The bacteria in figure 3.3 are viewed under an ordinary light microscope (*a*) and then under a phase contrast microscope (*b*).

Interference Microscope

The **interference microscope,** like the phase contrast microscope, depends for its functioning on changes in the speed of light as it passes through different materials. The most frequently employed microscope of this type is the **Nomarski differential interference contrast microscope.** This microscope has a device for separating light into two beams that travel close to each other (but separately) through the specimens and then recombine. The light waves are out of phase when they recombine, resulting in a three-dimensional appearance of the specimen (figure 3.3 *c*).

Dark-field Microscopy Highlights Specimens Against a Dark Background

A commonly used method for achieving a marked contrast between living organisms and the background is **dark-field microscopy.** In this technique, light is directed toward the specimen at an angle, so that only light that is scattered by the specimen enters the objective lens and is seen. The field is completely dark except for the objects being viewed, which are brilliantly illuminated (figure 3.3 *d*). In the dark-field microscope, the ordinary condenser of the light microscope is replaced by a dark-field condenser that does not allow the illuminating light to pass through the specimen directly. This method makes it possible to see objects and cells that are invisible by ordinary light microscopy. Indeed, van Leeuwenhoek was able to make such precise observations of bacteria probably because he discovered the technique of dark-field microscopy. When this technique is used, *Escherichia coli* is clearly visible at magnifications of only 100-fold, whereas the organism must be magnified approximately 450-fold to be seen by ordinary light microscopy. Dark-field microscopy is also useful for viewing very thin cells, such as the organism causing syphilis, *Treponema pallidum,* which is barely visible with ordinary light microscopy. Dark-field, phase contrast, and interference microscopy are all used to view living cells, allowing the observer the great advantage of being able to estimate the true size, shape, and motility of the bacteria.

Fluorescence Microscopy Is Used to Examine Materials that Fluoresce

Another type of microscope important in laboratories concerned with identifying microorganisms is the **fluorescence microscope.** This microscope is used to visualize objects that fluoresce—that is, that emit light when light of a different wavelength strikes them (figure 3.3 *e*). The fluorescence may be a natural property of the specimen being viewed, or it may result from the attachment of a fluorescent compound to a normally nonfluorescing material. For example, fluorescence microscopy is commonly used to identify molecules by their ability to bind to specific antibodies.[1] A fluorescent compound is attached to the antibodies that are not normally fluorescent. The antibodies then bind to the object of interest, thereby staining it with the fluorescent material, which can be seen when it is illuminated by ultraviolet light supplied by a special lamp. A special filter on the microscope allows the yellowish green light from the source of fluorescence to pass, but it blocks the passage of the ultraviolet light. Techniques such as this have numerous applications in medical microbiology.

The Electron Microscope Can Resolve Objects a Thousand Times Better than the Light Microscope

The light microscope allows the observer to identify only the major features of the bacterial cell—its size and shape and a few of its largest components. The light microscope was developed almost to the limit of its resolving power more than a century ago. To increase the resolving power significantly, a new type of microscope, the **transmission electron microscope (TEM),** was constructed by Knoll and Hruska in Berlin in 1931. In this microscope, electrons, rather than visible light, illuminate the specimen being viewed. Since the electrons have a wavelength about 1,000 times shorter than visible light, the resolving power increases about 1,000-fold to about 0.3 nanometers. Magnetic fields focus the beam of electrons and so function as a condenser lens. The TEM has an objective lens as does the light microscope. Some of the electrons pass through the specimen, others are scattered, and still others cannot pass. Instead of passing through an ocular lens, the electrons hit an electron-sensitive screen, thereby creating an image that is determined by the ability of the electrons to pass through various parts of the object being viewed (figure 3.3 *f*).

Preparation of Specimens for Transmission Electron Microscopy

Before the specimen can be viewed in the transmission electron microscope, it must be dried and attached to a supporting mesh. These procedures may distort the size of the organism and introduce other artifacts into the specimen. A major concern in electron microscopy is determining whether what is observed with the electron microscope is actually present in the living cell or is an artifact of the treatment involved in preparing the specimen.

[1]Antibodies—proteins produced by the body that react specifically with a foreign substance

• **Dark-field microscopy, p. 4**

• **Fluorescent antibodies, p. 387, 388**

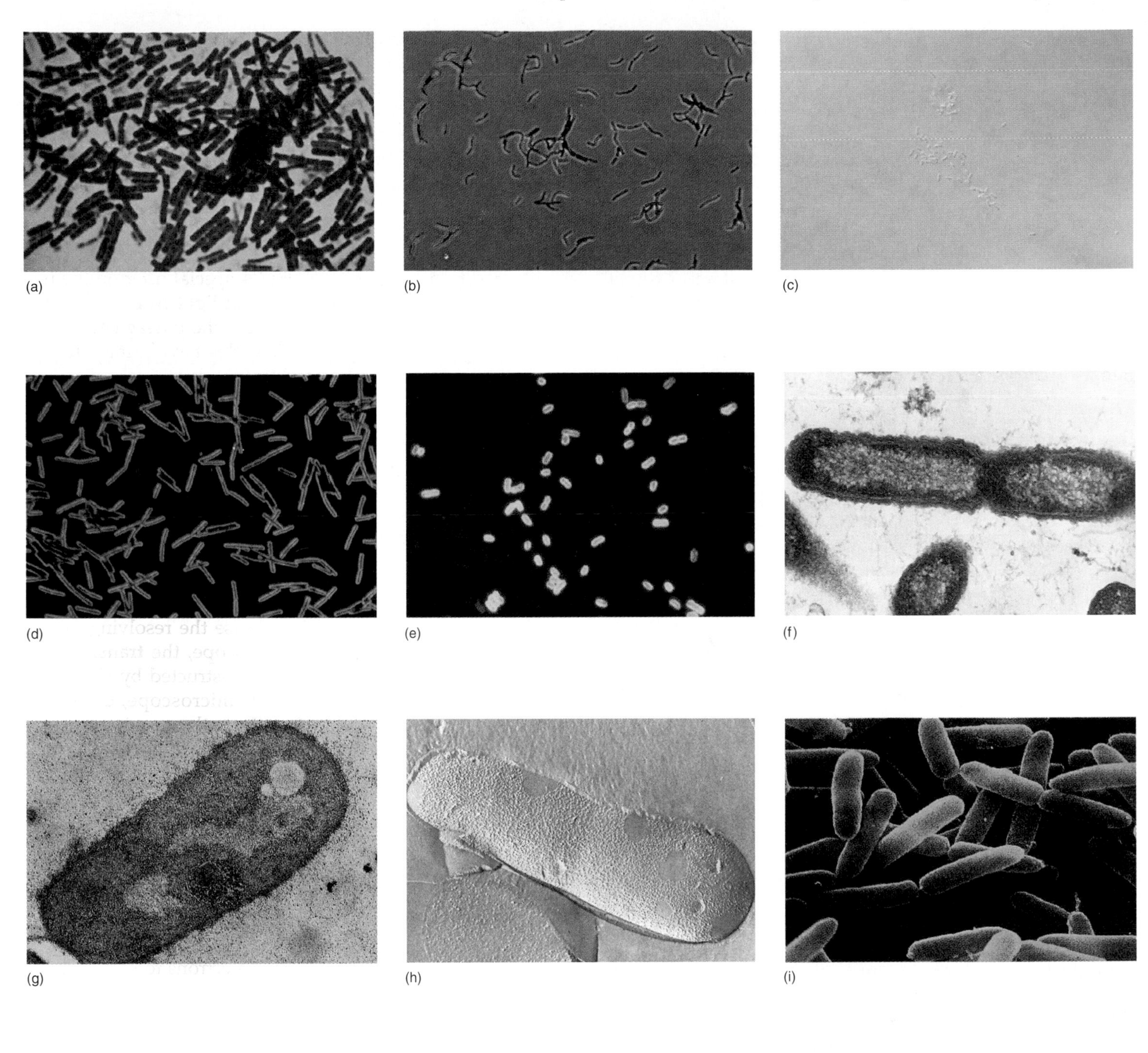

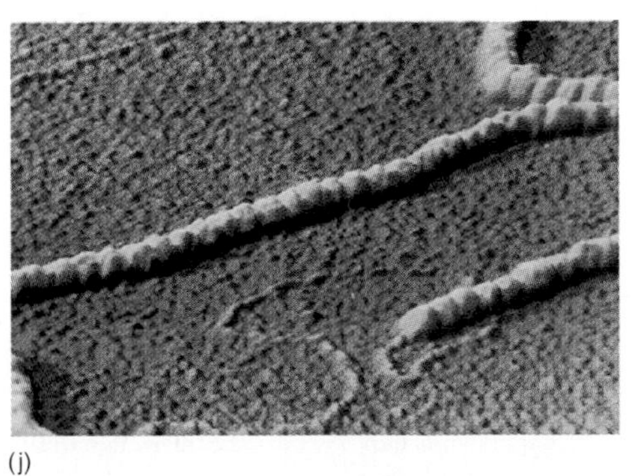

Figure 3.3
Photomicrographs (*a* through *i*) are of rod-shaped bacteria (bacilli). They illustrate the details that are possible with each type of preparation and microscope. (*a*) Bright-field (light) microscopy. The intracellular bodies are endospores. (*b*) Phase-contrast microscopy. (*c*) Bacilli as viewed through a Normarski microscope (X1,000). (*d*) Dark-field microscopy. (*e*) Fluorescence microscopy. (*f*) Transmission electron micrograph. (*g*) Thin section; electron photomicrograph. (*h*) Freeze-etched preparation. (*i*) Scanning electron photomicrograph. (*j*) Scanning tunneling electron micrograph of DNA-protein complex.

Although microorganisms can be viewed with the TEM, very little detail can be seen because of their thickness. Therefore, investigators commonly slice a specimen into very **thin sections** with a diamond or glass knife (figure 3.3 *g*).

Another preparation procedure is termed **freeze fracturing.** In this procedure, a frozen specimen is fractured by striking it with a knife blade. The surface of the section is then coated with a layer of carbon thin enough to be viewed with the electron microscope. This method is used primarily for viewing details in both internal and external surface structures (figure 3.3 *h*). Since this technique does not require fixation or thin sectioning of the specimen, artifacts are not as likely to arise. The photographs from such freeze-dried preparations can be very dramatic.

Scanning Electron Microscopes Allow a Three-dimensional View of Objects

The **scanning electron microscope (SEM)** operates on a different principle than the TEM. In the SEM, the beam of electrons scans back and forth over the surface of the specimen being viewed (hence the name of the instrument). The specimen is first coated with a thin film of metal. As the beam moves over the surface, electrons are released from the specimen and are reflected back into the viewing chamber. The SEM is especially useful for observing surface details and not the internal structures of cells, as is the TEM. Most scanning electron microscopes have resolving powers in the range of 1 to10 nm, which is not as good as the 0.3 nm of the TEM. However, relatively large specimens can be viewed and a dramatic three-dimensional effect is observed with the SEM (figure 3.3 *i*).

The Scanning Tunneling Microscope Can Observe Individual Atoms on the Surface of Samples

A new microscope was invented in 1981 by two scientists in Switzerland, Gerd Binning and Heinrich Rohrer. This microscope, termed the **scanning tunneling microscope (STM)** examines a sample in a new way. The microscope is focused at the surface of an object and produces a map showing the bumps and valleys of the atoms on the surface of the sample (figure 3.3 *j*). The mapping of the surface is done by "feel," much as a blind person explores the surface of the ground by tapping a cane. Its resolving power is much greater than the electron microscope and the samples do not have to be prepared in a special way as they do for electron microscopy. Binning and Rohrer received the Nobel Prize in Physics in 1986 for creating this instrument.

Microscopic Techniques: Staining

To overcome the difficulty of observing living, transparent, and often rapidly moving organisms with the light microscope, cells are frequently killed and then treated with **dyes** that have a special attraction for one or more cellular components. As a result, an entire organism, or specific parts of it, stands out in contrast to the unstained background.

In the staining procedure, a drop of liquid containing the organisms is placed on a glass slide, and the organisms are attached, or **fixed,** to the slide, usually by passing the slide over a flame. The dyes, or stains, are then applied to the fixed organisms (figure 3.4).

A wide variety of stains and staining procedures is currently employed, each with its own special use. Some dyes will stain only a particular cell component. Other staining procedures will stain one but not another group of organisms, thereby allowing the classification of organisms into groups based on their staining characteristics.

Stains can be divided into two major types on the basis of their affinity for cell components. **Positive stains,** which have a strong attraction for one or more cell components, color these components when added to cells fixed on a microscope slide. **Negative stains** cannot penetrate a cell and thus make it highly visible by providing a contrasting dark background. Negative stains are generally used on living cells to demonstrate surface structures, which are not stained well by positive stains. Examples of organisms stained with positive and negative stains are shown in figure 3.5 *a* and *b*.

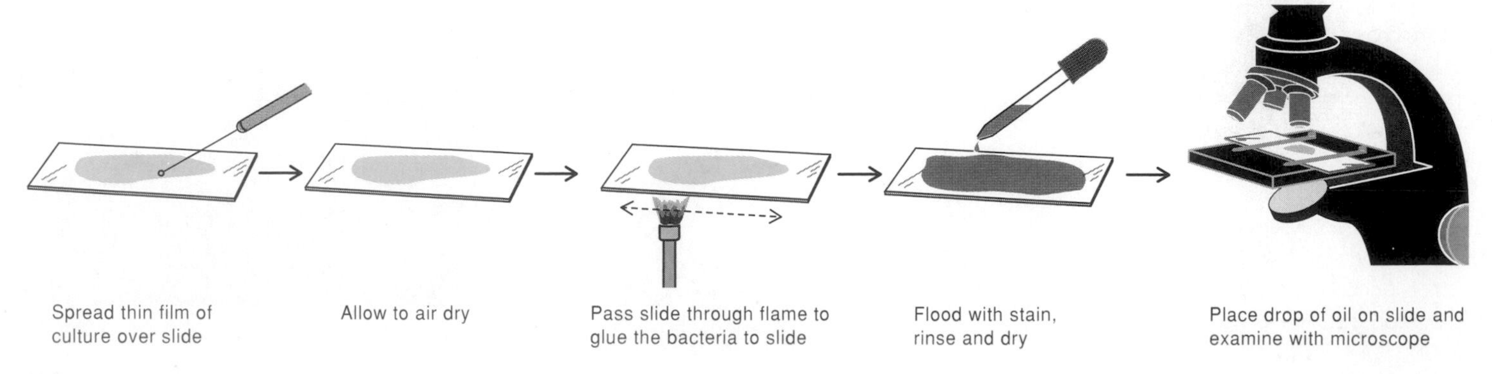

Figure 3.4
Steps in staining cells for microscopic observation.

Simple Staining Procedures Involve a Single Dye

In simple staining, only a single dye is applied to a cell. This technique may be used to stain the whole cell or specific structures in the cell. The dye methylene blue is frequently applied to fixed bacteria to stain entire cells blue without staining the background material (figure 3.5 *a*). This basic dye (which has a positive charge) binds primarily to the nucleic acids of the cell, both DNA and RNA (which have negative charges), so internal structures in the cell are not revealed. Acidic dyes (having negative charges), which include eosin, acid fuchsin, and Congo red, stain basic compounds in the cell, primarily proteins carrying a positive charge (those that are composed mainly of basic amino acids; figure 3.6). Sudan black is a dye that is very soluble in fat and is frequently used to identify and locate fat droplets and fatlike molecules in bacteria. These are all examples of positive staining.

Differential Stains Distinguish Various Kinds of Bacteria Because of Differences in Their Surface Structures

Several staining procedures known as **differential staining techniques** have been developed for identifying certain groups of bacteria. The two most frequently used are the Gram stain and the acid-fast stain. The surface structures of the bacteria determine their staining properties with these two procedures.

• **Basic amino acid, p. 32**

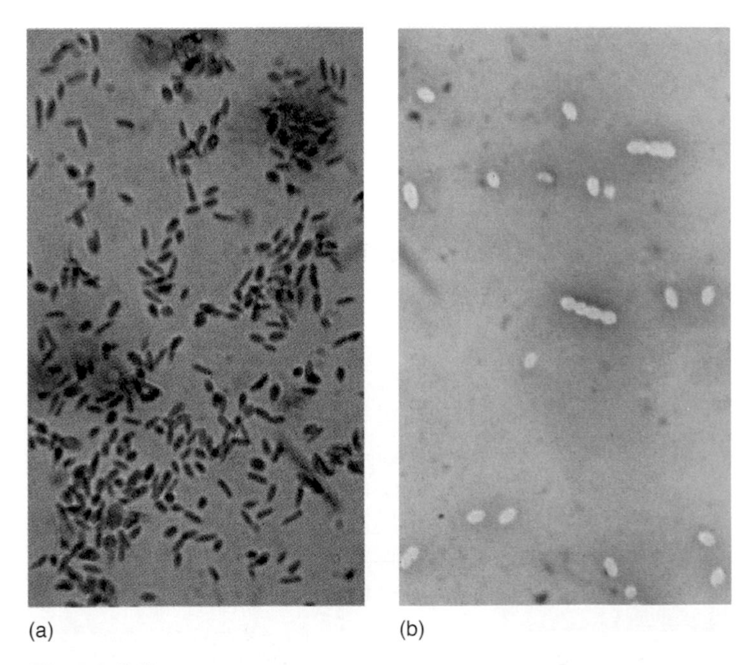

(a) (b)

Figure 3.5
(*a*) Positive staining of *Corynebacterium diphtheriae* (X1,000), using methylene blue; (*b*) negative staining of *Streptococcus pneumoniae*.

Gram Stain

Bacteria can be classified according to their ability to take up and retain certain dyes. The **Gram stain** is the most widely used staining procedure. It divides most bacteria into one of two groups. A Danish physician, Dr. Hans Christian Gram, working in a morgue in Berlin, developed this staining method in 1884 to distinguish between the bacteria that cause pneumonia and eukaryotic cell nuclei in infected mammalian tissues. The identity of the person who first had the idea to use the Gram stain as a means to separate bacteria into one or another group is not known. The procedure for separating bacteria using the Gram stain is as follows (figure 3.7).

A basic purple dye, usually crystal violet, is added to bacteria fixed on a microscope slide. All bacteria that are able to absorb this dye are stained. Next, a dilute solution of iodine is added, which decreases the solubility of the purple dye within the cell by combining with the dye to form a dye-iodine complex. Then, an organic solvent such as ethanol is added. This solvent readily removes the purple dye-iodine complex from some but not all genera of bacteria, thereby decolorizing the cells. A red dye, such as safranin, is then applied, which stains all bacteria. Bacteria that are decolorized by the ethanol appear red and are called **gram-negative bacteria.** Those that retain the purple dye appear purple, since the purple dye masks the red, and are called **gram-positive bacteria** (figure 3.7 *b*). The reason that some bacteria hold the basic purple dye and others do not is related to the chemical structure of their cell walls, which is discussed later in this chapter. The cell wall of gram-positive organisms is very different from that of gram-negative cells. These differences in cell wall structure serve as a major criterion for placing bacteria into different groups in their classification (see chapter 11).

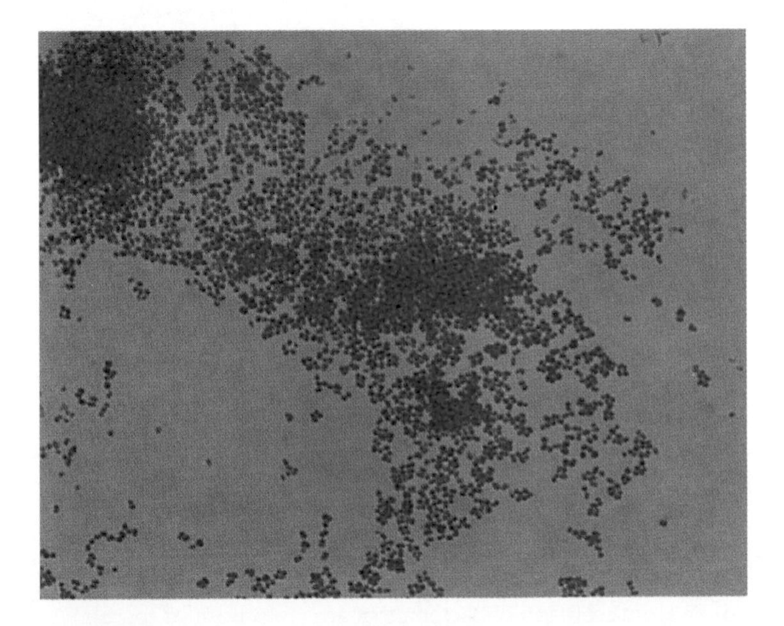

Figure 3.6
Acid stain (rose bengal) of micrococci (X1,000).

Steps in Staining	State of Bacteria
Step 1: Cells fixed by heating.	Shape of cells becomes distorted and cells shrink in size.
Step 2: Basic dye, crystal violet, applied.	All cells stain purple.
Step 3: Iodine added.	All cells remain purple.
Step 4: Alcohol added.	Gram-positive cells remain purple; gram-negative cells become colorless.
Step 5: Red dye such as safranin added.	Gram-positive cells remain purple; gram-negative cells appear red.

(a)

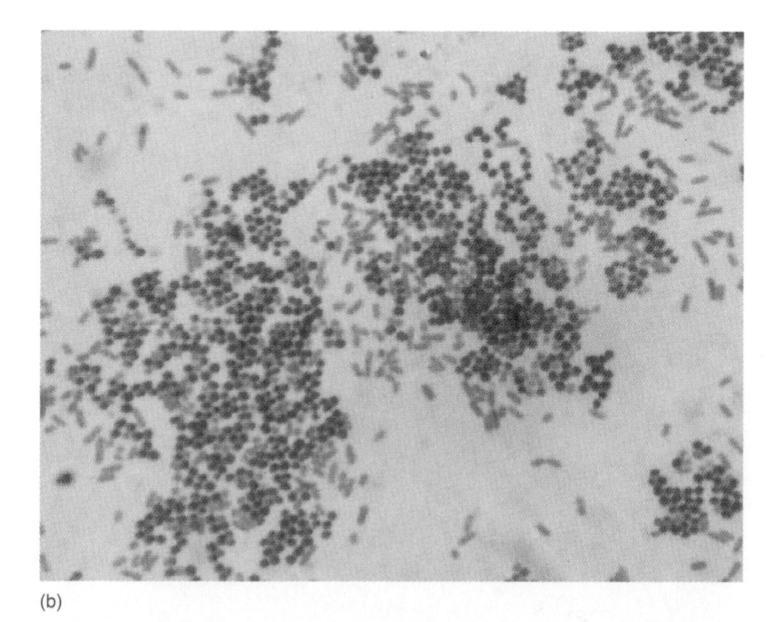

(b)

Figure 3.7
The Gram stain. (*a*) Steps in the Gram-staining procedure. (*b*) Photomicrograph of bacteria that are gram-positive (purple) and gram-negative (red). The species are *Staphylococcus aureus* and *Escherichia coli*, respectively.

Acid-Fast Stain

Another differential staining procedure is the **acid-fast stain** (figure 3.8).

This staining procedure is used to characterize a small group of organisms that stain only with great difficulty but, once stained, are very difficult to destain. These organisms are not decolorized with an acidic solution of alcohol. The bacteria are stained with the red dye basic fuchsin, which is applied with heat to bacteria fixed on a microscope slide. An acidic solution of alcohol is then added to decolorize the cells. Only a few groups of bacteria, **acid-fast bacteria,** retain the basic fuchsin under these conditions. Nonacid-fast organisms and the background of any host cells take up a second dye that is applied—methylene blue or another contrasting dye. Organisms are acid fast because they contain in their cell wall a unique lipid, **mycolic acid,** which binds to the basic fuchsin. Most acid-fast bacteria are in the genus *Mycobacterium,* which contains a species that causes tuberculosis and another that causes leprosy. Mycobacteria do not stain readily with the Gram stain because the lipid on their surface prevents the uptake of dyes. This acid-fast stain is extremely valuable in diagnostic laboratories concerned with detecting mycobacteria.

Staining of Specific Structures

Differential stains are also used to stain specific cell structures inside the cell. The staining procedure for each component of the cell is different, being geared to the chemical composition and properties of that structure. Some structures such as flagella (organelles of locomotion) are so thin (10 nm) that they can be observed with the light microscope only if their size is increased by dyes that stick to them (figure 3.9).

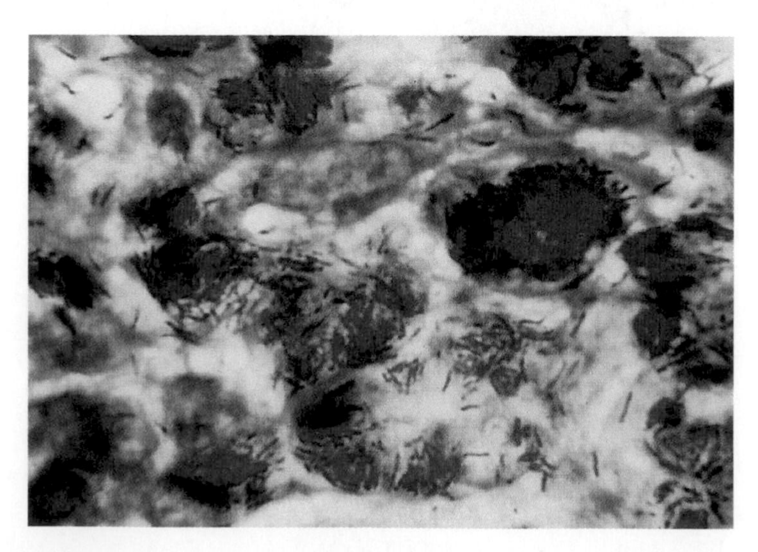

Figure 3.8
Acid-fast stain of *Mycobacterium tuberculosis.* The red-colored rods are *Mycobacterium.*

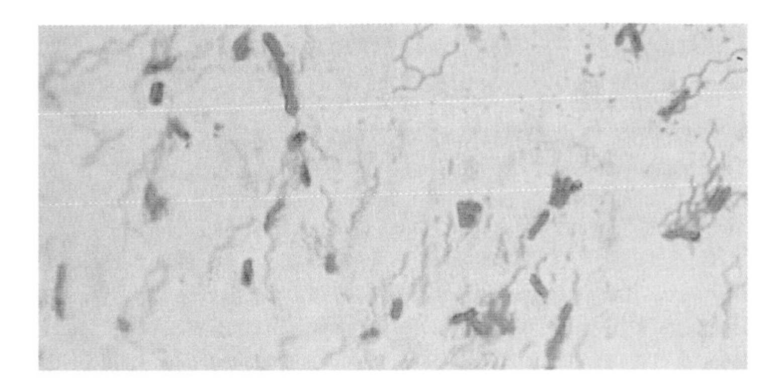

Figure 3.9
Stained bacterial flagella. The flagella not attached to the cells have been dislodgd from the cells during the staining procedure.

REVIEW QUESTIONS

1. Name the two lenses on a compound microscope. How does one calculate the magnification of the microscope?
2. Name two important properties of a microscope that determine its usefulness.
3. Why is the Gram stain so useful? What structure of the cell determines its Gram staining properties?
4. Some bacteria require unusual staining procedures. Give an example and discuss the reason. Why is this procedure useful in identification?

Shapes of Bacteria

Most bacteria have one of three shapes: spherical, commonly referred to as **cocci;** cylindrical, commonly referred to as *rod-shaped* or **bacilli;** and spiral, commonly referred to as **spirilla** (figure 3.10 *a–c*). Some bacteria have intermediate shapes such as short rods, termed **coccobacilli** (figure 3.10 *d*). In addition, a few species of bacteria have other shapes.

In 1980, square cells were isolated from the salty pools of the Sinai Peninsula in Egypt (figure 3.11 *a*). Other bacteria have extensions on their surface, termed **prosthecae,** which give the organisms a starlike appearance (figure 3.11 *b*).

Bacteria frequently do not occur as single cells but rather as independent cells attached to one another. Most bacteria divide by **binary fission,** a process in which one cell divides into two identical daughter cells. Following division, each cell lives independently of all other cells, even though the cells may not be separated. Cells adhering to one another form a characteristic arrangement that depends on the planes in which the bacteria divide. This is seen especially in the spherical cells (cocci). Cells that divide in one plane form chains (figure 3.12 *a*), while those dividing in several planes at random form clusters (figure 3.12 *b*). If all division occurs in sequence in two or three planes perpendicular to one another, cubical packets result (figure 3.12 *c*).

The bacteria that form long chains are frequently referred to as *streptococci.* Those that occur as two cells joined together are *diplococci,* while those that occur in clusters resembling bunches of grapes are called *staphylococci.* Note that these names are not specific genera. Other bacteria typically form packets of four or cuboidal packets of eight or more cells. The process of cell division will be covered in chapter 4.

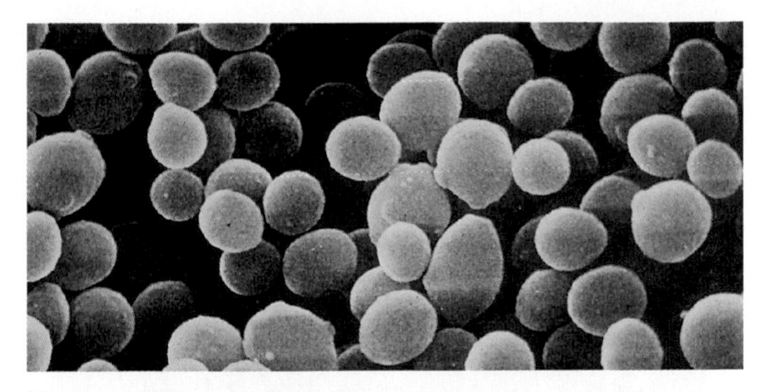

(a)

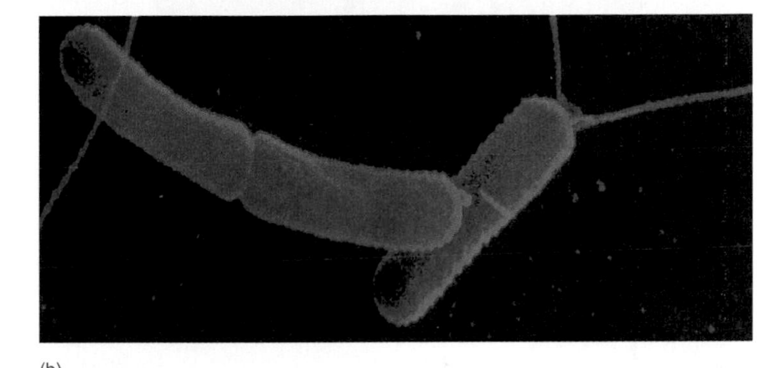

(b)

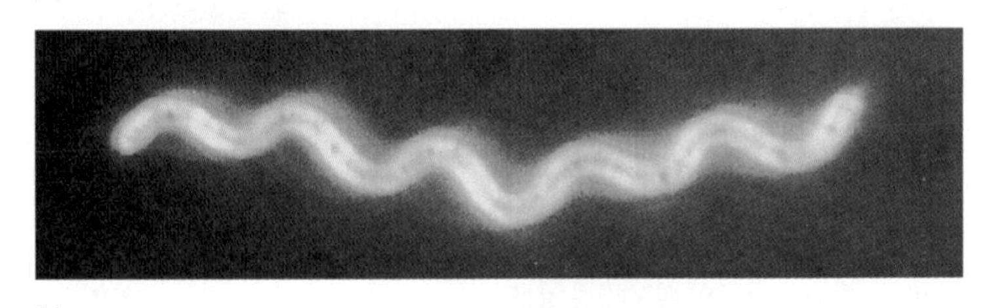

(c)

(d)

Figure 3.10
The three most common shapes of typical bacteria as viewed through a scanning electron microscope: (*a*) spherical (coccus); (*b*) cylindrical or rod-shaped (bacillus); and (*c*) spiral-shaped (spirillum). (*d*) Coccobacillus.

(a)

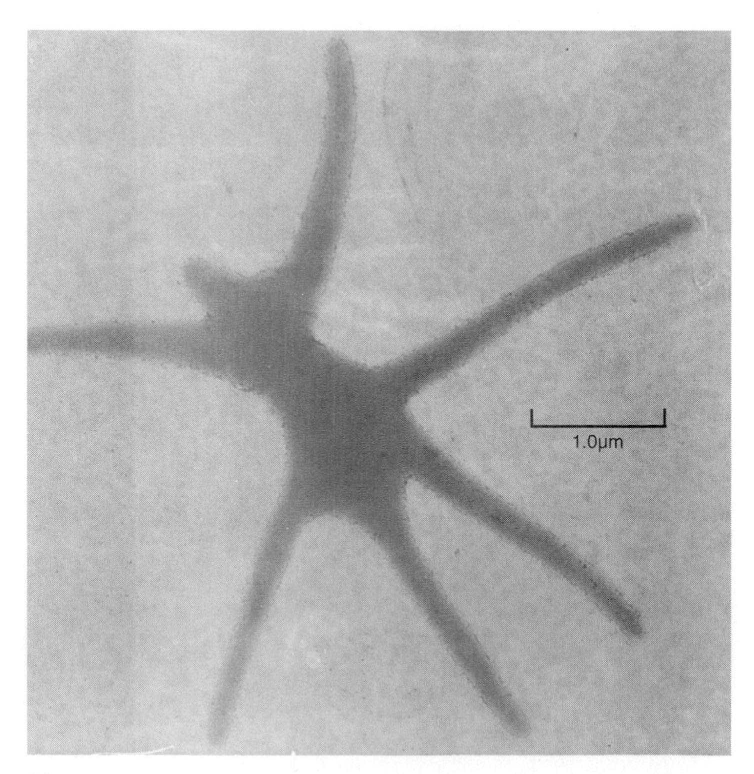

(b)

Figure 3.11
(*a*) Square bacterium. (*b*) *Ancalomicrobium*, an example of a prosthecate bacterium. Note that the cytoplasmic membrane and the cytoplasm are a part of each arm.

Functions Required in Prokaryotic Cells

The structures of the cell can be learned best in terms of their functions. Accordingly, we will discuss functions first and then the structures responsible for them. Because the emphasis of this course is on bacteria, we will emphasize the prokaryotic cell. Later in this chapter, we will discuss the structure of eukaryotic cells. Some structures are essential to the life of the cell, and others provide the cell with additional, useful capabilities.

The key features of prokaryotic anatomy must provide for

1. enclosure of the internal contents of the cell and separation from the external environment;
2. replication (duplication) of genetic information;
3. synthesis of all components of the cell;
4. generation, storage, and utilization of energy rich compounds; and
5. entry and exit of specific molecules.

Additional but nonessential functions include

1. cell movement;
2. transfer of genetic information to other cells;
3. storage of reserve materials such as carbohydrates, which later can be used for production of energy; and
4. formation of a different cell type called an *endospore.*

Prokaryotic Cells—Microscopic Appearance

An electron micrograph and diagrams of a typical prokaryotic cell, a bacterium cell of cylindrical shape, are given in figure 3.13.

This organism has structural features typical of many prokaryotes: a rigid cell wall and flagella giving it motility. Like all prokaryotes, it is unicellular and its internal structures are not enclosed in membranes.

We will now consider in some detail the structures that function to give bacteria their distinctive features.

Enclosure of the Cytoplasm in Prokaryotic Cells

Most bacteria have two structures, the **cell wall** and the **cytoplasmic membrane,** that surround the cell's gooey cytoplasm, which has the consistency of gelatin. Some bacteria have a third layer, the **capsule,** or **slime layer** (figure 3.14 *a* and *b*). Collectively, these structures are referred to as the cell's **envelope.** The discussion that follows will

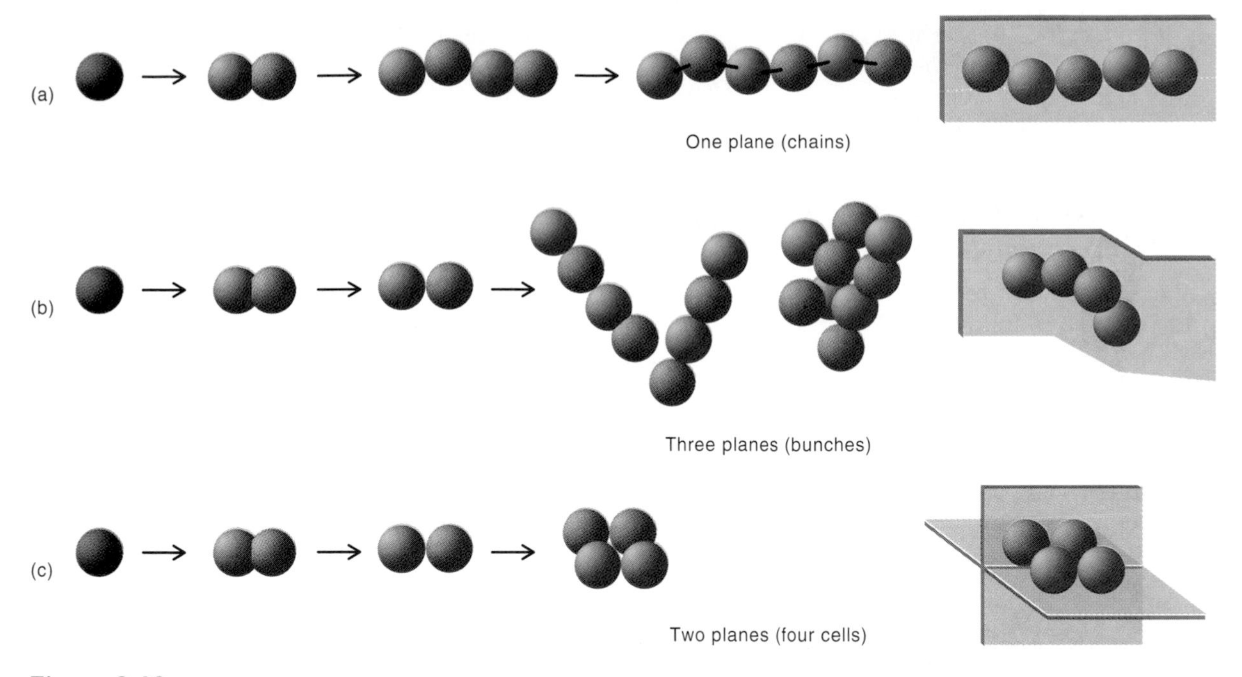

Figure 3.12
(*a–c*) Various cell arrangements. The reason for the different arrangements of cocci is that the orientation of planes in which the cells divide determines the arrangement of the cells. Arrangements of cells are specific for different cocci.

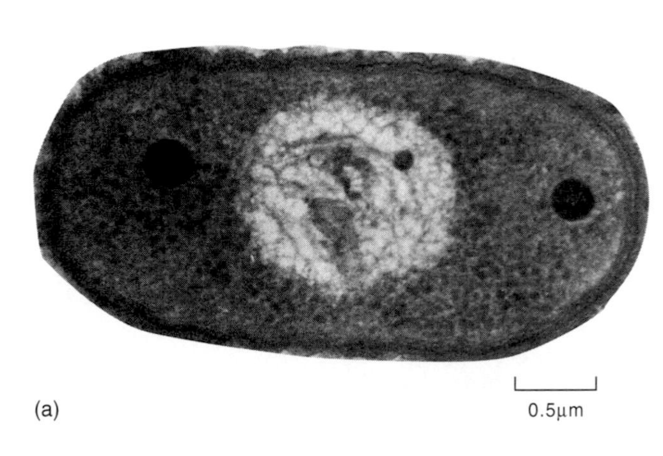

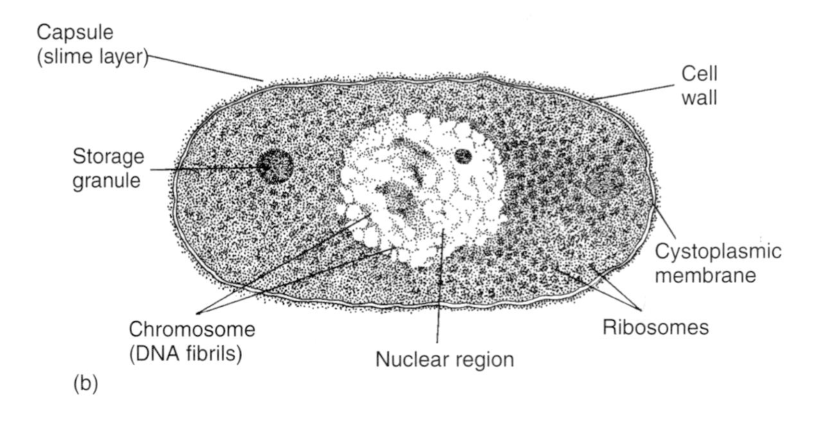

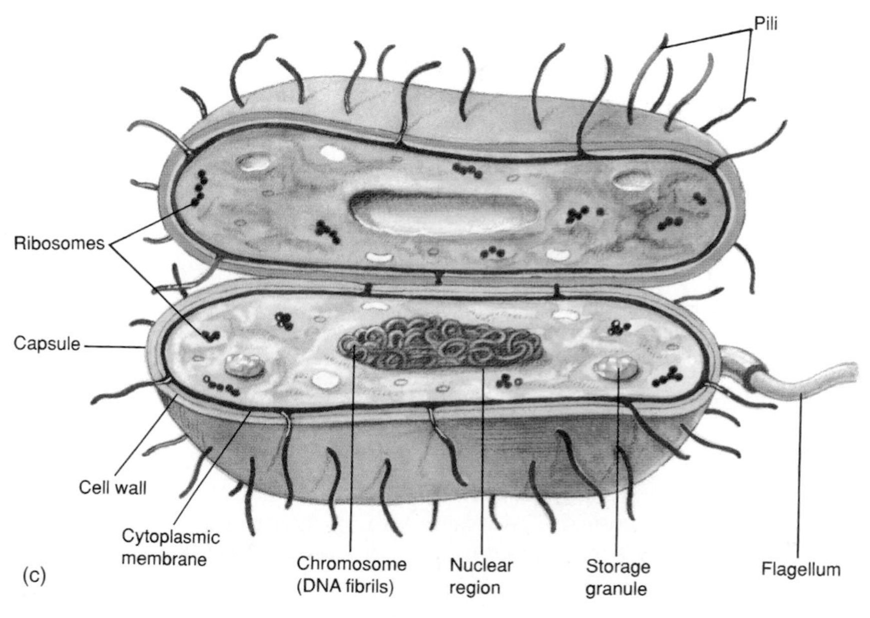

Figure 3.13
Typical prokaryotic cell. (*a*) Transmission electron photomicrograph of a prokaryotic cell, a rod-shaped bacterium, *Eikenella corrodens*, which is commonly found in the mouth. (The capsule is not well developed.) (*b*) Diagram of (*a*). (*c*) A three-dimensional diagrammatic representation of a typical bacterial cell.

proceed from the outermost (capsule) to the innermost (cytoplasmic membrane) layer. The cytoplasm membrane is often referred to as the **inner membrane** in cells that have an outer membrane.

The Capsule or Slime Layer Is a Gelatinous Structure that Can Attach Bacteria to Surfaces

The gelatinous capsule may be thick and regular or loose fitting and irregular in appearance. When loose fitting and irregular, it's usually referred to as a *slime layer.*

Capsules are most easily seen with negative staining, outlined as a light area against a darkened background (figure 3.15 *a*).

Colonies with capsules often appear moist, glistening, and slimy (compare *b* and *c* in figure 3.15). Capsules are produced only by certain bacteria and often only when certain nutrients that can be converted into these capsules are present.

In nature, the capsule may take the form of many tiny, short, hairlike structures or fibrils that often form on the

• **Negative stain, p. 50**

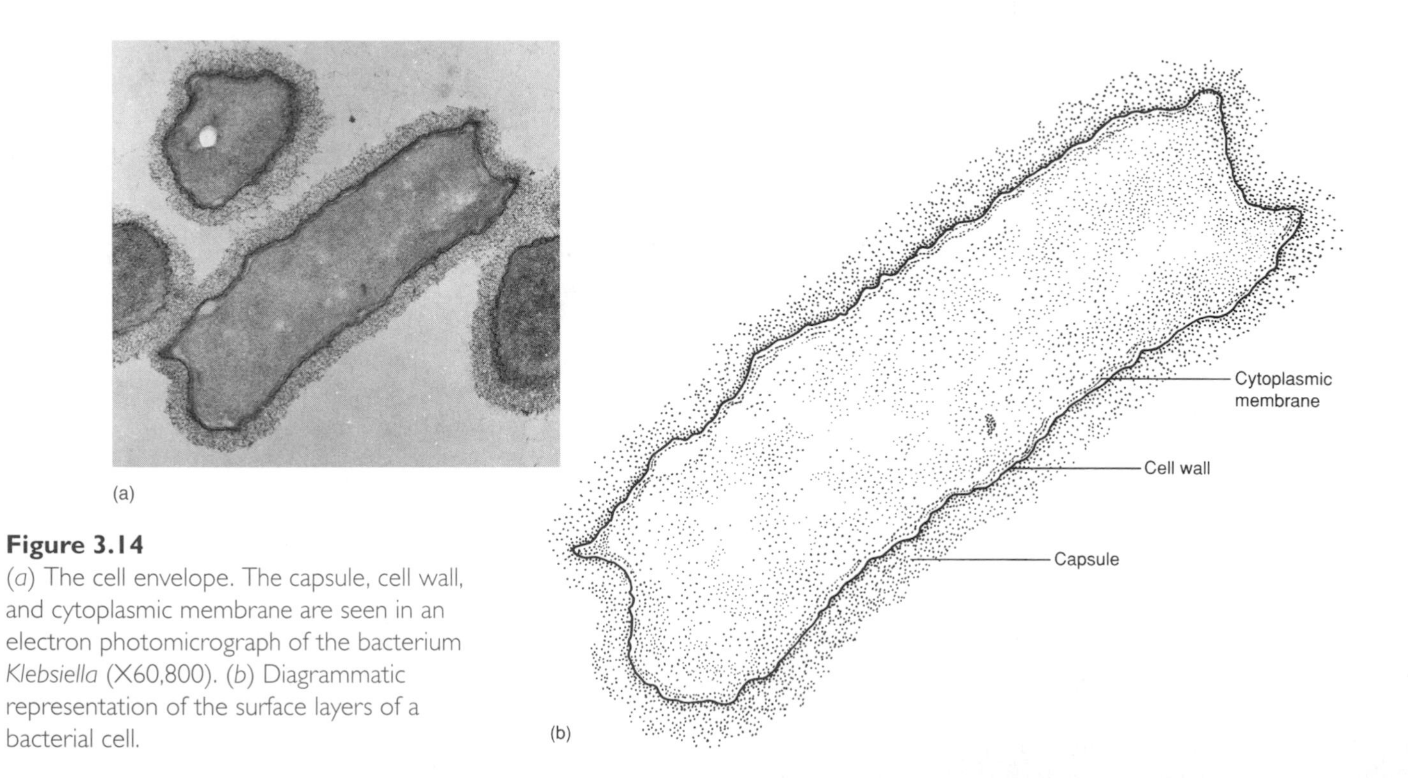

Figure 3.14
(*a*) The cell envelope. The capsule, cell wall, and cytoplasmic membrane are seen in an electron photomicrograph of the bacterium *Klebsiella* (X60,800). (*b*) Diagrammatic representation of the surface layers of a bacterial cell.

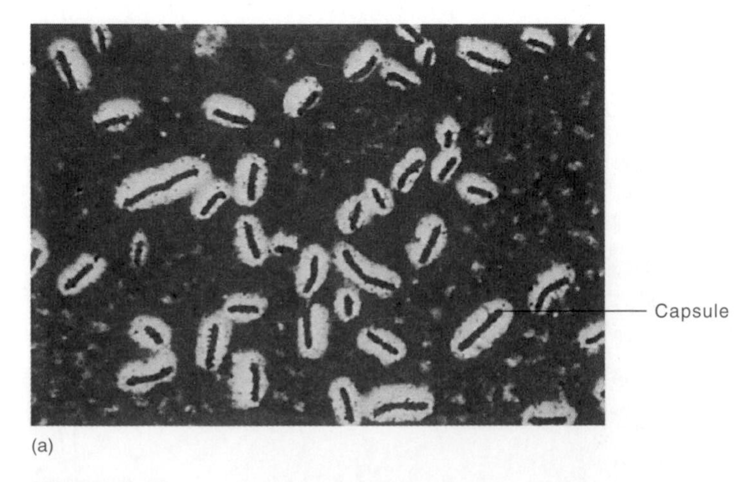

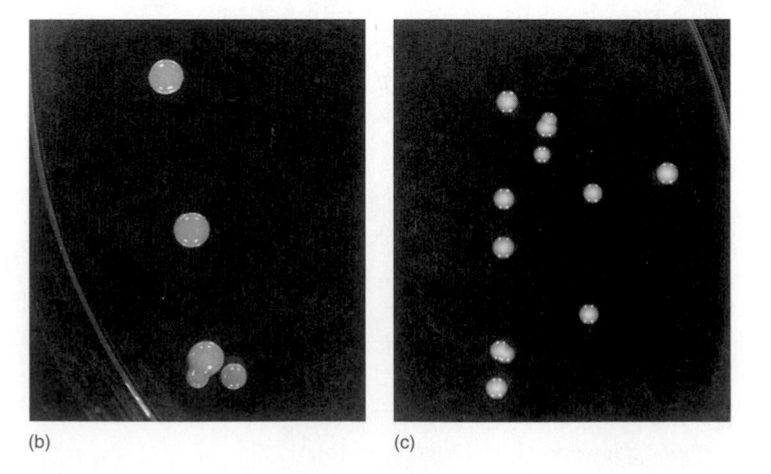

Figure 3.15
(*a*) Capsules as demonstrated by negative staining with India ink. The process of color enhancement has changed the usual black background of India ink to the red background seen here. The bacterial cytoplasm has been stained with a positive stain to increase the contrast (X410). Colonial morphology of *Streptococcus salivarius* growing on a medium with (*b*) and without (*c*) sucrose (X3,670). This organism forms a capsule if sucrose is present in the medium, and the capsule imparts a larger, gumdrop appearance to the colonies.

outside of the cell wall. These fibrils, frequently termed the **glycocalyx,** were only discovered relatively recently because bacteria grown under laboratory conditions generally do not synthesize them and they are not easily seen by microscopic techniques.

Chemistry and Function of Capsules (or Slime Layer or Glycocalyx)

Capsules vary in their chemical composition depending on the species of bacteria. Most are composed of polysaccharides, which are given the general name *exopolysaccharides* (*exo* means "outside"), but some consist of polypeptides made up of repeating subunits of only one or two amino acids. Of special interest is the fact that these amino acids are generally of the D-, or unnatural, form.

Capsules have several functions. First, they protect the bacteria under certain situations. This is best illustrated in the case of the organism that causes bacterial pneumonia, *Streptococcus pneumoniae.* Unless this organism synthesizes a capsule, it cannot cause pneumonia, since the unencapsulated organism is quickly destroyed by the defenses of the infected animal. Apparently, the phagocytic cells of the body, which engulf and then kill the microorganisms, have great difficulty surrounding the encapsulated organisms, whereas the unencapsulated organisms are quickly engulfed and destroyed.

Second, capsules serve to attach bacteria to a wide variety of surfaces, including human teeth, plant roots, the small intestine, rocks, and even other bacteria (figure 3.16 *a–c*). A familiar example of the role of the capsule (glycocalyx) in attachment is the formation of dental plaque. *Streptococcus mutans,*[1] the major organism causing dental caries, accumulates in large masses on the surface of the teeth (plaque). This attachment is greatly strengthened by a capsule made of glucan, a polysaccharide consisting of glucose subunits. The glucan is synthesized only from the disaccharide sucrose, composed of glucose and fructose after it is degraded into its subunits. This explains why the table sugar sucrose in the diet enhances dental decay, even though *S. mutans* can multiply in the laboratory on many other sugars.

The Bacterial Cell Wall Is Rigid and Is Composed of Unique Chemical Structures

The bacterial cell wall deserves special attention for several reasons: (1) It is composed of subunits found nowhere else in nature; (2) the cell wall and many of its component parts of many species of bacteria can produce symptoms of disease; (3) the cell wall is the site of action of some of the

[1] *Streptococcus mutans* is so named because it can change in shape from spherical to rodlike, in a process similar to a mutation.

• **Amino acid, p. 32**
• **Sucrose, p. 36**

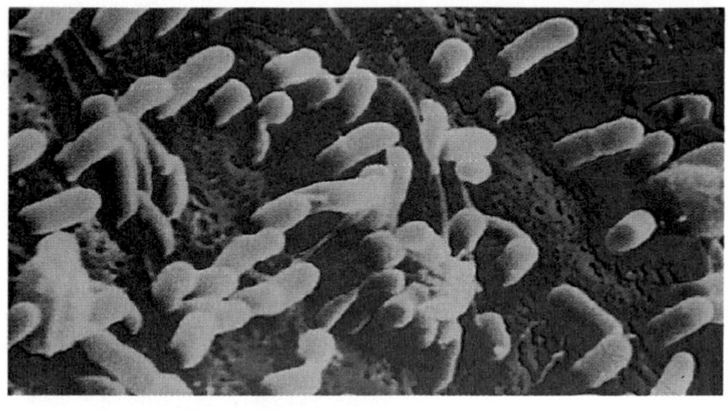
(a)

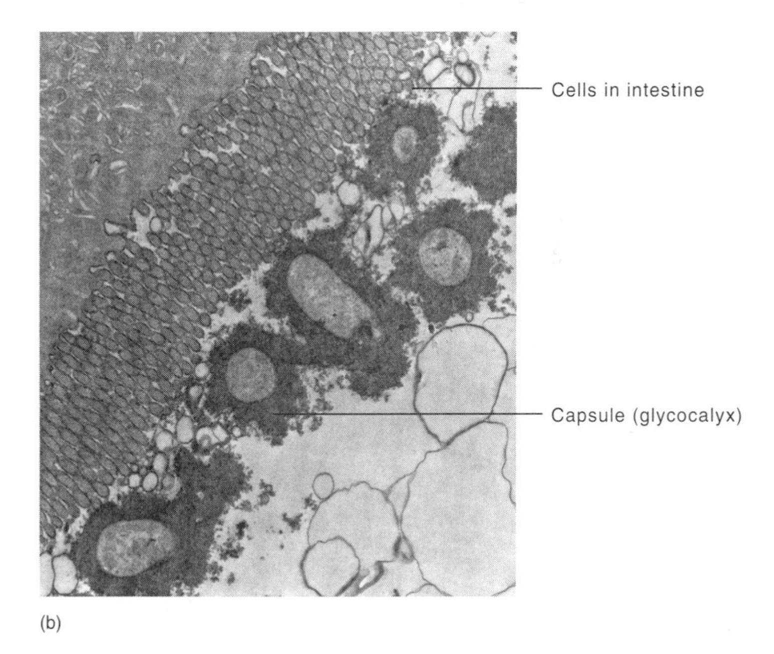

(b)

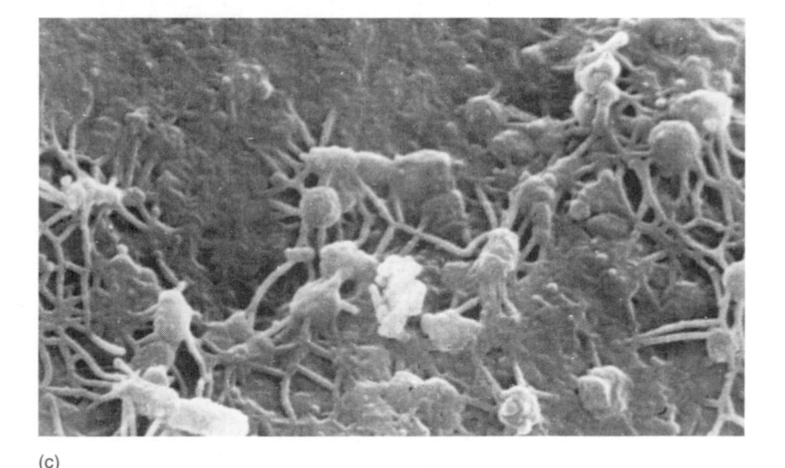
(c)

Figure 3.16
Attachment of bacteria to various surfaces. (*a*) Scanning electron micrograph of bacteria attached to the enamel surface of a tooth. (*b*) Cells of *Escherichia coli* attached to cells in the small intestine. (*c*) Masses of cells of *Eikenella corrodens* covered with slime, which joins the cells together.

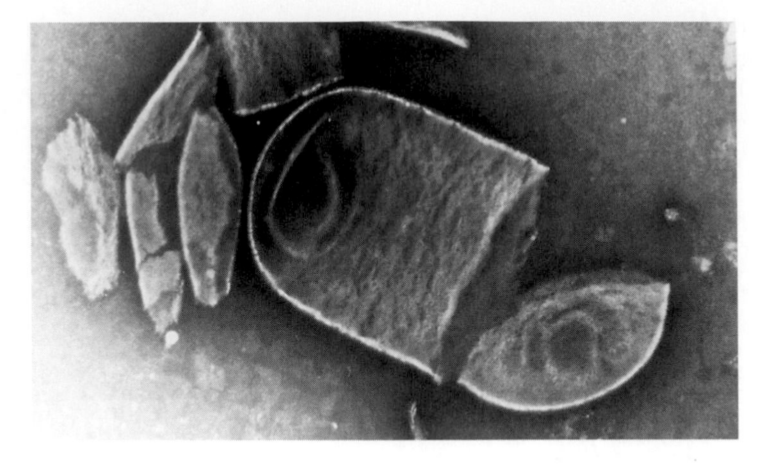

Figure 3.17
Rigidity of the bacterial cell wall. Note that the cell has split apart, but the cell wall maintains its original shape.

most effective antibiotics; and (4) differences in the chemical composition of the cell wall determine the gram-staining properties of the cell. The cell wall also determines the shape of the organism—for example, cylindrical cells have a cylindrical cell wall, and spherical organisms have a spherical cell wall. If the cell wall breaks, the cell wall does not collapse but maintains its rigid shape (figure 3.17).

Chemistry of the Bacterial Cell Wall

The chemical structure of the cell wall is responsible for its rigid nature. Although different species of bacteria have different cell wall structures, almost all eubacteria contain one particular macromolecule that is found only in bacteria. This macromolecule, known as **peptidoglycan,** consists of two major subunits. The backbone, or **glycan** portion, is composed of two different amino sugars, *N*-acetylmuramic acid and *N*-acetylglucosamine (figure 3.18). These sugars, which are related to glucose in their structure, alternate with one another through covalent bonds to form a high molecular weight polymer.

The peptide portion of the molecule consists of a chain of several different amino acids that are attached to each of the *N*-acetylmuramic acid molecules in the glycan portion of the molecule. These join adjacent glycan backbones, so that the peptidoglycan is in fact a single, very large molecule. Only a few of the 20 amino acids usually found in proteins occur in the peptidoglycan, and its exact composition varies among different bacteria. How the peptidoglycan forms a part of the cell wall will be considered shortly.

One unusual amino acid, diaminopimelic acid, which is related to the amino acid lysine, is found in no other place in nature except in certain bacterial cell walls. Also, many of the amino acids in cell walls are of the D-configuration, whereas those in proteins are of the L-configuration.

Not all prokaryotes have peptidoglycan in their walls. The Archaea do not contain either *N*-acetylmuramic acid or D-amino acids in their cell walls, differences that help distinguish them from members of the Eubacteria.

Although peptidoglycan is found only in bacterial cell walls, it is related chemically to cellulose found in plant cell walls and to chitin found in insect and crustacean exoskeletons,[1] as well as in fungal cell walls and the walls of certain protozoa.

Cell Walls of Gram-Positive Bacteria

The cell walls of most gram-positive bacteria consist mainly of numerous layers of glycan, up to 30, with amino acid bridges connecting each layer of glycan to other layers of glycan above and below it. In this way, a single large three-dimensional molecule is formed (figure 3.19 *a–c*).

Gram-positive bacteria usually have other components attached to the peptidoglycan layer. These are the **teichoic acids** (*teichos* means "wall" in Greek). The teichoic acids are polymers of subunits, which consist of gylcerol or ribitol bound to phosphate molecules as well as to various sugars and D-alanine. A typical subunit of a teichoic acid polymer is shown in figure 3.20. The teichoic acids are covalently bonded to the outside portion of the peptidoglycan layers and sometimes they are associated with the cytoplasmic membrane.

[1] Exoskeleton—external skeleton

• **Archaea, p. 13**
• **Eubacteria, p. 13**

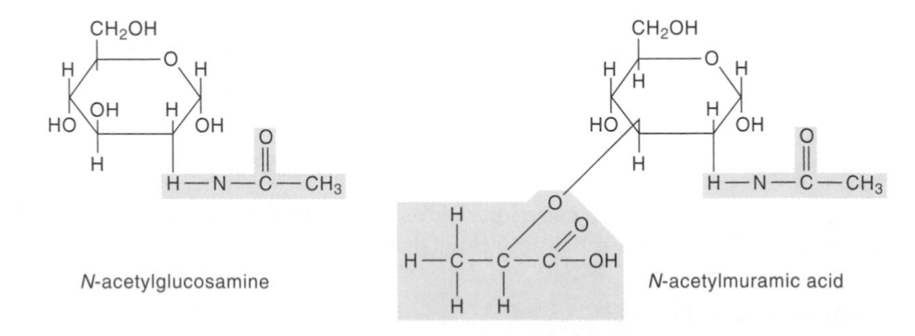

Figure 3.18
The chemical structure of *N*-acetylglucosamine and *N*-acetylmuramic acid. The nonshaded portions of the two molecules are glucose.

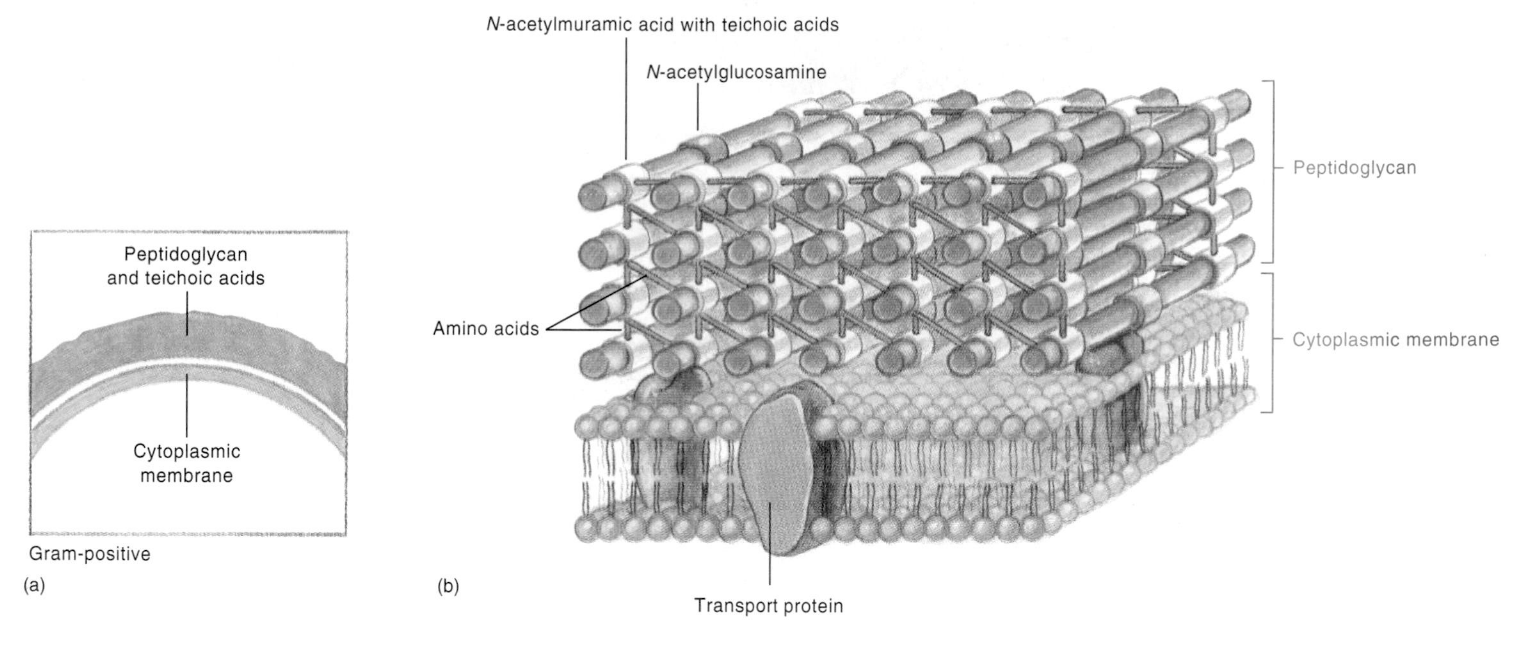

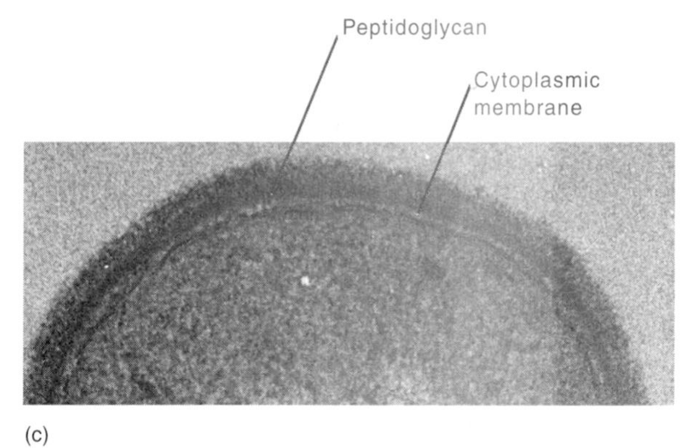

Figure 3.19 
Gram-positive cell wall and cytoplasmic membrane. (*a* and *b*) Schematic diagrams. Note that there are many layers of peptidoglycan held together by amino acid bridges. There is no periplasm. The teichoic acids associated with the outer layers of the peptidoglycan are not shown. (*c*) Transmission electron micrograph of a typical gram-positive cell wall (*Bacillus fastidiosis*, ×90,000).

D - glucose
D - alanine
D - alanine
Ribitol

Figure 3.20
A subunit of a typical teichoic acid. These acidic polysaccharides, which contain the sugar alcohol ribitol or glycerol, also contain glucose and D-alanine. The subunits are joined through ester bonds through the phosphate molecules.

Cell Walls of Gram-Negative Bacteria

The cell walls of gram-negative bacteria are more complex than the walls of gram-positive organisms. On the outside of the peptidoglycan layer is an **outer membrane** (figure 3.21 *a–c*). This consists of a lipid bilayer that has several other molecules embedded in it. These include **lipopolysaccharides,** or **LPS, lipoproteins,** and **porin proteins.** The lipopolysaccharides often have distinctive chains of sugar molecules useful in identifying the strain of organism.

The phospholipids form a bilayer with their hydrophobic tails oriented toward each other and away from the aqueous environment. The LPS is attached to the outer of the two layers, and the lipoprotein is covalently attached to the inner layer. Porins traverse both phospholipid layers of the outer membrane and serve as channels for the entrance and the exit of small molecules.

The LPS part of the outer membrane is of special interest because it is often toxic. Injection of purified lipopolysaccharides from gram-negative bacteria into the

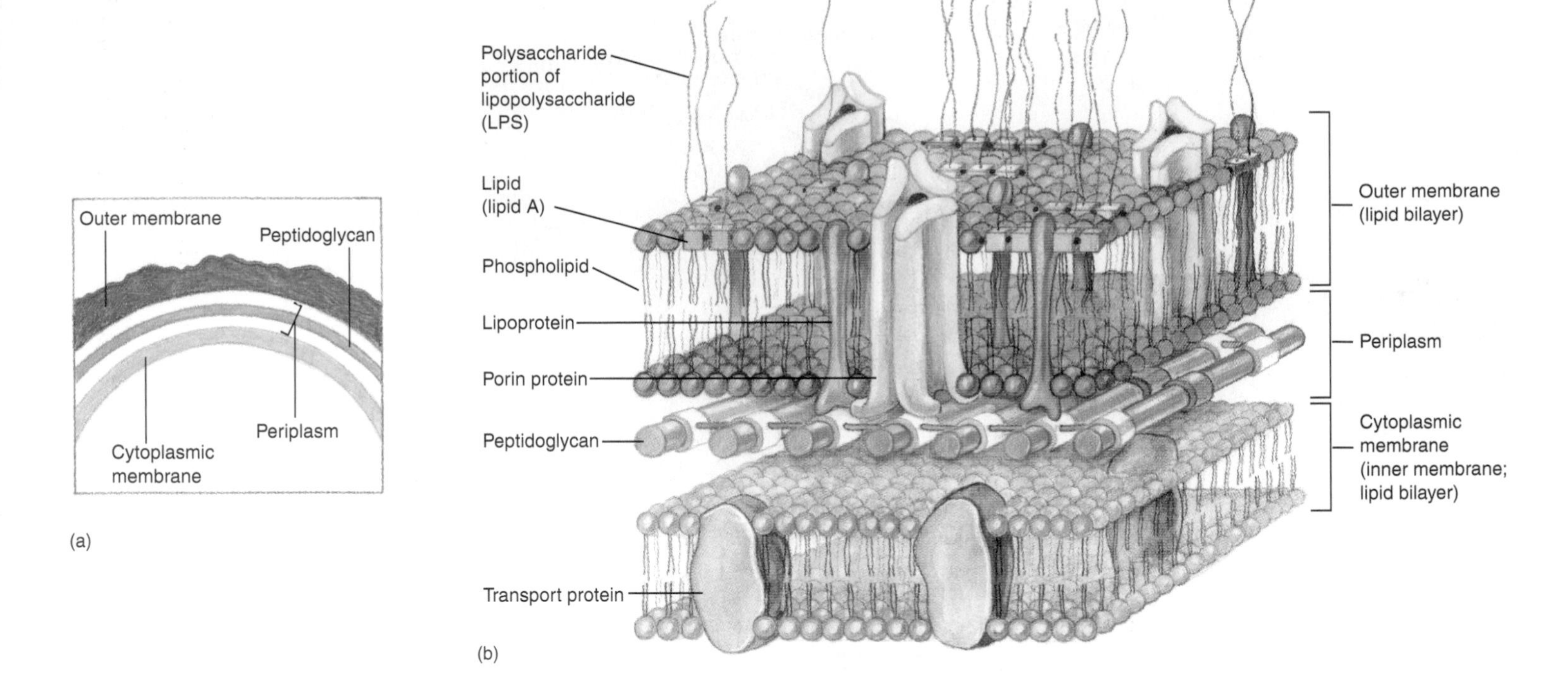

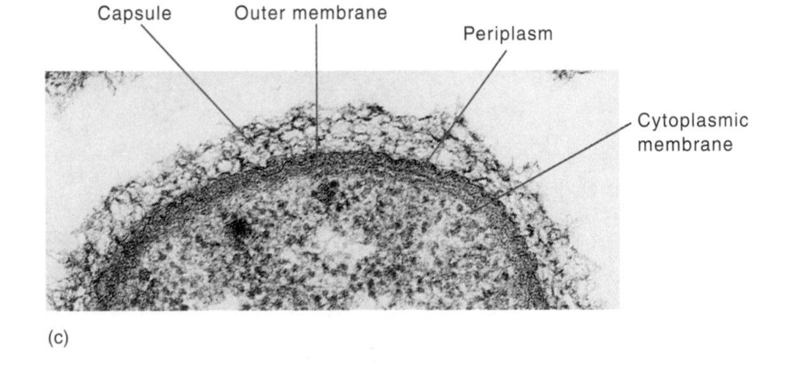

Figure 3.21
(*a* and *b*) Schematic diagrams of the wall of a gram-negative cell. Note that there is only a single layer of peptidoglycan. A periplasm and a complex outer membrane are both present. (*c*) A transmission electron micrograph of a typical gram-negative cell wall (*Pseudomonas aeruginosa*, ×120,000).

host often produces symptoms characteristic of the infections caused by whole organisms. The same symptoms occur regardless of the species of gram-negative organisms. The toxic element is the lipid portion of the lipopolysaccharide (lipid A), and is termed *endotoxin.*

The cell walls of gram-negative bacteria also contain one or more layers of peptidoglycan but far fewer than the numerous layers that occur in gram-positive organisms. Interestingly, the violent coughing characteristic of people with whooping cough is caused by subunits of the peptidoglycan released during growth of the gram-negative bacteria causing the disease. In gram-negative cells, the peptidoglycan layer is located in the **periplasm,** the region between the cytoplasmic membrane (inner membrane) and the outer membrane (figure 3.21 *a–c*). The periplasm, which also contains many different degradative enzymes such as deoxyribonucleases and proteases, is not found in gram-positive cells since these cells do not have an outer membrane. The peptidoglycan layer is joined to the outer membrane by means of lipoprotein molecules.

Functions of the Bacterial Cell Wall

The cell wall functions to hold the cell together and prevent if from bursting. Peptidoglycan is the major molecule that provides the rigid structure to the wall. Since gram-positive bacteria contain so many more layers of peptidoglycan, they are much tougher than gram-negative cell walls and more difficult to break. Bacteria normally grow, both in nature and in the laboratory, in water containing small amounts of some salts and other small molecules. However, the bacterial cytoplasm is a very concentrated solution of inorganic salts, sugars, amino acids, and various other small molecules, which collectively are termed the **solute.** Since the concentration of particles (the number of molecules and ions per unit volume) tends to equalize on the inside and outside of the cell, water flows from the medium into the cell, thereby reducing the concentration of molecules and ions inside the cell. This inflow exerts tremendous pressure, **osmotic pressure,** on all structures that enclose the cytoplasm, and it pushes the elastic cytoplasmic membrane against the cell wall. The rigid cell wall can withstand this pressure; without it, the cell would simply balloon out until it burst. If the bacteria are placed in an environment in which the salt concentration is higher on the outside than on the inside of the cell, then the cell wall retains its shape although water leaves the cell through the phenomenon of **plasmolysis.** This will be considered later in this chapter.

Often, the cell wall becomes less important if the bacterium is in an environment that contains a high concentration of low molecular weight compounds such as salt. Members of the Archaea found in the Great Salt Lake in Utah (30% NaCl) have a cell wall without a peptidoglycan layer, a common feature of this domain. These microorganisms, termed **halophiles** (salt loving), can live quite well at this high-salt concentration and have no need for a strong cell wall in their natural environment. However, if the salt concentration is lowered to 15%, water flows in and the cells burst.

The consequences of removing the cell wall can also be demonstrated in the laboratory. The wall can be destroyed by treating the cell with **lysozyme,** an enzyme that breaks the chemical bonds between the *N*-acetylmuramic acid and *N*-acetylglucosamine subunits. Figure 3.22 illustrates the results of this procedure. In the environment in which most bacteria grow, whether in nature or in the laboratory, the cells treated with lysozyme balloon and burst. However, in high-salt or high-sugar solutions, the treated cells do not burst but rather become spherical, the cells' most stable shape. These wall-less cells are termed **protoplasts,** or **spheroplasts.**[1]

[1]Strictly speaking, the term *protoplasts* describes gram-positive organisms in which the peptidoglycan has been removed. **Spheroplasts** are their gram-negative counterparts, lacking the peptidoglycan but containing other cell wall components.

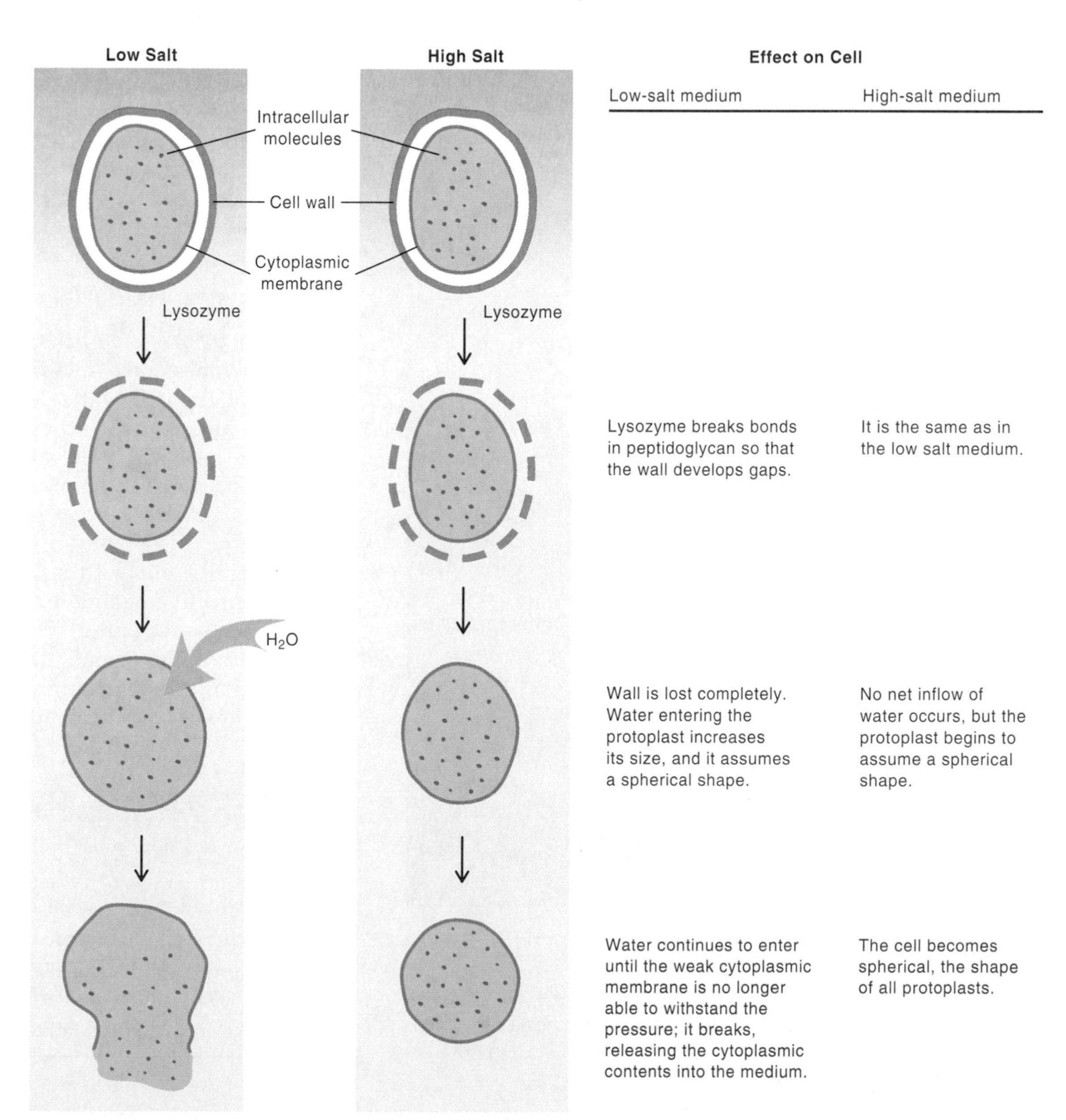

Figure 3.22
Effect of disruption of the cell wall by lysozyme in a low-salt and a high-salt medium. The concentration of solute molecules tends to equalize on the inside and outside of the cell. In a low-salt medium, water enters the cell in order to reduce the solute concentration.

The cell walls of both gram-positive and gram-negative bacteria are permeable to small molecules, of molecular weights of 5,000 and lower, but not to macromolecules. Even though the walls of gram-negative cells have a lipid bilayer, small hydrophilic molecules can pass because the protein porins that traverse this layer serve as channels for their entry (see figure 3.21). Some porins are specific for the molecule that pass through them; others are nonspecific and allow many different small molecules to pass. In part, the size of the porin determines the size of the molecule that can pass.

Wall-Deficient Organisms

Although the discussion thus far has emphasized the importance of the cell wall to bacteria in their natural environment, some microorganisms, such as members of the genus *Mycoplasma,* lack a cell wall. *Mycoplasma pneumoniae,* a cause of human pneumonia, grows slowly, and in the laboratory it requires a special medium, generally containing 20% serum,[1] to grow. To compensate for their lack of a wall, this group of bacteria has a stronger cytoplasmic membrane than most bacteria, apparently because the membrane contains **sterols,** which in some unknown way strengthen the membrane. Since they lack a rigid cell wall, mycoplasmas tend to be extremely variable in shape (figure 3.23).

[1]Serum—the amber-colored fluid that exudes from coagulated blood after blood clots

• **Sterols, p. 41**

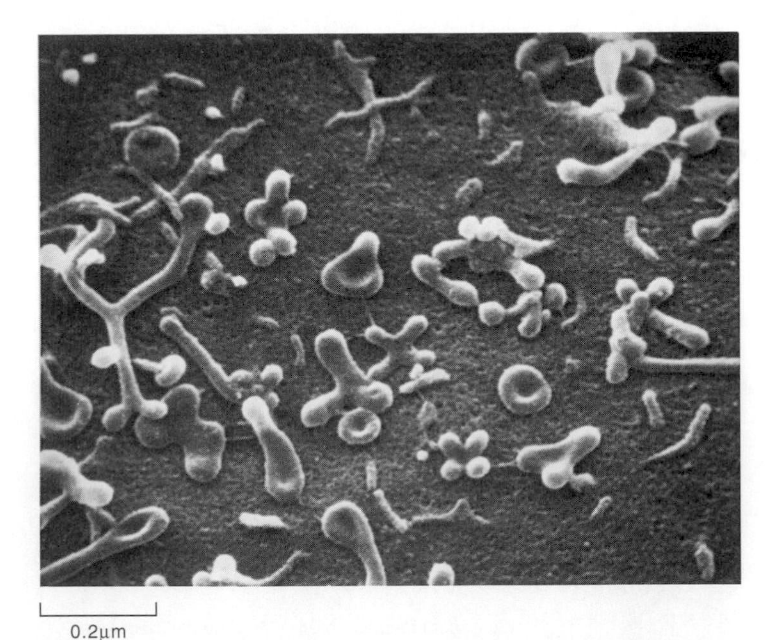

Figure 3.23
Mycoplasma pneumoniae (×16,300). This organism has a plastic shape because it lacks a rigid cell wall.

Consequences of Differences in Cell Wall Composition

As already pointed out, different bacteria stain differently when they are treated with the same dyes. On this basis, most bacteria can be separated into two groups: the gram-positive and the gram-negative. Differences in their cell wall composition account for their different staining characteristics, although it is not the wall but rather the inside of the cell that is stained. Gram-positive organisms retain the crystal violet stain after they are treated with iodine and washed with alcohol, whereas the alcohol washes out the crystal violet-iodine complex from gram-negative cells. How can differences in wall composition explain this? Apparently, the dye-iodine complex in alcohol becomes trapped in gram-positive organisms, whereas the gram-negative cell wall is permeable. This explanation also accounts for the observation that old cultures of gram-positive cells frequently become gram-negative in their staining properties. It is known that enzymes (termed *autolysins*) that damage the cell wall become active in cells of old cultures. Such damage increases the permeability of the cell wall to the dye-iodine complex, so that it dissolves and then is washed out by the alcohol.

Penicillin

The effectiveness of penicillin also depends on the composition of the cell wall. Penicillin is toxic to bacteria because it affects the synthesis of the peptidoglycan portion of the cell wall. The accepted explanation is that penicillin prevents the amino acid bridges from forming between adjacent glycan layers, resulting in a weak cell wall.

Generally, but with notable exceptions, penicillin is far more effective against gram-positive cells than gram-negative cells. Apparently, in gram-negative cells, the outer membrane prevents the penicillin from reaching its site of action, the peptidoglycan layer. (Other aspects of penicillin are covered in chapter 21.)

REVIEW QUESTIONS

1. What four functions must prokaryotic cells be able to carry out?
2. What three layers enclose the cytoplasm of bacteria?
3. What compound is unique to the cell walls of bacteria?
4. What two properties of bacteria are determined by the chemical composition of their cell walls?
5. What properties do capsules confer on bacteria?

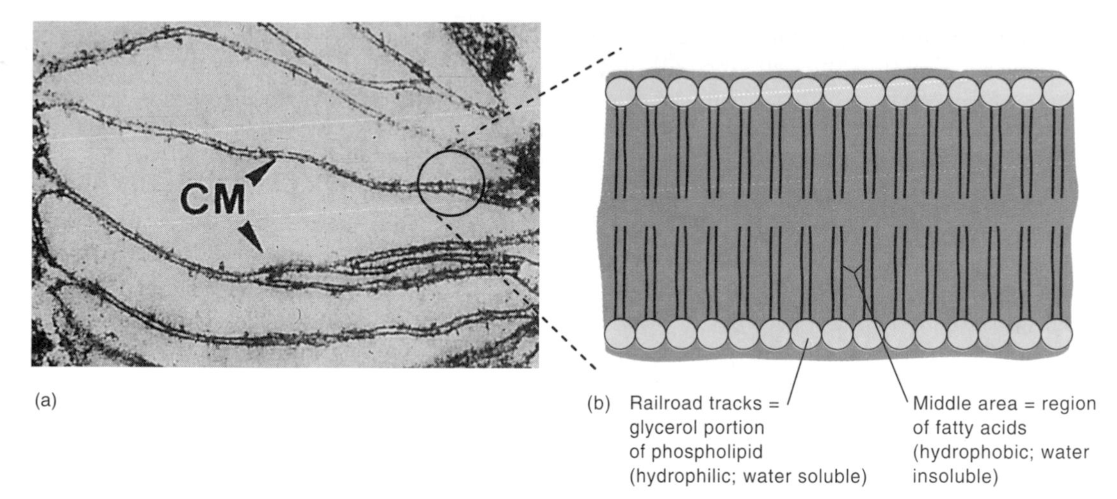

Figure 3.24 Cell membranes.
(*a*) Membranes in the cytoplasm (CM) of *Nitrosocystis oceanus.* The thin section was prepared from a partially disrupted cell (×36,000). The hydrophilic regions are the two dark "railroad tracks"; the light area between consists of hydrophobic fatty acids.
(*b*) Enlarged schematic view of membranes shown in (*a*).

The Cytoplasmic Membrane Is Flexible and Is Composed of a Lipid Bilayer in All Cells

The **cytoplasmic membrane,** sometimes simply called the **plasma membrane,** surrounds the cytoplasm both in cells that contain and those that lack a cell wall. In most bacteria, in dilute medium, the pressure on the inside of the cell forces this thin, delicate, elastic membrane against the peptidoglycan portion of the cell wall.

Chemical Composition

The cytoplasmic membrane isolated from eubacteria contains approximately 60% protein and 40% lipid, primarily in the form of a phospholipid bilayer. Whereas the cytoplasmic membrane of the Eubacteria contains fatty acids, those in the Archaea do not. They contain a different hydrophobic compound in their lipids. The presence or absence of fatty acids in the cytoplasmic membrane is a major distinguishing feature between these prokaryotes.

Structure

When viewed in the electron microscope, the cytoplasmic membrane appears virtually the same whether isolated from prokaryotic or eukaryotic cells: two dark bands separated by a light band (figure 3.24 *a*). The term **unit membrane,** or **bilayer membrane,** has been given to these membranes. Their appearance results from the chemical structure of the phospholipid component because of the ordered arrangement the phospholipids assume in an aqueous environment (figure 3.24 *b*).

Although all membranes have this same basic structure, membranes play so many different physiological roles that they must differ in certain ways. Different protein molecules are embedded in the phospholipid bilayer, which accounts for some of the different functions that membranes perform. These proteins can move sideways within the membrane, so that the structure of the membrane is changing constantly.

- **Lipid, p. 38**
- **Fatty acid, p. 40**
- **Protein, p. 30**
- **Phospholipid bilayer, p. 42**

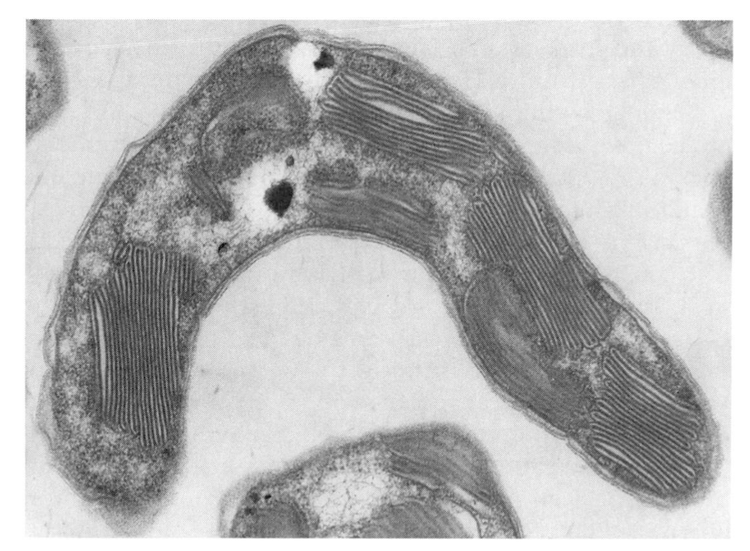

Figure 3.25
Extensive infoldings of the cytoplasmic membrane in the photosynthetic bacterium *Ectothiorhodospira mobilis* (×86,200).

Functions of the Cytoplasmic Membrane

The cytoplasmic membrane performs several functions vital to the life of the cell. In prokaryotic cells, many of the enzymes involved in obtaining energy from the breakdown of foodstuffs are attached to the cytoplasmic membrane **(membrane-bound enzymes).** Most other enzymes in bacteria are located in the cytoplasm **(soluble enzymes).** Some bacteria have an especially extensive array of membranes. For example, photosynthetic bacteria have an abundant array of membranes that are involved in converting light into chemical energy (figure 3.25).

The membrane also serves as a barrier to the entry of most molecules into the cell since all nutrients entering the cell (and all waste products of metabolism leaving the cell) must pass through the cytoplasmic membrane. Only small molecules whose molecular weights are no greater than several hundred can pass through freely without special mechanisms. Also, these molecules must be hydrophobic in order to pass through the hydrophobic membrane. Since most molecules in the cell's environment are polar and hydrophilic (they must be soluble in water), they require special mechanisms in order to pass through the membrane. However, there is one very important exception—the water molecule—that is small enough to pass between the phospholipid molecules. Therefore, the cytoplasmic membrane is **semipermeable,** allowing water but very few other molecules to pass through by themselves.

Since bacteria can multiply very rapidly, mechanisms must exist for nutrients to enter the cell fast enough to provide the growing cell with a source of energy and small molecules that can be converted into cellular components such as macromolecules. How do molecules get into the cell through this membrane barrier?

Many compounds enter into the bacterial cell by one of two distinct mechanisms: **diffusion** and **active transport.** Two distinct features separate these two mechanisms. First, in diffusion, the cell does not use energy to move molecules from the outside to the inside of the cell, whereas in active transport, energy is required. Second, in diffusion, the molecules pass through the membrane until their concentration is the same on both sides of the membrane. In active transport, however, the concentration of any molecule being transported can be much higher inside than outside the cell.

Diffusion There are two kinds of diffusion: **passive,** or **simple, diffusion** and **facilitated diffusion.** In passive diffusion, the molecules flow freely into and out of the cell (figure 3.26). Since only small hydrophobic molecules can pass through the hydrophobic lipid bilayer, passive diffusion is not an important transport mechanism in prokaryotes. Further, passive diffusion moves nutrients too slowly to allow rapid growth of bacteria.

Consequently, bacteria use **membrane transport proteins,** sometimes called **permeases,** to carry some small molecules into the cell. This process is an example of facilitated diffusion. The way in which this mechanism operates is not known, but one idea is that the membrane transport protein spans the phospholipid bilayer with one end protruding into the periplasm and the other end sticking out into the cytoplasm. The small molecule in the periplasm binds to the transport protein and may change its shape so that the small molecule can pass through the other end and on into the cell (figure 3.27).

As in simple diffusion, no energy is expended in this transport process, which only operates as long as the concentration of a particular molecule is higher on the outside

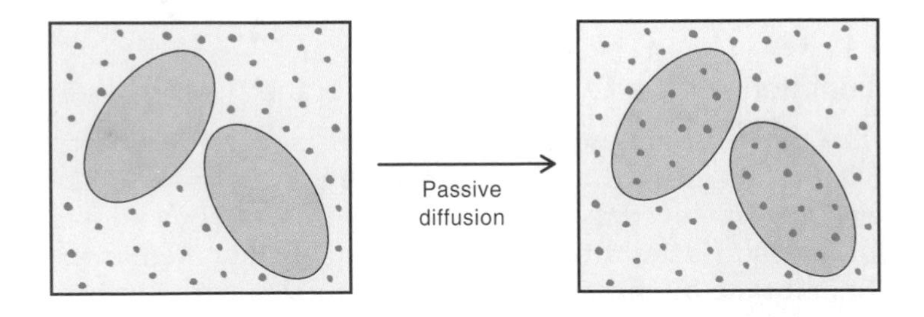

Figure 3.26
Passive diffusion. Following diffusion, the concentration of any particular molecule is the same on the inside as it is on the outside of the cell.

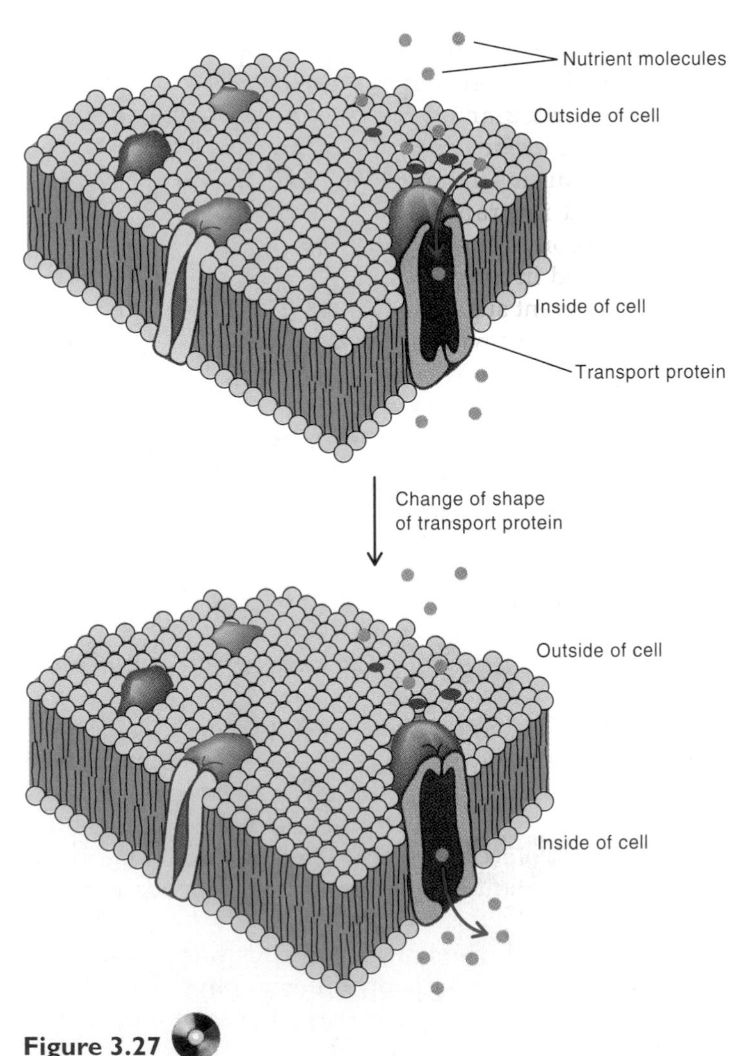

Figure 3.27
Facilitated diffusion of nutrient molecules from the outside to the inside of the cells by means of transport proteins. The schematic diagram shows how a conformational change in a transmembrane protein could serve to transport the nutrient molecules.

than on the inside of the cell. Further, the molecules are not concentrated to a higher level on the inside than they are on the outside of the cell, but they are transported to the inside very quickly.

The interaction of a particular molecule with the membrane transport protein is very specific. Generally, a separate membrane transport protein exists for each nutrient, and for example, one transport protein carries one compound, while another carries a different compound. In some cases, however, the same transport protein may be involved in carrying several compounds that have a similar structure.

Active Transport As previously mentioned, bacteria often live in environments in which nutrients are very dilute. In such environments, facilitated diffusion is not useful. Therefore, bacteria utilize a process for rapidly transporting nutrients that are in very low concentration on the outside of the cell to the interior of the cell, and concentrate them so that they can be used for cell multiplication. This process is called **active transport.** In it, the concentration of a particular nutrient may be a thousand times higher on the inside of the cell than on the outside. A number of mechanisms for active transport exist. In one mechanism, found in gram-negative bacteria, the compound being transported first binds to a specific **binding protein** that is located in the periplasm. Different binding proteins exist for different sugars and amino acids. The binding protein then interacts with a **membrane transport protein** located in the cytoplasmic membrane, which transports the molecule to the inside of the cell (figure 3.28). This process, like all other active transport processes, requires energy.

Many bacteria, called **oligotrophs** (*oligo* means "few"), live in an environment in which nutrients are very dilute. In such an environment, bacterial multiplication depends on very efficient mechanisms that pump in and concentrate these very dilute nutrients inside the cell. Some organisms that live in aquatic environments have exceptional abilities to concentrate very dilute nutrients, and some even grow in distilled water.

Table 3.1 summarizes the most important features of the various means by which molecules enter cells.

Degradation of Macromolecules Outside the Cell

The natural environments of bacteria often contain macromolecules such as nucleic acids and proteins whose subunits can be degraded to provide energy or that can be recycled for the synthesis of other macromolecules. Since such large molecules cannot pass through the cytoplasmic membrane, bacteria degrade these macromolecules into their subunits, which are small enough to be transported into the cell. In gram-positive bacteria, such enzymes as proteases and nucleases, termed **exoenzymes,** are secreted by the cell into the medium, where they act (figure 3.29).

In gram-negative cells, degradative enzymes are generally located in the periplasm (figure 3.30). Thus, any large molecules that pass through the porous cell wall will come in contact with these enzymes, which will break them down into smaller molecules that will then pass through the cytoplasmic membrane by means of the transport mechanisms previously discussed.

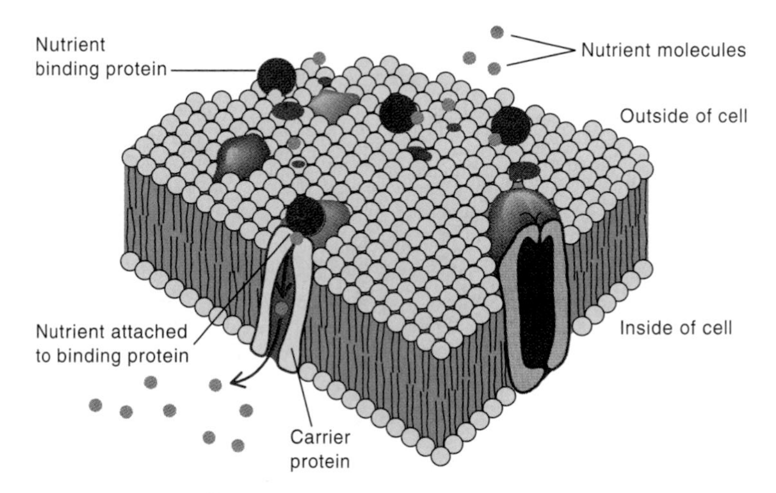

Figure 3.28 Active transport by means of specific binding proteins. These proteins are in the periplasm in gram-negative cells and are attached to the cytoplasmic membrane in gram-positive cells. The binding proteins interact with carrier proteins in the membrane. This process is rapid and results in the accumulation of the small molecules inside the cell.

Table 3.1 Features of Diffusion and Active Transport

Type of Transport	Energy Required	Concentration Outside and Inside of Cell	Mechanism of Transport
Passive diffusion	No	Same	Diffuses through cytoplasmic membrane
Facilitated diffusion	No	Same	Permeases in cytoplasmic membrane involved
Active transport	Yes	Higher on inside	Binds to protein in periplasmic space, which then interacts with a receptor protein in the cytoplasmic membrane

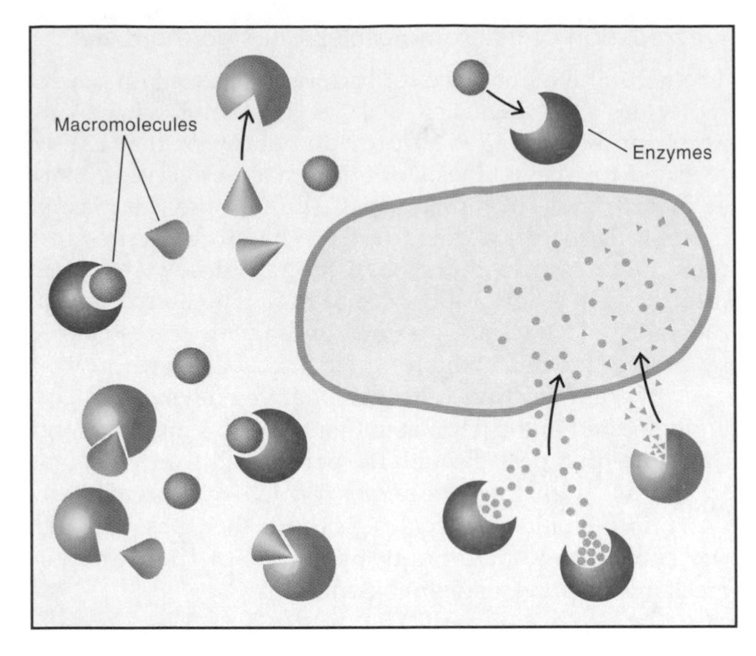

Figure 3.29
Degradation of large molecules by extracellular enzymes excreted by gram-positive bacteria. The breakdown of products then enter the cell by transport mechanisms.

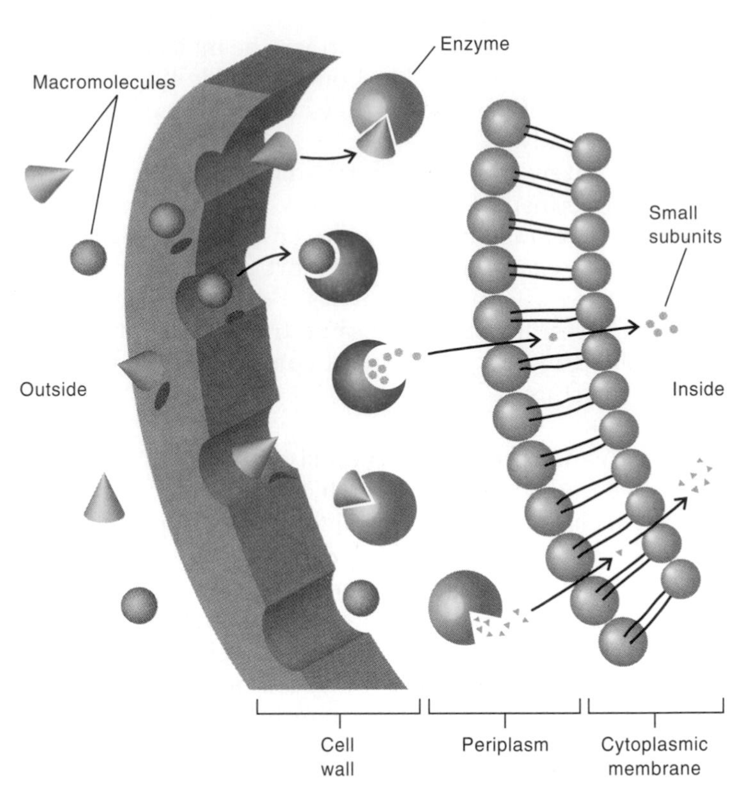

Figure 3.30
Degradation of large molecules by enzymes in the periplasm in gram-negative bacteria.

Table 3.2 Structures Surrounding the Cytoplasm—The Cell Envelope

Name of Structure	Present in All Cells?	Chemical Composition	Function	Notable Features
Capsule	No	Usually polysaccharides; sometimes polypeptides	Attachment; protection	Only synthesized under certain conditions
Cell wall	No	Peptidoglycan is unique structure; lipopolysaccharide; lipoprotein	Holds cells together	Site of penicillin action; rigid
Cytoplasmic membrane	Yes	Phospholipids, protein	Controls entrance and exit of molecules; site of certain enzymes	Structure similar in eukaryotic and prokaryotic cells

Important features of structures surrounding the cytoplasm (cell envelope) in prokaryotes are summarized in table 3.2.

Movement of Prokaryotic Cells

The major structures responsible for motility in prokaryotic cells are called **flagella.** However, certain bacteria have other organelles of locomotion and some microorganisms can glide over surfaces without any obvious organelles of locomotion. These organisms will be considered in chapter 11.

Flagella Have a Relatively Simple Structure in Bacteria

A number of structures are anchored in the cell envelope and then extend out of the cell. These structures include **flagella,** which are responsible for bacterial cell movement. The flagella are long, helical, protein appendages that are composed of three parts: the **filament, hook,** and **basal body** (figure 3.31).

The filament is composed of several identical protein chains twisted together into a helical structure with a hollow core. It is attached to the hook, which is fastened to the basal body, which contains several protein rings. The basal body anchors the flagellum to the cell wall and cytoplasmic membrane.

Figure 3.31 The structure of a flagellum in a gram-negative bacterium (*Escherichia coli*) showing the location of its component parts.

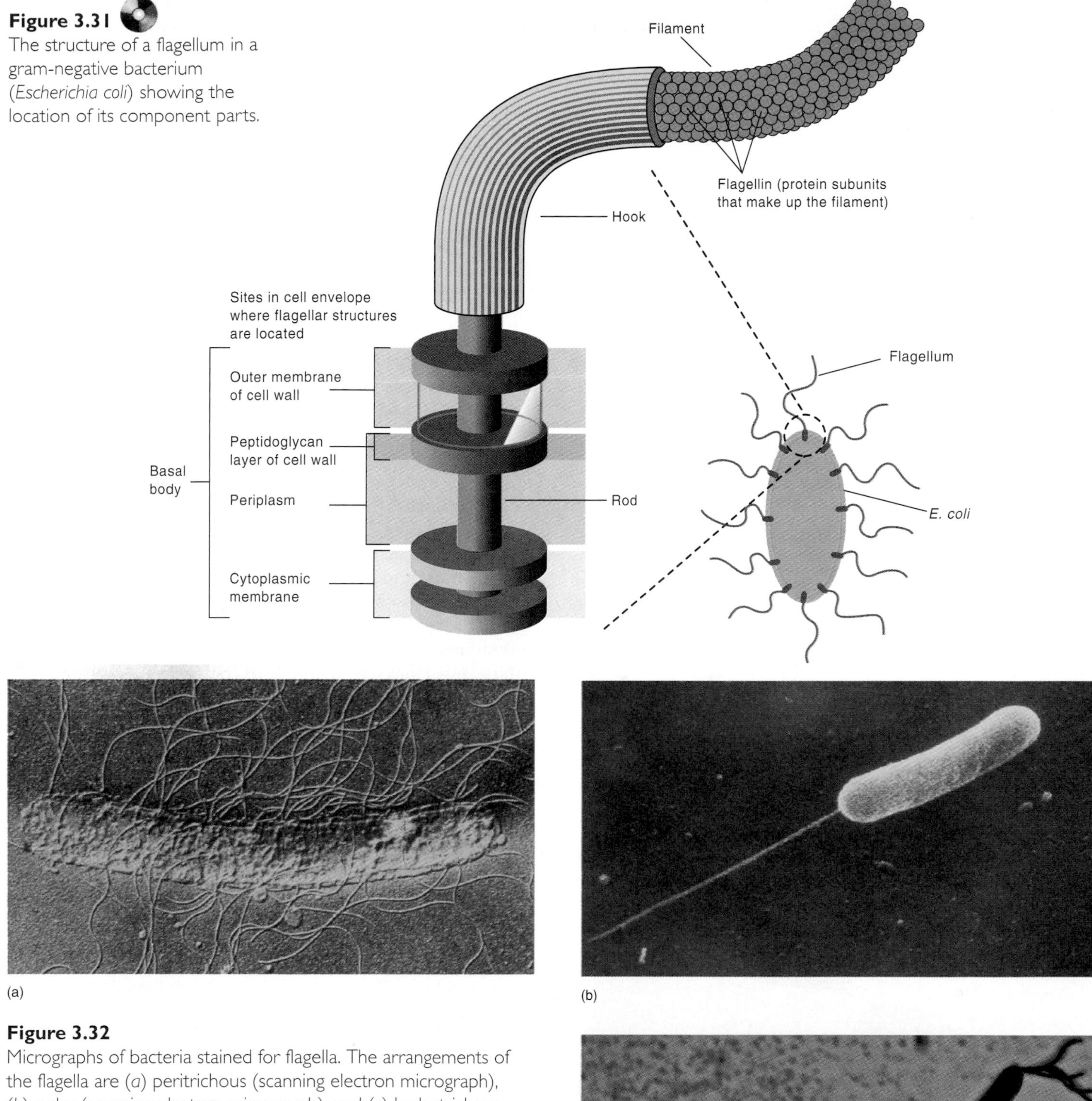

Figure 3.32 Micrographs of bacteria stained for flagella. The arrangements of the flagella are (*a*) peritrichous (scanning electron micrograph), (*b*) polar (scanning electron micrograph), and (c) lophotrichous (light micrograph).

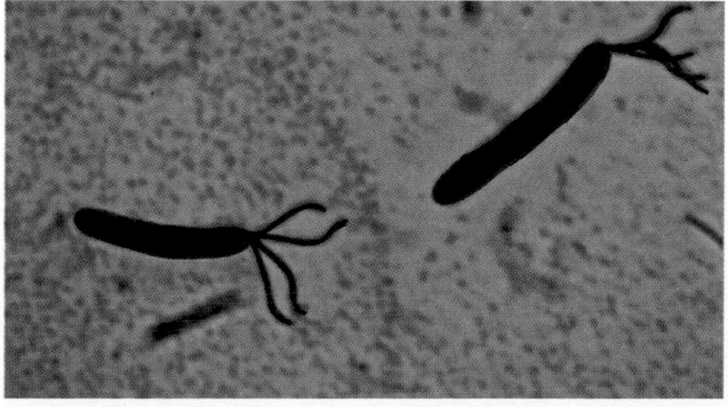

Bacteria have different arrangements of their flagella (figure 3.32 *a–c*). Some bacteria have a single flagellum at one end of the cell (known as *polar flagellaton*); others have a tuft (known as *lophotrichous flagellation; lopho* means "tuft"

and *trichous* means "hair"). Still others have flagella inserted at many points along the sides of the cell (known as *peritrichous flagellation; peri* means "around"). These arrangements, which are inherited, can be useful to help identify bacteria.

Mechanisms of Locomotion in Bacteria

Flagella push the bacterium through liquid much as a propeller pushes a ship through water. The flagella spin around their long axes, although little is known about the motor that drives them (figure 3.33). Flagella must work very hard to move a cell, since water has the same relative viscosity to bacteria as molasses has to humans. Nevertheless, their speed is quite phenomenal. *E. coli* can move at a rate of 20 body lengths per second, which is equivalent to a 6-foot man running 82 miles per hour. It does this by rotating the flagella more than a thousand revolutions per second.

Flagella may be important in the ability of an organism to cause disease. The organism that causes gastric ulcers has powerful multiple flagella at one end of its spiral-shaped cell. These flagella enable it to move through viscous gels that impede most flagellated organisms such as *E. coli.* The powerful flagella are likely important in allowing the organism to penetrate the mucous gel that coats the stomach epithelium. The ability of other bacteria to cause disease often is dependent on having flagella.

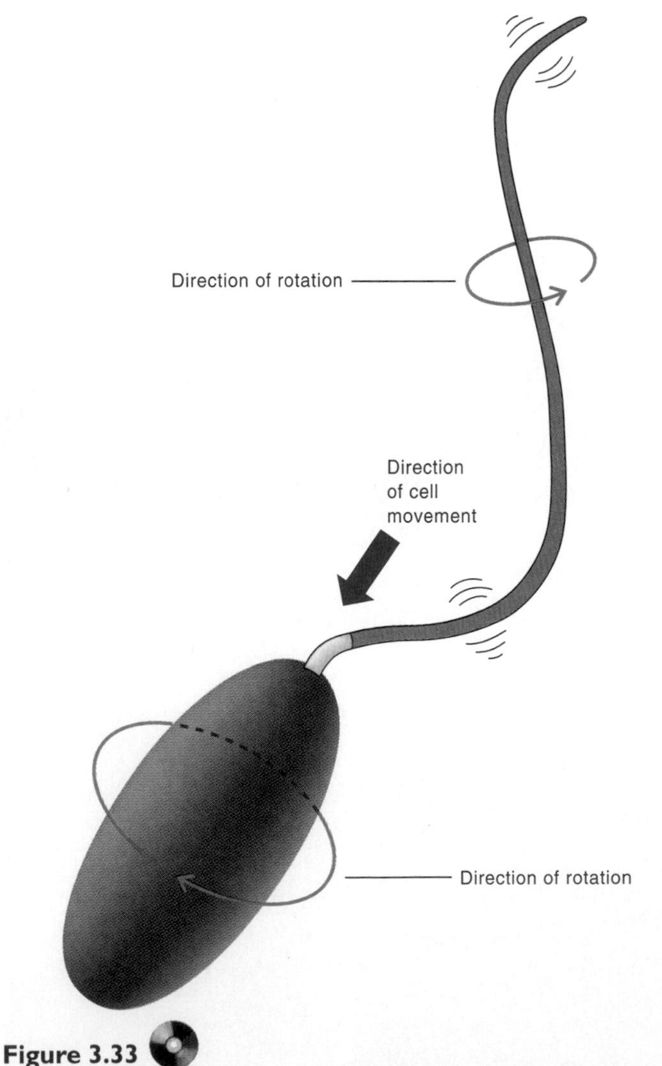

Figure 3.33
Flagellum pushing a bacterium. The flagellum and cell rotate in opposite directions.

Chemotaxis

Bacteria with flagella are able to move toward food sources or away from harmful materials in their environment. This phenomenon, called **chemotaxis,** can be explained as follows. In a uniform solution, bacteria move (swim) in one direction for about one second and then twiddle (tumble) for a fraction of a second, and then once again move in a straight line but not necessarily in the same direction. The direction of movement is random (figure 3.34 *a*). If, however, nutrients or toxic materials are in the vicinity, then movement is not random. The bacteria swim toward the nutrients, in a motion called *positive chemotaxis,* or away from the harmful chemicals, in a motion called *negative chemotaxis* (figure 3.34 *b* and *c*).

The cells sense the presence of nutrients, and this information is transmitted to the flagella, which then rotate either clockwise or counterclockwise. When the flagella rotate in a counterclockwise direction, the length of time the cell swims before it twiddles increases. When the flagella rotate in a clockwise direction, the time of swimming is very short before the cell twiddles. Thus, positive chemotaxis increases the swimming time between twiddles, and negative chemotaxis decreases the swimming time between twiddles. Consequently, in the presence of attractants, the cells swim for longer times as long as they can sense the attractants. If they move away from the attractants and no longer sense them, they will twiddle frequently. Thus, an attractant draws the cells toward itself as the time of swimming increases, whereas a repellent increases the frequency of twiddling and therefore reduces the time of swimming.

What is the situation if the cell has numerous flagella coming out of different parts? When such flagella rotate in a counterclockwise direction, they are all held together in a tight bundle that rotates like a single flagellum (figure 3.35). However, when this bundle rotates in a clockwise direction, the flagella no longer remain together and the bundle flies apart. Because the flagella do not rotate as a single unit, the cell tumbles.

In addition to reacting to chemicals, some bacteria can respond to variations in light (phototaxis), moving toward it, and others can respond to the concentration of oxygen (aerotaxis). Organisms that require oxygen for growth will move toward it, while bacteria that grow only in its absence tend to be repelled by it. Certain motile bacteria can react to the earth's magnetic field (magnetotaxis). They actually contain a row of magnetic particles that cause the cells to line up in a north-south direction much as a compass does

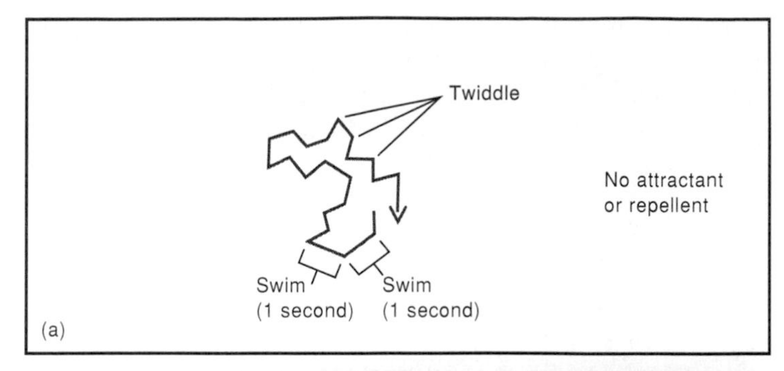

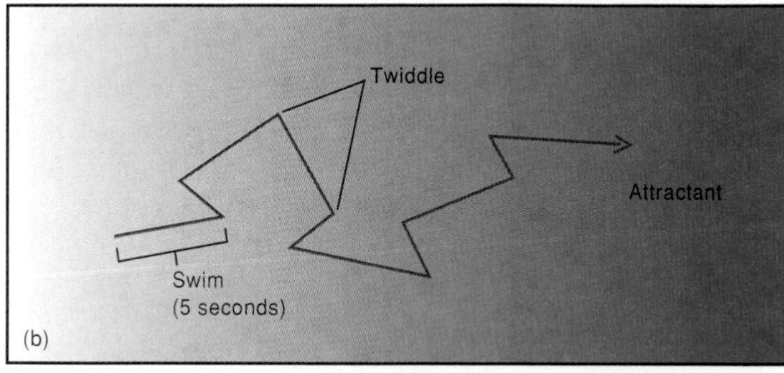

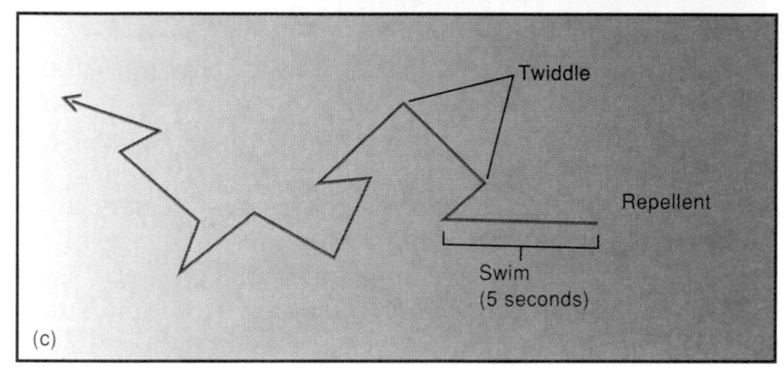

Figure 3.34
Diagrammatic representation of the movement of a bacterial cell. (*a*) Random movement of a cell when the attractant or repellent is uniformly distributed. Each swim is followed by a twiddle, and the twiddles occur fairly frequently, after which the cell moves in a new direction. (*b*) Directed movement toward a chemical attractant. The cell still swims in random directions, but when it swims toward the chemical gradient, the twiddles occur less frequently, and the time of swimming is longer. When the cell swims away from the chemical, twiddles occur more frequently. The result is that the cell moves toward the chemical. (*c*) Directed movement away from a chemical repellent. The cell swims for a longer time as it moves away from the repellent.

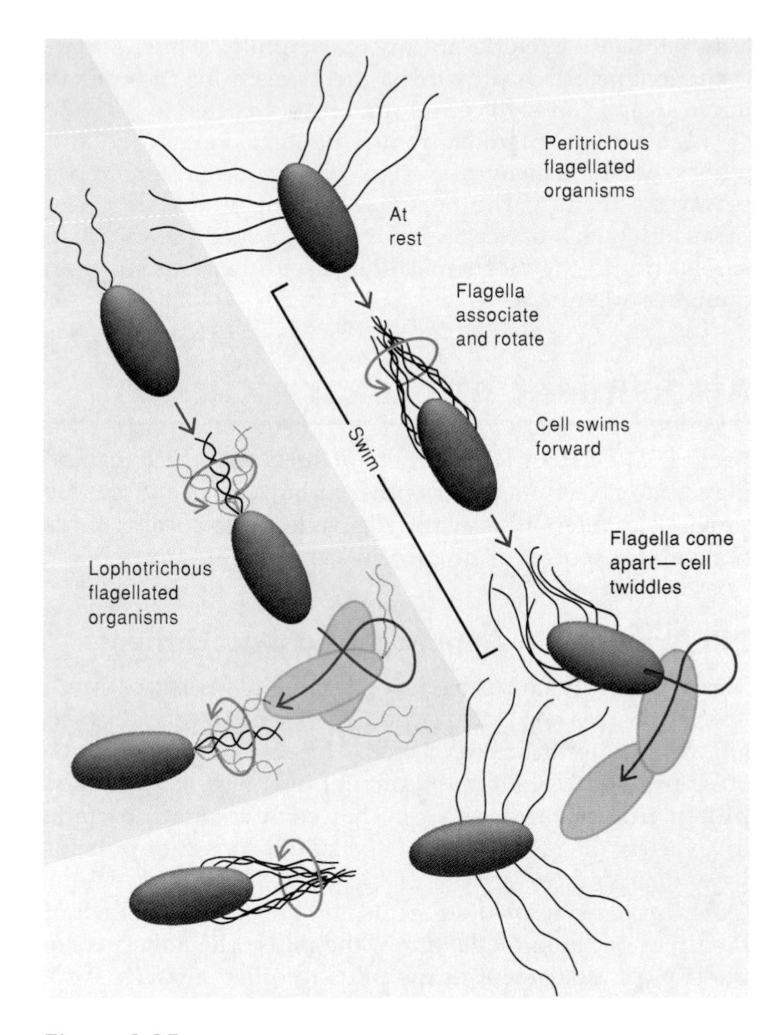

Figure 3.35
Flagellar association in bacteria with multiple flagella. The cell swims only when flagella form a single bundle.

(figure 3.36). Such bacteria in the northern hemisphere move toward the North Pole, and those in the southern hemisphere move toward the South Pole. Since the flagella are the organelles that push the cells forward, in the northern hemisphere the flagella are at the rear of the cell and the end of the magnetic particles seeking the North Pole at the other end. Thus, the flagella propel the cells in the direction of the North Pole. The magnetic forces of the earth attract the organisms so that they move downward and into sediments where the concentration of oxygen is low.

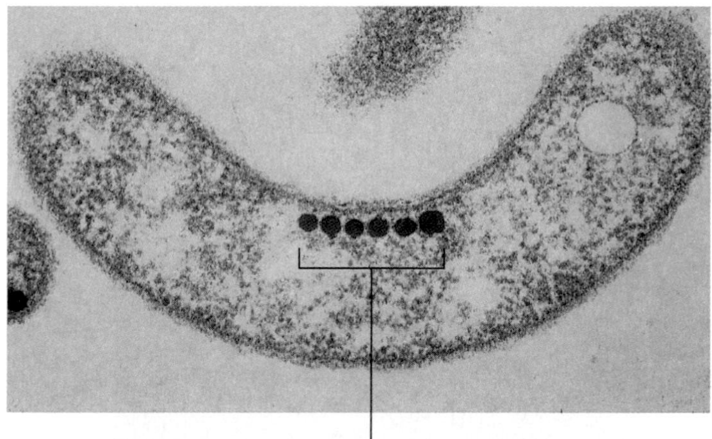

Figure 3.36
Negatively stained electron micrograph of a magnetotactic spirillum, *Aquaspirillum magnetotacticum.* This bacterium contains magnetic particles of Fe_3O_4 (magnetite) arranged in a chain; the particles serve to align the cell along geomagnetic lines. The organism was isolated from a water treatment plant in Durham, New Hampshire.

Magnetotactic bacteria are microaerophilic. Thus, apparently, magnetotaxis provides a mechanism for the cells to move away from high concentrations of oxygen.

The mechanism by which bacteria recognize and move toward or away from signals in the environment is a primitive form of the nervous system in animals. The means by which bacteria carry out chemotaxis bear certain similarities at the biochemical level with nervous function in higher organisms.

Attachment of Prokaryotic Cells

Bacteria attach to a wide variety of surfaces, which include inanimate objects as well as other living cells. As discussed previously, the polysaccharide capsule or glycocalyx functions in this process as do protein appendages termed *pili.*

Pili Are Protein Appendages of Attachment

Many gram-negative bacteria and some gram-positive species possess hairlike appendages of various sizes that extend from the surface of the cell. These appendages, which are considerably shorter and thinner than flagella, are called **pili,** or **fimbriae** (figure 3.37). They serve to attach bacteria to a variety of surfaces. Like flagella, pili consist of proteins wound around one another to form a hollow core structure. Other proteins, termed **adhesins,** are often located either at the tip of or along the length of the pilus. The adhesins are the sites of attachment of the pilus to other surfaces. Also, like flagella, pili can be removed without killing the cells by shaking the bacteria in a blender.

Pili of various sizes are located on the same cell and can be divided into a number of types based on their specific function as well as their size. Members of one group, called the *F,* or *sex, pili,* are especially long. They join bacteria prior to the transfer of DNA from one cell to the other. This process is discussed more fully in chapter 7.

The protein adhesins of other pili are necessary to attach bacteria to cells in the urinary or intestinal tracts of the host as the first step in infection (figure 3.38). Without this attachment, infection cannot occur. In many cases, pili attach only to specific glycoproteins or glycolipids in host tissues. For example, strains of *Neisseria gonorrhoeae* cause gonorrhea only if the bacteria have pili; nonpiliated strains fail to cause disease, apparently because they cannot attach to specific host cells. Thus, like the polysaccharide glycocalyx, pili attach bacteria to a variety of surfaces. One difference, however, is that pili attach either by their tip or by the side of the pilus only to specific cell receptors, whereas polysaccharide binding usually occurs on a wide variety of surfaces.

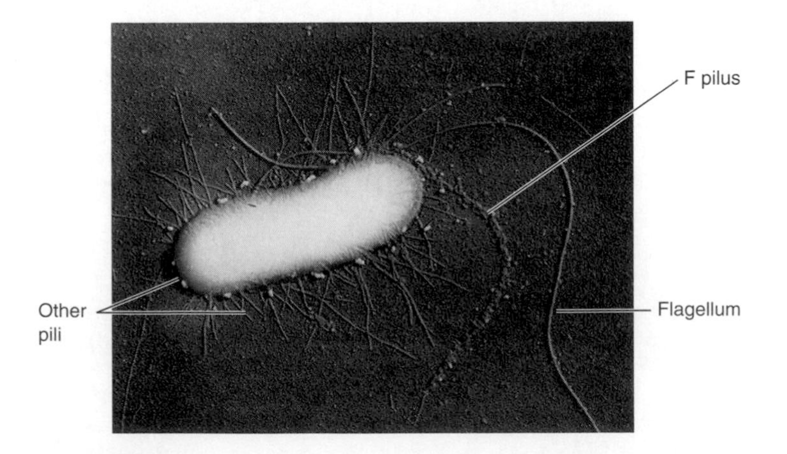

Figure 3.37
Pili of *Escherichia coli.* Note that there are many different kinds of pili. The pili that are involved in the transfer of DNA are the long F pili (×11,980).

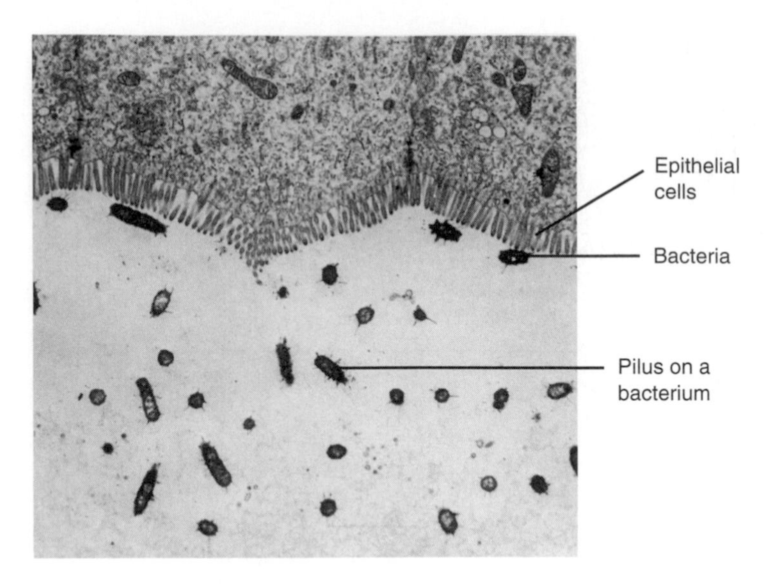

Figure 3.38
Escherichia coli attaching to epithelial cells in the small intestine of a pig. Note the pili on the bacteria.

• **Glycocalyx, p. 57**

REVIEW QUESTIONS

1. What two functions does the cytoplasmic membrane in bacteria perform?
2. How does the action of flagella propel prokaryotic cells?
3. Explain how bacteria move toward nutrients.
4. Name two appendages that protrude from the bacterial cell and give their function.
5. Name the structures that are involved in the attachment of bacteria to various surfaces.

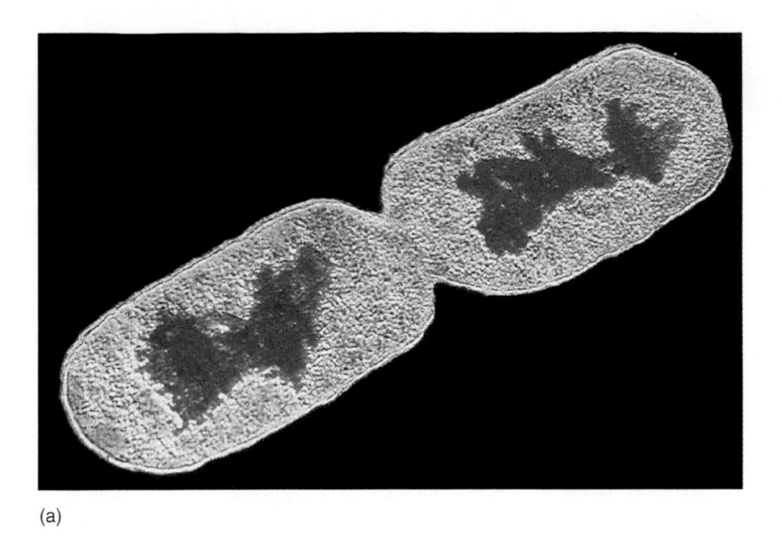

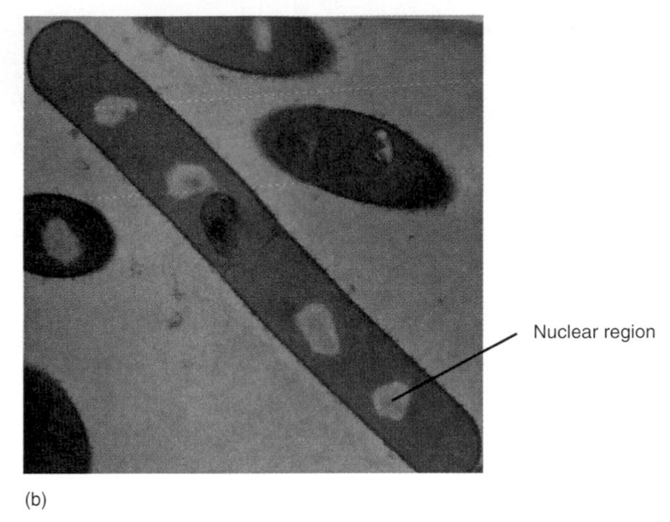

Figure 3.39
The bacterial nuclear region (nucleoid). (a) Transmission electron micrograph of a thin section of *Escherichia coli* with the DNA in red. (b) Light photomicrograph of cells of *Bacillus* with the nuclear region visible (X50,400).

Storage of Genetic Information in Prokaryotic Cells

The major structure in which the genetic information is stored is the **chromosome,** which is composed of DNA and is found in the **nuclear region** (nucleoid) of prokaryotic cells (figure 3.39 *a* and *b*).

The Bacterial Chromosome Is a Double-stranded, Circular DNA Molecule

In prokaryotic cells, only one or at most several identical chromosomes are present per cell and they are not surrounded by a nuclear membrane. The chromosome is composed of a single, circular, double-stranded molecule of DNA, which, when extended to its full length, is about 1 mm long, approximately 1,000 times longer than the cell itself (figure 3.40 *a*). It is tightly packed into about 10% of the total volume of the cell. To fit into a cell, the DNA must be twisted into a form that is called **supercoiled,** in which the ends are joined together by covalent bonds (figure 3.40 *b*).

Several basic proteins are bound to the DNA, but whether these proteins are involved in packing the DNA into the cell is not clear. Some of these proteins are called **histonelike** because they have properties similar to the basic proteins that bind to the DNA in eukaryotic cells.

• **DNA, p. 37**

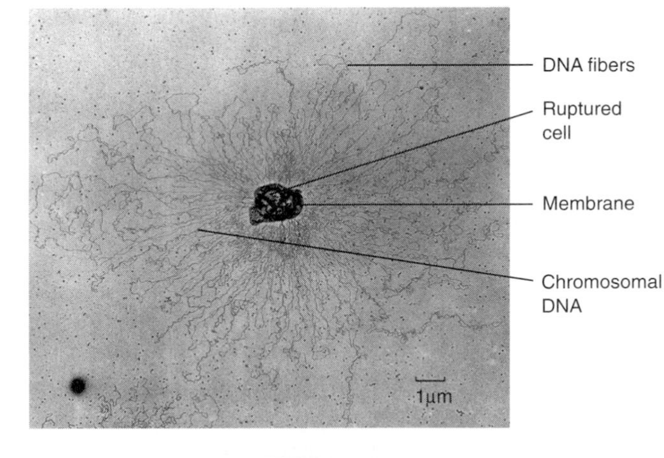

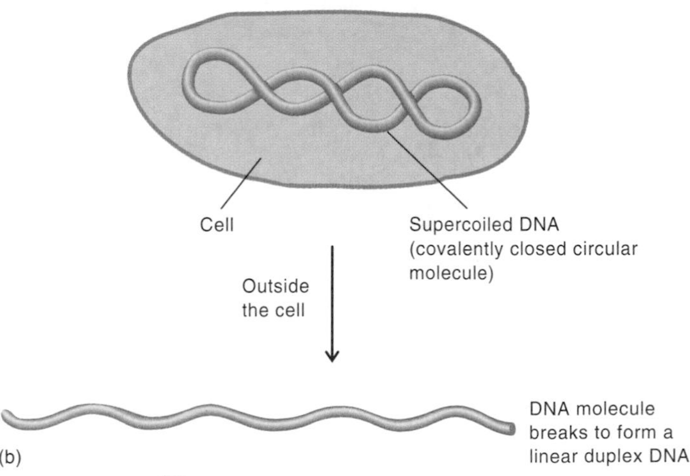

Figure 3.40
(*a*) Folded bacterial chromosome released from gently lysed cells of *Escherichia coli*. Note how tightly packed the DNA must be inside the bacterium. (*b*) Diagram of supercoiled DNA inside the cell and its breakage to form a linear molecule.

Perspective 3.1

Plasmids, Pasteur, and Anthrax

Many discoveries and observations made in the early days of microbiology can now be explained at the molecular level. A good example of recent molecular explanations for past discoveries is the work of Louis Pasteur on the disease anthrax.

In the late 1870s, an outbreak of anthrax in France resulted in a severe disease that killed flocks of sheep in the countryside. The serious nature of the disease convinced Louis Pasteur that he should study it. Since he had recently developed a method of eliminating the disease-causing ability of the organisms that caused chicken cholera by growing the cells at a high temperature, Pasteur decided to use a similar approach to the organism that caused anthrax. When he grew the bacteria at 42°C, they no longer caused disease.

How does heat eliminate the ability of the organism to cause disease? We now know that the parental disease-causing strain of the anthrax bacillus contains a plasmid, that is lost when the bacteria are grown at 42°C, rather than at their optimum temperature of 37°C. If the plasmid from the original disease-causing strain is reintroduced into the plasmidless strain, the bacteria regain their ability to cause disease. The plasmid is not required for growth of the organism, but in this case it codes for proteins required for disease production. Further, since the plasmid replicates independently of the bacterial chromosome, certain environmental conditions will inhibit its replication without having any effect on chromosome replication.

When Pasteur published his observations, he asked, "How does heat eliminate the disease-causing ability of a bacterium?" More than 100 years later, we finally know the answer.

Plasmids Are Small DNA Molecules that Code for Nonessential Information

In addition to the DNA that makes up the chromosome, most bacteria have **plasmids,** which also carry genetic information. Plasmids are DNA molecules that are also circular and double-stranded; they are approximately 0.1% to 10% of the size of the chromosome, and they replicate independently of it (figure 3.41 *a* and *b*).

Unlike chromosomes, many copies of a plasmid may be present in a cell. Whereas much of the information coded by the DNA in the chromosome is required for the life of the bacterium, the genetic information that plasmids contain is not, except under certain conditions. For example, the genetic information that confers resistance to a particular antibiotic often resides in plasmids. This property is not necessary to the life of the bacterium unless it encounters the antibiotic, in which case it becomes essential to the survival of the cell. Plasmids are discussed in more detail in chapter 7.

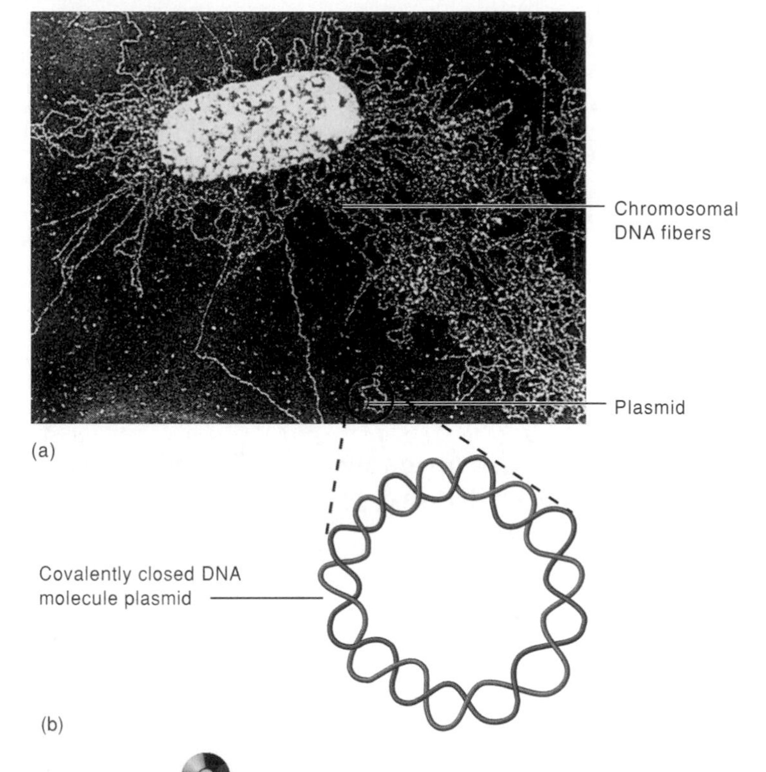

Figure 3.41 DNA leaking out of a cell of *Escherichia coli.* (*a*) Chromosomal and plasmid DNA. (*b*) Diagram of a covalently closed DNA molecule plasmid.

Synthesis of Protein in Prokaryotic Cells

Protein makes up more than 50% of the dry weight of a typical bacterial cell, and 90% of all the energy the cell uses goes into the manufacture of proteins.

Ribosomes Are Workbenches on Which Proteins Are Synthesized

In all cells, both prokaryotic and eukaryotic, the structures responsible for protein synthesis are the **ribosomes** (figures 3.13 and 3.42 *a* and *b*). Ribosomes are the structures on which amino acids are joined to form proteins. They are not merely passive workbenches but play an active role in a very complex process. Up to 15,000 of these small granules are found in the cytoplasm, the exact number depending on how rapidly the cell is multiplying. The faster the cell is

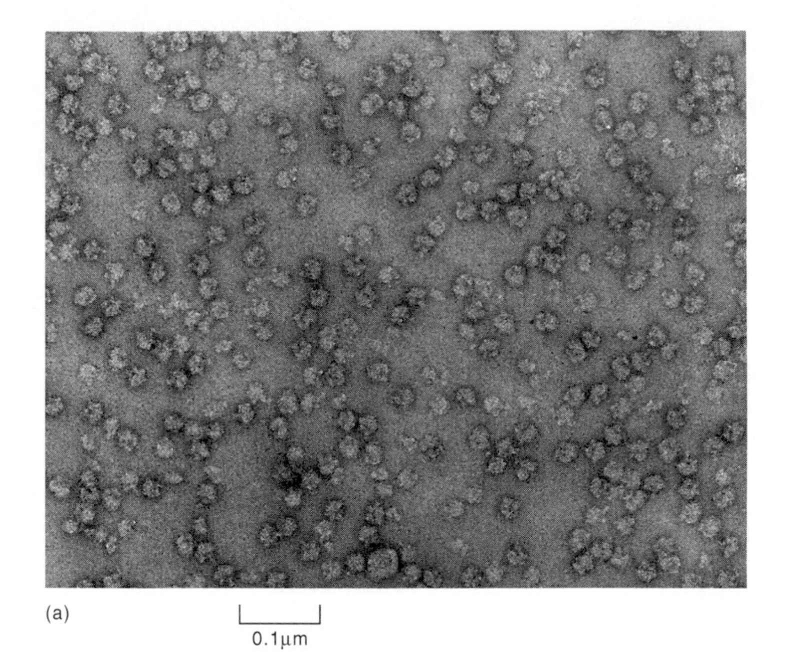

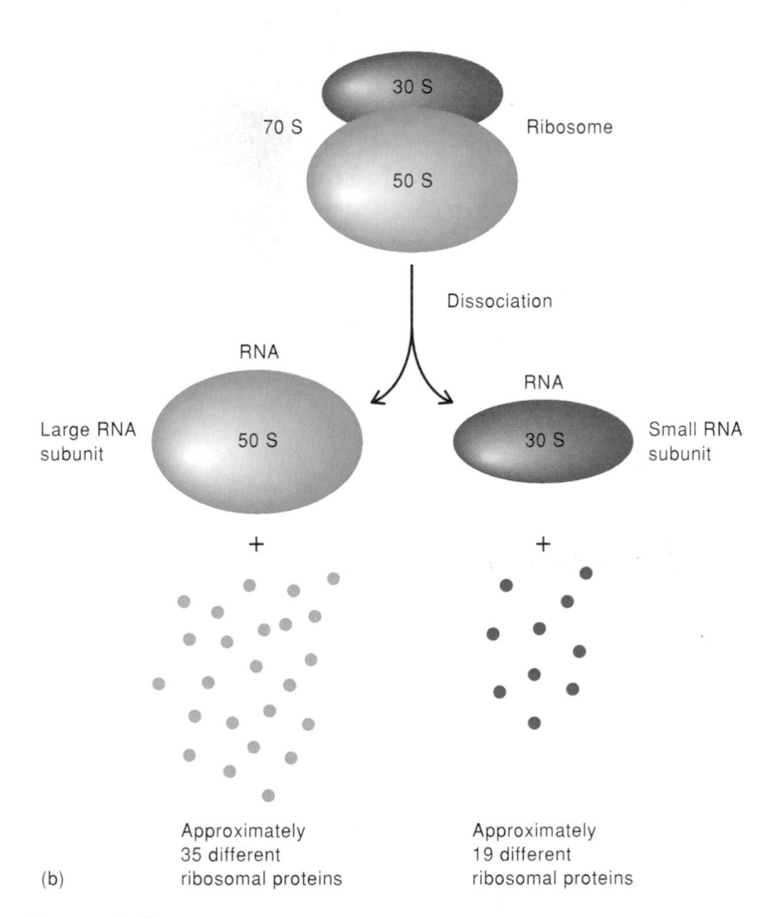

Figure 3.42
Bacterial ribosomes. (*a*) Electron micrograph of ribosomes from *Escherichia coli.* (*b*) The prokaryotic ribosome and its constituent parts. The 70S ribosome breaks down (dissociates) into a large (50S) and a small (30S) subunit. Each subunit consists of RNA (ribosomal RNA) and protein (ribosomal protein).

growing, the faster proteins are being synthesized and the greater the number of ribosomes.

Each ribosome is composed of two parts, each of which is made up of protein and RNA. These particular macromolecules are often called **ribosomal proteins** and **ribosomal RNA** to distinguish them from other types of proteins and RNA that also function in protein synthesis. The process of protein synthesis is covered in chapter 6.

Ribosomes are frequently described by how fast they move toward the bottom of a tube when they are centrifuged at very high speeds in an ultracentrifuge. The faster they move, the greater their density. Bacterial ribosomes are termed *70S ribosomes,* whereas in eukaryotic cells they are called *80S ribosomes.*

Storage Materials in Prokaryotic Cells

Bacteria often store a variety of materials that they later use as a source of nutrients. These storage products, which appear as large granules in the cells (figure 3.43), consist of

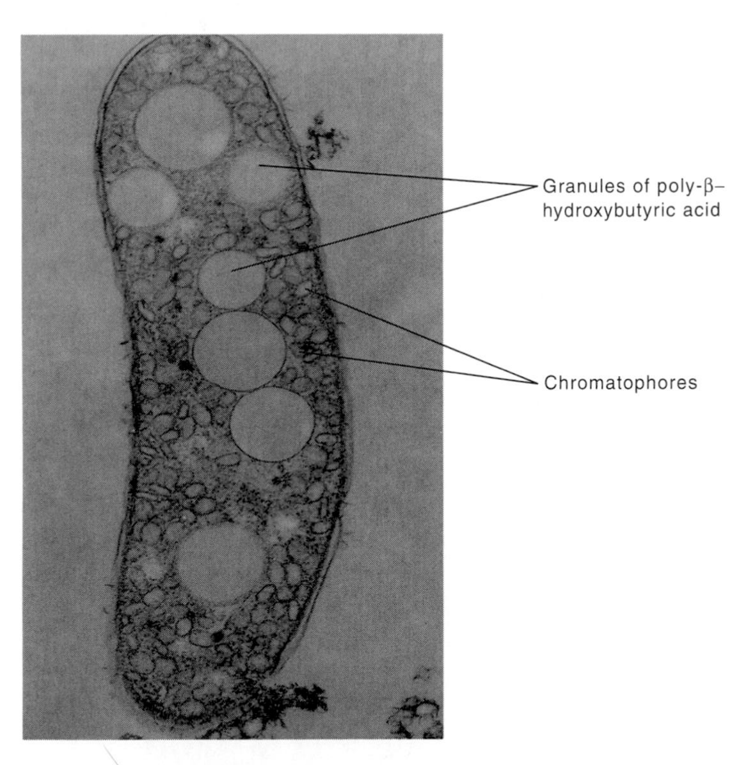

Figure 3.43
A thin section of a photosynthetic bacterium, *Rhodospirillum rubrum.* The cytoplasm is filled with chromatophores and inclusion granules of poly-ß-hydroxybutyric acid. Chromatophores are invaginations of the cytoplasmic membrane in photosynthetic bacteria in which enzymes and pigments concerned with using light energy are found.

high molecular weight polymers that are synthesized from small molecules for which the cell has no immediate need. Thus, if nutrients are provided in amounts greater than what the bacteria need at the moment, the bacteria store the nutrients for later use. However, they do not store the nutrients in the form in which they are provided because the accumulation of a large number of small molecules inside the cell would create osmotic pressure problems. Rather, they convert them into a few large storage granules. Depending on their composition, granules can serve as a source of energy and, frequently, as a source of materials for the synthesis of macromolecules. As a general rule, only one kind of reserve material is stored by a particular species.

Granules can be seen with special stains and are generally large enough to be readily detected by light microscopy. One common granule is **glycogen,** a polymer of glucose that serves as a storage form of both carbon and energy. Another storage unit supplying both carbon and energy is the polymer of β-hydroxybutyric acid (see figure 3.43). Phosphate is often stored as long chains of phosphate molecules called **volutin.** These granules stain red with certain blue dyes, such as toluidine blue, and thus are also called **metachromatic granules** (*meta* means "change" and *chromatic* means "color).

• **Glycogen, pp. 36, 37**

Endospores in Prokaryotic Cells

A very distinctive structure, the **endospore** (sometimes simply called a *spore*), is often seen inside certain species of gram-positive bacilli or in the medium in which the bacilli are growing (figure 3.44 *a* and *b*).

The endospore is not a storage granule, but rather a unique cell type that is formed within **vegetative cells,** the name given to the actively multiplying cell type. Vegetative cells develop into endospores when they are cultured in low amounts of carbon or nitrogen. Most species of endospore-formers fall within only two genera of bacteria: *Bacillus* and *Clostridium.*

Endospores Develop from Vegetative Cells when They Face Starvation

The process of spore formation, called **sporogenesis,** involves a complex, highly ordered sequence of morphological changes in the vegetative cell. In this process, which is illustrated in detail in figure 3.45 *a* and *b,* a vegetative cell develops into the endospore through a number of stages.

Endospores have a number of properties that distinguish them from vegetative cells. One property is that they are dormant and unable to multiply. Also, they do not degrade compounds to generate energy nor do they synthesize cell components.

Sporogenesis is an example of **cellular differentiation** (the conversion of one cell type into another) and involves very complicated regulation of protein synthesis as well as protein activity. This process has many features in common with the differentiation of cells in animals and is a model system for understanding how higher cells differentiate and develop. Some aspects of this regulation are considered in chapter 6.

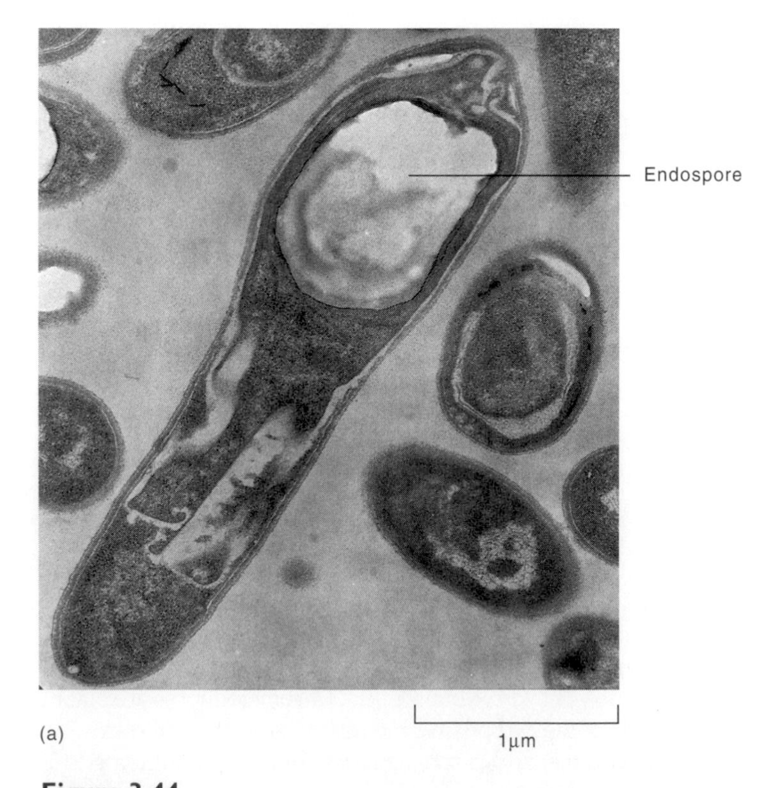

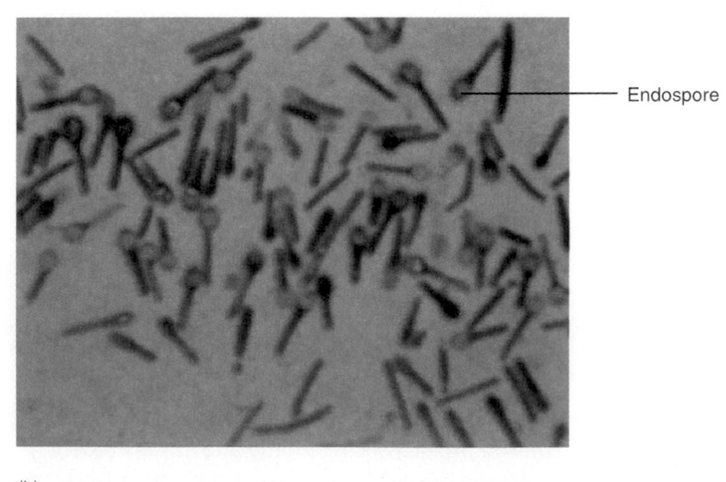

Figure 3.44
Endospores. (*a*) Endospore inside a vegetative cell of *Clostridium* sp. (*b*) Endospores inside cells of *Clostridium tetani.* Note the bulging of the cells at the end as a result of the endospore (×500).

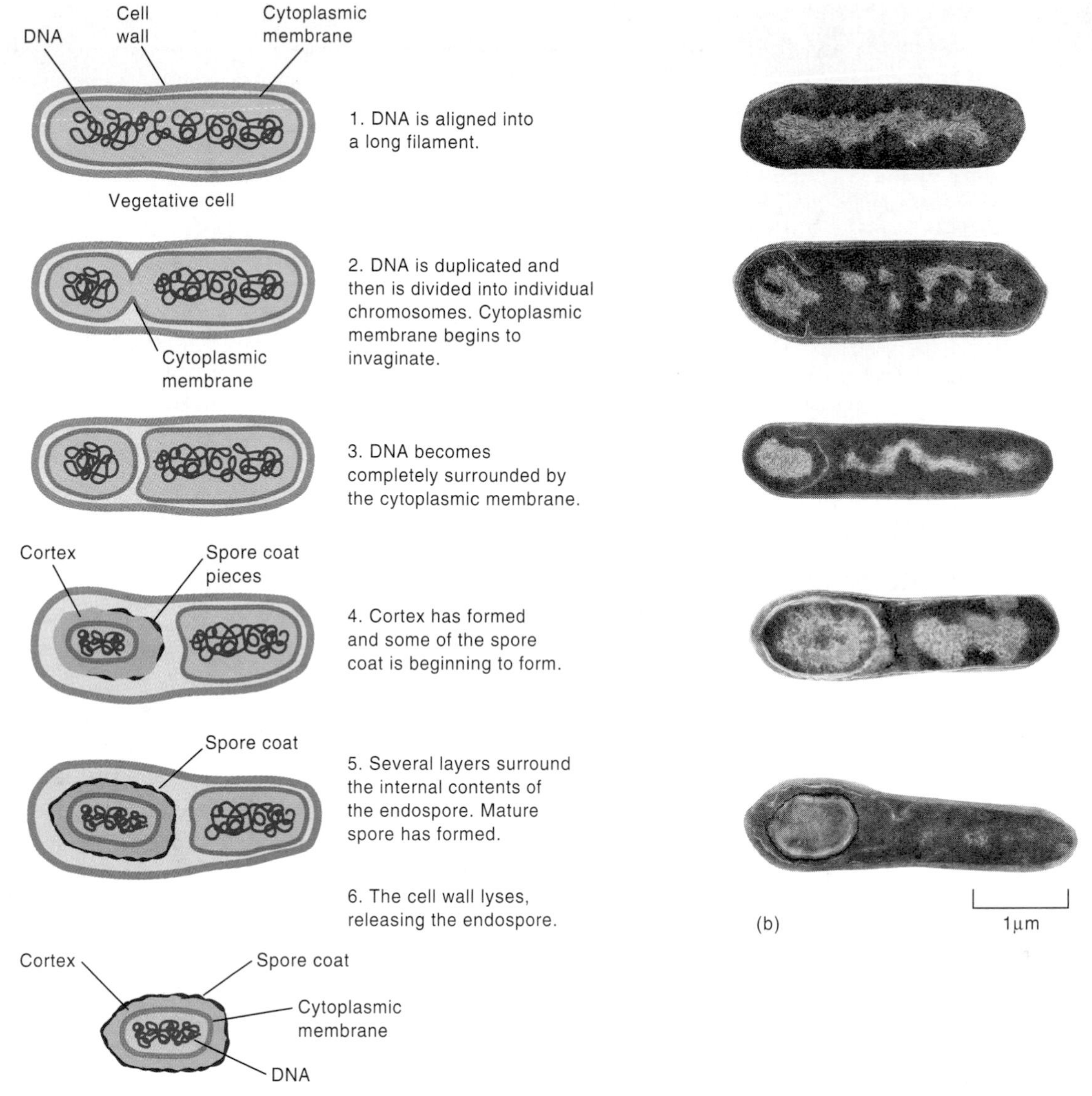

Figure 3.45
(*a*) Schematic representation of sporogenesis and (*b*) electron micrographs of each of the stages.

Endospores Can Develop into Vegetative Cells

Endospores can remain in the dormant state for long periods (at least as long as 100 years), but certain chemicals and heating to temperatures of 60°C to 70°C for a few minutes forces them to go through another developmental process termed **germination.** In this process, the endospore develops into a vegetative cell of the same kind from which it arose. Since one vegetative cell gives rise to one endospore, sporulation is obviously not a means of cell reproduction.

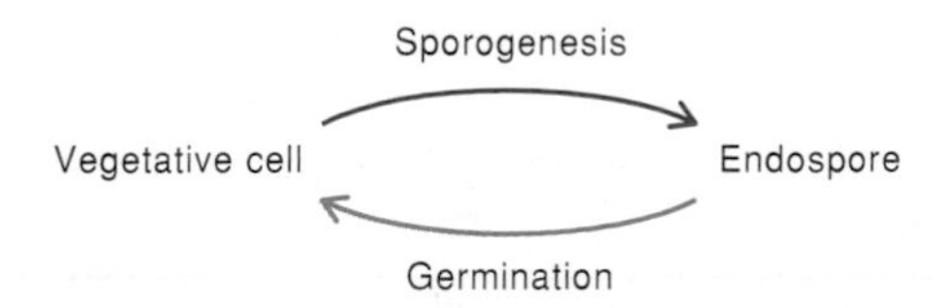

Endospores Have Unusual Staining and Resistance Properties

Endospores inside cells are readily observed under the light microscope because they do not take up stains through their thick spore coats, as do the rest of the cells (see figure 3.45 *b*). Staining spores requires that the cells be heated when the stain is applied. Since spores are located at characteristic positions inside the cell of different kinds of bacteria, their location can be helpful in identifying the particular species.

The most distinctive feature of the endospore is that it resists being killed by many different agents that kill the vegetative cell, such as heating, drying, freezing, toxic chemicals, and radiation. Thus, even boiling for hours may not kill endospores. This unusual resistance is due at least in part to the many coat layers that surround the cytoplasm (figure 3.46 *a*). One particular component of the innermost layer, termed the **core wall,** is **dipicolinic acid,** which is not present in vegetative cells (figure 3.46 *b*). This compound, in combination with Ca^{2+} ions, is important for the heat resistance of spores.

Perspective 3.2

Ancient Microbes

How long can living microorganisms survive in nature? This question has often been asked, but there is no accepted answer. Investigators have, in numerous studies, attempted to isolate living organisms from a variety of old specimens. The meaning of their results has often been questioned. For example, in the early 1930s, a scientist claimed to have recovered bacteria from the interior of rocks found in Canada, as well as from the Grand Canyon and from coal. However, the bacteria he isolated were identical to those present today. Could he have contaminated his samples? Another investigator found several species of bacteria in the frozen remains of a mammoth. Other investigators have failed to recover living microorganisms from ancient materials, including not only geological specimens, such as coal, but also a jar found in the tomb of Tutankhamen (1800 B.C.). However, several species of *Clostridium* were isolated from mummies found in the catacombs of a town in central Europe and estimated to be 180 to 250 years old. The temperature in the catacombs was 8°C to 10°C. None of the species isolated from the mummies was found in swabs from the skin of the mummies or from the coffins and the ground around them.

Reports continue to be published of endospores being brought back to life after many years. A recent report published in 1995 states that viable endospores could be extracted from the stomachs of ancient bees that had been embedded in amber for 30 million years, by far the oldest organisms believed to have survived. The organisms were identified as strains of *Bacillus sphaericus*, a common harmless bacterium that is found in the soil and in the stomachs of insects today. The investigators took scrupulous precautions to prevent contamination. However, since this particular species of bacteria is found so commonly in the environment, it is very difficult to prove and convince skeptics that the organisms are not modern contaminants. So the arguments continue about how long organisms and endospores can survive.

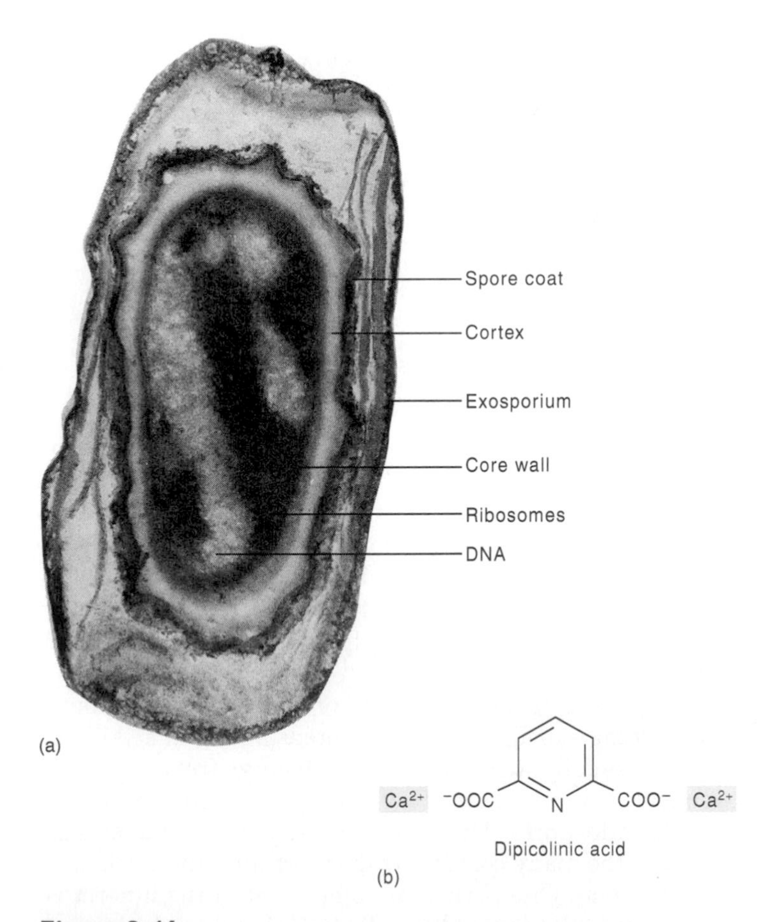

Figure 3.46
(*a*) Bacterial endospore. (*b*) Dipicolinic acid (DPA). Ca^{2+} ions associate with the carboxyl groups to form a complex.

Many Endospore-producing Bacteria Cause Disease

Endospores have a practical importance in microbiology, since they are formed by some bacteria that produce deadly toxins. Therefore, it is very important that foods that may contain these spores be processed adequately to insure that endospores are killed. Many people have died of botulism after ingesting improperly canned food that contained spores of *Clostridium botulinum.* In such improperly processed food, the heat-resistant spores germinate into vegetative cells under the appropriate conditions. These cells then synthesize the toxin, which when ingested causes botulism. However, since the toxin is destroyed by heat, boiling contaminated food for 15 minutes will destroy the toxin, although such treatment will not kill the endospores.

REVIEW QUESTIONS

1. Give two ways in which plasmids differ from chromosomes.
2. Describe the chemical composition of ribosomes.
3. Name two storage products found in bacteria and the function of each one in the cell.
4. Give three properties of endospores that distinguish them from vegetative cells.
5. What functions do the following serve: plasmids, chromosomes, ribosomes?

The Structure of Eukaryotic Cells

As pointed out in chapter 1, eukaryotic cells are distinguished from prokaryotic cells by their more complex structure and in general by their larger size (figure 3.47 *a–c*). Eukaryotic cells have a nucleus that is enclosed by a bilayer membrane. In addition, they have numerous other internal bilayer membrane-bound structures, such as mitochondria, all of which play important roles in the life of the cell. The microbial world contains numerous representatives, including algae, fungi, and protozoa, which are considered in chapter 12. In this chapter, we will include in our discussion the structure and function of the components of animal cells. Many of these components are important in the interactions of cells with infectious disease agents, specifically bacteria and viruses.

Figure 3.47
Typical eukaryotic cell. (*a*) Transmission electron photomicrograph of a eukaryotic cell, the protozoan *Polytomella agilis.* The outer structure is the cytoplasmic membrane because this protozoan does not have a rigid cell wall. (*b*) Diagram of (*a*). (*c*) Three-dimensional diagrammatic representation of a typical eukaryotic cell. Not all cells have all of the structures shown.

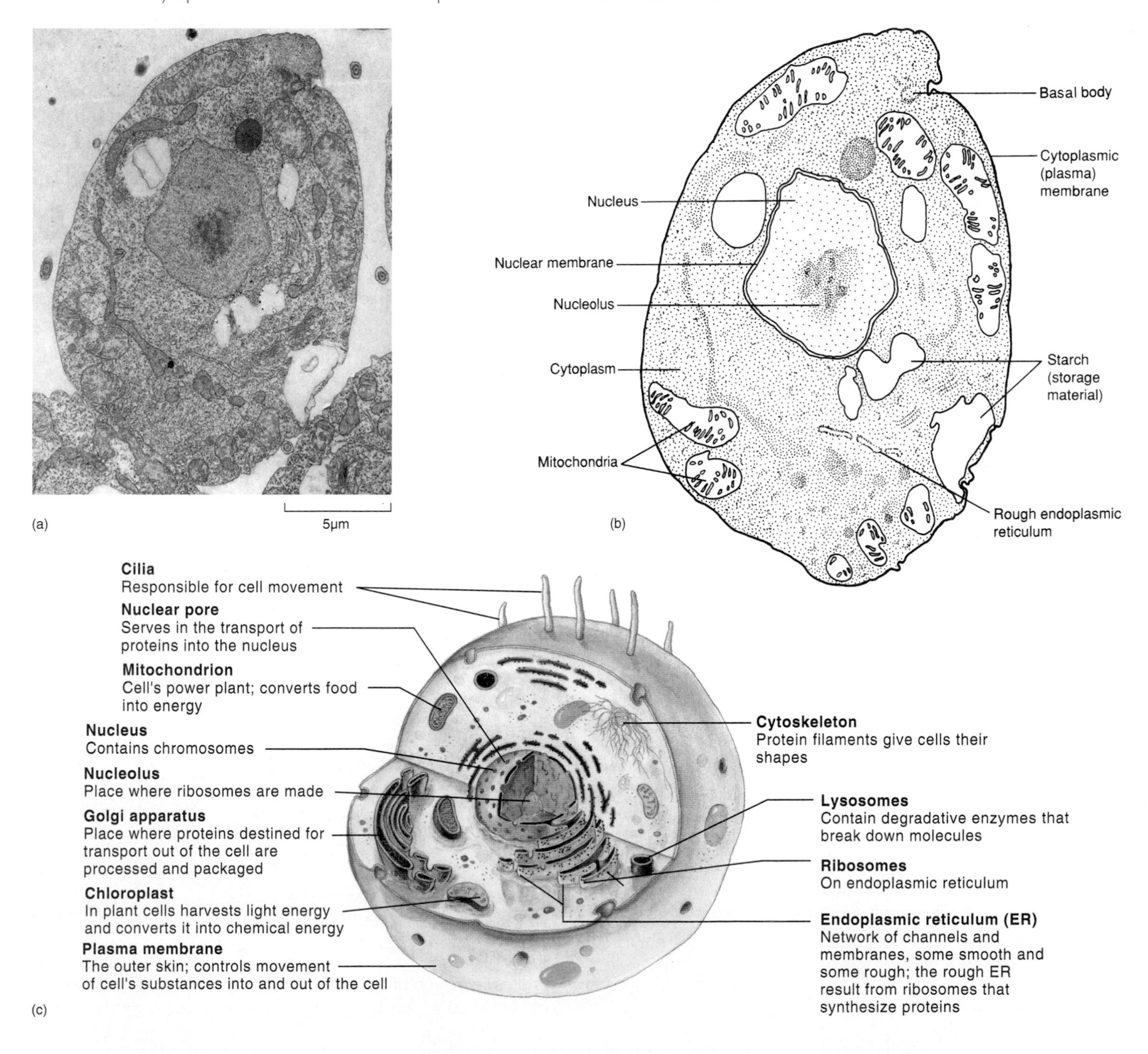

Enclosure of the Cytoplasm in Eukaryotic Cells

The cytoplasm in animal cells is enclosed by fewer structures than are found in prokaryotic cells. There is no rigid cell wall containing peptidoglycan and no capsule.

The Cytoplasmic Membrane is the Outer Layer in Many Eukaryotic Cells

Chemical Composition

The cytoplasmic membrane, more commonly termed the **plasma membrane** in animal cells, has a chemical composition quite similar to its prokaryotic counterpart, namely a bilayer of phospholipids with different proteins inserted into it. Various carbohydrates may also be attached to the proteins to form different glycoproteins. Specific glycoproteins frequently serve as sites of attachment for bacteria that cause disease. Such attachment is an early step in the infection process.

Function

All foodstuffs for biosynthesis and waste products of catabolism pass through the plasma membrane. As in prokaryotic cells, many of the proteins embedded in the membrane serve as channels for ions and pumps for specific molecules that are transported into and out of the cell. These channels are a site for action for many disease-causing agents. For example, a number of viruses can change the ion channels of cells they infect, resulting in the leakage of iron and calcium and consequently cell death. The organism causing cholera also alters ion channels of cells in the intestine, resulting in tremendous loss of water and possibly death.

Unlike the cytoplasmic membrane of prokaryotic cells, the eukaryotic membrane does not have enzymes of the electron transport chain embedded in it.

The entry of molecules into the cell often occurs through **endocytosis,** a process in which the flexible plasma membrane invaginates and pinches off to form a membrane-bound cytoplasmic vesicle. Some viruses and bacteria enter cells by this process which will be covered in later chapters. In fact, some bacteria that live inside animal cells actually induce the animal cells to take them up by this process. A special type of endocytosis carried out by certain cells of the body is termed **phagocytosis.** This process is an important part of the defense mechanism of the body. The process of **exocytosis** is the opposite of endocytosis. Membrane-bound vesicles inside the cell fuse with the plasma membrane and release their contents into the external medium. The processes of endocytosis and exocytosis result in the exchange of material between the inside and outside of the cell.

• **Glycoprotein, p. 35**

Internal Membrane Structures in Eukaryotic Cells

A basic feature of all eukaryotic cells is their extensive array of **internal membranes** (see figure 3.47 *a–c*). All of the internal structures are surrounded by membranes. These include the nucleus, Golgi apparatus, mitochondria, endoplasmic reticulum, lysosomes, and peroxisomes. These structures have a variety of different functions.

Several Membrane-enclosed Structures Are Involved in Biosynthesis and Transport

Several structures enclosed by membranes do not have a comparable counterpart in prokaryotic cells. The **endoplasmic reticulum (ER)** is a compartment formed by the meanderings of a single bilayer internal membrane that forms a meshlike network in all eukaryotic cells (figure 3.48). This compartment, which encloses a large intracellular space termed the **lumen,** is of two types.

The **rough endoplasmic reticulum (rough ER)** has ribosomes on its outer surface, which gives it a rough appearance. As in prokaryotic cells, the ribosomes are concerned with protein synthesis. However, their size (80S) as well as amino acid and nucleotide composition vary from their prokaryotic counterparts, which are 70S. Consequently, antibiotics that inhibit protein synthesis by interacting with the ribosome in prokaryotic cells may not affect protein synthesis in eukaryotic cells. The antibiotics that affect protein synthesis will be covered in chapter 21.

The other type of endoplasmic reticulum is called the **smooth endoplasmic reticulum (smooth ER)** because it lacks ribosomes. This structure is largely concerned with lipid metabolism. The **Golgi apparatus** (figure 3.49) consists of flattened sacs of a bilayer membrane that occurs in multiple layers. This structure is involved in the chemical modification of the molecules made in the endoplasmic

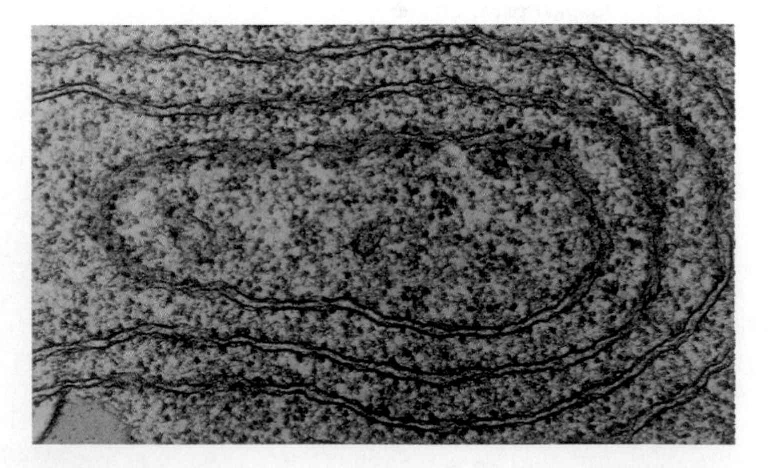

Figure 3.48
Photomicrograph of the endoplasmic reticulum of a mammalian cell (×151,000).

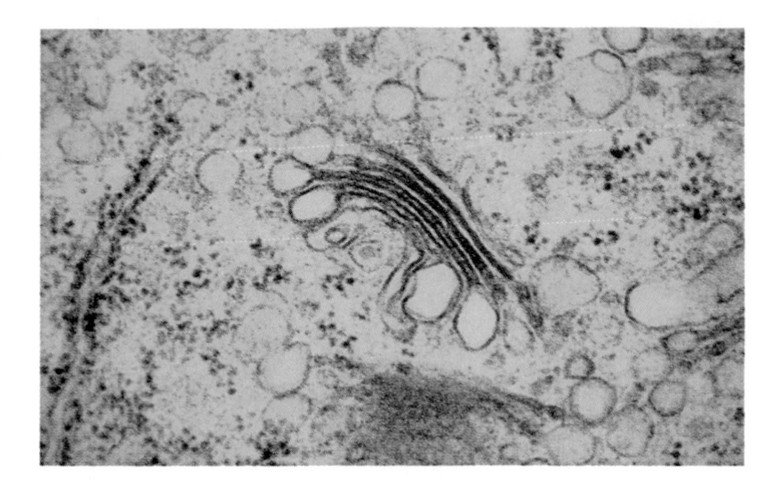

Figure 3.49
Electron photomicrograph of a Golgi apparatus from a plant cell (×21,000).

reticulum, such as proteins and lipids, and their transport to the outside of the cell or to other organelles within the cell. Once proteins and lipids are synthesized, they must be transported to their proper location in the cell in order to carry out their function. The Golgi apparatus is surrounded by many small vesicles that transport material from the Golgi apparatus to other parts of the cell.

Certain Membrane-enclosed Structures Are Involved in the Degradation of Molecules

In prokaryotic cells, degradative enzymes such as nucleases and proteases are secreted into the medium or into the periplasm. In eukaryotic cells, such enzymes are stored in sacs called **lysosomes,** which can be compared to a garbage disposal system. By being enclosed, they are prevented from destroying the structures in the cell. (Human beings who do not have their degradative enzymes enclosed in vacuoles suffer from a serious disease called *lysosomal storage disease,* which is often fatal in infancy.) Other sacs, called **peroxisomes,** contain enzymes that generate highly reactive molecules such as hydrogen peroxide and superoxide, a highly reactive form of oxygen. These dangerous, highly reactive molecules are also degraded in the peroxisome. Everything inside the plasma membrane that is not enclosed in membranes is termed the **cytosol.** It corresponds to the cytoplasm of prokaryotic cells.

Shapes of Eukaryotic Cells

Although animal cells do not have a rigid cell wall as do most bacteria, they nevertheless have definite shapes. All animal cells have an internal network of protein filaments, the **cytoskeleton,** that gives the cell its shape (figure 3.50).

• **Hydrogen peroxide, p. 93**
• **Superoxide, p. 93**

Not only is the cytoskeleton important in providing shape to the cell, but the filaments that make up the cytoskeleton are involved in movement of the cells and the transport of the internal organelles to various locations within the cells. One type of filament of the cytoskeleton is a protein called **actin** that is involved in the movement of individual cells. Bacteria that force animal cells to engulf them do so by attaching to the host cell and then sending chemical signals to the host, which results in a rearrangement of the actin filaments, which results in engulfment of the bacteria. This process will be discussed more fully in chapter 19.

Other kinds of cytoskeleton filaments are the **microtubules,** which consist of long, hollow cylinders (figure 3.51 *a*). These filaments are the main structure and energy-generating force for the **cilia** and **flagella,** the organelles of locomotion in certain eukaryotic cells (figure 3.51 *b*). Microtubules also form the **mitotic spindles,** the machinery that partitions chromosomes between two cells in the process of cell division. Without mitotic spindles, cells could not reproduce.

Cell Movement

Many motile eukaryotes are propelled by either flagella or cilia. Cilia also play an important role in host defense. For example, cilia on epithelial cells in the respiratory tract move mucus toward the mouth where it is swallowed or expelled, thereby removing any entrapped organisms.

Cilia in eukaryotic cells are similar in structure to eukaryotic flagella except that they are shorter and more numerous. However, the flagella of eukaryotes are far more complex than their prokaryote counterparts. Nine pairs of microtubules are arranged around a central pair, and the entire structure is enclosed in a membranous sheath (see

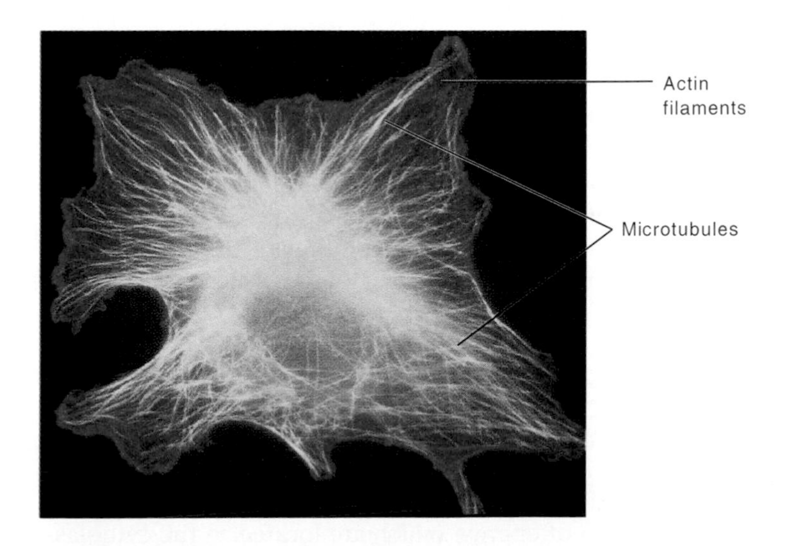

Figure 3.50
Photomicrograph of a eukaryotic cell cytoskeleton double-labeled with fluorescent antibodies to show actin filaments and microtubules.

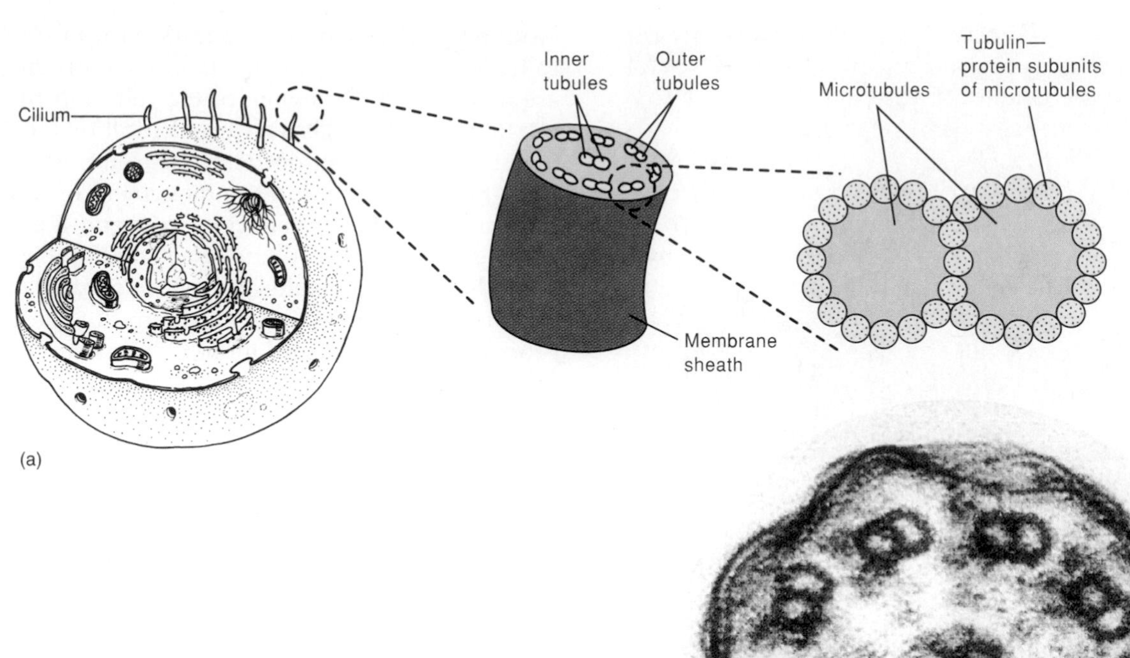

Figure 3.51
Cilium of a eukaryotic cell. (*a*) Diagram of a cilium showing the hollow nature of the microtubules that comprise the cilium. There is one pair of tubules in the center and nine pairs on the outside. Together with the membrane sheath, these comprise the eukaryotic cilium. (*b*) Electron photomicrograph of eukaryotic flagellum from the protozoan *Trichonympha* (×65,000).

figure 3.51). In eukaryotic cells, the flagella drive the cells forward by a whiplike action. In some cells, the flagella push the cell; in other cells, they pull the cell forward (figure 3.52). Recall that in prokaryotes, the flagella rotate like the propeller on a ship and push the cell.

Generation of Energy in Eukaryotic Cells

As pointed out already, many of the enzymes concerned with generating energy in prokaryotic cells are bound to the cytoplasmic membrane. In eukaryotic cells, distinct membrane-enclosed structures in the cytoplasm are involved in the generation of energy.

Energy Is Generated by Mitochondria

Mitochondria are highly complex structures about the size of a bacterial cell that have both an outer and an inner membrane (figure 3.53 *a* and *b*). The enzymes involved in the generation of energy, which are located in the cytoplasmic membrane of prokaryotic cells, are localized on the inner membrane.

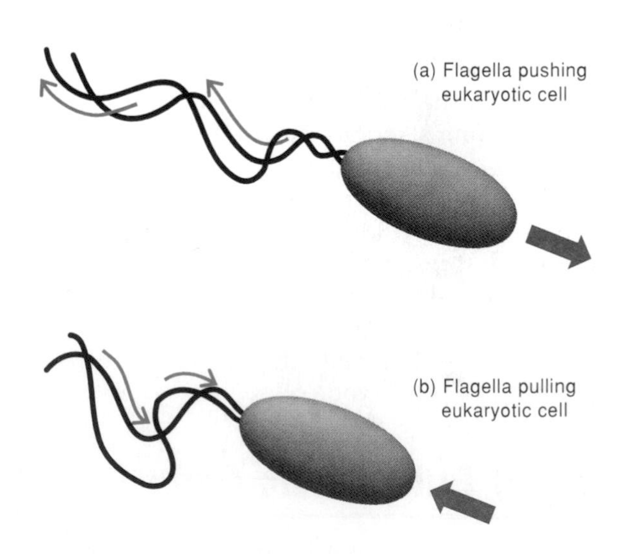

Figure 3.52
Manner in which flagella provide motility to eukaryotic cells. (*a*) Wave directed away from the cell pushes the cell in the opposite direction. (*b*) Wave directed toward the cell pulls the cell in the direction opposite to the wave.

Mitochondria also contain DNA, ribosomes, and the other materials necessary for protein synthesis. They do not have enough DNA to code for all the information needed for their own reproduction, so some of this information is coded by genes in the nucleus. However, mitochondria do elongate and divide in a fashion similar to bacteria. Further, the size of their ribosomes is more similar to that of bacterial ribosomes rather than that of eukaryotic ribosomes. These observations, together with the fact that mitochondria are similar in size to bacteria, have given rise to the popular theory that mitochondria have evolved from prokaryotic cells.

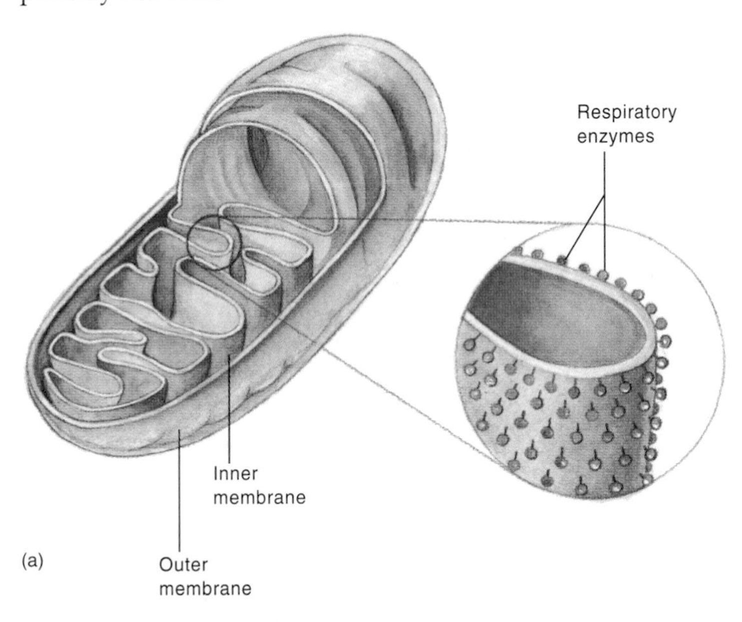

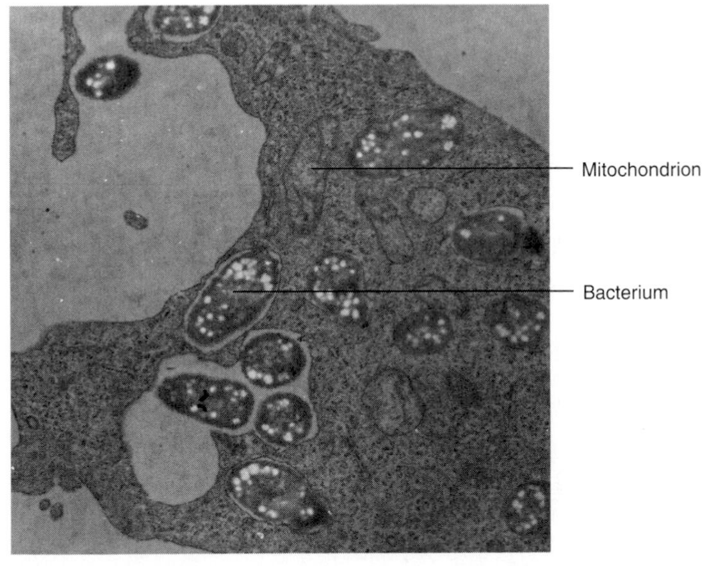

Figure 3.53
(*a*) Mitochondrion. The enzymes of energy metabolism (respiratory enzymes) are either located in the fluid inside or attached to the inner membrane. (*b*) Mitochondrion inside a mammalian cell that also contains bacteria. Note that the mitochondrion and bacterium are about the same size.

Storage of Genetic Information in Eukaryotic Cells

In eukaryotic cells, the genetic information is contained in chromosomes that consist of double-stranded DNA wound around basic proteins, called **histones,** to form **chromatin** (figure 3.54 *a, b*). The chromatin of each chromosome is arranged in thousands of loops. Each chromosome is divided into subunits termed **nucleosomes,** consisting of a histone core and its DNA (figure 3.54 *a, b*). The histones can be chemically modified through the action of enzymes. This modification loosens the grip of the histone on the DNA so that the genes in the DNA can be expressed. Unlike prokaryotic cells, eukaryotic cells contain multiple chromosomes each containing different genetic information. Each chromosome is a long, linear molecule (containing about 150 million nucleotide pairs in a human). The chromosomes are enclosed in a true **nucleus** formed by a bilayer membrane surrounding the

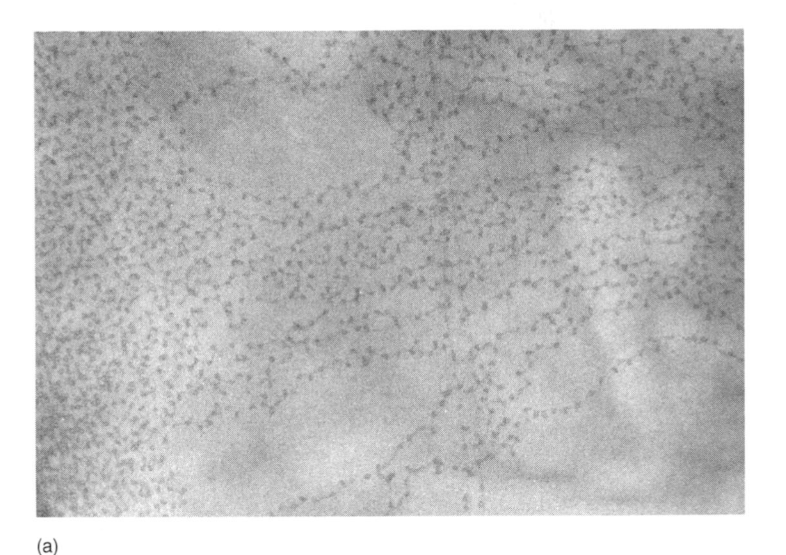

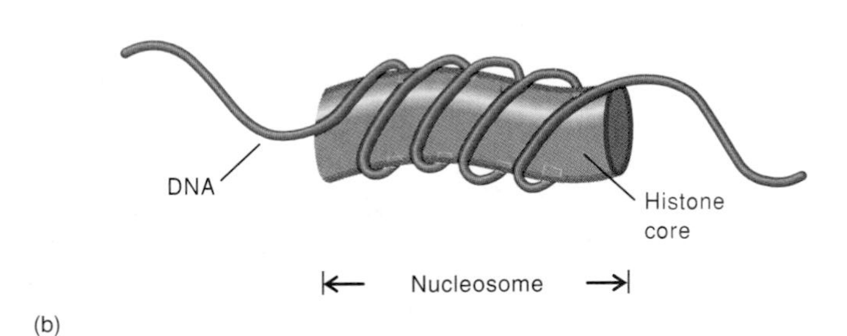

Figure 3.54
(*a*) Chromatin fibers streaming out of a nucleus of a red blood cell of a chicken. The threads as well as the "beads" on the threads consist of DNA wrapped around histone. These fibers and beads are packed tightly into the nucleus. (*b*) Diagram of a nucleosome. The histone is a basic protein that plays a role in the regulation of gene function in eukaryotic cells.

chromosomes (figure 3.55). The nuclear membrane is perforated with holes that allow large molecules such as RNA and proteins to be transported into and out of the nucleus (figure 3.56).

The nucleus also contains another structure, the nucleolus, a large, spherical body made up of loops of DNA that are not enclosed in a membrane (see figure 3.55). This structure plays an important role in the synthesis of ribosomal RNA and in the assembly of ribosomes. No comparable structure exists in prokaryotic cells.

The major structures of prokaryotic cells and their eukaryotic counterparts are compared in table 3.3.

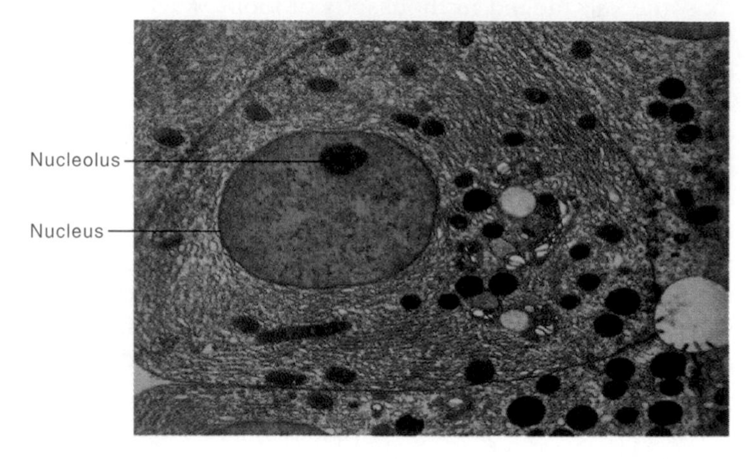

Figure 3.55
Transmission electron micrograph of an animal cell (rat pancreas) showing the nucleus and nucleolus (×1,875).

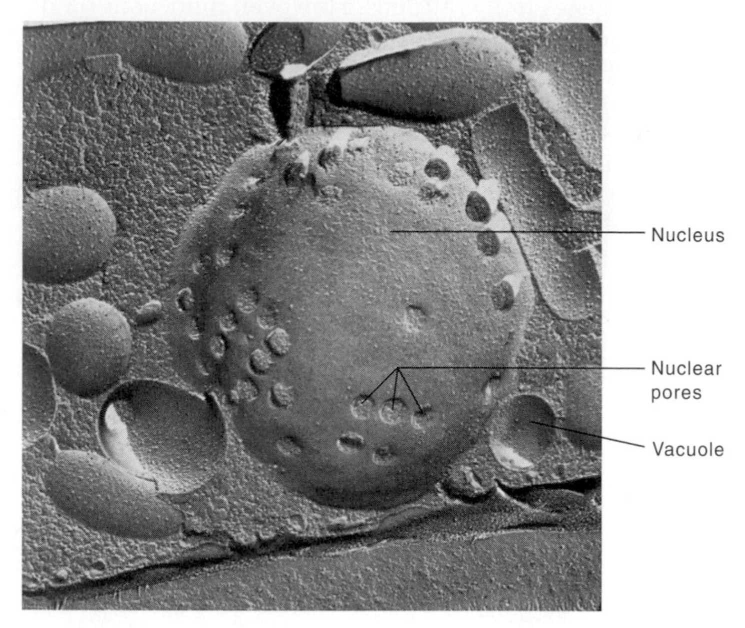

Figure 3.56
Electron micrograph of a yeast cell (*Geotrichum candidum*) by the freeze-fracture technique, showing a surface view of the nucleus and nuclear pores (×44,600).

Table 3.3 Comparison of Structural and Functional Features of Eukaryotic and Prokaryotic Cells

Structure	Distinguishing Features	
	Prokaryotic	***Eukaryotic***
Cell envelope		
capsule (glycocalyx)	generally complex polysaccharides	not present
cell wall	peptidoglycan for most bacteria	not present on animal cells
cytoplasmic membrane	phospholipid bilayer	phospholipid bilayer
Nucleus	no true nucleus, but a nuclear region with one or several chromosomes; identical; not surrounded by membrane; circular DNA; no histones	true nucleus; nuclear membrane surrounds many different chromosomes; DNA in long filaments associated with histones
Mitochondria	none; cytoplasmic membrane contains enzymes for generating energy	present; generate energy
Ribosomes	70S	80S
Flagella	simple structure composed of flagellin subunits; flagella rotate	complex structure composed of microtubules; flagella do not rotate

Summary

I. Microscopic Techniques: Instruments
   A. The simple microscope has one magnifying lens.
   B. The compound microscope has two sets of magnifying lenses, the objective lens and the ocular lens. The magnification is the product of the magnifications of the individual lenses. Resolving power determines the usefulness of a microscope and the amount of detail that can be seen. It depends on the wavelength of the illumination.
   C. The contrast between cells and the surrounding medium must be increased for bacteria to be seen clearly.
   D. Types of microscopes

1. The phase contrast microscope is a light microscope that increases the contrast between the bacteria and the surrounding medium.
2. The interference microscope enables the specimen to be viewed in three dimensions.
3. The dark-field microscope brightly illuminates the object being viewed against a dark background.
4. The fluorescence microscope is used for objects that fluoresce.
5. The electron microscope can resolve objects a thousand times better than the light microscope because it uses electrons rather than light to illuminate the specimen. (a) Specimens must be prepared for viewing in the electron microscope. Such preparation includes fixation, freeze-etching, and thin sectioning. (b) The scanning electron microscope sees surface details in a three-dimensional image. (c) The scanning tunneling microscope also views surface structures of specimens, but its resolving power is much greater than that of the electron microscope. Individual atoms can be seen with it.

II. Microscopic Techniques: Staining
- A. Simple staining procedures involve a single dye.
- B. Positive stains color cell components, whereas negative stains do not penetrate the cell.
- C. Differential staining techniques make use of a combination of dyes.
  1. The gram stain divides bacteria into two groups based upon their cell wall composition: gram-positive and gram-negative.
  2. The acid-fast stain is used on members of the genus *Mycobacterium*, which cause tuberculosis and leprosy.
- D. Specific cell structures can be stained through special techniques that depend on the structure of the material being stained.

III. Shapes of Bacteria
- A. Bacteria have three major shapes: cylindrical, or bacillus; spherical, or coccus; and spiral or spirilla.
- B. Individual cells, especially cocci, may remain attached to one another, forming particular arrangements that are useful in identifying genera.

IV. Functions Required in Prokaryotic Cells
- A. All cells must carry out the following functions:
  1. enclose the internal contents of the cell;
  2. replicate genetic information;
  3. synthesize cellular components;
  4. generate, store, and utilize energy-rich compounds; and
  5. control the entry and exit of specific molecules.
- B. In addition, some cells can move, transfer genetic information, store reserve materials, and form a new cell type, the endospore.

V. Enclosure of the Cytoplasm in Prokaryotic Cells
- A. The cytoplasm is enclosed by three layers: the capsule, the cell wall, and the cytoplasmic membrane.
- B. One layer is the capsule, or slime, layer. Some cells synthesize a capsule, or slime layer, when growing under certain nutritional conditions. Most capsules are composed of polysaccharides and are then called the *glycocalyx*. They are protective and also help attach bacteria to a variety of surfaces.
- C. Another layer is the cell wall.
  1. The bacterial cell wall has several important functions for the cell, and its composition confers important properties on it. The cell wall holds a cell together and gives the cell its shape.
  2. The unique structure in all bacterial cell walls that confers rigidity is glycan, which consists of repeating subunits of *N*-acetylglucosamine joined to *N*-acetylmuramic acid. These subunits are joined to glycan subunits by amino acid bridges. Each layer is termed a *peptidoglycan* layer.
  3. Cell walls of gram-positive bacteria consist of multiple layers of peptidoglycan along with teichoic acid. Cell walls of gram-negative bacteria consist of only one or a few layers of peptidoglycan with a complex outer membrane consisting of a phospholipid bilayer in which protein porins are embedded. Lipopolysaccharides stick out from the outer membrane. The peptidoglycan layer is located in the periplasm, the area between the outer and inner membrane.
  4. The Archaea have no peptidoglycan in their cell walls, and some bacteria have no cell wall.
  5. Differences in cell wall composition account for the gram-staining properties and the effectiveness of penicillin.
- D. Another layer is the cytoplasmic membrane. All cells, both prokaryotic and eukaryotic, have a cytoplasmic membrane that determines which molecules enter and leave the cell.
  1. The cytoplasmic membrane is composed of lipoprotein and phospholipids.
  2. The cytoplasmic membrane is a bilayer membrane.
  3. In bacteria, the cytoplasmic membrane is the site of enzymes of energy generation; it controls the exit and entrance of molecules through the mechanisms of diffusion and active transport.

VI. Movement of Prokaryotic Cells
- A. The major structures responsible for movement are flagella.
- B. Flagella have a relatively simple structure in bacteria. They consist of three parts: filament, hook, and basal body.
- C. Flagella push bacteria through liquids.
  1. Bacteria move toward food sources and away from harmful materials through the process of chemotaxis.
  2. Chemotaxis is a primitive nervous response to the environment and operates by determining the length of time the bacteria swim in one direction.

VII. Attachment of Prokaryotic Cells

Pili are protein appendages that attach bacteria to various surfaces or serve other functions.

VIII. Storage of Genetic Information in Prokaryotic Cells
- A. The major structure in which the genetic information is stored is the chromosome.
- B. The bacterial chromosome is a double-stranded, circular DNA molecule.
  1. In prokaryotic cells, only one or several identical chromosomes are present per cell, and the chromosomes are not surrounded by a nuclear membrane.
  2. Histonelike basic proteins are bound to the DNA.

C. Plasmids are small DNA molecules that code for nonessential information and replicate independently of the chromosome.

IX. Synthesis of Protein in Prokaryotic Cells
A. Ribosomes are structures on which proteins are synthesized. Ribosomes are composed of ribosomal proteins and ribosomal RNA molecules.

X. Storage Materials in Prokaryotic Cells
A. Bacteria often store a variety of materials that they later use as a source of nutrients. These include glycogen, volutin, and β-hydroxybutyric acid.

XI. Bacterial Endospores
A. The endospore is a unique cell type that develops from actively multiplying cells (vegetative cells) when they face starvation.
B. Endospores can develop into vegetative cells.
C. Endospores have unusual staining and resistance properties.
D. Many endospore-producing bacteria cause disease.

XII. The Structure of Eukaryotic Cells

XIII. Enclosure of the Cytoplasm in Eukaryotic Cells
A. The cytoplasmic (plasma) membrane is a lipid bilayer, similar in composition to the ones in prokaryotic cells. It encloses the cytoplasm. Various glycoproteins are embedded in it.
B. Animal cells do not have a rigid cell wall.

XIV. Internal Membrane Structures in Eukaryotic Cells
A. The endoplasmic reticulum (ER) encloses a large internal space and is of two types. The rough ER has ribosomes on its surface that are concerned with protein synthesis. The smooth ER, which lacks ribosomes, is concerned with lipid metabolism.
B. The Golgi apparatus is concerned with the chemical modification and transport of molecules made in the endoplasmic reticulum.
C. Lysosomes contain degradative enzymes such as nucleases and proteases.
D. Peroxisomes contain enzymes that both generate highly reactive toxic compounds in metabolism and also degrade these molecules.

XV. Shapes of Eukaryotic Cells
A. The cytoskeleton consists of a network of protein filaments that give the cell its shape and play other roles as well. Actin filaments are involved in the movement of cells. Microtubules, concerned with cell movement by flagella and cilia, form the mitotic spindles that participate in the separation of chromosomes in cell division.

XVI. Generation of Energy in Eukaryotic Cells
A. Mitochondria contain enzymes involved in the generation of energy. They function in ATP synthesis much like the cytoplasmic membrane of prokaryotic cells.
B. Mitochondria likely evolved from bacteria that were engulfed by cells. They contain DNA and ribosomes and are of the same size as bacteria.

XVII. Storage of Genetic Information in Eukaryotic Cells
A. Genetic information is stored in double-stranded DNA, which is associated with basic proteins (histones) to form chromatin. The chromatin makes up the chromosomes of the cell. Each cell contains multiple chromosomes that contain different genetic information. Each of the chromosomes replicates and divides into two during cell division.
B. The chromosomes are enclosed in a membrane bilayer, the nuclear membrane, to form the nucleus. This is one of the features that distinguishes eukaryotes from prokaryotes.
C. The nucleus also contains the nucleolus, the structure in which ribosomes are synthesized.

Critical Thinking Questions

1. A number of antibiotics that affect bacterial cell wall synthesis and ribosome function have proven useful in treating bacterial infections. However, very few useful antibiotics attack the cytoplasmic membrane. Explain why this is true.
2. What proteins would you expect to find in the periplasm? Why does this make sense in terms of cell function?
3. How do you explain the observation that old cultures of gram-positive bacteria become gram-negative?
4. Would penicillin kill membranes of the Archaea? Explain your answer.

Further Reading

Blakemore, R. P., and Frankel, R. B. 1981. Magnetic navigation in bacteria. *Scientific American* 245(6):58.

Bretcher, M. S. 1985. The molecules of the cell membrane. *Scientific American* 253(4):100.

Brock, T. D. 1988. The bacterial nucleus: a history. *Microbiol. Rev.* 52:397–411.

Ferris, F. G., and Beveridge, T. J. 1985. Functions of bacterial cell surface structures. *BioScience* 35(3):172–177.

Rietschel, E. T., and Brady, H. 1992. Bacterial endotoxins. *Scientific American* 267(2):54–61.

Rothman, J. E., and Orci, L. 1996. Budding vesicles in living cells. *Scientific American* 274(3):70.

Sharon, N., and Lis, H. 1993. Carbohydrates and cell recognition. *Scientific American* 268(1):82–89.

Vidal, G. 1984. The oldest eukaryotic cells. *Scientific American* 250(2):48.

4 Dynamics of Bacterial Growth

KEY CONCEPTS

1. Bacterial growth is measured by an increase in cell numbers; growth of most other organisms is measured by an increase in total size.
2. All bacteria require a source of carbon, hydrogen, oxygen, and nitrogen to synthesize components of cytoplasm as well as a source of energy in order to grow. Different bacteria use different materials for their nutrients.
3. The growth requirements of bacteria may change permanently when they are taken from their natural environment and are grown in the laboratory.
4. Representatives of different groups of bacteria can multiply over a very broad range of environmental and nutritional conditions.

Bacterial colonies growing on multicolored agar plates.

PREVIEW LINK

Microbes in Motion

The following books and chapters in the *Microbes in Motion* CD-ROM may serve as a useful preview or supplement to your reading:

Microbial Metabolism and Growth: Growth; Metabolism. *Control of Microorganisms* (all).

A Glimpse of History

From the days of Antony van Leeuwenhoek it was known that invisible organisms existed, but what they did, how they did it, and how they grew was a mystery. The efforts of a number of outstanding scientists in the late 1800s clarified some of the most important properties and activities of microorganisms. Louis Pasteur was one of these scientists.

Louis Pasteur was the first person to demonstrate that these invisible organisms could actually convert one substance into another. For example, he showed that bacteria present in a solution of glucose were responsible for the formation of butyric acid from the glucose. As butyric acid appeared, glucose was observed to disappear. Whenever Pasteur removed or destroyed the bacteria, no butyric acid was formed, and the glucose did not disappear.

In the course of these experiments, Pasteur made some other very important observations. The bacteria that formed butyric acid from glucose lived and multiplied in the absence of air:

These infusoria bacteria cannot only live in the absence of air, but air actually kills them. If a stream of carbon dioxide is passed through a medium in which they are multiplying, their viability and their reproduction are not affected in the least. On the contrary, under the same conditions, if one substitutes a stream of air for the carbon dioxide, in one or two hours they all die, and the butyric acid fermentation which requires their viability is stopped immediately. I believe this is the first example of an animal living in the absence of free oxygen.

Pasteur was correct. This report was the first to declare that not only is gaseous oxygen not required for life, but also that some organisms are killed by it. Prior to Pasteur's observations, people thought that all organisms died if air was removed for even a few minutes. This is only one example of the wide variety of conditions under which bacteria can grow. Other examples are given in this chapter.

In this chapter, it will become clear that, in contrast to humans, who thrive only in a limited range of environmental conditions, bacteria can exist within an extremely broad range. Some bacteria can grow at temperatures above 100°C (the boiling temperature of water) but not at room temperature. Other bacteria can grow in solutions of 30% salt or in the deep oceans where the pressure is 300 times greater than that at the surface of the earth. Other bacteria appear to be able to live on rocks alone! Any particular species can grow only under a limited set of environmental conditions, but as a whole, the members of the microbial world can grow in a wide range of different environments.

Pure Culture Methods

The variety of organisms of different sizes and shapes found in most environments represents a **mixed population** of various genera and species that have become adapted to a particular environment. In the 1860s, it was commonly believed that bacteria had the capacity to change their shapes and sizes and perhaps even their functions. However, this is not the usual case. Bacteria with different shapes and sizes generally represent different species. This was shown by separating the progeny (offspring) of a single bacterial cell from all other bacteria to obtain a **pure culture** and by demonstrating that all of the cells in the culture were approximately the same shape and size. The concept of a pure culture and the means of obtaining it in the laboratory are among the most significant advances in bacteriology. Without the techniques for obtaining pure cultures, this field of study would scarcely have progressed. With these techniques, the Golden Age of Bacteriology became possible.

The greatest contributor to pure culture methods was Robert Koch, a German physician who succeeded in combining a medical practice with a remarkably successful and productive research career for which he received a Nobel Prize in 1905. Koch was primarily interested in isolating and identifying disease-causing bacteria. However, early in his career in the late 1870s, he realized that it was necessary to have simple methods for obtaining pure cultures. Koch recognized that the isolation of pure cultures required a solid medium on which a single, isolated cell could multiply in a limited area. The population of cells that arises from a single bacterial cell in one spot is called a **colony.** About 1 million cells are required for a colony to be easily visible to the naked eye.

Koch initially experimented with growing bacteria on the cut surfaces of potatoes, but he found that a lack of nutrients in potatoes prevented growth of some bacteria. To overcome this difficulty, Koch realized that it would be advantageous to be able to solidify any liquid nutrient medium. He conceived the idea of adding gelatin as a solidifying agent. Since the gelatin-containing nutrient medium is liquid at temperatures above 28°C, it is only necessary to heat the medium above this temperature, pour it into a sterile container, and allow it to cool and harden. Once the medium hardens, a loopful of bacteria can be drawn lightly over the surface to deposit single bacterial cells at intervals. Each of those single cells divides, and after a day or two enough cells are present to form visible colonies. This **streak-plate method** is the simplest and most commonly used technique for isolating the progeny of a single bacterium (figure 4.1). A nutrient medium solidified with gelatin poses certain problems, however. The medium must be incubated at temperatures below 28°C if it is to remain solid. Furthermore, gelatin itself can be degraded by many microorganisms, resulting in the medium becoming liquid. Thus, its use as a hardening agent is severely restricted.

The perfect hardening agent came from a household kitchen. Frau Angelina Hesse, the wife of one of Koch's associates, suggested **agar,** a solidifying agent her mother had used to harden jelly, which remains a solid at warm temperature. Agar is an undigestible polysaccharide extracted from certain marine algae. In contrast to gelatin, a 1.5% agar gel must be heated above 95°C before it liquefies, and

• **Polysaccharide, p. 36**

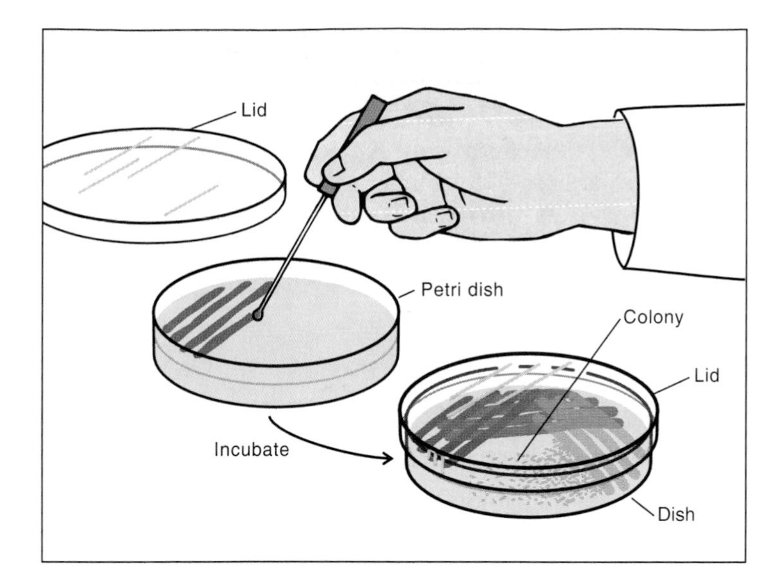

Figure 4.1
Streak plate method of producing isolated colonies. The loop is sterilized and then touched to bacteria, either in liquid or on a solid surface. The loop is then gently drawn across the surface of the agar several times, depositing bacteria in rows (streaks). The loop is sterilized again and touched to the agar surface at a location near the end of the first streaks. The loop is then drawn across the agar at an angle to the original streaks. This process is repeated twice more. Individual colonies should appear near the ends of the final streaks or in earlier streaks, depending on the number of bacteria initially present on the loop. The container in which the bacteria are growing is known as a petri dish, after its inventor.

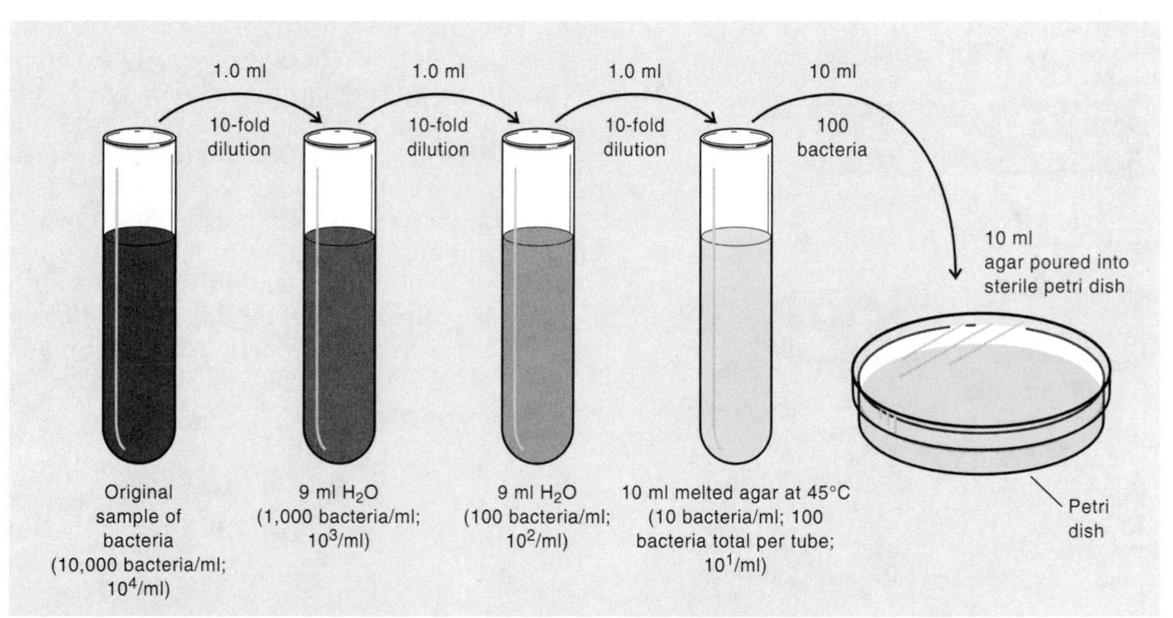

Figure 4.2
Pour-plate method of achieving isolated colonies.

it does not solidify until it is cooled to temperatures below 50°C. It is therefore solid over the entire range of temperatures at which most bacteria grow. Further, it is stable at high temperatures and can be sterilized by heating. Nutrients that might be destroyed at higher temperatures can be added before the agar hardens. Another advantage of agar is that very few bacteria can degrade it.

Another major technical advance was the development of a two-part dish, either glass or plastic, that can be readily sterilized and maintained in a sterile condition. This two-part dish, called the **petri dish,** named after its inventor, Julius Petri, who worked in Koch's laboratory, remains unmodified (figure 4.1). It serves as a convenient container for solid media, maintaining them free of any organisms in the air while still being exposed to the air.

An additional method for obtaining isolated single colonies that also takes advantage of the properties of agar is the **pour-plate method** (figure 4.2). In this method, one mixes an appropriate number of bacteria with melted agar at a temperature of about 45°C and then pours the mixture into a sterile dish. The agar solidifies and traps the organisms in a defined space; after incubation, colonies can be observed.

The significance of the pure culture technique in microbiology cannot be overestimated. This method proved crucial for the rapid isolation of pathogenic bacteria as well as other bacteria. Within 20 years of the development of the method, the causative agents of most of the major bacterial diseases of humans were isolated and characterized. Furthermore, the concept that bacteria divide to form

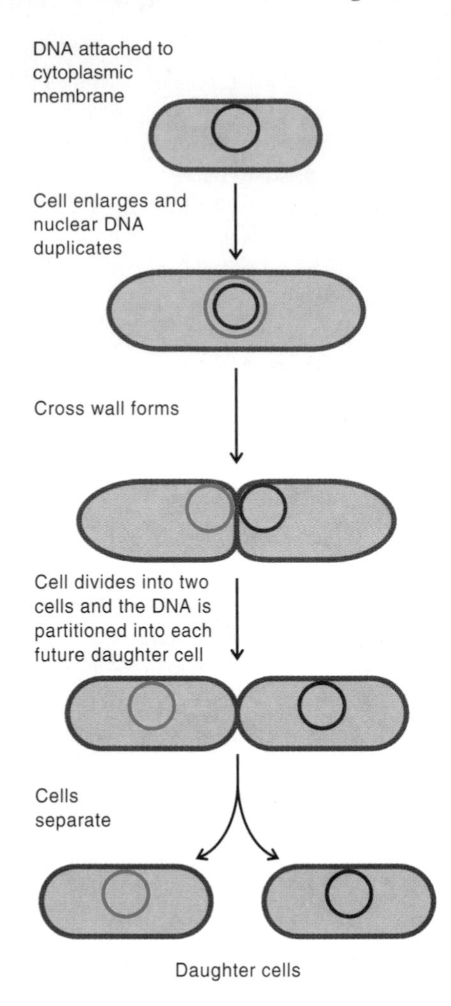

Figure 4.3
Schematic representation of cell division by binary fission. The chromosomal circular DNA is attached to the cytoplasmic membrane. As the cytoplasmic membrane increases in length, the DNA replicates and is partitioned into each of the two daughter cells.

organisms of similar shape and size was established once and for all. Koch recognized that bacteria of different shapes and sizes could be considered different species. However, we now know that the shape of some bacteria is not always constant; some bacteria can change their shape depending on the environment. The recognition that different bacteria had many distinguishable shapes with varying metabolic properties had a tremendous impact on the growth of the young science of bacteriology.

Scientists working with microorganisms in the laboratory are very careful to insure that they are working with pure cultures. Otherwise, it would not be possible to obtain reproducible results. However, it is also true that an understanding of the interactions of different organisms with their natural environment requires the use of mixed populations.

Measurement of Cell Growth

Growth is the orderly increase in the quantity of all components of the bacterial cell. After a bacterial cell has increased in size and doubled the amounts of each of its parts, it divides into two daughter cells through the process of **binary fission** (figure 4.3). The time required for one cell to divide into two cells is the **doubling,** or **generation, time.**

Bacterial Growth Is Defined in Terms of Population Size

Since the size of a single bacterial cell does not vary much, the growth of bacteria is measured by an increase in the number of cells. Thus, with bacteria, the rise in the number of cells indicates growth, whereas with multicellular organisms, growth is reflected primarily by an increase in the size of a single organism.

In a rapidly multiplying bacterial population, cell numbers increase exponentially because each cell gives rise to two cells, each of which then divides into two more, yielding a total of four, and so on. When the number of multiplying cells in a population is plotted against the time of incubation, a straight line results if the number of cells is plotted logarithmically and the time of incubation is plotted on an arithmetic scale (figure 4.4 line A). The

• **Logarithms, appendix I**

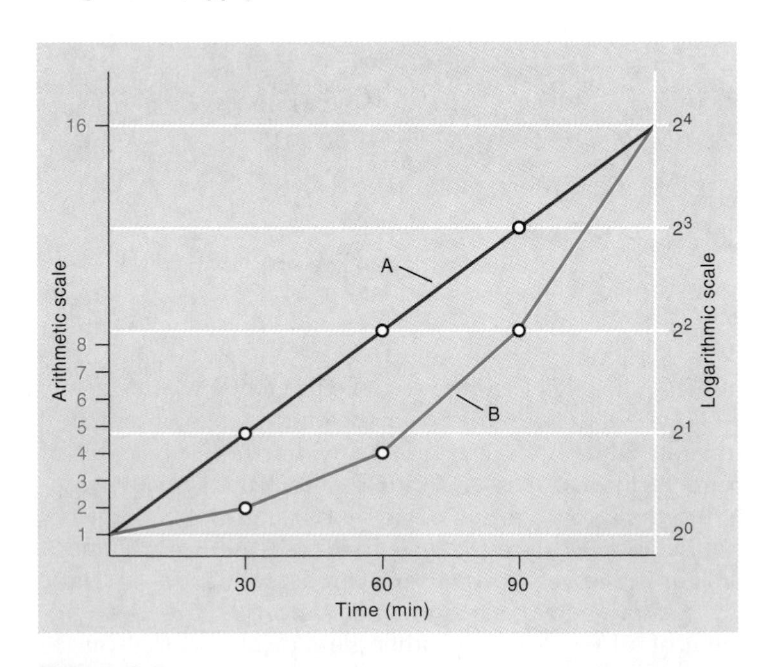

Figure 4.4
Relationship of cell number to number of generations plotted on arithmetic and logarithmic scales. Line A is the logarithmic scale, and line B is the arithmetic scale.

straight line indicates that the number of cells increases by the same percentage during each interval. If the ordinate on the graph representing cell numbers is arithmetic, rather than logarithmic, then the line is not straight, but rather becomes increasingly steep (figure 4.4 line B). If the number of cells increases from one to four in 1 hour (a generation time of 30 minutes), another fourfold increase in the population will occur in the next hour and yet another fourfold increase after each succeeding hour. Note, however, that the increase in the number of cells varies within the three intervals. The total number of cells at any given time depends on the number of cells present initially and the number of cell generations. Since many bacteria can divide every 30 minutes (and a few species even divide every 10 minutes), the number of cells in a population reaches very high levels in a relatively short time. If a given microorganism divides every 30 minutes, a single cell will result in a population of about 1 million cells (10^6) in 10 hours.

A simple equation expresses the relationship between the original number of cells (zero time), the number of cells in a population at any later time, and the number of divisions that the cells have undergone during this time. The latter value can be readily determined if the doubling (generation) time is known. If any two values are known, the third can be easily calculated from the equation

$$b = B2^n$$

where B is the number of cells at zero time, b is the number of cells at any later time, and n is the number of cell generations.

Growth Can Be Measured as an Increase in Mass (Weight) of the Population

The most convenient and simplest method of determining the mass of cells in a population is based on the fact that small particles, such as bacteria, scatter light that is passed through the cell suspension. The amount of light scattering is proportional to the mass of cells present, and because the bacterial cells are relatively uniform in size (mass), the number of cells can be measured from the amount of light that reaches a sensing device after passing through the suspension (figure 4.5). This method does not distinguish between living and dead bacteria.

One limitation to this method is that a medium containing 1 million bacteria (10^6) per milliliter is perfectly clear, and if it contains 10 million cells (10^7), it is barely turbid. Thus, although a turbid culture indicates that bacteria are present, a clear solution does not guarantee their absence. Not recognizing these facts can have serious consequences in the laboratory as well as outside the laboratory where bacteria are able to grow.

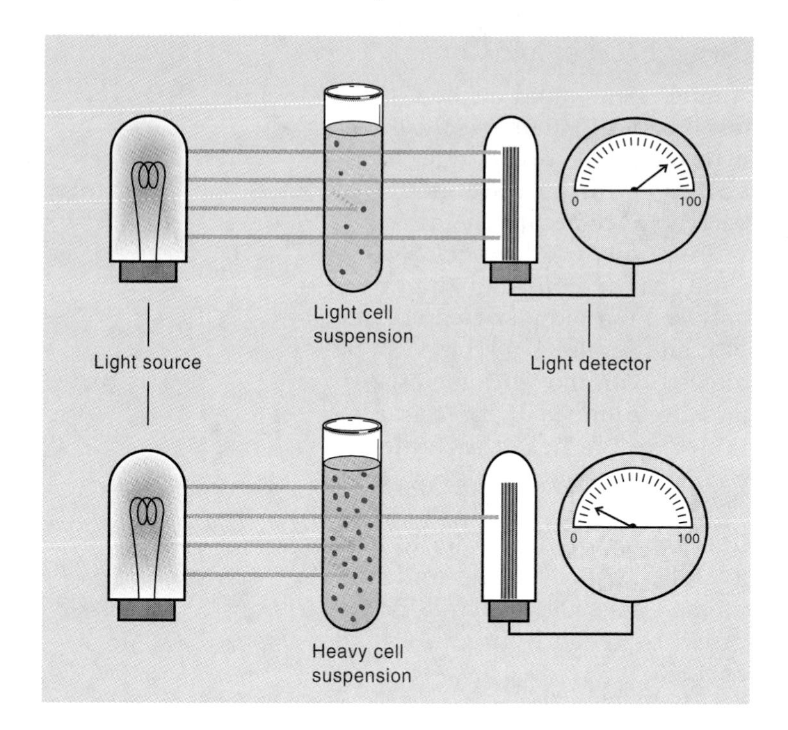

Figure 4.5
Measurement of bacterial mass by transmission of light. The amount of light that impinges on the light detector is inversely proportional to the number of bacteria in the suspension. The light detector is a spectrophotometer.

Growth Can Be Measured as an Increase in Cell Number

Two common laboratory techniques for directly measuring cell number are the plate count and the direct microscopic count.

Plate Count

The **plate count method** is based on the fact that, under the proper conditions, a bacterium will multiply to form a visible colony on solid nutrient medium. The number of colonies that appear after a suitable period of incubation thus represents the number of living cells in the original suspension (see figure 4.2). Because it is difficult and inaccurate to count more than about 300 colonies per petri dish, it is generally necessary to dilute the original culture before transferring a known volume onto the solid medium. If the cells do not separate from each other after cell division (a common situation with the chain-forming streptococci, for example), then the viable cell count measures the **colony-forming units** rather than individual cells. The plate count method measures only living cells, but it is extremely sensitive, since the presence of only a few cells per milliliter can be readily detected. However, the method is slow because the petri dish must be incubated at least 1 day before the bacteria have multiplied sufficiently to give rise to a visible colony.

Direct Microscopic Count

A much more rapid method of determining cell number is the **direct microscopic count,** which involves counting the number of bacteria in an accurately measured volume of liquid. Special glass slides called **counting chambers** are used to hold a known volume of liquid (figure 4.6). These can be viewed with the light microscope, and the number of bacteria contained in the liquid can thus be counted precisely. However, about 10 million bacteria (10^7) per milliliter are required in order to gain an accurate estimate, and this method does not ordinarily distinguish between living and dead bacteria.

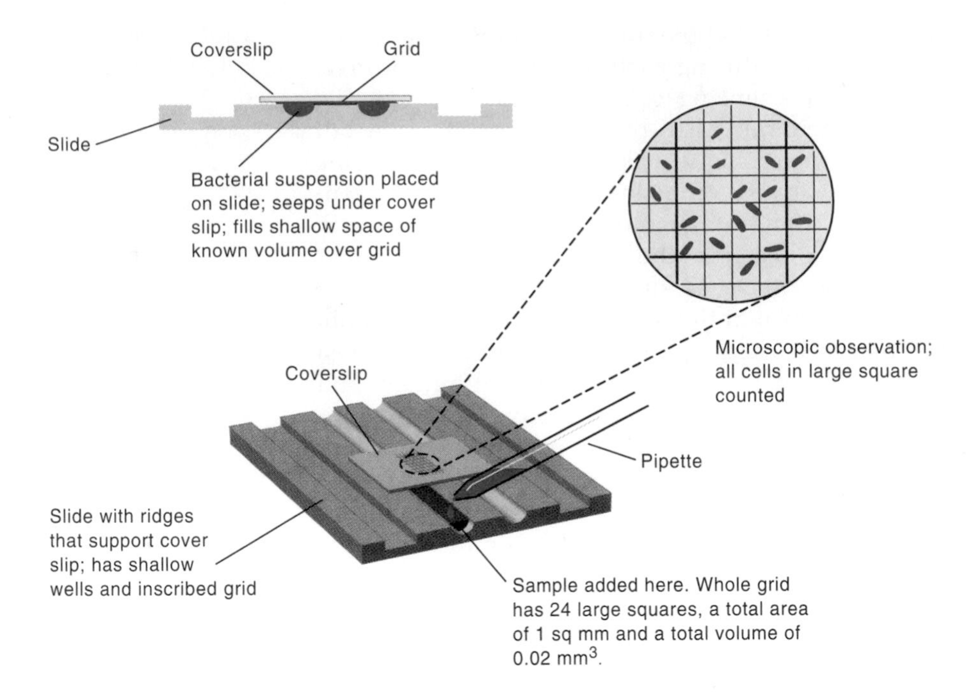

Figure 4.6
Direct microscopic counting procedure of bacteria in a counting chamber.

REVIEW QUESTIONS

1. List three useful properties of agar as a hardening agent in media.
2. By what process do bacteria multiply?
3. List three ways of measuring the number of bacteria in a liquid. Which method measures only living cells?
4. If bacteria divide every 30 minutes, how many cells will there be after 3 hours, beginning with a single cell?

Multiplication of Microorganisms–General Aspects

As a group, microorganisms multiply within an enormous range of environmental conditions, although no single species can multiply in every environment found in nature. The greatest number of species lives in the mild environmental conditions most commonly found on the earth's surface. However, some organisms live in very harsh environments, such as the Dead Sea (30% salt), the hot springs at Yellowstone National Park (temperatures of 90°C), and extremely acid mine waters. Such organisms usually do not grow when cultivated under the mild conditions under which most bacteria grow. These organisms are given the general name **extremophiles.**

An organism taken from its natural habitat often shows altered growth requirements when grown in the laboratory. For example, when initially isolated from its natural environment, *Brucella abortus,* the bacterium responsible for undulant fever in humans, needs 5% to 10% CO_2 in its environment to grow. However, after a few transfers on laboratory medium, this requirement is lost. Also, once an organism has been cultivated in the laboratory for a long time, it may no longer be able to survive in its original habitat. For example, the common, widely studied organism *Escherichia coli* K-12, originally isolated from a human intestinal tract, has been grown in the laboratory for years and cannot survive if it is reintroduced into the intestinal tract. Apparently, through the process of mutation and selection, the *E. coli* cells became adapted to their new environment in the laboratory. As a result, species of bacteria cultivated in the laboratory may differ markedly from their ancestors in the natural world.

Factors Influencing Microbial Growth

A variety of factors influence the multiplication of microorganisms. They can be divided into two broad categories: environmental factors and nutritional factors.

The Environmental Factors that Influence Growth Include Temperature, Oxygen, pH, and Osmotic Pressure

Temperature of Growth

Most bacteria grow within a temperature range of approximately 30°C, with each species having a well-defined upper and lower limit at which growth stops.

Bacteria are commonly divided into four groups based on their optimum growth temperatures (figure 4.7). The optimum temperature is the temperature at which the organism divides most rapidly. **Psychrophiles** have their optimum between –5°C and 20°C. These organisms are usually found in such environments as the Arctic and Antarctic regions and in lakes fed by glaciers. **Mesophiles,** which include the majority of bacteria, have their optimum within the range of 20°C to about 50°C. The optimal temperature for most disease-causing bacteria adapted to growth in the human body is between 35°C and 40°C. Microorganisms found in soil, which is a colder environment, generally have a lower optimum, close to 30°C. The third group, the **thermophiles,** have an optimum temperature between 50°C and 80°C. Another group, the **hyperthermophiles,** are organisms whose optimum growth temperature is above 80°C. They are frequently isolated from hot springs and are usually members of the Archaea. One such organism, isolated from an erupting volcano deep in the ocean, has its optimum at 105°C and a maximum temperature of growth of 110°C, which is among the highest growth temperatures yet recorded. Another member of the Archaea isolated from the same environment grows at temperatures between 90°C and 113°C, the highest recorded temperature for the growth of any organism.

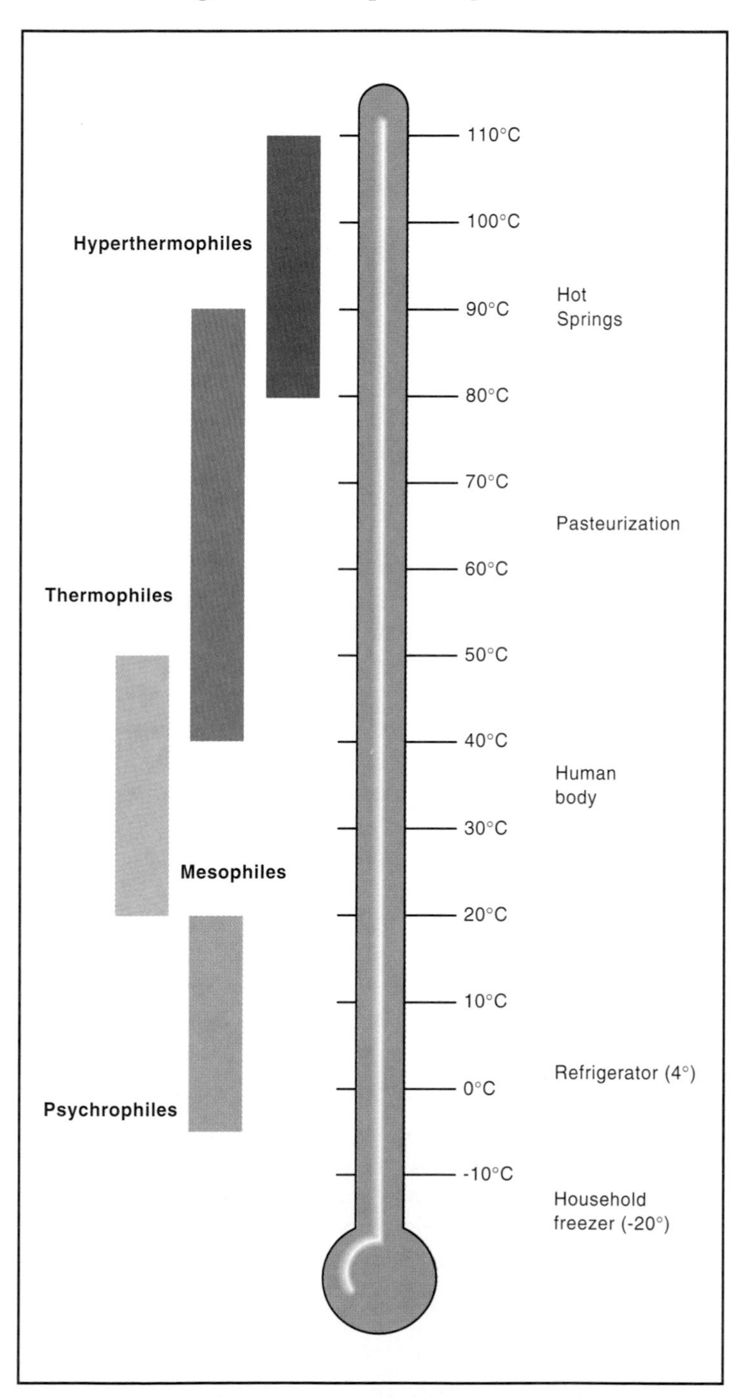

Figure 4.7
Temperature growth range of microorganisms. Organisms in each category cannot grow over the entire range for the group.

Why can some bacteria withstand such high temperatures but most cannot? As a general rule, protein molecules from thermophiles are not denatured at high temperatures. This thermostability is a property of the protein and is based on its composition and sequence of amino acids, which determine the strength of bonds that form within the protein and determine its three-dimensional structure. For example, the formation of many covalent bonds, as well as many hydrogen and other weak bonds, prevents denaturation of proteins.

Although bacteria can be classified conveniently into four groups according to their growth temperatures, no sharp dividing line exists between the temperatures at which organisms in each group will multiply (figure 4.8).

• **Denaturation, p. 35**

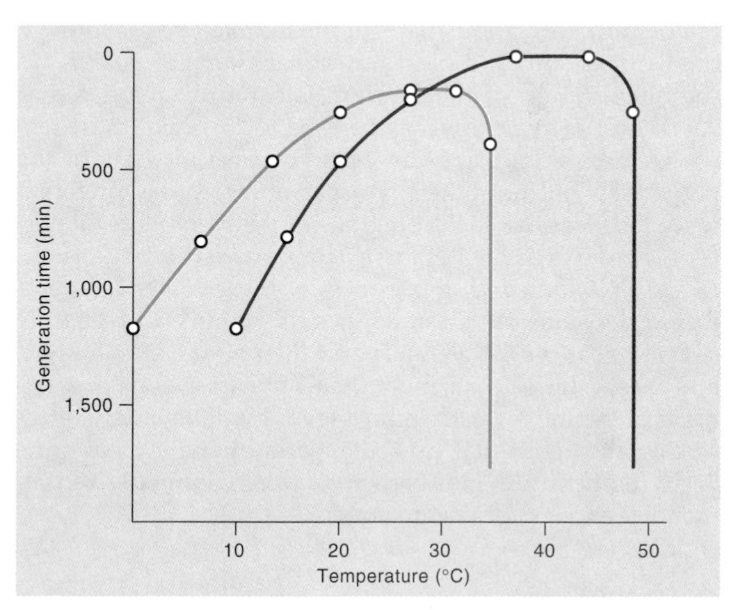

Figure 4.8
Effect of temperature on the generation time of bacteria with different optimum temperatures of growth. The generation time decreases as the temperature is raised, until a temperature is reached at which enzymes start to denature. Cell growth slows down sharply at this temperature. The organism whose growth is shown in blue has a lower optimum growth temperature than the organism whose growth is shown in red.

For example, the lower limit of growth of some thermophiles is below the upper limit of growth of some mesophiles. Furthermore, not every organism in a group can grow in the entire range indicated for its group. For example, some mesophiles will grow at a temperature of 45°C, but others will not. As a general rule, the optimum temperature for growth is close to the upper limit of its range (figure 4.8). This is because the speed of enzymatic reactions in the cell approximately doubles for each 10°C rise in temperature, and thus the cells grow more rapidly as the temperature rises. However, if the temperature becomes too high, enzymes required for the life of the cell are denatured and can no longer function, resulting in the slower growth or death of cells.

Temperature and Food Preservation Since the microorganisms that cause food spoilage in most cases are mesophiles, storage of fruits, vegetables, and cheeses at cool temperatures, approximately 15°C, retards food spoilage. Refrigerators, which maintain temperatures near 4°C, are very important in preserving foods. However, since psychrophiles can multiply at refrigerator temperatures, problems can arise in storing food and other materials, such as blood, that can supply enough nutrients to support microbial growth. To overcome this problem, long-term storage of substances that can withstand below-freezing temperatures must be frozen. Psychrophiles requiring liquid water will not multiply under these conditions.

Temperature and Disease Wide variations exist in the temperature of various parts of the human body. Although the heart, brain, and gastrointestinal tract are near 37°C, the temperature of the scrotum is carefully regulated at 32°C and the fingertips may be 20°C to 25°C. For these reasons, microorganisms can cause disease in some of the body parts but not others. For example, leprosy preferentially involves the coolest regions of the body (ears, hands, feet, and fingers) because the causative organism, *Mycobacterium leprae,* grows best at these lower temperatures. The same situation applies to syphilis, in which lesions appear on the genitalia and then on the lips, tongue, and throat. Indeed, for more than 30 years the major treatment of syphilis was to induce fever by deliberately introducing the agent that causes malaria, thereby raising the body temperature. Hot bath spas were commonly recommended as a less drastic treatment.

Oxygen Requirements for Growth

Bacteria can be classified into five major divisions based upon their requirements for gaseous oxygen, or air, which contains 20% oxygen. These differences depend on whether the bacteria have the necessary enzymes to detoxify toxic forms of oxygen as discussed in the next section and whether the bacteria require oxygen to degrade foodstuffs to provide energy. The various pathways of degradation will be discussed in chapter 5.

1. **Obligate (strict) aerobes.** These organisms have an absolute requirement for oxygen because they metabolize sugars through a pathway that requires oxygen. Since oxygen is not very soluble in liquids and it does not move quickly through them, aerobes grow best in flasks that are continuously shaken, which increases oxygen circulation. Members of this group include the genera *Bacillus* and *Pseudomonas.*

2. **Obligate (strict) anaerobes.** Members of this group cannot multiply if any oxygen is present. Some members are actually killed by traces of oxygen because they cannot modify the toxic forms of oxygen produced in metabolism. Furthermore, some of their enzymes are denatured by oxygen. The best known members of this group are in the genus *Clostridium.* It is now estimated that one-half of all cytoplasm on earth is in anaerobic bacteria!

3. **Facultative anaerobes.** These organisms can utilize oxygen in metabolic processes if it is available, but they can also grow in its absence. Growth is generally more rapid if oxygen is present, since more ATP, the readily utilizable storage form of energy, is generated by metabolic pathways that require oxygen. *Escherichia coli* and *Saccharomyces* (yeast) are common examples of facultative anaerobes.

4. **Microaerophilic organisms.** These organisms require small amounts of oxygen (2% to 10%), but higher concentrations are toxic. A few disease-causing organisms are included in this group, such as *Helicobacter pylori,* the agent that causes gastrointestinal ulcers. The natural environment of this organism, the mucous gel that coats the stomach epithelium, contains low levels of oxygen. Many microaerophiles grow more rapidly if the level of carbon dioxide normally present in the air is increased. Consequently, these organisms are often incubated in a candle jar, which consists of a jar with a tight lid and a lighted candle. Before extinguishing because of insufficient oxygen, the candle converts some of the O_2 in the air to CO_2, resulting in about 3.5% CO_2 and a small percent of oxygen. Enough oxygen remains in the jar that strict anaerobes cannot grow.

5. **Aerotolerant organisms.** These organisms grow in the presence or absence of oxygen, but unlike facultative anaerobes, they derive no benefit from the oxygen. In fact, they are said to be indifferent to oxygen. Many aerotolerant organisms grow better in containers used for anaerobic cultures because they contain higher levels of CO_2 and water vapor than is found in air. *Streptococcus pyogenes,* the cause of strep throat, is an aerotolerant bacterium.

It is often possible to determine in which class organisms belong by growing them in a nutrient agar medium

• **ATP, p. 114**

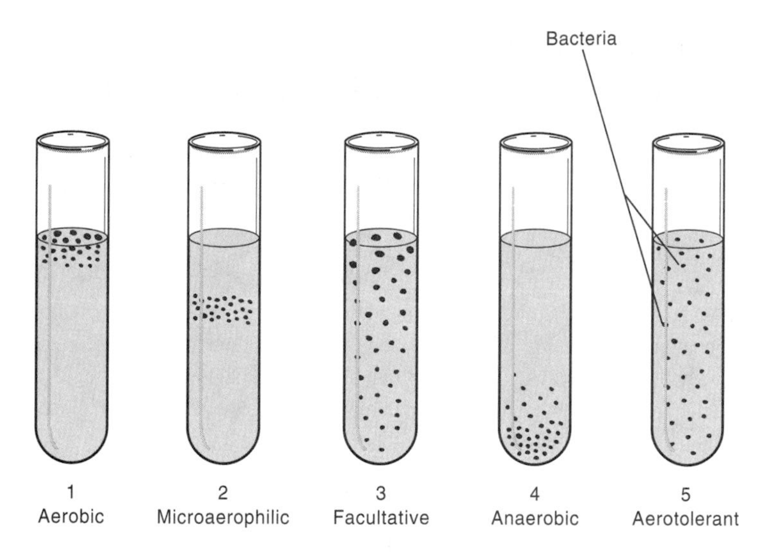

Figure 4.9
Shake tubes that demonstrate the oxygen requirements of different organisms. Note that the facultative bacteria growing with oxygen form larger colonies than do the same organisms growing in the absence of oxygen.

in **shake tubes** (figure 4.9). The tube with the agar not yet solidified is inverted several times, which disperses the bacteria throughout the tube. The agar is then allowed to harden. The top of the tube is highly aerobic and the bottom is anaerobic, since heating drives oxygen from the medium and solidified agar inhibits diffusion of oxygen. Therefore, where the bacteria are located in the tube indicates their requirement for oxygen.

Enzymatic Differences in Aerobes and Anaerobes

Aerobes require oxygen because they have metabolic pathways that require oxygen to generate energy. Anaerobes have other pathways that may generate energy but that do not require oxygen. Why are some anaerobes killed in the presence of oxygen, whereas others can tolerate it even though they cannot use it? The answer lies in the fact that oxygen can be converted into a number of forms that are highly toxic. Some of these toxic forms, such as hydrogen peroxide (H_2O_2), are formed by metabolic processes involving oxygen. Other toxic compounds, such as superoxide (O_2^-), are produced both as a part of normal metabolic processes and as chemical reactions in light. Cells that are not killed in the presence of oxygen contain enzymes that can convert these toxic compounds to nontoxic forms. For example, aerobic bacteria contain the enzyme **catalase,** which breaks down hydrogen peroxide to H_2O and O_2, and **superoxide dismutase,** which degrades superoxide. Although most strict anaerobes do not have superoxide dismutase, some do. Further, a few aerobes lack it. Therefore, this enzyme alone cannot protect organisms from the toxic forms of oxygen. The relationship between the ability of different bacteria to grow in the presence of oxygen and their content of enzymes of oxygen metabolism is shown in table 4.1.

Table 4.1 Enzyme Content of Bacteria with Different Requirements for Oxygen

Name	Enzymes in Cells for O_2 Detoxification
Strict aerobe	Catalase: $H_2O_2 \rightarrow H_2O + O_2$ Superoxide dismutase: $O_2^- + 2H^+ \rightarrow O_2 + H_2O_2 \rightarrow H_2O + O_2$
Facultative anaerobe	Catalase, Superoxide dismutase
Strict anaerobe	Neither catalase nor superoxide dismutase in most
Microaerophile	Small amounts of catalase and superoxide dismutase
Aerotolerant	Superoxide dismutase

pH scale	Common materials with pH indicated
0	Battery acid
1	Gastric juice
2	Lemon juice
3	Ginger ale, wine
4	Tomatoes, orange juice
5	Black coffee, beans
6	Milk begins to taste sour
7 (Neutral)	Distilled water, human saliva
8	Seawater
9	Bleach
10	Soap solution
11	Household ammonia
12	Lime
13	Drain cleaner
14	Lye

Increasing acidity (0–6); Increasing alkalinity (8–14)

Figure 4.10
pH values for various solutions.

pH

Most species of bacteria grow best in a medium with a pH of 7 (neutral), in which the number of acidic ions (protons, or H^-) is the same as the number of basic ions (hydroxyl, or OH^-). (See appendix II.) However, many bacteria can live and multiply at pH 5 (acidic) to pH 8 (basic). Yeasts and molds grow most rapidly in an acidic medium (figure 4.10).

Some prokaryotes as well as eukaryotes can tolerate very high acid concentrations, with the most acid-tolerant member of the Archaea able to grow at a pH of less than 1. Some bacteria that form vinegar (which contains acetic acid) can multiply in very strong acid. One acid-tolerant bacterium, *Helicobacter pylori*, causes ulcers in the human stomach, which is highly acidic. Members of the genus *Thiobacillus*, which accumulate sulfuric acid from their metabolic processes, can grow in media at a pH of 1 or lower. These bacteria are called **acidophilic** (acid loving). However, the pH inside the cell is considerably higher than this. *Thiobacillus* maintains its internal pH near neutral because it pumps out protons as quickly as they enter the cell. *Helicobacter* has a very powerful enzyme, urease, that splits urea in the stomach into carbon dioxide and ammonia. The ammonia neutralizes the stomach acid in the immediate surroundings of the bacteria.

In addition, amino acids and proteins can serve as buffers (see next section); thus, the cytoplasm of all bacteria serves as a natural buffer. The ability of organisms to grow within a broad pH range depends on their ability to maintain the inside of the cell at a neutral pH, rather than on the ability of their cytoplasmic proteins to withstand very high or very low pH.

Buffers Many bacteria produce enough acid as a by-product of their metabolism to inhibit their own growth (see chapter 5). To help prevent this, **buffers** are often supplied in the medium. Buffers are chemicals that limit an increase or decrease in pH. Buffers usually are a mixture of two salts of phosphoric acid—the sodium phosphates Na_2HPO_4 and NaH_2PO_4. These salts limit pH changes because they can combine chemically with the H^+ ions of strong acids and the OH^- ions of strong bases to produce neutral compounds. In an acidic medium, the HPO_4^{2-}, which is formed by the ionization of Na_2HPO_4, combines with the H^+ (a strong acid) to form a weak acid, $H_2PO_4^-$. This reaction can be written as follows:

$$H^+ + HPO_4^{2-} \rightarrow H_2PO_4^- \text{ (weak acid)}$$

The H^+ does not readily leave (ionize) the $H_2PO_4^-$, which makes $H_2PO_4^-$ a weak acid, so the solution is not as acidic. If an organism produces OH^- in its metabolism, the OH^- ion combines with $H_2PO_4^-$ to form another weak acid, HPO_4^{2-}, and water, HOH, by the following reaction:

$$OH^- + H_2PO_4^- \rightarrow HPO_4^{2-} + HOH$$

HPO_4^{2-} is a weak acid because the H^+ does not leave readily.

Some organisms grow only within a very narrow range of pH. For example, the organism causing Legionnaires' disease, *Legionella pneumophila*, could not be isolated from patients for many months after the disease was first recognized, partly because the organism only grows well at a pH of 6.85 to 7.0 and is killed if the pH changes much beyond these limits. To maintain the pH within these narrow limits, special buffers must be used.

• **Proton, p. 23**

Osmotic Pressure

Osmotic pressure can be defined as the pressure that is required to prevent the net flow of water across a semipermeable membrane, such as the cytoplasmic membrane. The osmotic pressure of a medium depends on the concentration of dissolved substances in it. The higher the concentration, the greater the osmotic pressure.

A substance commonly encountered in nature is sodium chloride (common table salt). Many bacteria can grow within a broad range of salt concentrations because the cell can maintain a slightly higher pressure on the inside than on the outside. If the concentration of NaCl increases on the outside of the cell, specific permeases pump K^+ ions from the medium into the cell so as to maintain the osmotic pressure slightly higher on the inside of the cell than on the outside. Consequently, water enters the cell and pushes the cytoplasmic membrane against the rigid cell wall (figure 4.11 *a*). Bacteria can also synthesize certain amino acids in high amounts that remain inside the cell, thereby increasing the internal osmotic pressure.

However, if the salt concentration on the outside of a cell is very high, the cell cannot compensate for it and the osmotic pressure becomes higher on the outside of the cell than on the inside. Consequently, water leaves the cell, and the cytoplasmic membrane shrinks from the cell wall, becoming damaged in the process (figure 4.11 *b*). As a result, cell growth is inhibited.

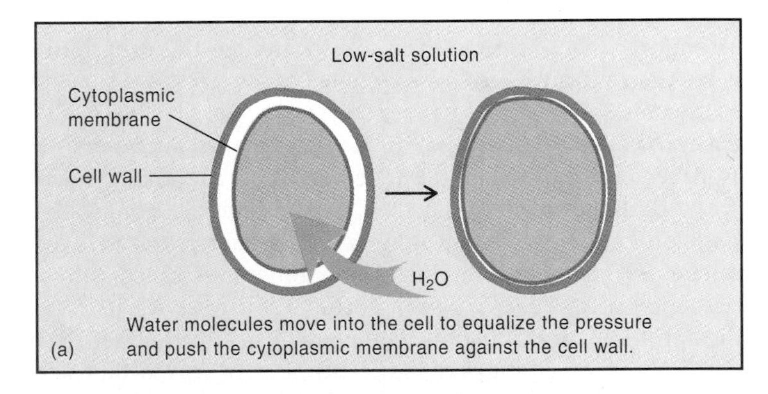

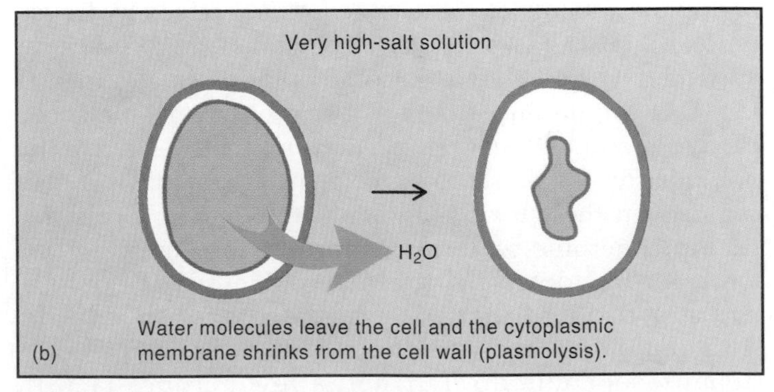

Figure 4.11
Movement of water into and out of cells. (*a*) Low- and (*b*) very high-salt solutions. The cytoplasmic membrane is semipermeable and allows only water molecules to pass through freely.

This growth-inhibiting effect of salt is taken advantage of in food preservation. High levels of salt are added to preserve such foods as bacon, salt pork, and anchovies. High concentrations of sugars can also inhibit the growth of bacteria. Many foods with a high sugar content, such as jams, jellies, honey, preserves, and sweetened condensed milk, are naturally preserved.

REVIEW QUESTIONS

1. List four environmental factors that influence the growth of bacteria.
2. List the categories into which bacteria can be grouped according to their optimum temperature of growth.
3. List two ways in which foods can be preserved.
4. List the categories into which bacteria can be classified according to their requirements for oxygen.
5. Explain how buffers maintain a constant pH.

Nutritional Aspects of Bacterial Growth

Cell Multiplication Requires a Source of Energy and Raw Material for Synthesis of Cell Components

For any organism to grow, not only must the physical environment be suitable, but foodstuffs must be available from which the organism can synthesize cell components and derive energy (see chapter 5).

Spectrum of Nutritional Types

Organisms can be divided into a number of groups based on (1) their source of energy and (2) their source of carbon for the synthesis of all the organic molecules that make up cell components (table 4.2).

The two major types of energy are light energy and chemical energy. The prefix **photo-** denotes organisms that can utilize light energy (*photo* means "light"), while the prefix **chemo-** indicates organisms that use chemicals as an energy source (*chemo* means "chemical"). If the energy source is inorganic, then the prefix **litho-** (which means "stone") is often used. Two sources of carbon can be utilized: organic materials and CO_2. Organisms that utilize organic materials are called **heterotrophs;** those that use CO_2 are called **autotrophs.** The combination of these two classification schemes leads to the following major nutritional groups, described in table 4.2: **photoautotrophs,** which use light energy and CO_2; **photoheterotrophs,** which use light energy and organic compounds; **chemoautotrophs,** which use chemical energy and CO_2; and **chemoheterotrophs,** which use chemical energy and an organic source of carbon.

Some chemoheterotrophs can utilize an amazingly large number of organic compounds. For example, certain members of the genus *Pseudomonas* can derive carbon and/or energy from more than 90 different organic compounds, including such unusual compounds as naphthalene (the active ingredient in mothballs). At the other extreme, some organisms can degrade only a single compound. One example is *Bacillus fastidiosus,* which can use only urea as a source of both carbon and energy. Most organic molecules found in nature, regardless of their complexity, can be degraded by at least one species of microorganism. However, a large number of human-made compounds, such as certain herbicides and plastics, are degraded very slowly, and some may not be degraded at all.

The nutritional requirements of certain bacteria vary, depending on the environment. For example, certain photosynthetic bacteria can grow anaerobically using light as an energy source and organic compounds as a carbon source (photoheterotrophs). They can also grow aerobically in the dark on organic sources of carbon and energy (chemoheterotrophs). Organisms that depend on a single mode of existence often have the name *obligate* appended to their nutritional description. The word *facultative* indicates that bacteria are flexible in their nutritional capabilities. For example, an obligate photoheterotroph can only grow with light as an energy source and with an organic compound as a carbon source. A facultative photoheterotroph can grow in the dark using chemical compounds as well as light as sources of energy. As already discussed, the terms *obligate* and *facultative* are also frequently used to describe an organism's requirement for oxygen.

Table 4.2 Classification of Bacteria According to Their Source of Energy and Carbon

Name	Energy Source	Carbon Source
Photoautotroph	Light	CO_2
Photoheterotroph	Light	Various organic compounds
Chemoautotroph (chemolithotroph)	Inorganic compounds such as H_2, NH_3, NO_2^-, Fe^{2+}, H_2S	CO_2
Chemoheterotroph	Organic compounds	Organic compounds

Table 4.3 Mineral Requirements

Name of Element	Common Source	Location or Function in Cell
Nitrogen (N)	NH_4^+, NO_3^-	Amino acids, nucleic acids
Sulfur (S)	SO_4^{-2}	Amino acids
Phosphorus (P)	PO_4^{-3}	Nucleic acids, ATP, phospholipids in membranes
Magnesium (Mg)	Mg^{+2}	Required for functioning of numerous enzymes
Potassium (K)	K^+	Stabilize ribosomes, cytoplasmic membrane, and nucleic acid
Iron (Fe)	Fe^{+3}	Part of certain enzymes
Trace elements* that include	Tap water	Parts of enzymes, required for enzyme activity
Cobalt (Co)		
Zinc (Zn)		
Molybdenum (Mo)		
Manganese (Mn)		

*Since trace elements are required in very small amounts, the tap water used in preparing the medium contains sufficient amounts to satisfy the needs of the bacteria.

Mineral Requirements

In addition to requiring carbon, organisms require that all other elements found in cell constituents, primarily nitrogen, sulfur, and phosphorus, be supplied in a form that they can utilize. Most microorganisms, both heterotrophs and autotrophs, can use inorganic salts as a source of each of these elements. The full list of the mineral requirements, their sources, and functions in the cell are given in table 4.3.

Growth Factors

Growth factors are small organic molecules, other than an energy or carbon source, that bacteria must be provided in order to grow. Examples include amino acids and vitamins. Microorganisms display a wide spectrum in their growth factor requirements, reflecting differences in their biosynthetic capabilities. At one extreme are the obligate intracellular parasites, such as members of the genera *Rickettsia* and *Chlamydia.* These organisms lack some of the enzymes or structures necessary for an independent existence and can only multiply inside living cells. At the other extreme are the photoautotrophs, such as the photosynthetic bacteria, which have enzymes that allow them to multiply with very simple nutrients—inorganic salts, carbon dioxide, and sunlight as a source of energy.

Biochemical studies on a wide variety of organisms have shown that most carry out a fundamental core of enzymatic reactions. These reactions include the ability to produce ATP as a useful form of energy and the ability to synthesize complex molecules, such as nucleic acids and proteins. Many heterotrophs, and a few autotrophs, do not have all of the enzymes necessary for synthesizing some of the small molecules such as amino acids, vitamins, purines, and pyrimidines required for growth. The fewer enzymes an organism has for the biosynthesis of small molecules, the more growth factors must be provided. One extreme example of this latter group is the lactic acid bacteria, which includes the bacterium that causes strep throat. This group must be supplied with as much as 95% of the different subunits that make up its macromolecules. These subunits include virtually all of the different amino acids, purines, and pyrimidines, as well as many of the vitamins.

Cultivation of Bacteria in the Laboratory

By knowing the environmental and nutritional factors that influence bacterial growth, it is possible to devise the appropriate conditions of temperature, oxygen, pH, and other conditions optimal for growth of most organisms. In some cases, it is impossible to devise a medium on which the bacteria will grow.

The Two Kinds of Media on Which Bacteria Are Grown Are Synthetic (Defined) and Complex (Undefined)

Microbiologists use two basic types of media for cultivating bacteria. One consists of measured amounts of pure chemical compounds and is referred to as a **chemically defined,** or **synthetic, medium.** The other type is called a **complex,** or **undefined, medium,** sometimes referred to as an **enriched medium.** It contains such complex components as digestion products of ground meat, plants, and yeast. Although the amount of each of these materials that is added can be weighed, exactly what is present in ground meat and plants is not known. Further, their composition varies with the type of meat or plant digested and the manner in which the digestion of these materials was carried out.

Perspective 4.1

Can Bacteria Live on Only Rocks and Water?

Within the past few years, bacteria have been isolated from environments that previously were thought to be incapable of sustaining life. For example, members of the Archaea have been isolated growing at a pH 10 times more acidic than that of lemon juice. Other archaea have been isolated from oil wells a mile below the surface of the earth at temperatures of 70°C and pressures of 160 atmospheres (at sea level, the pressure is 1 atmosphere). The isolation of these organisms suggests that thermophiles may be widespread in the earth's crust.

Perhaps the most unusual environment from which bacteria have been isolated are the volcanic rocks 1 mile below the earth's surface near the Columbia River in Washington state. What do these bacteria use for food? They apparently get their energy from hydrogen gas that is produced chemically in a reaction between iron rich minerals in the rock and groundwater. This groundwater also contains dissolved CO_2 which the bacteria can use as a souce of carbon. Thus, these bacteria apparently exist on nothing more than rocks and water.

Table 4.4 Synthetic Medium* for Growth of *E. coli:* Glucose-Inorganic Salts

Ingredient	Amount per Liter of Medium
Glucose†	5 g
Dipotassium phosphate (K_2HPO_4)	7 g
Monopotassium phosphate (KH_2PO_4)	2 g
Magnesium sulfate ($MgSO_4$)	0.08 g
Ammonium sulfate $(NH_4)_2SO_4$	1.0 g
Water	1,000 ml

*For preparing a solid medium, 1.5% agar (final concentration) is generally added.

†The glucose is generally sterilized separately from the salts, because heating the glucose with the salts converts glucose to materials that may be toxic to cell growth.

Synthetic Media (Defined)

The composition of a common defined medium used to grow *Escherichia coli* is given in table 4.4. This species, a chemoheterotroph, can synthesize all of its cellular constituents from glucose and the few inorganic salts that make up the medium.

This medium, commonly termed *glucose-salts,* can serve as a basic medium to which various growth factors can be added for growing other bacteria that require them. For example, some members of the *Neisseria* group (which includes the causative agent of gonorrhea) will only grow if the growth factors and inorganic substances listed in table 4.5 are added. Since the added growth factors are chemically pure, this medium is still defined.

Complex Media (Undefined)

Other bacteria with different nutritional requirements will not grow even in the highly enriched defined medium on which cells of *Neisseria* species can grow. In such cases, undefined nutrients, supplied by yeast, and meat or plant digests must be added to provide growth factors. The components of a common chemically undefined medium called **nutrient broth** are given in table 4.6.

Table 4.5 Synthetic Medium for Growth of *Neisseria*

Component	Component
NaCl	Uracil
K_2SO_4	Hypoxanthine
$MgCl_2$	
NH_4Cl	Spermine
K_2HPO_4	Hemin
KH_2PO_4	Nitrilotriethanol
$CaCl_2$	Polyvinyl alcohol
$Fe(NO_3)_3$	
Amino acids:	
20 different types	Sodium lactate Glycerine Oxaloacetate
Vitamins:	
7 different types	Sodium acetate Glucose

Source: Data from B. W. Catlin, *Journal of Infectious Diseases* 128:178.

Table 4.6 Complex Medium—Nutrient Broth

Ingredient	Amount per Liter
Peptone (0.5%)	5.0 g
Meat extract	3.0 g
Distilled water	1,000 ml

The peptones are degradation products of plant, yeast, or meat protein and consist primarily of chains of amino acids of various sizes. Beef extract, a water extract of lean meat, provides vitamins, minerals, and other nutrients. Yeast extract, an excellent source of vitamins, may be added to make the medium even more nutritious.

As a group, the causative agents of disease, the **pathogens,** require a large number of growth factors. These requirements probably result from their adaption over a long period to a particular environment—the animal body—that contains a variety of complex nutrients. In this environment, the bacteria do not need enzymes involved in the synthesis of these nutrients since they can simply use the nutrients provided by the animal body. In this way, bacteria save energy that they would otherwise expend in synthesizing these growth factors.

It is much simpler to grow a particular bacterial species in an undefined medium than it is to determine its exact nutrient requirements and then create a defined medium. This is especially true in laboratories in which many different organisms are grown. For this reason, clinical diagnostic bacteriology laboratories almost always use undefined media.

Problems Associated with Growth of Microorganisms in the Laboratory

Although it seems that it should be possible to devise a single medium to support the growth of any bacterium no matter how unusual its requirements, no such medium exists. One problem is that undefined media may contain toxic materials that must be removed or in some way detoxified. Many materials inhibit the growth of some species of pathogenic microorganisms. These materials include certain fatty acids, metal ions, and even traces of detergents from improperly rinsed glassware. For such reasons, some organisms, notably certain mycoplasmas, grow only in media containing a high concentration of **serum**[1] (20%). The serum supplies a protein (albumin) that binds a variety of toxic molecules. Charcoal, an absorbent, is also sometimes added to detoxify the medium.

Another problem in preparing media is that unusual growth factors must sometimes be provided. For example, members of the genus *Haemophilus* and certain other bacteria must be grown on **chocolate agar.** This medium is enriched with blood that has been carefully heated to release the growth factor **hematin** from hemoglobin, imparting a chocolate-brown color to the medium. Organisms that require this factor will grow on chocolate agar but not on unheated blood agar. The heating also denatures an enzyme present in blood that destroys the growth factor NAD, which is synthesized from the vitamin niacin, required by many bacteria. The importance of blood to this organism is indicated by its name, *Haemophilus*—*haema* means "blood" and *philus* means "loving."

Several biological supply companies manufacture hundreds of different types of media, each one specially formulated to permit the plentiful growth of one or several groups of organisms. However, even with the availability of all of these different media, some microorganisms still have never been cultivated in the absence of living cells. These include the rickettsias and chlamydias and the spirochete that causes syphilis, *Treponema pallidum.*

When organisms are removed from their natural habitat, and especially when they are separated from other organisms, their environment changes drastically. In nature, microorganisms often grow in association with many other kinds of organisms. For example, wounds frequently contain aerobes, facultative anaerobes, and anaerobes. The aerobes and facultative anaerobes multiply and use up the oxygen, creating an environment in which the anaerobes can multiply. Since it is very difficult to simulate the natural environment in the laboratory, it is also very difficult and sometimes impossible to grow certain organisms there. One group of organisms, the anaerobes, presents special problems that are considered next.

Anaerobes Require Special Cultivation Methods

Since the growth of anaerobes is inhibited by oxygen, special techniques are required for their cultivation. Two general methods of cultivation are available. One involves incubating cultures in **anaerobe jars,** from which the oxygen has been removed. A common type of anaerobe jar contains two main elements: a disposable packet that produces hydrogen and a catalyst that combines the hydrogen with any free oxygen to form water (figure 4.12 *a* and *b*).

A second method involves adding to the medium chemicals that react with oxygen, such as sodium thioglycollate, cysteine, and ascorbic acid. The medium is then often boiled to drive out the air before the bacteria are inoculated.

The cultivation of anaerobic organisms presents a great challenge to the microbiologist. Obligate anaerobes may be killed if they are exposed to oxygen for even a short time once they have been removed from their natural habitat. As techniques for cultivating anaerobes have improved in recent years, many more such organisms are being isolated from habitats, including wound and blood infections in which they had previously appeared to be absent.

From this discussion of the growth requirements for aerobes and anaerobes, it should be clear that the inability to demonstrate the presence of an unknown organism does not mean that the organism is absent. It may only be that the proper nutritional and environmental conditions for its cultivation have not been met. It is now believed that only 1% of the organisms in the environment have been cultivated in the laboratory.

[1] Serum—the liquid that remains after blood clots; contains many proteins.

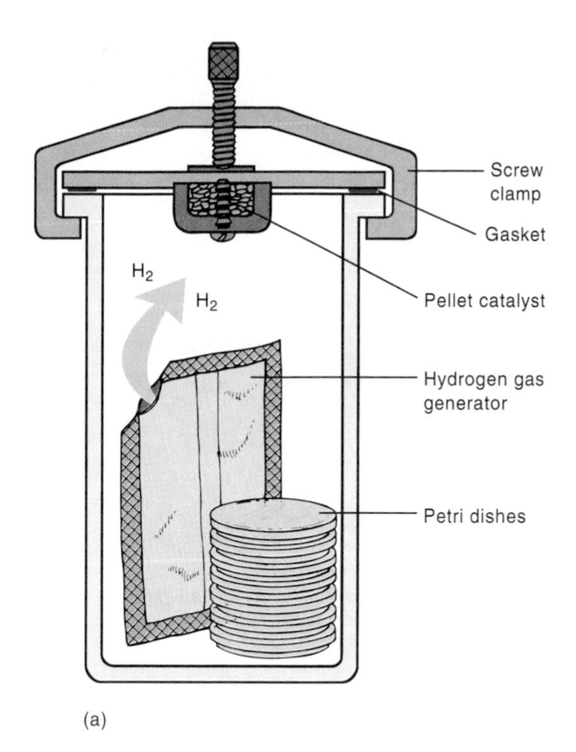

Figure 4.12
(a) Diagram of *(b)* anaerobe jar. The hydrogen released from the generator combines with any oxygen to form water, thereby forming an anaerobic environment.

Cultivation of Organisms that Occur as a Minor Part of the Microbial Population

In many cases, the organism that one is most interested in detecting or in isolating in pure culture is present as only a minor part of a mixed bacterial population. In order to isolate or even detect such organisms, it is necessary to resort to some clever techniques, which result in the bacteria becoming a much higher proportion of the population than they were originally.

Enrichment Cultures Enhance the Growth of Specific Organisms

An **enrichment culture** enhances the growth of one particular organism in a mixed population. This method is helpful in isolating an organism from natural sources when the bacterium is present in small numbers. The technique is based on the well-documented fact that, in any environment, the dominant organism is the one best able to grow in that particular environment. The steps in setting up an enrichment culture are illustrated in figure 4.13.

The sample from which the organism is to be isolated should be taken from an environment in which the organism is likely to be found. The sample is then inoculated into a liquid medium that contains nutrients that the desired organism can use but that other organisms are less likely to be able to use. The medium is incubated under conditions that preferentially promote the growth of the desired organism, taking advantage of any usual properties the organism may have. For example, can it grow using nitrogen gas in the air as a source of nitrogen? Can it use light as an energy source? Through appropriate culture methods, a skilled microbiologist can sometimes manipulate the nutritional and environmental conditions so effectively that a single species will predominate. It is then relatively simple to inoculate the culture onto agar medium and select a single colony to give a pure culture.

Selective Media Preferentially Inhibit the Growth of Certain Organisms

One problem frequently encountered in clinical laboratories is determining how to isolate a nutritionally fastidious (demanding) organism from other organisms in the environment that are less demanding. Since enrichment cultures are not useful in this situation, one technique is to use **selective media.** In this process, substances known to inhibit unwanted microorganisms are added to media to prevent their growth (of course, the organism being sought must be resistant to the inhibitor). For example, one procedure for diagnosing typhoid fever involves isolating the causative organism, *Salmonella typhi,* from the feces, a habitat

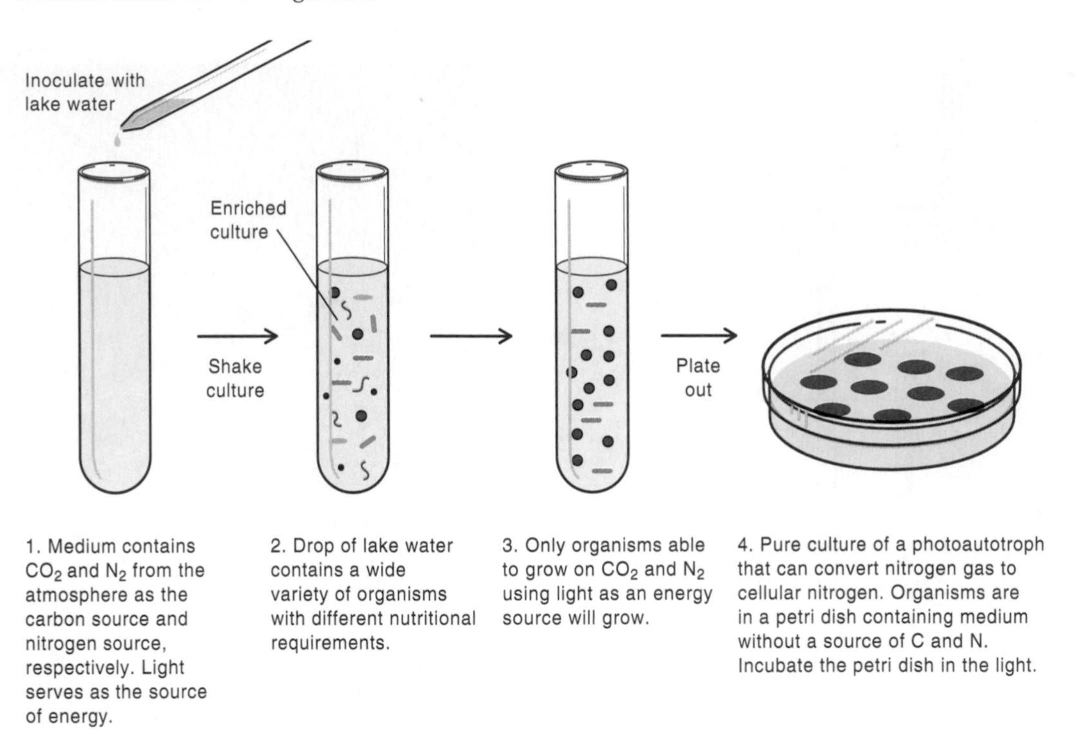

Figure 4.13
Enrichment culture—schematic diagram for isolating an organism that can utilize light as a source of energy, CO_2 as the source of carbon, and N_2 gas as the source of nitrogen.

containing many organisms closely related to *Salmonella*. The fecal specimens are first incubated in broth containing selenite or tetrathionate, two compounds that inhibit the growth of most bacteria normally found in stools but that do not generally inhibit the multiplication of *Salmonella typhi*. Therefore, when the cells from the inhibitory broth are then inoculated onto solid medium containing the same inhibitory compounds, colonies of *Salmonella typhi* make up a high proportion of the colonies that arise.

Differential Media Serve as Useful Diagnostic Aids

In many situations, it may be important to determine whether a particular disease-causing organism is present in a specimen before isolating the organism. It is sometimes possible to identify particular bacteria by their unusual growth patterns on **indicator,** or **differential, agar** media. These media contain a substance that certain bacteria change in a recognizable way. For example, clinical microbiologists use Hektoen enteric agar for the differentiation of *Salmonella* and *Shigella,* which are both gram-negative enteric microorganisms. Bacteria that produce hydrogen sulfide appear as colonies with black centers. *Salmonella* usually produce hydrogen sulfide (figure 4.14), while *Shigella* do not. In cases of suspected typhoid fever, clinical microbiologists may also use bismuth sulfite agar as a differential medium. *Salmonella typhi* produce hydrogen sulfide, which results in the formation of black colonies on bismuth sulfite agar.

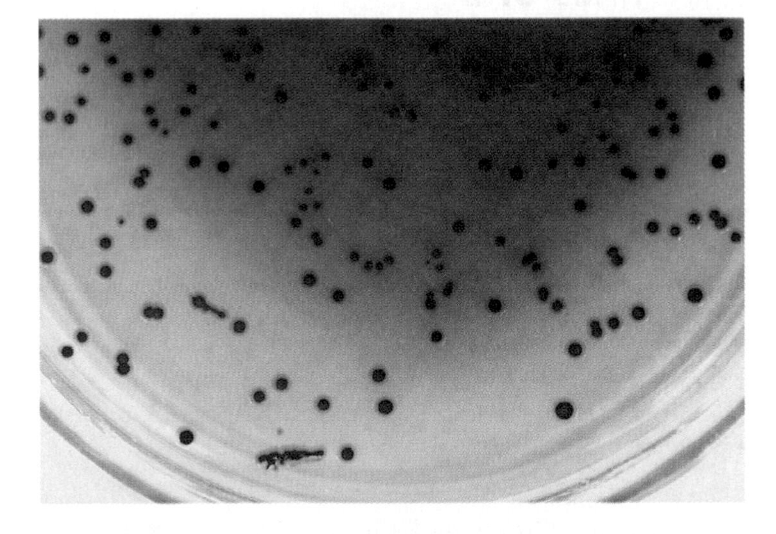

Figure 4.14
Salmonella sp. growing on hektoen enteric agar with the production of H_2S resulting in black colonies.

Another common indicator medium is blood agar. Colonies of the organism causing strep throat, *Streptococcus pyogenes,* are readily detected on this medium because they destroy the red blood cells and produce a clear zone around the colony (termed **beta hemolysis;** figure 4.15). In another example, certain indicator dyes in media will change color if the medium becomes acidic, which is useful in detecting any colonies of organisms that produce acidic substances from the sugar in the medium (figure 4.16).

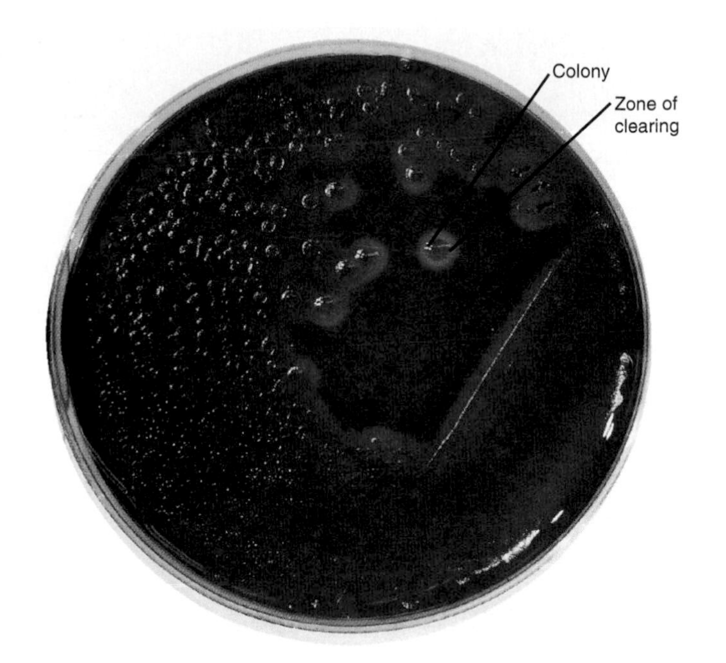

Figure 4.15
Photograph of petri dish containing *Streptococcus pyogenes* growing on blood agar, showing zones of clearing (beta hemolysis) around the colonies.

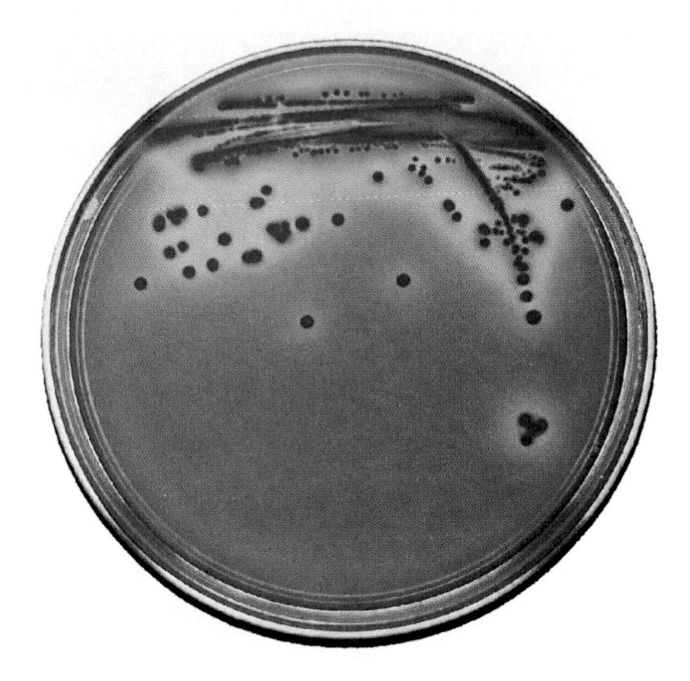

Figure 4.16
Photograph of *Escherichia coli* growing on MacConkey's agar containing lactose. Acid production from lactose by the colonies causes an indicator chemical to turn from colorless to red.

Dynamics of Bacterial Population Growth

If a pure culture of bacteria is transferred from one liquid medium into another and the number of viable (living) cells is measured at various times, the rate of increase in cell numbers of the population follows a predictable sequence (figure 4.17). Four readily distinguishable phases can be recognized: (1) the lag phase, in which the number of viable cells does not increase significantly; (2) the log, or exponential, phase in which the cell population increases exponentially with time; (3) the stationary phase, in which the total number of viable cells remains constant; and (4) the death phase, in which the total number of viable cells decreases exponentially. Each of these phases will now be discussed briefly.

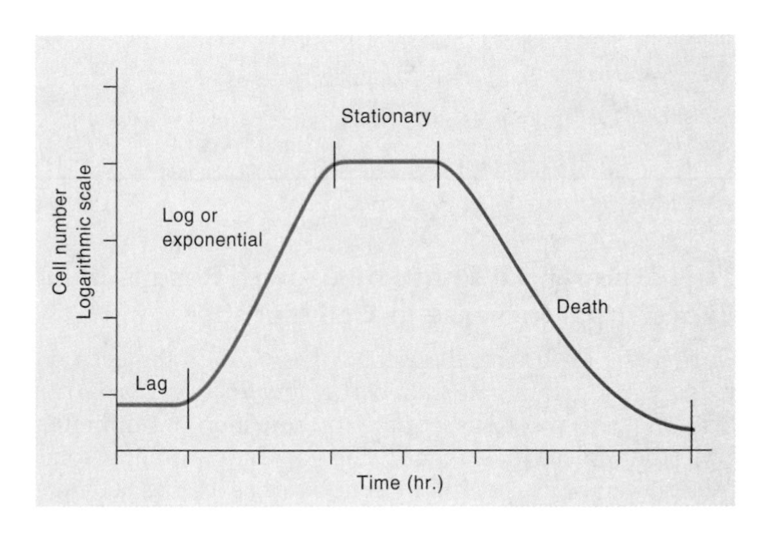

Figure 4.17
The phases of bacterial growth plotted on semilog graph paper. The vertical axis is logarithmic; the horizontal is arithmetic.

The Lag Phase of Growth Involves No Increase in Cell Numbers

When a culture of bacteria is diluted and then transferred into fresh medium, the number of viable cells does not increase immediately. Before cells can multiply, they increase in length and synthesize macromolecules such as the components of ribosomes needed for protein synthesis and the enzymes required for cell division. This phase is the **lag phase** of cell division.

The length of the lag phase depends in part on the conditions under which the bacteria were growing prior to their transfer into fresh medium and the medium into which they are transferred. The closer the two media are in composition and the closer the environmental conditions, such as temperature and pH, the shorter the time required for the cells to adapt to the fresh medium and, therefore, the shorter the lag time. Actively multiplying cells (log phase) diluted into the same medium and grown under the same environmental conditions will have a very short lag time. However, cells diluted from a rich complex medium into a simple, defined medium will have a longer lag time because they need to synthesize new enzymes before they can multiply.

Perspective 4.2

Bismuth Subsalicylate–A Medicine with Magical Properties

Few medicinal compounds have a longer history of use than do those that contain bismuth. Since the early 1500s, bismuth compounds have been used to treat a variety of ailments, including syphilis. Its most familiar form today is probably Pepto-Bismol, whose active ingredient is bismuth subsalicylate (BSS). Bismuth sulfite can be incorporated into agar to serve as an indicator agent for identifying *Salmonella typhi.* In this century, it has been largely used in the treatment of gastrointestinal ailments and in preventing traveler's diarrhea. This latter malady is well known to people who have traveled in such areas as Latin America, Africa, and southern Asia. Symptoms of traveler's diarrhea are an increase in the volume and number of stools. The most common cause is *Escherichia coli,* which produces a toxin (poison). A number of studies have shown that BSS is quite effective in preventing traveler's diarrhea, although reports are conflicting on how useful it is in treating it. It is even less certain how BSS acts, although considerable research is being carried out on it. BSS is probably effective in preventing traveler's diarrhea because it inhibits the growth of toxin-producing *E. coli.* Further, it has been shown that pathogens are isolated less frequently in patients treated with BSS than in patients treated with a placebo.* It is also possible that BSS may prove useful in curing gastric ulcers caused by *Helicobacter pylori.* At the present time, the four leading anti-ulcer drugs have enormous sales, but they can only treat rather than cure the condition.

*Placebo—a preparation containing no medicine but given for its psychological effect

Table 4.7 Doubling Times of Various Bacteria Under Optimal Conditions

Species	Doubling Time (min)	Time Required for One Cell to Grow to a Visible Colony (hr)
Clostridium perfringens	10	8
Beneckea natriegens	10	8
Escherichia coli	20	16
Mycobacterium tuberculosis	800	336 (2 weeks)

The Exponential Phase of Growth Results in an Exponential Increase in Cell Number

During the **exponential,** or **log,** phase, when the cells divide at a constant rate called the *doubling,* or *generation, time,* they are most susceptible to the action of antibiotics and other deleterious agents. The generation time of some bacteria can be as short as 10 minutes or as long as 800 minutes (table 4.7). Most bacteria have generation times between these two extremes and generally closer to 10 minutes.

The speed with which bacteria divide depends on all of the environmental and nutritional conditions previously mentioned in this chapter. Obviously, the optimal conditions vary for each particular organism. However, one important generalization can be made. The environmental and nutritional factors that influence the growth rate primarily affect the rate of ATP production by the cell. The more energy the bacteria can produce per unit of time, the faster they will multiply.

However, the division time of any particular species also depends on its genetic constitution. For example, *Mycobacterium tuberculosis* grows very slowly, doubling every 12 to 24 hours even under the most favorable environmental and nutritional conditions. Why some organisms have such a long generation time while others such as *E. coli,* which doubles every 20 minutes, can multiply rapidly is not clear.

• ATP, p. 114

The Stationary Phase Is Reached when the Number of Viable Cells Stops Increasing

Bacteria stop growing and enter the **stationary phase** once they have reached a population of about 5×10^9 cells per milliliter of medium. This occurs for one of two reasons: Either a nutrient is used up or toxic products accumulate from the cells' metabolism.

As cells of the genera *Bacillus* and *Clostridium* begin to run out of nutrients, they can form endospores and enter a dormant state. However, even bacteria that cannot form endospores can prepare for the time when food is in short supply and toxic compounds are in the environment. These cells reduce their size and increase their resistance to harmful chemicals. The relative constancy in the total number of viable cells in the stationary phase usually results from all cells in the population stopping growth and none dying.

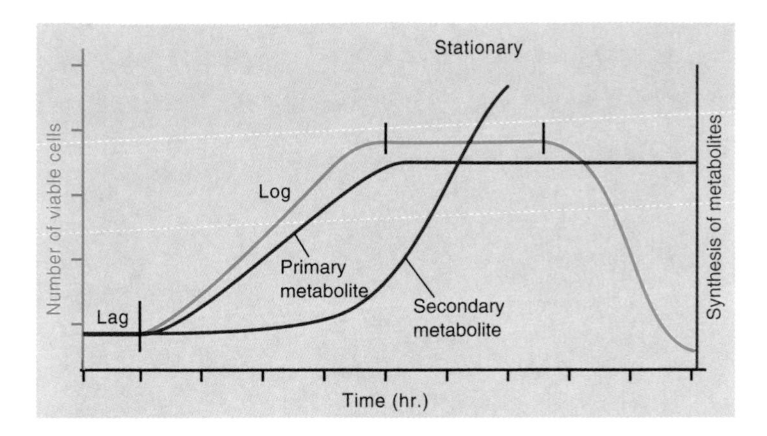

Figure 4.18
Synthesis of primary and secondary metabolites. Even though bacteria are not multiplying, they can still synthesize a variety of compounds.

Secondary Metabolites Are Synthesized at the End of the Log Phase of Growth

Whereas most compounds produced by bacteria are synthesized during the log phase of growth, a class of compounds is synthesized during the time the cells are not multiplying (see figure 4.18). The former class are termed *primary metabolites* and include such compounds as amino acids, lipids, polysaccharides, and other compounds that the cells must synthesize in order to grow. The second class are termed *secondary metabolites* and are not required for growth. They are synthesized from primary metabolites and include such compounds as the antibiotics. Indeed, most products synthesized by microorganisms of industrial importance are secondary metabolites.

The Death Phase Is Characterized by an Exponential Decrease in the Number of Viable Cells

The death phase always follows the stationary phase. The length of time cells remain in the stationary phase varies, depending on the species of organism and on environmental conditions. Some organisms remain in the stationary phase for only a few hours, whereas others remain for several days. Generally, if toxic products have accumulated as by-products of metabolism, the cells begin to die rapidly. Some cells, especially certain members of the genus *Bacillus,* lyse very quickly once they stop multiplying because an enzyme dissolves the cell wall. Consequently, even a few hours after they have reached the stationary phase, the number of viable cells in many cultures of *Bacillus* has decreased by 90%.

The total number of viable cells in the population begins to decrease at the onset of the **death phase.** By definition, a cell is dead if it can no longer multiply. A graph of the number of viable cells on a logarithmic scale versus time on an arithmetic scale results in a straight line, which indicates that the cell population dies exponentially (figure 4.17). Note that the death phase has a negative slope, whereas the exponential growth phase has a positive slope, although the rate of death bears no relationship to how fast the cells grow in the exponential phase. In both cases, the number of viable cells changes at a constant rate over any given period. The death phase is considered in more detail in chapter 9, which discusses the significance of exponential death in the deliberate killing of microorganisms.

Growth of a Bacterial Colony

The growth and development of a bacterial colony on solid media involve many of the same features as bacteria growing in liquid. Initially, the cells multiply exponentially. However, after a few divisions, the cells become very crowded and then must compete for the available nutrients. The cells on the edge of the colony can use oxygen from the air and obtain nutrients that are in the agar. Cells further toward the center deplete available oxygen and nutrients. A gradient of decreasing oxygen and nutrients develops from the outside to the center of the colony. The center tends to be anaerobic, with only a limited supply of nutrients available. Thus, colony enlargement results from cells multiplying at the outside edges of the colony.

The toxic end products of growth also influence the development of the colony. Any products, such as acids, may kill the cells in their immediate vicinity. Thus, within a single colony, cells may be in the exponential, stationary, or death phase, depending upon their location. Cells at the edge of the colony may be growing exponentially, whereas those in the center may be in the death phase.

Growth of Bacteria in Nature–An Open System

Thus far, we have discussed the growth of bacteria in the laboratory, in tubes or flasks of liquid or on agar plates. In these containers, nutrients are not renewed after they have been used up, and waste products are not removed. Such a system is termed a **closed system.** Under these conditions, bacteria grow at an exponential rate for only a relatively short time. In nature, however, such as in a running stream, nutrients may be continuously replaced, waste materials are removed, and cells are washed away from the solid surfaces to which they have attached. In this type of system, termed an **open system,** the bacteria may remain in the exponential phase of growth for a long period. However, as a general rule, bacteria multiply more slowly in their natural environment than in the laboratory under the most favorable conditions. In nature, nutrients are often in very short supply and the environment may not be optimal for rapid growth.

Microorganisms Commonly Grow in Communities in Nature

Members of the Eubacteria and Archaea domains in nature rarely live alone. Rather, the cells aggregate so as they almost appear to be a multicellular organism. These aggregations serve several functions. They may protect the organisms against harmful chemicals such as antimicrobial medicines. Other prokaryotes synthesize a glycocalyx so that they can attach to one another and grow as a mat over a solid surface. Surprisingly, this growth is not generally a haphazard mixture of microbes in a layer of slime, but rather a number of microcolonies enclosed in a slime but separated by open water channels (figure 4.19). This structure is termed a **biofilm.**

Bacteria frequently synthesize new structures when they are in their natural environment because chemical signals present in the environment are recognized by the bacteria and they tell the bacteria what structures and enzymes will be useful in that particular environment. Bacteria also can recognize signals emitted by other bacteria, telling them how many bacteria are present in the environment. Thus, bacteria may not attack tissue of a human or a plant unless enough organisms are present to insure that the attack is successful. From a large number of studies in a variety of systems, it is clear that bacteria can grow in complex communities that allow bacteria to do things they cannot do growing as single cells in a test tube and that give them properties that help them survive in a particular environment.

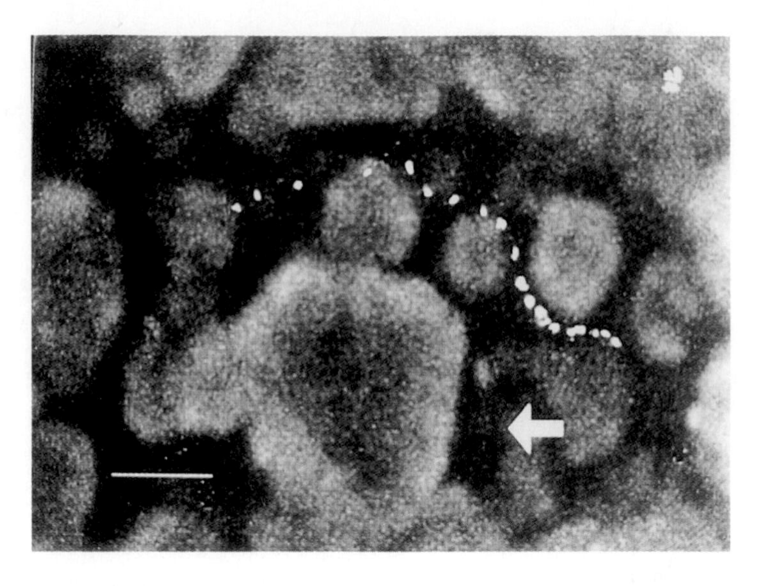

Figure 4.19
This slide shows a superimposed time sequence (2.3-second intervals) image from a Confocal Scanning Laser Microscope, showing a single bead moving through a biofilm channel (dark area). The direction of the bulk fluid is indicated by the arrow. Bar = 100 μm. The lighter gray shapes are biofilm clusters, containing microcolonies of bacteria. The image clearly shows how water oxygen nutrient and other chemicals can be transported through a biofilm between biofilm clusters.

Summary

I. Pure Culture Methods
- A. In nature, bacteria exist as a mixed population.
- B. Colonies represent the offspring of a single cell. The streak-plate method is used to separate cells.
- C. The properties of agar make it ideal for growing bacteria on a solid surface.

II. Measurement of Cell Growth
- A. Bacterial growth is defined in terms of population size, not size of individual cells.
  1. Bacteria divide by binary fission.
  2. The number of bacteria increases exponentially.
- B. Growth can be measured as an increase in the mass of the population.
  1. Scattering of light measures living and dead cells.
  2. Growth can be measured as an increase in cell number.
     - a. The plate count method measures the number of viable cells that can grow into a colony.
     - b. Direct microscopic count is rapid but it does not distinguish between living and dead bacteria.

III. Multiplication of Microorganisms—General Aspects
- A. Microorganisms multiply under an enormous range of environmental conditions.
- B. An organism taken from its natural habitat often shows altered growth requirements when grown in the laboratory.

IV. Factors Influencing Microbial Growth
- A. The environmental factors that influence growth include temperature, oxygen, pH, and osmotic pressure.
  1. Most bacteria grow at approximately 30°C. Psychrophiles grow optimally between –5°C and 20°C. Mesophiles grow optimally between 20°C and 50°C. Thermophiles grow optimally between 50°C and 80°C. Extreme thermophiles grow optimally above 80°C.
  2. Obligate (strict) aerobes have an absolute requirement for oxygen. Obligate (strict) anaerobes cannot multiply if oxygen is present, and it can kill the bacteria. Facultative anaerobes can utilize oxygen if it is available, but they can grow, although less well, in its absence. Microaerophilic organisms require small amounts of oxygen (2%–10%), but higher concentrations are toxic. Aerotolerant organisms grow in the presence or absence of oxygen, but they derive no benefit from the oxygen.
  3. Aerobes require oxygen because they have metabolic pathways that require oxygen to convert energy in foodstuffs into forms useful to the cell. Bacteria that require oxygen have the means to detoxify certain forms of oxygen, such as hydrogen peroxide. Detoxifying enzymes include catalase and superoxide dismutase.

4. Most bacteria grow best in a medium with a pH 7 (neutral). Buffers are added to media to maintain a constant pH. Bacteria can maintain their internal contents at neutral pH even though the external pH is below or above this.
5. Osmotic pressure is determined by the concentration of dissolved substances in the medium in which bacteria are growing. Under normal conditions, the osmotic pressure is slightly higher on the inside than on the outside of the cell. Bacteria can compensate for pressure differences to a certain extent if the external osmotic pressure is high.

V. Nutritional Aspects of Bacterial Growth
   A. Cell multiplication requires a source of energy and raw materials for synthesis of cell components.
      1. The spectrum of nutritional types includes (a) photoautotrophs, which use light energy and CO_2 as a source of carbon; (b) photoheterotrophs, which use light energy and organic compounds as a source of carbon; (c) chemoautotrophs, which use chemical energy and CO_2 as a source of carbon; (d) chemoheterotrophs, which use chemical energy and organic compounds as a source of carbon; and (e) lithoheterotrophs, which use an inorganic source of energy and organic compounds as a source of carbon.
      2. All organisms require a supply of nitrogen, sulfur, and phosphorus.
      3. Bacteria require small molecules that serve as subunits of biosynthetic macromolecules and cell components. Some bacteria can synthesize all of their growth factors; others that lack the necessary biosynthetic enzymes must have them provided.

VI. Cultivation of Bacteria in the Laboratory
   A. The two kinds of media on which bacteria are grown are synthetic (defined) and complex (undefined).
      1. Synthetic medium consists of chemically pure materials added in known amounts.
      2. Complex (undefined) medium consists of undefined nutrients, such as ground meat, added in known amounts.
   B. There are some problems associated with the growth of microorganisms in the laboratory.
      1. No single medium has been devised on which all bacteria will grow. This may result from several factors: (a) Toxic compounds may be present and must be removed. (b) Unusual growth factors may be needed. (c) Some bacteria can only be grown in a mixed culture, similar to their natural environment. One bacterium in the culture provides what the other one needs.
      2. Anaerobes require special cultivation methods. (a) An anaerobe jar may be used to remove all oxygen. (b) Chemicals may be added to react with oxygen.

VII. Cultivation of Organisms that Occur as a Minor Part of the Microbial Population
   A. Enrichment cultures enhance the growth of specific organisms.
   B. Selective media preferentially inhibit the growth of certain organisms.
   C. Differential media serve as useful diagnostic aids.

VIII. Dynamics of Bacterial Population Growth
   A. The lag phase of growth involves no increase in cell numbers—it is the "tooling up" phase.
   B. The exponential phase of growth is the time during which the cell number increases exponentially. Secondary metabolites are synthesized by cells at the end of the exponential phase.
   C. The stationary phase is reached when the number of viable cells stops increasing.
   D. The death phase is characterized by an exponential decrease in the number of viable cells.

IX. Growth of a Bacterial Colony
   A. Colony enlargement results from cells multiplying at the outside edges of the colony.

X. Growth of Bacteria in Nature—an Open System
   A. In the laboratory, bacteria grow in a closed system; nutrients are not renewed and waste products are not removed.
   B. An open system is the typical system found in nature. Nutrients are replenished and waste products are removed.

Chapter Review Questions

1. List the four nutritional types of bacteria.
2. Distinguish between a growth factor and a carbon source.
3. Give two reasons why it may not be possible to grow anorganism in the laboratory even though it grows in nature.
4. Distinguish between a selective medium and a differential medium. On what kind of medium would you inoculate a fecal specimen that you suspect might contain *Salmonella typhi*?
5. List the four stages of growth of bacteria in a liquid medium and describe what the population is doing in each stage.
6. What are the major differences between open and closed systems?
7. What did Rober Koch contribute to our understanding of mixed populations of microorganisms? What was the major contribution of Frau Hesse to studies on the growth of microorganisms?
8. What are secondary metabolites and when are they synthesized? Distinguish secondary from primary metabolites.
9. Name two enzymes that are involved in detoxifying harmful forms of oxygen. What reactions do they carry out?
10. Distinguish between an autotroph and a heterotroph in nutritional terms.

Critical Thinking Questions

1. You are attempting to isolate a spore-forming aerobic organism that can use N_2 gas as a source of nitrogen and CO_2 as a carbon source and that is a halophile. How would you proceed?
2. Many years ago, it was commonly believed that the reason bacteria stopped growing when they reached a high density of cells (i.e., 5×10^9) was that they became too crowded. Do you think this is a possible explanation? Give evidence from information provided in this chapter to support your explanation.

Further Reading

Atlas, R. 1993. Handbook of microbiological media. CRC Press, Boca Raton, Fla. A comprehensive reference to the formulations and applications of media commonly used in the laboratory.

Brock, T. D. 1985. Life at high temperatures. *Science* 230:132–138.

Difco Laboratories. 1984. Difco manual: Dehydrated culture media and reagents for microbiology. Difco Laboratories, Detroit.

Losick, R., and Kaiser, D. 1997. Why and how bacteria communicate. Scientific American 276(2):68–73.

Meyer, H. P., et al. 1985. Growth control in microbial cultures. Annual Review of Microbiology 39:299.

Metabolism: Energy Generation and Synthesis of Small Molecules

KEY CONCEPTS

1. Cells must carry out two metabolic functions: (1) the degradation of foodstuffs to generate energy and small molecules and (2) the use of the energy to convert the small molecules into macromolecules and cell structures.
2. Foodstuffs are degraded by three major pathways: glycolysis, the TCA cycle, and the pentose phosphate pathway.
3. The TCA cycle produces much more energy than glycolysis for the same amount of foodstuffs degraded.
4. Glycolysis and the TCA cycle have three functions: (1) to produce energy; (2) to synthesize precursor metabolites which are converted into the subunits of macromolecules; and (3) to generate reducing power to be used in biosynthetic reactions.
5. Bacteria as a group have a wide variety of enzymes and synthesize a large number of different products. Identifying these end products can help identify the species of bacteria.

World's largest redwood aging tank, Colony Wines, Asti Sonoma County, California.

Microbes in Motion

The following books and chapters in the *Microbes in Motion* CD-ROM may serve as a useful preview or supplement to your reading:

Bacterial Structure and Function: Cell Membrane. *Microbial Metabolism and Growth*: Metabolism; Growth.

A Glimpse of History

Louis Pasteur was trained as a chemist, and some of his major contributions to microbiology developed from his interest in the decomposition of natural products in the absence of air, the process of fermentation. Pasteur found that a large number of acids, including acetic, lactic, and butyric acids, as well as a large number of alcohols could be identified as the products of fermentation.

Two questions arose from his work: How did these products arise and what caused the fermentation? In 1855, Pasteur was asked by a wine maker to study the production of wine from grape juice. He jumped at the chance. A few years earlier, biologists had observed that yeast cells were always present in fermentation vats when alcohol and carbon dioxide were produced. Further, the yeast cells increased in number during the fermentation. These scientists argued that live yeast cells produced the alcohol and carbon dioxide from grape juice. Pasteur agreed with these views, but could not convince two very powerful and influential German chemists, Justus von Liebig and Friedrich Wöhler. These scientists believed that the breakdown of sugar was caused by substances released by the dying yeast cells. Both men lampooned the biologists' theory and attempted to discredit it by publishing illustrations of yeast cells depicted as miniature vertebrate animals taking in grape juice through one orifice and eliminating carbon dioxide and alcohol through the other.

To refute Liebig, Pasteur used a strategy commonly used by scientists today—simplifying the experimental system so that relationships can be more easily seen and identified. Pasteur prepared a clear solution consisting of sugar, ammonia, mineral salts, and trace elements, a synthetic medium on which yeast can grow. He then added a small amount of live yeast. As expected, the yeast grew and, as their total number increased, the sugar level decreased and the alcohol level increased. This suggested that the alcohol came from the sugar. Pasteur also found that other kinds of microorganisms (bacteria) produced different but characteristic kinds of fermentation products. If only one kind of microorganism was present in the sugar solution, the products of decomposition were always the same.

These experiments provided strong evidence that living cells caused the fermentation, but Liebig still could not believe that the fermentation was actually going on inside the microorganism. To convince him, Pasteur tried to extract something from the yeast cells that would convert sugar into alcohol and carbon dioxide. However, he failed, like many others before him. In 1897, Eduard Buchner, a German chemist, showed that ground-up yeast cells would convert sugar to ethanol and CO_2. These studies were the first to demonstrate that compounds could be transformed into other compounds by the cytoplasmic components from disrupted cells. We now know that the active ingredients of the ground-up cells were enzymes. For these pioneering studies, Buchner was awarded the Nobel Prize in 1907. He was the first of many investigators who received Nobel Prizes for studies on the biochemical pathways by which cells degrade sugars to other compounds. This father of biochemistry died in 1917 at age 57 as a result of wounds he received while serving in the German Army in World War I.

When a few cells of a common yeast are placed in a glucose-salts medium and incubated in a flask for several days, a cursory examination of the flask and a smell of its contents provide convincing evidence that obvious changes have taken place. Initially, the contents of the flask are clear. Later, they turn cloudy, and bubbles of gas appear at the surface. In addition, the odor of ethyl alcohol emanates from the liquid that was previously odorless. The increase in turbidity and the formation of ethyl alcohol and gas result from the metabolic activity of the yeast cells. The cells transform some of the glucose and inorganic salts into more yeast cells, which increase the turbidity of the medium. In addition, the cells convert some of the glucose into ethyl alcohol and CO_2. The following equation summarizes the process:

$$\underset{\text{glucose}}{C_6H_{12}O_6} + \text{yeast cells} \rightarrow \underset{\text{ethyl alcohol}}{2CH_3CH_2OH} + \underset{\text{carbon dioxide}}{2CO_2} + \text{more yeast cells}$$

This chapter will discuss how yeast cells accomplish this transformation. However, before discussing these biochemical processes, we will discuss the key players in the cell in these biochemical conversions: the enzymes.

Enzymes–Chemical Kinetics and Mechanisms of Action

Enzymes, the key proteins in the cell, determine in large part all of the properties of the cell. They are responsible for the cell's ability to carry out its functions necessary for cell multiplication. Therefore, before we can discuss the details of metabolism, we must understand something about how these key players in the process, the enzymes, function. This information allows us to better understand how cells gain energy from foodstuffs and use this energy to synthesize the molecules that comprise them.

Enzymes catalyze (speed up) the large number of chemical reactions in the cell. The conversion of one substance (the substrate) to another (the product) measured at varying intervals of time is shown in figure 5.1. In the absence of an enzyme, the substrate, the substance on which

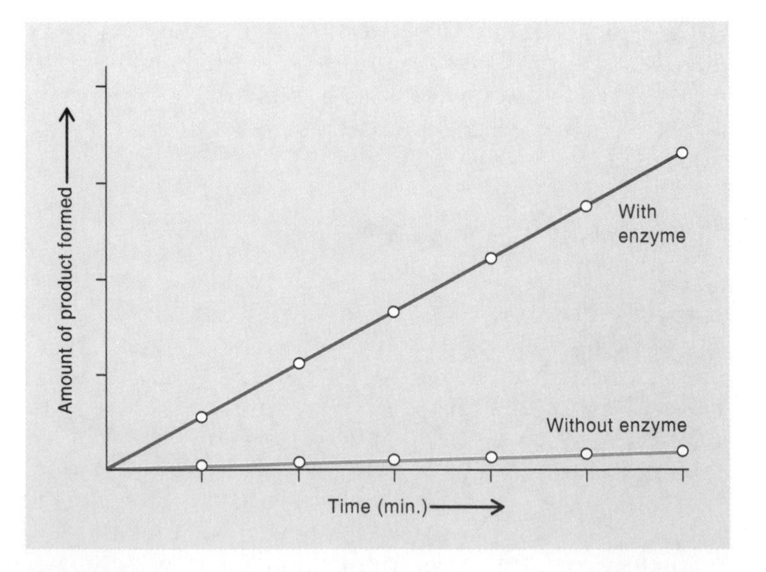

Figure 5.1
Effect of an enzyme on the conversion of a substrate to a product.

Perspective 5.1

Taking Enzymes to the Cleaners

In the past, marketing strategies for selling laundry detergents relied mainly on salesmanship and hype rather than on scientific principles. It now appears that science and technology may be entering the marketplace. Chemical companies and detergent makers are experimenting with adding new types of enzymes to detergents that will attack certain stains formerly resistant to removal, breaking the bonds the stains form with fibers in fabrics.

In May 1987, a large Japanese company, the equivalent of Procter and Gamble in the United States, introduced Attack, a detergent that uses a new commercial enzyme, alkaline cellulase, that degrades cellulose. Under alkaline conditions, this enzyme removes dirt embedded in cotton fibers. To counter this introduction, a competitor has introduced a different enzyme, lipase, that removes fats from fabrics. Other enzymes that are being added to detergents to break down proteins include heat-stable proteases, which are able to function at temperatures above 50°C and at an alkaline pH (pH 10).

In 1980, enzymes were virtually nonexistent in detergents, but today, roughly half of the detergents sold in the United States contain them. In Europe and Japan, nearly 100% of the detergents contain enzymes. Why is the United States lagging in this venture? In the 1960s, soap companies in the United States introduced enzymes in detergents, but factory workers and some homemakers had severe allergic reactions to the enzyme protein. Consequently, the companies voluntarily removed the enzymes from the detergents. To prevent these allergic reactions, the enzymes are now encapsulated in waxy balls.

There is considerable interest in finding enzymes that will work with bleaches and fabric softeners at low temperatures and short wash cycles. Psychrophilic bacteria will be a likely source of such enzymes. Once such technology develops, an increased use of new and improved enzymes in all detergents will certainly follow.

the enzyme acts, is changed into a product so slowly that it is impossible to measure the product's formation. However, an enzyme can covert substrate into a product in a short time. For example, if milk is left at room temperature for several days, it sours as a result of the action of bacterial enzymes on the substrate, lactose, to form the final product, lactic acid. If the milk did not contain microorganisms that could break down lactose, the conversion of lactose to lactic acid (souring) would still occur, but it would take many thousands of years. A single enzyme molecule can convert as many as 1 million substrate molecules per second to products. Reactions between molecules can occur rapidly in the absence of enzymes if the temperature or pressure is raised to very high levels as can be accomplished in a chemical laboratory. However, life cannot exist under these conditions. Thus, enzymes take the place of heat and high pressure, allowing organisms to live at the relatively mild temperatures that exist on and beneath the surface of the earth.

Enzymes Speed Up Reactions by Lowering Their Activation Energy

When oxygen (O — O) and hydrogen gas (H — H) are mixed together in a container, the molecules collide with one another and with the walls of the container. Each molecule has a certain level of energy of motion, or **kinetic energy.** At room temperature, this energy is not enough to allow the hydrogen atoms to break the strong covalent O — O bonds or to permit the oxygen atoms to push apart the H — H bonds. However, if the temperature is raised, the hydrogen and oxygen molecules move more rapidly and collide with enough energy to break apart the H — H and O — O bonds and to form O — H bonds.

In the example of heating a hydrogen-oxygen mixture, heat is necessary only to start the reaction. Once started, energy is given off by the few molecules that collide. This energy, in turn, raises the energy of motion of other molecules. Thus, a chain reaction is set in motion, and an explosion results as energy is released. This reaction can be written:

$$2H - H + O - O \rightarrow 2H - O - H + \text{energy}$$

The reason that energy must first be put into the system to start it can be explained by the analogy described in figure 5.2.

The hydrogen and oxygen molecules are represented as balls located near the top of a hill. If they can get over the top of the hill, the molecules will roll down and produce energy, which is necessary to break apart old bonds and form new ones. The energy required to lift the molecules to the top of the hill is called **energy of activation.** Enzymes act at this stage by lowering the activation energy hurdle. The activation energy barrier is crucial to life. Without it, atoms and molecules would undergo continual rearrangement, and life could not exist. Indeed, it would be difficult to live in a world in which wood and coal could burst into flames spontaneously.

The Substrate Fits the Shape of the Enzyme that Acts on It

An enzyme functions first by combining, through weak bonding forces, with its substrate. This interaction places a stress on the chemical bonds in the substrate, which weakens them enough to break them and then forms new bonds. Enzymes act in two steps. First, the substrate binds to a specific portion of the enzyme, the **active,** or **catalytic, site,** to

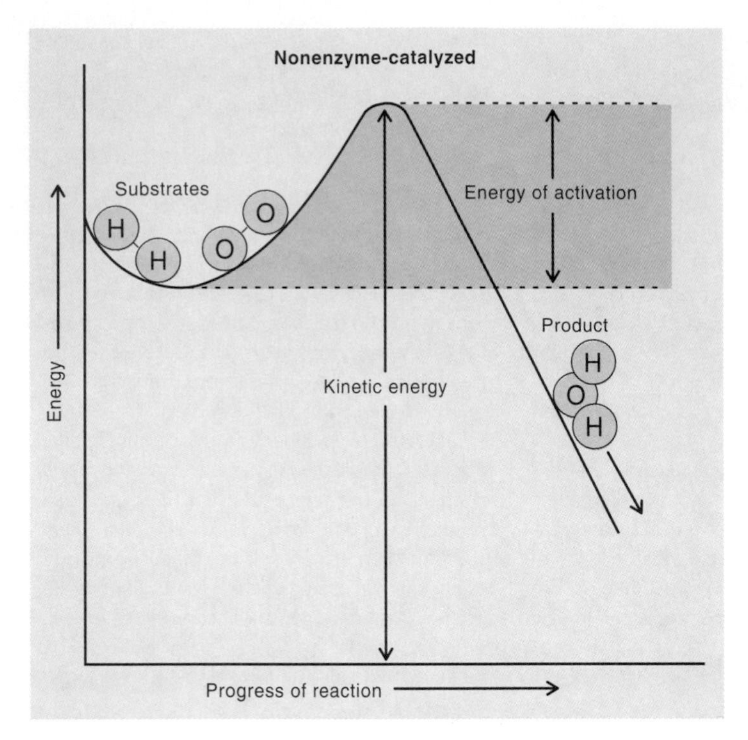

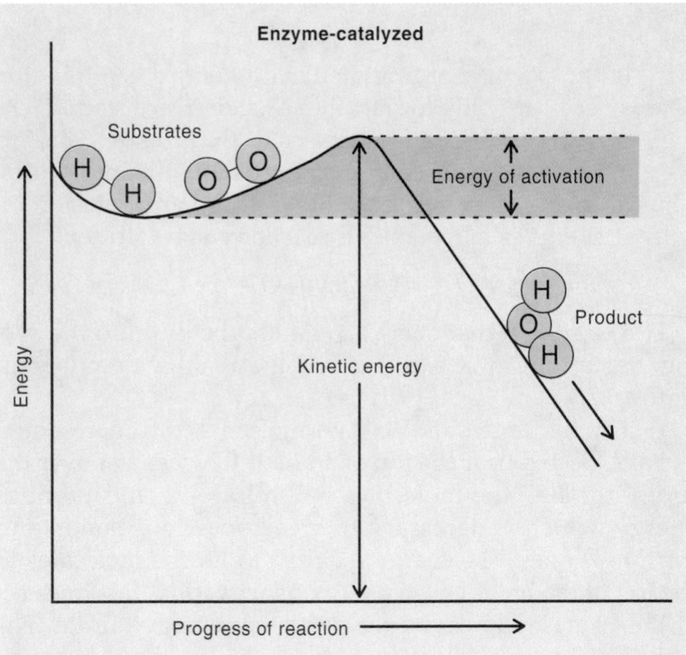

Figure 5.2
Energy barrier in chemical reactions. An enzyme lowers the energy of activation enough so that the energy of the H and O atoms alone is sufficient to lift them over the barrier.

form an enzyme-substrate complex (figure 5.3). Second, the products of the reaction are released, leaving the enzyme unchanged and free to combine with new substrate molecules. The arrangement of the enzyme and its substrate is commonly referred to as a "lock-and-key" arrangement; the substrate is the key and the enzyme the lock. Since the key must fit into the lock precisely, any particular enzyme will act on only one or a limited number of substrates, all of which must have a similar shape to fit the active site. This explains why

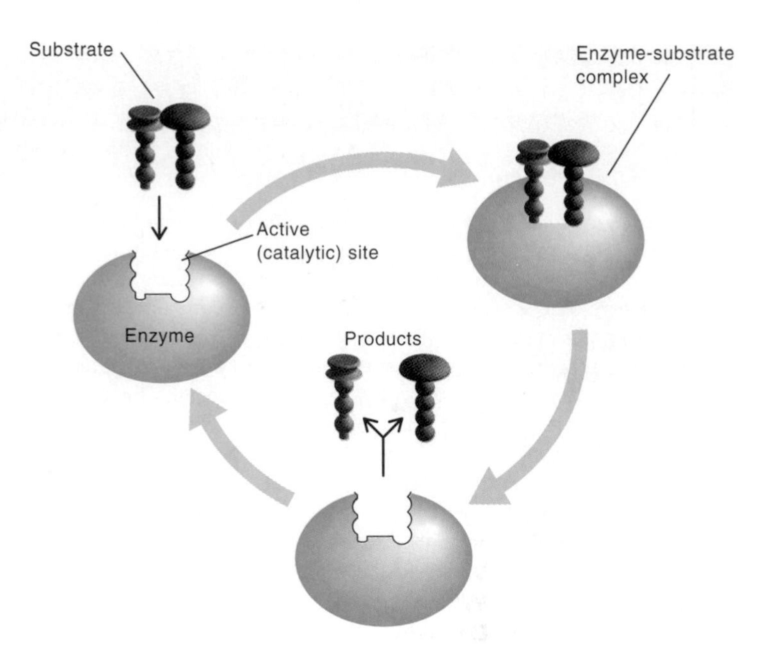

Figure 5.3
Enzyme catalysis, showing the requirement of a lock-and-key relationship between enzyme and substrate.

almost every reaction in a cell is catalyzed by a different enzyme. As a result, hundreds of different enzymes exist in the cell, but relatively few molecules of each enzyme are needed, since they can be used over and over again.

Some Enzymes Require Coenzymes for Their Function

Many enzymes cannot function unless they have another molecule, a **coenzyme,** bound to them. Coenzymes are small, nonprotein molecules such as nicotinamide adenine dinucleotide, abbreviated NAD, which is more than twice the size of a nucleotide. All coenzymes have the same general function—to transfer molecules, atoms, or electrons from one molecule to another. In some cases, this molecule is transferred when the coenzyme is still bound to the enzyme; in other cases, the coenzyme separates from the enzyme in the transfer process (figure 5.4). There are many different coenzymes but the same coenzyme can bind to different enzymes, so there are far fewer different coenzymes than there are enzymes. In some cases, the coenzyme is bound very tightly to the enzyme so that it is essentially a part of the enzyme as it carries out its function. In other cases, the coenzyme can readily separate from one enzyme and move to another enzyme.

Unlike enzymes, coenzymes do not take part in only a single reaction, but they associate with many different enzymes that catalyze different reactions. However, each coenzyme is specific in that it transfers only one type of small molecule, atom, or electron. Like enzymes, coenzymes are recycled and, therefore, are needed only in minute quantities in the human and bacterial diet.

• **Nucleotide, p. 37**

All coenzymes are synthesized from vitamins (table 5.1). If an organism lacks a vitamin, the functions of all the different enzymes whose activity requires that coenzyme are impaired. Thus, a single vitamin deficiency has serious consequences in animals.

Environmental and Chemical Factors Influence Enzyme Activity

The growth of any organism depends on the proper functioning of the enzymes that it contains. The enzymes, in turn, are influenced by the environment, which affects how rapidly cells multiply. The features of the environment that are most important in influencing enzyme activities are temperature, pH, and salt concentration (figure 5.5). Since a 10°C rise in temperature doubles the speed of enzymatic reactions, bacteria tend to grow more rapidly at higher temperatures. However, too high a temperature will denature proteins and the enzymes will become nonfunctional. Most enzymes function best at low salt concentrations and at pH values slightly above 7. Not surprisingly, the organisms from which these enzymes are taken grow the fastest under conditions of low salt and a pH of 7.

Table 5.1 Coenzymes and Their Function

Name of Coenzyme	Vitamin from Which It Is Derived	Molecule (Atoms) Transferred
Nicotinamide adenine dinucleotide (NAD)	Niacin	Hydrogen atoms (protons and electrons)
Flavin adenine dinucleotide (FAD)	Riboflavin	Hydrogen atoms
Coenzyme A	Pantothenic acid	Acetyl group (2-carbon molecule)
Thiamine pyrophosphate	Thiamine	Aldehydes
Pyridoxal phosphate	Pyridoxine	Amino group
Tetrahydrofolic acid	Folic acid	1-carbon atom

Enzyme Inhibition

In addition to being affected by extreme environmental conditions, enzymes are inhibited by a variety of compounds. Cyanide, arsenic, mercury, and nerve gases—all toxic substances—combine with and prevent the functioning of enzymes necessary for life processes. Some compounds inhibit in a reversible manner; others, in an irreversible manner. Whether reversible or irreversible depends on whether the enzyme functions after the inhibitor is removed. In reversible inhibition, the inhibitor binds to but readily separates (dissociates) from the enzyme. In irreversible inhibition, the inhibitor binds very strongly to the enzyme and does not dissociate readily. In both types, attachment often occurs to the active site of the enzyme, which prevents the substrate from binding to this site.

In the simplest type of reversible inhibition, **competitive inhibition,** the inhibitor has a chemical structure similar

• **Protein denaturation, p. 35**
• **pH, p. 93**

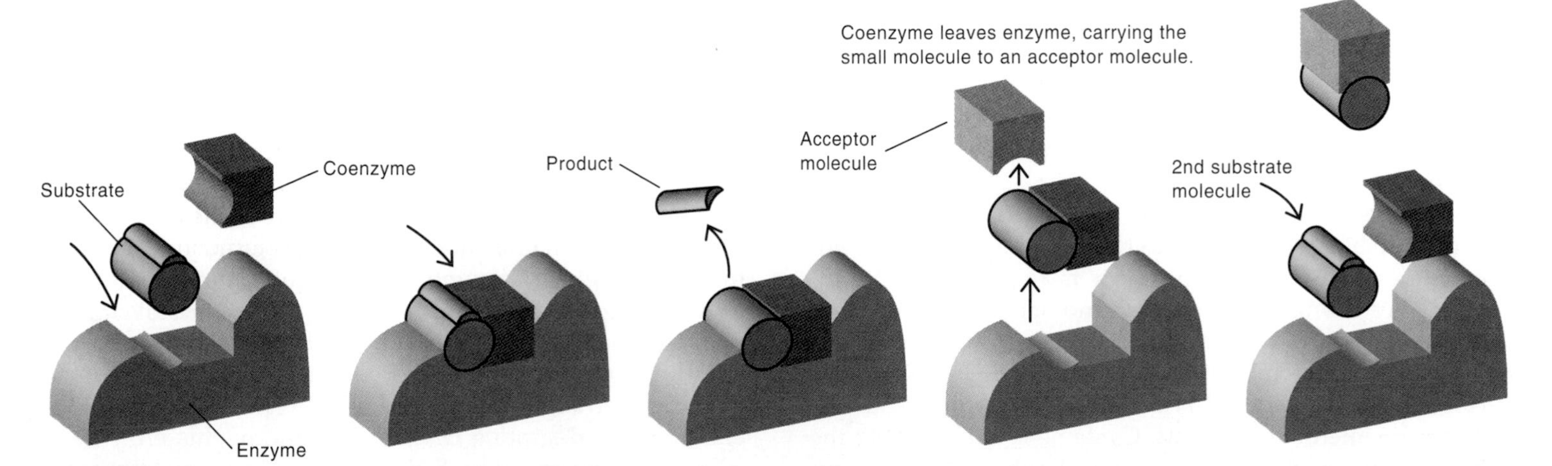

Figure 5.4
Interaction of a coenzyme with an enzyme and the small molecule transferred by the coenzyme. The coenzyme carries small molecules from an enzyme to the acceptor molecule, which may be a protein or another coenzyme.

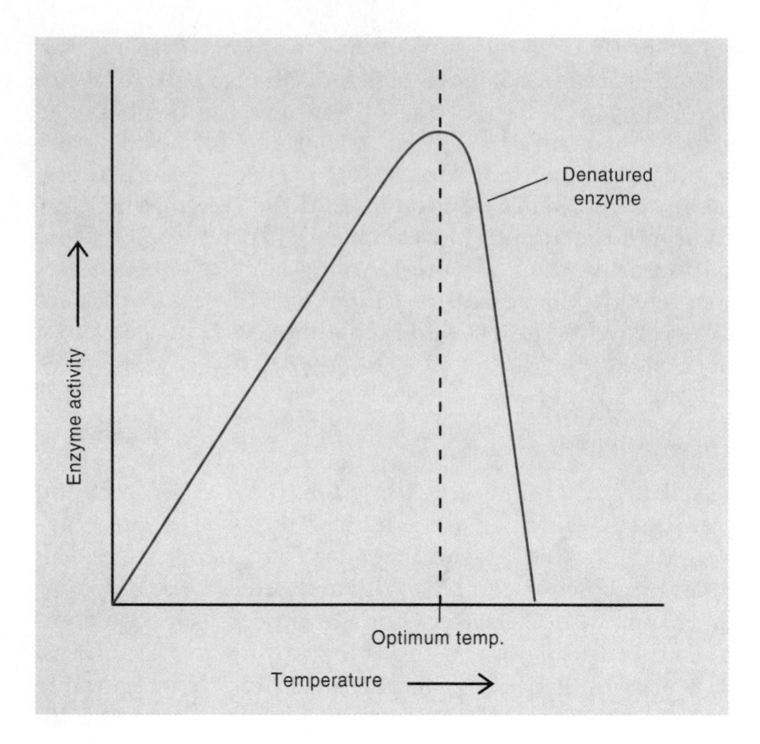

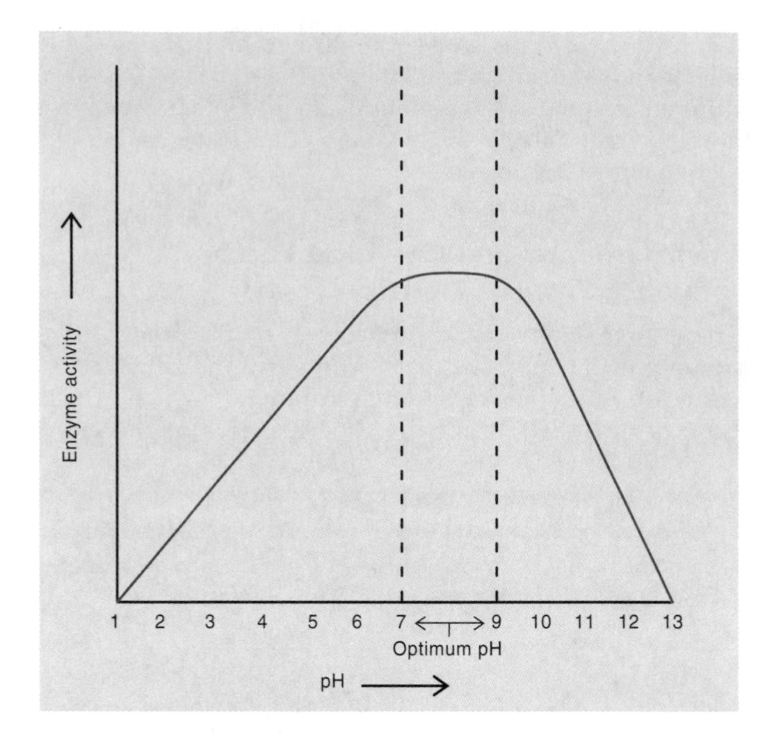

Figure 5.5
The effect of (*a*) temperature and (*b*) pH on the activity of an enzyme on its substrate.

to the normal substrate. A good example of reversible competitive inhibition is the action of the sulfa drugs, which have a similar structure to para-aminobenzoic acid, PABA (figure 5.6). In many bacteria, PABA is enzymatically converted to a required vitamin, folic acid, by a series of enzymes. Since PABA and sulfa drugs have similar structures, both can fit into the active site of one of these enzymes. When sulfa drugs combine reversibly with such an enzyme, the normal substrate, PABA, is prevented from combining with the active site of the enzyme (figure 5.7). The higher the ratio of sulfa molecules to PABA molecules, the more likely it is that the active site of the enzyme will be occupied by a sulfa molecule and, therefore, the greater the inhibition (see figure 5.7). Sulfa drugs and why they are effective in treating certain diseases are discussed further in chapter 21.

Irreversible inhibition has one of two causes. The inhibitor may bind so strongly to the enzyme that it does not separate even when the inhibitor is removed or the inhibitor may change important amino acids in the enzyme so that it can no longer function. For example, the mercury in mercurochrome inhibits the growth of microorganisms because it oxidizes the S—H groups of the amino acid cysteine in the microbial proteins. Cysteine is converted to the related amino acid cystine, which lacks the ability to form the important covalent disulfide bond (S—S). As a result, the protein cannot form its proper shape, the enzyme is unable to bind its substrate, and therefore, the enzyme is inactive.

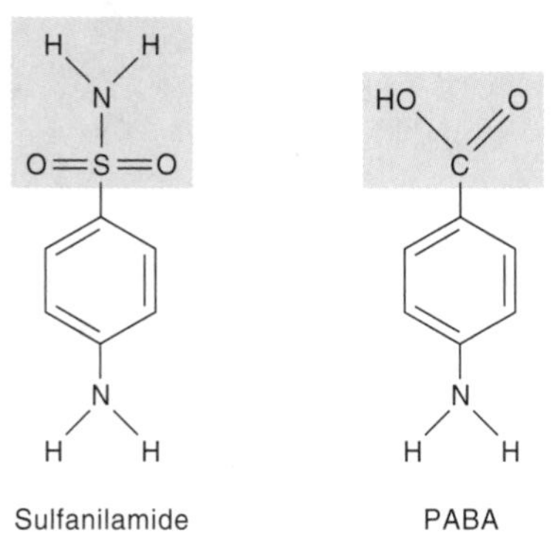

Figure 5.6
Structures of sulfanilamide (sulfa drug) and of para-aminobenzoic acid (PABA). The portions of the molecules that differ from each other are shaded.

Names of Enzymes Usually End in *-ase*

Over 2,000 enzymes have been discovered and named thus far. All enzymes can be assigned to one of six classes based on the general nature of the reactions that they catalyze. Many of these names are very cumbersome and have no meaning except to enzymologists. It is far more convenient to use the common names for enzymes. Frequently, an enzyme is named by simply adding the ending *-ase* to the substrate on which it acts. For example, penicillinase is an enzyme that degrades penicillin; deoxyribonuclease is an enzyme that degrades DNA (the name of this enzyme is frequently shortened to DNase). Sometimes, enzymes are named after the kinds of reactions they catalyze. Thus, the enzyme that bonds together the subunits of DNA (nucleotides) to form the DNA polymer is DNA polymerase. In this chapter, we will come across a variety of enzymes involved in anabolism and catabolism. The text will always refer to these enzymes by their common names. If their functions are not obvious from their names, then the text will explain their functions.

Overview of Metabolism

The collective biochemical reactions that take place within a cell are termed **metabolism,** and any compound produced in metabolism is a **metabolite.** In order to survive, microorganisms must multiply; that is, they must produce copies of themselves. Multiplication requires that the cells synthesize all of their cell components such as their chromosomes, ribosomes, and cell walls. The reactions involved in the synthesis of cell components are termed **anabolic reactions.** Collectively they are called **anabolism.** Anabolic reactions require energy. To gain this energy, cells must carry out other biochemical reactions termed **catabolic reactions,** which are *degradative reactions.* Degradative reactions yield energy. Thus, when yeast degrades glucose to the waste products ethanol and carbon dioxide with the production of more yeast cells, the yeast is carrying out both catabolic and anabolic processes. Both processes are *coupled* to one another; that is they occur simultaneously (figure 5.8).

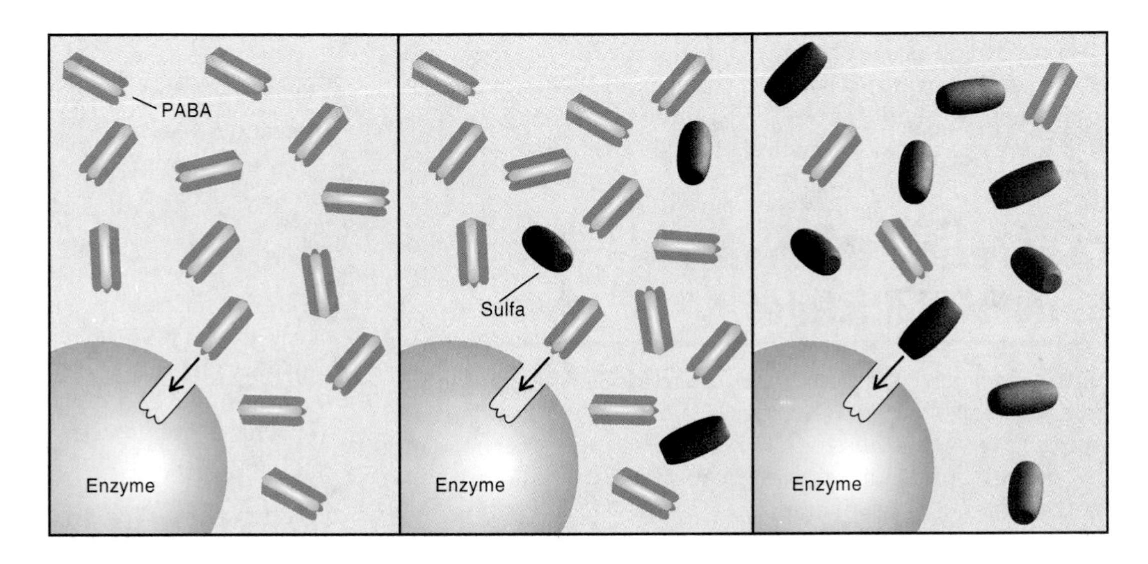

Figure 5.7
Reversible inhibition of folic acid synthesis by sulfa drug. The higher the concentration of sulfa drug molecules relative to PABA, the more likely that the enzyme will bind to the sulfa drug, and the greater the inhibition of folic acid synthesis.

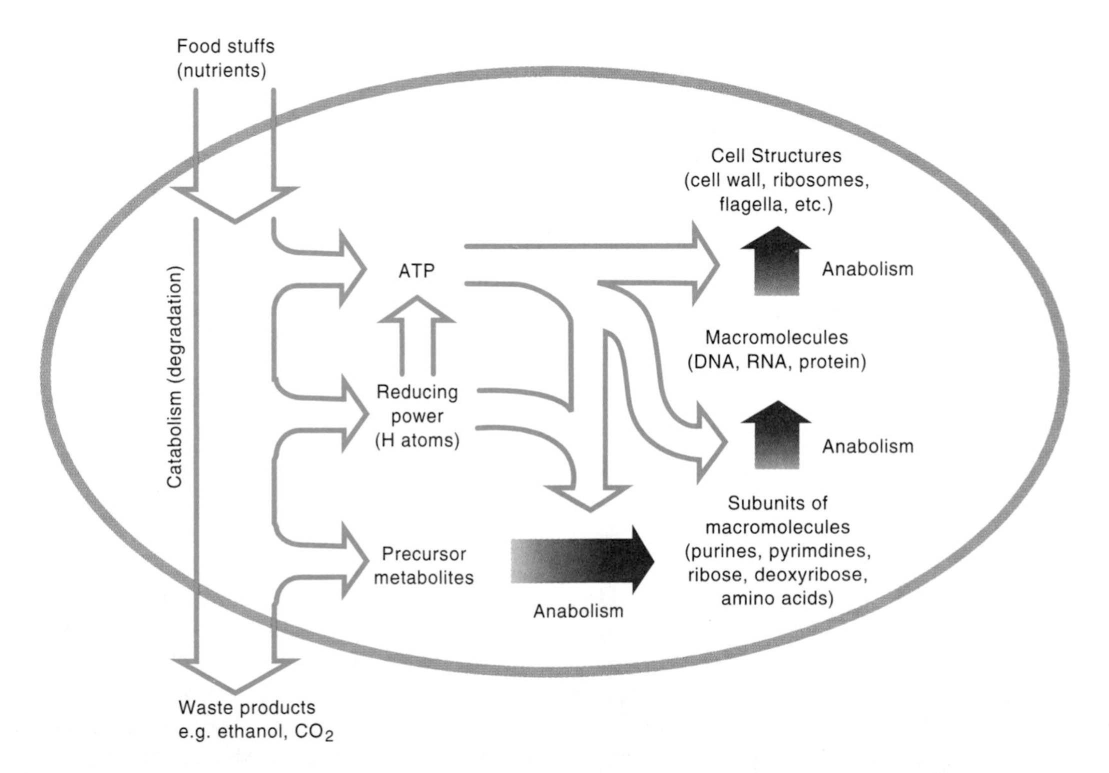

Figure 5.8
Overall view of metabolism.

In addition to generating ethanol and carbon dioxide, which are the waste products of catabolism, the degradation of glucose generates three kinds of molecules that are very important in anabolic processes. The first is **adenosine triphosphate, ATP,** the storage form of energy. Anabolic reactions generally require energy in order to proceed. The second is **reducing power** which consists of hydrogen (H) atoms; each atom is composed of an electron (e-) and a proton (H+). Many biosynthetic reactions require the input of reducing power. The third kind of molecules produced are **precursor metabolites,** the starting molecules from which the subunits of macromolecules are synthesized. An example of a precursor metabolite is pyruvic acid, which can be converted in one enzyme-catalyzed step to the amino acid, alanine. The synthesis of each of these groups of molecules will be emphasized in our discussion of catabolism. However, before we discuss some of the specific details of catabolism, we will discuss the general aspects of how cells obtain energy, a major function of catabolism.

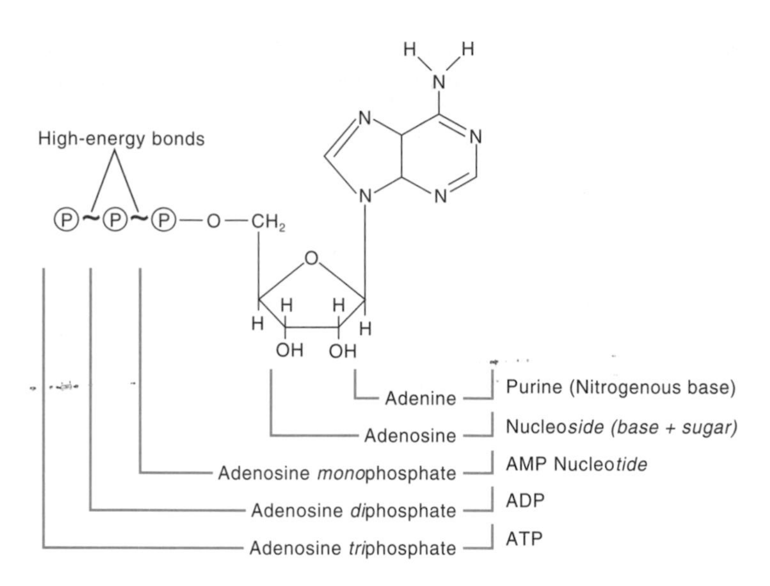

Figure 5.9
Formula of adenosine triphosphate (ATP). The two high-energy bonds are denoted by red wavy lines.

Generation of Energy—General Principles

Nutrients, such as glucose and other foodstuffs, contain a great deal of chemical energy in the covalent bonds that join the various atoms together. However, this energy is in low energy bonds and as such is not readily available to do work in the cells. As an analogy, a penny is not terribly useful for buying anything these days. However, if 100 pennies can be collected and exchanged for a dollar, then the dollar can be exchanged for a variety of goods and services. The pennies correspond to the many low energy covalent bonds, and the dollar corresponds to the high energy bond of adenosine triphosphate (ATP), the major usable form of energy in the cell (figure 5.9). This molecule has its energy concentrated in two high-energy bonds (indicated by the symbol ~). When either of these bonds is broken, a large amount of energy is released which the cell can use in biosynthetic (anabolic) reactions. Adenosine diphosphate (ADP) and inorganic phosphate (Pi) are formed when the terminal high energy bond of ATP is broken according to the following reaction:

ATP	⇄	ADP	+	Pi	+	energy
adenosine triphosphate		adenosine diphosphate				

Thus, the rearrangement of the atoms in the glucose molecule involves a series of enzyme catalyzed steps that result in the low energy bonds of glucose being converted into the high energy bonds of ATP. This rearrangement primarily involves two pathways, the glycolytic pathway and the tricarboxylic acid cycle. They will be discussed in greater detail shortly.

There are two mechanisms for generating these high energy bonds. One involves the transfer of electrons through a series of proteins and coenzymes from a high to a low energy state. This is the process of **oxidative phosphorylation.** The second process involves the transfer of a high energy phosphate bond intact to ADP to form ATP. This is the process of **substrate phosphorylation.**

Coenzymes Are Important for the Generation of Energy and the Storage of Reducing Power

Certain coenzymes are very important in the catabolism of glucose. They serve two functions. They (1) participate in the formation of ATP by the process of oxidative phosphorylation and (2) store reducing power, in the form of hydrogen (H) atoms, each hydrogen atom consisting of a proton and an electron. The two coenzymes that serve these two functions are **nicotinamide adenine dinucleotide (NAD)** and the closely related NADP, which has an extra phosphate group in the molecule. We shall consider these two molecules interchangeable and abbreviate them NAD. Both functions are important in biosynthetic reactions.

Formation of Reducing Power: Role of NAD

Reducing power is produced and stored in NAD following the removal of H atoms from certain compounds in the pathways of glucose degradation. These H atoms are removed by an enzyme (a dehydrogenase, a class of enzymes that take H atoms from a substrate) that has the coenzyme NAD bound to it. The compound from which the H atoms are removed becomes **oxidized** and the NAD that removes and then binds the H atom becomes **reduced.** Oxidation is a result of removal of H atoms; reduction is the gain of H atoms. One molecule of NAD removes a pair of H atoms at a time and thereby becomes reduced to NADH. Before it binds the H atoms, it is designated NAD^+, the oxidized form of NAD. The simplified reaction that illustrates this is

reduced substrate	+	NAD^+ →	oxidized substrate	+	NADH
H donor		H acceptor	oxidized		reduced acceptor

Note that oxidation must always be accompanied by a reduction and a reduction by an oxidation. H atoms cannot exist in solution inside the cell. The reduced form of NAD, which is NADH, is a major source of reducing power in the cell. NADH can transfer the H atom to a large number of different molecules in the cell, many of which are part of biosynthetic processes. Since the number of molecules of any coenzyme in the cell, such as an enzyme molecule, is very small, the NADH must be continuously recycled to NAD^+ so that the NAD^+ is available to once again participate in another oxidation—reduction reaction. Thus, the NADH must find an acceptor for the H atoms.

Formation of ATP: Chemiosmosis

In addition to being important as a source of reducing power in metabolism, NADH can transfer its H atom to another coenzyme, flavin adenine dinucleotide (FAD) which is on the path to generating ATP. ATP is generated by a series of oxidation and reduction reactions that involve the transfer of protons and electrons through the electron transport chain. In the process of this transfer, approximately 3 molecules of ATP are produced for every pair of H atoms that are transferred in bacteria. The process by which ATP is formed is called **oxidative phosphorylation.** The mechanism by which this occurs is **chemiosmosis.**

Chemiosmosis takes place in two steps in the cytoplasmic membrane of bacteria. In eukaryotic cells, the same process takes place in the inner membrane of mitochondria. In the first step (figure 5.10 *a*), hydrogen atoms are removed from a substrate (through an oxidation) by a dehydrogenase enzyme associated with the coenzyme NAD. The hydrogen atoms are then transferred along a chain of carriers, the electron transport chain, first to the coenzyme flavin adenine dinucleotide (FAD) and then to another small molecule called a *quinone.* The next carriers in the chain are a set of small proteins called **cytochromes.** However, these proteins can accept only electrons, so at this stage the protons separate from the electrons. The protons are transferred to the outside of the cytoplasmic membrane, whereas the electrons continue along the electron transport chain. The protons on the outside of the

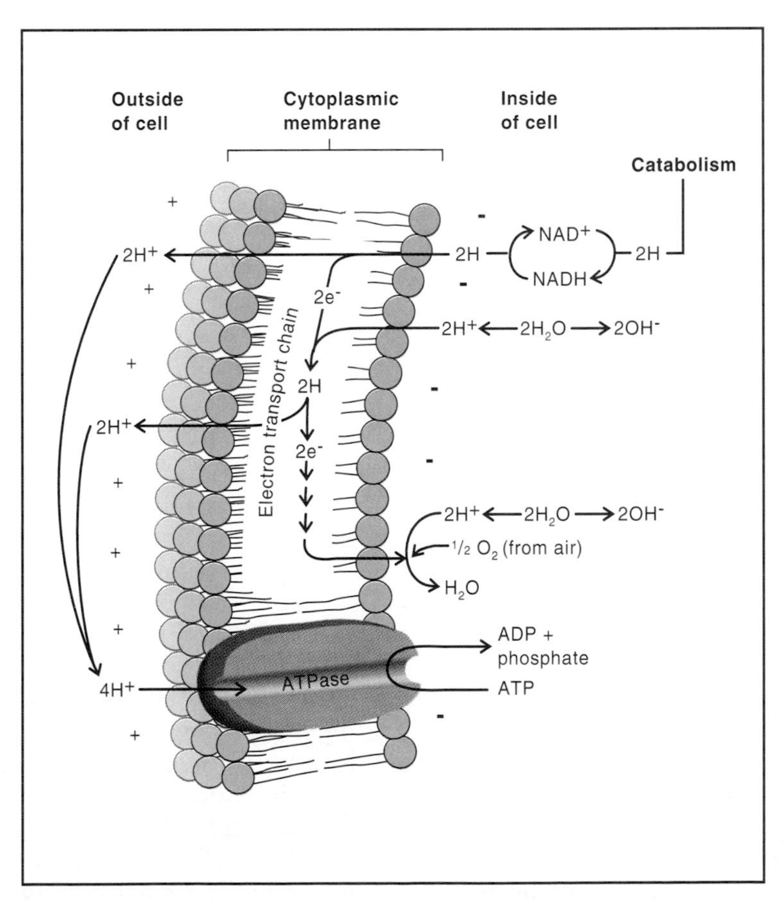

Figure 5.10 Chemiosmosis. Protons (H^+) are pumped out of the cell as electrons pass through the electron transport chain in the cell membrane to their final acceptor (O_2 in this illustration). A high concentration of protons develops at the cell surface relative to the cell interior. The force created by this proton gradient drives the reaction ADP + P → ATP, carried out by the enzyme ATPase.

membrane are at a much higher concentration than they are on the inside. The high concentration of H^+ creates a positive charge on the outside of the membrane, whereas on the inside of the membrane, a negative charge exists. This negative charge results from the dissociation of HOH molecules into H^+ and OH^-. The H^+ ions from this dissociation are also transported to the outside of the cell, whereas the OH^- remain inside (figure 5.10).

The protons in high concentration on the outside of the membrane generate a potential source of energy, much like water does when stored behind a dam. When released, the water flowing over the dam can be converted into electrical energy and can do work. The force created by the differential concentration of protons on either side of the membrane is termed the **proton motive force** and it can also do work. The energy of proton motive force is used in such energy-requiring processes as active transport and the movement of flagella.

The potential energy of proton gradient can also be converted into chemical bond energy of ATP by the following mechanism, the second step in chemiosmosis (figure 5.10). The production of ATP by this mechanism is termed **oxidative phosphorylation.** The protons tend to equalize their concentration on each side of the cytoplasmic membrane. This membrane is impermeable to the protons, but protons can pass to the inside of the cell by going through an enzyme in the cytoplasmic membrane, an adenosine triphosphatase or ATPase. As the protons pass through, this enzyme converts the energy of the proton gradient into ATP by catalyzing the following reaction:

$$3\,ADP + 3\,Pi \rightleftarrows 3\,ATP$$

Thus, for every pair of hydrogen atoms that passes through the electron transport chain, 3 molecules of ATP are generated. Oxidative phosphorylation is the method by which reducing power (NADH) is converted to energy in the form of high energy bonds of ATP.

Substrate Phosphorylation as a Mechanism to Generate ATP

Energy is generated in two ways in catabolic pathways. We have discussed one way, oxidative phosphorylation. The second mechanism is termed **substrate phosphorylation.** As mentioned previously, the energy in foodstuffs is diffused throughout the molecule and is not usable until it is converted into the high energy bonds of ATP. Many of the enzymatic reactions involved in catabolism concentrate the diffuse energy step by step into a few high energy phosphate bonds. The high energy phosphate bond can then be transferred intact to ADP to form ATP, in the process of substrate phosphorylation. The reaction can be diagrammed in the following equation, where X is an organic molecule and ~ symbolizes a high energy bond:

high energy bond

$$X \sim P + ADP \rightarrow X + ATP$$

Aerobic Respiration Involves Oxygen as a Final Hydrogen Acceptor

Since the concentrations of all coenzymes and proteins of the electron transport chain are at low levels, the reduced coenzymes and proteins must be continuously reoxidized. Therefore, a final H acceptor must be available in the environment that can become reduced by all of the H atoms that are removed from the substrate that is being oxidized. In the presence of air, this compound is usually oxygen and the product of the reaction is HOH (see figure 5.10). **Respiration** is the process of degradation in which the final electron acceptor is an inorganic compound. If the compound is oxygen, the process is called **aerobic respiration.** If the compound is some inorganic compound other than oxygen, the process is termed **anaerobic respiration.** The TCA cycle involves respiration.

Fermentation Involves an Organic Molecule as the H Acceptor

If the final H acceptor is an organic molecule, the catabolic pathway is termed a **fermentation.** In fermentations, the organic acceptors are generally produced as by-products of the fermentation pathway. Therefore, no external acceptor is necessary.

The catabolic pathways for degrading glucose, the TCA pathway, and the glycolytic pathway are considered next.

Catabolic Pathways—Overview

As previously stated, the degradation of glucose by bacteria as well as humans must result in the formation of three important classes of molecules: (1) ATP, (2) NADH as a source of reducing power for metabolic reactions, and (3) precursor metabolites that serve as the starting materials for the synthesis of the small molecules that are synthesized into cell structures. In order to generate all of these molecules, three catabolic pathways are necessary. These are **glycolysis,** in which glucose is converted to pyruvic acid; the **TCA cycle,** in which pyruvic acid is converted to CO_2 and water; and the **pentose phosphate pathway,** in which glucose is converted to compounds that enter the pathway of glycolysis. All three pathways generate reducing power and precursor metabolites. The pentose phosphate pathway does not generate any ATP, as do the other two pathways. Because of their importance, these three pathways are referred to collectively as the **pathways of central metabolism.** The pathways of central metabolism are summarized in outline form in figure 5.11. Not all cells have all three pathways and therefore not all cells can synthesize all precursor metabolites. Certain growth factors, such as amino acids, purines, pyrimidines, and vitamins, must be supplied in order for the cells to grow.

• **Growth factor, p. 96**

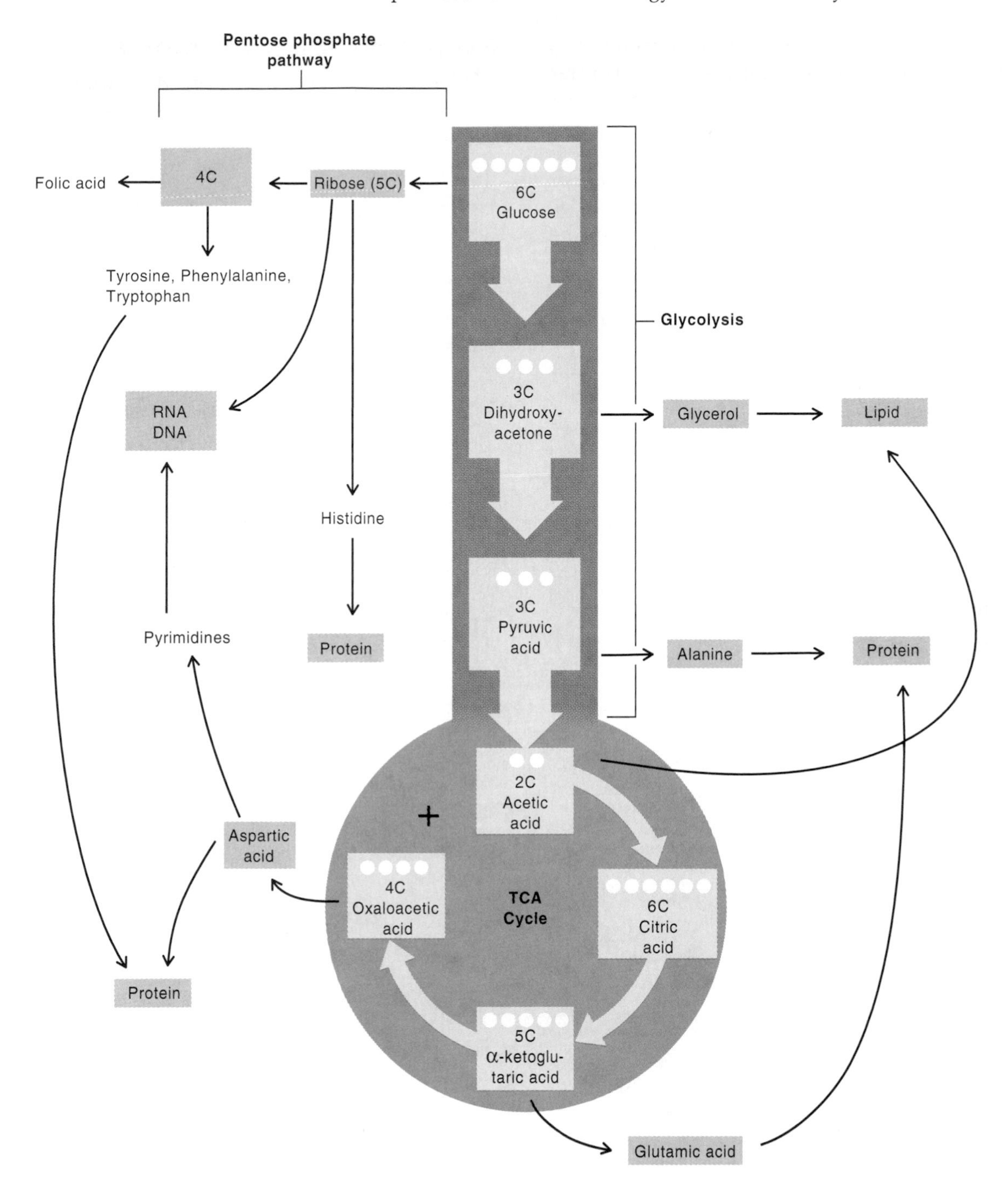

Figure 5.11
Summary of compounds originating in the glycolytic pathway and Krebs cycle leading to biosynthetic products. The diagram illustrates that most reactions involved in the degradation of a compound also serve to synthesize the beginning materials, the precursor metabolites, for the synthesis of macromolecules.

Glycolysis Is the Most Common Pathway for Degrading Sugars

The various species of microorganisms can degrade a tremendous variety of organic compounds to obtain energy, precursor metabolites, and reducing power. They do so through a variety of different pathways. The most common pathway is called **glycolysis,** (*glycos* means "sugar," and *lysis* means "dissolution") or the *Embden-Meyerhof pathway,* named after the two scientists who identified its major steps in the 1930s. Many bacteria and yeast use this pathway to degrade glucose and other sugars. For example, the yeast that produces alcohol from glucose uses this pathway, since glucose is converted to alcohol and CO_2.

The major steps in this pathway are diagrammed in figure 5.12, and the entire pathway with chemical formulas is illustrated in appendix IIIa. In a series of reactions, 1 glucose molecule (6-carbon atoms; $C_6H_{12}O_6$) is degraded to 2 molecules of pyruvic acid (3-carbon atoms each; $C_3H_4O_3$), as well as 2 pairs of H atoms of reducing power in the form of NADH. Much of the difference in bond energy between glucose and pyruvic acid is trapped in the form of high-energy bonds in 2 molecules of ATP (net).

Energy Balance Sheet in Glycolysis

For every 6-carbon molecule degraded, the following amounts of energy are used and gained:

energy expended	2 ATP molecules
energy gained	4 ATP molecules
net gain =	2 ATP molecules

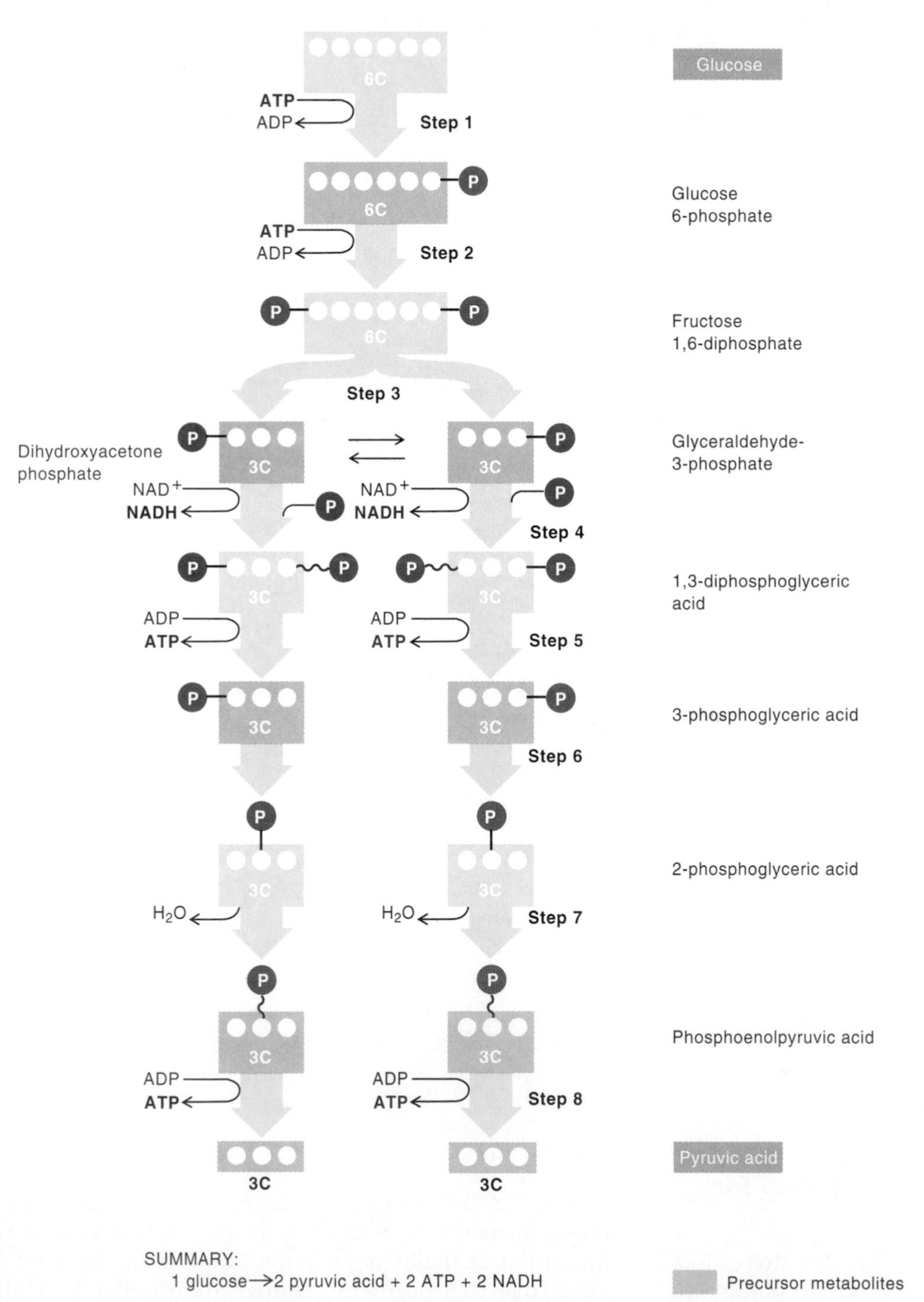

Step 1 Phosphate from ATP is added to glucose to form glucose-phosphate and ADP. A high energy phosphate bond is used up in this reaction.

Step 2 Another phosphate from ATP is added to the sugar so that the sugar (fructose) now carries two phosphate molecules. Another high energy bond is used up.

Step 3 The 6-carbon fructose diphosphate molecule splits into two molecules of three carbon atoms, each with an attached phosphate molecule. The two molecules are interchangeable and equivalent.

Step 4 Each 3-carbon compound is oxidized to another 3-carbon compound by the removal of a pair of H atoms by NAD^+ to form NADH. NADH can reduce organic compounds in the pathway (fermentation) or pass its H atoms through the electron transport chain (respiration). In this step, a phosphate molecule from the medium is added to the 3-carbon compound in the form of a high energy phosphate bond.

Step 5 The high energy phosphate bond is transferred to ADP to yield ATP by substrate phosphorylation. Thus far the cell has generated the same number of ATP molecules as it has produced.

Step 6 The phosphate molecule bound to the number 3-carbon atom is transferred to the number 2-carbon.

Step 7 The phosphate low energy bond is converted into a high energy phosphate bond by rearrangement of electrons in the 3-carbon molecule.

Step 8 The high energy phosphate bond is transferred to ADP to form ATP by substrate phosphorylation. Thus, a net gain of two molecules of ATP results for every molecule of glucose metabolized through the glycolytic pathway.

Figure 5.12
Glycolysis (Embden-Meyerhof) pathway. Note that gaseous oxygen is not used in any part of this pathway. Only a net of two ATP molecules are produced for every molecule of glucose degraded.

686,000 calories[1] of bond energy are present in each mole of glucose; each mole of pyruvic acid produced in glycolysis still contains 634,000 calories. Much of the 52,000-calorie difference can be accounted for in the 4 high-energy bonds of the ATP molecules formed in the degradation of glucose. Obviously, most of the energy originally present in the glucose molecule still remains locked up in the pyruvic acid produced in glycolysis. The cell can gain this energy by further metabolizing the pyruvic acid through the TCA cycle, which will be discussed shortly.

Generation of Precursor Metabolites in Glycolysis

Several compounds synthesized in the course of glycolysis serve as precursor metabolites as shown in figure 5.12. These include 3-carbon compounds that are converted to glycerol, a component of lipids, and pyruvic acid that is converted to alanine.

Reducing Power Balance Sheet in Glycolysis

The balance sheet for reducing power indicates that the cell generates 2 pairs of NADH molecules from each molecule of glucose degraded (see figure 5.12). However, since the NAD is in short supply in the cell, the NADH must be continuously recycled to NAD. Therefore, it must find a compound to accept the H atoms. The class of compound, organic or inorganic, that accepts the H atoms determines whether the particular compound that is degraded undergoes fermentation or respiration.

[1]Calorie—the amount of heat required to raise the temperature of 1 gram of water 1°C

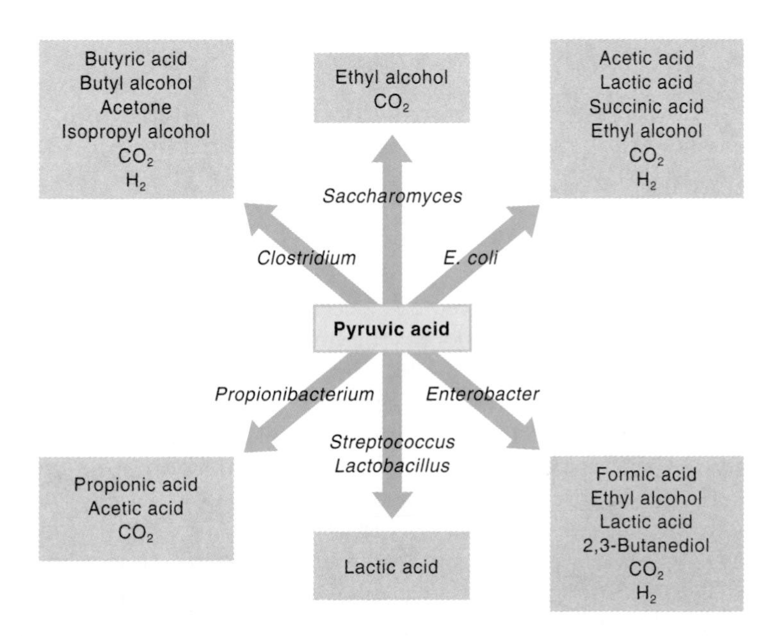

Figure 5.13
Conversions of pyruvic acid to different end products by different bacteria.

Metabolism of Pyruvic Acid in Fermentation

In addition to serving as a precursor metabolite for the synthesis of alanine, pyruvic acid can be metabolized further in the absence of oxygen, with the result that other products are formed. Depending on the microorganism, pyruvic acid is converted to a variety of different organic compounds through reactions in which the hydrogen from NADH is transferred to organic compounds. Products of fermentation include lactic acid, ethanol, and carbon dioxide. Note that in all fermentations, the NADH is not used to generate energy because the H is merely transferred to an organic compound. As in all fermentations, gaseous oxygen does not play a role. A number of different fermentations are possible depending on the organism, and several of the most common are discussed next. Many of these fermentations play an important role in food production.

Lactic Acid Fermentation

In this fermentation, 2 molecules of NADH, each containing a pair of H atoms, reduce the 2 molecules of pyruvic acid formed in glycolysis to form two molecules of lactic acid, regenerating NAD^+:

$$2\,CH_3\text{—}\overset{\overset{O}{\|}}{C}\text{—}\overset{\overset{O}{\|}}{C}\text{—}OH + 2\,NADH \rightarrow 2\,CH_3\text{—}\underset{\underset{H}{|}}{\overset{\overset{OH}{|}}{C}}\text{—}\overset{\overset{O}{\|}}{C}\text{—}OH$$

pyruvic acid lactic acid

Lactic acid fermentation plays important roles in nature. Lactic acid, probably the chief culprit in tooth decay, is produced from pyruvic acid. It is responsible for the pain that occurs in muscles during strenuous exercise, accumulating in them when oxygen is lacking. During rest, the lactic acid accumulated during exertion is metabolized by another pathway that requires oxygen (the TCA cycle, which is discussed later), and the pain disappears.

Lactic acid fermentation is also very important in both the spoilage, such as milk souring, and the production of a number of foods. Indeed, the production of foods by microorganisms can be looked upon as the useful application of controlled spoilage. For example, depending on the microorganisms present, fermentation of milk produces a variety of products such as yogurt, buttermilk, and cheeses. Although lactic acid is the major product responsible for the formation of cheeses, other fermentation products, often from organisms other than those that produce lactic acid, are important for their flavor and appearance. For example, the holes in Swiss cheese result from CO_2 formed from lactic acid by an organism that also produces the propionic acid that imparts the characteristic aromas and flavors to certain cheeses.

The bacteria that carry out the lactic acid fermentation are termed *lactic acid bacteria*. Some of these organisms carry out the following reaction:

$$1 \text{ glucose molecule} \rightarrow 2 \text{ lactic acid molecules}$$

Other lactic acid bacteria produce CO_2 and ethanol, as well as lactic acid. These latter organisms are largely responsible for the production of sauerkraut from cabbage and pickles from cucumbers.

Lactic acid bacteria are undesirable in beer and wine because the acid they produce can spoil these beverages. This example illustrates the fact that the production of a compound in one food is considered spoilage (lactic acid in beer) while its production in another food is responsible for its desirable properties (lactic acid in yogurt).

Alcoholic Fermentation

The production of alcohol from pyruvic acid involves two steps. In the first step, pyruvic acid is converted to acetaldehyde through the removal of CO_2. In the next step, the acetaldehyde is then reduced by the NADH produced in the glycolytic pathway to ethyl alcohol. These two reactions, which are carried out by yeast (*Saccharomyces*), form the basis for wine and beer production and for the rising of yeast breads. The overall reaction can be summarized as follows:

$$1 \text{ glucose} \rightarrow 2 \text{ ethanol} + 2\ CO_2$$

Other End Products of Fermentation

Humans, yeast cells, and bacteria all use the glycolytic pathway to degrade glucose. However, these organisms differ in the ways they metabolize pyruvic acid. As previously mentioned, humans may accumulate lactic acid in the absence of oxygen, while yeast forms ethanol and CO_2. Bacteria can form a wide variety of different organic end products. Some of these are given in figure 5.13. In each of these cases, the metabolism of pyruvic acid involves the same principles as have been discussed already—namely, that the reducing power gives up its hydrogen to an organic molecule, thereby becoming recycled to the NAD^+ form so it can again accept additional hydrogen atoms. This recycling does not generate energy. Since different acceptors of the hydrogen are used, different products result. These fermentations, called **mixed fermentations,** result in a variety of products. The types of products formed depend on the species of bacteria and some features of the environment, such as which hydrogen acceptors are available.

Closely related species of bacteria can often be distinguished from one another by their fermentation products. The products formed depend on enzymes, and even closely related bacteria often have different enzymes. For example, the two gram-negative rods, *Escherichia coli* and *Enterobacter aerogenes,* can be distinguished from each other because *E. aerogenes* synthesizes the fermentation product 2, 3-butanediol from glucose, whereas *E. coli* does not. Clinical microbiology laboratories determine the end products of fermentation routinely as an aid to identifying bacteria.

REVIEW QUESTIONS

1. List three environmental conditions that affect enzyme activity.
2. Distinguish between fermentation and respiration.
3. To what group of molecules are vitamins converted?
4. List three classes of molecules that cells must synthesize in order to multiply.

The TCA Cycle Oxidizes Pyruvic Acid to CO_2 and H_2O and Traps the Energy of Pyruvic Acid in ATP

Since the process of fermentation extracts only a very small fraction of the total energy available in the sugar molecule, any bacteria that can use this untapped energy would likely divide more rapidly. This energy can be tapped if the pyruvic acid is oxidized to compounds with less bond energy, namely CO_2 and H_2O. This process is accomplished by the enzymes in the tricarboxylic acid (TCA) cycle, sometimes called the *Krebs cycle* after Sir Hans Krebs, the German biochemist who worked out most of the steps in the pathway. The overall reactions beginning with glucose can be written as follows:

$$C_6H_{12}O_6 \xrightarrow[\text{(little energy gained)}]{\text{Glycolysis}} 2 \text{ pyruvic acid} + 2NADH \xrightarrow[\text{(much energy gained)}]{\text{TCA cycle}} 6\ CO_2 + 6\ H_2O + \text{cells}$$

In order to use this pathway, bacteria must have all of the necessary enzymes required in the pathway, as well as a plentiful source of final acceptor molecules for the hydrogen atoms from NADH. This acceptor molecule most commonly is oxygen, and it works through the process of aerobic respiration.

Summary of the TCA Cycle

Carbon Skeleton Transformation The key steps in this cyclic pathway can be described as follows (figure 5.14). The complete cycle is given in appendix III b.

Pyruvic acid entering this pathway is first converted to the 2-carbon compound acetic acid, which in the reaction attaches to a **coenzyme A molecule (CoA).** This coenzyme transfers acetic acid molecules (see table 5.1). In this oxidation reaction, NAD^+ picks up the hydrogen atoms (protons and electrons) to form NADH.

The acetyl CoA combines with a 4-carbon compound (oxaloacetic acid) to form the 6-carbon compound citric acid.[1] Citric acid is converted to the 5-carbon compound α-ketoglutaric acid by the release of CO_2 and a pair of hydrogen atoms. α-ketoglutaric acid is then converted to the 4-carbon compound succinic acid by the release of another CO_2 molecule and another pair of hydrogen atoms. After

[1]This molecule has three carboxyl groups, which explains how the tricarboxylic cycle got its name.

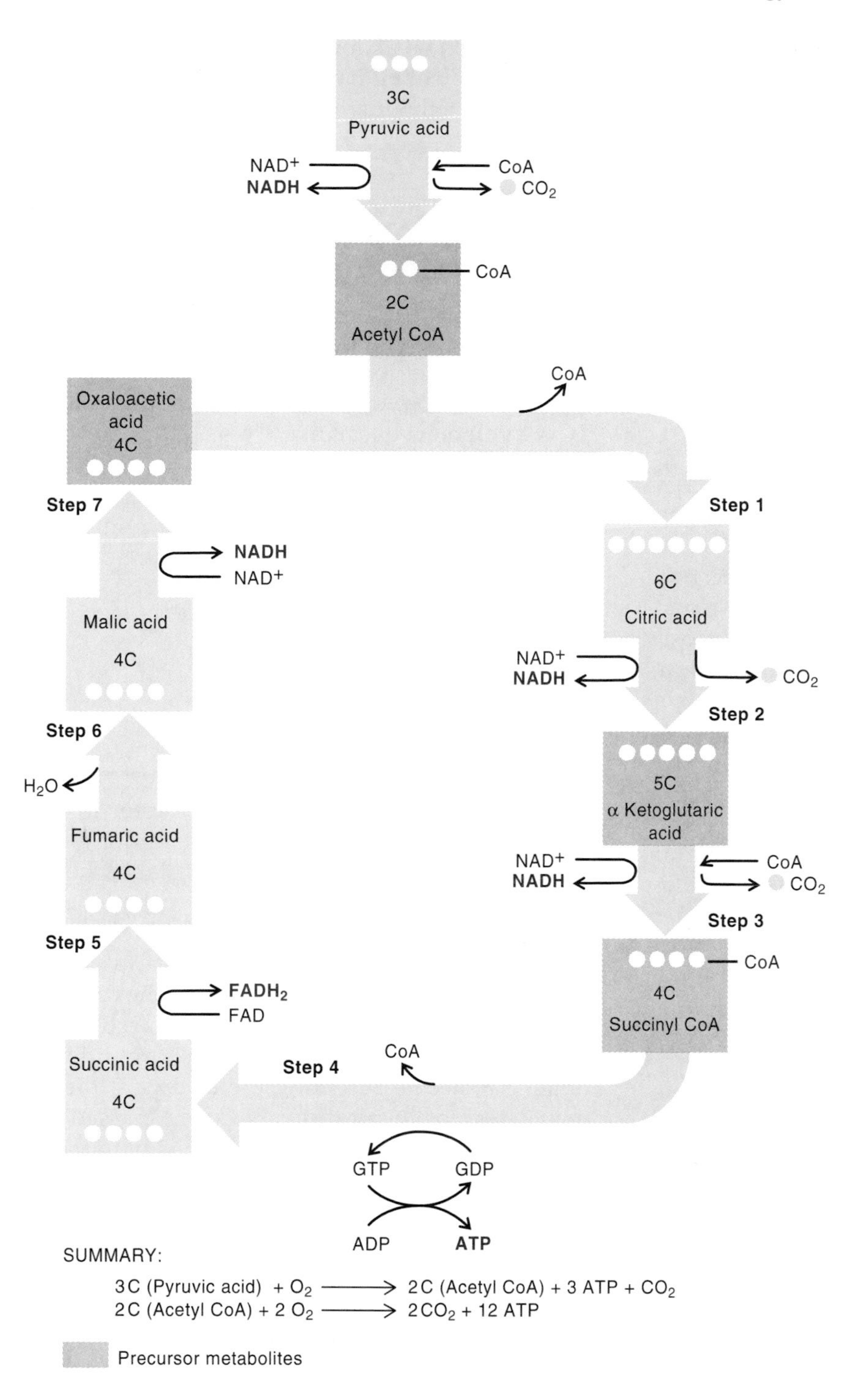

Step 1 The acetyl CoA generated from pyruvic acid in glycolysis combines with oxaloacetic acid to form the 6 carbon molecule, citric acid. This acid had 3 carboxyl (-COOH) groups, hence, the name of the cycle (citric acid or tricarboxylic acid).

Step 2 The citric acid is oxidized by the removal of a pair of H atoms by NAD^+. The NADH enters into the electron transport chain to yield 3 molecules of ATP (by oxidative phosphorylation and chemiosmosis), and if oxygen is available, it serves as the final H acceptor to produce water (aerobic respiration). CO_2 is also removed, with the formation of α ketoglutaric acid.

Step 3 α ketoglutaric acid is oxidized to succinic acid by the removal of a pair of H atoms; CO_2 is also removed. Coenzyme A (CoA) forms a high energy bond with the product of the reaction, succinic acid, to form succinyl CoA.

Step 4 The coenzyme A transfers its high energy bond to guanosine diphosphate (which is similar in structure to ADP) to form GTP and succinic acid. The high energy phosphate bond in GTP is transferred to ADP to form ATP by substrate level phosphorylation.

Step 5 The succinic acid is oxidized to fumaric acid (both 4 carbon atoms) by the removal of a pair of H atoms by the coenzyme FAD^+ to form $FADH_2$. This enters the electron transport chain to form two molecules of ATP.

Step 6 Fumaric acid is converted to another 4-carbon compound, malic acid, by the removal of water.

Step 7 Malic acid is converted to oxaloacetic acid by the removal of a pair of H atoms by NAD^+ to form NADH, which enters the electron transport chain to form 3 molecules of ATP. The oxaloacetic acid then combines with another molecule of acetyl CoA (Step 1) and the cycle starts over.

Figure 5.14
Tricarboxylic acid (TCA) cycle.

several more enzymatic steps, during which four more pairs of hydrogen atoms are released, the 4-carbon compound oxaloacetic acid is formed. This compound, which is continually regenerated in every turn of the cycle, then combines with another molecule of acetyl CoA, as it is generated from the glycolytic pathway, to again start the TCA cycle. Thus, in every turn of the cycle, two carbon atoms exit in the form of CO_2, and the 4-carbon molecule oxaloacetic acid is regenerated.

Balance Sheet of Reducing Power in the TCA Cycle

Three pairs of hydrogen atoms are removed by NAD^+ to form three NADH molecules. In addition, one pair is removed by another coenzyme flavin adenine dinucleotide (FAD) to form FADH. The NADH and FADH can either reduce compounds in biosynthetic reactions or pass the protons and electrons down the electron transport chain and generate ATP through oxidative phosphorylation.

Generation of Precursor Metabolites in the TCA Cycle

Several precursor metabolites are intermediates in the TCA cycle. These include α ketoglutaric acid, which is converted to the amino acid glutamic acid, in one step, and oxaloacetic acid, which is converted to aspartic acid in a single step (see figure 5.14).

Much More Energy Is Gained in Respiration Than in Fermentation

The energy balance sheet for the metabolism of glucose either through fermentation or respiration can now be summarized. Figure 5.15 illustrates the tremendous energy bonus a cell gains by oxidizing pyruvic acid to CO_2 and H_2O (respiration) compared with the energy it extracts if pyruvic acid or another organic compound is the electron acceptor (fermentation). Oxidative phosphorylation generates the greatest amount of energy available from glucose. Fermentation, through substrate-level phosphorylation, generates only about 5% as much. Consequently, to obtain the same amount of energy, a cell growing under anaerobic conditions must degrade about 20 times more glucose than a cell growing under aerobic conditions. Because the amount of work a cell can perform depends on its supply of ATP, a facultatively anaerobic cell can synthesize far more cell material per unit of time (and therefore can multiply more rapidly) under aerobic conditions than under anaerobic conditions.

Why do not all cells metabolize glucose through the TCA cycle if it is so advantageous? In order to use this cycle, the bacteria must have an inorganic electron acceptor available, usually oxygen. Under anaerobic conditions, the bacteria are only able to ferment, using pyruvic acid or another of its breakdown products as the final electron acceptor. Also, not all bacteria have all of the enzymes of the TCA cycle, without which they cannot use the cycle to generate energy.

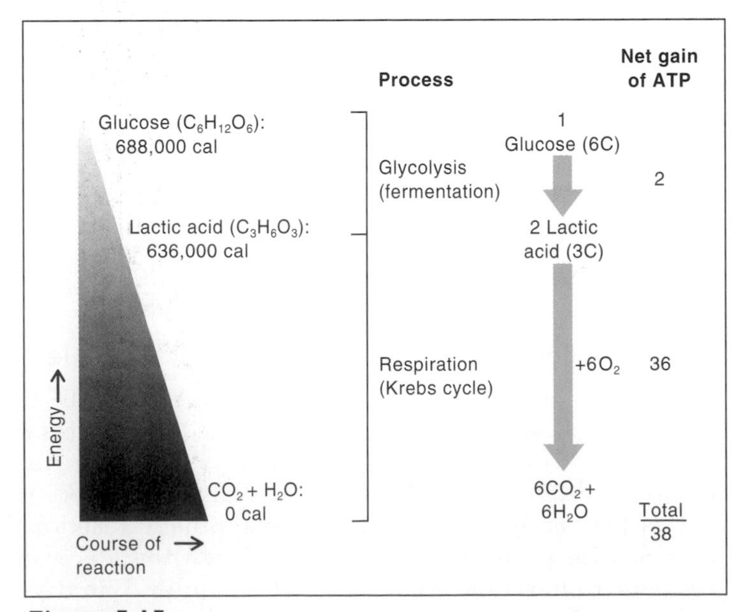

Figure 5.15
Energy gain in the breakdown of glucose.

Synthesis of ATP and Precursor Metabolites Depend on the Environment

Under aerobic conditions, the TCA cycle is primarily concerned with the generation of ATP, especially if the medium contains amino acids. However, under anaerobic conditions, in the absence of a terminal hydrogen acceptor, the TCA cycle serves primarily as a source of precursor metabolites for the synthesis of amino acids. Some anaerobes have some, but not all, of the enzymes of the TCA cycle. These bacteria can use these enzymes to synthesize precursor metabolites but not to generate energy. Energy generation requires a complete TCA cycle.

Respiration Under Anaerobic Conditions

Since the most common electron acceptor in respiration is oxygen, cells usually respire only under aerobic conditions. If gaseous oxygen is not available, however, some **facultative anaerobes** can utilize other inorganic compounds as electron acceptors. The most common acceptors are nitrate (NO_3^-), sulfate (SO_4^{2-}), and carbon dioxide (CO_2). Their reduction products are shown in figure 5.16. This type of metabolism is called **anaerobic respiration,** because it occurs without oxygen gas yet it involves an inorganic molecule accepting the electrons, the distinguishing feature of respiration. Electrons passed through the electron transport chain and then finally to these acceptors result in ATP formation through the same process described earlier for the TCA cycle. However, not all of the carriers of the electron transport chain participate in anaerobic respiration. Consequently, less ATP is generated in aerobic respiration.

The Pentose Phosphate Pathway

Another sugar metabolizing pathway present in most bacteria is the **pentose phosphate pathway** (figure 5.17 and appendix III c). This pathway is of great significance because its intermediates are 5-carbon and 4-carbon molecules that serve as precursor metabolites for nucleic acid and amino acid synthesis, respectively. Further, it serves to provide reducing power required in the biosynthesis of cell components. This pathway is especially important in organisms that carry out fermentations in which reducing

• **Facultative anaerobe, p. 92**

power is not available for biosynthetic reactions. This pathway, like glycolysis, can operate in the presence or absence of oxygen and is found in both aerobes and anaerobes.

The three pathways of central metabolism provide energy, reducing power, and precursor metabolites to the cells. These are summarized in table 5.2.

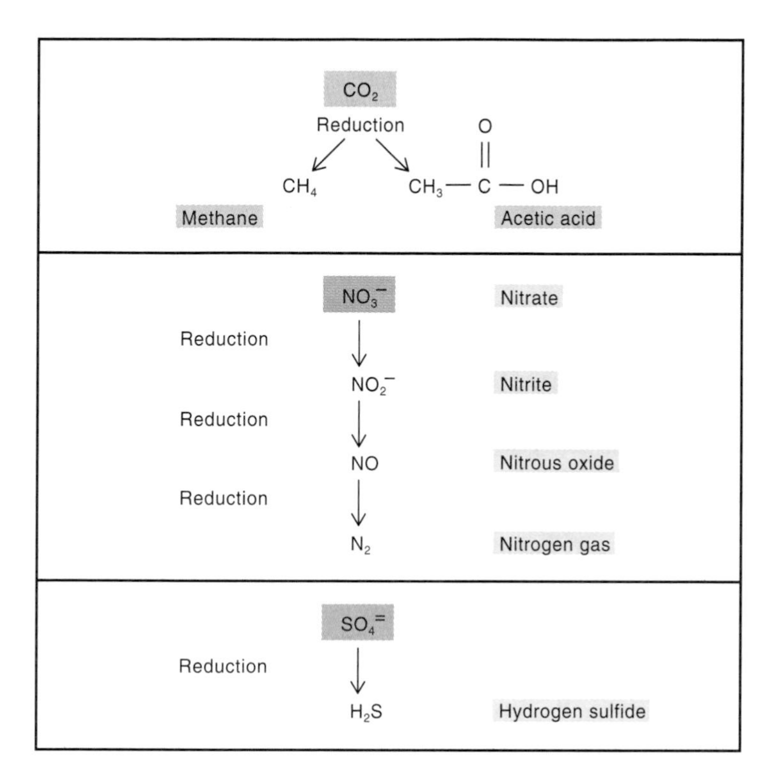

Figure 5.16
Various anaerobic respiration processes. All of these processes take place in the absence of oxygen.

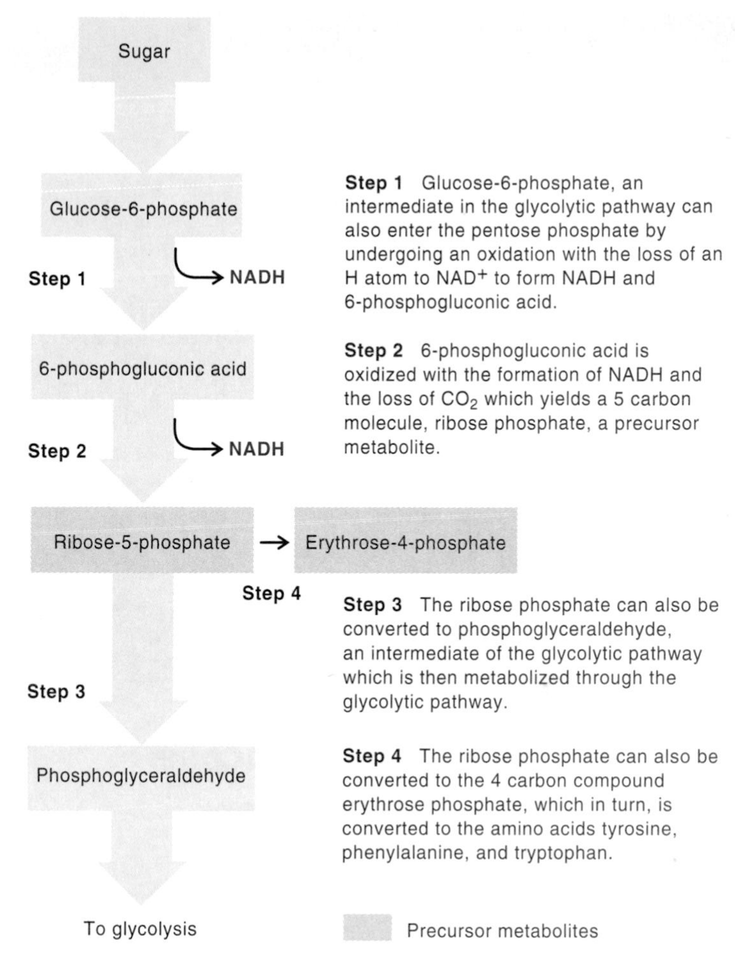

Figure 5.17
The pentose phosphate pathway. Note that oxygen is not required for any part of this pathway. Only two molecules of ATP are produced for every molecule of sugar degraded.

Table 5.2 Comparison of Pathways of Central Metabolism

Name of Pathway	Maximum Energy Generated (Net per Glucose Metabolized)	Precursor Metabolites Synthesized (Product)	Maximum Reducing Power Generated (Pairs of H Atoms per Glucose Molecule)	End Products	Environmental Conditions Under Which Pathway Operates
Glycolysis	2 ATP	Glucose-6-phosphate, fructose-6-phosphate, triose phosphate, 3 phosphoglyceric acid, 2 phosphoenolpyruvic acid	2 NADH	2 pyruvic acid	Aerobic and anaerobic
TCA cycle	~20	Acetyl CoA (lipids), oxacoacetic acid (aspartic acid), α-ketoglutaric acid (glutamic acid), succinyl CoA	8 NADH, 2 $FADH_2$	$CO_2 + H_2O$	Usually aerobic except for anaerobic respiration
Pentose phosphate pathway	~4	Ribose-5-phosphate (histidine), erythrose-4-phosphate (aromatic amino acids)	2 NADH	CO_2	Aerobic and anaerobic

Perspective 5.2

Plastics, Landfills, and Bacteria

More than a thousand new compounds are marketed every year, with their total annual world production exceeding 300 million tons. Although microorganisms can degrade a tremendous number of compounds, many compounds are totally resistant to degradation and others are degraded only very slowly, especially under conditions of low oxygen and no sunlight. These poorly degraded compounds include not only human-made compounds termed **xenobiotics,** to which microorganisms have not been exposed in the course of evolution, but also naturally occurring compounds, such as lignin, an important component of plant cell walls.

Plastics, which represent a major xenobiotic material, account for about 7% of municipal wastes and are its fastest growing component. Disposable plastic diapers represent 2% of the trash in the United States. The plastics made by current methods do not degrade readily and represent a major problem because of the nationwide shortage of landfill capacity and the increasing amounts of municipal wastes being generated. Biodegradable plastics are now being manufactured, which should lessen the problem. One approach to making plastics degradable is to mix 10% starch with the polymer that makes up the plastic. Bacteria and fungi then break down the starch, and the plastic falls apart into dust. Degradable shopping bags are being made from such a plastic.

Unfortunately, plastics containing starch do not degrade completely, especially in landfills where there is little sunlight and oxygen. Therefore, efforts continue to develop a completely biodegradable plastic. One recent attempt mixes polyhydroxybutyrate with plastic. This compound is an energy reserve storage product synthesized by *Alcaligenes eutrophus* when it is grown on high levels of glucose. The energy reserve can be readily harvested by breaking open the bacteria. It is then degraded into water, carbon dioxide, and complex organic materials.

Another Pathway of Glucose Degradation—the Entner-Doudoroff Pathway

The major scheme for the degradation of glucose begins with the glycolytic pathway. However, not all bacteria use this pathway; some produce pyruvic acid and derive energy through a different pathway, the **Entner-Doudoroff pathway,** named for the two scientists at the University of California, Berkelely, who discovered it in the early 1950s. It is presented in appendix III d.

Metabolism of Compounds Other than Glucose

Many bacteria have the enzymes necessary to convert a variety of naturally occurring compounds to low molecular weight sugars, which are then broken down by glycolysis. Indeed, if bacteria have the necessary enzymes, they will degrade any compound that has chemical bond energy and convert that energy into ATP. Most naturally occurring compounds are degradable by one mircoorganism or another, the end result being the production of ATP, although the particular catabolic pathways are in most cases unknown. Certain industries have taken advantage of this microbial versatility. Cheap and plentiful compounds, such as cellulose, sugarcane, molasses, and starch, are often used as the beginning materials for organisms to break down and derive energy and, in the process, produce useful products.

• **Cellulose, p. 36**

For example, molasses has been used as the starting material for the production of acetone and butanol as well as citric acid, and corn has been used as the starting material for ethanol production. Such ethanol has been widely used in combination with gasoline to make gasohol.

Most organic compounds that are catabolized, including amino acids, lipids, and carbohydrates, are converted into the compounds that are part of either the glycolytic pathway or the TCA cycle. In this way, they enter the pathways that lead to the formation of ATP, reducing power and precursor metabolites.

Oxidation of Inorganic Compounds

In addition to degrading organic compounds, many species of bacteria can oxidize a variety of inorganic compounds to gain energy. Table 5.3 presents important information about some of these organisms. (A number of these bacteria and their metabolic processes are considered in chapter 11.) Many of the organisms that derive energy from the oxidation of inorganic compounds can also gain energy by oxidizing organic compounds.

Note that the oxidation of inorganic compounds is quite different from the process of anaerobic respiration, which also involves the metabolism of inorganic compounds. The inorganic compounds give up electrons, thereby becoming oxidized and serving as a source of energy. Oxygen is the final electron acceptor and so the cells are aerobes. In anaerobic respiration, the inorganic compounds serve as electron acceptors and therefore are reduced. These acceptors take the place of oxygen and do not serve as a source of energy.

Table 5.3 Energy Metabolism of Some Aerobic Bacteria that Derive Energy from Inorganic Sources

Common Name of Organism	Source of Energy	Oxidation Reaction (Energy Yielding)	Important Features of Group	Common Genera in Group
Hydrogen bacteria	H_2 gas	$H_2 + \frac{1}{2}O_2 \rightarrow H_2O$	Can also use simple organic compounds for energy	*Hydrogenomonas*
Sulfur bacteria (nonphotosynthetic)	H_2S	$H_2S + \frac{1}{2}O_2 \rightarrow H_2O + S$ $S + 1\frac{1}{2}O_2 + H_2O \rightarrow H_2SO_4$	Some organisms of this group are able to live at a pH of less than 1	*Thiobacillus* *Beggiatoa* *Thiothrix*
Iron bacteria	Reduced iron (Fe^{2+})	$2Fe^{2+} + \frac{1}{2}O_2 + H_2O \rightarrow$ $2Fe^{3+} + 2\,OH^-$	Iron oxide present in sheaths of these bacteria	*Sphaerotilus* *Gallionella*
Nitrifying bacteria	NH_3	$NH_3 + 1\frac{1}{2}O_2 \rightarrow HNO_2 + H_2O$	Important in nitrogen cycle	*Nitrosomonas*
	HNO_2	$HNO_2 + \frac{1}{2}O_2 \rightarrow HNO_3$	Important in nitrogen cycle	*Nitrobacter*

Many Amino Acids Are Synthesized from the Same Precursor Metabolites

Although a number of different pathways are involved in synthesizing the 20 amino acids, it is useful to discuss some general principles that govern their synthesis. Certain amino acids such as glutamic acid, aspartic acid, and alanine are derived from precursor metabolites of the glycolytic pathway and TCA cycle. Other amino acids are synthesized from precursor metabolites generated in the pentose phosphate pathway.

The 20 amino acids can be grouped into families, the members of which are synthesized from the same starting material. A good illustration of this is the aromatic amino acid family, which consists of three amino acids—tyrosine, phenylalanine, and tryptophan (figure 5.18)—all of which have a benzene ring as part of their structure.

In the first step in their synthesis, the 4-carbon compound, which is produced by the pentose phosphate pathway, combines with a 3-carbon compound, which comes from the glycolytic pathway, to form a 7-carbon compound. The 7-carbon compound is then metabolized through a series of reactions, that finally end with the formation of the three aromatic amino acids. The entire sequence of reactions, which illustrates features common to many biosynthetic pathways, can be summarized as follows:

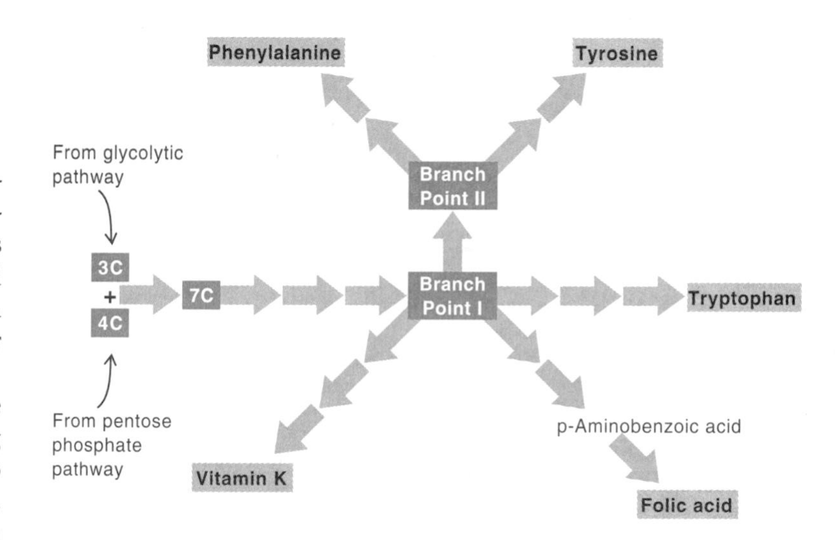

Figure 5.18
Biosynthetic pathway of aromatic amino acid biosynthesis—a model pathway. This is a complex biosynthetic pathway in which three different amino acids and several vitamins are synthesized. (The number of arrows does not reflect the number of steps actually involved.) The compound at branch point I is chorismic acid; at branch point II, the compound is prephenic acid.

1. A number of enzymes are common to the formation of all three amino acids. If any one of these enzymes is defective, none of the amino acids can be synthesized; then all three amino acids must be provided in the medium in order for the cells to grow.
2. Energy provided by the high energy bonds of ATP, must be supplied at several steps in the pathway. The high-energy compounds are formed in the glycolytic pathway and TCA cycle (see figure 5.14).
3. Several compounds in this pathway, termed *branch point compounds*, can be metabolized along different routes in the pathway.
4. The cell regulates the flow of compounds into the biosynthetic pathway at a rate that does not overtax the available energy of the cell. It makes no sense for the cell to channel the products of metabolism (metabolites) into biosynthetic pathways unless the energy required for the many biosynthetic reactions is available. Cells have elegant regulatory mechanisms to channel metabolites along the routes in which they are most needed. For example, if the energy sources of the cell are limited, the cell will metabolize the energy-rich compound phosphoenolpyruvic acid, a precursor metabolite, through the glycolytic pathway and generate ATP rather than use it for biosynthesis. The

Perspective 5.3

Mining with Microbes

Microorganisms have been used for thousands of years in the production of bread and wine. It is only in the past several decades, however, that they are being used with increasing frequency in another area of biotechnology, the mining industry. Although mining is a centuries-old industry, modern technologies have not yet been employed in mining as they have in medicine and agriculture. The mining process consists of digging crude ores from the earth, crushing them, and then extracting the desired minerals from the contaminants. The extraction process of such minerals as copper and gold frequently involves harsh conditions, such as smelting, and burning off the contaminants before extracting the metal, such as gold, with cyanide. Such activities are expensive and deleterious to the environment. With the development of biomining, some of these problems are being solved.

In the process of biomining copper, the ore is dumped outside the mine and then treated with sulfuric acid. The acid conditions encourage the growth of the acidophilic eubacterium, *Thiobacillus ferrooxidans*, which occurs naturally in the ore. This organism uses CO_2 as a source of carbon and gains its energy by oxidizing sulfides of iron first to sulfur and then to sulfuric acid. The sulfuric acid dissolves the insoluble copper and gold from the ore. Currently, about 25% of all copper produced in the world comes from this process of biomining. Similar processes are being applied to gold mining.

The current process of biomining employs only the bacteria that are indigenous to the ore. Many improvements should be possible. For example, the oxidation of the minerals generates heat to the point that the bacteria may be killed. Thus, the use of thermophiles should overcome this problem. Further, many ores contain heavy metals, such as mercury, cadmium, and arsenic, that are toxic to the bacteria. It should be possible to isolate bacteria that are resistant to these metals. Biomining is still in its infancy.

mechanisms that determine whether a precursor metabolite will be converted into a subunit for macromolecule synthesis or degraded further for energy are considered in chapter 6.

Utilization of Ammonium (NH_4^+) Salts

A very important reaction in the cells' metabolism is the reaction by which the cell incorporates ammonia (NH_3) or the ammonium ion (NH_4^+) into an organic molecule. One especially important example connects energy metabolism to biosynthesis. In this reaction, α-ketoglutaric acid, a precursor metabolite, synthesized in the TCA cycle, reacts with NH_4^+ to form glutamic acid. This reaction is reversible, and glutamic acid is readily converted to α-ketoglutaric acid:

$$\alpha\text{-ketoglutaric acid} + NH_4^+ \rightleftarrows \text{glutamic acid}$$

Glutamic acid can then be metabolized in several other ways. It may be incorporated into protein or it may be converted into another amino acid, glutamine, by the following reaction:

$$\text{glutamic acid} + NH_4^+ \rightarrow \text{glutamine}$$

Glutamine is an especially important amino acid because the ammonia that is part of the molecule can be transferred to a variety of different molecules, thereby synthesizing other molecules that are important in cell metabolism.

The amino group of glutamic acid can be transferred to other α-keto acids to yield other amino acids. These reactions, catalyzed by enzymes termed **transaminases,** are reversible.

Purines and Pyrimidines Are Synthesized Stepwise Along Different Pathways

The synthesis of purines and pyrimidines, parts of the subunits of nucleic acids, illustrates how compounds can be built up by the stepwise addition of a number of very simple molecules. This stepwise addition of simple molecules is common in biosynthetic processes. The carbon and nitrogen atoms originate from a variety of sources (figure 5.19). The

• **Purines and pyrimidines, p. 38**

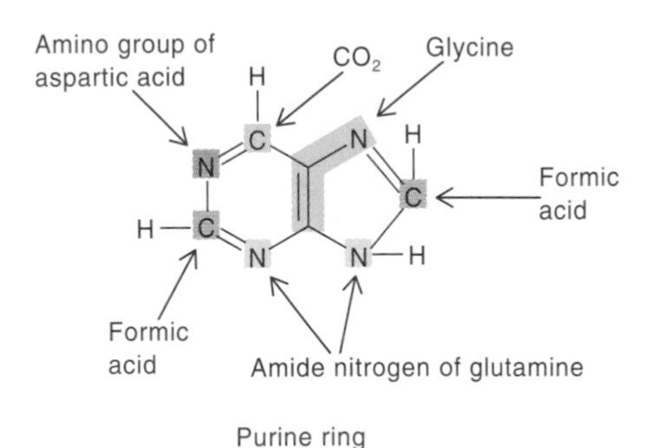

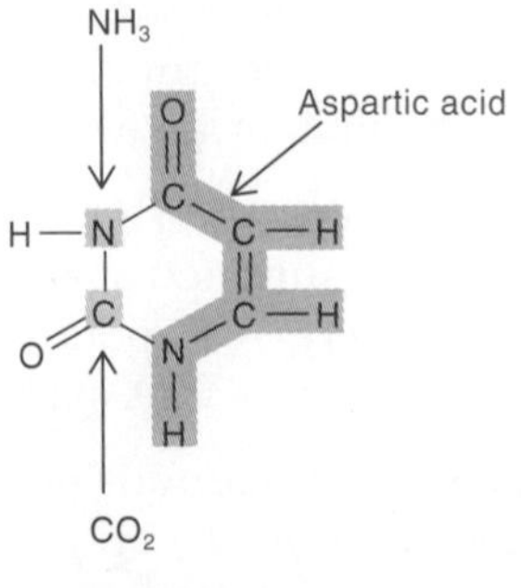

Figure 5.19
Origin of the ring structure of atoms for the biosynthesis of a purine molecule (top) and the ring structure of a pyrimidine molecule (bottom).

formation of purines and pyrimidines depends not only on the availability of the amino acids and other molecules from which they are formed, but also on about a dozen different enzymes, each one catalyzing a single step in the pathway. Purine and pyrimidine synthesis further illustrates the interrelationship between anabolic and catabolic reactions in the cell.

Bacterial Photosynthesis

Thus far, the discussion has focused on the pathways of degradation of organic molecules and biosynthesis of small molecules in bacteria. We have briefly discussed bacteria that can gain energy by oxidizing inorganic compounds. Another group of bacteria can absorb light energy, convert it into ATP, and multiply. These bacteria are highly visible in their natural habitats because they possess colored pigments, such as green, that absorb the light energy. These organisms are called **phototrophs** and **photoheterotrophs** if they use an organic source of carbon or **photoautotrophs** if they use CO_2 as a source of carbon.

Light Energy Absorbed by Chlorophyll Provides Energy to Reduce CO_2 to Cell Material

The overall reaction of photosynthesis carried out by green plants, algae, and cyanobacteria is summarized in the following equation:

$$6CO_2 + 6H_2O \xrightarrow{\text{light energy}} C_6H_{12}O_6 + 6O_2$$

The cell gains energy from the light, which it uses to convert CO_2 and H_2O into carbohydrates and other cellular material and to release oxygen. The formula $C_6H_{12}O_6$ represents all of the organic material that the cell synthesizes, and the reaction just illustrated represents all of the biosynthetic reactions carried out by the cell. The reverse of the reaction are the pathways by which glucose is oxidized completely via glycolysis and the TCA cycle:

$$C_6H_{12}O_6 + 6O_2 \rightarrow 6CO_2 + 6H_2O + \text{energy}$$

Using isotopes of carbon, hydrogen, and oxygen, scientists have been able to label the atoms in the important molecules and study their fates during photosynthesis. These experiments conclusively demonstrated that the carbohydrate $C_6H_{12}O_6$ is synthesized from CO_2 and the hydrogen atoms of water, while the O_2 originates from the oxygen atoms of water. Water is the source of the hydrogen atoms used to reduce CO_2 to $C_6H_{12}O_6$. This type of photosynthesis is termed **oxygenic** since oxygen is produced. The process of photosynthesis is more accurately shown by this balanced equation:

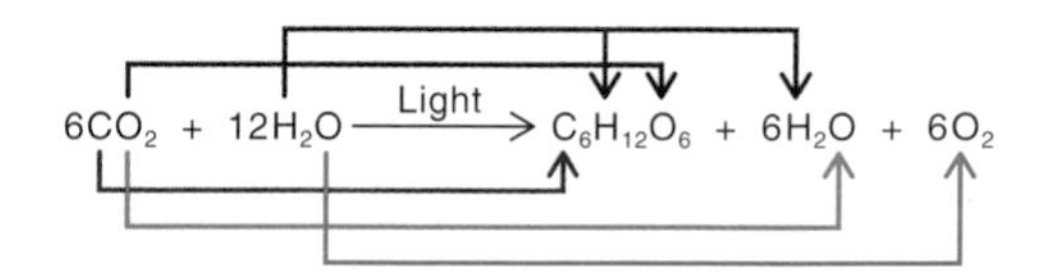

- **Photoheterotrophs, p. 95**
- **Photoautotrophs, p. 95**
- **Isotope, p. 24**

Compounds Other than H_2O Serve as the Reducing Agents in Photosynthetic Bacteria Other than Cyanobacteria

Whereas cyanobacteria, algae, and plants use H_2O as their source of hydrogen for the reduction of CO_2, no other photosynthetic bacteria use H_2O as a source of hydrogen. They utilize other reduced compounds, either inorganic or organic, depending on the particular organism. As a result, these bacteria, termed *purple* and *green bacteria* for the kind of chlorophylls they contain, never liberate oxygen but liberate other oxidized compounds. Consequently, this type of photosynthesis is termed **anoxygenic** (*anoxy* means "away from oxygen").

The green and purple sulfur bacteria are two groups of bacteria that utilize a reduced inorganic sulfur compound, such as H_2S, in place of water. The hydrogen in the $C_6H_{12}O_6$ comes from H_2S, and sulfur is produced. Other bacteria, the purple and green nonsulfur bacteria, can use reduced organic compounds as hydrogen donors if they grow aerobically. With this additional information, the general equation for photosynthesis, used by all bacteria, algae, and green plants, may be restated as follows:

$$CO_2 + H_2X \xrightarrow{\text{light}} C_6H_{12}O_6 + 6H_2O + X$$

Photosynthesis Requires Light for Energy Generation but Biosynthetic Reactions Do Not

The overall process of photosynthesis can be divided into two series of interrelated reactions: the "light" reactions, which occur only in the presence of light, and the "dark" reactions, which do not require light.

"Light" Reactions

The "light" reactions only occur in photosynthetic organisms, which are capable of harvesting the energy in light. In these reactions, the pigments absorb light energy, which is then transferred to electrons in the chlorophyll (figure 5.20).

These energized electrons are then transferred along an electron transport chain, and high-energy phosphate bonds of ATP are generated in the process. This process of ATP generation, **photophosphorylation,** involves the same sorts of carrier molecules (such as cytochromes) that participate in oxidative phosphorylation to generate ATP. The process of ATP synthesis is also the same. Some of the

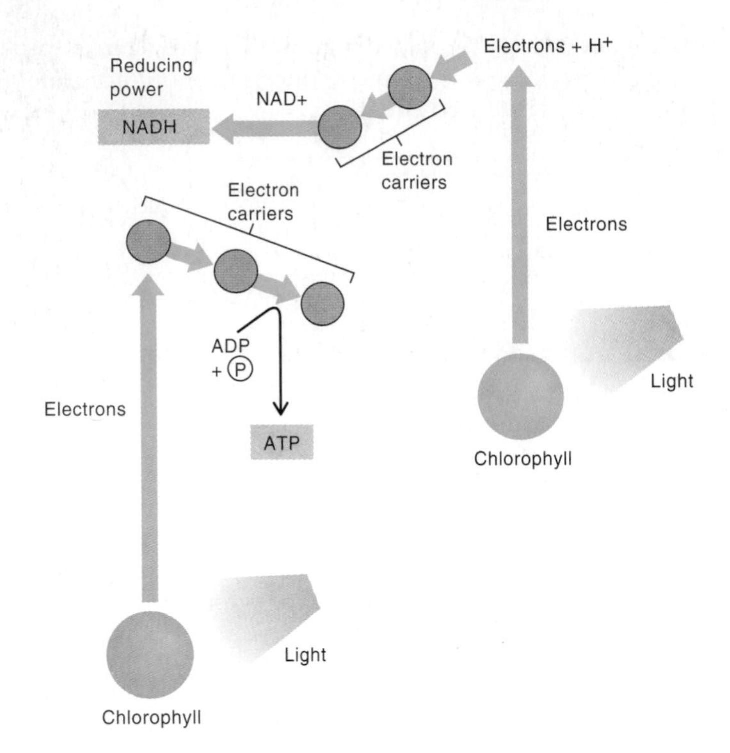

Figure 5.20
Generation of energy and reducing power in photosynthesis. The schematic diagram shows how light energy is absorbed by chlorophyll to generate both electrons (reducing power) and ATP.

electrons and protons produced in photosynthesis, which represent reducing power, also reduce small organic compounds, which are on the pathway from CO_2 and H_2O to $C_6H_{12}O6$. Thus, the absorption of light by photosynthetic pigments produces both chemical energy (ATP) and reducing power (NADH), as also happens in glycolysis and the TCA cycle.

"Dark" Reactions—Calvin Cycle

Once ATP and reducing power have been generated in the light, the actual formation of carbohydrate can occur either in the light or in the dark. In these reactions, frequently called "dark" reactions, CO_2, the source of carbon, is first incorporated into an organic molecule (figure 5.21). It is then converted to carbohydrate through additional enzymatic steps. One of the most important reactions occurring in all organisms able to use CO_2 as a major source of carbon is the incorporation of CO_2 into organic molecules. This is the same reaction that autotrophs that derive their energy from the oxidation of inorganic compounds use to convert CO_2 to organic molecules. The pathway, termed the *Calvin cycle,* is illustrated in figure 5.21. The CO_2 is added onto a 5-carbon compound (ribulose 1,5 bisphosphate[1]) and the resulting 6-carbon compound splits into 2 molecules of a 3-carbon compound. This compound (3-phosphoglyceric acid) is part of the glycolytic pathway. This reaction is limited to organisms that can use CO_2 as their major source of carbon. The glycolytic pathway functioning in reverse converts the 3-carbon compounds to sugar phosphates, which then are converted to the constituents of the cell. The cycle is named the Calvin cycle after the scientist Melvin Calvin who was most responsible for working it out.

[1]*Bisphosphate* indicates that there are 2 phosphate molecules in the molecule.

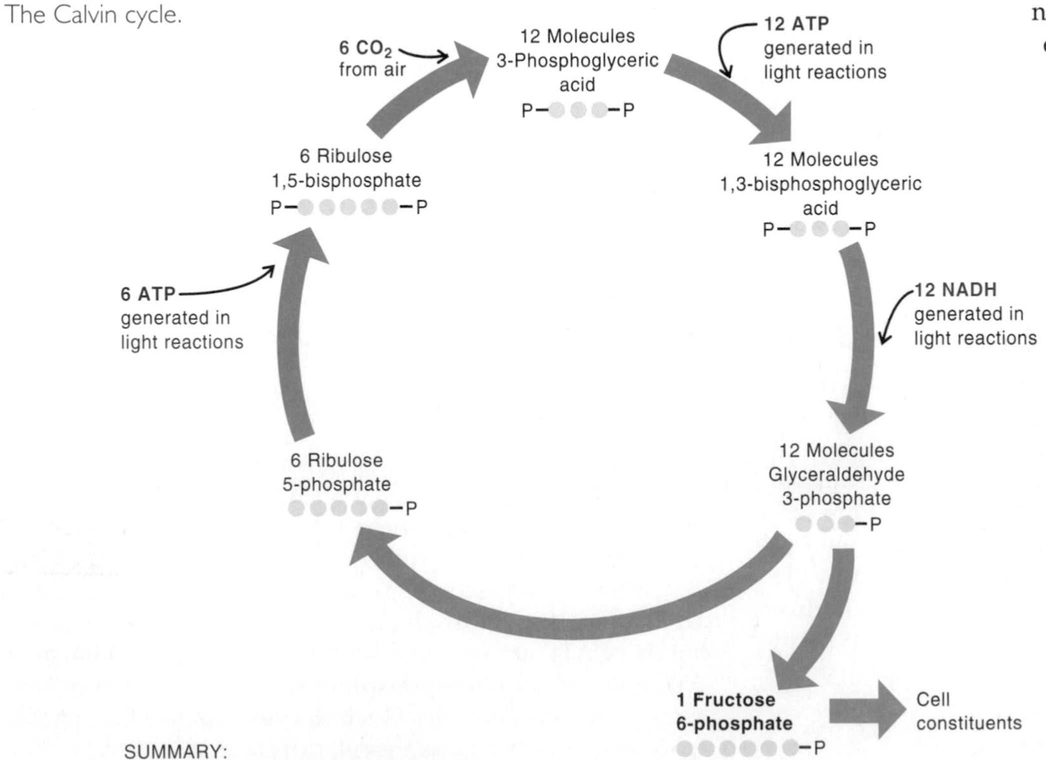

Figure 5.21
The Calvin cycle.

Summary

I. Enzymes—Chemical Kinetics and Mechanisms of Action
   A. Enzymes rearrange atoms under the mild conditions of temperature and pressure at which cells can live.
   B. Enzymes speed up reactions by lowering their activation energy.
   C. The substrate fits the shape of the enzyme that acts on it.
      1. An enzyme functions by combining with its substrate through weak bonding forces.
      2. Next, the products of the reaction are released, leaving the enzymes unchanged.
   D. Some enzymes require coenzymes for their function. Coenzymes are small molecules that transfer molecules, atoms, or electrons from one molecule to another.
   E. Environmental and chemical factors influence enzyme activity.
      1. Important features of the environment include temperature, pH, and salt concentration.
         a. A 10°C rise in temperature doubles the speed of enzymatic reactions.
         b. Most enzymes function best near a pH of 7.
         c. Most enzymes function best in low salt.
      2. Enzymes are inhibited by a variety of compounds that may result in a reversible or irreversible inhibition. An example of reversible inhibition is sulfa drugs, which compete with para-aminobenzoic acid for the active site of an enzyme used to synthesize the vitamin folic acid. An irreversible inhibitor may alter amino acids that give the enzyme its shape.
   F. Names of enzymes ususally end in -ase.

II. Overview of Metabolism
   A. To multiply, cells must degrade (catabolize) foodstuffs to gain energy. Foodstuffs also serve as a source of precursor metabolites and reducing power required for the synthesis of subunits of macromolecules. The synthesis of cell components requires energy, reducing power, and precursor metabolites.

III. Generation of Energy—General Principles
   A. The pathways of glucose degradation result in the concentration of the low energy in the chemical bonds to a few high energy bonds which can then be converted to the major energy currency of the cell, ATP.
   B. Certain coenzymes are important in the process, specifically NAD and FAD. These coenzymes are storage forms of reducing power. The reducing power can be converted to chemical bond energy in ATP. This latter process involves the transfer of electrons from a high energy level to a low energy level. This transfer occurs in the cytoplasmic membrane of bacteria, and ATP molecules are generated as protons flow back into the cell through an enzyme that synthesizes high energy bonds of ATP. This is the process of oxidative phosphorylation and the mechanism is termed *chemiosmosis*.
   C. ATP can also be synthesized by the transfer of a high energy phosphate bond on an organic molecule to ADP to form ATP, the process of substrate phosphorylation.

IV. Catabolic Pathways—Overview
   A. The three major pathways of catabolism are the pathways of central metabolism. They include glycolysis, the TCA cycle, and the pentose phosphate pathway. All produce ATP, precursor metabolites, and reducing power.
   B. Glycolysis is the most common pathway for degrading sugars.
      1. A glucose molecule metabolized through the glycolytic pathway results in the production of 2 molecules of ATP, 2 molecules of pyruvic acid, and 2 molecules of NADH.
      2. The pyruvic acid is metabolized to different compounds depending on the species of bacterium and whether oxygen is present. If oxygen is not present, pyruvate is converted to a variety of products, using the NADH generated in glycolysis. These include lactic acid in lactic acid fermentation, alcohol and CO_2 in the alcoholic fermentation, and a variety of other compounds in a mixed fermentation.
   C. The TCA cycle oxidizes pyruvic acid to CO_2 and H_2O and then traps the energy of pyruvic acid in ATP.
      1. In carbon skeleton transformation, the acetyl CoA is converted to 2 molecules of citric acid, which is converted to 2 molecules of CO_2 and 4 NADH molecules in every turn of the TCA cycle.
      2. In energy transformations, the electrons for the hydrogen atoms removed in the reactions are transferred through the electron transport chain, which generates the high-energy bonds of ATP. This process is called *oxidative phosphorylation*.
   D. Much more energy is gained in respiration than in fermentation. About 10 times more energy is gained if glucose is oxidized completely to CO_2 and H_2O, rather than if its metabolism stops at pyruvic acid.

V. Respiration Under Anaerobic Conditions
   A. Some bacteria can utilize other inorganic compounds besides oxygen as electron acceptors. The most common acceptors are nitrate, sulfate, and carbon dioxide. Less ATP is generated compared to aerobic respiration.

VI. The Pentose Phosphate Pathway

VII. Another Pathway of Glucose Degradation—the Entner-Doudoroff Pathway

VIII. Metabolism of Compounds Other than Glucose

IX. Oxidation of Inorganic Compounds
   A. Some bacteria can oxidize a variety of inorganic compounds to gain energy.
   B. Degradative and biosynthetic pathways may be part of the same pathway. Degradative pathways generate the compounds that serve as starting materials for the synthesis of cell components.

C. Many amino acids are synthesized from the same precursor metabolites.
D. A very important reaction in biosynthetic metabolism is the mechanism by which NH_4^+ is incorporated into organic compounds. An especially important reaction is the addition of NH_4^+ to α-ketoglutaric acid to form glutamic acid. This reaction is reversible and connects anabolism with catabolism. The NH_4^+ can be transferred to a variety of other α-keto acids to form other amino acids.
E. Purines and pyrimidines are synthesized stepwise along different pathways.

X. Bacterial Photosynthesis
A. Light energy absorbed by chlorophyll provides energy to reduce CO_2 to cell material.
B. Compounds other than H_2O serve as the reducing agents in photosynthetic bacteria other than cyanobacteria. If oxygen is released, then photosynthesis is termed oxygenic; if it is not released, it is termed anoxygenic.
C. Photosynthesis requires light for energy generation but biosynthetic reactions do not.

Chapter Review Questions

1. The end products of the TCA cycle are ____________ .
2. The two coenzymes that are involved in the removal of H atoms in the TCA cycle are ____________ and ____________ .
3. Compare the relative amounts of ATP generated in
   a. glycolysis;
   b. aerobic respiration; and
   c. anaerobic respiration.
4. Write out the reaction in which a compound synthesized in the TCA cycle is converted to an amino acid.
5. What is the energy of activation and how do enzymes influence it?
6. Define *precursor metabolites* and give two examples.
7. α-ketoglutaric acid is an especially important molecule in metabolism. In what cycle is it synthesized? To what two other compounds can it be converted by single enzymatic steps?
8. Write the chemical reaction by which phototrophs convert CO_2 and H_2O to glucose (organic material). Indicate the source of the atoms in the glucose molecule.
9. Explain why cells contain many more different enzymes than coenzymes.
10. The enzymes of the respiratory chain are located in different structures in eukaryotic and prokaryotic cells. Where are they located in each cell type?

Critical Thinking Questions

1. What are differences in the end products of photosynthesis by cyanobacteria and other photosynthetic bacteria?
2. Although not all bacteria have a complete TCA cycle, many have enzymes necessary for the synthesis of α-ketoglutaric acid. Speculate as to why this is true.

Further Reading

Anruckeu, Y. 1988. Bacterial electron transport chains. *Annual Review of Biochemistry* 57:101–132.
Battley, E. H. 1987. *Energetics of microbial growth.* John Wiley & Sons, New York.
Davis, E. A. 1986. *Microbial energetics.* Chapman, New York.
Gottschalk, G. 1986. *Bacterial metabolism,* 2nd ed. Springer-Verlag, New York.
Shiveley, J. M., and Burton, L. L. (ed.). 1991. *Variations in autotrophic life.* Academic Press, New York.

Informational Macromolecules: Function, Synthesis, and Regulation of Activity

KEY CONCEPTS

1. The general rule for the flow of information in cells is

   DNA $\xrightarrow[\text{the synthesis of}]{\text{codes for}}$ RNA $\xrightarrow[\text{amino acids in}]{\text{codes for}}$ protein

2. In the transfer of information, a sequence of nucleotides in DNA determines a sequence of amino acids in proteins.
3. DNA has two functions: storing genetic information and reproducing itself.
4. The sequence of nucleotides in one strand of DNA determines the sequence in the other strand.
5. The enzymes of bacteria may change with changes in the environment that turn on and off gene functions.
6. The synthesis of many small molecules, such as amino acids, is controlled by regulating the activity as well as the number of enzyme molecules involved.
7. Bacteria synthesize only the amounts of each cell component that they need in order to multiply most rapidly.

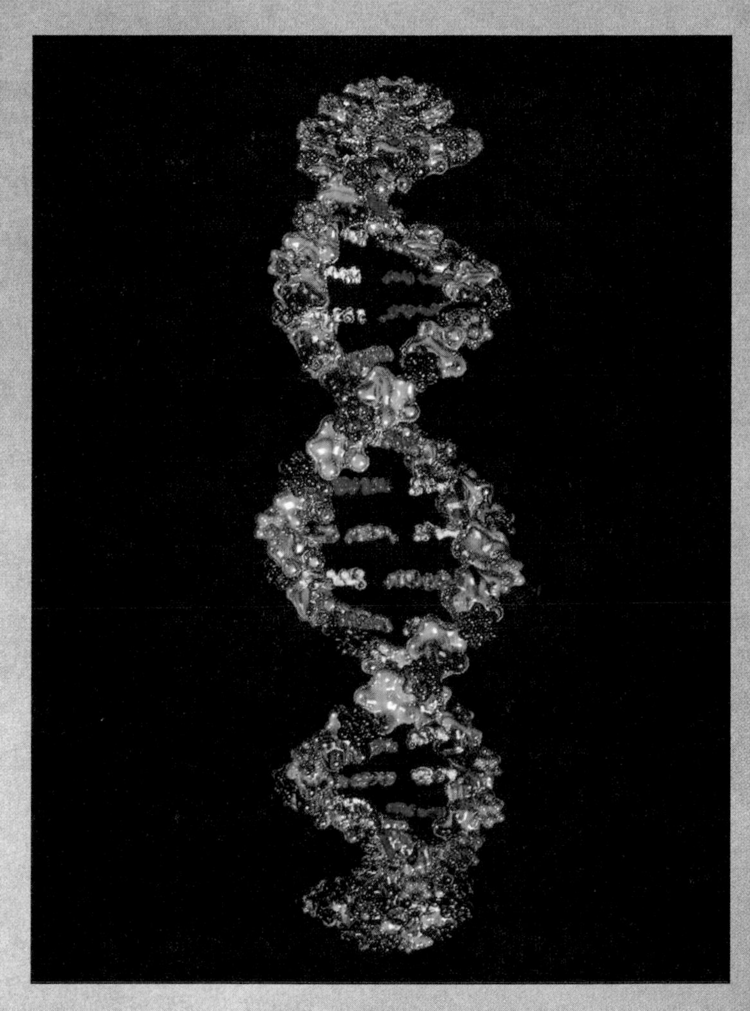

Computer-generated image of a portion of a DNA molecule.

Microbes in Motion

The following books and chapters in the *Microbes in Motion* CD-ROM may serve as a useful preview or supplement to your reading:

Bacterial Structure and Function: Internal Structures. *Microbial Metabolism and Growth:* Genetics. *Viral Structure and Function:* Viral Replication; DNA Replication.

A Glimpse of History

In 1866, the Czech monk Gregor Mendel showed that traits are inherited by means of physical units. It was not until 1903 that the Danish botanist Wilhelm Johannsen named these units *genes*, and it was not until 1940 that the function of *genes* became understood. In that year, George Beadle, a geneticist, and Edward Tatum, a chemist, published a scientific paper that reported that genes determine the structure of enzymes. These investigators answered the question of how genes function by studying the common bread mold *Neurospora crassa*. This organism can grow on a very simple medium containing sugar and simple inorganic salts. Beadle and Tatum believed that due to this property it should be possible to find offspring of *Neurospora* cells that would grow only if growth factors were added to the medium. This, in fact, was what they observed. Organisms with altered nutritional properties, **mutants,** presumably have an alteration in a gene, since genes determine the properties of organisms. Beadle and Tatum produced mutants by treating the organisms with X rays, which were known to alter genes. They then isolated organisms that could only grow if a growth factor were added to the glucose-salts medium. Each mutant presumably contained a defective gene. They then identified the specific biochemical defect of each mutant by determining which growth factor allowed the mutant to grow, thereby bypassing the function of the defective gene. By 1940, they had analyzed enough mutants to be able to conclude that each step in a biochemical pathway is catalyzed by a specific enzyme. Many biochemists had already shown that enzymes catalyze the conversion of one compound into another in a biochemical pathway. Furthermore, Beadle and Tatum showed that the requirement for each growth factor is inherited as a single gene. They concluded that in some way a single gene determines the production of each enzyme. Their results gave rise to an oft-cited statement in genetics, "one gene—one enzyme," which means that one gene is responsible for the synthesis of one enzyme (figure 6.1). In 1958, Beadle and Tatum shared the Nobel Prize in Medicine largely for these pioneering studies that ushered in the era of modern biology.

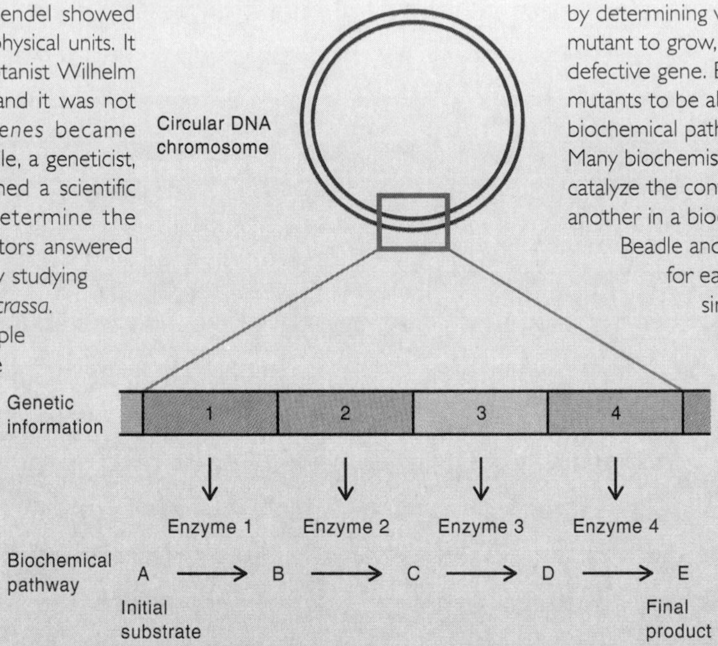

As so often occurs in science, the answer to one question raised many more questions. How do genes specify the synthesis of enzymes? What are genes made of? How do genes replicate? Numerous other investigators won more Nobel Prizes for answering these questions, many of which are covered in this chapter.

Circular DNA chromosome

Genetic information

1 2 3 4

Enzyme 1 Enzyme 2 Enzyme 3 Enzyme 4

Biochemical pathway

A ⟶ B ⟶ C ⟶ D ⟶ E

Initial substrate

Final product

Figure 6.1
One gene-one enzyme hypothesis. Four genes code for the synthesis of four enzymes, which by their sequential actions convert substrate A to the final product E.

Chapter 5 examined the mechanisms by which cells obtain energy and synthesize small molecules. In this chapter, we will consider how these small molecules are put together to form the three macromolecules: DNA, RNA, and protein. We then will discuss how the cell controls the function and synthesis of these macromolecules.

Chemistry of DNA and RNA

DNA and RNA have many features in common but they also differ in a number of ways. One difference is that DNA contains thymine; RNA contains uracil. Further, DNA usually occurs as a double-stranded molecule, whereas RNA is a single-stranded molecule.

DNA Is a Double-stranded, Helical Molecule

The composition and formation of the subunits of DNA (deoxyribonucleotides) and RNA (ribonucleotides) were discussed in chapter 2. Both subunits are referred to as **nucleotides.** The DNA in the cell usually occurs as a double-stranded, helical structure (figure 6.2) in which the two strands are held together by many weak hydrogen bonds between the nitrogenous bases in the two strands. Guanine (G) in one strand hydrogen bonds to cytosine (C) in the other, and adenine hydrogen bonds to thymine (T).

Only this hydrogen bonding arrangement is possible because of the shapes and chemical composition of the four purines and pyrimidines. The two nitrogenous bases that hydrogen bond together are said to be **complementary** to each other—as guanine is complementary to cytosine—so that one entire strand of DNA of the double helix is complementary to the other strand.

This complementarity explains a baffling observation made over 40 years ago: In double-stranded DNA, the num-

- **Nucleotides, pp. 37, 39**
- **Hydrogen bond, p. 26**
- **Purine and pyrimidine, pp. 37, 38**

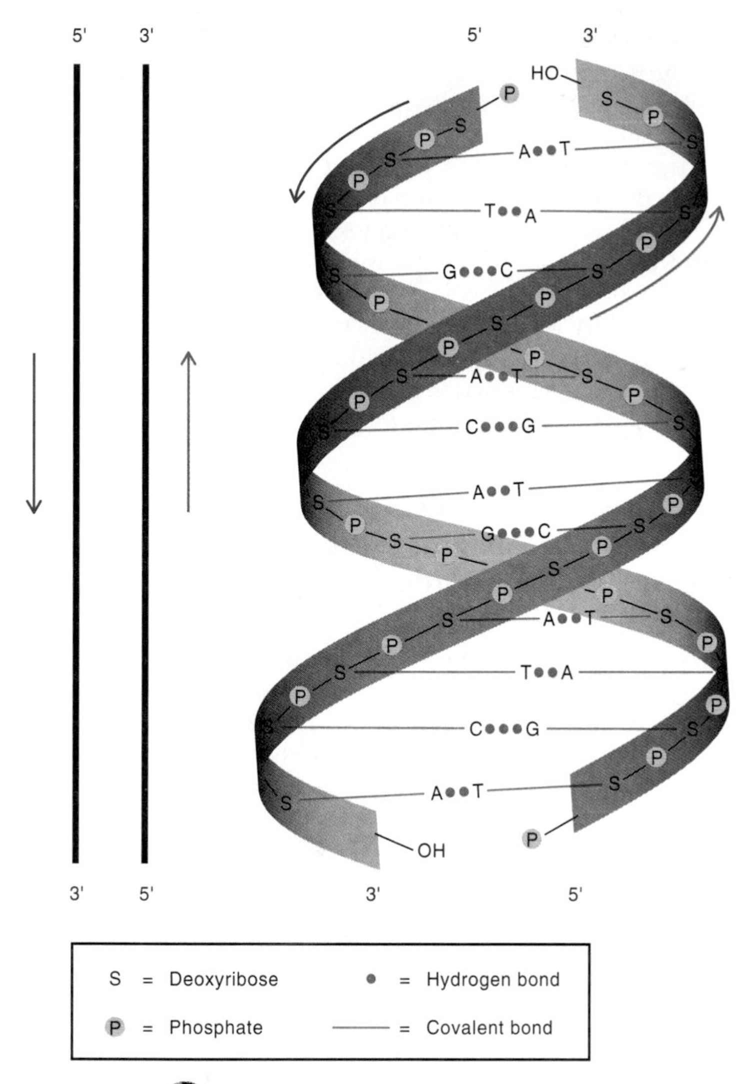

Figure 6.2
The double helix of DNA. Note the hydrogen bonding between the complementary base pairs (A-T and G-C). The chromosome of *Escherichia coli* consists of approximately 9,000 nucleotide pairs.

ber of molecules of thymine equals the number of adenine molecules, and the number of molecules of cytosine equals the number of guanine molecules, no matter from what organism the DNA is isolated. Only after James Watson and Francis Crick understood this hydrogen bonding between complementary bases could they solve the double-stranded structure of DNA in the early 1950s. The hydrogen bonding between bases in DNA illustrates how a large number of weak bonds can hold a very large molecule together.

The complementarity between adenine and thymine and guanine and cytosine is important for another reason: The sequence of bases on one DNA strand determines the sequence on the other strand, since thymine must always fit opposite adenine and guanine opposite cytosine. This ensures that when two new strands of DNA are synthesized prior to cell division, the two helices are identical to the old one (see figure 6.2).

The Two Strands of DNA Are Oriented in Opposite Directions

The two strands of the DNA helix wind around each other, with the two strands going in opposite directions. The two strands differ at their ends. One strand has a phosphate molecule attached to the number 5 carbon atom of the sugar, which is termed the *5 prime* (5') *end*. The other end has an OH group attached to the number 3 carbon atom of the sugar and is termed the *3 prime* (3') *end*. The 3' end is the growing end of the chain in DNA synthesis. The other strand, termed the *complementary strand*, goes in the opposite direction and has its 3' and 5' ends opposite to the 5' and 3' ends, respectively, of the other strand. The two strands are thus **antiparallel.**

RNA Is Usually Single-stranded and Not as Long as DNA

Ribonucleic acid also consists of a sequence of nucleotides, but unlike DNA, it usually exists as a single-stranded molecule. The uracil of RNA is equivalent to the thymine of DNA. Three different kinds of RNA function in the synthesis of protein and will be considered in the discussion of protein synthesis.

Information Storage and Transfer—Overview

DNA stores information in the sequence of its nucleotide subunits. This stored information is then transferred in cells by two different processes.

In one process, called **DNA replication,** the information in DNA is duplicated prior to cell division so that, after binary fission, both cells will have the same genetic information as the parental cell.

The second process is the transfer of information contained in the sequence of nucleotides in DNA into a sequence of amino acids in protein. The sequence of amino acids differentiates one protein from another. The functioning of a protein depends on its shape, which is determined by its amino acid sequence. Protein synthesis occurs in several steps. First, DNA determines the synthesis of a complementary strand of RNA, called **messenger RNA,** or **mRNA.** The mRNA, in turn, determines which amino acids are incorporated into protein. This flow of information from DNA

• **Binary fission, p. 88**
• **Protein structure, pp. 32, 33**

to RNA to protein is often referred to as the **central dogma of molecular biology** (figure 6.3). It was once believed that information flow proceeded only in this direction. Although this direction is the most common, it has now been shown that certain viruses, such as the one that causes AIDS, store their genetic information in RNA and transfer information from RNA to DNA. More recent studies have shown that this type of information flow also occurs in eukaryotic and prokaryotic cells to a limited extent.

The synthesis of a complementary strand of DNA as well as messenger RNA follows the same general pattern that is diagrammed in figure 6.4.

The sequence of nucleotides in the DNA serves as a **template,** or pattern, which is then copied into a complementary strand of either DNA, in the case of DNA replication, or RNA, in the case of messenger (mRNA) synthesis. The incorporation of nucleotides into newly synthesized strands requires the action of enzymes DNA polymerase and RNA polymerase in the case of DNA and RNA, respectively.

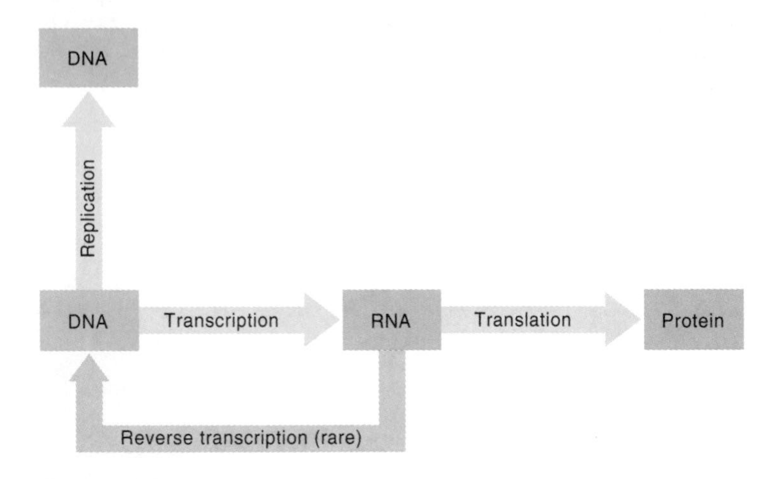

Figure 6.3
Flow of information in biological systems. None that the chemicals are not converted, but rather information is transferred in the process of transcription and translation. Reverse transcription is especially important in some viruses.

DNA Replication Begins at a Specific Site on the Bacterial Chromosome and Proceeds by the Sequential Addition of Nucleotides

To begin the replication of a chromosome, which is usually circular in bacteria, the two strands of DNA termed the **template strands** separate at a specific site termed the **origin of replication.** Synthesis of new strands begins at this site (figure 6.5 *a*). This separation of strands occurs at the **replication fork,** which progresses around the entire DNA molecule as the strands separate. As this fork moves, two new strands of DNA are synthesized, each one complementary to one of the original strands. In this process, the

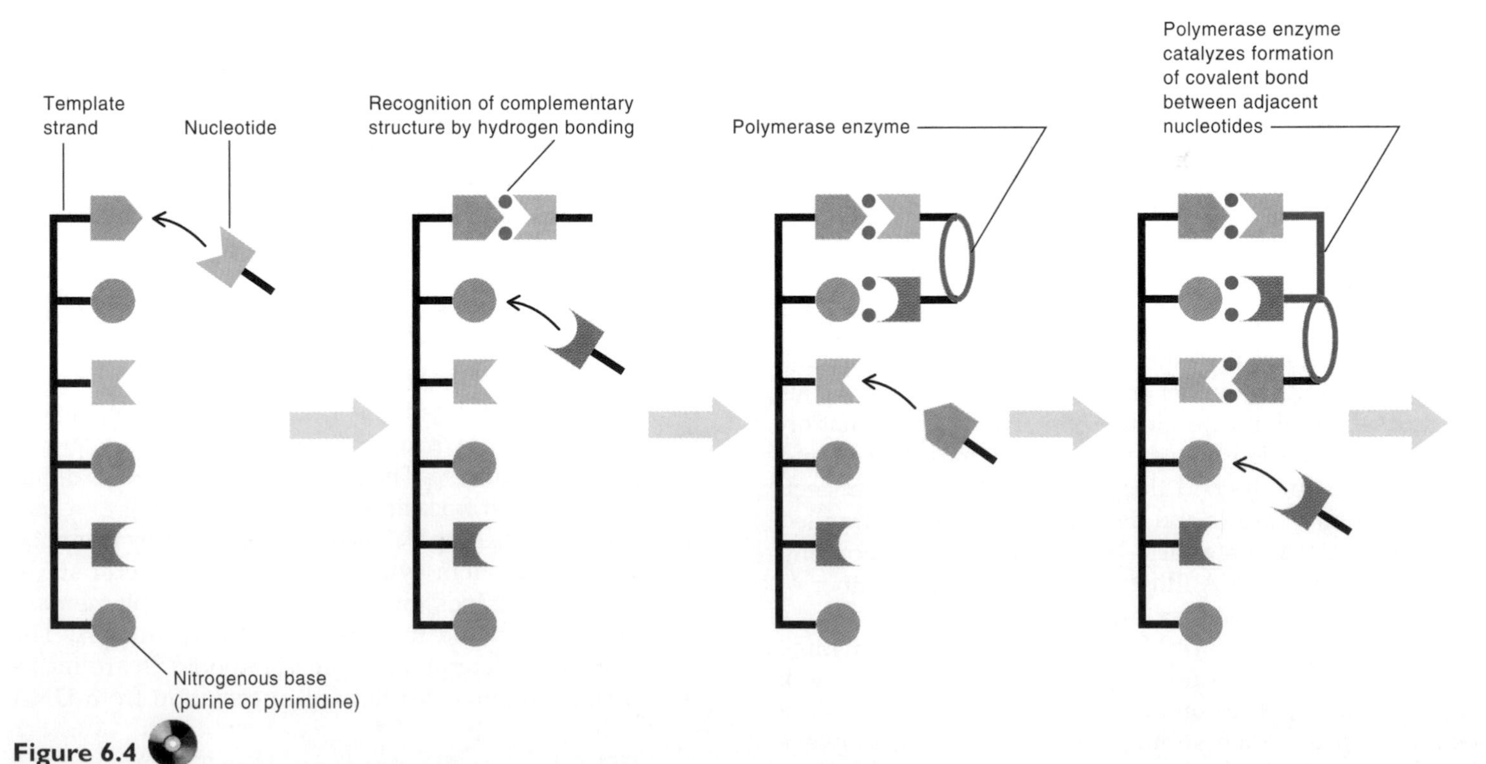

Figure 6.4
Sequence in template-directed nucleic acid synthesis. This general reaction mechanism diagrams the synthesis of DNA and RNA.

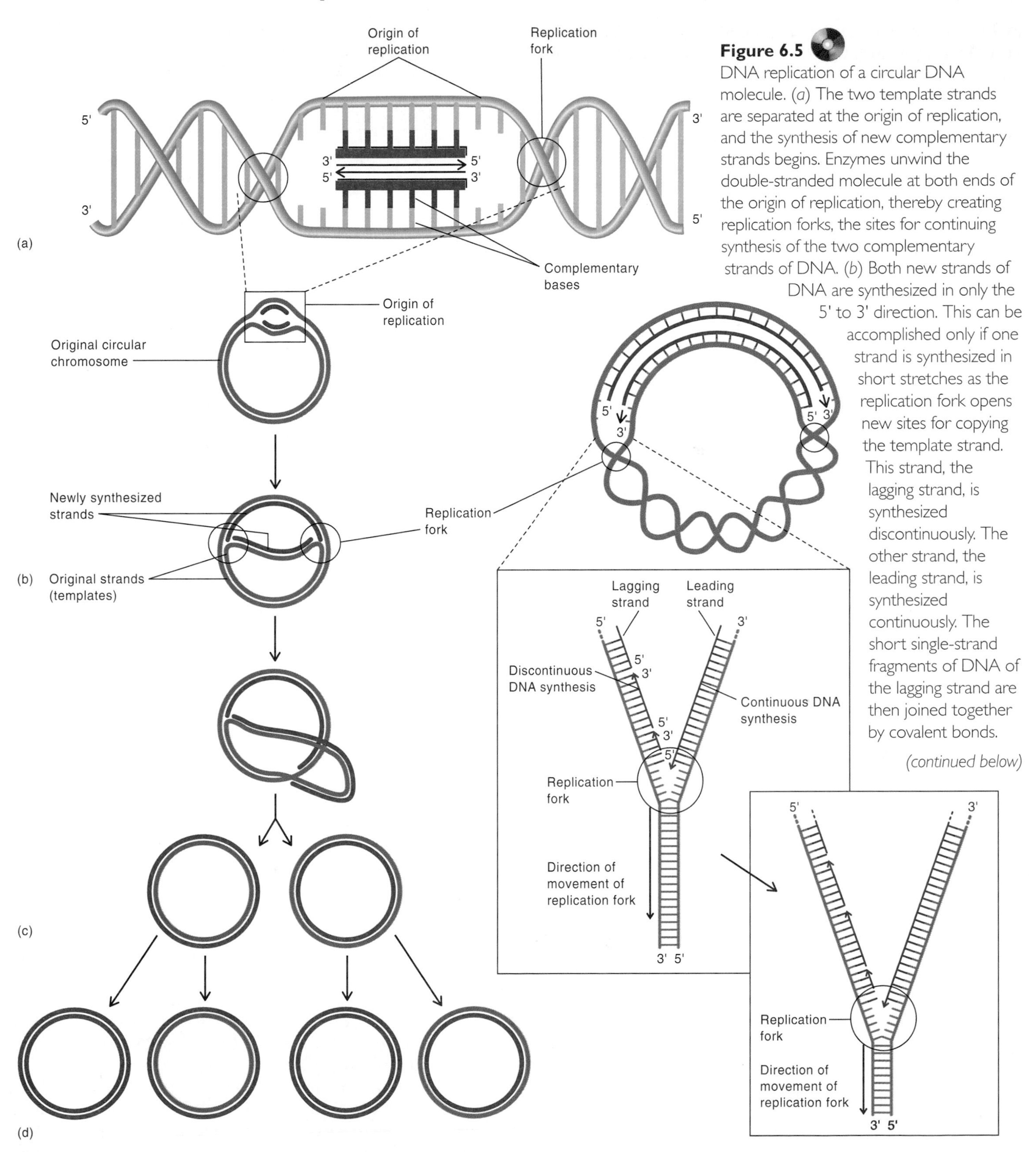

Figure 6.5 DNA replication of a circular DNA molecule. (*a*) The two template strands are separated at the origin of replication, and the synthesis of new complementary strands begins. Enzymes unwind the double-stranded molecule at both ends of the origin of replication, thereby creating replication forks, the sites for continuing synthesis of the two complementary strands of DNA. (*b*) Both new strands of DNA are synthesized in only the 5' to 3' direction. This can be accomplished only if one strand is synthesized in short stretches as the replication fork opens new sites for copying the template strand. This strand, the lagging strand, is synthesized discontinuously. The other strand, the leading strand, is synthesized continuously. The short single-strand fragments of DNA of the lagging strand are then joined together by covalent bonds.

(continued below)

(*c*) When the synthesis of two new strands of DNA has been completed, the two double-stranded molecules separate. Each double-stranded molecule is composed of one old strand and one newly synthesized strand. (*d*) Each DNA is replicated as indicated in steps *a* through *c*. The process involves many enzymes in addition to DNA polymerase. Only two of the molecules contain the original strands of DNA.

enzyme DNA polymerase sequentially adds nucleotides to the growing strand. The addition of each deoxyribonucleotide is always to the 3' end (containing the —OH groups). Thus, the DNA chain always grows in the 5' to 3' direction. The nucleotides that are added are determined by the fact that each nucleotide base must pair to its complement on the opposite pre-existing strand (figure 6.5 *b*).

When the replication fork has moved along the entire DNA molecule, two complete double-stranded DNA molecules are formed (figure 6.5 *c*). Since each of the two double-stranded molecules contains one of the original strands (the template strand) and one newly synthesized strand, this type of DNA replication is termed **semiconservative.** Because of this type of replication, only two of the daughter cells in a bacterial population have one of the original DNA strands that were present in the original parent cell, no matter how many cells have descended from this cell (figure 6.5 *d*).

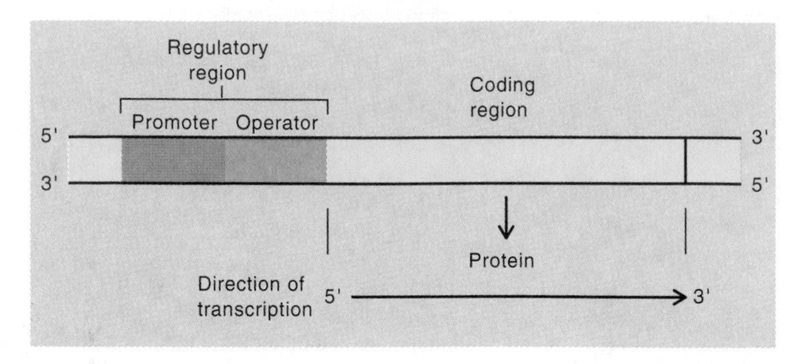

Figure 6.6
Diagram of a gene, showing its two regions. The coding region consists of the nucleotides that code for the sequence of amino acids that comprise the protein. The regulatory region plays a role in determining whether the gene is expressed.

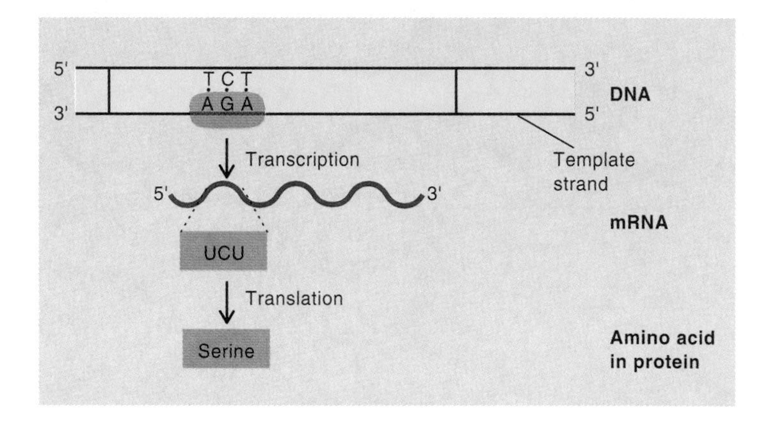

Figure 6.7
The transfer of information in a codon in the DNA to a codon in mRNA to an amino acid. Only the codons in the template strand of DNA are transcribed.

The Expression of Genes

The chromosome of *Escherichia coli* is divided into approximately 2,700 genes. Many of these genes consist of two major regions. One region determines, or codes for, the sequence of amino acids in one **polypeptide chain** or enzyme (the one gene–one enzyme hypothesis). Thus, the sequence of nucleotides determines the primary structure of the protein, which is important in conferring the distinctive properties on the protein, such as its shape. This region of the gene is commonly called the **coding region** (figure 6.6).

The second region determines whether the coding region will function in protein synthesis. This region contains nucleotide sequences to which **RNA polymerase,** a key enzyme involved in the information transfer from DNA to RNA, must bind in order to function. This region also contains a sequence of nucleotides to which other proteins bind; it thereby influences the function of RNA polymerase. It is called the **regulatory region** because it determines whether the gene will function; in other words, regulates gene expression. Two important sites exist in this region, the **promoter** and the **operator.** How they function in regulation will be considered shortly.

The Sequence of Three Nucleotides in mRNA Determines a Specific Amino Acid

The four different nucleotides in DNA and mRNA may be viewed as a four-letter alphabet whose sequences code for a message. Three nucleotides (letters) in a row spell out a word, which is one of the 20 common amino acid subunits of a protein molecule. The name **codon** is given to a set of three nucleotides.

• **Polypeptide, p. 32**
• **Primary structure, p. 32**

The sequences in DNA and RNA are equivalent because the first step in decoding the message in DNA is copying the sequence into a complementary strand of RNA, the **messenger RNA (mRNA),** the process of **transcription** (figure 6.7). The strand of DNA from which the mRNA is copied is the **template** strand.

The nucleotide code for each of the 20 amino acids is known. The codons and the amino acids they code for, termed the **genetic code,** are given in figure 6.8. The genetic code by convention is usually written as the sequence of nucleotides in messenger RNA rather than as the sequence in DNA. For example, in figure 6.8, the codon UCU in mRNA results in the amino acid serine being incorporated into the protein.

Note that no single codon specifies more than one amino acid although several codons specify the same amino acid. As this figure illustrates, three of the 64 triplets (UAA, UAG, and UGA) do not code for any amino acids. These codons, called **nonsense,** or **stop, codons,** serve to signal the end of the polypeptide chain and therefore are located at the end of every gene. They are transcribed into mRNA. A **start codon** in mRNA (AUG) signals the start of

	Middle Letter							
	U		C		A		G	
	5' 3'		5' 3'		5' 3'		5' 3'	
First Letter U	UUU	Phe	UCU	Ser	UAU	Tyr	UGU	Cys
	UUC	Phe	UCC	Ser	UAC	Tyr	UGC	Cys
	UUA	Leu	UCA	Ser	UAA	Nonsense	UGA	Nonsense
	UUG	Leu	UCG	Ser	UAG	Nonsense (Stop)	UGG	Trp
C	CUU	Leu	CCU	Pro	CAU	His	CGU	Arg
	CUC	Leu	CCC	Pro	CAC	His	CGC	Arg
	CUA	Leu	CCA	Pro	CAA	Gln	CGA	Arg
	CUG	Leu	CCG	Pro	CAG	Gln	CGG	Arg
A	AUU	Ile	ACU	Thr	AAU	Asn	AGU	Ser
	AUC	Ile	ACC	Thr	AAC	Asn	AGC	Ser
	AUA	Ile	ACA	Thr	AAA	Lys	AGA	Arg
	AUG	Met (Start)	ACG	Thr	AAG	Lys	AGG	Arg
G	GUU	Val	GCU	Ala	GAU	Asp	GGU	Gly
	GUC	Val	GCC	Ala	GAC	Asp	GGC	Gly
	GUA	Val	GCA	Ala	GAA	Glu	GGA	Gly
	GUG	Val	GCG	Ala	GAG	Glu	GGG	Gly

Figure 6.8
Genetic code. The dictionary of the mRNA genetic code words. The codons are read from left to right (5' → 3'). The three nonsense (stop) codons serve as punctuation marks between genes, and no tRNA recognizes them. The start codon codes for methionine (Met); thus, the first amino acid in almost all proteins is methionine. Note that many codons specify the same amino acid. The significance of this feature of the code is not known.

the protein. The people mainly responsible for "cracking" the genetic code were Marshall Nirenberg, an American biochemist working at the National Institutes of Health, and Har Gobind Khorana, working at the University of Wisconsin. In 1968, they were awarded the Nobel Prize for their extraordinary contributions to this area of molecular biology.

The First Step in Gene Expression Is Making an RNA Copy of the Genetic Information in DNA

Transcription involves three steps: (1) initiation of mRNA synthesis, (2) elongation of mRNA, and (3) termination of mRNA synthesis.

Initiation of Transcription

In the initiation step (figure 6.9), the enzyme RNA polymerase attaches to a specific sequence of nucleotides, the *promoter region,* in the regulatory region of the gene. This region contains the "start site" for RNA synthesis. Only one strand of the DNA, the **template** strand, is transcribed for any single gene. Which strand is transcribed is determined by the promoter region to which the RNA polymerase binds to start transcription. The sequence of nucleotides of the promoter is oriented and points the enzyme in one

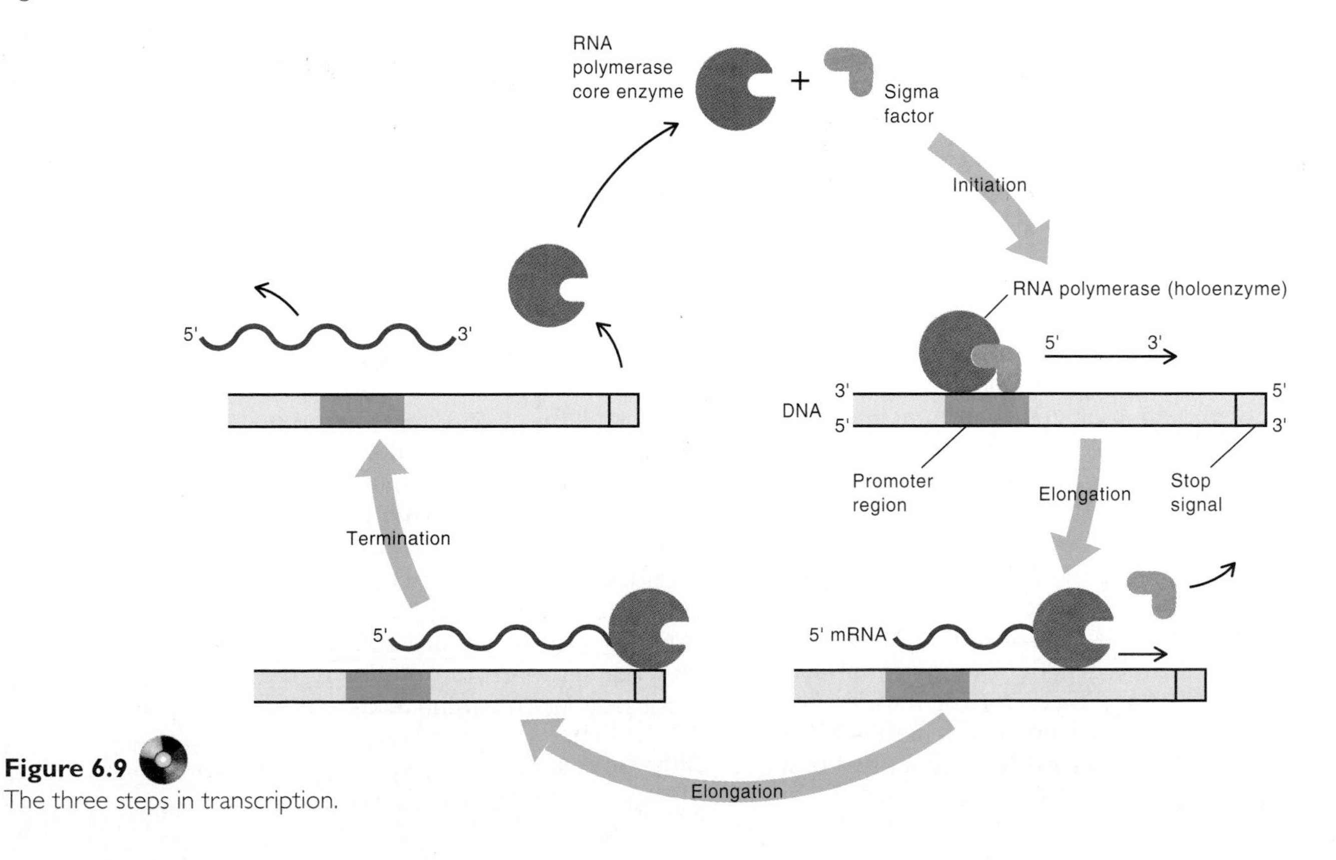

Figure 6.9
The three steps in transcription.

direction or the other. Because RNA strands are synthesized only in the 5' to 3' direction, the template strand must be oriented in the 3' to 5' direction. However, transcription of neighboring genes can occur on the opposite strand of DNA if the promoter is oriented in the opposite direction. Because genes are transcribed only in the 5' to 3' direction, the gene must be transcribed in the opposite direction.

RNA polymerase consists of several subunits that have different functions in transcription. The subunit termed *sigma* (σ) is especially important in the intiation of transcription. The σ subunit is a small protein that readily separates from the rest of the enzyme. When the enzyme lacks the σ factor, it is termed the **core** enzyme. When RNA polymerase contains the σ factor, it is termed the **holoenzyme.** The σ factor is responsible for the holoenzyme recognizing and binding to the promoter region through weak bonding forces. However, once transcription is intiated, the σ subunit dissociates from the RNA polymerase.

Elongation Phase

In the elongation phase (see figure 6.9), the RNA polymerase moves along the template strand of the DNA and synthesizes a single-stranded messenger RNA (mRNA) molecule, which is complementary to the strand of DNA along which it moves. After binding to the promoter, the RNA polymerase opens up a short stretch of double-stranded DNA to expose the nucleotides. The template strand then base pairs one at a time with complementary ribonucleotide triphosphate monomers, which are joined together by RNA polymerase. The enzyme moves stepwise along the DNA, unwinding the DNA and thereby exposing a new region of the template for complementary base pairing with mononucleotide triphosphates. Thus, the RNA chain elongates one nucleotide at a time in the 5' to 3' direction. The rate of polymerization is about 30 nucleotides per second. Therefore, synthesizing an RNA chain of 1,000 nucleotides, which codes for an average size protein, would take approximately 30 seconds.

Termination Phase

In the termination phase (see figure 6.9), the RNA polymerase comes to the end of the gene or genes and encounters a "stop" signal, a series of nucleotides that in some unexplained way cause the mRNA to drop from the DNA template.

Note that the "stop" signal, which causes the RNA polymerase and mRNA to separate from the DNA template is not the "stop" codon that is part of the genetic code. The stop signal in transcription is a sequence of nucleotides in the RNA that results in the mRNA forming a loop structure. The "stop" codon in the genetic code does not code for any amino acid (see next section on translation).

Genes vary in size, but their average length is about 1,000 nucleotides. Accordingly, the mRNA transcribed from a single gene of this length is also 1,000 nucleotides. Frequently in bacteria, several genes are transcribed as a single unit, termed an **operon,** and the stop signal occurs at the end of the unit. This mRNA is termed **polygenic mRNA** because it represents more than one gene. Polygenic mRNA is important in the regulation of gene expression and is discussed later in this chapter.

The Second Step in Gene Expression Is the Translation of mRNA into Protein

Once the message coded in a gene is transcribed into mRNA, the information contained in the mRNA is converted into a sequence of amino acids in the protein (polypeptide) molecule coded by the gene. About 90% of the total energy expended by a growing bacterial cell is devoted to protein synthesis. The process, termed **translation,** is very complex, involving several different enzymes, protein "factors," and two additional types of RNA, **transfer RNA (tRNA)** and **ribosomal RNA (rRNA).** These two types of RNA are also coded by the DNA and are transcribed by RNA polymerase. However, once transcribed, these two kinds of RNA are not translated because they do not code for any protein. The key molecule in translation is tRNA.

Transfer RNA (figure 6.10), is a molecule that has two important ends. One end, called the **anticodon,** has three nucleotides complementary to a particular codon in the mRNA. The other end has a specific amino acid covalently bonded to it. The tRNA is said to be "charged" if it carries an amino acid and "uncharged" if it does not. Each tRNA molecule has a different anticodon at one end and binds a different amino acid at the other end. Since most amino acids are coded for by more than one codon (see figure 6.8), a single tRNA molecule with its anticodon can recognize more than one codon that codes for the same amino acid. Note that any single codon codes for only one amino acid.

Initiation of Translation

Like transcription, translation also can be divided into three steps (figure 6.11): (1) initiation, (2) elongation, and (3) termination. During initiation, the small 30S subunit of the ribosome binds to a short nucleotide sequence termed the **ribosome binding site,** which is located in the mRNA near the site at which translation will begin. Three nucleotides that code for methionine serve as the **start signal** of translation, which explains why the first amino acid in virtually all proteins in bacteria is methionine (see figure 6.8). The series of nucleotides in the mRNA is translated into a sequence of amino acids, one codon at a time.

• **Ribosome, p. 72**

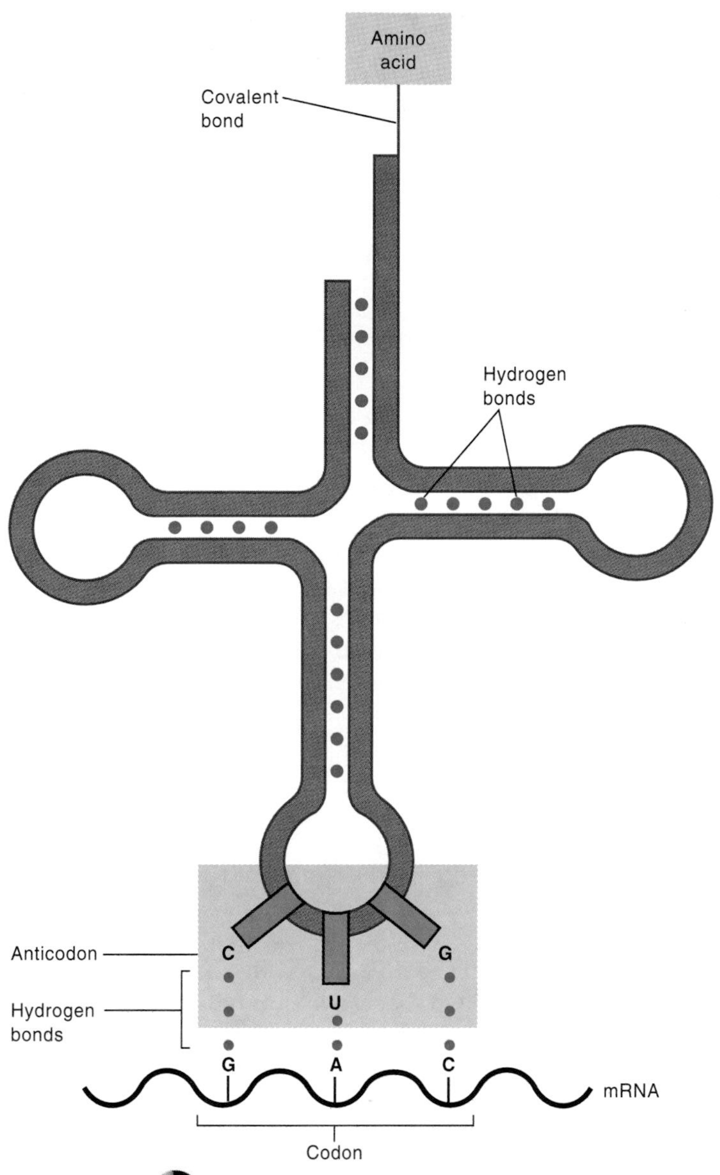

Figure 6.10 Transfer RNA. The different transfer RNA molecules that have different anticodons have different amino acids covalently bound to the other end. Because the genetic code is degenerate, transfer RNA molecules that have different codons can carry the same amino acid. However, all transfer RNA molecules that have the same anticodons bind the same amino acids.

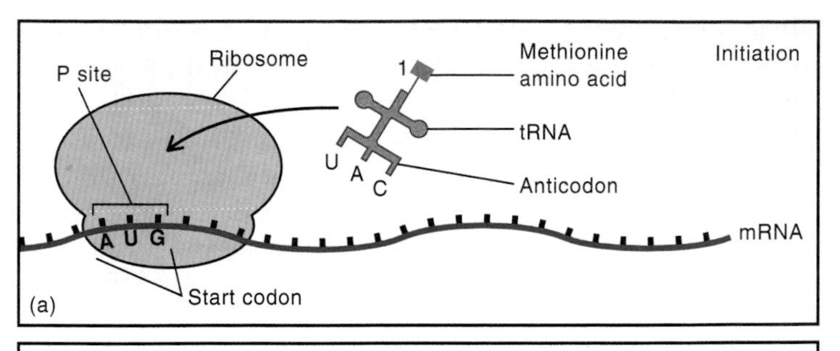

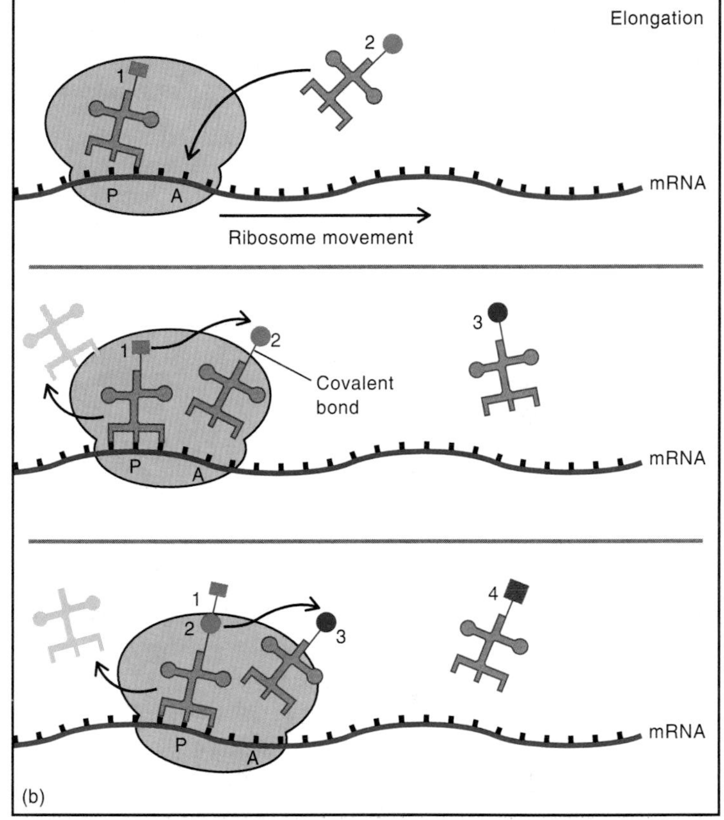

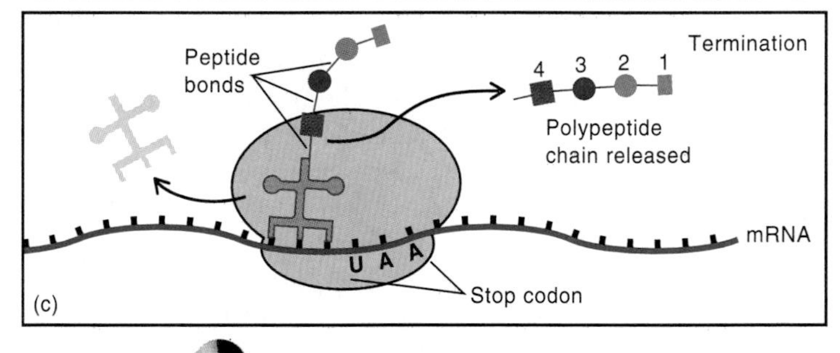

Figure 6.11 The three steps in translation of mRNA into amino acids in a polypeptide chain.

Elongation

During elongation (figure 6.11 *b*), the charged tRNA can bind to mRNA at two sites on the large 50S subunit of the ribosome. The first tRNA molecule involved in the synthesis of a polypeptide, charged with the amino acid methionine, binds to the **peptide,** or **P, site.** This binding occurs through hydrogen bonding between the three nucleotides in the mRNA codon and the three complementary nucleotides in the tRNA anticodon. The second site, termed the **acceptor,** or **A, site,** is initially unoccupied. However, once the methionine-charged tRNA molecule binds at the P site, a second charged tRNA molecule whose anticodon is complementary to the second codon in the mRNA binds to the

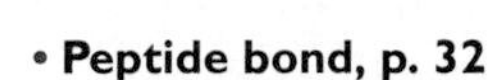
• **Peptide bond, p. 32**

A site. Next, the tRNA at the P site transfers its amino acid to the adjacent amino acid in the A site, and an enzyme joins the two amino acids through a peptide bond, leaving an uncharged tRNA in the P site. This tRNA leaves its P site on the ribosome and once again becomes charged with methionine. Note that in this entire process, the ribosome serves as the structure that brings the two amino acids to be joined into a favorable position in which the enzyme joining them can act easily to form a peptide bond. The tRNA acts only to transport amino acids to the ribosome. The amino acid that enters the A site is determined by the mRNA codon that is present.

A "ratchet" type of mechanism moves the mRNA along the ribosome a distance of one codon (three nucleotides) so that the second tRNA molecule, now carrying two amino acids bonded in a peptide bond, moves to the P site. Site A is now open to hydrogen bond another charged tRNA molecule whose anticodon is complementary to the codon on the mRNA. The two amino acids in the growing chain in site A form a peptide bond to the amino acid occupying site P. Once again, the uncharged tRNA at site P leaves the mRNA, becomes recharged, and returns to an open A site that has the proper codon to match its anticodon.

The peptide chain grows by the sequential addition of amino acids as the mRNA moves along the ribosome. The movement of the charged tRNAs to the A site and the movement of the growing peptide chain to the P site require the participation of a number of proteins called **elongation factors.** One of these factors, EF-2, is the target for the toxin synthesized by the bacterium that causes diphtheria. The toxin binds to EF-2 thereby inactivating it and thus stops protein synthesis. Consequently, the cell dies, resulting in the symptoms of diphtheria. At any one time, five or six ribosomes, termed **polyribosomes,** may be on the same mRNA molecule (figure 6.12), greatly increasing the speed with which mRNA is translated into many protein molecules. Note, however, that each ribosome is involved with the synthesis of only one protein molecule at any one time. The process of translation is very efficient. In one second, 20 amino acids can be added to the growing polypeptide chain. Since the first amino acid incorporated into the growing peptide chain has a free amino group (—NH_2), and the last amino acid has a free carboxyl (—COOH) group, the protein chain grows from the **amino terminal end** to the **carboxyl end.** Ribosomes closest to the beginning of the mRNA have the shortest polypeptide chain. The synthesis of a typical protein requires that each of the steps in elongation be repeated about 300 times, since there are about 300 amino acids in an average protein.

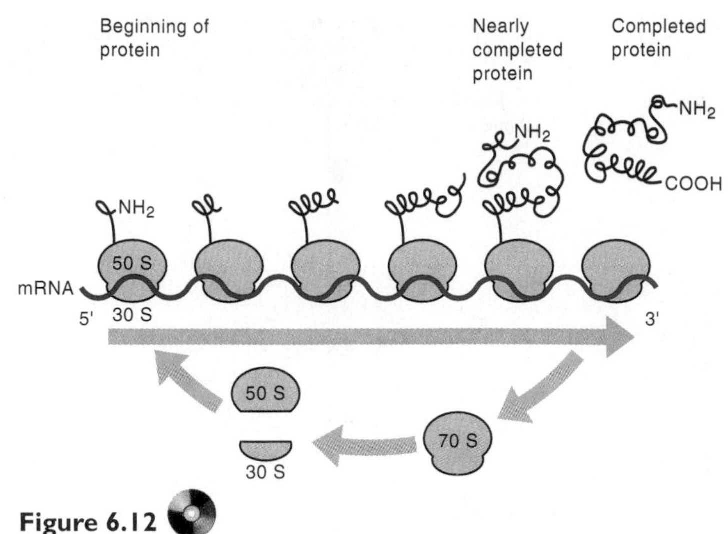

Figure 6.12
Polyribosomes. More than one ribosome is attached to the same mRNA molecule, but all ribosomes are involved in synthesizing the same protein as they move on the same RNA molecule. The closer to the end of the mRNA molecule the ribosome is, the more complete the protein along the mRNA that is being synthesized. The ribosomes move in the 5' to 3' direction.

Termination

During termination (see figure 6.11 *c*), the ribosome reaches a nonsense codon that serves as a "stop" signal (see figure 6.8). Recall that three codons of the genetic code do not code for any amino acid. At this "stop" signal, protein **release factors** promote the dissociation of the completed polypeptide chain from the tRNA molecule by breaking the covalent bond between the polypeptide and the tRNA at the P site. Also, the ribosome falls off the mRNA and dissociates into its two subunits, 50S and 30S. These two subunits join once again when the ribosome binds to the mRNA at the ribosome binding site.

Differences Between Prokaryotic and Eukaryotic Gene Transcription/Translation

Significant differences exist in transcription/translation in prokaryotes and eukaryotes. In prokaryotic cells, which have no nucleus, the processes of mRNA transcription and translation occur close together in time and space. As soon as the mRNA starts being synthesized, it binds to a ribosome, so that one end of an mRNA molecule is being translated while the other end is being synthesized from the DNA template. The processes of transcription and translation are coupled (figure 6.13).

The situation is very different in eukaryotic cells. The mRNA must be transported out of the nucleus, which is surrounded by a nuclear membrane, before it can be translated in the cytoplasm. Thus, the same mRNA molecule cannot be transcribed and translated at the same time.

Perspective 6.1

Wonder Foods and Virus Resistance

Learning how DNA replicates and transfers its genetic information into a sequence of amino acids in a protein molecule has been a remarkable achievement accomplished through the efforts of a large number of scientists. Knowing some of the details of this process has led to some unexpected practical applications, one of the most recent and exciting of which is **antisense technology.** This chapter explains that only one strand of DNA is transcribed into a single strand of mRNA. This mRNA strand that is synthesized is often termed the **sense** strand since it is translated into a sequence of amino acids.

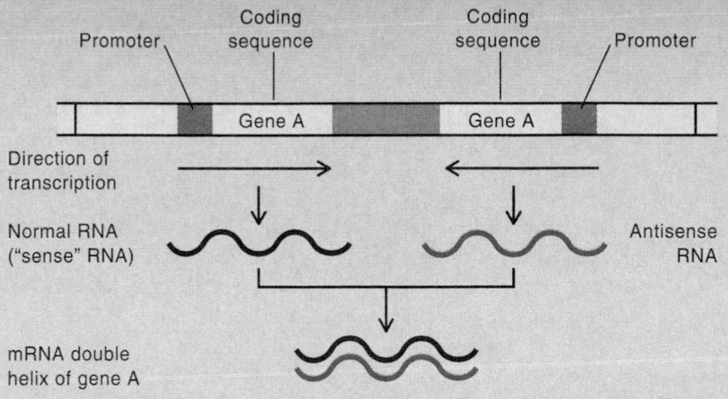

However, if the coding region of this same strand of DNA is flipped over, the same strand will be transcribed in the opposite direction. This mRNA strand, termed **antisense,** is complementary to the sense strand and therefore can associate with it to form a double-stranded RNA molecule (figure 1). The result of this process is that this molecule of mRNA cannot be translated, so the gene coding for this particular mRNA does not synthesize a protein.

This unusual control mechanism is actually used as a mechanism of control in a few bacteria. Investigators at biotechnology companies are also putting this mechanism to practical use. For example, a plant biotechnology company has genetically engineered tomato plants to synthesize antisense RNA of the gene that codes for the enzyme polygalacturonase. This plant enzyme breaks down plant cell walls and is responsible for the mushiness of ripe tomatoes a few days after they are picked. As a result, the tomatoes with antisense RNA to polygalacturonase do not get mushy for several weeks after they are picked since the antisense RNA does not allow the polygalacturonase to be synthesized.

This technology probably has many more applications. If it is possible to turn off any gene at will, it may be possible to turn off genes required for the multiplication of human viruses, such as those that cause AIDS, herpes, and certain leukemias. This technology has already been used to protect plants against certain viruses.

The use of antisense RNA is another example of how understanding fundamental principles of biological processes can lead to important practical applications.

Figure 1
Formation of sense-antisense RNA double helix. The formation of the double helix prevents the sense mRNA from being translated into protein.

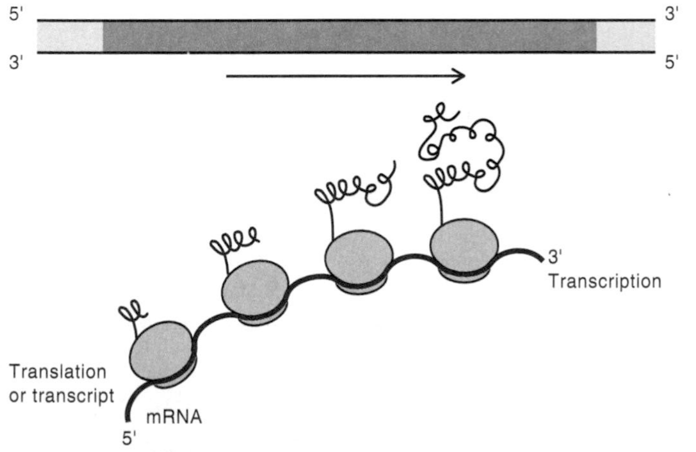

Figure 6.13
Polyribosomes. Many ribosomes can be bound to the same mRNA molecules. Binding is always at the 5' end of the mRNA. As the ribosome moves along the mRNA, another can bind. The ribosomes closest to the 3' end have synthesized the longest polypeptide chain.

Another major difference is that, in eukaryotic cells, most mRNA molecules are extensively modified in the nucleus after transcription is completed and before translation has begun. Apparently, most genes in eukaryotes contain blocks of DNA that do not code for amino acids in proteins, although this DNA is transcribed. However, the segments of mRNA coded by this DNA are removed from the mRNA before it leaves the nucleus and is translated on the ribosomes in the cytoplasm (figure 6.14). The piece of RNA that is removed is termed an **intron** (figure 6.14). In most cases, introns are junk DNA because they do not code for any proteins. After the removal of the introns, the ends of the mRNA molecules are rejoined, resulting in the mRNA molecule that is translated. This whole process is called **mRNA splicing.** The discoverers of RNA splicing, Phillip Sharp and Richard Roberts, were awarded the Nobel Prize in Medicine in 1993.

RNA splicing has been observed in a few genes of members of the Eubacteria and the Archaea as well as in viruses that infect bacteria, but it is extremely rare.

Perspective 6.2

RNA—The First Macromolecule?

The 1989 Nobel Prize in Chemistry was awarded to two Americans, Sidney Altman of Yale University and Thomas Cech of the University of Colorado, who independently made the surprising and completely unexpected observation that RNA molecules can act as enzymes. Before their studies, it was believed that only proteins had enzymatic activity. The key observation was made by Cech in 1982 when he was trying to understand how introns were removed from mRNA that coded for ribosomal RNA in the eukaryotic protozoan *Tetrahymena.* Since he was convinced that proteins were responsible for cutting out these introns, he added all of the protein in the cells' nuclei to the mRNA that still contained the introns.

As expected, the introns were cut out. As a control, Cech looked at the ribosomal RNA to which no nuclear proteins had been added, fully expecting that nothing would happen. Much to his surprise, the introns were also removed. It did not make any difference whether the protein was present—the introns were removed regardless. Thus, Cech could only conclude that the RNA acted on itself to cut out pieces of RNA.

The question of how widespread this phenomenon was remained. Did RNA have catalytic properties other than that of cutting out introns from rRNA? The studies of Altman and his colleagues, carried out simultaneously to and independently of Cech's, provided answers to these further questions. Altman's group found that RNA can convert a tRNA molecule from a precursor form to its final functional state. Additional studies have shown that enzymatic reactions in which catalytic RNAs, termed **ribozymes,** play a role are very widespread. Ribozymes have been shown to occur in the mitochondria of eukaryotic cells and to catalyze other reactions that resemble the polymerization of RNA. Whether catalytic RNA cuts out introns from mRNA in the nucleus is not known.

These observations have profound implications for evolution: Which came first, proteins or nucleic acids? The answer seems to be that nucleic acids came first, specifically RNA, which acted both as a carrier of genetic information as well as an enzyme. Billions of years ago, before the present universe in which DNA, RNA, and protein are found, probably the only macromolecule that existed was RNA. Once tRNA became available, these adapters could carry amino acids present in the environment to specific nucleotide sequences on a strand of RNA. Thus, in this scenario, the RNA nucleotides are the genes as well as the mRNA which codes for the sequence of amino acids in proteins.

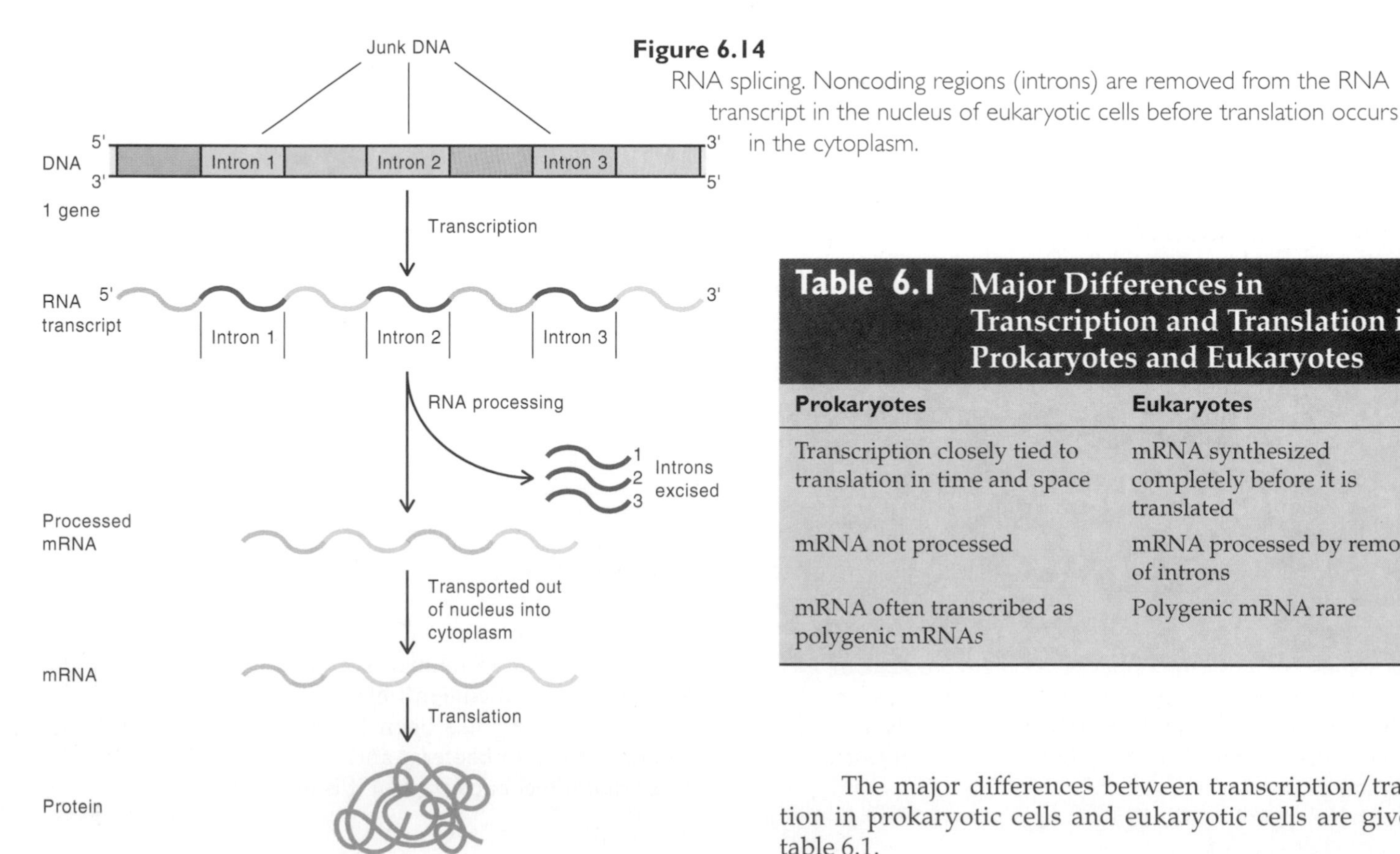

Figure 6.14
RNA splicing. Noncoding regions (introns) are removed from the RNA transcript in the nucleus of eukaryotic cells before translation occurs in the cytoplasm.

Table 6.1 Major Differences in Transcription and Translation in Prokaryotes and Eukaryotes

Prokaryotes	Eukaryotes
Transcription closely tied to translation in time and space	mRNA synthesized completely before it is translated
mRNA not processed	mRNA processed by removal of introns
mRNA often transcribed as polygenic mRNAs	Polygenic mRNA rare

The major differences between transcription/translation in prokaryotic cells and eukaryotic cells are given in table 6.1.

REVIEW QUESTIONS

1. Explain how the nucleotide sequence in one strand of DNA in a double-stranded DNA molecule determines the sequence of nucleotides in the other strand.
2. Name the two ways that DNA transfers information.
3. Name the three kinds of RNA. Which one is the decoder?
4. What is the region in DNA to which RNA ploymerase binds?
5. Give two ways in which transcription/translation differ in eukaryotic and prokaryotic cells.

The Environment and Control Systems—Overview

To survive in any environment, microorganisms must often reproduce more rapidly than other organisms in the same environment. The supply of available energy is generally the limiting factor in bacterial growth in nature, making it crucial that the bacterial cell synthesize the maximum amount of cell material from a limited supply of energy. Since biosynthetic reactions generally require energy, cells have developed very elaborate mechanisms for the control of their biosynthetic pathways. Specifically, cells shut down these pathways when the products do not need to be synthesized. Likewise, cells do not synthesize degradative enzymes unless the enzymes' substrates are present in the environment.

The control mechanisms that microorganisms have evolved take into account three fundamental principles related to their environment. First, if the organism can utilize any subunits of macromolecules that may be present in the environment, energy can be conserved. It takes far less energy for a cell to take in an amino acid than to synthesize it. This difference in energy output explains the reason that cells shut down biosynthetic pathways when the products of these pathways are available to them. Second, because the environment can change drastically, often in a matter of minutes, such control mechanisms must be reversible. Third, because of the great importance of the pathways of central metabolism, these pathways are not subject to the controls discussed in biosynthetic and degradative reactions.

Consider the situation of *Escherichia coli.* For over 100 million years, this organism has inhabited the gut of mammals, where it reaches concentrations of 10^6 cells per milliliter. Obviously, it is very well adapted to this particular niche in nature. In this same niche, however, it is subjected to alternating periods of feast and famine. For a limited time following a meal, the *E. coli* cells prosper, wallowing in a medium composed of amino acids, vitamins, purines, pyrimidines, and other end products of biosynthetic pathways. The cells actively take up these metabolites through their uptake systems, which require a minimal expenditure of energy. Simultaneously, the cells shut down their own biosynthetic pathways, conserving their energy in order to channel it into the rapid synthesis of macromolecules. Under these conditions, the cells divide at their most rapid rate. However, famine follows the feast. Between meals, which may be many days in the case of some mammals, the end products of metabolism are not available. Therefore, the cells' biosynthetic pathways must be activated to synthesize the different components of macromolecules. Cell division then markedly slows down because energy is drained off for the synthesis of these small component molecules. Thus, whereas cells can divide several times an hour in an environment composed of rich nutrients, they may divide only once every 24 hours in a famished mammalian gut.

• **Central metabolism, p. 65**
• **Uptake systems, p. 65**

Mechanisms of Control of Biosynthetic Pathways—Overview

Biosynthetic pathways are controlled by two independent mechanisms. One, termed **end product repression,** involves the control of gene expression. In this mechanism, the end product of a pathway prevents the synthesis of all of the enzymes of the pathway (figure 6.15). The second

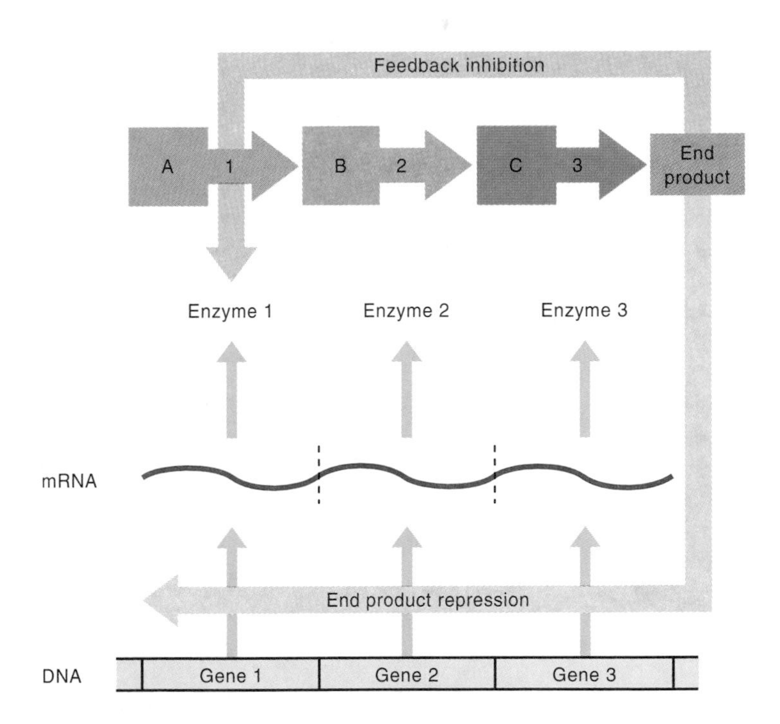

Figure 6.15
Two mechanisms by which biosynthetic pathways are controlled. The end product of the pathway can inhibit the expression of all genes of the pathway by inhibiting the synthesis of mRNA (end product repression). The end product can also inhibit the enzyme activity of the first enzyme of the pathway (feedback inhibition). In this diagram, the three genes form an operon, so a single repressor molecule in conjunction with the corepressor (end product) can inhibit synthesis of all three enzymes by inhibiting the transcription of the first enzyme. This assumes that there are no promoter regions between the three genes to which RNA polymerase can bind to initiate transcription.

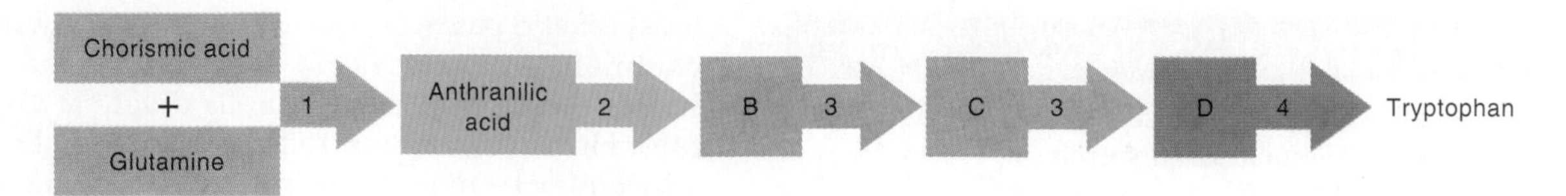

Figure 6.16
Pathway of tryptophan biosynthesis. The numbers indicate the enzymes, and the letters the intermediate compounds in the pathway. The same protein (enzyme 3) catalyzes two different sequential reactions, an unusual situation.

Table 6.2 Comparison of End Product Repression, Enzyme Induction, and Allosteric Control

Name of Control Mechanism	Mechanism of Action of Inhibitor or Activator	Pathways Involved	Inhibitor or Activator	Enzymes Involved	Speed of Control Mechanism
End product repression	Gene transcription inhibited	Biosynthetic	End product	All enzymes of pathway	Slow
Enzyme induction	Gene transcription activated	Degradative	First substrate	All enzymes of pathway	Slow
Allosteric control	Enzyme activity inhibited	Biosynthetic	End product	First enzyme of pathway	Rapid

mechanism, termed **feedback,** or **allosteric, inhibition,** involves the inhibition of only the first enzyme of the biosynthetic pathway by the end product of the pathway (see figure 6.15).

In the following discussion, we consider the biosynthesis of the amino acid tryptophan, which involves five enzymatic steps (figure 6.16). More is known about this pathway and its regulation in bacteria than about almost any other biosynthetic pathway. Therefore, it serves as a model system for understanding how small molecules such as amino acids, purines, pyrimidines, and vitamins control their own synthesis.

End Product Repression Controls the Start of Gene Transcription

Although the mechanism of end product repression may vary for different amino acids (table 6.2), the general principles are the same for all such mechanisms. To understand the mechanism of inhibition, it is necessary to know something about how the genes responsible for tryptophan biosynthesis are organized. In the case of tryptophan biosynthesis of *E. coli,* the five genes that code for the enzyme are next to one another on the chromosome (figure 6.17).

Transcription begins at the left end, transcribing gene 1 first. It ends after gene 5 is transcribed. Thus, the entire set of genes is transcribed into mRNA as a single unit, an example of a polygenic message. Another gene, a **regulatory gene,** codes for a protein that is not an enzyme. Rather, it codes for

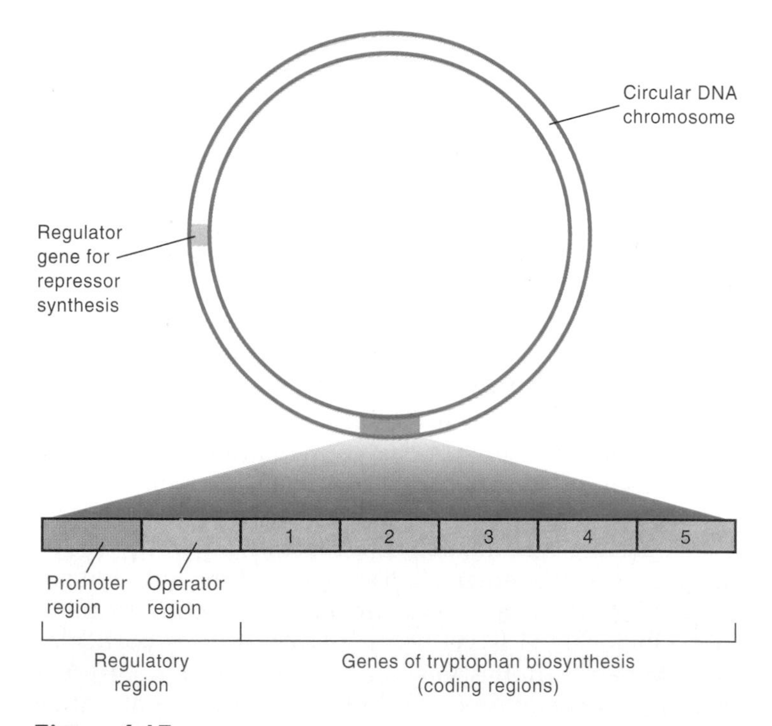

Figure 6.17
Genes of tryptophan biosynthesis. All of the genes that code for the enzymes of tryptophan biosynthesis are close together on the chromosome. The regulatory gene that codes for a repressor protein is not located close to these genes. The operator region and the promoter region are contiguous and next to the genes of tryptophan biosynthesis. In some pathways, the regulatory gene for repressor synthesis may be linked to the structural genes.

the synthesis of a **regulatory,** or **repressor, protein** that, under certain conditions, binds to a specific site on the DNA next to the promoter region, the site to which RNA polymerase binds to start transcription of the tryptophan genes. The DNA site to which the repressor protein binds the **operator region,** is part of the regulatory region of the gene.

In the absence of tryptophan, the repressor protein does not have the proper shape to bind to the operator region. RNA polymerase binds to the promoter region and then moves along the five genes, transcribing all of them and forming a polygenic mRNA (figure 6.18 *a*). However, if tryptophan is present in the medium, it attaches to the repressor protein, thereby changing its shape. The protein with its altered shape can now bind to the operator region. The repressor protein bound to the operator region gets in the way of RNA polymerase and prevents its binding to the promoter region (figure 6.18 *b*). Since transcription cannot start, none of the tryptophan biosynthetic enzymes are synthesized. Thus, a single repressor molecule can prevent the expression of all of the genes that are located next to one another and that are thus linked together. In bacteria, genes that are involved in the same biosynthetic pathway are often situated very close to one another. This arrangement allows a single repressor molecule to control transcription of all of the genes of the pathway. The name **operon** is given to a set of genes that are linked together and transcribed as a single unit. This clustering extends to many other genes that have related functions. For example, pathogenic bacteria often have their genes concerned with disease production in one or more operons termed **pathogenicity islands.** Although operons are very common in bacteria, they are not common in eukaryotic cells (see table 6.1).

The mechanism of regulation, like all other biologically important regulatory mechanisms, is reversible and depends on the presence of the end product. In addition, it is specific in that tryptophan regulates only the synthesis of tryptophan enzymes and not enzymes of other pathways. Each biosynthetic pathway has its own regulatory (repressor) protein that functions only in that pathway. Pathways controlled by end product repression are referred to as **repressible pathways.**

End Products Control the Activity of the First Enzyme of the Pathway Through Allosteric Control

To control synthesis of an end product through end product repression is a slow process. Although the synthesis of the enzyme stops immediately when an end product becomes available, the enzyme molecules in the cell prior to the addition of the end product still continue to function. Therefore, cells have an additional mechanism for controlling synthesis of end products that takes place immediately. This mechanism involves the inhibition of the first enzyme of the pathway rather than the synthesis of all enzymes (figure 6.19).

All enzymes have an active, or catalytic, site. In addition, the first enzyme of most biosynthetic pathways has a site that combines with the end product of the pathway. This site is called the **allosteric site,** and the end product is called a **feedback inhibitor.** The combination of the end product of the pathway, in this case tryptophan, with the allosteric site of the enzyme through weak bonding forces changes the shape (conformation) of the enzyme molecule so that it can no longer catalyze its reaction (see figure 6.19). Through this mechanism, the end product prevents the formation of the product of the first and any subsequent steps of the pathway, and the whole pathway is shut down immediately. Because the substrates on which the enzymes act are not available, the enzymes cannot function.

Mechanisms of Control of Degradative Pathways

In biosynthetic pathways, the end product turns off gene expression and inhibits the first enzyme of the pathway. In degradative pathways, the first compound to be metabolized often activates gene transcription but does not affect enzyme activity.

Metabolites that Can Be Degraded Activate the Genes Responsible for Their Degradation

The enzymes coded by genes in degradative pathways are called **inducible enzymes,** and the process of activating their transcription is called **enzyme induction.** The metabolites that activate gene transcription are called **inducers.** Inducers are usually, but not always, compounds that cells can use as a source of energy and carbon. In the absence of an inducer, the cells have no need to synthesize the enzymes required for the inducer's breakdown, just as they have no need to synthesize compounds provided to them in the environment.

The most thoroughly studied case of induced enzyme synthesis in bacteria involves the degradation of the di-saccharide lactose in *E. coli.* This system serves as a model for most other inducible systems. Lactose, the most abundant sugar in milk, induces two enzymes required for its breakdown: a permease that transports lactose into the cell and an enzyme (β-galactosidase) that breaks the bond joining the two component monosaccharides, glucose, and galactose. The monosaccharides can then be broken down through the glycolytic pathway to provide energy and precursor molecules.

The mechanism of enzyme induction (figure 6.20) is similar to enzyme repression. The same genetic elements are present: a regulatory gene that synthesizes a repressor

• **Enzyme catalytic site, p. 110**
• **Lactose, p. 36**

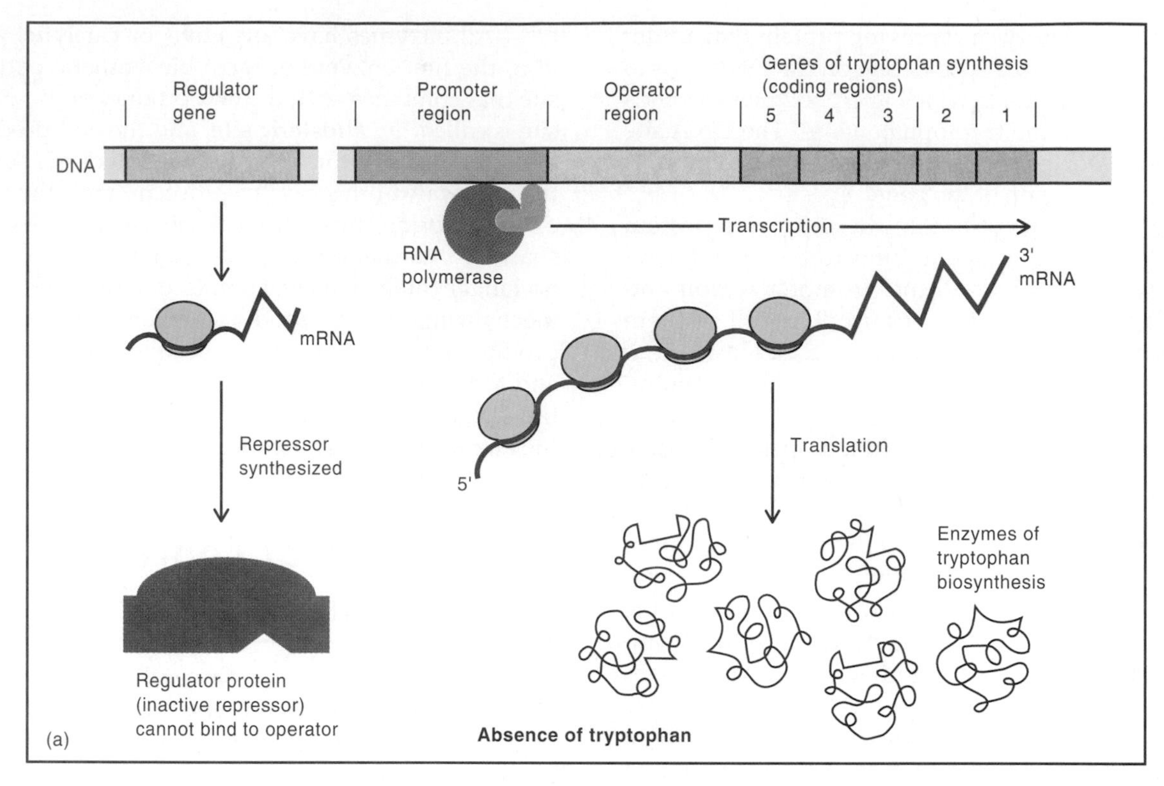

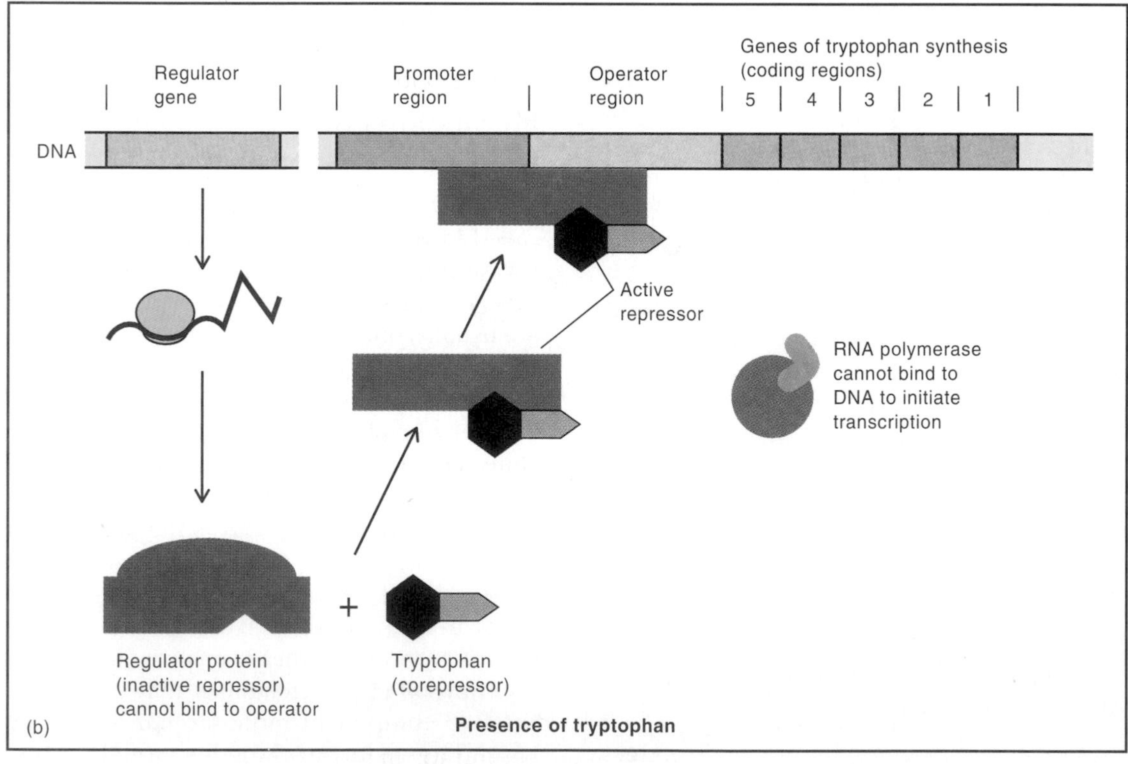

Figure 6.18
Regulation of tryptophan biosynthesis by the corepressor tryptophan. The genes of tryptophan biosynthesis are not expressed when tryptophan is in the environment. The active repressor bound to the operator region of the DNA prevents the RNA polymerase from binding to the promoter region. Thus, transcription is prevented.

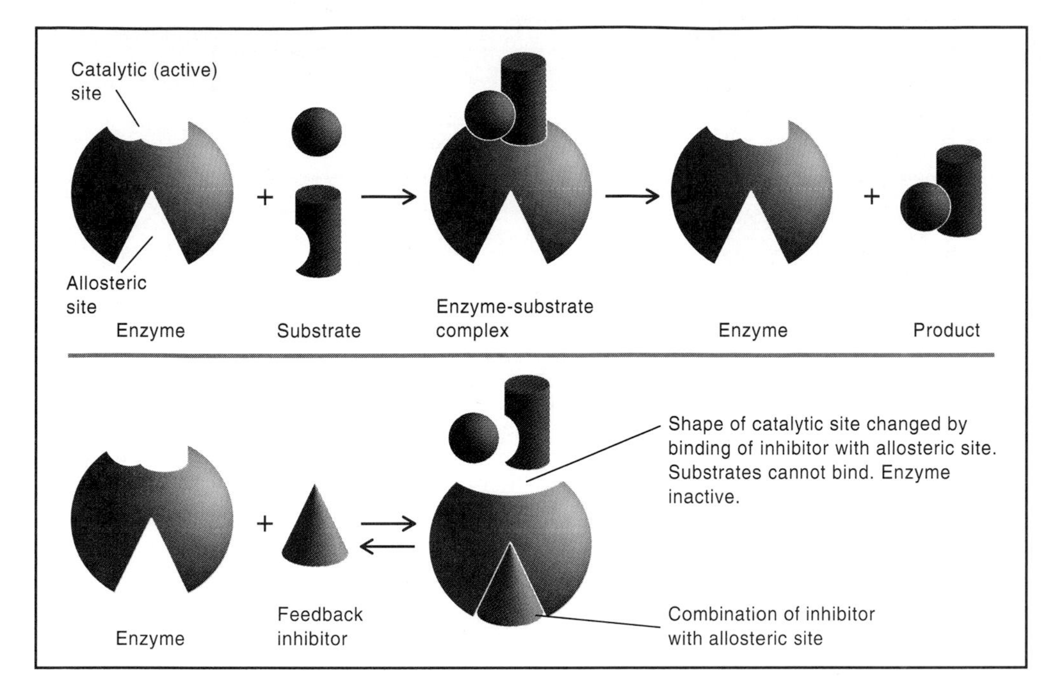

Figure 6.19
Inhibition of enzyme activity. The end product combines with the first enzyme of the pathwayconcerned with the synthesis of that product, thereby changing the shape of the catalytic(active) site and rendering the enzyme nonfunctional.

protein; an operator region next to the beginning of the first gene to be transcribed (that codes for β-galactosidase), to which the repressor protein binds; and an additional gene that codes for the β-galactosidase permease. These latter two genes form an operon. The single major difference between enzyme repression and induction is that, in the case of inducible enzymes, the repressor protein binds to the operator region in the absence of the inducer. When the inducer is present, it binds to the repressor protein, thereby changing its shape and inactivating it. Accordingly, the repressor does not attach to the operator region, and the genes of the lactose operon are transcribed. If the inducer is added to a bacterial culture, it can inactivate repressor molecules already bound to the operator, thereby causing the repressor to come off of the operator, allowing transcription to proceed.

Another inducible system is of great significance in medical microbiology. Many species of bacteria are resistant to the antibiotic penicillin because they can synthesize an inducible enzyme, **penicillinase,** that destroys the penicillin before it can kill the cell. Penicillin is the inducer in this system. As in the β-galactosidase system, the enzyme is synthesized only when it is needed.

The two mechanisms by which biosynthetic and degradative pathways are controlled are compared in table 6.2.

Catabolite Repression Ensures Efficient Use of Available Foodstuffs

The enzymes of the pathways of central metabolism are not subject to the controls discussed. Their levels remain essentially constant under varying nutritional conditions. Such enzymes are termed **constitutive.** As discussed, the breakdown of lactose requires enzyme induction. An interesting question arises: When both glucose and lactose are available to *E. Coli,* does the cell metabolize both sugars simultaneously or one after the other? The answer to this question, posed by the French scientist Jacques Monod in the early 1940s, revealed a new aspect of control in prokaryotes. It demonstrated the elegant mechanisms that these "simple" organisms employ in order to ensure efficient use of available foodstuffs to supply energy.

In *E. coli,* glucose is always metabolized first; after it disappears, lactose is degraded. Thus, in a medium containing both lactose and glucose, cells exhibit a two-step growth response termed **diauxic growth** (figure 6.21).

Glucose prevents growth on lactose by repressing the synthesis of enzymes of the lactose operon. Studies have shown that glucose represses synthesis of a large number of other inducible enzymes, in a phenomenon called **catabolite repression.** In contrast to tryptophan, which represses only the enzymes of its own pathway, glucose is said to be a **global repressor,** because it represses the synthesis of a large number of enzymes in many different pathways.

The overall features of glucose repression of the synthesis of many different enzymes are now partly understood. The repression by glucose results from the reduction in the level of the nucleotide **cyclic AMP (cAMP),** which is required for enzyme induction (figure 6.22), probably by inhibiting an enzyme involved in cAMP synthesis.

To function, cyclic AMP requires a protein called CAP (cyclic AMP–activating protein). Cyclic AMP binds to the CAP protein and changes its shape, thereby allowing the protein to attach to the promoter site of numerous genes coding for inducible enzymes. In some way, this facilitates the binding of RNA polymerase to the promoter region to transcribe the genes. Thus, transcription of the lactose operon in *E. coli* requires that not only the inducer lactose be present to inactivate the repressor protein, but that cAMP also be available (figure 6.23). The protein CAP is just one example of a *transcription factor,* proteins that bind to DNA and are necessary for the transcription of specific genes.

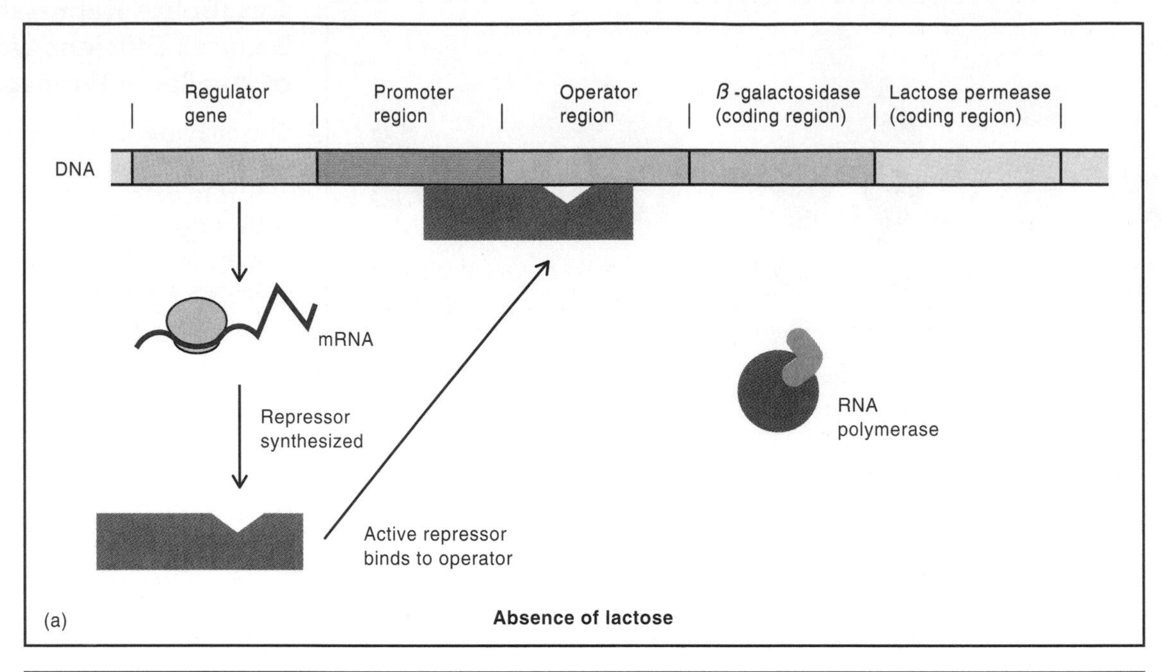

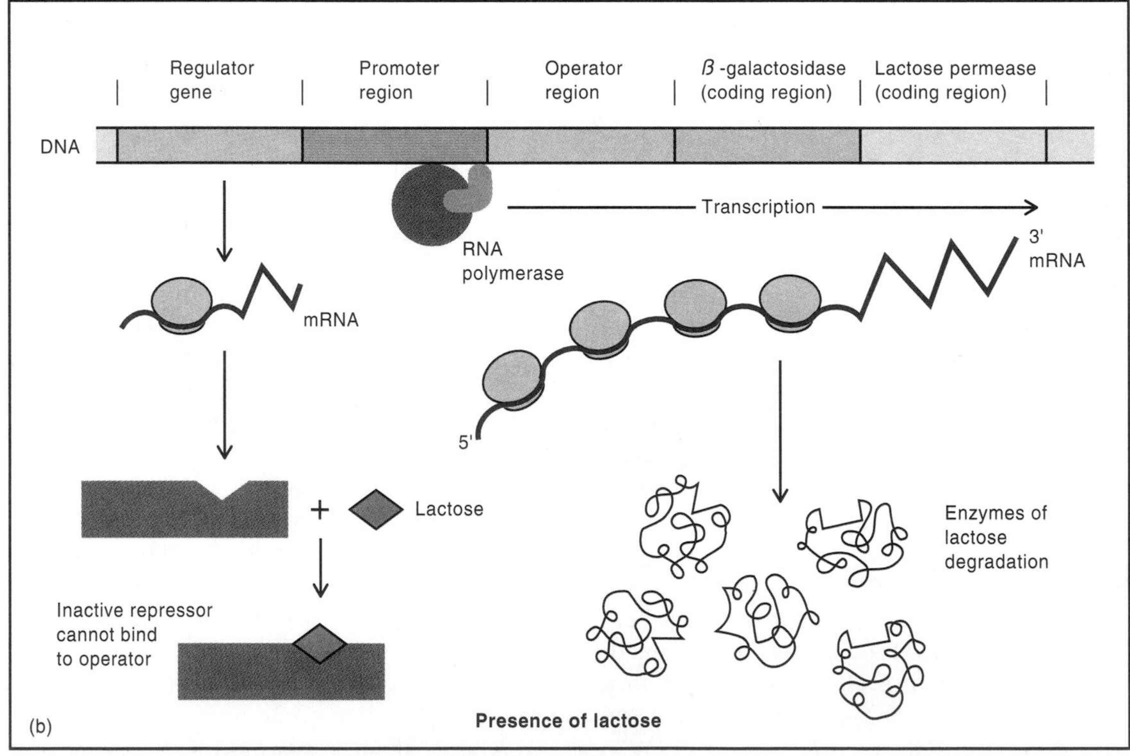

Figure 6.20
Regulation of the degradation of lactose by the substrate of the reaction, lactose. The corepressor lactose binds to the repressor protein and changes its conformation, so it cannot bind to the operator region. Thus, the RNA polymerase can bind to the promoter region and initiate transcription. However, unless the cyclic AMP–activating protein is also bound to the promoter (see figure 6.23), the RNA binds very weakly and transcription is very inefficient.

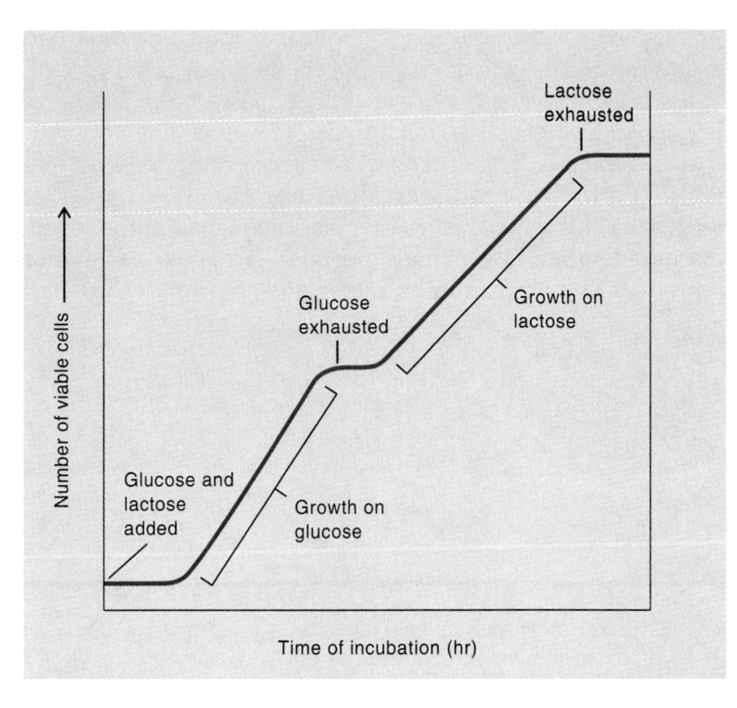

Figure 6.21
Diauxic growth curve. Note that the growth on lactose is slower than it is on glucose.

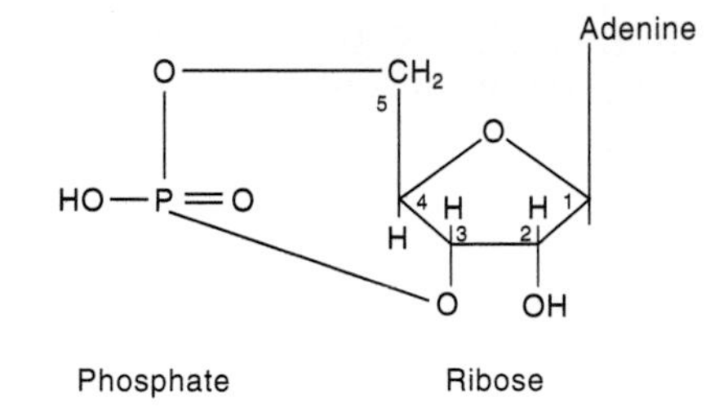

Figure 6.22
Cyclic AMP. The phosphate group is bound to carbon atoms 3 and 5 of the ribose molecule, thereby giving the molecule its name.

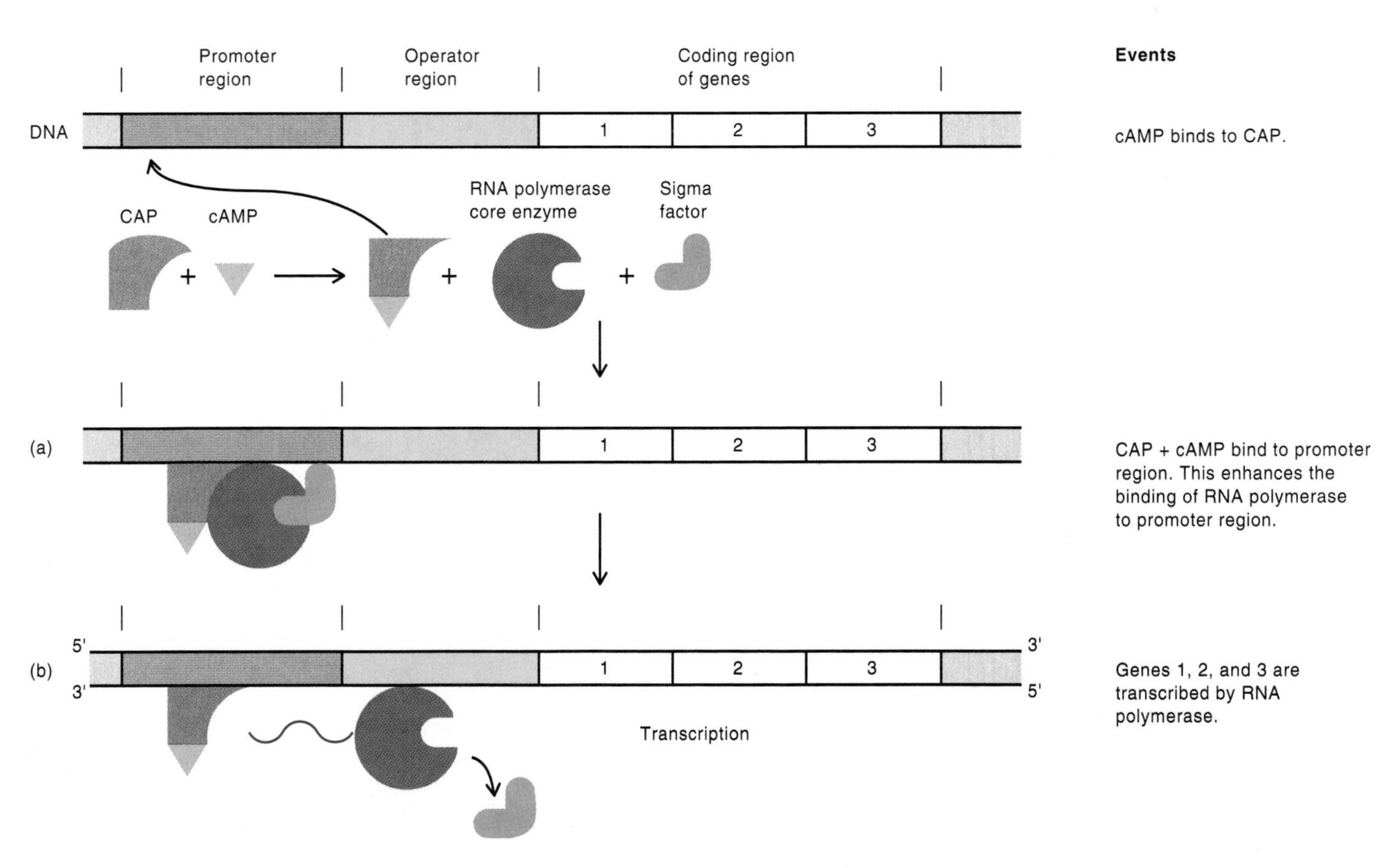

Figure 6.23
Cyclic AMP binding to CAP. The binding changes the shape of the protein so that it can now bind to the promoter region of the gene. This CAP protein bound to the DNA promotes the binding of RNA polymerase. The efficiency of transcription is greatly enhanced by CAP.

Since the CAP protein is required for transcription, it is called a **positive regulator.** In contrast, the repressor protein in tryptophan biosynthesis is called a **negative regulator** since its presence prevents transcription. Note that in all of the control systems in which a small molecule causes a change in gene expression, the molecule first binds to a protein molecule and changes the shape of the protein. This alters the ability of the protein to bind (or not bind) to a specific site on DNA. The small molecule never binds to DNA directly.

What is the importance of catabolite repression in regulating the metabolism of certain sugars? It ensures that a cell first uses the most rapidly metabolizable source of energy in its environment, which also supports the most rapid growth of the cell. Since glucose is metabolized faster than lactose in *E. coli,* it is used first, and lactose is only used when the glucose is depleted. If a particular bacterial species does not metabolize glucose readily, it does not act as a catabolite repressor. This mechanism of regulation, like all of the others discussed so far, has a single purpose—to allow bacteria to reproduce at their maximum rate in any environment.

Summary

I. Chemistry of DNA and RNA
   A. DNA occurs as a double-stranded helix held together by hydrogen bonds between adenine and thymine and between guanine and cytosine.
   B. The nucleotide sequences of the two strands run in opposite directions, one going in the 5' to 3' direction and the other in the 3' to 5' direction.
   C. RNA is usually single-stranded and shorter than DNA.

II. Information Storage and Transfer—Overview
   A. Information is stored in DNA and RNA in the sequence of their nucleotide subunits.
   B. Information in DNA can be transferred through two different processes.
      1. The sequence of nucleotides in DNA can be transferred to mRNA, which determines the sequence of amino acids in proteins.
      2. Replication of DNA begins at a specific site on the chromosome and proceeds by the sequential addition of nucleotides by DNA polymerase. Two chromosomes identical to the original are formed.

III. The Expression of Genes
   A. A bacterial chromosome consists of two major regions. One codes for proteins (coding region); the other determines whether the first region will be expressed (regulatory region).
   B. The sequence of three nucleotides in mRNA determines a specific amino acid, a triplet code.
   C. The first step in gene expression is making an RNA copy of the genetic information in DNA.
      1. The three steps of transcription are (a) initiation in which the sigma factor of RNA polymerase binds to a promoter region in the regulatory region of the gene; (b) elongation in which RNA polymerase moves along one of the strands of DNA (template strand) synthesizing a complementary single strand of mRNA; and (c) termination in which RNA polymerase comes to a stop signal at the end of a gene and is released.
   D. The second step in gene expression is the translation of mRNA into protein.
      1. The sets of three nucleotides (codons) that code for each amino acid are identified. Most amino acids have several codons.
      2. Translation involves three steps: (a) Initiation in which the ribosome binds to a region near the beginning of the mRNA to start translation. (b) Elongation in which a charged tRNA molecule attaches to its complementary codon of the mRNA positioned at a particular site on the ribosome. The ribosome moves along the mRNA three nucleotides at a time, so that another charged tRNA binds to the next complementary codon. The amino acids are joined together in a peptide bond. (c) Termination in which the protein dissociates from the tRNA and protein synthesis stops when the ribosome reaches a stop codon in the mRNA.

IV. Differences Between Prokaryotic and Eukaryotic Gene Transcription/Translation
   A. In prokaryotic cells, transcription and translation occur almost simultaneously; in eukaryotic cells, mRNA is transported out of the nucleus before translation begins.
   B. In eukaryotes, most mRNA molecules are modified before they are translated. Blocks of nucleotides are cut out, a situation very rare in prokaryotes.

V. The Environment and Control Systems—Overview
   A. Bacteria may be subjected to rapidly changing nutrients in the environment. They have the means to shut off or activate pathways of biosynthesis and degradation so as to synthesize only the materials they need in any particular environment.

VI. Mechanisms of Control of Biosynthetic Pathways—Overview
   A. End product repression controls the start of gene transcription. The end product binds to the repressor and changes its shape so that it can then bind to the operator region of the gene. This prevents the RNA from binding to the promoter.
   B. End products control the activity of the first enzyme of the pathway through allosteric control. The end product combines with the first enzyme of the pathway and changes its shape so it can no longer catalyze its reaction. The inhibition of the first enzyme shuts down the entire pathway.

VII. Mechanisms of Control of Degradative Pathways
   A. Metabolites that can be degraded activate the genes responsible for their degradation. The compound to be degraded turns on the pathway that can degrade it. The compound binds to the repressor protein and inactivates it, thereby allowing RNA polymerase to initiate transcription.
   B. Catabolite repression ensures efficient use of available foodstuffs.
      1. A catabolite repressor can reduce the level of cyclic AMP required for induction of certain pathways, so the catabolite repressor is metabolized first.
      2. Some very important enzymes (constitutive enzymes) are synthesized at the same level independent of the medium.

Chapter Review Questions

1. Name two mechanisms that bacteria use to control the synthesis of an end product.
2. In end product repression, which enzymes of a pathway are repressed? Which enzymes of a pathway are subject to allosteric control?
3. Name the two regions on the chromosome that interact with RNA polymerase and the repressor protein in an inducible pathway.
4. What nucleotide is important in catabolite repression?
5. Give examples of a pathway that is positively regulated and of one that is negatively regulated.
6. List three ways in which transcription and translation differ in eukaryotes and prokaryotes.
7. In what direction is DNA synthesized? mRNA? polypeptides?
8. Distinguish between the "stop" signals used in transcription and translation.
9. What is junk DNA as covered in this chapter?
10. Describe three processes discussed in this chapter in which complementarity between nucleotides plays a role.

Critical Thinking Questions

1. Discuss three specific examples given in this chapter of proteins recognizing and binding to specific nucleotide sequences in DNA.
2. Why might operons be so common in prokaryotes but rare in eukaryotes?
3. Why do small molecules, like cAMP, involved in gene regulation require another protein for their function?

Further Reading

Cech, T. R. 1986. RNA as an enzyme. *Scientific American* 255(5):64–75.

Darnell, J. E., Jr. 1985. RNA. *Scientific American* 253(4):68–78.

Dickerson, R. E. 1983. The DNA helix and how it is read. *Scientific American* 249(6):94–111.

Felsenfeld, G. 1985. DNA. *Scientific American* 253(4):59–67.

Freifelder, D. 1987. *Molecular biology,* 2d ed. Van Nostrand Reinhold, New York. A clearly written introductory textbook that emphasizes principles.

Radman, M., and Walker, R. 1988. The high fidelity of DNA duplication. *Scientific American* 259(2):40–46.

Söll, D., and Raj Bhandary, U. (ed.). 1994. tRNA structure, biosynthesis, and function. ASM Press, Herndon, Va.

Tijan, R. 1995. Molecular machines that control genes. *Scientific American* 272(2):54–61.

7 Bacterial Genetics

KEY CONCEPTS

1. Many chemicals as well as ultraviolet light cause changes in DNA which are passed on to future generations.
2. Cells have numerous mechanisms for repairing damaged DNA.
3. In bacteria, small pieces of the chromosome can be transferred from one cell to another.
4. Genes can move from one location to other locations in the DNA of the same cell.
5. Plasmids confer a large number of different but dispensable properties on bacteria and are readily transferred to other bacteria often unrelated to each other.
6. Bacteria are able to protect themselves from foreign DNA entering the cell.

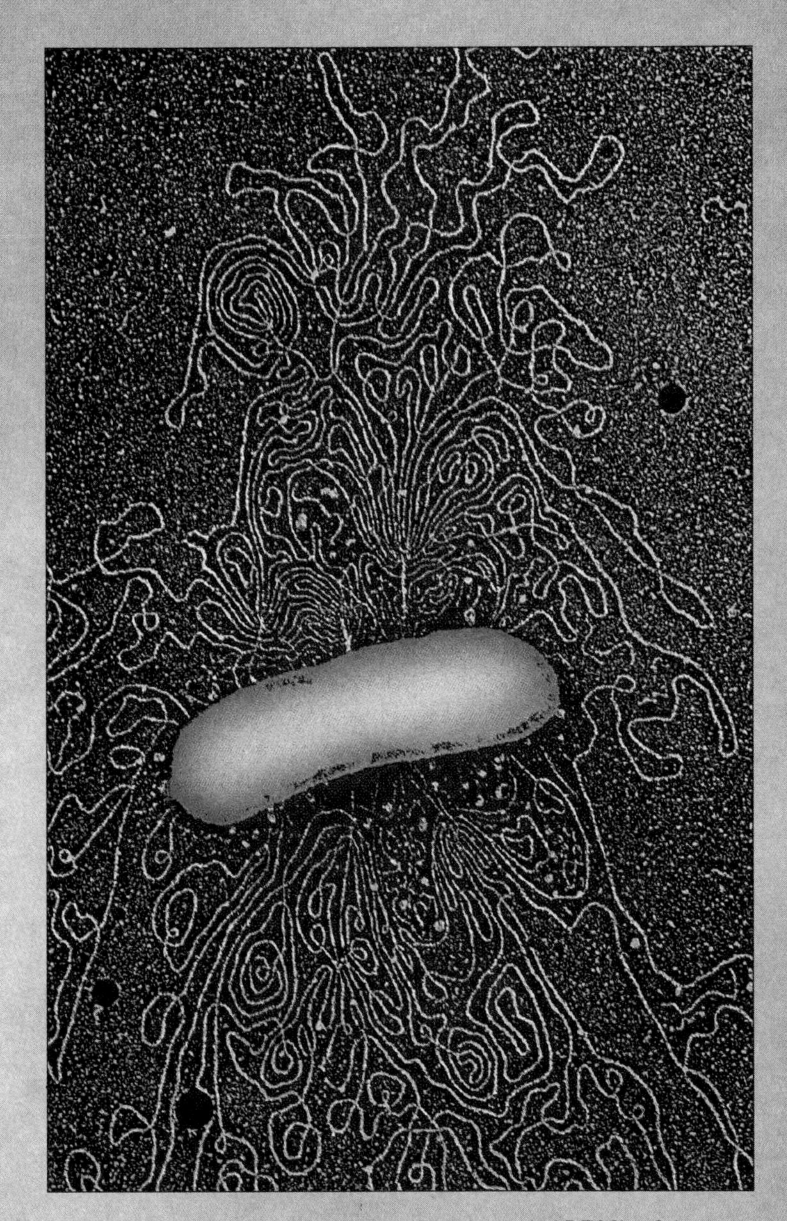

False-color transmission electron micrograph (X15,700) of the bacterium *Escherichia coli*. The bacterial cell in the center of the photo was treated with an enzyme to weaken its wall. Then it was placed in water, causing its DNA to leak out. The DNA is visible as the gold-colored fibrous mass lying around the bacterial shell. The length of the DNA is 1.5 millimeters, or 1,000 times the length of the cell from which it came. *E. coli* is usually a harmless inhabitant of the human intestinal tract. Much of the understanding of genetics has come from the study of this bacterium.

Microbes in Motion

The following books and chapters in the *Microbes in Motion* CD-ROM may serve as a useful preview or supplement to your reading:
Bacterial Structure and Function: Internal Structures. *Microbial Metabolism and Growth*: Genetics. *Viral Structure and Function;* Viral Replication.

A Glimpse of History

In the 1930s, it was well-known that DNA occurred in all cells, including bacteria. However, its function was a mystery. Since DNA consisted of only four repeating subunits, most scientists believed that it could not be a very important molecule. Its importance was discovered through studies conducted during a 20-year period, carried out by scientists in England and the United States.

In the 1920s, Frederick Griffith, an English bacteriologist, was studying the bacteria (pneumococci) that caused pneumonia. It was known that pneumococci could cause pneumonia only if they made a polysaccharide capsule. In trying to understand how they caused this disease, Griffith killed encapsulated pneumococci and mixed them with a strain of living pneumococci that could not manufacture a polysaccharide capsule. When he inoculated this mixture of organisms into mice, they developed pneumonia and died, a totally unexpected result (figure 7.1). Griffith was able to isolate living encapsulated pneumococci from the dead mice. When he injected the killed encapsulated organisms and living nonencapsulated organisms into separate mice, they did not develop pneumonia.

Two years after Griffith reported his findings, another investigator, M. H. Dawson, ruptured heat-killed encapsulated pneumococci so that their internal contents spilled out. He then passed the suspension of ruptured cells through a very fine filter, through which only the cytoplasmic contents of the bacteria could pass. When he mixed the filtrate (the material passing through the filter) with living bacteria that were unable to make a capsule, some of these bacteria were able to synthesize a capsule. Moreover, these bacteria passed on their ability to synthesize a capsule to all of their offspring. Thus, something in the filtrate was "transforming" the harmless, unencapsulated bacteria into bacteria with the ability to make a capsule.

What was this "transforming" principle? In 1944, after years of painstaking chemical analyses, Oswald T. Avery, Colin MacLeod, and Maclyn McCarty, three investigators from the Rockefeller Institute, submitted one of the most important papers ever published in biology. In it, they reported that the "transforming principle" was DNA, a chemical that could change (transform) a cell's properties.

The significance of their discovery was not appreciated at the time. Perhaps the discovery was premature, and therefore scientists were slow to recognize its significance and importance. None of the three investigators received a Nobel Prize, although many scientists believe that they deserved it. Their studies pointed out that DNA is a key molecule in the scheme of life and led to James Watson's and Francis Crick's determination of its structure, which they published in 1953. The understanding of the structure and function of DNA revolutionized the study of biology and ushered in the era of molecular biology. Microbial genetics serves as molecular biology's foundation.

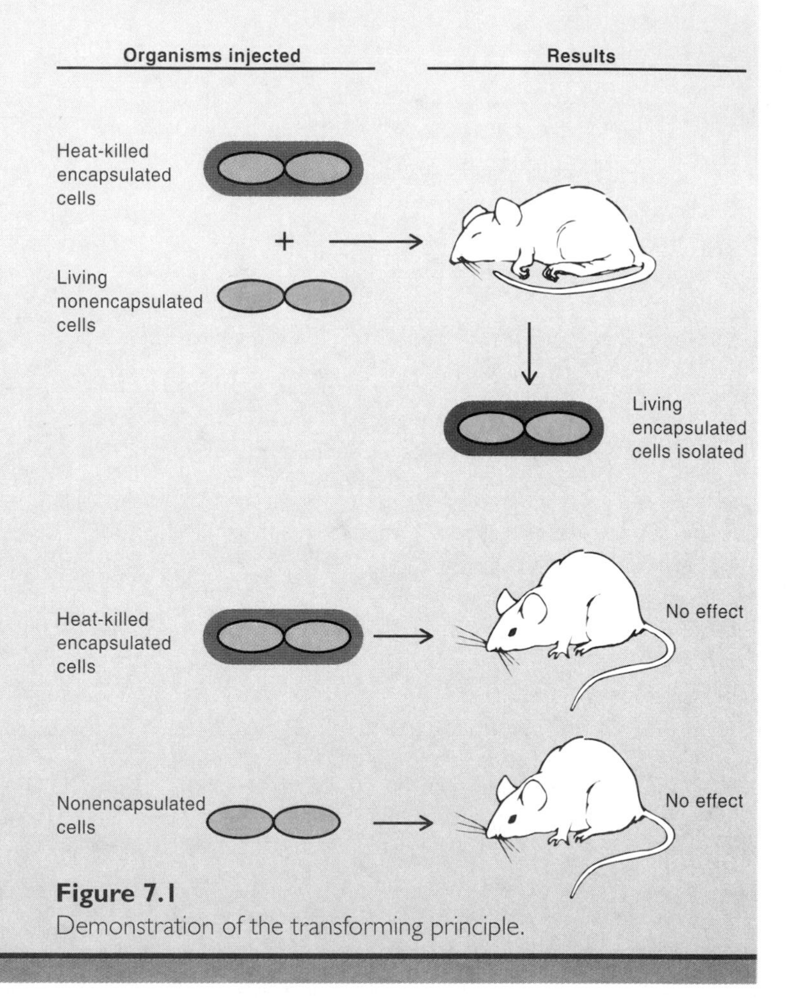

Figure 7.1
Demonstration of the transforming principle.

At the end of World War II, bacterial dysentery (caused by members of the genus *Shigella*) was a common disease in Japan. With the introduction of the sulfa drugs, the number of cases of dysentery decreased, as this antibacterial medicine proved highly successful. However, after 1949, the incidence of dysentery began to rise again, and many of the *Shigella* strains causing dysentery were resistant to sulfa. Fortunately, these organisms remained susceptible to such antibiotics as streptomycin, chloramphenicol, and tetracycline. However, a new problem arose. In 1955, a Japanese woman returning from Hong Kong developed *Shigella* dysentery that did not respond to any antibiotic treatment.

In the next several years, a number of dysentery epidemics developed in Japan. In some patients, the causative organisms were sensitive to antibiotics; in others, they were resistant. Curiously, bacteria isolated from certain patients who had been treated with a single antibiotic abruptly became resistant not only to that antibiotic but also to other antibiotics to which the patients had not been exposed. In addition, cells of *Escherichia coli* residing in the large intestine of many of the patients who carried antibiotic-resistant organisms were also resistant to the same antibiotics. How can these observations be explained? A reasonable conclusion is that the genes responsible for antibiotic resistance spread from one bacterium to another.

Antibiotics have reduced the incidence of many microbial infections. However, the development and spread of resistant strains of bacteria have created unique problems for the treatment of infectious diseases. Understanding the mechanisms by which organisms become resistant to antibiotics and how this resistance is transferred to sensitive cells is discussed in this chapter. With this knowledge, the medical community may be able to keep one step ahead of disease-causing bacteria and maintain the usefulness of antibiotics.

In order to study the mechanisms by which genes move in populations of bacteria, first it is necessary to recognize when genes have moved. To do this, genes must be altered in some way so that they change the properties of the cell into which they move.

First, we will discuss the source of variation in microorganisms and how genes are altered (mutated). Then, we will consider the mechanisms by which genes move from cell to cell as well as to different sites within the same cell.

Sources of Diversity in Microorganisms

The tremendous diversity seen in different microorganisms in the biological world stems from two major factors. First, the different sequences of nucleotides in DNA code for different proteins. Different proteins result in organisms with different properties. Second, the expression of genetic information that organisms possess is regulated so that genes are turned off and on, depending on the particular environment. Both of these concepts were discussed in chapter 6.

The total genetic constitution of an organism is its **genotype.** The genotype of a cell is determined by the genetic information contained in all of its DNA, or **genome.** This includes the chromosome as well as any plasmids in the cell. However, some scientists consider only genes in the chromosome to constitute the genome.

The characteristics displayed by an organism in any given environment comprise its **phenotype.** The environment can influence the expression of genes via induction and repression as well as by the activity of gene products, such as enzymes via allosteric control. Thus, even though *E. coli* has the genetic information to synthesize all of its amino acids, it does not do so if amino acids are present in the environment, for reasons discussed in the previous chapter. To summarize, the genotype represents the potential an organism has if all of its genes are expressed; the phenotype describes the actual properties of the organism at any one time. Thus, the interaction of the genome of the cell with its environment determines the properties of the cell; changes in either the DNA or its expression will alter the properties of the cell.

Whether a change in an organism is genotypic or phenotypic can be surmised from several properties. Genotypic changes are rare and involve changes in the nucleotide sequence of the DNA (mutations), so that only a few cells in a large population will be altered for a particular trait. In contrast, phenotypic changes involve almost all cells in the population. Further, phenotypic changes are readily reversible, in that all bacteria in the population can change back to their original property as the environment changes back to the original. In contrast, genotypic changes are relatively stable. In the next section, we discuss how genes can be altered.

Gene Alterations—Mutations

A change in the nucleotide sequence of DNA that results in a recognizable change in the organism is called a **mutation.** The properties of a cell depend largely on its proteins, which include both enzymes and proteins that form part of the cell's structures, such as the proteins in the cytoplasmic membrane and the ribosome. The substitution of even one amino acid for another in any given protein may cause the protein to be nonfunctional, thereby changing the properties of the cell. For example, if any gene of the tryptophan operon is altered so that the enzyme for which it codes no longer functions, the cells will grow only if this amino acid is added to the medium. However, some changes do not result in a recognizable change in which case they are termed **silent mutations.**

Mutations Can Occur by Substitution of One Base for Another

Genes mutate infrequently. The most common type of mutation involves a mistake during DNA synthesis, when an incorrect purine or pyrimidine base is incorporated into the DNA (known as *base substitution;* figure 7.2). Base substitution errors occur most frequently on the rare occasions when the hydrogen atoms on the molecules of A, T, G, or C change their locations, which alters the hydrogen-bonding properties of the base.

Thus, a shift in the hydrogen atom of adenine allows it to bond to cytosine. This mistake in incorporation then is passed on to the cell's progeny (offspring), resulting in an incorrect amino acid being incorporated into the protein coded by the gene (see figure 7.2).

Mutations Can Occur by the Removal or Addition of Nucleotides

The deletion (removal) or addition of one or several purine or pyrimidine nucleotides changes the nucleotide sequence, resulting in a mutation. Because translation of a gene begins at a specific codon and proceeds one codon (three bases) at a time, the deletion or addition of a purine or pyrimidine nucleotide shifts the codons of the DNA when it is transcribed into mRNA (figure 7.3). This type of mutation is termed a **frame shift mutation.**

• **Hydrogen bonding, DNA, p. 26**
• **Codon, p. 136**

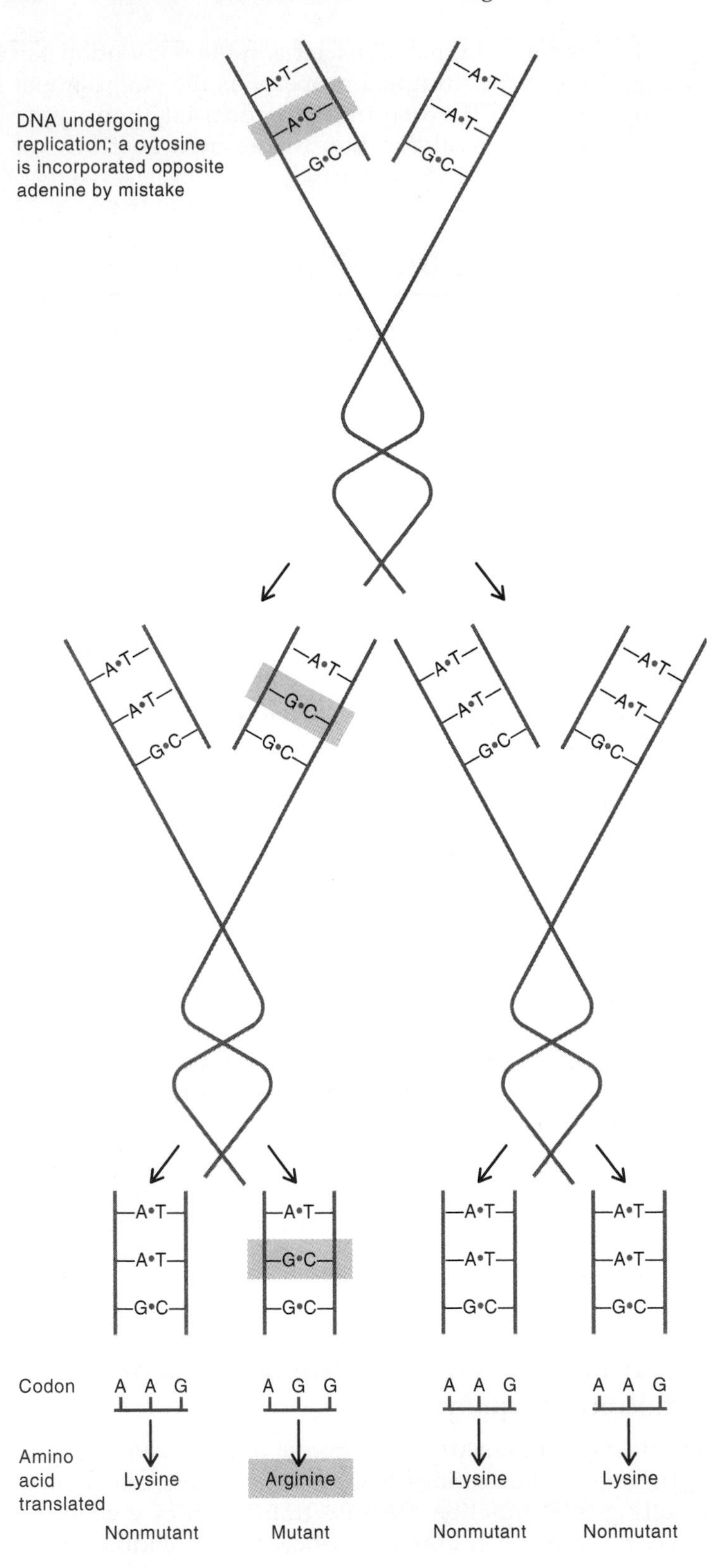

Figure 7.2
The generation of a mutant organism as a result of the incorporation of a pyrimidine base (cytosine) in place of thymine in DNA replication.

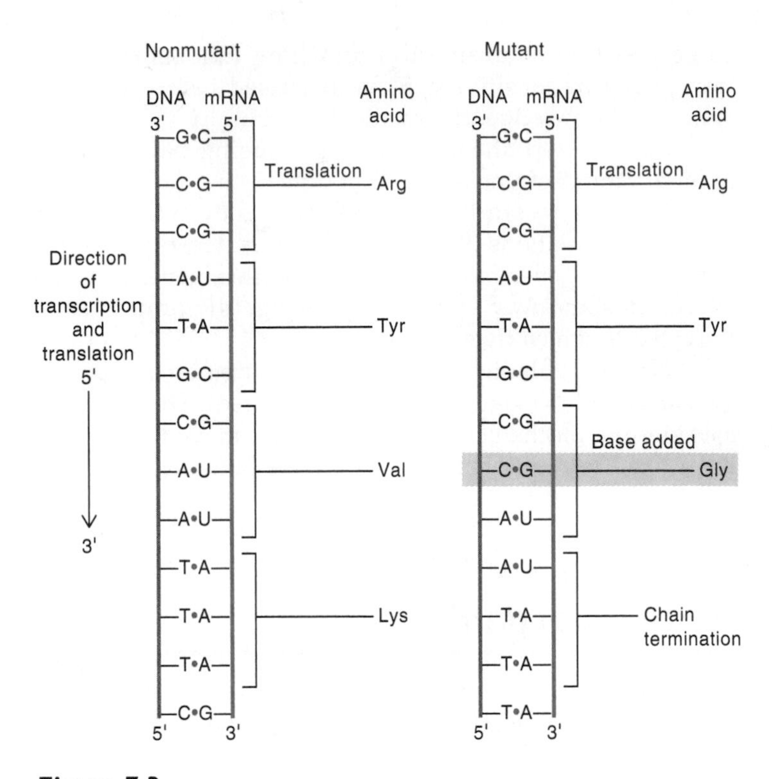

Figure 7.3
Production of mutation as a result of base addition. The addition of a nucleotide to the DNA results in a frame shift in the transcription of the DNA and a new triplet code word, which is translated as a new amino acid. The deletion of a nucleotide would have essentially the same effect. The protein chain terminates when a nonsense (stop) codon appears in the DNA.

Note that a deletion or addition of nucleotides affects most of the amino acids incorporated beyond the original site at which the addition or deletion occurred. These mutations change many more codons than are changed by the substitution of a base, thereby resulting in the alteration of many amino acids rather than just a single one.

Mutagenesis

Mutagenesis is the process by which a mutation is produced. Mutations occur at a frequency of about one in 10^9 (often stated merely as 10^{-9}) cells for any single gene, although the range is from one in 10,000 (10^{-4}) to one in 1,000 billion (10^{-12}). Because of mutations' rarity, bacteria in which mutations are being sought are often treated with **mutagens,** agents that increase the frequency of mutations 10-fold to 1,000-fold or more. Mutations that occur in the absence of a known mutagen are termed **spontaneous;** those that arise following treatment with a mutagen are termed **induced.** Mutagens can be divided into three broad categories: **chemical agents, transposable elements,** and **radiation.**

Chemical Mutagens Often Act by Altering the Hydrogen-bonding Properties of the Nitrogenous Bases

Any chemical treatment that alters the hydrogen-bonding properties of a purine or pyrimidine base already in the DNA will increase the frequency of mutations. One powerful mutagen that functions in this way is **nitrous acid** (HNO_2). This chemical primarily converts amino groups to keto groups—for example, by converting cytosine to uracil, which then pairs with adenine rather than guanine.

Alkylating Agents

The largest group of chemical mutagens consists of the **alkylating agents,** highly reactive chemicals that add alkyl groups (short chains of carbon atoms) onto the nitrogenous bases, altering their hydrogen-bonding properties. Many compounds used in cancer therapy, such as nitrogen mustard, are in this group. Evidence now exists that the use of some of these agents will cause new cancers to arise more than 10 years after they have been used successfully in treating the original cancer.

Base Analogs

Other chemical mutagens include **base analogs,** compounds that resemble the chemical structures of naturally occurring purine or pyrimidine bases closely enough that they are incorporated into DNA in place of the natural bases in the process of DNA replication (figure 7.4). However, base analogs do not have the same hydrogen-bonding properties as the natural bases. This difference in bonding properties increases the probability that, once incorporated into DNA, the base analog will pair with the wrong base as the complementary strand is being synthesized.

Certain analogs are also useful in treating diseases. For example, azidothymidine (AZT), also called zidovudine, an analog of thymidine, inhibits replication of the human immunodeficiency virus (HIV) and is used in treating AIDS. Note that the sugar deoxyribose and not the base thymine is modified in this base analog.

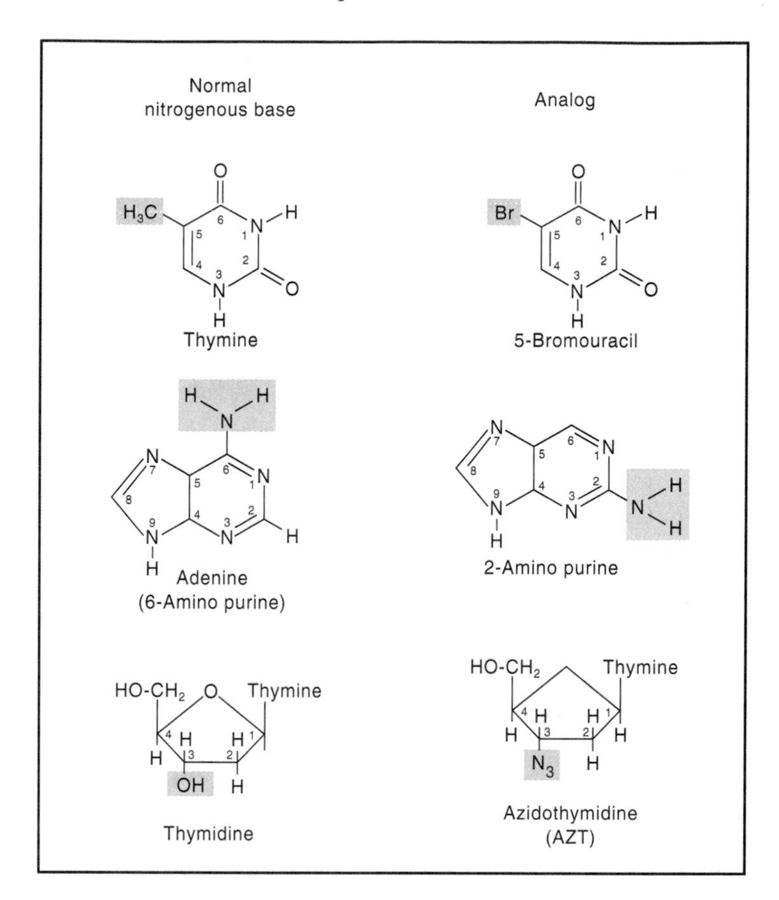

Figure 7.4
Common base analogs and the normal bases they replace in DNA. The important differences between the normal bases and the analogs are in boxes. The numbers refer to the carbon atoms in the ring which are not shown. Thymidine consists of thymine and deoxyribose. Because AZT does not have an OH group on C atom 3 of deoxyribose, replication of DNA stops when it is incorporated into the DNA chain being synthesized in the course of virus multiplication.

Intercalating Agents

A number of chemical mutagens, termed **intercalating agents,** are three-ringed molecules of about the same size as a pair of nucleotides in DNA (figure 7.5). These molecules do not alter hydrogen-bonding properties of the bases, but rather insert (intercalate) between two base pairs in both strands of the double helix and thereby move adjacent base pairs apart (figure 7.5). When the DNA containing an intercalating agent replicates, additional bases are inserted in the widened space between bases in the strand of DNA being synthesized, resulting in base additions. An intercalating agent commonly used in the laboratory to isolate plasmids is ethidium bromide. The manufacturer now warns users that it should be used with great care because it is a **carcinogen,** a cancer-causing agent.

Transposable Elements (Jumping Genes)

Transposable elements, sometimes called **transposons,** are special segments of DNA found in virtually all species of bacteria. They can move from one site to another in the same or different DNA molecules, in a process called **transposition** (figure 7.6). Any gene into which a transposable element inserts itself can no longer function. In fact, these movable pieces of DNA were first recognized as mutagens.

• **Plasmid, p. 72**

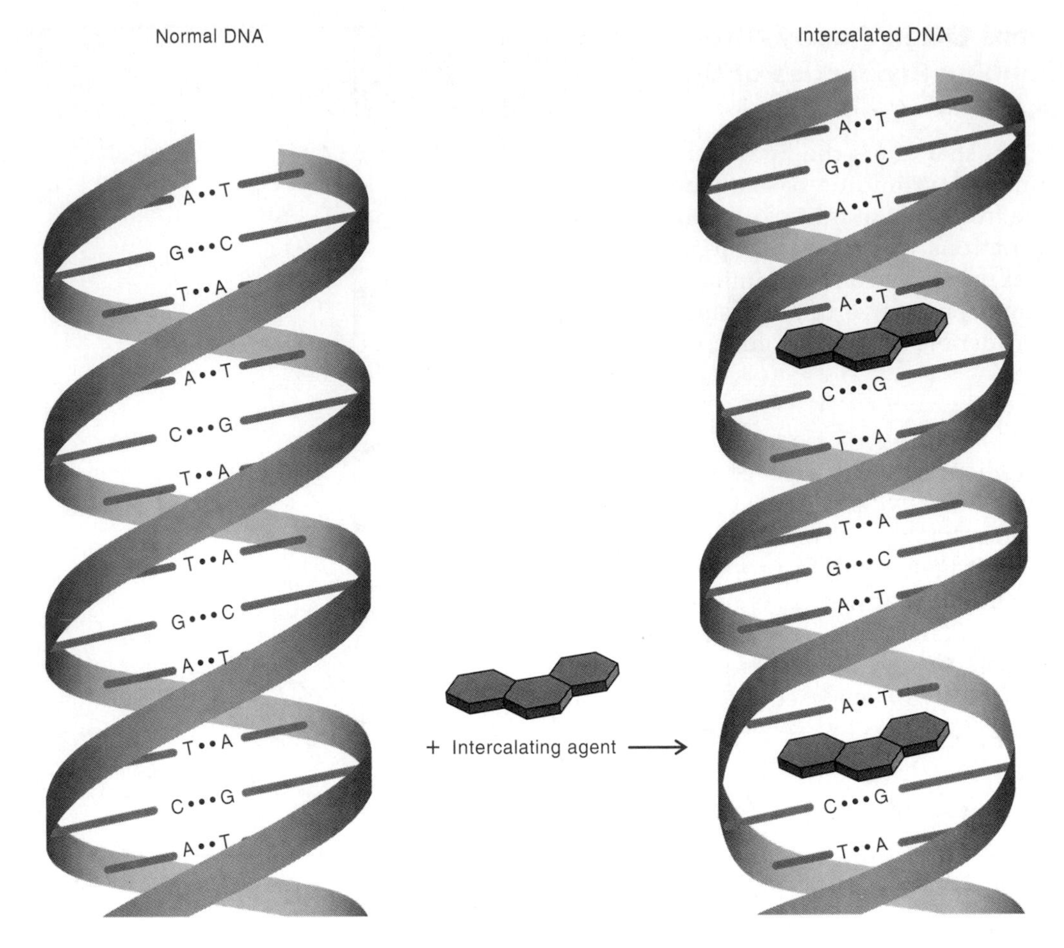

Figure 7.5
Intercalation. The insertion of the intercalating agent moves the adjacent base pairs farther from one another so base additions occur. This results in mutation.

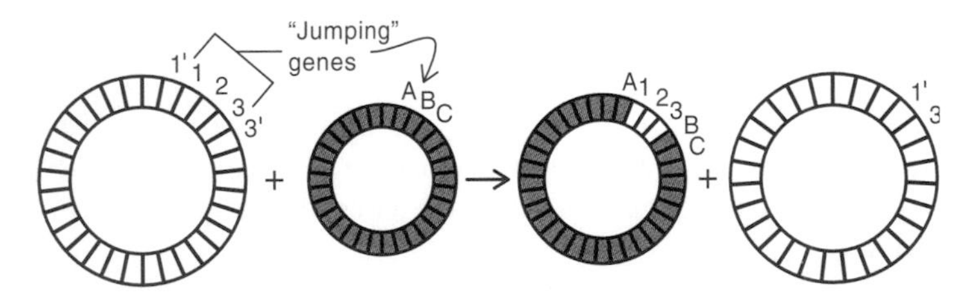

Figure 7.6
Transposition. The transposon has the ability to "jump" from one piece of DNA to another piece of DNA, where it becomes integrated. The transposon is not homologous to the region into which it integrates. The transposon, which includes genes 1, 2, and 3, leaves one piece of DNA and moves to another. The first piece of DNA loses the genes, and the second piece of DNA gains them. In some cases, the transposons are duplicated as they move so that there are two copies of them, one in the original location and another in the new location. The transposon may "jump" from the chromosome to a plasmid.

Several types of transposable elements exist, varying in the complexity of their structure. The simplest kind, termed an **insertion sequence (IS) element,** consists of a gene required for transposition bounded on either side by about 15 to 25 base pairs, which have the structure shown in figure 7.7 *a*. The important feature of this sequence of nucleotides is that the sequence in the top strand of DNA is the same as it is in the bottom strand but going in the opposite direction. Such a sequence is termed an **inverted repeat,** or **palindromic, sequence.** The *E. coli* chromosome contains more than a dozen different IS elements, labeled IS1, IS2, and so on. Several copies of each are present in the chromosome.

A more complex type of transposable element consists of a gene that is easily recognized, such as a gene coding for antibiotic resistance, flanked by IS elements (figure 7.7 *b*). Such a recognizable gene is capable of transposing. In theory, any gene or group of genes can move to another site if bounded by IS elements. It is not known how many mutants result from transposable elements inserting into genes, but they may account for many spontaneous mutants.

We will consider transposable elements further on in this chapter because they play an extremely important role in gene movement within the same cell and also between different cells.

Perspective 7.1

Barbara McClintock: The Discoverer of Jumping Genes

Barbara McClintock (1902–1992) was a remarkable scientist who made several very important discoveries in genetics dealing with chromosome structures. These studies were carried out at a time before the age of large interdisciplinary research teams and before the availability of sophisticated tools of molecular genetics. Her tools consisted of a clear mind, able to make sense of confusing and revolutionary observations, and a consuming curiosity that led her to work 12-hour days, six days a week in a small laboratory on Long Island, New York.

In 1983, at the age of 81, McClintock received a Nobel Prize in Medicine or Physiology in part for her discovery of transposable elements 40 years earlier. Her experimental system consisted of ears of corn; what she observed were the different colors of the kernels which were due to various pigments produced by certain enzymes. If the gene coding for the enzyme was inactivated, the kernel was not pigmented. If the enzyme was only partially inactivated, then the kernel was partially pigmented. By looking at kernel colors, McClintock was able to observe the results of mutations in genes of pigment production. McClintock concluded that fragments of DNA must be capable of moving from one site on the chromosome to another, thereby causing mutations (figure 1). A transposable element moving into a gene would inactivate it; when the element left a gene, it would reactivate it.

Figure 1
Variegation in color observed in the kernels of corn is caused by the insertion of transposable elements into genes involved in the synthesis of different pigments, thereby preventing the synthesis of the pigments.

At the time she published her results, most scientists believed that chromosomes were very stable. Consequently, McClintock's ideas were met with great skepticism by most geneticists, and she stopped publishing many of her observations. It was not until the late 1970s that her ideas became understood. By that time, transposable elements had been discovered in many organisms, including bacteria, and the techniques of molecular biology made it possible to actually isolate and identify these jumping agents.

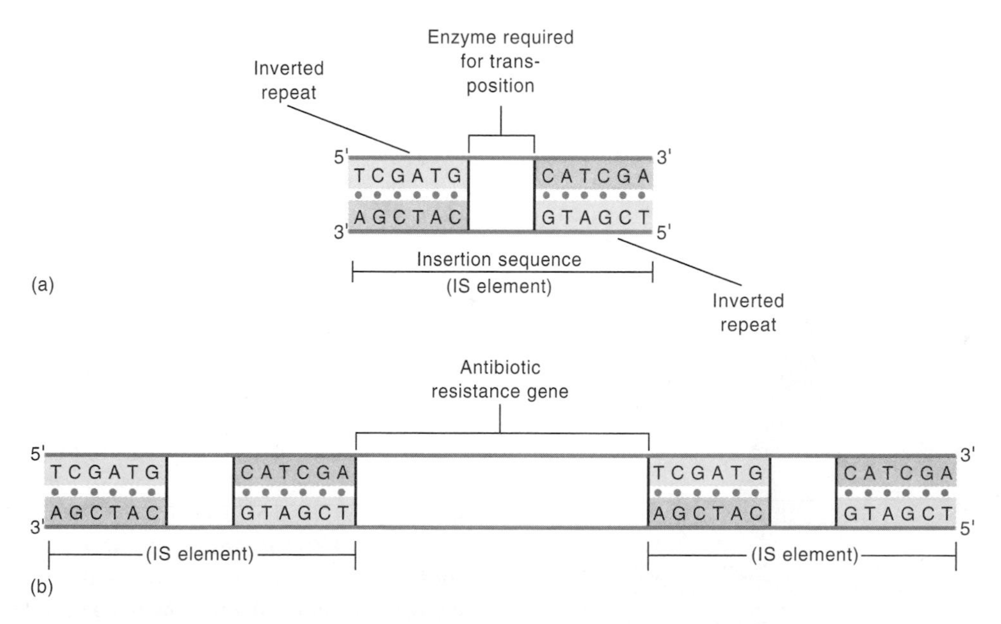

Figure 7.7
(*a*) Insertion sequence. Note that the sequence of bases at the end of the IS element in one strand is the same as the sequence in the opposite strand reading in the same direction (5' to 3'). Only 6 of the 15 to 25 base pairs are shown. (*b*) Antibiotic locus flanked by two insertion elements.

Ultraviolet Irradiation

Certain types of radiation, **X rays** and **ultraviolet (UV) light** in particular, are powerful mutagens. The wavelengths of X rays and UV light are shown in figure 7.8.

Irradiation of cells with ultraviolet light causes covalent bond formation between adjacent thymine molecules on the same strand of DNA (intrastrand bonding), resulting in **thymine dimers** (figure 7.9). A dimer is two identical molecules held together by covalent bonds.

The covalent bond distorts the DNA strand so much that the dimer cannot fit properly into the double helix, and the DNA is damaged. DNA cannot be replicated beyond this site of damage nor can genes be transcribed. However, the major mutagenic action of UV light does not result from the damaged DNA directly, but rather from the cell's attempt to repair the damage by a mechanism termed *SOS repair* (see the next section).

Table 7.1 summarizes some information on the major mutagens.

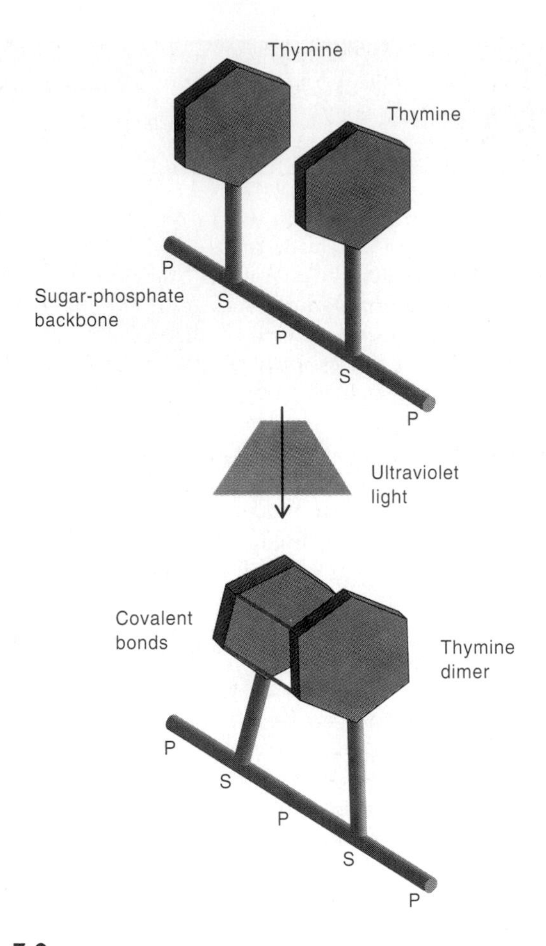

Figure 7.9
Thymine dimer formation. Covalent bonds form between adjacent thymine molecules on the same strand of DNA. This distorts the shape of the DNA and prevents replication past the dimer.

Repair of Damaged DNA

No molecule is more important to the cell than DNA. The amount of spontaneous and mutagen-induced damage to DNA that occurs in cells is enormous. Every 24 hours, the DNA in every cell in the human body is spontaneously damaged more than 10,000 times. This damage can lead to cell death and, in animals, cancer. If a cell is unable to repair damaged DNA, serious consequences may result. In humans, two breast cancer susceptibility genes code for enzymes that repair damaged DNA. Mutations in either one result in a high (80%) probability of breast cancer. Accordingly, it is not surprising that, in the course of millions of years of evolution, all cells, both prokaryotic and eukaryotic, have developed several different mechanisms for repairing any damage that DNA might suffer in a cell. Except for the light repair of UV dimers, the mechanisms that we will discuss operate on all types of damaged DNA such as misincorporated bases. However, the most-studied examples are the repair of UV dimers, so we will emphasize this type of damage.

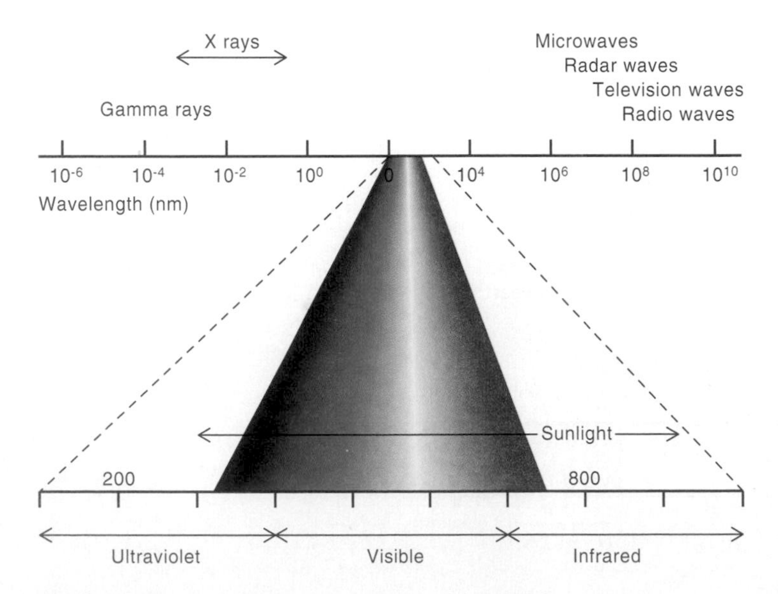

Figure 7.8
Wavelengths of radiation. The visible wavelengths include the colors of the rainbow.

Visible Light Activates an Enzyme that Breaks Covalent Bonds Forming Thymine Dimers

Many bacteria and other organisms, including people, have enzymes that can repair damage caused by UV light. Since UV light is a part of sunlight, cells are frequently exposed to this mutagenic agent in their natural environment, and they have developed several mechanisms to combat the harmful effects of these rays. For example, one particular enzyme can break the covalent bond of thymine dimers but only in the presence of visible light, in a phenomenon called **light repair** (figure 7.10).

Table 7.1 Chemical and Radiation Mutagenic Agents

Agent	Action	Result
Chemical		
Base analog		
5-bromouracil	Incorporated in place of thymine in DNA	AT base pair becomes GC base pair (base substitution)
Reactions with DNA		
Nitrous acid	Converts amino group to keto group in A and C	Base substitution
Alkylating agent		
Nitrogen mustard	Adds alkyl (CH_3 and others) to nitrogenous bases, such as guanine	Base substitution
Intercalating agent		
Ethidium bromide and acridine orange	Inserts between base pairs	Addition of base pairs
Radiation		
Ultraviolet (UV)	Thymine dimer formation	Misincorporation of bases due to SOS repair
X rays	Single- and double-strand breaks in DNA	Deletion of bases

Guanine
Thymine
Cytosine
Adenine
Covalent bonds
Thymine dimer results in a distortion of the sugar-phosphate backbone in both strands.
Visible light and photoreactivating enzyme
Light repair
Hydrogen bonds
An enzyme requiring visible light for activity breaks the two bonds joining the two thymine molecules together. The two strands of DNA assume their original shape.

Figure 7.10
Repair of a thymine dimer by a light-requiring enzyme.

Bacteria Can Excise Damaged DNA and Replace It with Undamaged DNA

Some bacteria also have enzymes that excise (cut out) the damaged segment from a single strand of DNA and other enzymes that then repair the resulting break by synthesizing a strand complementary to the undamaged strand. This process, termed **excision repair,** can cut out pieces of DNA that contain mismatched bases caused by chemical mutagens as well as DNA damaged by UV light. Because visible light is not required for the action of these enzymes, such repair is also called **dark repair** (figure 7.11).

This same repair mechanism is also important to the well-being of humans. People exposed to the sun for long periods have a higher incidence of skin cancer than do other people. However, people who have a defective dark repair enzyme have a more serious problem. The incidence of skin cancer in these people is increased markedly, illustrating both the damage UV light can cause to DNA and the importance of repair mechanisms in overcoming the damage.

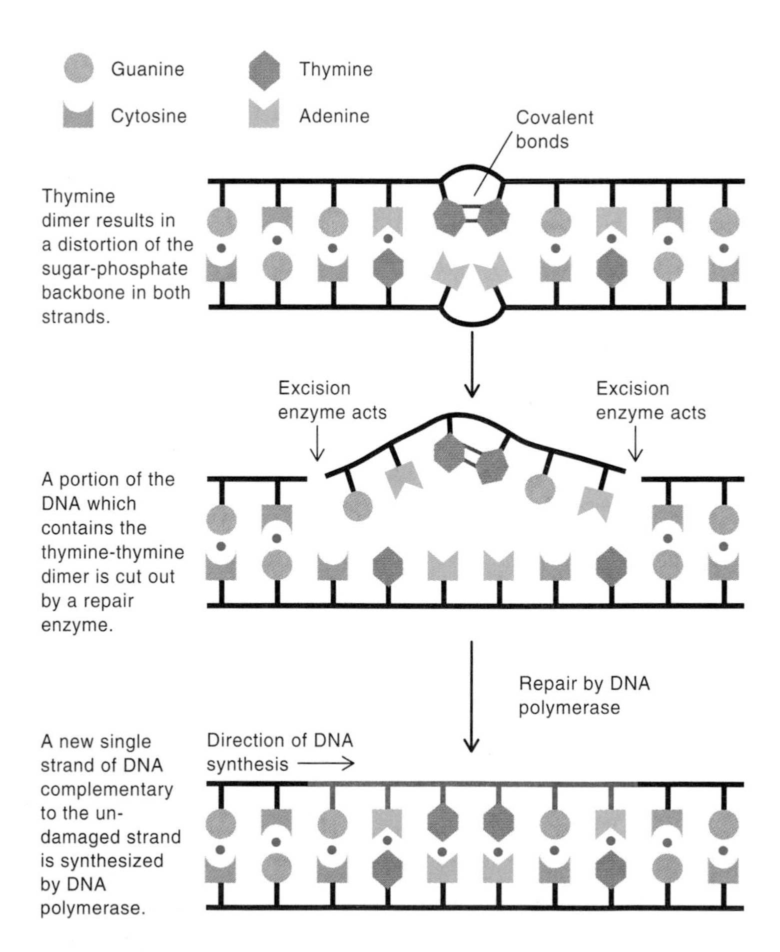

Figure 7.11
Dark, or excision, repair of a thymine dimer. The single strand of DNA containing the thymine dimer is removed and destroyed. The newly synthesized strand is joined to the end of the original strand by the enzyme DNA ligase.

Misincorporated Bases Can Be Removed by a Proofreading Function of DNA Polymerase

As mentioned already, a major cause of spontaneous mutations is the incorporation of the wrong base in the course of DNA replication by DNA polymerase. This enzyme selects the proper base that can hydrogen bond to the template strand and then polymerizes the nucleotide to the 3' OH group on the strand of DNA being synthesized. However, on rare occasions, the wrong base is selected so that the hydrogen bonding to the base in the template strand does not occur. If allowed to remain, this misincorporated base leads to a mutation. Cells have developed two ways of dealing with these errors of misincorporation. A major process is carried out by the DNA polymerase. This complex enzyme not only is involved in the synthesis of DNA but also has a **proofreading function** so that it can excise the base if it is not hydrogen bonded to the base in the template strand. The proper base is then selected and incorporated into the growing strand.

Misincorporated Bases Can Be Removed by a Mismatch Repair System

The proofreading function of DNA polymerase is very efficient. However, the precise replication of DNA is so important that cells have another repair system that recognizes bases that are not hydrogen bonded to the template strand but have been missed by the proofreading function of DNA polymerase. This repair mechanism is termed **mismatch repair.** In this process, the mismatch repair enzyme recognizes the pair of nonhydrogen-bonded bases and excises a short stretch of nucleotides from the newly synthesized strand (figure 7.12 *a*). The gap is then filled in with the proper nucleotides by DNA polymerase.

How does the mismatch repair enzyme recognize which strand of DNA should be excised? If the original strand of DNA were excised, the misincorporated base would not be removed and the mistake would not be repaired. Therefore, it is important that the repair enzyme be able to distinguish between the original strand and the newly synthesized strand. The original and newly synthesized strands differ in one respect—methylation. Certain bases in DNA are methylated by a process that occurs long after the DNA strand is synthesized. Therefore, the strand containing the misincorporated base which is newly synthesized is not methylated. The mismatch repair enzyme recognizes the difference in methylation of the two DNA strands and excises the DNA from the strand that is not methylated. The mismatch repair system also occurs in humans. Defects in this repair system lead to an increased incidence of colorectal cancers.

UV and Chemical Damage Can Be Overcome by an Inducible Enzyme System of DNA Replication that Makes Many Mistakes

As has been pointed out, replication of DNA by DNA polymerase cannot proceed past the block of thymine dimers. Thus, a cell with this damaged DNA would ordinarily die. However, a DNA replication repair system termed the **SOS system** can bypass the damaged DNA. This system differs from the usual DNA polymerase repair system. Two interesting and important features of this system require an explanation. First, the system is induced by damaged DNA, such as thymine dimers that, in effect, send a distress signal to the cells. This accounts for the

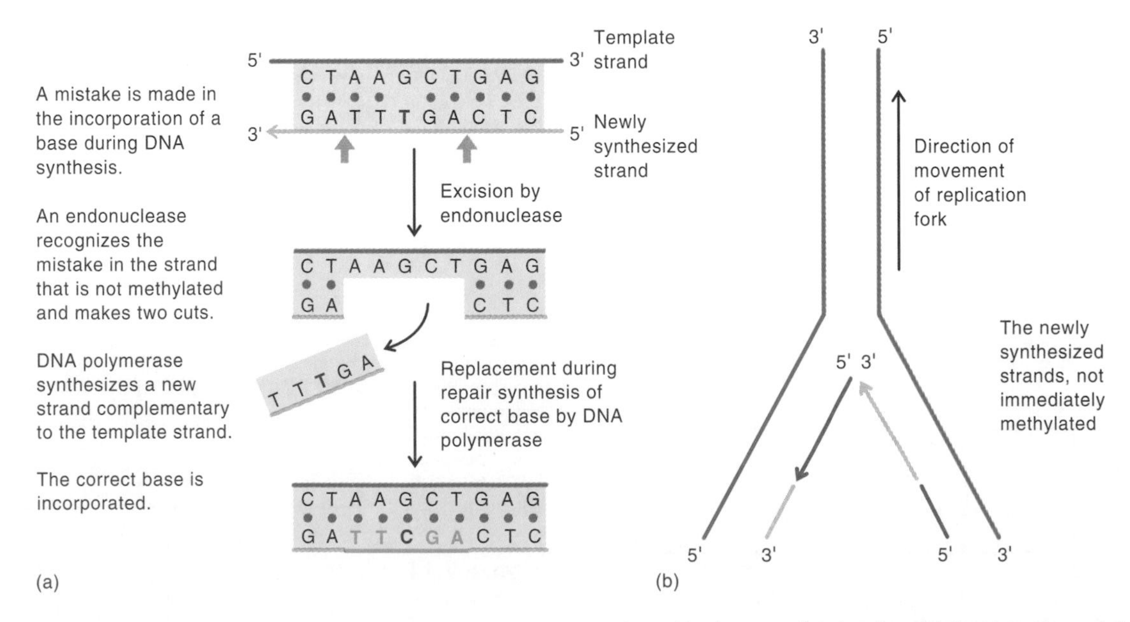

Figure 7.12
Mismatch repair. (*a*) The endonuclease excises a piece of DNA containing the misincorporated nucleotide. A new complementary strand is then synthesized and joined to the original strand by DNA ligase. (*b*) The endonuclease recognizes the strand of DNA that is not yet methylated as the strand to excise.

Table 7.2 Repair of Altered DNA

Mutagen	Type of Defect	Repair Mechanism	Biochemical Mechanism	Result
UV light	Thymine dimer formation	SOS repair	DNA synthesis by an inducible replication system bypasses site of damaged DNA	Mutations induced; cell survives
		Photoreactivation (light repair)	Breaking of covalent bond joining thymine dimers	Original DNA molecule restored
		Excision repair (dark repair)	Excision of a short stretch of single-strand DNA containing thymine dimer and synthesis of new strand by DNA polymerase	Mutation eliminated
Chemical	Chemical damage	Same mechanism as UV light except for photoreactivation	Same as for UV repair	Mutation eliminated or mutations induced by SOS repair
Spontaneous	Wrong base incorporated during DNA synthesis	Proofreading by DNA polymerase	Removal of mispaired base by DNA polymerase	Potential mutation eliminated
		Mismatch repair	Excision of a short stretch of unmethylated single-strand DNA and synthesis of new strand by DNA polymerase	Potential mutation eliminated

term *SOS*. Thus, the system does not function in normal cells which do not contain damaged DNA. Second, this DNA replication system introduces a high number of mispaired bases into the DNA strands that it synthesizes, thereby resulting in many mutations. This explains why UV is such a good mutagen. The SOS system saves cells from certain death by bypassing the site of damage. The cost is that the surviving cells have an increased frequency of mutations. Apparently it is better to be alive and mutant than not to be alive at all!

Table 7.2 summarizes the major features of the major DNA repair systems in cells.

Mutations and Their Consequences

The rate of mutation is defined as the probability that a mutation will occur in a given gene each time a cell divides and is generally expressed as a negative exponent per cell division. The mutation rate of different genes usually varies between 10^{-4} and 10^{-12} per cell division—that is, the chances that a gene will mutate are between one in 10,000 and one in 1,000 billion every time the cell divides. Thus, we can expect that many genes will be mutant in one in a million cells. Thus, the concept that all cells arising from a single cell are identical is not strictly true, since every large population will contain mutants. Even the cells in a single colony that contains about 1 million cells are not identical because of mutations. These mutations provide a mechanism by which organisms can respond to a changing environment. Thus, a spontaneous mutation to antibiotic resistance, even though it may be rare, will result in the mutants becoming the dominant organisms in a hospital environment where the antibiotic is present since only the resistant cells can survive. Note that the antibiotic is not a mutagen. It kills the sensitive cells and thereby allows the resistant cells to take over the population.

Since genes mutate independently of one another, the chance that two given mutations will occur within the same cell is very low. Indeed, the actual occurrence is the product of the individual rates of mutation of the two genes (the sum of the exponents). For example, if the mutation rate to streptomycin resistance is 10^{-6} per cell division and the mutation rate to penicillin resistance is 10^{-8} per cell division, the probability that both mutations will occur within the same cell is $10^{-6} \times 10^{-8}$, or 10^{-14}. For this reason, two or more drugs may be administered simultaneously in the treatment of some diseases such as tuberculosis. Any mutant cell resistant to one antibiotic is likely to be sensitive to the other and therefore will be killed by the combination of the two medicines.

Mutations are stable, so that the offspring of streptomycin-resistant cells will remain streptomycin resistant. However, occasionally the nucleotide will change back to its original state, resulting in the streptomycin-resistant cells becoming streptomycin sensitive. The process by which a cell phenotype changes back to its original state through mutation is termed **reversion.** Like the original mutation, it occurs spontaneously at low frequencies that can be increased by mutagens.

• **Exponents, appendix I**

Mutations Are Expressed Rapidly in Prokaryotic Cells

Recall that prokaryotic cells contain one or sometimes several identical chromosomes per cell but never pairs of different chromosomes, a feature characteristic of eukaryotic cells. For this reason bacteria are called **haploid,** and many eukaryotic cells are **diploid.**

If the prokaryotic cell contains only a single chromosome, with each gene present only as a single copy, any mutation will be observed right after it occurs. However, if several identical chromosomes are present in the prokaryotic cell, the mutation may be masked by a nonmutant gene present in the same cell. Such a mutation is termed **recessive** and the nonmutant gene is termed **dominant.** These terms are also used in eukaryotic genetics. In general, a gene that codes for the synthesis of any product is dominant to a mutation that results in the loss of the ability to synthesize that product. However, in a bacterium that may contain two identical chromosomes, one mutant and the other nonmutant, after one cell division the mutant and nonmutant genes separate (segregate) into different cells. Even if the mutation is recessive, it will be observed in one of the cells. However, dominant mutations will be seen immediately whether one or several identical chromosomes are present in the cell.

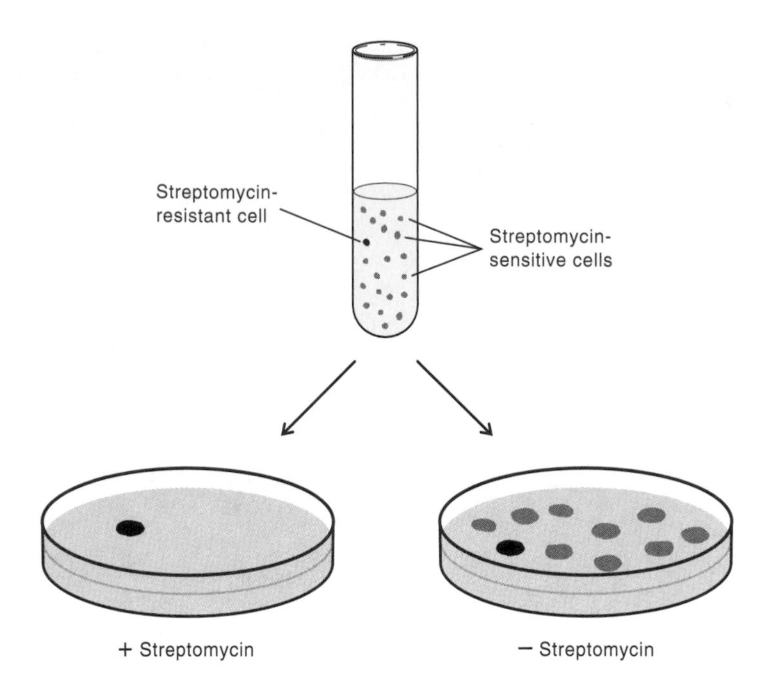

Figure 7.13
Direct selection of mutants. Only the streptomycin-resistant cells will grow on the streptomycin-containing medium. All cells will grow in the absence of streptomycin.

Mutant Selection

Genetic studies are most conveniently performed with very small organisms that can multiply on simple media and produce several billion cells per milliliter of medium in less than 24 hours. Every gene should be mutant in such a large population. However, the major problem is finding and then identifying them. Fortunately, some clever techniques have been devised that simplify the task of locating the rare mutant in a large population, the one cell in a million that has the mutation being sought.

Direct Selection Requires that the Mutant Grow on a Medium on Which the Parent Cannot

Depending on the mutant being sought, one of two techniques can be used, **direct** or **indirect selection.** Direct selection involves inoculating cells onto a medium on which the mutant, but not the parent, can grow. For example, mutants that are resistant to the antibiotic streptomycin can be easily selected directly by inoculating cells onto a medium containing streptomycin. Only the rare resistant cells in the population will form a colony; the others will be killed (figure 7.13). Antibiotic-resistant mutants are usually very easy to isolate, and most are dominant.

Indirect Selection Must Be Used to Isolate Mutants that Cannot Grow on a Medium on Which the Parent Can Grow

An example of a mutant that requires indirect selection is one that requires a growth factor to grow, while the parent strain does not. The term **auxotroph** (*auxo* means "increase," as in an increase in requirements) has been coined to describe a bacterium that requires a growth factor in order to grow. A **prototroph** (*proto* means "lowest," as in the least number of requirements) is a bacterium that does not require a growth factor to grow.

Replica Plating

Replica plating, a very clever technique for identifying mutants, was devised by the husband-and-wife team of Joshua and Esther Lederberg in the early 1950s (figure 7.14). In this technique, a master plate containing isolated colonies of all cells growing on an enriched complex medium is pressed onto sterile velvet, a fabric that has tiny threads that stand on end like tiny bristles.

A piece of every bacterial colony is transferred onto the velvet. Then two sterile plates, one containing a glucose-salts medium and the second a rich, complex medium, are pressed in succession onto the same velvet. In this way, cells imprinted on the velvet from the master plate

• **Growth factor, p. 96**

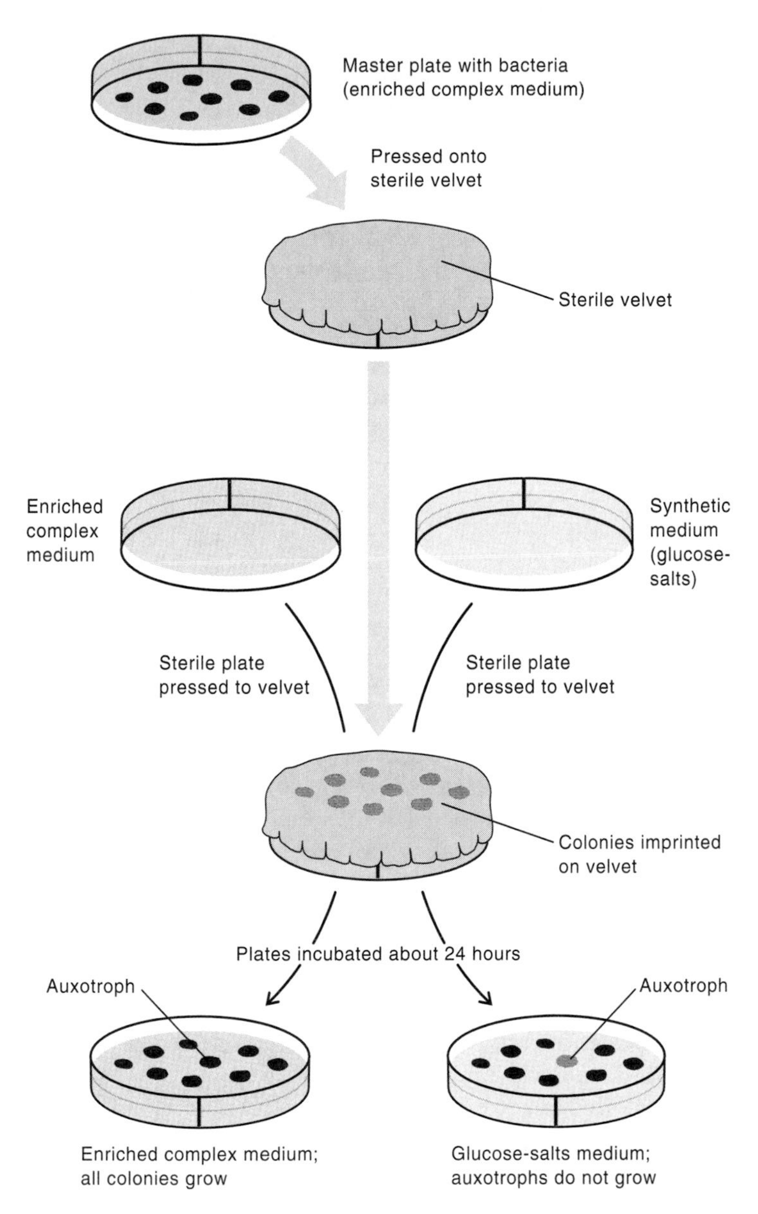

Figure 7.14
Indirect selection by replica plating, a method of identifying and isolating mutants that cannot be selected directly.

are transferred to both the glucose-salts medium and the complex medium. All prototrophs will form colonies on both the complex and the glucose-salts media, but auxotrophs will only form colonies on the complex medium. By keeping the orientation of the two plates constant as they touch the velvet, the investigator can identify any colony on the master plate that is able to grow on the complex medium but not on the glucose-salts medium. The cells in this colony must require a growth factor present in the complex medium to grow. The particular growth factor required by the organism can then be determined by adding the various factors individually to the glucose-salts medium and determining which one promotes cell growth. Before the invention of replica plating, each colony had to be individually transferred from the complex to the glucose-salts medium in order to identify the auxotrophs and determine their growth requirements, a very laborious job indeed.

Penicillin Enrichment

Even with mutagenic agents, the frequency of mutants in a population is low, ranging perhaps from about one in 1,000 to one in 1 million cells. To increase the proportion of mutants, a technique termed **penicillin enrichment** can be used (if the parent cells are killed by penicillin). In this technique, following treatment with a mutagen, the cells are grown first in a glucose-salts medium (to allow separation of a recessive mutation) and then in a glucose-salts medium containing penicillin. Since penicillin kills (by lysing) only growing cells, most of the growing prototrophs will be killed, while the nonmultiplying auxotrophs will survive (figure 7.15).

The enzyme penicillinase is then added to destroy the penicillin, and the cells are plated on an enriched complex medium. The practical result of using this enrichment technique is that far fewer colonies need to be examined to find the desired mutant.

• **Penicillin, p. 62**

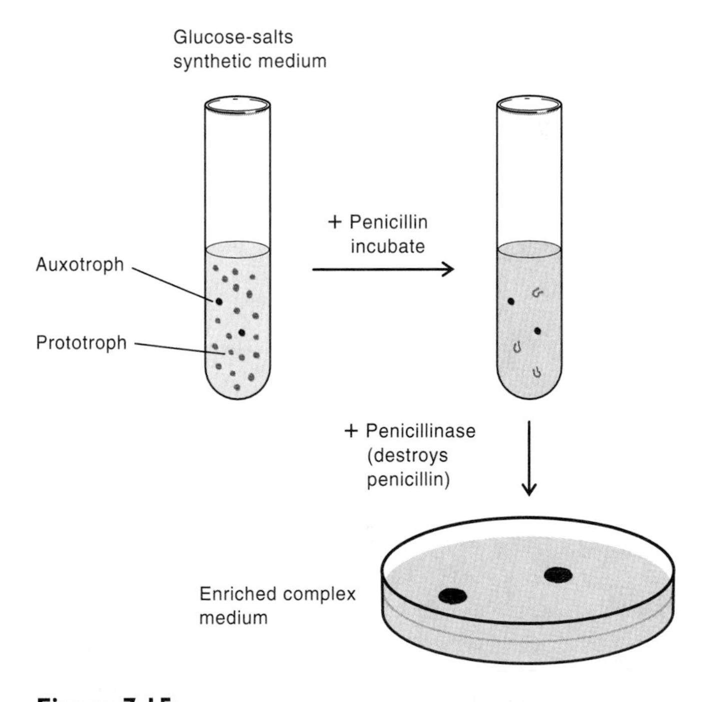

Figure 7.15
Penicillin enrichment of mutants. Since penicillin kills only growing cells, nonmultiplying auxotrophic mutants will not be killed in a glucose-salts medium, but the unmutated cells that can multiply will be lysed.

Conditional Lethal Mutants

The mutants discussed so far will grow if a growth factor they cannot synthesize is added to the medium. However, a large number of mutants may be defective in the enzymes required for macromolecule synthesis, such as factors involved in the translation of mRNA in protein synthesis or the enzymes of DNA replication. These mutations are ordinarily lethal to the cell because their defects cannot be overcome by adding small molecules to the growth medium. Therefore, they cannot be isolated by the techniques previously described. However, such mutants, given the name **conditional lethal mutants,** can sometimes be isolated, based on the fact that mutant proteins are often unstable and cannot function at the temperature at which the nonmutant enzyme catalyzes a reaction (37°C). However, they will function at a lower temperature (25°C). Thus, they are called **temperature-sensitive mutants,** one class of conditional lethal mutants. Such mutants can be isolated by the technique shown in figure 7.16 and described next.

In this technique, the master plate is incubated at a low temperature (25°C) at which all cells, both mutant and temperature-sensitive, will grow. The master plate is then replica plated onto two plates containing a complex medium. One is incubated at 37°C; the other at 25°C. Colonies growing at the low temperature but not the high one may be mutant in genes required for macromolecule synthesis.

• **Protein, pp. 30, 138**

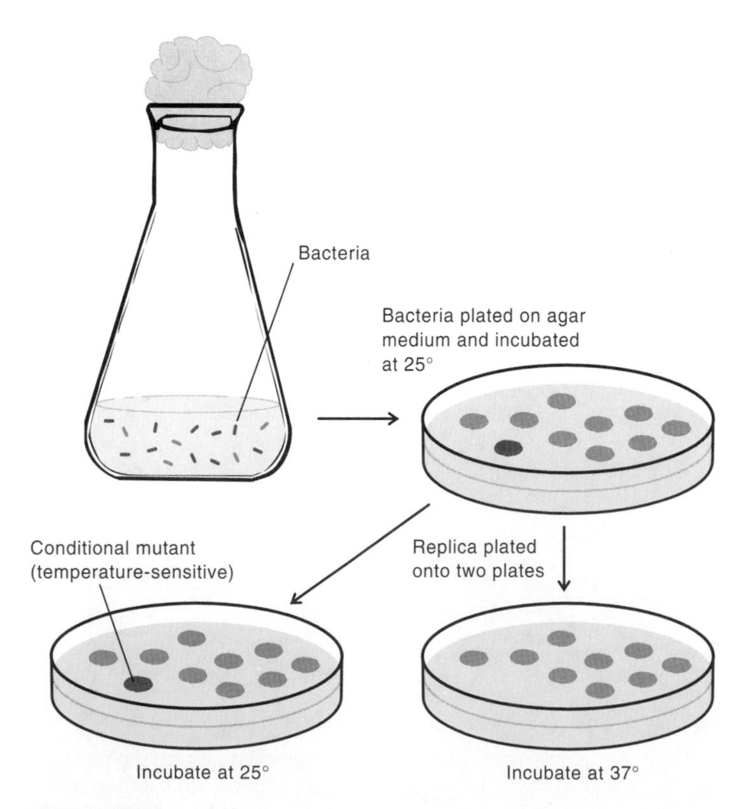

Figure 7.16
Isolation of temperature-sensitive (conditional) mutants. This procedure selects for mutants that, as a result of defective proteins, can grow only at their lower range of temperature.

Chemicals Can Be Tested for Their Cancer-causing Ability by Their Mutagenic Activity

Strong evidence exists that a substantial proportion of all cancers are caused by chemicals in the environment. How can the thousands of chemicals released into the environment, such as pesticides, herbicides, and the by-products of manufacturing processes, be tested inexpensively for their carcinogenic activity? Testing in animals takes two to three years and may cost $100,000 or more for the testing of a single compound. A relatively simple, inexpensive, and quick microbiological test for potential carcinogens was developed by Bruce Ames and his colleagues in the 1960s. The **Ames Test** takes only a few days and is based on three facts: (1) the reversion of a mutant gene in a biosynthetic pathway, such as histidine biosynthesis, can be readily measured; (2) the frequency of reversions is increased by mutagens; and (3) most carcinogens are mutagens. Specifically, the Ames Test measures the rate of reversion of a histidine auxotroph of *Salmonella* to prototrophy in both the presence and absence of the chemical being tested (figure 7.17). If the chemical is mutagenic, and by implication carcinogenic, it will increase the reversion rate. The test also gives some idea about how powerful a mutagen is and therefore how potentially hazardous a chemical it is by the number of revertants that arise.

The Ames Test fails to detect many carcinogens because some substances are not carcinogenic themselves but can be converted to active carcinogens by a metabolic reaction that occurs in animals but not in bacteria. Therefore, an extract of ground-up rat liver, which has the enzymes to carry out these conversions, is added to the petri plates containing the mutagen (carcinogen) being tested as part of the Ames Test. Thousands of chemicals have been tested for their mutagenic capabilities, a few of which are listed in table 7.3.

Additional testing must be done on any mutagenic agent identified in the Ames Test to confirm that it is actually carcinogenic in animals. Although data are not available on the percentage of mutagens that are carcinogens, it is clear that the Ames Test is useful as a rapid screening test to identify those compounds that have a high probability of being carcinogenic. Thus far, no compound that is negative in the Ames Test has been shown to be carcinogenic in animals.

REVIEW QUESTIONS

1. Name two factors that are important in changing the properties of organisms.
2. Distinguish between the phenotype and genotype of an organism.
3. List four ways in which mutations can occur.
4. How does light repair of DNA damage differ from dark repair?

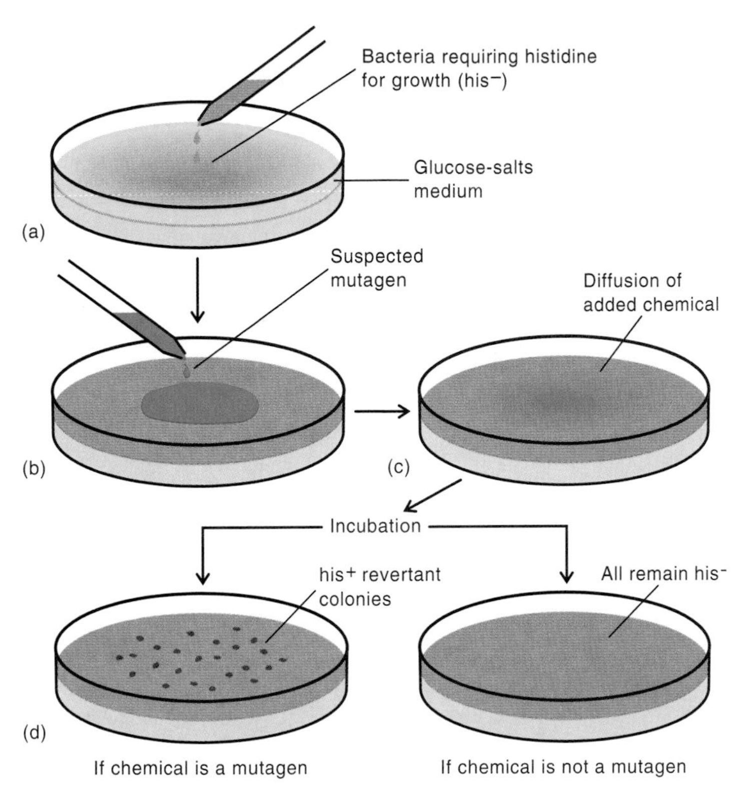

Figure 7.17
Ames test to screen for mutagens. The chemical will increase the frequency of reversion of his^- to his^+ cells if it is a mutagen and, therefore, most likely a carcinogen.

Table 7.3 Chemicals and Radiations Tested for Mutagenic and Carcinogenic Activity

Chemical or Radiation	Area of Use	Mutagenic	Carcinogenic in Animals
Vinyl chloride	Chemical industry	+	+
Captan	Fungicide	+	+
Sevin	Insecticide	0	0
Cigarette smoke condensate	Recreation	+	+
Nicotine	Tobacco	0	0
Caffeine	Coffee	0	Not tested
Caffeic acid	Coffee	Slightly	Not tested
Hydrogen peroxide		+	
UV	Sunlight	+	+
Radiation	Halogen lamps	+	+

Mechanism of Gene Transfer—General Aspects

The genetic information in a cell can be altered by, in addition to mutation, the transfer of sets of genes from other cells and of certain genes (transposable elements) from one location to another in the genome of the same cell.

The combination of intracellular and intercellular movement of DNA accounts for the rapid spread of resistance to antibiotics and heavy metals in bacterial populations, as described for *Shigella* earlier in this chapter.

Gene Transfer Between Bacteria

Mutants make it possible to determine whether **genetic recombination,** the combining of DNA (genes) from the same or from two different cells, occurs in bacteria. Recombination in the DNA from two different bacteria can be readily recognized because the resulting cells, termed **recombinants,** have properties of each of the bacteria. For example, when cells that are streptomycin resistant (str^R)[1] and require histidine (his^-)[2] and tryptophan (trp^-) to grow are mixed with cells that are killed by streptomycin (str^S) and require leucine (leu^-) and threonine (thr^-) to grow, rare recombinants appear that are resistant to streptomycin and grow on a glucose-salts medium—that is, that are his^+, trp^+, leu^+, thr^+ and str^R (figure 7.18).

These recombinants can arise only if DNA has been transferred from one cell (donor) to another (recipient). Reversion of the auxotrophs cannot account for the result because at least two genes would have had to revert in the same cell, an extremely rare event.

Genes are transferred between bacteria by three different mechanisms:

1. **DNA-mediated transformation,**[3] in which DNA is transferred as "naked" DNA. This is the mechanism discovered by Griffith (see "A Glimpse of History").
2. **Transduction,** in which bacterial DNA is transferred by a bacterial virus.
3. **Conjugation,** in which DNA is transferred between bacteria that are in contact with one another.

Although these processes differ from one another in the delivery system for DNA, they have certain features in common. First, only a part of the chromosome is transferred from the donor cell to the recipient cell. This contrasts with the

[1]str^R indicates that the cells are resistant (R) to streptomycin; str^S indicates the cells are sensitive.
[2]his^- indicates that the bacteria require histidine supplied in the medium in order to grow. By convention, only the growth factor requirements are designated. his^+ means that the cell can synthesize histidine, so it does not have to be supplied in the medium.
[3]DNA-mediated transformation—not to be confused with transformation of normal eukaryotic cells into cancer cells

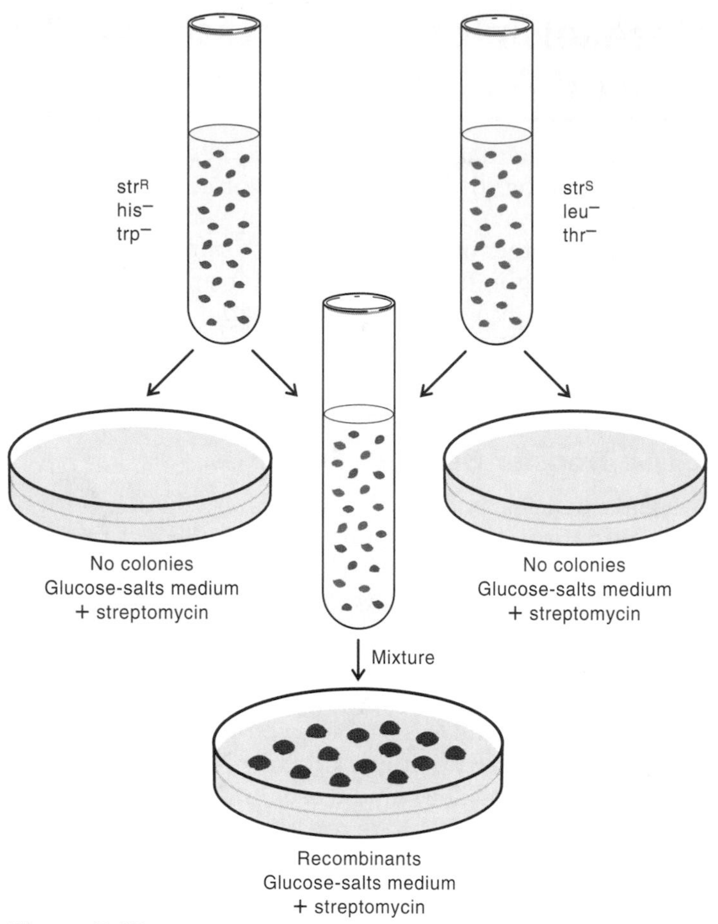

Figure 7.18
General experimental approach for detecting gene transfer in bacteria.

situation in eukaryotic cells in which cells, such as the egg and the sperm, fuse so that two sets of all chromosomes are present. Second, the DNA that is transferred probably undergoes the same process of integration in all three transfer schemes (figure 7.19).

Once inside the recipient cell, the piece of chromosomal donor DNA is positioned alongside the DNA of the recipient cell so corresponding (homologous) genes are adjacent. This type of recombination is therefore termed **homologous recombination.** Next, the donor DNA replaces a piece of the recipient chromosome DNA and becomes a part of the chromosome of the recipient cell. As a result, the chromosome of the recombinant cell contains DNA derived from both the donor cell and the recipient cell, and the information contained in this DNA is passed down to all progeny cells. Exactly how homologous recombination occurs is now beginning to be understood. Recombination of DNA molecules is extremely precise; there is no increase or decrease in the number of nucleotides in the recombinant, only an exact substitution of donor genes for recipient genes. This mechanism of recombination in prokaryotic cells is termed **breakage and reunion** and is very likely the mechanism of recombination in eukaryotic cells also.

Another feature common to all three mechanisms is that chromosomal DNA is generally transferred to no more than about one in 100 of the recipient cells. Therefore, to detect gene transfer it is helpful to be able to select directly for the recombinants, by inoculating cells on a medium on which only the recombinants will grow. Since several billion bacteria can be inoculated onto the agar contained in a single petri dish, rare recombinants can be detected readily.

DNA-mediated Transformation Involves the Transfer of Naked DNA from Donor Cells to Recipient Cells

If the cell walls of bacteria are ruptured, each long, circular molecule of chromosomal DNA that is tightly jammed into the bacterium breaks up into about 100 pieces, each of which consists of about 20 genes. This DNA can pass through the cell wall and cytoplasmic membrane of some of the recipient cells and can be integrated into the chromosome of the recipient cell (figure 7.20).

Natural Competence of Recipient Cells

One of the most interesting features of the DNA transformation process is the uptake of very large pieces of DNA by the recipient cell. As discussed already, only molecules whose molecular weight is no greater than a few hundred can usually pass through the cytoplasmic membrane. Therefore, it is still a mystery how DNA molecules whose molecular weight is a hundred thousand times greater can pass through. It is clear, however, that only under certain growth conditions do bacteria become able to take up large molecules of DNA. Such cells are termed **competent.** The manner in which a cell becomes competent is poorly understood, but it seems likely that a whole series of changes takes place, including modification of the cell wall and the synthesis of protein receptors on the cytoplasmic membrane that can bind DNA. Not all species of bacteria can become competent, and thus not all bacteria can be transformed. However, it has been observed that DNA-mediated transformation can occur in a wide variety of gram-positive and gram-negative bacteria. Some of these bacteria are listed in table 7.4. In addition, the eukaryote yeast *Saccharomyces cerevisiae* can be transformed.

Table 7.4 Some Species in Which DNA-Mediated Transformation Has Been Observed

Gram-Positive Bacteria	Gram-Negative Bacteria	Eukaryote
Bacillus subtilis	*Haemophilus influenzae*	*Saccharomyces cerevisiae*
Streptococcus pneumoniae	*Escherichia coli*	
Deinococcus radiodurans	*Rhizobium meliloti*	
Staphylococcus aureus	*Neisseria gonorrhoeae*	

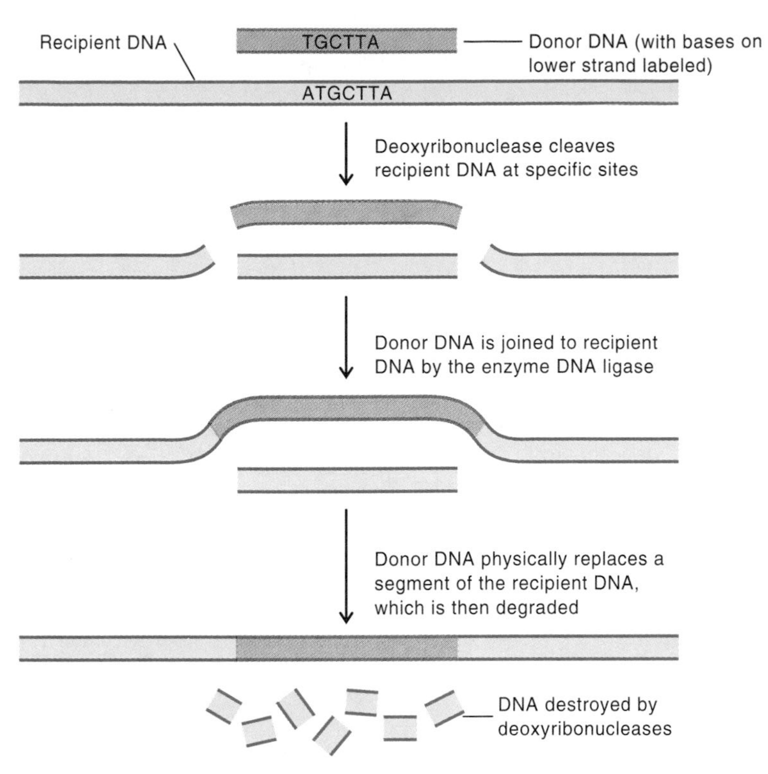

Figure 7.19
Integration of donor DNA with recipient cell DNA. Although this figure shows integration of both strands simultaneously, in many cases only one of the two strands of donor DNA is integrated. The complementary strand is then synthesized. The end result is the formation of recombinant DNA, which consists of a portion of the donor genome and most of the recipient cell genome.

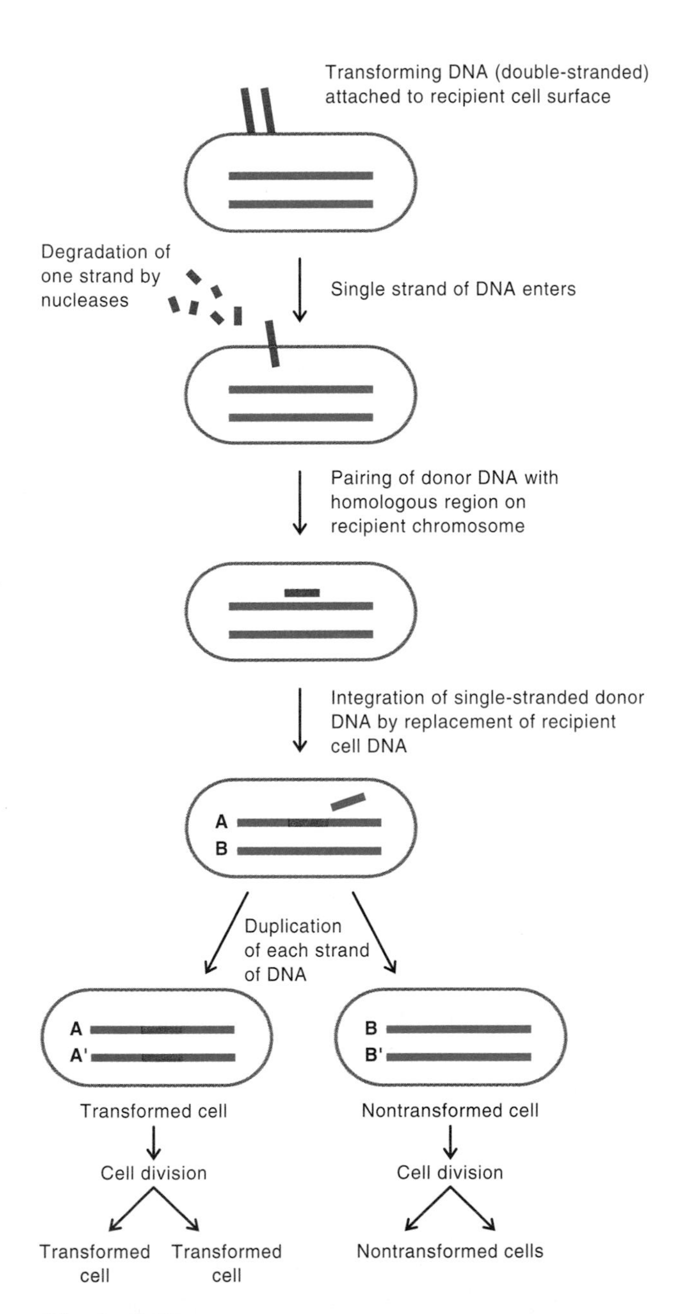

Figure 7.20
DNA-mediated transformation. The DNA is integrated as a single-stranded DNA molecule.

Artificial Competence

Although not all bacteria become competent under usual growth conditions, it is possible to introduce DNA into most cells (including bacteria, animals, and plants) through special techniques.

In one technique called **electroporation,** bacteria and DNA are mixed together and the mixture is treated with an electric current (figure 7.21). The current apparently makes holes in the bacterial cell wall and cytoplasmic membrane through which the DNA enters. Once inside the cytoplasm, the DNA becomes integrated into the recipient chromosome by homologous recombination.

Transduction Involves Viral Transfer of DNA

A bacterial virus called a **bacteriophage** (or simply **phage,** for short) can transfer genes from one bacterium to another (figure 7.22), which is the mechanism termed **transduction.**

In the most common type of transduction, after the phage infects the bacterium, the phage DNA codes for the enzyme deoxyribonuclease, which degrades the bacterial chromosome into about 100 pieces. One of these DNA fragments may accidentally become part of the phage particles as the phage multiplies. Once the phage, with the bacterial DNA inside it, is released from the infected cell, it may infect another cell, thereby transferring the bacterial genes. Once inside the recipient cell, the bacterial DNA is integrated by homologous recombination. This process of transduction is considered in more detail in chapter 13, in which the life cycles of phages are discussed.

• **Virus, p. 16**

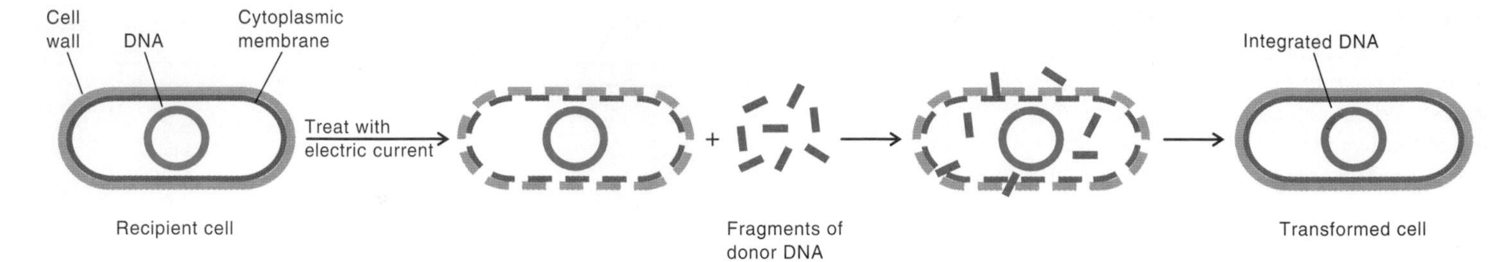

Figure 7.21
Electroporation. The electric current makes holes in both the cell wall and the cytoplasmic membrane through which the DNA can pass. These holes are then repaired by the cell, and the DNA becomes incorporated into the chromosome of the cell.

Table 7.5 Some Bacteria in Which Conjugation Has Been Observed

Gram-Positive Bacteria	Gram-Negative Bacteria
Enterococcus faecalis	*Escherichia coli*
Staphylococcus sp.	*Salmonella typhimurium*
	Pseudomonas aeruginosa
	Neisseria gonorrhoeae
	Escherichia coli (yeast in laboratory)

Conjugation Requires Cell-to-cell Contact

Another important and common mechanism of gene transfer is **conjugation.** This mechanism is observed primarily in gram-negative organisms, although some gram-positive organisms also transfer genes by this process. Some bacteria in which conjugation has been observed are listed in table 7.5.

Conjugation differs from transformation and transduction in that it requires contact between the donor cells and recipient cells. This requirement for contact can be shown through the following simple experiment. If two different auxotrophic mutants are placed on either side of a filter through which fluids, but not bacteria, can pass, recombination does not occur. However, if the filter is removed, allowing cell-to-cell contact, recombination takes place (figure 7.23).

This mechanism of transfer is far more complex than it appears from this simple experiment, but it is, in large part, understood. Two types of cells exist in any population of *E. coli.* One is the **male,** or **donor, cell;** the other is the **female,** or **recipient, cell.** DNA is transferred only in one direction, from donor cell to recipient cell. In the most common situation, the donor contains a special kind of plasmid, a small, circular, double-stranded DNA molecule that is not present in the recipient cell (figure 7.24).

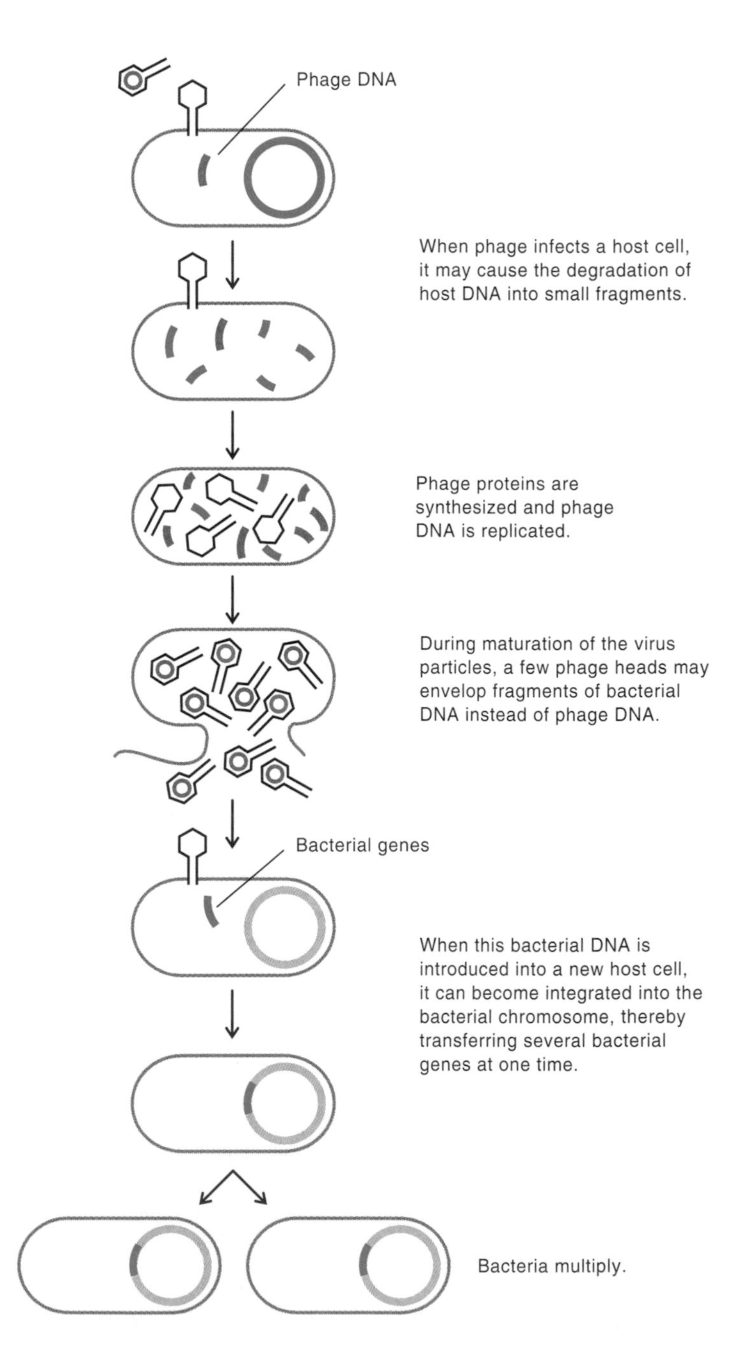

Figure 7.22
Transduction (generalized). Any piece of the chromosomal DNA of the donor cell can be transferred in this process. All of the DNA molecules, of the bacterial virus and of the bacteria, are double stranded.

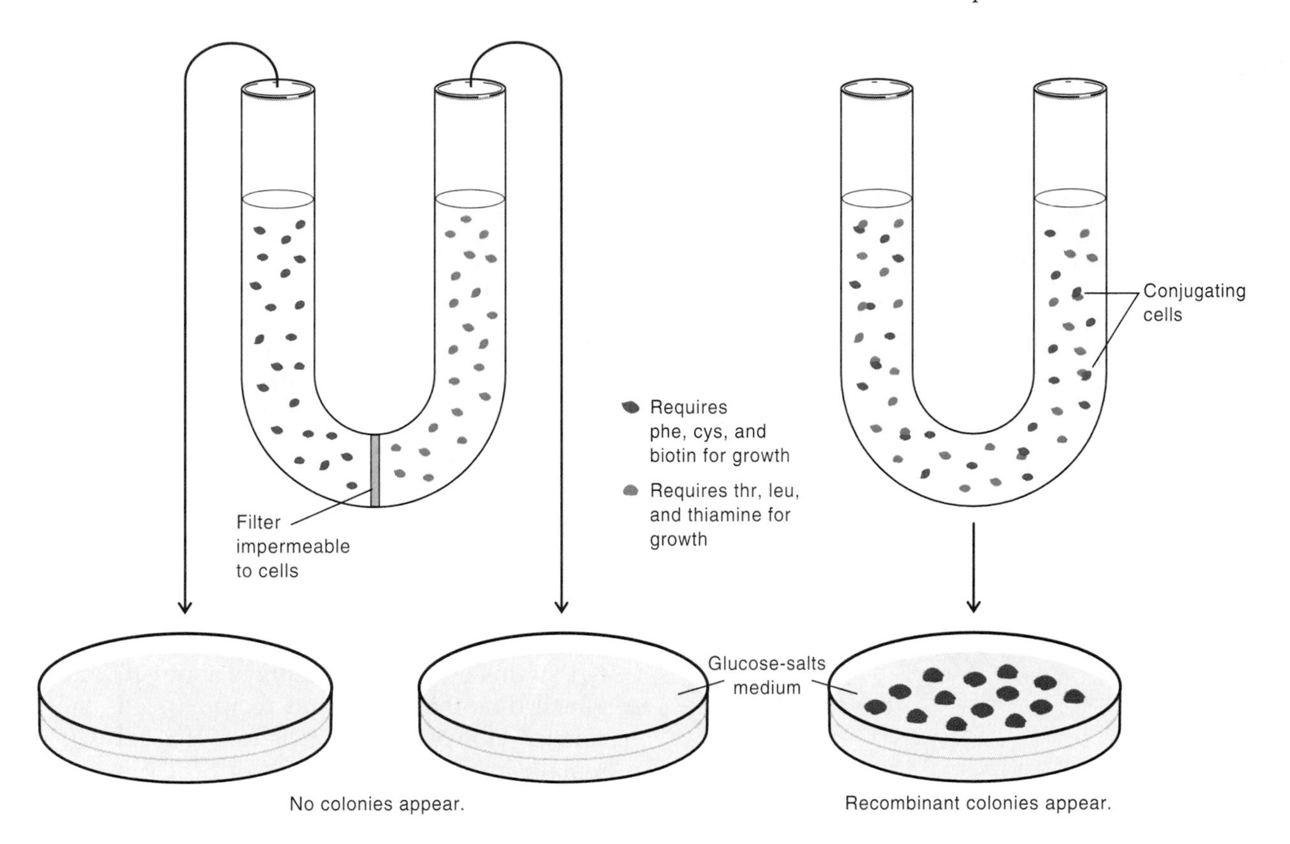

Figure 7.23
Demonstration of the need for cell-to-cell contact in conjugation. In order to grow on the glucose-salts medium, the bacteria must be able to synthesize all amino acids and vitamins.

Bacteria contain many different-sized plasmids that code for different properties of the cell. This particular plasmid is termed the **F plasmid** (for *fertility*). These donor cells are only able to transfer their F plasmid and not their chromosome.

The F plasmid in the donor cell carries the genetic information for the synthesis of the **sex pilus,** the organelle that attaches the donor to the recipient cell. It also carries additional genetic information required for the transfer of DNA. If the donor cell, which is also termed F^{+}, loses the F plasmid, it becomes F^{-} because it can no longer synthesize the sex pilus and attach to the recipient cell. In fact, it now acts as a recipient cell, which is F^{-}.

When donor and recipient cells are mixed together, the sex pili act as grappling hooks, pulling the two cells together (figure 7.25).

Within minutes of contact, the F plasmid from the donor cell enters the recipient cell. Because all F^{-} cells in the population receive the F plasmid, the entire population quickly becomes F^{+}. In addition, any other plasmids the donor cell contains, such as those coding for antibiotic resistance, may also be transferred. Only plasmids are transferred at a high frequency from F^{+} to F^{-} cells; the chromosome is rarely transferred (figure 7.26). Note that, once inside the recipient cell, the F plasmid is able to replicate and is not integrated into the DNA of the recipient cell. Thus, it does not undergo the process of homologous recombination discussed for the transfer of chromosomal DNA.

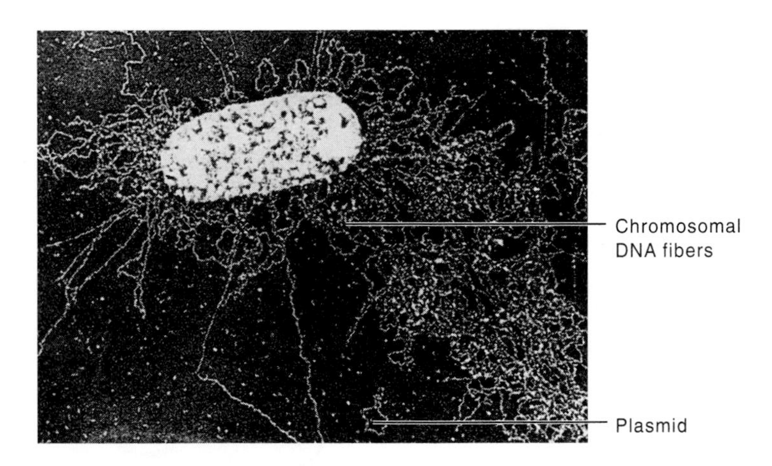

Figure 7.24
Chromosomal and plasmid DNA leaking out of a cell. Note the relative sizes of the two kinds of DNA.

• **Plasmid, p. 72**

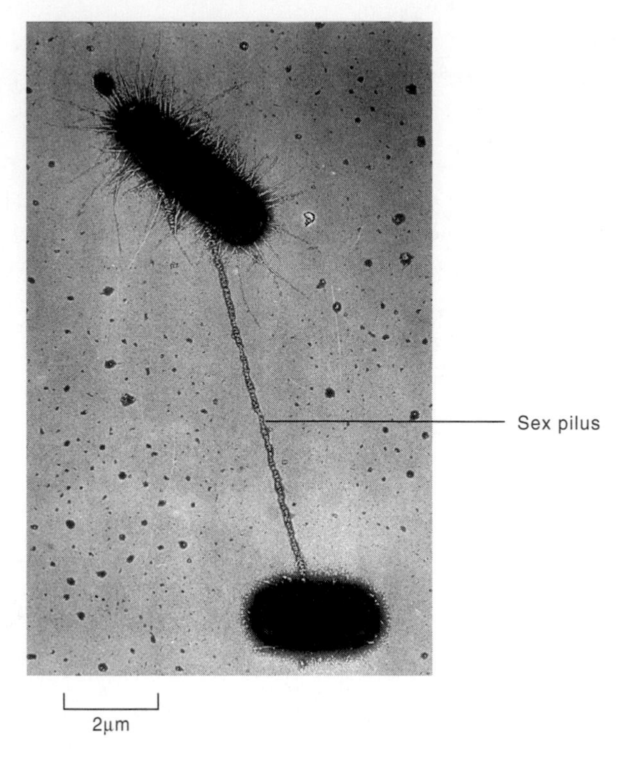

Figure 7.25
Sex, or F, pilus holding together donor and recipient cells of *E. coli* during DNA transfer. The tiny knobs on the pilus are bacterial viruses that have adsorbed to the pilus.

Hfr Formation

The bacterial chromosome, however, is sometimes transferred to F^- cells by a few cells in the F^+ population. These donor cells are not in fact F^+ but arise from F^+ cells as a result of the F plasmid becoming incorporated into the donor cell chromosome on rare occasions (figure 7.27). These cells are termed **Hfr** (for *h*igh *f*requency of *r*ecombination). In the Hfr cell, the F plasmid replicates as part of the chromosome, in contrast to the F^+ cell, in which the F plasmid replicates independently of the chromosome. Thus, the progeny of an Hfr cell are also Hfr.

The F plasmid must be incorporated into the chromosome for the donor chromosome to be transferred into the recipient cell. When the Hfr and F^- cells contact each other, the circular donor chromosome breaks at the site at which the F plasmid is integrated. The chromosome, now linear, is transported into the recipient cell (figure 7.28).

In contrast to the rapid transfer of the small F plasmid into F^- cells, the transfer of the entire chromosome from Hfr strains into F^- cells requires about 100 minutes, during which time the donor and recipient cells must remain in contact. However, since it is rare that the Hfr and F^- cells remain in contact for this length of time, generally only short segments of the chromosome are transferred. The chance that a gene will be transferred is directly related to its distance from the region that is transferred first, termed the **origin of transfer.** The genes closest to the origin, which include some of the F plasmid, are transferred first; genes farther away are transferred later and perhaps not at all if the cells separate before transfer takes place. The rest of the F plasmid integrated into the chromosome is transferred last from the donor cell, and in most cases the cells separate before the integrated plasmid is transferred. Therefore, the vast majority of the recipient cells remain F^- following chromosome transfer.

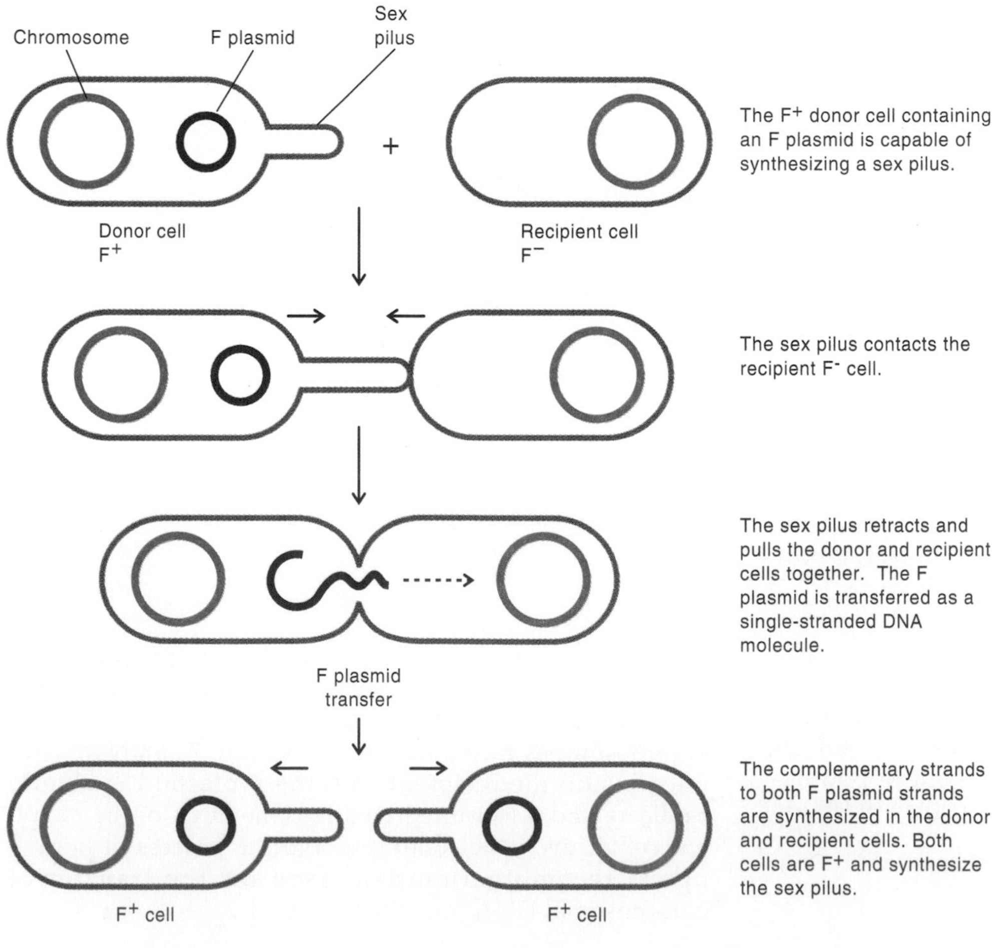

Figure 7.26
Conjugation—transfer of the F plasmid. The exact process by which the donor DNA passes to the recipient cells is not known.

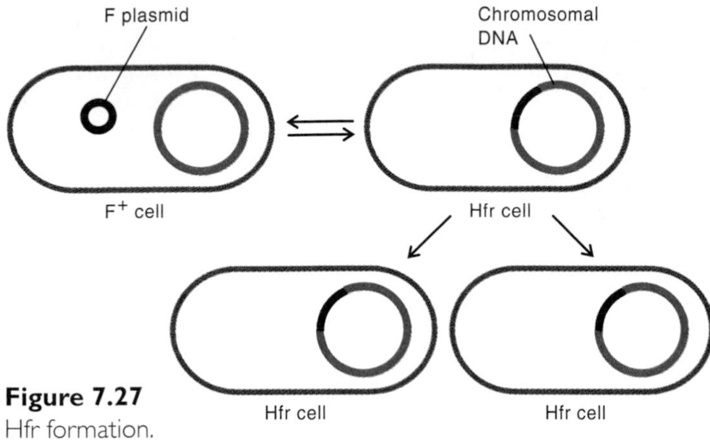

Figure 7.27
Hfr formation.
The F plasmid integrates at specific locations into the chromosome. This mobilizes the chromosome for transfer. The process is reversible.

F' Donors

The F plasmid in the Hfr strain can be removed (excised) from the chromosome; thus, the process of F plasmid incorporation is reversible. However, in some instances in the process of excision, a small piece of the bacterial chromosome remains attached to the F plasmid (figure 7.29). This altered F plasmid is called *F'* (F prime) and has the F plasmid's property of being rapidly and efficiently transferred to F$^-$ cells. Consequently, any chromosomal genes attached to the F plasmid are also transferred quickly. The F' plasmid usually remains extrachromosomal, (that is, not a part of the recipient cells' chromosome) so that, like the F plasmid, it multiplies independently of the chromosome. However, on rare occasions it can once again become incorporated into the chromosome of the cells.

The relationship between the donor chromosome and the extrachromosomal particles involved in gene transfer is illustrated in figure 7.30. The two people most responsible for unraveling the relationship between the F particle and the Hfr strains were the French microbiologists François Jacob and Elie Wollman. François Jacob developed a new concept in biology—that some DNA may exist in one of two forms, either extrachromosomally or as part of the chromosome and that these two states are reversible. Such a piece of DNA is termed an **episome.**

The three mechanisms of DNA transfer are compared in table 7.6.

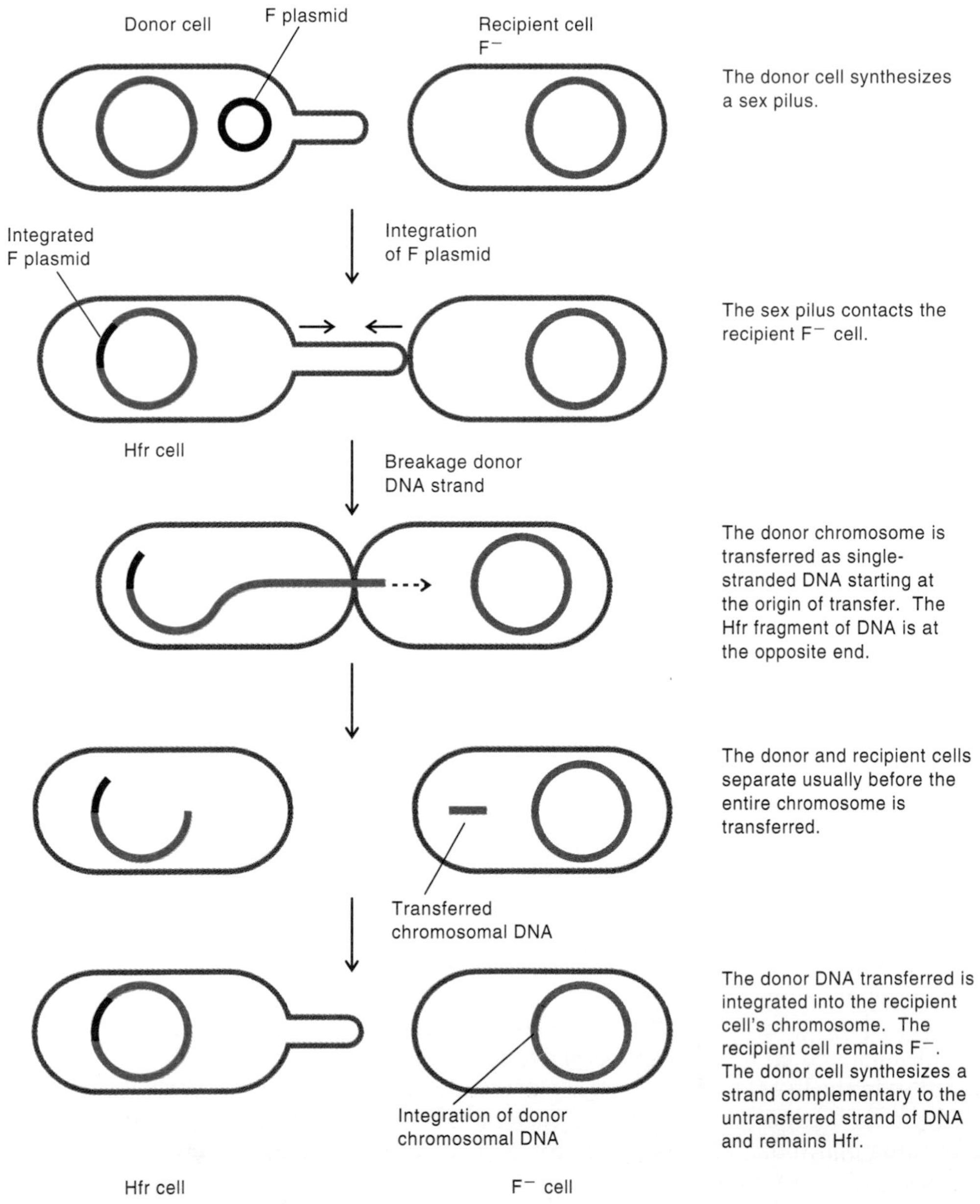

Donor cell
F plasmid
Recipient cell F$^-$

The donor cell synthesizes a sex pilus.

Integrated F plasmid
Integration of F plasmid

The sex pilus contacts the recipient F$^-$ cell.

Hfr cell
Breakage donor DNA strand

The donor chromosome is transferred as single-stranded DNA starting at the origin of transfer. The Hfr fragment of DNA is at the opposite end.

The donor and recipient cells separate usually before the entire chromosome is transferred.

Transferred chromosomal DNA

The donor DNA transferred is integrated into the recipient cell's chromosome. The recipient cell remains F$^-$. The donor cell synthesizes a strand complementary to the untransferred strand of DNA and remains Hfr.

Integration of donor chromosomal DNA

Hfr cell
F$^-$ cell

Figure 7.28
Conjugation—transfer of chromosomal DNA. The DNA is transferred as a single-stranded DNA molecule. The complementary strand to both the DNA that is transferred and the strand that remains behind is synthesized following the transfer process. Only a small fraction of the total chromosome is usually transferred.

Table 7.6 Comparison of Mechanisms of DNA Transfer

Mechanism	Main Feature	Frequency of Transfer	Size of DNA Transferred	Polarity?	Sensitivity to DN'ase?
Transformation	Naked DNA transferred	Rare	Approximately 20 genes	No	Yes
Conjugation					
Chromsomal DNA	Cell-to-cell contact required	Rare	All or part of chromosome	Yes	No
F plasmid	Cell-to-cell contact required	Frequent	Entire plasmid	Yes	No
Transduction	DNA enclosed in bacteriophage coat	Rare	Small fraction of chromosome	No	Yes

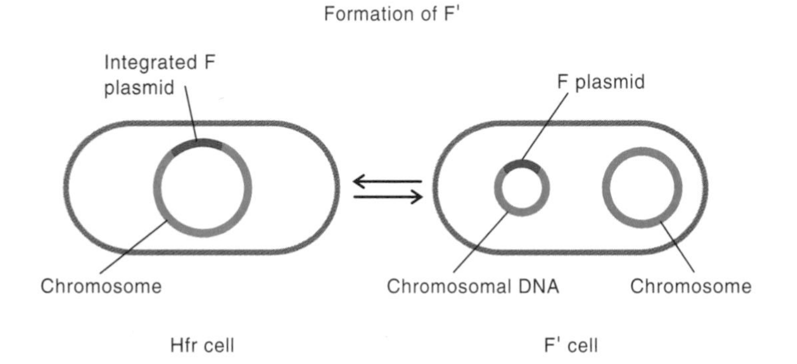

Figure 7.29
Formation of F' plasmid. This F' plasmid has properties of the F plasmid but carries chromosomal DNA. This process is reversible.

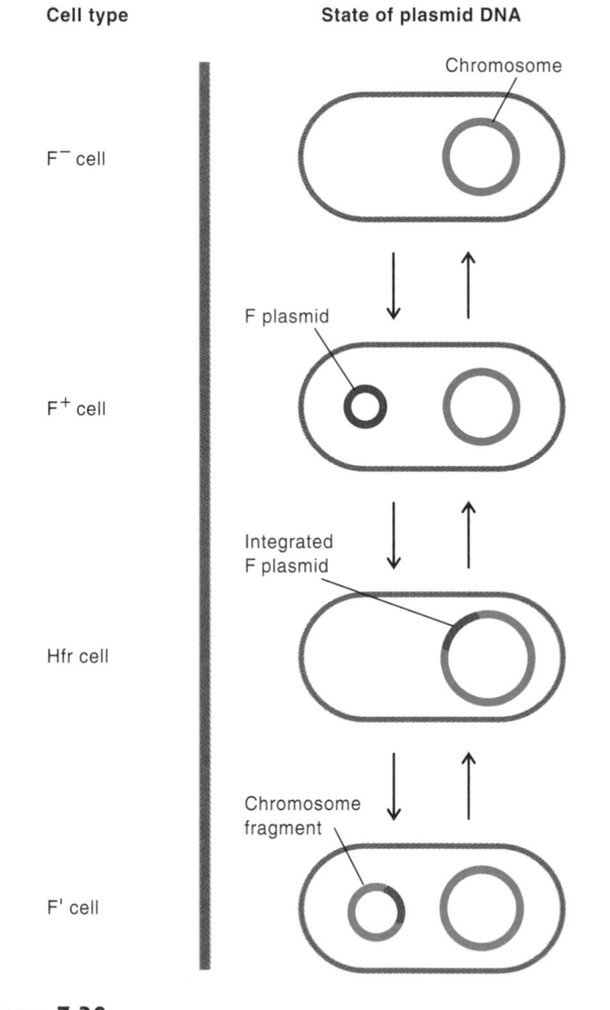

Figure 7.30
Interconversion of various cell types. The F^- cell receives the F plasmid from an F^+ cell to become F^+.

REVIEW QUESTIONS

1. In what mechanism of DNA transfer can the largest piece of DNA be transferred?
2. List four features of bacteria that make them ideal for genetic studies.
3. What distinguishes a donor cell of *E. coli* from a recipient cell in conjugation?
4. The term *competence* is associated with which mechanism of DNA transfer?
5. What mechanism(s) of DNA transfer involves polarity?

Plasmids

The F plasmid is only one example of a plasmid, an extrachromosomal piece of DNA that multiplies independently of the chromosome. Not only are plasmids very common in the microbial world, being found in eukaryotic algae, fungi, and protozoa, but they play very important roles in the lives of these organisms. Their role in the life of bacteria is especially well understood. One of the most interesting plasmids is the tumor-inducing (Ti) plasmid of *Agrobacterium*. It allows *Agrobacterium* to conjugate with plants (see perspective 7.2).

A study of a wide variety of plasmids in a large number of different bacteria has led to the following generalizations. First, most bacteria contain one or more different plasmids,

Perspective 7.2

Bacteria Can Conjugate with Plants

For over 40 years, scientists have known that DNA can be transferred between bacteria. In the last 20 years, it has been shown that a bacterium can even transfer its genes into plant cells, such as tobacco, carrots, and cedar trees, through a process analogous to conjugation. In the early years of this century, the causative agent of a common plant disease, termed *crown gall,* was shown to be the bacterium *Agrobacterium tumefaciens*. This disease is characterized by large galls or swellings that occur on the plant at the site of infection, usually near the soil line. When the diseased plant tissue was cultured on agar plates containing nutrients necessary for the growth of plant tissue, it displayed properties that differed from normal plant tissue. Whereas normal tissue requires several plant hormones for growth, the crown gall tissue grows in the absence of these added hormones. In addition, the crown gall tissue synthesizes large amounts of a compound termed *opine,* which neither normal plant tissue nor *Agrobacterium* synthesizes. The most surprising observation was that the plant cells maintain their altered nutritional requirements and the ability to synthesize the opine even after the bacteria are killed by penicillin. Indeed, the crown gall plant cells are permanently transformed. Although *Agrobacterium* is required to start the infection, the bacteria are not necessary to maintain the nutritional requirements and biosynthetic capabilities of the plant cells.

The explanation of the process by which *Agrobacterium* causes crown gall tumors and transforms plant cells was established in 1977. A team of microbiologists at the University of Washington showed that a specific piece of a **tumor-inducing,** or **Ti, plasmid** termed the **transferred,** or **T-DNA,** present in all crown gall-causing strains of *Agrobacterium* is transferred from the bacterial cell to the plant cell, where it becomes incorporated into the plant chromosome (figure 1).

This bacterial DNA acts like plant DNA because the promoter regions of the transferred genes resemble those of plants rather than bacteria. Therefore, the genetic information in the T-DNA is expressed in the plants but not in *Agrobacterium*. This DNA codes for enzymes for the synthesis of the plant hormones as well as of the opine. Thus, the expression of these genes supplies the plant cells with the plant hormones and explains why the transformed plant cells can grow in the absence of added hormones and are able to synthesize opine.

The *Agrobacterium*-crown gall system is of great interest for several reasons. First, it shows that DNA can be transferred across kingdoms, from prokaryotes to eukaryotes. Many people believed that such transfer between kingdoms would be impossible in nature and would only be possible in the laboratory, through such techniques as electroporation to deliver DNA into cells. Second, this system has spawned an entire plant biotechnology industry dedicated to improving the characteristics of higher plants. Thus, it is possible to replace the genes of hormone synthesis in the plasmid with any other genes, which will be transferred and incorporated into the plant. With this technology, genes conferring resistance to bacteria, viruses, insects, and different herbicides have been incorporated into a wide variety of plants. Genetic engineering of plants thus became a reality once scientists learned how a common soil bacterium, caused a well-known plant disease.

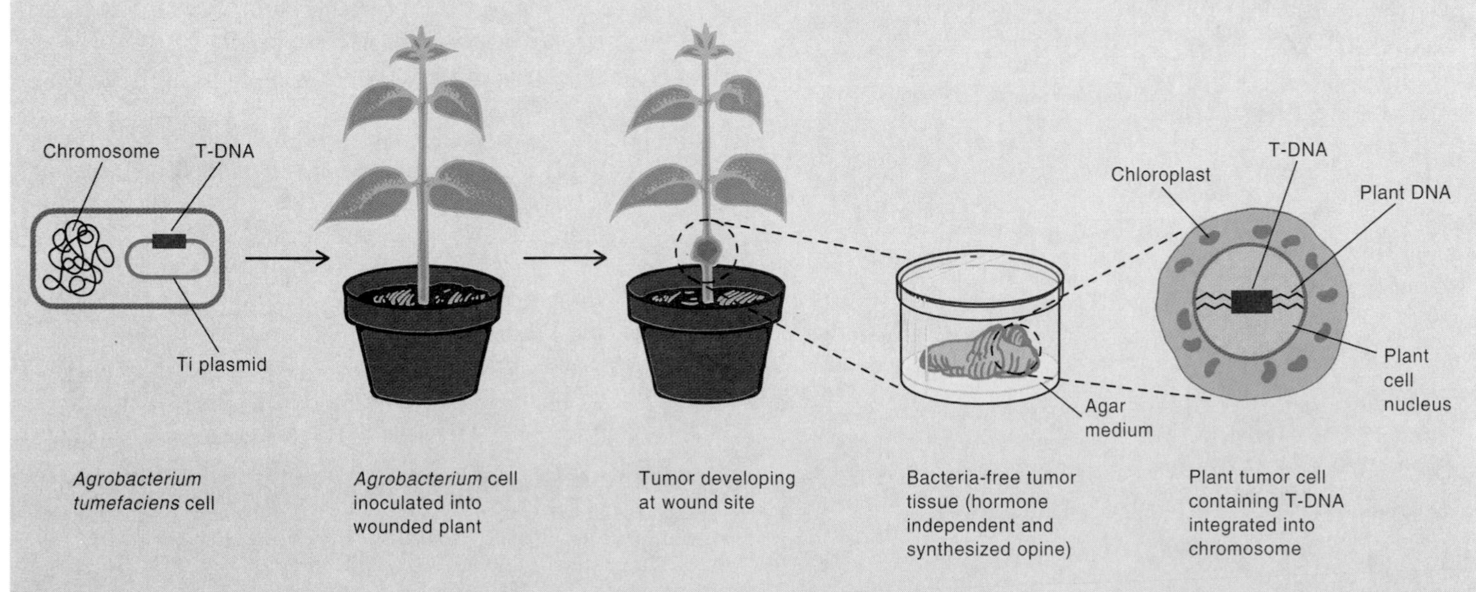

Figure 1
Agrobacterium sp. causes crown gall and transforms plant cells.

but their function is often unknown. Second, plasmids vary in size from a few genes to several hundred. Since the only function that all plasmids must carry out is replication, all of them are **replicons,** pieces of DNA that can replicate. Third, the traits coded by plasmids generally provide the bacterial cell with useful but not indispensable capabilities. For example, many members of the genus *Pseudomonas* can grow not only on many common compounds, such as glucose, but also on unusual compounds, such as camphor if it has a certain plasmid. If it does not have the plasmid, it can still grow on glucose. Thus, the plasmid extends the cells' biochemical capabilities for degrading certain compounds. The enzymes of the glycolytic pathway are never encoded on plasmids, since they are indispensable to the cell. However, plasmids code for a wide variety of other traits, some of which are listed in table 7.7. Fourth, the number of copies of a plasmid in a cell varies, depending on the plasmid. For example, the F plasmid is present in only one to two copies per cell; therefore it is a **low-copy number plasmid.** Other plasmids may be present in more than 100 copies per cell; these plasmids are called **high-copy number plasmids.**

Although plasmids can be transferred to other bacteria by the three methods we have discussed, conjugation is by far the most important means of transfer for many gram-negative species of bacteria. Most plasmids, termed **narrow host range plasmids,** can multiply only in one species of bacteria. The F plasmid is an example of a narrow host range plasmid. A few, however, termed **wide host range plasmids,** can multiply in a wide variety of different species of bacteria.

Table 7.7 Some Plasmid-Coded Traits

Trait	Organisms in Which Trait Is Found
Antibiotic resistance	*Escherichia coli, Salmonella, Neisseria, Staphylococcus, Shigella,* sp. and many other organisms
Pilus synthesis	*Escherichia coli, Pseudomonas* sp.
Tumor formation in plants	*Agrobacterium tumefaciens*
Nitrogen fixation (in plants)	*Rhizobium* sp.
Oil degradation	*Pseudomonas* sp.
Gas vacuole production	*Halobacterium* sp.
Insect toxin synthesis	*Bacillus thuringiensis*
Plant hormone synthesis	*Pseudomonas* sp.
Antibiotic synthesis	*Streptomyces* sp.
Increased virulence	*Yersinia enterocolitica*

Some plasmids, such as the F plasmid, carry all of the information required for transfer from one organism to another by conjugation; these are termed **conjugative plasmids.** Others do not have this information and are called **nonconjugative plasmids.** However, they can often be transferred by conjugation if the same cells also contain a conjugative plasmid.

Some important properties of plasmids are given in table 7.8.

R Factor Plasmids Confer Resistance to a Large Number of Antimicrobial Agents

Among the best-studied and most important groups of plasmids are the **resistance,** or **R factor, plasmids,** which confer resistance to many different antibiotics and heavy metals, such as mercury and arsenic. Heavy metals, like antibiotics, are also found in the hospital environment. Many of these plasmids are composed of two parts: a ***r**esistance **t**ransfer **f**actor* **(RTF),** which codes for the transfer of the plasmid to other bacteria by conjugation, and **resistance genes (R genes),** which code for the resistance trait (figure 7.31). The RTF portion contains genes that code for pilus synthesis and origin of transfer and other genes required for the transfer of the plasmid to the recipient cell. Such plasmids are conjugative.

The R genes code for resistance to widely used antibacterial medicines, which can include sulfanilamide, streptomycin, chloramphenicol, and tetracycline and a number of different heavy metals. Some R factors confer resistance to some, but not all, of these antimicrobial substances; yet others confer resistance to more than these particular antibacterial agents.

Perhaps the most important feature of many R factors is that they can be rapidly transferred to antibiotic- and heavy metal-sensitive bacteria, thereby conferring resistance on these recipient cells. Furthermore, many R factors have wide host ranges and can multiply in a wide variety of different genera, including *Shigella, Salmonella, Escherichia, Yersinia, Klebsiella, Serratia, Proteus, Vibrio,* and *Pseudomonas.* If some R factors are present in one species in a mixed population of these organisms, all of the cells will receive the R factor and thus become resistant to a variety of antimicrobial agents. This ready transfer of plasmids explains why so many different organisms in a hospital environment are resistant to many different antimicrobial medicines.

Table 7.8 Important Properties of Plasmids

Property	Range
Size (Kb)*	3–500
Number of copies per cell	1–500
Self transmissibilty from cell to cell	Most not
Host range	Narrow to broad

*Kb = 1,000 base pairs. An average-size gene is 1Kb.

Perspective 7.3

Do Dental Fillings Contribute to Antibiotic Resistance in Bacteria?

There is no question that the number of antibiotic-resistant organisms is increasing. The major reason for this is the expanded use of antibiotics both to treat a wide variety of diseases and to supplement the diet of farm animals to increase their weight.

The antibiotics select for those organisms in the environment that are antibiotic resistant. However, recall that resistance (R) plasmids often carry resistance genes to a variety of antibiotics. These same R plasmids often carry resistance to a variety of heavy metals such as arsenic and mercury. Thus, it is conceivable that bacteria exposed to mercury would become resistant not only to mercury but also to a number of antibiotics, the resistance to which is carried on the same R plasmid as the resistance to mercury. If this R plasmid were maintained in an environment because of the continuing presence of mercury, it could be transferred to other bacteria that would then become resistant to mercury as well as antibiotics.

Some experimental evidence suggests that this scenario might be happening, at least in monkeys. When researchers put amalgam dental restorations ("fillings"), which consist of 50% mercury (combined with silver), into the molars of six monkeys, they found that, within five weeks, bacteria in the intestines of the monkeys became resistant not only to the mercury but also to commonly used antibiotics including penicillin, streptomycin, kanamycin, chloramphenicol, and tetracycline.

How does the mercury get into the intestines? There is evidence that mercury vapor can seep out of the dental amalgam in the teeth. This vapor is inhaled into the lungs and enters the bloodstream. An enzyme in blood cells transforms it into a mercury ion, which is a poisonous form. From the blood it gets into the liver which excretes it in bile into the intestines. The bacteria that are mercury resistant contain an enzyme that converts the ionic form back into mercury vapor, a nontoxic form. Whether these observations of mercury and antibiotic resistance carried out in monkeys are relevant to antibiotic-resistant organisms in human beings requires further study.

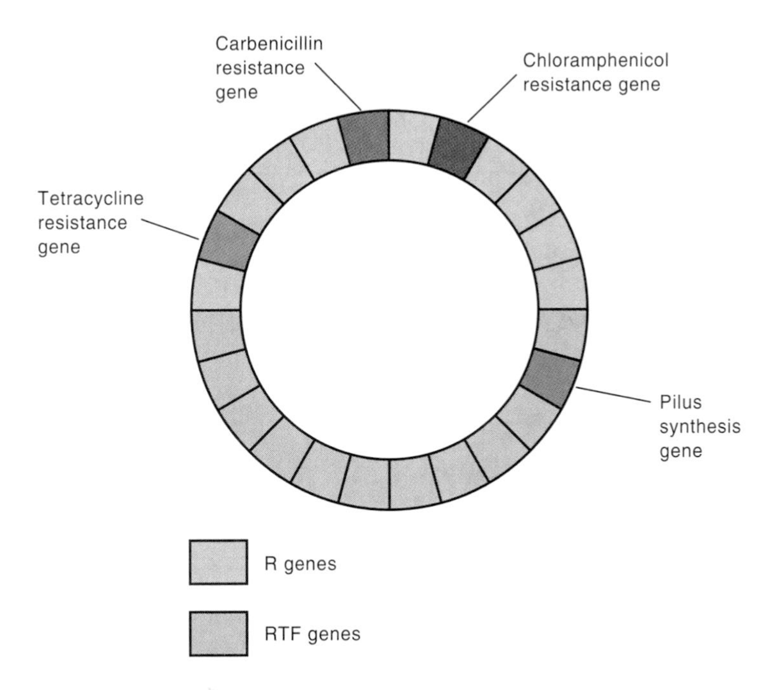

Figure 7.31
Diagrammatic representation of an R factor. Two regions are represented. The R genes code for resistance to various antibiotics; the RTF region codes for plasmid replication and the transfer of the plasmid to other bacteria.

Antibiotic-resistant organisms are most common, not surprisingly, in locations where antibiotics are in greatest use. This accounts for the fact that hospitals harbor many antibiotic-resistant bacteria. It is not surprising that antibiotic-resistant organisms are more common in certain parts of the world, particularly in developing countries. This probably results from the overuse of antibiotics in these countries, where antibiotics may be available without a doctor's prescription. Also, in areas that lack proper hygiene, bacteria are more readily transferred to people through contaminated food and water. Nondisease-causing bacteria that may be carried by healthy people can carry R factors, which can be transferred to disease-causing organisms.

Gene Movement Within the Same Bacterium—Transposable Elements

For many years, biologists thought that DNA was an extremely stable molecule and that genes did not move around inside a cell. Since DNA is the most important carrier of genetic information, it was believed that any rearrangements of DNA inside a cell would be harmful. However, as we have already discussed, genes flanked by inverted repeats often have the ability to transpose (move) from one site on a DNA molecule to a new position in the genome of the same cell. This movement occurs in both eukaryotes and prokaryotes. In bacteria, these jumping genes, called *transposable elements,* often code for proteins that confer resistance to antibiotics and heavy metals as well as for genes that code for toxin synthesis. Their movement can be very important to humans and animals. Indeed, the formation of R factor plasmids with their multiple antibiotic resistance genes probably results from the coming together of transposable elements coding for resistance to different antibiotics.

Transposable Elements Are Important for the Transfer of Genes to Unrelated Bacteria

Transposable elements can move from sites on the chromosome of cells into plasmids, which are rapidly transferred by conjugation to other cells (figure 7.32). The combination of inserting transposable elements into broad host range plasmids with their subsequent transfer into unrelated bacteria explains how the same antibiotic resistance genes can be found in nonrelated bacteria.

Restriction and Modification of DNA

Except for wide host range plasmids, a recipient bacterium will take up DNA by transformation, conjugation, or transduction only if the donor DNA is from the same species of bacteria. If other DNA, termed **foreign DNA,** enters the cell, a deoxyribonuclease enzyme in the recipient cell will degrade it. Such enzymes are called **restriction enzymes,** or **restriction endonucleases,** because they "restrict" the entry of foreign DNA into cells. The enzyme reacts with specific short sequences of nucleotides in foreign DNA and breaks the DNA at these sites. These enzymes are sometimes referred to as *molecular scissors*. Enzymes in different organisms "recognize" and cleave different nucleotide sequences (figure 7.33). Other deoxyribonucleases that degrade any kind of DNA then break down the DNA into nucleotides. Restriction enzymes have been found in virtually all species of prokaryotes but not in any eukaryotes. These enzymes, which have proven to be extremely valuable in genetic engineering of microorganisms, will be considered further in chapter 8.

Modification Enzymes Confer Resistance to Restriction Enzymes

How do bacteria prevent destruction of their own DNA? The partial answer to this question is that bacteria contain another enzyme, a **modification enzyme,** that adds methyl groups to adenine and cystosine in the fragments of DNA, which the restriction enzyme cleaves. Once the DNA is methylated the restriction enzymes are unable to recognize the specific base sequences as a substrate and so do not cleave the DNA. For example, if cytosine has a methyl (—CH_3) group, then it is not a substrate for the restriction enzyme. Since cytosine and methylcytosine have the same hydrogen-bonding characteristics, this alteration by the modification enzyme does not in-

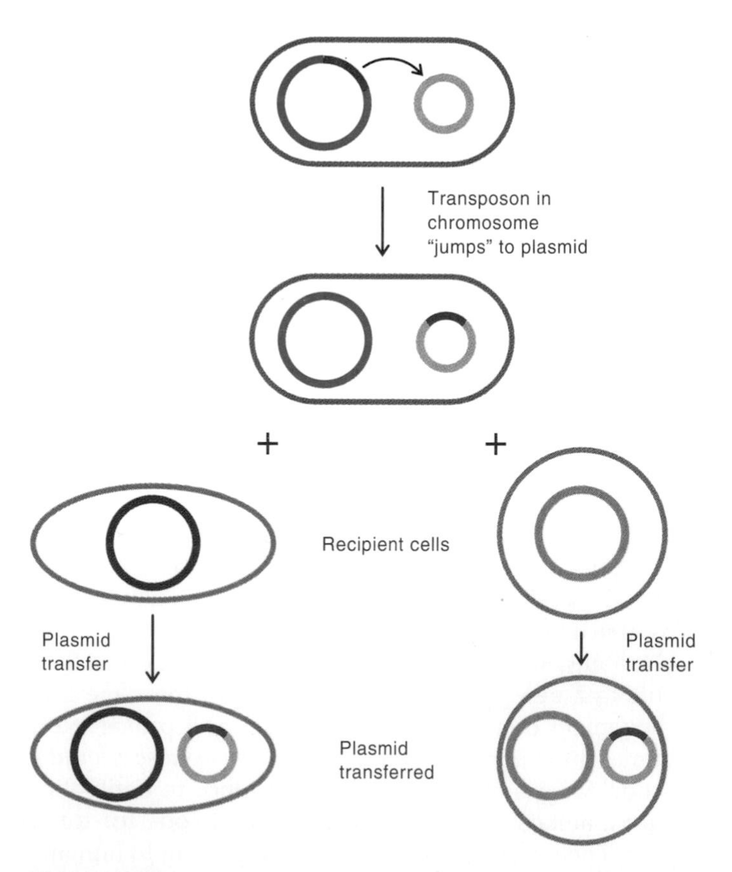

Figure 7.32
The movement of a transposon throughout a bacterial population. Note that transfer can be between bacteria belonging to different genera and between gram-negative and gram-positive bacteria.

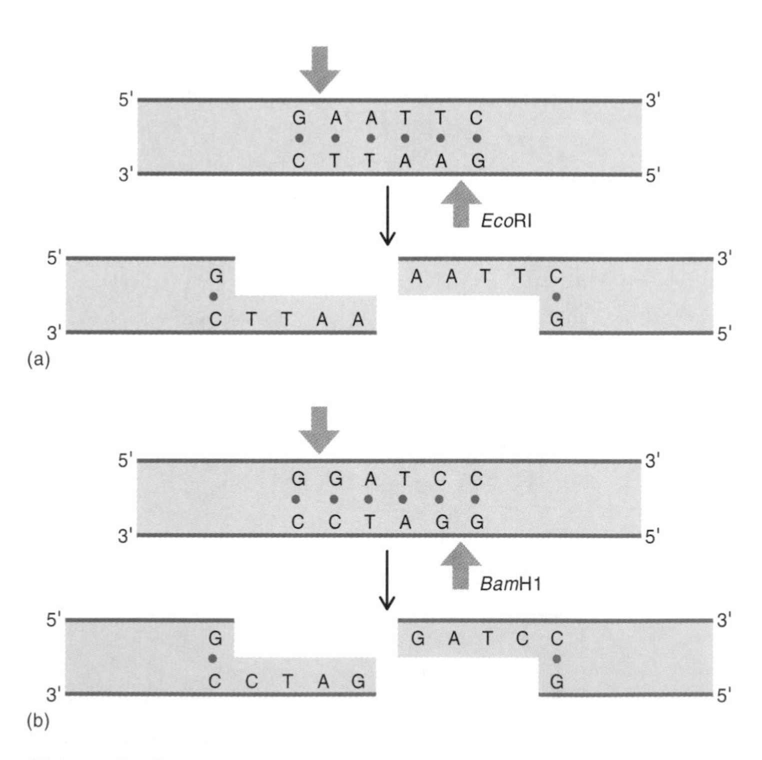

Figure 7.33
Action of restriction endonucleases. Cleaving of double-stranded DNA at specific sites indicated by the two arrows. The restriction endonuclease recognizes specific nucleotide sequences and cleaves the DNA at these sites. The few hydrogen bonds between the two cleavage sites are not able to hold the strands together. Consequently, the DNA splits into two molecules with overlapping ends. (*a*) Site of cleavage by the restriction enzyme *Eco*RI, isolated from *E. coli.* (*b*) Site of cleavage by the restriction enzyme *Bam*HI, isolated from *Bacillus amyloliquefaciens.*

duce mutations. Note that not all of the adenine and cytosine molecules are methylated, only those that are recognized by the restriction endonuclease of the cell.

Thus, when foreign DNA enters a cell, a race begins between methylation and restriction. If the DNA is methylated before it is degraded by the restriction endonuclease, it is no longer recognized as foreign by the restriction enzyme and will not be degraded. However, in most cases, it is degraded before it has a chance to be methylated (figure 7.34). Note that each species of bacteria has a different restriction enzyme and a modification enzyme that acts on the DNA sequence that the restriction enzyme cleaves.

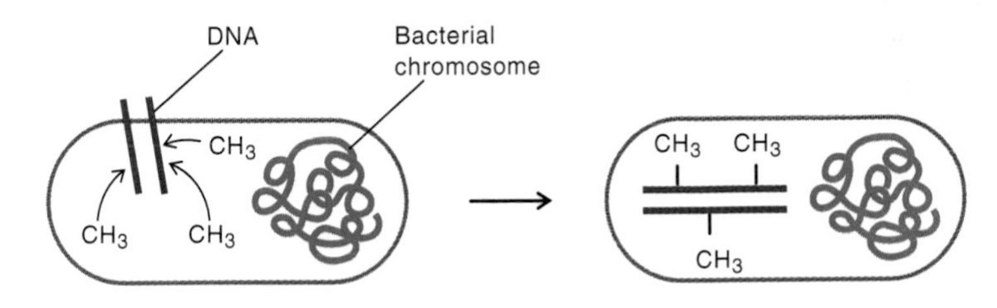

(a) The entering DNA is methylated by a modification enzyme and becomes resistant to degradation by the restriction endonuclease.

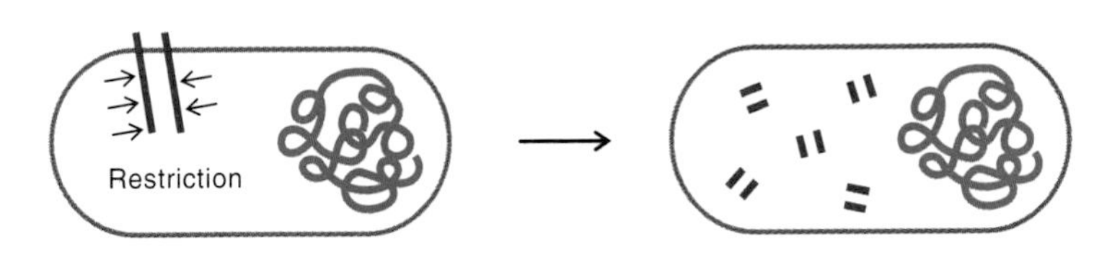

(b) The restriction endonuclease cleaves the DNA before it has a chance to be methylated.

Figure 7.34
Restriction and modification of entering DNA.

Importance of Gene Transfer to Bacteria

DNA-mediated transformation and conjugation have each been observed in quite a few species of bacteria (see tables 7.4 and 7.5). Conjugation has been shown to occur in only about a dozen species of mostly gram-negative bacteria, although it occurs in some very important gram-positive bacteria, in particular the streptococci and staphylococci. Plasmids are most commonly transferred in this process. Transformation has been demonstrated in about a dozen species of bacteria, both gram-positive and gram-negative. Transduction is probably the most widespread mechanism of genetic transfer, having been found in virtually all species in which it has been actively sought. The fact that conjugation and transformation have been found in relatively few species does not necessarily mean that these processes do not occur in other species. It may only mean that other species have not yet been investigated or that, if they have, the proper conditions have not yet been found under which gene transfer can occur. However, it is also true that the fact that transfer can be shown to occur in the laboratory does not mean that it occurs in nature.

Gene transfer provides new genetic information to microorganisms, which can allow them to survive changing environments. The major source of variation within a bacterial species is mutation. Unlike mutations in which only a single gene changes at any one time to provide new genetic information to the cell, in gene transfer many genes are transferred simultaneously, giving the recipient cell much more genetic information. For example, in a hospital environment in which a large variety of antibiotics are constantly used, a mutation to streptomycin resistance will confer a selective advantage on the mutant. The population will change very quickly from a largely streptomycin-sensitive population to one in which the streptomycin-resistant organisms predominate. However, only a single gene mutates at any one time and so the streptomycin-resistant organism will still be susceptible to other antibiotics in the environment. In contrast, in gene transfer, many genes can be transferred simultaneously and so a cell can become resistant to many different antibiotics if it receives a resistance plasmid.

Transposable elements also greatly expand the opportunity for new gene combinations to arise since their movement can inactivate and separate genes. They play other roles as well. As discussed already, insertion sequences are mobile genetic elements, and any gene or genes flanked by insertion sequences can move to different sites in the genome. It has been shown that a chromosomal gene flanked by insertion sequences can move to a broad host range plasmid in the same cell, which can be transferred to a large number of different genera of bacteria, both gram-positive and gram-negative. This process explains why the same antibiotic resistance genes occur in *E. coli* (gram-negative) and *Clostridium* (gram-positive).

Summary

I. Sources of Diversity in Microorganisms
   A. Variation in sequences of nucleotides that exist in DNA is one source.
   B. The environment, which regulates gene function and enzyme activity, is another source.

II. Gene Alterations—Mutations
   A. Mutations can occur by substitution of one base for another.
   B. Mutations can occur by the removal or addition of nucleotides.

III. Mutagenesis
- A. Spontaneous mutants occur in the absence of any known mutagen. These mutations are the same as those caused by mutagens.
- B. Three categories of mutagens are known: chemical agents, transposable elements, and radiation.
- C. Chemical mutagens often act by altering the hydrogen-bonding properties of the nitrogenous bases. They include alkylating agents, base analogs, and intercalating agents.
  1. Alkylating agents make up the largest group of chemical mutagens.
  2. Base analogs are incorporated into DNA in place of the natural base.
  3. Intercalating agents result in addition of nucleotides.

IV. Transposable Elements (Jumping Genes)
- A. An insertion sequence is the simplest transposable element. It is composed of two inverted repeat sequences flanking an enzyme required for transposition.
- B. More complex transposable elements often code for antibiotic resistance.

V. Ultraviolet Irradiation
- A. Ultraviolet light and X rays are mutagenic.
- B. UV causes the formation of thymine dimers.

VI. Repair of Damaged DNA
- A. Visible light activates an enzyme that breaks covalent bonds, forming thymine dimers, through the process of *light repair.*
- B. Bacteria can excise damaged DNA and replace it with undamaged DNA, which is called *excision repair* or *dark repair.*
- C. Misincorporated bases can be removed by a proofreading function of DNA polymerase.
- D. Misincorporated bases can be removed by a mismatch repair system.
- E. UV and chemical damage can be overcome by an inducible enzyme system of DNA replication that makes many mistakes.

VII. Mutations and Their Consequences
- A. Genes mutate independently of one another and mutations are stable.
- B. Mutations are expressed rapidly in prokaryotic cells because they have only one or two identical chromosomes.

VIII. Mutant Selection
- A. Direct selection requires that the mutant grow on a solid medium on which the parent cannot.
- B. Indirect selection must be used to isolate mutants that cannot grow on a medium on which the parent can grow.
  1. Replica plating involves transferring simultaneously all of the colonies on one plate to another plate.
  2. Penicillin enrichment increases the proportion of mutants in a population. It is based on the fact that penicillin kills growing cells only.

IX. Conditional Lethal Mutants
- A. These mutants result from mutations that ordinarily would be lethal.

X. Chemicals can be tested for their cancer-causing ability by their mutagenic activity.
- A. The Ames Test measures the frequency of reversion of his^- to his^+ cells.

XI. Mechanism of Gene Transfer—General Aspects
- A. Genes can be transferred between bacteria by three different mechanisms.
  1. DNA-mediated transformation involves the transfer of naked DNA. Recipient cells must be able to take up DNA through their envelope through natural competence. Cells can have holes punched in their envelopes if they are exposed to an electric current—artificial competence.
  2. Transduction involves viral transfer of DNA.
  3. Conjugation requires cell-to-cell contact. Only plasmids are usually transferred in F^+ to F^- cells. Hfr cells are formed if the F^+ plasmid integrates into the chromosome. Hfr cells can transfer chromosomal DNA. F' donor cells result when the F^+ plasmid integrated into the chromosome is excised and carries a piece of chromosomal DNA with the plasmid.
- B. Only short segments of DNA are transferred.
- C. Recombination between donor and recipient DNA occurs by a breakage and reunion mechanism.

XII. Plasmids
- A. R-factor plasmids confer resistance to a large number of antimicrobial agents. They can be transferred often to related and unrelated bacteria by conjugation.

XIII. Gene Movement Within the Same Bacterium—Transposable Elements
- A. Transposable elements are important for the transfer of genes to unrelated bacteria because they can move from a chromosomal site to a plasmid, which can then be transferred by conjugation.

XIV. Restriction and Modification of DNA
- A. DNA entering unrelated bacteria is recognized as being foreign and is degraded by deoxyribonucleases termed *restriction enzymes,* which cleave the DNA at specific sites.
- B. Modification enzymes confer resistance to restriction by altering the nucleotides in the DNA, at the sites where the restriction enzymes cleave the DNA.

XV. Importance of Gene Transfer to Bacteria
- A. Gene transfer allows microorganisms to survive changing environments by providing recipient cells with whole new sets of genes.

Chapter Review Questions

1. What is the one activity that all plasmids must be able to carry out?
2. What is the term that describes a plasmid that can transfer itself from *E. coli* into *Pseudomonas* where it can replicate?
3. What is the enzyme that is responsible for protecting cells against invasion of foreign DNA?
4. What are the two necessary features of an insertion sequence?
5. What enzyme has proofreading ability? How does it function in proofreading?
6. Give an example of gene transfer that involves homologous recombination. Give one that involves nonhomologous recombination.
7. Single-strand integration of DNA has been shown to be a feature of which mechanism of DNA transfer?
8. Methylation of DNA plays a role in two biological processes discussed in this chapter. What are they?
9. What environmental condition is useful in isolating conditional mutants? Explain.
10. Name three ways in which plasmids differ from bacterial chromosomes.

Critical Thinking Questions

1. Discuss the steps involved in transferring a gene from the chromosome of *E. coli* into *Pseudomonas* sp., an unrelated bacterium.
2. Devise a simple experiment to demonstrate that strepomycin does not cause mutation but selects for preexisting mutations. Hint: Replica plating is involved.
3. Why might an intestinal inhabitant like *E. coli* have a light repair system for repair of thymine dimers?
4. Can polarity of DNA transfer ever function in transduction? Explain.

Further Reading

Devoret, R. 1979. Bacterial tests for potential carcinogens. *Scientific American* 241(2):40–49.

Friedberg, E., Walker, G. and Silde, W. 1995. *DNA repair and mutagenesis.* ASM Press, Va. An advanced book for upper-division courses that covers DNA repair in prokaryotes and eukaryotes.

Leffell, D. J. and Brash, D. E. 1996. Sunlight and skin cancer. *Scientific American* 275(1):52–59.

Maloy, S. R., Cronan, J. E., Jr., and Freifelder, D. 1994. *Microbial genetics,* 2nd ed. Jones and Bartlett, Boston. A textbook for undergraduate and graduate courses in microbial genetics.

Novick, R. 1980. Plasmids. *Scientific American* 243(6):103.

Rennie, J. 1993. DNA's new twists. *Scientific American* 255(3):122–132.

Snyder, L., and Champness, W. 1997. *Molecular genetics of bacteria.* ASM Press, Va. A textbook for advanced undergraduate courses in bacterial genetics.

Stahl, F. W. 1987. Genetic recombination. *Scientific American* 256(2):91.

8 Microbiology and Biotechnology

KEY CONCEPTS

1. DNA is a macromolecule that is easily studied and manipulated in a test tube.
2. The discovery of restriction enzymes opened the door to gene cloning and genetic engineering.
3. Gene cloning can provide large amounts of specific DNA for study as well as large quantities of products coded by the cloned DNA.
4. In the laboratory, DNA from one organism can be introduced into other organisms, where it will replicate and be expressed.
5. The polymerase chain reaction is among the most important technical advances made in molecular biology in the past decade.

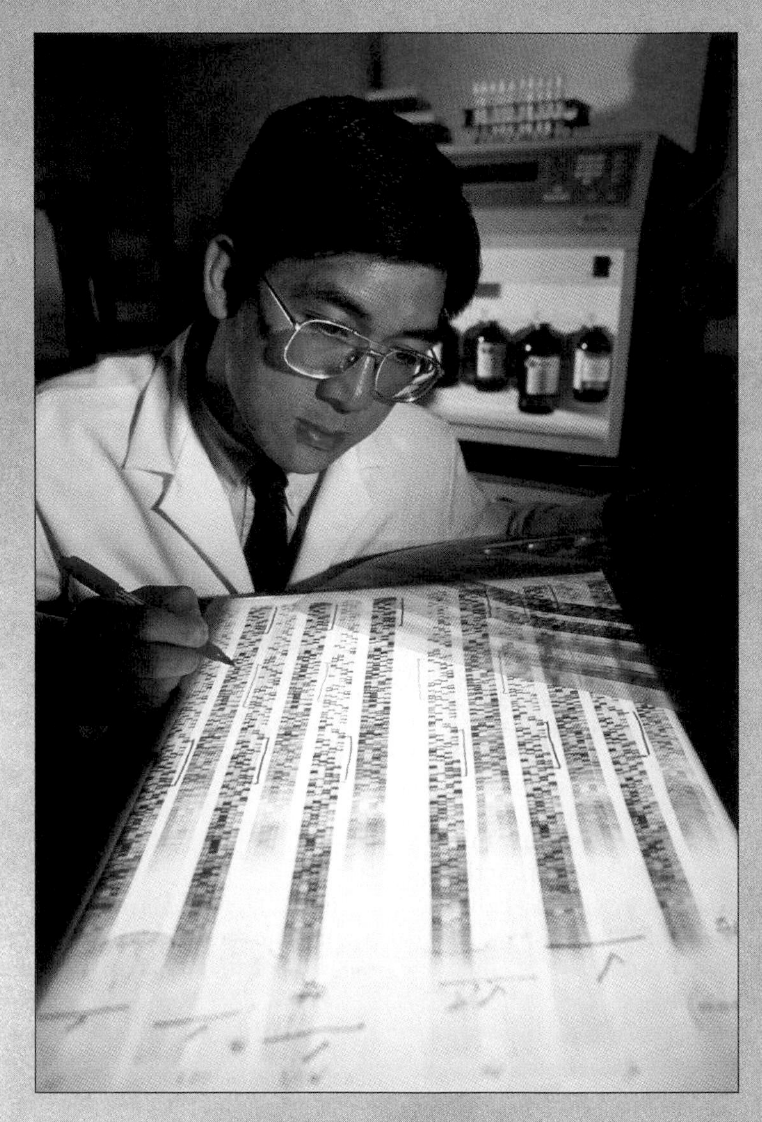

Scientist examining a film used to determine the sequence of nucleotides in a sample of DNA. The pattern of the bands on the film reflects the nucleotide sequence of the DNA.

Microbes in Motion

The following books and chapters in the *Microbes in Motion* CD-ROM may serve as a useful preview or supplement to your reading:

Microbial Metabolism and Growth: Genetics. *Viral Structure and Function*: Viral Detection. *Vaccines:* Vaccine Development.

A Glimpse of History

A revolution has occurred in biology over the past several decades—the science has transformed from a descriptive study of what cells are to a molecular study of how they function. In large part, this revolution coincided with the development of many important techniques for manipulating DNA outside the cell.

Certain technological developments were particularly important. One key development was the discovery of restriction enzymes, which are essential in genetic engineering. The story of restriction enzymes began in the early 1950s when Dr. Werner Arber observed that certain bacterial viruses could not infect certain strains of *Escherichia coli*. He later determined that bacteria had enzymes that degraded the DNA as it entered the cell. In 1970, Dr. Hamilton Smith and his colleagues purified and characterized the first of these enzymes, a *restriction endonuclease*, and discovered that the enzyme recognized a specific sequence of nucleotides and cleaved the DNA at these sites. Thus, using the enzymes, it became possible to isolate DNA fragments containing genes. Dr. Arber, Dr. Smith, and Dr. Daniel Nathans were awarded the Nobel Prize in 1978 for their studies on restriction enzymes.

Another crucial advance in the development of genetic engineering was pioneered by Dr. Stanley Cohen and Dr. Herbert Boyer in the early 1970s. These investigators engineered a small plasmid into which they could insert foreign genes. They also developed techniques for transforming bacteria with the plasmid. With these techniques and using restriction enzymes, it is possible to isolate a gene from one organism, insert it into a wide host range plasmid in a test tube, transfer the plasmid into *E. coli*, and have the foreign DNA replicate as part of the plasmid. This process is the technique of gene cloning.

Another technical breakthrough of major importance was the development of relatively simple ways to determine the arrangement (sequence) of nucleotides in DNA. These techniques were developed by two groups of investigators, Dr. Alan Maxam and Dr. Walter Gilbert in the United States and Dr. Frederick Sanger and his colleagues in England. Sequencing is now simple and rapid enough to perform that it is often taught to undergraduates in college. An entire gene (of over a thousand nucleotides) can be sequenced in a few days. The chromosomal DNA of humans is now being sequenced (in the human genome project). This sequence of nucleotides can provide very useful information about the protein for which it codes. For example, the sequence of amino acids in the protein coded by the gene can be determined and some idea of the shape of the protein can be deduced from its amino acid sequence. In addition, some idea of its function may be inferred if its sequence of amino acids is similar to the sequence of amino acids of other proteins whose functions are known. Known gene sequences are now stored in data banks that are available to all scientists. In 1980, the Nobel Prize in Chemistry was awarded to Dr. Paul Berg, Dr. Walter Gilbert, and Dr. Frederick Sanger for their contributions to the methodological revolution that is allowing researchers to examine the structure as well as the control of genes.

The most recently developed technique that is playing an increasingly prominent role in modern biology is the polymerase chain reaction (PCR). With this technique, based on an idea of Dr. Kary Mullis, it is possible to increase the quanity of the exact nucleotide sequence of any piece of DNA over 1 million-fold in about one hour. It has very wide applications and is currently being used in many areas of biomedical research. The 1993 Nobel Prize in Chemistry was awarded to Dr. Mullis for his development of PCR.

In this chapter, some examples of how these techniques are being employed in various areas of biotechnology are considered. Without any one of these techniques, biotechnology would not be the rapidly advancing field that it is today.

The use of modern microbiological and biochemical techniques to solve practical problems and to produce more useful products is called **biotechnology**. The field is actually a collection of many different technologies that have rapidly advanced over the last 30 years with the development of **recombinant DNA techniques.** These techniques enable researchers to isolate genes from one organism, manipulate the purified DNA in the laboratory, and then transfer the genes into another organism in a process called **gene cloning** (figure 8.1). These **genetically engineered** cells can act as biological factories and produce valuable proteins, serve as tools for researchers trying to unlock the mysteries of genes, or acquire economically useful traits such as pest resistance.

Biotechnology promises to bring advances to many fields. In medicine, the techniques are helpful in the development of safer, more effective vaccines and more rapid diagnoses of some diseases. The environment may be helped through the development of bacterial strains that can degrade toxic chemicals, while agriculture has already benefited from foods with improved ripening qualities. Meanwhile, scientific knowledge has increased tremendously through the use of recombinant DNA techniques. Much of the latest information in this textbook was discovered through human ability to manipulate, transfer, and express DNA in other cells in the laboratory.

Applications of Biotechnology

Recombinant DNA techniques provide a powerful tool for genetically engineering microorganisms specifically for industrial uses. For example, microorganisms can be altered so that they produce commercially important proteins more efficiently. If a microorganism normally makes a desired protein product in low amounts, the gene coding for the protein can be inserted into a plasmid present in multiple copies and then reintroduced into the same microorganism. The organism will then make increased amounts of the protein because the multiple genes can be transcribed and

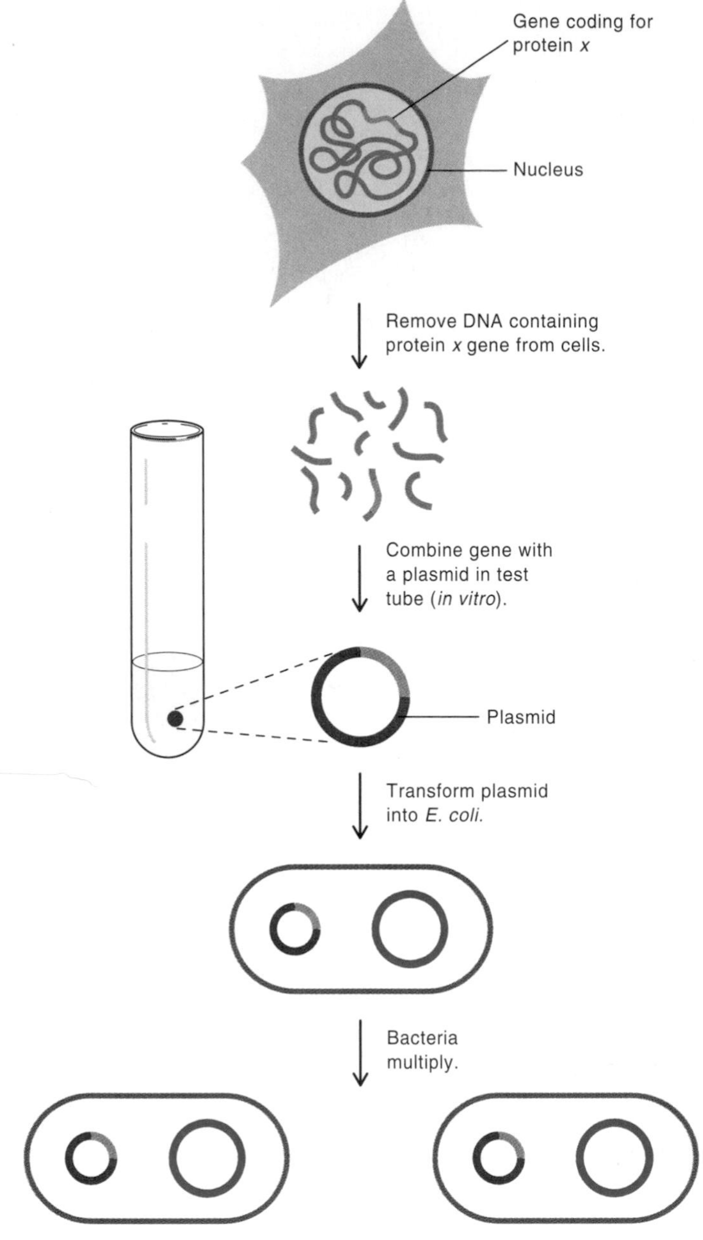

Figure 8.1
Overall aspects of gene cloning.

translated (figure 8.2). Likewise, an organism can be engineered to make a protein that is normally produced by a different organism. The same technique can also be used to give an organism new traits. For example, plants can be engineered to produce their own insecticide by introducing to them genes from *Bacillus thuringiensis* that encode proteins toxic to insect larvae.

Medically Important Proteins Can Be Produced by *E. coli* and Yeast

Mammalian proteins, such as insulin, human growth hormone, blood clotting factors, and β interferon, can be used to treat patients suffering from certain diseases (table 8.1). In the past, these proteins, which were extracted from living animals (including humans) or cadaver tissues, were expensive and often in short supply. The ability to clone the genes encoding these proteins and have them expressed in a microorganism ensures that the products can be produced in large quantities and with reliable quality controls.

Human insulin is an especially important pharmaceutical protein used in treating diabetics. It is in great demand, but the original commercial product, extracted from pancreatic glands of cattle and pigs, causes allergic reactions in 1 in 20 patients. However, the gene for human insulin has been cloned in *E. coli* and is now the major source of insulin sold in the United States. A 2,000 liter culture of *E. coli* yields 100 grams of purified insulin, an amount that would require 1,600 pounds of pancreatic glands. Thus, the bacterial production of insulin is safer and more economical than the use of animal glands.

Human growth hormone is used to treat approximately 20,000 patients in the United States for hypopituitary dwarfism, a disease caused by a deficiency of this hormone. Shortages of human growth hormone have now been alleviated by a product from a genetically engineered strain of *E. coli*. These genetically engineered products also eliminate the risk of infection from agents that might be found in the glands of cadavers.

Another class of proteins produced from genetically engineered bacteria are the interferons, substances found in human blood that have wide-ranging activity against viruses. In the past, when they still had to be extracted from human cells, which made them difficult to obtain in pure form, little was known about their clinical significance. In the 1990s, the genes for interferons were cloned into *E. coli*. The large-scale production of easily purified interferon by these genetically engineered microorganisms now permits tests of clinical activity. If interferon lives up to claims made by some optimistic researchers, it may one day be used to treat a variety of viral diseases, from the common cold to some types of cancer. The genes for three

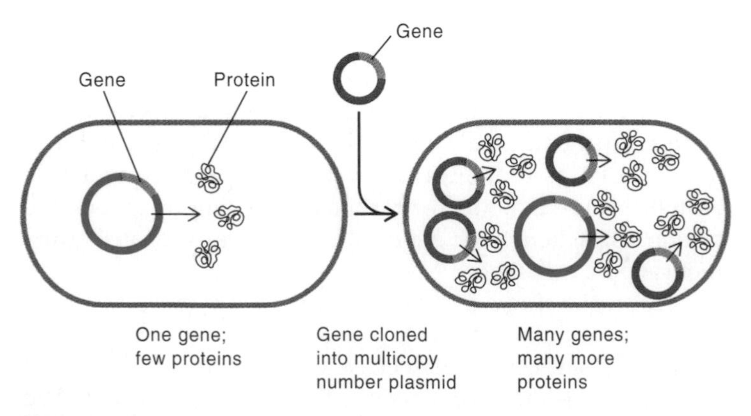

Figure 8.2
Use of a high copy number plasmid to obtain a strain with multiple copies of an important gene. This results in the synthesis of many more molecules of the desired protein.

Table 8.1 Genetically Engineered Products

Gene Cloned	Use
Factor VIII	Clotting blood (protein lacking in many hemophiliacs)
Insulin	Treating diabetes
Interferon α-2b	Treating hepatitis B infection
Growth hormone	Treating dwarfism
Gamma interferon	Treating cancer
α interferon	Treating cancer
Tissue plasminogen activator	Dissolving blood clots
Streptokinase	Dissolving blood clots
Deoxyribonuclease	Treating cystic fibrosis
β interferon	Treating multiple sclerosis

types of interferon have been cloned in *E. coli* and the yeast *Saccharomyces cerevisiae,* producing interferons that are now being used to treat patients with such diseases as AIDS, chronic viral hepatitis, and multiple sclerosis.

Many other medically useful proteins are being produced by bacteria containing recombinant DNA. These include human serum albumin (used to treat burn patients), blood clotting factors, streptokinase and tissue plasminogen activator (enzymes used to dissolve blood clots), erythropoietin (a blood protein that, in combination with AZT, may be useful in AIDS treatments), and even antibody fragments.

Safer Vaccines Can Be Made Using Genetic Engineering

Genetically engineered microorganisms are also being used in the production of a number of different **vaccines.** Vaccines protect against disease by harmlessly exposing a person's immune system to a killed or weakened attenuated form of the disease-causing agent or a part of the agent. A vaccine is often composed of whole bacterial cells or viral particles, but only specific proteins or parts of the proteins of the disease-causing agent are actually necessary to induce protection, or immunize, against the disease. The genes coding for these parts of proteins can be cloned in yeast or bacteria so that large amounts of the pure immunizing proteins can be produced. For example, genetically engineered vaccines are used in immunization against human hepatitis B and foot-and-mouth disease of domestic animals.

Bioremediation Involves the Use of Microorganisms to Degrade Harmful Chemicals

Bacteria have a voracious appetite and, as discussed already, can degrade a wide variety of compounds into nutrients. Because bacteria can break down a variety of compounds, including those that are considered harmful to the environment, they can be added to polluted areas such as oil spills to degrade the noxious materials. This application of microbiology is called **bioremediation.**

One of the most important large-scale bioremediation efforts was carried out on the oil-soaked beaches of Alaska, following the oil spill from the tanker *Exxon Valdez* in 1989. To promote growth of the bacteria already on the beaches, workers sprayed fertilizers that contained nitrogen and phosphorus, which are the limiting nutrients for microbial growth in coastal waters. The application of fertilizers promoted the growth of the natural population of bacteria already adapted to the cold Alaskan environment. Although it appeared that the oil-coated rocks were cleansed to a significant degree by the bacteria, some scientists feel that proper controls to determine the effectiveness of the process were lacking.

The use of bioremediation to clean up oil spills is difficult because oil is a complex mixture of hydrocarbons (compounds containing only carbon and hydrogen), some of which are readily degraded by bacteria and others that are not. As specific microorganisms grow, they utilize the hydrocarbons they can degrade. Later, other types of bacteria grow to high numbers and degrade other hydrocarbons. The result is that the oil slowly changes in its chemical composition over time and the population of bacteria changes in response.

Plants Can Be Genetically Engineered to Develop New Properties

One of the most exciting areas of biotechnology is the improvement of the qualities of plants through genetic engineering. Some plants have been engineered so that they are resistant to insect pests, viral and bacterial diseases, and a variety of herbicides. Further, as already mentioned, the properties of some fruits and vegetables such as the ripening properties of tomatoes, have been dramatically improved through biotechnology. These traits now have been introduced into a wide variety of plants, some of which are currently undergoing field testing in order to determine whether they behave in the field as they do in the laboratory (table 8.2).

Toxins from Bacteria Can Serve as Effective Biocontrol Agents

A wide variety of insecticides are used to control insect damage on plants. Most of them are chemically synthesized and cost farmers more than 3 billion dollars per year worldwide. There is increasing concern about their widespread use because many insecticides may be carcinogenic and harmful to many forms of life other than insects. For example, the pesticide DDT, a chlorinated hydrocarbon, does not degrade readily in the environment. In fact, it accumulates in the fat of predatory birds and results in fragile eggshell formation. Because of this detrimental effect, DDT

Table 8.2 Some Traits Genetically Engineered into Plants

Plant	Trait	Usefulness
Potato	Increased starch synthesis	Potato chips containing less oil
Rose	Blue flower	Pleasing appearance
Cotton	Resistance to various herbicides	Weeds killed without affecting engineered plants containing genes for herbicide resistance
Corn	*Bacillus thuringiensis* toxin synthesis	Lepidopteran (insect) resistance
Tomato	Synthesis of virus coat protein	Plants resistant to various viruses
Tomato	Antisense mRNA for polygalacturonase	Longer shelf life, spoilage resistance
Rape seed	Enzymes of oil synthesis	Modified oil composition of plant

is no longer used in the United States. In recent years, efforts have increased to identify and use natural biological control agents that are synthesized by organisms and do not persist in the environment. One of the most widely used of these natural insecticides is the insect toxin produced by *Bacillus thuringiensis* (B. T. toxin). It is synthesized as a protein crystal as the microorganisms form spores (figure 8.3).

Crystal/spore preparations have been applied to a wide variety of plants around the world for over 30 years without any apparent harm to the environment. The attractive features of these insecticides are that they are toxic to insects only and are degraded in the soil within a few days of application. Different strains of *B. thuringiensis* produce different toxins with different specificities; some are toxic to one group of insects, while others are toxic to another group. What is the basis of this specificity? Apparently, the toxin is synthesized as a protoxin, an inactive protein that must be processed in the gut of the insect to form the active toxin. The toxin binds to specific receptors in the insect's gut and causes the gut to dissolve. If the insect lacks the receptors, it is resistant to the toxin.

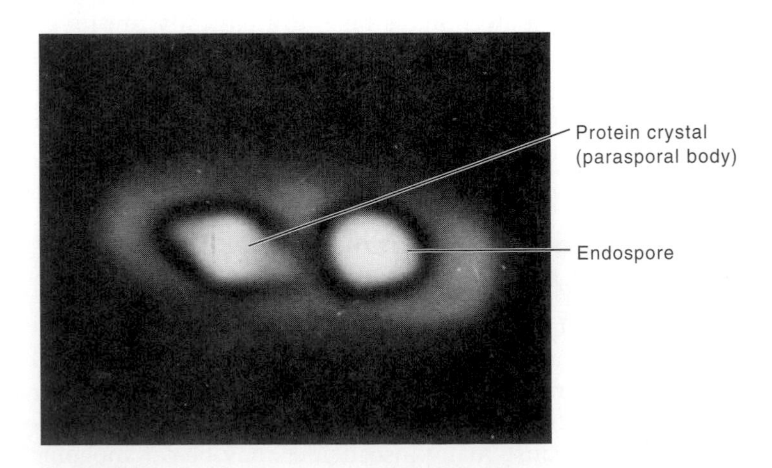

Figure 8.3
Protein crystal (parasporal body) and endospore in a cell of *Bacillus thuringiensis*.

A number of features have limited the commercial use of this biocontrol agent in the past. First, the toxin preparations are more expensive to produce than are chemical insecticides, and second, the crystal proteins are very unstable when they are applied in the field and may be degraded before they have a chance to kill the insects. Recently, innovative approaches have been taken to circumvent these problems. The stability of biocontrol agents has been increased by cloning the genes for toxin synthesis in cells of *Pseudomonas* in which they are expressed. The toxin-synthesizing *Pseudomonas* cells are grown in large vats and then are killed and treated in a way that stabilizes their cell walls and encapsulates the protein crystals. In these shells, the protein crystals remain active for at least one week.

Another popular and widely used approach for using *B. thuringiensis* toxins is to introduce the bacterial genes for toxin synthesis into the plant in which they will be expressed, thereby killing any insects that eat the leaves (figure 8.4). Many plants synthesizing these insecticidal toxins are now being field-tested.

REVIEW QUESTIONS

1. Give two beneficial uses of genetic engineering.
2. Name two proteins that are now produced by genetically engineered cells.
3. Name the process of using microorganisms to clean up polluted areas.
4. Name the organism that produces a protein crystal that is toxic to insects.

Cloning

Genetic engineering exploits the processes that naturally occur in cells and that result in new combinations of genetic information. Cloning a gene involves removing a

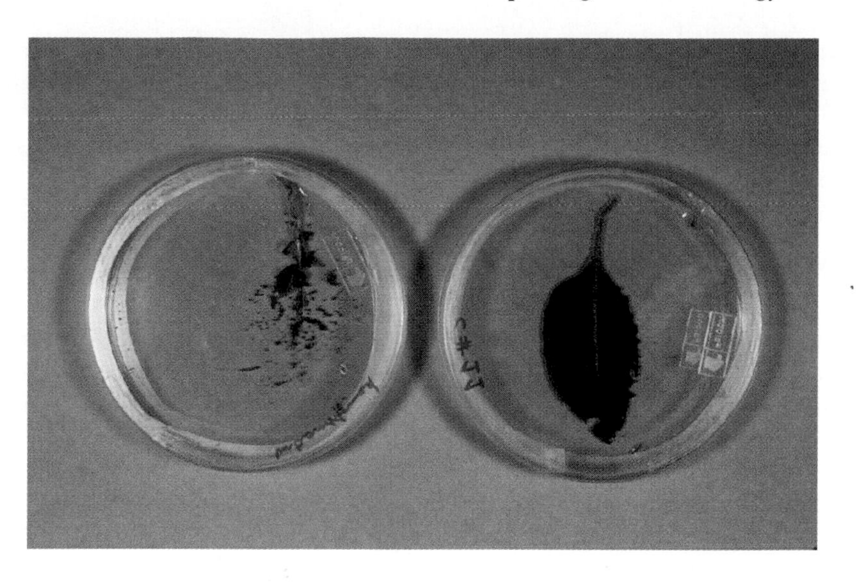

Figure 8.4
Genetic engineering of tobacco plants. The leaf on the left was taken from a tobacco plant that was not genetically engineered but was exposed to tobacco horn worms. The leaf on the right was taken from a tobacco plant that was genetically engineered with the gene for *Bacillus thuringiensis* toxin synthesis and then exposed to tobacco horn worms. Note that the leaf on the right has barely been eaten.

fragment of DNA from one organism and introducing it into another organism, most commonly *E. coli.* As the new host grows, the cloned fragment must replicate and be passed to daughter cells, thus generating a population of cells, each of which contains a DNA fragment identical to the original. If the original gene was in a human cell and it is cloned in *E. coli,* then the concentration of that gene can be increased dramatically because it is so easy to grow large quantities of bacteria. Further, the concentration can be increased 10- to 100-fold or more if the gene is cloned on a multicopy plasmid.

The cloning process appears relatively simple, but it is actually a complicated challenge for several reasons. As described in the previous chapter, linear pieces of DNA cannot replicate in *E. coli.* Even if the fragment of DNA is circularized, replication cannot occur unless the piece encodes an **origin of replication** (figure 8.5), which is highly unlikely because there is only one origin on a chromosome. Another problem is the fact that usually only a small portion of cells in an organism will actually take up the introduced DNA, so there must be a method to selectively isolate cells that have taken up the cloned fragment. These problems must all be overcome in the cloning procedure.

An even more formidable problem is the fact that a bacterial genome of over 1 million nucleotides long can encode over 1,000 different genes. Thus, a specific gene represents only a tiny fraction of the total genome. An approach frequently used to clone a specific gene is to first clone multiple genome fragments from an organism's chromosome into an *E. coli* host. The researcher must then identify which cells of the resulting population of *E. coli* cells contain the gene of interest. For example, if a researcher wants to clone the gene for luciferase, the enzyme that causes bioluminescence in *Vibrio fischerii,* he or she would first clone numerous DNA fragments that together contain all of the genes of an organism's chromosome and then identify the cells that contained the bioluminescence gene. This cloning technique is referred to as **shotgun cloning.** It results in a very large number of cells, each carrying different DNA fragments of the genome. Understandably, it is difficult to determine which of the cells contain the gene of interest unless the gene encodes a readily observable phenotype such as bioluminescence.

In order to overcome these problems, researchers planning cloning experiments need methods to (1) isolate and cut DNA into manageable fragments in order to obtain

• **DNA replication, p. 156**

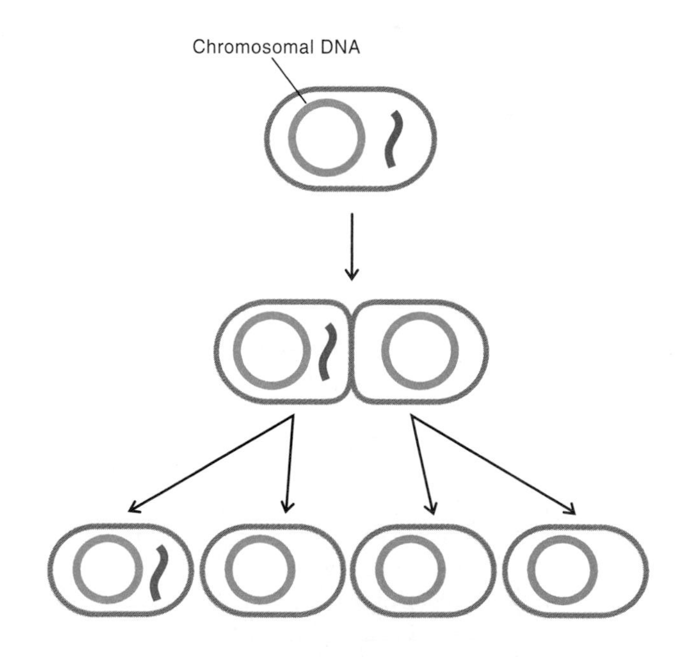

Figure 8.5
Fate of DNA fragments in a host cell.

pieces to be cloned; (2) introduce the fragments into the new host; (3) facilitate the cloned DNA's replication in the new host; (4) select for cells that have taken up a DNA fragment; and (5) determine which organisms contain the cloned gene (figure 8.6).

The Cells Are Lysed to Release All of Their DNA

Cells in an overnight broth culture are lysed by adding detergent to the suspension. The release of the long molecules of DNA from the lysed cells generates a characteristically viscous solution. During the process of release, the relatively fragile chromosome is inevitably sheared into long pieces (double stranded) of varying lengths. To purify the DNA, the cell lysate can be treated with specific enzymes such as protease and RNAase that degrade other cell contents such as protein and RNA.

Restriction Enzymes and DNA Ligase Are the Tools of Molecular Biology

Restriction enzymes act as molecular scissors that recognize and cleave specific sequences of DNA as discussed in chapter 7. Molecular biologists have found that these naturally occurring enzymes are remarkable tools to use in genetic engineering because they cut DNA in a predictable and controllable manner. "Cutting" or **digesting** long pieces of DNA with these enzymes generates smaller pieces called **restriction fragments.**

Hundreds of restriction enzymes have been discovered and are commercially available to scientists. Each enzyme has a seemingly peculiar name, but the name simply represents the bacterium from which the enzyme was initially isolated. The first letter of the name represents the first letter of the genus, the next two letters are derived from the species, and any other numbers or letters designate the strain and order of discovery. For example, a restriction enzyme from *E. coli* strain R is called *Eco*RI, and an enzyme from *Staphylococcus aureus* strain 3A is called *Sau*3A.

Most restriction enzymes recognize a sequence that is a palindrome. In other words, the sequence of one strand is identical to the sequence of the complementary strand when both are read in the 5' to 3' direction (table 8.3). The restriction enzyme recognizes and cuts the identical sequence on each strand of the double-stranded molecule which is why the sequences are palindromes.

Digestion of DNA with a restriction enzyme generates either a **sticky** or a **blunt** end, depending on where the enzyme cuts in the sequence. An enzyme such as *Alu*I cuts in the middle of the recognition sequence generating blunt ends, whereas an enzyme such as *Bam*HI produces a staggered cut, generating sticky or **cohesive** ends. DNA fragments with complementary sticky ends can form base pairs, or **anneal** to one another (figure 8.7).

If restriction enzymes are viewed as scissors that cut DNA into fragments, then the enzyme DNA ligase is the glue that pastes the fragments together. The annealing of complementary cohesive ends is due to the relatively weak hydrogen bonding, but DNA ligase is able to form permanent covalent bonds between the sugar-phosphate residues of adjacent nucleotides, thus joining two fragments. The combined actions of restriction enzymes and DNA ligase enable a researcher to easily remove fragments of DNA from diverse sources and then join them to generate a hybrid DNA molecule.

Vectors Act As Self-Replicating Carriers of Foreign DNA

A **vector,** usually a modified plasmid that can replicate independently of the chromosome, functions as a carrier of the cloned DNA. The DNA to be cloned is inserted into the

Table 8.3 Commonly Used Restriction Enzymes

Enzyme	Microbial Source	Sequence*
*Alu*I	*Arthrobacter luteus*	5′—A—G↓C—T— —T—C↑G—A—5′
*Bam*HI	*Bacillus amyloliquefaciens H*	5′—G↓G—A—T—C—C— —C—C—T—A—G↑G—5′
*Eco*RI	*Escherichia coli*	5′—G↓A—A—T—T—C— —C—T—T—A—A↑G—5′
*Hae*III	*Haemophilus aegyptius*	5′—G—G↓C—C— —C—C↑G—G—5′
*Hind*III	*Haemophilus influenzae*	5′—A↓A—G—C—T—T— —T—T—C—G—A↑A—5′
*Pst*I	*Providencia stuartii*	5′—C—T—G—C—A↓G— —G↑A—C—G—T—C—5′
*Sal*I	*Streptomyces albus*	5′—G↓T—C—G—A—C— —C—A—G—C—T↑G—5′
*Sma*I	*Serratia marcescens*	5′—C—C—C↓G—G—G— —G—G—G↑C—C—C—5′

*The arrows indicate the sites of cleavage on each strand.

• **Restriction enzymes, p. 178**

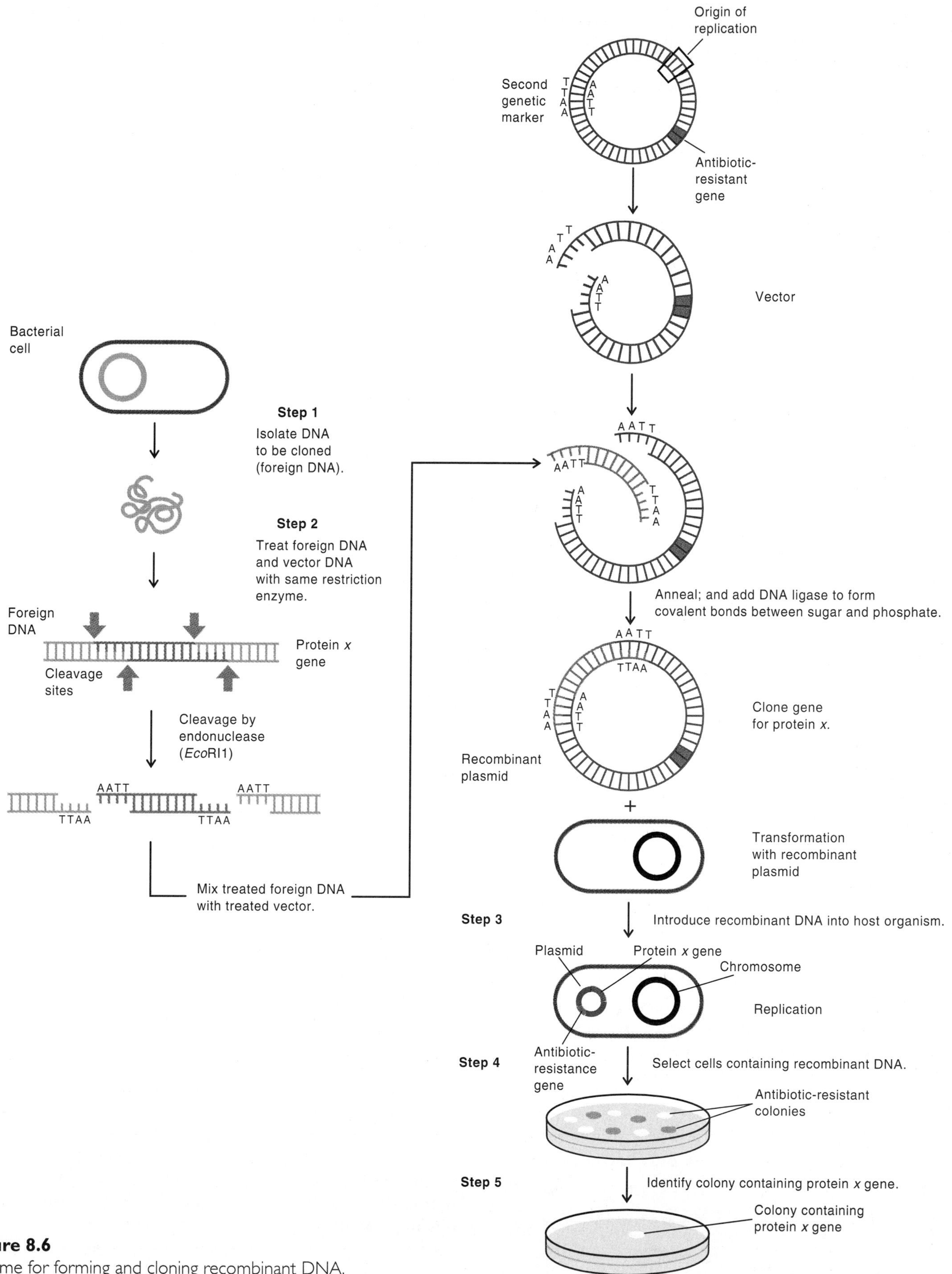

Figure 8.6
Scheme for forming and cloning recombinant DNA.

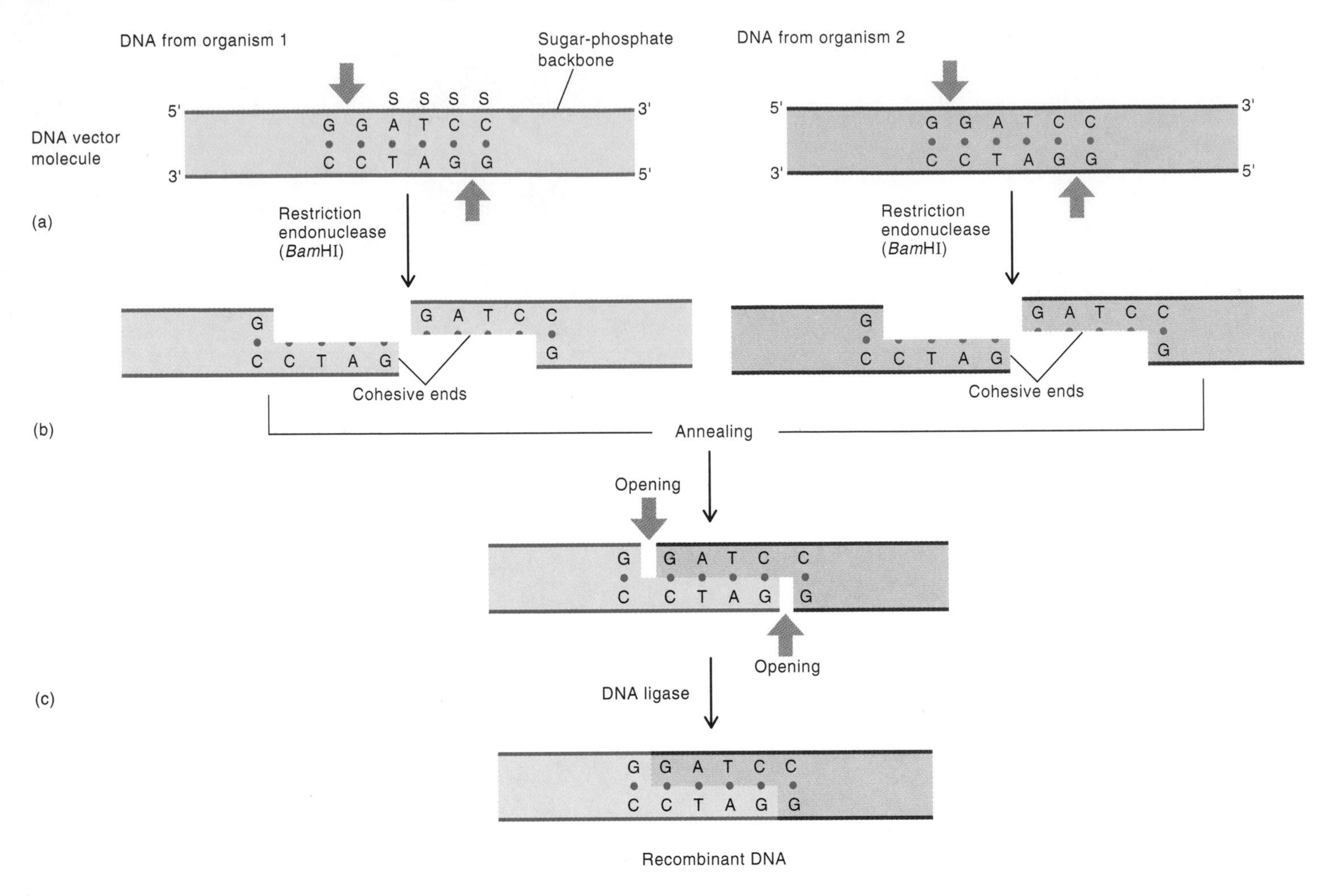

Figure 8.7
(*a*) Action of restriction endonucleases. Double-stranded DNA molecules from two different organisms are cut at specific base sequences. (*b*) Annealing—the association of DNA molecules cut with the same restriction endonuclease. (*c*) Sewing together of the two DNA molecules by the enzyme DNA ligase.

vector to form a hybrid molecule, or **recombinant plasmid,** that is part vector and part cloned DNA. The recombinant plasmid, like the vector alone, can replicate independently of the chromosome. Thus, it provides a mechanism for replication of the cloned DNA (figure 8.8)

In addition to having the ability to replicate, a vector usually can encode some type of **selectable marker** such as antibiotic resistance. The selectable marker enables the researcher to select for cells that contain the vector. This characteristic is important because when the DNA such as a vector is added to a host, usually by transformation, the majority of cells do not take up the DNA. The selectable marker is used to eliminate cells that have not been transformed with the vector. For example, if the selectable marker is antibiotic resistance, then only those cells that have taken up the vector are able to grow on antibiotic-containing medium.

• **Plasmid, p. 72**

Vectors also must have at least one restriction enzyme recognition site so that the circular vector can be cut to form a linear molecule. Many vectors have been engineered to contain a short sequence called a **multiple cloning site** that encodes several different restriction sites in the middle of a second genetic marker. This second genetic marker is later used to differentiate cells containing recombinant plasmids from cells that contain a vector alone. For example, one commonly used vector has a multiple cloning site in the middle of the *lacZ′* gene. The product of the *lacZ′* gene enables cells to break down a colorless compound called **x-gal** to form a blue-colored chemical. Thus, colonies that have the functional *lacZ′* gene encoded on an intact vector are blue (figure 8.9). The creation of a vector-insert hybrid disrupts the *lacZ′* gene so colonies harboring recombinant plasmids are white.

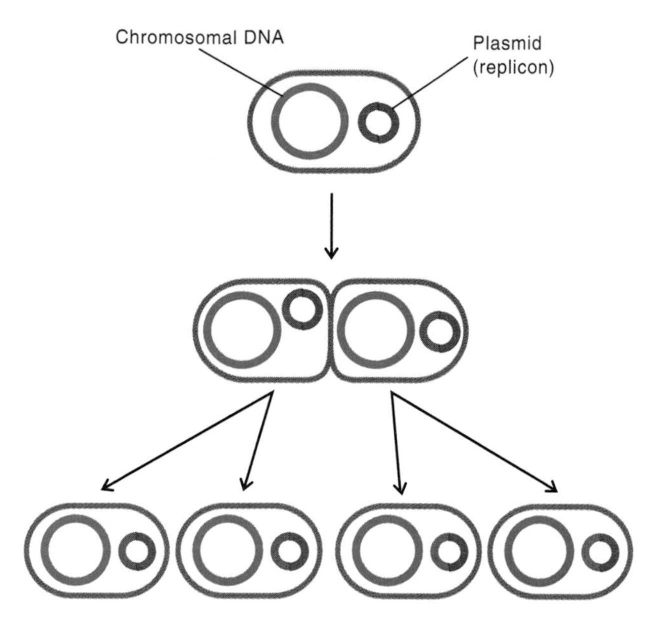

Figure 8.8
Fate of a replicon in a host cell.

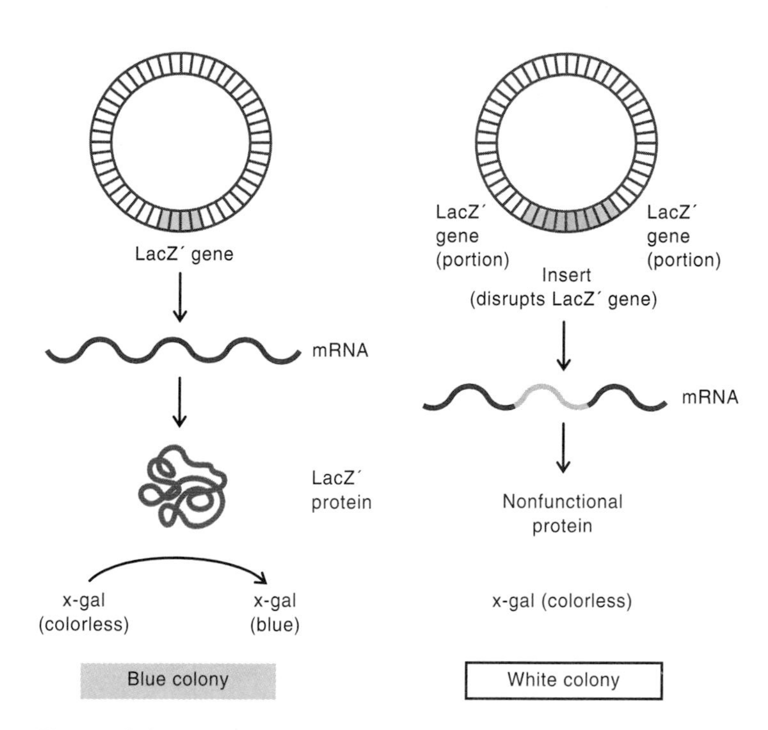

Figure 8.9
A second genetic marker such as the *lacZ′* gene can be used to differentiate cells that contain a recombinant plasmid from those that contain vector alone. A functional *lacZ′* gene results in blue colonies when the bacteria are plated on x-gal. Disruption of the gene by an insert generates a nonfunctional product so the resulting colonies are white.

Ligating the Vector and the Insert Together Creates a Recombinant Plasmid

When the vector and the DNA to be cloned (called the **foreign DNA** or **insert**) are both cut with the same restriction enzyme and the resulting DNA fragments are mixed together, a certain proportion of vector and insert will anneal together to form the desired circular recombinant plasmids. Other products, including a vector annealed to a vector, a vector annealed to itself, and an insert annealed to itself, will also occur, but these represent undesirable by-products of the reaction that are eliminated or ignored in future steps of the cloning procedure (figure 8.10). The annealed fragments are then joined permanently using DNA ligase which forms a covalent phosphodiester bond between the DNA molecules.

If the fragment to be cloned is one of the thousands of fragments obtained in a chromosomal digest, then a population of recombinant plasmids is generated, each of which carries a part of the chromosome. The set of cloned fragments, which together represent the entire genome, is called a **gene library.** Each recombinant plasmid containing a cloned fragment can be thought of as a "book" in the library.

The Recombinant Plasmid Must Be Introduced into a New Host

Once a group of recombinant plasmids is generated, they must be transferred into a suitable host where the molecules can replicate. Many well-characterized laboratory strains of *E. coli,* each with slightly different genetic characteristics, are readily available to be hosts for recombinant plasmids. A common method used to introduce DNA into a new host is DNA mediated transformation, which was discussed in chapter 7. Even cells that are not naturally competent can often be coaxed into taking up DNA under the proper conditions. For example, *E. coli* will become competent when suspended in a dilute calcium chloride solution. An alternative technique is to introduce the DNA by **electroporation,** a procedure that involves treating the cells with an electric current. This procedure works with *E. coli* as well as a variety of other bacteria.

After the DNA is introduced into the new host, the transformed bacteria are cultivated on medium that selects for cells containing vector sequences and differentiates those that carry recombinant plasmids. To select for cells containing vector sequences, the medium exploits the selective marker encoded on the vector to permit the growth of only those cells that have taken up plasmids that contain vector sequences. For example, if the vector encodes resistance to the antibiotic ampicillin, then the transformed cells are grown on ampicillin-containing

• **Transformation, p. 167**

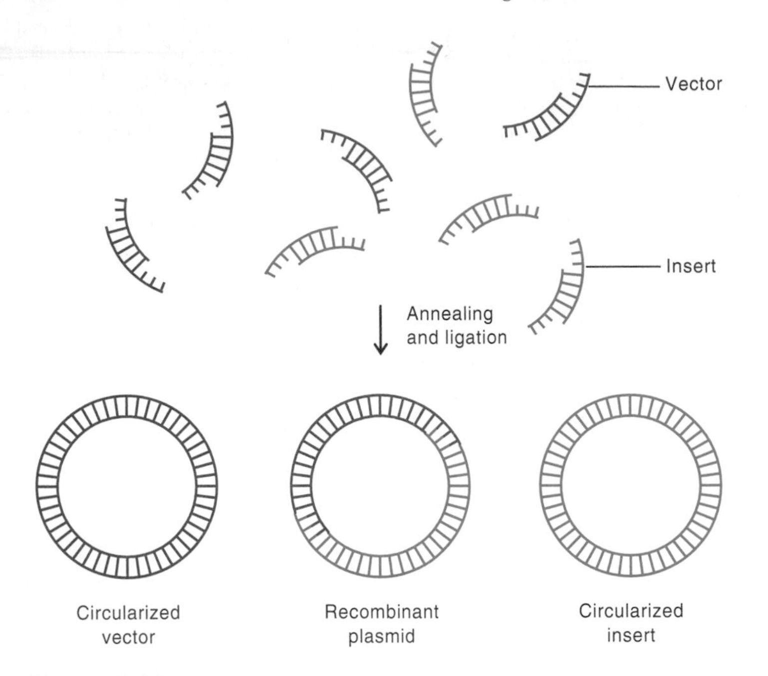

Figure 8.10
When linear fragments of vector and insert are mixed together, only a certain proportion of vector and insert will anneal to form recombinant plasmids.

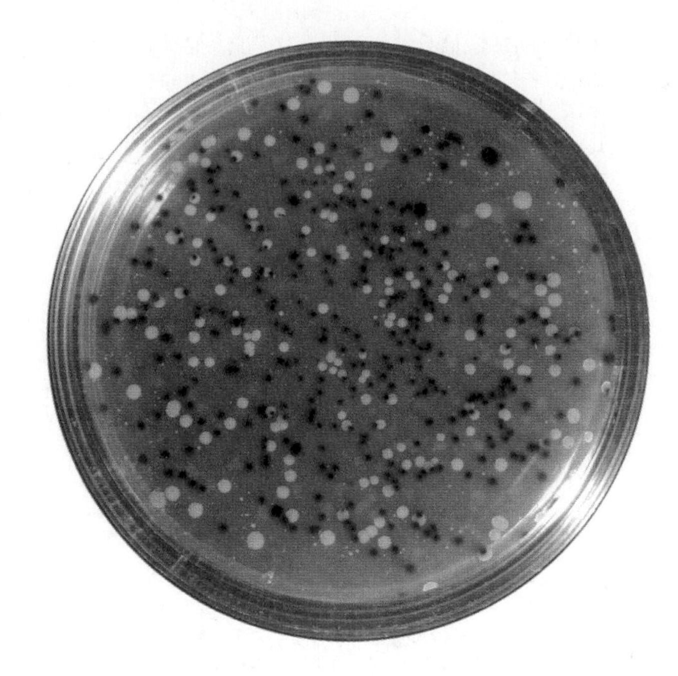

Figure 8.11
Colonies on an agar medium that contains x-gal. Colonies of bacteria that contain intact vectors are blue, whereas those that contain a recombinant plasmid are white.

medium. Cells that have not been transformed with a vector sequence are killed by the ampicillin. To differentiate cells that carry a recombinant plasmid, the medium exploits the second marker. In the case of vectors that use the *lacZ′* gene as a second marker, the chemical x-gal is added to the medium. Colonies of cells containing intact vector turn blue on the medium, whereas those containing recombinant plasmid are white. The white colonies are then further characterized to determine if they carry the gene of interest (figure 8.11).

Cells that Contain the Recombinant Clone Must Be Identified

When cloning genes that encode an observable phenotype, such as bioluminescence, it is relatively easy to identify cells that carry the DNA of interest. However, the task is a great deal more complex in most situations because there is no simple way to determine which cells carry the desired sequences. One common method is to use a **probe,** which is a piece of either DNA or RNA that has the same nucleotide sequence as the gene of interest, to seek out and bind to the cloned DNA. The use of a probe to seek out the DNA of interest relies on the high specificity of the base-pairing interactions between two strands of DNA. Double-stranded DNA, when exposed to high temperature or a high pH solution, will **denature,** or separate into two single strands. When the temperature is lowered or the pH is neutralized, the two strands will **renature** because of the base-pairing interactions of the complementary strands. Two complementary strands from different sources will also come together, but the process is then called **hybridization** to reflect the fact that the double-stranded DNA is a hybrid molecule. Because of the specificity of base pairing, if hybridization of DNA fragments occurs, it can be inferred that they have similar or identical nucleotide sequences and probably encode similar characteristics. Thus, hybridization can be used to identify or characterize unknown sequences of DNA (figure 8.12).

Obtaining a Probe

The probe can be obtained in a number of ways. In some cases, a similar gene has already been cloned so it can be used as a probe. For example, the toxin gene of *Shigella* can be used as a probe for similar toxin genes, such as the one encoded by *E. coli* 0157:H7. In other cases, the protein encoded by the gene has been isolated and its amino acid sequence determined. Knowing the amino acid sequence makes it relatively easy to deduce the potential nucleotide sequences using the genetic code. For example, if the protein has the amino acid methionine followed by a tryptophan, then one strand of the corresponding nucleotide sequence is ATGTGG. Machines are now available that can quickly chemically synthesize any short sequence of nucleotides, called an **oligonucleotide.** In practice, several

• ***Shigella*, p. 154**
• ***E. coli* 0157:H7, p. 153**

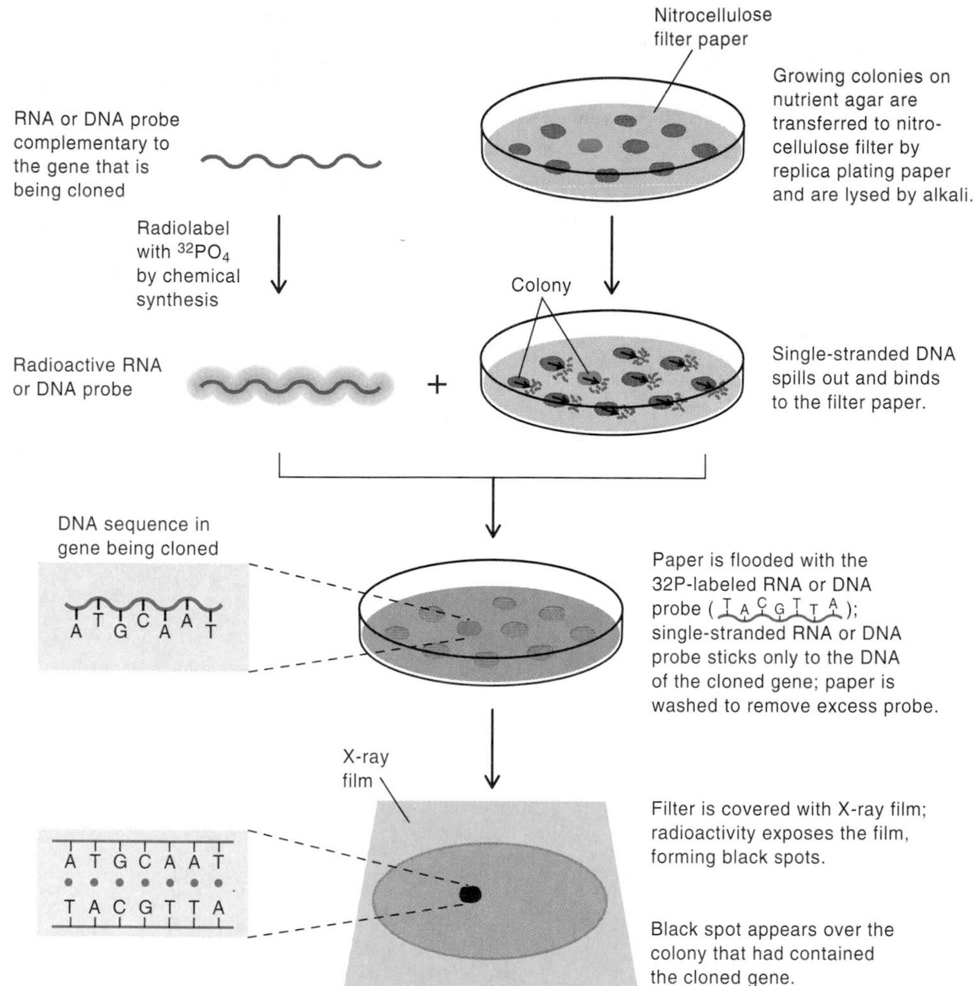

Figure 8.12
Identification of the proper colony in a shotgun cloning experiment using messenger RNA or a DNA probe.

different oligonucleotides are usually synthesized because the redundancy of the genetic code means that several slightly different oligonucleotides can encode the same short stretch of amino acids.

In order to facilitate detection in later steps, the single-stranded probe must be labeled before use, most commonly by adding a radioactive tag.

Preparing the Library

Before the probe can be used to seek out the DNA of interest, the cells that contain the library of recombinant plasmids must be prepared. The plate of colonies obtained after transformation of the recombinant plasmids is replica-plated on a nylon membrane, creating an identical pattern of colonies from the original plate. The membrane is then soaked in an alkaline solution, which simultaneously lyses the cells and denatures the DNA, thus generating single-stranded DNA. The radioactively labeled single-stranded probe is added to the membrane and incubated under conditions that allow the probe to hybridize to complementary sequences on the filter. The unbound probe is then washed off.

Detecting the Bound Probe

The location of the radioactively labeled hybridized probe is easily determined by placing the filter on a piece of X-ray film overnight. The radioactivity of the probe exposes the film, which, upon development, generates a black spot that can be correlated with a specific colony on the original plate. This technique for determining which colony contains the gene of interest is called a **colony blot.** It has many similarities to the Southern blot that will be discussed later in the chapter.

Successful Genetic Engineering Generally Requires Gene Expression

In some cases, an investigator clones genes in order to have large amounts of them available for study. However, in other cases, large quantities of the gene product are needed rather than the gene itself. For example, the insulin gene has been cloned to make it easier to produce large quantities of

• **Replica plating, p. 164**

insulin. The amount of product is proportional to the number of genes that code for the product; so, for example, the small cells of *E. coli* that contain a cloned gene on a high-copy number plasmid will produce much more of any product than will large eukaryotic cells each containing a single copy of the gene.

With human insulin or any other gene cloned for production purposes, cloning is only successful if the cloned gene is expressed in the new host. This can be a difficult problem, depending on the origin of the cloned gene and the relatedness of its new host. For expression to occur, the cloned gene must contain the appropriate signals that are recognized by the host machinery for the initiation and termination of transcription and translation. For example, the promoter of the cloned gene must be recognized by the host's RNA polymerase for transcription to occur, while the ribosome binding site must be recognized by the host's ribosomes for translation to occur. If the cloned gene comes from an organism that is not related to the host, such as a human insulin gene cloned in *E. coli,* then the original promoter and ribosome binding site must be replaced by the promoter and ribosome binding site of the new host (in this case *E. coli*) to ensure that the gene will be properly expressed (figure 8.13).

Another problem arises when genes of eukaryotic cells are cloned in bacteria. As discussed in chapter 6, most eukaryotic genes have intervening sequences, or **introns,** that are removed from precursor messenger RNA by the eukaryotic cell machinery before the mRNA is translated into protein. However, bacteria do not have introns themselves and do not have the means to remove them. Eukaryotic genes that have introns will generally not be expressed in *E. coli.* To overcome this problem, the eukaryotic genes are not cloned directly in the prokaryotic vector. Instead, mRNA from which the eukaryotic cell has already removed the introns is first isolated from the appropriate eukaryotic tissue. Then, a strand of DNA complementary to the mRNA is synthesized in vitro using **reverse transcriptase,** an enzyme encoded by retroviruses that is able to copy the sequence of mRNA into a complementary sequence of DNA. The resulting single strand of DNA is then converted into a double-strand DNA form, referred to as **cDNA** to reflect its origin (figure 8.14). The cDNA encodes the same protein as the original DNA, but it lacks the introns so it can be expressed when cloned in *E. coli.*

As a general rule, the more closely two organisms are related, the more likely the genes of one will be expressed in the cells of the other. For reasons that are not clear, *E. coli* seems to be especially able to express foreign genes from other bacteria even when their regulatory regions and ribosome binding sites are not modified. In addition, *E. coli* is a desirable host because it has been well-studied genetically and biochemically, However, like all gram-negative organisms, *E. coli* has an outer membrane that contains endotoxin, which is toxic when injected into humans and other animals. Therefore, great pains must be taken to purify any cloned gene product used intravenously, such as insulin, lest any outer membrane product contaminate the preparation.

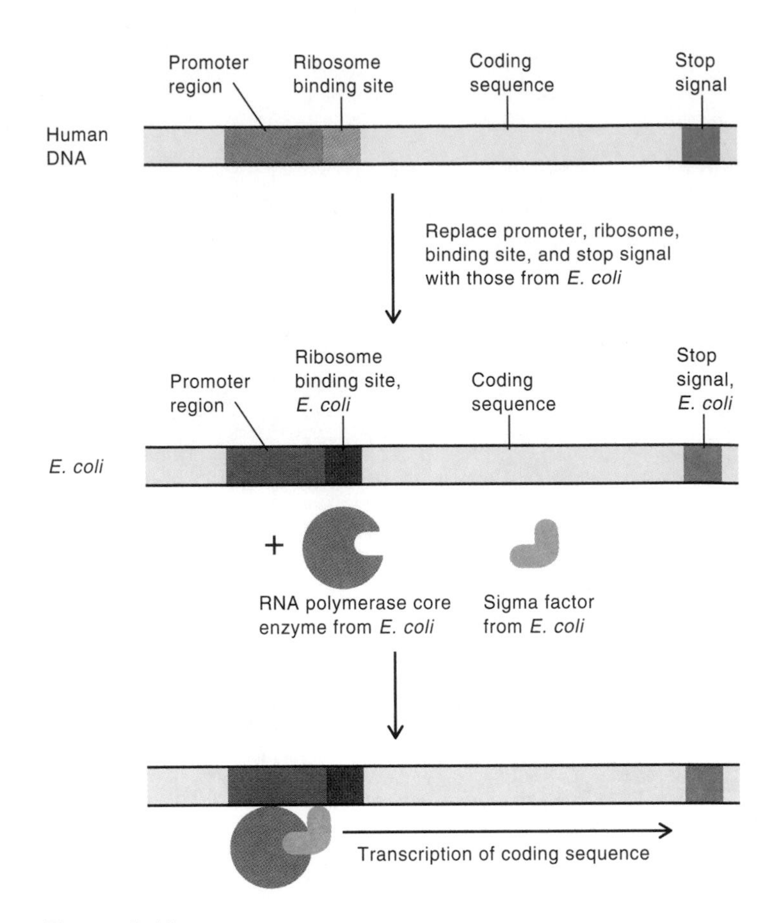

Figure 8.13
Some requirements for expression of human genes in *Escherichia coli.* The regulatory regions of the gene must be from *E. coli;* only the coding sequence is foreign. The excision of introns is not shown.

In Plant Genetic Engineering, Genes Are Introduced into Plant Cells by *Agrobacterium*

The workhorse of genetically engineered plants is the plant pathogen *Agrobacterium.* Recall that this bacterium has the capability to transfer a part of a large plasmid (the tumor-inducing, or Ti, plasmid) into a wide variety of plants, and its DNA is then incorporated (integrated) into the chromosome of the plant. In nature, this results in the important plant disease called *crown gall.* However, it is possible to remove the tumor-producing genes that are transferred to the plant and substitute useful genes such as the *B. thuringiensis* toxin genes. These useful genes are then

• **Promoter, p. 137**
• **Ribosome binding site, p. 138**
• **Intron, p. 141**

• ***Agrobacterium*, p. 175**

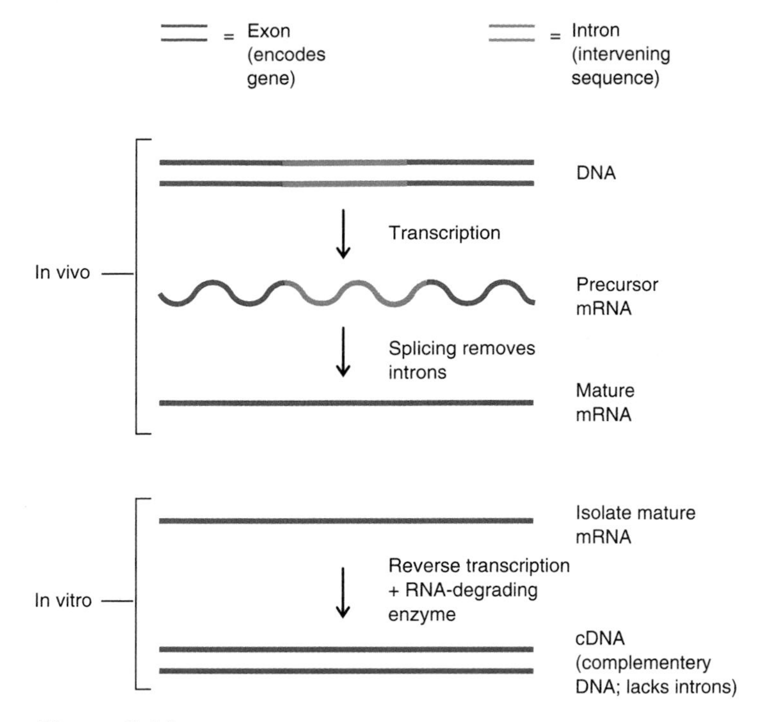

Figure 8.14
When cloning eukaryotic genes, the introns are removed by isolating the mRNA and using it as a template to make a new copy of DNA (cDNA) that lacks intron sequences.

transferred and integrated into the plant DNA. When the genes for *B. thuringiensis* toxin are transferred to a plant, the plant becomes toxic to many insect pests, thus conferring insect resistance (figure 8.15).

Resistance genes or any other useful genes transferred from animals or bacteria to a plant must be expressed if they are to alter the properties of the plant. Since the bacterial promoter is not recognized by plant RNA polymerase, it must be replaced by one that functions in the plant. The most commonly used promoter is one from a plant virus, the cauliflower mosaic virus, that infects plants and causes them to develop a serious disease. This promoter is recognized by the plant RNA polymerase and thus can be used as a regulatory region for the expression of any gene engineered into plants.

Southern Blot

The Southern blot technique, named after its developer Dr. Edward Southern, is a multistep procedure used to identify specific sequences of DNA and to determine the size of the restriction fragment that encodes the sequence (figure 8.16). For example, a Southern blot can be used to determine if a given strain of *E. coli* encodes the gene for Shiga-like toxin. The sizes of the restriction fragments can

• **Shiga-like toxin, p. 576**

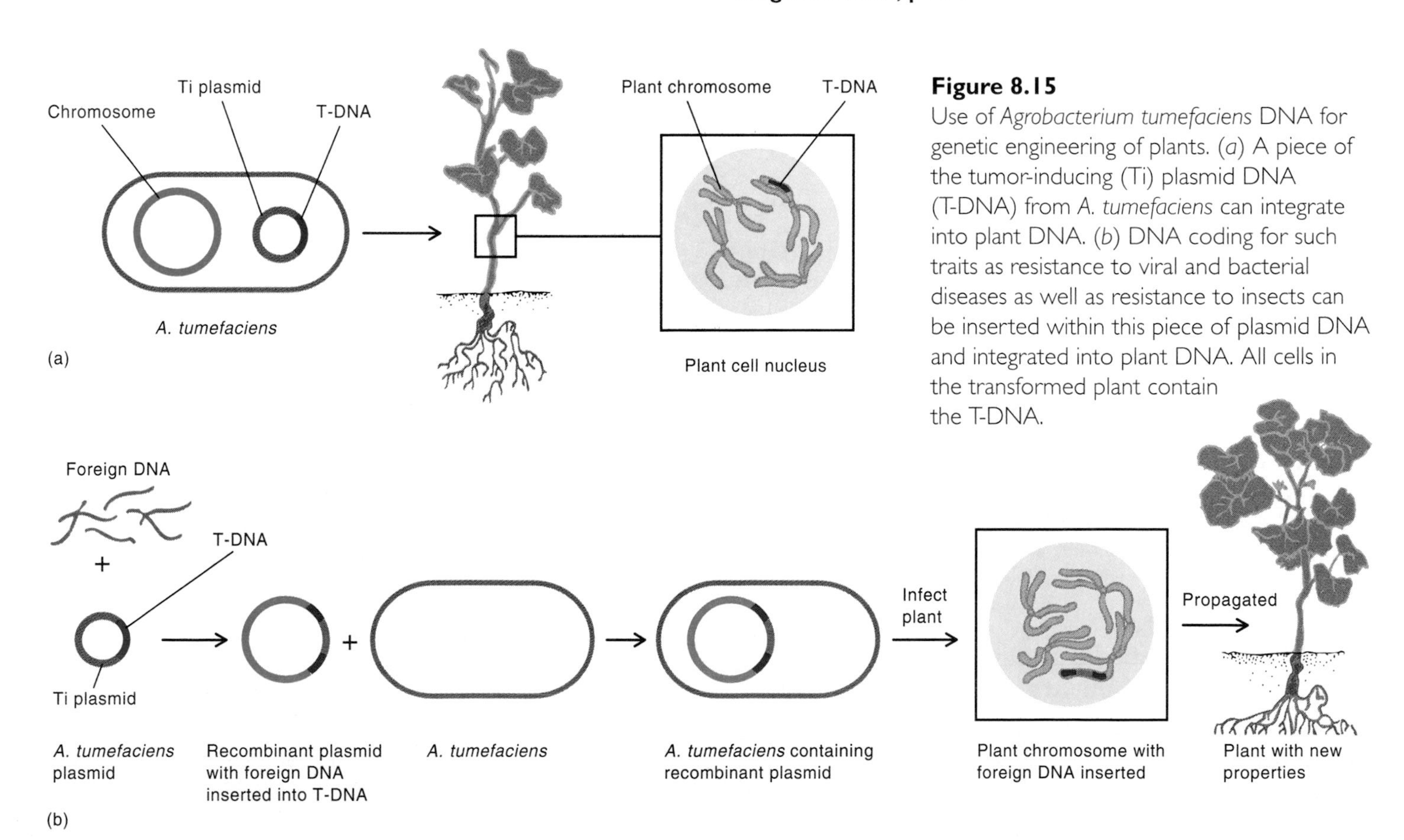

Figure 8.15
Use of *Agrobacterium tumefaciens* DNA for genetic engineering of plants. (*a*) A piece of the tumor-inducing (Ti) plasmid DNA (T-DNA) from *A. tumefaciens* can integrate into plant DNA. (*b*) DNA coding for such traits as resistance to viral and bacterial diseases as well as resistance to insects can be inserted within this piece of plasmid DNA and integrated into plant DNA. All cells in the transformed plant contain the T-DNA.

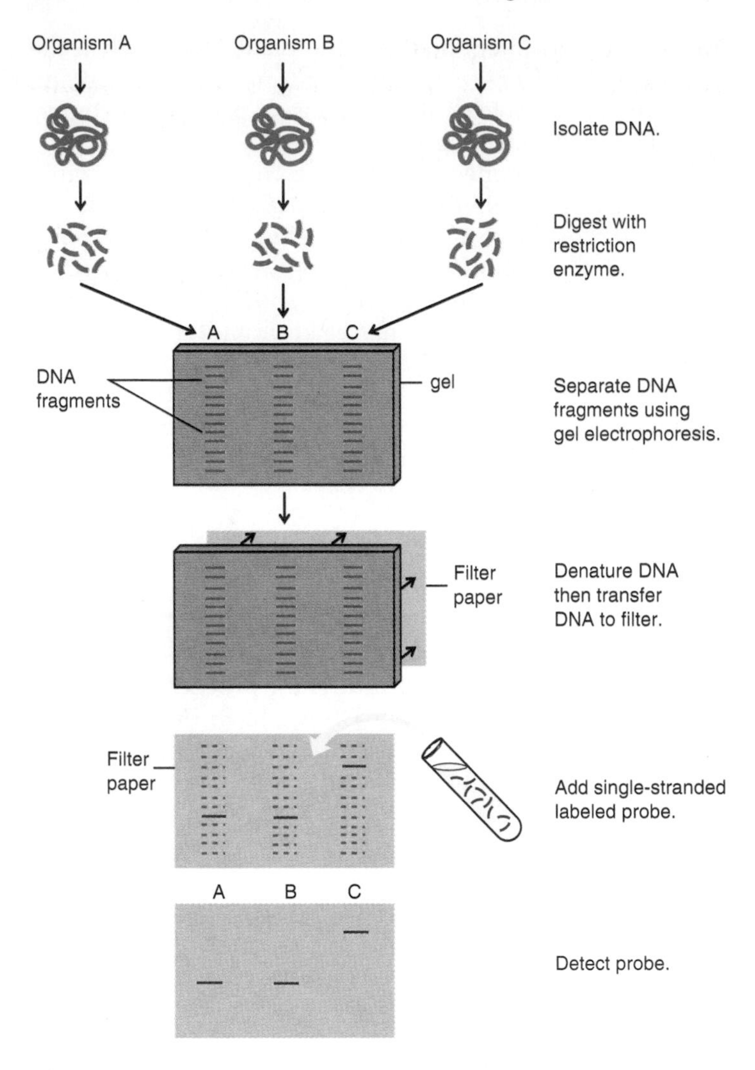

Figure 8.16
Steps of a Southern blot.

be used to determine relatedness of seemingly identical strains. For example, if a Southern blot is done using chromosomal DNA from three different isolates of an organism, and isolate A and B both have a 2,000 base pair fragment that hybridizes with the probe, while the fragment of isolate C that hybridizes to the probe is 3,000 base pairs, then isolate C is not the same strain as isolates A and B. While the data can be used to state which strains are different, they cannot be used to conclusively show that two strains are the same. The comparison of the sizes of fragments is fundamental to a procedure called **DNA fingerprinting** that has found its way into criminal trials and paternity suits.

The DNA Must Be First Digested and then The Fragments Separated According to Size

The first step of the Southern blot technique is the isolation and restriction enzyme digestion of the DNA to be studied. For example, a researcher studying the Shiga-like toxin gene of *E. coli* first isolates the DNA from several different strains of the organism and then separately digests each one with a restriction enzyme such as *Bam* HI. The DNA fragments from each strain are then separated according to their size using a technique called **agarose gel electrophoresis.** This technique uses a slab of gel made of agarose, a highly purified form of agar (which is the solidifying agent used in microbiological media). Like agar, agarose can be melted at high temperatures and poured as a liquid. When it cools, it is solid but flexible with the consistency of a very firm gelatin mixture. The digested DNA samples are put into slots or wells in the gel, and then the gel is placed in an electric field. DNA is negatively charged so it migrates toward the anode, the positively charged electrode (the anions migrate toward the anode). As the DNA fragments move in the electric field, the gel acts as a sieve, impeding the large fragments of DNA while allowing the smaller fragments to pass through more quickly. Because of the sievelike effect of the gel, the restriction fragments can be separated according to size.

The DNA is not directly visible on the gel so it must be stained by soaking the gel in a solution containing a dye called **ethidium bromide** that binds to nucleic acid molecules. Ethidium bromide–stained DNA is fluorescent when viewed with UV light and appears as bright bands on the gel. The bands of DNA in the gel can be viewed to ensure that separation of the fragments has occurred (figure 8.17).

The DNA Is Denatured and then Transferred to a Durable Membrane Support

The DNA in the gel is denatured by soaking the entire gel in an alkaline solution. Next, because the gel is too fragile to be easily used in future steps, the single-stranded fragments

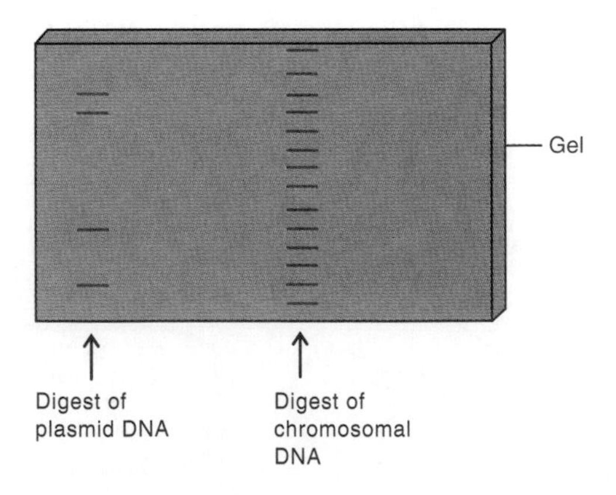

Figure 8.17
Each ethidium bromide stained band of DNA represents many copies of the same size fragment.

are transferred directly to a nylon membrane. This is done by placing the membrane directly on the gel so that the DNA is transferred to the membrane in the same relative position as it occupied on the gel.

Complementary Sequences Are Identified by Hybridizing a Probe

The DNA is now ready to be probed with the sequence of interest, which in this example is the Shiga toxin gene. Just as in the procedure for the colony blot hybridization, a radioactively labeled single-stranded DNA probe is used to seek out the sequence of interest. A liquid solution containing the probe is added to the membrane and incubated under conditions that allow the probe to hybridize to the complementary sequence. The unbound probe is then washed off, and the location of the bound probe is detected using X-ray film.

Polymerase Chain Reaction

The polymerase chain reaction (PCR) allows researchers to efficiently duplicate, or **amplify,** specific pieces of DNA. PCR can amplify extraordinarily small quantities of DNA. In fact, the amount of DNA in a single bacterial cell is sufficient to be amplified. In addition, even impure samples can be amplified using PCR, which is used in a wide variety of areas, from the medical laboratory, where it is used to detect pathogens in patient specimens, to the crime laboratory, where it is used to increase the amount of DNA in a small but critical blood stain or other sample. This powerful technique is currently revolutionizing research by making it possible to detect extremely small numbers of specific organisms, even those that cannot be cultivated. The basic techniques of PCR are elegant in their simplicity but ultimately depend on the existence of a heat-stable polymerase of a thermophilic bacterium, *Thermus aquaticus* (figure 8.18).

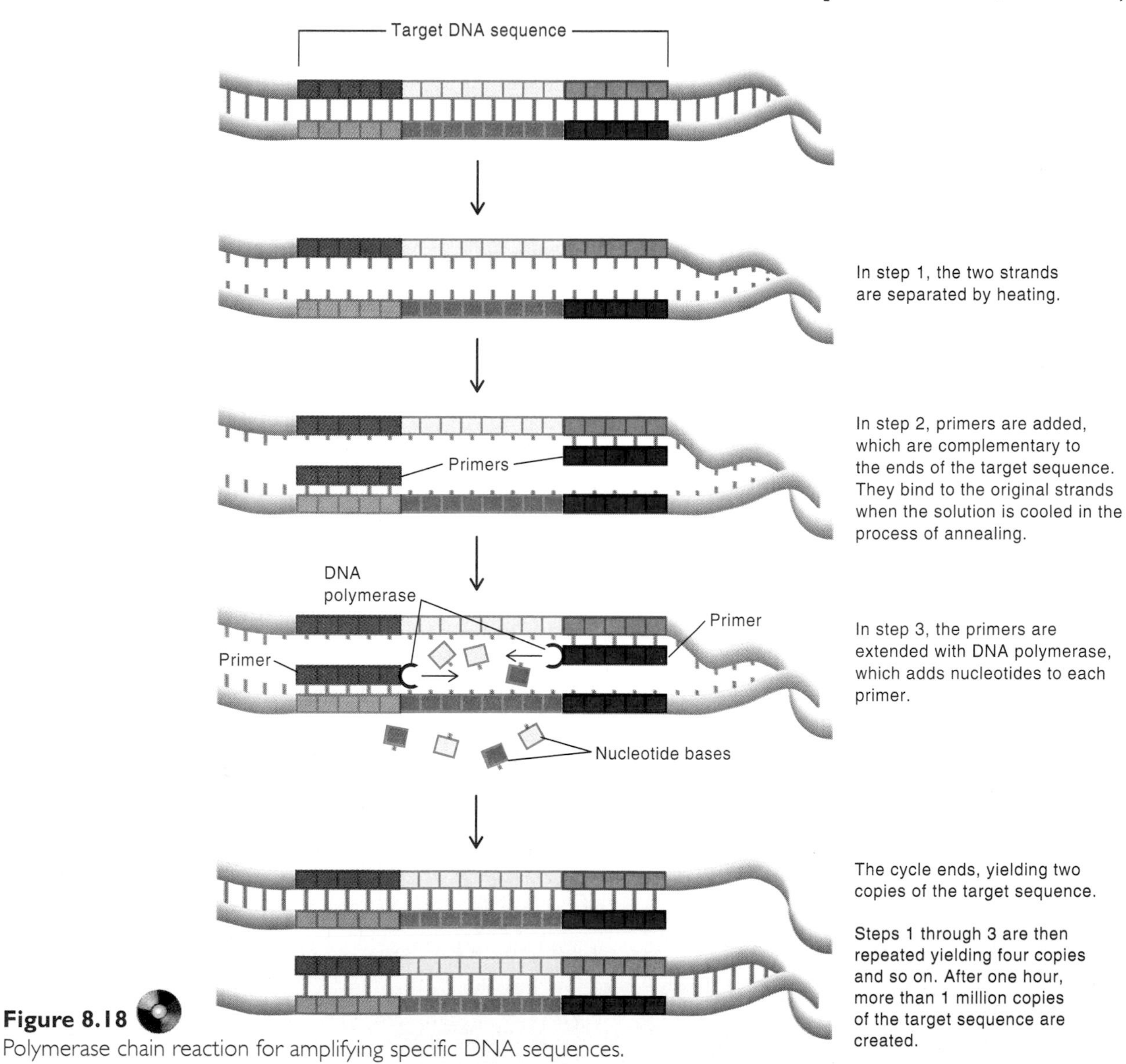

Figure 8.18
Polymerase chain reaction for amplifying specific DNA sequences.

Perspective 8.1

Science Takes the Witness Stand

Fingerprints have been used to identify criminals and solve crimes for over 100 years. More recently, DNA fingerprinting is gaining in prominence for its role in criminal investigations. If a criminal leaves a blood stain or semen at the scene of a crime, the DNA can be extracted and compared to the DNA from the blood of the suspect.

It is not possible to compare the entire nucleotide sequence of two samples; instead only certain regions of DNA that vary between different individuals are compared. These regions are located between genes and consist of a core sequence that is repeated a number of times; the exact number differs from one individual to the next. Depending on the number of times the sequence is repeated, different sizes of DNA fragments will be generated following digestion with a restriction enzyme that cuts outside of the core sequence. Variations in the size and number of these fragments can be detected by Southern blot hybridization using the core sequence as a probe (figure 1). By comparing the hybridization pattern of DNA from samples taken at the scene of the crime to that of DNA from the suspect, it is theoretically possible to determine the likelihood that the samples of DNA came from the same person.

It is essential that the results of DNA testing are interpreted with caution. For example, it is difficult to state with certainty that two bands are identical in size. Further, during electrophoresis DNA fragments of the same size can migrate at different rates because of variations in the preparation of the gel, concentration of the DNA, or other unknown variables. However, even with these technical problems, there seems little doubt that molecular biology will play an increasingly important role in criminal investigations as science takes the witness stand.

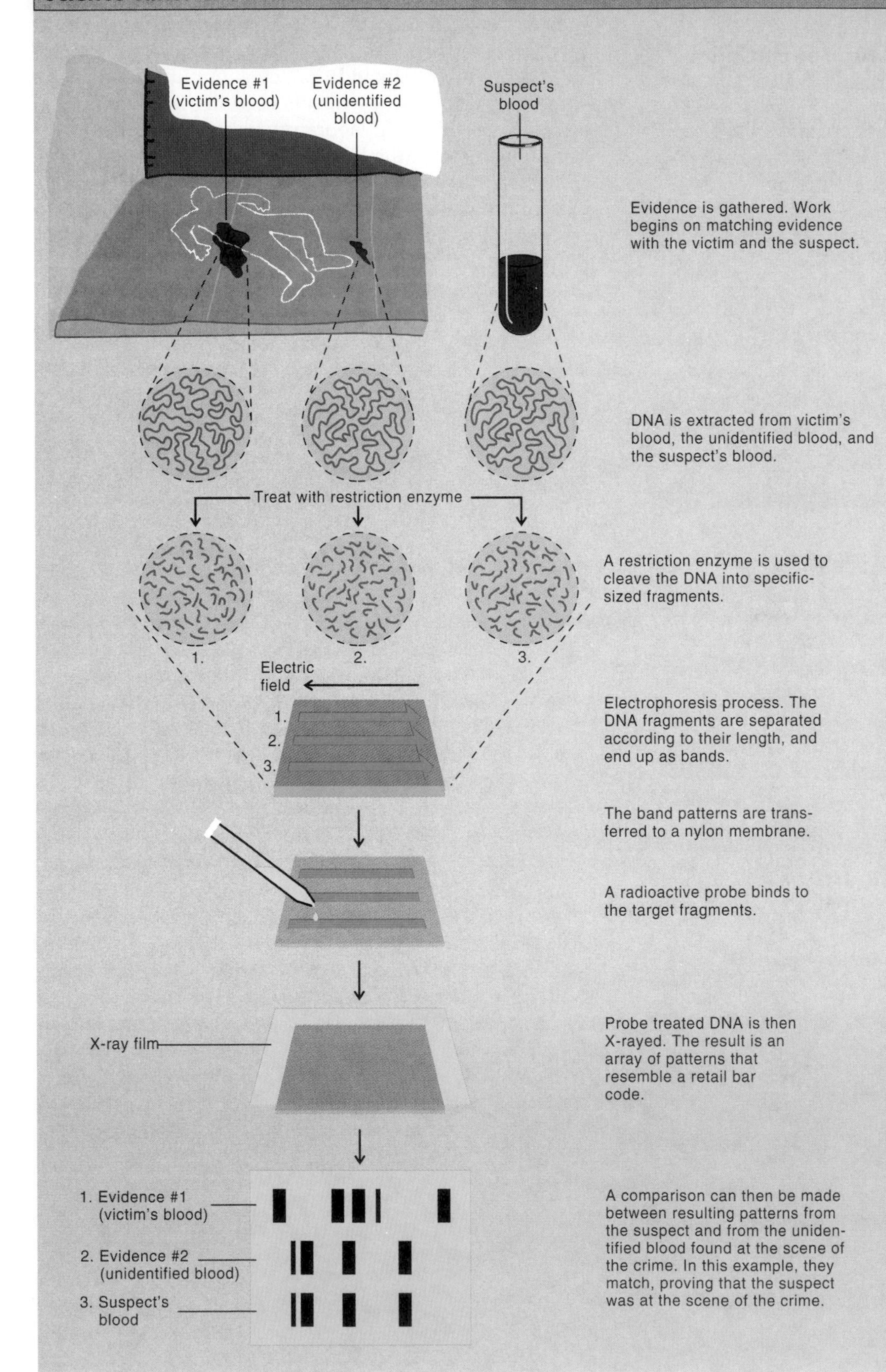

Figure 1
DNA fingerprinting (typing). The probe recognizes sequences in DNA that vary in number among people.

Repeated Temperature Cycles Amplify the Target DNA

Starting with a double-stranded DNA molecule, millions of copies can be produced in one hour through a reaction involving a repeating cycle of three steps. The ingredients in the reaction are: (1) the double-stranded DNA to be amplified, called the **target DNA;** (2) a heat-stable DNA polymerase; (3) each of the four nucleotides; and (4) small single-stranded segments of DNA called **primers.** The nucleotide sequence of the primers is critical because the primers dictate which portion of the double-stranded DNA is amplified. The primers must be complementary to the ends of the target DNA. If a researcher wants to amplify a DNA sequence that encodes a specific protein, then the researcher must first determine the sequence at both ends of the gene and synthesize appropriate primers.

The first step in PCR utilizes a temperature of approximately 95°C, which is just below boiling. At that high temperature, the hydrogen bonds between the A-T and G-C base pairs break, causing the two strands to come apart, or denature, to form two single-stranded pieces of DNA. In the second step, the temperature is lowered to approximately 50°C and within seconds the primers anneal to their complementary sequences on the denatured target DNA. The third step, DNA synthesis, occurs when the temperature is raised to the optimal temperature for *Taq* DNA polymerase, 70°C. As described in chapter 6, the DNA polymerase adds nucleotides to the 3' end of the DNA primer using the opposing strand as a template. The net result is the synthesis of two new strands of DNA, each of which is complementary to the other. In other words, the three-step cycle results in the duplication of the original target DNA. Since each of the newly synthesized strands can then serve as a template strand for the next cycle, the DNA is amplified exponentially. After a single cycle of the three-step reaction, there will be two double-stranded DNA molecules for every original double-stranded target; after the next cycle, there will be four, after the next cycle there will be eight, and so on. Since each three-step cycle takes only a few minutes, DNA can be amplified more than a millionfold in one hour.

A critical factor in PCR is the heat-stable DNA polymerase of the thermophile *Thermus aquaticus*. *Taq* polymerase, unlike the DNA polymerase of *E. coli*, is not destroyed at the high temperature used to denature the DNA in the first step of each amplification cycle. If the heat-stable polymerase were not used, new polymerase would be required for every cycle of the reaction.

• **Thermophile, p. 91**

REVIEW QUESTIONS

1. Name the two enzymes used to create a recombinant DNA molecule.
2. Define the term *gene library*.
3. Explain why cDNA is used when cloning a gene from a eukaryote.
4. Outline the steps of a Southern blot.
5. Explain why the enzyme from a thermophile is used in the PCR reaction.

Summary

I. Applications of Biotechnology
   A. Medically important proteins can be produced by *E. coli* and yeast.
   B. Safer vaccines can be made using genetic engineering.
   C. Bioremediation involves the use of microorganisms to degrade harmful chemicals.
   D. Plants can be genetically engineered to develop new properties.
   E. Toxins from bacteria can serve as effective biocontrol agents.

II. Cloning
   A. The cells are lysed to release all of their DNA.
   B. Restriction enzymes and DNA ligase are the tools of molecular biology.
   C. Vectors act as self-replicating carriers of foreign DNA.
   D. Ligating the vector and the insert together creates a recombinant plasmid.
   E. The recombinant plasmid must be introduced into a new host.
   F. Cells that contain the recombinant clone must be identified.
   G. Successful genetic engineering generally requires gene expression.
   H. In plant genetic engineering, genes are introduced into plant cells by *Agrobacterium*.

III. Southern Blot
   A. The DNA must be first digested and then the fragments separated according to size.
   B. The DNA is denatured and then transferred to a durable membrane support.
   C. Complementary sequences are identified by hybridizing a probe.

IV. Polymerase Chain Reaction
   A. Repeated temperature cycles amplify the target DNA.

Chapter Review Questions

1. Which of the following could be a restriction enzyme recognition site?
   A. AAAGGG
   B. ATCCTA
   C. ATGCAT
   D. AUCCUA
   E. ATCATC
2. An ideal vector has all of the following EXCEPT:
   A. an origin of replication
   B. a gene encoding a restriction enzyme
   C. a gene encoding resistance to an antibiotic
   D. a multiple cloning site
   E. the *lacZ'* gene
3. What should the first restriction enzyme isolated from *Serratia marcescens* be called?
4. What does the enzyme ligase do?
5. What is a gene library?
6. What is cDNA and why is it used when cloning eukaryotic genes?
7. The Ti plasmid is used to clone genes into plants. What is the name of the plant pathogen that naturally carries the plasmid?
8. On what basis does agarose gel electrophoresis separate DNA fragments?
9. What is the purpose of PCR (polymerase chain reaction)?
10. How many different temperatures are used in each cycle of the polymerase chain reaction?

Critical Thinking Questions

1. What might be the advantage to a bacterium to synthesize an insecticidal toxin?
2. Why is it advantageous for *Agrobacterium* to make a tumor on plants?
3. What features of plasmids make them good cloning vectors?
4. What are possible problems that might arise from genetically engineering a plant to synthesize B.T. toxin?

Further Reading

1983. Biotechnology. *Science* 219:611+. The entire issue of this weekly scientific publication is devoted to articles on various applications of biotechnology.

Brill, W. J. 1985. Safety concerns and genetic engineering in agriculture. *Science* 227:381.

Canby, T. Bacteria: Teaching old bugs new tricks. *National Geographic* 184(2). 1993.

Chilton, M. D. 1983. A vector for introducing new genes into plants. *Scientific American* 248(6):51–59.

Crueger, W., and Crueger, A. 1984. *Biotechnology: a textbook of industrial microbiology.* T. D. Brock (ed.) Sinauer Associates, Sunderland, Mass.

Demain, A. W., and Soloman, N. A. 1986. *Manual of industrial microbiology and biotechnology.* American Society for Microbiology, Washington, D.C.

Gasser, C., and Fraley, R. T. 1992. Transgenic crops. *Scientific American* 266(6):62–69.

Glick, B. R., and Pasternak, J. J. 1994. *Molecular biotechnology: principles and applications.* ASM Press, Herndon, Va. Well-illustrated textbook that is written for students and reference.

Mullis, K. B. 1990. The unusual origin of the polymerase chain reaction. *Scientific American* 262(4):56–65.

Persing, D., Smith, T. E., Tenover, F., and White, T. J. (ed). 1993. *Diagnostic molecular biology: principles and applications.* ASM Press, Herdon, Va. This book presents the principles and applications of molecular diagnostic methods, including nucleic acid probe technology and polymerease chain reaction.

Trevan, M. D., et al. 1987. Biotechnology: the biological principles. Open University Press, Milton Keynes, UK.

9 Control of Microbial Growth

Key Concepts

1. Sterilization is the process of removing or killing all microorganisms and viruses on or in a material. In contrast, disinfection only reduces the number of infectious agents to a point where they no longer present a hazard.
2. Time, temperature, growth stage of the organism, nature of the suspending medium, and the numbers of organisms present must all be considered when determining which sterilization or disinfection technique to employ.
3. Sterilization and disinfection can be accomplished by using heat, filtration, chemicals, or radiation.
4. Preservation techniques slow or halt the growth of organisms to delay spoilage.

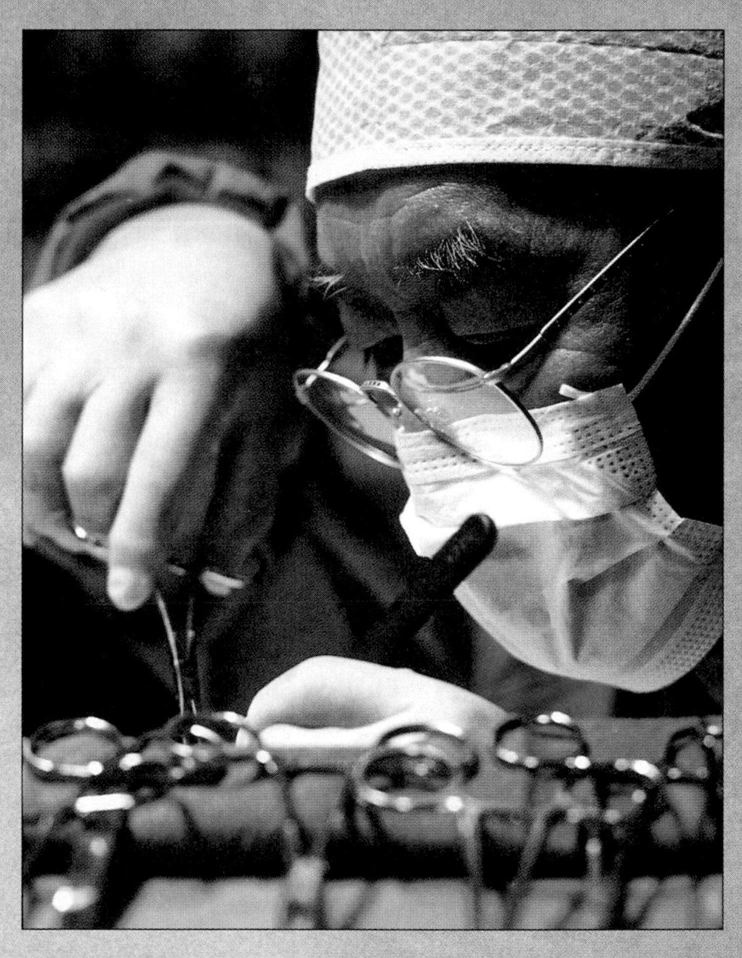

Surgeon performing an operation. Sterilization and disinfection techniques utilizing heat, irradiation, germicidal chemicals, and filtration can all be used to create an aseptic environment for surgical patients.

PREVIEW LINK

Microbes in Motion

The following books and chapters in the *Microbes in Motion* CD-ROM may serve as a useful preview or supplement to your reading:

Microbial Metabolism and Growth: Growth. *Control—Physical and Chemical:* Physical Control; Chemical Control. *Antimicrobial Action:* Cell Wall Inhibitors; Protein Synthesis Inhibitors; Agents Affecting Nucleic Acid; Agents Affecting Enzymes; Antibiotic Susceptibility Testing. *Antimicrobial Resistance;* Acquisition and Expression of Resistance Genes.

A Glimpse of History

Joseph Lister (1827–1912) was a British physician who revolutionized surgery by preventing surgical wounds from becoming infected. Impressed with Pasteur's work on fermentation (said to be caused by "minute organisms suspended in the air"), Lister wondered if "minute organisms" might be responsible also for the pus that forms in surgical wounds. He then experimented with a phenolic compound, carbolic acid, introducing it at full strength into wounds by means of a saturated rag. Lister was particularly proud of the fact that, after carbolic acid wound dressings became routine for his patients, the patients no longer developed gangrene. His work provided impressive evidence for the germ theory of disease, even though microorganisms specific for various diseases were not identified for another decade.

Later, Lister improved his antiseptic methods by introducing **aseptic surgery** that excluded bacteria from wounds by maintaining a clean environment in the operating room and by sterilizing instruments. These procedures were preferable to killing the bacteria after they had entered wounds because they avoided the toxic effects of the disinfectant on the wound.

Lister was knighted in 1883 and subsequently became a baron and a member of the House of Lords. The *British Medical Journal* stated that "he had saved more lives by the introduction of his system than all the wars of the 19th century together had sacrificed."

Once people recognized that microorganisms cause disease, they tried to discover ways of killing or inhibiting disease-causing bacteria and viruses. The control of microbial growth is also a concern in industries such as food production because microorganisms and the products they make during their growth cause spoilage and, in some cases, food-borne illness. This chapter will cover methods that have been developed to remove microorganisms from products and to kill them or inhibit their growth.

Approaches to Control

Routine control of undesirable microorganisms and viruses is achieved by mechanical removal such as washing and scrubbing with soaps and detergents. However, in many situations it is important to destroy all unwanted microorganisms and viruses.

Sterilization is the process of removing or killing all microorganisms and viruses on or in a product. Microorganisms and viruses must be rendered incapable of reproducing, even under the most favorable circumstances, so that they cannot reproduce even if they are transferred from the sterilized material to an ideal growth medium. This requirement is important because microorganisms that have presumably been killed can sometimes be "revived" by changing the conditions of growth. The word *inactivated* is sometimes used instead of the word *killed* in the context of viruses because, technically, viruses are not living organisms.

Microorganisms and viruses can be destroyed through some types of heat treatment, irradiation, or exposure to the gas ethylene oxide. Products that are commonly sterilized using these methods include bacteriological media, pipette and petri dishes that are used in microbiology laboratories, and medical instruments that are used in surgeries. Liquids such as antibiotic solutions can be sterilized by filtration to remove microorganisms and viruses.

Disinfection is the process that reduces the number of potential disease-causing bacteria and viruses on a material until they no longer represent a hazard. Unlike sterilization, disinfection permits some living microorganisms to persist. A **disinfectant** is a chemical used to disinfect inanimate objects, whereas an **antiseptic** is a disinfectant that is nontoxic enough to be used on human tissues. The term **decontamination** is often used interchangeably with *disinfection*, but it implies a broader role, including inactivation or removal of both microbial toxins and the living microbial pathogens themselves. In the fields of food technology and public health, the frequently used term **sanitize** means to reduce substantially a microbial population to meet accepted public health standards. Most people expect a sanitized object to be not only disinfected but also clean in appearance.

A disinfectant that kills microorganisms and viruses is called a **germicide** (*-cide* comes from *-cida* meaning "to kill"). Special terms, such as *bactericide, fungicide,* and *viricide,* are used to indicate killing action against specific microbial groups. Related terms, such as **bacteriostatic** and **fungistatic**, indicate that an antimicrobial agent is primarily inhibitory in its action, preventing growth without substantial killing (*-static* comes from a Greek word *statikas* meaning "causing to stand" and has the meaning here "to cease activity"). Growth can resume if the inhibitory agent is neutralized or diluted. An agent that is germicidal against one species of microorganism may be only inhibitory against another, and the nature of its action may change depending on its concentration, pH, and temperature.

Pasteurization involves a brief heat treatment that is used to reduce the number of spoilage organisms in products such as wine or to kill disease-causing organisms in milk and other foods. Pasteurization reduces the number of organisms in a product but does not eliminate them completely.

Preservation is the process that delays spoilage of food or other perishable products by inhibiting the growth of microorganisms. Common methods of preservation include storage at low temperature, lowering the pH, adding chemicals that inhibit microbial growth, and reducing the available water in the product being preserved.

Principles Involved in Killing or Inhibiting Microorganisms

Different microorganisms and viruses vary in their susceptibility to sterilizing and disinfecting procedures (table 9.1). For example, endospores are much more resistant than vegetative cells to killing. Thus, a product that may contain

Table 9.1 Ranking* of Different Types of Microorganisms and Viruses According to Resistance to Chemical Germicides

1. Bacterial endospores
2. Certain fungal spores
3. Mycobacteria such as *Mycobacterium tuberculosis*
4. Nonlipid-containing viruses (such as picornaviruses)
5. Vegetative fungi and most fungal spores
6. Vegetative bacteria
7. Lipid-containing and medium-sized viruses (such as influenza viruses, adenoviruses, HIV)

* Ranges from most resistant (bacterial endospores) to least resistant. The table depicts the general pattern, but there are exceptions.

and support the growth of the endospore-forming bacteria *Bacillus* and *Clostridium* must be subjected to much more rigorous sterilization or disinfection procedures than a product that does not. *Mycobacterium tuberculosis*, the causative agent of tuberculosis, is more resistant than other bacteria to many chemicals and conditions. Stronger disinfectants must be used to decontaminate environments that may contain *Mycobacterium*.

Numerous conditions alter the effectiveness of agents that kill microorganisms and viruses. For example, the phase of growth of the microorganism is important. Actively growing (log phase) cultures are the most susceptible to killing, whereas stationary phase cells are much more resistant. Another factor that strongly influences the action of disinfecting agents is the temperature. For example, sodium hypochlorite solution (household bleach) can kill a suspension of *M. tuberculosis* at 50°C in 150 seconds and one at 55°C in only 60 seconds. The nature of the suspending fluid (pH, viscosity, and the presence or absence of organic matter) also influences the rate of killing. For example, chlorine kills more readily at a low pH, but its killing power is neutralized by the presence of organic compounds.

When heat or other treatments are used to kill microorganisms, only a fraction of the microorganisms or viruses die during a given time interval. Moreover, this fraction tends to be constant under specific conditions. For example, if 90% of a bacterial population is killed during the first 3 minutes, then approximately 90% of those remaining will be killed during the next 3 minutes, and so on. In other words, the death curve is logarithmic. Recall that during the death phase of a cell population, cells die at a constant rate. However, the rate is less than what it would be if the culture were heated. The **D value**, or **decimal reduction time**, is defined as the time it takes to kill 90% of a population of bacteria under specific conditions (see perspective 9.6). Thus, if the D value for killing a specific organism is 2 minutes, then it would take 4 minutes (two D values) to reduce a population of 100 organisms to only one survivor (it takes 2 minutes to kill 90 organisms and 2 minutes more to kill 9 of the remaining 10), whereas it would take 20 minutes (10 D values) to reduce a population of 10^{10} cells to only one survivor. The concept of the D value reflects the fact that it takes more time to kill a large population of bacteria than it does to kill a small population. Thus, it explains why removing organisms by washing or scrubbing can decrease the time necessary to sterilize or disinfect a product.

Using Heat to Kill Microorganisms and Viruses

Even before the discovery of the microbial world, heat was used to control microorganisms. It kills organisms by coagulating (denaturing) the proteins of cells. Today, the use of heat is still one of the most generally satisfactory control methods because it is fast, reliable, and inexpensive relative to most other methods. In addition it does not introduce potentially toxic substances into the material being treated. However, the use of heat is limited to substances that can withstand the temperatures required for microbial killing and to objects small enough to fit into a heating chamber. Table 9.2 summarizes the methods, limitations, and uses of heat to control microorganisms.

Some types of heat treatment, such as pasteurization, are used to decrease the numbers of microorganisms, whereas other treatments, such as autoclaving, sterilize products. Cooking of foods is a familiar example of using heat to kill microorganisms. However, a "well-done" appearance does not always guarantee that a food is safe to eat because temperatures sufficient to kill potential pathogens must be maintained throughout the food and not just on its surface. In addition, some bacterial toxins are not destroyed by cooking.

• ***Staphylococcus aureus* exotoxin, p. 421**

Table 9.2 Using Heat to Control Microorganisms

Method	Limitations	Uses
Dry heat	Slow; high temperature required	Glassware, large metal objects
Boiling	Unreliable for killing endospores	Drinking water
Steam under pressure	Unreliable for sterilizing substances impenetrable to steam	Canned foods, surgical instruments
Pasteurization	Unable to kill many microorganisms	Foods and beverages

Perspective 9.1

Some Materials Can Be Sterilized by Temperatures Below Boiling

The physicist John Tyndall, one of the major contributors to the debate over spontaneous generation, realized that some bacteria could exist in a heat-resistant form (later shown to be the endospore) that could revert to a heat-susceptible vegetative form if growth conditions were favorable. Based on this concept, Tyndall developed a method of sterilizing media that required only modest amounts of heating (for example, 60°C). His process, now called **tyndallization,** involves heating the medium to a temperature sufficient to kill all the vegetative bacteria and then cooling the medium, allowing it to remain at incubator temperature until any spores present germinate. The medium is then reheated to kill the emerging vegetative cells. Robert Koch used this method to prepare a coagulated serum medium with which he was able to cultivate *Mycobacterium tuberculosis* and prove that it caused tuberculosis. Tyndallization is rarely used today because heat-sensitive nutrients can be sterilized more quickly and easily by filtration.

• **Tyndall, p. 5**

Dry Heat Requires More Time than Wet Heat to Kill Microorganisms

Heating in the absence of moisture is frequently used to kill microorganisms in materials other than food. For example, in the microbiology laboratory, the wire loops that are used to transfer bacterial cultures are sterilized between uses by direct flaming. To sterilize glass petri dishes and pipettes, the glassware is put into ovens with noncirculating air and left at temperatures of 160°C to 170°C for 2 to 3 hours. Alternatively, sterilizing ovens with a fan that circulates the hot air can sterilize glassware in half the time because of the more efficient transfer of heat. Even spacecraft have been sterilized using dry heat technology.

Dry heat requires much more time than wet heat to kill microorganisms. For example, 200°C for 1 1/2 hours of dry heat gives the killing equivalent of 121°C for 15 minutes of moist heat. The decreased effectiveness of dry heat is due to the fact that proteins do not denature as easily when they are dry.

Boiling Kills Most Vegetative Cells But Not Bacterial Endospores

The use of boiling to prevent disease extends back at least to the time of Aristotle (384–322 B.C.), who advised Alexander the Great to have his armies boil their drinking water. Under ordinary circumstances, with concentrations of microorganisms less than 1 million per milliliter, suspensions that do not contain bacterial endospores can be sterilized in boiling water at 100°C in about 10 minutes.

The drawback of boiling is that the temperature of boiling water (100°C) is not high enough to kill many bacterial endospores. For example, the endospores of the food-poisoning bacteria *Clostridium perfringens* and *C. botulinum* can survive many hours of boiling. When heat is applied to water at sea level, the temperature rises until it reaches 100°C (212°F). An additional amount of heat energy (heat of vaporization) is then taken up by the water without a change in temperature, whereupon boiling begins to occur. The temperature does not go above 100°C. At altitudes above sea level, water boils at temperatures considerably lower than this because of the reduced atmospheric pressure (table 9.3). Two percent sodium carbonate is sometimes added to boiling water to raise its pH and increase its killing power against bacterial endospores. However, even with such methods, boiling is not a reliable sterilizing technique because it cannot kill endospores of certain bacterial species with certainty.

Pressure Cookers and Autoclaves Are Effective Sterilizers

Pressure cookers and their commercial counterpart, the **autoclave** heat water in an enclosed vessel that achieves temperatures above 100°C. As increasing steam is formed from the heated water, the pressure inside the vessel increases beyond atmospheric pressure. At increased pressure inside the vessel, steam forms at increasing temperatures. Thus, while steam produced at atmospheric pressure never exceeds 100°C, steam produced at 15 pounds psi (pressure per square inch) is 121°C which is sufficient to kill bacterial endospores. Note that the pressure plays no direct role in the killing. The conditions usually used for sterilization are 15 pounds psi and 121°C (250°F) for 15 minutes.

Table 9.3 Relationship of Altitude to Boiling Point of Water

Locations	Approximate Altitude (in feet)	Boiling Point (H_2O)	
New Orleans	10	100°C	(212°F)
Atlanta	1,000	99°C	(210°F)
Tucson	2,500	97.5°C	(207.5°F)
Denver	5,300	94.7°C	(202.5°F)
Mt. Rainier	14,410	85.3°C	(185.5°F)
Mt. McKinley	20,320	79.3°C	(174.7°F)

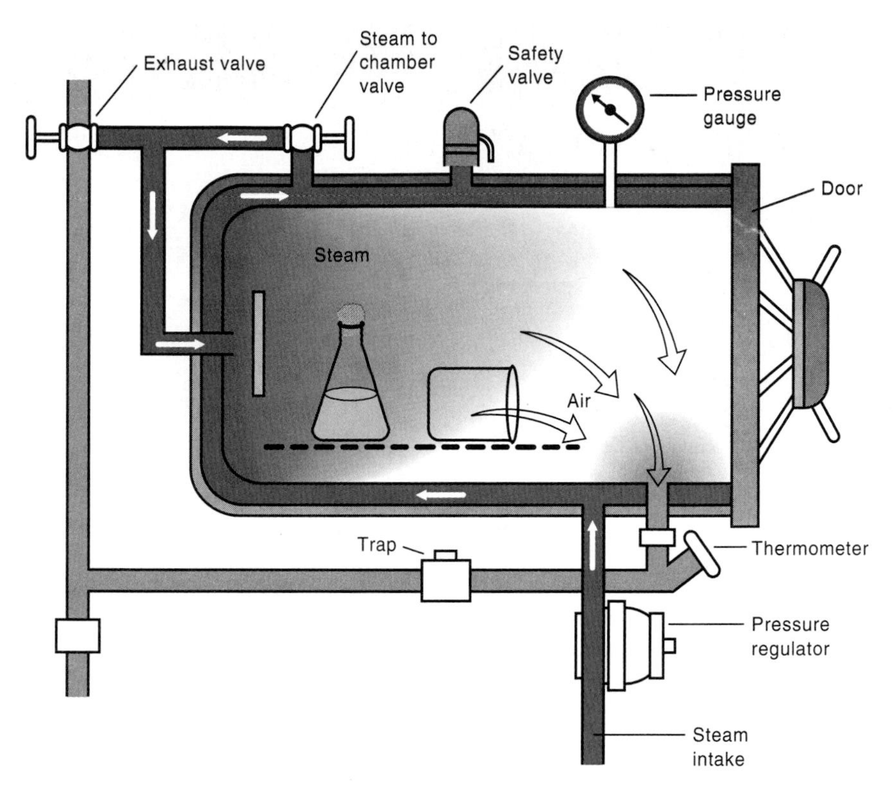

Figure 9.1
Steam-jacketed autoclave. Entering steam displaces air downward and out through a port in the bottom of the chamber. Dry objects are placed in the autoclave in a position to avoid trapping air. Watery liquids generate their own steam.

The main reason for the greater effectiveness of the autoclave over the hot air oven is related to the fact that heat coagulates proteins, including essential enzymes, more readily under moist conditions than dry conditions. The heat of vaporization is released when steam condenses on organisms and may play a role in the coagulation of proteins and in rapid killing. It is important, however, that the autoclave be used properly to ensure sterilization (see Technique 9.1 on using an autoclave).

The modern autoclave is, in principle, a sophisticated pressure cooker with mechanisms for regulating the steam pressure and for ensuring that air is completely evacuated from the chamber (figure 9.1). To be optimally effective, the steam should be at the temperature that water boils for any given pressure (figure 9.2). The presence of air allows the pressure to increase within the chamber without a corresponding increase in temperature (table 9.4). Thus, when autoclaving, it is important to always check the temperature as well as the pressure, because the latter is of little or no significance in killing microorganisms. Superheated steam (dry steam)[1] is likewise unsatisfactory and kills at slow rates, similar to hot air. In order to ensure that the autoclave is working properly, its sterilizing ability must be routinely checked using heat-sensitive chemical indicators, or biological indicators (figure 9.3). In operating rooms

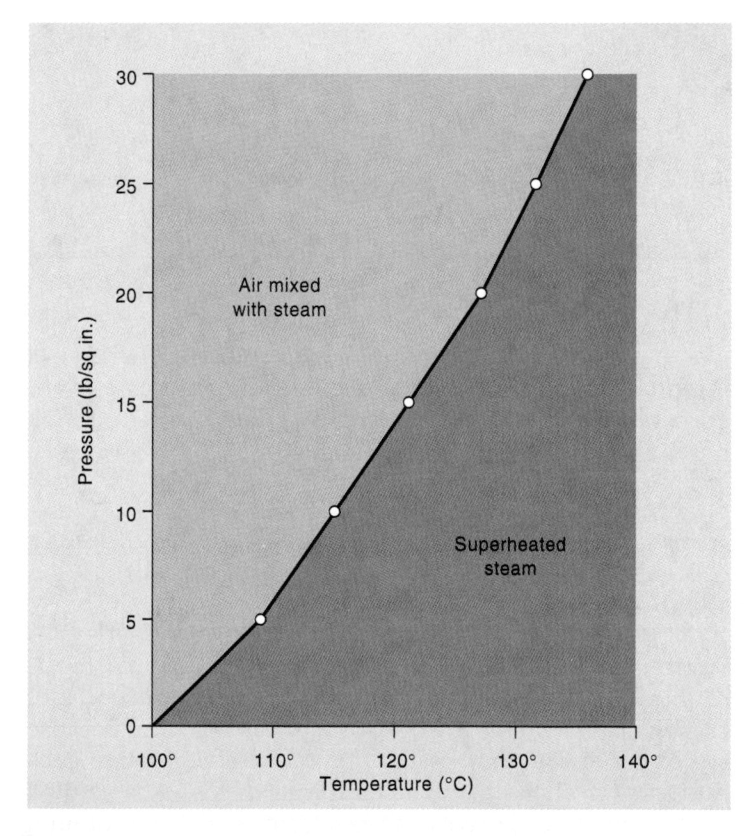

Figure 9.2
Relationship between pressure and the temperature at which steam is formed. Operating conditions that fall to the left or right of the curve are unsatisfactory for autoclaving.

[1]Superheated steam is steam heated to temperatures above the boiling point of water at a given pressure. This might happen if the autoclave water supply were exhausted but heat continued to be applied to the steam.

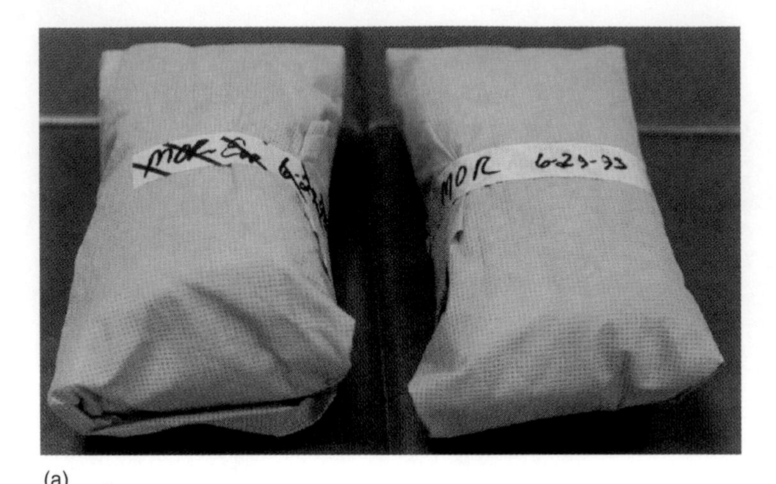

(a)

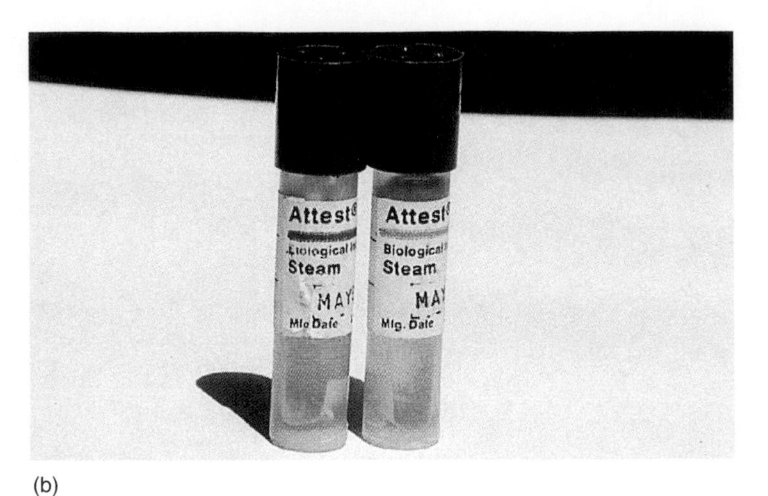

(b)

Figure 9.3
Indicators used in autoclaving. (*a*) Chemical indicators. The pack on the left has been autoclaved. Diagonal marks on the tape have turned black with heat, indicating that the object was exposed to heat but not necessarily that it is sterile. (*b*) Biological indicators. Heat-resistant endospores in tiny plastic tubes are placed near the center of the object being autoclaved. After autoclaving, the spores are mixed with the bacteriological medium by crushing a container within the tubes. Following incubation, a change of color from blue to yellow indicates growth of the spores and faulty autoclaving.

Table 9.4 Effect of Air Retained in an Autoclave

Pressure (pounds/square inch above atmospheric)	Temperature (degrees C)		
	Complete air removal	1/2 air removal	No air removal
5	109	94	72
10	115	105	90
15	121	112	100
20	126	118	109

where more rapid sterilization of instruments is sometimes important, a higher temperature is commonly employed. For example, at 135°C, sterilization is achieved in only 3 minutes. This is called **flash autoclaving**.

The Commercial Food-Canning Process

Some of the important steps in the commercial food-canning process involve ensuring that the endospores of *Clostridium botulinum* are destroyed. This is critical because surviving spores can germinate and the vegetative cells can grow in the anaerobic conditions of low-acid canned foods, such as vegetables and meats, to produce botulinum toxin, one of the most potent toxins known. Because ingestion of even minute amounts of botulinum toxin can be lethal, the canning process of low acid foods is designed to kill all endospores of *C. botulinum*. In doing so, the process also kills all other organisms capable of growing under normal storage conditions. Some thermophilic endospore formers that can spoil the canned food may survive the canning process, but they are usually of no concern because they are only capable of growing at temperatures well above those of normal storage. Thus, canned foods are called **commercially sterile** because they are not always free of all microorganisms. Figure 9.4 shows the steps involved in the commercial canning of foods.

Several factors dictate the time and temperature of the canning process. First, as discussed earlier, the higher the temperature, the shorter the time needed to kill all organisms. Second, the higher the concentration of bacteria, the longer the heat treatment required to kill all organisms. To provide a wide margin of safety, the commercial canning process is designed to reduce a population of 10^{12} *C. botulinum* endospores to only one. In other words, it is a 12 D process. Note that it is virtually impossible for a food to have this high a level of initial concentration of endospores, so the process has a wide safety margin.

An additional consideration in the treatment process is the type of food being canned. Generally, organisms are most resistant to heat at the pH that supports their optimum growth, usually around pH 7. Thus, extremes in pH usually promote more efficient killing. For example, acid foods (pH below 4.5) such as fruits, pickles, and most tomatoes do not need to be processed for very long times or at very high temperatures. Meats and vegetables that are nearly neutral in pH must be processed longer and at higher temperatures. Some new tomato varieties have a lower acid content than older varieties, so they must either be acidified before processing or processed using longer times and higher temperatures. Heat also kills organisms more readily in high concentrations of salt, such as in foods

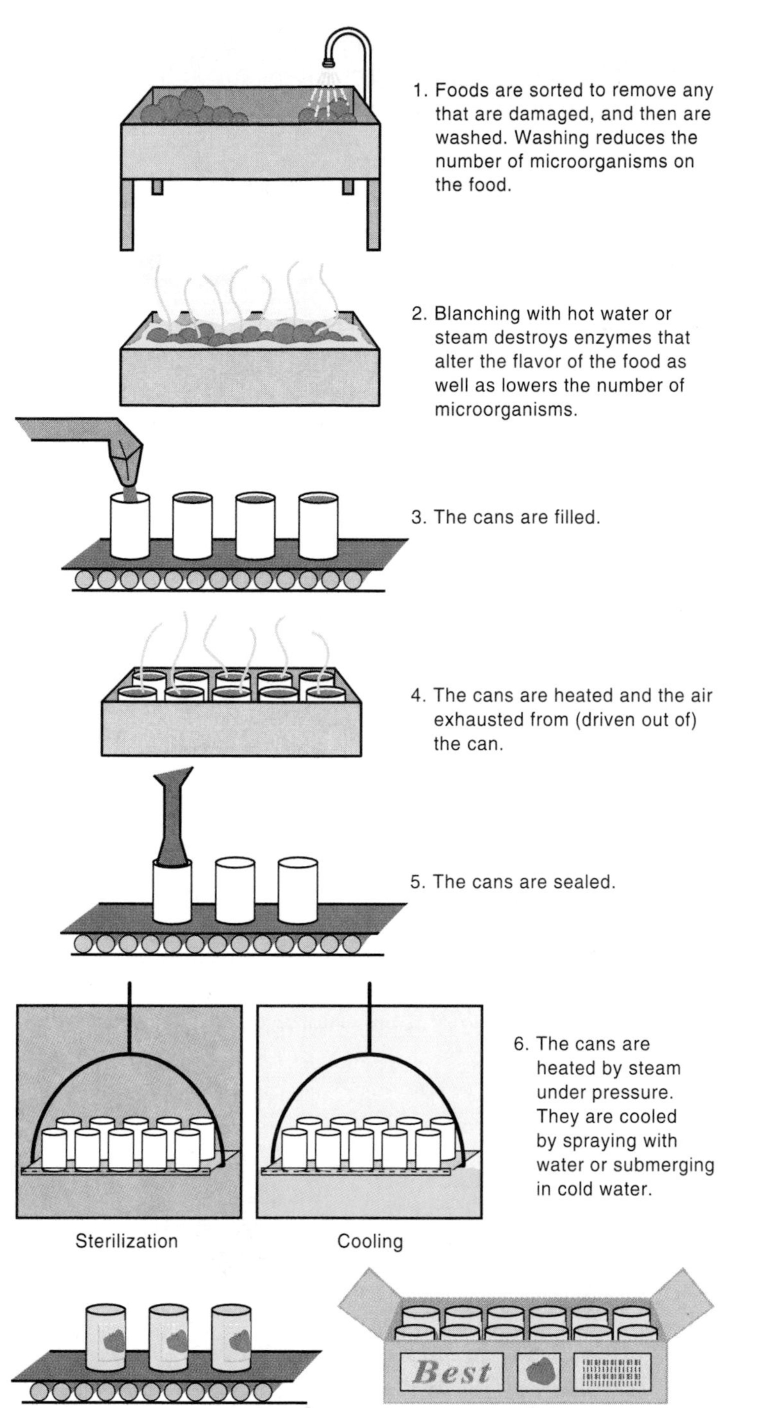

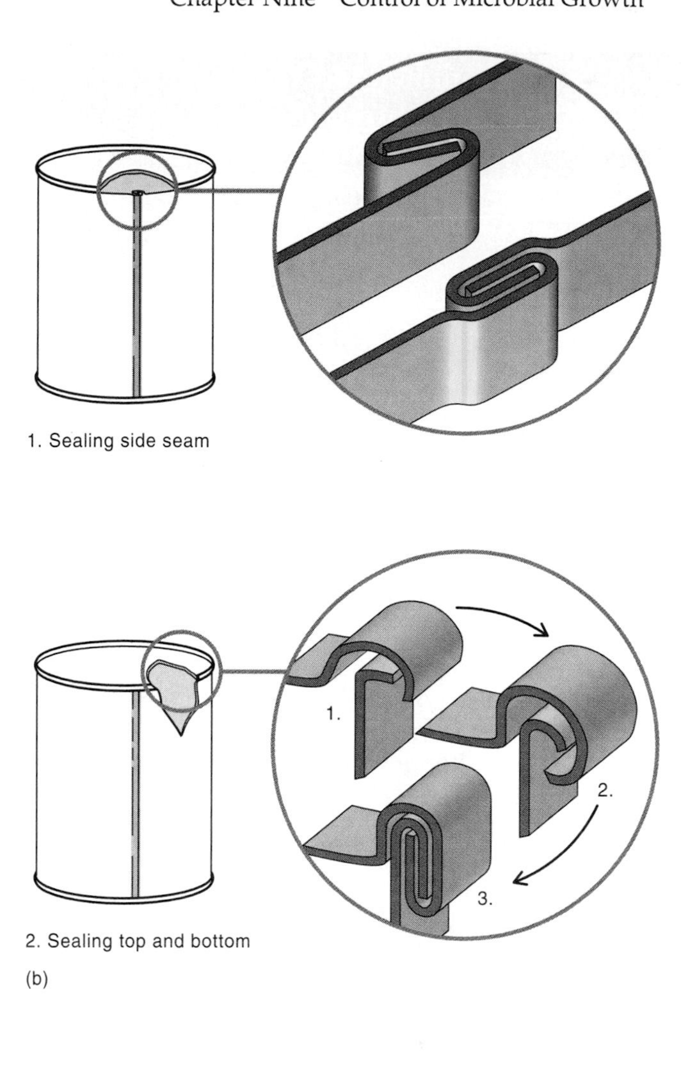

Figure 9.4
(*a*) Steps in the industrial canning of food. (*b*) Sealing metal cans.

in brine. Organisms are less heat sensitive in media containing proteins and fats, such as foods in cream sauces. Table 9.5 gives the time and temperature necessary for processing representative canned foods.

Pasteurization Eliminates Most Pathogens and Reduces the Number of Spoilage Organisms

In the 1800s, the French wine industry was threatened simultaneously by foreign competition and by a microbial invasion. Good wine results from the properly controlled conversion of fruit sugars to ethyl alcohol by yeast. Bacteria and molds from the environment may grow in the wine, degrading the alcohol and producing various bad-flavored metabolites. Louis Pasteur helped save the French wine industry when he found that moderate heating of wine, at just the right conditions of time and temperature, could kill these spoilage microbes without significantly changing its taste. Today, pasteurization is still used to destroy spoilage organisms in wine, vinegar, and a few other foods. The process does not sterilize substances but simply decreases the numbers of heat-sensitive organisms.

Table 9.5 Recommended Time and Temperature Necessary for Processing Several Foods*

Food	Temperature (Degrees F)	Time (Minutes)
Baked beans	240	105
Corn, cream style	240	90
Corn, in brine	240	50
Tomatoes	212	34

* Recommendations are based on bulletins issued by the National Canners Association. Source: The National Canners Association.

Pasteurization is widely used for ridding milk and other foods of disease-causing microorganism such as those that cause *Salmonella* food poisoning, tuberculosis, brucellosis, and typhoid fever. Long-term food preservation is not the primary goal in this pasteurization. Instead, milk and other foods are pasteurized to kill nonspore-forming pathogens without significantly altering the quality of the food.

The three methods of pasteurization in general use are (1) the **high-temperature–short-time method (HTST)**; (2) the **low-temperature–long-time method (LTLT)**, and (3) the **ultrahigh-temperature method (UHT)**. For example, the minimum requirement for milk using the HTST method is 72° C for 15 seconds; the LTLT method employs 62°C for 30 minutes; and the UHT uses 140°C for 15 seconds or 149°C for 0.5 seconds. Heat treatment for pasteurization must be adjusted to the individual food product. Ice cream, which is richer in fats than milk, requires a pasteurization process of 82°C for about 20 seconds or 71°C for 30 minutes. Aseptically boxed juices are processed using the newer UHT methods (140°C), but the procedure is designed to sterilize the product, so it is not considered pasteurization.

Although one does not ordinarily think of pasteurizing such things as cloth, this can easily be done by regulating the temperature of the water in a washing machine. The times and temperatures required may be different from those required for pasteurizing milk. For example, hospital anesthesia masks can be pasteurized at 80°C for 15 minutes. The temperatures and times used vary according to the organisms present and the heat stability of the material.

Sterilization by Filtration

Fluids that cannot tolerate heat can be sterilized by filtration. This procedure has many uses such as in space technology, the production of unpasteurized beer, and the sterilization of some heat-sensitive liquids. Filters capable of removing bacteria from fluids were developed during the last decade of the 19th century. They were made from materials such as porcelain, glass particles, diatomaceous earth[1], and asbestos. Filters of this type, called **depth filters**, are still used today. Their action is similar to that of a microscopic sieve, holding back microbes while letting the suspending fluid pour through the small holes in it. The passages that run through such filters are very tortuous; some have large open areas, and many others are blind (figure 9.5) The diameter of the passages is often considerably larger than that of the microbes they retain, and trapping of microbes is partly by electrical charges on the walls of the filter passages.

While the filters developed in the last century remove most species of bacteria from their suspending media, they have several drawbacks. For example, they cannot reliably remove viruses. They also cannot be used to separate bacteria from some desired products such as enzymes because they may retain proteins in addition to bacteria. The presence of proteins can also interfere with the filtering action by neutralizing electrostatic charges on the filter. Too much pressure applied to speed filtration can also overcome filtering action, allowing bacteria to pass through. Newer filters overcome some of these drawbacks.

Membrane Filters Have Multiple Uses

Membrane filters (figure 9.6) composed of compounds such as cellulose acetate, cellulose nitrate, polycarbonate, or polyvinylidene fluoride have been developed more recently and have largely replaced older types of filters. The filters are relatively inert chemically and absorb very little of the suspending fluid or its biologically important

[1] Diatomaceous—from diatoms, unicellular algae with silicon in their cell walls

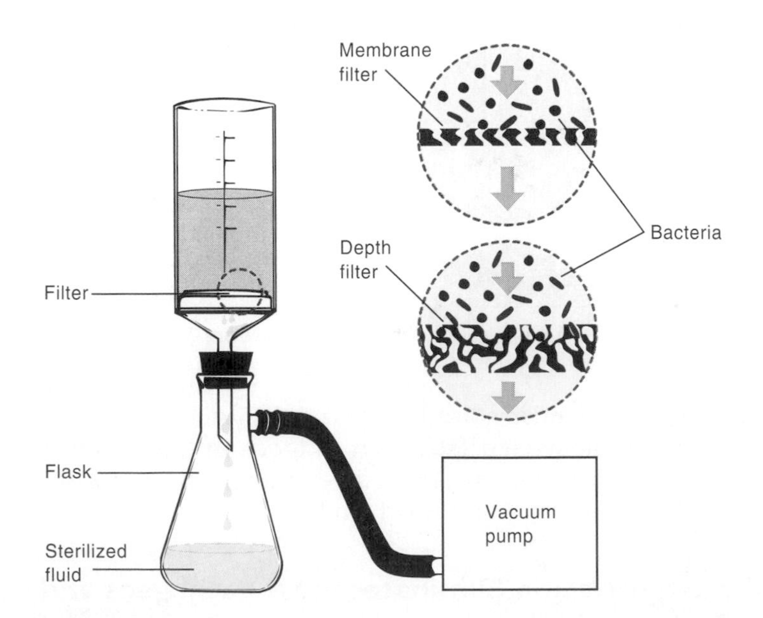

Figure 9.5
Filtration apparatus. The liquid to be sterilized flows through the filter on top of the flask in response to a vacuum produced in the flask by means of a pump.

(a)

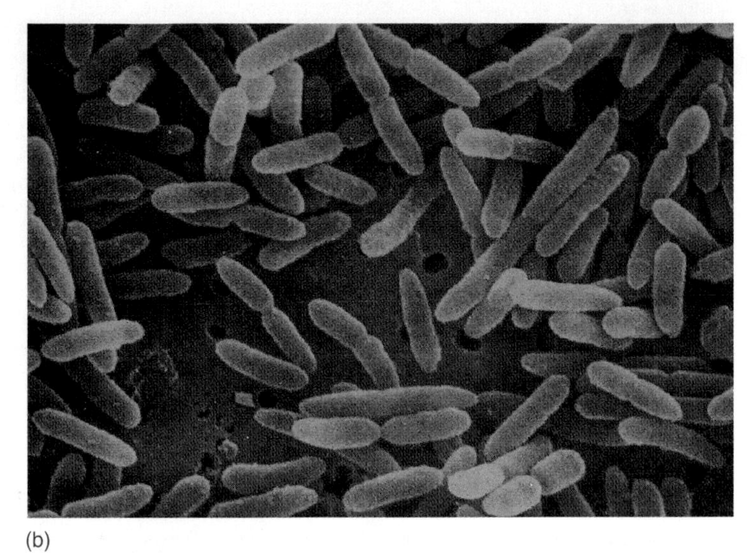

(b)

Figure 9.6
Membrane filters. (*a*) Scanning electron micrograph (X2,000) of a nylon filter retaining *Bacillus* sp. (*b*) Scanning electron micrograph (X5,000) of a polycarbonate nucleopore filter retaining *Pseudomonas* sp.

constituents such as enzymes. Paper-thin membrane filters are produced with graded pore sizes extending below the dimensions of the smallest known viruses. The most commonly used pore sizes for bacterial filters are 0.4 and 0.2 micrometers (μm). Membrane filtration of beer and wine results in a clear liquid free of spoilage organisms. Since pasteurization is unnecessary, filtered beer can be marketed as "draft" beer in bottles and cans. Some heat-sensitive medications that have no chemical preservatives are also sterilized by membrane filtration.

Only the smallest-pore membranes will sterilize fluids containing viruses. Bacteriophages and other viruses readily pass through most bacteriological filters and, if present in the original suspension, are present in the filtrate.

Membrane filters are also useful for detecting or enumerating organisms in a sample. For example, after filtering a suspension, the filter can be transferred to a sterile pad saturated with an appropriate medium. The bacteria retained on the filter surface will then grow into identifiable colonies. Counting the resulting colonies on the filter provides a good estimate of the number of bacteria in the volume of material filtered. Furthermore, specific pathogens deposited on membrane filters from measured volumes of air or liquids can be identified on the surface of the filters by using special techniques.

Chemicals Used for Sterilization, Disinfection, and Preservation

A number of chemicals are germicidal and can be used to sterilize and disinfect. Although generally less reliable than heat, these chemicals are suitable for treating large surfaces and many heat-sensitive items. In general, their effectiveness is impaired by organic material and enhanced by increased temperature. Some chemical germicides are sufficiently nontoxic to be used as antiseptics. Those that are only weakly germicidal but have a bacteriostatic action can be used as preservatives.

The Germicidal Activity of Chemicals Can Be Ranked

The activity of germicidal chemicals can be compared by testing them against different microorganisms and viruses (table 9.6). Germicides that kill the highly chemical-resistant bacterial endospores are called *high level.* If a germicide has little or no activity against endospores but is active against the chemically resistant nonspore-forming pathogen *Mycobacterium tuberculosis*, it is called *intermediate level.* Germicides are referred to as *low level* if they are unreliable against *M. tuberculosis* but are able to kill other vegetative bacteria. As a general rule, a disinfectant that kills an organism listed in table 9.1 will kill the microorganisms and viruses below it in the sequence of decreasing susceptibility. High-level germicides that kill large numbers of bacterial endospores within a reasonably short time (e.g., 6 to 12 hours) can be used for sterilizing. In more dilute solution, they are useful as disinfectants, acting relatively quickly (e.g., within 10 to 45 minutes).

To perform properly, germicides in each of these categories must be used strictly according to the manufacturers' directions, especially as they relate to dilution, temperature, and the amount of time they must be in contact with the object being sterilized or disinfected. Also, it is extremely important that the object be thoroughly cleaned and free of organic material before the germicidal procedure is begun. Such cleaning is necessary because organic material, such as dried blood or mucus, generally protects microbial pathogens from the action of chemical germicides.

Table 9.6 Usual Pattern of Germicidal Activity* of Disinfectants

Disinfectant Type	Bacteria			Fungi†	Viruses‡	
	Endospores	*M. tuberculosis*	Other vegetative bacteria		Small Nonlipid	Lipid and medium sized
High level	+	+	+	+	+	+
Intermediate level	–	+	+	+	–	+
Low level	–	–	+	+	–	+

* + = generally germicidal; – = generally unreliably germicidal
† Most fungal forms killed but some fungal spores resistant
‡ Methods of testing controversial

REVIEW QUESTIONS

1. Describe the difference between sterilization and pasteurization.
2. Describe three general methods used to preserve products.
3. Define the term *D value*.
4. Why does washing a product decrease the time necessary to sterilize the product?
5. Why does boiling not sterilize a product?

Germicidal Chemicals Used in Medical Fields Are Federally Regulated

Numerous chemical germicides are marketed for medical use under a variety of trade names. Frequently, they contain more than one germicidal chemical as well as other substances such as buffers that can influence their antimicrobial activity. The antimicrobial potency of these substances is difficult to assess and the assessment is subject to serious errors unless proper techniques are used. In the United States, the Environmental Protection Agency (EPA) has the main responsibility for ensuring that chemical germicides perform as advertised. To be registered with the EPA, their germicidal activity must be evaluated using its testing procedures (figure 9.7).

EPA-approved germicides capable of killing large numbers of bacterial endospores are labeled *sterilants* because they effectively kill all microbial types, including bacterial endospores, and therefore can be used for sterilization. Other germicides are unreliable sporicides but nevertheless are effective against *Mycobacterium tuberculosis*, the most resistant of the vegetative microorganisms. The EPA designates these formulations as "hospital disinfectants having tuberculocidal activity." A third category of germicides has even less antimicrobial activity, being unreliable against *M. tuberculosis*, but is active against other vegetative microorganisms. Chemicals in this category are labeled *hospital disinfectants* but do not claim to have tuberculocidal activity.

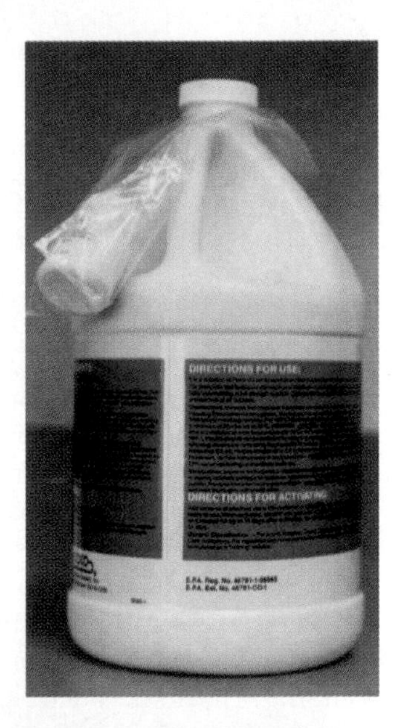

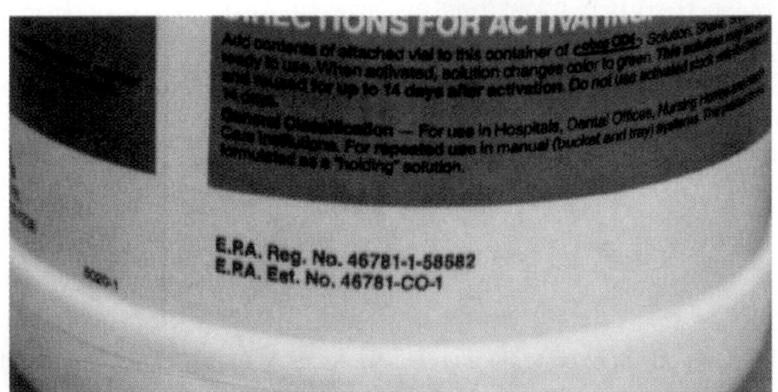

Figure 9.7
Aldehyde disinfectant that can be used for sterilizing. Before use, the disinfectant must be alkalinized by adding the contents of a vial in the plastic sack. Note the EPA registration number. Disinfectants registered with the Environmental Protection Agency must be tested according to specified procedures.

The type of germicidal procedure employed depends on the medical device, the material, and the intended use. Medical devices are divided into categories based on their risk of transmitting infectious agents. They are called "critical" if they enter body tissues, "semicritical" if they only come into contact with mucous membranes, and "noncritical" if they merely come into contact with human skin (table 9.7). In the United States, the manufacturer's recommendations for sterilizing or disinfecting critical and semicritical devices is supervised by the Food and Drug Administration.

Table 9.7 Risk Categories of Some Medical Devices

Category	Examples
Critical	Needles, scalpels; heart valves; artificial joints; biopsy forceps
Semicritical	Endoscopes; respiratory therapy equipment; vaginal specula
Noncritical	Stethoscopes; blood pressure cuffs; work surfaces

A Number of Chemical Families Provide Useful Germicides

Germicides are represented in a number of chemical families (table 9.8). Most chemical germicides react irreversibly with vital enzymes and other proteins, the cytoplasmic membrane, or viral envelopes. Some metallic compounds that act reversibly with microbial proteins are only bacteriostatic and are useful mainly as preservatives.

Alcohols

Ethyl and isopropyl alcohols at 50% to 80% concentration rapidly kill vegetative bacteria and fungi by coagulating essential proteins. For optimal results, the final concentration of alcohol in the material being disinfected should be about 75% by volume. Aqueous solutions are more effective than pure alcohol because proteins are not soluble in high concentrations of alcohol and, thus, they do not coagulate. Bacterial endospores do not take up water readily so alcohol is generally ineffective against them. Alcohols can be used to enhance the activity of other chemical agents, such as iodine, chlorhexidine, and quaternary ammonium compounds. The use of alcohols is limited to materials resistant to their solvent action. Another limitation of their use is that alcohols are volatile and may evaporate before adequate germicidal action has taken place.

Chlorine and Iodine

The use of chlorine as a disinfectant in municipal drinking water and in the water of swimming pools has undoubtedly saved hundreds of thousands of lives. The amount that must be added depends on the amount of organic matter in the water, since organic matter binds the chlorine, making it unavailable for action against microorganisms and viruses. Properly chlorinated drinking water requires about 0.5 parts per million (ppm) of free chlorine. Chlorine is widely used in disinfection in the form of free gas (Cl_2) and in solutions containing hypochlorite ions (OCl^-). Chlorine gas reacts with water to produce hypochlorite, which is thought to act on microorganisms by oxidizing essential proteins.

New, unopened containers of household bleaches (such as Clorox) consist of about a 5% sodium hypochlorite solution. Adding 3 ounces to a gallon of water results in a solution of approximately 1,000 ppm of available chlorine.[1] Although this concentration is several hundred times the amount required to kill most pathogenic microorganisms, it is usually necessary for fast killing. Also, organic material, which can neutralize much of the disinfectant activity, is often present. In locations where hepatitis B virus is a constant danger, such as in autopsy rooms on difficult-to-clean surfaces, even stronger sodium hypochlorite solutions have been recommended. The use of chlorine disinfectants is limited to materials resistant to their corrosive action. Certain kinds of rubber and metals, for example, are broken down by chlorine. Another drawback is that chlorine is irritating to the skin and mucous membranes. Chlorine may also react with some organic compounds to form trihalomethanes, which are potential carcinogens.

Iodine, like chlorine, is very active against most microorganisms, irreversibly oxidizing essential molecules in the cell so that they can no longer function. Like several other disinfectants, iodine has enhanced activity when dissolved in alcohol to produce a **tincture.** However, simple swabbing of the skin with tincture of iodine does not reliably kill bacterial endospores. Fortunately, high concentrations of the endospores of bacterial pathogens are not generally found on clean skin. Iodine stains skin and is painful when applied to damaged skin.

Compounds of iodine linked with carriers or surface-active agents such as detergents are called **iodophors.** The carriers slowly release the free iodine so iodophors do not irritate the skin or sting as much as tincture of iodine, and they are not as likely to stain. However, stock solutions of iodophores must be diluted properly because dilution affects the amount of free iodine. Some *Pseudomonas* species survive in the concentrated stock solutions for reasons that are unclear but possibly because there is inadequate free (unbound) iodine in these solutions, the iodine being released only with dilution. Nosocomial infections can result if a *Pseudomonas*-contaminated iodophore is erroneously used to disinfect instruments. Iodophores are not reliable for disinfecting instruments contaminated by *M. tuberculosis*. Iodophores and tincture of iodine are widely used as antiseptics but must not be used in individuals known to have an allergy to iodine (estimated at 1 in 10,000.)

Aldehydes

A solution of 8% formaldehyde in water is an extremely active disinfectant, killing most forms of microbial life very quickly. However, more dilute solutions may not kill bacterial

[1] Available chlorine—a measure of the antimicrobial activity of solutions containing chlorine compounds, determined by a chemical test; a 10% sodium hypochlorite solution contains about 9.52% available chlorine

• **Nosocomial infections, p. 442**
• **Aldehyde group, p. 29**

Table 9.8 Chemicals Used in Sterilization Disinfection, or Preservation

Chemical	Activity Level	Mechanism of Action	Advantages	Disadvantages	Common Uses
Alcohols	Intermediate	Denatures proteins	Leaves no residue; relatively nontoxic; easy to obtain; inexpensive	Volatile; evaporation limits contact time; does not kill endospores	Some laboratory instruments; skin disinfection
Halogens					
Chlorine	Intermediate	Oxidizes proteins and nucleic acids	Kills a wide variety of organisms; effective; convenient; inexpensive	Corrosive; inactivated by organics; reacts with organics to form carcinogenic compounds; high pH decreases effectiveness	Glassware; surfaces; drinking water; wastewater; food-processing equipment
Iodine tinctures	Intermediate	Oxidizes proteins	Not as quickly inactivated as chlorine by organics; effective at a wide pH range	Corrosive; inactivated by organics; stains; painful on damaged skin	Small-scale drinking water treatment; skin disinfection
Iodophors	Low to intermediate	Oxidizes proteins	Less corrosive and less staining than iodine tinctures	Proper dilution is essential; less effective than iodine tinctures; *Pseudomonas* can grow in some solutions	Skin disinfection
Phenolics	Low to intermediate	Destroys cell membranes; denatures proteins	Remains active in the presence of organics; kills a wide range of organisms; residual activity	Irritating to tissue; toxic residue accumulates; toxic vapors	Surfaces
Hexachlorophene				Neurotoxin	Hand scrubs
Chlorhexidine					Widely used as a skin antiseptic; hand scrubs; wound and burn treatment
Quaternary ammonium compounds	Low	Destroys cell membranes	Inexpensive; leaves germicidal residue	Readily inactivated by detergents, fibers, and other compounds; not very active against gram-negatives; *Pseudomonas* can grow even in concentrated solutions	Surfaces; food industry; general housekeeping
Metals (silver)	Low	Reacts with proteins		Toxic	Topical burn treatment
Aldehydes	Intermediate to high	Reacts with proteins, nucleic acids			
Formaldehyde			Wide range of activity; not corrosive	Irritating; toxic; potential carcinogen	Dialysis machines, vaccine production

(continued)

Table 9.8—*Continued*

Chemical	Activity Level	Mechanism of Action	Advantages	Disadvantages	Common Uses
Glutaraldehyde			More effective than formaldehyde against spores; can be used as a sterilant	Irritating	Very wide sterilant/ disinfectant
Ethylene oxide (gas)	High	Alkylates, proteins, and nucleic acids	Penetrates hard-to-reach places and fabrics	Explosive; toxic; carcinogenic	Heat sensitive devices such as disposable syringes and petri dishes; spices

Perspective 9.2

Formaldehyde Pollution of Household Air

Formaldehyde can be released from a number of substances in homes. Potential sources of it include particle board, plywood, paneling, upholstery, carpeting, foam insulation, gas and wood-burning stoves, and cigarette smoking. The release of formaldehyde gas often results when moisture from the air reacts with urea-formaldehyde glues and resins. Such a release generally continues for years, increasing or decreasing with temperature and humidity. Older mobile homes, constructed with large amounts of formaldehyde-releasing products, have been particularly problematic sites.

Ten to 20 percent of the population is highly sensitive to levels of formaldehyde that arise in some homes, especially under conditions of poor ventilation. Symptoms of formaldehyde sensitivity typically include chronic recurrent headaches, dizziness, nausea, and irritation of the eyes and respiratory system.

Newer building materials and strict building codes have helped to limit household formaldehyde buildup in some areas of the country. Local health departments can generally advise concerned people on how to control formaldehyde in homes.

endospores or inactivate some viruses. Failure to recognize this fact in the early days of vaccine development caused near disaster when dilute formaldehyde was used to inactivate poliovirus for use as a vaccine. Solutions of formaldehyde also have limited use because of their irritating vapors and suspected carcinogenicity.

In recent years, glutaraldehyde has largely replaced formaldehyde. It is now one of the most widely used high-level disinfectants. At higher concentrations, it can act as a sterilizing agent if a treatment time of 10 to 12 hours is employed. After treatment, glutaraldehyde is easily removed by rinsing in sterile water.

Phenolics

Phenolics (such as the ingredients in Lysol) include a large group of compounds such as cresols and xylenols that are chemically related to phenol (carbolic acid). They are effective germicides because they kill organisms by destroying the cytoplasmic membrane. Phenolics kill most vegetative bacteria, and in high concentrations (from 5% to 19%), many can kill *M. tuberculosis*. Five percent phenol is also effective against the major groups of viruses, but some of the more commonly used derivatives of phenol lack sufficient antiviral activity to be employed in viral disinfection. The major advantages of phenolics include their wide spectrum of activity, their reasonable cost, and the ability of many of the compounds to remain active in the presence of soaps, detergents, and organic compounds that would neutralize the effects of other disinfectants. Phenolic disinfectants are commonly used to wipe potentially contaminated surfaces such as those in hospital operating rooms. This practice effectively kills *Pseudomonas aeruginosa*, a gram-negative rod to which burn patients are particularly susceptible.

Hexachlorophene has been widely used as an antiseptic in hospitals. This chemical has substantial activity against *Staphylococcus aureus*, the leading cause of wound infections, and tends to be retained by the skin so that its antimicrobial activity increases with repeated use. It is not as effective against gram-negative organisms. The use of hexachlorophene during the 1950s and 1960s was of major importance in controlling infections caused by *S. aureus*. Since then, the use of hexachlorophene has been restricted in the United States because high quantities of the disinfectant were shown to cause brain damage in test animals. Hexachlorophene is now available with a prescription.

Perspective 9.3

Contamination of an Operating Room by a Bacterial Pathogen

A patient with burns infected with *Pseudomonas aeruginosa* was taken to the operating room for cleaning of the wounds and removal of dead tissue. After the procedure was completed, cultures of various surfaces in the room were performed to see the extent of contamination. *P. aeruginosa* was recovered from all parts of the room. Figure 1 shows how readily and extensively an operating room can become contaminated by an infected patient. Operating rooms and other patient care rooms are thoroughly cleaned and disinfected after use, in a process known as *terminal cleaning*.

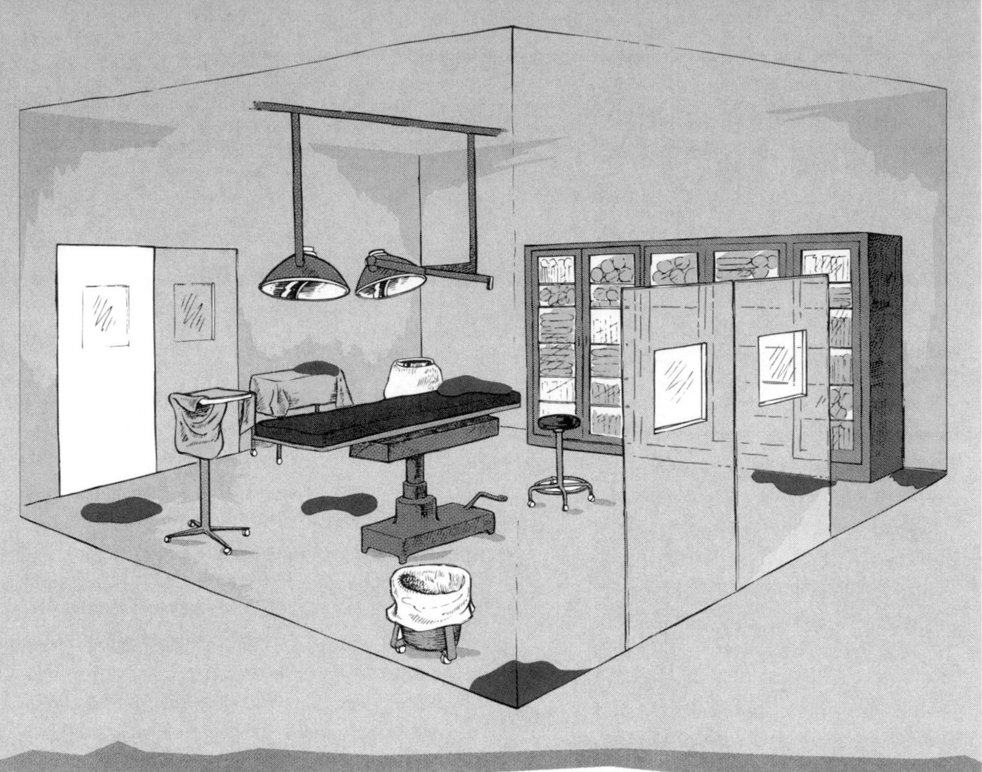

Figure 1
Diagram of an operating room in which dead tissue infected with *Pseudomonas aeruginosa* was removed from a patient with burns. Orange-colored areas indicate places where *P. aeruginosa* was recovered following the surgical procedure.

Another widely used phenol derivative, **chlorhexidine,** has largely replaced hexachlorophene as a routine hand and skin antiseptic because it is less toxic. In addition, it has the advantage of being active against gram-negative organisms while also possessing prolonged action against staphylococci. Chlorhexidine is widely used for surgeons' hand scrubs, for disinfecting patients' skin prior to an operation, and as a wound cleanser.

Quaternary Ammonium Compounds

Quaternary ammonium compounds, commonly called *quats*, represent a large group of compounds called *surface-active agents*, which reduce the surface tension[1] of liquids. Like the phenolics, they act against many vegetative bacteria by disrupting the cytoplasmic membrane. Quaternary ammonium compounds are widely used to disinfect clean inanimate objects and to preserve nonfood substances such as eyedrops. These uses have sometimes led to human infections because certain gram-negative bacteria, notably species of *Pseudomonas*, can grow in solutions of quaternary ammonium compounds. Unfortunately, most quats are easily inactivated by soaps, detergents, and organic materials such as gauze. Nevertheless, quaternary ammonium compounds are economical and effective sanitizing agents. They also enhance the effectiveness of other disinfectants.

[1]Surface tension—a property of liquids giving their surfaces the appearance of a thin elastic membrane under tension

Metal Compounds

Metals such as mercury compounds were widely used as disinfectants and antiseptics for many years because they were thought to have strong antimicrobial properties, including the ability to kill spores. Because of this mistaken idea, many people were infected by medical instruments that had been soaked in mercury compounds and were erroneously thought to be sterile.

Silver compounds have also fallen out of favor. For many years, doctors were required by law to instill drops of 1% silver nitrate into the eyes of newborn babies to kill any *Neisseria gonorrhoeae* that might have been acquired by the baby in its passage through the birth canal. This procedure is effective in preventing *N. gonorrhoeae* eye infection (gonococcal ophthalmia) but is not effective against another genital pathogen that can infect the eye, *Chlamydia trachomatis*. For this reason, and because silver nitrate is irritating to the eye, an antibiotic such as erythromycin is now used instead. Antibiotic resistance, increasingly common among *N. gonorrhoeae* strains, has not yet risen to the high antibiotic concentrations used in the eyes.

Perspective 9.4

Mercury Poisoning from Latex Paint

In August 1989, a four-year-old boy developed leg cramps, a generalized itchy rash followed by peeling of the skin, slight intermittent fever, rapid pulse and sweating, marked personality change, and marked weakness of his arms and legs. These symptoms suggested the diagnosis of acrodynia, a childhood form of mercury poisoning. This diagnosis was confirmed by testing a sample of his urine, which showed a high level of mercury. The boy slowly improved over four months of hospitalization and was able to walk at the time of discharge.

Investigation revealed that 10 days before the onset of symptoms, the child's home had been painted with interior latex paint. Samples of this paint were tested and showed more than twice the amount of mercury recommended by the EPA. The mercury was in the form of phenylmercuric acetate, added to the paint to inhibit the growth of molds.

After August 1990, it became illegal to add mercury compounds to interior latex paint. Prior to that time, it was legally permissible to add more mercury to latex paint than was recommended by the EPA, and there was no law that required the paint label to state the presence or concentration of mercury. There is no legal restriction on the sale of stocks of mercury-containing paints (which composed about one-third of all interior latex paints) manufactured before August 1990. Surfaces covered with these paints continue to release mercury in amounts that slowly decline over a number of years. Since the mercury from paints generally enters the body by inhalation, the best protection from toxicity is to ensure good ventilation.

Another silver compound, silver sulfadiazine, is widely used to prevent infection in burns. Silver ions act by binding to sulfhydryl groups of essential proteins, thereby preventing them from functioning. The combined silver and sulfadiazine lyses a wide variety of bacterial and fungal species.

Compounds of mercury, tin, arsenic, copper, and other metals were once widely used as preservatives and to prevent microbial growth in recirculating cooling water and industrial processes. Their extensive use has resulted in serious pollution of natural waters, which has promoted the issuing of strict controls on their use.

Ethylene Oxide

Ethylene oxide is an extremely useful gaseous sterilizing agent with many commercial and medical applications. It penetrates well into fabrics, equipment, and implantable devices such as pacemakers and artificial hips. In addition, it is one of the few chemical agents that can be relied on for sterilization. Disposable plastic dishes, syringes, rubber, and other heat-sensitive items used in laboratories or for medical purposes can be sterilized with ethylene oxide. Because it is explosive, it must be mixed with an inert gas such as carbon dioxide. The effectiveness of ethylene oxide depends on temperature and relative humidity, which must be carefully controlled. Therefore, it must be used in a special chamber in which these factors can be regulated. In practice, 3 to 12 hours are generally required to sterilize objects. The toxic ethylene oxide must then be eliminated from the treated material using heated forced air for 8 to 12 hours. Absorbed ethylene oxide must be allowed to dissipate because of its irritating effects on tissues and its undesired persistent antimicrobial effect. It should not be used for items employed for testing carcinogenicity, because traces of residual ethylene oxide can produce false positive results. Ethylene oxide is mutagenic and therefore potentially carcinogenic. Indeed, recent studies have shown a slightly increased risk of malignancies in long-term users of the gas.

Radiation

Electromagnetic radiation can be thought of as waves having energy but no mass. Examples include X rays, gamma rays, and ultraviolet and visible light rays. The energy that electromagnetic rays possess is proportional to the frequency of the radiation (the number of waves per second). Electromagnetic radiation of short wavelength, such as gamma rays, has much more killing power than that of long wavelength, such as visible light. Gamma rays and ultraviolet radiation are valuable tools for microbial control.

Gamma Rays Cause Biological Damage by Producing Hyperreactive Ions

Gamma rays are an example of ionizing radiation that causes biological damage by producing superoxide (O_2^-) and hydroxyl free radicals (OH·) when the rays transfer their energy to a microorganism. Gamma radiation from the radioisotope cobalt-60 is effective for controlling most microorganisms. Bacterial endospores are among the most radiation-resistant microbial forms, whereas the gram-negative rod-shaped bacteria, such as species of *Salmonella* and *Pseudomonas*, are among the most sensitive. Surprisingly, some bacteria such as *Deinococcus radiodurans* are more radiation resistant than bacterial endospores. They can often be recovered from foods that have been treated with radiation to prevent spoilage.

A number of biological materials (such as penicillin) and numerous disposable plastic items (such as hypodermic syringes) can be sterilized effectively with gamma irradiation. Sterilizing plastic with gamma rays is an alternative method to sterilizing with ethylene oxide, and it has the advantage that the sterilized objects can be used

• **Radiation, p. 160**

Perspective 9.5

Disinfection of Spilled Blood

For many years, people have been concerned about the effectiveness of disinfectants against blood-borne viruses, particularly hepatitis B virus because it can be present in enormous concentrations in the blood of unsuspecting carriers. It is now known that hepatitis B and C viruses and the human immunodeficiency virus (which causes AIDS) are all enveloped viruses that are susceptible to most high- and intermediate-level disinfectants. One study showed hepatitis B virus in highly infectious serum diluted about 1,000-fold to be inactivated in minutes by 80% ethanol or 0.1% glutaraldehyde.

It is unlikely that the same favorable results would occur from disinfecting dried blood on a tabletop or medical instrument. A reasonable approach for handling blood contamination is currently recommended by the Centers for Disease Control and Prevention. The blood is cleaned up using disposable gloves and towels and an EPA-registered tuberculocidal disinfectant. After all visible blood has been removed, disinfection is carried out under the conditions specified for the germicide. A cheaper and equally effective germicide is ordinary household 5.25% hypochlorite bleach, diluted to 3 ounces of bleach per 1 gallon of water. This procedure should substantially reduce or eliminate the risk of acquiring infection from blood-soiled objects.

immediately without the extended wait required after ethylene oxide sterilization. In addition, gamma irradiation has an advantage over other treatments in that it can be carried out after packaging. Large commercial firms sterilize many types of disposable medical equipment and supplies by irradiation.

Irradiating foods with gamma rays gives essentially the same results obtained by pasteurization. For example, it can be used to kill pathogens such as *Salmonella* in poultry with little or no change in the taste of the product. Foods can also be sterilized with irradiation, but undesirable flavor changes limit the usefulness of the procedure. In the United States, spices and herbs have been treated with irradiation for many years. The Food and Drug Administration (FDA) approved irradiated wheat and wheat flour in 1963, potatoes in 1964, fruits and vegetables in mid-1980, and poultry in 1990. Irradiation is considered a "food additive" and a statement such as "treated with gamma irradiation" must appear on the label.

The FDA and officials of the United Nations Food and Agriculture Organization and World Health Organization have endorsed food irradiation. Nevertheless, some people continue to question the safety of irradiated foods. They have lingering doubts about the possibility of irradiation-induced carcinogens being present in food even though available scientific evidence indicates that irradiated food is safe to eat. Their fears have led some foreign countries and states of the United States to ban the sale or manufacture of irradiated foods. Despite the controversy, many countries in the world are continuing to use irradiated foods.

Ultraviolet Radiation Damages Nucleic Acids

Certain wavelengths of radiation are much more effective antimicrobial agents than those with immediately shorter or longer wavelengths. An important example of this is the band of wavelengths from 200 to 310 nanometers in the ultraviolet (UV) zone. The absorption of these wavelengths damages the structure and function of nucleic acids by causing the formation of thymine dimers as discussed in chapter 7. Actively multiplying organisms are the most easily killed, while bacterial endospores are the most resistant. However, some organisms have mechanisms that can repair the damaged DNA. Damaged viruses that infect cells containing DNA-repair enzymes can also recover.

Ultraviolet rays penetrate very poorly. A thin film of grease on the UV bulb or extraneous materials covering the microorganism may markedly reduce effective microbial killing. Most types of glass and plastic also screen out ultraviolet radiation, so UV light is most effective at close range against exposed microorganisms in air or on clean surfaces. However, caution should be used because ultraviolet rays can also damage the skin and eyes and promote the development of skin cancers.

Microwaves Do Not Kill Organisms Directly

Microwaves do not affect microorganisms directly, but they can kill microorganisms with the heat they generate in a product. Indeed, organisms often survive microwave cooking because the food heats unevenly. Frequent stirring or rotating food in a microwave oven as well as observing the standing times in which the heat is allowed to penetrate all parts of the food are necessary. Some people have contracted trichinosis due to incomplete microwave cooking of pork sausage.

Preservation

Preventing or slowing the growth of microorganisms extends the shelf life of a product. The ingredients of food, soaps, medicines, deodorants, contact lens solutions, and many other common items often include preservative chemicals

• **Thymine dimers, p. 160**
• **Trichinosis, p. 798**

Perspective 9.6

Comparison of Sterilizing and Disinfecting Techniques

Different techniques for killing microorganisms can be compared in the laboratory by counting the number of surviving organisms after varying time intervals or by counting the survivors at a fixed time. The count must be made after neutralizing or removing the killing agent. Different dilutions of the suspension of microorganisms are inoculated on appropriate media, and the number of survivors that grow into visible colonies is determined.

When techniques of killing by heat are compared, the data can then be plotted (figure 1), and D or Z values can be determined. The **D value, or decimal reduction time,** is the time required for 90% of the organisms to be killed at a given temperature, whereas the **Z value** is the number of degrees of temperature change required to change the D value by a factor of 10. These values are more reliable than the time required for complete killing of a microbial population for evaluating different methods of microbial killing.

Disinfectants may be evaluated by comparing the ratio of their antimicrobial activity with that of pure phenol. The larger the resulting value—known as the *phenol coefficient**—the more active the agent under those conditions. The tests are carried out under rigorously controlled conditions of temperature, inoculum, growth medium, time, and so on and utilize specific strains of *Staphylococcus aureus*, *Salmonella typhi*, and *Pseudomonas aeruginosa*. The phenol coefficient has applicability in government control and labeling of phenolic disinfectants. However, it may be highly misleading in evaluating these substances for actual use because differences in killing rate depend on the species of microorganism, stage of growth, nature of the killing agent, suspending medium, pH, salt concentration, and many other factors. For these reasons, agents to be compared for use in a given situation (such as disinfection of a hospital floor or preparing the skin for an injection) should be evaluated under similar conditions to those under which they will be used. Widely conflicting information on the value of disinfecting agents has sometimes arisen because various and often inappropriate conditions of testing were used.

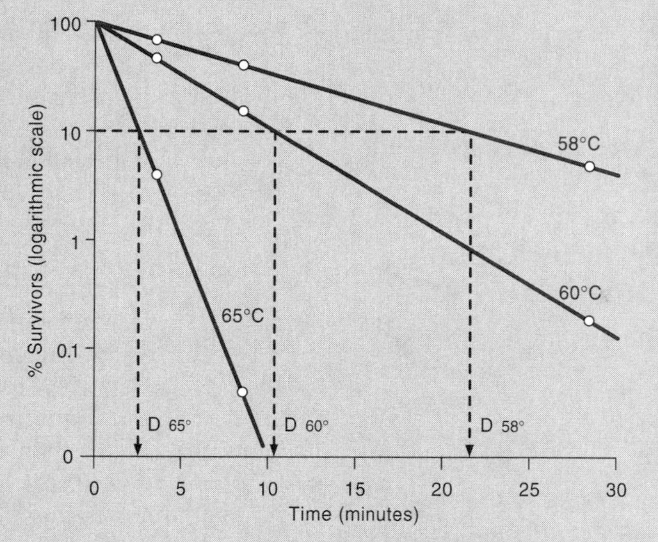

Figure 1
Effect of different temperatures on bacterial survival. D 58°, D 60°, and D 65° are decimal reduction times at 58°C, 60°C, and 65°C, respectively, for this bacterium. The Z value (change in temperature required to cause a 10-fold change in D value) is approximately 7°C.

*Pure phenol produces microbial killing at 1:90 dilution while the test disinfectant does the same at a dilution of 1:400. The phenol coefficient is 400 divided by 90, or 4.4.

(figure 9.8). The purpose of these agents is to prevent or slow growth of microbes that are inevitably introduced from the environment. Another common method of decreasing the rate of growth of microbes is using low-temperature treatments such as refrigeration or freezing. These methods are of particular importance in the preservation of foods.

Figure 9.8
Each of the items shown in this figure contains a preservative identified in its label.

Chemicals Are Used as Preservatives

Some of the germicidal chemicals previously described can be used to preserve nonfood items. For example, heparin (a medicine used to prevent blood clots) may contain a phenol derivative, contact lens solutions may contain a quaternary ammonium compound, and leather belts may be treated with one or more phenol derivatives. Food preservatives, however, must be nontoxic for repeated safe ingestion.

Benzoic, sorbic, and propionic acids are organic acids that are sometimes added to foods such as bread, cheese, and juices to prevent microbial growth. In the undissociated form that predominates at a low pH, these weak acids alter cell membrane functions and interfere with the generation of proton motive force. This reduces the energy-generating ability of the cell. The low pH at which they are most effective is itself sufficient to prevent the growth of most bacteria, so these preservatives are primarily added to acidic foods to prevent the growth of fungi. These organic acids also occur naturally in some foods such as cranberries and Swiss cheese.

Another preservative, nitrate, and its reduced form, nitrite, serve a dual purpose in processed meats. From a microbiological viewpoint, their most important function is inhibition of germination of endospores and subsequent growth of *Clostridium botulinum*. Without the addition of low levels of nitrate or nitrite to cured meats such as bologna, ham, bacon, and smoked fish, there is a risk that *C. botulinum* will grow and produce deadly botulinum toxin in these products. At higher concentrations than are required for preservation, nitrate and nitrite react with myoglobin in the meat to form a stable pigment that gives a desirable pink color associated with fresh meat. However, nitrates and nitrites also pose a potential hazard because they can be converted to nitrosamines by intestinal bacteria and by frying meats in hot oil. Nitrosamines have been shown to be potent carcinogens. Attempts are currently being made to reduce the levels of nitrates and nitrites in meats and to inhibit their conversion to nitrosamines.

Growth of Spoilage Microorganisms Is Slowed or Halted by Low-Temperature Treatment

The growth of microorganisms is temperature dependent. At low temperatures above freezing, many enzymatic reactions are very slow or nonexistent. Thus, low temperatures are extremely useful in preservation. However, household refrigerators slow microbial growth but do not always prevent it entirely. Unfortunately, psychrophilic microorganisms are able to grow at normal refrigerator temperatures 40°F or about 4°C; although the temperature may vary from 0°C to 10°C in different parts of the refrigerator.

Commercially, some fruits and vegetables such as apples and potatoes are held in cold storage for many months. In this method, products are stored at temperatures near 10°C, in the dark, and at appropriate humidity and oxygen concentrations. Prior to refrigeration, foods are sometimes irradiated in UV light to reduce the number of spoilage organisms.

Freezing is also an important means of preserving foods and other products. Water at freezer temperatures (–20°C) is not available for biological reactions because it is in the form of ice. Thus, freezing essentially stops all microbial growth. The formation of ice crystals can cause irreversible damage to microbial cell structures, killing up to 50% of the microorganisms. Foods can be stored frozen for many months with little loss of color, flavor, and other criteria of quality. However, caution should be exercised because remaining organisms can grow and spoil the foods after they are thawed.

• **Temperature of growth, pp. 90–92**
• **Psychrophiles, p. 91**

Desiccation (Drying) or Adding Sugar and Salt Prevent Microbial Growth by Reducing the Water Activity

For many centuries, drying has been used to preserve food. This process decreases the water activity of a food to below the limits required for microbial growth. Drying by natural means (sun drying) or by artificial means is often supplemented by other methods, such as salting or adding high concentrations of sugar or small amounts of chemical preservatives. Jams and jellies are made by adding sugar to fruit. Fish and meats are prepared by adding large amounts of salt or by soaking them in saltwater to preserve them. The salt or sugar draws out the water from the cells and essentially dries them, thereby preventing them from serving as food for spoilage microorganisms. Smoking of meats and fish not only reduces their water content but also coats the outer surfaces of the food with various compounds such as aldehydes or phenolic compounds that kill or inhibit the growth of microorganisms. Some caution should be exercised when using salt as a preservative because the food-poisoning bacterium *Staphylococcus aureus* is able to grow under quite high salt conditions.

Lyophilization (freeze-drying) is widely used for preserving some foods such as coffee, milk, meats, and vegetables. Freeze-dried foods' light weight and stability without refrigeration make them popular with hikers. In the process of freeze-drying, the frozen food is dried in a vacuum. When water is added to the lyophilized material, it reconstitutes, and the quality of the reconstituted product is often much better than that of products treated with ordinary freezing or drying methods.

Although drying stops microbial growth, it does not reliably kill bacteria and fungi in or on foods. For example, numerous cases of salmonellosis have been traced to dried eggs. Eggshells and even egg yolks may be heavily contaminated with *Salmonella* from the gastrointestinal tract of the hen. To prevent the transmission of such pathogens, some states now have laws requiring dried eggs to be pasteurized before they are sold.

• **Water activity, p. 761**

REVIEW QUESTIONS

1. How is a sterilant different from a hospital disinfectant?
2. Differentiate between a critical, semicritical, and noncritical medical instrument.
3. How does chlorine kill microorganisms?
4. How do microwaves kill microorganisms?
5. What is the risk of consuming nitrate or nitrite-free cured foods?

Technique 9.1

Using an Autoclave

The autoclave is the simplest and most consistently effective means of sterilizing most objects. The following are some precautionary steps in the use of an autoclave.

1. **Prevent air trapping.** It is critical that all air be removed from the autoclave in order to ensure adequate sterilization. Because steam displaces air downward in the chamber, long, thin containers should not be put into the autoclave in an upright position. Sterilization of materials in plastic bags will likewise fail, unless the bags are wide open and only partly filled so as to allow easy access of steam and escape of air.
2. **Ensure complete heat penetration.** A cold object placed in an autoclave takes time—over and above the time it takes the chamber temperature to reach 121°C—to warm up. For example, a 1-liter container of fluid must be autoclaved for 25 minutes after the chamber reaches 121°C, whereas a 4-liter container of fluid may require an hour.
3. **Ensure contact with steam.** Dry microorganisms protected from contact with steam (for example, within oil, plastic wrappings, or containers of talcum powder) cannot be killed reliably by autoclaving.
4. **Test with heat indicators and heat-resistant spores.** Test tubes and tapes containing a heat-sensitive chemical indicator are often included with objects when they are autoclaved. The indicator changes color during the autoclaving, giving a visual means of checking whether the objects have been subjected to heat. However, a changed indicator does not always indicate that the object is sterile because heating may not have been uniform.

   Tubes or envelopes containing large numbers of heat-resistant spores of the nonpathogenic bacterium *Bacillus stearothermophilus*, which acts as a biological indicator, are also frequently included with packs of materials being autoclaved. Death of these spores indicates adequate killing at the point where the tube or envelope was placed, which should be near the center of the pack.
5. **Cool before releasing pressure.** The elevated pressure in the autoclave prevents fluids from boiling, even though the temperature is well above their normal boiling point. If valves are opened to release this pressure at the end of the period of autoclaving, these fluids will immediately boil and may even explode their containers. The pressure must be maintained to prevent them from boiling until temperatures have dropped below the boiling point at atmospheric pressure. Tightly capped glass containers may collapse; therefore, cotton plugs or loose closures that can be tightened after cooling are used.
6. **Use different procedures for heat-sensitive materials.** Some media ingredients such as glucose are sterilized separately to prevent caramelization and interaction with other ingredients at high temperatures. Most autoclaves can operate at temperatures lower than 121°C. The use of lower temperatures is satisfactory for materials that can be tested for successful sterilization before being used. For example, certain heat-sensitive bacteriological media are sterilized at 115°C. This temperature is usually effective because heat-resistant microbial contaminants are rarely present in the media and because these media are generally tested for sterility before being used.
7. **Prevent recontamination.** Objects to be autoclaved are usually wrapped in paper or cloth to allow steam to penetrate during sterilization and to prevent recontamination thereafter. When wet, these coverings are readily permeable to bacteria, so they should be allowed to dry before they are removed from the autoclave chamber. After removal, they should be stored in a closed cupboard or drawer to help prevent the reintroduction of microorganisms.

Summary

I. Approaches to Control

II. Principles Involved in Killing or Inhibiting Microorganisms

III. Using Heat to Kill Microorganisms and Viruses
   A. Dry heat requires more time than wet heat to kill microorganisms.
   B. Boiling kills most vegetative cells but not bacterial endospores.
   C. Pressure cookers and autoclaves are effective sterilizers.

IV. The Commercial Food-Canning Process
   A. Pasteurization eliminates most pathogens and reduces the number of spoilage organisms.

V. Sterilization by Filtration
   A. Membrane filters have multiple uses.

VI. Chemicals Used for Sterilization, Disinfection, and Preservation
   A. The germicidal activity of chemicals can be ranked.
   B. Germicidal chemicals used in medical fields are federally regulated.
   C. A number of chemical families provide useful germicides. These include
      1. alcohols,
      2. chlorine and iodine,
      3. phenolics,
      4. quaternary ammonium compounds,
      5. metals,
      6. aldehydes
      7. ethylene oxide.

VII. Radiation
   A. Gamma rays cause biological damage by producing hyperreactive ions.
   B. Ultraviolet radiation damages nucleic acids.
   C. Microwaves do not kill organisms directly.

VIII. Preservation
   A. Chemicals are used as preservatives.
   B. Growth of spoilage microorganisms is slowed or halted by low-temperature treatment.
   C. Desiccation (drying) or adding sugar and salt prevent microbial growth by reducing the water activity.

Chapter Review Questions

1. What is the primary purpose of pasteurizing milk?
2. What is the primary purpose of pasteurizing wine?
3. What is the most chemically resistant nonspore-forming pathogen?
4. Why are low acid foods processed at higher temperatures than high acid foods?
5. Explain why it takes longer to kill a population of 10^9 cells than it does to kill a population of 10^3 cells.
6. How is an iodophore different from a tincture of iodine?
7. How does ultraviolet light kill bacteria?
8. How is preservation different from pasteurization?
9. How are heat-sensitive liquids sterilized?
10. Name two products that are commonly sterilized using ethylene oxide gas.

Critical Thinking Questions

1. Why might a solution start to boil as the pressure in an autoclave is lowered?
2. What are the limitations of chemical and biological indicators?
3. Why is it that laboratory comparisons of disinfectants may not reflect their relative value in everyday use as germicides?

Further Reading

Block, S. S. (ed). 1991. *Disinfection, sterilization and preservation*, 4th ed. Lea and Febiger, Philadelphia. A large and comprehensive reference.

Brown, Paul, et al. 1982. Chemical disinfection of Creutzfeldt-Jakob disease virus. *New England Journal of Medicine* 306:1279–1282.

Collins, C. H., et al. (ed). 1981. *Disinfectants: their use and evaluation of effectiveness.* The Society for Applied Bacteriology Technical Series no. 16. Academic Press, New York. Historical information and numerous home, medical, veterinary, and agricultural applications.

Davis, B. D., and R. Dulbecco. 1990. Sterilization and disinfection. In B. D. Davis et al. (ed.), *Microbiology*, 4th ed. J. B. Lippincott Company, Philadelphia. Concise, detailed, up-to-date discussion including radiation effects.

Favero, M. S., and Bond, W. W. 1991. Sterilization, disinfection, and antisepsis in the hospital. In A. Ballows et al. (ed.) *Manual of clinical microbiology*, 5th ed. American Society for Microbiology, Washington, D.C. Clear, up-to-date, concise, practically oriented discussion.

Kobayashi, H., Tsuzuki, M., Koshimizu, K., et al. 1984. Susceptibility of hepatitis B virus to disinfectants and heat. *J. Clinical Microbiology* 20:214–216.

Perkins, J. J. 1982. *Principles and methods of sterilization in health sciences*, 2nd ed. Charles C. Thomas, Springfield, Ill.

Sattar, S. A., and Springthorpe, V. S. 1991. Survival and disinfectant inactivation of the human immunodeficiency virus: a critical review. *Reviews of Infectious Diseases* 13:430–437.

Steenland, Kyle, et al. 1991. Mortality among workers exposed to ethylene oxide. *New England Journal of Medicine* 324:1402.

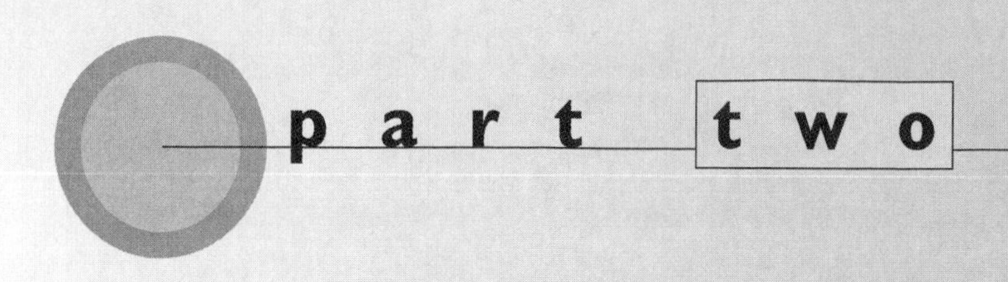

The Microbial World

Microscopic view of a drop of pond water showing myriad prokaryotic and eukaryotic species.

10 Classification and Identification of Prokaryotes

KEY CONCEPTS

1. The best classification schemes group organisms that are related through evolution and separate those that are unrelated.
2. At the present time, the classification scheme for bacteria is based not on evolution but on convenience and consists of readily determined characteristics such as shape and type of metabolism.
3. The most convenient method of determining the evolutionary relatedness of organisms is based on comparing the sequence of nucleotides in their DNA.
4. Unexpected relationships between various bacteria have been revealed by comparing their 16S ribosomal RNA sequences.
5. In clinical laboratories, identifying the genus and species of an organism is more important than understanding its evolutionary relationship to other organisms.

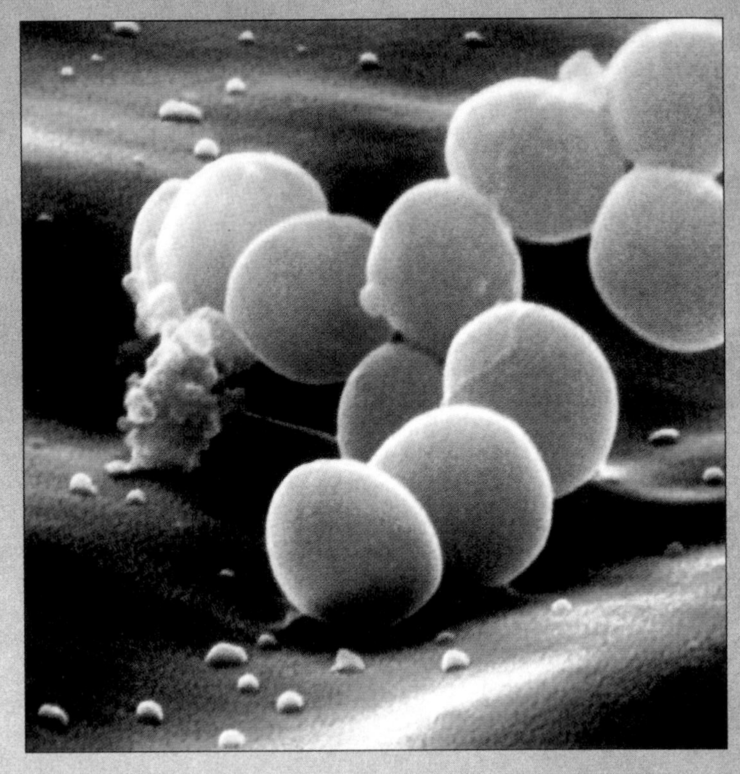

Staphylococcus aureus as viewed by the scanning electron microscope (×50,000).

PREVIEW LINK

Microbes in Motion

The following books and chapters in the *Microbes in Motion* CD-ROM may serve as a useful preview or supplement to your reading:

Miscellaneous Bacteria. *Gram Negative Organisms. Gram Positive Organisms. Bacterial Structure and Function:* External Structures. *Fungal Structure and Function:* Specific Fungal Structures; Metabolism and Growth. *Parasitic Structure and Function.*

A Glimpse of History

In the early 1870s, the German botanist Ferdinand Cohn published several papers on bacterial classification. Cohn grouped bacteria according to shape: spherical, short rods, elongated rods, and spirals. This classification scheme implied that bacteria are not plastic in shape but maintain a constant shape. Thus, spherical organisms give rise to spherical organisms; they do not become rods following binary fission. Cohn recognized that classification based solely on shapes was not adequate for classifying all of the different bacteria that he knew existed. There were too many bacteria and too few shapes.

The second major attempt at bacterial classification was initiated by Sigurd Orla-Jensen. Orla-Jensen's early training in Copenhagen was in the field of chemical engineering, but he soon became interested in microbiology and fermentations. In 1908, he proposed that bacteria be classified according to their physiological properties rather than their morphological properties. He considered organisms that gained their energy from inorganic sources as the most primitive.

A quarter of a century later, two Dutch microbiologists, Albert Kluyver and Jan van Neil, proposed two classification systems that were based on presumed evolutionary relationships. However, they recognized a very serious problem in their classification scheme: There was no way to distinguish between "resemblance" and "relatedness." The fact that two bacteria look alike does not mean they are related.

In 1970, Roger Stanier, a microbiologist at the University of California, Berkeley, pointed out that relationships could be determined at two different levels—the genetic level, by comparing the similarities of the DNA of organisms, and another level, by comparing the similarities in their gene products (such as proteins, cell walls, and ribosomal subunits). With the great advances made in analyzing nucleic acids, it became possible to directly compare the DNA sequences in various organisms at the genetic level and thereby gain insight into the relatedness between closely related organisms. The greater the similarity in nucleotide sequence of two organisms, the more closely related they are.

Stanier and all other microbiologists in the 1970s assumed that all bacteria are basically similar since they are all unicellular and prokaryotic in cell structure. However, when the chemical compositions of a wide variety of bacteria were examined in detail, it was found that many had features that differed from those present in the intensively studied *Escherichia coli*, considered a "typical" bacterium. These "unusual" features pertained to the chemical nature of the cell wall, cytoplasmic membrane, and ribosomal RNA.

In the late 1970s, Dr. Carl Woese and his colleagues at the University of Illinois took advantage of nucleic acid technologies, which allowed investigators to determine the sequence of bases in ribosomal RNA. When they compared such sequences from a wide variety of organisms, it became clear that prokaryotes could be divided into two major groups that differ from one another as much as they differ from the eukaryotic cell. Thus, the bacteria, which were once believed to be a part of the plant kingdom, have now been separated into two domains, each on the same level as the animals, plants and other eukaryotes.

This chapter discusses the problems in classifying prokaryotes, and how modern techniques of molecular biology and genetics are leading the way toward developing a classification scheme based on evolutionary relatedness. The use of these techniques has led to the surprising conclusion that not all prokaryotes are related to one another.

Problems in Classifying Microorganisms

Classification is the arranging of organisms into related groups, or **taxa; taxonomy** is the science of classification. Organisms are classified primarily to provide easy identification of the organisms. Thus, all classification schemes try to group organisms with similar properties and separate those that are different. The best classification schemes group organisms that are related through evolution and separate those that are unrelated. Unfortunately, this is more difficult in the case of bacteria than in plants and animals. Fossils of bacteria have been found that suggest that bacteria have existed on the earth for 3.5 billion years (figure 10.1). More recent data based on isotopic analysis of rocks in Greenland suggest that prokaryotes existed on this planet 3.85 billion years ago. However, the fossil record is incomplete and does not help in placing bacteria into their proper place in the evolution of living beings.

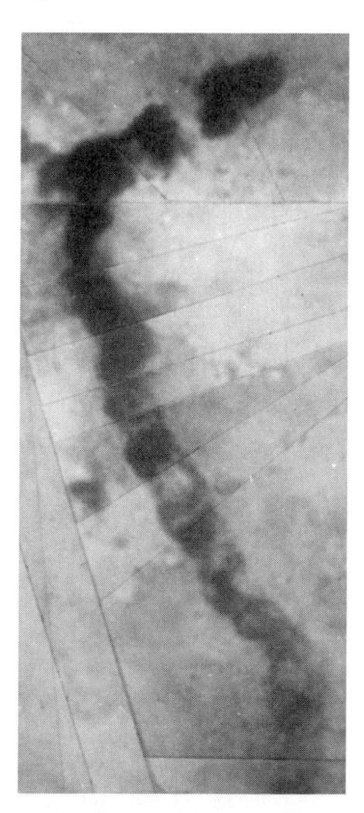
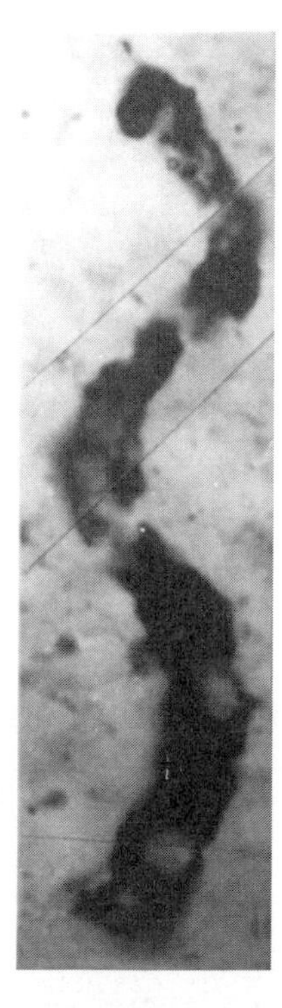

Figure 10.1
Carbonaceous microfossils shown in sections of rock from the Early Archaean era (~3,465 million years old). Apex chert of northwestern Western Australia.

Several basic distinctions exist between bacterial taxonomy and taxonomy of plants and animals. In both schemes, the **species** is the basic taxonomic unit. However, a species is defined differently for higher organisms than it is for bacteria. For higher organisms, a species is defined as a group that can transfer DNA between its members. In addition, members of one species have morphological characteristics that distinguish them from individuals of other species. These criteria cannot normally be applied to bacteria, because they have only a few different shapes, and genetic recombination (gene transfer) is relatively rare. Two species of bacteria differ from one another in several features that are determined by one or a few genes. The major problem in bacterial taxonomy is deciding how different two organisms must be in order to be classified as separate species. If two organisms differ only in minor phenotypic properties, then one organism may be considered a **subspecies** or **strain,** rather than another species. For example, many strains of *Escherichia coli* exist with few, but sometimes important, differences. Some strains have important features that result in serious consequences in particular settings. For example, *E. coli* strain 0157:H7 causes a very serious disease in children which common strains of *E. coli* do not do.

Present-Day Classification Schemes

As in the classification of higher organisms, several similar species of bacteria are grouped into a **genus,** and many similar genera are included within a **family.** At present, many families of bacteria are recognized. Families, in turn, are grouped into **orders;** many orders that have similar characteristics constitute a **class,** and several classes constitute a **division.** Similar divisions, in turn, are grouped into **kingdoms.** All prokaryotes, which include all members of the Eubacteria and Archaea, are officially classified in a single kingdom, *Prokaryotae,* because of their common prokaryotic cell structure. An example of how a particular species is classified in the various groups is illustrated in table 10.1.

The current taxonomic scheme for bacteria is practical and convenient, consisting of descriptions of organisms, including their morphology, staining characteristics, and nutritional and metabolic properties. The descriptions of bacteria are contained in the reference text *Bergey's Manual of Systematic Bacteriology* (first edition), as mentioned in chapter 1. This manual was published in four volumes, divided as follows:

Volume 1 Gram-negative organisms of general, medical, or industrial importance

Volume 2 Gram-positive organisms, other than actinomycetes

Volume 3 Archaebacteria, cyanobacteria, and the remaining gram-negative organisms

Volume 4 The Actinomycetes

In addition to containing descriptions of organisms, all four volumes contain information on the ecology, methods of enrichment, culture, and isolation of the organisms and methods for their maintenance and preservation. However, the heart of the book is a description of all of the prokaryotes and their groupings according to their shared properties. If the properties of a newly isolated organism do not agree with any descriptions in *Bergey's Manual,* then presumably a new organism has been isolated.

Although a major goal of bacterial taxonomists is to group organisms according to their evolutionary relationships, the present edition of the manual does not do this. The fact that organisms resemble each other in appearance, staining characteristics and metabolism does not necessarily mean they are related. As pointed out, the prokaryotes can be divided into two domains, the Eubacteria and Archaea, which *Bergey's Manual* does not do. Fortunately, several modern approaches to taxonomy indicate that evolutionary relationships between bacteria can be established more precisely than has been possible thus far. Future editions of this manual will no doubt take these more recent findings into account. Some of the modern techniques of bacterial taxonomy used to group organisms according to their evolutionary relationship are briefly considered here.

Table 10.1 Taxonomic Ranks of the Bacterium *Leptospira interrogans*

Formal Rank	Example
Kingdom	Prokaryotae
Division	Gracilicutes
Class	Scotobacteria
Order	Spirochaetales
Family	Leptospiraceae
Genus	*Leptospira*
Species	*interrogans*

Numerical Taxonomy

Numerical taxonomy is based on calculating the percentage of characteristics that two groups have in common. In this approach, the investigator conducts a large number of tests to determine whether certain features are present or absent in an organism. These tests include such characteristics as the ability to degrade lactose, the ability to form endospores, and the presence of flagella. Some characteristics, such as motility and spore-forming ability, depend on many genes and therefore are more fundamental and important properties of an organism than are a number of biochemical traits, such as the ability to degrade lactose into its monosaccharide components, which requires one or at

• Lactose, p. 36

most a few genes. However, there is no need to consider some features as more important than others if a large enough number of traits is examined. Thus, one of the advantages of numerical taxonomy is that the results are unbiased by subjective judgments of the investigator as to which characteristics are more important. Therefore, different investigators should arrive at the same conclusion.

The final result of classification by numerical taxonomy is expressed in terms of a **similarity coefficient,** defined as the percentage of the total number of characters tested common to two strains (figure 10.2). The greater the number of characteristics that two organisms have in common, the more closely they are related.

After a great many characteristics are examined in a large number of strains, a computer program is used to calculate similarity coefficients for each pair of organisms and to construct a similarity chart. On the basis of this chart, the strains can be ordered in such a way that those that have more than a 90% similarity coefficient are classified as a single species, while other more distinct groups are classified as separate species and perhaps even different genera.

(a)

Strain Number	1	2	3	4	5	6	7
1	100						
2	5	100					
3	10	95	100				
4	0	90	95	100			
5	80	15	35	15	100		
6	70	25	40	10	80	100	
7	95	10	20	10	90	75	100

(b)

	A				B		
Strain Number	1	7	5	6	3	2	4
A 1	100						
A 7	95	100					
A 5	80	90	100				
A 6	70	75	80	100			
B 3	10	20	35	40	100		
B 2	5	10	15	25	95	100	
B 4	0	10	15	10	95	90	100

Figure 10.2
Similarity coefficients. Seven strains of organisms were tested in 100 different ways, each test producing a positive or negative result. Each strain was then compared with the other strains by determining the similarity coefficient, the percentage of the total characters tested held in common in each of the strains. In (*a*), the strains are arranged randomly; in (*b*), they are arranged so that the organisms with similar properties are grouped together. Note that the seven organisms fall into two unrelated groups.

Molecular Approaches to Taxonomy

The more recent methods of classification are based on comparing more directly the information contained in the DNA of different groups of microorganisms. Because the genetic information of a cell is encoded in its DNA, the relatedness of two organisms is directly connected to the similarity of base sequences in their DNA. Differences in the base sequence are a measure of the number of mutations that have occurred since the two organisms diverged from a common ancestor. New proteins were formed as the base sequences of DNA molecules diverged in the course of evolution. Thus, modern approaches to an evolutionary taxonomy have generally involved (1) comparing DNA or RNA base sequences in one or more ways, as described in the next section and (2) comparing the amino acid composition of a specific protein or a group of proteins in different organisms.

DNA Base Compositions of Organisms Indicate Whether or Not They May Be Related

Since organisms that are related have DNA base compositions that are similar or identical, comparing the base composition can indicate whether two organisms are related. Such comparisons can be made by (1) analyzing the overall base composition and (2) comparing the sequence of bases.

As already discussed, the structure of double-stranded DNA is such that the number of molecules of adenine (A) equals that of thymine (T), and the number of molecules of guanine (G) equals that of cytosine (C). The base composition of an organism is usually expressed as the percent of guanine plus cytosine, termed the **GC content.** This is calculated from the following equation:

$$\frac{\text{moles G} + \text{moles C}}{\text{moles G} + \text{moles C} + \text{moles A} + \text{moles T}} \times 100$$

The GC content varies widely among different kinds of bacteria, with numbers ranging from about 22% to 78% (table 10.2). Organisms that are related by other criteria have DNA base compositions that are similar or identical. Thus, if the GC content of two organisms differs by more than a small percent, they cannot be closely related. For example, although *Proteus, Escherichia,* and *Pseudomonas* are all gram-negative, rod-shaped organisms, they cannot be closely related because their GC contents are so different. However, a similarity of base composition does not necessarily mean that the organisms are related since many arrangements of the bases are possible. For example, although the bacterium *B. subtilis* and humans have 40% GC in their DNA, obviously they are vastly different organisms. The arrangement of A, T, G, and C in their DNA differs greatly.

Table 10.2 DNA Base Composition of Different Bacteria

Bacteria	% G + C
Bacillus species (sp)	32–62
Clostridium sp	21–28
Proteus sp	39–42
Escherichia sp	50–53
Pseudomonas sp	58–70

The wide range of G + C content in some groups, such as *Bacillus* sp. and *Pseudomonas*, suggests that many organisms in these groups are unrelated to one another and that their members likely will be classified into other genera as more is learned about them.

Comparing Sequence of Bases in DNA Determines How Closely Organisms Are Related to One Another

Comparing the sequence of bases in DNA is probably the most widely used and accurate way to determine the relatedness of organisms. The technique for doing this, termed **nucleic acid hybridization,** is straightforward and relatively simple. It is based on the technique of nucleic acid hybridization discussed in chapter 8.

When double-stranded DNA is heated, the two complementary strands separate into two single strands (**DNA denaturation**). If this mixture of single-stranded molecules is then incubated over time at a lower temperature, complementary single-stranded molecules come together (renature) to form a double-stranded molecule whose properties are those of the original double-stranded DNA (figure 10.3). Alkali also denatures DNA.

Because the double-stranded DNA molecule can only form if the two strands of DNA have the complementary sequence of bases, this heating procedure can assess the extent of base sequence similarity in the DNAs of two organisms. In this procedure, single-stranded DNA molecules from the two different organisms are mixed together and incubated to allow the molecule to renature. If the two organisms are closely related but not identical, the sequence of purine and pyrimidine bases in their DNA will be similar, and the single strands will come together, not perfectly, to form double-stranded molecules (figure 10.4 *a*). If the organisms are unrelated, then none of the single-stranded molecules will come together (figure 10.4 *b*). This heating

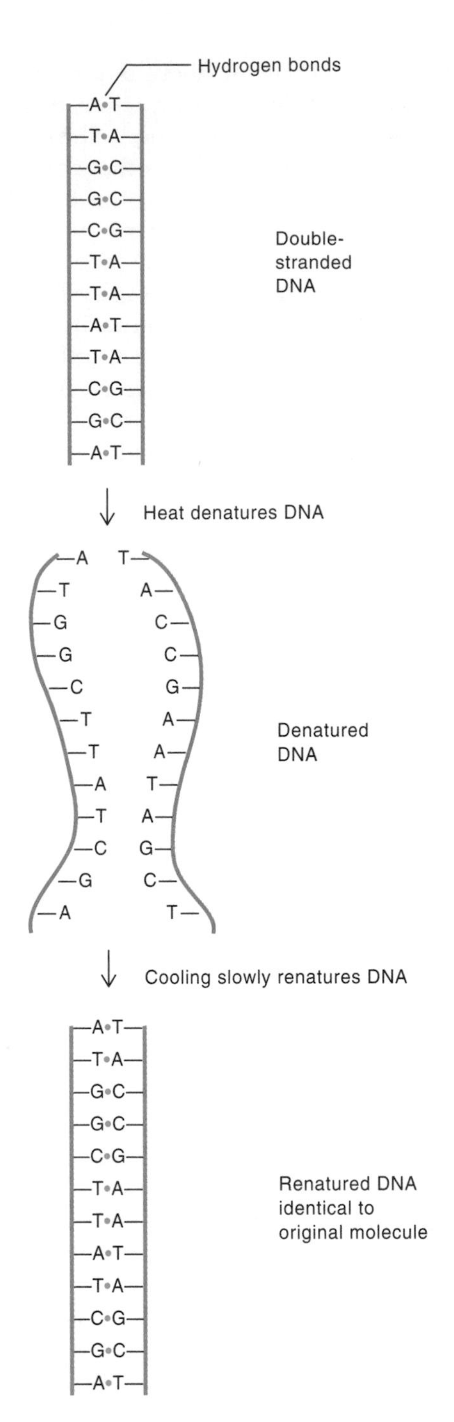

Figure 10.3
Denaturation and renaturation of DNA.

approach is a direct and powerful method for assessing relatedness between organisms. The more closely two organisms are related, the greater the degree of nucleic acid hybridization.

It is possible to estimate quantitatively the number of base changes that have accumulated by mutation in the evolution of two organisms since their descent from a common ancestor from the physical properties of the double

• **Nucleic acid hybridization, p. 192**
• **Purines and pyrimidines, p. 37**

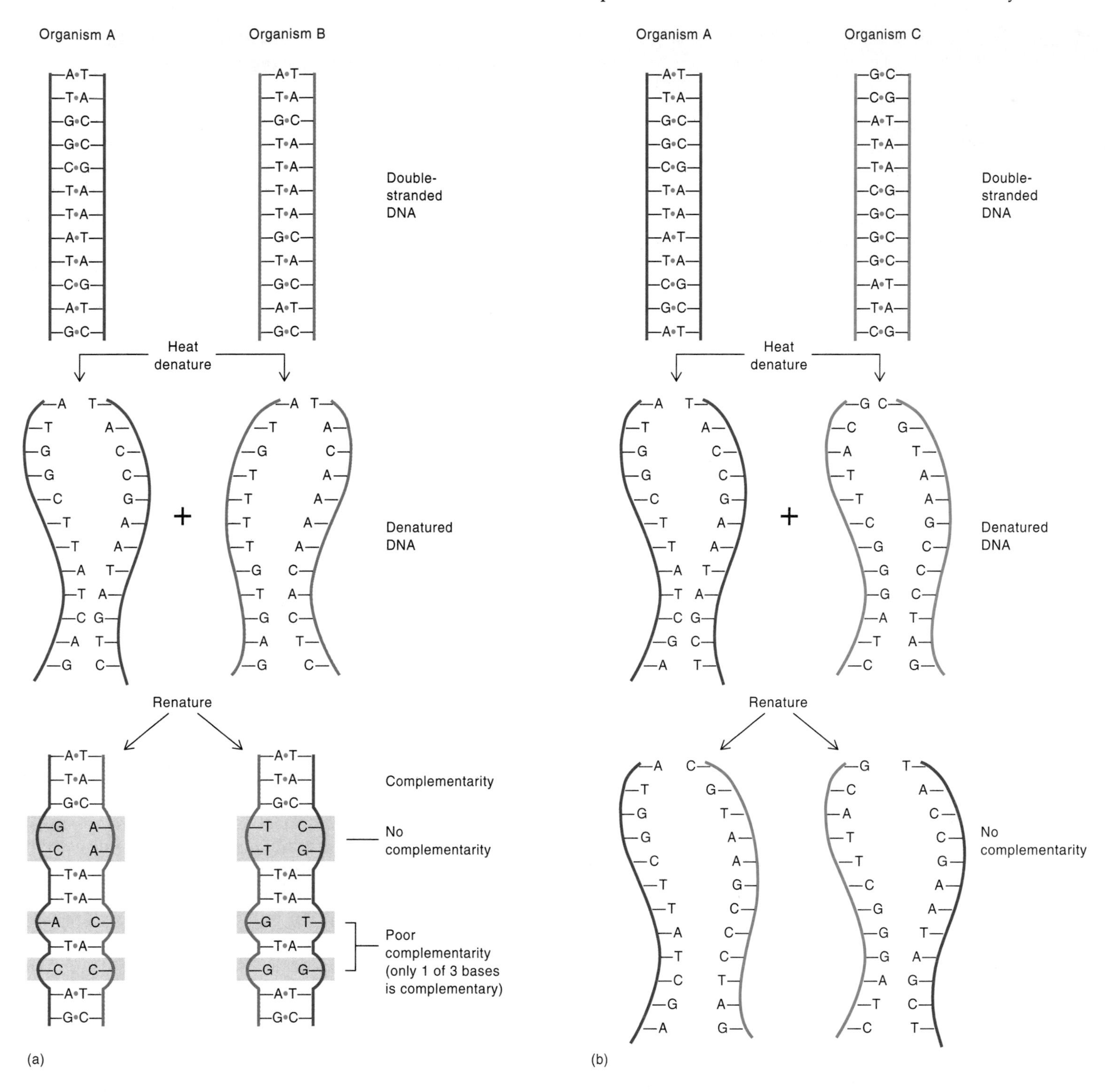

Figure 10.4
(*a* and *b*) Measuring relatedness between organisms by the similarity in their DNA base sequences. Whether the DNA of organism A will hybridize with organism B depends on the DNA base sequences being complementary to one another. The single-stranded DNA can be readily separated from double-stranded DNA on a filter. The DNA of one of the organisms is made radioactive and then mixed at very low concentrations with nonradioactive denatured DNA of the other organism, present in a 100-fold higher concentration. Only in related organisms will the radioactive DNA molecules form significant numbers of double-stranded DNA molecules, making the radioactivity in double-stranded DNA a measure of the degree of relatedness of the two organisms.

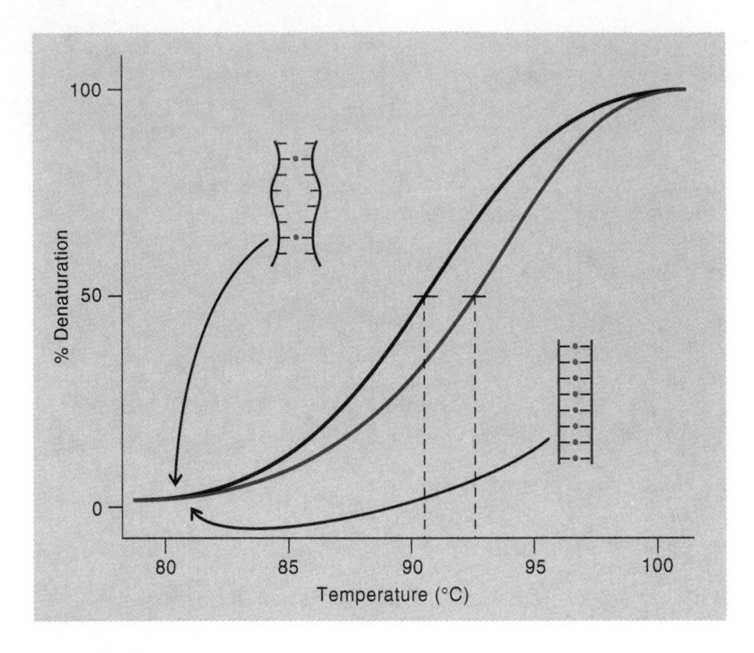

Figure 10.5
Measurement of the relatedness of two DNA molecules by determining the number of hydrogen bonds that form between complementary strands. The higher the relatedness of two organisms, the more hydrogen bonds that form between complementary strands and, thus, the higher the temperature required to separate the two strands.

strands formed by the renaturation of DNAs from the two different organisms. The more perfectly the complementary strands fit together, the more resistant the two strands are to dissociation by heating (figure 10.5). Minor differences in the two DNAs attributable to accumulated mutations will place two bases other than the normal A-T and G-C pairs opposite one another when DNA molecules renature (see figure 10.4 *a*). These will not pair, and the reduction in the number of hydrogen bonds reduces the renatured DNA's resistance to denaturation by heat.

Measuring the temperature to which the renatured DNA must be heated before it denatures, a measure of its stability to heat or **thermal stability,** provides a means of measuring the number of mismatched bases present, and thus the number of mutations that have accumulated in the DNA of the organisms since they diverged from a common ancestor. The reduction in thermal stability measured by the temperature at which 50% of the molecules are denatured directly measures the percentage of mismatched bases. With this method, the genera *Shigella* and *Escherichia* cannot be distinguished from one another and probably should be included in the same genus.

At the present time, nucleic acid hybridization is being used to assess the relationships of many groups of bacteria. Some groups of fungi have also been examined in this way, and the method has been applied to plants and animals.

Amino Acid Sequence Similarities in the Same Protein Can Indicate Relatedness Between Organisms

A gene can be described either by its base sequence or by the amino acid sequence of the protein molecule for which it codes. Thus, a comparison of the amino acid sequences of proteins with identical functions from two organisms gives some idea of the relationship between the genes that code for the protein. One problem with this approach is that one gene is only a minute portion of the total genetic information of the cell. This contrasts with the overall estimation of DNA similarity of the entire genome that can be made with DNA hybridization. Further, although the complete amino acid sequences have been determined for many proteins in a variety of eukaryotic organisms, not many have been determined from bacteria.

Determination of the Base Sequences in 16S Ribosomal RNA Reveals Evolutionary Relationships Between Diverse Groups of Bacteria

Another gene product that has been sequenced to determine genetic relatedness between organisms is 16S ribosomal RNA, a component of the small subunit of prokaryotic ribosomes. The 16S ribosomal RNA is isolated from the bacteria. Its nucleotide sequence is then converted into a complementary sequence in DNA, in vitro, using the enzyme reverse transcriptase. 16S ribosomal RNA consists of approximately 1,500 nucleotides (figure 10.6). There are several reasons this sequence was chosen to indicate genetic relatedness. As was already discussed, ribosomes play an indispensable role in protein synthesis, and the nucleotide sequence changes that can occur and still allow the ribosome to function are limited. Consequently, long after organisms have diverged in their nondispensable functions so much that they appear to be unrelated by nucleic acid hybridization, similarities between their ribosomal RNAs are still evident. Further, the number of nucleotides in 16S ribosomal RNA is large enough that numerous possible changes can occur and can be quantified as an indication of relatedness. Thus, this method makes it possible to determine the relatedness between organisms that appear unrelated by any of the other tests previously discussed. In addition, when the nucleotide sequences in 16S ribosomal RNA were determined in a wide variety of organisms, it was observed that specific base sequences, termed **signature sequences,** were almost always found in particular groups of organisms (see figure 10.6).

Furthermore, the presence of the specific signature sequence is correlated with the presence of other molecular

• **16S ribosomal RNA, p. 73**
• **Reverse transcriptase, p. 330**

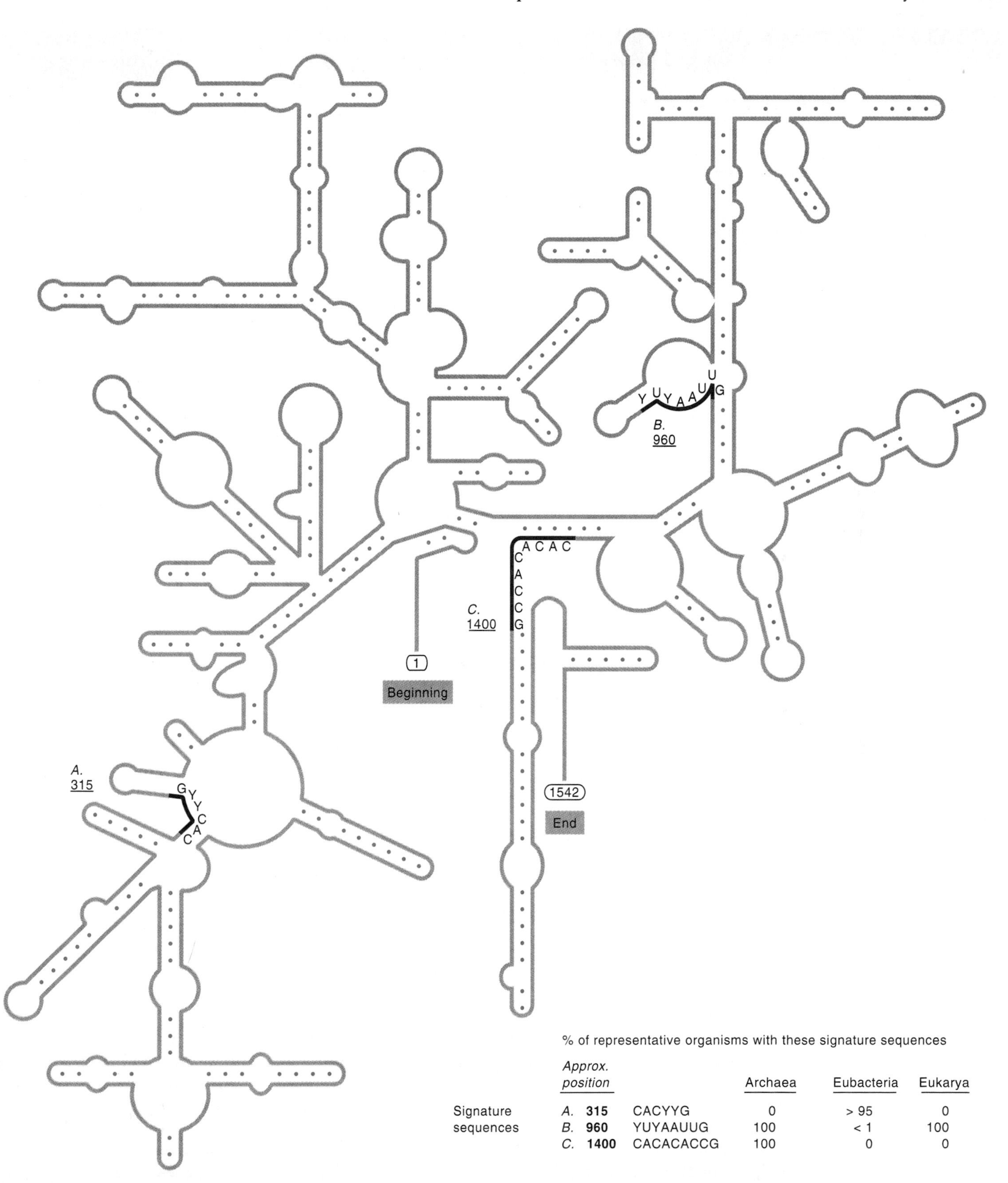

	Approx. position		Archaea	Eubacteria	Eukarya
Signature sequences	*A.* **315**	CACYYG	0	> 95	0
	B. **960**	YUYAAUUG	100	< 1	100
	C. **1400**	CACACACCG	100	0	0

Figure 10.6
Diagram of 16S ribosomal RNA. Some but not all of the signature sequences are shown in color. Although the RNA is single-stranded, many hydrogen bonds form between complementary bases, thereby leading to a double-stranded structure in much of the molecule. This is part of the 30S subunit of ribosomes.

Perspective 10.1

Surprising Relationships

Looking at relationships between organisms by comparing their nucleic acid sequences has resulted in some surprising conclusions. One such conclusion is about certain animal and plant pathogens. Members of the family *Rickettsiaceae* are small, gram-negative bacteria that, with a few exceptions, can only multiply inside animal cells. Members of this group live inside cells of insects and some live in cells of humans. They cause such diseases as Rocky Mountain spotted fever and Q fever. When the 16S ribosomal RNA of rickettsias was sequenced and compared to the 16S ribosomal RNA of other bacterial species, it was observed that the sequences were very similar to those of members of the family *Rhizobiaceae*. This family includes the species *Agrobacterium* and *Rhizobium*. The rhizobia induce root nodules in certain plants in which they live, and these nodules convert nitrogen from the air into ammonia, which the plant uses as a source of nitrogen fertilizer. *Agrobacterium* species form crown gall tumors on plants by transferring a part of their DNA into plant cells. Members of these two families, the *Rickettsiaceae* and the *Rhizobiaceae*, have always been considered to be totally unrelated to one another. However, they do have one obvious characteristic in common—an intimate, usually intracellular association with eukaryotic cells.

Woese and his colleagues made another observation, which may be a logical outcome of this association. It is now generally accepted from other studies that mitochondria in eukaryotic cells arose as bacterial **endosymbionts,** apparently after a progenitor to a eukaryotic cell had engulfed a bacterium. Woese compared the 16S ribosomal RNA of *Agrobacterium* with that of plant mitochondria and showed that they are closely related. This suggests that the chromosomal DNA of *Agrobacterium* and the mitochondrial DNA in the Eukarya have a common ancestor. This study is an excellent example of how a molecular approach to classification has given insights into relationships between organisms which otherwise would never have been suspected.

• **Endosymbiont, p. 248**

• **Agrobacterium, p. 175**
• **Rhizobium, p. 168**

Table 10.3 Some Properties of Archaea, Eubacteria, and Eukarya

Cell Feature	Archaea	Eubacteria	Eukarya
Cell wall	Has no peptidoglycan	Contains peptidoglycan	Has no peptidoglycan
Cytoplasmic membrane lipids	Hydrocarbons (not fatty acids) linked to glycerol by ether linkage*	Fatty acids linked to glycerol by ester linkage†	Fatty acids linked to glycerol by ester linkage†
Introns	Exist	Are rare	Exist
True nucleus	Does not exist	Does not exist	Exists

*Ether linkage - C - O - C -

†Ester linkage $-\overset{\overset{\displaystyle O}{\|}}{C}-O-$

properties of the organism. By determining the 16S ribosomal sequence of over 400 organisms, Dr. Carl Woese and his colleagues divided all organisms into one of three groups on the basis of the signature sequences and their related properties: the Eukarya, the Eubacteria, and the Archaea (table 10.3). They believe that these are the primary lines of evolutionary descent and represent three different domains. The domain is the highest level of organization in the biological world and should supplant the term *Kingdom* in the view of many microbiologists.

Figure 10.7 postulates the relationships among the three domains and their relationships to the "ancestral" cell, the last ancestor to all three domains. The first "organism" most likely was an RNA-containing "organism" without DNA. The length of time that it took for the RNA organism to develop into the ancestral cell is not known. Some scientists estimate that the time of divergence from the ancestral cell into its two branches was 1.3 to 2.6 billion years ago. The mitochondria and chloroplasts of the Eukarya arose from the engulfment of nonphotosynthetic bacteria and cyanobacteria, respectively, by ancestors of the Eukarya. In turn, eukaryotic cells were engulfed by other eukaryotic cells (see perspective 10.2).

In many ways, determining the base sequence is the most accurate and reliable procedure yet available for identifying evolutionary relationships of organisms. It is better than DNA hybridization, because it is far more quantitative. Organisms in which DNA hybridization can

Perspective 10.2

Captures and More Captures

There is general agreement that in the course of evolution eukaryotes "captured" cyanobacteria, which developed into chloroplasts and "captured" nonphotosynthetic prokaryotes, which developed into mitochondria. This endosymbiotic theory to explain the origin of these two organelles is based in part on the fact that they contain prokaryotic DNA and that both chloroplasts and mitochondria retain two outer membranes. Presumably the inner one is the outer membrane of the engulfed prokaryote and the outer one belongs to the cell that engulfed the victim. This capture of cells by other cells in the course of evolutionary history may be more common than previously suspected. The captured cells may even include eukaryotes. This was recently suggested by the surprising observation that a group of free-living parasites including those that cause malaria (*Plasmodium*) and toxoplasmosis (*Toxoplasma gondii*) contain chloroplasts, the organelles found in plants and green algae. Chloroplasts are the organelles that carry out photosynthesis, in which light energy is converted into the chemical energy of ATP. These chloroplasts do not function in the parasites but still contain their own DNA, as do mitochondria. What is remarkable is that they also have DNA nearly identical to algal DNA and that it has not undergone extensive mutations.

Many questions arise from these observations. What role, if any, do these chloroplasts play in the life of the parasite? Presumably they play an important role or they would have been eliminated in the course of evolution. Can infections such as malaria and toxoplasmosis caused by these parasites be cured with herbicides known to inhibit the growth of green plants? What other examples of cell and organelle engulfment exist in nature and what role do they play in evolution? The finding that some eukaryotic cells were captured and maintained by other eukaryotic cells has important implications for both clinicians and evolutionary biologists.

Table 10.4 Molecular Analytical Techniques for the Study of Relatedness Among Prokaryotes

I. Genetic level
   Comparison of nucleic acids in vitro
   A. DNA base ratio
      (% guanine + cytosine)
   B. Hybridization of nucleic acids
      (DNA/DNA; DNA/RNA)
   C. Comparison of 16S ribosomal RNA sequences
II. Protein level
   A. Comparison of amino acid sequence of the same protein in vitro

be barely detected are related, but how closely is not certain. Comparing their 16S ribosomal RNA sequences gives an accurate estimate. Indeed, this technique of determining the sequence of 16S ribosomal RNA has revealed relatedness between organisms that appear to be quite unrelated based on their phenotypic characteristics. For example, the mycoplasmas (cell wall deficient bacteria) are related to members of the genus *Clostridium*, anaerobic spore-forming rods, and the major photosynthetic bacteria are included in a group with many nonphotosynthetic species. For another example, see perspective 10.1.

Although this method may be able to reveal relationships between apparently unrelated organisms, at the present time it is not as easy to perform as nucleic acid hybridization. Therefore, its use is now restricted to relatively few research laboratories. The techniques used in determining relatedness between prokaryotes are summarized in table 10.4.

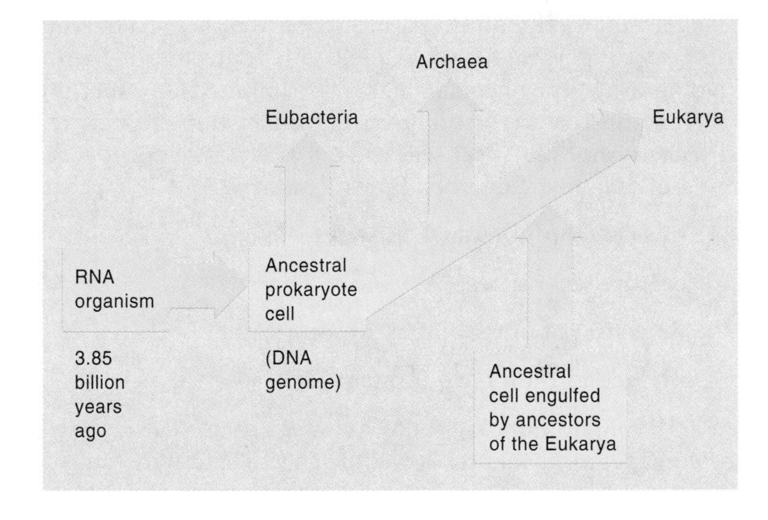

Figure 10.7
Probable evolution of the domains determined from 16S ribosomal RNA sequence comparisons.

REVIEW QUESTIONS

1. What is the basic unit of taxonomy?
2. What book attempts to classify the prokaryotes?
3. What technique is used to compare relatedness between organisms by comparing their sequences in DNA?
4. What are the three domains that have been identified through sequencing of 16S ribosomal RNA?

Methods of Identifying Bacteria

In practical terms, identifying the genus and species of a bacterium may be more important than understanding its genetic relationship to other bacteria. Therefore, the molecular analyses discussed thus far are carried out in research laboratories and not in clinical laboratories concerned with identifying infectious agents, the field of diagnostic microbiology. In clinical laboratories, it is critical to identify quickly the agents that are isolated from patients so the best possible treatment can be given. To do this, an organism is described in as many ways as is practically possible. The investigator identifies the organism genus and perhaps species by comparing the properties of the unknown organism with those of organisms that have already been identified. It is important to realize that the number of tests that can reasonably be performed in a clinical laboratory represents the gene products of only a small fraction of the total genome of the organism. The more tests that are done, the more accurate the identification.

In practice, diagnostic microbiology makes use of a wide assortment of technologies and strategies to accomplish its goals. The numbers and types of tests used depend upon the type of infectious agent to be detected or identified and the type of specimen being examined. Another important part of identifying an organism depends on the disease symptoms it causes. The major types of approaches used in diagnostic microbiology include

1. **direct techniques** such as stains;
2. **culture** techniques;
3. **detection of microbial** by-products; and
4. **molecular biological** techniques.

In addition, immunological techniques are used, which will be considered in chapter 16. While each of these methods is best suited for detecting certain kinds of organisms in specific types of specimens, combinations of the tests may provide the best identification of the infecting agents.

Direct Methods of Identification Involve Microscopic Examination Without Culturing the Organisms

Direct methods rely on direct microscopic examination of a specimen for microorganisms without the culturing of any organisms. These are some of the most versatile, inexpensive, and useful methods available to microbiologists because they do not require much equipment and laboratory space and they give information very quickly. Their greatest limitation is that they are relatively insensitive. A specimen must have 10^4 or 10^5 organisms per gram of specimen in order to be detected microscopically. Some direct methods are nearly as old as the invention of the microscope while others rely on new technologies.

Wet Mounts

Performing a simple **wet mount,** in which a liquid specimen is placed on a microscope slide and examined directly under a microscope, can provide all the information that is needed for diagnosis of some infections. For instance, a wet mount of vaginal secretions is routinely used to diagnose infection of the vagina caused by yeast (figure 10.8 *a*) and by the protozoan *Trichomonas* (figure 10.8 *b*). A wet mount of stool is examined for the eggs of parasites when certain roundworms are suspected (figure 10.8 *c*). The size, shape, and special features of the eggs are often sufficient to allow identification of the intestinal parasite.

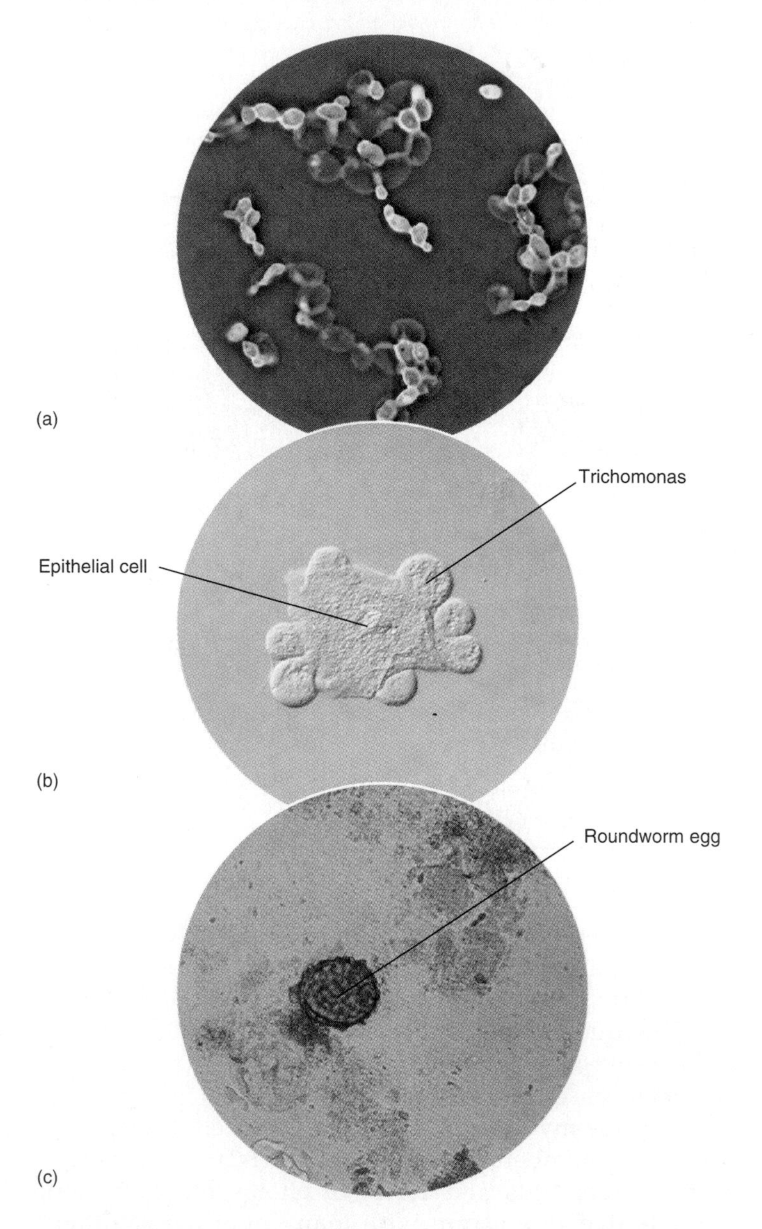

Figure 10.8
Photomicrographs showing wet mounts of (*a*) vaginal secretions containing yeast *(Candida albicans,* ×410); (*b*) *Trichomonas vaginalis* attached to squamous epithelium, as viewed through a Nomarski microscope (×400); and (*c*) roundworm (Ascaris) eggs in a stool (×400).

Gram Stain and Acid-Fast Stain

The differential stain named the Gram stain, which divides bacteria into two groups, gram-positive and gram-negative, was discussed in chapter 3. It provides rapid, suggestive information that can be used immediately by the clinician in choosing antimicrobial therapy. For example, if a Gram stain of a dried sample of urine on a microscope slide reveals more than one plump gram-negative rod per oil immersion field, the clinician will suspect that the patient has a urinary tract infection with a coliform-type[1] bacterium and will be able to start appropriate antimicrobial therapy. A gram stain from a male urethra secretion sample showing gram-negative diplococci in white blood cells is both a specific and sensitive indicator of infection with *Neisseria gonorrhoeae,* the causative agent of gonorrhea (figure 10.9 *a*). Likewise, a Gram stain of sputum showing numerous white blood cells and gram-positive encapsulated diplococci is highly suggestive of *Streptococcus pneumoniae,* an agent that causes pneumonia (figure 10.9 *b*).

While the direct Gram stain of clinical specimens by itself is not sensitive or specific enough to detect many types of infections, it is an extremely useful tool in clinical microbiology. The clinician can see both the Gram reaction as well as the shape and arrangement of the bacteria, whether they are growing in pure culture or with other bacteria and/or cells of the host.

If a patient has symptoms of tuberculosis, then an acid-fast stain will likely be done to determine whether *Mycobacterium tuberculosis* can be detected (figure 10.10 *a*). In a variation of the acid-fast stain, a fluorochrome such as auramine can be used to detect *M. tuberculosis*. These auramine-stained organisms fluoresce when viewed with a fluorescence microscope (figure 10.10 *b*). As with the conventional acid-fast stain, in auramine staining mycobacteria, unlike most other bacteria, retain the stain despite being washed with acid alcohol. Only a few specific types of organisms can be detected by the acid-fast stain.

[1]Coliform—gram-negative rods, including *E. coli* and similar species, that normally inhabit the colon

• **Gram stain, p. 51**

• **Acid-fast stain, p. 52**
• **Fluorescence microscopy, p. 48**

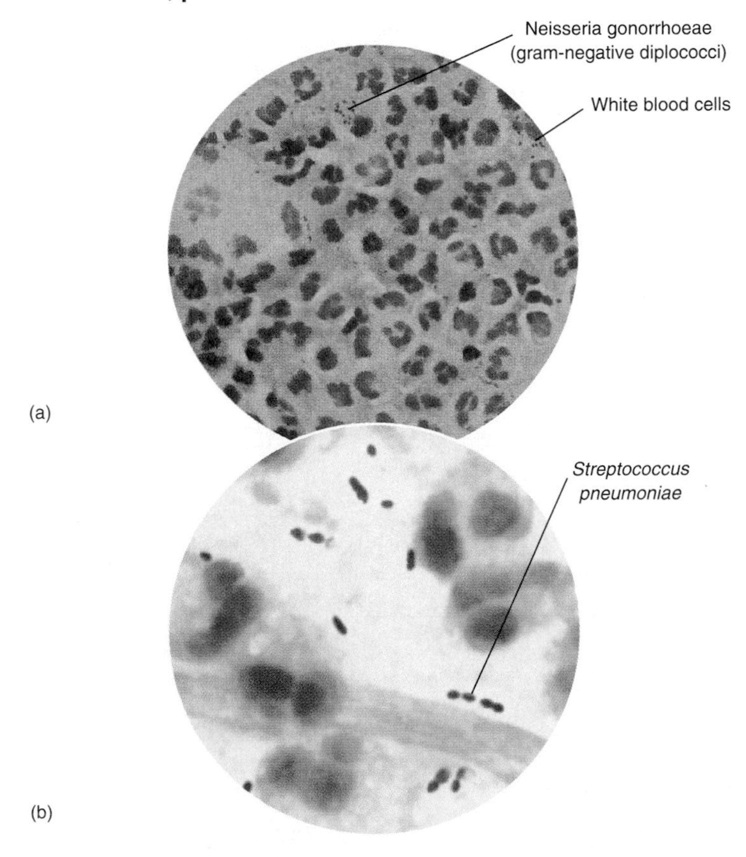

Figure 10.9
Gram stains of (*a*) male urethra secretions showing negatively stained diplococci *Neisseria gonorrhoeae* inside white blood cells and (*b*) sputum showing gram-positive diplococci *Streptococcus pneumoniae.*

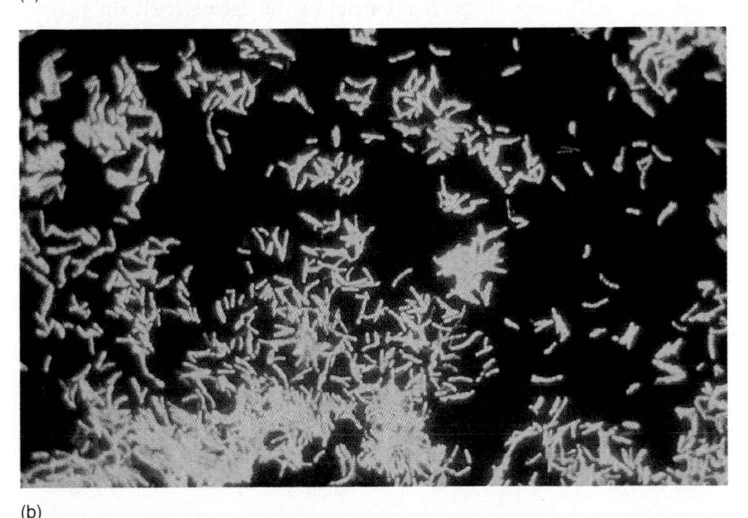

Figure 10.10
Photomicrographs of *Mycobacterium tuberculosis* showing (*a*) an acid-fast stain and (*b*) an auramine stain (fluorochrome; ×400).

Cultivation Techniques Involve Growing the Unknown Organisms in Pure Culture

The most specific and sensitive method for diagnosing some infections involves cultivating a specimen from the infected person. While direct staining methods are rapid and inexpensive, they are useful only in those cases in which relatively large numbers of organisms are present in the specimen. Normally sterile body fluids such as blood or cerebrospinal fluid do not have enough microorganisms to be detected by direct methods unless the infection is far advanced and the patient is near death. The only way to detect the infecting agent in the early stages of the infection is by cultivating the body fluid and observing bacterial growth. Two advantages of this approach are that (1) it allows precise identification of the organism and (2) the isolated organism is available for testing for susceptibility to antimicrobial medicines. However, some organisms such as *Mycobacterium tuberculosis* grow so slowly that physicians must begin treatment before the culture result is known. Treatment is then modified as indicated by the culture results. Also, some fastidious[1] organisms, such as members of the genera *Mycoplasma* and *Rickettsia*, require considerable time and expense to grow in vitro.

Other types of bacteria do not have distinctive shapes or staining characteristics that allow them to be identified by direct staining methods. For example, *Streptococcus pyogenes*, which causes strep throat, cannot be distinguished by staining from the other streptococci that are part of the normal flora of the throat. A Gram stain of a stool specimen cannot distinguish *Salmonella* from *E. coli*. Only specific tests following cultivation of the stool specimen can provide precise identification.

Specimens that are sent to the diagnostic laboratory are inoculated onto a set of media that has been chosen for its ability to support the growth of organisms that are likely to be present in that specimen type. For example, most organisms that cause urinary tract infections grow well on blood or a differential and selective medium containing lactose and bile salts such as MacConkey agar, which selects for coliform bacteria that cause primary infections. Bacteria that ferment lactose give rise to red colonies. Urine is routinely inoculated onto those types of media. As another example, organisms that cause bacterial meningitis may be extremely fastidious and present in very low numbers so that enriched media, such as chocolate agar, as well as an enriched broth will be used to cultivate that type of specimen.

The organisms of interest that grow on these agar plates are then isolated and grown in pure culture for identification. The genus and species of most bacteria that cause infection in humans can be identified using biochemical tests. These include simple tests for production of such enzymes as catalase (figure 10.11 *a*) and oxidase and the ability to degrade with the production of sugars, acid, and sometimes gas (figure 10.11 *b*). Some additional commonly employed biochemical tests are described in table 10.5.

[1] Fastidious—requiring many growth factors

• **Normal flora, p. 410**
• **MacConkey agar, p. 101**
• **Chocolate agar, p. 98**

Products of Metabolism May Be Detectable Before Visible Growth of a Microorganism

Even if a given organism can be cultured, detecting its growth may be difficult, especially if the organism grows very slowly. In such cases, bacterial growth can be detected by looking for by-products of metabolism rather than by isolating the organism on a culture plate. These by-products can be measured either with radioactively labeled com-

• **Catalase, p. 93**
• **Radioactivity, p. 24**

(a)

(b)

Figure 10.11
(*a*) Production of O_2 from H_2O_2 by bacteria producing catalase. $H_2O_2 \rightarrow H_2O + O_2$. A negative catalase is shown in the right tube. (*b*) Bacteria degrading three different sugars. The tubes on the left and right show acid (yellow color) and gas. The center tube shows no sugar utilization, therefore, no production of acid or gas. The tubes contain glucose (left), maltose (center) and fructose (right).

Table 10.5 Some Biochemical Tests for Identifying Bacteria

Name of Test	Reaction	Observed Reaction
Voges-Proskauer	Metabolism of glucose to 2,3-butylene glycol	A red color develops upon addition of chemicals that detect acetoin, a precursor of 2,3-butylene glycol.
Gelatinase	Enzyme breakdown of gelatin to polypeptides	The solid gelatin is converted to liquid.
Hydrogen sulfide production	Enzymes convert the S in sulfur-containing amino acids to H_2S	A black precipitate forms in the presence of iron salts.
Phenylalanine deaminase	Enzyme removes amino group from phenylalanine	The product of the reaction, phenylpyruvic acid, reacts with ferric chloride to give a green color.
Lysine decarboxylase	Carboxyl group removed from lysine	The medium becomes more alkaline, and a pH indicator changes color.
Indole	Amino group removed from tryptophan	The product, indole, changes the color of a chemical reagent to red.

pounds or with biochemical tests, which are reliable indicators that bacteria are present in a specimen.

When a blood specimen is taken for culture, some of the blood is generally inoculated into an enriched broth. If organisms are present in the blood specimen, they multiply and eventually cause the broth to become turbid. However, if the broth contains a carbohydrate such as glucose, which is radiolabeled with ^{14}C, the organisms will degrade this carbohydrate and release $^{14}CO_2$ as they multiply. This radiolabeled CO_2 indicating bacterial growth can be detected in much less time than the time it takes for the culture to become turbid, thereby allowing these patients to be treated earlier.

Such tests are especially useful for detecting *Mycobacterium tuberculosis,* a slow-growing organism that may not form visible colonies for three to four weeks. Sputum from patients suspected of having tuberculosis is often inoculated into a broth medium containing radiolabeled fatty acids. As these organisms degrade the fatty acids, $^{14}CO_2$ is released and, therefore, growth can be detected by the appearance of radioactive CO_2. Using this method, appropriate antimicrobial therapy can be started one to two weeks sooner than would be possible with waiting for discrete colonies to grow on culture media.

Another example of the use of isotopes to detect a bacterium by the products it forms is the use of the breath test for the ulcer-causing bacterium, *Helicobacter pylori.* The test depends on the ability of the organism to rapidly degrade urea to carbon dioxide. Patients drink a solution of urea that contains isotopic carbon that can be readily detected. If *H. pylori* is present in a patient's stomach, the organism will quickly break down the urea and release carbon dioxide containing the isotopic carbon atoms. The results are known within a day or two and, if they are positive, therapy can be started. This test is much cheaper and faster than analyzing a patient's stomach tissues for ulcer lesions.

Clostridium difficile, an organism associated with infections of the intestine, is an extremely fastidious anaerobe that is difficult to isolate on common culture media. If a stool specimen is inoculated into a selective broth, however, the organism multiplies and releases short chain fatty acids that are specific to this organism. These fatty acids can be easily identified by very sensitive chemical analysis of a portion of the broth without first isolating the organism on conventional media.

The methods described for detecting products of metabolism are relatively new to clinical microbiology laboratories. As the technology advances, more organisms will probably be detected by special biochemical tests and analyses without conventional culture techniques.

Molecular Approaches in Diagnostic Microbiology Usually Involve Analysis of DNA

Today, very few laboratories use nucleic acid technology as standard practice to identify bacteria. However, in the future, these methods will likely revolutionize the way most diagnostic laboratories identify infectious agents. While different strains of a given species can vary widely in their ability to utilize carbohydrates or produce such diagnostic enzymes as hemolysins, the DNA of these organisms are virtually identical to one another, somewhat like a fingerprint. Hence, comparing the base sequence of DNA from a known bacterium with that of an unidentified bacterium makes it possible to determine their relatedness and therefore perhaps to have some idea of the identity of the unknown organism. The techniques of nucleic acid hybridization were covered earlier in this chapter.

Another way of assessing whether the DNA of two organisms is similar is to compare the pattern of fragments produced when the same restriction enzyme is used to treat DNA samples from each organism. Recall that restriction endonucleases cleave DNA at specific sites, which are determined by the base sequences at the sites. However, the

• **Restriction endonuclease, p. 178**

entire chromosome of a bacterium is too large to result in a reasonable number of fragments when it is cleaved. Therefore, the fragmentation patterns of total DNA are seldom compared to assess the relatedness of organisms. However, a technique termed *restriction fragment length polymorphism,* which is based on cleavage of the total genome, is commonly used to establish whether two organisms are identical (perspective 10.3).

Looking at cleavage products of DNA directly is commonly used to assess possible relationships of much smaller DNA molecules such as plasmids. Identical plasmids will have the same restriction sites and will always yield the same number and size of DNA fragments after treatment with these enzymes and separation by electrophoresis on agarose gels (figure 10.12). DNA sequences that code for different proteins, however, will not yield the same size or number of fragments. Two pieces of DNA can thus be treated with the same restriction endonuclease and the resulting fragments analyzed to determine whether the two DNAs are similar or different. Using another restriction endonuclease gives futher evidence for the identity or non-identity of two plasmids.

Nucleic acid techniques have great power as diagnostic tools because they are very specific and provide information about the relatedness of one organism to another, which cannot be determined by any other method.

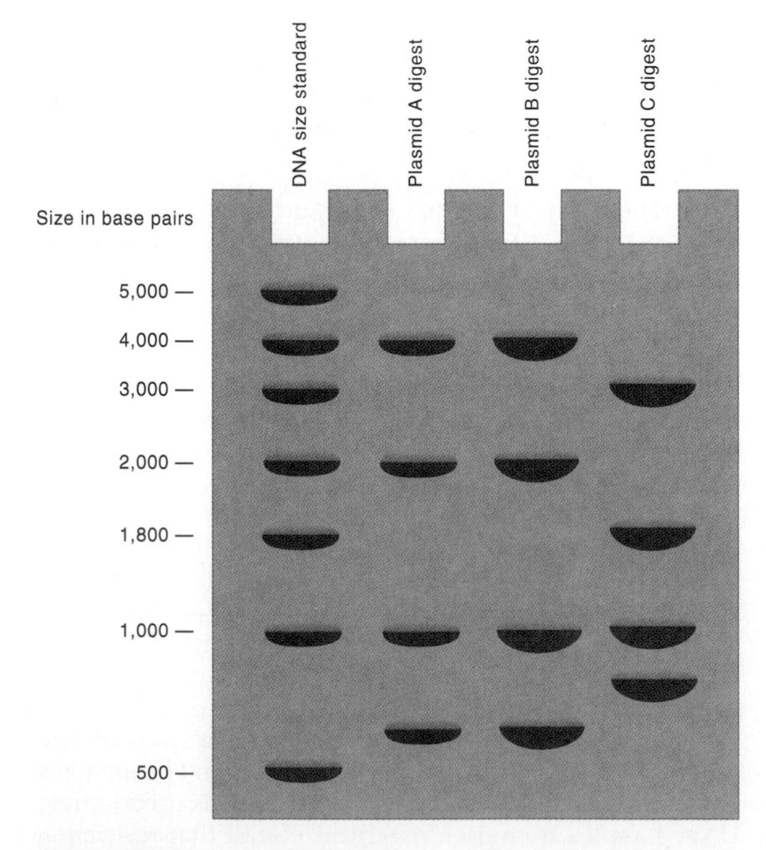

Figure 10.12
Restriction digestion and electrophoresis of three different plasmids. *A* and *B* are very similar and *C* is distinctly different.

DNA Probes

DNA probes, which are bits of nucleic acid labeled with either a radioactive isotope such as ^{32}P or a chemical such as biotin that can be easily detected, have been developed to detect and identify several different bacteria. These probes can be used to detect a single species from the numerous species present in a specimen from the throat, vagina, urethra, or sputum or to identify microorganisms in pure cultures. The general technique (figure 10.13) for using DNA probes was discussed in chapter 8.

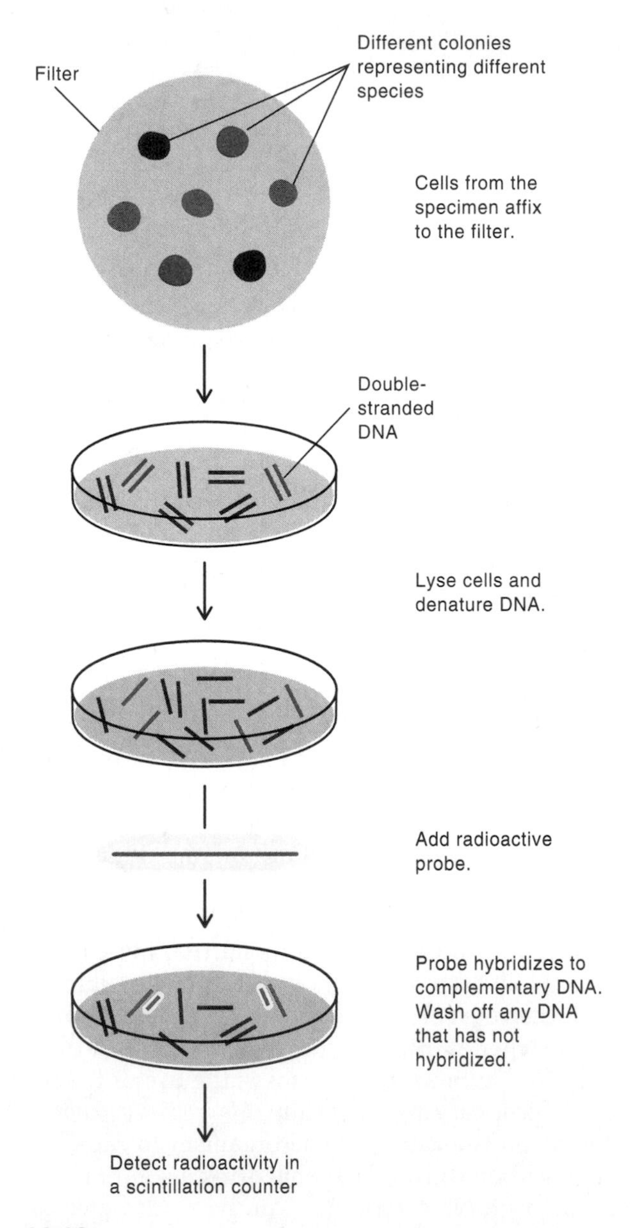

Figure 10.13
General procedure for detecting specific DNA sequence and thereby identifying a specific species of bacteria. This procedure uses a radioactive DNA probe that is complementary and therefore hybridizes to specific DNA sequences in the organism being studied. The probe should not hybridize to the DNA of any species other than the species being sought.

Perspective 10.3

Escherichia coli, Unpasteurized Apple Juice, and a Serious Disease Outbreak

An outbreak of food poisoning swept through western Washington in January 1993, affecting over 500 people and causing the deaths of three young children. The culprit was *Escherichia coli* strain 0157:H7. The bacteria apparently entered the body in undercooked contaminated hamburger eaten at a chain of fast-food restaurants. Hundreds of strains of *E. coli* exist, most of which live in the intestines of animals and pose no threat to humans. However, the strain 0157:H7 has characteristics that make it a killer of young children and the elderly. First, it has the ability to attach to walls of the intestine where it then synthesizes a potent toxin that destroys the walls leading to the symptoms of food poisoning—bleeding, stomach cramps, and diarrhea. Second, the toxin may enter the bloodstream, which can lead to damage of other organs, in particular the kidneys. In the summer of 1996, Japan suffered its worst outbreak of food poisoning in many years. Seven people died and almost 10,000 people were stricken. Again, *E. coli* 0157:H7, which had tainted hamburgers in the Pacific Northwest, was the causative agent. At about the same time, a new outbreak of the "*E. coli* disease" occurred in the Pacific Northwest, in California, and in Colorado. Again, the *E. coli* causing the disease was strain 0157:H7. This time, the source of the contamination was unpasteurized apple juice, apparently made from apples contaminated with manure.

How is this strain identified and separated from all of the hundreds of *E. coli* strains? The strain O designation is based on the antigenic structure of the lipopolysaccharide portion of the cell wall. The H designation refers to the composition of the flagella. Both structures are measured by their ability to react with specific antibodies that can detect small differences in chemical composition.

In any outbreak of the kind described, the immediate questions are "What is the source of the contamination?" and "Is there a single source or are they many different sources?" Once the source or sources have been identified, the offending products can be removed from the shelves or precautions can be taken to destroy the organisms before the food is eaten. To determine whether one or several different sources of contamination are the cause, it is necessary to determine whether the same *E. coli* 0157:H7 organism is causing all of the cases of food poisoning in the particular outbreak. To compare the two organisms more precisely, it is necessary to "fingerprint" the DNA of the organisms. A common method of fingerprinting organisms is a procedure called *restriction fragment length polymorphism*, abbreviated *RFLP* (see figure 1). This test is based on the fact that identical organisms have identical nucleotide sequences of their DNA and, therefore, identical numbers and sites at which restriction endonucleases will cleave the DNA. The number of DNA bands and their sizes which result from cleavage will be identical in both organisms. However, following cleavage by an endonuclease, too many bands are generated to separate them by gel electrophoresis. Nevertheless, it is possible to compare the sizes of specific bands in two DNA preparations following digestion of the DNA by the same restriction endonuclease by hybridizing a probe consisting of a known DNA fragment to the DNA digest after the bands have been separated by gel electrophoresis. If the organisms are identical, then the sizes of the bands produced by restriction endonuclease digestion should be identical and the bands will travel the same distance following electrophoresis. The probe used should be a gene of interest such as the toxin gene in the case of *E. coli* 0157:H7. To confirm identity of two strains, it is possible to use several different restriction endonucleases and demonstrate that the DNA bands are identical in size and number for each set of restriction enzymes in the two strains.

Using such techniques, it was possible to show that the *E. coli* 0157:H7 that caused the food poisoning in the Northwest was identical to the *E. coli* that caused the food poisoning in California and Colorado. Most of the affected people recalled drinking the same brand of unpasteurized apple juice shortly before becoming ill. This source of contamination was confirmed when the identical organism was isolated from an unopened bottle of the same juice.

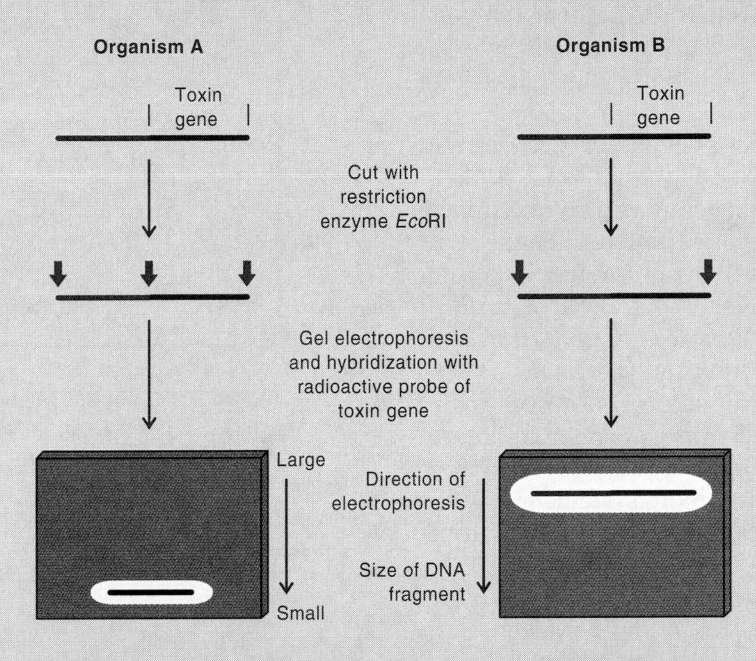

Figure 1

Restriction fragment length polymorphism (RFLP). This technique involves cleaving the total DNA of two organisms with the same restriction endonuclease, electrophoresing the cleaved DNA on an agarose gel, and then adding a radioactive probe complementary to a gene that is present in both organisms. Identical organisms will show the same size fragment that reacts with the probe, since both organisms have the same sites on the DNA that are cleaved by the restriction endonuclease.

• **Gel electrophoresis, p. 196**

Probe technology is extremely useful for three purposes: (1) to detect and identify an organism that is extremely difficult to cultivate by common methods (the labeled probe is directed against a gene that is characteristic for a certain species, such as ribosomal RNA genes); (2) to detect and quickly identify an organism if a reliable laboratory procedure is unavailable to cultivate the organism; and (3) to detect and identify an organism when the organism is easy to cultivate but the toxin or product of interest that is characteristic of the organism is difficult to detect by standard methods (a probe that is directed against the gene coding for toxin synthesis is used—if the gene is present, then it is assumed that it will function in the synthesis of the protein of interest).

Probes have been used successfully to detect *Mobiluncus*, a fastidious anaerobe that is difficult to cultivate and is found in vaginal specimens of women with vaginitis. They have been developed to detect *Mycoplasma* contaminating tissue culture systems in virology laboratories. DNA probes have been prepared from a plasmid of *Neisseria gonorrhoeae*, the causative agent of gonorrhea, and have been used to screen male urethral specimens. Probes have also been used to detect the toxin genes from toxin-producing *E. coli* strains isolated from patients with intestinal infections and to speed up identification of mycobacteria. As the techniques are refined and more systems are developed by commercial companies, these techniques will become more widely used.

At the present, probes have the potential for detecting the genes responsible for virulence in any pathogen. Such genes may be present in only a few strains within a species but also may be present in different species as a result of gene transfer. Since these genes are responsible for disease, probes may be more valuable in detecting specific important genes than in identifying organisms. In all these cases, it is very important that the DNA sequences in the probe be specific for the organism being sought or the gene product being tested for; otherwise, a false identification may be made.

Microorganisms that Cannot Be Cultivated Can Still Be Identified

In any environment, whether it be soil, the marine environment, or the human body, organisms exist that can be seen with the microscope but cannot be cultivated for any number of reasons, as already discussed in chapter 4. Some scientists estimate that every gram of fertile soil contains 4,000 to 5,000 species of bacteria, most of which have not been identified. Some organisms may be present in such minute numbers that they cannot even be seen. With PCR technology, it is possible not only to detect such organisms, but also to get some idea about their identity. The general procedure is illustrated in figure 10.14.

• **Probe technology, p. 192**
• **PCR technology, p. 197**

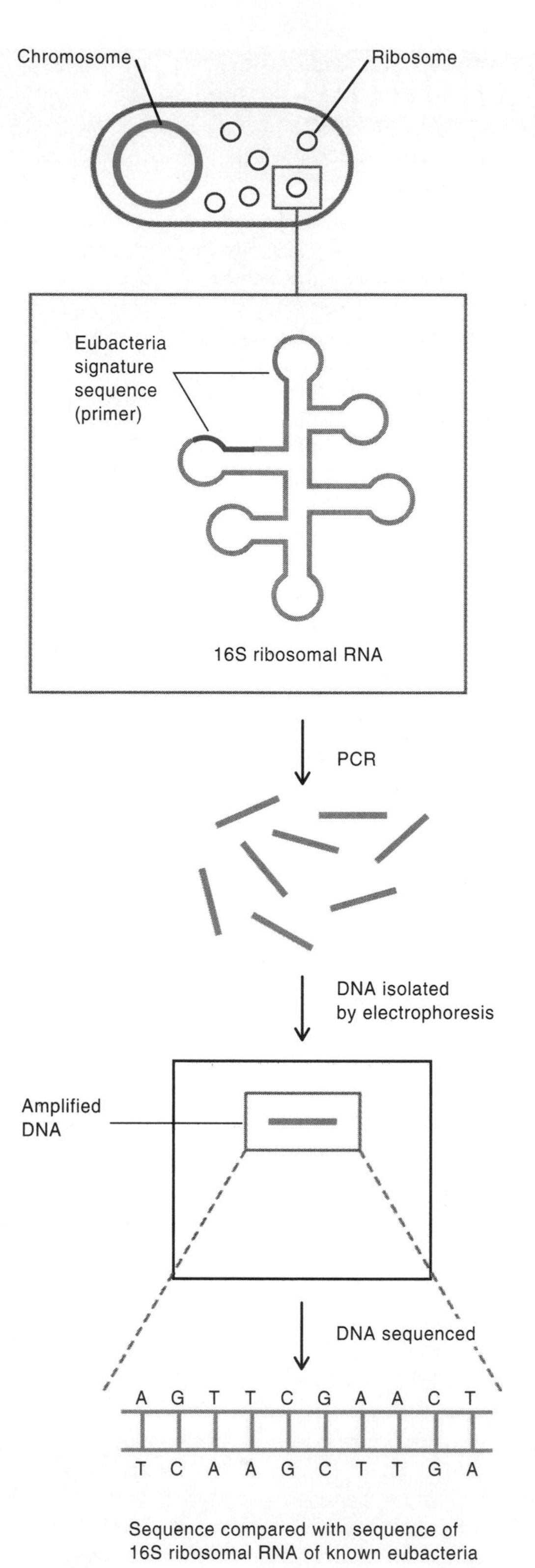

Figure 10.14
Identification of uncultivatable organisms. By using PCR to amplify sequences that are specific to a certain species of microorganism and then determining the nucleotide sequence of that specific sequence, one can compare that sequence with the sequences of bacteria that have already been identified. Bacteria having very similar sequences of nucleotides must be closely related.

To determine if eubacteria are present in any sample, investigators add the primers for the signature sequences in 16S ribosomal RNA that are specific for members of the Eubacteria (figure 10.6) to a reaction mixture containing the ingredients for the polymerase chain reaction. If eubacterial DNA is present, then these signature sequences will be amplified a millionfold and can be easily detected on gels following electrophoresis of the reaction mixture. The DNA on the gel is then eluted and the sequence of its nucleotides determined to confirm that it is indeed eubacterial DNA.

This technique can also be used to identify unknown organisms that cannot be cultivated in the laboratory. For example, although all members of the Eubacteria have the same signature sequence, the sequence of nucleotides between any two signature sequences varies between different eubacteria. Therefore, the sequences between the different signature sequences of the eubacteria should be amplified. This can be readily done by using two signature sequences as primers and amplifying the sequence between the primers using PCR. The amplified DNA is then isolated by electrophoresis, and its nucleotide sequence is determined. The sequence is then compared with that of known organisms, since DNA sequences between related organisms are similar. In this way, the organism may be identified without ever having been seen or cultivated.

This approach is being used to identify agents causing infectious diseases that cannot be cultured. For example, in one case, it was possible to show that the infectious disease, bacillary angiomatosis, was caused by a previously undescribed organism closely related to *Bartonella quintana,* the louse-borne agent that afflicted about 1 million military personnel in World War I and that causes trench fever. The same nucleotide sequences of 16S ribosomal RNA were shown to be present in all cases of bacillary angiomatosis, which were very similar to those found in *B. quintana.* Thus, even though the organism at that time was not cultured, its presence and identity could be determined using PCR and DNA sequencing technology. More recently, it has become possible to culture this organism in the laboratory.

Infectious Agents Present in Extremely Small Numbers Can Be Identified

A variation of this technique can be used to determine if a particular agent is present in any environment. For example, by using PCR technology, an investigator can determine whether HIV is present in people who cannot be identified as being HIV-positive by conventional means (figure 10.15). These individuals include infants without symptoms who, having been born to HIV-infected mothers, might very well contain very minute and therefore undetectable quantities of the virus. To determine whether even a few virus particles are present, it is only necessary to isolate DNA from mononuclear cells from blood, add the primers specific for a gene that codes for a protein specific

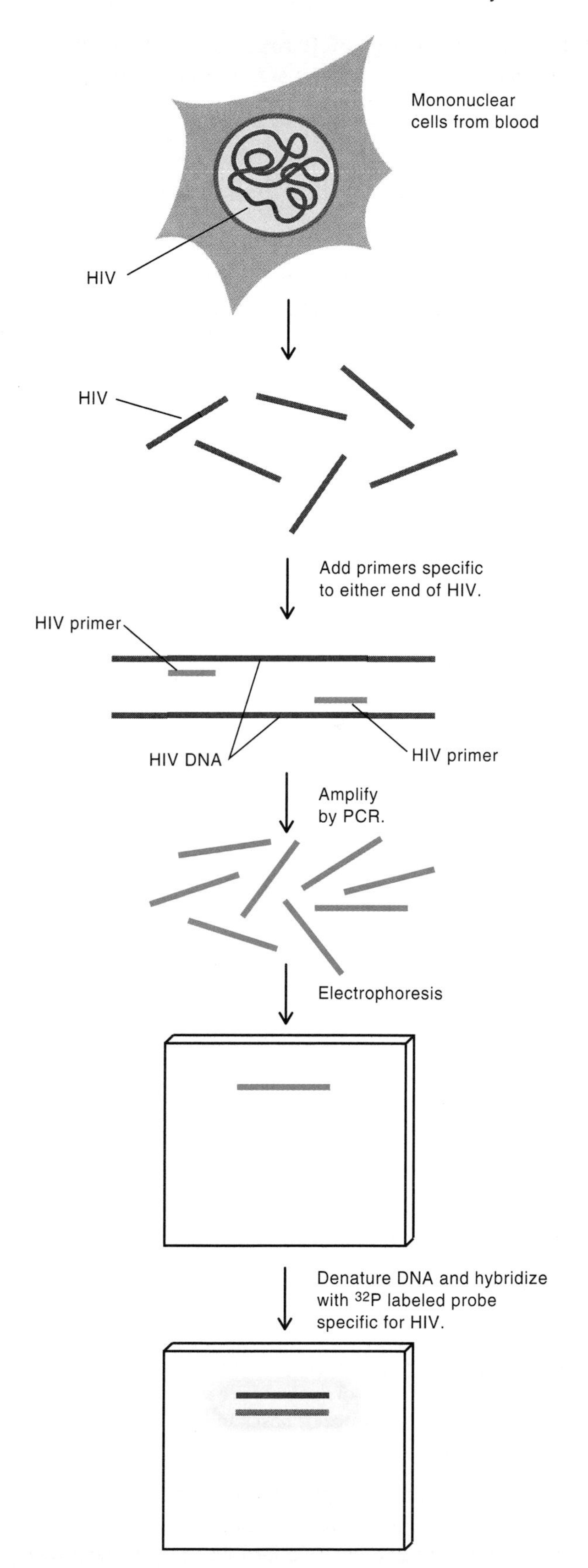

Figure 10.15
Identification of HIV provirus present in small amounts. This same technique can be used to identify any organism present in very small amounts in any environment.

to HIV, and allow the PCR reaction to amplify the sequence. The pieces of various sizes of DNA are then separated on a gel by electrophoresis, and the DNA is hybridized to a radioactive DNA fragment (probe) homologous to the amplified DNA. If HIV DNA is present in only one mononuclear cell, it will be amplified to the level that it can be readily detected.

Similar approaches are being developed to detect very small numbers of bacterial pathogens in food products.

The commonly used laboratory methods for identifying bacteria are summarized in table 10.6.

Table 10.6 Methods for Identifying Bacteria

Method
A. Colonial appearance
B. Microscopic analysis
1. Unstained preparations Shape, cell arrangement, inclusions, motility
2. Stained preparations Gram reaction, acid-fast properties
C. Biochemical tests
D. Growth requirements, including nutritional and oxygen requirements
E. Identification of products of metabolism
F. Radioactive probe technologies, including PCR amplification

Summary

I. Problems in Classifying Organisms
  - A. The primary purpose of any classification scheme is to provide easy identification of an organism.
  - B. The best classification schemes are based on evolutionary relatedness.
  - C. The basic taxonomic unit in bacteria is the species; two species differ from one another in several features, determined by genes.

II. Present-Day Classification Schemes
  - A. Descriptions of bacteria are contained in a reference text, *Bergey's Manual of Systematic Bacteriology*, published in four volumes.

III. Numerical Taxonomy
  - A. This technique measures the relatedness between organisms by determining how many characteristics the organisms have in common.

IV. Molecular Approaches to Taxonomy
  - A. Comparing the DNA base composition of organisms indicates whether they may be related.
    1. If the GC contents of two organisms differs by more than a few percentage points, the organisms cannot be related.
    2. Having the same GC content does not necessarily mean that two organisms are related.
  - B. Comparing sequence of bases in DNA determines the relatedness of organisms.
    1. Nucleic acid hybridization is the most widely used and accurate way to determine how organisms are related to one another.
  - C. Amino acid sequence similarities in the same protein can indicate relatedness between organisms.
    1. Only a minute portion of the total genetic information of the cell is measured by this method.
  - D. Determining the base sequence in 16S ribosomal RNA reveals evolutionary relationships between diverse groups of bacteria.
    1. This technique allows evolutionary relations to be seen between organisms that are not seen by nucleic acid hybridization.
    2. Certain sequences in 16S ribosomal RNA are always found in one of the three groups of organisms and identify an organism as belonging to the Eubacteria, Archaea, or Eukarya.

V. Methods of Identifying Bacteria
  - A. The more tests that are run, the more accurate the identification.
  - B. There are four major approaches: direct techniques, culture techniques, detection of microbial by-products, and molecular biological techniques.
  - C. Direct methods of identification involve microscopic examination of specimens without culturing the organisms.
    1. Wet mounts are made.
    2. Stains, especially gram stains, are useful if enough organisms are present.
    3. Many harmful bacteria cannot be distinguished from nonharmful bacteria by their appearance.
  - D. Culture techniques involve growing the organisms in pure culture.
    1. The choice of the identification method depends partly on the symptoms of disease.
    2. Culturing specimens is generally necessary because infectious agents are present in numbers too small to identify visually.
    3. Isolated organisms can be precisely identified and tests for antimicrobial susceptibility can be done.
  - E. Detecting products of metabolism may sometimes be easier and faster than culturing the organisms.

1. By-products can be measured by assaying for breakdown products of labeled energy sources.
2. This technique allows faster detection and therefore faster treatment.

F. Molecular approaches in diagnostic microbiology usually involve analysis of DNA.
1. Nucleic acid hybridization between the DNAs of two organisms—one known, the other unknown—will indicate whether they are related.
2. Comparison of the cleavage pattern of two DNA samples—one from a known, another from an unknown organism—by the same restriction enzyme will indicate whether they are related.
3. DNA probes can be used to identify species of bacteria in their natural environment or after culture.

G. Microorganisms that cannot be cultivated can still be identified.
1. One technique is to amplify by PCR a piece of DNA that is specific to one of the major cell domains (Eubacteria, Archaea, or Eukarya) and compare the sequence with the signature sequences from previously identified organisms.
2. PCR technology can be used to detect the presence of a single organism or virus particle.

H. Infectious agents present at extremely low levels can be demonstrated.

Chapter Review Questions

1. List two good reasons to directly examine specimens to identify disease-causing agents.
2. List two reasons that direct examination of specimens may not be successful.
3. What stain is most frequently used in diagnostic microbiology?
4. List two reasons for using DNA probes to detect and identify an organism.
5. What information must be available if a particular fragment is to be amplified using the polymerase chain reaction?
6. List three ways in which members of the Archaea differ from members of the Eubacteria.
7. Give two reasons 16S ribosomal RNA is a good choice for determining relatedness of organisms.
8. Give a reason that it is important to identify the causative agent of a disease as quickly as possible.
9. Define the term *signature sequence.*
10. List two end products that can serve as indicators that a particular carbohydrate has been metabolized.

Critical Thinking Questions

1. If a microbiologist suspects that a slow-growing organism is the cause of a disease and he or she wants to start antibiotic therapy as rapidly as possible, what approaches to identification of the organism can be used?
2. The GC content of an organism is frequently determined by measuring the temperature at which its DNA becomes denatured. Explain the basis of this measurement.
3. Why is the term *Archaea* preferred to *Archaebacteria?*

Further Reading

de Duve, C. 1996. The birth of complex cells. *Scientific American* 274(4):50–57.

Dyer, B. D. and Obar, R. A. 1994. *Tracing the History of Eukaryotic Cells: The Enigmatic Smile.* Columbia University Press, New York.

Holt, J. G. (ed). 1984–1989. *Bergey's Manual of Systematic Bacteriology,* 4 vols. Williams & Wilkins, Baltimore, MD.

Johnson, J. L. 1984. Nucleic acids in bacterial classification. In J. G. Holt and N. R. Krieg (eds.), *Bergey's Manual of Systematic Bacteriology,* vol. I, pp. 8–11. Williams & Wilkins, Baltimore, MD.

Jones, D. 1984. Genetic methods. In J. G. Holt and N. R. Krieg (eds.), *Bergey's Manual of Systematic Bacteriology,* vol. I, pp. 12–14. Williams & Wilkins, Baltimore, MD.

Knoll, A. 1991. End of the proterozoic eon. *Scientific American* 265(4):64–73.

Sneath, P. H. A. 1984. Numerical taxonomy. In J. G. Holt and N. R. Krieg (eds.), *Bergey's Manual of Systematic Bacteriology,* vol. I, pp. 5–7. Williams & Wilkins, Baltimore, MD.

Woese, C. R. 1987. Bacterial evolution. *Microbiological Reviews* 51(2):221–271.

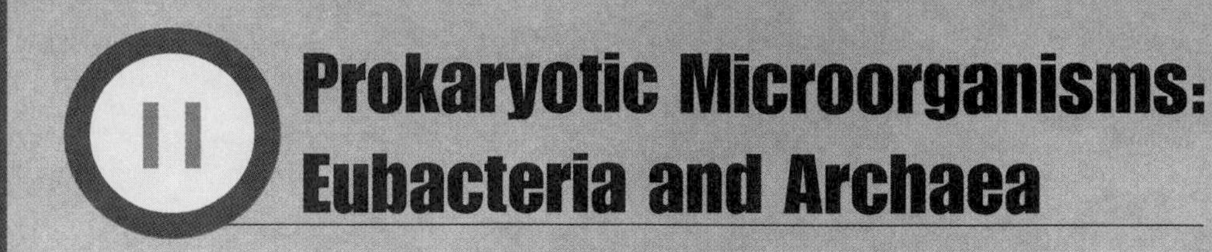

11 Prokaryotic Microorganisms: Eubacteria and Archaea

KEY CONCEPTS

1. Prokaryotes show extreme diversity in appearance, growth requirements, and metabolism.
2. Some prokaryotes prey on other prokaryotes.
3. Some prokaryotes can multiply only as symbionts within eukaryotic cells.
4. Prokaryotes are vitally important in the recycling of carbon, nitrogen, and sulfur.
5. Although they have a prokaryotic cell structure, archaea are probably as closely related to eukaryotes as to eubacteria.

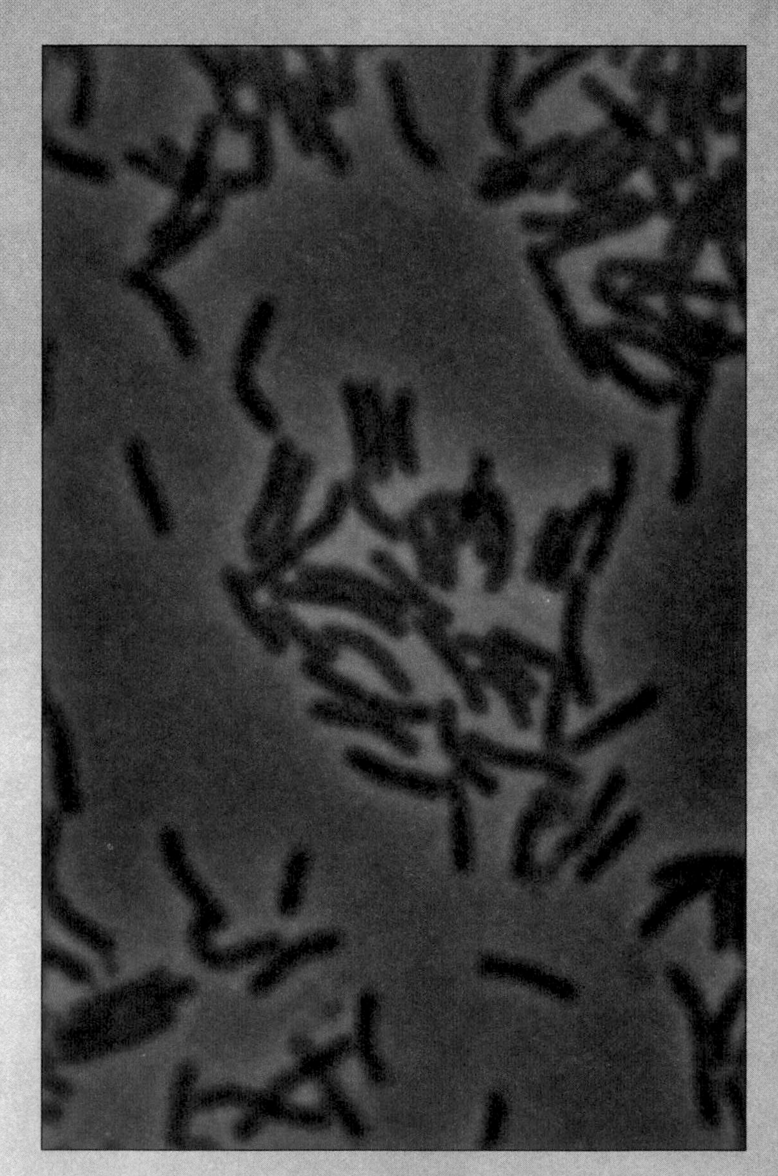

Methanobacterium species belong to the Archaea and are only remotely related to bacteria. Organisms of this genus reduce carbon dioxide to methane. These strict anaerobes are widely distributed, living in habitats as diverse as animal intestines and rice paddies.

PREVIEW LINK

Microbes in Motion

The following books and chapters in the *Microbes in Motion* CD-ROM may serve as a useful preview or supplement to your reading:

Bacterial Structure. Miscellaneous Bacteria: Zoonoses. *Gram Negative Organisms. Gram Positive Organisms.*

A Glimpse of History

Hans Christian Joachim Gram (1853–1938) was a Danish physician who developed a new method for staining bacteria. Strangely, his process stained some bacteria red and others blue. This still widely used staining process, now known as the **Gram stain,** efficiently identifies two large, distinct groups of bacteria, gram-positive (blue-staining) and gram-negative (red-staining) bacteria.

Christian Gram developed this process while working in Carl Friedlander's laboratory at the morgue of the City Hospital in Berlin. Friedlander was trying to identify the cause of pneumonia by studying patients who had died of it. Gram's task was to stain the infected tissue to make the organism easier to see under the microscope. However, his dyes revealed that two different kinds of organisms were causing pneumonia, one organism that stained blue and one that stained red.

For a long time, historians thought that Gram did not appreciate the significance of his discovery. In more recent years, however, several letters show that Gram did not want to offend the famous Dr. Friedlander under whom he worked, and therefore he played down the importance of his staining method. In fact, the Gram stain has been used as a key test in the initial identification of bacterial species ever since the 1880s. As discussed in chapter 3, the staining method separates eubacteria into two distinct groups, depending on the nature of their cell walls.

Prokaryotes inhabit virtually every ecological niche capable of supporting life. There are over 400 recognized genera of prokaryotes and hundreds of species not yet classified even though they have been grown in pure culture. The editors of *Bergey's Manual of Systematic Bacteriology* have proposed that all the prokaryotes together comprise a kingdom, the *Prokaryotae,* made up of four divisions based on cell wall type. These four divisions are

1. organisms with a gram-negative type of cell wall;
2. organisms with a gram-positive type of cell wall;
3. organisms lacking cell walls; and
4. organisms whose cell walls lack peptidoglycan.

Prokaryotes can also be grouped according to their genetic relatedness, and this has provided insight into how these microscopic creatures have evolved over millions of years. The genetic studies indicate that 11 kingdoms of prokaryotes exist among organisms of the first three divisions of *Bergey's Manual* alone!

Evidence from comparing their ribosomal nucleic acids shows that organisms of the fourth division are only distantly related to other prokaryotes. They appear to be a distinct life form equal in taxonomic rank to the eubacteria and eukaryotes. Since their discovery in 1977, these organisms have generally been called *archaebacteria* or *archaeobacteria*. Now many scientists prefer to call them the **archaea,** reserving the term *bacteria* for the eubacteria.

Bergey's Manual of Systematic Bacteriology is organized for convenience into 33 "sections," grouping prokaryotes according to shape, staining, metabolism and other easily determined characteristics.

The purpose of this chapter is to give a brief overview of the diversity of the prokaryotes by presenting features of some of these sections. We will discuss the sections in the order of the four divisions listed, starting with the prokaryotes that have a gram-negative type of cell wall and ending with those whose cell walls lack peptidoglycan. It should be recognized that a given section may contain totally unrelated species.

Prokaryotes with a Gram-Negative Type of Cell Wall

The "prokaryotes with a gram-negative type of cell wall" division is based on the electron microscopic appearance of the cell wall and not necessarily on its chemical composition or the Gram staining of the organism.

The Spirochetes Have a Distinctive Shape and Unusual Mechanism of Motility

The spirochetes comprise the first section in *Bergey's Manual* and are distinguished from other prokaryotes by their spiral shape, a flexible cell wall, and a unique motility mechanism. Movement of spirochetes occurs by means of structures called **axial filaments** (figure 11.1). Electron microscopic study of some of these organisms has revealed that each axial filament is actually made up of fibrils identical in structure to flagella, originating near each end of the organism. The fibrils extend toward each other between two layers constituting the cell wall and apparently overlap in the midregion of the cell. The layer of cell wall material covering the axial filament tends to separate easily in laboratory studies and is often referred to as a *sheath,* although it appears to be a typical gram-negative outer membrane. The molecular basis for movement with axial filaments is thought to be the same as for bacterial flagella.

The smaller spirochetes are very slender and therefore not easy to see by the usual microscopic methods, so that dark-field microscopy must be used (figure 11.2). Many are also difficult or impossible to cultivate, and their classification is based largely on their morphology and ability to cause disease. Some species are widespread in aquatic habitats, and others, such as members of the genus *Borrelia,* parasitize lice, ticks, and warm-blooded animals.

• **Gram stain, pp. 51–52**

• **Gram-negative cell wall type, pp. 59–60**

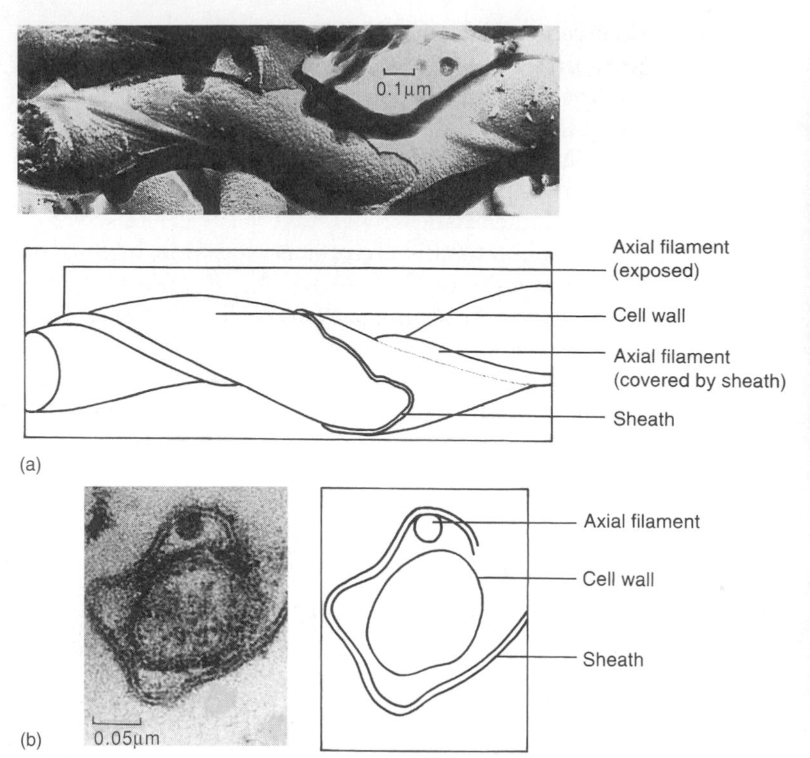

Figure 11.1
Electron micrographs showing the axial filament of a spirochete of the genus *Leptospira*. (*a*) Longitudinal view. (*b*) Cross section.

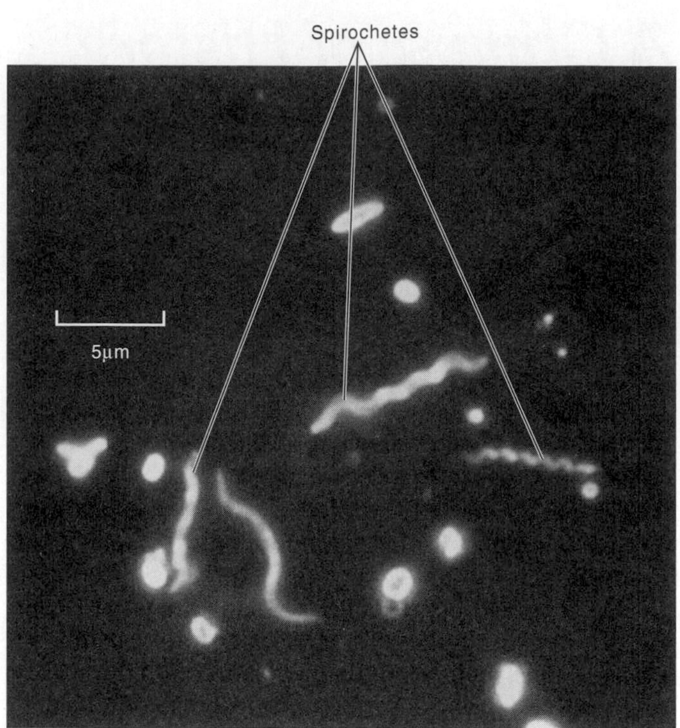

Figure 11.2
Dark-field photomicrograph of spirochetes from a monkey's mouth.

Table 11.1 Some Medically Important Spirochetes

Bacterium	Characteristics	Disease
*Treponema pallidum**	Slender, flexible, tightly coiled spirals best seen with dark-field illumination; has not been cultivated on cell-free laboratory media	Syphilis, a sexually transmitted disease that can generalize (spread throughout the body)
Borrelia burgdorferi†	Loose spirals; infects certain species of ticks; cultivable in vitro	Lyme disease; rash followed by generalized disease; acquired from tick bite
Leptospira interrogans‡	Long, tightly coiled spiral with hooked ends; lives in kidneys of animals; excreted in urine; cultivable in vitro	Leptospirosis; generalized disease transmitted by contaminated water

**Treponema pallidum*—from *trepo* for "turn" and *nema* for "a thread"and from *pallidum* for "pale"; a faintly visible turning thread
†*Borrelia burgdorferi*—named after A. Borrel and W. Burgdorfer
‡*Leptospira interrogans*—from *lepto* for "thin" and *spira* for "coil" and from *interrogans* for "interrogations"; a thin coil shaped like a question mark

Examples of medically important spirochetes are given in table 11.1. Some free-living species (genus *Spirochaeta*)[1] are extraordinarily long (figure 11.3), up to 500 µm in length. Spirochetes were among the organisms van Leeuwenhoek observed in saliva in 1683.

Not all spiral bacteria are spirochetes, since representatives of other groups may also have a spiral shape.

[1]*Spirochaeta*—from the Greek *spira* for "coil" and *chaete* for "hair"; a coiled hair

The Aerobic Motile, Helical, or Comma-Shaped Gram-Negative Bacteria Have Rigid Cell Walls and Flagella

The bacteria in this section, although curved, differ from spirochetes in possessing the nonflexible cell wall typical of most gram-negative eubacteria. Also, their movement occurs by means of one or more flagella located at one or both poles. Like the spirochetes, most are saprophytes residing in natural waters. Some species, including *Spirillum minus* and

Table 11.2 Some Medically Important Helical and Comma-Shaped Gram-Negative Aerobes

Bacterium	Characteristics	Disease
*Campylobacter jejuni**	Tiny curved rod, actively motile by polar flagella; requires 3% to 5% O_2 for growth; grows well at 42°C	Diarrhea and dysentery; from domestic animals
Spirillum minus†	Loose spiral with flagella at each end; not consistently cultivable in vitro; due for a name change	Rat bite fever; transmitted by bites of mice and rats or the animals that feed on them

**Campylobacter jejuni*—from *campylo* for "curved" and *bakter* for "rod"and from *jejuni* for "the mid portion of the small intestine"; a curved rod of the jejunum
†*Spirillum minus*—from *spirillum* for "small spiral" and *minus* for "small"; a small member of the genus *Spirillium* (differs from other members of the genus)

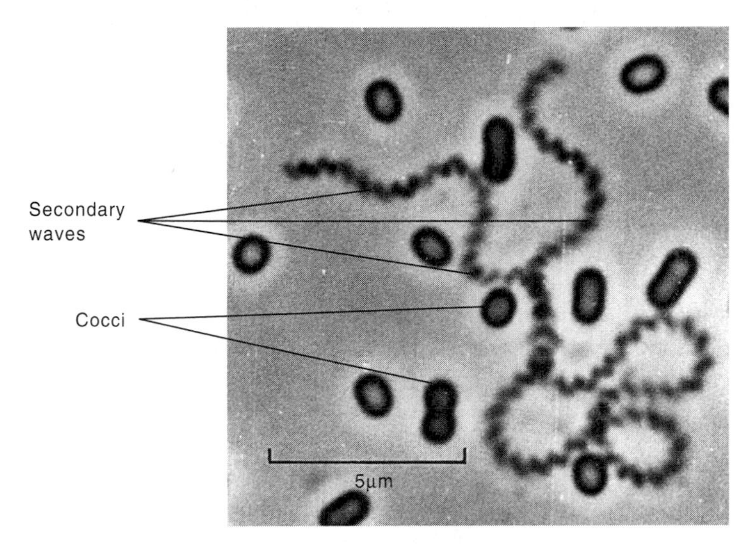

Figure 11.3
Phase contrast photomicrograph of a very long free-living spirochete showing secondary waves. Note the size relative to the cocci also present in the photograph.

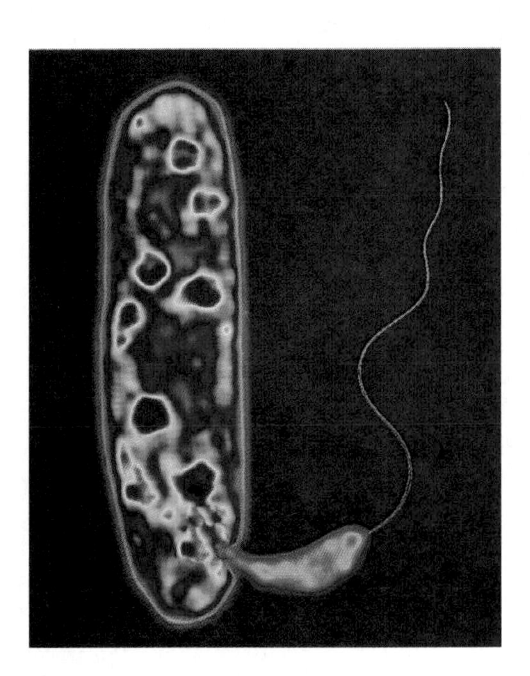

Figure 11.4
Color-enhanced transmission micrograph of *Bdellovibrio bacteriovorus*. The orange structure with the long flagellum is the bdellovibrio. Note its small size relative to its prey, the bacillus to which it is attached.

Campylobacter jejuni, cause disease in humans (table 11.2). All the bacteria in this group are aerobic or microaerophilic. *C. jejuni* is an example of a microaerophilic organism.

Bdellovibrios

One interesting and unusual aerobic group, the **bdellovibrios,** prey on other bacteria. In many instances, the bdellovibrio requires a specific host bacterium as its only food source. Bdellovibrios are unique in appearance—tiny gram-negative curved rods possessing a long, sheathed flagellum at one end of the cell. The generic name *Bdellovibrio* describes the organism's behavior and shape. It attaches to other bacterial cells (*bdello,* from the Greek for "leech") and has a comma shape like bacteria of the genus *Vibrio. Bdellovibrio bacteriovorus* (figure 11.4), the most intensively studied species of this group, is approximately 0.25 μm wide and 1 μm long. *Bdellovibrio* sp. attack the host cell by striking it at high velocity so that their nonflagellated end attaches to the host cell. Next, bdellovibrio penetrates the host cell wall through a hole made by the use of digestive enzymes and a spinning motion. The flagellum is left behind. The bdellovibrio then lodges between the host's cell wall and cytoplasmic membrane, growing in length at the expense of the cellular contents of the host bacterium, which is killed. Bdellovibrios utilize some components of the host after they are only partially degraded, thereby saving the energy that would be needed for complete resynthesis. After division, the bdellovibrio's progeny become motile and leave to find new hosts, repeating the growth and reproduction cycle (figure 11.5). Some strains of *Bdellovibrio* are able to grow heterotrophically in the absence of host cells.

• **Microaerophilic organisms, p. 92**

One surprising fact about *Bdellovibrio* is that it was not discovered until 1962, even though bdellovibrios can be easily isolated from soil, sewage, and natural waters. In contrast, viruses, which are not visible with the light microscope, were discovered in 1892, and rickettsias, which cannot be cultivated on cell-free media, were discovered in 1901.

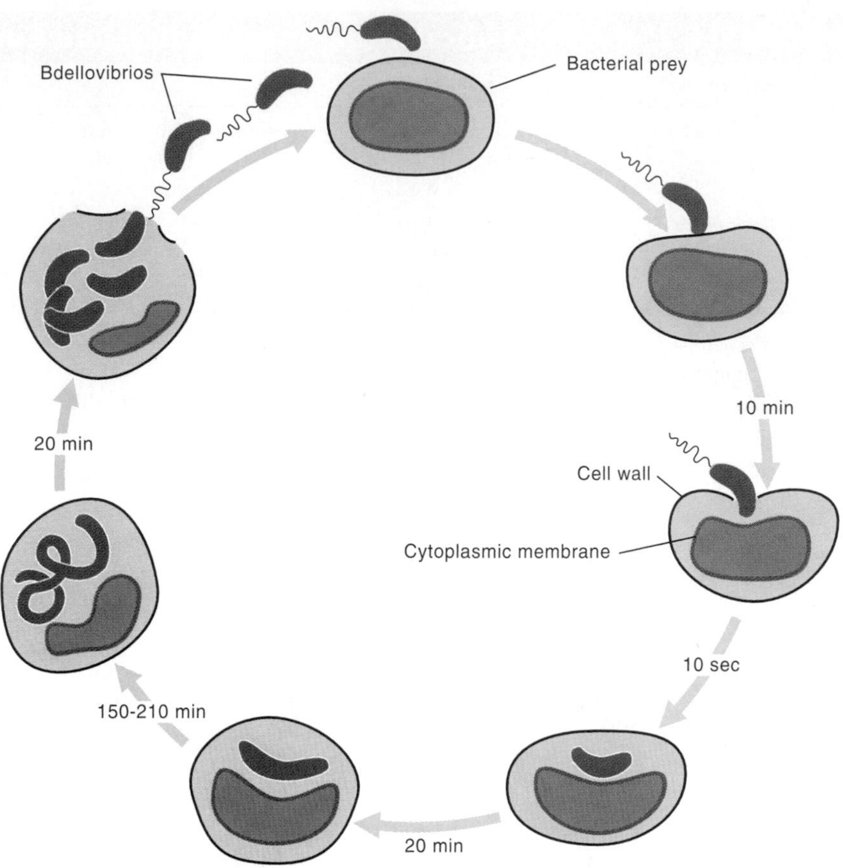

Figure 11.5
Life cycle of *Bdellovibrio bacteriovorus.* Note that the bdellovibrio multiplies in the space, much exaggerated here, between the cell wall and the cytoplasmic membrane of the bacterial cell upon which it preys.

Gram-Negative Aerobic Rods and Cocci Represent a Large and Diverse Group of Prokaryotes

One of the most important genera in this section is *Rhizobium,* members of which can live within the root cells of certain plants and fix atmospheric nitrogen (N_2) in a form (NH_4^+) that can be utilized by plants. Two other subgroups, the pseudomonads and the azotobacteria, are discussed briefly on the following page, and some medically important species are described in table 11.3.

Pseudomonads

The pseudomonads comprise members of the family *Pseudomonadaceae,* gram-negative rods that are motile by polar flagella and that often produce pigments (figure 11.6). Although most pseudomonads are strict aerobes, a few can be grown anaerobically in the presence of nitrate, which substitutes for oxygen as the final electron acceptor, in an example of anaerobic respiration. Many species of pseudomonads are free-living and harmless, whereas others can produce disease in plants and animals. There are a large number of genera in the family *Pseudomonadaceae.*

• **Anaerobic respiration, p. 122**

Table 11.3 Some Medically Important Gram-Negative Aerobes

Bacterium	Characteristics	Disease
Pseudomonas aeruginosa	Produces pigments that give the medium a green color; produces potent toxins	Burn, urinary, and bloodstream infections; often transmitted by aqueous solutions in which it can grow without obvious nutrients
Legionella pneumophila	Requires a medium containing increased iron and cysteine; organisms in infected tissue stain poorly; has facultative intracellular growth	Legionnaires' disease, mainly a lung infection; transmitted by contaminated water, in which it grows within protozoa
Neisseria gonorrhoeae	Diplococcus; colonies give positive oxidase test	Gonorrhea, a sexually transmitted disease of the urinary and genital tracts that can generalize
Bordetella pertussis	Tiny nonmotile coccobacilli; cultivable on medium heavily enriched with blood	Pertussis (whooping cough), a serious respiratory disease preventable by immunization
Francisella tularensis	Tiny nonmotile coccobacilli; special medium required	Tularemia; transmitted by skinning wild animals or by insect bites

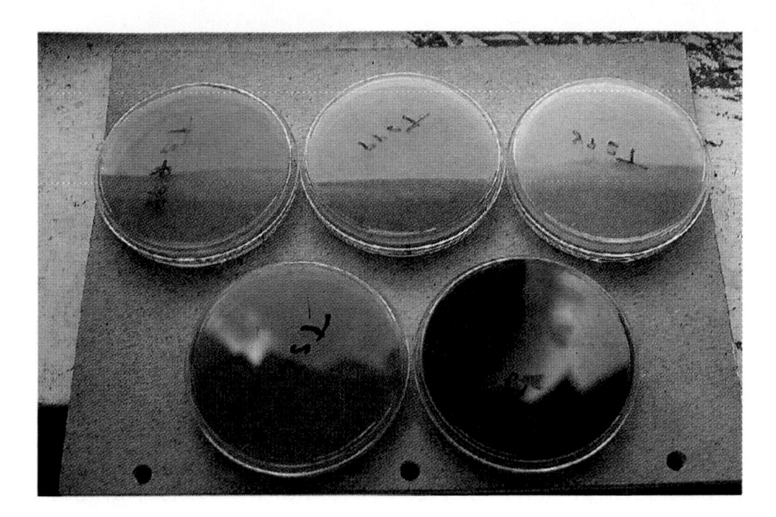

Figure 11.6
Cultures from different strains of *Pseudomonas aeruginosa* from infected people. Note the differing colors produced by water-soluble pigments.

Burkholderia and *Pseudomonas* are among the most important medically. The genus *Pseudomonas* contains species with remarkably diverse biochemical capabilities. Compounds used as nutrients by members of this genus include unusual sugars, amino acids, and far more complex compounds, such as those containing aromatic rings. Thus, *Pseudomonas* species play an important role in the degradation of many synthetic and natural compounds that resist breakdown by most other microorganisms. The ability to carry out some of these degradations is encoded by plasmids.

Azotobacteria

The azotobacteria (members of the genus *Azotobacter*) are unique in being able to fix atmospheric nitrogen under aerobic conditions in the absence of plants. **Nitrogenase,** the enzyme needed for nitrogen fixation, is ordinarily sensitive to traces of oxygen. However, azotobacteria have exceedingly high respiratory rates, and their metabolism apparently consumes oxygen so rapidly that an anaerobic environment is produced inside the cell and the nitrogenase enzyme is not inhibited.

Members of the genus *Azotobacter* live in alkaline soil. They represent one of the few groups of bacteria able to form a type of resting cell called a **cyst.** Each cyst is produced through division and shortening of a vegetative cell, followed by the elaboration of a thick protective wall that surrounds the cell (figure 11.7).

Cysts can withstand drying and ultraviolet radiation but are not highly resistant to heat; they differ from endospores in both their formation and their lesser degree of resistance to harmful agents. Furthermore, whereas endospore formation consists of a series of steps involving

• **Plasmids, pp. 174–177**

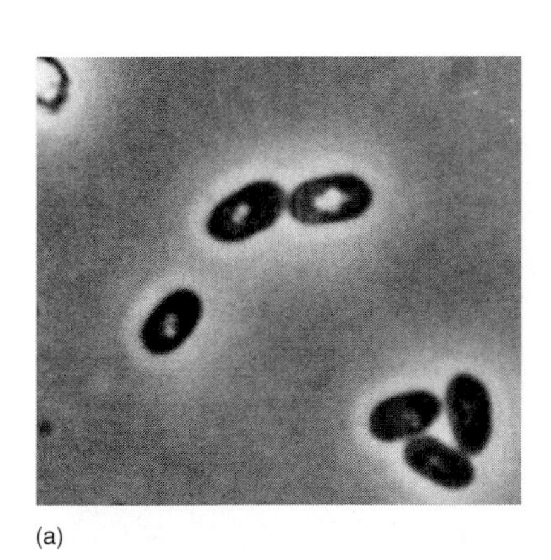

(a)

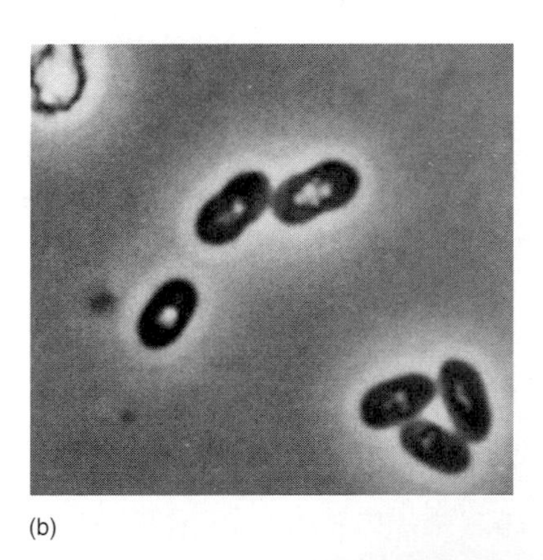

(b)

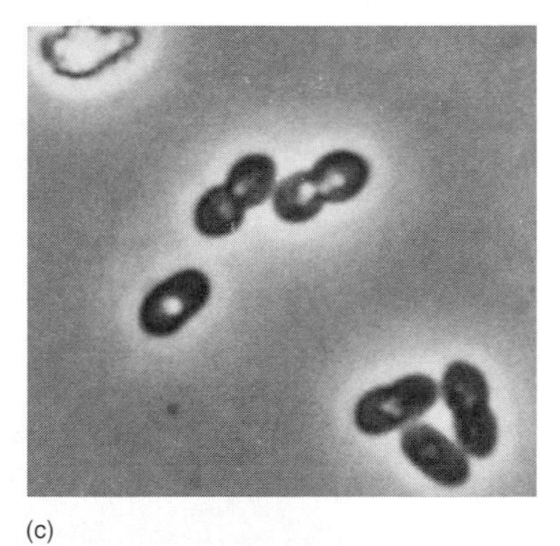

(c)

(d)

Cyst

5µm

Figure 11.7
Development of cysts of *Azotobacter,* (*a*) 30 minutes, (*b*) 2 1/2 hours, (*c*) 3 1/2 hours, and (*d*) 8 hours after inoculation.

the regulation of the expression of a large number of genes, the formation of *Azotobacter* cysts primarily concerns changes in the cell wall. The enzyme content of the cyst is similar to that of the cell from which it arises.

REVIEW QUESTIONS

1. What unique motility structure characterizes the spirochetes?
2. What illumination technique is especially useful in visualizing small spirochetes?
3. Give three locations where spirochetes might be found.
4. Do microaerophilic bacteria grow under strictly anaerobic conditions? Explain.
5. What is the name of a group of bacteria that preys on other bacteria?
6. Give two ways that bacteria in the aerobic, motile, helical, or comma-shaped group differ from spirochetes.
7. What important function do some members of the genus *Rhizobium* perform for plants?
8. What important function do members of the genus *Pseudomonas* perform?
9. What is unique about nitrogen fixation carried out by *Azotobacter* species?

Facultatively Anaerobic Gram-Negative Rods Are Represented by Enterobacteria and Vibrios

During anaerobic growth, facultative anaerobes gain their energy by a series of fermentation reactions. These bacteria are very widespread and diverse biochemically. Although most are harmless, some cause disease in plants and animals. Examples of medically important species are given in table 11.4.

Enterobacteria

The enterobacteria are so named because the characteristic species of the group, *Escherichia coli*, inhabits the intestine of humans and other animals (Greek *enteron* means "intestine"). However, some species of enterobacter live in the natural environment independent of animals. There are more than 25 recognized genera of enterobacteria; about half of them include medically important species. Sometimes hundreds of different strains can be distinguished within some species because of differences in their cell walls, flagella, and capsules (figure 11.8). Enterobacteria have peritrichous flagella, a feature important in distinguishing them from many other facultative anaerobes.

Vibrios

Vibrios and related bacteria constitute another large group of gram-negative facultative anaerobes. Some cause diseases in animals and human beings. Both curved and straight rods occur within this group, and some species are luminescent (figure 11.9). The latter are represented by certain species of *Vibrio* and *Photobacterium* that are salt-requiring marine organisms and that produce light in the presence of oxygen. The light-producing reaction is mediated by the enzyme

• **Fermentation, pp. 116–120**

Table 11.4 Some Medically Important Gram-Negative Facultative Anaerobes

Species	Characteristics	Disease
Enterobacteria		
Escherichia coli	Hundreds of different strains; plasmids often influence virulence	Newborn meningitis, diarrhea, urinary infections
Klebsiella pneumoniae	Nonmotile; large capsules	Pneumonia
Proteus mirabilis	Swarms across medium; urea metabolized to ammonia	Urinary infections
Salmonella typhi	Hydrogen sulfide producer; only one flagellar type	Typhoid fever; human source
Salmonella enteritidis	Hundreds of strains based on flagellar type	Gastroenteritis, "food poisoning"; animal source
Serratia marcescens	Red-pigmented colonies	Urinary and bloodstream infections
Yersinia pestis	Nonmotile; small colonies; infects rodents and their fleas	Bubonic plague; transmission by fleas
Vibrios		
*Vibrio cholerae**	Requires special high-salt medium	Cholera, a toxin-mediated small intestine disease
Vibrio parahaemolyticus	Requires special high-salt medium	Gastroenteritis; cramps and diarrhea last a day or less; from seafood

**Vibrio*—meaning to move rapidly to and fro

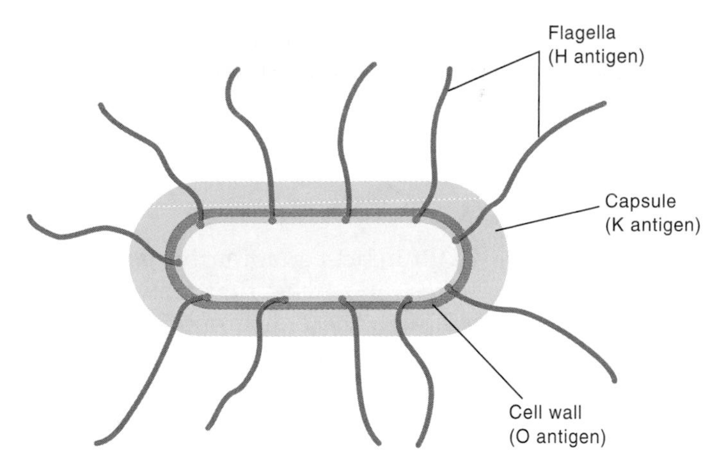

Figure 11.8
Schematic drawing of a typical enterobacterium.

bacterial luciferase. Some luminescent bacteria live in specialized pouches on certain species of fish and are their source of luminescence.

Most Fermentative Anaerobic Gram-Negative Straight, Curved, and Helical Rods Inhabit Animal Digestive Tracts

This section includes fermenting gram-negative species that are straight, curved, pleomorphic, or helical in shape. All are strict anaerobes and some are killed by brief exposure to oxygen. For the most part, these organisms inhabit body cavities such as the mouth and the intestinal tract of animals, including humans and insects. Because they require anaerobic growth conditions, they are more difficult to study than many other types of bacteria. Analysis of the acid products of metabolism by means of a device called a *gas chromatograph* has proved helpful in these species' identification. Table 11.5 shows the characteristics of some of the principal genera. The genus *Bacteroides*[1] is especially important medically because bacteroids can cause abscesses and bloodstream infections if introduced into body tissues by surgery or other trauma.

The Dissimilatory Sulfate- or Sulfur-Reducing Bacteria Generally Live in Mud Rich in Organic Material and Oxidized Sulfur Compounds

Dissimilatory sulfate- or sulfur-reducing bacteria utilize sulfur or oxidized sulfur compounds as a final election acceptor, thereby reducing these materials. The term *dissimilatory* was coined to contrast this process of sulfur utilization with *assimilatory* usage, wherein sulfur compounds are taken in as a nutrient, reduced by a different process, and incorporated mainly into the —SH groups of amino acids.

[1]*Bacteroides*—from *bakterum* for "rod" and *idus* for "shape"

(a)

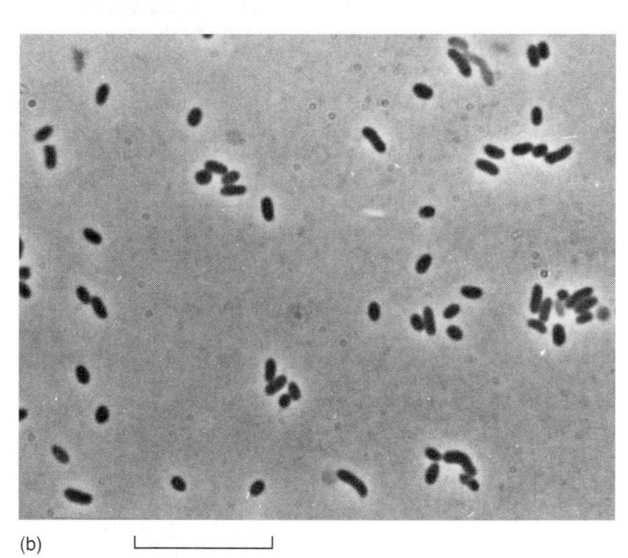

(b)

Figure 11.9
Photobacterium sp. (*a*) Photograph taken using light produced by the bacteria themselves. (*b*) Phase contrast photomicrograph of cells.

During growth of dissimilatory sulfur- or sulfate-reducing bacteria, oxidized sulfur receives the electrons generated by metabolism, and, as a result, large amounts of hydrogen sulfide are often produced. The H_2S is responsible for a foul odor and causes mud and water to turn black when it reacts with iron molecules. At least a dozen genera are recognized in this group, the most extensively studied of which is *Desulfovibrio*. Their important role in the sulfur cycle is discussed in chapter 32. A number of thermophilic archaea also reduce sulfur or sulfate.

• **Sulfur cycle, p. 737**

Table 11.5 Characterization of Anaerobic Gram-Negative Bacteria by Acid Products and Flagellation

Characteristics	Genus
Flagella peritrichous or absent	
Mainly butyric acid produced	*Fusobacterium**
Lactic acid sole acid produced	*Leptotrichia*†
Butyric, isobuteric, isovaleric, and succinic acids produced	*Bacteroides*
Polar flagella	
Butyric acid produced	*Butyrivibrio*
Succinic acid produced	
spiral shape	*Succinivibrio*
oval shape	*Succinimonas*
Flagellar tuft on side of crescent-shaped cells	*Selenomonas*‡

**Fusobacterium*—from *fusus* for "a spindle" and *bakterion* for "small rod"
†*Leptotrichia*—from *leptus* for "fine" and *thrix* for "hair"
‡*Selenomonas*—from *selene* for "moon" and *monas* "unit"

Most Rickettsias and Chlamydias Are Obligate Intracellular Parasites

Two orders of gram-negative bacteria, *Rickettsiales* and *Chlamydiales,* are composed of tiny, rod-shaped or coccoid (resembling cocci) organisms that are generally unable to reproduce outside a host cell. One advantage of living intracellularly is that the host cell supplies materials that the bacteria would otherwise have to synthesize for themselves. During evolution, most bacteria in this group lost the ability to synthesize substances needed for extracellular growth. However, medically important members of one genus, *Bartonella,* can be cultivated on cell-free media. Rickettsias and chlamydias are responsible for widespread and serious human diseases (table 11.6).

Rickettsias

Rickettsias (figure 11.10) infect certain arthropod species, including insects, mites, and ticks, that feed on warm-blooded animals. Although these arthropods are largely unaffected by the rickettsias, serious disease can result if they transmit the bacteria to their warm-blooded hosts. For most rickettsias, transmission from animal to animal depends on blood-sucking arthropods because rickettsias do not survive well in the environment. Rickettsias can

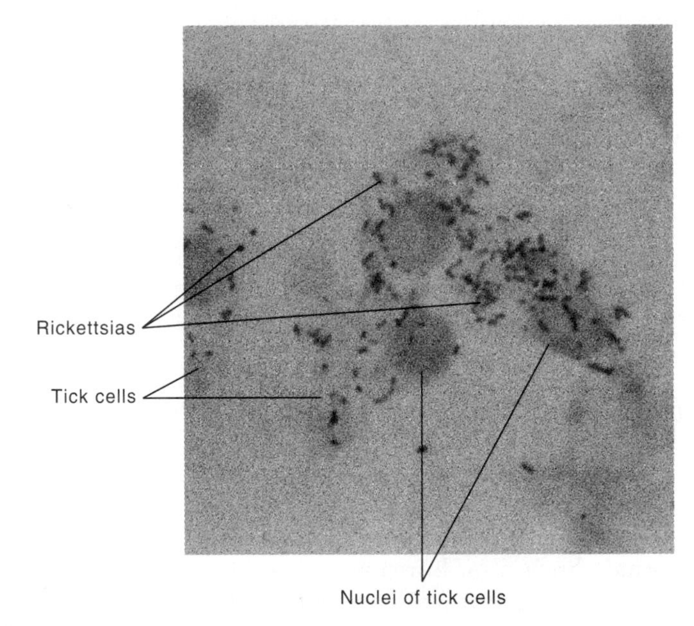

Figure 11.10
Rickettsia rickettsii in blood cells of an infected *Dermacentor andersoni* (wood tick). The tiny red objects are rickettsias.

Table 11.6 Some Medically Important Obligately Intracellular Bacteria

Bacterium	Characteristics	Disease
Rickettsia rickettsii	Short, gram-negative coccobacilli; survive poorly outside host cell	Rocky Mountain spotted fever; transmission by ticks
*Coxiella burnetii**	Form sporelike structure resistant to drying, heat, and disinfectants	Pneumonia, heart valve infections; usually acquired by inhalation
Chlamydia trachomatis	Multiple strains with differing disease potential; Glycogen-inclusion body at intracellular replication site	Trachoma, a serious eye infection; conjunctivitis of newborn; lymphogranuloma venereum; urethritis and genital tract disease; the most common sexually transmitted bacterial disease
Chlamydia pneumoniae	Pear-shaped elementary bodies; less than 10% DNA relatedness with other chlamydia	Mild pneumonia
Chlamydia psittaci	Chronically excreted by many bird species	Pneumonia (psittacosis or "parrot fever")

**Coxiella burnetii*—named after Herold R. Cox, who showed the organism could be cultivated in the chick embryo yolk sac, and Frank MacFarlane Burnet, who first described the organism's properties generally be cultivated in the laboratory in tissue cell cultures or embryonated eggs. Interestingly, rickettsias are closely related to *Agrobacterium* species, bacteria that associate with and transfer a piece of plasmid DNA into plant cells. Evidence suggests that rickettsias and plant pathogens may have evolved from a common ancestor.

One species of rickettsia, *Coxiella burnetii*, differs from the rest in being able to survive well outside the host and in being transmissible from animal to animal without necessarily involving a blood-sucking parasite. During its intracellular growth, *C. burnetii* forms endosporelike structures (figure 11.11) that allow it to survive in the environment, although the structures lack the extreme resistance to heat and disinfectants characteristic of most endospores. *C. burnetii* causes Q fever of humans. It is closely related in an evolutionary sense to *Legionella* species, facultative intracellular bacteria that parasitize protozoa and can cause Legionnaires' disease of humans.

• ***Agrobacterium*, pp. 175, 230**

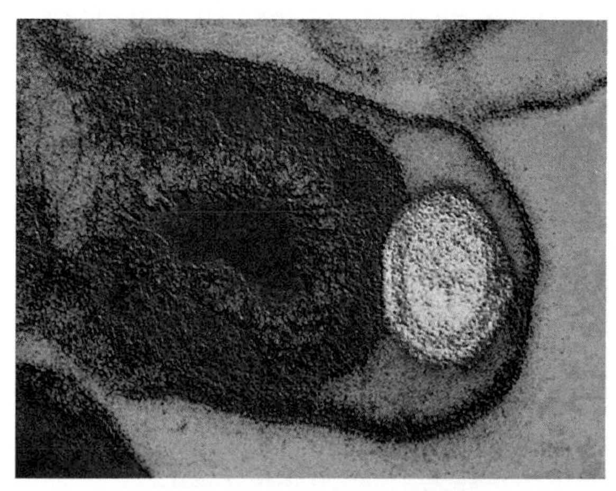

(a)

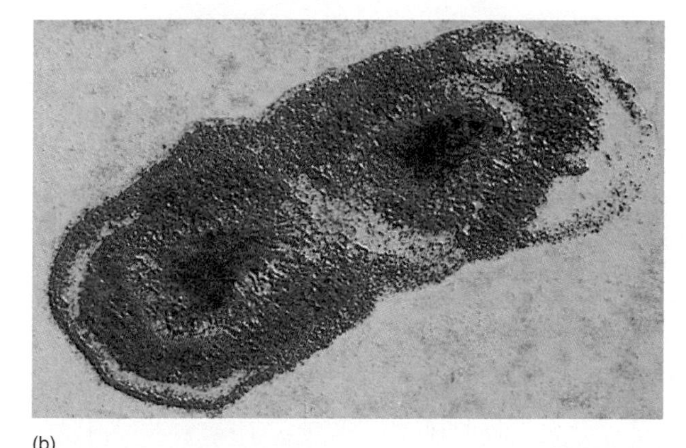

(b)

Figure 11.11
Color-enhanced transmission electron micrograph of *Coxiella burnetii*. (*a*) Endosporelike structure (oval, copper-colored object). Although not truly an endospore, this structure is probably the form of *C. burnetii* that is resistant to heat and drying. (*b*) *C. burnetii* cell lacking the endosporelike structure.

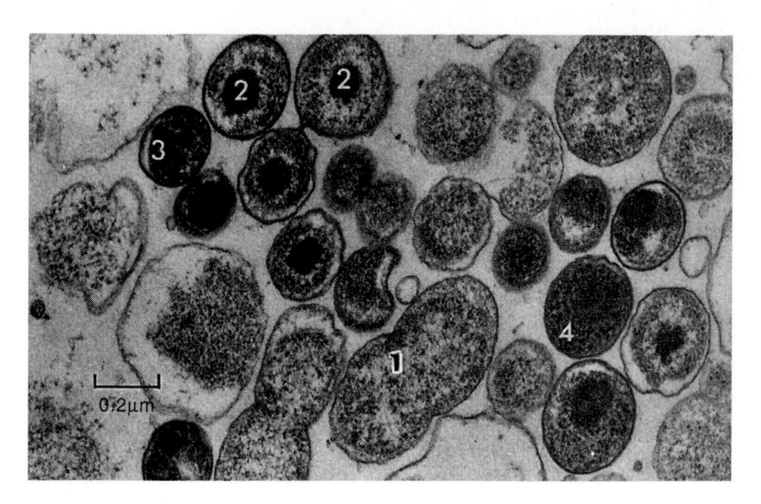

Figure 11.12
Chlamydias growing in a tissue cell culture. The numbers indicate development from a dividing form (*1*) to a mature, infectious bacterium (*4*).

Chlamydias

Chlamydias differ from rickettsias in that they are roughly spherical and do not parasitize blood-sucking arthropods. They show a unique growth cycle (figure 11.12). Inside the host cell, they initially exist as a fragile, noninfectious form that reproduces by binary fission. Later in the infection, the bacteria differentiate into a smaller, dense-appearing infectious form that is released upon death and rupture of the host cell. The cell wall of chlamydias is highly unusual among eubacteria in that it lacks peptidoglycan, although it has the general appearance of a gram-negative type of cell wall. The organisms lack an energy-generating mechanism and depend entirely on the host cell for ATP. This explains why they are obligate intracellular parasites.

REVIEW QUESTIONS

1. Why are enterobacteria important?
2. What bacterial structures help distinguish the different enterobacteria from each other?
3. How do the curved rods in the genus *Vibrio* differ from those in the aerobic, motile, helical, and comma-shaped group of bacteria?
4. What does luciferase do?
5. Why are the anaerobic gram-negative bacteria difficult to study?
6. What distinguishes the dissimilatory sulfur-reducing bacteria from other bacteria that utilize sulfur?
7. Where in nature are dissimilatory sulfur-reducing bacteria likely to be found?
8. What is an obligate intracellular bacterium?
9. How do rickettsias differ from chlamydias?

Endosymbionts Are Microorganisms that Have Established a Stable Intracellular Relationship with a Eukaryotic Host Cell

Many species of eukaryotic cells contain gram-negative bacterial endosymbionts (from *endo* for "inside" and *symbiosis* for "living together"). Generally, these endosymbionts cause no apparent harm to their host cells, and in a number of instances they are beneficial or even essential to survival of the host. In many cases, the endosymbionts cannot be cultivated outside the host cells and, when removed from them, cannot reinfect them. The endosymbionts simply multiply along with the host cell and are passed to daughter host cells when division occurs. There has been considerable speculation about what the ultimate evolutionary fate of such organisms might be. According to the **endosymbiotic theory,** eukaryotic subcellular organelles (such as mitochondria and chloroplasts) represent ancient prokaryotic endosymbionts. Studies comparing the 16S rRNA of mitochondria, bacterial endosymbionts, and other eubacteria indicate that the endosymbionts have been acquired much more recently in evolutionary history than mitochondria, and their genetic makeup is closer to that of extracellular eubacteria.

Endosymbionts of Protozoa

Prokaryotic endosymbionts of protozoa are common, but because few can be cultivated outside their host, characterization is difficult. Nevertheless, many species are recognized. Some clearly confer survival advantage to the host, as for example those that give the host the capability of killing other protozoa. In some cases, the mutually beneficial relationship changes with environmental conditions, resulting in excessive growth of the endosymbiont and death of its host. Some named and unnamed symbionts are given in table 11.7.

• **Mitochondria, chloroplasts, pp. 77, 80–81, 231**

Endosymbionts of Other Eukaryotes

Reference was made earlier in this chapter to *Rhizobium* species that grow in the root cells of some plants and fix atmospheric nitrogen that aids plant growth. Endosymbionts are also present in the cells of insects, fungi, sponges, various worms, and mollusks. Aphids become sterile or die if their endosymbionts are eliminated. One large tube worm discovered around hydrothermal vents deep in the ocean off the Galápagos Islands has no mouth or gut. Its main internal organ is a structure packed with prokaryotic endosymbionts that gain energy from hydrogen sulfide and metabolize nutrients diffusing into the worm from seawater. The worm appears to be entirely dependent on its endosymbionts. Another interesting example is *Neorickettsia helminthoeca,* an endosymbiont of certain parasitic flukes (leaf-shaped parasitic worms) that infest salmon from the Pacific Northwest. This endosymbiont readily infects dogs that feed on the fish, producing a fatal disease called "salmon poisoning." A bacterium similar to *N. helminthoeca* exists in the flukes that infest a marine fish. This bacterium has been reported to cause disease in people who eat the fish raw or partially cooked. Neorickettsias are presently grouped with the rickettsias and chlamydias. Another widespread endosymbiont, in the genus *Burkholderia* lives in large numbers in the cells of mycorrhizal fungi of land plants.

Nonphotosynthetic, Gliding Bacteria Are Widespread in the Environment and May Be Found in the Human Mouth

Certain single-celled and filamentous bacteria show slow progressive movement, termed *gliding motility,* ranging up to 600 μm per minute in liquids and across solid surfaces. Although they are capable of twisting and flexing movements, no motor organelles have been discovered on them. Although these bacteria produce a slimy material, it is not

• **Hydrothermal vents, p. 729**
• ***Burkholderia*, p. 247**
• **Mycorrhizas, pp. 288–289**

Table 11.7 Some Endosymbionts of Protozoa

Endosymbiont	Host	Cultivable?*	Role in Host
X-bacteria	*Amoeba proteus*	No	Required for survival of host
Lyticum flagellatum	*Paramecium aurelia*	No?	Produces a lysin that destroys sensitive paramecia; renders host nondependent on folic acid
Caedibacter varicaedens	*Paramecium aurelia*	No	Kills sensitive paramecia
"Bipolar bodies"	*Crithidia oncopelti*	No	Endosymbiont-free host crithidiae require nicotinamide

*Able to grow on cell-free media; one unconfirmed report of *L. flagellatum* cultivation responsible for pushing them. Possibly the same contractile material that causes these bacteria to twist and flex could produce wavelike contractions of their outer membrane, propelling them along as does the belly of a snake. Various other mechanisms have been suggested, and there may be different mechanisms in different species. The gliding bacteria in this section are not photosynthetic and produce no specialized structures for their movement.

A number of species of these gliding bacteria are able to break down complex polysaccharides such as cellulose and are therefore important degraders of organic material. Others, such as *Cytophaga columnaris,* are pathogens of fish. Pacific salmon are susceptible to heavy losses from this bacterium if water temperatures rise above 18°C.

Members of another genus, *Simonsiella,* are harmless residents of the mouths of humans and numerous other warm-blooded animals. Species of *Simonsiella* occur as short chains of flattened cells in a ribbonlike arrangement (figure 11.13). Interestingly, *Simonsiella* cells move in a direction perpendicular to the long axis of the ribbon.

Members of the genus *Saprospira* (figure 11.14) occur as long, multicellular helical filaments, measuring up to 500 μm in length. They show gliding motility and are found in both saltwater and freshwater.

Members of the genera *Beggiatoa* and *Thiothrix* (figure 11.15) are gram-negative, multicellular organisms that occur as unbranched filaments and live in sulfur springs, estuarine[1] sediments, and sewage-polluted water. These microorganisms obtain energy by oxidizing H_2S to granules of sulfur that are deposited intracellularly. Besides using H_2S, some strains can use certain organic chemicals for energy. The filaments of *Beggiatoa* strains are flexible and show gliding motility. The filaments of *Thiothrix* species are immobile and are fastened to rocks or other solid surfaces. *Thiothrix* cells exhibiting gliding motility detach from the ends of the filaments and disperse to new locations. Commonly, several such cells will then attach to a solid object in a rosette arrangement. As they multiply, these cells develop into new filaments.

[1]Estuarine—having to do with an estuary, the wide mouth of a river where it meets the sea

Myxobacteria Are Fruiting, Gliding Bacteria

Myxobacteria, such as species of *Chondromyces* and *Myxococcus,* are nonphotosynthetic, strictly aerobic gram-negative rods that employ gliding motility. The majority are unrelated to the organisms discussed in the previous section. They have a developmental cycle involving fruiting body formation, which is unique among prokaryotes (figure 11.16). At a certain time in the life of a culture of myxobacteria, which is apparently determined by exhaustion of nutrients, some of the cells begin to congregate. The chemical signal for this congregation is thought to be a guanosine polyphosphate, but the exact identity and the mechanism of signaling are not known. The congregated cells pile up, finally producing a mass of cells supported above the growth medium by a stalk. This structure, called

• **Inorganic compound oxidation, pp. 124–125**

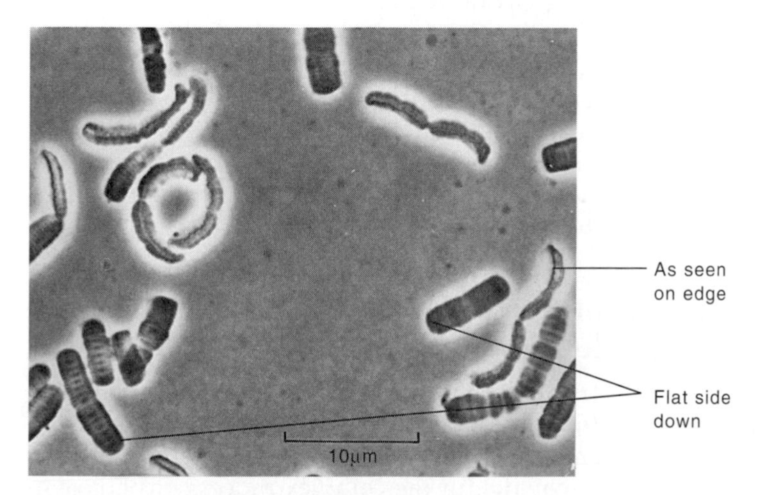

Figure 11.13
Simonsiella sp., a uniquely shaped, nonbranching, multicellular, gliding bacterium that uses organic energy sources.

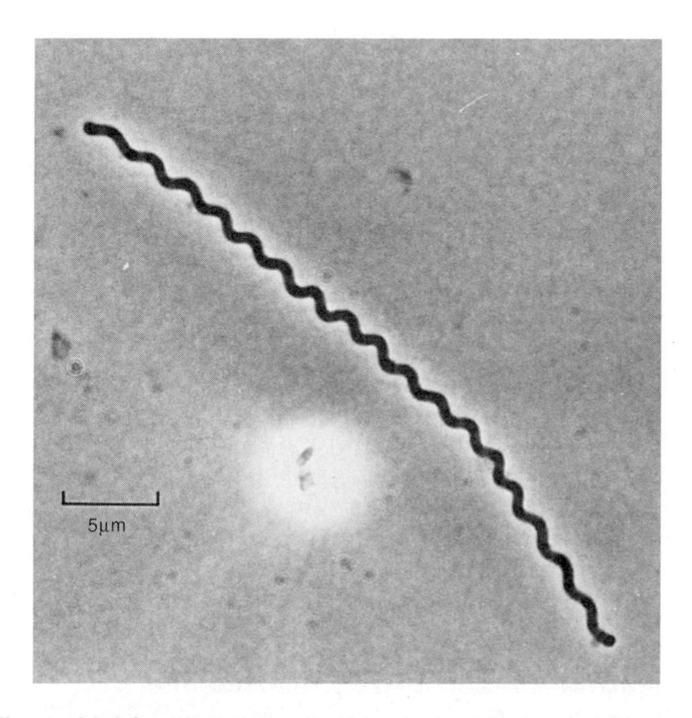

Figure 11.14
Saprospira sp. Large, multicellular, helical, gliding filaments characterize these bacteria.

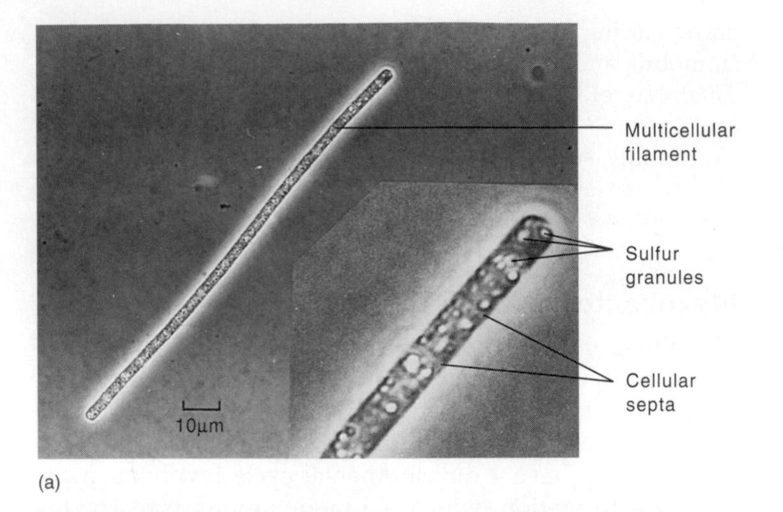

(a)

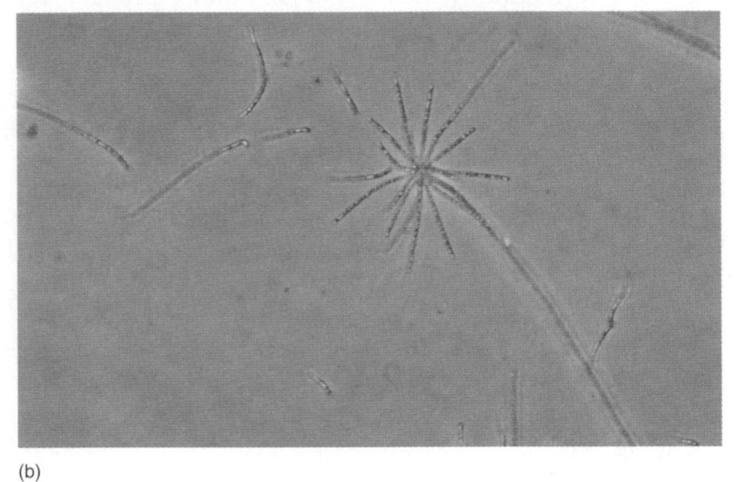

(b)

Figure 11.15
Phase contrast photomicrographs of gliding bacteria. (*a*) *Beggiatoa* sp. and (*b*) A rosette of *Thiothrix* species from a sulfur spring. As with *Beggiatoa* species, granules of elemental sulfur (bright areas) are present in the cells. Each of the multicellular filaments of *Thiothrix* is attached to the rosette by means of a holdfast.

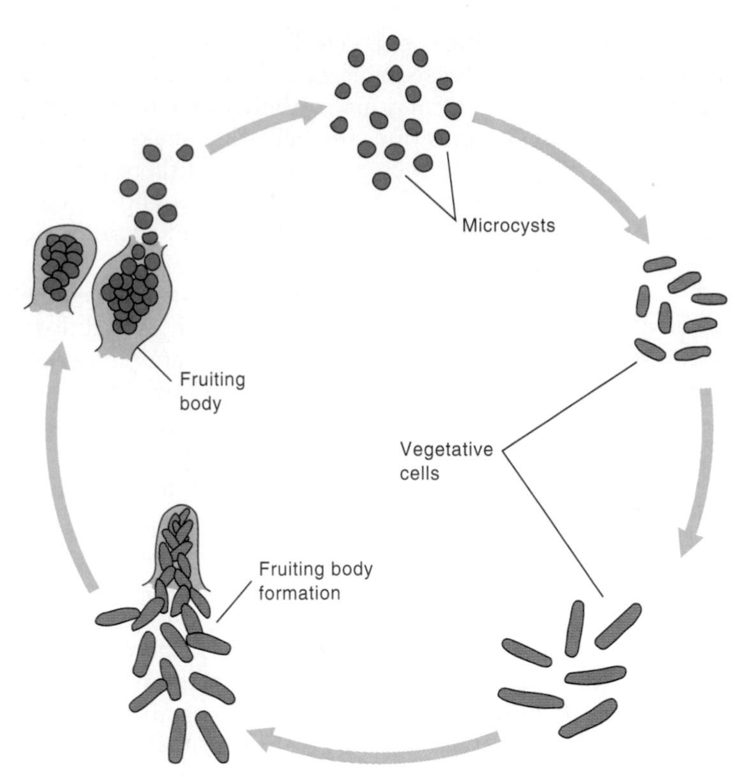

Figure 11.16
Life cycle of a fruiting myxobacterium.

the **fruiting body** (figure 11.17), may be brightly colored. The mass of cells on the stalk then differentiates into spherical cells called **microcysts,** which are very similar to the cysts of *Azotobacter* species. The microcysts are considerably more resistant to heat, drying, and radiation than are the vegetative cells of myxobacteria, but they are much less resistant than bacterial endospores. Myxobacteria are important in nature as degraders of complex organic substances; they can digest living or dead bacteria and certain algae and fungi.

Sheathed Bacteria Are Enclosed in a Lipoprotein-Polysaccharide Material

Sheathed bacteria constitute another group of multicellular filamentous organisms. They differ from other nonbranching filamentous forms in that their chain of bacterial cells is enclosed in a sheath composed of a lipoprotein-polysaccharide complex that is chemically distinct from the bacterial cell walls. The sheath helps attach the organisms to solid objects and protects them from attack by bdellovibrios and other organisms. *Sphaerotilus* (figure 11.18) is a widespread genus in this group. The filaments of members of this genus often produce masses of brownish scum that may be seen beneath the surface of polluted streams, and they frequently interfere with sewage treatment processes by plugging pipes. *Sphaerotilus* species reproduce by binary fission, in which polarly flagellated rods are released from the ends of the sheath. These rods then reproduce, secreting a new sheath. In freshwater habitats containing iron or manganese, ferric hydroxide or manganese oxide may be deposited on the sheaths of members of *Leptothrix,* the other common genus of sheathed bacteria.

Anoxygenic Phototrophic Bacteria Carry on Photosynthesis Without Releasing Oxygen

The **phototrophic bacteria** constitute a group of gram-negative rods that obtain their energy from sunlight, in contrast to **chemotrophs,** which obtain energy from organic or inorganic material. In the complex process of photosynthesis, light energy is converted to chemical energy, which is then used to synthesize organic materials.

• **Chemotrophic bacteria, p. 95**
• **Oxygenic and anoxygenic photosynthesis, p. 127**

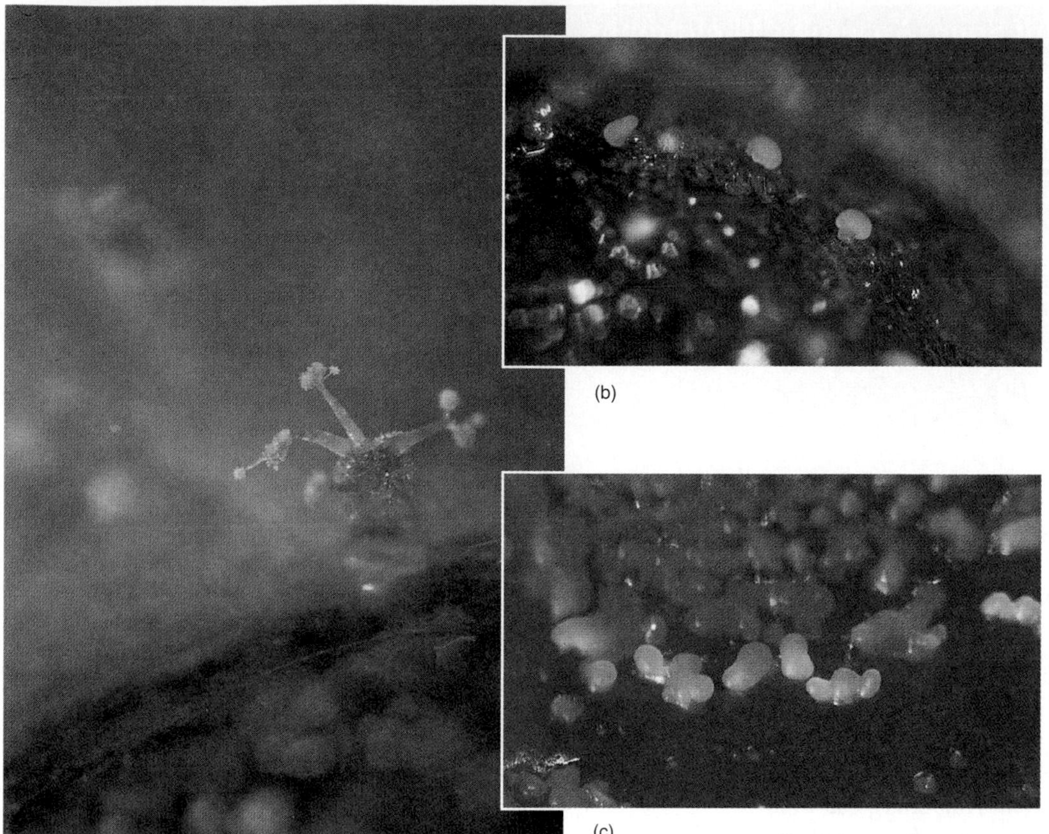

Figure 11.17
Myxobacterial colonies. (*a*) *Chondromyces crocatus,* (*b*) *Myxococcus fulvus,* and (*c*) *M. coralloides.*

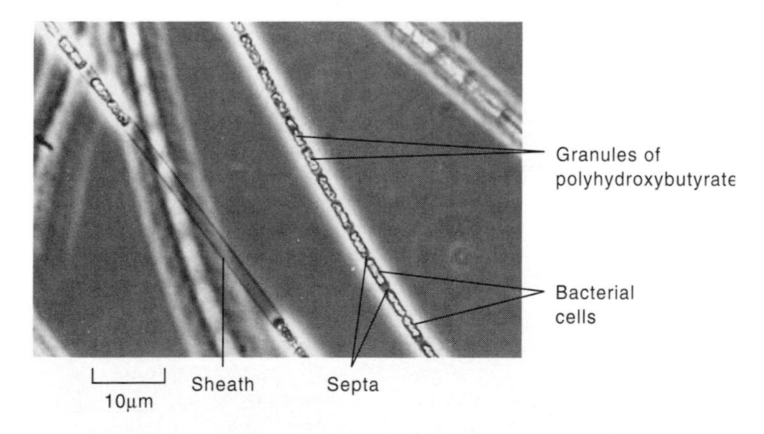

Figure 11.18
Phase contrast photomicrograph of *Sphaerotilus* sp., a sheathed bacterium.

Anoxygenic phototrophs grow photosynthetically only under anaerobic conditions. Since they do not use water as a source of hydrogen, they do not produce oxygen as a by-product of photosynthesis. Their chlorophyll pigments differ from those of the other groups of photosynthetic organisms, cyanobacteria, algae, and plants. Anoxygenic phototrophic bacteria inhabit a restricted ecological niche characterized by adequate light penetration and little or no oxygen. They require chemically reduced inorganic or organic compounds, such as hydrogen sulfide or fatty acids, as hydrogen donors. Thus, they may be found in certain wet soils and aquatic or marine habitats.

To achieve the location that provides optimum conditions for growth, some anaerobic phototrophs can change their position in a fluid environment by using their polar flagella or by altering the number of **gas vesicles** in their cytoplasm to give them more or less buoyancy. Gas vesicles are gas-filled cylindrical structures bounded by a single-layered protein membrane that inhibits the passage of water. Most of these bacteria can fix atmospheric nitrogen and, therefore, besides a hydrogen donor, require only light, water, carbon dioxide, molecular nitrogen, and some minerals for growth. Ancestors of these organisms probably predominated before the earth acquired atmospheric oxygen about 3 billion years ago. Today, their ecological role as the primary producers of organic materials from inorganic materials is insignificant in comparison with that of algae, except in a few special habitats.

Modern anaerobic phototrophs are taxonomically diverse and include rods, cocci, helices, and other forms. The characteristics of the four major groups—purple nonsulfur, purple sulfur, green nonsulfur, and green sulfur bacteria—as well as the names of representative genera, are shown in table 11.8. The four groups were named according to the colors of their photosynthetic pigments and whether they required reduced sulfur compounds to provide electrons for biosynthetic reactions. *Rhodomicrobium vannielii* (figure 11.19) is a budding anaerobic photobacterium that is also capable of aerobic growth under some conditions. Budding is a reproductive process in which a daughter cell develops asymmetrically on the parent cell. Budding bacteria are discussed further in a subsequent section of this chapter, "Budding and/or Appendaged Bacteria."

Oxygenic Photosynthetic Bacteria Liberate Oxygen as a By-Product of Their Photosynthesis

The oxygenic photosynthetic bacteria are called **cyanobacteria,** and like algae and plants, these gram-negative prokaryotes liberate oxygen during photosynthesis. Their

Table 11.8 Characteristics of Some Anaerobic Phototrophs

Bacterium	Motility	H_2S Used	Growth with Organic Compounds*
Purple Nonsulfur *Rhodospirillum* *Rhodopseudomonas* *Rhodomicrobium*	Polar or peritrichous flagella or nonmotile	–	+
Purple Sulfur *Chromatium* *Thiospirillum*	Polar flagella or nonmotile	+	–
Green Nonsulfur *Chloroflexus*	Gliding	–	+
Green Sulfur *Chlorobium* *Pelodictyon*	Nonmotile	+	–

*As an energy source

chlorophyll pigments and photosynthetic process are of the same type as those of algae and plants. However, cyanobacteria have unique accessory pigments called **phycobiliproteins** that absorb energy from light of wavelengths (shades of color) not well absorbed by chlorophyll. This energy is then transmitted to the photosynthetic apparatus, which converts light energy to chemical energy in the form of ATP.

Many cyanobacteria have the capacity to fix nitrogen, resembling the anaerobic photobacteria in this respect but differing from all algae and plants. As might be expected, since they obtain energy from sunlight and nitrogen from the air, their growth needs are simple. Cyanobacteria are therefore extremely widespread in nature and are found even where many other, more fastidious microbes are unable to grow. In natural waters polluted with nitrogenous compounds and phosphates, they often are found in enormous numbers, producing a colored scum on the surface. The color of such cyanobacterial masses, ranging from blue-green to red, violet, and black, depends on the types of light-trapping pigments that a particular species possesses.

All the prokaryotic morphological types are represented among the cyanobacteria: rods, cocci, and spirals, as well as nonbranching filamentous forms. Motility, when present, is of the gliding type, and some aquatic species can change their depth of flotation by changing the number of their intracytoplasmic gas vesicles. Two cyanobacterial structures, **thylakoids** and **heterocysts,** are unique among prokaryotes. Thylakoids are platelike membranes containing chlorophyll, and heterocysts (figure 11.20) are specialized cells whose primary function is thought to be nitrogen fixation. The genetically diverse cyanobacteria usually reproduce by binary fission, although some reproduce by budding, or formation of multiple progeny within a single large cell. Ancestors of cyanobacteria were probably responsible for the initial appearance of atmospheric oxygen on the earth, and there is evidence that as endosymbionts they may have evolved to become the chloroplasts of at least some algae. Of interest is the fact that red algae also contain phycobiliproteins, indicating that their chloroplasts are closely related to cyanobacteria.

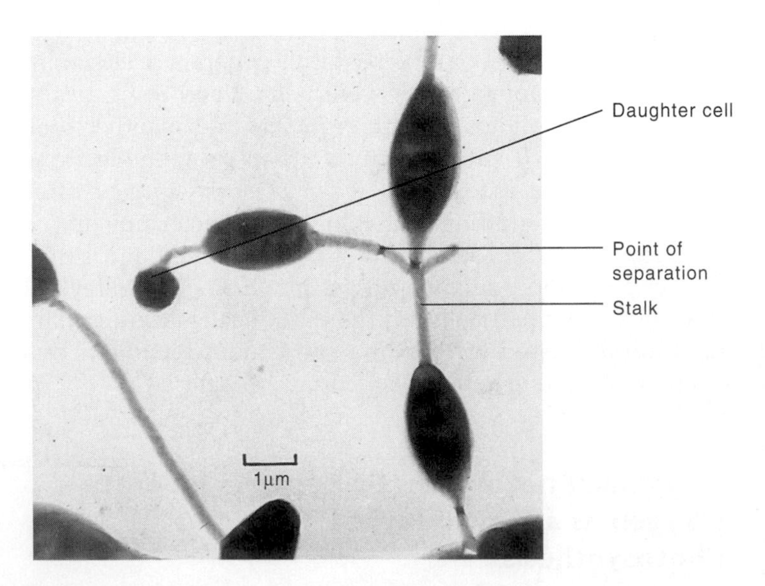

Figure 11.19
Rhodomicrobium vannielii, a budding photosynthetic bacterium.

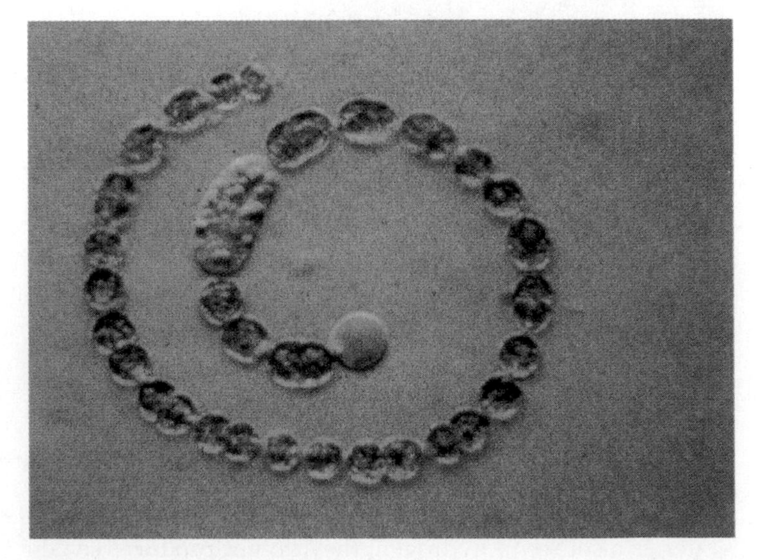

Figure 11.20
Anabaena spiroides. Note the terminal, spherical cell (heterocyst), site of nitrogen fixation (X800).

Budding and/or Appendaged Bacteria Are Characterized by Cytoplasm-Filled Projections from the Main Cell Body

A diverse group of gram-negative bacteria possess projections called **prosthecae,** which contain cytoplasm and have a cell wall. This is in contrast to spines and tubes found on other bacteria that do not contain cytoplasm and are not connected with the interior of the cell. In the case of *Caulobacter* species (figure 11.21), the prostheca (stalk or holdfast) serves to attach the organism to a solid substrate. Reproduction is asymmetrical (figure 11.22): The daughter cell is flagellated and, upon separation from the parent, swims away to attach and repeat the reproductive cycle. Other species of prosthecate bacteria, such as those of *Ancalomicrobium,* have two or more prosthecae (figure 11.23) whose function is not known. One theory is that they provide increased surface area to facilitate absorption of nutrients, since most prosthecate bacteria are capable of growth in water containing extremely dilute nutrients.

During reproduction of *Hyphomicrobium* species (figure 11.24), an enlargement (bud) is formed at the tip of the prostheca and subsequently reaches the size and shape of the parent cell. The duplicated genome is transferred from the mother cell down the prostheca to the developing bud. Formation of a septum in the prostheca allows separation of the daughter cell, although extensive networks of cells can arise if attachment is maintained. These bacteria are commonly found in natural waters and can thrive on very low concentrations of single-carbon organic substances, such as methanol or methylamine. Other budding species of bacteria do not form prosthecae.

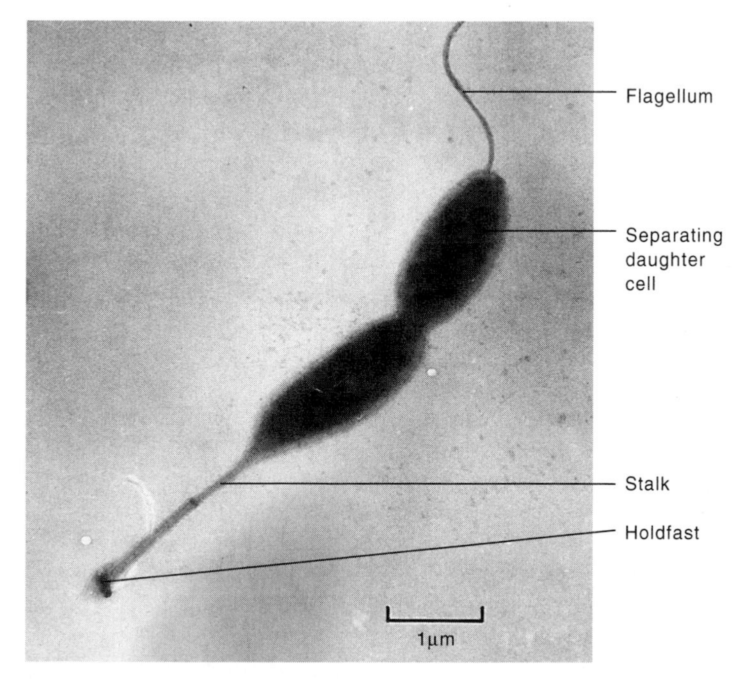

Figure 11.21
Caulobacter sp.

Many Aerobic Chemolithotrophic Bacteria Can Grow in the Dark on Media Containing Only Inorganic Substances

Bacteria in this section comprise a group of gram-negative organisms that can derive energy from oxidizing inorganic materials. The prefix *chemolitho-* is used in *Bergey's Manual* to indicate that the chemicals used for energy are inorganic. Many of these bacteria can use CO_2 as their sole carbon source and thus have the unique ability to grow in the dark on a medium containing only inorganic substances.

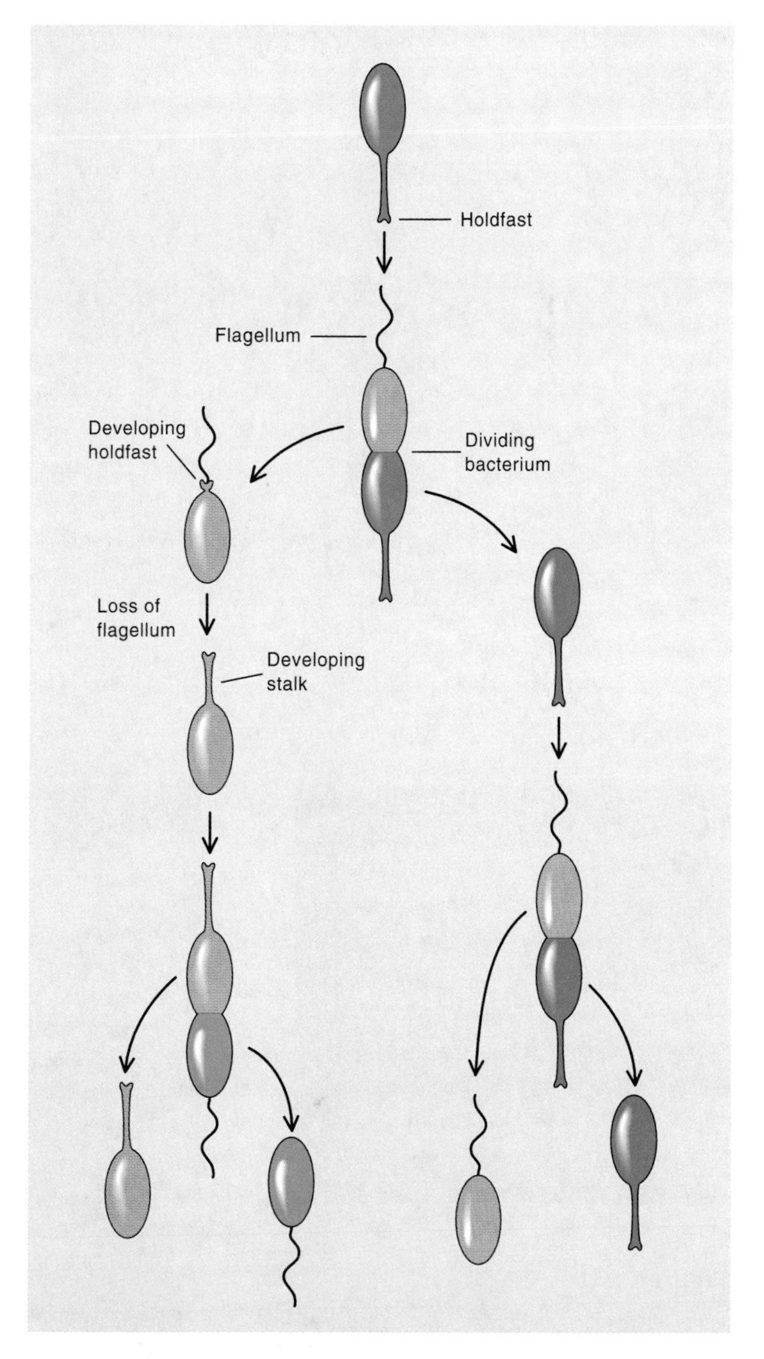

Figure 11.22
Life cycle of *Caulobacter* sp.

Members that fix carbon dioxide generally have well-defined intracytoplasmic structures called **carboxysomes,** which contain a portion of the CO_2-fixing enzymes. Some of these bacteria require one or more organic compounds in the form of vitamins or other growth factors. Although mentioned only briefly here, bacteria using inorganic energy sources are extremely important in the natural environment, as is explained in chapter 32.

Sulfur-Oxidizing Chemolithotrophs

Thiobacillus species are gram-negative, rod-shaped aerobic organisms that obtain energy by oxidizing sulfur or sulfur

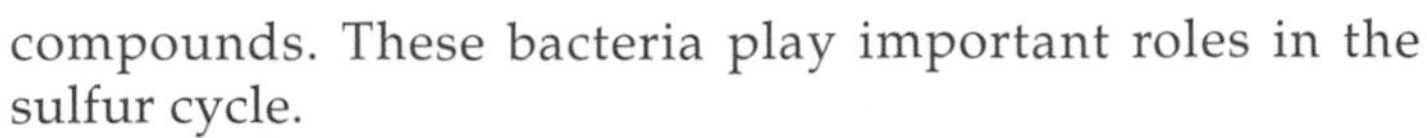

compounds. These bacteria play important roles in the sulfur cycle.

Oxidation of sulfur-containing minerals exposed by coal mining is carried out by certain acid-tolerant *Thiobacillus* species. The oxidation reactions produce sulfuric acid and convert toxic metals present in the minerals to a soluble form. Thus, the water that trickles from some coal mines (figure 11.25) can be highly acid and poisonous as a result of the bacterial action.

Thiomicrospira species are small, motile, spiral eubacteria that grow only on sulfur compounds and use CO_2 as a carbon source. They have been found in large numbers in areas surrounding deep-sea hydrothermal vents.

• **Lithotrophs, pp. 95, 124**

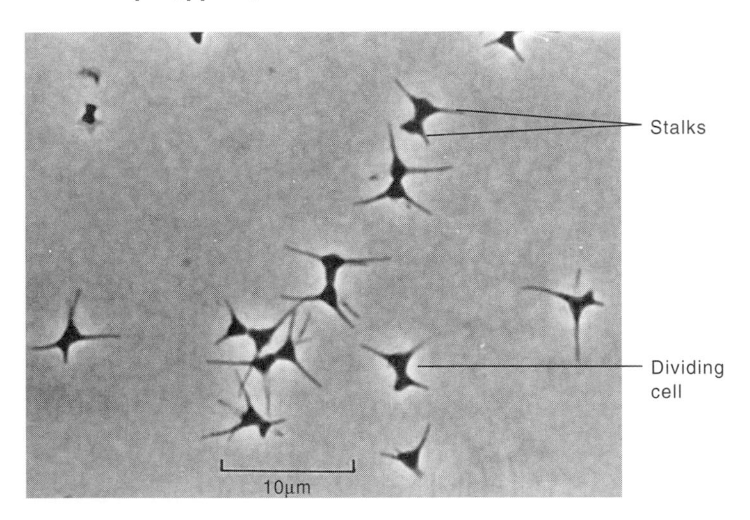

Figure 11.23
Ancalomicrobium sp.

Figure 11.24
Scanning electron micrograph of *Hyphomicrobium* sp. Note the developing buds at the ends of the hyphae.

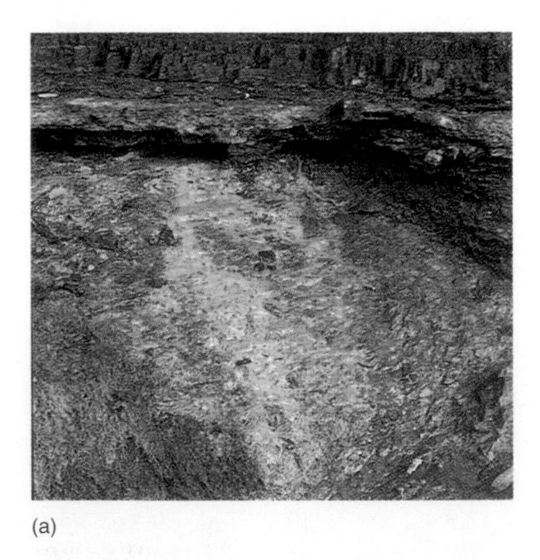

(a)

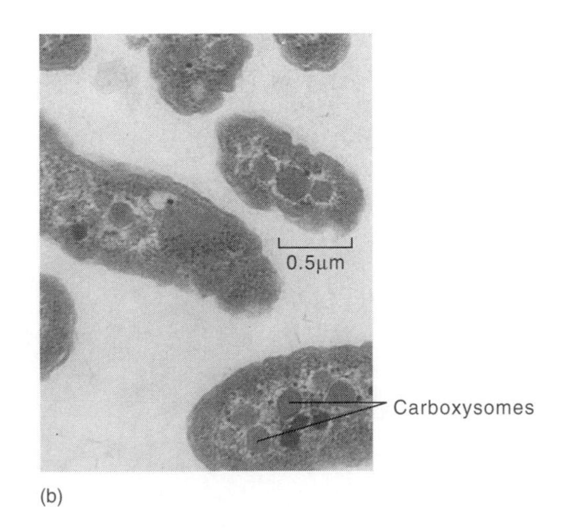

(b)

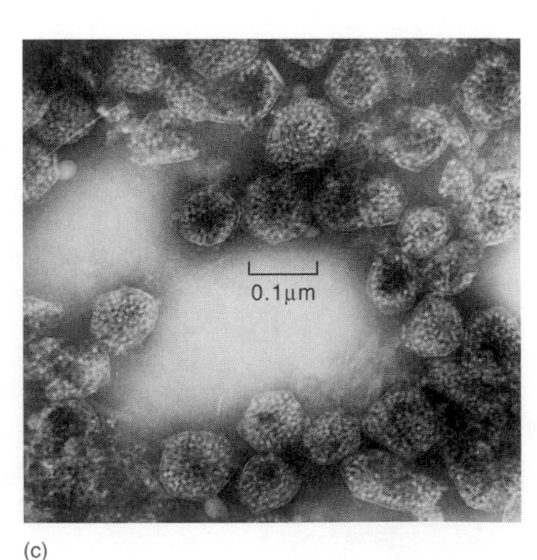

(c)

Figure 11.25
(*a*) Drainage from a coal seam exposed by mining. The drainage water is highly acidic and toxic due to the oxidation of sulfur compounds by *Thiobacillus* sp. (*b*) Electron micrograph of the internal structure of *Thiobacillus* sp. Note the presence of multiple spherical bodies called *carboxysomes.* (*c*) Higher power of the carboxysomes, the site of CO_2 fixation.

Chemolithotrophs that Oxidize Nitrogen Compounds

The nitrifiers are a group of gram-negative bacteria that obtain energy from the oxidation of inorganic nitrogen compounds such as ammonium or nitrite. Some, such as members of the genera *Nitrosomonas, Nitrosospira,* and *Nitrosolobus,* oxidize ammonium ion to nitrite. A second group, members of the genera *Nitrobacter, Nitrospira,* and *Nitrococcus,* oxidize nitrite to nitrate (table 11.9). These bacteria are important in the nitrogen cycle in both terrestrial and aquatic environments.

Species that Oxidize Hydrogen Gas

Some chemolithotrophs can obtain their energy from the oxidation of hydrogen gas (H_2), fixing CO_2 and producing cellular carbohydrates and H_2O in the process. The ability to grow under aerobic conditions using H_2 and CO_2 is found in diverse genera, including *Alcaligenes, Xanthobacter, Flavobacterium,* and some gram-positive genera. All these bacteria are also capable of growth utilizing organic sources of energy and are classified in various other sections of *Bergey's Manual.* In 1984, a new species, *Hydrogenobacter thermophilus,* which was isolated from hot springs in Japan, was found to be an obligate hydrogen oxidizer. It is also an aerobe and grows optimally at high temperatures, 70°C to 75°C.

Oxidizers of Reduced Metal Compounds

Acid waters and sediments of both freshwater and marine origin may contain dissolved iron and manganese in a reduced state (Fe^{2+}, Mn^{2+}). These reduced ions can be oxidized by some bacteria, and the energy gained may be used to fix carbon dioxide into cellular components. For example, some members of the genus *Thiobacillus* oxidize ferrous ions as an alternative to oxidizing reduced sulfur compounds. This reaction is actually used to obtain copper from low-grade ores. A solution containing the *Thiobacillus*-produced Fe^{3+} ions is trickled through the ore. The copper in the ore is oxidized by the iron ions and goes into solution; the dissolved copper is then recovered by a simple chemical process.

Table 11.9 Characteristics of the Nitrifying Bacteria

Organisms	Habitat	Gram Reaction	Shape
Ammonium oxidizers			
Nitrosospira	Marine, soil	–	Spiral
Nitrosomonas	Soil	–	Rod
Nitrosococcus	Marine, soil	–	Coccus
Nitrosolobus	Soil	–	Irregular coccus
Nitrite oxidizers			
Nitrobacter	Marine, soil	–	Rod
Nitrospira	Marine	–	Spiral
Nitrococcus	Marine	–	Coccus

REVIEW QUESTIONS

1. What are the endosymbionts? What significance might they have in the evolution of eukaryotic cells?
2. Name two habitats in which nonphotosynthetic gliding bacteria might be found.
3. What is a fruiting body?
4. What kind of bacteria might compose the subsurface scum of polluted streams?
5. What is a phototrophic bacterium?
6. Distinguish between oxygenic and anoxygenic phototrophs.
7. What are cyanobacteria?
8. What is a prostheca?
9. Give three examples of energy sources used by chemolithotrophs.

Prokaryotes with a Gram-Positive Type of Cell Wall

The second major division of prokaryotes, bacteria with a gram-positive type of cell wall, contains a large number of diverse organisms in about 90 different genera. They can be divided into two groups based on their content of guanine (G) and cytosine (C). Members of the first group, represented by the genera *Streptococcus, Bacillus,* and *Clostridium,* generally have less than 50% G and C, whereas members of the second group, represented by the actinomycetes, generally have more than 50% G and C.

Gram-Positive Cocci Comprise a Large Group, Subdivided on the Basis of Arrangement of Cells and Metabolic Differences

Two of the most important genera of gram-positive cocci are *Streptococcus* and *Staphylococcus,* which include species of great medical importance (table 11.10). Many other species of gram-positive cocci are harmless inhabitants of soil and water.

• **Gram-positive cell wall type, pp. 58–59**

Table 11.10 Medically Important Gram-Positive Cocci

Bacterium	Characteristics	Disease
Streptococcus pyogenes	β-hemolytic; group A cell wall carbohydrate; M-protein in cell wall responsible for virulence	Strep throat, rheumatic fever, wound infections, glomerulonephritis (kidney disease), toxic shock (one type)
Staphylococcus aureus	Positive test for catalase; protein A in cell walls; coagulases clot plasma	Leading cause of wound infections; food poisoning; toxic shock syndrome

Streptococci

Species that divide repeatedly in parallel planes produce chains of bacteria characteristic of the genus *Streptococcus,* which are either tolerant of oxygen or are anaerobic. Typically, streptococci give negative results in tests for catalase. Streptococci are characterized by the kinds of changes their colonies produce on blood agar (figure 11.26). If the red blood cells are partly destroyed around the bacterial colony, the effect is called **α-hemolysis.** If the red cells are completely destroyed, leaving a clear area around the colony, the effect is called **β-hemolysis.** Streptococci that produce no red cell destruction are referred to as **γ-streptococci.** Alpha-hemolytic streptococci that produce a green discoloration are called **viridans streptococci.**

Staphylococci

Repeated divisions in different planes produce clusters or packets of organisms characteristic of the genus *Staphylococcus.* These bacteria grow best in air but also can grow anaerobically. Staphylococci, unlike streptococci, are generally catalase positive.

Only a Few Genera of Gram-Positive Rods and Cocci Are Endospore-Forming

The few genera of bacteria known to produce endospores are very widespread in the environment as well as in the intestines of animals. Two such important genera of gram-positive, rod-shaped bacteria are *Bacillus* and *Clostridium* (figure 11.27).

Bacillus and *Clostridium* Species

Bacillus species are aerobic or facultatively anaerobic, whereas clostridia are obligate anaerobes, which in some cases are killed by even the faintest traces of oxygen. Their endospores represent the most resistant forms of life known; they tolerate extremes of heat and dryness, the presence of disinfectants, and radiation. Some members of *Bacillus* and *Clostridium* cause serious infectious diseases (table 11.11), and others play a role in fixing atmospheric nitrogen. Thermophilic strains of *Bacillus* can grow at temperatures above 80°C (158°F).

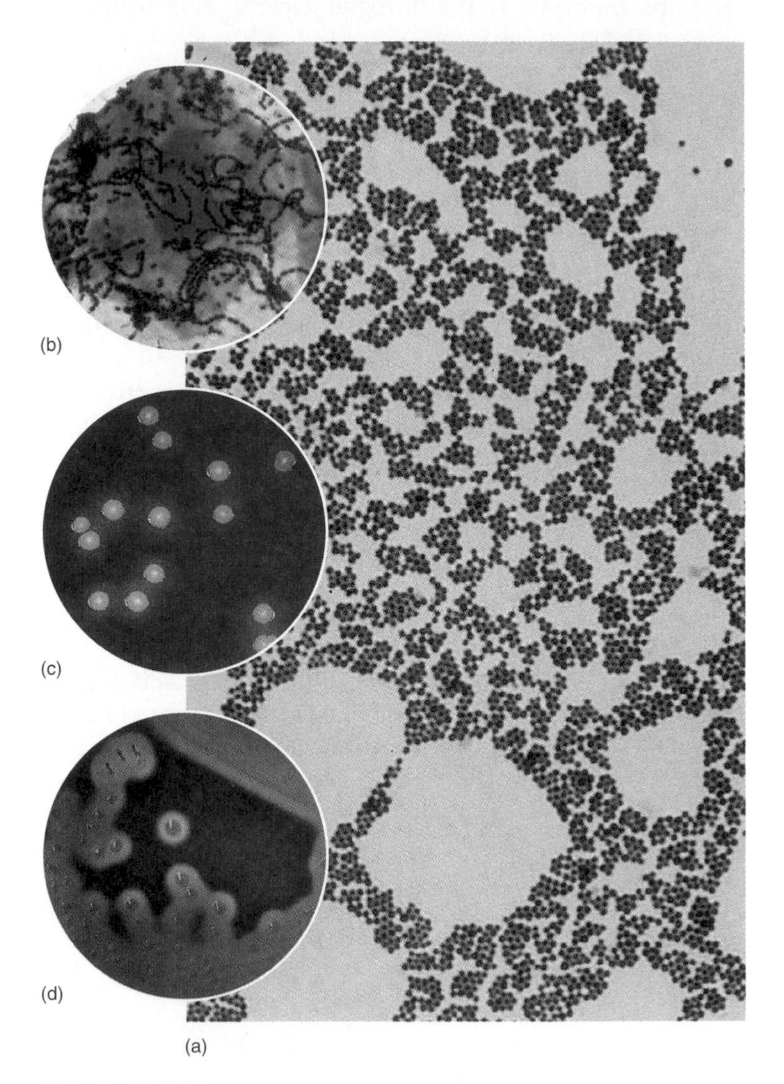

Figure 11.26
(*a*) Gram-positive cocci in clusters characteristic of staphylococci. (*b*) Gram-positive cocci in chains, characteristic of streptococci. (*c*) α-hemolysis. (*d*) β-hemolysis of streptococci.

• Aerotolerance, p. 92

• Anthrax, p. 72

Table 11.11 Some Medically Important Endospore Formers

Bacterium	Characteristics	Disease
Bacillus anthracis	Grows in air; produces toxin	Anthrax, disease acquired from spores in soil, or in hides or wool; in humans, usually severe wound infection; often fatal if spores are inhaled
Clostridium tetani	Anaerobic; toxin inhibitory for certain nerve impulses produced	Tetanus, often fatal type of paralysis, usually acquired when spores contaminate a wound; completely preventable by immunization
Clostridium botulinum	Anaerobic; toxin produced by organisms growing in food	Botulism, an often fatal type of paralysis; prevented by proper canning and cooking techniques

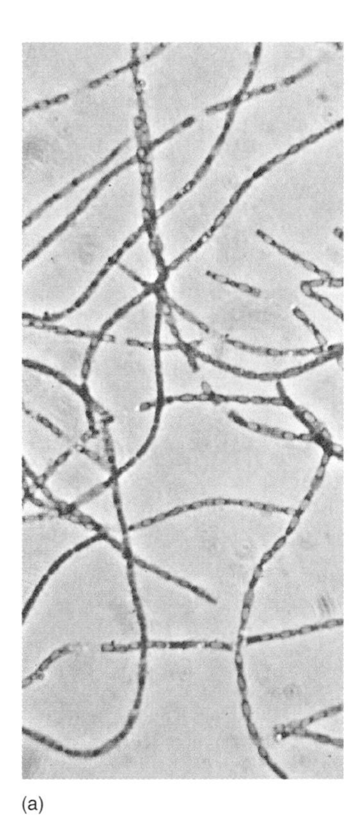

(a)

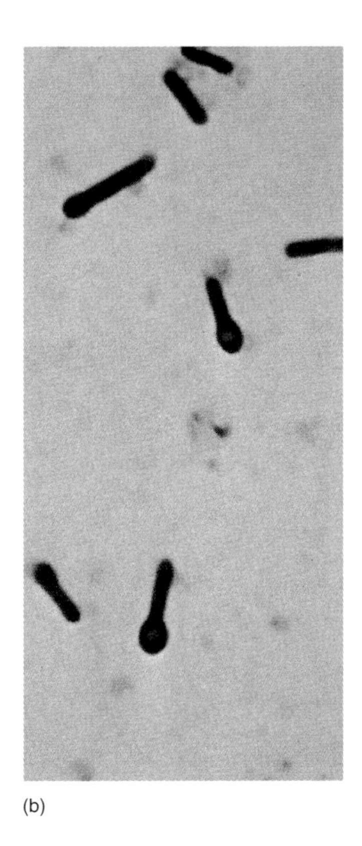

(b)

Figure 11.27
(*a*) Spores forming in the midportion of the cells in *Bacillus anthracis*, an aerobe. (*b*) Spores forming at the ends of the cells in *Clostridium tetani*. Both of these species can cause fatal disease in humans, but many other species of spore formers are harmless.

Some Cylindrical Gram-Positive Rods Do Not Form Endospores

This section is composed of gram-positive eubacteria that do not form endospores. The individual cells are straight-walled cylinders without bulges or appendages. The lactobacilli are important members of this group.

Lactobacilli

Lactobacillus sp. may occur as single cells or loosely associated chains of bacteria. They require enriched media to grow and produce lactic acid as a major end product of sugar metabolism. Lactobacilli are common among the microbial flora of the mouth and of the healthy human vagina during childbearing years. In the vagina, they metabolize glycogen, which has been deposited in the vaginal lining in response to the female sex hormone, estrogen. The resulting acidity helps the vagina resist infection. These bacteria have virtually no disease-producing ability except in the mouth where they may contribute to dental caries.

Lactobacilli are commonly present in decomposing plant material, milk, and other dairy products as well as their animal habitats. They are important in such processes as pickling and cheese making (figure 11.28).

Two medically important members of the group of cylindrical nonspore-forming gram-positive rods, *Listeria monocytogenes* and *Erysipelothrix rhusiopathiae*, are described in table 11.12.

The Irregularly Shaped, Nonspore-Forming, Gram-Positive Rods Include Many Unrelated Genera

Irregular, nonspore-forming, gram-positive rods may show misshapen cells, uneven staining, or short, branching filaments. A number of the species show **snapping division,** a fission process (figure 11.29) resulting in a side-by-side (palisade) arrangement or a pattern suggesting Chinese letters. Important genera include *Propionibacterium, Corynebacterium, Actinomyces,* and *Arthrobacter.* Two medically important species, *Corynebacterium diphtheriae* and *Actinomyces israelii,* are described in table 11.13.

• **Lactic acid, p. 119**

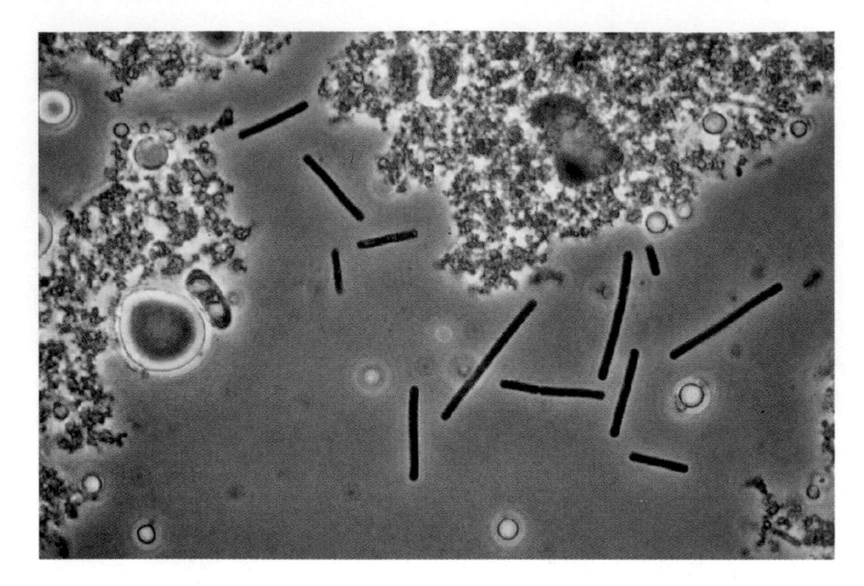

Figure 11.28
Lactobacillus bulgaricus from yogurt.

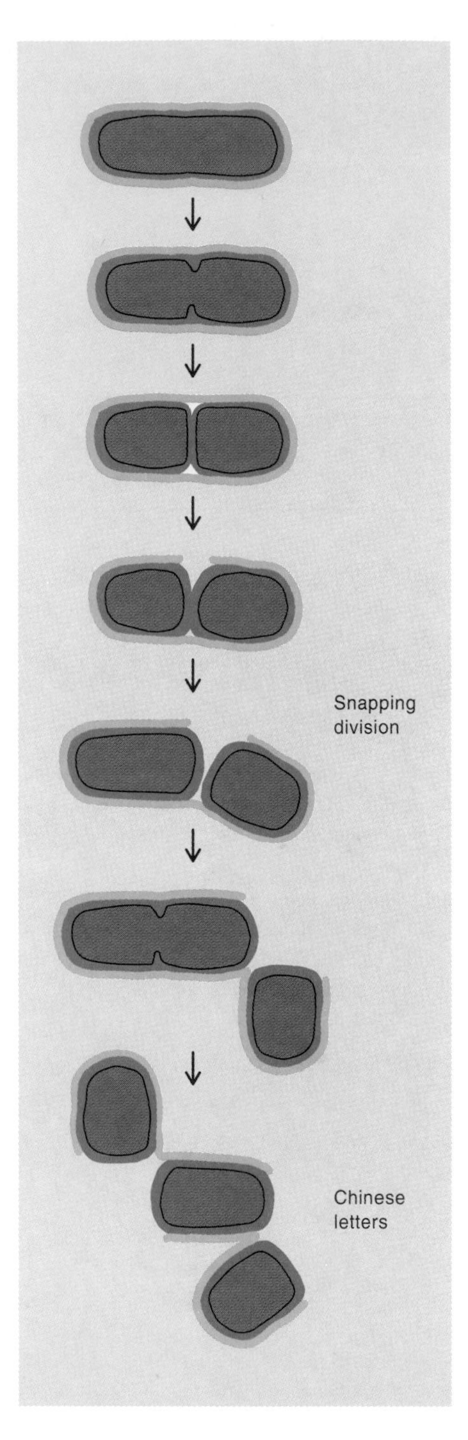

Figure 11.29
Snapping division, characteristic of *Corynebacterium* and a number of other gram-positive rod-shaped bacteria.

Table 11.12 Medically Important Straight-Sided Nonspore-Forming Gram-Positive Rods

Bacterium	Characteristics	Disease
Listeria monocytogenes	Motile; grows at refrigerator temperatures; positive test for catalase	Listeriosis-meningitis in those with AIDS or other immune defect; abortion and stillbirth; source often unpasteurized milk
Erysipelothrix rhusiopathiae	Nonmotile; negative test for catalase	Erysipeloid in humans, a painful chronic infection of fishhook wounds and other wounds; generalized acute and chronic infection of pigs and other animals

Table 11.13 Medically Important Misshapen Gram-Positive Rods

Bacterium	Characteristics	Disease
Corynebacterium diphtheriae	Powerful toxin that stops protein synthesis in certain cells; grows in air	Diphtheria, a frequently fatal throat infection completely preventable by immunization
Actinomyces israelii	Anaerobic; forms branching filaments that fragment into bacilli; dense colonies, "sulfur granules," form in tissue	Actinomycosis, a chronically draining abscess; "lumpy jaw" in cattle

Propionibacteria

Members of the genus *Propionibacterium* are mostly anaerobes or aerotolerant. They produce propionic acid as a fermentation product. One species, *P. acnes,* universally inhabits the tiny oil glands of the human skin, while another, *P. freudenreichii* is responsible for the holes and flavor of Swiss cheese.

Corynebacteria

Members of the genus *Corynebacterium* commonly inhabit plants, soil, and water. *C. diphtheriae* is an important human pathogen. Some members characteristically store energy as a phosphate polymer that is deposited in the cytoplasm in the form of granules. When a blue dye is used to stain these bacteria, the granules appear pink, an example of **metachromatic staining.**

Actinomyces Species

Members of the genus *Actinomyces* are pleomorphic rods that form from fragmentation of branching filaments. They are generally anaerobic or aerotolerant and prefer to grow on enriched media in the presence of 5% CO_2. Their usual habitat is the mouth or upper respiratory system of animals. One *Actinomyces* species is responsible for a chronic infection of cattle called "lumpy jaw" and may occasionally cause similar infections in humans. The term *actinomycete* is often used for any bacterium having the branching filamentous structure of *Actinomyces.*

Genus *Arthrobacter*

Strains of some species of *Arthrobacter* are largely responsible for the degradation of insecticides and herbicides in the soil. During growth, members of this genus may undergo transitions in which the shape of the cells is markedly altered. In the stationary phase, the cells are essentially spherical, but when such a culture is diluted with fresh medium, the spherical cells elongate and form the irregularly shaped rods characteristic of coryneform (club-shaped) bacteria. The rods once again revert to spherical cells when the culture reaches the stationary phase (figure 11.30). Interestingly, some species of *Arthrobacter* and *Micrococcus* are closely related, as revealed by the similarity of their ribosomal RNA. Thus, these organisms may be reclassified in the future.

The Mycobacteria Demonstrate the Acid-Fast Staining Property

Mycobacterium species are widespread in nature and include harmless saprophytes as well as organisms that produce disease in humans and domestic animals. One important feature of this group is that they take up dyes very poorly, but once they are stained, it is difficult to remove the stain. Although they have a gram-positive type of cell wall, mycobacteria stain poorly with the Gram-staining procedure. However, they can be stained by various dyes that are applied with heat.

Mycobacteria resemble many coryneform bacteria of the previous section except that they are strict aerobes. A waxy substance in their cell walls confers the acid-fast staining property (figure 11.31). Slightly branching rods but usually not long filaments occur in some species. They

• **Aerotolerant organisms, p. 92**

• **Acid-fast stain, p. 52**

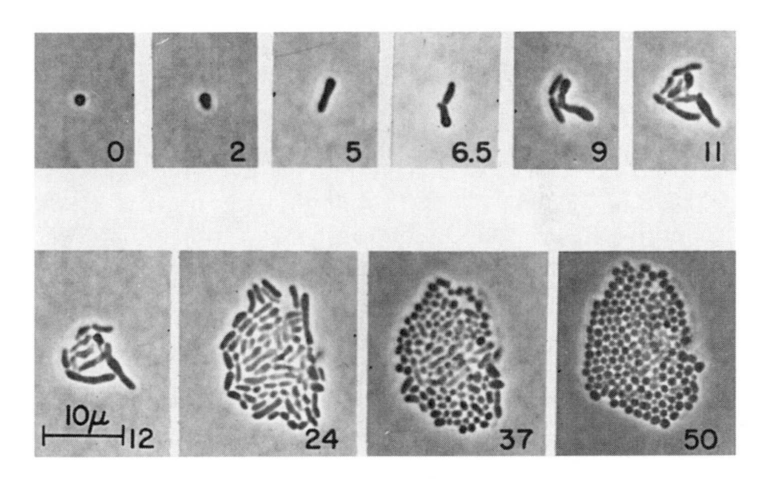

Figure 11.30
Phase contrast photomicrographs of *Arthrobacter crystallopoietes* colonial development from a single cell. Note the change in shape from coccus to rod to coccus. The numbers indicate time in hours. 10µ = 10µm.

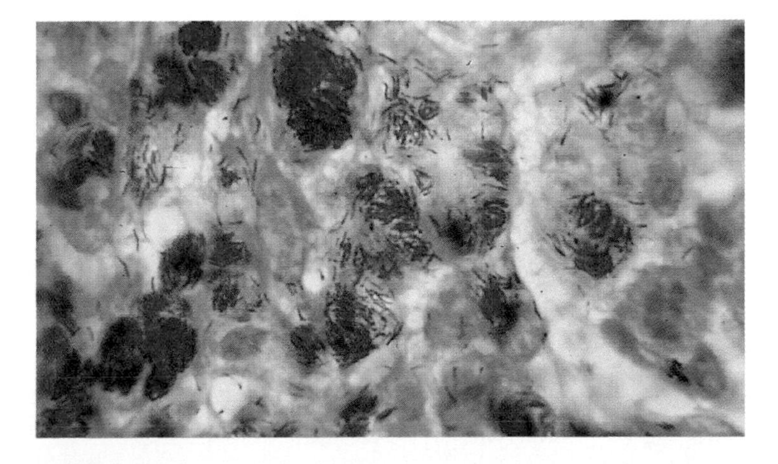

Figure 11.31
Acid-fast stain of *Mycobacterium leprae* in the tissues of a person with leprosy. The red material consists of masses of the bacteria.

are more resistant to disinfectants than most other vegetative bacteria and differ from them in their susceptibility to antibacterial medicines. Two species of mycobacteria pathogenic for humans, *M. tuberculosis* and *M. leprae*, are described in table 11.14.

Many Species of Branching, Gram-Positive Bacteria (Actinomycetes) Have Aerial and Specialized Hyphae

Rudimentary formation of branching cells and branched filaments sometimes occurs among the irregular gram-positive rods and the mycobacteria. The actinomycetes discussed next show more elaborate branching structures. As with fungi, the branching filaments are called **hyphae** (singular, *hypha*), and a tangled mass of them is called a **mycelium.** These hyphae, although analogous to those of fungi, are considerably narrower (generally 0.5 to 1.2 μm diameter) and are composed of prokaryotic cells. Prokaryotic mycelial growth occurs only among these gram-positive bacteria. The entire fourth volume of *Bergey's Manual of Systematic Bacteriology* is devoted to them. Actinomycetes can be divided into two groups: the nocardioforms and forms with specialized spores and hyphae.

Nocardioforms

The nocardioforms comprise nine genera of aerobic organisms that resemble members of the genus *Nocardia.* They show varying degrees of hypha formation but, unlike *Actinomyces nocardioforms* characteristically produce **aerial hyphae,** which project above the surface of the medium on which they are growing. Some nocardioforms produce rounded terminal cells that resemble spores (figure 11.32). Nocardioform hyphae, like those of *Actinomyces,* typically fragment into rods and cocci. Nocardioforms are widespread in nature and help degrade waxes, rubber, and petroleum pollutants. One species, *Nocardia orientalis,* produces the important antibiotic vancomycin, and others make part of the molecule for certain semisynthetic β-lactam antibiotics. A few species such as *N. asteroides* can cause lethal infections in patients with defective immunity due to cancer, AIDS, or other conditions.

Actinomycetes with Specialized Spores and Hyphae

The second group of actinomycetes with aerial hyphae are characterized by formation of distinctive spores **(conidia)** and spore-bearing structures. Some of the main patterns of growth and conidia formation of different actinomycetes are shown in figure 11.33. One group is characterized by **sporangia,** sacks of spores within or at the ends of hyphae that result from repeated divisions of hyphal cells in different planes. In some of these species, the spores released from the sporangia are motile by a tuft of flagella. The hyphae themselves do not usually break up into rod-shaped or coccoid cells.

• **β-lactam antibiotics, chapter 21**

Table 11.14 Medically Important Mycobacteria

Bacterium	Characteristics	Disease
Mycobacterium tuberculosis	Slender, irregularly staining, acid fast, slow growing; adhere side by side to form "cords" of organisms in suitable liquid medium	Tuberculosis, typically a chronic respiratory infection that can generalize
Mycobacterium leprae	Appearance similar to *M. tuberculosis;* not cultivable in cell-free media but grow in armadillos	Leprosy; peripheral nerve invasion characteristic

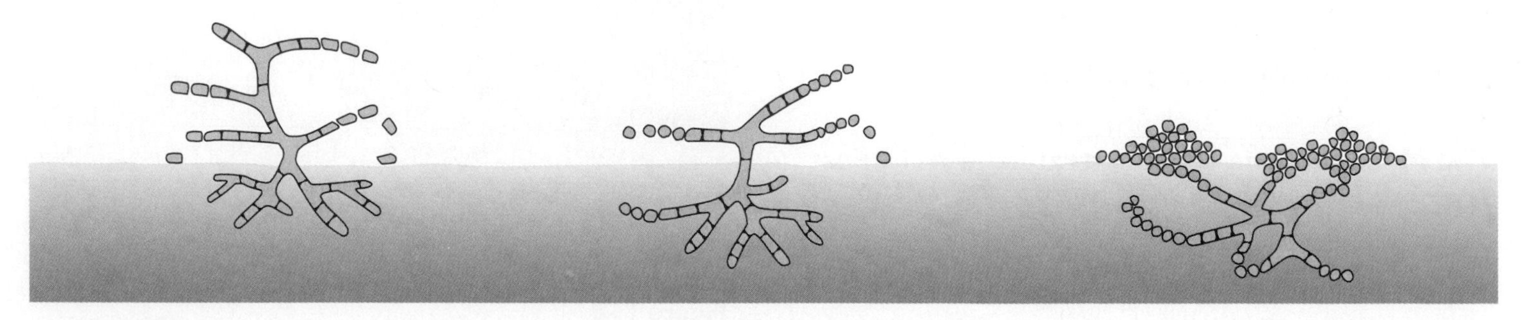

Figure 11.32
Nocardioforms. Most species produce filaments that project above the medium (aerial hyphae). The hyphae fragment into rods or spheres.

A common pattern in these specialized actinomycetes is the formation of one or more conidia at the ends of hyphae or on hyphal branches. These spores result from divisions of hyphal cells in planes perpendicular to the long axis of the hypha.

The spores of actinomycetes help distribute the organism, but, in contrast to bacterial endospores, they are generally not highly resistant to heat, radiation, and disinfectants. A notable exception to this rule occurs in the genus *Thermoactinomyces.* The spores of these bacteria are typical endospores, and although the organisms microscopically closely resemble other actinomycetes, nucleic acid studies relate them more closely with endospore-forming gram-positive rods of the genus *Bacillus. Thermoactinomyces* species are one of the causes of **farmers' lung,** a type of allergic pneumonia. The organisms grow in fermenting hay at temperatures as high as 65°C.

Other potential pathogens are described in table 11.15.

Frankia* and *Dermatophilus *Frankia* and *Dermatophilus* are two genera of sporangia-forming actinomycetes. *Frankia* species are nitrogen fixers and live symbiotically with certain plants, forming nodules on their roots. In contrast to species of *Rhizobium, Frankia* species have a wide host range, infecting a number of different plant species. One example is the alder tree, which is able to grow well on acid, rain-washed, nitrogen-poor soils because of *Frankia*-induced nitrogen-fixing root nodules. The nitrogenase enzymes of *Frankia* are apparently located in thick-walled swellings at the ends of their hyphae (figure 11.34).

Dermatophilus species also form sporangia, but they are pathogens of deer, domestic animals, and humans, causing an eczemalike skin infection. The sporangia release spores that are motile by means of a tuft of flagella.

Actinoplanes *Actinoplanes* species are representative of a group of aquatic actinomycetes that produce little or no

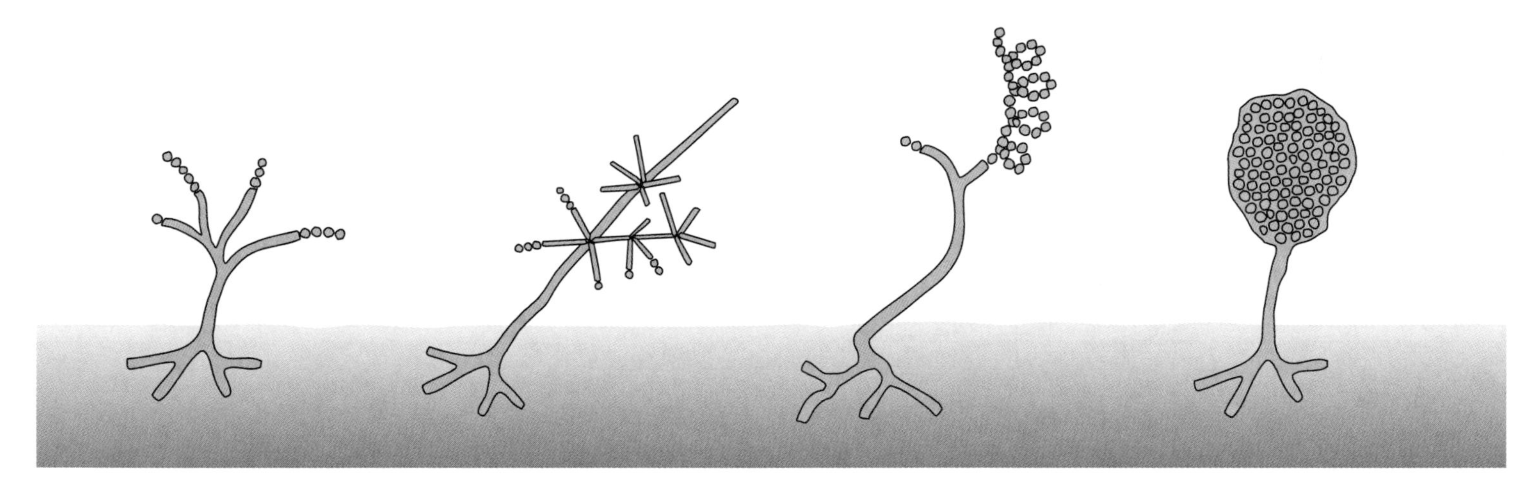

Figure 11.33
Spore-forming actinomycetes, showing some of the variations in structure.

Table 11.15 Potentially Pathogenic Actinomycetes

Group	Species	Type of Disease
Nocardioforms	*Nocardia asteroides*	Lung disease and generalized disease in AIDS and other immune deficiencies
	Nocardia brasiliensis	Mycetoma, a chronic draining abscess, usually of bones and tissues of the feet
	Rhodococcus equi	Wound and generalized infections
	Faenia rectivirgula	Allergic lung disease
Maduramycetes	*Actinomadura madurae*	Mycetomas
Thermomonosporas	*Nocardiopsis dassonvillei*	Mycetomas
Streptomycetes	*Streptomyces somaliensis*	Mycetomas
Thermoactinomycetes	*Thermoactinomyces vulgaris*	Allergic lung disease
Actinomycetes with multichambered sporangia	*Dermatophilus congolensis*	Eczemalike skin infection of wild and domestic animals and humans

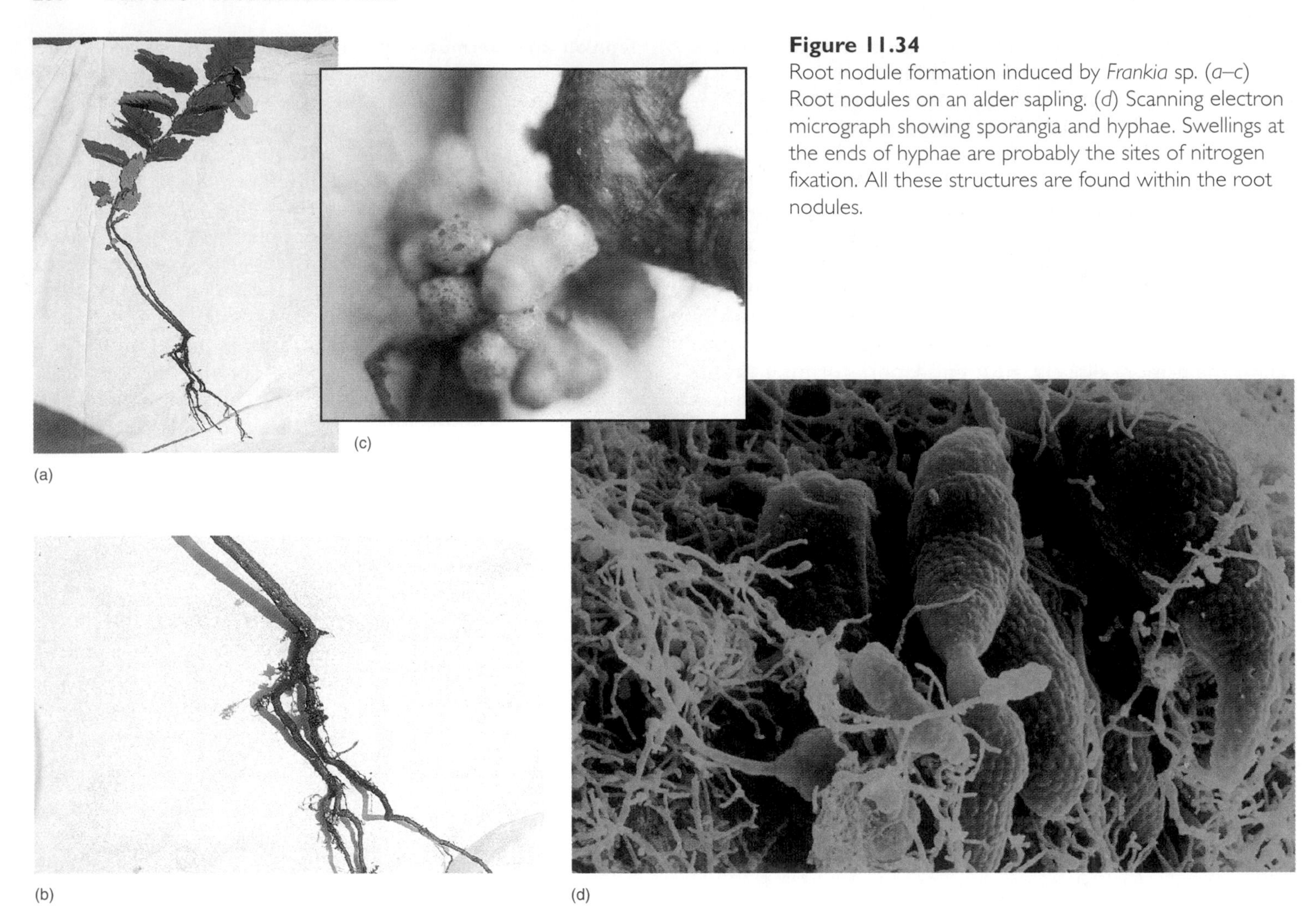

Figure 11.34
Root nodule formation induced by *Frankia* sp. (*a–c*) Root nodules on an alder sapling. (*d*) Scanning electron micrograph showing sporangia and hyphae. Swellings at the ends of hyphae are probably the sites of nitrogen fixation. All these structures are found within the root nodules.

aerial mycelium. They usually produce motile spores within a sporangium. Other members of this group are represented by the genus *Micromonospora,* which usually produces single thick-walled spores on short hyphae. *Micromonospora purpurea* produces the important antibiotic gentamicin.

Maduramycetes* and *Thermomonospora The maduramycetes compose a group of actinomycetes whose cell walls contain the sugar madurose.[1] Some members of this group produce sporangia, while others produce strings of spores by segmentation of their hyphae. Another diverse group of actinomycetes, the *Thermomonospora,* also vary considerably in appearance but have in common the same chemical cell wall structure. Both of these groups and the streptomycetes discussed next contain pathogenic species that characteristically cause **mycetomas,** chronic abscesses that involve skin, muscle, bone, and other tissues. The organisms typically infect the foot from contaminated soil entering small cuts or puncture wounds.

Streptomyces A distinctive feature of the genus *Streptomyces,* whose members are called *streptomycetes,* is the presence of strings of conidia at the ends of their hyphae (figure 11.35). The hyphal filaments are **coenocytic,** meaning that the cytoplasm of each cell is continuous with that of adjacent cells. *Streptomyces* species are mainly responsible for the characteristic odor of soil. A number of *Streptomyces* species are important antibiotic producers (see chapter 21), and others can infect humans.

Prokaryotes that Lack Cell Walls

The third of the four major divisions of the prokaryotes is composed of bacteria totally devoid of cell walls. Only three families are recognized, but a number of the organisms are as yet unclassified. Currently, all are placed in the order *Mycoplasma.*

[1]Madurose—3-O-methyl-D-galactose

• **Antibiotics, pp. 46, 62, 154**

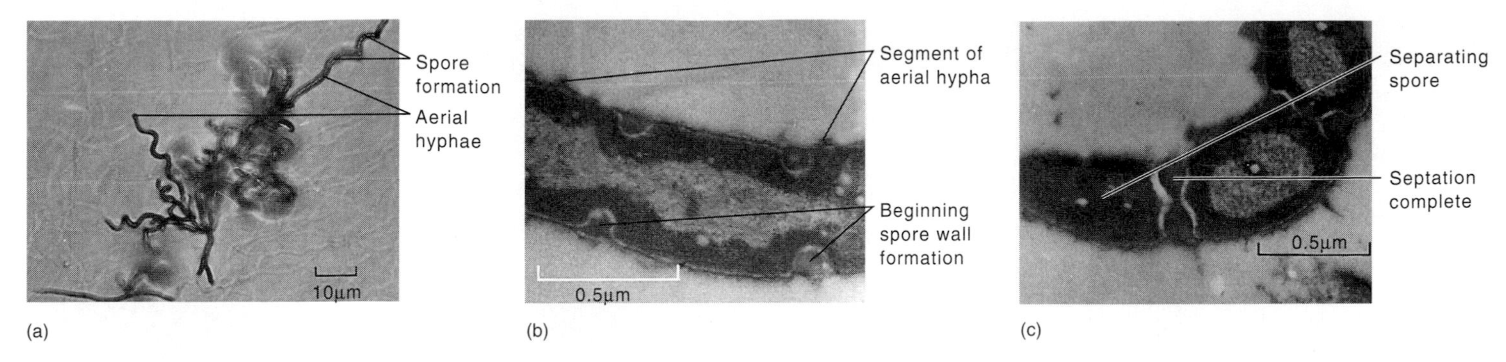

Figure 11.35
Reproduction of *Streptomyces* sp. (*a*) *Streptomyces fradiae.* (*b* and *c*) Electron micrographs showing steps in the formation of spores called *conidia.*

The Mycoplasmas Are Very Small Prokaryotes Bounded Only by a Plasma Membrane

The mycoplasmas are grouped together because they lack cell walls. Nevertheless, they are able to exist in relatively hypotonic environments because their cytoplasmic membranes are resistant to osmotic lysis. For the most part, studies of their ribosomal RNA indicate a close evolutionary relationship to gram-positive rods of the genus *Clostridium.* Members of the genus *Thermoplasma* represent an exception. They are free-living organisms that grow best at high temperatures under extremely acid conditions. The chemical nature of their lipids and RNA show them to be archaea.

Because they lack cell walls, mycoplasmas are easily deformed organisms (figure 11.36) that can pass through the pores of filters that retain other bacteria. Aside from some obligate intracellular bacteria, they generally contain less DNA than other prokaryotes. Most of them utilize carbohydrates as sources of energy, and both aerobic and anaerobic species are recognized. Characteristically, mycoplasmas growing on solid media produce colonies with a "fried egg" appearance (figure 11.37). Some species are found living harmlessly in soil and sewage, whereas others cause animal and plant diseases. Formerly, mycoplasmas were called *pleuropneumonialike organisms* (PPLO) because the earliest isolates were obtained from cattle with a fatal disease called **pleuropneumonia.** Medically important mycoplasmas are described in table 11.16. The genus *Mycoplasma* includes more than 65 species.

The Genus *Spiroplasma*

Organisms of the genus *Spiroplasma* (figure 11.38) first came to the attention of scientists in the late 1960s. Superficially, they resemble small spirochetes in size and helical shape, and they are similar also in their ability to spin and flex. However, they lack cell walls and axial filaments, form mycoplasmalike colonies, and pass through filters having very small pores. They cause certain plant and insect diseases. In 1976, a spiroplasma was shown to produce cataracts in baby mice. So far, however, they have not been implicated in human disease.

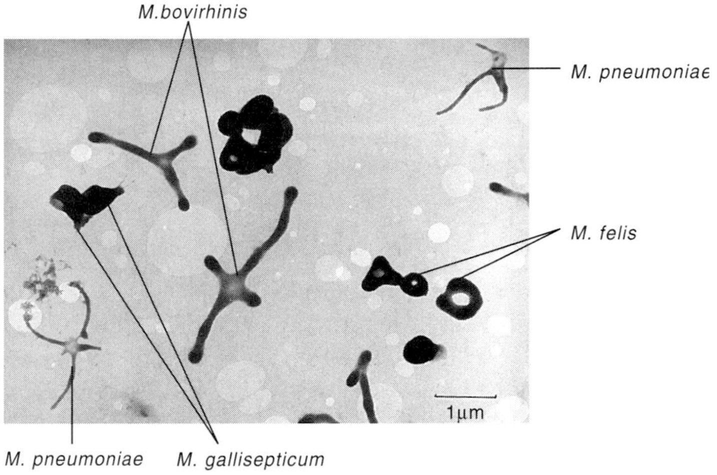

• **Mycoplasmas, p. 62**

M.bovirhinis / M. pneumoniae / M. felis / 1µm / M. pneumoniae / M. gallisepticum

Figure 11.36
Electron micrograph showing the morphology of different species of *Mycoplasma.*

Prokaryotes Whose Cell Walls Lack Peptidoglycan

Members of the fourth *Bergey's Manual* division of the prokaryotes lack peptidoglycan in their cell walls, a characteristic shared by very few other prokaryotes. Indeed, these organisms, members of the domain Archaea, are as distantly related to members of the domain Eubacteria as they are to humans. They are called archaea (from *archaios* for "ancient," "original") because evidence indicates that they evolved very early in the history of the earth and have

• **Peptidoglycan, p. 58**
• **Archaea, pp. 13, 14, 230**

Figure 11.37
Colonies of *Mycoplasma pneumoniae*, a cause of human respiratory disease. Note the dense central portion of the colonies characteristic of most mycoplasmas.

Table 11.16 Medically Important Mycoplasmas

Bacterium	Characteristics	Disease
Mycoplasma pneumoniae	Slow growing, aerobic; special medium required	"Walking pneumonia"
Ureaplasma urealyticum	Rapid aerobic or anaerobic growth; needs special medium	Part of normal vaginal flora in 60% of women; occasionally invades blood stream during childbirth, or causes urethritis

Figure 11.38
Scanning electron micrograph of *Spiroplasma* sp. growing on plant cells (X13,000).

retained some of their primitive traits. Multiple forms exist, including cocci, rods, spirals, multicellular filaments, and flat plates. Some, like mycoplasmas, lack a cell wall. Others move by means of flagella. They include aerobes, facultative anaerobes, and anaerobes. Some species grow at temperatures above 100°C and others in the icy waters of Antarctica. Determination of the complete sequence of bases of an archaeon genome was first reported in early 1996. While some of its genes resembled those of bacteria and others resembled eukaryotic genes, more than half of them were completely unknown to science.

Archaea Differ from Eubacteria in Many Respects

Generally, archaea grow in extreme environments and have therefore been termed **extremophiles.** They have unusual lipids in their cell membranes, contain unusual transfer RNA molecules, and have distinctive RNA polymerase enzymes. In short, the archaea share features that are quite different from those of eubacteria. They appear to be, if anything, more closely related to eukaryotes than eubacteria. There are three distinct groups of archaea: the **methanogens,** the **halobacteria,** and the **thermoacidophiles.**

Methanogens

Methane-producing archaea, termed *methanogens,* are strict anaerobes that grow in the digestive tracts of human beings and other animals and in sewage, swamps, and other places where organic material is decomposing under strongly anaerobic conditions. They have various shapes (figure 11.39); some stain gram-positive, and others stain gram-negative. All methanogens can obtain energy by oxidizing H_2, reducing CO_2 to methane gas (CH_4) in the process. Carbon is assimilated as CO_2 via a sequence of apparently unique reactions. Special enzymes not found in other bacteria participate in these and the methane-generating reactions. A list of genera of methanogens and some of their characteristics is given in table 11.17. Most of these bacteria can use formate (HCOOH), and the *Methanosarcina* can grow on methanol or acetate as well. More complex organic compounds do not serve as carbon and energy sources.

Methanogens are readily killed by oxygen, and, therefore, special techniques are required to grow them in the laboratory. The use of anaerobe jars as well as special anaerobic chambers (figure 11.40) is helpful, since even a brief exposure to O_2 can kill some methanogens. Cultures can be handled outside the jars and chambers by using the **Hungate technique** (figure 11.41). This technique involves the use of **roll tubes,** culture tubes that have been coated

• **Anaerobic jar, pp. 98–99**

with agar medium, leaving a hollow center. By flushing these continuously with gases such as nitrogen, having no O_2, cultures can be transferred and maintained with minimal exposure to O_2.

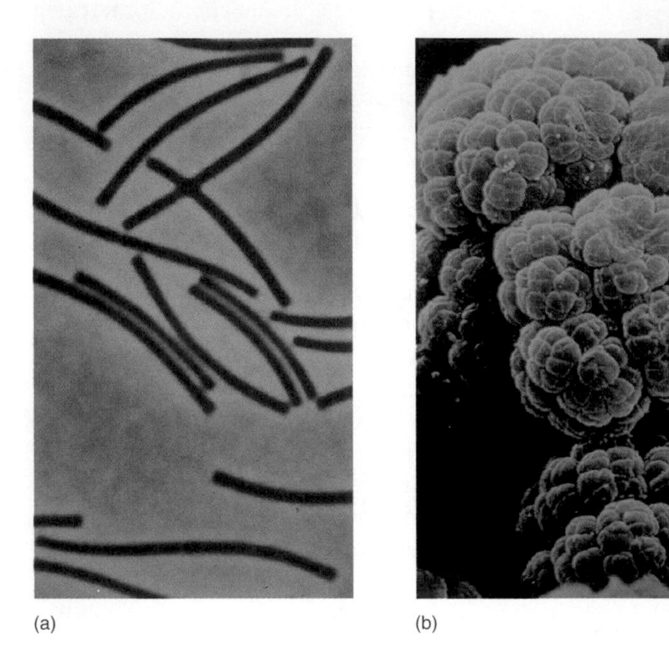

(a) (b)

Figure 11.39
Methanogenic bacteria. (*a*) Phase contrast photomicrograph of *Methanospirillum hungatei* (X2,000) and (*b*) scanning electron micrograph of *Methanosarcina* sp. (X6,000).

Figure 11.40
Chamber used for cultivating strict anaerobes.

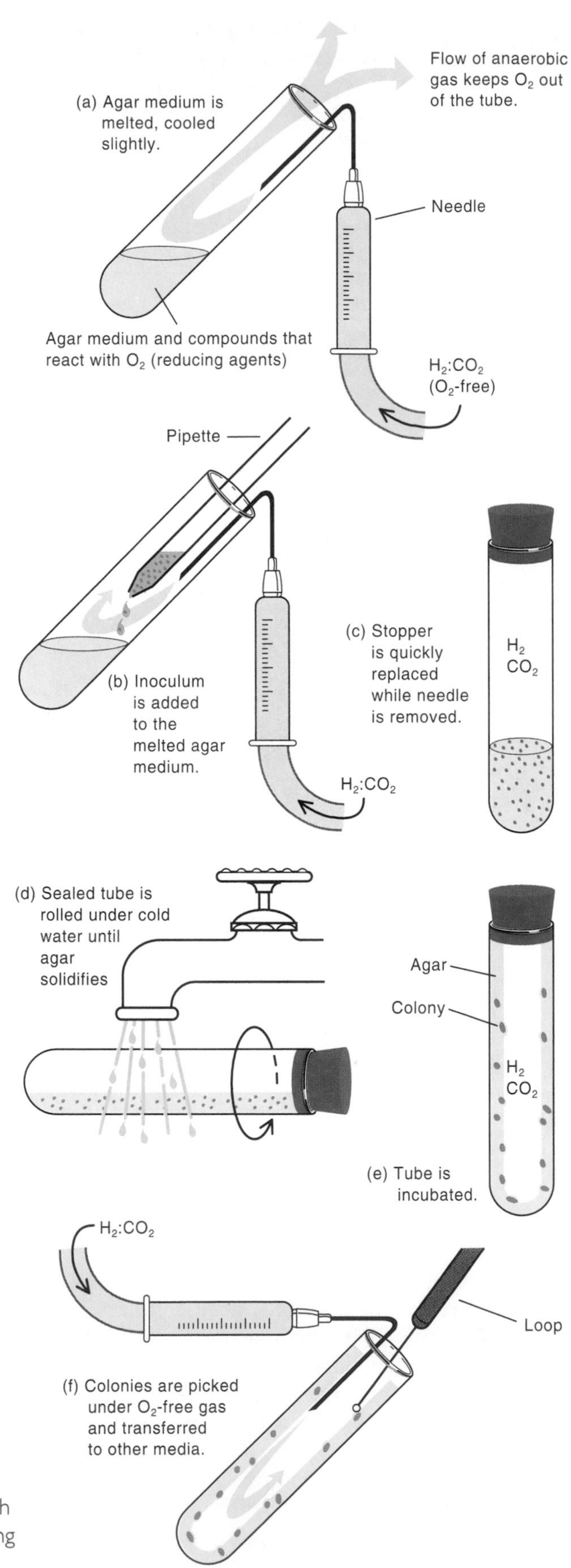

Figure 11.41
The Hungate technique. Tubes containing agar medium with a reducing agent are kept anaerobic after opening by flushing with gases free of O_2.

Table 11.17 Characteristics of the Methanogenic Bacteria

Genus	Shape	Growth Substrates
Methanobacterium	Long rod	H_2/CO_2
Methanobrevibacter	Short rod	H_2/CO_2
Methanococcus	Coccus	H_2/CO_2 or formate
Methanomicrobium	Short, straight, or curved, irregular rod	H_2/CO_2 or formate
Methanogenium	Irregular coccus	H_2/CO_2 or formate
Methanospirillum	Spiral	H_2/CO_2 or formate
Methanosarcina	Irregular coccus in packets	H_2/CO_2, methanol, or acetate

Halobacteria

A second group of archaea, the halobacteria, can live only in very strong salt solutions, such as salt ponds and brines used for curing fish, on which they may produce red patches of growth. They utilize chemical energy from organic material but are also capable of obtaining energy from light. Although they lack chlorophyll and the usual photosynthetic apparatus, they are able to absorb light energy by **bacteriorhodopsin,** a substance chemically related to rhodopsin, the light-sensitive pigment in the human retina. The absorbed light energy is used to generate the ATP needed for growth and motility.

Thermophilic Archaea

The last group of archaea includes more than a dozen genera that grow best under conditions of high temperature, usually in the presence of elemental sulfur. One type, called *Sulfolobus*, has been isolated from hot sulfur springs. These organisms have lobed, roughly spherical cells and obtain energy from oxidizing sulfur. They have temperature optima of 70°C to 75°C and pH optima of 2 to 3.

Another type is represented by *Thermoplasma* species, wall-less organisms that grow heterotrophically at a temperature optimum of 59°C and pH optima of 1 to 2. Although a number of other archaeal species grow in these extremely acid conditions, some prefer a neutral environment.

An extreme thermophile is *Pyrodictium occultum*, a dish-shaped archaeon that lives around volcanic vents in the floor of the sea. This species grows best at about 105°C, above the sea level boiling temperature of water. It is a strictly anaerobic lithotroph that uses hydrogen gas as an energy source and elemental sulfur as a final electron acceptor.

• **Hyperthermophiles, p. 91**

REVIEW QUESTIONS

1. Distinguish between α-hemolytic, β-hemolytic, γ, and viridans streptococci.
2. In what way does the metabolism of streptococci differ from that of staphylococci?
3. Name two genera of endospore-forming bacteria. How do they differ?
4. Give the principal metabolic by-product of lactobacilli.
5. Give two ways in which mycobacteria differ from most other vegetative bacteria.
6. What is an aerial hypha?
7. How do the two genera of nitrogen fixers, *Rhizobium* and *Frankia*, differ from each other?
8. How have species of *Streptomyces* contributed to the treatment of infectious diseases?
9. What distinguishes mycoplasmas and spiroplasmas from other bacteria?
10. Name three distinct groups within the domain Archaea.

Summary

I. Prokaryotes with a Gram-Negative Type of Cell Wall
   A. The spirochetes are flexible and have a spiral shape. Axial filaments contribute to their mobility.
   B. Aerobic, motile, helical, or comma-shaped gram-negative bacteria have a rigid cell wall, move by means of flagella, and require oxygen for growth.
      1. Microaerophiles require an atmosphere containing oxygen but at a much lower concentration than air.
      2. *Bdellovibrios* prey on other gram-negative bacteria, reproducing between the cell wall and cytoplasmic membrane.
   C. Gram-negative aerobic rods and cocci represent a large, diverse group of prokaryotes.
      1. *Rhizobium* species are important in converting atmospheric nitrogen (N_2) to a form usable by plants (nitrogen fixation). They live within the root cells of many plant species.
      2. Pseudomonads often produce nonphotosynthetic pigments that can color the medium green in some cases. They utilize many complex organic compounds. Some can use nitrate (NO_3^-) ion as a substitute for molecular oxygen (O_2), an example of anaerobic respiration.

3. Azotobacteria are soil organisms that can fix nitrogen under aerobic conditions.

D. Facultatively anaerobic gram-negative rods grow best aerobically but can also grow anaerobically.
   1. Enterobacteria include *Escherichia coli,* universally present in the human intestine.
   2. Vibrios and related bacteria are mostly marine organisms. Some possess luciferase and are luminescent.

E. Anaerobic gram-negative straight, curved, and helical fermentative rods have various shapes and are inhibited or killed by oxygen. Some species inhabit the body cavities of humans.

F. Dissimilatory sulfate- or sulfur-reducing bacteria utilize sulfur or oxidized sulfur compounds as a final electron acceptor. They are important in recycling sulfur.

G. The rickettsias and chlamydias are generally unable to grow outside the cells of a host animal but members of the genus *Bartonella* are an exception. Rickettsias are transmitted from one host to another by insects, mites, and ticks. Chlamydias multiply in a vegetative form and then differentiate into an infectious form. They are a unique group, lacking peptidoglycan in their cell walls.

H. Endosymbionts live within the cells of eukaryotic hosts in a stable association that can be beneficial to both organisms.
   1. Mitochondria and some chloroplasts of eukaryotes probably evolved from prokaryotic endosymbionts.
   2. Environmental factors can upset the endosymbiotic relationship in some protozoa, leading to death of both host and endosymbiont.
   3. Many types of endosymbionts of other eukaryotes are known, including endosymbionts of fungi, insects, and parasitic worms.

I. Gliding, nonphotosynthetic bacteria move slowly by an unknown mechanism when in contact with a surface. They can break down complex polysaccharides such as cellulose. One species causes a disease of fish.

J. Myxobacteria are gliding, fruiting bacteria. They can produce fruiting bodies, macroscopic aggregation of resting cells called *microcysts,* which are more resistant to heat, drying, and radiation than the vegetative cells. Chemical signals are involved in initiating and coordinating the action of the individual cells.

K. Sheathed bacteria are unbranching, filamentous forms in which the filament is enclosed in a sheath of lipoprotein-polysaccharide material. The sheath protects the bacteria and attaches them to solid structures. Sheathed bacteria may form brown scum in polluted streams, and they can plug pipes used in sewage treatment.

L. Anoxygenic phototrophic bacteria obtain their energy from sunlight but do not produce oxygen as a by-product of photosynthesis. They require anaerobic growth conditions and a hydrogen donor other than water. Their ancestors probably predominated in the early history of the earth.

M. Oxygenic photosynthetic bacteria
   1. Cyanobacteria are phototrophs that, like algae and plants, liberate oxygen during photosynthesis.
   2. Their chlorophyll type and photosynthetic process are essentially the same as those of algae and plants, but unlike these eukaryotes, many species of cyanobacteria can fix nitrogen.
   3. Accessory pigments called phycobiliproteins help trap light energy.

N. Budding and/or appendaged bacteria have projections of their cytoplasm into unusual structures. The structures function in attachment and increase the area for absorbing nutrients.

O. Aerobic chemolithotrophic bacteria can grow in the dark using an inorganic substance as an energy source and CO_2 for a carbon source. There are four types: nitrifiers, sulfur oxidizers, hydrogen oxidizers, and metal oxidizers.

II. Prokaryotes with a Gram-Positive Type of Cell Wall

A. Gram-positive cocci may occur in a chainlike arrangement or in clusters. Aerobic, facultative, and anaerobic forms exist. One group is remarkably radiation resistant.

B. Regular non-sporeforming gram-positive rods compose a group of mostly unicellular organisms with smooth, straight outlines. Some species of the genus *Lactobacillus* help protect against infectious disease; others have essential roles in making pickles, yogurt, and other foods.

C. Irregular shaped, nonspore-forming, gram-positive rods may show misshapen cells, uneven or metachromatic staining, or branching chains of bacteria. Snapping division is common.
   1. The corynebacteria are club-shaped and may show metachromatic staining. One species causes diphtheria.
   2. Arthrobacters help degrade pesticides in soil. Their life cycle includes changes in shape, from rod to coccus and back to rod.
   3. Members of the genus *Actinomyces* grow as branching filaments that fragment into rods.

D. Mycobacteria demonstrate the acid-fast staining property. Unusual cell wall lipids make them resistant to staining by other methods and also protect them against disinfectants. Certain species are responsible for human tuberculosis and leprosy.

E. Actinomycetes with aerial and specialized hyphae.
   1. Nocardioforms are filamentous organisms that usually form aerial mycelia and fragment into bacillary forms.
   2. Other actinomycetes develop spores called *conidia.* Conidia are not as resistant as endospores, except in the genus *Thermoactinomyces.* Conidia may occur in sacks, as in the *Frankia* and *Dermatophilus* genera. Streptomycetes and others include antibiotic-producing and disease-causing forms.

III. Prokaryotes that Lack Cell Walls
   A. The mycoplasmas
      1. The mycoplasmas lack cell walls. They appear to have evolved from an ancestor similar to gram-positive bacteria. One group, the spiroplasmas, has species with a spiral shape and motility similar to spirochetes.

IV. Prokaryotes Whose Cell Walls Lack Peptidoglycan
      1. Archaea are only distantly related to other prokaryotes. They lack peptidoglycan in their cell walls.
      2. They appear to have undergone less evolutionary change than eubacteria.
      3. They exist today mainly under conditions considered adverse for eubacteria.
      4. There are three kinds of archaea: methanogens, halobacteria, and thermophiles.

Chapter Review Questions

1. Why is it significant that some spirochetes parasitize lice and ticks?
2. What benefit do bdellovibrios get from attacking other bacteria?
3. What is the function of azotobacterial cysts? How do they differ from endospores?
4. Why were enterobacteria so named?
5. What bacterial structures can be used to distinguish one strain of *E. coli* from another?
6. Where in nature would bacteria of the genus *Vibrio* most likely be found?
7. Describe a characteristic of myxobacteria that distinguishes them from other gliding bacteria.
8. What kind of bacterium might be responsible for plugging the pipes in a sewage treatment facility?
9. What is the function of bacterial gas vesicles?
10. Name a genus of anaerobic endospore-forming bacteria.

Critical Thinking Questions

1. What sources of energy do different bacteria employ to carry out biosynthesis?
2. Can you think of any explanations for why archaea are found living in such adverse conditions as hot springs, strong salt solutions, and strict anaerobic environments?
3. Imagine that you are a scientist assigned the responsibility for establishing life on another planet. What are the tolerable limits of temperature, atmospheric and soil composition, and availability of water that would permit the use of naturally occurring (earth) bacteria?

Further Reading

Angert, E. R., Brooks, A. E., and Pace, N. R. 1996. Phylogenetic analysis of *Metabacterium polyspora:* clues to the evolutionary origin of daughter cell production in *Epulopiscium* species, the largest bacteria. *Journal of Bacteriology* 178:1451.

Baumann, Paul, et al. 1993. Origin and properties of bacterial endosymbionts of aphids, whiteflies, and mealy bugs. *ASM News* 59:21.

Bult, C. J., et al. 23 August 1996. Complete genome sequence of the methanogenic archaeon, *Methanococcus jannaschii. Science* 273:1058–1073.

Canby, T. Y. 1993. Bacteria. Teaching old bugs new tricks. *National Geographic* 184:36.

Earth's dominant life form is also its smallest: the microbe. 15 October 1996. *The New York Times,* p. B5.

Holt, J. G. (ed.). 1984, 1986, 1989. *Bergey's manual of systematic bacteriology,* vols.1–4.

Kluger, Jeffrey. 1992. Mars, in earth's image. *Discover* 13:70.

Morell, Virginia. 23 August 1996. Life's last domain. *Science* 273:1043–1045.

Murray, P. R. et al, (ed.). 1995. *Manual of clinical microbiology,* 6th ed. American Society for Microbiology, Washington, D.C. Detailed coverage of medically important bacteria.

Perry, J. J., and Staley, J. T. 1997. *Microbiology: dynamics and diversity.* Saunders College Publishing, New York. Excellent new text with an authoritative section on microbial evolution and divesity.

Prescott, L. M., Harley, J. P., and Klein, D. A. 1996. *Microbiology.* Wm. C. Brown Publishers, Dubuque, Iowa. Well-illustrated and readable general microbiology text.

Ruby, E. G., and McFall-Ngai, M. J. 1992. A squid that glows in the night: development of an animal-bacterial mutualism. *Journal of Bacteriology* 174:4865.

Shapiro, J. A. 1988. Bacteria as multicellular organisms. *Scientific American* 258(6):82–89.

Staley, J. T., et al. 1997. *The microbial world—foundation of the biosphere*. The American Society for Microbiology, Washington, D.C. A report based on an American Academy of Microbiology colloquium, January 19–21, 1996, concerning the three domains of life.

Stoeckenius, W. 1976. The purple membrane of salt-loving bacteria. *Scientific American* 234(6):38–46. An interesting paper on the role of bacteriorhodopsin in the energy metabolism of halobacteria, written by its discoverer.

Woese, C. R. 1981. Archaebacteria. *Scientific American* 244(6):98–122.

12 Eukaryotic Microorganisms: Algae, Fungi, and Protozoa

KEY CONCEPTS

1. Algae, fungi, and protozoa are eukaryotic organisms and as such have eukaryotic cell structures. Many of these organisms are microscopic.
2. Algae are very important in the food chain and carbon cycle but are of minor importance in causing disease in humans.
3. Fungi play an important role in the decomposition of organic materials. They also cause a great deal of damage to food crops as well as other materials, and they are responsible for some serious diseases in humans.
4. Protozoa are single-celled, often motile organisms that are an important part of zooplankton and the food chain. Some species cause diseases that affect a large part of the world's population.

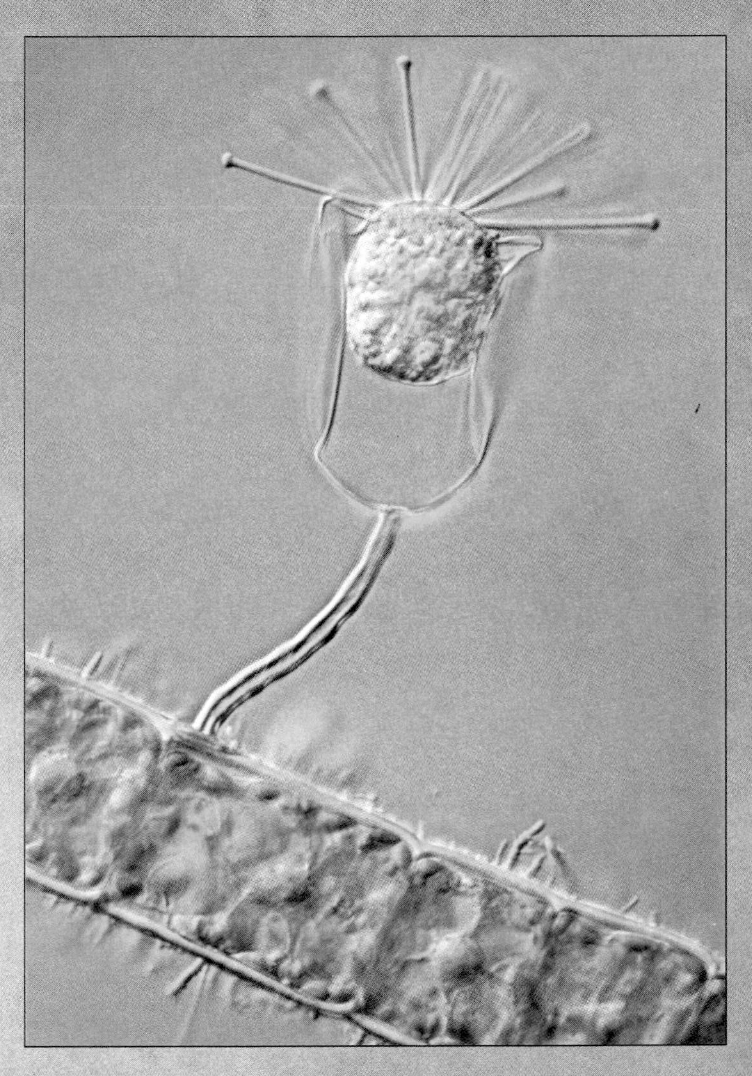

Protozoan (Paracineta) on green algae (spongomorpha, ×320).

PREVIEW LINK

Microbes in Motion

The following books and chapters in the *Microbes in Motion* CD-ROM may serve as a useful preview or supplement to your reading:

Fungal Structure and Function: Specific Fungal Structures; Metabolism and Growth. *Parasitic Structure and Function:* Protozoa Classification.

A Glimpse of History

Theobald Smith is a relatively unknown microbiologist yet his discovery that disease agents could be carried by insect vectors was one of the most important discoveries ever made in the history of science and in the control of disease. In 1884, when Smith became a staff member of the U.S. Bureau of Animal Industry, Texas cattle fever was a major problem among large cattle herds. Cattlemen would bring healthy cattle from the north and graze them beside healthy cattle in the south. In about 30 days, the northern cattle would come down with Texas cattle fever. The animals would stop eating and lose weight. Their urine would run red with blood and within a few days they would die. Surprisingly, the southern cattle remained unaffected.

In 1888 Dr. Salmon,[1] director of the Bureau of Animal Industry, directed Theobald Smith and F. L. Kilbourne to determine the cause of this disease. Local farmers suggested that ticks found on the cattle might be involved. To test this possibility, Smith and Kilbourne devised the following classic experiment.

Seven tick-infested cows were taken from a Texas fever area and four were put into a field with six healthy northern cows. The ticks were removed by hand from each of the three remaining cows, which were then placed in a separate field with four healthy northern cows. The healthy northern cows in the field with the tick-infested southern cows developed Texas fever and died. However, the healthy northern cows in the field with the southern tick-free cows remained healthy. Smith then took two of the healthy northern cows from the tick-free field and put them into the same field in which the northern cows had died. Although he had removed the southern cows, the ticks still remained in the grass. These healthy cows were bitten by the ticks, developed the symptoms of Texas cattle fever, and died.

Not only did Smith demonstrate that ticks were responsible for the spread of the disease, but he was also able to identify the causative agent of this disease as a pear-shaped protozoan, *Babesia bigemina,* carried by the ticks.

Why were northern cows susceptible and southern cows immune? The following explanation accounts for this observation. If the cattle contracted Texas fever when they were young, they suffered only a mild infection. These cattle recovered and were thereafter immune to the disease. However, cows that contracted the disease as adults developed the typical symptoms of Texas fever and usually died. This explained why the ticks affected only the northern adult cows who were moved to the south as adults. The southern cows exposed to the disease early in life were immune.

Smith's demonstration that disease agents could be transmitted by insect hosts opened the door to the discovery that other diseases could be transmitted in the same fashion. In later years, diseases such as malaria, yellow fever, and trypanosomiasis were shown to be transmitted in a similar fashion. When the insect vector was found, the spread of the disease could then be controlled by destroying or controlling the vector.

[1]The bacterium *Salmonella choleraesius,* the cause of hog cholera, was named for Dr. Daniel E. Salmon even though Theobald Smith was the actual discoverer of the organism.

Although algae, fungi, and protozoa are classified as different organisms, they have one feature in common: their eukaryotic nature. Their basic cell structure is the same as that of multicellular plants and animals and very different from that of prokaryotes. As such, they are not considered in *Bergey's Manual.* However, they are included in a textbook of microbiology because many of them are microscopic and are studied with techniques that are similar to those used to study bacteria. In addition, many of these organisms cause disease in humans as well as plants and animals.

We will now consider each of these groups, emphasizing especially those important in the health-related fields.

Algae

The algae are a diverse group of eukaryotic organisms that share some fundamental characteristics. Generally, algae contain chlorophyll *a* and carry out oxygenic photosynthesis but do not have the complex tissue development or multicellular reproductive structures found in higher plants (figure 12.1). This group includes both microscopic unicellular members and macroscopic multicellular members. Algae do not infect humans directly, but some produce toxins that cause paralytic shellfish poisoning. As primary producers, algae are extremely important in the food chains of the world.

Algae should not be confused with the cyanobacteria, which are prokaryotes and were formerly referred to as *blue-green algae.*

Algae Are Classified According to Their Photosynthetic Pigments

Some of the general characteristics of the algal groups are summarized in table 12.1. Classification is based on a number of properties including the principal photosynthetic pigments of each group, cell wall structure, type of storage products, mechanisms of motility, and mode of reproduction. The names of the different groups are derived for the most part from the major color displayed by most of the algae in that group.

Algae Occur in a Variety of Habitats and Can Be Either Microscopic or Macroscopic

Algae are found in freshwater and saltwater, in soil, and in symbiotic[2] relationships with other plants and animals. Since the oceans cover more than 70% of the earth's surface, aquatic algae are major producers of oxygen as well as important users of carbon dioxide. Unicellular algae make up

[2]Symbiotic—describes an intimate relationship between members of different species

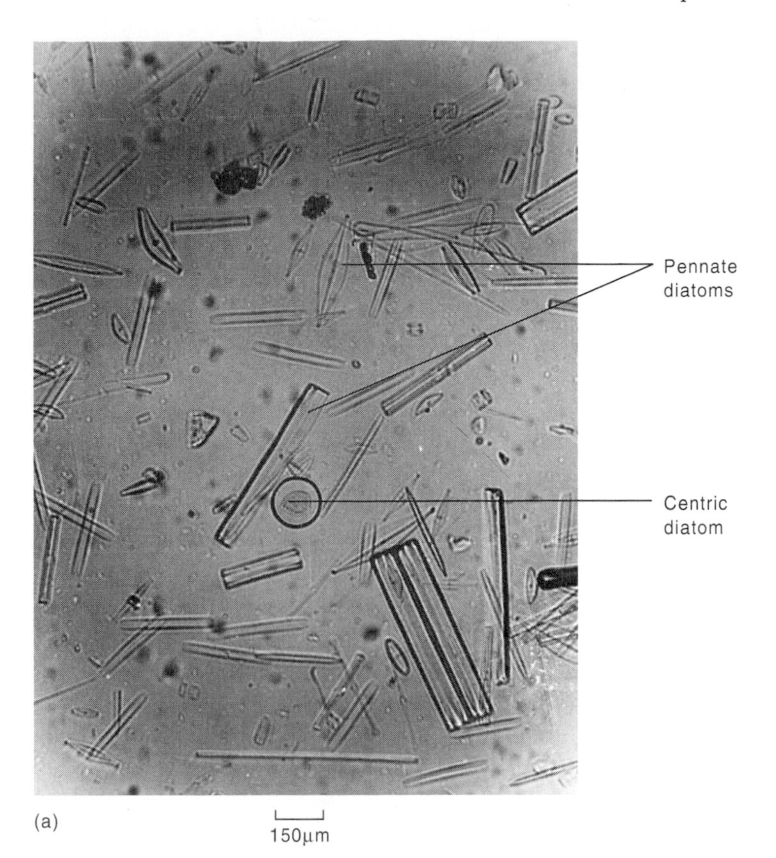

(a)

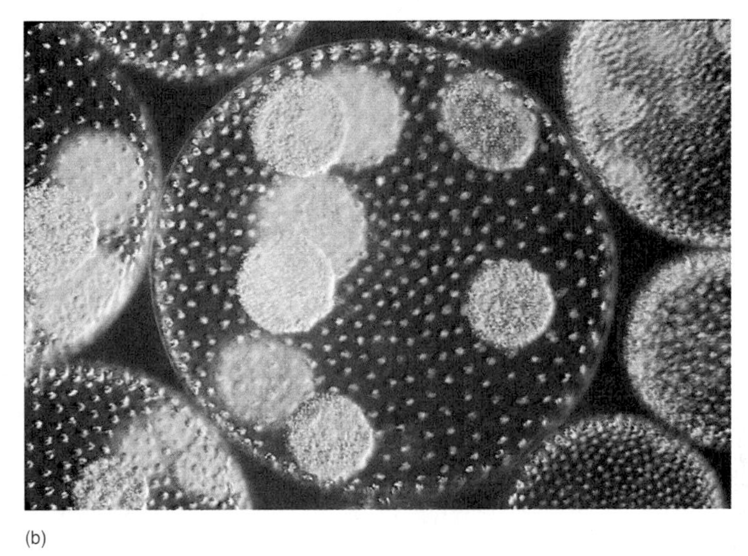

(b)

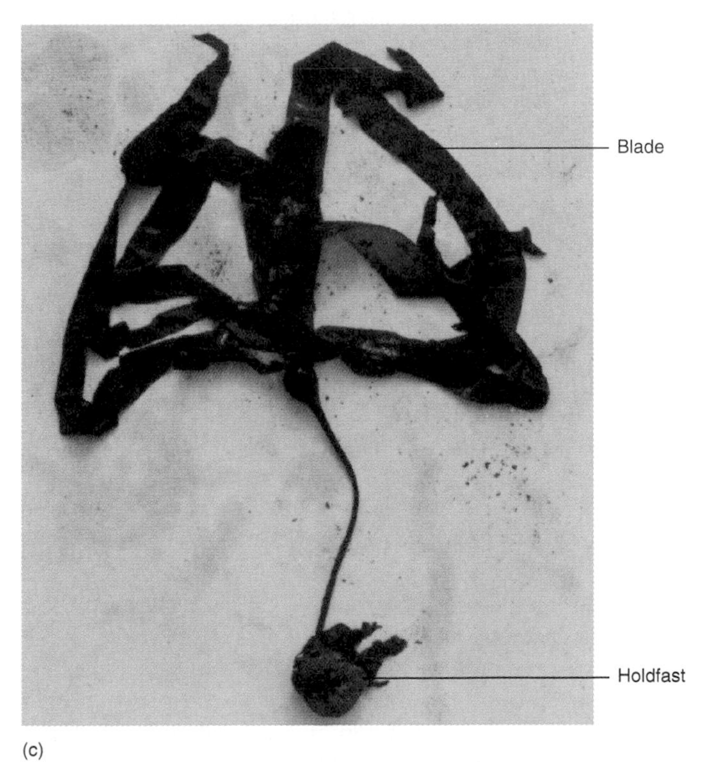

(c)

(d)

Figure 12.1
(*a*) *Various* diatoms, showing the variety of shapes and sizes. (*b*) *Volvox*, a colony of cells formed into a hollow sphere (×125). The green circles are reproductive cells that will eventually become new colonies. (*c*) *Laminaria saccharina* and (*d*) *Sargassum muticum* are examples of some of the large macroscopic species of algae found in the waters of Puget Sound, Washington. Note in (*d*) the small gas bladders that keep this alga afloat to take full advantage of the sun.

Table 12.1 Characteristics of Major Groups of Algae

Group and Representative Member	Usual Habitat	Principal Pigments*	Storage Products	Cell Walls	Mode of Motility (If Present)	Mode of Reproduction
Chlorophyta Green algae	Freshwater and saltwater; soil; tree bark, lichens	Chlorophyll *b*; carotenes; xanthophylls	Starch (α-1, 4-glucan)	Cellulose and pectin	Mostly nonmotile (except one order) but some reproductive elements flagellated	Asexual by multiple fission; spores or sexual
Phaeophyta Brown algae	Saltwater	Xanthophylls, especially fucoxanthin	Starchlike carbohydrates; annitol; fats	Cellulose and pectin; alginic acid	Two unequal, lateral flagella	Asexual, motile zoospores; sexual, motile gametes
Rhodophyta Red algae Corallines	Mostly saltwater; several genera in freshwater	Phycobilins including phycoerythrin and phycocyanin; carotenes; xanthophylls	Starchlike carbohydrates	Cellulose and pectin; agar; carrageenan	Nonmotile	Sexual, gametes; asexual spores
Chrysophyta Diatoms Golden brown algae	Freshwater and saltwater; soil, higher plants	Carotenes	Starchlike carbohydrates (β-1, 3-glucan); oils	Pectin, often impregnated with silica or calcium	Unique diatom motility; one, two, or more unequal flagella	Asexual or sexual
Pyrrophyta Dinoflagellates	Mostly saltwater but common in freshwater	Carotenes; xanthophylls	Starch; oils	Cellulose and pectins	Two unequal, lateral flagella in different planes	Asexual; rarely sexual
Euglenophyta Euglena	Freshwater	Chlorophyll *b*; carotenes; xanthophylls	Fats; starchlike carbohydrates	Lacking but elastic pellicle present	One to three anterior flagella	Asexual only, by binary fission
Cryptophyta Cryptomonads	Saltwater	Chlorophyll *d*; carotenes	Starch; oil	No cellulose	Two unequal flagella	Asexual by longitudinal division
Xanthophyta Yellow-green algae	Freshwater and saltwater	Carotenes	Chrysolamina; oils	Pectin	Two unequal, apical flagella	Asexual or sexual

*In addition to chlorophyll *a*, present in algae of all phyla

a significant part of the **phytoplankton,**[1] the free-floating, photosynthetic organisms that are found in marine environments. More oxygen is produced by the phytoplankton than by all forests added together. Phytoplankton is a major food source for many animals, both large and small. Microscopic zooplankton[2] graze on the phytoplankton. Zooplankton include the krill and multicellular but microscopic crustaceans that are food for the world's largest mammals (the benthic whales). The unicellular algae of the phytoplankton are well adapted to an aquatic environment. As single cells, they have large, adsorptive surfaces relative to their volume and can move freely about, thus effectively using the dilute nutrients available.

Because one or more algal species is capable of growing in almost any environment, algae often grow where other forms of life cannot thrive. Frequently, algae are among the first organisms to become established in barren environments, where they synthesize the organic materials necessary for the subsequent invasion and survival of other plants and animals. They are often found on rocks, preparing the surface for the growth of more complex members of the biological community.

[1]Phytoplankton—from the Greek *phyto* meaning "plant" and *plankton* meaning "drifting"

[2]Zooplankton—from the Greek *zoion* meaning "animal" and *plankton* meaning "drifting"

Perspective 12.1

Could the Algae Growing in the Ocean Be More Important to Global Warming than Tropical Rain Forests?

In January 1992, a research vessel left Seattle, Washington, with 33 scientists aboard to explore the Pacific Ocean between Hawaii and Tahiti. The object of the mission was to determine whether the oceans are capable of removing carbon dioxide from the atmosphere. By careful measurements, these scientists determined that of the 5.5 billion tons of CO_2 spewed into the air each year, 3.5 billion tons stay in the atmosphere. However, they did not know what happens to the other 2.0 billion tons. It is known that plants on the land use CO_2 during photosynthesis, which can account for some of the difference. Is it the algae in the oceans that can account for large amounts of CO_2 removal or is it something else? By analyzing the data collected on this excursion along with other data collected by scientists working worldwide, definitive answers may be found.

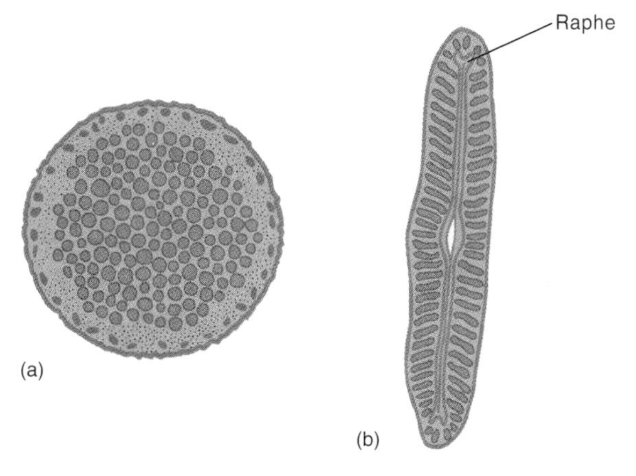

Figure 12.2
Diatom (*a*) Centric. (*b*) Pennate, with a central slit (raphe) through which mucilage is extruded. Diatoms with a raphe exhibit a characteristic gliding movement.

Figure 12.3
Construction of a diatom frustule. The hard coverings of diatoms (frustules) resemble covered boxes or dishes. When the cell divides, each daughter cell retains half of the parent frustule and subsequently forms another corresponding half.

Figure 12.4
Nereocystis luetkeana, the bladder kelp. The alga has a large bladder filled with carbon monoxide and air. The bladder keeps the blade (the active photosynthetic sites of the alga) floating on the surface of the water to expose them to sunlight. The holdfast anchors the kelp to rocks or other surfaces. The individual kelp in these pictures is young and relatively small. In a single season, these kelp can grow to lengths of 5 to 15 meters.

Microscopic Algae

Microscopic algae can be single-celled organisms floating free or propelled by flagella, or they can grow in long chains or filaments. Some microscopic algae such as *Volvox* form colonies of 500 to 60,000 biflagellated cells, which can be visible to the naked eye (see figure 12.1). Other microscopic algae such as the diatoms (figure 12.2) have three-dimensional structures called *frustules*. They are constructed in the same way as a petri dish, with one part of the cell fitting over the other (figure 12.3).

Macroscopic Algae

Macroscopic algae are multicellular organisms with a variety of specialized structures that serve specific functions (figure 12.4). Some possess a structure called a **holdfast,** which looks like a root system but in fact has a completely different function. A holdfast anchors the organism to a rock or some other firm substrate. Unlike a root, it is not used to obtain water and nutrients for the organism. The stalk of an alga, known as the **stipe,** usually has leaflike structures known as **blades** attached to it. The blades are the photosynthetic portion of an alga, and some also bear the reproductive structures. Many large algae have gas-containing bladders or floats that help them maintain their blades in a position suitable for obtaining maximum sunlight.

Algae Are Typical Eukaryotic Organisms

Algae have the typical eukaryotic cell wall structure with a membrane-bound nucleus as well as membrane-bound organelles.

Cell Walls

Algal cell walls are rigid and for the most part composed of cellulose (figure 12.5), often with pectin[1] associated with it. Additional materials such as silicon, calcium, alginic acid, mannan, xylan, agar, carrageenan, and some mucilaginous compounds may also be present. Some multicellular species of algae containing large amounts of compounds such as carrageenan and agar in their cell walls are harvested on a commercial scale. Carrageenan and agar both are used in food as stabilizing compounds. In addition, agar is important for growing microorganisms in the laboratory.

Diatoms have silicon dioxide incorporated into their cell walls. When these organisms die, their shells sink to the bottom of the ocean, and the siliceous[2] material does not disintegrate. Deposits of diatoms that formed millions of years ago are mined for a substance known as **diatomaceous earth,** used for filtering systems, abrasives in polishes, insulation, and many other purposes.

Genetic Material

As in all eukaryotes, the genetic information of algae is contained in a number of chromosomes that are tight packages of DNA with their associated basic protein. These chromosomes are enclosed in a nuclear membrane.

[1] Pectin—polygalacturonic acid, a water soluble polysaccharide found in a variety of algal and plant cell walls, often used to make jam or jelly

[2] Siliceous—containing silicon

- **Cell wall, pp. 54–62**
- **Chromosomes, p. 81**
- **Chloroplasts, p. 77**

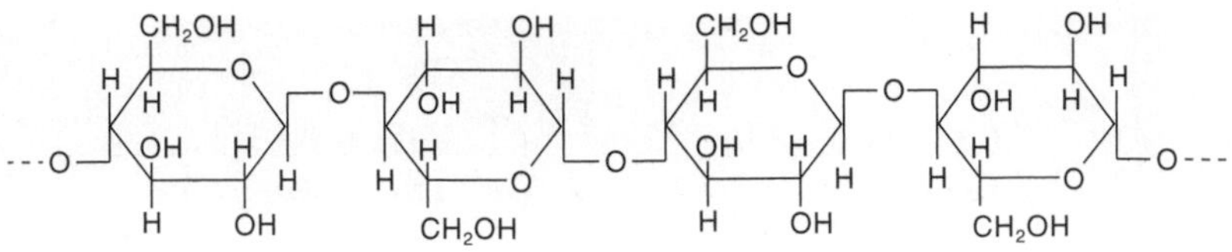

Figure 12.5
Chemical structure of cellulose.

Other Membrane-Bound Organelles

Algae have cytoplastic organelles called **chloroplasts.** These chloroplasts contain chlorophyll as well as other light-trapping pigments such as **carotenoids.** The chloroplast is the site in the cell where photosynthesis occurs. Mitochondria are the organelles in which respiration and oxidative phosphorylation occur.

Algae Reproduce by Both Asexual and Sexual Means

Most single-celled algae reproduce asexually by binary fission, as do most bacteria. The major difference between prokaryotic and eukaryotic fission involves events that take place in the nucleus. In prokaryotic fission, the circular DNA replicates and each daughter cell receives half the original double strand of DNA and a newly replicated strand. In eukaryotic organisms, after the DNA is replicated, the chromosomes go through a nuclear division process called **mitosis** (figure 12.6). This process ensures that the daughter cells receive the same number of chromosomes as the original parent.

Some algae, especially multicellular filamentous species, reproduce asexually by **fragmentation.** In this type of reproduction, portions of the parent organisms break off to form new organisms (figure 12.7), and the parent organism survives.

Sexual reproduction also regularly occurs in most algae. Sexual reproduction involves an actual fusion of the **haploid**[3] nuclei of two parental cells so that a **diploid**[4] cell is formed. The two cells that fuse are called **gametes,** and the product of their fusion is a **zygote.** The gametes are often flagellated and highly motile.

For an organism to reproduce sexually (that is, for its gametes to fuse to form a zygote), the number of chromosomes in the gamete must first be reduced from 2n (the diploid cell) to 1n (the haploid cell). When the gametes fuse, they will have a total of 2n chromosomes. Each species has a characteristic number of chromosomes. In humans, for

[3] Haploid—containing one complement of chromosomes

[4] Diploid—containing two complements of chromosomes, one from each haploid gamete

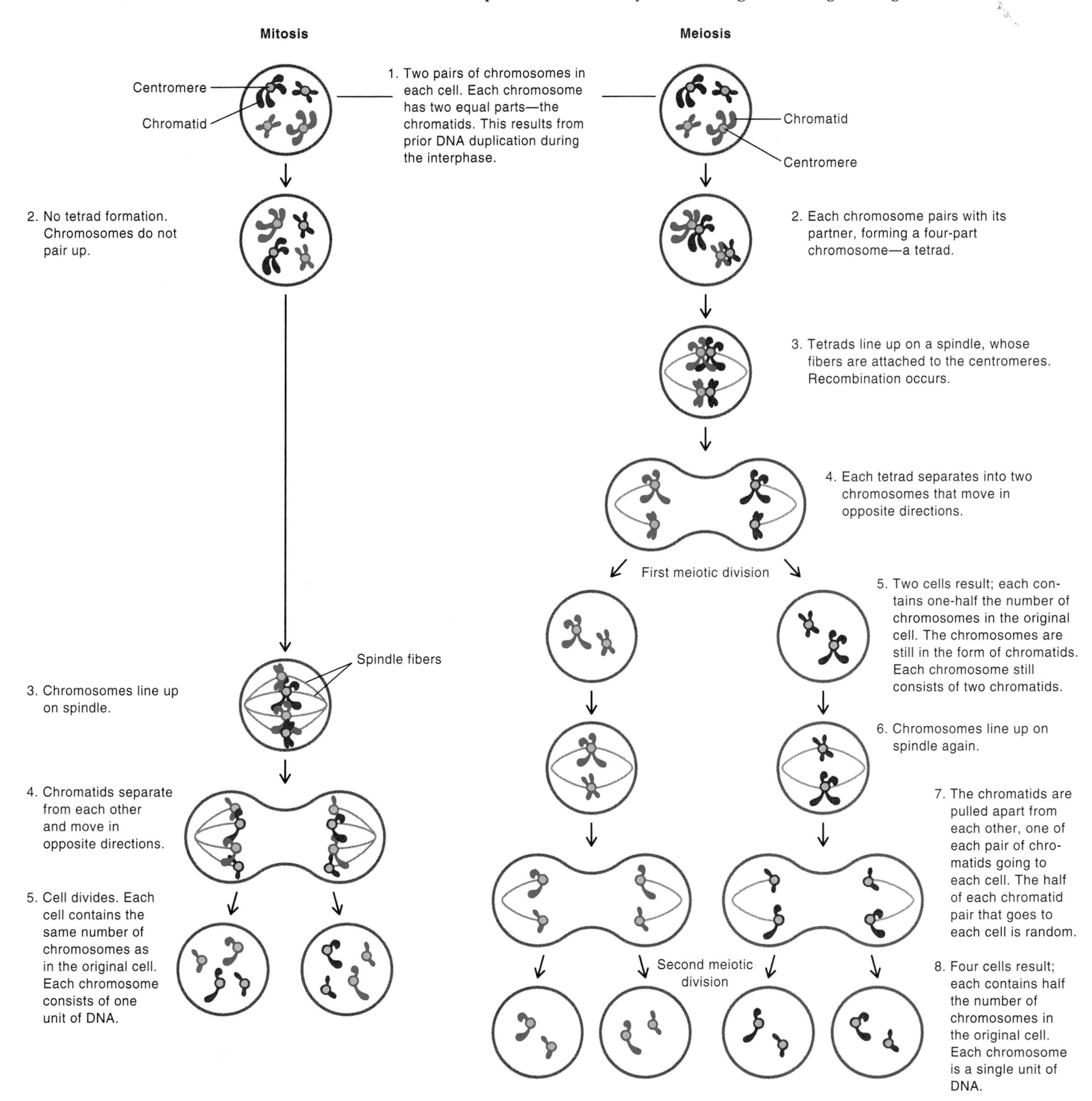

Figure 12.6
Diagrammatic representation of mitosis and meiosis.

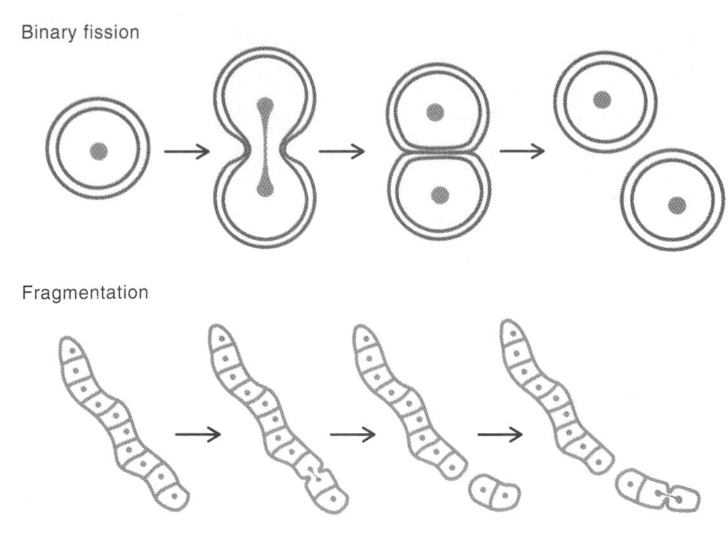

Figure 12.7
Binary fission is an asexual reproduction process in which a single cell divides into two independent daughter cells. Fragmentation is a form of asexual reproduction in which a filament composed of a string of cells breaks into pieces to form new organisms.

example, there are 23 pairs of chromosomes, for a total of 46. The process by which this reduction in chromosome number occurs is called **meiosis** (see figure 12.6).

Some Algal Species Produce Toxins that Cause Paralytic Shellfish Poisoning

Several dinoflagellates of the group Pyrrophyta cause "red tide," or algal blooms in the ocean. Red tide was reported in the Bible along the Nile and today seems to be spreading worldwide. In the warm waters of Florida and Mexico, the red tides result from an abundance of *Gymnodinium breve.* This dinoflagellate discolors the water about 10 to 45 miles from shore and produces brevetoxin which kills the fish that feed on the phytoplankton. It is unclear why algae suddenly grow in such large numbers, but it is thought that sudden changes in conditions of the water are responsible. There is some thought that the runoff of fertilizers along the waterways and coastlines may also be the cause of red tides. In addition, an upwelling of the water often brings more nutrients as well as the cysts[1] of the *G. breve* from the ocean bottom to the surface. When these cysts encounter warmer waters and additional nutrients at the surface, they are released from their resting state and begin to multiply rapidly. Persons eating fish that have ingested *G. breve* and thus contain brevetoxin may suffer a tingling sensation in their mouths and fingers, a reversal of hot and cold sensations, reduced pulse rate, and diarrhea. The symptoms may be unpleasant but are rarely deadly, and people recover in two to three days.

[1] Cysts—resting, resistant stage of the alga

Red tides caused by the dinoflagellate of the genus *Gonyaulax* are much more serious. *Gonyaulax* species produce neurotoxins such as saxitoxin and gonyautoxins, some of the most potent nonprotein poisons known. Shellfish such as clams, mussels, scallops, and oysters feed on these dinoflagellates without apparent harm and, in the process, accumulate the neurotoxin in their tissues. Then when humans eat the shellfish, they suffer symptoms of paralytic shellfish poisoning including general numbness, dizziness, general muscle weakness, and impaired respiration. Death can result from respiratory failure. *Gonyaulax* species are found in both the North Atlantic and the North Pacific. They have seriously affected the shellfish industry on both coasts over the years. At least 200 manatee died along the coast of Florida in the spring of 1996 as a result of red tide poisoning.

It should be noted that red tides may be present even when the water is not obviously discolored. Cooking the shellfish does not destroy the toxin.

Another algal toxin found in some species of diatoms has more recently been recognized to cause paralytic shellfish poisoning. This poison is domoic acid and is most often associated with mussels and other shellfish and crabs that accumulate the poison. Persons eating the shellfish suffer nausea, vomiting, diarrhea, and abdominal cramps as well as some neurological symptoms such as loss of memory. Cooking does not destroy this heat-stable poison.

A newly discovered dinoflagellate, *Pfisteria piscida,* kills fish by dispersing its poison directly into the water. This poison is so potent that researchers working with it in a laboratory were seriously affected. This organism now must be worked on in a level-three laboratory, the same level as HIV, the virus that causes AIDS.

State agencies constantly monitor for red tide toxins and domoic acid, and it is wise to check with the local health department before taking shellfish for human consumption.

REVIEW QUESTIONS

1. What is the major difference between bacteria and algae?
2. What is the primary characteristic that is used to distinguish the major groups of algae?
3. What disease or diseases do algae cause in humans?
4. Of what economic importance are algae?
5. What is the difference between mitosis and meiosis?

Fungi

The fungi are a diverse group of over 100,000 species of nonphotosynthetic, heterotrophic, eukaryotic organisms. Many are microscopic and can be examined using basic microbiological techniques; others are macroscopic (figure 12.8). Most fungi are saprophytes, obtaining their food from dead

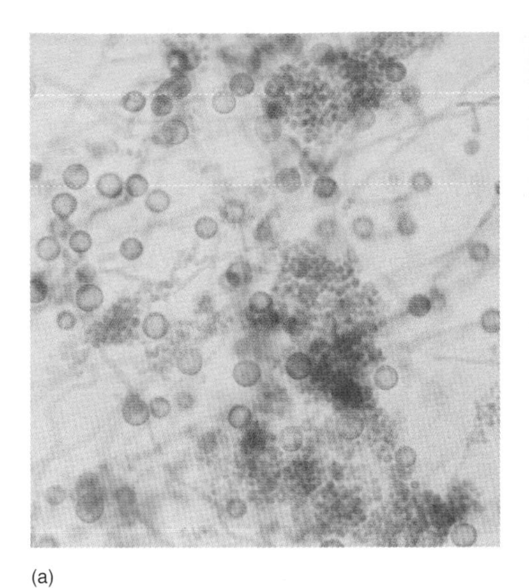

(a)

(b)

(c)

Figure 12.8
Fungi range in size from microscopic to macroscopic forms. (*a*) Microscopic *Candida albicans*, showing the chlamydospores (large, round circles) that are highly resistant to adverse conditions. (*b*) *Polyporus sulphureus* (chicken of the woods), a shelflike fungus growing on a tree. (*c*) *Amanita muscaria*, a highly poisonous mushroom, growing in a cranberry bog on the Oregon coast.

organic material. A large number of fungi cause disease in plants. Fortunately, only a few species cause disease in animals and humans. However, as modern medicine has advanced to treat once-fatal diseases, it has left many individuals with impaired immune systems. It is these immunocompromised individuals who are most vulnerable to the fungal diseases.

Some species are useful sources of food and industrial products, whereas others can spoil almost any organic material on which they grow. Along with bacteria, fungi are the principal decomposers of carbon compounds in the biosphere.[1] This decomposition releases carbon dioxide into the atmosphere and nitrogen compounds into the soil, which are then taken up by plants and converted into organic compounds. Without this breakdown of organic material, the world would quickly be overrun with organic waste.

Fungi Occur in Diverse Habitats

Fungi are found in nearly every habitat on the earth where organic materials exist. Whereas algae are primarily aquatic, the fungi are mainly terrestrial organisms. Some species occur only on a particular strain of one genus of plants, whereas others are extremely versatile in what they can attack and use as a source of carbon and energy. Materials such as leather, cork, hair, wax, ink, jet fuel, and even some synthetic plastics like the polyvinyls can be attacked by fungi. Some species can grow in concentrations of salts, sugars, or acids high enough to kill most bacteria. These fungi are often responsible for spoiling pickles, preserves, and other foods. Some fungi are resistant to pasteurization and others can grow at temperatures below the freezing point of water, rotting bulbs and destroying grass in frozen ground. Fungi are found in the thermal pools at Yellowstone National Park, in volcanic craters, and in lakes with very high salt content, such as the Great Salt Lake and the Dead Sea.

Fungal reproductive cells, or spores, are found throughout the earth. They also occur in tremendous numbers in the air near the earth's surface as well as at altitudes of more than 7 miles. Although not as resistant as bacterial endospores, fungal spores are generally resistant to the ultraviolet rays of sunlight. However, sunlight will sometimes kill fungal vegetative cells. Fungal spores are a major cause of asthma.[2]

Most fungi prefer a slightly moist environment with a relative humidity of 70% or more, and various species can grow at temperatures ranging from –6°C to 50°C. The optimal temperature for the majority of fungi is in the range of 20°C to 35°C.

The pH at which different fungi can grow varies widely, ranging from as low as 2.2 to as high as 9.6, but fungi usually grow well at an acid pH of 5.0 or lower. This explains why fungi grow well on fruits and many vegetables that tend to be acidic.

[1] Biosphere—that portion of the earth on which life exists

[2] Asthma—an immediate hypersensitivity (allergic) reaction that occurs in the lungs (see chapter 17)

Most Fungi Are Aerobic Heterotrophic Organisms

As heterotrophs, fungi secrete a wide variety of enzymes that degrade organic materials, especially complex carbohydrates, into small molecules that can be readily absorbed by other organisms. Most fungi are aerobic, but some of the yeasts are facultative anaerobes and carry out alcoholic fermentation. Facultatively anaerobic fungi live in the intestines of certain species of fish and help degrade algae. Some zoosporic fungi living in the rumen of cows and sheep are known to be obligately anaerobic. They are important in the digestion of the plant material that these animals ingest.

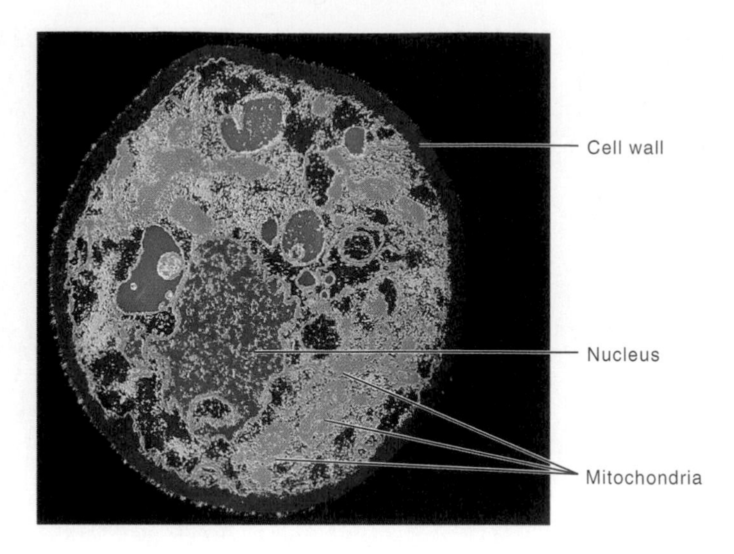

Figure 12.9
Morphology of a yeast cell as seen with an electron microscope.

The Terms *Yeast, Molds,* and *Mushrooms* Are Common Names that Refer to Various Fungal Forms

When people talk about fungi, they frequently use terms such as *yeast, mold,* and *mushroom.* These terms have nothing to do with the classification of fungi but instead indicate their morphological forms.

Yeasts are single-celled fungi. The morphology of a typical yeast cell can be seen in figure 12.9. Yeasts can be spherical, oval, or cylindrical and are usually 3 µm to 5 µm in diameter. They generally reproduce by budding, in which a small outgrowth on the cell produces a new cell (figure 12.10).

Molds are fungi that are filamentous. A single filament is known as a **hypha** (pl. hyphae), and a collection of hyphae growing together are known as a **mycelium.** Hyphae develop from fungal spores. A fungal reproductive spore is typically a single cell about 3 µm to 30 µm in diameter, depending on the species. When a fungal spore lands on a suitable substrate, it germinates and sends out a projection called a **germ tube** (figure 12.11). This tube grows at the tip and develops into a hypha. As the hypha grows, it branches into a tangle of hyphae, or a mycelium. The white mass seen on moldy bread (figure 12.12) is an example of a mycelium. Only a small portion of the mycelium is actually visible on the surface of the bread; the rest is buried deep within.

A mycelium is well adapted to absorb food. The narrow, threadlike hyphae with their high surface-to-volume ratio can absorb large amounts of nutrients. In addition, each hypha releases chemicals that repel the growth of other hyphae near it. As a result, hyphae spread throughout the food source, ensuring that each hypha will have access to adequate nutrients.

Parasitic fungi have specialized hyphae called **haustoria,** which are able to penetrate animal or plant cell walls to gain nutrients. Saprophytic fungi sometimes have specialized hyphae called **rhizoids,** which anchor them to the substrate.

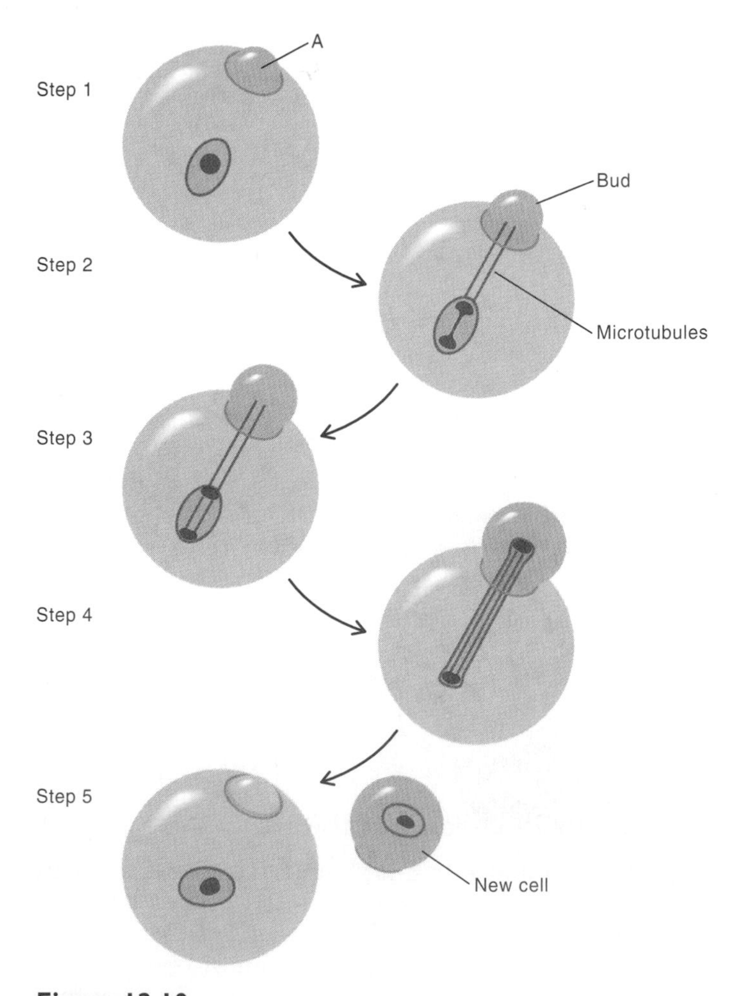

Figure 12.10
Budding in yeast. (*Step 1*) Cell wall softens at point A, allowing for the bulging out of the cytoplasm. (*Steps 2 and 3*) Nucleus divides by mitosis, and (*Step 4*) one of the nuclei migrates into the bud. (*Step 5*) Cell wall grows together, and the bud breaks off, forming a new cell.

Perspective 12.2

Pneumocystis carinii—Protozoan or Fungus?

As discussed in chapter 10, the classification of an organism is subject to change with each new development in classification techniques. As a result, organisms once thought to be related by purely phenotypic characteristics are often not related when the DNA or ribosomal RNA sequences are examined.

Since its discovery, the organism *Pneumocystis carinii,* a leading cause of pneumonia and death in patients with AIDS, has been classified as a protozoan. However, when the ribosomal RNA sequences of *P. carinii* were compared to the ribosomal RNA sequences of fungi, protozoa, animals, and plants, they were found to be most closely related to ustomycetus, the red yeast fungi, and not very closely related to protozoa.

Other characteristics of *P. carinii* also suggest that it is a fungus and not a protozoan. For example, pneumocystic infections generally occur only in lung tissue and appear to be acquired by inhalation in much the same manner as other fungal infections. Usually, protozoan infections require growth in an intermediate host and are passed from one individual to another by direct contact, ingestion of contaminated food, injection by an insect, or other similar means. This discovery has important implications for therapy leading to new and better medicines for treatment.

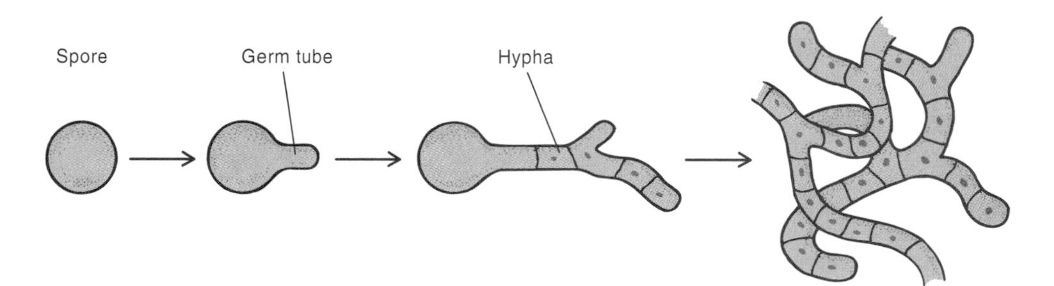

Figure 12.11
Spores of fungi germinate by forming a projection from the side of the cell called a *germ tube*, which elongates to form hyphae. As the hyphae continue to grow, they form a tangled mass called a *mycelium.*

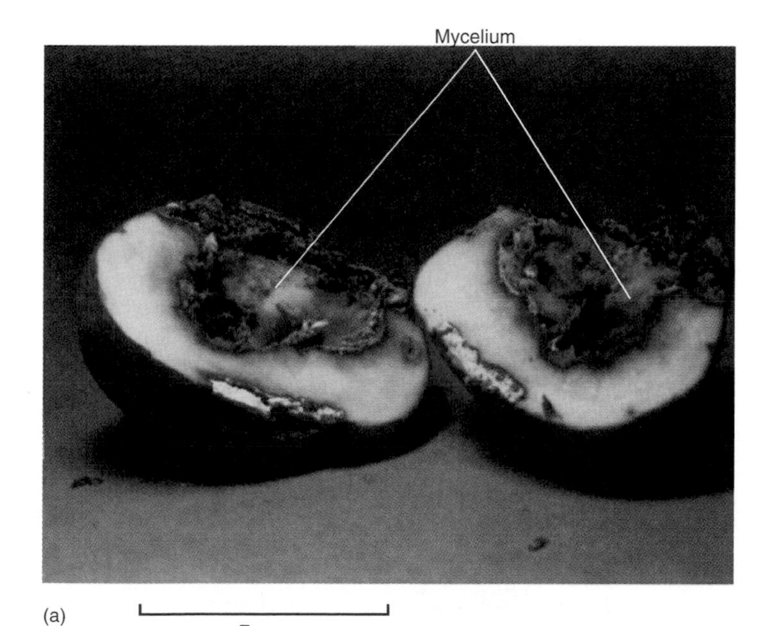

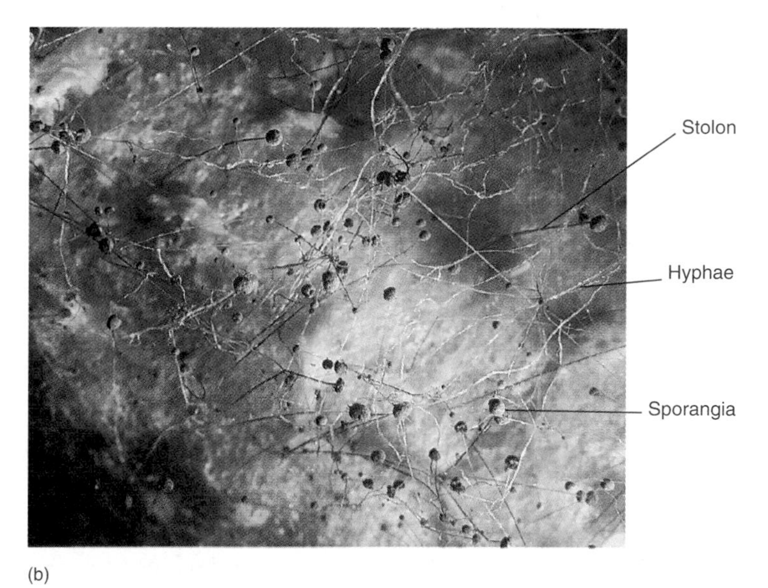

Figure 12.12
(*a*) A mycelium is the cottony white mass of hyphae seen on various foods, such as inside a potato. (*b*) The mycelium of *Rhizopus stolonifer* (black bread mold) is composed of three types of hyphae—arching hyphae called *stolons, sporangiophores,* which bear sporangia at their tips (black spots), and the *hyphae* that penetrate deep within the bread.

Filamentous fungi with large fruiting bodies that are frequently edible are called **mushrooms** (figure 12.13).

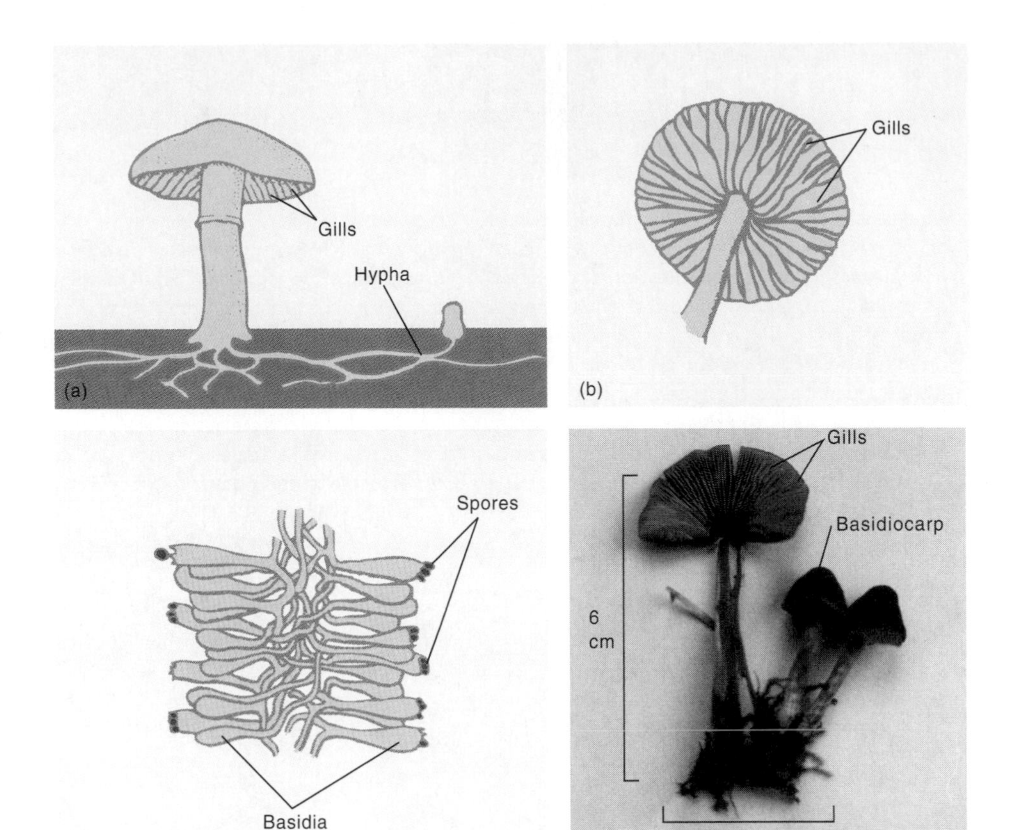

Figure 12.13
(a) The meadow mushroom, *Agaricus campestri*, has an extensive underground mycelium. The fruiting body emerges as a small button mushroom (right), which grows into a typical mushroom with a gilled cap (left). *(b)* The underside of the cap is composed of radiating gills. (c) Magnified view of the surface of a gill showing a mass of basidia, bearing spores. *(d) Lepiota rachodes* showing the fruiting body (basidiocarp) and gills.

Dimorphic Fungi

Species of fungi that are capable of growing either as yeastlike cells or as mycelia, depending on the environmental conditions, are known as the **dimorphic fungi.** This group is of great importance because many of the fungi that cause disease in humans are dimorphic. Certain fungi such as *Coccidioides immitis* grow in the soil as molds. When their spores, which are readily carried in the air, are inhaled into the warm, moist environment of the lungs, they develop into the yeast form of the organism and cause disease.

Classification of Fungi Is Revised Frequently

Classification of the fungi is still subject to debate. In this book, the fungi are placed into three divisions: the terrestrial fungi (*Amastigomycota*), the zoospore-producing fungi (*Mastigomycota*), and the slime molds (*Gymnomycota*; table 12.2). In this text, the slime molds are classified with the fungi, although other scientists classify them with the protozoa. They have characteristics of both groups.

Terrestrial Fungi

The terrestrial fungi **(Amastigomycota)** have both sexual and asexual spores and are divided into four subgroups based on their sexual spore-forming characteristics. Both sexual and asexual reproductive spores occur on specialized structures. The Zygomycetes form **zygospores** and include the common bread mold *(Rhizopus)* and other food spoilage organisms. The Ascomycetes form **ascospores** and include fungi that cause Dutch elm disease and rye smut (ergot). The Basidiomycetes form **basidiospores** and include the common mushrooms. The Deuteromycetes, or Fungi Imperfecti, do not produce sexual spores and include most of the disease-causing fungi such as those that cause athlete's foot and ringworm. However, as more is known about the Deuteromycetes, more of these fungi have been found to have sexual stages and are classified in the appropriate group. Table 12.2 gives additional characteristics of the various fungal groups.

Water Molds

The water molds **(Mastigomycota)** are found primarily in water or are associated with water. They have flagellated reproductive cells known as **zoospores.** The water molds cause some serious diseases of food crops. The late blight of potato (*Phytophthora infestans*) and downy mildew of grapes are included in this group.

Slime Molds

There are two kinds of slime molds, the plasmodial slime mold **(Myxomycetes)** and the cellular slime mold **(Acrasiomycetes).** Both are terrestrial organisms. Plasmodial slime molds are widespread and readily visible in their natural environment. The plasmodium is a multinucleated mass of protoplasm that oozes like slime over the surface of decaying wood and leaves. As it moves, it ingests microorganisms, spores of other fungi, and any other organic material with which it comes in contact.

The cellular slime mold has a vegetative form that is composed of single, amebalike cells. When cellular slime

Table 12.2 Characteristics of Major Groups of Fungi

Group and Representative Member	Usual Habitat	Some Distinguishing Characteristics	Cell Wall Composition	Asexual Reproduction	Sexual Reproduction
Amastigomycota					
Zygomycetes *Rhizopus stolonifer* (black bread mold)	Terrestrial	Multicellular, aseptate, and coenocytic mycelia, stolons, rhizoids; no flagellated cells	Chitin	Sporangiophores borne on sporangia	Zygospores
Basidiomycetes *Agaricus campestris* (meadow mushroom)	Terrestrial	Multicellular, uninucleated mycelia, with perforated septa; no flagellated cells	Chitin	Commonly absent	Produce basidiospores borne on basidia
Ascomycetes *Neurospora, Saccharomyces cerevisiae* (baker's yeast)	Terrestrial, on fruit and other organic materials	Unicellular and multicellular; mycelia septated and unicellular or multinucleated; no flagellated cells	Chitin	Budding; conidiophores borne on conidia	Ascospores born in an ascus
Deuteromycetes (Fungi Imperfecti) *Penicillium, Aspergillus*	Terrestrial	Unicellular and multicellular; septated mycelia; no flagellated cells	Chitin	Budding; conidiophores borne on conidia	Absent or unknown
Mastigomycota					
Oomycetes *Phytophthora infestans* (late blight of potato) and downy mildew of grapes	Water, soil	Unicellular or filamentous; nonseptated mycelia; flagellated cells	Cellulose	Flagellated zoospores	Oogonia
Chytridiomycetes *Allomyces, Synchytrium endobioticum* (potato black wart)	Water, soil	Unicellular, multicellular, filamentous, flagellated cells resembling protozoa	Chitin	Some have an alternation of generations; flagellated zoospores	Flagellated gametospores
Gymnomycota					
Myxomycetes (plasmodial slime molds)	Terrestrial	Plasmodium; multinucleated without rigid cell wall	Absent	Spores borne on sporangia	Gametospores
Acrasiomycetes (cellular slime molds)	Terrestrial	Unicellular; can congregate to form pseudoplasmodium	Cellulose	Sporocarps, spores	Probably absent

molds run out of food, the single cells congregate into a mass of cells that then forms a fruiting body and spores.

Slime molds are important links in the food chain in the soil. They ingest bacteria, algae, and other organisms and, in turn, serve as food for larger predators. Slime molds have been valuable as unique models for studying cellular differentiation during their aggregation and formation of their fruiting bodies.

Fungi Cause Disease in Humans

Fungi cause disease in humans in one of four ways. First, a person may develop an allergic reaction to fungal spores or vegetative cells. Second, a person may react to the toxins produced by fungi. Third, the fungi may actually grow on or in the human body, causing disease. Fourth, fungi can destroy the human food supply, causing starvation and death.

Allergic Reactions in Humans

Medical mycologists[1] study fungi that affect humans, including fungi that cause allergic reactions. Allergic diseases can result from exposure to fungi if exposed humans have become sensitized. For reasons discussed in chapter 17, immediate allergic reactions such as hay fever or asthma can result if the fungi or their spores are inhaled. Sometimes, severe, long-term allergic lung disease results from these allergic reactions.

[1] Mycologist—a person who studies fungi

• Immediate hypersensitivities or allergic reactions, pp. 396–403

Perspective 12.3

The Salem Witch Hunts

In Salem Village, Massachusetts, in 1692, a group of teenage girls was afflicted with symptoms of "bewitchment." They suffered convulsions; had sensations of being pinched, pricked, or bitten; and had feelings of being torn apart. Some became temporarily blind, deaf, or speechless and were sick to their stomachs. During the inquest into their bewitchment, the girls accused several women of being instruments of the devil. These women were tried and then executed as witches.

Recently, some historians have reexamined this event and, with the help of the medical and biological community, have concluded that the Salem witch hunts were a microbiological phenomenon, rather than a social phenomenon. A member of the microbial world may have been the culprit. The fungus *Claviceps purpurea*, the rye smut, produces the poison ergot, which when ingested produces symptoms similar to those experienced by the girls in Salem. Apparently, several conditions existed in New England at this time that would have predisposed the population to ergot poisoning. First, the years 1690 to 1692 were particularly wet and cool. Second, during those years, rye grass had replaced wheat in New England as the principal grain because the wheat was seriously affected by another fungus, the wheat rust. In addition, most of the victims were children and teenagers who would have been more affected by ergot poisoning than adults because they would have ingested more of the poison per body weight than adults.

This is only one highly publicized example of the way in which microbes have played a large role in the history of the world.

Effect of Fungal Toxins

For their hallucinogenic properties, certain mushrooms have long been used as part of religious ceremonies in some cultures. The lethal effects of many mushrooms have also been known for centuries. The poisonous effects of a rye smut[1] called **ergot** were known during the Middle Ages, but only recently has the active chemical been purified from these fungi to yield the drug ergot, which is now used to control uterine bleeding, relieve migraine headaches, and assist in childbirth.

Some fungi produce toxins that are carcinogenic. The most thoroughly studied of these carcinogenic toxins, produced by species of *Aspergillus*, are called **aflatoxins.** Ingestion of aflatoxins in moldy foods, such as grains and peanuts, has been implicated in the development of liver cancer (hepatoma) and thyroid cancer in hatchery fish. Governmental agencies monitor levels of aflatoxins in foods such as peanuts, and if a certain level is exceeded, the food cannot be sold.

Fungal Infections in Humans

Fungal diseases are called **mycoses.** The names of the individual diseases often begin with the names of the causative fungi. Thus, histoplasmosis, a disease seen worldwide, is a mycosis caused by the fungus *Histoplasma capsulatum.* Similarly, coccidioidomycosis is a mycosis caused by *Coccidioides immitis,* a fungus unique to certain arid regions of the western hemisphere. Diseases caused by the yeast *Candida albicans* are called *candidiasis* and are among the most common mycoses.

Infections can also be referred to by the parts of the body that they affect. **Superficial mycoses** affect only the hair, skin, or nails. **Intermediate mycoses** are limited to the respiratory tract or the skin and subcutaneous tissues. **Systemic mycoses** affect tissues deep within the body. Some diseases caused by fungi are given in table 12.3 and are discussed in the chapters dealing with the organ systems that are affected.

Some Fungi Form Symbiotic Relationships with Other Organisms

Fungi form several types of symbiotic relationships with other organisms. For example, lichens result from the association of an alga or a cyanobacterium with a fungus (figure 12.14). These associations are extremely close and, in some cases, the fungal hyphae actually penetrate the cell wall of the photosynthetic partner. **Mycorrhizas,** fungal symbioses of particular importance, are formed by the intimate association between fungi and the roots of certain plants such as the Douglas fir. The fungi occur only in association with the roots. By increasing their absorptive power, they often allow their plant partners to grow in soils where these plants could not otherwise survive. With the world

[1] Smut—plant fungal disease that produces black spores, giving the appearance of soot

• Symbiosis, p. 276

Table 12.3 Some Medically Important Fungal Diseases

Disease	Causative Agent	Chapter and Page with More Information
Candidial skin infection	*Candida albicans*	Chapter 22, p. 496
Coccidioidomycosis	*Coccidioides immitis*	Chapter 24, p. 545
Cryptococcal Meningoencephalitis	*Filobasidiella neoformans*	Chapter 28, p. 644
Histoplasmosis	*Histoplasma capsulatum*	Chapter 24, p. 547
Pneumocystosis	*Pneumocystis carinii*	Chapter 31, p. 711
Sporotrichosis	*Sporothrix schenckii*	Chapter 29, p. 669
Vulvovaginal candidiasis	*Candida albicans*	Chapter 27, p. 619

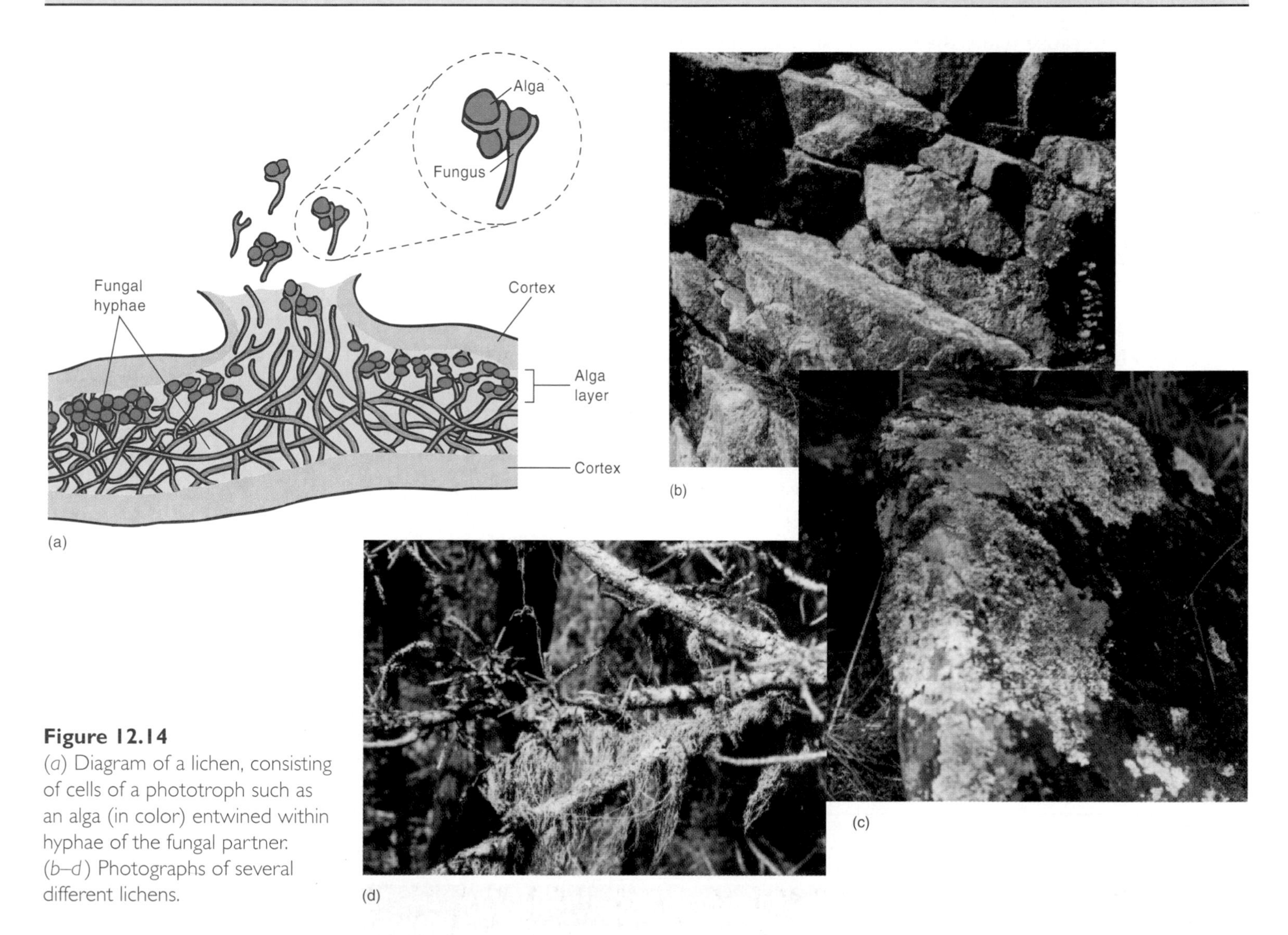

Figure 12.14
(*a*) Diagram of a lichen, consisting of cells of a phototroph such as an alga (in color) entwined within hyphae of the fungal partner. (*b–d*) Photographs of several different lichens.

facing major shortages of food and forest products, a better understanding of these symbiotic relationships is urgently needed. Many trees, including the conifers, are able to live in sandy soil because of their extensive mycelial mycorrhizas. In Puerto Rico, for example, the pine tree industry almost perished before the proper fungi were introduced to form mycorrhizal relationships with the trees. Now the industry is flourishing.

Certain insects also depend on symbiotic relationships with fungi. For example, some ants and termites grow fungi for food, nourishing the fungi with leaves that they carry into their nests.

Fungi Are Economically Important Organisms Both for the Good They Do and the Damage They Cause

Many fungi are important commercially. The yeast *Saccharomyces* has long been used in the production of wine, beer, and bread. Other fungal species are useful in making the large variety of cheeses that are found throughout the world. Penicillin, griseofulvin, and other antimicrobial medicines are synthesized by fungi.

Ironically, fungi are also among the greatest spoilers of food products, and large amounts of food are thrown away each year because they have been made inedible by species of *Penicillium, Rhizopus,* and others.

Fungi also cause many diseases of plants. Dutch elm disease caused by *Ceratocystis ulmi* is transmitted by beetles. It has destroyed the American elm trees that once shaded the streets in many cities. The wheat rust (*Puccinia graminis*) destroys tons of wheat yearly. Rust-resistant varieties have been bred to reduce the losses, but mutations in the rusts have made these advantages short lived. The genus *Phytophthora,* which contains about 35 species, is able to infect a large number of food crops. The most noteworthy, *P. infestans,* causes the disease late blight of potatoes, which led to the great famines in Ireland in 1845 to 1847. Thousands of Irish people emigrated to the United States during that time.

Fungi have been very useful tools for genetic and biochemical studies. *Neurospora crassa,* a common mold, has been widely used for modern genetic studies as well as for investigating biochemical reactions. More recently, yeasts have been genetically engineered to produce human insulin and the human growth hormone somatostatin, as well as a vaccine against viral hepatitis.

REVIEW QUESTIONS

1. What features distinguish the major groups of fungi?
2. What is the difference between a yeast, a mold, and a mushroom?
3. What are the four ways that fungi directly affect humans?
4. What are fungal diseases called?
5. What kind of symbiotic relationships do fungi form with other organisms? How are these relationships beneficial to each partner?

• **Food microbiology, p. 764**

Protozoa

Along with the algae and fungi, the protozoa constitute another group of eukaryotic organisms that have traditionally been considered part of the microbial world. The boundaries between these three groups is not always clearly delineated. In general, however, protozoa are microscopic, unicellular organisms that usually lack photosynthetic capability, are motile, and reproduce most often by asexual fission. Generally, protozoa are aerobic, although some protozoa live under anaerobic conditions in environments such as the human intestine or the rumen of animals.

Protozoan Habitats Are Diverse

A majority of protozoa are free-living and are found in marine, freshwater, or terrestrial environments; some species, however, are parasitic, living on or in other host organisms. The hosts for protozoan parasites range from simple organisms, such as algae, to complex vertebrates, including humans. All protozoa require large amounts of moisture, no matter what their habitat.

In marine environments, protozoa make up part of the zooplankton, where they feed on the algae of the phytoplankton. On land, protozoa are abundant in soil as well as in or on plants and animals. The specialized protozoan habitats include the guts of termites, roaches, and ruminants such as cattle.

Protozoan Classification Is Based Primarily on the Presence or Absence of Organelles of Locomotion

Protozoa are divided into groups based on their mode of locomotion (table 12.4).

The **Mastigophora** are protozoa that have one or more flagella. The flagella of eukaryotes are more complex than those of prokaryotes. Recall that nine pairs of hollow protein fibers, or **microtubules,** are arranged around a central pair (see figure 3.43 *b*), and the entire structure is enclosed in a membranous sheath. Mastigophora typically have one or two flagella, usually found at one end of the organism. Undulating waves of the flagellum pass from the base to the tip, driving the protozoan in the opposite direction.

Table 12.4 Some Properties of the Major Classes of Protozoa

Class	Mode of Motility	Mode of Reproduction*
Mastigophora	Flagella (one or more)	Longitudinal fission
Sarcodina	Pseudopodia (flagella also in some)	Binary fission
Sporozoa	Often nonmotile; may have flagella at some stages, creeping motion in some	Multiple fission; asexual reproduction in one host; sexual reproduction in a second host in some species
Ciliata	Cilia (multiple)	Transverse fission, asexual; also sexual reproduction involving micronuclei

*See figure 12.19.

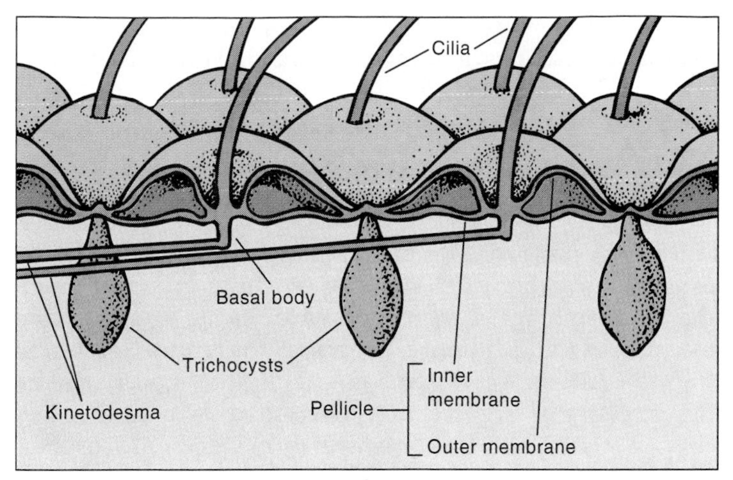

(a)

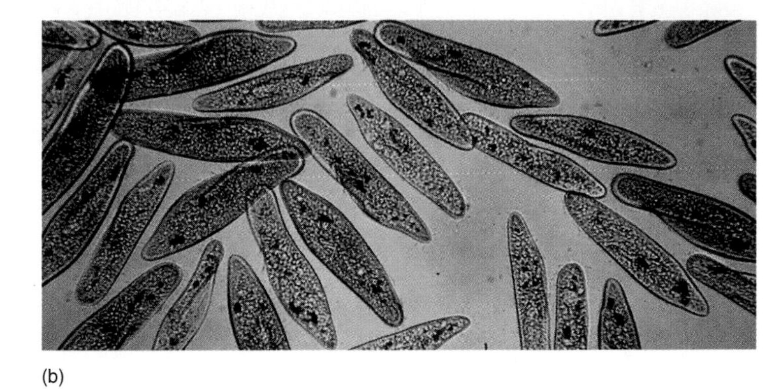

(b)

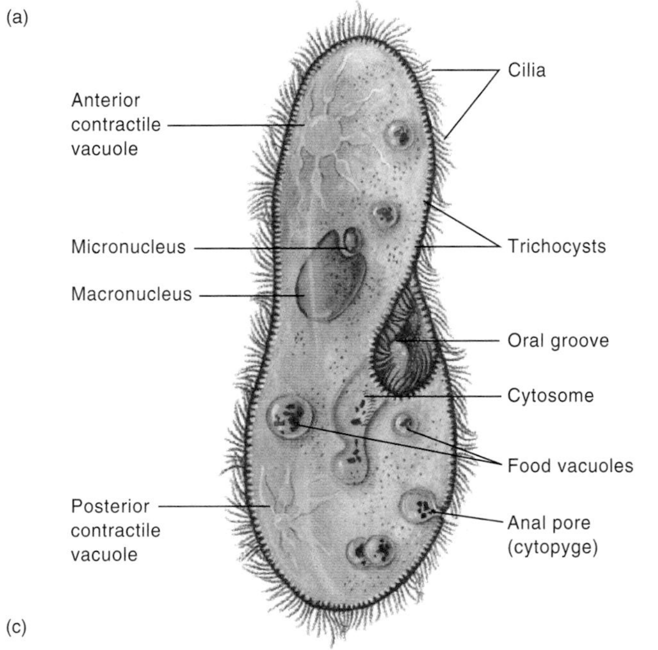

(c)

Figure 12.15
(*a*) Cross section of the pellicle of a *Paramecium* as seen by electron microscopy. The many cilia emerge through the inner and outer membranes of the pellicle. Undischarged trichocysts extend below the pellicle. (*b*) Paramecium (×50) as seen with a light microscope. (*c*) Morphology of a *Paramecium*. Drawing showing the major organelles of a typical paramecium.

The **Ciliata** are organisms that have cilia. The cilia are similar in construction to the flagella and usually completely cover the surface of an organism. Most often, they are arranged in distinct rows and are connected to one another by fibrils known as *kinetodesma* (figure 12.15). Cilia beat in a coordinated fashion in waves across the body of the protozoan. A beat of one cilium affects the cilia immediately around it, but there is no evidence that the connecting fibrils aid in this coordination. The cilia found near the oral cavity propel food into the opening.

Sarcodina are protozoa that move by means of pseudopodia. Sarcodina do not have rigid cell walls, and the cytoplasmic streaming within the cell results in amoeboid movement. During amoeboid movement, the cytoplasm moves in the direction in which there is least resistance. At the tip of the pseudopodium, the cytoplasm is less viscous and less concentrated. At the opposite end, the cytoplasm is under greater pressure (figure 12.16). Pseudopodia are involved in the process of phagocytosis.

• **Cilia and flagella, p. 66**

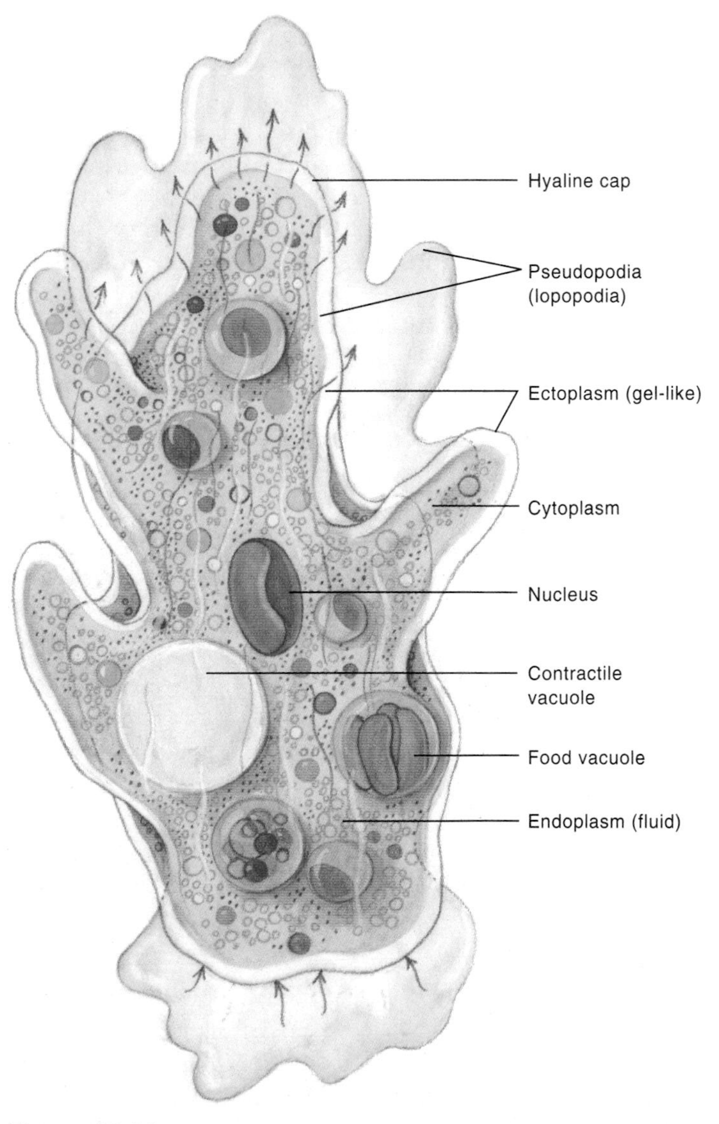

Figure 12.16
Amoeboid movement in a typical amoeba. The arrows indicate the direction of movement.

Protozoa Lack a Cell Wall

Protozoa lack the rigid cellulose or chitin cell wall found in algae or fungi. Most protozoa do, however, have a specific shape determined by the rigidity or flexibility of the ectoplasm lying just beneath the cell membrane. Foraminifera have distinct hard shells composed of silicon or calcium compounds (figure 12.17). The foraminifers, which secrete a calcium shell, have through the course of millions of years, formed limestone deposits such as the white cliffs of Dover on England's southern coast.

Polymorphism Is Often a Feature of the Protozoan Life Cycle

The life cycles of protozoa are sometimes complex, involving more than one habitat or host. Morphologically distinct forms of a single protozoan species can be found at different stages of the life cycle. Such organisms are said to be **polymorphic** (figure 12.18). This polymorphism is comparable in some respects to the differentiation of various cell types that form plant and animal tissues.

The ability to exist in either a vegetative (feeding, or **trophozoite**) form or resting **(cyst)** form is characteristic of many protozoa. Certain environmental conditions such as the lack of nutrients, moisture, and oxygen; low temperature; or the presence of toxic chemicals may trigger the development of a protective cyst wall within which the cytoplasm becomes dormant. Cysts provide a means for the dispersal and survival of protozoa under adverse conditions and can be compared to the bacterial endospore. However, protozoan cysts are not as resistant to heat and other adverse conditions as are bacterial endospores. When the cyst encounters a favorable environment, the trophozoite emerges. Thus, a number of parasitic protozoa are disseminated to new hosts during their cyst stage.

Protozoa Ingest Foodstuffs by Either Phagocytosis or Pinocytosis

Since most protozoa live in an aquatic environment, water, oxygen, and other small molecules readily diffuse through the cell membrane. Larger food particles are **phagocytized** by amoebae and other members of the group Sarcodina (figure 12.19 *a*). The organism engulfs the food particles, thus creating a large vacuole. This food vacuole then decreases in size, and its contents become acidic. Lysosomes bring enzymes that act in an acidic environment to the food vacuole, and the contents are digested. During digestion, the vacuole increases in size and its interior becomes alkaline. The products of digestion are then dispersed into the cytoplasm by pinocytosis and the undigested materials are excreted to the outside of the cell. Soluble substances that

• **Phagocytosis, p. 356**

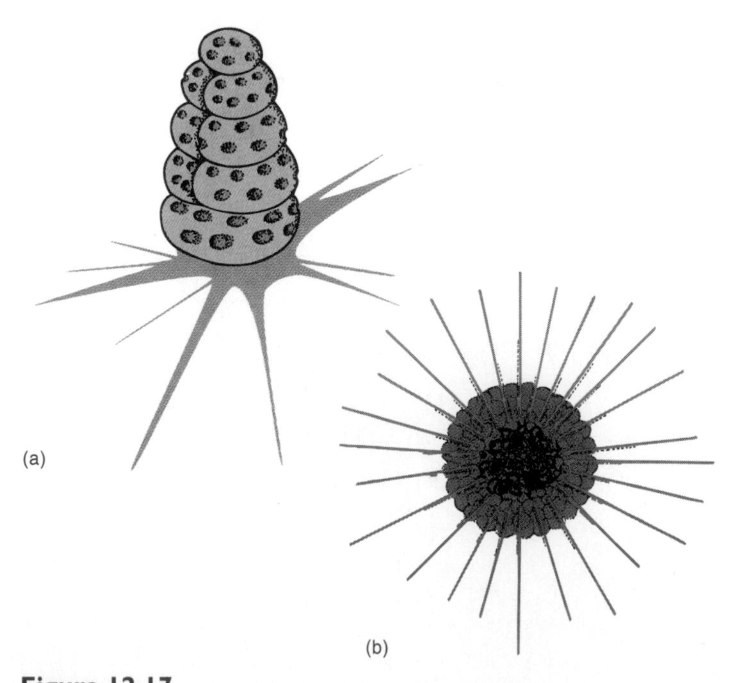

Figure 12.17
(*a*) Protozoan form with a multichambered shell, characteristic of some foraminifera. (*b*) Freshwater heliozoan. The marine radiolarians have forms similar to those of the heliozoans.

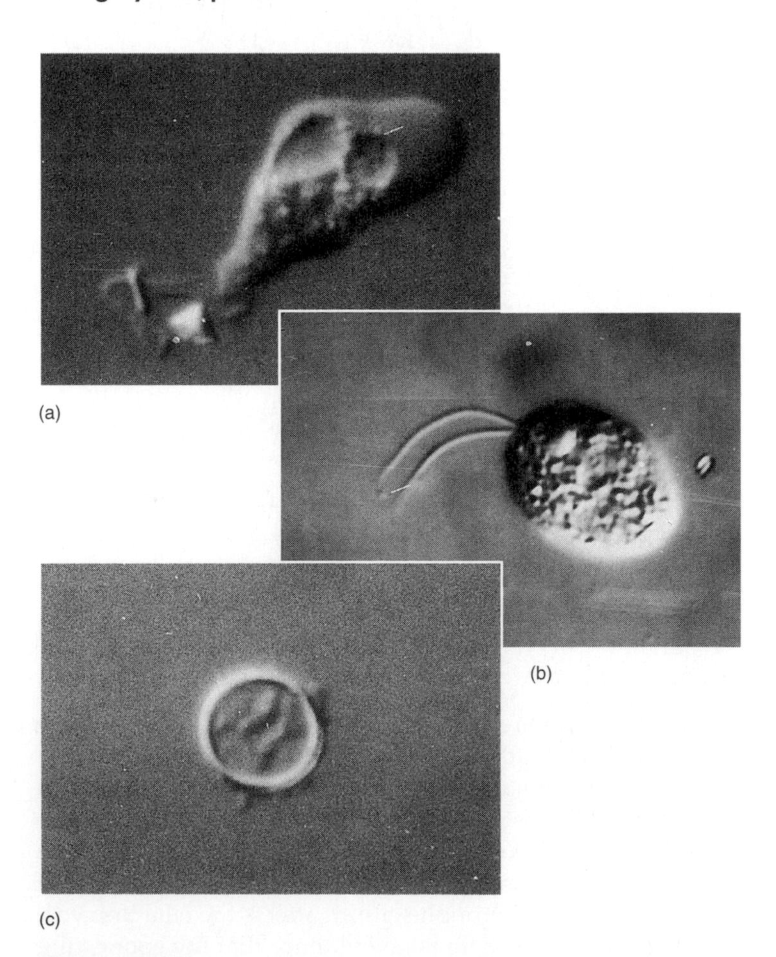

Figure 12.18
Polymorphism in a protozoan. This species of *Naegleria* may infect humans. (*a*) In human tissues, the organism exists in the form of an amoeba (10 to 11 μm at it widest diameter). (*b*) After a few minutes in water, the flagellate form appears. (*c*) Under adverse conditions, a cyst is formed.

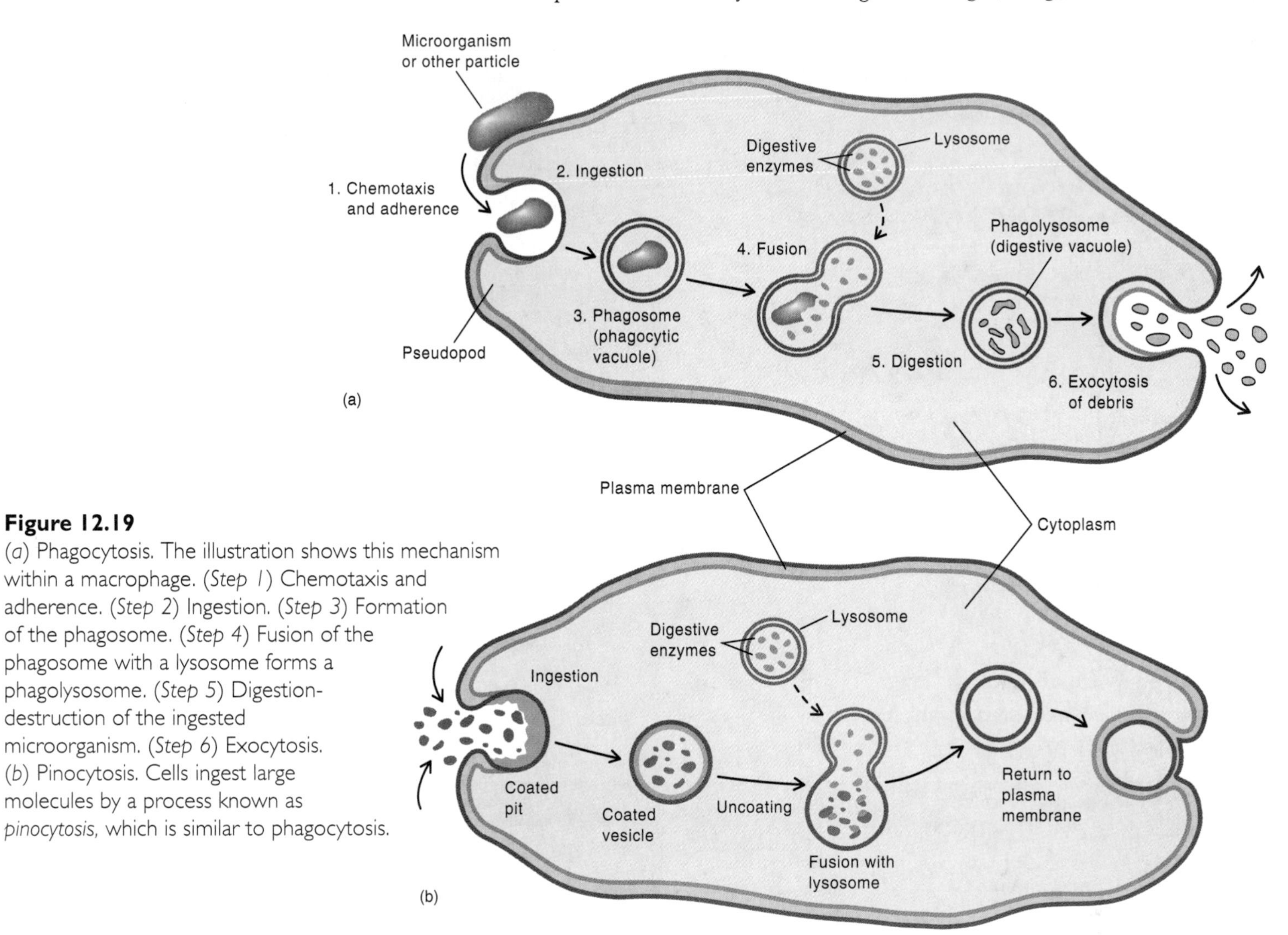

Figure 12.19
(*a*) Phagocytosis. The illustration shows this mechanism within a macrophage. (*Step 1*) Chemotaxis and adherence. (*Step 2*) Ingestion. (*Step 3*) Formation of the phagosome. (*Step 4*) Fusion of the phagosome with a lysosome forms a phagolysosome. (*Step 5*) Digestion-destruction of the ingested microorganism. (*Step 6*) Exocytosis. (*b*) Pinocytosis. Cells ingest large molecules by a process known as *pinocytosis*, which is similar to phagocytosis.

will not diffuse through the membrane are **pinocytized** (figure 12.19 *b*). A small vacuole[1] is formed around the molecules, thereby bringing them into the interior of the cell. From these vacuoles, the molecules then enter the cytoplasm of the organism.

Reproduction Can Be Either Asexual or Sexual

Both asexual and sexual reproduction are common in protozoa and may alternate during the complicated life cycle of some organisms. The methods of asexual reproduction serve as a second criterion for their classification (see table 12.3).

Binary fission takes place in many groups of protozoa (figure 12.20). In the flagellates, it usually occurs longitudinally, and in the ciliates, it occurs transversely. Since some protozoa possess both cilia and flagella, their method of asexual reproduction determines in which group they are classified. Some protozoa divide by multiple fissions, or **schizogony,** in which the nucleus divides a number of times and then the cell divides into a number of daughter cells—not just two, as in binary fission.

[1] Vacuole—a clear, bubblelike space in the cytoplasm of a cell

The alternation of sexual and asexual reproduction is well exemplified by the protozoan that causes malaria. Multiple fission of the asexual forms in the human host results in large numbers of parasites being released into the host's circulation at regular intervals, producing the characteristic cyclic symptoms of malaria. A more detailed account of this process is described in chapter 30.

Protozoa Are Important in the Food Chain and in the Control of Other Microorganisms

The protozoa are an important part of the food chain. They eat bacteria and algae and, in turn, serve as food for larger species. The protozoa help maintain an ecological balance in the soil by devouring vast numbers of bacteria and algae. For example, a single paramecium can ingest as many as 5 million bacteria in one day. Protozoa are important in sewage disposal because most of the nutrients they consume are metabolized to carbon dioxide and water, resulting in a large decrease in total sewage solids.

• **Waste treatment, pp. 750–758**

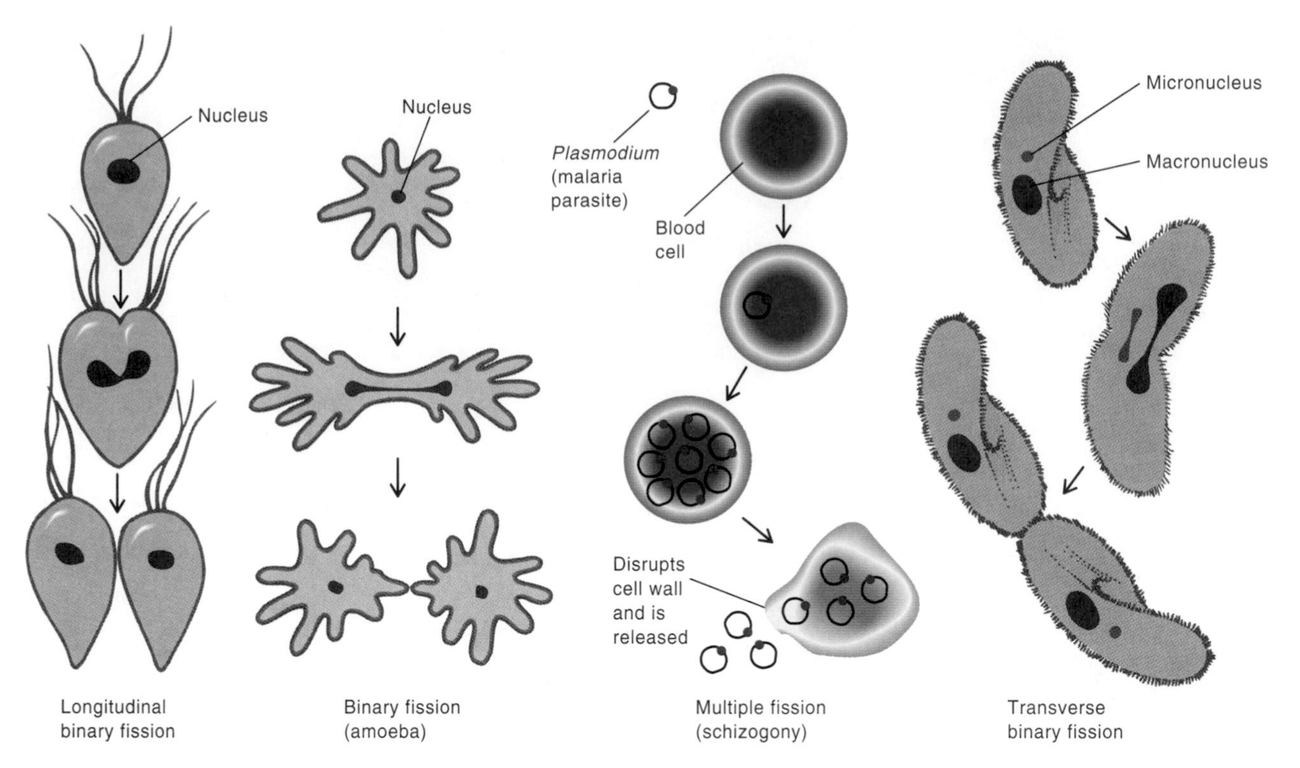

Figure 12.20
Various forms of asexual reproduction in protozoa.

Table 12.5 Some Medically Important Protozoan Diseases

Disease	Causative Agent	Chapter and Page with More Information
African sleeping sickness	*Trypanosoma brucei*	Chapter 28, p. 646
Amebiasis	*Entamoeba histolytica*	Chapter 26, p. 595
Giardiasis	*Giardia lamblia*	Chapter 26, p. 591
Malaria	*Plasmodium* sp.	Chapter 30, p. 688
Toxoplasmosis	*Toxoplasma gondii*	Chapter 30, p. 688
Trichomoniasis	*Trichomonas vaginalis*	Chapter 27, p. 620

Protozoa Cause a Number of Serious Diseases in Humans

The major threat posed by protozoa results from their ability to parasitize and often kill a wide variety of animal hosts. Human infections with the protozoans *Toxoplasma gondii*, which causes **toxoplasmosis,** *Plasmodium* species, which are responsible for **malaria,** *Trypanosoma* species, which cause **sleeping sickness,** and *Trichomonas vaginalis,* which causes **vaginitis,** are common in many parts of the world. Malaria alone has been one of the greatest killers of humans through the ages. At least 150 million people in the world contract malaria each year, and 1. 5 million die of it. Protozoan infections of animals are so common that it is difficult to estimate their extent or overestimate their economic importance. Some parts of tropical Africa are uninhabitable, due in large part to the presence of the tsetse fly, the carrier of the trypanosomes that cause African sleeping sickness. Humans have no natural defense mechanisms against this infection. Some animals, including cows and horses, are reservoirs for these organisms.

Some diseases caused by protozoa are given in table 12.5 and discussed more extensively in the chapters dealing with the organ systems that are affected.

Perspective 12.4

Hikers Beware! Is that Clear Mountain Stream Really Clear?

The water from a mountain stream tastes delicious on a hot summer hike, but that clear-looking water may hold an unpleasant surprise. Ingesting only a few cysts of *Giardia lamblia*,[1] a protozoan, can cause the disease **giardiasis** (figure 1). Symptoms of giardiasis include diarrhea, flatulence, abdominal cramps, and nausea. *Giardia* goes through its life cycle within the intestines of humans and other animals. The cysts are shed along with the feces into the water supply, where they await the next warm-blooded animal that takes a drink.

Giardia is found in nearly all water supplies that are untreated. The best way a hiker can prevent contracting giardiasis is to either bring bottled water from home or treat the water on the hike. The most reliable treatment is boiling the water. Chemicals are less reliable, in part because cysts are quite resistant to destruction in cold water. Iodine-based chemicals are usually more effective against *Giardia* than chlorine-based chemicals. Filters can be used as long as the pore size is small enough (under 5 micrometers).

So hikers beware—clear does not mean pure.

[1] *Giardia lamblia.* is also known as *Giardia intestinals.*

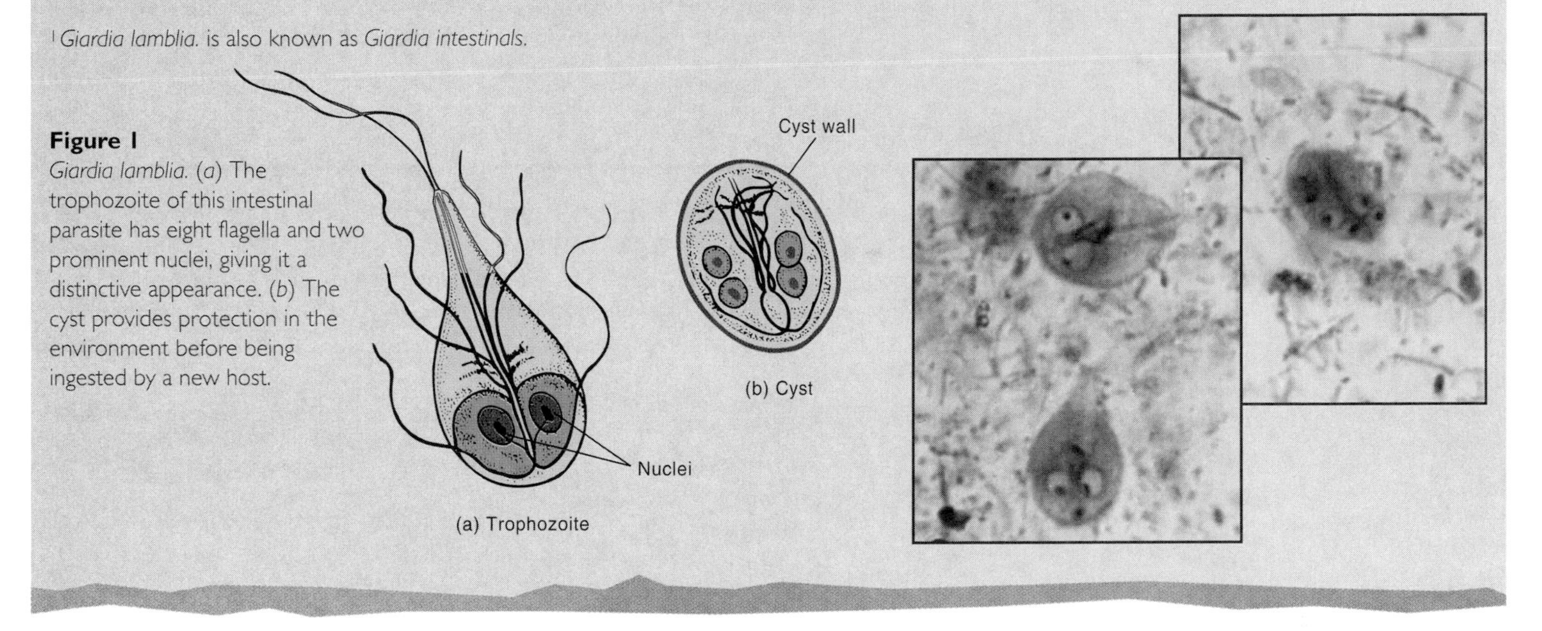

Figure 1
Giardia lamblia. (*a*) The trophozoite of this intestinal parasite has eight flagella and two prominent nuclei, giving it a distinctive appearance. (*b*) The cyst provides protection in the environment before being ingested by a new host.

Summary

I. Algae
   A. Algae are photosynthetic organisms that lack the complex tissue development of higher organisms.
   B. Algae are classified according to photosynthetic pigments.
   C. Algae are found in a wide variety of habitats.
   D. Algae are typical eukaryotic organisms.
      1. Algal cell walls are rigid and contain cellulose.
      2. Microscopic algae are usually single cells but can also live as colonies.
      3. Macroscopic algae are multicellular with a variety of structures such as holdfasts, blades, bladders, and stipes.
      4. Reproduction in algae can be asexual or sexual.
   E. Some dinoflagellates produce toxins that cause paralytic shellfish poisoning.

II. Fungi
   A. Fungi are nonmotile, nonphotosynthetic heterotrophic organisms.
   B. Fungi are primarily terrestrial and occur in a large variety of habitats.
   C. The terms *yeast, mold,* and *mushroom* refer to morphological forms of fungi.
      1. Yeasts are single cells.
      2. Molds are filamentous.
      3. Mushrooms are the large fruiting bodies of certain fungi and they are often edible.
      4. Dimorphic fungi exist as either yeast or mycelium, depending on conditions.
   D. Classification in fungi is frequently revised. In this text, they are considered in three groups:
      1. Amastigomycota—terrestrial fungi
      2. Mastigomycota—flagellated lower fungi and water molds
      3. Gymnomycota—slime molds
   E. Fungi cause disease in humans in four ways.
      1. Some people experience an allergic reaction to fungi.
      2. Ergot and aflatoxin are two fungal poisons that seriously affect humans.
      3. Fungal diseases are known as mycoses.
      4. Fungi destroy the human food supply.

F. Fungi form symbiotic relationships with a variety of other organisms. Lichens and mycorrhizas are examples.
G. Fungi are economically important organisms.
   1. They aid in the production of foods such as cheese, bread, beer, and wine.
   2. They are a source of antimicrobial medicines.
   3. They spoil foods and degrade organic materials.
   4. They cause many plant diseases.
   5. They are used for genetic and biochemical studies.
   6. Genetically engineered yeasts produce many useful compounds such as insulin and human growth hormone.

III. Protozoa
A. Protozoa are microscopic, unicellular organisms that lack photosynthetic capabilities.
B. Protozoan habitats are diverse but all protozoa require moisture in some form.
C. Classification of protozoa is based on their mode of locomotion.
   1. Mastigophora—flagella
   2. Sarcodina—pseudopodia
   3. Sporozoa—flagella or nonmotile
   4. Ciliata—cilia
D. Protozoa lack a cell wall but maintain their distinctive shape by other means.
E. Protozoa display polymorphism by taking on different forms in different environments.
F. Many protozoa take in nutrients by phagocytosis and pinocytosis.
G. Reproduction is by both asexual and sexual methods.
H. Protozoa are an important part of the food chain.
I. Protozoa cause a variety of serious diseases in humans.

Chapter Review Questions

1. Name the four groups of protozoa and give the main distinguishing features of each.
2. What is the difference between pinocytosis and phagocytosis?
3. What is polymorphism and how does it aid the protozoa in the spread of disease?
4. Of what economic importance are protozoa?
5. Describe and contrast the various modes of asexual reproduction in protozoa.
6. Describe and contrast the various modes of locomotion found in protozoa.
7. Discuss the differences in cell wall structure between the bacteria and the algae, fungi, and protozoa.
8. Discuss the differences in the way genetic information is organized in the bacteria and the algae, fungi, and protozoa.
9. Do algae cause disease in humans directly or indirectly? Why or why not?
10. Discuss the differences and similarities in the ways algae, protozoa, and fungi cause disease in humans.

Critical Thinking Questions

1. Compare the algae, fungi, and protozoa. What characteristics do all these organisms have in common? What features of each cause them to be separated into distinct classification groups?
2. What characteristics of algae account for their general lack of disease-causing properties?
3. What properties of fungi account for their ability to be so destructive?
4. Although we think most often of protozoa as disease-causing organisms, what would the world be like if there were no protozoa?

Further Reading

Alexopoulos, C. J., and Mims, C. W. 1979. *Introductory mycology,* 3d ed. Macmillan, New York. A classic textbook on the biology of fungi.

Anderson, D. M. 1994. Red tides. *Scientific American* 271(2):62–68.

Barnett, H. L., and Hunter, B. B. 1987. *Illustrated genera of imperfect fungi,* 4th ed. Macmillan, New York.

Bold, H. C., and Wynne, M. J. 1985. *Introduction to the algae: structure and reproduction.* Prentice-Hall, Englewood Cliffs, N.J.

Carefoot, T. 1977. *Pacific seashores: a guide to intertidal ecology.* University of Washington Press, Seattle.

Chapman, V. J., and Chapman, D. J. 1973. *The algae,* 2d ed. Macmillan, New York. Good introductory material.

Garcia, Lynne S., and Buckner, David A. 1996. *Diagnostic medical parasitology,* 3d ed. American Society of Microbiologists Press, Herndon, VA. Has chapters on protozoan diseases.

Kunzig, R. 1990. Invisible garden. *Discover.* Explains the importance of marine algae.

Larone, Davise H. 1995. *Medically important fungi,* 3d ed. American Society of Microbiologists Press, Herndon, VA.

Schmidt, G. D., and Roberts, L. S. 1985. *Foundations of parasitology,* 3d ed. Mosby, St. Louis. A comprehensive review of protozoa.

Sternberg, Steven. 1994. The emerging fungal threat. *Science* 266:1632–1634.

Sze, P. 1986. *A biology of the algae.* Wm. C. Brown Publishers, Dubuque, Iowa.

Walsh, J. J., and Pizzo, P. A. 1988. Nosocomial fungal infections: A classification for hospital-acquired fungal infections and mycoses arising from endogenous flora or reactivation. *Annual Review of Microbiology* 42:517–545.

13 The Nature and Study of Viruses

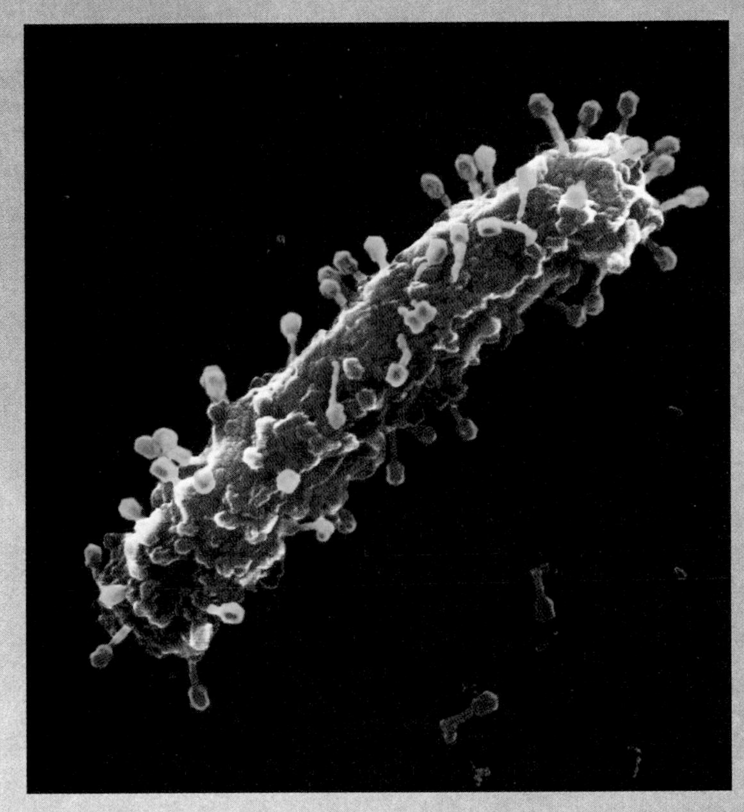

Color-enhanced scanning electron microscope image of T-phages attached to the cell wall of *Escherichia coli* (×36,000).

KEY CONCEPTS

1. Viruses are a set of genes enclosed in a protein coat.
2. In order to multiply, viruses depend to varying degrees on the enzymes of the cells they invade.
3. Viruses that infect bacteria serve as excellent models for viruses infecting animal and plant cells.
4. Viruses can replicate within cells that they subsequently lyse or they can become integrated into host cell DNA and confer new properties on the cells.
5. Different viruses can infect and replicate within all kinds of cells, both prokaryotic and eukaryotic.

PREVIEW LINK

Microbes in Motion
The following books and chapters in the *Microbes in Motion* CD-ROM may serve as a useful preview or supplement to your reading:
Viral Structure and Function: Structure; Replication; Assembly; Pathogenesis; Diseases.

A Glimpse of History

During the late 19th century, many bacteria, fungi, and protozoa were identified as infectious agents. Most of these organisms could be readily seen with the microscopes available at that time, and generally they could be cultured in vitro. In the 1890s, D. M. Iwanowsky and Martinus Beijerinck demonstrated that a disease of tobacco plants, *mosaic disease*, was also caused by an unusual agent. Early in the 20th century, F. W. Twort in England and F. d'Herelle in France demonstrated the existence of infectious agents that could destroy bacteria. Both groups of agents had similar properties that were different from anything that had been observed previously. The agents were so small that they could not be seen with the light microscope, and they passed through filters that retained almost all known bacteria. These same agents could not be grown in media free of living cells but would grow if intact, living host cells were present. The agents were called *filterable viruses* by Beijerinck (*virus* means "poison"), a term that once had been applied to all infectious agents. With time, the adjective *filterable* was dropped, and only the word *virus* was retained.

As more information became known, it appeared that viruses had certain features that were more characteristic of complex chemical substances than of cells. For example, tobacco mosaic virus could be precipitated from a suspension by the addition of ethyl alcohol and would still remain infective, whereas a similar treatment destroyed the infectivity of bacteria and other cells. Further, in 1935, Wendell Stanley of the University of California, Berkeley, crystallized tobacco mosaic virus, which showed that the physical and chemical properties of viruses differed from those of cells. Crystals are formed only by purified chemical compounds and not by cells or by mixtures of dissimilar molecules. The crystallized tobacco mosaic virus could still infect and multiply within cells.

The nature of these curious agents, some of which infect animals and others plants, became clearer by the study of viruses that infect only bacteria. Because bacterial cells can be grown easily and multiply rapidly, viruses that infect bacteria, termed **bacteriophages** or *phages* (*phago* means "to eat"), can be cultivated much more readily than the viruses that infect animals or plants. For this reason, bacteriophages have been studied extensively, and the knowledge gained from these studies has contributed enormously to an understanding of both viruses in general and the molecular biology of all organisms.

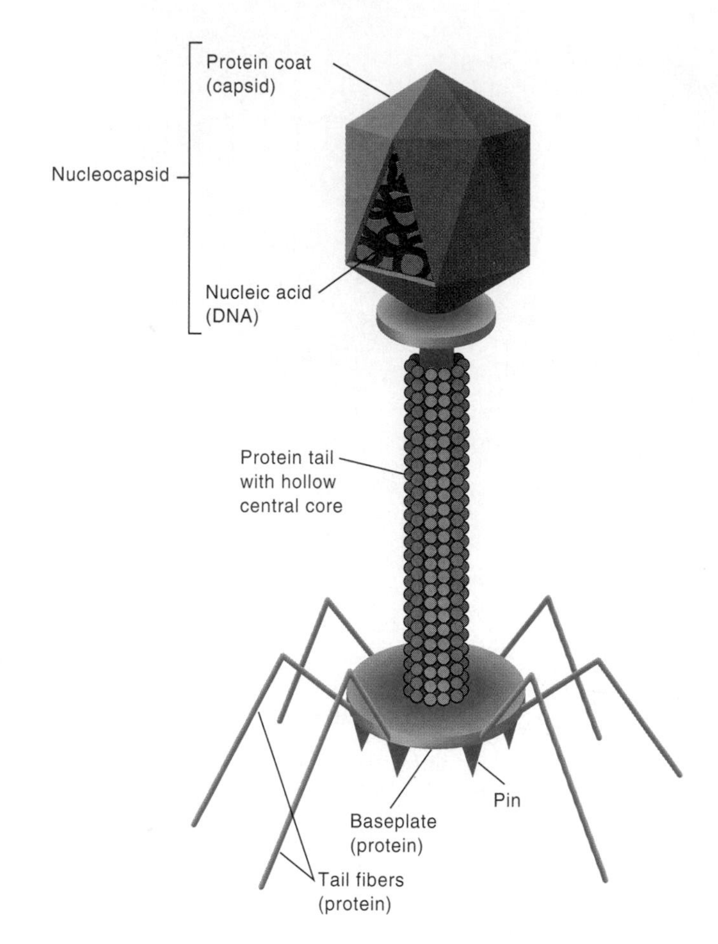

Figure 13.1
Morphology of a virus. Most viruses do not have a tail.

Table 13.1 Size and Genome Content of Viruses, *E. coli* and a Human Cell

	Size (Diameter)	Genome Content
Smallest virus	20 nm	4–5 genes
Largest virus	300 nm	150–200 genes
E. coli cell	1,000 nm	3,000 genes
Human cell	20,000 nm	<60,000 genes

Viruses—General Characteristics

Viruses are nonliving agents that are associated with all forms of life, including all members of the Eubacteria, Archaea, and Eukarya. Each virus particle, called a **virion,** consists of nucleic acid surrounded by a protective protein coat, termed a **capsid** (figure 13.1). Viruses have several features that distinguish them from cells. Perhaps the most obvious feature is their small size. Viruses are approximately 100-fold to 1,000-fold smaller than the cells they infect.

The smallest viruses are approximately 20 nm in diameter, while the largest are about the size of the smallest bacterial cells. Because of their small size, the smaller viruses contain very little nucleic acid and, therefore, very few genes (table 13.1).

The Nucleic Acids in Viruses Can Be DNA or RNA, but Never Both

The structure of the viral genome is also unusual compared to cells. Viruses contain either RNA or DNA but never both. Depending on the type of nucleic acid, they are frequently referred to as *RNA* or *DNA viruses*, respectively. The nucleic acids can occur in one of several different forms characteristic of the family of virus. These include double-stranded or single-stranded DNA and double-stranded or single-stranded RNA. Most single- and double-stranded DNA viruses contain all of their genetic information in a single, linear molecule. However, some

RNA viruses, termed **segmented viruses,** have several different RNA molecules in their capsids, each one carrying the same or different genetic information.

Viruses Can Grow Only Within Living Cells

Another feature that distinguishes viruses from cells is the viruses' lack of the cellular components necessary to generate energy and synthesize protein (for example, ribosomes). Because they contain so little nucleic acid, they do not have all of the genes necessary for the synthesis of many different proteins. Further, viruses contain very few if any enzymes, and the ones they do contain are generally involved with the entry of the virus into cells and the replication of their own nucleic acid. As a consequence, viruses can only multiply inside living cells, where they use cell structures and enzymes to support their own multiplication. However, all viruses must contain a minimum amount of genetic information. This includes the information required (1) to make its special protein coat, (2) to assure replication of its own chromosome, and (3) to move the virions into and out of the host cell.

Viruses Have Two Phases in Their Life Cycle

Viruses differ from cells in that they have two distinct phases in their life cycle. Since they can multiply only within living cells, outside of such cells they are metabolically inert; essentially, they are only macromolecules. However, in the second phase of the life cycle, inside susceptible cells, they can use the metabolic machinery and pathways of their host cells to produce new virions. Further, unlike cells, viruses can reproduce themselves if their genome only enters the host cell. Viruses are said to be "dead" if they cannot enter or multiply within living cells.

Differences between viruses and cells are summarized in table 13.2.

Virus Architecture

Different species of viruses have different shapes. Some are polyhedral (having many flat surfaces), although they appear to be spherical when viewed with the electron microscope; others are helical; and some are a combination of these two shapes (figure 13.2). Some viruses with these shapes are illustrated in figure 13.3. The viruses that infect bacteria, **bacteriophages** or **phages,** often are very complex in their structure. For example, a group of phages that infect *Escherichia coli* are composed of a polyhedral "head" containing double-stranded DNA, attached to a helical, hollow "tail" (figure 13.2 *c*).

The shape of a virus is determined by the shape of the protein capsid, either helical or spherical, that encloses the viral nucleic acid genome. Each capsid is composed of

Table 13.2 Differences Between Viruses and Cells

	Viruses	Cells
Growth	Grow only within cells; outside cell, metabolically inert	Most free-living
Nucleic acid content	Contain either DNA or RNA, never both	Always contain both DNA and RNA
Enzyme content	Have very few if any enzymes	Have many enzymes
Cell components	Lack ribosomes and enzymes to generate energy	Contains ribosomes and enzymes for generation of energy

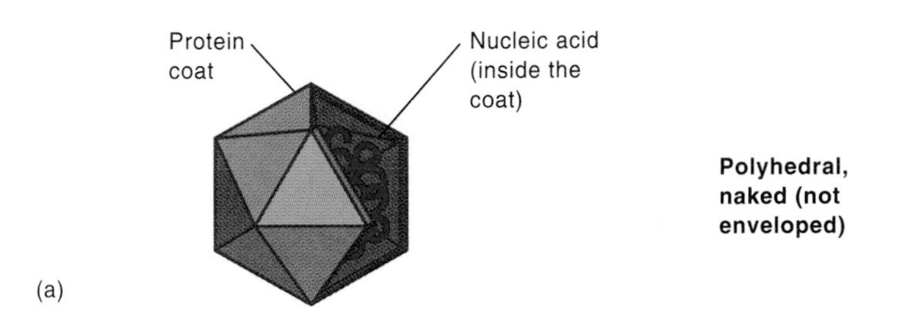

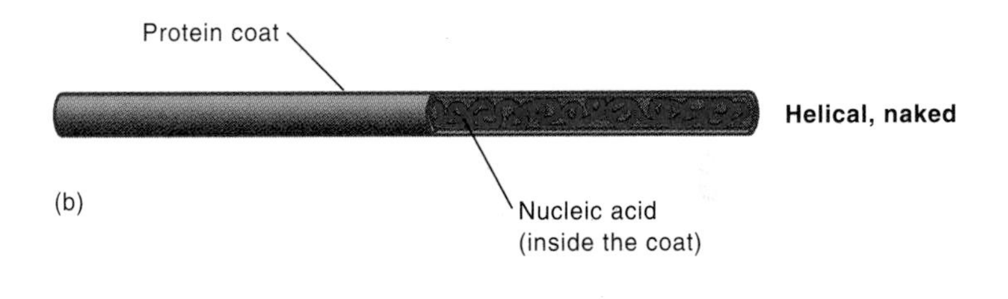

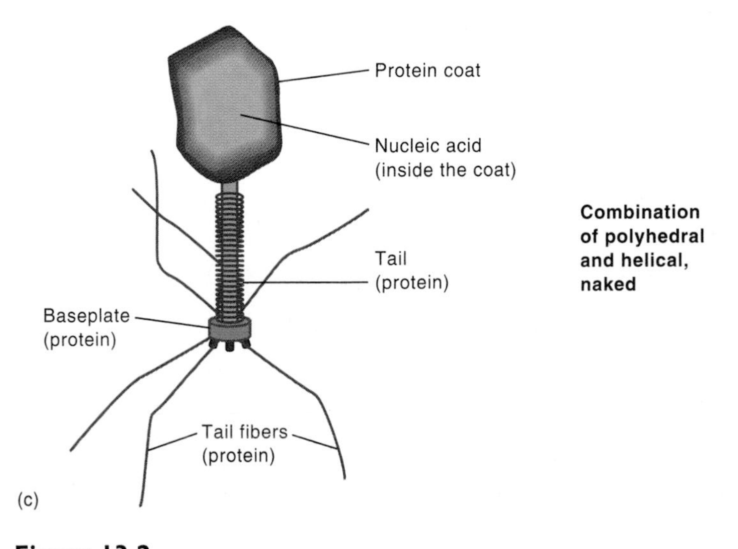

Figure 13.2
Different shapes of viruses. (*a*) Polyhedral. (*b*) Helical. (*c*) Combination.

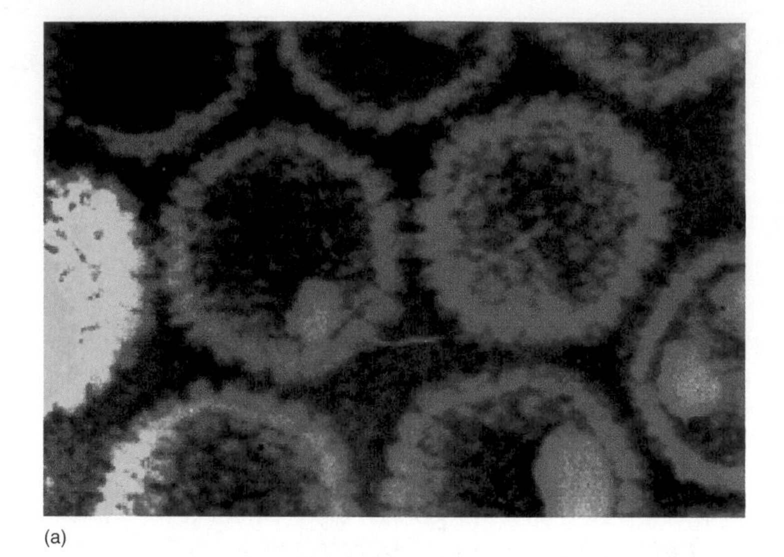
(a)

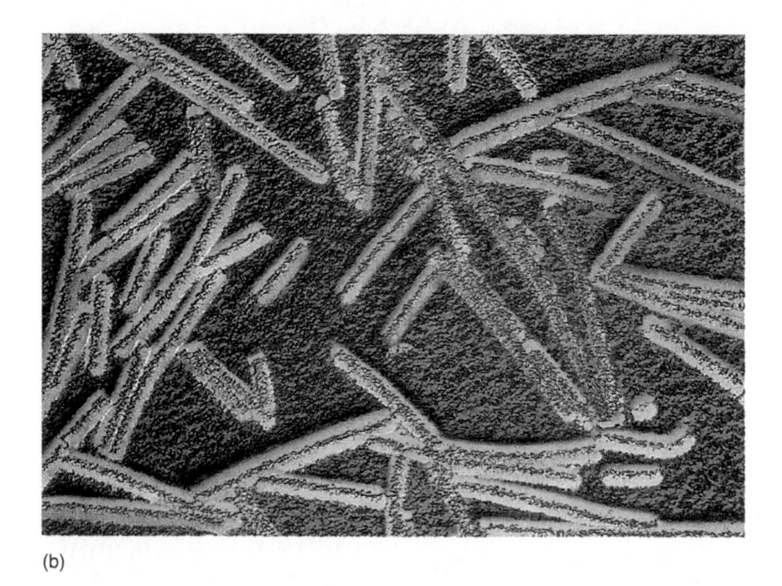
(b)

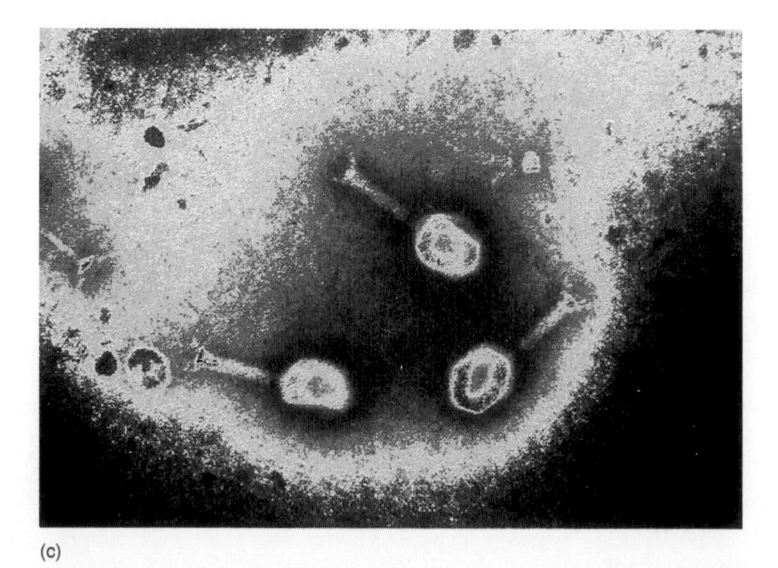
(c)

Figure 13.3
Virus morphology. (*a*) Herpes simplex viruses, (*b*) tobacco mosaic virus (×220,000), and (c) T4 bacteriophage.

many identical protein subunits called **capsomeres** (figure 13.4). The viral capsid together with the nucleic acid that is tightly packed within the protein coat is called the **nucleocapsid.** In many viruses, the capsid is the only outer coat and the virion is termed a **naked** or **nonenveloped capsid virus.** However, some viruses that infect humans and other animals have an additional lipid membrane or **envelope.** The envelope is usually acquired from the cytoplasmic membrane of the infected cell during the release of the virus from the cell (see figure 13.4). Such a virus is termed an **enveloped virus.** Thus, the structure of the viral envelope is similar to the cytoplasmic membrane of the cell, consisting of a double layer of lipids. Just inside the lipid envelope is a protein termed the **matrix protein.** Attachment protein or glycoprotein structures called **spikes,** which are involved in attaching the virion to the host cell, often project from the envelope or the capsid of nonenveloped virus particles. Even viruses without obvious spikes probably have short protein projections on their surface that function to attach the virus to the host cell.

• **Glycoprotein, p. 31**

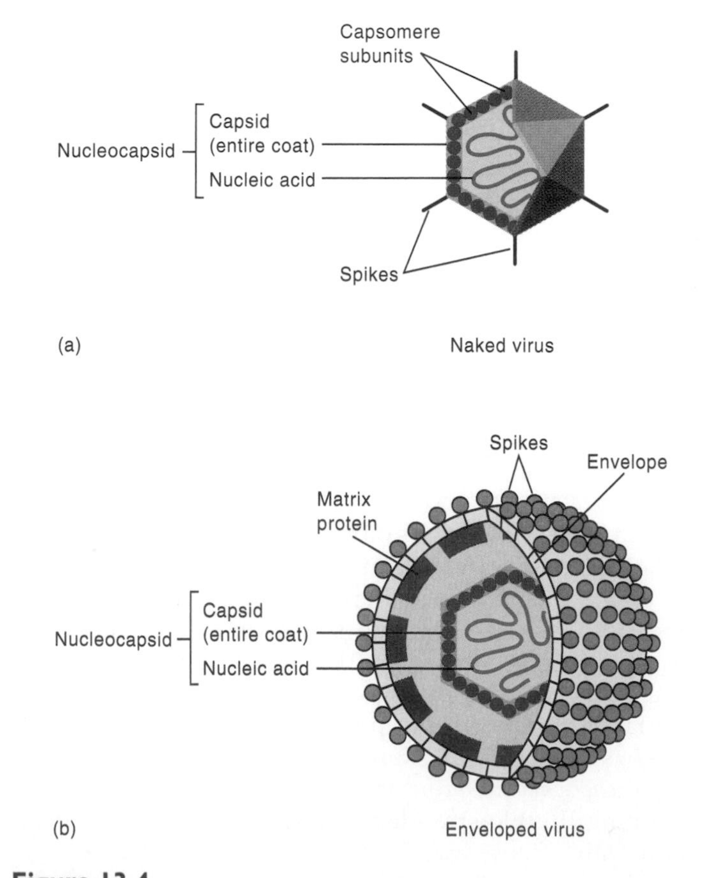

Figure 13.4
Schematic drawing of two basic types of virions—(*a*) naked, not containing an envelope around the capsid, and (*b*) enveloped, containing an envelope around the capsid.

Perspective 13.1

Mad Cows and Englishmen

Most scientists believe that infectious agents must contain nucleic acids in order to replicate. Until the structure of viroids was determined 25 years ago, most scientists believed that protein was also required. Now some, but not all, scientists are convinced that protein, without any nucleic acid, can be infectious and cause disease. This group of infectious agents, termed **prion protein** for *pro*teinaceous *in*fectious agents, has properties that distinguish it from both viruses and viroids. The agent causes diseases in a variety of animals and humans, all of which have similar symptoms of the nervous system. One disease, termed **scrapie,** was first recognized in sheep and goats over 250 years ago, but it was not until 50 years ago that it was shown to be transmissible from one animal to another. Scrapie was also shown to have an incubation period of up to several years. **Kuru,** a disease of people living in New Guinea, was shown in 1959 to be similar to scrapie. Kuru was shown to be transmitted from generation to generation of New Guinea natives by consumption of the brains of deceased elders. The incidence of the disease has dropped sharply since the practice of cannibalism has been abandoned.

In the animal world, similar diseases occur in mink, mule deer, and elk. In the United Kingdom, in 1986 a mysterious new disease, "mad cow disease" or bovine spongiform encephalopathy (BSE), appeared. It affected 600 head of cattle each month. In 1990, a BSE-like disease turned up in a pet cat in Great Britain. The disease symptoms in these animals are similar in all cases, The disease is a slowly progressive degenerative brain disease that makes the brain appear spongy and full of holes. All of the diseases have a very long period of incubation and death is the inevitable outcome.

There is no question that unconventional agents can be transmitted within the same species—for example, from humans to humans and from sheep to sheep. However, the evidence is now quite convincing that at least some of the agents can be transferred between species. BSE most likely was transferred to cattle by feed made from the flesh of scrapie-infected sheep, and it seems likely that the BSE-like disease in the cat also came from its food. Can BSE or scrapie be transmitted to humans who eat meat from infected cattle or from scrapie-infected sheep? The answer is not clear, but since human spongiform encephalopathies are very rare, and since scrapie has been known for about 250 years, it seems unlikely that there is much chance of humans contracting human spongiform encephalopathies by eating infected sheep. However, since BSE has been known for less than 10 years, it is not certain that transfer of this agent to humans is unlikely. Further, because of the lengthy incubation period between infection and appearance of symptoms, enough time has not elapsed to evaluate the possibility of transfer. Sales of beef have dropped significantly in Great Britain since the disease was publicized.

Virus-Host Interactions—General Features

Although viruses are found in all kinds of cells, the relationships between even the most extensively studied animal viruses and the mammalian host cells they invade are relatively poorly understood compared with the phage-bacteria host systems. This is not surprising in view of the differences in complexity of the two host cells, prokaryotic and eukaryotic. The volume of a typical animal cell is about 1,000-fold greater than that of a cell of *Escherichia coli,* and the animal cell genome contains 10,000 times more DNA than does the bacterium. Further, it takes an animal cell about 100 times longer to divide than it takes *E. coli,* and the time that viruses take to replicate is also much longer in animal cells. Thus, research on animal viruses has been relatively slow compared to results from studying bacterial viruses. Nevertheless, a great deal of progress has been made. Some aspects of this progress are considered in chapter 14. This chapter will focus on bacteriophages.

The relationships between certain phages and their bacterial cell hosts have been studied intensively for over 50 years now and a great deal is now known at the molecular level about their interactions. Of great importance is the fact that what is true for the interaction between phages and their bacterial hosts is also true, in general, for the interaction between viruses and the animals they infect. Thus, an understanding of phage-bacteria interactions provides the key to understanding virus-animal relationships, and accordingly phage will be discussed first.

Bacteriophages Can Be Conveniently Grouped on the Basis of Their Genome Structure and Their Interaction with Host Cells

Representative and important bacteriophages are listed with their properties in table 13.3. The important features include the cells that the phages infect, their nucleic acid composition (DNA or RNA; single- or double-stranded), and their relationship to the host cell. This latter topic is discussed here. Double-stranded RNA phages are extremely rare and will not be discussed. The classification of animal viruses is covered in chapter 14.

Viruses Can Have Different Relationships to Their Host Cells

How viruses affect the cells in which they multiply depends on the virus and the host cells. Some viruses multiply inside the cells they invade and actually cause their dissolution (lysis). Other viruses become a part of the genome of the cells they invade and modify the properties of the host cells. Other viruses leak out of host cells usually without killing them (figure 13.5).

Other less defined relationships undoubtedly exist. Each of the three major relationships can exist between phage

Table 13.3 Representative Important Bacteriophages

Bacteriophage	Host	Genome Structure	Relationship to Host Cell
T4	*Escherichia coli*	ds* DNA	Lytic
M13, fd	*Escherichia coli*	ss† DNA	Exit by extrusion
ΦX174	*Escherichia coli*	ss DNA	Lytic
Lambda (λ)	*Escherichia coli*	ds DNA	LATENT
MS2, Qβ	*Escherichia coli*	ss RNA	Lytic
Beta (β)	*Corynebacterium diphtheriae*	ds DNA	LATENT; codes for diphtheria toxin
Φ6	*Pseudomonas*	ds RNA	Lytic

*ds—double-stranded
†ss—single-stranded

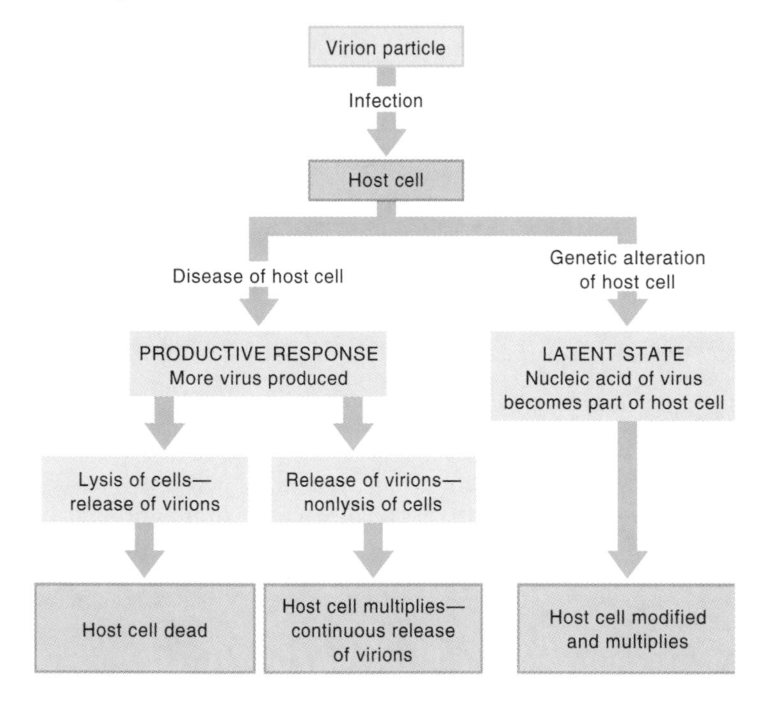

Figure 13.5
Major types of relationships between virus and the host cells they infect. These relationships are found in both bacterial and animal host cells.

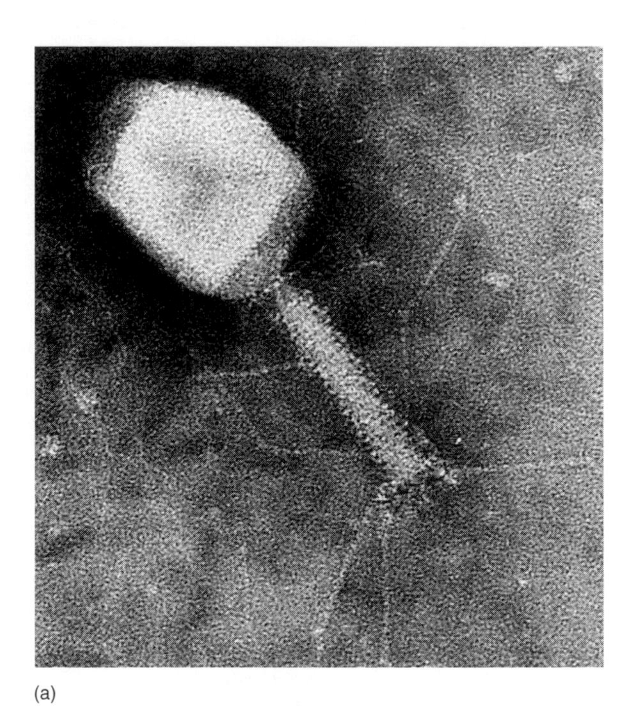

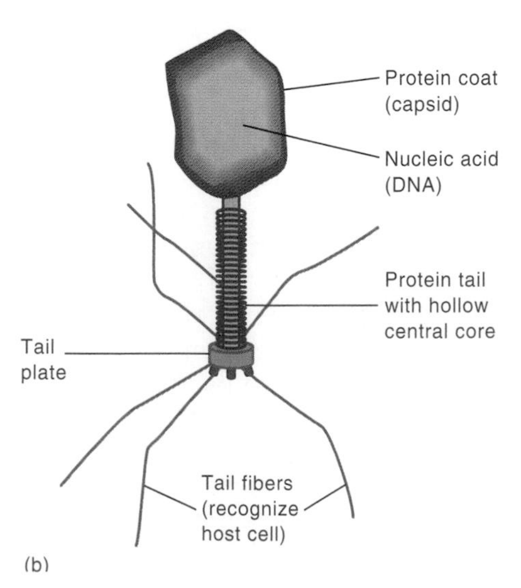

Figure 13.6
T4 bacteriophage. (*a*) Electron photomicrograph of one of the most complex of the bacterial viruses (×270,000). (*b*) Diagram showing the morphology of bacteriophage T4.

and bacteria, as well as between animal viruses and their host cells. However, since so much more is known in general about the phage-bacteria systems, we will emphasize these relationships in the following discussion. The relationships between animal cells and viruses is covered in chapter 14.

Virus Replication—Productive Infection

In a productive interaction, the phage nucleic acid enters the host cell and is replicated along with phage proteins. Some phages exit by lysing the bacterium. Phages that go through this productive life cycle are termed **virulent.** The virulent phage most intensively studied is the double-stranded DNA phage called **T4,** and its morphology is illustrated in figure 13.6. Each of the steps in its life cycle will now be discussed (figure 13.7).

Step 1: Phages Attach to Host Bacterial Cell Receptors

When a suspension of T4 phages is mixed with a susceptible strain of *E. coli*, the phages collide by chance with the bacteria (figure 13.7 *a*). Unlike bacteria, viruses do not have flagella and therefore cannot swim toward attractants. Protein fibers at the end of the phage tail attach to specific receptors on the bacterial cell wall.

Step 2: Viral Nucleic Acid Enters the Host Cell

After attachment, an enzyme **(lysozyme)** located in the tip of the phage tail degrades a small portion of the bacterial cell wall (figure 13.7 *b*). The tip of the tail then opens, and the linear viral DNA in the head of the phage passes through the channel of the phage tail. The DNA is literally injected through the cell wall and, by some unknown mechanism, passes through the cytoplasmic membrane, entering the interior of the cell. The protein coat of the phage remains outside the cell. The separation of nucleic acid from its protein coat is a key feature of all viral infections. However, many animal viruses enter the host cell intact, and then the nucleic acid separates from the capsid and envelope inside the cytoplasm (see chapter 14).

Step 3: Phage DNA Is Transcribed Leading to Production of Specific Proteins

Within minutes of the entry of phage DNA into a host cell, a part of the DNA is transcribed into mRNA, which is then translated into proteins that are specific for the infecting phage (figure 13.7 *c*). The first proteins produced are enzymes that are not normally

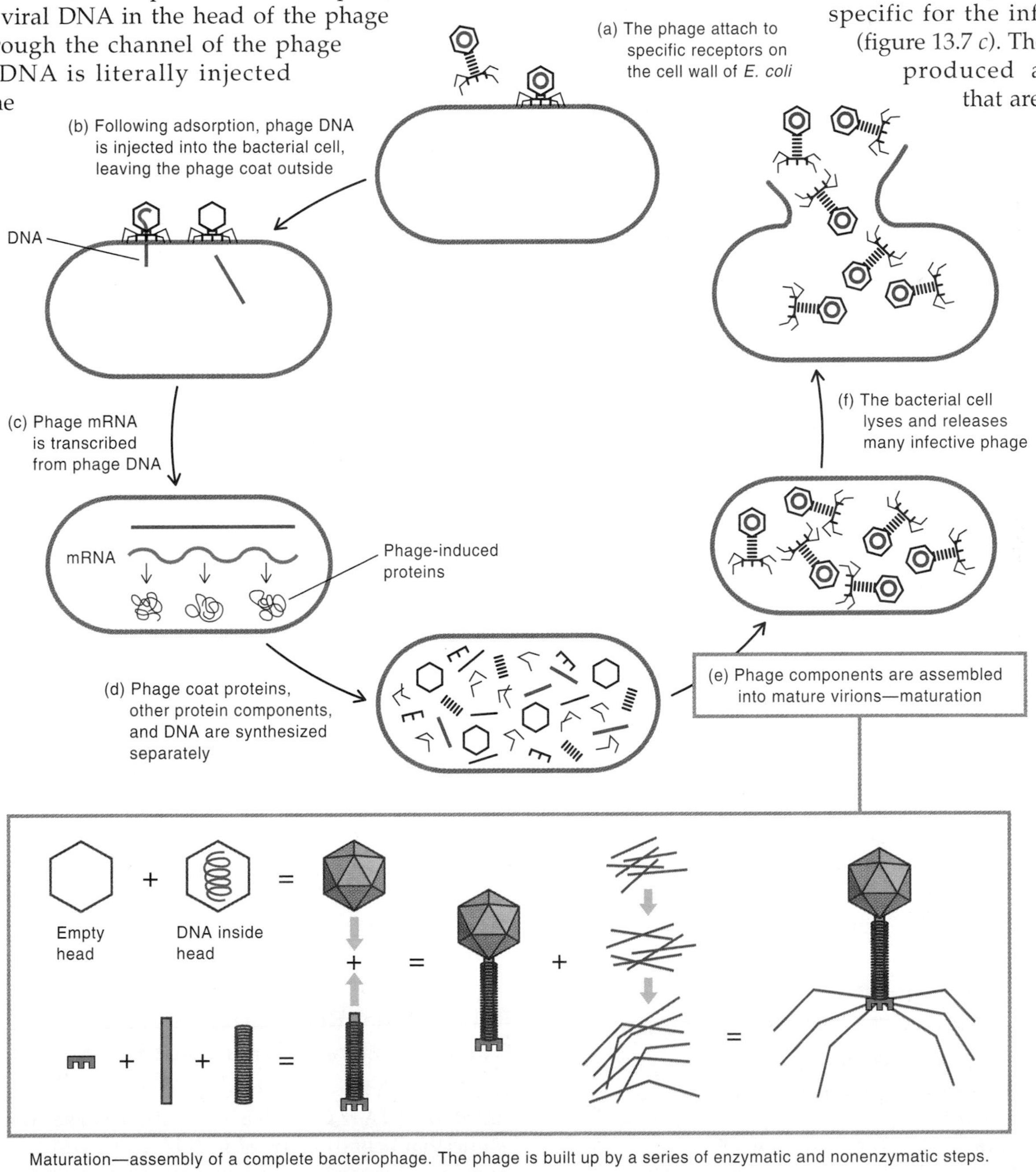

(a) The phage attach to specific receptors on the cell wall of *E. coli*

(b) Following adsorption, phage DNA is injected into the bacterial cell, leaving the phage coat outside

DNA

(c) Phage mRNA is transcribed from phage DNA

mRNA

Phage-induced proteins

(d) Phage coat proteins, other protein components, and DNA are synthesized separately

(e) Phage components are assembled into mature virions—maturation

(f) The bacterial cell lyses and releases many infective phage

Empty head

DNA inside head

Maturation—assembly of a complete bacteriophage. The phage is built up by a series of enzymatic and nonenzymatic steps.

Figure 13.7
Steps in the replication of T-4 phage in *Escherichia coli*.

present in the uninfected host cell. These are known as **phage-induced proteins** because they are coded by phage and not bacterial DNA. They are essential for the synthesis of phage DNA and phage coat protein. One phage-induced protein is a nuclease that degrades the DNA of the host cell; as a result, soon after infection host cell DNA cannot be transcribed and all mRNA synthesized is transcribed from the phage DNA. In this way, the phage alters the metabolism of the bacterial cell to its own purpose, namely the synthesis of more phages. However, existing host enzymes continue to function. They supply the energy necessary for phage replication through the breakdown of glucose; they also synthesize amino acids and nucleotides, the subunits of proteins and nucleic acids of the replicating phage. The host enzymes also play a role in the synthesis of phage nucleic acid and phage coat proteins, the latter being carried out on bacterial ribosomes.

Not all phage-induced enzymes are synthesized simultaneously. Rather, they are synthesized in a sequential manner during the course of infection. Those that are required early, such as enzymes of phage DNA synthesis, are synthesized early in infection; others that are not required until later in infection are synthesized later. This timing pattern is brought about because the phage DNA codes for the synthesis of different sigma factors of the RNA polymerase, which are synthesized at different times. These sigma factors are required for transcribing specific genes. An example of an early-induced enzyme is the nuclease that degrades the host chromosome. Late-induced enzymes include those concerned with the assembly of capsids and a type of lysozyme that lyses the bacteria and releases the newly formed virions.

Step 4: Phage DNA Replicates and Phage Structural Proteins Are Synthesized Separately

During the replication of all viruses, whether they are bacterial, animal, or plant, the viral protein and nucleic acid components replicate separately from one another (figure 13.7 *d*). The entering phage DNA has two independent functions: to serve as the template for replication of more phage DNA and to serve as the template for transcription of mRNA, which is then translated into phage-induced enzymes and the proteins that form the capsids. This mechanism of phage reproduction is termed **vegetative reproduction.**

Step 5: Phage DNA and Protein Assemble to Form Mature Virions—The Maturation Process

The **maturation,** or **assembly, process** involves the assembly of phage protein with phage DNA to form virions (figure 13.7 *e*). This is a complex, multistep process in the T4 phages. All of the structures of the phage, such as the heads, tails, tail spikes, and tail fibers, are synthesized independently of one another. Once the phage head is formed, it is packed with DNA; the tail is then attached, followed by the addition of the tail spikes. Some of the steps in the synthesis of these various components involve a **self-assembly process,** in which the protein components of the capsids come together spontaneously without any enzyme catalysis to form a specific structure. In other steps, certain proteins serve as scaffolds on which various protein components associate. The scaffolds do not themselves become a part of the final structure, much like scaffolding required to build a house does not become part of the house.

- **Transcription, p. 136**
- **Ribosomes, p. 72**
- **Sigma factor, p. 137**

Step 6: Virions Are Released from the Host Cell Following Lysis

During the latter stages of the infection period, the enzyme lysozyme coded by phage DNA is synthesized. This enzyme digests the host cell wall from within, resulting in cell lysis and the release of phage virions (figure 13.7 *f*). It is important that this enzyme not be synthesized early in the infection process, because it would lyse the host cell before any mature phage could be formed. This is why it is important that phage genes be expressed only at the time the enzymes for which they code are needed to function.

In the case of the T4 phages, the **burst size** or the number of phages released per cell is about 200. These virions then infect any susceptible cells in the environment, and the process of virus replication is repeated.

RNA Lytic Phages—General Features

Another group of lytic phages are the single-stranded RNA phages. The ones most intensively studied are those that infect *Escherichia coli.* These include MS2 and Qβ, which show many similarities to one another. These phages have several unusual properties. First, they infect only F^+ strains of *E. coli* because they attach to the sides of the sex pilus. Second, they replicate rapidly and, following lysis of the host cell, yield very high levels of progeny—approximately 10,000 per cell. They also all replicate by a similar mechanism.

RNA Phage Replication Involves Formation of Complementary Strands of RNA

The RNA phages contain only a single positive (+) sense strand of RNA (figure 13.8). A positive sense strand of RNA is one that acts directly as mRNA from which proteins are translated. In order to replicate more strands of positive sense RNA, the positive sense strands must first serve as a

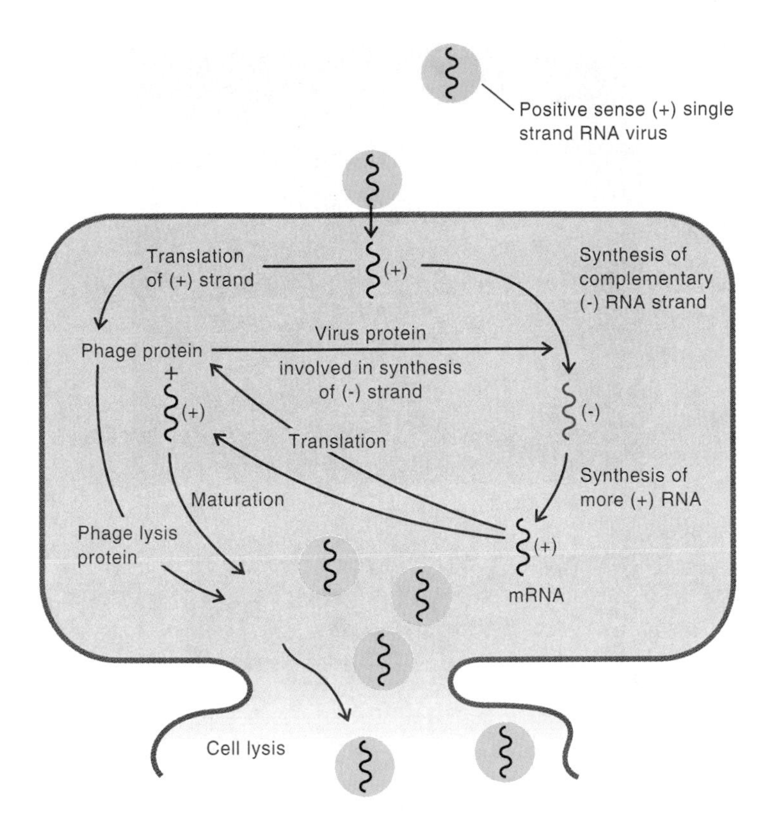

Figure 13.8
Replication of positive (+) sense strand RNA phage. The entering positive strand is translated to phage proteins and also is replicated by the synthesis of complementary negative strands by a phage-induced RNA polymerase. The negative RNA gives rise to more positive RNA strands, which are translated to more phage-induced proteins and are also incorporated into the virions, which then lyse the host cells.

template for the synthesis of negative sense strands by the phage induced enzyme RNA polymerase. The negative sense strands in turn serve as the template for positive sense strand synthesis, which are then incorporated into the virion in the maturation process. Note that the synthesis of a strand of RNA using another strand of RNA as the template is not an enzymatic reaction that occurs inside an uninfected cell. Therefore, the positive sense strand codes for the synthesis of this enzyme, an unusual RNA polymerase, which requires RNA as a template. RNA polymerase in the uninfected cell transcribes the template strand of DNA into RNA.

No RNA phages that have a single negative sense strand of RNA have been discovered and double-stranded RNA phages are extremely rare. However, viruses that infect animals do have these kinds of RNA and will be discussed in chapter 14. Another group of animal viruses are the retroviruses, which contain single positive sense strand RNA molecules and have a life cycle in which the RNA is copied into a single strand of DNA integrated into the DNA of the host cell. These retroviruses are not represented in any phages and will be discussed in chapter 14. A retrovirus causes AIDS, which is discussed in great detail in chapter 31.

Phage Growth–One-Step Growth Curve

The replication of bacteriophage from the time of attachment to the bacterium to the release of the virions can be conveniently measured by using a **one-step growth curve.** This procedure involves measuring the number of virions inside and outside the bacteria at various times after the phages have infected their hosts. The procedure is described in figure 13.9.

Enough phage particles are added to a suspension of bacteria so that every bacterium becomes infected by one or more phages. After the phages attach to the bacteria (within about 5 minutes), samples are removed at various times, and treated with chloroform. The chloroform lyses the bacteria, which releases any phages within the bacterium. The number of phages in the suspension of lysed cells at any particular time is determined by mixing samples containing phages with a bacterial suspension, which is then inoculated on an agar medium. The samples of phages are diluted so that only a single virion infects a bacterial cell. After incubation for 24 to 48 hours, this infection of bacteria by phages results in a **plaque,** an area of clearing in the bacterial lawn. The plaque is the site at which a single phage infected a cell and released its viral progeny to infect and lyse adjacent bacteria. Thus, counting the number of plaques gives the number of phage particles in the diluted sample (figure 13.10).

The curve (figure 13.11) derived from phage counts at different times shows that, shortly after the bacteria are infected, no intact phages remain inside the bacteria. This is termed the **eclipse period** and reflects the fact that the protein coat and DNA have separated from one another at the time of infection. After about 20 minutes, infectious phages can be detected inside the bacteria (by lysing the cells with chloroform). Their number rapidly increases, which indicates the time after infection that phages mature. The number of mature phages then remains constant, since all of the bacteria were infected initially and no uninfected cells are available to be infected.

If chloroform is not added to the bacterial suspension before the bacteria are lysed by the phage, then each infected cell will be counted as giving rise to a single phage because lysis will occur on the solid medium after the infected bacteria are plated. The phage particles released will only infect other bacteria in the immediate vicinity, thereby only giving rise to a single clear zone or plaque. The length of time between entry of DNA into the cell and phage release following lysis by the phage is termed the **latent period.** Note that this period is longer than the eclipse period, the length of time it takes for mature virions to be detected inside the cell.

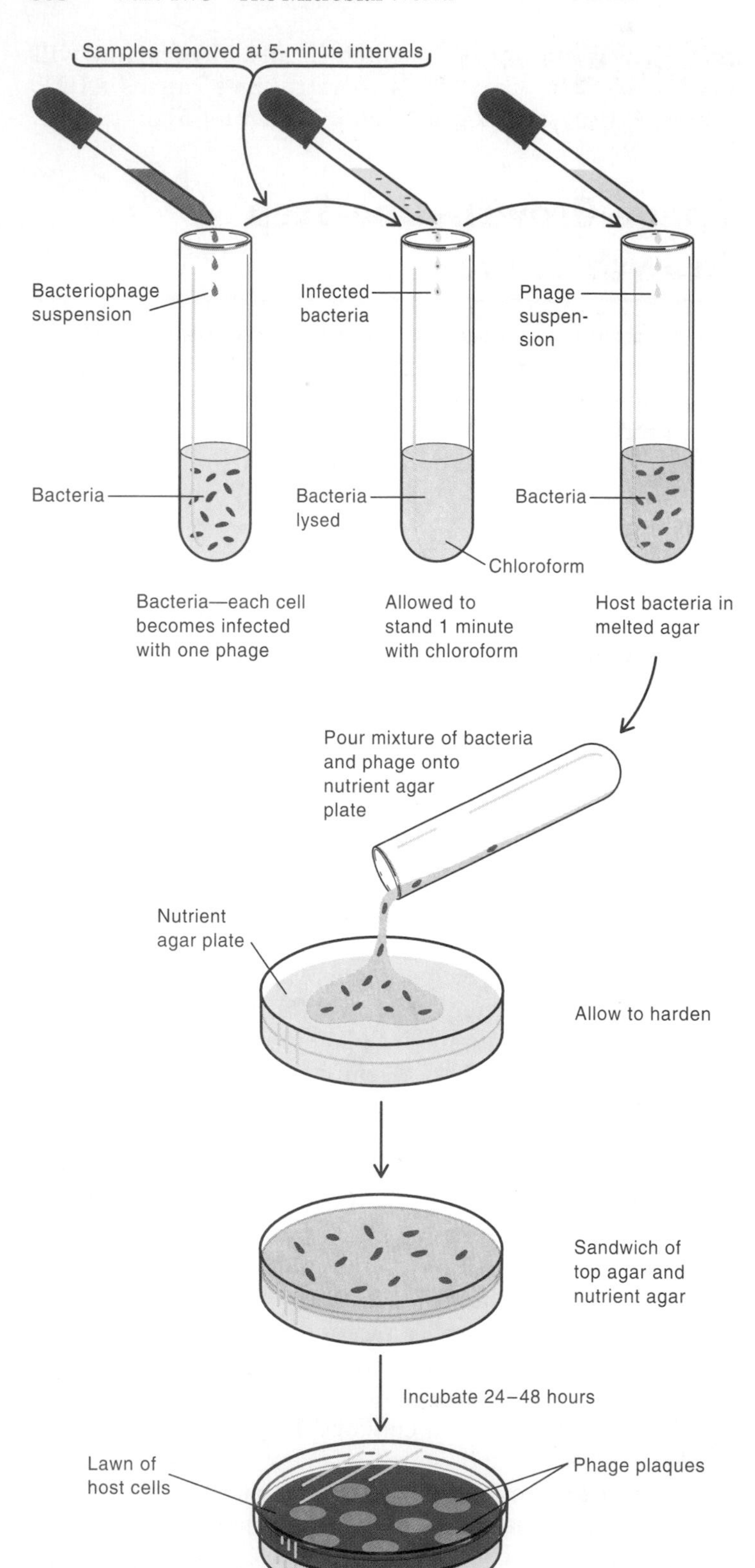

Figure 13.10
Bacteriophage plaques in petri dish. Each clear zone represents one phage particle that has infected a single bacterium, multiplied, lysed, and then infected surrounding cells.

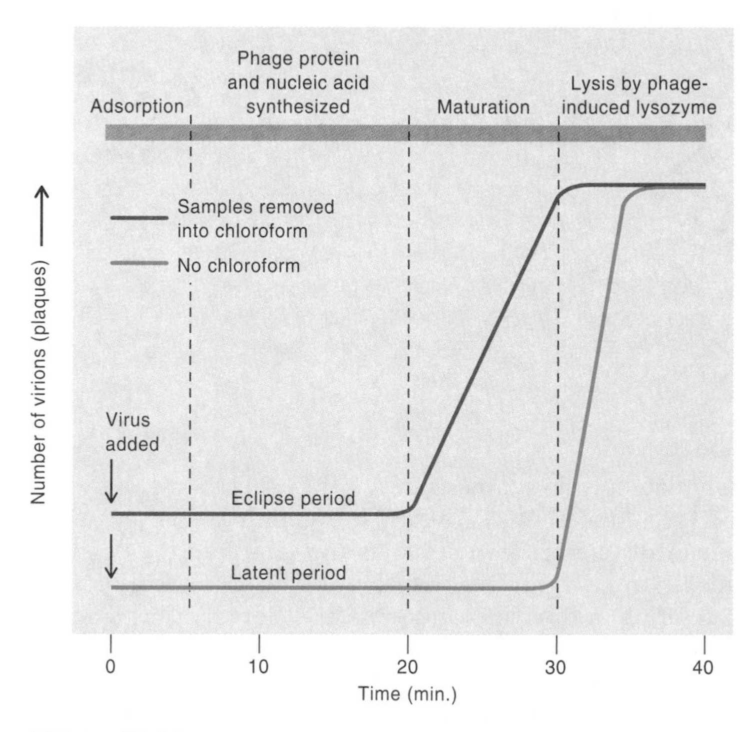

Figure 13.11
One-step growth curve of virus replication. Chloroform lyses cells, releasing any intact phages. Without chloroform, the phages are released as a result of lysis of the bacteria by phage induced lysozyme.

Figure 13.9
One-step growth curve. To assay the number of phages in a sample, a dilution of a suspension containing the virus material is mixed in a small amount of melted agar with the sensitive host bacteria, and the mixture is poured on the surface of a nutrient agar plate. The host bacteria, which have been spread uniformly throughout the top agar layer, begin to grow, and after overnight incubation, form a *lawn* of confluent growth. Each virus particle that attaches to a cell and reproduces may cause cell lysis. Then the virus particles released can spread to adjacent cells in the agar, infect them, multiply, and again lead to lysis and release. The size of the plaque formed depends on the virus, host, and conditions of culture.

Perspective 13.2

Viral Soup

It has been recognized for many years that viruses are important in nature because they cause many very deadly diseases of humans, animals, and plants. Within the past few years, scientists have discovered that they may play another important role in nature—that of being a major component of **bacterioplankton,** the bacteria found in the ocean. In the 1970s and 1980s, bacterial cells were found to be much more abundant than had been believed previously. Earlier estimates of the number of bacteria in aquatic environments were based on the numbers that could be cultured in the laboratory. However, when direct microscopic counts were made, the number of heterotrophic bacteria was estimated to be 10^5 to 10^7 bacteria per ml, much higher than had been previously estimated. Indeed, most carbon fixation in the ocean is carried out by bacteria. In addition, the concentration of cyanobacteria and single-celled eukaryotic algae is much greater than previously believed. However, what was most surprising was the number of bacteriophages in aquatic environments. To get an accurate estimate of their numbers, scientists centrifuged large volumes of natural unpolluted waters from various locations. They then counted the number of particles in the pellet. Much to their surprise, up to 2.5×10^8 phages per ml were counted, about 1,000 to 10 million times higher than previous estimates. Why are these new numbers important? First, they may answer the question of why bacteria have not saturated the ocean. The bacteria are most likely held in check by the phages. From the estimates of the numbers of bacteria and phages, it is calculated that one-third of the bacterial population may experience a phage attack each day. Second, the interaction of phages with bacteria in natural water implies that the phages may be actively transferring DNA from one bacterium to another, by the process of transduction.

Thus, the smallest agents in natural waters may have important consequences for the ecology of the aquatic environment. Perhaps viruses have a similar function for eukaryotic organisms.

• **Transduction, p. 169**

REVIEW QUESTIONS

1. Name the two important components of every virus.
2. List four features of viruses that distinguish them from cells.
3. What are the major shapes of viruses?
4. List the three major relationships that viruses can have with their host bacteria.

Virus Replication in a Latent State

Some viruses do not lyse their host cells, but rather "live" in harmony with them. In this relationship, the entering viral DNA becomes integrated into the chromosome of the host cell and often confers new properties on the host.

The DNA of Temperate Viruses Is Integrated into the Host Chromosome

The phage T4 is termed **virulent** because it can only replicate inside the host cell and then lyse its host. However, other phages exist that can either replicate and cause cell lysis or integrate their DNA into the host cell chromosome. These are known as **temperate phages.** The phage DNA when integrated into the host cell DNA is called a **prophage,** or a **provirus** if the virus is not a phage and the bacterial cell carrying a prophage is a **lysogenic cell,** or **lysogen.** Although viral latency has been studied most extensively in bacteriophages, many animal viruses also integrate their DNA into the host cell. The virus in this group being most actively studied today is probably the human immunodeficiency virus (HIV), which causes AIDS.

The most thoroughly studied temperate bacteriophage is the phage **lambda (λ),** which infects *E. coli.* Its size and shape are similar to T4 phage. The early stages in the infection of bacteria by T4 phage (see figure 13.6 *a* and *b*) are similar to the early stages of the life cycle of phage lambda, shown in figure 13.12. However, as shown in figure 13.12 *a,* once the phage DNA enters the host cell, one of two events can occur.

One possibility is that a productive infection can develop, resulting in cell lysis similar to that described for the T4 phage (see figure 13.7). The other possibility is that the phage DNA may become integrated into specific sites in the bacterial chromosome, becoming a part of the bacterial genome. In any culture infected with temperate phages, some of the bacteria will be lysogenized while others will be lysed. Whether the lambda phage lyses the majority of the cells or lysogenizes them depends largely on chance events, which can be tipped in favor of one or the other by modifying the external environment.

The Integration of Phage DNA into the Bacterial Chromosome Occurs Because of Identical DNA Sequences in Both

For a long time, how temperate phages could integrate into the bacterial chromosome was a mystery. However, during the last 25 years, the availability of sophisticated experimental techniques has made it possible to gain some detailed understanding of the mechanisms involved. It was learned that short nucleotide sequences in the DNA of phage and host are identical (homologous), allowing the phage and bacterial DNA to come together (**synapse**; figure 13.12 *b*). Following synapse, the phage DNA becomes integrated into the bacterial chromosome through the process of homologous recombination.

• **homologous recombination, p. 168**

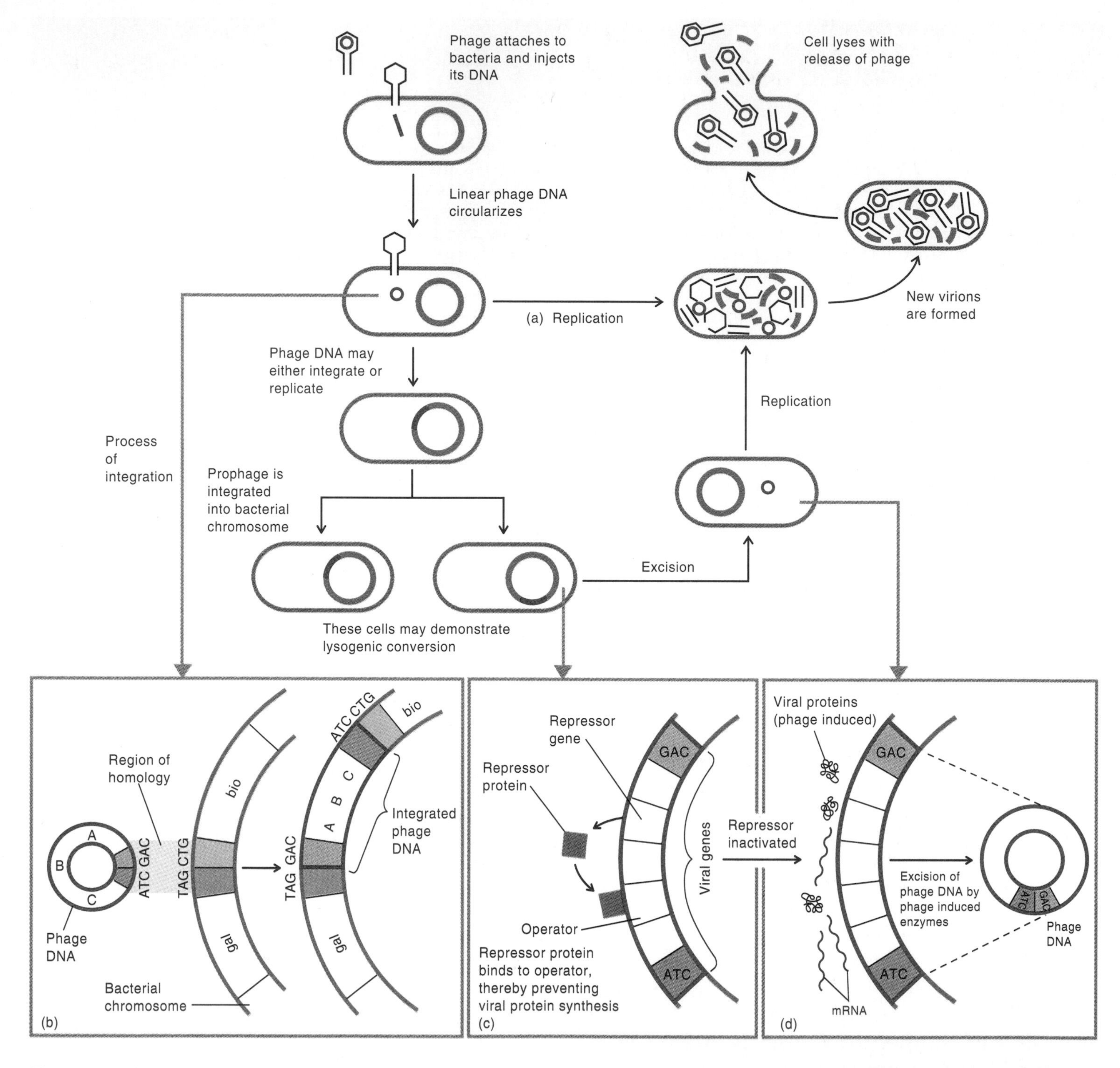

Figure 13.12
Life cycle of phage lambda (λ). One of two events can occur: (*a*) the lambda can go through a replication cycle like T4 or (*b*) the circular DNA can integrate into the bacterial chromosome because of regions of homology in the phage and bacterial chromosome. Once integrated, the phage DNA can remain in the prophage state indefinitely. (*c*) Repressor protein encoded by the phage DNA prevents the transcription of the gene necessary for the excision of the phage DNA. (*d*) If the repressor protein is inactivated, the prophage is excised from the bacterial chromosome. The phage DNA replicates and new virions are formed.

The Repressor Must Function Continuously to Keep the Prophage Integrated

Once integrated, the phage DNA can remain in the prophage state indefinitely. However, for the DNA to remain in an integrated state, a set of genes present on the integrated phage must be inhibited (repressed). These genes code for enzymes that remove (excise) the integrated DNA from the host chromosome as well as catalyze the ensuing steps in productive phage infection. How are these genes repressed? One gene in the integrated viral DNA codes for a **repressor** that prevents transcription of the integrated genes required for a productive infection (figure 13.12 c). As long

Table 13.4 Some Examples of Lysogenic Conversion

Microorganism	Property Affected
Corynebacterium diphtheriae	Synthesis of diphtheria toxin
Clostridium botulinum	Synthesis of botulinum toxin
Streptococcus (β-hemolytic)	Erythrogenic toxin responsible for scarlet fever
Salmonella	Modification of lipopolysaccharide of cell wall
Vibrio cholerae	Synthesis of cholera toxin

as this repressor is produced, the integrated phage chromosome is maintained as a prophage. However, if for any reason the repressor is no longer synthesized or is inactivated, the genes are transcribed and an enzyme that excises the viral DNA from the bacterial chromosome is synthesized. Once excised from the host chromosome, the viral DNA replicates, and codes for the synthesis of phage proteins, and the life cycle of a productive phage infection ensues.

Activation of The SOS Repair System Destroys the Repressor

Under ordinary conditions of growth of a lysogen, the phage DNA is released from the bacterial chromosome only about once in 10,000 divisions. However, if a lysogenic culture is treated with an agent that damages the DNA (such as ultraviolet light), the SOS system comes into play. This system activates a protease that destroys the repressor (figure 13.12 *d*). As a result, all of the prophages enter the lytic cycle, and a productive infection results. This process, termed **induction,** results in complete lysis of the culture. The term *induction* as used here should not be confused with induced enzyme synthesis as discussed in chapter 6 or with phage-induced enzymes discussed earlier in this chapter.

Lysogens Are Immune to Infection by the Same Phage

In addition to maintaining the prophage in the integrated state, the repressor protein produced by a prophage also prevents the infection of a lysogenic cell by phages of the type whose DNA is already carried by the lysogenic cell. Infections are blocked because the repressor binds to specific base sequences in the phage DNA as it enters the cell and inhibits its replication. Consequently, the cell is **immune** to infection by the same type of phage but not to infection by other phages, to whose DNA the repressor cannot bind.

- **Repressor, p. 146**
- **SOS system, p. 162**

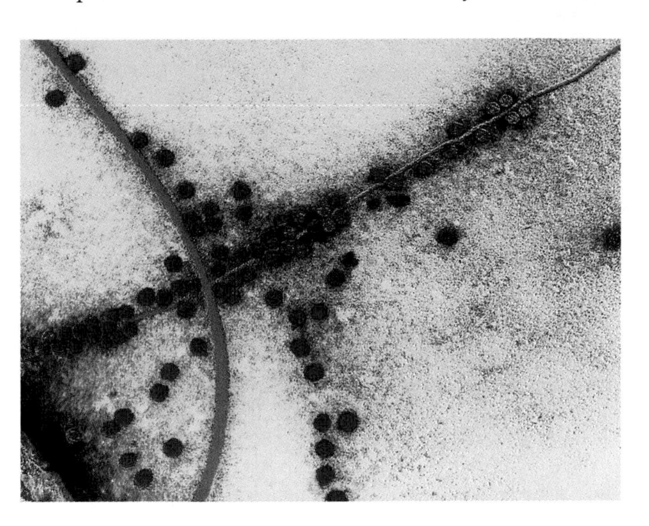

Figure 13.13
Colored transmission electron micrograph of bacteriophage f2 binding to the sex (F) pilus of *Escherichia coli.* The red arc across the bottom of the picture is a curved bacterial flagellum.

Lysogenic Conversion

In addition to having this immunity, lysogenic cells may differ from their nonlysogenic counterparts in other important ways. The prophage can confer new properties on the cell, a phenomenon called **lysogenic conversion.** For example, strains of *Corynebacterium diphtheriae* that are lysogenic for a certain phage (β phage) synthesize the toxin that causes diphtheria. Similarly, lysogenic strains of *Streptococcus pyogenes* and *Clostridium botulinum* manufacture toxins that are responsible for scarlet fever and botulism, respectively. In all of these cases, if the prophage is eliminated from the bacterium, the cells cannot synthesize the toxin. The genes that code for these toxins are phage genes, which are expressed only when the phage DNA is integrated into the bacterial chromosome. Some of the properties of bacteria for which prophage are responsible are given in table 13.4.

An analogous situation to lysogenic conversion can be found in the tumor virus animal system. Tumor causing viruses also have their DNA integrated into the host cell, which results in the gain of new properties by the host cell.

Filamentous Phages

In addition to developing productive (lytic) infections and latency, viruses can develop other relationships with their host cells. Both T4 and lambda phages lyse their host cells when they are released. However, a few bacterial viruses, the **filamentous phages** which include M13 and fd (figure 13.13) all of which are single-stranded DNA phages closely related to each other, are released by a process termed **extrusion,** which does not destroy the host cell. Indeed, these few phages are the only ones that appear to be incapable of causing a lytic infection. Their life cycle is diagrammed in figures 13.14 and 13.15.

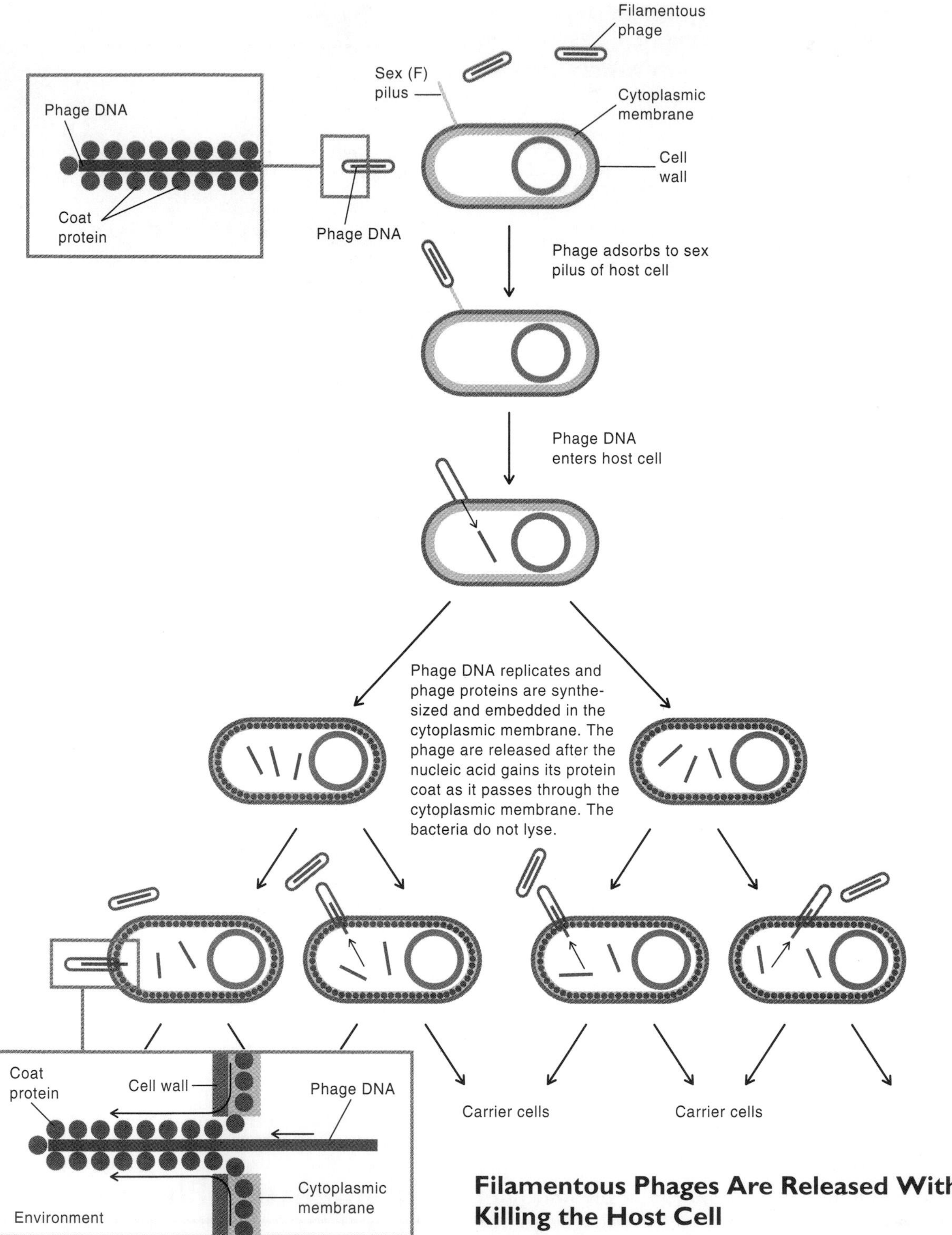

Figure 13.14
Life cycle of filamentous bacteriophages that are extruded from their host cell. These single-stranded DNA phages attach to the sex pili of *Escherichia coli.*

Filamentous Phages Are Released Without Killing the Host Cell

The filamentous phages have several features that distinguish them from the T4 virulent phages. First, their shape is very different (figure 13.13). The long thin fiber encloses a single molecule of circular, single-stranded DNA. Second, they infect only F^+ cells of *E. coli* because they attach to the tip of the sex pilus. Third, the filamentous phages do not completely take over their hosts' metabolism, because the infected bacteria continue to multiply. Fourth, neither mature filamentous phages nor their coats can be detected in the cytoplasm of the host cells. One explanation is that, after being synthesized, the phage coats are stored in the cytoplasmic membrane of the host cell, and the viruses are assembled as they are extruded from the cell. Fifth, viruses are continuously extruded rather than being released in a burst. Infected bacteria are termed **carrier cells.** They can be subcultured and stored in the same way as noninfected strains, and they continuously release virions into the medium. The next chapter discusses certain viruses that infect humans and exit by **budding,** a process that also does not necessarily kill the host cell. Examples of such viruses are the ones that cause influenza and mumps.

Single-Stranded DNA of Filamentous Phages Is Converted to a Double-Stranded Molecule in the Process of Replication

The DNA that enters the cell is a single-stranded molecule that is then converted to a double-stranded molecule by enzymes of the host cell. The double-stranded form, termed the **replicative form,** gives rise to the single-stranded positive(+) sense strand and is also transcribed into mRNA to give rise to viral proteins (figure 13.15). The double-stranded replicative form is also the form that replicates in much the same way as double-stranded DNA of bacteria replicates. The single-stranded DNA molecule that enters the cell is termed the **positive(+) sense strand** of DNA by convention because it is not the strand that can be transcribed into mRNA. By convention, the negative (–) sense strand of DNA is transcribed into mRNA which is the (+) strand and can serve in protein synthesis. The single-stranded DNA that is incorporated into the coat protein as the virus is extruded from the cell is derived from the double-stranded replicative form.

• **DNA replication, p. 156**

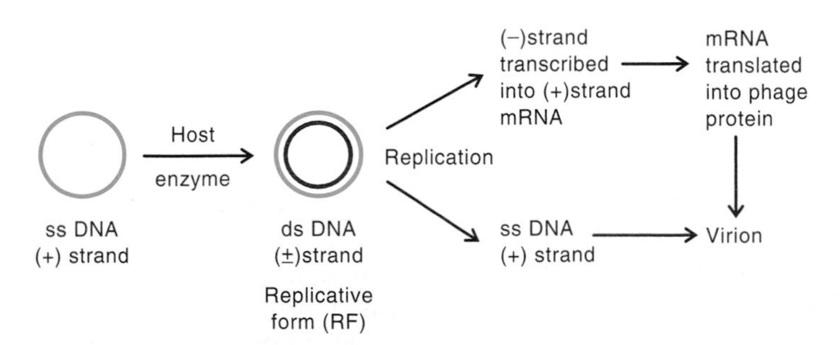

Figure 13.15
Macromolecule synthesis in filamentous phage replication.

Transduction

Bacteriophages play an important role in bacterial gene transfer from one bacterium to another. As briefly discussed in chapter 7, DNA can be transferred from one bacterial cell to another by phages, in the process termed **transduction.** There are two types of transduction. In one type, any bacterial gene can be transferred by the phages, through the phenomenon of **generalized transduction.** In this process, the phages are termed **generalized transducing phages.** The second type of transduction is termed **specialized transduction** since only a few specific genes can be transferred by the phages, which are called **specialized transducing phages.**

Generalized Transduction Is Carried out by Virulent and Temperate Phages

Virulent as well as temperate phages can serve as generalized transducing phages. Recall that some phages in their life cycle degrade the bacterial chromosome into many fragments at the beginning of a productive infection. Some of these DNA fragments can be incorporated into the phage head in place of phage DNA in the process of phage maturation. Once such a phage infects another bacterial cell, the bacterial DNA is transferred. Then the DNA can integrate into the recipient cell DNA by homologous recombination.

Why do the transducing virulent phages not lyse the cells they invade? The answer is that because the bacterial DNA replaces the phage DNA inside the phage's head, the genetic information necessary for the synthesis of many phage-induced proteins is lacking. The phage is termed **defective** because it lacks all of the DNA necessary to form complete virions and lyse the cell. Defective animal viruses are also known.

Specialized Transduction Is Only Carried out by Temperate Phages

Temperate bacteriophages are examples of **episomes,** genetic elements that can replicate as part of the bacterial chromosome (as a prophage) or independently of it (as a virulent phage). Their relationship to host cells is similar in many ways to the F^+-Hfr system, also an example of an episome, which was discussed in chapter 7. Just as the Hfr DNA can be removed from the bacterial chromosome (as a prophage) to become an F^+ plasmid, the prophage can be excised from the bacterial DNA and then replicate (figure 13.16). This process is the opposite of what happens when the temperate phage, such as lambda, integrates into

• **Generalized transduction, p. 169**

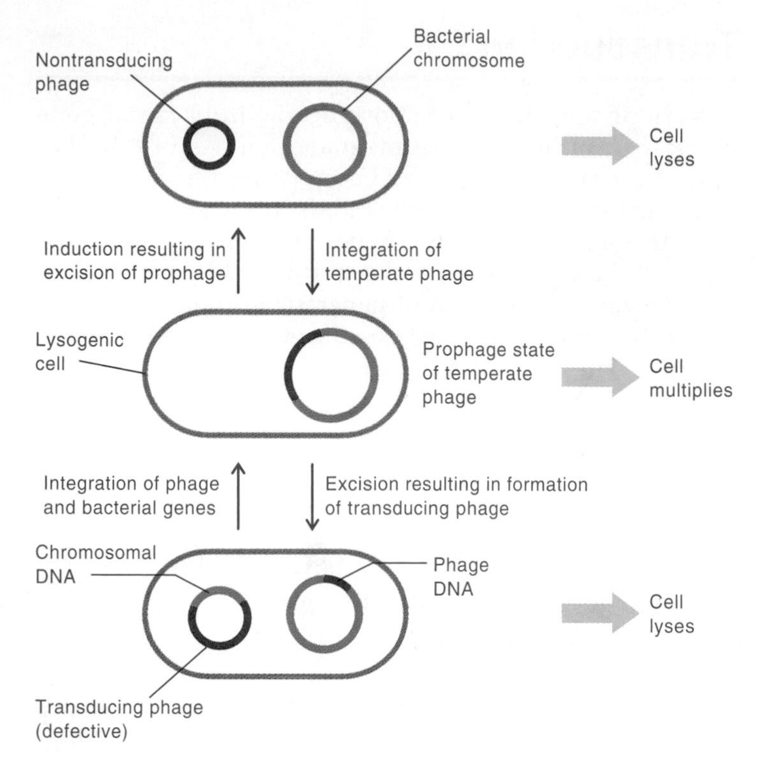

Figure 13.16
Integration and excision of temperate phage DNA. The excision may be imprecise and some of the chromosome may remain attached to the phage DNA, which is left behind in the chromosome.

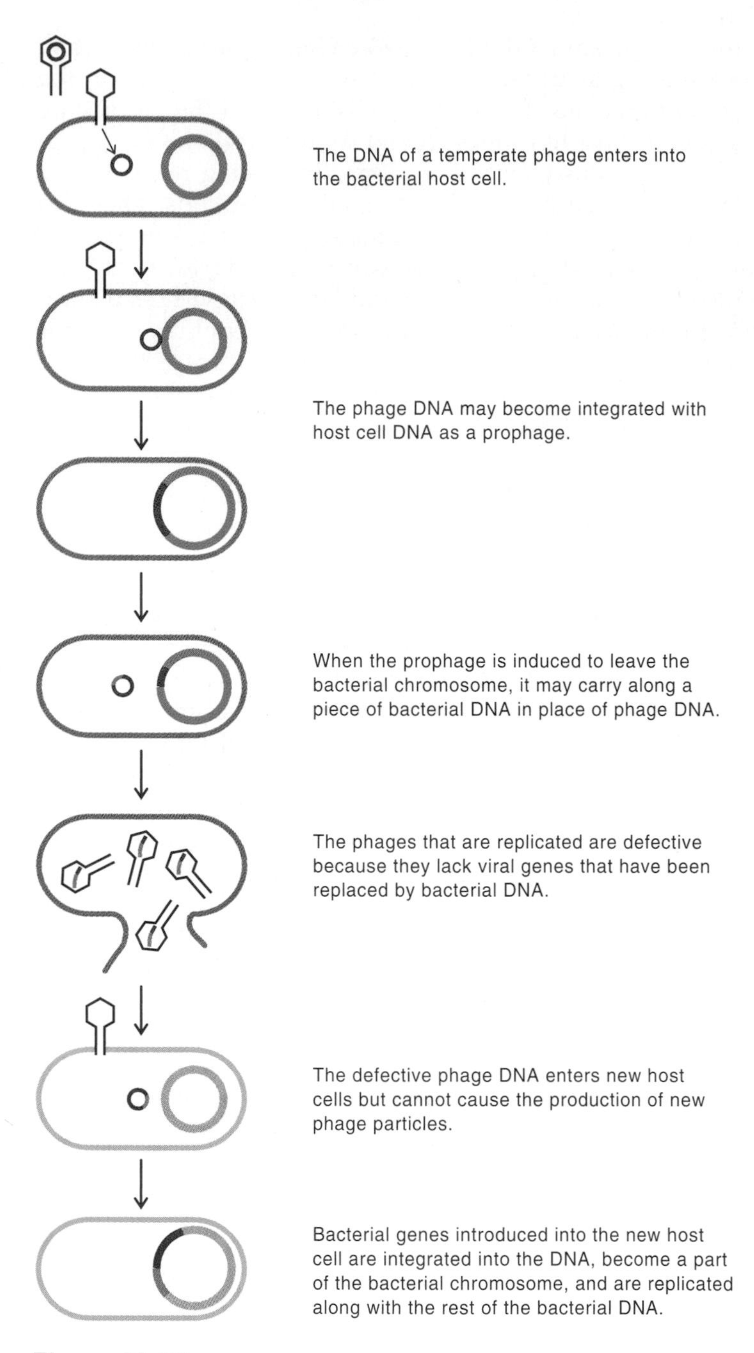

Figure 13.17
Specialized transduction by temperate phage. Only genes near the site where the prophage is integrated can be transduced.

the host chromosome (see figure 13.12 *b*). In the usual situations, only the phage DNA is excised. However, on rare occasions, a piece of bacterial DNA remains attached to the piece of phage DNA that is excised, which is also analogous to the formation of an F′ plasmid when Hfr DNA is excised from the bacterial chromosome. A piece of phage DNA is left behind in the bacterial chromosome, thereby creating a defective phage (figure 13.17).

The bacterial genes attached to the phage DNA replicate as part of the phage DNA and also become incorporated into mature phages in the maturation process. The phage, containing a piece of bacterial DNA, is then released from the lysed cells. When the phage infects another bacterial cell, both phage and bacterial DNA can again integrate into the chromosome. Thus, the resulting lysogen contains bacterial genes from the previously lysogenized cell (figure 13.17).

Only bacterial genes located near the site of integration of the phage DNA can be transduced. Because of the need for similarity (homology) between the DNA of the temperate phage and the site at which it integrates, DNA normally integrates at only one particular region of the host chromosome. Consequently, when the lambda DNA is excised from the chromosome, it can carry along only the limited group of genes that are near this region. Therefore, lambda phage is termed a **specialized transducing phage.**

Host Range of Viruses

The organisms that a virus can infect define its **host range.** The range of cells that can be infected by all the different viruses is very large. It is probable that any kind of cell can be infected by one virus or another. To date, viruses have been shown to infect cells of algae, fungi, and protozoa and most species of bacteria, plants, and animals. However, any given virus can generally infect cells of only one or a few species and often just a few strains of a species. Thus, the phage T4 will infect only certain strains of *E. coli,* and poliovirus naturally infects only certain cells of humans and a few other primates (the highest order of vertebrates, which includes humans, apes, and monkeys). However some viruses can infect widely different species.

Host Range Is Frequently Determined by Presence or Absence of Receptors

The host range is limited by the fact that animal and bacterial (but not plant) viruses must adsorb to specific receptor sites on the host cell surface in order to invade the cell. This adsorption requires that the attachment proteins on the virion be complementary to the attachment sites on the host cell. Since receptor sites vary in chemical properties and location, any particular virus can attach to only specific cells. Phages usually attach to the bacterial cell wall. This explains how prophages can confer resistance to similar phages by altering the surface of the host (lysogenic conversion). A few phages, however, attach to pili (figure 13.18 *a*), and a few others attach to sites on flagella (figure 13.18 *b*). The process by which the viral genome enters the cytoplasm of the host cell is unknown. Animal viruses bind to specific receptor sites on the cytoplasmic membranes of the host cells that they infect, and mutations in a host cell that alter or remove receptor sites confer resistance to that particular virus. Blocking absorption of viruses to their receptors is one obvious way to prevent viral infection.

Restriction-Modification System Limits Host Range of Phage

As discussed already, a recipient bacterium will take up DNA by transformation, conjugation, or transduction only if the donor DNA and recipient cells are of the same species or even subspecies. Cells are able to recognize and restrict the entry of foreign DNA if the methylation pattern of the entering DNA differs from its own. This restriction-modification system plays an important role in the host range of phage as illustrated in the following example.

Phages that infect, multiply, and lyse *E. coli* strain K-12 can infect the same strain of *E. coli* with a very high efficiency. However, these phages infect the B strain of *E. coli* with a very low efficiency. However, if the phages that are released from the few infected B cells are added again to the B strain, infection occurs at a high frequency. These observations can be explained by the presence of different restriction endonucleases in *E. coli* B and K-12 that recognize whether the entering DNA came from the B or K-12 strain of *E. coli* by its pattern of methylation (figure 13.19).

• **Antiviral medicines, p. 467**
• **Restriction modification, p. 178**
• **Restriction endonucleases, p.178**

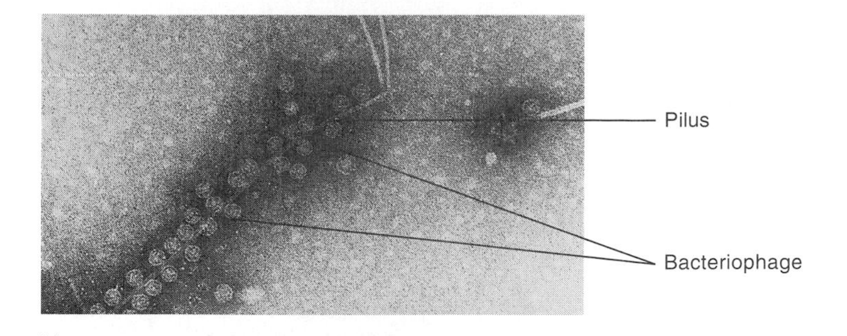

(a)

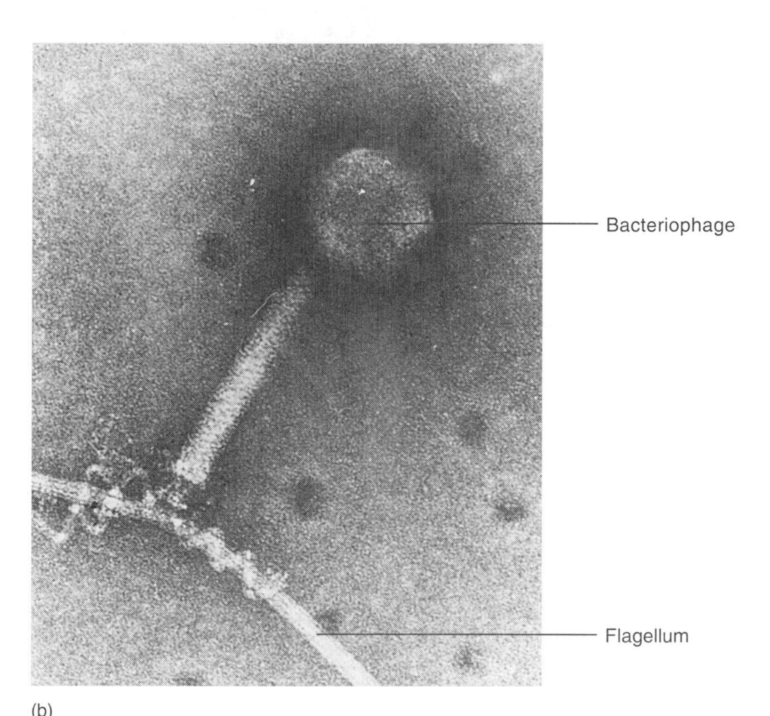

(b)

Figure 13.18
Adsorption of bacteriophage on various cell structures. (*a*) Bacteriophages on pilus of *Escherichia coli* (×19,200). (*b*) Bacteriophage tail fiber entwined around flagellum.

In rare cases, the phage DNA having replicated in a foreign cell can be methylated by the modification enzyme before the restriction endonuclease has an opportunity to degrade it. Once methylated in the specific pattern of the host, the "foreign" DNA is converted to "nonforeign" DNA. This DNA will replicate and give rise to phage proteins, resulting in the release of phages that can now infect the new host strain at a high efficiency. This process is not important in determining the host range of animal and plant viruses because eukaryotic cells do not contain restriction enzymes.

Methods Used to Study Viruses

Techniques are needed in the laboratory to recognize the presence of viruses, identify them, and grow them in large quantities.

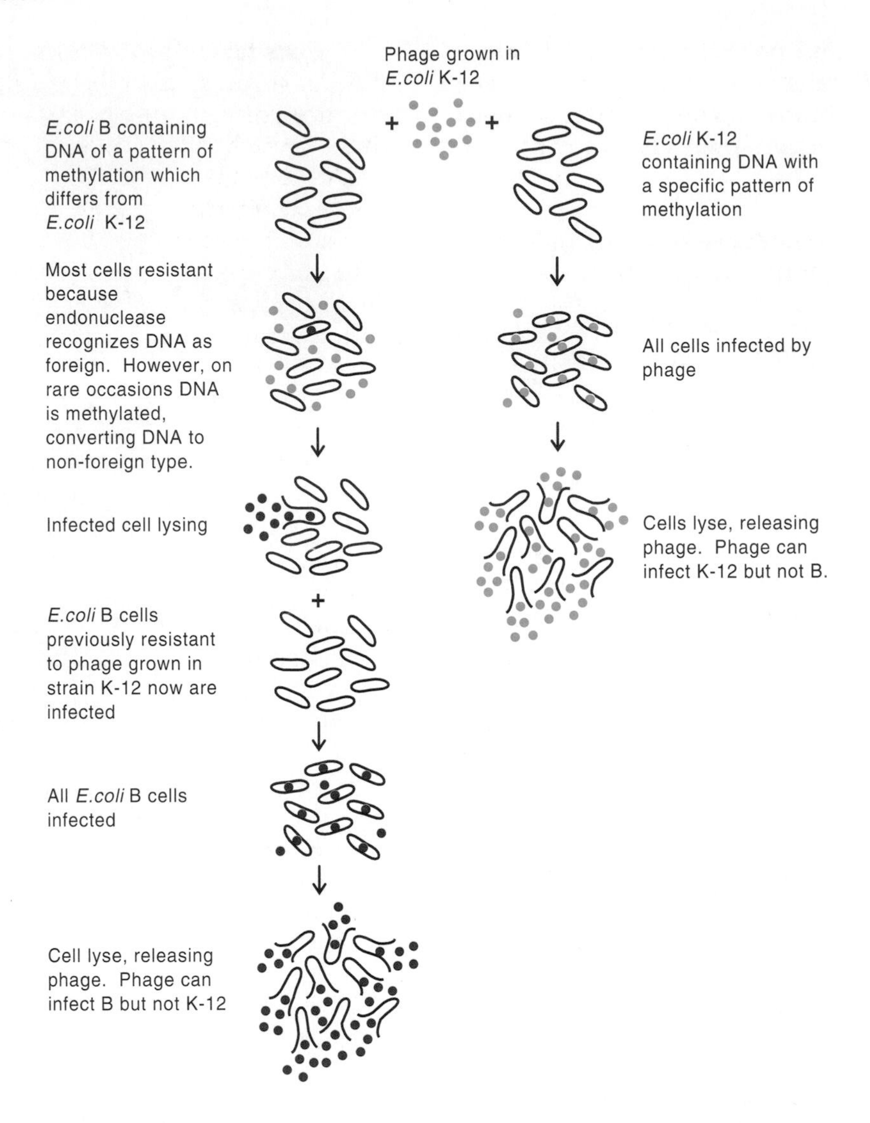

Figure 13.19
Limitation of phage host range on two strains of *Escherichia coli,* B and K-12. The restriction endonucleases in each strain recognize and cleave different sites in DNA. The modification enzyme in each strain converts the foreign DNA to non foreign DNA by methylating the site of cleavage of the restriction endonuclease, thereby making it resistant to cleavage.

Many Host Cells Can Be Cultivated in the Laboratory

Since viruses can multiply inside living cells only, their study requires that living cells be available for their growth. As previously mentioned, the study of bacterial viruses has advanced much more rapidly than investigations on animal and plant viruses, in large part because bacteria are easy to grow in large quantities in short time periods. With animal viruses, the primary difficulty is not so much in purifying the virions as it is in obtaining enough infected cells. Some viruses cannot be cultivated except by allowing them to multiply in living animals. Other animal viruses can be grown in cells taken from a human or another animal or in **embryonated chicken eggs** (those that contain developing chicks).

In the case of animal viruses that can be grown in animal cells, the host cells are cultivated in the laboratory by a technique called **cell,** or **tissue, culture** (figure 13.20).

To prepare cells for growth outside the body of the animal (in vitro), a tissue is removed from an animal and minced into small pieces. The cells are separated from one another by treating them with a protease enzyme, such as trypsin, that breaks down protein. The suspension of cells is then placed into a screw-capped flask in a medium containing a complex mixture of amino acids, minerals, vitamins, and sugars as a source of energy. The growth of animal cells also requires a number of additional growth factors present in blood serum, which have not yet been identified. Tissue cultures prepared directly from the tissues of an animal are termed **primary cultures.**

The cells bathed in the proper nutrients attach to the bottom of the flask and divide every several days, eventually covering the surface of the dish with a single layer of cells, termed a **monolayer.** When the cells become crowded, they stop dividing and enter a resting state. One can continue to propagate the cells by removing them from the primary culture, treating them with trypsin, diluting them, and putting the diluted suspension into another flask containing all of the required nutrients.

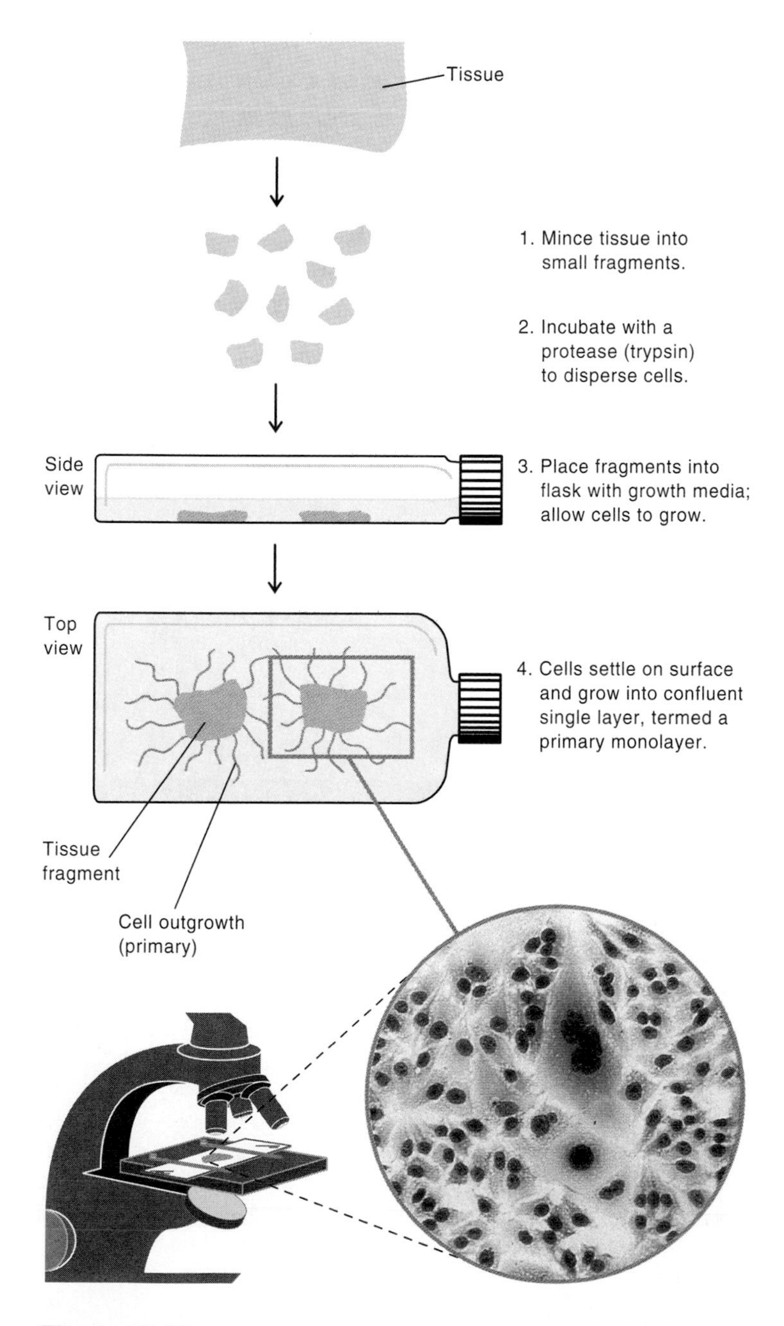

Figure 13.20
Preparation of primary cell monolayers.

Cells taken from a normal vertebrate tissue die after they have undergone a certain number of divisions in culture. For example, human skin cells divide 50 to 100 times and then die, even though they are diluted into fresh media. Accordingly, the cells must once again be taken from the animal and a new primary culture started. However, cells taken from a tumor can be cultivated in vitro indefinitely. Therefore, they are much easier to use for growing viruses than nontumorous tissue, and several tissue lines have been established from tumors. Further, they have several other properties that distinguish them from cell lines that have originated from normal tissue (see chapter 14).

Tissue culture is important in animal virology for several reasons. First, it is important for growing viruses in the laboratory. The virus is mixed with the susceptible cells, and the mixture is incubated until the infected cells are lysed. Following lysis, the unlysed cells and the cell debris are removed by centrifugation. The cells and debris sediment to the bottom of the centrifuge tube while the light, small virions remain in the liquid (the **supernatant**). This liquid containing the virions is also termed a **lysate.**

Second, tissue culture is used in virology to study the virus in both quantitative as well as qualitative terms. The commonly used method for detecting and quantifying the amount of virus present in any sample is the **plaque** assay.

Animal Viruses Can Also Form Plaques

We have already discussed the theory and technique of plaque assays in measuring the number of bacteriophages in a sample. The same theory and practice can be used to assay the number of animal viruses in a sample if the virus lyses the animal cells (figure 13.21).

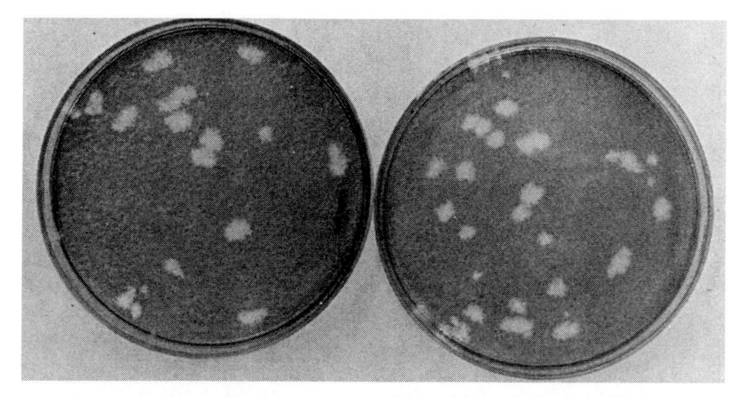

Figure 13.21
Plaques caused by infection of human embryonic tonsil cells with herpesvirus type 1.

Virus Infection May Change the Appearance of Cells

When a virus is propagated in tissue culture cells, it often changes the cells' appearance. Often, these changes are characteristic for a particular virus and are referred to as the **cytopathic effect** of the virus (figure 13.22). Thus, often the presence of a virus in an unknown sample and some idea of its identity can be gained by culturing the specimen on cells in tissue culture.

Counting of Virions with the Electron Microscope Can Determine the Number of Virions

If reasonably pure preparations of virions are obtainable, their concentration may be readily determined by counting the number of viral particles in a specimen prepared for the electron microscope (figure 13.23). Unfortunately, this method cannot distinguish between infective and noninfective viral particles.

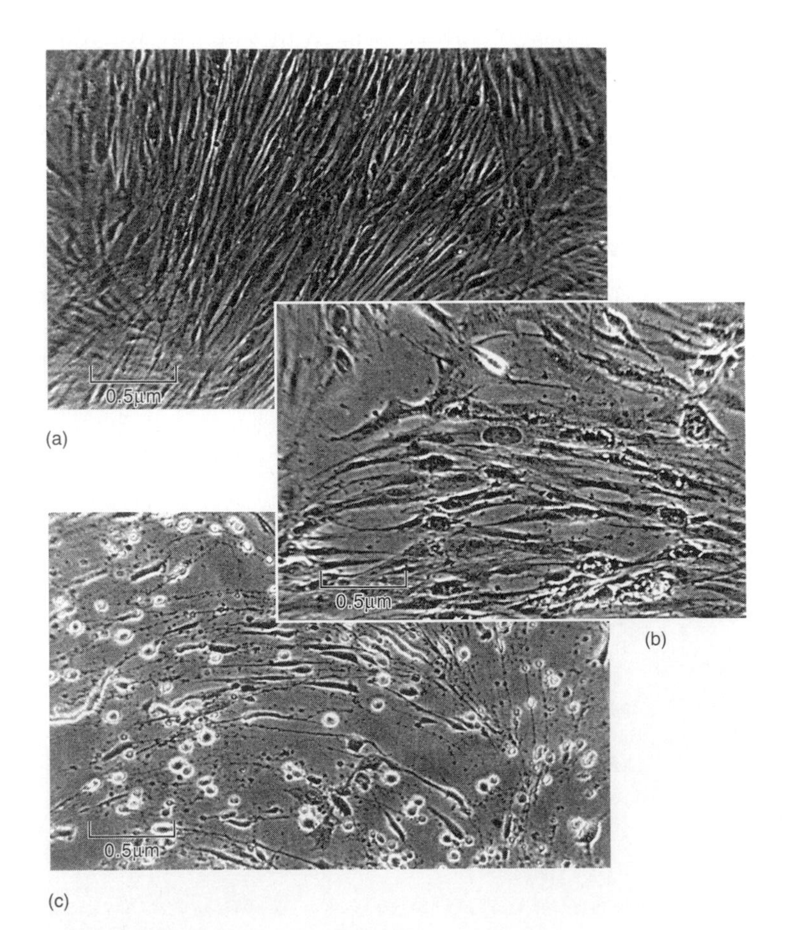

Figure 13.22
Cytopathic effects of virus infection on tissue culture. (*a*) Fetal tonsil diploid fibroblasts, uninfected. (*b*) Same cells infected with adenovirus. (c) Same cells infected with herpes simplex virus. Note that the monolayer is totally destroyed.

Quantal Assays Measure the Number of Infective Virions

Quantal assays are often used to give an approximate virus concentration. In this assay, several dilutions of the virus-containing preparation are made and administered to a number of animals, cell cultures, or chick embryos, depending on the host specificity of the virus. The **titer** of virus, or **endpoint,** is the dilution at which 50% of the hosts inoculated are infected or killed. This titer can be either the $\mathbf{ID_{50}}$ (infective dose) or the $\mathbf{LD_{50}}$ (lethal dose).

The Ability of Some Viruses to Clump Red Blood Cells Can Be Used to Measure Their Concentration

Many animal viruses, but not all, are able to clump (agglutinate) red blood cells because they interact with the surfaces of the red cells. This phenomenon is called **hemagglutination.** In this process, a virion attaches to two red cells simultaneously and causes them to clump (figure 13.24 *a*). Sufficiently high concentrations of virus cause large aggregates of red blood cells to form, which are readily visible (figure 13.24 *b*). Hemagglutination can be measured by mixing serial dilutions of the viral suspension with a standard amount of red blood cells. The highest dilution showing maximum agglutination is taken as the titer of the virus. One group of animal viruses that is able to agglutinate red blood cells is the myxoviruses, of which the influenza virus is a member.

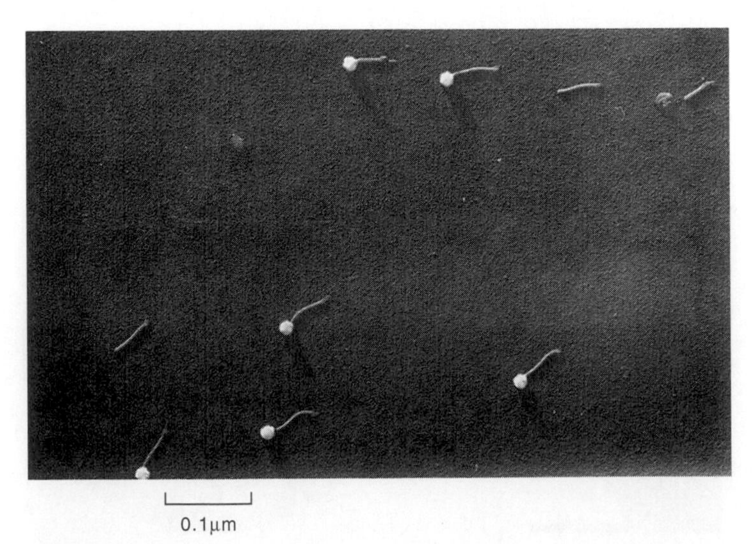

Figure 13.23
Bacteriophage particles. Note that not only can the number of virions be counted but the number of obviously damaged virions can be readily determined.

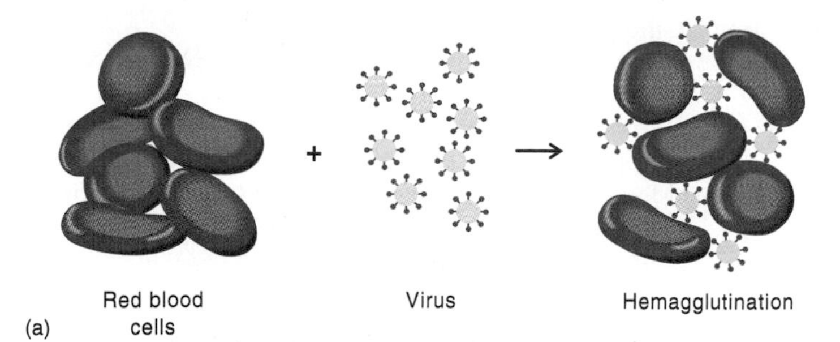

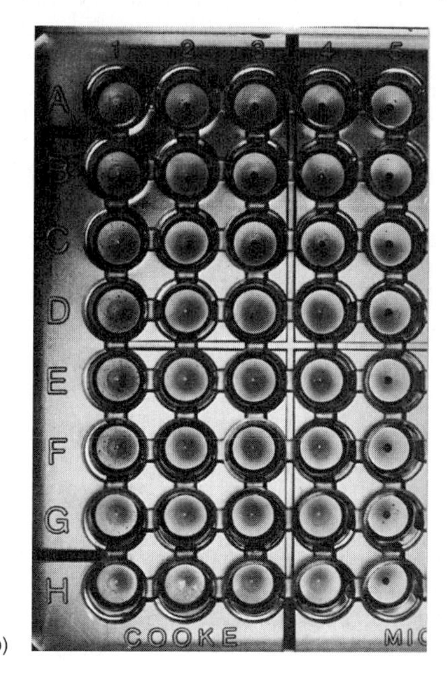

Figure 13.24
(*a*) Diagram showing antibodies combining with antigen to result in agglutination. (*b*) Assay of viral titer by hemagglutination. Each horizontal series of cups (A to H) represents serial twofold dilutions of different preparations of influenza virus mixed with a suspension of red blood cells. The various preparations have different titers. The diffuse precipitate at the bottom of each cup represents a positive hemagglutination reaction. The small button represents a settling of the red blood cells and a lack of hemagglutination.

REVIEW QUESTIONS

1. What is the name given to a phage whose DNA can be integrated into the host chromosome? What is the name of the host cell containing the phage DNA?
2. What is the process by which prophages confer new properties on their host cells?
3. Name the two types of transduction and explain how they differ from each other in the DNA that is transferred.
4. List two methods that can be used to determine the number of virions in a specimen.
5. Compare the number of plaques that would arise at the various stages in the one-step growth curve if chloroform were not added to the infected bacteria.
6. List the two mechanisms that allow bacteria to be resistant to phage infection. Which mechanism is more important in animal systems?
7. What is meant by *positive sense strand RNA? Positive sense strand DNA?*
8. Compare the lytic cycle of replication of double-stranded DNA phages with single-stranded DNA and single-stranded RNA phages in terms of cell lysis and number of phages produced.
9. How does UV light induce the synthesis of virions from lysogenic bacteria?
10. Why is it important that not all phage-induced enzymes be synthesized simultaneously? What general class of enzymes must be synthesized first? What class should be synthesized last?

Summary

I. Viruses—General Characteristics
Viruses are nonliving infectious agents that can be distinguished from cells in a number of ways.
   A. They are very small, 100 to 1,000 times smaller than the cells they invade.
   B. Most contain very few genes.
   C. They contain either DNA or RNA but never both.
   D. They contain no or very few enzymes.
   E. They contain none of the cell components such as ribosomes and enzymes to generate energy.

II. Virus Architecture
   A. Viruses can have different shapes. Some are spherical; others are rod shaped; and others are a combination.
   B. The shape of a virus is determined by the shape of its protein coat (capsid).
   C. Some animal viruses have a lipid membrane (envelope) surrounding the capsid (enveloped virus); others do not (naked virus).

III. Virus-Host Interactions—General Features
   A. Interactions of bacterial viruses with their hosts are much better understood than animal and plant viruses and serve as a model for the latter.
   B. Bacteriophages can be conveniently grouped on the basis of their interaction with their hosts.
   C. Bacteriophages can have different relationships with their host cells.
      1. In a productive relationship, more copies of the virus are synthesized and are released by lysis of the host cell.
      2. In a different productive relationship, the virus exits the cell without lysing it (extrusion).
      3. In a latent state, viral nucleic acid becomes integrated into the host chromosome.

IV. Virus Replication—Productive Infection
   A. In step 1, phages attach to specific sites on host bacteria.
   B. In step 2, phage DNA enters the host cell; the protein coat remains on the outside.

C. In step 3, phage genes are transcribed sequentially, leading to production of specific proteins (phage-induced enzymes) at different times.
D. In step 4, phage DNA is replicated and phage structural proteins are synthesized independently of one another.
E. In step 5, phage DNA and protein assemble to form mature virions in the maturation process.
F. In step 6, virions are released from the host cell following lysis of the bacterial cell wall by phage-induced enzymes.

V. RNA Lytic Phages—General Features
A. RNA phage contain only a single positive (+) sense strand of RNA. A positive sense strand codes for mRNA directly.
B. RNA phage replication involves formation of complementary strands, negative (–) sense strands of RNA.

VI. Phage Growth: One-step Growth Curve
A. This procedure involves assaying the number of virions inside and outside the bacteria at various times following adsorption of the phage to the bacteria.
B. Samples are removed at various times, one set is treated with chloroform, which lyses the bacteria; the other is not. Samples are inoculated onto a lawn of bacteria to lyse them, leading to formation of a plaque that represents the initial location of a single phage or single infected cell.

VII. Viral Replication in a Latent State
A. The DNA of temperate viruses is integrated into the host chromosome.
B. The integration of phage DNA into the bacterial chromosome occurs because of identical DNA sequences in the phage and the bacteria.
C. The repressor must function continuously to keep the prophage in the integrated state.
D. Activation of the SOS repair systems destroys the repressor; the phage DNA is excised and virions are produced that lyse the cells.
E. Lysogens are immune to infection by the same phage.

VIII. Filamentous Phages—Productive Infection
A. Filamentous phages are released without killing the host cell.
B. Single-stranded DNA of filamentous phages is converted to a double-stranded molecule in the process of replication.

IX. Transduction
A. Both virulent and temperate phages can transfer DNA from one host to another host.
1. Virulent phages can transfer any part of the genome through the process of generalized transduction.
2. Many temperate phages can transfer only a specific set of genes through the process of specialized transduction.

X. Host Range of Viruses
A. Any given virus can generally infect specific cells of only one or a few closely related species.
B. Limitation in host range most frequently results from the fact that animal and bacterial viruses must adsorb to specific receptors on the host cell surface. Plant viruses have no receptors on the host cells.
C. Restriction is a modification system that prevents entry of foreign DNA into bacteria.
1. Bacteria can recognize whether entering DNA is foreign by its pattern of methylation.
2. Foreign DNA is cleaved as it enters the cell by restriction endonucleases of the bacteria.
3. Modification enzymes of the bacteria can methylate (modify) the entering DNA and thereby convert it from "foreign" to "nonforeign." The "nonforeign" DNA is not cleaved.

XI. Methods Used to Study Viruses
A. Many eukaryotic host cells can be cultivated in the laboratory through the technique of tissue culture.
B. Cells from a normal vertebrate animal grow as a monolayer attached to the surface of a flask but only for a limited number of generations before they die.
C. Animal viruses, like bacteriophages, can infect, multiply, and lyse cells growing in a monolayer, resulting in a readily observable clearing (plaque).
D. Infected tissue culture cells undergo characteristic changes in their appearance depending on the virus.
E. Quantal assays measure the number of infective virions by their effect on different host cell systems.
F. The presence and concentration of some viruses can be detected by their ability to clump red blood cells (hemagglutination).

Critical Thinking Questions

1. Do you believe that viruses are living or dead? Explain your reason.
2. Why do phages form plaques and not a clearing of the entire surface?

Further Reading

Fields, B. N., and Knipe, D. M. (ed.). 1991. *Fundamental virology*, 2d ed. Raven Press, New York.

Karam, J. D. (ed.). 1994. *Molecular biology of bacteriophage T4*. ASM Press, Herndon, VA. This book emphasizes the value of the phage T4 as a research and teaching tool.

Levy, J. A., Fraenkel-Conrat, H., and Owens, R. 1994. *Virology*, 3d ed. Prentice-Hall, Englewood Cliffs, N.J.

Ptashne, M. 1986. *A genetic switch*. Blackwell Scientific Publications, Palo Alto, Calif. The story of phage lambda and how it's maintained in a prophage state, written by a major scientist in the field.

Scott, A. 1985. *Pirates of the cell: the story of viruses from molecule to microbe*. Basil Blackwell, New York.

14 Viruses of Animals and Plants

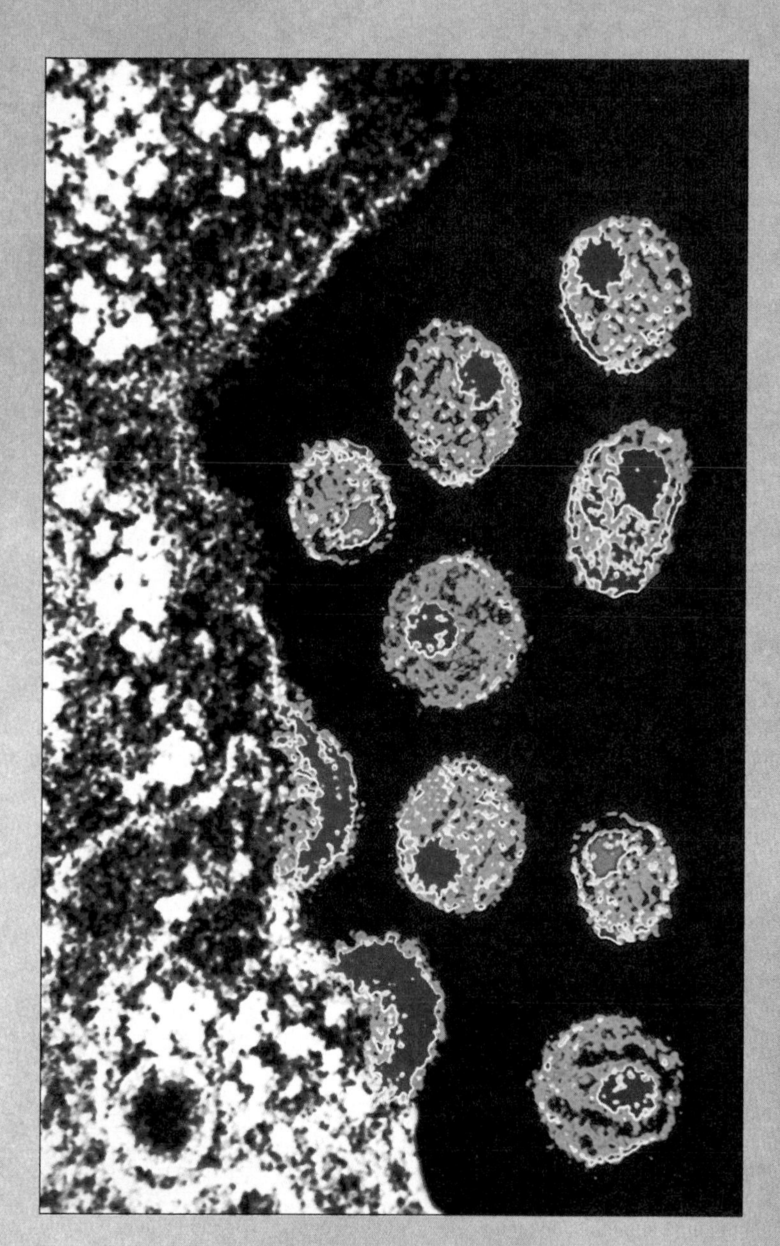
Rotaviruses as viewed through the transmission electron microscope (×575,000).

KEY CONCEPTS

1. Viruses are classified on the basis of their morphology, their size, the nature of their genome, and mode of replication.
2. The interactions of animal viruses with their hosts are similar to the interactions between bacteriophages and bacteria.
3. The replication of viral nucleic acid depends to varying degrees on the enzymes of nucleic acid replication of the host cell.
4. Symptoms of viral-caused disease result from tissue damage of the host.
5. Some tumors in animals result from viral DNA becoming integrated into the host genome and coding for the synthesis of abnormal proteins, which interfere with normal control of host cell growth.

Microbes in Motion

The following books and chapters in the *Microbes in Motion* CD-ROM may serve as a useful preview or supplement to your reading:
Viral Structure and Function: Structure; Replication; Assembly; Diseases

A Glimpse of History

Although scientific reports as early as the 1700s suggested that invisible agents might cause tumors, not until the early 1900s did the idea that viruses could be causative agents of tumors gain strong experimental support. At that time, Dr. Peyton Rous of the Rockefeller Institute showed that a suspension of ground-up cells from tumors of chickens that was filtered to remove all cells caused tumors when injected into healthy chickens. However, these studies were not taken very seriously at the time because of the generally accepted idea that viruses could not cause tumors. However, as the years passed, the idea that viruses caused certain tumors gained favor as several other investigators made supporting observations. In particular, in the early 1930s, Dr. Richard Shope's studies of skin tumors (papillomas) that appeared in wild rabbits in Iowa and Kansas lent support to the virus-tumor idea. He showed that these tumors were caused in wild rabbits by an agent that was not trapped in filters that removed bacteria. However, when Shope injected the filtrate into domestic rabbits rather than wild rabbits, he discovered that the agent, presumably a virus, often did not cause tumors but persisted in the rabbits without causing the disease. This phenomenon has been observed more recently with other viruses.

In 1957, almost 50 years after Peyton Rous's study, Dr. Sarah Stewart and Dr. Bernice Eddy were studying two tumor-causing viruses—polyoma, a DNA virus, and SV_{40}, an RNA virus. They noted that when the polyoma virus grown in monkeys or mouse embryo tissues was inoculated into a wide variety of animals such as rats, hamsters, and rabbits, it caused many different kinds of tumors. Eddy also pioneered the study of the RNA virus called SV_{40}, which could be isolated from certain monkeys, in which it did not cause any tumors. However, when injected into other animals, such as baby hamsters, it caused tumors. Thus, SV_{40} like the papillomavirus studied by Shope appears to be "silent" in its natural host but causes tumors when inoculated into unnatural or foreign hosts.

The study of tumor viruses has provided information that has helped in fighting cancer. In addition, studies have given scientists a great deal of basic information about how macromolecules such as DNA, RNA, and protein function in metabolic processes.

Much of the basic biology of bacteriophages also applies to animal and plant viruses; however, each viral group has certain properties that are unique. This chapter presents a general approach to the classification of animal viruses, followed by a discussion of their modes of replication and effects on host cells, including their role in causing certain tumors. A brief discussion of plant viruses is included.

Classification Scheme of Animal Viruses

The taxonomy of animal viruses has been constantly changing over the years as more is learned about their properties. For this reason, only the principles of taxonomy of viruses that infect animals are considered here. The most widely employed taxonomic criteria for animal viruses are based on five characteristics discussed in chapter 13:

1. The nature of the nucleic acid enclosed in the capsid—DNA or RNA, single-stranded or double-stranded, segmented or a single molecule.
2. The strategy of virus replication, which depends on whether the virus genome is single- or double-stranded DNA or RNA. Since the replication of some viruses depends on enzymes not present in the uninfected cell, the viruses carry their own enzymes that they introduce into the cell upon infection. Some viruses replicate within the nucleus; others in the cytoplasm.
3. Virus particle structure—polyhedral (spherical), helical (rod-shaped), or complex.
4. Presence or absence of a viral envelope. The viral envelope is composed of lipids of the host cell and proteins coded by the virus. The lipids make up from 20% to 35% of the dry weight of the viruses, and their composition depends on the host cell from which they come. The lipids are typical bilayers, into which the virus-coded glycoprotein spikes are embedded.
5. Size of the virion.

Beyond these physical characteristics, other criteria (immunological and the effect of the viruses on host cells) are used to subdivide the groups.

Based on these criteria, animal viruses are divided into a number of families whose names end in *-viridae,* (for example, Herpesviridae.) There are seven families of DNA-containing viruses and 15 families of RNA-containing viruses that infect vertebrates. The members of each family are derived from a common ancestor, related to one another as shown by nucleic acid hybridization results. Each family contains numerous genera whose names end in *-virus,* (for example, enterovirus). The species name is the name of the disease the virus causes (for example, polio or poliovirus). In contrast to bacterial nomenclature in which an organism is referred to by its genus and species name, viruses are commonly referred to only by their species name. The classification of a specific family of viruses is shown in table 14.1. The major families of DNA viruses and RNA viruses with some of their properties are given in tables 14.2 and 14.3. These tables need not be memorized but should serve as reference information.

Table 14.1 Classification of the Family Picornaviridae

Name of Group	Classification Scheme	
Family*	*Picornaviridae*	
Genus	*Enterovirus*	*Rhinovirus*
Species†	poliovirus coxsackievirus	cold virus

*The family and genus names are capitalized and written in italics.
†The species names are not capitalized or italicized.

Table 14.2 Classification of DNA Viruses that Infect Humans and Animals

Family	Virion Structure	Genome Structure*	Characteristics and Representative Pathogenic Members
Hepadnaviridae	Lipid-containing envelope	1 molecule, mainly ds DNA but with a single-strand gap	Hepatitis B virus
Parvoviridae	Nonenveloped	1 molecule, ss DNA	Outbreaks of gastroenteritis following eating of shellfish
Papovaviridae	Nonenveloped	1 molecule, circular ds DNA	Human papillomaviruses associated with genital and oral carcinomas
Adenoviridae	Nonenveloped	1 molecule, ds DNA	Some cause tumors in animals
Herpesviridae	Enveloped with surface projections	1 molecule, ds DNA	Herpes simplex virus, Cytomegalovirus
Poxviridae	Enveloped; large, brick-shaped	ds DNA; covalently closed ends	Smallpox virus, Vaccinia virus
Iridoviridae	Enveloped	1 molecule, ds DNA	No known human pathogens; only animal pathogens

*ds—double-stranded
ss—single-stranded

Table 14.3 Classification of RNA Viruses that Infect Humans and Animals

Family	Virion Structure	Genome Structure	Characteristics and Representative Pathogenic Members
Picornaviridae	Nonenveloped icosahedral	1 molecule, ssRNA*	Poliovirus Rhinovirus
Caliciviridae	Nonenveloped icosahedral	1 molecule, ssRNA	Norwalk virus; many members cause gastroenteritis
Togaviridae	Lipid-containing envelope	1 molecule, ssRNA	Many multiply in arthropods and vertebrates; encephalitis in humans
Flaviviridae	Lipid-containing envelope	1 molecule, ssRNA	Yellow fever virus; Dengue virus
Coronaviridae	Lipid-containing envelope	1 molecule, ssRNA	Colds and respiratory tract infections
Rhabdoviridae	Bullet-shaped; lipid-containing envelope	1 molecule, ssRNA	Rabies virus
Filoviridae	Long, filamentous; sometimes circular; lipid-containing envelope	1 molecule, ssRNA	Marburg virus Ebola virus
Paramyxoviridae	Pleomorphic; lipid-containing envelope	1 molecule, ssRNA	Mumps virus Parainfluenza virus
Orthomyxoviridae	Pleomorphic; lipid-containing envelope	7–8 segments of linear ssRNA	Influenza virus
Bunyaviridae	Lipid-containing envelope	3 molecules of ssRNA with hydrogen-bonded ends	Hantaan virus
Arenaviridae	Pleomorphic; lipid-containing envelope	2 molecules of ssRNA with hydrogen-bonded ends	Lassa virus
Reoviridae	Naked virion	Linear dsRNA divided into 10, 11, or 12 segments	Diarrhea in animals
Birnaviridae	Naked virion; icosahedral	2 segments of linear dsRNA†	No human pathogens; diseases in chickens and fish
Retroviridae	Icosahedral; lipid-containing envelope	2 identical molecules of ssRNA	HIV

*ssRNA—single-stranded RNA
†dsRNA—double-stranded RNA

In general, viruses with a similar genome structure replicate in a similar way. Indeed, animal viruses in general follow the same replication strategies as do phages whose genomes are similar. These strategies are covered in chapter 13. In this chapter, we will consider the replication of only one group of animal viruses, the *Retroviridae,* since there is no phage that has a similar life cycle.

Viruses Can be Grouped According to Routes of Transmission

Viruses that cause disease are often grouped according to their routes of transmission. These are not taxonomic groupings. The following groupings, summarized in table 14.4, provide examples of such a scheme.

The **enteric viruses** are usually ingested on material contaminated by feces. They replicate primarily in the intestinal tract where they usually remain localized. They often cause gastroenteritis, a disease of the stomach and intestine. However, some, such as the polio virus, replicate first in the intestines but do not cause gastroenteritis. Rather, they cause a generalized disease.

The **respiratory viruses** usually enter the body in inhaled droplets and replicate in the respiratory tract. The respiratory viruses include only those viruses that remain localized in the respiratory tract. Others that infect via the respiratory tract but then cause generalized infections are not considered respiratory viruses. These include the mumps and measles viruses.

Zoonoses are diseases that are transmitted from an animal to a human. Many viruses, such as rabies, are transmitted from animals to humans directly. Other viruses, such as canine distemper, can be transmitted from dogs to African lions. Another group of viruses, the **arboviruses,** are so named because they infect arthropods where they replicate. The viruses are transmitted to susceptible vertebrates through the bites of the arthropods, which include mosquitoes, ticks, and sandflies. Thus, the viruses are *arthropod-borne.* Note that, in many cases, viruses can invade and replicate in widely different species. The same arthropod may bite birds, reptiles, and mammals and transfer viruses among these different groups. Over 500 arboviruses are known; about 80 are capable of infecting humans; 20 of these viruses cause important diseases.

Sexually transmitted viruses cause lesions in the genital tract. These include herpes viruses and papillomaviruses. Other viruses are often transmitted during sexual activity but they cause generalized infections. They include certain viruses such as human immunodeficiency virus (HIV) and hepatitis viruses.

Table 14.4 Modes of Transmission of Human Viral Diseases

Mechanism of Transmission	Common Viruses Transmitted
Fecal-oral route	Enteroviruses (polio, coxsackie B); rotaviruses (diarrhea)
Respiratory or salivary route	Influenza; measles; rhinoviruses (colds)
Vector (such as arthropods)	Sandfly fever; dengue
Animal to human directly	Rabies; cowpox
Sexual contact	Herpes simplex virus-2 (genital herpes); HIV

Virus Life Cycle—Productive

The steps in the virus life cycle are the same in animal viruses as in phages. However, among the animal viruses, a number of phases differ markedly in different viruses and from the bacteriophage examples discussed in the previous chapter.

The First Step in Infection Is Adsorption of the Virus to the Host Cell

The process of adsorption (attachment) is basically the same in all virus-cell interactions except that the process in animal viruses seems far more complex than in phages. Animal viruses usually do not contain a single specific adsorption appendage, a tail with fibers, as do many phages. Rather, surface projections containing **attachment proteins** or **spikes** protrude all over the surface of a virion (figure 14.1). Frequently, there are several different attachment proteins. Antibodies formed against a virus by immune animals frequently bind to attachment proteins and thereby prevent viral adsorption and viral infection. The receptors to which the viral attachment proteins bind are usually glycoproteins located on the host cell plasma membrane. More than one receptor

- **Productive infection, p. 302**
- **Phage adsorption p. 303**
- **Glycoprotein, p. 35**
- **Plasma membrane, p. 78**

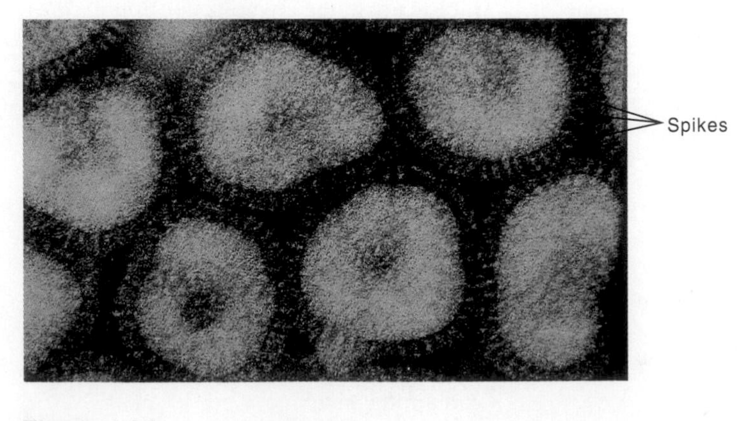

Figure 14.1
Electron micrograph of influenza virions showing attachment proteins or spikes (×350,000).

is frequently required for effective attachment. For example, the HIV virus must bind to two key molecules on the cell surface before it can enter the cell. The receptors number in the tens to hundreds of thousands per host cell. Interestingly, the function of these glycoproteins is completely unrelated to their role in virus adsorption. For example, many receptors are members of immunoglobulins; other viruses use hormone receptors and permeases. Different viruses may use the same receptor, and related viruses may use different receptors. Certain viruses can bind to more than one type of receptor and thus be able to invade different kinds of cells. The binding of the attachment proteins to their receptors often results in changes in the shape of viral proteins concerned with entry of the virion.

Because of this requirement to bind to specific receptors, frequently a particular virus can infect only a limited number of cell types, and most viruses can infect only a single species and some only a single cell type (e.g., a nerve cell) within a host species. This may account for the resistance that some animals have to certain diseases. For example, dogs do not contract measles from humans and humans do not contract distemper from cats. However, some viruses can infect unrelated animals such as horses and humans with serious consequences.

Viral Entry Can Occur by Two Different Mechanisms

The mechanism of entry of animal viruses into host cells depends upon whether the virion is enveloped or naked. In the case of enveloped viruses, two mechanisms exist. In one mechanism, after attachment to a host cell receptor, the envelope of the virion fuses with the plasma membrane of the host (figure 14.2 *a*). This fusion is promoted by a specific fusion protein on the surface of the virion. In some viruses, such as measles, mumps, influenza, and HIV, the protein, which recognizes the target protein on the cell, changes its shape when it contacts the host cell. Following fusion, the nucleocapsid is released directly into the cytoplasm where the nucleic acid separates from the protein coat. In another mechanism, enveloped viruses adsorb to the host cell with their protein spikes, and the virions are taken into the cell in a process termed **endocytosis** (figure 14.2 *b*). In this process, the host cell plasma membrane surrounds the whole virion in a vesicle. Then, the envelope of the virion fuses with the plasma membrane, after which the nucleocapsid is released into the host's cytoplasm.

In the case of naked virions, the virion also enters by endocytosis. However, since the virus has no envelope, the virion cannot fuse with the plasma membrane. Rather, the virions cause the vesicle to dissolve, resulting in their release into the cytoplasm, where the nucleic acid separates from its protein coat prior to replication. Note that the entire virion is taken into the cell whereas, in the case of T4 phages, the protein coat remains on the outside of the bacterium.

- **Enveloped virus, p. 300**
- **Naked virus, p. 300**

The Replication of Viral Nucleic Acid Depends on the Source of Enzymes and the Type of Nucleic Acid

In all virus systems, phage, animal, and plant, replication of viral nucleic acid depends to varying degrees on enzymes of the host cell. The extent to which viral nucleic acid replication depends on the enzymes of the host cell varies greatly. As a general rule, the larger the viral genome, the fewer host cell enzymes involved in replication. This is not surprising since enough DNA is present in large viral genomes to encode most enzymes of DNA synthesis. For example, the largest of the animal viruses, the poxviruses, like the T4 phages, are totally independent of host cell enzymes for the replication of their nucleic acid. The very small parvoviruses depend so completely on the biosynthetic machinery of the host cell that they require that the host cell actually be synthesizing its own DNA at the time of infection so that viral DNA can also be synthesized. Most animal viruses are between these two extremes. For example, small DNA viruses such as the papovaviruses code for a protein that is involved in initiating DNA synthesis, but once started, the rest of the process is carried out by host cell enzymes. This replication is similar to that of bacteriophage ΦX174, a single-stranded DNA virus that codes for proteins that are necessary for the initiation of phage DNA replication. However, once replication has started, the viral DNA is synthesized by the same enzymes, such as DNA polymerase, that replicate the host cell *E. coli* DNA.

The varying degrees of dependency of different DNA viruses on host cell enzymes for DNA replication are summarized in table 14.5.

Table 14.5 Relationship of Virus Size to Dependency on Host Cell Enzymes for DNA Replication

Virus*	Increasing Size of Virus	Dependence on Host Cell DNA Synthetic Enzymes
Parvovirus ssDNA	↓	Depends totally on host cell enzymes
Polyoma virus dsDNA	↓	Codes for a protein that is involved in start of DNA synthesis; rest depends on host enzymes
Adenovirus dsDNA	↓	Codes for proteins involved in initiation of DNA synthesis and also DNA polymerase
Poxvirus dsDNA	↓	Is totally independent of host cell enzymes

*ss—single-stranded; ds—double-stranded

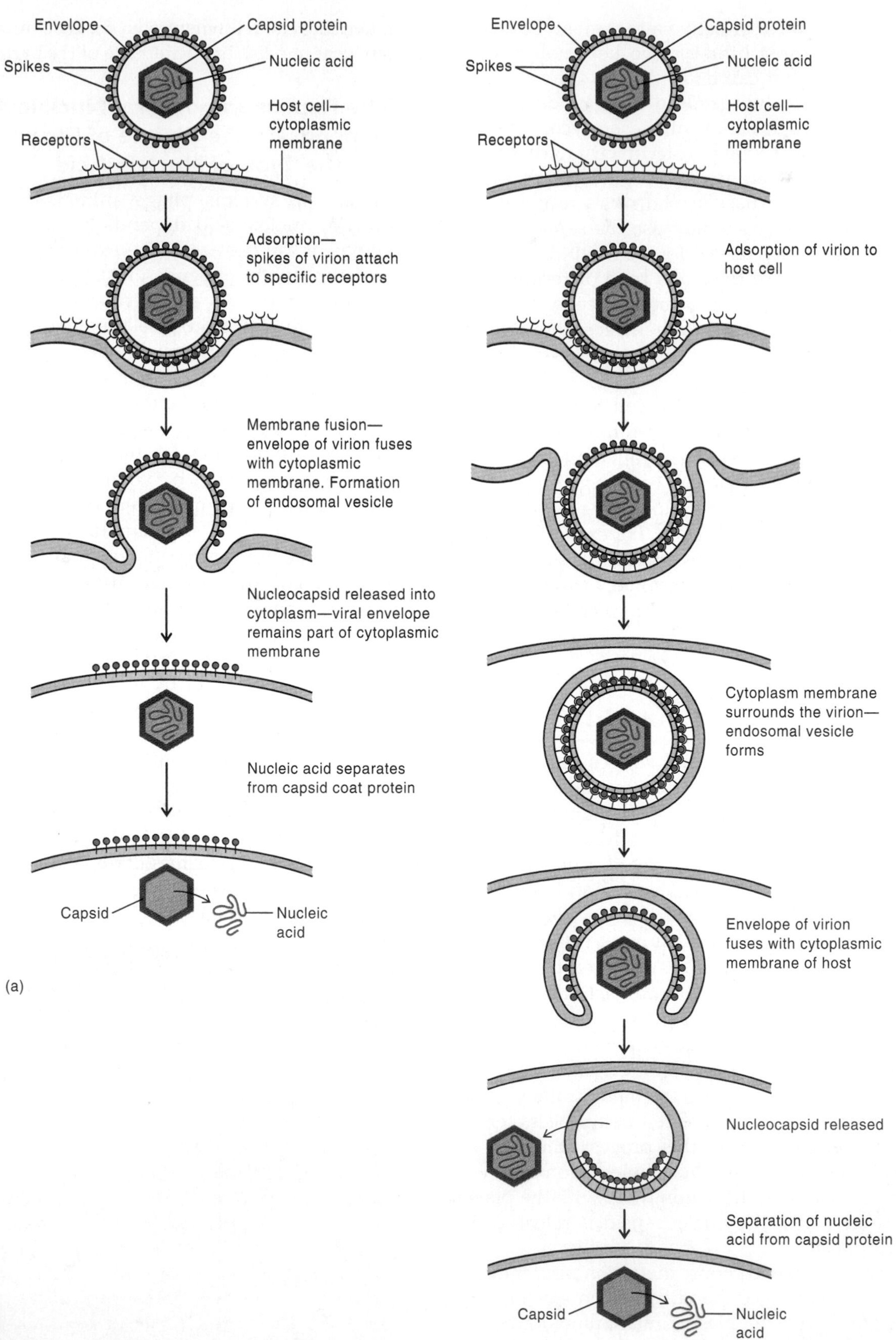

Figure 14.2
Entry of enveloped animal viruses into host cells. (*a*) Entry following membrane fusion and (*b*) entry by endocytosis.

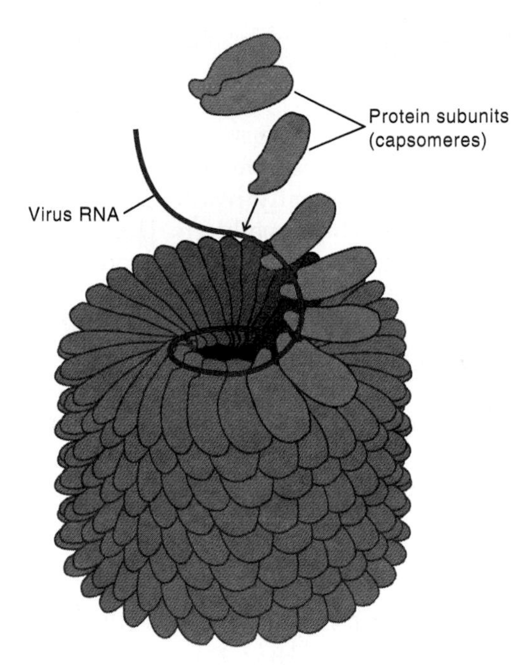

Figure 14.3
Tobacco mosaic virus maturation. The capsomeres are added, one by one, to the coat structure to enclose the viral nucleic acid (RNA).

Maturation of Virions Involves Self-Assembly of Viral Component Parts

The final assembly of the nucleic acid with its coat protein, the process of **maturation,** is preceded by formation of the protein capsid structure that surrounds the viral genome. The maturation process and multistep formation of the viral coat involve the same general principles in all kinds of viruses. This process has already been discussed for bacteriophage T4 (see chapter 13).

In the case of viruses that infect animal or plant cells, the maturation of the cylindrically shaped plant virus tobacco mosaic virus (TMV) has been studied extensively and serves as a model for both animal and plant viruses (figure 14.3). For TMV, many identical protein structural subunits (capsomeres) are first formed and then are added one by one to the growing coat structure that surrounds the viral RNA. The coat elongates in both directions, starting from a specific site on the single-stranded viral RNA. The RNA interacts with each protein disk as it is added, and when the end of the long RNA molecule is reached, the disks are no longer added. Enzymes are not required for the process, since it is a **self-assembly process.** Recall that the maturation of bacteriophage T4 is also, in part, a self-assembly process.

Virions Are Released as a Result of Lysis of Dead Cells or by Budding from Living Cells

Virions often are released following the death and lysis of infected cells. Unlike phages, animal viral nucleic acid does not code for enzymes that lyse the host cells. Infected cells usually die because viral DNA and proteins rather than host cell material are synthesized. Thus, functions required for cell survival are not carried out and cells die and then lyse as a consequence of cell death. In this way, virions are released. These virions may then invade any healthy cells in the vicinity.

Another mechanism for releasing enveloped virions in particular is **budding** (figure 14.4). As a first step in the process, the region of the host cell plasma membrane where budding is going to take place acquires the protein spikes that eventually are attached to the outside of the virion. Then the inside of the membrane becomes coated with the virion protein, the **matrix protein.** In the next step, the nucleocapsid becomes completely enclosed by the region of the plasma membrane into which the spikes and matrix protein are embedded. Almost all enveloped viruses obtain their envelopes as they exit the cell through budding. For these viruses, the budding process is part of the maturation process. The process of budding does not lead to cell death because the plasma membrane can be repaired following budding. As discussed in chapter 13, filamentous phages also are released from bacterial cells by budding or extrusion, which does not kill the bacterial cells either.

Interactions of Animal Viruses with Their Hosts—Overview

Discussion thus far has focused on the interaction of viruses with their host cells as studied in cell culture. In the case of bacterial viruses, the host organism is a single cell, so there can be no effects of other kinds of cells in the course of the infection. However, in the case of animals, including humans, the outcome of viral infection as measured by symptoms of disease depends on many factors that are independent of the infected cell. Of special importance are the defense mechanisms of the host, such as the presence of protective antibodies that can confer immunity against a virus that would ordinarily kill an individual without such immunity. The epidemics of measles and smallpox, which decimated the indigenous native population following the arrival of Europeans to the Americas, are good examples of the consequences of the lack of immunity to particular viruses.

Sudden epidemics causing widespread deaths are the most dramatic events of virus interactions with humans. However, the death of the host also means the death of the virus since viruses can only multiply in living cells. Probably most viruses develop a relationship with their normal host in which they cause no obvious harm or disease. Thus, the virus infects and persists within the host, in **balanced pathogenicity,** a state in which neither the virus nor the host is in serious danger. Indeed, most healthy animals, including humans, are **carriers** of a number of viruses

• **Capsomere, p. 300**

• **Extrusion, filamentous phages, p. 309**

as well as antibodies against these viruses without any ill effect. However, if a virus is transmitted to an animal that has no immunity to it, then disease may result.

Many viruses that are carried by one group of organisms without causing disease may cause serious disease when transferred to another group. For example, Lassa fever virus does not cause disease in rodents, but when the virus is transferred to humans, a large percentage of the infected population dies.

Improved hygienic precautions that protect children against certain viral infections can, in some instances, have harmful effects in the later life of the individual. For example, poliomyelitis typically results in a slight or nondetectable disease in young children. The virus replicates at low levels, and lifelong immunity is produced. However, if improved hygienic conditions prevent the child from contacting the virus at a young age, he or she may contract it as an adult. For unexplained reasons, this disease is often more serious when it occurs in adults.

The relationship between disease-causing viruses and their hosts can be divided into two categories based on the disease and the state of the virion in the host. These two categories of infection are *acute* and *persistent.*

Acute Infections Result in Readily Observed Disease Symptoms that Disappear Quickly

Acute infections are self-limited diseases in which the virus often remains localized (figure 14.5 *a*). Such diseases are usually of relatively short duration, and the host organisms may develop long-lasting immunity. Examples of acute infections are mumps, measles, influenza, and poliomyelitis. Viruses that cause acute infections result in productive infections, and the infected cells die and then lyse with the release of virions.

• **Productive infection, p. 302**

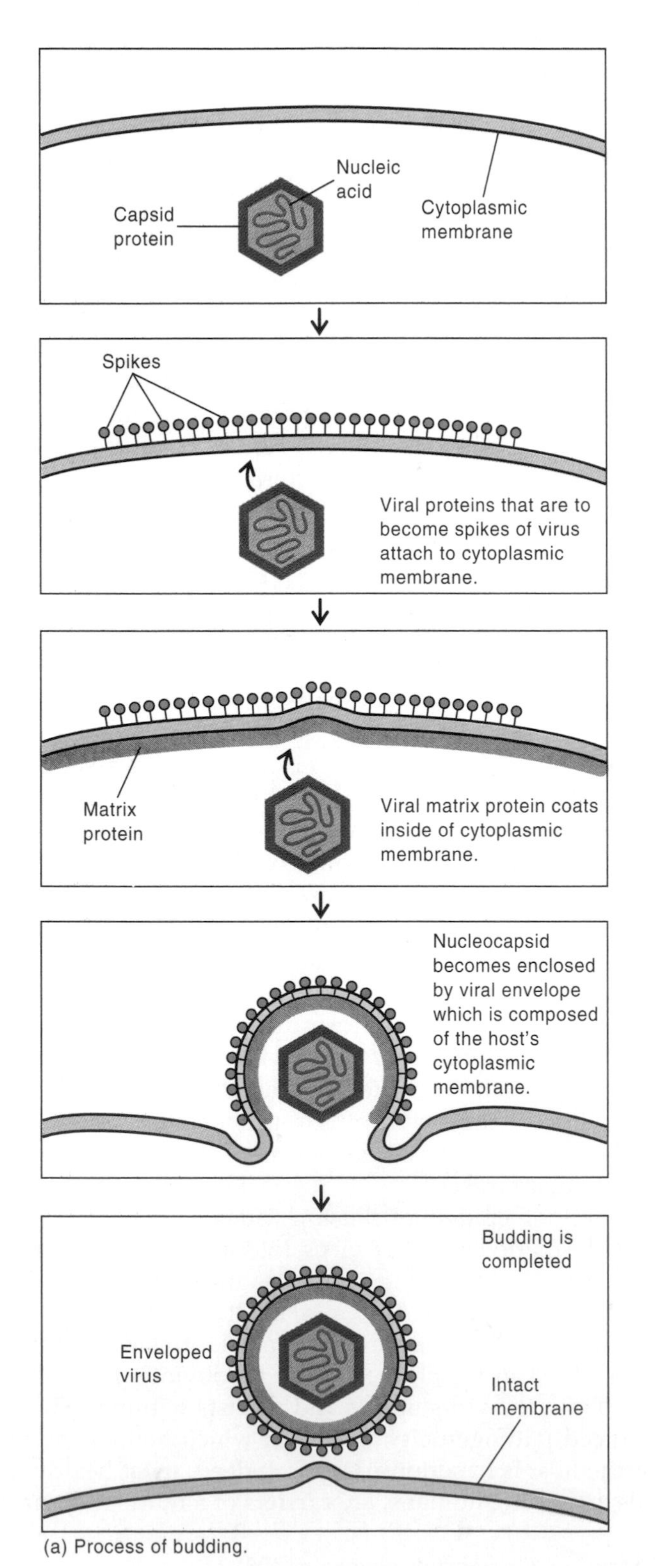

(a) Process of budding.

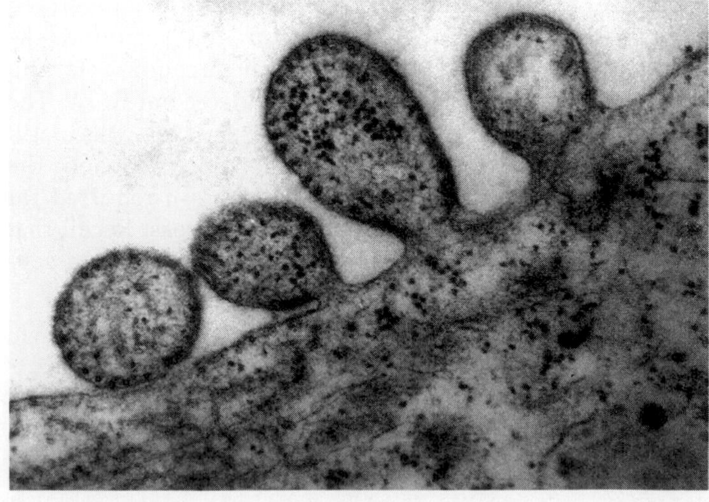

(b)

Figure 14.4
Mechanism for releasing enveloped virions. (*a*) Process of budding. (*b*) Electron micrograph of virus particles budding from the surface of a human cell. The virion on the left has completed the process. The other three are in various degrees of completion. It is clear from this photograph how the virions gain the plasma membrane of the host cell. Note that the membrane of the host remains intact after budding has been completed. (Courtesy of J. Griffith.)

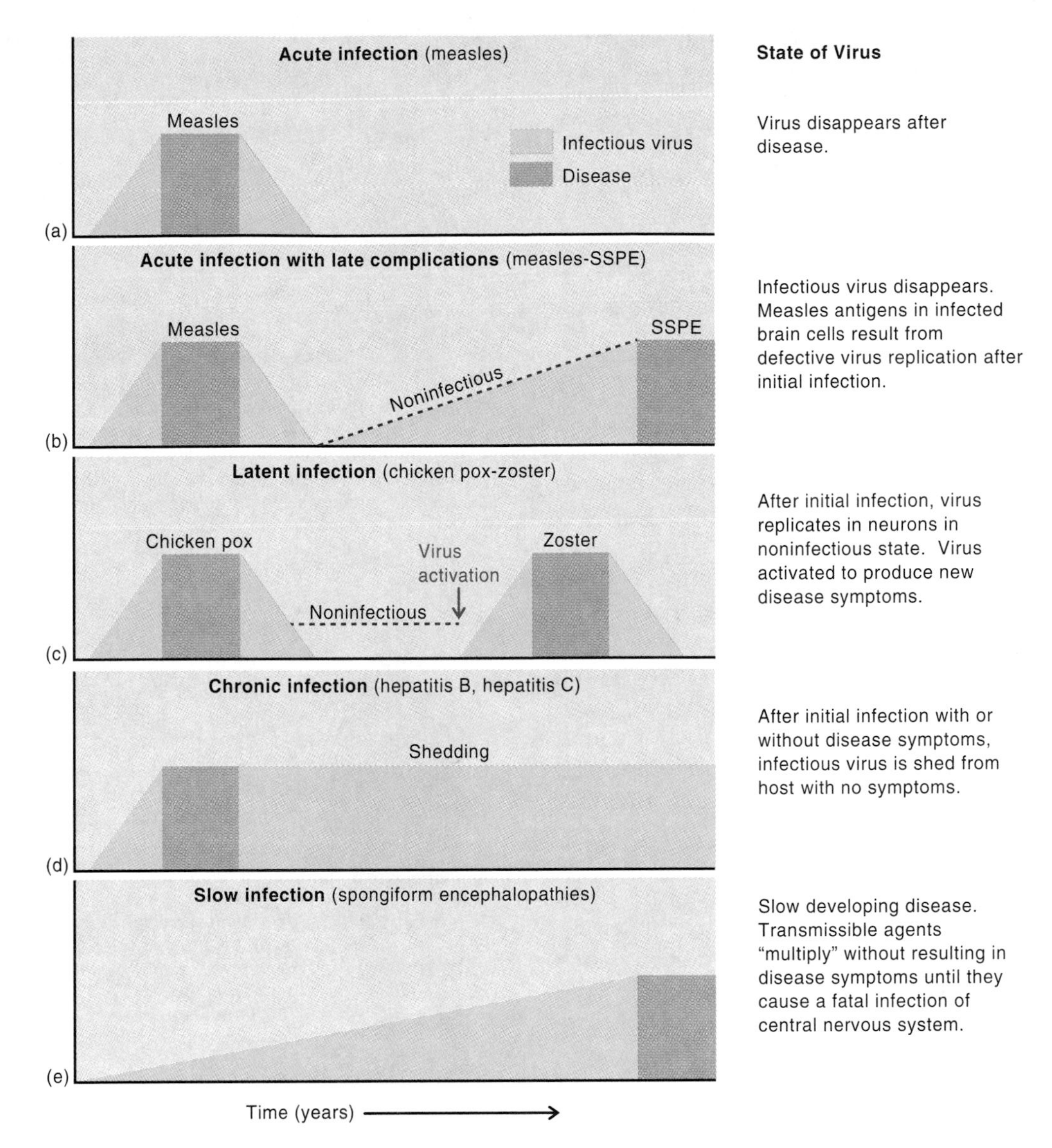

Figure 14.5
Time course of appearance of disease symptoms and infectious virions in various kinds of viral infections.

However, this does not mean that the hosts die. Disease symptoms result from localized or widespread tissue damage. With recovery, the defense mechanisms of the hosts gradually eliminate the virus over a period of days to months.

In Persistent Infections, the Virus Is Present for Many Years Often Without Any Disease Symptoms

Many viruses have the ability to establish infections that persist for years or even for life. In **persistent infections,** the viruses are continually present in the body. There are several kinds of persistent infections than can be divided conveniently into four categories. These are (1) **late complications following an acute infection,** (2) **latent infections,** (3) **chronic infections,** and (4) **slow infections.** In all categories, disease may or may not be associated with the persistent infection, but the fact that the infected person carries the virus makes the infected individual a potential source of infection to others.

The features of these four categories are shown in figure 14.5. The categories are distinguished from one another largely by whether a virus can be detected in the body during the long period of persistence. This raises the question of what the virus is doing if it persists in the body. Some persistent infections of all categories are associated with disease; others are not. However, in all cases, the person is a carrier and therefore able to spread disease. It is also important to realize that some persistent infections have features of more than one of these categories depending on such circumstances as the time after infection and the cell type in which the virus is located. For example, AIDS has features of latent, chronic, and slow infections.

Late Complications that Follow an Acute Infection

An example of a late complication that follows an acute infection is subacute sclerosing panencephalitis (SSPE), which follows an acute measles infection. This invariably fatal brain disorder occurs in about one in 300,000 people about one to 10 years after a person has had measles. At the time of onset of SSPE, very few infectious measles virions are found in the brain because nerve cells apparently prevent transcription of the viral RNA and, therefore, synthesis of viral proteins.

Latent Infections

Latent infections are those in which infectious virus particles cannot be detected until reactivation of the disease occurs. The viruses causing latent infections can be either DNA or RNA viruses, and the best-known examples of this type of persistent infections are caused by members of the herpes virus family (Herpesviridae), which are divided into two herpes simplex types, (HSV-1 and HSV-2). The latter, frequently called *genital herpes,* is an important and common sexually transmitted disease. Initial infection of young children with herpes simplex type 1 (HSV-1) may not lead to any symptoms, but cold sores and fever blisters often result. After this initial acute infection, the HSV infects the sensory nerve cells where it remains in a noninfectious form without causing symptoms of disease (figure 14.6). Replication of this virus in the nerve cells is repressed by some unknown mechanism but can be activated by such conditions as menstruation, fever, or sunburn. Following the initiation of replication, mature, infectious virions are produced and are carried to the skin or mucous membranes by the axons[1] of the nerve cells, once again resulting in cold sores. After these sores have healed, the virus and host cells once again exist in harmony with no virions being synthesized until the disease recurs.

Another example of a latent infection is provided by another member of the herpes virus family, the varicella-zoster virus, which causes chicken pox (varicella). Initial infection of normal children results in a rash termed *chicken pox.* This virus can remain latent for years without producing any disease symptoms. It can then be reactivated and produce the disease called *shingles* (herpes zoster). Thus, chicken pox and shingles are different diseases but are caused by the same virus.

Most herpes viruses, including HSV-2, tend to become latent under varying conditions. It appears that part or all of the viral DNA becomes integrated into the genome of the host, or copies of the nucleic acid of some herpes viruses replicate as plasmids in the host cell. Table 14.6 gives some examples of latent infections.

[1] Axons—long extensions from the main body of a nerve cell, responsible for transmitting sensory or motor signals

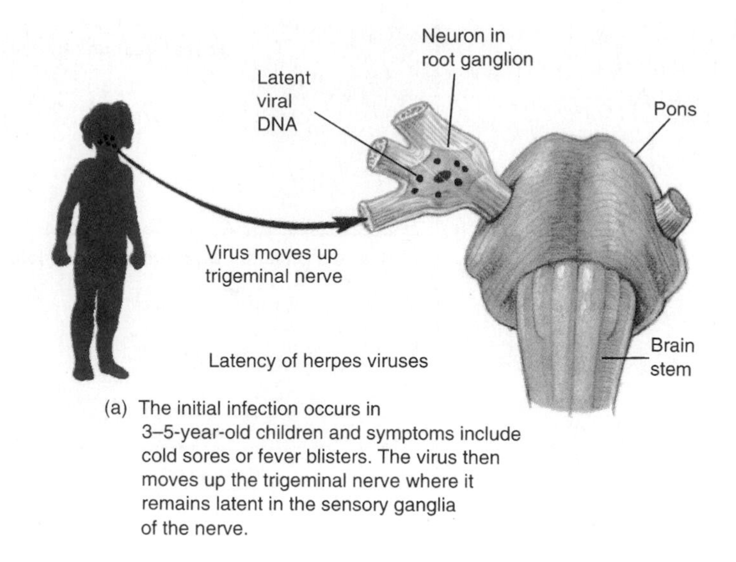

(a) The initial infection occurs in 3–5-year-old children and symptoms include cold sores or fever blisters. The virus then moves up the trigeminal nerve where it remains latent in the sensory ganglia of the nerve.

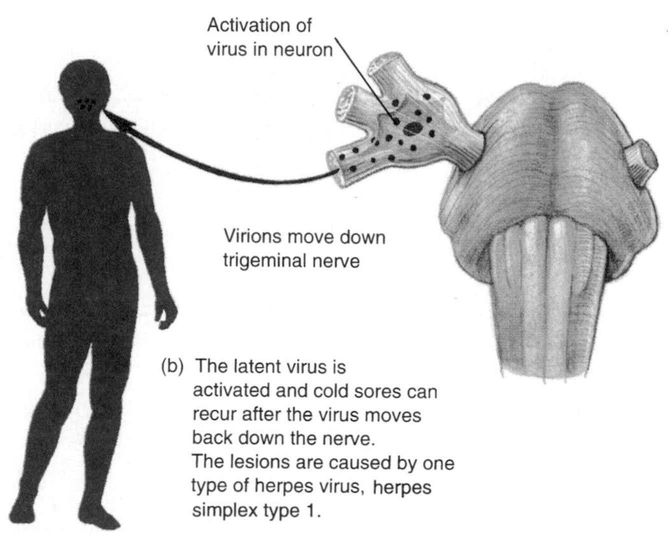

(b) The latent virus is activated and cold sores can recur after the virus moves back down the nerve. The lesions are caused by one type of herpes virus, herpes simplex type 1.

Figure 14.6
Life cycle of herpes simplex virus (HSV).

Chronic Infections

In chronic infections, the infectious virus can be demonstrated at all times. Disease may be present or absent during an extended period or may develop late, often with an immunopathologic[2] or neoplastic[3] basis. Perhaps the best-known chronic human infection is caused by the hepatitis B virus (serum hepatitis), which is transmitted sexually or from the blood of a chronic carrier who shows no symptoms. Some people who contract the virus develop an acute illness marked by nausea, fever, and jaundice. About 300 million people are carriers of the virus, and a significant number develop cirrhosis or cancer of the liver from which over 1 million people die each year.

[2] Immunopathologic—caused by immune cells
[3] Neoplastic—resulting from abnormal growth of cells, as in a tumor

Table 14.6 Examples of Latent Infections

Virus	Primary Disease	Recurrent Disease	Cells Involved in Latent State
Herpes simplex virus			
HSV-1	Primary oral herpes	Recurrent herpes	Viral DNA of virus in neurons of sensory ganglia
HSV-2	Genital herpes	simplex or Genitalis	Viral DNA of virus in neurons of sensory ganglia
Varicella-zoster virus (herpes-virus family)	Chicken pox	Herpes zoster (shingles)	Viral DNA in satellite cells of sensory ganglia
Cytomegalovirus (CMV; herpes-virus family)	Usually subclinical except in fetus or immunocompromised host	CMV pneumonia, eye infections, mononucleosislike symptoms	Viral DNA in salivary glands, kidney epithelium, leukocytes
Epstein-Barr virus (herpes-virus family)	Mononucleosis	Burkitt's lymphoma	Viral DNA in B cells*

*B cells—cells involved in antibody production

Table 14.7 Examples of Chronic Infections

Virus	Site of Infection	Location of Infectious Virions in Carrier State	Disease
Hepatitis B	Liver	Plasma, saliva, genital secretions	Hepatitis, cirrhosis, carcinoma
Hepatitis C	Liver	Plasma, saliva, genital secretions	Hepatitis, cirrhosis, carcinoma
Rubella virus	Many organs	Urine, saliva	Congenital rubella syndrome

In the carrier state, infectious virions of hepatitis B are continually produced and can be detected in the bloodstream, saliva, and semen. In carriers, the viral DNA genome may also occur in liver cells (hepatocytes) as a plasmid where it replicates and produces many infectious virions. Following replication in this plasmid state, the genome can integrate into the cells of the liver. In the integrated state, only portions of the protein components of the virion are synthesized, so infectious virions are not produced. Some examples of viruses that cause chronic infections are summarized in table 14.7.

Slow Infections

In slow infections, the infectious agent gradually increases in amount over a very long time during which no significant symptoms are apparent. Eventually, a slowly progressive lethal disease ensues. The term *slow infection* was originally used to describe slowly progressing retroviral diseases of sheep in Iceland. AIDS has features of slow virus infections.

Two groups of agents that cause slow infections have been identified. One group is the *Lentivirus* (*lenti* means "slow"), which are in the family *Retroviridae* (retroviruses). Members of this family cause tumors in animals. The second is the protein infectious agents called **prions.** Both groups cause diseases that have long preclinical phases and result in progressive, invariably fatal diseases. In both groups, the infectious agent can be recovered from infected animals during both the preclinical years when no symptoms are evident and the time that clinical symptoms are present.

AIDS Is a Slow Virus Disease The most significant slow viral disease caused by members of the *Lentivirus* group is AIDS. This disease results from the invasion and subsequent destruction of T lymphocytes and macrophages, important components in the immune system of the body. Without a healthy level of T lymphocytes and macrophages, the body becomes susceptible to a wide variety of infectious diseases. Like all retroviruses, the replication of HIV requires that its RNA genome can be converted into a double-stranded DNA copy. Two Americans, Howard Temin and David Baltimore, independently demonstrated in 1970 that retroviruses contain an unusual enzyme, **reverse transcriptase,** that enters the host cell at the time of infection. Reverse transcriptase is not found in uninfected cells and must enter into the host cells as part of the entering nucleocapsid. This enzyme copies the viral RNA into a complementary strand of DNA (figure 14.7). A second strand of DNA complementary to the first DNA strand is then synthesized. The DNA is integrated permanently into a chromosome of the host cell as a **provirus.** This provirus is superficially analogous to the phage lambda when it is present as a prophage in *E. coli.* The details of how the activity of the provirus is regulated are not nearly so clear as they are in the case of lambda. It is known that some of the infected cells are continuously synthesizing new virions that bud from the cell, but many more cells carry the provirus in the latent state in which no virions are produced. However, a variety of agents can activate the provirus so that it results in a productive infection in which the virions are released by budding. What these agents are in nature is not known. Small amounts of the virus are present continuously or intermittently in the blood and genital secretions, and carriers can transmit the infection through sexual contact.

Unlike the enzyme DNA polymerase, which makes very few mistakes in the replication of DNA because of its proofreading ability, reverse transcriptase makes many mistakes when copying RNA into DNA. This enzyme has no proofreading activity. As a result, the DNA that codes for a variety of different capsid proteins codes for mutant proteins. Many of these proteins are not recognized by antibodies that recognized the protein capsid of the original virus. These errors in copying help explain why the virus becomes resistant very quickly to antiviral drugs such as AZT and probably protease inhibitors. AIDS is covered in great detail in chapter 31.

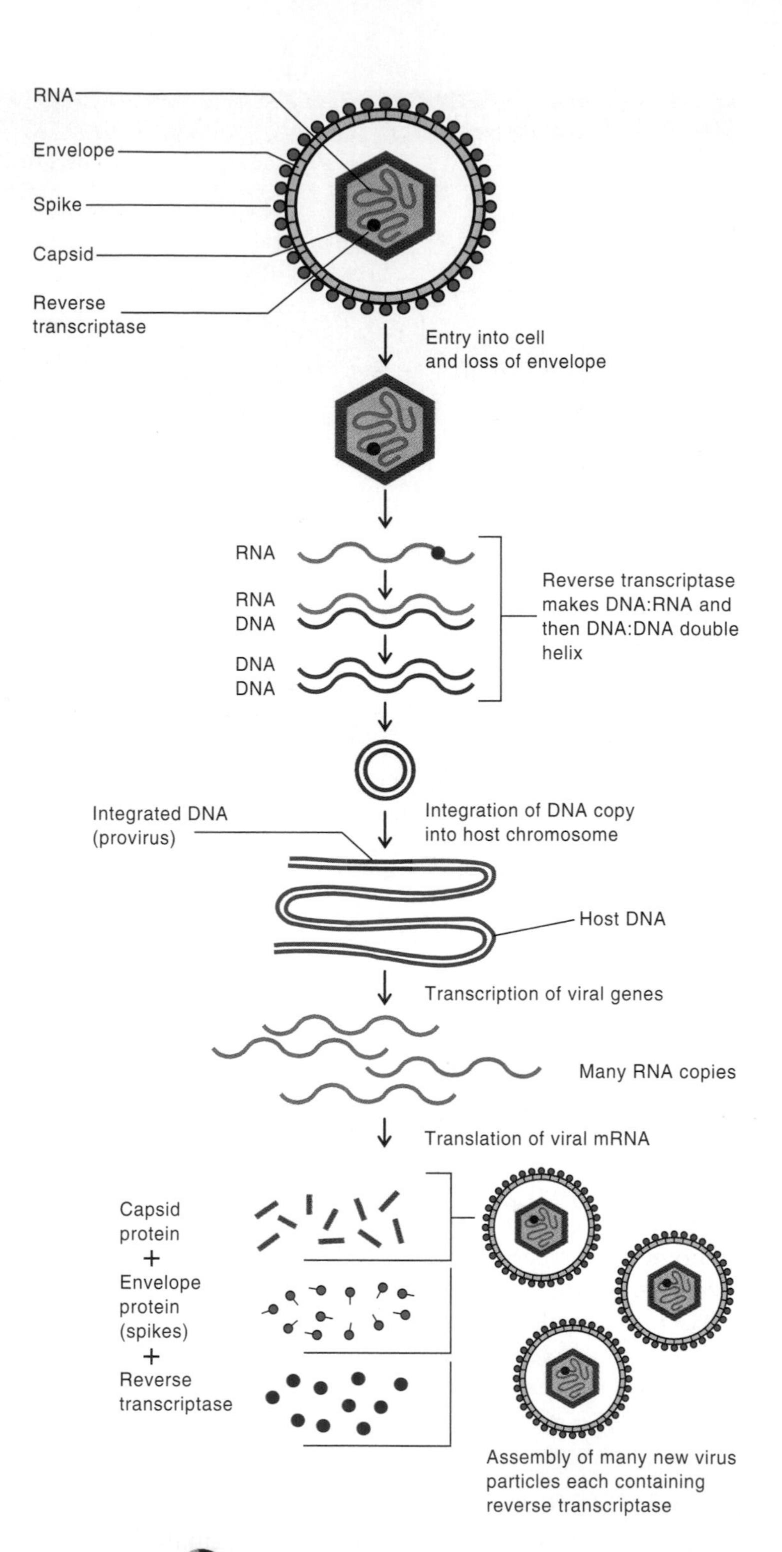

Figure 14.7 Life cycle of a retrovirus. A retrovirus is the cause of AIDS and many tumors in animals. The key features of the life cycle are the synthesis and integration of DNA into the host cell chromosomes. The DNA is copied from the RNA of the virion by viral reverse transcriptase. The proteins that are synthesized from mRNA are synthesized as a single long protein, which is then cleaved by a protease into different viral proteins. The inhibition of this protease by protease inhibitors prevents virus replication. This is proving to be a successful means for inhibiting the replication of HIV.

Prions Cause Fatal Neurodegenerative Diseases

The other group of agents that cause slow diseases are prions, infectious agents that apparently contain no nucleic acid. These agents have been linked to a number of fatal human diseases as well as diseases of animals. In all of these afflictions, brain function degenerates as neurons die, and brain tissue develops spongelike holes. Thus, the general term **transmissible spongiform encephalopathies** has been given to all of these diseases. Some of the slow infections that have been attributed to prions are listed in table 14.8. One major mystery of prions is how a protein molecule without any nucleic acid can cause an infectious disease. One current popular idea is illustrated in figure 14.8. The prion protein (PP) from a diseased animal, can convert a normal cellular protein (NP) to the prion protein conformation. The two proteins, one of them a prion and the other a normal protein of the host, have an almost identical amino acid sequence but differ significantly in their conformation. This hypothesis was put forth

- **AZT, pp. 452, 468**
- **Prophage, p. 307**
- **Proofreading, p. 162**
- **DNA polymerase, p. 134**

Table 14.8 Slow Infections Caused by Prions

Agent	Host	Site of Infection	Disease
Scrapie agent	Sheep	Central nervous system	Scrapie spongiform encephalopathy
Kuru agent	Humans	Central nervous system	Kuru spongiform encephalopathy
Creutzfeldt-Jakob agent	Humans	Central nervous system	Creutzfeldt-Jakob spongiform encephalopathy
Mad cow agent	Cow	Central nervous system	Mad cow spongiform encephalopathy

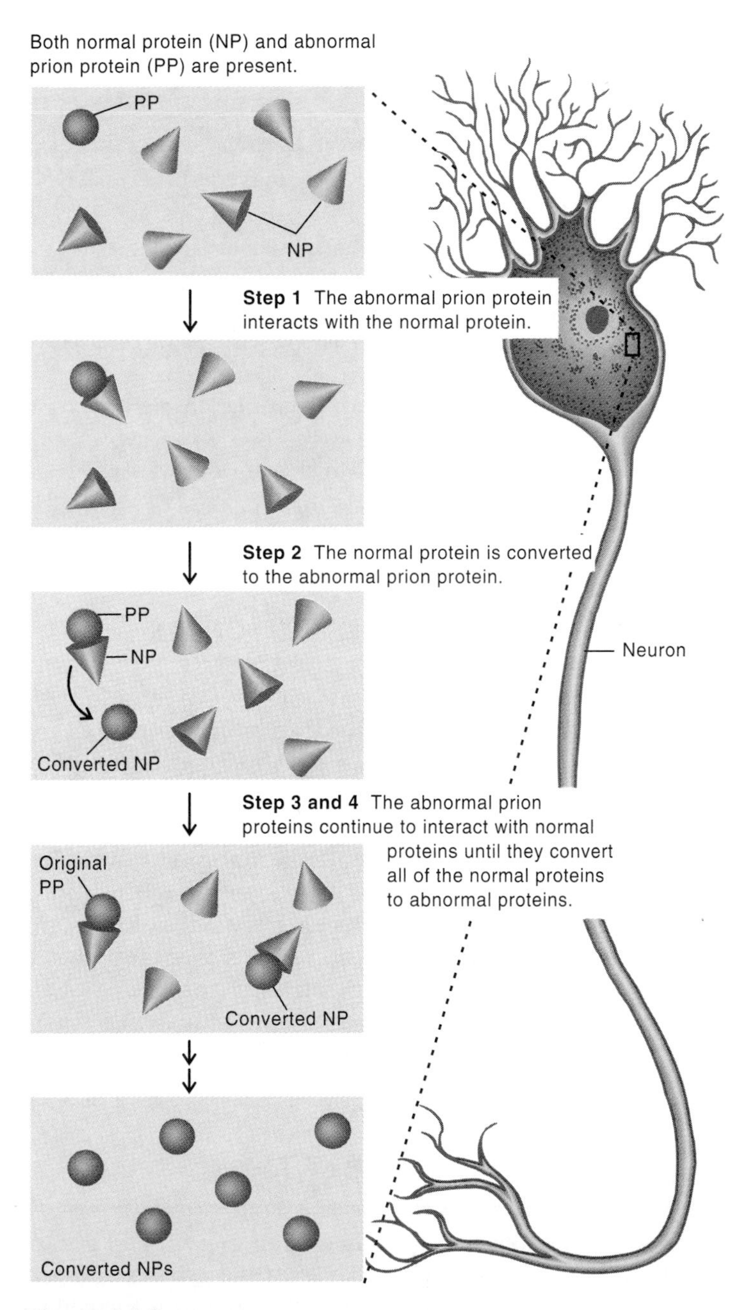

Figure 14.8
Proposed mechanism by which prions replicate. The difference between normal protein (NP) and prion protein (PP) is a single amino acid. This difference results in an altered conformation (shape) to the protein.

when it became known that normal humans and other vertebrates contain a protein similar in amino acid composition to prion protein in their neurons. However, the concept of the prion is not accepted by all scientists. Some scientists believe that these slow diseases are caused by a virus that has proven very difficult to isolate.

Modification of Animal Viruses—Role in Host Range

As discussed already, the bacteriophage genome can be modified by methylation enzymes of the host as the DNA enters the bacterium. This modification determines whether the phage DNA will be cleaved by the host's restriction endonuclease as it enters the cell. However, since eukaryotic cells do not have restriction endonucleases, this mechanism does not operate in animal virus systems. Viral infection can result in the modification of the viral genome in other ways that also play a role in the host range of the virus.

Viruses Can Exchange Their Protein Coats

Animal cells can be infected simultaneously by more than one virus, even when the viruses are from different genera. Different retroviruses can also infect the same cell. As the different viruses synthesize their different protein coats and replicate their nucleic acid in such multiply infected cells, an exchange of protein coats can occur (figure 14.9 *a*). This phenomenon is termed **transcapsidation.** Since the host range of any virus is determined in large part by its coat protein, a virus that was once unable to infect a certain cell type may gain the ability to do so with its new coat. Thus, its host range can be expanded by this mechanism. However, once inside the cell, the nucleic acid codes for its original coat protein and not the coat protein that it wore when it infected the cell.

Segmented Viruses Can Exchange Segments of Their Genome

Another mechanism for expanding the host range of viruses that have segmented genomes is termed **genetic reassortment** (figure 14.9 *b*). This phenomenon likely explains how

• **DNA methylation, p. 51**

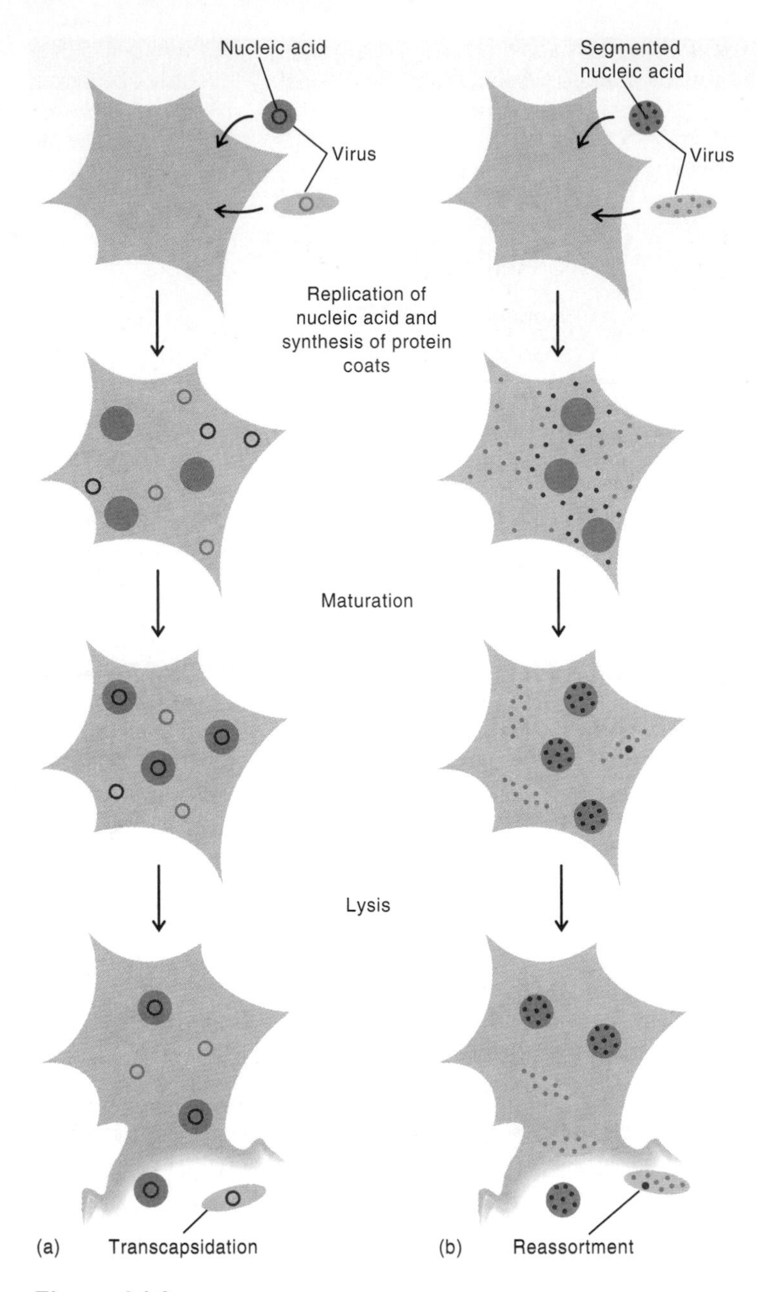

Figure 14.9
(*a*) Transcapsidation. Note that the exchange of protein coats is not permanent. (*b*) Genetic reassortment. Note that the exchange of segments is permanent.

the virulence of the human influenza virus changes so dramatically every 10 to 30 years. These changes result in deadly pandemics because the worldwide population does not have antibodies to the new subtype of the virus.

The genome of the influenza virus is divided into eight segments of RNA, each containing different genetic information. One of the segments codes for hemagglutinin, a protein that is very important in causing influenza. If a human has antibodies against a specific hemagglutinin, he or she is protected against the disease. However, if the structure of the hemagglutinin gene changes dramatically, then the antibodies will not recognize the protein and will not be able to protect against the disease. Experimental data suggest that avian and human influenza viruses can simultaneously infect a cell in a pig, which then serves as a mixing vessel for the 16 RNA segments of both virions. Only a virion that contains mainly RNA segments of the human influenza strain but also has a segment containing the hemagglutinin gene from the avian strain is able to infect humans and confer properties associated with a new subtype. The influenza virus thus undergoes an **antigenic shift** that results in a virus that can evade the host's antibody defense. In addition to experiencing this major change, the hemagglutinin gene, along with other viral genes, can undergo point mutations that result in only small changes in the protein. These types of changes are termed **antigenic drift.** Both of these processes will be discussed later in terms of the epidemiology of influenza.

REVIEW QUESTIONS

1. Name three kinds of interactions that animal viruses can have with their hosts.
2. Name three mechanisms by which animal viruses enter the host cell.
3. Describe one difference between the way animal viruses are released from cells and the way bacterial viruses are released in a productive infection.
4. Name an animal virus that has the same relationship to its host cells as the lambda phage has to *Escherichia coli.*
5. Name four criteria according to which animal viruses are classified.

Tumors—Definition of Terms

The word **tumor,** or **neoplasm,** indicates a swelling that results from the abnormal growth of cells. If the growth remains within a defined region and is not carried to other areas by the circulatory system, it is termed a **benign tumor.** If the abnormal growth spreads (metastasizes) to other parts of the body, it is termed a **malignant tumor** or **cancer.** It is estimated that less than one in 10,000 tumor cells that escape the primary tumor survives to colonize another tissue. Because the discussion in this chapter on basic virology is concerned only with primary tumors, we will use the term *tumor* exclusively as synonymous with *abnormal growth.* The term *cancer* will be used to denote a life-threatening tumor.

Tumors—General Aspects

Tumors result when the controls for cell growth and differentiation do not function properly. Normal growth and cell differentiation is a complex process that requires the cell to synthesize key proteins at the proper times and in the correct amounts. These proteins include those that are involved in communication between cells as well as those that regulate key activities within cells. Such synthesis requires the appropriate activation of genes at specific times in the cell cycle (up regulation) and their silencing at other times (down regulation). If these key genes do not function properly and at the correct times in the life of a cell, then abnormal growth resulting in tumor formation may result.

Two classes of regulatory genes are commonly involved in tumor formation. One class is compromised of **proto-oncogenes** which are activators of gene transcription. This class was identified by two microbiologists, Dr. Michael Bishop and Dr. Harold Varmus, who in 1989 were jointly awarded the Nobel Prize for their studies. Mutations in these genes are dominant since the altered protein causes the cells to grow abnormally. Fusing a normal cell with a tumor cell that contains an abnormal proto-oncogene will result in a cell that grows abnormally.

The second class of regulatory genes associated with tumor formation is composed of **tumor-suppressor genes.** These genes are responsible for putting the brakes on abnormal growth and keeping the cells growing normally. They prevent excessive growth of normal cells. Another class of tumor suppressor genes code for enzymes that repair damaged DNA. Mutations in these genes are recessive. Fusing a normal cell with a tumor cell that has a defective tumor-suppressor gene results in a normal cell. The normal cell supplies the wild type braking system required for normal cell growth and repairs damaged DNA.

Anything that causes the proto-oncogenes or the tumor-suppressor genes to not function properly can lead to abnormal growth and tumor formation. Proto-oncogenes and tumor-suppressor genes are most commonly altered in function through mutation. Such mutations can arise because of defects in the repair systems which operate in DNA replication as well as in mismatch repair. How these repair systems function has already been discussed for bacteria. Mutations can also arise as a result of mutagens in the environment mutating the genes. Recall that Bruce Ames pointed out that all carcinogens are mutagens because they affect the DNA of the organism. If the genes that are altered are essential for growth, such as the genes of macromolecule synthesis, then the cell will die. However, if the genes are one of the proto-oncogenes or tumor-suppressor genes, then a tumor may well develop. Thus it is not surprising that mutations in genes concerned with the repair of damaged DNA lead to an increased rate of tumor formation. It is estimated that 50% of all human cancers result from mutations in tumor-suppressor genes.

There are many proto-oncogenes and tumor-suppressor genes. Some human cancers, such as those of the colon, require that several of the proto-oncogenes and tumor-suppressor genes be mutant simultaneously. A mutation in only one of these genes leads to slightly abnormal growth; a second mutation results in somewhat more abnormal growth, and so on.

In addition to being caused by mutations of proto-oncogenes and tumor-suppressor genes, abnormal growth can be caused by viruses. Tumor-causing viruses, like mutations, alter the activity of the proto-oncogenes and the tumor-suppressor genes. However, viruses are not a frequent cause of tumors in humans—only 15% of all human tumors are estimated to be caused by viruses. However, this number will likely be revised upward because of AIDS. At least 30% of people with AIDS also develop cancer. There is the view that the majority of these cancers are caused by viruses. Viruses are recognized as a frequent cause of tumors in animals.

• **DNA repair, p. 160**
• **Mismatch repair, p. 162**
• **Ames test, p. 166**

Tumors Caused by Viruses in Animals

As pointed out in a "Glimpse of History" in this chapter, viruses were implicated in causing tumors in chickens in the early 1900s. Retroviruses are the most important tumor viruses in animals whereas, in humans, DNA viruses are the most important.

Tumor Viruses Can Transform Human and Animal Cells in Culture

Studies on the mechanism by which viruses cause tumors in animals were given a big boost when it was observed that tumor viruses could rapidly change the properties of human cells growing in culture. These changes are inheritable and are readily observed. Such changed cells are referred to as **transformed cells.** This phenomenon is quite different from what was discussed already for the transformation of bacteria by naked DNA.

A comparison of the properties of normal and tumor cells in culture is shown in table 14.9. The properties of the transformed cells in culture are similar to those of tumor cells in the body. Tumor cells do not respond to signals that limit their growth. Thus, while normal cells grow as a single layer (monolayer) on a glass surface, tumor cells grow in multiple layers. Also, normal cells go through only a limited number of cell divisions and then die, in the process of programmed cell death termed **apoptosis.** In contrast, tumor cells growing in tissue culture multiply indefinitely. They are immortal (see perspective 14.1). Further, normal cells stick tightly to the surface of the glass culture dish whereas transformed cells readily detach from the surface, in a phenomenon analogous to metastasizing in the body. Also, when transformed cells in culture are injected into animals, most, but not all, cause tumors.

Experimentation on tumor viruses became much simpler, cheaper, and faster once it was discovered that most tumor viruses such as the DNA viruses SV_{40} and polyoma as well as tumor retroviruses can transform normal cells in culture. The ability to assay the tumor-inducing ability of viruses quickly and cheaply by mixing the virions with tissue culture cells greatly expedited research on tumor viruses. Within a very short time, the viral genes responsible for cell transformation were identified.

• **DNA transformation, 167**

Perspective 14.1

The Immortal Henrietta Lacks

The most famous cells in culture outside of *Escherichia coli* are the eukaryotic cells HeLa. These cells originated from a very aggressive cervical cancer in the body of Henrietta Lacks, a 30-year-old mother of four from Baltimore, Maryland. In February 1951, while she was a patient in a Baltimore hospital, a sample of cells from her cancer was transferred to a petri dish, nutrients were added, and a cell culture of these cancer cells was established. Eight months later, Lacks died from the cancer, but her cells live on. These cells are frequently used as the host for growing animal viruses. They represent a uniform population of cells that grow rapidly to high cell numbers. They are very aggressive in their growth habits and they will readily take over another population of tissue culture cells if they are accidentally introduced as contaminants. Thus, Henrietta Lacks has achieved a kind of immortality, through the name given to these cells, which are being used in laboratories around the world.

Table 14.9 Some Properties of Normal Cells and Tumor Cells in Cell Culture

Normal Cells
1. Cells grow as a monolayer (single layer).
2. Cells grow attached to one another and to glass surfaces.
3. Cells grow for a limited number of generations and then die, even if diluted into fresh medium.
4. Cells do not form tumors in susceptible animals.
Tumor (Transformed) Cells
1. Cells grow in an unorganized pattern and in multiple layers.
2. Cells attach to surfaces less firmly.
3. Cells continue to grow indefinitely.
4. Cells may form tumors in susceptible animals.

Retroviruses Transform Animal Cells by Inserting a Piece of Their Genome into the Cells

Retroviruses transform cells by inserting the transforming gene of the virus into the genome of the host cell (see figure 14.7). The transformation requires that the genetic information in the entering single-stranded RNA molecule be converted into a double-stranded DNA molecule by the viral enzyme reverse transcriptase. This DNA is then integrated into the host cell genome where it is expressed. What is the nature of the transforming genes? These genes are termed **oncogenes,** from the Greek word *onkos,* which means "mass" or "lump." They are mutant forms of the normal cells' proto-oncogenes discovered by Bishop and Varmus. The mutation modifies the biochemical properties of the oncogene so that an oncogene integrated into the genome of a cell interferes with the normal intracellular control functions of the proto-oncogene. The disruption in these normal control functions results in tumor development. More than 60 oncogenes have been discovered in retroviruses thus far. Different tumor viruses contain different oncogenes that function somewhat differently. However, they are all involved in critical steps in regulating normal cell growth.

In summary, tumors develop from alterations in genes that play critical roles in regulating normal cell growth (figure 14.10). The mutations can be in the proto-oncogenes of the cell, in the oncogenes of the transforming virus, or in mutations in tumor-suppressor genes of cells. In all cases, the regulation of normal cell growth is disrupted and a tumor develops.

Oncogenes Arose from Proto-Oncogenes

The best available evidence suggests that oncogenes originated from proto-oncogenes which were captured by the retrovirus in the course of its excision from the host genome. Once inside the virus, the proto-oncogene underwent mutations that converted it into an oncogene. The temperate phage system in bacteria provides an excellent analogy for this capture hypothesis. Recall that temperate phages integrated into the bacterial host chromosome can incorporate a piece of bacterial DNA into their chromosome when the phage DNA is excised from the host chromosome. The bacterial DNA can then be transferred to other bacteria. By analogy, it is reasonable to hypothesize that retroviruses, which also can integrate into the chromosome of the host, could carry along a proto-oncogene from the host when excised from the host chromosome.

Only 15% of Human Cancers Are Caused by Viruses

Most human cancers are not caused by viruses, despite intensive efforts to prove otherwise. The majority of human cancers appear to be caused by mutations in either proto-oncogenes or tumor-suppressor genes. About 30% are estimated to be due to activation of a particular mutant proto-oncogene and 50% of all human cancers result from mutations in tumor-suppressor genes. The tumor-suppressor gene p53 (*p* stands for the *protein* for which it codes, and 53 is its molecular weight in thousands) seems especially

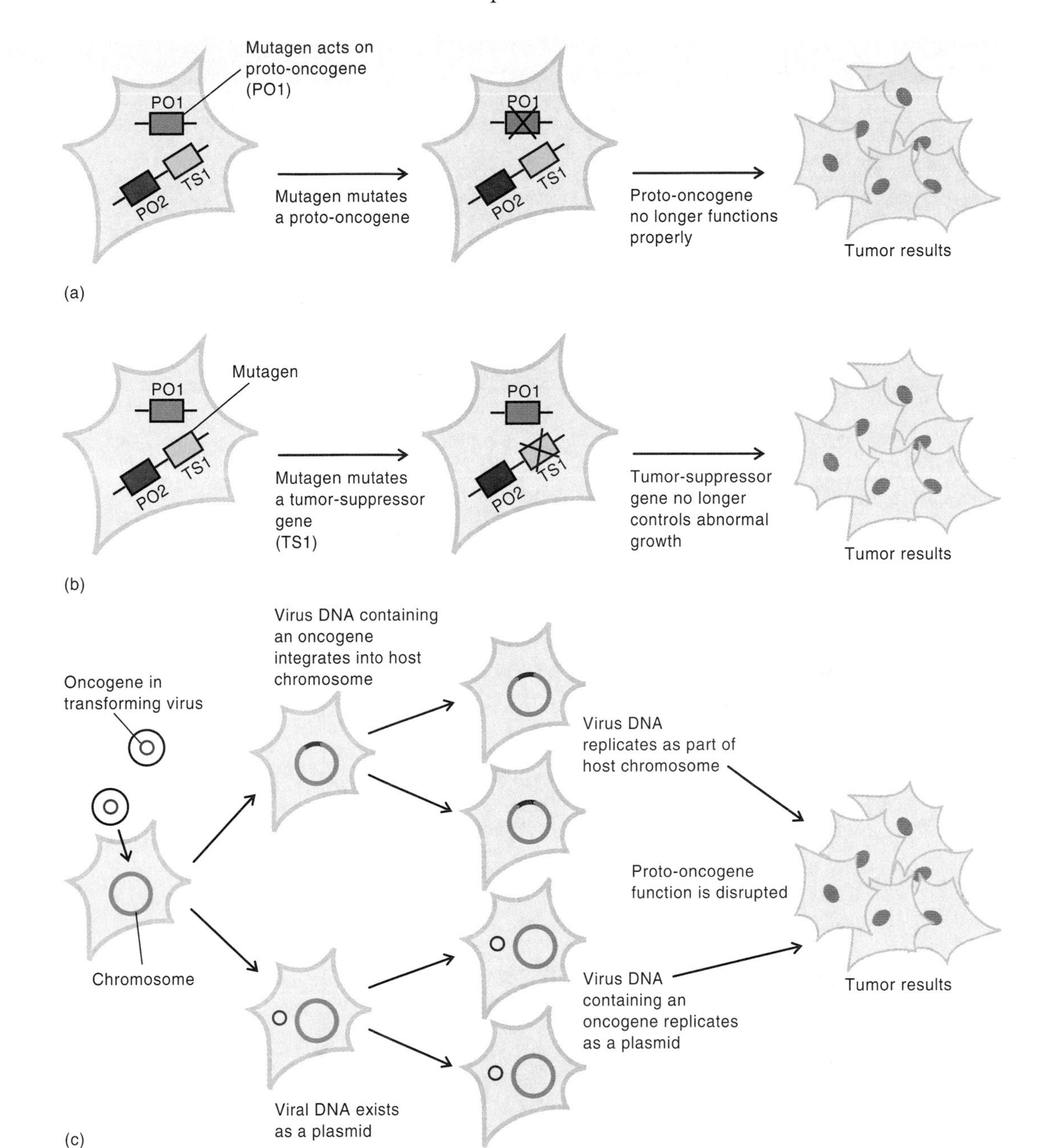

Figure 14.10
Tumor formation as a result of three different mechanisms. (*a*) Mutation in a proto-oncogene that disrupts the normal controls of growth regulation. (*b*) Mutation in a tumor-suppressor gene that removes the growth inhibition of abnormal cells. (*c*) The transforming virus inserts its oncogene into the chromosome of the host cell, or the DNA replicates as a plasmid. The expression of the oncogene disrupts normal functioning of the proto-oncogene of the host.

important in preventing cancer, since mutations in this gene may lead to cancer. The role of a defective p53 gene in causing lung and skin cancer is described in perspective 14.2 on page 339. The p53 gene product blocks the division of cells that have damaged DNA. Tumor viruses can inhibit the function of tumor-suppressor genes, and this is a major contributor to certain human cancers such as colorectal cancers.

Retroviruses are the main class of viruses causing tumors in animals. In 1980, a human leukemia was also shown to be caused by a retrovirus, named human T-cell lymphotrophic virus type L (HTLV-1). This rare tumor is restricted to certain geographic areas. The virus causes tumors by a somewhat different mechanism than discussed thus far. The virus carries a gene that, when integrated into the host cell genome, codes for an **activator protein** that activates regulatory genes of the cell. Thus, this activator gene functions as an oncogene.

Double-stranded DNA viruses are the main cause of tumors in humans (table 14.10). DNA tumor viruses interact with their host cells in one of two ways. They can go through a productive infection in which they lyse the cells or they can transform the cells without killing them. As in the case of retroviruses, the cancers caused by the DNA viruses result from the integration of all or part of the virion genome into the host chromosome. Following integration, the transforming genes are expressed, resulting in uncontrolled growth of the host cells. Thus, these cases of abnormal growth are analogous to lysogenic conversion observed in certain temperate phage infections of bacteria. In both cases, the expression of viral genes integrated into the host's chromosome confers new properties on the host cells.

• **Lysogenic conversion, p. 309**

Table 14.10 Viruses Associated with Cancers in Humans

Virus	Type of Nucleic Acid	Kind of Tumor
HTLV-1	RNA (retrovirus)	Adult T-cell leukemia
Human papillomaviruses (HPV)	DNA	Different kinds of tumors, including squamous cell and genital carcinomas, caused by different HPV types
Hepatitis B virus	DNA	Hepatocellular carcinoma
Epstein-Barr virus	DNA	Burkitt's lymphoma; Nasopharyngeal carcinoma; B-cell lymphoma
Hepatitis C virus	DNA	Hepatocellular carcinoma
Human herpes virus 8	DNA	Kaposi's sarcoma

In the case of some DNA viruses, such as papillomaviruses and herpesviruses, the viral DNA is not integrated but apparently replicates as a plasmid. The plasmid genes are expressed, leading to abnormal growth. Kaposi's sarcoma, a cancer of the skin and internal organs common in AIDS patients, is caused by a herpes virus. How the virus causes normal cells to become cancerous is not known. Note that in all cases of virus-induced cancers, the virus does not kill the host cell but rather changes its properties. The various interactions that viruses display with their hosts are illustrated in figure 14.11.

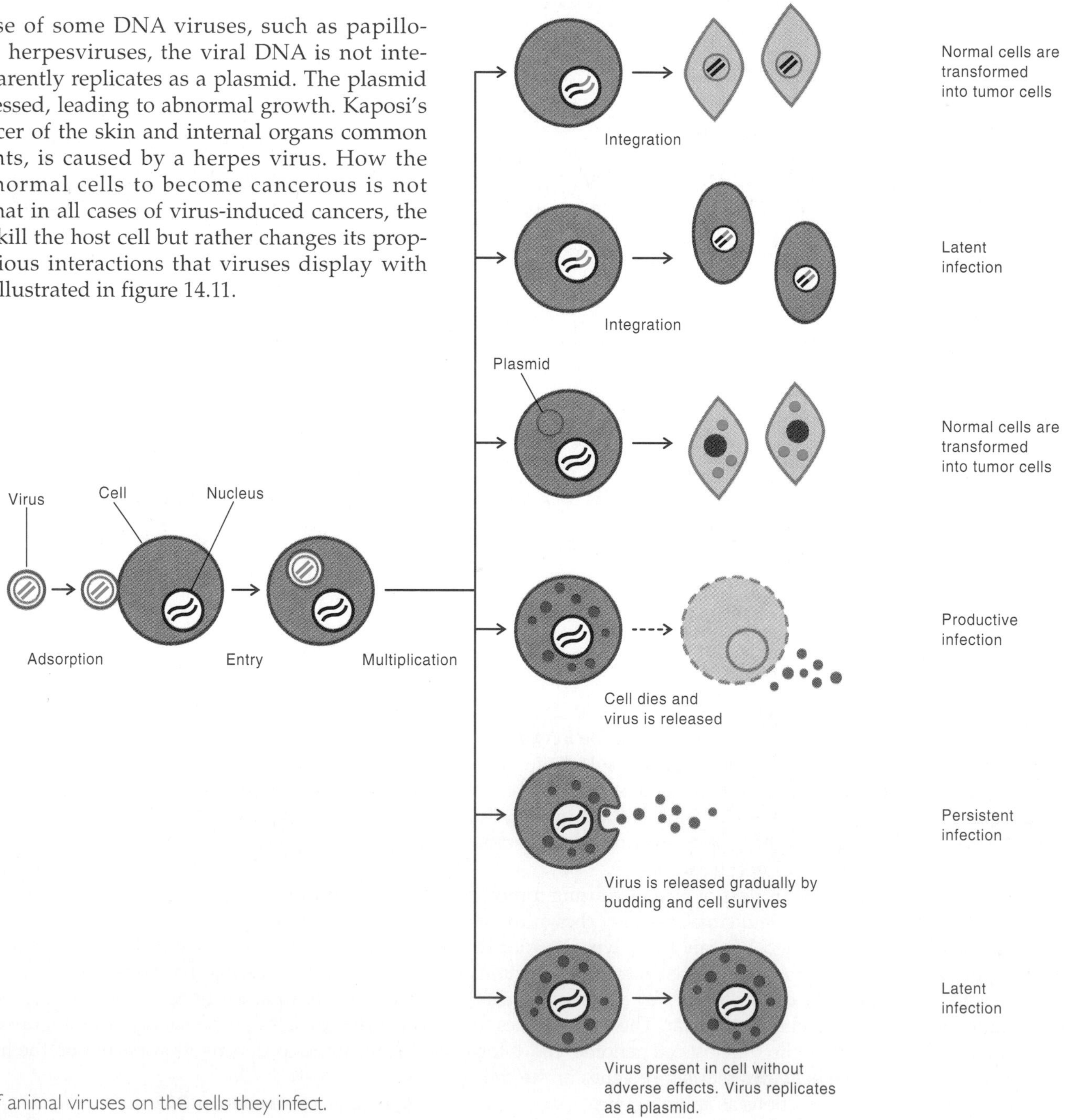

Figure 14.11
Possible effects of animal viruses on the cells they infect.

Viruses that Cause Diseases of Plants

A great number of plant diseases are caused by viruses, many of which are of major economic importance, particularly those occurring in crop plants. Virus infections are especially prevalent among perennial crop plants (those that live for many seasons) and those propagated vegetatively (not by seeds), such as potatoes. Other crops in which viruses cause considerable damage are wheat, soybeans, and sugar beets. A serious virus infection may reduce yields of these crops by more than 50%.

Infection of plants by viruses may be recognized through various outward signs. Localized abnormalities that result in a loss of green pigment may occur, and entire leaves may turn yellow (figure 14.12 *a*). In many cases, rings or irregular lines appear on the leaves and fruits of the plant (figure 14.12 *b*). Individual cells or specialized organs of the plant may die and tumors may appear. Usually, infected plants become stunted (figure 14.12 *c*), although in a few cases growth is stimulated, leading to deformed structures. In the vast majority of cases, plants do not recover from viral infections, for unlike animals, plants are not capable of developing specific immunity to rid themselves of invading viruses. On occasion, however, infected plants produce new growth in which visible signs of infection are absent, even though the infecting virus is still present. The reasons for this are not understood.

In severely infected plants, virions may accumulate in enormous quantities. For example, as much as 10% of the dry weight of a tobacco mosaic virus-infected tobacco plant may consist of virus.

In a few instances, plants have been purposely maintained in a virus-infected state. The best-known example of this involves tulips, in which a virus transmitted through the bulbs can cause a desirable color variegation of the flowers (figure 14.13). The infecting virus was transmitted through bulbs for a long time before the cause of the variegation was even suspected. The multiplication of viruses in plants is analogous to that of bacterial and animal viruses in most respects.

(a)

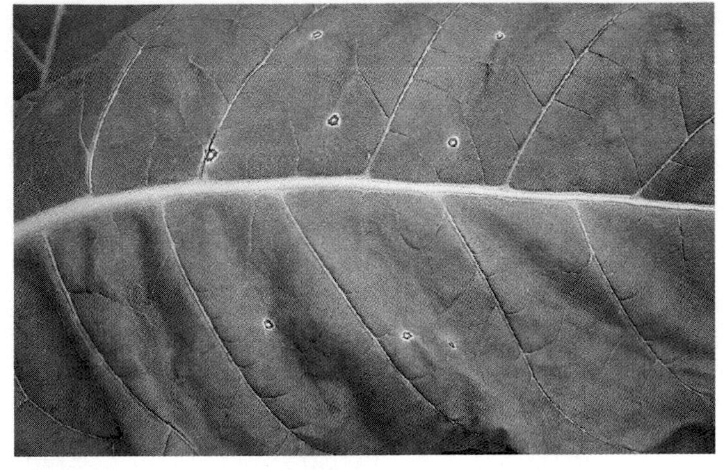

(b)

(c)

Figure 14.12
Various symptoms of viral diseases of plants. (*a*) A healthy wheat leaf can be seen in the center. The yellowed leaves on either side are infected with wheat mosaic virus. (*b*) Typical ring lesions on a tobacco plant leaf. (*c*) Stunted growth (right) in a wheat plant caused by wheat mosaic virus.

Figure 14.13
The variegated colors of tulips are due to viral infection.

Plant Viruses Are Spread by a Variety of Mechanisms

In contrast to phage and animal viruses, plant viruses do not attach to specific receptors on host plants. Instead, they enter through wound sites in the cell wall, which, unlike in animal cells, is very tough and rigid. Once started, infection in the plant can spread from cell to cell through openings that interconnect cells.

Many plant viruses are extraordinarily resistant to inactivation. Tobacco mosaic virus (TMV) apparently retains its infectivity for up to 50 years, which explains why it is usually difficult to eradicate the virions from a contaminated area. Smokers who garden can transmit TMV to susceptible plants from their cigarette tobacco. This stability of the virion is important in maintaining the virus because the usual processes of infection are generally very inefficient.

Some viruses are transmitted through soil contaminated by prior growth of infected plants. In some 10% of the known plant viruses, disease is transmitted through contaminated seeds or tubers or by pollination of flowers on healthy plants with contaminated pollen from diseased plants. Virus infections may also spread through grafting of healthy plant tissue onto diseased plants. Another more exotic transmission mechanism is employed via the parasitic vine dodder. This vine can establish simultaneous connections with the vascular tissues of two host plants, which serve as conduits for transfer of viruses from one host plant to the other.

Other important infection mechanisms involve vectors of various types. These include insects, worms, fungi, and humans. For example, tobacco mosaic virus is a serious disease of the tobacco crop, yet it has no known insect vectors. Instead, humans are the major vectors of this disease. Viruses are transmitted to healthy seedlings on the hands of workers who have been in contact with the virus from infected plants or by people who smoke. However, the most important plant virus vectors are probably insects; thus, insect control is a potent tool for controlling the spread of viruses.

Insect Transmission of Plant Viruses

Several distinct types of insect transmission of plant viruses are recognized. First, in external, or **temporary, transmission,** a virus is associated with the external mouthparts of the vector. The ability to transmit the virus lasts only a few days. Second, in **circulative transmission,** the virus circulates but does not multiply in the body of the insect and may be infective for the lifetime of the insect. Third, the transmission may involve actual multiplication of the virus within the insect. In this case, the virus is infectious for both an animal (insect) and a plant cell.

In many instances of insect infection with plant viruses, the viruses are passed from generation to generation of the insect and may be transmitted to plants at any time. The existence of insect-transmitted plant viruses raises several interesting questions about viral evolution. In particular, it seems that plant and animal viruses may not be as different as they first appeared. Several of these plant- and insect-infecting viruses show similarities to viruses that cause disease in higher animals, including humans. Potato yellow dwarf virus is very similar to animal myxoviruses, and wound tumor virus of plants and retroviruses of animals are similar morphologically.

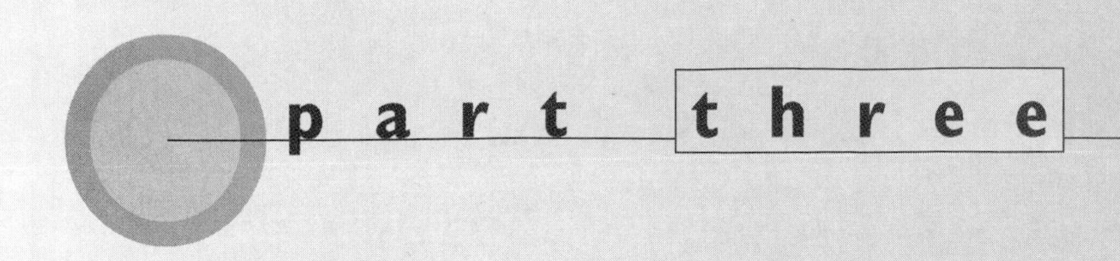

Microorganisms and Humans

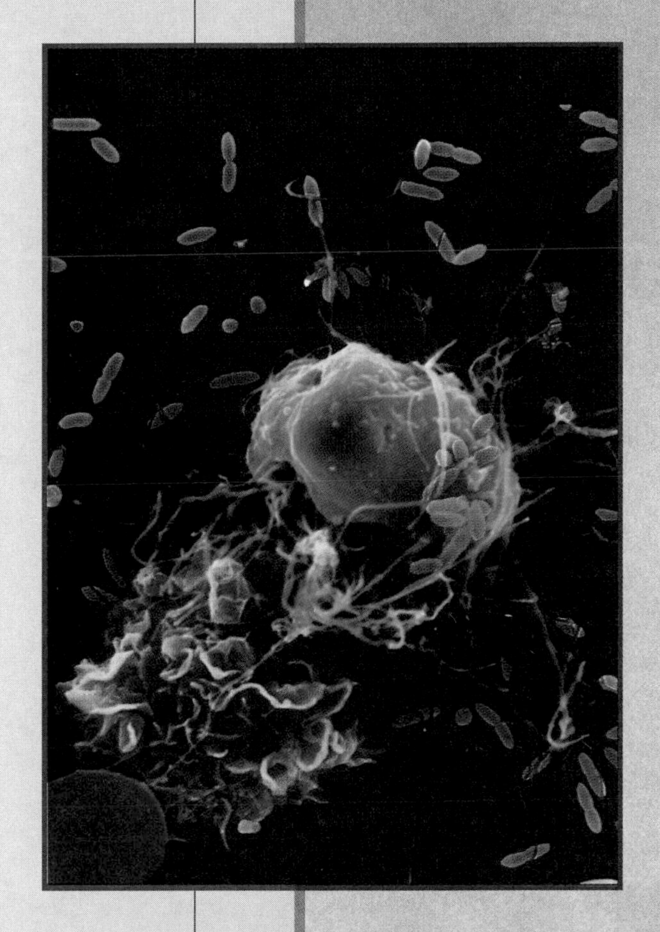

Color-enhanced scanning electron micrograph of a macrophage phagocytizing *Escherichia coli* and foreign debris.

15 Nonspecific Immunity

KEY CONCEPTS

1. The host defends itself with a large variety of both nonspecific and specific defense mechanisms.
2. Certain cells of the blood and tissues, including neutrophils, eosinophils, basophils, mononuclear phagocytes, and lymphocytes, are primarily responsible for body defense.
3. Innate nonspecific immunity is not affected by prior contact with the infecting agent.
4. Tissue barriers, such as skin and mucous membranes, and chemical factors, such as certain enzymes, the complement system, and cytokines, contribute to nonspecific immunity.
5. Inflammation, an early tissue response to injury, is nonspecific and very important in innate immunity.
6. Physiological changes, such as fever and metabolic changes, are also important in innate immunity.

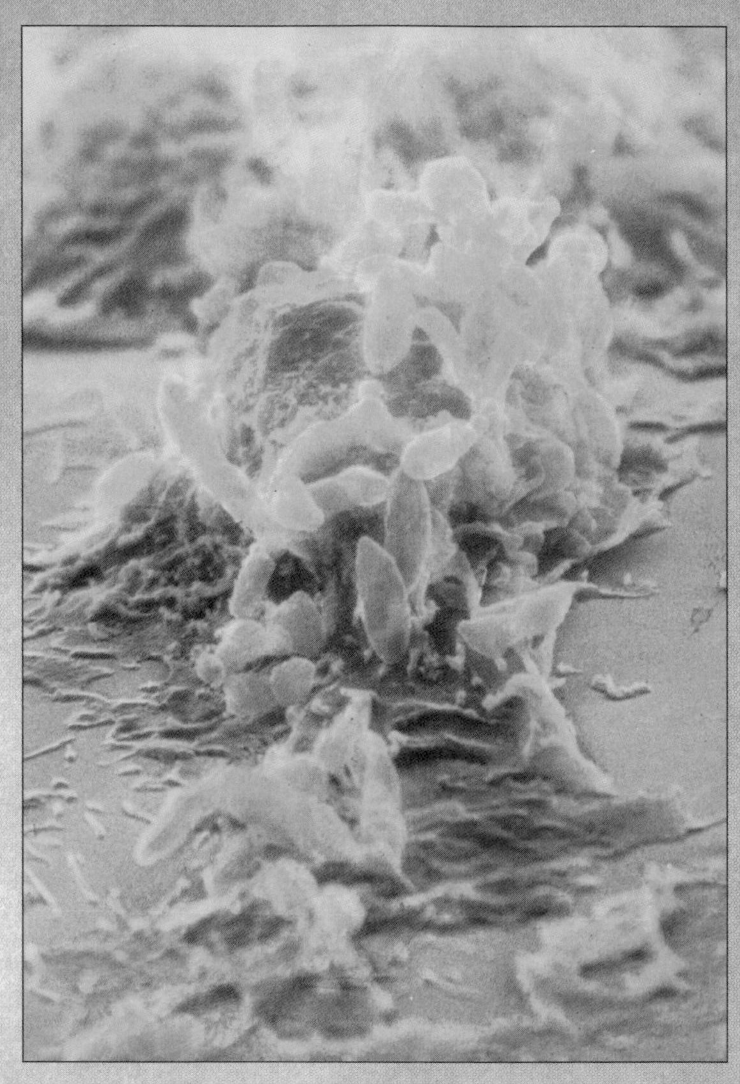

Color-enhanced scanning electron micrograph of macrophage ingesting bacteria (×3,200).

PREVIEW LINK

Microbes in Motion

The following books and chapters in the *Microbes in Motion* CD-ROM may serve as a useful preview or supplement to your reading:

Microbial Pathogenesis: Nonspecific Host Defense; Principles of Pathogenicity

A Glimpse of History

During the latter part of the 19th century, the studies of Louis Pasteur, Robert Koch, and others generated much interest in microorganisms and the diseases that they caused. Once microorganisms were shown to cause disease, scientists worked to explain how the body defended itself against invasion by microorganisms. Elie Metchnikoff, a Russian-born scientist, theorized that there were specialized cells within the body that could destroy invading organisms. His ideas arose from observations that he made while studying the transparent larvae[1] of starfish in Sicily in 1882. As he looked at the larvae in the microscope, he could see amoebalike cells within their bodies. He described his observations as follows:

. . . I was observing the activity of the motile cells of a transparent larva, when a new thought suddenly dawned on me. It occurred to me that similar cells must function to protect the organism against harmful intruders. . . . I thought that if my guess was correct a splinter introduced into the larva of a starfish should soon be surrounded by motile cells much as can be observed in a man with a splinter in his finger. No sooner said than done. In the small garden of our home . . . I took several rose thorns that I immediately introduced under the skin of some beautiful starfish larvae which were as transparent as water. Very nervous, I did not sleep during the night, as I was waiting for the results of my experiment. The next morning, very early, I found with joy that it had been successful.

Metchnikoff reasoned that certain cells present in animals were responsible for ingesting and destroying foreign material. He called these cells *phagocytes,*[2] and he proposed that these cells were primarily responsible for the body's ability to destroy invading microorganisms.

When Metchnikoff returned to Russia, he looked for a way to study phagocytosis. The water flea *Daphnia sp.*, which could be infected with a yeast, provided a way to do such studies. He observed phagocytes ingesting and destroying invading yeast cells within the experimentally infected transparent water fleas. In 1884, Metchnikoff published a paper that strongly supported his contention that phagocytic cells were primarily responsible for destroying disease-causing organisms. He spent the rest of his life studying phagocytosis and other biological phenomena and in 1908 was awarded the Nobel Prize.

[1]Larva (pl. larvae)—the immature stage in the development of an animal

[2]Phagocytes—from the Greek phago, "to eat," and cyto, "cells;" literally, cells that eat

From even before Metchnikoff's discoveries in the late 1800s to the present day, scientists have been studying ways in which humans and other vertebrates are able to protect themselves against invasion by foreign organisms or substances. The field of immunology is devoted to studying these many mechanisms of defense.

Although the science of immunology arose from studies of protection, or immunity, against infectious agents, it grew to encompass many other areas. Immune mechanisms operate not only against infectious agents, but also against cancers. They are responsible for rejection of transplanted cells and organs. Under some conditions, immune responses may be directed against the cells of one's own body, causing autoimmune diseases, and quite often they cause some damage as well as protection, thereby resulting in allergic or hypersensitivity reactions of various kinds. Immune reactions can be used in the laboratory to identify substances with a high degree of specificity. For example, immunological techniques are used to type blood and tissue cells, diagnose diseases, classify bacteria, and identify suspects in the investigation of criminal cases. Without effective immunological defense mechanisms, even the most innocuous microorganisms become deadly, as is seen in the current AIDS epidemic.

The Host Defends Itself: Nonspecific and Specific Defense Mechanisms

Vertebrate hosts, including humans, have two lines of defense against foreign invaders. The first, the **nonspecific immune response,** includes physical barriers such as the skin, mucous membranes and secreted mucus, and the flushing action of the urinary tract. The body also has physiological defense mechanisms, such as the increase in internal body temperature called *fever.* A very important defense is the nonspecific tissue response to injury called **inflammation**, which includes the release of chemical factors that attract certain cells to the area of injury and the actions of phagocytic cells that can ingest and destroy foreign material. All of these factors are termed *nonspecific* because they are directed against any organism that tries to invade the host. Many of these nonspecific defense mechanisms are found in most of the animal world. Recall that Metchnikoff observed his phagocytic cells in the invertebrate starfish and water fleas.

In addition to these nonspecific defense mechanisms, vertebrates have evolved a second line of defense, the **specific immune response.** This response is most highly developed in birds and mammals, and it depends upon specialized cells called **lymphocytes**. Birds and mammals have efficient blood and lymph circulation through which cells and antimicrobial fluids can travel to all tissues of the body. Lymphocytes and glycoprotein molecules called **antibodies** respond specifically to organisms or other foreign materials. When these same invaders are encountered again, there is an enhanced response.

The specific immune response takes time to develop and only becomes effective after the initial nonspecific reaction to the infection or injury. The lymphocytes of the specific response must multiply extensively in order to be effective.

The nonspecific and specific arms of the immune response involve many of the same cells and molecules that communicate between these cells. We will begin with a look at the cells and tissues involved in immune responses.

Cells and Tissues Involved in Immune Responses

Cells of the Blood and Lymphoid System Participate in Immune Responses

The bloodstream can be likened to an extensive highway system through which many cells and materials move from one part of the body to another. Those cells important in body defense are found in normal blood, but their numbers usually increase during infection. Fluids and some blood cells leave the bloodstream and enter the tissues through the lymphatic system. The **lymphatic system** removes proteins and other materials from the tissues in the fluid called **lymph.** The lymph is filtered through many **lymph nodes** along the way and empties back into the blood circulatory system at a large vein in the upper part of the duct. The structure of the lymphatic system and the lymphoid tissues of the system will be described in more detail in later chapters.

Some of the cells important in body defense may mature and **differentiate**[1] (gain different properties) after they leave the bloodstream and enter the tissues. All blood cells, whether white blood cells called **leukocytes,** red blood cells termed **erythrocytes,** or **platelets** (also called **thrombocytes** and really just fragments of cells), originate from the same type of cell, the **hematopoietic**[2] **stem cell,** found in the bone marrow. Under the influence of **colony-stimulating factors (CSF),** the specific types of blood cells are produced. Thus, each stem cell differentiates and matures through a series of intermediate stages to become one of the cell types of the blood and lymphatic system (figure 15.1).

Leukocytes are the cells primarily responsible for the defense of the body against microorganisms, viruses, multicellular parasites, and even tumor cells. There are several different subsets of leukocytes, each with special functions (table 15.1).

[1]Differentiate—undergo developmental changes to a more specialized form or function

[2]Hematopoietic—from the greek *haema,* "blood," *poien,* " to make," literally, blood-making tissue such as bone marrow

Table 15.1 Sets and Subsets of Human Leukocytes

Cell Type (% of Blood Leukocytes)	Morphology	Location in Body	Functions
Granulocytes			
Neutrophils (polymorphonuclear neutrophils or PMNs, often called polys; 55%–65%)	Lobed nucleus; granules in cytoplasm; ameboid appearance	Account for most of the leukocytes in circulation; few in tissues except during inflammation and in reserve locations	Phagocytize and digest engulfed materials
Eosinophils (2%–4%)	Large eosinophilic granules; nonsegmented or bilobed nucleus	Few in tissues except in certain types of inflammation	Participate in inflammatory reaction and immunity to some parasites
Basophils (0%–1%), mast cells*	Lobed nucleus; large basophilic granules in cytoplasm	Basophils in circulation; mast cells present in most tissues	Release histamine and other inflammation-causing chemicals from the granules
Mononuclear phagocytes			
Monocytes (3%–8%), macrophages†	Single nucleus; abundant cytoplasm	Macrophages present in all tissues and in lining of vessels; monocytes less-mature circulating forms	Phagocytize and digest engulfed materials; can participate in killing foreign cells that are not engulfed; play vital role in development of specific immunity
Lymphocytes			
Several types (25%–35%)	Single nucleus; little cytoplasm except with differentiation	In lymphoid tissues (such as lymph nodes, spleen, thymus, appendix, tonsils); also in circulation	Participate in specific immunological responses

*Mast cells—tissue cells similar in appearance and function to basophils
†Macrophages—mononuclear phagocytes found in tissues

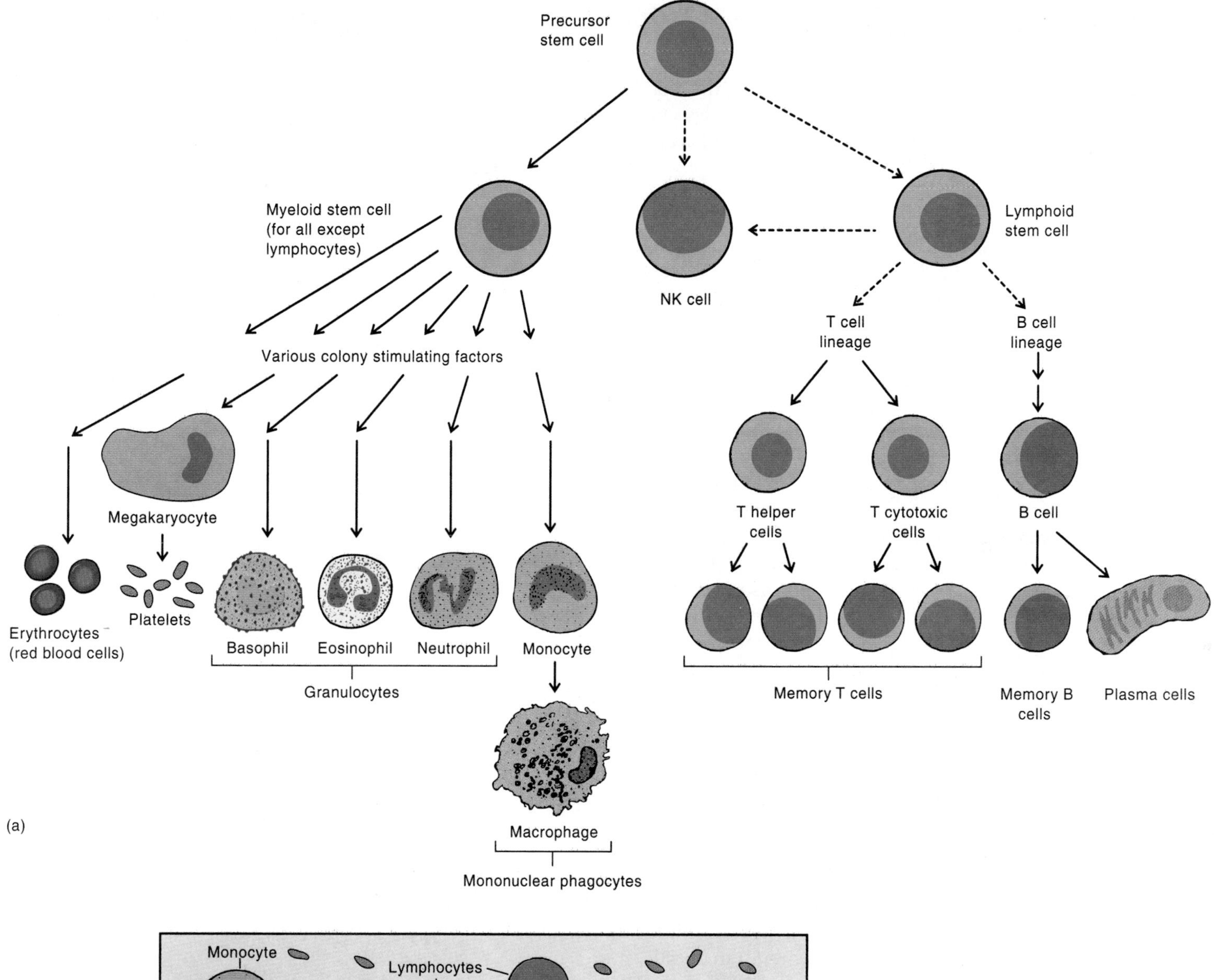

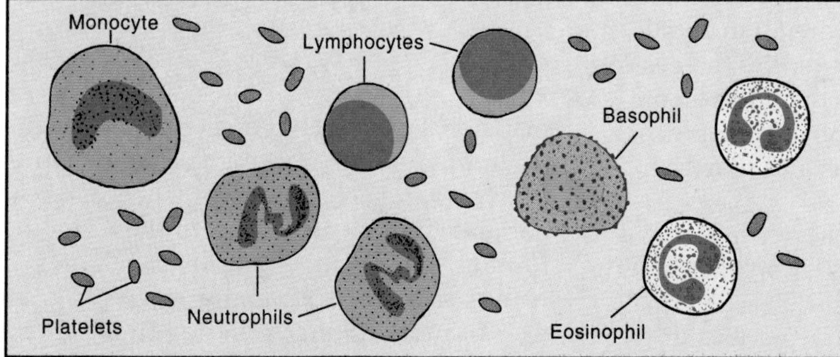

Figure 15.1
Cells of the blood and lymphoid system. (*a*) The family tree of cells of the blood and lymphoid system. This diagram shows that all these types of cells are derived from precursor stem cells found in the bone marrow. Some of the steps not yet clearly defined are indicated by dotted lines. Multiple steps occur between the stem cell and the final cells produced. (*b*) Leukocytes and platelets found in the normal circulating blood.

The **granulocytes** are so named because they contain characteristic cytoplasmic granules, which differ from one subset of granulocytes to another. The names of the various granulocytes reflect the staining properties of the cells when a certain mixture of dyes is used. The granules contain biologically active chemicals that are important in the function of each cell type.

The **neutrophil** granules stain poorly or not at all. These granules contain many antimicrobial substances, including peroxidase, lysozyme, and other degradative enzymes that are critically important in the destruction of engulfed bacteria and other materials. Mature neutrophils differ from other leukocytes in that their nuclei consist of several lobes. Although they each have a single nucleus, it is lobed and, thus, they are called **polymorphonuclear neutrophils (PMNs),** *neutrophils,* or simply *polys.* These are actively phagocytic cells and are extremely important in the first encounter with many foreign materials. They normally account for more than half the leukocytes in human blood, and their number increases with most acute (short and relatively severe) bacterial infections. It is estimated that for every PMN in the circulation, 100 are in reserve in the bone marrow, ready to be mobilized when needed. As they accumulate in diseased areas, they die. With some tissue fluid, the dead cells make up **pus.** A large collection of pus constitutes an abscess or boil.

Neutrophils are essential for human survival since, as Metchnikoff proposed, they play a vital role in defense by phagocytizing and destroying foreign or other unwanted materials, including microorganisms. Mature neutrophils are end cells that do not reproduce themselves. They contain a predetermined amount and variety of granules. Once the granules have been used, the cell dies. Although many types of cells in the body can ingest foreign material, polymorphonuclear neutrophils and mononuclear phagocytes (described next) are the major phagocytic cells of the body, the so-called professional phagocytes, and they are highly efficient at phagocytizing and destroying foreign materials.

The **eosinophil** granules are stained red by the eosin dye in the mixture of dyes. The granules contain basic proteins, peroxidases, and other antimicrobial substances. Eosinophils play a role in allergic reactions and in rejecting parasitic worms.

The **basophil** granules take up the basic dye in the mixture of dyes and stain a dark purplish blue (basophilic). They contain **histamine** and **serotonin,** substances that act on small blood vessels to increase blood flow and thus contribute to inflammatory reactions. A cell that is similar in appearance and function, the **mast cell,** is abundant in tissues all over the body. Mast cells do not circulate in the bloodstream.

Mononuclear phagocytes, which include **monocytes** and **macrophages,** are the other major group of phagocytic cells of the body. These are found in virtually all parts of the body and they constitute the **mononuclear phagocyte system (MPS),** (figure 15.2). Monocytes are found in normal blood where they represent about 3% to 8% of circulating leukocytes; they also migrate into tissues where they develop into tissue macrophages. The life span of tissue macrophages is between several months and several years. Within the tissues, macrophages develop or differentiate to perform various functions and, hence, take on slightly different appearances according to the tissue in which they are found. As a result, tissue macrophages are given special names, such as *histiocytes* for macrophages found in connective tissues, *Kupffer cells* for those in the liver, *alveolar macrophages* or *dust cells* in the lung, and *microglial cells* in the central nervous system. Cells with certain similar functions are the Langerhans-dendritic cells in the skin and lymphoid tissues. They can phagocytize and function during the process of antibody production much as macrophages do.

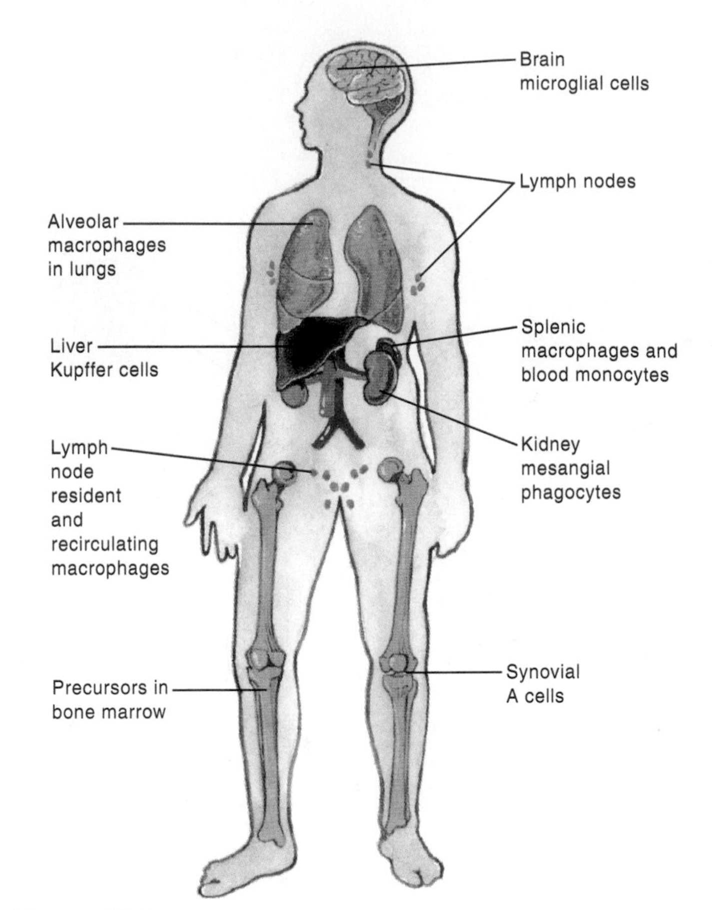

Figure 15.2
Monocytes and macrophages make up the *mononuclear phagocyte system,* formerly known as the reticuloendothelial system. These professional phagocytes are found in all parts of the body, and many have special names to denote their location; for example, Kuppfer cells are found in the liver, and alveolar macrophages are in the lung. The peritoneal body cavity that contains the abdominal organs is filled with peritoneal fluid, particularly rich in mononuclear phagocytes called *peritoneal macrophages.*

Macrophages are found in virtually all tissues. They are present in especially large numbers in the liver, the lungs, and the peritoneal cavity.[1] They are also abundant in the spleen and lymph nodes. Macrophages engulf and degrade foreign

[1] Peritoneal cavity—abdominal cavity enclosing the viscera

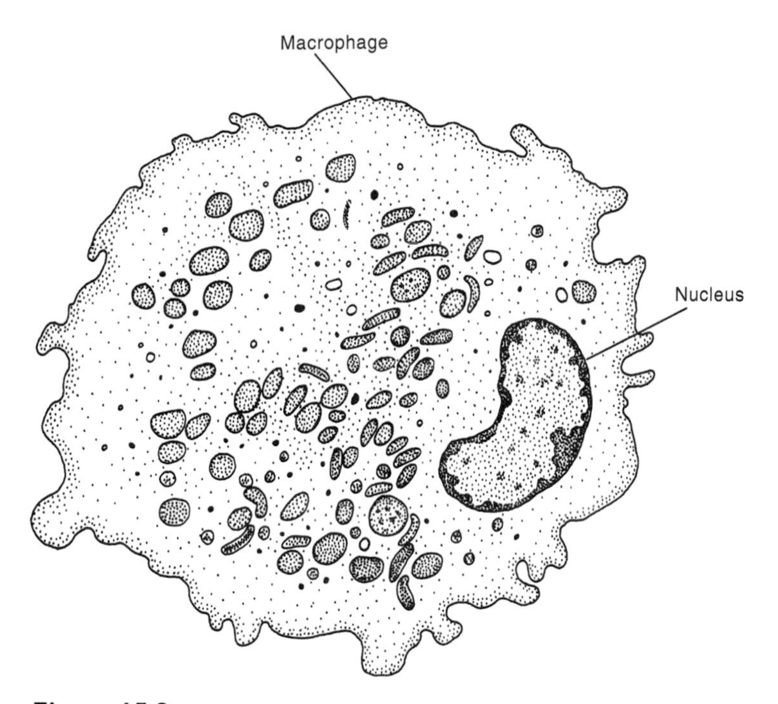

Figure 15.3
Schematic diagram of a macrophage.

materials, but unlike neutrophils, macrophages are not end cells. They can become activated (figure 15.3) and induced to make specific enzymes needed to break down engulfed particles and macromolecules. For example, when large amounts of fat are engulfed, lipases (enzymes that degrade fat) are made. If polysaccharides containing galactosides are ingested, large quantities of galactosidase (an enzyme that degrades galactosides) are produced to break down the large polysaccharide molecules. If these mechanisms of intracellular destruction fail, a large number of macrophages can fuse together to form a single **giant cell** that can retain particularly resistant engulfed organisms, such as tubercle bacilli, the organisms causing tuberculosis. Most importantly, macrophages can communicate with other cells via chemicals called *cytokines*. In addition to functioning in nonspecific immunity, macrophages play important roles in specific immunity.

Lymphocytes are the cells primarily responsible for the specific immune response. They will be discussed in more detail in the next chapter dealing with specific immunity.

REVIEW QUESTIONS

1. Of invertebrates and vertebrates, which have nonspecific and specific immune responses and why?
2. Describe the various kinds of granulocytes.
3. The number of leukocytes, especially PMNs, in the circulation often increases greatly during acute infections. Explain why this occurs.
4. What is the mononuclear phagocyte system and where is it found?
5. What kind of cells are giant cells?

Innate Nonspecific Immunity

When an infection occurs, the first defenses are those of innate, nonspecific immunity. Innate immunity can be defined as immunity that is not affected by prior contact with the infectious agent or other material involved. Innate immunity operates constantly to prevent the establishment of an infection. It may involve physical or chemical barriers. A number of physical barriers in the human body prevent the entry of microorganisms. Chemicals and enzymes normally produced by the body are potent nonspecific antimicrobial factors. The activities of many of the components of nonspecific defense are improved by specific immune mechanisms, but all of the nonspecific defenses function without a specific immune response.

Tissue Barriers and Nonspecific Factors Are Important in Nonspecific Immunity

Physical Barriers

The importance of skin and mucous membranes as **physical barriers** to infection cannot be overemphasized. Not only do these tissues form barriers, but both of them are endowed with antimicrobial secretions. Sweat and other secretions of skin are quite acidic and inhibit the growth of most disease-producing bacteria (the pH of the skin varies between 3.5 and 5.8). Saliva, tears, and mucus produced by mucous membranes are rich in a variety of antimicrobial substances (figure 15.4). Also, the movement of the mucus over the membrane surface helps to remove many trapped organisms. The flow of urine through the urinary tract is important in maintaining sterility there, as witnessed by the frequent occurrence of infection when urinary flow is impeded by anesthesia or other means. In order to counteract these mechanisms for preventing infection, microorganisms must be able to attach firmly to the tissues in order to colonize successfully.

An example of how some of the nonspecific defenses work together to protect the body against infection is found in the respiratory tract, where the nose removes up to 90% of inhaled microscopic particles, such as bacteria, fungal spores, and dusts. Mucus in the nose is rich in lysozyme and other antimicrobial substances. The motion of ciliated cells in the upper respiratory tract propels the blanket of mucus away from the lungs and prevents the entry of most extraneous material into the lungs. These and other defense mechanisms are discussed in more detail in the chapters dealing with each body system.

Nonspecific Antimicrobial Factors

Some important **nonspecific antimicrobial factors** are described in table 15.2. Two of these antimicrobial factors often found in body fluids, **lysozyme** and **beta-lysin**, are also found in high concentrations in white blood cells and

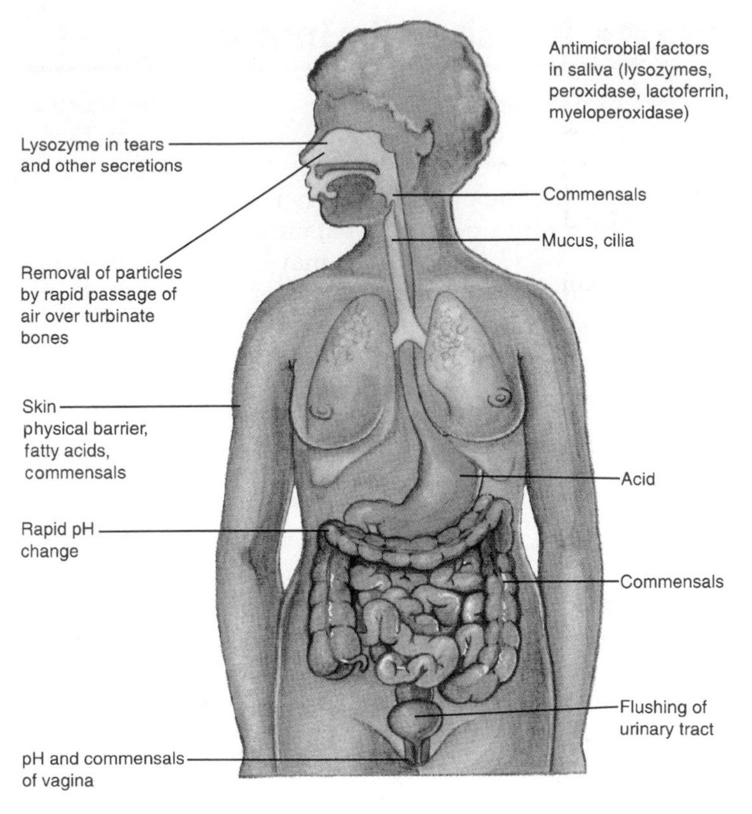

Figure 15.4
Nonspecific defense mechanisms in humans. Note the location of various nonspecific host defense mechanisms that work to prevent entry of microorganisms into the host's tissues.

platelets. They are released during inflammation, the nonspecific tissue reaction to injury, as when bacteria enter the body. Lysozyme degrades the peptidoglycan layer of the bacterial cell wall. Therefore, it is particularly effective against gram-positive bacteria whose peptidoglycan is more likely to be exposed and accessible to the enzymes than is that of gram-negative bacteria. Beta-lysin is also active against gram-positive bacteria.

Peroxidase enzymes, together with hydrogen peroxide and halide ions such as chloride and iodide, make up a very effective antimicrobial system. The peroxidase enzymes are found in neutrophil granules, in saliva, and in milk, and the peroxide is formed during oxygen metabolism, either by the host cell or by the invading organism. The interaction of hydrogen peroxide with peroxidase and chloride ions results in the formation of chlorine and hypochlorite, the active ingredient in bleach. It would seem that organisms would not have a chance of surviving a system that produces these powerful chemicals. However, bacteria that produce the enzyme catalase may be able to escape this mechanism of killing because the catalase breaks down the peroxide in a different manner than peroxidase does. Catalase-negative organisms are more sensitive to peroxidase killing.

• **Peptidoglycans, p. 58**

Table 15.2 Some Important Nonspecific Antimicrobial Factors

Antimicrobial or Antiviral Factor	Chemical Nature	Source	Effects
Lysozyme	Protein	Most body fluids; also within phagocytes	Destroys bacterial cell walls
Beta-lysin	Protein	Serum, leukocytes	Attacks cytoplasmic membrane; active against gram-positive bacteria
Peroxidase enzymes	Proteins	Leukocytes, saliva, milk	Kill a variety of microorganisms; important killing mechanism in saliva and within neutrophils
Complement system	Many distinct proteins	Produced by macrophages and other host cells	Proteins acting in special sequence to produce effects such as chemotaxis, opsonization, and cell lysis
Interferons	Glycoproteins	Produced by leukocytes and tissue cells	Interfere with the multiplication of viruses by causing the formation of antiviral proteins
Lactoferrin	Protein	Leukocytes, saliva, mucus	Successfully competes with intracellular bacteria for iron

Complement Is of Primary Importance, Not Only as an Antimicrobial Factor, but Also in Inflammation and a Number of Other Functions

Complement is a nonspecific reactant in the body's defenses because it can be activated by many different foreign invaders. It destroys invading bacteria and foreign cells by disrupting their cytoplasmic membranes. Complement is also involved in the inflammatory response by contributing to vascular permeability, stimulating chemotaxis, and enhancing phagocytosis.

Complement consists of a group of at least 26 different proteins. Some of these proteins are inactive enzymes that, once activated, act in sequence one after another in a cascade reaction. For example, one complement component is converted to an active enzyme that acts on the next component and converts it from an inactive to an active enzyme, and so on. Other complement components carry out functions described next, or act to regulate and control complement activity. The major components of complement are given a number along with the letter C. Major components are C1 through C9. These were numbered in the order in which they were discovered and not in the order in which they react.

• **Chemotaxis, pp. 68–70**

Complement can be activated in two ways, by either a pathway called the **classical pathway,** which requires a specific immune reaction for activation, or an **alternative pathway,** which does not require a specific immune reaction.

As can be seen in figure 15.5, the C3 molecule is the key component in the sequence of events occurring in either pathway. In the classical pathway, antibody that is combined with antigen interacts with the inactive enzyme C1 making it an active enzyme. This active enzyme then splits, or cleaves, C4 and C2, resulting in an enzyme called C3 convertase that cleaves C3 into C3a and C3b. The antigen-antibody activation of a single C1 molecule can ultimately cause the cleavage of millions of C3 molecules, making this a very efficient mechanism. In the alternative pathway, polysaccharides from the capsules and cell walls of microorganisms react directly with a tiny amount of C3b that is constantly available. It is available because native C3 contains an unstable (thioester) bond that for a few microseconds allows C3b to be split and active, before regulatory mechanisms inactivate it. This provides a

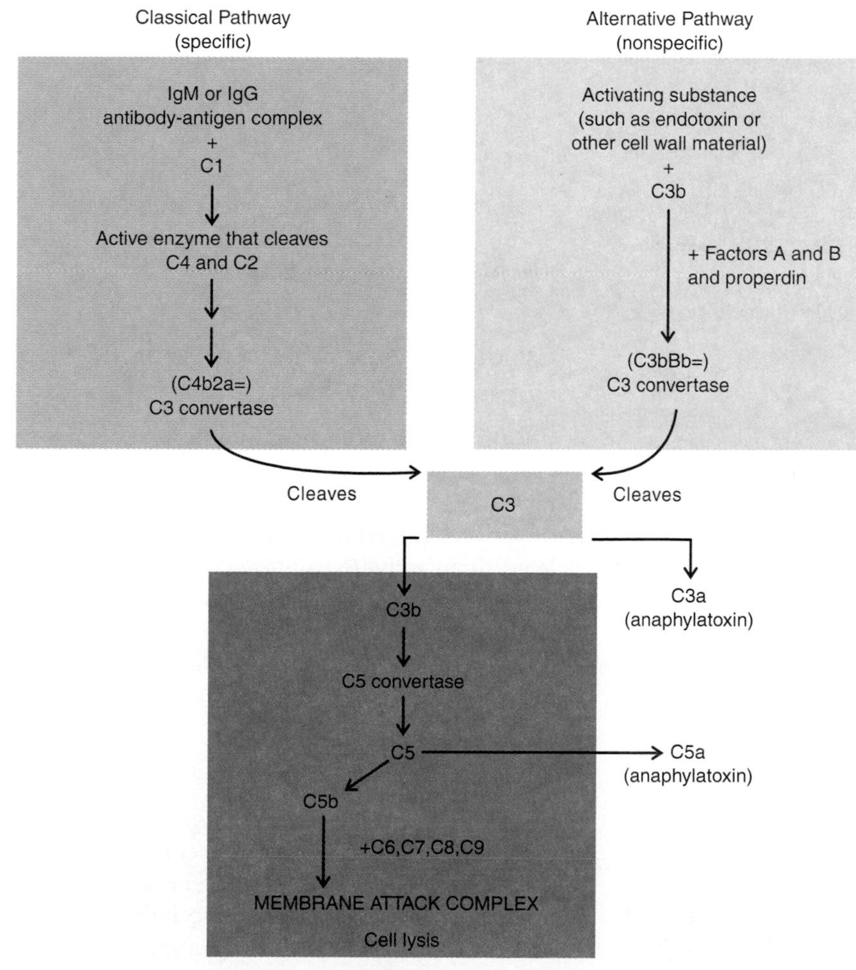

The complement system is made up of three sets of proteins. Two sets lead to the formation of C3 convertases, enzymes that cleave C3 molecules into two fragments, C3a and C3b.

- The blue background indicates one of the sets, the classical pathway proteins, which requires specific antibody complexed with antigen to start the reactions.
- The second set, the alternative pathway proteins indicated in the gold background, is nonspecific and is initiated by a variety of substances.
- The third set of proteins, in the red background, is concerned with the formation of the membrane attack complex that causes cell lysis. Fragments split off during the processes have important biological effects as indicated.

Figure 15.5
Simplified scheme of the activation of complement.

constant "trickle" of a few C3b molecules to start the alternative pathway. It is estimated that one C3b molecule deposited on a microorganism can become 4 million in four minutes by the activity of the alternative pathway. The classical pathway remains the more efficient, however, because many more C3 molecules are involved in the reactions.

C3b is also an enzyme, and it cleaves C5 into C5a and C5b. C3a and C5a are both small molecules that are important in inflammation. They trigger the release from mast cells of a number of chemicals that contribute to the vascular permeability of inflammation. Some of these chemicals and their biological effects are noted in table 15.3.

C3a and C5a also directly stimulate the metabolic activity of phagocytes and cause the phagocytes to produce more receptors for C3b on their surfaces. As figure 15.5 shows, C3b is formed when C3 is cleaved, and the C3b quickly attaches to the surface of the microorganism or other material that caused the cleavage. The C3b-coated material then adheres to the C3b receptors on phagocytes, which enhances uptake of the material into the phagocyte. An enhancement of phagocytosis is called **opsonization** and it can be caused in other ways besides C3b activity.

In addition to promoting phagocytosis, C5b molecules on the surface of bacteria or foreign cells bind together loosely and form a complex with C6 and C7. This complex together with C8 causes changes in C9 such that it polymerizes and forms a **membrane attack complex** that inserts through the membrane of the cell, resulting in cell lysis (figure 15.6). The multilayers of peptidoglycan in gram-positive bacteria may interfere with lysis, but C3b still increases phagocytosis and intracellular destruction of these bacteria. Gram-negative bacteria are more susceptible to complement lysis than are gram-positive organisms because their peptidoglycan occurs in a single layer.

Table 15.3 Chemicals Released from Mast Cells During Inflammation

Chemical	Effects
Histamine	Dilation and increase of permeability of small blood vessels; constriction of the bronchi
Chemotactic factors	Eosinophil and neutrophil (PMN) chemotaxis
Interleukins 3, 4, 5, and 6	Many interactions
Tumor necrosis factor alpha	Recruitment of granulocytes to area of inflammation; inducement of fever
Leukotrienes	Dilation of small blood vessels; constriction of bronchi; chemotaxis of leukocytes
Prostaglandins	Increase in vascular permeability; regulation of immune responses

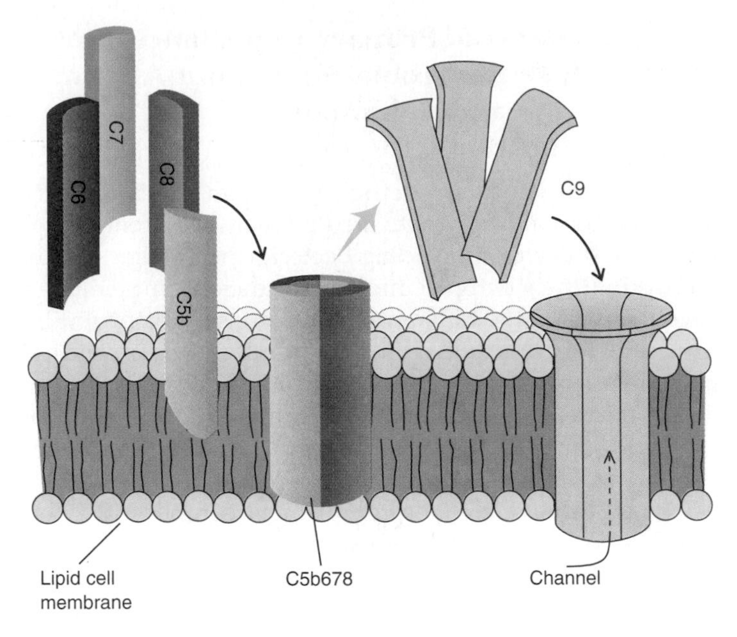

Figure 15.6
Membrane attack complex of complement (MAC). The MAC is formed after C5b, C6, and C7 form a complex on the cell surface of bacteria or other foreign cells. This complex, together with C8, causes changes in C9, allowing it to polymerize and form an MAC. The MAC inserts through the membrane, resulting in lysis of the cell. Varying numbers of C9 molecules, from 1 to 16, can polymerize in the MAC, depending on the cell surface involved.

As would be expected, stringent control mechanisms operate to control the complement system at various points under normal circumstances. Should these controls not function, disease results.

Cytokines Are Also Nonspecific Factors Important in Defense

Cytokines are molecular messages essential for communication between cells and for many other purposes besides defense. They were first discovered several decades ago, when it was recognized that cultured lymphocytes that were stimulated to an activated state secreted substances with biological activities into the culture medium. These substances were referred to as *lymphokines* because they were secreted by lymphocytes. Soon, macrophages and monocytes were also observed to secrete biologically active substances, which were called *monokines* after the cells that secreted them. Over the years, the number of cell types known to be capable of secreting such substances has increased, and the number of known substances has grown considerably. Currently, all of these substances are known collectively as *cytokines*. These are low molecular weight proteins and glycoproteins that act as messages between cells, especially immunologically active cells, and have many other biological activities. Table 15.4 lists some important cytokines, their sources, and their effects. We will return to these later as we discuss various facets of the immune response.

The groups of cytokines include interferons, interleukins, colony-stimulating factors, and others. They often act together or in sequence, in complex fashion.

Interferons are of special interest, as they are antiviral glycoproteins. Three types are known. One type, interferon alpha, is produced by various white blood cells; a second type, interferon beta, is made by tissue cells called fibroblasts; and a third type, interferon gamma, is made by T lymphocytes. In addition to being antiviral, gamma interferon also functions in development and regulation of the specific immune response.

The antiviral effects of all types of interferon are of value very early during a viral infection, before antibodies or immune lymphocytes are produced. Interferon is made soon after the start of viral infection, reaches a peak amount at about three days, and then decreases to almost nothing by a week after infection. It is nonspecific with respect to the virus; thus, it protects against most viral infections.

Virus infection of a cell results in the accumulation of double stranded RNA, which signals the cell to produce interferon (figure 15.7). The molecules of interferon act on other cells in the vicinity to prevent replication of the virus within those cells. First, the interferon molecules attach to receptors on the cell surface. Second, the cell responds by activating genes responsible for producing enzymes that degrade messenger RNA of both the virus and the host and inhibit protein synthesis in other ways. Viral replication is prevented by these antiviral proteins, but cellular protein synthesis is also inhibited, and the host cell dies in the process. Noninfected cells are not affected because the action of the enzymes requires the presence of double-stranded RNA formed in virus-infected cells, not normal cells. The end result is that virus replication is limited, at the sacrifice of some host cells. Other mechanisms of interferon activity against viruses are under study.

Clearly, it is of value to be able to use interferon in the treatment or prevention of viral infections, but for a time this was not possible because interferons, although nonspecific against viruses, are quite species-specific with regard to the host. That is to say, only human interferons are effective in human cells. Many other biologically active substances can be made in animal species and still be active in human cells, but not interferons. This problem has been overcome by using genetic engineering techniques. Human genes for interferons have been inserted into bacteria and yeast cells, which are then grown in huge quantities to produce large amounts of human interferons. These interferons can then be purified and used therapeutically, not just against virus infections but also against cancers and in various other ways. A number of examples of the therapeutic uses of interferons can be found in subsequent chapters.

Interleukins function in many different ways. They contribute to the inflammatory response, discussed next, and in various ways to the specific immune response. More than a dozen interleukins have been studied. Only a few examples are mentioned in table 15.4.

Colony-stimulating factors were mentioned previously as being important in the multiplication and differentiation of certain cells. During the immune response when leukocytes are needed in larger than normal numbers, colony-stimulating factors and other cytokines act to direct immature cells into the appropriate maturation pathways. Several different colony-stimulating factors are known, acting on different kinds of cells.

Tumor necrosis factors (TNF), aside from killing some tumor cells, act in a variety of ways to influence immune responses. For example, TNF alpha is stored in mast cell granules and released during inflammation. It recruits increased numbers of neutrophils into sites of inflammation, and it can be antiviral. Like interleukin-1, it can induce fever. Both TNF alpha and TNF beta act in many different ways.

Table 15.4 Some Important Cytokines

Cytokine	Source	Effects
Interferon alpha	Leukocytes	Antiviral
Interferon beta	Fibroblasts	Antiviral
Interferon gamma	T lymphocytes	Antiviral; macrophage activation; development and regulation of specific immune responses
Interleukin-1	Macrophages, fibroblasts	Proliferation of lymphocytes; macrophage production of cytokines; inducement of receptors for PMNs on blood vessel cells; inducement of fever
Interleukin-2	T lymphocytes	Changes in growth of lymphocytes; activation of natural killer cells
Interleukin-3	T lymphocytes, mast cells	Changes in growth of precursors of blood cells and also of mast cells
Colony-stimulating factors (CSFs)	Fibroblasts, endothelium, other cells	Stimulation of growth of different kinds of leukocytes
Tumor necrosis factors (TNF)		
Alpha	Macrophages, T lymphocytes, other cell types, mast cell granules	Cytotoxicity for some tumor cells; regulation of certain immune functions; inducement of fever; chemotacticity for granulocytes
Beta	T lymphocytes	Killing of target cells by T cytotoxic cells and natural killer cells

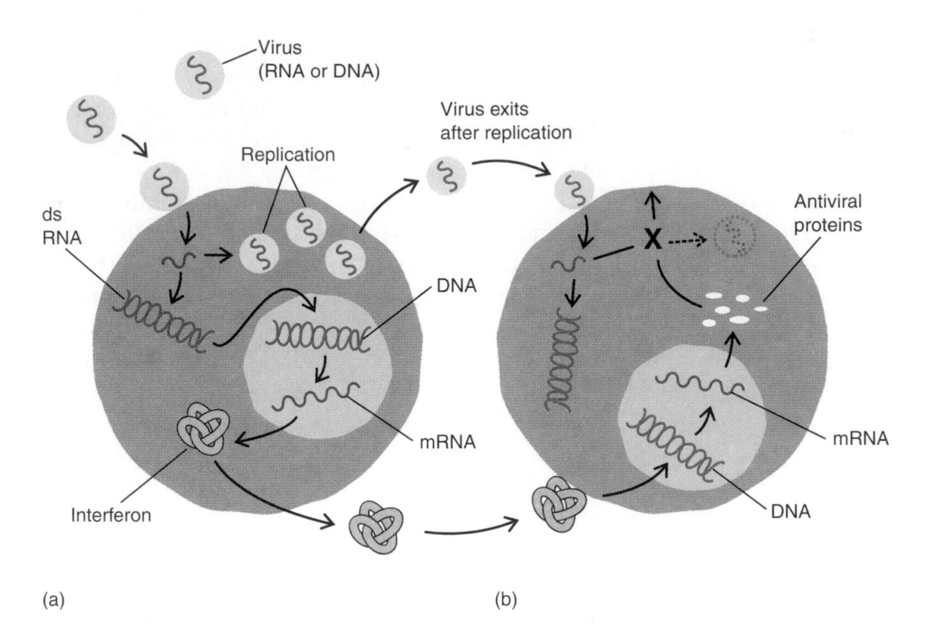

Figure 15.7
Mechanism of the antiviral activity of interferons. Interferons are small glycoproteins made by virus-infected cells that act on nearby cells causing them to produce antiviral proteins. These proteins are enzymes that inhibit virus replication in various ways. (*a*) Virus-infected host cell. Double-stranded RNA made during virus replication appears to initiate the production of interferon in the virus-infected cell. (*b*) The interferon reaches nearby cells and interacts with receptors on the cell membranes, inducing the cell to produce antiviral proteins, thereby preventing viral replication.

REVIEW QUESTIONS

1. What are the benefits of saliva in protection against infection?
2. Is lysozyme more effective against gram-positive bacteria, gram-negative bacteria, or viruses? Why?
3. How can the activation of a few molecules of C1 result in the formation of thousands of molecules of C3?
4. Are gram-positive or gram-negative bacteria more susceptible to complement lysis? Why?
5. Name at least two important cytokines and give their effects.

Inflammation Is an Early Reaction to Injury

Once an infectious agent has penetrated external barriers such as the skin or mucous membranes and has entered the tissues, the first host response is a nonspecific reaction to injury called the **inflammatory response,** or **inflammation.** Everyone is familiar with the signs of inflammation; in fact, the four cardinal signs were described by the Roman physician Celsus in the first century A.D. They are **redness, heat, swelling,** and **pain.** A fifth sign, loss of function, is sometimes present.

The same sequence of events occurs in response to any injury, whether caused by invading bacteria, burns, trauma, or other stimuli. Figure 15.8 and table 15.5 give the general sequence of events that takes place during inflammation. This response is protective in that useful substances and cells from the blood are delivered promptly to the area where they are needed. The extent of the inflammation varies, depending on the nature of the injury, but the response is local in the area of injury, begins immediately upon injury, and increases in intensity usually over a short period of time, although chronic (long-lasting) inflammation can occur.

The inflammatory response often activates coagulation proteins in the blood to produce clotting in the small vessels in the area. This helps to stop the spread of infection.

Table 15.5 Principal Events of the Inflammatory Process

Event	Effects
Tissue injury	Release of kinins,* prostaglandins,† and other chemicals that increase the permeability of adjacent small blood vessels
Blood vessel dilation and increased permeability to plasma, which may clot	Swelling of the tissues resulting from the leakage of plasma; elevated temperature of the region as a result of increased blood flow through the dilated vessels; redness for the same reason; pain from increased fluid in the tissues and from direct effect of chemicals on sensory nerve endings
Circulating white blood cells adherence to the walls of the altered blood vessels	White blood cell chemotactic migration through the vessel walls and to the area of injury to induce phagocytosis of foreign material and tissue debris and to initiate antibody production

*kinins—peptides cleaved enzymatically from certain proteins; act on blood vessels, causing dilation and increased permeability to plasma

†prostaglandins—long-chained fatty acids with effects on blood vessels similar to those of kinins and histamine

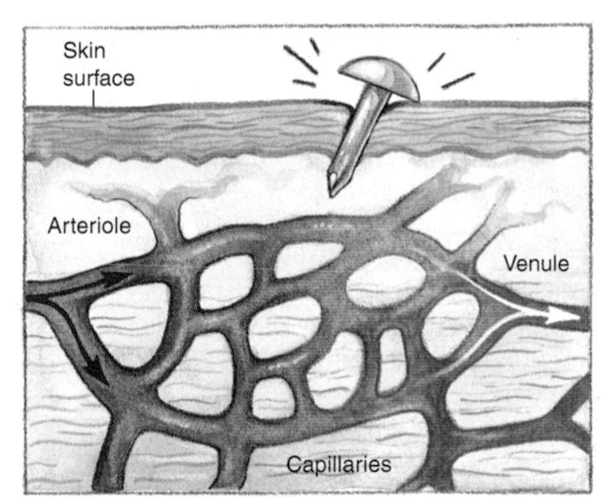

(a) Normal blood flow in the tissues as injury occurs.

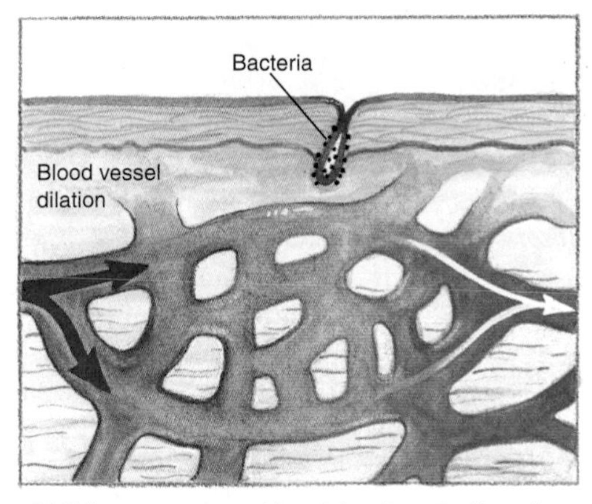

(b) Substances released from injured mast cells and as a result of the inflammatory reaction cause dilation of small blood vessels and increased blood flow in the immediate area.

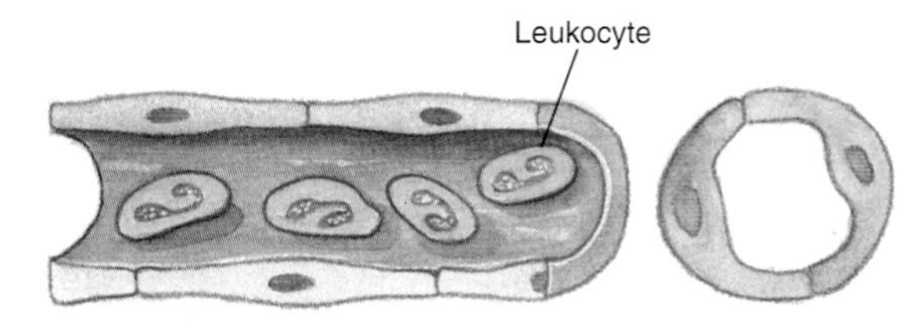

(c) Small blood vessels normally have tight junctions between the lining cells and blood stays within the vessels.

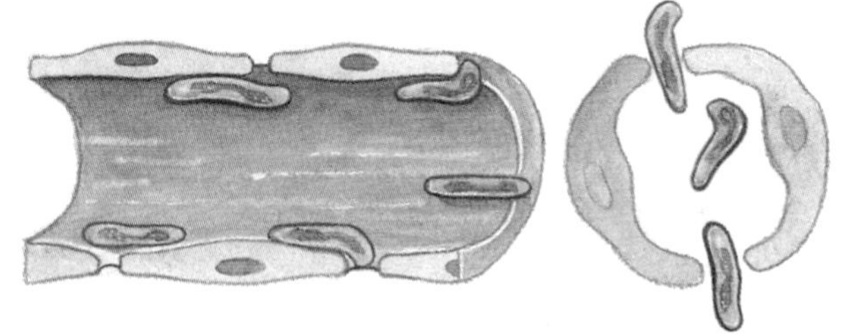

(d) During inflammation, the dilation the of vessels makes openings between cells lining the vessels. Leukocytes attach to the altered vessels and migrate through the openings between the cells into surrounding tissues. Some substances released during inflammation are chemotactic and attract certain leukocytes to the area of inflammation.

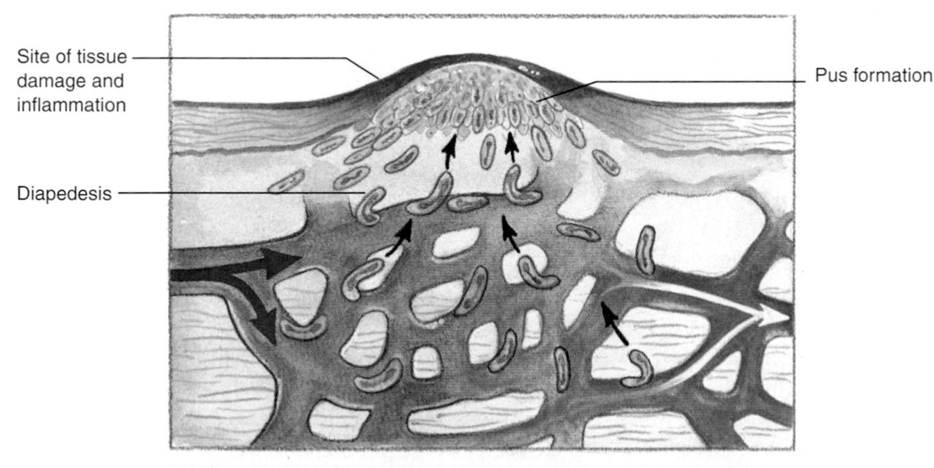

(e) The attraction of leukocytes causes them to move to the site of damage and inflammation. Collections of dead leukocytes and tissue debris make up the pus often found at sites of an active inflammatory response.

Figure 15.8
Inflammatory response.

During inflammation, C3a and C5a components of complement cause the release of chemicals from tissue mast cell granules. These chemicals, in turn, increase permeability of the small vessels, and this leads to increased blood flow and antimicrobial proteins in the area. Some of the chemicals cause the production of receptors for leukocytes on blood vessel walls in the area of the inflammation. Circulating leukocytes adhere to these receptors on the inner walls of the altered blood vessels and then migrate out of the blood vessels through the dilated permeable vessel walls in response to chemical attractants. This response is termed **chemotaxis.** The first kind of leukocyte to be lured from the circulation is the neutrophil (PMN). Soon after injury, neutrophils predominate in the area of inflammation. After the influx of neutrophils, monocytes and some lymphocytes accumulate in the area. Both the neutrophils and the monocytes (which mature into macrophages at the site of infection) actively phagocytize foreign materials.

Phagocytosis Is Essential to Nonspecific Defense

Phagocytosis involves a series of complex steps necessary for phagocytes to engulf and kill invading microorganisms (figure 15.9). First, the phagocyte must be attracted to the microorganism or virus.

Certain chemical products of microorganisms, components of the complement system (C5a), and phospholipids released by injured mammalian cell membranes act as chemoattractants to the phagocytes.

Second, once the phagocytic cell comes into contact with the microorganism, it must attach to it in order to engulf it efficiently. The C3b component of complement helps attach bacteria or other particles to phagocytes via C3b receptors on the phagocytic cells. The microorganism is then engulfed by the phagocyte into a vacuole known as a **phagosome.**

Other membrane-bound bodies within the phagocyte, called **granules** in neutrophils and **lysosomes** in mononuclear phagocytes, contain a variety of digestive enzymes including lysozyme and proteases. The granules or lysosomes come into contact with and then fuse with the phagosome and release their enzymes into the vacuole resulting from this fusion, known as a **phagolysosome.**

Within the phagolysosome, oxygen consumption increases enormously as sugars are oxidized via the TCA cycle, with the production of highly toxic oxygen products such as superoxide anions, hydrogen peroxide, singlet oxygen, and hydroxyl radicals. In mouse leukocytes, toxic products of nitrogen metabolism, such as nitric oxide, are also produced and are highly antimicrobial; however these products have not yet been found in human leukocytes. As the oxygen level declines, the metabolic pathway switches to fermentation with the production of lactic acid, which lowers the pH. Within the phagolysosome, various enzymes degrade the peptidoglycan of the bacterial cell walls and other macromolecular components of the cell. In this way, many microorganisms taken up by the phagocytic cell are killed and digested. Following the digestion of the microorganisms, the contents of the vacuole are excreted by exocytosis. As mentioned previously, once neutrophils have used their granules, they do not make more and the cells die, but mononuclear phagocytes can regenerate their lysosomes in order to meet the challenge of another bacterial attack.

Surprisingly, some microorganisms are not only able to resist all these defenses, but they can actually multiply within mononuclear phagocytes. This ability will be discussed in more detail later.

Physiological Change Affects the Immune Response

The host's metabolism is profoundly altered in response to infection, and most of the changes are geared to meet the needs of host defenses. The most significant changes include an alteration of temperature regulation leading to fever, redistribution of iron in the body, and changes in protein and carbohydrate metabolism.

• Phagocytosis, p. 292

Fever

Fever is one of the strongest indications of infectious disease, although significant infections can occur without it. There is abundant evidence that fever is an important host defense mechanism in a number of vertebrates, including humans. Within the human body, the temperature is normally kept within a narrow range by a temperature-regulating center in the hypothalamus[1] of the brain. The hypothalamus controls temperature by regulating the blood flow to the skin, the amount of sweating, and respiration. In this way, the heat produced by metabolism is conserved or lost in order to maintain a fairly constant temperature. In infection, the regulating center continues to function, but the body's thermostat is "set" at higher levels. This higher "setting" occurs because certain components of invading microorganisms attach to receptors on phagocytic cells thereby stimulating these cells to release the cytokine called **interleukin-1 (IL-1).** IL-1 is carried by the bloodstream to the hypothalamus, where it acts as a message to the hypothalamus that microorganisms have invaded the body. **Tumor necrosis factor alpha** can also induce fever. The temperature-regulating center responds by raising body temperature. The resulting fever inhibits the growth of many pathogens by at least two mechanisms: (1) elevating the temperature above the optimum growth temperature of the pathogen and (2) activating and speeding up a number of other body defenses.

The adverse effect of fever on pathogens correlates in part with their ideal growth temperature. Bacteria that grow best at 37°C are less likely to cause disease in people with fever. Recall from chapter 4 that the growth rate of bacteria often declines sharply as the temperature rises above their optimum growth temperature. A slower growth rate allows more time for other host defenses to destroy the invaders.

Recall also that a moderate increase in temperature increases the rate of enzymatic reactions. It is thus not surprising that fever has been shown to enhance the inflammatory response, phagocytic killing by leukocytes, the multiplication of lymphocytes, the release of substances that attract neutrophils, and the production of interferons and antibodies. Release of leukocytes into the blood from bone marrow reserves is also enhanced. Further, fever increases bacterial requirements for iron while decreasing the host's absorption of iron from food. For all these reasons, it is generally a good idea to avoid using drugs to reduce the fever of infectious diseases except when absolutely necessary, as when the fever is very high and sustained.

[1]Hypothalamus—part of the brain concerned with temperature regulation, regulation of metabolism, and certain other functions

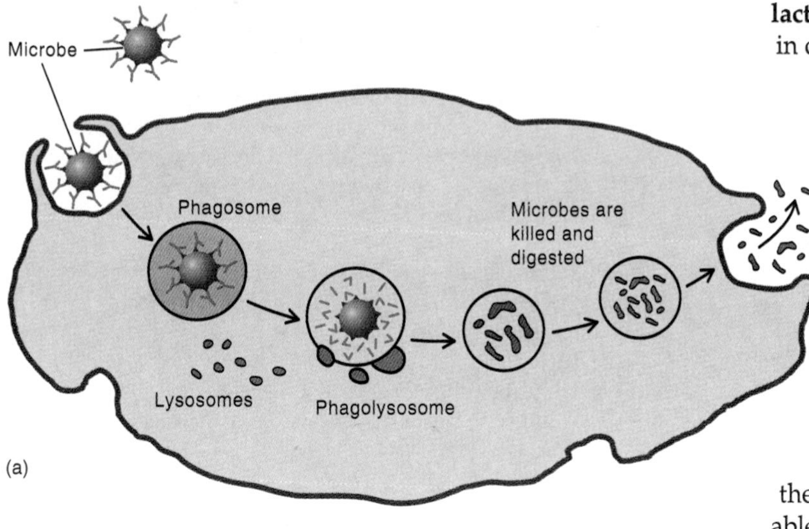

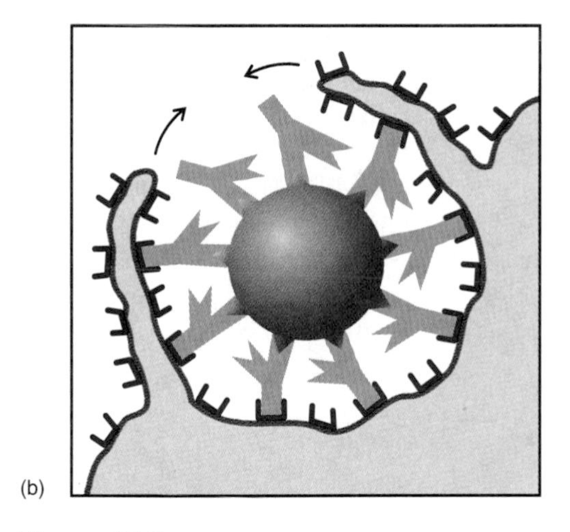

Figure 15.9

Phagocytosis and intracellular digestion. (*a*) Both neutrophils and macrophages take up particulate material in phagosomes, which then fuse with lysosomes or, in some cases, granules containing enzymes and antimicrobial substances. The fused phagosome and lysosome is the phagolysosome. If bacteria are phagocytosed, they may be killed by the contents of the phagolysosome. The resulting debris is removed from the cell by exocytosis. (*b*) In order to phagocytize efficiently, the cell membrane of the phagocyte must make multiple contacts with the particle to be ingested. This pulls the cytoplasmic membrane around the particle with what has been described as a "zipper effect." In this diagram, multiple contacts are made by Fc portions of attached antibody interacting with Fc receptors on the cell. Similarly, C3b on the microbe can react with C3b receptors on the phagocyte to facilitate phagocytosis.

Changes in Iron Metabolism

The ability to limit the amount of iron available to invading bacteria is a major nonspecific defense mechanism. Most bacteria as well as humans require iron for the functioning of certain important enzymes. Iron is found in human blood, but it is attached to specific proteins such as **transferrin** and **lactoferrin.** Bacteria must compete with these proteins for iron in order to obtain enough to grow. Some bacteria can successfully compete for the iron. For example, *Neisseria gonorrhoeae,* the cause of gonorrhea, can get iron directly from the carrier molecules, transferrin and lactoferrin. Unusually high concentrations of iron make it easier for the bacteria to obtain the iron they need.

Intracellular bacteria also must have iron. Iron bound to transferrin gets into cells of the host by means of transferrin receptors on the cell surface. These receptors bind the transferrin-iron complex and allow it into the cell. When mononuclear phagocytes ingest bacteria, they decrease the number of transferrin receptors on their surfaces, thereby decreasing the amount of iron available within the cell for use of the microorganisms. Also, lactoferrin in the phagolysosomes can compete successfully with most bacteria for intracellular iron. Iron administered in the form of medication can increase the susceptibility of people to infection, because it increases the amount of iron available to bacteria, as well as to the host.

Changes in Protein and Carbohydrate Metabolism

The response to infection requires that the host rapidly synthesize many different defensive proteins, including interferons and other cytokines, structural cell proteins and enzymes for cell division, enzymes involved in the lysis of microorganisms, and antibodies. These syntheses and the demands of phagocytosis require that cells of the host have abundant supplies of energy and raw materials for biosynthetic reactions. Energy is derived from both glycolysis and the TCA cycle as well as metabolism of fatty acids released during breakdown of fat tissue. Amino acids used in the synthesis of various proteins are generated primarily from hydrolysis of the proteins of muscle and other body tissue, as witnessed by the weight loss and weakness commonly seen in individuals with prolonged infectious diseases. The utilization of foodstuffs for energy increases the need for oxygen. Because vitamins are converted to coenzymes for the various enzymes of catabolism[1] and anabolism,[2] their levels decline in the blood. Severe infections can cause vitamin deficiency diseases, especially in people who are already malnourished.

REVIEW QUESTIONS

1. What are the four cardinal signs of inflammation?
2. How do white blood cells get into the tissues during an inflammatory response?
3. Explain some ways in which phagocytosis contributes to defense.
4. How does fever affect pathogens in the tissues?
5. Why is iron metabolism important in body defense?

[1]Catabolism—metabolic degradation of organic compounds
[2]Anabolism—biosynthetic reactions in a living organism

Summary

I. The Host Defends Itself: Nonspecific and Specific Defense Mechanisms

II. Cells and Tissues Involved in Immune Responses
   A. Cells of the blood and lymphoid system participate in immune responses.
   B. Leukocytes (white blood cells) are the cells primarily responsible for body defense. These include neutrophils, eosinophils, basophils, mononuclear phagocytes, and lymphocytes.

III. Innate Nonspecific Immunity
   A. Tissue barriers and chemical factors are important in nonspecific immunity.
      1. Physical barriers, such as skin and mucous membranes, are especially important in innate nonspecific immunity.
      2. Many nonspecific chemical antimicrobial factors, such as enzymes, the complement system, and cytokines, act together with physical barriers to prevent infection.
   B. Inflammation is an early reaction to injury. Inflammation is characterized by redness, heat, swelling, and pain.
   C. Phagocytosis is essential to nonspecific defense.
   D. Physiological change affects the immune response. Changes include
      1. fever;
      2. changes in iron metabolism; and
      3. changes in protein and carbohydrate metabolism.

Chapter Review Questions

1. Which occurs first, the nonspecific or the specific immune response?
2. Describe at least three kinds of leukocytes that participate in immune responses.
3. Discuss several important differences between PMNs and mononuclear phagocytes.
4. What is meant by innate nonspecific immunity?
5. How do the skin and mucous membranes function in nonspecific immunity?
6. Why are catalase-negative organisms usually more sensitive to killing by the peroxidase-peroxide-halide ion system than are catalase-positive organisms?
7. What is the difference between activation of complement by the classical pathway and activation by the alternative pathway? Which of these pathways functions in nonspecific immunity?
8. What effects do the small molecules C3a and C5a, formed during complement activation, have on inflammation?
9. What is the membrane-attack complex (MAC) of complement and what does it do?
10. Is fever deleterious or beneficial? Why?

Critical Thinking Questions

1. What are some differences between fever and the heat produced as a sign of inflammation?
2. Many paralyzed paraplegic persons develop recurrent urinary tract infections. Why might this occur?

Further Reading

Beck, G., and Habicht, G. S. 1996. *Scientific American:* Immunity and the invertebrates, 60–66.

Malaviya, R., et al. 1996. Mast cell modulation of neutrophil influx and bacterial clearance at sites of infection through TNF alpha. *Nature* 381: 77–79.

Roitt, Ivan M. 1994. *Essential Immunology,* 8th ed. Blackwell Scientific Publications, Oxford.

Salyers, Abigail A., and Whitt, Dixie D. 1994. Bacterial pathogenesis. A Molecular Approach. American Society for Microbiology, Washington, D.C.

Weiss, Alison. 1997. Mucosal immune defenses and the response of *Bordetella pertussis*. *ASM News* 63(1): 22–28.

16 Specific Immunity

KEY CONCEPTS

1. Specific immunity has memory and is more effective against a second or subsequent exposure to the same agent.
2. Antigens are substances that induce specific immune responses, either by cells or production of antibodies.
3. Antibodies are glycoproteins that react specifically with the antigens that induced their formation.
4. Lymphocytes are the cells primarily responsible for the specific immune response; B lymphocytes are concerned with making antibodies and T lymphocytes with cell mediated immunity.
5. The major histocompatibility antigens, which are genetically determined, allow cells to discriminate between self and foreign material and to communicate with each other during the immune response.
6. The antibody response involves recognition of antigen by immunoglobulin receptors on B cells, resulting in expansion of the B cell clone and antibody production.
7. Cell mediated immunity involves recognition of antigen by T cell receptors, resulting in expansion of the T cell clone and subsequent activities of the T cells produced.
8. Immunological tolerance denotes specific unresponsiveness to an antigen.
9. The immune response is carefully controlled by a multitude of mechanisms.

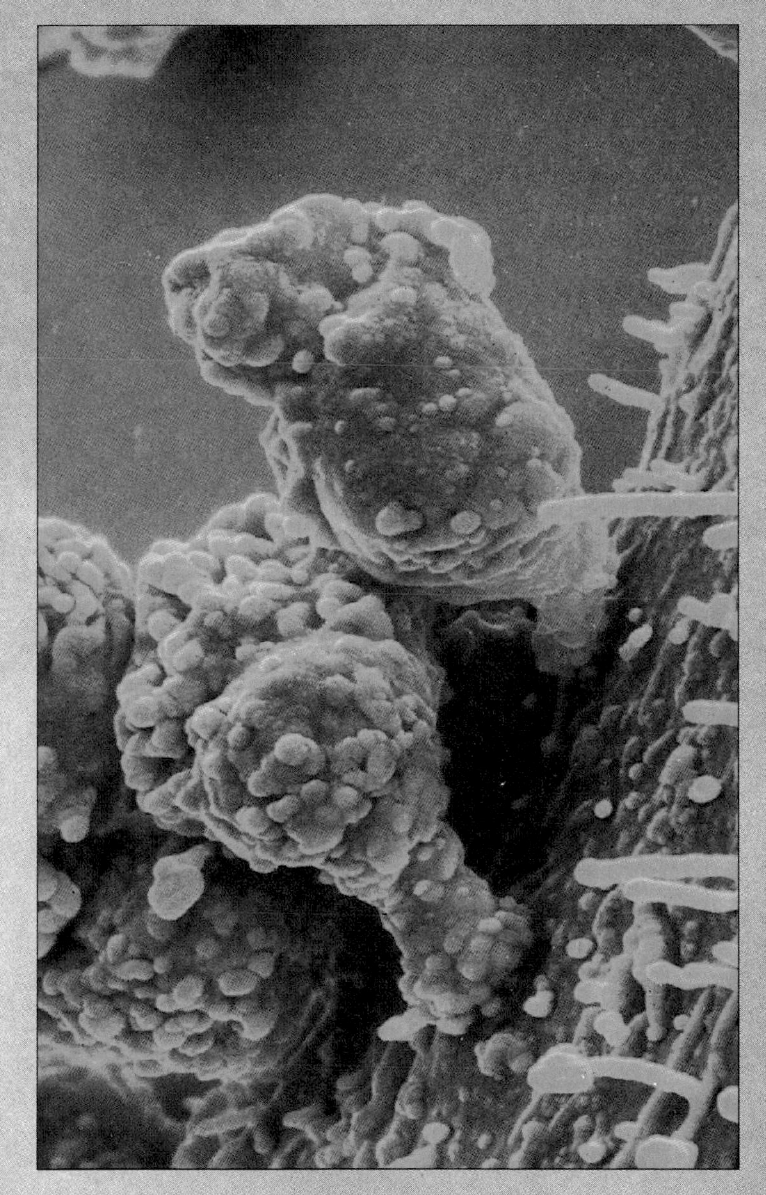

Cancer cell being attacked by a cytotoxic T cell.

Microbes in Motion

The following books and chapters in the *Microbes in Motion* CD-ROM may serve as a useful preview or supplement to your reading:

Viral Structure and Function: Viral Pathogenesis (Host Cell Response); Viral Detection.

A Glimpse of History

Near the end of the 19th century, diphtheria was a terrifying disease that killed many infants and small children. The first symptom of the disease was a sore throat, often followed by the development of a gray membrane in the throat that made breathing impossible. Death occurred rapidly, even in the absence of a membrane. Frederick Loeffler, working in Robert Koch's laboratory in Berlin, found club-shaped bacteria growing in the throat of people with the disease but not elsewhere in their bodies. He guessed that the organisms were making a poison that spread through the bloodstream. In Paris, at the Pasteur Institute, Emile Roux and Alexandre Yersin followed up by growing the bacteria in quantity and extracting the poison, or **toxin,** from strained culture fluids. The toxin injected into guinea pigs killed the animals.

Back in Berlin, Emil von Behring injected the diphtheria toxin into guinea pigs that had been previously inoculated with the bacteria and had recovered from diphtheria. These guinea pigs did not become ill from the toxin, suggesting to von Behring that something in their blood, which he called **antitoxin,** protected against the toxin. To test this theory, he mixed toxin with serum from a guinea pig that had recovered from diphtheria and injected this mixture into an animal that had not had the disease. The guinea pig remained well. In further experiments, he cured animals with diphtheria by giving them antitoxin.

The results of these experiments in animals were put to the test in people in late 1891, when an epidemic of diphtheria occurred in Berlin. On Christmas night in 1891, antitoxin was first given to an infected child who then recovered from the dread diphtheria. The substances in blood with antitoxin properties soon were given the more general name of **antibodies,** and materials that could generate antibodies were called **antigens.**

Emil von Behring received the first Nobel Prize in medicine in 1901 for this work on antibody therapy. It took many more decades of investigation before the biochemical nature of antibodies was elucidated. In 1972, Rodney Porter and Gerald Edelman were awarded the Nobel Prize for their part in determining the chemical structure of antibodies.

In the previous chapter, we discussed some mechanisms of **innate immunity**, present in all normal people, and ready to act whenever the body is infected or injured. These mechanisms are **nonspecific** with respect to the offending agent, and they are equally effective against a first attack or a subsequent attack by the invading organism. They provide a first line of defense, ever ready to protect.

While this early defense is at work, another even more effective response is being brought into play. This is the **specific acquired immune response,** so-called because it is directed only against the invading agent. In addition to being **specific,** this response has **memory,** so that it is more rapid and effective against a second or subsequent exposure to the same agent than it is against the first encounter with the agent. The memory response is also called an **anamnestic** or **secondary** response. For example, a first exposure to the measles virus causes a specific immune response that takes several weeks to develop fully and that usually halts the measles virus infection, but only after the virus has caused the disease known as *measles*. However, subsequent exposure to the same virus leads to a much faster and more effective memory response, so that a permanent state of immunity against measles exists. Similarly, infection with one strain of influenza virus induces immunity to that virus but not to other strains of influenza.

Substances that induce an immune response are called **antigens,** because they induce the formation of antibodies or, in other words, are *anti*body *gen*erators. **Antibodies** are glycoprotein molecules that react specifically with the antigen that induced them. Antigens may also cause the production of **activated (immune) T lymphocytes** that react specifically with the antigen. Antigens and antibodies or antigens and immune lymphocytes can only be defined in relation to each other. An **antibody response** is made by B cells, whereas specifically reactive T cells are responsible for the **cellular immune response.** Macrophages play important roles in both antibody and cellular immune responses.

Thus, the acquired immune response has the following fundamental features:

1. The response is highly specific.
2. The response has memory, in that once the host has responded to a specific antigen, a subsequent exposure to the same antigen is more rapid and effective, thereby quickly eliminating that antigen.
3. The response involves two sets of lymphocytes, the B and the T cells, acting together with macrophages, and it may be predominately an antibody (B cell) response or a cellular (T cell) immune response.

These and other features of the specific immune response will be discussed in more detail next.

The Nature of Antigens

Antigens and Haptens Have Special Properties

Antigens not only induce an immune response, but also react specifically with the antibodies or immune cells produced during the response. Antigens are usually **foreign** to the host and are recognized as such by the immune system. Most antigens are macromolecules of high molecular weight, such as proteins, polysaccharides, and occasionally lipids or nucleic acids. Molecules differ in their effectiveness in stimulating an immune response. Proteins and polysaccharides are generally good antigens, whereas lipids and nucleic acids are rarely antigenic alone. When complexed, as in a lipoprotein, lipids become better antigens. Substances of less than 10,000 daltons[1] do not make good antigens.

Although antigens are large molecules, the immune response is not directed toward the entire molecule but rather to particular chemical groups on the molecule known as **antigenic determinants** or **epitopes** (figure 16.1).

[1]Dalton—unit of weight equal to the weight of a single hydrogen atom

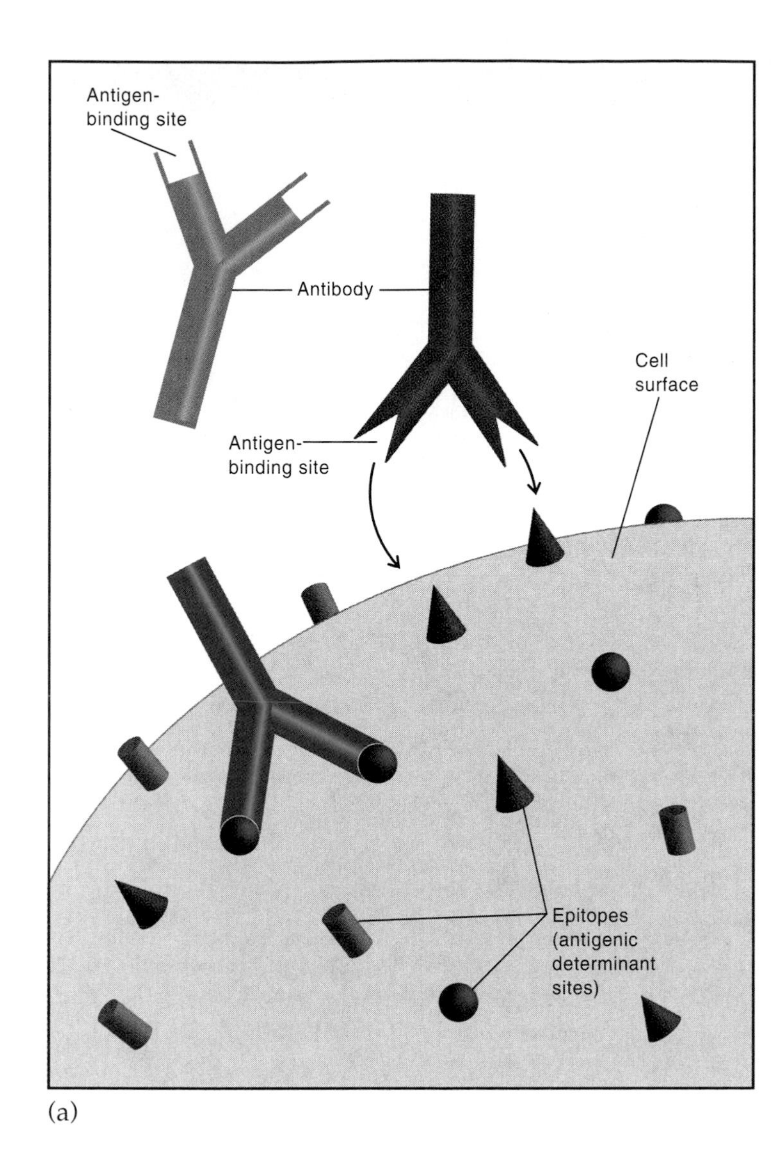

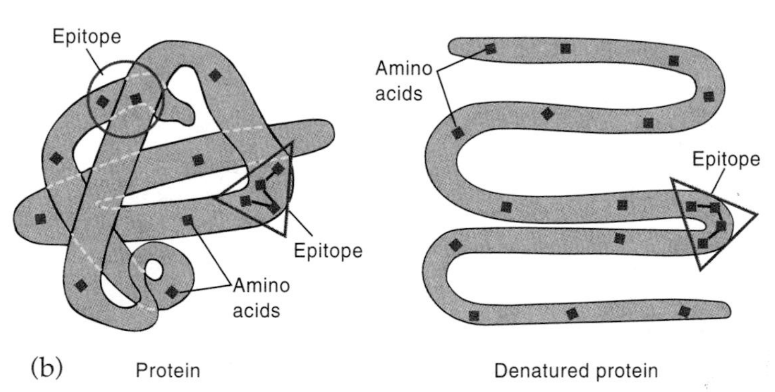

Figure 16.1
Schematic drawings of antigen and antibody. (*a*) Antibodies react with specific epitopes on an antigen. There must be a close complementary fit between the antigen-binding site on the antibody and the epitope on the antigen. Note that there may be a number of different epitopes, indicated by shape, and a number of the same epitopes on a single antigen molecule. (*b*) Protein antigen made of folded peptide chains. Two different epitopes are suggested by the colored shapes. An epitope may consist of a number of amino acids from the same part of the molecule (red) or widely separated amino acids from different parts of the chains that come together when folding occurs (blue). When the protein is denatured and loses its folded characteristic, the blue epitope is lost.

On large protein molecules, sequences of 10 to 25 amino acids act as epitopes, which are recognized by the antibodies. Complex structures, such as most bacterial cell walls, may have 100 or more epitopes.

While low molecular weight molecules by themselves are not antigenic, some small molecules called **haptens** can combine with large "carrier" molecules such as protein or polysaccharide to become antigenic and induce an immune response. The hapten alone can then react with the antibodies produced. Penicillin, for example, has a molecular weight of about 350 daltons and by itself is incapable of causing antibody formation. In the body, however, it is changed by enzymes so that it can combine with large protein carrier molecules. The penicillin then acts as an epitope and induces the formation of antibodies specific for it. The reaction of these antibodies with the hapten penicillin can result in allergic reactions in a low percentage of the population, precluding the further use of penicillin in these reactive individuals.

The Nature of Antibodies

Antibodies Are Immunoglobulins

Antibodies are glycoprotein molecules called **immunoglobulins** that have special properties that permit them to recognize antigenic epitopes specifically and to interact with antigens in particular ways. Immunoglobulins all have a similar molecular structure consisting of two heavy molecular weight polypeptide chains (H chains) and two light polypeptide chains (L chains) joined together to form a single molecule, a monomer. There are five classes of human immunoglobulin (Ig): IgM, IgG, IgA, IgD, and IgE. Of these, IgM is a pentamer, composed of five monomeric subunits, IgA is often a dimer of two subunits, and the others are monomers. Each class of antibody differs in some characteristics of its heavy chains and in its corresponding biological properties. Table 16.1 gives some properties of the different classes of immunoglobulins.

Since **IgG** has been most extensively studied, we will first examine the structure of IgG molecules. They are Y-shaped molecules consisting of two identical halves. Each half contains an H polypeptide chain and an L polypeptide chain. These two chains are held together by disulfide bonds as are the two halves of the molecule (figure. 16.2).

The Y shape of the antibody molecule diagrammed in figure 16.2 can actually be seen in electron micrographs. The very high magnification reveals a flexible or "hinge" region of the molecule at the fork of the Y. Antibody molecules combined with antigen molecules form different angles around the hinge, and the arms of the Y move. The arms can move out far enough to give the molecule a T shape. In

Table 16.1 Functions of the Various Classes of Human Immunoglobulins

Class	Percent of Total Antibody	Functions	Properties
IgG	80%–85%	Protection within the body; causes lysis or removal of antigens; neutralizes viruses and toxins	Specific attachment to phagocytes; complement fixation; ability to cross the placenta
IgM	5%–10%	Protection within the body, as with IgG; especially effective in agglutinating antigens and in activating complement	Complement fixation
IgA	10%	Protection of mucous membranes	Secretion into saliva, milk, mucus, and other external secretions
IgD	1%–3%	Facilitating the development and maturation of the antibody response	Antigen receptor on B lymphocytes
IgE	0.05%	Involved in hypersensitivity and allergic reactions	Specific attachment to mast cells and basophils

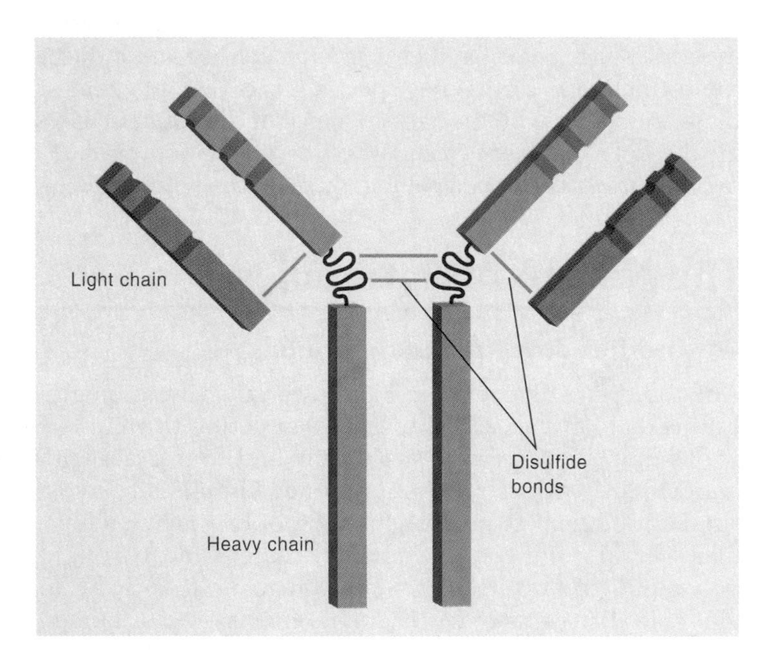

Figure 16.2
Immunoglobulin molecule. The molecule consists of two identical light chains and two identical heavy chains held together by disulfide bonds.

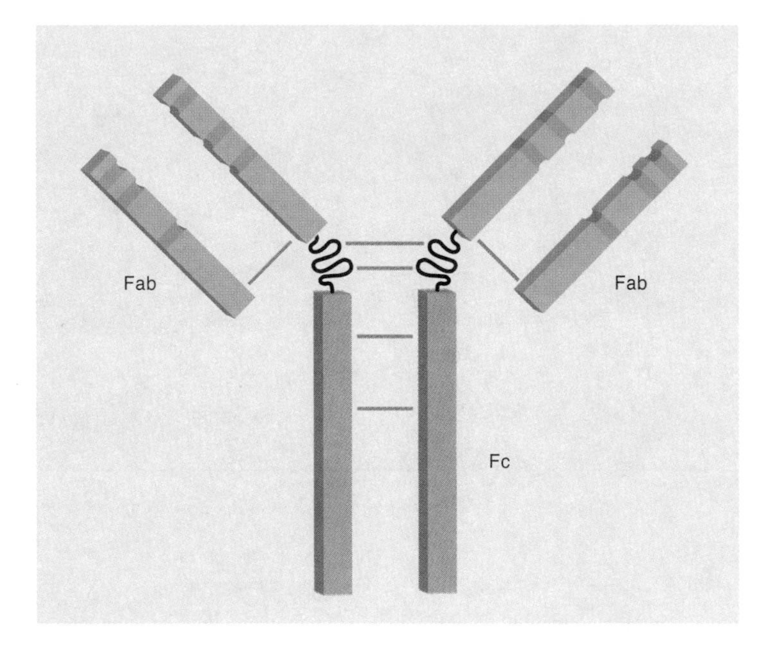

Figure 16.3
The Fc region of immunoglobulin is part of the constant amino acid component. The proteolytic enzyme papain cleaves the immunoglobulin molecule into two fragments: 2 Fab (fragment antigen binding) and 1 Fc (fragment crystallizable). Fc is crystallizable because it consists of constant amino acid sequences, virtually the same in all molecules of the same class. The Fc region of immunoglobulins is responsible for the special characteristics of each class, such as the ability to fix complement, to cross the placenta, and to interact with Fc receptors on cells.

fact, recent evidence indicates that antibody molecules are not only flexible in the hinge region, but they may be quite pliable in other places as well, rather like clay. For example, some antibodies have been shown to bind to proteins inside crevices on the surface of rhinovirus virions.

Each polypeptide chain in immunoglobulin molecules consists of a variable region and a constant region. The constant regions of H chains for all molecules of the same immunoglobulin class have virtually the same amino acid sequence and are responsible for the characteristic functions and properties of that class. Part of this area of the immunoglobulin molecule is called the **Fc region** (figure 16.3). The different classes of immunoglobulins differ in their constant regions.

The variable regions on the outer end of each arm of the Y on both the H and L chains contain the section of the molecule that interacts with the epitopes of antigens. Since each epitope differs from other epitopes, the variable regions of the antibody also differ in their amino acid sequences. This difference accounts for the specificity of antibodies.

An IgG molecule has two identical **antigen-binding sites,** one at the end of each arm of the Y. The differences in the variable region are primarily found in three small **hypervariable regions.** Because of the folding of the polypeptide chains, these areas lie near the end of the arms, where they can easily interact with antigen (figure 16.4).

The interaction of antigens and antibodies depends on a **close complementarity** between the antigen-binding sites and the specific antigenic epitopes (figure 16.5). The fit must be close and complementary because the bonds that bind the antibody with the antigen are weak and of short range, such as hydrogen bonds. Many such bonds must be formed in order to hold the two together. As a consequence, the antigen-antibody reaction is always reversible. No chemical changes occur in either the antigen or the antibody when the interaction between them occurs.

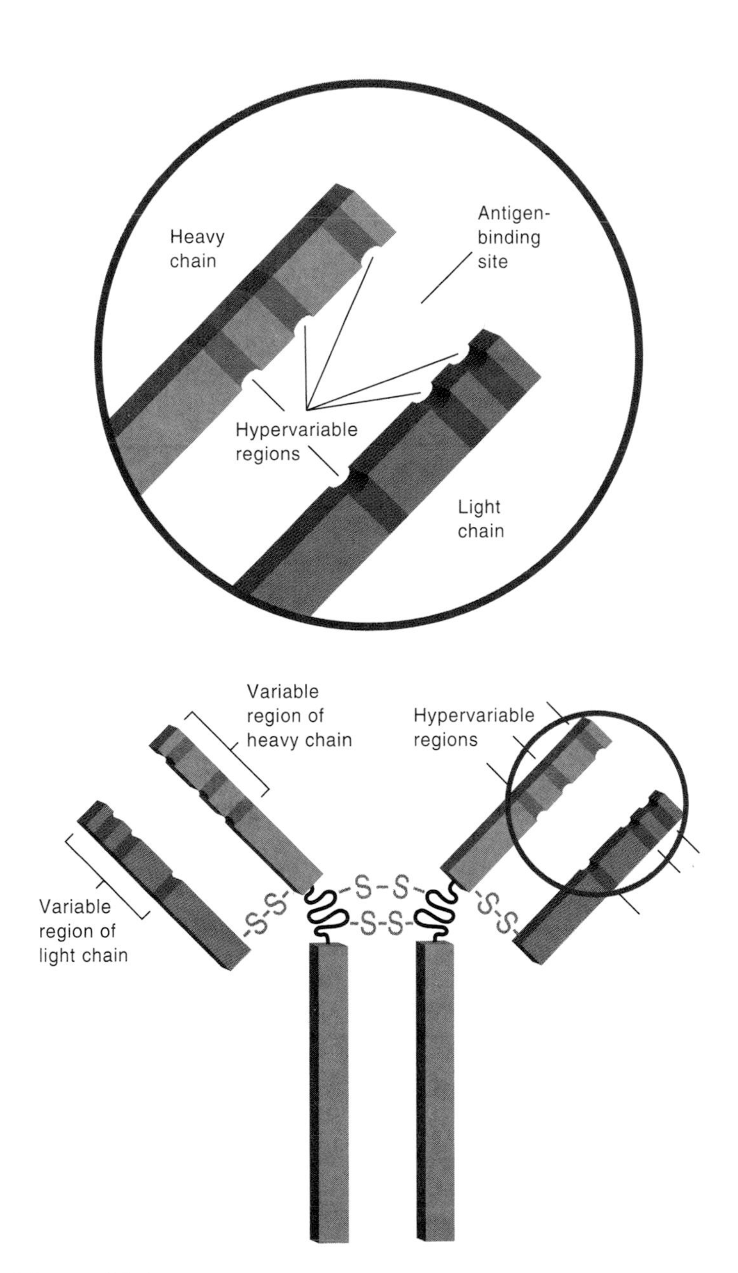

Figure 16.4
Antigen-binding sites of an IgG molecule. Hypervariability of amino acids in the terminal regions of both the heavy and the light chains accounts for the specificity of antibodies.

Immunoglobulin Classes Differ in Their Properties

As noted, all five classes of immunoglobulin molecules have the same basic structure: two identical light polypeptide chains connected by disulfide bonds to two identical heavy polypeptide chains. The amino acid sequences and carbohydrate content of the heavy chains vary with each class. Polymers of the basic subunit can form in IgM and IgA. Some properties of each immunoglobulin class are outlined next. The functions of immunoglobulin antibodies are discussed more fully in the next chapter.

Immunoglobulin G (IgG)

IgG accounts for about 80% of the total immunoglobulin in normal immunocompetent people over the age of two years. Most of it is in the circulation. It is the only class of immunoglobulin that can cross the placenta from the

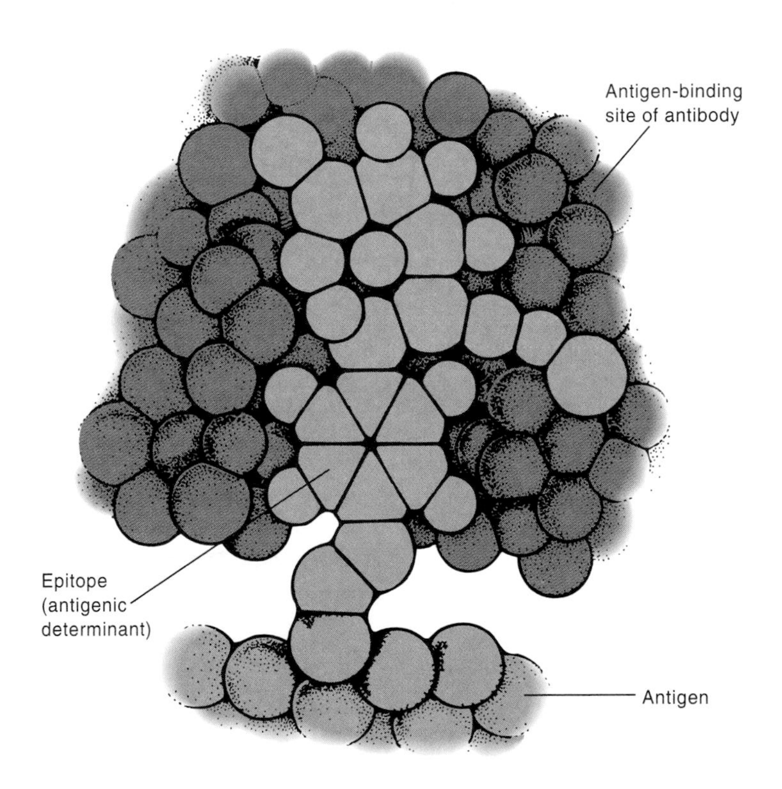

Figure 16.5
Antigen epitope with a complementary antigen-binding site of an antibody. The schematic drawing shows how a protruding part of the antigen molecule, the epitope, fits into the antigen-binding site of the antibody molecule to bring the atoms of each into close proximity. With such closeness, weak chemical bonds such as hydrogen bonds can form and hold the antigen and antibody molecules together. The closer the fit, the more bonds that are formed and the greater the affinity (strength of binding) of the reaction.

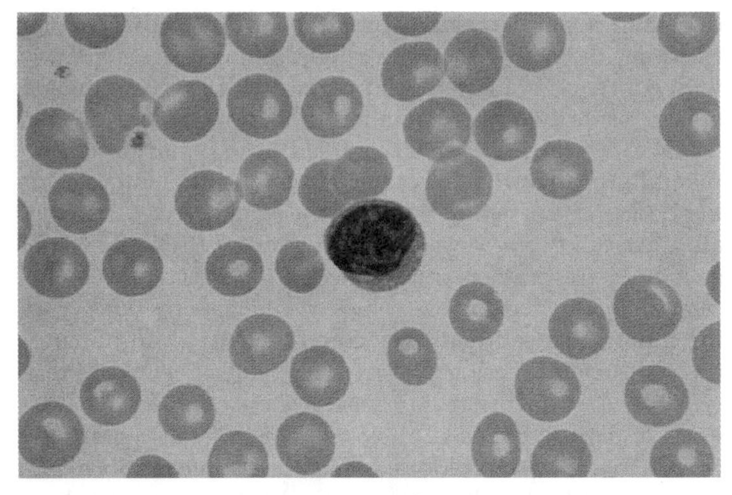
(a)

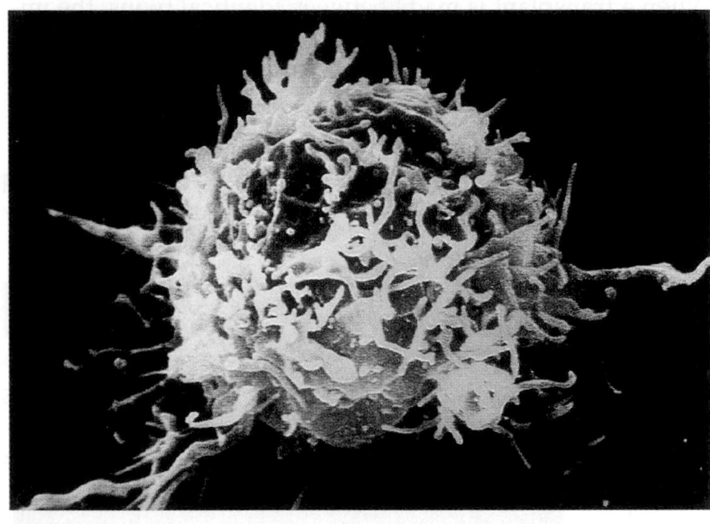
(b)

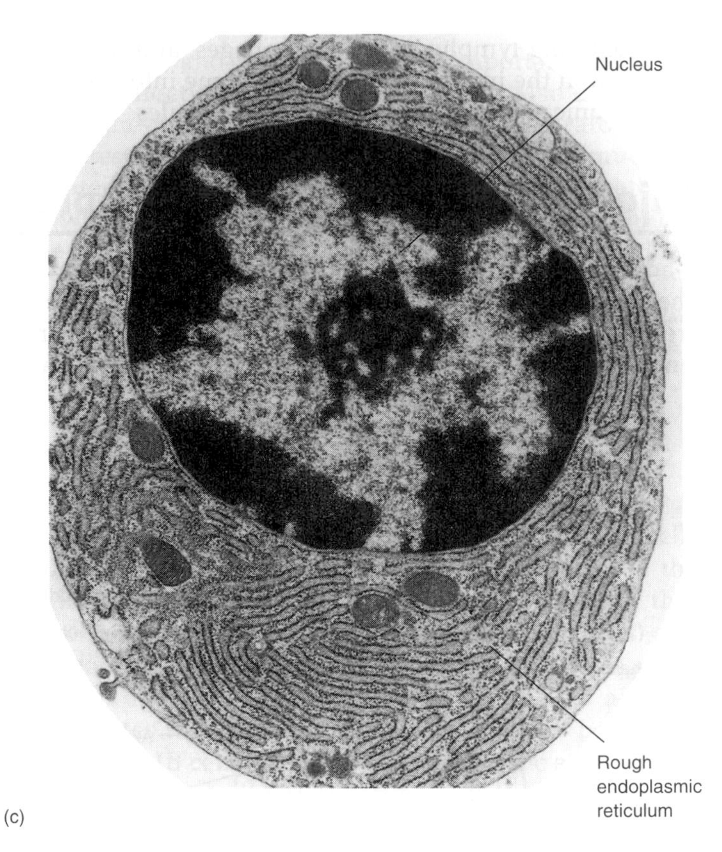

(c)

Figure 16.12
Lymphocytes and plasma cells. (*a*) Light micrograph of a T lymphocyte. The morphology is the same as that of a B lymphocyte. (*b*) Scanning electron micrograph of a T lymphocyte. T and B lymphocytes cannot be distinguished by microscopy. (*c*) Plasma cell, a form of B cell that is highly differentiated to produce large amounts of antibody. Note the extensive endoplasmic reticulum, the protein-synthesizing equipment of the cell. All of the antibodies produced by a single plasma cell have the same specificity.

cells are transplanted between nonidentical people, the immune system recognizes the histocompatibility antigens on the transplanted cells as foreign (nonself) and makes a vigorous immune response, leading to rejection of the tissue. These observations seemed strange at the time of discovery. Since transplanting tissues is very unnatural, why should the body have such a complex system for rejecting the transplant? A great deal of research in many areas of immunology has uncovered some reasons. The MHC gene products are important in many ways that do not involve tissue transplantation. They are important in cell-to-cell contact, essential for the functioning of the immune system. The MHC molecules on one cell interact with those on the other cell to allow responding cells to recognize self or foreign substances (nonself), accepted if self and disposed of if recognized as foreign.

Two classes of MHC molecules, class I and class II, are particularly important. Class I molecules are found on all nucleated cells and thus virtually all human cells except mature red blood cells, which lack a nucleus. They are called *leukocyte antigens* because they are found in large quantities on white blood cells, but they are also present on almost all other cells. The CD8 markers previously discussed function as receptors for MHC class I substances. Class II molecules are found principally on macrophages and B cells, along with class I molecules. CD4 markers act as receptors for MHC class II molecules. The markers indicated in table 16.2 can be used to distinguish between T and B lymphocytes.

Development of the Antibody Response

We will now examine the events that occur during the development of an antibody response. Later we will discuss the second type of specific immunity, cell-mediated immunity, and consider the similarities and differences between the antibody and cell-mediated responses.

Complex Events Occur During the Antibody Response

In the antibody response, the antigen may be soluble molecules or antigen epitopes on cells. For example, a soluble antigen could be a protein toxin secreted by bacteria, while the protein flagella found on bacteria could be an example of antigen that is part of a cell. In either case, the protein antigen, being a large molecule, will probably have a number of different epitopes on its surface. Individuals who have never been exposed to this antigen will have a few B lymphocytes with surface antibodies that recognize each of the foreign epitopes of the antigen. Each B lymphocyte will interact with the epitope it recognizes.

Protein antigens are **thymus-dependent antigens** (also referred to as *T-cell-dependent*); that is, they require the cooperation of T helper (Th) cells in order for the B cells to respond to the antigen by making antibodies.

Extracellular antigens, such as proteins secreted by invading bacteria, must be taken up and **processed** by macrophages, dendritic cells, or other cells, collectively referred to as **antigen presenting cells.** The antigenic epitopes must be processed and **presented** to the Th cells in special ways. For example, Th cells recognize antigen only when it is associated with major histocompatibility complex class II molecules. Both class I and class II antigens have a groove in one side of the molecule where fragments of antigen are inserted (figure 16.13 *a*). When the Th cell recognizes the fragment presented in the class II molecule groove, it secretes cytokines that activate the B cell to multiply and begin to secrete specific antibodies (figure 16.13 *b*). B cell recognition of an antigen alone does not result in antibody formation. The signal from the Th cell or elsewhere is necessary for the cells to multiply and to synthesize antibodies.

B cells can produce antibodies to two kinds of **thymus-independent antigens** in addition to thymus-dependent antigens. Bacterial lipopolysaccharides are examples of type 1 thymus-independent antigens. These large molecules have O antigens that are recognized by B cells. The lipid A part of the lipopolysaccharide molecule stimulates the B cells to multiply, thus substituting for the function of the Th cells to stimulate the multiplication of the B cells. Type 2 thymus-independent antigens are large molecules with the same antigenic determinant repeated many times, such as the polysaccharide capsules of pneumococci. At least 12 to 16 evenly spaced repeating epitopes on the antigen are needed to combine with antibodies on the B cell surface to induce a good response. The cross-linking of surface antibodies activates enzymes that lead to cell multiplication, substituting for the Th stimulus given for thymus-dependent antigens.

• **Bacterial lipopolysaccharides, pp. 360, 366**

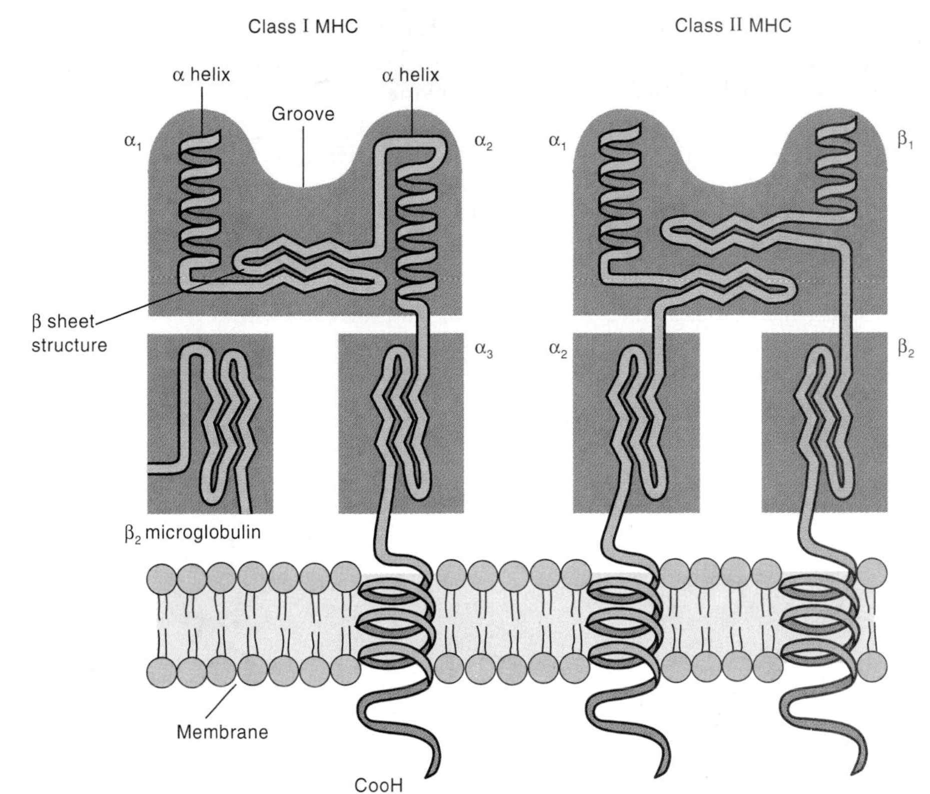

Figure 16.13
Structure of major histocompatibility (MHC) antigens. The lower part of the MHC antigen is inserted into the cell membrane. The upper part of the class I MHC molecule forms a distinct groove or cavity that usually contains a fragment of an antigen, a peptide of 8 to 10 amino acids. Peptides of antigens are recognized by T cytotoxic cells only when they are complexed with class I MHC molecules in this manner. Class II MHC molecules are also characterized by a groove that contains fragments of antigen, but the class II grooves can accommodate peptides 12 to 25 amino acids in length. T helper cells recognize antigen fragments only when they are complexed with class II MHC molecules.

Clonal Selection Occurs During Immune Responses

Thus, during an antibody response, B lymphocytes recognize the antigen epitope by means of specific antibody molecules on the surface of the B cells. This recognition alone does not initiate antibody production but, with specific signals given by the Th cells or in other ways, it induces the B cells to multiply. Each cell divides to give two cells with the same antibody specificity, then four, then eight, and so on to form large **clones** of B lymphocytes producing the same antibody. As the cells divide, they also differentiate, resulting finally in plasma cells, highly specialized antibody-producing cells. Plasma cells are capable of producing thousands of molecules of antibody per second. Also, some of the B cells develop into small lymphocytes called **memory cells** with the same specificity as the plasma cells. These memory cells persist in the body for years and are present in numbers sufficient to give a prompt and effective **anamnestic,** or **memory, response** when the same antigen is encountered again at a later time. The activities of the memory cells are responsible for the resistance to a particular disease once a person has recovered from it and also for the effectiveness of vaccines. Plasma cells have a limited life span. After they die, just the memory cells survive.

During antibody formation, antigen **selects** the B cells with specific antibody for that antigen on their surfaces. With the help of T helper cells, the selected cells form large **clones** of cells making the same specificity of antibody. This sequence of events is known as **clonal selection** (figure 16.14).

The Secondary (Memory) Response Is Faster and More Efficient than the Primary Response

If the amount of antibody present in the blood, the **antibody titer,** is measured over a period of time, it is seen that the first time a particular antigen is introduced into the body, about 10 days to two weeks is required before a substantial amount of antibody is detected in the blood. The titer rises until the antigen is removed and the response wanes (figure 16.15). This is typical of antibody production during the **primary antibody response.** If, however, the same antigen is introduced at a later date, the level of antibodies rises much faster and higher, due to the immediate activation of

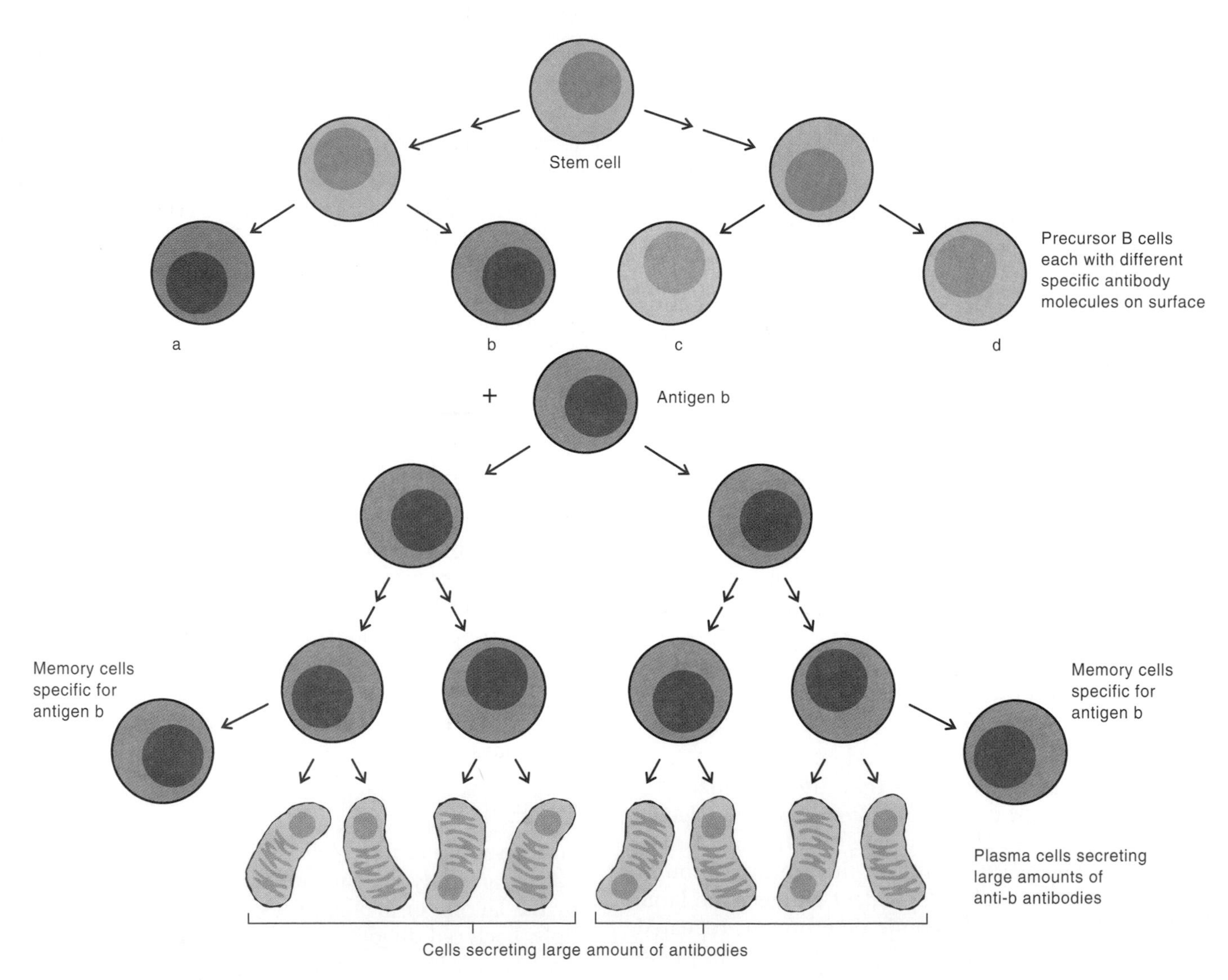

Figure 16.14
Differentiation of B cells through clonal expansion of the antigen-stimulated B lymphocyte. Antigen selects lymphocytes that bear antibodies specific for that antigen on their surface. In the diagram, antigen b selects lymphocyte b with specific anti-b antibody on its surface. This leads to formation of a clone of cells by the multiplication of these lymphocytes and differentiation of the lymphocytes to plasma cells producing anti-b antibodies and memory cells specific for antigen b. All cells of a clone produce the same specificity of antibodies. Plasma cells die after producing copious quantities of antibodies. Memory cells remain as small lymphocytes with the same specificity, living for many years.

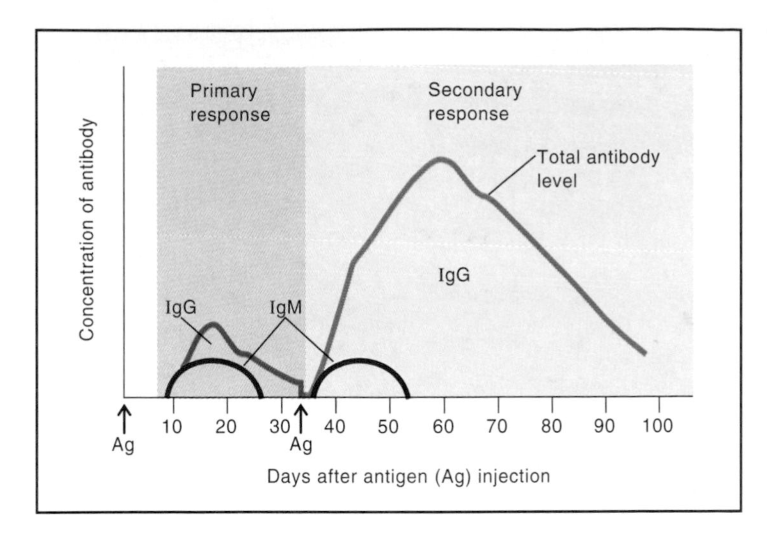

Figure 16.15
A memory response is a characteristic of the antibody response. The first exposure to antigen causes a primary response during which small amounts of antibodies are produced for a short time. Subsequent doses of the same antigen stimulate a secondary (anamnestic) response during which much more antibody is produced over a longer period.

the clones of memory cells present. This is typical of the **secondary (memory) antibody response.**

IgM is the first class of antibody produced during an immune response, but with thymus-dependent antigens there is a switch to other classes of immunoglobulin as the response develops, as shown in figure 16.15. In animals, it has been shown that cytokines produced by Th cells play a major role in causing the switch. To date, little is known about how such switching occurs in humans.

In the secondary antibody response, essentially the same sort of IgM response occurs as in the primary response, but the IgG response is greatly accelerated, as shown in figure 16.15. In other words, there is memory for IgG production but not for IgM.

Measuring the **rise in titer,** or increase in amount of antibody, seen in the antibody response is helpful in determining whether an infection is present or the antibodies remain from a previous infection. Also helpful is determination of the class of antibodies present. During the primary response to an infection, IgM antibodies are made, then IgG is made, and both rise in titer. If the antibodies remain from a previous infection or immunization, the antibodies will be mostly IgG and the amount will not increase substantially.

Immune Specificity Permits Millions of Different Antigens to Be Recognized

Normal humans can respond immunologically to millions of different antigen epitopes. Each epitope is recognized by specific antibodies on the surface of **B lymphocytes,** or specific antigen-receptors on **T lymphocyte** surfaces. During development, as lymphocytes differentiate from the bone marrow stem cell into either B cells or T cells, a small number will make surface receptors for each antigenic epitope. For example, a small number of lymphocytes will produce surface receptors for epitope a and another few lymphocytes will have receptors for epitope b, and so on for about a million different epitopes.

The development of the millions of specificities needed to respond to antigens in the environment results from the rearrangement of genes in the B and T lymphocytes. Perspective 16.1 diagrams some of the ways in which genes are rearranged in B lymphocytes. Similar diversity occurs in T cell receptors.

REVIEW QUESTIONS

1. Why are lymphocytes important in specific immunity?
2. Outline some major differences between B and T lymphocytes.
3. What is the importance of the major histocompatibility complex?
4 What is the function of T helper lymphocytes?
5. Why is the secondary (memory) response faster and more efficient than the primary response?

Development of the Cell-Mediated Immune Response

The Cell-Mediated Immune Response Involves T Lymphocytes

In **cell-mediated immunity,** a population of T cells develops that can kill the antigen-bearing **target cells;** hence, cells of the developing population are called **T cytotoxic (Tc) cells.** In contrast to antibody formation, in which antibodies are produced to antigens that are either soluble or on cells, the Tc cell response is directed only to antigens that are on whole cells, such as virus-infected cells or tumor cells. This is because the Tc cells only recognize antigen plus class I MHC antigen in a complex on the host cell membrane. Cell-mediated immunity is initiated by interaction of the antigen with a few T lymphocytes capable of recognizing that specific antigen.

Recall that, in antibody formation, the antigen interacts with specific receptors, antibodies, on the surface of B cells. In cell-mediated immunity, antigen is recognized by specific molecules called **T cell receptors** on the T cell surface. The T cell receptors for antigen are not immunoglobulins, but they have certain similarities to antibodies (figure 16.16). There are two kinds of T cell receptors: TCR1 and TCR2. Each kind consists of two polypeptide chains, each with variable and constant regions. It is most likely that the variable regions supply the antigen-reaction sites. The genes that code for T cell receptors are rearranged genes, of the same kind as in antibodies (perspective 16.1).

Perspective 16.1

Each Different Kind of Antibody Molecule Results from Rearrangements of Genes in Antibody-Producing Cells

One of the mysteries of antibody formation was the way plasma cells, using the limited genetic information available to them, produce antibody molecules that are effective against the millions of antigenic molecules known to exist. Dr. Susumu Tonegawa won the Nobel Prize for Medicine in 1988 for solving the mystery.

As the lymphocyte divides during maturation, its genes are rearranged. DNA sequences for each light chain are assembled by the random selection of a V (variable) gene segment and a J (joining) gene segment. DNA sequences for each heavy chain are assembled by the random selection of a V gene segment, a J gene segment, and a D (diversity) gene segment. The constant region of the molecule is encoded by a single C (constant) gene segment (figure 1).

Because of the large number of ways in which these genes can be rearranged, the various B lymphocytes together have the capacity to produce a similarly large number of antibody proteins. Also, mutation can occur in immunoglobulin V genes, and several other means exist to increase the variability. In mice, for example, in which extensive studies have been made, it is known that different arrangements of these DNA sequences can code for at least 10^8 different antibody molecules. This number of different antibody molecules is ample enough to account for all antigens that might be encountered in the environment.

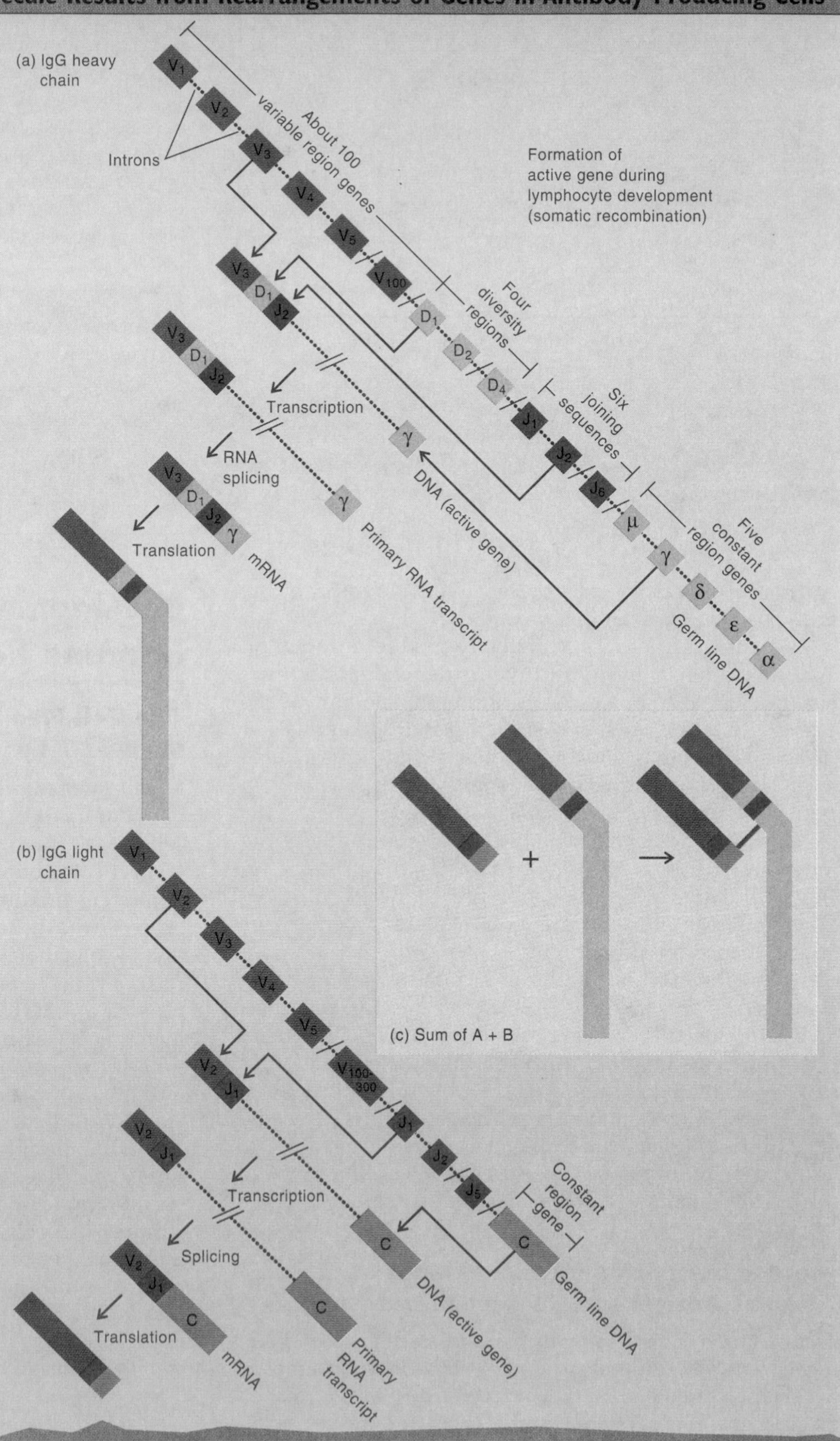

Figure 1
Immunoglobulin (IgG) gene arrangement in an immature lymphocyte and the mechanism of active gene formation. (*a*) Heavy chain; (*b*) light chain; and (*c*) formulation of one-half an antibody molecule.

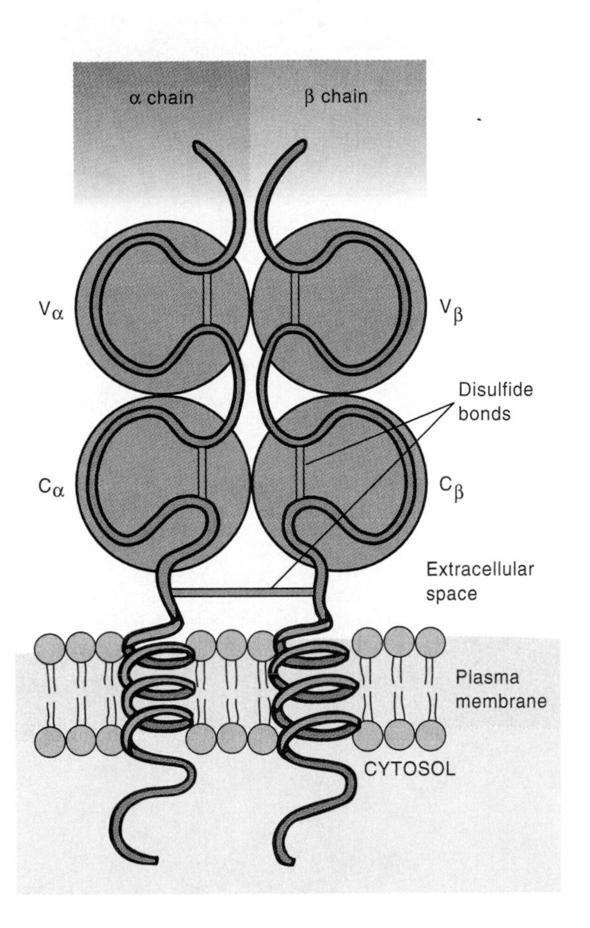

Figure 16.16
T cell receptor composed of an α and a β polypeptide chain. Each chain has one variable (V) and one constant (C) region. Unlike antibodies, which have two binding sites for antigen, T cell receptors have only one.

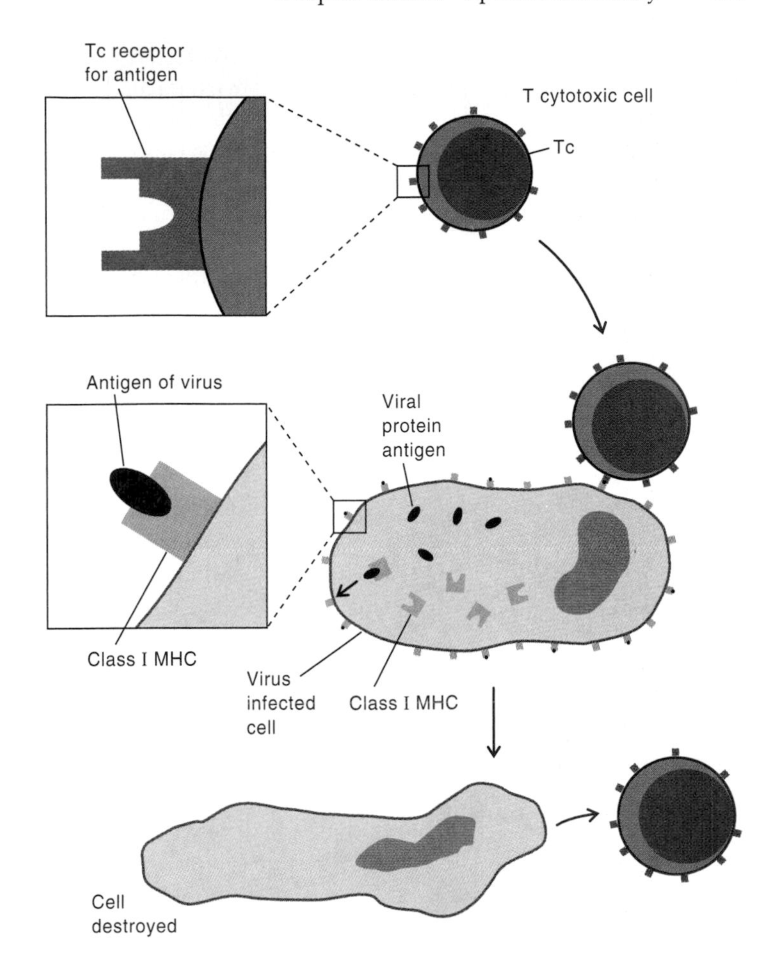

Figure 16.17
T cytotoxic (Tc) cell activity. Tc (CD8) cells can kill antigen-marked target cells. In this case, the target is a virus-infected cell. Proteins of the virus, synthesized with the host cell, are taken into a vacuole and degraded by enzymes into small peptide fragments. These peptides complex with the class I MHC molecules being formed there. The peptide, within the cavity of the MHC class I antigen, then moves to the surface where it is inserted into the cell membrane. The peptide epitope, complexed in the MHC class I molecule, is recognized by the T cell receptor. The CD8 molecules on the Tc cell also interact with CD8 receptors on the target cell, making more bonds to hold the two cells together. Perforin or other mechanisms cause the destruction of the target cell. The Tc cell survives and can kill other specific target cells.

Cytotoxic T Cells Kill Virus-Infected Cells and Other Cells Recognized as Foreign

As in antibody formation, during cell-mediated immunity the antigen selects cells that can recognize it. In antibody formation, the T helper cells only recognize antigen presented in combination with MHC class II molecules; during the cell-mediated response, T cytotoxic cells only recognize antigen combined with MHC class I molecules.

All nucleated cells routinely degrade the proteins they have synthesized, and the resulting peptides are inserted into the grooves of the class I molecules and displayed on their surfaces. Peptides of self, under normal circumstances, do not induce an immune response, but others are recognized by T cytotoxic cells, which then can cause the destruction of the cell.

For example, during cell-mediated immunity, a virus-infected cell becomes a **target** cell to be destroyed. The virus causes the host cell to synthesize proteins of the viral coat. Just as with normal cell proteins, some of these are taken into a vacuole and enzymatically degraded into short peptide sequences (figure 16.17). Then, within the cell, they complex with MHC class I molecules that are being formed there. The complex of MHC class I with antigen peptide is then moved to the cell surface where it can be presented to specific responding Tc cells. The Tc cells also have receptors for MHC class I markers.

Interaction of the epitope and the T cell receptor and interaction between the MHC class I molecules and the CD8 molecules that serve as receptors for class I are sufficient to hold the Tc and the target cell together. Contents of granules in the Tc cell are released in close contact with the target cell membrane, and the target cell is killed. The Tc cell survives and can go on to kill other targets.

Only protein antigens can induce cell-mediated immunity, and the epitopes recognized are short peptides. It would be quite a disadvantage if Tc cells destroyed a cell that was not really infected but only had a molecule of foreign antigen attached to its surface. The fact that the antigen is produced and cleaved to peptides within the infected cell and is only recognized when presented on the cell surface together with MHC class I molecules prevents uninfected cells from being recognized as foreign and destroyed.

Immunity to Most Cancers and to Tissue Transplants Is Cell-Mediated

Since tumor cells arise from host tissue, it might be expected that they would not incite an immune response. However, in laboratory animals it has clearly been shown that they do. When tumors are induced by certain chemicals, each tumor has its own unique antigen not shared by other tumors produced in the same way. When tumors are induced by viruses, however, they have shared antigens; that is, all tumors induced by one virus will have the same surface antigen. This lends hope that it may be possible eventually to immunize against some tumors. While it would be difficult, if not impossible, to immunize against tumors that do not have common antigens, it should be feasible to immunize against tumors with a common antigen, such as those caused by viruses. In fact, this has already been done for one tumor of chickens, Marek's disease, caused by a virus.

Both tumor cells and transplanted cells are destroyed largely by Tc cell activity. NK cells (described next) and other mechanisms also act against these foreign cells.

Cell-Mediated Immunity Controls Microbial Growth in Macrophages

Cell-mediated immunity also operates against some pathogens that multiply in host cells, especially macrophages, where they are shielded from antibodies and many other host defenses. Cytokines produced by Th cells, notably interferon gamma, activate macrophages, including those that contain the intracellular pathogens. The activated macrophages can then kill the intruders. Examples of this kind of infection are tuberculosis, leprosy, listeria infection, and some fungal infections. Figure 16.18 shows the changes in macrophages that permit them to become more actively antimicrobial and thereby kill the pathogens living inside them.

Natural Killer Cells Kill Foreign Cells but Are Not Antigen-Specific

Like Tc cells, **natural killer (NK) cells** also act against foreign cells. NK cells are large, granular lymphocytes, lacking

• **Transplantation, p. 403**

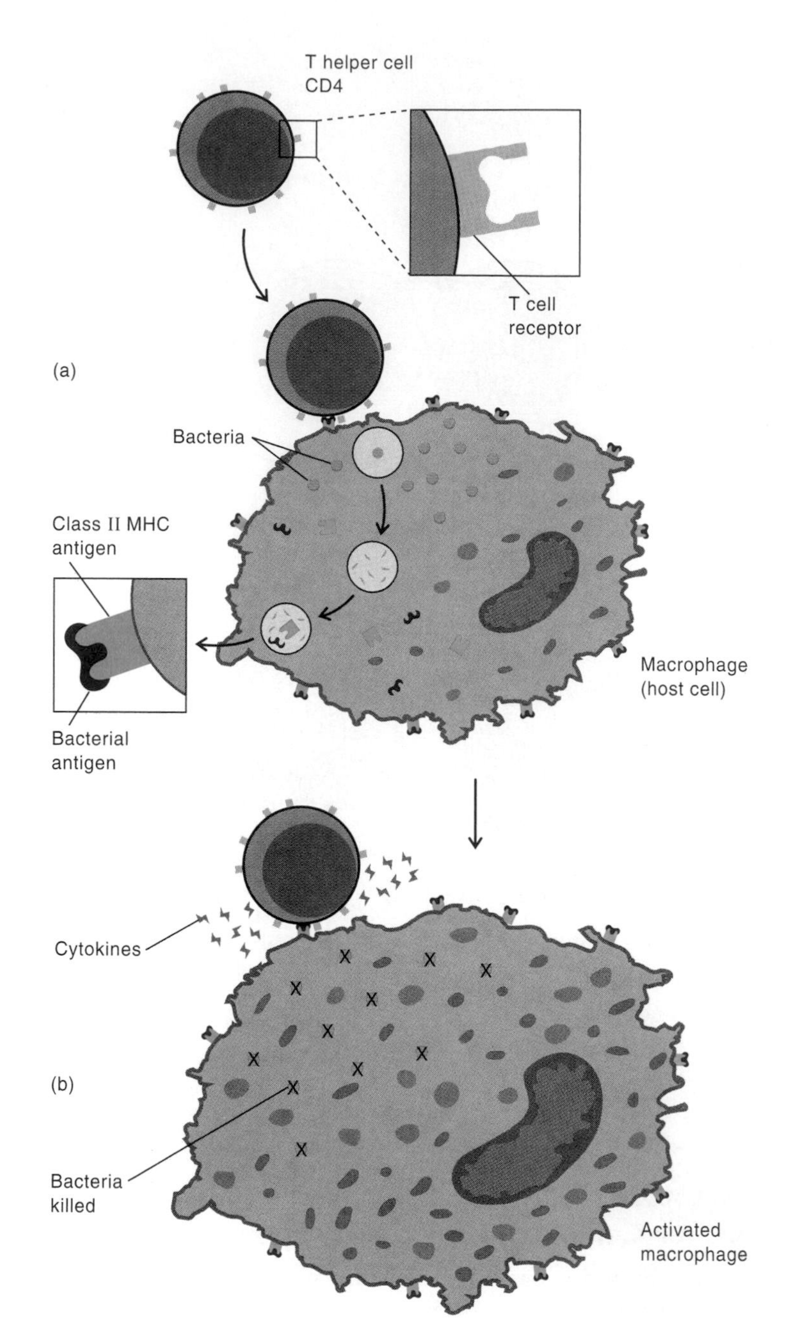

Figure 16.18
T cell activity in the killing of intracellular pathogens within macrophages. Th (CD4) cells function in the killing of intracellular pathogens by causing the activation of macrophages. The macrophages become larger and contain greatly increased amounts of enzymes and degradative materials. There is an increase in surface receptors and greater phagocytic capability. The activated macrophage is a cell with the capacity to kill many intracellular pathogens. (*a*) In this example, antigens of bacteria or other organisms living within the host cell are degraded to fragments, complexed with class II MHC antigens, and inserted into the cell membrane. They are recognized by the T cell receptor on the CD4 T cell, and contact is made. (*b*) The T cell then releases large amounts of cytokines at the cell surface, causing activation of the host cell and resulting in killing of intracellular organisms.

markers characteristic of T or B lymphocytes, but they are probably related to T cells. They do not have T cell receptors for specific antigen, but they are able to recognize abnormal cells, such as virus-infected or tumor cells, as being different and unwanted. They can attach to target cells, release some of their granule contents close to the target cell membrane, and thereby kill the targets. This is similar to the way in which Tc cells kill, but NK cells lack antigenic specificity.

NK cells may also act together with specific antibodies to kill target cells in **antibody-dependent cellular cytotoxicity.** Antibodies attach to specific target cells, leaving their Fc portions exposed. NK cells have receptors for the Fc portions of antibodies, so they can come in close contact with the target cells, resulting in target cell killing (figure 16.19).

Immunological Tolerance and Control

Immunological Tolerance Denotes Specific Unresponsiveness to Antigen

A state of specific unresponsiveness is called **immunological tolerance.** Lack of response to self antigens is an example of immunological tolerance. It is possible under some circumstances to develop immunological tolerance to antigens other than self. Various mechanisms account for the development of tolerance.

One well-established mechanism for inducing immunological tolerance is the destruction of cells that recognize and respond to an antigen. This is known as **clonal deletion.** Immature T cells that interact with the specific antigen they recognize during development are deleted, usually in the thymus. Some immature B cells are also deleted in the bone marrow.

Another means of inducing immunological tolerance, called **clonal anergy,** occurs when clones of potentially responsive cells cannot actively respond. This can happen when T helper cell function is lacking, so that B cells do not get the signal to multiply. Also, active cells can be suppressed by various means, possibly by T suppressor cells.

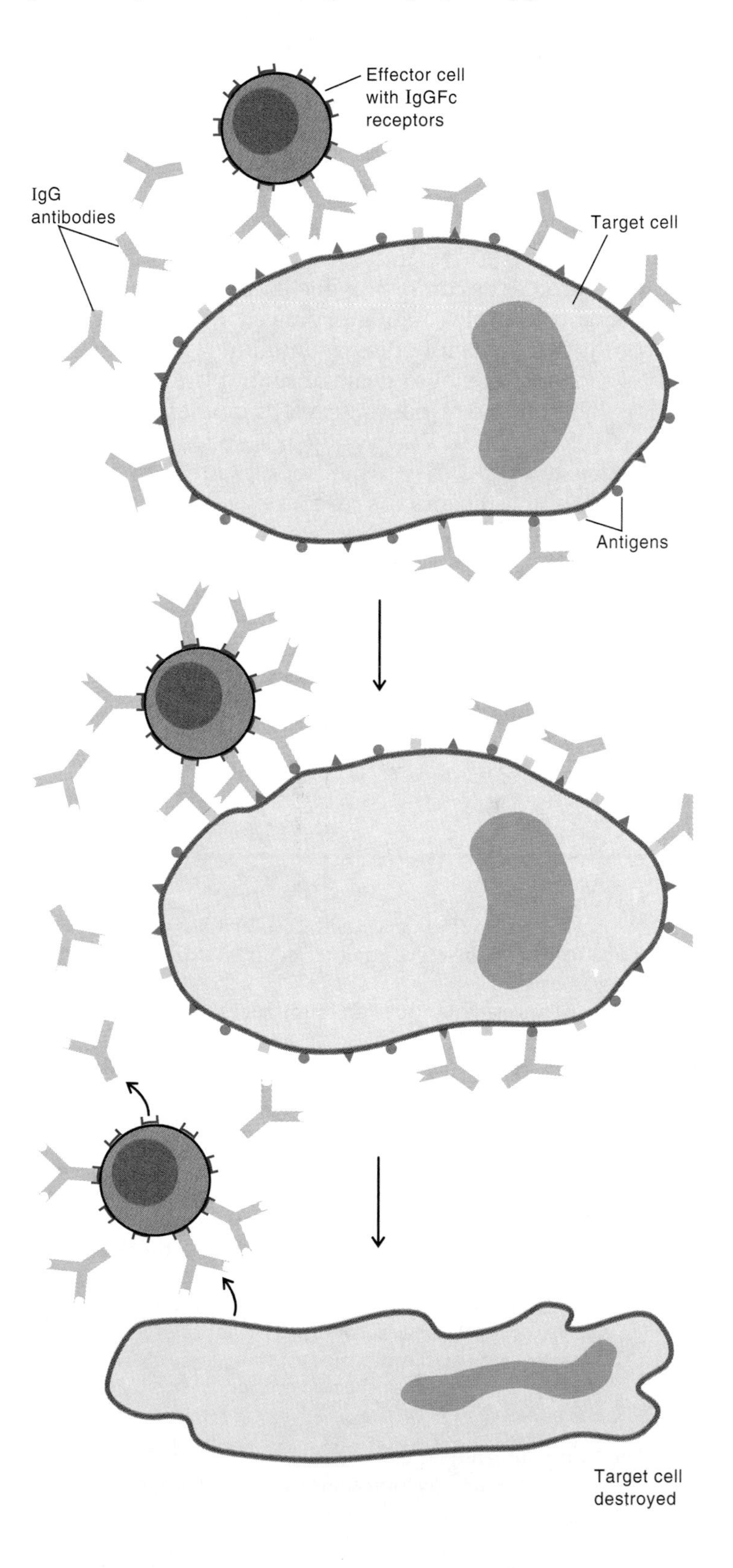

Figure 16.19
Target cell killing by antibody-dependent cellular cytotoxicity (ADCC). Various cells can function as effector cells in ADCC, including certain granulocytes, monocytes, and also natural killer (NK) cells, all of which have Fc receptors for immunoglobulins. NK cells are not specific for antigens, but they do have surface Fc receptors. Antibody molecules, free or attached to epitopes on the target cell membrane, act to bring effector cells and target cells in close contact by means of the Fc portion of the antibody reacting with the Fc receptors on the cells. The close contact permits killing to occur via a number of different mechanisms. For example, NK cells, like Tc cells, produce perforin, which makes lesions in the target cell membrane, leading to lysis and cell death. Some of the granulocytes kill by means of their granule contents.

Experiments reported recently have challenged the idea that the immune system distinguishes between antigens of self and foreign antigens. It has been suggested that the immune response is activated only when an antigen is associated with causing damage or harm. Thus, this theory holds that, following recognition of the antigen, a danger signal must be given by damaged cells in order for the T cell to deliver the stimulus to multiply and become active. In the absence of the danger signal, the T cells must be unable to respond, resulting in immunological tolerance. If such mechanisms do operate, they might function with clonal deletion.

The Immune Response Is Carefully Controlled

Clearly, a process as complex as the immune response must have equally complex regulatory mechanisms to prevent overproduction of antibodies or immune T lymphocytes. One of the more obvious agents of control is the antigen itself. As the immune response increases in intensity, antigen is effectively removed. When antigen is no longer available, the response wanes. Also, as antibodies are formed, they can act to inhibit the specific response, not only by removing antigen, but in several other ways as well. Antibody is used clinically to prevent destructive immune responses to Rh-positive antigen in Rh-negative mothers (see chapter 18).

Immune responses are also regulated by T cell activities. Some of the cytokines produced by T cells stimulate and others inhibit other immunologically active cells. As mentioned earlier, some CD8 T lymphocytes, T suppressor cells, have been shown to have the ability to suppress antibody production and Tc cell development. Current evidence suggests they produce cytokines and other soluble factors that suppress the other cells.

The genetic makeup of an individual has an influence over immune responses. Some individuals lack the genetic capability to respond or they respond poorly to certain antigens, while others have the genetic makeup to respond well to those antigens. It is possible to breed animals to produce offspring that are either high- or low-antibody producers to various antigens. Genes that influence the presentation of antigens also help to regulate the response.

It is well-established that general body fitness, good nutrition, well-balanced endocrine function, and lack of stress are important in maintaining optimal immune function. This is seen, for example, in the failure of immunity in severely malnourished persons.

Thus, some important control mechanisms include antigen, antibody, T cell activities, genetic influences, and general health and body function.

REVIEW QUESTIONS

1. What are the capabilities of T cytotoxic cells?
2. How does the T cell receptor for antigen differ from the B cell receptor for antigen?
3. Why is T cell killing of target cells a specific reaction?
4. Is there reason to hope for a vaccine to protect against all cancers or any cancers? Explain.
5. Define *immunological tolerance.*

Summary

I. Specific Immunity
   A. Specific immunity has memory so that it is more effective against second or subsequent exposure to the same agent.
   B. Substances that induce an immune response are called *antigens.*
   C. Glycoprotein molecules that react specifically with the antigen that induced them are called *antibodies.*
   D. Specific immune responses involve two sets of lymphocytes, the B and the T lymphocytes.

II The Nature of Antigens
   A. Antigens are usually foreign to the host.
   B. Most antigens are large molecules, such as proteins and polysaccharides.
   C. Small chemical groups on the antigen molecule, called *epitopes,* are the areas recognized by antibodies.
   D. Haptens are small molecules that can complex with large molecules to induce a specific immune response against the haptens.

III. The Nature of Antibodies
   A. Antibodies are glycoproteins called *immunoglobulins.*
   B. There are five classes of immunoglobulins, each with different properties, but antibodies of all five classes can combine specifically with antigen.
   C. The five immunoglobulin classes are IgG, IgM, IgA, IgD, and IgE.

IV. The Role of Lymphocytes in Specific Immunity
   A. Lymphocytes are the cells primarily responsible for the specific immune response.
   B. Lymphocytes fall into two major categories, the T lymphocytes and the B lymphocytes.

V. Major Histocompatibility Complex
   A. The major histocompatibility complex antigens are genetically determined and allow cells to recognize each other as part of self or as foreign.
   B. Class I and class II major histocompatibility antigens are important in the immune response.

VI. Development of the Antibody Response
   A. Complex interactions occur during the immune response.
   B. Clonal selection occurs during immune responses.
   C. The secondary (memory) response is faster and more efficient than the primary response.

D. Immune specificity permits millions of different antigens to be recognized.

VII. Development of the Cell-Mediated Immune Response
A. The cell-mediated immune response involves T lymphocytes.
B. Immunity to most cancers and to tissue transplants is largely cell mediated.
C. Cell-mediated immunity controls microbial growth in macrophages.
D. Natural killer cells kill foreign cells but are not antigen-specific.

VIII. Immunological Tolerance and Control
A. Immunological tolerance denotes specific unresponsiveness to antigen.
B. The immune response is carefully controlled by a combination of mechanisms.

Chapter Review Questions

1. Diagram and compare the structures of IgG and IgA.
2. Which immunoglobulin classes can activate complement? How does the ability to activate complement relate to function of these immunoglobulins?
3. Why is IgE so scarce in circulation?
4. Describe the cells that produce antibodies.
5. How can T lymphocytes be distinguished from B lymphocytes?
6. What is the major difference between responses to thymus-dependent antigens and responses to thymus-independent antigens?
7. What is meant by *clonal selection?*
8. How do T cytotoxic cells function in cell-mediated immunity?
9. How does cell-mediated immunity operate against intracellular pathogens in macrophages?
10. How do natural killer cells differ from T cytotoxic cells?

Critical Thinking Questions

1. What kinds of diseases would be expected to occur as a result of lack of T or B lymphocytes?
2. How might it be possible to have effective antibody-mediated-immunity and yet lack cell-mediated immunity?
3. Why might one wish to induce immunological tolerance?

Further Reading

Paul, William E., ed. 1993. *Fundamental immunology,* 3rd ed. Raven Press, New York.
Rajewsky, Klaus. 1996. Clonal selection and learning in the antibody system. *Nature* 381:751–758.
Roitt, Ivan M. 1994. *Essential immunology,* 8th ed. Blackwell Scientific Publications, Oxford.
1993. *Scientific American* 269(3). The entire edition is devoted to immunology.
Zinkernagel, Rolf M. 1996. Immunology taught by viruses. *Science* 271:173–178.

17 Functions and Applications of Immune Responses

KEY CONCEPTS

1. Antibodies and immune T cells protect the human body by diverse mechanisms.
2. In vitro immunological reactions are the basis of many laboratory tests.
3. Immunization, or vaccination, can control or even eliminate some (but not all) serious infectious diseases.
4. Immunizing agents have been dramatically improved with new technology.

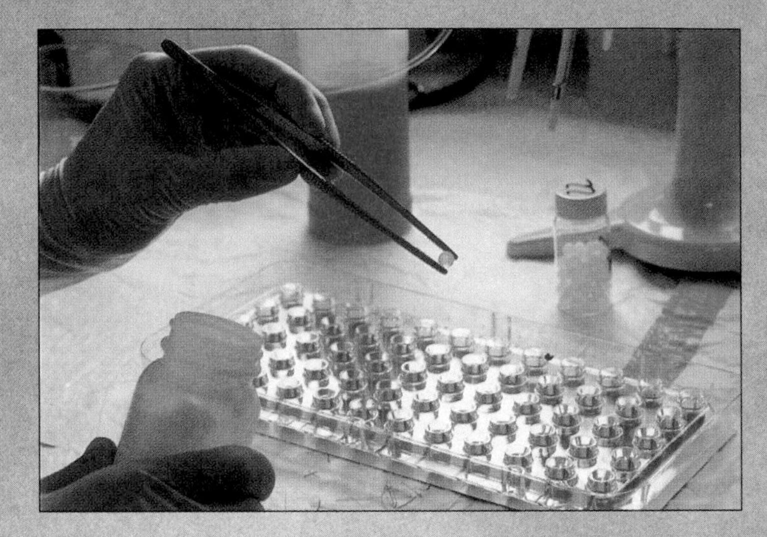

Test for AIDS antigen.

PREVIEW LINK

Microbes in Motion

The following books and chapters in the *Microbes in Motion* CD-ROM may serve as a useful preview or supplement to your reading:

Viral Structure and Function: Viral pathogenesis; Viral Detection, *Vaccines:* Vaccine Development; Vaccine Targets (Bacterial Targets).

A Glimpse of History

Vaccination, the practice of deliberately stimulating the immune system in order to protect individuals against a disease, has become routine in most countries of the world. As a result, many of the diseases that once caused widespread death and disability, such as smallpox, measles, whooping cough, mumps, and poliomyelitis, are now nonexistent or much less common.

Even before people knew that microorganisms caused disease, it was recognized that individuals who came down with and eventually recovered from a certain disease such as smallpox rarely contracted it a second time. A procedure known as **variolation** is described in old Chinese writings dating from the Sung dynasty (960–1280 A.D.). A person would deliberately inhale the powdered crusts of smallpox pustules or place a small amount of this material into a scratch made in the skin. Usually the resulting disease was mild, and the person would recover and be immune to smallpox for the rest of his or her life. Occasionally, a severe case of the disease developed, which often resulted in death. Nevertheless, the risk of getting a serious case of the disease using this procedure was rare enough that people were willing to take the risk.

Although variolation was practiced in China and the Mideast a thousand years ago, it was not until the early part of the 1700s that European doctors learned of the practice. Even then, it was not widely used until after 1719 when variolation was popularized by Lady Mary Wortley Montagu, the wife of the British ambassador to Turkey. She had their children immunized against smallpox in this way. Since smallpox was often fatal and was disfiguring, producing deep scars ("pocks"), variolation subsequently became very popular in Europe.

Although a person exposed to smallpox through variolation would usually completely recover and so was not at great risk, he or she would become contagious. Because of the danger of contagion and because the procedure was reasonably expensive, large segments of the population in Europe were left unprotected.

As an apprentice physician, Edward Jenner noted that milkmaids who had had cowpox infections rarely contracted smallpox. Cowpox was a disease of cows that caused few or no symptoms when contracted by humans. In 1796, long before bacteria and viruses had been discovered, Jenner conducted a classic experiment in which he deliberately transferred material from a cowpox lesion on the hand of a milkmaid, Sarah Nelmes, to a scratch on the arm of a young boy named James Phipps. Six weeks later, he exposed Phipps to pus from a smallpox victim. Phipps did not develop the disease. The boy had apparently been made immune to smallpox when he was inoculated with pus from the cowpox lesion. Using the less dangerous cowpox material in place of the pustules from smallpox cases, Jenner and others worked to spread the practice of variolation. Later, Pasteur used the word **vaccination** (from the Latin *vacca* for "cow") to describe any type of protective inoculation. By the 20th century, most of the industrialized world was generally free of smallpox as the result of routine vaccination of large populations.

In 1967, the World Health Organization (WHO) initiated a program of intensive vaccination wherever smallpox appeared. Since there were no animal hosts and no nonimmune humans to which it could be spread, the disease died out. The last case of naturally contracted smallpox occurred in Somalia, Africa, in 1977, and in 1979 in a ceremony in Nairobi, Kenya, WHO declared the world free of smallpox.

It is now known that the smallpox and cowpox viruses are closely related. It is of interest that the virus used in recent years for vaccination against smallpox is neither the cowpox nor the smallpox virus. It is called the **vaccinia** virus and it is not entirely clear where this virus originated. Some scientists believe that the vaccinia virus is in fact a hybrid of the cowpox and smallpox viruses. Quite possibly it developed during Jenner's time, when calves injected with cowpox virus to produce vaccine material accidentally received the smallpox virus at the same time. In addition to its importance in defeating smallpox, the vaccinia virus has been genetically engineered in recent years to make experimental vaccines against other diseases such as AIDS and viral hepatitis. Thus, Edward Jenner's legacy extends beyond the eradication of smallpox.

In the last two chapters, we discussed basic defense mechanisms, both nonspecific and specific, and became acquainted with antibodies and immune cells. This chapter will explore in more detail the functions of antibodies and immune cells, both in the body (in vivo) and in the laboratory (in vitro). The manipulation of the immune system in order to provide protection through immunization procedures will also be addressed. We will consider how vaccination such as that pioneered by Jenner can be used to enhance the immune response and how immunization procedures have advanced remarkably in recent years to become even safer and more effective.

Functions of Antibodies in Vivo

Antibodies are referred to by their class, such as IgM or IgG, but they may also be known by their function. Thus, an IgG antibody that causes the antigen to precipitate is a precipitin; one that causes it to agglutinate is an agglutinin; one that coats the antigen so that it is more readily phagocytized is an opsonin; one that leads to complement lysis of the antigen is a complement-fixing antibody; one that neutralizes the effects of toxin is an antitoxin; and one that neutralizes the ability of a virus to infect cells is a neutralizing antibody. Some of the important functions carried out by the different classes of antibodies within the human body are listed in table 16.1, in the previous chapter.

The first antibodies made during infection are of the class IgM. These large molecules have up to 10 sites that bind antigen, which makes them very efficient at agglutinating particles of antigen. The agglutinated clumps of antigen and antibody are rapidly removed by phagocytes. IgM is the class of antibody that is most efficient at activating complement. The complement products formed through this action increase phagocytosis, cause inflammation, and can lyse and destroy some cells. The existence of IgM antibodies in circulation early during infection

As mentioned earlier, cells infected with viruses gain new surface antigens as a result of infection. These antigens are made and their peptides inserted together with class I histocompatibility markers into the cell membrane of the virus-infected cell (figure 17.3). The membrane-bound antigens serve as targets for direct killing of the virus-infected cell by Tc cells. This killing can occur before complete new viruses are made. The immune system needs to prevent the replication of the viruses, even if it means sacrificing some of the body's own cells.

Although tumor cells arise from host tissue, the cells of many tumors have new surface antigens that induce immune responses. Whether caused by a virus or in other ways, tumor cells are usually destroyed by specific Tc cell killing (figure 17.3). Also, nonspecific killing by natural killer cells aids in antitumor protection.

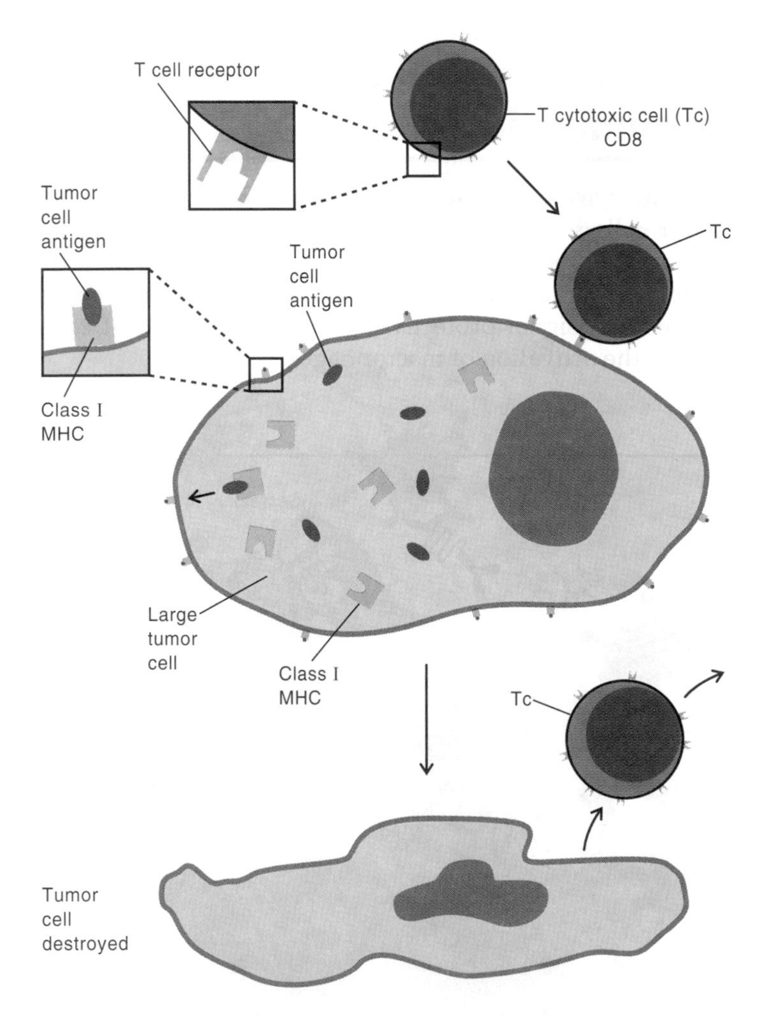

Figure 17.3
T cell killing of target cells. Tc (CD8) cells kill target cells recognized as different from cells of self. In this example, the target cell is a tumor cell. Tumor antigens are made and degraded to peptides by enzymes within the tumor cell. These peptide fragments are taken into vesicles and complexed with class I MHC molecules being formed there. The peptide, within the groove of class I molecule, is moved to the tumor cell surface and inserted into the cell membrane. The peptide epitope, complexed with MHC, is recognized by the T cell receptor. Perforin or other substances are released, and they destroy the target cell. The Tc cell survives and can kill other specific target tumor cells.

There are a number of mechanisms for killing by Tc cells. When the Tc cell comes in close contact with the target cell membrane, contents of granules in the T cells are released. One theory is that the contact between Tc cell and target cell triggers intracellular events that lead to programmed cell death (apoptosis), characterized by enzyme degradation of the DNA of the target cell. Another mechanism that has received much study is the action of a protein called **perforin.** Some Tc cells and some natural killer cells secrete perforin, which makes a hole in the target cell's membrane, causing the cell to lyse and die. Perforin has some similarities to the C9 component of complement that makes up the membrane attack complex. Recall that the membrane attack complex of complement also makes holes in cell membranes, resulting in cell lysis. Another mechanism of killing is the action of the cytokine tumor necrosis factor beta, found in the granules of both Tc cells and natural killer cells.

Different mechanisms prevail in cell-mediated immunity against intracellular organisms. It is not useful for the immune system to kill the cell in which intracellular organisms are multiplying, as this would result in the infection of many more cells by the released organisms. It is much more advantageous to activate the host cells in such a way that the intracellular microbes are killed or inhibited in growth. This is achieved by inserting antigens from the intracellular organisms into the cell membrane with class II major histocompatibility markers. The class II molecules do not interact with Tc cells but instead recognize Th cells. The Th cells respond by making cytokines, such as interferon gamma, that cause activation of the host macrophages and consequently efficient killing of intracellular organisms (figure 16.18 *b*).

This form of cell-mediated immunity is exemplified by the disease tuberculosis, caused by *Mycobacterium tuberculosis.* These bacteria usually are inhaled and infect via the lung. Like all other extraneous inhaled material, the tubercle bacilli are engulfed by the many alveolar macrophages in the lungs. Unfortunately, the tubercle bacilli are hard for the macrophages to handle. The chemical composition of their cell walls, with a very high concentration of lipids complexed with peptides and polysaccharides, makes the bacilli resistant to destruction by enzymes within normal macrophages. In fact, once inside, the bacilli happily multiply for a time. Then, as cell-mediated immunity to the tubercle bacilli develops, changes occur in the macrophages. Th cells stimulated by antigens of the bacilli produce abundant cytokines that dramatically change the macrophages, which become highly active cells called **activated macrophages** (figure 15.3). They increase in size, in membrane activity, and in content. Their lysosomes increase in number and contain greatly increased amounts of degradative enzymes.

Perspective 17.1

Monoclonal Antibodies

In 1975, an exciting breakthrough occurred in immunology. Georges Kohler and Cesar Milstein developed techniques that fused normal antibody-producing B lymphocytes with myeloma[1] tumor cells, resulting in clones of cells they termed **hybridomas.** Since these hybridomas are clones, they produce antibodies with a single specificity that are, therefore, known as **monoclonal antibodies.**

Usually, when an animal is injected with an immunizing agent, it responds by making a variety of antibodies directed against different epitopes on the antigen. Therefore, even though there is a single antigen, the result is a mixture of different antibodies. When these antisera are used in immunological tests, standardizing the results is difficult, since there are differences each time the antiserum is made. Monoclonal antibodies, however, will be of the same immunoglobulin class and have the same variable regions and thus the same specificity and affinity.[2] With such specificity, tests can be standardized much more easily and with greater reliability.

It was hoped that monoclonal antibodies would also become a useful therapeutic tool. In theory, it should be possible to make monoclonal antibodies that are specific for, say, a cancer cell. Radioactive materials that destroy cancer cells could be attached to the monoclonals making a sort of "magic bullet." These antibodies would search and find the cancer cells and attach specifically to them, allowing the radioactive material to destroy the cells. In reality, many problems have arisen when the monoclonal antibodies have been used in vivo to treat human patients. One problem was that human patients reacted to the mouse antigens on the hybridoma antibodies, causing them to be rapidly removed from the body. This limited the effectiveness of the monoclonal antibodies to one or a few doses. This and many other unforeseen difficulties are being addressed now, with some encouraging results.

In vitro, monoclonal antibodies form the basis of a number of diagnostic tests. For example, monoclonal antibodies against a hormone can detect pregnancy only 10 days after conception. Specific monoclonal antibodies are used for rapid diagnosis of hepatitis, influenza, herpes simplex, and chlamydia infections. Kohler and Milstein won the Nobel Prize in 1984 for their work.

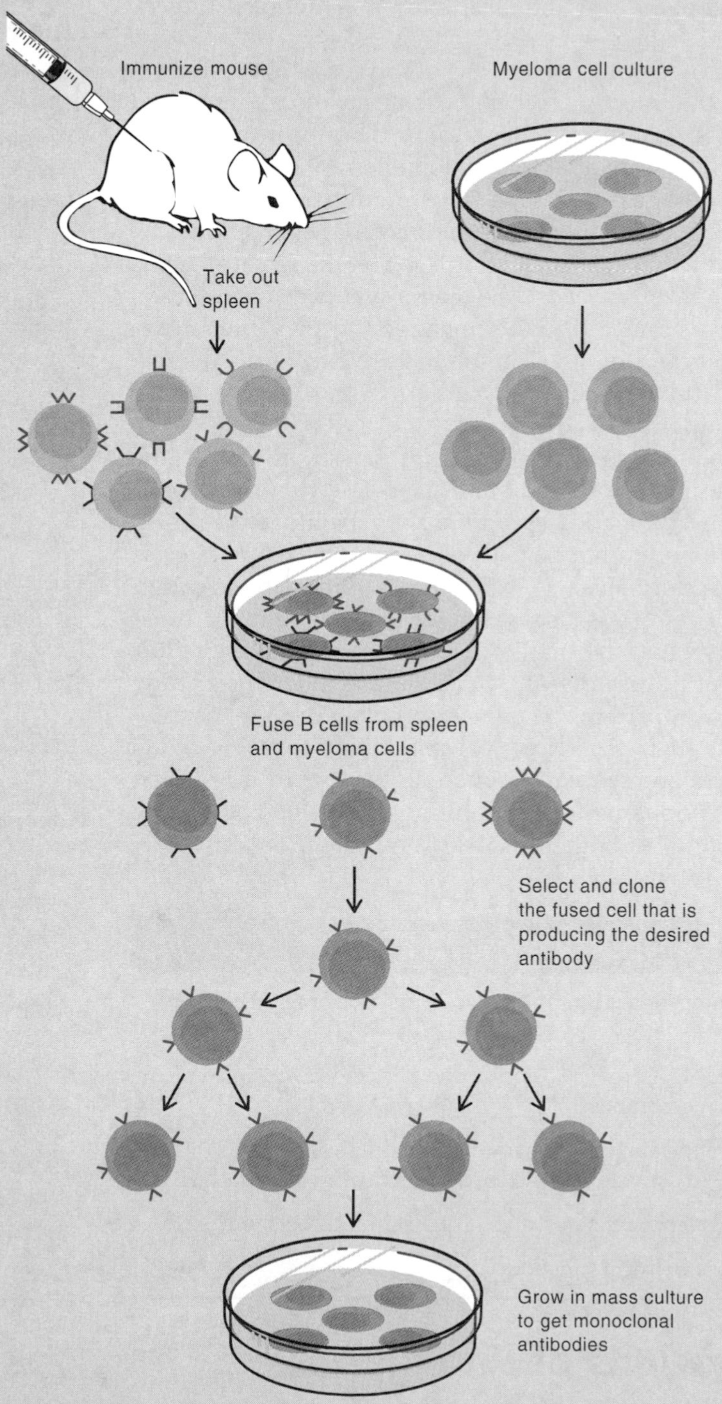

Immunize mouse
Myeloma cell culture
Take out spleen
Fuse B cells from spleen and myeloma cells
Select and clone the fused cell that is producing the desired antibody
Grow in mass culture to get monoclonal antibodies

[1] Myeloma—tumor of malignant plasma cells
[2] Affinity—strength of binding between an antigen-binding site on an antibody and the corresponding epitope on an antigen

New enzymes are also induced. All of this activity is directed toward destruction of the intracellular bacilli. If the immune response does not keep up with the growth of the bacteria, disease results. However, this kind of cell-mediated immunity is often enough to control the infection and prevent progression to a disease state. A few organisms usually remain in cells, enough to maintain a state of responsiveness to the antigens, known as **delayed hypersensitivity** (chapter 18).

If the macrophage response described is not sufficient to control the infection, other mechanisms come into play. Activated macrophages that cannot destroy intracellular microorganisms are able to fuse together. As many as 50 to 100 macrophages may fuse their membranes to form a giant cell. These giant cells retain bacteria, such as the tubercle bacilli, that cannot be destroyed by the macrophages, and the giant cells prevent the microorganisms from escaping to infect other cells. Macrophages and giant cells form foci, called **granulomas.** Granulomas are part of the disease process in tuberculosis and similar diseases, sometimes called **granulomatous diseases.**

If the immune system is compromised, as in AIDS or other immunosuppressed states, the few tubercle bacilli that remain in the body can multiply, causing active tuberculosis. This often happens in older people, who were infected at an early age and were able to contain the infection without developing disease. Later in life, when immune function declines, the small number of bacilli resume multiplication and cause disease. Reactivation tuberculosis, due to immunosuppression, is a major problem in some parts of the world, such as Africa, where inactive tuberculosis is common (in some areas estimated to occur in more than 80% of the population), and immunosuppression caused by AIDS, malaria, and other diseases is rampant.

REVIEW QUESTIONS

1. Diagram opsonization that occurs by means of IgG Fc receptors.
2. How does IgA prevent colonization of microorganisms on mucous membranes?
3. Compare the ability of the immunoglobulin classes IgG, IgA, and IgM to activate complement and explain your analysis.
4. How are tumor cells destroyed by the immune system?
5. How does perforin function?

Applications of Immune Responses in Vitro

The same functions that occur in vivo can be used in vitro in diagnostic tests or other applications. Many laboratory tests involve serum antibodies and, thus, this branch of immunology is known as **serology.** Some in vitro techniques are described next. It is important to remember that one of the reactants in an antigen-antibody reaction has to have known specificity in order to identify the other, unknown reactant. Thus, with a known antibody, the corresponding antigen can be identified. Conversely, with a known antigen, the unknown antibody can be found.

Precipitation Reactions Form the Bases for Many Immunological Tests

Antibodies that combine with soluble antigens in vitro to form a visible precipitate are called **precipitins.** This phenomenon is the basis for a number of diagnostic tests. The precipitin reaction takes place in two stages. First, within seconds after they are mixed, antigen and antibody molecules collide and form small **primary complexes.** Second, over a period lasting from minutes to hours, large precipitating complexes form as the result of latticelike cross-links between the small primary complexes (figure 17.4).

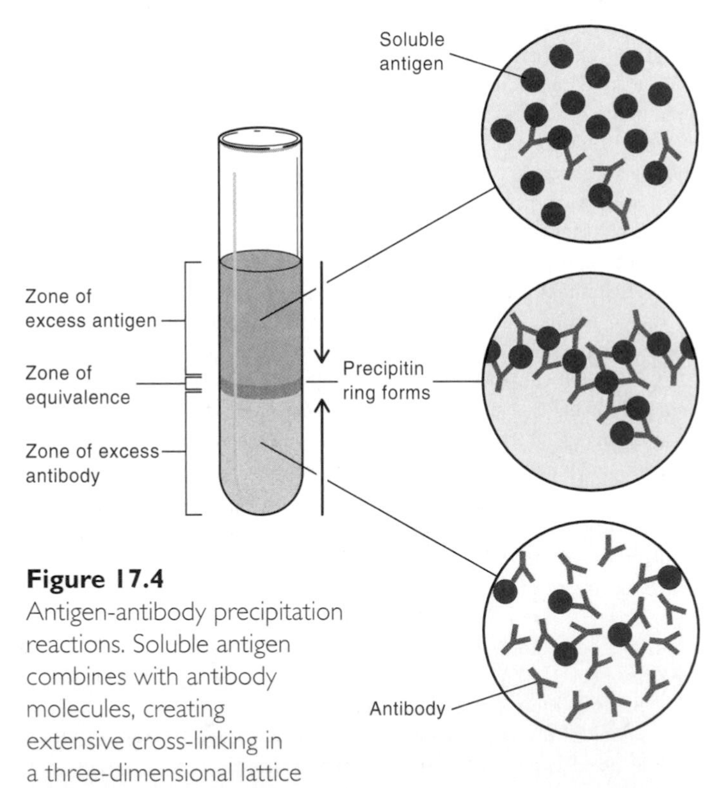

Figure 17.4
Antigen-antibody precipitation reactions. Soluble antigen combines with antibody molecules, creating extensive cross-linking in a three-dimensional lattice formation. When a large number of molecules have linked, the lattice precipitates to give a visible precipitation reaction. When excess antigen is present, all the antibody molecules are complexed to antigen so that cross-linking cannot occur efficiently. Similarly, when excess antibody is present, antigen is completely complexed and cannot form an effective lattice. In the zone of equivalence, all the antigen and antibody molecules are linked to give the maximum amount of precipitation. In this example, serum containing specific antibody was put in the test tube and antigen was carefully layered on top. A line of visible precipitate formed at the interface between antigen and antibody, in the zone of optimal proportion. No precipitate was observed above, in the zone of antigen excess, or below, in the zone of antibody excess.

Precipitation occurs in a **zone of optimal proportion;** that is, where the amount of antibody equals or nearly equals the amount of antigen (figure 17.5). If there is a great excess of either, a cross-linking lattice cannot form; consequently, no precipitate is formed.

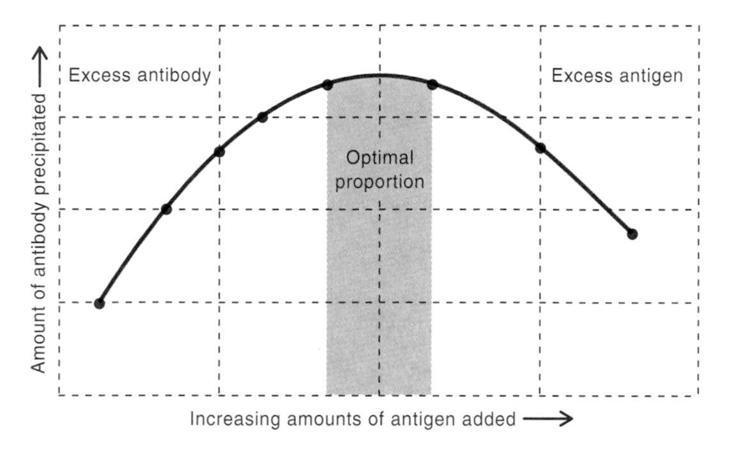

Figure 17.5
Precipitin curve. The maximum amount of precipitate forms near the zone of optimal proportion where no free antigen (Ag) or antibody (Ab) exists.

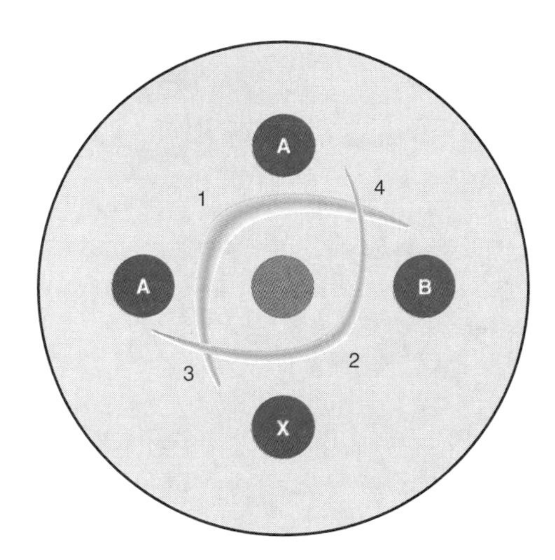

Figure 17.6
Antigen-antibody precipitation in agar gels. Using the Ouchterlony method of double diffusion in agar, antigen is placed in one well cut in the agar, and antibody is placed in another well. Each diffuses through the agar, and a line of precipitate is formed where the two meet in proper proportions. This method can be used to determine whether an unknown substance is identical to or shares antigenic determinants with a known substance. If the known and unknown substances are identical in terms of antigenic determinants, the lines of precipitate fuse (a reaction of identity); if they are not the same, the lines do not fuse but cross each other (a reaction of nonidentity). In the center well are anti-A antibodies and anti-B antibodies; antigen A and antigen B are in the outer wells as shown. X indicates an unknown antigen. Fusion of the lines at 1 and 2 shows reactions of identity (antigen A = antigen A; antigen B = antigen X). Reactions of nonidentity at 3 and 4 indicate that antigen A and antigen B are not the same and that antigen A and antigen X are not the same.

A number of tests use precipitation reactions. Immunodiffusion tests are precipitation reactions carried out in agar or other gels. One such test, the **Ouchterlony technique** (also called the *double diffusion technique*) involves diffusion of both antigen and antibody in agar (figure 17.6). Antigen and antibody are placed in separate wells cut in the agar and are allowed to diffuse toward each other. When a specific antigen and its antibody meet at optimal proportions between the wells, a line of precipitate forms. Since there is often more than one antigen present, more than one line can form, each line in its area of optimal proportions. This method is useful for identifying unknown substances and can be used to identify more than one antigen or antibody at a time.

Immunoelectrophoresis is a variation of the precipitation in gel technique, combining precipitation with electrophoresis. Electrophoresis is used to separate mixtures of proteins on the basis of their movement through an electrical field (figure 17.7). Once the proteins have been separated, the antibodies are placed in a trough and allowed to

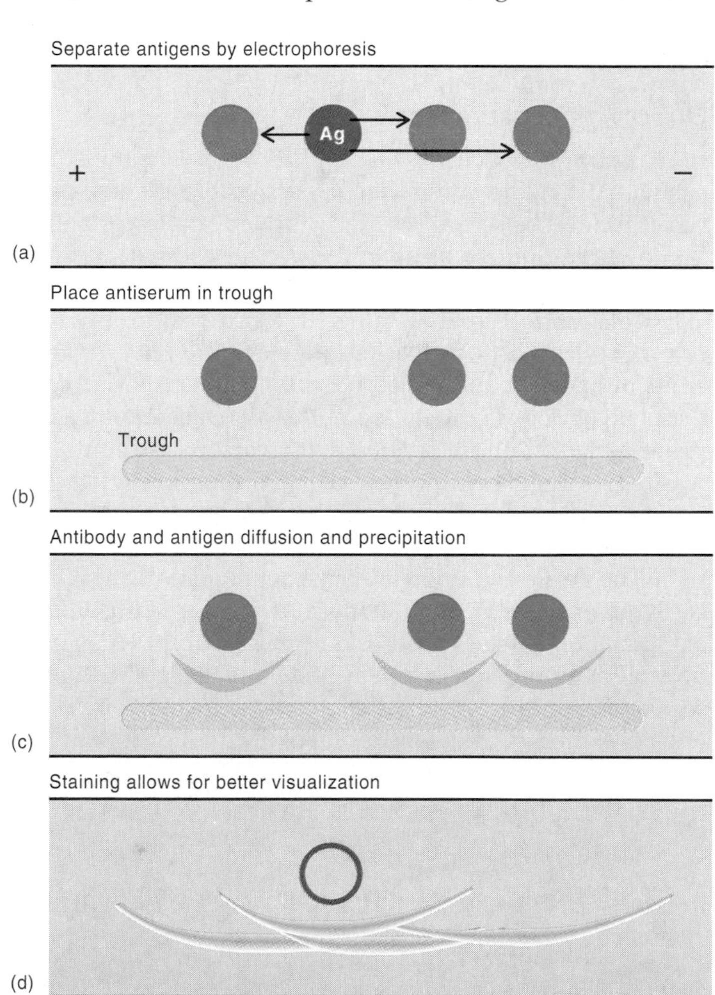

Figure 17.7
Immunoelectrophoresis. (*a*) Antigens are separated in an agar gel by an electrical charge. (*b*) Antibody (antiserum) is then placed in a trough cut parallel to the direction of antigen migration. (*c*) The antigens and antibodies diffuse through the agar and form precipitin arcs. (*d*) After staining, better visualization is possible.

diffuse toward the separated proteins. A line of precipitate forms at the location of each antigen that is recognized.

Countercurrent immunoelectrophoresis (CIE) is immunodiffusion carried out in an electrophoretic field, so that antigen and antibody are quickly brought together into a zone of optimal proportion, making the test more rapid and sensitive. CIE is used to diagnose some diseases such as bacterial meningitis. This test is useful in determining unknown infections when a known serum antibody is used.

Agglutination Reactions Are Similar to Precipitin Reactions Except That the Antigens Are Particles, as Opposed to Soluble Antigens

Agglutination reactions are similar in principle to precipitin reactions; however, in agglutination reactions, the antigen is relatively large particles rather than molecules, so much larger aggregates of antigen and antibody are formed. Figure 17.8, which illustrates an agglutination reaction between red blood cells and specific antibodies (agglutinins), indicates how readily the agglutinates can be seen.

Direct Agglutination Tests

In direct agglutination tests, antibodies are measured against particulate antigens, such as red blood cells, bacteria, or fungi. The serum is serially diluted and tested against a known amount of antigen to give an estimation of the amount of antibody present (the **antibody titer**). The direct agglutination test is done either in test tubes or now more commonly in plastic **microtiter plates** (figure 17.9). Proteins, either antigens or antibodies, adsorb to the plastic surface of the plate strongly, so they are not easily washed out. These proteins can then be treated with various reagents and washed without being removed from the plate surface. A standard microtiter plate has 96 wells, allowing for a large number of tests to be carried out at the same time.

The greatest dilution at which agglutination can be seen easily gives the **titer** of antibody in the serum. Thus, if clear agglutination is seen at serum dilutions of up to 1:256, questionable agglutination is seen at 1:512, and no agglutination is seen at a dilution of 1:1,024, the titer of serum antibody is 256. (Titer is expressed as the reciprocal of the dilution.)

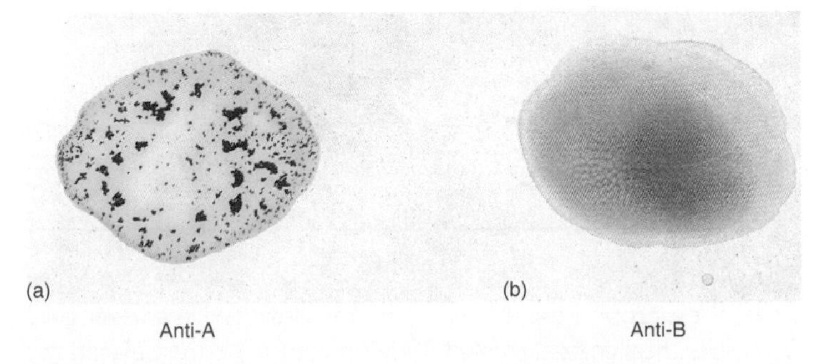

Figure 17.8
Agglutination of erythrocytes. (*a*) Agglutinated red blood cells and antibody. (*b*) Nonagglutinated red blood cells (control).

A **rise in titer** of antibodies to an organism over the course of an illness usually indicates an infection with that organism. As the infection progresses, more and more antibodies are formed and the amount of antibody in the blood increases.

Indirect Agglutination Tests

It is possible for soluble antigen to be used in an agglutination test if the antigen is first attached to red blood cells, latex beads, or other particles. Then, when antibody is added in the test, agglutination occurs and the reaction is much more readily visible than a precipitate of soluble molecules would be. A frequently used indirect agglutination test is the test for rheumatoid factor, commonly found in patients with rheumatoid arthritis. These patients make autoantibodies, known as *rheumatoid factor*, against their own

Figure 17.9
Microtiter plates used for immunoassays. Each plate has 8 rows of 12 wells each, so there are 96 reaction wells per plate. The wells are very small so a minimum of reagents is needed. Special equipment allows rapid dilution and mixing of specimens. For example, in doing a test to measure specific antibodies in a serum sample, the serum must be diluted and each dilution mixed with antigen to look for an antigen-antibody reaction. Undiluted serum is placed in the first well and diluted with a saline solution in twofold dilutions, so that well 2 contains a serum dilution of 1:2; well 3 contains 1:4; well 4 1:8; and so on to a final dilution of 1:512 in well number 10. The last two wells are used for a positive control and a negative control, to be sure the test is working properly. To each dilution of serum is added an equal volume of antigen, making the final serum dilutions in the range from 1:2 through 1:1,024. After mixing, the plates are incubated and then each well is examined for a visible reaction. The titer of antibodies is determined by the highest dilution of serum that gives a definite positive reaction. If there is a positive reaction in the dilution of 1:256 but a negative one in the dilution of 1:512, the titer is the reciprocal of the dilution, or a titer of 256. The plates are small, but eight tests such as the one described can be done on a single plate.

IgG. When IgG (the antigen) is coated on latex particles, the rheumatoid factor (the antibody) in the patient's serum causes easily visible agglutination of the latex.

Hemagglutination Inhibition

Hemagglutination is the clumping of red blood cells. Some viruses such as those causing infectious mononucleosis, influenza, measles, and mumps are able to agglutinate red blood cells by an interaction between surface components of the virus and the red cell which is not a specific immunological reaction. Since specific antiviral antibodies inhibit the agglutination, antibodies to the virus can be measured by **hemagglutination inhibition.** Serum is serially diluted and mixed with the virus and red blood cells. Dilutions of serum containing specific antibodies inhibit the usual agglutination of virus and red cells. The titer of antibodies is determined by the highest dilution of serum in which agglutination does *not* occur.

Neutralization Tests Can Be Used to Test for Virus Infections

Neutralization tests are frequently used to test for virus infections. In the body, viral infections result in the production of antibodies directed against the viral antigens. Thus, the tests involve mixing serum containing the antiviral antibody with a known viral suspension. If antibodies exist to a particular virus, they will bind to it and neutralize the virus, preventing its attachment to and subsequent infection of cells. When the virus is then put onto cell culture or injected into chicken embryos, it is unable to replicate and cause cell damage (figure 17.10).

Neutralization tests can also be used to test for toxin or antitoxin. Specific antibodies will neutralize the effects of the toxin.

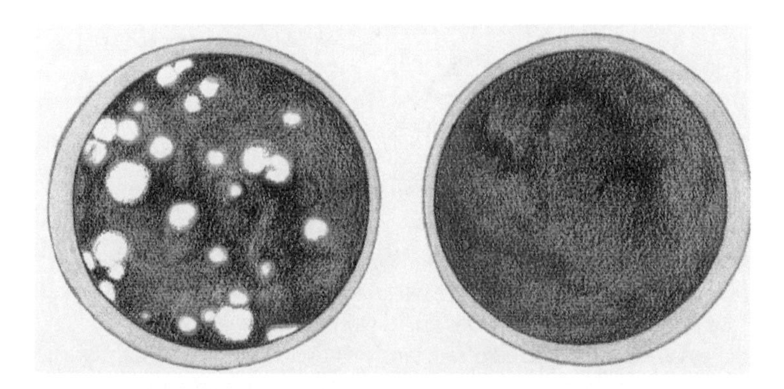

Figure 17.10
Virus neutralization test. (*a*) When a susceptible cell culture is exposed to poliovirus particles, cells become infected and are destroyed, producing visible clear areas (plaques) in a lawn of cells. (*b*) If specific neutralizing antibodies are mixed with polioviruses before exposure to the cells, the viruses cannot combine with host cell receptors. As a result, the cells are not infected and no plaques appear in the lawn of cells.

Immunofluorescence Tests Use a Visible Tag to Detect Antigen or Antibody

Fluorescent dyes such as fluorescein isothiocyanate (FITC) or rhodamine can be attached to known specific antibodies that are then used in tests to detect the presence of a specific antibody in serum or microorganisms in a clinical sample. The tests are quick and sensitive. There are two basic versions of these tests.

Direct Fluorescent Antibody Test

In this test, the microorganism to be tested is fixed to a slide. Known antibodies labeled with fluorescein isothiocyanate are added to the slide and incubated and then the slide is washed. Antibody that binds to the microorganism will not wash away but will stay on the slide. The slide is then examined under a microscope using ultraviolet light. If the microorganisms have bound the antibody, they will glow a yellow-green color (figure 17.11 *a*). If rhodamine is used instead of FITC, the microorganisms will glow a red color. By using various fluorescent dyes, it is possible to locate different antigens in the same cell or preparation.

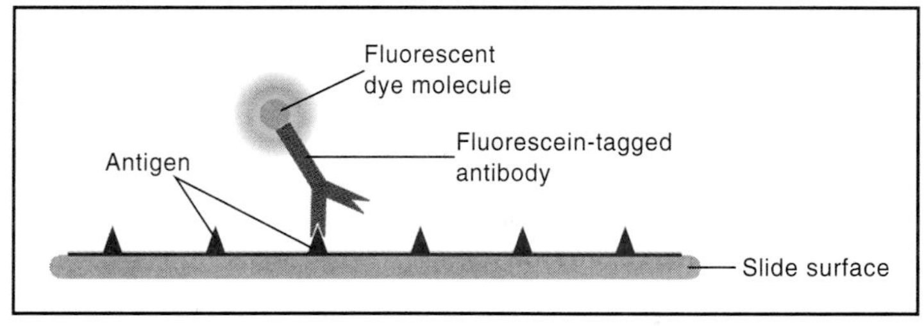

(a) Direct testing

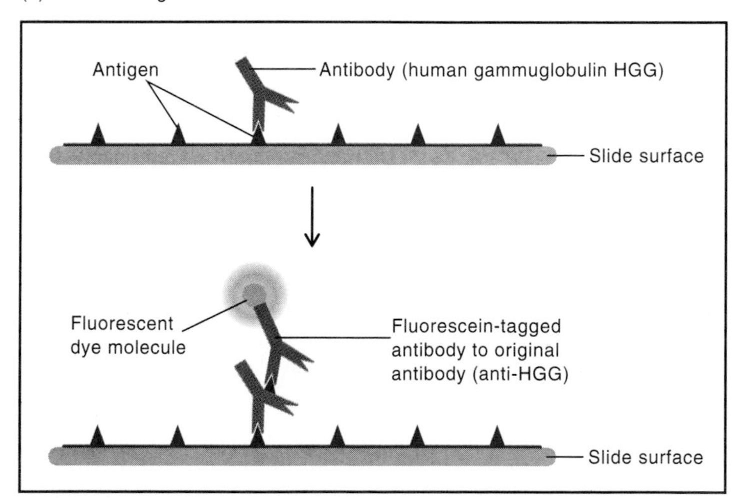

(b) Indirect testing

Figure 17.11
Direct and indirect immunofluorescence. (*a*) In direct testing, antigen is fixed to the slide. The antibody to which fluorescent dye (FITC) has been attached is added. When viewed with UV light, antibodies appear yellow-green. (*b*) In indirect testing, a known antigen reacts with a known antibody (HGG). The fluorescent antibody (anti-HGG) recognizes the HGG antibodies.

Indirect Fluorescent Antibody Test

This test is used to find specific antibody in human serum following an infection. A known organism is fixed to the slide, a sample of the unknown serum is added, and the mixture on the slide is incubated. In order for the antigen-antibody complexes formed to be seen, any unfixed material is washed off and an anti–human-gamma-globulin (anti–human-antibody) antiserum that has been labeled with FITC dye is added to the slide. The anti–human-gamma-globulin attaches to any antibodies remaining on the slide from the unknown serum. Then the slide is examined with ultraviolet light. Again, the organisms appear as yellow-green (figure 17.11 *b*).

The Complement Fixation Test Involves an Indicator System in Which Complement Causes Visible Cell Lysis

Bacteria, red blood cells, or other cells may sometimes undergo lysis as the result of an interaction between antigenic components of the cell, specific antibody, and complement. Recall that complement is a complex system of interacting proteins that can bind to cells and cause them to lyse. This phenomenon is the basis for the **complement fixation test.** Once widely used, this test has been largely superseded by newer, easier techniques, such as ELISA, described next.

RIA and ELISA Use Radioactive Labels or Enzyme Reactions to Visualize Antigen-Antibody Reactions

RIA and **ELISA** are widely used because these tests are extremely sensitive and can be used to test a large number of samples in a short period of time using very little reagents. For example, the ELISA technique is now routinely used to test donated blood for HIV (the virus that causes AIDS) before the blood is used for transfusion. RIA and ELISA apply similar technology, using an easily measurable label to detect either antigen or antibody.

In both **RIA and ELISA**, the known antigen is put into a plastic test tube or on a plastic microtiter plate. Some of the antigen adsorbs to the plastic surface and the excess is then washed away. The serum to be tested is added to the tube or plate and incubated for a short time. If antibodies are present, they bind to the antigen.

In the **radioimmunoassay (RIA),** any excess antibody is washed away and a radioactively labeled material that can react with the antibody is added. The excess radioactive materials are washed away and the tubes are then counted using a gamma counter. This method can detect as little as 1 to 50 ng/ml of antibody; however, the gamma counter is an expensive piece of equipment not found in all laboratories.

The **enzyme-linked immunosorbent assay (ELISA)** is done in a similar manner except, instead of using radioactive material to label, it uses an enzyme such as peroxidase as a label. This binds to the test antibody, excess label is washed away, and chromogen is added. Chromogen is a colorless substrate that produces a colored end product when acted upon by an enzyme such as peroxidase. This color change can then be observed and measured (figure 17.12).

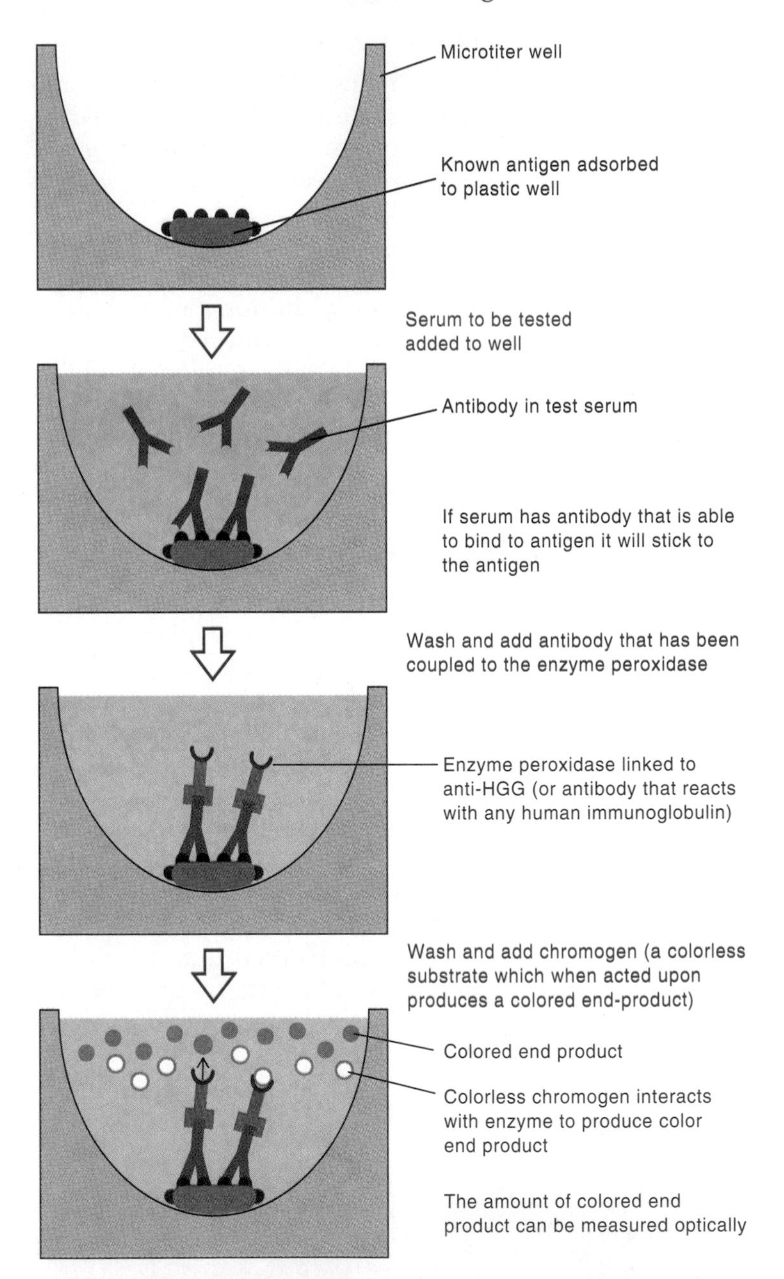

Figure 17.12
Enzyme-linked immunosorbent assay (ELISA). Antibody, which is human gamma globulin, reacts with known specific antigen adsorbed to a plastic surface. After washing away excess reagents, it is necessary to add a labeled reactant that allows the antigen-antibody reaction to be detected. In ELISA, the label used is an enzyme that will react with substrate to give a visible color reaction. For example, an antibody against human gamma globulin (anti-HGG) with an enzyme label is used. The labeled anti-HGG combines with antibodies that have reacted with antigen. After again washing away excess reagents, the substrate for the enzyme is added and the visible colored end product can be measured optically.

Western Blot Combines ELISA with Electrophoresis

Western blot, or immunoblot, is a technique that combines ELISA with electrophoresis to separate antigens. It is used to diagnose infections and screen blood to reduce the risk of transmitting human immunodeficiency virus and in a number of other applications. In this technique, a mixture of antigens is separated by electrophoresis in a polyacrylamide gel. The resulting bands of separated antigens can be identified using specific antibodies, but antibodies do not diffuse well into these gels. Therefore, it is necessary to transfer the bands from the gel onto paper which is then incubated with known antibodies. In order to see the antigen-antibody combinations formed, a detectable label is used, usually an enzyme-labeled anti-antibody, as in ELISA. Western blotting is useful clinically as a test to confirm results obtained from other simpler tests. For example, ELISA is often used to screen for HIV disease (AIDS), and a positive screening test is confirmed with a Western blot test.

REVIEW QUESTIONS

1. Name at least five types of serological tests.
2. What is the titer of an antiserum?
3. Explain the hemagglutination inhibition test used for detecting antibodies against influenza virus.
4. Why is anti–human-gamma-globulin antiserum labeled with detectable substances used in some immunological tests?
5. Why is ELISA much more widely employed than RIA?

Vaccination and Immunization

Immunity Can Be Active or Passive

Immunity can be natural or artificial and active or passive (table 17.1). Natural immunity occurs through exposure to infectious agents or other immunizing agents that induce an immune response; the immunity that results following an attack of measles is an example. Artificial immunity results from deliberate exposure of the host to antigen in order to protect, as with well-known children's "shots." Active immunity is produced by the individual in response to an antigenic stimulus, whereas passive immunity results from the transfer of immune serum or cells produced by other individuals or animals. This section will focus on artificial immunity deliberately induced, both passively and actively.

Passive Immunity Usually Involves the Transfer of Preformed Antibodies

Chapter 16 explored the use of antitoxin to prevent the disease diphtheria that is caused by a toxin. This is an example of passive immunity. Von Behring experimented with antibodies produced in horses to treat diphtheria in an early test of passive immunity. Similarly, antibodies produced in animals have been used to treat tetanus and other diseases. Unfortunately, complications have often arisen following such treatments. Patients have been found to respond to foreign antigens in horse serum with hypersensitivity reactions (these reactions will be discussed in the next chapter). In order to preclude such reactions, antibodies are now obtained from humans who are immune to the antigens because of previous infection or immunization or from hybridomas (perspective 17.1). Human antibodies obtained from different people are similar in structure and therefore much less likely to cause a serious reaction than are antibodies from other species.

Passive immunity also occurs naturally. During pregnancy, the mother's IgG antibodies that cross the placenta protect the fetus and the infant. Consequently, a number of infectious diseases normally do not occur until a baby is three to six months of age, by which time the maternal antibodies have been degraded. Obviously, there is no memory for passive immunity. Once the transferred antibodies are degraded, the protection is lost.

Table 17.1 Examples of Innate and Acquired Immunity

Innate	Inborn as a result of the genetic constitution of a species or an individual; independent of previous experience	For example, immunity of humans to distemper viruses of cats and dogs; protection provided by phagocytes and protective substances such as lysozyme
Acquired	Naturally acquired; active	Antibodies and specialized lymphocytes acquired after natural exposure to a foreign agent; long-lasting; specific
	Naturally acquired; passive	Transfer of immunity to disease from mother to fetus through the placenta; temporary
	Artificially acquired; active	Acquired following immunization (for example, with poliovirus vaccine); long lasting; specific
	Artificially acquired; passive	Acquired by administration of protective antibodies (for example, the transfer of preformed antibodies or gamma globulin for hepatitis); temporary

In many parts of the world, expectant mothers may not be immunized against tetanus, and sometimes contaminated instruments are used to cut the umbilical cord. As a result, newborns develop fatal tetanus. Immunization of a mother before or even during pregnancy allows protective antibodies to cross the placenta and protect her infant. A concerted effort by the World Health Organization to immunize all antenatal mothers against tetanus has helped eliminate this problem.

Table 17.2 lists some passive immunization preparations currently in use. One example is **immune serum globulin,** or gamma globulin, which is the immunoglobulin G (IgG) fraction of pooled plasma from many donors. This contains a variety of antibodies that the various donors have made because of infections and immunization procedures. This may be given to travelers who visit areas where sanitation is minimal, to offer some protection against hepatitis A and other common diseases associated with poor hygiene. In addition, **hyperimmune globulin** is prepared from the sera of donors with high titers of antibodies to certain diseases, and these are used to prevent or treat specific diseases. Examples are human tetanus immune globulin, rabies immune globulin, and hepatitis B immune globulin. These preparations given during the incubation period, after exposure but before disease develops, can often prevent severe diseases from developing.

Active Immunity Is Induced Artificially by Giving Vaccines

A variety of vaccines or immunizing preparations are used to induce active immunity. Sometimes these preparations contain **adjuvants,** substances that help to induce a better response. Alum is the only adjuvant currently used in vaccines for humans. Antigens adhere to the alum and induce a stronger immune response than they would without the adjuvant.

Living, attenuated agents (either organisms or viruses) cause an infection with undetectable or mild disease and usually a solid and long-lasting immunity. *Attenuated* means that the agent, although viable, has been modified so as to be incapable of causing disease under ordinary circumstances. A single dose of an attenuated agent is generally sufficient to induce immunity. Since the agent multiplies in the body, the antigen is present for a longer period and in greater amounts than with nonliving agents. The disadvantage of using living agents to immunize is that they have the potential to cause disease in immunosuppressed people, and rarely they can revert or mutate to strains that cause serious disease. Such a living preparation has the added potential of being spread from an individual being immunized to other nonimmune people. If the virus reaches a recipient who is immunosuppressed, the results can be tragic, but normal individuals can become immunized by the inadvertent spread of the attenuated organisms.

Examples of living attenuated vaccines in widespread use are measles, mumps, rubella, and yellow fever. The Sabin poliomyelitis vaccine is another living, attenuated virus. This has given protection against polio to millions of people over many years. The attenuated polio vaccine has to be given in a series of three doses rather than one because of interactions among the three types of virus included in the vaccine. Currently, it is recommended that a dose of killed vaccine be given before the living attenuated vaccine to give a degree of protection against the spread of the virus. Table 17.3 lists some living attenuated vaccines in current use.

Care must be taken to avoid giving living vaccines to pregnant women, because the vaccines may cross the placenta and cause damage to the developing fetus.

Inactivated immunizing agents include killed bacteria and viruses and inactivated bacterial toxins. These are the **whole agent vaccines.** Also included with the inactivated

Table 17.2 Passive Immunization Procedures

Disease	Route	Preparation
Cytomegalovirus (CMV)	IV*	CMV immune globulin
Diphtheria	IM† or IV	Antiserum made in horses
Hepatitis A	IM	Human immune gamma globulin
Hepatitis B (HB)	IM	Human HB immune globulin
Measles	IM	Human immune serum globulin
Rabies	IM and locally	Human rabies immune globulin
Rubella	IM	Human immune serum globulin
Tetanus	IM	Human tetanus immune globulin
Varicella-zoster	IM	Human zoster immune globulin

*IV—intravenous †IM—intramuscular

Perspective 17.2

A Vaccine Success Story

Haemophilus influenzae type b is a tiny gram-negative rod that is a common pathogen affecting young children. This organism living in the upper respiratory tract can invade the bloodstream and reach other tissues. Of particular concern are serious invasive infections, such as meningitis in babies under two years of age and pneumonia or other invasive infection in children two to five years old. In the past, it was estimated that about 1 in 200 children developed serious *H. influenzae* infections by five years of age. The component of the bacteria principally responsible for the ability to cause disease was identified as the polysaccharide type b capsular material, polyribitol phosphate.

A vaccine made of the polyribitol phosphate was not effective in children under two years of age, the age of the most dangerous infections. Recall that polysaccharides are usually thymus-independent antigens that induce the formation of IgM antibodies and do not induce a memory response or the production of IgG antibodies. Children under two years old do not respond well to thymus-independent antigens. For this reason, a vaccine made of the polyribitol phosphate was not effective in the younger chldren who were most at risk for *H. influenzae* type b infections. Consequently, a new vaccine, in which the polyribitol phosphate was conjugated with proteins from the diphtheria toxoid (universally used for immunization) and some other proteins, was developed. This vaccine was licensed for use in 1989, and in 1990 it was recommended for routine immunization of infants, beginning at two months of age. The efficiency of this conjugate subunit vaccine can be seen in the statistics. The incidence of invasive *H. influenzae* infections has dropped more than 94% since this vaccine was introduced into general use.

Table 17.3 Live Attenuated Vaccines

Disease	Route and Number of Doses Used
Adenovirus	Oral; single dose (virus types 4 and 7; (for military only)
Measles	SC*; single dose
Mumps	SC; single dose
Poliomyelitis	Oral; three doses following primary dose of killed vaccine
Rubella	SC; single dose
Tuberculosis	SC or ID†; single dose of Bacille Calmette-Guerin (in endemic areas)
Typhoid	Oral; dose repeated every other day for four doses (for travelers)
Varicella	IM‡; single dose
Yellow fever	SC; single dose every 10 years (for travelers)

*SC—subcutaneous †ID—intradermal ‡IM—intramuscular

Table 17.4 Inactivated Whole Agent Vaccines

Disease	Route and Number of Doses Used
Cholera	SC*; several doses (limited effectiveness)
Influenza	IM†; once yearly
Plague	SC; several doses (for those at high risk)
Poliomyelitis	SC; several doses or single dose followed by live attenuated vaccine
Rabies	Depends on exposure (for at-risk individuals)

*SC—subcutaneous †IM—intramuscular

vaccines are the **subunit vaccines,** made of products or portions of an agent.

Inactivated whole agent vaccines include those against cholera, plague, influenza, and rabies and the Salk vaccine against polio (table 17.4). The agents are often killed or inactivated by treatment with formalin, a chemical that does not significantly change the surface epitopes, leaving the agent antigenic even though it cannot reproduce. These vaccines are advantageous in that they cannot cause infections or revert to dangerous forms. They also have several disadvantages. Since they do not replicate, it is usually necessary to give several doses of the vaccine to induce solid immunity. Parts of the agents that are not concerned with the immune response are included in the vaccines and can sometimes cause unwanted reactions. For example, the whooping cough (pertussis) killed vaccine that was previously used routinely for immunizing babies and young children often caused reactions such as pain, tenderness at the site of the injection, fever, and occasionally, convulsions. Currently, this killed whole cell vaccine is being replaced by a subunit (acellular) vaccine, made of fragments of the bacteria, that does not cause these side effects. Another success story of a subunit vaccine is given in perspective 17.2, which examines immunization against infections with *Haemophilus influenzae* type b.

The **toxoids** of diphtheria and tetanus bacteria are the bacterial toxins, treated so as to destroy the toxic part of the molecules but retain the antigens. These toxoids, along with alum adjuvant, make up the vaccines against diphtheria and tetanus.

Subunit vaccines are made by isolating the antigens or antigenic fragments of the agent and administering only a subunit of the total agent. Some subunit vaccines in use are those directed against meningococci, pneumococci, *Haemophilus influenzae* type b, and pertussis (table 17.5).

Genetic engineering is used to produce another kind of subunit vaccine called **recombinant vaccine.** An example is the vaccine against the hepatitis B virus. Genes for a part of the viral protein coat are inserted into yeast cells, which then produce the proteins. These make a safe vaccine but one that requires several doses in order to be effective. Genetically engineered bacteria have been used to produce vaccines, and genetically altered vaccinia virus is also being used for experimental vaccines. Much remains to be done in developing vaccines against some serious and widespread infectious diseases, such as malaria and AIDS.

In addition to recombinant vaccines, some other new approaches to vaccine development are being intensively studied. Table 17.6 lists some of these experimental approaches.

Among the new vaccines are **peptide vaccines.** The peptides from pathogenic organisms are either isolated or synthesized in the laboratory and made into vaccines. Such vaccines are stable to heat and other environmental influences, do not contain extraneous materials to cause unwanted hypersensitivity reactions, and have several other desirable features. However, peptide vaccines are weakly immunogenic and are very expensive to prepare. Although under active investigation, no peptide vaccines are yet approved for human use.

Another new approach is the use of DNA alone as an immunizing agent. Segments of DNA from infectious organisms can be introduced in various ways into human beings so as to replicate and code for production of microbial antigens that then induce an immune response. This procedure eliminates the possibility of infection with the immunizing agent, as only a small portion of the microbe's DNA is used. This approach is under active investigation.

Another promising means of immunization, especially for developing nations, is the use of edible vaccines produced in plants. Preliminary studies have shown that genes from various pathogenic organisms can be introduced into plants. For example, potato plants were transformed with a gene for part of an *Escherichia coli* enterotoxin. The potatoes expressing the enterotoxin fragment were then fed to mice, inducing the production of serum and secretory antibodies against the toxin. It is hoped that appropriate genes can be introduced into common plants, such as bananas, resulting in very low-cost immunization for whole populations. However, these investigations are in their early stages.

While vaccination is widely used to protect against infectious diseases, it is also under study as a means to control fertility and hormone activity.

The Importance of Routine Immunizations for Children Cannot be Overemphasized

Although there is some risk associated with almost any procedure, including a very slight risk with routine immunizations, the benefits greatly outweigh the risks. Consider that before vaccination was available for common childhood diseases, thousands of children died or were left with permanent disability from these diseases. Pertussis (whooping cough) decreased markedly in incidence in the United States after the introduction of routine immunization.

Table 17.5 Some Subunit Vaccines in Current Use

Disease or Infection	Vaccine
Haemophilus influenzae type b	Polyribitol phosphate-protein conjugate
Meningococcus (groups A and C)	Capsular polysaccharide; used for military and other high-risk groups
Pertussis	Acellular vaccine
Pneumococci	Capsular polysaccharides of a number of types; used in elderly populations and for people at high risk

Table 17.6 Some Experimental Approaches to Vaccine Development

Vaccine	Characteristics and Challenges
Peptide vaccines	Peptides isolated or synthesized, stable but expensive and weakly immunogenic
DNA	Genes for antigens introduced into nonpathogenic viruses or microbes and then into humans
Edible vaccines	Genes for antigens introduced into foods, leading to immune responses
Cholera B-subunit with killed cholera vibrios	Oral vaccine effective in trials; may be able to use other antigens along with cholera B-subunit as oral vaccine
Human fertility control	Trials aimed at immunizing against sperm, egg, or reproductive hormones; far from ready for human use
Modulation of hormone action	Enhancement or depression of activity of hormones, such as growth hormone

Table 17.7 Schedules of Common Immunizations*

Age	HBV†	DTP††	TOPV§	HbCV‖	MMR‡	Td**
2 months	X	X	X	X		
4 months	X	X	X	X		
6 months	X	X		X		
12 months				X		
15 months		X	X	X	X	
4–6 years		X	X		X	
14–16 years						X

*This is the schedule of immunizations recommended by the Centers of Disease Control and Prevention in the United States.
†HBV—hepatitis B vaccine
††DTP—diphtheria, tetanus, and pertussis vaccines
§TOPV—trivalent oral poliovirus vaccine; first dose of polio vaccine at 2 months uses killed vaccine; others are live attenuated poliovirus
‖HbCV—haemophilus b conjugate vaccine
‡MMR—measles, mumps, and rubella vaccines
**Td—tetanus and diphtheria toxoids; given every 10 years after adolescence

Because of some adverse reactions, many parents refused to allow their babies to get this vaccine. By 1990, this refusal of vaccination resulted in the highest incidence of pertussis cases in 20 years and the deaths of some children, mostly under one year of age. In recent years, the recommended protocol for whooping cough immunization has used killed whole cell vaccine given at two, four, and six months of age, because it induces a strong response, and a subunit pertussis vaccine that gives fewer side reactions for booster doses after 15 months. Several recent large-scale studies carried out in Europe of the use of acellular pertussis vaccines for primary immunization have shown the acellular vaccines to be more effective and have fewer side effects than the whole cell vaccines. In 1996, the Federal Drug Administration approved the acellular pertussis vaccine for use in the United States as the primary immunizing agent against whooping cough. We can expect that soon only the acellular vaccines will be used in this country, without the need for whole cell preparations.

The recommendations of the United States Public Health Service for childhood and adolescent immunizations are shown in table 17.7.

REVIEW QUESTIONS

1. What is immune serum globulin? Give several examples.
2. What is the purpose of adjuvants used in vaccines?
3. Give four examples of living, attenuated vaccines currently in use and four examples of killed vaccines also in use.
4. What makes up the vaccines used to protect against diphtheria and tetanus?
5. What is a subunit vaccine?

Summary

I. Functions of Antibodies in Vivo
   A. IgM antibodies are especially efficient in agglutination and the activation of complement, which leads to inflammation and increased phagocytosis.
   B. IgG antibodies are the only ones that can cross the placenta to protect the fetus and newborn. IgG Fc receptors cause increased and more effective phagocytosis, as does complement activation by IgG. This class of antibodies is also effective in agglutination, precipitation, and neutralization.
   C. IgA antibodies inhibit adherence of microorganisms and prevent colonization. IgA is of primary importance in the protection of mucous surfaces.
   D. IgE can function in antibody-dependent cellular cytotoxicity with eosinophils, and it functions in immunity to parasites as well as in allergies.

II. Functions of Cell-Mediated Immunity in Vivo
   A. Cytotoxic T cells kill virus-infected cells, tumor cells, and cells of tissue transplants.
   B. Cell-mediated immunity limits the growth of intracellular organisms within cells, especially macrophages.

III. Applications of Immune Responses in Vitro
   A. Precipitation reactions involve the cross-linking of soluble antigen molecules by specific antibody molecules to form visible precipitates. This is the basis of many laboratory tests, such as the double-diffusion Ouchterlony test and immunoelectrophoresis.
   B. Agglutination reactions involve the cross-linking of particles of antigen by specific antibody molecules to form easily visible clumps. Agglutination tests may be direct, in which the antigen is part of the clumped particles, or indirect, in which the antigen is attached to larger particles.

C. The titer of antibodies in a sample is the greatest dilution at which a visible reaction occurs when the sample is tested. A rise in titer during an illness usually indicates an active infection.
D. Neutralization tests are used to test for virus infections. Specific antibodies bind to the virus, preventing its attachment to and infection of cells, thereby neutralizing it. Toxins may also be neutralized by specific antibodies, rendering them nontoxic.
E. Immunofluorescence tests use a visible fluorescent dye attached to either antigen or antibody to visualize the antigen-antibody reaction.
F. The complement fixation test involves a specific antibody acting with antigen and an indicator system containing complement to give a visible reaction.
G. Radioimmunoassay and ELISA use radioactive or enzyme reactions to visualize antigen-antibody reactions.
H. Immunoblot techniques combine ELISA with electrophoresis to separate antigens, providing highly specific and sensitive tests.

IV. Vaccination and Immunization
A. Passive immunity involves the transfer of preformed antibodies, as from mother to fetus, gamma globulin injections to protect against hepatitis or other diseases, or antitoxin injections to counteract toxins.
B. Active immunity is induced by active immune responses against agents or by giving living, attenuated or killed vaccines.
C. Other forms of immunizing agents are toxoids, subunit vaccines, recombinant vaccines, and peptide vaccines.
D. The United States Public Health Service regularly issues recommended schedules for immunizations suitable for residents of the United States.

Chapter Review Questions

1. Why is IgM especially efficient in agglutination and complement activation?
2. What class of antibodies is found in milk and what is the function of these antibodies?
3. Which class(es) of antibodies can attach by Fc receptors to neutrophils and which to eosinophils?
4. How does cell-mediated immunity limit the growth of organisms in macrophages?
5. What is the significance of a rise in titer of specific antibodies during an illness?
6. How can a soluble antigen be used in an agglutination test?
7. Draw a diagram to explain why no precipitate is found if there is a great excess of antigen.
8. In immunoelectrophoresis, why are there usually multiple lines of precipitation?
9. List some advantages and disadvantages of killed vaccines and of living, attenuated vaccines.
10. How has genetic engineering improved available vaccines?

Critical Thinking Questions

1. List some possible immune defects that could interfere with effective cell-mediated immunity (such as lack of Th cells) and note how these would interfere with the response (lack of cytokines to stimulate and produce activated macrophages).
2. Suppose that a new and dangerous virus has been isolated and that the virus and its soluble antigens are available. Suggest at least two tests that could be developed to test for antibodies against the virus. What are some of the advantages and disadvantages of each test?
3. In a precipitation reaction, what happens when more antibody is added to a tube in which there were optimal proportions of antigen and antibody and, hence, maximum precipitation?

Further Reading

Hall, Stephen S. 1995. Monoclonal antibodies at age 20: promise at last? *Science* 270: 915–916.

1996. Acellular pertussis primary vaccine approved. *ASM News* 62(10):518–519.

1996. Vaccine side effects, adverse reactions, contraindications, and precautions. *Morbidity and Mortality Weekly Report* 45(RR-12):2.

18 Immunologic Disorders

KEY CONCEPTS

1. Hypersensitivities result from the immune system reacting to certain antigens in such a way as to lead to undesirable consequences.
2. Immunodeficiency results when the immune system cannot respond adequately to antigenic stimulation; overproduction of immune factors can also occur.
3. Autoimmune disease can result when the immune system responds to tissues in its own body.

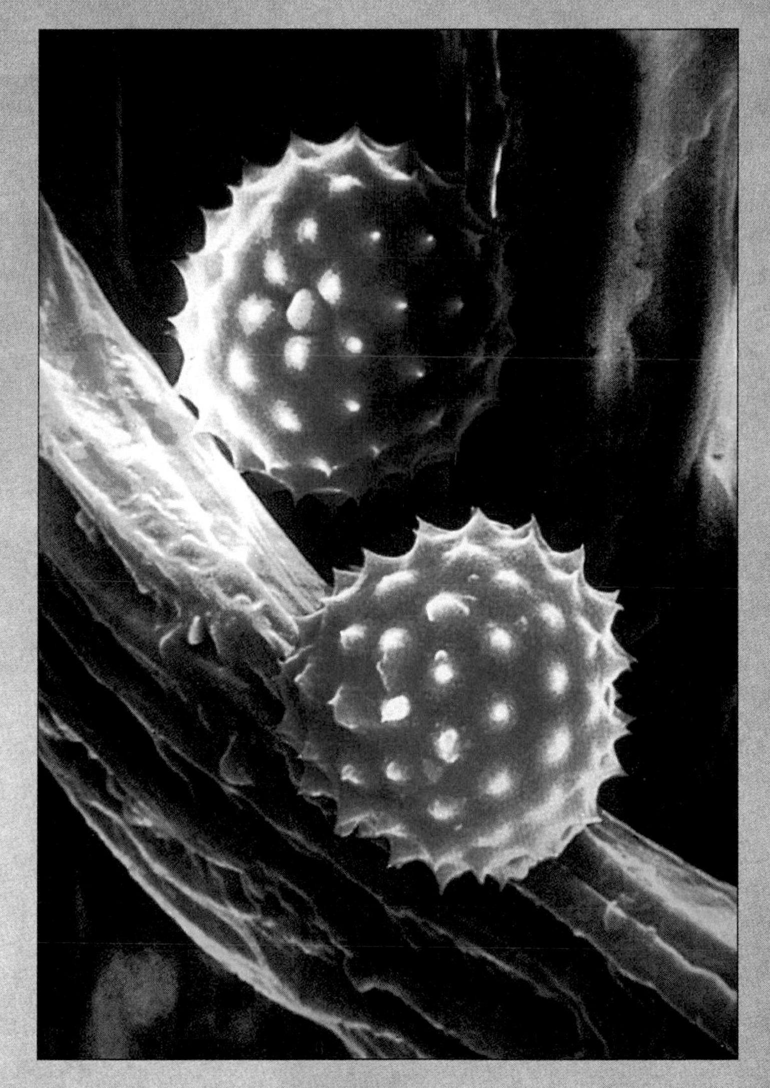

Scanning electron microscopy of ragweed pollen. Abundant in many parts of the United States during the late Summer, ragweed pollen is one of the most common causes of hay fever (allergic rhinitis).

A Glimpse of History

Pasteur is widely quoted as saying, "Chance favors the prepared mind." This was certainly the case with Charles Richet in his discovery of hypersensitivities near the end of the 19th century. Richet was a well-known physiologist who had discovered, among other things, that the acid in the stomach is hydrochloric acid and who had done early experiments with antisera. He and his colleague, Paul Portier, while cruising in the South Seas on Prince Albert of Monaco's yacht, decided that the Portuguese man-of-war jellyfish must have a toxin responsible for its ugly stings. They made an extract of the tentacles of one such jellyfish and showed that it was very toxic to rabbits and ducks.

Upon returning to France, they were no longer able to study the man-of-war, so they switched to the sea anemone, testing in dogs extracts of its tentacles. Some dogs survived the toxic dose, but when these animals were tested with the toxin again, they died a few minutes after receiving a small dose. One very healthy dog survived the first challenge with the toxin. When given a second dose 22 days later, the dog became very ill within a few seconds. It could not breathe, lay on its side, had violent diarrhea, vomited blood, and died within 25 minutes.

Richet and Portier, because of their early work with antisera, had prepared minds. They recognized that the dogs' reactions were probably immunological but represented the opposite effect of prevention of disease (prophylaxis) resulting from antisera. Therefore, they named this development of hypersensitivity to relatively harmless substances **anaphylaxis** (the extreme opposite of prophylaxis). For this and other outstanding contributions to medicine, Richet received the Nobel Prize in 1913.

For the most part, the immune system does a superb job protecting the body from invasion by various microorganisms and viruses; however, the same mechanisms that are so effective in protecting can, under some circumstances, be detrimental. First, the immune system can act in an inappropriate manner, giving rise to hypersensitivity reactions that cause tissue damage. Second, the immune system can respond too little or too much, resulting in immunodeficiency or overproduction. Third, immune responses against self-antigens, substances of self, can lead to autoimmune diseases.

Hypersensitivities

The term **hypersensitivity** is used to describe immune responses that occur in an exaggerated or inappropriate fashion, causing tissue damage. Hypersensitivities are categorized according to which parts of the immune response are involved and how quickly the response occurs. Most hypersensitivity reactions fall into one of the four major types: types I, II, III, and IV. Their main characteristics are shown in table 18.1.

Type I, or Immediate, Hypersensitivities Are Mediated by IgE

Type I hypersensitivities are characterized by an immediate allergic[1] reaction upon exposure to antigen. Some people develop such a reaction when exposed to substances such as dust, pollens, animal dander, and molds, which do not evoke a response in nonallergic people.

The following events occur during a Type I hypersensitivity response (figure 18.1). First, the antigen makes contact with some part of the body (pollen grains contact the mucosa of the nose, for example). The antigen is taken up and processed by antigen-presenting cells and with the Th cells is presented to B cells. Although B cells making all the classes of immunoglobulin are probably induced to make

[1]Allergic—from the Greek *allos* for "altered" and *ergeai* for "energy"

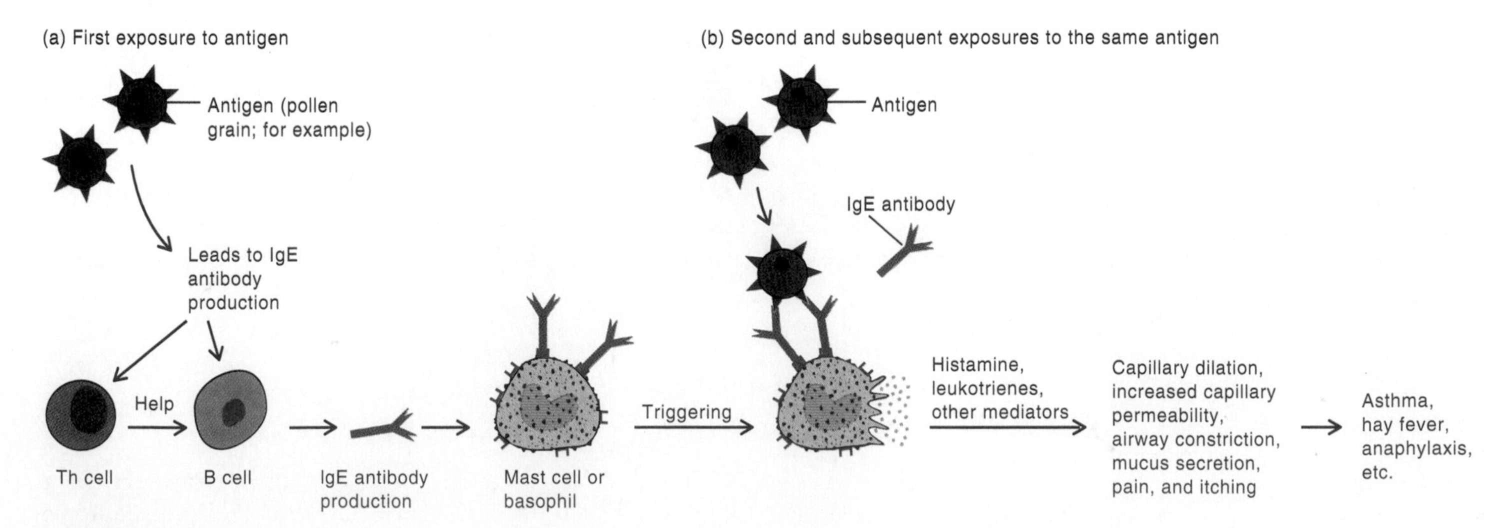

Figure 18.1
Mechanism of immediate (type I) hypersensitivity, or anaphylaxis.

Table 18.1 Some Characteristics of the Major Types of Hypersensitivities

Characteristic	Type I hypersensitivity	Type II hypersensitivity	Type III hypersensitivity	Type IV hypersensitivity
Descriptive name of response	Immediate; IgE-mediated	Cytotoxic	Immune-complex	Delayed
Cell type responsible	B cells	B cells	B cells	T cells
Location of antigen	Soluble	Cell-bound	Soluble	Soluble or cell-bound
Type of antibody	IgE	IgG, IgM	IgG	None
Other cells involved	Basophils, mast cells	Red blood cells, white blood cells, platelets	Various host cells	Various host cells
Mediators	Histamine, serotonin, leukotrienes	Complement, ADCC	Complement, neutrophil proteases	Cytokines
Transfer of hypersensitivity	By serum	By serum	By serum	By T cells
Time of reaction after challenge with antigen	Immediate, up to 30 minutes	Hours to days	Hours to days	Peaks at 48 to 72 hours
Skin reaction	Wheal and flare	None	Arthus	Induration, necrosis
Examples	Anaphylactic shock, hay fever, hives	Transfusion reaction, hemolytic disease of newborns	Serum sickness, farmer's lung, malaria	Tuberculin reaction, contact dermatitis, tissue transplant rejection

antibodies, the tissues under the mucous membranes are rich in B cells committed to IgA and IgE production, and IgE-producing cells are more abundant in allergic individuals than in others. These cells have been switched from making other classes of immunoglobulin to making IgE, probably by the action of cytokines. As IgE is produced in this area that is rich in mast cells, the IgE molecules attach via their Fc portion to receptors on the mast cells and also on circulating basophils. Mast cells and basophils contain many large cytoplasmic granules packed with histamine, leukotrienes, and other potent **chemical mediators** of the type I hypersensitivity reaction.

Once attached, the IgE molecules can survive for many weeks with their reaction sites waiting to interact with antigen. People with a type I hypersensitivity will have many IgE antibodies fixed to mast cells throughout their bodies. When such a person is not exposed to antigen, these antibodies are harmless. Upon exposure, however, the antigen readily combines with the cell-fixed IgE antibodies. At least two cell-bound IgE molecules must react with an antigen, cross-linking the IgE molecules, in order for a reaction to occur. Within seconds, the IgE-antigen attachment and cross-linking of IgE in the cell membrane cause the mast cell to release histamine, leukotrienes, and other mediators from its cytoplasmic granules. These chemical mediators are the direct causes of the diseases seen, including **urticaria (hives), allergic rhinitis (hay fever), asthma, anaphylactic shock,** and other allergic manifestations.

One might wonder why this type of harmful reaction has persisted throughout evolutionary development. There is evidence that IgE antibodies are of value in immunity against parasitic worms. During a worm infestation, soluble worm antigens react with specific IgE antibodies, causing mast cells to degranulate and release proteins that cause an increase in vascular permeability and attract inflammatory cells. Also, IgE antibodies and eosinophils together can kill some parasites (chapter 17). Thus, IgE seems to have a protective role.

The tendency to have type I allergic reactions is inherited. The reactions occur in a relatively small percentage of people. Some statistics suggest the figure may be about 20% to 30% of the population of the United States.

Localized Anaphylaxis

Anaphylaxis, the name given by Richet and Portier to the condition they observed, may be generalized or local. By far, the most usual allergic reactions are examples of localized anaphylaxis. Urticaria (hives) is an allergic skin condition characterized by the formation of wheals, itchy red swellings that have a general resemblance to mosquito bites. They may occur, for example, when a person allergic to lobster eats some of the seafood. Lobster antigen absorbed from the intestinal tract enters the bloodstream and is carried to tissues such as skin where it reacts with mast cells that have antilobster IgE antibody attached to them. Reaction between antibody and antigen on the mast cell surface releases histamine, which in turn causes dilation of tiny blood vessels and the leaking of plasma into skin tissues. Because histamine is a major mediator in this situation, the reaction is blocked by antihistamines. Life-threatening respiratory obstruction is possible during a reaction if a giant hive involves the throat and larynx.

Allergic rhinitis (hay fever), marked by itching, teary eyes, sneezing, and runny nose, occurs when allergic persons inhale an antigen such as ragweed pollen to which they are sensitive. The mechanism is similar to that of hives and is also blocked by antihistamines.

Asthma is another type of immediate respiratory allergy. Antigen-induced release of chemical mediators from IgE-sensitized mast cells causes spasms of the bronchi, markedly interfering with respiratory function. Mediators other than histamine, mainly leukotrienes, are responsible for the bronchospasm and increased mucus production. Antihistamines are therefore of no value in treating asthma, but several other drugs are available to block the reaction.

Generalized Anaphylaxis

Generalized anaphylaxis is a rare but serious form of IgE-mediated allergy. This is the form of anaphylaxis described by Richet. Instead of involving localized areas of the skin or respiratory tract, it affects almost the entire body. Loss of fluid from the blood vessels into tissues causes not only swelling, but also a condition called **shock.** Shock is the result of such extensive blood vessel dilation and loss of fluid from the blood that blood pressure falls dramatically and there is insufficient blood flow to vital organs such as the brain. Anaphylactic reactions may be fatal within minutes. Bee stings and penicillin injections probably account for most cases of generalized anaphylaxis. As mentioned in chapter 16, penicillin molecules are changed in the body to a haptenic form that can react with body proteins. In a percentage of people, this hapten protein complex causes the formation of IgE antibodies that react with penicillin. Fortunately, only a tiny fraction of people who make the antibodies are prone to anaphylactic shock. Anyone who has had a urticarial reaction to bees, penicillin, or other substance is at risk of anaphylactic shock if exposed to the same substances and should wear a medical-alert bracelet and carry emergency medications.

Immunotherapy

In view of the danger of severe generalized anaphylaxis upon exposure to even tiny amounts of the offending antigen, it might seem paradoxical that one way of preventing type I hypersensitivity reactions is to inject the sensitive person with extremely dilute solutions of the antigen. The concentration of antigen in the injected solution is very gradually increased over a period of months in a series of injections. The patient gradually becomes less and less sensitive to the antigen and may even lose the hypersensitivity entirely. Following treatment, there is an increase in the levels of IgG and Ts cell activity and a decrease in the IgE response to the antigen. The IgG antibodies produced are thought to protect the patient by binding to the offending antigen, thus preventing its attachment to IgE (figure. 18.2).

Also, Ts cells are known to develop and these may suppress the production of IgE antibody.

Type II Hypersensitivity Reactions, or Cytotoxic Reactions, Are Caused by Humoral Antibodies that Can Destroy Normal Cells

Complement-fixing antibody that reacts with cell surface antigens may cause injury or death of the cell. This phenomenon is especially striking in the case of red blood cells because the reaction is **cytolytic,** resulting in rupture of the cell and release of hemoglobin. The reaction can occur as a result of foreign antigenic or haptenic material such as a drug that attaches to erythrocytes, but more common examples are transfusion reactions and hemolytic disease of the newborn. In these cases, normal red cells are lysed as a result of antibodies reacting with erythrocyte epitopes.

Transfusion Reactions

Normal erythrocyte surfaces are characterized by a large number of antigens, and these antigens differ from individual to individual. When a person receives transfused red blood cells that are antigenically different from his or her own, immune cytolysis often results. This cytolytic reaction is especially likely when the antigens involved are those of the ABO blood group system.

A locus on one of the 23 pairs of human chromosomes, chromosome number 9, determines which of the ABO antigens will be present on an individual's red blood cells. Since one gene of the chromosome pair is originally derived from the mother and the other is from the father, the genes of the pair may differ. In the case of the ABO system, there are three possible types of genes for the chromosomal locus: A, B, and O. A and B genes each code for a different glycosyltransferase responsible for attaching A or B polysaccharide antigen to the red cell surface. The O gene does not code for either, indicating a lack of A or B glycosyltransferases. When both an A and a B gene are present, they are codominant, meaning that both antigens are expressed on the red blood cell surface. The possible ABO

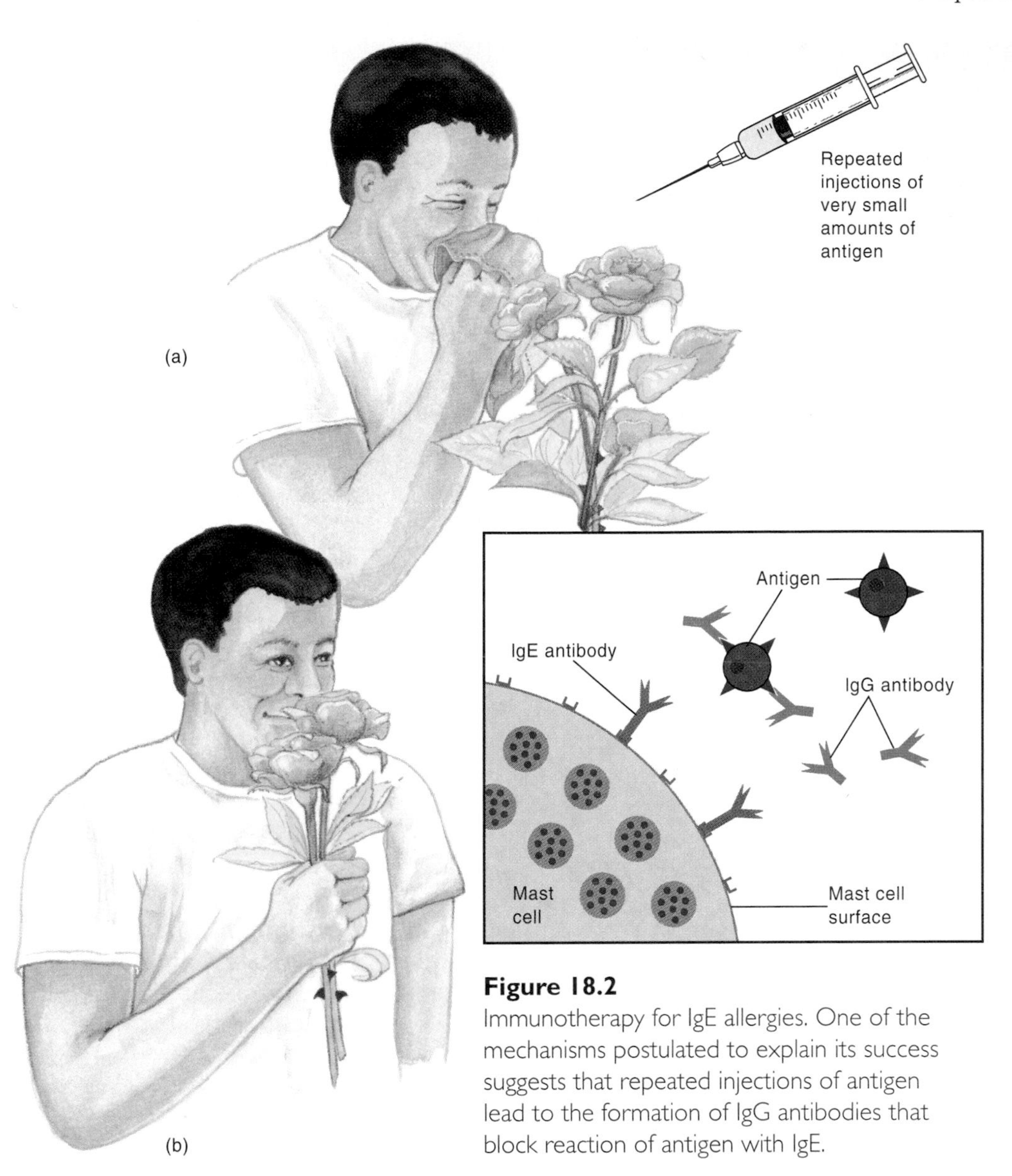

Figure 18.2
Immunotherapy for IgE allergies. One of the mechanisms postulated to explain its success suggests that repeated injections of antigen lead to the formation of IgG antibodies that block reaction of antigen with IgE.

blood types are shown in table 18.2. People are designated as having blood type A, B, AB, or O, depending on which, if any, ABO polysaccharide is present.

One important and unexpected feature of the ABO system is that people who lack A or B erythrocyte polysaccharides have antibodies against the lacking antigen. Thus, people with blood type O have both anti-A and anti-B antibodies; those of blood type A have anti-B; and those of type B have anti-A. Type AB people have neither anti-A nor anti-B antibodies. These antibodies are called **natural antibodies** because they are present without any obvious or deliberate stimulus (the anti-A and anti-B antibodies are not present at birth but generally appear before the age of six months). These natural antibodies are mostly of the class IgM and are capable of fixing complement and lysing red cells. Since they are IgM, they cannot cross the placenta. One theory is that they arise because of multiple exposures to small amounts of substances similar to the blood type antigens, substances known to be found in many environmental materials such as bacteria, dust, and foods.

In addition to these natural antibodies, anti-A and anti-B antibodies of IgG class arise after exposure to A and B antigens through mismatched blood transfusions or during pregnancy and delivery of a baby whose blood type differs from the mother's. These IgG hemolysins are generally more potent than the corresponding natural antibodies.

This discussion is a simplified explanation of just one of the red cell antigen systems; there are a number of others. Except in dire emergencies, it is the practice to test donor and recipient bloods for possible incompatibilities in addition to ABO. Cross-matching the bloods and other techniques are used to ensure compatibility of donor and recipient. A transfusion reaction can occur if a patient receives erythrocytes differing antigenically from his or her own during a blood transfusion. In the case of ABO incompatibility, IgM antibodies cause a type II hypersensitivity reaction. The symptoms of a typical transfusion reaction include fever, low blood pressure, pain, nausea, and vomiting. These symptoms occur because the foreign erythrocytes are agglutinated by the recipient's antibody, complement is activated, and red blood cells are lysed.

However, when death is imminent from blood loss and there is no time to test the patient's blood, type O blood cells can be transfused if they lack the Rh antigen. This is a relatively safe option because most severe transfusion reactions involve the A, B, and Rh antigens. In an emergency, people of blood type AB (who lack anti-A and anti-B) can receive any Rh-negative ABO type blood with reasonable safety.

Hemolytic Disease of the Newborn

The Rh (Rhesus) blood group system is complex and involves various antigens. Only the one with the greatest antigenic potency, the D antigen or simply the Rh antigen, will be discussed here. If the Rh antigen is present on a person's erythrocytes, he or she is Rh positive; if it is lacking, the person is Rh negative.

The Rh-positive adult is generally not at risk for most Rh blood group problems. As long as other antigens are compatible, such a person can receive either Rh-positive or

Table 18.2 Antigens and Antibodies in Human ABO Blood Groups

Blood Type	Antigen Present on Erythrocyte Membranes	Antibody in Plasma	Incidence of Type in U.S. Among Whites	Incidence of Type in U.S. Among Blacks
A	A	Anti-B	41%	27%
B	B	Anti-A	10%	20%
AB	A and B	Neither anti-A nor anti-B	4%	7%
O	Neither	Anti-A and anti-B	45%	46%

Rh-negative blood transfusions, since Rh-positive blood is generally compatible with other Rh-positive blood, whereas Rh-negative cells lack the Rh antigen. It is the Rh-negative adult who may experience an Rh blood transfusion reaction as a result of being immunized against the Rh antigen, such as through a transfusion, organ graft, or pregnancy that might lead to introduction of the antigen and consequent development of anti-Rh antibodies.

Anti-Rh antibodies that are formed by a pregnant woman may damage her offspring. The resulting disease is called **hemolytic disease of the newborn** or simply *Rh disease* (figure 18.3).

While an Rh-negative mother is carrying an Rh-positive fetus, few fetal red blood cells enter the mother's circulation via the placenta and usually not enough to provoke a primary antibody response. However, at the time of birth, enough of the Rh-positive baby's erythrocytes may enter the mother's circulation to incite a vigorous immune response. Induced or spontaneous abortions may also be responsible for sensitizing an Rh-negative mother to the Rh antigen.

The anti-Rh antibodies formed by the mother cause her no harm because her erythrocytes lack the Rh antigen. However, there is an increasing chance of problems with every subsequent Rh-positive fetus the mother might carry. With each such pregnancy, even a few Rh-positive cells that might enter the mother's circulation from the fetus are enough to provoke a secondary response. The result is production by the mother of large quantities of anti-Rh antibodies of the class IgG. Like other IgG antibodies, anti-Rh antibodies readily cross the placenta, enter the circulation of the fetus, and cause extensive disease from fetal red cell damage. Although miscarriage with loss of the fetus can result, the disease often is not apparent until shortly after birth. Blood incompatibilities other than Rh may be responsible for this disease, but such cases are generally less severe.

The reason the fetus survives the Rh antibody attack while still in utero is that harmful products of red cell destruction can be eliminated from its system by enzymes of the mother. Since these enzymes are present in only scant amounts in the newborn, the toxic effects quickly appear after birth. As early as 36 hours after birth, jaundice may appear and result in permanent brain damage or even death. Not only does the baby become seriously ill from the toxic products of red cell destruction, but it also develops a severe anemia. In this critical situation, it may be necessary to withdraw a portion of the baby's blood and replace it with Rh-negative blood. Since Rh-negative cells are not harmed by the maternal anti-Rh antibody, they can supply the infant's needs for a number of weeks until the maternal antibody is largely eliminated from the baby's circulation. The baby's production of its own Rh-positive cells can then supply its own needs. In severe cases, tranfusions can even be given to unborn babies.

Exchange transfusions as described previously are rarely needed these days because new methods, such as **light treatment,** exist to detoxify the erythrocyte breakdown products. Irradiation with light at 420 to 480 nm makes the red blood cell breakdown products more readily excretable. In addition, most cases of hemolytic disease caused by Rh incompatibility are preventable by injecting the Rh-negative mother with anti-Rh antibodies within the first 24 to 48 hours and preferably within the first few hours of abortion or delivery, thus passively immunizing her against Rh antigen. These administered antibodies block the development of anti-Rh antibodies by the mother to Rh-positive cells from the fetus.

Cells can be destroyed in type II reactions by, in addition to complement lysis, antibody-dependent cellular cytotoxicity (ADCC) or a mechanism involving macrophage "nibbling" of the red cell membrane until the red cell is destroyed. This sort of red cell destruction occurs when the antigen with antibody attached is not in the proper quantity or position for complement lysis to occur, as is the case with the Rh antigens.

Type III Hypersensitivity, or Immune Complex-Mediated Reactions, Activate Inflammatory Mechanisms

When antibody reacts with a specific soluble antigen, an **immune complex** results. Often, these complexes fix complement and thus contain complement components. Immune complexes usually adhere to Fc receptors on cells of the mononuclear phagocyte system and are engulfed and destroyed intracellularly. Thus, under ordinary circumstances, they are rapidly removed from circulation. However, under conditions where a moderate excess of antigen over antibody exists, the complexes are not quickly removed and destroyed but persist in circulation or at their sites of formation in tissue. Immune complexes possess considerable biological

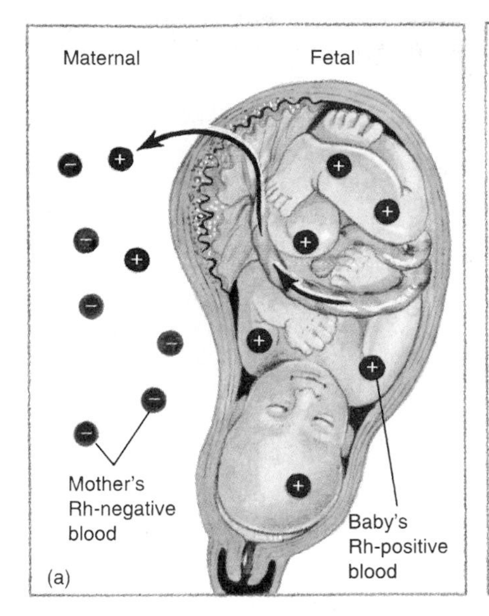

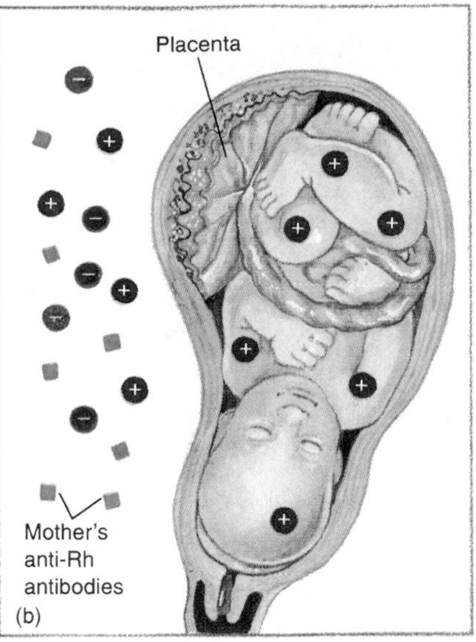

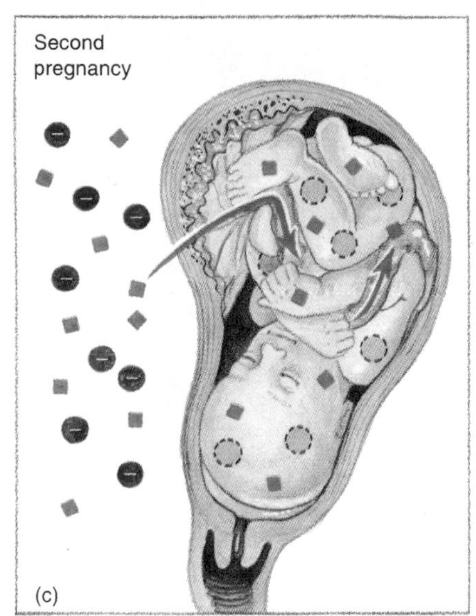

Figure 18.3
Hemolytic disease of the newborn. (*a*) Rh-positive blood from the fetus reaches the mother. (*b*) The mother makes a small primary response to Rh antigen. (*c*) During a subsequent pregnancy with an Rh-positive fetus, a secondary anti-Rh response is made, and the IgG antibodies enter the body's circulation and destroy its erythrocytes.

activity; they activate the blood clotting mechanism, and they cause the production of the C5a component of complement, which attracts neutrophils and contributes to inflammation (figure 18.4). Proteases released from the neutrophils are especially active in inducing tissue damage. Circulating immune complexes are commonly deposited in the kidneys, joints, and the skin, where they are responsible for the rashes, joint pains, and other symptoms seen in a number of diseases. Immune complexes can also precipitate a devastating condition, **disseminated intravascular coagulation,** described in chapter 30, in which bleeding develops at multiple sites.

Immune complex effects may arise during a variety of bacterial, viral, and protozoan infections, as well as from inhaled dusts or bacteria and injected medications such as penicillin. The steps in the formation of immune complex pathology are summarized in table 18.3.

Immune complex formation is also responsible for the localized injury or death of tissue **(Arthus reaction)** that occurs if antigen is injected into the tissue of a previously immunized animal or person with high levels of circulating specific antibody. Neutrophils predominate in these immune complex reactions. A few examples of disease states in which immune complexes play a prominent role are presented in table 18.4. They include **farmer's lung,** in which the foreign antigen consists of thermophilic actinomycetes inhaled with dust from spoiled hay; the skin rash of rubella measles; and the kidney damage of malaria. In addition, some forms of **glomerulonephritis**[1] result from immune complex deposition in the kidney. Another immune complex disease is **serum sickness,** caused by injecting antiserum from an animal into humans to prevent or treat a disease such as diphtheria.

[1]Glomerulonephritis—inflammation of small tufts of tiny blood vessels (glomeruli) within the kidney (nephros)

Table 18.3 Pathogenesis of Immune Complex Disease

1. Antibody combines with excess soluble antigen.
2. The antibody-antigen combination reacts with complement.
3. Complexes are deposited in sites such as skin, kidney, and joints.
4. Fragments of complement cause release of histamine and other mediator substances from mast cells or basophils and also attract neutrophils.
5. Release of the mediators causes increased permeability of blood vessel walls.
6. Immune complexes penetrate or form in blood vessel walls.
7. Neutrophils enter the vessel walls chemotactically.
8. Neutrophils release lysosomal enzymes, especially proteases, that induce tissue injury.

Type IV Hypersensitivity Reactions, or Delayed Hypersensitivity Reactions, Involve T Lymphocytes

Untoward effects produced by the mechanisms of cell-mediated immunity are referred to as *delayed hypersensitivity.* The name reflects the fact that the response to antigen appears

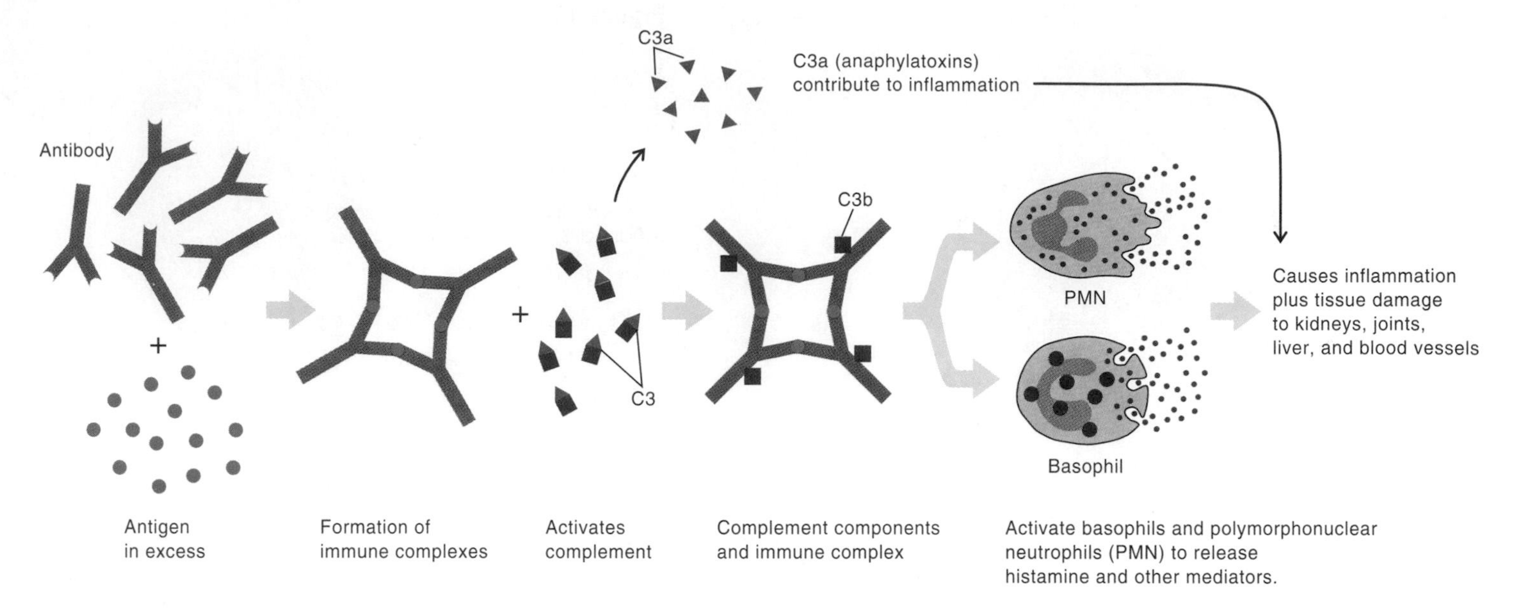

Figure 18.4
Type III (immune complex) hypersensitivity. The antibody and antigen combine to form immune complexes that activate complement. This causes basophils and platelets to degranulate, releasing histamine and other mediators that increase vascular permeability. The increased permeability allows the immune complexes to be deposited in the blood vessel wall. This induces platelet aggregation to form blood clots on the vessel wall. Polymorphonuclear neutrophils, or PMNs, stimulated by complement degranulate, causing enzymatic damage to the blood vessel wall.

Table 18.4 Some Disease States in Which Immune Complexes Play a Role

Disease	Antigen	Effect of Complexes
Farmer's lung	Inhaled hay bacteria	Lung damage
Bacterial endocarditis	Bacteria from infected heart	Kidney damage
German measles (early)	Rubella virus	Skin rash
Malaria	Malaria protozoa	Kidney damage

more slowly than in immediate hypersensitivity. As would be expected with cell-mediated responses, T cells are responsible and antibodies are not involved. Delayed hypersensitivity reactions can occur anywhere in the body and are wholly or partly responsible for **contact dermatitis** (such as from poison ivy), tissue damage in a variety of infectious diseases, rejection of tissue grafts, and some autoimmune diseases.

Tuberculin Skin Test

A familiar example of a delayed hypersensitivity reaction is the positive reaction to a tuberculin skin test that occurs in most people who have been infected with *Mycobacterium tuberculosis*. This test involves the introduction of very small quantities of protein antigens of the tubercle bacillus into the skin. In those with delayed hypersensitivity to *M. tuberculosis,* the site of injection reddens and gradually becomes **indurated** (thickened) within 6 to 24 hours. The reaction reaches its peak at two to three days. There is no wheal formation, as would be seen with IgE-mediated reactions. The redness and induration of delayed skin hypersensitivity reactions are the result mainly of the reaction of sensitized T cells with specific antigen, followed by the release of lymphokines and the influx of macrophages to the injection site.

Contact Hypersensitivities

Familiar examples of contact hypersensitivity (contact allergy or contact dermatitis) are poison ivy (figure 18.5) and allergic reactions to the nickel of metal jewelry or the chromium salts in certain leather products.

In the case of poison ivy, the antigen is an oily product of the poison ivy plant. In the case of a metal, a soluble salt of the metal acts as a hapten. Contact hypersensitivity is commonly detected by **patch tests,** in which the suspect substance is applied to the skin under an adhesive bandage. Positive reactions are maximum in about three days and consist of redness, itching, and blisters of the skin.

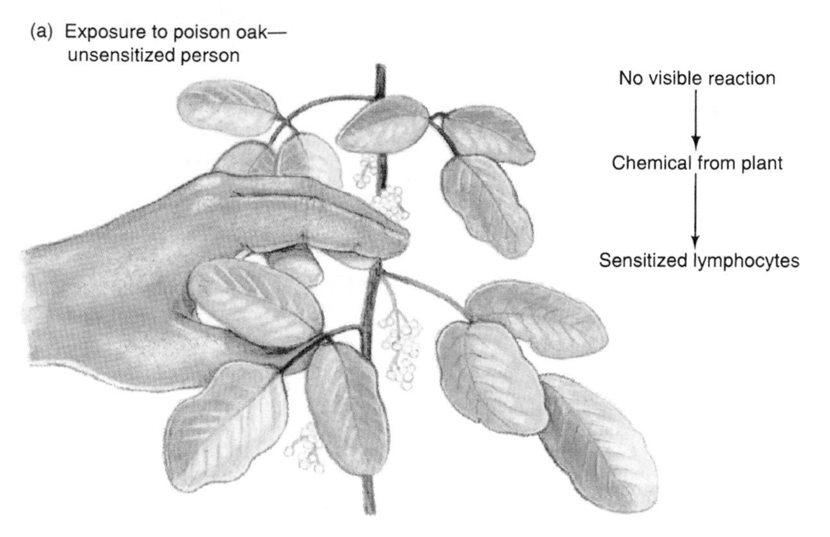

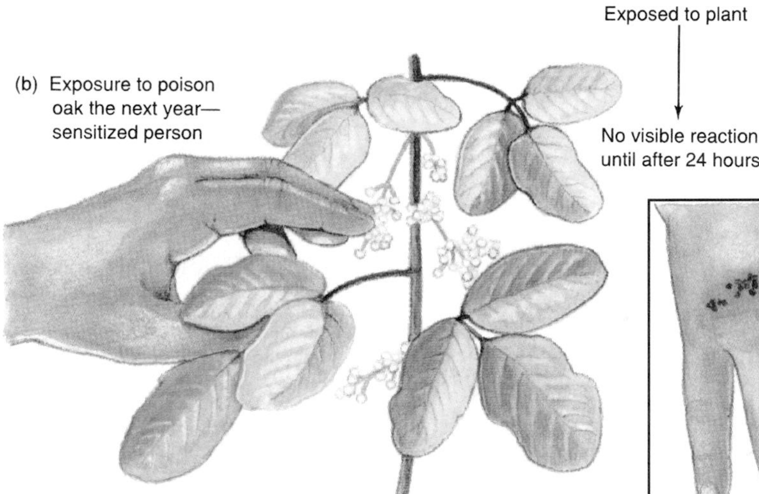

Figure 18.5
Poison oak is an example of delayed (type IV) hypersensitivity. This type of reaction is mediated by T cells. T cells that have become sensitized to a particular antigen release lymphokines when they come in contact again with the same antigen. These lymphokines cause inflammatory reactions that attract macrophages to the site. The macrophages then release mediators that add to the inflammatory response.

Delayed Hypersensitivity in Infectious Diseases

The role of cell-mediated immunity in combating intracellular infections through the cell-destroying activity of activated macrophages and T lymphocytes was discussed previously. These infections may be caused by viruses, mycobacteria and certain other bacteria, protozoa, and fungi. They include leprosy, tuberculosis, leishmaniasis, and herpes simplex, among many others. In particularly slowly progressing infections, such as some cases of leprosy and hepatitis B infections, delayed hypersensitivity causes cell destruction and progressive impairment of tissue function, such as the damaged sensory nerves in leprosy and the progressive liver damage of chronic hepatitis.

Transplantation Rejection

Like red blood cells, other body cells vary antigenically from individual to individual, and differences between donors' and recipients' tissues can cause immunological reactions. There are many different tissue antigens, but those of the MHC system are the ones most commonly involved in transplantation graft rejections.

MHC tissue typing is done in an effort to ensure that no major tissue incompatibility exists between a prospective tissue donor and the recipient patient. Antigenic incompatibility leads to graft rejection, a process consisting principally of a delayed hypersensitivity reaction to the grafted tissue. Killing of the graft cells occurs through a complex combination of mechanisms, including direct contact with sensitized T lymphocytes and NK cells.

In addition to carefully matching donor and recipient tissue antigens, in transplantation it is necessary to employ **immunosuppressant drugs** indefinitely to prevent graft rejection. These drugs are needed because it is almost impossible to find a donor compatible in all the many tissue antigens. Only genetically identical siblings, such as identical twins, have tissues that match in every way.

Cyclosporine, produced by a fungus, is an effective immunosuppressant drug. Cyclosporine specifically suppresses T cells but does not kill them. It has no effect on B cells. Use of cyclosporine has boosted the success rate (one-year graft survival) of kidney transplants to 90% to 95% in patients with kidneys from living related donors. The drug has fewer side effects than other immunosuppressant medications primarily because it selectively suppresses T cell activity but it does not inhibit other parts of the immune system.

• **MHC antigens, pp. 367, 368**

REVIEW QUESTIONS

1. How do generalized and localized anaphylactic reactions differ?
2. Describe an example of a type II hypersensitivity reaction and explain its cause.
3. What type of hypersensitivity reaction may result when antitoxin is given to destroy toxin?
4. Itching of eyelids developing a couple of days after use of eyeshadow is likely a hypersensitivity reaction of which type?
5. How does hypersensitivity contribute to the rejection of transplanted organs?

Immunodeficiency Disorders

In immunodeficiencies, the body is incapable of making or sustaining an adequate immune response. There are two basic types of immunodeficiency diseases: **primary,** or **congenital,** and **secondary,** or **acquired.** Primary immunodeficiency can be inborn as the result of a genetic defect or can result from developmental abnormalities. Secondary immunodeficiency can be acquired as the result of infection or other trauma such as starvation. People with either type of immunodeficiency are subject to repeated infections. The types of these infections will often depend on which part of the immune system is absent or malfunctioning.

Primary Immunodeficiencies Result from Genetic or Developmental Abnormalities

Primary immunodeficiencies occur because of a defect in some component of the immune system, a defect that is not secondary to environmental factors such as viral infections, malnutrition, drugs, or malignancies. Primary immunodeficiencies may result from genetic abnormalities or developmental abnormalities. For example, when certain cell lines or organs, such as the thymus, fail to develop normally, the result is that lymphocytes cannot respond to antigenic stimuli or there may be a metabolic defect that causes the cells to function improperly.

Primary immunodeficiencies may affect B cells or T cells or both. They may also affect natural killer (NK) cells, phagocytes, or complement components. In general, immunodeficiencies are characterized by an increased susceptibility to infection. The type of infection reflects the particular deficiency.

Primary immunodeficiencies are rare. Agammaglobulinemia[1] occurs in a ratio of about 1:50,000, and severe combined immunodeficiency[2] has a ratio of about 1:500,000 live births. An exception is selective IgA deficiency, which has been reported in different studies to occur in ratios as high as 1:333 to 1:700.

Primary deficiencies may occur in various components of the complement system. For example, the few individuals who lack C3 are prone to develop severe, life-threatening infections with encapsulated bacteria. Patients with deficiencies in the late components of the classical pathway of C activation (C5, C6, C7, C8) have recurrent infections caused by *Neisseria.* The effects of defects in other parts of the C system are reflected by various symptoms, depending on the deficient component.

Some of the more important immunodeficiency diseases are listed in table 18.5.

In people with **DiGeorge's syndrome,** there is a failure of the thymus to develop in the embryo. As a result, T cells do not differentiate and are absent. Affected people have other developmental defects as well, such as heart and blood vessel abnormalities, and a characteristic appearance of the upper lip and ears. As expected from a lack of T cells, affected people are very susceptible to infections by eukaryotic pathogens, such as fungi, as well as viruses, obligate intracellular bacteria, and *Pneumocystis carinii* (see chapter 30).

Boys with **infantile X-linked agammaglobulinemia** lack B cells and therefore do not produce immunoglobulins. They generally are healthy babies, protected by maternal antibodies that have crossed the placenta. But by six to nine months of age, they become highly susceptible to streptococcal and staphylococcal infections.

Selective IgA deficiency, in which very little or no IgA is produced, is the most common primary immunodeficiency known. Although people with this disorder may appear healthy, many have repeated bacterial infections of areas where secretory IgA is normally protective, such as the respiratory, gastrointestinal, and genitourinary tracts.

Severe combined immunodeficiency (SCID) results when neither T nor B lymphocytes are produced from bone marrow stem cells. People with SCID die of infectious diseases at an early age unless they are sucessfully treated, which may mean that they receive a bone marrow transplant to reconstitute the bone marrow with healthy cells.

There are a variety of causes of SCID. One cause is defects in an enzyme necessary for V, D, and J chain recombination to form B and T cell receptors for antigen. Without these receptors, there is a lack of functioning B and T cells. Other individuals with SCID have been found to lack adenosine deaminase, an enzyme important in the proliferation of B and T cells, and a number of these people have responded well to repeated replacement of the adenosine deaminase enzyme. It has been possible to correct this condition in a few children by collecting their own defective T cells, transfecting[3] them with the adenosine deaminase gene linked to a retrovirus vector, and returning the cells to them. There is much excitement about the possibility of treating other severe disorders with gene therapy.

Chronic granulomatous disease involves the phagocytes, which fail to produce hydrogen peroxide and certain other active products of oxygen metabolism, due to a defect in an oxidase system normally activated by phagocytosis. Hence, the phagocytes are defective in their ability to kill some organisms, especially the catalase-positive *Staphylococcus aureus.*

[1]Agammaglobulinemia—deficiency or absence of gamma globulin in the blood

[2]Severe combined immunodeficiency—deficiency or failure of both B cell mediated and T cell mediated immunity

[3]Transfection—transfer of genetic information by introducing a vector capable of replicating (such as a virus) that has had DNA sequences for the desired genetic information inserted into its genome

Table 18.5 Immunodeficiency Disorders

	Disease	Part of the Immune System Involved
Primary immunodeficiencies		
	DiGeorge's syndrome	T cells (deficiency)
	Congenital agammaglobulinemia	B cells (deficiency)
	Infantile X-linked agammaglobulinemia	Early B cells (deficiency)
	Selective IgA deficiency	B cells making IgA (deficiency)
	Severe combined immunodeficiency (SCID)	Bone marrow stem cells (defect)
	Chediak-Higashi disease and chronic granulomatous disease	Phagocytes (defect)
Secondary immunodeficiencies		
	Acquired immune deficiency syndrome (AIDS)	T cells (destroyed by virus)
	Monoclonal gammopathy	B cells (multiply out of control)

Chediak-Higashi disease also affects the phagocytes. The lysosomes of affected people's phagocytes are deficient in certain enzymes, and thus phagocytized bacteria are not destroyed. These people suffer from recurring pyogenic[1] bacterial infections.

Secondary Immunodeficiency Results from Malnutrition, Immunosuppressive Agents, Infections, and Some Malignancies

Secondary or acquired immunodeficiency diseases are the result of malignancies, advanced age, certain infections (especially viral infections), immunosuppressive drugs, or malnutrition. Often, an infection will cause a depletion of certain cells of the immune system. The measles virus, for example, causes the death of many lymphoid cells, leaving the body temporarily open to other infections. Syphilis, leprosy, and malaria affect the T cell population and also macrophage function, causing defects in cell-mediated immunity.

Some immunodeficiency diseases result from cells overproducing a single immunoglobulin. For example, multiple myeloma is a malignancy arising from a single plasma cell that proliferates out of control and produces large quantities of its specific immunoglobulin. This overproduction of a single kind of molecule results in the body using its resources to produce useless immunoglobulins at the expense of those needed to fight infection. It is of interest that scientists have used multiple myeloma cells in vitro to produce monoclonal antibodies (perspective 17.1).

[1]Pyogenic—pus-producing

Malnutrition causes decreased immune responses, especially the cell-mediated response. Malignancies involving the lymphoid system often decrease effective antibody-mediated immunity. In addition to multiple myeloma, these lymphoid disorders include macroglobulinemia (overproduction of IgM) and some forms of leukemia. In Hodgkin's lymphoma, cell-mediated immunity is decreased.

One of the most serious of the secondary immunodeficiencies is AIDS (acquired immune deficiency syndrome), caused by human immunodeficiency virus (HIV). This RNA virus of the retrovirus group infects and destroys helper T cells, leaving the affected person highly susceptible to infections, especially with opportunistic agents. AIDS and opportunistic infections are covered in detail in chapter 30.

REVIEW QUESTIONS

1. Why do the symptoms of infantile X-linked agammaglobulinemia appear at about six months of age instead of at birth?
2. What is the defect in severe combined immunodeficiency and how might this defect be corrected?
3. Multiple myeloma is a plasma cell tumor in which a clone of malignant plasma cells produces large amounts of immunoglobulin. With all this excess immunoglobulin, how can a person with multiple myeloma be immunodeficient?

Autoimmune Diseases

Usually, the body's immune system recognizes its "self" antigens and does not attack its own tissues. However, a growing number of diseases are suspected of being caused by an autoimmune process, meaning that the immune system of the body is responding to the tissues of the body as foreign. Some of these diseases are listed in table 18.6.

Some bacterial and viral agents try to evade destruction by the immune system by developing amino acid sequences that are similar to some self antigens. As a result, the immune system is unable to discriminate between the agent and self. The immune system may then destroy the substances of self as well as those of the bacteria or virus. It has been found that the likeness of amino acid sequences need not be exact for this self destruction; as slight as 50% likeness may lead to an autoimmune response.

Autoimmune responses may occur after tissue injury in which antigens are released from the injured organ, as in the case of a heart attack. The autoantibodies formed react with heart tissue and cause further damage. A variety of mechanisms account for the development of autoimmune responses.

Autoimmune reactions occur over a spectrum ranging from organ-specific reactions, such as several kinds of thyroid disease, to widespread responses not limited to any one tissue, such as lupus erythematosus, in which autoantibodies are made against nuclear constituents of all body cells, and rheumatoid arthritis, in which autoantibodies are made against IgG and collagen.[1]

Some autoimmune diseases are mediated by antibodies and others by delayed hypersensitivity. **Myasthenia gravis** is an example of a disease in which humoral antibodies are involved. **Sympathetic ophthalmia** is an example of disease in which delayed hypersensitivities are involved. As mentioned, autoantibodies are commonly demonstrated against gamma globulins in **rheumatoid arthritis** and against cell nuclei in **lupus erythematosus.** Two examples illustrating humoral and cell-mediated immune diseases are given next.

Table 18.6 Some Diseases Suspected of Being Autoimmune

Disease	Main Areas Affected
Rheumatoid arthritis	Joints
Lupus erythematosus	Joints, skin, heart, kidneys
Graves' disease	Thyroid
Diabetes mellitus	Pancreas
Myasthenia gravis	Muscles
Sympathetic ophthalmia	Eyes
Multiple sclerosis	Brain, spinal cord
Hemolytic anemia	Blood

Some Autoimmune Diseases Are Caused by the Production of Antibodies to Structures Within the Body: Myasthenia Gravis

An individual with myasthenia gravis experiences muscle weakness. The cause of the disease is known to be the production of antibodies to the acetylcholine receptor proteins that are present on muscle membranes. These receptors are located where the nerve contacts the muscle. Normally, transmission of the impulses from the nerve to the muscle takes place when acetylcholine is released from the end of the nerve and crosses the gap to the muscle fiber.

When these areas were examined by immunochemical techniques, IgG antibody molecules and complement were evident. The antibody binds to the acetylcholine receptors, thereby blocking access of acetylcholine to these receptors. Further evidence that IgG antibodies are involved comes from babies born to mothers with myasthenia gravis. The babies also experience muscle weakness since IgG antibodies cross the placenta. Fortunately, the effect is not permanent as these IgG antibodies decay within a few months and the babies are no longer affected. Treatment of myasthenia gravis includes the administration of drugs that inhibit the enzyme cholinesterase, allowing acetylcholine to accumulate so some contact with receptors can occur. Immunosuppressive medications and thymectomy are helpful in many cases. The role of the thymus in this disease is not understood.

Some Autoimmune Diseases Are Mediated by Mechanisms of Delayed Hypersensitivity: Sympathetic Ophthalmia

Sympathetic ophthalmia is a condition in which a pentrating wound of one eye produces blindness in the other eye. For this to occur, some of the darkly pigmented tissue present inside the eye must be affected by the penetrating wound, and there must be a prolonged inflammatory response stimulated by foreign material. Under these conditions, instead of normal resolution of the inflammation and healing, the eye tissues become infiltrated with sensitized lymphocytes and macrophages. Delayed hypersensitivity against the pigmented eye tissue can be demonstrated at this point, and it leads to progressive loss of vision and eventual shrinkage of the eye.

Unfortunately, the destructive process is not limited to the injured eye but soon appears in the healthy eye. If the injured eye is surgically removed promptly at this point, the process in the healthy eye resolves without detectable damage. If there is a delay, the process progresses in the healthy eye whether or not the injured eye is removed.

[1]Collagen—main supportive protein of skin, tendon, bone, cartilage, and connective tissue.

The phenomenon of sympathetic ophthalmia can be explained as follows (figure 18.6): Injury and inflammation in one eye cause alteration of pigmented eye tissue so that it becomes antigenic and is recognized as foreign by the immune system. Sensitized lymphocytes that arise in response to this antigen are carried by the bloodstream to the healthy eye, where they attach to epitopes in the pigmented tissue. If the attack is intense and prolonged, tissue in the healthy eye can be destroyed, and the process becomes self-perpetuating.

Autoimmune diseases result from a loss or lack of immunological tolerance to substances that normally would not induce an immune response. To correct this abnormal response, it is desirable to produce an immunological tolerance to the antigens in question. An interesting experimental approach is being tested, with some preliminary success, in some autoimmune diseases. Rheumatoid arthritis patients often have an active immune response to collagen, a protein prominent in the joints and surrounding tissues, causing extreme pain in the joints. The new treatment, which has given encouraging results, involves feeding solutions of animal collagen daily to the patients. The rationale depends on a well-known phenomenon called **feeding tolerance.** Antigen introduced by the oral route can sometimes induce tolerance instead of the expected immunity, for reasons that are not entirely clear. It will be interesting to see if this approach will prove effective in the treatment of a variety of autoimmune diseases.

REVIEW QUESTIONS

1. Distinguish between primary and secondary immunodeficiency diseases.
2. Which occur more often, primary or secondary immunodeficiency diseases?
3. Give an example of a secondary immunodeficiency disorder.
4. How can infectious agents cause autoimmune disease?
5. What is the spectrum of autoimmune diseases from organ-specific to nonorgan specific?

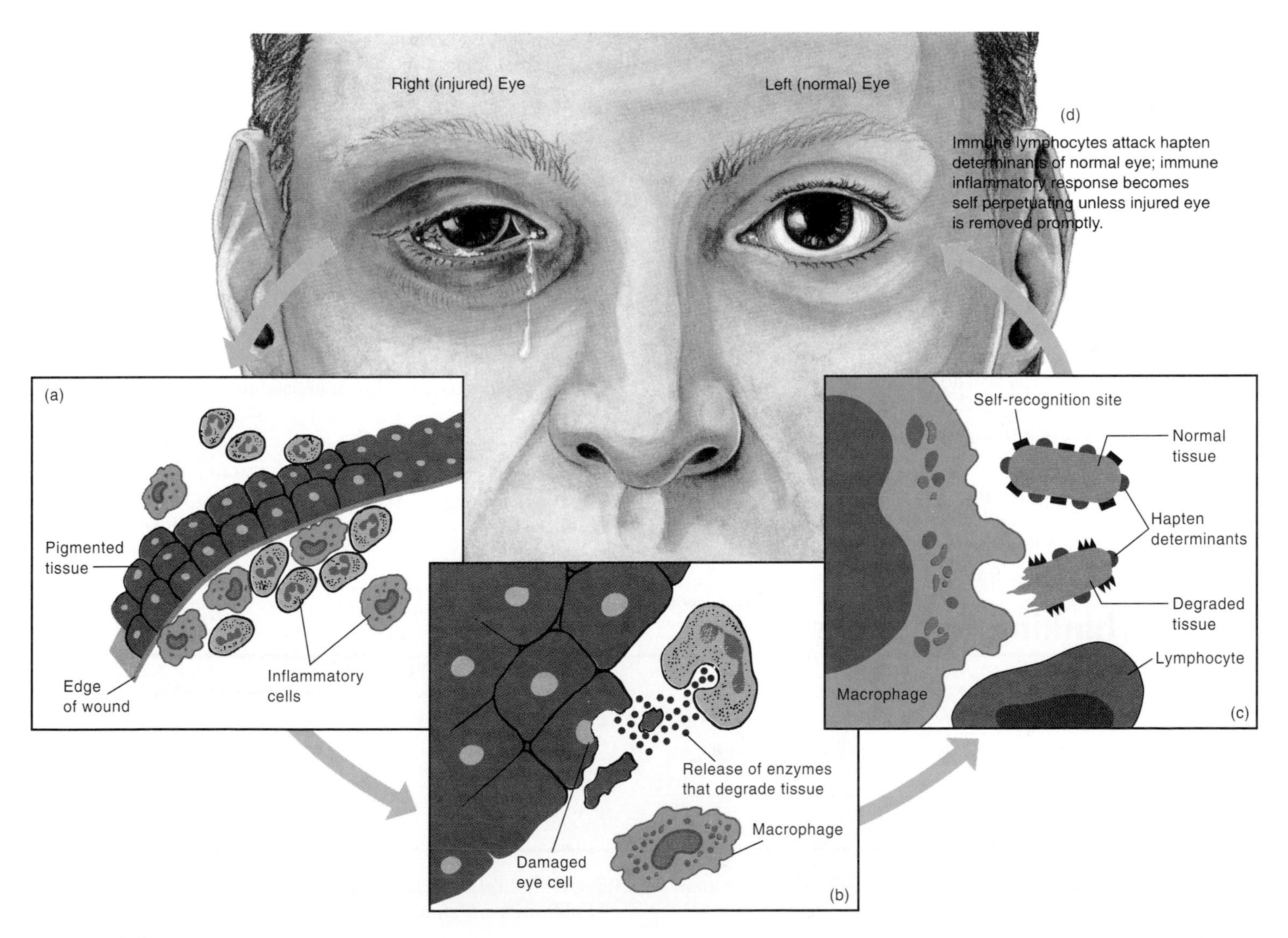

Figure 18.6
One mechanism postulated to explain the autoimmune process of sympathetic ophthalmia.

Summary

I. Hypersensitivities
   A. Type I, or immediate, hypersensitivities (allergies) are mediated by IgE.
      1. IgE attached to mast cells or basophils reacts with specific antigen, resulting in the release of powerful mediators of the allergic reaction.
      2. Localized anaphylactic (type I) reactions include urticaria (hives), allergic rhinitis (hay fever), and asthma.
      3. Generalized anaphylaxis is a rare but serious reaction that can lead to shock and death.
      4. Immunotherapy is often effective in decreasing the type I hypersensitivity state.
   B. Type II hypersensitivity reactions, or cytotoxic reactions, are caused by antibodies that can destroy normal cells.
      1. Transfusion reactions are examples of type II reactions.
      2. Hemolytic disease of the newborn is also a type II reaction.
   C. Type III hypersensitivity reactions, or immune complex-mediated reactions, activate inflammatory mechanisms.
   D. Type IV hypersensitivity reactions, or delayed hypersensitivity reactions, involve T lymphocytes.
      1. The tuberculin skin test provides a familiar example of a delayed hypersensitivity reaction.
      2. Contact hypersensitivities (contact allergy, contact dermatitis) occur frequently to substances such as poison ivy, nickel in jewelry, and chromium salts in leather products.
      3. Delayed hypersensitivity is important in responses to many chronic, long-lasting infectious diseases.
      4. Transplantation rejection is caused largely by mechanisms of delayed hypersensitivity.

II. Immunodeficiency Disorders
   A. Primary immunodeficiencies result from genetic or developmental abnormalities.
      1. B cell immunodeficiencies result in diseases involving a lack of antibody production, such as agammaglobulinemias and selective IgA deficiency.
      2. T cell deficiencies result in diseases such as DiGeorges's syndrome.
      3. Lack of both T and B cell functions results in combined immunodeficiencies, which may be severe.
      4. Defective phagocytes are found in chronic granulomatous disease and Chediak-Higashi disease.
   B. Secondary immunodeficiencies can result from malnutrition, immunosuppressive agents, infections (such as AIDS), and malignancies.

III. Autoimmune Diseases
   A. Some autoimmune diseases are caused by antibodies produced to body components.
   B. Some autoimmune diseases are mediated by mechanisms of delayed hypersensitivity.

Chapter Review Questions

1. Why are antihistamines useful for treating many type I allergic reactions but not effective in treating asthma?
2. Penicillin is a very small molecule, yet it can cause several types of hypersensitivity reactions, especially type I. How can this occur?
3. Note several ways that immune complexes can contribute to an inflammatory reaction.
4. What are some major differences between a type I skin reaction, such as hives, and a type IV reaction, such as that revealed by a tuberculin skin test?
5. Why might malnutrition lead to immunodeficiencies?
6. What is the most common primary immunodeficiency disorder?
7. Give at least two lines of evidence showing that myasthenia gravis results from antibody activities.
8. How could a bacterial infection lead to an autoimmune disease?
9. Can genetic deficiencies leading to immunodeficiency disorders be corrected? If so, give an example.
10. Give an example of an organ-specific autoimmune disease and one that is widespread, involving a variety of tissues and organs.

Critical Thinking Questions

1. Hypersensitivity reactions, by definition, lead to tissue damage. Can they also be beneficial? Explain.
2. How might immunotherapy operate to help patients with type I allergies?
3. Why does blood need to be cross-matched before transfusion?
4. Patients with advanced leprosy do not give positive type IV reactions to any of a variety of antigens. Why might this be?

Further Reading

Auchincloss, Hugh, Jr., and Sachs, David H. 1993. Transplantation and graft rejection. *Fundamental Immunology,* 3rd ed., pp. 1099–1141. Raven Press, New York.

Lachmann, P. J., et al. 1993. *Clinical Aspects of Immunology,* 5th ed. Blackwell Scientific Publications, Oxford.

Roitt, Ivan. 1994. *Essential Immunology,* 8th ed. Blackwell Scientific Publications, Oxford.

19 Interactions Between Microorganisms and Humans

KEY CONCEPTS

1. Establishing and maintaining the normal flora as a dynamic human-microbe system is important for good health.
2. Microorganisms communicate with their human hosts and force the host cells to do things they would not normally do.
3. Pathogens have a variety of virulence factors responsible for their ability to cause disease. These factors are only synthesized when the pathogens need them.
4. The outcome of infection depends upon the host response as well as the activities of the infecting agent.

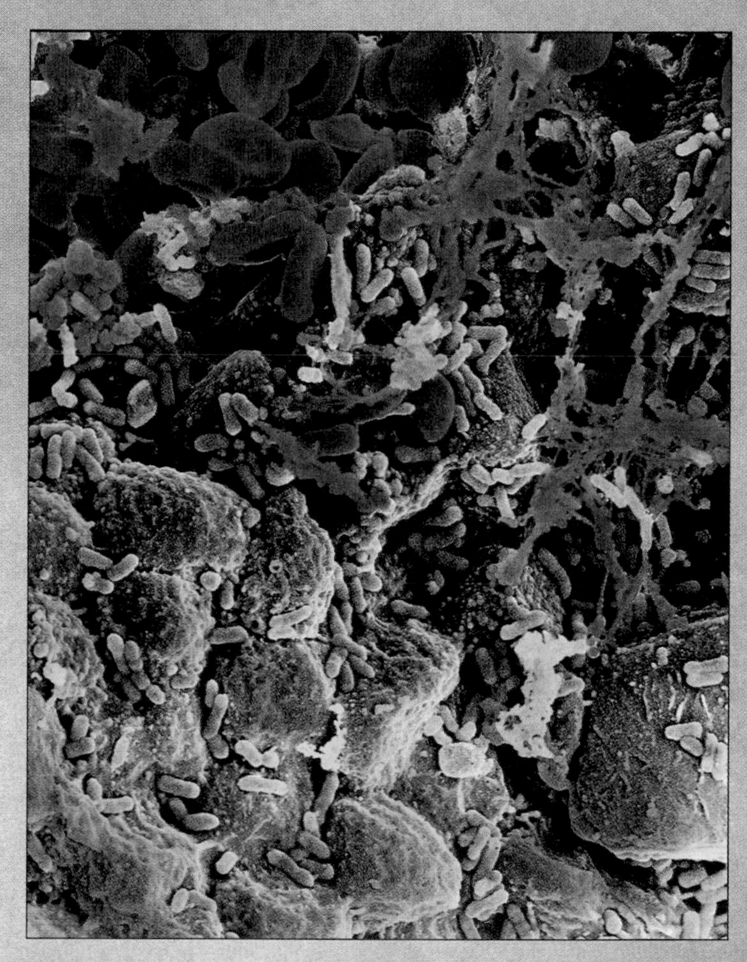

Color-enhanced scanning electron micrograph of *Escherichia coli* infection of the bladder. Rod-shaped *E. coli* (yellow) are attached to epithelial cells (blue). The epithelium has secreted thick mucus filaments (orange). The inflammation has caused some bleeding (top left, red blood cells).

Microbes in Motion

The following books and chapters in the *Microbes in Motion* CD-ROM may serve as a useful preview or supplement to your reading:
Bacterial Structure and Function: External Structures. *Viral Structure and Function:* Viral Structure. *Gram-Positive Organisms in Human Hosts:* Normal Flora. *Gram-Negative Organisms in Human Hosts:* Normal Flora. *Microbial Pathogenesis of Infectious Disease:* Principles of Pathogenicity.

A Glimpse of History

The concept that diseases are infectious and caused by unknown entities appeared in human history long before the discovery of the microbial world. Indeed, Fracastoro proposed a "germ theory of disease" in 1546. With the discovery of microorganisms in the late 1600s by Leeuwenhoek, people began to suspect that microorganisms might cause disease. However, it was not easy to prove this. Because techniques were neither very sophisticated nor well understood, scientists often produced conflicting results. It was not until 1876 that Robert Koch offered convincing proof of the germ theory of disease when he showed that *Bacillus anthracis* is the cause of anthrax, a serious and often fatal disease of humans, sheep, and other animals. With his microscope, he observed *B. anthracis* bacteria in the blood and spleen of dead sheep. He then inoculated mice with this infected sheep blood and was able to recover *B. anthracis* from the blood of the mice. In addition, he was able to grow the *B. anthracis* in pure culture and show that it caused anthrax when injected into healthy mice.

From his experiments with *B. anthracis* and later with *Mycobacterium tuberculosis*, Koch formalized a group of criteria now known as *Koch's Postulates* (perspective 19.1). The main importance of Koch's Postulates is that they allowed proof of the germ theory of disease, but it is often not possible to use them in establishing the cause of infectious diseases of humans.

After learning some of the many ways in which human beings can protect themselves against organisms and other foreign materials, it may seem surprising that microorganisms are actually able to colonize and infect people. However, microbes and viruses have a vast multitude of ploys for gaining entry into people, living in or on the human body, and countering body defenses. Viruses are arguably the most efficient of entities in that they are able to enter host cells, take over the cellular metabolism, and use it for their own reproduction. In this chapter, we will explore some of the ways in which microorganisms and viruses colonize the human host and either live in harmony with the host in the normal flora or cause disease. We will also see how the host responds to some of the disease-producing strategies of microbes.

As indicated in previous chapters, microorganisms are marvelously adaptive, able to adjust quickly to changing conditions by regulating the expression of genes. These capabilities are put to the test when microorganisms enter the hostile environment of the human body. Consider, for example, potentially pathogenic bacteria living in a household environment where they have adapted to a particular environment, including room temperature and humidity, aerobic conditions, and available nutrients. Suddenly, a passing breeze propels the bacteria into a human respiratory tract. Now the bacteria must battle against innate host defenses, such as the blanket of mucus being moved outward by the ciliated epithelium, and against noxious chemicals in the mucus. The bacteria must immediately express new sets of genes necessary to allow them to survive in this new environment with its higher temperature, different nutrients, and other conditions. It is urgent that genes coding for virulence factors (factors that enhance disease production) be expressed. The ability to regulate all these genes is essential for the pathogenicity of these bacteria.

In recent years, the study of human-microbial interactions has become very exciting, as much more has been learned about ways in which microorganisms and hosts interact. Techniques developed for genetics, molecular biology, and immunology have been used to determine how microorganisms cause disease. This study at the molecular level reveals complicated mechanisms and poses new questions. Use of these new techniques has also made possible a molecular version of Koch's Postulates (perspective 19.1).

Methods for Establishing the Cause of an Infectious Disease

Molecular Postulates Are Proving to Be Useful

These new postulates deal with the **virulence** of microorganisms, their properties that enhance disease production. The postulates are outlined in perspective 19.1. As with the traditional Koch's Postulates, it is not always possible to apply all of these postulates, but they provide many new approaches to determining the causal agents of infectious diseases. Note that the gene becomes more important than the species in identifying potential pathogens.

Normal Flora of the Human Body

In nature, all organisms, including humans and microbes, interact with one another as part of a biological community called an **ecosystem.** Because microorganisms live in a wide range of natural habitats, it is not surprising that some of them grow abundantly on various surfaces of the human body, such as the skin and parts of the genitourinary, respiratory, and gastrointestinal tracts. The microorganisms that grow on the external and internal surfaces of the body without producing obvious harmful effects to these tissues are known as the **normal microbial flora,** or **normal flora.** Figure 19.1 shows some of the organisms that are part of the normal flora and their locations on the human body. Microorganisms that inhabit the body only occasionally are termed **transient microbial flora.**

Microorganisms and Humans Form a Variety of Symbiotic Relationships

Organisms that live within biological communities interact both positively and negatively with each other. The term **symbiosis** (for "living together") describes interactions that occur between different organisms that live close

Perspective 19.1

The Traditional Koch's Postulates and the Modern Molecular Postulates for Determining the Causative Agents of Infectious Diseases

The Traditional Version of Koch's Postulates

1. The microorganisms must be present in every case of the disease.
2. The microorganisms must be grown in pure culture from the diseased host.
3. The same disease must be reproduced when a pure culture of the microorganism is inoculated into a healthy, susceptible host.
4. The microorganisms must then be recovered from the experimentally infected host.

The Molecular Postulates

1. The gene or its disease-causing products (virulence factors) should be found in pathogenic strains but not in nonpathogenic strains of the suspected organism.
2. Disrupting the function of the virulence genes should change an organism from pathogenic to nonpathogenic or, alternatively, introduction of the cloned gene should change a nonpathogenic organism to a pathogenic one.
3. The genes for virulence must be expressed during the disease process.
4. Antibodies or immune cells specific for the virulence gene products should be protective.

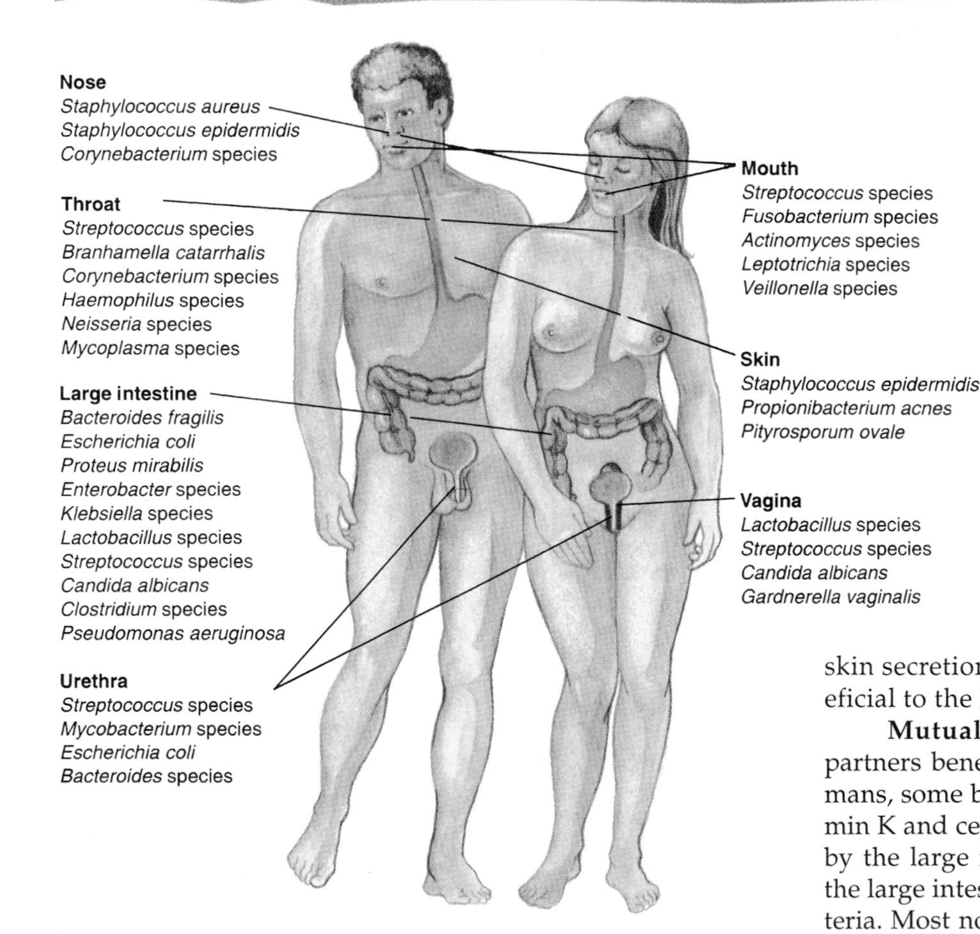

Figure 19.1
Location of many of the organisms that are part of the normal flora in and on the male and the female human body.

together on more or less a permanent basis **(symbionts).** Microorganisms have a variety of symbiotic relationships with the human body. These relationships can take on different characteristics depending upon the closeness of the association, the relative advantage to each partner in the association, and the dependence of partners upon each other. Symbiotic associations can be described as either commensal, mutualistic, or parasitic.

Commensalism refers to an association in which one partner benefits but the other remains unaffected. Many microorganisms on the surfaces of the human body are commensals. For example, the *Propionibacterium* sp. found on the surface of skin live off skin secretions but are generally neither harmful nor beneficial to the human host.

Mutualism refers to an association in which both partners benefit. For example, in the large intestine of humans, some bacteria such as *Escherichia coli* synthesize vitamin K and certain B vitamins. These nutrients are absorbed by the large intestine, which benefits the human. In turn, the large intestine provides nutrients and shelter to the bacteria. Most normal flora organisms are commensals or mutualists. They are either mutually beneficial or they do not live at the expense of the host.

Parasitism refers to an association in which one organism, the **parasite,** benefits at the expense of the other organism, the **host.** All living organisms can act as hosts for parasites. Most often, the parasites are members of the microbial world: the bacteria, fungi, protozoa, and viruses. Some invertebrates such as round worms and flatworms can also be parasitic. Usually, the host is a macroscopic organism, and the parasite is a microscopic one.

Some parasites do very little damage to the host, but others do great damage, often killing the host. Most disease-causing organisms are considered parasites, but their parasitic nature depends on the state of host defenses.

Establishing and Maintaining the Normal Flora as a Dynamic Human-Microbe Ecosystem Is Important for Good Health

Establishment of Normal Flora

Normally, a human fetus has no resident microbial population. During the passage of the fetus through the vagina, the mother's microorganisms and viruses take up residence on the skin and in the gastrointestinal tract of the baby. Further, various microorganisms in food or on other humans and in the environment may establish themselves as permanent residents on the newborn. These microorganisms then become part of the newborn's normal flora and are said to have "colonized" the host. **Colonization** simply implies establishment of microbial growth on or within the host.

Importance of Normal Flora

The normal flora is important to the overall health of the human host. One of the most beneficial effects of these organisms is prevention of the growth of other potentially more harmful organisms. They do this in several possible ways. They may provide a surface incompatible for the attachment of the invader. Frequently, the growth of one organism is prevented or slowed down by the growth of another organism living close by. Such mutual inhibition often occurs among members of the normal flora found on the human body as well as between organisms of the established normal flora and potentially harmful organisms that appear.

One microorganism may inhibit the growth of another by competing for essential nutrients, such as vitamins, amino acids, and iron. Other mechanisms of inhibition are also known. For example, *Corynebacterium diphtheriae,* the bacterium causing diphtheria, is inhibited by hydrogen peroxide produced by certain groups of streptococci, organisms found in the normal flora of the nasopharynx and oral cavities. Additionally, *Staphylococcus epidermidis* and *Propionibacterium* sp. that grow all over the skin break down lipids to fatty acids that are harmful to most other bacteria. Some bacteria produce antibiotics or bacteriocins that inhibit the growth of other bacteria. Some microorganisms consume large amounts of oxygen, reducing the supply available, thereby inhibiting the growth of aerobic microorganisms. Microorganisms may produce toxic metabolic by-products such as acids and peroxides. When organisms of the normal flora are removed or their growth suppressed, as can happen during treatment of the host with antibiotics, potentially harmful antibiotic-resistant organisms can colonize and possibly cause disease.

In addition, organisms of the normal flora can stimulate the immune system. They frequently cause the production of antibodies that cross-react with potentially pathogenic organisms and prevent the harmful organisms from colonizing or invading.

Finally, in some animals such as cows and deer, members of the normal flora of the digestive tract are essential for degrading cellulose and providing vitamins and other nutrients for the animal. In humans lacking a well-balanced diet, the intestinal flora may also be a significant source of essential nutrients.

The Normal Flora: Subject to Changes

The number of different species of microorganisms that comprise the normal microbial flora is very large. Figure 19.1 lists only a few representative species of the many that inhabit the human body. At any one time, the composition of this complex ecosystem represents a dynamic balance of many forces that may alter the bacterial population quantitatively as well as qualitatively. The microbial composition is constantly changing, both in response to external influences and also as a direct result of the activities of the human host. Further, each member of this ecosystem is influenced by the other members and, in turn, exerts its influence upon others in the system. For example, any activity that increases perspiration and, thereby, moisture and nutrients on the skin, increases multiplication of skin organisms causing body odor and even skin disease.

Intestinal microbial populations change with variations in diet, acidity of the stomach, ingestion of antibiotics, and peristalsis.[1] For example, people who have a substantial amount of meat in their diets will have more *Bacteroides* and other gram-negative rods in their large intestine than will people on vegetarian diets. Also, many antibiotics will not only kill or inhibit pathogens, but also destroy some of the normal flora, allowing unwanted organisms to proliferate. In hospitalized patients, the suppressed normal flora may be replaced by antibiotic-resistant organisms such as *Pseudomonas* or *Staphylococcus,* that can cause serious infections.

Infectious Disease Terminology

A Number of Terms are Used in the Study of Infectious Disease

Many host and microbial factors interact to determine the effects of microorganisms on the human body, regardless of whether these organisms are established members of the normal flora or new arrivals from the environment or from other hosts. The terms used to describe the most common interactions or results of these interactions are given in table 19.1.

[1]Peristalsis—progressive wavelike contractions of the intestinal wall that move the contents in the proper direction

Table 19.1 Definition of Terms Used in the Study of Infectious Diseases

Term	Definition
Avirulent	refers to microorganisms that lack the ability to cause disease
Colonization	the establishment of microbial growth on a body surface
Disease	any condition in which there is a noticeable impairment of body function; may be caused by a microorganism or virus but can be from other causes such as autoimmune disease
Infection	result of a microorganism penetrating the body surfaces, entering the body tissue, multiplying, and causing the host to react by making an immune or other type of response
Infectious disease	any condition in which there is a noticeable impairment of body function caused by an infecting microorganism or virus
Opportunist	a microorganism that is able to produce disease only in hosts with impaired defense mechanisms
Pathogen	any disease-producing microorganism
Pathogenic	disease-causing
Pathogenicity	the ability to cause disease
Virulence	enhanced disease-producing capabilities of microorganisms

One outcome of the interactions, **colonization,** implies establishment of microbial growth on or within the host. If the microorganism that colonizes a surface penetrates the surface, enters body tissues, and multiplies, **infection** is said to have occurred. The term **disease** describes a noticeable impairment of body function. A disease caused by a microorganism or virus is an **infectious disease.** It is important to remember that infection does not always lead to disease; disease only occurs when body function is significantly impaired.

A **pathogen** is any disease-causing microorganism or virus. Two other commonly encountered terms are **pathogenic,** meaning "disease-causing" and **pathogenicity,** meaning the "ability to cause disease." Common usage has sometimes restricted the term *pathogen* to those species of microorganisms that most commonly cause disease in normal people.

Opportunistic pathogens, or **opportunists**, are organisms that cause disease only in immunologically suppressed individuals. Table 19.2 gives examples of bacteria that are pathogenic.

The **opportunists** are often members of the normal flora or occur in the environment. They are able to cause disease in hosts with impaired defense mechanisms, such as those with wounds, genetic defects, alcoholic or drug dependencies, and AIDS or those who are undergoing irradiation therapy or surgery. These opportunists can sometimes cause infection if introduced into an unusual location. For example, *Escherichia coli,* normally harmless in the large intestine, causes infection if it gets into the urinary bladder.

Infections do not necessarily result in disease. Even when disease occurs, its severity may vary. As previously mentioned, the term **virulence** describes attributes of a microorganism or virus that promote pathogenicity. Different strains within a species may vary in virulence. For example, pneumococci that have a capsule are virulent and often cause disease; those lacking a capsule are **avirulent** and are much less likely to cause disease. The word *virulent* is also used in a quantitative sense, indicating that an organism has more disease-promoting attributes than do other less virulent strains of the same species. This implies that such an organism is more likely to cause disease, particularly severe disease, than is a less virulent strain. In some instances, factors responsible for virulence are known, such as the capsule of pneumococci or toxin produced by bacteriophage-containing *Corynebacterium diphtheriae*. In other cases, virulence factors are not known. Neither virulence nor pathogenicity should be viewed as absolutes. Both depend on factors within the host as well as in the microorganism and on their relationship to each other.

REVIEW QUESTIONS

1. How were Koch's Postulates used to prove the germ theory of disease?
2. Name three types of symbiotic relationships microorganisms form with humans and describe the differences.
3. When and how is the normal flora of a human established?
4. List three ways in which the normal flora of humans helps to protect the body against invasion by microorganisms.
5. Are opportunists pathogens? Explain.

Mechanisms of Pathogenesis

A variety of factors involving both the host and microorganisms determine the effects that microorganisms have on the human body. This interaction is independent of whether the organisms are established members of the normal flora or new arrivals from the environment. In the broadest sense, the total capabilities of both the host and the microorganism are called into play.

Table 19.2 Examples of Pathogenic Bacteria

Pathogenic in Normal People*	Commonly Opportunistic†	Occasionally Opportunistic††
Bacillus anthracis	*Actinomyces israelii*	*Acinetobacterium calcoaceticus*
Bordetella pertussis	*Bacteroides fragilis*	*Bacillus cereus*
Borrelia recurrentis	*Clostridium perfringens*	*Eikenella corrodens*
Brucella abortus	*Clostridium tetani*	*Enterobacter aerogenes*
Chlamydia psittaci	*Escherichia coli*	*Flavobacterium meningosepticum*
Clostridium botulinum	*Haemophilus influenzae*	*Propionibacterium acnes*
Corynebacterium diphtheriae	*Klebsiella pneumoniae*	*Pseudomonas cepacia*
Coxiella burnetii	*Neisseria meningitidis*	*Staphylococcus epidermidis*
Escherichia coli	*Pasteurella multocida*	*Streptococcus agalactiae*
Francisella tularensis	*Proteus mirabilis*	*Enterococcus faecalis*
Leptospira interrogans	*Pseudomonas aeruginosa*	
Mycobacterium tuberculosis	*Serratia marcescens*	
Mycoplasma pneumoniae	*Staphylococcus aureus*	
Neisseria gonorrhoeae	*Streptobacillus moniliformis*	
Salmonella typhi	*Streptococcus pneumoniae*	
Shigella dysenteriae		
Streptococcus pyogenes		
Treponema pallidum		
Vibrio cholerae		
Vibrio parahaemolyticus		
Yersinia pestis		

*Strains of these species cause disease in normal, nonimmune people.

†Commonly pathogenic only in cases of burns, wounds, viral infections, alcoholism, or similar conditions of impaired resistance

††Commonly pathogenic only in cases of immaturity, birth trauma, defective immunity, or large infecting dose or when other factors overwhelmingly favor the microbe

Successful Pathogens Carry Out a Sequence of Activities

The **successful pathogen** has the ability to carry out the following steps: (1) **transmission** of the causative agent to a susceptible host, (2) **adherence** of the agent to a target tissue, (3) **colonization** and usually **invasion,** (4) **damage** to the host by the activities of **toxins** or other mechanisms, (5) **exit** from the host, and (6) **survival** outside the host long enough for step 1 to occur.

Pathogenic bacteria often have a variety of factors that contribute to their ability to cause disease or, in other words, to be virulent. These are known as **virulence factors.** Some important virulence factors are listed in table 19.3 and discussed next.

Table 19.3 Some Important Virulence Factors of Bacteria

Pili and other ligands that facilitate attachment of bacteria to host cells
Ability to trigger actin rearrangement in nonphagocytic cells, causing endocytosis of the bacteria
Capsules, protein A, and enzymes that interfere with phagocytosis by professional phagocytes
Exotoxins
Endotoxin
Enzymes with toxic effects
Superantigens
Proteases that specifically break down secretory IgA
Ability to vary antigens, negating the antibody response of the host

The Sources of Infectious Agents and Their Modes of Transmission Are Varied and Numerous

The specific sources of particular infectious agents and their transmission to humans will be discussed in more detail in the chapters on disease. In brief, infectious agents can originate from soil (*Coccidioides immitis*); water (*Legionella pneumophila*); numerous animals (*Salmonella enteritidis, Brucella abortus*) including insects (*Rickettsia rickettsii, Plasmodium vivax*); and human beings, including a person's own normal flora (*Streptococcus pneumoniae, Staphylococcus aureus*).

Infectious disease agents are transmitted or communicated from one individual to another by **direct contact,** such as touching, handshaking, or sexual intercourse, or **indirectly** through the **air,** through **food** and **water,** or by **insect bites.** Transmission is successful if enough organisms are transferred to colonize and infect. The number of organisms that are necessary to insure infection, termed the **infectious dose,** varies with the pathogen. For example, it takes only 10 to 100 *Shigella* sp. but as many as 10^6 *Salmonella* sp. to establish infection in humans. The establishment of infection does not necessarily mean that disease develops, but it is an essential early step in the disease process.

Adherence Is Usually a Necessary Step in the Establishment of Infection

Humans as well as plants and animals have very effective mechanisms to keep most microorganisms and viruses from entering the interior organs of the body and causing damage.

All surfaces of the body exposed to the outside world are covered with **epithelial cells.** These cells are packed tightly together and rest on a thin layer of fibrous material, the **basement membrane** (figure 19.2). The parts of the body that are exposed to the outside world include not only the skin, which is in direct contact, but also the mucous membranes of the genitourinary tract, the respiratory tract, and alimentary

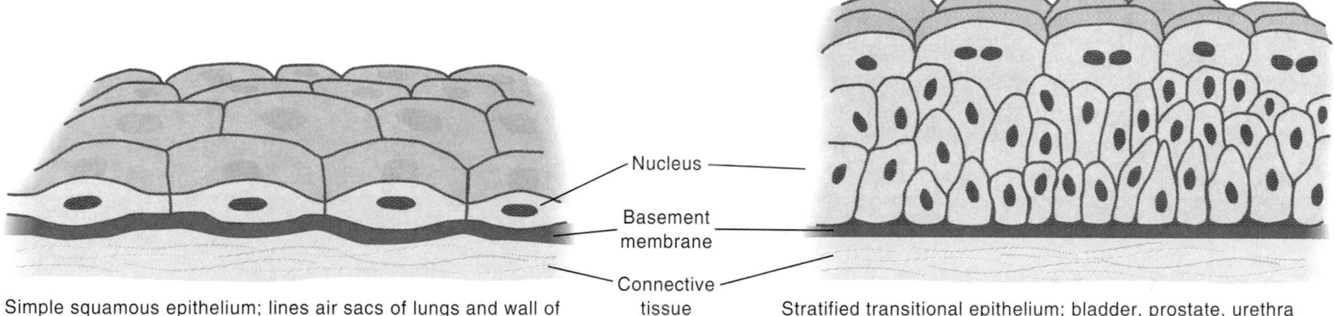

Simple squamous epithelium; lines air sacs of lungs and wall of capillaries; covers membranes that line body cavities

Stratified transitional epithelium; bladder, prostate, urethra

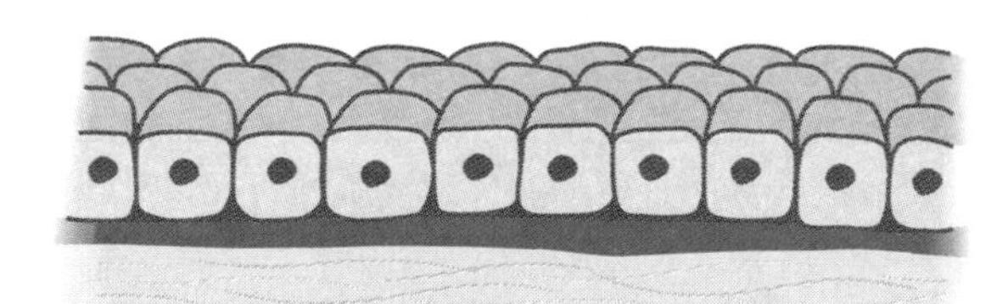

Simple cuboidal epithelium; covers ovaries; lines the kidney tubules, ducts of various glands (salivary, pancreas, liver)

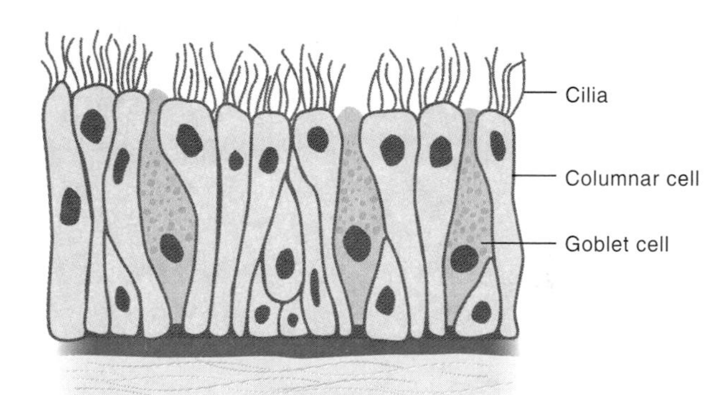

Pseudostratified columnar epithelium; passages of respiratory system, various tubes of the reproductive systems

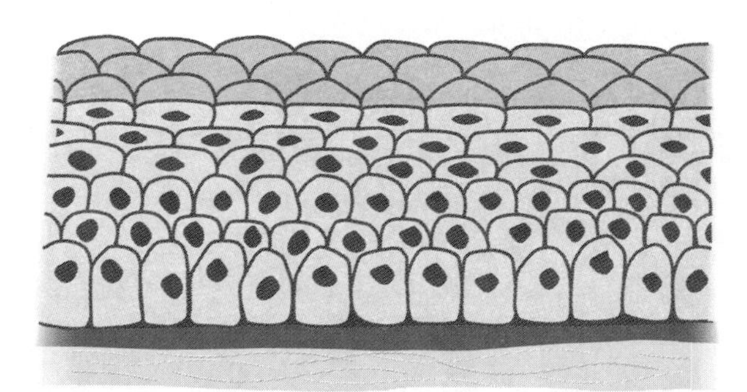

Stratified squamous epithelium; mouth cavity, esophagus, most of the female urethra, vagina, anal canal

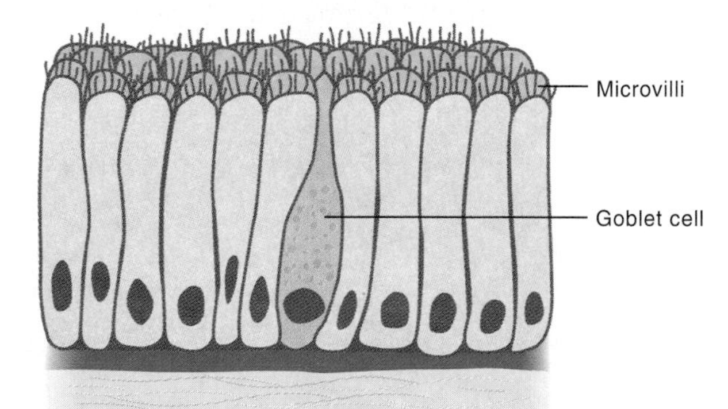

Simple columnar epithelium; intestines, stomach, lining of the uterus

Figure 19.2
Various types of epithelium. Common features of the different types of epithelium are the tight junctions between cells that make entry of microorganisms virtually impossible under normal conditions and the presence of basement membranes that also discourage entry of materials through the epithelium. Note the cilia on some epithelial cells—for example, the respiratory system—that function to sweep materials away from the epithelium.

tract. Although they are "inside the body," they are nevertheless exposed to the external environment through the intake of food and air.

Each region of the body has, in addition to the tight packing of the epithelial cells, other mechanisms to keep microorganisms from colonizing and gaining entry. For example, skin cells are shed regularly. Ciliated cells of the respiratory tract transfer mucus-entrapped microorganisms out of the lungs and into the throat where they are first swallowed and then destroyed by stomach acid. The peristaltic movement in the intestine moves potential pathogens out of the body, and the flushing action of the urinary tract is its main defense against potential pathogens (table 19.4).

Because these mechanisms are very effective in removing organisms, a pathogen must adhere or attach to host cells as a necessary first step in the establishment of infection. As a general rule, both pathogens and host cells are negatively charged and therefore tend to repel each other. For attachment to occur, these repulsive forces must be circumvented. In many instances, projections from the surface of the microorganism or virus, termed **ligands,** bind specifically to surface components of the host cells, termed **receptors** (figure 19.3).

The ligand and receptor molecules are held together by short-range, weak bonds that include hydrogen bonds. In most cases studied, the ligands are proteins and the receptors are glycoproteins. For example, the ligands of *Escherichia coli, Vibrio cholerae,* and *Mycoplasma pneumoniae* are proteins, and their receptors are glycoproteins containing mannose, fucose, and sialic acid, respectively. More than one kind of ligand may occur on the same pathogen, and the cells of different tissues of the same host may have different receptors. Some pathogens, such as the streptococci that are responsible for tooth decay, possess polysaccharides that form a sticky network of fibrils called the **glycocalyx** that allow the bacteria to attach to surfaces nonspecifically.

Among the most common attachment ligands are the **pili** found on many bacteria, especially on gram-negative organisms (figure 19.4). Tips of the pili attach to the host cell receptors, holding the bacteria to the host cell. The host responds by producing antibodies that combine specifically

Table 19.4 Barriers to Entry of Microorganisms

Body Site	Defense Mechanism	Conditions that Foster Entry
Skin	Dryness, acidity, toxicity, constant shedding	Wounds, excess moisture, serous discharge
Respiratory tract	Ciliated cells that constantly move mucus to throat	Reduced movement of ciliated cells as in smoking, chilling, narcotics, viral infection
Gastrointestinal tract	0.2% hydrochloric acid, enzyme pepsin	Reduced stomach acid, ingestion of antacid
Vagina	Lactobacilli during childbearing years	Reduced numbers of lactobacilli from douching, soaps, menopause, antibiotic therapy
Urinary tract	Flushing action of urination	Short urethra in women, incomplete or infrequent urination, sexual intercourse

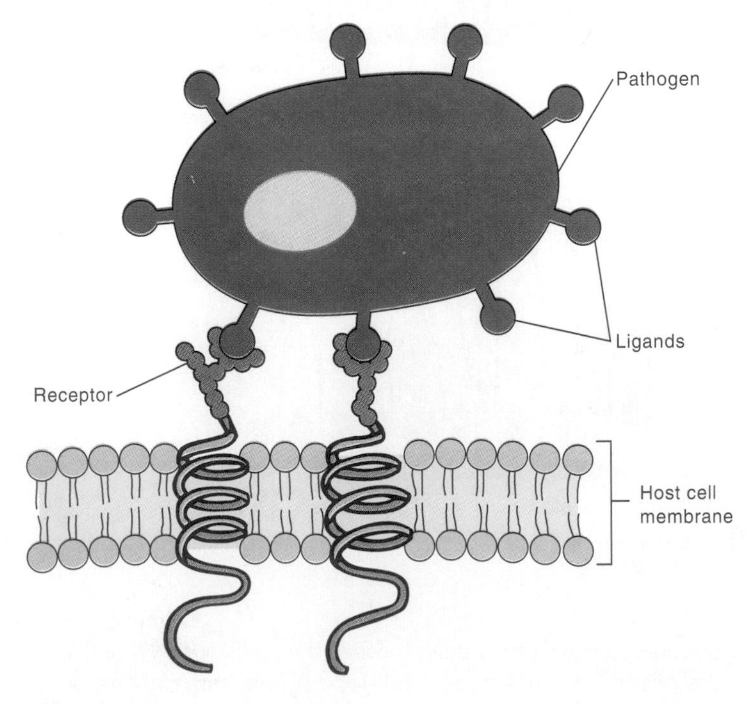

Figure 19.3
Attachment of pathogen to host cells.

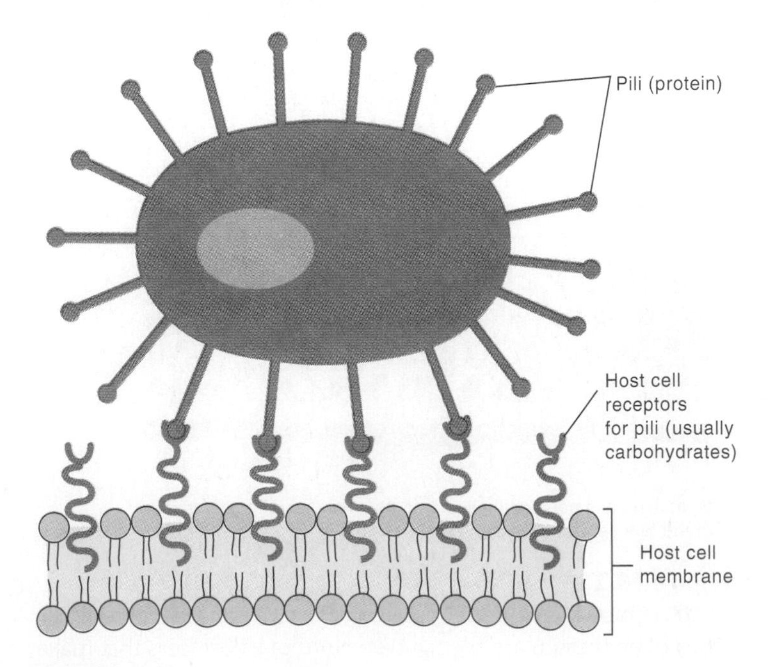

Figure 19.4
Pili attachment to host cell.

with the tips to prevent adherence. The bacteria often counter by switching to a promoter gene which activates the synthesis of different pili, thus negating the specific antibody response. This phenomenon is an example of **phase variation.**

Different strains of the same species of pathogen may have different ligands. This helps account for the variety of diseases caused by different strains of the same organism. For example, certain strains of *Haemophilus influenzae* selectively infect the eye and cause conjunctivitis ("pink eye"), whereas most *H. influenzae* strains infect the upper respiratory tract. The ligands on the strains causing conjunctivitis are specific for receptors found on the eye. Similarly, certain *Streptococcus pyogenes* strains attach to throat epithelial cells causing "strep throat," whereas others attach to the skin.

In some cases, the genes that code for particular ligands are carried on plasmids and can be transferred from one strain to another. For example, *E. coli*, usually a benign member of the normal flora of the large intestine, can acquire a plasmid containing the genes that code for the colonization factor antigen (CFA) as well as the production of an enterotoxin.[1] The CFA antigen is a pilus protein that allows the *E. coli* to attach to and colonize the human small intestine. Once the cells have attached, they produce enterotoxin which results in the severe diarrhea caused by these strains. Many other plasmid-transferable virulence factors account for the pathogenicity of various *E. coli* strains.

The presence of host cell receptors often varies with the host's age. They also vary in number and kind among individuals and among species. This variety accounts in part for differences in susceptibility to infection by various microorganisms and viruses. For example, most humans are not susceptible to diseases of cats and dogs, such as the viral disease distemper. The cell receptors are not compatible with the ligands on the surface of the virus and, therefore, the virus cannot attach.

Just because a microorganism attaches to cells does not mean it will cause disease. Many members of the normal flora of the mouth and intestine attach strongly to cells yet do not cause disease. Other factors such as toxin production and invasive ability of the agents must come into play before disease results.

The Establishment of Infection Involves Invasion of Host Tissues

Most pathogens must invade tissues in order to cause disease. However, since cells on the body surfaces lie tightly together, entry into the host is not easy. When a bite, abrasion, wound, or burn damages the epithelial layer, microorganisms are able to gain entry. In addition, some microorganisms gain entry by producing substances that destroy the basement membrane; others disrupt the cell membranes or degrade the carbohydrate-protein complexes between cells, thereby allowing the epithelial cells to separate and create a passageway for invasion.

[1]Enterotoxin—microbial toxin or poison that affects the intestine

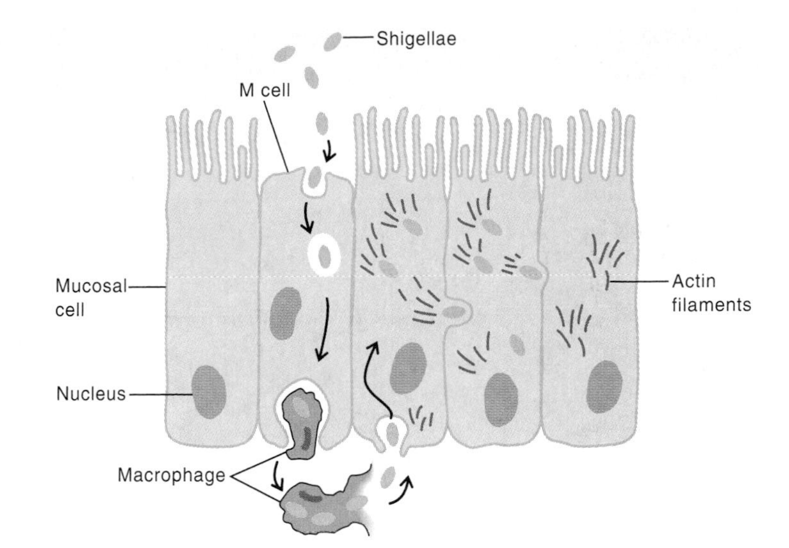

Figure 19.5
Mucosa-associated lymphoid tissue (MALT) in the gastrointestinal tract includes M cells. MALT in the gastrointestinal tract is often referred to as GALT (gastrointestinal-associated lymphoid tissue). Peyer's patches are part of this lymphoid tissue. They contain some M cells that are capable of phagocytizing but not destroying organisms. Shigellae, for example, can cross the mucosal barrier by being ingested by M cells and then egested into the internal tissue. There they can invade mucosal cells and move from cell to cell by being pushed along actin filaments.

Within the Peyer's patches of lymphoid tissue in the gastrointestinal tract (part of the MALT) there are M cells (figure 19.5) that are capable of phagocytosis. They function to take up antigens in the intestines and convey them across the mucosa to the antibody-producing lymphoid tissues of the Peyer's patches. Some bacteria, such as *Shigella* sp., can cross the mucosal barrier by being phagocytosed by M cells and then released on the other side of the mucous membrane, where they cause disease by means of various virulence factors.

Following attachment, microorganisms and viruses may be ingested by host cells via **endocytosis,** a process that resembles phagocytosis. Table 19.5 gives some examples of pathogens that can enter host cells by this process. It is of interest that the microbe actually signals the eukaryotic cell and induces the cell to take it up. The mechanism of this cell-cell communication is now being actively studied. It has been found that eukaryotic cells normally polymerize soluble actin molecules in the cytoplasm into insoluble filaments that give the cell its shape; when professional phagocytes engulf materials, the actin filaments rearrange to make the pseudopods that surround materials and bring them into the cell. During the process of endocytosis, the pathogen attaches to the normally nonphagocytic cell and is able to induce the cell to rearrange its actin and become phagocytic (figure 19.6). If the pathogen cannot induce the cell to take it up, it cannot cause disease. Some bacteria with the capability

• **MALT, p. 769**

Table 19.5 Some Pathogens that Enter Epithelial Cells by Endocytosis

Pathogen	Disease	Characteristics
Rhinovirus	Common cold	Infected cells of the respiratory tract lose contact with the basement membrane and are carried away.
Chlamydia sp.	Conjunctivitis, urethritis, pneumonia	Obligate intracellular bacteria infect the eye, urethra, or other surface, depending on the strain.
Escherichia coli (some strains), *Shigella* sp., *Campylobacter jejuni*	Dysentery	Infection is generally confined to the intestinal mucosa and supporting tissues.
Salmonella sp. (some strains)	Blood infections with or without diarrhea	Bacteria may pass through the intestinal epithelial cells but cause damage mostly to other parts of the body.
Neisseria gonorrhoeae	Gonorrhea	As with *Salmonella* sp., the bacteria are discharged from the host cell by exocytosis and may then spread to other tissues via the bloodstream.

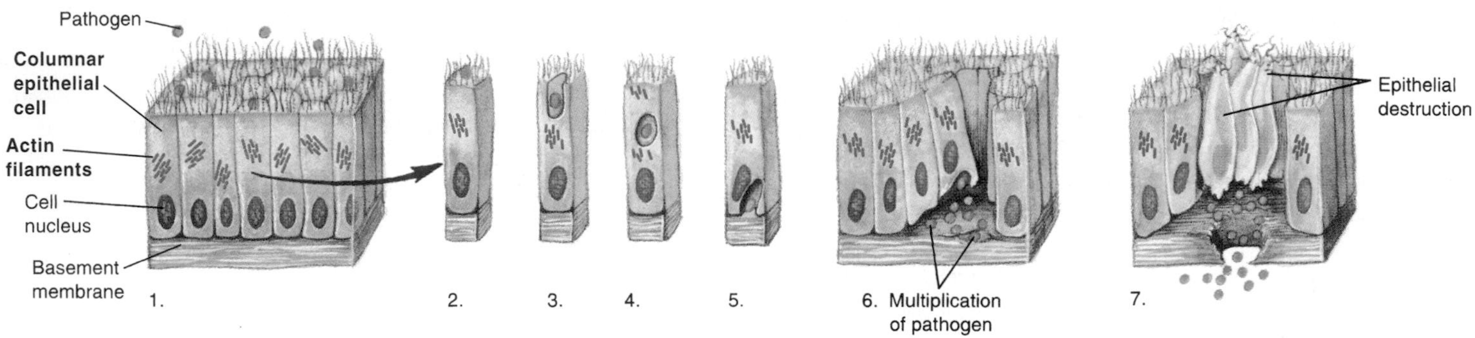

Figure 19.6
Infection of epithelium involving endocytosis and exocytosis. (*Steps 1 to 4*) Endocytosis by a process closely resembling phagocytosis. (*Step 5*) Exocytosis. (*Steps 6 and 7*) Destruction of the epithelial cells following multiplication of the pathogen.

of inducing endocytosis, such as the nonmotile *Shigella* sp., can also be pushed by actin filaments within and between host cells, as indicated in figure 19.5.

Once organisms enter the cell by means of endocytosis, they may reproduce intracellularly or they may exit the cells by the process of **exocytosis** and invade other cells.

Some pathogens actively penetrate cells using their enzymes and their own motility. The protozoan *Toxoplasma gondii* seems to make a small hole in the host cell and then enter. As mentioned earlier, certain enveloped viruses enter cells by a process involving fusion of the viral envelope with the cell membrane.

Tissue Invasion Is Fostered by Large Concentrations of Pathogens

Microorganisms in high concentration are more likely to penetrate surrounding tissues than those in low concentration. It is now known that many bacteria can sense their cell density by cell-cell communication. As mentioned earlier, when harmless microorganisms and viruses of the normal flora are eliminated or suppressed by antibiotic treatment, antibiotic-resistant members of the normal flora rapidly reach high concentrations and may cause infection. For example, the yeast *Candida albicans* is commonly found in relatively small numbers in the vagina. However, if an antibiotic suppresses the competing normal bacterial flora of the vagina (mostly *Lactobacillus* and *Streptococcus* sp.), the *C. albicans*, which is not affected by the antibiotic, may vastly increase in number. The yeast is then likely to invade the tissues of the host, producing vaginitis.[1]

Large numbers of normally harmless organisms can sometimes invade the eye, with serious consequence. Small numbers of organisms of the genera *Bacillus* and *Pseudomonas* frequently come in contact with the eye with no evidence of any infection. However, if contaminated eye medications in which these bacteria have grown to large numbers are introduced into the eyes, the organisms can invade and destroy the eyes with their protein-destroying enzymes (proteases).

Colonization Is a Necessary Step in Establishing an Infection

Following attachment and entry, an organism must multiply in order to establish itself on or within the host; that is, it must **colonize** the host. Colonization is a prerequisite for the production of an infectious disease. In order to colonize, bacteria must not only overcome many obstacles but must also find conditions that allow them to multiply. They must compete with the normal flora inhabiting the site for space and nutrients and overcome toxic products that the resident flora may

[1]Vaginitis—inflammation of the vagina

produce such as certain fatty acids or bacteriocins. Nonspecific and specific components in the host's immune response may also play a role. Also essential for the organisms to multiply are a supply of nutrients and a physical environment that includes the proper moisture, gaseous content, and acidity.

Microorganisms Interact with Their Hosts in Many Ways

Extracellular, Facultatively Intracellular, and Obligately Intracellular Pathogens Cause Damage to Hosts in Different Ways

Pathogens can be described as **extracellular, facultatively intracellular,** and **obligately intracellular.** An extracellular parasite reproduces in the spaces and fluids surrounding the cells and tissues. In cholera, the most extreme example, the toxin-producing organism never penetrates the outer layer of the intestinal epithelium and causes no visible damage to it. The symptoms are produced entirely by the toxin the bacterium produces. For the most part, extracellular parasites invade and cause damage to host tissues ranging from the superficial to the profound but all the while grow and reproduce outside of cells.

In sharp contrast, obligate intracellular parasites, such as rickettsia, chlamydia, viruses, and certain protozoan parasites, can only reproduce inside cells. To varying degrees, these infectious agents lack the capability for independent growth and replication. They depend on host cells for the missing components or capabilities. Members of a third category, facultative intracellular parasites, have the ability to multiply both intracellularly and extracellularly.

Species of bacteria that multiply within host cells can evade the immune responses of the host. For example, the rickettsias produce a phospholipase that helps them penetrate the cell and probably also helps them escape from the endocytic vacuole into the cytoplasm where they multiply. *Mycobacterium tuberculosis* changes the phagosome surface and thereby prevents lysosomal fusion. Other bacteria such as *Yersinia pestis,* the cause of plague, can actually reproduce inside the phagolysosome despite its unfriendly environment. Microorganisms living within macrophages remain viable and may be carried to other parts of the body. By remaining within the phagocyte, they are protected from some antimicrobial drugs and the immune system of the host.

REVIEW QUESTIONS

1. List the steps usually necessary for a pathogen to produce an infectious disease in a susceptible host.
2. What is a virulence factor?
3. List at least four sources of microorganisms that cause disease in humans.
4. What is meant by the term *colonization?*
5. How do extracellular, facultatively intracellular, and obligately intracellular pathogens differ?

Pathogens Can Interfere with Phagocytosis and Thereby Establish Infection

Once a pathogen has entered the body, in order to establish an infection it must cope with phagocytes. Pathogens may interfere with the activities of the phagocytes. For example, *Streptococcus pneumoniae* has a large capsule that inhibits the attachment of phagocytes and thus inhibits killing (figure 19.7). These capsules, which are important virulence factors, induce an antibody response. When specific antibodies are present, they coat the capsule, activate complement, and permit the organism to be ingested. Also effective in preventing phagocytosis are the hyaluronic acid capsules of *Streptococcus pyogenes* and the sialic acid capsules of some strains of meningococci. These capsular substances are also found in human cells and so are not recognized as foreign, and they do not induce opsonizing antibodies.

Staphylococcus aureus produces a protein, coagulase, that causes fibrin[1] to clot. This weblike clot traps and impedes the movement of phagocytes, thus protecting the microorganisms. *S. aureus* also produces proteins called **leukocidins** that kill phagocytes. One particular leukocidin damages the cytoplasmic membrane of the phagocyte, causing the contents of the cell to leak out, resulting in cell death.

Recall that phagocytes move toward the C5a component of complement and this is important in the inflammatory process. *S. pyogenes* produces an enzyme called *C5a peptidase* that breaks down C5a and interferes with its activity. As a result, phagocytes do not migrate effectively into the area and phagocytosis is impeded. Also some gram-negative bacteria are able to add on to the O antigen side chains of their endotoxin molecules, making them long enough that complement membrane attack complexes cannot penetrate the bacterial cell to cause lysis (figure 19.8). All of these ploys interfere with phagocytosis and complement action.

[1]Fibrin—insoluble protein that forms the essential portion of the blood clot

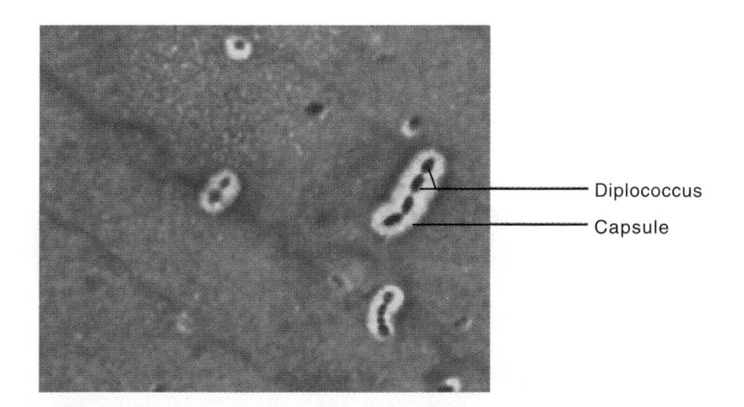

Figure 19.7
Capsules can interfere with phagocytosis. The large polysaccharide capsule of the diplococci *Streptococcus pneumoniae* effectively prevents phagocytosis. Specific antibody and the C3b portion of activated complement can combine with the capsule and allow phagocytosis to occur.

Toxins May Be Responsible for the Ability of Microorganisms to Cause Disease

A number of bacterial pathogens produce harmful substances called toxins. These toxins can be subdivided on the basis of their chemical nature into **exotoxins** (proteins) and **endotoxins** (lipopolysaccharides) See table 19.6 for a description of the properties of toxins.

Exotoxins Are Proteins Usually Secreted by Microorganisms

Exotoxins are diverse in their effects and some are powerful poisons. As a group, they can affect a wide variety of different cellular structures and functions, although generally each has a specific action (table 19.7) Both gram-positive and gram-negative species produce exotoxins. Exotoxins are either actively secreted by the organism or leak into the surrounding fluid following lysis of the bacterial cell wall. Most exotoxins are inactivated by heating at 60°C to 100°C for 30 minutes, a notable exception being the staphylococcal toxins that cause food poisoning. Generally, exotoxins are carried by the bloodstream from the point of infection to distant parts of the body where they can cause dramatic effects. In the case of botulism poisoning, ingestion of minute amounts of the toxin in food is sufficient to cause the paralysis that characterizes the disease. In this case, bacterial colonization and infection are not necessary prerequisites.

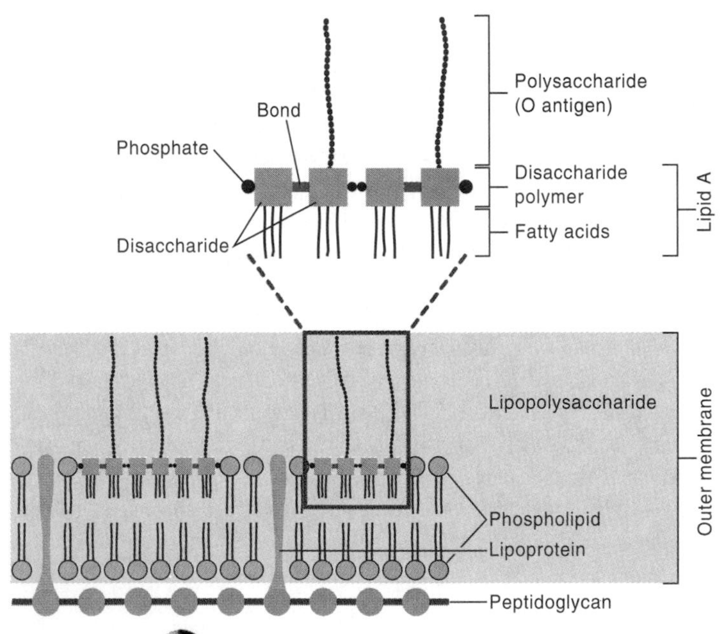

Figure 19.8
Endotoxin of the gram-negative cell wall. The lipid A portion of the lipopolysaccharide endotoxin is responsible for its toxic properties and is the same in all endotoxin molecules. Being a lipid, it is nonantigenic. The polysaccharide portion varies from one organism to another and is antigenic (the O antigens).

Some exotoxins are called **A-B toxins** because they have two parts, A and B. The A portion is an enzyme constituting the toxic part, and the B portion binds to host cell receptors, allowing the toxic enzyme to enter the cell (figure 19.9). Examples of the A-B toxins are those causing diphtheria, cholera, tetanus, and *Shigella* dysentery. A portion of many of these toxins are enzymes that remove the ADP-ribosyl group from the coenzyme NAD and attach it to some host cell protein, inhibiting the function of that protein. For example, the cholera toxin acts on a protein that normally regulates cyclic AMP levels in the cell, causing uncontrolled water and ion loss into the bowel and resulting in a severe watery diarrhea that is often fatal.

A second type of exotoxin acts by disrupting cell membranes. This type includes, for example, a cell-lysing cytotoxin of *S. aureus* and a phospholipase toxin of *Clostridium perfringens*. The alpha toxin of *S. aureus* is a protein that inserts into host cell membranes, forming pores that allow leakage from the bacteria and allow water to enter, disrupting the cell. The *C. perfringens* toxin is an enzyme that disrupts cell membranes. Some of these membrane-disrupting toxins and toxic enzymes are described in table 19.8.

Superantigens comprise a third type of exotoxin. They are described in perspective 19.2. These antigens are not processed intracellularly as other antigens are, but instead bind directly and indiscriminately to the major histocompatibility class II antigen on helper T cells. Rather than binding to a small number of specific T cells, the superantigen binds to a great many T cells. The result is the release

Table 19.6 Important Properties of Bacterial Toxins

Property	Exotoxins	Endotoxins
Bacterial source	Gram-positive and gram-negative species	Gram-negative species only
Location in bacterium	Synthesized in cytoplasm and released from cell	Component of cell wall
Chemical nature	Protein	Lipopolysaccharide containing lipid A
Ability to form a toxoid	Present	Absent
Stability	Generally heat labile, 60°C–100°C for 30 minutes	Heat stable
Action	A distinctive effect from each	Same effect: fever, circulatory collapse, and characteristic damage to tissues

Table 19.7 Examples of Bacterial Exotoxins

Bacterium	Gram Stain	Disease	Exotoxin Effect
Corynebacterium diphtheriae	+	Diphtheria	Enzyme; blocks protein synthesis; cells most severely affected are those of nerves, heart, and kidney
Clostridium botulinum	+	Botulism	Inhibits acetylcholine release from motor nerve endings, causing paralysis; resists gastrointestinal proteases
Clostridium tetani	+	Tetanus	Blocks function of certain nerves that normally prevent muscle spasm; spastic paralysis results
Bacillus anthracis	+	Anthrax	Causes swelling and hemorrhage of tissue and circulatory failure
Staphylococcus aureus	+	Scalded skin syndrome; food poisoning; toxic shock syndrome	Causes skin layers to separate, causing blisters and skin sloughing; toxin produced by *S. aureus* growing in food; heat resistant; diarrhea results from small intestinal hypersecretion; stimulates vomiting center of brain by acting on visceral receptors
Streptococcus pyogenes	+	Scarlet fever	Toxin absorbed from "strep throat" acts on skin
Bordetella pertussis	–	Pertussis (whooping cough)	Damages cilia of respiratory mucosal cells
Escherichia coli	–	Diarrhea	Two kinds of toxin, one similar to cholera toxin, the other heat resistant; production controlled by plasmid genes
Pseudomonas aeruginosa	–	Septic shock	Enzyme; acts like diphtheria toxin but affects different tissues
Vibrio cholerae	–	Cholera	Binds to cell receptors of the small intestinal epithelium and causes abnormally high cAMP production with consequent copious diarrhea

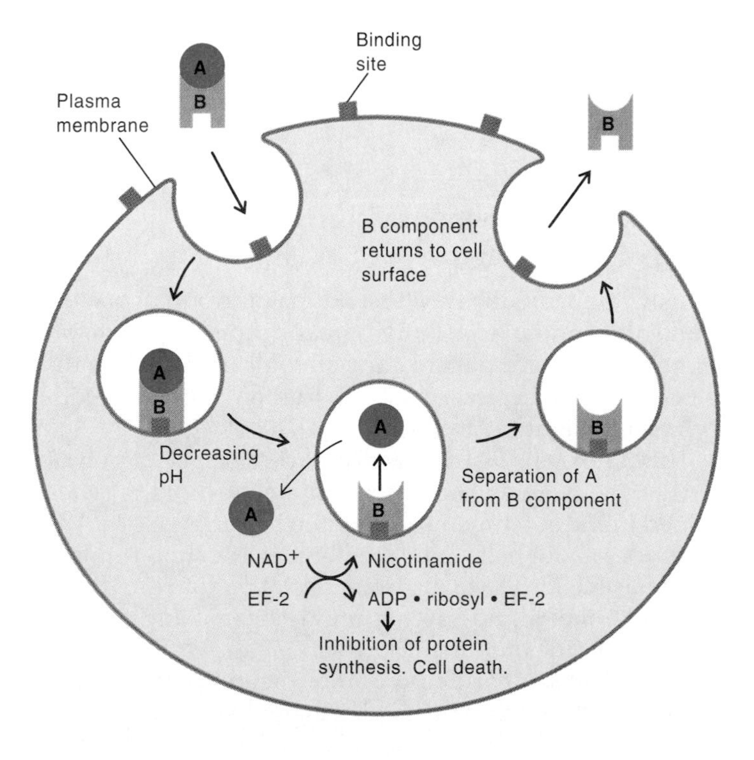

Figure 19.9
The mode of action of A-B exotoxins. The A-B exotoxins have two components, A and B. The A portion is the toxic part, and the B portion functions to get the toxin into host cells. To do this, the B portion binds to receptors on host cells and thus determines which cells will be affected by the toxin. After binding, the A-B molecule is taken into a vacuole by endocytosis. After endocytosis, the contents of the vacuole become more acidic, causing the A and B portions to separate. The B portion remains in the vacuole and is removed by exocytosis, while the A portion enters the cytoplasm of the cell and exerts its toxic effect. For example, the diphtheria toxin A portion catalyses the removal of ADP-ribosyl from NAD and its attachment to elongation factor 2 (EF-2), making EF-2 inactive. This interferes with protein synthesis and results in cell death. The A portions of most A-B exotoxins act in a similar manner, by attaching ADP-ribosyl to an important host cell protein, thereby interfering with the action of that protein. Some A-B toxins enter cells by a different mechanism, in which the B portion binds to the host cell membrane and the A portion enters the cell cytoplasm, leaving the B part on the outside. Possibly a pore forms through the membrane, through which the A part enters, but this has not been proven.

Table 19.8 Some Bacterial Membrane Disrupting Toxins and Toxic Enzymes that May Contribute to Virulence

Product	Description	Function	Producing Bacterium
Phospholipase	Enzyme	Breaks down lecithin, a lipid component of mammalian cell membranes	*Clostridium perfringens*
Hemolysin	Oxygen-sensitive protein	Destroys red blood cells	*Streptococcus pyogenes*
Coagulase	Enzyme	Clots plasma	*Staphylococcus aureus*
Collagenase	Enzyme	Breaks down collagen, tissue fiber	*Clostridium perfringens*
Hyaluronidase	Enzyme	Breaks down hyaluronic acid, a tissue component	*Staphylococcus aureus*
Lipase	Enzyme	Breaks down fat	*Staphylococcus aureus*
Deoxyribonuclease	Enzyme	Breaks down DNA	*Staphylococcus aureus*
Alpha toxin	Protein	Forms pores in membranes	*Staphylococcus aureus*

and activity of excess amounts of cytokines that enter the bloodstream instead of only local activity. This leads to a variety of symptoms and sometimes shock, with failure of many organ systems, circulatory failure, and often death.

Transfer of the Production of Toxins and Other Virulence Factors from One Bacterium to Another Temperate phages may carry the genes coding for exotoxin production in some bacteria, notably those bacteria that cause diphtheria, scarlet fever, and botulism. Plasmids may also carry genes for exotoxins, such as the genes for the production of certain enterotoxins in *E. coli.* The phages and plasmids can transfer the ability to make toxins from one strain of bacterium to another by transduction, transformation, and conjugation. The modes of action of various exotoxins and enterotoxins and the effects of these in diseases such as diphtheria, botulism, and tetanus are discussed in more detail in later chapters.

Genes for toxins and other virulence factors are found in many different locations on the bacterial chromosome. However, in gram-negative enteric rods, several large groups of genes have been associated with pathogenicity. These large gene regions are called **pathogenicity islands**, and each can be inserted into a particular location on the bacterial chromosome. For example, some strains of *E. coli* are known to cause urinary tract infections, while other strains are enteropathogenic; that is, cause gastrointestinal disease. These strains have been found to have different pathogenicity islands at the same chromosome location. The pathogenicity island, in the strains causing urinary tract infections codes for a cytolysin, while the pathogenicity island in the enteropathogenic strains codes for secreted proteins that lead to intestinal cell damage. Usually, a pathogenicity island contains genes for a number of different virulence factors. A single bacterial strain may contain several different pathogenicity islands. Obviously, pathogenicity islands are transferred between bacteria and are inserted into the chromosome, but the mechanisms of transfer and insertion are not yet known. It is clear that the virulence of microbial strains can be changed rapidly and dramatically by the exchange or loss of these genetic transfers.

Use of Toxoids Most exotoxins of gram-positive bacteria can be converted to toxoids by mild heat treatment or reaction with chemicals such as formalin. As discussed in chapter 17, a toxoid lacks the toxic properties but resembles the toxin closely enough to stimulate antibody production when injected into a person. Toxoids have been extremely effective as vaccines in the prevention of diseases such as diphtheria and tetanus, caused by toxins of gram-positive bacteria. For reasons that are not clear, even though antibody production is stimulated by them, toxoids prepared from the exotoxins of gram-negative bacteria have generally been disappointing as immunizing agents in disease prevention.

Endotoxins Are Lipopolysaccharides, a Part of Gram-Negative Cell Walls

As shown in table 19.6, the second major group of bacterial toxins, the endotoxins, differs greatly from the exotoxins. All endotoxins are part of the cell wall of gram-negative bacteria and are necessary for cell division in gram-negative bacteria (see figure 19.8).

Endotoxins consist of a two-part molecule, a polysaccharide and a lipid. The lipid component (lipid A) is embedded in the outer membrane of the bacterial cell wall and constitutes about half of the membrane. The lipid A portion is responsible for the toxic properties of endotoxin. Lipid A alone, like most lipids, is not immunogenic and does not induce a specific immune response.

The polysaccharide portion of the molecule, or the O antigen, sticks out into the environment like little hairs. The O chains are different for each bacterial strain. The specific immune response is made to the O antigen.

Perspective 19.2

Superantigens

Usually when the immune system encounters a foreign substance such as a virus or microoganism, only a relatively small number (1 in 10,000) of T cells reacts. This small number of cells is able to stimulate the rest of the immune system for an attack that eliminates this foreign invader. **Superantigens** are proteins that cause the immune system to produce large numbers of T cells that are usually useless in fighting the infection. Superantigens most often are toxins such as those produced by *Staphylococcus aureus* in food poisoning or toxic shock syndrome.

Superantigens such as the toxin A in *S. aureus* do not need to be processed by antigen-presenting cells. Instead, they bind directly with the MHC molecule, attaching themselves to the outside of it instead of to the interior pocket (figure 1), and cause the cells to secrete interleukin-2 at very high levels. This causes a massive production of T cells with many different antibody configurations. Excess interleukin-2 causes the symptoms of fever, nausea, vomiting, diarrhea, and shock. When many T cells are activated, some may recognize self antigens, provoking an attack on healthy tissues of the self **(autoimmunity).** When so many of the T cells are directed to replicate, many of them will die, leaving the body unable to combat other infections.

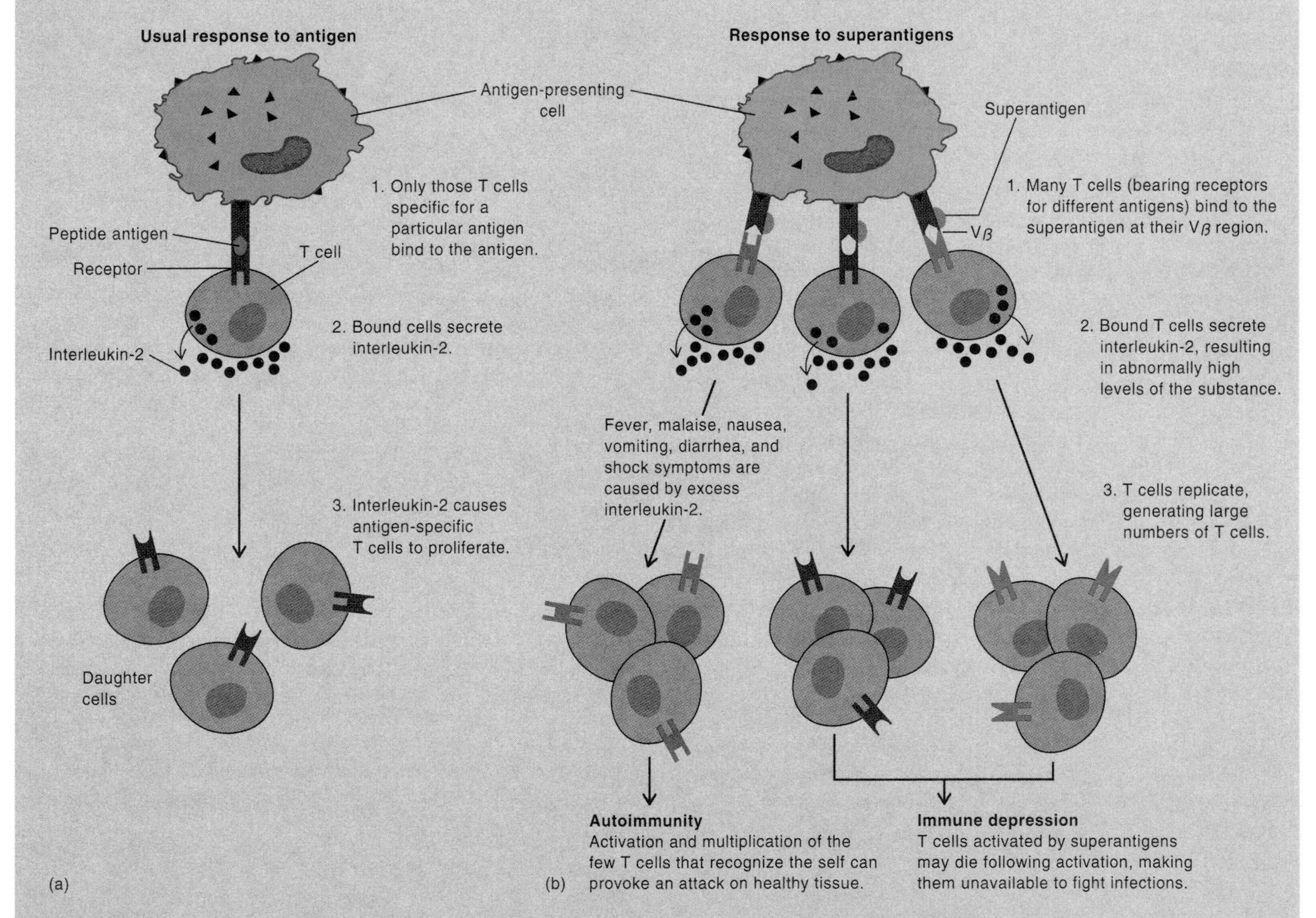

Figure 1
How superantigens cause disease. (*a*) Usually, foreign antigens activate only those helper T cells whose receptors can bind the antigens tightly. (*b*) Superantigens indiscriminately activate any T cells whose receptors include selected V_β segments. They thereby activate millions of helper cells, sometimes with dire consequences.

Endotoxins do not affect the host unless they are released from the bacterial cell, which occurs when the cell divides or dies. After the endotoxin is released, it binds to macrophages and causes them to release certain proteins, including interleukin-1, which act in a number of ways including to induce fever. These proteins are also responsible for the other symptoms of infection: weakness, generalized aching, and in the case of an overwhelming infection, shock through the release of large amounts of endotoxin.

Nearly all gram-negative bacteria, whether pathogens or not, possess endotoxins. By contrast, only a relatively small number of species of gram-positive and gram-negative bacteria produce exotoxins. Except in certain cases of septic shock in which large numbers of bacteria enter the bloodstream, endotoxins generally play a contributing rather than a primary role in pathogenesis.

Endotoxins are resistant to heat, cannot be converted to toxoids, and have the same mode of action regardless of the species of gram-negative bacterium from which they were derived.

Since endotoxins are heat stable, they are not affected by autoclaving. Consequently, solutions for intravenous administration must not only be sterile, but must be free of endotoxin as well. Disastrous results including death have resulted from the administration of intravenous fluids contaminated with minute amounts of endotoxin. A very sensitive test known as the *Limulus* assay is used to verify that these fluids are not contaminated. This test uses material from the horseshoe crab *Limulus polyphemus* that precipitates in the presence of endotoxin. The precipitate is examined with a spectrophotometer, and as little as 10 to 20 picograms[1] of endotoxin per milliliter can be detected.

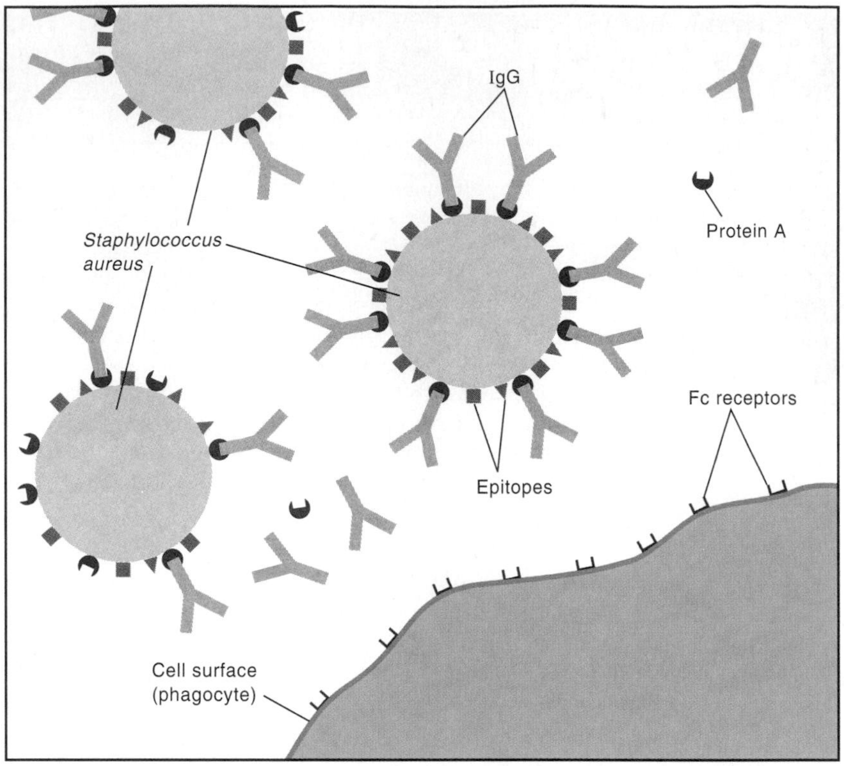

Figure 19.10
Protein A of *Staphylococcus aureus* inhibits opsonization by antibodies. Bacteria that are coated with IgG antibody molecules are usually opsonized and efficiently phagocytized by interaction of the Fc portions of antibodies and Fc receptors on phagocytes. However, strains of *S. aureus* that produce protein A are not opsonized by IgG because protein A coats the staphylococci and binds to the Fc region of antibodies. This prevents the interaction of antibody Fc regions with Fc receptors on phagocytes. Also, because the antigen-binding sites of antibodies may not combine with epitopes, protein A inhibits other antibody activities as well.

Evasion of the Host's Immune Response Bacteria and viruses are adept at evading host defense mechanisms. They often can make genetic variations from one substance to another; for example, they may vary from one type of pilus to another (phase variation). Thus, as antibodies are produced to the old pili, the new pili are still able to cause adherence and colonization. *S. aureus* produces protein A that binds to the Fc portion of immunoglobulins and inhibits opsonization via the Fc portion of immunoglobulin (figure 19.10).

The protein G of *Streptococcus pyogenes* functions in a similar manner. As mentioned earlier, some bacterial capsules resemble host tissue components and so do not induce an immune response. Several pathogens that successfully colonize mucous membranes make a protease that specifically breaks down secretory IgA, permitting the bacteria to attach to cells. Microorganisms have evolved many such ways to evade the immune response. Some of these mechanisms are summarized in table 19.9.

Exit from the Host and Survival in the Environment Are Essential Steps in the Infection Process

Exit from the Host

Microorganisms may leave the host from the respiratory tract, the alimentary tract, the genitourinary tract, the skin, or the conjunctiva of the eye. Coughing and sneezing, diarrhea, pus, and other discharges all offer microorganisms routes of exit from the host. Some pathogens exit via insect bites. For example, *Rickettsia prowazekii*, the cause of typhus, is taken into lice when they bite an infected individual. The organisms multiply within the lice and, after a

[1] Picogram—1 picogram = 10^{-12} grams

Table 19.9 Some Mechanisms Microorganisms Use to Evade Host Responses

Evasion of Complement Activity	Evasion of Phagocytic Activity	Evasion of Antibody Activity
Some capsules prevent complement activation. Some gram-negative rods can lengthen O chains in their lipopolysaccharide to prevent complement membrane attack.	C5a peptidase production destroys C5a and inhibits the attraction of phagocytes to the area of infection. Leukocidins produced are toxic for phagocytes. Some microorganisms can escape from phagosomes before they fuse with lysosomes. Some microorganisms can prevent fusion of phagosomes with lysosomes. Some microorganisms resist antimicrobial lysosomal contents.	Production of proteases destroys secretory IgA. Staphylococcal protein A and similar proteins interfere with antibody-mediated opsonization. Variation in surface antigens of bacteria, including pili, makes antibodies useless. Some microorganisms produce capsules that are not antigenic.

period of days, appear in large numbers in the feces. Lice defecate when they feed, so the rickettsia are deposited on the skin of humans and rubbed into the bite wound to enter a new human host. The organisms are also able to survive drying in the feces and can reach new hosts when the dried feces reach the mucous membranes of the eye or respiratory tract.

Survival in the Environment

Once outside the host, the organism must activate new genes to allow it to survive in the different environment. Some bacteria, fungi, and eukaryotic parasites are able to maintain an independent existence in soil, water, or other environmental niches until they reach new hosts. Others live in animals and are transmitted to humans via meat, milk, or other animal products. For many if not most pathogens, infection of the human host is an accidental occurrence. For those transmitted by direct contact between humans, their inability to survive very long outside the human host makes it imperative that they immediately find a new host.

The Outcome of Infection Depends upon Host Responses as Well as the Activities of the Infecting Agent

Impaired Body Defenses Promote Infectious Disease

The nonspecific and specific immune responses, discussed in previous chapters, are vitally important in maintaining the health of the host. Therefore, impairment in any one of these mechanisms makes the host more susceptible to invasion by microbes or viruses (chapter 18). For example, some people are born unable to produce functional phagocytes or antibodies. Others lose these capabilities later in life. Still others, such as those receiving heart or kidney transplants from other people, may be given drugs that impair their inflammatory response or immune response or both, in order to prevent rejection of the transplant. Certain conditions, including cancer, diabetes, AIDS, and malnutrition, may cause defects in protective mechanisms of the host. In all of these instances, microorganisms of the normal flora can invade, and latent infections are reactivated more frequently than they would be in hosts having normal defenses.

Genetic Factors Are Important in Determining Susceptibility to Infections

Different strains of laboratory animals have marked differences in susceptibility to certain infections. For example, some strains of inbred mice are much more susceptible to *Salmonella* infections than are other strains. Humans also show genetic differences in susceptibility. For example, persons with genes for the sickle cell trait are more resistant to malaria but more susceptible to *Salmonella* infections than is the general population.

Age and Stress Also Influence Susceptibility to Infection

Infants and the aged have less effective immune systems; thus, both of these populations are more susceptible to infections. It has long been recognized that stress can adversely affect the outcome of infection. In response to stress, high levels of corticosteroid hormones are produced and these are well known to be immunosuppressive. Also, lymphoid tissue has been shown to have receptors for catecholamines, compounds that act on nerve cells, and catecholamines can suppress T cell activity in lab animals. These and other findings suggest that messages go, not only between immune cells, but also between the immune system and the nervous system, to affect the overall immune response.

Some Potential Pathogens May Live for Years in Host Tissues, Held in Check by Host Defenses

In some instances, viruses and microorganisms live in the host's tissues without causing any symptoms. Such agents may remain latent for many years, only revealing their pathogenic capabilities when the balance of the host-parasite ecosystem shifts in their favor. Such is often the case with the virus of herpes simplex, the cause of "cold sores" and genital herpes. It also occurs with chicken pox virus (varicella-zoster virus) and the intracellular bacterium *Rickettsia prowazekii,* the cause of typhus. Another example is *Mycobacterium tuberculosis,* the cause of human tuberculosis. The mycobacteria are often confined within a small area of tissue by host defense mechanisms, but they may begin growing and destroying tissue when the host suffers from malnutrition, the effects of anti-inflammatory medicines, AIDS, or other factors that influence his or her ability to combat the pathogen.

In subsequent chapters, a variety of infectious diseases along with the organisms causing them are discussed. Mainly, these organisms are the pathogens that frequently cause diseases in normal persons who have not had previous exposure to them. Keep in mind, however, that many other microorganisms have the capacity to cause disease in abnormal or normal hosts under special conditions favoring the microbes. Some of the opportunistic infections are discussed in chapter 31.

REVIEW QUESTIONS

1. What is an important difference between the polysaccharide capsules of *Streptococcus pneumoniae* and the hyaluronic acid capsules of *S. pyogenes?*
2. What are the chemical natures of exotoxins and of endotoxins?
3. Distinguish between the functions of the A and B portions of A-B exotoxins.
4. Give two examples of membrane-disrupting toxins.
5. What are toxoids and of what use are they?

Summary

I. Methods for Establishing the Cause of an Infectious Disease
   A. A molecular version of Koch's Postulates is proving to be useful.

II. Normal Flora of the Human Body
   A. Microorganisms and humans form a variety of symbiotic relationships.
      1. Commensalism
      2. Mutualism
      3. Parasitism
   B. Establishing and maintaining the normal flora as a dynamic human-microbe ecosystem is important for good health.
      1. Establishment of normal flora
      2. Importance of normal flora
      3. The normal flora: subject to changes

III. Infectious Disease Terminology
   A. A number of terms are used in the study of infectious disease.
      1. Colonization
      2. Infection, disease
      3. Pathogenic organisms
      4. Opportunistic organisms
      5. Virulence

IV. Mechanisms of Pathogenesis
   A. Successful pathogens usually carry out a sequence of activities.
      1. Transmission to a susceptible host
      2. Adherence
      3. Colonization and invasion
      4. Damage to the host
      5. Exit from the host
      6. Survival outside the host until transmission can occur again
   B. The sources of infectious agents and their modes of transmission are varied and numerous.
   C. Adherence is usually a necessary step in the establishment of infection.
   D. The establishment of infection involves invasion of host tissues.
      1. Tissue invasion is fostered by large concentrations of pathogens.
      2. Colonization is a necessary step in establishing an infection.
   E. Microorganisms interact with their hosts in many ways.
      1. Extracellular, facultatively intracellular, and obligately intracellular pathogens cause damage to hosts in different ways.
      2. Pathogens can interfere with phagocytosis and thereby establish infection.
   F. Toxins may be responsible for the ability of microorganisms to cause disease.
      1. Exotoxins are proteins usually secreted by microorganisms.
         a. Transfer of the production of toxins and other virulence factors from one bacterium to another
         b. Use of toxoids
      2. Endotoxins are lipopolysaccharides, a part of gram-negative cell walls.
   G. Exit from the host and survival in the environment are essential steps in the infection process.
   H. The outcome of infection depends upon host response as well as the activities of the infecting agent.
      1. Impaired body defenses promote infectious disease.
      2. Genetic factors are important in determining susceptibility to infections.
      3. Age and stress also influence susceptibility to infection.
      4. Some potential pathogens may live for years in host tissues, held in check by host defenses.

Chapter Review Questions

1. List at least five ways in which the normal flora helps protect against infectious diseases.
2. How can intensive antibiotic therapy affect the normal flora?
3. Why is attachment of bacteria to host cells important in establishing infection and how does attachment occur?
4. Describe several ways in which microorganisms or viruses gain entry into the body through the gastrointestinal tract.
5. What advantage is gained for pathogens by the ability to live and reproduce intracellularly?
6. Discuss at least five ways in which organisms can evade phagocytosis and/or complement attack.
7. Compare and contrast exotoxins and endotoxins.
8. Superantigens "turn on" T cells to become immunologically active. Why is this harmful rather than protective?
9. What is a pathogenicity island? Give an example.
10. Give examples of three general mechanisms by which microorganisms can evade host defenses.

Critical Thinking Questions

1. Explain why *E. coli* is or is not a pathogen.
2. Does having an infection mean that an infectious disease is present?
3. Explain how and when normal microbial flora can become a threat to human health.

Further Reading

Mecsas, Joan, and Strauss, Evelyn J. 1996. Molecular mechanisms of bacterial virulence: Type III secretion and pathogenicity islands. A synopsis of these novel but widespread mechanisms of virulence. *Emerging Infectious Diseases* 2:271–288.

Ochman, Howard, *et al.* 1996. Identification of a pathogenicity island required for *Salmonella* survival in host cells. A report of virulence genes for *S. typhimurium* clustered in a chromosomal pathogenicity island. *Proc. Natl. Acad. Sci. USA* 93: 7800–7804.

Salyers, Abigail A. and Whitt, Dixie D. 1994. Bacterial pathogenesis. A molecular approach. American Society for Microbiology, Washington. An in-depth look at pathogenesis at the molecular level.

20 Epidemiology and Public Health

KEY CONCEPTS

1. Identifying the reservoir and mode of transmission of a disease agent can provide a means of preventing the disease.
2. Diseases in which symptomatic humans are the only reservoir are the easiest to control.
3. Worldwide travel and distribution of foods increases the global threat of disease.
4. Identifying the origin of an epidemic depends on a careful comparison of the characteristics and activities of those affected and unaffected by the epidemic.

A variety of diseases are spread through contaminated water. Unfortunately, many parts of the world lack adequate sources of clean drinking water.

PREVIEW LINK

Microbes in Motion

The following book and chapter in the *Microbes in Motion* CD-ROM may serve as a useful preview or supplement to your reading:
Epidemiology: Disease Acquisition (Communicable).

A Glimpse of History

For centuries, it was accepted that many mothers developed fever and died following childbirth. Puerperal fever, an illness resulting from bacterial infection of the uterus following childbirth, had been known since the time of Hippocrates (460–377 B.C.), but it was not until the 18th century, when it became popular for women to deliver their babies in hospitals, that the incidence of the disease rose to epidemic proportions. In Prussia between 1816 and 1875, over 363,600 women died of puerperal fever. In the middle of the 19th century in the hospitals of Vienna (the major medical center of the world at that time), about one of every eight women died of puerperal fever following childbirth.

In 1841, Ignaz Semmelweis, a Hungarian, traveled to Vienna to study medicine. After finishing medical school, he became the first assistant to Professor Johann Klein at the Lying-in Hospital. This was about 30 years before Pasteur and Koch established the germ theory of disease.

There were two sections of the Lying-in Hospital. The first section was under the management of Professor Klein and the medical students. The second section was served by midwives and midwifery students. Being an astute observer, Semmelweis soon noticed that the incidence of puerperal fever in the first section often rose as high as 18%, four times that in the second section. When he examined the conditions in the two sections, they appeared to be the same except in terms of their management. Semmelweis was dismissed from his post when he implicated Professor Klein and the students in the spread of the disease, but he was reinstated a few months later when friends intervened on his behalf.

An important clue in the source of the disease came from the death of Semmelweis's friend Kolletschka, who had died of an infection as the result of a scalpel wound he incurred while doing an autopsy. The symptoms were the same as those seen in the women suffering from puerperal fever. Semmelweis reasoned that the same "poison" that had killed his friend was probably on the hands of the medical students who did autopsies and then attended the women in the first section. Since midwives did not perform autopsies, there was a lower incidence of puerperal fever in the second section. Semmelweis instituted the practice of having doctors and students wash their hands with a solution of chloride of lime, a strong disinfectant, before attending their patients. The result was a drop in the incidence of puerperal fever to one-third its previous level. Instead of accepting these findings and the new techniques that Semmelweis employed, his colleagues in Vienna refused to accept responsibility for the deaths of so many patients. The work of Semmelweis was so fiercely attacked that he was forced to leave Vienna and return to his native Hungary. There he was again able to use disinfection techniques to achieve a remarkable reduction in the number of deaths from puerperal fever.

Semmelweis became increasingly outspoken and bitter, finally becoming so deranged he was confined to a mental institution in 1865. Ironically, he died one month later of a generalized infection similar to the kind that had killed his friend Kolletschka and the many women who had contracted puerperal fever following childbirth. The infection originated from a finger wound received before his confinement. Some said he deliberately infected himself from a cadaver while doing an autopsy.

The accomplishments of Ignaz Semmelweis would be the envy of today's hospital epidemiologists. Without a knowledge of microbiology, he rigorously defined the affected population, located the source of the causative agent and its mode of transmission, and showed how to stop the spread of the disease.

Epidemiology is the study of the frequency and distribution of disease. Recognizing that diseases do not occur at random but instead have risk factors that influence their distribution makes it possible for authorities to set guidelines for the prevention and control of certain diseases. **Epidemiologists** are the "disease detectives." They study the patterns of disease occurrence and attempt to identify the cause, the source, and the route of transmission of disease. Once these factors have been determined, recommendations for prevention can be instituted.

Several features of today's world underscore the importance of epidemiology: (1) the burgeoning world's population, which increasingly leads to crowding and malnutrition, facilitating the spread of infectious diseases; (2) social unrest, which interferes with the prevention, detection, and control of disease; (3) rapid transportation of disease, which makes a disease outbreak anywhere in the world the urgent concern of the rest of the world; and (4) mass distribution of food and supplies, sometimes contaminated with pathogens, which can cause widespread disease when items are shipped to several states or nations.

Descriptive Terms in Epidemiology

Diseases that are constantly present in a given population are **endemic.** For example, both the common cold and influenza are endemic in the United States. An unusually large number of cases in a population constitute an **epidemic.** Epidemics may be caused by diseases that are normally endemic, such as influenza and pneumonia (figure 20.1), or by diseases that are not normally present in a population, such as the recent outbreaks in Africa of the Ebola virus. An epidemic that spreads worldwide such as AIDS is a **pandemic.**

Epidemiologists are concerned not only with the number of people affected by the disease, but also and most importantly with the **rate** of disease; that is, the fraction of people who have the disease in a given population (diseased population/total population). For example, if 100 people in a large city (population 500,000) develop disease X, and 100 people in a rural community (population 1,000) develop the same disease, then the rate is much higher in the rural community and, thus, of greater concern to the epidemiologist. A related rate is the **attack rate,** which is the number of cases developing the infection per 100 people exposed.

Several terms, each of which is most often expressed as a rate, are important in describing the epidemiology of disease. **Morbidity rate** is calculated as the number of cases divided by the population at risk. Contagious diseases such as influenza often have a high morbidity rate because each infected individual may transmit the infection to several others. **Mortality rate** is the fraction of people who die from the disease. Diseases such as plague, AIDS, and Ebola are feared because of their very high mortality rate. **Incidence rate** reflects the number of new cases in a specific time period in a given population, whereas **prevalence rate** reflects the number of total existing cases in a given population. The

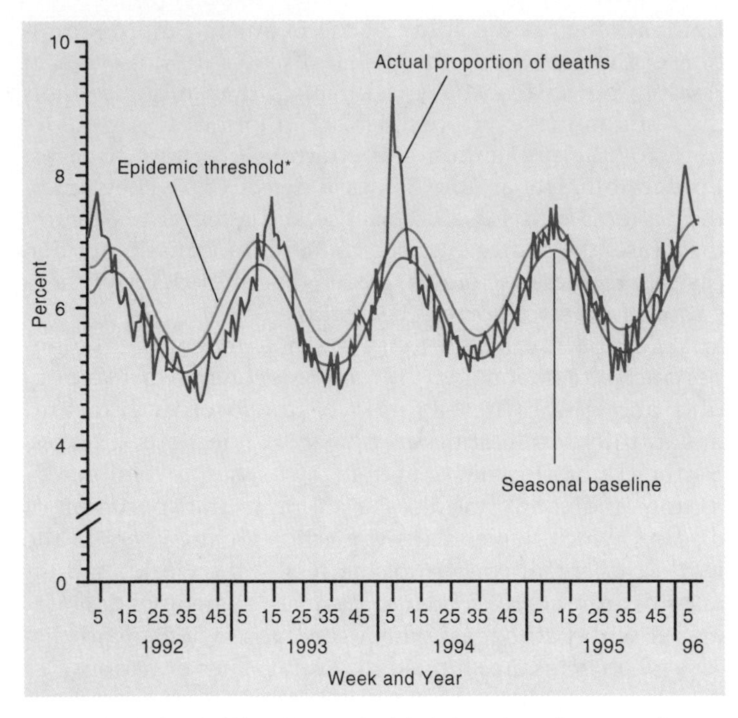

*The epidemic threshold is 1.645 standard deviations above the seasonal baseline calculated using a periodic regression model applied to observed percentages since 1983. The baseline was calculated using a robust regression procedure.

Figure 20.1
Weekly pneumonia and influenza mortality as a percentage of all deaths for 121 cities of the United States, January 1, 1992, to February 3, 1996.

Source: *Morbidity and Mortality Weekly Report.* 45(6)135. 1996.

prevalence rate is thus useful to assess the overall impact of the disease on society, whereas the incidence rate provides a means of measuring the risk of an individual contracting the disease at a certain time. Diseases that are typically of long duration have a higher prevalence than incidence because the total number of existing cases continues to climb.

Diseases that can be transmitted from one person to another, such as measles, colds, and influenza, are **communicable diseases.** In contrast, diseases that are nontransmissible, such as pneumonia resulting from the inhalation of normal throat flora or botulism that results from the ingestion of foodborne botulinum toxin, are **noncommunicable diseases.** Some illnesses, such as measles and colds, are typically **acute diseases,** which means the symptoms develop and then subside fairly rapidly, whereas others are **chronic diseases** in which the symptoms may persist for years. With some diseases, all infected people have symptoms and are said to be **symptomatic.** With other diseases, a portion of the population may not have obvious symptoms; these **asymptomatic** people are said to have **subclinical infections.** A **latent disease** is one in which the disease agent can remain inactive for an extended period after which the disease may reappear. For example, cold sores caused by herpesvirus remain latent but reactivate when a person is stressed.

Spread of Disease

In order for disease to spread, an infectious agent must have (1) a suitable habitat in which it normally lives and multiplies, (2) a mode of transmission to the next host, and (3) an appropriate route to enter the next host (figure 20.2). By determining the habitat, transmission, and entry route, an epidemiologist can recommend steps to eliminate or interrupt the spread of the disease.

Humans, Animals, and Nonliving Environments Can Be Reservoirs for Disease

The natural habitat or source of an organism is called the **reservoir.** In the case of communicable diseases, infected humans are the most significant reservoir. Obviously, people who have symptomatic illness are a source of infection, but ideally they understand the importance of taking precautions to avoid the transmission of their illness to others. Diseases in which symptomatic humans are the only reservoir are theoretically among the easiest to control. Smallpox, for example, was successfully eradicated because humans were the only reservoir and they were always symptomatic. Thus, widespread vaccination and the successful isolation of all remaining cases eradicated the disease. A more problematic source of infection occurs in those diseases in which some people harbor organisms with no obvious clinical symptoms. These asymptomatic people unknowingly act as **carriers** of the disease and spread the pathogen as they move freely about. Sometimes, carriers harbor the disease-causing organism for only a short time, but sometimes they may become chronic carriers who continue to excrete the organisms intermittently or constantly for months, years, or even a lifetime. After a disease such as typhoid fever has been controlled in a population, for example, the presence of chronic carriers poses a threat of disease recurrence for a period of many years. Fortunately, most carriers can be successfully treated with antibiotics. Asymptomatic infections are also a complicating factor in the control of sexually transmitted diseases such as gonorrhea. Up to 50% of women infected with *Neisseria gonorrhoeae* have no obvious symptoms, which means they often unknowingly transmit the disease to their sexual partners. Treatment of gonorrhea infections is easily accomplished through antibiotic therapy, but tracking down sexual contacts of infected people is difficult and costly.

Nonhuman animal reservoirs exist for many pathogens including *Salmonella, Giardia,* and the rabies virus. Rodents, for example, are a reservoir for *Yersinia pestis,* the causative agent of plague, which killed one fourth of the population of Europe in the 14th century. Occasional transmission to humans is still reported in the Southwestern states where *Yersinia pestis* is endemic in prairie dogs and other rodents. While controlling the rodent reservoir is an obvious and acceptable solution for the prevention of plague, other diseases have more exotic animal reservoirs such as the wild game animals of Africa that are even less desirable to eliminate.

Diseases such as plague and rabies that can be transmitted to humans but primarily exist in other animals are called

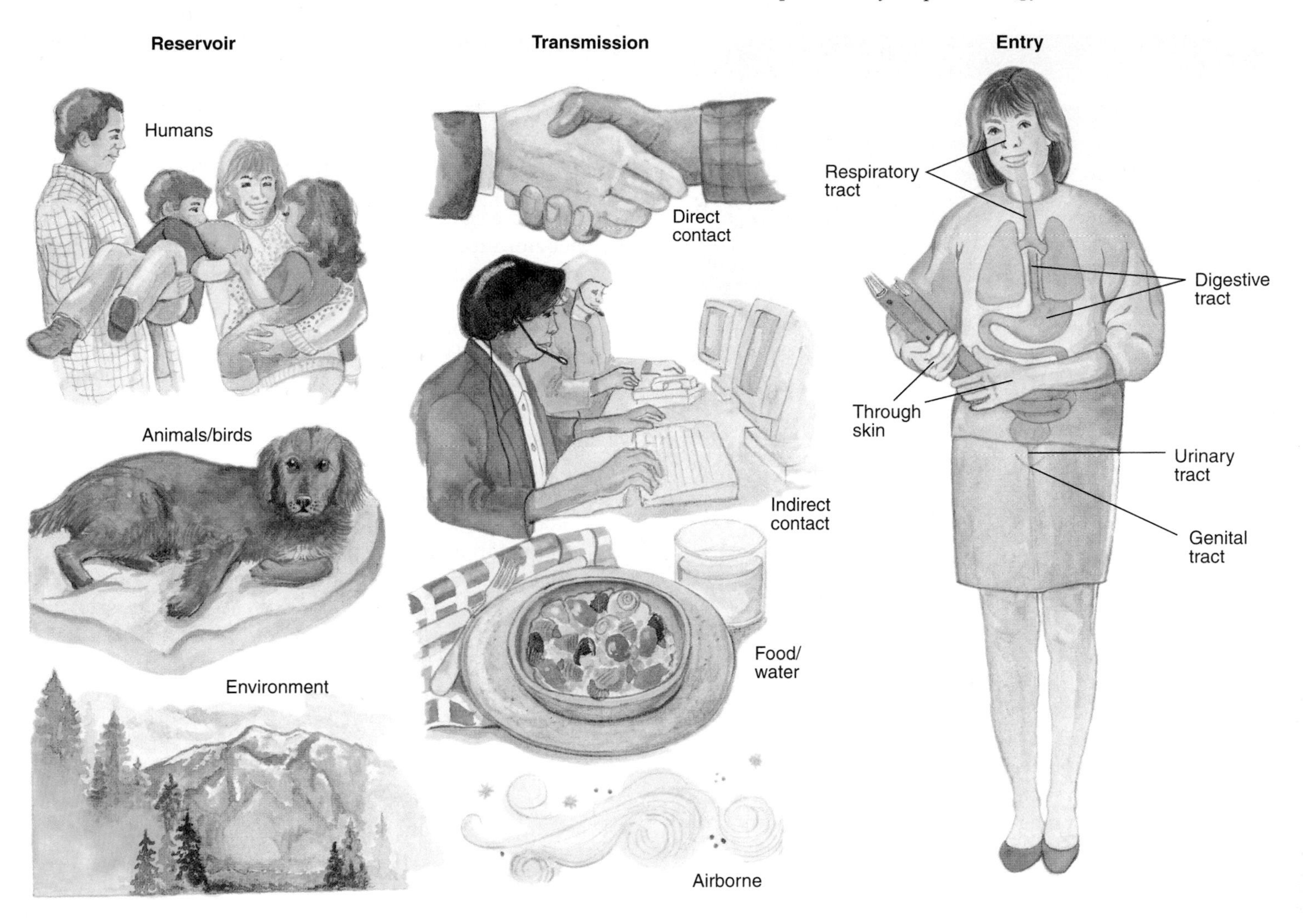

Figure 20.2
Some habitats (or reservoirs), transmission modes, and entry routes for diseases.

zoonotic diseases. Zoonotic diseases are often more severe in humans than in the normal animal host because the infection in humans is an accident; there has been no evolution toward the balance that normally exists between a host and parasite.

Some pathogens have environmental reservoirs. For example, both *Clostridium botulinum,* which causes botulism, and *Clostridium tetani,* which causes tetanus, are widespread in soils. Unfortunately, pathogens that have environmental reservoirs, as well as those in which animals are a reservoir, are probably impossible to eliminate.

Diseases Can Be Transmitted from Person to Person Through Direct Contact, Droplet Transmission, or Indirect Contact via Inanimate Objects

A successful pathogen must somehow be transmitted from its reservoir to the next susceptible host. Transmission of a pathogen from a human reservoir generally requires direct person-to-person contact, inhalation of droplets of respiratory secretions or saliva, or indirect contact by way of a nonliving object.

Direct Transmission

Pathogens that cannot survive for extended periods in the environment must, because of their fragile nature, be transmitted through **direct contact.** For example, *Treponema pallidum,* which causes syphilis, and *Neisseria gonorrhoeae,* which causes gonorrhea, both require intimate sexual contact for their transmission; otherwise, they readily die when exposed to the environment. Other organisms are not as fragile, but the numbers required to initiate infection, called the **infectious dose,** are low enough to make direct transmission possible. For example, it is estimated that the infectious dose of the intestinal pathogen *Shigella* is only 100 organisms, which makes it easily transmitted via a handshake. Hand washing, a fairly simple routine that physically removes organisms, is important in preventing this type of spread of infection. Even washing in plain water reduces the numbers of potential pathogens on the hands which, in turn, decreases the possibility of transferring a dose of an organism sufficient to establish an infection.

Spread of respiratory diseases over distances of less than 1 meter (approximately 3 feet) through inhalation of large respiratory droplets is referred to as **droplet transmission.** People continually discharge microorganisms into the air

Figure 20.3
Photograph of droplet transmission from a sneeze.

when sneezing, coughing, talking, and singing (figure 20.3). Many organisms are enclosed within large droplets of saliva or mucus, most of which fall quickly to the ground no farther than 1 meter from release. Although it does not require physical contact, droplet transmission is considered direct transmission because it requires close proximity. Droplet transmission is particularly important as a source of contamination in crowded locations such as schools and military barracks. Thus, desks or beds in such locations ideally should be spaced more than 4 feet and preferably 8 to 10 feet apart to minimize the transfer of infectious agents in this manner. Another way to minimize the spread of these diseases is to educate people about the importance of covering their mouths when they cough or sneeze.

These examples of disease that are transmitted person to person are considered **horizontal spread,** in contrast to diseases passed directly from a pregnant woman to the fetus or a mother to a newborn, which are **vertical spread.** Some agents, such as those that cause diseases such as syphilis and rubella, can cross the placenta and damage a developing fetus; other disease agents, such as the herpes virus and Group B streptococcus, can infect the newborn as it passes through the birth canal.

Indirect Transmission

Indirect transmission involves transfer of pathogens via inanimate objects, or **fomites,** such as clothing, tabletops, doorknobs, or drinking glasses. For example, a carrier of *Staphylococcus aureus* may contaminate his or her hands when touching a skin infection or his or her colonized nose. Organisms on the hands can then easily be transferred to a fomite that serves as a source of infection when handled by another person. Again, hand washing is an important control measure. The hand scrub performed by nurses and doctors before participating in an operation or when working in a newborn nursery or intensive care or isolation unit takes 10 minutes and uses a strong disinfectant.

Pathogens Can Be Transmitted in Food or Water

Pathogens, particularly those that infect the gastrointestinal tract, can be transmitted through contaminated food or water. Attack rates of foodborne infections tend to be higher than those of waterborne infections because dilution is less. Moreover, food is often a nutrient for infectious agents, and a sizable increase in numbers of the agents may occur before food is eaten. Waterborne pathogens, generally originating from sewage, can be concentrated by filter feeding shellfish such as clams and oysters. The source of the contamination varies with the food product. Foods can be contaminated at their source, as with *Salmonella* in hens' eggs and *Brucella* in cows' milk. Pathogens can also be added during food preparations. For example, bacteria such as *Salmonella typhi* and *Staphylococcus aureus* can be transmitted if food handlers who harbor the organisms inadvertently inoculate food with their contaminated hands.

Food and waterborne disease outbreaks have the potential of involving large numbers of people because modern technology has made widespread distribution a reality. For example, the 1993 waterborne outbreak of *Cryptosporidium,* an intestinal parasite, in Milwaukee, Wisconsin, was estimated to have involved approximately 400,000 people. An *Escherichia coli* O157:H7 outbreak traced to contaminated hamburger patties served at a fast-food restaurant chain was spread to people in four states. The contaminated meat was supplied to all the restaurants from a single distributor. Prevention of these large-scale outbreaks can be accomplished through treatment of drinking water and sewage as well as the sanitary production, adequate cooking, and proper storage of foods.

Pathogens Can Be Transmitted Through Air

Disease transmission can also be airborne. As mentioned earlier, people continually discharge microorganisms in liquid droplets created during talking and sneezing. Large droplets quickly fall to the ground, whereas smaller droplets dry, leaving one or two organisms attached to a thin coat of the dried material, creating **droplet nuclei.** In the presence of even slight air currents, the droplet nuclei can remain suspended indefinitely. In addition, dust particles, including household dust and particles from soil disturbed by the wind, may carry respiratory pathogens. The inhaled dust particles and droplet nuclei may enter airways. Only those smaller that 10 μm can enter the lungs because their small size allows them to avoid entrapment in the mucus lining of the nose and throat. Fortunately, under usual conditions of good ventilation, the air movement rapidly dilutes the droplet nuclei so that only highly infectious agents, such as the viruses that cause chicken pox and measles, pose a risk.

The design of hospitals and other large buildings must take into account the fact that pathogens can be airborne. Skin pathogens also enter the air on skin cells. Understandably, airborne transmission of pathogens is the most difficult transmission to control.

Perspective 20.1

Escherichia coli O157:H7 in Unpasteurized Apple Juice

In October 1996, physicians at the Children's Hospital and Medical Center in Seattle, Washington, noticed a slight increase in the number of cases of children with severe bloody diarrhea caused by *Escherichia coli* O157:H7. Because isolated cases of *E. coli* O157:H7 regularly occur at low levels, physicians at the hospital were not certain whether the increase was due to a greater number of unrelated infections or to the start of a single source epidemic.

Using recombinant DNA technology, microbiologists at the University of Washington quickly analyzed DNA of the *E. coli* O157:H7 isolates from each patient and showed that most of the isolates had the same "genetic fingerprint," indicating that the outbreak originated from the same contaminated source. Meanwhile, epidemiologists from the Seattle–King County Health Department were rushing to identify the source of the outbreak so that the public could be alerted and steps could be taken to prevent more people from becoming ill. Using a checklist of hundreds of different possible foods, epidemiologists interviewed the parents of stricken children to determine what foods had been consumed by the children in the days preceding their illness. When it became apparent that most of the ill children had consumed apple juice, the parents were again questioned to determine the brand. Many parents recalled purchasing the same premium brand of apple juice manufactured by a company that specializes in nonpasteurized juices and is known for its freshness and healthful image.

When the company that produced the apple juice learned that its product was linked to the outbreak, it agreed to immediately recall the juice and related products from all stores in the distribution area, which spanned seven states and British Columbia. The preparedness and prompt response of the physicians, microbiologists, and public health officials that led to the immediate recall of the contaminated juices provide an excellent example of how efficient infectious disease surveillance can halt epidemics and ultimately save lives.

Figure 20.4
The air sampler directs measured amounts of air against the surface of the microbiological medium. After incubation of the medium, the numbers and kinds of microorganisms per cubic foot of air can be determined. (*a*) Air sampling machine. (*b*) Air sample culture from a clean, empty hospital room showing no growth. (*c*) Air sample culture from a small room containing 12 people. In both (*b*) and (*c*), 5 cubic feet of air were sampled.

The number of organisms in air can be estimated by using a machine that pumps a measured volume against the surface of microbiological medium in petri dishes. (figure 20.4). This technique has shown that the number of bacterial colonies that develop per cubic foot of air sampled rises in proportion to the number of people in a room. To prevent the buildup of airborne pathogens, architects design modern public buildings with ventilation systems that constantly change the air. In many hospitals, the airflow can be regulated so that it is supplied to the operating room under slight pressure, thereby preventing contaminated air in the corridors from flowing into the room. Hospital microbiology laboratories can be kept under a slight vacuum so that air flows in from the corridors but microorganisms and viruses cannot escape to other parts of the building. Special contamination problems may arise if the pumping action of elevators changes the pressure relationships or if the chimneylike effect of laundry chutes carries pathogens in the warm air from the dirty laundry areas in the basement to other areas of the building. Air conditioning systems may be sources of infectious agents, since their air intake may draw in contaminated air or dust. In addition, some pathogens can grow in the water of an air conditioning system.

The survival of organisms in air varies greatly with the type of organism and with air conditions such as relative humidity, temperature, and exposure to light. In general, gram-positive organisms survive longer in air than do gram-negative organisms, and survival in dim light is greater than that in bright light. Organisms in still air and those attached to large dust particles eventually fall to the floors of buildings but readily resuspend in the air if the floor is swept. For this reason, a dampened mop or a floor washer is used to clean floors in hospitals. Although the use of ultraviolet light or air filtration can markedly reduce the number of viable organisms in air, these methods are generally expensive and unnecessary except in some special types of hospital units where the risk of microbial contamination is great. Instead, satisfactory control of airborne infection is usually achieved by good ventilation and dust control.

Perspective 20.2

Tuberculosis Epidemic in a Hospital

Adequate dilution of airborne pathogens may not occur under conditions of inadequate ventilation as is demonstrated by the following example of a hospital epidemic of tuberculosis. A patient with an unsuspected case of tuberculosis was treated for pneumonia in a large hospital room (ward) for two and a half days. During this time, 13 of the 44 people present in the ward became infected with the tubercle bacillus. As is often the case, the air conditioning for the ward merely recirculated the room's air without filtering out airborne pathogens. Fortunately, none of the 13 individuals developed active tuberculosis since, soon after tuberculosis was diagnosed in the **index case** (the person who first brought the disease to the ward), tuberculin skin testing was performed and the individuals were given antituberculosis treatment.

Arthropod Vectors May Transmit Disease

Some diseases, such as malaria, plague, and Lyme disease, are transmitted through an arthropod **vector** such as a flea, tick, or mosquito. The vector either injects the infectious agent while taking a blood meal or deposits it on a person's skin where it can then be inadvertently inoculated when the individual scratches the itchy bite. In the case of malaria, caused by the eukaryotic pathogen *Plasmodium*, the mosquito is not only the injector of the organism, but also serves as an essential part of the reproductive life cycle of the parasite.

Prevention of vectorborne disease relies on mosquito, tick, and insect control. The success of such control was demonstrated in the late 1940s when malaria, which was once endemic in the continental United States, was successfully eradicated from the nation. This was accomplished through a combination of mosquito elimination and prompt treatment of infected patients. Unfortunately, worldwide eradication efforts, which initially showed great promise, later failed, in part due to the decreased vigilance that accompanied the dramatic but short-lived decline of the disease.

The Portal of Entry May Determine the Outcome of Disease

In order to cause disease, a pathogen must not only be transmitted from its reservoir to a new host, but also somehow enter the new host. For example, *Shigella* that is transmitted via a handshake will only cause disease when the new host inadvertently ingests the pathogen, giving it entry so that it can establish itself in the intestinal tract. Respiratory pathogens that are released into the air during a cough will only cause disease when they are inhaled by the next host. Many organisms that cause disease if they enter one body site are harmless if they enter another. For example, *Enterococcus* may cause a bladder infection if it enters the normally sterile urinary tract, but it is harmless in the intestine where it frequently resides as a member of the normal flora. An interesting example of the importance of the route of entry in disease development is illustrated in the case of plague transmission. When fleas that normally reside on the rodent reservoir become infected and an infected flea bites a person, *Yersinia pestis* is injected and bubonic plague develops. The bacteria multiply rapidly inside lymph nodes, resulting in a disease that is not contagious but has a mortality rate of 50% to 75% unless treated promptly. In a small percentage of people with bubonic plague, however, the organism spreads to the lungs, resulting in a different disease called **pneumonic plague.** Pneumonic plague presents a much more severe situation because it is transmitted person to person through respiratory droplets. The mortality rate of untreated pneumonic plague is nearly 100%. Plague epidemics are due to the pneumonic form of the disease because of the ease of transmission.

Other Factors Influence the Spread and Outcome of Disease

The Incubation Period of Disease Influences the Extent of the Spread of the Disease

The extent of the spread of an infectious agent is influenced by the length of time between exposure to the agent and the onset of disease symptoms, which is called the **incubation period.** Diseases with long incubation periods such as AIDS can spread extensively before the first cases appear. An excellent example of the importance of this factor was the spread of typhoid fever from a ski resort in Switzerland in 1963. As many as 10,000 people had been exposed to drinking water containing small numbers of *Salmonella typhi*, the causative agent of the disease. The long incubation period of the disease, 10 to 14 days, allowed widespread dissemination of the organisms by the skiers, since they flew home to various parts of the world before they became ill. As a result, there were more than 430 cases of typhoid fever in at least six countries.

The Intensity of Exposure to an Infectious Agent Influences Frequency, Incubation Period, and Severity of Disease

The **dose** (amount) of the infecting agent received by susceptible individuals is also important in epidemiology. The probability of infection and disease is generally low if an individual is exposed only to small numbers of a potential pathogen. The direct relationship between the dose of a potential pathogen and the likelihood of developing disease is explained by the fact that there must be a certain minimum number of pathogens in the body to produce enough damage to cause disease symptoms. For example, if only 30 *Salmonella typhi* are ingested in contaminated drinking water yet 1,000,000 are required to produce typhoid fever symptoms, then it will take some time for the bacterial population to increase to that number. Because host defenses are being mobilized at the same time and are racing to eliminate the bacteria before they can multiply sufficiently to cause disease, small doses often result in higher percentages of asymptomatic infections—the immune system sometimes has time to eliminate the organism before symptoms appear. On the other hand, there are few if any infections for which immunity is absolute. An unusually large exposure to a pathogen such as can occur in a laboratory accident may produce serious disease in a person who has immunity to ordinary doses of the pathogen. Therefore, even immunized persons should take precautions to minimize exposure to infectious agents. This principle is especially important for medical workers who attend patients who have infectious diseases.

The magnitude of the dose also affects the length of the incubation period. It takes longer for disease symptoms to appear when the dose of the infecting agent is small because it takes time for the organism to grow to high enough numbers to cause disease. As would be expected, a large dose results in a shorter incubation period.

Immunization, or Prior Exposure, Influences the Susceptibility to Infectious Agents

The number of people who become ill from a disease is influenced by previous exposure or immunization of the population to that disease agent or antigenically related agents. A disease is unlikely to spread very widely in a population in which 90% of the people have previously been immunized or infected by the same strain and thus have acquired immunity to the disease agent. If humans are the only reservoir, then a continuous source of susceptible people is required or the disease will disappear from the population. As mentioned earlier, this requirement for a susceptible host allowed smallpox to be eradicated. When an infectious disease cannot spread in a population because it lacks a critical concentration of nonimmune hosts, a phenomenon called **herd immunity** results. The nonimmune individuals are essentially protected by the lack of a reservoir of infection. Unfortunately, many infectious agents have the ability to dramatically alter their antigens so that they are able to continue to propagate even in previously exposed populations. For example, antibodies produced in response to a *Neisseria gonorrhoeae* infection are usually directed against the pili, which are the most prominent surface antigens. However, the pilin gene of *N. gonorrhoeae* can undergo a chromosomal rearrangement that gives rise to a new type of pili not recognized by the original antibodies.

Nutrition, Crowding, and Other Environmental and Cultural Factors Can Shape the Character of Epidemics

Malnutrition, overcrowding, and fatigue increase the susceptibility of people to infectious diseases and enhance the diseases' spread. Thus, infectious diseases have generally been more of a problem in poor areas of the world where individuals are crowded together without proper food or sanitation. Crowding, stress, and fatigue also appear to have contributed to epidemics of meningococcal meningitis in military recruits in the United States. Adequate rest, nutrition, freedom from stress, and other factors that promote good general health result in increased resistance to infections such as tuberculosis. When infection does occur in a healthy individual, it is more likely to be asymptomatic or to result in mild disease.

The distribution of diseases is also influenced by religious and cultural practices. For example, infants who are breast fed are less likely to have diarrhea caused by pathogenic strains of *Escherichia coli,* presumably because of the protective effects of IgA in the mother's milk. The effects of cultural practices on the distribution of disease is also illustrated by the observation that Scandinavian, Japanese, and Jewish people who eat traditional dishes made from raw freshwater fish are more likely to acquire tapeworm, which would normally be killed by cooking the fish. In the waterborne epidemic of typhoid fever in Switzerland discussed earlier, illness occurred primarily in tourists because the local people rarely drank water, preferring wine.

The Genetic Background of an Individual Can Influence Susceptibility to Infection

It is known that natural immunity can vary with genetic background, but it is usually difficult to determine the relative importance of genetic, cultural, and environmental factors. In a few instances, however, the genetic basis for resistance to infectious disease is known. For example, many people of black African ancestry are immune to malaria caused by *Plasmodium vivax* because they lack a specific red blood cell receptor for the organism. Studies of the incidence of tuberculosis in identical twins suggest that genetic factors play a role, and there is also evidence of a genetic influence in the development of paralytic poliomyelitis.

REVIEW QUESTIONS

1. Describe the difference between the terms *epidemic, endemic,* and *pandemic.*
2. What are diseases that develop and subside rapidly called?
3. What is the epidemiological significance of people having asymptomatic infection?
4. Why are zoonotic diseases often severe when transmitted to humans?
5. Name an important control measure for preventing person-to-person transmission of a disease.
6. Define the term *herd immunity.*

Epidemiological Studies

Epidemiologists investigate disease outbreaks to determine the causative agents, as well as their reservoirs, routes of transmission, and portals of entry in order to recommend ways to minimize the spread. They are concerned with not only the spread of newly emerging diseases such as that caused by *Escherichia coli* O157:H7, but also the spread of diseases that have been recognized for centuries, such as tuberculosis and cholera. The studies epidemiologists conduct are in many ways similar to criminal investigations. After a disease outbreak, investigators conduct a **descriptive study** to describe the characteristics of the ill persons involved and the place and the time of the outbreak. This information gives clues as to the possible cause, reservoir, and transmission of the illness. Once the occurrence of the outbreak has been fully described, an **analytical study** is done to identify specific factors, the risk factors, that result in high frequencies of disease.

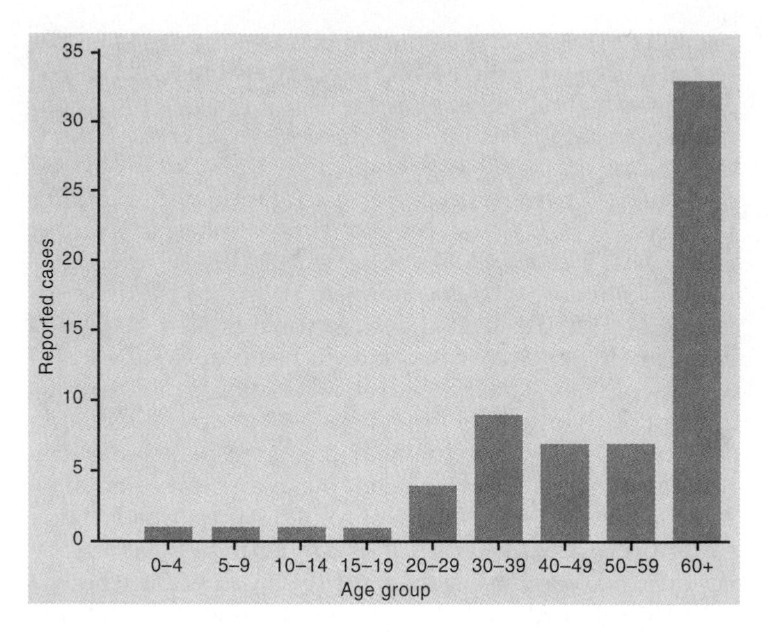

Figure 20.5
Tetanus by age group, United States, 1990.

Source: *Morbidity and Mortality Weekly Report.* 39(53)43: 1990.

Descriptive Studies Examine the Persons Affected by, the Place of, and the Time of a Disease Outbreak

Determining the personal characteristics of those who become ill in an outbreak is a critical step in defining the population at risk. Variables such as age, sex, race/ethnicity, occupation, personal habits, previous illnesses, and socioeconomic class may all yield clues about risk factors for developing the disease. For example, the observation that the neurological disease kuru was transmitted among New Guinea populations that practiced cannibalism was important in finally establishing that the disease was probably transmitted during rituals in which the brains of dead relatives were handled and ingested (perspective 13.1). The fact that people over the age of 60 are more likely to develop tetanus may indicate that older adults are not receiving adequate immunizations (figure 20.5), which are recommended once every 10 years.

The geographic location of a disease helps identify the probable site of contact between the infected person and the infectious agent. The location may also give clues about potential disease reservoirs, vectors, or geographical boundaries that may affect disease transmission. An example of the importance of determining the place of the outbreak as well as the people involved was illustrated in 1976 when over 200 mostly elderly people around the United States were hospitalized with pneumonia and 34 of them died. Epidemiologists determined that those affected had attended the same American Legion convention in Philadelphia several days earlier, and most had stayed at the same hotel. Studies further showed that most who contracted the disease were smokers and had spent time in the lobby of the hotel. Curiously, hotel employees and other nonconvention attendees who also spent time in the lobby did not get the disease. Eventually, it was determined that a previously undescribed bacterium, later named *Legionella pneumophila*, was contaminating the water of the air conditioning unit of the hotel. Presumably, the air conditioner released airborne organisms into the lobby. Further epidemiological studies have shown that *Legionella* rarely causes serious disease in normal, healthy people but is instead an opportunist that causes disease in people who fit the profile of many of those attending the convention: older smokers with impaired host defenses. As this outbreak at the Legionnaires' convention illustrates, determining the correct geographic location of disease acquisition as well as factors that predispose people to contracting the disease can be essential in unraveling the epidemiology of an illness.

The timing of the outbreak may also yield helpful clues. If there is a rapid rise and fall in the numbers of people who became ill, then it is likely they were all exposed to a single common source of the infectious agent, such as contaminated chicken at a picnic. This type of outbreak is called a **common source epidemic.** If, on the other hand, the numbers of ill people rise gradually and finally fall, then the pattern indicates that the disease is contagious, with one person transmitting it to several others, who each then transmit it to several more, and so on. This type of disease is called a **propagated epidemic** (figure 20.6).

The season of occurrence of the epidemic may also be significant. Respiratory diseases such as influenza, respiratory syncytial virus infections, and the common cold are more easily transmitted in crowded indoor conditions during the winter (figure 20.7). Conversely, vectorborne and foodborne diseases are more likely to be transmitted in the summer when people are outside, exposed to mosquitoes and ticks and eating picnic food that has incubated all afternoon in the warm sun (figure 20.8).

The data obtained through descriptive studies allow the formulation of a hypothesis regarding the potential risk factors involved in the spread of disease.

Perspective 20.3

A Legionnaires' Disease Outbreak Originating in a Grocery Store

In October 1989, two physicians reported an outbreak of serious lung infections to the Louisiana Department of Health and Hospitals. An investigation showed the outbreak to be Legionnaires' disease and to involve 33 people. Investigators selected a control group—people from the same general area who did not become ill and who were matched with the patients according to age, physician, and the presence of conditions (such as chronic bronchitis) that might predispose the person to Legionnaires' disease. When the infected and control groups were compared, it became apparent that all the people with Legionnaires' disease had visited a certain grocery store within two weeks of the start of their illness, whereas only about half of the control group had done so. Moreover, the people with Legionnaires' disease were more likely to have spent a long time in the store and to have selected foods located close to a mist machine used to keep the produce fresh. The mist machine was shown to produce water droplets less than 5 μm in diameter, tiny enough to be inhaled, and to contain *Legionella pneumophila*, the cause of Legionnaires' disease. Further study showed that this strain of *L. pneumophila* was the same serotype as the one that infected the people. No further cases of the disease occurred after use of the mist machine was discontinued.

The mist machine used in this grocery store contained a reservoir of relatively stagnant water and therefore was well-suited to colonization of *Legionella* species. (Most mist machines used in grocery stores around the country have a different design and are unlikely to be contaminated with the organisms.) Since this episode occurred, the Food and Drug Administration and Consumer Product Safety Commission have publicized measures to help ensure that all grocery store mist machines remain free of *Legionella*.

Source: *Morbidity and Mortality Weekly Report* 39:108, February 1990, Centers for Disease Control and Prevention, Atlanta, Ga.

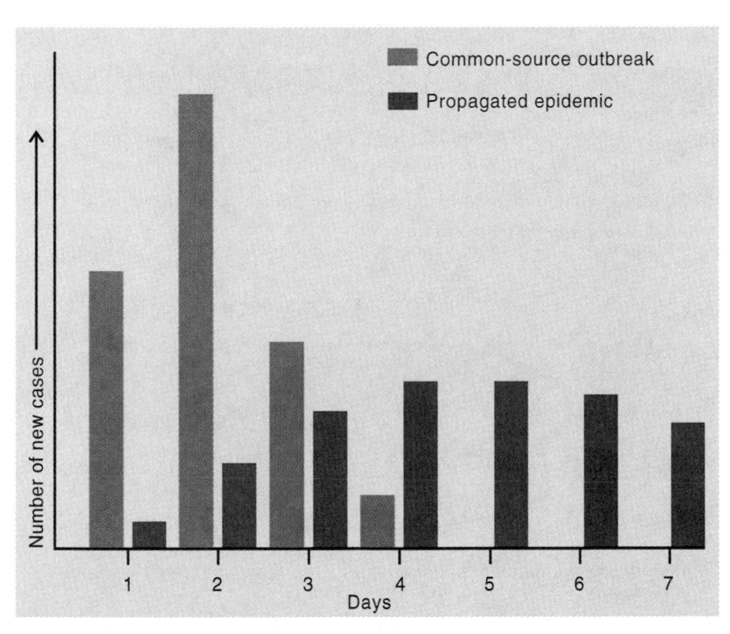

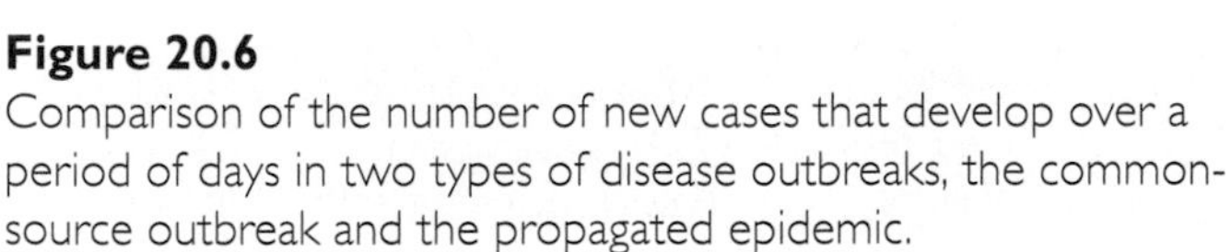

Figure 20.6
Comparison of the number of new cases that develop over a period of days in two types of disease outbreaks, the common-source outbreak and the propagated epidemic.

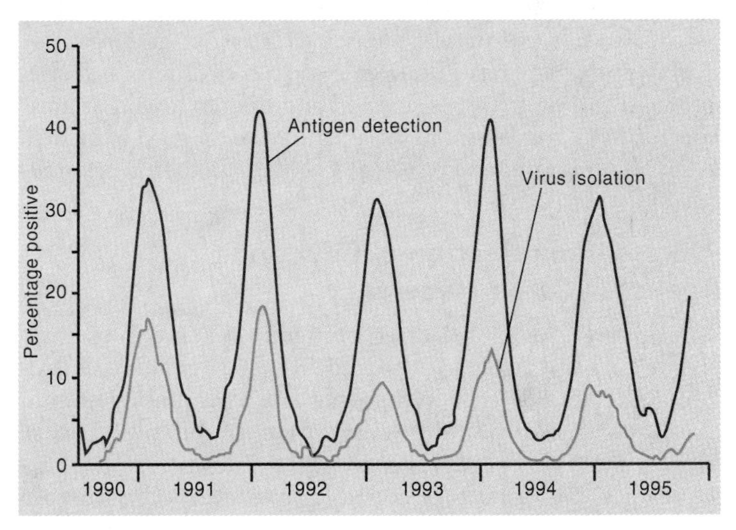

Figure 20.7
Percentage of specimens positive for respiratory syncytial virus by method of confirmation and month in the United States, 1990–1995.

Source: *Morbidity and Mortality Weekly Report.* 44(48)901. 1995.

Figure 20.8
Reported cases of arboviral infections, United States, 1985–1994.

Source: *Morbidity and Mortality Weekly Report.* Summary of Notifiable Diseases, 43(53)18: 1994.

Analytical Studies Compare Risk Factors for Developing Disease in Cases and Controls

Analytical studies are designed to determine which of the potential risk factors identified by the descriptive studies are actually relevant. A **retrospective study** is done following a disease outbreak. This type of study looks back to compare the actions of and events surrounding those who developed the disease with those of people who did not develop the disease (controls). The activity or event that was common among the cases but not the controls, is likely to have been a factor in the development of the disease. It is important to select controls that match the cases with respect to variables such as age, sex, and socioeconomic status to ensure that all had equal probability of coming in contact with the disease agent. Table 20.1 shows the retrospective data that linked the development of bloody diarrhea in a Washington State outbreak of *E. coli* O157:H7 in January 1993, to the consumption of hamburgers at a specific fast-food restaurant chain. The controls in this case were the neighborhood friends of the ill patients.

A **prospective study** looks ahead to see if the identified risk factors can predict a tendency to develop the disease. A study group that has a known exposure to the risk factor, termed a **cohort group,** is selected and then followed in time. The incidences of disease of those who were exposed to the risk factor and of those who were not exposed are then compared.

Precise Identification of Infectious Agents Is Often Necessary

Determining the source of an epidemic is sometimes complicated by the facts that different disease agents may produce the same disease symptoms and that a single agent may sometimes produce a wide variety of different disease symptoms. For example, more than 100 different virus strains can cause the common cold, whereas poliovirus can produce an illness whose symptoms range from those of a cold to fatal paralysis. Thus, to determine if potentially related illnesses are indeed connected, it is often necessary to identify the causative agents of diseases precisely.

Table 20.1 Frequency of Exposures Among Cases of Bloody Diarrhea and Controls, Washington, 1993

Exposure	Percent of cases	Percent of controls
Chain A (hamburgers)	75	0
Chain B	20	20
Chain C	21	29
Pork	11	79
Chicken	79	79
Hot dogs	67	42

Source: Bell, et al. 1994. *JAMA* 272(17):1349–1353.

The sources of epidemics that are due to common normal flora or environmental organisms are particularly difficult to trace because of the ubiquity of the organisms. In these cases, it is sometimes necessary to identify strains of the organisms in order to identify the actual origins of the outbreaks. Individual bacterial strains can be identified using a procedure called **bacteriophage typing** in which various strains are tested for their sensitivity to a series of different types of bacteriophage (figure 20.9). A strain can also be characterized and differentiated by its **antibiogram,** or antibiotic susceptibility pattern (figure 20.10). Newer techniques to distinguish individual strains utilize genetic engineering technology to detect differences in DNA sequences (figures 20.11 and 20.12).

(a) An inoculum of *S. aureus* is spread over the surface of agar medium.

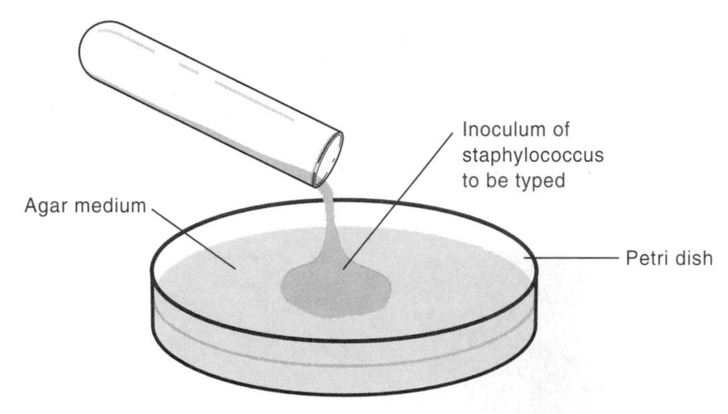

(b) 31 different bacteriophage suspensions are deposited in a fixed pattern.

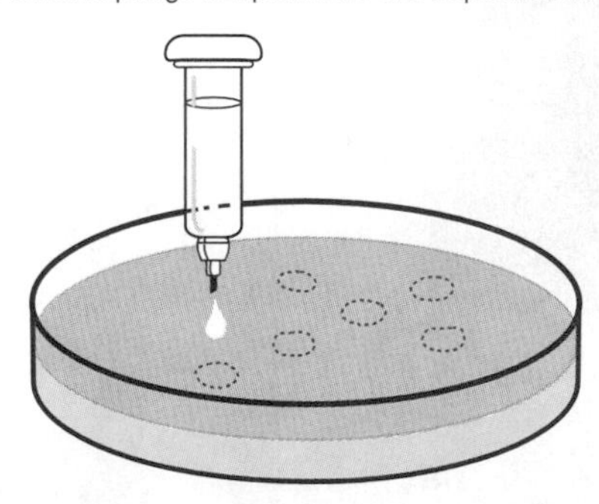

(c) After incubation, different patterns of lysis are seen with different strains of *S. aureus*.

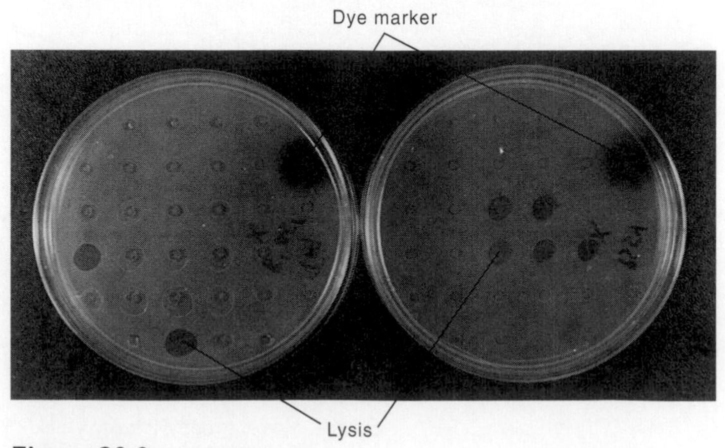

Figure 20.9
Bacteriophage typing of *S. aureus*.

Perspective 20.4

Ebola Hemorrhagic Fever in Africa

In 1995, an outbreak of Ebola hemorrhagic fever in Zaire captured international headlines and claimed over 230 lives. Not since the disease was first recognized in 1976, when two simultaneous outbreaks struck 550 people and claimed 340 lives in Sudan and neighboring Zaire, had any large outbreaks been reported (a small outbreak of 34 occurred in Sudan in 1979).

The Ebola virus is one of the most virulent viruses known, with an incubation period of 2 to 21 days and a mortality rate of 50% to 90%, depending on the strain. The disease, named for a river near its discovery, is characterized by the abrupt onset of fever, weakness, muscle pains, sore throat, and headache. Initial symptoms soon progress to include vomiting, diarrhea, rash, both internal and external bleeding, and often death due to organ failure and shock. There is no vaccine or cure for Ebola hemorrhagic fever; there is only supportive therapy.

Transmission of the virus is through direct person-to-person contact with contaminated blood, secretions, or bodily fluids. Because of this type of transmission, hospital personnel and family members caring for infected patients are at particular risk for contracting the virus. For example, the 1995 epidemic in Zaire appears to have started when members of a surgical team became infected while operating on a patient whose symptoms were later recognized as hemorrhagic fever. Nosocomial transmission is even more likely in developing areas of the world where adequate medical supplies such as protective gowns, masks, and gloves are not readily available and needles and syringes are reused.

The rapid mortality and rare occurrence of the Ebola hemorrhagic fever suggest that the disease is zoonotic, and humans are merely an accidental host. Nonhuman primates are also susceptible to Ebola, but because of the severity of their disease symptoms, they too are thought to be accidental hosts. In fact, two decades after the discovery of the virus, the thousands of animal and plant samples collected from areas near the outbreaks have yielded no clues as to the original source. The natural reservoir of the deadly Ebola virus remains a mystery.

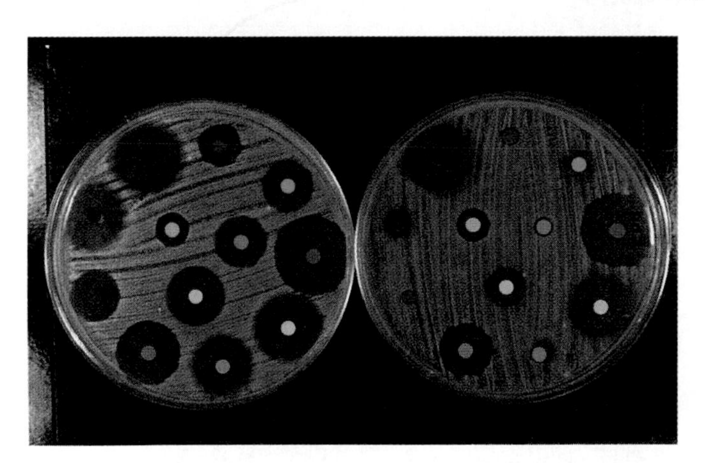

Figure 20.10
Distinguishing strains of *Staphylococcus aureus* by their antibiotic susceptibility patterns. In this photograph, 12 different antimicrobial drugs incorporated in paper disks have been placed on the two staphylococcal cultures. Clear areas represent zones of inhibited growth.

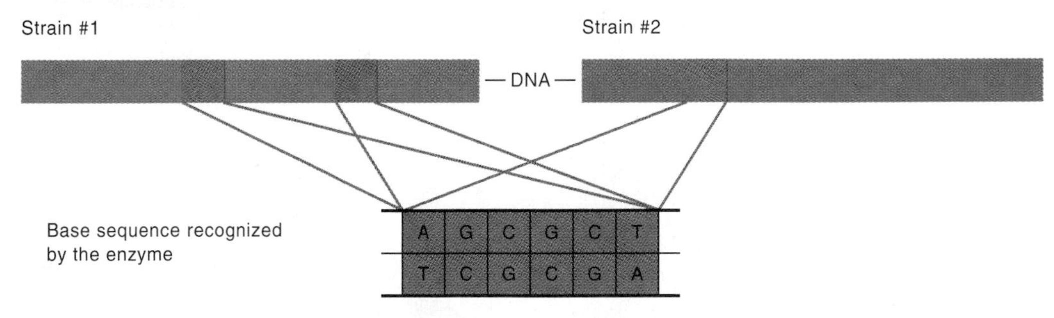

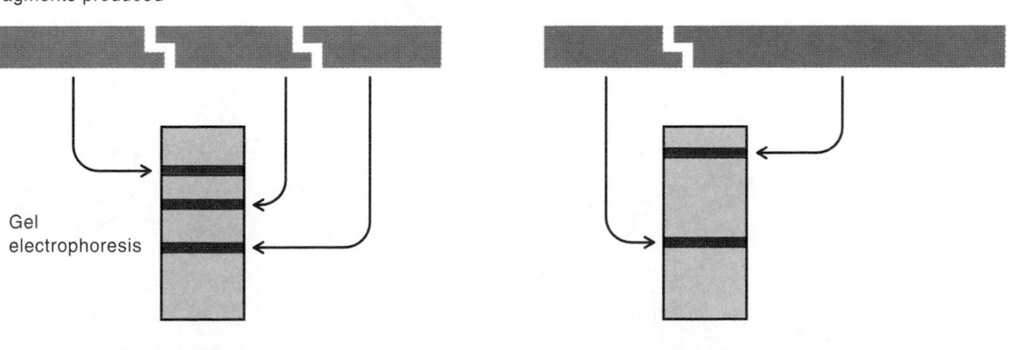

Figure 20.11
Use of restriction nucleases to identify different strains of herpes simplex virus. The base sequence recognized by the nuclease occurs at different locations in the DNA strands of the two viral strains. Therefore, the fragments produced by the enzymatic digestion of the two DNAs are of different lengths and move different distances in a gel when an electric current is applied.

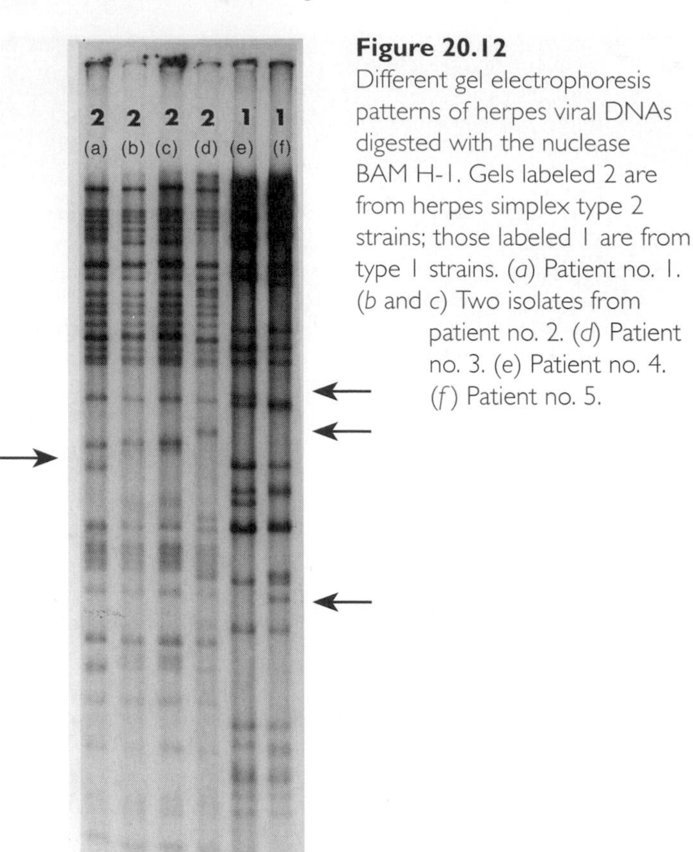

Figure 20.12
Different gel electrophoresis patterns of herpes viral DNAs digested with the nuclease BAM H-1. Gels labeled 2 are from herpes simplex type 2 strains; those labeled 1 are from type 1 strains. (*a*) Patient no. 1. (*b* and *c*) Two isolates from patient no. 2. (*d*) Patient no. 3. (*e*) Patient no. 4. (*f*) Patient no. 5.

The value of precisely identifying microorganisms is shown by the following case study. An unusually large number of *Streptococcus pyogenes* infections appeared over the course of several months, but because they were reported from different hospitals and involved patients attended by many different medical and other personnel, it was not clear that an epidemic existed. When the surface antigens of the streptococci were identified by using antisera, it was found that three-quarters of the patients had been infected with the same strain of *S. pyogenes*, whereas the remainder had been infected by miscellaneous strains of the organism. Furthermore, all patients infected with the first type of streptococcus had been attended by the same physician, who was then found to be a carrier of that strain of *S. pyogenes* (figure 20.13). Only by identifying the precise strains of the same species was it possible to trace the origin of the strain back to a single carrier.

Figure 20.13
Use of antisera and antibiotic susceptibility tests to distinguish different strains of *S. pyogenes*. No common factor of exposure could explain all the streptococcal cases in the five hospitals, but cases due to the T9, tetracycline-resistant strain could all be traced to an asymptomatic carrier.

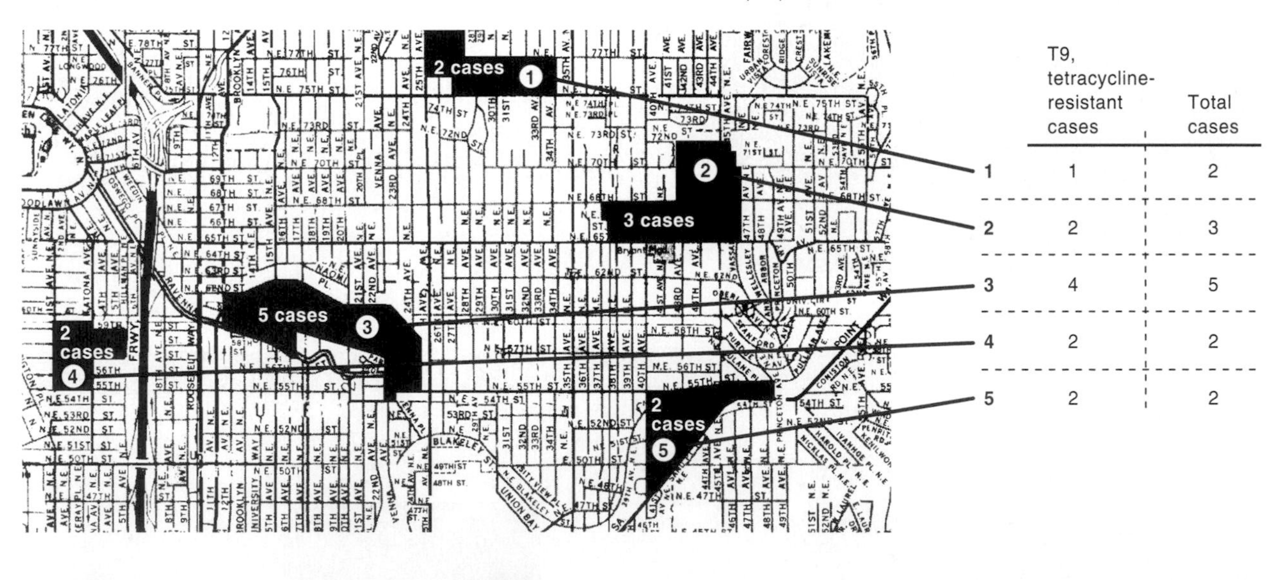

	T9, tetracycline-resistant cases	Total cases
1	1	2
2	2	3
3	4	5
4	2	2
5	2	2

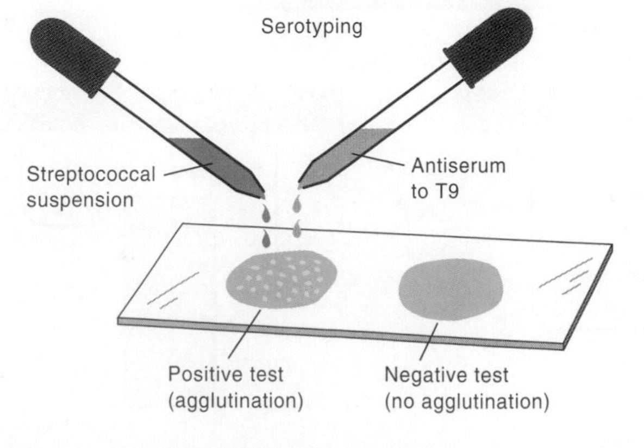

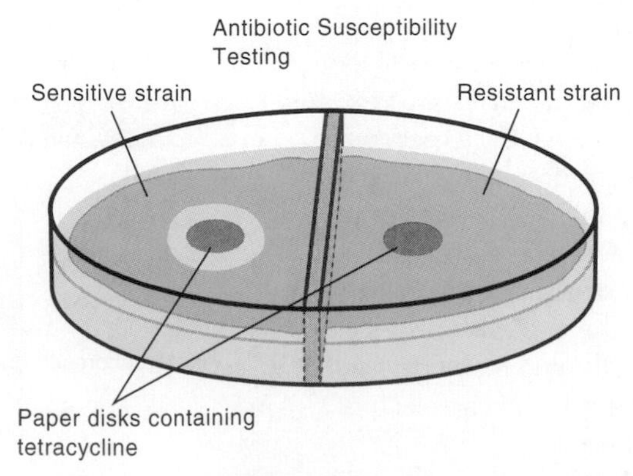

Infectious Disease Surveillance

Infectious disease control nationwide depends heavily on a network of people and agencies across the country that monitor disease development. It is partly because of the success of this network that infectious diseases do not claim more lives in the United States.

The Public Health Network Includes Government Agencies, Hospital Laboratories, and News Media

The National Centers for Disease Control and Prevention (CDC) is part of the United States Department of Health and Human Services and is located in Atlanta, Georgia. It provides support for infectious disease laboratories in the United States and abroad and collects data on diseases of public health importance. Each week, the CDC publishes a pamphlet, *Morbidity and Mortality Weekly Report* (MMWR), that summarizes the status of a number of diseases (figure 20.14). The number of new cases of 52 diseases that are considered nationally notifiable is voluntarily reported to the CDC by individual states. The numbers collected by the CDC are published in the MMWR along with historical numbers to reflect any trends. Included in the 52 diseases are tuberculosis, hepatitis, sexually transmitted diseases, rabies, chicken pox, measles, mumps, and malaria. Potentially significant case reports, such as the 1981 report of a cluster of opportunistic infections in young gay men which heralded the AIDS epidemic and the 1995 report on the Ebola outbreak in Zaire, are also included in the MMWR. This publication, now available on-line, is an invaluable aid to physicians, public health agencies, teachers, students, and anyone else studying infectious disease or public health. In fact, many of the epidemiological charts and stories in this textbook are taken from the MMWR. The CDC also conducts research relating to infectious diseases and can dispatch teams to various states or to other parts of the world to assist with controlling epidemics. The CDC also works with state laboratories to provide refresher courses that update the knowledge of laboratory and infection control personnel.

Each state has a public health laboratory that is involved in infection surveillance and control as well as other health-related activities. Individual states have the authority to mandate which diseases must be reported by physicians to the state laboratory. The list of reportable diseases varies slightly from state to state but generally includes the diseases that are considered nationally notifiable by the CDC.

It is interesting to note that although the *Escherichia coli* O157:H7 outbreak described in table 20.1 appeared in other states, Washington was the first state to recognize the outbreak and intervene to stop its spread. The prompt response was due in part to the fact that, at that time, Washington was one of the few states that had active surveillance and reporting measures for *E. coli* O157:H7 cases.

State laboratories also provide specialized laboratory services for the examination of specimens or cultures submitted by physicians, local health departments, hospital laboratories, epidemiologists, sanitarians, and others. Examples of these laboratory services are listed in table 20.2. Public health laboratories also deal with environmental health matters. They test water supplies for potential pathogens, give advice on laboratory safety and design, and assist in handling outbreaks of infectious disease. Additional programs enhance

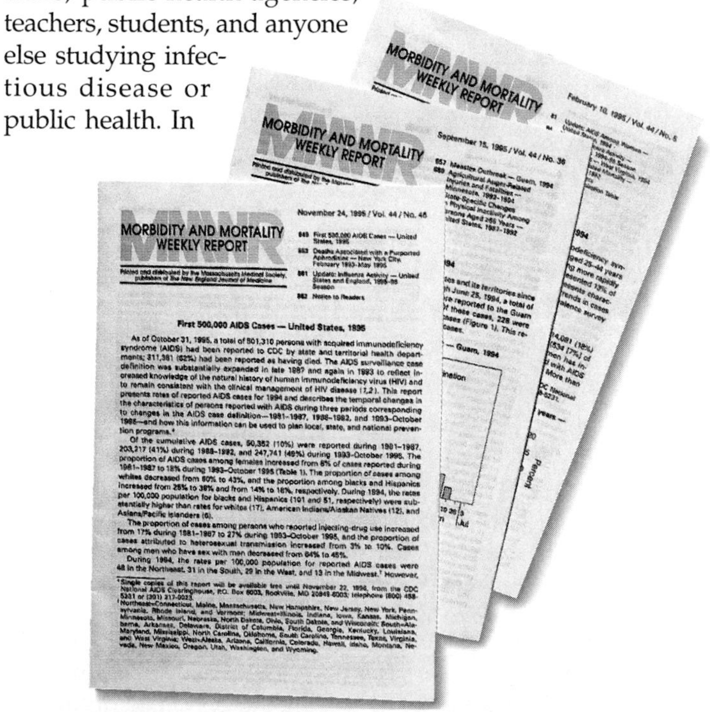

MORBIDITY AND MORTALITY WEEKLY REPORT

MORBIDITY AND MORTALITY WEEKLY REPORT

MORBIDITY AND MORTALITY WEEKLY REPORT

Printed and distributed by the Massachusetts Medical Society, publishers of The New England Journal of Medicine

November 24, 1995 / Vol. 44 / No. 46

849 First 500,000 AIDS Cases — United States, 1995
853 Deaths Associated with a Purported Aphrodisiac — New York City, February 1993–May 1995
861 Update: Influenza Activity — United States and England, 1995–96 Season
862 Notice to Readers

First 500,000 AIDS Cases — United States, 1995

As of October 31, 1995, a total of 501,310 persons with acquired immunodeficiency syndrome (AIDS) had been reported to CDC by state and territorial health departments; 311,381 (62%) had been reported as having died. The AIDS surveillance case definition was substantially expanded in late 1987 and again in 1993 to reflect increased knowledge of the natural history of human immunodeficiency virus (HIV) and to remain consistent with the clinical management of HIV disease (1,2). This report presents rates of reported AIDS cases for 1994 and describes the temporal changes in the characteristics of persons reported with AIDS during three periods corresponding to changes in the AIDS case definition—1981–1987, 1988–1992, and 1993–October 1995—and how this information can be used to plan local, state, and national prevention programs.*

Of the cumulative AIDS cases, 50,352 (10%) were reported during 1981–1987, 203,217 (41%) during 1988–1992, and 247,741 (49%) during 1993–October 1995. The proportion of AIDS cases among females increased from 8% of cases reported during 1981–1987 to 18% during 1993–October 1995 (Table 1). The proportion of cases among whites decreased from 60% to 43%, and the proportion among blacks and Hispanics increased from 25% to 38% and from 14% to 18%, respectively. During 1994, the rates per 100,000 population for blacks and Hispanics (101 and 51, respectively) were substantially higher than rates for whites (17), American Indians/Alaskan Natives (12), and Asians/Pacific Islanders (6).

The proportion of cases among persons who reported injecting-drug use increased from 17% during 1981–1987 to 27% during 1993–October 1995, and the proportion of cases attributed to heterosexual transmission increased from 3% to 10%. Cases among men who have sex with men decreased from 64% to 45%.

During 1994, the rates per 100,000 population for reported AIDS cases were 48 in the Northeast, 31 in the South, 29 in the West, and 13 in the Midwest.† However,

*Single copies of this report will be available free until November 22, 1996, from the CDC National AIDS Clearinghouse, P.O. Box 6003, Rockville, MD 20849-6003; telephone (800) 458-5231 or (301) 217-0023.

†Northeast=Connecticut, Maine, Massachusetts, New Hampshire, New Jersey, New York, Pennsylvania, Rhode Island, and Vermont; Midwest=Illinois, Indiana, Iowa, Kansas, Michigan, Minnesota, Missouri, Nebraska, North Dakota, Ohio, South Dakota, and Wisconsin; South=Alabama, Arkansas, Delaware, District of Columbia, Florida, Georgia, Kentucky, Louisiana, Maryland, Mississippi, North Carolina, Oklahoma, South Carolina, Tennessee, Texas, Virginia, and West Virginia; West=Alaska, Arizona, California, Colorado, Hawaii, Idaho, Montana, Nevada, New Mexico, Oregon, Utah, Washington, and Wyoming.

Figure 20.14
Morbidity and Mortality Weekly Report, a publication of the Centers for Disease Control and Prevention.

Table 20.2 Public Health Laboratory Services

1. Enteric pathogen laboratory: isolation and identification of *Salmonella, Shigella, Yersinia, Campylobacter, Vibrio,* and other species
2. Reference laboratory: identification or confirmation of identification of pure cultures submitted by personnel from other laboratories who are unable to be certain about the identity of a microorganism
3. Serology laboratory: testing for antibody against syphilis, Lyme disease, brucellosis, leptospirosis, or tularemia in specimens of serum from patients
4. Virology-rickettsiology laboratory: testing for poliomyelitis, influenza, rickettsiae, and chlamydiae
5. Fluorescence microscopy laboratory: identification of certain species, including *Legionella pneumophila* and *Bordetella pertussis,* by fluorescent antibody tests
6. Parasitology laboratory: identification of parasitic protozoa, worms, and others
7. Mycobacteria and fungus laboratory: identification of mycobacteria and testing of their antimicrobial susceptibility; identification of pathogenic fungi
8. Food microbiology laboratory: examination of foods for food-poisoning bacteria, botulism toxin, and algal (*Gonyaulax*) toxins of shellfish

Table 20.3 Diseases Considered Eradicable

Disease	Current Annual Toll Worldwide	Chief Obstacles to Eradication	Conclusion
Dracunculiasis (Guinea worm disease)	<2 million persons infected; few deaths	Lack of public and political awareness; inadequate funding	Eradicable
Poliomyelitis	100,000 cases of paralytic disease; 10,000 deaths	No insurmountable technical obstacles; increased national/international commitment needed	Eradicable
Lymphatic filariasis	800 million cases	Need for better tools for monitoring infection	Potentially eradicable
Mumps	Unknown	Lack of data on impact in developing countries; difficult diagnosis	Potentially eradicable
Taeniasis/cysticercosis (pork tapeworm)	50 million cases; 50,000 deaths	Need for simpler diagnostics for humans and pigs	Potentially eradicable

Source: Centers for Disease Control and Prevention. 1993. *Morbidity and Mortality Weekly Report* 42(no. RR-16):8–9.

the reliability of laboratories in the state and ensure that they meet standards for certification. This assessment of reliability often involves sending cultures of known pathogens to laboratories to see whether they can identify the cultures properly.

The public health network also includes public schools, which report absentee rates, and hospital laboratories, which report on the isolation of pathogens that have epidemiological significance for the community. In conjunction with these local activities, the news media help inform the general public of the presence of infectious diseases and the best preventive measures.

The World Health Organization Has Targeted Some Diseases for Eradication

Diseases in which the epidemiology is well understood, and for which there is a reliable vaccine and/or treatment, have the potential to be eliminated, as smallpox was. The World Health Organization, in conjunction with the Centers for Disease Control, has developed a list of diseases that are targeted for eradication (table 20.3). One of the first diseases on the list, dracunculiasis, which is caused by the eukaryotic parasite *Dracunculis medinenisis* (appendix V), was targeted for elimination by the end of 1995. While dracunculiasis has not been eliminated yet, progress is being made. An interesting effect of the effort is that warring parties in Sudan, which has been embroiled in a civil war for 12 years, agreed to a temporary cease fire in order to allow the disease eradication program to proceed.

Infectious Disease Control in Special Situations

Day-care centers, where the young clientele cares little about sanitation, and hospitals, where patients are often predisposed to illness, can present unique problems with respect to infectious disease control.

Special Attention to Good Hygiene Is Essential in Day-Care Centers

Day-care centers, where infants in diapers who are oblivious to sanitation and hygiene mingle, are a relatively new component of American society. For obvious reasons, the centers can be hotbeds of contagious diseases. Young children have not acquired immunity to many common communicable diseases, so illnesses such as colds and diarrhea are readily transmitted among that susceptible population. Some intestinal pathogens such as *Giardia* and *Shigella* have low infectious doses, and diapered infants can serve as prolific reservoirs for these illnesses. The day-care staff must be well-trained in sanitation procedures in order to avoid transmission of infectious diseases.

Up-to-date immunizations should logically be a requirement for admission to any day-care facility. Table 20.4 presents some questions that help in assessing day-care centers. Unfortunately, it is costly to carry out good infectious control techniques in this setting, and few day-care centers can come close to the ideal.

Nosocomial Infections Are Those that Are Acquired in a Hospital

An infection that is acquired during hospitalization is called a **nosocomial infection.** It is estimated that 2% to 10% of all hospitalized patients in this country acquire nosocomial infections. This amounts to roughly 1 million patients each year in the United States and results in $6 billion in additional hospital costs. In fact, nosocomial infections make up at least half of all cases of infectious disease treated in hospitals in the United States.

Nosocomial infections can result from diagnostic or therapeutic procedures. For example, bacteria may be inadvertently introduced into the body by catheterization of the bladder or a blood vessel or during surgery. The organisms may come from the hospital environment, from contact with medical personnel, or even from the patient's own normal flora. The illness following the infection may range

Table 20.4 Assessment of Infection Control in Day-Care Centers

1. Does the center require a health history and immunization record for each child?
2. Is there adequate staff—one adult for every five children under 12 months?
3. Is there adequate space—35 to 50 square feet per child?
4. Does the staff routinely look for the signs and symptoms of contagious disease? Can and do staff members arrange for separation of sick children from other children until they can be sent home?
5. Is the diaper-changing area kept clean and disinfected?
6. Are diapers and contaminated tissues disposed of properly?
7. Do the staff members wash their hands after changing diapers and wiping noses?
8. Are children's hands washed in a sink distant from the food preparation area before they eat?
9. Are formula bottles and food jars properly refrigerated, and are they properly labeled with name and date?
10. Are cribs and toys designed for easy cleaning, and are they disinfected daily?

from being mild enough to go unnoticed to being fatal. Sometimes, because of a long incubation period, the illness may appear after the patient has been discharged from the hospital. Many factors determine which microorganisms or viruses are responsible for nosocomial infections. These include the virulence and number of organisms, the length of time the person is exposed to the agent, and particularly important, the state of the patient's host defenses.

Nosocomial infections have been a problem since hospitals began, but the nature of the problem has changed because of modern medical practices. A hospital can be regarded as a high-density population made up of unusually susceptible people, into which the most virulent and antibiotic-resistant microbial pathogens are continually introduced. One of the differences between hospitals today and the hospitals of 100 years ago is that antimicrobial drugs are now used extensively. Their use results in the selection of mutant microbial strains that resist antibiotic therapy. Members of the hospital staff may become carriers of these strains. Furthermore, hospitalized patients often have impaired host defenses. For example, surgery causes a breach of the skin barrier, allowing opportunistic pathogens to reach underlying tissues. Patients taking antibiotics have their normal microbial flora suppressed, thereby allowing colonization by resistant pathogens. In addition, certain medications impair the inflammatory and immune responses.

The increasing problem of nosocomial infections led in 1976 to the requirement that all hospitals accredited by the Joint Commission of Accreditation of Healthcare

Table 20.5 Responsibilities of the Hospital Epidemiologist

1. Review results of cultures from hospitalized patients and check admissions records for infections acquired in the community that might be transmitted to other patients
2. See that proper precautions and isolation methods are used with suspected and known infections in hospital patients
3. Teach patients, their families, hospital staff, and new employees the proper measures for control of the spread of infectious disease
4. Conduct infection control orientation programs for new employees
5. Check any infections that could have originated from hospital staff
6. Report infections to the regional health department
7. Keep in contact with supervisors in departments such as housekeeping, dietary, and X ray to ensure their procedures are consistent with accepted infection control methods
8. Investigate outbreaks of infections in the hospital to determine their source, their mode of spread, and effective control measures
9. Prepare and present information to the infection control committee, including (a) presence of community-acquired infections that might threaten to spread through the hospital; (b) nosocomial infections—numbers, causes, locations in the hospital, types of infection (wound, respiratory, septicemia, for example); (c) potential problems such as faulty ventilation or improper disposal of contaminated materials; and (d) new information on infection control from other hospitals and state and national agencies

Organizations have on staff a hospital epidemiologist (or "infection control practitioner") responsible for identifying and controlling nosocomial infections. The hospital epidemiologist generally reports to an infection control committee composed of representatives of the various professionals in the hospital who are concerned with infection control. These include nurses, doctors, dietitians, engineers, housekeeping staff, and microbiology laboratory personnel (table 20.5).

REVIEW QUESTIONS

1. What are three important factors that a descriptive epidemiological study attempts to determine?
2. Draw a representative graph (time versus number of people ill) depicting a propagated epidemic and a common source epidemic.
3. What two groups of people does an analytical study compare?
4. Name four public agencies that help control epidemics and describe how they contribute to minimizing the spread of infectious disease.
5. What measures can be taken to prevent nosocomial infections?

Perspective 20.4

Control of Hospital Infections

Although many hospital infections arise from a patient's own microbial flora, others are acquired from some other source in the hospital environment. An estimated one-third of these infections are preventable. Traditionally, prevention has depended on diagnosing the infectious disease and then instituting isolation measures appropriate to stopping the spread of the causative agent. For example, according to this approach, AIDS and chicken pox would be treated as shown in table 1.

In the early 1980s, it became increasingly apparent that health-care workers were acquiring hepatitis B, and, later, AIDS from contact with the blood or other body fluids of patients who were not suspected of having bloodborne disease. Because of this, the Centers for Disease Control recommended **universal blood and body fluid precautions** for all patients, regardless of diagnosis. These precautions were applicable to all situations in which contact with blood was likely, as indicated in table 2.

In recent years, many hospitals have broadened the concept of universal precautions, making the traditional diagnosis-dependent isolation procedures largely unnecessary. The newer approach is referred to as **body substance isolation.** As with universal precautions, body substance isolation does not depend on a patient's diagnosis, and it helps protect both patients and others. There are six basic components to this approach:

1. Disposable gloves are worn whenever there is possible contact with blood, secretions, mucous membranes, skin wounds, or moist body substances. Gloves are changed between patients. Hand washing is unnecessary unless the hands become soiled.
2. Thorough hand washing is practiced between patients when gloves are not used.
3. Gowns, masks, and eye protection are used whenever secretions, blood, or other body fluids are likely to splash onto the face or soil the skin or clothing.
4. Soiled, reusable items are placed in protective bags to prevent leaking and contamination.
5. All needles and sharp objects are discarded in a rigid, puncture-proof container without touching them or replacing needle caps.
6. Private rooms are used for patients with airborne diseases and those who might soil other people or their surroundings. Health-care workers not known to be immune to airborne diseases such as measles and chicken pox should avoid entering the rooms of patients with these diseases even with face masks.

Table 1 Isolation Measures for AIDS and Chicken Pox

Isolation measures	AIDS	Chicken Pox
Private room	Yes, if hygiene is poor	Yes
Face mask	Generally not necessary	Yes
Protective gown	Yes, if soiling is likely	Yes
Disposable gloves	Yes, if touching infected material	Yes
Infective material	Blood and body fluids	Respiratory and skin lesions
Duration of precautions	Duration of illness	Until all lesions are crusted
Comment	Use special care when handling blood or blood-soiled instruments and body substances.	Persons known to be immune do not need a face mask; others should avoid entering the room.

Table 2 Universal Blood and Body Fluid Precautions

Activity	Gloves	Gown	Mask	Goggles or Face Shield
Controlling spurting blood	Yes	Yes	Yes	Yes
Assisting childbirth	Yes	Yes	Yes	Yes
Starting intravenous fluid	Yes	No	No	No
Taking a patient's temperature	No	No	No	No

Summary

I. Descriptive Terms in Epidemiology

II. Spread of Disease
   A. Humans, animals, and nonliving environments can be reservoirs for disease.
   B. Diseases can be transmitted from person to person through direct contact, droplet transmission, or indirect contact via inanimate objects.
      1. Direct transmission
      2. Indirect transmission
   C. Pathogens can be transmitted in food or water.
   D. Pathogens can be transmitted through air.
   E. Arthropod vectors may transmit disease.
   F. The portal of entry may determine the outcome of disease.
   G. Other factors play a role in disease outcome and spread.
      1. The incubation period of a disease influences the extent of the spread of the disease.
      2. The intensity of exposure to an infectious agent influences frequency, incubation period, and severity of disease.
      3. Immunization, or prior exposure, influences the susceptibility to infectious agents.
      4. Nutrition, crowding, and other environmental and cultural factors can shape the character of epidemics.
      5. The genetic background of an individual can influence susceptibility to infection.

III. Epidemiological Studies
   A. Descriptive studies examine the persons affected by, the place of, and the time of a disease outbreak.
   B. Analytical studies compare risk factors for developing disease in cases and controls.
   C. Precise identification of infectious agents is often necessary.

IV. Infectious Disease Surveillance
   A. The public health network includes government agencies, hospital laboratories, and news media.
   B. The World Health Organization has targeted some diseases for eradication.

V. Infectious Disease Control in Special Situations
   A. Special attention to good hygiene is essential in day-care centers.
   B. Nosocomial infections are those that are acquired in a hospital.

Chapter Review Questions

1. Which of the following is an example of a fomite?
   a. table
   b. flea
   c. *Staphylococcus aureus* carrier
   d. water
   e. air
2. Which of the following would be the easiest to eventually eradicate?
   a. a pathogen that is common in wild animals but sometimes infects humans
   b. a disease that occurs exclusively in humans, always resulting in obvious symptoms
   c. a mild disease of humans that often results in no obvious symptoms
   d. a pathogen that is found in marine sediments
   e. a pathogen that readily infects both wild animals and humans
3. Give one example of a disease that is transmitted vertically.
4. Why is pneumonic plague spread more easily than bubonic plague?
5. Name three things that a descriptive epidemiological study seeks to determine.
6. Give one possible reason respiratory diseases are more common in the winter.
7. What is an antibiogram?
8. Explain the role of the Centers for Disease Control (CDC) in disease prevention.
9. What precautions should be taken in day-care centers to avoid disease transmission?
10. Define *nosocomial infection.*

Critical Thinking Questions

1. How might hand washing help prevent the spread of infectious diseases, even though it fails to remove all potential pathogens from the hands?
2. Why might an epidemic disease occurring in a Laotian village be of any concern to us?
3. How would you go about choosing the next epidemic disease to be eliminated from the planet?

Further Reading

Bell B.P., Goldoft, M., Griffin, P. M., Davis, M. A., Gordon, D-C., Tarr, P. I., Bartleson, C. A., Lewis J. H., Barrett, T.J. and Wells J. G. 1994. A multistate outbreak of *Escherichia coli* O157:H7 associated bloody diarrhea and hemolytic uremic syndrome from hamburgers. The Washington experience. *JAMA* 272(17)1349–1353.

Bendiner, Elmer. 1987. Semmelweis: lone ranger against puerperal fever. *Hospital Practice* 22:194–225. Excellent account, interesting and readable.

Bennett, J. V., and Brachman, P. S. 1986. *Hospital Infections.* Little, Brown & Co., Boston.

Berkelman, R. L., Bryan, R. T., Oserholm, J. W., LeDuc, J. W., Hughes, J. M. 1994. Infectious disease surveillance: a crumbling foundation. *Science* 264:368–370.

Brandt, S. L., Simon, H. L., and Dechairo, D. C. 1978. *Epidemiology for the infection control nurse.* C.V. Mosby Co., St. Louis.

Centers for Disease Control, [http://www.cdc.gov/]. Publications including *Morbidity and Mortality Weekly Report* and *Emerging Infectious Diseases* can be accessed from this Web site.

Cliff, Andrew, and Haggett, Peter. 1984. Island epidemics. *Scientific American* 250:138. Fascinating studies of the spread of measles epidemics.

Donowitz, L. G. 1990. Hospital acquired infections in children. *New England Journal of Medicine* 323:1836. Editorial outlining nosocomial infection problems unique to children.

Geldreich, E. E. 1986. Potable water: new directions in microbial regulations. *AMS News* 52:530. A succinct review of water contamination problems.

Kaiser, A. B. 1991. Surgical wound infections. *New England Journal of Medicine* 324:123. Nice discussion of current problems.

Linnan, M. J., et al 1988. Epidemic listeriosis associated with Mexican-style cheese. *New England Journal of Medicine* 319:823.

MacMahon, B., Trichopoulos, D. 1996. *Epidemiology, Principles and Methods,* 2d ed. Little, Brown and Co., New York.

Nightingale, F. 1969. *Notes on nursing.* Dover Publications, New York. (Unabridged republication of the first American edition as published by D. Appleton & Co., 1860).

Outbreak of *E. coli* O157:H7 infections associated with drinking unpasteurized commercial apple juice. 1995. *Morbidity and Mortality Weekly Report* 45(44)975.

Pfaller, J. A., 1991. Typing methods for epidemiologic investigations. *In* Balows, A., et al. (ed.), *Manual of Clinical Microbiology,* 5th ed. American Society for Microbiology, Washington, D.C.

Roueche, Berton 1980. *The medical detectives.* The New York Times Book Company, New York. Entertaining, mostly true accounts of solving epidemics, written in a whodunit style.

Sanchez, A., Ksiazek, T. G., Rollin, P. E., Peters, C. J. , Nichol, S. T., Khan, A. S., and Mahy, B. W. 1995. Reemergence of Ebola virus in Africa. *Emerging Infectious Diseases* (July-September).

21 Antimicrobial Medicine

KEY CONCEPTS

1. Certain chemicals selectively kill or inhibit the growth of one organism while sparing another.
2. Many antimicrobial drugs exploit the differences that exist between prokaryotic and eukaryotic cell structure and metabolism.
3. Viruses and eukaryotic pathogens have fewer targets for selective toxicity.
4. Mutations and transfer of genetic information have allowed microorganisms to develop resistance to each new antimicrobial drug that has been developed.
5. Laboratory tests of susceptibility of microorganisms to antimicrobial drugs help predict the effectiveness of the drugs in treating disease.

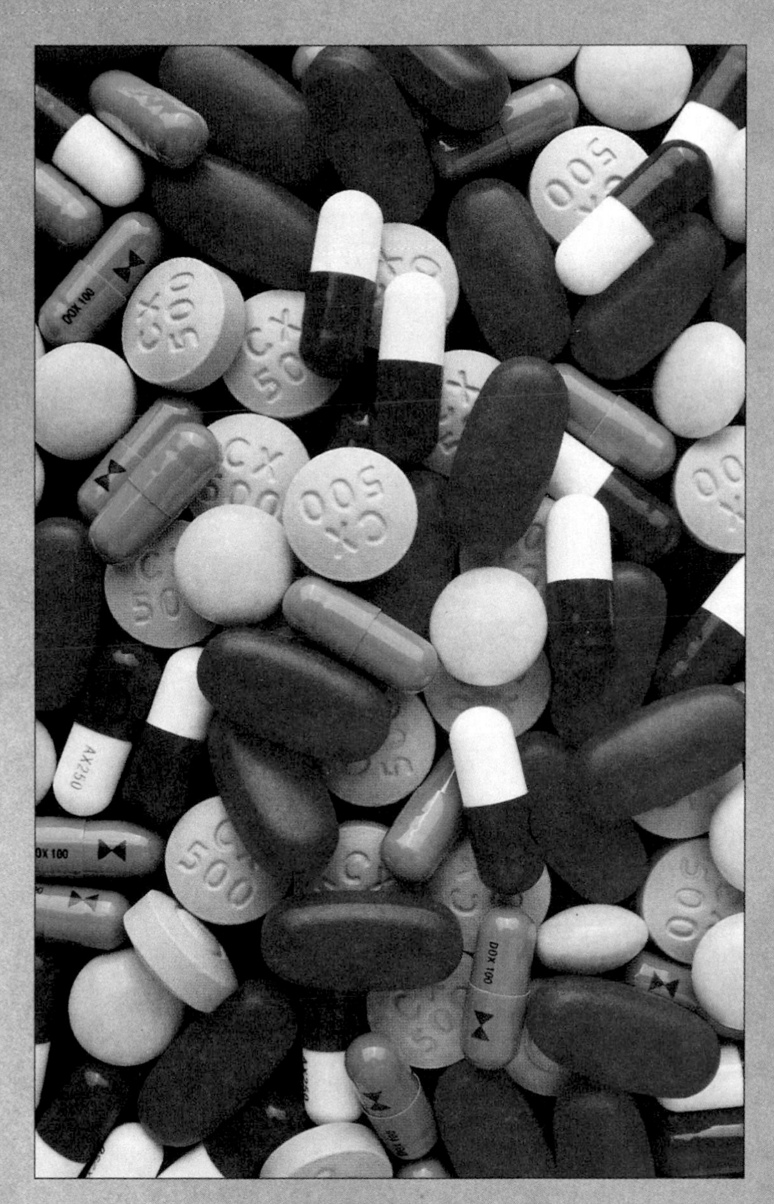

A wide variety of antimicrobial drugs are available to treat infections.

Microbes in Motion

The following books and chapters in the *Microbes in Motion* CD-ROM may serve as a useful preview or supplement to your reading:
Control of Microbes: Terminology; Physical Control; Chemical Control. *Antimicrobial Action:* Cell Wall Inhibitors; Protein Synthesis Inhibitors; Agents Affecting Enzymes; Agents Affecting Nucleic Acids.

A Glimpse of History

Paul Erlich (1854–1915), a German physician and bacteriologist, was born into a wealthy family, the only son after many daughters. Understandably, his interests were indulged by family and servants, even though his large collections of frogs and snakes occasionally entered the laundry room. As an adult, he was rarely without a good cigar and habitually scribbled notes on his shirt cuffs. He received a degree in medicine in 1878 and then became intrigued with the way various types of body cells differ in their ability to take up dyes and other substances. After observing that certain dyes stain bacterial cells but not animal cells, indicating that the two cell types are somehow fundamentally different, it occurred to him that it might be possible to find a dye or chemical that selectively harms bacteria without affecting human cells.

Erlich began a systematic search, synthesizing many chemicals in an attempt to find what he termed a "magic bullet," a drug that would kill microbial pathogens without harming the human host. One disease he was particularly interested in treating was the sexually transmitted disease syphilis which is caused by the bacterium *Treponema pallidum*. Much of the mental illness during his time resulted from tertiary syphilis, a late stage of the disease. Since an arsenic compound had shown some success in treating a protozoan disease of animals, Erlich and his colleagues began synthesizing hundreds of different arsenic compounds to use in treating experimental infections of laboratory animals. The 606th compound, arsphenamine, proved to be highly effective in treating syphilis in animals. When the drug was administered to patients, it killed the spirochete of syphilis at concentrations that usually did not kill the patients (although it made them very sick). Many patients whose infections were previously considered hopeless were cured by this new medicine. The name given to the drug, *Salvarsan*, was derived from the words *salvation* and *arsenic*. The use of Salvarsan to cure syphilis proved that chemicals could selectively kill pathogens without harming the human host.

The term **antimicrobial** is a general term used to define a diverse group of naturally occurring or laboratory-synthesized chemicals that, at very low concentrations, are able to kill or inhibit the growth of microorganisms. Salvarsan was the first example of an antimicrobial drug synthesized in the laboratory. **Antibiotics** are a subset of antimicrobials that are naturally produced by the biosynthetic processes of molds and bacteria. For example, the antimicrobial penicillin is produced by the mold *Penicillium*, so it is called an antibiotic. In contrast, Salvarsan and other chemically synthesized drugs are not called antibiotics because they are not natural products of microorganisms. However, the distinction is not so clear because many antibiotics are now at least partially synthesized in the laboratory.

Many diseases that were considered life threatening before World War II are now readily treated with antibiotics and other antimicrobials. Before the discovery of penicillin, which initiated the antibiotic era, the prognosis for people with diseases such as bacterial pneumonia and tuberculosis or with staphylococcal infections was grim. Physicians were able to identify the cause of the disease, but were generally unable to recommend treatments other than bed rest. Today, however, antimicrobials are routinely prescribed, and the simple cure they provide for so many bacterial diseases is often taken for granted (table 21.1). Unfortunately, the overuse and misuse of these life-saving drugs, coupled with the bacterial world's amazing ability to adapt under selective pressure, has led to an increase in the number of organisms that are resistant to the effects of antimicrobials. The increase in antibiotic resistance among a wide range of bacteria has caused some people to speculate that we are in danger of seeing an end to the antibiotic era. In response to the increased resistance, the pharmaceutical industry is scrambling to develop new antimicrobial drugs.

The usefulness of antibiotics in medicine stems from the fact that their toxicity is often due to their ability to interfere with essential biochemical structures or processes that are unique to prokaryotes. Consider, for example, the synthesis of peptidoglycan. By interfering with a target that is unique to prokaryotes, the antibiotics cause relatively little harm to the human host, in a phenomenon called **selective toxicity.** That is, the drugs are toxic to bacteria but not the human host.

The use of Salvarsan to treat syphilis was an early example of **chemotherapy,** using a chemical as a therapeutic treatment. Antimicrobial chemotherapy took a giant step forward in 1929, when a British scientist, Alexander Fleming, discovered the antibiotic penicillin which is produced by the mold *Penicillium* ("A Glimpse of History," chapter 3). Ten years after Fleming's discovery, two other British scientists, Ernst Chain and Howard Florey, were successful in their attempts to isolate penicillin. In 1941, the antibiotic was tested for the first time on a police officer who had a life threatening *Staphylococcus aureus* infection. He improved so dramatically that, within 24 hours, his illness seemed under control. Unfortunately, the supply of purified penicillin ran out, and the man eventually died of the infection. Later, with greater supplies of the drug, the experiment was repeated and two deathly ill patients were successfully cured. Because England was fully involved in World War II, Chain and Florey found two American companies willing to produce large quantities of the drug so that it could be used to treat infected soldiers and workers.

Soon after the discovery of penicillin, Selman Waksman observed that a bacterium he isolated from soil, *Streptomyces griseus*, produced an antibiotic he called streptomycin. The realization that bacteria as well as molds could produce medically useful antimicrobial drugs prompted researchers to begin laboriously screening hundreds of thousands of different strains of microorganisms for antibiotic production. Even today, pharmaceutical companies examine soil samples from around the world for organisms that produce novel antibiotics.

Perspective 21.1

A Red Dye and Good Luck Produce a Breakthrough

The first real breakthrough in the developing science of antimicrobial chemotherapy came almost 20 years after the death of Ehrlich in 1915. Stimulated by his work, the German chemical industry began screening the effectiveness of thousands of dyes for the treatment of streptococcal infections in animals. One of them was a red dye called Prontosil, which was found to be dramatically effective without producing toxic effects in the infected animals. Surprisingly, Prontosil had no effect on streptococci in test tubes. The reason for these findings soon became apparent when it was discovered that, once inside the animal body, the Prontosil molecule was split apart by enzymes in the blood, and a smaller molecule called *sulfanilamide* was produced. It was the sulfanilamide that acted against the infecting streptococci. Thus, the discovery of sulfanilamide, the first sulfa drug, was based on luck as well as persistence. If Prontosil had only been screened against bacteria in test tubes and not given to infected animals, its effectiveness might never have been discovered.

Table 21.1 Antibacterial Therapy

Bacterium	Disease	Medication
Streptococcus pyogenes	Strep throat	Penicillin V
Staphylococcus aureus	Boil	Dicloxacillin
Haemophilus influenzae	Meningitis	Ceftriaxone
Legionella pneumophila	Legionnaires' disease	Erythromycin
Rickettsia rickettsii	Rocky Mountain spotted fever	Tetracycline
Mycobacterium tuberculosis	Tuberculosis	Isoniazid + rifampin
Pseudomonas aeruginosa	Septicemia	Imipenem
Escherichia coli	Bladder infection	Sulfa drug

Overview of Antibiotics

Antibiotics are examples of **secondary metabolites** of microorganisms, meaning that they are not directly involved in the growth of the producing organisms. While a wide variety of antibiotics has been discovered, not all of the antibiotics are selectively toxic, which is why only a small percentage can be safely used for treating humans. Hundreds of tons and many millions of dollars' worth of antibiotics are now produced each year. Most antibiotics come from three main groups of microorganisms: the branching filamentous prokaryotic actinomycetes (*Streptomyces*); spore-forming, gram-positive rods (*Bacillus*); and eukaryotic molds (*Penicillium* and *Cephalosporium*). In commercial production, a carefully selected strain of the appropriate producing species (e.g., of *Penicillium chrysogenum* for penicillin) is inoculated into a broth medium and incubated in a huge vat. As soon as the maximum antibiotic concentration is reached, the drug is extracted from the medium and extensively purified. In many cases, the antibiotic is chemically altered after synthesis and purification in order to impart to it new characteristics such as increased stability. Antibiotics that have been chemically altered are called **semisynthetic.** In some cases, scientists are able to synthesize the entire antibiotic in the laboratory. By convention, these partially or totally synthetic chemicals are called *antibiotics,* since microorganisms can make them in whole or in part.

- **Secondary metabolite, p. 103**
- ***Streptomyces*, p. 266**
- ***Bacillus*, p. 19**
- ***Penicillium*, pp. 46, 448**

The Suitability of a Specific Antimicrobial Depends on the Clinical Situation

When deciding which antimicrobial drug to prescribe for a given clinical situation, a physician must take several factors into account: the cause and site of the infection as well as host factors such as age, immune status, and kidney function. Each of these characteristics can make some antimicrobial drugs more appropriate than others in specific clinical situations.

Antimicrobials Differ in Their Toxicity and Spectrum of Activity

While the ideal antimicrobial harms bacteria but not the human host, most in reality are at least slightly toxic to humans as well. The relative toxicity is expressed as the therapeutic ratio or **therapeutic index,** which is defined as the

Perspective 21.2

A Near-Fatal Human Infection Is Cured with Prontosil

On May 17, 1933, at the meeting of the Dusseldorf (Germany) Dermatological Society, Drs. Schreus and Foerste gave an astonishing report. They had had under their care a 10-month-old boy with a skin infection caused by staphylococci. Despite repeated efforts to cure him using all known methods, the bacteria invaded his bloodstream, his condition deteriorated, and he was near death. By chance, the two doctors received a small supply of Prontosil from the I. G. Farben Chemical Company, with the information that it had been shown to be effective for treating infections in mice. The new medicine was given to the child in a last-ditch effort to save his life. His fever promptly subsided and he returned rapidly to good health. There were no ill effects whatsoever from the medicine.

dose toxic to the patient divided by the dose toxic to bacteria. Antibiotics that have a high therapeutic index are less toxic to the patient often because they act against a vital biochemical process of bacteria that does not exist in human cells. For example, penicillin, which interferes with peptidoglycan synthesis, has a very high therapeutic index. When an antimicrobial that has a low therapeutic index is given, the concentration of the drug in the patient's blood must be carefully monitored to ensure that it does not reach a level that is toxic to the patient.

Antimicrobial drugs that inhibit the growth of bacteria without killing them are **bacteriostatic.** These drugs depend on the normal host defenses to kill or eliminate the pathogen after its growth has been inhibited. For example, sulfa drugs, which are frequently prescribed for urinary tract infections, simply inhibit the growth of bacteria in the bladder until they are cleared during the normal process of urination. Drugs that kill bacteria are **bactericidal.** These drugs are particularly useful in situations in which the normal host defenses cannot be relied upon to remove or destroy pathogens.

Some antimicrobials inhibit or kill a narrow range of microorganisms, such as only gram-positive bacteria, while others affect a wide range, generally including both gram-positive and gram-negative organisms. Antimicrobials that affect a wide range of bacteria are called **broad-spectrum antimicrobials.** They are very important in the treatment of acute life-threatening diseases when immediate antimicrobial therapy is essential and there is no time to culture and identify the disease-causing agent. The disadvantage of broad-spectrum antimicrobials lies in the fact that, by affecting a wide range of organisms, they disrupt the normal flora that play an important role in excluding pathogens. This, in turn, leaves the patient predisposed to other infections. Antimicrobials that affect a narrow range of bacteria are **narrow-spectrum antimicrobials.** Their use requires that the pathogen be identified, but they do not disrupt the normal flora as much as broad-spectrum antimicrobials.

• **Normal flora, pp. 410–12**

Antimicrobials Differ in How They Are Distributed, Metabolized, and Excreted

Antimicrobials differ not only in their action and activity, but in how they are distributed, metabolized, and excreted by the body. For example, only some drugs are able to cross the blood-brain barrier into cerebral spinal fluid, an important factor for a physician to consider when prescribing a drug to treat meningitis. Likewise, drugs that are actively transported into tissues are most appropriate for treating tissue infections. Drugs that are unstable in acid are destroyed by stomach acid when taken orally, so these drugs must instead be administered through intravenous or intramuscular injection.

Another important characteristic of individual antimicrobials is their rate of elimination, which is expressed as their **half-life.** The half-life of a drug is the amount of time it takes for the body to eliminate one-half of the original effective dose in the serum. The half-life of a drug dictates the frequency of doses required to maintain a specific level of the drug in the body. For example, penicillin, which has a half-life of only 1 1/2 hours, needs to be taken four times a day, whereas azithromycin, which has a half-life of over 24 hours, is taken only once a day or less. Patients who have kidney or liver dysfunction often excrete or metabolize drugs more slowly, so their drug dosages must be adjusted accordingly.

New antimicrobials must be evaluated extensively to determine their usefulness, safety, and individual characteristics. Their spectrum of activity against a wide variety of potentially pathogenic microorganisms must be determined. Their toxicity is evaluated by giving increasing doses to laboratory animals to see how much can be given before tissues such as the kidneys, liver, bone marrow, or central nervous system are damaged. Doses are then given to human volunteers to identify how the drugs are changed and excreted by the host. By measuring the concentrations of antimicrobials in urine, mother's milk, spinal fluid, blood, and other body fluids, it can be determined which sites of infections can be treated, what doses are needed, and how often the drug must be given.

Combinations of Antimicrobial Drugs Can Be Antagonistic, Synergistic, or Additive

Combinations of antimicrobials are sometimes used to treat infections, but care must be taken when selecting the combinations because some drugs will counteract the effects of others. For example, bacteriostatic drugs interfere with the bactericidal effects of penicillin because penicillin only kills actively dividing cells. The combination of one drug that interferes with the activity of another is called **antagonistic.** In contrast, drugs that are more effective when taken together are called **synergistic.** For example, the combination of penicillin and streptomycin is more lethal to bacteria than is the sum of their individual activities because penicillin interferes with cell wall synthesis which, in turn, allows streptomycin to more easily enter cells. This synergistic effect sometimes allows lower doses of toxic drugs to be used. Drug combinations that are neither antagonistic nor synergistic are called *additive*. An **additive** combination results in a total effect that is neither greater nor less than the sum of the effects of the two individual drugs.

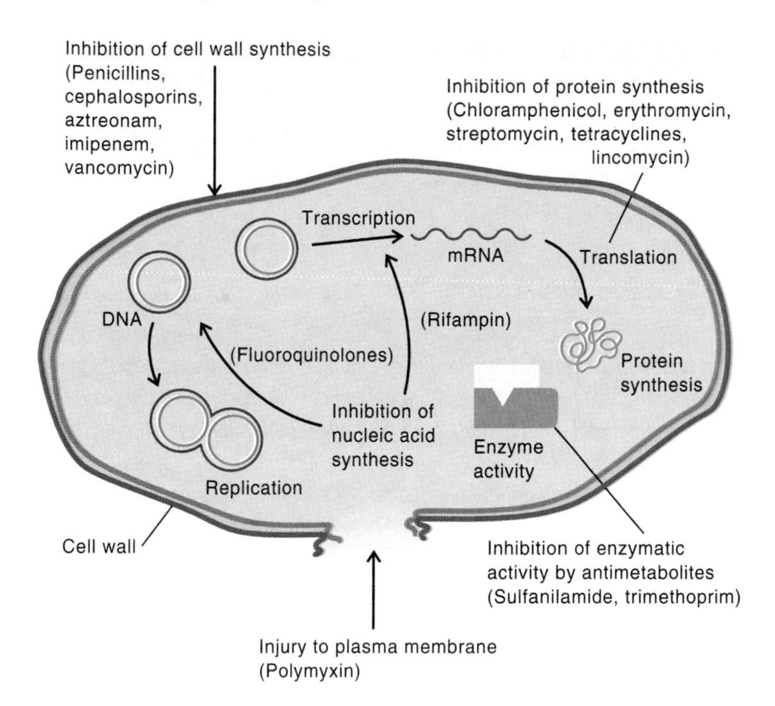

Figure 21.1
Bacterial cell showing the targets of antibiotic action and the antimicrobials used against those targets.

Targets of Antimicrobial Drugs

As stated earlier, medically useful antibiotics and other chemotherapeutic drugs interfere with a process that is unique to the disease-causing agent. Otherwise, the drugs would be toxic to the patients as well as the pathogens. Several differences between prokaryotes and eukaryotes can be targets for selective toxicity, but fewer such targets exist for effective antifungal, antiparasitic, and antiviral drugs.

Differences Between Prokaryotic and Eukaryotic Cells Are Useful Targets of Antibacterial Drugs

Consider the differences between eukaryotic and prokaryotic cell structure/function, which were discussed in earlier chapters. It is precisely because of these differences that antibacterial drugs are medically useful (figure 21.1).

Cell Wall Synthesis

The target of several medically useful antimicrobial drugs is cell wall biosynthesis. Because the peptidoglycan cell wall is unique to bacteria, eukaryotic cells are not affected by these drugs. Thus, antimicrobials that target peptidoglycan synthesis usually have a very high therapeutic index. Recall that peptidoglycan is a three-dimensional structure composed of sheets of alternating subunits of *N*-acetylglucosamine (NAG) and *N*-acetylmuramic acid (NAM), which are interconnected through a peptide bridge between the amino acid side chains of NAM (chapter 3). The intact structure provides the cell the rigidity to avoid bursting.

Penicillin, the first antibiotic discovered, interferes with the formation of the peptide bridge by binding to proteins, called **penicillin-binding proteins (PBP),** that are involved in cell wall biosynthesis. The resulting lack of cross-linking weakens the structural integrity of the cell wall, ultimately leading to cell lysis. Since peptidoglycan is only synthesized in actively growing cells, penicillin is only effective against multiplying bacteria. Other antibiotics such as bacitracin and vancomycin also interfere with peptidoglycan synthesis, but their action is not restricted to the cell wall and, therefore, their therapeutic index is low.

Protein Synthesis

Several groups of antibiotics including streptomycin, chloramphenicol, and tetracycline exert their effect on bacteria by interfering with steps of protein synthesis. However, only a few of these antibiotics are selective enough in their toxicity to be administered safely to patients, which is not surprising since protein synthesis is a feature of all cells, not just those of pathogenic microorganisms. Fortunately, 80S eukaryotic ribosomes are different

• **Peptidoglycan, p. 58**

• **Ribosome, p. 72**

enough in their structure that some selectivity in the action is possible. Even so, bacterial-type (70S) ribosomes are present in mitochondria. This may partially account for the side effects often observed in patients undergoing treatment with antimicrobial agents that interfere with prokaryotic protein synthesis.

Nucleotide Replication/Transcription

Several enzymes involved in DNA replication and transcription in prokaryotes are sufficiently different from those of eukaryotes to act as selective targets. Rifampin, for example, binds strongly to bacterial RNA polymerase and thereby inhibits mRNA synthesis. The family of synthetic antibacterial drugs called *quinolones* are also inhibitors of nucleic acid synthesis, but they act against the enzyme DNA gyrase, thereby stopping DNA synthesis. DNA gyrase mediates the breaking and reunion of DNA strands and is required for replication of DNA as well as transcription of DNA to mRNA.

Cell Membrane Function

A few antimicrobial drugs damage the bacterial cytoplasmic membranes. For example, polymyxin B binds to the membrane of prokaryotic cells and alters the permeability. This, in turn, leads to leakage of the cells' contents and, ultimately, cell death. Unfortunately, these drugs also bind to eukaryotic cells, though to a lesser extent, which limits their medical usefulness.

Metabolic Pathways

Some antimicrobial drugs interfere with metabolic pathways common in prokaryotes but not humans. One such pathway is the synthesis of folic acid, which is an important precursor of an essential coenzyme. Many types of bacteria are able to synthesize folic acid through a multistep pathway, but they cannot take it up from their external environment. Humans, on the other hand, lack this pathway, so folic acid must be provided in their diet. Drugs such as sulfonamide and trimethoprim that inhibit essential enzymes in the pathway of folic acid synthesis are selectively toxic to bacteria because humans do not have the target enzymes.

Eukaryotic Pathogens Have Many Similarities to Human Cells so They Offer Fewer Targets for Selective Toxicity

Human cells more closely resemble eukaryotic pathogens such as fungi and protozoa than they do bacteria. It is not surprising, therefore, that far fewer drugs are available for use against eukaryotic pathogens, and the ones that are available tend to be more toxic.

The target of antifungal drugs is often the chemical compound **ergosterol** which is found in the cytoplasmic membrane of fungal but not human cells. By interfering with the function or synthesis of ergosterol, these drugs are selectively toxic to fungi. Unfortunately, these drugs sometimes have a low therapeutic index because they interfere with compounds found in eurkaryotic cells, which limits their internal use to the treatment of life-threatening infections. However, some can be applied topically to treat skin, hair, and nail infections.

Most antiparasitic drugs are thought to interfere with either nucleic acid synthesis of protozoan parasites or the neuromuscular function of worms. Unfortunately, relatively little research and development goes into antiparasitic drugs because these diseases are concentrated in the poorer areas of the world where people simply cannot afford to spend money on expensive medications.

Viruses Have Few Targets for Selective Toxicity

Viruses rely almost exclusively on the host cell's metabolic machinery for their replication which makes them extremely difficult targets for selective toxicity. Recall from earlier reading that viruses are acellular, which means they have no cell wall, no ribosomes, nor any other structure targeted by commonly used antibiotics. Thus, viruses are completely unaffected by these antibiotics. In fact, no drug has been developed yet that will actually cure a viral infection, although a few drugs slow down specific virus replication enough to enable the normal host defenses to bring the infection under control. Most antiviral drugs, such as acyclovir and AZT, take advantage of the error-prone virally encoded enzymes involved in replication or transcription of viral nucleic acid (figure 21.2). Many researchers and pharmaceutical companies are currently trying to develop more effective antiviral drugs.

Concerns in Using Antimicrobial Medications

As with any medication, several concerns and dangers are associated with antimicrobial drugs. It is important to remember, however, that antimicrobials are extremely valuable drugs that have saved countless lives when properly prescribed and used.

• **DNA replication, p. 156**
• **Transcription, p. 136**
• **AZT, p. 157**
• **Acyclovir, p. 467**

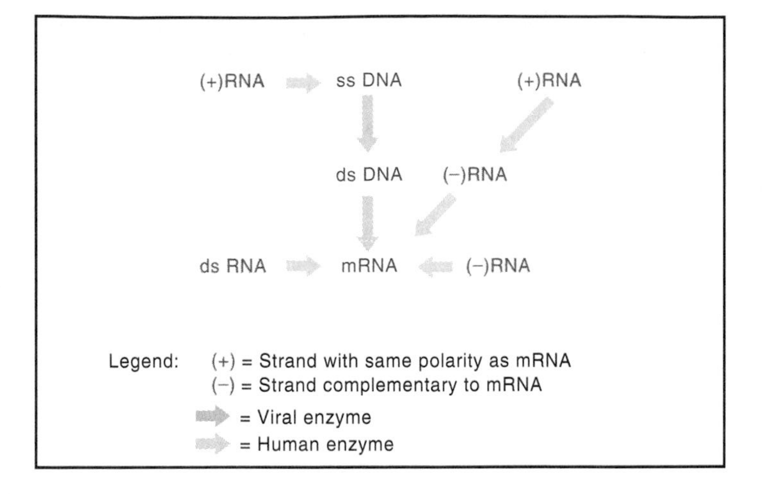

Figure 21.2
Differing replication strategies of RNA and DNA viruses.

Some Antimicrobials May Elicit Allergic or Other Adverse Reactions

Some people develop hypersensitivities or allergies to certain antimicrobials. An allergic reaction to penicillin and other related drugs may result in a fever or rash and can sometimes lead to life-threatening anaphylactic shock. For this reason, it is important that people who have allergic reactions to antimicrobials alert their physicians and pharmacists so that alternative drugs can be prescribed.

Several antimicrobials are toxic at high concentrations or occasionally cause adverse reactions. For example, streptomycin can damage kidneys and impair sense of balance, but its most serious toxic effect is irreversible deafness. Patients taking this drug must be closely monitored because it has a very low therapeutic index. Some antimicrobials have such severe potential side effects that they are reserved for only life-threatening conditions. For example, in rare cases, chloramphenicol causes the potentially lethal condition *aplastic anemia* in which the body is unable to make white and red blood cells. For this reason, chloramphenicol is used only when no other alternatives are available such as for treating a penicillin-allergic patient who has bacterial meningitis.

Antimicrobials Suppress the Normal Flora

As discussed in chapter 19, the normal flora play an important role in excluding pathogens. When the composition of the normal flora is altered, which happens when a person takes antimicrobials, pathogens that would not normally be able to compete may grow to high numbers. Patients who take broad-spectrum antibiotics orally are at risk of developing the life-threatening disease **antibiotic associated colitis (pseudomembranous colitis)** caused by the growth of toxin-producing *Clostridium difficile* in the intestine. This organism is not usually able to establish itself in the intestine due to competition with other bacteria. However, when the growth of the normal flora is inhibited or flora are killed, *C. difficile* can flourish and cause disease.

• **Hypersensitivity reaction, pp. 396–403**
• **Anaphylactic shock, pp. 397, 398**

REVIEW QUESTIONS

1. Describe the difference between the terms *antimicrobial* and *antibiotic.*
2. Define *therapeutic index* and explain its importance.
3. Why is the bacterial cell wall a good target for selective toxicity?
4. Why is it difficult to develop antiviral drugs?
5. List two adverse reactions associated with the use of antimicrobials.

Bacterial Sensitivity Testing

For some pathogenic species, susceptibility to antimicrobial agents is predictable. For example, members of the genus *Mycoplasma* lack a cell wall, so they are resistant to the effects of penicillin. In contrast, different strains of *Staphylococcus aureus* vary in antibiotic sensitivity because they frequently have R plasmids that encode antibiotic resistance. Thus, there is no way of knowing which antibiotic is likely to be effective in a given *S. aureus* infection. Unfortunately, it has often been the practice to try one drug after another until a favorable response is observed or, if the infection is very serious, to give several antibiotics together. Both approaches are undesirable. With each unnecessary drug given, needless risks of toxic or allergic effects arise, and the normal flora may be altered, which may permit the overgrowth of pathogens that are resistant to the antimicrobial. A better approach is to determine the susceptibility of the specific pathogen to various antimicrobial drugs and then choose the drug that acts against the offending pathogen but against as few other bacteria as possible.

Susceptibility Can Be Quantified by Determining Minimal Inhibitory Concentrations

To measure the susceptibility of a bacterium to an antimicrobial, decreasing concentrations of the drug are first prepared in a suitable growth medium. Then, a suspension of the microorganism is added to test tubes containing each concentration of the drug. Following incubation to allow growth of the bacteria, the tubes are examined for the

• **R plasmids, p. 176**

presence of visible growth or turbidity. The lowest concentration capable of preventing growth is called the **minimum inhibitory concentration** (MIC; figure 21.3). The organism is said to be "sensitive" to that concentration. The **minimum bactericidal concentration (MBC)** can then be determined by subculturing a sample from each tube in which the antibiotic prevented growth to fresh antibiotic-free medium. If no growth occurs in the antibiotic-free medium, then no living organisms remained, indicating that the antibiotic was bactericidal at that concentration.

The Terms *Sensitive* and *Resistant* Refer to Whether a Microorganism Is Likely to Be Treatable

Although often used in a quantitative sense, the word *sensitive* is also used as a qualitative term to describe organisms susceptible to the concentration of the antimicrobic that is attainable in a patient who is taking the drug. The word *resistant* is applied to microorganisms requiring substantially higher concentrations. For example, an organism with an MIC of only 20 micrograms per milliliter of polymyxin would nevertheless be called *resistant* since levels of this drug that can be achieved in the blood are usually lower than 5 micrograms per milliliter. Microorganisms requiring inhibitory concentrations on the borderline between sensitive and resistant are often called *intermediate*.

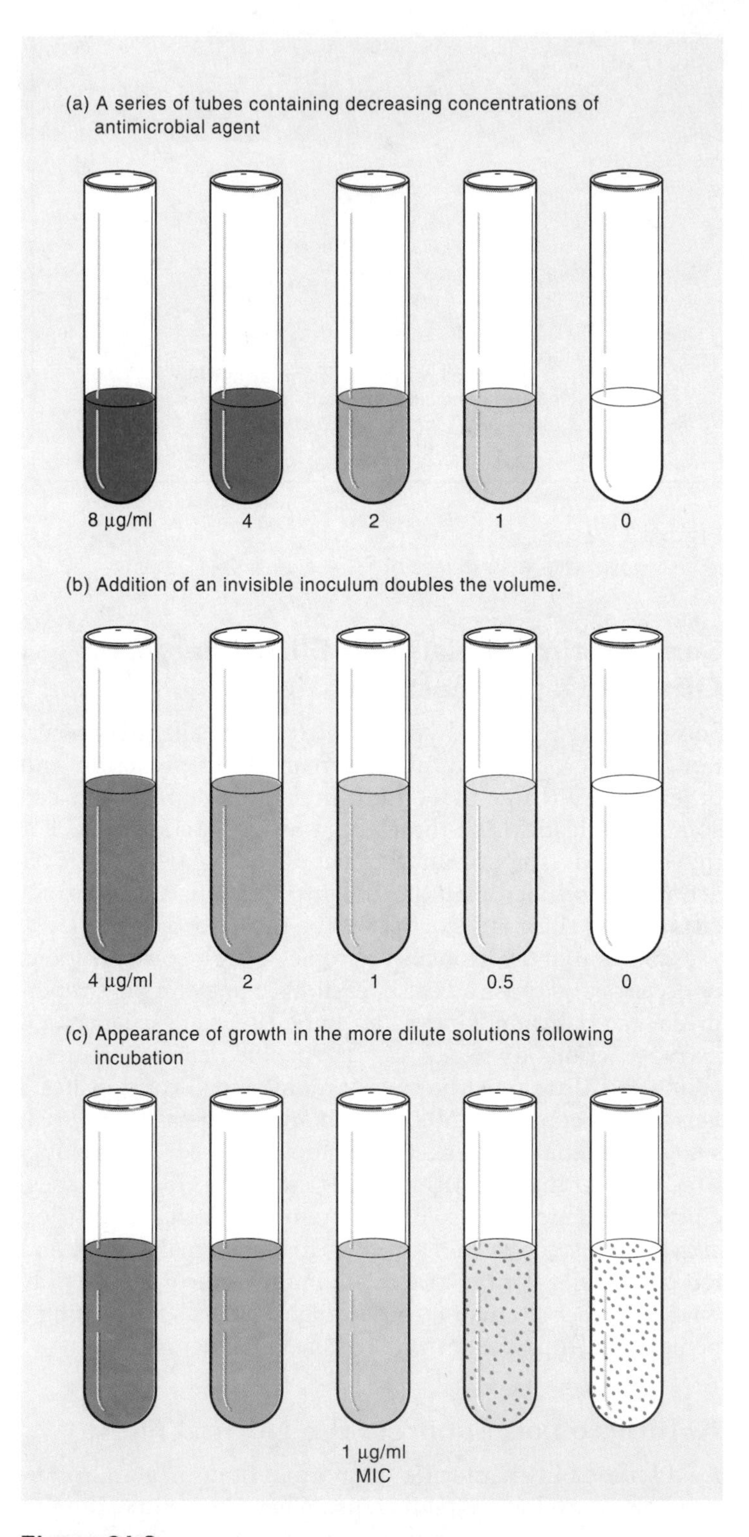

Figure 21.3
Minimal inhibitory concentration (MIC). In this example, the MIC is 1 µg/ml. The tube containing medium without antibiotic is a growth control. In actual practice, an antibiotic control using an organism of known MIC is also included.

Development of Resistance

A few years after the introduction of sulfa drugs and penicillin, there was great hope that such drugs would soon eliminate most bacterial diseases. It is now recognized that the development of drug resistance limits the usefulness of all known antimicrobials. Understanding the mechanisms and the spread of antimicrobial resistance is an important step in curtailing the problem, and it may also allow pharmaceutical companies to develop new drugs that foil common resistance mechanisms.

Some Organisms Are Innately Resistant to Certain Antibiotics

Organisms that naturally lack a target for a specific antibiotic are innately resistant to that antibiotic. For example, members of the genus *Mycoplasma* lack a cell wall, so not surprisingly, they are resistant to penicillin and other drugs that target peptidoglycan. Additionally, many gram-negative organisms are inherently resistant to penicillin because the selective permeability of their outer membrane excludes the drug from the cell wall. Innate resistance is consistent and predictable because it reflects the natural composition of an organism.

Perspective 21.3

Measuring the Amounts of Antimicrobial Medicines in Body Fluids

Merely knowing the MIC and MBC is not enough to tell whether a drug will be effective in treating an infection. It is also necessary to know whether these concentrations are likely to occur in a patient's infected tissues. To determine this, each new drug is given to human subjects in doses known to be safe and nontoxic. Samples of the blood, urine, and other body fluids are then collected at different time intervals following administration of the drug. A very sensitive organism is then used to measure the amount of the anitmicrobial drug present in the fluids (figure 1).

This can be done in the following way. A culture of the organism is spread over the surface of agar medium. Holes are then punched out of the agar, and some are filled with fluid containing known concentrations of the drug to be assayed, while others are filled with the body fluid being tested. Following incubation, zones of inhibition form around the agar wells; the higher the concentration of antimicrobial, the larger the zone of inhibition. By measuring the zone sizes and plotting them against the corresponding concentrations, a curve relating zone size to concentration is obtained, from which the concentration of the antimicrobial medicine in the body fluid can be read. Thus, the concentrations present in body fluids at different times following administration of a drug can be determined, as shown in figure 2.

(a) Holes punched in agar medium inoculated with a very sensitive bacterium

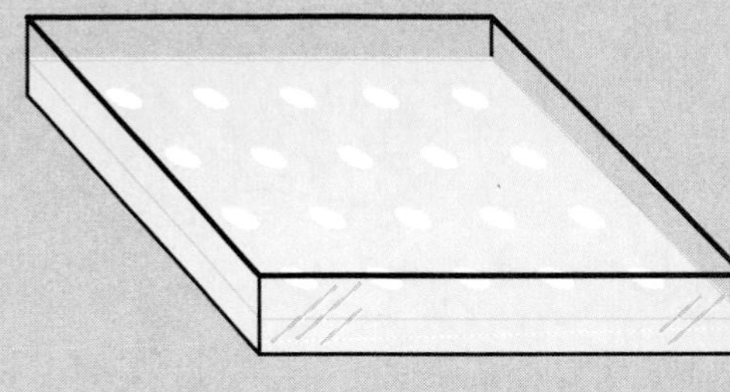

(b) Three tubes containing solutions of penicillin in known concentration and a fourth tube with the patient's serum

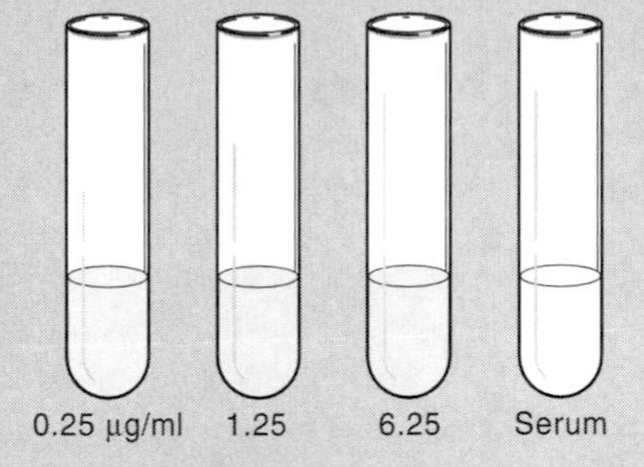

(c) The holes have been filled with fluid from the tubes and the culture plate incubated overnight; the sensitive bacterium grew everywhere except in areas around the holes in the agar.

(d) The diameters of the areas of inhibited growth are averaged for each solution and plotted semi-logarithmically; the diameter **D** of the zone of inhibition produced by the serum is used to determine the concentration **C** of antibiotic in the serum.

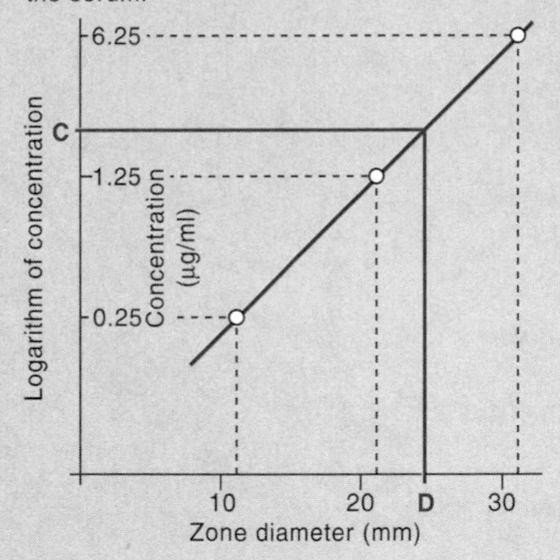

Figure 1
Biological assay to determine the concentration of penicillin in a patient's serum.

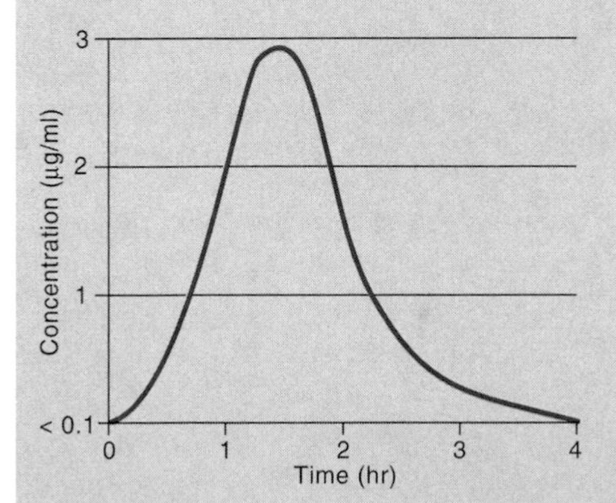

Figure 2
Concentration of penicillin in a patient's blood at different times after an oral dose of the antibiotic.

Perspective 21.4

Distinguishing Sensitive and Resistant Organisms by Using Dry Paper Antimicrobial-Containing Disks

Automated and miniaturized equipment is now available to determine MICs, but it is expensive. Therefore, most medical laboratories determine qualitative susceptibilities of bacterial pathogens by instead using filter paper disks impregnated with antimicrobial agents (Kirby-Bauer procedure, figure 1). Thus, to determine whether the *S. aureus* strain isolated from a boil is sensitive to penicillin, a culture of the organism is spread over the surface of an agar medium, and an antibiotic disk containing a precise amount of penicillin is placed on top of it. Following incubation, the zone of inhibited growth is measured, and if it is large enough, the organism is called *sensitive*.

• ***S. aureus*, p. 260**

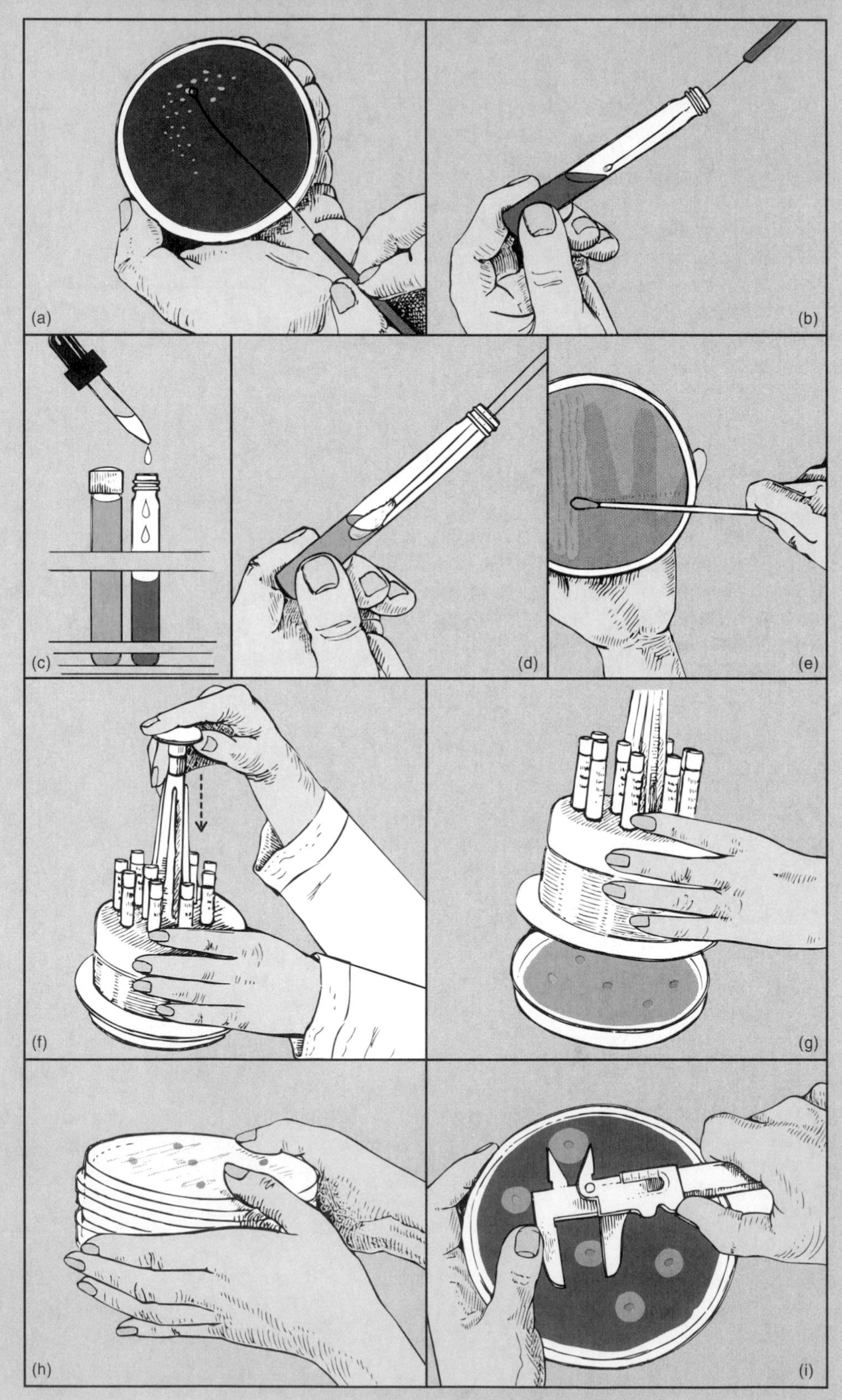

Figure 1
Steps involved in the Kirby-Bauer procedure for sensitivity testing using filter paper disks impregnated with antimicrobial medicines. (*a*) A microbiological loop is used to remove a colony from a culture of infected material. (*b*) The colony is transferred to a tube of liquid broth and incubated for a short time. (*c*) The growth of the bacteria causes the fluid to become turbid; this turbidity is adjusted to standard turbidity by adding more sterile fluid. (*d*) A sterile swab is dipped into the culture. (*e*) The swab saturated with a suspension of bacteria is used to inoculate the entire surface of the sensitivity testing medium contained in a petri dish, and (*f* and *g*) a mechanical dispenser is used to deposit filter paper disks containing different antimicrobial medicines. (*h*) The petri dishes are incubated overnight at 35°C to 37°C. (*i*) The next day, the diameters of the inhibited growth around the disks are measured. Sensitivity and resistance to the different antimicrobial medicines are determined by comparing the diameters with the range of diameter sizes for sensitive and resistant bacteria as given in special tables.

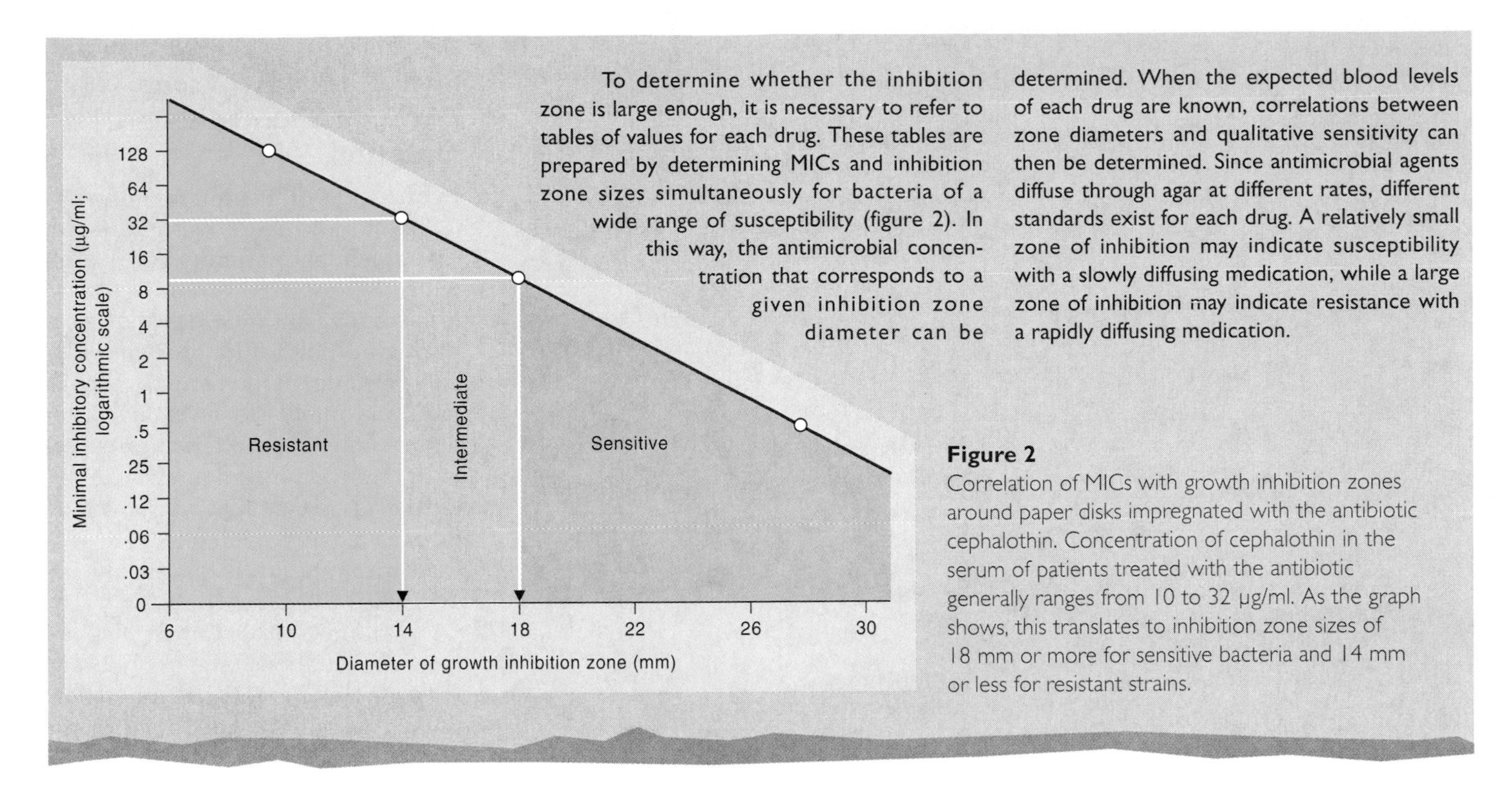

To determine whether the inhibition zone is large enough, it is necessary to refer to tables of values for each drug. These tables are prepared by determining MICs and inhibition zone sizes simultaneously for bacteria of a wide range of susceptibility (figure 2). In this way, the antimicrobial concentration that corresponds to a given inhibition zone diameter can be determined. When the expected blood levels of each drug are known, correlations between zone diameters and qualitative sensitivity can then be determined. Since antimicrobial agents diffuse through agar at different rates, different standards exist for each drug. A relatively small zone of inhibition may indicate susceptibility with a slowly diffusing medication, while a large zone of inhibition may indicate resistance with a rapidly diffusing medication.

Figure 2
Correlation of MICs with growth inhibition zones around paper disks impregnated with the antibiotic cephalothin. Concentration of cephalothin in the serum of patients treated with the antibiotic generally ranges from 10 to 32 µg/ml. As the graph shows, this translates to inhibition zone sizes of 18 mm or more for sensitive bacteria and 14 mm or less for resistant strains.

Sensitive Bacteria Can Acquire Antibiotic Resistance

Unlike innate resistance, acquired antibiotic resistance is ever changing. As antibiotics are increasingly used and misused, the bacterial strains that are resistant to their effects have a selective advantage over their sensitive counterparts when the antimicrobial is in the environment. For example, when penicillin was first introduced, less than 3% of *Staphylococcus aureus* strains were resistant to its effects. Heavy use of the drug, measured in hundreds of tons per year, progressively eliminated sensitive strains, so that now 85% or more are resistant (table 21.2). This development is understandably of great concern to health professionals because of the impact on the cost, complications, and outcomes of treatment. Even more ominous is the recent report of an *S. aureus* strain isolated from a patient in Japan that is resistant to normal levels of all antimicrobials.

Bacteria have evolved diverse and remarkable ways to avoid the effects of antimicrobials (figure 21.4). In several cases, resistance is due to a minor structural alteration in the target so that it is no longer bound by the drug yet still functions. For example, streptomycin normally binds to a part of the prokaryotic 30S ribosomal subunit that is critical for protein synthesis. A slight alteration in the structure of the ribosome results in a distortion, so that streptomycin is no longer able to bind but the ribosome can still functionally translate mRNA. Similarly, changes in the penicillin-binding proteins (PBPs) do not alter their function but prevent the binding of penicillin.

Table 21.2 *Staphylococcus Aureus* **Isolates Resistant to Penicillin G**

Year	Percentage Resistant
1940	<3
1946	14
1948	58
1960	83
1995	>90

Some bacteria have evolved the ability to overproduce the target as a way of avoiding the effects of an antimicrobial drug. The increased quantity of target molecules overwhelms the drug. For example, sulfa drugs normally interfere with folic acid synthesis by acting as a decoy substrate for the enzyme, thus competitively inhibiting the enzyme. When an organism produces excess enzyme, enough uninhibited enzyme will be available to complete the synthesis of folic acid.

An entirely different mechanism of drug resistance involves the destruction or inactivation of the antibiotic. Some organisms produce specific enzymes that can cleave or chemically modify the essential portion of an antibiotic to destroy its activity. For example, the enzyme **penicillinase**

• **Sulfa drugs, p. 154**

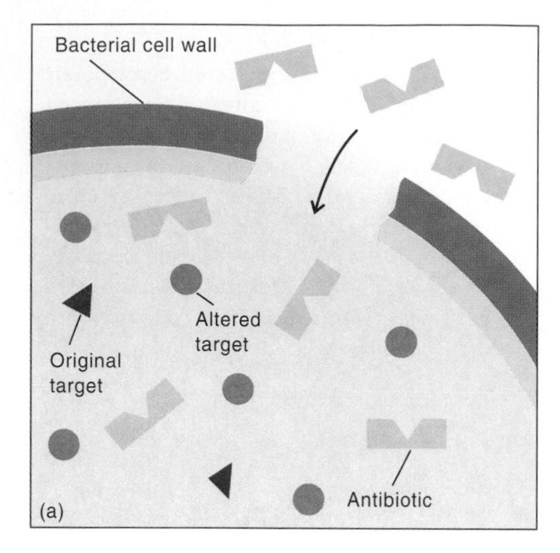

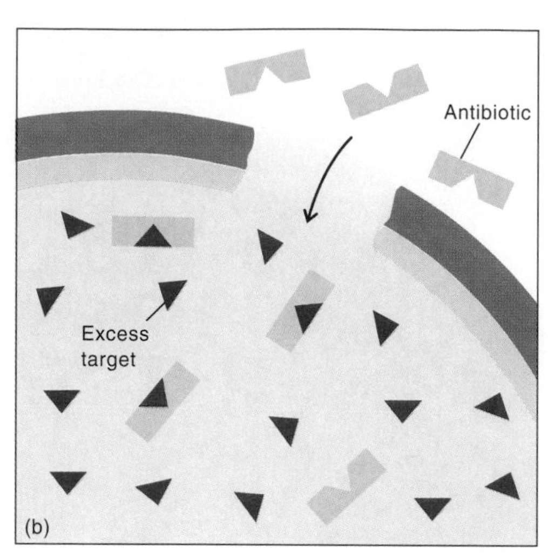

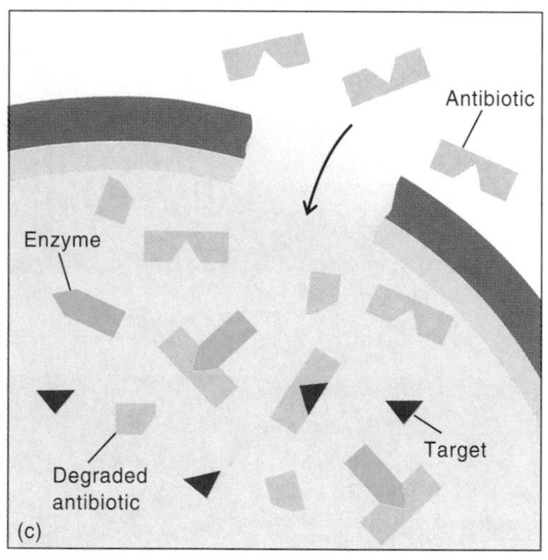

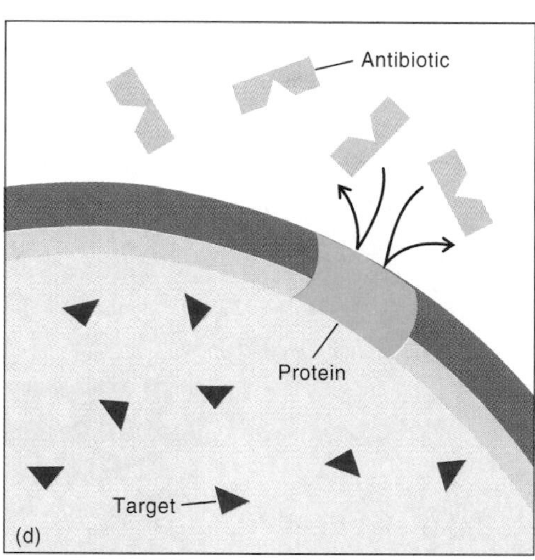

Figure 21.4
Some alternative mechanisms of bacterial resistance to antimicrobial drugs. (*a*) Minor structural alterations in the protein target can result in a functional protein that is no longer recognized by the antimicrobial. (*b*) Over-production of the target overwhelms the binding capacity of normal achievable levels of the antimicrobial drug. (*c*) Bacteria can produce specific enzymes that cleave or chemically modify specific antibiotics. (*d*) Mutational alterations in membrane proteins can prevent drugs from entering the cell.

is one of a group of enzymes called **β-lactamases** that destroy the activity of penicillin and some other similar drugs by cleaving an essential portion, the β-lactam ring. Similarly, another enzyme chemically modifies chloramphenicol by adding an acetyl group. The modified form of chloramphenicol is not toxic to bacteria.

Alterations in membrane permeability or its other function may also confer antibiotic resistance. In some cases, mutational alteration of a membrane protein responsible for maintaining selective permeability prevents the drug from entering the cell. In the case of tetracycline resistance, however, the drug is actively pumped back out of the cell.

Acquisition of Antimicrobial Resistance May Occur Through Spontaneous Mutation

Acquisition of antimicrobial resistance may be due to spontaneous mutations that naturally occur during cell growth. For example, streptomycin resistance is acquired through a point mutation that, like all spontaneous mutations, occurs only very rarely. However, given the high numbers of bacteria associated with an active infection and the selective advantage that resistant mutants have when antibiotics are used, the rare mutation is significant indeed (figure 21.5).

Antimicrobials to which spontaneous mutation frequently occurs are sometimes given in combination with a second antimicrobial drug. For example, streptomycin alone is never used to treat tuberculosis but is instead used in combination with other drugs such as rifampin. The chance that an organism will simultaneously develop resistance to both drugs is extremely low, and so an organism that develops resistance to one drug will still be killed by the other drug.

Acquisition of Antimicrobial Resistance May Occur Through DNA Transfer

The most troublesome source of resistance is the transfer of genes from resistant organisms to those that are sensitive by the mechanisms discussed in chapter 7. The resistance genes that code for enzymes that inactivate the antimicrobials are often found on conjugative plasmids called *resistance plasmids* or **R plasmids.** A single R plasmid frequently encodes on several different genes resistance to several different antimicrobial drugs, thus enabling an organism to simultaneously gain resistance to several completely different drugs. Unfortunately, the mobility of these genes makes the possibility of widespread resistance a grim reality (table 21.3). For example, extensive use of antibiotics selects for normal flora such as *Staphylococcus epidermidis* that carry R plasmids. Normally, this would not be

• **R plasmid, p. 176**

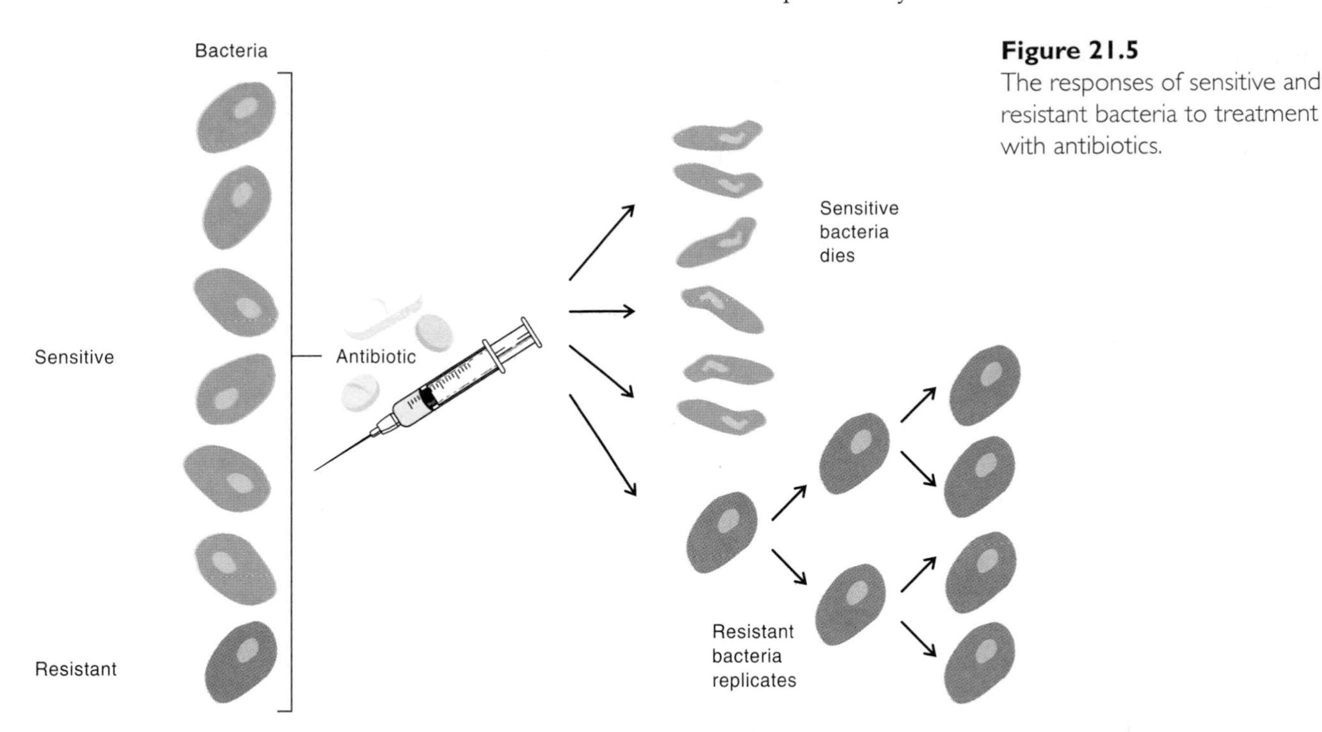

Figure 21.5
The responses of sensitive and resistant bacteria to treatment with antibiotics.

Table 21.3 Plasmid-Mediated Resistance in *Neisseria Gonorrhoeae*

Date	Event
1944	Penicillin is introduced for treatment of gonorrhea.
1949	Tetracycline is introduced for treatment of gonorrhea.
1974	Penicillin resistance plasmids are detected in *Haemophilus influenzae.*
1976	Penicillin resistance plasmids appear in *N. gonorrhoeae.*
1982	Tetracycline resistance plasmids appear in *N. gonorrhoeae.*
1989	Neither penicillin nor tetracycline is useful in the primary treatment of gonorrhea.

threatening, except for the fact that *S. epidermidis* is capable of transferring its plasmids to the common pathogen *S. aureus*. Because of the ease with which antibiotic resistance on R plasmids can be transferred through conjugation, many isolates of *S. aureus* are now resistant to all antibiotics except vancomycin. Vancomycin is a drug usually reserved for life-threatening conditions. Even more serious is the fact that some strains of *Enterococcus*, a common opportunistic pathogen that is part of the normal flora, are resistant to all known antimicrobial agents including vancomycin (figure 21.6). These **vancomycin-resistant enterococci,** or **VRE,** are particularly a problem in intensive care settings in which patients are prone to opportunistic infections. Infections caused by these strains are untreatable with conventional drug therapy.

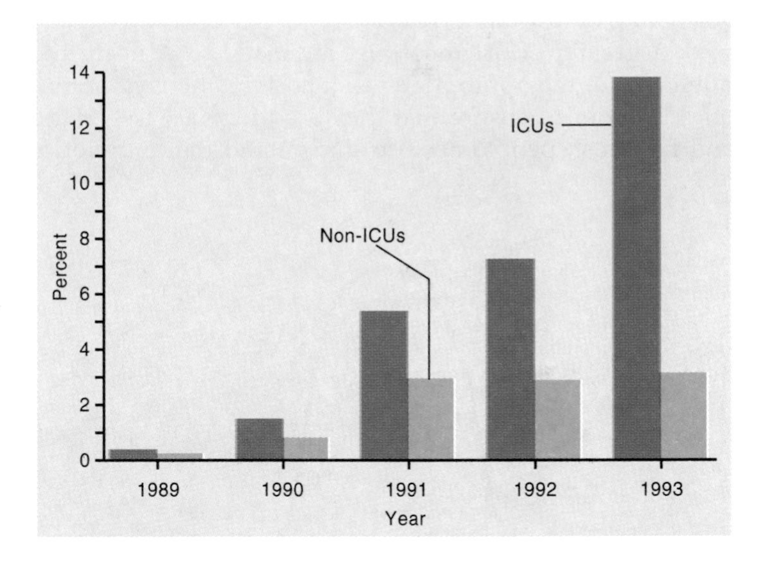

Figure 21.6
Emergence of vancomycin-resistant enterococci. The chart shows the percentage of resistant enterococci isolated from infections in people in intensive-care units (ICUs) and people not in ICUs, by year.

Source: Morbidity and Mortality Weekly Report. 42(30) 598. 1993

Control of Antibiotic Resistance Relies on the Cooperation of Patients and Physicians

In order to reverse the alarming trend of increasing antimicrobial resistance, physicians as well as the general public must take more responsibility for the appropriate use of these life-saving drugs. Physicians need to increase their efforts to identify the causative agent of infectious diseases and, only if appropriate, prescribe suitable antimicrobials.

They must also educate their patients about the proper use of prescribed drugs in order to increase patient compliance. While these efforts may be more expensive in the short term, they will ultimately save both lives and money.

Patients, meanwhile, need to carefully follow the instructions that accompany their prescriptions, even if those instructions seem inconvenient. It is very important to maintain the concentration of the antimicrobial in the blood at the required level for a specific time period. This is important because in any population of sensitive bacteria, some organisms will be more sensitive to the effects of the drug than others. Initially, the most sensitive cells in the population are killed or inhibited, but over several days in the presence of the appropriate level of antimicrobial, the least sensitive are eventually eliminated too (figure 21.7). When a patient skips a scheduled dose of a drug, the blood level of the drug may not remain high enough to inhibit the growth of the least sensitive of the population. Likewise, failure to complete the prescribed course of treatment may leave the least sensitive organisms to multiply. Misusing antimicrobials by skipping doses or failing to complete the prescribed duration of treatment promotes the gradual emergence of resistant organisms.

A greater effort must also be made to educate the public about the appropriateness and limitations of antibiotics in order to ensure that they are utilized wisely. First and foremost, people need to understand that antibiotics are not effective against viruses. Taking antibiotics will not cure the common cold nor any other viral illness. A few antiviral drugs are available, but their use is limited. Unfortunately, surveys indicate that far too many people erroneously believe that antibiotics are effective against viruses and often seek prescriptions to "cure" viral infections. This misuse only changes the normal flora by selecting for antibiotic-resistant bacteria. Even though these organisms are not pathogenic themselves, they can serve as a reservoir for R plasmids, eventually transferring their resistance genes to an infecting pathogen.

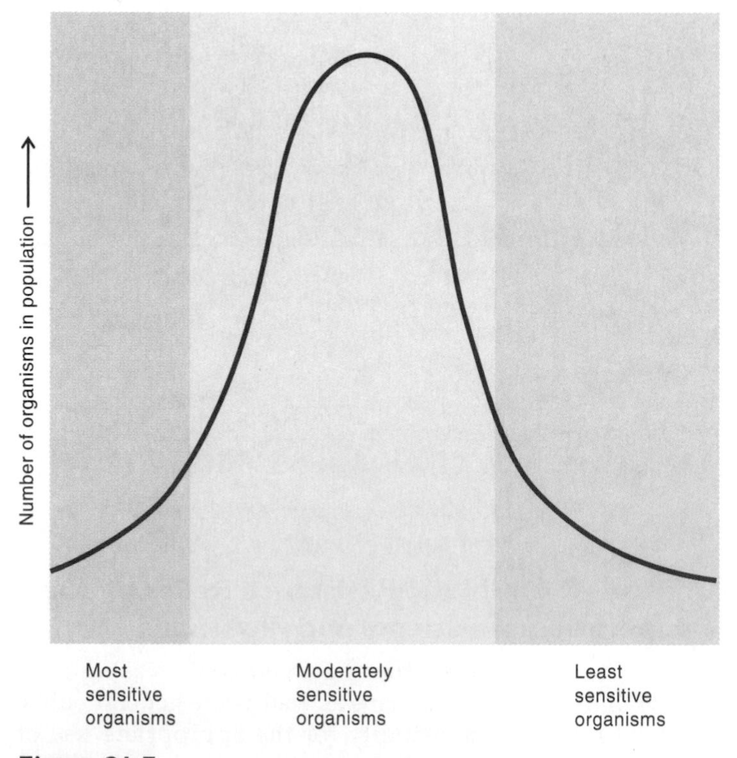

Figure 21.7
Population curve of most sensitive, moderately sensitive, and least sensitive organisms.

Survey of Antimicrobial Drugs

A wide variety of antimicrobial drugs from different sources is currently available. Each drug has individual characteristics that make it more useful in some clinical situations than in others.

Antibacterial Drugs

A summary of the properties of selected antibacterial drugs is presented in table 21.4.

Penicillins

Penicillin G, the first antibiotic commercially produced, remains the drug of choice for treating susceptible organisms because of its high therapeutic index. As mentioned previously, the drug interferes with cell wall synthesis by binding to enzymes, known as *penicillin-binding proteins*, that are essential in the synthesis of peptidoglycan. Penicillin G has drawbacks however, including its instability in the acidic conditions of the stomach, which requires that it be injected rather than taken orally. It also has a narrow spectrum of activity, which limits its use primarily to infections caused by gram-positive organisms and a few gram-negative cocci.

Since the discovery of penicillin G, scientists have discovered they can alter the chemical structure of penicillin to give it new properties including acid stability, a broader spectrum of activity, and resistance to enzymatic destruction. Figure 21.8 shows the structural formulas of what is now known as the family of penicillins. Notice that all the penicillins have a common portion, the β-lactam ring, but each has a slightly different side chain. The altered side chains are what give each penicillin derivative a slightly different property. The β-lactam ring, on the other hand, is common to all of the derivatives, and its integrity is essential for antibacterial activity. Unfortunately, some organisms produce an enzyme, penicillinase or β-lactamase, that breaks the β-lactam ring, thereby destroying the activity of the antibiotic. Over 30 different β-lactamases with slightly different specificities exist, many of which are encoded on transferable R plasmids. Penicillin derivatives such as

Table 21.4 Properties of Selected Antibacterial Drugs

Target	Drug	Action	Original Source	Representative Mechanism of Resistance
Cell wall synthesis	Penicillins	Binds to proteins essential for cell wall synthesis	*Penicillium*	Enzymatic inactivation by β-lactamase
	Cephalosporins	Same as penicillin	*Cephalosporium*	Enzymatic inactivation by β-lactamase
	Imipenem	Same as penicillin	*Streptomyces*	Prevention of entry into cell
	Aztreonam	Same as penicillin	*Chromobacterium*	Enzymatic inactivation not by β-lactamase
	Vancomycin	Inhibits assembly of peptidoglycan	*Streptomyces*	Altered target
Cell membrane function	Polymyxin	Binds to membrane protein and alters permeability	*Bacillus*	(Resistance is rare)
Protein synthesis	Streptomycin	Binds to 30S ribosomal subunit	*Streptomyces*	Enzymatic inactivation
	Chloramphenicol	Binds to 50S ribosomal subunit	*Streptomyces*	Enzymatic inactivation
	Erythromycin	Binds to 50S ribosomal subunit	*Streptomyces*	Enzymatic inactivation
	Tetracyclines	Binds to 30S ribosomal subunit	*Streptomyces*	Prevention of entry into cell
	Lincomycin	Binds to 50S ribosomal subunit	*Streptomyces*	Altered target
Nucleic acid synthesis	Rifampin	Binds to RNA polymerase	*Streptomyces*	Altered target
	Fluoroquinolones	Interferes with DNA gyrase	Chemically sythesized	Altered target
Folic acid synthesis	Sulfonamides	Competitively inhibits enzyme	Chemically synthesized	Altered target
	Trimethoprim	Competitively inhibits enzyme	Chemically synthesized	Altered target

methicillin have been modified to make them resistant to many, but not all, β-lactamases.

Because the action of the penicillins is directed strictly against bacterial cell walls, the penicillins are almost completely nontoxic. However, some people develop rare but life-threatening allergies to the drug, which is why doctors, nurses, and pharmacists routinely ask patients about penicillin allergies before giving these drugs.

Cephalosporins

Eukaryotic fungi of the genus *Cephalosporium* produce antibiotic substances that are similar in structure to penicillin and have the same mode of action that results in bacterial cell lysis. Like the penicillins, the **cephalosporins** have been chemically modified to produce a family of antibiotics that differ in their properties. These are grouped into categories—the first, second, and third generation cephalosporins—to reflect their order of development. The first generation drugs are relatively resistant to the β-lactamases produced by *Staphylococcus*, whereas the later versions are more resistant to the β-lactamases produced by gram-negative rods. Interestingly, the later versions are less effective against gram-positive cocci. Some of the cephalosporins are destroyed by stomach acid and must be given by injection, whereas others can be given orally. About 5% of people who are allergic to penicillin also react to the cephalosporins, so they are generally not used to treat penicillin-sensitive patients. Some examples of cephalosporins and their uses are given in table 21.5.

Other β-lactam Antibiotics (Carbapems and Monobactams)

After the discovery of penicillin and then cephalosporin, other antibiotics that possess the characteristic β-lactam ring structure were found or synthesized in the laboratory. Like the penicillins, these drugs interfere with cell wall synthesis.

Streptomyces produces a category of β-lactam antibiotics called **carbapenems,** but only one derivative, imipenem, is currently in use. Imipenem is very resistant to inactivation by β-lactamases, a characteristic that gives it the broadest spectrum of activity of any of the β-lactam antibiotics. However, people who are allergic to penicillin are often also sensitive to imipenem. In addition, the drug is readily inactivated by one of the body's kidney enzymes, so it must be given in combination with another drug called *cilastatin* that inhibits the kidney enzyme. Resistance to imipenem is likely due to mutations in outer membrane proteins that would normally transport the drug into the cell to its site of action.

While most β-lactam antibiotics have a double ring structure, one group has only a single ring as reflected in its name, the monobactams. Aztreonam is the first monobactam licensed for medical use. The single ring structure of monobactams makes them resistant to most β-lactamases. In addition, they can safely be administered to patients who are allergic to other β-lactam antibiotics. Their mode of action is similar to other β-lactamases, but they have little activity against gram-positive

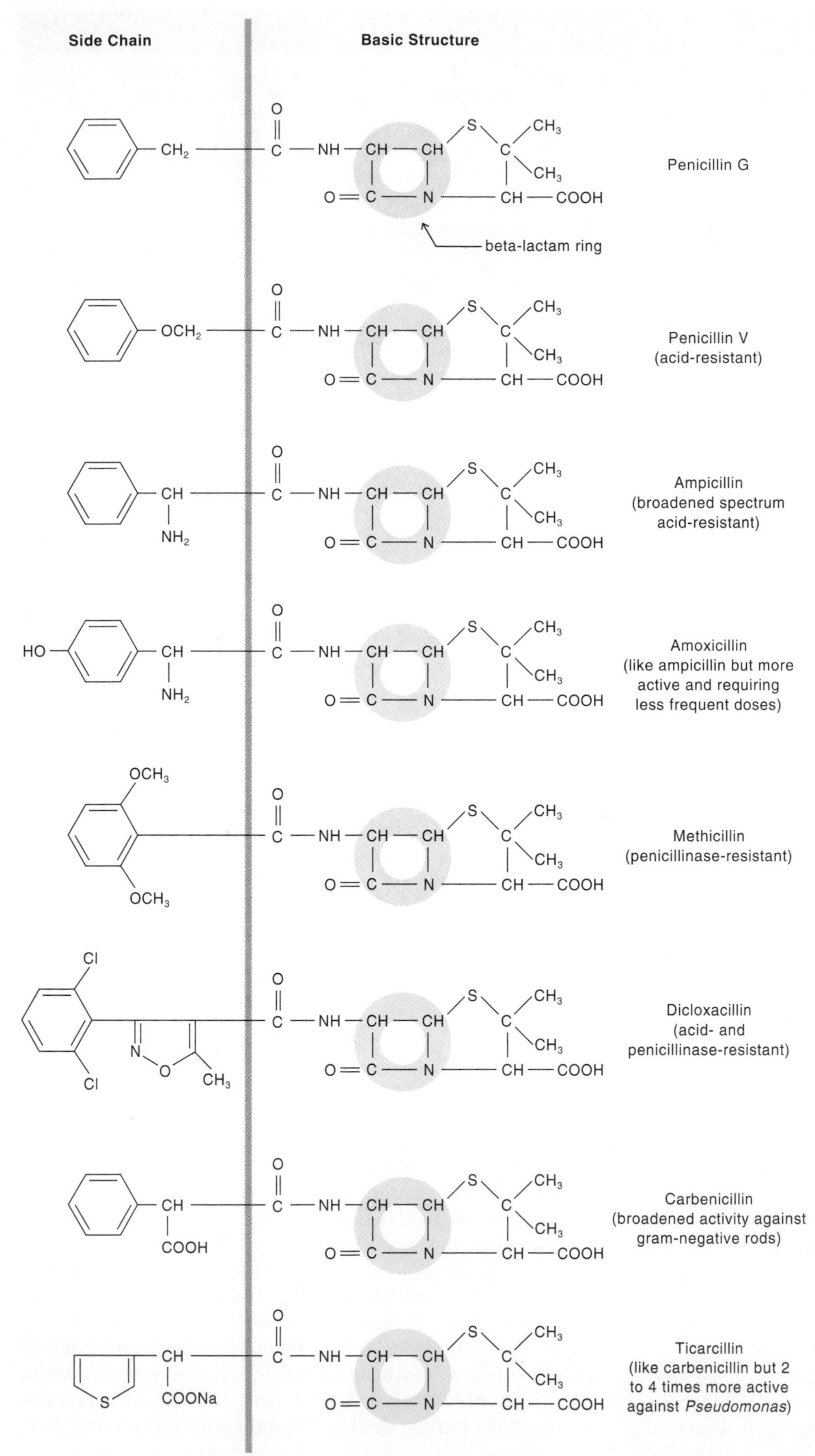

Figure 21.8 
Some members of the penicillin family. The chemical groups to the left of the vertical line are responsible for changes in properties of the antibiotic. The β-lactam rings are marked by the shaded circles.

Table 21.5 Cephalosporin of Choice for Different Infections

Cephalosporin	Infection
Cephalexin	Oral treatment of staphylococcal and some *Escherichia coli* and *Proteus mirabilis* infections
Cephalothin	Intravenous treatment of staphyloccocal infections
Cefuroxime	Intravenous treatment of most pneumonias
Ceftriaxone	Intravenous or intramuscular treatment of childhood meningitis, gonorrhea

organisms, apparently because they do not bind as well to their penicillin-binding proteins. The spectrum of activity of imipenem is broader than that of aztreonam (table 21.6)

Table 21.6 Comparisons of Antibacterial Spectrum of Aztreonam and Imipenem

Species	Aztreonam	Imipenem
Staphylococcus aureus	R*	S**
Streptococcus pneumoniae	R	S
Streptococcus pyogenes	R	S
Streptococcus faecalis	R	S
Haemophilus influenzae	S	S
Neisseria gonorrhoeae	S	S
Escherichia coli	S	S
Enterobacter spp.	S	S
Pseudomonas aeruginosa	S (≥80%)	S (≥80%)
Bacteroides fragilis	R	S
Clostridium perfringens	R	S

*R = resistant
**S = sensitive

Glycopeptides

Vancomycin and teicoplanin are **glycopeptides,** short chains of amino acids attached to a sugar molecule. Both of these antibiotics interfere with peptidoglycan synthesis, resulting in bactericidal activity against gram-positive organisms. They are poorly absorbed when given orally and must thus be administered by injection except when used to treat intestinal infections. They have been considered a last resort for treating infections caused by bacteria that are resistant to other antibiotics. As mentioned previously, vancomycin-resistant strains of the opportunistic pathogen *Enterococcus* are becoming increasingly common, and a vancomycin-resistant strain of *Staphylococcus aureus* was recently isolated in Japan.

Tetracyclines

The tetracyclines are a group of bacteriostatic antibiotics that interfere with protein synthesis in prokaryotes by binding to the 30S ribosomal subunit, thus blocking the attachment of tRNA. Unlike the penicillins and cephalosporins, which are of fungal and therefore eukaryotic origin, the tetracyclines are produced by prokaryotes of the genus *Streptomyces*. The newer tetracyclines, including doxycycline, demeclocycline, and minocycline, are derivatives that vary primarily in their achievable blood levels and duration of activity.

The tetracyclines are actively transported into bacterial cells but not mammalian cells, which effectively concentrates the drug in bacteria. The drugs inhibit both gram-positive and gram-negative bacteria as well as *Mycoplasma* and intracellular *Rickettsia* and *Chlamydia*. Because of their broad spectrum of activity, the tetracyclines are widely used to treat infections in which the causative agents have not been determined. They are also used to prevent secondary bacterial infection in people suffering from colds, influenza, measles, or surgical wounds.

In reality, their use sometimes actually *increases* the chance of secondary infection probably because they alter the normal flora. Gastrointestinal disturbances including diarrhea and nausea as well as vaginal yeast infections are common side effects. Tetracyclines should be avoided by young children because they bind strongly to developing teeth, causing a permanent yellowish discoloration. Calcium, magnesium, iron, and aluminum interfere with the absorption of tetracyclines. Therefore, these drugs should not be taken with milk, antacids, or iron-containing medications.

Resistance to tetracycline is becoming increasingly common. It often involves an active transport mechanism that ejects the drug from the cell. In some cases, resistance is due to the synthesis of a protein that adheres to the tetracycline and blocks the drug from binding to the ribosome.

Chloramphenicol

Like the tetracylines, chloramphenicol is a broad spectrum antibiotic originally isolated from *Streptomyces*. It interferes with the action of the 50S ribosomal subunit. It readily diffuses into cells and spinal fluid to act on bacterial pathogens beyond the reach of other antibiotics. However, it is toxic at high doses, mainly because it interferes with the normal development of erythrocytes (red blood cells). Much more serious, however, is a reaction that occurs in about 1 of every 50,000 patients who receive chloramphenicol. This complication, *aplastic anemia*, is not related to dose and is characterized by the inability of the body to form white and red blood cells. Because aplastic anemia is often fatal, the use of chloramphenicol is generally reserved for life-threatening infections for which no effective alternative treatment is available. Nevertheless, chloramphenicol is used extensively in some countries because of its low price.

Resistance to chloramphenicol is due to acquisition of a plasmid that encodes the enzyme that chemically inactivates the drug.

Aminoglycosides

The aminoglycosides are a group of bactericidal antibiotics that share a similar chemical structure. They all bind to and disrupt the function of the prokaryotic ribosome. Like the tetracyclines, they are actively transported into prokaryotic cells but not eukaryotes, which partially accounts for their selective toxicity. They are active against a variety of aerobic and facultative gram-positive and gram-negative bacteria.

Streptomycin, produced by *Streptomyces griseus*, was one of the first antibiotics and the earliest aminoglycoside discovered. It binds to a ribosomal protein of the 30S ribosomal subunit, causing distortions in the ribosome that prevent proper protein synthesis. Unfortunately, organisms rapidly develop resistance to the drug due to spontaneous mutation in the gene encoding the ribosomal protein. Streptomycin is still used to treat tuberculosis and other mycobacterial infections, but it is used in combination with other antimicrobial drugs since an organism is less likely to simultaneously develop mutational resistance to two different targets.

The newer aminoglycosides gentamycin, tobramycin, netilmicin, and amikacin can each bind to several different sites on the bacterial ribosome, which lessens the likelihood of the development of mutational resistance. They are also less susceptible than streptomycin to enzymes that chemically inactivate drugs. Amikacin is the least susceptible to enzymatic inactivation. Unfortunately, all aminoglycosides can cause serious side effects, including kidney damage and loss of hearing. The severe toxic effects occur mainly when prolonged high concentrations are present in the bloodstream as a result of kidney diseases that interfere with normal excretion of the drug in the urine.

Another aminoglycoside, neomycin, is very toxic when injected, but because it is poorly absorbed from the gastrointestinal tract, it can be taken orally to reduce the numbers of normal intestinal flora. It is often prescribed by surgeons preparing to operate on a patient's intestine. Following oral neomycin treatment, fewer bacteria are left to cause infection of the surgical wound. Neomycin is also used topically to treat superficial infections.

Macrolides

Erythromycin, produced by *Streptomyces erythreus*, is the most commonly prescribed of a group of chemically similar antibiotics called **macrolides.** Erythromycin interferes with prokaryotic protein synthesis by binding to the 50S ribosomal subunit. It is active against gram-positive organisms as well as the most common causes of atypical pneumonia, *Mycoplasma* and *Chlamydia*. It is often the drug of choice for patients who are allergic to penicillin. Serious side effects are rare, although it can cause gastric distress and sometimes reversible liver damage. Resistance can result from mutations in chromosomal genes or plasmids transferred from resistant bacteria.

Two new macrolides have recently been introduced, azithromycin and clarithromycin. They have several advantages over erythromycin but cost more. They both are well absorbed and last longer in the serum so they only need to be taken once or twice a day. In addition, they are active against some of the opportunistic pathogens that commonly infect AIDS patients.

Lincosamides

Lincomycin, originally derived from a species of *Streptomyces* isolated from soil near Lincoln, Nebraska, is chemically distinct from other antibiotics, but it has an activity and antimicrobial spectrum similar to that of erythromycin. Clindamycin, a more biologically active derivative of lincomycin, is the form in general use. It is generally bacteriostatic against many gram-negative and gram-positive rods. Unfortunately, while all people on antibiotics are at risk for developing pseudomembranous colitis caused by the intestinal overgrowth of *Clostridium difficile*, the risk is greater for those on clindamycin because the organism is generally resistant to the drug.

Polypeptide Antibiotics

Polypeptide antibiotics, so called because they contain many peptide bonds, include bacitracin and polymyxin, both of which are produced by members of the genus *Bacillus*. Bacitracin interferes with peptidoglycan synthesis, resulting in lysis of gram-positive bacteria. In contrast, the polymyxins are active against gram-negative organisms and act by binding to and altering the permeability of their cytoplasmic membrane. The resulting loss of cytoplasmic contents is bactericidal. Both bacitracin and polymyxin are poorly absorbed from the intestine so they must be administered by injection. Unfortunately, they have a low therapeutic index so their use internally is reserved for the treatment of rare infections caused by bacteria resistant to all other antimicrobial drugs. However, they are frequently used as a topical treatment for superficial infections. Commercial preparations containing a combination of bactitracin, polymyxin, and neomycin are readily available without a prescription.

Rifamycins

The most widely used rifamycin, rifampin, is a derivative of a compound produced by *Streptomyces*. Rifampin blocks prokaryotic RNA polymerase from initiating transcription and is active against many gram-positive and gram-negative organisms as well as members of the genus *Mycobacterium*. It is used primarily to treat tuberculosis and leprosy and to prevent meningitis in specific situations. Generally, little toxicity has been noted, but liver damage can occur on rare occasions. In some patients, a reddish-orange pigment appears in urine and tears, and soft contact lenses may be stained by it. Rifampin's main drawback is that highly resistant mutants commonly arise. To avoid the outgrowth of resistant mutants, rifampin is often used in combination with other drugs such as streptomycin.

Sulfa Drugs

Sulfa drugs are concentrated in the urine so they are widely used to treat bladder infections, most of which are caused by *E. coli*. The chemically synthesized drugs are structurally similar to PABA, a substrate in the pathway for folic acid biosynthesis (figure 21.9). As discussed in chapter 5, they inhibit the growth of bacteria by competitively inhibiting the enzyme for which PABA is a substrate. The bacteriostatic action is effective against both gram-positive and gram-negative bacteria, although resistance is common.

• **Competitive inhibition, p. 111**

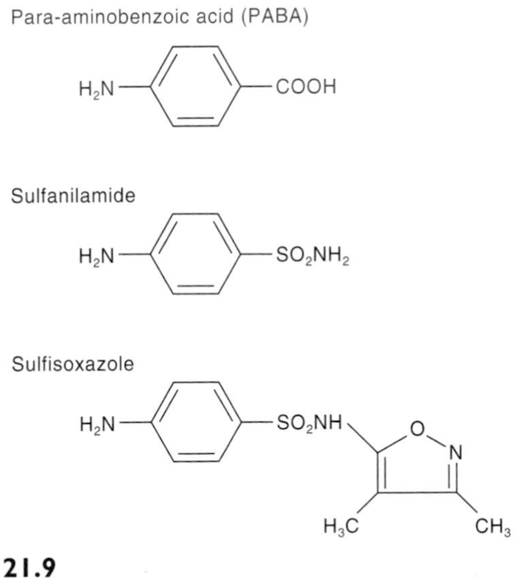

Para-aminobenzoic acid (PABA)

H_2N—(benzene ring)—COOH

Sulfanilamide

H_2N—(benzene ring)—SO_2NH_2

Sulfisoxazole

H_2N—(benzene ring)—SO_2NH—(isoxazole ring: O, N, H_3C, CH_3)

Figure 21.9
Structural formulas of para-aminobenzoic acid (PABA) and two sulfa drugs, sulfanilamide and sulfisoxazole. The different chemical structure of sulfisoxazole makes it much more soluble than sulfanilamide and, therefore, less likely to precipitate in the kidneys and damage them.

Trimethoprim

Trimethoprim blocks folic acid synthesis but at a different step than sulfa drugs (figure 21.10). When used in combination with sulfa drugs, the two are markedly more effective than either is alone. The combination of medications has a wide spectrum of activity against gram-positive and gram-negative bacteria as well as *Pneumocystis carinii*, an opportunistic fungus that infects AIDS patients.

Quinolones

The quinolones are bactericidal drugs that interfere with nucleic acid synthesis by inhibiting gyrase, an enzyme involved in the breaking and ligation of DNA strands during replication (figure 21.11). The first quinolone developed, naladixic acid, is active primarily against gram-negative rods. Unfortunately, drug resistance due to a single point mutation in the gyrase gene develops rapidly, which limits the use of the drug. Newer derivatives of the drug, such as norfloxacin, ciprofloxacin, ofloxacin, enoxacin, and lomefloxacin, appear to bind to several sites on the gyrase enzyme. The presence of several alternative binding sites decreases the likelihood of the development of resistance due to spontaneous mutations. The newer quinolones also have a wider spectrum of activity that includes both gram-

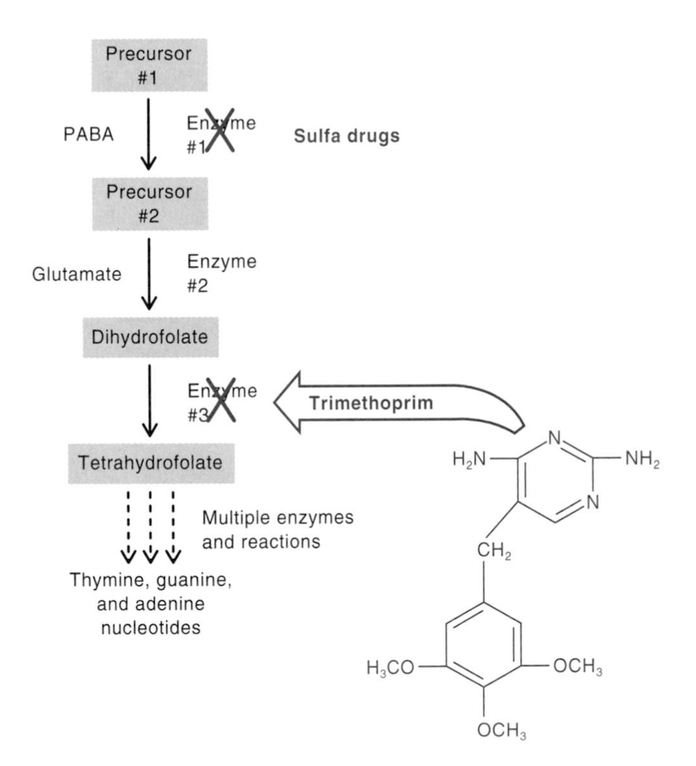

Figure 21.10
Timethoprim. When this drug is given with a sulfa drug, the effectiveness of both is enhanced because they act at two different steps in folic acid metabolism.

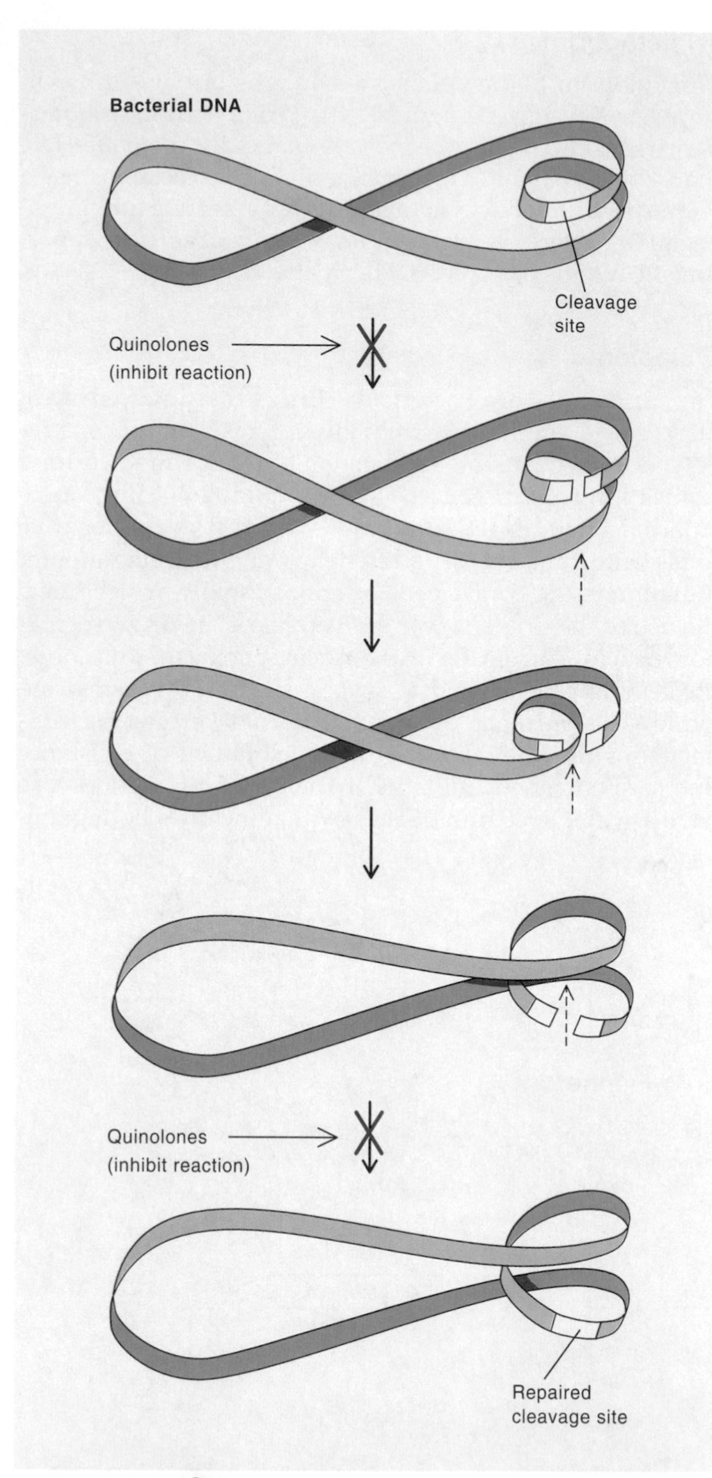

Figure 21.11
The enzyme DNA gyrase allows for the breaking and rejoining of the DNA strand as it coils or uncoils. Quinolone antimicrobials inhibit this enzyme.

negative and gram-positive organisms. Ciprofloxacin, which has the widest spectrum of activity, is effective against *Pseudomonas aeruginosa*, an organism that is resistant to many other antimicrobials.

Antifungal Drugs

Relatively few antifungal drugs are available because these eukaryotic pathogens have many of the same structures and metabolic pathways as human cells, so there are few targets for selective toxicity. Antifungal drugs range from the relatively nontoxic drugs used to treat superficial skin infections to the highly toxic drugs reserved for treating life-threatening systemic fungal infections (table 21.7).

Griseofulvin

Griseofulvin, produced by a species of *Penicillium*, is used to treat infections caused by dermatophytes, the fungi that invade keratin-containing structures such as skin and nails. Griseofulvin is taken orally because it can then be absorbed from the intestinal tract and concentrated in the dead keratinized layers of the skin. There, griseofulvin is taken up by the invading fungi and inhibits their growth by interfering with cell division. Treatment with griseofulvin takes many months.

Azoles

The azoles are a large family of chemically synthesized drugs, some of which are antifungal in their activity. The antifungal azoles, including clotrimazole, ketoconazole, miconazole, itraconazole, and fluconazole, act by interfering with the synthesis of ergosterol, an essential component of the fungal cytoplasmic membrane. Ergosterol is not found in mammalian or prokaryotic cells. The azoles result in defective fungal membranes that leak cytoplasmic contents. The azoles are active against a wide variety of fungi.

Polyenes

Amphotericin B is the most important of the polyene antibiotics, all of which are produced by *Streptomyces*. The polyenes bind to ergosterol in the cytoplasmic membrane of fungal cells. The binding disrupts the membrane, causing leakage of cytoplasm, which eventually results in cell death. Amphotericin B is quite toxic to humans but, nevertheless, it is the most effective drug for use against life-threatening systemic fungal infections. It is not absorbed from the gastrointestinal tract and must be given by intravenous infusion. Another polyene called *nystatin* is even more toxic, which limits its use to topical application.

Flucytosine

Flucytosine is a synthetic derivative of cytosine, a natural pyrimidine found in nucleic acids. Enzymes within fungal cells convert flucytosine to 5-fluorouracil, which inhibits an enzyme required for nucleic acid synthesis. Flucytosine is

Table 21.7 Antifungal Medicines for Deep Infections

Drug (How Given)	Origin	Principal Genera Used Against	Side Effects
Amphotericin B	Antibiotic from *Streptomyces*	*Aspergillus, Blastomyces, Candida, Coccidioides, Cryptococcus, Histoplasma, Mucor, Paracoccidioides*	Commonly, fever, nausea, vomiting, phlebitis,* kidney damage, and anemia**
Flucytosine (oral)	Synthetic	*Candida, Phialophora, Cladosporium, Cryptococcus*	Occasionally, diarrhea, skin rash, liver injury, decreased leukocytes
Ketoconazole (oral)	Synthetic	*Coccidioides, Candida, Paracoccidioides, Blastomyces, Cryptococcus, Histoplasma*	Gastrointestinal symptoms and itching
Potassium iodide (oral)	Synthetic	*Sporothrix*	Frequently, swollen saliva glands and acnelike rash
Hydroxystilbamidine	Synthetic	*Blastomyces*	Commonly, loss of appetite and nausea; occasionally, liver injury, skin rash, numbness, and tingling

*Phlebitis—inflamed veins, in this case the veins into which the medicine is injected
**Anemia—insufficient number of normal erythrocytes

Table 21.8 Antiviral Therapy

Virus	Disease	Drug
Cytomegalovirus	Eye, nervous system, generalized disease, in infants or those with immunodeficiency	Ganciclovir; foscarnet for ganciclovir-resistant strains
Hepatitis B or C (HBV or HCV)	Chronic hepatitis	α-interferon
Herpes simplex (HSV)	Encephalitis, genital herpes	Acyclovir, foscarnet, or vidarabine
Human immunodeficiency (HIV)	HIV disease, AIDS	Zidovudine, dideoxyinosine, protease inhibitors
Influenza A	Influenza (prevention)	Amantadine
Respiratory syncytial (RSV)	Pneumonia	Ribavarin

less toxic than amphotericin. Unfortunately, resistant mutants are common and, therefore, flucytosine is used mostly in combination with amphotericin or as an alternative drug for patients who are unable to tolerate amphotericin.

Antiviral Drugs

Antiviral drugs can be used to prevent or shorten the duration of a viral disease. Thus far, however, no drug has been developed that can cure a viral infection (table 21.8).

Acyclovir

Acyclovir is used to limit herpesvirus infections. The drug is taken in its inactive form, but it is modified in infected cells to its active form by a virally encoded enzyme. The active form of acyclovir then inhibits the viral DNA polymerase. Because the activating enzyme is found only in virally infected cells, acyclovir has little toxicity against uninfected host cells. Herpesviruses, however, can remain latent in cells, so active infection and symptoms can recur.

Amantadine and Rimantadine

Two similar drugs, amantadine and rimantadine, block the uncoating of influenza A virus after it enters a cell and, thus, can prevent or reduce the severity and duration of the disease. Use of amantadine or rimantadine prophylaxis is particularly useful in preventing or curtailing influenza outbreaks in high-risk situations such as nursing homes. Generally, the drugs are only used until active immunization can be achieved, since immunization is important for long-term protection. In addition, amantadine resistance develops frequently and may eventually limit the usefulness of the drug.

Nucleotide Analogs

A growing number of antiviral drugs are nucleotides or **base analogs,** which are chemicals structurally similar to the nucleotides of DNA and RNA. The virally encoded polymerases are more prone to incorporating these analogs than are host polymerase. They mistakenly incorporate the base analogs during the synthesis of viral nucleic acid. As a result, viral nucleic acid production is either halted or the resulting end product is nonfunctional. An important base analog is zidovudine, also called *azidothymidine* (*AZT*), which was the first drug useful in treating AIDS patients. The HIV-encoded enzyme reverse transcriptase occasionally incorporates AZT rather than the normal deoxynucleotide. When this occurs, viral DNA elongation is terminated. Anti-HIV drugs that act in an analogous manner include dideoxyinosine (ddI) and dideoxycytidine (ddC).

Drugs Active Against Protozoan Parasites

Table 21.9 presents some of the drugs used in treating diseases caused by protozoa. Most of these have a low therapeutic index, and their mode of action is not completely known.

• **AZT, p. 157**

REVIEW QUESTIONS

1. What is the difference between innate and acquired resistance to antibiotics?
2. List three different mechanisms that microorganisms may use to resist the effects of an antimicrobial drug.
3. Explain the significance of R plasmids.
4. What is the target of β-lactam drugs?
5. Explain the significance of vancomycin-resistant enterococci.

Table 21.9 Some Medicines Used in Treating Infections Caused by Protozoa

Drug	Disease	Causative Genus of Protozoa	Side Effects
Amphotericin	Amebic meningitis, American leishmaniasis	*Naegleria, Leishmania*	Frequently, kidney damage, fever, nausea, phlebitis
Chloroquine	Amebic abscess, malaria	*Entamoeba, Plasmodium*	Occasionally, itching, vomiting, weight loss, vision problems
Dehydroemetine and emetine	Amebic abscess, amebic dysentery	*Entamoeba*	Frequently, cardiac irritability, chest pain, muscle weakness
Iodoquinol	Amebic dysentery	*Entamoeba*	Occasionally, rash, nausea, diarrhea; rarely, loss of vision in children
Melarsoprol	African sleeping sickness	*Trypanosoma*	Frequently, heart, kidney, nerve, and brain damage; abdominal pain; vomiting
Metronidazole	Vaginitis, giardiasis, amebic dysentery, amebic abscess	*Trichomonas, Giardia*	Frequently, nausea, headaches, metallic taste; occasionally, vomiting, diarrhea, dizziness
Primaquine	Malaria (liver stage)	*Plasmodium*	Occasionally, blood disorders, gastrointestinal upset
Quinine	Malaria (chloroquine-resistant forms)	*Plasmodium*	Frequently, headache, ringing ears, vision abnormalities

Summary

I. Overview of Antibiotics
   A. The suitability of a specific antimicrobial depends on the clinical situation.
   B. Antimicrobials differ in their toxicity and spectrum of activity.
   C. Antimicrobials differ in how they are distributed, metabolized, and excreted.
   D. Combinations of antimicrobial drugs can be antagonistic, synergistic, or additive.

II. Targets of Antimicrobial Drugs
   A. Differences between prokaryotic and eukaryotic cells are useful targets of antibacterial drugs.
      1. Cell wall synthesis
      2. Protein synthesis
      3. Nucleotide replication/transcription
      4. Cell membrane function
      5. Metabolic pathways
   B. Eukaryotic pathogens have many similarities to human cells so they offer fewer targets for selective toxicity.
   C. Viruses have few targets for selective toxicity.

III. Concerns in Using Antimicrobial Medications
   A. Some antimicrobials may elicit allergic or other adverse reactions.
   B. Antimicrobials suppress the normal flora.

IV. Bacterial Sensitivity Testing
   A. Susceptibility can be quantified by determining minimal inhibitory concentrations.
   B. The terms *sensitive* and *resistant* refer to whether a microorganism is likely to be treatable.

V. Development of Resistance
   A. Some organisms are innately resistant to certain antibiotics.
   B. Sensitive bacteria can acquire antibiotic resistance.
   C. Acquisition of antimicrobial resistance may occur through spontaneous mutation.
   D. Acquisition of antimicrobial resistance may occur through DNA transfer.
   E. Control of antibiotic resistance relies on the cooperation of patients and physicians.

VI. Survey of Antimicrobial Drugs
   A. Antibacterial drugs
      1. Pencillins
      2. Cephalosporins
      3. Other β-lactam antibiotics (carbapems and monobactams)
      4. Glycopeptides
      5. Tetracyclines
      6. Chloramphenicol
      7. Aminoglycosides
      8. Macrolides
      9. Lincosamides
      10. Polypeptide antibiotics
      11. Rifamycins
      12. Sulfa drugs
      13. Trimethoprim
      14. Quinolones
   B. Antifungal drugs
      1. Griseofulvin
      2. Azoles
      3. Polyenes
      4. Flucytosine
   C. Antiviral drugs
      1. Acyclovir
      2. Amantadine and rimantadine
      3. Nucleotide analogs
   D. Drugs active against protozoan parasites

Chapter Review Questions

1. Which of the following targets would you expect to be the MOST selectively toxic?
   A. cytoplasmic membrane function
   B. DNA synthesis
   C. glycolysis
   D. peptidoglycan synthesis
   E. 70S ribosome
2. Are antibiotics effective against viruses as well as bacteria?
3. Penicillin has been modified to make derivatives that differ in all of the following EXCEPT?
   A. spectrum of activity
   B. resistance to stomach acid
   C. resistance to penicillinases (β lactamases)
   D. potential for allergic reactions
4. Which of the following is the target of β-lactam antibiotics?
   A. peptidoglycan synthesis
   B. DNA synthesis
   C. RNA synthesis
   D. protein synthesis
   E. glycolysis
5. What is an antibiotic that affects a wide range of microorganisms is called?
6. Why does a drug with a long half-life need to be taken less frequently than a drug with a short half-life?
7. What is the significance of R plasmids with respect to antibiotic resistance?
8. Why are bacteria less likely to develop mutational resistance to antibiotics that bind several sites of a target than to those that bind to only one site?
9. Why are sulfa drugs often used in combination with trimethoprim?
10. Why are there relatively few antiviral drugs available?

Critical Thinking Questions

1. Would you expect a bacteriostatic antibiotic given along with penicillin to have an increased or decreased effect or no effect on its bactericidal action? Explain.
2. What is the basis for selective toxicity, and what is the meaning of the therapeutic ratio?
3. How are bacteria tested for susceptibility to different antibacterial medicines? How can laboratory tests predict effectiveness *in vivo?*
4. How have scientists dealt with resistance to antibacterial medicines?
5. What can be done about the problem of increasing microbial resistance to antimicrobial medicines?

Further Reading

Davies, Julian. 1996. Bacteria on the rampage. *Nature* 383:219–220.

Davies, Julian. 1986. Life among the aminoglycosides. *ASM News* 52:620.

Dixon, Bernard. 1992. Covert activities. *Biotechnology* 10:343. Speculations about secondary metabolites.

Donowitz, G. R., and Mandell, G. L. 1988. Beta-lactam antibiotics. *The New England Journal of Medicine* 318:419–423 and 490–496.

Elsbach, Peter. 1990. Antibiotics from within: antibacterials from human and animal sources. *Trends in Biotechnology* 8:26.

Franklin, T. J., and Snow, G. A. 1989. *Biochemistry of antimicrobial action*, 4th ed. Chapman and Hall.

Hare, R. 1970. *The birth of penicillin and the disarming of microbes.* George Allen and Unwin, London. Fascinating account of the discovery of penicillin.

Hirsch, M. S., and Kaplan, J. C. 1987. Antiviral therapy. *Scientific American* 256:76.

Hooper, D. C., and Wolfson, J. S. 1991. Fluoroquinolone antimicrobial agents. *The New England Journal of Medicine* 324:384–394.

Jacoby, G. A., and Archer, G. L. 1991. New mechanisms of bacterial resistance to antimicrobial agents. *The New England Journal of Medicine* 324:601–612.

Levy, S. B. 1990. Starting life resistance free. *The New England Journal of Medicine* 323:335–337.

Mandell, G. L., Douglas, R. G., Bennett, J. E., and Dolin, R. 1995. *Principles and practice of infectious diseases antimicrobial therapy 1995/1996*. Churchill Livingstone Inc.

Neu, H. C. 1992. The crisis in antibiotic resistance. *Science* 257:1064.

Wright, Karen. 1990. Bad news bacteria. *Science* 249:22–24. Discussion of evolving patterns of virulence and resistance to antimicrobials.

1994. Frontiers in biotechnology: resistance to antibiotics. *Science* 264:360–388. Several excellent articles about the increasing problem of antibiotic resistance.

22 Skin Infections

KEY CONCEPTS

1. The skin is a large, complex organ covering the external body surfaces.
2. Properties of the skin cause it to resist microbial colonization, provide a physical barrier to the entry of potential pathogens, and assist the body's regulation of temperature and fluid balance.
3. The changes in the skin in an infectious disease commonly reflect disease involvement of other parts of the body.

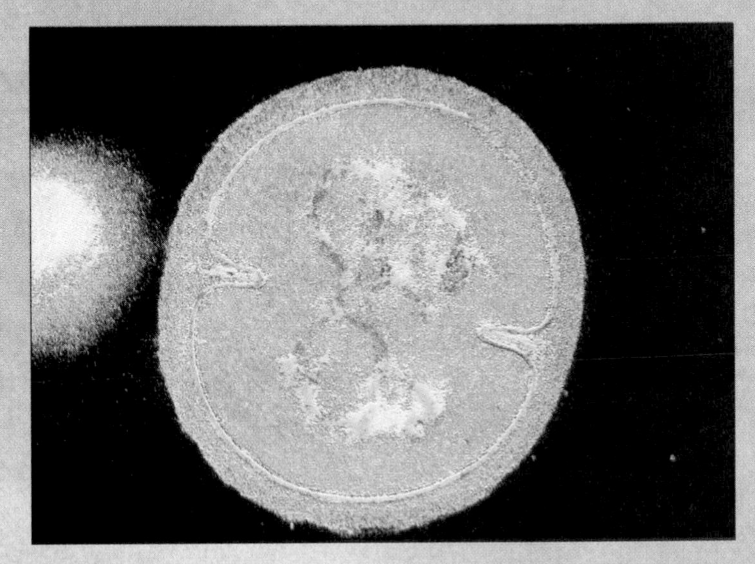

Color-enhanced transmission election micrograph of a *Staphylococcus epidermidis* cell undergoing division. This bacterium is almost universally present among the normal flora of the human skin. However, it can be a life-threatening pathogen because of its ability to colonize plastic objects and its resistance to many antibacterial medications.

Microbes in Motion

The following books in the *Microbes in Motion* CD-ROM may serve as a useful preview or supplement to your reading: *Gram-Positive Organisms. Gram-Negative Organisms. Miscellaneous Bacteria Microbial Pathogenesis in Infectious Diseases:* Toxins. *Vaccines:* Vaccine Targets. *Viral Structure and Function:* Viral Diseases. *Fungal Structure and Function:* Diseases. *Parasitic Structure and Function.*

A Glimpse of History

Howard Ricketts (1871–1910), a pathologist, conducted studies on Rocky Mountain spotted fever. In 1906, he showed that the disease is caused by a tiny, obligate, intracellular coccobacillus. He was able to infect guinea pigs by injecting them with the blood of people suffering from the disease. Other experiments showed that the bite of certain kinds of ticks transmit the disease to the guinea pigs. Through microscopic examination he could detect the causative coccobacillus in the blood of these diseased animals. Further experiments showed that not all species of ticks were infected with this bacterium, later to be known as *Rickettsia rickettsii*, but that certain ticks of the genus *Dermacentor* were prime carriers. Ricketts also proved that ticks acquiring *R. rickettsii* by feeding on an infected person or animal transmit the infection from one generation of ticks to the next. This phenomenon occurs when the infected female tick transmits the organism to her young through her eggs.

The study of Rocky Mountain spotted fever was carried on by other scientists when Howard Ricketts turned his attention to typhus, another rickettsial disease. Unfortunately, during his experiments with typhus, he contracted the disease and died shortly afterwards, a young man in the prime of his career.

The skin is a large, complex organ that covers the external surface of the body. Its most important functions are the control of body temperature and prevention of loss of fluid from body tissues. It also plays an important role in the synthesis of vitamin D, which is needed for normal teeth and bones. Numerous sensory receptors of various types occur in the skin, providing the central nervous system with information about the environment. The skin also produces hormonelike substances that aid the development and function of cell-mediated immunity.

The major part of the body's contact with the outside world occurs at the surface of the skin. As long as skin is intact, this tough, flexible outer covering is remarkably resistant to infection. Because of its exposed state, however, it is frequently subject to cuts, punctures, burns, or chemical injury as well as hypersensitivity reactions. These skin injuries provide a way for pathogens to enter and infect the skin and underlying tissues. Skin infections also occur when microorganisms or viruses are carried to the skin by the bloodstream after entering the body from another site, such as the respiratory or gastrointestinal systems.

Anatomy and Physiology

The **epidermis,** the surface layer of the skin (figure 22.1 and figure 22.2), ranges from 0.007 to 0.12 mm thick. The outer portion is composed of scaly material made up of flat cells containing **keratin,** a durable protein also found in hair and nails. The cells on the skin surface are dead and, along with any resident organisms, continually peel off, replaced by cells from deeper in the epidermis. These cells, in turn, become flattened and die as keratin is formed within them. This process results in a complete regeneration of the skin about once a month.

The epidermis is supported by the **dermis,** a second, deeper layer of skin cells through which many tiny nerves, blood vessels, and lymphatic vessels penetrate. The dermis adheres in a very irregular fashion to the fat and other cells that make up the subcutaneous tissue.

Fine tubules of sweat glands and hair follicles traverse the dermis and epidermis (figure 22.1). The hair follicles commonly provide passageways through which microorganisms can penetrate the skin to reach deeper body tissues. One or more pilosebaceous glands that produce an oily secretion called **sebum** feed into the sides of the hair follicles. This secretion flows up through the follicles and spreads out over the skin surface, keeping the hair and skin soft, pliable, and water repellent. **Acne** is a disorder of the sebaceous glands in which sebum secretion is excessive, while at the same time the glands become plugged and inflamed. The possible role of bacteria in the causation of acne is controversial.

The secretions of the sweat and sebaceous glands are very important to the microbial population of the skin because they supply water, amino acids, and lipids, which serve as nutrients for microbial growth. The normal pH of skin ranges from 4.0 to 6.8. Breakdown of the lipids by the microbial residents of normal skin results in fatty acid byproducts that inhibit the growth of many potential disease-producers. In fact, the normal skin surface is a rather unfriendly terrain for most pathogens, being too dry, unstable, acidic, and toxic for their survival.

Factors that interfere with the normal functioning of the skin tend to neutralize its antimicrobial properties. For example, the allergic condition **eczema,**[1] which results in the oozing of plasma onto the surface of the skin, neutralizes the antimicrobial action of the skin's fatty acids.

Normal Flora of the Skin

Large numbers of microorganisms live on and in the various components of the normal human skin. For example, depending on the body location and amount of skin moisture, the number of bacteria on the skin surface may range from only about 1,000 organisms per square centimeter on the back to more than 10 million on the scalp and in the armpit. Most of these organisms can be categorized in three groups (table 22.1).

[1]Eczema—an itchy, chronic skin eruption with various causes; often IgE-mediated or a delayed hypersensitivity reaction

• **Rickettsias, pp. 230, 250**
• **Cell-mediated immunity, pp. 371, 373–74**
• **Normal flora, pp. 410, 412**

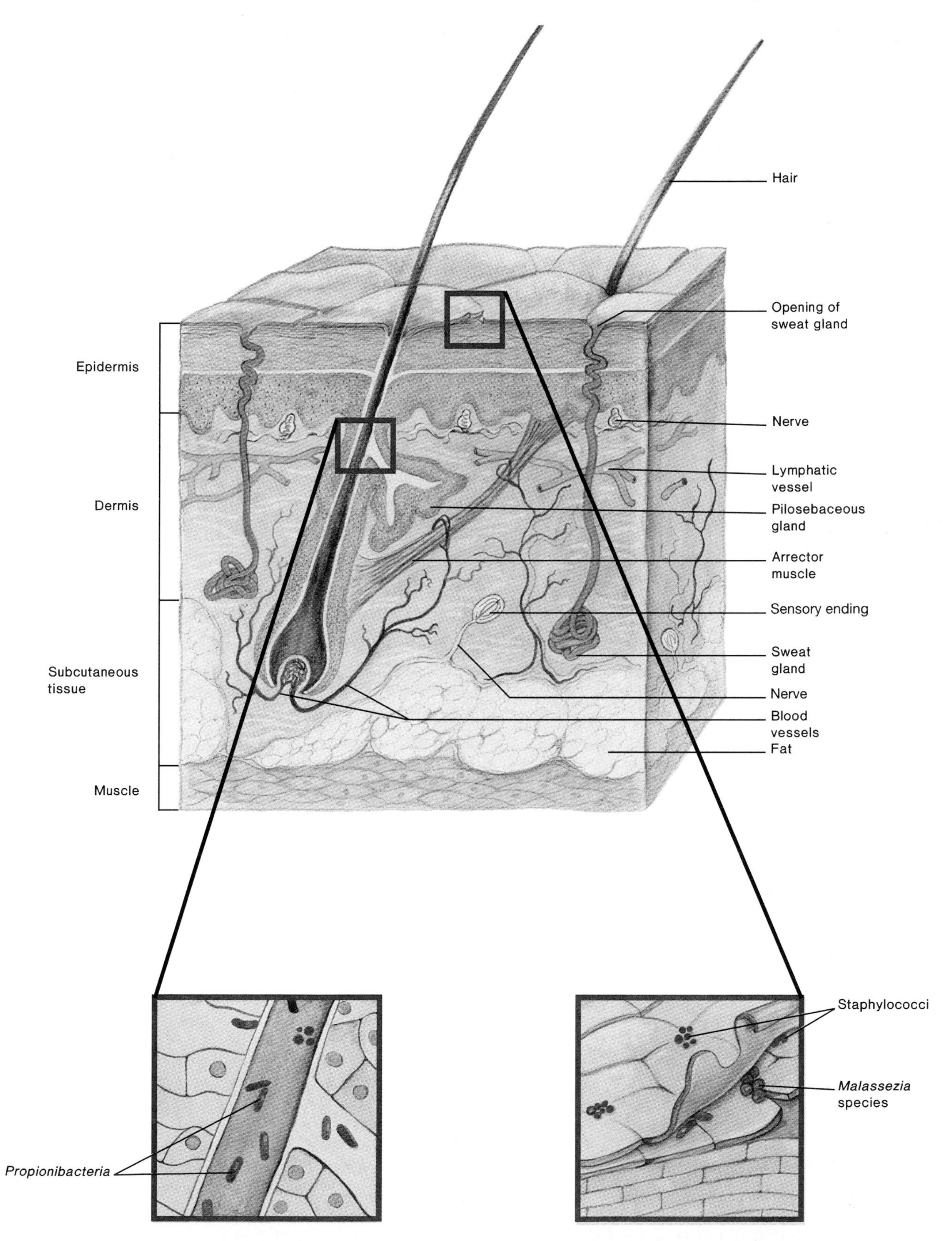

Figure 22.1
Microscopic anatomy of the skin. Notice that the pilosebaceous unit, composed of the hair follicle and attached sebaceous gland, almost reaches the subcutaneous tissue.

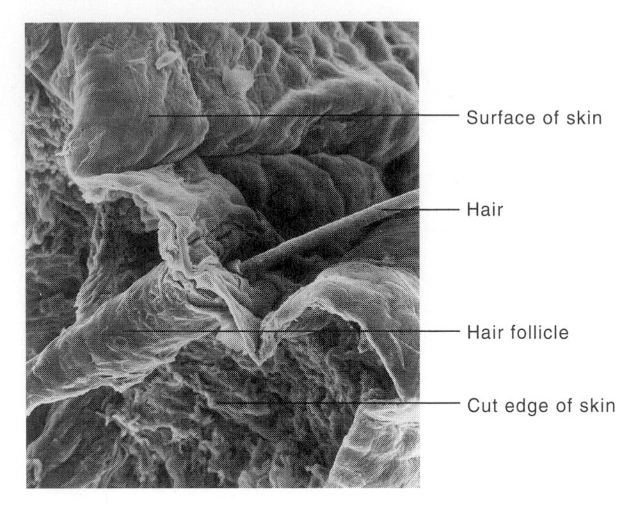

Figure 22.2
Scanning electron micrograph of the surface and cut edge of normal human skin. Notice that one of the surface cells near the upper edge of the photograph is peeling. Also note the pit at the entrance of the hair follicle, the site of entry for skin bacteria.

Normal Skin Flora Includes the Diphtheroids, a Group of Gram-Positive Nonspore-Forming Rods

The diphtheroids are a group of bacteria named for their resemblance to the diphtheria bacillus, *Corynebacterium diphtheriae*. Diphtheroids are gram-positive, pleomorphic ("many shaped") rods of low virulence. They are nonmotile, do not form spores, and are distinguished from *C. diphtheriae* by their different morphology, carbohydrate fermentation patterns, and lack of toxin production.

The diphtheroid species found on the skin in the largest numbers is *Propionibacterium acnes*, which is present on virtually all humans. Surprisingly, most strains of *P. acnes* are anaerobic, although some strains are aerotolerant. It grows primarily within the hair follicles where conditions are anaerobic. Growth of *P. acnes* is enhanced by the oily secretion of the sebaceous glands, and the organisms are usually present in large numbers only in areas of the skin where these glands are especially well-developed, on the face, upper chest, and back. These are also the areas of the skin where acne develops, and the frequent association of *P. acnes* with acne inspired its name, even though most people who carry the organisms do not have acne. It was once thought that fatty acids released from oily skin secretions by the action of *P. acnes* enzymes contributed to the severity of acne. However, it has since been shown that certain medicines that completely suppress fatty acid production do not cure acne. The reason that many patients with acne appear to be helped by antibiotic treatment is not understood. It certainly is not due to the elimination of *P. acnes* since this is rarely, if ever, accomplished by such treatment.

Table 22.1 Principal Members of the Normal Skin Flora

Name	Characteristics
Diphtheroids	Pleomorphic nonmotile, gram-positive rods of the *Corynebacterium* and *Propionibacterium* genera
Staphylococci	Gram-positive cocci arranged in packets or clusters; coagulase negative; facultatively anaerobic
Malassezia sp.	Small yeasts that require oily substances for growth

Diphtheroids are significant for several reasons. Since they are common in cultures of clinical material, they must be distinguished from more virulent species. Their production of fatty acids undoubtedly helps defend the skin from infection. However, they occasionally produce abscesses or other infections of the liver, bone, or heart, especially in people with AIDS or other immunodeficiencies.

Staphylococci Are Gram-Positive Cocci that Are Generally Arranged in Clusters

The second large group of microorganisms universally present on the normal skin is the gram-positive cocci, principally species of *Staphylococcus*. As with the diphtheroids, these bacteria have little virulence although they certainly can cause serious disease if host defenses are impaired. Generally, staphylococci are the most common of the skin bacteria able to grow aerobically. The principal species is *Staphylococcus epidermidis*.

The chief importance of the skin's staphylococci is probably in preventing colonization by pathogens and in maintaining the balance among the skin's various flora. These gram-positive cocci have been shown to produce antimicrobial substances highly active against *P. acnes* and other gram-positive bacteria, which may be a factor in maintaining a balance among the various species of microorganisms on the skin.

Species of Single-Celled Fungi (Yeasts) Are Generally Represented Among the Skin Flora

Tiny yeasts almost universally inhabit the normal human skin from late childhood onward. Their shape varies with different strains, being round, oval, or sometimes short, fat rods. The yeasts can be cultivated on laboratory media containing fats such as olive oil, and they belong to the genus *Malassezia* (formerly *Pityrosporum*). They are generally harmless, although in some people they cause skin conditions such as a scaly face rash, dandruff, and **tinea versicolor**

• **Aerotolerance, p. 92**

• **Staphylococci, p. 259**

(figure 22.3). The latter is a common skin disease that causes a patchy scaliness and increase in pigment in light-skinned persons or a decrease in pigment in dark-skinned people. Scrapings of the affected skin show large numbers of *Malassezia furfur* both in its yeast form and as short hyphae. Antifungal medications that eliminate the organism cure the rash. Unknown factors, probably relating to the host, are important in these diseases since most people carry *Malassezia* sp. on their skin without any disease. AIDS patients often have a severe rash with pus-filled pimples caused by *Malassezia* yeasts, and the organisms may even infect internal organs in patients receiving fat-containing intravenous feedings.

REVIEW QUESTIONS

1. Give three routes by which microorganisms invade the skin.
2. Give four characteristics of skin that help it resist infection.
3. Describe the three kinds of microorganisms usually encountered on the normal skin.

• **Yeasts, p. 284**

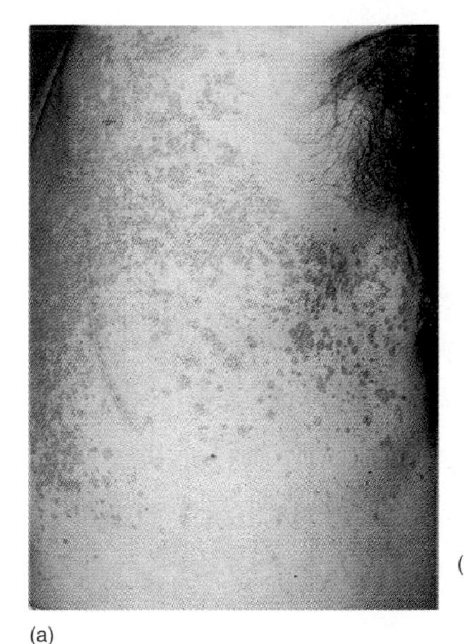

(a)

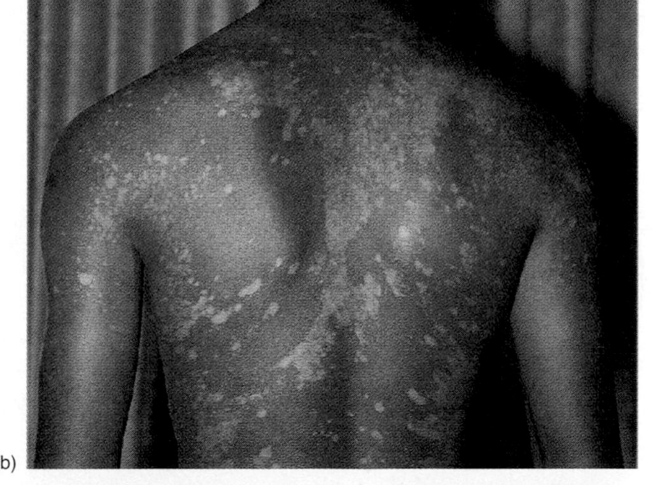

(b)

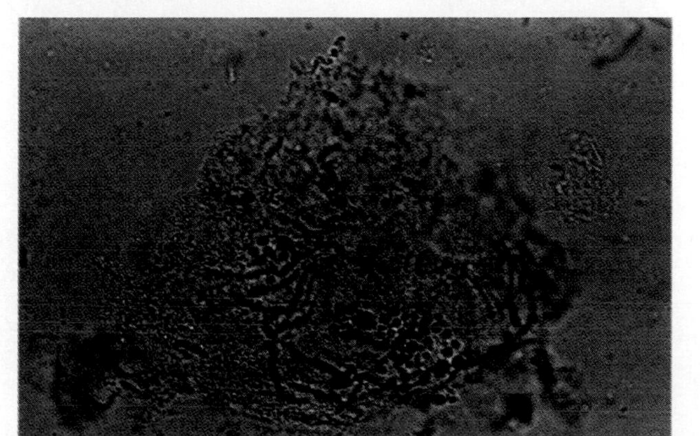

(c)

Figure 22.3
Tinea versicolor. (*a*) Fair-skinned individual and (*b*) dark-skinned individual. (*c*) Microscopic examination of skin scraping showing *Malassezia furfur* spores and hyphae.

Skin Diseases Caused by Bacteria

Only a few species of bacteria commonly invade the intact skin directly, which is not surprising in view of the anatomical and physiological features discussed. Hair follicle infections exemplify direct invasion. However, many bacterial pathogens invade the skin through minor injuries such as scrapes, scratches, and insect or tick bites. Others invade the skin from the bloodstream.

Infections of the Hair Follicles Cause Folliculitis, Furuncles, and Carbuncles

Infections originating in hair follicles commonly clear up without treatment. In other instances, however, they progress into severe or even life-threatening disease.

Symptoms

Folliculitis, furuncles, and carbuncles represent different outcomes of hair follicle infections. In **folliculitis**, a small red bump develops at the site of the involved hair follicle. Often, the hair can be pulled from its follicle, sometimes accompanied by a small amount of pus, and then the infection goes away without further treatment. However, if the infection extends from the follicle to adjacent tissues, causing localized redness, swelling, severe tenderness, and pain, the lesion is called a **furuncle** (boil). Pus may drain from the boil along with a plug of inflammatory cells and dead tissue. A **carbuncle** is a large area of redness, swelling, and pain punctuated by several sites of draining pus. Carbuncles develop in areas of the body where the skin is thick. Fever is often present, along with other signs of a serious infection.

Causative Agent

Most furuncles and carbuncles, as well as many cases of folliculitis, are caused by *Staphylococcus aureus*,[1] a facultatively anaerobic, gram-positive coccus (perspective 22.1).

[1] *Staphylococcus aureus*—*staphyle* means "bunch of grapes"; *aureus* means "golden"

Perspective 22.1

Laboratory Identification of *Staphylococcus aureus*

Staphylococcus grows readily on usual laboratory media such as blood agar; well-developed cream-colored colonies are present after 18 hours of incubation. Most strains of *S. aureus* produce a small zone of clearing (β-hemolysis) around their colonies on blood agar, caused by one or more hemolysins produced by the bacteria. Gram stains show that the organisms generally occur in clusters (figure 1 *a*).

Unlike most other medically important species of bacteria, *S. aureus* can grow in high concentrations of sodium chloride (table salt). In fact, a selective agar medium containing 7.5% salt is commonly used to recover staphylococci from samples containing other bacteria. The ability of *S. aureus* to ferment mannitol is also an important identifying feature, and mannitol with a pH indicator is usually included in the salt agar (figure 1 *b*).

Another important aid in identifying staphylococci is the coagulase test (figure 1 *c*). Virtually all strains of *S. aureus* produce a protein called **coagulase** that causes plasma to coagulate (clot), whereas none of the other medically important cocci do this under the conditions of the test. The coagulase test is the most important way to distinguish *S. aureus* from *S. epidermidis*, which is much less likely to cause disease.

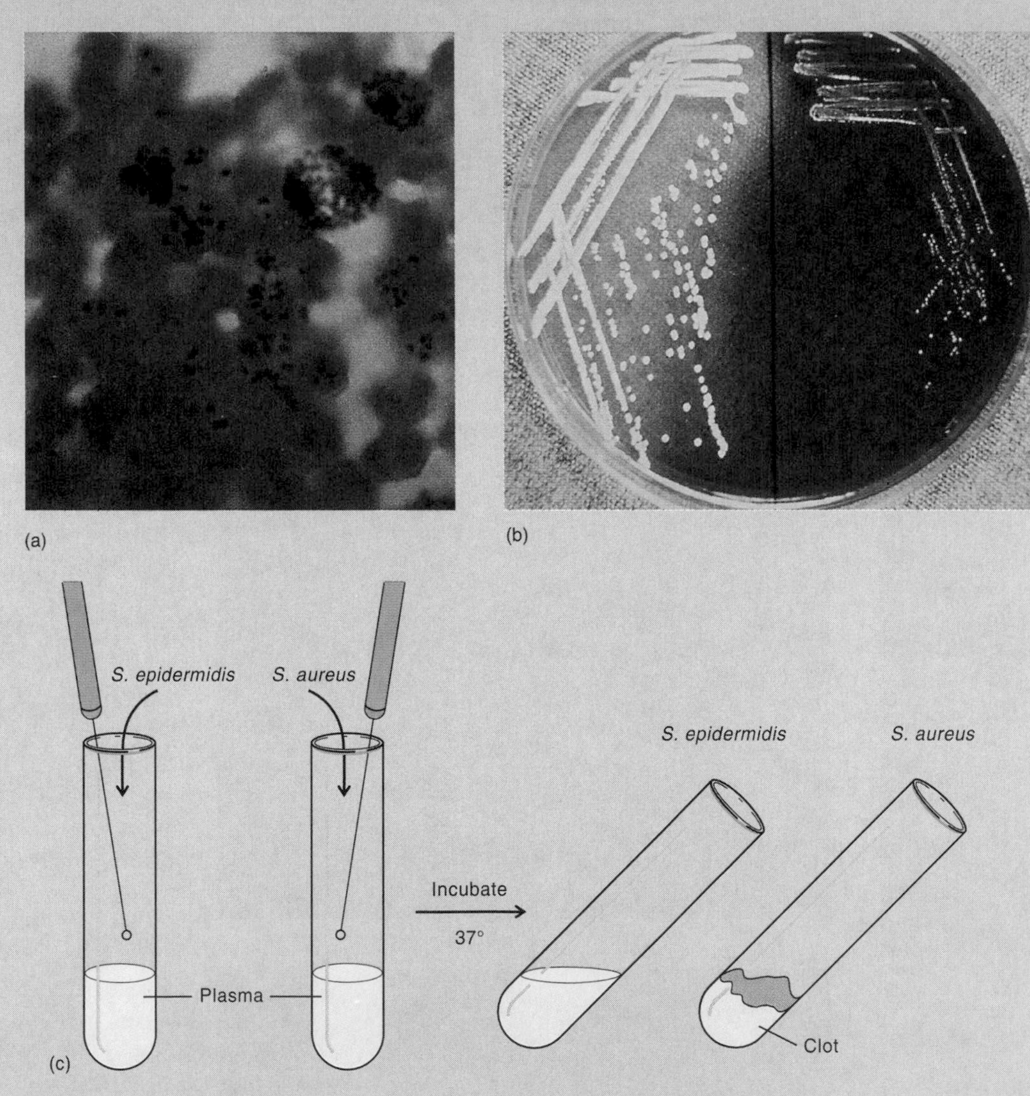

Figure 1
Tests used in identifying *Staphylococcus aureus*. (*a*) *S. aureus* in pus, on a Gram-stained smear. The red objects are leukocytes, and the dark-staining dots are staphylococci. (*b*) Mannitol-salt agar. The left half of the petri dish shows a yellow color because of acid produced by *S. aureus* fermentation of the mannitol present in the medium. The right side of the petri dish is inoculated with *S. epidermidis*, which does not ferment mannitol. (*c*) Coagulase test plasma is clotted by *S. aureus*.

Pathogenesis

Infection is initiated when virulent staphylococci attach to the cells of a hair follicle, multiply, and spread downward to involve the follicle and pilosebaceous glands. The infection induces an inflammatory response with swelling and redness, followed by congregation of polymorphonuclear leukocytes. If the infection continues, the follicle becomes a plug of inflammatory cells and necrotic tissue. The infectious process spreads deeper, reaching the subcutaneous tissue where an abscess[1] forms. Without effective treatment, pressure within the abscess increases, causing it to expand to other hair follicles (figure 22.4). If organisms enter the bloodstream, the infection can spread to other parts of the body, such as the heart, bones, or brain.

The properties of *S. aureus* that can contribute to its virulence are shown in table 22.2. Virtually all strains possess an unusual cell wall component called **protein A.** This protein, some of which is released from the cell, binds to the Fc region of antibody molecules, thereby preventing the antibody from attaching to Fc receptors on phagocytes.

[1]Abscess—localized collection of pus

• **Fc receptors, p. 362**

• **Polymorphonuclear cells, pp. 346–348**

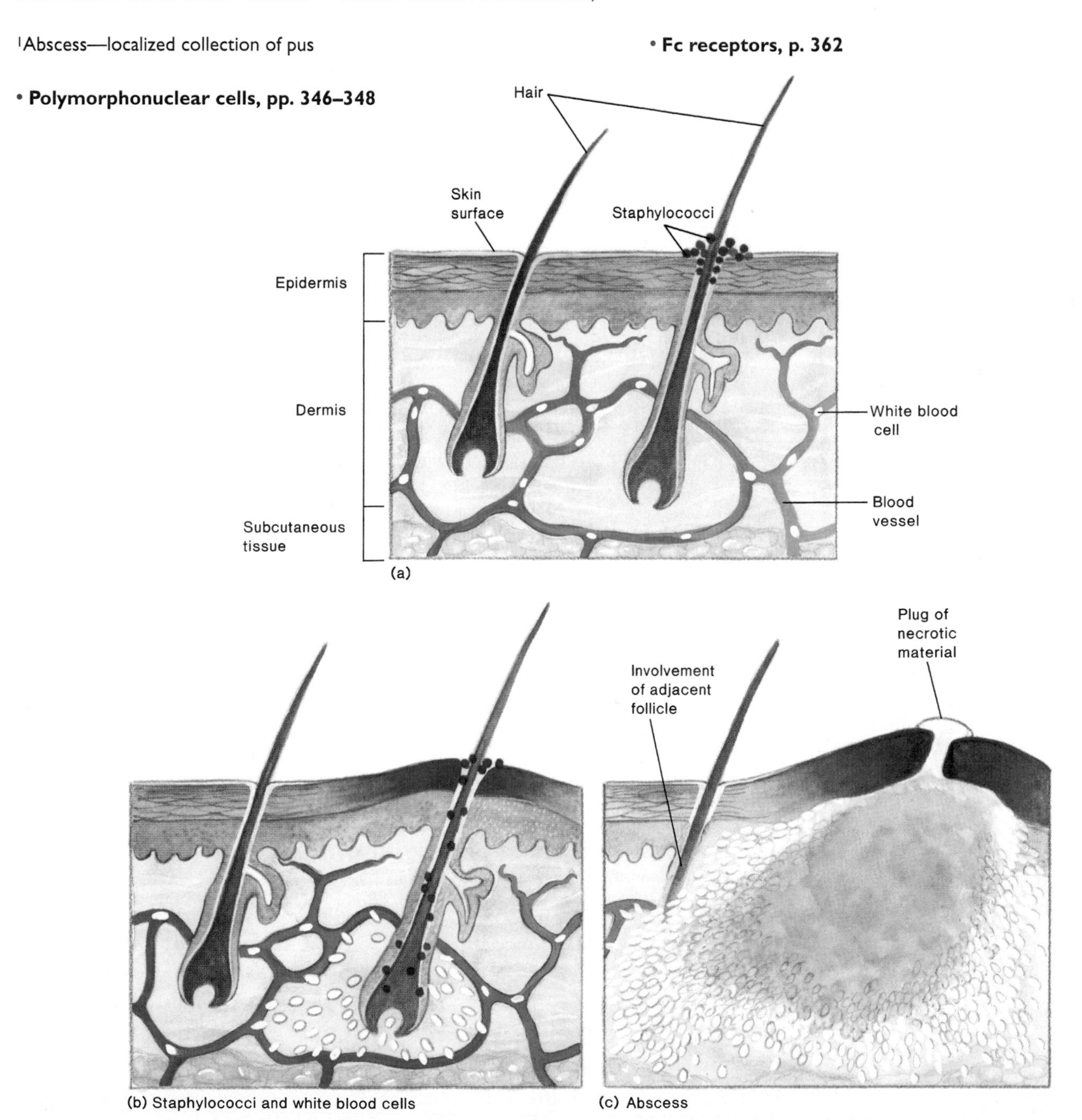

Figure 22.4
Pathogenesis of a boil. (*a*) *S. aureus* infects a hair follicle through its opening on the skin surface. (*b*) The infection spreads deeper along the hair follicle to the subcutaneous tissue, where an abscess develops. (*c*) The infection in the skin results in a small abscess, or more commonly, a plug of necrotic material. Most of the abscess is below the skin.

Table 22.2 Some *Staphylococcus aureus* Products that May Contribute to Its Virulence

Product	Effect
Leukocidin	Kills white blood cells by producing holes in their cytoplasmic membrane
Coagulase	May impede progress of leukocytes into infected area by producing clots in the surrounding capillaries
Epidermolytic toxin (exfoliatin)	Separates layers of epidermis, causing scalded skin syndrome
Toxic shock syndrome toxin	Causes rash, diarrhea, shock
Protein A	Binds to Fc portion of antibody, inhibiting phagocytosis (blocks attachment to Fc receptors on white blood cells)
Capsule	Inhibits phagocytosis of nonopsonized bacteria
Lipase	Breaks down fats by hydrolyzing the bond between glycerol and fatty acids
Proteases	Degrade collagen and other tissue proteins
Hyaluronidase	Breaks down hyaluronic acid component of tissue, thereby promoting extension of infection

Thus, a major effect of protein A is to interfere with phagocytosis. Many strains of *S. aureus* growing in body tissues synthesize a polysaccharide capsule that also inhibits phagocytosis. Interestingly, the same cells generally do not synthesize the capsule when growing in the laboratory.

Besides producing these cellular components, *S. aureus* produces numerous extracellular products that might contribute to virulence. These products include leukocidins, which kill white blood cells; hyaluronidase, which degrades hyaluronic acid, a component of host tissue; proteases, which degrade collagen[1] and other host proteins; and lipases, which degrade lipids. The lipases may assist colonization of the oily hair follicles since strains of *S. aureus* that cause follicle infections tend to produce more lipase than do other strains.

Epidemiology

Staphylococcus aureus inhabits the nostrils of virtually everyone at one time or another, reaching counts as high as 10^8 bacteria. About 20% of healthy adults have continually positive nasal cultures for a year or more, while over 60% will be colonized at some time during a given year. The organisms are mainly disseminated to other parts of the body and to the environment by the hands. Thus, hand washing is an important control measure. Although the nasal chambers seem to be the preferred habitat of *S. aureus*, moist areas of skin are also frequently colonized. People with boils and other staphylococcal infections shed large numbers of *S. aureus* and should not work near patients with surgical wounds or chronic illnesses. Staphylococci survive well in the environment, which favors their transmission from one host to another. Various domestic animals carry *S. aureus*, but animal strains do not commonly cause human infections.

Techniques for characterizing strains of *S. aureus* include the pattern of susceptibility to multiple antibiotics, bacteriophage typing, and plasmid identification. However, all of these methods are inherently unreliable. A more reliable method is to compare the electrophoretic patterns of the DNA fragments produced by treatment with a restriction enzyme. Since *S. aureus* is so commonplace and there are many different strains of it, epidemics of staphylococcal disease can generally be traced to their sources only by precise identification of the epidemic strain.

Treatment

Effective treatment of boils and carbuncles requires that the pus be drained from the lesion and an antistaphylococcal medicine be given. Since the passage through the skin is often narrow or completely blocked, a surgical incision into the deep subcutaneous abscess is often necessary. Squeezing lesions, especially those on the face, is dangerous because it can spread the infection to the brain, bones, or other tissue. Antibiotic treatment is complicated by the fact that about 90% of *S. aureus* strains produce the penicillin-destroying enzyme **penicillinase** (β-lactamase), so penicillin cannot routinely be used in treatment. Strains of *S. aureus* vary widely in their sensitivity to the various antimicrobial medicines. Some are resistant to multiple antibiotics including β-lactamase–resistant penicillins and cephalosporins. Therefore, each isolate of *S. aureus* must be tested against multiple antimicrobial medicines to determine which one is best for treatment.

Scalded Skin Syndrome Is Also Caused by *Staphylococcus Aureus*

Staphylococcal scalded skin syndrome is a potentially fatal toxin-mediated disease that occurs mainly in infants but can also occur in children and adults.

Symptoms

As the name suggests, the skin appears to be scalded (figure 22.5). SSSS begins as a generalized redness of the skin affecting 20% to 100% of the body. Other symptoms,

[1]Collagen—a white fibrous protein present in skin, tendons, cartilage, connective tissue, and so on

- **Bacteriophage typing, p. 438**
- **Penicillinase, p. 147**
- **Cephalosporins, pp. 449, 451**

Figure 22.5
Staphylococcal scalded skin syndrome (SSSS). Toxin produced by certain strains of *S. aureus* causes the outer layer of skin to separate.

such as malaise,[1] irritability, and fever are also present. The nose, mouth, and genitalia may be painful for one or more days before the typical features of the disease become apparent. Within 48 hours after the redness appears, the skin becomes wrinkled, and large blisters filled with clear fluid develop. The skin is tender to the touch and looks like sandpaper.

Causative Agent

Staphylococcal scalded skin syndrome (SSSS), or Ritter's disease, is caused by epidermolytic[2] toxins (exfoliatins) produced by about 5% of strains of *Staphylococcus aureus*. At least two types of epidermolytic toxin exist: one is coded by a plasmid gene, while the other is chromosomal. Some strains of *S. aureus* produce both kinds of toxins.

Pathogenesis

Staphylococci growing at a site of infection, usually in the skin, produce toxin that is absorbed and carried by the bloodstream to large areas of the skin. In the skin, it causes a split in the cellular layer of the epidermis just below the dead keratinized outer layer. The toxins have esterase activity which is thought to break chemical bonds that hold these cells together. *Staphylococcus aureus* is usually not present in the blister fluid but may be cultured from the primary site of infection or from the blood, skin, or nose.

Epidemiology

The disease can appear in any age group but occurs most frequently in newborn infants, the elderly, and immunocompromised adults. Staphylococcal scalded skin syndrome usually appears in isolated cases, although small epidemics in nurseries sometimes occur. Rapid diagnosis is important to prevent death. Because the outer layers of skin are lost as in a severe burn, mortality can range up to 40%, depending on a patient's age and whether he or she is weakened by another disease.

Treatment

Patients suspected of having SSSS are placed in protective isolation.[3] Initial therapy includes a bactericidal antistaphylococcal antibiotic such as methicillin, a penicillinase-resistant derivative of penicillin. All dead skin and other tissue are removed to help prevent secondary infection[4] with gram-negative bacteria such as *Pseudomonas* sp. and fungi such as *Candida albicans* that can easily colonize the damaged skin.

Table 22.3 describes the main features of this disease.

Streptococcal Pyoderma Can Result in Acute Glomerulonephritis

A skin infection characterized by pus production is called **pyoderma.** Pyodermas can result from infection of an insect bite, burn, or other wound. There is generally no obvious wound in **impetigo,** the most common type of pyoderma (figure 22.6).

Symptoms

Impetigo is a superficial skin infection, involving patches of epidermis just beneath the dead, scaly outer layer. Thin-walled blisters first develop, then break, and are replaced by yellowish crusts that form from the drying of plasma that weeps through the skin. Usually, little fever or pain develop, but lymph nodes near the involved areas often enlarge, indicating that bacterial products have entered the lymphatic system.

[1]Malaise—a feeling of general discomfort or uneasiness, often the first indication of an infection
[2]Epidermolytic—destroying the material that binds the layers of skin cells together
[3]Isolation—procedures that help keep environmental pathogens from infecting the damaged skin; usually, use of a private room, careful hand washing, and use of a protective gown and gloves by medical staff
[4]Secondary infection—invasion by another organism of tissues damaged by an earlier infection

• **Plasmids, p. 72**
• ***Candida*, pp. 232, 283**

Causative Agent

Although *Staphylococcus aureus* can cause impetigo, many cases are due to *Streptococcus pyogenes*. These gram-positive, chain-forming cocci are β-hemolytic (figure 22.7) and are frequently referred to as "β-hemolytic group A streptococci" because their cell walls contain a polysaccharide called group A carbohydrate. (A more detailed description of *Streptococcus pyogenes* is found in chapters 23 and 29.) Table 22.4 compares *Staphylococcus aureus* and *Streptococcus pyogenes*.

Pathogenesis

Many different strains of *S. pyogenes* exist, some of which are able to colonize the skin. Infection is probably established by scratches or other minor injuries that introduce the bacteria into the deeper layer of epidermis.

As with *Staphylococcus aureus*, a number of extracellular products may contribute to the virulence of *Streptococcus pyogenes*. These products include enzymes that break down protein (proteases), nucleic acids (nucleases), and the hyaluronic acid that is a component of most host tissues (hyaluronidase). As with staphylococci, it is probable that such enzymes contribute to streptococcal pathogenicity. However, none of them appears to be essential since antibody to them fails to protect experimental animals. The surface components of *S. pyogenes*, notably a hyaluronic acid capsule and a cell wall component known as the **M-protein,** are nevertheless very important in enabling this organism to cause disease because they interfere with phagocytosis.

• ***S. pyogenes*, p. 506**

Table 22.3 Staphylococcal Scalded Skin Syndrome (SSSS, or Ritter's Disease)

Symptoms	Tender red rash with sandpaper texture, malaise, irritability, fever, large blisters, peeling of skin
Incubation period	Variable, usually days
Causative agent	Strains of *Staphylococcus aureus* that produce epidermolytic toxin
Laboratory diagnosis	Gram-stained smears and cultures from blood, skin, nose, or sputum
Pathogenesis	Toxin produced by staphylococci at an infection site, usually of the skin, and carried by the blood stream to the epidermis where it causes a split in a cellular layer; loss of body fluid and secondary infections contribute to mortality
Epidemiology	Person-to-person transmission; seen mainly in newborns but can occur in any age group
Prevention	Careful hand washing between patients by hospital personnel to prevent transfer of infection
Treatment	Isolation of patient to protect from environmental potential pathogens; penicillinase-resistant penicillin; removal of dead tissue

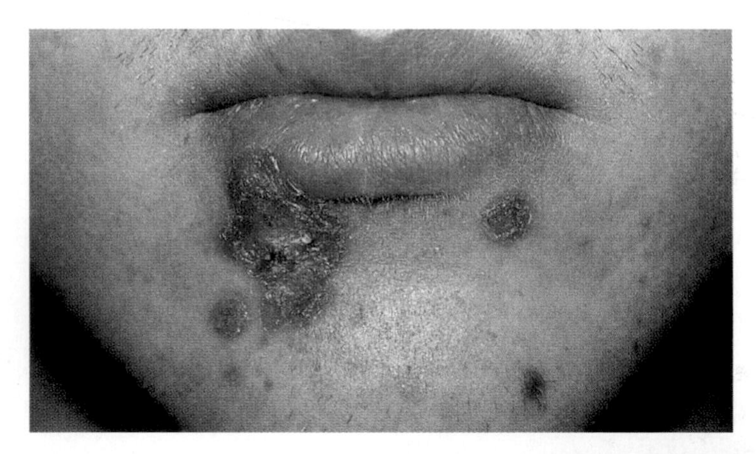

Figure 22.6
Impetigo. This type of pyoderma is often caused by *Streptococcus pyogenes* and may result in glomerulonephritis.

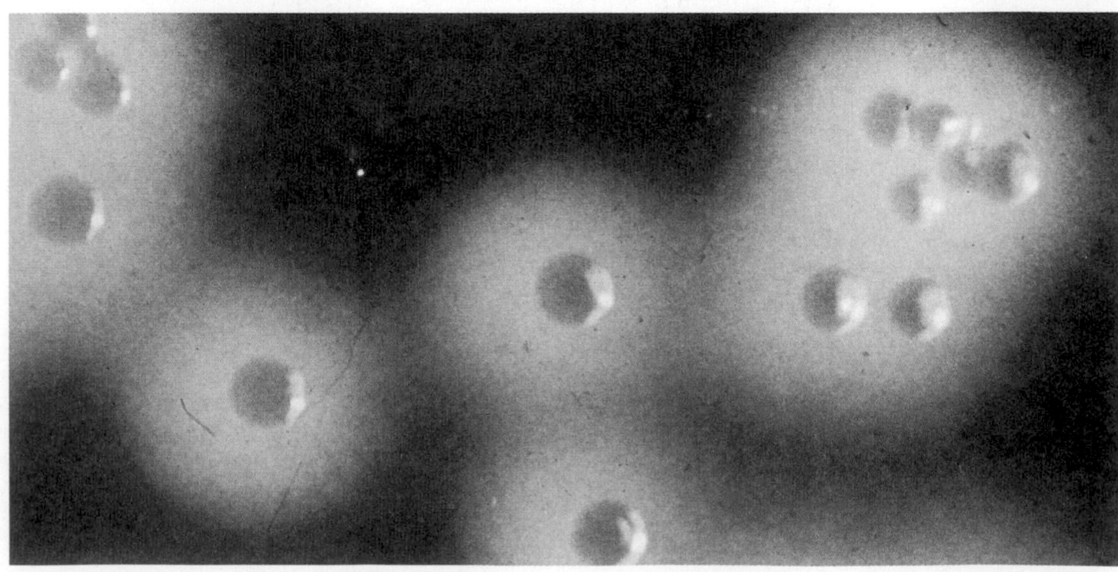

Figure 22.7
Streptococcus pyogenes growing on blood agar. The colonies are small and are surrounded by a wide zone of β-hemolysis.

Nester/Roberts/Nester—Microbiology: A Human Perspective, 2/e

Chapter 22 Study Card—**Skin Infections**

Disease	Cause	Significant Features
Furunculosis (boils)	*Staphylococcus aureus*	The organisms invade the skin via hair follicles. Protein A plays a role in pathogenesis. The bacteria generally originate from the nose.
Scalded skin syndrome	*S. aureus* strains that produce exfoliatin	Resembles a severe burn. Most cases are in newborn infants.
Impetigo	*Streptococcus pyogenes*, aerotolerant gram-positive cocci in chains; β-hemolytic; possess Group A carbohydrate and M protein	This and other streptococcal skin infections can result in acute glomerulonephritis.
Rocky Mountain spotted fever	*Rickettsia rickettsii*, an obligate intracellular, small gram-negative rod	*Dermacentor* ticks transmit the disease to humans. Blood vessel damage causes hemorrhages throughout the body. Treatment can be successful if given early in the disease, but diagnosis is difficult.
Lyme disease	*Borrelia burgdorferi*, a spirochete transmitted by *Ixodes* ticks	A rash, erythema chronicum migrans, characterizes the disease, but it can involve the heart, joints, nervous system, and other tissues as well.
Varicella (chicken pox)	Varicella-zoster virus (herpesvirus family)	The virus persists as a latent infection of nerve ganglia. Reactivation may occur as herpes zoster (shingles).

Nester/Roberts/Nester—Microbiology: A Human Perspective, 2/e

Chapter 23 Study Card—**Upper Respiratory System Infections**

Disease	Cause	Significant Features
Streptococcal pharyngitis	*Streptococcus pyogenes*	Immunity is strain specific; there are many strains, identified by specific M proteins. Rheumatic fever and acute glomerulonephritis are potential complications.
Diphtheria	*Corynebacterium diphtheriae* (nonmotile, club-shaped, nonsporing gram-positive rods that show metachromatic granules)	A potent toxin is produced as a result of lysogenic conversion. Toxin taken up by tissues such as the heart and nerves irreversibly halts protein synthesis.
Bacterial otitis media (middle ear infection)	Usually *Streptococcus pneumoniae* (bile-soluable α-hemolytic gram-positive diplococci) or *Haemophilus influenzae* (small gram-negative rods requiring x and v factors)	Viral respiratory infections commonly precede the bacterial infection.
Bacterial sinusitis	Usually *S. pneumoniae, H. influenzae* or *Bacteriodes* sp. (small anaerobic gram-negative rods)	It is often preceded by a cold. Meningitis is a feared complication.
Bacterial conjunctivitis (pink eye)	Usually *S. pneumoniae or H. influenzae*	Conjunctivitis caused by viruses generally shows less pus and does not respond to antibacterial medicines.
Acute afebrile infectious coryza (common cold)	Mostly rhinoviruses (picornavirus family)	Cold temperatures do not promote colds. Cold remedies are of little or no value. Colds can be transmitted by fingers as well as by coughing. Colds can pave the way for more serious infections.
Adenoviral pharyngitis	Adenoviruses	It resembles "strep throat" but does not respond to penicillin.

Nester/Roberts/Nester—Microbiology: A Human Perspective, 2/e

Chapter 24 Study Card—**Lower Respiratory System Infections**

Disease	Cause	Significant Features
Pneumococcal pneumonia	*Streptococcus pneumoniae*	Immunity is strain specific. There are many strains distinguished by their differing capsule antigens. The capsules are responsible for virulence.
Klebsiella pneumonia	Usually *Klebsiella pneumoniae*, a nonmotile, encapsulated enterobacterium	Usually occurs in persons with impaired host defenses. The endotoxin and the antiphagocytic action of the capsules enhance virulence. Permanent lung damage often results.
Mycoplasma pneumonia	*Mycoplasma pneumonia*, a slow-growing aerobic bacterium lacking a cell wall	Patients with this common type of "walking pneumonia" often develop cold agglutinins in their blood and this assists diagnosis.
Pertussis (whooping cough)	*Bordetella pertussis*, an encapsulated, aerobic gram-negative rod	It can be fatal for infants. Because of waning immunity, adolescents and adults may develop a mild case of the disease that spreads to others.
Tuberculosis	*Mycobacterium tuberculosis*, a slow-growing acid-fast rod that resists drying, disinfectants, and many antibacterial medicines	Granuloma formation is characteristic of the disease. Delayed hypersensitivity contributes to pathogenesis. Reactivated disease can be a lifelong threat.

Nester/Roberts/Nester—Microbiology: A Human Perspective, 2/e

Chapter 26 Study Card—**Lower Alimentary System Infections**

Disease	Cause	Significant Features
Cholera	*Vibrio cholerae*, a curved, facultatively anaerobic gram-negative rod	Choleragen, an exotoxin produced by *V. cholerae*, acts on the small intestinal mucosa, causing an outpouring of fluid without producing visible damage to the intestinal cells.
Shigellosis	*Shigella* sp., nonmotile, generally lactose nonfermenting enterobacteria	A cause of dysentery, some produce Shiga toxin.
Salmonellosis	*Salmonella choleraesuis*, lactose nonfermenting, citrate-utilizing, H_2S producing enterobacteria of over 2,000 serotypes; many exhibit phase variation	Gastroenteritis is the commonest form. Wild and domestic animals are the usual sources of human infection. Eggs and poultry are often contaminated with salmonellas.
E. coli gastroenteritis	*Escherichia coli*, a lactose-fermenting enterobacterium	There are four main groups of pathogenic *E. coli* strains: ETEC produce toxins that cause excessive water and electrolyte secretion, a common form of traveler's diarrhea; EPEC cause alteration of intestinal cell structure and are a common cause of nursery epidemics; EIEC invade the intestinal epithelium and can cause dysentery; EHEC possess a "Shigalike" toxin and can cause the hemolytic-uremic syndrome.
C. jejuni diarrhea	*Campylobacter jejuni*, a curved gram-negative microaerophilic bacterium cultivated on a special medium	A leading cause of bacterial diarrhea and dysentery, it usually originates from a domestic animal.

Nester/Roberts/Nester—Microbiology: A Human Perspective, 2/e

CHAPTER 25 Study Card—**Upper Alimentary System Infections**

Disease	Cause	Significant Features
Dental caries	Mostly *Streptococcus mutans* and similar bacteria that convert sucrose to insoluble polysaccharides	Dental plaques containing these bacteria produce acid in response to dietary sucrose.
Periodontal disease	Dental plaque bacteria growing at the gingival crevice	Bacterial products and the host's response to them cause loosening of the teeth in a chronic process.
Acute necrotizing ulcerative gingivitis ("ANUG," "Trenchmouth," "Vincent's angina")	*Fusobacterium* (spindle-shaped gram-negative anaerobes) and *Treponema* sp. (spirochetes)	An acute infection, it results in necrosis of gingival tissue, pain, bleeding, and fever.
Bacterial endocarditis	Viridans streptococci and other oral bacteria	Oral bacteria forced into the bloodstream during dental procedures colonize a structurally abnormal heart (chapter 30). Properly timed antibacterial treatment may prevent the disease.
Herpes simplex ("cold sore," "fever blister")	Herpes simplex virus	After the initial infection, the virus becomes latent in nerve ganglia. It can be present in saliva in the absence of symptoms and infect areas (such as the finger) of another person.
Mumps	Mumps virus (Paramyxovirus family)	Contracted from saliva, it infects the upper respiratory system and spreads throughout the body, infecting parotid glands, ovaries, testicles, and other tissues.
Infectious esophagitis	Herpes simplex virus; the fungus *Candida albicans*	Infection of the esophagus is unusual except in immunodeficiency conditions including AIDS.
Infectious gastritis	*Helicobacter pylori*, an acid-tolerant, urease-producing curved rod with sheathed flagella	Infection can predispose a person to gastric and duodenal ulcers and may be a factor in stomach cancer.

Nester/Roberts/Nester—Microbiology: A Human Perspective, 2/e

CHAPTER 22 Study Card (*cont'd.*)—**Skin Infections**

Disease	Cause	Significant Features
Rubeola (measles)	Rubeola virus (Paramyxovirus family)	Measles virus can cause pneumonia and encephalitis. Measles is also commonly complicated by bacterial pathogens.
Rubella (German measles)	Rubella virus (Togavirus family)	Generally, there are mild symptoms. The most important complication is the congenital rubella syndrome.
Erythema infectiosum (Fifth disease)	Parvovirus B-19	The rash has a "slapped cheek" appearance and often comes and goes.
Exanthem subitum (roseola infantum)	Herpesvirus, type 6	It is seen in infants. Several days of high fever are followed by a rash.
Warts	Human papilloma viruses (Papovavirus family)	There is a long incubation period. Removal of the wart may not eliminate the infection.
Dermatophytosis	Molds of the genera *Trichophyton*, *Microsporum*, *Epidermophyton*, and others	It invades outer layers of the skin, hair, and nails (keratin-containing structures), particularly if moist.

Nester/Roberts/Nester—Microbiology: A Human Perspective, 2/e

CHAPTER 26 Study Card (*cont'd.*)—**Lower Alimentary System Infections**

Disease	Cause	Significant Features
Typhoid Fever	"Salmonella typhi," a serological type of *S. choleraesuis* with unusual biochemical features, pathogenic only for humans	It causes "enteric fever," with fever, headache, and shock.
Staphylococcal food poisoning	Enterotoxin-producing strains of *S. aureus*	Staphylococcal enterotoxins are resistant to destruction by cooking.
Clostridial food poisoning	*Clostridium perfringens*, an anaerobic, encapsulated gram-positive rod	Toxin is released from the bacteria when they sporulate in the intestine. Usually, a meat is the source of the bacteria.
B. cereus food poisoning	Toxin-producing strains of *Bacillus cereus*, an aerobic gram-positive spore-forming rod	Different toxins produce differing symptoms. Commonly, a rice is the source of the bacteria.
Botulism	*Clostridium botulinum*, an anaerobic, gram-positive spore-forming rod	*C. botulinum* toxins cause a severe paralysis that commonly results in death.
Fungal food poisoning	Various fungi, including *Aspergillus* sp.	Ergot poisoning can result from moldy grain; aflatoxins can be carcinogenic.
Viral hepatitis	Mostly HAV, HBV, HCV	Hepatitis A virus (HAV) is transmitted by the fecal-oral route; HBV, by blood and semen; HCV, by blood. Carriers of HBV are common.
Viral gastroenteritis	Rotaviruses, Norwalk viruses	In the United States, rotaviruses are the most common cause of severe diarrhea in young children.
Giardiasis	*Giardia lamblia*, a flagellated protozoan with an adhesive disk	In the United States, it is the most commonly identified waterborne illness.
Amebiasis	*Entamoeba histolyica*, an ameba	Intestinal ulceration and extraintestinal infections are features of this illness.

Nester/Roberts/Nester—Microbiology: A Human Perspective, 2/e

CHAPTER 24 Study Card (*cont'd.*)—**Lower Respiratory System Infections**

Disease	Cause	Significant Features
Legionnaire's Disease	*Legionella pneumophila*, a gram-negative rod that stains poorly in infected tissue; requires a special medium	Water aerosols are the usual source. Smokers and others with impaired host defenses are susceptible.
Influenza	Mainly influenza A and B viruses (Orthomyxovirus family)	Viral strains are identified by their H and N antigens. Antigenic drift and antigenic shift play a big role in epidemiology.
Coccidiodomycosis (Valley Fever)	*Coccidioides immitis*, a dimorphic soil fungus of hot dry areas of the Americas.	The infectious form is a barrel-shaped arthrospore; spherules develop in infected tissue. The disease can mimic tuberculosis.
Histoplasmosis ("Spelunkers disease")	*Histoplasma capsulatum*, a dimorphic fungus associated with soil containing bird or bat droppings.	The highest U. S. incidence is in the Mississippi River drainage states and South Atlantic states. It can mimic tuberculosis.

Nester/Roberts/Nester—Microbiology: A Human Perspective, 2/e

CHAPTER 27 Study Card—**Genitourinary Infections**

Disease	Cause	Significant Features
Bladder infections	Mostly *E. coli* or other enteric bacteria	Burning pain with urination is characteristic. Kidney infections can result from bladder infections.
Leptospirosis	*Leptospira* sp., a spiral bacterium	Water contaminated by the urine of various animals is the usual source of infection.
Gonorrhea	*Neisseria gonorrhoeae*, a gram-negative coccobacillus with specific attachment pili	In the United States, the most common of the reportable bacteria diseases, it is a leading cause of pelvic inflammatory disease. A progressive increase in resistance to antimicrobial medicines has occurred.
Chlamydial genital disease	*Chlamydia trachomatis* (certain strains), a spherical obligate intracellular bacterium	Mimics gonorrhea and may even be more common.
Syphilis	*Treponema pallidum*, a slender, spiral bacterium best seen with dark-field illumination	It is a chronic progressive disease transmissible during its primary and secondary stages. Transplacental infection of fetuses can occur in asymptomatic mothers.
Chancroid	*Haemophilus ducreyi*, a small gram-negative rod that requires X factor	It is one cause of painful genital sores.
Genital herpes simplex	Usually herpes simplex virus type 2	Painful recurrent genital rash. There is no cure.
Genital warts	Various human papilloma viruses	There is probably no cure for infection, although warts can be removed. Some viral strains are associated with cancer of the cervix. Asymptomatic infections are common.

Nester/Roberts/Nester—Microbiology: A Human Perspective, 2/e

CHAPTER 28 Study Card—**Nervous System Infections**

Disease	Cause	Significant Features
Meningococcal meningitis	*Neisseria meningitidis*, a gram-negative diplococcus	The disease can occur at any age but is most common in infants. It can occur in epidemics. Endotoxic shock can occur and be rapidly fatal.
Haemophilus meningitis	*Haemophilus influenzae*, a small gram-negative rod that requires X and V factors and demonstrates satellitism	It causes meningitis mainly in children. Type B is usually responsible.
Pneumococcal meningitis	*Streptococcus pneumoniae*, an encapsulated diplococcus	It is the leading cause of meningitis in adults.
Leprosy	*Mycobacterium leprae*, an acid-fast rod that can be grown in armadillos, but not *in vitro*	Tuberculoid-type disease occurs in those with cell-mediated immunity; lepromatous disease, in those whose cell-mediated immunity is overcome.
Sporadic viral encephalitis	Herpes simplex, mumps, measles, and other human viruses	Cases occur at low frequency in a population.
Epidemic viral encephalitis	La Crosse, St. Louis, and other arborviruses	The viruses are maintained in the wild in various animals and arthropods such as mosquitoes that bite them. The viruses spread to humans through the bites of infected arthropods.

Nester/Roberts/Nester—Microbiology: A Human Perspective, 2/e

CHAPTER 29 Study Card—**Wound Infections**

Disease	Cause	Significant Features
Surgical wound infections	*Staphylococcus aureus*, other staphylococci, enterobacteria, and others	Factors that determine the development and type of infection and duration of surgery, presence of foreign material, the status of host defenses, and resistance to antimicrobial medicines.
Infections of burns	*Pseudomonas aeruginosa* and others	These bacteria produce greenish pigments and are resistant to many antimicrobials. They also release a potent toxin.
Tetanus	*Clostridium tetani*, a motile, anaerobic gram-positive rod with a terminal spore	Spastic paralysis results from an exotoxin; the bacteria generally do not invade tissue.
Gas gangrene	*Clostridium perfringens* and other clostridia	Muscle tissue is invaded and destroyed by the bacteria. Dirty wounds with dead and poorly oxygenated tissue are especially susceptible.
Actinomycosis	*Actinomyces israelii*, an anaerobic, branching bacterium	Chronic condition characterized by abscesses that burrow to the surface and drain pus containing "sulfur granules."
Bite wound infections	*Pasteurella multocida*, a facultatively anaerobic, gram-negative rod; various other species	Infections are often severe but can usually be successfully treated with penicillin. Human bites can result in synergistic infections involving mouth streptococci and gram-negative anaerobes.
Cat scratch disease	*Bartonella henselae*, a newly described bacterium	Pus-filled lymph nodes are a common feature.
Rat bite fever	*Streptobacillus moniliformis*, a facutative anaerobe that produces L-forms	Fevers, joint pain, and a rash can be fatal if untreated.
Sporotrichosis	*Sporothrix schenckii*, a soil fungus	The organisms are introduced by thorns or splinters and cause chains of draining skin ulcers.

Nester/Roberts/Nester—Microbiology: A Human Perspective, 2/e

CHAPTER 30 Study Card—**Blood and Lymphatic Infections**

Disease	Cause	Significant Features
Subacute bacterial endocarditis	*Viridans streptococci* from the mouth or *Staphylococcus epidermidis* from the skin; various other bacteria of low virulence	It is a chronic illness dependent on a structural abnormality of the heart.
Acute bacterial endocarditis	*Staphylococcus aureus*, *Streptococcus pneumoniae*, and various other virulent bacteria	A complication of infection elsewhere in the body, it can occur in the absence of a structural heart abnormality. It can rapidly destroy heart valves.
Gram-negative septicemia	Usually enterobacteria such as *Escherichia coli* or *Pseudomonas aeruginosa* or gram-negative anaerobes such as *Bacteroides* sp.	Release of endotoxin from the organisms can cause shock and D.I.C.
Tularemia ("rabbit fever")	*Francisella tularensis*, an aerobic gram-negative rod that requires a special growth medium	Bacteria from infected wild animals enter the body through mucous membranes, through scratches, or with the bites of flies or ticks. Ulcer formation at the entry site and enlargement of regional lymph nodes are characteristic. Hunters and trappers are at risk for the disease.
Brucellosis ("undulent fever," "Bang's disease")	Species of *Brucella*, aerobic gram-negative rods	It is generally acquired from chronically infected domestic animals. Workers in the meat-packing industry are at particular risk for the disease.
Plague ("black death")	*Yersinia pestis*, a nonmotile enterobacterium that, when stained, resembles a safety pin	It exists in wild rodents, in a number of states. Transmission is mainly by the rat flea, causing "bubonic plague." Lung infection can spread directly person to person ("pneumonic plague").

Nester/Roberts/Nester—Microbiology: A Human Perspective, 2/e

CHAPTER 28 Study Card (*cont'd.*)—**Nervous System Infections**

Disease	Cause	Significant Features
Poliomyelitis	Polioviruses 1, 2, 3	The viruses selectively infect and destroy motor nerve cells, causing paralysis. A WHO program aims to eradicate this disease through vaccination by the year 2,000.
Rabies	Rabies virus	The virus is widespread in wild animals. The disease in humans is almost always fatal. However, because of the long incubation period, the disease can often be prevented by vaccination after exposure to the virus.
Cryptococcal meningitis	*Cryptococcus neoformans*, a fungus commonly present in soil contaminated by pigeon droppings	Infection begins in the lung and then spreads to the meninges and brain. Pathogenicity depends on the presence of a large capsule on the fungal cells.
African sleeping sickness	*Trypanosoma brucei*, slender flagellated protozoa	Transmission is via the bite of tsetse flies. The trypanosomes evade host immunity by changing their surface antigens.
Spongiform encephalopathy	Unknown; possibly a new type of infectious agent	Transmission among humans has occurred in the past through cannibalism and more recently with corneal and dura grafts and growth hormone prepared from human cadavers.

Nester/Roberts/Nester—Microbiology: A Human Perspective, 2/e

CHAPTER 27 Study Card (*cont'd.*)—**Genitourinary Infections**

Disease	Cause	Significant Features
AIDS	Human immunodeficiency virus (HIV)	Immune deficiency is a late stage of HIV disease. Transmission can occur during an asymptomatic period that can last for years.
Bacterial vaginosis	Unknown; associated with an overgrowth of *Gardnerella vaginalis* and other vaginal anaerobes	It is signaled by a characteristic odor and the presence of "clue cells" in vaginal secretions.
Vulvovaginal candidiasis	Candida albicans, a kind of fungus	It is a common result of treatment with antimicrobial medicines that suppress normal vaginal bacterial flora.
Trichomoniasis	*Trichomonas vaginalis*, a flagellated protozoan	It usually is manifested as vaginitis. Asymptomatic carriers are common, especially among men.
Puerperal (childbirth) fever	Usually *Streptococcus pyogenes*	It is a severe infection of the female genital system following childbirth. Medical attendants who are asymptomatic carriers can be a source.
Toxic shock syndrome	*Staphylococcus aureus* strains that produce toxic shock syndrome toxin (TSST)	It can result from infection by these bacteria anywhere in the body. Highly absorbent vaginal tampons predispose a woman to the condition.

Nester/Roberts/Nester—Microbiology: A Human Perspective, 2/e

CHAPTER 30 Study Card (*cont'd*)—**Blood and Lymphatic Infections**

Disease	Cause	Significant Features
Infectious mononucleosis ("kissing disease," "mono")	Epstein-Barr virus (a member of the herpesvirus family)	The virus preferentially infects B lymphocytes which causes them to produce immunoglobulins and to proliferate. T cells bring the infection under control, but latent infection persists in some B cells. These B cells are immortalized by the infection. There are many asymptomatic carriers. The virus is transmitted by kissing.
Yellow fever	Yellow fever virus	Jungle primates are reservoirs of the disease. Mosquitoes transmit the infection to and among humans. The name *yellow fever* relates to the fact that many victims suffer severe liver damage and become jaundiced.
Malaria	Four species of *Plasmodium,* protozoa with a complex developmental cycle in humans and certain mosquitoes	Attacks of chills and fever correlate with the release of merozoites from infected red blood cells. Only gametocytes are infectious for the mosquito hosts, in which sexual reproduction of the plasmodia occurs. Infections with *P. falciparum* are the most severe since red blood cells of all ages are infected. Resistance to medications is widespread. Global warming could cause extension of malarious areas.

Nester/Roberts/Nester—Microbiology: A Human Perspective, 2/e

CHAPTER 31 Study Card—**HIV Disease and Complications of Immunodeficiency**

Disease	Cause	Significant Features
HIV disease and AIDS	Human immunodeficiency virus, type 1 (most cases)	Flulike symptoms followed by an asymptomatic period of months or years during which the disease progresses and is transmissible. Malignant tumors can lead to the diagnosis. Unusual infectious diseases such as pneumocystosis herald the development of AIDS. An increasing proportion of cases result from heterosexual transmission. Well over 300,000 AIDS deaths have occurred in the U.S. The epidemic can be halted by changes in human behavior.
Kaposi's sarcoma	(?) Kaposi's sarcoma associated virus (herpesvirus 8)	Malignant tumor of blood or lymphatic vessels. Rare occurrence in elderly men without HIV disease or organ transplantation. Much more common in immunodeficiencies.
Lymphomas	(?) Epstein-Barr virus (a herpes virus)	Malignant tumors of lymphoid tissues. Commonly develop in the brain of people with AIDS.
Carcinomas of the anus and uterine cervix	(?) Papilloma-virus types 16 and 18	Malignant tumors of epithelial cells. Screening with Pap tests helps early diagnosis.
Pneumocystosis	*Pneumocystis carinii*	A lung disease. *P. carinii* probably an unclassified fungus. Medications help prevention and treatment.
Toxoplasmosis	*Toxoplasma gondii*	Cause of serious brain infection in people with immunodeficiencies. *T. gondii* is a protozoan that reproduces sexually only in feline intestinal epithelium. Acquired from materials contaminated with cat feces and from eating rare meat. Medications help prevent and treat the disease.
Cytomegalovirus disease	Cytomegalovirus (a herpesvirus)	Widespread virus. Causes encephalitis and blindness in people with immunodeficiencies. Antiviral medications helpful in some cases.

Table 22.4 Comparison of *Staphylococcus aureus* and *Streptococcus pyogenes*

Identification	Extracellular Products	Pathogenic Potential
S. aureus		
Gram-positive cocci arranged in clusters; cream-colored colonies produce catalase and coagulase; ferment mannitol; cell wall contains protein A; strain identification using bacteriophages and other methods	Hemolysins, leukocidin, hyaluronidase, nuclease, protease, lipase, penicillinase, and others	A prominent cause of boils, wound infections, abscesses, impetigo, food poisonings, and toxic shock syndrome
S. pyogenes		
Gram-positive cocci in chains; small β-hemolytic colonies catalase-negative; cell wall contains group A polysaccharide and M-protein; strains distinguished by cell wall proteins (M and T antigens)	Hemolysins (streptolysins) O and S, streptokinase, DNAase, hyaluronidase, and others; certain lysogenic strains produce erythrogenic toxin; some produce exfoliatins	Causes impetigo, pharyngitis, wound infections, scarlet fever, puerperal fever, toxic shock, fasciitis; late complications of infection include glomerulonephritis and rheumatic fever

Acute glomerulonephritis is a serious complication of *S. pyogenes* pyoderma, discussed in the next chapter. Rheumatic fever, a serious complication of "strep throat," is not generally a complication of streptococcal pyoderma.

Epidemiology

Impetigo is most prevalent among poor children of the tropics or elsewhere during the hot, humid season. Children two to six years are mainly afflicted. Person-to-person contact spreads the disease as possibly do fomites, flies, and other insects. Impetigo patients often become throat and nasal carriers of *S. pyogenes*.

Prevention

General cleanliness and avoiding people with impetigo help prevent the disease. Prompt cleansing of wounds and application of antiseptic probably also decrease the chance of infection.

Treatment

So far, *S. pyogenes* strains remain susceptible to penicillin. In patients allergic to penicillin, erythromycin can be substituted.

Table 22.5 summarize the main features of impetigo.

Table 22.5 Impetigo

Symptoms	Blisters that break and "weep" plasma and pus; formation of golden colored crusts; lymph node enlargement
Incubation period	2–5 days
Causative organisms	*Staphylococcus aureus, Streptococcus pyogenes*
Laboratory diagnosis	Fluid from lesion cultured on blood agar
Pathogenesis	Initiated by organisms entering the skin through minor breaks; certain strains of *S. pyogenes* prone to cause impetigo; some *S. aureus* strains that make epidermolytic toxin produce large blisters (bullae)
Epidemiology	Spread by direct contact with carriers or patients with impetigo
Prevention	Cleanliness; care of skin injuries
Treatment	Oral penicillin if cause is known to be *S. pyogenes*; otherwise, an antistaphylococcal antibiotic orally or topically

Rocky Mountain Spotted Fever Is a Serious Tickborne Disease Caused by an Obligate Intracellular Bacterium

Rocky Mountain spotted fever is representative of a group of serious rickettsial diseases that occur worldwide and are transmitted by certain species of ticks, mites, or lice.

- **Glomerulonephritis, p. 509**
- **Rheumatic fever, p. 510**

Symptoms

Rocky Mountain spotted fever generally begins suddenly with a headache, pains in the muscles and joints, and fever. Within a few days, a rash consisting of faint pink spots appears on the palms, wrists, ankles, and soles. This rash spreads up the arms and legs to the rest of the body and becomes raised and hemorrhagic[1] (figure 22.8). Bleeding may occur at various sites, such as the mouth and nose. Involvement of the heart, kidneys, and other body tissues can result in shock and death unless treatment is given promptly.

[1]Hemorrhagic—due to blood breaking out of blood vessels

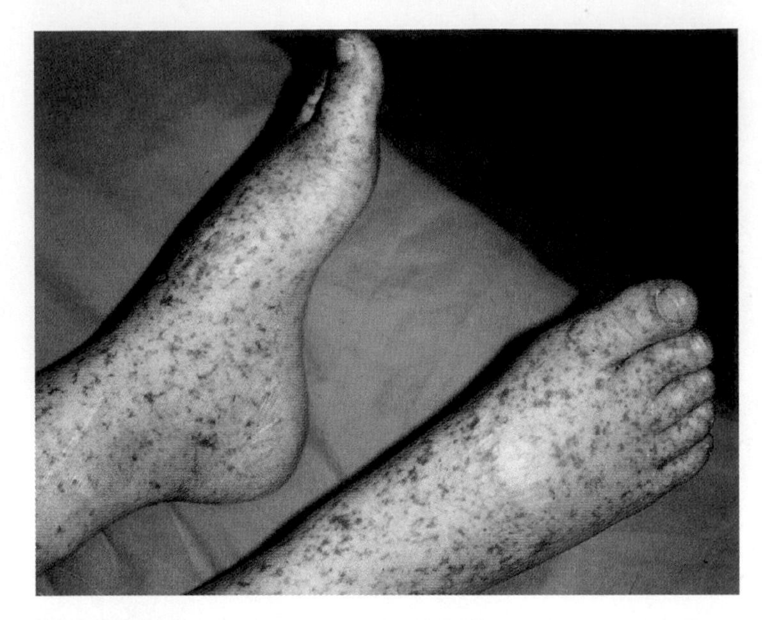

Figure 22.8
Rash caused by Rocky Mountain spotted fever. Characteristically, the rash begins on the arms and legs and, as shown in this photo, becomes hemorrhagic.

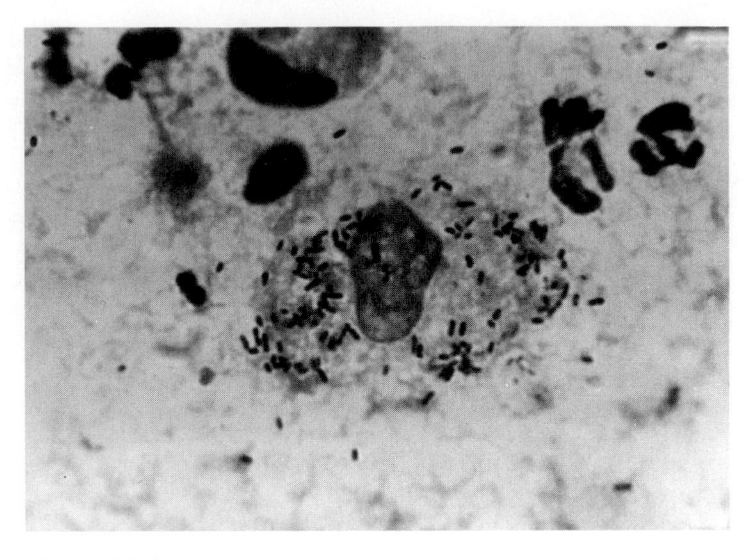

Figure 22.9
Rickettsia rickettsii growing within a cell of a rodent.

Causative Agent

Rocky Mountain spotted fever is caused by *Rickettsia rickettsii* (figure 22.9), an obligate intracellular bacterium. The organisms are tiny, gram-negative, nonmotile coccobacilli. Their cell walls contain a lipopolysaccharide that cross-reacts immunologically with various other bacteria, including enterobacteria of the genus *Proteus*. *Rickettsia rickettsii* is difficult to see well in Gram-stained smears but can be seen using special stains. The organism can be cultivated in embryonated hens' eggs and a variety of tissue cell cultures.

Following attachment to host cells, *R. rickettsii* is taken into the cells by endocytosis. Inside the cell, the bacteria leave their phagosome and multiply in both the cytoplasm and nucleus without being enclosed in vacuoles. Some exit the cell early in the infection by entering, and then lysing, fingerlike host cell cytoplasmic projections. Eventually, the integrity of the cell membrane suffers from this process, the cell takes in water, it lyses, and it releases the remaining rickettsias. Very few laboratories cultivate *R. rickettsii* because maintaining the tissue cell cultures is expensive, and laboratory workers may accidently become infected.

The only sure way of establishing the diagnosis of Rocky Mountain spotted fever early in the illness is to demonstrate *R. rickettsii* in biopsies[1] of skin lesions using specific antibody labeled with a fluorescent dye or peroxidase enzyme, a technique available in certain large medical centers and public health laboratories. Most other laboratories rely on detecting antibodies to *R. rickettsii* in the patient's blood. The best test available at present is an enzyme-linked immunoassay using antigen from killed rickettsias. This test only becomes positive about one week after illness begins. Because of the need to make the diagnosis sooner than this, specific tests for *R. rickettsii* DNA have been developed but are not yet generally available.

Pathogenesis

Rocky Mountain spotted fever is acquired from the bite of a tick infected with *R. rickettsii*. The bite is usually painless and unnoticed; the tick remains attached for hours while it feeds on capillary blood. Rickettsias are not immediately released into tick saliva from the tick's salivary glands. Therefore, the infection is not usually transmitted until the tick has fed for 4 to 10 hours. When the organisms are released into capillary blood with the tick saliva, they are taken up preferentially by the cells lining the small blood vessels. In the blood vessel cells, they multiply and seed the bloodstream, thus infecting even more cells. Infection can also extend into the walls of the small blood vessels, causing an inflammatory reaction, clotting of the blood vessels, and small areas of necrosis. This process is readily apparent in the skin as a hemorrhagic rash but, more ominously, occurs throughout the body, resulting in damage to vital organs such as the kidneys and heart. Potentially even more serious is the release of endotoxin into the bloodstream from the rickettsial cell walls, causing shock and generalized bleeding because of disseminated intravascular coagulation.

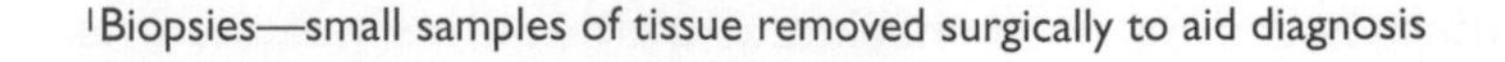

[1]Biopsies—small samples of tissue removed surgically to aid diagnosis

• **Obligate intracellular bacteria, p. 250**
• **Enzyme-linked immunosorbent assay, p. 388**
• **Endotoxin, pp. 60, 422–424, 676**
• **Disseminated intravascular coagulation, p. 401**

Epidemiology

Rocky Mountain spotted fever is an example of a **zoonosis,** a disease that exists primarily in animals other than humans. It occurs in a spotty distribution across the contiguous United States and extends into Canada, Mexico, and a few countries of South America. The involved areas change over time, but in the United States the highest incidence has generally been in the South Atlantic and South Central states (figure 22.10). Sources of infection have even included certain parks in New York City. Rocky Mountain spotted fever is maintained in nature in various species of ticks and mammals. Generally, little or no illness develops in the natural host, but humans, being an "accidental" host, often develop severe disease. Several species of ticks transmit the disease to humans. The main vector in the western United States is the wood tick, *Dermacentor andersoni* (figure 22.11), while in the East it is the dog tick, *Dermacentor variabilis.* Once infected, ticks remain infected for life, transmitting *R. rickettsii* transovarially[1] from one generation to the next, as well as to their mammalian hosts. Ticks are most active from April to September, and it is during this time period that most cases of Rocky Mountain spotted fever occur.

Prevention

No vaccine against Rocky Mountain spotted fever is currently available to the public, although promising vaccines that are genetically engineered are under development. The disease can be prevented if people take the following precautions: (1) avoid tick-infested areas when possible; (2) use protective clothing; (3) use tick repellents such as dimethyltoluamide; (4) carefully inspect their bodies for ticks several times daily; and (5) remove attached ticks carefully to avoid crushing them and thereby contaminating the bite wound with their tissue fluids. Gentle traction with tweezers applied near the mouth parts is the safest method of removal. Touching the tick with gasoline or whiskey may help to loosen its attachment. After removal of the tick, the site of the bite should be treated with an antiseptic.

[1]Transovarially—by way of infected eggs

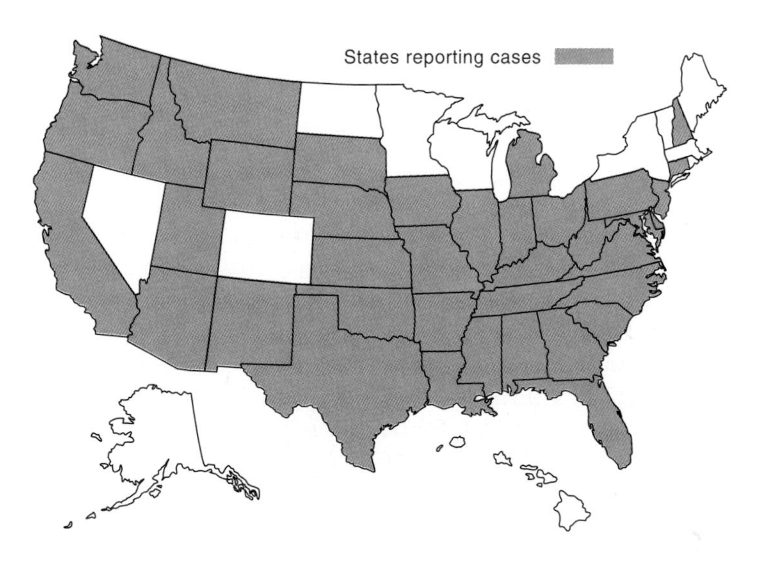

Figure 22.10
States that reported cases of Rocky Mountain spotted fever during 1990. The higher concentrations of cases are generally in the South Atlantic and South Central states.

Source: Centers for Disease Control and Prevention. 1990. *Morbidity and Mortality Weekly Report* 39:281.

Treatment

The antibiotics tetracycline and chloramphenicol are highly effective in treating Rocky Mountain spotted fever if given early in the disease before irreversible damage to vital organs has occurred. Without treatment, the overall mortality from the disease is about 20%, but it can be considerably higher in older patients. With early diagnosis and the treatments now available, the mortality rate is less than 5%.

The main features of Rocky Mountain spotted fever are summarized in table 22.6.

Lyme Disease Is a Chronic Tickborne Spirochetal Disease

In the mid-1970s, studies of a group of cases in Lyme, Connecticut, led to the first recognition in the United States of Lyme disease as a distinct entity. It was not until 1982 that the cause was finally identified by Dr. Willi Burgdorfer.

• **Antiseptic, p. 202**
• **Chloramphenicol, pp. 154, 463**

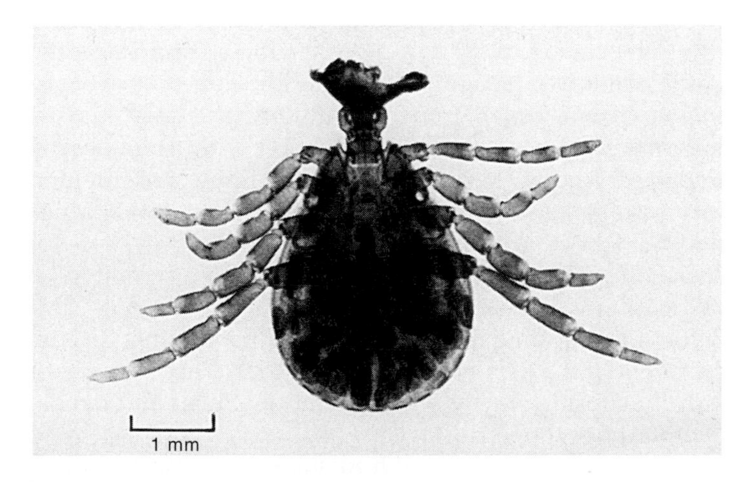

Figure 22.11
Dermacentor andersoni, the wood tick, principal vector of Rocky Mountain spotted fever in the western United States. This specimen still holds a portion of its victim's skin, pulled away when the tick was removed.

Table 22.6 Rocky Mountain Spotted Fever

Symptoms	Headache, pains in muscles and joints, and fever, followed by hemorrhagic rash
Incubation period	4–8 days
Causative organism	*Rickettsia rickettsii*
Laboratory diagnosis	Fluorescent antibody stain of skin biopsy; ELISA test of patient's serum for antibody against rickettsial antigen
Pathogenesis	Organisms multiply at site of bite; bloodstream is invaded and endothelial cells of blood vessels are infected; vascular lesions and endotoxin account for pathologic changes
Epidemiology	Transmitted by bite of infected tick, usually *Dermacentor* spp.
Prevention	Avoidance of tick-infested areas, use of tick repellent, removal of ticks within four hours of exposure
Treatment	Tetracycline or chloramphenicol

Symptoms

Lyme disease is another tickborne bacterial disease. It is caused by *Borrelia burgdorferi*, a spirochete. As with another spirochetal disease, syphilis, symptoms of Lyme disease mimic a wide variety of other illnesses.

Symptoms of Lyme disease can be divided roughly into three stages, although individual patients may lack symptoms in one or more of the three. The first stage typically begins a few days to several weeks after a bite by an infected tick. It is characterized by a skin rash called **erythema chronicum migrans** (figure 22.12). The rash begins as a red spot or bump at the site of the tick bite and slowly enlarges to a median diameter of 15 cm (about 6 inches). The advancing edge is bright red, while the redness of the central portion fades as the lesion enlarges. About half of these cases develop satellite lesions that behave similarly but are smaller. Erythema chronicum migrans is the hallmark of Lyme disease but is present in only two-thirds of the cases. Most of the other symptoms that occur during this stage are influenzalike—fatigue, chills, fever, headache, stiff neck, joint and muscle pains, and backache. Without treatment, these symptoms slowly subside, commonly overlapping the next stage.

The characteristic symptoms of the second stage generally begin two to eight weeks after the appearance of erythema chronicum migrans and involve the heart and the nervous system. Electrical conduction within the heart is impaired, leading to dizzy spells or fainting, and a temporary pacemaker is sometimes required to maintain a normal heartbeat. Involvement of the nervous system can cause one or more of the following symptoms: paralysis of the face, severe headache, pain on moving the eyes, difficulty concentrating, emotional instability, fatigue, and impairment of the nerves of the legs or arms. These symptoms usually subside slowly without any treatment and overlap symptoms of the third stage.

The symptoms of the third stage are characterized by joint pain, swelling, and tenderness, usually of a large joint such as the knee. These symptoms develop in 60% of untreated cases, beginning on the average six months after the skin rash, and slowly disappear over subsequent years.

• Spirochetes, p. 243

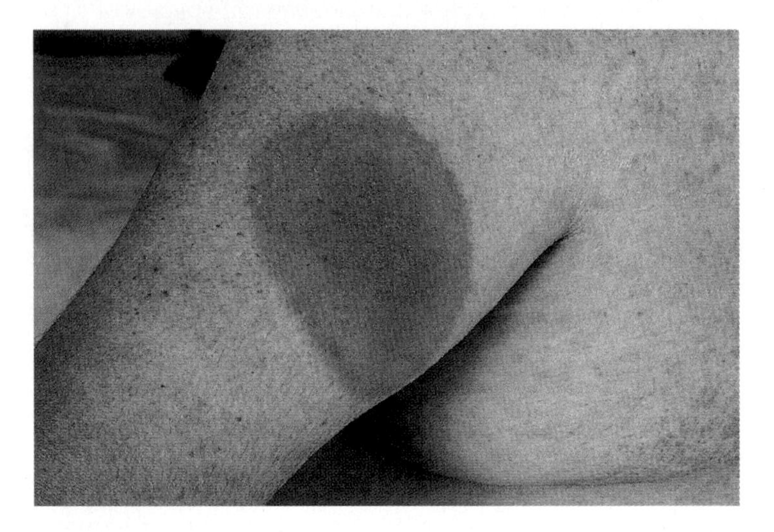

Figure 22.12
Erythema chronicum migrans. This expanding rash often takes on a bull's-eye appearance. The rash usually causes little or no discomfort. It is highly suggestive of Lyme disease, but not every case has it.

Causative Agent

Borrelia burgdorferi is a large, spiral organism (figure 22.13) 11 to 25 μm in length, with a number of axial filaments wrapped around its body and enclosed in the outer sheath of the cell wall. The organism can be cultivated in the laboratory on a special medium under microaerophilic conditions. Different strains can be distinguished using monoclonal antibodies. Surface antigens remain stable during infection. Cultures are generally not fast or sensitive enough for diagnosis. Usually, serological methods, such as ELISA followed by Western blot, are used.

Pathogenesis

The spirochetes are introduced into the skin by an infected tick, multiply, and migrate outward in a radial fashion. The cell walls of the organisms cause an inflammatory reaction in the skin, which produces the expanding rash. The host's immune response is initially suppressed, allowing continued

• Inflammation, p. 366

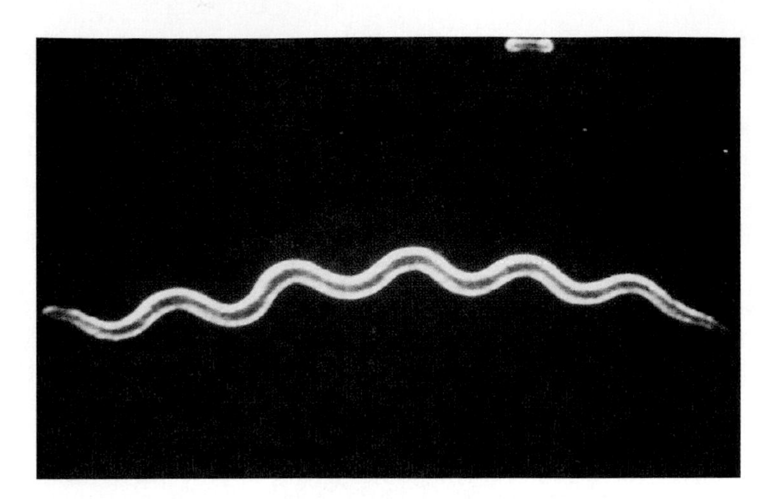

Figure 22.13
Scanning electron micrograph of *Borrelia burgdorferi*, the cause of Lyme disease.

multiplication of the spirochete. The organisms then enter the bloodstream and become disseminated to all parts of the body but generally do not cross the placenta of pregnant women. Wide dissemination of the organisms accounts for the influenzalike symptoms of the first stage. After the first few weeks, an intense immune response occurs, and thereafter, it becomes very difficult to recover *B. burgdorferi* from blood or body tissues. The immune response against the bacterial antigens is probably responsible for the symptoms of the second stage. The third stage of Lyme disease is characterized by arthritis, and the affected joints have high concentrations of highly reactive immune cells and immune complexes. Thus, the joint symptoms of the third stage also probably result from immune responses against persisting bacterial antigens.

Epidemiology

Lyme disease is now known to occur throughout the world. In the United States, it is the most commonly reported vectorborne disease, with cases occurring in more than 40 states.

Like Rocky Mountain spotted fever, Lyme disease is a zoonosis, and humans are accidental hosts. Several species of ticks have been implicated as vectors, but the most important in the eastern United States is the deer tick *Ixodes scapularis* (figure 22.14). In some areas of the East Coast, 80% of these ticks are infected with *Borrelia burgdorferi*. Because of their small size (1–2 mm before feeding; 3–5 mm when fully engorged with blood), these ticks often feed and drop off their host without being detected, so that two-thirds of Lyme disease patients are unable to recall a tick bite. The ticks mature during a two-year cycle (figure 22.15); the nymph stage[1] is an aggressive feeder and mainly responsible for transmitting Lyme disease. The preferred host of *I. scapularis* is the white-

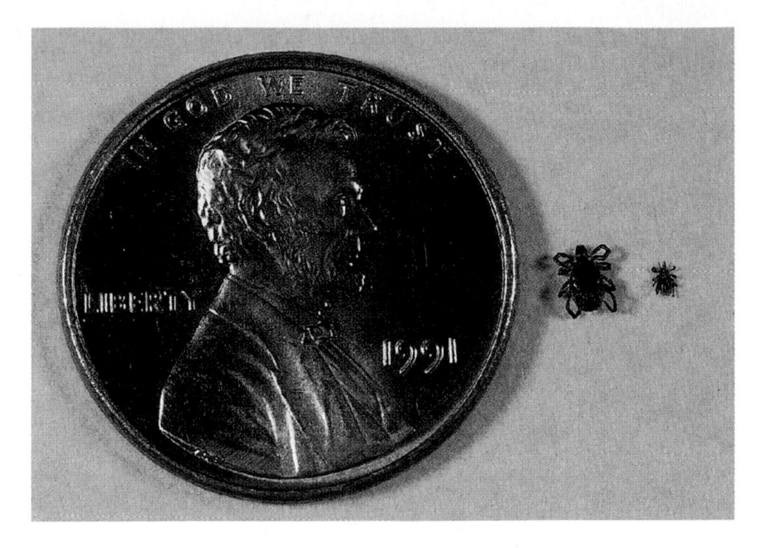

Figure 22.14
Ixodes scapularis, adult and nymph. This tick is the most important Lyme disease vector in the eastern United States.

footed mouse, which acquires *Borrelia burgdorferi* from an infected tick and develops a sustained bacteremia[2] with the organism, thus becoming a source of infection for other ticks. There is a low frequency of transovarial passage of the spirochete that helps sustain the organism in nature. Infected ticks and mice constitute the main reservoir of *B. burgdorferi*, but deer, while not a significant reservoir, are important because they can quickly spread the disease over a wide area and are the preferred winter host of the adult ticks. Tick nymphs are the most active from May to September, corresponding to the peak occurrence of Lyme disease cases. Adult ticks sometimes bite humans late in the season and transmit the disease. Infectious ticks can be present in well-mowed lawns as well as in wooded areas.

Prevention and Treatment

Preventive measures for Lyme disease are the same as those for Rocky Mountain spotted fever. No vaccine is yet available for general use in humans although one is being used for preventing Lyme disease in dogs. A biological control program for *Ixodes* ticks using a tiny parasitic wasp is being evaluated.

Antibiotics such as doxycycline, amoxicillin, and erythromycin are effective in patients with early disease. In late disease, the response to treatment is less satisfactory, presumably because the spirochetes are not actively multiplying and antibacterial medicines are usually ineffective against nongrowing organisms. Nevertheless, prolonged treatment with intravenous ampicillin or ceftriaxone has been curative in many cases.

Table 22.7 summarizes some features of Lyme disease.

[1]Nymph stage—a developmental stage in the life of a tick: a six-legged larval form emerges from the egg; after growing, it sheds its outer coat (molts) and becomes an eight-legged form called a *nymph*; after further molts, the nymph becomes the larger, sexually mature adult form

[2]Bacteremia—bacteria in the bloodstream

• **Immune complexes, p. 400**
• **Reservoir, p. 430**
• **Antibiotics, pp. 62, 154, 448–453**
• **Erythromycin, p. 464**

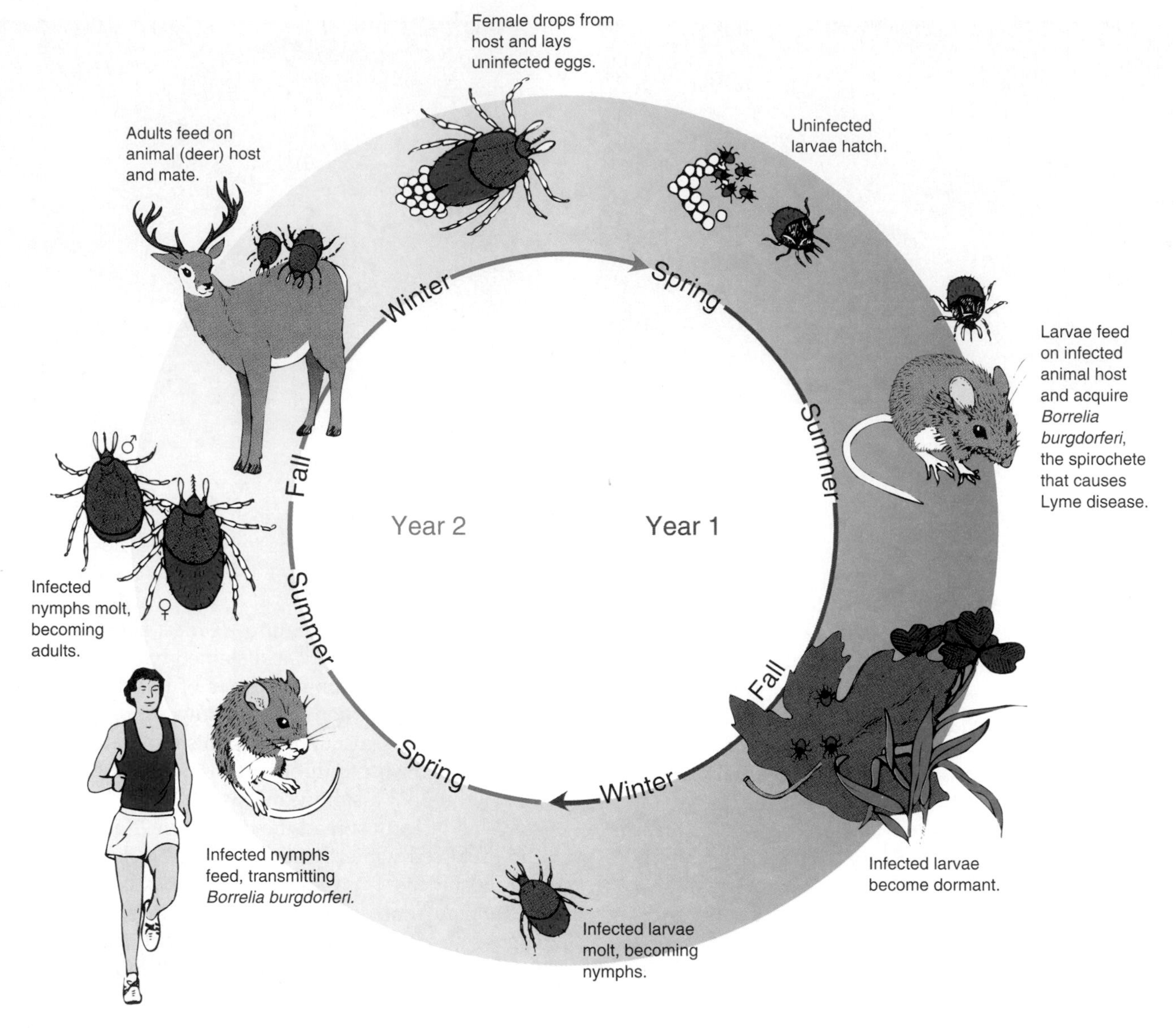

Figure 22.15
Life cycle of *Ixodes scapularis*, a vector of *B. burgdorferi*. The life cycles of other tick vectors involve different hosts and may, therefore, require different control measures.

Table 22.7 Lyme Disease	
Symptoms	Enlarging, red, ringlike lesion at the site of the bite; fever, malaise, headache, general achiness, enlargement of lymph nodes near bite, arthritic pain
Incubation period	Approximately one week
Causative agent	*Borrelia burgdorferi*, a spirochete
Laboratory diagnosis	ELISA, Western blot
Pathogenesis	Spirochetes injected by bite of infected tick, multiply and spread radially in the skin; spirochetes disseminate to body tissues by the bloodstream; organisms persist in tissue in quiescent state; immune reaction to organisms produces tissue damage
Epidemiology	Spread by the bite of ticks, *Ixodes* sp., usually found in association with animals such as white-footed mice and white-tailed deer living in wooded areas
Prevention	Protective clothing; tick repellents
Treatment	Early treatment with doxycycline, amoxicillin, or erythromycin

Perspective 22.2

Lyme Disease in California

There are probably many different ways for *Borrelia burgdorferi* to maintain itself in natural settings around the world. For example, in the United States, the natural history of the spirochete differs between the Northeast and the Far West. In the Northeast, *B. burgdorferi* is primarily maintained in white-footed mice, transmitted from generation to generation by *Ixodes scapularis*.[1] The same tick will then feed on humans, causing Lyme disease. In California, the pattern is quite different. The spirochete is maintained in nature in a wood rat, transmitted from rat to rat by a species of tick that does not bite humans. However, another tick species, *I. pacificus*, sometimes feeds on rats and becomes infected with *B. burgdorferi*. Since these ticks will also bite humans, they transmit Lyme disease. The fact that only 1% to 5% of *I. pacificus* ticks are infected with *B. burgdorferi* helps explain why Lyme disease is less common in California than in Connecticut. Since wood rats (also called *pack rats*) like to collect various items and take them to their nests, scattering insecticide-treated cotton balls in Lyme disease areas has been proposed. If the rats take the cotton balls to their nests, the insecticide will kill the *B. burgdorferi*-carrying ticks.

[1] *I. scapularis* was reported in 1993 to be the same species as *I. dammini*, the black-legged tick.

REVIEW QUESTIONS

1. What bacterium is generally found in cultures of furuncles and carbuncles? Give three features that characterize this organism.
2. List the events leading to bone infection by *Staphylococcus aureus* starting with infection of a hair follicle.
3. How does protein A of *S. aureus* interfere with phagocytosis of the antibody-coated organisms?
4. List four extracellular products of *S. aureus* that contribute to its virulence.
5. Why is the blister fluid of staphylococcal scalded skin syndrome free of staphylococci?
6. Give three ways in which *Streptococcus pyogenes* resembles *Staphylococcus aureus* and three ways in which they differ.
7. Compare and contrast the causative agents and epidemiology of Rocky Mountain spotted fever and Lyme disease.

Skin Diseases Caused by Viruses

Several childhood diseases are characterized by distinctive skin rashes (exanthems) caused by viruses carried to the skin by the blood from sites of infection in the upper respiratory tract. This group of diseases is usually diagnosed by inspection of the rash and other clinical findings. However, when the disease is not typical, tests can be performed to identify specific antibody against the virus, or the virus can be cultivated in the laboratory from skin lesions or upper respiratory secretions. In some cases, the virus causes characteristic changes in skin or respiratory cells, which can be identified under the microscope. For example, in chicken pox, cold sores, and measles, clinical specimens show large cells called **giant cells** formed by the fusion of two or more infected cells. Virus-infected cells can also be examined by electron microscopy or by using fluorescent antibodies against suspected viral pathogens.

Varicella (Chicken pox) Is Caused by a Virus that Remains Latent in the Body After Recovery

Varicella is currently the most widespread of the childhood exanthems (rashes) in the United States. The causative virus is a member of the herpesvirus family and, like others in that group, produces latent infection that can reactivate after recovery from the initial illness has occurred.

Symptoms

Most cases of varicella (chicken pox) are mild, sometimes unnoticed, and recovery is usually uncomplicated. The typical case has a rash, which is diagnostic. It begins as small, red macules (spots), papules (bumps), and vesicles (blisters) surrounded by a narrow zone of redness. The lesions can erupt anywhere on the body, but usually they first appear on the back of the head, then the face, mouth, main body, and arms and legs, ranging from only a few lesions to many hundreds. The lesions appear at different times and within a day or so go through a characteristic evolution from macule to papule to vesicle to pustule (pus-filled blister). After the pustules break, leaking viral-laden fluid, a crust forms, and then healing takes place. At any time during the rash, lesions are at various stages of evolution (figure 22.16). The lesions are pruritic (itchy), and scratching may lead to secondary infection by *Streptococcus pyogenes* or *Staphylococcus aureus*.

Symptoms of varicella tend to be more severe in older children and adults. In about 20% of adults, pneumonia develops, with rapid breathing, cough, shortness of breath, and a dusky skin color. The pneumonia subsides with the rash, but respiratory symptoms often persist for weeks. Varicella is also a major threat to newborn babies if the mother develops the disease within five days before delivery to two days

• **Latent infections, pp. 327–29**

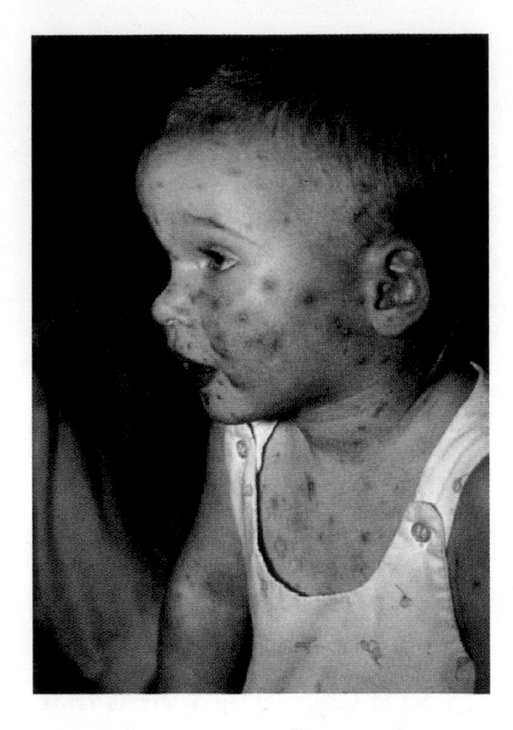

Figure 22.16
Child with varicella (chicken pox). Characteristically, lesions in various states of evolution—macules, papules, vesicles, and pustules—are present.

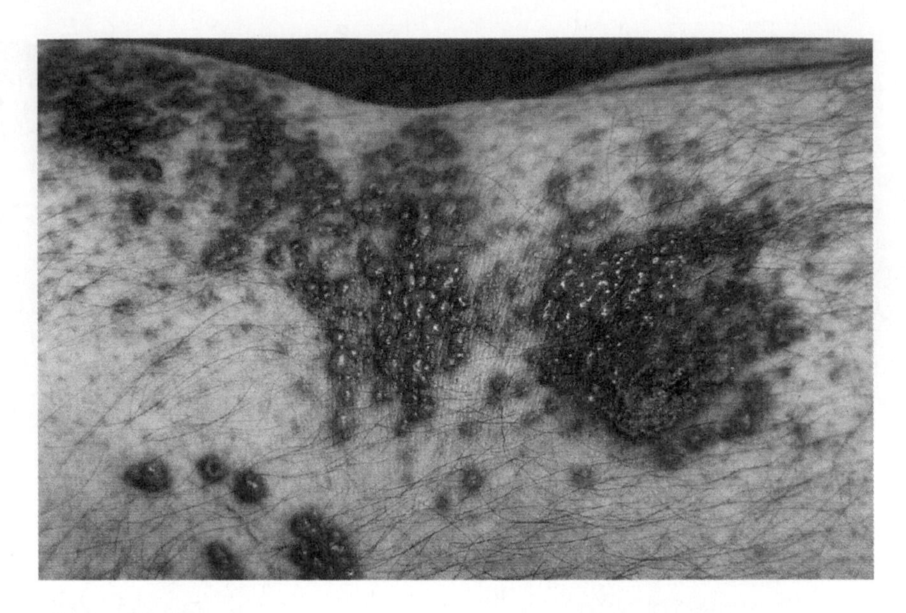

Figure 22.17
Shingles. The rash mimics that of chicken pox, except that it is limited to a sensory nerve distribution on one side of the body.

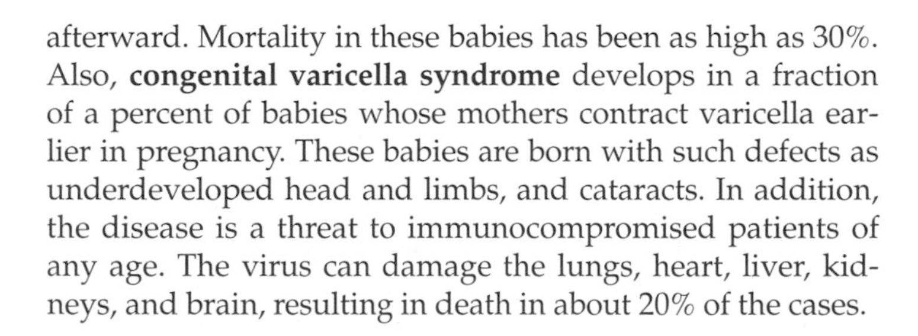

afterward. Mortality in these babies has been as high as 30%. Also, **congenital varicella syndrome** develops in a fraction of a percent of babies whose mothers contract varicella earlier in pregnancy. These babies are born with such defects as underdeveloped head and limbs, and cataracts. In addition, the disease is a threat to immunocompromised patients of any age. The virus can damage the lungs, heart, liver, kidneys, and brain, resulting in death in about 20% of the cases.

Herpes zoster Reactivation of varicella is called *herpes zoster*[1] (shingles). It can occur at any age but becomes increasingly common with advancing age. It begins with pain in the area supplied by a nerve of sensation, often on the chest or abdomen but sometimes on the face or an arm or leg. After a few days to two weeks, the characteristic rash of varicella appears along the course of the nerve (figure 22.17). The rash generally subsides within a week, but pain may persist for weeks or months.

Causative Agent

Varicella is caused by the varicella-zoster virus, a member of the herpesvirus family. It is an enveloped, medium-sized (150–200 nm) double-stranded DNA virus, indistinguishable from other herpesviruses in appearance. It can be cultivated in tissue cell cultures and identified using specific antiserum.

[1]Herpes zoster is often mistakenly considered to be the name of a virus.

• **Viral size, p. 18**

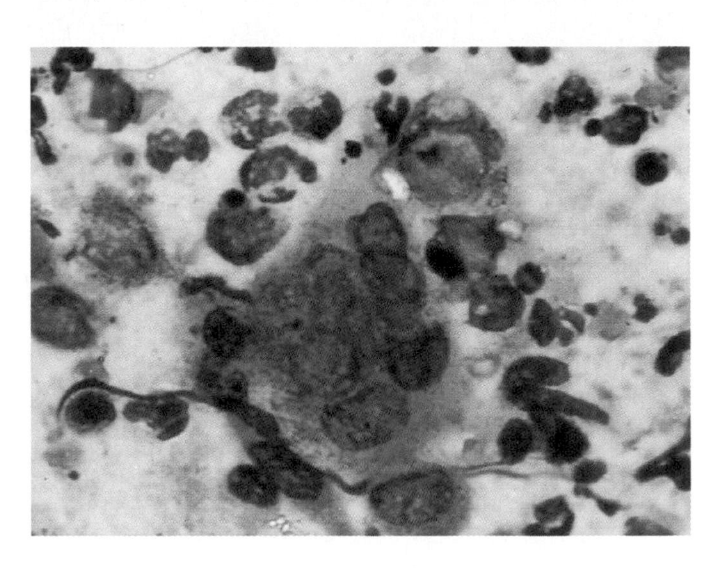

Figure 22.18
Tzanck test (named for a Russian dermatologist, Arnault Tzanck 1886–1954). Stained material from the base of a vesicle often shows large, multinucleated cells like the one shown. Finding these cells (a positive Tzanck test) is not specific for varicella-zoster virus since infections by other members of the herpesvirus family produce the same result.

Pathogenesis

The virus enters the body by the respiratory route, establishes an infection, replicates, and disseminates to the skin via the bloodstream. After the living layers of skin cells are infected, the virus spreads directly to adjacent cells, and the characteristic skin lesions appear. Stained preparations of infected cells show **intranuclear inclusion bodies** at the site of viral replication in the nucleus. Some infected cells fuse together forming multinucleated giant cells, the basis of a positive Tzanck test (figure 22.18). The infected cells swell and ultimately lyse.

Perspective 22.3

Shingles Patient Starts Chicken Pox Epidemic

A 29-year-old woman complained to her physician of a painful rash that appeared on one side of her chest. The doctor examined it and noticed that it consisted of small, red bumps; blisters; and scabs that followed the branches of one of the nerves of skin sensation. He informed her that she had "shingles," also known as herpes zoster and that it represented a reactivation of the chicken pox infection she had years ago in elementary school. This seemed somewhat unlikely to the patient, but her skepticism lessened appreciably when one of her four children developed chicken pox two weeks later, followed by the other three, and then their friends, until half of the younger children in the immediate community had developed the disease.

The virus enters the sensory nerves, presumably when an area of skin infection advances to involve a sensory nerve ending. Conditions inside the nerve cell do not permit full expression of the viral genome; however, viral DNA is present in the ganglia[1] of the nerves and is fully capable of coding for mature infectious virus. The mechanism of suppression of viral replication within the nerve cell is not known but is probably under the control of immune cells. Thus, the occurrence of herpes zoster correlates with a decline in cell-mediated immunity. With the decline, infectious virus is presumably produced in the nucleus of the nerve cells and is carried to the skin by the normal circulation of cytoplasm within the axon[2] of the nerve cell. With the appearance of the skin lesions, a prompt, intense anamnestic boost of both cellular and humoral immunity ensues. A marked inflammatory reaction occurs in the ganglion with an accumulation of immune cells, and the herpes zoster quickly disappears, although sometimes leaving scars and chronic pain.

Epidemiology

Usually less than 200,000 cases of varicella are reported each year in the United States, but the true incidence is estimated at 3.7 million. Most cases go unreported, and many are so mild as to go unnoticed. As with many diseases transmitted by the respiratory route, most cases occur in the winter and spring months. Humans are the only reservoir, and because of the high infectivity of varicella-zoster virus, about 90% of humans are infected by the age of 15. The incubation period of the disease averages about two weeks, with a range of 10 to 21 days. Cases are infective from one to two days before the rash appears until all the lesions have crusted (usually four days after the onset).

The mechanism by which the varicella-zoster virus persists allows it to survive in small, isolated populations. When a highly infectious virus such as measles is introduced into such a population, it spreads quickly and infects most of the susceptible individuals, who either become immune or die. If susceptible victims are unavailable, the measles virus will disappear from the community. By contrast, when varicella-zoster virus is introduced into such a community, it can persist indefinitely and cause recurrent epidemics of chicken pox whenever sufficient numbers of susceptible children have been born. Cases of shingles, which are the source of infectious virus, may occur in as many as 1% of elderly people.

Prevention

In March 1995, a live, attenuated varicella vaccine was licensed in the United States. The virus was originally isolated from a varicella case in Japan in 1970 and was attenuated by subculturing it many times in various tissue cell cultures. It has proven safe with use in over 2 million people in various countries, starting in about 1984. The vaccine is recommended for all healthy persons age 12 months or older who do not have a history of varicella or who lack laboratory evidence of immunity to the disease. Immunization should be done sometime before one's 13th birthday because of the increased likelihood of serious complications from varicella in older children and adults. The vaccine should not be administered to people with malignant or immunodeficiency diseases until studies of its safety in such individuals are complete.

There are increasing numbers of patients with impaired immunity who are at risk of severe disseminated varicella-zoster virus infections. These patients include those with cancer, AIDS, and organ transplants and newborn babies whose mothers contracted varicella near the time of delivery. Patients with impaired immunity can be partially protected from severe disease if they are passively immunized by injecting them with zoster immune globulin (ZIG) derived from the blood of recovered herpes zoster patients. In the future, most of these cases ideally will be prevented by active immunization.

[1]Ganglia—plural of *ganglion*, a small nodule near the spine that contains the nuclei and cell bodies of sensory nerves

[2]Axon—a long, cytoplasm-filled extension of a nerve cell, such as from the cell body in the ganglion all the way to a fingertip

- **Passive immunization, p. 389**
- **Attenuated vaccine, p. 390**

The antiviral drug acyclovir, when given intravenously, has been fairly effective in treating people who develop disseminated disease. The same medication taken orally in high doses shortens the illness and decreases the chronic pain of herpes zoster that often occurs in older age groups.

The main features of varicella are summarized in table 22.8.

Rubeola (Measles) Is Often Complicated by Secondary Bacterial Infections

One of the great success stories of the last half of the 20th century has been the dramatic reduction of measles cases through the use of a live vaccine against the disease. There is hope that the disease can be entirely eliminated from the United States.

Symptoms

Rubeola (measles, figure 22.19) begins with fever, runny nose, cough, and swollen, red, weepy eyes. Within a few days, a fine red rash appears on the forehead and spreads outward over the rest of the body. Unless complications occur, symptoms generally disappear in about one week.

Unfortunately, many cases are complicated by secondary infections caused by bacterial pathogens, mainly *Staphylococcus aureus*, *Streptococcus pneumoniae*, *Streptococcus pyogenes*, and *Haemophilus influenzae*. These pathogens readily invade the body because rubeola damages the normal body defenses. Secondary infections most commonly involve the middle ear and the lungs, causing earaches and pneumonia, respectively.

In about 5% of cases, the rubeola virus itself causes pneumonia, with rapid breathing, shortness of breath, and dusky skin color from lack of adequate oxygen exchange in the lungs. Another serious complication is brain involvement (encephalitis), which is marked by fever, headache, confusion, and seizures. This complication occurs in about one out of every 1,000 cases of measles. Permanent brain damage, with mental retardation, deafness, and epilepsy, commonly results from measles encephalitis.

Very rarely, measles is followed 2 to 10 years later by a disease called **subacute sclerosing panencephalitis,** which is marked by slowly progressive degeneration of the brain, generally resulting in death within two years. A defective measles virus can be detected in the brains of these patients, and high levels of measles antibody are present in their blood. This is an example of a "slow virus" disease. It has all but disappeared from the United States with widespread vaccination against measles.

Measles occurring during pregnancy results in an increased risk of miscarriage, premature labor, and low birth weight. However, birth defects are generally not seen.

• **Slow virus, pp. 329–30**

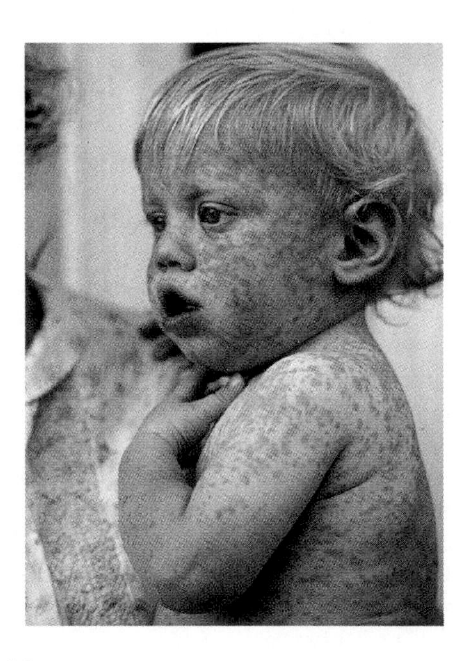

Figure 22.19
Child with rubeola (measles). The rash is accompanied by fever, runny nose, and a bad cough.

Table 22.8 Chicken Pox (Varicella)

Symptoms	Small, itchy bumps and blisters over skin and mucous membranes, fever; latent infection may become manifest as herpes zoster (shingles)
Incubation period	10–21 days
Causative agent	Varicella-zoster virus; double-stranded DNA, enveloped virus of the herpesvirus family
Laboratory diagnosis	Fluorescent antibody to detect viral antigen in vesicle fluid; viral isolation in tissue cell cultures
Pathogenesis	Upper respiratory virus multiplication followed by dissemination by bloodstream to the skin; cytopathic effect of virus in epidermis; vesicles formed containing serum, polymorphonuclear leukocytes; multinucleated giant cells result from fusion of infected cells
Epidemiology	Highly infectious; transmission by respiratory route; humans the only source (cases of chicken pox and herpes zoster)
Prevention	Passive immunization with zoster immune globulin (ZIG) for immunocompromised individuals; attenuated vaccine
Treatment	Nonaspirin medicines to relieve itching, control fever; acyclovir for severe varicella cases and for older people with herpes zoster

Causative Agent

Measles is caused by a pleomorphic, medium-sized (120–200 nm diameter), single-stranded, negative sense RNA virus of the paramyxovirus family. The virus can be cultivated from the patient's nasal secretions, blood, and urine until the second or third day of the rash. A number of tissue cell cultures can be used to grow the virus. Rubeola virus is unusual for RNA viruses in that it replicates in the host cell nucleus; stains of infected cells show both cytoplasmic and intranuclear inclusion bodies. The viral envelope has two biologically active projections. One, H, is responsible for viral attachment to host cells and causes hemagglutination *in vitro.* The other, M, is responsible for fusion of the viral outer membrane with the host cell. The M antigen also causes adjacent infected host cells to fuse together, producing multinucleated giant cells.

Pathogenesis

Rubeola virus is acquired by the respiratory route. It presumably replicates in the upper respiratory epithelium, spreads to lymphoid tissue, and following further replication, eventually spreads to all parts of the body. Mucous membrane involvement is responsible for an important diagnostic sign, **Koplik's spots** (figure 22.20), which are usually best seen in the mouth opposite the molars.[1] Koplik's spots look like tiny grains of salt lying on an oral mucosa that is red and rough, resembling red sandpaper. Damage to the respiratory mucous membranes partly explains the markedly increased susceptibility of measles patients to secondary bacterial infections, especially infection of the middle ear (otitis media) and lung (pneumonia). Involvement of the intestinal epithelium may explain the diarrhea that sometimes occurs in measles and contributes to high measles death rates in impoverished countries.

The skin rash of measles results from the cytopathic effect of rubeola virus replication in skin cells and the cellular immune response against the viral antigen in the skin. It is not known why the rash characteristically spreads outward, often clearing on the face before it reaches the lower extremities.

The measles virus temporarily suppresses cellular immunity, which can cause reactivation of latent herpes simplex (cold sores) and tuberculosis.

In the United States, deaths from measles occur in about one to two of every 1,000 cases, mainly from pneumonia and encephalitis.

Epidemiology

Humans are the only natural host of rubeola virus. Before vaccination became widespread in the 1960s, 4 million cases of measles per year were estimated to occur in the United States, and probably less than 1% of the population escaped infection. Continued use of measles vaccine resulted in a progressive decline in cases, reaching a low of less than 1,500 reported cases in 1983. There was hope in the 1970s that measles would be quickly eradicated in the United States, but epidemics of the disease continue to occur. For example, more than 27,000 cases were reported in 1990; the actual incidence was estimated to be several times higher, based on known underreporting of cases. Interestingly, many of the cases were acquired in doctors' offices and emergency departments.

The resurgence of measles in the United States was due to the continued presence of nonimmune populations (figure 22.21) and the influx of immigrants and travelers from other countries. Nonimmune individuals include

[1]Molars—the teeth that grind food, located in the back of the jaw

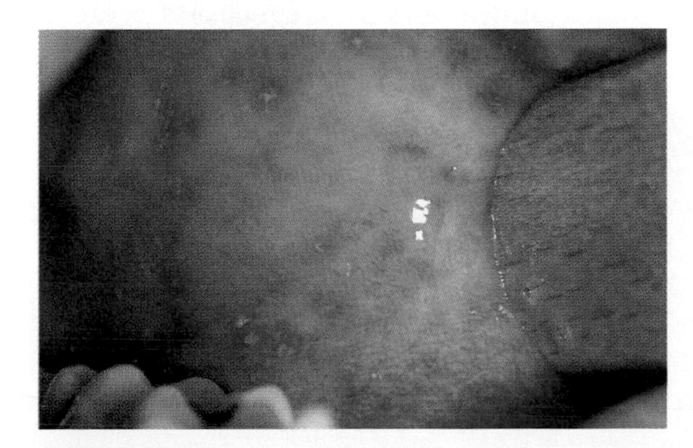

Figure 22.20
Koplik's spots, characteristic of rubeola, are often transitory. They have been described as grains of salt on a red base.

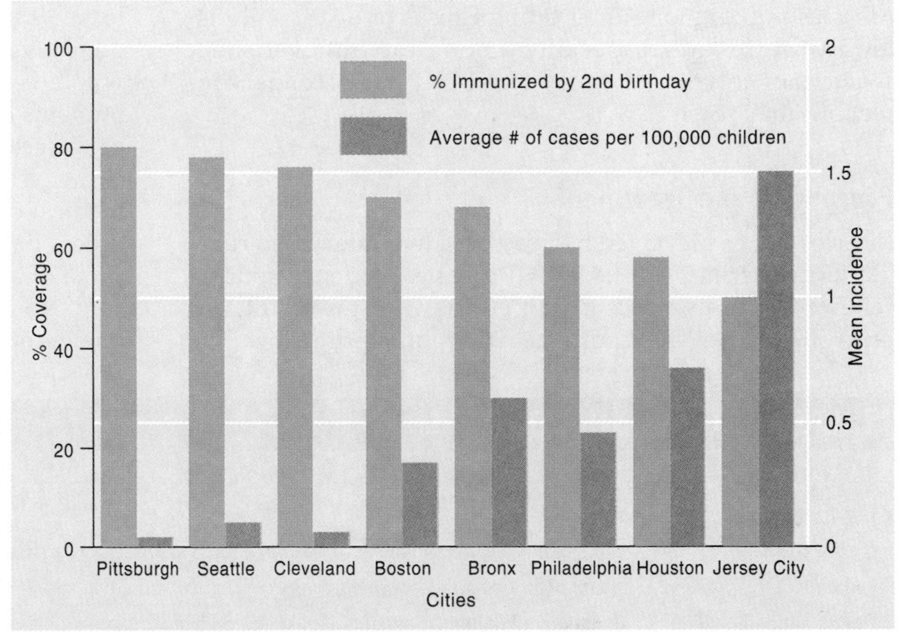

Figure 22.21
The incidence of measles correlates inversely with the immunization rates for different cities.

Source: Centers for Disease Control and Prevention. 1991. *Morbidity and Mortality Weekly Report* 40:37.

Perspective 22.4

Measles Sparks a Massacre

In February 1836, Marcus and Narcissa Whitman set out to cross the continent with fur traders for their honeymoon. After a long journey marked by considerable hardship, they arrived in Oregon territory and settled among the Cayuse people near what is now Walla Walla, Washington. They worked hard to learn the Cayuse language, to establish farming among the people, and to convert the people to Christianity. Other settlers arriving from the disease-ridden Mississippi Valley began to stop at the mission, bringing with them typhoid fever, malaria, cholera, and other diseases. Marcus, who was a physician, worked hard to treat the ill, but none of his medicines were effective. Many Cayuse people died, while the white people seemed to have a natural resistance to the disease and generally survived. The Cayuse, none of whom converted to Christianity, suspected the missionaries were evil shamans who used their power to kill people. One man noted that his wife developed spots on her skin and died after taking the doctor's medicine. Another man decided to test the doctor by pretending to be sick. Marcus gave him medicine and later the man contracted measles and died. The leaders of the Cayuse met and agreed that this test proved the doctor's guilt, and the missionaries would have to be killed for the Cayuse to be saved. On November 29, 1847, the Cayuse attacked the mission and killed many of the inhabitants, Marcus and Narcissa being among the first to die. Angry settlers and militiamen subsequently drove the Cayuse people out of the Walla Walla Valley.

(1) children too young to be vaccinated; (2) preschool children never vaccinated; (3) children and adults inadequately vaccinated; and (4) persons not vaccinated for religious or medical reasons. Of almost 250 outbreaks that occurred in 1989 and 1990, the majority were in people 5 to 19 years old, but many outbreaks occurred among those less than five years of age and those 20 years old or older. Some cases occurred in people born before 1957, presumably because they escaped natural measles and therefore did not become immune. About 3% of the cases either were acquired abroad or resulted from contact with such cases. Measles ranks among the leading causes of death and disability worldwide, especially among the impoverished, where the mortality rate may reach 15% and secondary infections may reach 85%.

Prevention and Treatment

Measles can be prevented by injecting a live, attenuated rubeola virus vaccine. The 1989–1991 measles epidemic spurred renewed intensive vaccination efforts with the result that only 309 cases were reported in 1995, an all-time low. The vaccine is usually given together with mumps and rubella vaccines (MMR). Until 1976, measles vaccine was generally given at 9 to 12 months of age. However, because many of these vaccinated infants failed to develop adequate immunity, the age of administration of the vaccine was raised to 15 months. At this age, the immune system is more responsive, and there is no danger that maternal placentally transferred antibody will interfere with the vaccine. Immunity develops in at least 95% of the recipients. Since 1989, a second injection of vaccine is recommended at entry into elementary school. The two-dose regimen has resulted in at least 99% of the recipients becoming immune. In an epidemic, vaccine is given to babies as young as six months, who are then reimmunized before their second birthday. Students entering high school or college are also advised to get a second dose of vaccine if they have not received one earlier. Those at special risk of acquiring rubeola, such as medical personnel, should be immunized regardless of age, unless they definitely have had measles or have laboratory proof of immunity.

No antiviral treatment exists for rubeola at present. Some features of rubeola are summarized in table 22.9.

Table 22.9 Measles (Rubeola)

Symptoms	Rash, malaise, fever, conjunctivitis, cough, and nasal discharge
Incubation period	10–12 days
Causative agent	Rubeola virus, a single-stranded, negative sense RNA virus, paramyxovirus family
Laboratory diagnosis	Complement fixation or hemagglutination inhibition tests; cultivation of virus in tissue cell cultures
Pathogenesis	Virus multiplies in respiratory tract; carried by blood to various parts of body, notably skin, lungs, and brain; damage to respiratory tract epithelium leads to secondary infection of ears and lung
Epidemiology	Acquired by respiratory route; highly contagious; humans only source
Prevention	Attenuated virus vaccine administered to infants, preferably at 15 months of age, with a second dose at kindergarten age or adolescence
Treatment	None but symptomatic; antibiotics for secondary infections

Case Presentation

The patient was a 20-year-old asymptomatic man who was immunized against measles as a requirement for starting college. He had received his first dose of measles vaccine at approximately one year of age.

Past medical history revealed that he was a hemophiliac and had contracted the human immunodeficiency virus from clotting factor (a blood product given to control bleeding) contaminated with the virus.

Laboratory tests showed that he had a very low CD4+ lymphocyte count, indicating a severely damaged immune system.

About a month after his precollege immunization, he developed pneumocystosis, a lung infection characteristic of AIDS, was hospitalized, had a good response to treatment, and was discharged. Ten months later, he was again hospitalized for symptoms of a severe lung infection. He had no rash. Multiple laboratory tests to determine the cause of his infection were negative. Finally, a lung biopsy was performed and revealed "giant cells," very large cells with multiple nuclei. Cytoplasmic and intranuclear inclusion bodies were also present. This picture was highly suggestive of measles pneumonia, and measles virus subsequently was recovered from tissue cell cultures of the biopsy material. Subsequent studies showed it to be the measles vaccine virus. The patient received intravenous gamma globulin and an experimental antiviral medication, ribavirin, and improved. However, subsequently, his condition deteriorated, and he died of presumed complications of AIDS.

1. Is measles immunization a good idea for people with immunodeficiency?
2. Is it surprising that the vaccine virus was still present in this patient 11 months after vaccination? Explain.
3. Despite the severe infection there was no rash. Why?

Discussion

1. Measles is often disastrous for persons with AIDS or other immunodeficiencies. They should be immunized as soon as possible in their illness, before the immune system becomes so weakened it cannot respond effectively to the vaccine. Also, as this and other cases have shown, the vaccine virus can itself be pathogenic when immunodeficiency is severe. With the worldwide effort to eliminate measles, the risk of exposure to the wild measles virus is declining, but outbreaks in colleges and other institutions still occur. A severely immunodeficient individual can be passively immunized against measles with gamma globulin if exposure to the wild virus occurs.
2. Measles is often given as an example of a persistent viral infection, meaning that, following infection, the virus can persist in the body for months or years in a slowly replicating form. It has been suggested but not proven that this explains the lifelong immunity conferred by measles infection in normal people. In rare presumably normal individuals and more commonly in malnourished or immunodeficient individuals, persistent infection leads to damage to the brain, lung, liver, and, possibly, the intestine. Subacute sclerosing panencephalopathy can follow measles vaccination but at a much lower rate than after wild virus infection.
3. Following acute infection, the measles virus floods the bloodstream and is carried to various tissues of the body, including the skin. The rash of measles is caused by T lymphocytes attacking measles virus antigen lodged in the skin capillaries. In the absence of functional T lymphocytes, the rash does not occur.

The Most Serious Consequence of Rubella Is Birth Defects

In contrast to varicella and rubeola, rubella (also called *three-day measles* or *German measles*) is typically a mild, often unrecognized disease that is difficult to diagnose. Nevertheless, infection of pregnant women can have tragic consequences.

Symptoms

Characteristic symptoms of rubella are slight fever, mild cold symptoms, and enlarged lymph nodes behind the ears and on the back of the neck. After about a day, a faint rash consisting of innumerable pink spots appears over the face, chest, and abdomen (figure 22.22). Unlike with rubeola, there are no diagnostic mouth lesions. Adults commonly develop painful joints, with pain generally lasting three weeks or less. Other symptoms generally last only a few days. However, the significance of rubella lies not with these symptoms but rather with rubella's threat to the fetuses of pregnant women, which will be discussed in detail next.

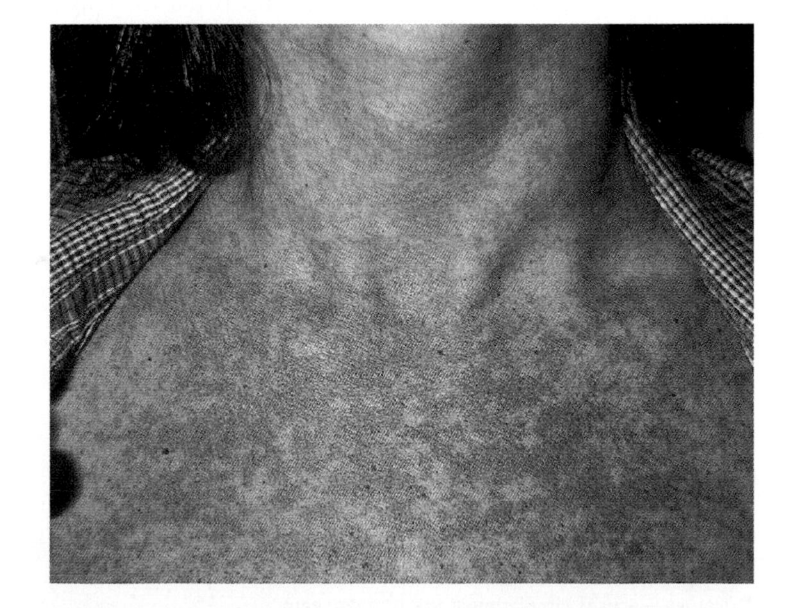

Figure 22.22
Adult with rubella (German measles). Symptoms are often very mild, but the effects on a fetus may be devastating.

Causative Agent

Rubella is caused by the rubella virus, a member of the togavirus family. It is a small (about 60 nm in diameter), enveloped, single-stranded, positive sense RNA virus that can readily be cultivated in a variety of tissue cell cultures. Surface glycoproteins give the virus *in vitro* hemagglutinating ability for some species of red blood cells. Hemagglutination is inhibited by specific antibody, allowing serological identification of the virus.

Pathogenesis

The rubella virus enters the body by the respiratory route. It multiplies in the nasopharynx and enters the bloodstream, causing a sustained viremia.[1] The blood transports the virus to various body tissues, including the skin and joints. Humoral and cell-mediated immunity develop against the virus, and the resulting antibody-antigen reactions probably account for the rash and joint symptoms and then for the rapid disappearance of the disease.

In pregnant women, the placenta becomes infected during the period of viremia. Early in pregnancy, the virus readily crosses the placenta and infects the fetus, but as the pregnancy progresses, fetal infection becomes increasingly less likely. Virtually all types of fetal cells are susceptible to infection; some cells are killed, while others develop a persistent infection in which cell division is impaired and chromosomes are damaged. The result is a characteristic pattern of fetal abnormalities that is referred to as the **congenital rubella syndrome**. The abnormalities include cataracts and other abnormalities of the eyes, brain damage, deafness, heart defects, and low birth weight despite normal gestation. Babies may be stillborn, but more commonly they are born alive. Those that live continue to excrete rubella virus in throat secretions and urine for many months. The viral persistence occurs despite the presence of neutralizing antibody. The babies produce antiviral IgM and IgG antibodies, as well as a cellular immune response that, in time, eliminates the persistently infected cells. The likelihood of the syndrome varies according to the age of the fetus when infection occurs. Infections occurring during the first six weeks of pregnancy result in almost 100% of the fetuses having a detectable injury, most commonly minor deafness. However, even infants who are apparently normal excrete rubella virus for extended periods and thus can infect others.

Epidemiology

Humans are the only natural host for rubella virus. The disease is highly contagious although less so than rubeola; it is estimated that in the prevaccine era, 10% to 15% of people reached adulthood without being infected. Complicating the epidemiology of rubella is the fact that over 40% of infected individuals fail to develop symptoms. People with few or no symptoms continue their usual activities and can spread the virus. People who develop typical rubella can be infectious for as much as seven days before the rash appears until seven days afterwards. Before widespread use of the vaccine began in 1969, periodic major epidemics arose. One epidemic in 1964 resulted in about 30,000 cases of congenital rubella syndrome. Past estimates of the average cost of lifetime medical care of patients with the syndrome exceeded $200,000. Undoubtedly, the estimates would be considerably higher today.

[1]Viremia—viruses circulating in the bloodstream

• **Persistent viral infection, p. 327**

Prevention and Treatment

Prevention of rubella depends on subcutaneous injection of a live attenuated rubella virus administered to babies at 12 to 16 months of age. The vaccine produces long-lasting immunity in about 95% of recipients. Unfortunately, many poor and inner-city infants fail to receive the vaccine. To reduce the risk of rubella transmission, most states require documentation of vaccination before entry into school, but many preschool children are underimmunized. Among women of childbearing age, 6% to 25% of different groups lack rubella immunity. The vaccine is not given to pregnant women for fear that it might result in congenital defects. As an added precaution, women are advised not to become pregnant for three months after receiving the vaccine.

Use of the vaccine has markedly reduced the incidence of rubella in the United States (figure 22.23). Nevertheless, outbreaks of the disease can continue to occur because of the presence of underimmunized populations. For example in 1990, 10 outbreaks occurred in prisons, four in colleges, and four in religious groups that refuse vaccinations. Since then, intensified vaccination programs have reduced the incidence of rubella to generally less than 250 cases per year.

After a person has been exposed to rubella, the administration of gamma globulin does not prevent the disease. No specific antiviral therapy is available.

Some features of rubella are summarized in table 22.10.

There Are Many Other Childhood Viral Exanthems

The kinds of viruses that can cause childhood rashes probably number in the hundreds. One group alone, the enteroviruses, has about 50 members that have been associated with skin lesions. The causes of two common childhood exanthems, erythema infectiosum and exanthem subitum, have only been established in recent years.

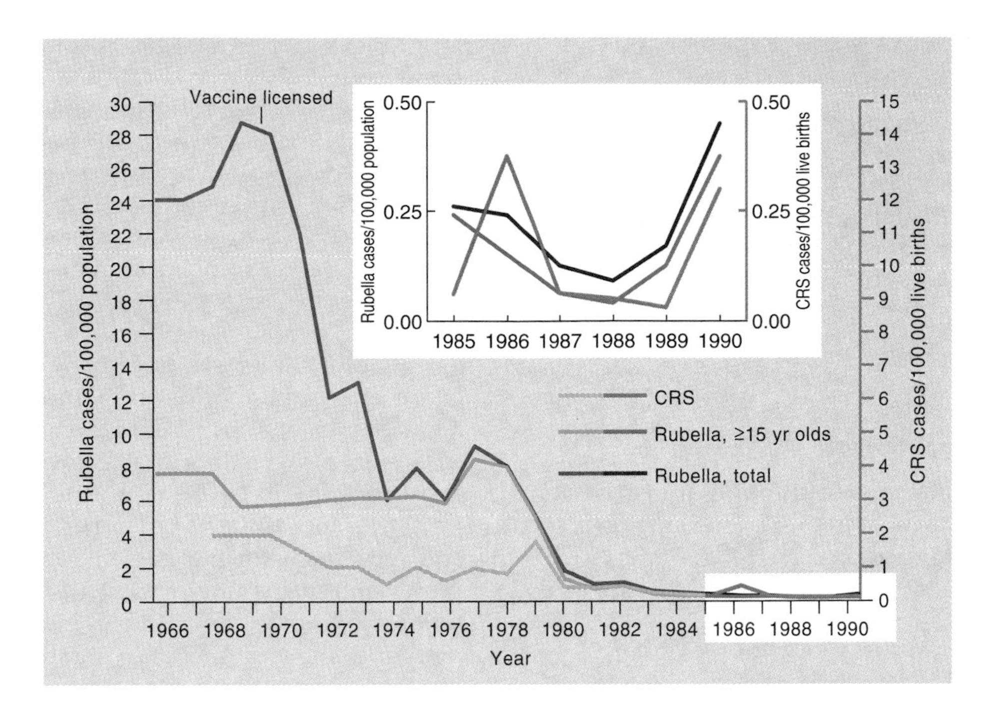

Figure 22.23
Incidence of rubella and congenital rubella syndrome (CRS).

Source: Centers for Disease Control and Prevention. 1991. *Morbidity and Mortality Weekly Report* 40:93.

Table 22.10 Rubella ("Three-day Measles" or "German Measles")

Symptoms	Malaise, headache, fever, mild conjunctivitis, rash beginning on forehead and face, enlarged lymph nodes behind the ears
Incubation period	14–21 days
Causative agent	Rubella virus (an RNA virus, togavirus family)
Laboratory diagnosis	Inoculation of tissue cultures with throat, blood, or urine specimens; hemagglutination inhibition, complement fixation, and neutralization tests
Pathogenesis	Following replication in the upper respiratory tract, virus disseminates to all parts of the body and crosses the placenta; surviving fetuses may develop abnormally and excrete virus for months after birth
Epidemiology	Virus possibly present in nose and throat from one week before rash to one week after; acquisition by respiratory route; humans only source
Prevention	Live attenuated rubella virus vaccine administered to children at 15 months of age
Treatment	None but symptomatic

Erythema Infectiosum

Erythema infectiosum (fifth disease)[1] occurs in both children and young adults. The illness begins with nonspecific symptoms—fever and malaise, head and muscle aches. Then a diffuse redness appears on the cheeks, giving the appearance of the face if it is slapped. The rash commonly spreads in a lacy pattern to involve other parts of the body, especially the extremities. The rash may come and go for two weeks or more before recovery. Joint pains are a prominent feature of some adult infections. The disease is caused by parvovirus B–19, a small (18 to 28 nanometers), nonenveloped, single-stranded DNA virus of the parvovirus family. Equal numbers of virions contain negative as contain positive sense DNA. The virus preferentially infects certain bone marrow cells and is a major threat to persons with sickle cell and other anemias because the infected marrow sometimes stops producing blood cells, a condition known as **aplastic crisis.** Also, about 10% of women infected with the virus during pregnancy suffer spontaneous abortion.

[1]Fifth disease—in the early 1900s, common childhood exanthems were numbered 1–6 as follows: (1) rubeola, (2) scarlet fever, (3) rubella, (4) uncertain, (5) erythema infectiosum, and (6) exanthem subitum

Exanthem Subitum

Exanthem subitum (roseola infantum or sixth disease) occurs chiefly in infants six months to three years old. It begins abruptly with fever that may reach 105°F and cause convulsions. After several days, the fever vanishes and a transitory rash appears, mainly on the chest and abdomen. The patient has no symptoms at this point, and the rash vanishes in a few hours to two days. This disease is caused by herpesvirus, type 6.

Warts Are Skin Tumors Caused by Papillomaviruses

A few viruses can infect the skin through abrasions and other injuries. One of the most interesting examples is the group of viruses that cause warts. Warts are in fact small tumors called **papillomas**[1] that are caused by wart virus infection. These tumors usually remain benign, although some genital warts are associated with cancer of the uterine cervix (which is discussed in chapter 27). In more than half the cases, warts on the skin disappear within one to two years without any treatment. Warts are caused by papillomaviruses (figure 22.24), a group of viruses within the papovavirus family. They are small (about 50 nm diameter), nonenveloped, double-stranded DNA viruses. Over 45 different papillomaviruses are known to infect humans.

Papillomaviruses have been very difficult to study because they fail to grow in tissue cell cultures or experimental animals. Wart viruses can survive on inanimate objects such as wrestling mats, towels, and shower floors, and infection can be acquired from such contaminated objects. The virus infects the deeper cells of the epidermis and reproduces in the nuclei. Some of the infected cells grow abnormally and produce the papilloma. The incubation period ranges from 2 to 18 months. Infectious virus is present in the wart and can contaminate fingers or objects that pick or rub the lesions. Like other tumors, warts can only be treated effectively by killing or removing all of the abnormal cells. This can usually be accomplished by freezing the wart with liquid nitrogen, by cauterization[2] with an electric arc, or by surgical removal. However, a persistent papillomavirus infection of normal appearing adjacent skin can subsequently lead to the formation of additional papillomas.

[1]Papillomas—benign tumors consisting of nipplelike protrusions of tissue covered with skin or mucous membrane

• Tissue cell culture, pp. 314, 315

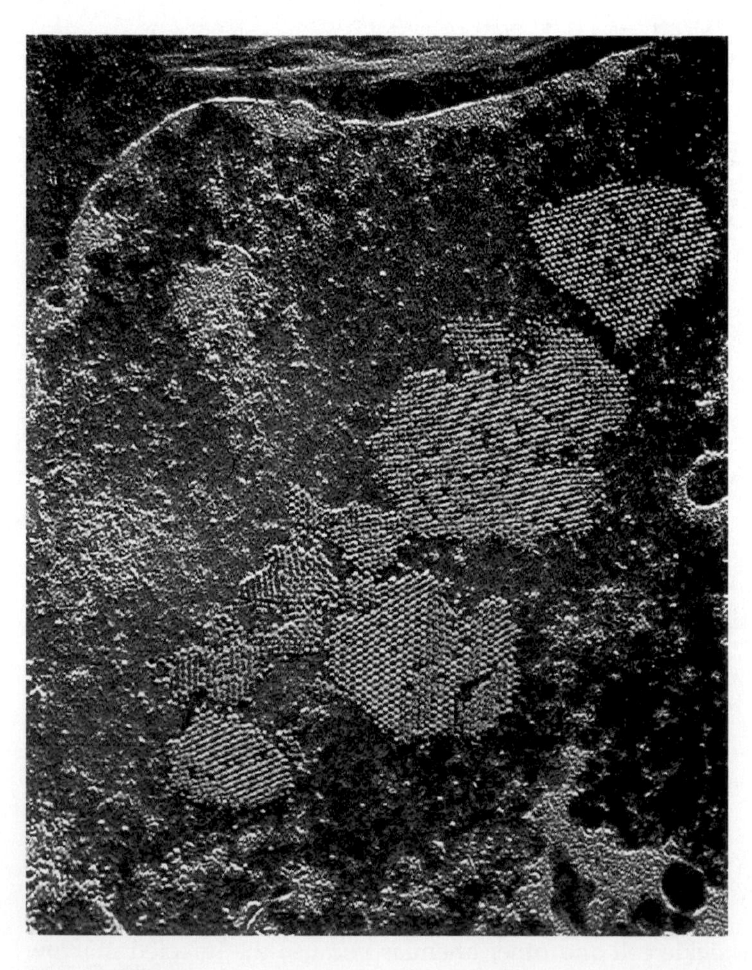

Figure 22.24
Wart virus. The virions appear yellow in this color-enhanced transmission electron micrograph.

Skin Diseases Caused by Fungi

Earlier in this chapter, we mentioned the role of yeast of the genus *Malassezia* in causing mild skin diseases, such as tinea versicolor. Other fungi are responsible for more serious infections of the skin, although even in these cases the condition of the host's defenses against infection is often crucial. The yeast *Candida albicans* (figure 22.25) may live harmlessly among the normal flora of the skin, but in some people it invades the deep layers of the skin and subcutaneous tissues. In many people with candidal skin infections, no precise cause for the invasion can be determined. Certain molds also cause skin disease, but they do not invade the deep skin layers.

Fungi Cause Ringworm

Certain species of molds can invade hair, nails, and the keratinized portion of the skin. The resulting conditions have colorful names such as *jock itch, athlete's foot,* and *ringworm.*

[2]Cauterization—burning of tissue, usually with an electrically heated needle

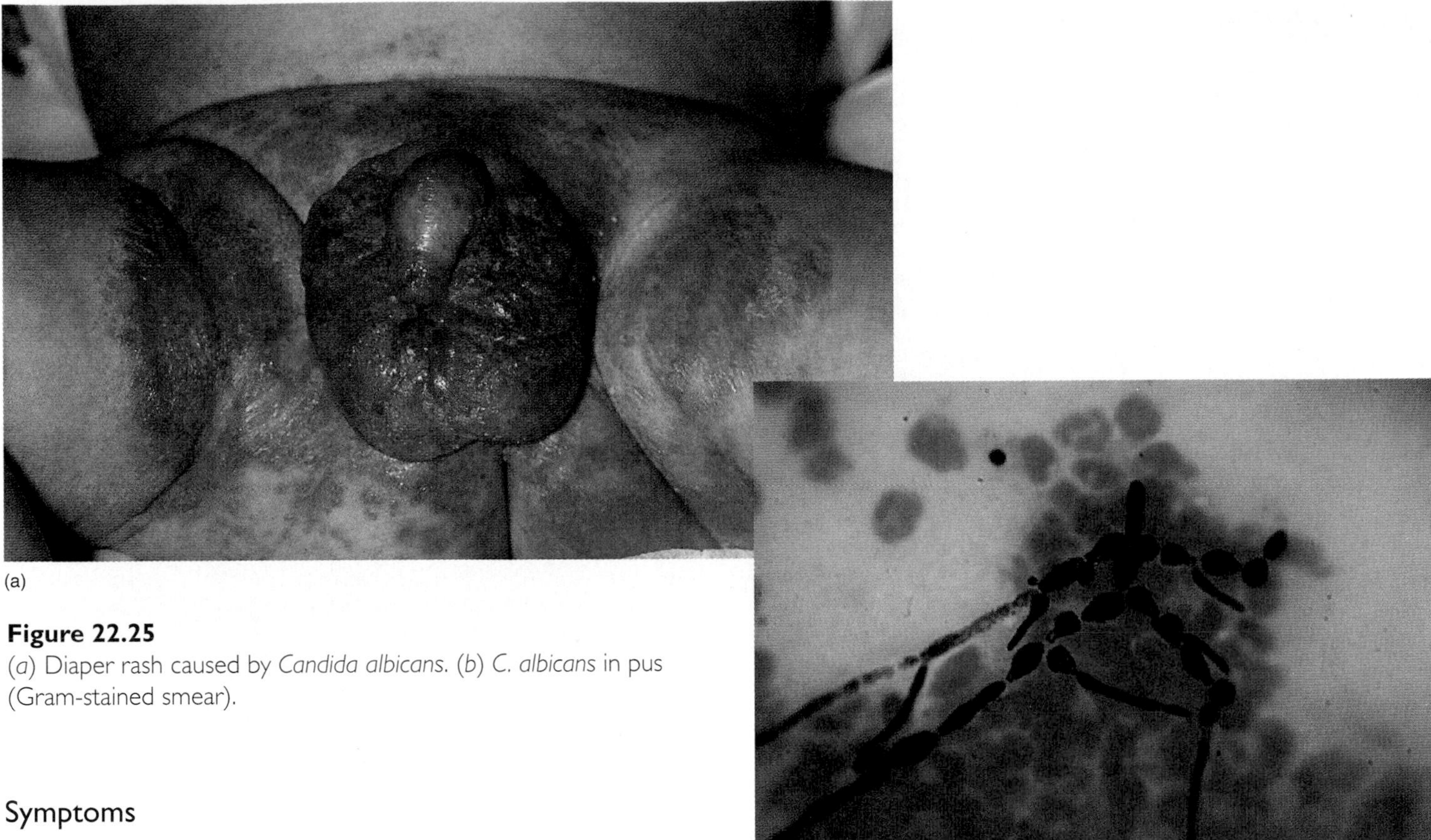

(a)

(b)

Figure 22.25
(*a*) Diaper rash caused by *Candida albicans*. (*b*) *C. albicans* in pus (Gram-stained smear).

Symptoms

Most people colonized by these molds have no symptoms at all. Others complain of itching, a bad odor, or a rash. The rash usually occurs at the site of the infection and consists of a scaley area surrounded by redness at the outer margin, producing irregular rings or a lacy pattern on body skin. On the scalp, patchy areas of hair loss occur, with a fine stubble of short hair left behind. Involved nails become thickened and brittle and may separate from the nailbed. Sometimes a rash consisting of fine papules and vesicles develops distant from the infected area. This rash is referred to as a *dermatophytid* ("id") reaction.

Causative Agents

The skin-invading molds belong mainly to the genera *Epidermophyton*, *Microsporum*, and *Trichophyton* and are collectively termed **dermatophytes** (figure 22.26). They can be cultivated on media especially designed for molds and are identified by their colonial and microscopic appearance, their nutritional requirements, and biochemical tests.

Pathogenesis

The normal skin is generally resistant to invasion by dermatophytes. However, some species are relatively virulent and can even cause epidemic disease, especially in children. In conditions of excessive moisture, dermatophytes can invade keratinized structures, including the epidermis down to the level of the keratin-producing cells. A keratinase enables them to dissolve keratin and use it as a nutrient. Hair is invaded at the level of the hair follicle because the follicle is relatively moist. Fungal products diffuse into the dermis and provoke an immune reaction, which probably explains why adults tend to be more resistant to infection than children. It also explains why some people develop allergic ("id") reactions.

Epidemiology

As mentioned, age, virulence of the infecting strain of mold, and excessive moisture are important factors in causing infections. Common causes of excessive moisture are obesity causing folds of skin to lie together, tight clothing, and plastic or rubber footware. Potentially pathogenic molds may be present in soil and on pets such as young cats and dogs.

Prevention

Attention to cleanliness and maintenance of normal dryness of the skin and nails effectively prevent most dermatophyte infections. Powders, open shoes, changing of socks, and application of alcohol after bathing may help prevent toenail infections.

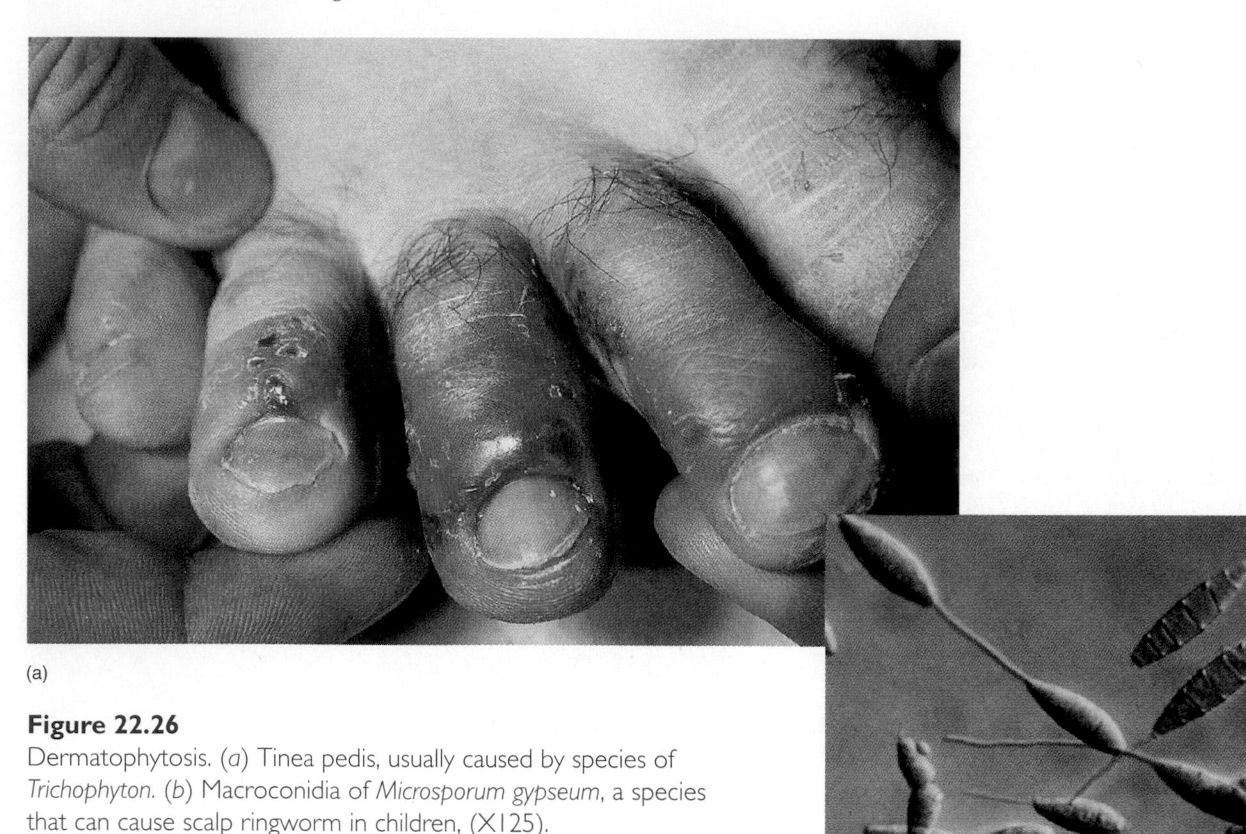

(a)

(b)

Figure 22.26
Dermatophytosis. (*a*) Tinea pedis, usually caused by species of *Trichophyton*. (*b*) Macroconidia of *Microsporum gypseum*, a species that can cause scalp ringworm in children, (X125).

Treatment

Numerous prescription and over-the-counter medications are promoted for treating dermatophytoses. Many of the medications contain undecylenic acid, an 11-carbon, unsaturated compound that is fungistatic in action. Naftifine and numerous -azole compounds (miconazole, clotrimazole, econazole, itraconazole, etc.) are fungicidal, and some act against both *Candida albicans* and dermatophytes. They act by inhibiting synthesis of ergosterol, an essential component of the fungal cell wall. Nystatin, an antibiotic produced by *Streptomyces noursei*, was discovered by the New York State Health Laboratory (hence its name). It acts by attaching to ergosterol in the fungal membrane and causing the formation of pores from which cellular contents are lost. Griseofulvin, another antibiotic, is produced by *Penicillium griseofulvum* and is fungistatic. Some stubborn toenail infections respond to griseofulvin, but it may take months to achieve a cure.

REVIEW QUESTIONS

1. What is the relationship between varicella and herpes zoster?
2. What is the epidemiological significance of shingles?
3. Why do so many people suffer permanent damage or die from rubeola?
4. What important diagnostic sign is often present in the mouth of rubeola victims?
5. What is the significance of rubella viremia during pregnancy?
6. How does a person contract warts?
7. What is the cause of ringworm?

Summary

I. Anatomy and Physiology
   A. The skin repels potential pathogens by shedding and being dry, acidic, and toxic.

II. Normal Flora of the Skin
   A. Diphtheroids are gram-positive, pleomorphic, rod-shaped bacteria. The growth of *Propionibacterium acnes* is enhanced by oily secretion of sebaceous glands. It produces fatty acids from the gland secretion. It has anaerobic metabolism, but some are able to tolerate growth in air.

B. Staphylococci are gram-positive cocci that help prevent colonization by potential pathogens and maintain the balance among flora of the skin. *Staphylococcus epidermidis* is universally present and pathogenic at times.
C. *Malassezia* sp. are single-celled yeasts found universally on the skin. They probably cause some cases of dandruff as well as tinea versicolor and more serious rashes in AIDS patients.

III. Skin Diseases Caused by Bacteria
   A. Furuncles and carbuncles are caused by *Staphylococcus aureus*, which is coagulase positive and often resists penicillin and other antibiotics. A carbuncle is more dangerous because the infection is more likely to be carried to the heart, brain, or bones.
   B. Staphylococcal scalded skin syndrome results from exotoxins produced by certain strains of *Staphylococcus aureus*.
   C. Impetigo is a superficial skin infection caused by *Staphylococcus aureus* and *Streptococcus pyogenes*.
      1. The capsule and the M-protein of the cell wall of *S. pyogenes* interfere with phagocytosis, thus aiding pathogenesis.
      2. Penicillin or erythromycin is effective in most *S. pyogenes* infections.
      3. Acute glomerulonephritis, a kidney disease, is a complication of *S. pyogenes* infections.
   D. Rocky Mountain spotted fever, caused by the obligate intracellular bacterium *Rickettsia rickettsii*, is transmitted to humans by the bite of an infected tick. Tetracycline or chloramphenicol is used in treatment.
   E. Lyme disease can imitate many other diseases. It is caused by a spirochete, *Borrelia burgdorferi*, and is transmitted to humans by ticks. Erythema chronicum migrans is the hallmark of the disease. Doxycycline, amoxicillin, and erythromycin are effective in treating early stages of the disease.

IV. Skin Diseases Caused by Viruses
   A. Chicken pox is a common disease of childhood caused by the varicella zoster virus, a member of the herpesvirus family.
      1. Herpes zoster, or "shingles," can occur months or years after chicken pox and represents a reactivation of the varicella virus infection in the distribution of a sensory nerve.
      2. Herpes zoster cases can be responsible for chicken pox epidemics.
   B. Rubeola (measles) is a potentially dangerous viral disease that can be fatal in some cases and can lead to serious secondary bacterial infections in others. Rubeola can be controlled by immunizing young children and susceptible adults with a live attenuated vaccine. A two-dose regimen is now employed in an effort to decrease the size of the nonimmune population.
   C. Rubella (three-day measles, German measles) contracted by a woman in the first eight weeks of pregnancy results in at least a 90% chance of congenital rubella syndrome. Immunization with a live virus protects against this disease, but many women still reach childbearing age without proper vaccination.
   D. Numerous viruses can cause rashes. In recent years, the viral cause of two exanthems has been determined.
      1. Erythema infectiosum is characterized by a "slapped cheek" rash. It is caused by parvovirus B-19. It can be fatal to people with various anemias.
      2. Exanthem subitum is marked by several days of high fever and a transitory rash appearing as the temperature returns to normal. The disease is caused by human herpesvirus, type 6.
   E. Warts are skin tumors caused by a number of papillomaviruses. They are generally benign, but some genital warts have been associated with cancer of the uterine cervix.

V. Skin Diseases Caused by Fungi
   A. Invasive skin infections are sometimes caused by the yeast *Candida albicans*.
   B. Athlete's foot and ringworm are caused by certain species of mold-type fungi that feed on keratin-containing cells.

Chapter Review Questions

1. Under what circumstances is *Malassezia furfur* likely to be pathogenic?
2. What is the difference between a furuncle and carbuncle?
3. In what age group is Ritter's disease most likely to occur?
4. What does *transovarial passage* mean?
5. In which countries of North America might Rocky Mountain spotted fever be contracted?
6. What is the causative agent of Lyme disease?
7. What is erythema chronicum migrans?
8. Describe the hemagglutination inhibition test and its use in diagnosing rubella.
9. What is characteristic about the rash of varicella?
10. What childhood exanthem might be associated with an aplastic crisis?

Critical Thinking Questions

1. Why is it a good idea to immunize little boys against rubella?
2. Why might it be more difficult to eliminate a disease like Lyme disease or Rocky Mountain spotted fever from the earth than rubeola or rubella?

Further Reading

Barinaga, Marcia. 1992. California rides its own bi-cycle. *Science* 256: 1385. Describes the epidemiology of Lyme disease in California.

Fischetti, V. A. 1991. Streptococcal M protein. *Scientific American* 264: 58–65.

Kelley, P. W., et al. The susceptibility of young adult Americans to vaccine-preventable infections: a national survey of U.S. Army recruits. *Journal of the American Medical Association* 266:2724.

Klein, J. O. 1990. From harmless commensal to invasive pathogen. Coagulase-negative staphylococci. *The New England Journal of Medicine* 323:339.

Mahalingam, R., et al. 1990. Latent varicella-zoster viral DNA in human trigeminal and thoracic ganglia. *The New England Journal of Medicine* 323:627–631.

Oliver, J. H., et al. 1993. Conspecificity of the ticks *Ixodes scapularis* and *I. dammini* (Acari: Ixodidae). *Journal of Medical Entomology* 30:54.

1996. Prevention of varicella. *Morbidity and Mortality Weekly Report* 45(no. RR-11):1.

1994. Reported vaccine-preventable diseases—United States, 1993, and the childhood immunization initiative. *Morbidity and Mortality Weekly Report* 43(4):57.

Szer, J. S., Taylor, E., and Steere, A. C. 1991. The long term course of Lyme arthritis in children. *The New England Journal of Medicine* 325:159.

23 Upper Respiratory System Infections

KEY CONCEPTS

1. Untreated skin or throat infections by *Streptococcus pyogenes* can occasionally result in serious kidney or heart damage even though these organs are not infected. The damage is probably caused by the immune system.
2. Some diseases arise because an exotoxin is absorbed from a localized infection. The toxin then circulates throughout the body, causing damage to tissues some distance from the site of infection.
3. Circulating toxins may not affect all tissues equally. A toxin may target specific organs because it attaches to receptors on the cells composing the organs.
4. Many different kinds of infectious agents can produce the same symptoms and signs of disease.

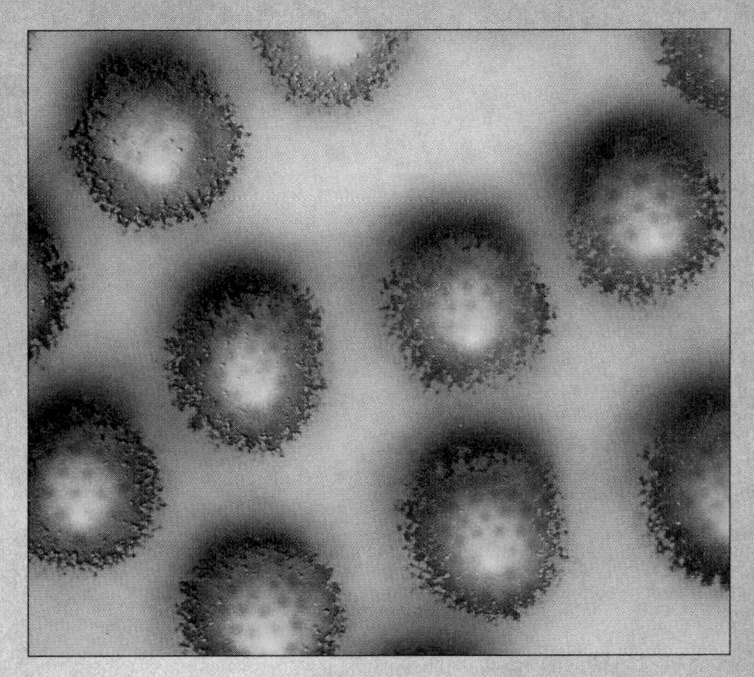

Scanning electron micrograph of an adenovirus. Adenoviruses are nonenveloped, double-stranded DNA viruses that can cause a variety of respiratory syndromes, sometimes mimicking "strep" throat, colds, or "pinkeye."

Microbes in Motion

The following books in the *Microbes in Motion* CD-ROM may serve as a useful preview or supplement to your reading: *Gram-Positive Organisms. Gram-Negative Organisms. Miscellaneous Bacteria. Microbial Pathogenesis in Infectious Diseases:* Toxins. *Vaccines:* Vaccine Targets. *Viral Structure and Function:* Viral Diseases. *Fungal Structure and Function:* Diseases. *Parasitic Structure and Function.*

A Glimpse of History

Arataeus, a Cappadocian,[1] first described diphtheria in the second century A.D. In 1826, Pierre Bretonneau, a French physician, named the disease *diphtheria*, derived from the Greek word for leather or membrane, because of the loosely adhering membrane that characteristically forms in the throats of infected people. In 1883, Edwin Klebs, a professor of pathology at the University of Zurich in Switzerland, first described the organism that causes the disease. The following year, Friedrich Loeffler presented a report to the German Imperial Health Office describing his comprehensive studies of the disease, including numerous animal experiments. In recognition of the work done by Klebs and Loeffler, the organism that causes diphtheria became known as the *Klebs-Loeffler bacillus*. (It was later named *Corynebacterium diphtheriae*.) After reviewing Loeffler's work and carrying out studies of their own, Emile Roux and Alexandre Yersin, two co-workers of Louis Pasteur, published a paper in 1888 that proved that a toxin circulates in the blood of laboratory animals infected with the diphtheria bacillus. This explained how diphtheria, usually a localized infection in the nose or throat, could cause damage throughout the body.

In 1890, Emil von Behring, Paul Ehrlich, and Shibasaburo Kitasato discovered that treating the diphtheria toxin with dilute formalin[2] destroys its toxic properties. This modified toxin is called a **toxoid.** When the toxoid is injected into animals, antibodies are formed that neutralize the natural toxin. In 1892, Emile Roux solved the problem of producing large amounts of the antitoxin by immunizing horses. The resulting horse serum was then injected into children with diphtheria, and the severity of the disease was reduced in many cases. Roux reported the success of the new treatment at the International Congress of Hygiene in Budapest in 1894. Today, almost all children in the United States begin immunization with diphtheria toxoid when they are about six weeks old. As a result of immunization, diphtheria, once common, is now a rare disease.

This story illustrates how discoveries in science are made. Rarely are important discoveries the work of a single person; rather, they most often depend upon the work of many scientists over long periods. Each discovery makes the next one possible in a step-by-step progression.

[1] Cappadocia—an ancient province in Asia Minor
[2] Formalin—a 37% solution of formaldehyde in water

• **Alexandre Yersin, p. 673**

Most upper respiratory infections are uncomfortable but not life threatening and go away without treatment in about a week. However, because they are so common, they far outweigh other infections in terms of the cumulative misery and loss of time from work they cause. Some infections such as the childhood rashes discussed in the last chapter produce minor upper respiratory symptoms but progress to involve the skin, lung, bloodstream, or central nervous system, where they can cause major damage. This chapter focuses on infections that produce major symptoms involving the nose, throat, middle ear, sinuses, and eye.

Anatomy and Physiology

The structures commonly involved in infections of the upper respiratory system include the moist surfaces of the eyes and eyelids, the nasolacrimal ducts, the middle ear, the air-filled chambers of the skull (sinuses and mastoid air cells), and the main respiratory passage of the nose and throat as far as the epiglottis and vocal cords (figure 23.1). The tonsils are composed of lymphoid tissue, strategically located to come into contact with and respond immunologically to incoming microorganisms. The tonsils are important producers of immunity to infectious agents, but paradoxically they can also be the sites of infection. Enlargement of the pharyngeal tonsil ("adenoids") can contribute to ear infections by interfering with normal drainage from the eustachian tubes. These tubes, which extend from the middle ear to the nasopharynx, are 3 to 4 cm long. They serve to equalize the pressure in the middle ear and to drain normal mucous secretions.

A person normally breathes about 16 times per minute, inhaling about 500 ml of air with each breath, or more than 11,500 liters of air per day with accompanying microbes. The air enters the respiratory system at the nostrils (a common site of colonization by *Staphylococcus aureus*), flows into the nasal cavity, and is deflected downward through the throat. The air then enters the lower respiratory tract[3] below the epiglottis, a muscular fold of tissue that closes off the lower respiratory tract during swallowing. The nasal cavity is a chamber above the roof of the mouth, incompletely divided into right and left halves by a vertical wall extending almost back to the throat. Spongy masses of tissue bulge into each half of this chamber from its outside walls. These tissues are similar to other erectile tissues of the body in that they expand and contract with alterations in blood flow controlled by nervous reflexes. If they are enlarged, these spongy tissues contribute to nasal congestion or obstruction of the airways. Many stimuli can cause these changes, including variations in temperature and humidity, emotional factors, and the irritating effects of smoke and infections.

The Movement of Mucus by Cilia Is Important in the Removal of Microorganisms and Other Foreign Material

The cells lining the nasal cavity, tear ducts, eustachian tubes, sinuses, mastoid air cells, and middle ear all have tiny, hairlike projections called **cilia** along their exposed

[3] Lower respiratory tract—that part of the respiratory system that begins with the trachea and ends with the lungs

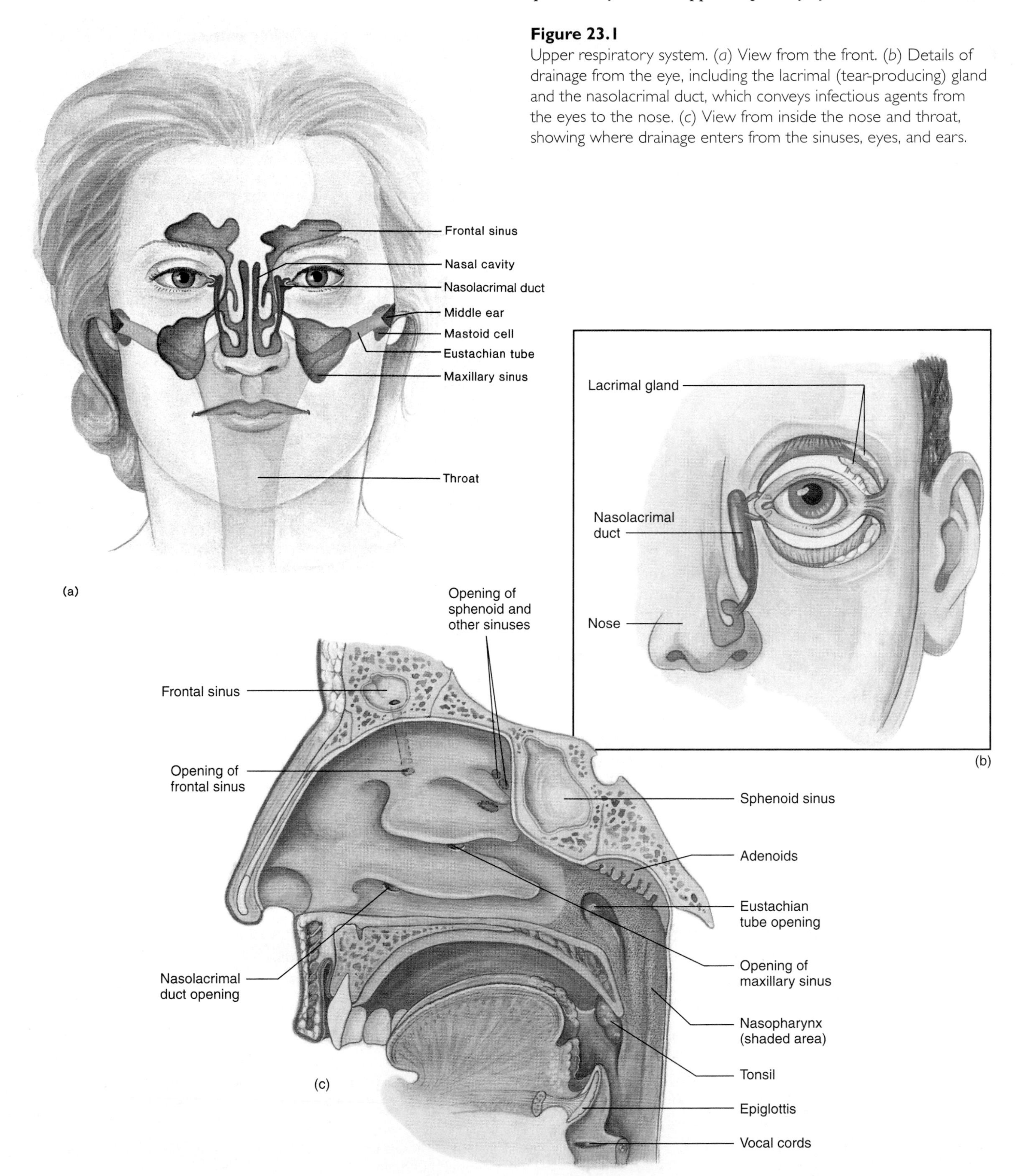

Figure 23.1
Upper respiratory system. (*a*) View from the front. (*b*) Details of drainage from the eye, including the lacrimal (tear-producing) gland and the nasolacrimal duct, which conveys infectious agents from the eyes to the nose. (*c*) View from inside the nose and throat, showing where drainage enters from the sinuses, eyes, and ears.

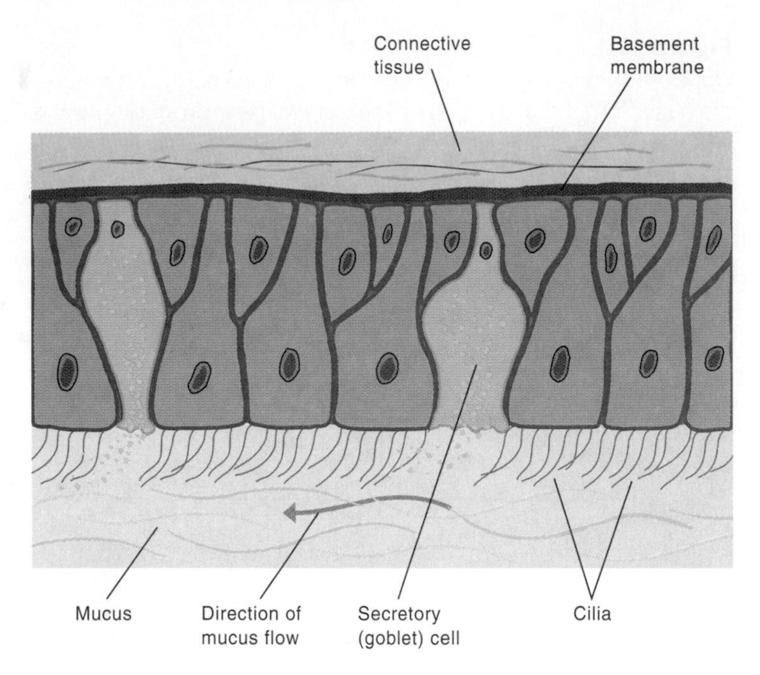

Figure 23.2
Ciliated mucous membrane, an important defense mechanism for the respiratory system. It lines the middle ears and eustachian tubes, mastoids, sinuses, nasolacrimal ducts, and nasal cavity.

free border (figure 23.2). These cilia beat synchronously at a rate of about 1,000 times a minute and continually propel a film of mucus (secreted by cells called **goblet cells** that line the respiratory system) into the nose and throat. The mucus is then swallowed, and any entrapped microorganisms and viruses are exposed to the killing action of stomach juices. The nasal hairs produce turbulence that helps bring the larger inhaled particles into contact with mucus. Only about 50% of particles the size of bacteria escape to pass through the upper airways. The ciliary action of the respiratory epithelium also normally keeps the middle ears and the sinuses free of microorganisms that may have entered them from the nose or throat. Viral infection as well as the use of tobacco, alcohol, and narcotics all impair ciliary movement and thereby promote infection.

Upper Respiratory Tract Functions Include Regulation of the Temperature and Humidity of Inspired Air

The primary function of the upper respiratory tract is not only to remove microorganisms and other foreign material from the inspired air, but to regulate its temperature and water content. When cold air enters the upper tract, nervous reflexes immediately increase the blood flow to the spongy tissues in the nose, thereby transferring heat to the air. This mechanism usually adjusts the temperature to within 2 to 3 degrees of body temperature by the time it reaches the lungs. Inspired air also becomes saturated with water vapor, the nasal tissues giving up as much as a quart of water per day to keep the air humidified. The warmth and moisture provide optimum conditions for the ciliary action, mucus flow, and macrophages that act together to defend the lungs against infection.

Normal Flora of the Upper Respiratory System

Although parts of the upper respiratory system are normally sterile, other parts are colonized by numerous species of bacteria. Aerobes, facultative anaerobes, aerotolerant organisms, and anaerobes are all represented. Although generally harmless, members of the normal flora can sometimes cause disease.

The Normal Bacterial Flora Include Potentially Pathogenic Species

The principal types of bacteria inhabiting the upper respiratory system are presented in table 23.1. Surprisingly, even though the eyes are constantly exposed to a multitude of microorganisms, about half of all normal healthy people have no bacteria on the conjunctiva, the moist membrane covering the eye and inner surfaces of the eyelids. Presumably, this results from the frequent automatic washing of the eye with lysozyme-rich

Table 23.1 Normal Flora of the Upper Respiratory System

Genus	Characteristics	Comments
Staphylococcus	Gram-positive cocci in clusters	Commonly includes the potential pathogen *S. aureus*
Corynebacterium	Pleomorphic, gram-positive rods; nonmotile; nonspore-forming	Nonpathogenic species collectively referred to as *diphtheroids*
Moraxella	Gram-negative diplococci	Resemble *Neisseria* species
Haemophilus	Small, pleomorphic, gram-negative rods	Commonly includes the potential pathogen *H. influenzae*
Bacteroides	Small, pleomorphic, gram-negative rods	Strict anaerobes
Streptococcus	Gram-positive cocci in chains	α (especially viridans streptococci), β, and γ types; commonly includes the potential pathogen *S. pneumoniae*

tears and from the eyelid's blinking reflex, which cleans the eye surface. Thus, viruses and microorganisms adhering to the moist membrane covering the eye are almost immediately swept into the nasolacrimal duct and thence to the nasopharynx. Indeed, viruses that cause colds can infect the nose by this route. Organisms recovered from the normal conjunctiva are usually few in number and consist of species normally found on the skin.

The secretions of the nasal entrance usually contain large numbers of diphtheroids, micrococci, and staphylococci and smaller numbers of other bacteria that normally inhabit the environment, such as *Bacillus* sp. About 20% of all normal humans also carry *Staphylococcus aureus* in the nose, and an even higher percentage of hospital personnel are likely to be carriers of these opportunistic pathogens.[1] Farther inside the nasal passages, the microbial population increasingly resembles that of the nasopharynx. The nasopharynx contains large numbers of microorganisms (figure 23.3), mostly α-hemolytic[2] streptococci of the viridans group, nonhemolytic streptococci, *Moraxella* (*Branhamella*) *catarrhalis*, and diphtheroids. Anaerobic gram-negative bacteria, including species of *Bacteroides*, are also normally present in large numbers in the nasopharynx. In addition, commonly pathogenic bacteria such as *Streptococcus pneumoniae*, *Haemophilus influenzae*, and *Neisseria meningitidis* are often found in this area, especially during the cooler seasons of the year. More virulent species such as *Streptococcus pyogenes*, *Mycoplasma pneumoniae*, *Corynebacterium diphtheriae*, and *Bordetella pertussis* are sometimes found, but in only a small percentage of healthy humans.

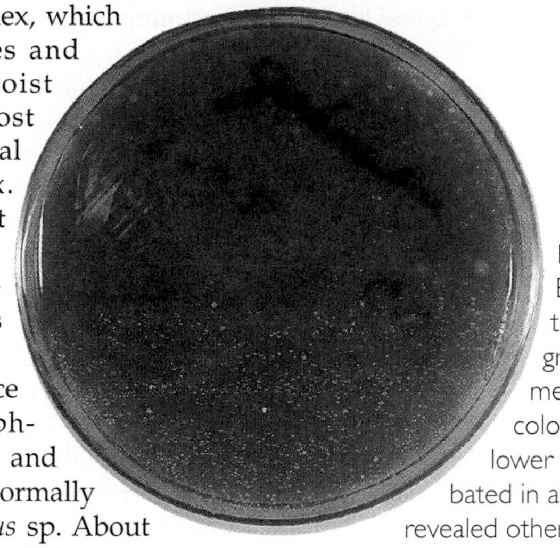

Figure 23.3
Bacterial flora of the nasopharynx. Notice the enormous numbers of colonies and the greenish discoloration of the blood-agar medium. Non-hemolytic and β-hemolytic colonies are visible on the upper half but not the lower half of the medium. This culture was incubated in a candle jar. An anaerobic culture would have revealed other members of the normal flora.

REVIEW QUESTIONS

1. What potentially pathogenic bacterium commonly colonizes the nostrils?
2. Explain how the mucous membranes of the upper respiratory system defend against infection.
3. How does contamination of the eye lead to upper respiratory infection?
4. Give the names of three commonly pathogenic species of bacteria that may colonize the upper respiratory system during the cooler months of the year.

[1] Opportunistic pathogens—organisms that generally can cause disease only when normal body defenses are impaired

[2] α-hemolytic—producing partial red blood cell destruction around colonies grown on blood agar medium; many α-hemolytic streptococci produce a greenish color as a result of hemoglobin breakdown and are called viridans streptococci (*viridans* means "green")

• **Streptococci, p. 259**

Bacterial Infections of the Upper Respiratory System

Many different bacterial infections involve the upper respiratory system. Some, such as sore throats caused by *Haemophilus influenzae* and β-hemolytic streptococci of groups C and G, generally do not require treatment because they are usually quickly eliminated by the body's immune system. The sections that follow will focus on infections that require treatment because they are not easily eliminated by the body and can cause serious complications.

Streptococcal Pharyngitis (Strep Throat) Can Cause Serious Complications, Including Rheumatic Fever and Glomerulonephritis

Sore throat is one of the most common reasons that people in the United States seek medical care, resulting in about 27 million doctor visits per year. Many of these visits are due to a justifiable fear of streptococcal pharyngitis.

Symptoms

Typically, the throat is red, with patches of adhering pus and scattered tiny hemorrhages, and the lymph nodes in the neck are enlarged and tender. Abdominal pain or headache may be prominent in older children and young adults. *Not* usually present are red, weepy eyes, cough, runny nose, or generalized lymph node enlargement. Although abscess[3] formation and other local complications may prolong the illness, most patients with streptococcal

[3] Abscess—localized collection of pus, composed of living and dead leukocytes, components of tissue breakdown, and infecting organisms

sore throat recover spontaneously after about a week. In fact, many infected people have only mild symptoms or no symptoms at all.

Causative Agent

The cause of strep throat is *Streptococcus pyogenes*, the group A β-hemolytic streptococcus (figure 23.4) described in chapter 22 that is also a cause of impetigo. Like many other streptococci, *S. pyogenes* is aerotolerant.

In infected material, *S. pyogenes* usually occurs as diplococci or short chains, whereas in culture, longer chains are characteristic (figure 23.5). *Streptococcus pyogenes* is easily cultivated on various laboratory media, but it grows preferentially on a medium enriched with blood, incubated in an atmosphere containing more carbon dioxide and water vapor than are generally present in air. Suitable atmospheric conditions can be achieved using a candle or anaerobic jar. Some strains have prominent hyaluronic acid capsules and form large mucoid colonies.[1] The majority of strains of *S. pyogenes* also show a zone of β-hemolysis around their colonies where complete erythrocyte destruction and clearing of the medium has occurred. The β-hemolysis results from the activity of two toxins, **streptolysin O** and **streptolysin S,** that are released by *S. pyogenes* cells. If *S. pyogenes* is cultivated with other species, as from the throat, its growth may be inhibited. However, anaerobic incubation tends to suppress the growth of competing flora and enhances β-hemolysis (streptolysin O is oxygen-labile), thus making it much easier to detect *S. pyogenes* colonies. The growth of *S. pyogenes* is more likely to be inhibited by low concentrations of the antibiotic bacitracin than is growth of other β-hemolytic streptococci, which aids in its identification (figure 23.6).

• **Impetigo, p. 479**
• **Aerotolerance, p. 92**

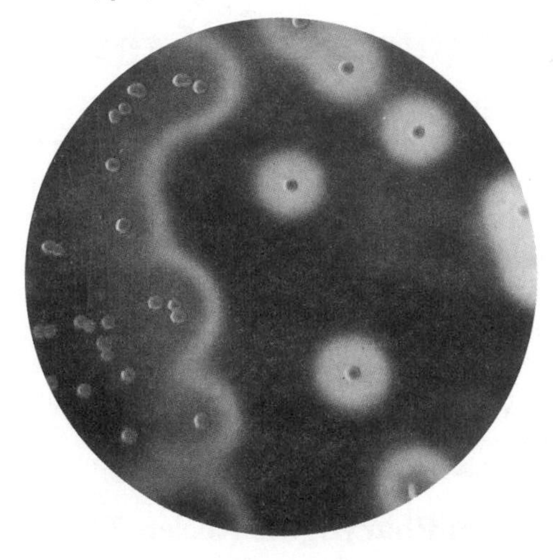

Figure 23.4
Streptococcus pyogenes culture on blood agar.

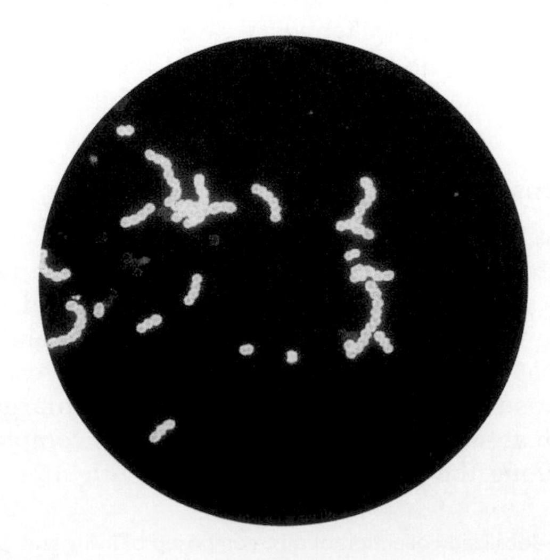

Figure 23.5
Chain formation as revealed by fluorescence microscopy of organisms from a fluid culture of *Streptococcus pyogenes*.

Some of the components of *S. pyogenes* cells are diagrammed in figure 23.7. The **M protein,** to be discussed further, varies from one strain to another and is therefore useful for differentiating between strains. **Protein F** is responsible for adherence to the throat epithelium. **Protein G** has a function similar to that of protein A of *Staphylococcus aureus* in binding the Fc segment of immunoglobulin G.

The cell wall carbohydrates of streptococci are important in their identification (figure 23.8). The group A cell wall carbohydrate of *S. pyogenes*, a polysaccharide polymer of *N*-acetylglucosamine and rhamnose, differs antigenically from the cell wall carbohydrates of most other streptococci. *S. agalactiae*, for example, can be β-hemolytic, but it possesses

• **Candle jar, p. 92**
• **Anaerobic jar, p. 99**
• ***S. pyogenes* epidemic, pp. 440, 506–512**

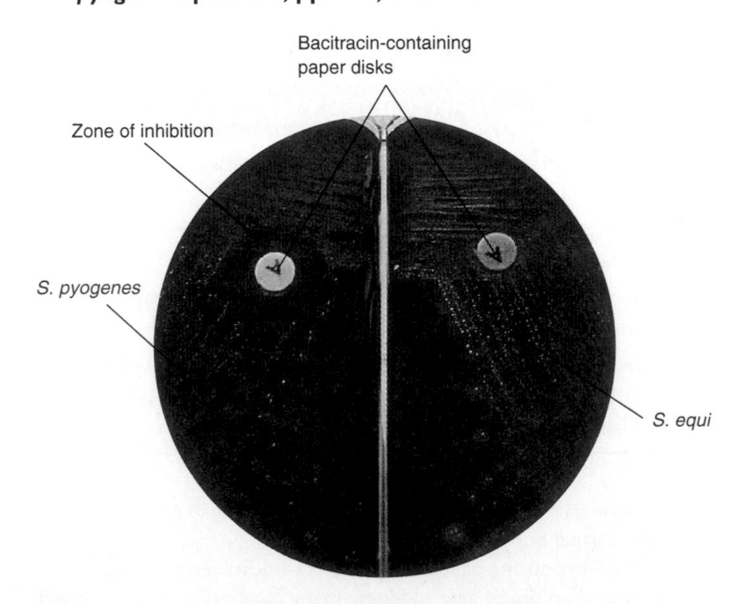

Figure 23.6
Bacitracin susceptibility of *Streptococcus pyogenes*. Growth of *S. pyogenes* is inhibited by bacitracin diffusing from a filter paper disk, while growth of most other β-hemolytic streptococci is not.

[1]Mucoid colonies—resembling a drop of mucus.

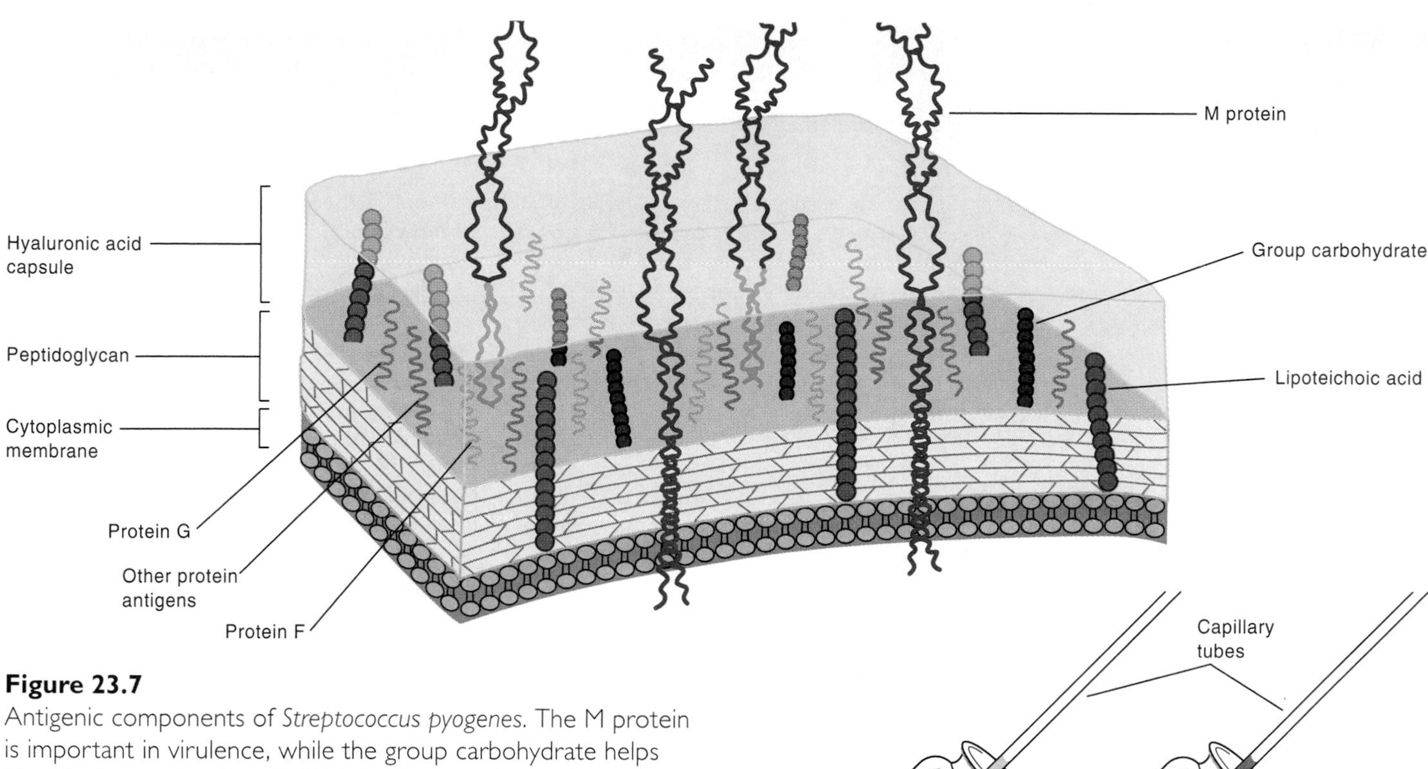

Figure 23.7
Antigenic components of *Streptococcus pyogenes*. The M protein is important in virulence, while the group carbohydrate helps identify the species. Protein F binds the organism to fibronectin on epithelial cells of the throat, while protein G binds immunoglobulin G by its Fc portion.

the group B carbohydrate, which is antigenically different. The various carbohydrates of streptococci are identified by the use of antisera, in a process called **Lancefield grouping.** Antisera are also used to identify M proteins.

Pathogenesis

Protein F mediates attachment of *S. pyogenes* by adhering to fibronectin[1] on the surface of epithelial cells of the throat. M protein interacts with certain components of the complement system, the end result being the breakdown of C3b, an opsonin. Thus, M protein helps the streptococcus evade destruction by phagocytes.

M protein is of central importance to the virulence of *S. pyogenes*. Immunity depends on production by the host of antibody to the M protein of the infecting strain. Immunity is "strain specific," meaning that antibody to one strain protects against infection with that strain only and not to the other 80 or more strains of *S. pyogenes*. The hyaluronic acid capsules produced by this bacterium also inhibit phagocytosis. Hyaluronic acid is a component of human tissue, and its presence as a capsule may impair recognition of the streptococci by the immune system. Protein G may

[1] Fibronectin—an adhesive glycoprotein found on cell surfaces, in connective tissue, and circulating in the bloodstream

• **Complement, pp. 28, 351–352, 397**
• **Opsonins, p. 364**

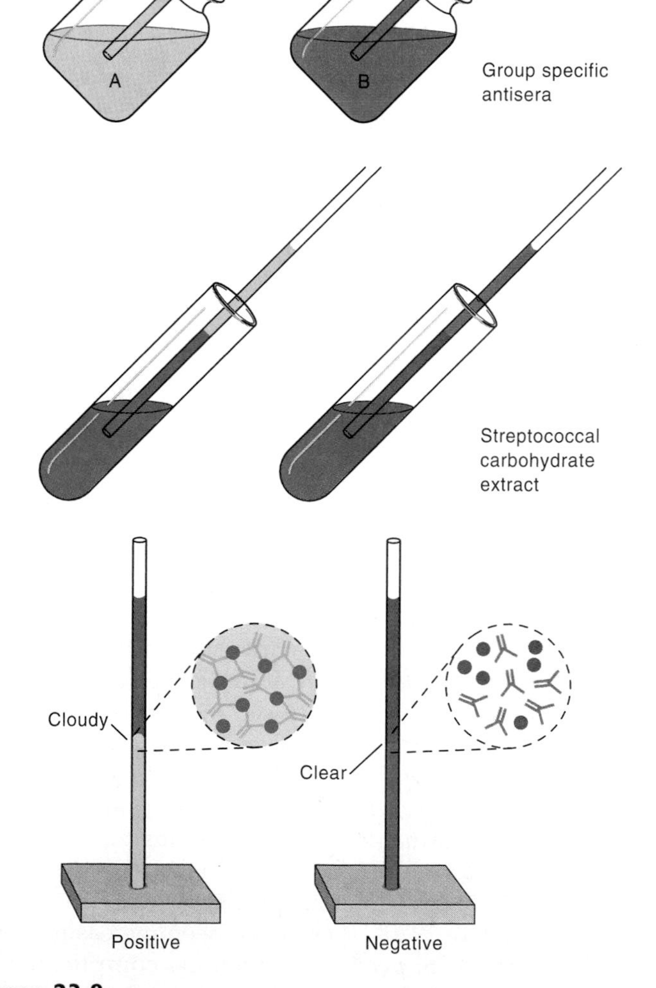

Figure 23.8
Serogrouping of *Streptococcus pyogenes*. Lancefield grouping uses antisera to identify the Group A carbohydrate of the cell wall.

Perspective 23.1

Scientist of Distinction: Rebecca Lancefield

Rebecca Lancefield was born Rebecca Craighill on a Staten Island, New York, army base in 1895. Her mother was a direct descendent of Lady Mary Wortley Montagu, who promoted smallpox vaccination in England more than 75 years before Jenner. Lancefield was educated in various schools as her parents moved from one army base to another. In 1912, she entered Wellesley College, became interested in biology, and obtained her degree in this field. She was awarded a scholarship to Columbia University and studied under the famous microbiologist Hans Zinsser. After receiving her Master's degree, she married Donald Lancefield, a graduate student in Zoology, who was drafted into the armed services in World War I. He was assigned to the Rockefeller Institute, where Rebecca got a job as a laboratory technician for the distinguished microbiologists O. T. Avery and A. R. Dochez.

Well known for their studies of pneumococci, Avery and Dochez had been commissioned to study streptococcal cultures from personnel at army camps. At that time, classification of the numerous kinds of streptococci was based largely on whether their colonies on blood agar produced β-hemolysis or a green discoloration. Years of study by various scientists would be needed before streptococci could be grouped more precisely and their pathogenicity defined. Studies of the army camp cultures at the Rockefeller Institute were among the first to use immunological methods to classify streptococci. By using different streptococcal strains to immunize mice, Avery and Dochez showed that four different antigenic types were represented among their cultures.

When funding for this work was discontinued, the Lancefields returned to Columbia where Rebecca received her Ph.D. in 1925 for her studies of α-hemolytic streptococci. At that time, α-hemolytic streptococci were considered a possible cause of rheumatic fever, but her studies showed that antibodies against the various antigens of these streptococci occurred as often in healthy people as in those with rheumatic fever. Thereafter, Lancefield returned to the Rockefeller Institute where she spent the remainder of her scientific career studying streptococci. Building on the discoveries of others, she showed that almost all the strains of β-hemolytic streptococci from human infections had the same cell wall carbohydrate "A," which could easily be extracted from the cells by hot acid. Streptococcal cultures from other sources were shown to have different carbohydrates: "B" from certain cattle infections; "C" from cattle, horses, and guinea pigs; "D" from cheese and human normal flora; and so forth. The grouping of streptococci by their cell wall carbohydrates proved to be a much more useful prediction of pathogenic potential than hemolysis on blood agar. Lancefield went on to establish a typing scheme for group A streptococci based on their M and T protein antigens, which was later adopted worldwide. She received numerous honors for her work and, in 1960, became the first woman president of the American Association of Immunologists. In 1970, Lancefield was elected to the prestigious National Academy of Sciences. She died in 1981 at the age of 86.

• **Mary Wortley Montagu, p. 379**

have a similar function in that it coats the organisms with immunoglobulin molecules.

S. pyogenes has yet another way of defending itself against phagocytosis, **C5a peptidase.** This enzyme destroys C5a, the complement component that is a chemoattractant for phagocytes.

S. pyogenes produces a number of extracellular substances that probably supply it with nutrients. Their role in pathogenesis is debatable because antibody against them fails to give protection. The substances include streptolysins O and S, which lyse leukocytes and other cells by making holes in their cell membranes; hyaluronidase (spreading factor), which allows the bacteria to advance into tissue by degrading its hyaluronic acid; NADase, which breaks down nicotinamide adenine dinucleotide and is toxic to leukocytes; streptodornases, which degrade DNA; streptokinases, which activate a normal host substance causing blood clots to dissolve; and a protease of uncertain function associated with highly invasive strains of *S. pyogenes*.

Since the late 1970s, there have been increasing numbers of reports of *S. pyogenes* infections complicated by shock, rash, and a highly mortality. So far, only a small percentage of the cases have resulted from throat infections, most having been associated with infections elsewhere in the body. The increased virulence of these strains of *S. pyogenes* is thought to be due in large part to **streptococcal pyrogenic exotoxins (SPEs)**. The mode of action of SPEs is controversial. They are superantigens, causing massive T cell activation and release of cytokines. However, this does not fully explain how they cause shock and a rash. At least one of these exotoxins is coded on a chromosomal gene, while others result from lysogenic conversion. SPEs have been incriminated in shock associated with some cases of nongroup A streptococcal infections, which suggests the possibility that SPE genes move readily among different species of streptococci. The rash of scarlet fever (discussed next) is caused by an SPE known as **erythrogenic toxin.**

Scarlet Fever Some strains of *S. pyogenes* produce an erythrogenic (*erythro* means "red," and *-genic* means "development," as in *genesis*) toxin that is absorbed and carried by the bloodstream throughout the body, causing a red rash and whitish coating of the tongue. The rash subsides within a week. This condition is called **scarlet fever.** Only *S. pyogenes* strains lysogenic for a specific bacteriophage produce the toxin, an example of lysogenic conversion.

• **Lysogenic conversion, pp. 307, 309**

Except for the effects of erythrogenic toxin on the skin and mucous membranes, the clinical picture of infection with these strains is similar to that with toxin nonproducing strains. However, in the 19th century, scarlet fever was much more severe, sometimes killing entire families. This perhaps represented a third variation in the spectrum of SPE-related disease.

Glomerulonephritis Streptococcal throat infections, like skin infections, may lead to **acute glomerulonephritis**, a disease of the kidney probably caused by antibody reacting with streptococcal products to form immune complexes. This condition appears during convalescence from untreated *S. pyogenes* infections and is characterized by the abrupt appearance of fever and high blood pressure and the occurrence of blood and protein in the urine. Acute glomerulonephritis is caused by inflammation of small tufts of tiny blood vessels (glomeruli) within the kidney (nephros; figure 23.9). It is unusual to find *S. pyogenes* or other bacteria in the urine or diseased tissues, since the bacteria have been eliminated by body defenses by the time symptoms of glomerulonephritis appear. Damage to the kidney is caused by immune complexes that are deposited in the glomeruli and provoke an inflammatory reaction.

Only a small percentage of the many different strains of *S. pyogenes* cause acute glomerulonephritis. Complete recovery without permanent damage is the norm, although fatalities occasionally occur as the result of kidney failure, and chronic kidney disease sometimes complicates adult cases. Recurrences are rare and depend on being infected with another glomerulonephritis-causing *S. pyogenes* strain.

• Immune complexes, pp. 401, 610–611

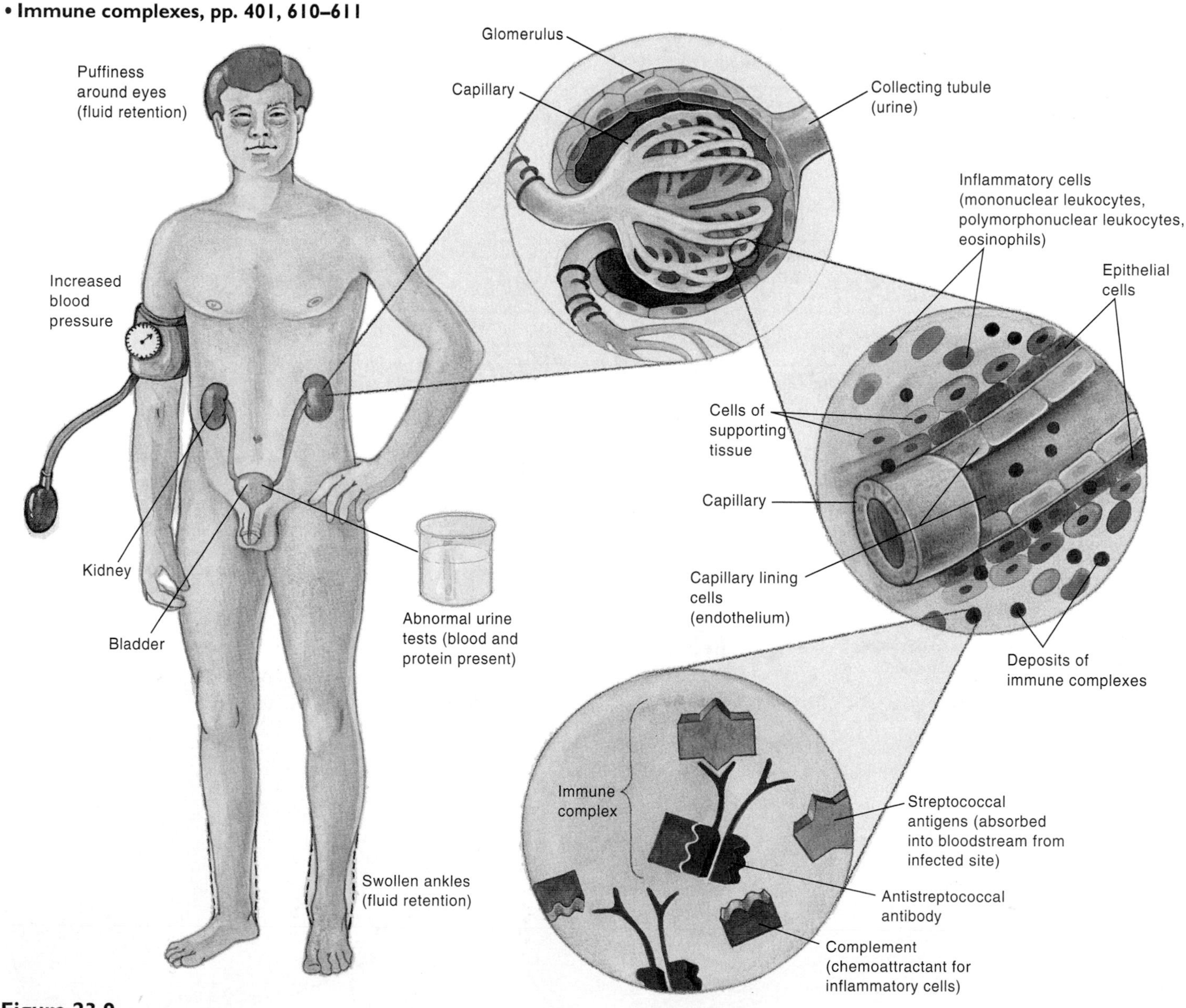

Figure 23.9
Pathogenesis of poststreptococcal acute glomerulonephritis. Immune complexes are deposited in the kidney glomeruli, inciting an inflammatory response.

Rheumatic Fever Rheumatic fever occurs only after *S. pyogenes* throat (not skin) infections. Symptoms begin about three weeks after the onset of streptococcal pharyngitis and reflect inflammation of various tissues, including the joints, heart, skin, and brain. Pain in one or more joints is the most common symptom, and it may move from one joint to another. Serious heart involvement can be asymptomatic or manifest by chest pain and shortness of breath. Painless nodules may develop under the skin, and there may be a flat lacy painless rash. After a delay sometimes as long as six months, **chorea** may develop, manifest by jerky, purposeless movements involving the face and other muscles. Heart failure and death can occur during the acute illness, but usually symptoms subside with rest and a medication such as aspirin. Nevertheless, permanent damage to the heart valves can occur, causing them to leak and leading to heart failure later in life. The deformed valves are also susceptible to an infectious complication, subacute bacterial endocarditis (figure 23.10).

• **Subacute bacterial endocarditis, p. 675**

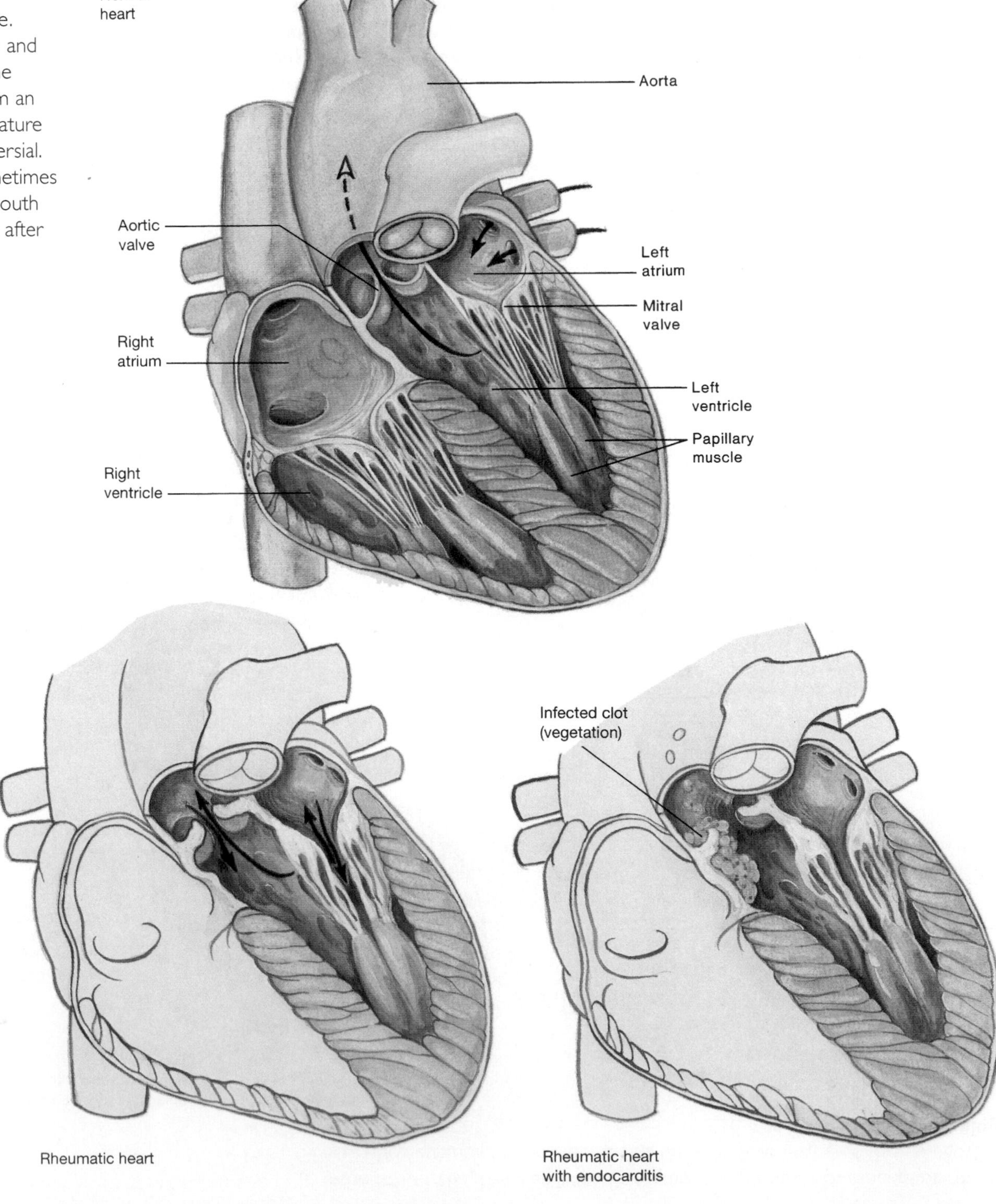

Figure 23.10 Rheumatic heart disease. Views of the left atrium and ventricle. Damage to the heart valves results from an immune reaction, the nature of which is still controversial. The damaged area sometimes becomes infected by mouth or skin organisms years after acute rheumatic fever.

After recovery from acute rheumatic fever, people may develop a recurrence of the disease if they contract another strep throat and do not treat it. Estimates of the risk of a recurrence after an untreated throat infection range from 5% to 50%. Recurrent rheumatic fever does not develop any faster but follows the same pattern as the first episode.

Usually, by the time symptoms of rheumatic fever appear, *S. pyogenes* can no longer be cultured from the throat.

However, epidemiological evidence indicates that only certain strains of *S. pyogenes* cause rheumatic fever and that they differ from those that cause glomerulonephritis. For example, using M protein typing, types 3, 5, 6, 14, 18, 19, 24, and 29 have been associated with rheumatic fever, while types 4, 12, 49, 55, 57, and 60 have been linked to glomerulonephritis. M protein types 1 and 18 have been associated with both conditions, but there are undoubtedly many differences among strains that have the same M protein. People with severe untreated *S. pyogenes* throat infections generally have up to a 3% risk of developing acute rheumatic fever, although evidence is accumulating that only certain strains of *S. pyogenes* can initiate the disease. Rheumatic fever may also develop in individuals with untreated mild infections, although the risk is only about 0.1%. However, mild infections do result in many cases of rheumatic fever because they are common and much more likely to go untreated.

Cultures of the blood, heart, and joint tissue are negative for *S. pyogenes* in rheumatic fever. Also, the risk that a patient will develop rheumatic fever following untreated strep throat rises with the magnitude of antibody response to *S. pyogenes* antigens. These facts suggest that rheumatic fever, while a consequence of the initial infection, is due to an immunological response that takes time to develop. The mechanism involved in this process is still not understood. Although far from proven, a hypothesis popular for many years is that some type of autoimmune reaction is responsible for the tissue damage. Antibodies, increasing in response to some streptococcal antigen, may cross-react with an infected individual's tissues. Indeed, various streptococcal components show antigenic cross-reactivity with the human heart and other tissues. Of these antigens, the one considered most likely to cause rheumatic fever is a segment of the M protein molecule that is held in common by a number of *S. pyogenes* strains. The many different M proteins are composed of both conserved segments, which are the same for different M proteins, and variable segments unique to each strain.

The trend in incidence of rheumatic fever in the United States has been generally downward for many years. The decline in rheumatic fever is due most likely to the practice of administering penicillin for sore throats and partly to a relative scarcity of rheumatic fever-causing strains of *S. pyogenes*. In 1994, only 112 cases of rheumatic fever were reported, compared with over 10,000 cases in 1961. In the United States during the 1980s, the reported incidence of rheumatic fever was less than two cases per million population. However, there have been a number of localized outbreaks of the disease in Ohio, Pennsylvania, Missouri, Utah, and California. These outbreaks have mostly involved unusually virulent strains of *S. pyogenes* possessing both abundant M protein and large capsules. The fact that these virulent strains were once prevalent years ago when rheumatic fever was common, and have now reappeared in association with outbreaks of the disease, suggests that they are especially likely to cause rheumatic fever.

People who recover from rheumatic fever are treated with penicillin continuously for at least five years, sometimes for life if they have heart damage, to prevent infection with a new strain of *S. pyogenes* and to minimize the risk of recurrent rheumatic fever. In addition, those who have rheumatic heart disease are given antibacterial medication immediately before any procedure likely to cause bacteremia. This is to reduce the risk of subacute bacterial endocarditis.

Epidemiology

Streptococcal infections spread readily by respiratory droplets, especially in the range of about 2 to 5 feet. Following untreated strep throat, a person may be an asymptomatic carrier for weeks. People who carry the organism in their nose spread the streptococci more effectively than do pharyngeal carriers. Anal carriers are not common but can be dangerous because they are usually unsuspected. Food contaminated by a carrier or a symptomatic case can be responsible for epidemics. Some people become long-term carriers of *S. pyogenes*. In these cases, the infecting strain usually becomes deficient in M protein and is not a threat to the carrier or to others. The peak incidence of strep throat occurs in winter or spring and is highest in grade school children. Among students visiting a clinic because of sore throat at a large West Coast university, less than 5% had strep throat. However, with some groups of military recruits, the incidence has been above 20%.

Prevention

Adequate ventilation and avoidance of crowding help to control the spread of streptococcal infections. Because so many strains exist, preparing a vaccine has been impractical. However a basis for a vaccine is being sought by identifying conserved epitopes among the different M proteins. Persons with fever and sore throat should have a throat culture for *S. pyogenes* because diagnosis cannot be reliably made from people's symptoms or the appearance of their throats. It is important that the diagnosis be made so that treatment can be prescribed, thereby halting spread of the disease and preventing complications of *S. pyogenes* pharyngitis. Newer tests that detect *S. pyogenes* antigens in throat secretions obtained with a swab give reasonably reliable results in much less time than a culture

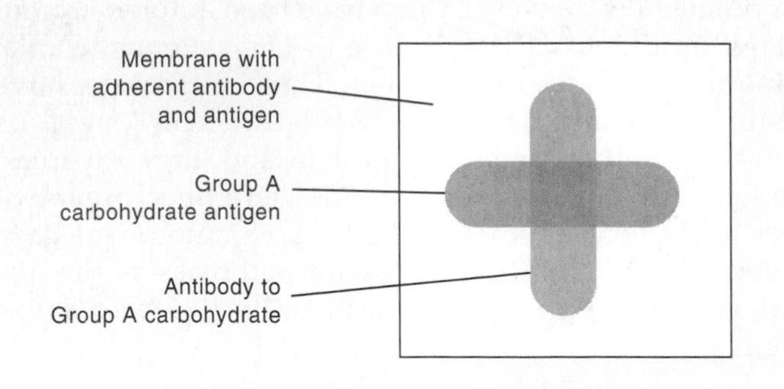

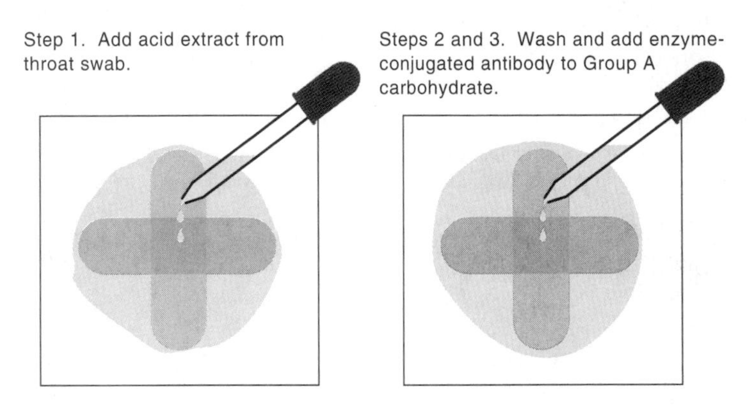

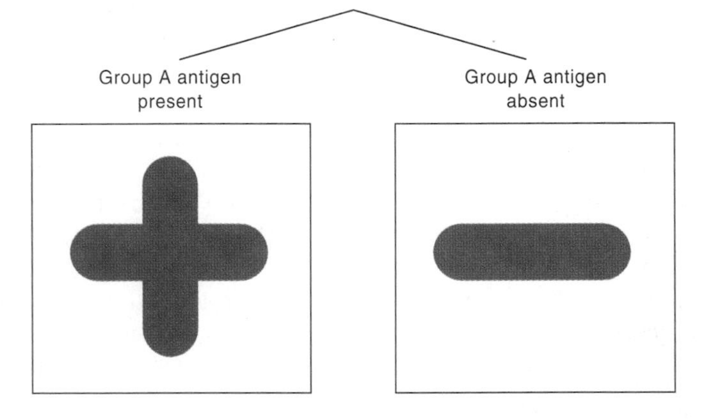

Figure 23.11
Rapid diagnosis of strep throat using enzyme-labeled antibody against Group A carbohydrate of *S. pyogenes*.

requires (figure 23.11). However, negative rapid tests need to be confirmed by culture because the rapid tests give too many false negative results. These tests generally utilize latex particles or a pad coated with specific antibody to detect the streptococcal antigen. The antigen is extracted from the bacteria in throat secretions by treatment with acid or an enzyme. Newer tests using a DNA probe are commercially available but also lack sensitivity.

Treatment

Confirmed *S. pyogenes* pharyngitis is treated with full ten days of therapy with penicillin or erythromycin, which eliminate the organism in about 90% of the cases. Treatment given as late as nine days after the onset prevents rheumatic fever from developing. Abscesses generally require surgical drainage in addition to antibiotic therapy. Table 23.2 summarizes some important facts about streptococcal pharyngitis.

• DNA probes, pp. 192–193, 236

Table 23.2 Streptococcal Pharyngitis (Strep Throat)

Symptoms	Red throat, often with exudate and tiny hemorrhages, enlargement and tenderness of lymph nodes in the neck; less frequently, abscess formation involving tonsils; occasionally, rheumatic fever and glomerulonephritis as sequels
Incubation period	2–5 days
Causative agent	*Streptococcus pyogenes;* isolation of organism by culture of material from throat; rapid immunological methods
Pathogenesis	Virulence associated with hyaluronic acid capsule and M protein both of which inhibit phagocytosis; protein F for mucosal attachment
Epidemiology	Direct contact and droplet infection; ingestion of contaminated food
Prevention	Avoidance of crowding; adequate ventilation; daily penicillin to prevent recurrent infection in those with a history of rheumatic heart disease
Treatment	Penicillin, erythromycin

Diphtheria Is a Deadly Toxin-Mediated Disease

Diphtheria is now rare in the United States; only a few cases are reported each year. However, recent events in other parts of the world are a reminder of what can happen when public health is neglected. In 1990, a diphtheria epidemic began in the Russian Federation. Over the next five years, it spread to all the newly independent states of the former Soviet Union. By the end of 1995, 125,000 cases and 4,000 deaths had been reported. The social and economic disruption following the breakup of the Soviet Union had allowed diphtheria to escape after being well controlled over the previous quarter of a century.

Symptoms

Diphtheria usually begins with a mild sore throat and slight fever, accompanied by a great deal of fatigue and malaise. Swelling of the neck is often dramatic. A whitish gray membrane forms on the tonsils and the throat or in the nasal cavity. Heart and kidney failure and paralysis may follow these symptoms.

Perspective 23.2

Diphtheria Takes the Life of a 62-year-old Man

A 62-year-old man was seen by a doctor because of a sore throat. The man had exudate on his epiglottis and one of his tonsils, and the doctor prescribed an antibiotic without taking a specimen for culture. Ten days later, it was learned that the man had been exposed to diphtheria. Specimens were obtained and cultured specifically for isolation of *C. diphtheriae*. These specimens were negative, probably because of the prior antibiotic treatment received by the patient. The man was given diphtheria toxoid immunization since he could not remember ever being immunized against diphtheria. Six weeks after his sore throat started, the man experienced difficulty in swallowing, had double vision, and became very weak. Two days later, he was dead from diphtheria.

This case illustrates the treachery of diphtheria, a disease that is currently so rare that physicians do not consider the possibility of its diagnosis. The exact cause of death is not known in this case, but most likely it was due to a cardiac arrhythmia (rapid, ineffective heartbeat) of a heart damaged by diphtheria toxin. Perhaps the arrhythmia was precipitated by choking on food or liquid as the result of throat paralysis. Injection of antitoxin might have saved the man's life if it had been given early in the disease while the toxin was still circulating free in the bloodstream and before it entered body cells. However, it is certainly far better to prevent the disease by immunization with toxoid than to try to treat it with antitoxin.

• **Toxoid, p. 422**

Table 23.3 Usual Biochemical Reactions of *Corynebacterium Diphtheriae*

				Fermentations:		
Catalase	Nitrate reduction	Urease	Gelatin hydrolysis	Glucose	Maltose	Sucrose
+	+	–	–	+	+	–

Causative Agent

The cause of diphtheria is *Corynebacterium diphtheriae*, a club-shaped, nonmotile, gram-positive rod of variable size that often stains irregularly. Slide preparations of *C. diphtheriae* commonly show the organisms arranged in "Chinese letter" patterns (figure 23.12) or side by side in "palisades." Their cytoplasm frequently contains metachromatically stained granules.

Corynebacterium diphtheriae grows aerobically on many media. Growth is generally rapid, although some strains do not develop visible colonies for several days. Different colony types can be identified, a property that is useful in tracing epidemic spread of the organisms. These bacteria require special media and microbiological techniques to aid their recovery from the mixture of bacteria present in throat exudate or other material. It is unlikely that they would be recovered by the usual methods for culturing *S. pyogenes* from sore throats. Media useful in identifying *C. diphtheriae* include Loeffler's, which is composed largely of solidified serum, and a medium containing potassium tellurite, a chemical that inhibits the growth of many other bacterial species. Tellurite forms a black compound within *C. diphtheriae*, causing the colonies to appear gray or black.

• **Corynebacteria, p. 263**
• **Metachromatic granules, p. 74**

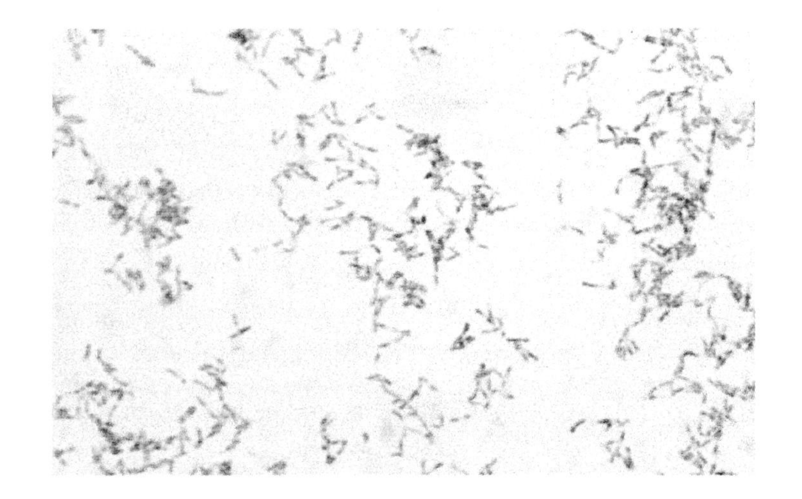

Figure 23.12
Corynebacterium diphtheriae. Note the "Chinese letter" pattern. Metachromatic granules (pink) are difficult to discern in this photograph.

Identification of this organism depends on its appearance after staining with the dye methylene blue, on colonial appearance, on biochemical activity such as the breaking down of carbohydrates and other substances (table 23.3), and on demonstrating the diphtheria toxin. The identification of different strains of *C. diphtheriae* by differences in the results of these tests can aid in tracing epidemics.

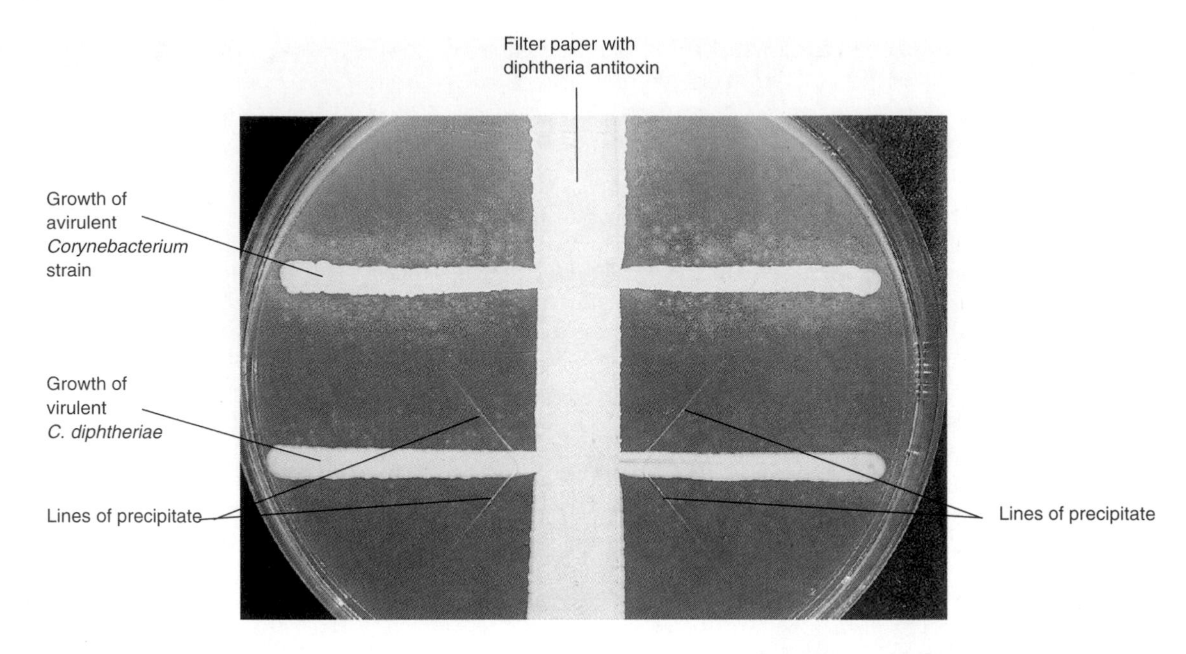

Figure 23.13
Test for diphtheria toxin. The top horizontal streak is an avirulent *Corynebacterium* strain; the bottom horizontal streak is virulent *C. diphtheriae*. The vertical band is filter paper soaked with diphtheria antitoxin. Note the fine lines of precipitate at 45° angles to the bottom streak, formed when antitoxin meets toxin diffusing into the medium.

Most but not all strains of *C. diphtheriae* secrete a powerful exotoxin (figure 23.13) that diffuses from the organism and is responsible for the seriousness of diphtheria infections. Production of this toxin requires that the bacterium be lysogenized by one of a certain group of bacteriophages, in another example of lysogenic conversion. Strains that are nontoxigenic become toxin producers when infected with the bacteriophage. Toxin production can be demonstrated *in vitro* or *in vivo* by its effect on guinea pigs. The toxin is produced only when the medium on which *C. diphtheriae* is growing has too little iron for optimal growth of the bacterium.

Pathogenesis

C. diphtheriae has very little invasive ability, rarely entering the blood or tissues, but the powerful exotoxin it releases is absorbed by the bloodstream. The gray-white membrane that forms in the throat of people with diphtheria is made up of clotted blood along with cells of the host mucous membrane and inflammatory cells[1] that have been killed by exotoxin from growing diphtheria bacteria. This membrane may come loose and obstruct the airways, causing the patient to suffocate. Absorption of the toxin by body cells results in damage to the heart, nerves, and kidneys.

The diphtheria toxin is a 58,000 molecular weight protein secreted by the bacterium in an inactive form. It is cleaved extracellularly into two chains, A and B, joined by a disulfide bond. The B chain attaches to specific receptors on a host cell membrane, and the toxin molecule is taken into the cell by endocytosis (figure 23.14 *a*). Cells that lack the appropriate receptors do not take up the toxin and are unaffected

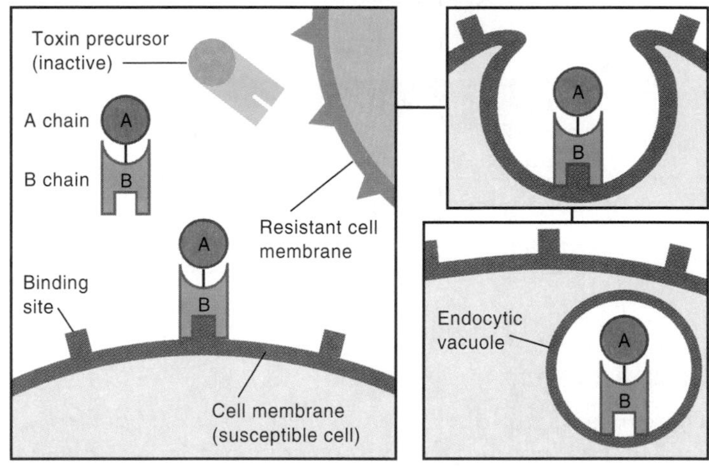

(a) Endocytosis

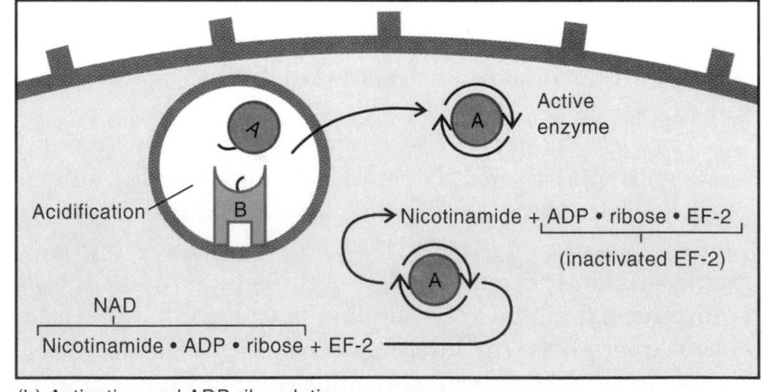

(b) Activation and ADP ribosylation

Figure 23.14
Mode of action of diphtheria toxin. (*a*) Toxin precursor splits into A and B chains joined by a disulfide bond; the B chain attaches to a specific receptor on the cell membrane of a susceptible cell. The toxin enters the cell by endocytosis. (*b*) With acidification of the endocytic vacuole, the A chain separates from the B chain, enters the cytoplasm as an active enzyme and catalyses the inactivation of EF-2 by ADP-ribosylation.

[1] Inflammatory cells—cells from blood and tissue that congregate in the area as part of the inflammatory response

• **Inflammatory response, pp. 345, 354–355**

by it. This receptor specificity explains why some tissues of the body are not affected in diphtheria, while others such as the heart and nerves are severely damaged. The need for specific receptors also explains why some animals such as mice are resistant to the toxin, while others such as guinea pigs are very susceptible.

Once the toxin molecule is inside the cell, the A chain separates from the B chain and becomes an active enzyme. This enzyme catalyses a chemical reaction in which part of the NAD molecule, ADP-ribose, is transferred to elongation factor 2 (EF-2). This reaction is known as **ADP ribosylation** (figure 23.14 *b*) and is essentially irreversible. Since it inactivates EF-2, a substance required for movement of the ribosome on mRNA, it halts protein synthesis, and the cell dies. Since the toxin A chain is an enzyme, it is not consumed in the process, so one or two molecules of toxin can inactivate essentially all the cell's elongation factor 2. This explains the extreme potency of diphtheria toxin. However, in vitro it is easily inactivated by heat and certain chemicals. Interestingly, the mode of action of diphtheria's toxin is similar to those of the toxins of some other bacterial pathogens, including *Vibrio cholerae, Pseudomonas aeruginosa, Escherichia coli,* and *Bordetella pertussis*. The toxins produced by these organisms also act by ADP ribosylation, transferring ADP-ribose to EF-2 or some other important component of the cells.

Epidemiology

Humans are the primary reservoir for *C. diphtheriae*. Sources of infection include carriers who have recovered from an infection, new cases not exhibiting symptoms, and people with active disease. The organisms are inhaled and establish infection in the upper respiratory system, usually in the nose or throat.

Prevention

Toxoid, prepared by formalin treatment of diphtheria toxin, is used for immunization against diphtheria. Toxoid administration results in the production of antibodies that specifically neutralize the diphtheria toxin. Since the disease results primarily from toxin absorption rather than microbial invasion, its control can be accomplished most effectively by immunization with toxoid. The well-known childhood vaccination DPT consists of diphtheria and tetanus toxoids and pertussis vaccine, all three generally given together. Unfortunately, these immunizations have often been neglected, particularly among socioeconomically disadvantaged groups, and serious epidemics of the diseases have occurred periodically. Since the 1980s, there has been an active campaign in most of the United States to ensure that children who are entering school are immunized against diphtheria. As a result, today only a few cases of diphtheria are annually reported in the United States, as compared to the 30,000 cases reported in 1936.

- **Protein synthesis, pp. 72, 138–40**
- **NAD, pp. 114–115**
- **ADP ribosylation, p. 114**

Treatment

Effective treatment of diphtheria depends on giving antiserum against diphtheria toxin to the patient as soon as possible. If the disease is suspected, administration of antiserum must be given without waiting for proof of the diagnosis, since a delay of the several days needed to obtain confirmation from the culture results can be fatal. The bacteria are sensitive to antibiotics such as erythromycin and penicillin, but such treatment stops only the transmission of the disease; it has no effect on the toxin that has been absorbed already. Even with treatment, on average, about 1 of 10 diphtheria patients dies (range 3.5% to 22% depending on age, severity of disease, promptness of treatment, etc.).

What has become apparent from analysis of some recent diphtheria epidemics around the world is that widespread childhood immunization against diphtheria has shifted susceptibility to the disease to adolescents and adults. Immunity wanes after childhood because of lack of exposure to diphtheria bacilli to boost immunity. In order to prevent buildup of a large population of susceptible older individuals, booster injections should be given every 10 years following childhood immunization.

Table 23.4 summarizes some important facts about diphtheria.

Table 23.4 Diphtheria

Symptoms	Sore throat, fever, fatigue, and malaise; membrane forms on tonsils and throat; paralysis, heart and kidney failure
Incubation period	2–6 days
Causative agent	*Corynebacterium diphtheriae*; cultured from nose and throat on special media; in vivo or in vitro tests for toxin
Pathogenesis	Infection in upper respiratory tract; exotoxin released and absorbed by bloodstream; toxin kills cells by interfering with protein synthesis; effect is on cells that have receptors for the toxin—mainly, heart, kidney, and nerve tissue
Epidemiology	Inhalation of infected droplets; direct contact with patient or carrier; indirect contact with contaminated articles
Prevention	Immunization with diphtheria toxoid; given to children at 6 weeks, 4 months, 6 months, 18 months, and 4–6 years; boosters every 10 years
Treatment	Antitoxin, erythromycin

REVIEW QUESTIONS

1. Which group of β-hemolytic streptococci can cause rheumatic fever? What component of the streptococcal cell wall determines the group?
2. Why is the hemolysis of *S. pyogenes* enhanced by anaerobic growth conditions?
3. Name the protein component of *S. pyogenes* cell walls that is mainly responsible for the organism's disease-causing ability.
4. What is scarlet fever?
5. Describe two serious complications that may develop after recovery from strep throat.
6. Why is diphtheria often fatal even though the causative organism usually only infects the nose or throat? How is the disease prevented?

Otitis Media and Sinusitis Typically Arise from Nasopharyngeal Infections

Otitis media (infection of the middle ear) occurs in three-quarters of children by the time they reach two years of age. It is responsible for 30 million doctor visits and costs over $1 billion annually.

Sinusitis (infection of the air-filled cavities of the skull) is common in both children and adults and shares similar etiology, pathogenesis, epidemiology, and treatment.

Symptoms

Otitis Media In a typical case, a young child with cold symptoms wakes from sleep with excruciating pain in one ear. Fever is generally mild or absent. Vomiting often occurs at the height of the pain. Sometimes the pain ends abruptly, and drainage of fluid appears in the external ear canal.

Sinusitis Symptoms of sinusitis also commonly accompany cold symptoms. Characteristically, pain and pressure sensation occur in the region of the involved sinus. Tenderness is also present over the sinus in many instances. Maxillary sinusitis is usually accompanied by a headache in the area of the forehead, and often an adjacent tooth aches. When a sphenoid sinus is involved, headache can be in the back of the head. Sinus infections are often associated with severe malaise.

Causative Agents

Otitis Media *Streptococcus pneumoniae* and *Haemophilus influenzae* are the most important bacterial pathogens in otitis media. Less commonly, *Mycoplasma pneumoniae, Streptococcus pyogenes*, and *Staphylococcus aureus* are the cause. In many if not most cases of otitis media, a respiratory virus initiates the infection and then respiratory bacteria colonize the middle ear. Some of the bacteria, such as *Moraxella catarrhalis*, appear to have little virulence. In 1992, a study was reported in which middle ear fluid from a large number of otitis media cases was examined for bacteria and viruses. Slightly more than one-third of the cases showed only bacteria in the fluid, while an equal number showed both a bacterium and a virus. The remaining cases had either a virus alone or negative results for both viruses and bacteria. Such studies help explain why some otitis media cases fail to respond to antibacterial medicines, which have no effect on viruses.

Sinusitis Sinusitis can also be caused by viruses or bacteria. A viral sinus infection is usually part of the common cold and subsides along with the cold without treatment. However, bacterial sinus infections (table 23.5) with opportunistic nasopharyngeal inhabitants such as *S. pneumoniae, H. influenzae*, or *Bacteroides* sp. can arise as a complication of colds.

Pathogenesis

Typically, otitis media is preceded by infection of the nose and throat. Presumably, infection spreads upward through the eustachian tube (figure 23.15), moving cell-to-cell along the epithelium, undoubtedly assisted at times by changes in airway pressure that force infected secretions upward in the tube. The infection damages the ciliated cells, resulting in inflammation and swelling. Since the damaged eustachian tube cannot move secretions from the middle ear, pressure builds up from fluid and pus collecting behind the eardrum. The throbbing ache of a middle ear infection is produced by pressure on nerves supplying the middle ear. Bacterial infections of the middle ear can cause the eardrum to perforate, giving discharge of blood or pus from the ear and immediate relief from pain. With treatment, the eardrum perforations usually heal promptly.

Almost all patients with otitis media have some fluid behind the eardrum that impairs movement of the drum.

• **Mycoplasmas, pp. 266–267**

Table 23.5 Principal Bacterial Causes of Sinus Infections

S. pneumoniae	*H. influenzae*	*Bacteroides sp.*
Gram-positive diplococci, often encapsulated; bile soluble; α-hemolytic colonies	Gram-negative pleomorphic rods; facultative anaerobes; require X and V factors* for growth	Gram-negative, pleomorphic rods; strict anaerobes

*X and V factors—X is a fraction of hemoglobin called *hematin*; V is NAD

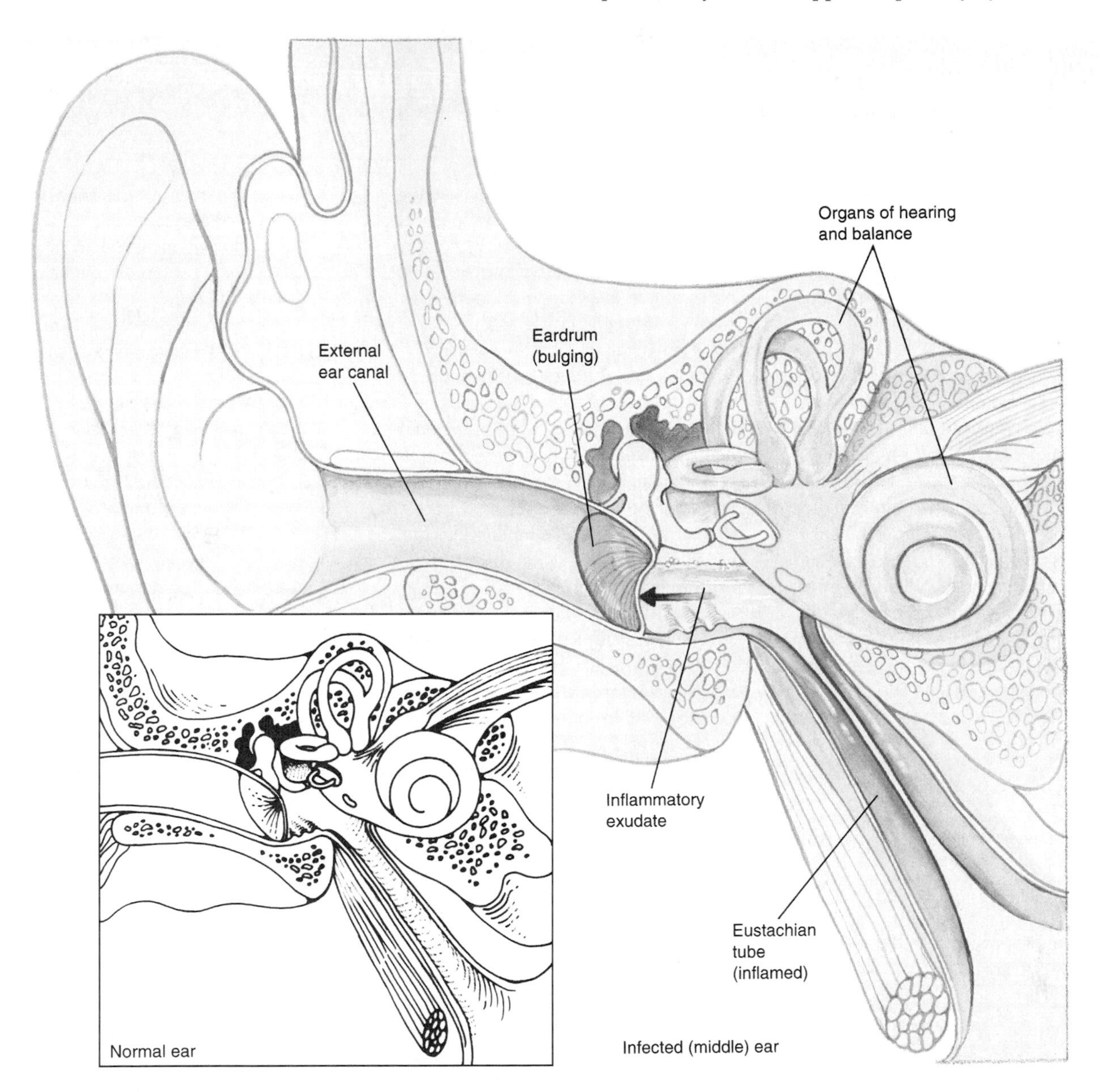

Figure 23.15
Otitis media. A stylized view of the infected middle ear showing accumulation of fluid, swelling of the eustachian tube, and bulging outward of the eardrum.

This fluid disappears within three months of treatment in about 90% of the cases, but it may persist for months in the remainder. Because of the fluid, some children cannot hear clearly and show a delay in normal speech development.

In the absence of timely treatment, bacterial otitis media can spread to the mastoid air cells, which communicate with the middle ear cavity, or even into the bone surrounding the middle ear cavity and mastoids. This can result in damage to the nerves that normally pass through the bone, causing paralysis of an eye or facial muscle. Further extension of the infection can result in meningitis or, more rarely, brain abscess. All these complications require intensive treatment in the hospital. Fortunately, they are now generally rare in the United States.

Epidemiology

Otitis Media Otitis media is rare in the first month. It becomes very common in early childhood, especially during winter and spring when viral respiratory diseases are frequent. Use of pacifiers beyond the age of two years is associated with a substantial increase in otitis media. Other conditions that cause inflammation of the nasal mucosa besides viral infections appear to play a role in some cases. Nasal allergies and exposure to air pollution and cigarette smoke are examples of nasal irritants. Older children develop immunity to *Haemophilus influenzae*, and it becomes an increasingly uncommon cause of otitis media after about five years.

Sinusitis Sinusitis tends to involve older children in whom the sinuses are more fully developed and adults. The incidence follows the same seasonal pattern and relationship to allergy and irritants as otitis media.

Case Presentation

The patient was a 19-month-old boy who woke up screaming with left ear pain shortly after midnight. He vomited one time. He was previously well except he had had a slight cold. One older sibling attended a day-care facility. One parent smoked cigarettes but only outside the house.

On examination, the patient was cranky and alert and had a temperature of 100.2°F. His left eardrum was red and bulged outward. His neck was not stiff and he had no rash.

The doctor diagnosed acute otitis media, probably bacterial. He prescribed medicine for pain and amoxicillin (a kind of penicillin).

Over the next 72 hours, the patient failed to improve. Using a needle and syringe, the doctor withdrew fluid from the middle ear, giving relief of pain. A Gram-stained smear of the fluid revealed gram-positive diplococci. Culture media were inoculated. The boy's antimicrobial medication was changed, and he made a rapid recovery.

1. What factors predisposed this child to contract otitis media?
2. What was the most likely causative agent in this case?
3. What was the likelihood that the causative agent was resistant to amoxicillin?
4. What are some approaches to the prevention of otitis media?

Discussion

1. For unknown reasons, being male increases the risk of having otitis media. The patient had cold symptoms, indicating a viral respiratory infection that impaired body defenses against bacterial middle ear infection. Both attendance at a day-care facility and attendance of a family member at a day-care facility increase the risk of otitis media. In day-care centers, potential respiratory pathogens spread readily and multiply to high numbers in their young hosts, who then transfer them to family members and others. Cigarette smoking by household members probably did not play a role in this case because no one smoked inside the house.
2. In infants older than three months, *Streptococcus pneumoniae* is the most common cause of otitis media, accounting for 35% to 40% of cases. The gram-positive diplococci seen on the stained smear of this patient's middle ear fluid were confirmed to be *S. pneumoniae* by culture. In the United States, pneumococci account for an estimated 7 million cases of acute otitis media each year.
3. Up until the late 1980s, only very rare strains of *S. pneumoniae* were resistant to treatment with the penicillins. Since then, a progressive increase in resistance has been documented. In some parts of the United States, more than 25% of pneumococcal isolates from middle ear fluid now show in vitro resistance to the penicillins. Other otitis media pathogens, including *Haemophilus influenzae*, are also more likely to be resistant. The penicillin resistance of pneumococci results from mutant penicillin-binding proteins to which penicillins attach poorly. This type of resistance increases in small steps, so that merely increasing the dose of medication is sometimes effective at first. This contrasts with the plasmid-coded resistance of *H. influenzae*, which cannot be overcome by increasing the dose of antibiotic. Pneumococcal resistance to other antibacterial medications is also increasing. This means that doctors can no longer assume that every case of otitis media will be cured by any given medication. They must be prepared to obtain middle ear fluid for culture in an increasing number of cases.
4. General preventive measures include promotion of breast feeding, avoidance of bottle feeding when supine, avoiding pacifiers beyond age two, and avoiding cigarette smoke. Influenza vaccine helps prevent otitis media by preventing a predisposing viral illness. *Haemophilus influenzae* type b vaccine is an excellent vaccine against the invasive type b strains but not the strains of *Haemophilus influenzae* that cause otitis media. The currently licensed pneumococcal vaccine consists of a mixture of the capsular polysaccharides of 23 of the 90 or more known *S. pneumoniae* strains. Besides being only weakly immunogenic in infants, the polysaccharides are T cell independent antigens, stimulating B cells directly, without the participation of T cells. Since the polysaccharides are poor stimulators of immunologic memory, natural exposure to pathogenic pneumococci fails to provoke an anamnestic response, and immunity wanes with time. A number of conjugate vaccines are under development and some are in phase III trials, meaning that they have already been shown to be safe and immunogenic and are now being tested in humans for their effectiveness in preventing pneumococcal disease. These vaccines are composed of pneumococcal polysaccharides to which a protein such as diphtheria toxoid has been attached. The conjugated vaccines are T cell dependent antigens and provoke immunologic memory. It is hoped they will stimulate better production of IgG antibody, which could cross the placenta of pregnant women and, therefore, protect young infants. They also may cause better production of IgA antibodies, which are secreted onto mucous membranes, giving good protection against otitis media and ensuring faster elimination of the carrier state.

Prevention

Otitis Media Administration of influenza vaccine to infants in day-care facilities substantially decreases the incidence of otitis media during the "flu" season. Pneumococcal vaccine as presently constituted has proved disappointing in preventing otitis media because infants under two years of age fail to get an adequate immune response. New, presently experimental, vaccines will probably be better. Ampicillin or sulfasoxazole given continuously over the winter and spring are useful preventives in people who have three or more bouts of otitis media within a six-month period. Surgical removal of enlarged adenoids improves drainage from the eustachian tubes and can be helpful in preventing recurrences in certain patients. In patients with chronically malfunctioning eustachian tubes, plastic ventilation tubes are sometimes installed in the eardrums so that pressure can equalize on both sides of the drum. In the United States, this is the most common operation performed on children, but its value in preventing acute otitis media has been questioned.

Sinusitis The approaches used in preventing otitis media probably apply in large part to sinusitis, but few studies have been done.

Treatment

Antibacterial therapy with ampicillin is generally effective. In communities where β-lactamase producing *H. influenzae* strains are present, ampicillin plus clavulanic acid is appropriate. Erythromycin plus sulfasoxazole, cefaclor, and cefixime are among the numerous alternative medications that are used. Since it is usually difficult or impossible to identify the causative agent, several different antibacterial medications are sometimes tried until one "works" or, more commonly, the patient gets better on his or her own. Occasionally, pus must be drained surgically by making an opening in the eardrum in otitis or by inserting a large needle through the bone in the case of sinusitis. In general, antibacterial treatment has been highly successful in preventing the serious complications of these two conditions. Decongestants and antihistamines generally are ineffective and can be harmful.

Haemophilus influenzae and *Streptococcus pneumoniae* Are Common Causes of Bacterial Conjunctivitis (Pink Eye)

Eye infections are included in this discussion of upper respiratory infections because the eyes connect directly to the upper respiratory system by the nasolacrimal ducts, providing one of the major routes for pathogens to enter the system. Also, almost all the common eye infections are caused by respiratory pathogens. Finally, symptoms and signs of eye infection commonly accompany other upper respiratory infections.

Symptoms

The symptoms of acute bacterial conjunctivitis include excessive tears, redness of the conjunctiva, swelling of the eyelids, sensitivity to bright light, and large amounts of pus. These symptoms, seen mainly in children, generally last a few days to three weeks. Considerable variation in symptoms exists, depending on the causative agent. Acute bacterial conjunctivitis must be distinguished from conjunctivitis caused by a large number of viruses. In the latter cases, eyelid swelling and pus are usually minimal.

Causative Agents

The most common bacterial pathogens are *Haemophilus influenzae* and *Streptococcus pneumoniae*. Other bacteria, including *Moraxella lacunata*, *Chlamydia trachomatis*, some enterobacteria, and *Neisseria gonorrhoeae* also infect the conjunctiva. Conjunctival cultures and microscopic examination of stained smears of the eye discharge are important aids in diagnosis.

Pathogenesis

Few details are known about the pathogenesis of acute bacterial conjunctivitis. Presumably, the organims are inoculated directly onto the conjuctiva from airborne respiratory droplets or rubbed in from contaminated hands. The organisms must have a mechanism for adhering to the conjunctiva and resisting the constant sweeping action of the eyelids. Adhesins are present on the subgroup of *H. influenzae* strains that infect the conjunctiva. Conjunctival adhesins are also demonstrable for some enterobacteria. An adhesin for epithelial cells has been demonstrated for *S. pneumoniae*, but its role in conjunctivitis is not yet known. Bacterial conjunctivitis may be accompanied by otitis media.

Epidemiology

Epidemics of bacterial conjunctivitis sometimes occur among groups of schoolchildren. The carrier rates for *H. influenzae* and *S. pneumoniae* can sometimes reach 80% in the absence of disease. The factors involved in the appearance and spread of conjunctivitis-causing strains are unknown.

Prevention

Suspected cases are removed from school or day-care settings for diagnosis and start of treatment. General measures include hand washing, avoiding rubbing or touching the eyes, and no sharing of towels.

Treatment

Bacterial conjunctivitis is effectively treated with eyedrops or ointments containing an antibacterial medicine to which the infecting strain is sensitive.

Viral Infections of the Upper Respiratory System

Viral infections of the eyes, ears, sinuses, and throat are discussed in more detail in the following section.

Colds Are Caused by a Large Number of Different Viruses

Medical practitioners have been taught for generations that the symptoms of the common cold, "if treated vigorously, will go away in seven days, whereas if left alone, they will disappear over the course of a week." So far, little has occurred to change this assessment.

Causative Agents

The all-too-familiar symptoms of the common cold ("acute afebrile infectious coryza") result from upper respiratory

Perspective 23.3

Laboratory Diagnosis of Rhinovirus Infections

The rhinoviruses were initially difficult to cultivate in the laboratory, since most failed to infect laboratory animals or tissue cell cultures. Researchers in England, however, discovered that if tissue cell cultures of monkey or human origin were incubated at 33°C instead of at body temperature (37°C) and at a slightly acid pH instead of at the alkaline pH of body tissues, many rhinovirus strains would infect them. It is noteworthy that these conditions of lower temperature and pH normally exist in the upper respiratory tract. Rhinoviruses, however, are killed if the pH drops below 5.3 and, therefore, they are usually destroyed in the human stomach. Their susceptibility to acid is used to distinguish them from other picornaviruses, such as coxsackievirus and poliovirus.

tract infections by any of about 200 known viruses and a few bacteria. The cold-causing viruses include members of the orthomyxoviruses (influenza A and B), paramyxoviruses (parainfluenza and respiratory syncytial viruses), coronaviruses, adenoviruses, and enteroviruses (Coxsackie- and ECHO viruses). However, 30% to 50% of colds are caused by the 100 or more types of rhinoviruses (*rhino* means "nose," as in rhinoceros, or "horny nose"), members of the picorna group (*pico* is Italian for "small" and *rna* is *r*ibo*n*ucleic *a*cid [RNA]; thus, "small RNA viruses"; figure 23.16). Their RNA is single-stranded, positive sense.

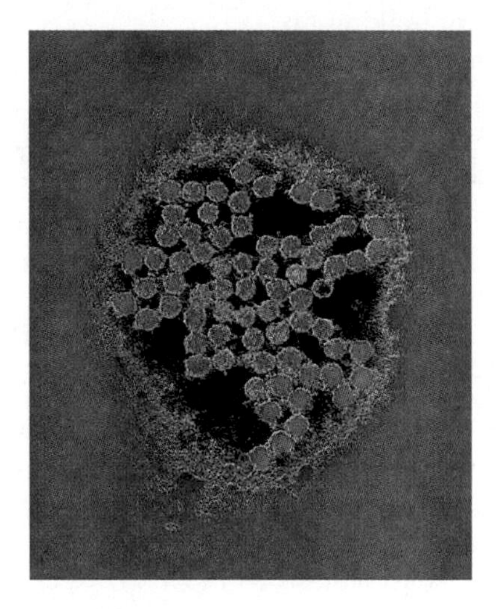

Figure 23.16
Rhinovirus. Transmission electron micrograph. The red spheres are the rhinovirus virions. These virions range from 20 to 27 nm in diameter and contain single-stranded, positive sense RNA. The blue color is the cytoplasm of the host cell in which they replicate.

Pathogenesis

After the rhinoviruses lodge on the respiratory epithelium cells, they infect a few of them. The virus then replicates, is shed from the cells, and infects adjacent cells. The injury causes the release of inflammatory mediators and stimulates nervous reflexes. The result is an increase in nasal secretions, sneezing, and swelling of the mucosa and nasal erectile tissue. This swelling partially or completely obstructs the airways. Later, in the inflammatory response, dilation of blood vessels, oozing of plasma, and congregation of leukocytes in the infected area occur. Secretions from the area may then contain pus and blood. The infection is eventually halted by the inflammatory response, interferon release, and cellular and humoral immunity, but it can extend into the ears, sinuses, or even the lower respiratory tract before it is stopped.

• **Mediators of inflammation, pp. 351, 397, 400, 415**

Epidemiology

Humans are the only source of cold viruses, and close contact with an infected person is generally necessary for viruses to be transmitted. A person with severe symptoms early in the course of a cold is much more likely to transmit a cold than is someone whose symptoms are mild or who is in the late stage of a cold. Very high concentrations of virus are found in the nasal secretions and often on the hands of infected people during the first two or three days of a cold. By the fourth or fifth day, virus levels are often undetectable, but low levels can be present for two weeks. A few virions are sufficient to infect the nasal mucosa, but the mouth is quite resistant to infection. In adults the disease is usually contracted when airborne droplets containing virus particles are inhaled. Transmission can also occur when virus-containing secretions are unwittingly rubbed into the eyes or nose by contaminated hands. Virus introduced into the eye is promptly transmitted to the nasal passage via the nasolacrimal duct. With reasonable caution, however, colds are not highly contagious. In a study of nonimmune adults exposed in a family or dormitory setting, less than half contracted colds. Young children, however, transmit colds and other respiratory viruses very effectively because they are often careless with their respiratory secretions.

Prevention

Since such a large number of different viruses cause colds, vaccines are impractical. Rhinoviruses, like all other viruses, are resistant to antibiotics and other medications

Perspective 23.4

Does a Cold Temperature Cause Colds?

Many people believe that colds occur more often when the body is exposed to cold temperatures, and scientific studies have been undertaken repeatedly to define this relationship. An important study reported in 1933 showed that colds disappeared from the Arctic island of Spitzbergen during the long winter when no ships came, and they reappeared when ships arrived in late spring. This finding indicated that new sources of infectious virus are necessary to produce colds, regardless of the temperature of a region. Other studies have shown that the incidence and severity of colds were the same when a rhinovirus was administered to two groups of nonimmune volunteers and one group was exposed to chilling and the other was not. Finally, it has been shown that, in semitropical areas, a sharp increase in colds occurs with the onset of the rainy season, even though the mean temperature stays about the same. Thus, the influence of season on the incidence of colds and other respiratory diseases has been clearly demonstrated, but the mechanism for this phenomenon has not. One important factor is the tendency for people to congregate indoors when the weather is bad, increasing the opportunities for infectious viruses to spread.

Although studies have failed to show that exposure to cold temperatures increases susceptibility to colds, solid evidence exists that psychological stress increases the risk of colds. A study of 394 healthy people intentionally exposed to cold viruses showed that the incidence of colds rose progressively from 27% to 47% according to the level of psychological stress as revealed by a stress-measuring questionnaire.*

* Excerpted from information appearing in Cohen, Sheldon, et al. 1991. New England Journal of Medicine, 325:606.

that control bacterial infections. Prevention of the spread of rhinoviruses includes washing the hands, even in plain water, which readily removes rhinoviruses, and keeping the hands away from the face. In addition, one should avoid crowds and crowded places such as commercial airplanes when respiratory diseases are prevalent and especially avoid people with colds during the first couple of days of their symptoms.

Experimentally, interferon, produced by genetic engineering techniques, helps prevent colds when it is sprayed into the nose at the time of exposure to a cold and for several days thereafter. However, interferon is currently too expensive to use for preventing colds.

Table 23.6 summarizes some facts about the common cold.

Table 23.6 Common Cold

Symptoms	Scratchy throat, nasal discharge, malaise, headache, cough
Incubation period	1–2 days
Causative agent	Mainly rhinoviruses—more than 100 types; many other viruses
Pathogenesis	Viruses lodge on respiratory epithelium, starting infection that spreads to adjacent cells; ciliary action ceases and cells slough; mucus secretion increases, and inflammatory reaction occurs; infection stopped by interferon release and antibody production
Epidemiology	Inhalation of infected droplets; transfer of infectious mucus to nose or eye by contaminated fingers; children initiate many outbreaks in families because of lack of care of nasal secretions
Prevention	Hand washing; avoidance of people with colds and touching face
Treatment	None except for symptom relief

Adenoviral Pharyngitis Can Resemble Strep Throat

Adenoviruses are widespread and, throughout the year, produce a spectrum of illnesses ranging from coldlike ailments to pneumonia or diarrhea. However, most commonly they cause upper respiratory ailments.

Symptoms

Adenoviral infections often cause coryza (profuse runny nose) but, unlike with the common cold, fever is usually also present. Typically, the throat is sore, and grey-white pus is present on the pharnyx and tonsils. The lymph nodes of the neck enlarge and become tender. Often, conjunctivitis is present, and there may be a mild cough. Less commonly, a severe cough develops, indicating possible pneumonia. With or without treatment, recovery occurs in one to three weeks.

Causative Agent

More than 45 antigenic types of adenoviruses infect humans. The viruses are nonenveloped, 70 to 90 nanometers in diameter, with double-stranded DNA. They can be cultivated in a

• **Interferon, pp. 184–185, 560**

Perspective 23.5

Cold Medicines Are of Doubtful Value and Some May Be Harmful

Commonly used cold remedies have little or no beneficial effect on colds and some may be harmful. In a study of 56 healthy subjects with intentionally induced rhinovirus colds, subjects treated with aspirin or acetaminophen had significantly more nasal swelling and obstruction than did those given only a placebo. Moreover, aspirin and acetaminophen partly suppressed the antibody response to the rhinovirus. These results indicate that treating colds with aspirin or acetaminophen is best avoided unless required to relieve significant headache or other discomfort.

variety of tissue cell cultures where they produce a characteristic cytopathic effect. They can remain infectious in the environment for long periods of time and can be transferred readily from one patient to another on medical instruments. However, they are easily inactivated by heat at 56°C, chlorine, and various other disinfectants.

Pathogenesis

Little is known about the pathogenesis of adenoviral disease. Humans are the only source of infection; however, related viruses infect animals. Once inside the cells of the host, the virus multiplies in the nuclei. In severe infections, extensive cell destruction and inflammation occur. Different types of adenoviruses vary in the tissues they affect. For example, adenovirus types 4, 7, and 21 typically cause illness characterized by sore throat and enlarged lymph nodes, whereas type 8 is likely to cause extensive eye infection with few other symptoms. Adenoviruses 1, 2, 5, and 6 produce a mild throat infection in young children and then, in half the cases, become latent in the tonsils and other lymphoid tissues, where they remain for years. Adenoviral throat infections can resemble infectious mononucleosis and strep throat. Some human adenoviruses cause tumors when injected into baby hamsters but do not appear to cause tumors in humans.

Epidemiology

Adenoviral illness is prevalent among schoolchildren, causing outbreaks of respiratory sickness in winter and spring. Summertime epidemics occur as a result of transmission of the viruses in inadequately chlorinated swimming pools, resulting in **pharyngoconjunctival fever.** These cases typically have fever, conjunctivitis, pharnygitis, enlargement of the lymph nodes of the neck, a rash, and diarrhea. Adenoviral disease is common in any congregation of young people and has been a big problem in military recruits. Epidemic spread is fostered by a high percentage of asymptomatic infections. The viruses are shed from the upper respiratory tract during the acute illness and continue to be excreted in the feces for months thereafter.

• **Cytopathic effect, pp. 315, 401**

Table 23.7 Adenoviral Pharyngitis (Pharyngoconjunctival Fever)

Symptoms	High fever, very sore throat, severe cough, swollen lymph nodes of neck, exudate over tonsils and throat, and conjunctivitis.; less frequently, pneumonia
Incubation period	5–10 days
Causative agent	Adenoviruses—more than 45 types
Pathogenesis	Virus multiplies in epithelial cells; cell destruction and inflammation occur; latent infections occur with some strains
Epidemiology	Inhalation of infected droplets; possible spread from gastrointestinal tract
Prevention	Live virus vaccine used by military
Treatment	None except for relief of symptoms

Prevention and Treatment

No vaccines are marketed for civilian use because adenovirus infections are ordinarily not serious or frequent enough to require widespread vaccination. However, military recruits can be immunized with attenuated or inactivated vaccines containing several adenovirus types. Antibiotic treatment is of no value in treating adenovirus infections and sometimes can even be harmful because it suppresses some of the body's normal bacteria, thereby allowing unchecked growth of resistant opportunists that might be present in the body.

Table 23.7 summarizes some important facts about adenoviral pharyngitis.

REVIEW QUESTIONS

1. What role do viruses play in bacterial infections of the middle ear and sinuses?
2. What causes the common cold?
3. How are colds transmitted?
4. How can colds be prevented?
5. What kinds of diseases are caused by adenoviruses?

Summary

I. Anatomy and Physiology
   A. The moist lining of the eyes (conjunctiva), nasolacrimal duct, middle ears, sinuses, mastoid air cells, nose, and throat comprise the main structures of the upper respiratory system.
   B. An important defense system is the movement of mucus by the ciliated cells that line much of the respiratory system. The middle ears, mastoids, and sinuses are usually kept free of microorganisms by this mechanism.
   C. The functions of the upper respiratory tract include temperature and humidity regulation of inspired air and removal of microorganisms.

II. Normal Flora of the Upper Respiratory System
   A. Secretions of the nasal entrance are often colonized by diphtheroids, micrococci, and *Staphylococcus aureus* (coagulase-positive staphylococci).
   B. The nasopharyngeal flora include α-hemolytic streptococci, nonhemolytic streptococci, *Moraxella catarrhalis*, diphtheroids, and anaerobes of the genus *Bacteroides*.

III. Bacterial Infections of the Upper Respiratory System
   A. *Streptococcus pyogenes* causes strep throat, a very important bacterial infection. Infection may lead to scarlet fever, rheumatic fever, toxic shock, or glomerulonephritis.
   B. Diphtheria, caused by *Corynebacterium diphtheriae*, is a toxin-mediated disease that can be prevented by immunization.
   C. Otitis media and sinusitis develop when infection extends from the nasopharynx.
   D. Conjunctivitis (pink eye) is usually caused by *Haemophilus influenzae* or *Streptococcus pneumoniae* (pneumococcus). Viral causes, including adenoviruses and rhinoviruses, usually result in a milder illness.

IV. Viral Infections of the Upper Respiratory System
   A. The common cold can be caused by many different viruses, rhinoviruses being the most common.
   B. Adenoviruses cause illnesses varying from mild to severe, which can resemble a common cold or strep throat.

Chapter Review Questions

1. After you recover from "strep throat," can you get it again? Explain.
2. Why do streptococcal pyrogenic exotoxins (SPEs) cause a release of cytokines?
3. After a person's recovery from untreated streptococcal pharyngitis, what symptoms would indicate a diagnosis of chorea?
4. Name the causative agent of diphtheria.
5. Where is the gene for diphtheria toxin production located?
6. Describe the process by which diphtheria toxin enters host cells and kills them.
7. Describe the two bacterial species responsible for most cases of bacterial sinusitis, otitis media, and conjunctivitis.
8. Describe the group of viruses responsible for most colds.
9. What are two ways to decrease the chance of contracting a cold?
10. What is the causative agent of pharyngoconjunctival fever?

Critical Thinking Questions

1. Explain how it is that diphtheria toxin affects some tissues and spares others.
2. What possible advantage might there be for *C. diphtheriae* to be lysogenized by a bacteriophage that codes for toxin production when there is little iron in the medium?

Further Reading

Bisno, A. L. 1991. Group A streptococcal infections and acute rheumatic fever. *The New England Journal of Medicine* 325:783.

Clements, D. A., et al. 1995. Influenza A vaccine decreases the incidence of otitis media in 6- to 30-month-old children in day care. *Arch Pediatr Adolesc Med* 149:1113–7.

Collier, R. J. 1990. Corynebacteria. In B. D. Davis, R. Dulbecco, H. N. Eisen, and H. S. Ginsberg (ed.), *Microbiology*, 4th ed. J. B. Lippincott Company, Philadelphia, 507–514.

1996. Defining the public heath impact of drug-resistant *Streptococcus pneumoniae:* report of a working group. *Morbidity and Mortality Weekly Report* 45(no. RR-1):1–20.

1996. Diphtheria outbreak-Saraburi Province, Thailand, 1994. *Morbidity and Mortality Weekly Report* 45(no. 13):271.

Fischetti, V. A. 1991. Streptococcal M protein. *Scientific American* 264:58–65.

Hierholzer, J. C. 1995. Adenoviruses. In P. R. Murray (ed.), *Manual of clinical microbiology*, 6th ed. ASM Press, Washington, D. C., 947–955.

Kleinman, L. C., et al. 1994. The medical appropriateness of tympanostomy tubes proposed for children younger than 16 years in the United States. *Journal of the American Medical Association* 271:1250–1255.

Niemela, M., et al. 1995. A pacifier increases the risk of recurrent acute otitis media in children in day care centers. *Pediatrics* 96:884–888.

Salyers, A. A., and Whitt, D. D. 1994. *Bacterial pathogenesis. A molecular approach.* ASM Press, Washington, D. C.

1996. Update: diphtheria epidemic—new independent states of the former Soviet Union, January 1995–March 1996. *Morbidity and Mortality Weekly Report* 45(no. 32):693.

24 Lower Respiratory System Infections

KEY CONCEPTS

1. Very efficient mechanisms protect the lower respiratory system; bacterial lung infections usually occur only in people whose defenses are impaired.
2. Most pneumonias are bacterial or viral, but eukaryotic microorganisms, chemicals, and allergies can also cause pneumonias.
3. A vaccine composed of killed organisms can confer protection against some gram-negative bacterial diseases such as pertussis.
4. Most *Mycobacterium tuberculosis* infections become latent, posing a risk of reactivation throughout life.
5. Most deaths from influenza are caused by secondary bacterial infections.
6. The host's immune system can play a key role in the pathogenesis of lung disease.

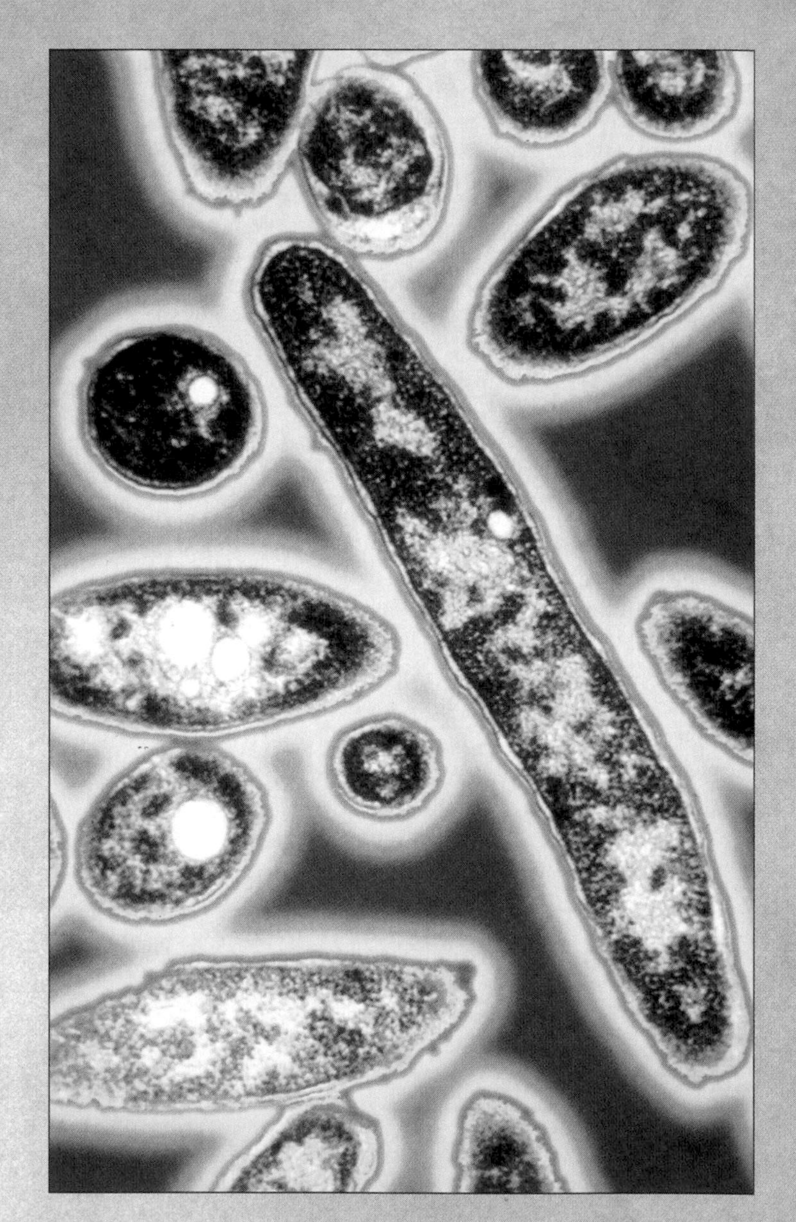

Color enhanced electron micrograph of *Legionella pneumophila*, the cause of Legionnaires' disease.

PREVIEW LINK

Microbes in Motion

The following books in the *Microbes in Motion* CD-ROM may serve as a useful preview or supplement to your reading: *Gram-Positive Organisms. Gram-Negative Organisms. Miscellaneous Bacteria. Microbial Pathogenesis in Infectious Diseases:* Toxins. *Vaccines:* Vaccine Targets. *Viral Structure and Function:* Viral Diseases. *Fungal Structure and Function:* Diseases. *Parasitic Structure and Function.*

A Glimpse of History

Tuberculosis is an ancient disease as evidenced by tubercular damage found in skeletons of Stone Age humans. With increasing urbanization in the 1300s, the incidence of pulmonary tuberculosis rose markedly. Many people died rapidly from the disease or had chronic debilitating symptoms, while others appeared to conquer the infection and lead normal lives.

In 1804, René Laënnec, after performing microscopic examination of tissues from nearly 400 autopsies, stated that the many different manifestations of tuberculosis, previously considered separate diseases, are in fact a single disease. This disease was named *tuberculosis* by a Swiss physician in 1839 for the characteristic nodules, or tubercles, that are formed in infected tissue. A French army surgeon, Jean Antoine Villemin, subsequently proved that tuberculosis is contagious by inducing the disease in rabbits with injections of material from human cases. However, it was Robert Koch who proved that tuberculosis is actually caused by a bacterium. In 1881, he performed experiments with the tubercle bacillus using guinea pigs. Six months later, he presented a preliminary report to the Berlin Physiological Society, in which he demonstrated the unusual staining methods required to identify the tubercle bacillus. Koch's presentation was exceptionally well received by the audience, including the highly respected scientist Paul Ehrlich who remarked that the presentation represented his "greatest experience in science." In 1882, Koch published "The Etiology of Tuberculosis," which is considered his major work, and in 1905, he received the Nobel Prize.

Albert Calmette and Camille Guérin developed vaccines against tuberculosis in cows. In 1924, after 20 years of studies, they introduced a vaccine against tuberculosis in humans, called BCG (Bacille Calmette-Guérin). BCG was used in trials worldwide, and its use increased as its success became more apparent. Ironically and tragically, at the height of BCG's success, a culture of BCG became contaminated with a virulent strain of the tubercle bacillus at the Public Health Laboratories at Lübeck, Germany in 1929. Within a few weeks, 71 of 252 children who had received the contaminated vaccine died from tuberculosis. BCG was initially blamed for the disaster, but later proof of the contamination became widely known and the vaccine was used again. Today, vaccination with BCG is still widely used in some parts of the world where tuberculosis is uncontrolled.

Compared to the upper respiratory tract, the lower respiratory tract is well protected from colonization by microorganisms. However, some very serious diseases, such as tuberculosis and pneumonia, can affect the lower respiratory tract and are explored in detail in this chapter.

Anatomy and Physiology

The lower respiratory system consists of the windpipe (trachea) and its various branching divisions and subdivisions (bronchi and bronchioles) ending in the tiny, thin-walled air sacs (alveoli) that make up the lungs (figure 24.1). The lungs are surrounded by two membranes: One adheres to the lung and the other to the wall of the chest and diaphragm. These membranes, termed **pleura,** normally slide against each other as the lung expands and contracts. Thus, a potential space exists between the two membranes in which products of infection can accumulate and compress the lungs.

As with the upper respiratory tract, much of the lower respiratory system is lined with ciliated cells and a film of mucus, the **mucociliary escalator.** This film is constantly swept upward from the bronchioles and bronchi toward the throat at the rate of about 2 centimeters per minute under normal conditions. It traps and removes microorganisms and other foreign materials, although extraneous factors such as tobacco smoke and chilling may decrease the ciliary action of the lower tract, and prolonged exposure to irritants such as tobacco smoke leads to loss of cilia and flattening of the lining cells. The cough reflex, which also helps to expel foreign materials, is activated by irritants and excessive secretions, but it can be depressed by external factors, such as alcohol and narcotics. Finally, macrophages are numerous in the lung tissues and readily move into the alveoli and airways to engulf infectious agents and other foreign substances. These protective mechanisms are very efficient, especially against bacteria, and it is unusual to see bacterial lung infections except in people whose defenses are impaired.

Normal Flora of the Lower Respiratory System

In contrast to their presence in portions of the upper respiratory system, microorganisms normally are absent from the entire lower tract.

Bacterial Pneumonias

Approximately 2 million cases of pneumonia occur in the United States each year, accounting for 40,000 to 70,000 deaths. Pneumonia tops the list of infectious killers in the general population and of infections acquired in hospitals. There are many different causes of pneumonia, but most commonly the cause is bacterial. In this section, we will discuss three common causes of bacterial pneumonia.

Streptococcus pneumoniae Is One of the Most Common Causes of Bacterial Pneumonia

Streptococcus pneumoniae accounts for over 60% of the bacterial pneumonias of adults who require hospitalization.

- **Macrophages, pp. 348, 349, 382, 677–78**
- **Streptococci, p. 260**

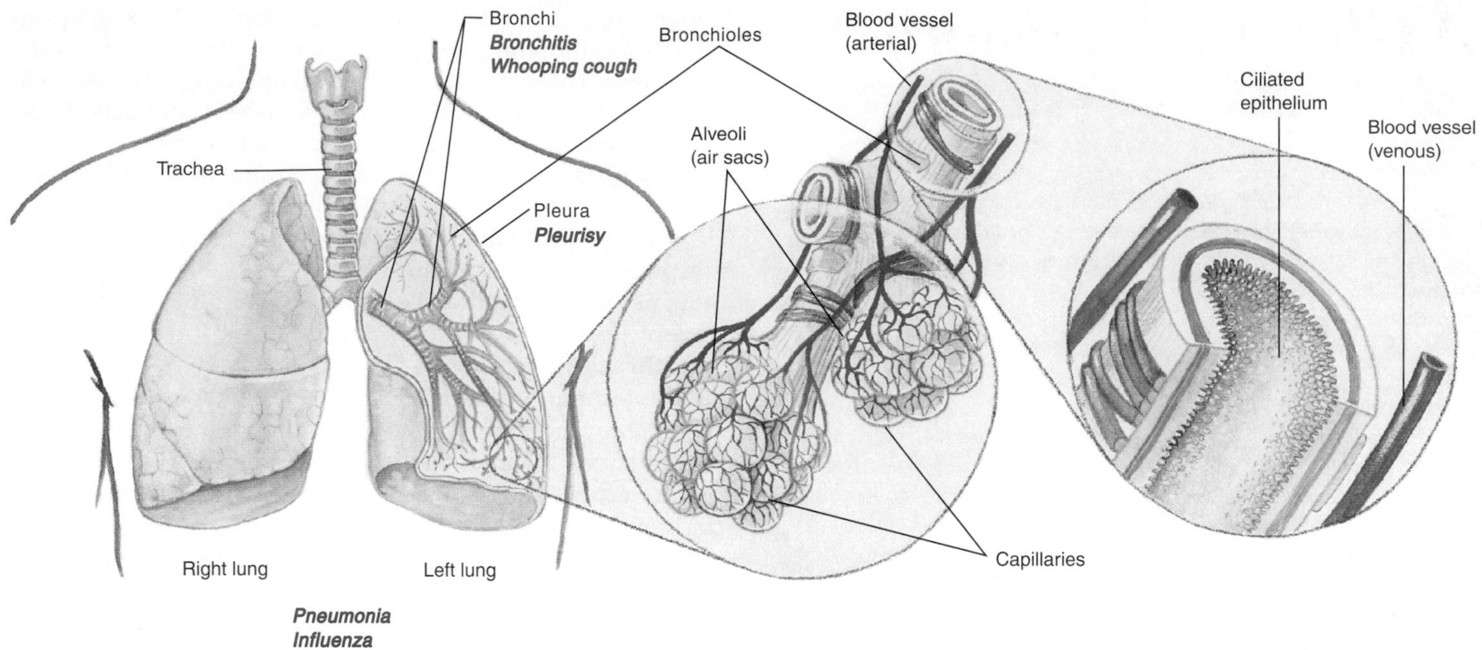

Figure 24.1
Lower respiratory system.

Symptoms

The typical symptoms of pneumococcal pneumonia are cough, fever, chest pain, and sputum[1] production. These symptoms are usually preceded by a day or two of runny nose and upper respiratory congestion, ending with an abrupt rise in temperature and a single body-shaking chill. The sputum quickly becomes pinkish or rust colored, and severe chest pain develops, aggravated by each breath or cough. As a result of the pain, breathing becomes shallow and rapid. The patient becomes short of breath and develops a dusky color because of poor oxygenation. Without treatment, after 7 to 10 days people who survive show profuse sweating and a rapid fall in temperature to normal.

Causative Agent

Streptococcus pneumoniae (also known as the *pneumococcus,* figure 24.2) is a gram-positive diplococcus. The most striking characteristic of *S. pneumoniae* is its thick capsule, which is responsible for the organism's virulence. There are over 80 different types of *S. pneumoniae,* as determined by differing capsular antigens.

[1]Sputum—thick discharge that accumulates in respiratory diseases, usually consisting of mucus, microorganisms, pus, sloughed respiratory cells, and sometimes blood

• **Capsules, pp. 58–61**

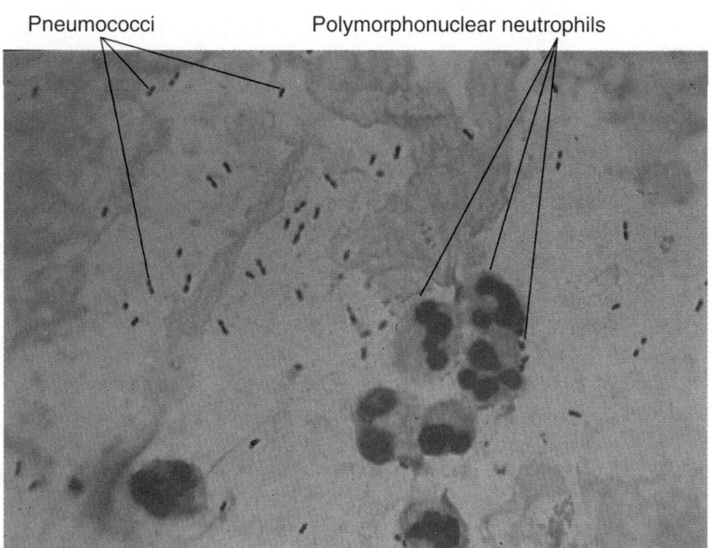

Figure 24.2
Streptococcus pneumoniae. Gram stain of sputum from a person with pneumococcal pneumonia showing gram-positive diplococci and PMNs.

Pathogenesis

Pneumococcal pneumonia develops when encapsulated (virulent) pneumococci are inhaled into the alveoli of a susceptible host, multiply rapidly, and cause an inflammatory response. Serum and phagocytic cells pour into the air sacs of the lung, causing difficulty in breathing and sputum production. This increase in fluid produces abnormal shadows on X-ray films of the chest in patients with pneumonia (figure 24.3).

The inflammation can involve nerve endings, causing pain; when this pain comes from the pleura, the result is **pleurisy.** The bacteria commonly enter the bloodstream from

• **Inflammatory response, pp. 345, 354–55**

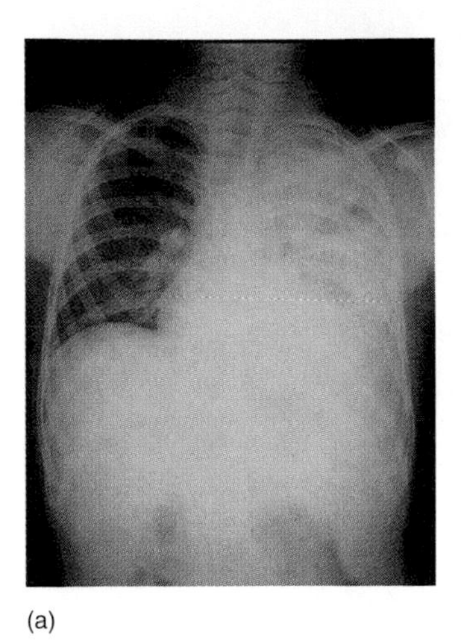
(a)

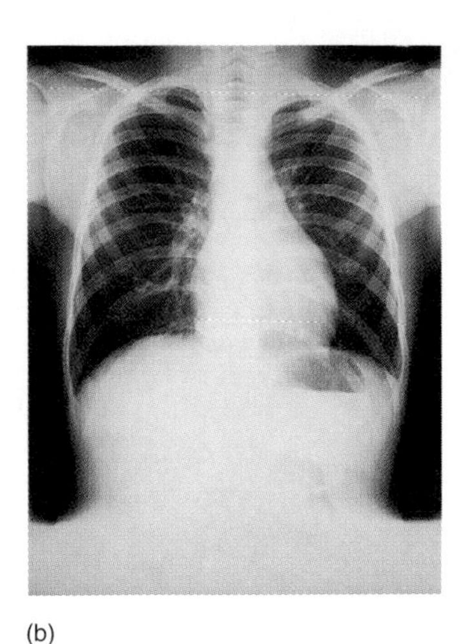
(b)

Figure 24.3
Pneumococcal pneumonia, X-ray appearance. (*a*) Pneumonia. The left lung (right side of figure) appears white because fluid-filled alveoli stop the X-ray beam from reaching the X-ray film and turning it black. (*b*) Normal X ray after recovery. The X-ray beam passes through air-filled alveoli and turns the film black.

the inflamed lung and occasionally produce septicemia, a symptomatic infection of the bloodstream; endocarditis, an infection of the heart valves; or meningitis, an infection of the membranes covering the brain and spinal cord. Infected people who do not have such complications usually develop sufficient specific antibodies within a week or two to permit trapping and destruction of the infecting organisms by lung phagocytes. Complete recovery usually results. Most pneumococcal strains cannot destroy lung tissue.

Prevention

A vaccine is available that stimulates production of anticapsular antibodies and gives immunity to pneumococcal disease. Of the more than 80 known capsular types of *S. pneumoniae,* relatively few types account for most of the serious pneumococcal infections in the United States. A commercially available vaccine consists of a mixture of the purified capsular polysaccharides of 23 of the most commonly encountered strains. Immunization is especially important for certain high-risk patients, such as those suffering from chronic lung disease or alcoholism and those who lack a functioning spleen or suffer from AIDS. Unfortunately, some of the people in these high-risk groups fail to develop protective antibody responses to the current vaccine, probably because their immune systems are impaired. Immunization would probably be of great benefit in countries that lack well-developed facilities for diagnosis and treatment of pneumococcal diseases.

Treatment

Most pneumococcal infections can be cured if treatment is given early in the illness. Most strains of pneumococci are highly susceptible to penicillin, erythromycin, and tetracycline, although resistant strains are increasingly being encountered. Strains resistant to several antibiotics have been reported from various countries, including the United States. The mechanism that accounts for the rising incidence of antibiotic resistance in pneumococci is unknown. Since this genus is transformable, perhaps the resistance genes are acquired by DNA transformation. About 25% of hospitalized patients with blood cultures positive for *S. pneumoniae* die, largely because of alcoholism, underlying disease of the heart or lung, or not seeking medical attention soon enough.

Table 24.1 describes the main features of pneumococcal pneumonia.

Table 24.1 Pneumococcal Pneumonia

Symptoms	Cough; fever; rust-colored sputum; shortness of breath; chest pain
Incubation period	1–3 days
Causative agent	*Streptococcus pneumoniae*, encapsulated strains
Laboratory diagnosis	Stained smear of sputum shows pus and gram-positive diplococci; cultures of sputum and blood show α-hemolytic colonies of gram-positive diplococci that are lysed by bile
Pathogenesis	Colonization of the alveoli incites inflammatory response; plasma, blood, inflammatory cells fill the alveoli.
Epidemiology	Inhalation of infected droplets; predisposing factors are alcoholism, chronic disease of heart or lung, and influenza
Prevention	Vaccine available contains 23 capsular antigens
Treatment	Penicillin, erythromycin, and others

Pneumonia Caused by *Klebsiella* Sp. Can Result in Permanent Lung Damage

Most enterobacteria can cause pneumonia, especially if host defenses are impaired. These pneumonias attack the very old, the very young, nursing home patients, those debilitated by other diseases, and immunocompromised people. Most fatal nosocomial infections are pneumonias caused by enterobacteria and other gram-negative rods. *Klebsiella* pneumonia is a classic example of this group of diseases.

Symptoms

In general, the symptoms of *Klebsiella* pneumonia—cough, fever, and chest pain—are indistinguishable from pneumococcal pneumonia. However, *Klebsiella* pneumonia patients

• **DNA transformation, pp. 154, 168–69**

Perspective 24.1

Laboratory Diagnosis of Pneumococcal Pneumonia

Streptococcus pneumoniae generally occurs as oval diplococci whose adjacent sides are flattened (figure I *a*). Colonies (figure I *b*) growing aerobically on blood agar are surrounded by a zone of α-hemolysis, which is often greenish in color due to the effect of the bacteria on the hemoglobin released from the lysed red blood cells. Unlike any of the many other α-hemolytic streptococci, *S. pneumoniae* is strongly inhibited by optochin, a chemical relative of quinine (figure I c) and is lysed by bile and some other detergents. These properties are useful in identifying *S. pneumoniae*.

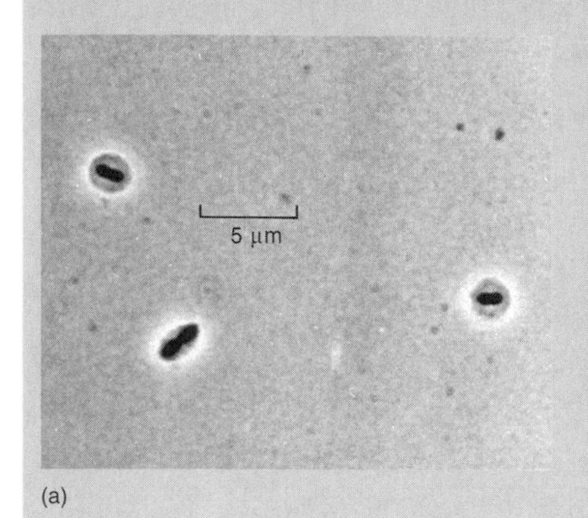

(a)

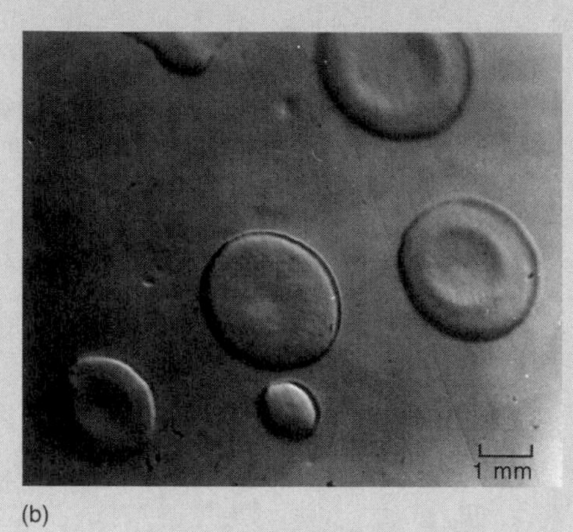

(b)

(c)

Figure I
Streptococcus pneumoniae. (*a*) Capsules are made more distinct by their reaction with specific capsular antiserum in the surrounding fluid. (*b*) Umbilicated colonies. The centers of the colonies tend to collapse due to autolysis of the bacteria. (c) Inhibition of growth by optochin. Note α-hemolysis.

typically have repeated chills, and their sputum is red and gelatinous, resembling currant jelly. Their mortality rate is higher, and they tend to die sooner than other pneumonia patients.

Causative Agent

Several species of *Klebsiella* cause pneumonia, but *Klebsiella pneumoniae* (table 24.2, figure 24.4) is the best known. It has a large capsule and big, strikingly mucoid colonies. Like other klebsiellas, *K. pneumoniae* is nonmotile at 37°C.

Pathogenesis

The organisms first colonize the mouth and throat and then are carried to the lung by inspired air or a ball of aspirated mucus. Evidence indicates that specific adhesins aid colonization, but their exact nature is not yet known. The capsule is an essential virulence factor, probably functioning like the pneumococcal capsule in limiting the availability of C3b. The C3b component of complement is a critically important opsonin for the body early in infection before the appearance of antibody. Unlike *S. pneumoniae, K. pneumoniae* causes death of tissue and rapid formation of lung abscesses. Therefore, even with effective antibacterial medication, the lung is permanently damaged. The cause of the tissue destruction is unknown. The organisms are more likely than pneumococci to invade the bloodstream and establish abscesses elsewhere in the body.

Epidemiology

Klebsiellas are part of the normal flora of the intestine in a small percentage of normal people. Colonization of the mouth and throat is common in debilitated individuals, especially in an institutional setting. In hospitals and nursing homes, the organisms are often resistant to antimicrobial medications, and spread is fostered by the use of these medications. There are more than 70 capsular serotypes, and serotyping is sometimes useful in tracing the sources of institutional epidemics.

Prevention

There are no specific preventive measures. Disinfection of the environment, use of sterile respiratory equipment, and use of antimicrobial medications only when necessary help control the organisms in hospitals.

Table 24.2 Some Identifying Characteristics of *Klebsiella pneumoniae*

Fermentations		Voge-Proskauer*	Methyl red**	Citrate	Motility
Glucose	Lactose	reaction	reaction	utilization	(37°C)
+	+	+	–	+	–

*Voge-Proskauer: test for 2,3-butylene glycol production from glucose
**Methyl red: a pH indicator; – means failure to produce enough acid from glucose to turn the indicator red

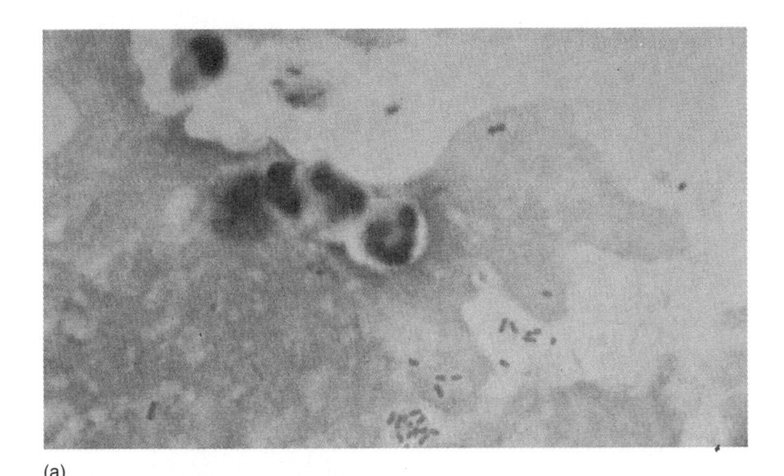
(a)

(b)

Figure 24.4
Klebsiella pneumoniae (*a*) in infected sputum and (*b*) in colonies. When growth is touched with a microbiological loop, it "strings" out.

Treatment

Klebsiella resistance to antimicrobial medications can be chromosomal or plasmid-mediated. The organisms easily acquire and are a source of R factors. Patients are usually treated by injecting both an aminoglycoside, such as amikacin, gentamicin, or tobramycin, and a cephalosporin or cefamycin.

Table 24.3 describes the main features of this disease.

• **R factors, p. 176**

Table 24.3 *Klebsiella* Pneumonia

Symptoms	Chills, fever, cough, chest pain, and bloody, mucoid sputum
Incubation period	1–3 days
Causative agent	*Klebsiella pneumoniae*
Laboratory diagnosis	Stained smears of sputum showing pus and large, gram-negative, rod-shaped bacteria; cultures of sputum and blood showing mucoid enterobacterial colonies
Pathogenesis	Colonization of the alveoli incites inflammatory response; plasma, blood, inflammatory cells fill the alveoli; destruction of lung tissue and abscess formation common; infection spreads by blood to other body tissues.
Epidemiology	Inhalation of infected droplets, aspiration of mucus
Prevention	No vaccine available
Treatment	A cephalosporin with an aminoglycoside

Mycoplasma pneumoniae Is a Common Cause of Walking Pneumonia

Mycoplasmal pneumonia is the leading cause of pneumonia in college students and is also common among military recruits. The disease is commonly mild and difficult to diagnose, hence the popular name *walking pneumonia*.

Symptoms

The onset of mycoplasmal pneumonia is typically gradual. The first symptoms are fever, headache, muscle pain, and fatigue. After several days, a dry cough begins, but later mucoid sputum may be produced. About 15% of cases have middle ear infections. Hospitalization is rarely necessary.

Causative Agent

The causative agent of mycoplasmal pneumonia is *Mycoplasma pneumoniae*, a small (0.2 μm diameter), easily deformed bacterium lacking a cell wall. Distinguishing characteristics are slow growth, aerobic metabolism, and

Perspective 24.2

Laboratory Diagnosis of *Klebsiella* Pneumonia

Laboratory diagnosis of this disease is complicated by the possible presence of *Klebsiella* strains in the throat of normal people, which may contaminate sputum specimens. Thus, merely finding colonies of klebsiellas in a sputum culture is not sufficient evidence with which to diagnose *Klebsiella* pneumonia. Other evidence is needed, such as finding massive numbers of large, plump gram-negative rods and pus cells in a smear of the patient's sputum, a positive blood culture, and a chest X ray consistent with this disease.

binding of red blood cells (detected by washing a red blood cell suspension over colonies growing on solid medium). As with other mycoplasmas, the central portion of *M. pneumoniae* colonies grows down into the medium, producing a fried egg appearance (figure 24.5).

Pathogenesis

Only a few inhaled organisms are necessary to start an infection with *M. pneumoniae.* The organisms attach to specific receptors on the respiratory epithelium (figure 24.6), interfere with ciliary action, and cause the ciliated cells to slough off. An inflammatory response characterized by infiltration of lymphocytes and macrophages causes the walls of the bronchial tubes and alveoli to thicken.

Epidemiology

The mycoplasmas are spread by aerosolized[1] droplets of respiratory secretions. Transmission from person to person is aided by the long time period in which *M. pneumoniae* is

[1]Aerosolized—having tiny liquid or solid particles suspended in air

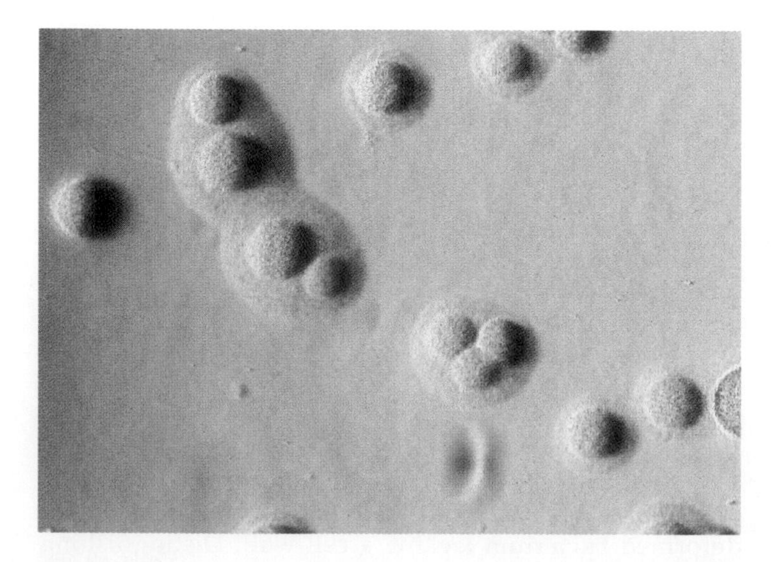

Figure 24.5
Mycoplasma pneumoniae. A typical colony has a fried egg appearance.

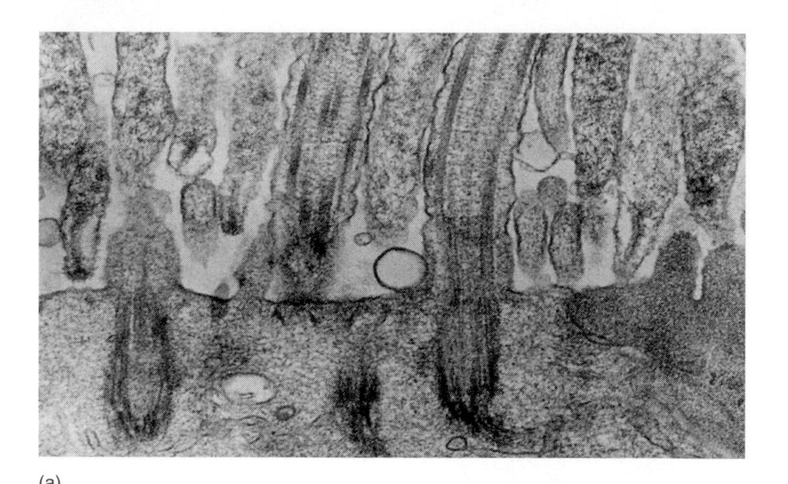

Figure 24.6
Electron micrograph (*a*) and diagram (*b*) showing attachment of *Mycoplasma pneumoniae* to respiratory epithelium. Notice the distinctive appearance of the tips of the mycoplasmas adjacent to the host epithelium. The tips probably represent a site on the microorganism that is specialized for attachment.

Perspective 24.3

Laboratory Diagnosis of Mycoplasmal Pneumonia

Diagnosis is generally made by testing the patient's blood for complement-fixing antibodies to *M. pneumoniae*. Cultures can confirm the diagnosis but are often too slow to aid in early treatment. However, a fast and widely available test is one for **cold agglutinins,** which are IgM antibodies that react to the I antigens of human red blood cells, causing agglutination at 0°C to 4°C. Cold agglutinins arise in the blood of about two-thirds of the cases of *Mycoplasma* pneumonia, presumably because of a cross-reacting antigen of the bacterium. Cold agglutinins lack the diagnostic significance of culture or complement fixation tests because other infectious diseases can sometimes cause their appearance. A nucleic acid probe could be very helpful for diagnosis, but the ones available so far have not been reliable.

present in respiratory secretions, ranging from about one week before symptoms begin to many weeks afterward. *Mycoplasma* pneumonia accounts for about one-fifth of bacterial pneumonias, and has a peak incidence in the 5 to 15 year age group, and is common in college students. It is much less common over the age of 40. Immunity after recovery is not permanent, and repeat attacks have occurred within five years.

Prevention

No practical preventive measures exist for *Mycoplasma* pneumonia, except avoiding crowding in schools and military facilities.

Treatment

Antibiotics that act against bacterial cell walls (penicillins, cephalosporins) are, of course, not effective. Tetracycline and erythromycin shorten the illness if given early, but they are bacteriostatic and do not readily eliminate *M. pneumoniae* from the respiratory secretions.

The main features of *Mycoplasma* pneumonia are presented in table 24.4.

Other Bacterial Infections of the Lower Respiratory Tract

Some lower respiratory bacterial diseases have features that readily distinguish them from the bacterial pneumonias discussed earlier. Three examples of such diseases—pertussis, tuberculosis, and Legionnaires' disease—are examined next.

Pertussis Is Caused by *Bordetella pertussis*

Fortunately, pertussis (whooping cough) is now uncommon in the United States, although it remains a threat to people who lack immunity. Worldwide, it causes 500,000 deaths each year.

Symptoms

Pertussis typically begins with a number of days of rhinorrhea (runny nose) followed by prolonged, violent bouts of coughing, vomiting, and sometimes seizures.

Causative Agent

The disease is caused by *Bordetella pertussis* (figure 24.7), a tiny, encapsulated, strictly aerobic, gram-negative rod that agglutinates red blood cells in vitro. These organisms do not tolerate drying or sunlight and die quickly outside the host.

Pathogenesis

Bordetella pertussis enters the respiratory tract with inspired air and attaches specifically to ciliated cells of the respiratory epithelium. A number of surface proteins may be responsible for the attachment, but evidence is strongest for

Table 24.4 *Mycoplasma* Pneumonia

Symptoms	Cough, fever, sputum production, headache, fatigue, and muscle aches
Incubation period	2–3 weeks
Causative Agent	*Mycoplasma pneumoniae*
Laboratory diagnosis	About two-thirds of cases develop cold agglutinins; complement fixation test using *M. pneumoniae* antigen reveals a rise in antibody titer in about 60% of cases; culture often takes a week or more
Pathogenesis	Cells attach to specific receptors on the respiratory epithelium; inhibition of ciliary motion and destruction of cells follow
Epidemiology	Inhalation of infected droplets; mild infections common
Prevention	No vaccine currently available; Avoidance of crowding in schools and military facilities advisable
Treatment	Tetracycline or erythromycin

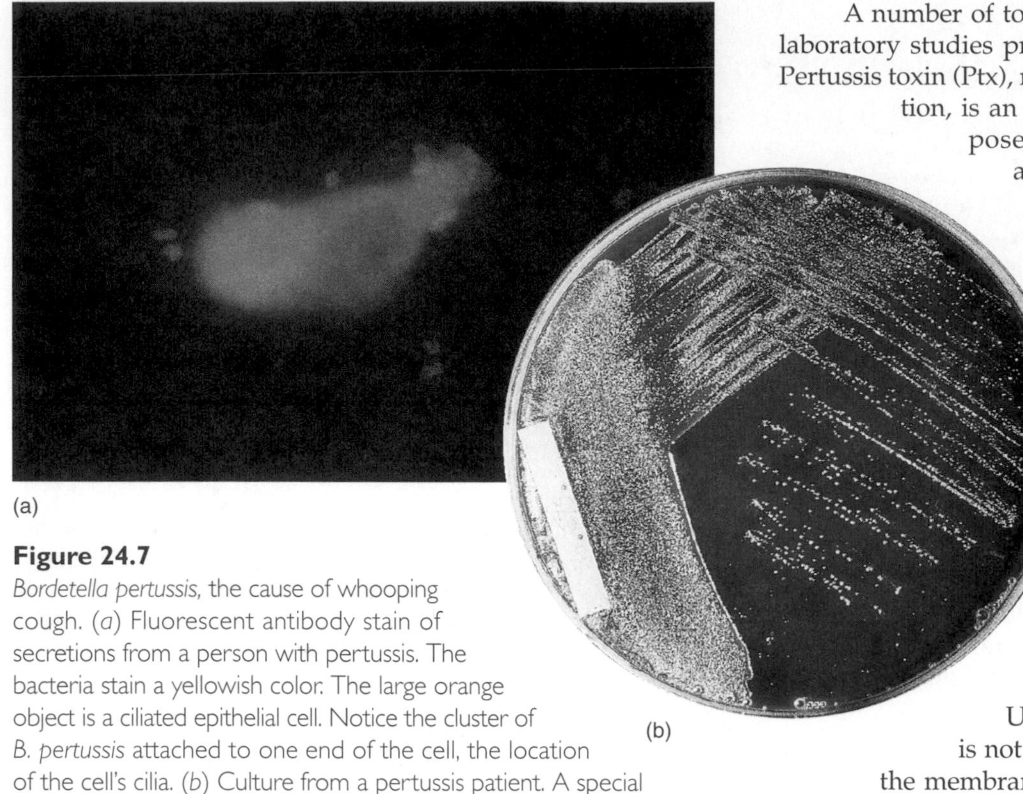

(a)

(b)

Figure 24.7
Bordetella pertussis, the cause of whooping cough. (*a*) Fluorescent antibody stain of secretions from a person with pertussis. The bacteria stain a yellowish color. The large orange object is a ciliated epithelial cell. Notice the cluster of *B. pertussis* attached to one end of the cell, the location of the cell's cilia. (*b*) Culture from a pertussis patient. A special medium heavily enriched with fresh blood is required.

pertussis toxin (Ptx) and **filamentous hemagglutinin (Fha).** Pertussis toxin acts as both adhesin and toxin. Some of it is bound to the surface of the bacteria and attaches to receptors on the surface of ciliated cells of the respiratory epithelium. Experimentally, antibodies to Ptx prevent both attachment to the epithelium and infection. Ptx will be discussed in more detail later.

Filamentous hemagglutinin (Fha) consists of long protein strands extending from the bacterial surface. These strands bind to certain glycolipids present on ciliated cells, thereby acting as adhesins. *B. pertussis* mutants deficient in Fha show reduced ability to colonize the respiratory epithelium. Therefore, both Ptx and Fha appear to have a role in colonization of the respiratory tract, at least in experimental animals.

The areas colonized by *B. pertussis* are shown in figure 24.8. Only involvement of the passageways of the lower respiratory tract appears to be important. The organisms grow in dense masses on the epithelial surface, but they do not invade. Mucus secretion increases markedly, while ciliary action declines precipitously, and patches of ciliated cells slough off. Only the cough reflex remains for clearing the secretions. The coughs come in sequences of five or more followed by an intense inspiratory effort, the "whoop," as the body attempts to get air into the lungs. Some of the bronchioles become completely obstructed, resulting in small areas of collapsed lung. Others, because of spasm or mucus plugging, let air enter but not escape, causing hyperinflation. Hemorrhages occur in the eyes and brain, and seizures can occur. Pneumonia due to *B. pertussis* or, more commonly, secondary infection is the chief cause of death.

A number of toxic products of *B. pertussis* identified in laboratory studies probably play a role in its pathogenesis. Pertussis toxin (Ptx), mentioned earlier for its role in colonization, is an A-B type of toxin. The B portion, composed of five subunits, is responsible for attachment to specific receptors on the host cell. The A portion is the sixth subunit. It becomes an active ADP-ribosylating enzyme when it enters the host cell. Its mode of action closely resembles the toxin of a completely unrelated bacterium, *Vibrio cholerae*, discussed in chapter 26. Figure 24.9 shows the steps involved in the toxin's action. First, the B portion attaches to receptors on the host cell surface. Second, the A portion is transferred through the cytoplasmic membrane of the host cell and becomes activated in the process. Unlike with diphtheria toxin, endocytosis is not involved in the entry of Ptx. Just inside the membrane is a G protein responsible for initiating down-regulation of cyclic adenosine monophosphate (cAMP). The Ptx enzyme inactivates the G protein by ADP-ribosylation, thereby allowing maximum production of cAMP.

Invasive adenylate cyclase, another *B. pertussis* toxin, also causes an increase in cAMP production by the host cell. It is probably an A-B toxin similar in structure to diphtheria toxin but having a different action. Upon entry into a host cell, the potentially enzymatic portion of the toxin binds to calmodulin[1] and becomes the active enzyme, adenylate cyclase. Adenylate cyclase catalyses the conversion of AMP to cAMP. Invasive adenylate cyclase is probably important in the pathogenesis of pertussis since mutants lacking the toxin can colonize but not kill experimental animals.

It is not clear how the increased output of cAMP caused by these two toxins contributes to the pathogenesis of pertussis. However, findings in the disease that could be explained by unregulated cAMP production include marked increase in mucus output in the respiratory tract, decreased killing ability of phagocytes, massive release of lymphocytes into the bloodstream, ineffectiveness of natural killer cells, and low blood sugar.

Tracheal cytotoxin is another toxin that probably plays an important role in the pathogenesis of pertussis. Chemically, it resembles a fragment of peptidoglycan. It kills ciliated epithelial cells and causes them to slough off. It also causes the release of interleukin-1 (Il-1), a fever-causing cytokine.

• **cAMP, p. 147**
• **Diphtheria toxin, pp. 514, 515**

[1]Calmodulin—a calcium-binding protein found in all nucleated cells

Epidemiology

Pertussis spreads through aerosolized respiratory secretions. Patients are most infectious during the period of rhinorrhea, the numbers of expelled organisms decreasing substantially with the onset of paroxysmal[1] coughing. The disease is typically mild in older children and in adults and is often overlooked as a minor cold, thus fostering transmission.

[1]Paroxysmal—sudden, violent, uncontrollable

Prevention

Intensive vaccination of infants with killed *B. pertussis* cells can prevent the disease in about 70% of individuals. Its widespread administration to young children is responsible for the drop from 235,239 reported cases of pertussis in the United States in 1934 to only about 1,000 in 1981. Since then, reported cases have generally ranged from 2,000 to 6,000 per year (figure 24.10).

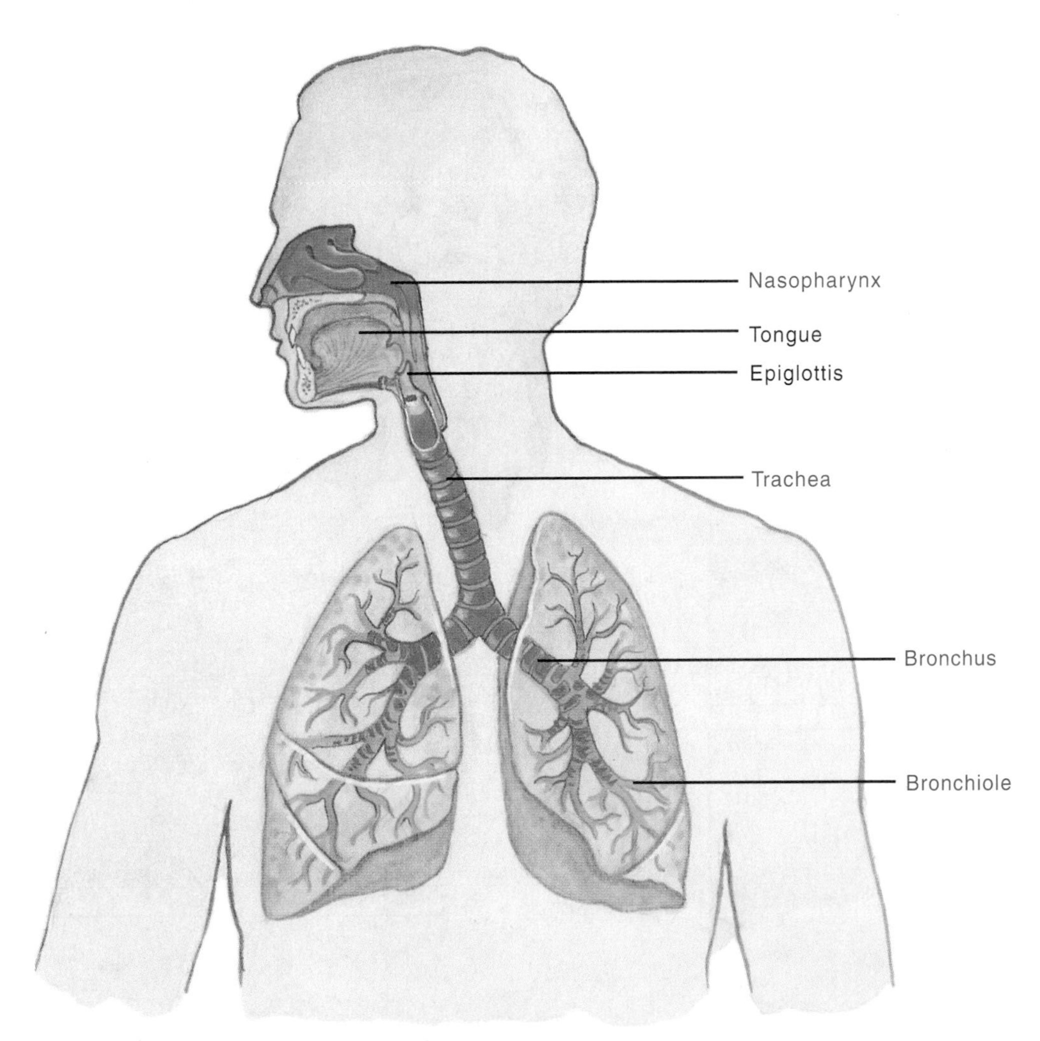

Figure 24.8
Whooping cough. Areas colonized by *Bordetella pertussis* are shown in red.

Perspective 24.4

Laboratory Diagnosis of Pertussis

In the early stage of pertussis, the organism is present in enormous numbers in the patient's respiratory secretions and is readily detected in smears using specific fluorescent antibody. Pertussis toxin can be demonstrated in these secretions using a monoclonal antibody in an enzyme-linked immunosorbent assay (ELISA). *Bordetella pertussis* can also be grown on a special medium enriched with fresh blood, but it often takes three or more days before colonies appear. A small amount of penicillin (to which *B. pertussis* is resistant) is incorporated into the medium to inhibit the growth of other organisms that may be present in the patient's secretions.

• **Monoclonal antibody, pp. 369–70**

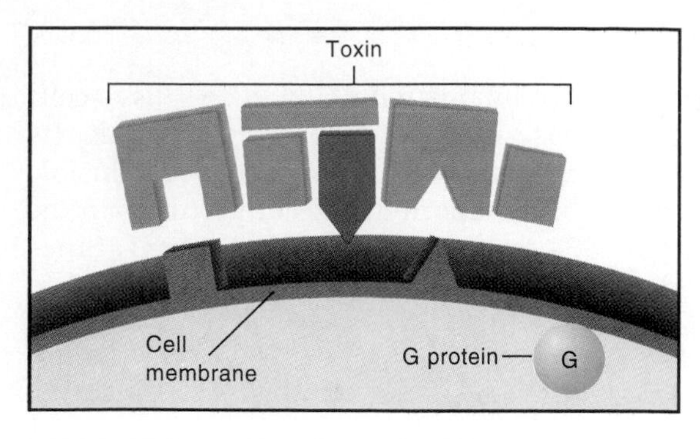

Attachment

Penetration

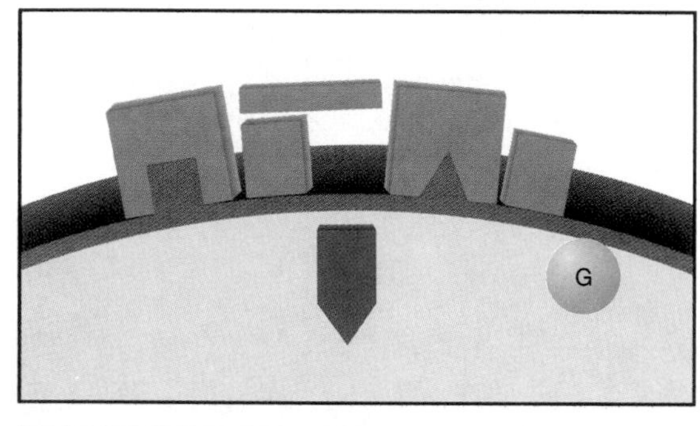

Separation

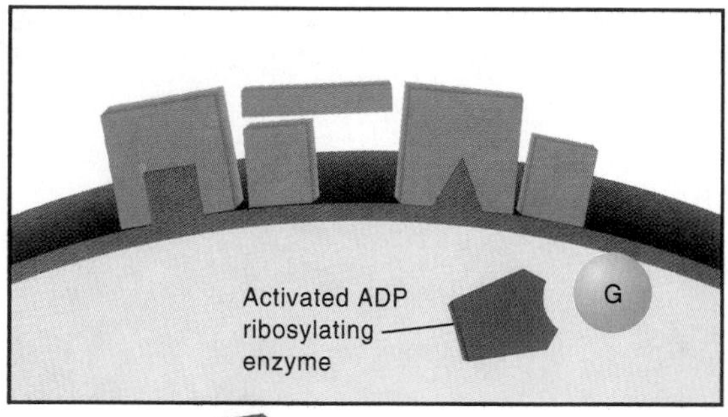

Activation

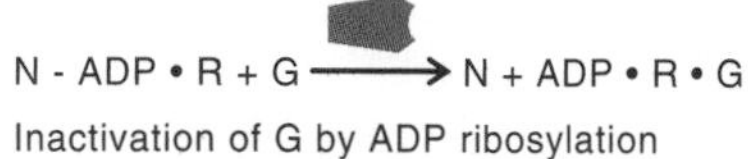

Figure 24.9
Mode of action of *Bordetella pertussis* toxin.

Table 24.5 Pertussis (Whooping Cough)

Symptoms	Nasal congestion and mild cough followed by spasms of violent coughing; possible vomiting and convulsions
Incubation period	7–21 days
Causative agent	*Bordetella pertussis*
Laboratory diagnosis	Smears of nasopharyngeal secretions stained by fluorescent antibody; ELISA test for toxin in respiratory secretions
Pathogenesis	Colonization of the surfaces of the upper respiratory tract and tracheobronchial system; ciliary action slowed; toxin released when organisms disintegrate causes death of epithelial cells and a rise in the number of lymphocytes in the bloodstream
Epidemiology	Inhalation of infected droplets; older children and adults have mild symptoms
Prevention	Whole cell and acellular vaccines, for immunization of children
Treatment	No cure; erythromycin given to limit spread

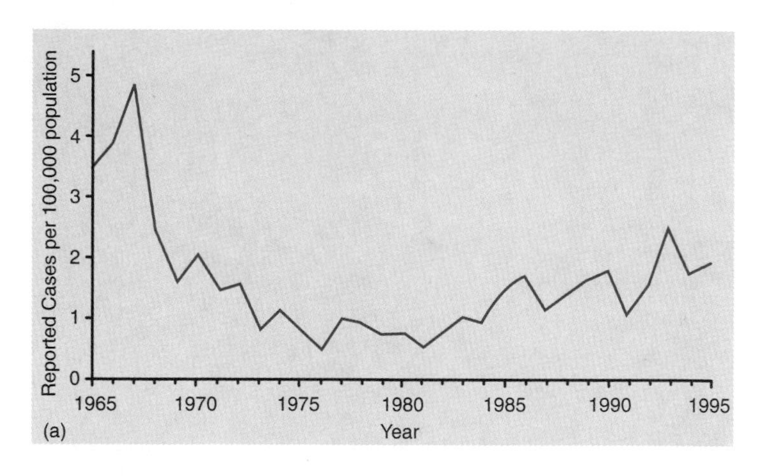

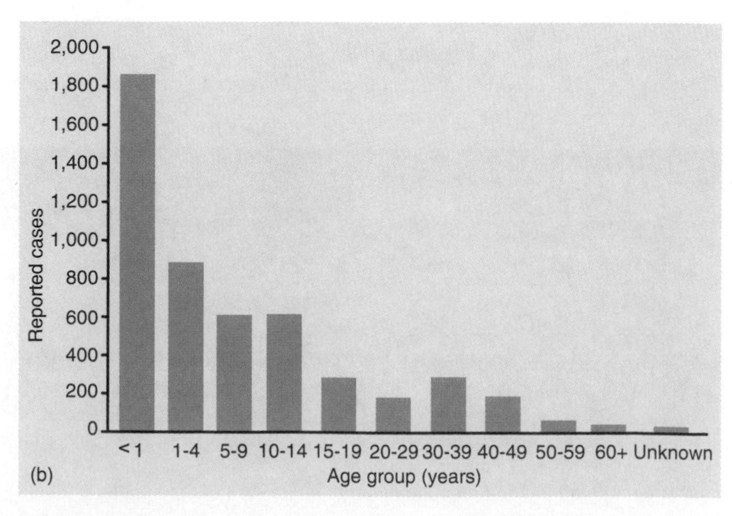

Figure 24.10
Pertussis (whooping cough) in the United States. (*a*) Incidence from 1965 to 1995 and (*b*) age distribution in 1994.

Source: Centers for Disease Control and Prevention. 1996. Summary of notifiable diseases, United States, 1995. *Morbidity and Mortality Weekly Report* 44(53):46.

Pertussis vaccine is given along with diphtheria and tetanus toxoids (DPT) by injection at 6 weeks, and 4, 6, and 18 months of age, with a booster dose between 4 and 6 years. Common reactions to the vaccine include pain and tenderness at the injection site and fever, occasionally high enough to cause convulsions. Acellular vaccines,[1] first approved for use in the United States in 1991, and licensed for all four doses in infants in 1996, give as good or better immunity with a much lower incidence of side effects. Newer vaccines hopefully will be safe enough to give as booster doses to adolescents and adults who are now a major reservoir of *B. pertussis* because of their waning immunity. The main features of this disease are shown in table 24.5.

Treatment

Treatment with the antibiotic erythromycin early in the disease, before the paroxysmal cough, is somewhat effective and decreases the number of *B. pertussis* in the secretions.

REVIEW QUESTIONS

1. What feature of the *S. pneumoniae* cell is responsible for its virulence?
2. How might *S. pneumoniae* acquire the genes to make it resistant to multiple antibiotics?
3. How do alcoholism and cigarette smoking predispose a person to pneumonia?
4. Give a mechanism by which *Klebsiella* sp. become antibiotic resistant.
5. Why are people with *Mycoplasma* pneumonia not treated with penicillin?
6. Why do doctors use cold agglutinin tests rather than culture to diagnose *Mycoplasma* pneumonia?
7. Why start immunization against *B. pertussis* at two months or less?
8. Why does the incidence of whooping cough rise promptly when pertussis immunizations are stopped?

Tuberculosis Is Typically a Chronic Disease

During most of this century, with improved living standards and chemotherapy, tuberculosis has shown a marked decline in the industrialized countries of the world (figure 24.11).

In the United States, for example, the decline averaged about 6% per year. However, in 1985, the decline slowed, and then the incidence of tuberculosis actually began to rise. The rise continued for the next six years in association with the expanding AIDS epidemic and increases in treatment-resistant cases. Fortunately, starting in 1993, with better funding of control measures and more aggressive treatment, tuberculosis once again appeared to be retreating.

Symptoms

Tuberculosis is a chronic bacterial infection of the lungs (and often other body organs) characterized by slight fever, weight loss, sweating at night, and chronic cough productive of blood-streaked sputum.

Causative Agent

Most cases of tuberculosis are now caused by *Mycobacterium tuberculosis* (figure 24.12), although in the years before widespread milk pasteurization, the cattle-infecting species *M. bovis* was a common cause. *M. tuberculosis*, or the "tubercle bacillus," is a slender, strongly acid-fast, rod-shaped bacterium that stains in an irregular ("beaded") fashion. The organism is a strict aerobe that grows very slowly, thus making it difficult to diagnose tuberculosis quickly. In comparison to most other bacteria, the tubercle bacillus is unusually resistant to drying, disinfectants, and strong acids and alkali. However, it is easily killed by pasteurization. There are no toxins, and its virulence depends on its ability to survive within host macrophages.

M. tuberculosis resists staining, drying, disinfectants, and extremes of pH because of its unique cell wall. The bacterium has a typical gram-positive cell wall, but bound covalently to the cell wall peptidoglycan is a thick layer composed of complex glycolipids. The peripheral (lipid) portion of this layer is bound in turn to other lipids composing the bacterial surface. Up to 60% of the dry weight of the *M. tuberculosis* cell wall consists of lipids, a much higher percentage than that in most other bacteria. These lipids include waxes[2] and **mycolic acids.** The mycolic acids are extremely long chained fatty acids unique to corynebacteria, nocardias, and mycobacteria.

Pathogenesis

Infection with *M. tuberculosis* usually occurs by inhaling aerosolized organisms from a person with tuberculosis. The microorganisms lodge in the lungs, where they are ingested by pulmonary macrophages. They resist destruction and may multiply within the macrophages, which carry them to lymph nodes in the region. At this stage, there is little inflammatory response to the infection, the organisms continue to multiply, and with lysis of the macrophages, they are released into the lymphatics and blood vessels. Thus, dissemination to other parts of the

[1]Acellular vaccines—vaccines that contain the antigens that stimulate immunity but lack other components of the bacterial cell

[2]Wax—ester of fatty acid with fatty alcohol

• **Mycobacteria, pp. 263–64**

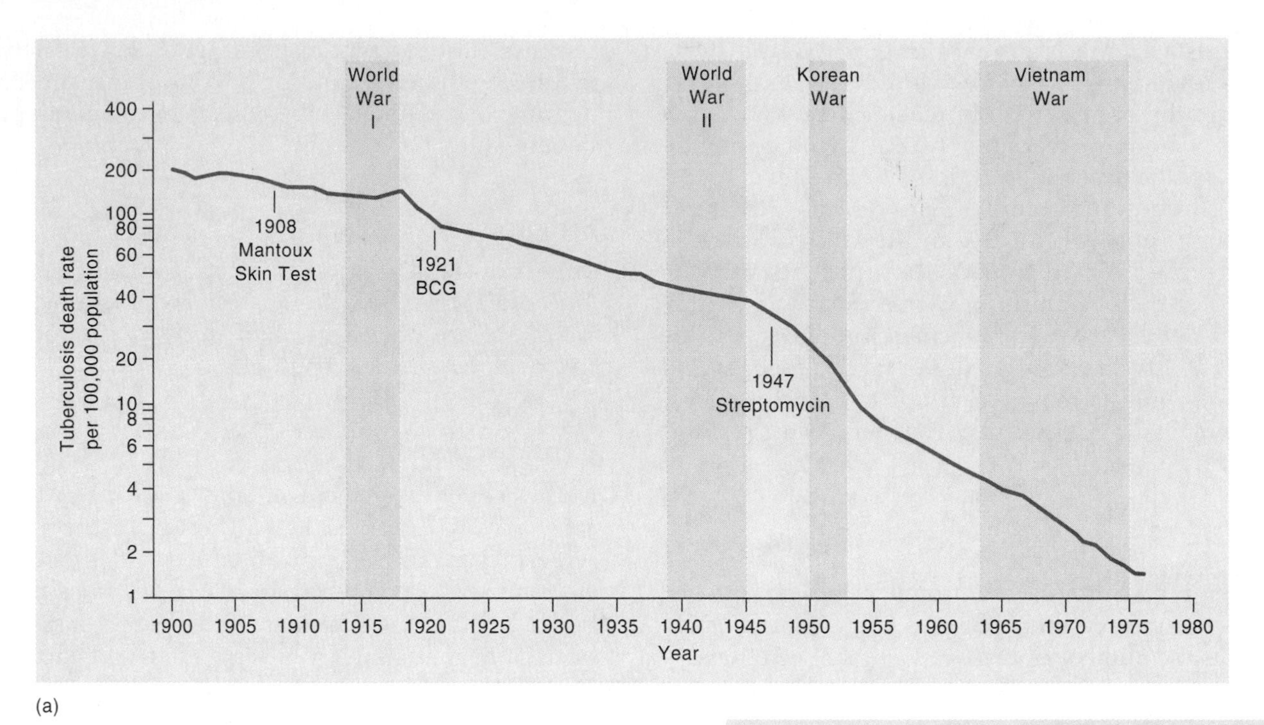

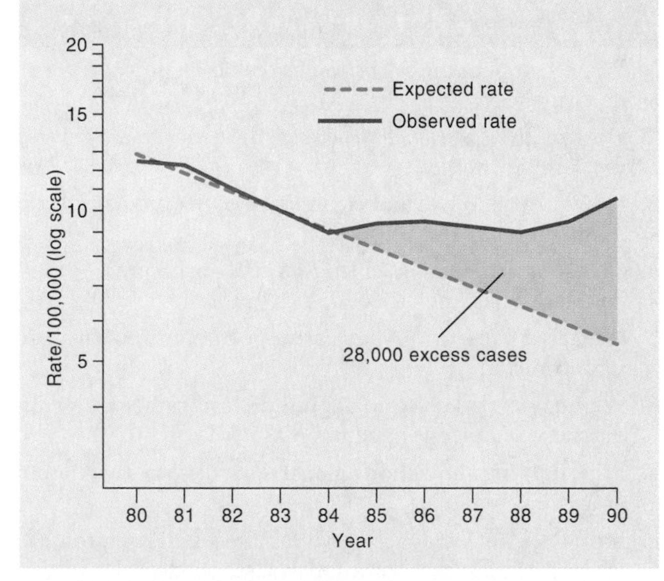

Figure 24.11
Incidence of tuberculosis. (*a*) After many decades of decline associated with improved living standards, (*b*) tuberculosis showing a resurgence. Much of the increase can be attributed to the AIDS epidemic.

Source: (*a*) Data from Mandell, G. L., et al. 1979. *Principles and practice of infectious diseases*, p. 1909. John Wiley & Sons, Inc., New York. (*b*) Centers for Disease Control and Prevention. 1991. Surveillance Summaries 40(no. SS-3):27.

body can be widespread. After about two weeks, delayed hypersensitivity to the tubercle bacilli develops. An intense reaction then occurs at sites where the bacilli have lodged. Macrophages, now activated, collect around the bacilli, and some macrophages fuse together to form large multinucleated giant cells. Lymphocytes and macrophages then collect around these multinucleated cells and wall off the infected area from the surrounding tissue. The localized collection of inflammatory cells is called a **granuloma** (figure 24.13), which is the characteristic response of the body to microorganisms and other foreign substances that resist digestion and removal. The granulomas of tuberculosis are called **tubercles.** In most cases, growth of the mycobacteria is halted by granuloma formation and no significant illness develops, although *M. tuberculosis* may remain alive in the tubercles for many years. In areas of intense struggle between activated macrophages and *M. tuberculosis,* many of the macrophages lyse and release their enzymes into the infected tissue. The result is death of tissue, with the formation of a cheesy material (the process called **caseous necrosis**). If this process involves a bronchus, the dead material may discharge into the

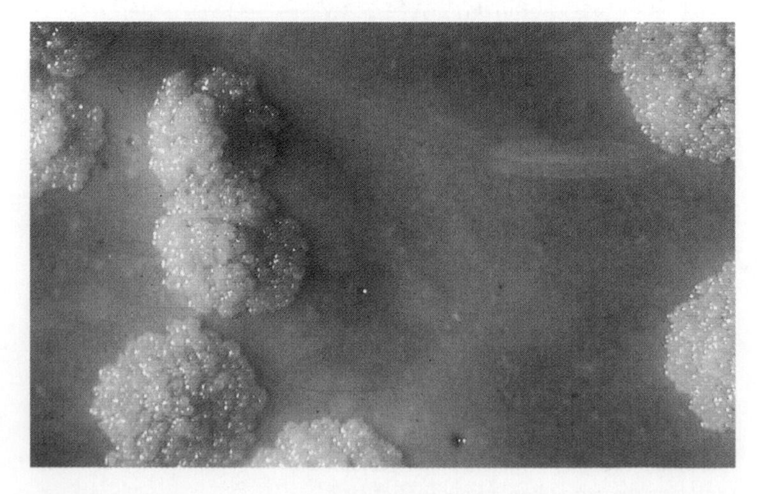

Figure 24.12
Colonies of *Mycobacterium tuberculosis* on Lowenstein-Jensen medium.

• **Delayed hypersensitivity, pp. 401–03, 636**

Perspective 24.5

Laboratory Diagnosis of Tuberculosis

Acid-fast stains of sputum (figure 1) or other infected material provide a quick way of detecting the presence of slow-growing *M. tuberculosis*. Cultures must be done to confirm the diagnosis because nonpathogenic mycobacteria are normally present on the human body and in certain foods. They are also frequently present in tap water and air and, therefore, contaminate stains and other reagents used in microbiology.

M. tuberculosis can generally be recovered from specimens, such as sputum containing multiple other organisms, by treating them with strong acid or alkali. If conditions such as time and temperature are carefully controlled, the other organisms are all killed, leaving most of the tubercle bacilli intact. They can then be inoculated onto laboratory culture media. *M. tuberculosis* has simple growth requirements, but it fails to grow on most microbiological media because they contain substances that inhibit its growth. Special media therefore are required. Under moist conditions, the organisms characteristically form long, tangled cords of cells attached side by side (figure 2). Other characteristics that help identify *M. tuberculosis* are their production of niacin (a B vitamin readily detected in the laboratory), reduction of nitrate, and susceptibility to certain antimicrobial medicines.

- **Acid-fast stain, pp. 52, 233**
- **Niacin, pp. 117, 118**

Due to the slow growth of *M. tuberculosis*, laboratory diagnosis of tuberculosis can take weeks to accomplish. However, the process can be speeded up considerably by using a liquid medium containing radioactive palmitic acid. As mycobacteria multiply, they metabolize the palmitic acid, releasing radioactive carbon dioxide. As soon as significant radioactivity is detected in the container, an acid-fast stain is done to verify the presence of mycobacteria. The organisms in samples of the culture are then lysed, and nucleic acid probes are used to identify the species. These probes react with species-specific segments of ribosomal RNA and, therefore, readily identify pathogenic mycobacteria, including *M. tuberculosis* and *M. avium*. Both of these species commonly infect AIDS patients, *M. avium* generally being highly resistant to antitubercular medicines.

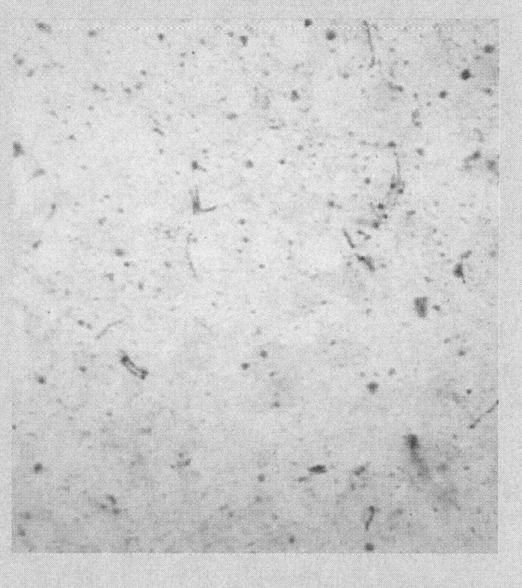

Figure 1
Mycobacterium tuberculosis, as seen in digested sputum. The pink-red organisms are long and slender and often lie side by side or have a beaded appearance.

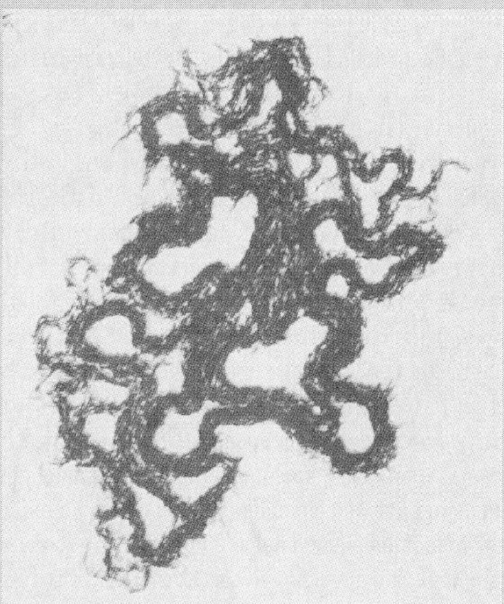

Figure 2
Acid-fast stain of *Mycobacterium tuberculosis*. On moist media, organisms in a developing colony adhere to each other side by side, resulting in a mass resembling a tangled cord. This cording phenomenon is characteristic of *M. tuberculosis*.

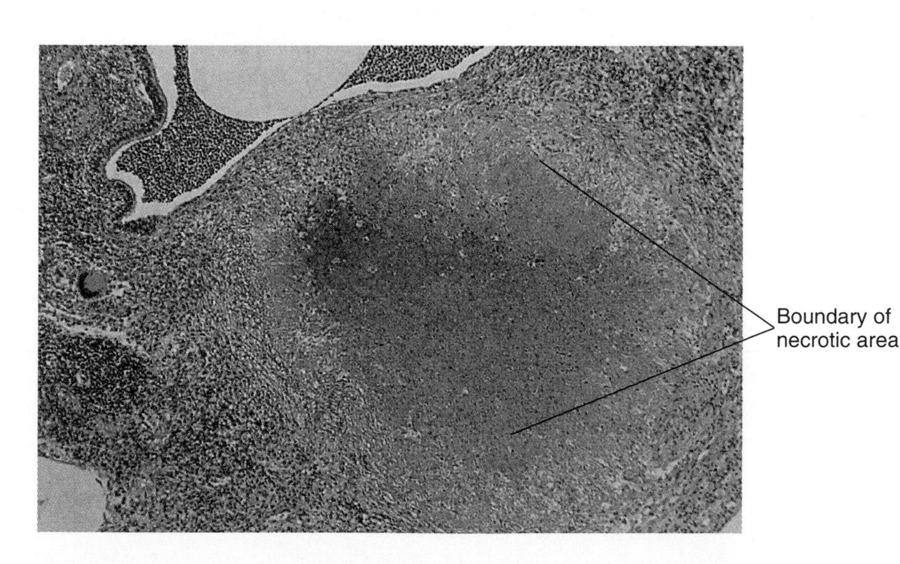

Figure 24.13
Low-power view of stained lung tissue showing a tubercle, a kind of granuloma caused by *Mycobacterium tuberculosis* (X16). The innumerable dark dots around the outer portion of the picture are nuclei of lung tissue and inflammatory cells. Centrally and extending to the right of the photograph, most of the nuclei have disappeared because the cells are dead and the tissue has begun to liquefy.

airways, causing a large defect (**cavity**) in the lung and the spread of the organisms to other parts of the lung. Coughing and spitting transmit the organisms to other people. Indeed, lung cavities characteristically persist, slowly enlarging and shedding tubercle bacilli into the bronchi. In other areas of caseous necrosis, multiplication of *M. tuberculosis* ceases, but wherever living organisms persist, there is the potential for their renewed growth if immunity becomes impaired as with stress, advanced age, or AIDS. Disease resulting from renewed growth of the organisms is called **reactivation tuberculosis.**

Tests to Determine Infection with *M. Tuberculosis*

Hypersensitivity to the tubercle bacilli is detected by injecting very small amounts of **tuberculin,** a sterile fluid obtained from a culture of *M. tuberculosis* or, more commonly, by injecting a purified fraction of the fluid **(purified protein derivative, PPD)** into the skin. (Similar preparations are available to test for hypersensitivity to other species of pathogenic *Mycobacterium.*) People who are sensitive develop redness and a firm swelling at the injection site, reaching a peak intensity after 48 to 72 hours. This test is called the **tuberculin,** or **Mantoux,**[1] **test.** A strongly positive reaction to the test generally indicates that living *M. tuberculosis* bacilli are present somewhere in the body of the person tested. Any impairment to the general health or treatments that suppress the immune response of individuals who are tuberculin-positive can result in renewed multiplication and actively progressing tuberculosis. This result may happen years after the initial infection. A positive tuberculin test implies that the person is resistant to acquiring a new infection but, because of the possibility of later reactivation, is probably more likely to develop active tuberculosis than a tuberculin-negative individual.

Epidemiology

About one-third of the world's population is infected with *M. tuberculosis,* and 3 million people die from tuberculosis each year. Over 20,000 cases of tuberculosis were reported each year in the United States from 1980 to 1995, with new cases occurring at a rate of about 9 per 100,000 population. An estimated 10 million Americans are infected by *M. tuberculosis* as evidenced by positive tuberculin tests. Infection rates are highest among nonwhites and elderly poor people. Transmission of tuberculosis occurs almost entirely by the respiratory route; 10 or even fewer inhaled organisms are enough to cause infection. Factors important in transmission include the frequency of coughing, the adequacy of ventilation (transmission is unlikely to occur outdoors), and the degree of crowding. Immunodeficiency can result in activation of latent tuberculosis, and over 5% of AIDS patients have developed active tuberculosis. Small epidemics are commonly seen in families, schools, nursing homes, or hospitals, usually starting with an unsuspected case of reactivated tuberculosis. Bacteriophage typing has been helpful in tracing the epidemic strain of *M. tuberculosis.*

Resistance to Medications

Drug-resistant mutants occur with high frequency among sensitive *M. tuberculosis* strains. Since mutants simultaneously resistant to more than one drug occur with a very low frequency, two or more drugs are usually given together in treating tuberculosis. Because of the long generation times of *M. tuberculosis* and its resistance to destruction by body defenses, drug treatment of tuberculosis must generally be continued for a minimum of six months to one or more years to cure the disease. Even then, nonmultiplying bacilli enclosed within old caseous tubercles may not be killed by treatment.

Unfortunately, during the prolonged treatment, many patients have no symptoms and become careless about taking their medications, or they discontinue them prematurely. Such negligence can lead to high rates of relapse, but of even greater concern is that up to two-thirds of the *M. tuberculosis* strains obtained from inadequately treated patients are resistant to one or more antitubercular medications (the record is a strain resistant to 11 drugs). These resistant strains can infect others in the community and are increasingly responsible for new infections. The problem of drug-resistant tuberculosis strains has reached alarming proportions in some areas. For example, in New York City as many as 45% of *M. tuberculosis* cultures are resistant to one or more antitubercular medicines. New cases are treated with four antituberculosis medicines pending determination of the susceptibility of their infecting strains.

Prevention

Vaccination against tuberculosis has been widely used in Scandinavia and many other parts of the world and is of proven value. The vaccinating agent, a living attenuated mycobacterium known as Bacille Calmette-Guérin (BCG), is derived from *M. bovis.* Repeated subculture in the laboratory over many years resulted in selection of this strain of *M. bovis,* which has little virulence in humans but does result in some immunity. At best, BCG vaccination is about 80% effective (the incidence of tuberculosis being one-fifth that occurring in unvaccinated persons) and lasts at least several years. People receiving the vaccine also develop delayed hypersensitivity to tuberculin. Public health authorities have discouraged routine use of the BCG vaccine in the United States because, by causing a positive Mantoux test, BCG vaccination prevents early diagnosis of tuberculosis. BCG is not safe to use in severely immunocompromised patients because the bacillus can spread throughout the body and cause disseminated disease.

[1]Pronounced "Man-too"

• **Bacteriophage typing, pp. 438, 478**

Treatment

Control of tuberculosis is aided by identifying unsuspected cases using skin tests and X rays. Individuals with active disease are then treated, thus interrupting the spread of *M. tuberculosis.* People whose Mantoux tests have changed from negative to positive are also treated, even when no other evidence of active disease exists. Treatment reduces the risk (estimated to be about 12%) that these people will develop active disease later in life.

Antimicrobial medicines useful in the treatment of infections with *M. tuberculosis* include isoniazid (isonicotinic acid hydrazide, or INH), ethambutol, streptomycin, pyrazinamide, and rifampin (chapter 21). The combination of rifampin and INH is favored because both drugs are bactericidal against actively growing organisms in cavities as well as relatively inactive intracellular organisms. Active tuberculosis is always treated with at least two antimicrobial medicines.

A Plan to Eliminate Tuberculosis

In 1989, *A Strategic Plan for the Elimination of Tuberculosis in the United States* was published by the Centers for Disease Control and Prevention. The goal of the plan is to reduce the infection rate from the approximately 9.5 new cases per 100,000 population in 1989 to 3.5 per 100,000 by the year 2000 and to less than one case per million population by 2010. The plan depends largely on increased efforts in case finding and treatment among the high-risk groups, particularly nonwhite poor people, people with AIDS, prisoners, and immigrants from countries with high rates of tuberculosis. The high rate of tuberculosis in these groups and the alarming increase in *M. tuberculosis* strains resistant to multiple medications have put the strategic plan in jeopardy.

The main features of tuberculosis are shown in table 24.6.

Table 24.6 Tuberculosis

Symptoms	Fever, weight loss, cough, sputum production
Incubation period	2–10 weeks
Causative agent	*Mycobacterium tuberculosis*
Laboratory diagnosis	Sputum treated with a strong alkali to kill other organisms; smears examined for acid-fast organisms; special media inoculated; biochemical testing and nucleic acid probes for identification
Pathogenesis	Colonization of the alveoli incites inflammatory response; ingestion by macrophages follows; organisms survive ingestion and are carried to lymph nodes, lungs, and other body tissues; tubercle bacilli multiply; granulomas form
Epidemiology	Inhalation of infected droplets; latent infections can reactivate
Prevention	BCG vaccination, which is not useful in the United States; Mantoux test for detection and early therapy of cases; treatment of young people with positive tests and those whose skin test converts from negative to positive
Treatment	Isoniazid (INH), ethambutol, pyrazinamide, rifampin, and streptomycin

Legionnaires' Disease Is a Bacterial Lung Infection Acquired by Inhaling Aerosols of Contaminated Water

Legionnaires' disease was unknown until 1976, when a number of people attending the American Legion Convention in Philadelphia developed a mysterious pneumonia that was fatal in many cases. Months of scientific investigation eventually paid off when the cause was discovered to be a bacterium commonly present in the natural environment.

Symptoms

Legionnaires' disease typically begins like most bacterial pneumonias with headache, malaise, rapid rise in temperature, and shaking chills. A dry cough develops that later produces small amounts of sputum, sometimes streaked with blood. About one-fourth of the cases also have some alimentary tract symptoms such as diarrhea, abdominal pain, and vomiting. Shortness of breath is common and oxygen therapy is often needed. Recovery is slow, and weakness and fatigue last for weeks.

Causative Agent

The causative agent is *Legionella pneumophila* (figure 24.14), a rod-shaped bacterium that stains poorly with the Gram stain and fails to grow on the usual laboratory culture media, which explains why it escaped detection for so long. The organism is usually motile by one polar flagellum, and the structure of the *L. pneumophila* cell wall resembles that of gram-negative bacteria. The poor staining of *L. pneumophila* probably results from the presence of some unusual fatty acids and other chemical characteristics of the cell wall.

Pathogenesis

Legionella pneumophila infection is acquired by breathing aerosolized water contaminated with the organism. Normal, healthy people are quite resistant to infection, but smokers and those with impaired host defenses from

• ***Legionella*, pp. 94, 246**

Perspective 24.6

Laboratory Diagnosis of Legionnaires' Disease

Legionella pneumophila can be cultivated on a special medium designed to meet its growth requirements, mainly an amino acid supplement rich in cysteine and a source of iron. Charcoal is included to neutralize inhibitory substances present in agar, and the medium is buffered to maintain a pH of 6.9 because more acid or alkaline conditions impair growth of the bacterium. Growth occurs under aerobic conditions and is slow, colonies often taking several days to appear. Antibiotics that inhibit other organisms but not *L. pneumophila* are included in the medium.

chronic diseases such as cancer and heart or kidney disease are susceptible. The organisms lodge in and near the alveoli of the lung and are taken up by macrophages. They are not killed by the phagocytes but instead multiply within them and are released upon cell death to infect other tissues. Necrosis (tissue death) results, followed by formation of multiple small abscesses. Respiratory failure followed by death occurs in about 15% of hospitalized cases. Curiously, *L. pneumophila* infections remain confined to the lung in most cases.

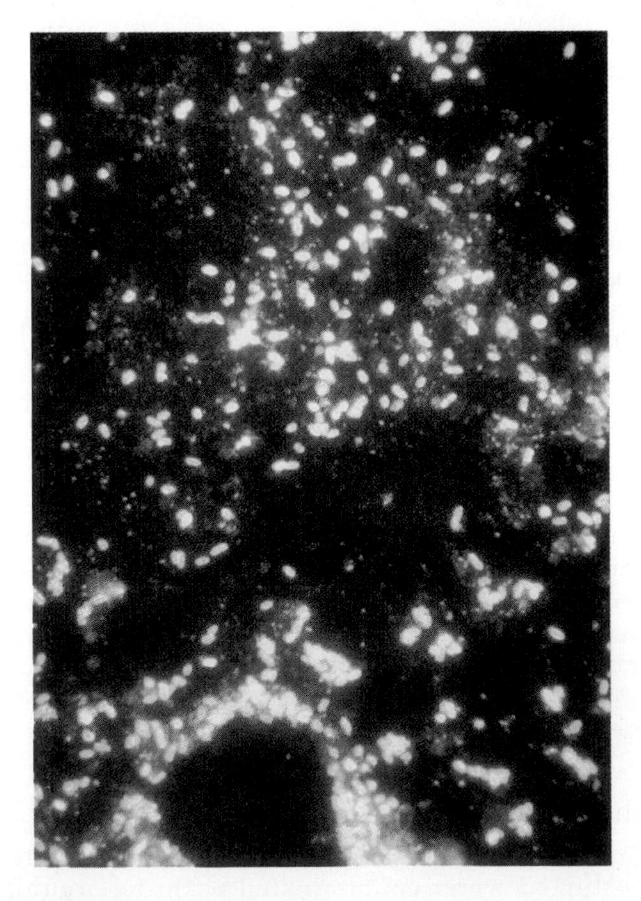

Figure 24.14
Legionella pneumophila stained with fluorescent antibody. When present in tissue or sputum, *L. pneumophila* fails to stain with most other microbiological stains.

Epidemiology

The organisms are widespread in natural waters, but fortunately most sources are noninfectious, probably because many *L. pneumophila* strains have little virulence and their concentrations are low. The organism is able to grow in water when other microorganisms are present to supply its growth requirements. In natural waters, *L. pneumophila* is taken up and multiplies within amebas in a manner similar to the way it grows in human macrophages. It is considerably more resistant to chlorine than the usual gram-negative bacterial pathogens and survives well in the water systems of buildings, particularly in hot water systems (where chlorine levels are generally low). Legionnaires' disease cases have originated from contaminated aerosols from air conditioner cooling towers, nebulizers,[1] and even from showers and water faucets. However, direct person-to-person spread does not occur. Individual strains of *L. pneumophila* can be identified using monoclonal antibodies, which helps trace the source of epidemics.

Prevention

Most efforts at control have focused on designing equipment to minimize the risk of infectious aerosols and on disinfecting procedures once the source of an epidemic has been identified. Environmental surveillance is not practical because of the lack of a simple method for detecting virulent strains.

Treatment

Legionnaires' disease is treated with high doses of erythromycin, sometimes concurrently with rifampin. *Legionella pneumophila* produces the enzyme β-lactamase, which makes it resistant to many penicillins and cephalosporins (chapter 21).

The main features of Legionnaires' disease are given in table 24.7.

[1]Nebulizer—device for producing a fine spray of liquid

Perspective 24.7

Hantavirus Pulmonary Syndrome

In the spring of 1993, the Indian Health Service, the Navaho Nation Division of Health, and the Health Departments of Arizona, Colorado, New Mexico, and Utah began investigating an outbreak of a mysterious flulike illness that rapidly progressed to severe respiratory failure and often death. Most of the victims were otherwise healthy adults. Subsequent studies revealed that affected individuals developed antibodies that reacted with a group of enveloped RNA rodent viruses called *hantaviruses*, of the family *Bunyaviridae*. The name *hantavirus* comes from the Hantaan River in Korea. In the 1950s during the Korean War, thousands of the United Nations' troops developed fever, hemorrhages, and kidney failure. Their disease, **Korean hemorrhagic fever,** was finally shown by Professor Ho Wang Lee of Seoul National University College of Medicine to be caused by a hantavirus. Similar hantavirus diseases exist in Scandinavia and other parts of the world, but before 1993, no hantavirus disease was known to occur in the United States.

The new American disease, now known as the **hantavirus pulmonary syndrome,** is characterized mainly by pulmonary symptoms, thereby differing from previous hantavirus diseases. Fluid leaks from the blood into the pleural space that surrounds the lungs, into lung tissue, and into the alveoli themselves. Many mononuclear cells infiltrate the lung tissues in response to the infection. The result of these changes is that oxygen exchange is severely limited. Sixty percent of infected people have died from the disease.

Initially, cases were identified using cross-reacting antigen from one of the known hantaviruses. Later, a monoclonal antibody was prepared against epitopes shared by the known hantaviruses and used to detect the presence of the unknown hantavirus in tissue. The polymerase chain reaction was then used to amplify hantavirus RNA extracted from autopsy specimens.

This RNA from the unknown hantavirus was cloned in *E. coli*, providing a better antigen for diagnostic serological tests. Finally, in November 1993, the virus itself was cultivated in vitro by scientists at the Centers for Disease Control and Prevention and the Army Medical Research Institute. The name for the new virus is **Sin Nombre,** which in Spanish means "nameless."

By the spring of 1996, a total of 131 cases from 24 states had been reported, with an overall mortality of 50%. Deer mice are the main source of the virus in the West, but other hantaviruses from other rodent hosts may be responsible for the pulmonary syndrome in the eastern United States and in other parts of the world. Infected mice show no illness but excrete the virus in saliva, urine, and feces. Humans are mainly infected by inhaling aerosols containing the virus, and extreme caution is advisable to ensure good ventilation and to avoid inhaling dust from areas inhabited by mice and rats.

Table 24.7 Legionnaires' Disease

Symptoms	Muscle aches, headache, fever, cough, shortness of breath, chest pain, diarrhea
Incubation period	2–10 days
Causative agent	*Legionella pneumophila*
Laboratory diagnosis	Special stains used to visualize the organisms in lung tissue; medium containing cysteine and iron required for in vitro cultivation
Pathogenesis	Organism multiplies within phagocytes; death of these and other cells followed by sloughing of cells lining the alveoli, tissue necrosis, and formation of microabscesses
Epidemiology	Originates mainly from warm water contaminated with other microorganisms, such as found in air conditioning systems
Prevention	Avoidance of contaminated water aerosols; regular cleaning and disinfection of humidifying devices
Treatment	Erythromycin and rifampin

Viral Infections of the Lower Respiratory Tract

The DNA viruses cytomegalovirus, varicella-zoster virus, and the adenoviruses can all cause serious pneumonias, some adenovirus infections mimicking pertussis. However, the paramyxovirus and orthomyxovirus families are of greater overall importance. Of the paramyxoviruses, measles (rubeola), parainfluenza, and respiratory syncytial viruses (RSV) are common causes of lower respiratory tract disease. RSV is responsible for 25% of pneumonia and 75% of bronchitis hospitalizations in infants and children. Of the orthomyxoviruses, influenza A and B are the most important.

Influenza Is a Lung Infection that Occurs in Pandemics About Every 10 Years

Influenza is a good example of the constantly changing interaction between hosts and infectious agents. Changes in influenza virus antigenicity result in frequent major epidemics of the disease. Similarly, epidemics occur in nonhuman animals, and the possibility exists that antigenic variability of influenza viruses is partly due to movement of viral genes from one animal species to another.

Symptoms

After a short incubation period (on average, two days), influenza typically begins with headache, fever, and muscle pain, which reach a peak in 6 to 12 hours. A cough without production of sputum appears, worsening over a few days. The acute symptoms then go away within a week, leaving the patient with a lingering hacking cough, fatigue, and generalized weakness for additional days or weeks.

Causative Agent

Influenza viruses belong to the orthomyxovirus family (chapter 14), single-stranded RNA viruses with an envelope derived from the host cell membrane (figure 24.15). Three major serotypes, designated A, B, and C, based on nucleoprotein antigens, can be identified. Types A and B are the most important clinically. Projecting from their envelope are two kinds of glycoprotein spikes, **hemagglutinin (H)** and **neuraminidase (N).** Hemagglutinin functions to attach the virus to specific receptors on the host cell and thus initiate infection. Antibody to hemagglutinin confers immunity to influenza. As the name implies, hemagglutinin adheres to red blood cells, causing them to agglutinate. This is used for in vitro tests but has a role in pathogenesis. Neuraminidase, on the other hand, is an enzyme that destroys the cell receptor to which the hemagglutinin attaches. This aids in the release of newly formed virions from the infected host cell. It may also foster the spread of the virus to uninfected host cells. The viruses are easily cultivated in embryonated hens' eggs and a variety of tissue cell cultures such as monkey kidney cells. Growth of the viruses is detected by adding a suspension of chicken red blood cells to amniotic fluid from the egg or the medium overlying all cultures. If an influenza virus is present, the red cells will agglutinate. Infected tissue culture cells will also adsorb red blood cells because of the viral hemagglutinin present on the surfaces of the tissue cells. Neutralization of hemagglutination or hemadsorption by specific antisera identifies the type of hemagglutinin present on the virus.

Pathogenesis

Influenza is acquired by inhaling aerosolized respiratory secretions from a person with the disease. The virions attach by their hemagglutinin to specific receptors on ciliated epithelial cells, their envelope fuses with the cell membrane, and the virus enters the cell. Host cell protein and nucleic acid synthesis cease, and rapid synthesis of viral nucleoproteins begins. Regions of the cell membrane become embedded with viral glycoproteins, specifically hemagglutinin and neuraminidase. Within six hours, mature virions bud from the host cell, receiving an envelope of cell membrane containing viral hemagglutinin and neuraminidase as they are released. The virus spreads rapidly to nearby cells, including mucus-secreting cells and cells of the alveoli. Infected cells ultimately die and slough off, thus destroying the mucociliary escalator and severely impairing one of the body's major defenses against infection. The immune response of the body quickly controls the infection in the vast majority of cases, although complete recovery of the respiratory epithelium may take two months or more. Only a small percentage of people with influenza die, but epidemics are often so widespread that the number of deaths is high. Influenza virus infection alone can kill apparently normal, healthy people. However, more often death occurs because of bacterial secondary infections, usually by *Staphylococcus aureus*, *Streptococcus pyogenes*, or *Haemophilus influenzae*. Influenza also takes a heavy toll on people whose hearts or lungs are on the borderline between adequate and inadequate, the infection causing these organs to fail.

• **Enveloped viruses, p. 300**
• **Segmented viruses, p. 299**
• **Serological tests, pp. 384–87**

Epidemiology

Outbreaks of influenza occur every year in the United States and are associated with an estimated 10,000 to 40,000 deaths. A moderately severe influenza epidemic in the United States imposes an economic burden estimated at $10 billion to $12 billion. Figure 20.1 shows the variations in the percentage of total deaths attributed to pneumonia and influenza. Pandemics occur periodically over the years (table 24.8).

• **Figure 20.1, p. 430**

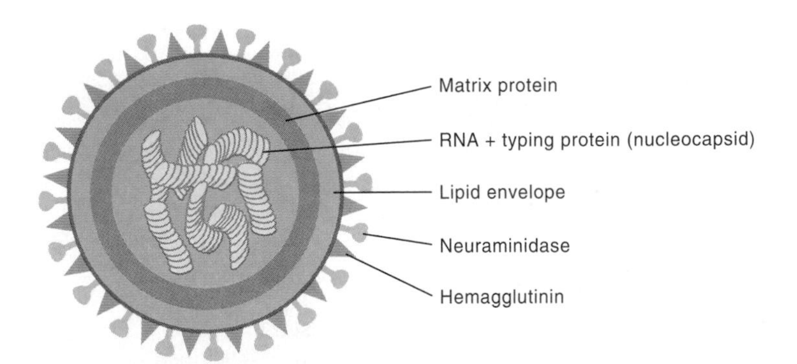

Figure 24.15
Influenza virus: diagrammatic view showing a segmented genome and types of antigens.

Table 24.8 Pandemic Influenza

Year	Influenza A Type	Severity
1889–1890	H2N8*	very high morbidity, mortality
1918–1919	H1N1	"
1957–1958	H2N2	"
1968–1969	H3N2	moderately severe
1977–1978	H1N1	relatively mild

*H—hemagglutinin, N—neuraminidase

Although other factors are involved in the spread of influenza viruses, major attention in recent years has focused on their antigenic changeability. Figure 24.16 shows how the antigenic composition of influenza viruses has varied from year to year. Two types of variation are seen: **antigenic drift** and **antigenic shift** (figure 24.17). Antigenic drift is seen in interepidemic years and consists of minor mutations in the hemagglutinin (H) antigen that makes immunity developed during prior years less effective and ensures that enough susceptible people are available for the virus to survive. This type of change is exemplified by A/Texas/77(H3N2) and A/Bangkok/79(H3N2), wherein there have been mutations that have altered the H antigen slightly. (The geographic names are the places where the virus was first isolated.) The antibody produced by people who have recovered from A/Texas/77(H3N2) is only partially effective against the mutant H3 antigen of A/Bangkok/79(H3N2). Thus, the newer Bangkok strain might be able to spread and cause a minor epidemic in a population previously exposed to the Texas strain.

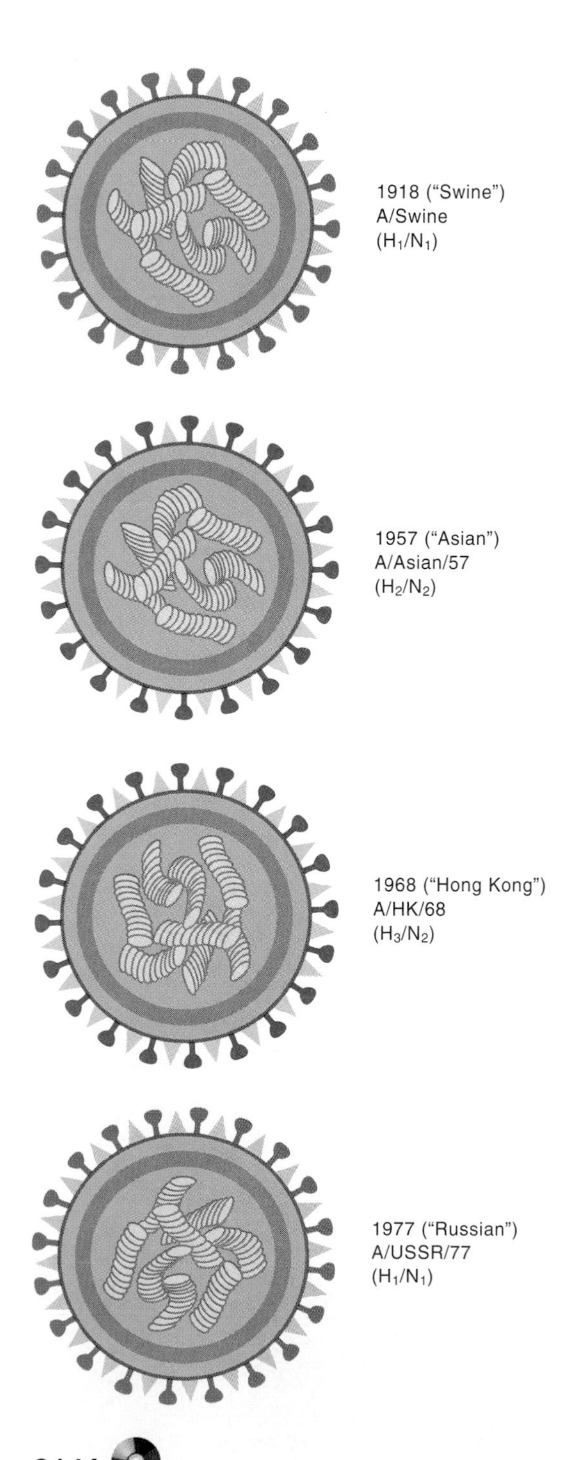

Figure 24.16
Influenza virus: antigenic changes with different epidemics.

Antigenic shifts are represented by more dramatic changes, and virus strains appear that are markedly different antigenically from the strains previously seen. The manner in which they arise is a matter of speculation, but most likely they are a result of rare events in which two different viruses infect a cell at the same time. As can easily be demonstrated in the laboratory, dual infections result in an exchange of large segments of the viral genomes, so that totally different viruses result. It is probable that genes from animal strains of influenza virus can on rare occasions infect humans and cause dual infections with human strains. Obviously, if one of these newly created viruses were infectious, were virulent, and possessed a hemagglutinin for which a population had no immunity, it could cause widespread disease before the pool of susceptible people was exhausted.

These major shifts are characteristic of influenza A viruses, whereas influenza B viruses tend to change more slowly. This difference in the rate of change may explain why influenza A viruses generally cause more severe and widespread disease than do influenza B viruses.

Reye's Syndrome

A curious affliction known as **Reye's syndrome** occasionally occurs in association with influenza B infections (figure 24.18), usually within 2 to 12 days of the onset of the infection. The syndrome occurs predominantly in children between 5 and 15 years old and is characterized by liver and brain damage. The death rate has generally been about 30%. The incidence of the syndrome has varied between nine and less than one case per million people in the United States, with a general trend to declining incidence. Reye's syndrome is also seen in association with a number of other viral infections including influenza A and chicken pox. Epidemiologic evidence suggesting that aspirin therapy increases the risk of Reye's syndrome has led physicians to use this drug sparingly in children with fever.

Prevention

Influenza can be prevented by using killed vaccines that are 80% to 90% effective in preventing disease when the vaccine is produced from the epidemic strain. Split vaccines consisting mostly of hemagglutinin are now widely used instead of whole virus vaccines. Because of antigenic drift, vaccines produced from earlier strains having the same H type may be decreasingly effective, so that new vaccines must be produced each year. A serious reaction was observed during a nationwide immunization program with the swine

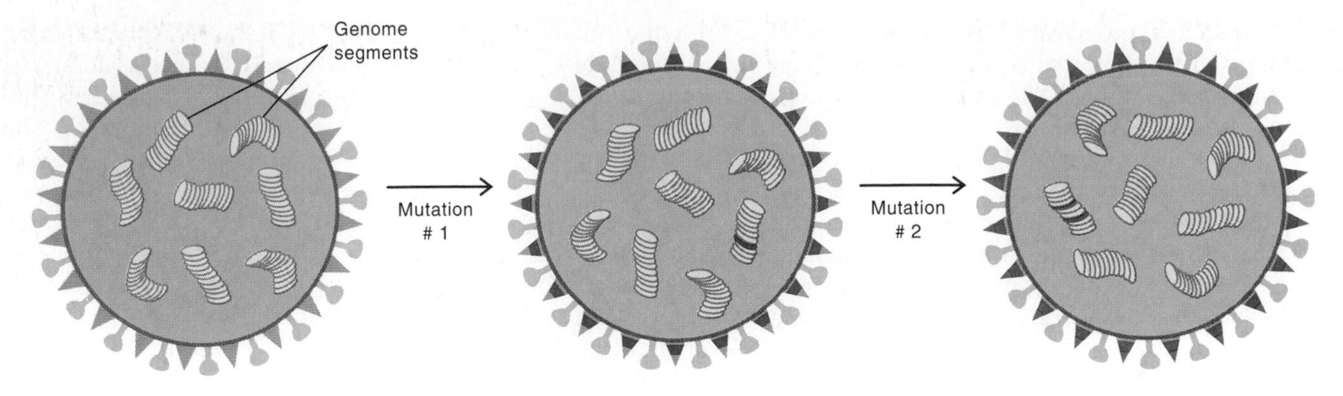

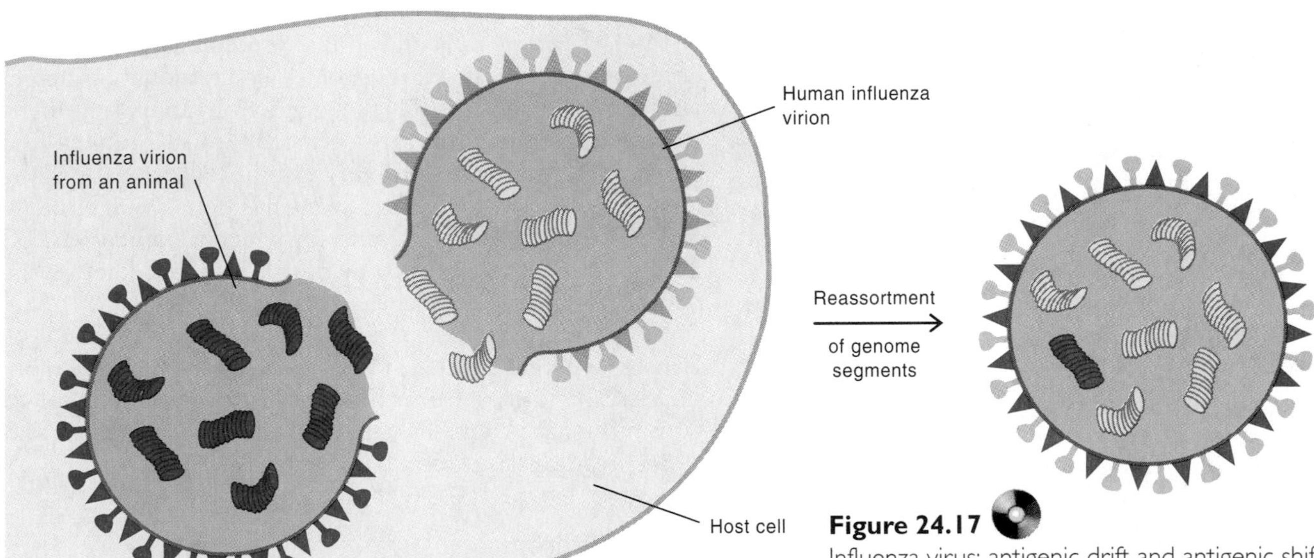

Figure 24.17
Influenza virus: antigenic drift and antigenic shift. (*a*) With drift, there is a gradual change in the hemagglutinin antigens so that antibody against the original virus becomes progressively less effective. (*b*) With shift, there is a major change in the hemagglutinin antigens because the virus acquires a new genome segment, which in this case codes for hemagglutinin. Changes in neuraminidase could occur by the same mechanism.

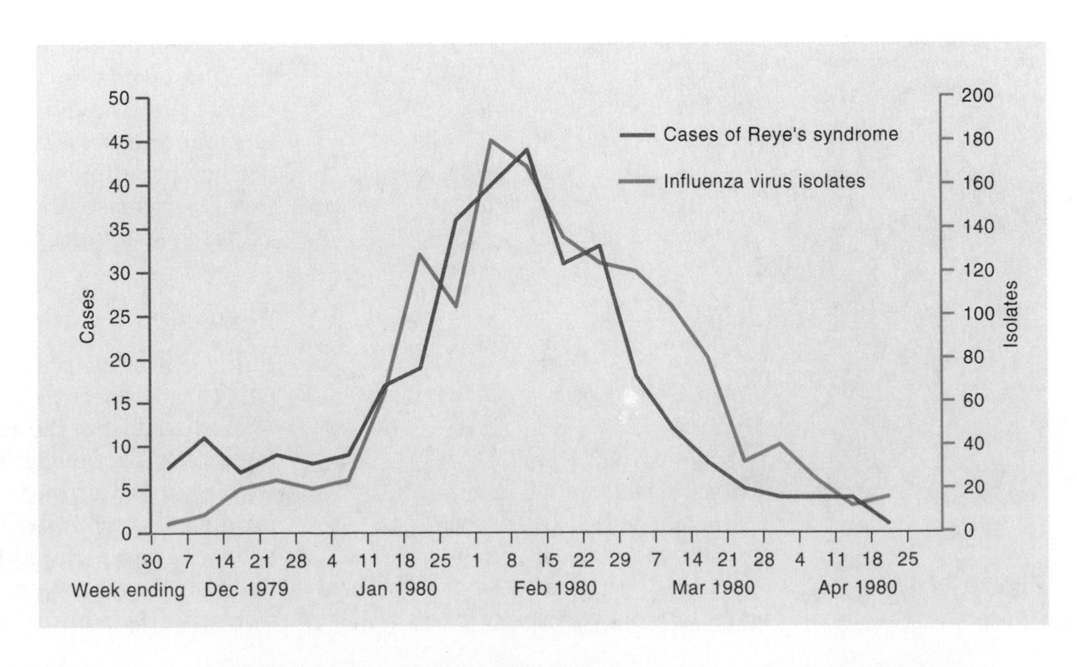

Figure 24.18
Association between influenza B and cases of Reye's syndrome. The syndrome probably results from aspirin given to control the symptoms of influenza.

Source: Centers for Disease Control and Prevention. 1980. *Morbidity and Mortality Weekly Report* 29:27.

influenza[1] vaccine produced against the 1976 Fort Dix H1N1 influenza A strain. This reaction, called the **Guillain-Barré syndrome,** occurred in about 1 of every 100,000 persons vaccinated. It is characterized by severe paralysis, and although most people recover completely, about 5% die of the paralysis. The Guillain-Barré syndrome occurs following a variety of viral illnesses. The reason for its apparent association with the swine influenza vaccine is not known.

It takes six to nine months from the appearance of a new influenza strain before adequate amounts of vaccine can be manufactured. Amantidine and a related antiviral medicine, rimantadine, can be employed for short-term prevention of influenza A disease when vaccine is not available. These medicines are usually used in conjunction with vaccination to protect exposed individuals until they can develop immunity. They do not prevent disease caused by influenza B viruses.

The main features of influenza are summarized in table 24.9.

Table 24.9 Influenza

Symptoms	Fever, muscle aches, lack of energy, headache, sore throat, nasal congestion, cough
Incubation period	1–2 days
Causative agent	Influenza virus, an orthomyxovirus
Laboratory diagnosis	Recovery of virus from pharyngeal or nasal secretions; tests for antibody carried out in blood samples from acute and convalescent stages of illness
Pathogenesis	Infection of respiratory epithelium; cells destroyed and virus released to infect other cells
Epidemiology	Inhalation of infected droplets
Prevention	Amantidine and rimantadine effective for type A but not type B viruses; vaccines
Treatment	Above medications marginally effective when given early in the disease

Fungal Infections of the Lung

Serious lung diseases caused by fungi are quite unusual in healthy, immunocompetent individuals. However, symptomatic and asymptomatic infections that subside without treatment are common. Coccidioidomycosis and histoplasmosis are two mycoses widespread in parts of North America.

Coccidioidomycosis Is Caused by a Soil Fungus Occurring in Certain Hot, Dry Regions of the Americas

In the United States, most cases of coccidioidomycosis occur in California, Arizona, Nevada, New Mexico, Utah, and West Texas. People who are exposed to dust and soil, such as farm workers, are most likely to become infected, but only 40% develop symptoms.

Symptoms

Flulike symptoms such as fever, cough, chest pain, and loss of appetite and weight are common symptoms of coccidioidomycosis. About 1 of 10 people suffering the illness experiences hypersensitivity to the fungal antigens, manifested by a rash on the shins or on other parts of the body, and pain in the joints. The majority of people afflicted with coccidioidomycosis recover spontaneously within a month and, in contrast to people with tuberculosis, have little risk of later reactivation of the disease. In a few people, the disease closely resembles tuberculosis.

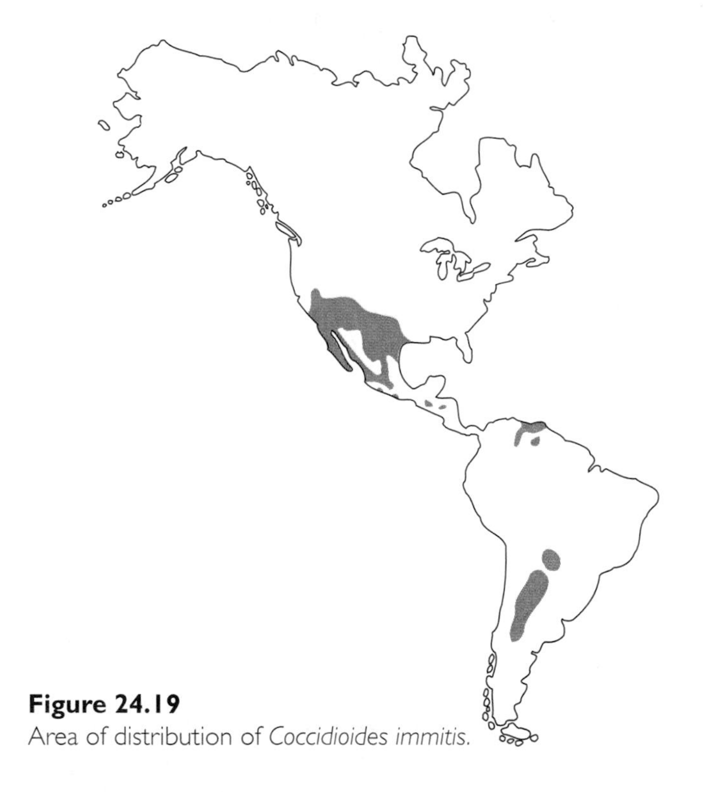

Figure 24.19
Area of distribution of *Coccidioides immitis.*

Causative Agent

The causative agent of coccidioidomycosis, *Coccidioides immitis,* is a dimorphic fungus living in the soil. It grows only in hot, dry, dusty areas of the Western Hemisphere (figure 24.19).

[1]Swine influenza—the name given to disease caused by certain H1N1 influenza viruses. These viruses have an H1 antigen (H_{SW}) that matches the H1 of a pig influenza virus. (The same term, *swine influenza,* is the name of an influenzal disease of pigs. In this disease, the influenza virus can become latent in earthworms and in a parasitic worm of pigs. The virus replicates and causes respiratory disease when the pig is stressed by cold or infection with *Haemophilus influenzae.*)

• **Mycoses, p. 288**

• **Dimorphic fungi, pp. 286, 545**

Laboratory Diagnosis

The "tissue phase" of *C. immitis* is the form present in infected tissues and can be identified by microscopic examination of sputum or pus. It is characterized by a thick-walled sphere called a **spherule,** which in contrast to most other fungi growing in tissue, never buds. The larger spheres of the fungus contain several hundred small cells called "endospores" (figure 24.20). These "endospores" have little resemblance to the endospores of bacteria, which are smaller and considerably more resistant to heat and disinfectants. The mold form of the organism grows readily on most laboratory media at room temperature and is the form that grows in soil. On Sabouraud's medium (a specialized medium for growing fungi), the mold form of *C. immitis* usually grows in three to five days. Such cultures are extremely infectious, since most of the hyphae of the organism develop numerous barrel-shaped **arthrospores** (figure 24.21), which separate easily from the hyphae and become airborne. Since other molds can resemble *C. immitis,* cultures should be converted to the spherule form to complete the identification. This conversion occurs when cultures are grown under conditions similar to those in body tissues.

Pathogenesis

Arthrospores enter the lung along with inhaled air and develop into spherules. Each spherule matures and ruptures, spilling its numerous endospores. The endospores mature into spherules and repeat the process. Each rupture of a spherule provokes an inflammatory response that initially includes a large number of polymorphonuclear neutrophils. As acquired immunity develops, the cellular response changes, showing mainly lymphocytes, plasma cells, and macrophages. Symptoms and tissue injury are caused mainly by the host response to coccidioidal antigens. Usually, the infection is eliminated, but in a few cases tissue death leads to cavities throughout the body, involving the skin, mucous membranes, brain, and other organs.

Delayed hypersensitivity to the organism can be detected by skin testing with **coccidioidin,** a liquid derived from the fluid portion of a *C. immitis* culture in a manner similar to that used for tuberculin. Most people show a positive coccidioidin skin test within three weeks of inhaling the fungal spores of *C. immitis,* and they retain the capacity to produce a positive skin test for life if they remain in the area where the fungus occurs. Patients who move away for several years and those whose disease spreads throughout their bodies often lose their skin reactivity to coccidioidin. Within the first month of illness, most people with coccidioidomycosis develop precipitating antibodies, which generally drop to low titers by the third month, regardless of the progress of the disease. Complement-fixing antibodies to the organism arise later and often remain detectable for years, although the titer of these antibodies falls after recovery from the illness. However, titers of complement-fixing antibodies continue to rise in disseminated infections. This laboratory finding indicates that treatment must be given to save an infected person's life.

Epidemiology

People become infected with *C. immitis* by inhaling the airborne spores of the fungus, which are easily dispersed in dust from soil in which the organism is growing. Infectious spores have unknowingly been transported to other areas (such as the southeastern United States), but there is no evidence that the fungus can establish itself in other climates. In the areas where *C. immitis* is endemic, infections occur only during the hot, dusty seasons when its spores are airborne. Dust arising during the January 1994 California earthquake was blamed for one recent epidemic. However, periodic rainfall may promote growth of the soil fungus, as witnessed by the dramatic rise in California cases when rains in 1992 ended in a prolonged drought. Virtually all individuals who

• **Precipitins, p. 384**

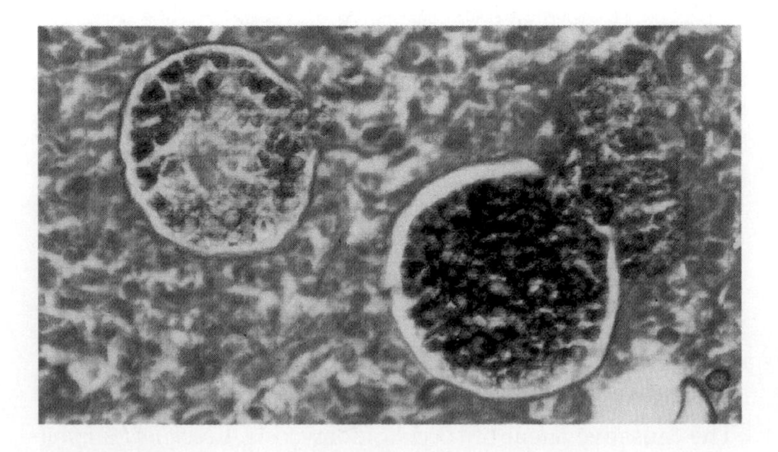

Figure 24.20
Coccidioides immitis. Spherules containing "endospores" in tissues of an experimentally infected mouse (X400).

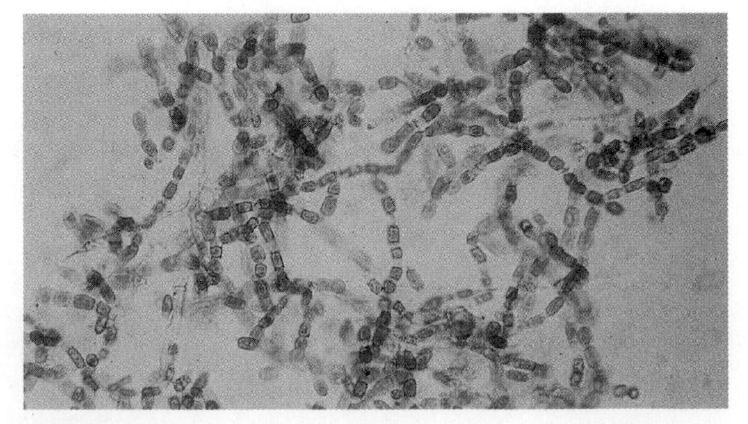

Figure 24.21
Coccidioides immitis. Mold phase mycelium fragmenting into barrel-shaped arthrospores (X460). The arthrospores are the infectious form of *C. immitis.*

live in affected areas show evidence of having been infected. Indeed, people have become infected with *C. immitis* by simply traveling through contaminated areas on a bus.

Prevention

Preventive measures include avoiding dust in the endemic areas. Watering and planting vegetation aid in dust control.

Treatment

Without antifungal treatment, about half of all patients with disseminated coccidioidomycosis die. Medications approved for treatment include amphotericin B and ketoconazole. They markedly improve the prognosis but must be given for long periods of time and cause troublesome side effects. Even with treatment, the disease can reactivate months or years after completion of treatment.

Table 24.10 describes the main features of this disease.

Histoplasmosis Is Associated with Bird and Bat Droppings

Like coccidioidomycosis, histoplasmosis is usually benign but occasionally mimics tuberculosis. The distribution is more widespread than that of coccidioidomycosis and is associated with different soil and climate.

Symptoms

Most infections are asymptomatic. Fever, cough, and chest pain are common, sometimes with shortness of breath. Mouth sores may develop, especially in children.

Table 24.10 Coccidioidomycosis (Valley Fever)

Symptoms	Fever, cough, chest pain, loss of appetite and weight; less frequently, rash, pain in joints; skin, mucous membranes, brain, and internal organs sometimes involved
Incubation period	2 days to 3 weeks
Causative agent	*Coccidioides immitis*, a dimorphic fungus
Laboratory diagnosis	Microscopic examination of sputum or pus for spherules; blood tests for antibody
Pathogenesis	After lodging in lung, arthrospores develop into spherules that mature and discharge their endospores, each of which can then develop into another spherule
Epidemiology	Inhalation of spores from soil
Prevention	Dust control methods such as grass planting and watering
Treatment	Amphotericin B or ketoconazole

Causative Agent

The cause of histoplasmosis is the dimorphic fungus *Histoplasma capsulatum*. This organism prefers to grow in soils contaminated by bat or bird droppings; thus, bat caves and chicken coops are notorious sources of the infection. In its tissue phase, *Histoplasma capsulatum* is an oval yeast that grows within host macrophages (figure 24.22). The mycelial (mold) form of the organism shows **macroconidia,** which are characterized by numerous projecting knobs (figure 24.23).

Laboratory Diagnosis

Diagnostic measures include microscopic examination of stained smears of infected material and cultures on laboratory media or in laboratory mice. Delayed hypersensitivity to the organism is detected by skin testing with **histoplasmin,** an antigen analogous to coccidioidin and tuberculin. Precipitating and complement-fixing antibodies develop in

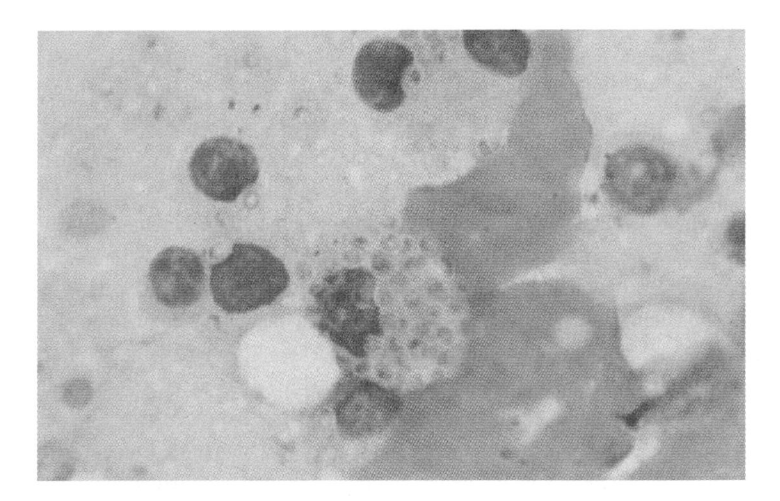

Figure 24.22
Histoplasma capsulatum. Yeast phase organisms in macrophage of an infected mouse.

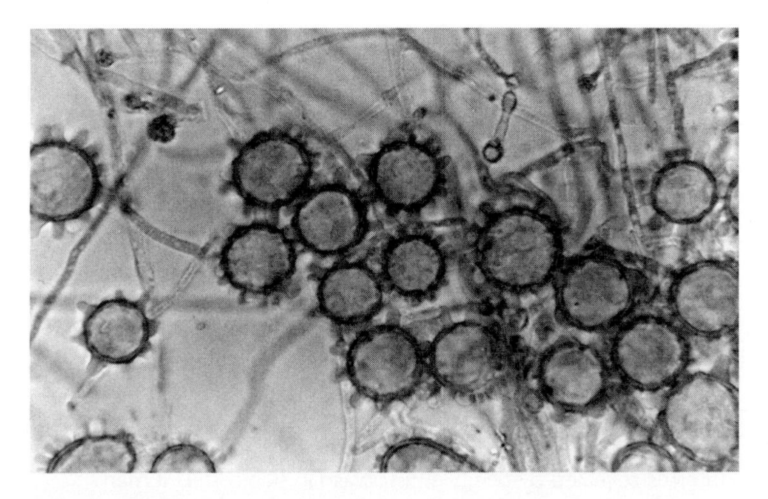

Figure 24.23
Histoplasma capsulatum. Characteristic tuberculate macroconidia in mold phase culture (X1,000).

histoplasmosis as they do in coccidioidomycosis. These antibodies often cross-react with antigens of *Coccidioides* and *Blastomyces dermatitidis*, another pathogenic fungus. Skin tests and antibody determinations therefore help to establish the cause of a fungal infection only if antigens from all three fungi are tested at the same time. The antigen corresponding to the infecting fungus usually shows the largest skin test reaction and greatest antibody response.

Pathogenesis

Few details of the pathogenesis of histoplasmosis are known. Spores are inhaled with dust from contaminated soils and develop into the yeast form. The organisms probably grow intracellularly. Granulomas develop in the infected areas, closely resembling those seen in tuberculosis, sometimes even showing caseation necrosis. Eventually, the lesions are replaced with scar tissue, and many calcify. In rare cases, the disease is not controlled and spreads throughout the body.

Epidemiology

Histoplasmosis is very similar to coccidioidomycosis, except in its distribution. Figure 24.24 shows the distribution of histoplasmosis in the United States, but the disease also occurs in tropical and temperate zones scattered around the world. Since cave explorers (spelunkers) are at risk for contracting the disease, it has been called **spelunkers disease.**

About half of the reported cases of histoplasmosis have occurred in the United States, primarily in the Mississippi River drainage area and in the South Atlantic states such as Maryland and North Carolina. Skin tests reveal that millions of people living in these areas have been infected. However, as with *Coccidioides*, only a small fraction of people infected with *Histoplasma* develop serious illness.

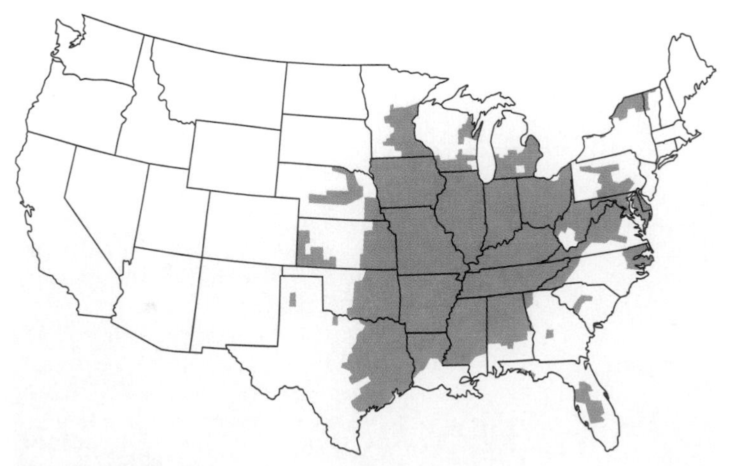

Figure 24.24
Geographic distribution of *Histoplasma capsulatum* in the United States as revealed by positive histoplasmin skin tests. Human histoplasmosis has been reported in more than 40 countries besides the United States, including Argentina, Italy, South Africa, and Thailand.

Prevention

No proven preventive measures are known. It is a good idea to avoid areas containing soil heavily enriched with bat, chicken, and bird droppings, especially if they have been left undisturbed for a long period. Some researchers have recommended that several inches of clay soil be placed over soils containing large quantities of old droppings.

Treatment

Treatment of histoplasmosis is similar to that of coccidioidomycosis. Amphotericin B, ketoconazole, and itraconazole are approved for treating histoplasmosis.

Table 24.11 describes the main features of histoplasmosis.

REVIEW QUESTIONS

1. Why is tuberculosis due to *Mycobacterium bovis* much less common since the implementation of routine milk pasteurization?
2. How can *M. tuberculosis* be cultivated from mixtures with rapidly growing microbes?
3. Why are two or more antitubercular medicines used together to treat tuberculosis?
4. Why did it take so long to discover the cause of Legionnaires' disease?
5. Why are there so many deaths from influenza when it is generally a mild disease?
6. What are the differences between antigenic drift and shift of influenza virus?
7. Why does the epidemiology of coccidioidomycosis differ from that of histoplasmosis?

Table 24.11 Histoplasmosis

Symptoms	Mild respiratory symptoms; less frequently, general malaise, weakness, fever, chest pain, cough
Incubation period	5–8 days
Causative agent	*Histoplasma capsulatum*, a dimorphic fungus
Laboratory diagnosis	Stained smears of ulcer exudates, bone marrow, sputum, or blood; blood tests for antibody
Pathogenesis	Spores undergo metamorphosis to tissue phase; granuloma formation follows
Epidemiology	Inhalation of airborne spores from soil; grows in soil contaminated by bird or bat droppings
Prevention	Avoidance of contaminated chicken coops and bat caves
Treatment	Amphotericin B, ketoconazole, or itraconazole for serious infections

Case Presentation

The patient was a two-month-old boy brought to the hospital because he vomited a large quantity of dark red material. Previously well, a few days earlier he had abruptly interrupted a crying episode and then appeared to have difficulty breathing.

There were no recognized illnesses and no history of trauma or abuse. He was bottle fed. Both parents smoked in the home, which was located in an area of the city characterized by substandard housing in poor repair. Three weeks before admission, their house became flooded during a storm. Cleanup had not yet been accomplished.

On examination, the baby was well developed and nourished, and there was no evidence of trauma or infection. His breathing was labored. A test of his vomitus showed it to be blood. His white blood count was within normal limits, but he was moderately anemic. His blood clotted normally, but his blood oxygen was low. His chest X ray showed patchy shadows consistent with fluid in the alveoli.

The patient was initially placed on a respirator and treated in the intensive care unit for five days. In two weeks, he appeared to be normal and was discharged to his grandmother's house.

Over the next few days, infants with similar illnesses were admitted to the hospital. All came from the same area of the city as the first patient. Some of them died. Some were sent home well but were later readmitted because of recurrent symptoms.

1. Which body system was primarily involved in this illness?
2. Would you expect this illness to be infectious, allergic, or toxic in origin?
3. Why was this patient discharged to his grandmother's house rather than his own home?
4. How would you go about establishing the cause of the illness?

Discussion

1. This is a pulmonary problem. Blood from the lungs was swallowed by the baby and then vomited.
2. Numerous possibilities had to be considered. Infection seemed least likely because of the normal white blood count, lack of evidence of contagion, and the good response to treatment with a respirator. Allergy seemed unlikely because bleeding occurred without other signs of hypersensitivity. An inhaled toxic substance seemed most likely.
3. The patient was sent to his grandmother's house because of the doctor's suspicion that something in the air of the patient's home caused the illness. This idea was supported by the fact that some other infants had recurrence of symptoms when they went home.
4. A case-control study was carried out by investigators from the Air Pollution and Respiratory Health Branch of the Center for Environmental Health, part of the Centers for Disease Control and Prevention. Ten cases of the bleeding lung condition were matched according to age with 30 control infants who had no illness but were from the same part of the city. Among the information gathered about the two groups were presence or absence of toxic chemicals in the home, whether the infants were breast fed, and how many homes had cigarette smokers. The most striking finding was that 100% of the case homes had sustained major water damage within one month of the onset of the illness, whereas this was true of only 23% of the control homes. Moreover, the water damage had been cleaned up in the control homes but not in the case homes. Inspection of the homes of cases revealed molds growing on materials such as wet wood, newspapers, and cloth, and air samples showed higher levels of mold spores in the homes of cases. One of the mold species, *Stachybotrys atra* (*S. chartarum*), is known to cause intestinal bleeding in animals that eat moldy grain. A toxin from this mold was a leading suspect as the cause of the babies' illness.

Summary

I. Anatomy and Physiology
   A. The lower respiratory system includes the trachea, bronchi, bronchioles, and alveoli. Pleural membranes cover the lungs and line the chest cavity.
   B. Ciliated cells line much of the respiratory tract and remove microorganisms by constantly propelling mucus out of the respiratory system.

II. Normal Flora of the Lower Respiratory System
   Viruses and microorganisms are normally absent from the lower respiratory system.

III. Bacterial Pneumonias
   A. *Streptococcus pneumoniae* is one of the most common causes of pneumonia. The bacterium is also known as the *pneumococcus*.
   B. *Klebsiella pneumoniae* can cause permanent damage to the lung. Serious complications are more common than with other bacterial pneumonias. Treatment is more difficult, partly because klebsiellas often contain R factor plasmids.
   C. *Mycoplasma* pneumonia is often mild, and often called *walking pneumonia*. Serious complications are rare. Penicillins and cephalosporins are not useful in treatment because *M. pneumoniae* lacks a cell wall. Cold agglutinins often help confirm the diagnosis.

IV. Other Bacterial Infections of the Lower Respiratory Tract
   A. Pertussis (whooping cough) is characterized by violent spasms of coughing and gasping. The cause is a gram-negative rod, *Bordetella pertussis.* Childhood immunization prevents the disease.
   B. Tuberculosis is generally slowly progressive or heals and remains latent, presenting the risk of reactivation. The cause is an acid-fast rod, *Mycobacterium tuberculosis.*
   C. Legionnaires' disease occurs when there is a high infecting dose of the causative microorganisms or an underlying lung disease. The cause, *Legionella pneumophila,* is a rod-shaped bacterium common in the environment. It requires special stains and culture techniques for diagnosis.

V. Viral Infections of the Lower Respiratory Tract
   A. Serious epidemics are characteristic of influenza A viruses. Antigenic shifts and drifts are responsible.
   B. Deaths are usually but not always caused by secondary infection.
   C. Antibody against hemagglutinin gives protection.

D. Reye's syndrome may rarely occur during recovery from influenza B and other viral infections but is probably not caused by the virus itself.

VI. Fungal Infections of the Lung
   A. Coccidioidomycosis occurs in hot, dry areas of the Western Hemisphere. The airborne spores of the dimorphic soil fungus *Coccidioides immitis* initiate the infection.
   B. Histoplasmosis is similar to coccidioidomycosis but occurs in tropical and temperate zones around the world. The causative fungus, *Histoplasma capsulatum*, is dimorphic and found in soils contaminated by bat or bird droppings.

Chapter Review Questions

1. Give three complications of pneumococcal pneumonia.
2. What preventive measure decreases the chance of developing pneumococcal pneumonia?
3. Give an important difference between the pathogeneses of of pneumonia due to *Streptococcus pneumoniae* and that due to *Klebsiella pneumoniae.*
4. Which group of bacteria accounts for most cases of nosocomial pneumonia?
5. Give the name of one of the surface proteins of *Bordetella pertussis* that facilitates its attachment to respiratory epithelium.
6. At what stage in their illness are pertussis patients the most infectious?
7. Name a cell wall component of *Mycobacterium tuberculosis* that helps make it resistant to staining and to disinfectants.
8. On what characteristic of *Mycobacterium tuberculosis* does its virulence depend?
9. Why do AIDS and active tuberculosis often go together?
10. How long does it take after a new strain of influenza appears for a vaccine against it to become available?

Critical Thinking Questions

1. Would you expect a person who recovered from pneumococcal pneumonia to be immune to further bouts of the disease? Explain.
2. What caused the incidence of tuberculosis to increase after a long decline?
3. If we completely stop all transmission of tuberculosis from one person to another, how long will it take for the world to be rid of the disease?

Further Reading

1990. Bovine tuberculosis—Pennsylvania. *Morbidity and Mortality Weekly Report* 39:201–203.

1993. Coccidioidomycosis—United States, 1991–1992. *Morbidity and Mortality Weekly Report* 42:21.

Culliton, B. J. 1992. Drug-resistant TB may bring epidemic. *Nature* 356:473.

Davis, S. F., et al. 1992. Pertussis surveillance—United States, 1989–1991. *Morbidity and Mortality Weekly Report*: 41(no. SS-8):11–19.

1996. Defining the public health impact of drug-resistant *Streptococcus pneumoniae:* report of a working group. *Morbidity and Mortality Weekly Report*:45(no. RR-1):1–20.

Fields, B. S. et al. 1990. Amoebae and Legionnaires' disease. *ATCC Quarterly Newsletter* 10(1):1.

Fincher, J. 1989. America's deadly rendezvous with the Spanish lady. *Smithsonian*: 130.

1996.Food and Drug Administration approval of an acellular pertussis vaccine for the initial four doses of the diphtheria, tetanus, and pertussis vaccination series. *Morbidity and Mortality Weekly Report*: 45(31):665–671.

1996. Hantavirus Pulmonary Syndrome—United States, 1995 and 1996. *Morbidity and Mortality Weekly Report* 45(14):291–295.

Lieberman D., et al. 1996. *Legionella* species community-acquired pneumonia: a review of 56 hospitalized adult patients. *Chest* 109:1243–1249.

1996. Multidrug-resistant tuberculosis outbreak on an HIV ward Madrid, Spain, 1991–1995. *Morbidity and Mortality Weekly Report* 45(16):325–330.

Nennig, M. E. et al. 1996. Prevalence and incidence of adult pertussis in an urban population. *JAMA* 275:1672–1674.

Pennisi, Elizabeth. 1996. Experts wary of ever-changing influenza A virus. *ASM News* 62(7):356–360.

1996. Prevention and control of influenza. *Morbidity and Mortality Weekly Report* 45(no. RR-5).

1996. Prevention and control of tuberculosis in correctional facilities. *Morbidity and Mortality Weekly Report* 45(no. RR-8).

1993. Resurgence of pertussis—United States, 1993. *Morbidity and Mortality Weekly Report* 42:952.

Straus, W. I. et al. 1996. Risk factors for domestic acquisition of legionnaires disease. *Arch Intern Med* 156:1685–1692.

1996. The role of BCG vaccine in the prevention and control of tuberculosis in the United States. *Morbidity and Mortality Weekly Report* 45(no. RR-4).

1996. Tuberculosis Morbidity—United States, 1995. *Morbidity and Mortality Weekly Report* 45(18):365–370.

25 Upper Alimentary System Infections

KEY CONCEPTS

1. The alimentary tract is a major route for the entry of pathogenic microorganisms into the body.
2. The pathogenesis of tooth decay depends on both diet and acid-forming bacteria which are adapted to colonize hard, smooth surfaces within the mouth.
3. Infections of the esophagus are so unusual in normal people that their occurrence suggests immunodeficiency.
4. A properly functioning stomach destroys most microorganisms before they reach the intestine.

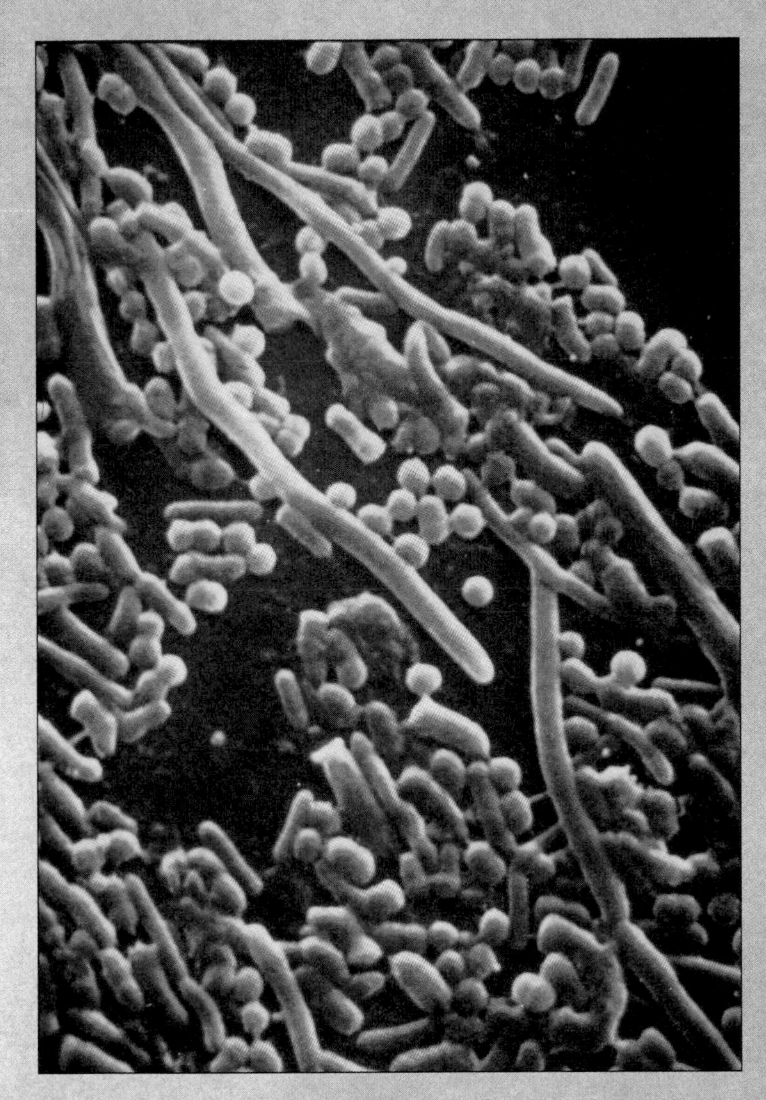

Scanning electron micrograph of dental plaque showing diverse bacteria. Cocci, bacilli, and branching filaments are seen.

PREVIEW LINK

Microbes in Motion

The following books in the *Microbes in Motion* CD-ROM may serve as a useful preview or supplement to your reading: *Gram-Positive Organisms. Gram-Negative Organisms. Miscellaneous Bacteria. Microbial Pathogenesis in Infectious Diseases:* Toxins. *Vaccines:* Vaccine Targets. *Viral Structure and Function:* Viral Diseases. *Fungal Structure and Function:* Diseases. *Parasitic Structure and Function.*

A Glimpse of History

The presence of **dental caries** (tooth decay) in prehuman skeletons more than 500,000 years old demonstrates the antiquity of tooth decay. In Europe, through the medieval period, less than 10% of human teeth had cavities, and they were mostly on parts of the teeth exposed by the receding gums of older people. However, in the 17th century, an increase in the incidence of tooth decay occurred, and the location of cavities changed to include the sides and biting surfaces of the teeth, particularly of young people. This change in the frequency and location of the cavities is related to the introduction of two new substances into the European diet: refined flour and sugar (sucrose). We now know that, although refined flour can produce dental caries, it is of minor importance compared with sucrose. Initially, only the royalty and other rich people had rampant tooth decay, since only they could afford the price of sugar. By the middle of the 19th century, however, the development of ocean shipping and use of slave labor on huge sugarcane plantations brought sugar (and dental caries) within the financial reach of most western populations. Sugar in the diet has spread throughout the world since then, with dental caries progressively increasing in developing countries.

The alimentary tract is the passageway running from the mouth to the anus. Like the skin, it is one of the body's boundaries with the environment. Most of it is colonized by a normal microbial flora, and it is exposed to other organisms from food and drink. It is one of the major routes into the body for invading microbial pathogens.

The upper portion, consisting of the mouth, esophagus, and stomach, prepares food for digestion and absorption in the lower or intestinal portion of the tract. The alimentary tract has important appendages, such as the salivary glands of the mouth, that are connected to the tract by ducts (tubes). These appendages produce digestive juices that pour into the tract and aid in the processing of food. The alimentary tract and its appendages together compose the alimentary system, whose purpose is to provide nourishment for the body. Figure 25.1 illustrates the relationships among the various parts of the alimentary system.

There are relatively few infectious diseases of the upper alimentary system, but one—dental caries (tooth decay)—is exceedingly widespread and economically important. Others such as herpes simplex and mumps cause considerable discomfort and some deaths. Infection of the stomach, once thought to be rare, is now known to be commonplace.

Anatomy and Physiology

Most foods are composed of large, complex molecules that cannot be absorbed into the body until they are broken down into their simple components.

The Initial Breakdown of Complex Foods Occurs in the Mouth Through the Action of Teeth and Saliva

The teeth (figure 25.2) are made up largely of hydroxyapatite[1] crystals but also contain some protein especially in the inner (**dentin**) portion. The outer portion, the **enamel,** comprises a hard, protective layer of densely packed crystals, yet certain bacteria are able to penetrate it and produce cavities and tooth destruction. Pits and crevices normally present on the surfaces of the teeth collect food particles and offer protected sites for microbial colonization. The **gingival crevice** (space between the tooth and gum) is important because both **gingivitis**[2] and **periodontal disease**[3] originate there.

Saliva-producing glands are located under the tongue, laterally in the floor of the mouth, and on each side of the face below the ears. The saliva produced by these glands is conveyed by ducts that discharge it through openings beside the upper second molar and under the tongue. About 1,500 ml of saliva is secreted into the mouth from the salivary glands each day. Saliva keeps the mouth clean and lubricated and helps maintain a neutral pH. Since it is saturated with calcium, saliva tends to prevent the calcium phosphate of the teeth from dissolving in acid foods and the acids produced by certain mouth bacteria as a product of their metabolism. Also, saliva contains the enzyme **amylase,** which begins the breakdown of starches.

The Esophagus Functions to Convey Food Through the Chest Cavity to the Stomach

The tongue and throat muscles work together in swallowing to propel food from the mouth and throat into the esophagus, a collapsible tubelike structure about 10 inches long, located behind the trachea (windpipe). In the process of swallowing, a muscular flap (epiglottis) folds down to cover the opening of the trachea to prevent any material from entering the respiratory system. The esophagus has a muscular wall that contracts rhythmically (in a process called *peristalsis*), actively propelling food or liquid into the stomach. Once the swallowed material is in the stomach, the lower esophagus normally closes, preventing regurgitation of stomach contents. Occasionally, the esophagus can rupture during violent vomiting, causing bleeding and infection of the underlying tissues. Defective lower esophageal closure results in regurgitation of stomach acid, which inflames the esophagus and causes pain under the sternum (breastbone), commonly referred to as *heartburn*.

[1] Hydroxyapatite—repeating units of $Ca_{10}(PO_4)_6(OH)_2$
[2] Gingivitis—inflamed gingiva (gums)
[3] Periodontal disease—abnormality of the tissues surrounding the base of the tooth

Perspective 25.1

Role of Saliva in Preventing Tooth Decay

People with poor production of saliva because of disease of their salivary glands or medication often have rampant dental caries. Experimentally, rats whose salivary glands have been removed show a 10-fold increase in caries. Saliva helps to maintain the teeth because of several important properties: (1) it has a high buffering capacity, meaning it can neutralize acids; (2) it is supersaturated with calcium and phosphate ions, which can crystallize in early dental lesions and slow the advance of dental caries; and (3) it contains antibacterial substances, including lysozyme (which degrades peptidoglycan of bacterial cell walls), lactoferrin (which binds iron ions, critically required by bacteria), and specific IgA antibodies (which inhibit bacterial attachment).

• **Iron, pp. 28, 259, 357, 411**

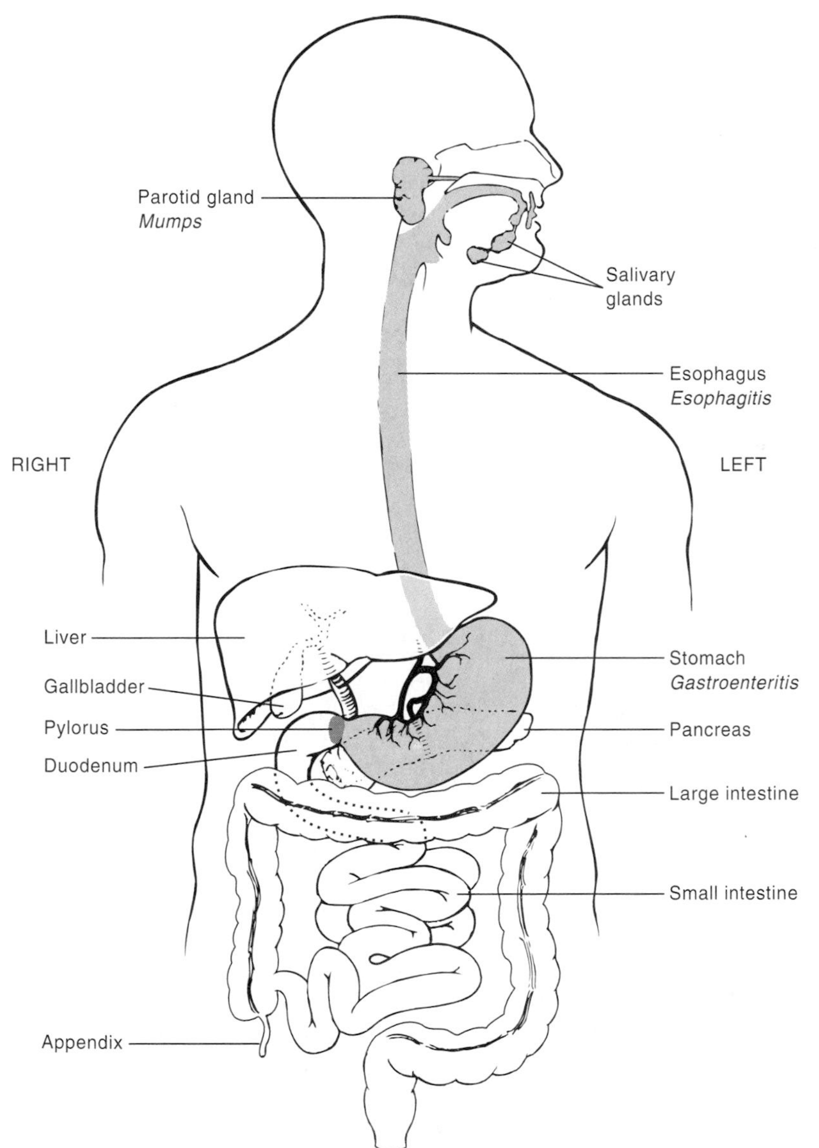

Figure 25.1
Upper alimentary system, shown in pink. This part of the alimentary system mainly prepares food for final breakdown and absorption in the lower alimentary system. Disorders associated with some parts of the system are indicated.

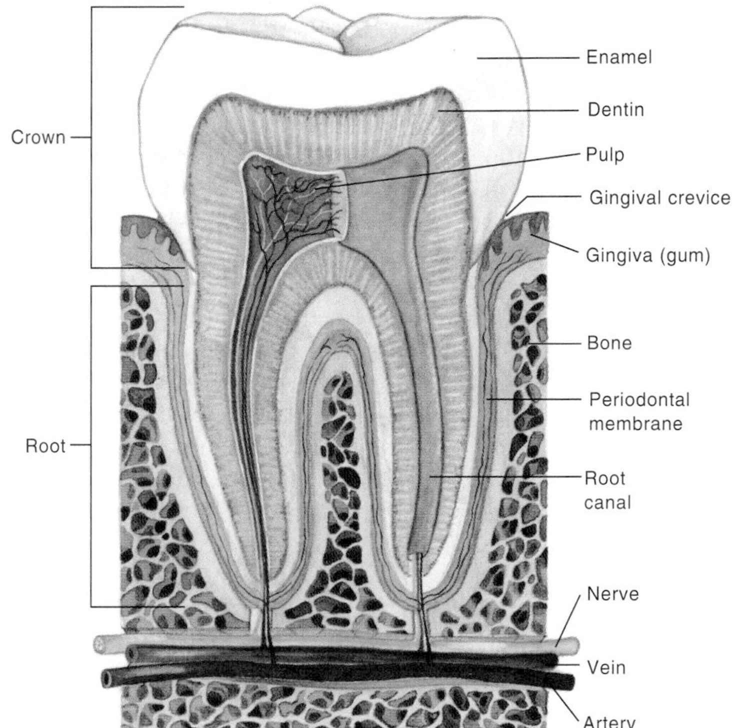

Figure 25.2
Structure of a tooth and its surrounding tissues.

Hydrochloric Acid and the Protein-Splitting Enzyme Pepsin Are Secreted by the Stomach and Continue the Breakdown of Food

The stomach is an elastic saclike structure with a muscular wall. Some of the cells that line it produce hydrochloric acid, while others produce pepsinogen, which becomes the protein-splitting enzyme pepsin upon contact with acid. In the presence of food, the stomach begins a mixing action, bringing acid and enzymes into contact with the ingested material. The stomach itself is protected from the acid and enzymes by a thick layer of mucus secreted by a third kind of lining cell. Stomach emptying is controlled by a complex system of nerve impulses and hormones acting on a muscular valve, the **pylorus,** which determines the rate at which the stomach contents enter the intestine.

If food stays too long in the stomach, particularly if it is inadequately chewed and mixed with stomach juices, organisms present in the food cause it to rot, causing halitosis (bad breath). Food that passes through the stomach too quickly may not be easily digested by the small intestine, resulting in excess intestinal gas or diarrhea.

Normal Flora

This section focuses on the oral cavity (mouth) flora, since the esophagus has only a relatively sparse population consisting of species represented in the mouth and upper respiratory system. The normal fasting[1] stomach is devoid of microorganisms because of the killing action of acid and pepsin.

A variety of bacterial species inhabit the mouth, occurring on its mucous membranes, in tooth deposits, and in saliva. Two features of this oral bacterial population are especially noteworthy. First, of all the species of bacteria introduced into the mouth from the time of birth onward, relatively few are able to colonize and persist in the oral cavity. Second, the distribution of these bacterial species is far from uniform; certain members colonize one area, and others colonize another. The most prominent species of bacteria in the mouth are of the genus *Streptococcus*, gram-positive, chain-forming cocci that metabolize carbohydrates with the production of lactic acid. *Streptococcus salivarius* preferentially colonizes the upper part of the tongue, *S. sanguis* colonizes the teeth, and *S. mitis* colonizes the mucosa of the cheek. Collections of bacteria adhere to the smooth surfaces and crevices of the teeth. These masses of bacteria, called **dental plaque** (figure 25.3), are composed chiefly of streptococci intermixed with various species of filamentous, gram-positive, branching and nonbranching bacteria. The gingival crevice characteristically is populated by strict anaerobes of the genera *Bacteroides, Porphyromonas,* and *Fusobacterium*, and by smaller numbers of spirochetes. Dental deposits of bacteria range in number up to 100 billion organisms per gram of plaque, and salivary organisms number up to 1 billion per milliliter of saliva.

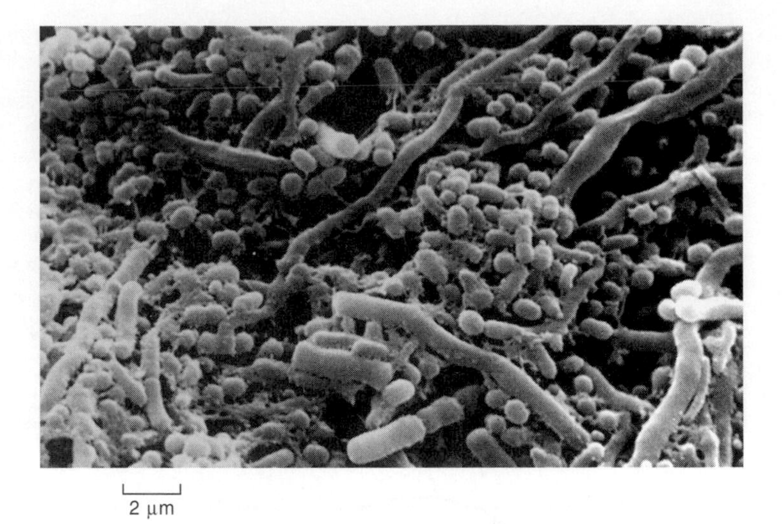

Figure 25.3
Scanning electron micrograph of dental plaque. The many different kinds of bacteria composing the plaque exhibit specific attachments to the tooth and to each other.

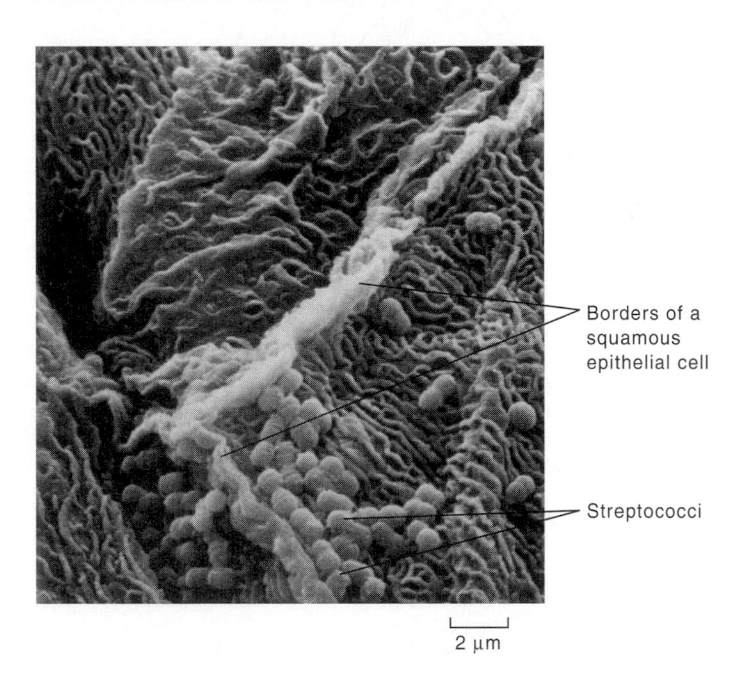

Figure 25.4
Scanning electron micrograph of streptococci adhering to oral mucosa.

Colonization of the Mouth Depends upon the Ability of Microorganisms to Adhere to Oral Surfaces

Certain strains of streptococci and other bacteria have surface filaments (fimbriae) that attach specifically to receptors on host tissues, allowing microorganisms to resist the scrubbing action of food and the tongue and the flushing action of salivary flow (figure 25.4). The fact that the lining cells of different mouth tissues have differing receptors accounts for the distribution of the various species of streptococci in the mouth. The host limits the extent of bacterial colonization of its mucous membranes by constantly shedding the superficial layers of cells and replacing them with new ones; the rate of this shedding correlates with the number of microorganisms present on the surface. Furthermore, the antibodies and mucins[2] in saliva may tend to cover up the bacterial attachment filaments and thus prevent or weaken adherence to the oral tissues.

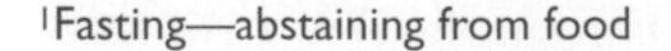

[1]Fasting—abstaining from food

[2]Mucins—the glycoproteins in saliva, mucus, and body tissues

• Gram-negative anaerobes, pp. 249, 250

Colonization of the mouth by strictly anaerobic flora requires that teeth be present because only the teeth can provide a sufficiently anaerobic habitat to allow the growth of these organisms. The pockets created by the gingival crevices and fissures in the teeth are one such habitat. Tooth surfaces also represent a nonshedding tissue in which enormous bacterial concentrations are able to build up. The metabolic activity of the various organisms growing on teeth consumes oxygen and thus provides another area in which anaerobic species can grow.

Infectious Diseases Caused by Oral Bacteria

For the most part, bacterial diseases of the mouth are caused by normal mouth flora. These organisms can cause serious tooth and gum disease and may also cause disease elsewhere in the body. Infections caused by bites or sharp objects that have been in the mouth are discussed in chapter 29.

Dental Caries Is a Chronic Infectious Disease of the Teeth Caused by Mouth Bacteria

Dental caries is probably the most common infectious disease of human beings. The cost to restore and replace teeth damaged by this disease is estimated to be about $20 billion per year in the United States.

Causative Agent

Dental caries is an infectious disease caused principally by *Streptococcus mutans* and closely related species. The bacteria live only on the teeth and cannot colonize the mouth in the absence of teeth. These gram-positive, chain-forming cocci grow readily on a variety of laboratory media, generally preferring anaerobic conditions. The colonies are usually nonhemolytic on sheep blood agar, although many strains are β-hemolytic and some are α-hemolytic. They are distinguished from other streptococci by their abilities to ferment the sugar alcohols mannitol and sorbitol and to produce extracellular **glucan** from sucrose (table sugar). Glucans are insoluble polysaccharides composed of repeating subunits of glucose.

• ***S. mutans*, p. 57**

Formation of Dental Plaque

The first step in the formation of human dental plaque is the adherence of streptococci or actinomycetes to the tooth **pellicle,** a thin film of proteinaceous material adsorbed on the tooth from the saliva. The streptococci that adsorb to this film (usually *S. sanguis*) have a specific affinity for the tooth pellicle and are not the predominant streptococci in the saliva. Other species of bacteria attach specifically to earlier arrivals, including species of streptococci, *Haemophilus* and *Veillonella*. Two species that might not attach to each other may attach to a third. If dietary sucrose is present, *S. mutans* soon produces sticky extracellular polysaccharides (glucans) through the action of extracellular enzymes called glucosyltransferases. The glucans bind the organisms together and to the tooth, and provide attachment sites for additional species.

Initially, these may include aerobic organisms, such as members of the genera *Neisseria* and *Rothia*. However, metabolic activity of these early colonizers consumes oxygen, thus making the plaque a suitable location for anaerobic species. Production of lactic acid and other acids by streptococci in the plaque provides a suitable acidic medium for growth of lactobacilli. In fact, lactobacilli colonize dental plaque but not other parts of the mouth, and they can sometimes be cariogenic. Plaque-colonizing anaerobes of the genus *Veillonella* use as a nutrient some of the lactic acid produced from the metabolism of streptococci and lactobacilli. Filamentous organisms of the genus *Actinomyces* colonize the plaque and, by the end of a week, make up most of its bulk. The characteristics of some of these dental plaque bacteria are shown in table 25.1.

• **Lactobacilli, pp. 119–120, 261**

Table 25.1 Characteristics of Some of the Bacteria Found in Dental Plaque

Genus	Characteristics
Actinomyces	Gram-positive, branching filaments; generally prefer or require anaerobic growth conditions; ferment sugars, producing acid end-products
Lactobacillus	Gram-positive rods that are nonmotile and generally prefer anaerobic growth conditions; produce lactic acid
Neisseria	Aerobic, gram-negative diplococci; possess cytochrome C oxidase, an enzyme easily identified by a simple laboratory test
Rothia	Gram-positive; aerobic; mostly lactic acid produced during sugar fermentation; may show spherical, rod-shaped, or filamentous forms depending on growth conditions and age of the colony
Streptococcus	Gram-positive cocci in chains; usually prefer anaerobic growth conditions or are indifferent to oxygen; produce mostly lactic acid from sugars
Veillonella	Gram-negative, anaerobic cocci that do not ferment carbohydrates; can use lactic acid as a nutrient

Perspective 25.2

Sealants

Toothbrush bristles cannot remove plaque from the pits (fissures) normally present in children's teeth because the fissures are too deep and narrow. Caries arising from these plaques is enhanced by *S. mutans* and sucrose but is dependent on neither. Fortunately, fissure-related dental caries can be prevented by using sealants. In this procedure, the teeth are professionally cleaned and treated with an epoxy material that coats them with a polymerized methacrylate.* The coating seals the fissures, eventually kills the plaque, and prevents bacterial recolonization. Since dental procedures and wearing down of the tooth surface eliminate many fissures, older people have fewer fissure-related caries.

*Methacrylate—an ester of methacrylic acid, $C_4H_6O_2$

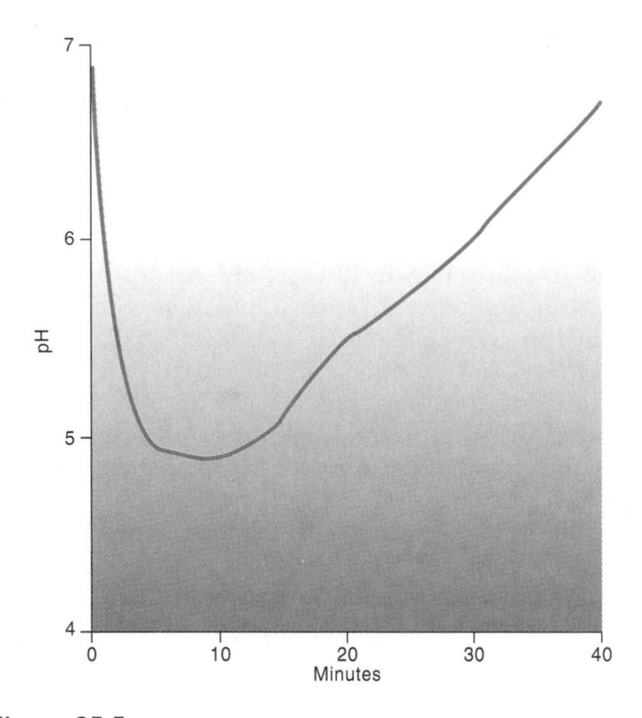

Figure 25.5
Increase in acidity in dental plaque after rinsing the mouth with a glucose solution. Tooth enamel begins to dissolve at about pH 5.5.

Pathogenesis

Scientists have placed tiny electrodes into dental plaque and have studied the change in pH when sugar enters the mouth. The response is dramatic (figure 25.5). The pH of the plaque drops from its normal value of about 7 to below 5 within minutes, more than a 100-fold increase in the acidity of the plaque, to a level at which the calcium phosphate of teeth dissolves. The length of this acidic state depends on how long the teeth are exposed to sugars and on the concentration of the sugars. After food leaves the mouth, the pH of the plaque rises slowly to neutrality. The delay in return of the pH to neutrality is due to the ability of *S. mutans* to store a portion of its food as an intracellular, starchlike polysaccharide that is later metabolized to acid. Plaque thus acts as a tiny, acid-soaked sponge closely applied to the tooth. Note that both *S. mutans* and a suitable sugar-rich diet are required to produce dental caries. Species of streptococci exist that can colonize teeth and produce plaque even if sugar is not present, but sucrose is generally required to establish a cariogenic plaque on smooth tooth surfaces.

The process of tooth decay is usually slow. The outer portion of the tooth (enamel) is very hard and only slowly dissolves in acid. However, acid diffuses into the tooth from overlying cariogenic plaque, penetrates the enamel, and attacks the more easily dissolved dentin. After a period of about six months to two years (or as little as one month in those with poor mouth care), a white or brown spot is visible on the tooth under the plaque, which represents an area of weakened enamel. Beneath this spot is the beginning of a small cavity where the dentin has dissolved. Eventually, the weakened enamel gives way, exposing the underlying cavity. The protein-rich dentin (having 20 times more protein than enamel) allows a number of bacterial species to grow, and the disease progresses rapidly. Finally, the root canal with its nerve and blood vessels is invaded and an abscess forms.

Prevention

Studies undertaken years ago demonstrated that some populations had a far lower incidence of dental caries than other populations receiving the same diet. This led to the discovery that people who drank water with trace amounts of fluoride had fewer caries than those who drank water without it. These trace amounts of fluoride are now known to be required for teeth to resist the acid of cariogenic plaques. Fluoride enters spaces among the hydroxyapatite crystals that compose the enamel, making it harder and more resistant to dissolving in acid. In the United States, over 100 million people are currently supplied by fluoridated public drinking water, which has resulted in a 60% reduction in dental caries in those populations. In areas where fluoridated drinking water is not available, fluoride tablets or solutions are often used, although they are considerably more expensive. To have optimum effect, children should begin receiving fluoride before the permanent teeth erupt. Fluoride applied to tooth surfaces in the form of mouthwashes, gels, or toothpaste is generally less effective, but it may have antibacterial action and can be a valuable supplemental treatment for people especially prone to developing dental caries.

Receding gums of older people can expose root surfaces that become sites for dental caries. Fortunately, the incidence of root caries decreases when water fluoridation is introduced. However, fluoride ingestion has relatively little effect (33% reduction) on dental caries on the biting surfaces of the molars, where over half of carious infections develop. These surfaces normally have tiny pits and fissures where food becomes impacted and caries becomes established.

Another method to control dental caries is restricting sugar in the diet and thereby reducing the ability of *S. mutans* to adhere to teeth and produce acid. Various studies have shown that caries can be reduced by 90% if sweets are eliminated from the diet. However, it is not simply the quantity of sugar in the diet that is important. The frequency of eating and the length of time food stays on the teeth may be more critical than the actual quantity of sugar ingested. Foods low in sugar content may be more likely to cause caries than high-sugar foods if they are eaten as frequent, between-meal snacks and in a form slowly cleared from the mouth. Since it leaves the mouth more quickly, sugar in liquids is less likely to cause caries than is sugar in food. Interestingly, chewing paraffin or sorbitol-sweetened gum tends to reduce the acidity that develops in dental plaques in response to sugar-containing food, probably because it increases the flow of saliva.

Another important preventive measure against dental caries is mechanical removal of plaque by toothbrushing and use of dental floss. Careful cleaning of teeth once a day for 5 to 10 minutes followed by flossing can reduce the incidence of dental caries by 50%. These measures in prevention have resulted in a significant reduction in dental caries in industrialized nations since the 1970s.

It has also been proposed that mothers' tooth decay be vigorously treated before their children's teeth appear since this decreases the quantity of *S. mutans* in mothers' saliva. Maternal saliva is usually the source of a child's *S. mutans*.

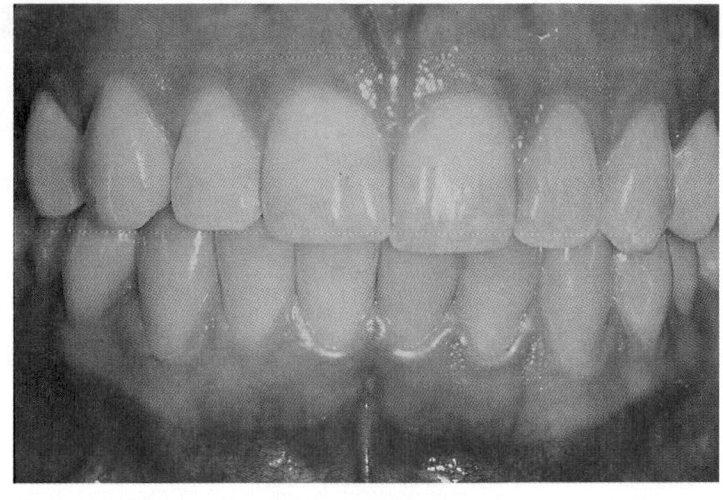

(a)

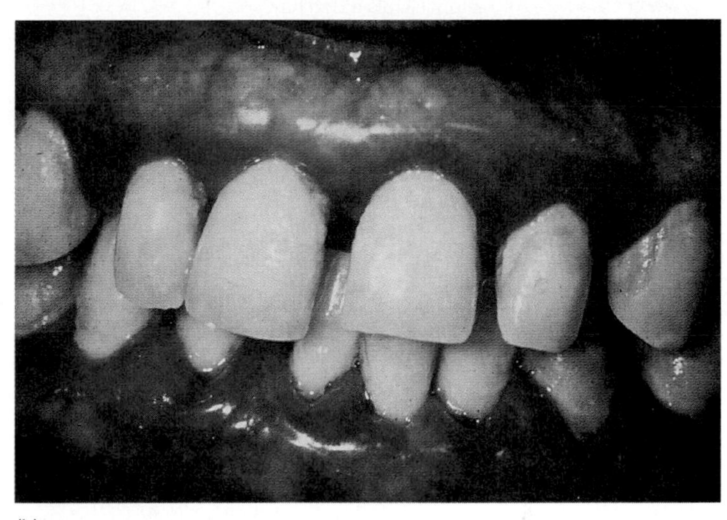

(b)

Figure 25.6
Periodontal disease. (*a*) Normal gingiva. (*b*) Periodontal disease with plaque and inflammation.

Dental Plaque Causes Periodontal Disease, the Chief Cause of Tooth Loss in Older People

Periodontal disease is a chronic inflammatory process involving the tissues around the roots of the teeth. It usually develops very slowly and is the chief cause of loss of teeth from middle age onward. Formation of dental plaque is necessary for the development of periodontal disease, but little is known about how the disease progresses. Plaque and **calculus** (calcified plaque) in the region of the gingival crevice cause an inflammatory process that at first produces swelling, redness, and the appearance of inflammatory cells in the gingiva (figure 25.6). This gum inflammation is known as **gingivitis.** When gingivitis is chronic, the gingival crevice becomes widened and deepened, allowing a marked overgrowth of bacteria. An abscess then develops and penetrates through the base of the gingival crevice. The large population of microorganisms in the abscess produces the enzymes collagenase, protease, and hyaluronidase[1] along with endotoxins. These products penetrate the tissues. The host's immune response to the bacterial products undoubtedly contributes to the inflammatory process. The membrane that attaches the root of the tooth to the bone also becomes inflamed, and the calcium phosphate composing the bone surrounding the tooth gradually dissolves. Periodontal disease thus eventually leads to the

[1]Collagen is a tough, fibrous protein found in skin, bone, cartilage, and ligaments; hyaluronic acid is a gelatinous material that helps tissue cells adhere.

Perspective 25.3

U.S. Dental Health Objectives

Among the health objectives for the United States published by the U.S. Public Health Services in 1980 were a number of objectives dealing with dental caries and periodontal disease:

1. The proportion of nine-year-old children who have experienced dental caries in their permanent teeth should be decreased to 60%.
2. Prevalence of gingivitis in children 6 to 17 years should be decreased to 18%.
3. In adults, the prevalence of gingivitis and destructive periodontal disease should be decreased to 20% and 21%, respectively.
4. No public elementary or secondary school (and no medical facility) should offer highly cariogenic foods or snacks in vending machines or in school breakfast or lunch programs.
5. At least 95% of schoolchildren and their parents should be able to identify the principal risk factors related to dental diseases and be aware of the importance of fluoridation and other measures in controlling these diseases.
6. At least 75% of adults should be aware of the necessity for both thorough personal oral hygiene and regular professional care in the prevention and control of periodontal disease.
7. At least 95% of the population on community water systems should be receiving the benefits of optimally fluoridated water.
8. At least 50% of schoolchildren living in fluoride-deficient areas that do not have community water systems should be served by an optimally fluoridated school supply.
9. At least 65% of schoolchildren should be proficient in personal oral hygiene practices and should be receiving other needed preventive dental services in addition to fluoridation.

Only goals 1 and 6 were met over the subsequent decade, and even with these goals caries decline has not been uniform. Some groups of black, Hispanic, and Native American children still have high rates of dental decay.

Source: *Morbidity and Mortality Weekly Report* 39:646, 1990, Centers for Disease Control and Prevention, Atlanta, Ga

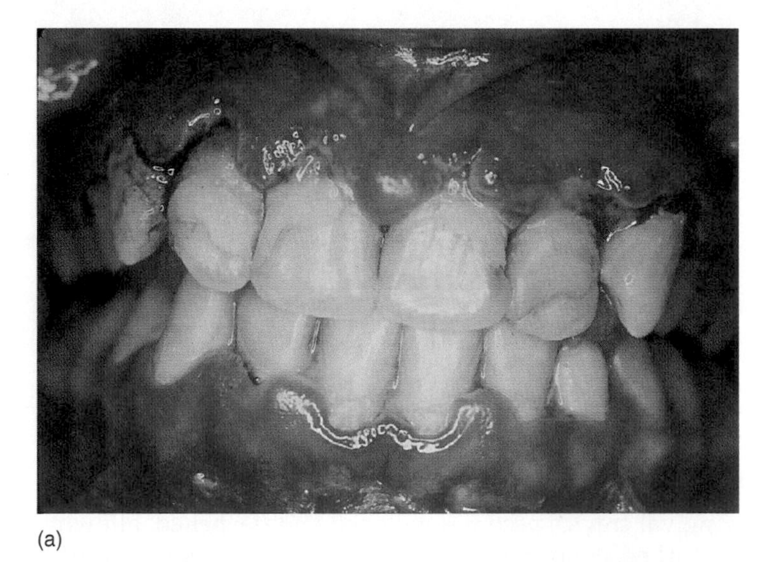

(a)

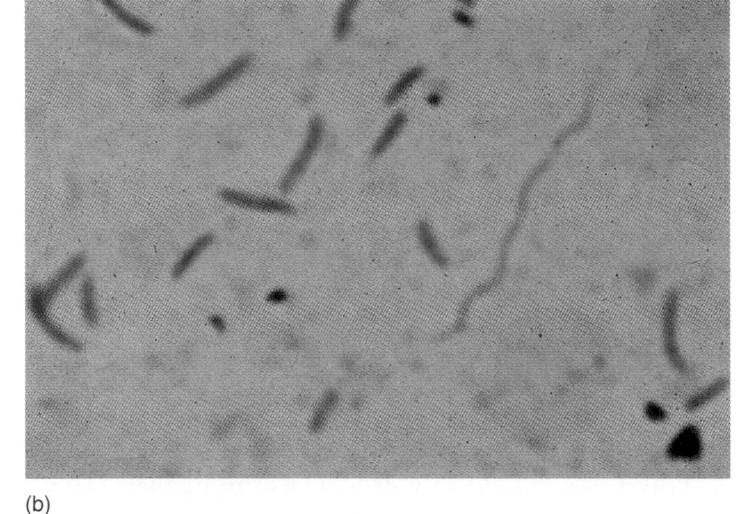

(b)

Figure 25.7
Acute necrotizing ulcerative gingivitis (ANUG). (*a*) Appearance of gingiva. (*b*) Gram stain of exudate.

loosening and loss of teeth, but it can be halted in its earlier stages by cleaning out the inflamed crevice and removing plaque and calculus. Careful flossing and toothbrushing can help prevent periodontal disease. In the future, a vaccine may become available to help prevent the disease. A vaccine against *Porphyromonas gingivalis*, one of the suspected causes of periodontal disease, completely prevented the disease in germ-free rats. *P. gingivalis* is an anaerobic, glucose nonfermenting, gram-negative rod formerly assigned to the genus *Bacteroides*.

Trenchmouth Is an Acute Form of Gingivitis with Bacterial Invasion

Trenchmouth (Vincent's disease, acute necrotizing ulcerative gingivitis, ANUG) is a severe, acute form of gingivitis (figure 25.7) characterized by tissue necrosis at the edge of the teeth, bleeding, pain, a foul odor, and fever. Trenchmouth is characterized by a marked overgrowth of plaque fusiforms and spirochetes (*Fusobacterium* and *Treponema* sp.) and an actual invasion of the tissue by spirochetes. It can occur at any age in association with poor mouth care, malnutrition, and stress. Trenchmouth responds to dental care and antibacterial medicines and is not contagious.

Oral Bacteria Can Cause Endocarditis, a Serious Heart Infection

Dental procedures such as cleaning or pulling teeth usually introduce oral bacteria into the bloodstream. Normally, these organisms are quickly removed by the body's mononuclear phagocyte system, but in individuals with abnormal heart valves, the bacteria can colonize the valves and cause bacterial endocarditis. To prevent endocarditis, dentists treat patients with known or suspected abnormal heart valves with an antibiotic to decrease the numbers of bacteria living around the teeth and to kill or inhibit the growth of any bacteria that enter the blood as a result of the dental procedure. An antibiotic such as penicillin results in a temporary drop in the numbers of oral bacteria, but in a few days the oral flora return to their normal levels even if the antibiotic treatment is continued. The rise in numbers of oral bacteria is due to growth of strains resistant to penicillin. This explains why, in attempting to prevent endocarditis, antibiotic medication is started only a few hours before a dental procedure, continued a day or two afterward, and then promptly discontinued.

REVIEW QUESTIONS

1. What role does saliva play in preventing dental caries?
2. Why is the distribution of bacterial species in the mouth not uniform?
3. Why must teeth be present in order for anaerobes to colonize the mouth?
4. What two attributes of *Streptococcus mutans* make it a good dental caries producer?
5. Give three methods for preventing dental caries.
6. What is periodontal disease?
7. What is acute necrotizing ulcerative gingivitis?

Viral Diseases of the Mouth and Salivary Glands

Some viral diseases commonly involve the mouth but produce more dramatic symptoms elsewhere in the body. For example, measles produces Koplik's spots in the mouth and a dramatic skin rash and respiratory symptoms; chicken pox causes oral blisters and ulcers and a characteristic skin rash; infectious mononucleosis can cause multiple oral ulcers and bleeding gums and also enlargement of lymph nodes and spleen. Coxsackieviruses can cause **hand, foot, and mouth disease,** with painful mouth ulcers and a rash of the palms and soles; **herpangina,** with painful oral ulcers alone; or **coxsackie parotitis,** with swollen, painful salivary glands.

In this section, we focus on **herpes simplex,** with its characteristically painful oral ulcers, and **mumps,** with its enlarged, painful parotid glands.

• **Endocarditis, pp. 510, 676–678**
• **Koplik's spots, pp. 491**

The Mouth and Lips Are Common Sites of Herpes Simplex Virus Infections

Herpes simplex is extremely widespread. Although usually insignificant, the disease can have tragic consequences. The virus can be transmitted by asymptomatic carriers.

Symptoms

Herpes simplex is a disease that commonly begins during childhood with fever and the formation of vesicles (small blisters) and ulcers on the mouth and throat that can be so painful that it is difficult to eat or drink. The characteristic lesion is a group of vesicles surrounded by an area of redness. The vesicles break within a day or two, leaving superficial painful ulcers that heal without treatment within about 10 days. Thereafter, the infection becomes latent and the affected person may have recurrent disease (herpes simplex labialis, "cold sores, fever blisters"), usually appearing on the lips (figure 25.8). The symptoms of most

• **Latent infection, pp. 327–328**

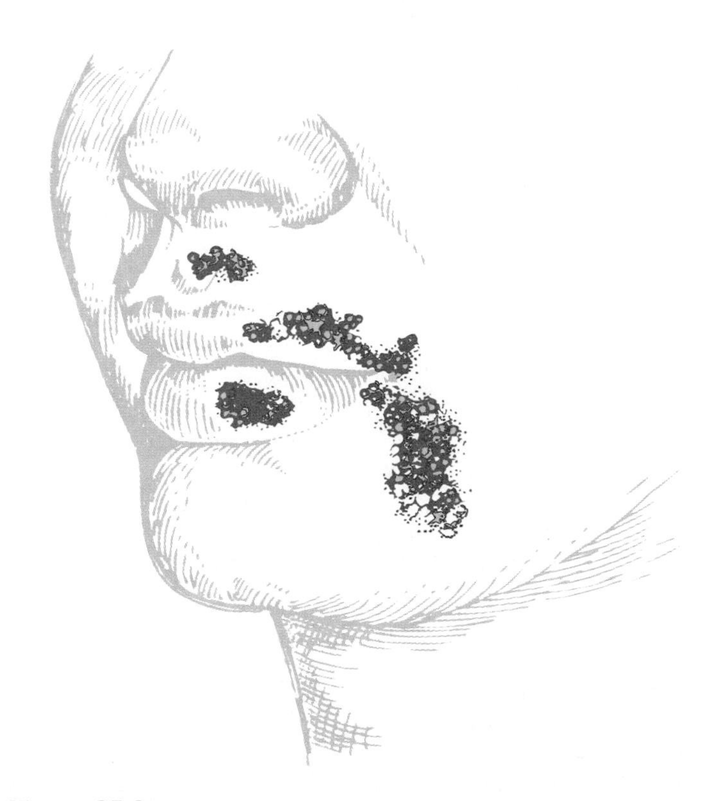

Figure 25.8
Herpes simplex labialis, also known as *cold sores or fever blisters.*

recurrences begin on the lips and include a tingling, itching, and burning sensation or pain that commonly lasts a few hours but sometimes two days or more. Sometimes no further symptoms develop, but usually these abnormal sensations are followed by the appearance of vesicles that generally ulcerate and then crust over within a couple of days. Pain is most severe during the first few days and generally is gone within a week. Healing occurs within 7 to 10 days.

Causative Agent

Herpes simplex is caused by the herpes simplex virus (HSV), a medium-sized, enveloped virus containing double-stranded, linear DNA. HSV can readily be cultivated in a variety of tissue cell cultures and is detected by using specific anti-HSV antibody. It is a member of the herpesvirus family, which includes varicella-zoster virus (the cause of chicken pox and shingles) and Epstein-Barr virus (the cause of infectious mononucleosis).

Pathogenesis

Following infection of the mouth or throat, the virus multiplies in the epithelium and destroys cells. Vesicles form that contain large numbers of infectious virions. Other epithelial cells fuse together, producing large, multinucleated giant cells. Nuclei of infected cells characteristically contain a deeply staining area called an **intranuclear inclusion body** (figure 25.9), which is the site of earlier viral replication. Giant cells and intranuclear inclusion bodies can be seen in stained matter on a microscope slide that has been pressed against an unroofed vesicle. Some of the virus particles are

• **Herpesvirus family, pp. 487–489**

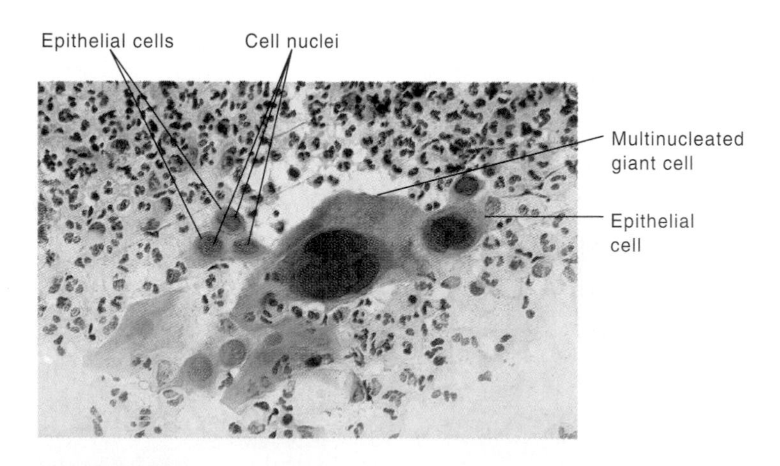

Figure 25.9
Stained smear of material from a herpes simplex lesion, showing a multinucleated giant cell and intranuclear inclusion bodies. The pink areas within the epithelial cell nuclei are intranuclear inclusion bodies.

carried by lymph vessels to the nearby lymph nodes. The immune response of the host quickly limits the infection by producing interferon as well as humoral antibodies and a cell-mediated immune response.

However, as with other herpesviruses, HSV establishes latent infections. Some of the virions produced by the acute infection enter the nerves that are responsible for sensation in the area. These sensory nerve cells are extremely long and thin. The nucleus and main cell body are located in the ganglion near the spine, while long extensions reach the tissues of the body and the spinal cord (figure 25.10). Viral DNA persists in these nerve cells in a noninfectious form. The normal circulation of material within the nerve cells can convey the latent virus to the nerve endings in the epithelium. There, the virus can revert to its infectious form and, if immunity is weak, destroy epithelium, producing recurrent disease. Certain stresses such as sunburn, illness

• **Interferon, pp. 184–185**

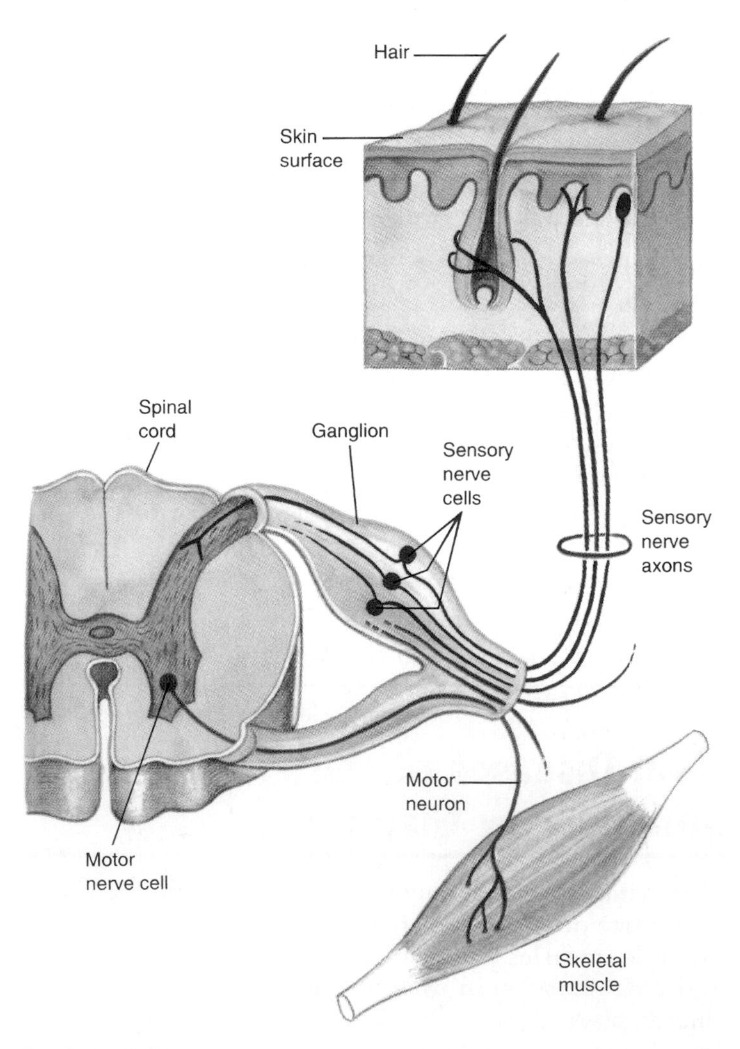

Figure 25.10
Ganglion of a sensory nerve, the site of latent herpes simplex virus infections.

with fever, and menstruation can trigger recurrences of herpes simplex.

Epidemiology

HSV is extremely widespread and infects up to 90% of some U.S. inner-city populations, usually resulting in mild, if any, symptoms. An estimated 20% to 40% of Americans suffer recurrent herpes simplex. The virus is transmitted primarily by close physical contact, although it can survive for several hours on plastic and cloth. The greatest risk of infection is from contact with lesions or saliva from patients within a few days of disease onset, because at this time large numbers of virions are present. However, small quantities of virus can often be demonstrated in the saliva of people with no symptoms who represent a risk to therapists such as nurses and dentists. HSV can infect almost any body tissue. For example, **herpetic whitlow,** a painful finger infection is not uncommon, and wrestlers can develop infections at almost any skin site because herpes-containing saliva can contaminate wrestling mats and get rubbed into abrasions. Recurrent disease involving the eye is the most common cause of corneal[1] blindness in the United States, and although uncommon, HSV is the most frequently identified cause of nonepidemic viral encephalitis, a serious brain disease.

Treatment and Prevention

Acyclovir inhibits HSV DNA polymerase and halts the progress of HSV disease, but it does not affect the latent virus and thus cannot rid the body of HSV infection. The medication is useful for treating severe cases and for preventing disabling recurrences. Since the ultraviolet portion of sunlight can trigger recurrent disease, sunscreens are sometimes helpful.

Mumps Is a Viral Disease that Characteristically Involves the Parotid Glands

Mumps results from an acute viral infection that spreads throughout the body, preferentially attacking certain types of glands.

Symptoms

The early symptoms of fever, loss of appetite, and headache are similar to many other viral diseases, but in the typical case of mumps painful swelling develops in one or both parotid glands (figure 25.11). Spasm of the underlying muscle makes it difficult to chew or talk. The word *mumps* probably derives from the verb *to mump*, meaning "to mumble" or "whisper." Symptoms of mumps usually disappear in about a week.

Although painful parotid swelling is the most common symptom of mumps, other symptoms can arise elsewhere in the body with or without parotid swelling. For example, headache and stiff neck indicative of meningitis are common symptoms, while up to half of the cases of mumps show no obvious parotid involvement. In these cases, the diagnosis of mumps virus infection needs to be made by cultures and serological tests.

The symptoms of mumps are generally much more severe in patients past the age of puberty. For example, about one-quarter of cases of parotitis in postpubertal men and boys are complicated by rapid swelling of one or both testicles to three to four times their normal size, accompanied by intense pain. Although shrinkage (atrophy) of the involved testicle commonly occurs after recovery from the acute infection, sterility is rare. In women and postpubertal girls, ovarian involvement occurs in about 1 of 20 cases, manifested by pelvic pain. Pregnant women with mumps commonly miscarry, but birth defects do not result from mumps as they do from rubella. Serious consequences of mumps are rare and are most likely to occur in older people. These consequences include deafness and death from encephalitis (brain infection).

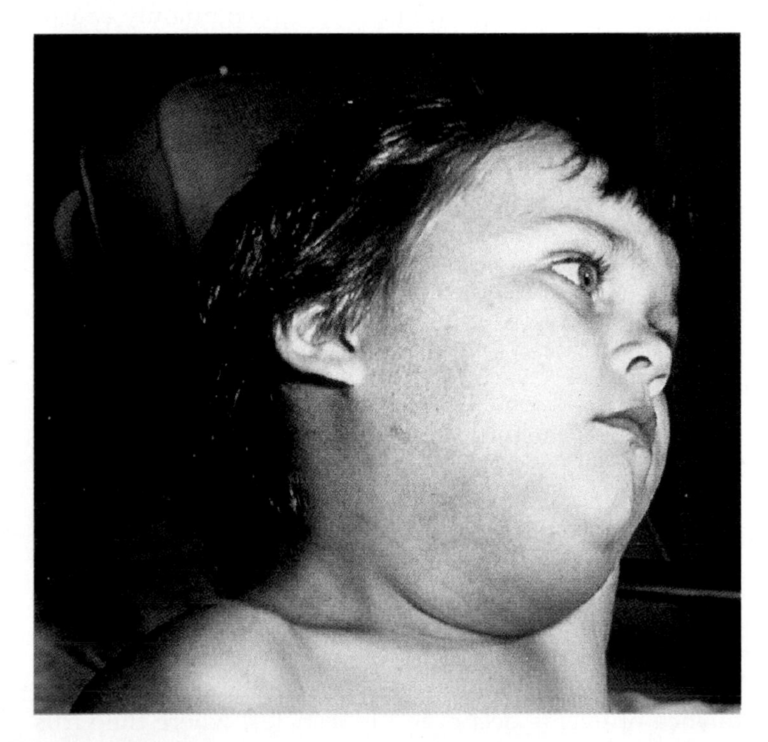

Figure 25.11
Child with mumps. The swelling directly below the earlobe is the enlarged parotid gland.

[1] Corneal blindness—blindness resulting from damage to the cornea, the clear anterior portion of the eyeball

• **Acyclovir, pp. 467, 490**

Perspective 25.4

Mumps Vaccine

A live attenuated mumps virus vaccine grown in chicken embryos was introduced in the United States in 1967. In that year, 185,691 cases of mumps were reported. Thereafter, with gradually increasing use of the vaccine, the incidence of mumps generally declined so that, by 1985, only 2,982 cases were reported. During this time, vaccine use was inconsistent and mainly involved children two to four years of age. Since there were few cases of mumps, many children reached adolescence and adulthood without exposure either to vaccine or an infectious case and therefore lacked immunity to mumps. Consequently, mumps epidemics began to appear in high schools and colleges. From 1985 to 1987, there was more than a sixfold increase in the incidence of mumps in 10 to 14 year olds and more than an eightfold increase in 15 to 19 year olds. The current recommendation is that everyone who has not had mumps, has not received the vaccine, or is unsure should be immunized. The exceptions are people born before 1957 (who are assumed to have had naturally acquired mumps infection); children less than 12 months of age (because transplacental mumps antibody might still be present and interfere with the vaccine); women who are pregnant; people with extreme allergy to egg protein or neomycin (because traces of these substances are present in the vaccine); and people with severe immune deficiencies, because their immune systems might not be able to restrain the growth of the vaccine virus. However, mumps vaccine is recommended for those infected with the AIDS virus, because the benefit far outweighs the risk from the vaccine.

Causative Agent

Mumps virus is similar in size to herpes simplex virus (about 200 nm diameter) and also has a lipid-containing envelope. Otherwise, the viruses are quite different. Mumps virus is classed in the paramyxovirus family, a group of single-stranded RNA viruses that includes rubeola (measles) virus; Newcastle virus, which causes diarrhea; and respiratory syncytial virus, which commonly causes lung infections in infants. Mumps virus is readily cultivated in tissue cell cultures and embryonated hens' eggs, where it can be easily detected by adding some red blood cells. The virus causes the red blood cells to clump (hemagglutinate) or adhere to infected tissue cell cultures. Only one antigenic type (serotype) of the mumps virus is known.

Pathogenesis

Infection occurs when virus-laden droplets of saliva are inhaled by a person who lacks immunity to mumps. The incubation period tends to be rather long (generally 15 to 21 days) because the virus reproduces in the respiratory tract, spreads throughout the body by the bloodstream, and produces symptoms only after infecting other tissues such as the parotid glands, meninges, or testicles. In the salivary glands, the virus multiplies in the epithelium of ducts that convey saliva to the mouth. This destroys the epithelium, which releases enormous quantities of virus into the saliva. The body's inflammatory response to the infection is responsible for the severe swelling and pain. A similar sequence of events occurs in the testicles, where the virus infects the system of tubules that convey the sperm. The marked swelling and pressure often impair the blood supply, leading to hemorrhages and death of testicular tissue. Not much is known about infection of other body structures, but kidney tubules are presumably infected since the virus can be cultivated from the urine for 10 or more days following the onset of illness. The immune system of the host eliminates the infection, and only rarely does significant permanent damage result.

• **Paramyxovirus family, p. 541**

Epidemiology

Humans are the only natural host of mumps virus, and natural infection confers lifelong immunity. Individuals sometimes claim to have had mumps more than once, probably because other infectious and noninfectious diseases can cause parotid swelling. The virus is spread by the high percentage of people (about 30%) who are asymptomatic and continue to mingle with other people while infectious. In symptomatic patients, the virus can be present in saliva from almost a week before symptoms appear to two weeks afterward, although peak infectivity is from one to two days before parotid swelling until the gland begins to return to normal size. The mumps virus does not cause latent infections.

Control

An effective live attenuated mumps vaccine has been available in the United States since 1967. Figure 25.12 shows how the incidence of mumps has generally declined, although it increased in the 1980s because of a decline in funding for vaccinations. Since its host range is limited to humans, there is only a single viral serotype, and latent recurrent infections do not occur, mumps should be a candidate for eradication just like smallpox, which was eliminated in 1977.

Perspective 25.5

Helicobacter Pylori and Stomach Cancer

Although the incidence of stomach cancer ranks 12th among cancers in the United States, it is second only to lung cancer worldwide. The cancer generally appears many years after initial exposure to unknown environmental factors during the early decades of life. More than one factor is involved in causing stomach cancer.

Considerable evidence exists that *H. pylori* infection increases the risk of stomach cancer. Areas of the United States and elsewhere that have a high incidence of stomach cancer also have a high incidence of *H. pylori* infection, including among the younger age groups. Compared to age-, sex-, and race-matched controls, a significantly higher percentage of stomach cancer patients were infected years earlier with *H. pylori.* Since only a small percentage of people with *H. pylori* infection develop stomach cancer and it can occur in the absence of infection with the bacterium, *H. pylori* is not the only factor involved in the development of stomach cancer. Nevertheless, *H. pylori* may be responsible for as many as 60% of stomach cancers. Proof of this possibility awaits studies showing that treatment of *H. pylori* infections reduces the risk of stomach cancer.*

*Excerpted from information appearing in 1991. *New England Journal of Medicine* 325:1127, 1132, 1170.

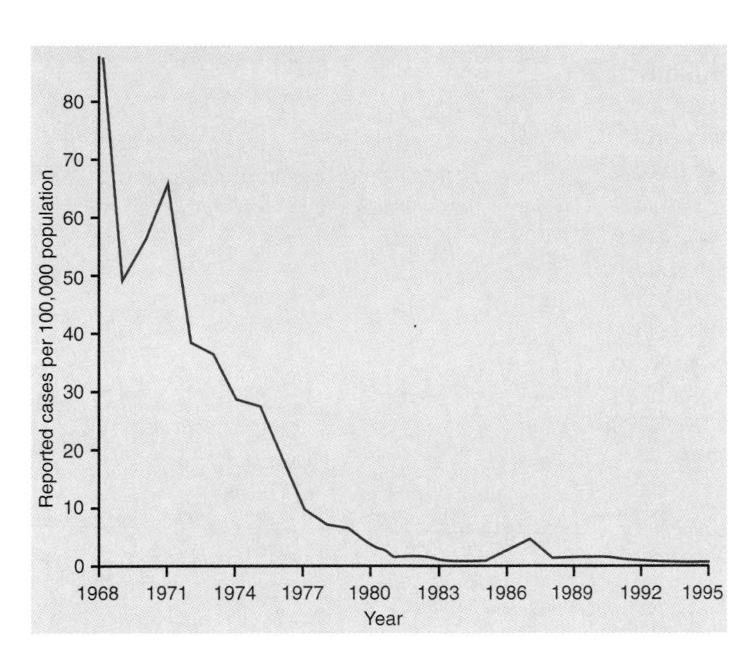

Figure 25.12
Cases of mumps in the United States, 1968–1995. Mumps vaccine was licensed in 1967.

Source: Centers for Disease Control and Prevention. 1996. Summary of Notifiable Diseases—United States, 1995. *Morbidity and Mortality Weekly Report* 44:45.

Infections of the Esophagus and Stomach

Less than two decades ago, infections of the esophagus and stomach were considered to be rare curiosities. Now esophageal infections are common because they are a complication of AIDS and the immunodeficiency of patients with organ transplants. Moreover, new techniques have revealed stomach infections with an unusual bacterium, *Helicobacter pylori,* to be common in otherwise healthy people.

Esophageal Infections Are Sometimes the First Indication of AIDS

Inflammation of the esophagus **(esophagitis)** is common and usually the result of acid reflux from the stomach. The symptoms range from discomfort after swallowing to severe heartburn or painful spasm. The same symptoms occur when the esophagus becomes infected in patients with immunodeficiency caused by AIDS or by the medications used to treat cancer or organ transplant patients. The two main infectious agents are herpes simplex virus and a common yeast of the normal flora, *Candida albicans.* Diagnosis is made by endoscopy[1] and culture.

Helicobacter Pylori Gastritis Is a Common Stomach Ailment

Stomach inflammation **(gastritis)** can be asymptomatic or produce symptoms ranging from belching and mild indigestion to severe mid- and left-upper abdominal pain. Various substances such as alcoholic beverages and aspirin can cause gastritis and, surprisingly enough, so can a species of bacterium, *Helicobacter pylori.*

[1]Endoscopy—inspection of an internal area of the body employing a flexible fiber-optic instrument or similar device

• *C. albicans,* pp. 232, 283

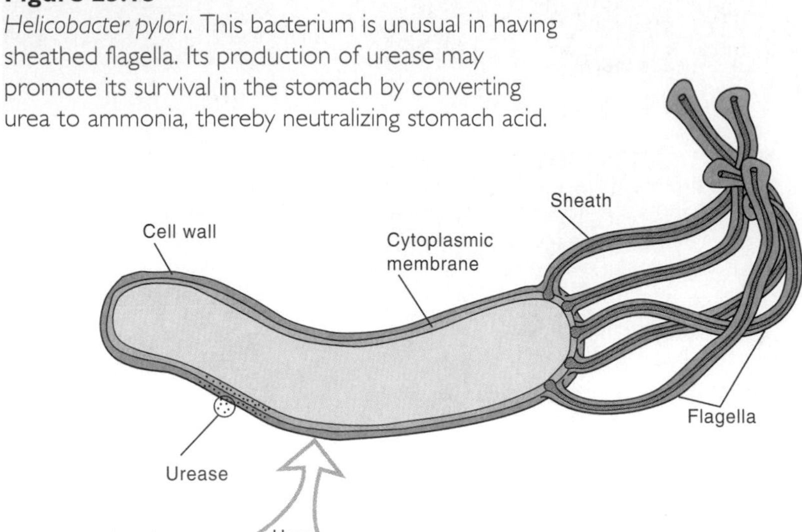

Figure 25.13
Helicobacter pylori. This bacterium is unusual in having sheathed flagella. Its production of urease may promote its survival in the stomach by converting urea to ammonia, thereby neutralizing stomach acid.

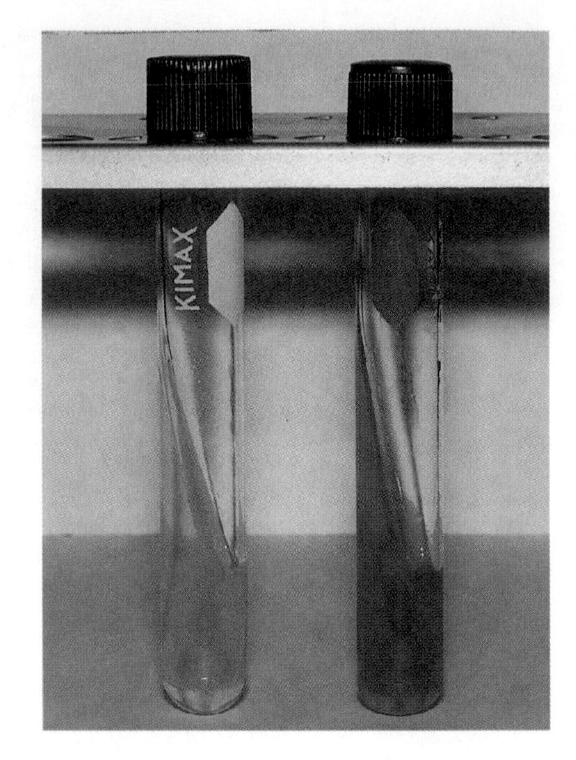

Figure 25.14
Positive urease test. A urease-producing organism converts urea in the medium to ammonia. The resulting alkalinity turns an indicator in the medium a red color.

Causative Agent

Helicobacter pylori had been seen in stained samples of stomach tissue obtained at autopsy many years ago, but it remained almost unknown to the medical world until its successful cultivation in 1982. It was first named *Campylobacter pyloridis* but later was placed in a new genus, *Helicobacter,* when marked differences from *Campylobacter* species were observed. The organisms are gram-negative and appear S-shaped on stained material, but they are actually short spirals. They have multiple polar flagella, which are unusual in that they possess a sheath (figure 25.13). The organisms are readily cultivated on specialized laboratory media, which suppress growth of other bacteria that might be present. *Helicobacter pylori* produces the enzyme urease, which converts urea to ammonia (figure 25.14). Urea is a waste product of protein catabolism by body cells and is present in gastric mucus.

Pathogenesis

These remarkable organisms use their flagella to corkscrew their way through the thick layer of gastric mucus and then attach to the epithelial cells of the stomach lining. Perhaps because of some toxic effect of the bacteria or the body's inflammatory response to them, mucus production decreases, causing a thinning or absence of mucus at the site of infection (figure 25.15). This decrease in protective mucus may be a factor in the development of peptic ulcers of the stomach and duodenum. Indeed, more than 75% of patients with gastric ulcer and 95% of those with duodenal ulcer are infected with *H. pylori.* Infection is associated with a rise in gastric pH, probably as a result of damage to the stomach lining and the organisms' degradation of urea to ammonia in the stomach. This series of events could result in spoilage of food retained in the stomach and account for the halitosis

Figure 25.15
Gastric ulcer formation associated with *Helicobacter pylori* infection.

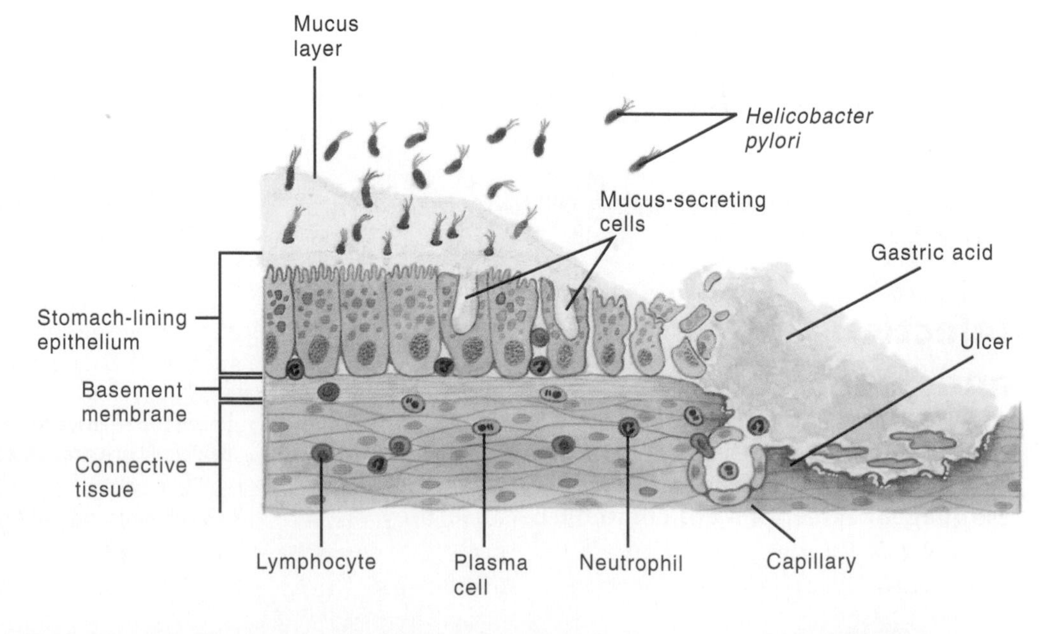

Case Presentation

The patient was a 35-year-old man who consulted his physician because of upper abdominal pain. The pain was described as a steady burning or gnawing sensation, like a severe hunger pain. Usually it came on one and one-half to three hours after eating, and sometimes it woke him from sleep. Generally, it was relieved in a few minutes by food or antacid medicines.

On examination, the patient appeared well, without evidence of weight loss. The only positive finding was tenderness slightly to the right of the midline in the uppermost part of the abdomen.

A test of the patient's feces was positive for blood. The remaining laboratory tests were normal.

Endoscopy showed a patchy redness of parts of the stomach lining. A biopsy was taken. The endoscopy tube was passed through the pylorus and into the duodenum. About 2 centimeters into the duodenum there was a lesion 8 millimeters in diameter that lacked a mucous membrane and appeared to be "punched out." The base of the lesion was red and showed adherent blood clot. After the endoscopy, a biopsy portion was placed on urea-containing medium. Within a few minutes, the medium began to turn color, indicating a developing alkaline pH.

1. What is the patient's diagnosis?
2. What would you expect microscopic examination and culture of the gastric mucosa biopsy to show?
3. Outline the pathogenesis of this patient's disease.
4. Why did it take so long for doctors to accept that this condition had an infectious etiology?

Discussion

1. This patient had a duodenal ulcer. The ulcer had penetrated deeply beyond the mucosa, involving small blood vessels and causing bleeding. This was apparent from the clot that was visualized at endoscopy and the positive test for blood in the stool.
2. Microscopic examination of the biopsy showed curved bacteria, confirmed by culture to be *Helicobacter pylori.*
3. The bacteria probably enter the gastrointestinal tract by the fecal-oral route. In the stomach, they escape the lethal effect of gastric acid because they produce urease. Highly motile, they enter the gastric mucus and follow a gradient of acidity ranging from pH 2 at the gastric lumen to pH 7.4 at the epithelial surface. Mutant strains that lack the ability to produce urease are only infectious if they are introduced directly into the mucus layer. Multiplication occurs just above the epithelial surface, but some of the bacteria attach to the epithelial cells and cause a loss of microvilli and thickening at the site of attachment. An inflammatory reaction develops beneath the affected mucosa. The products of two other genes, *vacA* and *cagA*, correlate with virulence. VacA is a cytotoxin similar to the adenylate cylcase of *Bordetella pertussis*. CagA is not itself cytotoxic but provokes a strong immune response. Once established, *H. pylori* infections persist for years and often for a lifetime. It is not known why some people develop gastric or duodenal ulcers and others do not. Both host and bacterial factors are almost certainly involved. For example, strains of *H. pylori* isolated from peptic ulcer patients tend to be more virulent than those from patients who just have gastritis; patients with blood group O have more receptors for the bacterium and a higher incidence of peptic ulcers than do other people. Stomach acid and peptic enzymes probably play a role in ulcer formation by acting on damaged epithelium unprotected by normal mucus.
4. Claude Bernard, a scientist of Pasteur's time, put it this way: "It is that which we do know which is the greatest hindrance to our learning that which we do not know." In 1983, when Dr. Barry J. Marshall proclaimed before an international gathering of infectious disease experts that a bacterium caused stomach and duodenal ulcers, everyone "knew" it could not be true because no organism could survive stomach acidity and enzymes. Indeed, almost everyone already "knew" the cause of ulcers to be psychosomatic. In the same year as Dr. Marshall's presentation, one scientist reported, "The critical factor in the development of ulcers is the frustration associated with the wish to receive love. When this wish is rejected, it is converted to a wish to be fed, (ultimately leading) to an ulcer." There may be other reasons Dr. Marshall's findings did not occur sooner and were not accepted more quickly, but the fact is that the spiral bacteria had been seen and largely ignored by a number of investigators many decades earlier.

that commonly occurs in infected people. *H. pylori* undoubtedly possesses other as yet unknown factors contributing to its virulence since the numerous other bacterial species that produce urease are unable to colonize the normal stomach.

Epidemiology

Infections by *H. pylori* appear to cluster in families. If one member is infected, then it is highly likely that the others are infected also. Overall, about 20% of the adult U.S. population is infected with the bacterium, and the incidence increases with age. About half the population over age 50 has antibody to *H. pylori*, indicating current or past infection. Organisms with a similar appearance occur in other animals, but their relationship to *H. pylori* is not yet defined.

Diagnosis

Infection is generally diagnosed by using a fiber-optic endoscope introduced into the stomach through the mouth. The instrument pinches off a small piece of the stomach lining, a portion of which is tested for the presence of urease. This test can be done quickly and easily and, if it is positive, the presence of the organism can be verified by culture and microscopic methods.

Treatment

H. pylori is susceptible to a number of antimicrobial medications when tested *in vitro*. However, as might be expected, the organisms respond poorly to treatment perhaps because the medicines are inactive in the harsh gastric environment. Interestingly, bismuth compounds, long used to treat gastric distress, are among the most effective treatments. They cause the organisms to be released from the epithelial cells and degraded by the stomach juices. Effective treatment of *H. pylori* infection substantially reduces the chance of recurrence of peptic ulcers.

REVIEW QUESTIONS

1. What risk might contact with the saliva from an apparently normal person hold for another person?
2. Why is the incubation period of mumps so long?
3. If a person is extremely allergic to eggs, should he or she be vaccinated against mumps? Explain.
4. The development of *Candida* esophagitis suggests the possibility of which underlying disease?
5. Name two of the stomach or upper small intestine diseases in which *Helicobacter pylori* is suspected of playing a causal role.
6. Give two distinctive characteristics of *H. pylori.*

Summary

I. Anatomy and Physiology
  - A. The upper alimentary system is composed of the mouth and salivary glands, esophagus, and stomach.
  - B. Degradation of complex foods begins in the mouth when the food is ground up by teeth and mixed with saliva. Saliva is produced by glands in the floor of the mouth and on the sides of the face. It contains the enzyme amylase that begins the degradation of starch. It protects the teeth by several mechanisms.
  - C. Three different kinds of cells line the stomach. One produces hydrochloric acid; another, enzymes; and another, mucus. One of the important stomach enzymes is pepsin, which begins the degradation of proteins under acid conditions.

II. Normal Flora
  - A. The species of bacteria that inhabit the mouth colonize different locations depending on their ability to attach to structures such as the tongue, cheeks, teeth, or other bacteria.
  - B. Dental plaque consists of enormous quantities of bacteria of various species attached to teeth or each other. Streptococci and actinomycetes are generally responsible for the initiation of plaque formation.
  - C. The presence of teeth allows for colonization by anaerobic bacteria.

III. Infectious Diseases Caused by Oral Bacteria
  - A. Dental caries is an infectious disease caused mainly by *Streptococcus mutans* and related species.
  - B. Pathogenesis of cariogenic dental plaque involves formation by *S. mutans* of extracellular glucans from dietary sucrose.
  - C. Penetration of the calcium phosphate tooth structure depends on acid production by cariogenic dental plaque. Prolonged acid production is fostered by the ability of *S. mutans* to form intracellular polysaccharide.
  - D. Control of dental caries depends mainly on supplying adequate fluoride and restricting dietary sucrose. Dental sealants are important in children.
  - E. Periodontal disease is mainly responsible for tooth loss in older people.
  - F. Endocarditis in patients with abnormal heart valves is commonly due to oral bacteria. Entry of these bacteria into the bloodstream frequently occurs with dental procedures.

IV. Viral Diseases of the Mouth and Salivary Glands
  - A. Herpes simplex is caused by an enveloped DNA virus that can be readily cultivated in tissue cell cultures. Infected cells show intranuclear inclusion bodies.
  - B. Latent infections are characteristic of herpesviruses. HSV persists inside sensory nerves in a noninfectious form that can become infectious and produce active disease when the body undergoes certain stresses.
  - C. Mumps is caused by an enveloped RNA virus that is prone to infect the parotid glands, meninges, testicles, and other body tissues.
  - D. Mumps virus generally causes more severe disease in persons beyond the age of puberty.
  - E. Mumps can be prevented using a live vaccine.

V. Infections of the Esophagus and Stomach
  - A. Esophageal infections are seen mainly in immunocompromised individuals. The main causes are herpes simplex virus and the yeast *Candida albicans.*
  - B. *Helicobacter pylori* gastritis is a common infectious disease of the stomach.
    1. *H. pylori* predisposes the stomach and upper duodenum to peptic ulcers.
    2. Treatment is difficult even though the organism is susceptible to antimicrobial medications *in vitro*. However, effective treatment prevents peptic ulcer recurrence.

Chapter Review Questions

1. What acid is produced when streptococci metabolize carbohydrates?
2. What component of tooth dentin is a nutrient for oral cavity bacteria?
3. What trace element is important for resistence of a tooth to dental caries?
4. What is the usual source of a child's *Streptococcus mutans?*
5. Name a heart disease caused by normal bacterial mouth flora.
6. What common viral infection of the mouth and throat causes intranuclear inclusion bodies?
7. Give three examples of stressful events that can reactivate herpes simplex.
8. At what stage of maturation of boys is mumps likely to be complicated by orchitis?
9. Give two infections of the esophagus that suggest immunodeficiency.
10. What enzyme produced in large quantity by *Helicobacter pylori* might contribute to a rise in gastric pH?

Critical Thinking Questions

1. Compare the difficulty of eliminating herpes simplex virus and that of eliminating mumps virus from the world.
2. Why is timing important in giving prophylactic antibiotics to persons with abnormal heart valves receiving dental treatment?

Further Reading

Briss, P. A., et al. 1994. Sustained transmission of mumps in a highly vaccinated population: assessment of primary vaccine failure and waning vaccine-induced immunity. *Journal of Infectious Diseases* 169:77–82.

1993. Dental health of schoolchildren—Oregon, 1991–92. *Morbidity and Mortality Weekly Report* 42(46):887–891.

Drumm, Brendan, et al. 1990. Intrafamilial clustering of *Helicobacter pylori* infection. *New England Journal of Medicine* 322:359–363.

Dubois, Andre. 1995. Spiral bacteria in the human stomach: the gastric helicobacters. *Emerging Infectious Diseases* 1(3):79–85.

Ginsberg, H. S. 1990. Herpesviruses. In B. D. Davis, R. Dulbecco, H. N. Eisen, and H. S. Ginsberg (ed.), *Microbiology,* 4th ed. J. B. Lippincott Co., Philadelphia, 929–945.

1996. Guidelines for school health programs to promote lifelong healthy eating. *Morbidity and Mortality Weekly Report* 45 (no. RR-9):1–41.

1990. Herpes gladiatorum at a high school wrestling camp—Minnesota. *Morbidity and Mortality Weekly Report* 39:69–71.

Kaplan, K. M., et al. 1988. "Mumps in the workplace: further evidence of the changing epidemiology of a childhood vaccine-preventable disease." *Journal of the American Medical Association* 260:1434.

Kolenbrander, P. E. 1988. Intergeneric coaggregation among human oral bacteria and ecology of dental plaque. *Annual Reviews of Microbiology* 42:627.

1995. Mumps surveillance—United States, 1988–1993. *Morbidity and Mortality Weekly Report* 44 (no. SS-3):1–14.

Richardson, Sarah. 1995. Ulcers from drinking? *Discover* 16 (10):38.

Robinovitch, M. R. 1994. Dental and periodontal infections. In K. J. Ryan (ed.) *Medical microbiology: an introduction to infectious diseases.* Appleton and Lange, Norwalk, Conn.

Shaw, J. H. 1987. Causes and control of dental caries. *New England Journal of Medicine* 317:996–1004.

Soll, A. H. 1990. Pathogenesis of peptic ulcer and implications for therapy. *New England Journal of Medicine* 322:909–916.

26 Lower Alimentary System Infections

KEY CONCEPTS

1. Some microbial toxins alter the secretory function of cells in the small intestine without killing or visibly damaging them.
2. Unsuspected human carriers can remain sources of enteric infections for many years.
3. Viruses, bacteria, and protozoa are important causes of human intestinal diseases.
4. Food poisoning can be caused either by infectious microorganisms consumed with the food or by toxic products of microbial growth in food.

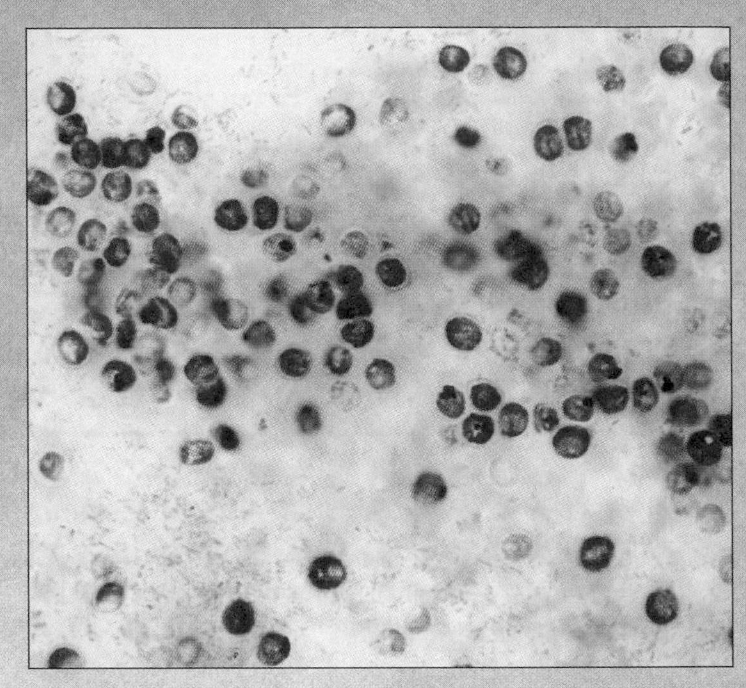

Cryptosporidium parvum oocysts in the feces of an infected individual (acid-fast stain). This protozoan can spread person to person in day-care centers, or infect thousands when municipal water supplies become contaminated. Persons with AIDS are especially prone to severe illness.

Microbes in Motion

The following books in the *Microbes in Motion* CD-ROM may serve as a useful preview or supplement to your reading: *Gram-Positive Organisms. Gram-Negative Organisms. Miscellaneous Bacteria. Microbial Pathogenesis in Infectious Diseases:* Toxins. *Vaccines:* Vaccine Targets. *Viral Structure and Function:* Viral Diseases. *Fungal Structure and Function:* Diseases. *Parasitic Structure and Function.*

A Glimpse of History

"The face was sunken as if wasted by lingering consumption,* perfectly angular, and rendered peculiarly ghastly by the complete removal of all the soft solids, in their places supplied by dark lead-colored lines. The hands and feet were bluish white, wrinkled as when long macerated in cold water; the eyes had fallen to the bottom of their orbes, and envinced a glaring vitality, but without mobility, and the surface of the body was cold."

This vivid description of cholera was written by Army surgeon S. B. Smith in 1832 when the disease first appeared in the United States. Cholera is a very old disease and is thought to have originated in the Far East thousands of years ago. Sanskrit writings indicate that it existed endemically in India many centuries before Christianity. With the increased shipping of goods and mobility of people during the 19th century, cholera spread from Asia to Europe and then to North America. Cholera was a major epidemic disease of the 19th century, appearing in almost every part of the world.

In 1854, John Snow, a London physician, demonstrated that cholera was transmitted by contaminated water. He observed that almost all people who contracted cholera got their water from a well on Broad Street. When the handle of the Broad Street pump was removed, people were forced to obtain their water elsewhere and the cholera epidemic in that area subsided. Snow's explanation was not generally accepted by other doctors mostly because disease-causing "germs" had yet to be discovered. It was not until 1883 that Robert Koch isolated *Vibrio cholerae*, the bacterium that causes cholera.

In the United States, cholera epidemics occurred in 1832, 1849, and 1866. The disease seemed to infect poorer people, those unfortunate enough to be crowded together in cities where cleanliness was impossible to maintain. The prevailing attitude was that the cholera victims themselves were at fault. In fact, the theme of many church sermons was that these people had sinned against God. The medical community was not any more enlightened. Many doctors were of the opinion that not only were the personal habits of these people "rash and excessive," but also that they insisted on taking the "wrong" medicines. However, by 1866, it was evident that where cholera appeared, the lack of sanitation rather than the morality of the victims was at fault. Public health agencies then played a major role in the elimination of epidemic cholera.

*Consumption—wasting of the body due to tuberculosis

The stomach guards the lower alimentary tract against infection by producing hydrochloric acid and pepsin which kill most microorganisms. However, a few pathogens can escape destruction particularly if large numbers of them are ingested. This is especially likely if the organisms enter the tract with food, which tends to protect them by partially neutralizing the acid and binding the digestive enzymes. Also, some microorganisms are quite resistant to digestion, and others bypass the stomach altogether, attacking the alimentary system via the bloodstream.

Anatomy and Physiology

The lower alimentary system is shown schematically in figure 26.1, along with its relationship to the upper alimentary tract. It consists of the small intestine, liver, pancreas, and large intestine (colon). Its most important functions include (1) degrading the macromolecules of food to simpler subunits; (2) absorbing these subunits, essential vitamins, and minerals into the bloodstream; (3) recycling the enormous quantities of water in the digestive secretions by reabsorption into the bloodstream; and (4) concentrating waste products for excretion. Two very important appendages, the liver and the pancreas, discharge their secretions into the upper part of the small intestine (duodenum).

The Liver Assists Digestion, Detoxifies Poisons, and Degrades Medications

The liver has many functions, one of which is to remove the breakdown products of hemoglobin from the bloodstream. These yellow-green products are excreted with the bile, the fluid produced by the liver. Interference with this excretory process, for example by infection or obstruction of bile ducts, can produce **jaundice** (a yellow color of the skin and eyes caused by bile pigments). The gallbladder is a saclike structure in which bile is concentrated and stored. **Cholecystitis,** infection of the gallbladder, is caused by several species of intestinal bacteria and is often life-threatening. Surprisingly, one feared pathogen, typhoid fever-causing *Salmonella typhi,* can colonize the gallbladder for years without causing symptoms. Since bile inhibits the growth of many bacteria, it can be put into media used to select for bile-resistant pathogens, such as *S. typhi*. Substances in the bile called **bile salts** act as detergents and help emulsify[1] fats in the intestine, which aids the absorption of fats and the fat-soluble vitamins A, D, E, and K. The normal brown color of feces results from the action of intestinal bacteria on bile.

The liver also inactivates potentially toxic substances that enter the bloodstream. For example, the ammonia produced by intestinal bacteria could poison the body if it were not detoxified by the liver. The liver also chemically alters many medications, such as the antibiotic erythromycin. The normal liver removes it from the bloodstream and excretes it in the bile in an inactive form. If the liver is severely damaged, as it might be by a viral infection, the dosage of such medications might have to be decreased to avoid the buildup of toxic levels.

[1]Emulsify—to produce an emulsion, which is a mixtue of two liquids not mutually soluble

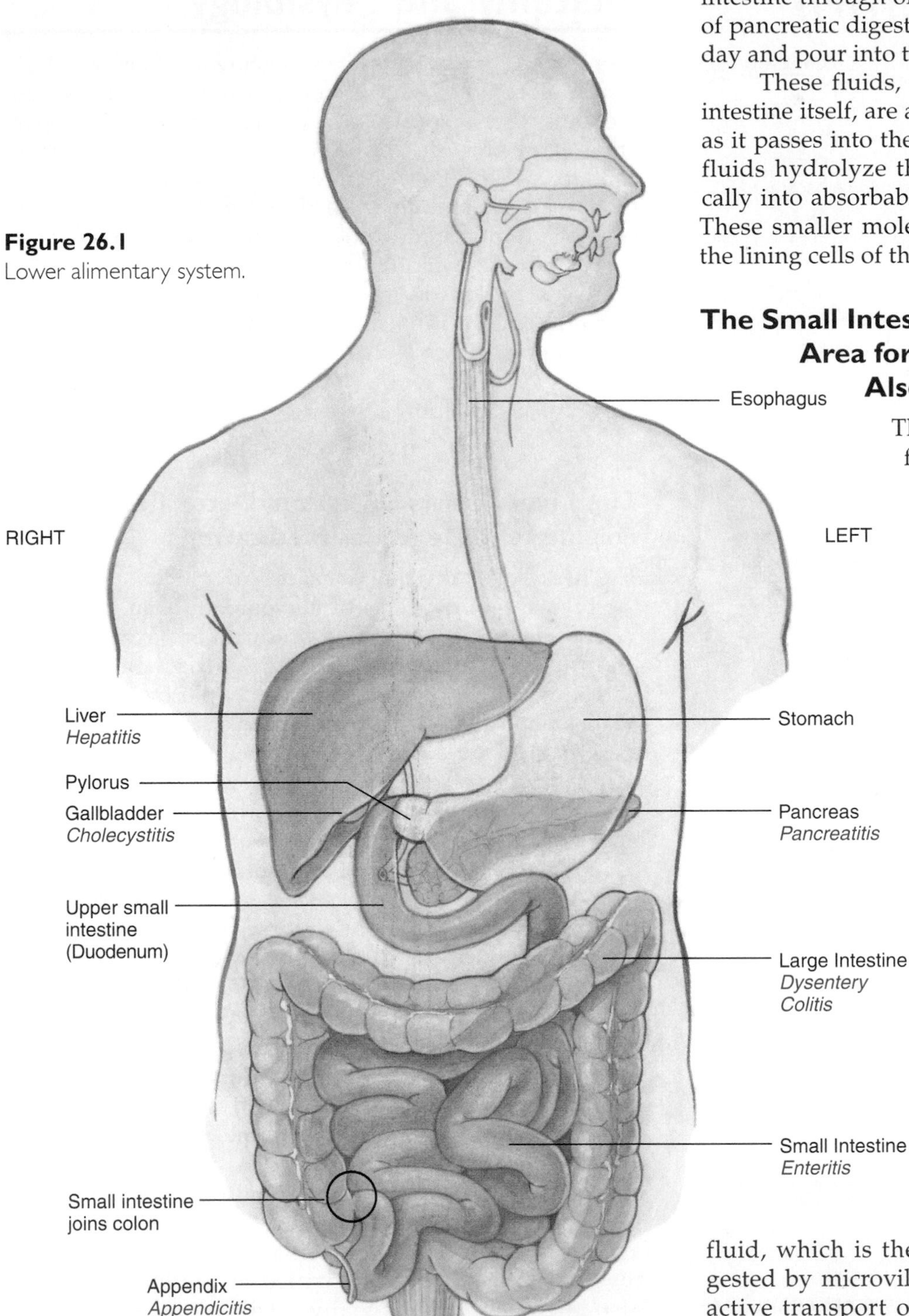

Figure 26.1
Lower alimentary system.

The Pancreas Produces Insulin and Digestive Enzymes

Some pancreatic cells release hormones such as insulin[1] into the bloodstream, while others produce enzyme-containing digestive secretions that are discharged into the intestine through one or more tubes. About 2,000 milliliters of pancreatic digestive juices join 500 milliliters of bile each day and pour into the upper portion of the small intestine.

These fluids, as well as digestive juices of the small intestine itself, are alkaline and neutralize the stomach acid as it passes into the intestine. The pancreatic and intestinal fluids hydrolyze the macromolecules of foods enzymatically into absorbable amino acids, fats, and simple sugars. These smaller molecules are then selectively taken up by the lining cells of the small intestine.

The Small Intestine Has an Enormous Surface Area for Absorbing Nutrients, but It Also Has Great Secretory Ability

The pyloric valve (pylorus) controls the flow of the well-mixed gastric contents from the stomach into the small intestine. The rate of passage slows down if there is a considerable amount of fat in the food or if it is very concentrated or dilute. As the food enters the small intestine, digestive enzymes and bile are added from the pancreas and liver.

Although it is only 8 or 9 feet long, the small intestine has tremendous surface area to facilitate nutrient and fluid absorption. The inside surface of the intestine is covered with many small, fingerlike projections called **villi,** each 0.5 to 1 millimeters long. Each of these villi are, in turn, covered with cells that have cytoplasmic projections called **microvilli** (figure 26.2). The total surface area of the small intestine is about 30 square feet. The microvilli contain most of the intestinal digestive enzymes. This large epithelial surface has both secretory and absorptive functions that go on simultaneously. Tiny intestinal glands continuously secrete large amounts of fluid, which is then reabsorbed along with the food digested by microvilli enzymes. These functions depend on active transport of nutrients and electrolytes across the plasma membranes of the epithelial cells. For example, sodium ion (Na^+) is taken up by the cell when glucose or amino acids are absorbed, while hydrogen ions (H^+) and bicarbonate ions (HCO_3^-) are secreted to adjust the pH of the intestinal contents. Movement of these substances is accompanied by water molecules and is controlled by hormones and nerve signals.

The intestinal epithelial cells play an indispensable role in digestion. Final digestion takes place as the cells secrete juices rich in enzymes such as dipeptidases (which hydrolyze peptides to amino acids) and disaccharidases (which similarly convert disaccharides such as sucrose and

[1]Insulin—hormone released into the blood from certain pancreatic cells; insulin deficiency results in diabetes mellitus

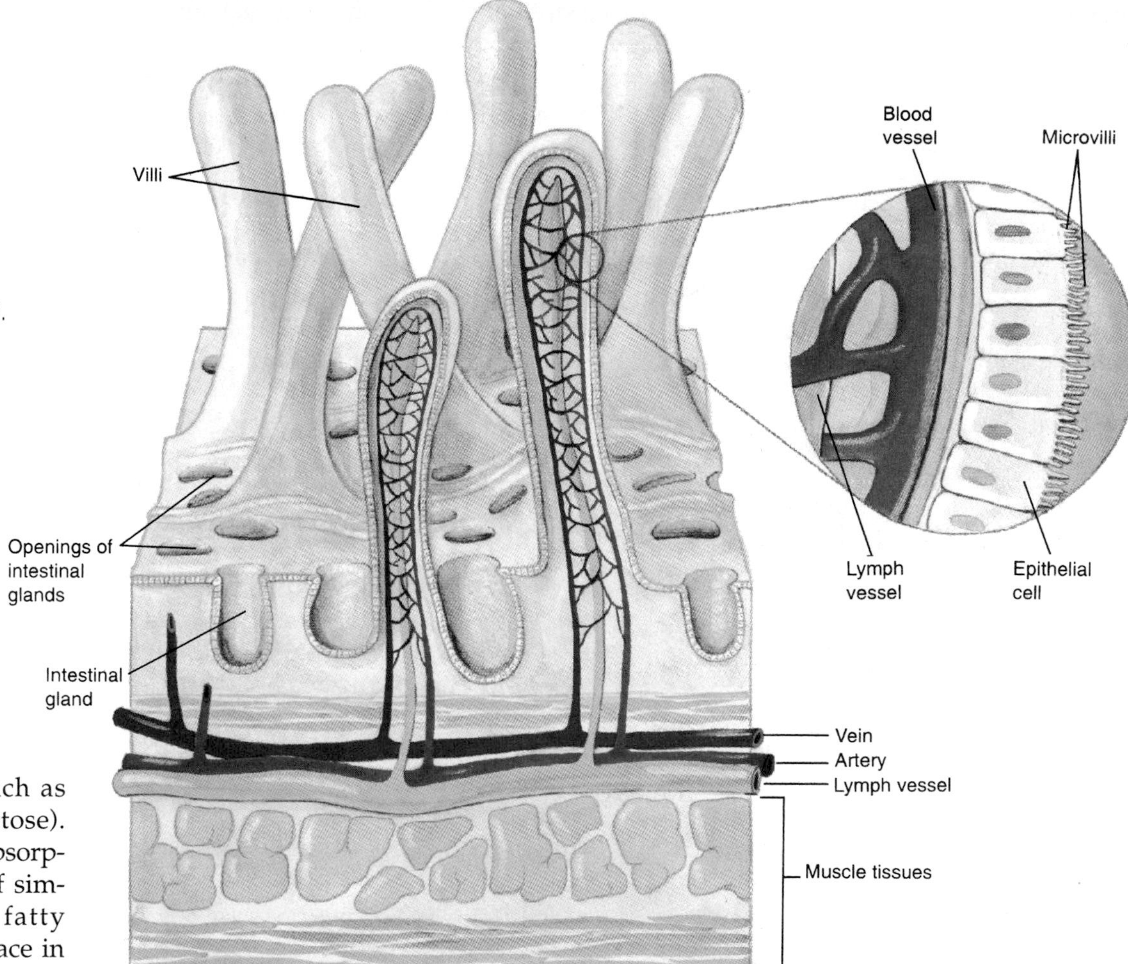

Figure 26.2
Small intestine; villi and microvilli.

lactose to simple sugars such as glucose, galactose, and fructose). The major part of nutrient absorption, including absorption of simple sugars, amino acids, fatty acids, and vitamins, takes place in the small intestine. Some nutrients, such as iron, can only be absorbed here. In addition, the intestine reabsorbs approximately 9 liters of fluid per day. The lining cells of the intestine are continuously shed and replaced by new cells, so that cells poisoned by a microbial toxin are soon replaced. Intestinal epithelial cells are completely replaced every nine days, one of the fastest turnover rates in the body.

Microbial diseases interfere with the functions of the small intestine by injuring or destroying the lining epithelium or by altering its ability to absorb and secrete. Some bacterial toxins markedly increase secretion without apparently damaging the epithelium. Failure of absorption or too much secretion can cause diarrhea and dehydration or, if chronic, vitamin deficiency and malnutrition. Small patches of lymphoid tissue (called Peyer's patches) are found on the walls of the small intestine. They are responsible for the immune response to invading microorganisms, but paradoxically they are selectively invaded by *S. typhi*.

A Major Function of the Large Intestine Is to Recycle Intestinal Water and Absorb Nutrients

Due to the great reabsorptive capacity of the small intestine, only 300 to 1,000 milliliters of gastrointestinal fluid normally reach the large intestine per day. From this volume of fluid, the large intestine absorbs water, electrolytes,[1] vitamins, and amino acids from foods. After this absorption, the semisolid feces, composed of indigestible material and bacteria, remain. The feces are an important source of opportunistic pathogens, particularly for the urinary tract and bloodstream. Infection of the large intestine often interferes with absorption and stimulates peristalsis,[2] resulting in diarrhea.

Normal Flora of the Digestive Tract

Only small numbers of bacteria live in the upper small intestine. The predominant organisms are usually aerobic and facultatively anaerobic, gram-negative rods and some streptococci. Lactobacilli are commonly found in small numbers, as are yeasts such as *Candida albicans*. The low concentrations of microorganisms result from the flushing

[1] Electrolytes—substances that when dissolved separate into ions that make the liquid electrically conducting; e.g., NaCl → Na^+ and Cl^-

[2] Peristalsis—progressive, wavelike contraction of intestinal walls that normally moves the contents in the proper direction but can result in painful abdominal cramps during intestinal infections

action produced by the rapid passage of digestive juices. There is a general increase in the bacterial population as the intestinal contents move toward the colon.

In contrast to the relatively scanty numbers of organisms in the stomach and small intestine, the colon (large intestine) contains very high numbers of microorganisms, approximately 10^{11} bacteria per gram of feces. In fact, bacteria make up about one-third of the fecal weight. The numbers of anaerobic organisms, notably including members of the genus *Bacteroides,* generally exceed other organisms by about 100-fold. Facultatively anaerobic, gram-negative rods, particularly *Escherichia coli* and members of the genera *Enterobacter, Klebsiella,* and *Proteus,* predominate among fecal microorganisms able to grow in the presence of air.

Intestinal Microorganisms Produce Vitamins and Inhibit Growth of Pathogens

Intestinal bacteria synthesize a number of useful vitamins, including niacin, thiamine, riboflavin, pyridoxine, cyanocobalamin (vitamin B_{12}), folic acid, pantothenic acid, biotin, and vitamin K. These vitamins are of enormous importance in the nutrition of some animals, but for humans, they are of doubtful importance except when the diet is inadequate. Thus, when an individual is poorly nourished and not getting adequate dietary vitamins, oral antibiotics can cause vitamin deficiency by suppressing growth of the intestinal flora.

The normal flora of the intestine also play an important role in preventing colonization of the colon by potential pathogens. In fact, a sometimes life-threatening disease called **antibiotic-associated colitis** (or pseudomembranous colitis) can follow the use of certain broad-spectrum antibiotics. The disease is caused by the growth of a toxin-producing anaerobic bacterium, *Clostridium difficile,* which readily colonizes the intestine of patients whose normal intestinal flora has been reduced by antimicrobial chemotherapy. The toxins of *C. difficile* are lethal to intestinal epithelium and cause small patches (pseudomembranes) of cell debris, inflammatory cells, and clotted serum to form on the colon's lining. Fever, abdominal pain, and profuse diarrhea result. Suppression of intestinal flora with antibacterial medicines can increase susceptibility to other pathogens, such as *Salmonella enteritidis* (discussed next).

Metabolic Effects of Intestinal Flora May Contribute to Human Discomfort and Disease

Some people suffer from a condition called the **blind loop syndrome** in which a pocket of the small intestine becomes relatively isolated from the rest of the organ as the result of disease or surgery. The flow of intestinal juices is reduced in the pocket and large populations of the intestinal bacteria flora build up. Affected individuals become anemic and lose weight because the large numbers of bacteria that accumulate in the loop deplete vitamins and degrade bile. Repairing the structural abnormality or giving antibiotics to reduce the numbers of organisms in the loop corrects the condition.

The enormous population of colon bacteria comprises more than 300 species that collectively can enzymatically change numerous materials. The currently recommended high-fiber diets contain substances that are indigestible by the gastric and intestinal juices but are readily degraded by intestinal bacteria, often with the production of large amounts of carbon dioxide, hydrogen, and methane gas. Abdominal discomfort and the passage of gas (flatus) from the rectum is the inevitable result. Also, bacterial enzymes can convert various substances to carcinogens and therefore may be involved, along with diet, in the production of intestinal cancer. Bacterial enzymes even have an important role in cholesterol production, and treatment with a variety of antibiotics can result in a temporary 30% drop in blood cholesterol.

- ***Bacteroides*, p. 249**
- **Vitamins, p. 110–11**
- ***Clostridium*, p. 260**

Infectious Intestinal Disease

Infectious intestinal disease is a major cause of death worldwide. In fact, in developing countries, one out of five children dies of diarrhea before the age of five. Each year, more than 5 million people die of diarrhea, 80% of whom are less than one year of age. Diarrhea also contributes significantly to malnutrition by impairing the absorption of nutrients. The situation becomes cyclical since malnourished individuals are more susceptible to diarrhea.

Microorganisms can cause diarrhea by at least two different mechanisms: toxin production and invasion of the intestinal epithelium. Some microorganisms use both mechanisms. The site of infection can influence the diarrheal symptoms. Generally, involvement of the small intestine results in a very watery diarrhea with resulting dehydration, whereas damage to the large intestine causes a bloody, pus-filled diarrhea known as **dysentery.**

In cases of severe watery diarrhea, up to 1 liter of fluid may be excreted per hour. This loss of fluids can reach 15% of the body weight, resulting in rapid dehydration. In some cases, much of the body's water and electrolytes are lost so that the blood becomes thickened and reduced in volume, producing insufficient blood flow to keep vital organs, such as the kidneys, working properly. Severe watery diarrhea can be rapidly fatal unless the lost fluid can be replaced promptly, ideally by intravenous treatment. This solution is often highly impractical when diarrheal epidemics occur in developing countries during the monsoon season, when it is difficult or impossible for most patients to get to a center for medical treatment. Fortunately, the ability of the small intestine to absorb fluid remains intact in most diarrheal diseases. An **oral rehydration solution (ORS)** was developed that consists of water, glucose, and electrolytes. With adequate instruction, administration of ORS is safer and cheaper than intravenous fluids. The glucose in the solution is essential for the absorption of the electrolyte solution by the intestinal cells. Packets of dehydrated ORS are distributed in developing countries with the cooperation of organizations such as UNICEF and the World Health Organization in order to decrease the mortality due to diarrhea. In the 1980s, evidence was presented that starchy substances such as rice powder can substitute for glucose in oral rehydration solutions. Rice powder is simpler to use, has a lower cost, and has improved caloric and nutritional value. The newer solutions can be readily absorbed because most diarrhea patients have sufficient enzymes to degrade the starch to simple sugars. Proteins and peptides are also present in grain products and can be degraded to amino acids, which also aid electrolyte absorption.

Bacterial Diseases of the Digestive Tract

With knowledge developed over the past few decades it has been possible to identify the pathogen in 80% or more of the infectious diarrhea cases. Species of bacteria often lead the list of causes, especially in locations where sanitation is poor.

Vibrio cholerae Causes Cholera, a Life-Threatening, Toxin-Mediated Diarrhea

There have been seven cholera pandemics since the early 1800s. Between 1832 and 1836, more than 200,000 Americans died when the second and fourth pandemics of cholera involved North America. The seventh pandemic began in 1961 in Indonesia, spreading to South Asia, the Middle East, and parts of Europe and Africa. By 1989, three fourths of the cases of cholera worldwide were African. South America had been cholera-free for a century when, in January, 1991, the disease abruptly appeared in Peru. The disease was probably introduced by a Chinese freighter's bilgewater discharged into a Lima harbor. The municipal water supply was not chlorinated and quickly became contaminated. The disease then spread rapidly, and by the end of 1992, 731,312 cases had been reported with 6,323 deaths through much of South and Central America. Cases appeared in 14 states in the United States in travelers and persons who ate seafood imported from Latin America.

Symptoms

Cholera is the classic example of a severe form of diarrhea. In cholera, the diarrheal fluid can amount to 20 liters a day. Because of its appearance, this watery fluid has been described as "rice water stool." Vomiting also occurs in most people at the onset of the disease, and many people suffer muscle cramps caused by loss of fluid and electrolytes.

Causative Agent

The cause is *Vibrio cholerae,* a curved, gram-negative, rod-shaped bacterium with a flagellum at one end. It is facultatively anaerobic and oxidase positive. It is able to tolerate strong alkaline conditions and high salt concentrations, allowing the use of selective culture media that inhibit other bacteria but not *V. cholerae.*

Pathogenesis

Vibrio cholerae adheres to the small intestinal epithelium by means of pili and probably also by other surface proteins. The organisms multiply on the epithelial cells (figure 26.3) but do no visible damage to them. However, the bacteria produce a potent enterotoxin[1] **choleragen** that is responsible for the symptoms of cholera. As with a number of other bacterial exotoxins, choleragen is heat labile,[2] and its protein

[1] Enterotoxin—exotoxin that acts on the intestine
[2] Heat labile—inactivated by heat

• **Vibrio, p. 248**

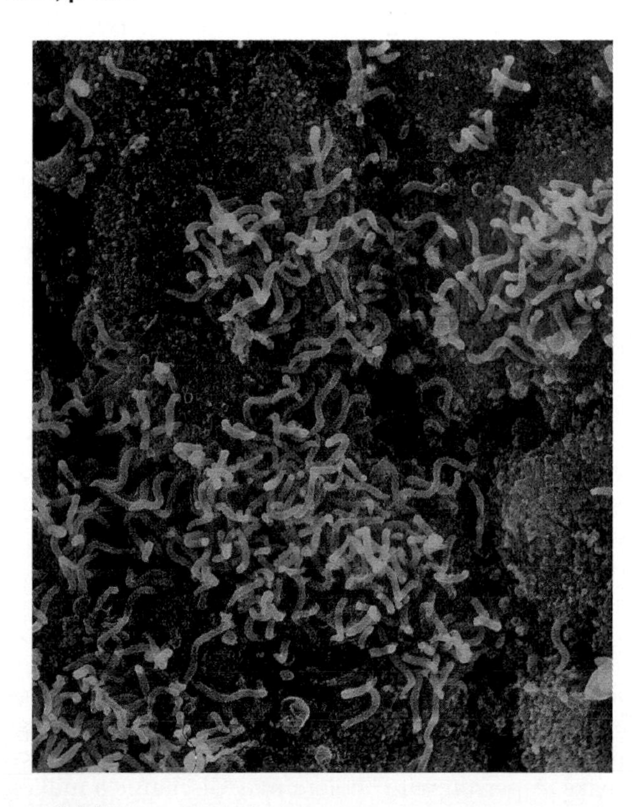

Figure 26.3
Scanning electron micrograph of *Vibrio cholerae* attached to intestinal mucosa (X1,200). Attachment occurs by means of pili.

Perspective 26.1

Toxigenic *Vibrio cholerae* O1 Infection Acquired in Colorado

On August 17, 1988, a middle-aged man was treated at a Colorado emergency department for profuse watery diarrhea, vomiting, and dehydration. Two days earlier he had eaten about a dozen raw oysters purchased from an oyster processing plant in the town. About 26 hours later, his symptoms suddenly began, and he passed 20 stools before seeking medical attention. A stool culture yielded toxigenic *V. cholerae* O1, biotype El Tor, serotype Inaba. The patient had no underlying illness, was not taking medications, and had not traveled outside the region during the preceding month.

Investigations revealed that the oysters were harvested August 8, 1988, off the coast of Louisiana and shipped by refrigerator truck to Colorado where they were maintained in disinfected artificial seawater baths before being packaged for sale. These oysters were probably the source of the man's infection. Bacteriophage typing showed the strain of *V. cholerae* to be identical to others recovered from infections originating in the Gulf of Mexico. Over the next three months, five additional cholera cases related to eating oysters were identified in five other states. Thorough cooking of shellfish is recommended to prevent acquiring infectious diseases.

Source: Centers for Disease Control and Prevention. 1989. *Morbidity and Mortality Weekly Report* 38:19.

molecule is composed of two parts, A and B. The B fragment has no toxic activity but binds irreversibly to specific receptors on the microvilli of the epithelial cells. The A fragment, responsible for toxicity, causes the activation of the enzyme adenylate cyclase, which converts ATP to cyclic adenosine monophosphate (cAMP). This is another example of ADP-ribosylation. In this case, the target of the toxin is a G protein that under normal conditions switches back and forth from active to inactive to regulate adenylate cyclase activity. ADP-ribosylation of this G protein makes it permanently active, thus causing maximal cAMP production by the cell. Accumulation of cAMP in the cell causes a markedly increased secretion of water and electrolytes. Although the colon is not affected by the toxin, it cannot absorb the huge volume of fluid that rushes through it, and diarrhea results. The normal shedding of intestinal cells eventually gets rid of the toxin.

In mid-1996, scientists at Harvard Medical School showed that choleragen was encoded by a filamentous bacteriophage that infects *V. cholerae* via the same pili involved in colonization of the intestine. The bacteriophage can integrate into the bacterial chromosome or replicate independently as a plasmid. Expression of both choleragen and the pili is regulated by the same bacterial gene, and expression is enhanced by in vivo conditions. Production of choleragen is another example of lysogenic conversion.

Epidemiology

Fecally contaminated water is the most common source of cholera infection, although foods such as crab and vegetables fertilized with human feces have also been implicated in outbreaks. A person with cholera may discharge a million or more *V. cholerae* organisms in each milliliter of feces. Although cholera is relatively common worldwide, no cases were known to have been acquired in the United States from 1911 to 1973. However, since 1973, sporadic cases have occurred along the Gulf of Mexico. Most of these infections have been traced to coastal marsh crabs that had been eaten. Bacteriophage typing showed that these cases were due to a strain of *V. cholerae* different from the pandemic strain, although both belong to the same (O1) serogroup. It is not known how long this strain has been in the United States or how it got here. It probably lives independent of humans in association with the crabs of coastal waters. Cholera cases due to this strain of *V. cholerae* average about three per year. On occasion, the pandemic strain has been introduced into waters along the Gulf Coast of the United States by ballast water discharged by freighters and taken up by oysters. In order to prevent transmission to humans, these oyster beds are monitored and closed down until negative for *V. cholerae.*

Ominously, in late 1992, a new strain of *V. cholerae* appeared in India and spread rapidly across South Asia, attacking even those people with immunity to the seventh pandemic strain. This new strain belongs to a different O group (O139), and its rapid spread suggests it could initiate another pandemic.

Treatment

Treatment of cholera depends on the rapid replacement of salts and water before irreversible damage to vital organs can occur. The prompt administration of intravenous or oral rehydration fluid decreases the mortality of cholera to less than 1%. Antibiotic therapy limits the duration of the illness somewhat, but the primary concern is replacement of lost fluids.

- **ADP-ribosylation, pp. 515, 532**
- **Lysogenic conversion, pp. 309, 508**
- **Bacteriophage typing, pp. 438, 478**

Prevention

Control of cholera is largely dependent on adequate sanitation and the availability of safe, clean water supplies. Travelers to areas where cholera is occurring are advised to cook food immediately before eating it. Crab should be cooked for no less than 10 minutes. No fruit should be eaten unless peeled personally by the traveler, and ice should be avoided unless it is known to be made from boiled water. A vaccine of killed *V. cholerae* is available that stimulates bactericidal and toxin-neutralizing antibodies in its recipients, but it has been only about 50% effective in preventing symptomatic cholera for three to six months. Well-nourished people living under sanitary conditions appear to be resistant to cholera. Cases are rare among the millions of Americans who travel in areas of the world where cholera is endemic. Cholera vaccination is not required for travelers entering the United States.

The main features of cholera are summarized in table 26.1.

Table 26.1	Cholera
Symptoms	Abrupt onset of massive, painless diarrhea, generally followed by vomiting without nausea, muscle cramps, and when severe, dehydration and shock; spontaneous recovery if fluid replacement is adequate
Incubation period	Short, generally 12–48 hours
Causative agent	*Vibrio cholerae*, a curved, gram-negative, rod-shaped, motile bacterium
Laboratory diagnosis	Recovery of *V. cholerae* from cultures of feces or vomitus on special media
Pathogenesis	Heat-labile enterotoxin causes excessive secretion of water and electrolytes by the intestinal epithelium
Epidemiology	Ingestion of fecally contaminated food or water; some natural sources, associated with marine crustaceans; asymptomatic carriers
Prevention	Purification of water and careful hand washing; vaccination gives short-lived, partial protection
Treatment	A solution of electrolytes and glucose, given intravenously in severe cases; or similar electrolyte solution containing a glucose source given by mouth in milder cases; orally administered tetracycline antibiotic

Epithelial Invasion and Destruction Characterize Dysentery Caused by *Shigella* Sp.

Shigellosis, a disease caused by enterobacteria of the genus *Shigella*, is distributed worldwide and is an important diarrheal illness wherever sanitary practices are lacking. Reported cases in the United States have averaged about 21,500 per year.

Symptoms

The classic symptoms of shigellosis are frequent liquid stools containing blood and pus, abdominal cramps, and a painful but unproductive urge to defecate. Other symptoms are headache, stiff neck, convulsions, and joint pain.

Causative Agent

Species of *Shigella* are gram-negative, nonmotile, rod-shaped, facultatively anaerobic enterobacteria. All virulent strains contain a large plasmid that is essential for the attachment and entry of *Shigella* into host cells. The bacteria typically do not ferment lactose, produce H_2S, or use citrate as a substrate. In addition, they are able to convert the amino acid tryptophan to indole. These characteristics (figure 26.4) help distinguish *Shigella* sp. from other lactose nonfermenting enterobacteria such as *Salmonella* sp.

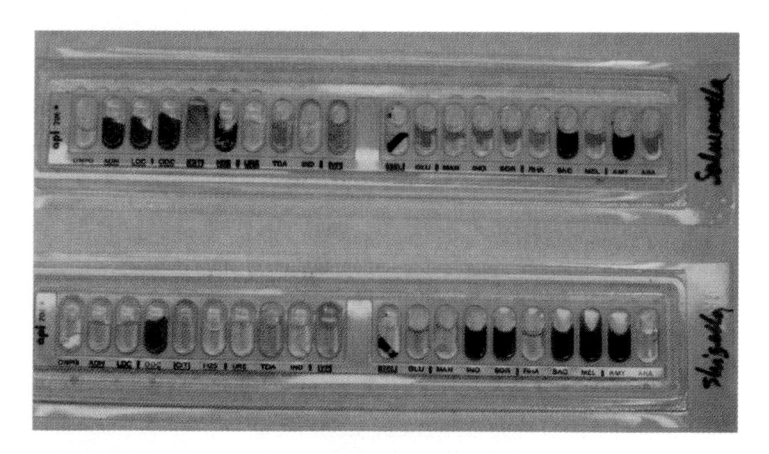

Figure 26.4
Comparison of biochemical reactions of *Salmonella* and *Shigella* on different substrates. The tests from left to right are ß-galactosidase; arginine dihydrolase; lysine and ornithine decarboxylase; citrate utilization; H_2S production; urease; tryptophan deaminase; indole; Voge-Prokauer; gelatin hydrolysis; and fermentations of glucose, mannitol, inositol, sorbitol, rhamnose, sucrose, melibiose, amygdalin, and arabinose.

Pathogenesis

Shigellas enter the mouth on fecally contaminated materials, are swallowed, survive passage through the stomach and small intestine, and establish infection in the colon. The colon tends to be progressively more heavily infected toward the rectum. Evidence indicates that the first step in

Perspective 26.2

Laboratory Diagnosis of Salmonellosis

The causative organisms of salmonellosis are usually present in the feces in large numbers. Since many species of *Salmonella* are relatively resistant to toxic chemicals, selective media containing such substances as bismuth sulfite or brilliant green dye can be used to suppress the normal fecal flora and allow the salmonellae to grow. Antisera are used to determine their cell wall (O) and flagellar (H) antigens when identifying serotypes.

Most *Salmonella* cultures can express two different kinds of flagella, a phenomenon known as *phase variation,* which is important in identifying individual strains of the bacteria. In any given *Salmonella* culture, almost all the bacterial cells have the same type of flagella, which can be called *H-1*. If an inoculum of the culture is mixed with anti-H-1 antiserum and placed in a localized area of a soft agar medium, then cells with H-1 flagella are immobilized by the antiserum, while cells with the alternative flagellar antigen (H-2) disperse throughout the medium and multiply. A subculture of these cells with H-2 flagella can be converted back to one with H-1 flagella by a similar technique. The two types of culture, called *phases,* along with somatic antigens help identify a *Salmonella* strain. For example, *S. choleraesuis* serotype *typhimurium* has O antigen 4, phase 1 antigen i, and phase 2 antigen 1, 2. *S. choleraesuis* serotype *choleraesuis* has O antigen 6, 7, phase 1 antigen c, and phase 2 antigen 1, 5.

Phase variation can be explained by the presence of two flagellar genes, *H-1* and *H-2,* on the *Salmonella* chromosome. The first gene to be transcribed (*H-1*) is linked to a repressor gene that blocks expression of *H-2*. Thus, if *H-1* is expressed, *H-2* is not. However, the *H-1* promoter region is flanked by inverted repeats that allow the promoter to be inserted in the bacterial chromosome in a reverse direction. When oriented in this way, the promoter is nonfunctional, preventing *H-1* expression and allowing expression of *H-2*.

The phenomenon of phase variation occurs in other species besides *Salmonella* and with other than flagellar antigens. It may provide pathogens a way of escaping the initial immune response of the host animal.

- **Inverted repeats, pp. 158**
- **Promoter region, pp. 136, 194**

the infection is phagocytosis of the bacteria by antigen-presenting cells associated with pockets of lymphoid tissue in the colon. These cells transport the bacteria beneath the epithelium where they are discharged by exocytosis and taken up by macrophages. Il-1 released by the infected macrophages causes an intense inflammatory reaction, with an outpouring of PMNs. The shigellas attach to specific receptors located on the colon epithelial cells at or near their bases. Attachment induces the process of endocytosis, which takes the bacteria into the epithelial cells. Intracellularly, they multiply at a high rate and kill the cells. Even though *Shigella* sp. are nonmotile, they are moved efficiently from cell to cell by a process first described in 1995. A chromosomal gene of the bacterium codes for a protein that causes the host cell's actin filaments to form progressively lengthening tail attached to one end of the bacterium. Actin tails push the bacteria from one cell to another.

The final result of the infectious process described is sloughed areas of epithelium surrounding Peyer's patches and other lymphoid tissues. The denuded areas are intensely inflamed, pus-covered, and bleeding, the source of blood and pus in the diarrheal stool.

Some strains of *Shigella dysenteriae* produce a potent toxin known as the **Shiga toxin.** Experimentally, this enterotoxin causes increased fluid secretion in the small intestine and produces paralysis. It is a chromosomally coded A-B toxin structurally similar to cholera toxin but otherwise unrelated. The toxin does not cause ADP-ribosylation. Instead, the A subunit reacts with the host cell ribosome, thereby halting protein synthesis. Mutants that lack ability to produce Shiga toxin are fully virulent. The importance of this toxin is that it has a strong association with the **hemolytic uremic syndrome,** a serious condition of humans that follows dysentery and often causes death from kidney failure. Shiga toxin is also produced by some strains of *Escherichia coli.*

Epidemiology

As with other diarrhea-causing bacteria, *Shigella* sp. are transmitted most readily in overcrowded populations with poor sanitation. The disease is highly communicable since ingestion of as few as 100 organisms can initiate infection. Humans are the only natural host of *Shigella,* but the organisms can easily be transmitted by direct person-to-person contact. In the United States, spread of *Shigella* is a common problem in day-care centers. Fecally contaminated food and water have also been implicated in outbreaks. Fortunately, the most virulent species, *Shigella dysenteriae,* is also the least prevalent.

Prevention

The spread of *Shigella* sp. is controlled by sanitary measures and surveillance of food handlers and water supplies.

Treatment

Antimicrobial medicines such as ampicillin and trimethoprim-sulfa are useful because they shorten the symptomatic period and the time during which shigellas are excreted. Plasmid-mediated antibiotic resistance is common in some areas of the world and requires that sensitivity testing be

Table 26.2 Shigellosis

Symptoms	Fever, diarrhea, vomiting, pus, and blood in feces; less frequently, headache, stiff neck, convulsions, and painful joints
Incubation period	3–4 days
Causative agent	Species of *Shigella*, gram-negative, nonmotile enterobacteria
Laboratory diagnosis	Isolation of organisms from cultures of feces
Pathogenesis	Invasion of and multiplication within intestinal epithelial cells; death of cells, followed by intense inflammation and ulcerations of intestinal lining
Epidemiology	Source: infected persons via fecal-oral route; sometimes transmitted by fecally contaminated food or water; asymptomatic carriers occur; humans generally the only source
Prevention	Sanitary precautions including careful hand washing
Treatment	Antibacterial medicines such as ampicillin and trimethoprim-sulfa shorten period of symptoms and time shigellas are excreted; multiple resistances due to R factors may be present

done to help choose the best antimicrobial drug. Two specimens of feces, collected at least 48 hours after stopping antimicrobial medicines, must show negative cultures for *Shigella* before a person is allowed to return to a day-care center or food-handling job.

Table 26.2 describes the main features of shigellosis.

Domestic Animals Are a Common Source of the *Salmonella* Strains that Cause Gastroenteritis

Salmonellosis, a disease caused by bacteria of the genus *Salmonella*, can originate from animals as diverse as chickens, rats, and turtles. On the average, about 47,500 cases are reported each year in the United States. However, most cases are not reported, and the actual number is estimated at 2,000,000 per year.

Symptoms

Symptoms including diarrhea, abdominal pain, nausea, and vomiting are indistinguishable from those of many other diarrheal illnesses commonly called *gastroenteritis*. The symptoms vary considerably depending on the virulence of the strain of *Salmonella* and the number of infecting organisms.

Causative Agent

Salmonella strains are gram-negative, facultatively anaerobic, rod-shaped enterobacteria. Generally, they do not ferment lactose, but they usually use citrate as their sole carbon source and produce H_2S. These features help distinguish *Salmonella* from other enterobacteria.

Although it is generally recognized that there is only one species of *Salmonella*, *S. choleraesuis*, it may be subdivided into over 2,000 serotypes. Many of these serotypes were formerly assigned species names that often reflected the place where they were first isolated, such as Dublin, Derby, and Heidelberg, to name but a few. These names are still commonly used to indicate specific serotypes.

Pathogenesis

Virulent salmonellas that survive passage through the stomach invade the lining of the colon and lower small intestine, but unlike shigellas they commonly penetrate the epithelium to the underlying tissues. The resulting inflammatory response increases fluid secretion and decreases absorption by the intestine. Exotoxins are produced by some strains, but their role in pathogenesis is not yet fully understood. A few strains tend to invade the bloodstream and tissues, producing abscesses, fever, and shock, sometimes with little or no diarrhea.

Epidemiology

Salmonellas generally enter the gastrointestinal tract with food, although fecally contaminated water, fingers, and other objects may also be sources. The food products most commonly contaminated with *Salmonella* strains are eggs and poultry. One of the largest outbreaks, traced to contaminated milk, occurred in 1985 with over 17,000 cases reported. Other outbreaks have resulted from contaminated brewer's yeast, protein supplements, and dry milk and even from a medication used to help diagnose intestinal disease. The reason for such a large variety of substances being contaminated with salmonellas is that salmonellas infect a wide variety of animals, from cows and baby ducks to insects and snakes. Generally speaking, in contrast to shigellosis, *Salmonella* gastroenteritis has an animal source rather than a human source.

Prevention and Treatment

Control of *Salmonella* infections depends on a reporting system for salmonellosis, tracing of sources by careful identification of individual strains, and routine sampling of animal products for contamination. Adequate cooking is very important, especially of frozen fowl, since heat penetration to the center of the carcass may be inadequate to kill salmonellas even when the outside appears done. Unfortunately, the number of reported cases of salmonellosis has continued to rise, and the organisms have also shown increasing plasmid-mediated resistance to antimicrobial medicines.

Perspective 26.3

Salmonellosis Due to *Salmonella Enteritidis* in Raw Eggs

Salmonella enteritidis is the most frequently reported *Salmonella* serotype in the United States. One of the outbreaks of *S. enteritidis* infection reported in 1991 involved 15 patrons and 23 of the 78 employees of a restaurant. The main symptom was diarrhea, associated in most cases with intestinal cramps, fever, chills, and nausea. The 13 patrons who sought medical care had fecal cultures positive for *S. enteritidis*. Development of the illness was strongly associated with eating Caesar salad. The dressing for this salad was made from raw eggs, olive oil, anchovies, garlic, and warm water and was kept in a cool container near the salad during the 8 to 12 hours the restaurant was open. Bacteriophage typing indicated that the epidemic strain of *S. enteritidis* was the same as one isolated from the restaurant egg supply.

From 1985 to 1991, 375 outbreaks of salmonellosis caused by *S. enteritidis* were reported, totalling 12,784 cases, of which 1,508 were hospitalized and 49 died. Many of the cases originated from raw or inadequately cooked eggs, which are commonly used in foods such as homemade ice cream, hollandaise sauce, and Caesar salad dressing. Salmonellosis tends to be severe in infants and the elderly, who should avoid food prepared from eggs unless they have been pasteurized or thoroughly cooked. Since early 1990, any flock of chickens that produces eggs associated with an *S. enteritidis* epidemic is examined by the U.S. Department of Agriculture, and if cultures are positive for the organism, the flock is destroyed or their eggs are pasteurized. Also, starting in 1991 all eggs must be refrigerated during interstate shipping.

Source: Centers for Disease Control and Prevention. 1992. *Morbidity and Mortality Weekly Report* 41: 369.

Table 26.3 Salmonellosis

Symptoms	Diarrhea and vomiting; rarely, abscesses, fever, and shock
Incubation period	6–72 hours
Causative agent	*Salmonella* sp., motile, gram-negative, lactose nonfermenting, enterobacteria
Laboratory diagnosis	Culture of feces; special media used to suppress normal fecal flora and allow salmonellas to grow
Pathogenesis	Invasion of the lining cells of colon and lower small intestine, with penetration to underlying tissues; body's inflammatory response causes increase in fluid secretion and decrease in intestinal fluid absorption
Epidemiology	Ingestion of food contaminated by animal feces, especially poultry
Prevention	Adequate cooking and handling of food
Treatment	Usually no antimicrobial needed unless invasion of tissues or blood occurs

This resistance is partly due to selection of resistant strains of *Salmonella* by the widespread, ill-advised use of antibiotics such as tetracycline in animal feeds. Fortunately, most people with salmonellosis recover spontaneously and do not require treatment with antimicrobial medicines.

Table 26.3 summarizes some of the features of salmonellosis.

Gastroenteritis Is Commonly Caused by Certain Strains of *Escherichia coli*

Escherichia coli, a lactose-fermenting, gram-negative enterobacterium, is an almost universal member of the normal intestinal flora of humans (and a number of other animals). Long ignored as a possible cause of gastrointestinal disease, it was not until the early 1960s that strains of *E. coli* were shown to cause life-threatening epidemic gastroenteritis in infants. Certain *E. coli* strains also cause gastroenteritis in adults, notably as the agents responsible for many cases of traveler's diarrhea (also called "Delhi belly," "Montezuma's revenge," "Turkey trots," and the like).

Symptoms

Symptoms depend largely on the virulence of the infecting *E. coli* strain. They range from mild diarrhea to a cholera-like picture to frank dysentery.

Causative Agent

There are at least four groups of *E. coli* implicated in diarrhea: **enterotoxigenic, enteroinvasive, enteropathogenic,**

Table 26.4 *Escherichia coli* **Gastroenteritis**

Symptoms	Vomiting and diarrhea; sometimes dysentery
Incubation period	2 hours–6 days
Causative agent	Certain strains of *E. coli*, typically lactose-fermenting, motile enterobacteria
Laboratory diagnosis	Culture of feces followed by serologic identification of *E. coli* isolates
Pathogenesis	Various mechanisms; attachment to small intestinal cells allows colonization; some strains produce one or more enterotoxins; some strains, resembling shigellas, invade colon epithelium; others cause host cell membrane thickening and loss of microvilli
Epidemiology	Common in travelers; can be foodborne or waterborne; fecal-oral route transmission
Prevention	Sanitary precautions including careful hand washing; bismuth compounds
Treatment	Replacement of fluid loss; treatment with antibiotics such as gentamicin or polymyxin for infants

and **enterohemorrhagic** *E. coli.* In contrast to other strains of *E. coli,* strains in those groups possess virulence factors that allow them to be intestinal pathogens.

Enterotoxigenic *E. coli* (ETEC) is a common cause of traveler's diarrhea and diarrhea in infants. Some members of this group are responsible for significant mortality in young livestock due to diarrhea. ETEC strains usually possess adhesins that allow them to adhere to and colonize the intestinal epithelium where they secrete one or more toxins. One such toxin is heat-labile and similar to cholera toxin in action and antigenicity.

Enteroinvasive *E. coli* (EIEC) infections result in a disease closely resembling that caused by *Shigella* sp.

Enteropathogenic *E. coli* (EPEC) strains cause diarrheal outbreaks in hospital nurseries and chronic diarrhea in children. They possess plasmid-dependent adhesins and cause loss of microvilli and a thickening of the cell surface at the site where the organisms attach. Enteropathogenic *E. coli* strains are easily distinguished from other *E. coli* strains by using fluorescent antibodies to identify their somatic and capsular antigens.

The fourth group of diarrhea-producing strains, enterohemorrhagic *E. coli,* was discovered in 1982. Almost all of these organisms belong to a single serological type O 157:H7. They often produce severe illness including bloody diarrhea. This strain has been shown to produce a potent group of toxins that cause the death of intestinal epithelium by interfering with protein synthesis. The toxins are closely related to Shiga toxins found in some *Shigella* strains. However, unlike shigellas, enterohemorrhagic *E. coli* generally do not penetrate intestinal epithelial cells. Their toxin production depends on lysogenic conversion by a distinct bacteriophage. Many infected patients develop **hemolytic-uremic syndrome,** marked by lysis of red blood cells and kidney failure. Fatalities are common in infants and the elderly.

Outbreaks of O 157 *E. coli* disease have occurred commonly in Canada and Great Britain, as well as in a number of states in the United States. A large 1993 epidemic in Washington state was traced to inadequately cooked hamburgers served by a fast-food restaurant chain. Cattle have been the source of a number of epidemics.

Pathogenesis

Hundreds of different strains of *E. coli* exist, but only those possessing certain virulence factors cause gastrointestinal disease. Some of these factors were discussed in the previous section; there is strong evidence that not all factors have been identified. Two important virulence factors, enterotoxin production and the ability to adhere to the small intestine, are coded by plasmids. These plasmids can be transferred to other *E. coli* organisms by conjugation, through which virulence is conferred on the recipient strain. Gastroenteritis-producing strains of *E. coli* often have more than one type of virulence plasmid. It is interesting that the heat-labile toxin of *E. coli* is antigenically closely related to the choleragen of *Vibrio cholerae.* This similarity suggests that the genes responsible for the two toxins have a common ancestry.

Treatment

Treatment of *E. coli* gastroenteritis includes replacing the fluid loss resulting from vomiting and diarrhea. In addition, infants may require antibiotics such as gentamicin or polymyxin for a few days. Traveler's diarrhea can be prevented with bismuth preparations (such as Pepto-Bismol) or an antibiotic such as tetracycline if the *E. coli* strains in the geographic area visited are sensitive to the antibiotic. However, the widespread use of an antibiotic to prevent diarrhea promotes the appearance of resistant strains by fostering the spread of R plasmids and should not be used routinely.

Some features of *E. coli* gastroenteritis are summarized in table 26.4.

- **Conjugation, p. 174**
- **Plasmids, p. 173**
- ***E. coli* O157: H7, p. 237**

In the United States, *Campylobacter jejuni* Is One of the Common Bacterial Causes of Diarrhea

It was not until 1972 that *Campylobacter jejuni* was isolated from a diarrheal stool. It took another five years for a suitable culture medium to be developed and widely used, which led to its recognition as a common pathogen.

Symptoms

Symptoms are similar to those with *Shigella* sp.: fever, abdominal cramps, and diarrheal stools containing blood and pus.

Causative Agent

C. jejuni (figure 26.5) is a small, gram-negative, comma- or S-shaped microaerophilic bacterium that requires a special selective medium for its isolation from stool specimens.

• ***Campylobacter*, p. 245**

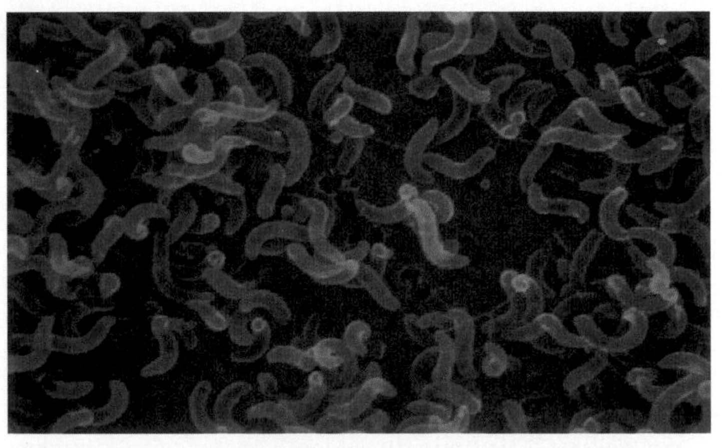

(a)

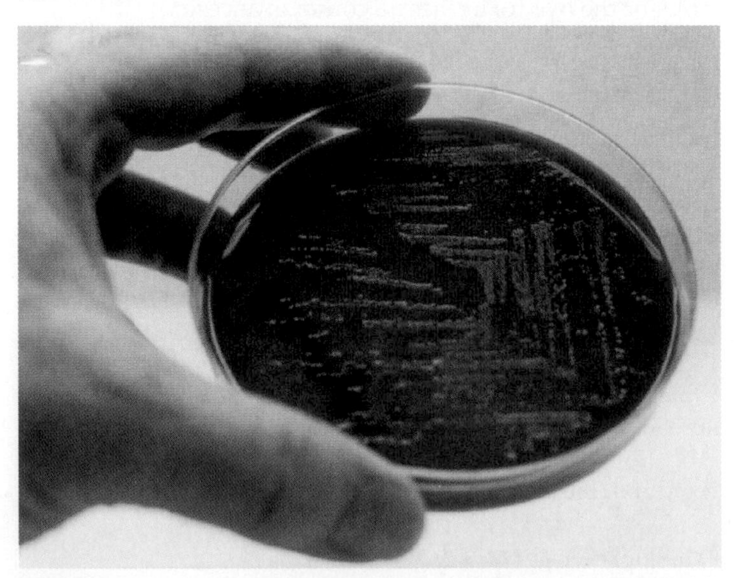

(b)

Figure 26.5
Campylobacter jejuni. (*a*) Scanning electron micrograph (×10,000). (*b*) Growth on a special medium under microaerophilic conditions.

Pathogenesis

The number of organisms required to produce symptomatic infection is 500 or less. The bacteria penetrate the intestinal epithelial cells of the small and large intestine, multiplying within and beneath the cells, causing an inflammatory reaction. As with most strains of *Shigella* and *Salmonella*, penetration into the bloodstream is uncommon.

Epidemiology

Numerous foodborne and waterborne outbreaks of *C. jejuni* have been reported; one, in 1978, involved an estimated 3,000 people. Like the salmonellas, *C. jejuni* lives in the intestines of a variety of domestic animals. Poultry is a common source of infection, and as many as 89% of raw poultry products can harbor the organisms. Cats have also frequently been implicated as a source. Large epidemics have resulted from ingesting unpasteurized cow and goat milk and from drinking nonchlorinated surface water. Person-to-person transmission is uncommon.

Prevention

Prevention of outbreaks depends mostly on pasteurization of milk and on chlorination of water. Proper cooking and handling of raw poultry so as to avoid contamination of hands and kitchen surfaces can prevent most of the other cases.

Treatment

Most cases of *C. jejuni* diarrhea subside without antimicrobial treatment in 10 days or less. The antibiotic erythromycin is recommended for severe cases.

Typhoid Fever Is an Example of "Enteric Fever," a Serious Form of *Salmonella* Infection

Typhoid fever occurs only in humans. It is an example of "enteric fever," which can sometimes be caused by a few other *Salmonella* serotypes. Once very common in the United States, reported cases of typhoid fever now average about 450 per year.

Symptoms

The disease is characterized by fever that increases over several days and by severe headache and abdominal pain. In some cases, the intestine ruptures, causing bleeding and shock. If treatment is not given, about one in five affected people dies of the disease.

Causative Agent

Typhoid fever is caused by *Salmonella typhi*, a serotype that has distinctive biochemical features. It differs from other salmonellas in that it does not produce gas from glucose and forms little or no H_2S. It has only one phase antigen and possesses a capsular antigen called *Vi* (figure 26.6). Different strains are distinguished by their susceptibility to various bacteriophages.

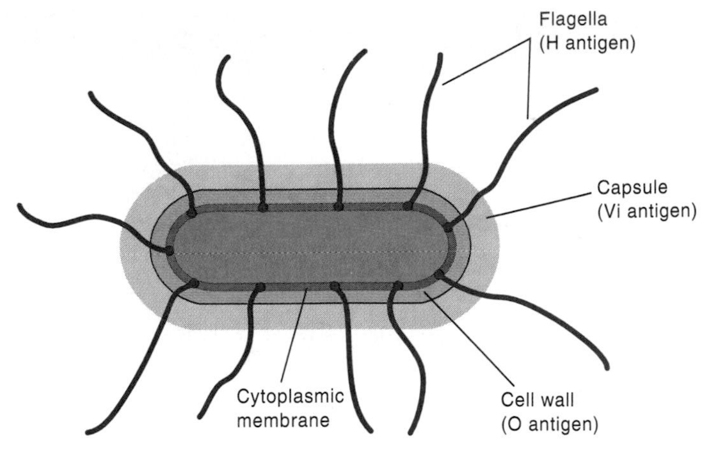

Figure 26.6
"Salmonella typhi" principal antigens. The Vi antigen is associated with the virulence of this *Salmonella* serotype, a cause of enteric fever.

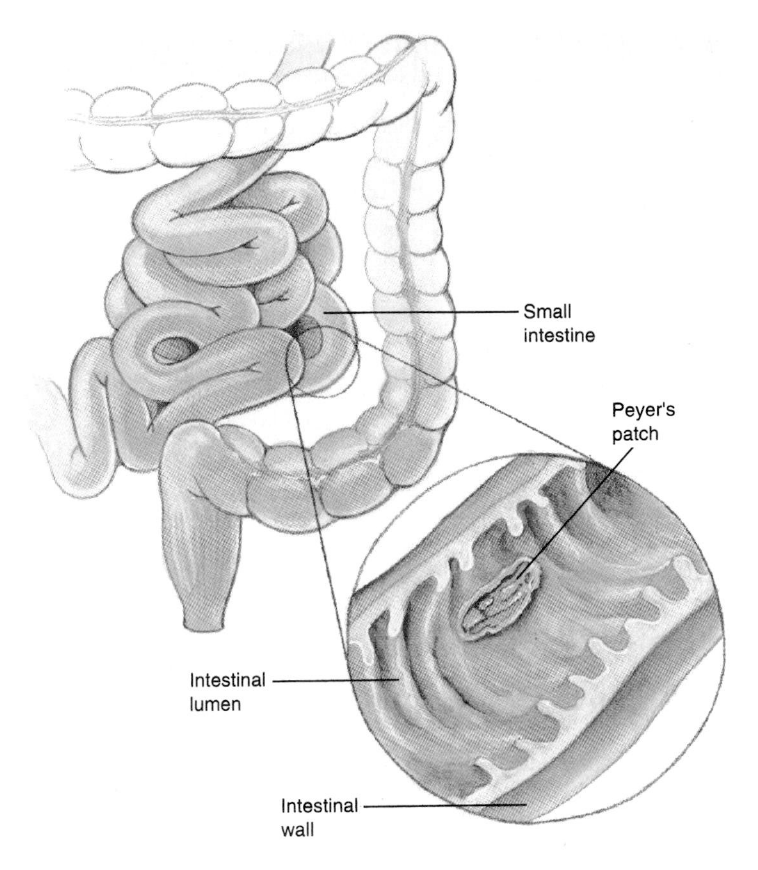

Figure 26.7
Peyer's patch, a collection of lymphoid tissue in the intestinal wall and potential site of intestinal perforation in typhoid fever.

Pathogenesis

Under natural conditions, *S. typhi* infects only humans. The organisms usually enter the gastrointestinal tract in fecally contaminated food or water, as do other salmonellas, and, like them, readily penetrate the intestinal epithelial lining. Phagocytic cells appear in the gut wall and ingest the invaders, but the typhoid bacilli are not destroyed and multiply inside these cells. These infected cells are then carried by the bloodstream to other parts of the body where they release the bacteria, causing infection of the bones, kidneys, brain, or other tissues. *S. typhi* also localizes in Peyer's patches, collections of lymphoid cells in the intestinal wall (figure 26.7); destruction of Peyer's patches by *S. typhi* can lead to intestinal rupture and hemorrhage. Typhoid organisms can be recovered from the blood or bone marrow of an infected person and, later, from their feces by using selective enrichment media. Over a one- to five-week period, a rise in antibody titer against the somatic, flagellar, and capsular (O, H, and K) antigens of the organisms can often be demonstrated in the blood of an infected person. This rise in antibody titer may be helpful in identifying the cause of the illness, but since *S. typhi* shares antigens with other salmonellas, the antibody rise is not specific.

Epidemiology

S. typhi is maintained in nature by human carriers who appear perfectly well but excrete as many as 10 billion typhoid bacilli per gram of their feces. The source of these organisms is almost always the gallbladder, where the organisms multiply in concentrations of bile that kill or inhibit most other bacteria. Since only a few organisms are necessary to cause infection, it is easy to see how dangerous typhoid carriers can be. One of the most notorious carriers was "Typhoid Mary," a young Irish cook living in New York state in the early 1900s. She is known to have been responsible for at least 53 cases of typhoid fever transmitted during a 15-year period. At the time, about 350,000 cases of typhoid occurred in the United States each year. Improved sanitation and public health surveillance measures are responsible for the low incidence of typhoid fever in the United States today.

• **Enrichment media, pp. 89, 90**

Prevention

Two vaccines are currently available for preventing typhoid fever. One is a killed vaccine that is administered by injections. Protection is partial, reducing the incidence of disease by 51% to 76% in different studies. The second vaccine is a live attenuated strain of *S. typhi* that became generally available in the United States in 1990. It is given by mouth. Its effectiveness is similar but longer lasting than that of the injectable vaccine.

Treatment

Typhoid carriers can be detected by culturing feces and by serological tests for antibodies against the Vi (capsular) antigen of *S. typhi.* Carriers are given prolonged treatment with antimicrobial medicines such as ampicillin or trimethoprim combined with a sulfa drug. For many carriers, surgical removal of the gallbladder is necessary to rid them of infection.

Table 26.5 describes the main features of typhoid fever.

Table 26.5 Typhoid Fever

Symptoms	Fever, severe headache, abdominal pain; less frequently, rupture of intestine and shock
Incubation period	7–21 days
Causative agent	*Salmonella typhi,* a motile, gram-negative enterobacterium with distinctive antigens and biochemical features that distinguish it from other salmonellas
Laboratory diagnosis	Organisms recovered from the blood, bone marrow, or feces; by growth on enrichment and selective media
Pathogenesis	Organisms colonize small intestine and penetrate its lining with little or no apparent damage, rapidly appearing in bloodstream; phagocytic cells ingest invaders but typhoid bacilli are not destroyed and instead multiply inside cells; infected cells transport bacteria to other parts of body via blood
Epidemiology	Usually ingestion of food contaminated by feces of human carriers
Prevention	Oral attenuated or killed injected vaccines give partial protection
Treatment	Antibiotics, such as ampicillin, chloramphenicol, or trimethoprim with sulfamethoxazole, depending on susceptibility of the infecting strain

REVIEW QUESTIONS

1. What is cholecystitis?
2. Why is bile included in some selective microbiological media?
3. How do pathogens resist the flushing action of intestinal fluids?
4. What is *Clostridium difficile?* Name a disease it causes.
5. What causes the "blind loop syndrome"?
6. What is the source of flatus?
7. Explain how *V. cholerae* causes cholera without apparent intestinal damage.
8. Contrast the epidemiology and pathogenesis of shigellosis and salmonellosis.
9. Why did it take so long to identify *C. jejuni* as a common cause of diarrhea?
10. How does the epidemiology of *S. typhi* differ from that of *S. enteritidis?*

Food Poisoning

Microbial growth in food can cause illness chiefly by two different mechanisms: (1) the contaminating microorganisms can infect the person who ingests the food, or (2) products of microbial growth in the food can be poisonous. Gastroenteritis caused by *Salmonella* sp. of animal origin exemplifies the first mechanism, whereas a variety of illnesses caused by species of *Clostridium, Bacillus,* and *Staphylococcus* result from the second mechanism. Although all illnesses arising from microbial growth in ingested food are popularly considered "food poisoning," the following discussion deals with the illnesses caused by microbial products rather than those resulting from infection.

Staphylococcal Food Poisoning Is the Result of a Heat-Resistant Exotoxin

One of the most common forms of food poisoning is due to *Staphylococcus aureus,* the same species of bacterium responsible for boils and other infections. The illness begins with nausea, vomiting, and diarrhea a few hours after a person eats contaminated food. Recovery generally occurs within another few hours. This type of food poisoning is caused by enterotoxins (exotoxins that affect the intestine) produced by certain strains of *S. aureus* growing in a suitable food, usually one high in carbohydrates. This toxin may be produced by staphylococci in food at room temperature, often within only one or two hours. Moreover, the toxin is relatively heat-stable, so subsequent cooking often kills the staphylococci but leaves their toxin intact. One epidemic reported in 1989 caused by staphylococcal toxin in canned mushrooms involved more than 100 persons in three states. Several antigenically distinct varieties of enterotoxins are produced by the various food-poisoning strains of *S. aureus* and can be identified in extracts of contaminated foods by using specific antisera.

Food Poisoning Can Be Caused by *Clostridium perfringens* and *Bacillus cereus*

Clostridium perfringens is commonly considered a cause of food poisoning, although most of its toxin is probably released in the intestine. Symptoms appear typically 6 to 12 hours after eating and consist mainly of diarrhea and abdominal cramps, although half the victims also have nausea and vomiting. *C. perfringens* is an encapsulated, gram-positive, nonmotile anaerobic rod that is commonly found in human and animal intestines as well as in soil. It is also the primary cause of gas gangrene (chapter 26). One

• ***Staphylococcus aureus,* p. 476**

Case Presentation

The patient was a 51-year-old woman who was admitted to a California hospital because of fever of 105°F and swelling, redness, and tenderness of both legs. Two days before admission, she had eaten raw oysters at a party. She was well until one day before admission when she developed fever, nausea, and muscle aches. Past history included a diagnosis of breast cancer 14 years before admission and chronic hepatitis C 10 years before admission.

In the hospital, her condition deteriorated rapidly. Bloody blisters appeared on her skin, and she went into shock. Despite being transferred to the intensive care unit, and receiving multiple antibiotics, she died the day after admission.

1. What aspects of this case presentation point to the correct diagnosis?
2. What is the natural history of the causative agent?
3. What persons are at especially high risk of contracting this condition?
4. What preventive measures can be taken?

Discussion

1. Persons who develop high fever, cellulitis (the thickened, red, tender skin), and hemorrhagic bullae (the bloody blisters) after eating raw oysters should be suspected of having a serious *Vibrio vulnificus* infection. This bacterium, like most other vibrios, is a curved or straight, oxidase positive, facultatively anaerobic gram-negative rod.
2. The natural habitat of *V. vulnificus* is warm saltwater, especially along the coast of the Gulf of Mexico. The bacterium may be present in as many as 50% of the oyster beds along the Gulf Coast during the warmer months of the year.
3. People, like this patient, who have liver disease are at very high risk of fatal bloodstream infection with this bacterium. Those with AIDS and other immunodeficiencies as well as those with diabetes, cancer, and certain inherited blood diseases are also at high risk for serious infection.
4. This organism can infect through open wounds. Therefore, those with open wounds are advised to avoid contact with seawater and water used to rinse seafood. About half the infections are acquired from eating raw or partly cooked seafood and raw oysters in more than 95% of the cases. Approximately one-third of these cases die of the infection. Stores that sell raw oysters post signs warning of the dangers, and health officials of states along the Gulf of Mexico maintain surveillance of the problem. Also, states outside the area often require that oysters sold in their states be tagged with the place of origin. This patient's infection was traced to a Louisiana oyster bed.

outbreak of food poisoning by *C. perfringens* was traced to bean-filled burritos, but more commonly meat and poultry dishes are the sources of clostridial poisoning. When large numbers of *C. perfringens* are ingested, they reach the intestine where they release their enterotoxin upon sporulation. The illness rarely lasts more than 24 hours and usually no treatment is needed.

Bacillus cereus, a gram-positive, aerobic spore-former, also produces exotoxins. Some strains of *B. cereus* cause mainly diarrhea and have a heat-labile toxin similar to that of *E. coli.* Other strains cause mainly vomiting and produce a toxin similar to that of *S. aureus.* The incubation period is two to six hours, and symptoms clear within 12 hours. Most cases have resulted from ingestion of contaminated rice dishes.

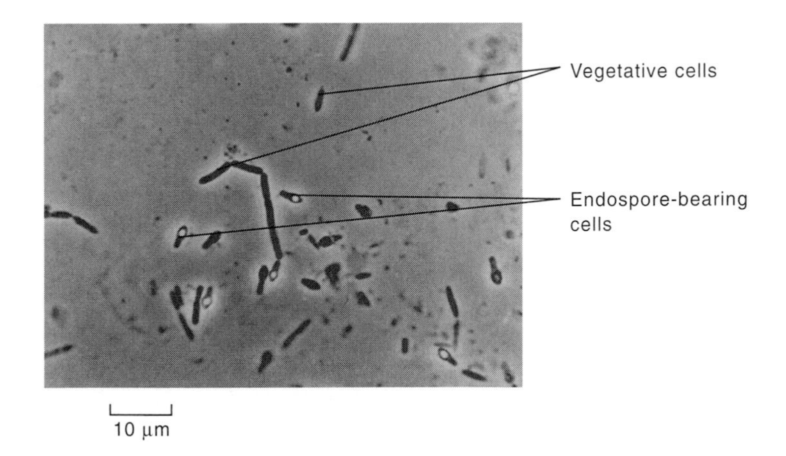

Figure 26.8
Clostridium botulinum. Note the spores that form near the ends of the organism. These spores are not reliably killed by the temperature of boiling water.

Botulism Is the Most Feared Type of Food Poisoning Because It Can Result in Paralysis and Death

Botulism, one of the most feared of all diseases, is a microbial poisoning that results in paralysis. It is caused by an exotoxin synthesized by the anaerobic, gram-positive, spore-forming, rod-shaped bacterium, *Clostridium botulinum,* which is widely distributed in soils around the world. Endospores are formed near the ends of the cells (figure 26.8). The name *botulinum* comes from the word for "sausage" and was chosen because some of the earliest recognized cases occurred in people who had eaten contaminated sausage. The vast majority of serious botulism cases

Table 26.6 Botulism

Symptoms	Blurred or double vision, weakness, nausea, vomiting, and diarrhea; nerve involvement leads to generalized paralysis and respiratory insufficiency
Incubation period	12–36 hours
Causative agent	*Clostridium botulinum,* an anaerobic, gram-positive, spore-forming, rod-shaped bacterium
Laboratory diagnosis	Identification of toxin in blood and food
Pathogenesis	During growth of *C. botulinum,* exotoxin released into food; neurotoxin survives stomach acid and pepsin, is absorbed into the bloodstream, and is carried to nerves; toxin acts by blocking the transmission of nerve signals to the muscles, producing paralysis; occasionally, *C. botulinum* colonizes intestine or wounds and releases enough toxin to cause botulism
Epidemiology	Ingestion of contaminated food, often home-canned food that was not heated enough to kill *C. botulinum* spores; spores widespread in soil and dust
Prevention	Education to improve home canning methods; heating food to boiling for 15 minutes just prior to serving; immunization with toxoid not generally available
Treatment	Antitoxin given intravenously

are caused by eating food in which this organism has grown and released its toxin. Some cases occur when the organisms colonize the intestine or a wound, but they are generally milder.

Foodborne Botulism

Foodborne botulism was first recognized in the late 18th century and is the most common and most severe of the three types of botulism. In the five years preceding 1922, there were 83 outbreaks of botulism in the United States, many traceable to commercially canned foods. The work of the distinguished microbiologist Karl F. Meyer and his colleagues showed that canning methods then in use were inadequate to kill the heat-resistant spores of *C. botulinum.* Strict controls were then placed on commercial canners to ensure adequate sterilizing methods. Since then, outbreaks caused by commercially canned foods have been infrequent. The occasional lapses in proper canning technique highlight the potential for disaster when defective processing occurs in large companies that produce and distribute food products.

Symptoms Twelve to 36 hours (or sometimes longer) after a person has eaten toxin-containing food, blurred or double vision occurs, indicating eye muscle weakness. Progressive paralysis then ensues, generally involving all voluntary muscles, but respiratory paralysis is the most common cause of death. Despite treatment, about one-fourth of the victims of botulism die. Table 26.6 describes the main features of this disease.

Epidemiology and Pathogenesis Like other clostridia, *C. botulinum* produces endospores that are often highly resistant to heat and can thus persist in foods such as vegetables, fruit, meat, seafood, and cheese despite cooking and canning processes. These spores later germinate if the environment is favorable—such as having anerobic conditions, a pH above 4.6, and a temperature above 39°F (4°C)—and growth of the bacteria results in the release of exotoxin into the food. When a person eats food in which *C. botulinum* has grown, the exotoxin is absorbed into the bloodstream, where it may continue to circulate for as long as three weeks. This exotoxin is a neurotoxin, meaning that it acts against the nervous system, and is one of the most powerful poisons known. This circulating toxin is then carried in the blood to the various nerves of the body, where it blocks the transmission of nerve signals to the muscles, thus producing paralysis. A few milligrams of the toxin would be sufficient to kill the entire population of a large city. Indeed, cases of botulism have resulted from a person eating a single contaminated string bean and another person licking a finger contaminated with toxin. Fortunately, the toxin is heat-labile, and even high concentrations are completely inactivated by boiling contaminated food for 15 minutes. Seven antigenically different variants of the neurotoxin, designated A, B, C, D, E, F, and G, are synthesized by different strains of *C. botulinum.* Types A, B, and E are responsible for most human cases, while types C and D affect only birds and other animals. In some strains, toxin production results from lysogenic conversion.

Like a number of other bacterial exotoxins, botulinal toxin is composed of two portions, A and B. The B portion attaches to specific receptors on motor nerve endings, while the A portion enters the nerve cell. The A portion then becomes an active peptidase enzyme that prevents release of

• **Heat sterilization, pp. 203–8**

• **Lysogenic conversion, pp. 309, 508**

a neurotransmitter,[1] probably by destroying a receptor on the vesicles containing the neurotransmitter.

C. botulinum toxin can be used to treat people with certain chronic spastic conditions. Minute amounts of the toxin give prolonged relief of symptoms when injected into the area of spasm.

Treatment and Prevention Botulism is treated by administering the antitoxin specific for the causative type of *C. botulinum.* However, the antitoxin only neutralizes toxin circulating in the bloodstream, and the nerves already affected by it recover slowly, over weeks or months. Prevention of the disease depends on proper sterilization and sealing of food at the time of canning and on adequate heating prior to serving. One cannot rely on a spoiled smell, taste, or appearance to detect contamination, because such changes are not always present.

Intestinal Botulism

Proper food handling cannot completely eliminate the danger of botulism since *C. botulinum* can occasionally colonize the human intestine, especially of infants six months of age or less, and produce a mild form of the disease. Botulism is characterized by constipation followed by generalized paralysis. It can vary greatly from mild lethargy to respiratory insufficiency. In 1976, when infant botulism was first recognized, four previously healthy infants were hospitalized in California because they became paralyzed. *C. botulinum* organisms and toxin were present in their feces, but the toxin levels were too low to be detectable in their blood. Although one infant required respiratory support and another required tube feeding, all recovered without receiving antitoxin treatment. Recovery probably resulted from their own antibody production. *C. botulinum* organisms and toxin persisted in the infants' feces for a time but were gradually replaced by competing normal intestinal flora. The source of these infants' infections was undoubtedly *C. botulinum* spores, which are commonly present in dust and contaminate various foods such as honey. Indeed, ingestion of honey has been associated with this illness in about one-third of the cases. Generally, 25 to 100 cases of infant botulism are reported in the United States per year.

Intestinal botulism also occurs in adults, particularly in immunodeficient patients whose normal intestinal flora have been suppressed by antibiotic treatment.

Wound Botulism

C. botulinum occasionally colonizes dirty wounds and releases enough toxin to cause botulism. Some of the cases have been due to abuse of injected drugs. All eleven cases of wound botulism reported in 1994 were due to injected-drug abuse.

[1]Neurotransmitter—chemical that conveys nerve impulses from nerve to muscle or nerve to nerve

Poisoning by Moldlike Fungi

In earlier ages, ergot poisoning ("St. Anthony's Fire") resulted from eating bread prepared from grain on which a certain kind of mold had grown. The victims of this poisoning had agonizingly painful convulsions, and some developed gangrene of the hands and feet because of spasms of the arteries. Epidemics of ergot poisoning still occur occasionally, but effective medicines are now available to counteract the ergot. Proper standards of agricultural practice and public health surveillance of food grains now make it unlikely that ergot-producing fungi will contaminate food and produce ergot poisoning.

Other fungal poisons, called **aflatoxins,** are produced by common molds of the genus *Aspergillus.* Aflatoxins cause acute poisoning in many species of animals, especially the young, producing liver damage. A few milligrams, for example, can kill a dog within 72 hours. Interest in aflatoxins rose with the observation that tiny traces of it in the food of certain animals caused tumors. Feeding aflatoxins to laboratory animals results in liver cancers in a number of species.

Climatic conditions influence aflatoxin levels in foods. For example, in the summer of 1988, the midwestern section of the United States experienced severe drought conditions. Many corn kernels cracked, allowing fungal spores to enter and grow. The resulting production of aflatoxin made much of this corn unsuitable for either animal feed or human consumption, and it had to be destroyed. In developing countries in the tropics, aflatoxin concentrations in human foods are generally much higher than in industrialized countries and correlate with a higher incidence of human liver cancer. However, several factors appear to be involved in the development of human liver cancers. Most likely, a metabolite of aflatoxin known to bind covalently to nucleic acids is involved, perhaps acting together with hepatitis B virus infection and recurrent malnutrition. Cancer arising in liver cells is much less common in the Western world.

Viral Infections of the Lower Alimentary System

A number of viruses can infect the gastrointestinal tract or its appendages. Two common illnesses resulting from these infections are viral hepatitis and viral gastroenteritis.

• **Ergot, p. 288**
• **Aflatoxins, p. 288**

Table 26.7 Hepatitis Viruses

Virus (Size)	Nucleic Acid	Incubation Period	Transmission	Laboratory Diagnosis	Treatment	Prevention
HAV (27 nm)	RNA, single-stranded	3–5 weeks	Fecal-oral	IgM and IgG antibody determination	Treatment of symptoms only	Hand washing; cooking food; gamma globulin; vaccine
HBV (42 nm)	DNA, double-stranded	10–15 weeks	Blood, semen	Determine presence of viral antigens and antibodies	Interferon for chronic hepatitis	Vaccine; caution in handling blood; safe sex; blood and organ donors screening; hyperimmune globulin (HBIG)
HCV (60 nm)	RNA, single-stranded	5–9 weeks	Blood	Determine presence of antibody to viral proteins; polymerase chain reaction	Same as HBV	Same as HBV except no vaccine yet
HDV (Delta, 36 nm)	RNA, single-stranded	With co-infection, same as HBV; HDV superinfection may exacerbate hepatitis B	Blood, semen	Determination of HDV antibody; HDV antigen also possible in some laboratories	Like HBV	Vaccination against HBV; same as HBV
HEV (29 nm)	RNA, single-stranded	6 weeks	Fecal-oral	Determination of HEV antibody; also HEV RNA by polymerase chain reaction	Like HAV	Sanitary measures; boiling drinking water
HGV	RNA, single-stranded	?	Blood	HGV RNA by polymerase chain reaction	None	?

Various Viruses Infect the Liver, Causing Acute and Chronic Disease

Hepatitis, or inflammation of the liver, can be caused by allergic reactions to medicines, certain toxic chemicals including anesthetic gases and cleaning fluids, and various microorganisms and viruses. Typically, the symptoms of hepatitis are loss of appetite and vigor, fever, and jaundice. Hepatitis A virus (HAV) and hepatitis B virus (HBV) account for most cases of hepatitis, although there are other viral causes. The principal characteristics of HAV and HBV are given in table 26.7. The two viruses are completely unrelated and differ in the way in which they spread and in their incubation period. Although they formerly were thought to be unusually resistant to killing by heat or chemical disinfectants, more recent studies indicate they are not much different in susceptibility than many other small viruses.

Hepatitis A Virus (HAV) Disease

Hepatitis A virus is a small single-stranded RNA virus of the picornavirus family. HAV differs from other members of the family and has been given the name *hepatovirus.* It can be cultivated in a variety of tissue cell cultures. Hepatitis A virus disease (formerly called *infectious hepatitis*) spreads in epidemic fashion, principally through fecal contamination of hands, food, or water. Eating raw shellfish is a frequent source of infection since these animals concentrate the hepatitis A virus from fecally polluted seawater. Following ingestion, the virus reaches the liver by an unknown route. The liver is the main site of replication and the only tissue known to be damaged by the infection. The virus is excreted in the feces two weeks into the incubation period until the liver starts to recover. Most cases of the disease are mild and self-limited, and many are asymptomatic. However, some cases are severe, requiring many weeks of bed rest for recovery. The infection is sufficiently widespread in the population that many people have hepatitis A virus antibody in their blood. For example, in New York City, 20% to 80% of the residents have HAV antibody, the higher percentages occurring in the lower socioeconomic groups because of crowding and inadequate sanitation. Since the antibody to HAV is so common, gamma globulin obtained from the donated

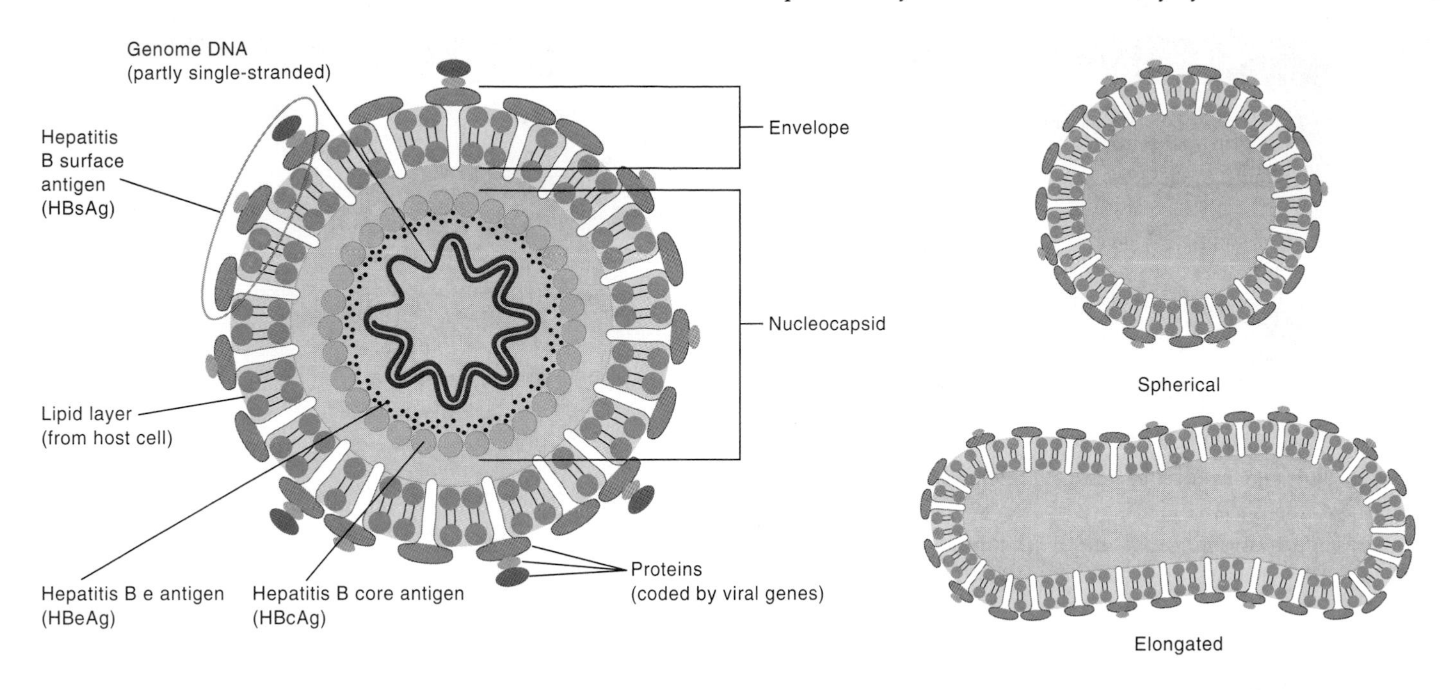

Figure 26.9
Hepatitis B virus components found in the blood of infected patients. (*a*) Complete infectious virion. (*b*) The small spherical and elongated envelope particles lack DNA.

blood of many people is pooled and administered intramuscularly to susceptible exposed people in order to provide passive immunity to the disease. Also, a highly effective formalin inactivated vaccine has been available since 1995.

Hepatitis B Virus (HBV) Disease

Cause and Symptoms Hepatitis B, formerly known as *serum hepatitis*, tends to be more severe than hepatitis A, causing mortality from liver failure in 1% to 10% of hospitalized cases. However, many people with hepatitis B virus (HBV) infection show few or no symptoms. Indeed, for every case with jaundice, there are at least three without. Hepatitis B is unique in that enormous quantities of viral material appear in the patient's blood.

HBV is a member of the hepadnavirus family (chapter 14), some members of which (but not HBV) infect certain rodents and ducks. Unlike HAV, HBV has not been cultivated *in vitro;* however, its genome has been cloned in bacteria, yeast, and mammalian cells. The virus contains double-stranded DNA and has a lipid-containing outer envelope. Three important HBV antigens are (1) surface antigen (HBsAg), (2) core antigen (HBcAg), and (3) e antigen (HBeAg), a soluble component of the viral core. HBsAg is produced during viral replication in amounts far in excess of that needed for viral envelope production. It occurs in the bloodstream on small spheres and filaments (figure 26.9) in quantities often 1,000 or more times greater than complete virions. HBsAg is responsible for the ability of the virus to attach to and infect its hosts; antibody to surface antigen (anti-HBsAg) confers immunity. HBcAg represents the outer covering of the nucleocapsid. The presence of IgM antibody to HBcAg indicates acute rather than chronic hepatitis B. The presence of HBeAg in the blood indicates a strong likelihood that the blood is infectious.

Pathogenesis The virus enters the body with contaminated blood or semen or from mother to fetus across the placenta and is carried to the liver by the bloodstream. The mechanism by which HBV causes liver injury is not known. Following entry of HBV into the liver cell, the virus replicates by an unusual mechanism involving reverse transcriptase (figure 26.10). HBsAg appears in the bloodstream days or weeks after infection, often long before laboratory or clinical signs of liver damage are evident. In more than 90% of HBV-infected persons, antibodies arise to the viral antigens, causing them to be removed from the blood as the infection subsides

• **Passive immunization, p. 389**

• **Reverse transcriptase, pp. 329, 330**

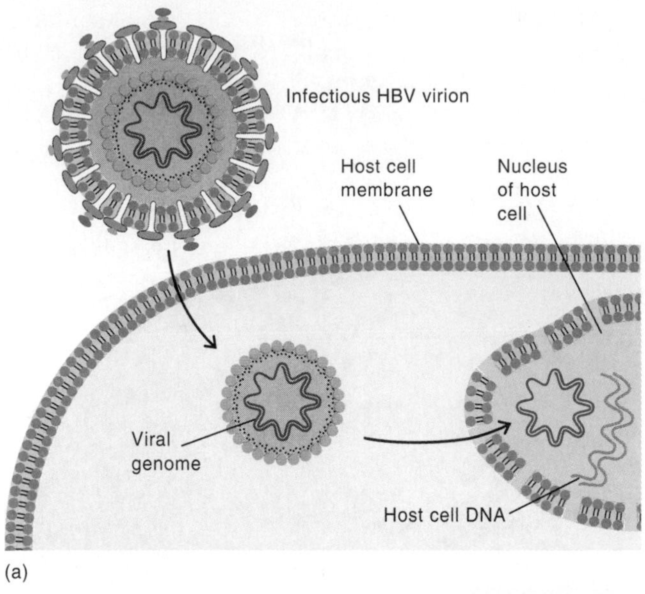

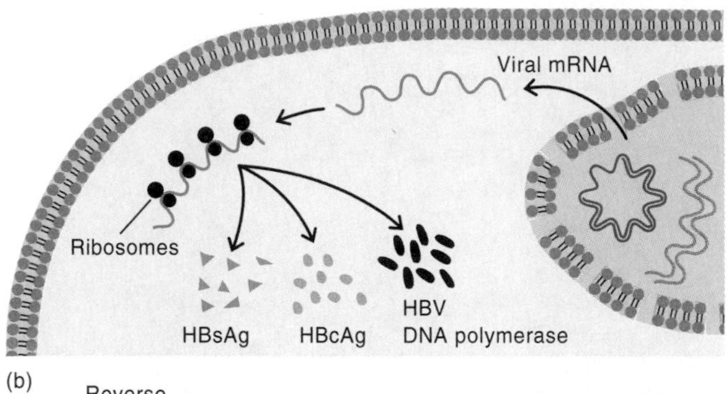

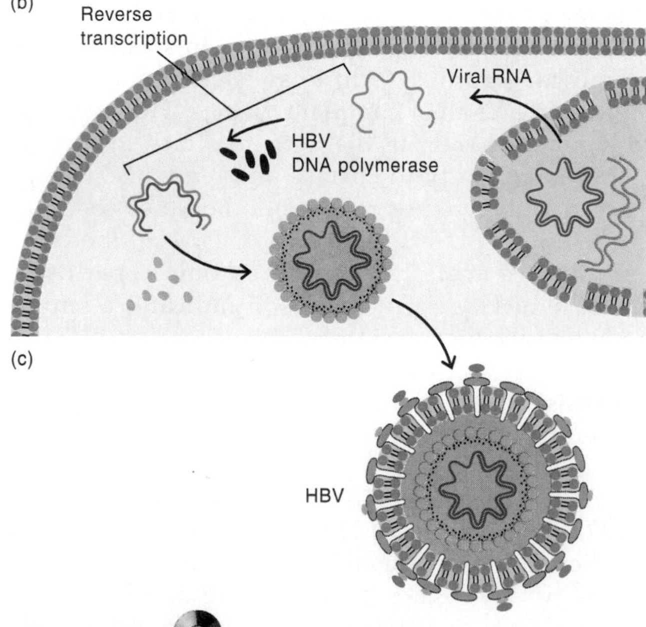

Figure 26.10
Replication of hepatitis B virus. Note that this DNA virus employs reverse transcriptase in its replication cycle.

(figure 26.11). Liver cell destruction most likely results from the body's immune system attacking the infected liver cells.

In a small percentage of cases (2% to 10%), the infection smolders unsuspected for years, releasing infectious virus into the person's blood and body secretions. About

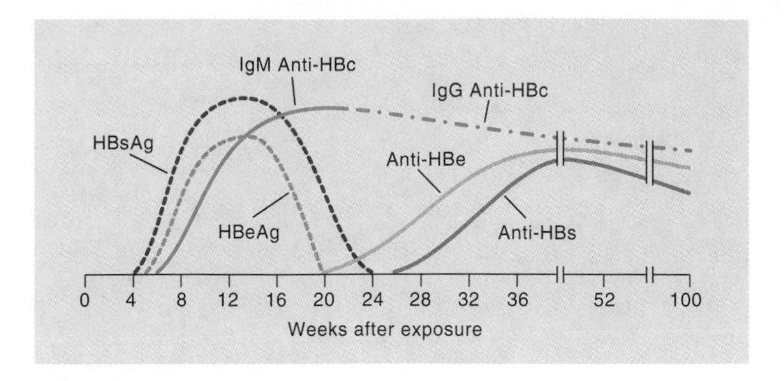

Figure 26.11
Blood test results at different times during acute hepatitis B virus infection. Antigen decreases in the blood as antibody is produced. Antibody becomes detectable when it exceeds antigen.

40% of chronically infected people eventually die from cirrhosis (scarring) of the liver or from liver cell cancer. Indeed, in one study, chronic HBV infection was shown to increase the risk of liver cell cancer more than 100-fold. Evidence indicates that these cancers can result from HBV transformation of liver cells.

Epidemiology From 1965 to 1985, a progressive rise in reported hepatitis B cases occurred. Since then, the incidence of the disease has appeared to plateau or decline somewhat (figure 26.12). Currently, an estimated 300,000 new cases of HBV occur in the United States each year. The number of asymptomatic HBV carriers in the United States is estimated to be 750,000 to 1,000,000, with as many as 30,000 new carriers arising each year. Hepatitis B infects at least 5% of the world's population and is the ninth leading cause of death.

HBV is spread mainly by blood, blood products, and semen. Persisting viremia[1] can follow both symptomatic and asymptomatic cases, and the virus may continue to circulate in the blood for many years. Carriers are of major importance in the spread of hepatitis B because they are often unaware of their infection. If only a minute amount of blood from an infected person is injected into the bloodstream or rubbed into minor wounds, infection can result. A small amount of plasma from a patient with the disease, when given experimentally by mouth to another subject, has also produced the disease, the virus probably infecting the recipient through small scratches or abrasions in the mouth. Many hepatitis B virus infections result from sharing of needles by drug abusers. Unsterile tattooing and ear-piercing instruments and shared toothbrushes, razors, or towels can also transmit HBV infections. Blood transfusions and improperly sterilized dental instruments were a common source of HBV infection in the past. At some blood banks, hepatitis B virus antigen was demonstrated in

[1]Viremia—presence of viruses in the circulating blood

• **Viral transformation, pp. 333, 334**

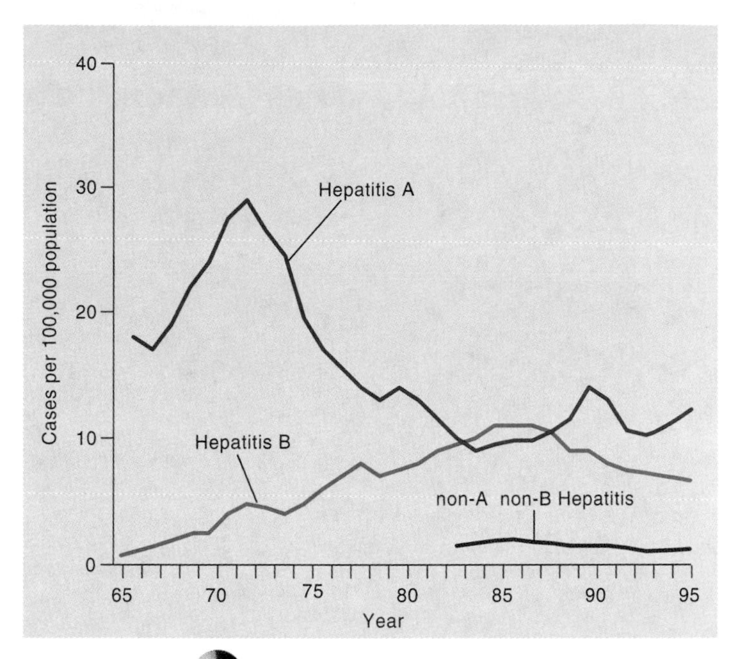

Figure 26.12
Incidence of viral hepatitis in the United States, 1965–1995. Much of the non-A and non-B hepatitis is caused by the hepatitis C virus. The first hepatitis B vaccine was licensed in 1962. An inactivated hepatitis A vaccine was licensed in 1995. A test for hepatitis C antigen became available in 1990.

Source: Centers for Disease Control and Prevention. 1996. Summary of notifiable diseases—United States, 1995. *Morbidity and Mortality Weekly Report* 44(53):37.

as many as 5 of every 1,000 prospective donors. However, tests now used to screen the blood are very sensitive and, when they indicate the presence of HBV, the blood is not used for transfusions. In addition, dental instruments are routinely cleaned and sterilized between patients, and disinfectants and plastic coverings are used to prevent the handles of instruments, trays, and tabletops from transmitting the disease.

Besides being present in the blood, HBV antigen is often present in saliva and breast milk, but these fluids are not important in spreading the disease, probably because the quantity of infectious virus is low. However, semen can transmit the disease during sexual intercourse.

Another very important mode of spread of HBV is from a mother to her baby. Five percent or more of pregnant women who are carriers transmit the disease to their babies at the time of delivery as do more than two-thirds of women who develop hepatitis late in pregnancy or soon after delivery. While infected babies may sometimes die of liver failure, most have asymptomatic infections and become long-term carriers.

Prevention and Treatment An effective vaccine is available against hepatitis B. The first vaccine to be released for general use consisted of HBsAg obtained from the blood of chronic carriers and treated to remove infectious viruses. The serum of many chronic carriers contains enough viral antigen in one milliliter of blood serum to immunize eight adults. However, following the cloning of the gene for HBsAg in the yeast *Saccharomyces cerevisiae,* a genetically engineered vaccine was developed, tested, and released for use in 1986. Both of these vaccines are safe and produce long-lasting immunity. Since hepatitis B is a strictly human disease, vaccines provide a way to eventually eliminate it, just as a vaccine eliminated smallpox in the late 1970s. Currently, prevention has been directed toward immunizing infants born to mothers with HBV infection and other high-risk groups such as health care workers exposed to blood. Active immunization is of course most effective when administered before exposure to HBV. If a person is already known to be exposed, such as an infant born to an infected mother, hepatitis B immune globulin (HBIG) containing a high titer of antibody to HBV is administered promptly at the same time as the vaccine. Passive immunization with HBIG offers immediate, partial protection against HBV infection and does not prevent the development of active immunity from the vaccine.

Worldwide, there are an estimated 500,000 to 1 million new liver cell cancers each year, many of them associated with HBV infection. Studies are currently under way to determine the effect of hepatitis B vaccination on the incidence of primary liver cancer.

Measures designed to educate high-risk groups about the epidemiology of hepatitis B virus appear to be helpful in some instances. People likely to be exposed to blood are taught to treat all human blood as though it were infectious, wash hands contaminated with blood, wear gloves and protective clothing, and handle contaminated sharp objects such as needles and scalpel blades carefully. Teaching the use of condoms also helps limit spread of the infection. These educational efforts now have a greater impact because the epidemiology of HBV is identical to that of the AIDS virus.

A generally effective antiviral treatment is not currently available for curing hepatitis B, although remarkable improvement has occurred in about one-third of the chronic cases treated with genetically engineered interferon.

Hepatitis C Virus (HCV) Disease

When blood that tested positive for HBV was excluded from transfusions, post-transfusion hepatitis remained common and was believed to be caused by a "non-A, non-B" virus. A number of years passed as researchers unsuccessfully tried to isolate and identify this hepatitis virus. Finally, in 1989, scientists were able to clone a part of the genome of the transfusion-associated virus, now known as hepatitis C virus (HCV). They were then able to produce a viral antigen that was shown to react with antibody in the blood of patients with hepatitis C. HCV is enveloped and

• **Genetically engineered vaccines, pp. 183–185**
• **Interferon, pp. 184–85, 560**

contains single-stranded RNA. It represents a new group of viruses in the flavivirus family.

Hepatitis C virus is responsible for most cases of post-transfusion hepatitis. Since many people infected with HCV fail to develop detectable antibody for many weeks or months, it is difficult to detect the infection in many blood donors. However, screening tests for HCV in donated blood have become progressively more sensitive since their introduction in 1990. Evidence indicates that sexual transmission of HCV is uncommon. Fifty% to 80% of infected people become chronic carriers and develop cirrhosis. Interferon is an effective treatment for many of these patients.

Delta Hepatitis Virus (HDV)

Delta hepatitis virus is a defective RNA virus that depends on hepatitis B virus for replication. It has a worldwide distribution, with relatively high frequency in Italy, the Middle East, and parts of Africa and South America. Its structure is shown schematically in figure 26.13. The outer envelope of HDV is apparently identical to its co-infecting HBV. Patients can be infected with the two viruses simultaneously, or HDV can be introduced in someone previously infected with HBV. In the latter instance, there may be a marked worsening of the patient's disease in some cases. In other cases, HDV appears to accelerate the course of liver scarring.

Hepatitis E Virus (HEV) Disease

Hepatitis E virus (HEV) is a small, nonenveloped single-stranded RNA virus probably of the calicivirus family. Like HAV, it is transmitted by the fecal-oral route. Epidemics of hepatitis E have occurred in many places around the world, including Mexico, India, Pakistan, Burma, China, and parts of Africa. The disease is often waterborne, tending to occur in the cities of underdeveloped countries and rural areas of developed countries. Mortality is generally low except in infants and pregnant women, in whom it reaches 20%. Long-term carriers and chronic liver disease are not known to occur.

Hepatitis G Virus (HGV)

Discovery of another "non-A, non-B" virus, hepatitis G virus (HGV), was reported in early 1996. HGV is a positive sense, single-stranded RNA virus transmitted by blood transfusions. It can cause chronic hepatitis leading to cirrhosis, and long-term carriers exist. Studies of its epidemiology are under way. Development of screening tests for HGV should help make blood transfusion still safer.

A Number of Viruses Cause Vomiting and Diarrhea (Gastroenteritis)

At least five different groups of viruses produce epidemic gastroenteritis ("stomach flu"). Two of the most important groups are the rotaviruses and the Norwalk viruses. Table 26.8 gives the similarities and differences between these viruses and the illnesses they produce. Both groups of viruses infect the small intestinal epithelium and interfere with absorption of intestinal fluid. Rotaviruses have a 70-nanometer diameter, double-walled capsid, and RNA genome typical of the reoviruses. About 50% to 70% of the strains can be cultivated *in vitro,* although it requires a laborious and time-consuming process and is therefore of little use diagnostically. Rotaviruses are the most common cause of severe diarrhea in children, causing about 3,500,000 cases in the United States per year. Generally, by the age of four years, children have acquired some immunity and, if exposed to these viruses, develop at most a mild illness. However, rotaviruses can cause traveler's diarrhea, being implicated in about 25% of cases. In general, childhood epidemics tend to occur in winter in temperate climates, perhaps because children are more apt to be confined indoors in groups where the viruses can spread easily. The illness generally lasts five to eight days. Diagnosis can be made by electron microscopy of fecal specimens (figure 26.14), although more sensitive and easier immunological methods are now generally available. An effective live attenuated vaccine has been developed.

By contrast, Norwalk viruses are only 27 nanometers in size, and they have not been grown in culture nor has their genome been completely characterized. Like HEV, they are members of the calicivirus family. Immunity to Norwalk viruses is not acquired until adulthood. About half of the outbreaks of viral gastroenteritis in the United States are due to Norwalk viruses.

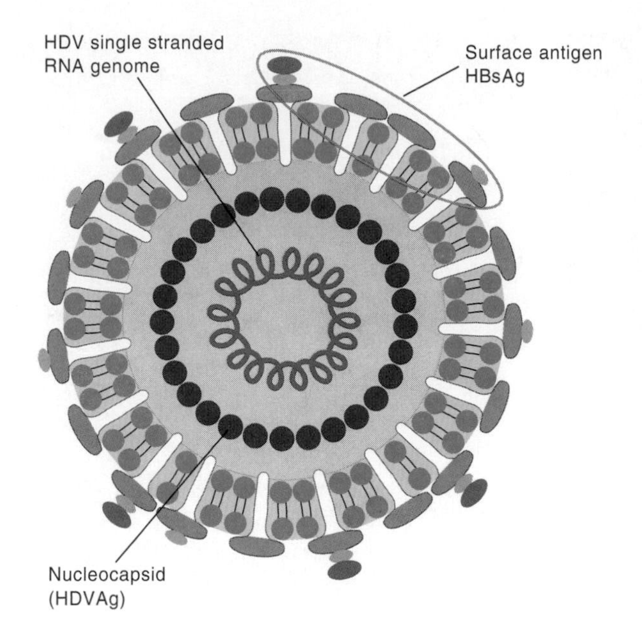

Figure 26.13
Delta hepatitis virus (HDV). Its surface antigens (Ag) are identical to those of hepatitis B virus, on which it is dependent for replication.

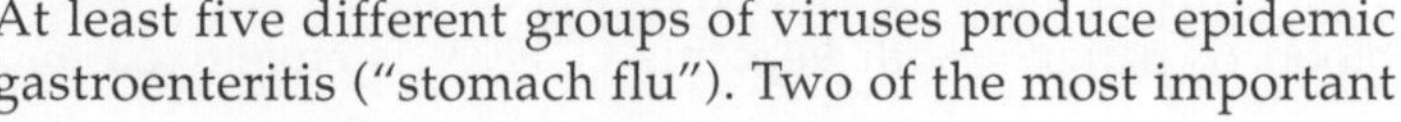
• **Defective viruses, pp. 321–25**

Table 26.8 Viral Gastroenteritis

Causative Agent	Characteristics	Incubation Period	Symptoms	Laboratory Diagnosis	Treatment
Rotavirus	70 nm diameter, double-walled capsid; double-stranded RNA; many strains cultivable	24–48 hrs	Diarrhea, abdominal cramps, vomiting; 5–8 days	Electron microscopy or ELISA of diarrheal stool for virus	No antiviral; occasionally intravenous fluid replacement for infants
Norwalk virus	27 nm diameter, single-stranded, positive sense RNA; not cultivable	18–72 hrs	Abdominal cramps, vomiting, muscle- and headaches; 24–48 hrs	Same	Same

Table 26.9 Pathogenic Intestinal Protozoa of Humans

Protozoan	Main Characteristics
Ameba:	
Entamoeba histolytica	Medium size; trophozoite shows distinctive mobility; quadrinucleate cysts; may infect liver and other tissues outside the colon
Ciliate:	
Balantidium coli	Large size; rare; may be contracted from pigs; cyst form; inhabits colon; rapid rotating movement of trophozoite
Flagellate:	
Giardia lamblia	Unmistakable pear-shaped, flattened, binucleate trophozoite with ventral sucking disk; cyst form; inhabits small intestine
Sporozoan:	
Cryptosporidium parvum	Tiny oocysts with four sporozoites; prolonged symptoms in those with AIDS or other immuno-compromised state; acute, self-limiting illness very common in normal people; waterborne epidemics have occurred, as have epidemics in hospital and day-care settings; inhabits small intestinal mucosa
Cyclospora cayetanensis	Symptoms similar to those of *C. parvum;* oocysts larger; not infectious when passed in feces; need warm, moist conditions to mature; waterborne and food-related epidemics; can be treated with trimethoprim-sulfamethoxazole

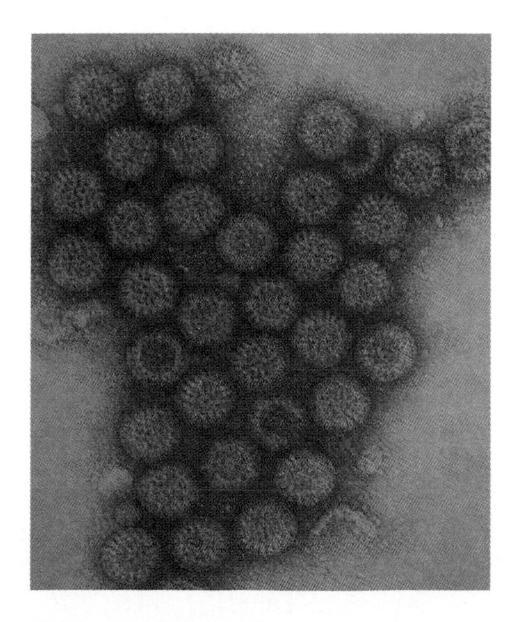

Figure 26.14
Transmission electron micrograph of rotavirus in human feces (X93,000).

Protozoan Diseases of the Lower Alimentary System

Protozoa are important causes of human intestinal disease (table 26.9). All the major protozoan types are represented: flagellates, ciliates, sporozoa, and amebas. Four important protozoan diseases are discussed here.

Giardiasis Is a Common Widespread Disease of the Small Intestine

Giardiasis, a disease caused by *Giardia lamblia* (also known as *G. intestinalis*), is the most commonly identified waterborne illness in the United States. It can be contracted from clear mountain streams and chlorinated city water and can be transmitted by person-to-person contact. Giardiasis occurs worldwide and is responsible for many cases of traveler's

• **Protozoa, pp. 290–94**
• ***Giardia*, p. 295**

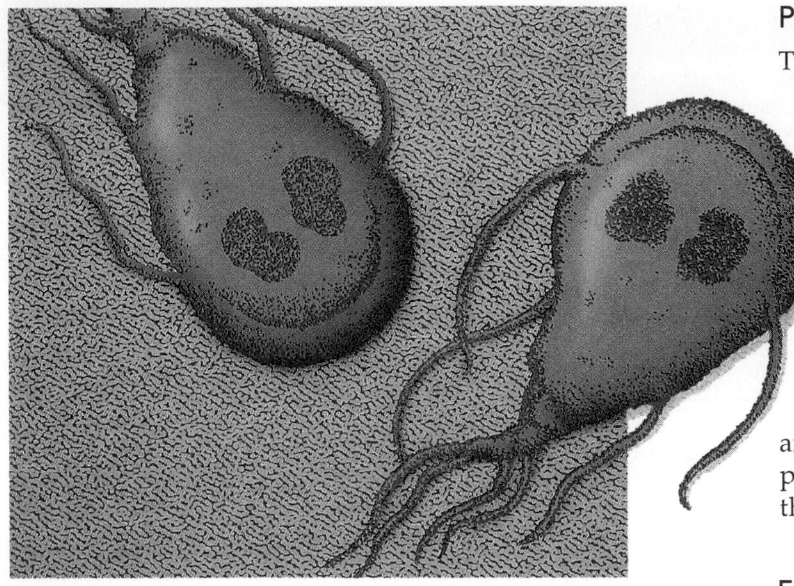

Figure 26.15
Scanning electron micrograph of *Giardia lamblia* trophozoites.

diarrhea. In Americans, *G. lamblia* is the most commonly identified intestinal parasite, and in economically underdeveloped areas of the world, its incidence often exceeds 10%.

Cause and Symptoms

G. lamblia (figure 26.15) is a flagellated protozoan shaped like a pear cut lengthwise, with two side-by-side nuclei that resemble eyes. These features along with an adhesive disk on the undersurface give the organism an unmistakable appearance. *G. lamblia* can exist in two forms: as a vegetative trophozoite or as a resting form (cyst). The trophozoite is the actively feeding form, and it colonizes the upper part of the small intestine. When trophozoites are carried by intestinal juices toward the colon, they form cysts that have thick walls composed of the tough, flexible, nitrogen-containing polysaccharide **chitin.**[1] The cyst wall protects the organism from harsh environmental conditions. As mentioned in chapter 1, *Giardia* lacks mitochondria, and its ribosomal RNA resembles that of prokaryotes.

In the usual epidemic of giardiasis, about two-thirds of the cases develop symptoms. The usual incubation period is generally 6 to 20 days. Symptoms can range from mild (indigestion, "gas," and nausea) to severe (vomiting, explosive diarrhea, abdominal cramps, fatigue, and weight loss). The symptoms usually disappear without treatment in one to four weeks, but some cases become chronic. Both symptomatic and asymptomatic cases can become long-term carriers, unknowingly excreting infectious cysts with their feces.

Pathogenesis

The cyst form is responsible for infection because, unlike the trophozoite, it is resistant to stomach acid. Two trophozoites emerge from each cyst that survives passage through the stomach and reaches the intestine. Some of these trophozoites attach to the epithelium by their adhesive disk, while others move freely in the intestinal mucus using their flagella. Some may even migrate up the bile duct to the gallbladder and cause crampy pain or jaundice. The protozoa do not destroy the host epithelium, but in severe infestations[2] they may entirely cover the epithelial surface. They interfere with the ability of the intestine to absorb nutrients and secrete digestive enzymes. Some of the impairment is probably a side effect of the host immune system attacking the parasites.

Epidemiology

Since *G. lamblia* occurs worldwide and in all types of climates, it is a common cause of diarrhea in travelers. Transmission is usually by the fecal-oral route, especially via fecally contaminated water. Known or suspected sources of *G. lamblia* include beavers, raccoons, muskrats, dogs, cats, and humans. A single human stool can contain 300 million *G. lamblia* cysts; only 10 cysts are required to establish infection. The cysts can remain viable in cold water for more than two months. In the United States, *G. lamblia* is responsible for most of the waterborne disease epidemics in which a cause is identified. Usual levels of chlorination of municipal water supplies are ineffective against the cysts, and water filtration is necessary to remove them. Giardiasis has occurred in city areas considered safe because of failure of their water filtration systems. Hikers who drink from streams, even in remote areas thought to contain safe water, are at risk of contracting giardiasis.

Although waterborne outbreaks are the most common, person-to-person contact can also transmit the disease. This mode of transmission is especially likely in day-care centers where hands become contaminated in the process of diaper changing. People who promiscuously engage in anal intercourse and fellatio are also prone to contracting the disease. Transmission by fecally contaminated food has also been reported. Good personal hygiene, especially hand washing, decreases the chance of passing on the infection.

Diagnosis and Treatment

Giardiasis can generally be diagnosed by microscopically identifying *Giardia* cysts in the patient's feces. In chronic cases, it is often necessary to examine several specimens

[1]Chitin—a polymer of N-acetyl glucosamine subunits linked in the same way as the glucose subunits in cellulose; a component of the cell walls of some eukaryotic microorganisms and the exoskeletons of arthropods; second to cellulose as the world's most abundant organic material.

[2]Infestation—living as a parasite on or in a host

• **Day-care centers, pp. 442, 443**

because the cysts may be excreted intermittently and at a given moment may be few in number. A more sensitive test for the presence of *G. lamblia* involves using an antibody to detect *Giardia*-specific antigen in the feces (figure 26.16).

Several medicines including quinacrine (Atabrine) and metronidazole (Flagyl) are effective in treating giardiasis.

The best way to make drinking water safe from giardiasis is to boil it for one minute. Using a few drops of household sodium hypochlorite bleach or tincture of iodine per quart of water or commercial water-purifying tablets is also effective. However, as in all sterilization procedures, time and temperature are important. Only an hour may be necessary to treat warm water, but many hours are required to kill cysts in cold water.

Table 26.10 describes the main features of giardiasis.

• **Antibody, see ELISA, pp. 345, 388**
• **Sterilization, pp. 201–5**

Coccidial Diarrhea, Now Common, Was Almost Unknown Two Decades Ago

Coccidia comprise a subgroup of protozoa in the class *Sporozoea*. Some coccidia multiply intracellularly in the small intestinal epithelium, their entire life cycle occurring in a single host. This section discusses two diseases caused by such coccidia that were unknown to most American physicians until very recently. In otherwise healthy individuals, the symptoms of these diseases are troublesome but rarely fatal. Infection in immunodeficient subjects, on the other hand, can be serious.

• **Sporozoa, p. 290**

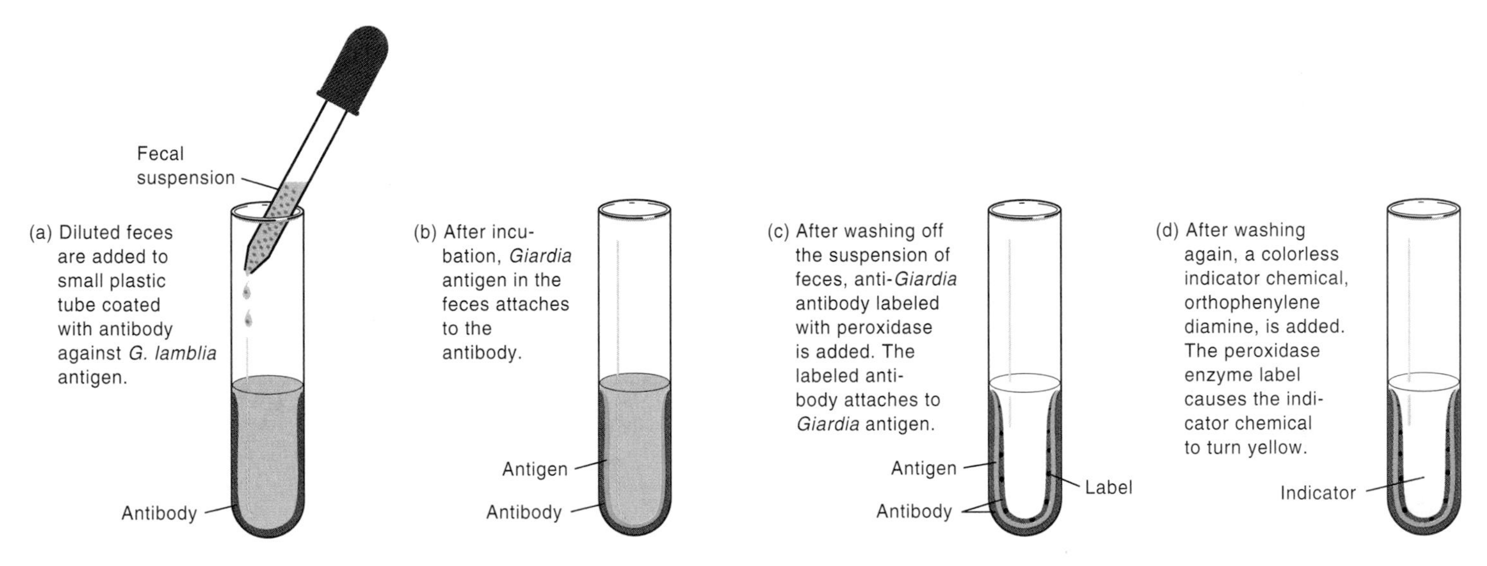

Figure 26.16
Detection of *Giardia lamblia* in stools by using an enzyme-labeled antibody against an antigen specific for *G. lamblia.*

Table 26.10 Giardiasis	
Symptoms	Mild illness: indigestion, flatulence, nausea; severe: vomiting, diarrhea, abdominal cramps, weight loss
Incubation period	6–20 days
Causative agent	*Giardia lamblia* (also known as *G. intestinalis*)
Laboratory diagnosis	Examination of feces for cysts and trophozoites or tests for specific antigen
Pathogenesis	Ingested cysts survive stomach passage to small intestine, where trophozoites are "hatched" from cysts; some trophozoites attach to epithelium, others are free moving; mucosal damage is seen, which is reversible after parasite has been eradicated by medication
Epidemiology	Ingestion of fecally contaminated water; person-to-person, in day-care centers
Prevention	Boiling or disinfecting drinking water; filtration of community water supplies
Treatment	Quinacrine hydrochloride (Atabrine) or metronidazole (Flagyl)

Symptoms

Disease caused by the coccidium *Cryptosporidium parvum* is characterized by profuse watery diarrhea, abdominal cramps, nausea, and loss of appetite, beginning after an incubation period of 4 to 12 days. The symptoms generally last 10 to 14 days, but in people with AIDS and other immunodeficiency diseases, they can last for months.

In disease caused by *Cyclospora cayetanensis*, the symptoms begin with malaise, slight fever, watery diarrhea, fatigue, loss of appetite, vomiting, and weight loss, after an incubation period of about one week. The diarrhea subsides in three to four days, but relapses occur for up to four weeks.

Causative Agents

The oocyst stage of *Cryptosporidium parvum* can be detected by microscopic examination of the patient's stool. It is a tiny (4–5 micrometers) acid-fast sphere that contains four banana-shaped sporozoites. *Cyclospora cayetanensis* oocysts appear similar, except they are larger, about 8 to 10 μm in diameter, and the organisms from feces do not contain sporozoites. For a number of years, *C. cayetanensis* was known only as a "cyanobacteriumlike body" because its true nature was unknown, and it resembled certain cyanobacteria. Later, the oocysts were found to produce two sporocysts, each containing two sporozoites. Thus, the cyanobacteriumlike body was shown to be a protozoan.

Pathogenesis

Few details are known yet about the pathogenesis of these diseases. Following ingestion, the organisms establish infection in the epithelial cells of the small intestine. The infection results in some deformity of the epithelium and intestinal villi and an inflammatory response beneath the cells. Cell-meditated immunity is important to the body's defense against the infections. Spread to organs such as the lung can occur with *Cryptosporidium parvum* in AIDS patients or others with immunodeficiency.

Epidemiology

C. parvum is widespread in natural waters and infects pets and other domestic animals. Its small size makes it very difficult to filter from drinking water, and it is also highly resistant to chlorine. Not suprisingly, there have been a number of epidemics traced to municipal water supplies. A single such outbreak in Milwaukee, Wisconsin, in 1993, involved hundreds of thousands of cases. The oocysts are fully infective when they are passed in diarrheal stool, so person-to-person spread occurs readily under conditions of poor sanitation. Epidemics have been spawned by day-care centers and contaminated swimming pools. The organism is responsible for many cases of travelers' diarrhea.

The epidemiology of *Cyclospora cayetanensis* differs in important ways from that of *Cryptosporidium parvum*. The oocysts of *C. cayetanensis* are immature when excreted in the stool and are noninfectious, so person-to-person spread does not occur. Under warm, moist conditions, it takes the oocysts five or more days to become infectious. Most infections are waterborne and occur in spring and summer. Travelers to tropical areas are more likely than others to become infected. So far, there are no proven animal sources. Fresh fruits have been implicated in a number of epidemics. Outbreaks have occurred in Canada, the United States, Nepal, Peru, Haiti, and various other countries of South and Central America, Southeast Asia, and Eastern Europe.

Prevention

The only known preventive measures are those that interrupt fecal-oral spread of disease, including sanitary disposal of feces, hand washing before eating, and sanitary preparation of food. *Cryptosporidium parvum* contamination of municipal water supplies despite strict adherence to existing standards has stimulated new research on how to improve the quality of domestic water supplies.

Treatment

No effective treatment exists for *C. parvum* infections. Trimethoprim-sulfamethoxazole is effective against *Cyclospora cayetanensis*.

Amebiasis Is a Disease of the Large Intestine Caused by *Entamoeba histolytica*

Disease resulting from *Entamoeba histolytica* infection is called **amebiasis.** In the United States, cases occur mainly among male homosexuals who have many sexual partners, in poverty-stricken areas of the South and on some Indian reservations, and among migrant farm workers. Although usually a mild disease, worldwide it causes about 30,000 deaths per year, most of which occur in Mexico, parts of South America, Asia, and Africa. Life-threatening disease occurs in these areas because of the presence of virulent *E. histolytica* strains and crowded, unsanitary living conditions.

Causative Agent

Amebiasis is commonly asymptomatic, but symptoms ranging from chronic mild diarrhea to acute dysentery and death can occur. The causative organism is *Entamoeba histolytica,* an ameba that characteristically moves in a direct ("purposeful") path across a microscope slide (figure 26.17). Its size generally ranges from about 20 to 40 μm in diameter, and like *Giardia,* it has a cyst form with a chitin-containing cyst wall. Immature cysts have a single nucleus, a large mass of glycogen, and a prominent, elongated structure called a **chromatoidal body.** The glycogen serves as an energy store and the chromatoidal body is a collection of ribosomes. As the cyst matures, the nucleus divides twice, and the glycogen and chromatoidal bodies disappear. The mature cyst, now with four nuclei (quadrinucleate cyst), is the infectious form for the next host. The life cycle of *E. histolytica* is shown in figure 26.18. The cysts of *E. histolytica* can be identified by direct microscopic examination of the feces of infected persons. If only small numbers are present, they must first be concentrated from fecal material by various laboratory methods in order to be detected.

• **Glycogen, p. 37**

E. histolytica can be grown anaerobically in pure cultures or aerobically in fluid cultures containing bacteria. The organisms grow better when certain bacteria are added to the medium, and a mixture of bacteria promotes better growth than does a single bacterial species. Indeed, in studies with some germ-free animals, *E. histolytica* caused little or no injury to the intestinal lining unless bacteria were also present.

Pathogenesis

Ingested quadrinucleate cysts of *E. histolytica* survive passage through the stomach. The organisms are released from their cysts in the small intestine, whereupon the cytoplasm and nuclei divide, yielding eight trophozoites. Upon

Table 26.11 Amebiasis

Symptoms	Diarrhea, abdominal pain, blood in feces
Incubation period	2 days–several months
Causative agent	*Entamoeba histolytica*
Laboratory diagnosis	Trophozoites seen in fecal sample; in chronic cases, cysts seen in fecal sample
Pathogenesis	Injested cysts liberate trophozoites in the small intestine; in the large intestine, trophozoites feed on mucus and cells lining intestine; digestive enzymes are produced and allow penetration of the intestinal epithelium, sometimes intestinal wall and blood vessels and thence to the liver and other organs, resulting in abscesses
Epidemiology	Ingestion of fecally contaminated water
Prevention	Good sanitation and personal hygiene
Treatment	Metronidazole, paromomycin

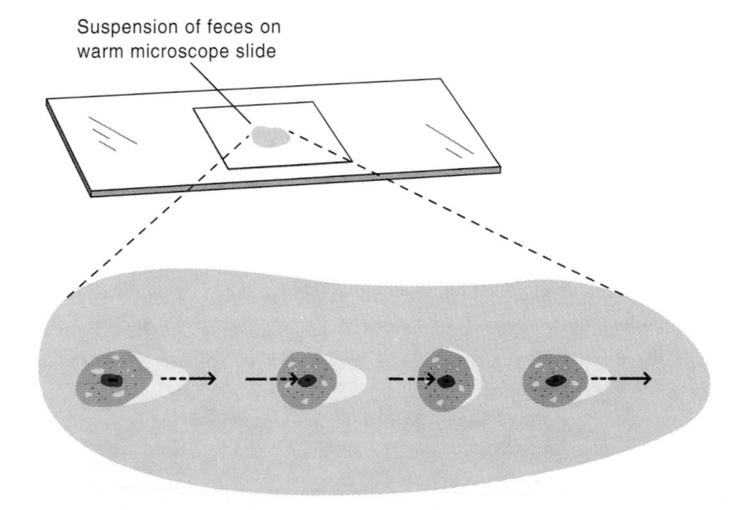

Figure 26.17
Progressive movement characteristic of *Entamoeba histolytica.*

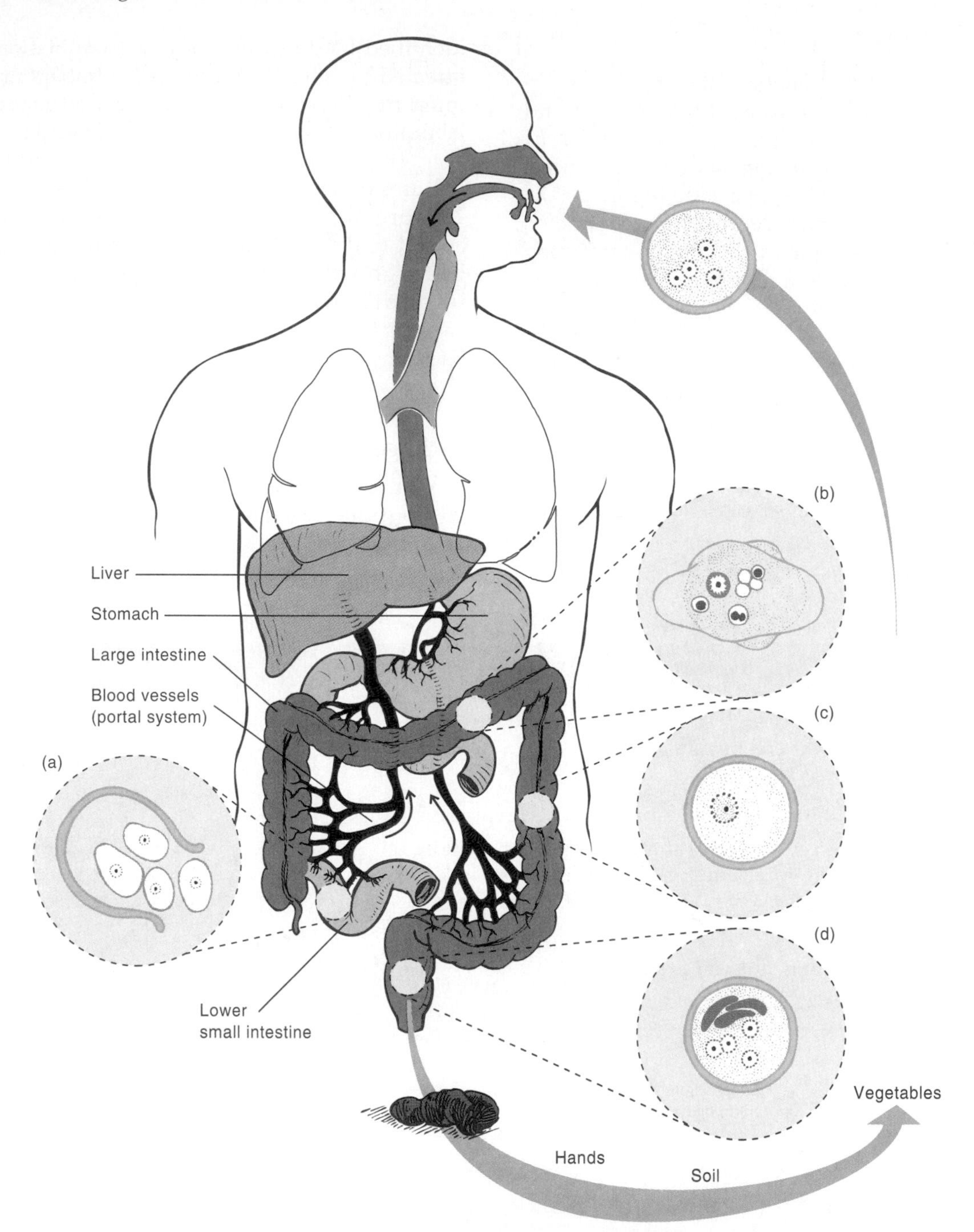

Figure 26.18
Life cycle of *Entamoeba histolytica.* An infectious quadrinucleate cyst enters the mouth on contaminated food and passes through the stomach to the lower small intestine. (*a*) Four daughter protozoa are released from each cyst and develop into the feeding form (*b*). (*c*) Dehydration of intestinal contents in the lower portion of the large intestine stimulates progressive stages in cyst development. (*d*) Mature cysts are passed in the feces to contaminate soil, water, and hands. Trophozoites that burrow into blood vessels of the intestine can be carried by the portal system to the liver, causing abscesses.

reaching the lower intestine, these trophozoites begin feeding on mucus and intestinal bacteria. Many, but not all, strains produce a cytotoxic enzyme that kills intestinal epithelium on contact, allowing the organisms to penetrate the lining cells and enter deeper tissues of the intestinal wall. Sometimes, they penetrate into blood vessels and are carried to the liver or other body organs. Multiplication of the organisms and tissue destruction in the intestine and in other body tissues can result in amebic abscesses. The irritating effect of the amebas on the cells lining the intestine causes intestinal cramps and diarrhea. Due to intestinal ulceration, the diarrhea fluid is often bloody, and the condition is referred to as **amebic dysentery.**

Table 26.11 gives the main features of amebiasis.

REVIEW QUESTIONS

1. Give two mechanisms by which bacteria cause food poisoning.
2. Explain how *S. aureus* causes food poisoning from cooked food that has no living bacteria.
3. What cultural feature distinguishes *B. cereus* from *C. perfringens?*
4. What symptom of botulism distinguishes it from most other food poisoning?
5. What is the source of aflatoxin in foods? Why is there concern about it?
6. Contrast the cause and epidemiology of hepatitis A and B.
7. Contrast the cause and epidemiology of giardiasis and amebiasis.

Summary

I. Anatomy and Physiology
   A. Important functions include breakdown of food macromolecules to their subunits, absorbing nutrients, and recycling fluids.
   B. Bile, produced by the liver, is inhibitory for many bacteria and aids digestion of fats and vitamins. It gives feces its brown color when acted on by intestinal bacteria.
   C. The pancreas produces insulin and alkaline fluid containing digestive enzymes.
   D. The small intestine secretes digestive juices and absorbs nutrients. It has a large surface area.
   E. The large intestine contains large numbers of microorganisms, many of which are opportunistic pathogens. It helps recycle body water.

II. Normal Flora of the Digestive Tract
   A. The small intestine has few microorganisms.
   B. Microorganisms make up about one-third of the weight of feces.
   C. Anaerobes of the *Lactobacillus* and *Bacteroides* genera are the most prevalent.
   D. The biochemical activities of microorganisms in the digestive tract include synthesis of vitamins, degradation of indigestible substances, competitive inhibition of pathogens, production of cholesterol, chemical alteration of medications, and carcinogen production.

III. Infectious Intestinal Disease

IV. Bacterial Diseases of the Digestive Tract
   A. Diarrhea is a major cause of death worldwide.
   B. Cholera is a severe form of diarrhea caused by a toxin of *Vibrio cholerae* that acts on the small intestinal epithelium.
   C. Species of *Shigella* are common causes of dysentery because they invade the colon epithelium.
   D. Gastroenteritis is often caused by salmonellas of animal origin. The organisms often enter the intestinal tract with food, usually eggs and poultry.
   E. Gastroenteritis can be caused by certain strains of *Escherichia coli*. However, only strains possessing virulence factors cause disease. These factors often depend on plasmids. Plasmids and, therefore, virulence can be transferred to other enteric organisms.
   F. *Campylobacter jejuni* is a common bacterial cause of diarrhea in the United States. Like the salmonellas, it usually originates from domestic animals. Like the shigellas and certain strains of *E. coli*, it can cause dysentery.
   G. Typhoid fever is caused by *Salmonella typhi*.
      1. The disease is characterized by high fever, headache, and abdominal pain. Untreated, it has a high mortality rate.
      2. Human carriers maintain the disease.
      3. A new oral attenuated vaccine helps in prevention.

V. Food Poisoning
   A. Illness is caused by toxic products of microbial growth in food.
   B. Staphylococcal food poisoning, caused by certain strains of *Staphylococcus aureus*, is one of the most common forms. The toxin is heat-stable.
   C. Food poisoning is also caused by the spore-forming, gram-positive rods *Clostridium perfringens* and *Bacillus cereus*.
   D. Botulism is characterized by paralysis and occurs in three forms: foodborne, intestinal, and wound.

VI. Poisoning by Moldlike Fungi
   A. Growth of certain species of fungi on grain can cause ergot poisoning.
   B. Aflatoxins are fungal poisons produced by molds of the genus *Aspergillus*. Liver damage and certain cancers can result from ingestion of aflatoxins.

VII. Viral Infections of the Lower Alimentary System
   A. Various viruses infect the liver, causing acute and chronic disease.
      1. Hepatitis A virus disease (HAV) is spread by fecal contamination of hands, food, or water. Most cases are mild and may even be asymptomatic. Antibody to the virus is widespread in the population.
      2. Hepatitis B virus (HBV) is spread mainly by blood and blood products but also by sexual intercourse and from mother to infant. HBV is generally more severe than HAV. Chronic infection is fairly frequent and can lead to scarring of the liver and liver cancer. Carriers are common and can have infectious virus in their bloodstream for years without knowing it.
      3. Other hepatitis viruses are hepatitis C virus (HCV), hepatitis delta virus (HDV), hepatitis E virus (HEV),

and hepatitis G virus (HGV). HCV causes most of the cases of transfusion-associated hepatitis. HDV is a defective RNA virus that requires the presence of HBV to replicate. HEV, an RNA virus, is transmitted by the fecal-oral route and can cause fatalities, especially in pregnant women. HGV causes some cases of transfusion-associated hepatitis.

B. A number of viruses cause vomiting and diarrhea (gastroenteritis). Rotaviruses and Norwalk viruses are the most important causes of this disease. Rotaviruses affect mainly children but also can cause traveler's diarrhea. Norwalk viruses cause about half of the U.S. gastroenteritis outbreaks.

VIII. Protozoan Diseases of the Lower Alimentary System

A. Giardiasis is caused by *Giardia lamblia,* which is usually transmitted by drinking water contaminated by feces of humans and wild animals. It is distributed worldwide and is a common cause of traveler's diarrhea.

B. Coccidial diarrhea is mainly caused by *Cryptosporidium parvum* and *Cyclospora cayetanensis*.

C. *Entamoeba histolytica* is an important cause of dysentery. Amebiasis is often chronic, and infection can spread to the liver and other organs.

Chapter Review Questions

1. What might the lack of a brown color of feces indicate?
2. What is ADP-ribosylation?
3. How can shigellas move from one host cell to another even though they are nonmotile?
4. Contrast the sources of *Shigella* and most *Samonella* strains pathogenic for humans.
5. Name four different groups of *Escherichia coli* based on their pathogenic mechanisms.
6. What is the hemolytic-uremic syndrome? Name an organism that can cause it. Give one documented source of this organism.
7. What is the usual source of *Campylobacter jejuni* infections?
8. Name the causative agent of a kind of food poisoning that results in paralysis without necessarily infecting the victim.
9. Name two kinds of hepatitis for which inactivated vaccines are available.
10. Name two causative agents of coccidial diarrhea. How does treatment differ for the two?

Critical Thinking Questions

1. One reason given by Peruvian officials for not chlorinating their water supply was that chlorine can react with substances in water to produce carcinogens. How do you assess the relative risks of chlorinating or not chlorinating drinking water?
2. HBV vaccine could be considered an anticancer vaccine. Since liver cell cancer probably has multiple causes, how would you measure the sucess of an anticancer vaccination program?

Further Reading

Blacklow, N. R., and Greenberg, H. B. 1991. Viral gastroenteritis. *The New England Journal of Medicine* 325:250.

1995. Cholera's new coat. *Discover* 16:17.

Colwell, R. 1996. Global climate and infectious disease: the cholera paradigm. *Science* 274:2025.

1996. Cryptosporidium and enteric disease: a case in point. *ASM News* 62:14.

1995. Foodborne botulism—Oklahoma. *Morbidity and Mortality Weekly Report* 44(11):200–202.

Glausiusz, Josie. 1996. How cholera became a killer. *Discover* 17: 28.

Hirschhorn, Norbert, and Greenough III, W. B. 1991. Progress in oral rehydration therapy. *Scientific American* 264:50.

Huttner, W. B. 1993. Snappy exotoxins. *Nature* 365:104–105.

Linnen, Jeff, et al. 1996. Molecular cloning and disease association of hepatitis G virus: a transfusion-transmissible agent. *Science* 271:505–508.

1996. Outbreak of *Escherichia coli* 0157:H7 infection—Georgia and Tennessee, June 1995. *Morbidity and Mortality Weekly Report* 45(12):249–251.

1996. Outbreaks of *Salmonella* serotype enteritidis infection associated with consumption of raw shell eggs—United States, 1994–1995. *Morbidity and Mortality Weekly Report* 45(34):737–742.

Pereira, B. J. G., et al. 1991. Transmission of hepatitis C virus by organ transplantation. *The New England Journal of Medicine* 325:454.

Portnoy, D. A., and Smith, G. A. 1992. Devious devices of salmonella. *Nature* 357:536. Gives new information on how salmonellas enter host cells.

1996. *Shigella sonnei* outbreak associated with contaminated drinking water—Island Park, Idaho, August 1995. *Morbidity and Mortality Weekly Report* 45(11):229–231.

Tiollais, Pierre, and Buendia, Marie-Annick. 1991. Hepatitis B virus. *Scientific American* 264:116.

Waldor, M. K., and Mekalanos, J. J. 1996. Lysogenic conversion by a filamentous phage encoding cholera toxin. *Science* 272:1910–1914.

Waters, Tom, and Winters, Dan. 1992. The fine art of making poison. *Discover* 13:29. Discusses some uses of *Clostridium botulinum* toxin for treating certain conditions.

27 Genitourinary Infections

KEY CONCEPTS

1. The flushing action of urination is a key defense mechanism against bladder infections.
2. During the childbearing years, a woman's hormones are important in the vagina's resistance to infection because estrogen influences the vagina's normal flora.
3. Transmission of genital tract infections usually requires direct human-to-human contact.
4. Unsuspected sexually transmitted infections of pregnant mothers pose a serious threat to fetuses and newborn babies.
5. Asymptomatic infections can cause serious genital tract damage and are transmissible to other people.

Scanning electron micrograph of *Treponema pallidum*, the cause of syphilis. During 1990, the incidence of syphilis was ten times higher in the United States than any other industrialized country. Syphilis, serious in itself, also increases the risk of contracting AIDS.

Microbes in Motion

The following books in the *Microbes in Motion* CD-ROM may serve as a useful preview or supplement to your reading: *Gram-Positive Organisms. Gram-Negative Organisms. Miscellaneous Bacteria. Microbial Pathogenesis in Infectious Diseases:* Toxins. *Vaccines:* Vaccine Targets. *Viral Structure and Function:* Viral Diseases. *Fungal Structure and Function:* Diseases. *Parasitic Structure and Function.*

A Glimpse of History

Although some historians argue that syphilis was transported to Europe from the New World by Columbus's crew, others find convincing evidence, including biblical references, that syphilis existed in the Old World for many years before Columbus returned from the New World. History and literature record that kings, queens, statesmen, and heroes degenerated into madness or mental incompetence or were otherwise seriously disabled probably as a result of the later stages of syphilis. Henry VIII (King of England, 1509–1547), Ivan the Terrible (Czar of Russia, 1547–1584), Catherine the Great (Empress of Russia, 1762–1796), and Benito Mussolini (Premier of Italy, 1883–1945) are a few of the famous people who probably suffered from syphilis.

Syphilis was first named the "French pox" or the "Neapolitan disease" because it was believed to have come from France. In 1530, Girolamo Fracastoro, an Italian physician who suggested the germ theory long before the discovery of microorganisms, wrote a poem about a shepherd named Syphilis who had ulcerating sores covering his body. The description matched the symptoms of syphilis, and from that time on, the disease was known by the shepherd's name.

Initially, the method of transmission was unknown, but gradually it became generally recognized that the disease was sexually transmitted. The symptoms of syphilis were very severe in the early years of the epidemic, often causing death within a few months. Treatment, consisting of doses of mercury and guaiacum,* had little beneficial effect. As the decades went by, mutations and natural selection resulted in more resistant hosts and a less virulent microbe, and the disease evolved into the chronic illness known today.

By the late 1800s, the disease was shown to be transmissible to laboratory animals, but the various bacteria readily seen in syphilitic sores all proved to be contaminating organisms not related to the cause of the disease.

In 1905, Fritz Schaudinn, a German protozoölogist, examined some material from a syphilitic sore and saw a faintly visible organism "twisting, drilling back and forward, hardly different from the dim nothingness in which it swam." When Schaudinn used dark-field illumination, he was able to see the organism much more clearly. It appeared very thin and pale, similar to a corkscrew without a handle. The spirochetes appeared in specimens from other cases of syphilis, and Schaudinn later succeeded in staining them. By this time, he felt certain he had discovered the cause of syphilis, but the organism could not be cultivated on laboratory media. Schaudinn named the organisms *Spirochaeta pallida*, "the pale spirochete." (This organism is now called *Treponema pallidum*, from *treponema* for "a turning thread.")

*Guaiacum—the resin from a tropical American tree of the genus *Guaiacum*, formerly used to treat a variety of diseases

• Mercury, pp. 177, 214–215

The human body is constructed to prevent microbial invasion, a fact particularly evident in studying the anatomy of the genitourinary tract. Under certain circumstances, however, the genitourinary tract can be invaded by a variety of opportunistic pathogens from the normal flora as well as by other species.

Anatomy and Physiology

The urinary tract consists of the kidneys (in the upper urinary tract), the bladder (in the lower urinary tract), and accessory structures. The kidneys act as a specialized filtering system to clean the blood of many waste materials, selectively reabsorbing substances that can be reused. Waste materials are excreted in the urine, which is usually acid (about pH 6) because of excess hydrogen ions from foods and metabolism. The normal pH can range from about 4.8 to 7.5, but a consistently alkaline urine suggests infection with a urease-producing bacterium that converts the urea in urine to ammonia. Figure 27.1 shows that each kidney is drained by a tube called the **ureter,** which connects it with the urinary bladder. The bladder acts as a holding tank. Once filled, it empties through the **urethra.**

The Urinary Tract Is Generally Well-Protected from Infection

Infections of the urinary tract occur far more frequently in women than in men because the female urethra is short (about 1.5 inches, compared with 8 inches in the male) and is adjacent to the genital and intestinal tracts. Special groups of muscles near the urethra keep the system closed most of the time and help prevent infection. The downward flow of urine also helps clean the system by flushing out microorganisms before they have a chance to multiply and cause infection.

The urinary tract is protected from infection by a number of mechanisms besides its anatomy. Normal urine contains antimicrobial substances such as organic acids and small quantities of antibodies. During urinary tract infections, larger quantities of specific antibodies can be found in the urine. There is also evidence that antibody-forming lymphoid cells infiltrate the infected kidneys or bladder and form protective antibodies locally at the site at which they are needed. In addition, during infection an inflammatory response occurs in which phagocytes are of the utmost importance in engulfing and destroying the invading microorganisms.

Many kidney infections are difficult to eradicate, are often chronic, and can destroy normal kidney functions. Failure of both kidneys results in death unless the patient is fortunate enough either to be able to use an artificial kidney or to receive a kidney transplant.

The Female Genital Tract Is More Often Involved in Infection than Is the Male Genital Tract

The anatomy of the female and male genital tracts is shown in figure 27.2. In women, ova are expelled from the ovary and swept into the adjacent fallopian tube and then into the

• Phagocytes, p. 419

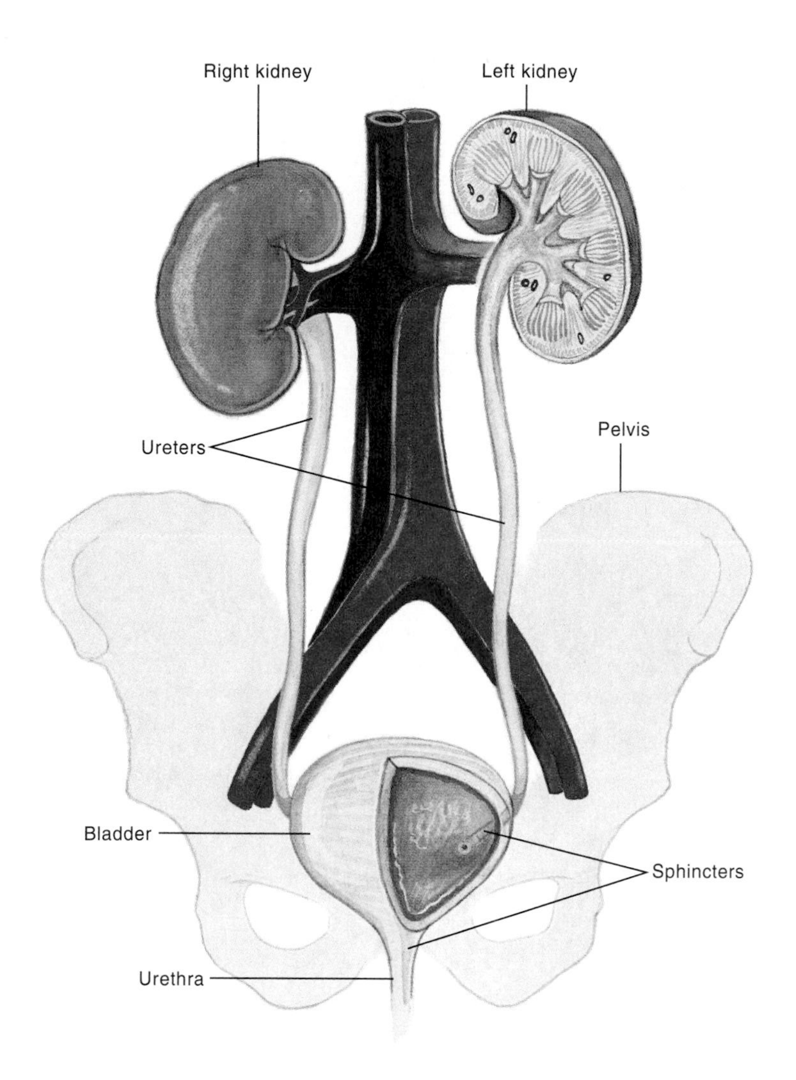

Figure 27.1
The anatomy of the urinary tract. Urine flows from the kidneys, down the ureters, and into the bladder, which empties through the urethra. Sphincter muscles help to prevent any contaminants from ascending.

uterus, by the action of ciliated cells (figure 27.2 *b*). However, such cells can be destroyed by infection. In addition, the vagina is subject to colonization by potential pathogens from the anus.

Normal Flora of the Genitourinary Tract

Normally, the urine and urinary tract above the entrance to the bladder are essentially free of microorganisms; the lower urethra, however, has a normal resident flora. Species of *Lactobacillus*, *Staphylococcus* (coagulase-negative), *Corynebacterium*, *Haemophilus*, *Streptococcus*, and *Bacteroides* are common inhabitants of the normal urethra.

The normal flora of the genital tract of women is influenced by the action of estrogen hormones on the lining (epithelial) cells of the vaginal mucosa. When estrogens are present, glycogen is deposited in these epithelial cells. The glycogen is then converted to lactic acid by lactobacilli, resulting in an acidic pH that inhibits the growth of many potential pathogens. Thus, the normal flora and resistance to infection of the female genital tract vary considerably with a woman's hormonal status. For several weeks after birth, the vagina of newborn girls remains under the influence of maternal hormones and has an acidic pH. During this time, lactobacilli predominate in the normal flora. As the influence of the maternal hormones decreases, however, the pH of the infant vagina increases to neutrality and remains neutral until puberty. During this childhood period, the normal flora of the vaginal tract consist of a variety of cocci and rod-shaped organisms, and are highly susceptible to a variety of bacterial pathogens including *Streptococcus pyogenes* and *Neisseria gonorrhoeae*. At puberty and throughout the childbearing years, lactobacilli again become predominant, although smaller numbers of yeasts and other bacterial species are also present, and resistance to infection is high. After the menopause, the tract returns to a neutral pH and to the mixed flora typical of childhood and is again susceptible to infection.

Urinary Tract Infections

A number of factors can predispose the urinary tract to infection. Any situation in which the urine does not flow naturally increases the chance of such infection. After anesthesia and major surgery, for example, the reflex ability to void urine may be inhibited for a time. In this circumstance, urine accumulates and distends the elastic bladder. Even a few bacteria that manage to evade the urinary tract defenses and enter the bladder can multiply to high levels during this time, causing infection. Urine provides abundant nutrients for many species of bacteria. Paraplegics[1] are almost always afflicted with urinary infections. Since they lack nerve control, paraplegics are unable to void normally and require a catheter (tube) that stays in the bladder to carry urine to an outside container. Many other medical conditions also require insertion of a bladder catheter for periods ranging from several days to months. This connection to the outside makes it easier for pathogens to reach the bladder and cause urinary tract infections. In the United States, about 0.5 million hospitalized patients develop bladder infections each year, mostly after catheterization.

Bladder Infections in Women Are Sometimes Related to Sexual Intercourse

Since the vagina is adjacent to the urethra in women, the rubbing action associated with sexual intercourse may introduce organisms from the lower urethra into the urinary

[1]Paraplegic—individual with paralysis of the lower half of the body

• **Lactobacilli, pp. 261–62**

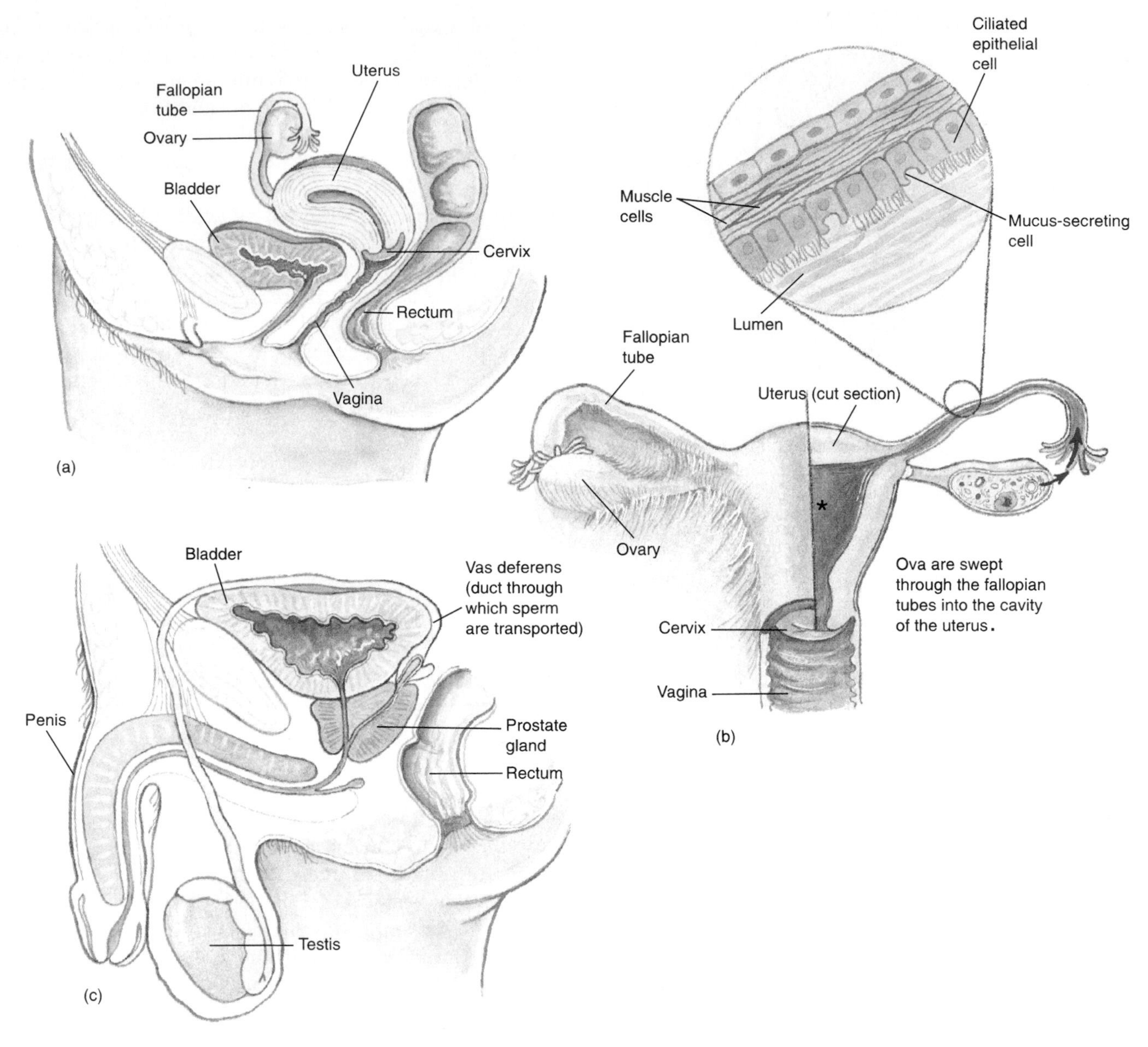

Figure 27.2
The anatomy of the genital tract. (*a* and *b*) Female and (*c*) male.

bladder. In fact, many women develop their first bladder infections following their first sexual intercourse. The main symptoms of bladder infection are frequent urination associated with a burning pain. For reasons that are not clear, using a diaphragm or taking birth control pills increases colonization of the vagina by urinary pathogens and the frequency of bladder infections.

Kidney Infections May Follow Infections of the Bladder

Unless infection of the lower urinary tract is overcome either by antimicrobial treatment or by natural defense mechanisms, it can ascend through the ureters and involve the kidneys. Thus, most kidney infections develop after a bladder infection. However, bacteria circulating in the bloodstream can sometimes infect the kidneys.

Bladder Infections Usually Originate from the Normal Intestinal Flora

More than 90% of urinary tract infections are caused by bacterial species that are part of the normal fecal flora and consequently can readily contaminate the genital area and invade the urinary tract when its natural defenses are weakened. Studies of infecting organisms have shown that they are usually strains from the patient's own intestinal bacteria rather than from the environment. In these studies,

Perspective 27.1

Urine Specimens Are Used to Diagnose Urinary Infections

The organisms causing urinary tract infections can usually be recovered easily and in large numbers from the urine of infected patients. Careful collection of such specimens is necessary in order to avoid external contamination from the patient's genitalia or intestinal tract. Avoiding specimen contamination is more difficult with women than with men. Normal urine usually contains less than 10,000 (10^4) viable bacteria per milliliter, and concentrations of bacteria greater than 100,000 (10^5) per milliliter indicate infection. Lower counts are usually due to normal flora bacteria that enter the urine as it passes from the lower urethra, but they may indicate infection if they are composed of a single species of recognized urinary pathogen. Since many bacterial species multiply exponentially in urine, cultures for colony counts must be done promptly after the urine is collected in order to obtain meaningful counts of urinary bacteria.

• **Colony counts, p. 89**

for example, *Escherichia coli* isolated from the urine during urinary tract infections was almost always of the same serotype[1] as *E. coli* from the intestine of the patient. The *E. coli* strains that cause urinary infections have adhesins that attach specifically to receptors on bladder and kidney epithelium. Using genetic engineering, adhesins can be produced in large quantities for studies. When the adhesins are injected into experimental animals, the animals produce antiadhesin antibodies that protect them against bladder infections.

E. coli accounts for the majority of urinary tract infections in young, otherwise healthy women, but other gram-negative rods of different genera, such as *Proteus* and *Enterobacter*, can also be the cause (figure 27.3), particularly in hospitalized patients or those with a predisposing condition such as bladder catheterization or diabetes. *Pseudomonas aeruginosa*, an aerobic, gram-negative rod, is a particularly troublesome urinary tract pathogen because its resistance to antimicrobial medicines makes it difficult to treat successfully. *Enterococcus faecalis* (commonly called *enterococcus*), found in the normal bowel flora, is the most frequently isolated gram-positive organism responsible for urinary tract infections.

Urine specimens are cultured on media such as MacConkey agar especially suitable for the growth of enterobacteria such as *E. coli* and on blood agar, which will also support the growth of enterococci and some other gram-positive pathogens (figure 27.4).

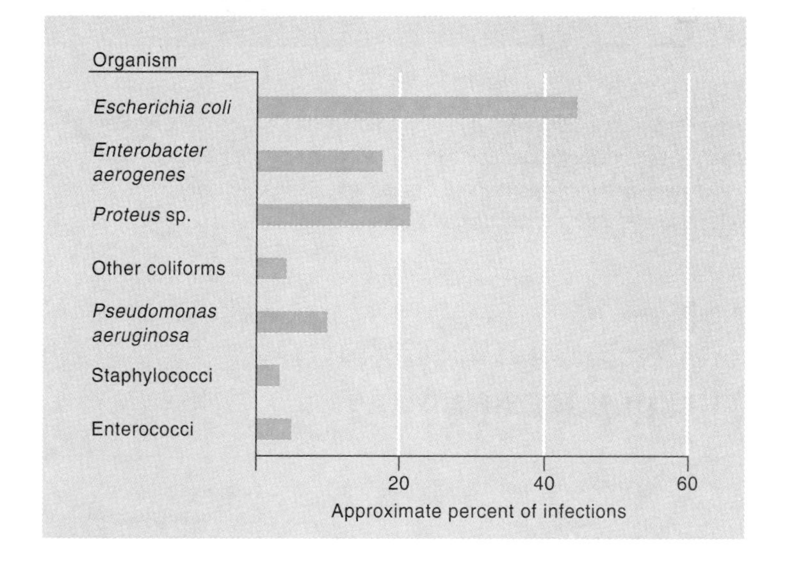

Figure 27.3
Common causes of urinary tract infections. The actual percentages vary according to the population studied.

Microorganisms from Disseminated Disease Can Be Excreted in Urine

The causative organisms of disease involving areas of the body other than the urinary tract can also be found in the urine. Thus, in typhoid fever, *Salmonella typhi* organisms become disseminated throughout the body and are excreted in the urine about one to two weeks after infection, sometimes establishing a chronic urinary tract infection. In like manner, slender spirochetes of the genus *Leptospira* (the cause of leptospirosis) may become established in the kidneys. Water contaminated with urine from infected humans or animals can be responsible for leptospirosis in swimmers. The *Leptospira* organisms may continue to be excreted in the urine for many months after recovery from the illness. Leptospirosis is a common infection of animals and is one of the diseases against which dogs are commonly immunized.

[1]Of the same serotype—having the same antigens, as detected by using known antisera

• ***Pseudomonas*, pp. 246–47**

• ***S. typhi*, p. 580**
• ***Leptospira*, p. 244**

Figure 27.4
Appearance of *Escherichia coli* on blood (left) and MacConkey agar, (right) two media commonly used in the diagnosis of urinary infections. Individual colonies on the MacConkey medium are red, indicating fermentation of the lactose in the medium.

Genital Tract Infections

The genital tract is the portal of entry for numerous infectious diseases. Many of these diseases seem to affect women more severely than men. Some of these infectious diseases are transmitted by sexual intercourse. Serious genital tract infections can follow childbirth and spontaneous or induced abortions. The normal vagina can also be colonized by various pathogens, producing symptoms that range from annoying to life-threatening.

Sexually transmitted diseases have become increasingly widespread over the past several decades in association with relaxation of social taboos, the availability of oral contraceptives, inadequate education of adolescents about the responsibilities associated with sexual intercourse, and a decrease in funding for tracking contacts of diagnosed cases. A rise in male and female prostitution due to poverty or to support drug addiction has been another important epidemiological factor, especially in the spread of syphilis and AIDS. The health cost to the nation is alarming, more so since the wide dissemination of the human immunodeficiency virus which causes AIDS.

The list of diseases that can be transmitted sexually is very long and includes shigellosis, giardiasis, viral hepatitis, and many others that have nonsexual modes of transmission as well. The appearance of a sexually transmissible disease in an individual suggests the possibility that he or she has more than one such disease from unprotected sexual intercourse with numerous partners or with one partner who has had numerous partners (figure 27.5). Some symptoms that suggest a sexually transmitted disease and warrant clinical evaluation are listed in perspective 27.2. In theory, prevention is simple: abstinence or a monogamous relationship with a noninfected person. Proper use of latex condoms and spermicides, while not an absolute guarantee of protection, probably markedly reduces the risk of acquiring a sexually transmitted disease.

Some of the important sexually transmitted diseases are discussed in the following sections. Additional information is given in the discussion of HIV infection in chapter 31.

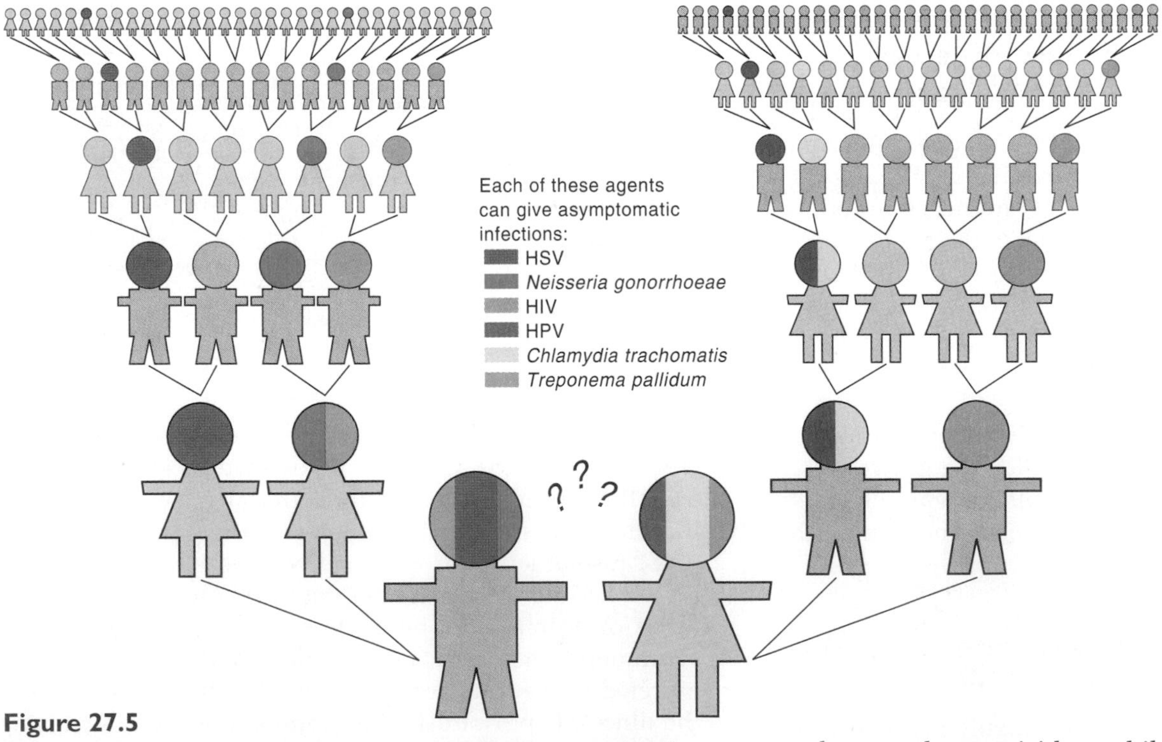

Figure 27.5
The possible risk of having an asymptomatic sexually transmitted disease in two individuals contemplating unprotected sexual intercourse. Each partner had two previous sexual partners, and each of these partners had two previous partners, and so on. (HSV—herpes simplex virus, type 2; HIV—human immunodeficiency virus; HPV—human papilloma virus.)

Perspective 27.2

Symptoms Suggestive of a Sexually Transmitted Disease

The following symptoms can indicate the possibility of a sexually transmitted disease, even if they go away by themselves, particularly if they appear within a few weeks of a person having sex with a new partner.

1. Abnormal discharge from the vagina or penis
2. Pain or burning sensation with urination
3. Sore or blister, painful or not, on the genitals or nearby; swellings in the groin
4. Abnormal vaginal bleeding or unusually severe menstrual cramps
5. Itching in the vaginal or rectal area
6. Pain in the lower abdomen in women; pain during sexual intercourse
7. Skin rash or mouth lesions

Sexually Transmitted Diseases: Bacterial

Most of the bacteria that cause sexually transmitted diseases survive poorly in the environment. Transmission from one person to another usually requires intimate physical contact and is highly unlikely to occur via a handshake or contact with a contaminated toilet seat.

Gonorrhea Is High on the List of Reportable Bacterial Diseases in the United States

Gonorrhea is not a new problem. Thirteen % of World War I draftees were found to be infected with syphilis or gonorrhea. Today's concerns are its prevalence and rapidly increasing resistance to antibacterial treatment.

Symptoms

The incubation period of gonorrhea is generally short—symptoms appear two to five days after exposure. In men, gonorrhea is characterized by inflammation of the urethra, with pain during urination and a thick, pus-containing discharge. Complications are rare since most infections cause symptoms that inspire prompt treatment. The disease in men can be self-limiting and disappear on its own accord. However, if the disease is not treated, there is a risk that the infection may spread to the prostate gland and testes, producing complications. For example, extensive inflammatory reaction to the infection can lead to the formation of scar tissue that partially obstructs the urethra, predisposing the man to future urinary tract infection. If scar tissue blocks the tubes that carry the sperm or if tissue in the testes is destroyed by inflammation, sterility may result.

Gonorrhea follows a different course in women. Gonococci thrive in the cervix and fallopian tubes, as well as in vaginal glands and other areas of the female genital tract. Symptomatic infections in women usually begin with painful urination (dysuria) and vaginal discharge. However, the infection can also be asymptomatic or mild, and often the victim is unaware of the disease. Thus, women are especially likely to be unknowing carriers of gonorrhea. If the disease is not limited at this stage, 15% to 30% of infections progress upward through the uterus into the fallopian tubes, causing **pelvic inflammatory disease (PID).** Scar tissue formed as a result of the infection in the fallopian tubes may block normal passage of the ova through the tubes, causing sterility.

Scarring of a fallopian tube can also lead to a dangerous complication—**ectopic pregnancy**—in which the ovum is fertilized and develops in the fallopian tube or even in the abdominal cavity outside the uterus. Ectopic pregnancy commonly leads to life-threatening internal hemorrhaging.

The areas of the body affected by gonorrhea discussed in the preceding paragraphs are illustrated in figure 27.6.

Causative Agent

The causative organism of gonorrhea is *Neisseria gonorrhoeae* (gonococcus), a gram-negative diplococcus that shows flattening of adjacent sides. The organisms are typically found on and within the leukocytes in urethral pus (figure 27.7). They grow under aerobic conditions but only on special media such as chocolate agar (an agar medium that has a chocolate color because it contains heated blood). Growth of the gonococci either requires or is enhanced by carbon dioxide, depending on the strains. The organisms are oxidase positive, meaning that their colonies give a positive reaction to the enzyme cytochrome C oxidase (figure 27.8). Glucose is metabolized by the organisms, yielding acid by-products. Gonococci are parasites of humans only, preferring to live on the mucous membranes of their host. Most strains are susceptible to cold and drying and, hence, do not survive well outside the host. For this reason, gonorrhea is transmitted primarily by direct

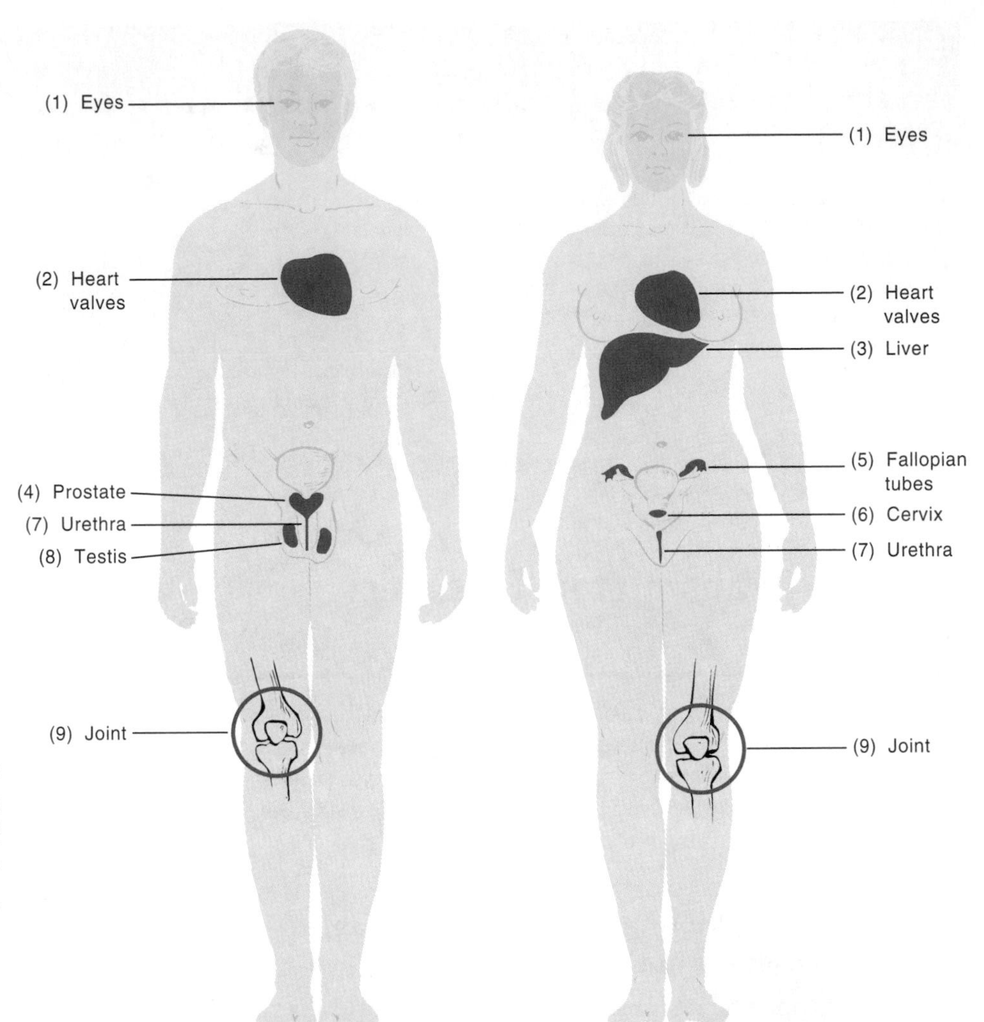

Figure 27.6
Complications of gonorrhea. (*1*) Eyes of both adults and children are susceptible to the gonococcus; serious infections leading to loss of vision are likely in the newborn. (*2*) Organisms carried by the bloodstream affect the heart valves and cause damage. (*3*) The outer covering of the liver is infected when gonococci enter the abdominal cavity from infected fallopian tubes. (*4*) Prostatic gonococcal abscesses may be difficult to eliminate. (*5*) Infection of the fallopian tubes results in scarring, which can lead to sterility or ectopic pregnancy. (*6*) The cervix is the usual site of primary infection in women. (*7*) Urethral scarring from gonococcal infection can predispose to urinary tract infections by other organisms. (*8*) Scarring of testicular tubules can cause sterility. (*9*) Gonococcal arthritis results from organisms carried by the blood.

contact, and since the bacteria mainly live in the genital tract, this contact is almost always sexual. An increasing percentage of strains contain R plasmids that render them resistant to antibiotic medicines, including penicillin and tetracycline.

Pathogenesis

Gonococci selectively attach to certain epithelial cells of the body, notably those of the urethra, uterine cervix, pharynx, and conjunctiva. Initial attachment is by pili (fimbriae) that project from the surface of the cocci (figure 27.9) and attach specifically with receptors on host cells. These pili, as well as certain other surface proteins involved in attachment, can either be expressed or not expressed (phase variation), but *N. gonorrhoeae* is noninfectious when it lacks pili, since it cannot attach without them. A single strain of gonococcus can genetically express many different kinds of pili by chromosomal rearrangements within the pili genes. This variation in pili expression explains why cultures from different body sites in an infected individual and those from his or her sex partner often yield *N. gonorrhoeae* with different types of pili. Presumably, this ability of the gonococcus to genetically express different surface antigens (pili) allows it to attach to many different kinds of cell receptors. It also allows the organism to escape the effects of antibody formed against one kind of pilus. The ability to express a variety of surface antigens has complicated the task of

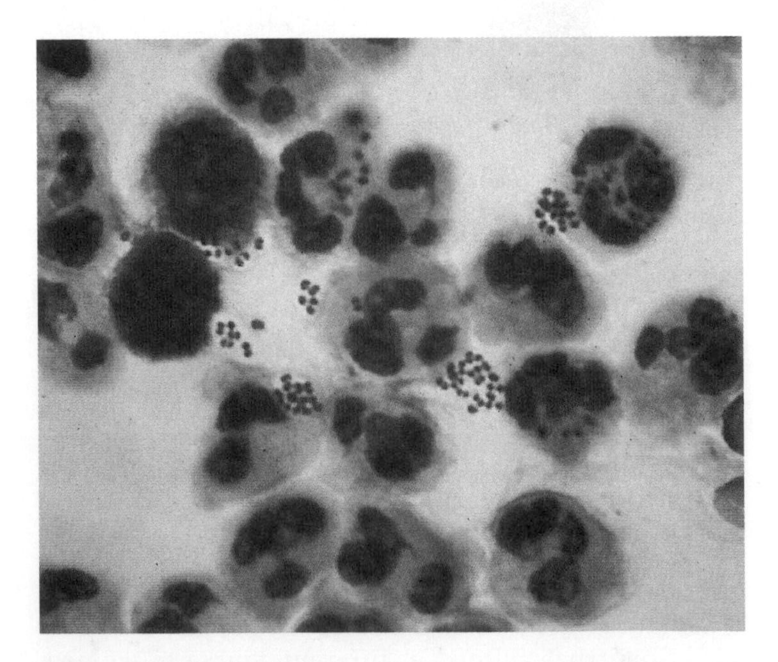

Figure 27.7
Appearance of *Neisseria gonorrhoeae* in pus from the urethra.

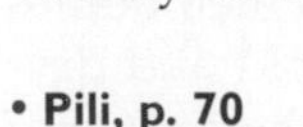

• **Pili, p. 70**
• **Phase variation, pp. 575, 576**

Perspective 27.3

Plasmid-Mediated Antibiotic Resistance of *Neisseria gonorrhoeae*

A 30-year-old man was seen by a physician in late November 1979 because of symptoms of acute urethritis (pain on urination and discharge of pus from the urethra). The diagnosis of gonorrhea was made by a stained smear of the urethral pus showing gram-negative, intracellular diplococci. The presence of *N. gonorrhoeae* was confirmed by culture. The patient was treated with a single, large dose of ampicillin given by mouth, but symptoms persisted, and in mid-December he was treated again, this time with 2.4 million units of penicillin injected into each buttock. This treatment also failed, and in mid-January he was again treated, this time with large doses of tetracycline. When his symptoms failed to improve after a week, another culture was obtained and the infecting strain of *N. gonorrhoeae* was sent to a reference laboratory to be tested for susceptibility to antimicrobial medicines.

The results showed that the patient's strain of *N. gonorrhoeae* was highly resistant to the drugs he had received. Moreover, resistance to penicillin and ampicillin resulted from the enzyme penicillinase, indicating the presence of a resistance plasmid. Immediately, the Public Health Department began the task of identifying the man's sexual partners, interviewing them, and learning the names of their other sexual partners. Cultures from all these people and from all other gonococcal cultures that came into the Public Health laboratory were tested for penicillinase-producing *N. gonorrhoeae*. By the end of April, 25 cases of gonorrhea were identified as resulting from the resistant strain. Unfortunately, because of the long time lapse before the case was investigated, the person who originated this epidemic could not be traced.

The first case of gonorrhea due to a penicillinase-producing, R plasmid–containing gonococcus was reported in March 1976. The disease was contracted in southeast Asia. By 1986, 2% of all cases reported in the United States contained the plasmid, and by 1990, almost 1 out of 12 had it. As in the previous account, these plasmids often conferred resistance to other antibiotics beside the penicillins, such as tetracycline. Obviously, if this trend continues, gonorrhea will become increasingly more difficult to treat.

• **Penicillinase, p. 147**

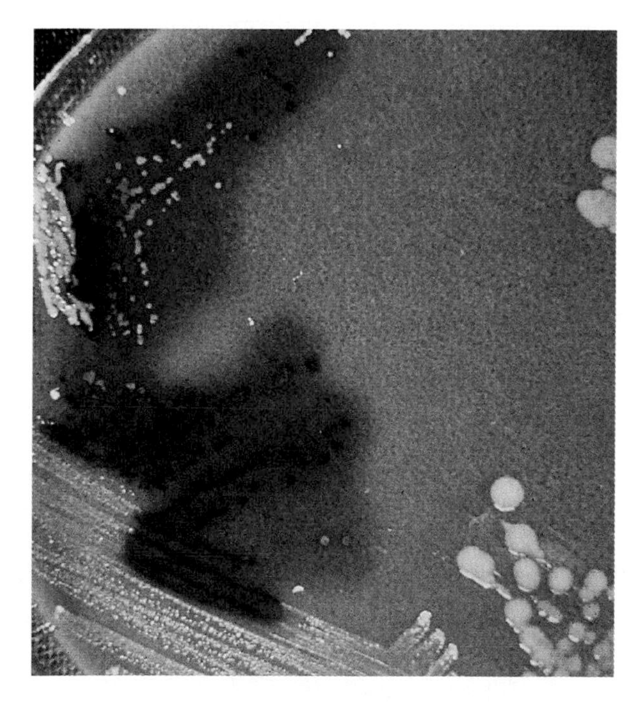

Figure 27.8
Oxidase test. The dark-colored colonies are showing a positive oxidase reaction to a reagent that has been dropped on the plate.

developing a vaccine for the prevention of gonorrhea. Besides playing a role in attachment, pili interfere with phagocytosis by macrophages.

It is not clear how *N. gonorrhoeae* (which is nonmotile) can traverse the uterus to reach the fallopian tubes. One possibility is that the organisms hitch a ride on sperm, to which the bacteria are known to attach.

Infrequently, *Neisseria gonorrhoeae* can also cause disseminated gonococcal infection (DGI). Curiously, this complication is not usually preceded by urogenital symptoms. Disseminated gonococcal infections are characterized by one or more of the following: rash, toxemia, and arthritis caused by growth of gonococci within the joint spaces. Any joint may be affected, especially the larger ones. The heart valves can also become infected. Some strains of gonococci have a greater tendency than others to disseminate to the joints or other areas away from the original site of the infection.

Epidemiology

Gonorrhea is among the most prevalent of the sexually transmitted diseases. In the United States, its incidence is the highest of any reportable bacterial disease other than *Chlamydia* infections. An increase from 400,000 cases of gonorrhea in 1967 to close to 1 million cases in 1976 was an alarming development. From 1976 through 1980, over 1 million cases were reported each year. Over the last decade, the number of reported cases has declined, but many cases go unreported. The widespread use of oral contraceptives (such as "the pill") has contributed to the increased incidence of gonorrhea and other venereal (transmitted by

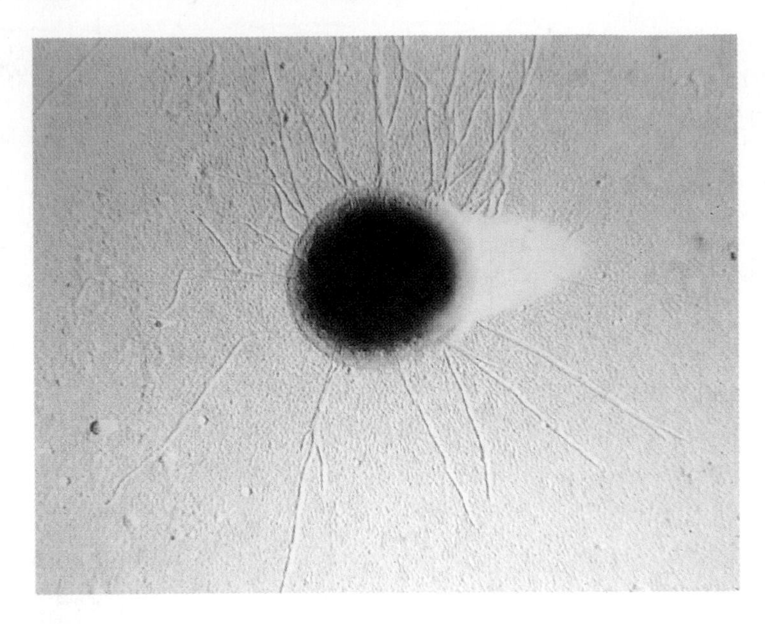

Figure 27.9
Electron micrograph of *Neisseria gonorrhoeae* showing fimbriae (pili).

sexual contact) diseases for a variety of reasons, both social and medical. Generally, women taking oral contraceptives begin having intercourse at an earlier age and have more sex partners than was the case before "the pill" became available. In addition, long-term use of oral contraceptives can lead to migration of gonorrhea-susceptible epithelial cells from upper areas of the cervix into more exposed areas of the outer cervix. Oral contraceptives also tend to increase both the pH and the moisture content of the vagina. All these factors favor infection, not only with gonococci but also with several other agents of sexually transmitted diseases. Women taking oral contraceptives are thus especially vulnerable to gonorrheal infection; they are also more likely to develop serious complications if they contract the disease.

Several other factors contribute to the high incidence of gonorrhea. Carriers of gonococci, both male and female, can unknowingly transmit these bacteria over months or even years. Identifying carriers is therefore important in controlling gonorrhea. Another factor contributing to the incidence of gonorrhea is the lack of immunity following recovery from the disease. The individual who has recovered is susceptible to reinfection. It is probable that repeated infection results from the multiplicity of immunologically distinct strains of the gonococcus.

Since about 1980, the incidence of gonorrhea has declined (figure 27.10). Perhaps increased condom use to control the spread of AIDS may have the added benefit of controlling the spread of gonorrhea and other sexually transmitted diseases.

Ophthalmia Neonatorum

One form of very serious gonococcal disease not transmitted by direct sexual contact is an eye infection, **ophthalmia neonatorum**, that can be transmitted at birth from mother to infant during the infant's passage through an infected birth canal. This form of gonococcal infection has mostly been controlled in the United States by laws requiring the use of a 1% silver nitrate solution or an antibiotic ointment such as 1/2% erythromycin[1] placed directly into the eyes of all newborn infants. This treatment must be given within one hour of birth. Some mothers who feel certain that they do not have gonorrhea have challenged these laws because they do not want their babies to have the unpleasant experience of receiving irritating eye drops. However, the long-standing and often asymptomatic nature of gonorrhea in many women (and men) makes omission of prophylactic treatment of a baby's eyes risky.

Treatment and Control

Years ago, all gonococci were susceptible to penicillin, but they have gradually become more resistant over the years. Thus, in 1944, 100,000 units of penicillin were usually sufficient to cure a case of gonorrhea, whereas by the mid 1970s, a dose of 4.8 million units given intramuscularly was required. This increased resistance of gonococci to penicillin was mainly due to chromosomal mutations that changed the protein of the cell wall to which penicillin normally bound. In 1976, gonococcal strains with a penicillinase-coding plasmid appeared. These highly penicillin-resistant gonococci have become progressively more common. To make matters worse, in 1982, strains with plasmid-mediated tetracycline resistance also appeared and spread, so that within a few years neither penicillin nor tetracycline could be used reliably to treat gonorrhea. For a time, the antibiotic spectinomycin was used as a reliable alternative; however, resistance to it is increasing and one of the semisynthetic cephalosporins such as ceftriaxone must be used. Despite an intense effort, no effective immunization is available against gonorrhea since either the antigens are poorly immunogenic or antibodies arising from them are not protective. Contributing to the problem is the fact that virtually all pathogenic *N. gonorrhoeae* produce an enzyme that destroys the protective IgA antibody found on mucosal surfaces.

Table 27.1 describes the main features of this disease.

[1]Erythromycin—Besides being less irritating to the eye than silver nitrate, erythromycin is effective against *Chlamydia trachomatis* infections, which are even more common than gonorrhea and are the most common cause of neonatal conjunctivitis. Silver nitrate does not prevent this infection.

• **Plasmids, p. 174**
• **Cephalosporins, pp. 449, 461–63**

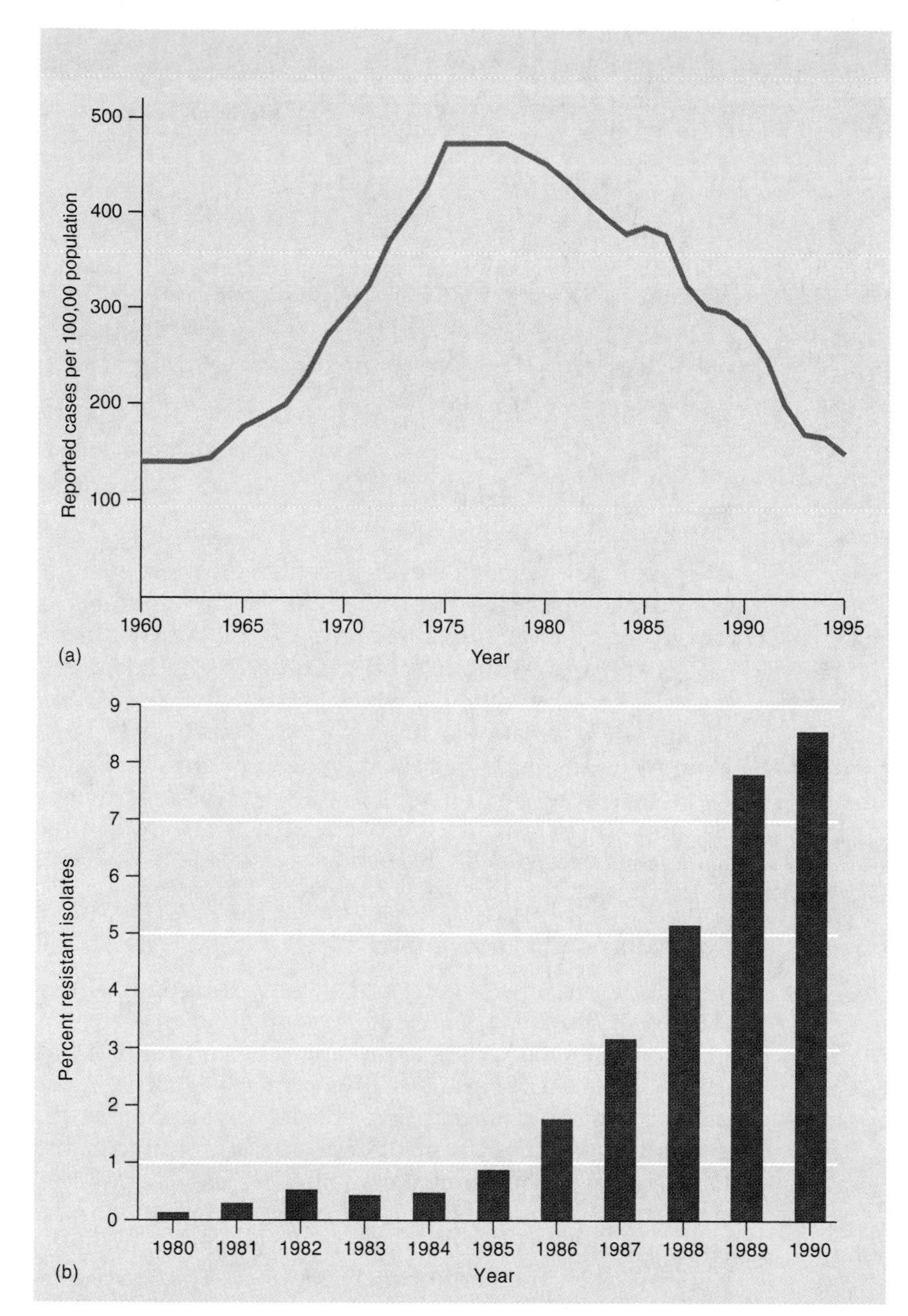

Figure 27.10

Gonorrhea strains in the United States, 1960–1995. (*a*) Incidence and (*b*) rise in antibiotic-resistant strains.

Source: Centers for Disease Control and Prevention. 1990, 1991, 1996. Summary of notifiable diseases—United States, 1989, 1990, 1995. *Morbidity and Mortality Weekly Report* 38(54):54, 39(53):22, 24, 44(53):73.

Chlamydial Infections Rival Gonorrhea in Frequency

Eighty % to 90% of college men with symptoms suggesting gonorrhea actually have other sexually transmitted infections. Twenty-five % to 40% of these students are infected with *Chlamydia trachomatis*, a common cause of sexually transmitted disease of men and women that mimics *N. gonorrhoeae* infections in several ways, including in its production of urethritis, epididymitis, and salpingitis.[1] Like *N. gonorrhoeae, C. trachomatis* attaches to sperm, and it has been suggested that the bacteria reach the fallopian tubes by hitching a ride on sperm cells. The main importance of the infection is that it can produce pelvic inflammatory disease in women (figure 27.11), damaging the fallopian tubes and promoting sterility or ectopic pregnancy; tubal damage can occur with or without symptoms. The infection in men can produce sterility by infecting the testicles.

Symptoms

The symptoms of *C. trachomatis* infection generally appear 7 to 14 days after exposure. In men, the main symptom is a thin, grey-white discharge from the penis, while women most commonly develop an increased vaginal discharge, sometimes accompanied by painful urination, abnormal vaginal bleeding, or lower abdominal pain. Many infections of men and women are asymptomatic.

Causative Agent

Chlamydia trachomatis is a spherical, obligate intracellular bacterium. Diagnosis is most reliably made by inoculating material from a clinical specimen onto tissue cell cultures. After two to three days of incubation, growth of *C. trachomatis* is detected in the cells by using a fluorescent antibody. For routine diagnosis in clinics and hospitals, a faster, less expensive method is now available that uses a monoclonal antibody to detect *C. trachomatis* antigen directly in pus or other clinical specimens.

Epidemiology

Asymptomatic carriers of *C. trachomatis* are fairly common, as shown by one study of college women that revealed that almost 5% had positive genital cultures in the absence of any symptoms. Nonsexual transmission of this agent also occurs, for example, in nonchlorinated swimming pools. Moreover, newborn babies of infected mothers may experience eye infections or pneumonia from infections contracted during passage through the birth canal. In 1995, genital chlamydial infections led all reportable infectious diseases.

[1]Urethritis, epididymitis, and salpingitis—inflammation of the urethra, epididymis (collection of tubules in the testes), and fallopian tubes, respectively

- ***Chlamydia*, p. 250**
- **Fluorescent antibody, pp. 387–88**

Table 27.1 Gonorrhea

Symptoms and complications	In men, inflammation of urethra, pain on urination and discharge; may lead to urinary tract infection, sterility, and arthritis; in women, possibly no symptoms in early stages; later, infection of fallopian tubes, sterility, ectopic pregnancy, and arthritis
Incubation period	3–5 days
Causative agent	*Neisseria gonorrhoeae*
Laboratory diagnosis	Gram-stained smears of purulent discharge; smears often reveal gram-negative diplococci, but in order to identify them as *N. gonorrhoeae* cultures, biochemical tests have to be done
Pathogenesis	Organisms attach to certain epithelial cells by pili, which also interfere with phagocytosis; variation in surface proteins allows attachment to different host cells and escape from immune mechanisms
Epidemiology	Transmitted by sexual contact
Prevention	Education, condoms, early treatment of sexual contacts
Treatment	Intramuscular ceftriaxone; penicillin or tetracycline if strain proven susceptible

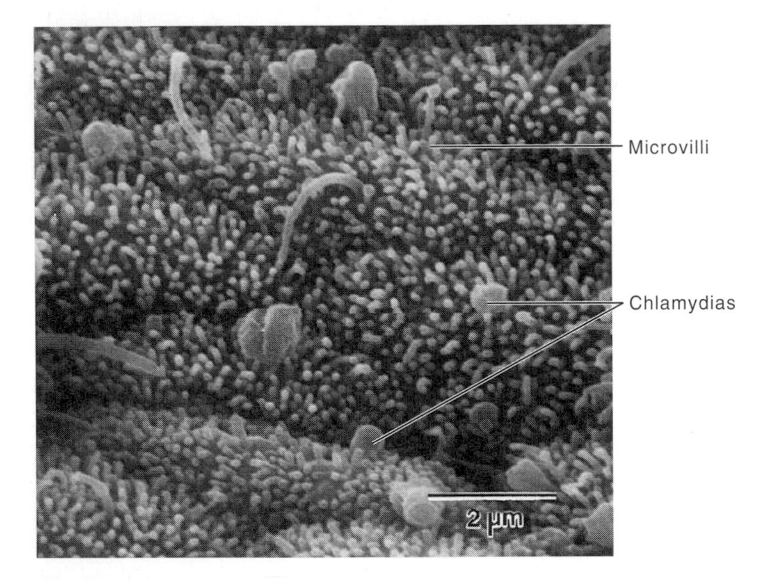

Figure 27.11
Scanning electron micrograph of *Chlamydia trachomatis* attached to fallopian tube mucosa.

Treatment

Infections caused by *C. trachomatis* can usually be treated effectively with a variety of antimicrobials including tetracycline, erythromycin, and azithromycin but not with the penicillins.

Syphilis Is the "Great Imitator" of Other Diseases

Syphilis, another sexually transmitted disease of major importance, is caused by a spirochete, *Treponema pallidum*.

Syphilis was very common in the United States after World War II, but by the mid 1950s, it was almost eradicated. This was accomplished by aggressively locating syphilis cases and their sexual contacts and then treating them with penicillin. However, in contrast to most other industrialized countries, the United States saw a number of factors combine to cause a resurgence of the disease. Inner-city poverty, prostitution, and drug use were linked to a high incidence of syphilis, which exceeded 100 new cases per 100,000 population in at least seven cities in 1990. Since then, renewed efforts in education, case finding, and treatment have caused a dramatic drop in new cases, spawning the hope that transmission of the disease in the United States can once again be stopped.

Symptoms and Pathogenesis

Syphilis occurs in so many forms that it is easily confused with other diseases and, therefore, is often called "the great imitator." Generally, its manifestations occur in three clinical stages. In the first stage, called **primary syphilis,** *T. pallidum* grows and multiplies in a localized area of the genitalia, spreading from there to the lymph nodes and bloodstream. About three weeks after infection, a painless, red ulcer with a hard rim called a **hard chancre** (pronounced "shanker") appears at the site of infection (figure 27.12). The hard chancre represents an intense inflammatory response of the body's white blood cells to the bacterial invasion. Examination of a drop of fluid squeezed from the chancre reveals that it is teeming with infectious *T. pallidum*. Whether or not treatment is given, the chancre disappears within four to six weeks, and the patient may mistakenly believe that the disease is cured.

About 2 to 10 weeks later, the manifestations of **secondary syphilis** may appear. These symptoms usually include runny nose and watery eyes, aches and pains, sore throat, and a rash that involves the palms and soles. Many of these manifestations of secondary syphilis are due to the reaction of circulating *T. pallidum* with specific antibodies to form **immune complexes.** By this time, the spirochetes

• **Immune complexes, pp. 400–1**

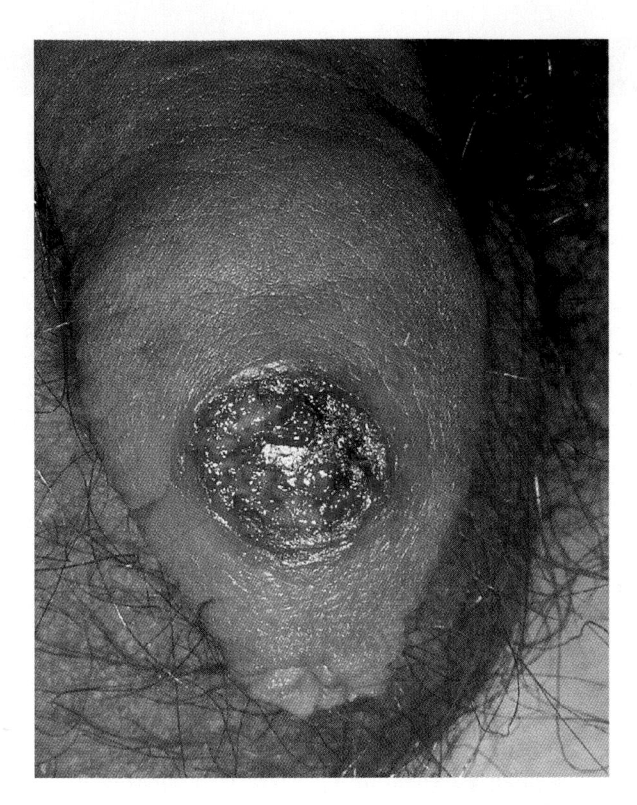

Figure 27.12
Syphilitic chancre on the foreskin of an uncircumcised man. This is the site where *Treponema pallidum* entered the man's body.

have spread throughout the body, and infectious lesions occur on the skin and mucous membranes in various locations, especially in the mouth. Syphilis can be transmitted by kissing during this stage. The secondary stage lasts for weeks to months, sometimes as long as one year, and then gradually subsides. About 50% of untreated cases never progress past the secondary stage; however, after a latent period of from 5 to 20 years or even longer, some people with the disease develop **tertiary syphilis.**

Tertiary syphilis, or the third stage of syphilis, represents a hypersensitivity reaction to small numbers of *T. pallidum* that grow and persist in the tissues. In this stage, the patient is no longer infectious. The remaining *T. pallidum* organisms may be present in almost any part of the body, and the symptoms of tertiary syphilis depend on where the hypersensitivity reactions occur. If they occur in the skin, bones, or other areas not vital to existence, the disease is not life-threatening. However, if they occur within the walls of a major blood vessel such as the aorta, the vessel may become weakened and even rupture, resulting in death. Hypersensitivity reactions to *T. pallidum* in the eyes cause blindness; central nervous system involvement most commonly manifests itself as a stroke. A characteristic pattern of symptoms and signs called **general paresis** develops an average of 20 years after infection. Typical findings include personality change, emotional instability, delusions, hallucinations, memory loss, impaired judgment, abnormally active tendon reflexes,[1] reduction in the size and ability of the pupils to react to light, and speech abnormalities. The main characteristics of the three stages of syphilis are summarized in table 27.2.

Table 27.2 Manifestations of Syphilis

Stage of Disease	Main Characteristics	Infectious?
1. Primary	Firm ulcer (hard chancre) at site of inoculation; lymph node enlargement	Yes
2. Secondary	Rash, aches, and pains; mucous membrane lesions	Yes
3. Tertiary	Damage to large blood vessels, eyes, nervous system; insanity	No

Congenital Syphilis

During pregnancy, *T. pallidum* readily crosses the placenta and infects the fetus. This can occur at any stage of pregnancy, but damage to the fetus does not generally develop until the fourth month. Therefore, if the mother's syphilis is diagnosed and treated before the fourth month of pregnancy, the fetus will be treated too and will not develop the disease. Without treatment, the risk to the fetus depends partly on the stage of the mother's infection. Three-fourths or more of the fetuses become infected if the mother has primary or early secondary syphilis; risk decreases with the duration of the mother's infection but is still significant into the latent period. Fetal infections can occur in the absence of any symptoms of syphilis in the mother. About two out of every five infected fetuses are lost through miscarriage or stillbirth. The remainder are born with congenital syphilis. Although these infants frequently appear normal at the time of birth, some of them develop secondary syphilis, which is often fatal, within a few weeks of birth. Others develop characteristic deformities of their face, teeth (figure 27.13), and other body parts later in childhood.

Causative Agent

Syphilis is caused by *T. pallidum*, an extremely slender, motile organism with tightly wound coils, which ranges up to 20 μm in length (figure 27.14). Dark-field microscopy is used to see *T. pallidum* and to observe its characteristic slow

[1]Tendon reflex—sudden involuntary contraction of a muscle caused by momentarily stretching its tendon; e.g., in the knee jerk, the leg kicks forward when the tendon beneath the kneecap is tapped

• **Dark-field microscopy, p. 48**

Perspective 27.4

Laboratory Diagnosis of Syphilis

Laboratory tests are extremely important because it is often difficult to diagnose syphilis by symptoms and physical examination alone. If a chancre fails to develop, the primary stage of the disease generally goes unnoticed, and symptoms that develop later may be blamed on other diseases. Furthermore, congenital syphilis may go unrecognized if the disease is not apparent in the mother. Pregnant women who have no symptoms of the disease commonly transmit *T. pallidum* across the placenta to the fetus. The diagnosis may only be suspected when characteristic deformities of the teeth, legs, and facial bones appear later in childhood.

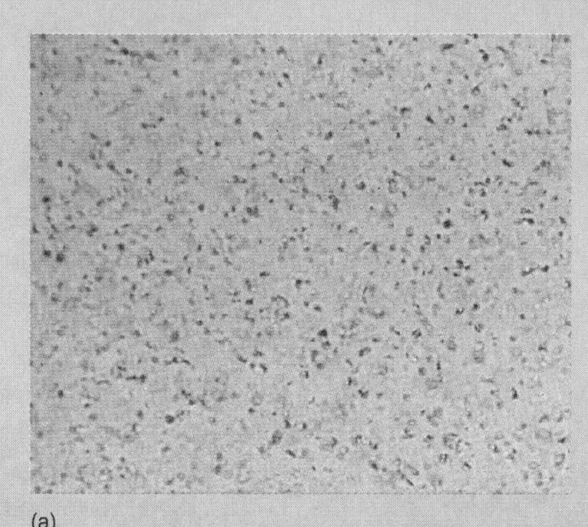
(a)

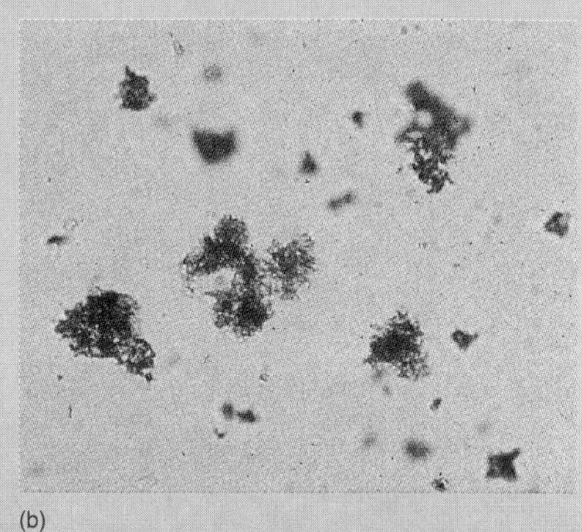
(b)

Figure 1
Flocculation test for syphilis. (*a*) Negative test. (*b*) Positive test.

Primary and secondary syphilis can be diagnosed by finding typical-looking spirochetes by dark-field examination of fresh material from lesions on the skin. Mucous membrane lesions of the genitalia and mouth pose more of a problem because identical-looking spirochetes sometimes exist among the normal flora, especially in the mouth. *T. pallidum* can also be detected in dry smears of specimens by using fluorescent antibody. In tertiary syphilis, internal lesions contain so few organisms that they generally cannot be detected by these means.

During all stages of syphilis, serological tests are helpful in diagnosis. The most common test is based on the accidental finding that patients with syphilis develop antibodylike protein in their blood that will react *in vitro* with a lipid material extracted from beef heart. This chance discovery led to the development of tests for *T. pallidum* infection using the beef heart substance as the "antigen." Most laboratories currently detect this protein by mixing a drop of the patient's plasma with a suspension of a purified form of the beef heart lipid. If the protein is present in the plasma, flocculation (formation of a coarse precipitate) occurs (figure 1). Note that this test does not employ an antigen from *T. pallidum* and so is called **nontreponemal.** Such tests are useful for preliminary testing of large numbers of people, but a positive test does not necessarily indicate that an individual has syphilis, since false positive reactions occur in various other diseases. Besides their use in diagnosing syphilis, the tests are of great help in evaluating the success of treatment, since the titer of the antibodylike protein generally falls fourfold within three months when treatment is effective.

Thus, other serological methods that use *T. pallidum* antigens must be employed to confirm the diagnosis of syphilis. The fluorescent treponemal antibody absorbed test (FTA-ABS) is widely used and employs the entire treponeme. In this indirect **immunofluorescence test** (figure 2), the patient's serum is mixed with nonpathogenic spirochetes to remove (absorb) any antibodies that might cross-react with similar antigens on *T. pallidum*. The serum is then allowed to interact under carefully controlled conditions on a microscope slide with killed *T. pallidum* organisms that have been grown in the testes of rabbits. If specific antibodies to *T. pallidum* are present in the serum, they attach to the treponemes; excess serum proteins can then be washed from the slide. To detect the reaction, fluorescent dye-tagged antibody against human gamma globulin is applied to the slide. If specific antitreponemal gamma globulin antibodies have combined with them, the treponemes will fluoresce when examined under ultraviolet light with the fluorescence microscope (figure 3).

A hemagglutination test is also widely used to diagnose syphilis. It is much simpler to perform than the FTA-ABS test and uses antigen from disrupted *T. pallidum* attached to specially treated sheep or turkey red blood cells. As with the FTA-ABS test, the patient's serum is absorbed with nonpathogenic treponemes to remove cross-reacting antibodies. It is easy to determine if specific anti–*T. pallidum* antibodies remain because they will attach to the antigen on the red blood cells and cause the cells to agglutinate.

These tests are very helpful in diagnosing syphilis when used to confirm a positive nontreponemal test. However, despite inducing the absorption of the patient's serum with nonpathogenic treponemes, the tests are not completely specific for *T. pallidum* infection. Also, the tests are not sensitive enough to detect many early syphilitic infections, such as in babies born to mothers who acquired syphilis very late in pregnancy or in patients with AIDS who produce antibody poorly. In addition, these tests generally remain positive for life and, therefore, do not distinguish active infections from previous infections that have been treated.

Better tests are now being developed and evaluated in laboratories that specialize in studies of syphilis. For example, with **immunoblot techniques,** the many antigens present in *T. pallidum* are separated from each other and allowed to react with the patient's serum. This technique gives a highly specific

• **Fluorescent antibody, pp. 387–88**
• **Serological tests, p. 384**

Figure 2
Fluorescent treponemal antibody absorbed (FTA-ABS) test. This is a commonly used method for detecting *Treponema pallidum* antibody in a person's blood.

(a) Killed *Treponema pallidum* spirochetes, fixed to the slide, are incubated with the subject's serum. If specific antibodies are present they combine with the spirochetes.

(b) The slide is washed to remove excess serum, and fluorescent-labeled anti-human gamma globulin (anti-HGG) anti-serum is added. The anti-HGG antibodies combine with any human antibodies already on the spirochetes.

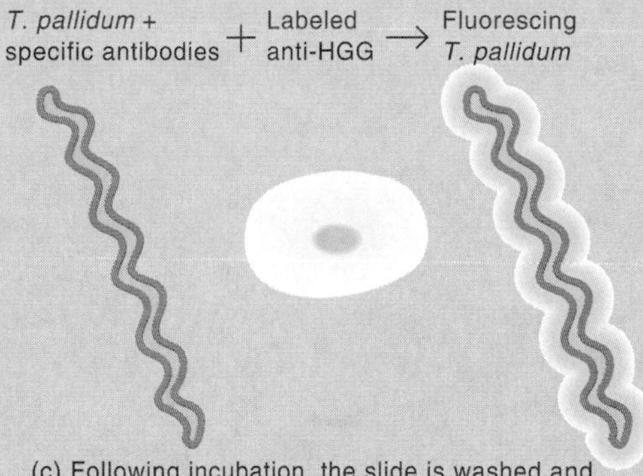

(c) Following incubation, the slide is washed and examined microscopically using ultraviolet light. The *T. pallidum* organisms coated with antibodies fluoresce and can be visualized easily.

test for syphilis because it shows whether antibodies in the serum react with antigens specific for *T. pallidum* and not just the antigens it shares with other bacteria. The same kind of test has been important in diagnosing AIDS.

Another type of test, the **IgM capture method,** uses a surface coated with a highly specific antibody against the heavy chain portion of IgM. When the patient's serum is applied to the surface, its IgM attaches by heavy chains and remains when other serum proteins are washed away. If the patient has syphilis, the presence of IgM antibody can be detected by applying specific *T. pallidum* antigen. IgM antibody is more indicative of active infection than is IgG antibody, and it does not cross the placenta. This type of test is therefore especially helpful in diagnosing congenital syphilis. It should be noted that, in the past, these serological tests depended on relatively crude antigens derived from *T. pallidum* grown in laboratory rabbits. Now, however, many of the genes of *T. pallidum* have been cloned in *E. coli*, making it possible to produce unlimited quantities of purified and highly specific antigen.

The inability to cultivate *T. pallidum* may no longer impede diagnosis of syphilis. The polymerase chain reaction can be used to amplify markedly the DNA present in only a few *T. pallidum* organisms to a level easily detected with a specific *T. pallidum* nucleic acid probe.

• **IgM, pp. 361–64**

• **Polymerase chain reaction, pp. 183, 197**

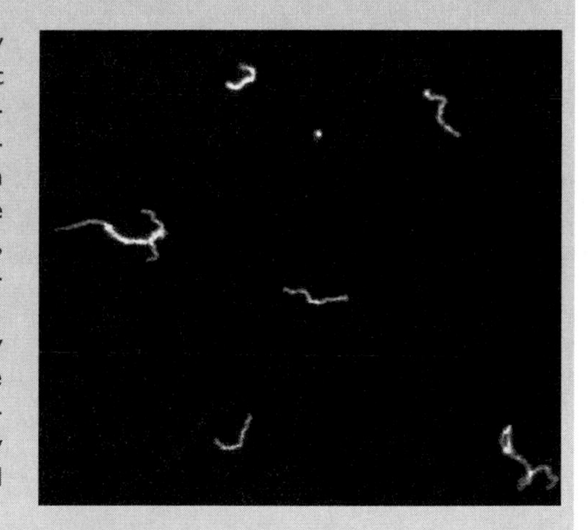

Figure 3
FTA-ABS test. Appearance of the antigen in a positive test.

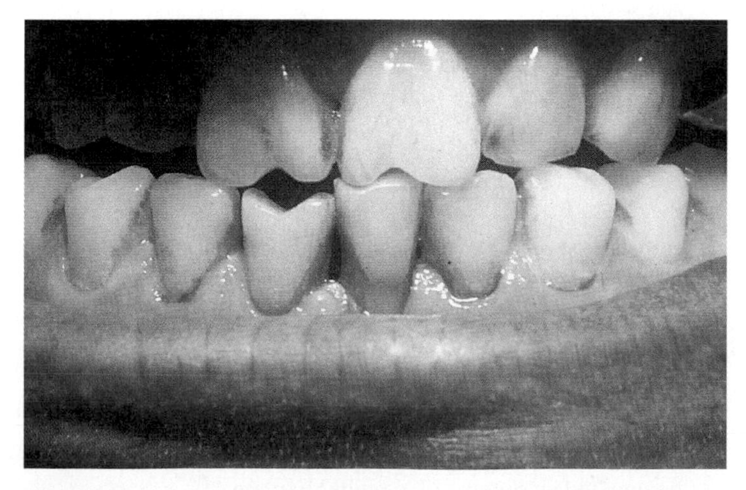

Figure 27.13
Hutchinson's teeth. Notice the notched, deformed incisors, a late manifestation of congenital syphilis.

rotational and flexing motions. The organism is difficult to study because it cannot be cultivated *in vitro* and must be grown in the testicles of laboratory rabbits. Although the organism can be maintained *in vitro* and its metabolism studied, it either does not multiply on culture or does so to a very limited extent. Like many strains of the gonococcus, *T. pallidum* is killed by drying and chilling and is therefore transmitted almost exclusively by sexual or oral contact. Normally, *T. pallidum* is only a parasite of humans.

Treatment

Primary and secondary syphilis are effectively treated with an antibiotic such as penicillin. Treatment must be continued for a longer period for tertiary syphilis, however, probably because many of the organisms are not actively multiplying.

Table 27.3 summarizes the main features of syphilis.

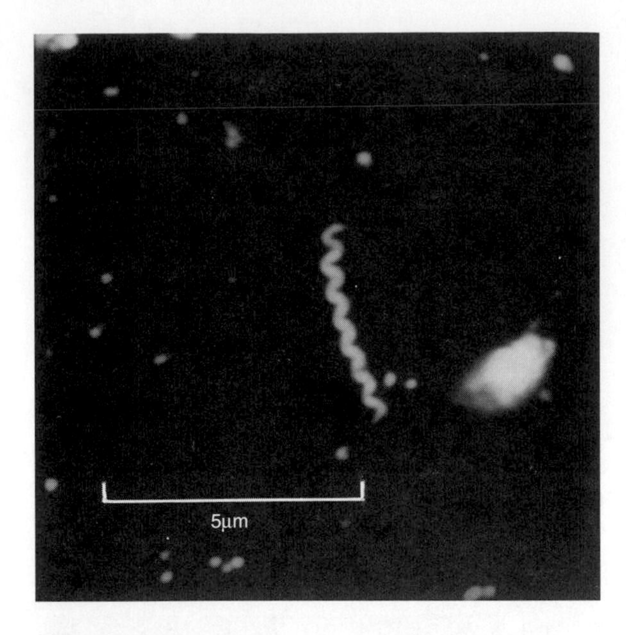

Figure 27.14
Appearance of *Treponema pallidum* with dark-field illumination. This technique readily detects the organism in the skin and mucous membrane lesions of syphilis.

Table 27.3 Syphilis

Symptoms	Chancre, fever, rash, stroke, nervous system deterioration; can imitate many other diseases
Incubation period	10–90 days
Causative agent	*Treponema pallidum*
Laboratory diagnosis	Dark-field illumination; direct fluorescent antibody staining techniques; serological tests, treponemal and nontreponemal
Pathogenesis	Primary lesion (chancre) appears at site of inoculation, heals after 2–6 weeks; *T. pallidum* invades vascular system and is carried throughout the body, causing fever, rash, mucous membrane lesions; damage to brain, arteries, and peripheral nerves appear years later
Epidemiology	Sexual contact with infected partner
Prevention	Education, use of condoms, treatment of sexual contacts, reporting cases
Treatment	Penicillin

Chancroid Is Characterized by Painful Genital Ulcers

Chancroid is another bacterial sexually transmitted disease that showed a marked increase in frequency followed by a progressive decline over the last several decades. It is widespread in the United States but is not commonly reported. The states of Florida, Louisiana, New York, and Texas each reported over 200 cases during 1991, while 18 other states reported fewer numbers. Reported cases declined more than fivefold between 1991 and 1995.

Symptoms

The disease is characterized by single or multiple genital sores that are painful, unlike the chancre of syphilis.

Causative Agent

The causative organism of chancroid is *Haemophilus ducreyi*, a small, pleomorphic, gram-negative rod that can only be cultivated on special media containing X-factor (a blood component called *hematin*).

Epidemiology and Treatment

Epidemics in American cities are associated with prostitution; in some tropical countries, chancroid is second only to gonorrhea among the sexually transmitted diseases. Chancroid and other ulcerating diseases of the genitalia appear to increase the risk of HIV infection.

Chancroid usually responds well to treatment with the antibiotics erythromycin, azithromycin, or ceftriaxone, but the effectiveness of therapy is often sharply reduced if the patient also has AIDS.

• ***Haemophilus*, p. 98**

REVIEW QUESTIONS

1. Describe the mechanisms that defend the urinary system against infection.
2. What does estrogen have to do with the resistance of the vagina to infection?
3. What is the leading cause of bladder infections in otherwise healthy women? Where does this organism come from?
4. How is gonorrhea diagnosed? What complications occur with the disease?
5. Why is a person who has recovered from gonorrhea not immune to the disease?
6. Why is antimicrobial medicine put into the eyes of newborns?
7. How are chlamydial infections diagnosed?
8. Why is syphilis called the "great imitator"?
9. Give the characteristics of the three stages of syphilis.
10. Can a baby have congenital syphilis without its mother ever having had symptoms of syphilis? Explain.
11. What, if any, is the relationship between chancroid and development of AIDS?

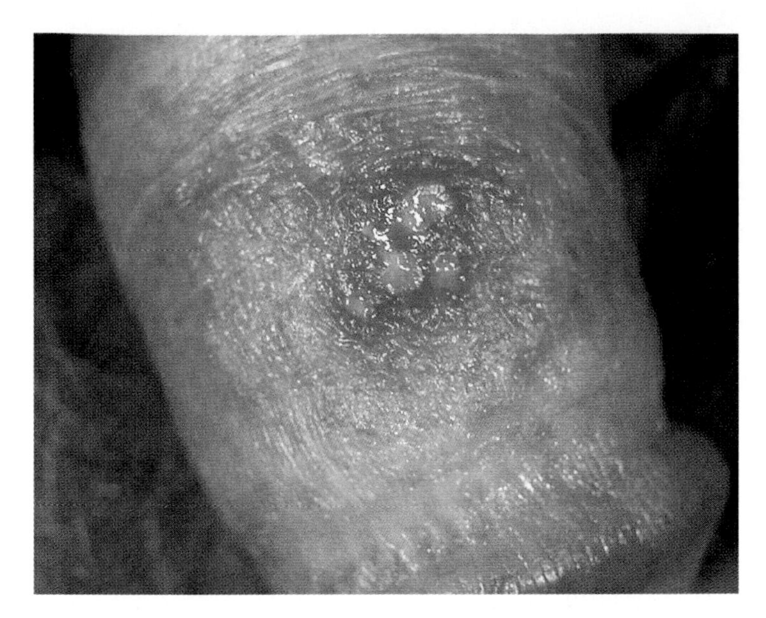

Figure 27.15
Genital herpes simplex. The lesions typically begin as a group of small vesicles surrounded by redness. Within three to five days, they break, leaving the small, superficial ulcers as shown. The lesions are often very painful, especially in women.

Sexually Transmitted Diseases: Viral

Sexually transmitted diseases caused by viruses are probably as common or more common than the bacterial diseases, but they are incurable. Their effects can be severe and long lasting. Genital herpes simplex can give recurrent symptoms for years, papilloma virus infections can lead to cancer, while HIV infections usually cause AIDS.

Genital Herpes Simplex Is Characterized by Small, Painful Genital Blisters and Ulcers

Genital herpes simplex (figure 27.15) is among the top three or four most common sexually transmitted diseases. An estimated 30 million Americans are infected with the causative virus.

Symptoms

Symptoms of genital herpes simplex begin 2 to 20 days (but usually about 6 days) after exposure, with genital itching, burning, and in women, often severe pain. Clusters of small red bumps appear on the genitalia, which then become blisters surrounded by redness. The blisters break in three to five days, leaving an ulcerated area. The ulcers slowly dry and become crusted and then heal without a scar. Symptoms are the worst in the first one to two weeks and disappear within three weeks. They recur in an irregular pattern; about 75% of patients will have at least one recurrence within a year, and the average is about four recurrences a year. The recurrent symptoms are usually not quite as severe as those of the first episode, and recurrences generally decrease in frequency with time. However, some individuals may have recurrences for life.

• **Latent infections, pp. 327–28**

Causative Agent

The cause of the disease is usually herpes simplex virus type 2, a medium-sized, enveloped, double-stranded DNA virus closely resembling the herpes simplex virus type 1 that commonly causes cold sores. This disease, like cold sores, can recur because the virus becomes latent. Either virus can infect the mouth and the genitalia, but the type 2 virus causes more severe genital lesions whose frequency of recurrence is greater than that of the type 1 virus. During initial infection and for as long as one month after, the virus is found in genital secretions. During recurrence, the virus is usually present in large numbers for less than a week. However, the virus can be transmitted at any time, even if symptoms are absent.

Pathogenesis

Pathogenesis of the disease is poorly understood, but viral DNA exists within nerve cells in a circular form during times when there are no symptoms. The virus in this state is not infectious, and only a small portion of its genome is transcribed. However, at times the entire viral chromosome can be transcribed and form complete infectious virions. The mechanisms by which the latent infection is maintained or reactivated are not known.

Treatment and Prevention

There is no cure for genital herpes, although medications such as acyclovir and famciclovir can decrease the severity of the first attack and the incidence of recurrences. The disease is important not only because of the symptoms it produces, but also because if the primary infection in a pregnant woman occurs near the time of delivery, the baby acquires the infection and often dies from it. Genital herpes is associated with cancer of the uterine cervix, since women with genital herpes have more than five times the risk of precancerous changes of the cervix compared with women who do not have herpes. However, the association of herpes simplex type 2 infection with cervical cancer does not necessarily mean that the virus causes the cancer. Women with recurrent genital herpes should have a Papanicolaou smear every 12 months to detect cervical cancer in its early (curable) stage. The "Pap test" consists of removing cells from the cervix with a small brush or spatula, staining them on a microscope slide, and examining them for any abnormal cells. Further, infected men and women should alter their sexual activity to avoid spreading the virus.

Once infected, there is a lifelong risk of transmitting the virus to another person. Avoidance of sexual intercourse during active symptoms and the use of condoms with a spermicide reduce but do not eliminate the chance of transmission. Spermicidal jellies and creams can inactivate herpes simplex viruses.

Some Papillomaviruses Cause Genital Warts and Others Probably Cause Cervical Cancer

Human papillomaviruses are probably the most common of the sexually transmitted disease agents. Although they cannot be grown in the laboratory, their presence in human tissues can be studied by using nucleic acid probes.

Causative Agents

Human papillomaviruses (HPV) are small, nonenveloped, double-stranded DNA viruses of the papovavirus family (chapter 14). There are over 60 HPVs, more than a dozen of which are sexually transmitted. Similar viruses cause warts on domestic animals, but they are not infectious for humans. HPVs that cause the common warts of the hands and plantar warts of the feet generally do not infect the genitalia.

HPV and Genital Warts

Genital warts (figure 27.16 *a*) are caused by sexually transmitted HPV. They usually appear one to three months (sometimes longer) after infection and can occur on the penis, around the anus and vagina, or on the cervix of the uterus. Only a small percentage of people infected with the viruses develop warts, but the virus genome can nevertheless be detected in normal looking skin or mucous membrane. This latent infection is long lasting, perhaps for years. Evidence indicates that in warts, the virus genome exists in multiple copies as circular DNA. Babies born to mothers with genital warts occasionally develop warts of their respiratory tract.

Epidemiology of Genital Warts

The causative viruses are widespread, and it is possible that transmission can occur by means other than sexual intercourse. One study of women university students seeking routine gynecological examinations detected HPV DNA in 46%. The risk of acquiring genital warts increases with the number of sex partners.

Prevention and Treatment of Genital Warts

Warts can be removed by laser treatment or freezing with liquid nitrogen, but these measures do not cure the infection. Whether treating genital warts decreases the risk of transmitting the virus to others is not yet known.

• **Nucleic acid probes, pp. 192–93**

HPV-infected individuals are advised to employ condoms to decrease the chance of transmission.

HPV and Cancer

Worldwide, cancer of the cervix is second in frequency only to breast cancer among the malignant tumors of women. Fortunately, cervical cancer is generally preceded by visible changes—small, flat areas of abnormal cell growth (dysplasias, figure 27.16 *b*). These abnormal cells can be detected by the Papanicolaou test and then removed, thereby preventing development of the cancer.

HPVs are probably an important contributing factor in cervical cancer, apparently causing cervical dysplasias that can then as a rare event interact with carcinogens absorbed by the body from cigarette smoking or other sources to produce cancer. The HPVs that have been associated with these cancers are different from the HPVs that cause genital warts.

The HPVs associated with cervical cancer have also been found in anal cancers in homosexual men and in enlarged prostate glands with and without cancer. The occurrence of such HPVs in men raises the possibility that these viruses may be a factor in the development of other types of cancers.

AIDS Is a Late Manifestation of Human Immunodeficiency Virus Infection

The AIDS epidemic was first recognized in the United States in 1981. By the end of 1995, the total number of people with AIDS had risen to well over 500,000, of whom more than 300,000 had died of the disease. AIDS occurs because human immunodeficiency virus (HIV) slowly destroys a person's immune system until it is no longer able to fight infections or cancers. When this happens, the individual is already in the late stage of a disease that has been going on for a long time, mostly without symptoms, but that is still transmissible to others.

In the last few years, a number of developments have occurred that give hope that AIDS can be defeated. First, new methods for quantifying the virus have given a clearer understanding of how the disease progresses, pointing to new approaches to treatment. Second, new medications such as protease inhibitors have been developed that attack the virus in different ways than the older reverse transcriptase inhibitors. This has allowed the use of combinations of medications that are many times more effective than zidovudine (AZT), the first medication to be used. Third, failures in earlier attempts at vaccine development have provided a better understanding of what must be done to make an effective vaccine. Fourth, some individuals apparently fight off the virus on their own, leading to the hope that someday human virus-fighting subtances can be produced and used as medicine to fight the infection.

Despite the emerging optimism among scientists, the AIDS pandemic will not be conquered soon. Newer treatments may only control, not cure the disease, and they are far too expensive for general use.

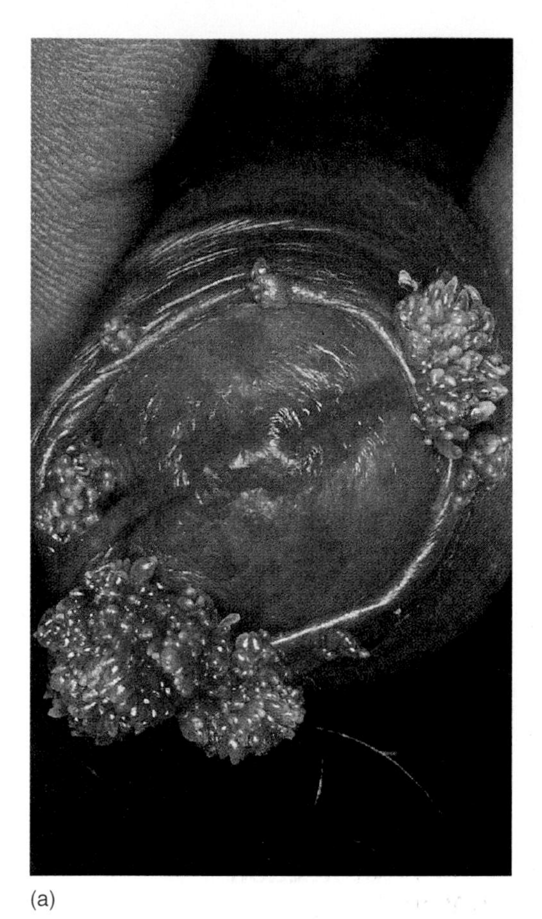

(a)

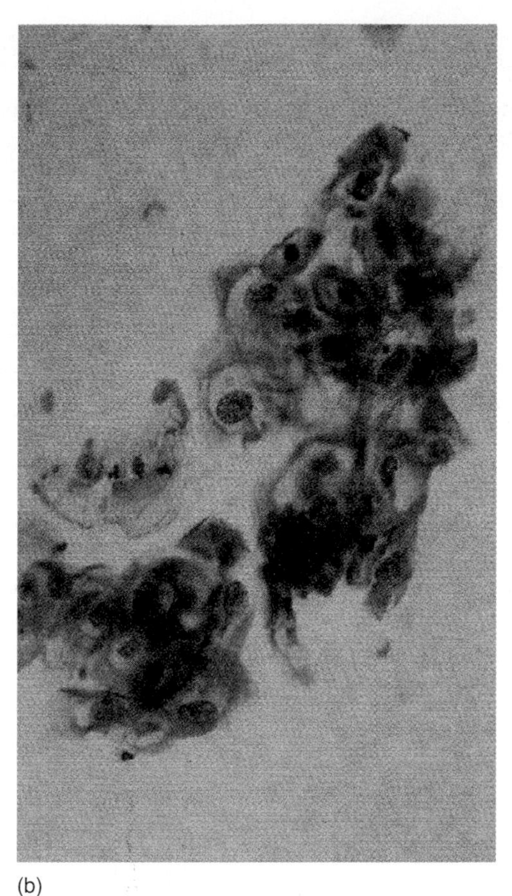

(b)

Figure 27.16
Papillomavirus infections. (*a*) Genital warts on a penis. Cauterization of the warts does not cure the papillomavirus infection. (*b*) Appearance of cervical epithelial cells infected with HPV. Such infections are probably a factor in the development of cervical cancer.

Everyone still has an important role in controlling the pandemic by studying the disease and doing his or her part in preventing spread from one person to another. Those who, since the late 1970s, have engaged in injected drug use or risky sexual behavior and anyone having had unprotected sexual intercourse with them should probably obtain a blood test to rule out HIV infection. This can now be done at home using commercially available test kits.

HIV and AIDS are discussed in detail in chapter 31.

Other Genital Tract Infections

A variety of other genital tract infectious diseases are not generally sexually transmitted. Bacterial vaginosis is very common, but its exact cause is unclear. Two other common vaginal infections are caused by eukaryotes—candidiasis by a fungus and trichomoniasis by a protozoan. Puerperal sepsis results from infection of tissues traumatized by childbirth or abortion, while toxic shock syndrome is often associated with menstruation.

Bacterial Vaginosis Is Characterized by an Unpleasant Odor and Large Numbers of *Gardnerella Vaginalis*

In the United States, symptoms of vaginal infection in women of the childbearing years are most commonly due to **bacterial vaginosis**. Pregnant women with this condition have a sevenfold increase in the risk of having a premature baby when compared with those without bacterial vaginosis. Nevertheless, the risk is small (less than 10%).

Symptoms

Bacterial vaginosis is characterized by a homogeneous, thin, grayish white, slightly bubbly vaginal discharge having a characteristic pungent odor, pH elevated above the normal of 4.5, and the presence of **clue cells** (figure 27.17). No inflammation occurs unless there is another, concurrent vaginal infection. For many patients, the odor is more distressing than the vaginal discharge, which is often slight.

Causative Agent

The cause of bacterial vaginosis is unknown. *Gardnerella vaginalis*, a small, variable in Gram staining, nonmotile aerotolerant bacterium with a gram-positive type of cell wall, is consistently present in large numbers. Also commonly present are species of *Mobiluncus*, an anaerobic curved rod with pointed ends, motile by subpolar flagella. These bacteria also have a gram-positive type of cell wall, and nucleic acid studies indicate they are related to one of the actinomycetes. Other anaerobes are also increased in number, notably species of *Bacteroides* and anaerobic streptococci. Lactobacilli, which predominate in normal women, are markedly decreased in number. Clue cells are epithelial cells that have sloughed off the vaginal wall and are covered with *G. vaginalis* and *Mobiluncus* sp. The strong odor is due to metabolic products of the anaerobes, such as the amines putrescine and cadaverine. Pure cultures of these various species do not consistently produce bacterial vaginosis in normal people voluntarily inoculated, although the secretions from a patient with the condition do produce the infection. All these species of bacteria occur in

• ***Mobiluncus*, p. 238**

Perspective 27.5

HIV Infection on College Campuses

Currently, the North American AIDS epidemic is advancing most rapidly among the poor minorities of the larger cities. Although these populations are not well represented among the more than 12 million students attending colleges and universities, there is little room for complacency when considering the possibility of HIV spreading among students.

Is HIV present on college campuses? One study reported in 1990 found that of 19 colleges surveyed, 9 had one or more students infected with HIV. Other studies indicated that most of the HIV-infected students were homosexual men. Most students had a good understanding of how AIDS is transmitted, but nevertheless, many engaged in high-risk sexual practices or intravenous drug abuse.

Will college students experience the rapid increase in heterosexual transmission of HIV now occurring in the general population? Various surveys over the past few years suggest that this is a real possibility. One study found that, of first-year students less than 25 years old, three-fourths of the men and two-thirds of the women had had sexual intercourse. Of these students, about 1 out of 5 men and 1 out of 12 women reported having more than 10 sexual partners during their lifetime. More than 1 out of 6 sexually active women had engaged in anal intercourse, a major risk factor for acquiring HIV infection. Most sexually active students, especially those with multiple partners, did not consistently use condoms.

Almost 1 out of 10 students at one university reported having sexual intercourse with a person known to have HIV infection or with someone at high risk for HIV infection (homosexual or bisexual men, intravenous drug abusers, prostitutes). Students engaging in sexual intercourse with high-risk people had a much higher incidence of other sexually transmitted diseases (chlamydial infection, gonorrhea, anogenital warts, herpes simplex, syphilis) compared to students not exposed to high-risk partners. The risk of sexually transmitted disease correlated with the number of sexual partners. Sexually transmitted diseases that cause ulcerations, such as herpes, chancroid, and syphilis, further increase the risk of acquiring HIV infection.

The prevalence of HIV infection on college campuses is probably still low. Clearly, to keep it that way, some students will need to modify their behavior. Also, students who have ever engaged in high-risk behavior or who have had a sexually transmitted disease are advised to be tested for HIV infection. If they are found to be infected with HIV, they can receive treatment to help delay development of AIDS, and they can take steps to avoid transmitting the virus to others. HIV disease can be very sneaky. One woman (not a college student) is reported to have discovered her HIV infection when her baby was diagnosed as having AIDS. She then found out she acquired the infection from her asymptomatic husband who, unknown to her, had become infected when he experimented with intravenous drugs years earlier.

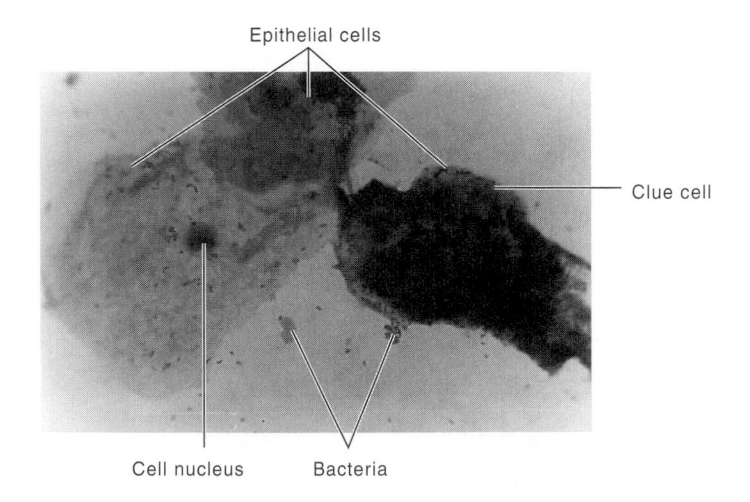

Figure 27.17
Clue cells in a patient with bacterial vaginosis. The cells in the photograph are sloughed vaginal epithelial cells, one of which (the clue cell) is completely covered with adherent anaerobes.

vaginal secretions of some normal women, although usually in smaller numbers. They may act synergistically to cause bacterial vaginosis.

Epidemiology

Since the cause of bacterial vaginosis is not known, its epidemiology is incomplete. It has not been established that it is a sexually transmitted disease. The disease is limited to sexually active women and to some children who have been sexually abused. Sexual promiscuity and a new sex partner also increase risk of the disease.

Treatment

Most cases of bacterial vaginosis respond dramatically to treatment with the antimicrobial medicine metronidazole, which kills anaerobes and permits growth of the lactobacilli normally present. Since metronidazole is not safe to use during pregnancy, the antibiotic clindamycin is used instead.

Vulvovaginal Candidiasis Often Results from Antibacterial Therapy

Candidiasis is the second most common cause of vaginal symptoms after bacterial vaginosis.

Symptoms

The most common symptom of vulvovaginal candidiasis is itching. Vaginal discharge is scanty and whitish, often

Case Presentation

The patient was a 27-year-old married waitress seen in the clinic because of her first pregnancy.

The physical examination showed a healthy appearing black woman. The examination was unremarkable except for the pregnancy. The baby's gestational age was estimated at 16 weeks, consistent with the date of the woman's last menstruation.

Her blood count was normal; however, her test for hepatitis B surface antigen (HBsAg) was positive.

There was no history of receiving blood transfusions or donating blood, sexual contact with bisexual men, multiple sexual partners, abuse of injected drugs, or employment in the health care field. Her husband of three years had a tattoo on his left shoulder.

1. How could the patient have acquired hepatitis B?
2. Is her husband likely to be infected also?
3. What steps need to be taken at the birth of her baby?
4. What groups of people are high risk for this disease?

Discussion

1. Hepatitis B is transmitted most efficiently by blood or blood products and by semen. However, in more than 30% of the cases, the mode of transmission cannot be identified, even with intense questioning.
2. It is likely that her husband was infected, too. Vaginal secretions can transmit the infection but, more likely, her husband contracted the infection from a tattoo needle and transmitted it to her during sexual intercourse.
3. As many as 90% of babies born to hepatitis B virus (HBV) infected women will become infected. Ninety % of infected babies will become chronic HBV carriers, and 25% of them will die of cirrhosis of the liver or liver cancer. However, 90% of these babies can be protected from infection by giving them hepatitis B immune globulin and starting active immunization at the time of birth.
4. Groups of people at high risk of carrying HBV include homosexual and bisexual men, heterosexual people with multiple sexual partners, abusers of injected drugs, laboratory and health care workers exposed to blood, family members of HBV carriers, refugees from Southeast Asia, Australian Aborigines, and Alaskan natives.

occurring in curdlike clumps. There is often a rash on the vulva, the external female genitals.

Causative Agent

The condition is caused by *Candida albicans* (figure 27.18), a yeast that is part of the normal flora of the vagina in about 35% of women. Since *C. albicans* is a fungus, it has a eukaryotic cell structure. Normally, it causes no symptoms. The interaction between large numbers of the normal bacteria and small numbers of these fungi results in a balance between the groups probably based mostly on competition for nutrients. However, when this balance is upset, most often during intensive antibacterial treatment with antibiotics such as ampicillin and tetracycline, *C. albicans* multiplies in large numbers, and the symptoms of vulvovaginitis (inflammation of the vulva and vagina) occur. Other predisposing factors to *Candida* infection are pregnancy, the use of oral contraceptives, and sugar diabetes.

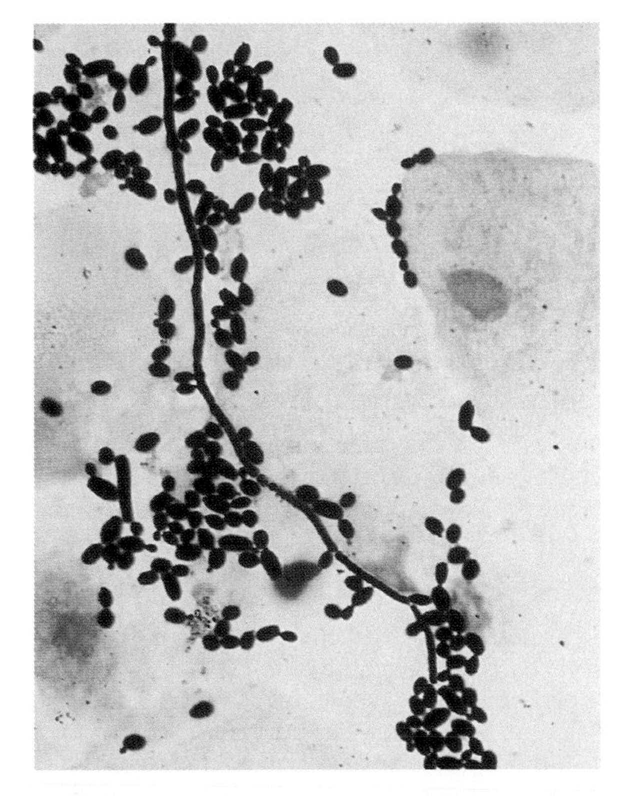

Figure 27.18
Candida albicans in the vaginal secretions of a woman with vaginal candidiasis.

Treatment

Vaginal treatment of *C. albicans* infections with antifungal medicines such as nystatin or clotrimazole creams are usually effective. Simultaneous treatment of any male sex partners is usually unnecessary unless a *Candida* infection of the penis is present.

Perspective 27.6

Laboratory Diagnosis of *Candida albicans*

Candida albicans is an oval, budding yeast that forms smooth, creamy colonies when grown at 37°C on enriched media. Under less favorable conditions, such as a less nutritious medium at room temperature, **pseudohyphae** are formed along with the budding yeast cells. These pseudohyphae consist of very elongated budding cells in chains, resembling hyphae in a mycelium. Frequently, round, thickened spores **(chlamydospores)** form at the ends of its pseudohyphae (figure 1). These morphological characteristics along with characteristic carbohydrate fermentation patterns are used to distinguish *C. albicans* from other yeasts. In addition, this organism produces tubelike projections called **germ tubes** in human serum (figure 2).

• **Yeast, p. 15**

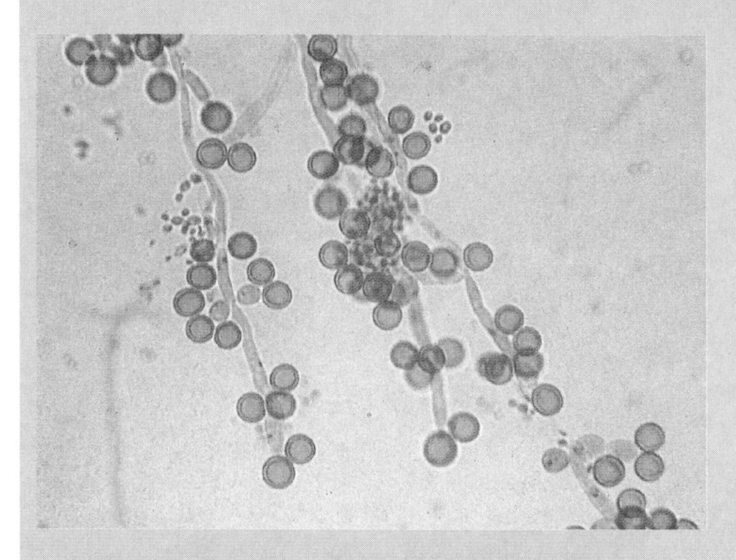

Figure 1
Candida albicans. Chlamydospore formation with growth on cornmeal agar. Besides the large, spherical chlamydospores, tiny, oval blastospores, and pale, slender pseudohyphae are visible.

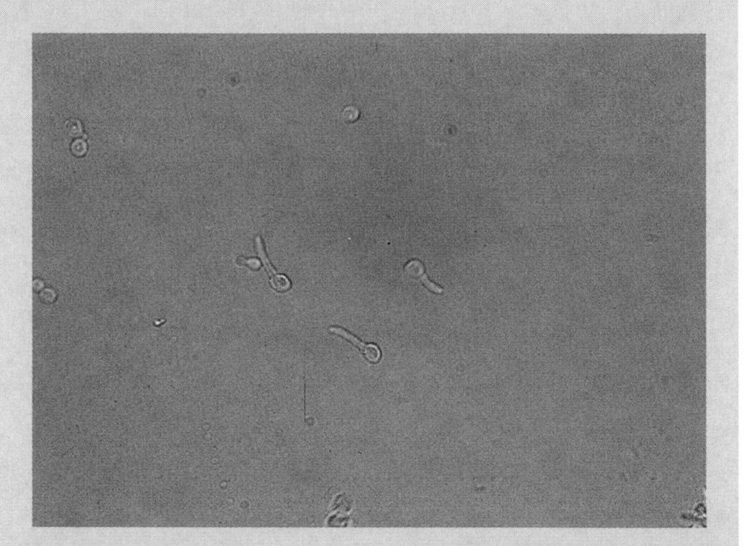

Figure 2
Candida albicans. When incubated in serum, the blastospores develop tubelike projections called *germ tubes*.

The Protozoan *Trichomonas Vaginalis* Can Cause Urogenital Infections in Men and Women

Trichomonas vaginalis infections (trichomoniasis) are the third most common cause of vaginal symptoms. Infected men usually are asymptomatic carriers of the organism.

Symptoms

Most symptomatic *T. vaginalis* infections occur in women. The protozoan causes itching of the vulva and inner thighs, itching and burning of the vagina, a frothy, yellowish vaginal discharge, and sometimes burning discomfort with urination. Although most infected men are free of symptoms, some have burning pain with urination, painful testicles, or discomfort arising from involvement of the prostate gland.

Causative Agent

Trichomonas vaginalis measures about 10 × 30 micrometers, has four anterior flagella, and has a long posterior flagellum attached to an undulating membrane (figure 27.19). It also has a slender posteriorly protruding rigid rod called an **axostyle.** The organism is a eukaryote, but like some other flagellated aerobic or microaerophilic flagellates, it lacks mitochondria. *T. vaginalis* has interesting cytoplasmic organelles called **hydrogenosomes.** Enzymes within these double membrane–bounded organelles remove the carboxyl group (COOH) from pyruvate and transfer electrons to hydrogen ions, thereby producing hydrogen gas. The molecular structure of one of these enzymes closely resembles the corresponding enzyme in an aerobic bacterium, *Pseudomonas putida*, suggesting that the organelle and the bacterium evolved from a common ancestor.

• **Protozoa, pp. 285, 290–95**

• **Mitochondria, pp. 80–1**

Perspective 27.7

Laboratory Diagnosis of *Trichomonas Vaginalis*

When large numbers of *T. vaginalis*, about 100,000 per milliliter, are present in vaginal secretions, they can be easily identified microscopically by their appearance and by their characteristic jerky motility. Although this identification method is fast, it detects only about half of the cases of active infections. The organisms can be cultivated in laboratory media, a much more sensitive method of diagnosis, but one that takes several days. Another option is to use a mixture of fluorescent antibodies against the many antigenic strains of the organism to detect *T. vaginalis* in dried smears on a microscope slide. This method is faster than culture and can detect fewer organisms than microscopic inspection of vaginal secretions.

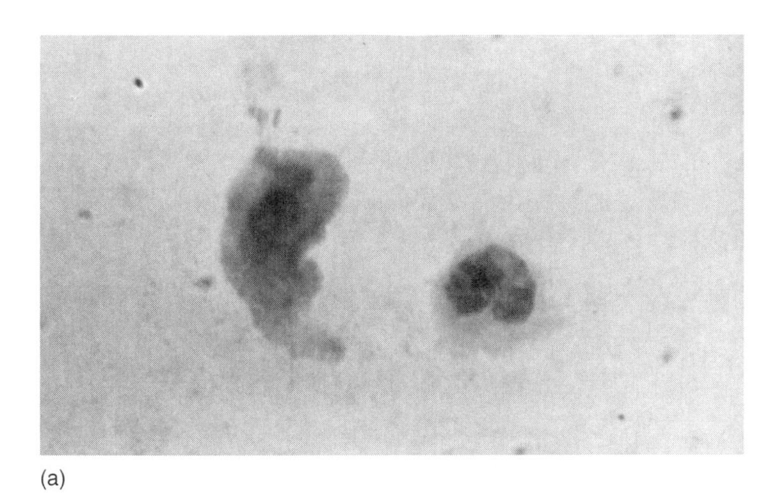

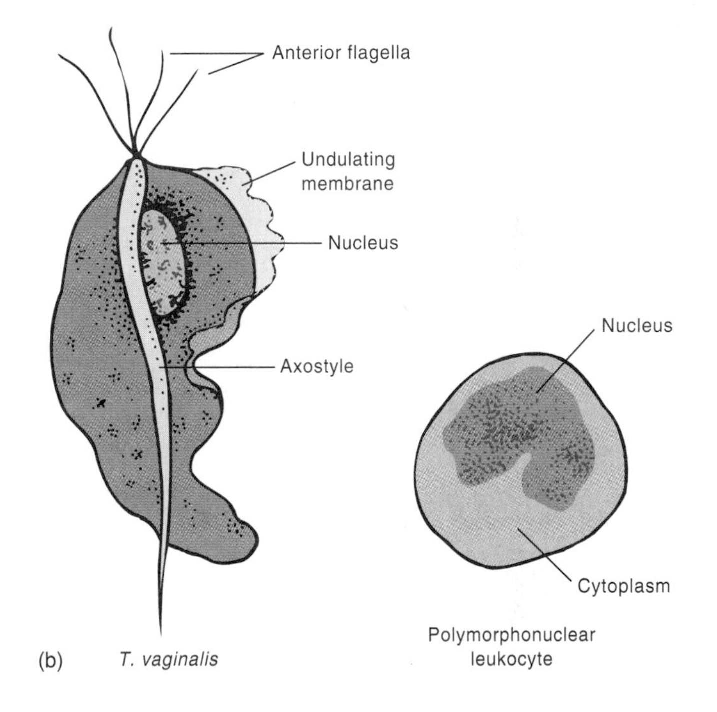

Figure 27.19
Trichomonas vaginalis, one of the common causes of vaginitis. (*a*) Stained smear of vaginal secretions showing the relative sizes of *T. vaginalis* and a polymorphonuclear leukocyte. (*b*) Diagrammatic representations of (*a*). Although it is a flagellated protozoan, *T. vaginalis* often exhibits pseudopodia.

Epidemiology

T. vaginalis is distributed worldwide as a human parasite and has no other reservoirs. It lacks a cyst form and is easily killed by drying. Therefore, transmission is usually by sexual contact. The organism can survive for a time on moist objects such as towels and bathtubs, so it can be transmitted nonsexually. Nevertheless, *T. vaginalis* infections in children should at least raise the question of sexual abuse and possible exposure to other sexually transmitted diseases. There is a high percentage of asymptomatic carriers, especially among men and this fosters transmission of the disease. Infection rates are highest in men and women with multiple sex partners.

Prevention and Treatment

Transmission is prevented by the use of condoms. Most strains of *T. vaginalis* respond quickly to metronidazole treatment, but a few are resistant. These resistant strains sometimes respond to higher doses of the medication given for a longer time.

Puerperal Fever Is Now Rare Because of Aseptic Birthing Practices and Antibiotics

Historically, *Streptococcus pyogenes* was a major cause of puerperal fever (childbirth fever), with mortality rates of 20% or higher. The organisms were commonly transmitted by direct contact at childbirth with contaminated instruments or with the physicians' hands. Today, such outbreaks occasionally occur, usually traced to an asymptomatic carrier of *S. pyogenes*, but antibiotic treatment is usually curative.

• ***S. pyogenes*, p. 506**

Organisms from dirt and dust and from a woman's own fecal bacteria can also attack the uterus following childbirth or abortion. One of the most feared bacteria, *Clostridium perfringens,* causes gas gangrene and has been responsible for many fatalities following abortions induced under unclean conditions.

Toxic Shock Syndrome Is Often Caused by Certain Strains of *S. Aureus* Growing in the Vagina

Toxic shock syndrome occurs as the result of infection by certain strains of *Staphylococcus aureus*. The infection can be located in any part of the body and can affect men, women, and children. It was not until the 1980s that staphylococcal colonization of the vagina was shown to be a major cause of the syndrome.

Cause and Symptoms

Abnormally heavy growth of certain strains of *S. aureus* within the vagina leads to fever, diarrhea, muscle aches, low blood pressure, and a sunburnlike rash. This collection of symptoms is called the **toxic shock syndrome,** which occurs mainly among women who use highly absorbent intravaginal tampons during their menstrual periods, especially when the tampons are left in place for extended periods. The symptoms of the disease result from absorption of one or more staphylococcal toxins, a situation analogous to staphylococcal scalded skin syndrome. One of the *S. aureus* exotoxins has been designated TSST-1 (toxic shock syndrome toxin-1).

Epidemiology

Although cases of toxic shock syndrome occur in menstruating women who are not tampon users, some tampons are especially likely to promote toxin production, thus increasing the risk. Laboratory studies indicate that the toxin production is enhanced when the magnesium ion content of the medium is reduced. It appears that some tampon fibers bind magnesium ions, thus reducing their concentration and promoting toxin synthesis. Brands that were incriminated in the disease were removed from the market, and the incidence of TSS subsequently fell dramatically (figure 27.20).

• **S. aureus, p. 476**
• **Scalded skin syndrome, p. 478**
• **Exotoxins, pp. 420–21**

Prevention

From 1980 to 1984, 2,683 cases of toxic shock syndrome were reported in the United States, resulting in 114 deaths. Beginning in 1980, widespread dissemination of knowledge about toxic shock syndrome promoted earlier diagnosis and prompt treatment, resulting in a drop in mortality to about 3%. The percentage of cases associated with menstruation dropped from about 93% to 70%, perhaps in large part because of greater care in the use of tampons. Since March 1990, the U.S. Food and Drug Administration has required that the absorbency of tampons be indicated on packages so that users can choose the lowest absorbency possible for their needs. In addition, packages of tampons contain an insert detailing the causes and symptoms of toxic shock syndrome.

Table 27.4 describes the main features of toxic shock syndrome.

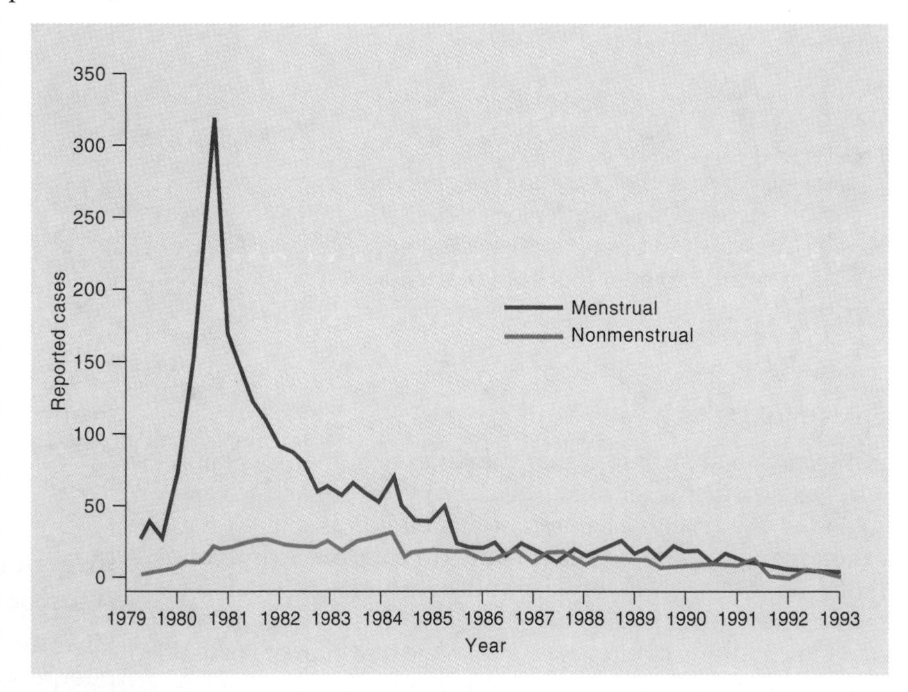

Figure 27.20
Toxic shock syndrome in the United States, 1979–1992. Awareness of the role of high-absorbency vaginal tampons resulted in a sharp decline in menstruation-related toxic shock.

Source: Centers for Disease Control and Prevention. 1993. Summary of notifiable diseases—United States, 1992. *Morbidity and Mortality Weekly Report* 41(55):58.

Table 27.4 Toxic Shock Syndrome

Symptoms	Fever, diarrhea, muscle aches, low blood pressure, and rash
Incubation period	3–7 days
Causative agent	*Staphylococcus aureus*
Laboratory diagnosis	Cultivation of large numbers of *S. aureus* from the vagina
Pathogenesis	Toxin (TSST-1) produced by certain strains of *S. aureus* in vagina may be enhanced by low magnesium ion concentration due to the absorption of the ion by tampon fibers
Epidemiology	Absorption of staphylococcal toxin often from the vagina during tampon use; also as a result of infection of other parts of the body such as skin, bone, lung
Prevention	Prompt treatment of serious *S. aureus* infections; frequent change of tampon in menstruating women
Treatment	Antimicrobial medication based on laboratory tests of susceptibility of the *S. aureus* strain

REVIEW QUESTIONS

1. Why should a person be concerned about genital herpes simplex—is it not just a cold sore on the genitals? Explain.
2. What are the possible consequences of papillomavirus infections for men, women, and newborns?
3. What, if any, effect does laser removal of genital warts have on a genital HPV infection?
4. What kinds of sex partners present a high risk of transmitting human immunodeficiency virus?
5. In what ways has the epidemiology of AIDS changed since the early years of the epidemic?
6. Describe the symptoms of and laboratory findings in bacterial vaginosis.
7. Describe the epidemiology of bacterial vaginosis.
8. What conditions are associated with vaginal candidiasis?
9. What is interesting about the cytoplasm of *Trichomonas vaginalis?*
10. What possible significance does *T. vaginalis* infection of a child have?
11. What is puerperal fever?
12. Why is toxic shock syndrome commonly associated with menstruation?

Summary

I. Anatomy and Physiology
   A. The main structures of the urinary tract—the kidneys, bladder, ureters, and urethra—are generally well-protected from infection.
   B. Urinary tract infections occur more frequently in women than in men because of the shortness of the female urethra and its closeness to the genital and intestinal tracts.
   C. The vagina is the site of entry for a number of infections.

II. Normal Flora of the Genitourinary Tract
   In women, the normal flora is dependent on the action of estrogen hormones promoting deposition of glycogen in the lining cells of the vagina. Lactobacilli normally help prevent colonization by pathogens.

III. Urinary Tract Infections
   A. Bladder infections occur when frequent, complete emptying fails to take place. This allows accumulation of urine, a nutritious medium for many bacteria, and permits enormous multiplication of any pathogenic bacteria that are present.
   B. Kidney infection may complicate untreated bladder infection when pathogens ascend through the ureters and involve the kidneys.
   C. Ninety % of urinary tract infections are caused by enterobacteria from the person's own normal intestinal flora.
   D. *Escherichia coli* is the cause of the majority of urinary tract infections in otherwise healthy women.
   E. *Pseudomonas aeruginosa* infections are difficult to treat because these bacteria are generally resistant to many antibiotics.
   F. Proper collection and culturing of urine specimens are critically important in identifying the cause of urinary infections because during urination, normal flora bacteria are introduced into the specimen from the lower urethra and vagina.
   G. Causative organisms of diseases involving areas of the body other than the urinary tract may be found in urine.

IV. Genital Tract Infections

V. Sexually Transmitted Diseases: Bacterial
   A. Gonorrhea, caused by *Neisseria gonorrhoeae*, is among the most prevalent venereal diseases. Ophthalmia neonatorum, a very destructive eye disease of the newborn, can be acquired during passage through the mother's infected birth canal. Infections in men may extend to the prostate gland and testes. In women, infection can spread upward through the uterus into the fallopian tubes, resulting in pelvic inflammatory disease. The resulting scar tissue formation can block the passage of ova through the fallopian tubes, causing sterility or ectopic pregnancy.

B. Symptoms and complications of chlamydia infections are very similar to those of gonorrhea. Asymptomatic carriers are common.

C. Syphilis, caused by *Treponema pallidum,* can imitate many other diseases as it progresses. An infected asymptomatic pregnant woman can transmit the organism to her fetus.

D. Chancroid is caused by *Haemophilus ducreyi,* an X-factor requiring bacterium. In some tropical countries, it is second only to gonorrhea among the sexually transmitted diseases. Like other ulcerative sexually transmitted diseases, it can increase the risk of acquiring infection with HIV.

VI. Sexually Transmitted Diseases: Viral

A. Genital herpes simplex, caused by herpes simplex virus type 2, is among the most common venereal diseases. Recurrences are common. The viral chromosome exists in a noninfectious form within nerve cells during asymptomatic periods. However, the disease can be transmitted, indicating that mature virions can be produced. The disease is incurable but can be suppressed by medication.

B. Genital warts are caused by human papillomaviruses (HPV). Some HPV strains are strongly associated with cervical cancer. Removal of warts does not necessarily cure the infection.

C. HIV infection is currently pandemic and ultimately fatal in most cases. No vaccine or medical cure is yet available, but spread of infection can be halted by applying current knowledge.

VII. Other Genital Tract Infections

A. Bacterial vaginosis is the most common cause of vaginal symptoms. It is characterized by a decrease in lactobacilli and an increase in *Gardnerella vaginalis* and anaerobic bacteria.

B. Vulvovaginal candidiasis is caused by *Candida albicans,* a yeast that is commonly a normal part of the vaginal flora. This organism becomes pathogenic when the normal bacterial flora is reduced by antibiotic treatment.

C. Vaginitis can be caused by the protozoan *Trichomonas vaginalis.* Infections in men are usually asymptomatic.

D. Puerperal fever can be caused by *Streptococcus pyogenes* or anaerobic bacteria. Contaminated instruments or hands can spread the disease.

E. Toxic shock syndrome is caused by the growth of certain strains of S. *aureus* and the resulting absorption of the toxins produced. This may occur in the vagina after high-absorbency tampons have been left in place too long.

Chapter Review Questions

1. List four things that favor the development of infection of the urinary bladder.
2. What is generally the source of bacteria that infect the urinary bladder?
3. Name two genera of bacteria that infect the kidneys from the bloodstream.
4. Name two bacterial causes of pelvic inflammatory disease (PID).
5. What is ophthalmia neonatorum?
6. Which bacterium leads all others as a cause of reportable infectious disease?
7. Name and describe the causative agent of syphilis.
8. Name the virus linked to cancer of the anus in homosexual men and of the uterine cervix and anus in women.
9. Name three diseases of the intestine or liver that can be transmitted by sexual intercourse.
10. What is the reservoir of *Trichomonas vaginalis?* Why does transmission generally depend on close physical contact?

Critical Thinking Questions

1. Discuss the factors involved in the high incidence of sexually transmitted diseases.
2. Discuss the implications of sexually transmitted diseases for the unborn.
3. Discuss the difficulties in the laboratory diagnosis of syphilis in a newborn infant.

Further Reading

Aral, S. O., and Holmes, K. K. 1991. Sexually transmitted diseases in the AIDS era. *Scientific American* 264:62.

Bauer, H. M., et al. 1991. Genital human papillomavirus infection in female university students as determined by a PCR-base method. *Journal of the American Medical Association* 265:472–477.

Bloom, B. R. 1996. A perspective on AIDS vaccines. *Science* 272:1888–1890.

Cohen, Jon. 1996. The marketplace of HIV/AIDS and protease inhibitors: a tale of two companies. *Science* 272:1880–1883.

Cook, R. L., et al. 1992. Clinical microbiological and biochemical factors in recurrent bacterial vaginosis. *Journal of Clinical Microbiology* 30:870. A study showing that vaginal abnormalities persist after treatment in many women with bacterial vaginosis.

Dorfman, D. H., and Glaser, J. H. 1990. Congenital syphilis presenting in infants after the newborn period. *New England Journal of Medicine* 323:1299.

Gayle, H. D., et al. 1990. Prevalence of the human immunodeficiency virus among university students. *New England Journal of Medicine* 323:1538.

Hook III, E. W., and Marra, C. M. 1992. Acquired syphilis in adults. *New England Journal of Medicine* 325:1060. A comprehensive readable review of the current status of syphilis.

1995. Increasing incidence of gonorrhea—Minnesota, 1994. *Morbidity and Mortality Weekly Report* 44(14):282–286.

Kotloff, K. L., et al. 1991. Assessment of the prevalence and risk factors for human immunodeficiency virus type 1 (HIV-1) infection among college students using three survey methods. *American Journal of Epidemiology* 133:2–8.

MacDonald, N. E., et al. 1990. High risk STD/HIV behavior among college students. *Journal of the American Medical Association* 263:3155–3159.

1996. Outbreak of primary and secondary syphilis—Baltimore City, Maryland, 1995. *Morbidity and Mortality Weekly Report* 45(8): 166–169.

Piot, Peter. 1996. AIDS: a global response. *Science* 272:1855.

28 Nervous System Infections

KEY CONCEPTS

1. Infectious diseases of the central nervous system are uncommon compared to infections elsewhere in the body because pathogens usually cannot cross the blood-brain barrier.
2. The presence or absence of receptors on the surface of different kinds of nervous system cells determines their susceptibility to infection and thus the symptoms of an infectious disease.
3. The nervous system is infected primarily from the bloodstream, via the nerve cells, or through cranial bone.
4. Antimicrobial medicines are effective in central nervous system infections only if they are able to cross the blood-brain barrier.

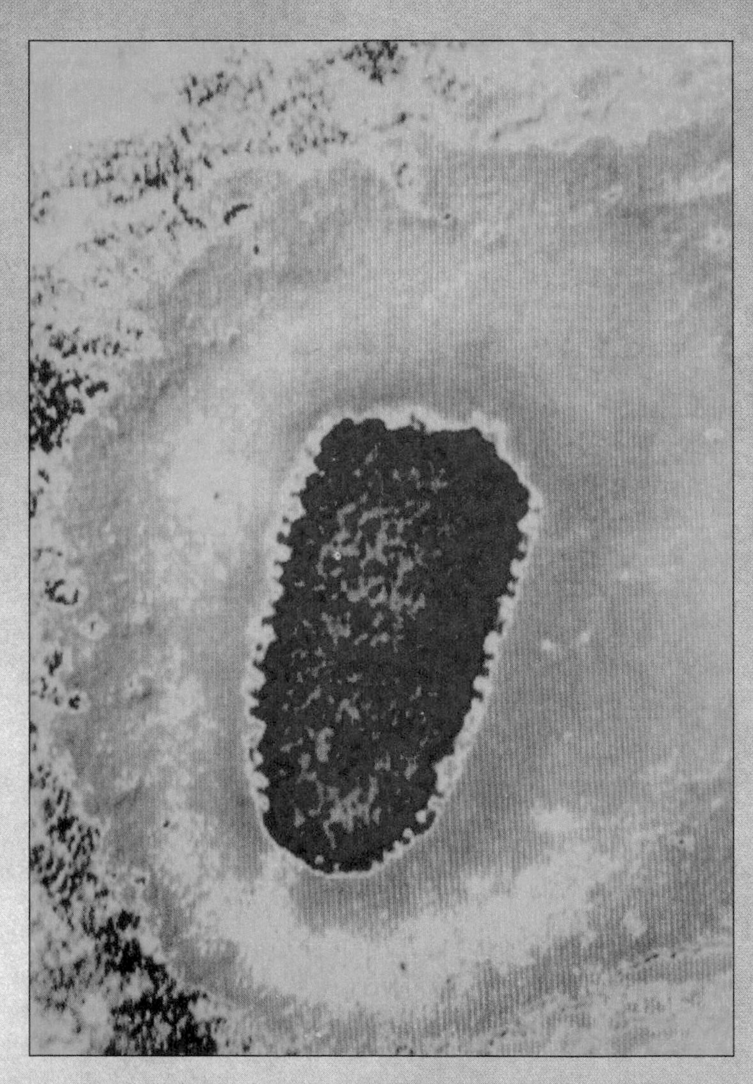

Color enhanced transmission electron micrograph of a rabies virion. Rabies is rampant in wildlife, and therefore a constant threat to humans.

PREVIEW LINK

Microbes in Motion

The following books in the *Microbes in Motion* CD-ROM may serve as a useful preview or supplement to your reading: *Gram-Positive Organisms. Gram-Negative Organisms. Miscellaneous Bacteria. Microbial Pathogenesis in Infectious Diseases:* Toxins. *Vaccines:* Vaccine Targets. *Viral Structure and Function:* Viral Diseases. *Fungal Structure and Function:* Diseases. *Parasitic Structure and Function.*

A Glimpse of History

Poliomyelitis (or polio, as it is more commonly known) was first recognized as a distinct disease early in the 19th century, although based on ancient Egyptian inscriptions, it probably existed long before that. The first polio epidemic was not reported until 1887 in Sweden. For the next 30 years, polio epidemics were mostly confined to the Scandinavian countries, Canada, New Zealand, Australia, and the northeastern United States. Until the late 1950s, polio was a terrifying threat, especially in economically advanced nations. Fear of the disease was widespread, and hundreds of thousands of people in the United States contributed money to the National Foundation for Infantile Paralysis (March of Dimes) to help victims and to support research for the solution to this problem. Former president Franklin Delano Roosevelt, himself a polio victim, helped dramatize the need for this support.

One of the most important developments toward defeating polio was demonstrated in 1949. Dr. John Enders and his colleagues at Harvard University showed that polioviruses could grow readily in monkey kidney cell cultures, producing clear-cut cytopathic effects. Dr. Enders, Dr. Thomas Weller, and Dr. Frederick Robbins were awarded the Nobel Prize in Medicine in 1954 for this work. In this same year, Dr. Jonas Salk perfected a vaccine consisting of formalin-inactivated polio virus that had been grown in monkey kidney cell cultures. This inactivated vaccine was widely employed, and the incidence of paralytic poliomyelitis was lowered dramatically. At about this same time, Dr. Albert Sabin selected mutant poliovirus strains that did not attack nervous tissue but could stimulate antibody formation against the wild-type, virulent virus. Orally administered vaccines consisting of living attenuated viruses were quickly developed,; they proved more effective than the killed virus vaccines. The results of vaccine programs using these vaccines has been dramatic. In 1952, just before the development of the killed vaccine by Salk, 57,879 cases of polio were reported in the United States. Ten years later, the CDC reported 910 cases; by 1972, there were 72 cases and in 1986 there were only 3 reported cases. By 1991, poliomyelitis had been eliminated from the Western Hemisphere. Global eradication by the year 2000 appears feasible.

• **Cytopathic effect, p. 315**
• **Attenuated virus, p. 390**

Infections of the nervous system, although relatively uncommon, are apt to be serious because they threaten a person's ability to move, to feel, and to think normally. These infections are caused by bacteria, viruses, fungi, and even some protozoa.

Anatomy and Physiology

The brain and spinal cord make up the central nervous system. Both are enclosed by bone—the brain by the skull and the spinal cord by the spinal column (figure 28.1). The network of nerves throughout the body is connected with the central nervous system by large bundles of nerve fibers that penetrate the protective bony covering at intervals. These larger nerves can be damaged if the bones are infected at sites of nerve penetration. Two kinds of nerves are involved in the body's nerve network: **motor nerves,** which cause different parts of the body to act, and **sensory nerves,** which transmit sensations, such as of heat, pain, light, and sound. All nerves are made up of special cells with very long, thin extensions **(axons)** that transmit electrical impulses. Bundles of these axons make up a nerve. They can sometimes regenerate if severed or damaged, but they cannot be repaired if the nerve cell of which they are part is itself killed, as may occur, for example, in poliomyelitis. Some viruses and toxins can move through the body within the cytoplasm of the nerve fibers, and some herpesviruses can remain latent in nerve cells for many years.

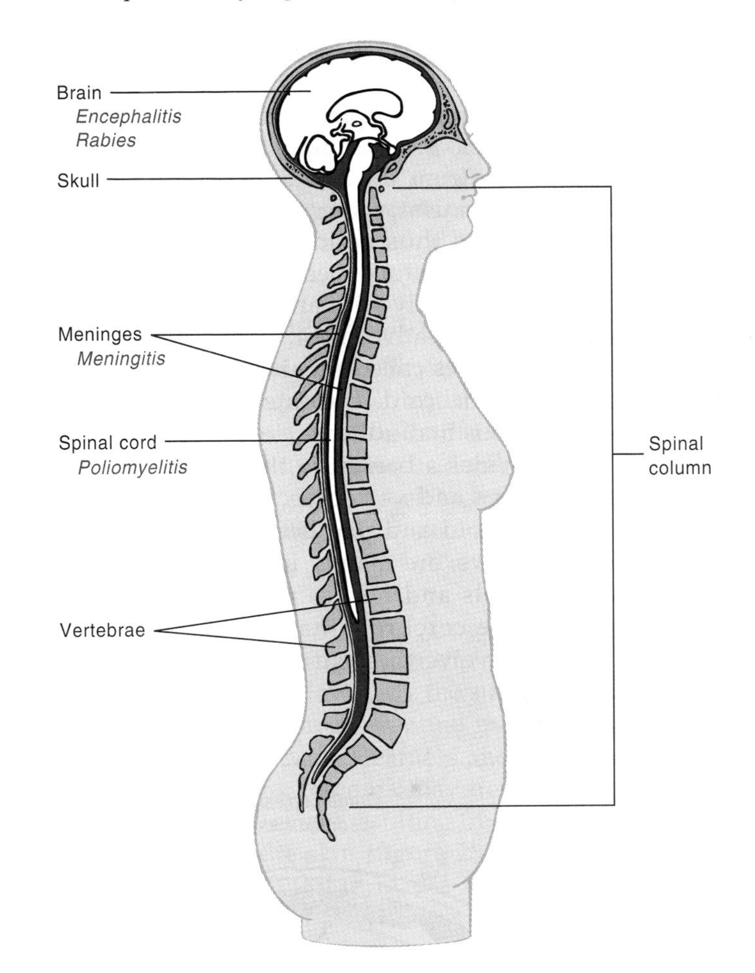

Figure 28.1
The central nervous system.

Deep inside the brain are several cavities called *ventricles*, which are filled with a clear fluid called **cerebrospinal fluid.** This fluid is continually produced in these cavities and then flows from them through small openings at the base of the brain and spreads over the surface of the brain and spinal cord. The cerebrospinal fluid then flows into the bloodstream through specialized structures between the

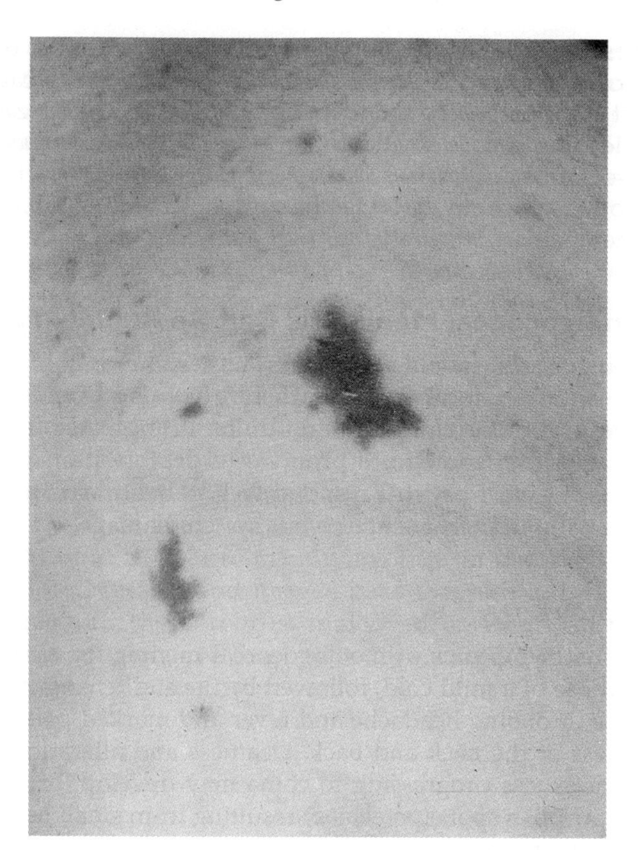

Figure 28.3
Petechiae of meningococcal disease. Meningococci are present in the lesions.

They are then taken in by the respiratory tract epithelial cells, pass through them, and then invade the bloodstream. The blood carries the organisms to the meninges and cerebrospinal fluid, where they multiply faster than they can be engulfed and destroyed by the leukocytes that enter the fluid in large numbers. The bacteria and leukocytes degrade glucose and thereby cause a marked drop in the glucose concentration in the cerebrospinal fluid. The inflammatory response, with its formation of pus and clots, may cause brain swelling and infarcts.[1] It can also lead to obstruction of the normal outflow of cerebrospinal fluid, squeezing the brain flat against the skull with a buildup of internal pressure. The infection may damage the nerves of hearing or vision or the motor nerves, producing paralysis. Moreover, *N. meningitidis* circulating in the bloodstream releases endotoxin, thereby causing a drop in blood pressure that can lead to shock. The smaller blood vessels of the skin are also damaged by the circulating meningococci, causing petechiae.

The organism's endotoxin also increases the body's sensitivity to repeated endotoxin exposure. This strikingly increased sensitivity, called the **Shwartzman phenomenon,** can be demonstrated in experimental animals following repeated injections of meningococci. Thus, it may well be that intermittent "showers" of meningococci entering the blood from infected meninges are responsible for the rapidly fatal outcome in some human cases of meningitis. The tendency of meningococci to autolyse (rupture spontaneously) may enhance their release of endotoxin.

Epidemiology

Meningococcal meningitis can spread rapdily in crowded populations such as military recruits. However, more typically it appears at widely separate locations and occurs throughout the year (figure 28.4). Humans are the only source of the infection, and transmission can occur with exposure to a person with the disease or an asymptomatic carrier of *N. meningitidis*. The majority of meningococcal infections are limited to the nose and throat and pass unnoticed. If a selective medium is used to suppress the growth of other nasopharyngeal flora, meningococci can be recovered from the throats of 5% to 15% of apparently normal people.

Over the years, the frequency of meningitis caused by different strains of *N. meningitidis* has varied markedly. Factors that might play a role in these changing frequencies include (1) relative immunity of the population to strains involved in previous epidemics; (2) widespread use of sulfa drugs, which resulted in the selection of rare sulfa-resistant strains known to be present among groups B and C even before the introduction of sulfa drugs; (3) crowding of susceptible populations, as in military barracks, which allow rapid spread of virulent strains; and (4) transfer of virulence genes among different strains of meningococcus, possibly by DNA-mediated transformation. Numerous types can be identified within the serogroups, which helps trace individual strains during study of an epidemic.

Prevention and Treatment

Opsonins specific for the different group antigens protect the body against meningococcal disease. A vaccine composed of purified capsular polysaccharides from the four major disease-causing groups is available. The vaccine is effective in controlling meningitis caused by strains of these capsular types and is being used to control epidemics and to immunize people at high risk for the disease. Attempts are in progress to develop an effective vaccine against group B organisms, which possess a poorly immunogenic polysaccharide. Since meningococcal strains from patients with active disease are likely to be more virulent than other strains, people intimately exposed to cases of meningococcal disease are given prophylactic treatment

[1]Infarct—death of tissue resulting from obstructed blood supply, such as might occur with a clot in a blood vessel

• **Endotoxin, pp. 60, 420, 422, 424**

• **Sulfa drugs, pp. 154, 465**
• **DNA-mediated transformation, pp. 154, 168–69**

Perspective 28.1

Laboratory Diagnosis of Meningococcal Disease

In patients with meningococcal meningitis, *Neisseria meningitidis* can usually be found on smears made from the sediment of centrifuged cerebrospinal fluid or in scrapings of the skin petechiae characteristic of the disease. Meningococci are gram-negative cocci occurring in pairs (diplococci); each coccus has a flattened side where it faces its neighbor. Many of the organisms seen in stained cerebrospinal fluid smears occur within leukocytes (figure 1), which have entered the fluid in response to the infection. *N. meningitidis* grows readily on media enriched with blood, especially if incubated in an atmosphere with an increased carbon dioxide concentration. The colonies are oxidase positive. Some meningococcal strains die quickly at room temperature, which may account for some failures to recover the organism in cultures from patients. For this reason, it is important that specimens be transmitted to the laboratory promptly.

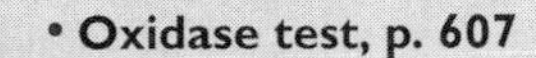
• **Oxidase test, p. 607**

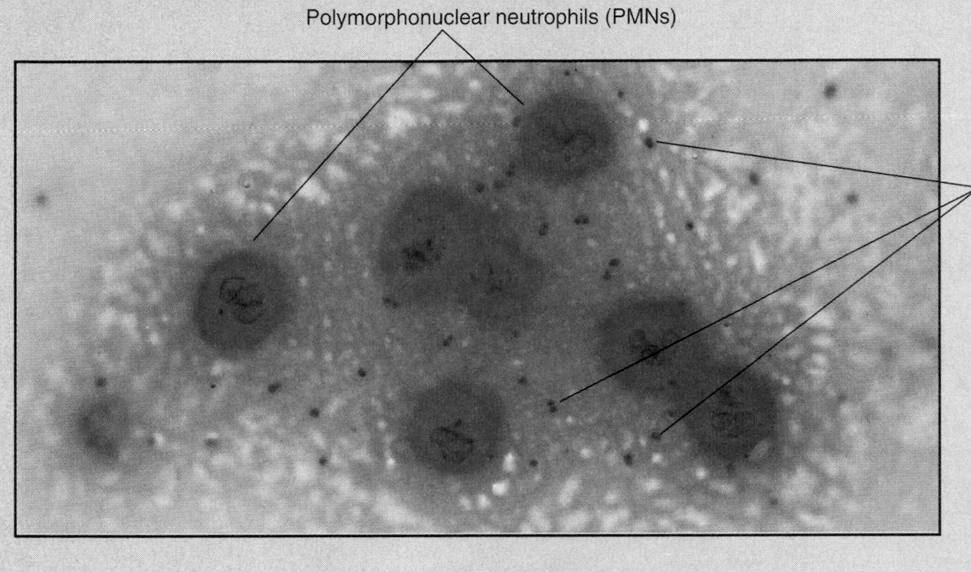

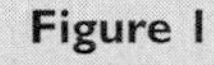
Figure 1
Polymorphonuclear leukocytes and meningococci in the spinal fluid of a person with meningococcal meningitis. This case is somewhat unusual because so many of the meningococci can be seen outside of the leukocytes.

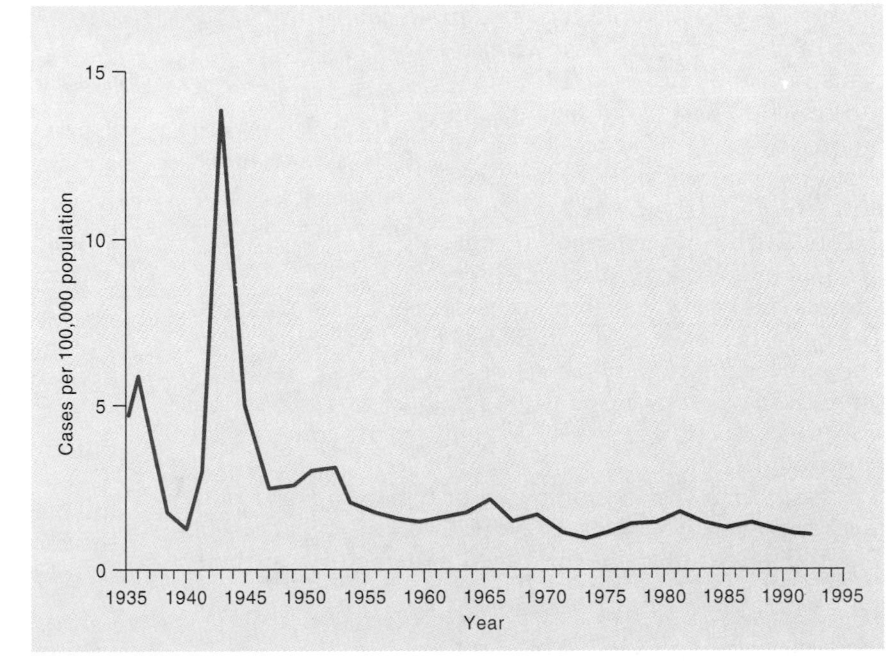

Figure 28.4
Meningococcal disease in the United States, 1935–1992. Note the high incidence during World War II when epidemics occurred among military recruits crowded in barracks.

Source: Centers for Disease Control and Prevention. 1993. Summary of notifiable diseases—United States, 1992. *Morbidity and Mortality Weekly Report* 41(55):41.

with the antibiotic rifampin. Early cases of meningococcal meningitis can usually be cured by penicillin or chloramphenicol unless severe brain injury or shock is present. The mortality rate is less than 10% in treated cases.

Table 28.1 gives the main features of this disease.

• **Opsonins, pp. 379, 380**

Table 28.1 Meningococcal Meningitis

Symptoms	Mild cold followed by headache, fever, pain, stiffness of neck and back, and purplish spots on skin (petechiae)
Incubation period	1–7 days
Causative agent	*Neisseria meningitidis*
Laboratory diagnosis	Smears and cultures of centrifuged sediment of cerebrospinal fluid, scrapings of petechiae
Pathogenesis	Meningococci adhere by pili and colonize upper respiratory tract and then enter bloodstream; carried to the meninges and spinal fluid; inflammatory response obstructs normal outflow of this fluid; increased pressure caused by obstructed flow and edema impairs brain function; damage to motor nerves produces paralysis; endotoxin released from bacteria causes shock
Epidemiology	Close contact with a case or carrier; inhalation of infectious droplets; crowding and fatigue appear to predispose to the disease
Prevention	Polysaccharide vaccine against types A, C, W135, and Y used to immunize high-risk populations; rifampin given to those exposed
Treatment	Penicillin, ceftriaxone, or chloramphenicol

Numerous Other Species of Bacteria Can Cause Meningitis

Besides *N. meningitidis*, a number of other bacterial species can infect the meninges. In many cases, just as with *N. meningitidis*, these organisms can be carried by healthy people, can be transmitted by inhalation, and can only produce meningitis in a small percentage of people who are infected. Organisms in this category include the two other leading causes of bacterial meningitis, the respiratory pathogens *Streptococcus pneumoniae* and *Haemophilus influenzae*. *H. influenzae* is a tiny, gram-negative, facultatively anaerobic coccobacillus that requires two components of blood, designated X and V, for growth. These requirements can be met by certain other species of bacteria, resulting in the phenomenon of **satellitism** (figure 28.5). Except during rare periods of epidemic meningococcal disease, *H. influenzae* has been the leading cause of bacterial meningitis. The peak incidence occurs in infants 6 to 12 months of age. There are six antigenic types of *H. influenzae*, labeled a through f; type b is responsible for most cases of serious disease. Antibody against the type b capsule confers resistance to infection.

In 1985, a vaccine composed of type b capsular polysaccharide was licensed for use in the United States. This T-independent antigen produced satisfactory antibody responses only in children older than 24 months and thus could not be used to protect the age group at greatest risk for meningitis. In 1987, the first "conjugate vaccine," consisting of *H. influenzae* b polysaccharide joined covalently to diphtheria toxoid, was licensed. Linking the diphtheria protein to the *H. influenzae* polysaccharide made the vaccine T cell dependent and much more immunogenic, and in areas where it was intensively used in healthy children there was a decline in *Haemophilus* meningitis in those older than 18 months. Newer conjugate vaccines licensed in 1990 produce an immune response in two-month-old infants and have already produced a dramatic decline in meningitis caused by *H. influenzae*.

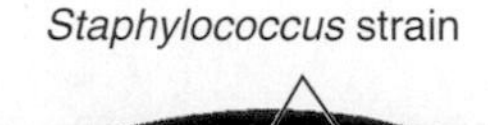

Figure 28.5
Satellitism by *Haemophilus influenzae*. The whitish streak in the center of the blood agar medium is confluent growth of a *Staphylococcus* strain. The small colonies that border the streak are *H. influenzae*.

Streptococcus pneumoniae (figure 28.6) is the leading cause of adult meningitis. The organism is a prominent cause of middle ear infections (otitis media), sinusitis, and pneumonia (chapters 23 and 24), which precede many cases of pneumococcal meningitis.

Table 28.2 compares the three leading bacterial causes of meningitis.

S. pneumoniae, *H. influenzae*, and *N. meningitidis* seldom cause meningitis in newborn babies because most mothers have antibodies against them, which cross the placenta to the baby. Instead, gram-negative rods such as *E. coli* from the mother's intestinal tract, and *Streptococcus agalactiae* (serological group B streptococcus), which colonizes the

• ***Haemophilus*, pp. 98, 516**
• ***S. pneumoniae*, pp. 525–27, 528**
• **T-dependent antigen, p. 369**
• **Otitis media, p. 516**
• **Group B streptococci, pp. 506–7**

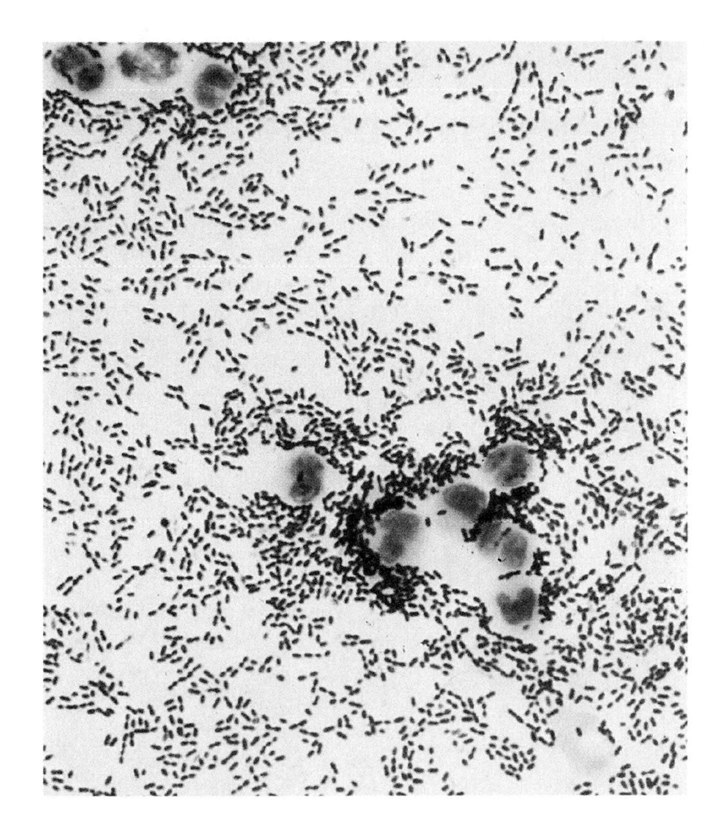

Figure 28.6
Streptococcus pneumoniae in the spinal fluid of a person with pneumococcal meningitis (Gram stain). The large, red objects are polymorphonuclear leukocytes that have entered the spinal fluid in response to the infection.

vagina in 15% to 40% of pregnant women, are more likely to cause meningitis in the newborn. The reason that the newborn is susceptible to gram-negative rod infections is that its mother's bactericidal antibodies to the organisms are IgM, which are unable to cross the placenta and confer immunity.

A much less common cause of meningitis in newborn infants is the nonspore-forming, motile, gram-positive rod, *Listeria monocytogenes*, which can also be carried among the vaginal flora of some women. Epidemics have sometimes occurred because of dairy products contaminated with *Listeria* (chapter 20), but it generally causes only a small percentage of meningitis cases in the United States.

Table 28.3 shows the relationship between the age of the infected individual and the main causes of bacterial meningitis.

Table 28.2 The Leading Causes of Bacterial Meningitis

Causative Agent	Distinguishing Characteristics
Neisseria meningitidis	Gram-negative, oxidase-positive diplococcus; grows readily aerobically on blood agar; a number of different antigenic groups
Haemophilus influenzae	Tiny, gram-negative coccobacillus; requires X- and V-factors for growth; forms satellite colonies around colonies of bacteria that supply the growth factors; serious disease caused by type b organisms
Streptococcus pneumoniae	Encapsulated gram-positive diplococci; grow readily on blood agar in the presence of air; colonies are alpha hemolytic and are lysed by bile; a large number of serotypes, based on differing capsular antigens

Table 28.3 The Main Causes of Bacterial Meningitis at Different Ages

Age Group	Principal Species
Newborn infants	Group B streptococci and *E. coli* account for 70% of cases; other gram-negative enteric bacilli, *Listeria monocytogenes*, and *S. pneumoniae* are occasional causes; *H. influenzae* and *N. meningitidis* are rare
Children	*H. influenzae* 40%–60% of cases; *N. meningitidis* 25%–40%; *S. pneumoniae* 10%–20%.
Adults	*S. pneumoniae* 30%–50%; *N. meningitidis* 10%–35%; *H. influenzae* 1%–3%; increasing risk of enteric gram-negative rods and *Listeria monocytogenes* in people over 50

Leprosy Is Caused by *Mycobacterium Leprae*, an Acid-Fast, Slow-Growing Rod that Has Not Been Cultivated *in Vitro*

Although leprosy (Hansen's disease) is now a relatively minor problem in the Western world, it was once common in Europe and America. Like tuberculosis, leprosy began to recede for unknown reasons before an effective treatment was discovered. Today, it is chiefly a disease of tropical and economically underdeveloped countries, with an estimated worldwide occurrence of 11 million cases, about 60% of which are in Asia. In recent years, reported new cases in the United States have averaged about 250 per year, mostly acquired outside the country.

• **IgM antibody, pp. 361–64**

Case Presentation

The patient was a 31-month-old girl admitted to the hospital because of fever, headache, drowsiness, and vomiting. She had been previously well until 12 hours before admission when she developed a runny nose, malaise, and loss of appetite.

Her birth and development were normal. There was no history of head trauma.

On examination, her temperature was 40°C (104°F), her neck was stiff, and she did not respond to verbal commands.

Her white blood count was elevated and showed a marked increase in the percentage of polymorphonuclear neutrophils (PMNs). Her blood sugar was in the normal range.

A spinal tap was performed, yielding cloudy cerebrospinal fluid under increased pressure. The fluid contained 18,000 white blood cells per microliter (normally, there are few or none), a markedly elevated protein, and a markedly low glucose. Gram stain of the fluid showed many tiny gram-negative coccobacilli, most of which were outside the white blood cells.

1. What is the diagnosis and what is the causative agent?
2. What is the prognosis in this case?
3. What age group is most susceptible to this illness?
4. Compare the pathogenesis of this disease in children and adults.

Discussion

1. The patient had bacterial meningitis caused by *Haemophilus influenzae*, serotype b.
2. With treatment, the fatality rate is approximately 5%. Formerly, ampicillin was effective in most cases, but beginning in 1974, an increasing number of strains possessed a plasmid coding for β-lactamase. So far, however, these strains have been susceptible to the newer cephalosporin-type antibiotics. Unfortunately, about one-third of those who are treated and recover from the infection are left with permanent damage to the nervous system, such as deafness or paralysis of facial nerves. Prompt diagnosis and correct choice of antibacterial treatment minimize the chance of permanent damage.
3. The peak incidence of this disease is in the age range of 6 to 18 months, corresponding to the time when protective levels of transplacental antibody are lacking and acquired active immunity has not yet developed. Since 1987, vaccines consisting of type b capsular antigen conjugated with a protein such as the outer membrane protein of *Neisseria meningitidis* have been used to immunize infants and thus eliminate the immunity gap. As a result, meningitis caused by *H. influenzae* type b is now rare. It is important to know, however, that strains other than type b can cause meningitis. So far, such strains are uncommon.
4. *Haemophilus influenzae* type b strains are referred to as "invasive" strains because they establish infection of the upper respiratory tract, pass the epithelium by some means, and enter the bloodstream or lymphatic vessels. In this way, they gain access to the general circulation and are carried to the central nervous system. Most strains of *H. influenzae*, unlike type b, are noninvasive; although they can cause infections of respiratory epithelium, they usually do not enter the circulation. Most adults are immune to type b strains, but some adults develop meningitis from noninvasive *H. influenzae* strains that gain access to the nervous system because of a skull fracture or by direct extension to the meninges from an infected sinus or middle ear.

Symptoms of Leprosy

Leprosy begins gradually, usually with the onset of increased or decreased sensation in certain areas of skin. These areas typically have either increased or decreased pigmentation. Later, they enlarge and thicken, losing their hair, sweating ability, and all sensation. The nerves of the arms and legs become visibly enlarged with accompanying pain, later changing to numbness, muscle wasting, ulceration, and loss of fingers or toes (figure 28.7). In the lepromatous type of leprosy, the skin lesions are not sharply defined but slowly thicken and spread. Changes are most obvious in the face, with thickening of the nose and ears and deep wrinkling of the facial skin. Collapse of the supporting structure of the nose occurs with accompanying congestion and bleeding.

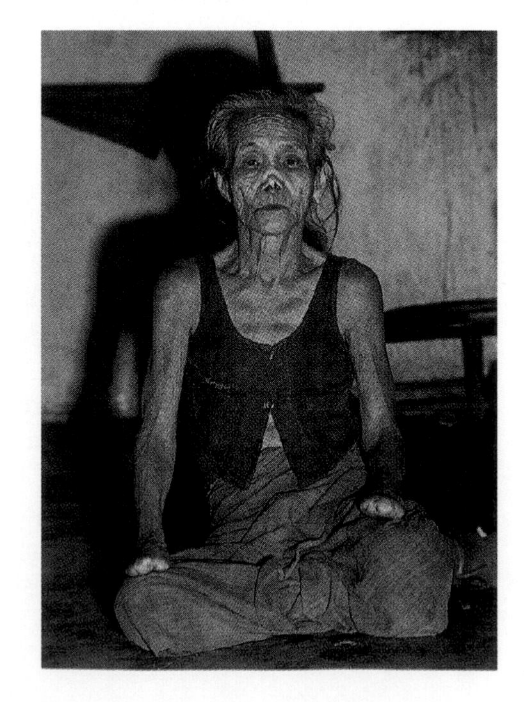

Figure 28.7
Person with leprosy. Note the absence of her fingers as a result of the disease.

Perspective 28.2

Laboratory Diagnosis of Leprosy

Leprosy is diagnosed by microscopically inspecting scrapings or biopsies of infected tissues that have been stained to detect acid-fast bacteria. Finding acid-fast bacilli with typical leprous nerve involvement establishes the diagnosis. Tuberculoid cases are identified by the presence of typical nerve and skin involvement, with or without acid-fast bacilli. The important finding in 1971 that the nine-banded armadillo (figure 1), which is prevalent in Texas and Louisiana, was susceptible to infection by *M. leprae* led to the production of large quantities of the organism in the tissues of the infected animal. These organisms can be used to test a patient's blood for antibody against the leprosy bacilli, thus identifying individuals who may have been infected with *M. leprae*.

• **Acid-fast stain, pp. 52, 233**

Figure 1
Armadillo. The discovery that *Mycobacterium leprae* bacteria from people with leprosy grows well in the nine-banded armadillo has made it possible to obtain large quantities of the bacteria for scientific studies.

Causative Agent

In 1868, Dr. Armauer Hansen, a Norwegian physician, demonstrated the causative bacterium *Mycobacterium leprae* in the tissues of leprosy patients. This discovery was the first to causally link a bacterium to a human disease. *Mycobacterium leprae* (figure 28.8), which is morphologically indistinguishable from *M. tuberculosis*, is an aerobic, acid-fast rod that typically stains with a beaded appearance. Despite many attempts, *M. leprae* has still not been grown in the absence of living cells. In the 1960s, it was successfully cultivated to a limited degree in the footpads of laboratory mice, and more recently, generalized infections were established in armadillos and certain monkeys. The growth of *M. leprae* is very slow, with a generation time in mice of about 12 days or more. A clone bank of the genome of *M. leprae* has been made in *E. coli*, thus making available for study large quantities of the organism's antigens.

Pathogenesis

The earliest detectable finding in human infection with *M. leprae* is generally the invasion of the small nerves of the skin demonstrated by biopsy of skin lesions. Indeed,

Figure 28.8
Masses of *Mycobacterium leprae* in a biopsy specimen from a person with lepromatous leprosy.

• ***Mycobacterium*, p. 402**
• **Cloning, pp. 183, 186–88**

M. leprae is the only known human pathogen that preferentially attacks the peripheral nerves. The bacterium also apparently grows very well within macrophages. The course of the infection depends on the immune response of the host. In most cases, cell-mediated immunity and delayed hypersensitivity develop against the invading bacteria. Immune macrophages limit the growth of *M. leprae*, and the bacteria therefore do not become numerous. However, the attack of immune cells against chronically infected nerve cells can progressively damage the nerves, which leads to disabling deformity. The disease spontaneously stops progressing in some cases and thereafter the nerve damage, although permanent, does not worsen. This limited type of leprosy in which cell-mediated immunity suppresses proliferation of the bacilli is called the **tuberculoid type.** People with tuberculoid leprosy rarely, if ever, transmit the disease to others. However, in many cases of tuberculoid leprosy, the immune system is gradually overwhelmed by the growth of *M. leprae*, leading to the more serious lepromatous form of the disease discussed in further detail next. Early treatment is important to prevent the disease from progressing.

In some infected people, cellular immunity and delayed hypersensitivity to *M. leprae* either fail to develop or are suppressed, and unrestricted growth of *M. leprae* occurs in the cooler tissues of the body, notably in skin macrophages and peripheral nerves. This relatively uncommon form of leprosy is called the **lepromatous type.** The tissues and mucous membranes contain billions of the leprosy bacteria, which produce almost no inflammatory response. The mucus of the nose and throat is highly contaminated with *M. leprae*, which can be transmitted to others. The lymphocytes present in the lesions are mostly T suppressor cells, and there is little or no evidence of macrophage activation.

In lepromatous leprosy, cell-mediated immunity to *M. leprae* is absent, although immunity and delayed hypersensitivity to other infectious agents are usually normal. Normal immune function tends to return when the disease is controlled by medication.

The very long generation time of this bacterium most likely accounts for the long incubation period of leprosy, usually more than 2 years and often 10 years or longer.

Epidemiology

Transmission of leprosy is by direct human-to-human contact. The source of the organisms is mainly nasal secretions of a lepromatous case, which transport *M. leprae* to another individual via the mucous membranes or skin abrasions. Although leprosy bacilli readily infect people exposed to a case of lepromatous leprosy, the disease develops in only a tiny minority, being controlled by body defenses in the rest. Leprosy is no longer a much feared contagious disease, and the morbid days when lepers were forced to carry a bell or horn to warn others of their presence are fortunately past.

No animal or environmental reservoir for *M. leprae* is known to exist. Some of the wild nine-banded armadillos in Louisiana are infected with a bacterium similar to the human leprosy bacillus, but there is no evidence that these animals are a source of human leprosy.

Prevention and Treatment

No proven vaccine to control leprosy is available at present, although with the cloning of *M. leprae* genes in *E. coli* its antigens are available for study as possible immunizing agents. Tuberculoid leprosy can be arrested by treatment with the antimicrobials dapsone and rifampin administered in combination for six months. Lepromatous leprosy is generally treated for a minimum of two years, with a third drug, clofazimine, included in the treatment regimen. As in tuberculosis, multiple drug therapy is required to minimize the development of strains resistant to the antimicrobials.

Some features of leprosy are summarized in table 28.4.

• **Macrophage, pp. 348–49**
• **Delayed hypersensitivity, pp. 384, 403**
• **T suppressor cells, p. 366**
• **Generation time, p. 102**

• **Reservoir, p. 430**
• **Drugs for leprosy, p. 465**

REVIEW QUESTIONS

1. Why can an infection in the brain's ventricles usually be detected in spinal fluid obtained from the lower back?
2. What are the main entry routes for infections of the central nervous system?
3. What is meningitis?
4. How does the epidemiology of meningococcal meningitis differ from those of other common types of meningitis?
5. What cell wall component of *N. meningitidis* is probably responsible for the shock and death that sometimes occur with infections by this bacterium?
6. Name and describe the organism that has generally been responsible for most cases of bacterial meningitis.
7. Why are most newborn babies unlikely to contract meningococcal, pneumococcal, or *Haemophilus* meningitis?
8. Describe the only bacterial pathogen that prefers to invade peripheral nerves.

Table 28.4 Leprosy (Hansen's Disease)

Symptoms	Progressive nerve damage, chronic skin lesions, ulcerative lesions of mucous membranes, deformed face, loss of fingers or toes
Incubation period	3 months to 20 years; usually 3 years
Causative agent	*Mycobacterium leprae*
Laboratory diagnosis	Microscopic inspection of skin scrapings or biopsies of infected tissues stained to detect acid-fast bacteria
Pathogenesis	Invasion of small nerves of skin; multiplication in macrophages; course of disease depends on immune response of host; immune macrophages limit growth of organism; attack of immune cells against infected nerve cells produces nerve damage leading to deformity; immunity sometimes overwhelmed by accumulating bacterial antigen, allowing unrestrained growth of *M. leprae.*
Epidemiology	Direct contact with organisms from lesions of mucous membranes or skin
Prevention	Disinfection of contaminated articles, hand washing; vaccines under evaluation
Treatment	Dapsone plus rifampin for months or years; clofazimine added for lepromatous disease

Viral Diseases of the Nervous System

Many viruses can infect the central nervous system, including the Epstein-Barr virus of infectious mononucleosis; the mumps, rubeola, varicella-zoster, and herpes simplex viruses; and arboviruses, a diverse group of viruses transmitted by insects, mites, or ticks. In most cases, nervous system involvement occurs in only a very small percentage of people infected with the viruses.

The following section discusses four kinds of illness resulting from viral central nervous system infections: sporadic encephalitis, epidemic encephalitis, poliomyelitis, and rabies.

Cases of Sporadic Encephalitis Are Widely Scattered in Time and Place

Viral encephalitis is an inflammatory disease of the brain caused by a virus. Typical symptoms include fever, headache, stiff neck, disorientation, seizures, coma, and impairment of one or more motor or sensory nerves. Most people recover from the disease but are left with permanent impairment, such as epilepsy, paralysis, deafness, or difficulty thinking. **Sporadic encephalitis** occurs at a fairly constant low frequency in a given population, with cases widely scattered in time and place. Herpes simplex virus is the most common cause of sporadic encephalitis, but other widespread human viruses, including those that cause mumps, measles, and infectious mononucleosis, can occasionally cause encephalitis.

• **Herpes simplex virus, pp. 559–61**

Epidemic Encephalitis Generally Results from the Spread of Arboviruses to Humans

Arboviruses (arthropodborne viruses) can cause **epidemic encephalitis.** LaCrosse encephalitis virus is the leading cause of epidemic encephalitis in the United States. Like most other arboviruses, the LaCrosse virus maintains itself in nature, humans being only accidental hosts. The virus infects *Aedes* mosquitoes, transmitted from one mosquito to another by the male mosquito's semen and from one mosquito generation to another through the female mosquito's eggs. These mosquitoes feed on and infect squirrels and chipmunks, which thereby acquire the infection and pass it on when bitten by other mosquitoes. Humans then acquire the virus from the bite of an infected mosquito, human bites occurring mostly when there is a shortage of natural hosts or an abundance of mosquitoes. Another common cause of epidemic encephalitis is St. Louis encephalitis virus, maintained in nature in birds and *Culex* mosquitoes.

Poliomyelitis (Infantile Paralysis) Is Caused by Poliovirus Infection of the Motor Nerves

The characteristic feature of poliomyelitis ("polio") is selective destruction of motor nerve cells, usually of the spinal cord, resulting in permanent paralysis of one particular group of muscles, such as those of an arm or leg. One goal of the World Health Organization is to eliminate this disease from the earth by the year 2000.

Symptoms

Poliomyelitis usually begins with the symptoms of meningitis: headache, fever, stiff neck, and nausea. In addition, pain

• **Mosquitoes, p. 434**

Perspective 28.3

Arthropodborne Encephalitis Epidemic, Florida, 1990

In Florida, midsummer 1990, blood tests on sentinel chickens showed that increasing numbers were developing antibody to St. Louis encephalitis virus. This result indicated that many of the mosquitoes in the area had become infected with the virus. Since *Culex* mosquitoes tend to shift their feeding from birds to humans in the fall, various health agencies were alerted to the possibility of an epidemic of St. Louis encephalitis among humans. When cases were identified, public warnings and intensive mosquito control measures were undertaken. Due to such measures, only 38 cases of St. Louis encephalitis were documented, much lower than the 110 cases in the previous epidemic in 1977.

When faced with an arbovirus encephalitis epidemic, people can take preventive measures in addition to using insecticides. These measures include limiting time outdoors at night, using insect repellents, mending window and door screens, and eliminating places where mosquitoes reproduce, such as old tires and other containers that collect water.

Source: Centers for Disease Control and Prevention. 1990. *Morbidity and Mortality Weekly Report* 39:756.

and spasm of some muscles generally occur, later followed by paralysis. Over the ensuing weeks and months, muscles shrink and bones do not develop normally in the affected area (figure 28.9). In the more severe cases, the muscles of respiration are paralyzed, and the victim requires an artificial respirator (a machine to pump air in and out of the lungs). Some recovery of function is the rule if the person survives the acute stage of the illness. The nerves of sensation (touch, pain, temperature) are not affected.

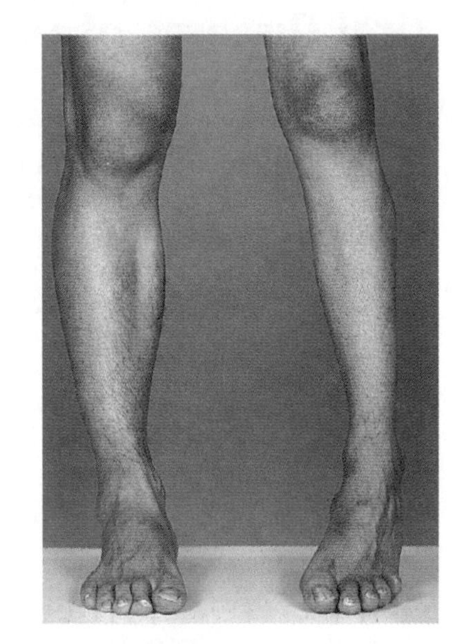

Figure 28.9
Person with an atrophied leg due to poliomyelitis.

Causative Agent

Poliomyelitis is caused by three types of poliovirus, designated 1, 2, and 3, which are distinguished by using antisera. These small, nonenveloped, single-stranded, positive sense RNA viruses are members of the enterovirus subgroup of the picornavirus family (chapter 14). They can be grown in vitro in tissue cell cultures, causing cell destruction. With low concentrations of virus, the areas of cell destruction, termed *plaques*, are separated from one another and are readily seen with the unaided eye (figure 28.10).

Pathogenesis

Polioviruses enter the body orally, infect the throat and intestinal tract, and then invade the bloodstream. In most people, the immune system conquers the infection and recovery is complete. Only in a small percentage of people does the virus enter the nervous system and attack motor nerves. The viruses can infect a cell only if the surface of that cell possesses specific receptors to which the virus can attach. This specificity of receptor sites helps explain the fact that polioviruses selectively infect motor nerve cells of the brain and spinal cord, while sparing many other kinds of cells. The infected cell is destroyed when the mature virus is released.

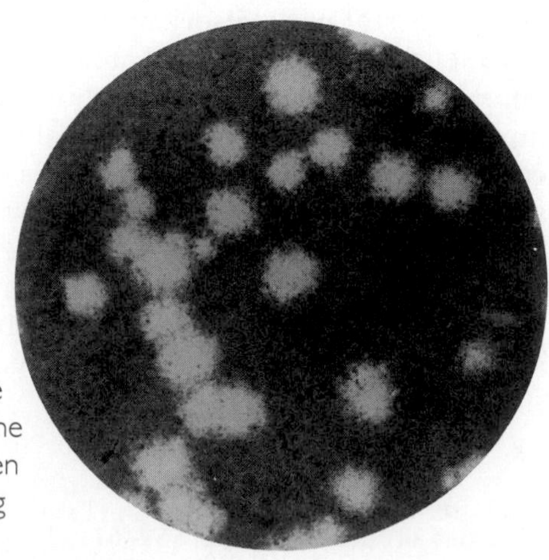

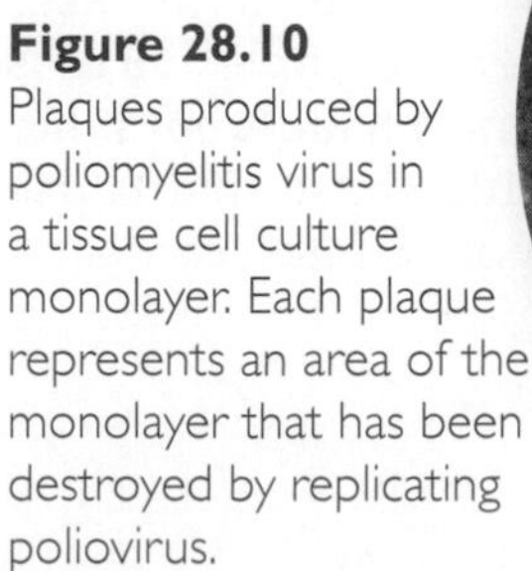

Figure 28.10
Plaques produced by poliomyelitis virus in a tissue cell culture monolayer. Each plaque represents an area of the monolayer that has been destroyed by replicating poliovirus.

Perspective 28.4

Laboratory Diagnosis of Poliomyelitis

Poliovirus can be recovered from the patient's throat and feces early in the development of the illness, later from the blood, and finally, in some patients, from the cerebrospinal fluid. Excretion of the virus in the feces continues for weeks or months but is transitory from other sites. When introduced into appropriate tissue cell cultures (usually monkey kidney cells), the virus produces a cytopathic (cell-damaging) effect in which the cells become spherical, no longer adhere to the surface of the container, and eventually lyse. The areas of viral multiplication are thus visible microscopically. To identify the virus, known antisera are added to samples of the virus, which are then added to tissue cell cultures. The antiserum that prevents the cytopathic effect identifies the virus.

Epidemiology

Poliomyelitis has had the greatest impact in the most economically advanced countries. The disease occurs in poorer, more crowded nations (the World Health Organization estimates 450,000 cases per year), but epidemics of paralytic poliomyelitis generally do not occur there. This characteristic can be explained by the fact that, in areas where sanitation is poor, the polioviruses spread throughout the population, transmitted by the fecal-oral route. Thus, very few people escape childhood without becoming infected and developing immunity to the disease. Therefore, most babies receive antipolio antibodies transplacentally from their mothers. Newborn infants in these nations thus are partially protected against bloodstream infection and nervous system invasion by poliomyelitis virus for as long as their mothers' antibodies persist in their bodies—usually about two or three months. During this time, because of exposure to the poliovirus through crowding and unsanitary conditions, infants are likely to develop mild infections of the throat and intestine, thereby achieving lifelong immunity. Even if exposure is delayed beyond two or three months, it is likely to occur at an early age, when the disease is less likely to cause severe paralysis.

In contrast, in areas with efficient sanitation, the poliomyelitis virus sometimes cannot spread to enough susceptible people to sustain itself. If its reintroduction from another community is delayed sufficiently, some people of all ages, including older children and adults, will lack antibody and will be susceptible to infection. When the virus is finally reintroduced, a high incidence of infection and paralysis will result. This situation occurred in the United States in the 1950s, resulting in many cases of paralysis and death. However, with most people now routinely immunized against the disease, this scenario no longer occurs.

Prevention

Like other enteroviruses, polioviruses are quite stable under natural conditions (including the conditions of contaminated swimming pools) but are inactivated by pasteurization and properly chlorinated drinking water.

Control of poliomyelitis using vaccines represents one of the greatest success stories in the battle against infectious diseases (figure 28.11). Due to vaccination, not a single case of naturally acquired poliomyelitis occurred in the entire Western Hemisphere between 1991 and 1996. Ironically, the recent cases of paralytic polio in the United States have been caused by the live oral polio vaccine that has been used since 1961, which can, on rare occasion, produce paralysis in a recipient. Paralytic polio can develop in infants and children receiving the oral vaccine and in nonvaccinated people who change the diapers of vaccinated infants. Since the vaccine virus multiplies in the intestinal cells, large concentrations may be present in the feces of a recently vaccinated infant or child. Vaccine-related paralytic poliomyelitis occurs mostly with the first dose of vaccine, with an incidence of about one case in every 520,000 first doses.

In 1978, a killed injectable vaccine more highly purified and concentrated than the 1955 killed (Salk) vaccine became available. The newer killed vaccine is more immunogenic than the earlier one. The live (Sabin) oral vaccine produces better local immunity in the throat and intestine, is less expensive, and does not require an injection. However, based on current evidence, it is probably safer to use killed vaccine for a child's first vaccine doses. The live vaccine can safely be given as the fourth and fifth doses at 18 months and five years, respectively. In 1996, this approach was formally recommended by the Centers for Disease Control and Prevention but challenged on the grounds of increased cost. Parents who have not been fully immunized against polio should also be immunized with killed vaccine before their children receive the oral form.

Postpolio Syndrome

Even though poliomyelitis transmission no longer occurs in the United States, about 300,000 people, most of whom recovered years ago, have survived the disease. Since the

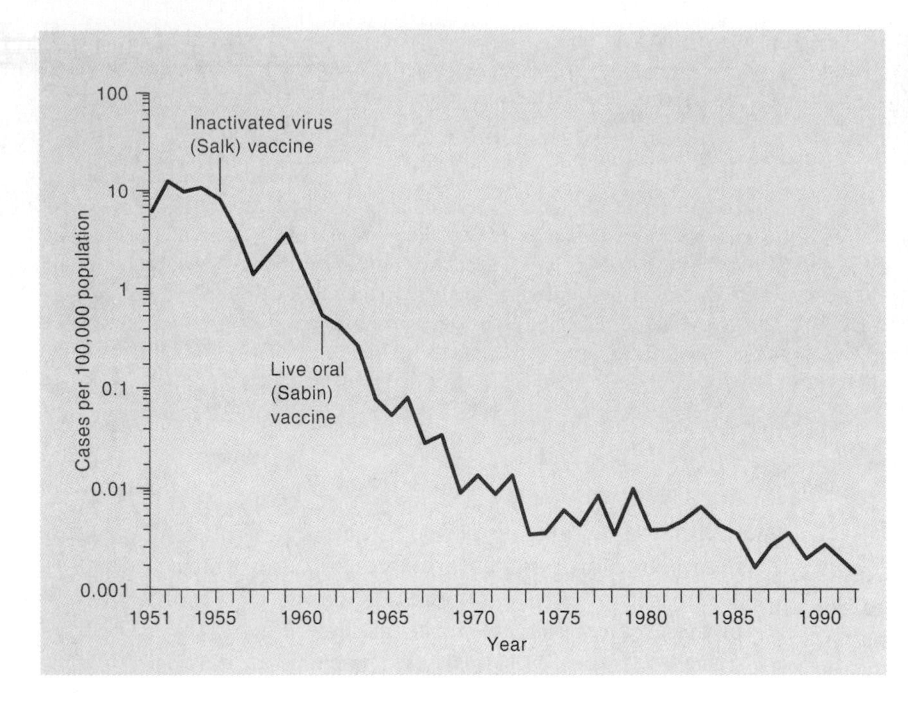

Figure 28.11
Incidence of poliomyelitis in the United States, 1951–1992. Ironically, since 1980 all cases acquired in the United States have been caused by the oral vaccine.

Source: Centers for Disease Control and Prevention. 1993. Summary of notifiable diseases—United States, 1992. *Morbidity and Mortality Weekly Report* 41(55):46.

Table 28.5 Poliomyelitis (Infantile Paralysis)

Symptoms	Headache, fever, stiff neck, nausea, pain, and spasm of muscles followed by paralysis
Incubation period	7–14 days
Causative agents	Polioviruses 1, 2, 3
Laboratory diagnosis	Culture of poliovirus from throat and feces
Pathogenesis	Virus infects the throat and intestine, circulates via the bloodstream, and enters some motor nerve cells of the brain and spinal cord; infected nerve cells lyse upon release of mature virus
Epidemiology	Fecal-oral route; asymptomatic and nonparalytic cases common
Prevention	Oral live polio vaccine or injected killed polio vaccine
Treatment	Supportive; artificial ventilation for respiratory paralysis; physical therapy and rehabilitation

early 1970s, there have been reports that some of these survivors can develop muscle pain, increased weakness, and muscle degeneration 15 to 50 years after they had acute poliomyelitis. This condition, called the *postpolio syndrome,* sometimes involves muscles not obviously affected by the original illness. The late progression of muscle weakness is not due to recurrent multiplication of polio viruses. During recovery from acute poliomyelitis, surviving nerve cells branch out to take over the functions of the killed nerve cells. The late progression is thus most likely due to the death of these nerve cells that have been doing "double duty" for so many years. The slow loss of these nerve cells produces no noticeable change to most victims of poliomyelitis who have relatively good muscle function, but in those who already have moderately severe impairment, further deterioration can cause severe disability.

Poliomyelitis is summarized in table 28.5.

Rabies Has a Long Incubation Period and Is Generally Fatal

In the United States, immunization of dogs against rabies has practically eliminated them as a source of human disease. However, the rabies virus remains rampant among wildlife, a constant threat to nonimmunized domestic animals and humans. Many questions remain about the pathogenesis of rabies, and no effective treatment exists for the disease.

Symptoms of Rabies

Rabies is one of the most feared of all diseases because its terrifying symptoms almost invariably end with death. Like many other viral diseases, it begins with fever, head and muscle aches, sore throat, fatigue, and nausea. The characteristic symptom that strongly suggests rabies is a tingling or twitching sensation at the site of viral entry, usually an animal bite. These early symptoms generally begin one to two months after viral entry and progress rapidly to symptoms of encephalitis, agitation, confusion, hallucinations, seizures, and increased sensitivity to light, sound, and touch. The body temperature then rises steeply, and increased salivation combined with difficulty swallowing result in frothing at the mouth. **Hydrophobia,** painful spasm of the throat and respiratory muscles provoked by swallowing or even seeing liquids, occurs in half the cases.

Perspective 28.5

Human Rabies, Texas, 1990

A 22-year-old man was bitten on the right index finger by a bat while in a Texas tavern. He did not get medical attention and remained well until about six weeks later when his right hand became weak. Over the next four days, he developed numbness in his right arm, and then 10- to 15-second episodes of staring and unresponsiveness, muscle rigidity, hallucinations, and throat spasms causing him to refuse liquids. His temperature rose to 107°F. Rabies was suspected, and cerebrospinal fluid and skin biopsies were sent for testing. He went into a coma and died five days after his symptoms began. Fluorescent antibody demonstrated rabies virus in his brain tissue, which was obtained postmortem. The rabies virus antigens were then identified using monoclonal antibodies, which showed the viral strain to be similar to a strain known to be carried by a common bat species in the area. More than 50% of this species of bat have proved to be rabies-positive after they demonstrate abnormal behavior. In fact, three of the four cases of human rabies acquired in the United States from 1980 to 1990 were a result of exposure to bats.

• Monoclonal antibody, p. 383

To help avoid contracting rabies, people should avoid contact with any animal showing abnormal behavior, dogs and cats should be vaccinated against the disease, and, if a person is bitten, he or she should immediately clean the wound thoroughly and seek medical attention. Even a trivial physical contact with a bat warrants medical evaluation because bites are not recognized in most cases.

Source: Centers for Disease Control and Prevention. 1991. *Morbidity and Mortality Weekly Report* 40:132

Coma develops, and about 50% of patients die within four days of the first appearance of symptoms.

Causative Agent

The cause of rabies is the rabies virus (figure 28.12 *a*), a member of the rhabdovirus family (see chapter 14). This virus is about 180 × 75 nanometers in size, has a striking bullet shape, is enveloped, and contains single-stranded negative sense RNA. It buds from the surface of infected cells.

Pathogenesis

The principal mode of transmission of rabies to humans is via the saliva of a rabid animal (figure 28.12 *b*) introduced into bite wounds or abrasions of the skin. Rabies can also be contracted by inhaling aerosols containing the virus, such as from bat feces. Only a few details of the events that follow introduction of the virus into the body are known. During the incubation period of the disease, the virus multiplies in muscle cells and probably other cells at the site of infection. Knoblike projections on the viral surface attach to receptors in the region where the nerve joins the muscle. At some point, the virus enters an axon and is carried along the course of the nerve by the normal flow of the axon's cytoplasm, eventually reaching the brain. The long incubation period, usually one to two months but sometimes exceeding a year, is partly determined by the length of the journey to the brain. Patients with head wounds into which the virus is introduced tend to have a shorter incubation period than those with extremity wounds. Severe wounds and those with a large amount of introduced rabies virus also generally have short incubation periods. The virus then multiplies extensively in brain tissue, causing the symptoms of encephalitis. Characteristic **inclusion bodies,** called **Negri bodies** (figure 28.13), form at the sites of viral replication in the brain, but the cells are not lysed. The virus spreads outward from the brain via the nerves to various body tissues, notably the salivary glands, eye, and fatty tissue under the skin, as well as to the heart and other vital organs. The presence of the virus in the eye is of some practical significance, since cases can be diagnosed before death by stained smears made from the surface of the eye. Moreover, several cases of rabies have occurred in patients who received corneal transplants from donors who died of an atypical form of rabies that escaped diagnosis.

The immune response of the host probably plays an important role in pathogenesis since viral antigens are expressed on the surface of infected cells, which are therefore subject to attack by antibody and complement as well as cell-mediated immunity.

Epidemiology

Rabies is widespread in wild animals; about 5,000 cases are generally reported each year in the United States. This represents an enormous reservoir from which infection can be transmitted to domestic animals and humans. In the United States, skunks, raccoons, and bats constitute the chief reservoir hosts. The virus can remain latent in bats for long periods, and healthy looking bats can have the virus in their salivary glands.

• Complement, p. 351

Perspective 28.6

Laboratory Diagnosis of Rabies

Rabies is usually diagnosed most quickly by preparing smears from the conjunctiva or fatty tissue of the neck or from biopsy of the brain. These smears are stained with a combination of dyes called **Seller's stain** or, better yet, with fluorescent antirabies antibody and are inspected for Negri bodies. Presence of the rabies virus can be verified by cultivating it in laboratory mice or tissue cell cultures and using a known antirabies antiserum for final identification.

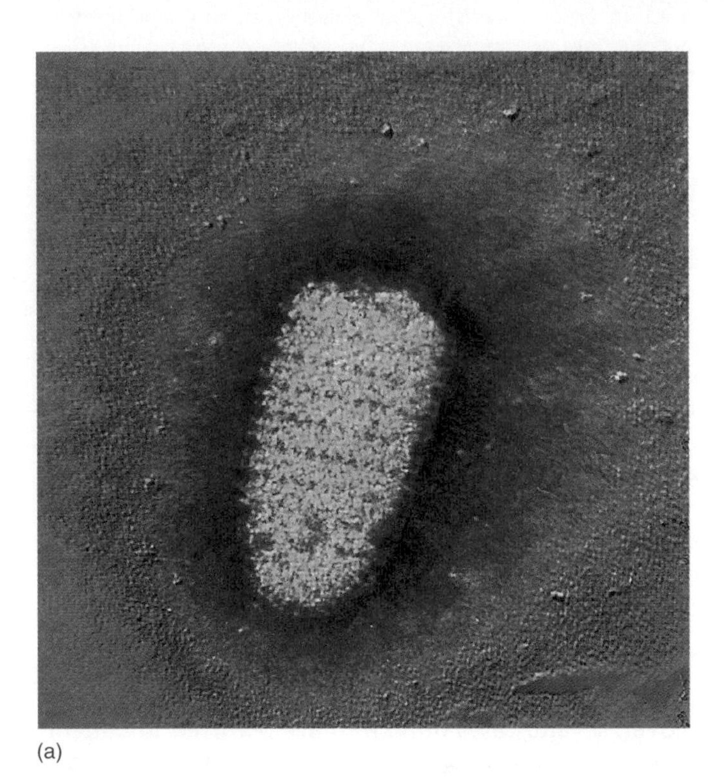

(a)

(b)

Figure 28.12
In most cases, rabies is transmitted by the saliva of a biting animal. (*a*) Color-enhanced transmission electron micrograph of rabies virus. Note the bullet shape. (*b*) Rabid raccoon caged in a Virginia animal shelter.

Until the mid-20th century, most rabies cases in the United States resulted from dog bites, which is the primary mode of transmission in the nonindustrialized world. The dog population in the United States is about 25 million, and an estimated 1 million people are bitten each year. Fortunately, since World War II, the incidence of dog rabies has dropped dramatically, and now more than 85% of reported rabies infections are in wild, as opposed to domestic, animals. The incidence of rabies in people has declined because dogs and cats have been immunized against rabies infection. In effect, this immunization creates a partial barrier to the spread of the rabies virus from wild animal reservoirs to humans. Since pets vary in their tolerance to rabies vaccines depending on their age and species, several kinds of vaccines are available.

About three-quarters of dogs that develop rabies excrete rabies virus in their saliva, and about one-third of these begin excreting it one to three days before they get sick. Therefore, when a person is bitten by a nonvaccinated, apparently healthy dog, the animal should be confined for 10 days to see if symptoms of rabies appear. Some dogs become irritable and hyperactive with the onset of rabies, produce excessive saliva, and attack people, animals, and inanimate objects. Perhaps more common is the "dumb" form of rabies, in which an infected dog simply stops eating, becomes inactive, and suffers paralysis of throat and leg muscles. Obviously, one should not be tempted to try to remove a suspected foreign body from the throat of a sick, choking, nonvaccinated dog! The reported number of rabies cases in humans now generally ranges from zero to

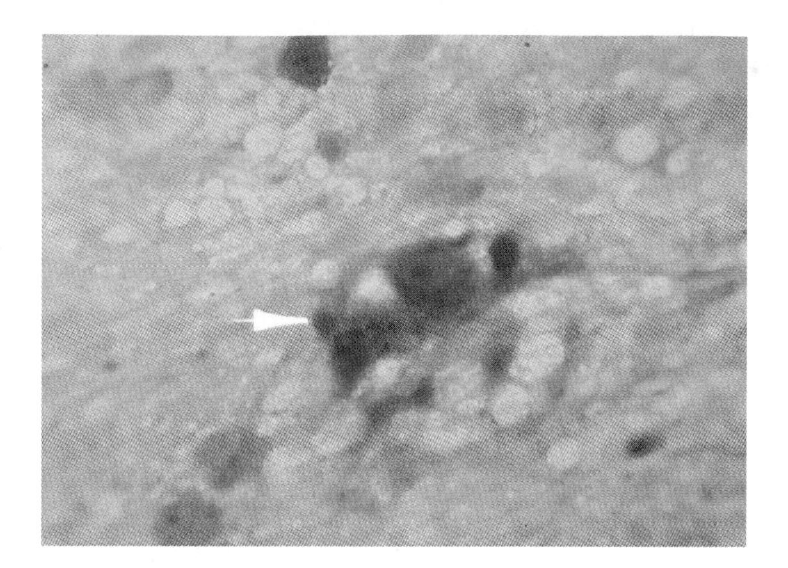

Figure 28.13
Smear of brain tissue from a rabid dog (Seller's stain). The arrow points to one of several Negri bodies within the triangular-shaped nerve cell. The Negri bodies represent the sites of rabies virus replication.

four per year in the United States. Some of these cases have occurred in immigrants apparently infected as long as six years before developing rabies.

Pasteur's Work with the Rabies Virus

In people bitten by dogs having rabies virus in their saliva, the risk of developing rabies is about 30%. Louis Pasteur discovered that this risk can be lowered considerably by administering rabies vaccine as soon as possible after exposure to the virus. Presumably, the vaccine provokes an immune response much sooner than the natural infections, inactivating free virus and killing infected cells during the long incubation period when the number of infected cells is relatively small.

Pasteur's vaccine was made from dried spinal cords of rabies-infected rabbits. Unfortunately, the rabbit nervous system tissue in the vaccine sometimes stimulated an immune response against the patient's own brain, causing **allergic encephalitis.** An inactivated (killed) rabies virus grown in human cell cultures (human diploid cell vaccine) is now used, and the risk of serious side effects is very low.

Prevention and Treatment

A person who has been bitten by an animal should first wash the wound immediately and thoroughly with soap and water and then apply an antiseptic. If there is a possibility that the animal is rabid, the patient should then receive a series of five injections of human diploid cell[1] vaccine intramuscularly. Antirabies antibody is also injected at the wound site and intramuscularly. Formerly, the sources of this antibody were horses that had been immunized by injection with rabies vaccine. However, since horse antiserum is a foreign protein, many human recipients quickly formed antibodies against it. The reaction between the patient's antibody and the horse protein often resulted in **serum sickness** or anaphylactic shock. Today, antirabies globulin obtained from humans who have been immunized against rabies is used instead of horse serum. In the United States, about 30,000 people annually receive rabies vaccine to prevent rabies after having been bitten by suspected rabid animals.

Studies are underway to evaluate different methods of immunizing wildlife against rabies, but no proven techniques are available yet.

There is no effective treatment for rabies, and only three people are known to have ever recovered from the disease. Interestingly, all three people recovered completely.

Some features of rabies are summarized in table 28.6.

• **Anaphylactic shock, pp. 397–98**
• **Serum sickness, p. 401**

[1]Diploid cell—one having the normal duplicated set of chromosomes, as opposed to tumor cells, which tend to have abnormal chromosomal arrangements

Table 28.6 Rabies

Symptoms	Fever, headache, nausea and vomiting, sore throat, and cough at onset; later, spasms of the muscles of mouth and throat, coma, and death
Incubation period	Usually 30–60 days; sometimes many months or years
Causative agent	Rabies virus; has an unusual bullet shape
Laboratory diagnosis	Fluorescent antibody-stained smears from conjunctiva, fatty tissue of neck, or brain show Negri inclusion bodies in infected cells
Pathogenesis	During incubation period, virus multiplies at site of bite, then travels via nerves to the central nervous system; here it multiplies and spreads outward via multiple nerves to infect heart and other organs
Epidemiology	Bite of rabid animal, usually dog, cat, skunk, raccoon, or bat
Prevention	Avoid suspect animals; immunization of pets; immunization of wildlife
Treatment	No effective treatment once symptoms begin; effective post-exposure measures: wash wound with soap and water and apply antiseptic; injections of human diploid cell vaccine and human rabies antiserum as soon as possible after exposure

Fungal Diseases of the Central Nervous System

Fungi very rarely invade the central nervous system of normal people, but patients with cancer, sugar diabetes, and AIDS as well as those receiving immunosuppressing medications risk serious infection. Such infections are dangerous because they are difficult to treat with antimicrobial medication. Common fungi from the soil and decaying vegetation sometimes infect the nose and sinuses of these patients; from there they penetrate to the brain and cause the patient's death. Cryptococcal meningoencephalitis differs from this general pattern, since about half the cases occur in apparently normal people, and many cases are cured by medication.

Cryptococcal Meningoencephalitis Is Caused by the Yeast Form of the Fungus *Filobasidiella Neoformans*

Cryptococcal meningoencephalitis (inflammatory disease of the meninges and brain) was an uncommon disease before the AIDS epidemic, but since its onset, a large increase has occurred in AIDS patients. The disease is among the top four life-threatening infectious complications in AIDS.

Symptoms of Cryptococcal Meningoencephalitis

Symptoms of cryptococcal meningoencephalitis develop very gradually in most cases and generally consist of difficulty in thinking, dizziness, intermittent headache, and slight fever. After weeks or months of slow progression of these symptoms, vomiting, weight loss, stupor, seizures, and impairment of one or more nerves may appear. The disease occurs frequently in people with immune deficiency and generally progresses much faster than in other patients. Without treatment, death can occur in as little as two weeks.

Causative Agent

Cryptococcal meningoencephalitis is an infection of the meninges and brain by the encapsulated yeast form of *Filobasidiella neoformans*, an organism previously placed in the genus *Cryptococcus*. However, in 1975, discovery of a sexual form was reported, showing that the fungus was closely related to a group of pathogenic basidiomycetes that infect plants. Due to this discovery, the organism was renamed *F. neoformans*, but the earlier genus name is still generally used. As can be seen in infected material from patients, the organism is a small, spherical yeast generally 3 to 7 micrometers in diameter surrounded by a large capsule (figure 28.14).

• Basidiomycetes, p. 287

Pathogenesis

The *F. neoformans* fungus becomes airborne with dust, enters the body by inhalation, and establishes an infection first in the lung. This infection causes mild or no symptoms in most people and is usually eliminated by body defenses, chiefly lung phagocytes. However, phagocytic killing of *F. neoformans* is slow and inefficient in nonimmune individuals. In some cases, the organisms continue to multiply, enter the bloodstream and are then distributed throughout the body. The capsule is essential to pathogenicity since nonencapsulated strains do not cause disease. Capsular material inhibits phagocytosis and migration of leukocytes and diffuses from the organism, neutralizing opsonins. Progressive infection and dissemination are much more likely to occur when a person's cell-mediated immunity is impaired, particularly in AIDS and certain cancers. Macrophage activation by immune lymphocytes, lacking in many immunodeficiencies, is essential for rapid phagocytic killing of *F. neoformans*. Meningoencephalitis is the most common infection outside of the lung, but organisms spread by the bloodstream can also infect skin, bones, or other body tissues. In meningoencephalitis, the organisms typically cause thickening of the meninges, sometimes impeding the flow of cerebrospinal fluid, thereby increasing the pressure within the brain. They also invade the brain tissue, producing multiple abscesses.

Epidemiology

F. neoformans is distributed worldwide in soil and vegetation but is especially numerous in areas where pigeon droppings accumulate. For every case of cryptococcal meningoencephalitis, millions of people are infected by the organism without harm. About half of the patients with cryptococcal meningoencephalitis prove to have an underlying disorder that impairs their cell-mediated immune system. The infection is often the first indication of AIDS. Person-to-person transmission of the disease does not occur.

Treatment and Prevention

Treatment with the antibiotic amphotericin B is often effective, particularly if given concurrently with flucytosine (5-fluorocytosine) or newer oral antifungal medicines such as itraconazole. Amphotericin B must be given intravenously and the dose carefully regulated to minimize the toxic effects of the antibiotic, mainly against the kidneys. Since amphotericin B does not reliably cross the blood-brain barrier, it is often necessary to administer it through a plastic tube inserted through the skull into a lateral ventricle of the brain. Except in AIDS patients, treatment is successful in about 70% of cases. AIDS patients respond poorly to treatment, most likely because they lack T cell dependent

• Opsonins, p. 352
• Antifungal medicines, pp. 466–67

Perspective 28.7

Laboratory Diagnosis of Cryptococcal Meningoencephalitis

During laboratory diagnosis, cerebrospinal fluid is inoculated onto culture media and is examined under the microscope. The encapsulated yeasts are readily visualized microscopically when a few drops of India ink are added to the fluid to outline their capsules (figure 28.14). The organisms are also easily demonstrated by staining dried smears of infected material. On culture media, they grow readily at 37°C and produce dark, water-insoluble pigments if the media has an appropriate substrate. The organisms produce the enzyme urease, which is also helpful diagnostically. At least four different serotypes exist, which can be identified by using known antisera against their capsules.

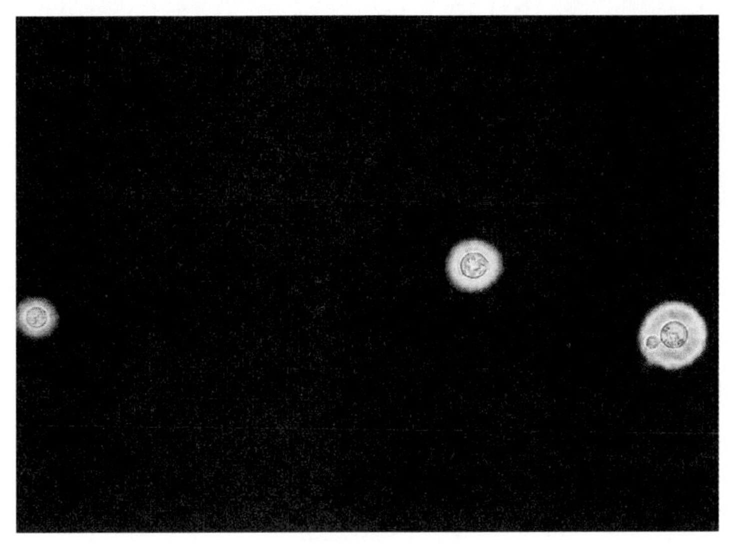

Figure 28.14
Filobasidiella (*Cryptococcus*) *neoformans* in the spinal fluid of a patient with cryptococcal meningitis (×450). India ink has been added to the fluid to outline the organism's capsule.

killing of *F. neoformans* that normally assists the action of the antifungal medications. There is no vaccine or other preventive measure available.

The main features of cryptococcal meningoencephalitis are summarized in table 28.7.

Table 28.7 Cryptococcal Meningoencephalitis

Symptoms	Headache, vomiting, confusion, and weight loss; fever often absent; symptoms may progress to seizures, paralysis, coma, and death
Incubation period	Widely variable, few to many weeks
Causative agent	*Cryptococcus (Filobasidiella) neoformans*
Laboratory diagnosis	Cerebrospinal fluid examined under the microscope with India ink; cultures
Pathogenesis	Infection starts in lung; in nonimmune individuals, encapsulated organisms multiply, enter bloodstream and are carried to various parts of the body; phagocytosis inhibited and opsonins neutralized; meninges and adjacent brain tissue become infected; half of the patients have underlying disease that interferes with immunity
Epidemiology	Inhalation of dust containing dried pigeon droppings contaminated with the fungus; most people resistant to the disease
Prevention	None
Treatment	Amphotericin B with flucytosine or itraconazole

Protozoan Diseases of the Central Nervous System

Only a few species of protozoa are important central nervous system pathogens for humans. One interesting example, *Naegleria fowleri*, causes meningoencephalitis after penetrating the skull along the nerves responsible for the sense of smell. Less than 200 cases have been reported, but the disease is almost always rapidly fatal. It is usually acquired by swimming in warm freshwater, with or without chlorine. For every case of *Naegleria* meningoencephalitis, many millions of people are exposed to the organism without harm. *N. fowleri* can exist in three forms: ameba, flagellate, and cyst. It is one of only a few free-living protozoa pathogenic for humans. In contrast, African sleeping sickness is a quite different protozoan disease.

• **Protozoa, pp. 285, 290–94**

Perspective 28.8

Laboratory Diagnosis of African Sleeping Sickness

Microscopic inspection of blood or other material withdrawn from the lymph node with a needle and syringe usually reveals the actively motile protozoa. The finding of strikingly high levels of IgM antibody strongly supports the diagnosis of African sleeping sickness.

Table 28.8 African Sleeping Sickness

Symptoms	Fever, sleepiness, rash; enlargement of lymph nodes, liver, and spleen; later, involvement of the central nervous system, uncontrollable sleepiness, headache, poor concentration, and unsteadiness
Incubation period	3 weeks
Causative agent	*Trypanosoma brucei*
Laboratory diagnosis	Microscopic inspection of lymph node material reveals the protozoa, high levels of IgM antibody
Pathogenesis	After bite of fly, organisms multiply at site and then enter blood and lymphatic circulation; as new cycles of parasites are released, their surface protein changes and the body is required to respond with a new antibody
Epidemiology	Bites of the infected tsetse fly transmit organism through fly saliva; wild animal reservoir for *T. brucei rhodesiense*
Prevention	Protective clothing, insecticides, clearing of brush where flies breed
Treatment	Suramin; if central nervous system is involved, melarsoprol is used

African Sleeping Sickness Is Acquired from the Bite of a Tsetse Fly

African sleeping sickness is important because residents of and visitors to Africa can contract the disease in a wide area across the middle of the African continent. The causative protozoan has a clever way of protecting itself against the host's immune system.

Symptoms of Sleeping Sickness (African Trypanosomiasis)

The first symptoms of this disease appear within a week after a person is bitten by a tsetse fly. A tender nodule develops at the site of the bite. The regional lymph nodes may enlarge, but symptoms may all disappear spontaneously. Weeks to several years later, recurrent fevers develop that may continue for months or years. Involvement of the central nervous system is marked by gradual loss of interest in everything, decreased activity, and indifference to food, progressing to slurred speech, coma, and eventually death.

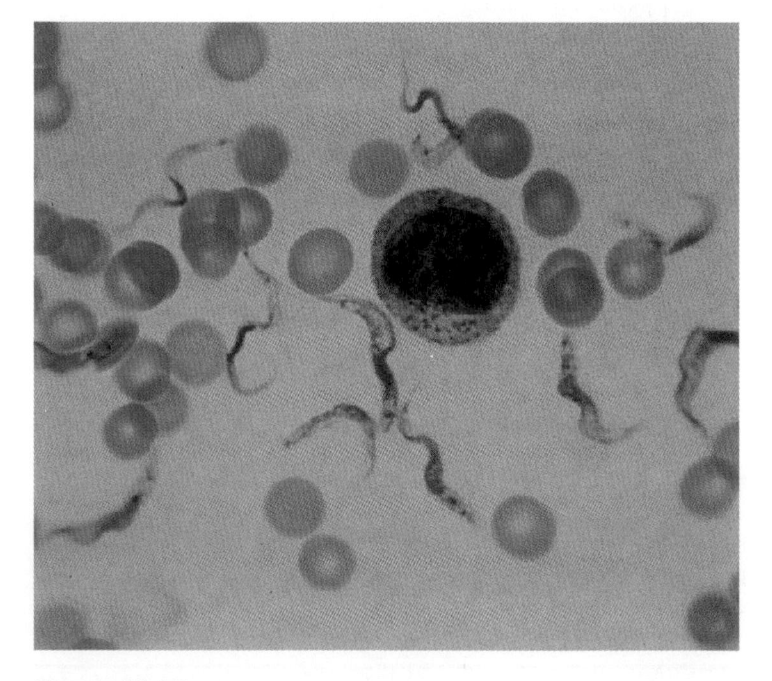

Figure 28.15
Trypanosoma brucei rhodesiense in the blood smear of a person with African trypanosomiasis (×500).

Causative Agent

African sleeping sickness is caused by the flagellated protozoan, *Trypanosoma brucei*. The organisms are slender, with a wavy, undulating membrane and an anteriorly protruding flagellum (figure 28.15). There are two subspecies that are morphologically identical, *T. brucei rhodesiense* and *T. brucei gambiense*. The *rhodesiense* subspecies occurs mainly in the cattle-raising areas of East Africa, while the *gambiense* subspecies occurs mainly in forested areas of Central and

Perspective 28.9

Spongiform Encephalopathies

A mysterious group of chronic degenerative brain diseases involves wild and domestic animals and humans. Microscopically, brain tissue affected by these diseases has a spongy appearance due to the loss of neurons and other changes—this is why these diseases are collectively referred to as transmissible *spongiform encephalopathy*. Brain and other tissues from affected animals can transmit the disease to normal animals. The incubation period is often long, measured in months, years, or decades.

Some of these encephalopathies have been transmitted to small laboratory animals including mice and hamsters. Cow-to-human transmission probably occurs on rare occasions following peoples' consumption of flesh from cattle with mad cow disease. Human-to-human transfer of one type of spongiform encephalopathy called Creutzfeldt-Jakob disease has resulted from corneal transplants (the cornea is the clear middle part of the eye in front of the lens). Another human spongiform encephalopathy, kuru, was probably transmitted by cannibalism, which was formerly practiced by New Guinea natives.

Table 1 Characteristics of Spongiform Encephalopathy Agents

1. They increase in quantity during the incubation period of the disease.
2. They resist inactivation by ultraviolet and ionizing radiation.
3. They resist inactivation by formaldehyde and heat.
4. They are not readily destroyed by proteases.
5. They are not destroyed by nucleases.
6. They are much smaller than the smallest virus.
7. They are composed of protein coded by a normal cellular gene but modified after transcription.

The main signs and symptoms of spongiform encephalopathies are dementia and unsteadiness.

The causative agents of these diseases (widely known as proteinaceous infectious particles, or "prions") may constitute a new class of infectious agents that differ from bacteria, viruses, and viroids. Their nature is still under active investigation, but there is strong evidence that they are unusual proteins, devoid of nucleic acid and more resistant to proteases than normal proteins are. Prions have an amino acid sequence identical to a normal brain protein but are folded differently. The normal protein contains helical segments. Whereas the corresponding segments of the prion protein are pleated sheets.

Little is known about the pathogenesis of spongiform encephalopathies, and there are no inflammatory or immune responses to them. Also, there is no treatment for these diseases, which, fortunately, are rare in humans.

• **Mad cow disease, p. 301**

• **Prions, p. 16**

• **Protein folding, p. 33**

West Africa. Both are transmitted by tsetse flies, a group of biting insects of the genus *Glossina*.

Pathogenesis

During the bite of an infected tsetse fly, the protozoan enters the bite wound in the fly's saliva. The organism multiplies at the skin site and within a few weeks enters the lymphatics and blood circulation. The patient responds with fever and production of IgM antibody against the protozoa, and symptoms improve. However, within about a week and at roughly weekly intervals thereafter, there are recurrent increases in the number of parasites in the blood. Each of these bursts of increased parasitemia (parasites in the circulating blood) coincides with the appearance of a new protein on the surface of the trypanosomes. By activating genes transposed to certain sites of the protozoan chromosome, a single strain of *T. brucei* can produce hundreds of different surface proteins. The patient's immune system must then respond with production of a new antibody. These recurrent cycles of parasitemia and antibody production occur until the patient is treated or dies.

In infections due to *T. brucei rhodesiense*, the disease tends to progress rapidly, with the heart and brain invaded within six weeks of infection. Irritability, personality changes, and mental dullness result from brain involvement, but the patient usually dies from heart failure within six months.

In *T. brucei gambiense*, progression of infection is much slower, and years may pass before death occurs, often from secondary infection. Much of the damage to the host is due to immune complexes formed when antibody reacts with complement and high levels of protozoan antigen.

Epidemiology

African sleeping sickness occurs on the African continent within about 15° of the equator, with 10,000 to 20,000 new cases each year, including a number of American tourists.

• **Transposons, p. 177**

• **Immune complexes, pp. 400–1**

The distribution of the disease is determined by the distribution of the tsetse fly vectors. The main reservoirs of the severe (Rhodesian) form of the disease are wild animals; for the milder (Gambian) form, humans are the main reservoir. Bites of the infected tsetse fly transmit the disease, but less than 5% of the flies are infected.

Treatment and Prevention

If the disease has not progressed to involve the central nervous system, the drug suramin is effective for treatment. It must be given intravenously, and toxic effects of the drug, including itching, vomiting, and nerve damage, are common. If central nervous system involvement is present, then the arsenic compound melarsoprol must be used because it is able to cross the blood-brain barrier, whereas suramin does not. A promising new drug, difluoromethylornithine, may replace melarsoprol, which can cause fatal brain damage.

Preventive measures directed against tsetse fly vectors include insect repellents and protective clothing to prevent bites, insecticides, and clearing brush that provides breeding habitats for the flies. Populations can be screened for *T. brucei* infection by examining blood specimens. Treatment of infected people helps reduce the reservoir of *T. brucei gambiense*. A single intramuscular injection of the medication pentamidine prevents the Gambian form of the illness for a number of months, although it will not necessarily prevent an infection that could progress at a later time.

• **Suramin, melarsoprol, p. 468**

REVIEW QUESTIONS

1. Why is it important that people keep their polio immunizations up-to-date?
2. What is the postpolio syndrome?
3. Why is rabies now rare in humans when it is so common in wildlife?
4. What is a Negri body?
5. How does *T. brucei* evade host immunity?
6. How likely is it that a person who sniffs dust containing dried pigeon droppings or swims in warm freshwater will become ill?

Summary

I. Anatomy and Physiology
   A. The brain and spinal cord make up the central nervous system.
   B. There are two kinds of nerves.
      1. Motor nerves cause parts of the body to act.
      2. Sensory nerves transmit sensations, such as heat, pain, touch, light, and sound.
   C. Cerebrospinal fluid is produced in cavities inside the brain and flows out over the brain, spinal cord, and nerves. Samples can be removed and tested for the cause of a central nervous system infection.
   D. Meninges are the three membranes that cover the surface of the brain and spinal cord. Cerebrospinal fluid flows between the innermost two.
   E. Infectious organisms can reach the brain and spinal cord by way of
      1. the bloodstream (the blood-brain barrier acts as protection);
      2. the nerves; and
      3. direct extensions through the skull of infections elsewhere, such as from the middle ear or sinuses.

II. Normal Flora

III. Bacterial Infections of the Nervous System
   A. *Neisseria meningitidis* is the cause of meningococcal meningitis. It is a gram-negative diplococcus. Release of endotoxin from the organism can cause shock and death. It can cause epidemics.
   B. *Haemophilus influenzae* is a leading cause of bacterial meningitis. It is a tiny, gram-negative coccobacillus and requires X and V factors for growth. It shows satellitism. Some strains have acquired a penicillinase-coding plasmid.
   C. *Streptococcus pneumoniae* is the cause of pneumococcal meningitis. It is a gram-positive, encapsulated diplococcus. Pneumonia and ear or sinus infection often accompanies meningitis. The most common cause of meningitis in adults is *S. pneumoniae*.
   D. *Escherichia coli* is an enteric, gram-negative rod; it is one of leading causes of meningitis in the newborn.
   E. *Streptococcus agalactiae*, another leading cause of meningitis of the newborn, has a group B cell wall polysaccharide.
   F. *Listeria monocytogenes*, a motile, aerobic, gram-positive rod, causes meningitis in newborns and patients with other diseases.
   G. Leprosy (Hansen's disease) is characterized by invasion of peripheral nerves by the acid-fast bacillus, *Mycobacterium leprae*, which has not been cultivated *in vitro*. The disease occurs in two main forms, tuberculoid and lepromatous, depending on the immune status of the patient.

IV. Viral Diseases of the Nervous System
   A. Herpes simplex virus is the most important cause of sporadic encephalitis.

B. Epidemic encephalitis is usually caused by arboviruses.
   1. LaCrosse encephalitis virus is maintained in *Aedes* mosquitoes and rodents.
   2. St. Louis encephalitis virus is maintained in *Culex* mosquitoes and birds.

C. Poliomyelitis involves the motor nerve cells of the brain and spinal cord; it is caused by three picornaviruses, polioviruses 1, 2, and 3.
   1. Motor nerve destruction leads to paralysis, muscle wasting, and failure of normal bone development.
   2. Postpolio syndrome occurs years after poliomyelitis, and it is probably caused by the death of nerve cells that had taken over for the ones killed by poliomyelitis virus.

D. Rabies is transmitted mainly through the bite of an infected animal.
   1. The incubation period is long, often measured in months and sometimes years.
   2. The virus multiplies in muscle and then travels in nerve cells to the brain. After multiplying in the brain, it spreads outward via the nerves to infect other body tissues, rapidly causing death.

V. Fungal Diseases of the Central Nervous System
   A. Cryptococcal meningoencephalitis is an infection of the meninges and brain caused by *Filobasidiella* (*Cryptococcus*) *neoformans*. Infection originates in the lung after a person inhales dust laden with the fungi. The organism is often associated with old pigeon droppings. This disease is a common, serious complication of AIDS.

VI. Protozoan Diseases of the Central Nervous System
   A. African sleeping sickness is caused by *Trypanosoma brucei*.
      1. During infection, the organism shows bursts of growth, each appearing with different surface proteins.
      2. Each of these variants requires that the body respond with a new antibody since the prior one is ineffectual.

Chapter Review Questions

1. Describe the Shwartzman phenomenon.
2. What measures can be undertaken to prevent meningococcal meningitis?
3. Describe the organism that is the leading cause of bacterial meningitis in adults.
4. Name two causes of bacterial meningitis in newborn infants.
5. Describe the differences between tuberculoid and lepromatous leprosy.
6. How is leprosy transmitted?
7. What are arboviruses?
8. What is the difference between sporadic encephalitis and epidemic encephalitis? Name one cause of each.
9. Why does poliomyelitis virus attack motor and not sensory neurons?
10. Can a dog infected with rabies virus transmit the disease while appearing well? Explain.

Critical Thinking Questions

1. Give the pros and cons for using live or inactivated polio vaccine.
2. Why is it important to learn about rabies when only a few cases occur in the entire United States each year?

Further Reading

Dalakas, M. C., et al. 1986. A long-term follow-up study of patients with post-poliomyelitis neuromuscular symptoms. *The New England Journal of Medicine*, 314:959–963.

Donelson, J. E., and Turner, M. J. 1985. How the trypanosome changes its coat. *Scientific American*, February.

Fox, J. L. 1990. Rabies vaccine field test undertaken. *ASM News* 56(11):579–583. An interesting study to evaluate a genetically engineered vaccinia virus to immunize wildlife against rabies. Also a brief discussion of the theoretical risks of employing vaccinia virus to immunize wildlife.

1997. Human rabies—Kentucky and Montana, 1996. *Morbidity and Mortality Weekly Report* 46(18):397.

Lieberman, J. M., et al. 1996. Safety and immunogenicity of a serogroups A/C *Neisseria meningitidis* oligosaccharide-protein conjugate vaccine in young children: a randomized controlled trial, *Journal of the American Medical Association* 275:1499–1503. Describes evaluation of a T cell dependent vaccine against groups A and C meningococci.

Mestel, Rosie. 1996. Putting prions to the test. *Science* 273:184–189. A nice review of current knowledge about prions.

Pinner, R. W., et al. 1992. Epidemic meningococcal disease in Nairobi, Kenya, 1989. *The Journal of Infectious Diseases* 166:359. A study of a recent epidemic including strain identification and control using vaccine.

1996. Progress toward global eradication of poliomyelitis, 1995. *Morbidity and Mortality Weekly Report* 45(26):565–568.

Roueche, Berton. 1984. *The incurable wound*. In *The Medical Detectives*. Truman Talley Books, New York. A story about rabies.

Scully, R. E., et al. 1985. Case records of the Massachusetts General Hospital. *The New England Journal of Medicine*, 313:1464–1472. A case of leprosy in a 25-year-old man.

1995. Serogroup B meningococcal disease—Oregon, 1994. *Morbidity and Mortality Weekly Report* 44(7):121–124.

Smith, J. S., et al. 1991. Unexplained rabies in three immigrants in the United States. A virological investigation. *The New England Journal of Medicine* 324:205–211. Presents evidence for very long incubation periods.

Whitley, R. J. 1990. Viral encephalitis. *New England Journal of Medicine 323(4):242–250.*

Winkler, W. G., and Bögel, Konrad. 1992. Control of rabies in wildlife. *Scientific American* 266:86. An interesting description of efforts to control rabies in wildlife by using vaccine-containing baits.

29 Wound Infections

KEY CONCEPTS

1. Bacteria with little invasive ability can produce serious disease if they synthesize toxins that are absorbed and carried to other parts of the body.
2. Since a wide variety of pathogens infect wounds, widely differing diagnostic and treatment methods may be required.
3. *Staphylococcus aureus* is the most important cause of wound infections because it is so commonly carried by humans.
4. Bacterial exotoxins from different species of bacteria may have the same mode of action but cause distinctly different diseases because the toxin molecules enter different body cells.
5. The prevalence of certain wound infections is determined by factors such as geography and occupation.
6. Bacteria with little invasive ability when growing alone can invade tissue when growing with other bacteria.

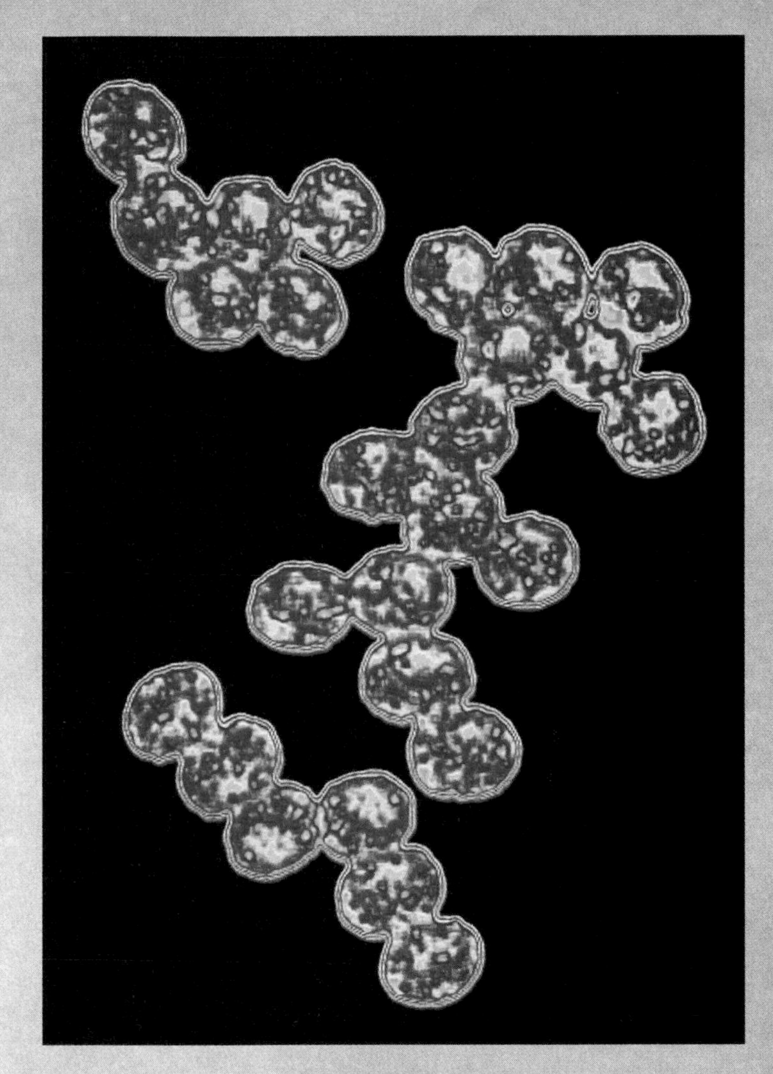

Color-enhanced transmission electron micrograph of *Staphylococcus aureus*. This resident of the human nose is not only a leading cause of wound infections, but also a cause of conditions as diverse as boils, toxic shock, and food poisoning.

Microbes in Motion

The following books in the *Microbes in Motion* CD-ROM may serve as a useful preview or supplement to your reading: *Gram-Positive Organisms. Gram-Negative Organisms. Miscellaneous Bacteria. Microbial Pathogenesis in Infectious Diseases:* Toxins. *Vaccines:* Vaccine Targets. *Viral Structure and Function:* Viral Diseases. *Fungal Structure and Function:* Diseases. *Parasitic Structure and Function.*

A Glimpse of History

Clostridium tetani, the cause of the disease **tetanus,** is found in soil and dust and, thus, virtually everywhere. Before the development and widespread use of tetanus toxoid immunization, this often lethal disease was common because of the widespread occurrence of the organisms. In many economically developing nations, tetanus continues to be a serious health problem.

Dr. Shibasaburo Kitasato (1856–1931), while working in Dr. Koch's laboratory in Germany, discovered how to cultivate this organism and recognized that a toxin produced by *C. tetani* caused tetanus. His findings paved the way for the development of immunization against the disease.

Kitasato was born in a mountainous village in southern Japan in 1856. He was sent by his family to medical schools in both Kumamoto and Tokyo. Following his graduation in 1883, Kitasato went to work for the Central Hygienic Bureau in Japan. The Japanese government needed to control epidemics of typhoid, cholera, and blackleg,* so in 1885 they sent Kitasato to study with Koch in Germany.

In Koch's laboratory, Kitasato was given the tetanus problem to study. In the course of his experiments, he discovered that *C. tetani* would only grow under strictly anaerobic conditions. Once he had isolated the organism in pure culture, he was able to show that it produced tetanus in laboratory animals. However, he was puzzled by the finding that although the animals died of generalized disease, there was no *C. tetani* at sites other than the site where the organism had been injected. By doing experiments in which he injected the tails of mice and then removed the injected tissue at hourly intervals after injection, he showed that the animals only developed tetanus if the organisms were allowed to remain for more than an hour. He further showed that the organisms remained at the site of inoculation; at no time were they found in the rest of the body. Kitasato reasoned that something other than bacterial invasion was causing the disease.

About this time, another investigator, Emil von Behring, was busy investigating how *Corynebacterium diphtheriae* caused the disease diphtheria. Together, Kitasato and von Behring were able to show that both diseases were caused by toxins that were produced by the bacteria. The concept that a bacterial toxin could cause disease in the absence of bacterial invasion was an extremely important advance in the understanding and control of certain infectious diseases.

In 1890, Kitasato published his studies. In 1892, even though he was urged to stay in Germany, he returned to Japan. The Japanese government was not prepared to support basic research at this time, so Kitasato established his own institute for infectious diseases where he worked and trained Japanese scientists for the rest of his life. In 1908, Koch paid a visit to Kitasato in Japan and a Shinto shrine was built in Koch's honor.

During his later years, Kitasato was instrumental in establishing laws regulating health practices in Japan. He died in 1931 at age 75. To honor him, a shrine was erected next to Koch's.

*Blackleg—often fatal disease affecting mainly calves, caused by a species of *Clostridium*

• **Bacterial toxins, pp. 420, 422**

Most people occasionally sustain wounds that produce breaks in the skin or mucous membranes. Almost always, microorganisms contaminate these wounds from the air, fingers, or the object causing the wound. Whether these microorganisms cause trouble depends on how virulent they are, the size of the inoculum, the status of host defenses, and the presence or absence of foreign material. Ordinarily, wounds heal uneventfully despite microbial colonization, but sometimes severe, even fatal, infection results from even a trivial wound.

Burns represent one large category of accidental wounds. In the United States each year, more than 2.5 million people sustain burns severe enough to seek medical attention. Of these, more than 100,000 require hospitalization, and 12,000 die from their burns.

Intentional wounds inflicted by surgery represent about 23 million cases annually in the United States. More than 900,000 of these become infected, the percentages varying according to the type of surgery and the status of the patients' host defenses. These infections increase the cost of hospitalization by approximately $1.5 billion each year.

Anatomy and Physiology

Wounds expose components of tissue normally protected from the outside world by skin or mucous membranes. These components include substances such as fibronectin, fibrin, and fibrinogen that provide receptors to which potential pathogens specifically attach. Fibronectin is a glycoprotein that occurs both as a circulating form and as component of tissue where it serves to tie cells and other tissue substances together. Shortly after the occurrence of a wound, the soluble protein, fibrinogen, is converted to the fibrous material, fibrin, thereby forming clots in the severed blood and lymphatic vessels. This stops the flow of blood and lymph as the first step in the repair process. Wound healing begins with the outgrowth of connective tissue cells (fibroblasts) and capillaries from the surfaces of the wound, producing a nodular, red, translucent material called **granulation tissue.** In the absence of dirt or infection, granulation tissue fills the void created by the wound, contracts, and eventually is converted to scar tissue covered by overlying skin or mucous membrane. Sometimes in the presence of dirt or infection, the granulation tissue overgrows, bulging from the wound to form a **pyogenic granuloma.**

Wounds can be classified as incised, as when produced by a knife or other sharp object; puncture, as with penetration of a small sharp object such as a nail; lacerated, when the tissue is torn; or contused, as when caused by a blow that crushes tissue. All wounds should be considered potentially dangerous because of the risk of serious infection. However, the wounds most likely to be complicated by infection are those that contain devitalized (dead) tissue, dirt, or other foreign material and those that have a small opening that impairs drainage of dirt and pus. Burns present special problems. Although initially sterile, thermal burns are often extensive, with a large area of weeping devitalized tissue highly susceptible to infection.

Wound Abscesses Are Difficult to Cure Using Antimicrobial Medicines Alone

An **abscess** (figure 29.1) is a localized collection of pus, composed of living and dead leukocytes, components of tissue breakdown, and infecting organisms. There are no blood vessels in abscesses because they have been destroyed or pushed aside. A surrounding area of inflammation separates the abscess from normal tissue. Consequently, abscess formation helps to localize an infection and prevent its spread. Microorganisms in abscesses often are not killed by antimicrobial medicines because many of the microorganisms cease multiplying, and active multiplication is generally required for microbial killing by the medicines. In addition, the chemical nature of pus interferes with the action of some antibiotics, while some diffuse poorly into abscesses because they lack blood vessels. Microorganisms in abscesses are a potential source of infection of other parts of the body if they escape the surrounding area of inflammation and enter the blood or lymph vessels. Generally, abscesses must be drained in order to effect a cure.

Anaerobic Conditions Are Present in Many Wounds

Another important feature of many wounds is that they are relatively anaerobic, thus allowing anaerobic pathogens to multiply. Such conditions are especially likely in dirty wounds, wounds with crushed tissue, and puncture wounds. Puncture wounds caused by nails,

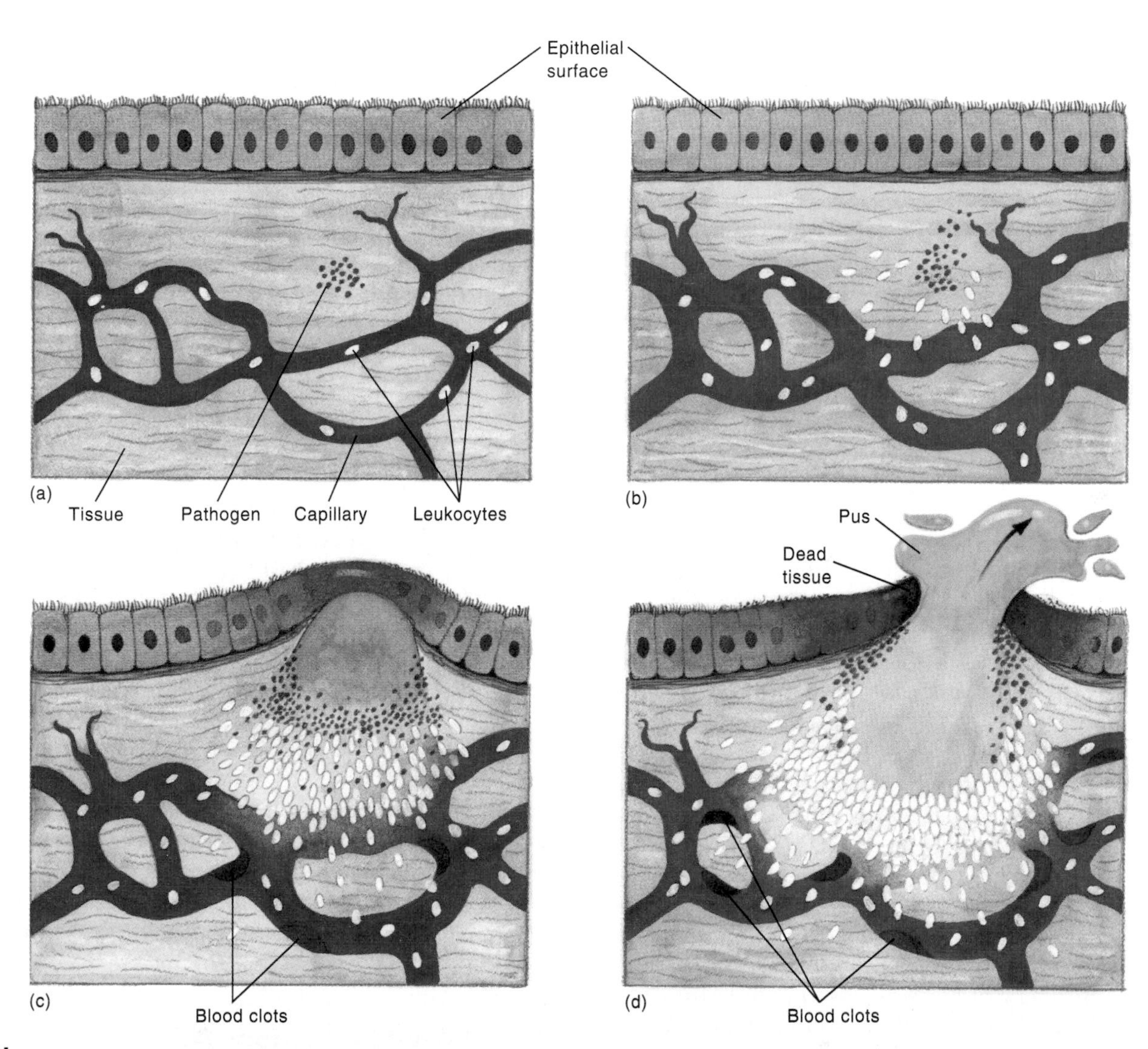

Figure 29.1
Abscess formation. (*a*) A pathogenic microorganism is deposited in the tissue from a wound or from the bloodstream, for example. (*b*) Blood vessels dilate, and leukocytes migrate chemotactically from them to the area of the developing infection. (*c*) Pus, composed of the products of tissue cell breakdown, leukocytes, and bacteria, forms. Clotting occurs in the adjacent blood vessels. (*d*) Buildup of pressure causes the abscess to expand in the direction of least resistance. If it reaches a body surface, it may rupture and discharge its contents.

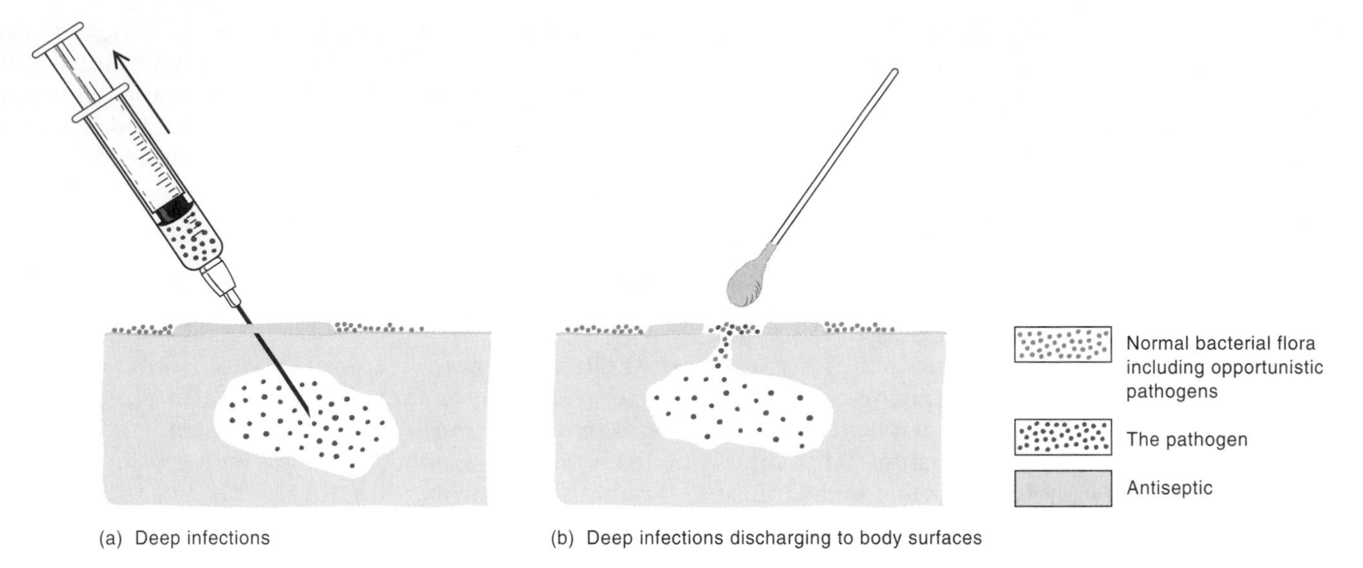

Figure 29.2
Techniques used to culture wounds. In the case of (*b*), it would be easy to mistake opportunists from the normal flora for the pathogen responsible for the infection, especially when the pathogen grows more slowly than the opportunists.

thorns, splinters, and other sharp objects can introduce foreign material and microorganisms deep into the body. Bullets and other projectiles that enter at high speed can carry contaminated fragments of skin or cloth into the tissues. Because of the force with which they enter, projectiles cause relatively small breaks in the skin that may close quickly, thereby masking areas of extensive tissue damage.

Diagnosing Wound Infections Can Be Difficult

A number of generally harmless bacteria can sometimes cause serious wound infections, depending on the kind of wound and the immune status of the wounded person. For example, *Aeromonas hydrophila*, a facultatively anaerobic, gram-negative rod that lives in soil and water, can multiply in a clean minor wound but not interfere with its healing. In a person who is immunodeficient or has a crushing wound with dirt and dead tissue, the same organism can cause a life-threatening infection.

Merely cultivating an organism from a wound does not prove that the wound is infected. Things that favor the diagnosis of wound infection are (1) a heavy growth of the organism, (2) its isolation in pure culture, (3) evidence of the host's response to it in the form of a polymorphonuclear neutrophil response and other signs of acute inflammation, and (4) recovery of a species of known pathogenic potential. Some important causes of wound infections are shown in table 29.1.

Depending on the type of wound, it may be difficult to identify the cause of an infection because of the presence of normal body flora that are known to have pathogenic potential. Figure 29.2 illustrates how the normal flora can mislead the diagnosis.

Table 29.1 Important Causes of Wound Infections

Type of Wound	Causative Agent
Wound from accidental trauma	*Staphylococcus aureus*, *Streptococcus pyogenes*, *Clostridium* sp., *Sporothrix schenckii*
Surgical wound	*S. aureus*, *S. pyogenes*, *Pseudomonas aeruginosa*, enterobacteria, *Enterococcus* sp.
Burn	*P. aeruginosa*, *S. aureus*, *S. pyogenes*, enterobacteria
Animal-inflicted wound	*Pasteurella multocida*, *S. aureus*, *Bartonella henselae*

Bacterial Wound Infections

Almost all wounds sooner or later become contaminated with bacteria from the environment or a person's own body. Whether they can multiply and cause disease depends on the species of bacterium, its virulence, the numbers of organisms introduced into the wound, the status of host defenses, and other factors. The following sections present some of the bacterial species successful in causing wound infections.

Most Wound Infections Are Caused by Staphylococci

Staphylococci lead the causes of wound infections, both surgical and accidental (figure 29.3). As discussed in chapter 22, staphylococci are almost universally present on the human skin or in the nose. Of the more than 30 recognized species of

Perspective 29.1

The Coagulase Test

The coagulase test is used to identify *S. aureus* because it is easy to perform and almost all strains of the bacterium give a positive reaction, while all but a few other staphylococcal species give a negative result. Despite its *-ase* ending, coagulase is not an enzyme. It is a largely extracellular protein product of *S. aureus* that reacts with a blood-clotting substance called *prothrombin*. The resulting complex, called *staphylothrombin*, acts on fibrinogen to produce the fibrin strands that cause blood to clot. There is no conclusive evidence that coagulase is a virulence factor for *S. aureus*. However, besides being extracellular, some coagulase is tightly bound to the surface of the bacterium, and it is known that fibrin-coated staphylococci resist phagocytosis.

The test for **clumping factor,** sometimes referred to as the *slide coagulase test*, is often used in place of the coagulase test because it is faster and the two tests correlate well with each other. The clumping factor test is performed by mixing growth of staphylococci from an agar medium with a drop of plasma on a microscope slide. In a positive result, the staphylococci promptly aggregate in coarse clumps. The clumping is caused by a cell-fixed protein that attaches specifically to fibrinogen in the plasma. The clumping factor protein is a virulence factor for *S. aureus* because it attaches to both fibrinogen and fibrin, thus allowing colonization of wound surfaces. Plastic devices, such as intravenous catheters and heart valves, become coated with fibrinogen shortly after insertion, thus making them too a target for *S. aureus* colonization. The gene for clumping factor is distinct from the one controlling coagulase.

• **Coagulase test, p. 476**

staphylococci, only two, *S. aureus* and *S. epidermidis*, are thought to account for most human wound infections. However when better ways of identifying the coagulase-negative staphylococci become available, many isolates formerly labeled *S. epidermidis* may prove to be other species. Generally, coagulase-positive staphylococcal isolates have been called *S. aureus* and coagulase-negative isolates have been called *S. epidermidis*.

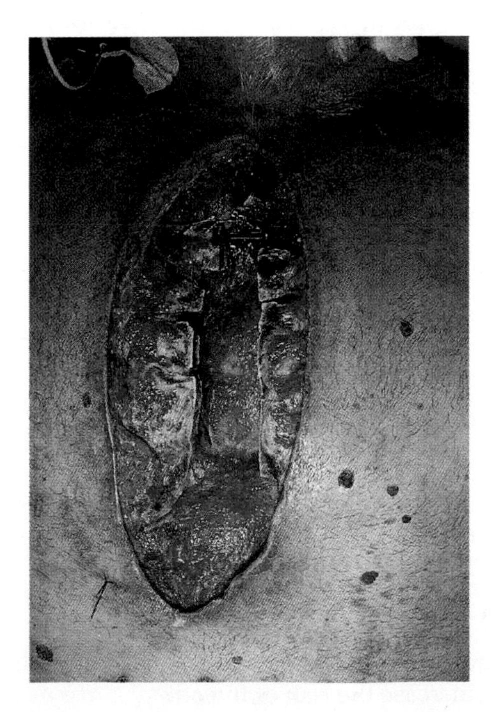

Figure 29.3
Surgical wound infection due to *Staphylococcus aureus*. The stitches pull through the infected tissue, causing the wound to open.

Symptoms

Staphylocci are **pyogenic cocci,** meaning that they characteristically cause the production of a purulent discharge (pus). An inflammatory reaction is generally elicited, causing swelling, redness, and pain. If the infected area is extensive or if the infection has spread to the general circulation, fever is a prominent symptom. Wound infections by some strains of *S. aureus* produce the **toxic shock syndrome,** with high fever, muscle aches, and life-threatening shock, sometimes accompanied by a rash and diarrhea. In general, symptoms produced by *S. epidermidis* are minor unless the infection involves the circulatory system.

Causative Agents

Staphylococci are gram-positive cocci arranged in clusters that grow readily aerobically and anaerobically on the usual laboratory media. They are catalase positive and salt tolerant. *S. aureus* ferments mannitol and is coagulase and clumping factor positive. These tests are negative for *S. epidermidis*.

Staphylococcus aureus *S. aureus* possesses a number of traits that make it much more virulent than *S. epidermidis*. Colonization of wounds is aided by a number of adherence factors, including clumping factor and binding proteins for fibronectin, fibrin and fibrinogen, and collagen. **Protein A** is another component of the bacterial surface that aids virulence. It has the property of binding IgG by the Fc portion of the immunoglobulin molecule. Thus, protein A not only competes with the Fc receptors of phagocytes for antibody molecules, but it also causes staphylococci to be coated with host protein. Most strains of *S. aureus* also produce

α-toxin, which kills cells by attaching to specific receptors on the cell membranes of susceptible cells, causing holes to form. Platelets and monocytes are especially susceptible to α-toxin.

All of these properties contribute to the virulence of almost all strains of *S. aureus*. However, a relatively small percentage of strains also produce one or more additional toxins. One type of toxin, a **leukocidin,** kills neutrophils in the higher concentrations of toxin or causes them to release their enzymes at lower concentrations. Other toxins are responsible for toxic shock. These include the six **enterotoxins** responsible for food poisoning and **toxic shock syndrome toxin-1 (TSST-1).**

Table 29.2 summarizes these virulence properties of *S. aureus*. The bacterium also produces a variety of extracellular enzymes that probably do not contribute directly to virulence but supply it with nutrients by degrading blood cells and components of damaged tissue. Some of these are discussed in chapter 22.

Staphylococcus epidermidis Most strains of *S. epidermidis* have little or no invasive ability for normal people. However, most strains can adhere to the plastic catheters and other devices employed in modern medicine. Following adherence, the bacteria produce a slime (glycocalyx) that cements the growing colony to the plastic and protects it from attack by phagocytes and other host defense mechanisms as well as antibacterial medications.

Table 29.2 Properties of *S. Aureus* Implicated in Virulence

Virulence Factor	Action Site	Action
Fibronectin binding protein	Bacterial surface	Attachment to acellular tissue matrix, clots, indwelling plastic devices, endothelium, epithelium
Clumping factor	Bacterial surface	Attachment to fibrin, fibrinogen, plastic devices
Protein A	Bacterial surface, extracellular	Competes with Fc receptors of phagocytes; coats bacterium with host's immunoglobulin
α-toxin	Extracellular	Makes holes in cell membranes
Leukocidin	Extracellular	Kills or degranulates neutrophils
Enterotoxins	Extracellular	If systemic, cause toxic shock
Toxic shock syndrome toxin-1	Extracellular	If systemic, causes toxic shock

Pathogenesis

Staphylococcus aureus Many years ago, it was shown that it takes more than 100,000 staphylococci injected into the skin to produce a small abscess. However, when injected along with a suture, only about 100 *S. aureus* are required to produce the same lesion. These studies, which used students as guinea pigs, dramatized the effect of foreign material in the pathogenesis of staphylococcal infections.

More recent studies have involved cloning the genes controlling suspected virulence factors and then deleting or reinserting the genes to see what effect they have on pathogenesis in laboratory animals. It is likely that multiple virulence factors act together to produce the usual wound infection. Clumping factor and the other binding proteins attach the organisms to clots and tissue components, fostering colonization. Clumping factor, coagulase, and protein A serve to coat the organisms with host proteins, giving them a disguise that hides them from attack by phagocytes and the immune system. This may explain why immunity to staphylococcal infection is generally weak or nonexistent. Some protein A is released from the bacterial surface and reacts with circulating immunoglobulin. The resulting complexes activate complement and probably contribute to the intense inflammatory response and accumulation of pus.

Colonization of plastics and other foreign materials occurs because they quickly become coated with fibrinogen and fibronectin to which the staphylococci attach. Systemic spread of wound infections can lead to abscesses in other tissues, such as the heart and joints. Staphylococcal toxins that enter the circulation act as superantigens, causing the widespread release of TNF and other cytokines from T cells, thereby producing toxic shock.

Staphylococcus epidermidis Wound infections by *S. epidermidis* in normal people are frequently cleared by host defenses alone. Organisms colonizing indwelling plastic catheters come loose and are carried by the bloodstream to the heart and other tissues. This can result in subacute bacterial endocarditis or multiple tissue abscesses in people with impaired host defenses as from cancer, diabetes mellitus, or other causes.

Epidemiology

The epidemiology of *S. aureus* is discussed in chapter 22. Various studies have shown that, in the case of surgical wound infections, nasal carriers have a two to seven times greater risk of infection than do those who are not nasal carriers. Thirty % to 100% of the infections are due to a patient's own staphylococcus strain. Advanced age, poor general health, immunosuppression, prolonged preoperative hospital stay, and infection at a site other than the site of surgery increase the risk of infection.

• *S. aureus*, extracellular products, p. 478

• Superantigens, pp. 420, 422–423

Prevention and Treatment

Cleansing and removal of dirt and devitalized tissue from accidental wounds minimizes the chance of infection as does prompt closure of clean wounds by sutures. Trying to eliminate the nasal carrier state with antistaphylococcal medications is occasionally successful but incurs the risk of acquiring a staphylococcal strain resistant to the medications. Surgical infections in critical cases can be reduced by half by administering an effective antistaphylococcal medication immediately before surgery. For unknown reasons, the infection rate is actually increased if the medication is given more than three hours before or two hours after the surgical incision.

Treatment of staphylococcal infections has been problematic because of the development of resistance to antibacterial medications. When penicillin was first introduced, more than 95% of the strains of *S. aureus* were susceptible to it. These strains soon largely disappeared with the widespread use of the antibiotic. The remaining strains are resistant by virtue of plasmid encoded β-lactamase. Treatment of these strains became much easier with the development of penicillins and cephalosporins resistant to β-lactamase. However, soon thereafter, strains of *S. aureus* appeared that were resistant to the new antibiotics.

The resistance is due to a chromosomal gene that codes for a peptidoglycan transpeptidase that binds penicillins and cephalosporins poorly. Fortunately, these strains, usually referred to as MRSAs (methicillin-resistant *Staphylococcus aureus*), have not spread as rapidly or widely as other resistant strains. Although many of the MRSAs have R plasmids, making them resistant to most antistaphylococcal medications, vancomycin has generally been effective against them. Unfortunately, plasmidborne vancomycin-resistance is increasingly common among *Enterococcus* sp., which are members of the normal intestinal flora of humans and which conjugate with staphylococci in vitro. This resistance among enterococci raises the specter of vancomycin resistance among staphylococci as use of the antibiotic increases. Partly, this increased use is against *S. epidermidis* infections in immunocompromised patients. *S. epidermidis* is commonly resistant to methicillin and other β-lactamase–resistant antibiotics.

Group A Streptococci Can Be "Flesh Eaters"

Streptococcus pyogenes is introduced in earlier chapters as the cause of "strep throat," rheumatic fever, and glomerulonephritis. It is also a common cause of wound infections, which have generally been easy to treat since penicillin became available. However, over the last decade or so concern has increased about *S. pyogenes* causing death and massive tissue destruction; *S. pyogenes* is the "flesh eating bacterium" of the tabloid press.

• ***S. pyogenes*, p. 506**

Symptoms

At the site of a surgical wound or of accidental trauma, sometimes even without an obvious break in the skin, pain develops acutely. Within a short time, swelling becomes apparent, and the injured person develops fever and confusion. The overlying skin becomes tense and discolored because of the swelling. Unless treatment is initiated promptly, shock and death usually follow in a short time.

Causative Agent

As described in earlier chapters, *S. pyogenes* is a gram-positive coccus arranged in chains, is catalase negative, is hemolytic on blood agar, and has the Lancefield group A cell wall polysaccharide. Current evidence indicates that the strains of *S. pyogenes* that cause the symptoms described are more virulent than other strains by virtue of at least two extracellular products. The first of these is **pyrogenic exotoxin A** which is a superantigen and causes **streptococcal toxic shock.** The second product, **exotoxin B,** is a cysteine protease that destroys tissue by breaking down protein.

Pathogenesis

Like *Staphylococcus aureus, Streptococcus pyogenes* has a fibronectin binding protein that aids colonization of wounds. The organisms multiply in the wound and release their enzymes and toxins. The subcutaneous fascia and fatty tissue are destroyed, in a condition called **necrotizing fasciitis.** In some cases, the fascia[1] surrounding muscle is penetrated, and muscle tissue is destroyed, in a condition known as **streptococcal myositis.** Intense swelling occurs as fluid is drawn into the area because of increased osmotic pressure from the breakdown of tissue into small molecules. The organisms continue to multiply and produce toxic products in the mass of dead tissue, using the breakdown products as nutrients. In most cases, toxic products and organisms enter the bloodstream. Superantigens and other streptococcal products cause shock by releasing TNF from lymphocytes and probably by other as yet unproven mechanisms.

Epidemiology

These invasive infections by *S. pyogenes* are known to have occurred over a span of many years, but there is insufficient data to prove a recent increase in the disease. One estimate of the current incidence is 10 to 20 cases per 100,000 population. Cases have been associated with injected drug abuse, diabetes, liposuction, hysterectomy, and bunion surgery, but most cases begin at the site of minor injuries.

[1] Fascia—bands of fibrous tissue under the skin and surrounding muscle and other organs

Prevention and Treatment

There are no proven preventive measures. Urgent surgery is necessary to release the pressure of the swollen tissue and to remove dead tissue; amputation is sometimes necessary to promptly rid the patient of the source of toxins. Penicillin is the drug of choice for early infection, but it has little or no effect on streptococci in necrotic tissue and no effect on toxins.

Pseudomonas aeruginosa Infections Contribute to Deaths Among Burn Patients

Burned areas with their damaged skin are ideal sites for infection by bacteria from the environment or normal flora. Almost any opportunistic pathogen can infect burns, but one of the most common and hardest to treat is the gram-negative rod *Pseudomonas aeruginosa* (figure 29.4).

Symptoms

In burns and other wounds, *P. aeruginosa* can actually color the tissues green from the formation of its pigments, pyocyanin and fluorescein. It is especially dreaded because it often invades the bloodstream and causes chills, fever, and shock.

Causative Agent

P. aeruginosa is a common cause of hospital-acquired infections. It is motile by means of a single flagellum at the end of the cell (figure 29.5). The organisms grow readily and rapidly on many media at 37°C. Some strains of this organism can also grow in a variety of aqueous solutions, even in distilled water and some disinfectant solutions. *P. aeruginosa* generally utilizes oxygen as the final electron acceptor in the breakdown of nutrients. However, it can also grow well anaerobically on many media containing nitrate, which can substitute for oxygen as a final electron acceptor, in the process of anaerobic respiration. As mentioned, most strains of *P. aeruginosa* produce a greenish discoloration of growth media resulting from the formation of two pigments.

Pathogenesis

P. aeruginosa is one of the more common causes of wound infection. It can even infect clean surgical wounds, especially in people with impaired humoral immunity, and is especially troublesome in those with extensive burns. Several of its attributes have been linked with virulence, at least in experimental animals. Its overall effect is to produce tissue damage, prevent healing, and increase the risk of septic shock. Most strains produce an extracellular enzyme called **exoenzyme S** that in vitro catalyses the transfer of an ADP-ribose fragment of NAD to G proteins. The signaling mechanism from host cell surface receptors to the appropriate genes is thereby impaired. In other words, exoenzyme S functions in vitro in the same way as the toxins of *Vibrio cholerae* and *Bordetella pertussis*. As with the cholera toxin, exoenzyme S is inactive when released by the bacterium and requires a host cell protein for its activation. The target cells for exoenzyme S and its mode of entry into host cells have not yet been identified. Strains of *P. aeruginosa* having an experimentally impaired gene for the enzyme show marked decrease in virulence.

• **Anaerobic respiration, p. 122**

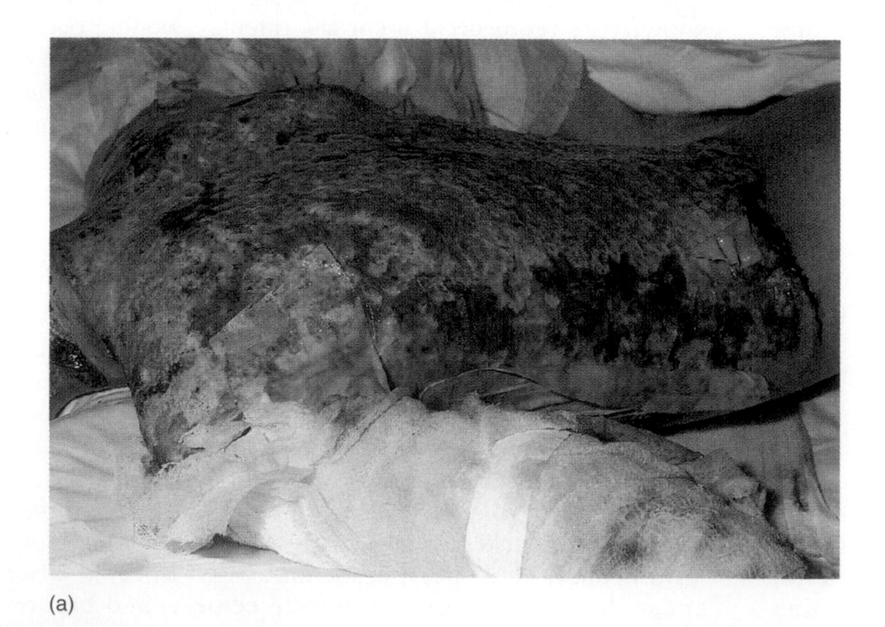

(a)

(b)

Figure 29.4
Pigment production by *Pseudomonas aeruginosa*. (*a*) Extensive burn infected with *P. aeruginosa*. Note the green discoloration. (*b*) Culture. Note the green discoloration of the medium, the result of water-soluble pigment diffusing from the *P. aeruginosa* colonies.

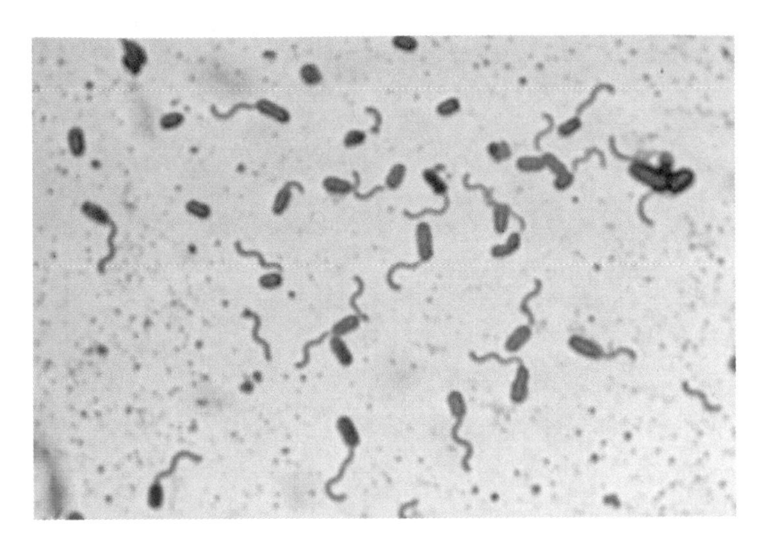

Figure 29.5
Pseudomonas aeruginosa has a single polar flagellum.

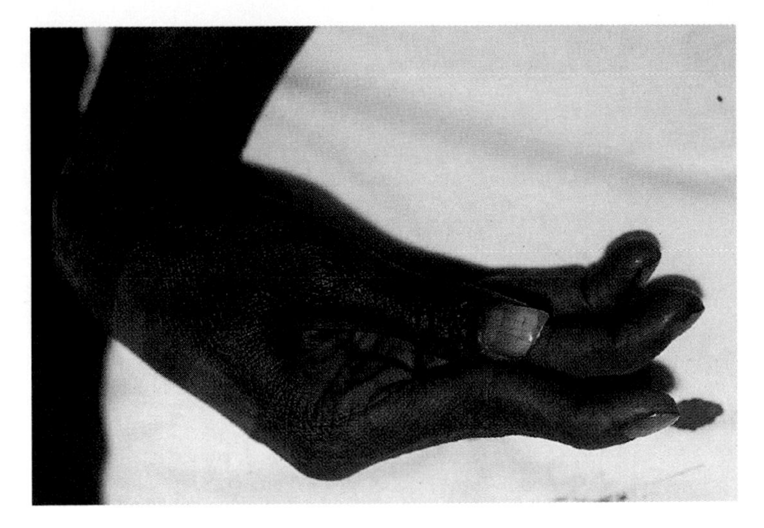

Figure 29.6
Marked muscular spasm of the wrist and hand of a person with tetanus. In many tetanus cases, the muscular spasms are generalized (they involve all the body's muscles).

Most strains of *P. aeruginosa* produce **toxin A,** another enzyme that requires activation after it is released from the bacterium. This toxin has a mode of action identical to that of *Corynebacterium diphtheriae* toxin, although the two toxins are antigenically distinct. These toxins catalyze the transfer of ADP-ribose to elongation factor-2, thereby halting protein synthesis by the host cell. Like the diphtheria toxin, toxin A production is enhanced by lowering the concentration of iron. However, the cell receptors for *P. pseudomonas* toxin A and the host cells that it targets differ from those of diphtheria toxin. Toxin A deficient mutants of *P. aeruginosa* show a marked decrease in virulence.

Most strains of *P. aeruginosa* produce extracellular **proteases** that cause localized hemorrhages and tissue necrosis but not the death of experimental animals. It is not known what role they play in human infections.

Many strains of *P. aeruginosa* produce a heat-labile hemolysin, a **phospholipase C,** identical in mode of action to the principal toxin of the gas gangrene bacillus, *Clostridium perfringens*. Phospholipase C hydrolytically degrades lecithin, an important lipid component of cell membranes. Its role in the pathogenesis of *P. aeruginosa* infections has not yet been defined.

Epidemiology

P. aeruginosa is widespread in nature, on plants and in soil and water. It is introduced into hospitals on shoes, on ornamental plants and flowers, and on produce. It can persist in most places where there is dampness or water, even in some soaps and disinfectants.

Prevention and Treatment

Prevention involves elimination of potential sources of the bacterium and prompt care of wounds. Careful removal of dead tissue from burn wounds, followed by application of a cream containing one of the sulfa drug derivatives, mafenide acetate or silver sulfadiazine, is effective in preventing infection with *P. aeruginosa*. Established infections are notoriously difficult to treat because *P. aeruginosa* is usually resistant to multiple antibacterial medications. Antibiotic susceptibility tests must be done to guide the selection of an effective regimen. Usually, one of the newer β-lactam antibiotics (carbenicillin, ticarcillin, ceftazidime, imipenem, azthreonam) is given along with one of the newer aminoglycocidelike medications (gentamicin, tobramycin, amikacin). Ciprofloxacin is also sometimes effective. All these medications generally require intravenous administration in high doses.

Tetanus Has a High Mortality but Is Easily Prevented by Immunization

Pathogenic clostridia are primarily soil organisms. They are commonly present in the human intestinal tract because food is contaminated with soil and dust. A few species of *Clostridium* can also colonize the intestine. Although they are anaerobes, they can multiply on wound surfaces because pus and dead tissue cells consume available oxygen and wound exudates supply nutrients. Clostridia are frequently present in cultures of wounds, but in only a small minority of cases do they produce disease. Tetanus is a clostridial disease that can be fatal even when the causative infection is so mild as to be unnoticed.

Symptoms

Tetanus is characterized by uncontrollable contraction of one or more muscles of the body (figure 29.6), often beginning with the jaw muscles, giving the disease the popular

• ***Clostridium*, p. 260**

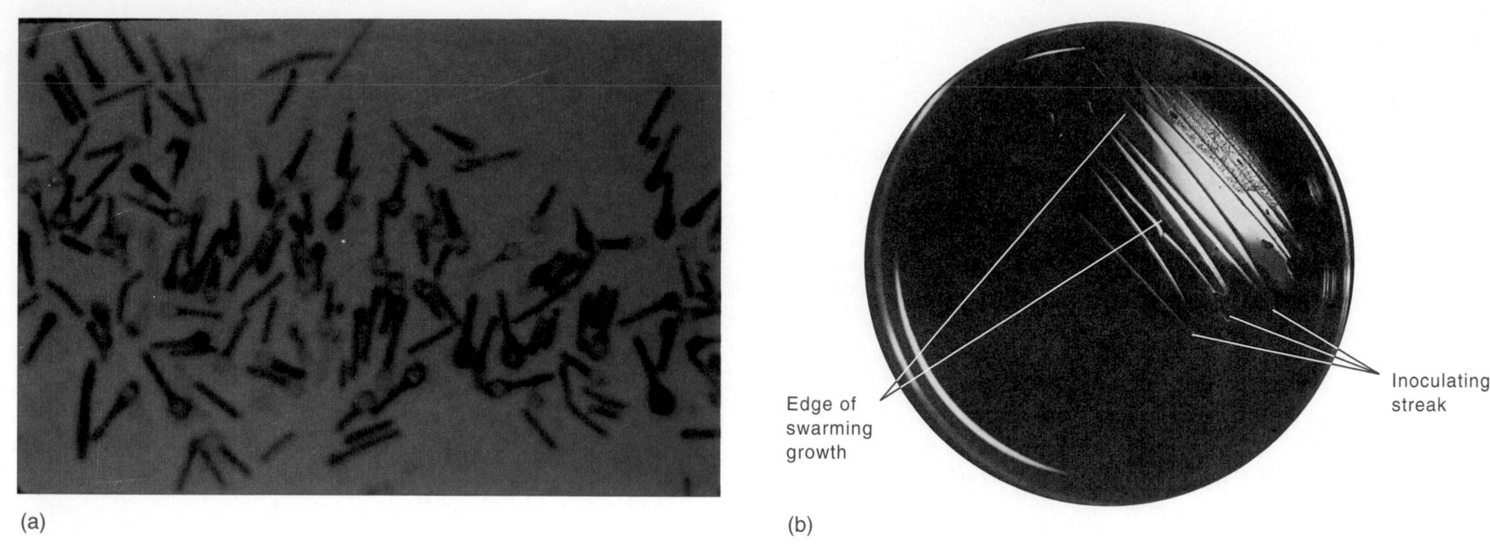

(a) (b)

Figure 29.7
Clostridium tetani. (*a*) Terminal endospores. (*b*) Swarming growth of *C. tetani* on agar medium after a four-hour incubation. Rapid spreading from the site of inoculation provides a way of obtaining a pure culture from mixtures with other species.

name "lockjaw." The disease begins with restlessness, irritability, stiffness of the neck, contraction of the muscles of the jaw, and sometimes convulsions, particularly in children. As more muscles tense, the pain grows more severe and is similar to that of a severe leg cramp. Breathing becomes labored, and after a period of almost unbearable pain, the infected person often dies of pneumonia or from aspiration of stomach contents regurgitated into the lung.

Causative Agent

Tetanus is caused solely by the action of a powerful exotoxin produced by vegetative cells of *C. tetani*. This organism is a motile, anaerobic, spore-forming, gram-positive rod. *C. tetani* shows two striking features (figure 29.7). The first feature is the spherical endospore that forms at the end of the bacillus, in contrast to the oval endospore that develops near the center of the rod in other pathogenic species of *Clostridium*. The second feature is its swarming growth that spreads to cover the entire surface of solid media. Final identification of *C. tetani* depends on identifying its toxin, which is coded by a plasmid.

Isolation of *C. tetani* from wounds does not prove that a person has tetanus. Conversely, failure to find the organism in a person's wound cultures does not eliminate the possibility of tetanus. The reasons are that tetanus endospores may contaminate wounds that are not sufficiently anaerobic to allow germination and toxin production, or the person may simply be immune to the toxin by prior vaccination. Only vegetative cells, not endospores, synthesize toxin. Spores can remain dormant in the tissues for a long time and then germinate. The resulting vegetative cells then produce the toxin after a wound has healed. One study in the United States found that *C. tetani* could not be isolated from 7% of the tetanus cases studied, probably because the number of organisms present was too small or because little or no wound drainage was available.

• **Exotoxin, pp. 420–422**

Pathogenesis

C. tetani is not invasive and colonization is generally localized to a wound. Its pathologic effects are entirely the result of a 150,000 molecular weight toxin called **tetanospasmin,** released from the organism during the stationary phase of growth. The toxin is composed of two chains joined by a disulfide bond. The heavier chain attaches specifically to receptors on motor neurons (motor nerves), which then take up the toxic, lighter chain by endocytosis. This uptake of the toxin occurs at neuromuscular junctions (sites where motor nerves join muscle) near the infected wound. The toxin is carried by the cytoplasm of the neuron's axon to its cell body in the spinal cord. There the motor neuron is in contact with other neurons that control its action. Some of them cause stimulation of the motor cell, thereby producing a muscle contraction. Others make the motor neuron resistant to stimulation, thereby inhibiting muscle contraction. Toxin that has been carried from the wound to the spinal cord by the nerve's axon is released into inhibitory neurons where the neurons contact the motor nerve. Like most other neurons, inhibitory neurons exert their effect on other nerve cells by releasing chemicals called **neurotransmitters.** The effect of the toxin is to prevent the release of neurotransmitters such as the amino acid **glycine** and thereby block the action of the inhibitory neurons. This leaves the stimulatory neurons unopposed.

The toxin generally spreads across the spinal cord to the side opposite the wound and then downward. Thus, typically, spastic muscles first appear on the side of the wound, then on the opposite side, and then downward, depending on the amount of toxin. However, in many cases, the amount of toxin released from the infected wound is very large. It is quickly taken up by the blood vessels and lymphatics, enters the general circulation, and is carried throughout the central nervous system. In these cases, inhibitory neurons of the brain are first affected, and the muscles of the jaw are among the first to become spastic. Like other neurotransmitters, glycine and its main counterpart in the brain, **γ-aminobutyric acid,** exist in the inhibitory neurons in tiny membrane-bound sacks called *vesicles*. In order for the neurotransmitters to be released to another neuron, the vesicles must first attach specifically to receptors on their cell's membrane to initiate the process of exocytosis. Tetanospasmin is a peptidase, and current evidence indicates that it acts by removing from the vesicles the ligand responsible for attaching it to the cell membrane. Thus, exocytosis does not occur, and the inhibitory effect of the neuron is blocked.

Epidemiology

C. tetani organisms are widely distributed in soil and enter wounds with dirt and dust. Even trivial puncture wounds, surface abrasions, or burns can result in tetanus. They are also commonly found in the gastrointestinal tract of humans and other animals that have eaten soil-contaminated foods. Contrary to popular opinion, horse manure and other animal manures do not represent the major source of the organism.

Prevention

Prophylactic immunization with tetanus toxoid is by far the best weapon against tetanus. Three injections of the toxoid are given to stimulate development of antibody-producing cells. Immunization is usually begun during the first year of life. For infants and young children, the tetanus toxoid is usually given in combination with diphtheria toxoid and pertussis vaccine. The three together are commonly known as DPT. A "booster" dose is given a year later and again when children enter school. Once immunity has been established by this regimen, "booster" doses of tetanus toxoid given at about 10-year intervals will maintain an adequate level of protection. More frequent doses of toxoid are not recommended because danger of an allergic reaction following intensive immunization exists. This type of hypersensitivity reaction results when antigen (the toxoid) is introduced into an individual who already has large quantities of antibodies against the antigen.

Even though tetanus is easily prevented by immunization with tetanus toxoid, inadequately vaccinated people in the United States contract the disease each year, and many die from it. Ordinarily, 50 to 100 cases are reported annually, with a mortality rate around 30%. Age is an important factor (figure 29.8), probably because immunity is inadequate in many older people. About 97% of the people who developed tetanus in recent years had never been immunized with the toxoid. Adequate immunization would have prevented their disease. To help prevent cases of tetanus, active and passive immunization are often indicated when a wound is sustained (table 29.3).

• **Toxoid, p. 344**

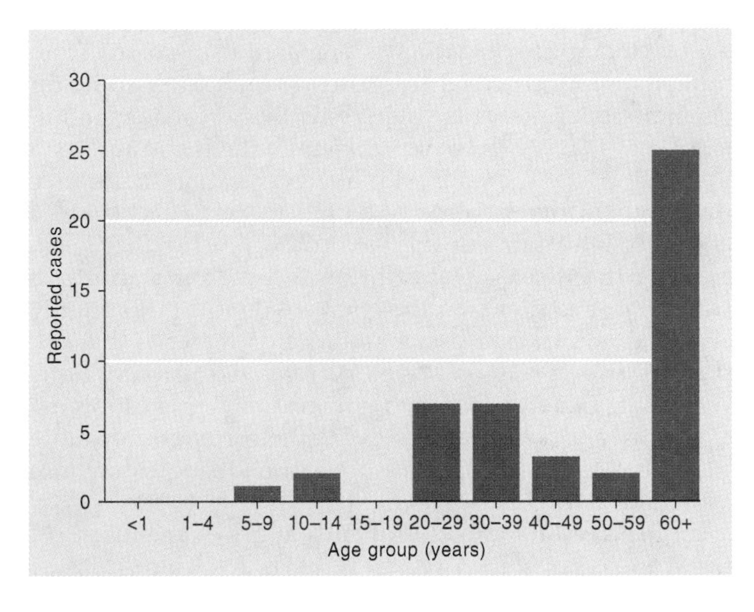

Figure 29.8
Tetanus by age group in the United States, 1992.

Source: Centers for Disease Control and Prevention. 1993. Summary of notifiable diseases—United States, 1992. *Morbidity and Mortality Weekly Report* 41(55):57.

Table 29.3 Use of Tetanus Toxoid and Immune Globulin in Caring for Wounds

Immunization History	Clean Minor Wounds		All Other Wounds	
	Toxoid	Immune globulin	Toxoid	Immune globulin
Unknown, or fewer than 3 injections	Yes	No	Yes	Yes
Fully immunized (3 or more injections of toxoid)				
1) 5–10 years since last dose	No	No	Yes	No
2) more than 10 years since last dose	Yes	No	Yes	No
3) 5 years or less since last dose	No	No	No	No

Tetanus occurs much more frequently on a global basis than it does in the United States due to lack of immunization and proper care of wounds. In some parts of the world, newborn babies commonly die of tetanus as a result of their umbilical cord being cut with unsterilized instruments contaminated with *C. tetani*. If the mothers of these babies had been immunized, their infants would have been protected by antibody crossing the placenta. In 1989, the World Health Organization began an effort to eliminate neonatal tetanus. In 1994, there were an estimated 490,000 neonatal deaths from tetanus, but 730,000 neonatal deaths were prevented by immunizing mothers in developing countries.

Treatment

Tetanus is treated by administering tetanus antitoxin to neutralize any toxin not yet attached to motor nerve cells. The treatment of choice is **tetanus immune globulin (TIG),** a gamma globulin with a high antibody titer against tetanus toxin, which is prepared using blood from humans immunized with tetanus toxoid. This antitoxin generally does not cause hypersensitivity reactions in humans and is much more effective than the horse antitoxin formerly used. The antitoxin, however, cannot neutralize the exotoxin that is already bound to nerve tissue. This explains why tetanus antitoxin is often ineffective in treating the disease. In addition to involving antitoxin treatment, treatment requires that a wound be thoroughly cleaned and all dead tissue and foreign material that can provide anaerobic conditions be removed. An antibiotic such as penicillin is given to kill any actively multiplying clostridia and thereby to prevent the formation of more exotoxin. Antibiotics would not kill endospores or nongrowing bacteria.

Table 29.4 describes the main features of tetanus.

• **Immune globulin, pp. 361–365**

REVIEW QUESTIONS

1. Give two reasons why an abscessed wound might not respond to antibiotic treatment.
2. What kinds of wounds are subject to infection by anaerobic bacteria?
3. What situations favor infection by microorganisms normally considered harmless?
4. What is the most common cause of surgical wound infections? Where does it come from?
5. Why do wound infections due to *Pseudomonas aeruginosa* sometimes produce green pus?
6. Why are *P. aeruginosa* infections so hard to treat?
7. Describe the symptoms of tetanus. How is the disease prevented?
8. Why are cultures not helpful in diagnosing tetanus?
9. Why do many tetanus cases fail to respond to tetanus antitoxin?

Gas Gangrene Is a Complication of Neglected, Dirty Wounds

Almost every sample of soil and dusty surface has endospores of the gas gangrene bacillus. However, only rarely does contamination of a wound result in gas gangrene.

Symptoms

Gas gangrene (figure 29.9) begins abruptly, with pain rapidly increasing in the infected wound. Increased swelling occurs in the area, and a thin, bloody or brownish fluid leaks from the wound. The fluid may have a frothy appearance due to gas formation by the organism. The overlying skin becomes stretched tight and mottled with blue. The patient appears very ill but remains alert until late in the illness when, near death, he or she becomes delirious and lapses into a coma.

Table 29.4 Tetanus (Lockjaw)

Symptoms	Restlessness, irritability, headache; muscle pain and spasm in jaw, abdomen, or back; difficulty swallowing
Incubation period	3 days to 3 weeks; average 8 days
Causative agent	*Clostridium tetani*
Laboratory diagnosis	Of limited value; wound tissue may be cultured for organisms, but false negative results occur in up to two thirds of patients; *C. tetani* occurs in wounds without tetanus
Pathogenesis	Tetanus results from action of exotoxin produced by organism; toxin enters the circulation and is carried to brain and spinal cord; toxin acts by attaching to nerve cells and preventing them from releasing a chemical that normally inhibits contraction of opposing muscle groups; uncontrolled spread of nervous impulses results in spasms of muscles
Epidemiology	Organisms common in soil; contamination of any wound, especially those showing anaerobic conditions, particularly dirty or puncture wounds
Prevention	Immunization of children at ages 2 months, 4 months, 6 months, 18 months; booster dose at time of entering school and at 10-year intervals after that
Treatment	Tetanus immune globulin (TIG), cleaning of wound, penicillin

Perspective 29.2

Laboratory Diagnosis of Gas Gangrene

Diagnosis of gas gangrene depends on the patient's symptoms, supported by a stained smear of the brownish wound fluid showing near absence of leukocytes (they are killed by the toxin), bits of partly digested muscle, and plump gram-positive rods, usually without spores. All *C. perfringens* types produce toxins. Alpha (α) toxin is responsible for a wide zone of partial α-hemolysis around colonies growing anaerobically on blood agar (figure 1). A smaller zone of complete β-hemolysis is due to theta (θ) toxin. The enzymatic activity of α toxin can be demonstrated by growing the organism on a medium containing egg yolk. α toxin hydrolyses lecithin in the egg yolk and changes its color (figure 2). Cultures of the infected tissue help to distinguish clostridial causes of gas gangrene from other, rarer causes. However, since the disease can progress rapidly, the disease must generally be treated before a definitive diagnosis can be made.

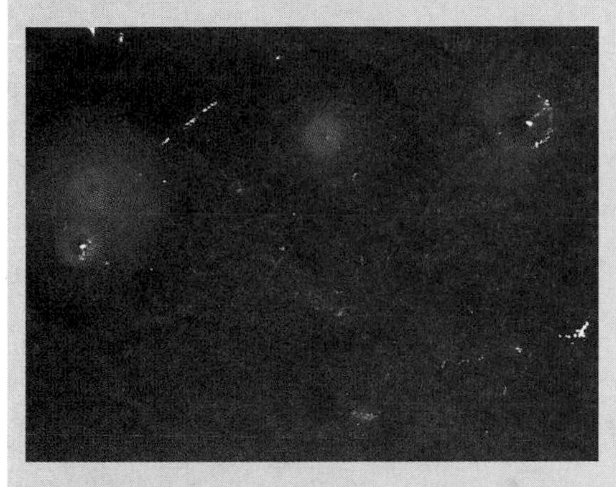

Figure 1
Clostridium perfringens on blood agar. The inner zone of β-hemolysis immediately surrounding each colony is due to θ toxin; the outer zone of α-hemolysis is due to α toxin.

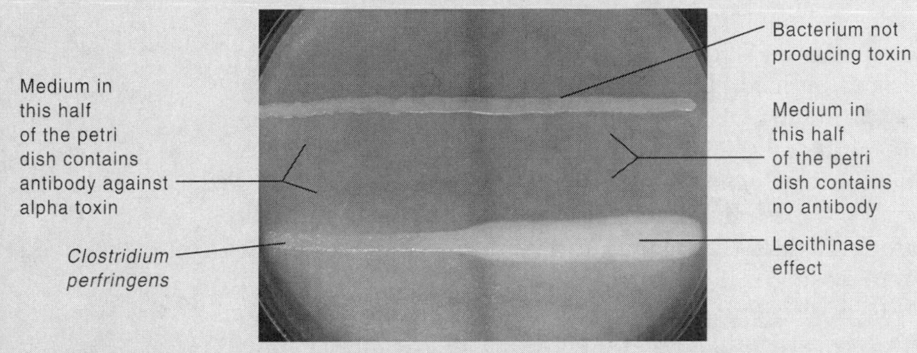

Figure 2
Effect of clostridial α toxin (a lecithinase) on egg yolk incorporated into the growth medium.

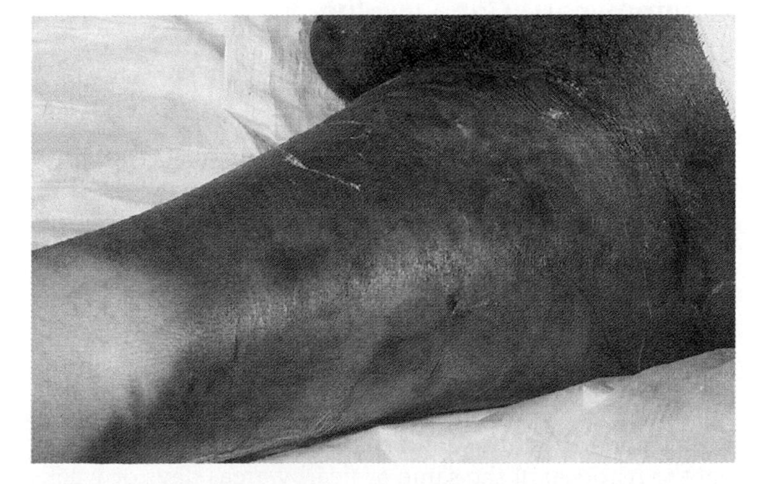

Figure 29.9
Person with clostridial myonecrosis (gas gangrene). Fluid seeping from the involved area typically shows bits of muscle digested by *Clostridium perfringens*. White blood (pus) cells are absent because the clostridial toxin kills them.

Causative Agent

Several species of *Clostridium* can produce life-threatening gas gangrene when they invade injured muscle, but by far the most common offender is *C. perfringens*. Five different types—A, B, C, D, and E—are recognized by the toxins they produce; most cases are caused by *C. perfringens*, type A.

Pathogenesis

Two main factors foster the development of gas gangrene: (1) the presence of large amounts of dirt in the wound and (2) long delays before the wound is carefully cleaned of dirt and dead tissue. *C. perfringens* is unable to infect healthy tissue but grows readily in dead and poorly oxygenated tissue, releasing its toxin. Growth is fostered by anaerobic conditions, as well as by the presence of peptides and amino acids in such tissue. This toxin diffuses from the area of infection, killing leukocytes and tissue cells. Several tissue-attacking enzymes produced by the pathogen, including collagenase and hyaluronidase, probably promote this process.

Curiously, these infections are usually localized and produce few symptoms unless the bacteria invade muscle. Diffusion of toxin produces necrosis (death of tissue) adjacent to the infected wound. The organisms then grow readily in the fluids of the dead tissue, producing hydrogen and carbon dioxide from fermentation of amino acids and muscle glycogen. The hydrogen gas and some carbon dioxide then accumulate in the tissue and raise the pressure, thereby fostering spread of the infection. The reason for the rapid onset of severe toxicity when muscle becomes involved is unclear. It is not reversed by administering antibody to toxin.

Epidemiology

Gas gangrene is quite unusual except on battlefields where wounds cannot be treated promptly. However, it occurs occasionally as a result of surgery, induced abortion performed by inadequately trained personnel, and accidents. According to one series of reports in the medical literature, only 1.8% of 187,936 major open wounds of violence resulted in gas gangrene. Impaired oxygenation of tissue, as occurs, for example, when blood vessels are damaged by arteriosclerosis (hardening of the arteries) or diabetes, is a predisposing factor. Cancer patients also seem to have increased susceptibility to gas gangrene. The natural habitat of *C. perfringens* includes both the soil and the human intestine, and its endospores commonly contaminate clothing, skin, and wounds.

Prevention

Neither toxoid nor vaccine is available for immunization against gas gangrene. Prompt cleaning and debridement of wounds, the surgical removal of dead and damaged tissue, dirt, and other foreign material, is highly effective in preventing the disease.

Treatment

Treatment of gas gangrene depends primarily on the prompt surgical removal of all dead and infected tissues. Some authorities recommend hyperbaric oxygen treatment[1] because it inhibits growth of the clostridia, thereby stopping release of toxin, and it also improves oxygenation of injured tissues. Some people appear to improve dramatically with hyperbaric oxygen, but others show no benefit and develop complications from the treatment. Antibiotics such as penicillin are also used to help stop bacterial growth and toxin production, but when used alone, they are ineffective in treating the disease because they do not diffuse well into large areas of dead tissue and they do not inactivate toxin.

Table 29.5 describes the main features of this disease.

[1]Hyperbaric oxygen treatment—regimen in which the patient is placed in a special chamber and breathes pure oxygen under three times normal air pressure

Table 29.5 Gas Gangrene (Clostridial Myonecrosis)

Symptoms	Severe pain, gas and fluid seeping from wound, blackening of overlying skin; shock and death commonly follow
Incubation period	Usually 1–5 days
Causative agent	*C. perfringens;* other clostridia less frequently
Laboratory diagnosis	Diagnosis depends mostly on patient's symptoms; stained smear of wound fluid showing digested muscle, gram-positive rods, and few or no leukocytes
Pathogenesis	Organism grows in dead and poorly oxygenated tissue and releases alpha toxin; toxin kills leukocytes and normal tissue cells by degrading the lecithin component of their cell membranes; involvement of muscle causes shock by unknown mechanism
Epidemiology	Dirt contamination of wounds, especially crushing injuries resulting in impaired circulation to muscle tissue
Prevention	Prompt cleaning and debridement of wounds.; no vaccine available
Treatment	Surgical removal of infected tissues of primary importance; hyperbaric oxygen sometimes of possible value; penicillin to kill vegetative *C. perfringens*

Actinomycosis Can Develop Following Dental Surgery

Many wound infections are caused by anaerobes other than clostridia. The main offenders are members of the normal flora of the mouth and intestine. Actinomycosis is an interesting example. It is not a mycosis since it is not caused by a fungus. Its name persists from former times when it was thought to be a fungal disease.

Symptoms

Actinomycosis is characterized by slowly progressive, sometimes painful swellings under the skin that eventually open and chronically drain pus. The openings may heal, only to reappear at the same or nearby areas days or weeks later. Most cases involve the area of the jaw and neck; the resulting scars and swellings gave rise to the popular name **lumpy jaw** (figure 29.10). In other cases the process develops on the chest or abdominal wall or even in the genital tract of women.

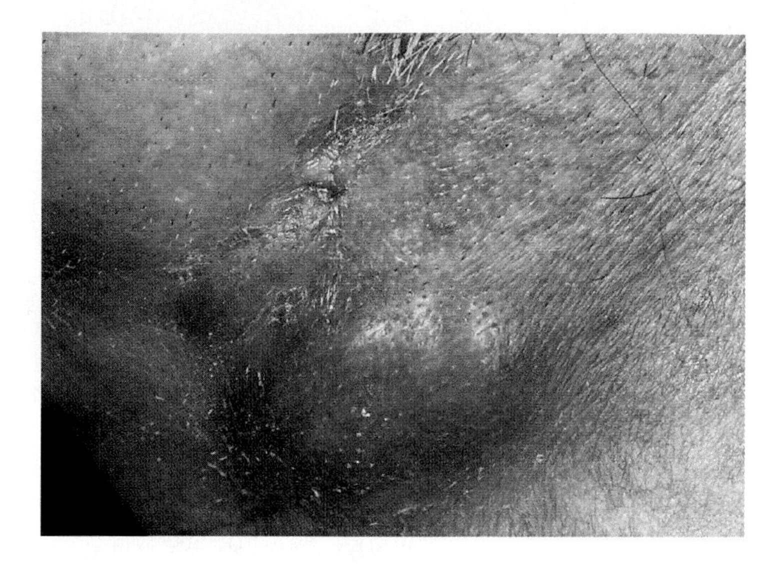

Figure 29.10
Actinomycosis on the face ("lumpy jaw"). Abscesses, such as the one shown, form, drain, and heal and then recur at the same location or somewhere else. Sulfur granules, colonies of *Actinomyces israelii*, can be found in the drainage.

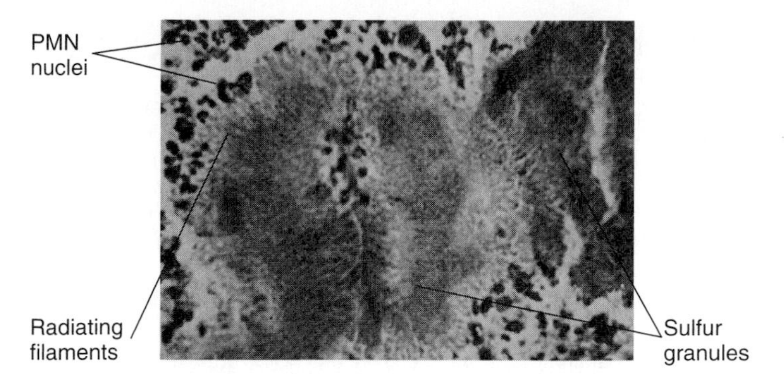

Figure 29.11
Actinomyces israelii sulfur granule. Microscopic view of stained smear of pus from an infected person. Filaments of the bacteria can be seen radiating from the edge of the colony.

Causative Agent

Most cases of actinomycosis are caused by *Actinomyces israelii*, a gram-positive, filamentous, branching, anaerobic bacterium that grows slowly on commonly employed laboratory media such as blood agar. A number of similar species can cause the disease in animals and humans.

Pathogenesis

A. israelii cannot penetrate the normal mucosa, but it can establish an infection in association with other organisms if introduced into tissue by wounds. The infectious process is characterized by cycles of abscess formation, scarring, and formation of sinus tracts (passages) to a surface where pus is discharged. In the tissue, *A. israelii* grows as dense yellowish colonies called **sulfur granules** (figure 29.11), which must be searched for in the sinus drainage and cultivated in order to establish the diagnosis. Most cases originate from wounds in the mouth or intestine as may occur from dental or other surgery, but some begin with a lung infection following aspiration of material from the mouth. The detailed mechanisms by which the organism produces the disease have not yet been established.

Epidemiology

A. israelii normally inhabits the mucosal surfaces of the mouth, upper respiratory tract, intestine, and sometimes the vagina. It is common in the gingival crevice, particularly with poor dental care. Other species responsible for actinomycosis of dogs, cattle, sheep, pigs, and other animals generally do not cause human disease. The disease is sporadic and is not transmitted person to person.

• **Gingival crevice, p. 553**

Prevention and Treatment

There are no preventive measures. Actinomycosis responds to treatment with a number of antibacterial medications, including penicillin and tetracycline. However, to be successful, treatment must be given over an extended period of time. This need for prolonged therapy probably is due to the slow growth of the organisms and their tendency to grow in dense colonies.

Bite Wounds Are Commonly Infected with *Pasteurella multocida*

Each year, over 3 million animal bites occur in the United States, and many of them become infected with bacteria from the mouth of the biting animal. *Pasteurella multocida*, a common cause of these infections, can follow bites from dogs, cats, monkeys, humans, and other animals.

Symptoms

There are no reliable symptoms or signs that distinguish among bacterial bite wound infections. Spreading redness and tenderness and swelling of tissues adjacent to the wound, followed by the discharge of pus, are early indications of infection.

Causative Agent

P. multocida is a small, nonmotile, gram-negative, facultatively anaerobic, coccobacillus. It grows on blood agar, is oxidase positive, and ferments various sugars. Most isolates are encapsulated. Antigenically, there are at least four capsular and over a dozen cell wall types. Based on DNA hybridization studies, some of the animal isolates may prove to be *Pasteurella* species other than *P. multocida*.

Perspective 29.3

Infection Caused by a Human "Bite"

A 26-year-old man injured the knuckle of the long finger of his right hand during a tavern brawl when he punched an assailant in the mouth. Due to his inebriated condition, a night spent in jail, and the insignificant early appearance of his knuckle wound, the man did not seek medical help until more than 36 hours later. At that time, his entire hand was massively swollen, red, and tender. Furthermore, the swelling was spreading to his arm. The surgeon cut open the infected tissues, allowing the discharge of pus. He removed the damaged tissue and washed the wound with sterile fluid. Smears and cultures of the infected material showed aerobic and anaerobic bacteria characteristic of mouth flora, including species of *Bacteroides* and *Streptococcus*. The patient was given antibiotics to combat infection, but the wound did not heal well and continued to drain pus. Several weeks later, X rays revealed that infection had spread to the bone at the base of the finger. To cure the infection, the finger had to be amputated.

Pathogenesis

The details of pathogenesis are not yet known. There is evidence for one or more adhesins. The capsules of *P. multocida* are antiphagocytic. When opsonized by specific antibody, the organisms are ingested and killed by phagocytes.

Epidemiology

P. multocida is widespread among the normal oral and upper respiratory flora of animals. Usually, the animals are asymptomatic, but epidemics of fatal pneumonia and bloodstream infection occur in rabbits, cattle, sheep, domestic fowl, and mice. Young animals under stress from crowding or shipping are especially susceptible. At least some of these animal strains can colonize the mouth and upper respiratory tract of humans.

Prevention and Treatment

Experimental vaccines for use in domestic animals have given mixed results so far. In humans, immediate cleansing of bite wounds and prompt medical attention usually prevent the development of serious infection and possible permanent impairment of function, as can occur, for example, in the infected hand. Unlike many species of gram-negative rods, *P. multocida* is quite susceptible to penicillin and ampicillin. Usually, before the cultural diagnosis is known, ampicillin and a β-lactamase-blocking substance (such as Augmentin, which is ampicillin plus clavulanic acid) are administered. This is because there is a high risk of a concomitant infection with penicillinase-producing *Staphylococcus aureus* in bite wounds.

Cat Scratch Disease Is Common, and It Can Be Serious

In the United States, cat scratch disease is the most common cause of chronic lymph node enlargement at one body site in young children. Typically the infection is mild and localized, lasting from a few weeks to a few months. However, of the estimated 24,000 annual cases, about 2,000 require hospitalization.

Symptoms

In about 9 of 10 cases, the disease begins within a week of a scratch or bite with the appearance of a papule or pustule (pimple) at the site of injury. Enlargement of the lymph nodes of the region develops in one to seven weeks. About one-third of the patients develop fever, and in about half that number the lymph nodes become pus-filled and soften. The disease generally disappears without treatment in two to four months. Of the remaining 10% of cases, some develop irritation of an eye with enlargement of the lymph node in front of the ear, while others abruptly develop epileptic seizures and coma, heralding the development of encephalitis. Still others develop acute or chronic fever associated with bloodstream or heart valve infection. Persons with AIDS are prone to develop these more serious complications. **Peliosis hepatis** and **bacillary angiomatosis** are two complicating conditions seen in people with HIV disease, but they are not necessarily associated with a cat scratch or bite. In peliosis hepatis, the liver enlarges because of blood-filled cysts that can rupture, causing massive hemorrhage. In bacillary angiomatosis, nodules composed of proliferating blood vessels develop in the skin, lymph nodes, and other parts of the body.

Causative Agent

Cat scratch disease is caused by *Bartonella henselae*, a tiny (0.6 by 1.0 micrometers) slightly curved gram-negative rod that can be cultivated on a special medium containing 5% fresh rabbit blood incubated in 5% CO_2 at high humidity. Most of the usual biochemical tests are negative with this organism, including sugar fermentations

Perspective 29.4

Laboratory Diagnosis of Sporotrichosis

In nature or in laboratory cultures incubated at 25°C, *S. schenckii* forms a fluffy mold mycelium, tan or brown in color. In body tissues and laboratory cultures at 35°C, it grows as a budding yeast. In the mold form, the hairlike hyphae of *S. schenckii* are more slender than those of most fungi, and spores are formed along them. The spores are oval or rounded and occur in clusters on branches at right angles to the hyphae, producing a flowerlike appearance (figure 1). In cultures grown at 35°C and in tissues, the yeast phase is characterized by elongated, cigar-shaped cells with one to three buds at each end of the cell (figure 2).

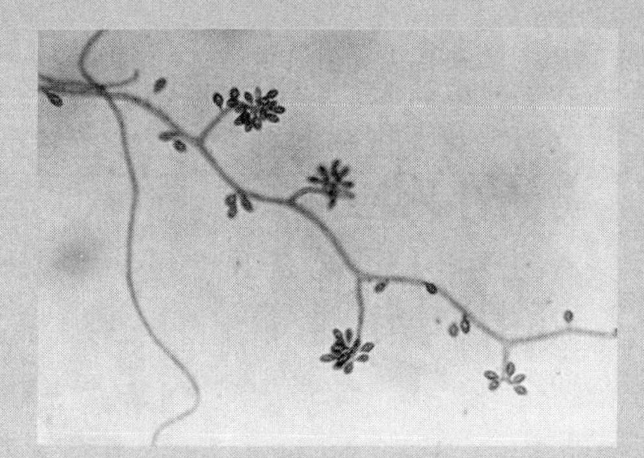

Figure 1
Mold form of *Sporothrix schenckii.*

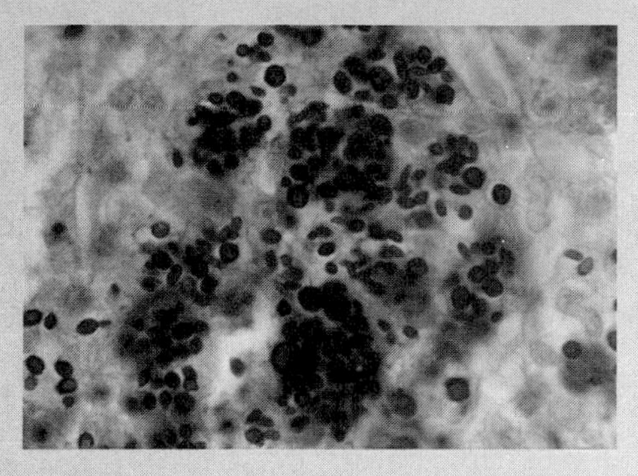

Figure 2
Microscopic view of a stained smear of the yeast phase of *Sporothrix schenckii* as seen in infected tissue.

and catalase, oxidase, urease, and nitrate reduction. *B. quintana* (the cause of trench fever, a disease not associated with cats) can cause symptoms and signs similar to those of *B. henselae*.

Pathogenesis

The virulence factors of *B. henselae* and the process by which it causes disease are not yet known. The organisms can be identified in the lesions of typical cat scratch disease, as well as in the cyst walls of peliosis hepatis and the nodular vascular lesions of bacillary angiomatosis.

Epidemiology

There is clear evidence that the organisms are present in cats and transmitted to humans by their bites and scratches. Other modes of transmission are likely because many cases occur after people handle cats but are not bitten or scratched. The disease occurs worldwide.

Prevention and Treatment

There are no proven preventive measures for cat scratch disease. However, it is prudent to avoid handling stray cats, especially young ones and those with fleas. Any cat-inflicted wound should be promptly cleaned with soap and water and then treated with an antiseptic. Severely immunodeficient persons should avoid cats if possible and, if not, control fleas and abstain from rough play. Severe *B. henselae* infections can be treated with a variety of antibacterial medications, including ampicillin, tetracycline, rifampin, and trimethoprim-sulfasoxazole. Some strains are resistant to β-lactams and tetracycline, so susceptibility tests are indicated.

Streptobacillary Rat Bite Fever Is a Hazard to Those Who Work with Rats

Rat bites are fairly common among the poor people of large cities and among workers who handle laboratory rats. In the past, as many as 1 out of 10 bites resulted in rat bite fever, but fortunately, the disease is now rare in the United States.

Symptoms

Usually, the bite wound heals promptly without any problem noted. However, about 10 days later, chills, fever, head and muscle aches, and vomiting develop. A rash appears in most cases after a few days, followed by pain on motion of one or more of the large joints. Later complications result in brain abscesses and infection of the heart valves. Without treatment, about 1 in 10 victims dies.

Causative Agent

The cause of streptobacillary rat bite fever is *Streptobacillus moniliformis*, a highly pleomorphic, facultatively anaerobic gram-negative rod that is catalase and oxidase negative but ferments some sugars. Stained smears from cultured material show a multiplicity of forms ranging from small coccobacilli to granule-containing, unbranched filaments more that 100 micrometers long. *S. moniliformis* can be cultivated using media containing blood or serum. The

Perspective 29.5

An Epidemic of Sporotrichosis in the United States

During a five-week period in the summer of 1988, 84 cases of sporotrichosis were diagnosed, scattered over 14 different states. In all of the cases, people had handled evergreen tree seedlings packed in sphagnum moss or the moss alone. The seedlings had been packed in the moss in Pennsylvania, using moss obtained from Wisconsin. After packing, the seedlings were shipped to nurseries in the various states, where they were unpacked and distributed. Sphagnum moss is a known potential source of *Sporothrix schenkii*, having been associated with another sporotrichosis epidemic in 1976.

To help prevent sporotrichosis, gloves and a long-sleeved shirt should be worn when handling evergreen seedlings and sphagnum moss. This chronic disease is often misdiagnosed, leading to delayed and inappropriate treatment.

Source: Centers for Disease Control and Prevention. 1988. *Morbidity and Mortality Weekly Report* 37:652.

organism is unique in that it spontaneously develops **L-forms.** L-forms are cell wall deficient variants, first identified at the famous Lister Institute (hence, the *L* in L-form). As might be expected, L-form colonies resemble those of mycoplasmas.

Pathogenesis

The details of how this bacterium causes streptobacillary rat bite fever are not yet known.

Epidemiology

Both wild and laboratory rats, as well as mice and other rodents can carry the organism in their nose and throat. Rat bites are the usual source of human infections and can occur while the victim is sleeping. However, epidemics of the disease have arisen from ingesting milk or food contaminated with *S. moniliformis*. The foodborne disease is called **Haverhill fever** from a 1926 epidemic in Haverhill, Massachusetts. A more recent epidemic occurred in an English boarding school in 1983, when 208 students acquired the disease from drinking raw milk.

Prevention and Treatment

Wild rat control and care in handling laboratory rats are reasonable preventive measures. Penicillin, given by injection, is the treatment of choice for cases of rat bite fever.

Human Bites Can Cause Disastrous Infections

Wounds caused from human bites or striking the teeth of another person are common and can result in very serious infections. Rarely, diseases such as syphilis, tuberculosis, and hepatitis B are transmitted this way. Much more commonly, it is the normal mouth flora that cause trouble.

• **Mycoplasmas, p. 267**

Symptoms

The wound may appear insignificant at first but then becomes painful and swells massively. Most of the wounds are on the extensor surface on the hand, but the swelling may soon involve the palm also, and movement of some or all of the fingers becomes difficult or impossible. Discharged pus often has a foul smell.

Causative Agents

Strains and cultures usually show members of the normal mouth flora, including anaerobic streptococci, fusiforms, spirochetes, and *Bacteroides* sp., often in association with *Staphylococcus aureus*.

Pathogenesis

The crushing nature of bite wounds provides suitable conditions for anaerobic bacteria to establish infection. Although most members of the mouth flora are harmless alone, together they produce an impressive number of toxins and destructive enzymes. There include leukocidin, collagenase, hyaluronidase, ribonuclease, various proteinases, neuraminidase, and complement and antibody-destroying enzymes. Capsules of some species inhibit phagocytosis. Facultative organisms reduce available oxygen and thus encourage the growth of anaerobes. The result of all these factors is a **synergistic infection,** meaning that the sum effect of all the organisms acting together is greater than sum of their individual effects. Irreversible destruction of tissues such as tendons and permanent loss of function can be the result.

Prevention and Treatment

Prevention involves avoiding situations that lead to uncivilized behavior such as biting and hitting. Prompt cleansing of wounds followed by application of an antiseptic may be helpful, but most important is immediate medical attention if there

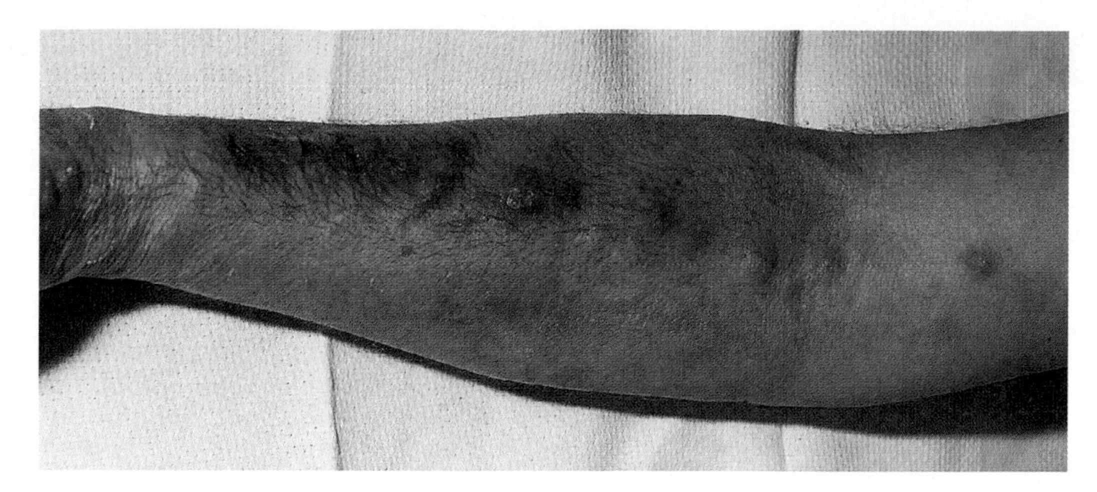

Figure 29.12
Person with sporotrichosis. Notice the multiple abscesses along the course of the lymphatic drainage from the hand.

is any suspicion of developing infection. Treatment of infected wounds consists of opening the infected area widely with a scalpel, washing the wound thoroughly with sterile fluid, and removing dirt and dead tissue. The initial choice of antibacterial medication should include medication effective against anaerobes and should avoid aminoglycosides, trimethoprim-sulfamethoxazole, and fluoroquinolones because they are generally ineffective against anaerobic bacteria.

Fungal Infections of Wounds

Fungal infections of wounds are unusual in the United States, although one type of infection, sporotrichosis, occasionally causes small epidemics.

Sporotrichosis Is a Chronic Fungal Disease of People Whose Work Exposes Them to Splinters and Thorns

Cause and Symptoms

Sporotrichosis is caused by the dimorphic fungus *Sporothrix schenckii*, which lives in soil and on vegetation. The organism is introduced into wounds by thorns and splinters and results in a chronic, slowly progressive infection. Typically, a chronic ulcer forms at the wound site followed by a chain of ulcers along the lymphatic drainage from the wound (figure 29.12). Lymph nodes in the region of the wound enlarge, but patients generally do not become ill. However, if they are immunodeficient, the disease can be severe, spreading throughout the body.

• **Dimorphic fungus, p. 286**

Epidemiology

Sporotrichosis is distributed worldwide, mostly in the warmer regions but extending into temperate climates. In the United States, most cases occur in the Mississippi and Missouri river valleys.

Treatment

Surprisingly, unlike other infections, sporotrichosis can usually be cured by oral treatment with the simple chemical compound potassium iodide (KI). The antibiotic amphotericin B is used in cases in which the disease has spread throughout the body.

REVIEW QUESTIONS

1. What is the most common cause of gas gangrene? Where does it come from?
2. What factors favor the development of clostridial myonecrosis?
3. Describe actinomycosis and its cause. From what kinds of wounds is the disease likely to arise?
4. What gram-negative organism commonly infects wounds caused by animal bites?
5. What is the most common cause of chronic localized lymph node enlargement in young children?
6. What are the symptoms of rat bite fever?
7. What kind of variants are seen in *Streptobacillus moniliformis* cultures?
8. Why are human bites dangerous?
9. What fungus infection is likely to result from thorn or splinter injuries?

Case Presentation

The patient was a 24-year-old woman, a surgical nurse, seen in the clinic for evaluation of a needle puncture wound to the hand. Earlier in the day, while assisting in a frantic attempt to revive a man with cardiac arrest, she sustained a deep puncture wound to her right palm from a needle that had accidentally dropped into the bedclothes. The needle was visibly contaminated with blood. She immediately washed her hand thoroughly with soap and water, applied an antiseptic, and dressed the puncture site with a loose adhesive bandage.

She was married, with one 14-month-old child. There was no history of blood transfusion or injected drug abuse. She had donated blood the previous month and it was not rejected. Her tetanus immunization was up-to-date, but she had not been immunized against hepatitis B.

Two days after the clinic visit, a test of the cardiac arrest patient's blood was reported as positive for a viral infection.

1. What were the main diagnostic considerations?
2. What risk of infection did the patient face?
3. What measures could be taken to reduce the risk? How much time could expire before preventive measures became ineffective?
4. What was this patient's prognosis?

Discussion

1. The viruses of concern are hepatitis B virus (HBV), human immunodeficiency virus (HIV), and hepatitis C virus (HCV). Each of these could be transmitted to the nurse by a needle stick and cause serious illness.
2. There are an estimated 750,000 to 1 million carriers of HBV in the United States. They typically have large amounts of circulating infectious virus, so that even a tiny amount of their blood can transmit the disease. The risk of infection from a needle puncture wound when the blood originates from a hepatitis B virus carrier is estimated to be 10% to 35%.

   The AIDS-causing human immunodeficiency virus infects approximately 1 million Americans. The blood of these persons is also potentially infectious, but the risk of transmission by a needle stick is considerably lower than the risk for hepatitis B, averaging about 0.4%. The lower risk results from smaller amounts of circulating infectious virus in HIV-infected individuals. The risk is probably higher early in HIV disease, during the acute infection, and later, when AIDS develops, because much higher levels of circulating infectious virus are then present.

   Hepatitis C virus transmission by blood accounts for most cases of post-transfusion hepatitis. Transmission from surgeon to patient has been documented, presumably by the multiple pricks from surgical needles that often penetrate the surgeons' gloves during major surgery. The risk of transmission by needle stick from an HCV positive individual is about 1.8%. The number of new hepatitis C virus infections in the United States each year has been estimated at between 150,000 and 170,000, but the mode of transmission is unknown in most cases.

   Other viruses, such as cytomegalovirus (CMV) and Epstein-Barr virus (EBV) can be transmitted by blood. The risk from a needle stick injury is unknown but is probably much lower than from the viruses already mentioned. Obviously, all blood should be considered potentially infectious.
3. In the case of needle puncture wounds that expose a person to HBV, hepatitis B immune globulin (HBIG) is given as soon as possible after the wound occurs. HBIG is gamma globulin obtained from individuals that have a high titer of antibody against HBV. At the same time, active immunization is started with hepatitis B vaccine. These measures must be initiated within seven days of the injury to be effective. This nurse, as with all persons at high risk of blood exposure, should have already been immunized with hepatitis B vaccine and no other preventive measures would need to be taken.

   Those exposed to HIV by needle punctures should be given zidovudine (AZT), plus one or more other anti-HIV medications, immediately and for four weeks. There is probably little protective effect if therapy is delayed beyond two hours.

   There is no proven preventive measure for HCV exposure. Approaches similar to those for HBV may become available in the future.
4. Preventive measures for hepatitis B exposure are highly effective, reducing the risk of infection by 75% or more. Also, the already relatively low risk from needle puncture wound for HIV exposure can probably be reduced by 75% to 80% with preventive medication. New medications on the horizon may decrease the risk further.

   The patient's prognosis for remaining free of infection was good. However, the small chance of becoming infected, and the long incubation period of these diseases, add up to considerable worry. Every effort should be made to avoid needle puncture wounds in the first place.

Summary

I. Anatomy and Physiology
   A. Wound abscesses are localized collections of pus. They lack blood vessels, and inflammation surrounds the abscess. The microorganisms present may not be actively growing and therefore not affected by antibiotics. Microorganisms in abscesses are potential sources of infection elsewhere in the body.
   B. Anaerobic conditions in wounds permit growth of anaerobic pathogens. Puncture wounds, dirty wounds, and those with dead tissue are likely to be anaerobic.
   C. Even microorganisms usually considered harmless can cause serious infections if the tissue is crushed or the host is immunodeficient.

II. Bacterial Wound Infections
   A. *Staphylococcus aureus* (coagulase-positive staphylococcus) is the most common cause of surgical wounds.
   B. Some strains of *Streptococcus pyogenes* can destroy tissue and produce shock.
   C. *Pseudomonas aeruginosa* is a common cause of burn infections. It is usually resistant to most antibiotics.

D. Clostridial wound infections include tetanus and gas gangrene.
   1. Tetanus (lockjaw) is caused by a toxin produced by *Clostridium tetani*. Colonization of even minor wounds by *C. tetani* can result in tetanus. The disease is often fatal but can be prevented by proper immunization and care of all wounds.
   2. Gas gangrene is marked by necrosis of muscle tissue. It is usually caused by the anaerobe *C. perfringens*.

E. Nonclostridial anaerobic infections of wounds are caused by *Bacteroides, Fusobacterium*, and *Peptostreptococcus* sp. Actinomycosis is caused by *Actinomyces israelii*, a member of the normal human mouth and intestinal flora. It can originate from or follow dental and intestinal surgery.

F. Bite wounds are commonly infected.
   1. Infection of bite wounds may result from a single species of bacterium or from multiple species acting synergistically.
   2. *P. multocida* infection commonly complicates animal bite wounds.
   3. Cat scratch disease is characterized by localized skin papules, followed by large and often pus-filled lymph nodes. It is caused by *Bartonella henselae*.
   4. Rat bite fever can be caused by *Streptobacillus moniliformis*. *S. moniliformis* characteristically has cell wall deficient variants called *L-forms*.
   5. Human bites often result in severe synergistic infections. Prompt treatment is important.

III. Fungal Infections of Wounds
   A. Sporotrichosis is a chronic fungal disease mainly of people who work with vegetation. The cause is *Sporothrix schenckii*, usually introduced into wounds caused by thorns and splinters.

Chapter Review Questions

1. List three findings that indicate infection in the microbiological examination of a wound.
2. Give an example of a pyogenic coccus and explain why it is called *pyogenic*.
3. What property of *Staphylococcus epidermidis* aids its ability to colonize plastic materials used in medical practice?
4. What is the role of clumping factor in the pathogenesis of infections due to *Staphylococcus aureus?*
5. What is the relationship between the superantigens of *Staphylococcus aureus* and the organism's production of toxic shock?
6. Why is there major concern about vancomycin-resistant enterococci?
7. Give two sources of *Pseudomonas aeruginosa*.
8. What organism causes "lockjaw"? Outline the pathogenesis of the disease.
9. What is the causative agent of cat scratch disease? Why is it a threat to patients with AIDS?
10. What is a synergistic infection? How might one be acquired?

Critical Thinking Questions

1. Why does the etiology of wound infections change?
2. Could colonization of a wound by a noninvasive bacterium cause disease?
3. Contrast actinomycosis and sporotrichosis.

Further Reading

1990. *Aeromonas* wound infections associated with outdoor activities—California. *Morbidity and Mortality Weekly Report* 39:334. *Aeromona* sp. are common in the environment and can colonize wounds.

1994. Encephalitis associated with cat scratch disease—Broward and Palm Beach Counties, Florida, 1994. *Morbidity and Mortality Weekly Report* 43(49):909.

Huttner, W. B. 1993. Snappy exocytotoxins. *Nature* 365:104. Gives evidence of a striking similarity in the actions of tetanus and botulism toxins.

Nowak, Rachel. 1994. Flesh-eating bacteria; not new, but still worrisome. *Science* 264:1665.

Perl, T. M., and Roy, M. C. 1995. Postoperative wound infections: risk factors and role of *Staphylococcus aureus* nasal carriage. *Journal of Chemotherapy* 7(suppl. no. 3):29.

1996. Progress toward elimination of neonatal tetanus—Egypt, 1988–1994. *Morbidity and Mortality Weekly Report* 45(4):89.

1995. Recommendations for preventing the spread of vancomycin resistance. *Morbidity and Mortality Weekly Report* 44(RR-12):1.

Stevens, D. L. 1995. Streptococcal toxic shock syndrome: spectrum of disease, pathogenesis, and new concepts in treatment. *Emerging Infectious Diseases* 1(3):69.

1993. Tetanus fatality—Ohio, 1991. *Morbidity and Mortality Weekly Report* 42:148. Fatal tetanus in an 80-year-old woman with a splinter. Despite many past visits to physicians' offices over the years, she apparently had never been immunized.

Zangwill, K. M., et al. 1993. Cat scratch disease in Connecticut. Epidemiology, risk factors, and evaluation of a new diagnostic test. *The New England Journal of Medicine* 329:8.

30 Blood and Lymphatic Infections

KEY CONCEPTS

1. A systemic infection represents failure of the body's mechanisms for keeping infections localized to one area.
2. The inflammatory response, although vitally important in localizing infections, can be life threatening if generalized.
3. Systemic infections threaten the transportation of oxygen and nutrients to body tissues and removal of waste products.
4. The circulatory system can expose all the body's tissues to infectious agents and their toxins.

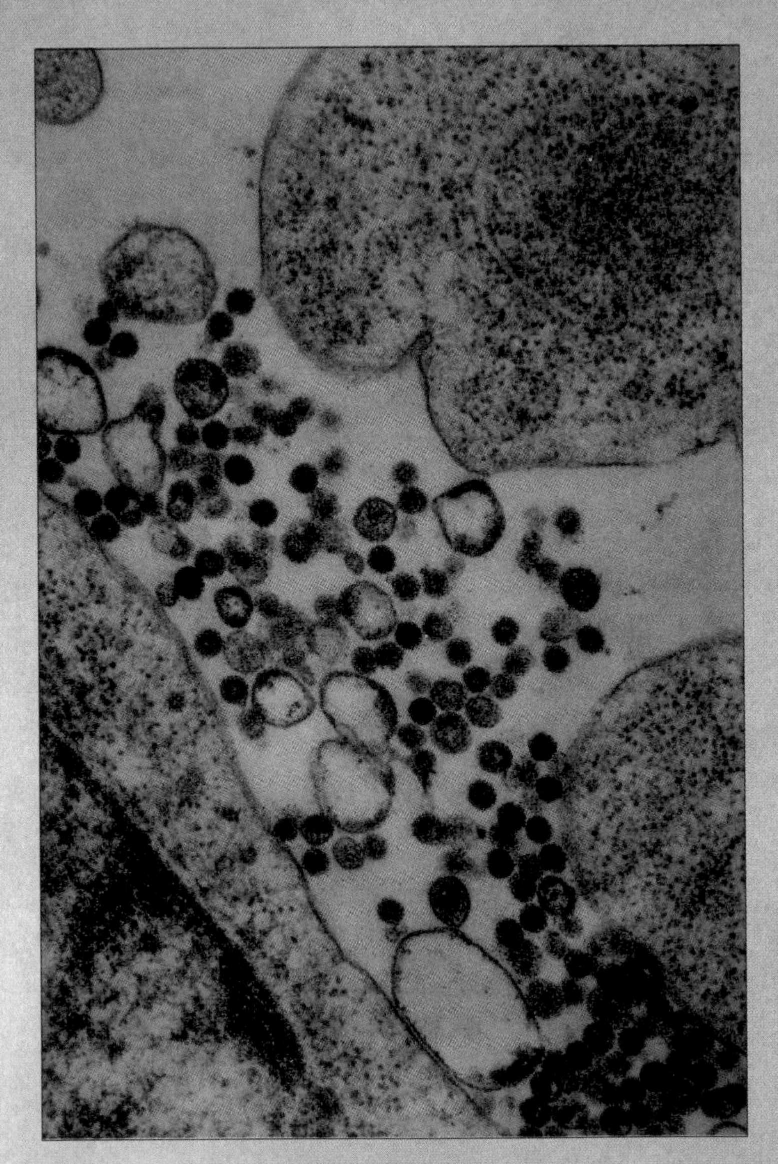

Transmission electron micrograph of Epstein-Barr virus (black spheres), the cause of infectious mononucleosis. This member of the herpesvirus family becomes latent in B lymphocytes and probably plays a role in causing certain malignancies.

Microbes in Motion

The following books in the *Microbes in Motion* CD-ROM may serve as a useful preview or supplement to your reading: *Gram-Positive Organisms. Gram-Negative Organisms. Miscellaneous Bacteria. Microbial Pathogenesis in Infectious Diseases:* Toxins. *Vaccines:* Vaccine Targets. *Viral Structure and Function:* Viral Diseases. *Parasitic Structure and Function.*

A Glimpse of History

Alexandre Emile John Yersin is one of the most interesting though relatively unknown contributors to the conquest of infectious diseases. He was born in 1863 in the Swiss village of Lavaux, where his family lived in a gunpowder factory. His father, director of the factory and a self-taught insect expert, died unexpectedly three weeks before Yersin's birth. Alexandre developed an interest in science when he discovered in an attic trunk his father's microscope and dissecting instruments. A local public health physician influenced him to study medicine, and at age 20 he began premed studies in Lausanne. After a year, Yersin transferred to the University of Marburg, Germany. He continued his medical studies in Paris at a hospital for the poor while working in a laboratory where he translated German scientific literature and performed autopsies on patients who had died of rabies. Despite the pressures of classroom and clinical responsibilities, he volunteered to help at the Pasteur Institute. In his fourth year of medical training, he was hired by Emile Roux, a co-worker of Pasteur. Yersin's clinical experience with diphtheria and tuberculosis influenced Roux to study these diseases.

Yersin became increasingly recognized for his work at the Institute, but he was bored with research, and he also felt it was wrong for physicians to make a living from sickness. He abruptly left the Pasteur Institute and was employed as a physician on a ship sailing from Marseilles to Saigon. Most of Indochina was under French control and was largely unexplored and undeveloped. Yersin became enchanted with the land and its peoples, exploring crocodile-infested rivers in a dugout canoe and making expeditions on elephant into jungles where tigers roared. For the rest of Yersin's life, Vietnam was his home. His many contributions to the region included the introduction of improved breeds of cattle, cultivation of rubber and quinine-producing trees, and the establishment of a medical school.

Yersin also studied the diseases that affected the Vietnamese people. One of them, plague, was a devastating and recurring problem. The disease had killed millions in Europe in medieval times and had struck France as recently as 1720. It was a greatly feared, unpredictable, incurable disease, and its cause and transmission were unknown. In 1894, Yersin went to study a Hong Kong outbreak of plague because the facilities for studying the disease were better in Hong Kong than in Vietnam. Unfortunately for Yersin, Shibasaburo Kitasato, the famous colleague of Robert Koch and Emil von Behring, had arrived three days before he did with a large team of Japanese scientists. British authorities had given them access to patients and laboratory facilities. Since Yersin did not speak English, it was difficult for him to communicate with the authorities. Moreover, although both he and the Japanese scientists spoke fluent German, the Japanese team treated him coolly, perhaps as a reflection of the intense rivalry between Pasteur and Koch. Yersin was reduced to setting up a laboratory in a bamboo shack. He bribed British sailors responsible for disposing of the bodies of plague victims to allow him to get samples of material from their buboes (swollen, pus-filled lymph nodes).

A week after his arrival in Hong Kong in the summer of 1894, Yersin reported to the British authorities his discovery of a bacillus of characteristic appearance, invariably present in the buboes of plague victims. The bacterium could be cultivated, and it caused plaguelike disease when injected into rats. Later, he showed that the disease was transmitted from one rat to another. It was not until three years later that another Pasteur Institute scientist was able to show that the rat flea was crucial in transmission. Yersin's plague bacillus (now known as *Yersinia pestis*) was used to make an antiplague vaccine, and in 1896, antiserum prepared against the organism provided the world's first cure of a patient with plague.

Kitasato erroneously claimed to have isolated the causative agent of plague shortly after his arrival in Hong Kong. Nevertheless, because of his great prestige, Kitasato was credited with co-discovery of the plague bacillus.

• **Emile Roux, p. 502**
• **Robert Koch, pp. 86, 525**
• **Shibasaburo Kitasato, p. 652**

The circulation of blood and lymph fluids supplies nutrients and oxygen to the body's tissue cells and carries away the cells' waste products. The circulatory system also heats and cools body tissues to maintain an optimum temperature. Infection of the system can have devastating effects because infectious agents become systemic, meaning they can be carried to all parts of the body, producing disease in one or more vital organs or causing the circulatory system itself to stop functioning. Thus, even a small scratch that introduces an infectious agent into the blood or lymphatic circulation can cause considerable harm if the agent is not localized.

This chapter discusses some examples of the many infectious diseases that involve the circulatory system.

Anatomy and Physiology

The Heart Supplies the Force that Moves Blood

The general structure and function of the body's circulation system are discussed in this section. The heart, a muscular pump enclosed in a sac called the **pericardium,** supplies the force that moves the blood. As shown schematically in figure 30.1, the heart is divided into a right and a left side, separated by a septum (wall of tissue) through which blood normally does not pass after birth. The right and left sides of the heart are each divided into two chambers, one that receives blood **(atrium)** and another that discharges it

(ventricle). Blood from the right ventricle flows through the lungs and into the atrium on the left side of the heart. This blood then passes into the left ventricle and is pumped through the **aorta** to the arteries and capillaries that supply the tissues of the body. Although not common, infections of the heart valves, muscle, and pericardium may be disastrous because they affect the vital function of blood circulation.

Veins Return Blood From the Tissues to the Heart

The veins collect the blood from the tissue capillaries and carry it back to the right atrium of the heart. Since pressure in the veins is low, one-way valves help keep the blood flowing in the right direction. Veins are easily compressed, so the action of muscles aids the flow of venous blood. Thus, as the blood flows around the circuit, it alternately passes through the lungs and through the tissue capillaries.

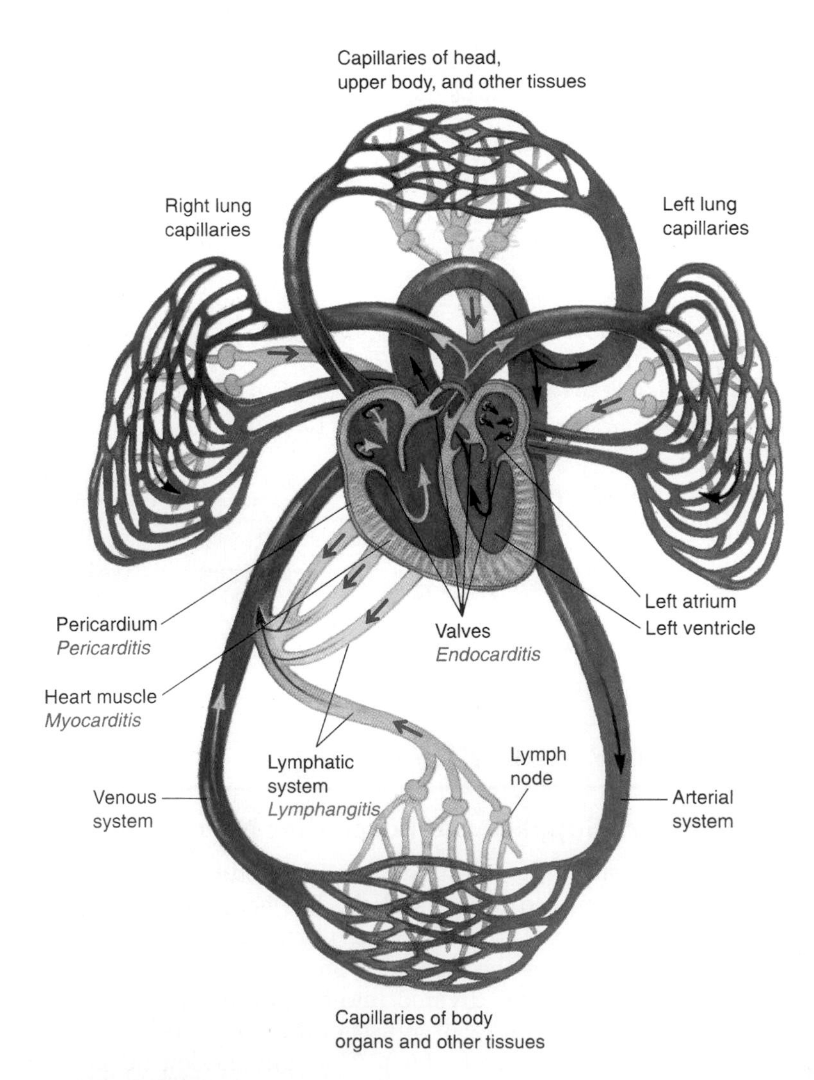

Figure 30.1
The blood and lymphatic systems. For simplicity, the spleen is not shown. It is a large, blood-filled lymphoid organ located under the ribs on the left side of the abdomen.

During each circuit, a portion of the blood passes through organs such as the spleen, liver, and lymph nodes, all of which, like the lung, contain phagocytic cells of the mononuclear phagocyte system. These phagocytes help cleanse the blood of foreign material, including infectious agents, as the blood passes through these tissues.

The Lymphatics Convey Lymph From the Tissue Cells

The system of lymphatic vessels begins in tissues as tiny tubes that resemble blood capillaries but differ from them in having closed ("blind") ends and in being somewhat larger. Lymph, an almost colorless fluid, is conveyed within these lymphatic vessels. It originates from **plasma** (the noncellular portion of the blood) that has oozed through the walls of the blood capillaries to become the **interstitial fluid** that surrounds tissue cells. This fluid bathes and nourishes the tissue cells and then enters the lymphatics. Unlike the blood capillaries, the readily permeable lymphatic vessels take up foreign material such as invading microbes and their products, including toxins and other antigens. The tiny lymphatic capillaries join progressively larger lymphatic vessels. Many one-way valves in the lymphatic vessels keep the flow of lymph moving away from the lymphatic capillaries. Both contraction of the vessel walls and compression by the movements of the body's muscles force the lymph fluid along.

At many points in the system, lymphatic vessels drain into small, bean-shaped bodies called **lymph nodes.** These nodes are constructed so that foreign materials such as bacteria are trapped in them; the nodes also contain phagocytic cells and antibody-producing cells. Lymph flows out of the nodes through vessels that eventually unite into one or more large tubes that then discharge into a large vein, usually behind the left collarbone and thus back into the main blood circulation.

When a hand or a foot is infected, a visible red streak may spread up the limb from the infection site (figure 30.2).

• **Phagocytic cells, pp. 356, 419**

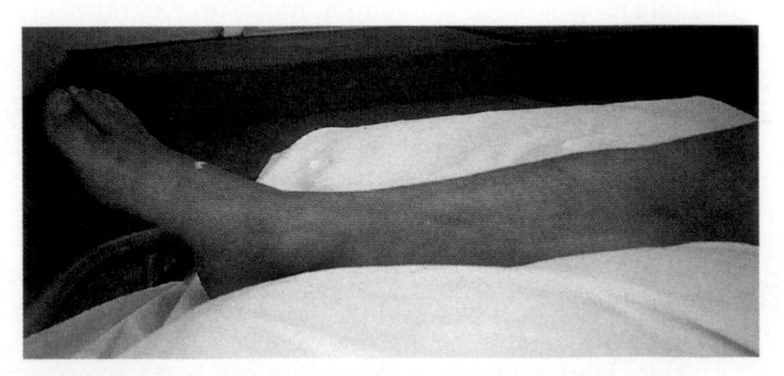

Figure 30.2
Lymphangitis. Notice the red streak extending up the leg from the infected toe. The streak represents an inflamed lymphatic vessel.

This streak represents the course of lymphatic vessels that have become inflamed in response to the infectious agent. This condition is called **lymphangitis.** It may stop abruptly at a swollen and tender lymph node, only to continue later to yet another lymph node. This pause in the progression of lymphangitis demonstrates the ability (even though sometimes temporary) of the lymph nodes to clear the lymph of an infectious agent.

As indicated earlier, the blood and lymph carry infection-fighting leukocytes and such antimicrobial proteins as antibodies, complement, lysozyme, β-lysin, and interferon. An inflammatory response may cause lymph and blood to clot in vessels that are close to areas of infection or antibody-antigen reactions, one of the ways the body has to localize an infection from the rest of the body.

Normal Flora

As in the nervous system, the blood and lymphatic systems lie deep within the body and are normally sterile.

Bacterial Diseases of the Blood Vascular System

Bacterial infections of the vascular system can be rapidly fatal, or they can smolder for months, causing a gradual decline in health. They are not common, but they are always dangerous.

Subacute Bacterial Endocarditis Involves the Valves and Lining of the Heart Chambers

Subacute bacterial endocarditis (SBE) is an infection of the inner lining of the heart, usually localized to one of the heart valves. It commonly occurs in hearts that are abnormal as a result of a birth defect, rheumatic fever, or some other disease.

Symptoms

People with subacute bacterial endocarditis usually suffer from marked fatigue and slight fever. They typically become ill very gradually and slowly lose energy over a period of weeks or months.

Causative Agent

The causative organisms of endocarditis are usually members of the normal bacterial flora of the mouth or skin, notably viridans streptococci and *Staphylococcus epidermidis.* The infecting organisms are usually shed from the infected heart valve into the circulation and can be found by culturing samples of blood drawn from an arm vein.

Pathogenesis

The bacteria that cause SBE can gain entrance to the bloodstream during dental procedures, tooth brushing, or other trauma. In an abnormal heart, a thin blood clot may form in areas where there is turbulent blood flow around a deformed valve or other defect. This clot traps circulating organisms and makes them inaccessible to phagocytic killing. People with bacterial endocarditis often have high levels of antibodies, which are of little value in eliminating the bacteria and may even be harmful because they make the bacteria clump together and adhere to the clot. As the organisms multiply, more clot is deposited around them, gradually building up a fragile mass. Bacteria continually wash off the mass into the circulation, and pieces of infected clot may break off. If large enough, these clots may block important blood vessels and lead to death (infarction) of the tissue supplied by the vessel (as in a stroke) or to its weakening and ballooning out (aneurysm formation). Circulating immune complexes may lodge in the skin, eyes, and other body structures. In the kidney, they produce a kind of **glomerulonephritis** by their inflammatory effects. Even though the organisms normally have little invasive ability, great masses of them growing in the heart are sometimes able to burrow into heart tissue to produce abscesses or damage valve tissue, resulting in a leaky valve. This is a good example of how pathogenicity depends on both host factors and the virulence of the microorganism. With certain host factors, even bacteria of low virulence can cause serious, even fatal, infections.

In 5% to 15% of cases of infective endocarditis, culturing the blood from an arm vein fails to yield the causative bacteria. This is especially likely to occur when the infection is on the right side of the heart, and blood from the infected site must pass through the lung and spleen, both richly supplied with phagocytes, before reaching the arm vein.

Epidemiology

In recent years, viridans streptococci have accounted for a smaller percentage of SBE cases than previously, partly because most dentists are careful to give antibiotic treatment to patients with prominent heart murmurs[1]. Also, nowadays, SBE cases occur in abusers of intravenous drugs, in hospitalized patients who have plastic

[1]Heart murmur—abnormal sound the physician hears when listening to the heart; may indicate a deformed valve or other structural abnormality

- **Antimicrobial substances, p. 763**
- **Inflammatory response, pp. 354–55**
- **SBE, pp. 510, 559**
- **Rheumatic fever, p. 510**
- **Viridans streptococci, p. 260**
- **Immune complexes, pp. 400–1**
- **Virulence, p. 413**

intravenous catheters for long periods, and in those with artificial heart valves. These people are usually infected with *S. epidermidis* or a wide variety of species other than viridans streptococci.

Prevention

No scientifically proven preventive methods are available. However, in people with known or suspected defects in their heart valves, the accepted practice is to give an antibacterial medication shortly before dental or other bacteremia-causing procedures. The medication is chosen according to the expected species of bacterium and its likely antimicrobial susceptibility.

Rigorous attention to sterile technique when inserting plastic intravenous catheters, changing the catheters to new locations every several days, and discontinuing their use as soon as possible probably helps prevent colonization of the catheters and consequent bacteremia.

Treatment

Antibacterial medications are chosen according to the susceptibility of the causative organism. Only bactericidal medications are effective, and usually two or more antimicrobials are used together. Treatment usually requires a number of weeks. If foreign material such as an artificial heart valve is present, it must generally be replaced to effect a cure. Sometimes, it is necessary to perform surgery to remove the mass of infected clot or to drain abscesses.

The main characteristics of SBE are presented in table 30.1.

Table 30.1 Subacute Bacterial Endocarditis (SBE)

Symptoms	Fever, loss of energy over a period of weeks or months
Incubation period	Poorly defined, usually weeks
Causative agents	Usually oral α-hemolytic viridans streptococci or *Staphylococcus epidermidis*
Laboratory diagnosis	Blood drawn from arm vein and cultured for the organism
Pathogenesis	Organisms of normal flora gain entrance to bloodstream through dental procedures or other trauma; in an abnormal heart, turbulent blood flow causes formation of a thin clot that traps circulating organisms and makes them inaccessible to phagocytic killing; pieces of clot break off and may block important blood vessels, leading to tissue death
Epidemiology	Persons at risk are mainly those with hearts that have congenital defects or are damaged by disease such as rheumatic fever; situations that cause bacteremia
Prevention	Administration of an antibiotic at time of anticipated bacteremia, such as dental work
Treatment	Bactericidal antibiotics such as penicillin and gentamicin

Acute Bacterial Endocarditis Is Caused by Bacteria More Virulent than Those that Cause SBE

Acute bacterial endocarditis differs from SBE in the suddenness and severity of its onset. An infection is usually evident somewhere else in the body, or there is evidence of abuse of intravenous drugs.

Causative Agents

Infections of the heart can be caused by bacteria more virulent than those that cause SBE, such as *Staphylococcus aureus* or *Streptococcus pneumoniae*. These organisms can produce a rapidly progressing acute endocarditis and are more likely to invade the heart and its valves and permanently destroy tissue than are the organisms that cause subacute bacterial endocarditis. Acute bacterial endocarditis often results as a complication of drug addiction, when unsterile needles are used to inject substances intravenously. Many other cases occur as an occasional complication of streptococcal pneumonia, gonorrhea, and other acute infectious diseases marked by bacteremia (bacteria circulating in the bloodstream).

• **Oral bacteria and endocarditis, p. 559**
• ***Staphylococcus aureus*, pp. 475–79, 655**
• ***Streptococcus pneumoniae*, pp. 525–28, 633**

The Seriousness of Gram-Negative Septicemia Relates to Endotoxin

Septicemia, an acute illness caused by infectious agents or their products circulating in the bloodstream, or "blood poisoning," is a major cause of illness among hospitalized patients. An estimated 400,000 cases occur in the United States each year, about 30% of which are caused by gram-negative bacteria. Endotoxin release by gram-negative bacteria can lead to shock and lack of sufficient oxygen.

• **Endotoxin, pp. 60, 420, 422, 424**

Perspective 30.1

Endotoxin Contamination of Cardiac Catheters

Thirty-two patients undergoing cardiac catheterization* experienced fever, shaking chills, and a drop in blood pressure, suggesting the presence of endotoxin in their blood, but blood cultures showed no bacteria. A detailed epidemiological investigation revealed that the catheters had been thoroughly washed, rinsed in distilled water, and sterilized after each use. A sample of the distilled water used to rinse the catheters was found to contain large numbers of the gram-negative bacterium *Acinetobacter calcoaceticus*. This organism was the source of the endotoxin, contaminating the catheters when they were rinsed before sterilization. Since even distilled water becomes contaminated with gram-negative organisms when it is exposed to air, it is difficult to obtain water free of endotoxin. Species of *Pseudomonas* and *Acinetobacter* as well as a variety of other gram-negative bacteria are widely found in samples of water.

*Cardiac catheterization—process of inserting a long plastic tube (catheter) into the heart via one of the blood vessels of the leg; the purpose is usually to measure pressures in the heart

Symptoms

The symptoms of septicemia include violent shaking chills, and fever, often accompanied by anxiety and rapid breathing.

Causative Agents

Septicemia is caused by gram-negative bacteria, gram-positive bacteria, viruses, and fungi. Probably because they possess endotoxin, gram-negative bacteria tend to cause more serious septicemias than do gram-positive bacteria. Shock is common, and despite treatment, only about half of all people afflicted with this kind of infection survive. Cultures of blood from these patients usually reveal enterobacteria such as *Escherichia coli*, *Enterobacter aerogenes*, *Serratia marcescens*, and *Proteus mirabilis*. Among the aerobic, gram-negative, rod-shaped bacteria encountered in septicemia are organisms commonly found in the natural environment, such as *Pseudomonas aeruginosa*. This organism has such extraordinary biochemical capabilities that it can metabolize a variety of organic compounds, resist antibiotics and disinfectants, and grow under conditions unfavorable to many of the usual bacterial pathogens.

Some of the gram-negative organisms causing septicemia are anaerobes. For example, *Bacteroides* sp., which make up a major percentage of the normal flora of the large intestine and the upper respiratory tract, cause a significant percentage of septicemia cases.

Pathogenesis

Septicemia almost always originates from an infection somewhere in the body other than the bloodstream (e.g., boil, kidney infection, abscess, pneumonia) that has become uncontrolled. In addition, alterations in normal body defenses as the result of medical treatment (e.g., surgery, catheters, and drugs that interfere with the immune response) may allow microorganisms that normally have little invasive ability to infect the blood.

Endotoxin is released from the outer cell walls of gram-negative bacteria growing in a localized infection or in the bloodstream. Unfortunately, antibiotics that act against the bacterial cell wall can also enhance the release of endotoxin from the organisms. These antibiotics are typically used in treating gram-negative bacterial infections.

Many of the body's cell types, especially tissue macrophages and circulating leukocytes, respond defensively to endotoxin, just as they do to many other foreign substances, but the response to endotoxin is particularly intense. Quite likely, this exaggerated response to endotoxin is a type of hypersensitivity, which develops at an early age in response to endotoxin that diffuses into the circulation from the normal intestinal flora.

The response of the body cells to endotoxin is appropriate for localizing gram-negative bacterial infections in tissues and killing the invaders. However, when localization fails and endotoxin enters the circulation, it causes the nearly simultaneous triggering of defense cells throughout the body, which can be fatal to the host. Macrophages are of central importance in body defense, but they also play a key role in septic shock. The interaction of endotoxin with macrophage cell membrane causes the cell to synthesize and release cytokines.

One cytokine, **tumor necrosis factor (TNF),** is released from macrophages within minutes of exposure to endotoxin. Levels rise quickly and then fall off.

TNF has diverse effects, one of which is a change in the setting of the body's thermostat, causing the temperature to rise. TNF also causes circulating polymorphonuclear leukocytes to adhere to capillary walls, leading to large

• **Enterobacteria, p. 248**
• ***Pseudomonas*, pp. 246–47, 658**
• **Hypersensitivity, pp. 396–402**
• **Cytokines, p. 352**

Perspective 30.2

Laboratory Diagnosis of Gram-Negative Septicemia

Laboratory diagnosis of gram-negative septicemia is made by scrubbing the skin over a vein with iodine and alcohol and then carefully (without contaminating the area) withdrawing about 10 milliliters of blood from the vein with a sterile needle and syringe. The blood is then added to culture media. Most enterobacteria and species of *Pseudomonas* are easy to cultivate and identify. Anaerobic bacteria sometimes present difficulties in cultivation and testing of antimicrobial susceptibility because of their requirements for strict anaerobiosis. However, gas chromatography is a useful tool for identifying this group of bacteria. In this technique, volatile (gaseous) microbial products from a culture are passed through a device that separates them from each other in a sequence determined by their chemical properties. The time of appearance of each substance is automatically recorded and compared with the times of appearance of known substances to identify the substance.

accumulations of these inflammatory cells in tissues such as the lung, which have large populations of macrophages. Experimentally, antibody against TNF gives substantial protection against endotoxic shock. Interleukin-1 (IL-1) is another cytokine released from macrophages. Besides acting with TNF to cause fever and the release of leukocytes from bone marrow, it has many other effects. One potentially harmful action is to cause the release of enzymes from polymorphonuclear leukocytes.

Macrophages also synthesize and secrete complement, which is activated by endotoxin. Components of activated complement attract leukocytes and cause them to release tissue-damaging lysosomal enzymes. Activated complement also causes capillaries to leak excessive amounts of plasma.

The circulating proteins responsible for blood clotting are also activated by endotoxin. Activation causes small clots to form, which plug capillaries, cutting off blood supply and causing tissue necrosis. This condition, called **disseminated intravascular coagulation,** is often accompanied by hemorrhage[1] because not enough clotting proteins are left to stop bleeding in the injured tissues.

These bad effects of endotoxemia (endotoxin circulating in the bloodstream) are made worse by the hypotension (low blood pressure) that usually is present. The hypotension is caused by decreased muscular tone of the heart and blood vessel walls, and the low blood volume that results from leakage of plasma from the blood vessels. Current evidence indicates that the release of cytokines and their effects on the heart and blood vessels play a key role in causing hypotension in gram-negative septicemia. Shock results when the blood pressure falls so low that vital organs are no longer supplied with adequate amounts of blood to maintain their function.

Although multiple organs are affected by endotoxemia, the lung is particularly vulnerable to serious, irreversible damage. This damage often results in death despite successful cure of the infection and correction of shock.

[1]Hemorrhage—bleeding

• **Complement, p. 351**

Prevention

Prevention of septicemia depends largely on the prompt identification and effective treatment of localized infections, particularly in people whose host defenses are impaired. Also, conditions such as bedsores and bladder infections, which commonly lead to septicemia in patients with cancer and diabetes, can usually be prevented.

Treatment

Antimicrobial medications, directed against the causative organism, are given. Measures are taken to correct shock and poor oxygenation. Despite these treatements, the mortality rate remains high, generally 30% to 50%, partly because most of the patients have serious underlying conditions. Treatment with monoclonal antibody directed against endotoxin has been disappointing, showing some benefit when given before shock occurs but no decrease in mortality at one month. Similarly, monoclonal antibody against TNF has shown a 40% decrease in mortality at three days but no significant decrease in deaths one month after treatment. Evaluating this kind of therapy is difficult because many of the patients have underlying life-threatening illnesses in addition to their septicemia.

The principal pathological events that can occur in gram-negative bacterial septicemia are outlined in table 30.2.

REVIEW QUESTIONS

1. What is a "systemic" infection?
2. Which side of the heart, right or left, do bacteria in infected lymph reach first? Where do they go from there?
3. How is the bacteriological diagnosis of SBE usually made?
4. What is the significance of immune complex formation in SBE?
5. Contrast acute and subacute bacterial endocarditis.
6. What is the difference between bacteremia and septicemia?
7. How does circulating endotoxin cause septicemic shock?

Table 30.2 Principal Events in Gram-Negative Bacterial Septicemic Shock

Event	Result
Gram-negative bacterial infection (e.g., pyelonephritis, pneumonia, peritonitis from ruptured appendix)	The bacteria or their endotoxins enter the bloodstream
Endotoxin triggers response from many kinds of the body's cells, notably macrophages	The cells release their cytokines, resulting in fever, decreased muscle tone of heart and blood vessel walls, increased leakage of plasma from blood vessels, and increased adhesiveness of PMN; blood pressure drops and oxygen exchange in the lung is impaired
Endotoxin activates complement	Components of the complement system attract leukocytes, increase capillary permeability to plasma, cause release of lysosomal enzymes from leukocytes; tissue is damaged, especially the lungs
Endotoxin activates the blood-clotting mechanism	Clots form in the capillaries, causing necrosis of tissue and hemorrhage.

Bacterial Diseases Involving the Lymph Nodes and Spleen

The enlargement of the lymph nodes and spleen is a prominent feature of certain infectious diseases that involve the mononuclear phagocyte system. Three examples of these diseases—tularemia, brucellosis, and plague—are discussed next. All three are now uncommon human diseases in the United States. However, they represent a constant threat because they exist as **zoonoses** (diseases of animals transmissible to humans).

Tularemia Is a Dangerous Zoonosis Caused by an Unusual Gram-Negative Rod

Tularemia is widespread in the United States, involving many different animal species including rabbits, muskrats, and bobcats. Many human cases are acquired when people are skinning animals that appear to be free of disease. The causative organism enters through unnoticed scratches or by penetration of a mucous membrane. The disease can also be acquired from the bites of flies and ticks and by inhalation of the causative organism.

Symptoms

Tularemia (figure 30.3) is characterized by development of a skin ulceration and enlargement of the regional lymph nodes two to five days after a person is bitten by a tick or insect or handles a wild animal. The usual symptoms of fever, chills, and achiness that occur in many other infections are also present in tularemia.

Causative Agent

Tularemia is caused by *Francisella tularensis*, a pleomorphic, nonmotile, aerobic, gram-negative rod that derives its name from Edward Francis, an American physician who studied tularemia in the early 1900s, and from Tulare County, California, where it was first studied. The organism is unrelated to other common human pathogens and is unusual in that it requires a special medium enriched with the amino acid cysteine in order to grow.

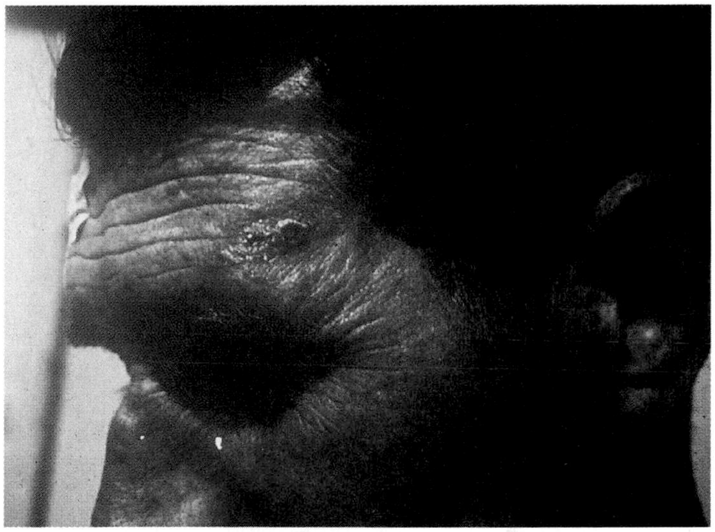

• **Mononuclear phagocytes, p. 348**

Figure 30.3
Ulceroglandular form of tularemia in a muskrat trapper. The healing ulcer above the patient's left eyebrow is the site of entry of *F. tularensis*.

Pathogenesis

Typically, *F. tularensis* causes a steep-walled ulcer where it enters the skin. Lymphatic vessels draining the area carry the organisms to the regional lymph nodes. These nodes then become large and tender and may become filled with pus and drain spontaneously. Later, the organisms spread to other parts of the body via the lymphatics and blood vessels. Pneumonia, caused when the organisms reach the lung by the bloodstream or by inhalation, has a mortality rate as high as 30%. *F. tularensis*, like *Mycobacterium tuberculosis*, is ingested by phagocytic cells and grows within them. This may explain why tularemia persists in some people despite

Perspective 30.3

Laboratory Diagnosis of Tularemia

The organism that causes tularemia can be cultivated from blood and pus, but if the disease is suspected, generally no attempts are made to isolate the organism because of the danger to laboratory technicians. Instead, fluorescent antibody can be used to detect the organisms in smears of infected material. If clinical specimens are contaminated with other more rapidly growing bacteria from the environment or the normal flora, the material can be injected into a laboratory animal such as a guinea pig or mouse. The defense mechanisms of the animal destroy the contaminating flora, while *F. tularensis* invades the animal's tissues. Fluorescent antibody or other serological techniques can then be used to identify *F. tularensis* from the infected tissues. Most cases of human tularemia are diagnosed by demonstrating a rise in antibody titer to the organisms in people's blood.

• Serological techniques, p. 384

the high titers of antibody in their blood. Cell-mediated immunity is probably responsible for ridding the host of this infection, as it is with other organisms that can live intracellularly. Both delayed hypersensitivity and serum antibodies quickly arise during infection, so that even without treatment over 90% of infected people survive.

Epidemiology

Tularemia occurs in many areas of the Northern Hemisphere, including all the states of the United States except Hawaii. In the eastern United States, human infections usually occur in the winter months, as a result of people skinning rabbits (hence, the common name, "rabbit fever"). However, hunters and trappers in various parts of the country have contracted the disease from muskrats, beavers, squirrels, deer, and other wild animals. The animals generally appear to be free of illness. In the West, infections mostly result from the bites of infected ticks and deer flies and, thus, usually occur during the summer. Generally, 150 to 250 cases of tularemia are reported each year in the United States (figure 30.4).

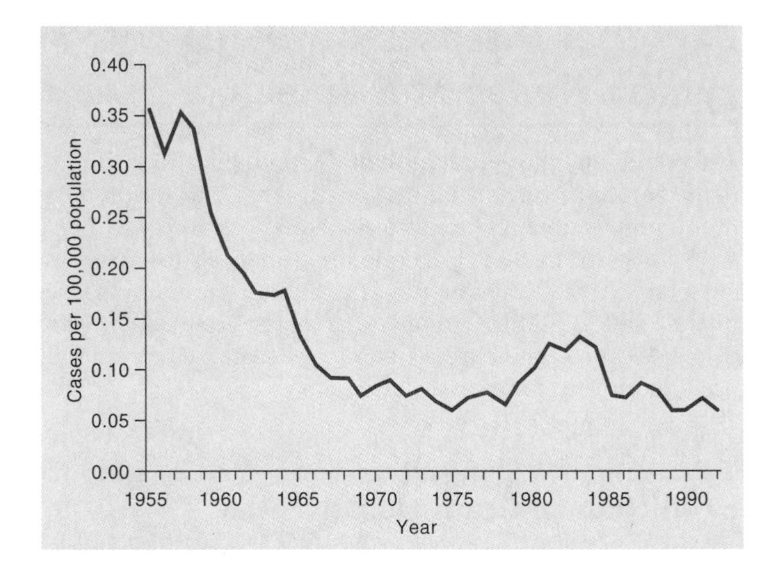

Figure 30.4
Incidence of tularemia, United States, 1955–1992.

Source: Centers for Disease Control and Prevention. 1993. Summary of notifiable diseases, 1992. *Morbidity and Mortality Weekly Report* 41(55):62.

Prevention

Rubber gloves and goggles or face masks are advisable for people skinning wild animals. Insect repellants and protective clothing help guard against insect and arachnid vectors. It is a good practice to inspect routinely for ticks after exposure to the out-of-doors and to remove them carefully. A vaccine is available for laboratory workers and others at high risk of infection.

Treatment

Most cases of tularemia are effectively treated with tetracycline or gentamicin.

The main features of this disease are summarized in table 30.3.

• Cell-mediated immunity, p. 371
• Delayed hypersensitivity, pp. 384, 403

Brucellosis Is a Zoonosis Acquired from Domestic Animals

Only about 150 cases of human brucellosis are reported each year in the United States. However, 10 to 20 times that number of cases go unreported annually. Brucellosis is often called *undulant fever* or *"Bang's disease,"* after Frederik Bang (1848–1932), a Danish veterinary professor who discovered the cause of cattle brucellosis.

Symptoms

The onset of brucellosis is usually gradual, and the symptoms are vague. Typically, patients complain of mild fever, sweating, weakness, aches and pains, and weight loss. The recurrence in some cases of fevers over weeks or months gave rise to the alternate name "undulant fever."

Table 30.3 Tularemia

Symptoms	Ulcer at site of entry, enlarged lymph nodes in area
Incubation period	1–10 days; usually 2–5 days
Causative agent	*Francisella tularensis*
Laboratory diagnosis	Cultures using cysteine-containing media; fluorescent antibody stain of pus; detection of rise in antibody titer
Pathogenesis	Organisms spread throughout body, are ingested by phagocytic cells, and grow within these cells
Epidemiology	Mucous membrane or broken skin contact with organism (skinning rabbits, for example); bite of infected insect or tick
Prevention	Vaccination only in special cases; avoiding bites of insects and ticks; wearing rubber gloves when skinning rabbits; taking safety precautions when working with organisms in laboratory
Treatment	Gentamicin or tetracycline

Causative Agent

Four varieties of the genus *Brucella* cause brucellosis in humans. DNA studies show that all members of the genus fall into a single species, *Brucella melitensis*, but traditionally, the different varieties were assigned species names depending largely on their preferred host: *B. abortus* invades cattle; *B. canis*, dogs; *B. melitensis*, goats; and *B. suis*, pigs. *Brucella* sp. are strictly aerobic, nonmotile, pleomorphic, small, gram-negative rods and have complex nutritional requirements. *B. melitensis* variety *abortus* differs from the other human pathogens in requiring an atmosphere of 5% to 10% CO_2 when it is first isolated. Growth on the dyes basic fuchsin and thionin, production of urease and hydrogen sulfide, antisera, bacteriophage typing, and DNA probes are used to distinguish the named varieties and more than a dozen subtypes. The distinctions between the various strains are mainly useful epidemiologically and generally have little pathogenic significance.

Pathogenesis

As with tularemia, the organisms responsible for brucellosis penetrate mucous membranes or breaks in the skin and are disseminated via the lymphatic and blood vessels to the heart, kidneys, and other parts of the body. Like *F. tularensis*, *Brucella* sp. are not only resistant to phagocytic killing but also can grow intracellularly in phagocytes, where they are inaccessible to antibody and some antibiotics. Up to 15% of cases persist for several months, but mortality, generally due to endocarditis, is about 2%. Bone infection (osteomyelitis) is the most frequent serious complication.

Epidemiology

Brucellosis is typically a chronic infection of domestic animals involving the mammary glands and the uterus, thereby contaminating milk and causing abortions in the affected animals. Abortion is not a feature of human disease. Sixty % of the cases of brucellosis occur in workers in the meat-packing industry; less than 10% arise from ingesting raw milk or other unpasteurized dairy products. Worldwide, brucellosis is a major problem in animals used for food, causing yearly losses of many millions of dollars. In the United States, infections have been acquired by hunters from elk, moose, bison, caribou, and reindeer.

Prevention

The most important control measures against brucellosis are pasteurization of dairy products and inspection of domestic animals for evidence of the disease. The use of goggles and rubber gloves helps protect veterinarians, butchers, and slaughterhouse workers. A live attenuated vaccine effectively controls the disease in domestic animals.

Treatment

Treatment with the antibiotics tetracycline and streptomycin, gentamicin, or rifampin is usually effective.

Table 30.4 gives the main features of brucellosis.

Table 30.4 Brucellosis

Symptoms	Fever, body aches, weight loss, enlargement of lymph nodes and spleen; symptoms may subside without treatment but then recur
Incubation period	5–21 days
Causative agent	*Brucella melitensis*
Laboratory diagnosis	Culture from blood, bone marrow, or urine; complement fixation test
Pathogenesis	Organisms penetrate mucous membranes and are carried to heart, kidneys, and other parts of the body via the blood and lymphatic system; are resistant to phagocytic killing and grow within these cells
Epidemiology	Sources are domestic animals and wild animals such as moose, caribou, and bison; butchers, farmers, veterinarians, and people who drink unpasteurized milk at risk
Prevention	Eradication of disease in animals; pasteurization of milk and milk products; gloves and face protection
Treatment	Tetracycline with rifampin

• **Pasteurization, pp. 207–8**

Plague, Once Pandemic, Persists Endemically in Many Parts of the World

Plague, once known as the "black death," was responsible for the death of approximately one-fourth of the population of Europe between 1346 and 1350. Crowded conditions in the cities and a large rat population undoubtedly played major roles in the spread of the disease.

Symptoms

The symptoms of plague develop one to six days after an individual is bitten by an infected flea. The person characteristically develops markedly enlarged and tender lymph nodes (buboes; hence, the name "bubonic plague") in the region that receives lymph drainage from the area of the flea bite. High fever, shock, delirium, and patchy bleeding under the skin quickly develop, and in many cases, there may also be cough and bloody sputum if the lungs become involved ("pneumonic plague").

Causative Agent

Plague is caused by the facultatively intracellular enterobacterium *Yersinia pestis*, a small, oval, pleomorphic, nonmotile, gram-negative rod that grows best at 28°C. The organism resembles a safety pin in stained preparations of material taken from infected lymph nodes because the ends of the bacterium stain more intensely than does the middle (figure 30.5).

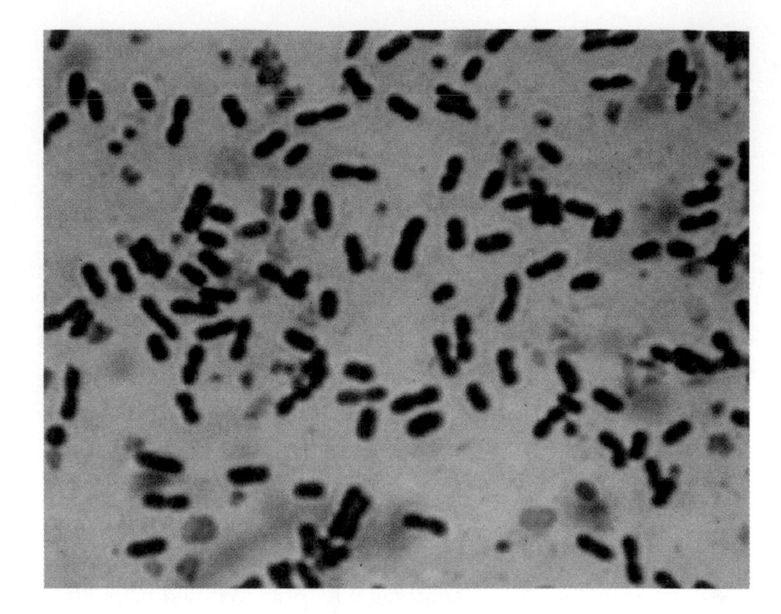

Figure 30.5
Yersinia pestis. Each pole of the organism is stained intensely by certain dyes, producing a safety pin appearance.

Extensive study has revealed some of the complex mechanisms by which *Y. pestis* achieves its impressive virulence. The bacterium has three kinds of plasmids distinguishable by their size. The smallest has 9.5 kilobase pairs (9,500 base pairs) and codes for an important protease called *Pla* because it activates plasminogen activator. Plasminogen activator, in turn, converts plasminogen to plasmin, a potent dissolver of blood clots. (Plasminogen activator, plasminogen, and plasmin are proteinaceous substances normally found in the human body.) Another activity of the Pla protease is that it destroys the C3b and C5a components of complement.

The middle-sized plasmid has 72 kilobase pairs, and almost its entire genome codes for (1) a group of proteins that interfere with phagocytosis and (2) regulators of their expression. These proteins are referred to as *Yops*, for Yersinia outer-membrane proteins. Although they associate with the *Yersinia* outer membrane at 37°C, they can also be released from the bacteria. While the role of Yops in actual human infections is uncertain for the most part, *Y. pestis* loses its virulence if this plasmid is lost. Experimentally, several kinds of actions can be demonstrated. For example, Yop E destroys microfilaments of host cells, while Yop H interferes with the ability of phagocytes to receive environmental signals. Yop M acts by preventing the release of cytokines from platelets. These cytokines would normally call forth an inflammatory response to infection. Another of the proteins, **LcrV** (formerly known as *V antigen*), is important in virulence because antibody against it protects against *Y. pestis* infection. However, its mode of action is uncertain. The sum of the effects of Yops is to interfere with phagocytes that normally would ingest and kill *Y. pestis* and initiate an immune response to the bacterium.

The size of the largest plasmid is 110 kilobase pairs, and it codes for **Fra 1.** This protein is transported to the *Y. pestis* surface to become part of an antiphagocytic capsule. The stimulus for capsule production is the relatively increased temperature of a mammalian host (37°C for humans) compared to that of a flea (about 26°C.)

The codes for some other *Y. pestis* virulence factors reside on the bacterial chromosome. Included in these properties are resistance to the lytic action of activated complement, mechanisms for storing hemin and using it as an iron source, and production of a pilus adhesin, probably as a result of intracellular growth.

A summary of these potential virulence factors is presented in table 30.5.

Pathogenesis

Masses of *Y. pestis* partially obstruct the digestive tract of infected rat fleas. Consequently, the flea is not only ravenously hungry and bites repeatedly but also regurgitates

• **Plasmids, p. 174**
• **Complement, p. 351**
• **Filaments, p. 79**
• **Iron requirement, p. 357**

infected material into the bite wounds. Chronically infected fleas do not have their digestive tract obstructed but excrete *Y. pestis* in their feces; the organisms can then be introduced into human tissue when a person scratches the flea bite.

The *Y. pestis* protease Pla is essential for the spread of the organisms from the site of entry of the bacteria by clearing the lympatics and capillaries of clots. The organisms are carried to the regional lymph nodes where they are taken up by macrophages. The bacteria are not killed by the macrophages and instead multiply and elaborate Fra 1 capsular material. The capsules make them resistant to phagocytosis, and various proteins are synthesized in response to the intracellular growth conditions. After several days, an acute inflammatory reaction develops in the nodes, producing enlargement and marked tenderness. At this stage, the bacteria are quite resistant to destruction because of capsules and other virulence factors. The lymph nodes then become necrotic, allowing large numbers of *Y. pestis* to spill into the bloodstream. This stage of the disease is called **septicemic plague,** and endotoxin release results in shock and disseminated intravascular coagulation (DIC). Based on experimental evidence, it might be expected that the bacteria released from dead macrophages are encapsulated and express Yops, pilus adhesin, complement resistance, and heme storage. Infection of the lung from the bloodstream occurs in 10% to 20% of the cases, resulting in pneumonic plague. Organisms transmitted to another person from a case of pneumonic plague are already fully virulent and, therefore, especially dangerous. The dark hemorrhages into the skin from DIC and the dusky color of skin and mucous membranes probably inspired the name *black death* for the plague.

The mortality rate for persons with untreated bubonic plague is between 50% and 80%. Untreated pneumonic plague progresses rapidly and is nearly always fatal within a few days.

Epidemiology

Plague is endemic in rodent populations of all continents except Australia. In the United States, the disease is mostly confined to wild rodents in about 15 states in the western half of the country (figure 30.6). Prior to 1974, only a few cases of plague were reported each year. However, over the last few decades, the number of cases has generally been higher, averaging about 15 reported cases per year, as towns and cities expand into the countryside. Prairie dogs, rock squirrels, and their fleas constitute the main reservoir, but rats, rabbits, dogs, and cats are potential hosts. Hundreds of species of fleas can transmit plague, and the fleas can remain infectious for a year or more in abandoned rodent burrows. Epidemics in humans, initiated by infected rodent fleas, can spread from person to person by household fleas, as well as by aerosols produced by coughing patients with pneumonic plague. *Yersinia pestis* can remain viable for weeks in dried sputum and in flea feces and for months in the soil of rodent burrows.

• **Reservoir, p. 430**

Table 30.5 Factors Associated with Virulence of *Yersinia Pestis*

Factor	Code	Action
Pla (protease)	9.5 kbp plasmid	Activates plasminogen activator; destroys C3b, C5a
Yop (proteins)	72 kbp plasmid	Interferes with phagocytosis and the immune response by differing mechanisms
Fra 1	110 kbp plasmid	Forms antiphagocytic capsule at 37°C
PsaA (adhesin)	Chromosome	Has maximal expression at intracellular pH
Complement resistance	Chromosome	Protects against lysis by activated complement
Iron acquisition	Chromosome	Traps hemin and other iron-containing substances; stores iron compounds intracellularly

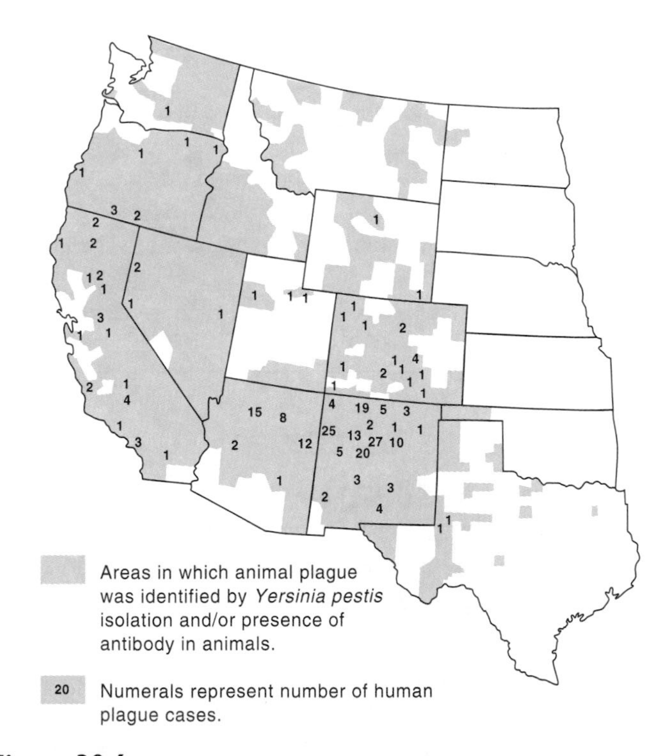

Figure 30.6
Distribution of plague in the United States, 1956–1987. Plague continues to be present among the wild animals of the states indicated, with occasional sporadic spread to humans.

Source: Centers for Disease Control and Prevention. 1988. *Morbidity and Mortality Weekly Report* 37(no. SS-3):12.

Prevention

Plague epidemics can be prevented by rat control measures such as proper garbage disposal, constructing rat-proof buildings, installing guards on the ropes that moor ocean-going ships, and extermination programs. The latter must be combined with the use of insecticides to prevent the escape of possibly infected fleas from dead rats.

A killed vaccine that gives short-term partial protection against plague is available to control epidemics and for those who are at high risk in laboratories or endemic areas.

The antibiotic tetracycline can be given as a preventive and is useful in controlling epidemics because of its immediate effect.

Treatment

Treatment with gentamicin or tetracycline is effective, especially if given early in the disease.

REVIEW QUESTIONS

1. Give the two main ways in which tularemia is contracted.
2. What role does cell-mediated immunity play in recovery from tularemia?
3. How is tularemia usually diagnosed?
4. Workers in what industry are especially likely to contract brucellosis? How can they protect themselves from the disease?
5. Why is pasteurization of milk so important in the control of brucellosis?
6. What is the difference between bubonic and pneumonic plague?
7. What changes occur in *Yersinia pestis* when it enters the human body from a rat flea?
8. In the United States, what constitutes the main reservoir of plague?

Viral Diseases of the Lymphoid System

Infectious Mononucleosis Is Followed by a Latent Infection of B Lymphocytes

Infectious mononucleosis ("mono") is a disease familiar to many students because of its high incidence among people between the ages of 15 and 24 years. The term *mononucleosis* refers to the fact that people afflicted with this condition have an increased number of mononuclear leukocytes in their bloodstream.

• Leukocytes, pp. 346–47

Symptoms

Typically, symptoms of infectious mononucleosis appear after a long incubation period, usually 30 to 60 days. They consist of fever, sore throat covered with pus, marked fatigue, and enlargement of the spleen and lymph nodes of the neck.

Causative Agent

Infectious mononucleosis is caused by the Epstein-Barr (EB) virus, named after its discoverers, M. A. Epstein and Y. M. Barr. It is a double-stranded DNA virus of the herpesvirus family (chapter 14), and although identical in appearance, it is not closely related to any of the other known herpesviruses that cause human disease. This interesting virus was unknown until the early 1960s when it was isolated from **Burkitt's lymphoma,** a malignant tumor derived from B lymphocytes. Subsequent studies showed that EB virus is the cause of infectious mononucleosis.

Pathogenesis

Primary infection with EB virus is analogous to throat infections with the herpes simplex virus in that both viruses initially infect the mucosal cells and then become latent in another cell type. The probable sequence of events is shown in figure 30.7. The Epstein-Barr virus initially replicates in the epithelial cells of the mouth and throat, and then infects B lymphocytes, which have specific surface receptors for the virus. During the illness, up to 20% of the circulating B lymphocytes are infected with the virus. For most of these cells, the infection is nonproductive, since the viral genome replicates as an episome[1] within the cell. However, profound changes in the cells result from the infection. The virus activates the B cells, causing multiple clones of B lymphocytes to proliferate and produce immunoglobulin. The infected cells are also "immortal," meaning that they are able to reproduce indefinitely in laboratory cultures. The T lymphocytes then respond actively to the infection. The abnormal-appearing lymphocytes characteristically seen in smears of the patient's blood are activated helper T cells. The proliferating lymphocytes are responsible for the large numbers of mononuclear cells that give the disease its name. Their numbers and appearance sometimes mistakenly suggest the diagnosis of leukemia, which is disproved when the patient spontaneously recovers.

In many cases, a consequence of B cell infection is the appearance of an IgM antibody that will react with an antigen on the red blood cells of certain animal species, notably

[1]Episome—any genetic element that can replicate either as integrated into the chromosome of the cell or independently in the cytoplasm; some plasmids are episomes, for example

• B lymphocytes, pp. 366–67

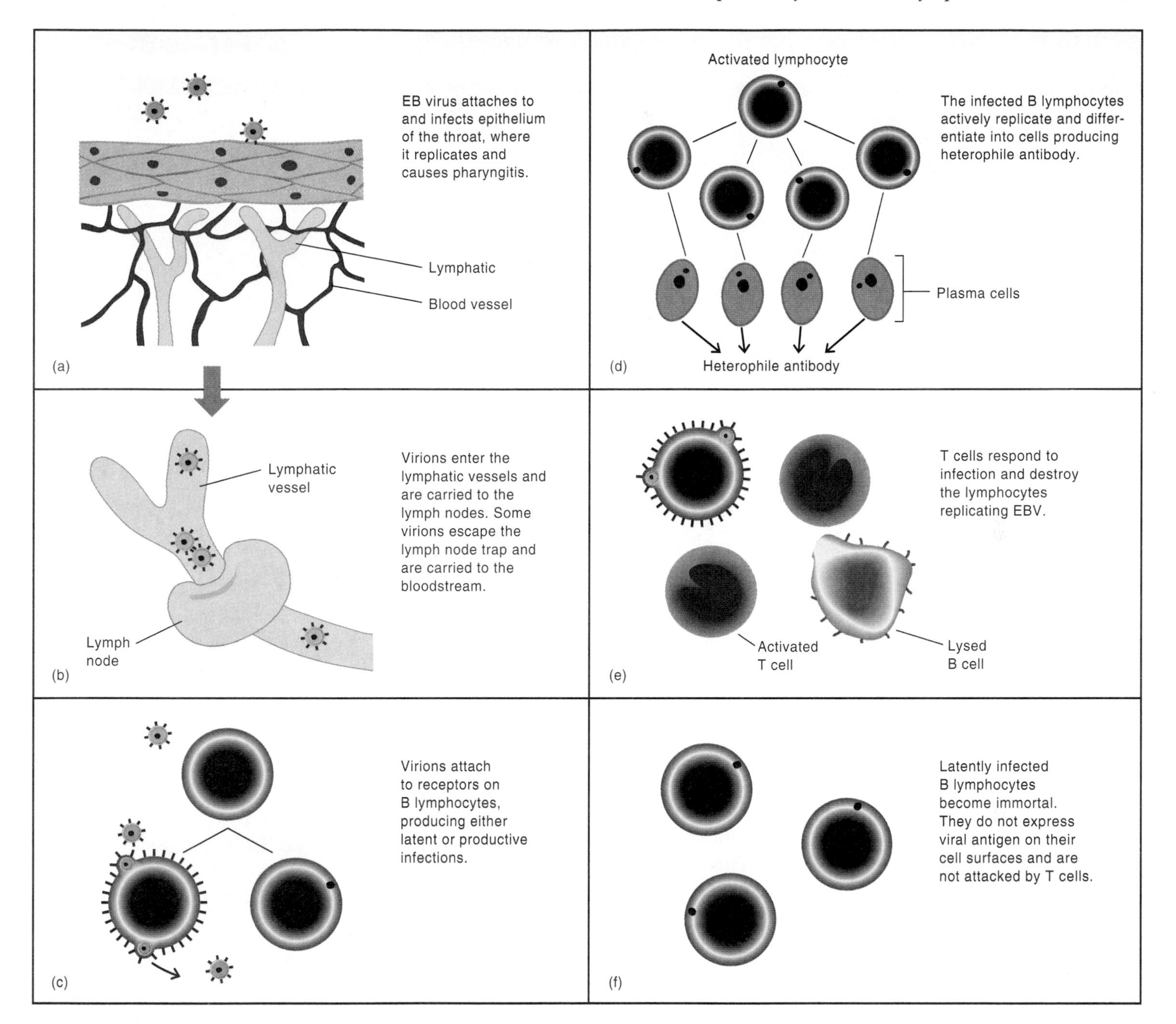

Figure 30.7
Pathogenesis of infectious mononucleosis.

sheep, horse, and ox. This kind of antibody arising against antigens of another animal is called a **heterophile antibody.** It generally has no pathologic significance, but its presence helps in diagnosing infectious mononucleosis. The antibody does not react specifically with EB virus.

The possibility that EB virus could play a role in causing certain malignant tumors has been intensely investigated. The malignancies most closely related to EB virus infection—Burkitt's lymphoma and nasopharyngeal carcinoma—cluster dramatically in certain populations. This suggests that factors other than simple EB viral infection must be present in order for these malignancies to develop. EB virus-associated Burkitt's lymphoma cases occur mainly in children in East Africa and New Guinea, whereas nasopharyngeal carcinoma is common in Southeast China. The EB virus genome is detectable in 90% to 100% of these two malignancies. Infectious mononucleosis in otherwise healthy American college students carries no increased risk of malignant tumor. However, evidence suggests that EB virus may be a factor in causing some malignancies that develop in patients with defective immunity (for example, as a result of AIDS or the use of immunosuppressive medications in organ transplants).

Perspective 30.4

Laboratory Diagnosis of Infectious Mononucleosis

Laboratory diagnosis of infectious mononucleosis usually depends on demonstrating characteristically abnormal lymphocytes on smears of the peripheral blood (figure 1) and a heterophile antibody that agglutinates sheep or horse red blood cells. More precise diagnostic techniques include demonstrating specific IgM antibody against EB virus, viral isolation, and use of nucleic acid probes.

- **IgM antibody, pp. 361–64**
- **Probes, p. 236**

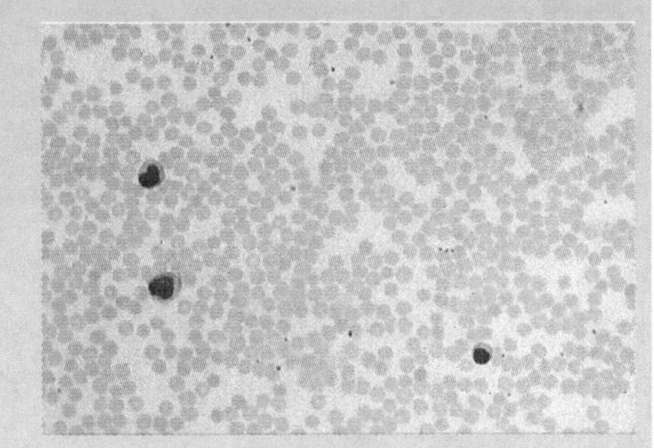

Figure 1
Blood smear in infectious mononucleosis. The two cells on the left are characteristic for this disease. The remaining cell is a normal-appearing lymphocyte.

Epidemiology

The EB virus is distributed worldwide and infects individuals in crowded, economically disadvantaged groups at an early age without producing significant illness. In such populations, the characteristic infectious mononucleosis syndrome is quite rare. More affluent populations such as students entering college in the United States often have escaped past infection with the agent and lack immunity to EB virus. Infectious mononucleosis occurs almost exclusively in adolescents and adults who lack antibody to the virus. However, even in the age group of 15 to 24 years, only about half of the EB virus infections produce infectious mononucleosis; the remainder develop few or no symptoms.

Virus can be demonstrated in the saliva of about one out of five healthy persons who have previously been infected with EB virus as shown by the presence of antibody in their blood. The salivary shedding of virus is even more common in people whose immune systems have been compromised by medication or a disease such as AIDS. Kissing is an important mode of transmission of infectious mononucleosis in young adults, giving the disease the name "kissing disease." The donor of the virus generally is asymptomatic. The virus is not highly contagious, rarely spreading within households; nevertheless, by middle age, most people have demonstrable antibody to the virus, indicating past infection.

Prevention

Prevention is aided by avoiding the use of objects such as toothbrushes or drinking glasses contaminated with another person's saliva. No vaccine for infectious mononucleosis is available.

Treatment

Most people recover within a few weeks without specific treatment. The antiviral medication acyclovir inhibits productive infection by the virus and is of some value in rare serious cases; however, it has no activity against the latent infection.

Table 30.6 gives the main features of infectious mononucleosis.

Table 30.6 Infectious Mononucleosis ("Mono")

Symptoms	Fatigue, fever, sore throat, enlargement of lymph nodes and spleen; complications are myocarditis, meningitis, hepatitis, or paralysis
Incubation period	Usually 1–2 months
Causative agent	Epstein-Barr (EB) virus
Pathogenesis	Productive infection of epithelial cells of throat and salivary ducts; latent infection of B lymphocytes; activation of B and T lymphocytes
Epidemiology	Spread by saliva; lifelong recurrent shedding of virus into saliva of asymptomatic, latently infected individuals
Prevention	Avoid sharing of articles such as toothbrushes and drinking glasses, which may be contaminated with the virus from saliva
Treatment	Usually none needed; acyclovir of possible benefit in rare cases

REVIEW QUESTIONS

1. What characteristic change occurs in the blood of patients with infectious mononucleosis?
2. What is the name of the tumor from which EB virus was first isolated?
3. What type of leukocytes does EB virus infect and immortalize?
4. What is a heterophile antibody?
5. How is infectious mononucleosis transmitted?

Yellow Fever Persists as a Zoonosis in Tropical Climates

Yellow fever was first recognized in Central America in 1648, probably introduced there from Africa. One of the worst outbreaks of yellow fever in this century occurred in Ethiopia in the 1960s, producing 100,000 cases and 30,000 deaths. In 1989, an epidemic of yellow fever occurred in Bolivia among poor people who moved into the jungle to try to make a living growing coca.

Symptoms

The symptoms of yellow fever can range from very mild to severe. Symptoms of mild disease may be only fever and a slight headache lasting a day or two. Patients suffering severe disease, however, may experience a high fever, nausea, bleeding from the nose and into the skin, "black vomit" (from gastrointestinal bleeding), and jaundice (hence, the name *yellow fever*). The mortality rate from yellow fever can reach 50% or more.

Causative Agent

Yellow fever is caused by an enveloped single-stranded, positive sense RNA arbovirus of the flavivirus family (chapter 14). The virus multiplies in species of mosquitoes, apparently without harming them, and the mosquitoes transmit the infection to humans.

Pathogenesis

The yellow fever virus is introduced into humans by the bite of an *Aedes* mosquito. It is then carried by blood and lymph to all parts of the body. Viral liver damage results in jaundice, and injury to small blood vessels produces petechiae throughout the body. The virus affects the circulatory system by directly damaging the heart muscle, by causing bleeding from blood vessel injury in various tissues, and by causing disseminated intravascular coagulation (DIC).

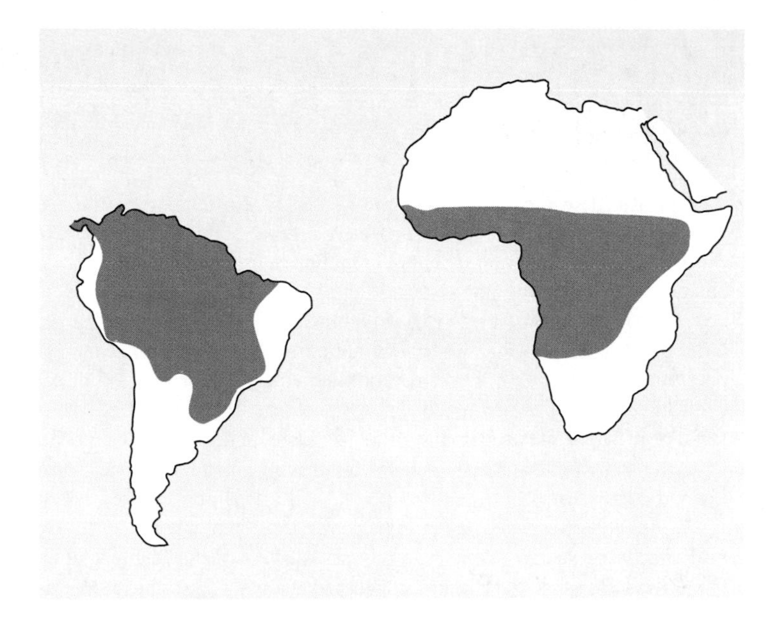

Figure 30.8
Distribution of yellow fever. Extensions may sometimes also occur into Central America and the island of Trinidad.

Epidemiology

The reservoir of the disease is mainly infected mosquitoes and primates living in the tropical jungles of Central and South America and in Africa (figure 30.8). Periodically, the disease spreads from the jungle reservoir to urban areas where it is transmitted to humans by *Aedes* mosquitoes.

Prevention

In urban areas, control of yellow fever is achieved by spraying insecticides and eliminating the breeding sites of its principal vector, *Aedes aegypti*. In the jungle, control of yellow fever is almost impossible because the mosquito vectors live in the forest canopy and transmit the disease among canopy-dwelling monkeys. A highly effective live attenuated vaccine is available to immunize people who might become exposed, including foreign travelers to the endemic areas.

Protozoan Infections

Protozoa infect the blood vascular and lymphatic systems of millions of people worldwide. Malaria is one of the most important and widespread of the protozoal diseases.

• **Mosquito, p. 434**
• **DIC, pp. 401, 678**
• **Attenuated vaccine, p. 390**
• **Protozoa, pp. 285, 290–94**

Perspective 30.5

Walter Reed and Yellow Fever

Walter Reed was born in Virginia in 1851. After receiving two medical degrees, he entered the Army Medical Corps. His superior abilities earned him an appointment as professor of bacteriology at the Army Medical College in 1893. In 1900 Reed was appointed president of the Yellow Fever Commission of the United States Army. Although improved sanitation had eliminated many other diseases in Cuba, the incidence of yellow fever remained high.

Reed suspected that an insect was the real vector of yellow fever. As early as 1881, Dr. Carlos Finlay of Havana had suggested that a mosquito might be the carrier of this disease, but he was unable to prove his hypothesis.

Beginning with mosquitoes raised from eggs, Reed and the others in his commission proceeded with a series of experiments. First, the laboratory-raised mosquitoes were allowed to feed on patients with yellow fever and, second, on members of the commission. After initial failures, one member of the commission, Dr. James Carroll, came down with classic yellow fever three days after an experimental mosquito bite. Dr. Jesse Lazear, another commission member, noted that the yellow fever patient on which the mosquito had originally fed was in his second day of the disease and that 12 days had elapsed before it bit Carroll. This timing was critical for the transmission of the disease. Later, Lazear was bitten by a mosquito while working in a hospital yellow fever ward, developed yellow fever, and died. He was the only yellow fever fatality among commission members, although other volunteers lost their lives before the mysteries of this highly fatal disease were resolved.

The commission then had a mosquito-proof testing facility constructed, called Camp Lazear in honor of their deceased colleague, to house volunteers for their experiments. Some subjects lived with the soiled clothing and bedding from yellow fever victims in mosquito-free housing. Other volunteers were subjected to the bite of mosquitoes that had bitten persons ill with yellow fever, and still others were bitten by mosquitoes that had not fed on yellow fever patients. As a result of these and other experiments, Dr. Reed and his colleagues made the following discoveries: (1) mosquitoes are the vectors of the disease; (2) an interval of about 12 days must elapse between the time the mosquito ingests the blood of an infected person and the time it can transmit the disease to an uninfected person; (3) yellow fever can be transmitted from a person acutely ill with the disease to a person who has never had the disease by injecting a small amount of the ill person's blood; (4) yellow fever is not transmitted by soiled linens, clothing, or other items that have come into contact with infected persons; and (5) yellow fever is caused by an infectious agent that passes through a filter that excludes bacteria.

Armed with these findings, Major William C. Gorgas, the chief sanitary officer for Havana, instituted mosquito control measures in Havana that resulted in a dramatic reduction in yellow fever cases. Later, Gorgas used mosquito control in the Panama Canal Zone to allow construction of the Panama Canal. Previously, French workers had tried to build the canal but had failed because of heavy losses from yellow fever and malaria.

It is an ancient scourge, as evidenced by early Chinese and Hindu writings. During the fourth century B.C., the Greeks noticed its association with exposure to swamps and began drainage projects to control the disease. The Italians gave the disease its name, *mal aria,* which means "bad air," in the 17th century. In early times, malaria ranged as far north as Siberia and as far south as Argentina. In 1902, Ronald Ross received a Nobel Prize for demonstrating the life cycle of the protozoan cause of malaria.

Malaria Is the Most Common Serious Infectious Disease Worldwide

After initial successes in eliminating the disease, the incidence of malaria rose dramatically in many areas. There are now 400 to 490 million people infected worldwide, with 2.2 to 2.5 million deaths annually.

Symptoms

The first symptoms of malaria are flulike, with fever, headache, and pain in the joints and muscles. These symptoms generally begin about two weeks after the bite of an infected mosquito, but in some cases they can begin many weeks afterward. After two or three weeks of these symptoms, the pattern changes and becomes highly suggestive of malaria. The patient abruptly feels cold and develops shaking chills that can last for as much as an hour (cold phase). Following the chills, the temperature begins to rise steeply, often reaching 40°C (104°F) or more (hot phase). After a number of hours of fever, the temperature falls, and drenching sweating occurs (wet phase). Except for fatigue, the patient feels well until 24 or 48 hours later (depending on the causative species) when the pattern of symptoms repeats.

Causative Agent

Human malaria is caused by protozoa of the genus *Plasmodium*. Four species are involved—*P. vivax, P. falciparum, P. malariae,* and *P. ovale*—and evidence for a fifth species similar to *P. vivax* has been presented. These species differ in microscopic appearance and, in some instances, life cycle, type of disease produced, severity, and treatment. In recent years, the majority of patients diagnosed in the United States have been infected with *P. vivax*, but up to

Case Presentation

The patients were two American boys, seven and nine years old, living in Thailand, who almost simultaneously developed irritated eyes and slightly runny noses, progressing to fever, headache, and severe muscle pain. It hurt them to move their eyes, and they refused to walk because of pain. The symptoms subsided after a couple of days, only to recur at lesser intensity. No treatment was given, and they were completely back to normal within a week. Their illness was diagnosed as **dengue** (pronounced DEN-gay), also known as "breakbone fever."

1. What causes dengue and where does it occur?
2. Why is dengue important?
3. What is known about the pathogenesis of dengue?
4. What can be done to prevent dengue?

Discussion

1. Dengue is caused by any of four closely related flaviviruses, dengue 1, 2, 3, and 4. The disease occurs in large areas of the tropics and subtropics around the world. In the Americas, the disease is transmitted mainly by *Aedes aegypti* mosquitoes, which have staged a comeback after being almost eradicated in the 1960s. This vector now occurs year-round along the Gulf of Mexico. Cases of dengue were diagnosed in 20 states and the District of Columbia in 1995. Most of these cases were contracted in Mexico, Central America, or the Caribbean region, but some were contracted in the southern United States.
2. The two patients presented here had a mild form of the disease that characteristically involves newcomers to an area where the disease commonly occurs. In endemic areas, the disease is characterized by fever, headache, muscle aches, rash, nausea, and vomiting. In a small percentage of cases, however, the effects of dengue infection are much more serious, resulting in **dengue hemorrhagic fever.** This form of the disease is characterized by bleeding and leakage of fluid from the capillaries. An important result is that the blood pressure drops, and the blood thickens. With expert treatment, the mortality is about 1% to 2%. **Dengue shock syndrome** is another potential development. It is characterized by profound shock and disseminated intravascular coagulation (DIC) and a mortality above 40%.
3. About 90% of the cases of hemorrhagic fever or shock occur in subjects who have previously been infected by a dengue virus and have antibody to it. The remaining cases occur largely in infants who still have transplacentally acquired maternal antibody against dengue viruses. These antibodies, whether from earlier infection or from the mother, attach to the infecting dengue virus strain and thereby promote its uptake by macrophages. The virus, instead of being killed, reproduces in the macrophage. This results in the death of the macrophage and release of chemicals that cause leaky capillaries and shock. T lymphocytes may also play a role in this process in older children and adults. Evidence also exists that the virus causes a depression of the bone marrow, accounting for the very low white blood cell and platelet counts seen in dengue.
4. Scientists at Mahidol University in Bangkok and other research centers around the world are working on vaccines designed to bring dengue under control. At present, control efforts are largely directed at killing mosquitoes and their larvae and eliminating water containers around houses where the mosquitoes breed. Scientists at Colorado State University have successfully genetically engineered mosquitoes to render their cells incapable of reproducing dengue virus. This was accomplished by using another virus to introduce an antisense segment of the dengue genome into the mosquito cells. This was a dramatic accomplishment, but years of work remain before it can play a role in the control of dengue.

• **Antisense technology, p. 141**

30% have been infected with the more dangerous species *P. falciparum*.

The *Plasmodium* life cycle is complex (figure 30.9). The parasite grows and divides in the erythrocytes of the host (figure 30.10). After a time, the infected erythrocytes break open, and the offspring of the division, called **merozoites,** are released into the plasma. Most of the merozoites then enter new erythrocytes and multiply, repeating the cycle. Some of them, however, develop into **gametocytes,** which are specialized sexual forms different from the other circulating plasmodia in both their appearance and susceptibility to antimalarial medicines. These sexual forms do not rupture the red blood cells. They cannot develop further in the human host and are not important in causing the symptoms of malaria. They are, however, infectious for certain species of *Anopheles* mosquitoes and are thus ultimately responsible for the transmission of malaria from one person to another.

Shortly after entering the intestine of the mosquito and stimulated by the drop in temperature, the male gametocyte produces tiny, whiplike bodies that unite with the female gametocyte in much the same way as the sperm and ovum unite in higher animals. The resulting **zygote** burrows into the wall of the midgut of the mosquito and forms a **cyst,** which enlarges as the zygote undergoes meiosis, dividing asexually into numerous offspring. The cyst then ruptures into the body cavity of the mosquito, and the released parasites, called **sporozoites,** find their way to the mosquito's salivary glands and saliva, from which they may be injected into a new human host.

In the human host, the sporozoites are carried by the bloodstream to the liver, where they infect liver cells. In these cells, each parasite enlarges and subdivides, producing thousands of daughter cells. The daughter cells are then released into the bloodstream and establish the cycle involving the erythrocytes.

Pathogenesis

The characteristic feature of malaria, recurrent bouts of fever followed by feeling healthy again, results from the erythrocytic cycle of growth and release of merozoites. Interestingly, the infections in all the millions of different red blood cells become nearly synchronous. Thus, cell rupture and release of daughter protozoa occur at roughly the same time for all infected cells, and each release causes a fever. For *P. malariae,* the growth cycle takes 72 hours, so that fever recurs every third day. For the other species, fevers generally occur every other day. Infections by *P. falciparum* tend to be very severe, probably because all erythrocytes are susceptible to infection, whereas other *Plasmodium* species infect only young or old erythrocytes. Thus, very high levels of parasitemia (parasites in the bloodstream) can develop with *P. falciparum* infections. The infected red blood cells become rigid, in contrast to normal red cells, which are flexible. Also, they adhere to each other and to the walls of capillaries. These tiny blood vessels therefore become plugged, and the affected tissue becomes deprived of oxygen as a result. Involvement of the brain, or **cerebral malaria,** is particularly devastating, but almost any organ can be severely affected. *P. vivax* and *P. ovale* malaria often relapse after treatment of the blood infection because treatment-resistant forms of the organisms continue to infect the liver **(exoerythrocytic cycle)** and from there can initiate new erythrocytic cycles of infection. The spleen, which is a kind of filter for the blood, characteristically enlarges in malaria to cope with the large amount of foreign material and abnormal red blood cells, which it removes from the circulation. The parasites can cause anemia by destroying red blood cells, but the spleen also contributes to it by removing malaria-damaged cells. The large amount of foreign material in the bloodstream strongly stimulates the immune system. In some cases, the overworked immune system fails and immunoincompetence results. Malaria-induced immunoincompetence is thought by some to be a factor in the development of Burkitt's lymphoma following EB virus infection since the tumors tend to occur in areas where malaria is endemic.

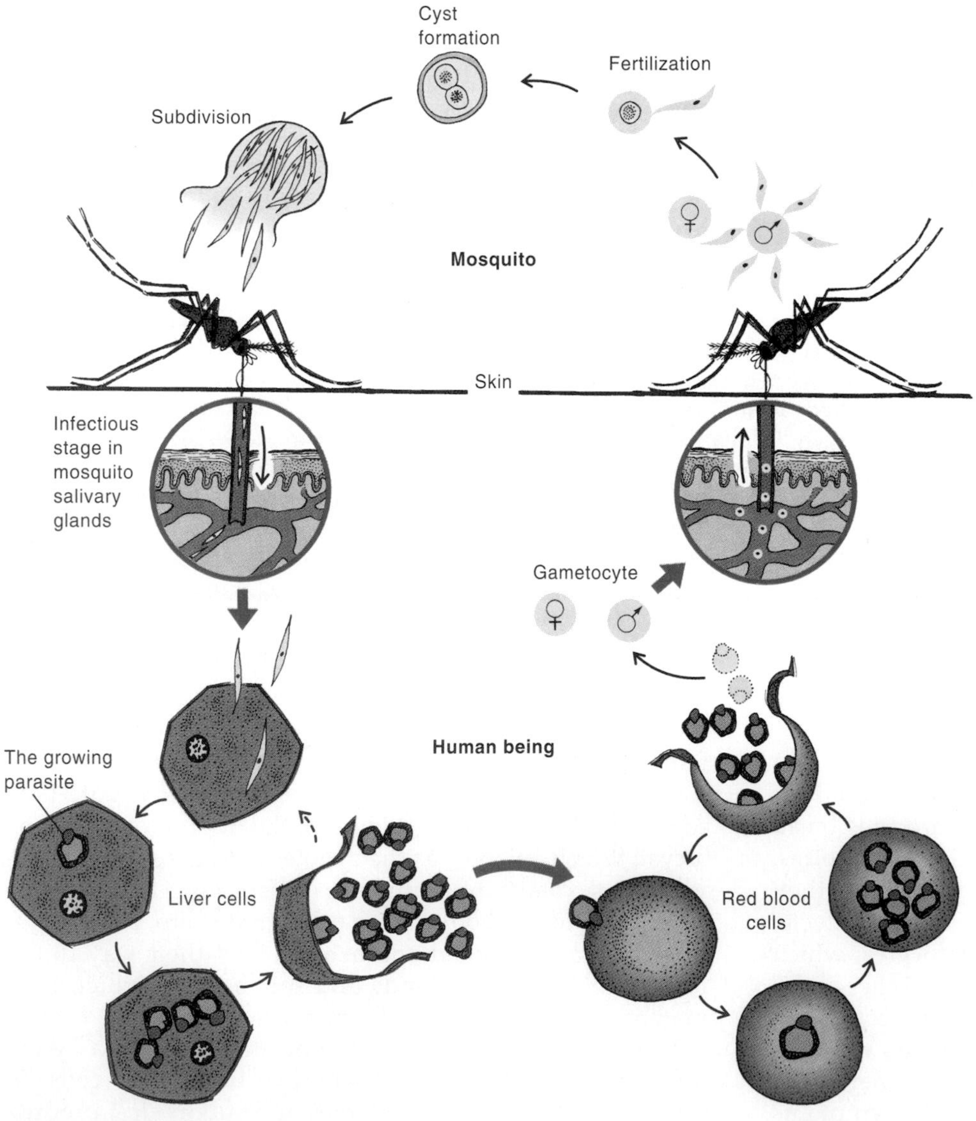

Figure 30.9
Life cycle of *Plasmodium vivax.*

Epidemiology

Malaria was once common in both temperate and tropical areas of the world, and endemic malaria was only eliminated from the continental United States in the late 1940s. Today, malaria is predominantly a disease of warm climates (figure 30.11), but almost half the world's population lives in endemic areas. Certain mosquito species of the genus *Anopheles* are vectors of human malaria. Since suitable vectors are abundant in North America, the potential exists for the spread of malaria whenever it is introduced. Malaria can be transmitted by, besides mosquitoes, blood transfusions or the sharing of syringes among drug users. Malaria contracted in this manner is easier to treat since it involves only red blood cells and not the liver. Only sporozoites from mosquitoes can infect the liver. Some

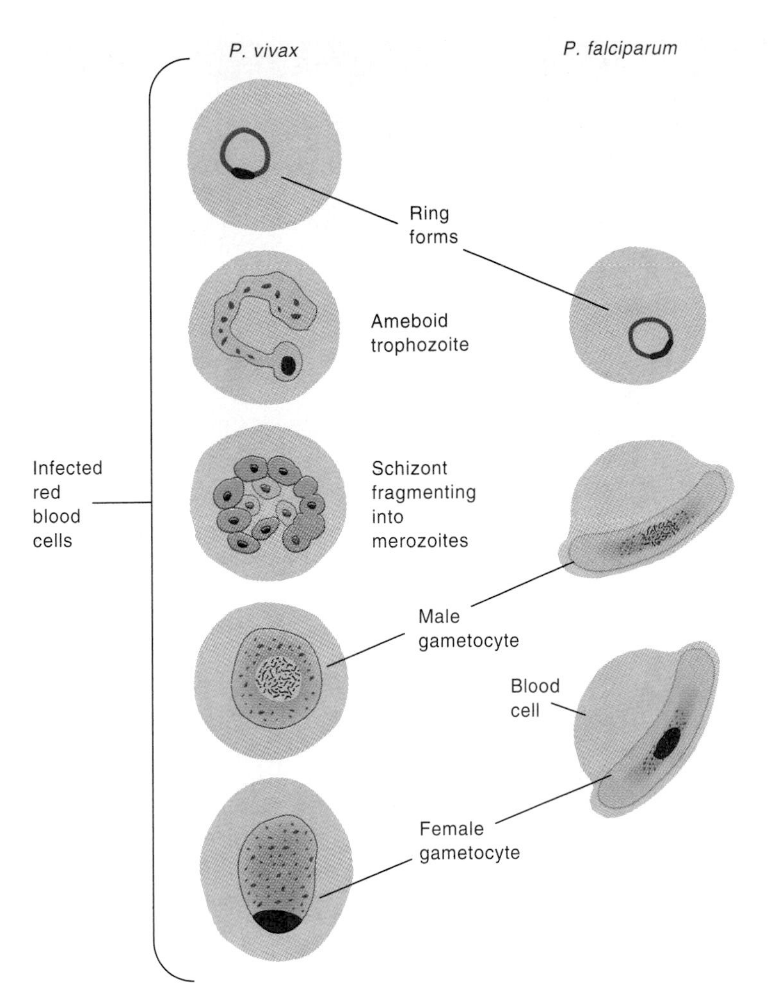

Figure 30.10
Developmental forms of *Plasmodium vivax* and *P. falciparum* in red blood cells.

black people are genetically resistant to *P. vivax* malaria because their red blood cells lack the receptors for the parasite's ligands.

Prevention and Control

Travelers to malarious areas can generally prevent symptomatic malaria but not necessarily infection by the plasmodia, with weekly doses of chloroquine. Using insect repellants and mosquito netting impregnated with insecticide, avoiding the outdoors from dusk to dawn, and wearing protective clothing all help avoid infection. After leaving malarious areas, people take primaquine to eliminate any possible exoerythrocytic infection.

• **Receptors, ligands, pp. 416–17**

Areas of the world with resistant strains of *P. falciparum* present an increasingly vexing problem since the only effective preventive medications can cause serious, even fatal reactions in some people. Travelers to such areas should consult with a health officer familiar with local conditions before deciding on a preventive regimen.

In 1955, the World Health Organization began a program for the worldwide elimination of malaria. Initially there was great success, as WHO employed insecticides against the mosquito vector, detected infected patients by obtaining blood smears, and provided treatment of infected cases. Fifty-two nations undertook control programs and, by 1960, 10 of them had eradicated the disease. Sri Lanka, with about 3 million cases in the 1940s after World War II, had only 29 cases in 1964. Unfortunately, strains of *Anopheles* mosquitoes resistant to insecticides began to appear, and in cooperation with bureaucracy and complacency, malaria began a rapid resurgence. By 1975, Sri Lanka had more than 400,000 cases, and figures from other tropical countries were equally discouraging. In 1976, the World Health Organization acknowledged that the eradication program was a failure.

Unfortunately, the malaria problem is still unsolved and is likely to worsen. The world population is presently 5.7 billion, with 250 new births occurring every minute. Moreover, various computer models project the climate to warm significantly due to increasing atmospheric CO_2, resulting in an increase in the areas where malaria is likely to occur. One estimate gives an increase in malarious areas of about 25% 50 years from now. The combination of increasing population, increasing areas where malaria can be easily transmitted, and the development of medication-resistant plasmodia and insecticide-resistant mosquitoes underscores the need for an effective antimalarial vaccine.

Scientists have been trying to perfect a vaccine against malaria for many years. The first major breakthrough was reported in 1976 from the laboratory of Dr. William Trager at the Rockefeller University. Trager described a method for the continuous in vitro cultivation of *P. falciparum*, allowing for the production of antigens from different stages of the organism's life cycle. Later on, other scientists using genetic engineering techniques cloned the antigens and identified which ones were important in immunity to malaria. The most promising antigens were attached to other proteins (such as tetanus toxoid) to make them more immunogenic and then tested for their ability to give an immune response when injected into experimental animals. Finally, trials were undertaken in human beings to see whether the vaccine would protect them from malaria. Results so far have been encouraging for a vaccine against the plasmodial sporozoites, although only low titers of antibody have been produced and an effective cell-mediated response is lacking. Estimated protection in one trial in African children was about 30%.

Perspective 30.6

Transplacental Transmission of Malaria by an Asymptomatic Mother

Malaria can be transmitted through the placenta to the fetus, resulting in a spontaneous abortion or a child born with active malaria. A child with malaria was born to Kampuchean parents who arrived in the United States two months before his birth. After two days of fever and vomiting, the child was admitted to the hospital, where blood smears revealed *Plasmodium vivax*. Treatment was effective, and he was then sent home. Many people (like this child's mother) recover from an acute attack of malaria and live with few or no symptoms despite the presence of parasites in their blood. These parasites can often be detected by examining blood smears. Although there were no symptoms of malaria during the mother's pregnancy, sufficient organisms were present in her blood to infect the unborn child.

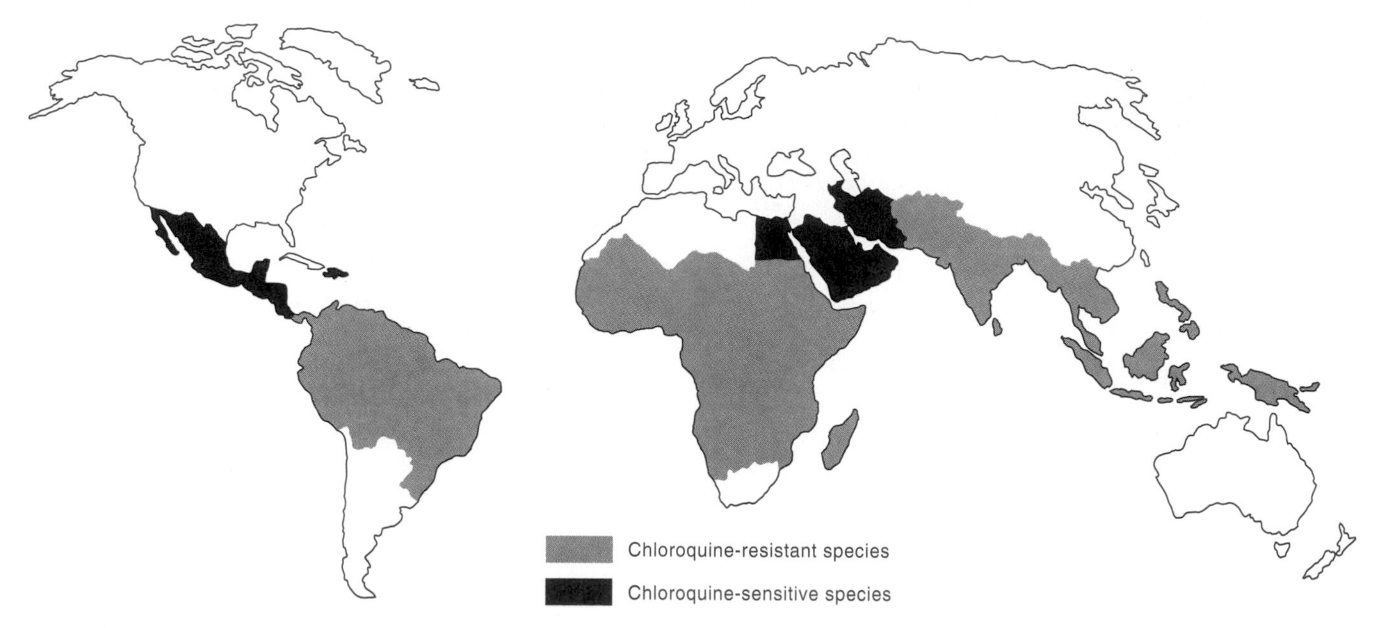

Figure 30.11
Distribution of malaria in 1990. The malaria range is expected to expand greatly with global warming.

Source: Centers for Disease Control and Prevention. 1990. *Morbidity and Mortality Weekly Report* 39(RR-3):8.

Present hopes for malaria control depend on vaccines currently under development and evaluation. Current strategy for containment of the disease focuses on improving access of populations to treatment facilities. Destruction of mosquito habitat, along with earlier diagnosis, faster treatment, and earlier detection of epidemics should help contain malaria, presently responsible for 20% to 30% of the deaths of African children under five years of age.

Treatment

Treatment of malaria is complicated by the fact that different stages in the life cycle of the parasite respond to different medications. Chloroquine, for example, is generally effective against the erythrocytic stages, although it will not cure the liver infection with *P. vivax* or kill the gametocytes of *P. falciparum*. Primaquine is generally effective against the exoerythrocytic stage and the *P. falciparum* gametocytes. Unfortunately, strains of chloroquine-resistant *P. falciparum* are now common in many parts of the world, and resistant strains of *P. vivax* have appeared in some areas. Some patients infected with chloroquine-resistant strains respond to mefloquine, while others require treatment with intravenous quinine or quinidine together with an oral medication such as tetracycline or a sulfa drug.

Table 30.7 gives the main features of malaria.

Table 30.7 Malaria

Symptoms	Recurrent bouts of violent chills and fever alternating with feeling healthy
Incubation period	Varies with species; 6–37 days
Causative agent	Four species of protozoa of the genus *Plasmodium*
Laboratory diagnosis	Blood smears, serologic tests
Pathogenesis	Cell rupture and release of protozoa cause fever; infected red blood cells adhere to each other and to walls of capillaries; vessels become plugged and tissue is then deprived of oxygen; spleen enlargement in response to its removing large amount of foreign material and abnormal blood cells from the circulation
Epidemiology	Transmitted from person to person by bite of infected anopheline mosquito
Prevention	Weekly doses of chloroquine while in malarial areas; after leaving, primaquine is given; other medicines for resistant strains; eradication of mosquito vectors; vaccine under development
Treatment	Dependent upon species involved; medicines used include one or more of chloroquine, primaquine, quinine, mefloquine, tetracycline, and sulfa drugs

REVIEW QUESTIONS

1. What gives yellow fever its name?
2. How is yellow fever transmitted?
3. Name a viral disease that can be complicated by disseminated intravascular coagulation.
4. Travelers to and from which areas of the world should have certificates of yellow fever vaccination?
5. Which microorganism causes the most dangerous form of malaria?
6. Which form of the malaria parasite, present in the saliva of infected mosquitoes, infects the human host's liver?
7. Why is it simpler to eradicate a malaria infection contracted from a blood transfusion than one contracted from a mosquito bite?
8. Why is malaria increasing despite many years of attempted eradication?
9. What is the current strategy for containing malaria?

Summary

I. Anatomy and Physiology
   A. The blood vascular system includes the heart, arteries, veins, and capillaries.
   B. The lymphatic system consists of the lymphatic vessels and lymph nodes. The blind-ended lymph vessels collect extracellular fluid and carry it back to the bloodstream.

II. Normal Flora

III. Bacterial Diseases of the Blood Vascular System
   A. Subacute bacterial endocarditis (SBE) involves the valves and lining of the heart chambers and is commonly caused by oral streptococci or *Staphylococcus epidermidis*. Infection usually begins in structurally abnormal heart chambers on valves previously damaged by disease or on congenital abnormalities of the valves or chambers.
   B. Acute bacterial endocarditis is usually caused when virulent bacteria such as *Staphylococcus aureus* or *Streptococcus pneumoniae* enter the bloodstream from a focus of infection elsewhere in the body or when contaminated material is injected by drug abusers. The normal heart is commonly infected.
   C. Septicemia usually results from bacteremia. It is often caused by gram-negative organisms, mainly enterobacteria and anaerobes of the genus *Bacteroides*.

IV. Bacterial Diseases Involving the Lymph Nodes and Spleen
   A. The mononuclear phagocyte system is involved in these diseases.
   B. Tularemia is often transmitted from wild animals to humans by insects and ticks. The cause is the gram-negative aerobe *Francisella tularensis*.
   C. Brucellosis is usually acquired from cattle or other domestic animals and is caused by species of *Brucella*. Pasteurization of milk is an important control measure.
   D. Plague, once pandemic, is now endemic in rodent populations. It is caused by *Yersinia pestis*, which is transmitted to humans by fleas or persons suffering from pneumonic plague.

V. Viral Diseases of the Lymphoid System
   A. Infectious mononucleosis is a viral disease principally of B lymphocytes. The cause is the Epstein-Barr virus, which establishes a lifelong latent infection of

lymphocytes, making them "immortal." The virus may be partly responsible for Burkitt's lymphoma and nasopharyngeal carcinoma.

B. Yellow fever is caused by an arbovirus and is characterized by fever, jaundice, and hemorrhage.

VI. Protozoan Infections

Malaria is caused by four species of *Plasmodium* and is transmitted by the bite of the *Anopheles* mosquito.

Chapter Review Questions

1. Where do the organisms that cause subacute bacterial endocarditis originate?
2. Give two examples of enterobacteria that can cause septicemia.
3. What is disseminated intravascular coagulation (DIC)?
4. What activity of humans is likely to expose them to tularemia?
5. In the microbiological attempt to diagnose undulant fever, in what atmosphere should the medium be incubated?
6. Describe three virulence factors of *Yersinia pestis* that are coded by plasmids.
7. Why might rodent burrows be a source of plague months after they are abandoned?
8. Why is brucellosis a threat to big game hunters?
9. What is the most frequent serious complication of brucellosis?
10. Why might the *Yersinia pestis* bacteria from a patient with pneumonic plague be more dangerous than the ones from fleas?

Critical Thinking Questions

1. Why do dentists prescribe an antibiotic for patients with damaged heart valves such a short time before performing dental work?
2. Why is the mortality from gram-negative septicemia so high even though effective medicines are available to kill the causative bacteria?
3. What risk is there in playing with your dog after it has sniffed a dead prairie dog?
4. Distinguish between EB virus infection and infectious mononucleosis.
5. Why was timing critical in the transmission of yellow fever from a patient to Dr. James Carroll?

Further Reading

Abraham, E., et al. 1995. Efficacy and safety of monoclonal antibody to human tumor necrosis factor alpha in patients with sepsis syndrome. A randomized, controlled, double-blind, multicenter clinical trial. *Journal of the American Medical Assn.* 273:934–941.

Bone, R. C., et al. 1995. A second large controlled clinical study of E5, a monoclonal antibody to endotoxin: results of a prospective, multicenter, randomized, controlled trial. *Critical Care Med.* 23:994–1006.

1994. Brucellosis at a pork processing plant—North Carolina, 1992. *Morbidity and Mortality Weekly Report* 43(7):113–116.

Harrison, G. 1978. *Mosquitoes, malaria and man: a history of the hostilities since 1880.* E. P. Dutton, New York. Fascinating historical account of the steps in unraveling the mysteries of malaria.

1994. Human Plague—United States, 1993–1994. *Morbidity and Mortality Weekly Report* 43(13):242–246.

Local Transmission of *Plasmodium vivax* malaria—Houston, Texas, 1994. *Morbidity and Mortality Weekly Report* 44(15):295.

McEvedy, Colin. 1988. The bubonic plague. *Scientific American* 258:118.

1996. Mosquito transmitted malaria—Michigan, 1995. *Morbidity and Mortality Weekly Report* 45(19):398–400.

1992. Pneumonic plague—Arizona, 1992. *Morbidity and Mortality Weekly Report* 41:737. Describes the death of a 31-year-old man from pneumonic plague.

Sixbey, J. W., et al. 1994. Epstein-Barr virus replication in oropharyngeal epithelial cells. *New England Journal of Medicine* 310:1225.

Teutsch, S. M., et al. 1979. Pneumonic tularemia on Martha's Vineyard. *New England Journal of Medicine* 301:826.

1994. Update: human plague—India, 1994. *Morbidity and Mortality Weekly Report* 43:761–762.

1994. White, N. J. Tough test for malaria vaccine. *Lancet* 344:1172–1174. A disappointing trial of a synthetic polypeptide vaccine.

31 HIV Disease and Complications of Immunodeficiency

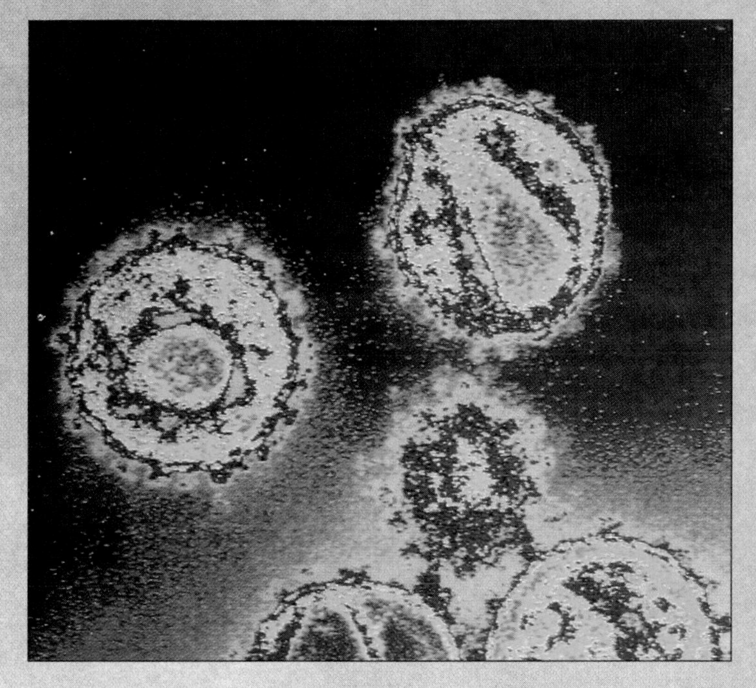

Color-enhanced transmission electron micrograph of the human immunodeficiency virus (HIV), the enveloped, single-stranded RNA virus that causes AIDS. The cone-shaped appearance of the nucleocapsid, as seen in the upper right virion, is characteristic.

KEY CONCEPTS

1. The signs and symptoms of persons with AIDS are mainly due to the opportunistic infections and tumors that complicate HIV disease.
2. AIDS is not highly contagious.
3. The AIDS pandemic could be stopped by changes in human behavior.

Microbes in Motion

The following books in the *Microbes in Motion* CD-ROM may serve as a useful preview or supplement to your reading:

Viral Structure and Function: Viral Diseases; Viral Replication; Viral Invasion (Adsorption, Penetration). *Fungal Structure and Function:* Diseases.

A Glimpse of History

It was disturbing news when, in 1981, five published reports described an illness appearing in previously healthy young homosexual men, characterized by unusual opportunistic infections, certain malignant tumors, and immunodeficiency. The constellation of symptoms and signs associated with this illness came to be known as the acquired immunodeficiency syndrome (AIDS). Initially, there was wild speculation about what might be the cause of AIDS, but by 1982, the Centers for Disease Control had convincing epidemiological evidence that AIDS was caused by a previously unknown infectious agent. By 1984, three laboratories had independently recovered and partly characterized the causative virus.

These accomplishments occurred in a remarkably short period of time due to the skill and effort of scientists in different parts of the world. They were able to build on some important scientific advances of the 1960s and 1970s—specifically, the discovery of subsets of lymphocytes along with the functions of T cells and the role of lymphokines. In 1975, the Nobel Prize was awarded to R. Dulbecco, H. Temin, and D. Baltimore for discovering reverse transcriptase and its function in making a DNA copy of retrovirus genomes. Moreover, in early 1978, Dr. Robert Gallo's laboratory at the National Cancer Institute discovered the first human retrovirus, human T lymphotrophic virus (HTLV). Dr. Gallo developed a technique for cultivating the retroviruses in normal lymphocytes activated with IL-2. Viral growth in the cultures was detected by an assay for reverse transcriptase and by electron microscopy.

There were troubling problems facing the scientific community when AIDS was recognized in 1981. In that year, Ronald Reagan, a popular former movie star, began his first term as president of the United States, elected with a mandate to reduce taxes and expand defense spending. This posed a threat to funding for science and sparked intensified competition for research dollars among different groups. Some scientists became more reluctant to exchange information, reagents, and cultures, especially when their work had commercially valuable potential. They raced to patent discoveries ahead of competing laboratories because the winner could receive royalties from commercial exploitation of his or her work.

At first, the federal government failed to recognize the seriousness of the AIDS epidemic and had difficulty formulating a unified policy on how to deal with it. There was little political support for research on a disease acquired by homosexual men through anal intercourse and by drug addicts through the sharing of needles. Politicians feared offending their constituencies by appearing to support gay men and intravenous drug users. Others in the political spectrum urged mandatory testing for AIDS and quarantining those with evidence of infection. Conflicts that arose over which government agency would coordinate AIDS research were only partly resolved when the National Institute of Allergy and Infectious Disease (NIAID) was made the lead agency. Important research on sexual behavior was not funded until the early 1990s.

In January 1983, an important scientific breakthrough occurred in the laboratory of Dr. Luc Montagnier at the famous Pasteur Institute. Using techniques developed earlier by Gallo for cultivating HTLV, Montagnier recovered a new virus from a patient with lymphadenopathy syndrome, an AIDS-associated condition. Because the IL-2 activated lymphocyte cultures died out quickly, he was only able to obtain small quantities of the new virus (named LAV) but enough to develop a blood test that showed that AIDS cases were infected with the virus. He sent a sample of this virus to Gallo at the National Cancer Institute, and the two exchanged material from other patients. He also made application to the U.S. Patent Office for a patent on his blood test.

Meanwhile, Gallo's laboratory began recovering a virus from AIDS patients and reported the finding in the same issue of *Science* in which Montagnier reported his finding. A number of these viral isolates were introduced together into continuous cell cultures (as opposed to IL-2 activated normal lymphocytes) to see if a strain of the virus could replicate in the cells. One did replicate well, enabling Gallo to obtain large quantities of the virus. Gallo named the virus HTLV-III because of a resemblance to the two human T lymphotrophic viruses discovered earlier. Using HTLV-III, Gallo perfected a blood test for AIDS and, like Montagnier, applied for a patent. These conflicting patent claims caused a bitter scientific and legal battle that raged for years, the subject of charges and counter charges and investigations by scientific and congressional committees costing hundreds of thousands of dollars.

Under pressure from President Reagan, a settlement was negotiated in 1987 wherein Gallo and Montagnier were named co-discoverers of the test. Eighty % of the royalties were to go to an AIDS foundation, the remainder to be divided equally between the National Institutes of Health (NIH) and the Pasteur Institute. The settlement did not end the conflict, and genetic analysis showed that HTLV-III was actually LAV, apparently introduced into Gallo's cultures in a laboratory mixup. Later, the Pasteur Institute suffered its own embarrassment, acknowledging that due to a laboratory mixup of its own LAV actually came from a different patient than stated in its publications. In 1994, the earlier settlement was modified, NIH admitting that HTLV-III and LAV were the same and giving the French side a larger share of the royalties. Even so, by the end of 1995, Gallo's share of the royalties had generated $40 million for the NIH over the years.

The value of the blood tests far exceeded any monetary figure. They helped prove beyond reasonable doubt that HTLV-III/LAV (renamed HIV) causes AIDS. They also showed that many asymptomatic people are infected with the virus and can transmit it and that the epidemic is far more extensive than previously suspected. Thus, they helped spur public support for controlling the disease so that, despite the legal and political turmoil, the federal government spent over $900 million to control AIDS in just the first four years following discovery of the causative virus and much larger amounts thereafter. Finally, by 1985, the blood test was generally available for routine testing of donated blood, thus markedly improving the safety of blood transfusion and products prepared from pooled blood.

People are commonly exposed to microorganisms such as *Pneumocystis carinii* that are of such low virulence that they can only cause disease in individuals with immunodeficiency disorders. Therefore, the appearance of a disease due to one of these opportunistic microorganisms strongly suggests that the victim is immunodeficient. In the United States in 1981, a number of cases of *P. carinii* pneumonia in previously healthy homosexual men led to the recognition of the disease now generally known as AIDS (acquired immunodeficiency syndrome). In the decade subsequent to its recognition, the disease killed more United States citizens than the Korean and Vietnam wars combined. By 1994, in the United States it had become the leading cause of death among those 25 to 44 years of age. Currently, worldwide, it is estimated that a new infection occurs every 13 seconds, and a person dies of AIDS every nine minutes. This chapter will cover AIDS and some of the opportunistic infections and malignant tumors that complicate the disease.

AIDS

AIDS is arguably the worst pandemic the world has seen in the last half of the 20th century. Without question, it has been studied more intensely than any other disease in history with techniques at the very cutting edge of science. Although the cause of AIDS is known, many details of its pathogenesis are still not understood. New hypotheses continually appear as older ones are disproven. The following sections summarize current knowledge of the disease.

• **Immunodeficiency disorders, pp. 396, 404–5**

AIDS Is a Late Manifestation of Human Immunodeficiency Virus Infection

The acronym AIDS, for acquired immunodeficiency syndrome, was first used in 1982 by the Centers for Disease Control for diseases that were at least moderately predictive of a defect in cell-mediated immunity in subjects with no apparent cause for low resistance to the disease. These AIDS-defining diseases (table 31.1) were useful in studying the AIDS epidemic, especially before the cause was known. It is now known that immunodeficiency is the end stage of a disease with many other manifestations. AIDS is therefore more appropriately called *HIV disease,* after its causative agent, human immunodeficiency virus (HIV).

Symptoms

The first symptoms of HIV disease appear after an incubation period of six days to six weeks and usually consist of fever, headache, sore throat, muscle aches, enlarged lymph nodes, and a generalized rash. Some subjects develop central nervous system symptoms ranging from moodiness and confusion to seizures and paralysis. These acute symptoms

Table 31.1 AIDS-Defining Conditions Employed for Surveillance*
Candidiasis involving the esophagus, trachea, bronchi, or lungs
Cancer of the uterine cervix, invasive
Coccidioidomycosis, of tissues other than the lung
Cryptococcosis, of tissues other than the lung
Cryptosporidiosis of duration greater than one month
Cytomegalovirus disease of the retina with vision loss or other involvement outside liver, spleen, or lymph nodes
Encephalopathy (brain involvement with HIV)
Herpes simplex virus causing ulcerations lasting one month or longer or involving the esophagus, bronchi, or lungs
Histoplasmosis of tissues other than the lung
Isosporiasis (a protozoan disease of the intestine) of more than one month's duration
Kaposi's sarcoma
Lymphomas, such as Burkitt's, or arising in the brain
Mycobacterial diseases, including tuberculosis
Pneumocystosis (pneumonia due to *Pneumocystis carinii*)
Pneumonias occurring repeatedly
Progressive multifocal leukoencephalopathy (a brain disease caused by the JC polyomavirus)
Salmonella infection of the bloodstream, recurrent
Toxoplasmosis of the brain
Wasting syndrome (weight loss of more than 10% due to HIV); also known as *slim disease*

*Additional conditions used in surveillance of childhood AIDS

Perspective 31.1

Origin of AIDS Viruses

Where did the AIDS viruses come from? Genetic evidence indicates that the HIV-1 virus mutated to its present form fairly recently, between 50 and 150 years ago. Although it first appeared in the United States in the 1970s, serological evidence indicates that it was present in Africa in rare individuals in the 1950s.

Viruses similar to the AIDS viruses exist in a number of wild and domestic animals including cats, milk cows, and monkeys. A virus closely related to HIV-1 has been found in several captive chimpanzees, and another virus, closely related to HIV-2, is present in a kind of large monkey called a *sooty mangabey*. One popular theory is that the AIDS viruses "jumped" to humans from these simian relatives, presumably through contact with their blood. Indeed, in Africa, chimps and mangabeys have sometimes been killed for food, exposing humans to their blood. Also, a number of humans were intentionally injected with chimpanzee and mangabey blood in the course of research on malaria carried out between 1922 and 1955.

As plausible as this idea is for the origin of human AIDS, there is no compelling scientific evidence that human HIV arose from these animals. There is also little evidence for other popular theories including one that the viruses were transferred to humans from monkeys by inadequately sterilized Salk polio vaccine grown in monkey kidney tissue cell cultures or that the viruses resulted from botched biological warfare experiments by Russia or the United States.

Leading AIDS researchers believe it likely that the AIDS viruses may have existed in people living in isolated Central African villages perhaps for centuries. With population increases and migration to big, crowded cities, the viruses began to spread rapidly, becoming more virulent in the process.

The answer to the question about the origin of AIDS may never be known, but the question continues to be intriguing.

Table 31.2 Examples of Agents Active Against HIV

Most commercially available high-level disinfectants
Many detergents such as Triton X and Nonidet P-40
Household bleach (5% sodium hypochlorite) diluted 1:10
Alcohol 70%–90%
Heat at 56°C, 30 minutes (blood serum, aqueous solutions)

typically subside within six weeks. However, most HIV infections are asymptomatic, or the symptoms are mild and attributed to the "flu." Following the acute illness (if any), there is an asymptomatic period that typically lasts for years even though the disease advances in the infected person and can be transmitted to others. The asymptomatic period may end with persistent enlargement of the person's lymph nodes, a condition known as *lymphadenopathy syndrome* (*LAV*). Other symptoms heralding immunodeficiency include fever, weight loss, tiredness, and diarrhea, referred to as the *AIDS-related complex* (*ARC*). The first symptoms in many cases are those due to tumors or opportunistic infections resulting from severe immunodeficiency. As one might expect, symptoms at this stage of the disease vary widely according to the kind of infection. For example, a frequently encountered symptom is a fuzzy white patch on the tongue, **hairy leukoplakia,** a result of latent EB virus reactivation. Severe skin rashes, cough and chest pain, stiff neck, confusion, blindness, and diarrhea are manifestations of other infections.

Causative Agent

In the United States, and most other parts of the world, AIDS is generally caused by human immunodeficiency virus type 1 (HIV-1), a single-stranded RNA virus of the retrovirus family. There are many kinds of retroviruses, naturally infecting hosts as diverse as fish and humans. HIV-1 belongs to the lentivirus subgroup of the retroviruses, which characteristically infect mononuclear phagocytes. Outside the host, HIV-1 is easily inactivated by various disinfectants and modest amounts of heat (table 31.2). However, virus present in dried blood or pus may be more difficult to inactivate. Infectious HIV-1 persists in samples of blood plasma for at least one week after they are taken from AIDS patients.

HIV-1 viruses can be classified into subtypes based on nucleic acid sequences of certain genes. Nine subtypes have been described, designated A through I. The different subtypes vary in their geographic distribution. For example, subtype B dominates in North and South America and Western Europe, whereas subtype E dominates in Thailand. These subtypes may also vary in their ability to infect by different routes. Infection by two different subtypes can give rise to "hybrids" having properties of both.

Although other viruses cause AIDS, they will not be considered further. For the remainder of this chapter, we will use the term *HIV* to indicate HIV-1. More than a hundred cases of acquired immunodeficiency syndrome have been described that were not caused by any of the known retroviruses. These cases are called ICL (Idiopathic CD4+ T lymphopenia). ICL has multiple causes and does not represent a new epidemic threat.

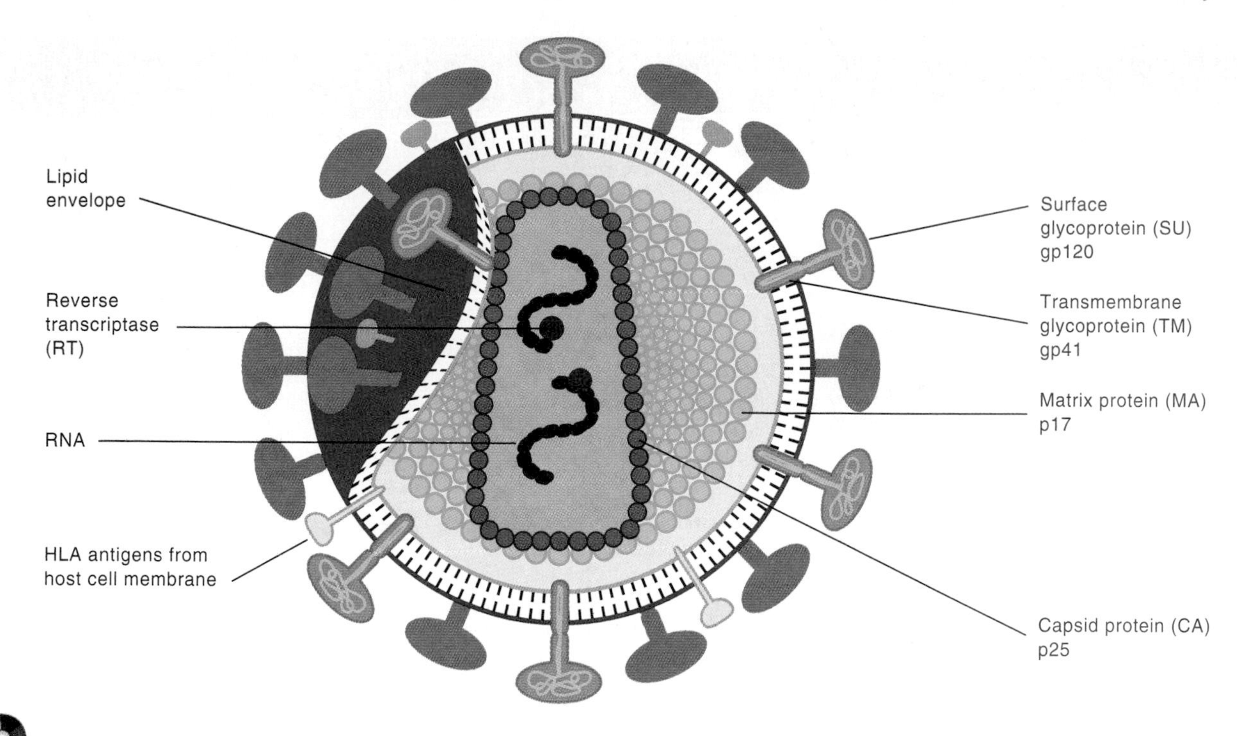

Figure 31.1
Diagrammatic representation of human immunodeficiency virus type 1 (HIV-1). The integrase and protease enzymes are not shown.

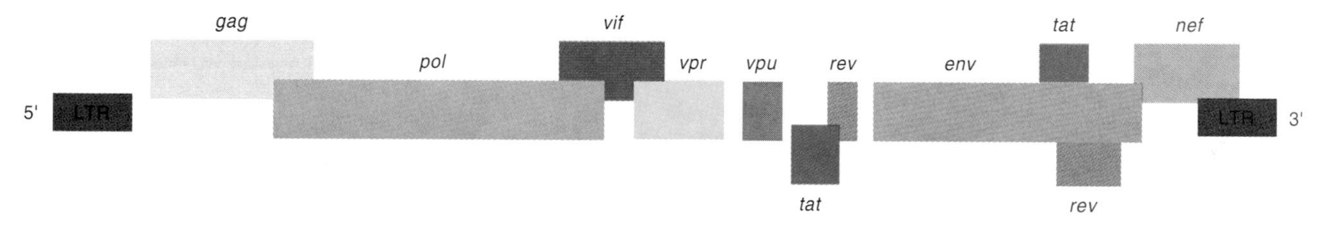

Figure 31.2
Map of the human immunodeficiency virus (HIV) genome showing its nine genes and flanking long terminal repeats (LTRs). The LTRs contain regulatory sequences recognized by the host cell.

Viral Structure The structure of HIV is shown schematically in figure 31.1. The locations of its important antigens are indicated. The numerous knobs projecting from the surface of the virion represent the gp120 (SU) antigen mainly responsible for attachment to the host cell. The gp41 (TM) antigen traverses the viral envelope and is closely associated with SU; it probably plays a role in entry of the virus into the host cell. The p17 (MA) matrix protein, located inside the viral envelope, helps maintain viral structure, transport the viral genome to the host cell nucleus, and assemble new virions. The core of the virion is composed of the p25 (CA) capsid antigen, two copies of the single-stranded RNA viral genome, and various proteins, including three important viral enzymes, reverse transcriptase (RT), protease (PR), and integrase (IN).

The HIV Genome The HIV genome is shown schematically in figure 31.2. The first two genes, *gag* (from "group antigen") and *pol* (from "polymerase"), are translated as a unit from the full-length viral messenger RNA. The resulting protein is split into four segments by viral protease, yielding three enzymes: protease, reverse transcriptase, and integrase. The fourth segment is split into a number of other proteins including p25 (CA) capsid and p17 (MA) matrix. Reverse transcriptase and protease are the targets of the medicines currently available for treating AIDS.

The *env* gene is translated from a spliced messenger RNA, yielding a precursor protein that is processed by host cell enzymes to give the gp120 (SU) surface glycoprotein and the gp41 (TM) transmembrane glycoprotein.

There are at least six additional genes, known as *accessory genes*, all translated from spliced messenger RNAs.

• **RNA splicing, pp. 141, 142**

Case Presentation

The patient was a young woman from West Africa presenting with the complaint of generalized enlargement of her lymph nodes.

She had recently moved to the United States. She had no history of drug abuse or of receiving blood transfusions. She had had three sex partners in her life. Her single pregnancy the year before her arrival was delivered by emergency cesarean section. She and the baby's father had routine tests for HIV at that time, and both tests were negative. The baby and the father remained well. Ten years before her evaluation, she was treated for a fever by scarification, and this was repeated for an unrelated complaint four years before her evaluation. The native healer performing the scarification used a razor blade, the sterility of it unknown to the patient.

Her initial test results included nondiagnostic lymph node biopsies and a negative HIV test (enzyme-linked immunoassay, ELISA). A repeat HIV test some months later was weakly positive by ELISA, and the confirmatory Western blot showed only questionable reactions of the patient's serum with gp41 and two other HIV antigens. A test for HIV-2 was negative. A CD4+ cell count was very low. A test to detect HIV-1 DNA using the polymerase chain reaction was negative.

1. Could this patient have had HIV disease? If so, how could the negative tests be explained?
2. Does the history give any possible ways in which she could have contracted HIV disease?
3. How could the diagnosis be established?
4. Does this case suggest that any changes should be made in the way HIV disease is diagnosed?

Discussion

1. This patient could well have had HIV disease/AIDS because of her persistent generalized lymphadenopathy and very low CD4+ cell count. Her HIV strain might have differed enough from the pandemic group M HIV-1 strains so that the usual laboratory tests did not detect antibody to it.
2. An AIDS-causing virus could have been contracted during sexual intercourse, from unsterile instruments during her emergency cesarean, or from a blood-contaminated razor blade used for scarification.
3. The Centers for Disease Control and Prevention have a global surveillance system designed to detect cases like this because they raise the possibility of new or rare AIDS-causing viruses being introduced into a population. In the present case, samples of the patient's blood were examined by the CDC and she was shown to be infected with a group O HIV-1 virus. The patient's serum reacted with peptides specific to group O strains, and the virus was isolated from her blood. Nucleic acid sequences of the *env*, *gag*, and *pol* genes matched those of previously isolated group O strains. Group O strains of HIV-1 were first found in Cameroon, a country on the West African coast, where they account for 6% of HIV infections.
4. This patient was the first case of group O HIV disease identified in the United States. Studies indicate that previous introductions, if any, were not accompanied by spread of the virus. Nevertheless, federal agencies are working with manufacturers of HIV tests to try to increase sensitivity of the tests to unusual HIV variants.

Table 31.3 Products of HIV Accessory Genes

Gene Product	Function
Tat (transactivating protein)	Interacts with a cellular protein to increase HIV replication
Rev (regulator of viral expression)	Affects pattern of mRNA splicing
Nef (negative factor)	Interacts with a cellular protein to suppress transcription of proviral DNA
Vif	Plays a role in virion maturation during budding from host cell
Vpr (viral protein R)	Stops final step in division of HIV-infected cell
Vpu	Plays a role in release of virus from HIV-infected cell

These genes, *tat, rev, nef, vif, vpr,* and *vpu,* code for the proteins Tat, Rev, Nef, Vif, Vpr, and Vpu. The gene products and their known or assumed functions are listed in table 31.3.

The functions of these HIV accessory genes are exceedingly complex, as they interact in different ways with host cell substances and vary with the type of cell infected. Developing knowledge of their gene products promises to reveal new ways to attack AIDS. Suggestive evidence exists that a strain of HIV with defects in *nef* could be made into an AIDS vaccine. Tat may influence the development of Kaposi's sarcoma by stimulating the growth of blood vessel cells. Perhaps Vpr could be used therapeutically to stop the growth of cancer cells.

Epidemiology

Promiscuity Initially, promiscuous homosexual men were the hardest hit by the epidemic. An estimated 1.4% to 10% of American men are active homosexuals. Before the arrival of AIDS, a survey found that 33% to 40% of gay men had more than 500 lifetime sexual partners, and another 25 percent had 100 to 500. By 1978, 4.5% of a group of gay men in San Francisco were infected with HIV; by 1984, two-thirds were infected and almost one-third had developed AIDS.

The pace of the epidemic among gay men has been slowing, but there were almost 35,000 cases reported among

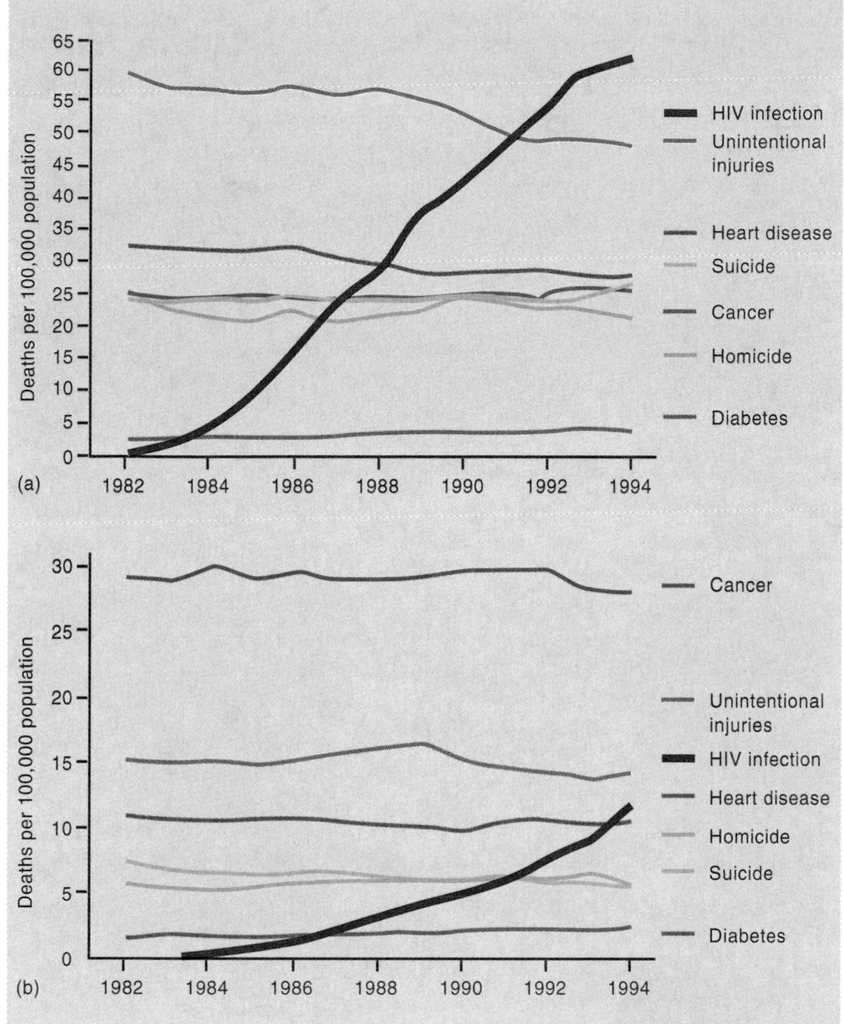

Figure 31.3
Death rates per 100,000 population attributed to AIDS and other conditions, United States, 1982–1994. (*a*) Men and (*b*) women, 25 to 44 years of age.

Source: Centers for Disease Control and Prevention. 1996. *Morbidity and Mortality Weekly Report* 45(6):122.

male homosexuals in 1994, and sex between men is still the leading mode of transmission of HIV.

Drug Abuse In the United States, the population of abusers of intravenous drugs is estimated to be 1.5 million. Many of them, both men and women, support their drug habits with prostitution. Because HIV is acquired and spread through sexual intercourse and through the sharing of hypodermic needles, drug users have been hard hit by AIDS. It has been very difficult to find ways to stop the spread of HIV by and among drug abusers.

Blood and Blood Products Individuals infected with HIV who donated blood before a screening test became available in 1985 unknowingly infected thousands of transfusion recipients. One of the products from donated blood, clotting factor VIII, was used to treat bleeding episodes among hemophiliacs. By 1984, over half of the hemophiliacs in the United States and 10% to 20% of their sexual partners were HIV positive. The first introduction of HIV into Japan was through commercial clotting factor VIII administered to hemophiliacs. Fortunately, the risk of HIV transmission by factor VIII, very low since 1985, was eliminated in 1992 when recombinant factor VIII was licensed.

Current Status of the AIDS Epidemic Well over 500,000 cases of AIDS have been reported in the United States since the epidemic began, resulting in excess of 300,000 deaths. Almost half of the AIDS cases occurred in just the last few years. AIDS became the leading cause of death among persons aged 25 to 44 years (figure 31.3). The epidemic is increasing more rapidly among injected drug abusers and those infected through sexual intercourse between men and women.

AIDS in Women and Children Women represent an increasing percentage of AIDS cases (figure 31.4). At least 38% of the more than 14,000 cases reported in women in 1994 were acquired through heterosexual contact with an HIV positive partner. This is only a crude reflection of the extent of the epidemic among women. An estimated 80,000 women were HIV positive in 1992, and they can leave as many as 150,000 children if they die of AIDS over the rest of the century.

About 1 out of 10 pregnant HIV positive women will miscarry, and of live born babies, 15% to 40% will develop AIDS unless treatment is given to the mother. However, remarkably, in a study of 219 babies positive for HIV by culture or PCR, almost 3% cleared their infection without any treatment.

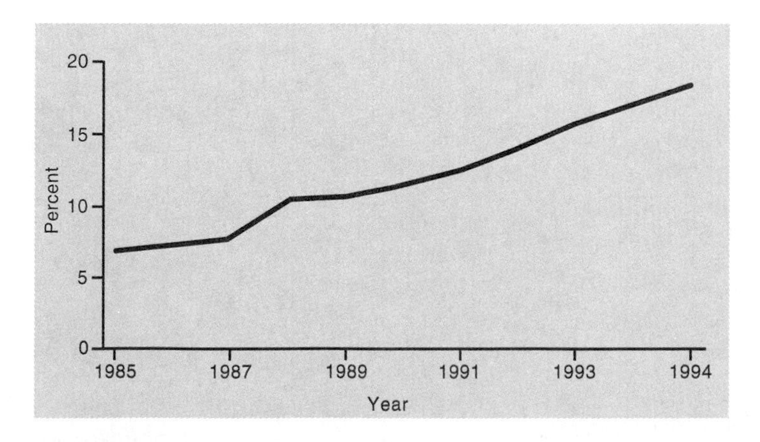

Figure 31.4
Rising percentage of AIDS cases in women 13 years or older, United States, 1985–1994.

Source: Centers for Disease Control and Prevention. 1995. *Morbidity and Mortality Weekly Report* 44(5):81.

Figure 31.5
Some mechanisms of pathogenesis in HIV disease.

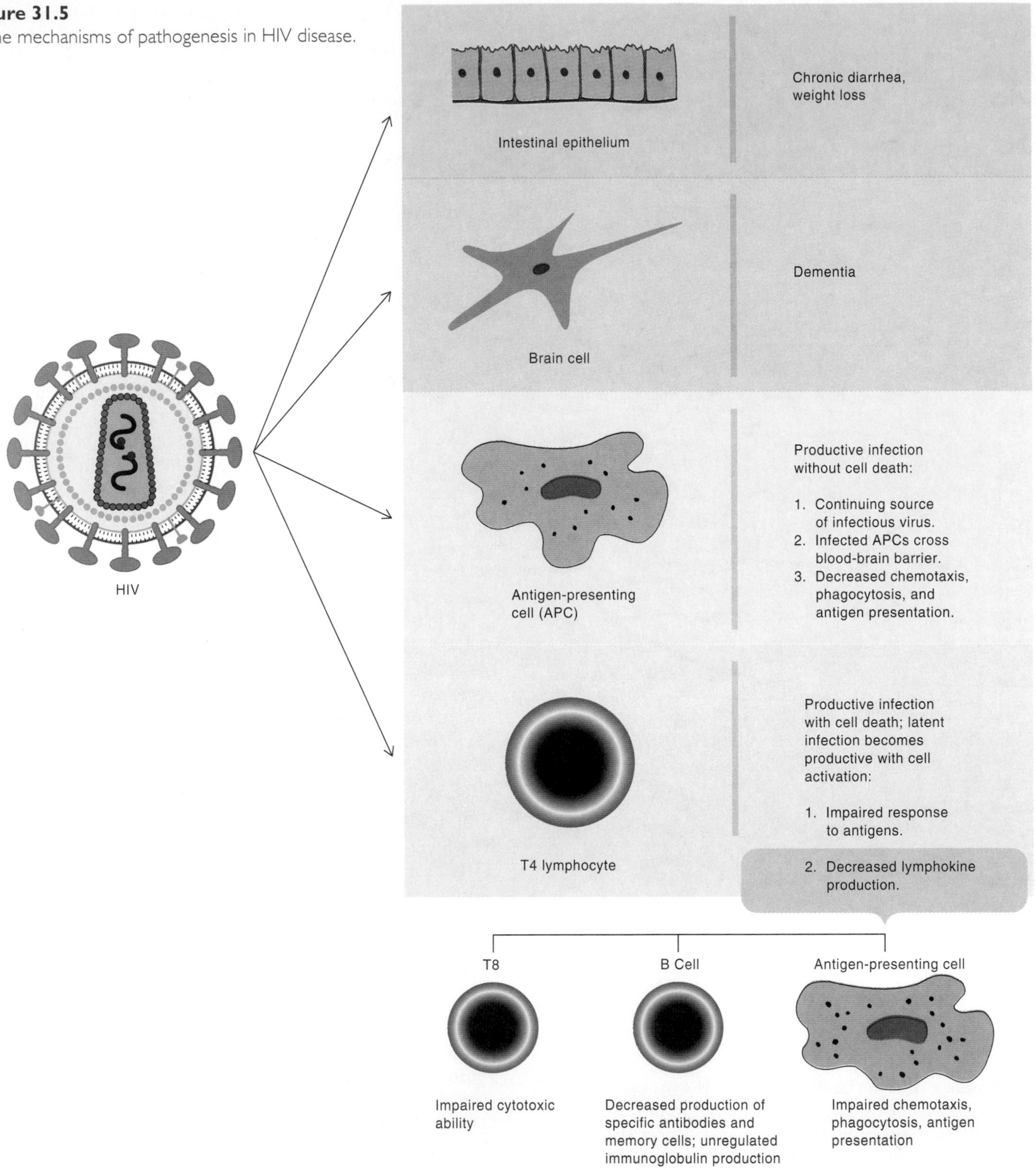

HIV

Intestinal epithelium — Chronic diarrhea, weight loss

Brain cell — Dementia

Antigen-presenting cell (APC) — Productive infection without cell death:

1. Continuing source of infectious virus.
2. Infected APCs cross blood-brain barrier.
3. Decreased chemotaxis, phagocytosis, and antigen presentation.

T4 lymphocyte — Productive infection with cell death; latent infection becomes productive with cell activation:

1. Impaired response to antigens.
2. Decreased lymphokine production.

T8 — Impaired cytotoxic ability

B Cell — Decreased production of specific antibodies and memory cells; unregulated immunoglobulin production

Antigen-presenting cell — Impaired chemotaxis, phagocytosis, antigen presentation

Pathogenesis

When HIV virions enter the body, they attach to and infect certain types of cells. Which cells are affected and their response to infection vary with the particular viral strain and the type of host cell. Some of the cell types and the consequences of infection are shown in figure 31.5.

Two prime targets of HIV are the T4 lymphocytes and the macrophages. The importance of these cell types lies with their central role in the body's specific immune response.

Attachment and Entry Like most other cells susceptible to HIV infection, T4 lymphocytes and macrophages are

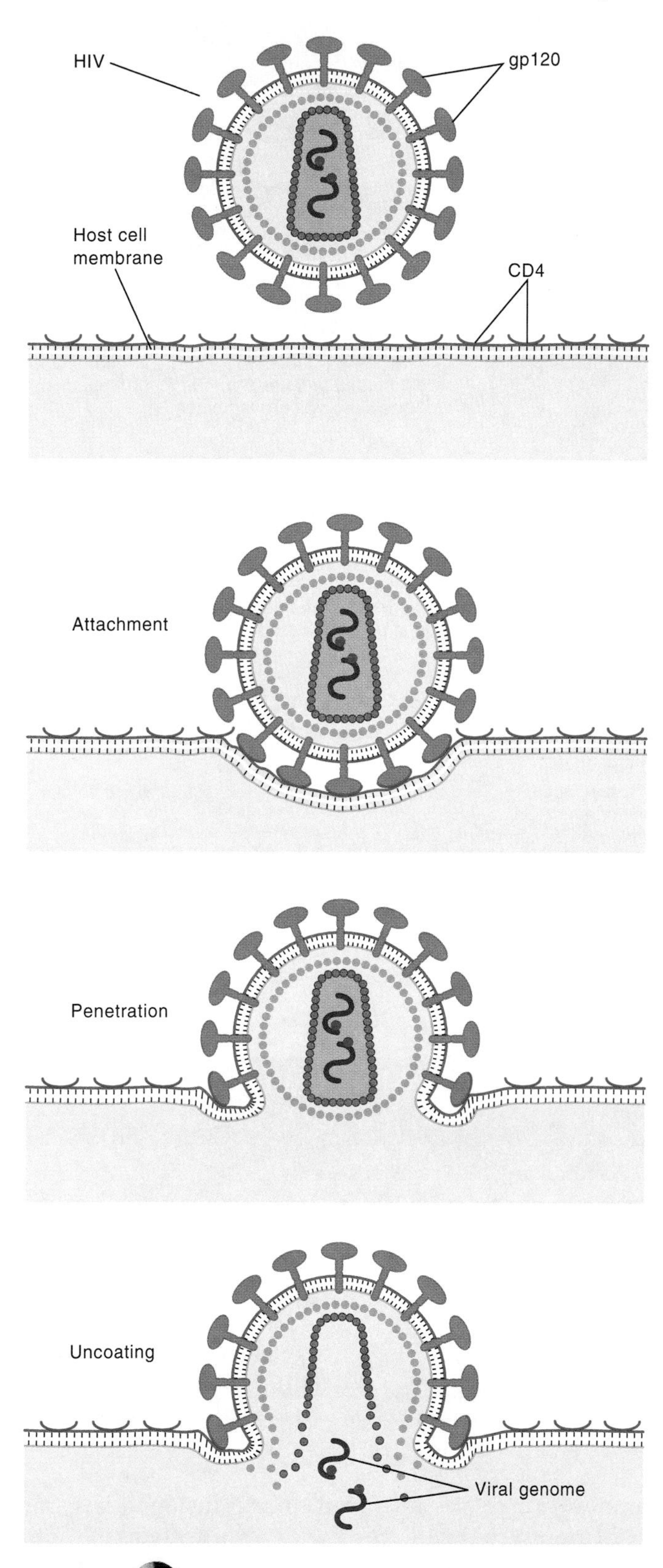

Figure 31.6 Attachment and entry of HIV into a host cell. The chemokine receptors required for entry are not diagrammed.

CD4+ (they possess the CD4 surface antigen). HIV attaches tightly to this structure by its surface gp120 (SU; figure 31.6). The presence of the CD4 antigen is by itself not sufficient to cause entry of HIV. In addition, there must be a host cell coreceptor specific for HIV in order for viral entry to occur. The coreceptor differs for different cell types. In early 1996, after a 10-year search, identification of a coreceptor for T lymphocytes was accomplished at the National Institutes of Health. Initially called *fusin*, it was later identified as one of the cell's chemokine receptors. An HIV coreceptor of macrophages has also been identified and shown to be a chemokine receptor. These important findings have opened up several new avenues of research in the battle to defeat HIV disease.

Events Following HIV Entry Once inside the host cell, the HIV reverse transcriptase enzyme makes a DNA copy of the viral RNA genome. A complementary DNA strand is then added, and the ends of the resulting double-stranded DNA segment are joined noncovalently. The resulting circular DNA is then moved to the nucleus and is inserted into the host cell chromosome by the viral integrase (IN) enzyme. The viral segment thus becomes a **provirus,** a step necessary for efficient replication of HIV. No specific location is required for the insertion into the host chromosome. Following integration, spliced mRNA segments appear, mainly transcribed from *tat, rev,* and *nef*, and are translated into proteins that regulate HIV gene expression. High levels of Tat are associated with replication of infectious virions, whereas Nef protein depresses virion production, favoring viral latency. In the case of T4 (helper) lymphocytes, replication of mature virions results in lysis of the cells and release of infectious HIV. Macrophages, on the other hand, can release infectious virus over long periods of time without death of the cells. The life cycle of HIV is shown schematically in figure 31.7, but in an infected person, at any one time, the vast majority of infected cells show neither lysis nor latency. Most of the HIV-infected cells show accumulations of various viral products that probably impair the normal functioning of the cells.

HIV Variability HIV characteristically produces a wide range of variant virions during replication, especially during the time when immunity of the infected person is still functional. The variant viruses show differences in their preferred host cell, rates of replication, response to host immunity, and other characteristics. The variability is due to a high rate of "error" when reverse transcriptase copies the viral genome. This could result from that unusual HIV characteristic, the double genome. Having two genomes gives reverse transcriptase the "choice" of copying segments from each. Some regions of the genome are highly conserved, meaning they do not change much from one strain of virus to another. Other regions, V regions, are highly variable. The V3 variable region of the gene that codes for gp120 (SU) is important because it plays a role in the virulence of HIV. This antigenic variability of HIV enormously complicates the task of developing an effective vaccine against the virus.

Mechanisms of Cell Destruction by HIV It is not yet known precisely how HIV causes death of a person's helper lymphocytes and other host cells. Lysis of CD4+ lymphocytes

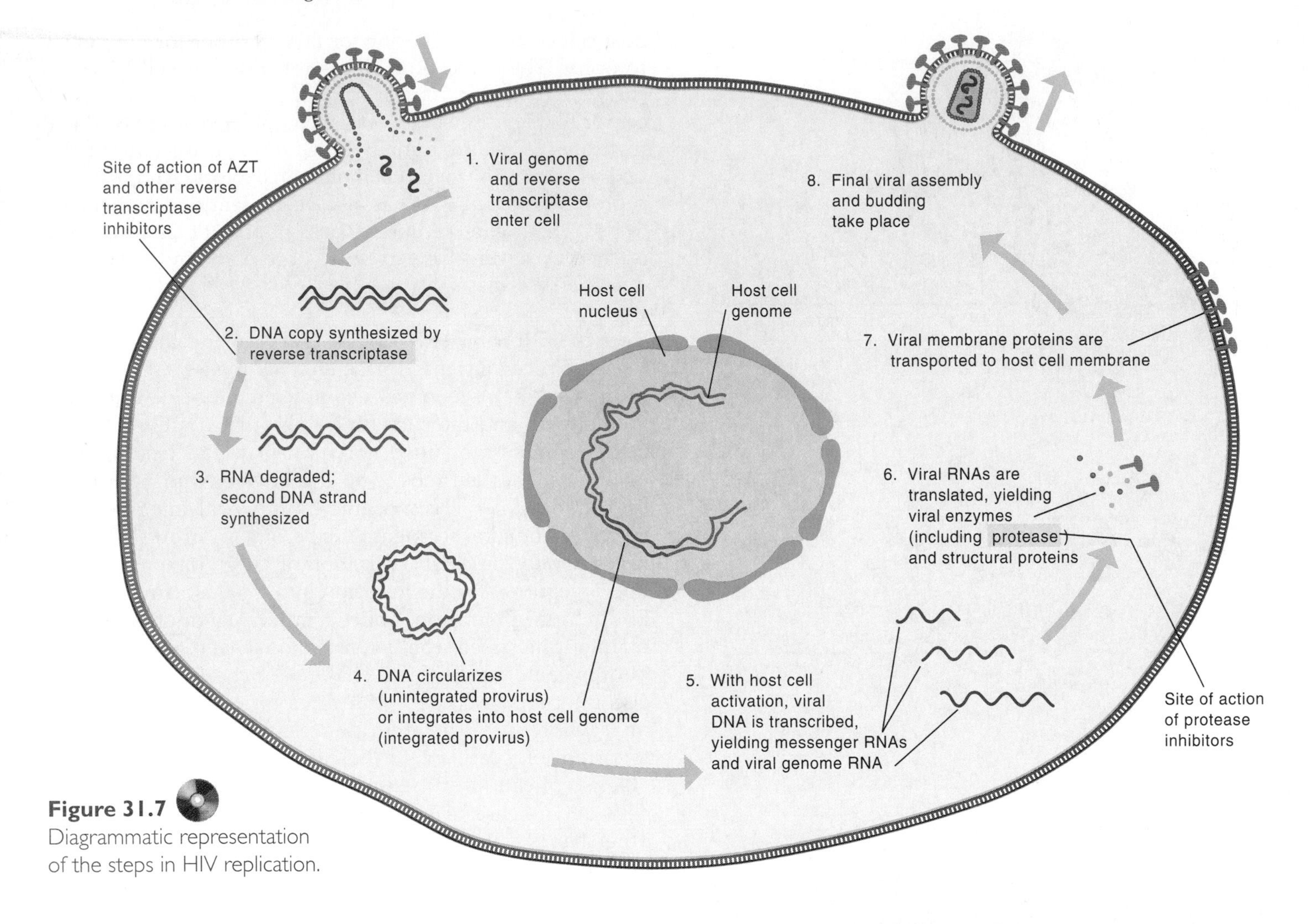

Figure 31.7 Diagrammatic representation of the steps in HIV replication.

during replication of infectious virus is certainly one mechanism, but by itself it cannot account for the devastation of the immune system that results from HIV disease. The host's immune system is itself responsible for the destruction of cells expressing viral antigens. The earliest detectable immune response to HIV infection involves HIV-specific CD8+ T lymphocytes, which attack and lyse infected cells. Natural killer cells probably also play a role in cell destruction. Immune cell attack may be broadened if gp120 (SU) antigen is released from infected cells and attaches to uninfected CD4+ cells, thus marking them for attack by immune cells. Humoral antibody may also play a role through antibody-dependent cellular cytotoxicity. There is also some evidence that autoimmunity could be responsible since HIV envelope proteins show some homology with MHC-II antigen. Fusion of host cells with each other and accumulation of intracellular viral products represent two other probable mechanisms by which cells of the immune system are destroyed. In cell fusion, a large number of uninfected cells fuse with an infected cell, whereupon the resulting syncytium is destroyed. There is also evidence that the accumulation of viral RNA and unintegrated viral DNA inside the cytoplasm of the infected cell can result in its death.

Table 31.4 summarizes some proposed mechanisms of cell destruction.

- **Natural killer cells, pp. 374–75**
- **Antibody-dependent cellular cytotoxicity, p. 375**
- **Major histocompatibility complex, type II (MHC-II), pp. 367–68**

Table 31.4 HIV-Related Destruction of Immune System Cells
Lysis following HIV replication
HIV-specific CD8+ lymphocytes
Natural killer cells
Antibody-dependent cellular cytotoxicity
Autoimmune process
Fusion of infected and uninfected cells
Apoptosis (acceleration of death due to aging)

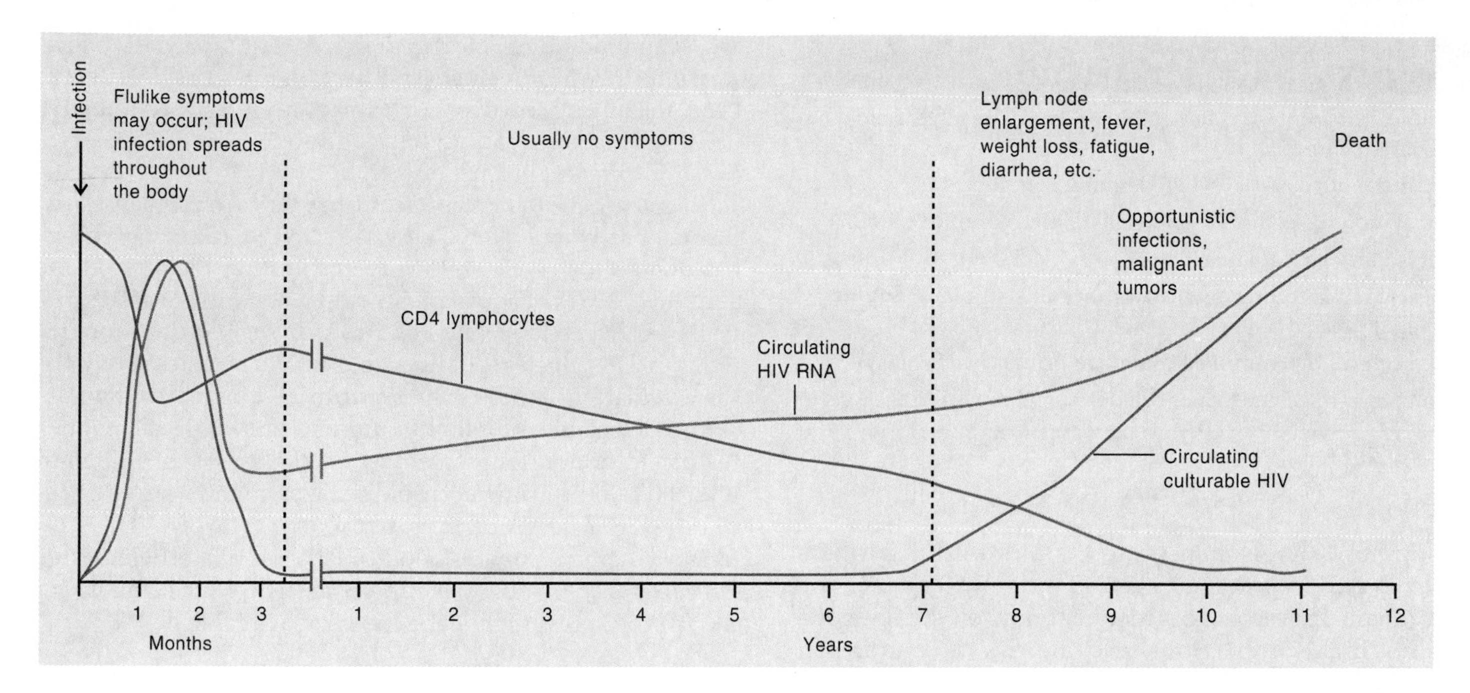

Figure 31.8
Natural history of HIV disease. Blood levels of infectious virus are very high at the beginning and at the end of the disease. Antibody tests for diagnosing the disease are negative in the early stage of the disease even though infected people are highly infectious. The disease steadily progresses in the absence of symptoms as shown by the rising levels of viral RNA and falling CD4+ cell count.

Clinical Features Following the initital infection, the virus spreads throughout the body via the blood and lymphatic systems, quickly infecting the lymphoid tissues. The concentration of HIV in the blood rises to high levels as the newly infected cells release their progeny virus. About half of newly infected subjects develop sore throat, fever, and muscle and headaches. The lymph nodes enlarge and a rash often appears, probably due to circulating antibody-antigen complexes. The marked viremia usually subsides promptly as the supply of uninfected cells dwindles and anti-HIV MHC-I CD8+ cytotoxic T cells appear. The CD4+ T cell level falls initially but then slowly recovers. Following the acute episode, the HIV-infected individual generally enters an asymptomatic period ranging from months to many years. The levels of HIV virus as measured by circulating and lymph node viral RNA and DNA are good predictors of the course of the illness, high levels pointing to a more rapid progression to AIDS. Regardless of the lack of symptoms, a silent struggle goes on between HIV and the immune system for the rest of the person's life. Infectious virus, a threat to others, continues to be present in the blood and body secretions.

The Typical Case In approximately three-fourths of HIV-infected people, the immune system slowly loses ground to HIV even though the body can normally replace over a billion CD4+ cells per day. The peripheral blood CD4+ count (normally about 1,000 cells per microliter) steadily falls about 50 cells per year, symptoms of AIDS usually appearing when CD4+ counts fall below 300 cells per microliter. Levels of infectious virus and viral nucleic acid again rise dramatically. At this point in the illness, the pathology is dominated by malignant neoplasms or opportunistic infections. Half of the patients reach this stage of the illness within 9 to 10 years; the rest take longer. The typical course of HIV disease is summarized in figure 31.8.

Atypical Progression to AIDS In contrast to the typical cases discussed, roughly 10 percent of persons with HIV disease have high virus levels with the acute infection that do not fall dramatically within a few weeks. These subjects progress rapidly to AIDS within a few years. Another 5% to 10% of HIV-infected individuals show no fall in CD4+ cells. They maintain high levels of anti-HIV antibody and HIV-specific CD8+ cytotoxic T cells. Estimates suggest that, with these cases and the more slowly progressing typical cases, 10% to 17% of HIV-infected persons will be free of AIDS 20 years after their infection.

Malignant Tumors and AIDS Viruses are suspected to have a causative role in Kaposi's sarcoma, lymphomas, and cervical and anal carcinoma, the three main kinds of malignancies in AIDS. These are discussed in more detail later in this chapter.

REVIEW QUESTIONS

1. What is the leading cause of death among 25 to 44 year olds in the United States?
2. What is the purpose of "AIDS defining conditions"?
3. What symptoms comprise the AIDS related complex (ARC)?
4. What is hairy leukoplakia?
5. Name two disinfecting agents active against human immunodeficiency virus (HIV).
6. Which cells of the immune system are prime targets of HIV?
7. Outline three mechanisms of cell destruction in HIV disease.

Prevention

Knowing how HIV is transmitted is a powerful weapon against the AIDS epidemic. This weapon can be far more effective than any vaccine or treatment now on the horizon. HIV is not highly contagious, and the risk of contracting and spreading it can be eliminated or markedly reduced by assuming a lifestyle that prevents transmission of the virus (table 31.5).

Some groups of gay men lowered the incidence of new HIV infections from 10% to 20% annually to only 1% to 2% annually by avoiding practices that favor HIV transmission. The improvement in infection rates has not always been sustained, however. Attempts are underway to better define the factors leading to the adoption of safe sex practices for different groups of people.

Since 1985, the risk of acquiring HIV from blood transfusions has been lowered dramatically by screening potential donors for HIV risk factors and testing their blood for antibody to HIV. The risk is now estimated to be less than 1 in 400,000 transfused units and possibly could drop more if tests for HIV nucleic acid become practical. Infectious HIV can circulate in a person's blood for as long as six months before detectable antibody is present. Screening techniques for blood donors have also reduced the risk of infection with HIV from artificial insemination or from organ transplantation. The blood product clotting factor VIII is no longer obtained from donated blood. Since late 1992, factor VIII produced in a baby hampster tissue cell line containing the appropriate human gene has been available.

Education Education about how HIV is transmitted can be a very effective tool in bringing the epidemic of HIV disease under control. Because videotapes and written material developed for one group of people are often completely ineffective for and even offensive to another group, additional research is needed to determine how best to target groups in which the epidemic is advancing rapidly, especially teenagers and economically disadvantaged Hispanic and black people.

All persons unsure of their HIV status and especially those at increased risk of HIV disease (table 31.6) are advised to get tested for HIV. More than one of five people presenting with AIDS has no knowledge of his or her HIV status. By knowing, those with HIV disease can receive prompt preventive treatment for infections and cancers that complicate the disease. They can also do their part in preventing transmission. Home test kits are now commercially available.

Vaccines Efforts toward development of a vaccine against HIV disease began not long after the discovery of the causative agent. In theory, the vaccine could be used in either

Table 31.5 Some Ways to Eliminate or Decrease the Risk of HIV Infection

1. Do not engage in sexual intercourse.
2. Stay with one faithful sexual partner. Put your energy into a close relationship with one person rather than into multiple superficial relationships.
3. Have sexual intercourse only with a person you know very well.
4. Have sexual intercourse only with a person who is known not to be at risk of HIV infection (table 31.6).
5. Avoid trauma to the genitalia and rectum. Small breaks in the skin and mucous membranes allow HIV to infect.
6. Do not engage in sexual intercourse when sores from herpes simplex or other causes are present. They represent sites where HIV can infect.
7. Do not engage in anal intercourse. Receptive anal intercourse carries a high risk of HIV infection.
8. Use latex condoms with a spermicide from beginning to end of sexual intercourse. Do not use other kinds of condoms. Do not use oil-based lubricants.
9. Postpone pregnancy indefinitely if you are a woman infected with HIV. If you are not sure of your HIV status, get blood tests to rule out HIV disease before considering pregnancy.
10. Use extreme care to avoid needles, razors, toothbrushes, etc., that could be contaminated with someone else's blood.

Table 31.6 People at Risk for HIV Disease

1. Homosexual and bisexual men.
2. Intravenous drug abusers who have shared needles.
3. Persons who received blood transfusions or pooled blood products between 1978 and 1985.
4. Sexually promiscuous men and women, especially prostitutes and drug abusers.
5. People with hepatitis B, syphilis, gonorrhea, or other sexually transmitted diseases.
6. People who have had blood or sexual exposure to any of the people listed.

of two ways. One would be to immunize uninfected individuals against the disease. The other would be to boost the immunity of those already infected with HIV before they became severely immunodeficient. The latter approach would be similar to the postexposure treatment of rabies infections. A successful vaccine must induce mucosal in addition to bloodstream immunity because HIV disease is primarily sexually transmitted. The vaccine must also get around the problem of HIV antigenic variability and stimulate cellular and humoral responses against virulence determinants. Moreover the vaccine has to be safe. A live attenuated agent must not be capable of reverting to virulence and must not be oncogenic. Additionally, the vaccine must not stimulate an autoimmune response, and it must not cause production of "enhancing antibodies" that aid the passage of HIV into the body's cells. The vaccine should induce neutralizing antibodies against cell-free virions and also prevent direct spread of HIV from a cell to neighboring cells.

Despite many difficulties, HIV vaccine research continues as the best hope for controlling an epidemic that every year produces more than 50,000 new HIV infections in the United States alone. Findings that have encouraged vaccine researchers include failure of some African prostitutes to develop AIDS although repeatedly exposed to HIV. Also, HIV-infected pregnant women that have high titers of HIV-neutralizing antibodies are less likely to transmit the infection to their fetuses than women with low antibody titers. Another finding showing the potential power of the immune system is that some babies born with HIV infection appear to clear the infection on their own and do not progress to AIDS. Finally, infection by some strains of HIV apparently progresses to AIDS very slowly or not at all. Development of vaccines against HIV disease takes time and costs enormous amounts of money. At present, there is no vaccine in sight that could prevent 80% to 90% of HIV infections, but the epidemic is expanding so rapidly that even a poor vaccine quickly put into use could save more lives than would be saved by researchers' waiting for a highly effective vaccine to be perfected.

Research on Potential Vaccines Candidate vaccines undergo an extensive evaluation in experimental animals before they are tested in humans. Some of the substances in this "preclinical" stage of development are live attenuated HIV, killed HIV, adenovirus and poliomyelitis vaccine viruses carrying parts of the HIV genome, BCG and *Salmonella* bacteria carrying parts of the HIV genome, HIV peptides and proteins, HIV proteins with adjuvants, naked DNA, and various particles containing antigens of HIV.

Vaccines Being Tested in Humans Vaccine trials in people progress from phase I (testing safety and ability to provoke an immune response) to phase II (determining optimum dose and kinds and duration of immune response) to phase III (testing effectiveness in preventing HIV infection or AIDS). Substances that have entered phase I trials include purified glycoprotein products of the *env* gene, live vaccinia virus containing HIV genes, synthetic peptides that are copies of epitopes associated with virulence, HIV antigens in viruslike particles, and naked DNA representing parts of the HIV genome.

Glycoprotein Products of *env* Using recombinant technology, the code for *env* is inserted into a baculovirus, an insect virus that produces a large amount of protein as part of its normal life cycle. The recombinant virus is allowed to grow in an insect tissue cell culture, and HIV gp160 (gp120 and gp41 combined) instead of the normal viral protein is expressed in large amounts. The gp160 is then purified for use as a candidate vaccine. Other methods used for preparing purified gp160 or gp120 use a genetically engineered vaccinia virus (the one formerly used to stamp out smallpox) to express genetic information from *env* in mammalian tissue cell cultures such as hampster ovary cells. These purified preparations of HIV surface glycoproteins have been disappointing because the immune response to them has been weak and short lived when they are used alone. Also, the HIV virus used in their preparation has been cultivated in the laboratory for years and differs markedly from "wild" HIV viruses.

Vaccinia Virus Vaccines Vaccinia, a pox virus, has a large genome with multiple sites where foreign genes can be spliced without interfering much with its ability to infect and multiply. One of the most promising HIV vaccines is a vaccinia virus that has been genetically engineered to express both HIV surface and core antigens. When injected into volunteers, it produces a localized infection in which the HIV antigens are expressed and the volunteers develop both cell-mediated and humoral responses to the antigens.

Synthetic Peptides The amino acid sequence of the surface proteins of HIV is known, so that peptides that exactly mimic immunologically important segments such as V3 (the "V3 loop") can be synthesized. These peptides can be used effectively for booster injections after primary immunization with the modified vaccinia virus already discussed.

Viruslike Particles When both *gag* and *env* HIV genes are introduced into vaccinia and the recombinant virus is allowed to propagate in tissue cell cultures, particles form that resemble viruses. These viruslike particles have gp120 antigen on their surfaces, but they lack the HIV genome. The same phenomenon occurs with other viruses, such as baculovirus, and with **Ty,** an interesting transposon of yeast, which resembles a retrovirus except that it is not infectious. Viruslike particles make excellent antigens because their shape makes them easily recognized by the body's immune system.

Naked DNA Segments of DNA containing the HIV *env* gene have been made that are taken up and expressed by body cells when introduced into body tissues. This is the technique used in gene therapy. The expressed HIV proteins

Table 31.7 Vaccines Against HIV Disease: Status of Clinical Trials

Vaccine Type	Year Trial Started		
	Phase I	Phase II	Phase III
env gene product	1989	1993	1997 (Thailand)
Live recombinant vaccinia	1989	1996	? (before 2000)
Synthetic peptide	1993	1996	?
Viruslike particle	1995	?	?
Naked DNA	1996	?	?

are in a form readily recognized by the immune system, and cell-mediated immunity is elicited. Tests in experimental animals suggested that this immunization method could prevent syncytium formation.

Table 31.7 shows the status of some of these vaccines.

Antiviral Treatment

The last few years have witnessed the dawning of a much more hopeful period in the antiviral treatment of HIV disease. For the first time, therapy can significantly lengthen the asymptomatic stage of the disease and prolong life. Advances in antiviral treatment mainly result from the development of new medications and their use in combination. Presently available medications fall into two groups: inhibitors of reverse transcriptase and inhibitors of viral protease.

Reverse Transcriptase Inhibitors Medications that interfere with reverse transcriptase include zidovudine (AZT), stavudine (D4T), didanosine (ddI), zalcitabine (ddC), and lamivudine (3TC). All are nucleoside analogs, but they differ chemically and in terms of the site of action on the enzyme. Use of these and newer nucleoside analogs in combination with each other can markedly improve the extent and duration of their antiviral effect. For example, AZT used in combination with ddC and ddI significantly raised blood levels of CD4+ T cells and lowered the death rate by 40% compared with AZT used alone.

These substances owe their effectiveness to their resemblance to the normal purine and pyrimidine building blocks of nucleic acids. During nucleic acid synthesis, the viral reverse transcriptase enzyme incorporates the medication molecule into the growing DNA chain, thereby blocking completion of the DNA strand. This process for zidovudine is illustrated in figure 31.9.

Protease Inhibitors The protease inhibitor saquinavir was approved for therapy in December 1995, the first in a series of powerful new anti-HIV medications. The impact of these inhibitors on the treatment of people with AIDS has in some cases been immediate and dramatic. For example, one 30-year-old woman with AIDS had exhausted all available AIDS therapies. She was a wasted 85 pounds, her hair had fallen out, and she coped with her fatigue by napping every two hours. She was given indinavir (a newer protease inhibitor) together with D4T and 3TC and, within a month, her hair returned, she gained weight, she began skating, and her CD4+ cell count rose from 3 to over 200 per microliter of blood.

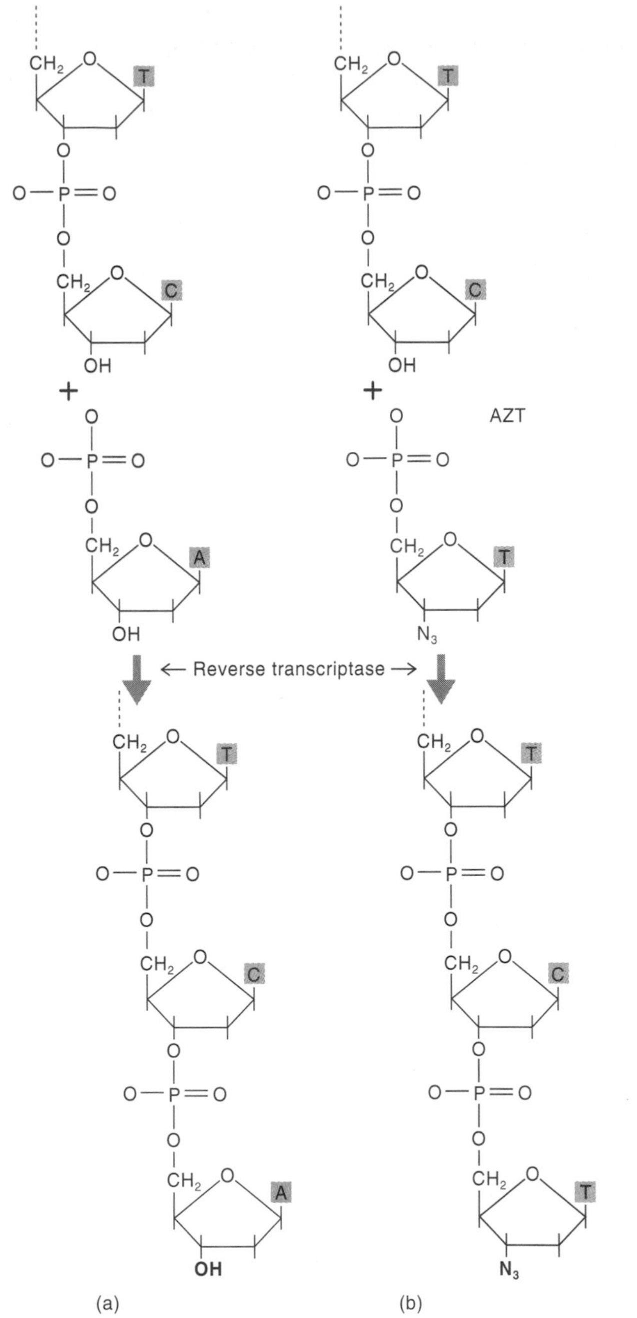

Figure 31.9
Mode of action of zidovudine (AZT). (*a*) Normal elongation process of DNA in which reverse transcriptase catalyses the reaction between the OH group on the chain with the phosphate group of the nucleotide being added. (*b*) Reverse transcriptase catalyses the reaction of zidovudine with the growing DNA chain. Since zidovudine lacks the reactive OH group, no further additions to the chain can occur, and DNA synthesis is halted.

The protease inhibitors do not cure AIDS, but they do provide another good weapon in the fight against HIV disease. Unlike the reverse transcriptase inhibitors, which prevent HIV from entering the genome of the body's cells, the protease inhibitors prevent the virus from reproducing. Of one group of advanced AIDS cases, all of whom had more than 20,000 copies of the HIV genome per milliliter of their blood, 86% had no detectable copies after six months of combined treatment with indinavir, AZT, and 3TC. One of the questions yet to be answered is whether combination therapy given during acute HIV disease can prevent or delay the development of AIDS by markedly decreasing the amount of circulating virus at the beginning of the disease.

Limitations of Anti-HIV Therapy Viral resistance limits the effectiveness of all the anti-HIV medications. Some resistant HIV strains existed even before the medications were introduced, and resistance frequently develops during therapy with the medications. The development of HIV resistance during treatment has been associated with worsening of the disease. In one study of saquinavir, almost half the patients developed HIV resistance to the medication within one year. High doses and combined therapy delay the development of HIV resistance to medications.

Toxic effects of the medications are another limitation of anti-HIV therapy. For example, zidovudine (AZT) can cause anemia, low white blood count, vomiting, fatigue, headache, and muscle and liver damage. Painful peripheral nerve injury and inflammation of the pancreas can occur with ddI therapy, and ddC can cause rash and mouth and esophagus sores and fever. Also, indinavir may be responsible for kidney stone formation, and ritonavir (another protease inhibitor) causes nausea and diarrhea. Diabetes mellitus may also be a side effect of protease inhibitors.

Finally, a big limitation of the use of anti-HIV medications is their cost. The cost of a combination of these medications can easily exceed $1,000 per month, with the total cost of therapy being much higher because of the need for prevention and treatment of opportunistic infections and other conditions. The cost of medications puts this form of therapy beyond the reach of millions of the world's impoverished HIV victims.

Other Possible Treatments Being Explored Many hundreds of substances are being evaluated as potential anti-HIV medications. These include substances that block attachment of HIV to host cells, new inhibitors of reverse transcriptase, new inhibitors of viral protease, substances that preferentially kill HIV-infected cells, cytokines and substances that neutralize cytokines, and naturally occurring anti-HIV products of CD8+ lymphocytes. Other experimental techniques include some that enhance the body's immune response, bone marrow grafts, and gene therapies that introduce anti-sense HIV RNA or a modified *rev* gene that prevents HIV replication. Some of these approaches to anti-HIV therapy are presented in table 31.8.

Table 31.8 Some Experimental Approaches to HIV Disease Therapy

Preventing Attachment of HIV to Host Cells	
Examples:	1. Soluble CD4 produced using recombinant technology and modified to make it more stable is introduced into the patient's bloodstream where it adheres to infectious virions and prevents their attachment to host cells. 2. Highly negatively charged sulfonated polysaccharides block attachment of HIV to host cells.
Destruction of HIV-Infected Cells	
Examples:	1. *Pseudomonas* or another toxin is attached to soluble CD4, or monoclonal antibody is guided to cells presenting HIV surface antigen (gp120) and kills them. 2. Tricosanthin, a Chinese abortion drug, preferentially destroys HIV-infected macrophages.
Gene Therapy	
Example:	Peripheral blood CD4+ T cells are withdrawn from an AIDS patient and inoculated with a genetically engineered virus containing a modified *rev* gene. The cells are allowed to multiply outside the body in the presence of anti-HIV medications and then injected back into the patient. These cells survived more than four times longer than other CD4+ T cells because the modified *rev* did not permit replication of HIV.
Attack Other HIV Proteins	
Examples:	1. The gene product of *gag* is essential for assembling infectious HIV virions. It must associate with a host cell substance that reshapes proteins to make them biologically active. 2. HIV integrase binds to the DNA copies of the HIV genome and assists their entry into the host chromosome.
Identify Anti-HIV Substances Normally Produced by the Body	
Example:	Three polypeptides produced by CD8+ T cells when acting together strongly supress HIV replication. These substances could probably be produced in large quantity through genetic engineering. A distinguished AIDS researcher has predicted that these substances in conjunction with other medications will make it possible to arrest HIV disease indefinitely.

Malignant Tumors Complicating Acquired Immunodeficiencies

This section will discuss certain malignant tumors associated with HIV disease, organ transplantation, and other acquired immunodeficiency states. Most of these malignancies fall into one of only three types: Kaposi's sarcoma, lymphomas, or carcinomas arising from anal or cervical epithelium. They tend to metastasize (jump to new areas) quickly and be difficult to treat. Evidence indicates that viruses are a factor in their causation. A popular theory is that certain viral antigens, perhaps with the aid of cytokines, cause rapid multiplication of a host cell type. A mutation to malignancy then occurs among the rapidly dividing cells, the result of insertion of viral DNA into their genome or other carcinogen. Finally, the malignant cell escapes destruction because of defective cellular immunity, and it multiplies without restraint.

Kaposi's Sarcoma, Once Rare, Became Common with the Arrival of the AIDS Epidemic

Kaposi's sarcoma (figure 31.10) is a malignant tumor arising from blood or lymphatic vessels. Formerly, it was a rare tumor occurring mainly in elderly men. However, coincident with the spread of HIV, the tumor began to appear in young men, and the incidence rose dramatically. Among a group of never married San Francisco men, the incidence of Kaposi's sarcoma was 2,000 times higher in 1984 than it was in the period before HIV became widespread. The tumor has been so common among AIDS patients that it has become an AIDS-defining condition despite the fact that it generally appears before the development of severe immunodeficiency. Kaposi's sarcoma is often associated with, besides HIV disease, the use of medications to suppress the immune system in organ transplant patients. Indeed, the incidence of Kaposi's sarcoma is 70 times higher in transplant patients than in the general public. One of the peculiarities of the tumor is that it can disappear if a patient's organ graft is rejected. Another oddity is that Kaposi's sarcoma in women often regresses (gets smaller) if the woman becomes pregnant. This regression appears to be due to the high levels of gonadotropic hormone (the one tested for in pregnancy tests) during pregnancy. These observations indicate that the tumor is subject to both immune and hormonal controls. In patients with HIV disease, other factors may operate. Tat, the protein product of the HIV *tat* gene, causes increased production of a host cell growth factor. Finally, epidemiologic evidence indicates that a sexually transmissible infectious agent is important in causing Kaposi's sarcoma.

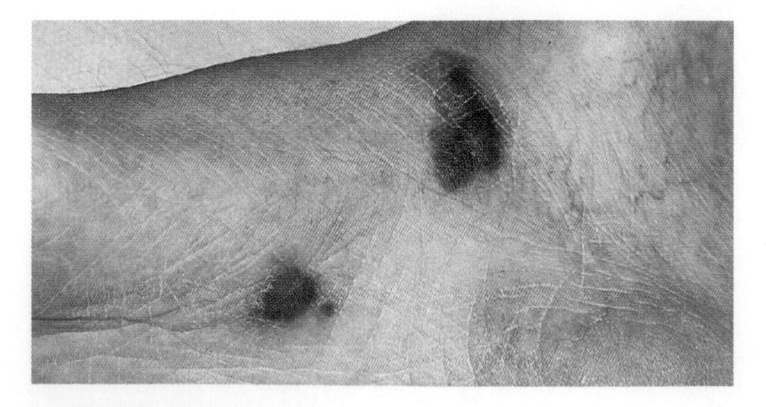

Figure 31.10
Kaposi's sarcoma occurring on a person's foot.

In 1994, scientists at Columbia University detected a new herpes virus in Kaposi's sarcomas, tentatively named *KS associated virus* or *herpesvirus 8*. The virus was subsequently cultivated at the University of California, San Francisco, from the tumors. The role this virus plays in Kaposi's sarcoma and other conditions should become clearer over the next few years.

With Acquired Immunodeficiencies, B Lymphocytic Tumors Can Develop in the Brain

Lymphomas are a group of malignant tumors that arise from lymphoid cells. They occur in 5% to 10% of AIDS patients and 60 times more frequently in organ transplant patients than in the general public. Most of these tumors arise from B lymphocytes, although T lymphocytic lymphomas also occur. Intense, sustained replication of lymphoid cells is a constant feature of HIV disease. As stated at the beginning of this chapter, HIV was first isolated from a patient with the lymphadenopathy syndrome, a condition in which the lymph nodes are markedly enlarged for three months or more, now known to be a prelude to AIDS. The lymph node enlargement reflects proliferation of lymphoid cells in response to cytokines and to the HIV gp41 (TM) antigen. Also in HIV disease, sustained replication of T cells occurs as billions of new cells are produced each day to replace those destroyed by HIV. Unlike in the case of Kaposi's sarcoma, with B lymphocytic tumors there is no epidemiologic evidence of involvement of a sexually transmitted infection. However, EB virus may play a role in some of the lymphomas. The EB virus-related tumor, Burkitt's lymphoma, is at least 1,000 times more frequent among AIDS patients than in the general public, representing about one-fourth of the lymphomas associated with AIDS. In general, lymphomas rarely arise in the brain, but they do in approximately 25% of AIDS patients with lymphomas. EB virus is found in almost all the AIDS-associated brain tumors and in about 60 percent of all AIDS-associated B cell lymphomas. Among organ transplant–associated cases, almost all lymphomas are EB virus related. There is no evidence that EB virus is a direct cause of B cell lymphomas, but it may well play an indirect role. HIV infection causes activation of latent EB virus infection, and this, in turn, causes polyclonal B cell proliferation and increased life span. These effects on B cells could be a factor in lymphoma development, but the exact role of EB virus is as yet undefined.

There Is an Increased Rate of Cervical and Anal Carcinoma in People with HIV Disease

Carcinoma (cancer) of the uterine cervix in women and carcinoma of the anus in women and gay men are strongly associated with human papillomaviruses (HPV) types 16 and 18. The cells involved in these cancers are epithelial cells and, therefore, differ from those in Kaposi's sarcomas and lymphomas. HPV is transmitted during sexual activity and infects the cervical and anal epithelium, appearing to cause increased replication of the cells, blocking expression of a cellular gene responsible for controlling cell growth. In HIV disease, organ transplantation, and other conditions, HPV replication increases with the decline of the host's cellular immunity. Precancerous changes can be demonstrated in the anal cells of HIV-positive gay men twice as frequently as in gay men who are HIV negative and six times more frequently among HIV-positive gay men with low CD4+ T cell counts than in HIV-positive gay men with high CD4+ T cell counts. Interestingly, even before the arrival of HIV, the incidence of anal carcinoma in gay men exceeded the incidence of cervical carcinoma in women. One important implication of these findings is that both women and gay men should be screened for precancerous lesions of the cervix and anus at frequent intervals, especially if they are HIV positive.

REVIEW QUESTIONS

1. List five groups of people at increased risk for HIV infection.
2. How is clotting factor VIII produced now that it is no longer obtained from donated blood?
3. What are four requirements of an acceptable HIV vaccine?
4. Describe three observations that give encouragement to developers of HIV vaccines.
5. Give three limitations of current anti-HIV medications.
6. What are the three main types of malignant tumors that complicate HIV disease?

Infectious Complications of Acquired Immunodeficiency

Immunodeficient individuals are susceptible to the same infectious diseases as other people. In addition, infections that pose no threat to normal people can cause severe, even fatal disease in those with impaired immune systems. Bacteria, viruses, fungi, protozoa, and even parasitic worms such as *Strongyloides stercoralis* can be life-threatening. Geographical area may determine which infectious complications are most important. For example, the mold *Penicillium marneffei* is the third most common opportunistic agent among AIDS patients in Thailand but is almost unheard of elsewhere.

This section will present some common examples of opportunistic infections in patients with immunodeficiency.

Pneumocystosis, Once the Most Common Cause of AIDS Fatalities, Is Now Responsible for Less than 15% of Deaths

Pneumocystosis, a severe infectious lung disease, was recognized just after World War II in Europe when it caused deaths among hospitalized, malnourished, premature infants. Subsequently, scattered cases were recognized among immunodeficient patients until the onset of the AIDS epidemic when the incidence soared. By 1995, almost 128,000 cases had been reported to the Centers for Disease Control and Prevention.

Symptoms

The symptoms of pneumocystosis typically begin slowly, with gradually increasing shortness of breath and rapid breathing. Fever is usually slight or absent, and only about half of the patients have a cough. A dusky appearance of the skin and mucous membranes gradually worsens, a reflection of poor oxygenation of the blood, which can become fatal.

Causative Agent

Pneumocystosis is caused by *Pneumocystis carinii* (figure 31.11), a tiny fungus of uncertain classification. Formerly considered a protozoan, *P. carinii* has a number of

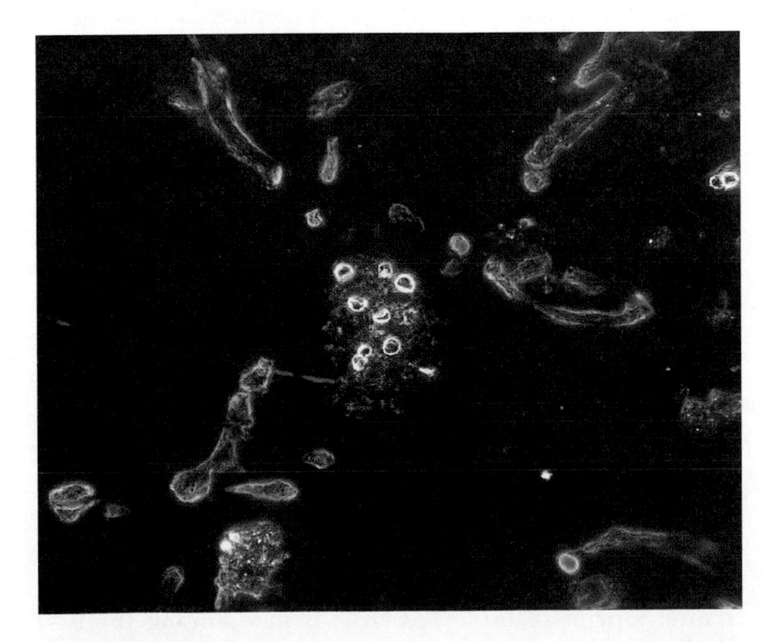

Figure 31.11
Fluorescent antibody stain of *Pneumocystis carinii.*

characteristics that support its inclusion among the fungi. These features include the fungal-like ultrastructure of its spore case wall and mitochrondria, spore case contents that resemble ascospores, a 16S rRNA that shows homology with ascomycete 16S rRNA, and an elongation factor (EF 3) that is homologous with that of an ascomycete. It differs from many fungi in the chemical makeup of its cell wall; consequently, it is resistant to the medications often used against fungal pathogens.

Pathogenesis

The infectious form of *P. carinii* is not known, but its small spores, measuring only 1 to 3 micrometers, could easily be inhaled into lung tissue. In experimental infections, the spores attach to the alveolar (air sac) walls, and the alveoli fill with fluid, mononuclear cells, and *P. carinii* cells in various stages of development. Later, the alveolar walls become thickened and scarred, preventing the free passage of oxygen.

Epidemiology

Various strains of *P. carinii* are widespread among animals, including dogs, cats, horses, and rodents, persisting in their lungs as a latent infection. Serological tests indicate that most children are infected by age two and a half. The infection is asymptomatic and apparently remains latent for long periods. The organisms are known to persist in the lungs of people without any underlying diseases. The source and transmission of human infections are unknown. Animal strains are not thought to be infectious for humans. Almost all cases of pneumocystosis occur in persons with immunodeficiency, but it is uncertain whether their disease is caused by activation of latent infection or by a newly acquired infection. Most cases are presumed to be due to reactivation of latent infection, but epidemics have occurred in nursing home residents, suggesting a single environmental source or person-to-person spread.

Prevention

Formerly, pneumocystosis eventually occurred in about four-fifths of all AIDS patients. Nowadays, physicians give a preventive medication such as trimethoprim-sulfamethoxazole to patients with HIV disease as soon as their CD4+ T cell count falls below 200 per milliliter or when other hallmarks of immunodeficiency appear, such as prolonged unexplained fever or *Candida* infection of the mouth and throat (thrush). An inhaled medication, pentamidine, is also effective in preventing pneumocystosis for those who cannot tolerate oral medications.

Treatment

Trimethoprim-sulfamethoxazole is also among the best medications for treating pneumocystosis and, along with oxygen and other measures, can reduce mortality from nearly 100% to about 30%. Alternate medications are available for treating those who cannot tolerate trimethoprim-sulfamethoxazole because of its side effects—mainly rash, nausea, and fever. For unknown reasons, people with HIV disease are more likely to develop these side effects than are other groups of patients. After treatment, patients must receive preventive medication indefinitely.

The Eyes and Brain Are Commonly Attacked in Toxoplasmosis, a Protozoan Disease

Toxoplasmosis is generally a rare disease among normal people but can be a serious problem in people with malignant tumors, recipients of organ transplantation, or people with HIV disease. It can also be a serious congenital disease.

Symptoms

Despite the fact that infection of normal people is quite common, only about 10% of infected people develop symptoms. The disease presents itself differently in the three main categories of patients: the immunologically normal, unborn children, and those with immunodeficiency.

Toxoplasmosis of Immunologically Normal People Symptomatic toxoplasmosis can occur in normal individuals who eat undercooked meat or who are exposed to cat feces. The typical symptoms are similar to those of infectious mononucleosis. They usually consist of sore throat, fever, enlarged lymph nodes and spleen, and sometimes a rash. These symptoms subside over weeks or months and do not require treatment. Rarely, a severe, even life-threatening illness develops, due to involvement of the heart or central nervous system.

Congenital Toxoplasmosis Symptomatic or asymptomatic toxoplasmosis acquired by the mother during pregnancy involves the fetus in about 50% of cases. Fetal toxoplasmosis during the first trimester of pregnancy is the most severe, often resulting in miscarriage or stillbirth. Babies born live may have severe birth defects including small or enlarged heads, and their lungs and liver may also be damaged by the disease. Later, these babies may develop seizures or manifest mental retardation.

About two-thirds of fetal toxoplasmosis cases occur during the last trimester of pregnancy, and the effects on the fetus are usually less severe. Most of these infants appear normal at birth, although later in life, **retinitis** can be a big problem for them. This manifests itself as recurrent episodes of pain, sensitivity to light, and blurred vision, usually involving only one eye. Less common late consequences of congenital toxoplasmosis include mental retardation and epilepsy.

Toxoplasmosis Complicating Immunodeficiency Toxoplasmosis is a common life-threatening illness in patients with AIDS and other acquired immunodeficiencies. Brain involvement (encephalitis) occurs in more than half the cases, manifest by confusion, weakness, impaired coordination, seizures, stiff neck, paralysis, and coma. Involvement of the brain, heart, and other organs often results in death.

Causative Agent

Toxoplasmosis is caused by *Toxoplasma gondii*, a tiny (3 by 7 micrometers) banana-shaped protozoan (figure 31.12 *a*) that has a worldwide distribution and infects many vertebrate hosts, including household pets, pigs, sheep, cows, rodents, and birds. The life cycle of *T. gondii* is shown in figure 31.12 *b*. The definitive host (one in which sexual reproduction occurs) is the cat or other feline (ocelot, puma, bobcat, and so on). The organism reproduces in the cat's intestinal lining, resulting in numerous offspring, some of which spread throughout the body and others that differentiate into male and female gametes (sexual forms). Union of gametes results in the formation of a thick-walled oval structure (oocyst) about 12 micrometers in diameter that is shed in the cat's feces. Millions of the oocysts are excreted each day generally over a period of one to three weeks. In the soil, the oocysts undergo further development over one to five days into an infectious form containing two sporocysts, each containing four sporozoites. These may remain viable for up to a year, contaminating soil and water and, secondarily, hands and food. In general, the cats recover from the acute infection and do not excrete oocysts again, although a few might do so following infection with another pathogenic species. The oocysts are infectious for other animals including humans.

When nonfeline animals ingest the oocysts, usually from contaminated food or water, the sporozoites emerge from the oocysts and invade the cells of the small intestine, especially its lower part (ileum), but there is no sexual cycle. The intestinal infection spreads throughout the tissues of the host, infecting cells of the heart, brain, and muscles. As host immunity develops, multiplication slows, and a tough fibrous capsule forms, surrounding large numbers of a smaller form of *T. gondii*. These capsules packed with organisms are called **cysts** (figure 31.12 *c*), and they remain viable for months or years. The life cycle is completed when a cat becomes infected by eating an animal with cysts in its tissues.

Pathogenesis

Entry into the host cell is aided by an enzyme produced by the organism, which alters the host cell membrane. Proliferation of *T. gondii* in cells of the host causes destruction of the cells. This process is normally brought under control by the immune response of the host. However, in patients with immunodeficiency, infection can be widespread and uncontrolled, producing many areas of tissue necrosis. The disease process can result from a newly acquired infection, or declining immunity can allow reactivation of latent infection with escape of *T. gondii* from cysts in a person's tissues.

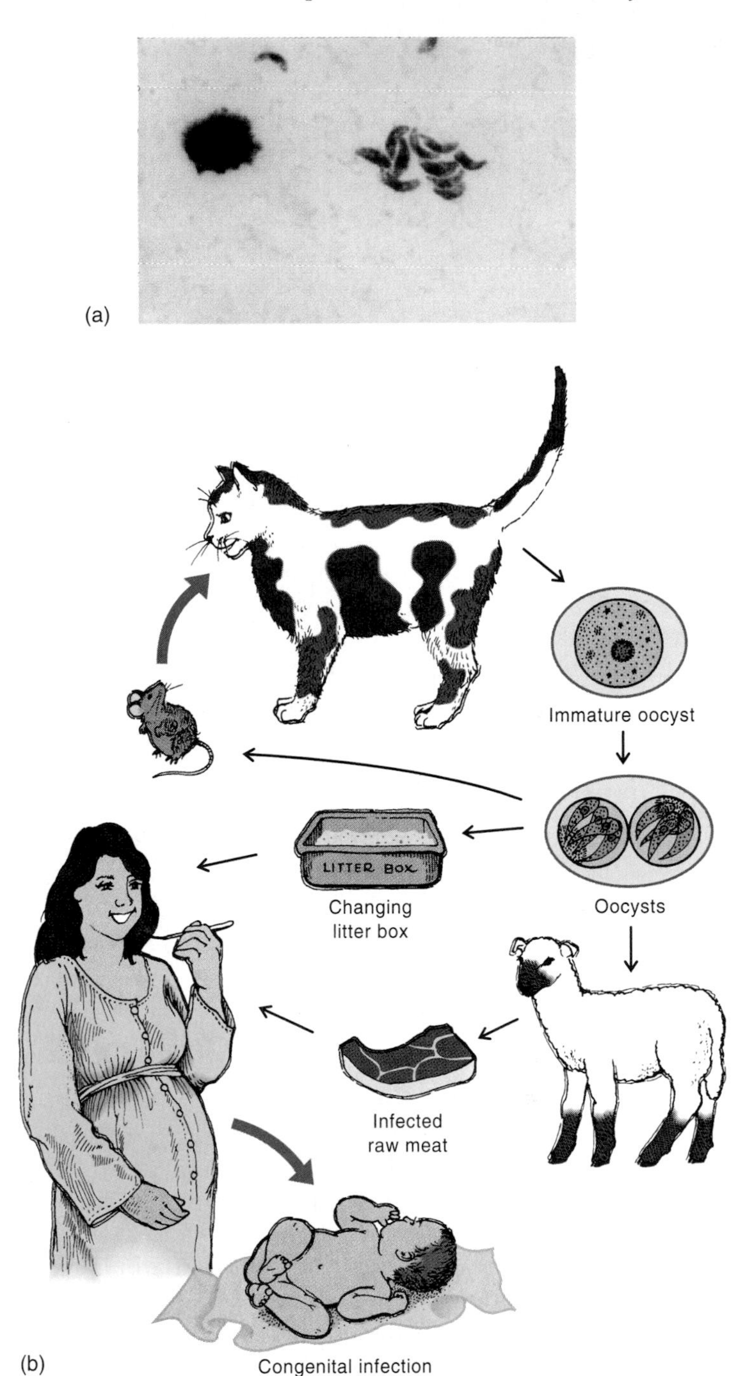

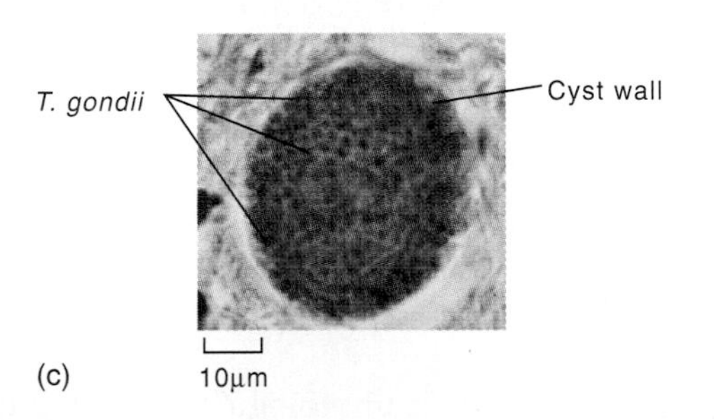

Figure 31.12
Toxoplasma gondii. (*a*) Invasive forms. (*b*) Life cycle. Oocysts from cat feces and cysts from raw or inadequately cooked meat can infect humans and many other animals. (*c*) Cyst in tissue.

Epidemiology

T. gondii is distributed worldwide, but it is less common in cold and in hot, dry climates. Infection of humans is widespread. Serological surveys show the infection rate increases with age, ranging from 10% to 67% among those over 50. Most infections are acquired from ingesting oocysts. Small epidemics have occurred from drinking water contaminated with oocysts. Eating rare meat poses a definite risk. Tissue cysts are present in about 25% of pork, in 10% of lamb, and less commonly in beef and chicken.

Prevention

General measures for preventing *T. gondii* infection include washing hands after handling raw meat, coming in contact with soil, or changing cat litter. Raw or undercooked meat should not be eaten. Meat, especially lamb, pork, and venison, should be cooked until the pink color is lost from its interior. Fruits and vegetables should be washed before eating. Cats should not be allowed to hunt birds and rodents. Children's sandboxes should be kept tightly covered when not in use. These measures are especially important for pregnant women and persons with immunodeficiency.

HIV-infected patients and those about to receive immunosuppressant medications are tested for antibody to *T. gondii*. A positive test indicates they have latent infection. Therefore, those with positive tests are given prophylactic trimethoprim-sulfamethoxazole if they have fewer than 100 CD4+ T cells per microliter.

Cytomegalovirus Disease Is a Common Cause of Blindness in AIDS Patients

Cytomegalovirus (CMV) is a member of the herpesvirus family, which includes herpes simplex virus, EB virus, and varicella-zoster virus, any of which can cause troublesome symptoms in patients with immunodeficiency. CMV, like other herpesviruses, is commonly acquired early in life and then remains latent. With impairment of the immune system, the infection activates and can cause severe symptoms.

Symptoms

Symptoms of cytomegalovirus disease follow a pattern similar to that of toxoplasmosis. Severe damage can occur to the fetus if the mother develops an acute infection during pregnancy. This condition is known as *cytomegalo inclusion disease* and is characterized by jaundice, large liver, anemia, eye inflammation, and birth defects. The vast majority of infected infants appear normal at birth, but almost one of five manifests hearing loss or mental retardation later in life.

Acute infections in immunologically normal individuals are usually without symptoms, but adolescents and young adults sometimes develop fever, fatigue, and enlarged lymph nodes and spleen for weeks or months.

In immunodeficient individuals, CMV infection causes fever, loss of appetite, painful joints and muscles, rapid, difficult breathing, ulcerations of the gastrointestinal tract with bleeding, mental dullness, and blindness.

Causative Agent

Human cytomegalovirus is an enveloped, double-stranded DNA virus that looks like other herpesviruses on electron micrographs but has a larger genome. Its name (*cyto* for "cell" and *megalo* for "large") derives from the fact that cells infected by the virus are two or more times the size of uninfected cells. Infected cells show a large intranuclear inclusion body surrounded by a clear halo, inspiring its description as an "owl's eye." The envelope is acquired as the virion buds from the nuclear membrane.

There are many different strains of the virus detected by the patterns of endonuclease digests, and antigenic differences also occur. Like other herpesviruses, CMV can cause lysis of the infected cell or become latent and subject to later reactivation.

Pathogenesis

A wide variety of tissues are susceptible to infection, but the exact host cell receptors and mode of entry into cells are uncertain. Once cell entry has occurred, the viral genome can exist in latent (noninfectious) form, in a slowly replicating form, or as a fully productive infection. Control of viral gene expression depends partly on the

• **Endonuclease typing, p. 178**

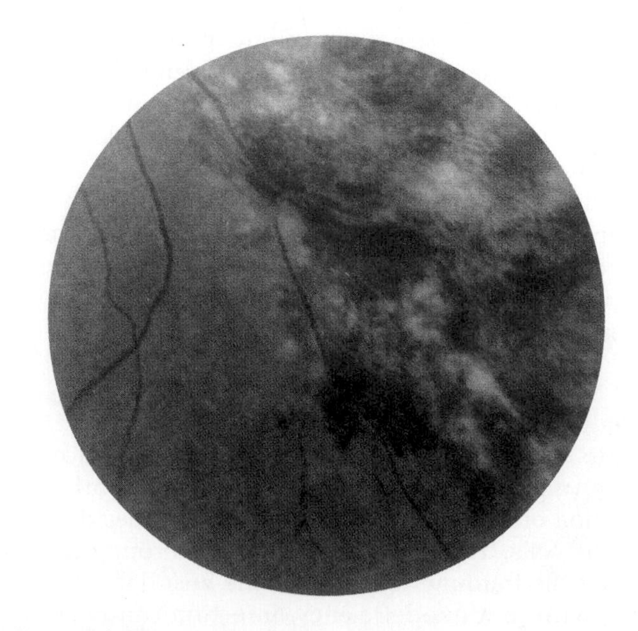

Figure 31.13
Cytomegalovirus (CMV) retinitis. Photograph of a CMV-infected retina as it appears when viewed through the eye's pupil using an ophthalmoscope.

type of cell infected. Monocytes allow low levels of infectious virus production. The CMV genome is ordinarily quiescent in T and B lymphocytes, expressing some viral genes but not producing viral DNA. Integration of viral DNA into the host cell genome probably does not occur. If CMV-infected T cells are also infected with HIV, productive CMV infection occurs. Fully productive infection of a wide variety of tissues occurs during acute infections, causing tissue necrosis. Whether a latent state occurs in tissue cells is not known. Transplanted organs and blood transfusions can transmit the disease, indicating that virus-producing cells are present, or that nonproducing cells in the blood or organ start producing infectious CMV under conditions of immune suppression. Cell-mediated immunity probably plays a role in suppressing production of infectious virus as well as in lysis of infected cells. However, in cells infected with CMV, transfer of MH antigen to the cell surface is impaired and, therefore, CMV antigens are not recognized as being "foreign." CMV infection is associated with an increase in CD8+ cells and a decrease in CD4+ cells, thus enhancing the effect of HIV infection. Latent infections activate with AIDS, organ transplants, and other immunodeficient states. Also, immunodeficient subjects are highly susceptible to newly acquired infection.

Epidemiology

CMV is found worldwide. One U.S. study found that more than 50% of adults 18 to 25 years old and more than 80% of people over 35 years had been infected. Infection is lifelong. Infants born with CMV infection and those who acquire it shortly after birth excrete the virus in their saliva and urine for months or years. Virus is found in semen and cervical secretions in the absence of symptoms, and sexual intercourse is a common mode of transmission in young adults. Up to 15% of pregnant women secrete the virus, and 1% of newborn infants have CMV in their urine. Breast milk may contain CMV and be responsible for transmission. Almost all prostitutes and promiscuous gay men are infected with CMV. CMV spreads readily in day-care centers.

Prevention

There is no approved vaccine yet available. Use of condoms is effective in decreasing the risk of sexual transmission. People with HIV disease or other immunodeficiency syndromes or who lack CMV antibody should avoid contact with day-care centers if possible and, in any case, should wash their hands if exposed to saliva, urine, or feces. Tissue and blood donors can be screened for antibody to CMV. Those who have anti-CMV antibody are assumed to be infected and should not donate to those lacking antibody to CMV. The antiviral medication ganciclovir, given orally, halves the incidence of CMV retinitis (infection of the retina, the light-sensitive part of the eye, figure 31.13) in HIV disease patients with low CD4+ cell counts and positive tests for CMV antibody. A ganciclovir implant designed to release the medication into the eye over a long period also delays the progression to CMV retinitis.

Treatment

The antiviral drugs ganciclovir and foscarnet can reduce the severity of CMV disease. They inhibit CMV DNA polymerase at different sites. Both medications have serious side effects, mainly bone marrow suppression with ganciclovir and kidney impairment with foscarnet.

Mycobacteria Lead the List of Opportunistic Bacteria Complicating Immunodeficiencies

Mycobacterium tuberculosis, the cause of tuberculosis, usually causes an asymptomatic infection that is controlled by the immune system and becomes latent. As might be expected, defects in cellular immunity are associated with reactivation of latent tuberculosis, which commonly results in unrestrained disease in AIDS and other immunodeficiencies. *M. tuberculosis* is not the only mycobacterium that causes disease in immunodeficient people. Disease has been reported due to *M. kansasii, M. scrofulaceum, M. xenopi, M. szulgai, M. ganavense, M. haemophilum, M. celatum,* and others. This section discusses disease caused by organisms of *Mycobacterium avium* complex, next to *M. tuberculosis* the most common mycobacterial opportunists complicating immunodeficiency diseases.

Symptoms

The vast majority of *Mycobacterium avium* complex (MAC) infections in normal people are asymptomatic. Elderly people, especially those with underlying lung diseases, develop a chronic cough productive of sputum and sometimes lesions in the lungs resembling tuberculosis. Children sometimes develop chronic enlargement of lymph nodes on one side of their neck, easily treated by surgical removal of the affected nodes. Patients with immunodeficiencies have symptoms ranging from chronic cough productive of sputum to fever, drenching sweats, marked weight loss, abdominal pain, and diarrhea.

Causative Agent

Mycobacterium avium complex is a group of mycobacteria consisting of two closely related species, *M. avium* and *M. intracellulare*. More than two dozen strains of mycobacteria fall into the MAC, distinguishable by serological tests, optimum growth temperature, and host range. Their growth rate is almost as slow as that of *M. tuberculosis*, but they are easily distinguished by using biochemical tests and nucleic acid probes.

• **Tuberculosis, pp. 535–39**

Pathogenesis

MAC organisms enter the body via the lungs and the gastrointestinal tract. They are taken up by macrophages by phagocytosis, but they resist destruction because they inhibit acid production in the phagosome. Surviving organisms multiply within phagocytes and are carried by the bloodstream to all parts of the body. When cellular immunity is intact, most organisms are destroyed, and disease is localized. Disease spreads throughout the body with profound immunodeficiency, as in AIDS patients with CD4+ cell counts below 50 per micrometer. In these patients, there is persistent bacteremia, with counts as high as 1 million per milliliter of blood. The small intestinal lining contains macrophages packed with the bacteria, and infected tissues may resemble leprosy, containing up to 10 billion of the acid-fast bacteria per gram of tissue. Despite the presence of enormous numbers of the bacteria, there is little or no inflammatory reaction, and the clinical effect is a slow decline in the patient's well-being rather than a quickly lethal effect. One cause of deterioration may be activation of HIV replication when HIV-infected macrophages are infected with MAC.

Epidemiology

MAC organisms are widespread in natural surroundings and have been found in food, water, soil, and dust. In the United States, they are most prevalent in the Southeast, parts of the Pacific Coast, and the North Central region. Some strains are important pathogens of chickens and pigs. In AIDS patients, they are the most common bacterial cause of generalized infection. It is not known whether infection in AIDS patients is mostly newly acquired or a reactivation of latent disease. Presumably, most infections are from environmental sources rather than from person-to-person spread.

• **Leprosy, pp. 633–37**

Prevention

No generally effective measures are available to prevent exposure to MAC organisms. For HIV patients whose CD4+ count is below 75 cells per microliter, the antimycobacterial medication **rifabutin** is recommended to help prevent MAC infections. It is necessary to check for the development of MAC bacteremia because it may occur despite rifabutin, and it requires intensive treatment.

Treatment

Once MAC bacteremia develops, patients must be given two or more medications together, such as clarithromycin plus rifabutin, for life.

REVIEW QUESTIONS

1. Bacteria, fungi, protozoa, and viruses can all cause life-threatening diseases in AIDS patients. Name a fifth kind of infectious agent that can do the same.
2. The disease pneumocystosis mainly involves which part of the body?
3. How do physicians prevent pneumocystosis in AIDS patients?
4. How is toxoplasmosis acquired?
5. In AIDS patients with toxoplasmosis, which part of the body is affected in more than half the cases?
6. What is a feared complication of cytomegalovirus infection in AIDS patients?
7. Where in an AIDS patient's surroundings might MAC organisms be found?

Summary

I. Introduction
   A. AIDS was first recognized when a group of previously healthy young men developed life-threatening infections by organisms generally considered harmless.
   B. Worldwide, it is estimated that a new infection with an AIDS-causing virus occurs about every 13 seconds.

II. AIDS
   A. Acquired immunodeficiency syndrome (AIDS) is a late manifestation of human immunodeficiency virus (HIV) disease.
      1. Symptoms of HIV disease are varied.
         a. "Flulike" symptoms can occur six days to six weeks after infection by the virus.
         b. Asymptomatic persons can transmit the disease.
         c. The disease progresses in the absence of symptoms.
         d. The onset of immunodeficiency may be marked by LAV, ARC, tumors, or opportunistic infections.
         e. Hairy leukoplakia is probably the result of Epstein-Barr virus reactivation.
      2. Human immunodeficiency virus type 1 (HIV-1) is the main cause of AIDS in the United States. It is a single-stranded RNA virus of the retrovirus family. There are nine subtypes, A through I, based on distinctive nucleic acid sequences. It is easily inactivated by many disinfectants.
         a. Important structural components include the gp120 (SU) surface antigen, the gp41 (TM) transmembrane antigen, the p17 (MA) matrix protein, and the p25 (CA) capsid antigen.

b. The virion contains three important enzymes: reverse transcriptase (RT), protease (PR), and integrase (IN).
c. There are two copies of the viral genome.
d. Two genes, *gag* and *pol*, are translated as a unit. The resulting large protein is then split into PR, RT, IN, CA, MA, and other viral proteins.
e. The *env* gene is translated from a spliced messenger RNA, the resulting protein being processed by host cell enzymes to give SU and TM.
f. Six additional genes are involved mainly with regulating viral reproduction.

3. The disease is spread mainly by sexual intercourse, by blood, and from mother to fetus or newborn.
   a. Early on, the epidemic involved mainly promiscuous homosexual men. Transmission has slowed among this group but still represents a leading source of new infections.
   b. Abusers of intravenous drugs who share needles represent an important factor in the epidemic because they spread HIV by both blood and prostitution. There has been little response to control efforts.
   c. Blood transfusions and blood products are now a relatively minor mode of transmission because of screening tests of donated blood and use of recombinant clotting factor for hemophiliacs.
   d. In the United States, AIDS is the leading cause of death in the 25 to 44 age group. The percentage of cases due to heterosexual transmission is increasing as is the percentage of cases in women and children.
4. A number of different kinds of body cells can be infected by HIV, including those of the brain, those of the intestinal epithelium, T4 lymphocytes, and macrophages.
   a. In T4 lymphocytes, macrophages, and some other cells, the CD4 surface antigen is an important receptor for the virus, binding to its gp120 (SU) antigen.
   b. Entry of the virus into T lymphocytes requires that the virus attach specifically to a chemokine receptor (formerly called *fusin*) of the lymphocyte.
   c. Entry into macrophages requires that the virus attach to a different chemokine receptor.
   d. Inside the host cell, a DNA copy of the viral genome is made through the action of reverse transcriptase, a complementary DNA strand is made, and the double-stranded DNA is inserted into the host genome as a provirus by viral integrase (IN).
   e. Replication of mature virions results in the death of infected lymphocytes, whereas macrophages can release infectious HIV without cell death.
   f. HIV mutants readily appear because reverse transcriptase (RT) is prone to make errors in copying the viral genome.
   g. Mechanisms of T4 lymphocyte destruction include viral replication, accumulation of viral products, host immune attack, and fusion of cells with lysis of the resulting syncytium.
   h. Following infection, large amounts of HIV are released into the bloodstream whether or not symptoms occur. Thus, tissues throughout the body become infected. There is a transitory drop in the number of CD4+ lymphocytes.
   i. The immune system largely clears infectious virus from the bloodstream, and CD4+ lymphocytes return nearly to normal levels.
   j. The battle between the immune system and the virus continues, with huge numbers of CD4+ lymphocytes and virions destroyed each day. The immune system gradually falls behind, finally becoming unable to limit the infection. Viremia recurs and AIDS develops in half the cases within 9 to 10 years and in the rest of cases over a longer period.
5. Prevention of AIDS depends on changes in human behavior. HIV disease is not highly contagious.
   a. Persons unsure about whether they might be infected with HIV are advised to get a blood test.
   b. Educational techniques must be designed specifically for each group at high risk.
   c. It is unlikely that an effective vaccine will become available in the near future, although a number of possibilities are being examined.
6. Treatment of AIDS depends on antiviral medications and prevention and treatment of opportunistic infections and tumors.
   a. Combinations of different antiviral medications in conjunction with therapy directed at infectious complications have significantly improved the life expectancy and quality of life of AIDS patients.
   b. The effectiveness of treatment is limited by the expense and side effects of medications and by the development of medication resistance by HIV and other infectious agents.

III. Malignant Tumors Complicating Acquired Immunodeficiencies
  A. Kaposi's sarcoma is a malignant tumor arising from blood or lymphatic vessels.
    1. The tumor is a complication of organ transplantation, AIDS, and other acquired immune deficiencies.
    2. The tumor is common enough to be one of the surveillance conditions for AIDS, although it may appear before severe immune deficiency develops.
    3. Accumulating evidence implicates a previously unknown herpesvirus as a causative agent.
  B. Lymphomas are malignant tumors that arise from lymphoid cells. They have a markedly increased incidence in AIDS and transplant patients.
    1. Both B and T lymphocytes can give rise to lymphomas, but B cell lymphomas are more common in patients with immune deficiencies.
       a. Strong evidence exists that EB virus plays a causative role in these tumors.
  C. There is an increased rate of anal and cervical carcinoma in people with HIV disease.
    1. These tumors arise from epithelial cells and, therefore, differ from Kaposi's sarcoma and lymphomas.
    2. Cancers of the uterine cervix in women and of the anus in women and homosexual men are strongly

associated with human papillomavirus (HPV) types 16 and 18.
   a. The viruses are transmitted by sexual intercourse.
   b. Replication of the viruses increases in the presence of immunodeficiency.
   c. Precancerous changes are detectable before the development of cancer.

IV. Infectious Complications of Acquired Immunodeficiency
  A. Pneumocystosis is a serious lung disease first recognized in malnourished premature infants.
    1. The disease symptoms develop slowly, with gradually increasing shortness of breath and rapid breathing. Fever and cough may or may not be present. Patients can die from lack of oxygen.
    2. The causative agent is *Pneumocystis carinii*, a tiny unclassified organism, most likely a fungus.
    3. Experimentally, the tiny spores are inhaled into the lung, and they attach to the air sacs, which fill with fluid, the multiplying *P. carinii*, and mononuclear cells. The alveolar walls thicken, impeding the passage of oxygen into the blood.
    4. The organisms are widespread in humans and other animals, generally causing asymptomatic infections that become latent. Reactivation of latent infection is probably the source of pneumocystosis in most immunodeficient patients, but an environmental source or person-to-person spread are probable in some cases.
    5. The disease can be prevented in most cases by giving trimethroprim-sulfamethoxazole orally as soon as evidence of immunodeficiency develops. An inhaled medication, pentamidine, is also effective.
    6. Trimethoprim-sulfamethoxazole and other medications, along with oxygen therapy, are available for treatment of pneumocystosis. For unknown reasons, people with HIV disease develop intolerance to trimethoprim-sulfamethoxazole more often than do other patients.
  B. Toxoplasmosis is rare among normal people but can be a serious problem for those with malignant tumors, organ transplantation, and HIV disease. The disease can also be congenital.
    1. Symptoms differ for the immunologically normal, unborn children, and the immunodeficient.
      a. Infection is common in normal people, but only about 10% develop symptoms. Typically, symptoms consist of sore throat, fever, and enlarged lymph nodes and spleen, sometimes with a rash. The clinical picture can be confused with that of infectious mononucleosis. Symptoms disappear over weeks or months and generally do not require treatment.
      b. Infection of the fetus occurs about 50% of the time when the mother develops symptomatic or asymptomatic infection during pregnancy. Early in pregnancy, miscarriage can occur or babies can be born with birth defects or damage to the brain or other organs. Infections later in pregnancy are usually milder but can result in epilepsy, mental retardation, or recurrent retinitis in the child.
      c. Toxoplasmosis is common and life threatening in those with immunodeficiencies. More than half of the cases with toxoplasmosis and AIDS develop encephalitis, and death often occurs because of involvement of the brain, heart, and other organs.
    2. The causative agent is *Toxoplasma gondii*, a tiny banana-shaped protozoan that undergoes sexual reproduction in the intestinal epithelium of cats but can infect humans and many other vertebrates. Oocysts excreted in the feces of acutely infected cats become infectious in soil and contaminate food, water, and fingers. The flesh of animals that have ingested oocysts contain cysts filled with *T. gondii*, also infectious for humans and other carnivores.
    3. An enzyme produced by *T. gondii* aids penetration of the host cell membrane. Proliferation of the protozoan results in destruction of the host cell and release of the microorganisms to infect other cells. Normally, host immunity brings the process under control, and *T. gondii* is confined to cysts. With immunodeficiency, the microorganisms are released from the cysts and cause generalized infection.
    4. *T. gondii* is present worldwide. Most people become infected from oocysts. Epidemics have resulted from contaminated drinking water. Eating rare meat, especially pork and lamb, can also cause infection.
    5. Preventive measures include hand washing after contact with soil, cat litter, or raw meat; washing fruits and vegetables; thorough cooking of meats; and preventing cats from eating birds and rodents. Patients with HIV disease and those scheduled to receive immunosuppressive therapy are tested for antibody to *T. gondii*. A positive test indicates the presence of cysts in body tissues. Prophylactic medication is given when CD4+ cells decline to low levels. Trimethoprim-sulfamethoxazole is effective.
  C. Cytomegalovirus is a common cause of blindness in people with AIDS. Like other herpesviruses, it is often acquired early in life and then becomes latent, only to reactivate with the development of immunodeficiency.
    1. As with toxoplasmosis, symptoms differ among normal people, fetuses, and those with acquired immunodeficiency. Normal individuals generally have asymptomatic infections; fetuses may develop cytomegalo-inclusion disease or appear normal at birth but show mental retardation or hearing loss later; immunodeficient individuals develop fever, gastrointestinal bleeding, mental dullness, and blindness.
    2. Cytomegalovirus (CMV) is an enveloped, double-stranded DNA virus. Infected cells are enlarged and have an "owl's eye" appearance because of a large intranuclear inclusion body surrounded by a clear halo. Multiple strains can be identified by their differing endonuclease digests.

3. A wide variety of tissues is susceptible to infection. The virus can exist in a latent form, a slowly replicating form, or a fully replicating form. Co-infection with HIV results in fully productive infection and tissue death.
4. The virus occurs worldwide, and most adults have been infected. Infants with CMV secrete the virus in saliva and urine for months or years. The virus may also be found in semen and cervical secretions and is commonly transmitted sexually. Breast milk may also contain the virus.
5. No vaccine is available. Condoms decrease the risk of sexual transmission. Those with immunodeficiency are advised to avoid day-care centers and to wash hands after contact with saliva, urine, or feces. Tissue donors are screened for antibody to CMV, and if positive, their tissues are not used for persons negative for the antibody. The antiviral medication ganciclovir can be given to prevent CMV retinitis.
6. Ganciclovir and foscarnet are helpful in treating CMV disease but can cause serious side effects.

D. Mycobacteria commonly cause severe infections in immunodeficient patients. *Mycobacterium tuberculosis* and *Mycobacterium avium* complex (MAC) organisms are most commonly responsible.
1. Normal people usually get asymptomatic or mild infections with MAC organisms, but immunodeficient patients may have fever, drenching sweats, severe weight loss, diarrhea, and abdominal pain.
2. *Mycobacterium avium* complex is comprised of *M. avium* and *M. intracellulare* strains. They can be distinguished from *M. tuberculosis* by biochemical tests and nucleic acid probes.
3. MAC organisms enter the body via the lungs and gastrointestinal tract. They are taken up by macrophages but resist destruction and are carried to all parts of the body. In immunodeficiency, MAC organisms multiply without restriction, producing massive numbers of the organisms in blood, intestinal epithelium, and tissues. There is little or no inflammatory response to the bacteria.
4. MAC bacteria may be present in water, food, and dust. In AIDS patients, they are the most common cause of generalized bacterial infection.
5. No generally effective measures are available for preventing exposure to MAC bacteria. Prophylactic antimycobacterial medication is advised for severely immunodeficient patients, but it can fail to prevent infection.
6. Treatment requires two or more antimycobacterial medications used together for life.

Chapter Review Questions

1. About how many people have died of AIDS in the United States since the epidemic of HIV disease began?
2. What role do asymptomatic people with HIV disease play in the epidemiology of AIDS?
3. What is the main symptom of patients with lymphadenopathy syndrome (LAV)?
4. Why is HIV protease so important to replication of HIV?
5. What is a function of HIV accessory genes?
6. Why might the infant son of a hemophiliac man develop AIDS when the son's parents were strictly monogamous nonabusers of drugs?
7. What are two consequences of the rising percentage of AIDS cases in women?
8. Why is reverse transcriptase needed in order for HIV to become a provirus?
9. Give two reasons it is a good idea to know whether you are infected with HIV.
10. What is a phase III vaccine trial?

Critical Thinking Questions

1. Why is it important to focus on HIV-1 infection rather than AIDS?
2. What differences might have been expected in the response to the AIDS epidemic if HIV-1 had been introduced into the United States in 1928 instead of 1978?

Further Reading

Balter, Michael. 1996. New hope in HIV disease. *Science* 274:1988.

Bloom, B. R. 1996. A perspective on AIDS vaccines. *Science* 272:1888.

Bovsun, Mara. 1993. Probability of polio vaccine tie to AIDS deemed extremely low. *ASM News* 59(2):56.

Chang, Y., et al. 1994. Identification of herpesvirus-like DNA sequences in AIDS-associated Kaposi's sarcoma. *Science* 266:1865.

Cocchi, F., et al. 1995. Identification of RANTES, MIP-1-α and MIP-1-β as the major HIV-suppressive factors produced by CD8+ T cells. *Science* 270:1811.

1995. Case-control study of HIV seroconversion in health-care workers after percutaneous exposure to HIV-infected blood—France, United Kingdom, and United States, January 1988–August 1994. *Morbidity and Mortality Weekly Report* 44:929.

1995. Doubts cast on 1959 AIDS claim. *Science* 268:35. Suggests it was unlikely that the diagnosis of AIDS in an English sailor who died in 1959 was correct.

Feng, Yu, et al. 1996. HIV-1 entry cofactor: functional cDNA cloning of a seven-transmembrane, G protein-coupled receptor. *Science* 272:872.

1995. First 500,000 AIDS cases—United States, 1995. *Morbidity and Mortality Weekly Report* 44(46):849.

Gayle, H. D., et al. 1990. Prevalence of the human immunodeficiency virus among university students. *The New England Journal of Medicine* 323:1538.

Hoffman, Michelle. 1994. AIDS. Solving the molecular puzzle. *American Scientist* 82(2):171. Nice discussion and illustrations.

Lackritz, E. M., et al. 1995. Estimated risk of transmission of human immunodeficiency virus by screened blood in the United States. *The New England Journal of Medicine* 333:1721.

Levy, J. A. 1994. *HIV and the pathogenesis of AIDS*. American Society for Microbiology, Washington, D.C. A nice summary of information about AIDS as of 1994.

Mills, John, and Masur, Henry. 1990. AIDS related infections. *Scientific American* 263:50.

Perelson, A. S., et al. 1996. HIV-1 dynamics in vivo: virion clearance rate, infected cell life span, and viral generation time. *Science* 271:1582.

Piot, Peter. 1996. AIDS: a global response. *Science* 272:1855.

Richardson, Sarah. 1997. AIDS. Crushing HIV. *Discover* 18:28.

Weiss, Robin. 1994. Of myths and mischief. *Discover* 15:36. Clears up some of the misinformation about AIDS.

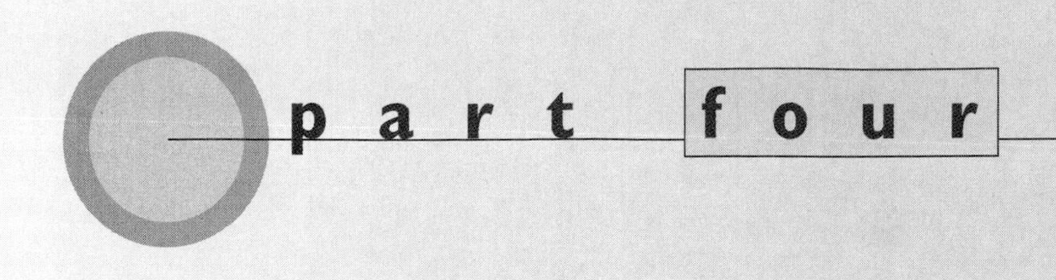

Applied Microbiology

Polluted river. Fertilizers rich in nitrates and phosphates are washed out of farmland by rain and carried into streams and rivers, where they stimulate overgrowth of algae and weeds (eutrophication). The overabundant vegetation blocks out light and, as it decays, exhausts the oxygen supply, allowing growth of anaerobic bacteria that produce obnoxious odors.

32 Environmental Microbiology

KEY CONCEPTS

1. For the microorganism, the environment immediately surrounding it, its microenvironment, is most important for its survival and growth.
2. Different microorganisms are capable of growing in a wide variety of environments, but each will grow best in those environments for which it is well adapted.
3. Life supported by solar energy and photosynthetic primary production is very important, but life sustained by chemosynthetic primary production may actually be more common.
4. Microorganisms are essential for recycling the biologically important elements oxygen, carbon, nitrogen, sulfur, and phosphorus.
5. Microorganisms play important roles in bioremediation, the biological cleanup of pollution.

Soil microorganisms play an important role in maintaining farmland productivity. Elements essential for growth, such as oxygen, nitrogen, carbon, sulfur, and phosphorus, are recycled by soil microorganisms.

A Glimpse of History

For centuries, farmers have understood that they could not continue growing the same crop on the same piece of land year after year without reducing the crop's yield. They know that allowing a field to lie fallow* will enable it to recover its productivity. The wild plants that grow on the field for a year or two appear to rejuvenate the soil. It was not until the late 1800s that scientists began to discover just why this was so. They isolated soil microorganisms that could take nitrogen from the air and chemically transform it into forms of nitrogen that plants could use. This process is called **nitrogen fixation.** Although many scientists have worked for years to understand how microorganisms fix nitrogen, one scientist from the Netherlands, Martinus Willem Beijerinck, stands out as an early contributor in these studies.

In 1887, Beijerinck reported on the properties of the root nodule bacterium, which he called *Bacterium radicicola*. (The genus name of this bacterium was later changed to *Rhizobium*.) He showed in 1890 that root nodules were formed when *B. radicicola* was incubated with certain plant seedlings. Russian microbiologist Sergei Winogradsky then showed that bacteria formed a symbiotic relationship with the roots of certain plants, the legumes. These nodules, which contain bacteria, could fix nitrogen.

Beijerinck also made major contributions to other areas of microbiology. He worked on yeasts, plant viruses such as tobacco mosaic virus, *Azotobacter*, and plant galls. *Beijerinckia*, a group of gram-negative aerobic rods, is named for him. The genera *Azotobacter* and *Beijerinckia* include bacteria that can fix nitrogen under aerobic conditions in the absence of plants.

Beijerinck was described as a "keen observer," a person who was able "to fuse results of remarkable observations with a profound and extensive knowledge of biology and the underlying sciences." This ability was undoubtedly responsible for the great success of his work.

*Fallow—left unplanted for one or more seasons

Microorganisms play an essential role in the environment, in cycling of elements, maintenance of fertile soil and usable water supplies, and decomposition of wastes and other pollutants. Earlier chapters in this text have been concerned mainly with principles of studying microorganisms in the laboratory; the study of microorganisms in their natural environments is much more complex. Instead of growing as **pure cultures** under controlled conditions, organisms in nature exist as members of heterogenous *communities* under poorly defined and often changing environmental conditions. In the laboratory, the investigator provides the nutrients necessary to ensure the optimal growth of the culture. In nature, nutrient supplies are generally limiting. In the laboratory, organisms are often grown in liquid culture, whereas in nature, most microorganisms grow in association with solid surfaces, on plants or animals.

Some principles that govern the growth and development of populations in different environments as well as the role of microorganisms in the cycling of key biological components are discussed in this chapter. Bioremediation, the cleanup of pollutants such as pesticides, oil spills, and metals, will also be discussed.

The Principles of Microbial Ecology

All the living organisms in a given area, the **community,** interacting with the nonliving environment of that area, form an ecological system, or **ecosystem.** The major ecosystems of the world include the sea, rivers and lakes, deserts, marshes, grasslands, forests, and tundras. Each possesses its own particular physical environment and spectrum of living organisms.

Microorganisms play a major role in most ecosystems, and many ecosystems are populated with unique microorganisms. The role that an organism plays in its particular ecosystem as well as the physical space it occupies are called its **ecological niche. Ecology** is the study of the relationship of organisms to each other and to their environment.

The region of the earth inhabited by living organisms is called the **biosphere.** Within the biosphere, wide diversity exists, both in the number of different species of organisms inhabiting particular ecosystems and in the total quantity of biomass[1] present in any given area. The fundamental requisite for all organisms in any environment is that they be capable of multiplying, at least from time to time. In an environment conducive to life, as for example, the top few inches of fertile soil, there is likely to be a large biodiversity (variety of different species) as well as a large quantity of biomass. It has been estimated that the top 6 inches of fertile soil may contain more than 2 tons of bacteria and fungi per acre. When a sample of such fertile soil is cultured, millions of organisms per gram of soil are isolated. However, the number of organisms cultured is extremely misleading since only about 1% of the organisms in many environments can be cultured in the laboratory. Both the number and the diversity of organisms are much greater than ever imagined.

In harsh environments, exposed, for example, to extremes of temperature or moisture, relatively few species are capable of multiplying and, therefore, of existing for a significant period. Microorganisms survive such environments by producing resting forms such as endospores.

The external environment has a more profound effect on conditions inside microbial cells than it does on cells of higher organisms. This is especially obvious in the case of temperature. Some multicellular organisms contain sensitive controls in specialized organs; these maintain reasonably constant internal conditions despite wide fluctuations in the external environment. In microorganisms, the external and internal environments may be similar, although even bacteria have the ability to concentrate essential foodstuffs and to keep out some potentially harmful chemicals.

Because microorganisms are so small, the true environment of a microorganism is often difficult to identify and measure. The environment immediately surrounding an individual cell, the **microenvironment,** is most relevant.

[1]Biomass—total weight of all organisms in an environment

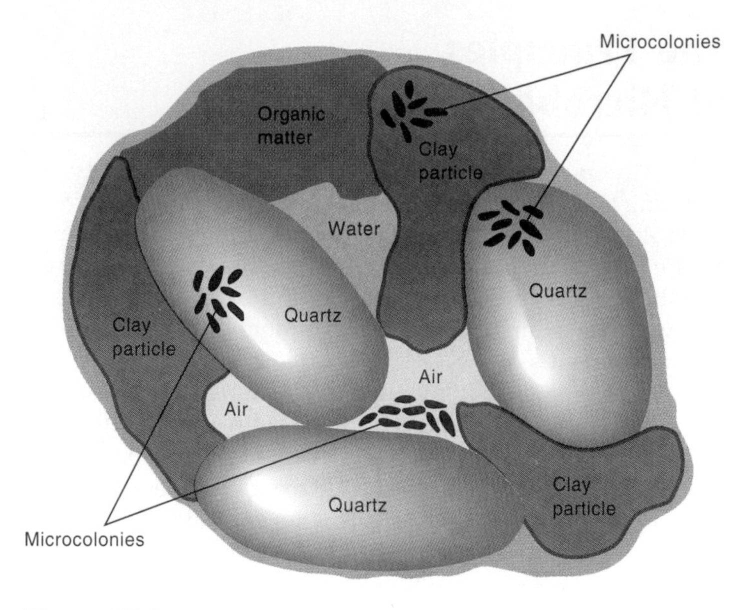

Figure 32.1
Soil microorganisms usually form microcolonies within soil aggregates or crumbs. The nature of the surrounding microenvironments determines where specific microorganisms are localized.

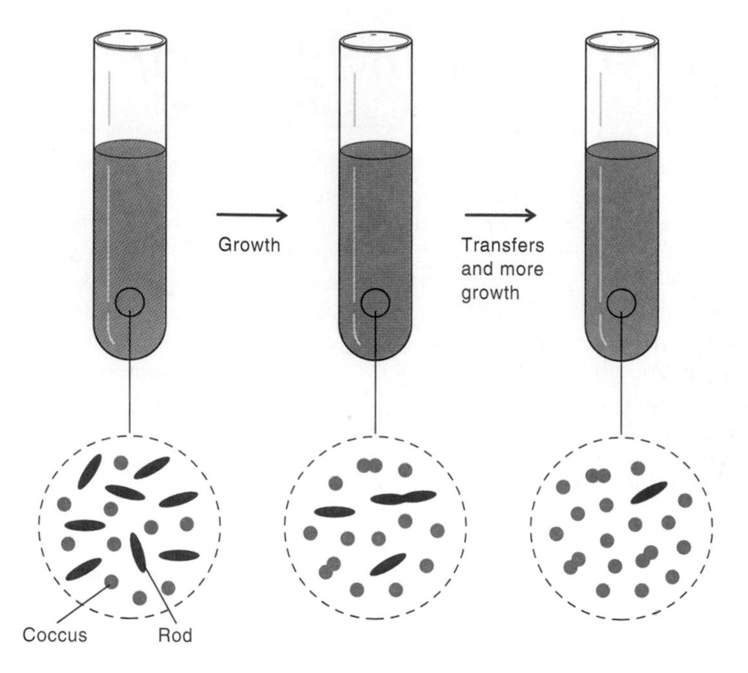

Figure 32.2
Competition between two bacteria. The coccus has a generation time of 99 minutes, while the rod divides every 100 minutes.

The more readily measured gross environment, or macroenvironment, may be very different from the microenvironment. To a bacterial cell living within a crumb of soil (figure 32.1), the environment inside this crumb is more important than that outside it. Not only may the two environments differ radically from each other, but a variety of microenvironments may actually be found within a single crumb. For example, crevices can rapidly become anaerobic while outer surfaces remain aerobic.

Microbial Competition Demands Rapid Reproduction and Efficient Nutrient Use

Although many different species may be capable of living in the same microenvironment, only a few species, sometimes only one, actually occupy a specific microenvironment. To return to a previous example, a large number of different species are found within a crumb of soil. If the crumb is carefully dissected into squares of about 70 micrometers per side, only a single species will be found living as a microcolony within any one of these squares. The species best adapted to live in a particular microenvironment is the one that inhabits it to the exclusion of all others.

Perhaps nowhere in the living world is competition more fierce and the results of competition more quickly evident than among microorganisms. The fitness of an organism, as measured by its ability to compete successfully for its ecological niche, is generally related to the rate at which the organism multiplies, as well as to its ability to withstand adverse environmental conditions. The complete takeover of one species by another is especially likely when two species have similar nutrient requirements and one of these nutrients is in limited supply. Thus, the organism that multiplies faster will yield the greater cell population, utilizing more of the limited nutrient supply. Because cells are often able to divide at least once every several hours, any small differences in their generation times will result in a very large difference in the total number of cells of each species after a relatively short time. For example, if 10 cells each of a rod and a coccus are growing in the same natural environment, under conditions in which the generation time of the rod is 100 minutes and that of the coccus is 99 minutes, and if the number of cells reaches about 10^9, there will be 1.35 times more cocci than rods. If the cells are then diluted so that they can continue to multiply, the ratio of cocci to rods will increase to the point where there are very few rods left (figure 32.2).

These considerations probably explain the observation that the species of bacteria living in the human intestine tend to remain quite stable with time. Despite the fact that humans swallow large populations of bacteria, including some closely related to those already present, the new strains have great difficulty becoming established. The resident population not only competes successfully for a limited food supply, it also produces toxic substances that inhibit the growth of newcomers. The resident microbial population thus provides a natural defense mechanism against intestinal pathogens.

The cells of colonies growing close together hamper one another's growth by competing for the same nutrients. This phenomenon is readily demonstrated in the laboratory on agar plates. Colonies close together are much smaller than those that are well separated (figure 32.3).

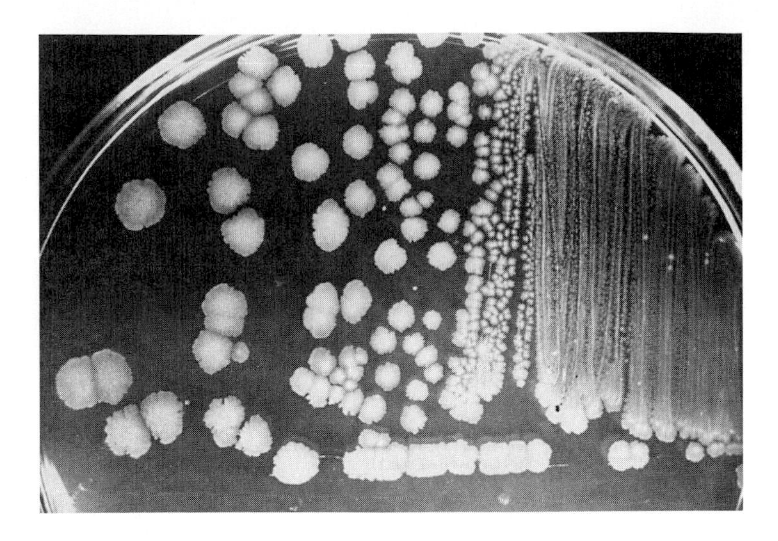

Figure 32.3
Competition for nutrients among bacteria. The closer the cells are to one another, the smaller the colonies are.

Microbial Populations Both Cause and Adapt to Environmental Changes

Environmental changes often result in population changes. Those organisms that have adapted to live several inches beneath the surface of an untilled field will probably not be well adapted if the field is plowed up, fertilized, and irrigated. If the organisms are to survive under these new conditions, they must adapt to the altered environment using one or more mechanisms. For example, the same cells may be able to synthesize new enzymes that help them cope with the new environment. Generally, cells synthesize only the enzymes they absolutely require for growth. Under changed environmental conditions, however, additional or different enzymes may become important. The cells will then manufacture these enzymes and stop synthesizing others that have lost their usefulness. For example, some mercury-resistant bacteria can produce an enzyme that inactivates mercury, a toxic metal. The enzyme is only formed when mercury is present. In other words, cells have the ability to sense changes in the environment and respond to these changes by altering the enzymes that they synthesize.

Another type of adaptation takes place when the pre-exisiting population changes. Thus a mutant within a pre-existing population may be especially well adapted to the new environment, even though it is in the minority. It may then become the dominant organism. The emergence of antibiotic-resistant organisms in hospital environments illustrates this type of adaptation (figure 32.4).

Similarly, a minor species in a population may become dominant owing to a changed environment. In addition to external sources creating environmental changes, the growth and metabolism of organisms themselves may alter the environment dramatically. Nutrients may become depleted, and a variety of waste products, many of which are toxic, may accumulate. In some environments, the changing conditions bring about a highly ordered and predictable succession of bacterial species: First one species becomes dominant, then another, and then a third. An example of such a microbial succession is the one that occurs in unpasteurized milk (figure 32.5).

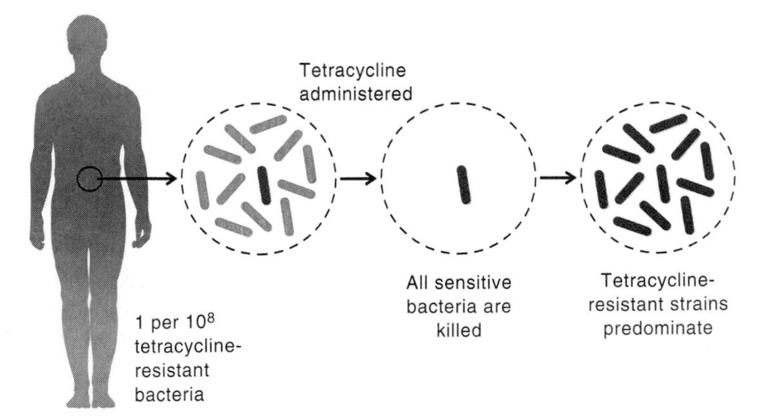

Figure 32.4
Selection of an antibiotic-resistant mutant in a population by the application of antibiotic therapy.

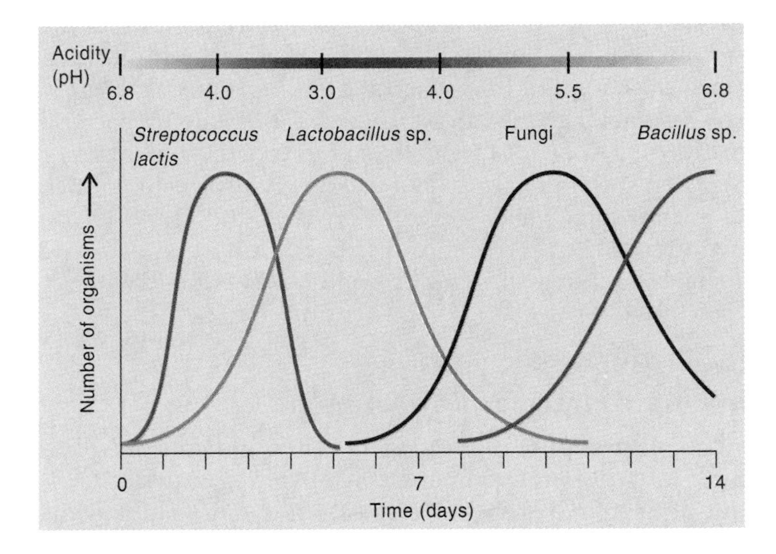

Figure 32.5
Growth of microbial populations in raw milk at room temperature. Unpasteurized milk naturally contains *Streptococcus lactis* and other bacteria that multiply efficiently at room temperature, producing enough lactic acid and other acids to decrease the pH from near neutral (6.8) to acid (4.0 to 4.5) within a few days. As the acidity becomes greater, the growth of lactobacilli, which are also normally present in milk, is favored. The lactobacilli produce even more acid, further decreasing the pH to a very acidic 3.0 to 3.5. This acid pH discourages the growth of most bacteria but permits many fungi to grow. The fungi actually oxidize the acids, releasing energy and producing carbon dioxide and water. As the acidity is lost, species of *Pseudomonas*, some spore-forming bacteria, and other organisms digest the proteins and fats that putrefy the milk.

Unpasteurized milk often contains numerous different organisms, including various species of bacteria, yeasts, and molds, which are derived mainly from the immediate environment around the cow. Initially, the dominant organism is usually the bacterium *Streptococcus lactis*, which breaks down the milk sugar lactose to lactic acid. The resulting acid inhibits the growth of most other organisms in the milk, and eventually enough acid is produced to repress even the further growth of the *S. lactis*. The acid sours the milk and also curdles it, a result of denaturation of the milk proteins. Bacterial species such as *Lactobacillus casei* and *L. bulgaricus* are capable of multiplying in this highly acidic environment. These species metabolize any remaining lactose, forming more acid until their growth is also inhibited. Yeasts and molds, which grow very well in this highly acidic environment, then become the dominant group and convert the lactic acid into nonacidic products. Because most of the milk sugar has already been used, the streptococci and lactobacilli cannot resume multiplication since they require this substrate. Milk protein (casein) is still available, and it can be utilized for energy by bacteria of the genus *Bacillus*, which secrete proteolytic enzymes that digest the protein. This breakdown of protein, known as **putrefaction,** yields a completely clear and very odorous product. The milk thus goes through a succession of changes with time, first souring and finally putrefying.

Some Bacteria Can Grow in a Low-Nutrient Environment

Since **oligotrophic** (low-nutrient) environments are common in nature, microorganisms capable of growth in dilute aqueous solutions are also common. These organisms are by no means restricted to lakes, rivers, and streams. Indeed, microorganisms even grow in distilled-water reservoirs such as are found in research laboratories and mist therapy units used in hospitals to treat patients with lung diseases. In these environments, the microbes are able to extract trace amounts of volatile nutrients absorbed by the water from the air and to grow slowly, reaching concentrations as high as 10^7 per milliliter. Since this cell concentration is not high enough to result in a cloudy solution, the growth is usually unnoticed. This can have serious consequences for the health of hospitalized patients and for the success of laboratory experiments that depend on water purity. The fact that bacteria can grow in such a dilute medium indicates they are able to concentrate nutrients very efficiently inside themselves. Generally, organisms that grow in dilute environments have been found to contain highly efficient transport systems for moving nutrients inside the cell.

• **Contamination of dilute solutions, p. 431**

Microorganisms and Soil

The many microorganisms that live in the soil play an indispensable role in maintaining life on this planet. The soil contains a large variety of microorganisms able to degrade or chemically modify organic and inorganic molecules. It also contains organisms that are indispensable to life on earth because they transform chemical substances that cannot be readily utilized into others that serve as nutrients for a variety of other organisms. For example, although we live in a virtual ocean of air that is 79% nitrogen, this element cannot be used by humans unless it is first converted into amino acids. Certain soil microorganisms such as those of the genus *Rhizobium* are able to fix atmospheric nitrogen by reducing it to ammonia (NH_3). This ammonia is then available to the plants that use it to synthesize amino acids and eventually protein.

Considerable human interest in soil organisms, such as *Streptomyces* sp., stems from the fact that some of them produce antibiotics (table 32.1). The pharmaceutical industry has tested many thousands of soil microorganisms to

Table 32.1 Some Useful Antibiotics Synthesized by Species of *Streptomyces*

Antibiotic	Produced by	Active against*
Streptomycin	*S. griseus*	*Mycobacterium tuberculosis, Yersinia pestis, Brucella* sp.
Spectinomycin	*S. spectabilis*	Penicillinase-producing *Neisseria gonorrhoeae* (most strains)
Neomycin	*S. fradiae*	Bowel flora, except anaerobes; generally used only in ointments applied to body surfaces due to toxicity
Tetracycline	*S. aureofaciens*	*Franciscella tularensis, Brucella* sp., *Yersinia pestis,* rickettsias and chlamydias, *Mycoplasma*
Erythromycin	*S. erythreus*	*Bordetella pertussis, Corynebacterium diphtheriae, Mycoplasma pneumoniae, Chlamydia trachomatis, Legionella pneumophila*
Clindamycin	*S. lincolnensis*	Effective against obligate anaerobes, especially *Bacteroides fragilis*
Nystatin	*S. noursei*	Used on body surfaces against fungi, especially *Candida albicans*
Amphotericin B	*S. nodosus*	Used against invading fungi such as *Histoplasma capsulatum* and *Coccidioides immitis*
Chloramphenicol	*S. venesuelae*	*Salmonella typhi* and *Haemophilus influenzae* (strains resistant to less toxic antibacterials)

*Most antibiotics are effective against several different species of bacteria. The entries in this column refer to the usual applications of a given antibiotic. See appendix VIII, p. 756.

find ones that produce useful antibiotics. There is also an ongoing search to find organisms capable of synthesizing other useful compounds. Probably in no other habitat can one hope to find a greater range of synthetic capabilities than are represented in the soil.

Not all soil microorganisms are beneficial. The soil is also the habitat of a number of fungi and bacteria that are potentially pathogenic for plants and animals. Among those pathogenic for humans are members of the bacterial genera *Clostridium* and *Nocardia* and the fungal genera *Coccidioides, Histoplasma,* and *Blastomyces.*

Soil Teems with a Variety of Life Forms

Most soils can support a variety of life ranging from the submicroscopic viruses of bacteria and plants to the macroscopic nematodes, earthworms, insects, and other larger forms.

Bacteria

The physiological diversity of bacteria allows them to colonize the various types of soil, with many different genera being present. Although bacteria are the most numerous soil inhabitants, the biomass of fungi is greater. Since soil contains aerobic and anaerobic microhabitats, both aerobes and anaerobes are present in it. In acid or alkaline soils, specifically adapted organisms will be present. In desert or arctic soils, bacteria with resistant forms, such as endospores or cysts, will predominate.

In general, the largest group of bacteria in the soil are the **actinomycetes.** Actinomycetes are often resistant to dry conditions and can survive in desert soils. Soil actinomycetes are aerobes and use organic compounds obtained by decomposing dead plant and animal matter. The most common genera are *Nocardia, Arthrobacter,* and *Streptomyces.* Bacteria of the genus *Streptomyces,* in addition to producing antibiotics, produce metabolites called **geosmins,** which give soil its characteristic odor.

Another common group of heterotrophic soil bacteria is the **myxobacteria.** Many of these are pigmented and form brightly colored masses that can sometimes be seen on forest debris. Under starvation conditions, myxobacteria will swarm together to form complex macroscopic fruiting bodies. These fruiting bodies contain spores that can survive nutrient-poor conditions.

Soil bacteria are important in cycling the biologically important elements nitrogen, oxygen, carbon, sulfur, and phosphorus. Each of these cycles is discussed later in this chapter. Bacteria are also involved in controlling the availability of some key metals such as iron and manganese that are necessary trace elements for living cells.

Fungi

Many types of fungi can be found in soils, but the specific genera will vary with the type of soil and with its physical and chemical properties. Some soil fungi are free living, and some, the mycorrhizae, occur in symbiotic relationship with plant roots. Species of the genera *Candida, Aspergillus, Penicillium, Rhizopus,* and *Mucor* are common. Some fungi are capable of degrading and using components of soil that cannot be attacked by other microorganisms. Other fungi are important plant pathogens attacking wheat and corn among others.

Fungi in soil generally exhibit active metabolism during periods when environmental conditions are favorable and become dormant when conditions change for the worse or when nutrient supplies diminish. Since they are aerobes, they are usually found in the top 10 centimeters of soil. The soil fungi are extremely important in the decomposition of plant matter, degrading and utilizing complex macromolecules such as cellulose and lignin.[1]

Algae

A number of algal types may be found in the soil. Because they depend on sunlight and photosynthesis to provide their energy needs, they mostly live on or near the soil surface. Algal populations are quite sensitive to environmental conditions, declining during periods of drought or low temperature. Algae are a major nutrient source for a number of soil inhabitants, including protozoa, fungi, earthworms, and nematodes. In barren soils, lichens, which are algal-fungal symbionts, are common.

Protozoa

Protozoa are found in most soils at a density of about 10^4 to 10^5 organisms per gram of soil. They generally occur near the surface, since most require oxygen. Protozoa are predators of soil bacteria and algae.

Other Eukaryotic Organisms in Soil

Larger forms of life are present in soil in vast numbers. They vary in size from the large moles and gophers through the smaller earthworms down to the barely visible nematodes and tiny insects. The numbers of these barely visible forms can be staggering. A single acre of soil may contain 1 billion insect eggs and larvae and up to 1 billion mites.

The Environment Influences the Bacterial and Fungal Flora in Soil

Environmental conditions affect the density and composition of the bacterial and fungal flora of the soil. The primary environmental influences include moisture, acidity, temperature, and nutrient supply.

Moisture in the Soil Affects the Oxygen Supply

Wet soils are unfavorable for most bacteria simply because the pore spaces in the soil fill up with water, thus diminishing

[1]Lignin—complex, aromatic polymer of phenylpropane; major cell wall component of woody plants

the amount of air in the soil. However, when the moisture content of soil drops to a low level, as in a desert environment, the metabolic activity and number of soil bacteria are markedly reduced. Many species of soil bacteria, both those that form spores and those that do not, are markedly resistant to drying. Representatives of these groups can survive drought for years.

Because the filamentous fungi are strict aerobes, they are found close to the surface of the soil. In waterlogged soils, the numbers of vegetative cells of these fungi are also markedly reduced, but their spores often persist. Once the water in the soil diminishes and the level of soil oxygen increases, the molds resume multiplication and become an important part of the population.

Acidity Influences the Relative Proportion of Fungi

Highly acid or alkaline conditions generally inhibit the growth of common bacteria. However, many species of fungi can multiply over a broad pH range—from alkaline above pH 9 to acid almost as low as pH 2. Thus, fungi will dominate the soil population in areas of low pH. For this reason, adding lime to a soil reduces the abundance of fungi and treating the soil with acid-forming fertilizers, such as those containing ammonium, increases the fungal abundance. Some common bacteria in soil can oxidize the ammonium to nitric acid, thereby creating an acid environment. This is why mushrooms may appear in abundance in a lawn fertilized with an acid-producing fertilizer such as ammonium chloride. The increased abundance of fungi at low pH results primarily from the fact that the bacteria are suppressed and thus do not compete with the fungi for nutrients.

Temperature Regulates Microbial Activities

Temperature governs the rates of biochemical processes through its effect on the rates of enzyme reactions. A warming trend favors the biochemical changes in soil that are brought about by its microbial population. The bulk of the soil bacteria are **mesophiles** (bacteria that grow best at temperatures between 20°C and 50°C). True **psychrophiles** (bacteria that grow below 20°C) are rare or absent. In winter, the bacteria in soil are usually cold-tolerant mesophiles rather than psychrophiles. **Thermophiles** (bacteria that can grow above 50°C) are ubiquitous, occurring especially in such environments as compost heaps, which generate a great amount of heat.

Availability of Nutrients Limits the Size of the Microbial Population

The great majority of soil bacteria and fungi are heterotrophic. The organic matter that serves as their source of energy is produced largely by the photosynthetic activity of higher plants. The size of the microbial population is limited by the organic matter available. The addition of organic material, such as occurs when manure or crop residues are plowed into the soil, dramatically increases the number of bacteria and also fungi in the soil.

In soils in which plants are growing, a variety of organic materials exude from the plant roots. The zone that surrounds the roots and contains these exudates is termed the **rhizosphere.** The exudates provide an abundant energy source for bacteria, fungi, and other heterotrophic organisms. These organisms tend to concentrate on or near the root surfaces of plants.

Energy Sources for Ecosystems

All ecosystems must have access to a source of energy. The energy trapped in chemical bonds cannot be totally recycled, since a portion is always lost as heat. Thus, energy is continually lost from biological systems. To compensate for this outflow, energy is continually added to most ecosystems in the form of solar or light energy. Before this energy is useful, however, it must be converted into chemical bond energy. The conversion process, called **photosynthesis,** is carried out by chlorophyll-containing plants and microorganisms. These organisms, the producers, make the light energy available.

The requirement for solar energy has traditionally been used to explain why life is not equally abundant everywhere in and on the earth. Solar energy is confined primarily to a thin shell in the upper levels of the soil and water and to the lower levels of the atmosphere. It does not penetrate to deep soil, rocks, or the depths of the oceans. Until recently, it was thought that any organism present in these zones must depend for its food on organisms that have access to light. However, the discovery of different types of ecosystems has dramatically changed these ideas.

First, it was found that large communities of organisms exist in the hot water around vents on the ocean floor near the Galápagos Islands, in the Juan de Fuca vent off the northwest coast of the United States, and in other areas. Thousands of feet below the surface, a number of undersea hot springs (or vents from the interior of the earth) spew out vast quantities of minerals and gases, including hydrogen sulfide (figure 32.6). The communities that grow on the dark ocean floor near these hot (40°F to 70°F), undersea geysers are supported not by photosynthesis, but by chemoautotrophy or **chemosynthesis.** Large numbers of sulfur-oxidizing *Thiomicrospira* sp. have been found in the area, both free living and in association with the large tube worms and clams discovered there. The chemoautotrophs obtain energy from oxidation of hydrogen sulfide, and they reduce CO_2 to synthesize organic compounds, which the tube worms and clams then feed upon. This type of

Perspective 32.1

Life on Mars?

The world listened with extreme interest in 1996 when NASA scientists reported finding what could be life forms in a meteorite from Mars. The meteorite, called ALH 84001, is a 4-pound potato-shaped rock found in Antarctica in the 1980s. It is estimated to have formed on the surface of the red planet 4.5 billion years ago and to have been launched into space by an asteroid collision with Mars about 15 million years ago, finally falling on Antarctica 13,000 years before it was found there. Chemical analysis has established the Martian origin of the rock. The evidence for living forms within the rock is not so conclusive.

Four types of evidence were put forward to support the thesis that life once existed within the rock. Carbonate globules, mixtures of calcium and magnesium carbonates, were present along fractures and pores in the rock. These could have been deposited by water flowing along the rock, water being a prerequisite for life. Magnetite and iron sulfide particles were present. These are often formed by bacteria and other living organisms. Polycyclic aromatic hydrocarbons (PAHs) were present. These are compounds left behind when living organisms decay. Very tiny microscopic fossils, resembling bacteria but smaller, were found inside the carbonate globules. It is proposed that these were tiny bacteria or bacterialike organisms that lived on Mars long ago, died, and left their remains within the rock.

In 1997, at the next large gathering of planetary scientists, those who made the first report were more confident that they were right, but other researchers doubted their conclusions. The latter suggested that the carbonate globules were formed by high-temperature impacts on the surface of Mars, rather than by water. Some thought the magnetite and iron sulfide particles looked like the trail left by gases from volcanic eruptions. Still others felt that the PAHs were contaminants from the Antarctic ice and snow. The supposed bacterial fossils were about 100 times smaller than any known bacteria on Earth, bolstering the skeptics.

The opinions and data accumulate. One reason for disbelief has been the extreme conditions under which these purported organisms must have survived. The recent discoveries of extremophiles thriving under previously unbelievable conditions on Earth has made the reports of Martian microbes easier to accept or at least to imagine. The series of spacecraft being sent to Mars during the next decade may bring back more evidence to enable scientists to answer the question, "Life on Mars?"

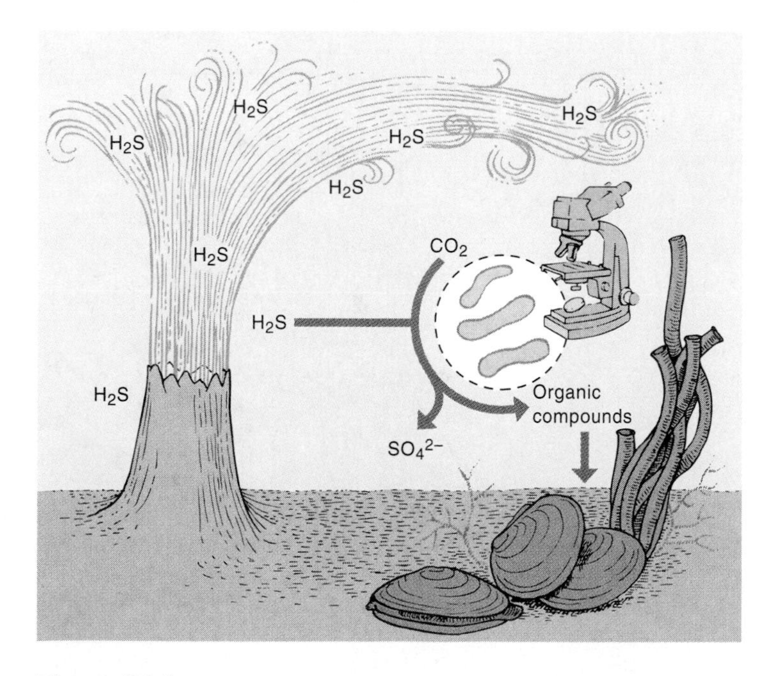

Figure 32.6
A hydrothermal vent community—an ecosystem in which the producers are thought to be chemoautotrophic bacteria. Water escaping from the vent is full of minerals and dissolved gases, including hydrogen sulfide.

symbiosis is now thought to be widespread among species that live in environments where oxygen, carbon dioxide, and hydrogen sulfide are present.

Subsequently, another type of seafloor vent was studied, the conical mounds up to 30 feet high, called **smokers.** The smokers spout superheated waters at temperatures up to 600°F or even higher. Water can remain liquid at this temperature because the underwater pressure at these depths is so high, reaching 265 atmospheres. The large numbers of microbes isolated from the smokers were found to be **anaerobic hyperthermophiles** as well as **barophiles.** Organisms that live under such extreme conditions are known as **extremophiles.** The finding of large numbers of organisms living under extreme conditions suggested that the growth of microorganisms is limited not by high temperatures, but by a requirement for liquid water, as long as other conditions are suitable for growth. If this is so, it follows that conditions suitable for life exist in many environments on the earth and in the universe that were previously thought uninhabitable by living organisms.

In 1994, there were reports of microbial populations living more than 9,000 feet under the ground. Then, in 1995, scientists found large populations of microbes thriving in volcanic rocks called *basalts* from more than

Perspective 32.2

Acid Rain: The Canadian Lakes Experiment

As the world becomes more industrialized, the atmosphere becomes increasingly filled with the by-products of that industrialization, chemicals that, when combined with rainwater, form solutions of acids. When this rain, called *acid rain*, falls to earth, it can cause serious damage to trees and lakes.

For many years, companies around the world did not want to believe that the pollutants they were spilling into the atmosphere were in any way connected to the phenomenon of acid rain and damaged lakes. They felt that other things must be accounting for the dying lakes.

To answer the questions of whether acid rain could be harmful to lakes, Canadian scientists undertook a carefully controlled scientific study. In 1976, in a remote area of Ontario Province, they deliberately added acids and other pollutants to several small lakes and observed the effects on the lakes and their inhabitants. These lakes were in an area that was not otherwise affected by acid rainfall. In this controlled situation, the following observations were made:

1. Most lakes have some capacity to neutralize acid so that, in the first two years, not much change was observed.
2. However, there were some subtle changes that occurred even during the early stages of acidification. At a pH of 5.9, the freshwater shrimp disappeared, and fathead minnows failed to reproduce. The phytoplankton remained abundant but with a different mix of species.
3. Lake trout stopped reproducing as the pH approached 5.6. The crayfish numbers declined as they also stopped reproducing. Adult trout were starving and the shrimp population declined.
4. The biomass remained fairly constant, but the composition changed as the lakes became more acid. The number of species of insects dropped from 70 to 35. Large clumps of sticky, filamentous algae were seen floating just above the lake floor as the pH dropped below 5.6.
5. At a pH of 4.5, a level that has been reached by some Scandinavian lakes due to acid rain, the total biomass declined and the lakes began to die.

The good news was that the study showed that the acidification process could be reversed, provided the lake had not become too acidic. As the pH returned to 7.0, species began to reproduce again. In addition, the plankton increased and the algal balls declined. Nevertheless, indications are that it would take many years for lakes to fully recover after acid rain ceases.

These studies helped convince industrial polluters that it is essential that they find ways to avoid emitting materials into the air that will form acids.

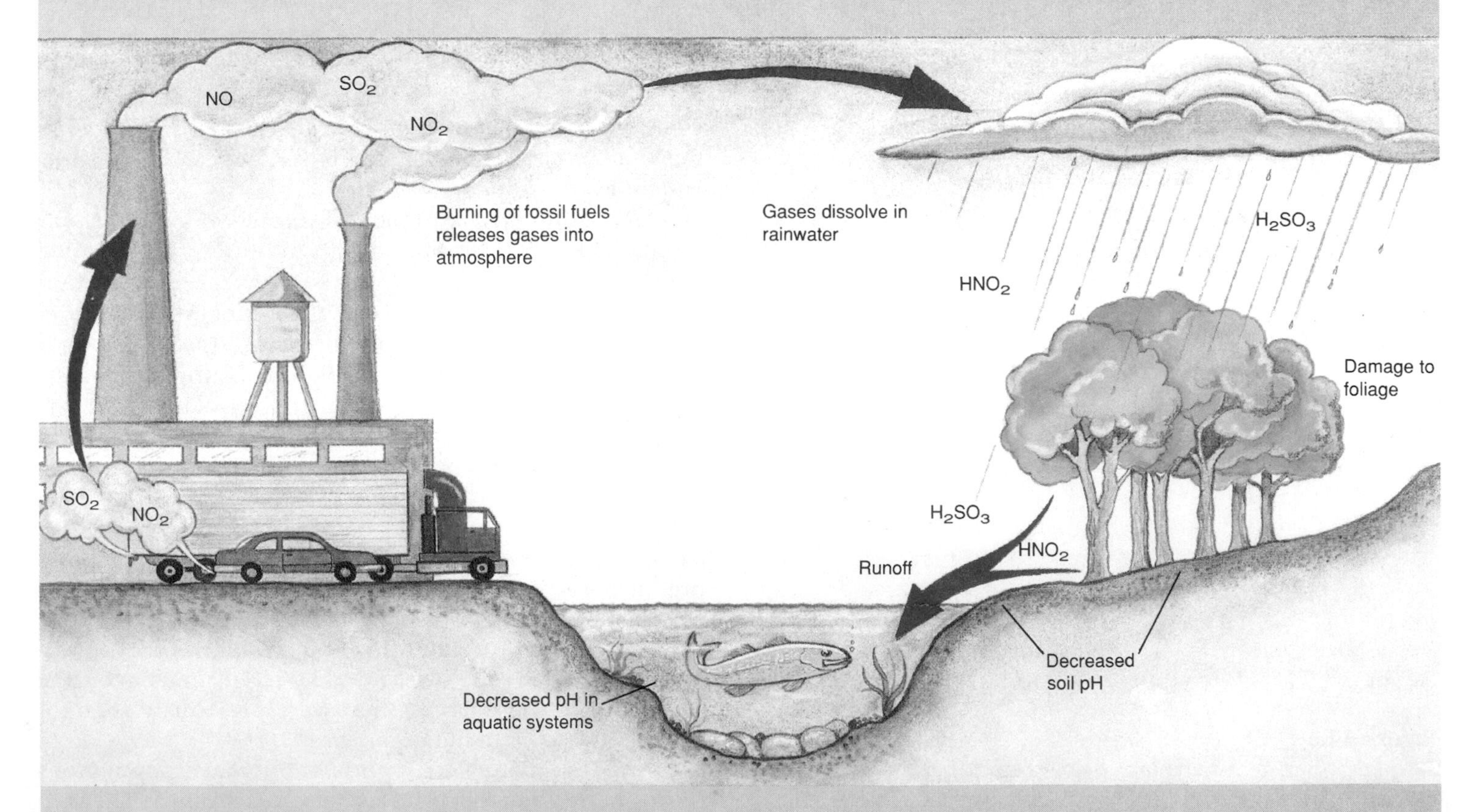

Generation of acid rain. Emissions of sulfur and nitrogen oxides from the burning of fossil fuels result in dilute solutions of sulfuric and nitric acids reaching the earth in the form of rain, snow, or particles.

3,000 feet below the surface of the Columbia River. These organisms gain energy from hydrogen produced by reactions between the basalt and water seeping through the rocks. It has been estimated that *if* (and it is a big if at this point) most appropriate rocks contain microbes, there could be as much as 2×10^{14} tons of underground microorganisms, equivalent to a layer 5 feet thick over the entire land surface of the earth.

Thus, evidence is pointing to the conclusion that life based on solar energy and photosynthetic primary production may be the exception and life sustained by chemosynthetic primary production the more common means of existence.

Extremophiles found in superheated smokers require very high temperatures in order to multiply. For example, the archaen *Pyrobolus fumarii* can multiply at temperatures as high as 221°F, but it cannot grow below 194°F. Other extremophiles isolated from Antarctic sea ice require opposite extremes of cold. The bacterium *Polaromonas vacuolata* grows best at about 39°F and does not multiply at temperatures above 50°F. The extremophile archaens and bacteria have some enzymes that function at extremes of environmental conditions, and these enzymes are proving to be very useful in industrial applications.

Aquatic Environments

Water has several unique properties that help make it such a necessary part of the environment for many microorganisms and that contribute to its essential functions within all living cells. For example, water can absorb light of certain wavelengths, an important property for photosynthetic aquatic microbes. Also, it is an extremely efficient solvent, a property that is of paramount importance for all organisms, whether they are terrestrial or aquatic. Virtually all water contains living microorganisms, the numbers and varieties varying greatly with different aquatic environments.

Rainwater normally contains very small amounts of dissolved nutrients obtained from material suspended in the air. However, increasing world industrialization has led to a serious problem known as **acid rain** (perspective 32.2)

Freshwater may be either surface waters, such as rivers and lakes, or underground water, called *groundwater*. Groundwater close to the surface may break through to form springs, while deeper groundwater is accessed by digging or drilling wells. Water underground gains dissolved nutrients, and it varies in temperature depending on its depth. The water temperature increases about 1°F for every 50 feet of depth. Thus, the temperature of natural waters can vary from nearly freezing to nearly boiling at surface atmospheric conditions.

As water follows its course to the ocean, it becomes progressively more enriched with dissolved salts. Seawater contains about 3.5% salt, compared with about 0.05% for freshwater. Phosphate and nitrate are scarce in seawater. Thus, seawater usually contains fewer organisms than freshwater, and **halophilic** organisms thrive in it.

Specialized aquatic habitats include salt lakes, such as the Great Salt Lake in Utah, which has no outlet. Water in the lake evaporates, leaving high concentrations of salt. Only extremophile halophilic organisms live in this environment. Other specialized habitats include iron springs that contain large quantities of ferrous ions; these springs are the habitat for species of *Gallionella* and *Sphaerotilus*. Also, sulfur springs support the growth of both photosynthetic and nonphotosynthetic sulfur bacteria.

• **Industrial applications of extremophile products, pp. 90, 269**

REVIEW QUESTIONS

1. List several ways in which the resident population in the intestines helps to prevent infection by pathogenic bacteria.
2. Why do antibiotic-resistant bacteria become common in a hospital environment?
3. Why do people not very often observe the souring and putrefaction of milk?
4. What is the largest group of bacteria found in most soils?
5. What kind of organisms live in the Great Salt Lake and why?

Biochemical Cycling

All living organisms are involved in the important process of cycling elements. Plants are ingested by animals that in turn die and are then decomposed through the action of numerous microorganisms. A number of microorganisms are capable of degrading complex cell constituents to inorganic substances such as ammonia, sulfates, phosphates, and carbon dioxide. Microorganisms play a major role in those processes owing to their ubiquity, their unique metabolic capabilities, and their high rates of conversion.

A fixed and limited amount of the elements that make up living cells exists on the earth and in the atmosphere. For an ecosystem to sustain its characteristic life forms, these chemical elements must eventually be recycled. Only a relatively small percentage of the total amount of these elements is found in available biomass; the rest is unavailable to the biological community. It is either buried too deep in the earth's crust or present in forms that are not useable. Therefore, some portion of the elements must be continuously recycled from nonutilizable to utilizable forms if the growth of new organisms is to continue. Thus, the recycling of oxygen, carbon, nitrogen, phosphorus, and sulfur is essential.

Recycling processes require the activities of organisms in three ecological roles: **production, consumption,** and **decomposition.** On the basis of these roles, organisms can be divided into three groups: (1) the **producers,**

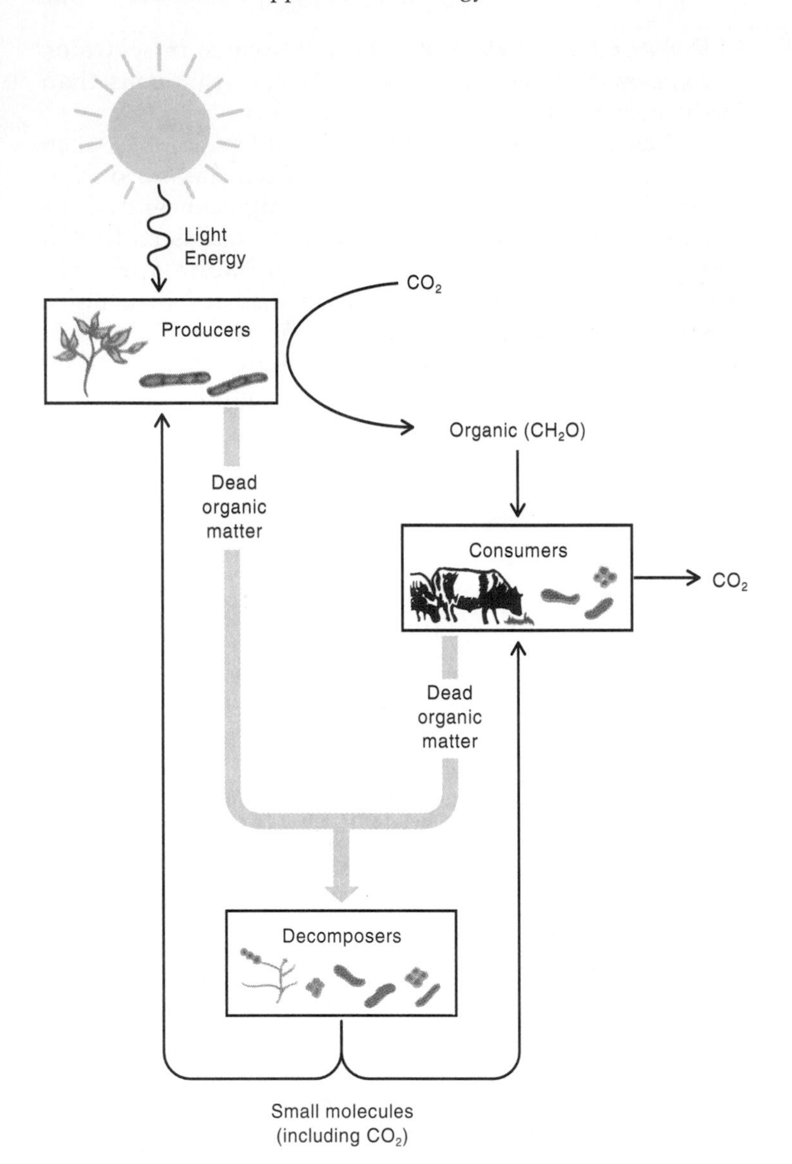

Figure 32.7
Relationship between producers, consumers, and decomposers in an ecosystem.

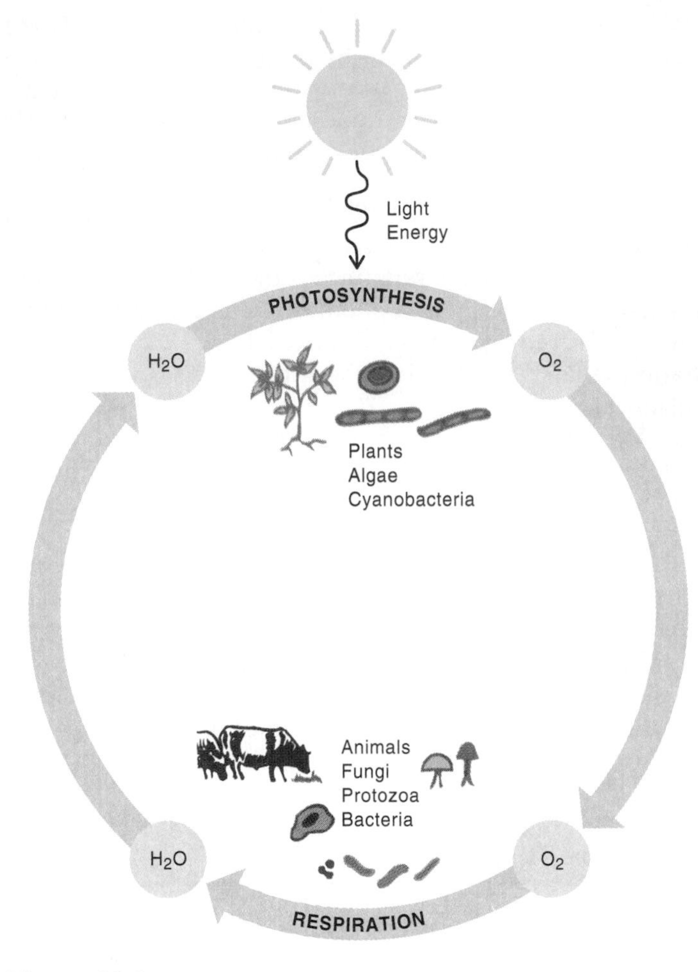

Figure 32.8
Oxygen cycle.

which convert carbon dioxide or other inorganic materials into organic material, (2) the **consumers,** which metabolize the organic matter synthesized by the producers, and (3) the **decomposers,** which digest and convert dead plant and animal material into small molecules that can be used by both the consumers and producers (figure 32.7). For each chemical element used in cell structures, recycling will occur only if members of all three functional groups are present; the loss of any group breaks the cycle.

Oxygen Cycles Between Respirers and Photosynthesizers

Oxygen is cycled by the processes of respiration and photosynthesis (figure 32.8). During photosynthesis, cyanobacteria, algae, and higher plants produce oxygen from water. It is believed that the earth's early atmosphere contained little or no oxygen and that, as cyanobacteria developed, with chlorophyll and other compounds that permitted them to carry out photosynthesis, they increased the level of oxygen in the atmosphere. This made possible the development of organisms that respire using oxygen. The current level of oxygen in the atmosphere is kept in balance by chemical reactions in the upper atmosphere, aerobic respiration, and photosynthesis.

Carbon Cycles Between Organic Compounds and Carbon Dioxide

All organisms are composed of organic molecules built from carbon "skeletons." The carbon cycle revolves around carbon dioxide; its fixation into organic compounds by primary producers, mainly green plants, algae, and cyanobacteria; and its regeneration, primarily by microorganisms. The overall cycle is depicted in figure 32.9. Consumer organisms utilize the organic compounds and release the carbon in the carbon dioxide generated by respiratory metabolism.

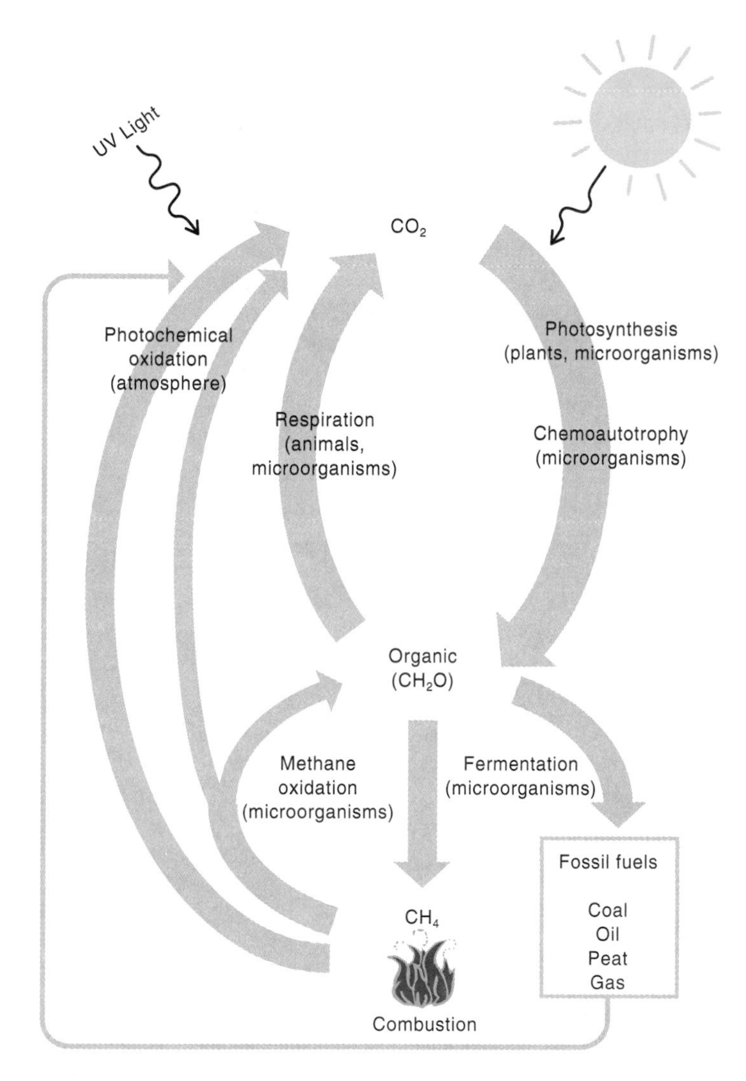

Figure 32.9
Carbon cycle.

The decomposition of organic matter is carried out by "decomposers"—chiefly, bacteria, fungi, protozoa, and small animals. The decomposition of organic debris involves the combined activity of a large number of organisms. Fungi and many bacteria utilize the more readily decomposable organic substances, such as sugars, amino acids, and proteins. Members of one group of bacteria, the *Cytophaga* are capable of degrading cellulose, a glucose polymer found in plants that is highly resistant to degradation by most other bacteria. Other major plant constituents, such as lignin and pectins, are generally decomposed initially by fungi. Some of the breakdown products of these constituents can then be further metabolized by bacteria. Lignin, however, is relatively resistant to degradation but can be degraded aerobically by some basidiomycetes.

As a general rule, bacteria appear to play the dominant role in the decomposition of animal flesh; fungi appear to be the most important in the initial degradation of plant material. Aerobic decomposition proceeds by aerobic respiration. Anaerobic decomposition, an important process in nature, occurs by fermentation and anaerobic respiration.

The main gas evolved in the aerobic decomposition of organic matter is carbon dioxide:

$$(CH_2O)n + 6O_2 \rightarrow \underset{\text{enzymes}}{\text{bacterial}} \rightarrow CO_2 + H_2O$$

When the level of oxygen is low, as is the case in wet rice-paddy soil, marshes, swamps, and manure piles, methane (CH_4) production may be great. The biosynthesis of methane is limited to a specialized group of archaens, the **methanogens,** all of which are strict anaerobes that live in poorly aerated environments. These organisms have the ability to gain energy from the oxidation of hydrogen gas and the reduction of carbon dioxide, according to the following reaction:

$$4H_2 + CO_2 \xrightarrow{\text{anaerobic conditions}} CH_4 + 2H_2O$$

This reaction is an excellent example of an anaerobic respiration in which CO_2 is reduced. The methanogens cannot utilize common organic compounds, as can most heterotrophs. The carbon cycle is completed by the methane oxidizers, a group able to oxidize methane to CO_2 and organic compounds under aerobic conditions. Methane that enters the atmosphere is oxidized photochemically (by ultraviolet light and chemical ions) to carbon monoxide (CO) and carbon dioxide (CO_2).

A proportion of the organic matter produced is buried under anaerobic conditions, such as in mud or ocean sediments. Under these conditions, it is not decomposed and eventually forms fossil fuels such as coal, oil, and peat. When these fuels are burned, their carbon is released into the atmosphere as CO_2.

Nitrogen Cycles Between Organic Compounds and a Variety of Inorganic Compounds

Next to carbon and oxygen, the element that organisms require in greatest quantity is nitrogen, an important constituent of proteins and nucleic acids. Although the atmosphere consists of 80% nitrogen, relatively few organisms can use this gaseous form of the element. In most microorganisms and plants, inorganic nitrogen is taken up as nitrate (NO_3^-) or ammonium (NH_4^+) ions. These compounds of nitrogen are called **fixed nitrogen** because the nitrogen is combined with hydrogen (NH_4^+) or oxygen (NO_3^-). They are provided to soils and water mainly by the action of the few microorganisms that can convert nitrogen gas into these forms. Only prokaryotic cells have developed the ability to fix nitrogen. Alternatively, the fixation of molecular nitrogen can be carried out industrially by the energy-expensive Haber-Bosch process which is used to synthesize nitrogen fertilizers. Fertilizers produced synthetically by humans are assuming an increasingly larger role in the nitrogen cycle. However, on a global scale, the bulk of the nitrogen added to soil is in the form of organic materials, largely as crop waste.

To be used by plants and many microorganisms, the organic nitrogen in these materials must be converted first to ammonium ions **(ammonification)** and then to nitrate **(nitrification).** Nitrogen can be lost from the soil because some species of bacteria can convert nitrate to gaseous nitrogen by using nitrate as a metabolic electron acceptor in place of oxygen. This process, another example of anaerobic respiration, is called **denitrification.** The result is a loss of soil fertility, since gaseous nitrogen is only available after it is fixed.

All forms of nitrogen undergo a number of transformations. The entire series of reactions involving nitrogen comprises the nitrogen cycle (figure 32.10), and microorganisms are essential at each stage of this cycle. The cycle consists of several individual transformations of nitrogen, in which the nitrogen becomes progressively oxidized from its highly reduced state (NH_4^+) to the highly oxidized state (NO_3^-). Some of the important features of each individual step are considered next.

Ammonification

Almost all of the nitrogen found in the upper surface of the soil and in aquatic sediments exists in organic molecules—primarily, amino acids, purines, and pyrimidines.

A wide variety of microorganisms, including aerobic and anaerobic bacteria as well as fungi, are capable of decomposing protein. They do this initially through the action of extracellular proteolytic enzymes that convert protein into chains of amino acids of varying lengths. Other enzymes then proceed to degrade these long chains into smaller chains, ammonium and sulfate ions (derived from the sulfur-containing amino acids), and water.

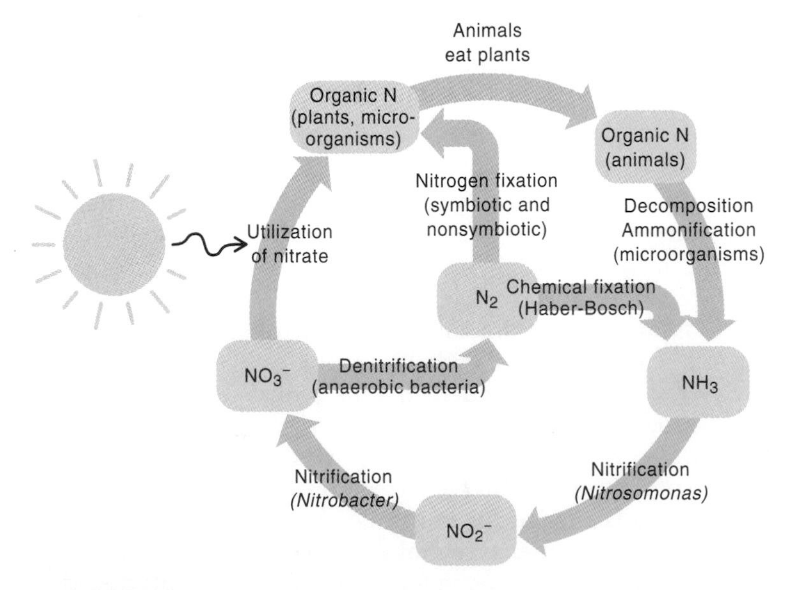

Figure 32.10
Nitrogen cycle.

Nitrification

During the process of ammonification, some nitrogen may be released into the atmosphere as ammonia (NH_3, a gas) from nitrogen-containing compounds. This is likely to happen only in alkaline environments such as heavily limed soils in which large amounts of organic nitrogen are being metabolized. Usually, however, any ammonia formed reacts with water to give NH_4^+; the ammonium ion is then oxidized to nitrite (NO_2^-) and then to nitrate (NO_3^-). It is the process that converts NH_4^+ to NO_3^- that is called **nitrification.**

It might be helpful to consider the conversion of ammonium ion to nitrate in terms of the oxidation levels of the various elements involved. The ammonium ion (NH_4^+) is the most reduced form of nitrogen. As such, it can be oxidized by bacteria with the release of energy. Nitrification can be carried out only by members of a few genera. This oxidation occurs in two steps: (1) the oxidation of NH_4^+ to NO_2^-, carried out by one group of bacteria, and (2) the further oxidation of NO_2^- to NO_3^-, performed by a second group. In soils, the dominant bacteria oxidizing NH_4^+ to NO_2^- are members of the genus *Nitrosomonas;* the conversion of NO_2^- to NO_3^- is carried out by *Nitrobacter* sp. In marine environments, bacteria from a variety of genera, including those mentioned previously, are involved in nitrification.

The reactions involved can be written as follows:

$$NH_4^+ + 1/2\ O_2 \rightarrow NO_2^- + H_2O + 2H^+ + \text{energy}$$

$$NO_2^- + 1/2\ O_2 \rightarrow NO_3^- + \text{energy}$$

As might be expected from looking at these reactions, they occur only under aerobic conditions, and therefore no nitrification takes place in waterlogged soils or in anaerobic regions of aquatic environments. The nitrifiers are generally autotrophs, utilizing CO_2 as their sole source of carbon. The energy required for the reduction of CO_2 to the constituents of cytoplasm is obtained from the oxidation of NH_4^+ to NO_2^- to NO_3^-.

Nitrite ion (NO_2^-) is highly toxic to plants, and it is therefore important that the organisms oxidizing NO_2^- to NO_3^- be in the same environment as those converting NH_4^+ to NO_2^-. Nitrate is soluble and can be leached from soil by rainwater. If it accumulates in groundwater, it can be dangerous to humans who drink the water. Nitrate itself is not highly toxic, but it can be converted to nitrite by intestinal bacteria. Nitrite is toxic because it combines with hemoglobin of the blood, causing a reduction in the oxygen-carrying capacity of the blood. Infants and children, more likely to be affected than are adults, may develop respiratory distress and a bluish skin color from their poorly oxygenated blood.

The Centers for Disease Control and Prevention reported outbreaks of miscarriages of pregnancies in a small number of women in Indiana during 1993 and 1994. All drank water from private wells that were subsequently shown to have high levels of nitrate. Some of the wells tapped into the same aquifer,[1] which was apparently contaminated by a leaking waste pit on a hog farm in the area. In another case, only 12 miles away, the contamination seemed to come from the septic system on a particular property.

Denitrification

Certain facultatively anaerobic bacteria that are active in proteolysis and ammonification can use nitrate as a final metabolic electron acceptor under anaerobic conditions (anaerobic respiration). The end results of this process are the reduction of nitrate with the liberation of nitrogen gas and the net loss of nitrogen from the ecosystem.

Nitrogen Fixation

Biologically available nitrogen is continually being removed from the soil through the leaching action of water, which removes nitrate through the incorporation of nitrogen into plants, which are harvested, and through denitrification. Usable nitrogen must, therefore, be continually added to the soil to maintain its nitrogen status. Growth in many ecosystems, including agricultural ecosystems, is limited by the availability of usable nitrogen.

The natural supply of fixed nitrogen (nitrogen in nongaseous compounds) is limited, limiting the capacity of world agriculture. None of the processes described thus far explains how fixed nitrogen is added to the soil. One source is the small amounts of fixed nitrogen, in the form of ammonium and nitrate ions, that occur in rainwater. Fertilization of soil also adds fixed nitrogen. However, the combination of these two processes usually does not compensate for the amount of nitrogen lost from the soil by denitrification. Nitrogen must, therefore, be fixed in other ways, and nitrogen-fixing bacteria represent the most important means for this fixation. No eukaryotic organisms have been shown to fix nitrogen without the aid of prokaryotes.

The two types of organisms that can fix nitrogen are those that are **free-living** and those that live in **symbiotic** association with various plants. A variety of these bacteria are capable of reducing nitrogen gas to the amino group of the amino acids that make up proteins. This process, carried out by the enzyme nitrogenase, occurs in several steps. The nitrogen gas is first activated and then reduced to the ammonium ion. The overall reaction requires a tremendous expenditure of energy, about 15 molecules of ATP for every molecule of nitrogen fixed. The ammonium formed inside the cell is then incorporated into amino acids by transamination.[2]

[1] Aquifer—underground lake

[2] Transamination—reaction in which an ammonium ion is transferred from one amino acid to an α-keto acid to make a different amino acid

Free-Living Nitrogen Fixers A variety of free-living microorganisms are capable of fixing nitrogen. Among these free-living forms of eubacteria are members of the genus *Azotobacter* (figure 32.11). These heterotrophic, aerobic, gram-negative rods may be the chief suppliers of fixed nitrogen in grasslands and other ecosystems lacking plants with nitrogen-fixing symbionts. Azotobacters are rare in soils with a pH below 6.0. Members of the genus *Beijerinckia*, another genus containing heterotrophic nitrogen-fixing species, are found in highly acid soil. For reasons that are not clear, members of the genus *Beijerinckia* occur principally in the tropics and rarely in temperate zones. The dominant free-living, anaerobic, nitrogen-fixing organisms of the soil are members of the genus *Clostridium*

In aquatic environments, the most significant nitrogen fixers are cyanobacteria. They are especially important in flooded soils. Rice has been cultivated successfully for centuries without the addition of nitrogen-containing fertilizer because of the cyanobacterium *Anabaena* that grows on the aquatic fern *Azolla* in the water of paddy fields. Before planting rice, the farmer allows the flooded rice paddy to overgrow with *Azolla* ferns. As the rice grows, it eventually crowds out the fern, which dies and releases the nitrogen into the water. Sufficient fixed nitrogen is produced in this way to fertilize the rice plants, and the farmer does not need to add any chemical fertilizers. Other nitrogen-fixing bacteria are found in genera that include photosynthetic, methane-oxidizing, sulfate-reducing, and sulfur-oxidizing organisms.

Symbiotic Nitrogen Fixers Although free-living bacteria are potentially capable of adding a considerable amount of fixed nitrogen to the soil, symbiotic nitrogen-fixing organisms are far more significant in benefiting plant growth and

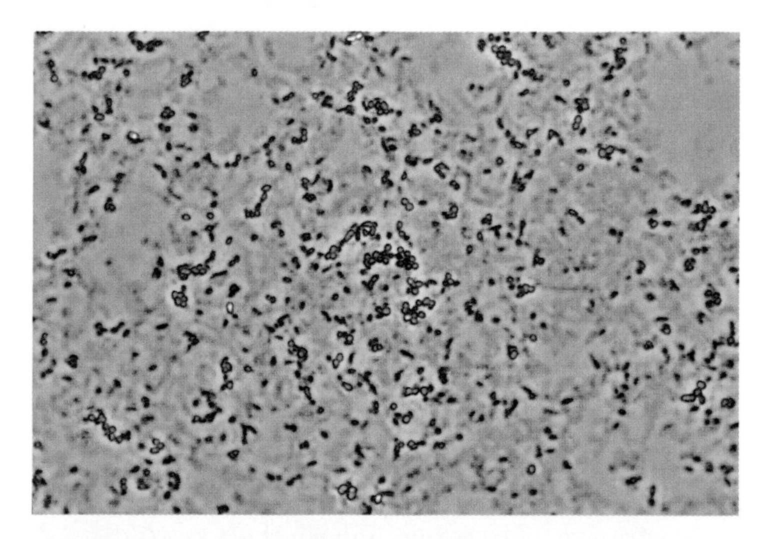

Figure 32.11
Phase-contrast photomicrograph of *Azotobacter chroococcum*, a nitrogen-fixing bacterium (×410).

crop production. The **rhizobia,** found in the three genera *Rhizobium, Bradyrhizobium,* and *Azorhizobium,* are the most agriculturally important symbiotic nitrogen-fixing bacteria. The plants with which these organisms associate are all leguminous plants and include alfalfa, clover, peas, beans, peanuts, and vetch (table 32.2).

The input of soil nitrogen from the microbial symbionts of leguminous plants such as alfalfa may be roughly 10 times the annual rate of nitrogen fixation attainable by nonsymbiotic organisms in a natural ecosystem. The amount of nitrogen fixed by nitrogen-fixing bacteria depends primarily on two factors, the organism and the habitat (table 32.3).

The symbiotic association of the bacteria with the plant results in the formation of nodules on the roots of the plant. Both the leguminous plants and the bacteria can be cultivated separately without each other; however, in soils lacking leguminous plants, rhizobia compete poorly with other bacteria and tend to disappear from the soil if leguminous plants are not grown in them for long periods of time. Generally speaking, rhizobia are capable of fixing nitrogen only when they are growing in the root nodules of the plants (figure 32.12). In order to encourage plant growth, people often plant the seeds of certain legumes together with inocula of the appropriate rhizobia.

Table 32.2 Examples of Legume Hosts Infected by Various Rhizobia

Legume Host	Bacterial Species
Alfalfa	*R. melioti*
Clovers	*R. leguminosarum* biovar *trifolii*
Pea	*R. leguminosarum* biovar *viciae*
Phaseolus bean	*R. leguminosarum* biovar *phaseoli*
Soybean	*Bradyrhizobium japonicum*

Table 32.3 Quantities of Nitrogen Fixed by Microorganisms

Group	Species or Habitat	N_2 Fixed per Acre per Year (lbs)
Nodulated legumes	Alfalfa	113–297
	Soybean	57–105
	Red clover	75–171
Nodulated nonlegumes	Alder tree	200
Cyanobacteria	Arid soil in Australia	3
	Paddy field in India	30
Free-living heterotrophs	Soil under wheat	14
	Soil under grass	22
	Rain forest in Nigeria	65

How Roots Are Infected with Rhizobia The symbiotic relationship between rhizobia and leguminous plants has been the object of study for years. Many of the details have been worked out and are outlined briefly here.

The infection process of legumes by rhizobia consists of several stages. First, the bacteria recognize and attach to specific molecules on the root hairs of the legume. Next, the root hair curls and forms an infection thread through which the bacteria invade. Once inside the plant, the bacteria become changed in appearance to a form called *bacteroids* and begin to fix nitrogen. Finally, the repeated division of both plant cells and bacteroids leads to the formation of complete root nodules (figure 32.13).

The genes that code for all these activities are distributed among the plant, the bacterial chromosome, and a large bacterial plasmid called *Sym.* A special oxygen-carrier called *leghemoglobin* has the heme portion coded for by genes in the bacteria and the globin coded for by plant genes. The genes for nodulation, the *nod* genes, are largely on the *Sym* plasmids. Products of the *nod* genes are called **NOD factors.** Also on the plasmid are some *specificity* genes that account for the host specificity.

The growth of rhizobia in the area immediately surrounding the roots of the plant (the rhizosphere) and subsequent nodulation are host specific, as indicated in table 32.2. For example, nodules are formed on alfalfa roots by *Rhizobium melioti* and on the roots of clovers by *Rhizobium leguminosarum* biovar *trifolii.* Some of these bacteria have a broad host range, and others are able to nodulate only one or a few host plant species. The reason for the host specificity is that the bacteria make surface molecules that specifically bind to surface molecules (lectins) on the legume.

Some NOD factors, secreted by the bacteria in the vicinity of the root hairs, act as signals to the plant to switch on the nodulation process. NOD factors cause changes in the host root hairs, with branching and curling of the hairs, and development of the cellulose infection

Figure 32.12
Nitrogen-fixing bacterium *Rhizobium* on clover.

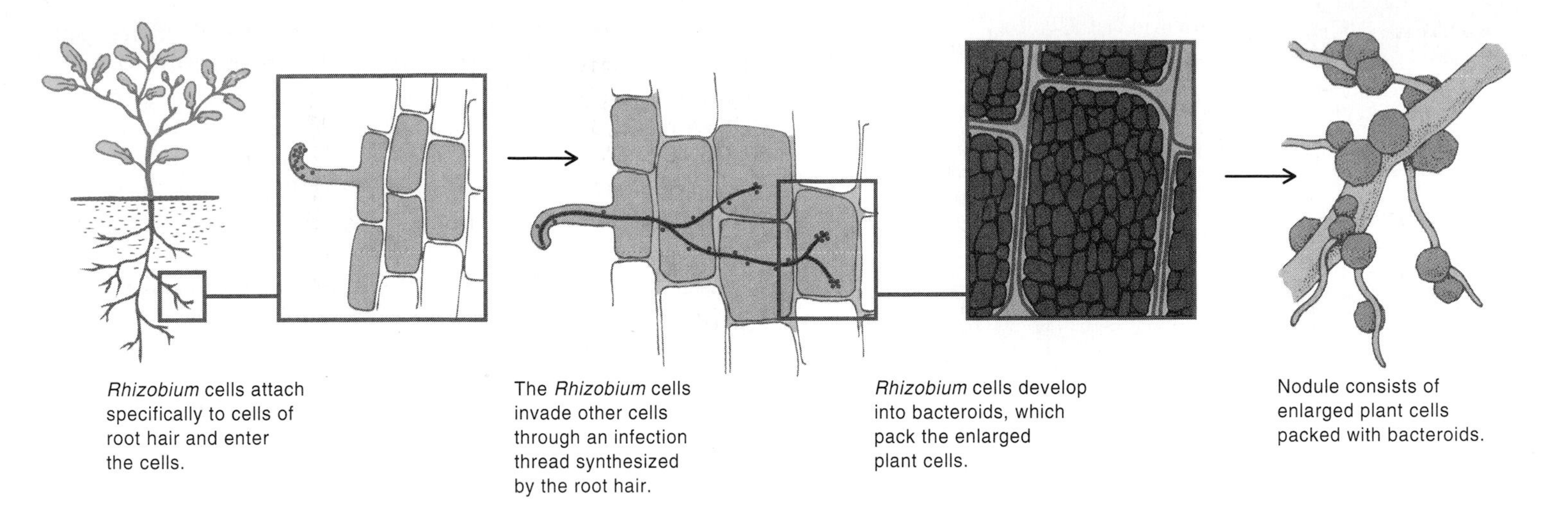

Figure 32.13
Symbiotic nitrogen fixation. The major steps leading to the formation of a root nodule in leguminous plants by *Rhizobium*.

threads. Bacteria move through the infection thread into the roots, where they enter plant cells. The bacteria multiply rapidly within the host cells and become surrounded by a membrane synthesized by the host plant. They then undergo further morphological changes into the swollen and irregularly shaped **bacteroids,** and their formation is apparently induced by several chemicals in the plant tissue. Both bacteroids and plant cells multiply repeatedly, resulting in the formation of the nodule.

Leghemoglobin regulates oxygen concentration in the nodule to keep free oxygen at a very low level and maintain anaerobic conditions. Nitrogen fixation is carried out by the rhizobia in the anaerobic environment of the nodule. The fixed nitrogen then diffuses out of the cells into the root environment where it is used by the plant for the synthesis of proteins.

Nodules are not found on all species of legumes, and not all species of rhizobia are capable of nodulating leguminous plants. Moreover, the ability to develop a symbiotic relationship resulting in nitrogen fixation varies markedly among strains of rhizobia. Transfer of the *Sym* plasmid can sometimes transfer host specificity.

In addition to legumes, several genera of nonleguminous trees, including alder and gingko, possess nitrogen-fixing root nodules at some stages in their life cycle. The microorganisms involved in symbiosis with the alder tree are actinomycetes of the genus *Frankia* (chapter 11).

Phosphorus Also Cycles Between Organic and Inorganic Forms

Phosphorus occurs in the soil in both organic and inorganic forms. Organic phosphorus-containing compounds occur almost exclusively near the surface. They are derived mostly from surface vegetation and are accordingly found as constituents of plant tissue, animals, and microorganisms. The inorganic forms of phosphorus occur largely as insoluble salts of calcium, iron, and aluminum. Plants generally cannot use these inorganic and organic forms readily, but **orthophosphate (PO_4^{3-})** is readily used by most plants and microorganisms, being part of such biologically important molecules as ATP and nucleic acid. Microorganisms play three major roles in phosphorus transformation: They mineralize[1] organic phosphorus, they convert insoluble forms of inorganic phosphorus to soluble forms, and they immobilize[2] inorganic phosphorus.

Overall Transformations of Phosphorus

Organic phosphate is first converted to inorganic phosphate, which can then be utilized by plants or immobilized by bacteria into their own organic forms of phosphorus. Some minerals containing phosphate may also be converted into orthophosphate by microorganisms.

Mineralization of Organic Phosphates

Bacteria and fungi are largely responsible for rendering organic phosphorus available to succeeding generations of microorganisms as well as to plants. Most microorganisms do this by synthesizing an enzyme, phosphatase, which separates the organic phosphate from a wide variety of compounds in which it is found. The breakdown of phosphate-containing nucleic acids is initiated by the action of

[1] Mineralize—convert an element from an organic form to an inorganic form
[2] Immobilize—convert an element from an inorganic form to an organic form

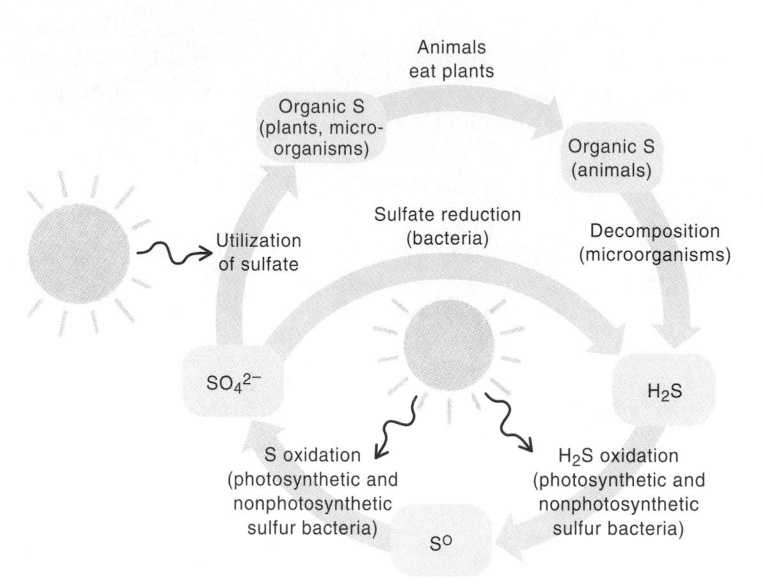

Figure 32.14
Sulfur cycle.

nucleases, which degrade DNA and RNA into their individual subunits (nucleotides). The phosphate is then separated from these nucleotides by the enzyme phosphatase to form nucleosides:

$$\text{nucleo}\textbf{tide} \rightarrow \text{nucleo}\textbf{side} + PO_4^{3-}$$

Dissolution of Minerals

A wide variety of commonly occurring soil organisms are capable of dissolving calcium phosphate [$Ca_2(HPO_4)_2$ and $CaHPO_4$], which is much more soluble than the tribasic phosphate salt [$Ca_3(PO_4)_2$]. The amount of phosphate that dissolves depends primarily on the amount of acid that these organisms produce in their metabolic processes.

In many aquatic habitats, growth of algae and cyanobacteria (the primary producers) is often limited because concentrations of phosphorus are too low. Addition of phosphates to such habitats from agricultural fields, for example, can result in algal blooms and deterioration of water quality.

The Sulfur Cycle Involves Several Important Types of Organisms

Sulfur also occurs in all living matter, chiefly as a component of the amino acids methionine, cystine, and cysteine. The sulfur cycle bears many similarities to the nitrogen cycle (figure 32.14).

Like nitrogen, sulfur is present in the soil chiefly as a part of proteins, and it is taken up by plants in its oxidized form, sulfate (SO_4^{2-}). The sulfur-containing proteins are first degraded into their constituent amino acids by proteolytic enzymes excreted by a large variety of soil organisms. The sulfur of the amino acids is generally converted to hydrogen sulfide (H_2S) by a variety of soil microorganisms, a process analogous to NH_4^+ formation. The H_2S formed would be highly toxic to many biological systems if it were allowed to accumulate. However, under aerobic conditions, H_2S oxidizes spontaneously to sulfur and is then converted to its most readily utilized form, sulfate, by sulfur bacteria. This process is analogous to nitrification. Under anaerobic conditions, the counterparts of the denitrifying organisms operate—these sulfate-reducing bacteria reduce SO_4^{2-} to H_2S.

The microbial flora concerned with the sulfur cycle have a number of unique features. The oxidation of H_2S to SO_4^{2-} is carried out principally by two major groups of organisms, the nonphotosynthetic autotrophs, members of the genera *Thiobacillus* and *Beggiatoa,* and less commonly, by the photosynthetic autotrophs (green and purple sulfur bacteria).

The reduction of sulfate to sulfide (sulfate reduction) is carried out by a small group of anaerobic bacteria that are capable of utilizing sulfate as the final electron acceptor in their anaerobic respiration. These anaerobes include organisms of the genus *Desulfovibrio* as well as a few anaerobic spore-forming rods. Hydrogen sulfide is thus likely to accumulate in waterlogged soils. Since hydrogen sulfide is toxic to higher plants, sulfate reduction is undesirable from an agricultural point of view.

Bioremediation: The Biological Cleanup of Pollutants

Pollutants

Traditionally, pollutants from domestic and industrial wastes have been dumped into the environment, with the hope that they will be degraded and removed through natural recycling. Most organic compounds of natural origin can be degraded by one or more species of soil or aquatic organisms, a fact that led some people to believe that no organic compound would ever persist in the environment. However, this is not the case with a large number of chemicals synthesized industrially. People have increased the number of slowly degraded or nonbiodegradable compounds in the soil and in aquatic systems of many parts of the world by adding thousands of tons of pesticides, herbicides, nonbiodegradable detergents, and plastics. The chemical nature of nonbiodegradable synthetic compounds covers an extremely broad range of organic compounds—organic acids, nitrophenols, chlorinated organic acids, and other organic substances. Although certain of these chemically synthesized compounds disappear from an ecosystem within the reasonably short time of a few days or weeks, many remain undegraded for years. Chemically synthesized compounds are most likely to be biodegradable if they have a chemical composition similar to that of naturally occurring compounds. If a synthetic chemical compound is totally different from any that occurs in nature, organisms are less likely to have the enzymes necessary for degrading it rapidly, and the compound is likely to persist for long periods. Specific alterations of a compound, such as the addition of a halogen atom, often result in the need for totally new mechanisms to degrade it.

It is highly desirable that toxic compounds be degradable, because it is becoming apparent that most herbicides and insecticides not only are toxic to their target weeds or insects, but also have far-ranging deleterious effects on fish, birds, and other animals. DDT is an example of such a compound. This nonbiodegradable pesticide accumulates in the fat of predatory birds. Figure 32.15 illustrates the phenomenon of **biological magnification** that occurs.

The continuing ingestion and reingestion of DDT, which accumulates in fat, results in an increasingly greater concentration of the DDT as it is passed upward through the food web. DDT, as well as other chlorinated hydrocarbon insecticides, also interferes with the reproductive process of birds, leading to the production of fragile eggs, which break before the young can hatch. Furthermore, DDT and other chlorinated hydrocarbons are carcinogenic. Although banned in the United States, DDT is still heavily used in other countries where its effects can be far reaching. For example, DDT use in South America caused a depletion of predatory birds that normally summer in Greenland. The result was an overpopulation of Greenland's rodents. Thus, although a biodegradable compound must be applied repeatedly to maintain its effectiveness, it is preferred over a toxic, nonbiodegradable one that may have widespread and long-term effects.

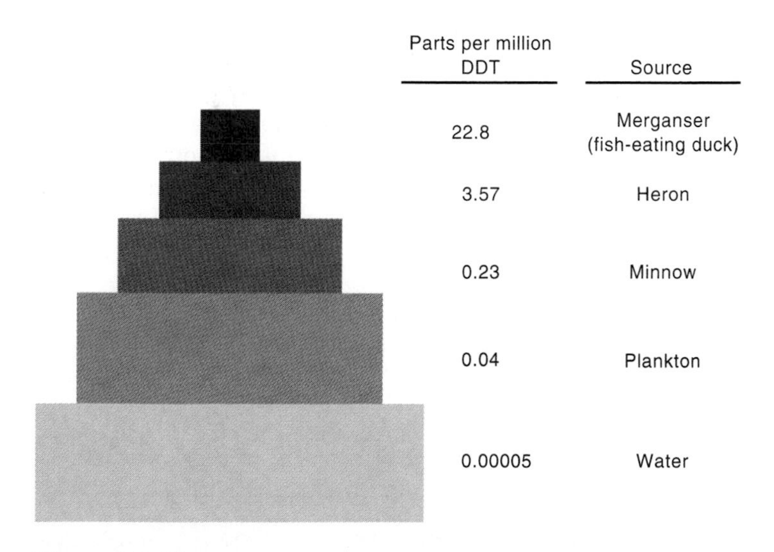

Figure 32.15
Biological magnification of DDT, a fat-soluble toxic compound used as a pesticide.

Relatively slight molecular changes markedly alter the biodegradability of a compound. Perhaps the best-studied example involves the herbicides 2,4-dichlorophenoxyacetic acid (2,4-D) and 2,4,5-trichlorophenoxyacetic acid (2,4,5-T). The only difference between these two compounds is the additional chlorine atom on 2,4,5-T (figure 32.16).

When 2,4-D is applied to the soil, it completely disappears within a period of several weeks, as a result of its degradation by a segment of the microbial population in the soil. However, when 2,4,5-T is applied, it is often still present more than a year later (figure 32.17). Its persistence is apparently due to the third chlorine atom, which blocks the enzyme that makes the initial attack on 2,4-D. The persistence of a variety of herbicides is shown in table 32.4.

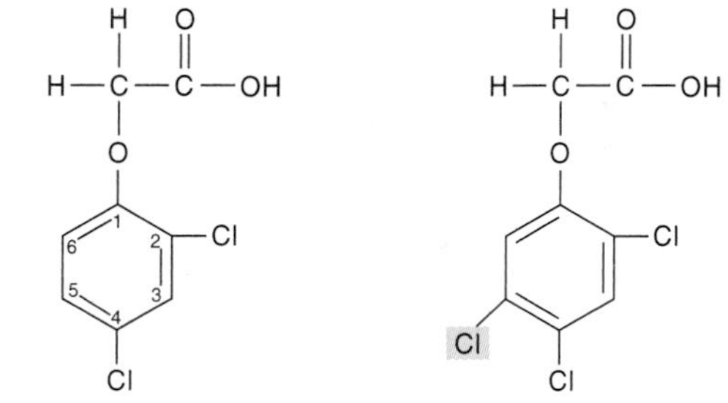

Figure 32.16
Formulas of 2,4-D (left) and 2,4,5-T (right). Note that 2,4,5-T has an additional chlorine atom attached to carbon atom 5.

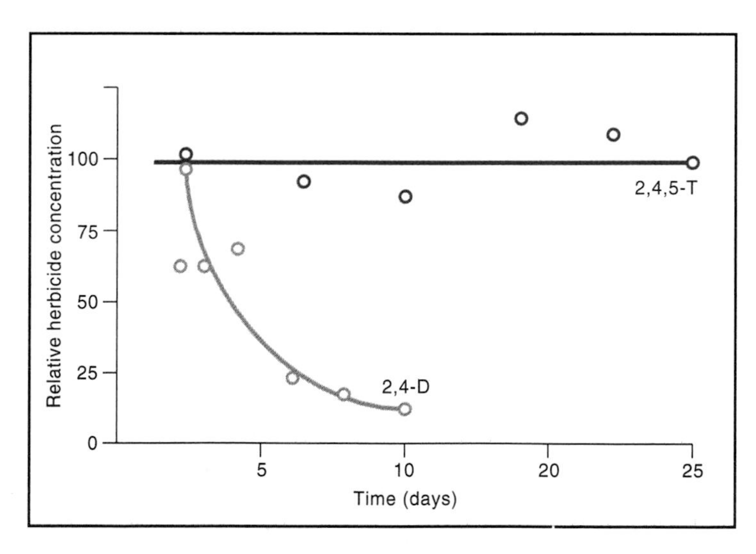

Figure 32.17
Comparison of the rates of disappearance of two structurally related herbicides, 2,4-D and 2,4,5-T.

Source: Alexander, M. 1961. Introduction to soil microbiology. John Wiley & Sons, Inc., New York.

The estimates in table 32.4 are only approximations, since many factors influence the rate of degradation of compounds by the soil population. As a general rule, any practice that favors the multiplication of microorganisms will increase the rate of degradation. Thus, raising the temperature, maintaining the pH near neutrality, and providing an optimal amount of moisture are all likely to increase the rate of degradation of most materials added to the soil.

Another class of dangerous compounds comprises the polychlorinated biphenyls (PCBs, figure 32.18), which have been used in many manufacturing operations. These aromatic molecules with chlorine atoms attached to their benzene ring are not biodegradable, and their concentration in the environment increased steadily for several years. In 1968, cooking oil contaminated with PCBs poisoned 1,000 people in Japan. PCBs have also been blamed for reproductive failure in birds and in mink and are carcinogenic. The use of PCBs was banned in the United States in 1978, but even so, one-third of the United States population is estimated to contain PCBs in concentrations greater than 1 part per million in their tissues.

It seems unlikely that organic chemical compounds can be created that will selectively affect only certain undesirable groups of plants or insects. Therefore, greater emphasis is being placed on developing and using synthetic materials, including pesticides, detergents, and plastics, which can be broken down in the natural environment.

In addition to human-made materials, some natural materials may create problems, as in large oil spills. Petroleum accidentally spilled into the ocean can result in the death of large numbers of aquatic organisms, from microscopic ones to birds and mammals. Bacteria have proven their worth in helping to remedy pollution from oil spills and many other kinds of pollution.

Table 32.4 Persistence of Some Herbicides and Insecticides

Name of Compound	Time of Persistence
Insecticides	
Parathion	1 week
Malathion	1 week
Diazinon	3 months
Heptachlor	2 years
Aldrin	3 years
Lindane	3 years
DDT	4 years
Chlordane	5 years
Herbicides	
2,4-D (2,4-dichlorophenoxyacetic acid)	4 weeks
Dalapin	8 weeks
2,4,5-T (2,4,5-trichlorophenoxyacetic acid)	20 weeks
Atrazine	40 weeks
Simazine	48 weeks
Propazine	1.5 years

Means of Bioremediation

The problem of toxic chemicals in natural environments has prompted a great deal of research into the development of means of rapidly degrading the polluting compounds. One goal of such research is developing strains of microorganisms that can be used in biological pollution control, either by large-scale inoculation of polluted areas or by introduction of the necessary degradative capabilities into the natural population of a given area. Pseudomonad strains capable of growth on 2,4,5-T have been used successfully to remove 2,4,5-T from soil samples in the laboratory, and dual cultures of an *Acinetobacter* and a pseudomonad have been shown to degrade PCBs (figure 32.19).

Many of the bacteria that have been isolated by virtue of their ability to grow on toxic chemicals contain the genes required for degrading these compounds on self-transmissible plasmids. Therefore, it should be possible either by selective breeding or by using recombinant DNA techniques to combine several characteristics on one plasmid and produce a strain that has multiple degradative capacities. With this method, a pseudomonad that is capable of rapidly degrading certain fractions of petroleum has been bred. Naturally occurring microorganisms are responsible for removing much of the oil that is regularly

Cl Cl Cl Cl

2,2',5,5'-Tetrachlorobiphenyl

Figure 32.18
Example of a polychlorinated biphenyl. More than 200 different chlorinated varieties are possible.

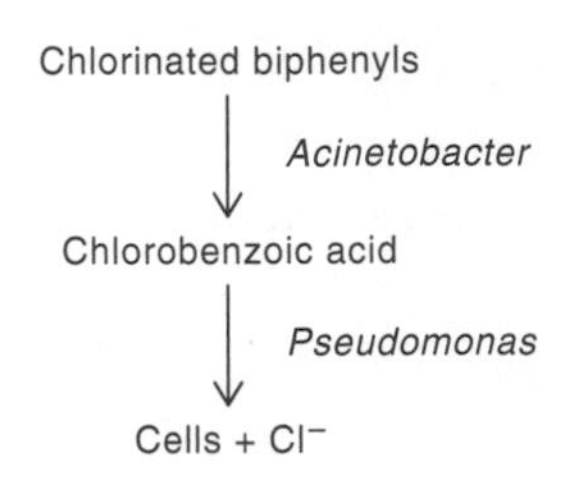

Figure 32.19
Utilization of two bacteria to degrade PCBs.

spilled into the oceans; the use of genetically engineered organisms can help to speed the recovery from ecological accidents and disasters.

All of the petroleum-degrading bacteria are limited in their growth by the low concentrations of nitrogen and phosphorus in seawater. The development of a nitrogen- and phosphorus-containing fertilizer that adheres to the spilled oil has greatly improved the results of bacterial treatment of oil spills. Use of the fertilizer has led to better growth of the bacteria and thereby increased the effectiveness of degradation by the bacteria at least threefold. Bioremediation is becoming ever more important in maintaining a healthy environment.

REVIEW QUESTIONS

1. Why are microorganisms well suited to recycle elements?
2. How is the decomposition of organic matter carried out?
3. How are proteins decomposed in natural environments?
4. What is the importance of nitrogen fixation?
5. What is meant by bioremediation?

Summary

I. The Principles of Microbial Ecology
   A. Microbial competition demands rapid reproduction and efficient nutrient use.
   B. Microbial populations both cause and adapt to environmental changes.
   C. Some bacteria can grow in a low-nutrient environment.

II. Microorganisms and Soil
   A. Soil teems with a variety of life forms.
      1. Bacteria
      2. Fungi
      3. Algae
      4. Protozoa
      5. Other eukaryotic organisms in soil
   B. The environment influences the bacterial and fungal flora in soil.
      1. Moisture in the soil affects the oxygen supply.
      2. Acidity influences the relative proportion of fungi.
      3. Temperature regulates microbial activities.
      4. Availability of nutrients limits the size of the microbial population.

III. Energy Sources for Ecosystems

IV. Aquatic Environments

V. Biochemical Cycling
   A. Oxygen cycles between respirers and photosynthesizers.
   B. Carbon cycles between organic compounds and carbon dioxide.
   C. Nitrogen cycles between organic compounds and a variety of inorganic compounds.
      1. Ammonification
      2. Nitrification
      3. Denitrification
      4. Nitrogen fixation
         a. Free-living nitrogen fixers
         b. Symbiotic nitrogen fixers
         c. How roots are infected with rhizobia
   D. Phosphorus also cycles between organic and inorganic forms.
      1. Overall transformations of phosphorus
      2. Mineralization of organic phosphates
      3. Dissolution of minerals
   E. The sulfur cycle involves several important types of organisms.

VI. Bioremediation: The Biological Cleanup of Pollutants
   A. Pollutants
   B. Means of Bioremediation

Chapter Review Questions

1. On a plate of blood agar in the lab, why are colonies smaller when they are close together than when they are widely separated? What is the significance of this when organisms are growing in the body?
2. List at least four genera of soil microorganisms that are pathogenic for humans and note whether each genus is bacterial, fungal, or something else.
3. Why is there a high concentration of microbes in the rhizosphere of plants?
4. Contrast ecosystems supported by photosynthesis with those that depend on chemosynthesis.
5. What are some differences between sea vents and smokers, and how do these differences affect the microbial flora found in each location?
6. How can apparently barren basalt rocks deep under the surface of the earth support microbial growth?
7. Give examples of free-living and symbiotic nitrogen-fixing microorganisms. Are these prokaryotic or eukaryotic?
8. Outline the symbiotic relationship between rhizobia and leguminous plants.
9. What is meant by *mineralization* and *immobilization*, as the terms are used in discussing the cycling of elements?
10. Describe the use of bioremediation in the cleanup of oil spills.

Critical Thinking Questions

1. Elemental sulfur is deposited around the large tube worms and clams growing near underwater hot vents. What is the source of the sulfur?
2. Describe the kind of life forms one would expect to find on Mars or the moon, given conditions there.
3. Recent reports suggest that human activities, such as the generous use of nitrogen fertilizers, have doubled the rate at which elemental nitrogen is fixed, raising concerns of environmental overload of nitogen. What problems could arise from too much fixed nitrogen, and what could be done about this situation?

Further Reading

Chung, King-Thom, and Ferris, Deam Hunter. 1996. Martinus Willem Beijerinck (1851–1931). Pioneer of general microbiology. *ASM News* 62(10):539–543.

Dixon, Bernard. 1996. Bioremediation is here to stay. *ASM News* 62(10):527–528.

Madigan, Michael T., and Marrs, Barry L. 1997. Extremophiles. *Scientific American:* 82–87.

33 Water and Waste Treatment

KEY CONCEPTS

1. Waterborne diseases result from contamination of water with a variety of microorganisms and chemicals.
2. Adequate water treatment and regular testing assure safe drinking water and recreational waters.
3. Proper sewage treatment, necessary to ensure the health of a community, depends on the stabilization of wastes by microorganisms.
4. Pathogenic bacteria are usually eliminated by the secondary sewage treatment process.
5. Methods such as trickling filters and the production of artificial wetlands are good solutions to small-scale sewage treatment problems.
6. Both backyard and commercial composting reduce the need for large landfills for disposal of solid waste.

Maraine Lake, near Lake Louise, Alberta, Canada.

PREVIEW LINK

Microbes in Motion

The following books in the *Microbes in Motion* CD-ROM may serve as a useful preview or supplement to your reading:

Miscellaneous Bacteria. Chlamydia (Growth and Metabolism). *Anaerobic Bacteria:* Normal Flora. *Gram-Negative Organisms:* Normal Flora. *Gram-Positive Organisms:* Normal Flora. *Microbial Pathogenesis:* Toxins; Exotoxins.

A Glimpse of History

Delivering freshwater to urban areas and removing human wastes has been practiced at least since Roman times. Ruins of aqueducts used to deliver fresh water long distances can be seen today in many parts of Europe. Ridding cities of human wastes has been more difficult, and the sewers that were used until the mid-19th century were not much more than large, open cesspools.

As early as the 1840s, Edwin Chadwick, an English activist, championed a new idea on how wastes could be removed. His idea was to construct a system of narrow, smooth ceramic pipes through which water could be flushed along with solid waste materials. This system would carry the waste materials away from the inhabited part of the city to a distant collection site. There he hoped to collect the waste materials, turn them into fertilizer, and sell it to farmers. The system he envisioned required the installation of new water and sewer pipes along with pumps to deliver water under pressure to houses. With the water under pressure and smooth narrow pipes, the system could be kept well flushed.

In 1848, with the threat of a cholera epidemic imminent, the Board of Health in England instituted widespread reforms and began the installation of a sewage system along the lines envisioned by Chadwick. New York City did not establish its Board of Health and a proper sewage disposal system until 1866, again in response to a threatened cholera epidemic. It was usually the threat of a cholera epidemic that prompted cities to install adequate water and sewage treatment systems. By the end of the 19th century, most large European and United States cities had established water-sewer systems to deliver safe drinking water and remove and treat waste materials. Cholera in the industrialized nations of Europe and North America disappeared.

Although Chadwick's idea for cleaning up the cities was successful, his hope of turning the wastes into fertilizer and selling it was a failure. In the late 1800s, inexpensive chemically synthesized fertilizers were introduced. Today, the problem of disposing of the large amounts of sludge that accumulate in sewage treatment plants is a serious one. Some cities, such as Seattle, are collecting and treating sewage sludge. This treated sludge is then used as fertilizer in certain circumstances, such as to fertilize large groves of Douglas fir trees.

Long before the discovery of the microbial world, it was recognized that some diseases are associated with water supplies. Cholera epidemics occurred years before the cholera-causing *Vibrio cholerae* was identified, yet, even at that time, it was obvious that the epidemics were associated with drinking water. The desire for clean, clear water led to the use of a sand filtration system in London and elsewhere in the early 19th century. Late in that century, Robert Koch showed that this kind of filtration not only yielded clear water, but also removed more than 98% of bacteria from the water.

Even the clearest water is not really "pure," because it contains dissolved substances and likely some microorganisms. The term **potable water** refers to water that is safe to drink but far from pure in the chemical or microbiological sense.

Water Pollution by Microorganisms and Chemicals

Pollution of water occurs from a variety of sources (figure 33.1). Contamination of water with pathogenic organisms remains a major cause of epidemics of disease. The Centers for Disease Control and Prevention (CDC) has released a summary of surveillance for waterborne disease outbreaks in the United States during 1993 and 1994, including unreported outbreaks in 1992. The report covered outbreaks associated with both drinking water and recreational water, such as swimming pools, whirlpools, hot tubs, water parks, rivers, ponds, and lakes. A total of 30 outbreaks, in 17 states and one territory, were associated with drinking water. Of these, 22 were either known or suspected to be caused by infectious microorganisms, and 8 had chemical contaminants, such as lead, copper, fluoride, and nitrate (table 33.1).

More than 405,000 persons became ill as a result of these outbreaks, 403,000 of them from cryptosporidiosis in Milwaukee. Fewer than 100 cases were associated with chemical contamination, the remainder being caused by known organisms or microbial infection.

Table 33.1 Causes of Waterborne Disease Outbreaks Associated with Drinking Water in the United States, 1993–1994

Causative Agent	Number of Outbreaks	Number of Cases
Cryptosporidium parvum	5	403,271
AGI*	5	495
Giardia lamblia	5	385
Campylobacter jejuni	3	223
Lead	3	3
Fluoride	2	43
Nitrate	2	4
Salmonella serotype Typhimurium	1	625
Shigella sonnei	1	230
Copper	1	43
Shigella flexneri	1	33
Non-01 *Vibrio cholerae*	1	11
Total	30	405,366

*AGI—acute gastrointestinal illness

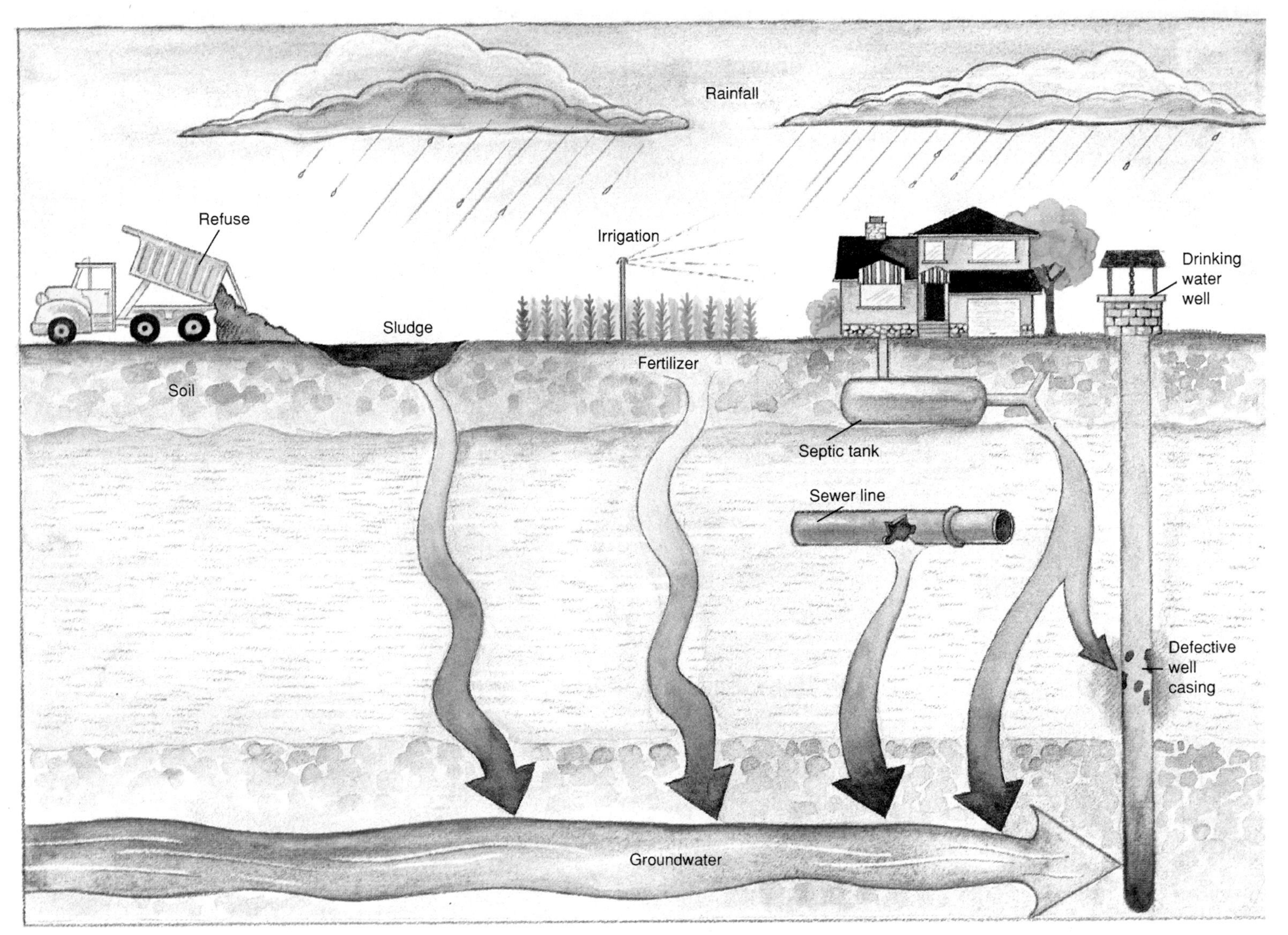

Figure 33.1
Ground water environment and relationships between pollution sources and drinking water quality. Some important soil factors influencing the fate of chemical and microbiological contaminants in water are texture, organics (humic acids), ionic strength, adsorption, structure, pH, and permeability.

The CDC study of diseases associated with recreational water included outbreaks of gastroenteritis[1] (table 33.2) and dermatitis[2] (table 33.3). In addition, one case of meningoencephalitis[3] caused by the ameba *Naegleria fowleri* was reported, associated with water from a pond or river.

Among the late reports from 1992 was notice of an outbreak of 40 cases of dermatitis, apparently caused by *Schistosoma* sp. and occurring in people who had gone swimming in a lake in New Jersey.

[1]Gastroenteritis—inflammation of the stomach and intestines
[2]Dermatitis—inflammation of the skin
[3]Meningoencephalitis—inflammation of the brain and meninges

The CDC reports that every year, despite federal and state standards for water systems, many outbreaks of disease occur as a result of water pollution. Figure 33.2 shows that the number of outbreaks in recent years reached a peak in the late 1970s and then, in general, decreased since then; 647 separate outbreaks were reported between 1971 and the end of 1993.

Chemicals contaminating water supplies caused more than a quarter of the reported outbreaks (figure 33.3); the numbers affected in these outbreaks of disease were low. When people drink water contaminated by feces, breakdown of large amounts of organic matter in the feces leads to accumulation of nitrates in the water. A number of cases of

Table 33.2 Waterborne Outbreaks of Gastroenteritis Associated with Recreational Water—United States, 1993–1994

Causative Agent	Number of Cases per Outbreak	Source of Outbreak	Setting
Giardia lamblia	80	Pool	Community
	12	Lake	Park
	43	Lake	Swim club
	6	River	River
Shigella flexneri	35	Lake	Park
Shigella sonnei	242	Lake	Park
	160	Lake	Park
Cryptosporidium parvum	101	Pool	Motel
	418	Lake	Park
	51	Pool	Motel
	5	Pool	Community
	54	Pool	Community
	64	Pool	Motel
Escherichia coli O157:H7	166	Lake	Camp

Table 33.3 Waterborne Outbreaks of Dermatitis Associated with Recreational Water—United States, 1993–1994

Causative Agent	Number of Cases per Outbreak	Source of Outbreak	Setting
Pseudomonas aeruginosa	4	Hot tub	Bed & breakfast
	2	Pool/whirlpool	Resort
	10	Hot tub	Ski resort
	14	Hot tub	Hotel
	6	Hot tub	Hotel
	15	Pool	School
	53	Pool	School
	30	Pool	Hotel
	113	Pool/hot tub	Lodge
Chemical (probable)	9	Pool	Hotel
	20	Pool	Hospital

disease caused by nitrate contamination of water have been reported in recent years. Lead and copper leaching from plumbing systems have caused a number of outbreaks.

Another example of chemical contamination of water that has proved fatal in the past is mercury contamination from industrial wastes. In this type of contamination, metallic mercury, thought to be harmless, was discharged into water. In this environment, the mercury was converted into methyl mercury, a powerful nerve toxin for humans. Fish concentrated the methyl mercury in their tissues. Many humans who ate contaminated fish died, and many more suffered severe mercury poisoning.

Water Testing and Treatment

In order to ensure its quality, water is treated in various ways. The principal concern is detecting and removing contamination, especially by pathogenic microorganisms. A typical treatment procedure is diagrammed in figure 33.4.

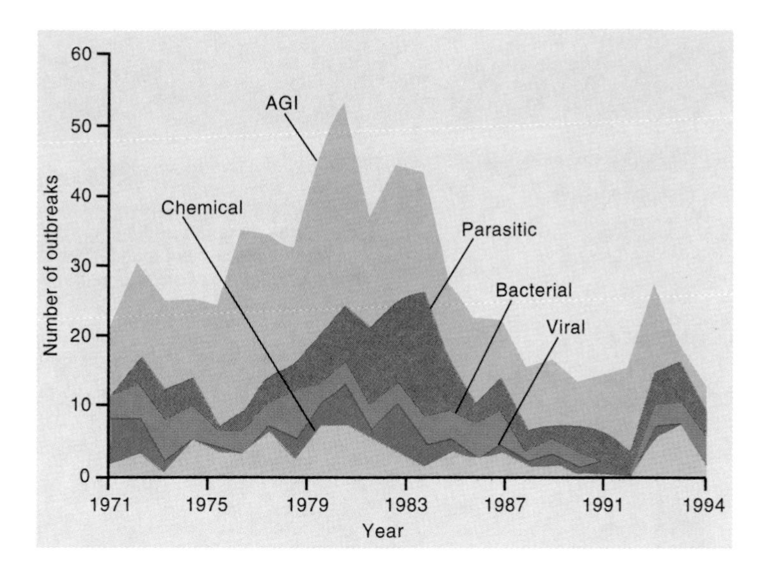

Figure 33.2
Number of waterborne disease outbreaks associated with drinking water, United States, 1971–1994. A total of 647 cases were reported to the United States Public Health Service (USPHS) during the 23-year period. The causes are indicated by the colors: pink—acute gastrointestinal illness of unknown cause; green—parasitic; brown—bacterial; blue—viral; orange—chemical. Because they are acute and short-lived diseases, the causes of many of these illnesses are never established with certainty. Many more episodes of disease occur but are not reported to the USPHS.

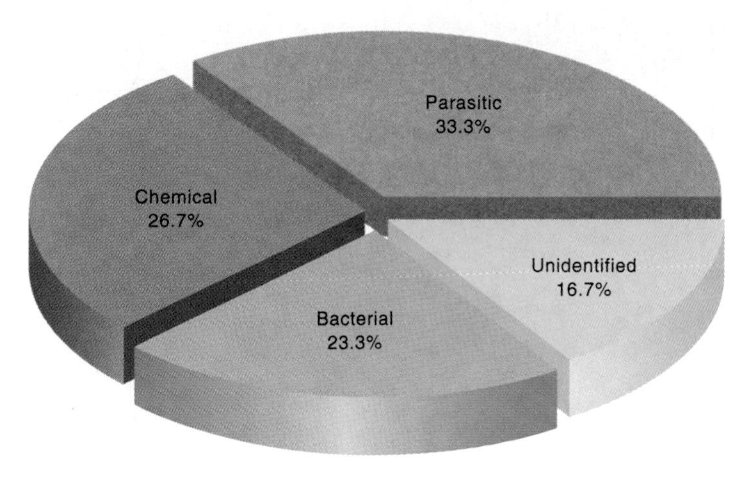

Figure 33.3
Causes of waterborne disease outbreaks associated with drinking water, United States, 1993–1994. There were 30 outbreaks reported to the USPHS during the two-year period, with an estimated 403,366 people affected. Of these outbreaks, 33.3% were caused by parasites, 26.7% by chemicals in the water, 23.3% by bacteria, and 16.7% by unidentified causes. Five outbreaks of cryptosporidiosis caused illness in over 403,000 people and some deaths. Seven people infected with *Salmonella typhimurium* died.

First, the water is allowed to stand long enough for particulate matter to settle. The water is then removed from the sediment and placed in a tank where it is mixed with a flocculent chemical, such as alum. Alum causes a flocculation (fluffy clumping) of materials still suspended in the liquid. The clumps settle, thereby removing unwanted materials from the water, including some bacteria and viruses. Following the flocculation, the water is filtered through sand or diatomaceous earth. Sometimes, filtration through an activated charcoal filter is also used, especially when toxic chemicals may remain. As a last step, the water is disinfected with chlorine or other treatments to kill any harmful organisms that might remain.

Regular testing must be done to be sure that the water is safe. It is not feasible to test for pathogens in the water on a regular basis, so the accepted method tests for coliform bacteria, a sign of fecal contamination. Coliforms are defined as a group of gram-negative, rod-shaped, nonsporeforming, aerobic and facultatively anaerobic bacteria that ferment lactose, forming acid and gas within 48 hrs at 95°F (35°C). These bacteria are commonly found in soil and in the gut and feces of warm-blooded animals. Their presence in water may indicate contamination with human and/or animal feces. For municipal water supplies, a maximum of 1 coliform organism per 100 milliliters is considered safe for potable water. Higher numbers are acceptable in water used for other purposes, such as recreational waters. Figure 33.5 outlines two procedures used to test for coliforms in water.

Public water systems in the United States are regulated under the Safe Drinking Water Act of 1974, since amended. In 1996, a bill was enacted to help Americans know what is actually in the water they drink. The law requires local water agencies to issue annual reports disclosing the chemicals and bacteria the tap water contains. This information is required to be sent directly to water users, along with their water bills. Also, water suppliers are required to provide 24-hour public notification when a contaminant poses a significant risk. It is hoped that public awareness will offer even more protection against waterborne diseases.

REVIEW QUESTIONS

1. What is potable water?
2. What is the benefit of filtering water through sand?
3. Name at least five diseases associated with the contamination of drinking water.
4. What kind of chemical contaminants in drinking water have caused problems?
5. How are public water systems in the United States regulated?

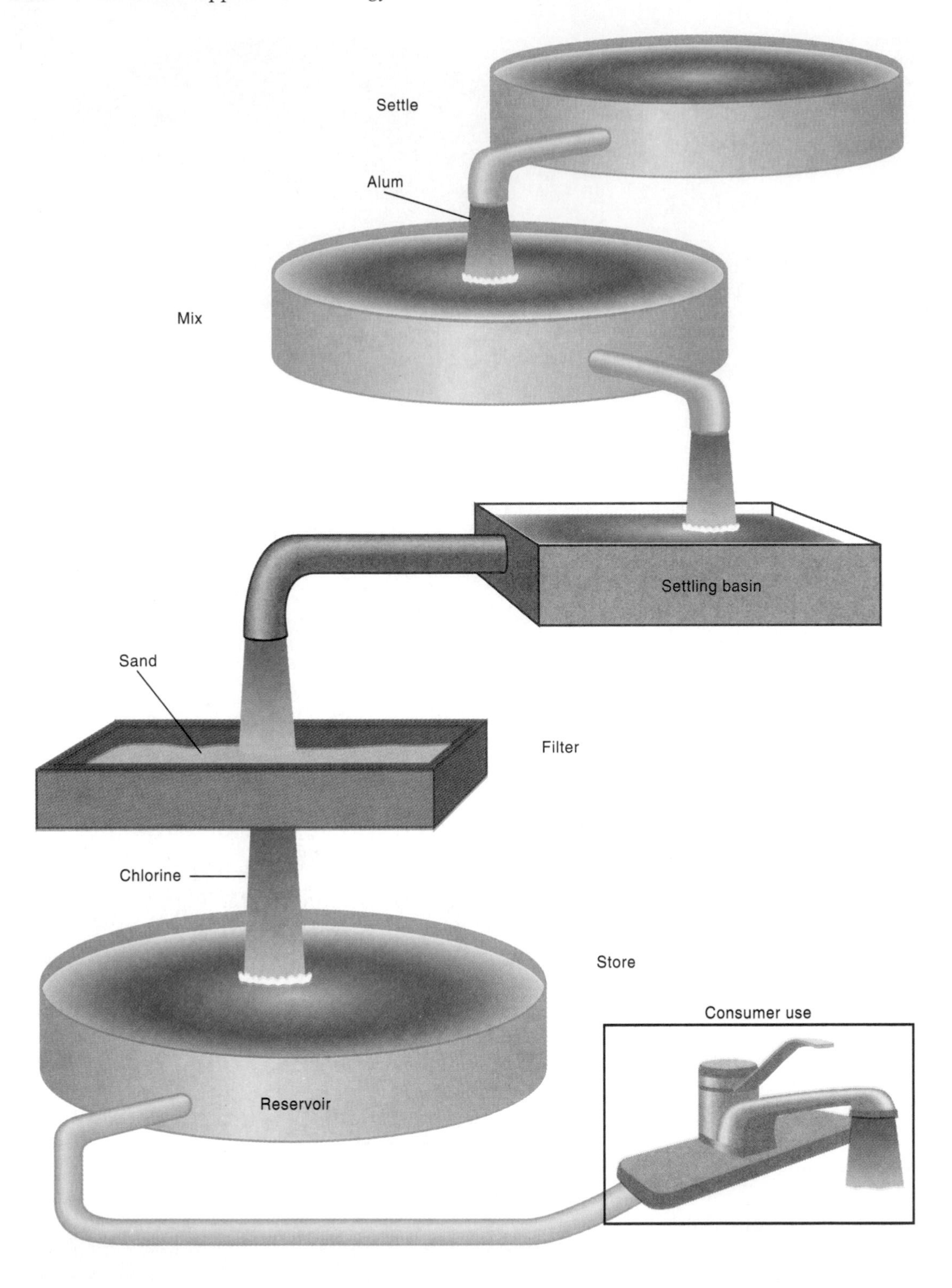

Step 1 Water is held in reservoirs, where large materials sediment. Aluminum sulfate may be added to cause flocculation of organic matter, which then settles out.

Step 2 The water is then filtered through beds of sand which removes almost all of the bacteria. Filtration through activated charcoal may be used to remove toxic or objectionable organic materials.

Step 3 Finally, chlorination is used to disinfect the water, killing any pathogens that might remain. It is possible to treat drinking water with ultraviolet irradiation or with ozone so that it does not have to be chlorinated. Las Vegas will be the first city to use such methods in place of chlorination. Some cities also add fluoride to protect against dental cavities.

Figure 33.4
Steps in the treatment of metropolitan water supplies. (*a*) Water is held in reservoirs, where large materials sediment. Aluminum sulfate may be added to cause flocculation of organic matter, which then settles out. (*b*) Water is then filtered through beds of sand, removing almost all of the bacteria. Filtration through activated charcoal may be used to remove toxic or objectionable organic materials.
(*c*) Finally, chlorination is used to disinfect the water, killing any pathogens that might remain. Some cities also add fluoride to protect against dental caries.

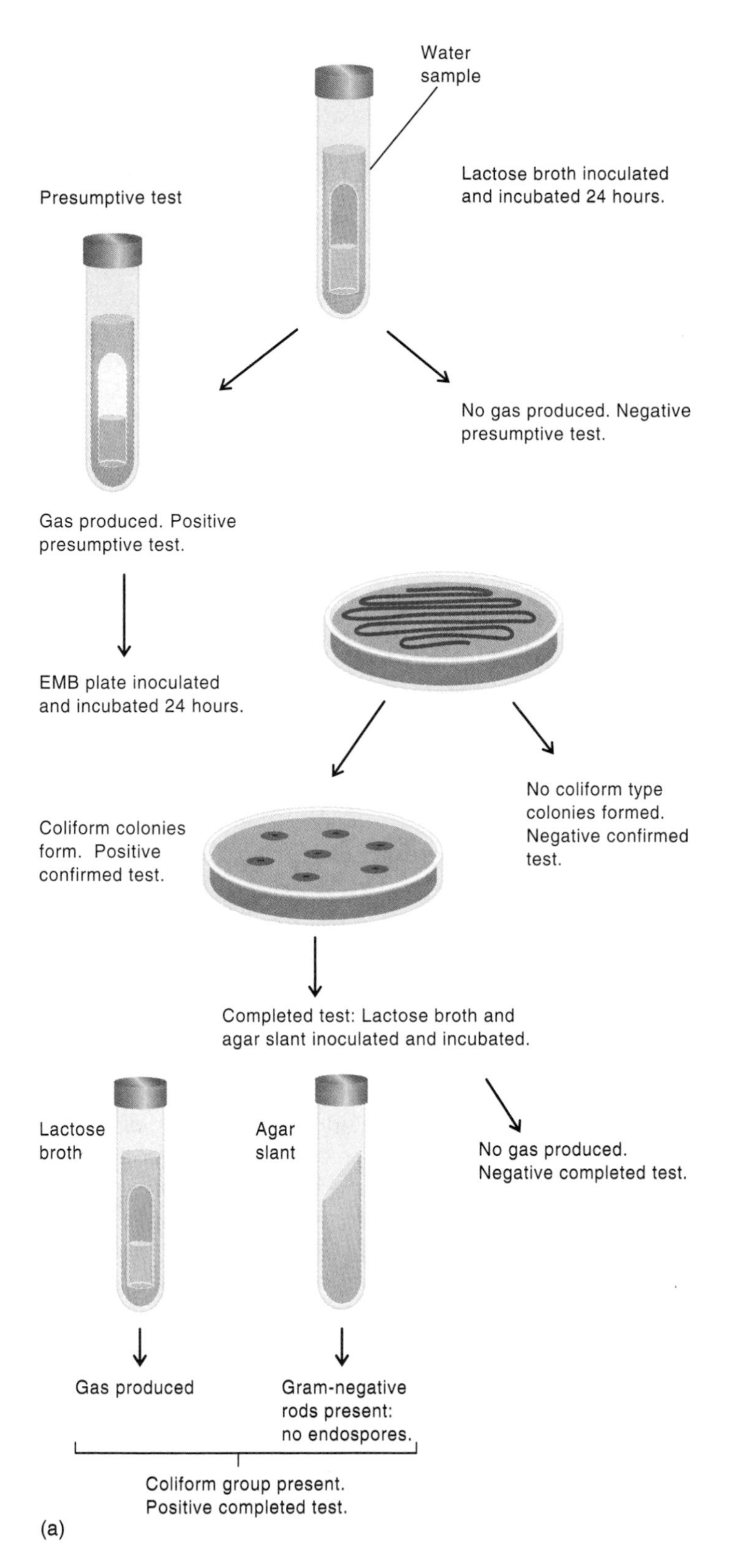

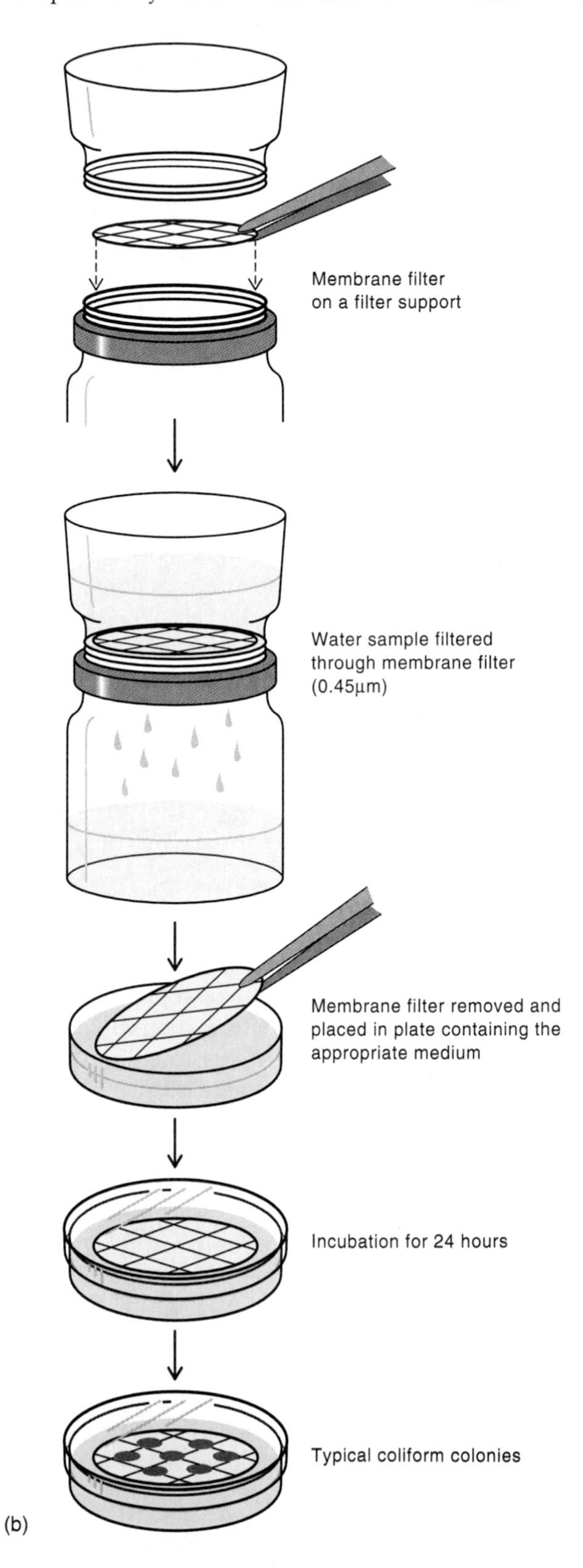

Figure 33.5

Methods used for testing water. Fecal contamination of water is indicated by the presence of the coliform *Escherichia coli*. (*a*) Traditional method for detecting *E. coli*. *E. coli* ferments lactose, producing acid and gas. In the first step, the water sample is added to lactose broth and incubated for 24 hours. The production of gas is a positive presumptive test. The next step is to confirm that the gas-forming organism in the sample is *E. coli*. To do this, the 24-hour culture is streaked onto an EMB (eosin-methylene blue) plate and incubated for 24 hours. *E. coli* grown on EMB gives characteristic colonies with a metallic sheen. The presence of such colonies is taken as a positive confirmed test. To complete the investigation, lactose broth and an agar slant are inoculated. The presence of gram-negative rods and the absence of endospores on the agar slant and the presence of gas production in the lactose broth constitute a positive completed test. (*b*) The membrane filter procedure is used for direct recovery of indicator bacteria from water.

Microbiology of Waste Treatment

A large majority of the population of the United States is concentrated in cities that cover a very small percentage of the land. To understand what this means in terms of waste treatment, consider that every day the average American uses about 150 gallons of water, 4 pounds of food, and 19 pounds of fossil fuel, which are converted into some 120 gallons of sewage, 5 pounds of trash, and almost 2 pounds of air pollutants! This means that a city with only 1 million inhabitants is faced with the disposal of 120 million gallons of domestic sewage daily, to say nothing of industrial and other wastes.

Microbial activity is extremely important in the recycling of all these waste materials. Most of the methods used to treat wastes depend on the conversion, by microorganisms, of organic materials to inorganic forms—the process of **mineralization** or **stabilization.** In general, the **primary treatment** of sewage involves the removal of large objects and much of the particulate matter through the physical processes of screening and sedimentation. **Secondary treatment** involves the chemical and biological processes of converting the remaining materials in sewage into odorless inorganic substances that can be reused. **Tertiary treatment** involves the removal from sewage of phosphates and nitrogen compounds that could cause eutrophication.

Microorganisms Degrade the Components of Sewage to Inorganic Compounds

During the **aerobic** treatment of sewage, microbial oxidations of organic compounds yield carbon dioxide and inorganic nitrogen-containing nutrients for plants. During the **anaerobic** decomposition of sewage, similar changes occur, except that anaerobic microbes ferment organic compounds. The products of this fermentation are subsequently utilized through aerobic respiration. The methanogens are important in anaerobic degradation, transforming the small breakdown products formed by other bacteria from organic carbon compounds into CO_2 and CH_4 (methane). In the presence of hydrogen, methanogens can further convert CO_2 into methane as well. The remaining CO_2 can be metabolized by photosynthetic organisms and plants, and the CH_4 generated by the methanogens can either be discarded, conserved for fuel, or oxidized to CO_2 by the methane-oxidizing bacteria.

Microorganisms Reduce the Biochemical Oxygen Demand of Sewage

The term **biochemical oxygen demand (BOD)** designates the oxygen-consuming property of a wastewater sample; it is roughly proportional to the amount of degradable organic material present in the water sample. To measure the BOD, a well-aerated sample of water is incubated in a sealed container under standard conditions of time and temperature (usually five days at 20°C), and the amount of oxygen consumed by the microorganism growing in the sample is determined. The higher the BOD, the more oxygen is being used in biological degradation processes during the incubation. Thus, high BOD values reflect large amounts of degradable organic materials in a sample of wastewater or other material. For example, rich nutrient media used for laboratory cultivation of bacteria have BOD values of from approximately 2,000 to 7,000 milligrams of oxygen per liter of solution, as compared with 100 to 300 milligrams per liter for unpolluted natural waters. The decrease in BOD of wastewater during treatment reflects the effectiveness of the treatment in converting organic wastes in the water to inorganic materials.

Effective treatment decreases the BOD of sewage as much as possible and removes toxic materials and other objectionable matter from the sewage. The methods of sewage treatment chosen depend on a number of factors, including the amount of sewage material, the BOD, the presence of toxic materials, and the nature of the receiving waters (that is, the bodies of water into which the sewage treatment products are emptied).

Methods for Waste Treatment Include Primary, Secondary, and Tertiary Treatment

The steps in the treatment of metropolitan wastes are outlined in figure 33.6 and discussed next.

Primary Treatment

Primary treatment of sewage is designed to remove materials that will settle or sediment out. During this step, sewage is passed through a series of screens to remove large objects such as sticks, rags, and trash and then allowed to settle for a period of from 90 minutes to two hours. Sometimes, aluminum sulfate, ferrous sulfate, or other chemicals are added to flocculate particles so they settle more rapidly. After the settling period, the remaining fluid is given secondary treatment, and the sedimented material from the primary tanks is usually either sent to a large tank called a **digester** for further treatment or is incinerated.

Secondary Treatment

Most municipal treatment plants in the United States now use secondary treatment because the Clean Water Act requires that all municipal treatment plants provide for it. Secondary treatment of sewage is designed to stabilize most of the organic materials and reduce the BOD of the sewage.

The mechanisms by which populations of aerobic organisms stabilize sewage at most secondary treatment

Perspective 33.1

Now They're Cooking with Gas

Farmers in rural China have historically been very frugal when it came to using any waste materials. However, in rural areas, fuel for cooking and heating are often in short supply, and up to 80% of the hay that should go for animal feed is used for fuel. To alleviate this problem, many of the farmers build methane-producing tanks on their farms (figure 1). Near the farmhouse is an underground cement tank that is connected to the latrine and pigpen. Human and animal wastes along with water and some other organic materials such as straw are added to the tanks. As the natural process of fermentation occurs, methane gas (CH_4) is produced. This gas rises to the top of the tank and is connected to the house with a hose. Enough gas is produced to provide lights and cooking fuel for the farm family. By producing its own gas, a family also saves money because it does not need to buy coal for cooking.

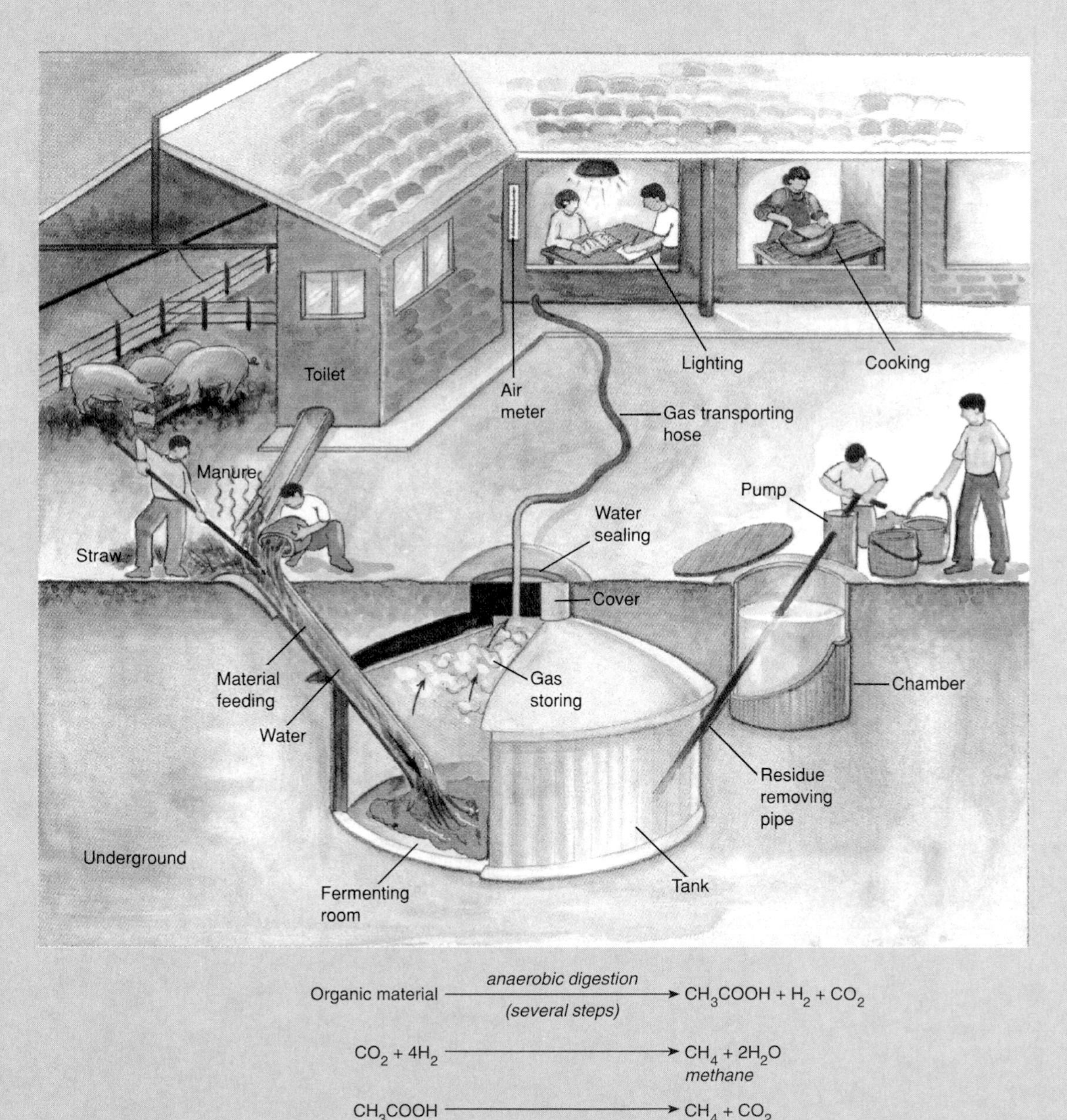

Figure 1
Production and use of methane on a small scale.

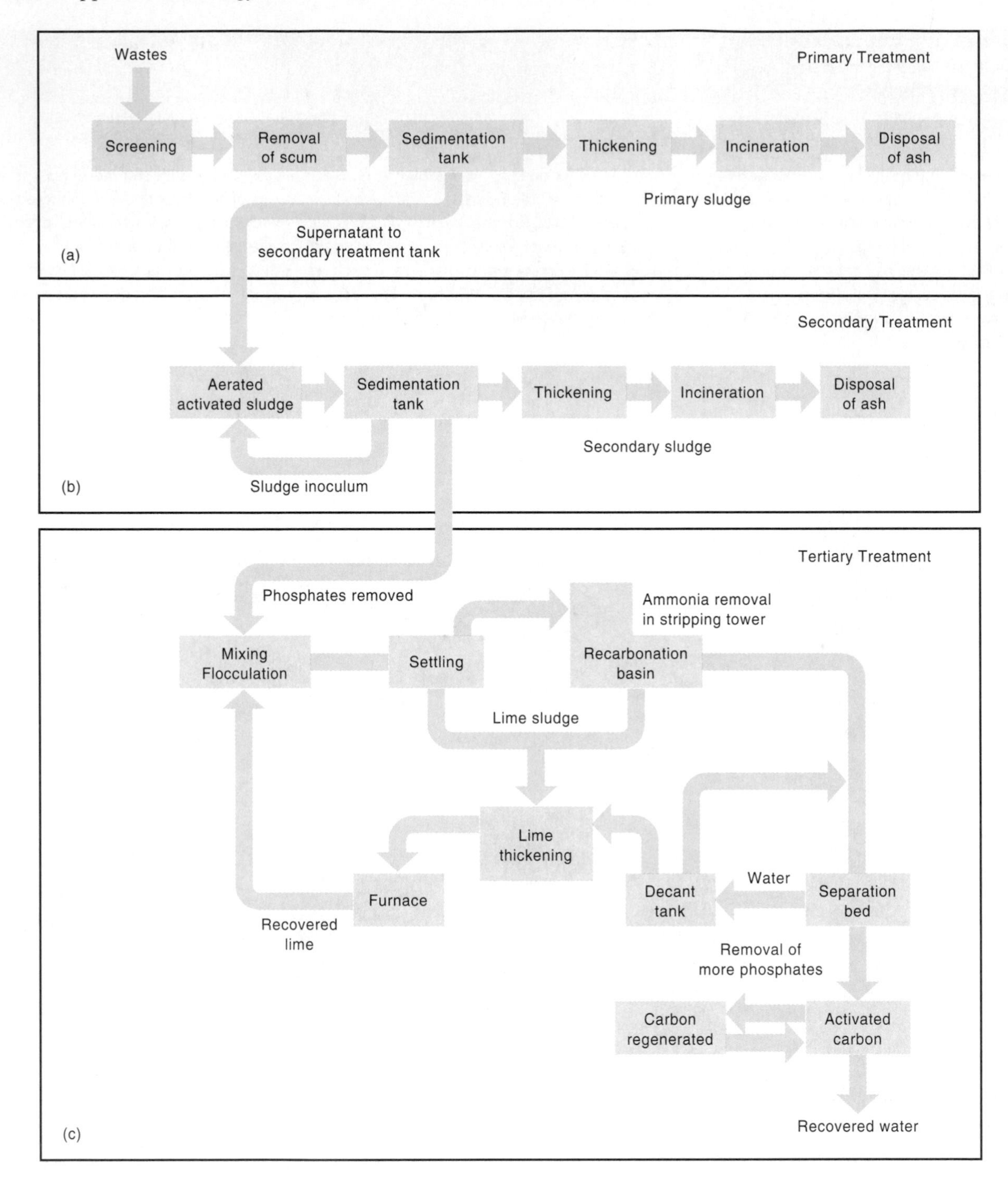

Figure 33.6
Metropolitan waste treatment. The most advanced waste treatment schemes utilize tertiary as well as primary and secondary treatments. (*a*) Primary treatment; (*b*) secondary treatment; and (*c*) tertiary treatment.

plants is called the **activated sludge method.** In this treatment, the sewage serves as a nutrient source for mixed populations of aerobic organisms adapted to grow in it. An abundance of oxygen is supplied by mixing the sewage in an aerator. Most of the biologically degradable organic material is then converted into gases or oxidized products, and a very small percentage is incorporated into the cell material of the organisms growing in the sewage. This process is often the most satisfactory way to treat domestic sewage, but it can be rendered totally ineffective by the presence of toxic industrial wastes, which poison the microbial population.

After the bacteria and fungi in sewage treatment have used certain nutrients, they serve as the food for ciliates, protozoa, nematodes, and other forms of life. The end result is a small increase in the mass of organisms present in a given amount of treated sewage and a large decrease in the amount of biodegradable organic materials. The increased microbial mass in the sewage is then removed to the digester; a portion is left behind as the inoculum to act on a new load of waste materials.

Tertiary Treatment

At present, many sewage treatment plants discard the effluent from secondary treatment into lakes, rivers, and other bodies of water, often causing eutrophication. The large quantities of phosphates or nitrates remaining in sewage after secondary treatment may increase the growth of microorganisms that gradually deplete the oxygen and thus threaten other forms of aquatic life. The tertiary treatment of sewage, to remove nitrates and phosphates, can greatly alleviate this problem.

In some designs for the tertiary treatment of sewage, chemical precipitation of phosphates has been combined with biological removal of nitrates. Certain bacteria (particularly species of *Pseudomonas* and *Bacillus*) can completely reduce nitrates (NO_3^-) to N_2 (denitrification). The N_2 gas is inert, nontoxic, and easily removed.

Digester Treatment Anaerobically Stabilizes Sewage

Within a sewage digester, anaerobic organisms act on the solids remaining in sewage after its aerobic treatment. The digester provides **anaerobic stabilization** and removes water from the sewage so that a minimum of solid matter remains in it. Various populations act sequentially. In the sewage digester, the anaerobic methane-forming organisms can perform their role of converting the simple organic acids in sewage into the useful end product methane (CH_4). Many sewage treatment plants are equipped to use their methane, thereby avoiding the cost of using other sources of energy to run their equipment.

Pathogens May Survive During Sewage Treatment

Pathogenic bacteria are generally eliminated from sewage during secondary treatment, but disease-producing viruses may survive. Pathogens account for only a small proportion of the total number of bacteria in feces, and they are greatly diluted by the water in sewage. During secondary treatment of sewage, these pathogens must compete for nutrients with the huge mass of bacteria that have been adapted to grow best at the temperature and conditions provided. As a result, most pathogenic bacteria are rapidly overgrown and eliminated by their competitors. Animal viruses lack appropriate host cells in sewage and cannot replicate there, although they may survive for long periods. If, therefore, large quantities of virus particles are present in raw sewage, some may be recovered after secondary treatment. Furthermore, although sewage effluents are often chlorinated before being discharged into receiving waters, chlorine treatment at this stage does not kill viruses, because virus particles are commonly protected from this chemical by their enclosure within small aggregates of effluent materials. Because the viruses do adhere to larger particles, they can be removed along with the other solid materials.

Sewage Can Also Be Treated on a Small Scale

In recent years, considerable progress has been made in developing better methods for the disposal of urban wastes. At present, however, these methods are not practical for small communities or for isolated dwellings. In such settings, other means of waste disposal that employ much the same basic principles of microbial degradation but in different ways must be used.

Small towns sometimes depend for their waste disposal on a process called **lagooning,** in which sewage is channeled into shallow ponds, or lagoons, where it remains for several days to a month or more, depending on the design of the lagoon. During this period, settling occurs and sewage materials are stabilized by anaerobic or aerobic organisms or both. Pathogenic bacteria are usually eliminated by competition, as described previously.

Trickling filters (figure 33.7) are frequently used for smaller sewage treatment plants; they are also sometimes used in place of activated sludge digesters for secondary treatment. The rotating arm of the trickling filter sprays controlled amounts of sewage over a bed of coarse gravel and rocks. The pebbles and rocks in the bed then become coated with a film of organisms that aerobically degrade the sewage. The film, about 2 millimeters thick, consists of an outer layer in which fungi predominate; a middle layer of fungi, algae, and cyanobacteria; and an inner layer composed largely of bacteria, fungi, and algae. Protozoa, especially some of the ciliates, are also present. The rate of sewage flow over the rocks can be adjusted so that waste materials are maximally degraded. As is the case with the activated sludge method, different populations act in turn to degrade various compounds. Nematodes, rotifers, ciliates, bacteria, and other organisms cooperate during sewage stabilization with trickling filters and in lagoons.

Isolated dwellings or very small communities customarily rely on **septic tanks** for sewage disposal. In theory, the septic tank makes sense; in practice, however, it

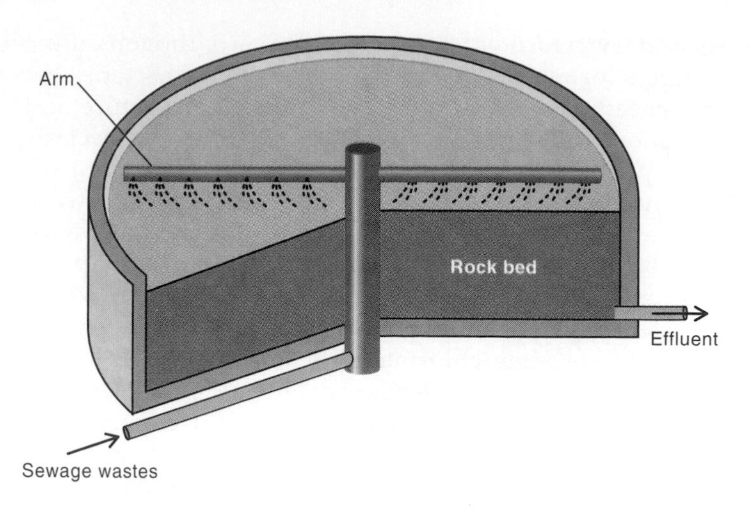

Figure 33.7
Diagrammatic representation of a trickling filter. Sewage wastes are channeled into the revolving arm and trickle through holes in the bottom of the arm onto a gravel and rock bed. The rocks are coated with microorganisms that stabilize the sewage as it trickles through the bed so that the effluent has a greatly reduced load of degradable organic materials.

often does not work correctly. Figure 33.8 illustrates the design of a septic tank. Sewage is collected in a large tank in which much of the solid material settles and is degraded by anaerobic microorganisms. The fluid overflow from the tank has a high BOD and must be passed through a drainage field of sand and gravel. Theoretically, stabilization should occur in the drainage field in the same manner described for the trickling filter. However, stabilization depends on adequate aeration and sufficient action by aerobic organisms associated with sand and gravel in the drainage field, conditions that are often not met. For example, clay soil under a drainage field may prevent adequate drainage, allowing anaerobic conditions to develop, or toxic materials may inhibit microbial activity in the drainage field. The possibility exists that the drainage from a septic tank contains pathogens; therefore, the tank must never be allowed to drain where it can contaminate water supplies.

Better ways of dealing with small quantities of sewage are constantly being sought. One method that has proven satisfactory in some small communities is the use of **artificial wetlands.** With an artificial wetlands system, individual septic tanks and drainage fields are not needed. Instead, land can be subdivided into more lots of smaller size, with a common portion of the land set aside for a series of ponds (figure 33.9). Sewage is channeled into

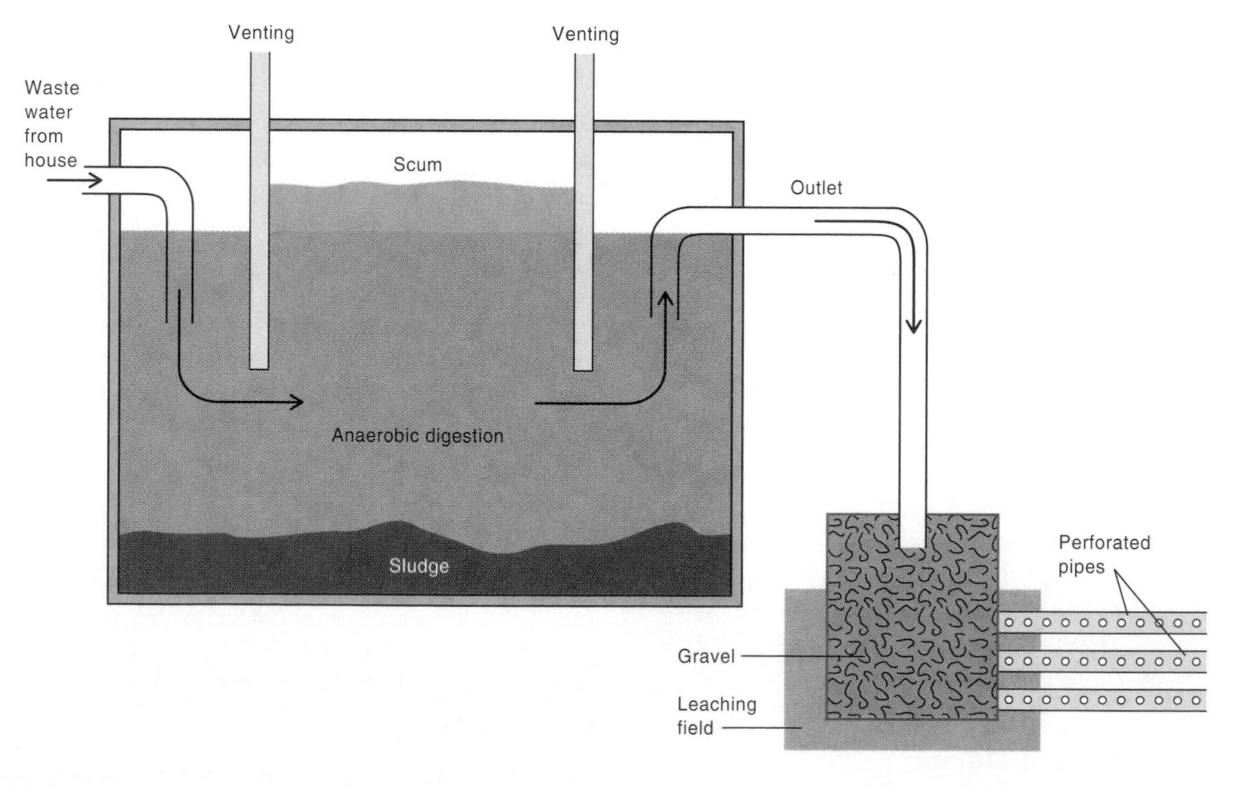

Figure 33.8
Septic tank. Sewage wastes enter the tank from the house. Within the tank, solid materials settle and undergo anaerobic stabilization. Materials that do not settle exit through the outlet pipe, which permits seepage into the drainage field. Conditions in the drainage field must be aerobic so that materials remaining can be degraded by the activities of aerobic microorganisms. If the drainage area is not properly designed, contaminated materials readily enter adjacent surface waters.

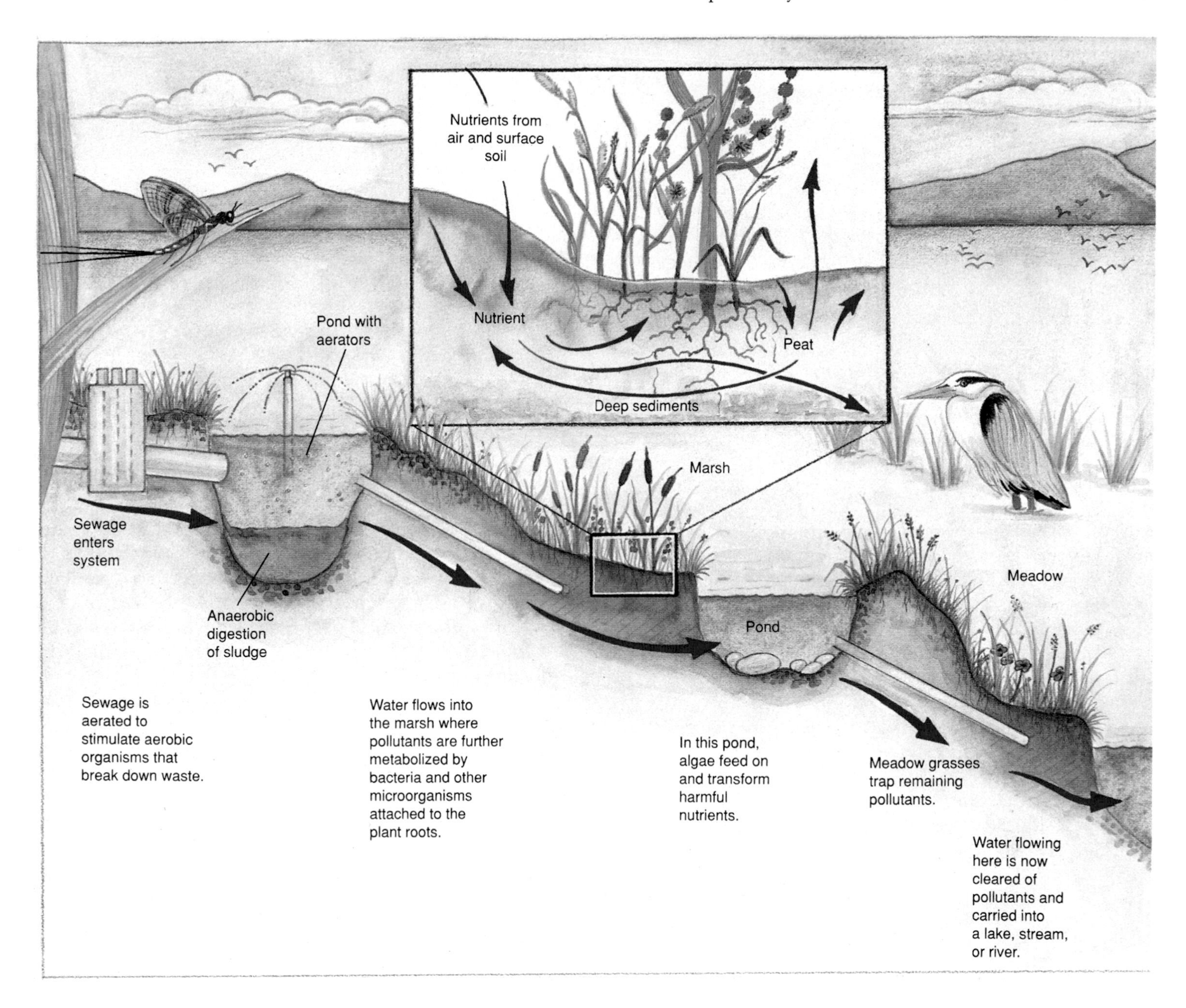

Figure 33.9
Artificial wetland.

successive ponds, which are carefully designed to carry out both aerobic and anaerobic stabilization. By the time the water reaches the final pond, it is suitable for decorative or recreational use. This type of system depends on proper design, adequate maintenance, and careful monitoring. In communities where it has been used it has provided safe and efficient waste disposal with the added bonus of a park area around the pond system.

Progress is also being made in many other areas of waste treatment. For example, soil filters are being developed to remove gases, such as H_2S, that are formed during sewage treatment. If the area is kept moist so that bacteria can grow, the foul-smelling and toxic hydrogen sulfide is removed during passage through the soil and converted by bacteria (such as *Thiobacillus* sp.) to sulfate ions.

Use of Treated Waste Residues Is Still a Problem

One area of waste treatment in which research and development are greatly needed is in the utilization of treated waste residues. Clearly, receiving waters deteriorate following the addition of treated wastes, as a result of the nutrients added to them as well as from toxic materials and changes in their pH and temperature. Even when the best available methods of tertiary sewage treatment are used,

Perspective 33.2

Sludge for Trees

Sludge is a by-product of sewage treatment that can be difficult to dispose of. If it is trucked to the landfills, it takes up valuable space. If it is discharged into waterways or incinerated, it can become a major source of pollution.

Scientists from the Forestry Department at the University of Washington in Seattle are experimenting with at least one strategy for dealing with the problem of sludge disposal. They have taken the sludge from the city's sewage treatment plant to their experimental forest and used it as fertilizer for Douglas fir trees. They have found that the sludge boosts the growth of these trees by over 400%.

There are two major problems with using sludge as fertilizer. The first problem is that heavy metals and other similar pollutants remain in the sludge following digestion of the sewage. The second problem is that pathogenic microorganisms and viruses also may be left in the residues. Because of these two problems, sludge cannot be used as fertilizer on food crops or in other situations in which humans might be contaminated. However, the sludge can be used to grow trees that will be processed for pulp or lumber. It is important that the sites where sludge is used be carefully monitored to be certain that heavy metals, organic pollutants, large amounts of nitrogen, and pathogenic microorganisms do not present public health problems in the runoff, or groundwater leaching. Whether large-scale use of sludge for forest production is an economical or practical solution to the disposal of sludge is still open to debate.

the quality of receiving water decreases. For instance, although the recovered water at some treatment plants is pure enough to drink, it still contains some microbial nutrients. For this reason, the purified water is collected in reservoirs rather than being rechanneled into a lake.

The use of sewage sludge, the solid residue of waste treatment, is also a difficult problem. When this waste is burned, polluting gases are often formed and must be removed. In some areas, the treated sludge is used for fertilizer, but this is possible only if toxic compounds are not present in the sludge.

The treatment of wastes and utilization of waste effluents and residues currently constitute one of the most challenging areas of research for microbiologists and other scientists working in this area.

Microbiology of Solid Waste Treatment

In addition to ridding our environment of wastes in water, we must take care of the solid wastes that are generated each day. In the United States, more that 150 million tons of solid waste are produced each year. Some of this is industrial waste and some is from households. Some of it is made up of inert materials such as glass, metals, and plastics, but a large amount is made up of organic materials that are affected by microorganisms. Eliminating waste products from the environment has become an increasingly complex problem.

Use of Landfills Has Been a Low-Cost Method of Disposing of Solid Wastes

Landfills are used to dispose of solid wastes near towns and cities. Usually, the area of land that is chosen for the landfill is not particularly valuable. Because just piling the solid wastes up on the ground attacts insects and rodents, causing both aesthetic and public health problems, **sanitary landfill** methods are most often used. At the end of each day, a layer of dirt is put over the wastes. When the landfill is completed, it can be used for recreation and eventually as a site for construction.

There are several disadvatanges to this type of waste management. First, there are only a limited number of sites that are available for use near urban and suburban areas. Second, the organic content of landfills anaerobically decomposes very slowly, over a period of at least 50 years. During this time, the methane gas that is produced must be removed. If buildings are constructed before most of the methane is removed, disastrous explosions can be induced by the gas. Pollutants such as heavy metals and pesticides can leach from landfill sites into the underground aquifers. It is very difficult to purify these aquifers once they have become contaminated.

Although sanitary landfill has traditionally been a low-cost method of handling large quantities of solid waste, because of increased costs and decreased availability of land, many cities are looking for ways to decrease the amount of solid waste. In Seattle, the fees charged to people for garbage collection are based upon the size of the container collected. The smaller the can, the lower the cost. Waste disposal companies are now considering weighing garbage and charging people accordingly to further reduce the amount of solid waste generated. Both of these methods are intended to raise people's awareness of how much solid waste they are generating as well as to offer an incentive to recycle. Companies in some areas offer lower costs for garbage collection to people who recycle. Programs to recycle paper, plastics, glass, and metal are being implemented in many cities and counties with great success. Through these programs, landfill areas can be expected to be available for a longer period of time.

Backyard and Commercial Composting Is an Attractive Alternative to Landfill

Commercial and home composting programs are becoming popular in many areas and are succeeding in reducing the amount of solid waste generated. Composting is the natural decomposition of organic solid material.

Backyard composting involves mixing garden debris with kitchen organic waste (excluding meats and fats). If proper amounts of sunlight, water, and air are provided to the mix, in a matter of months, the bulk of the waste is reduced by two-thirds. The black organic material generated by composting can then be used as fertilizer in garden beds. This material is rich in organic compounds and improves the soil for growing plants. Figure 33.10 shows the steps in managing a backyard compost heap.

A compost pile usually starts with a supply of organic material such as leaves and grass clippings. Often, some soil

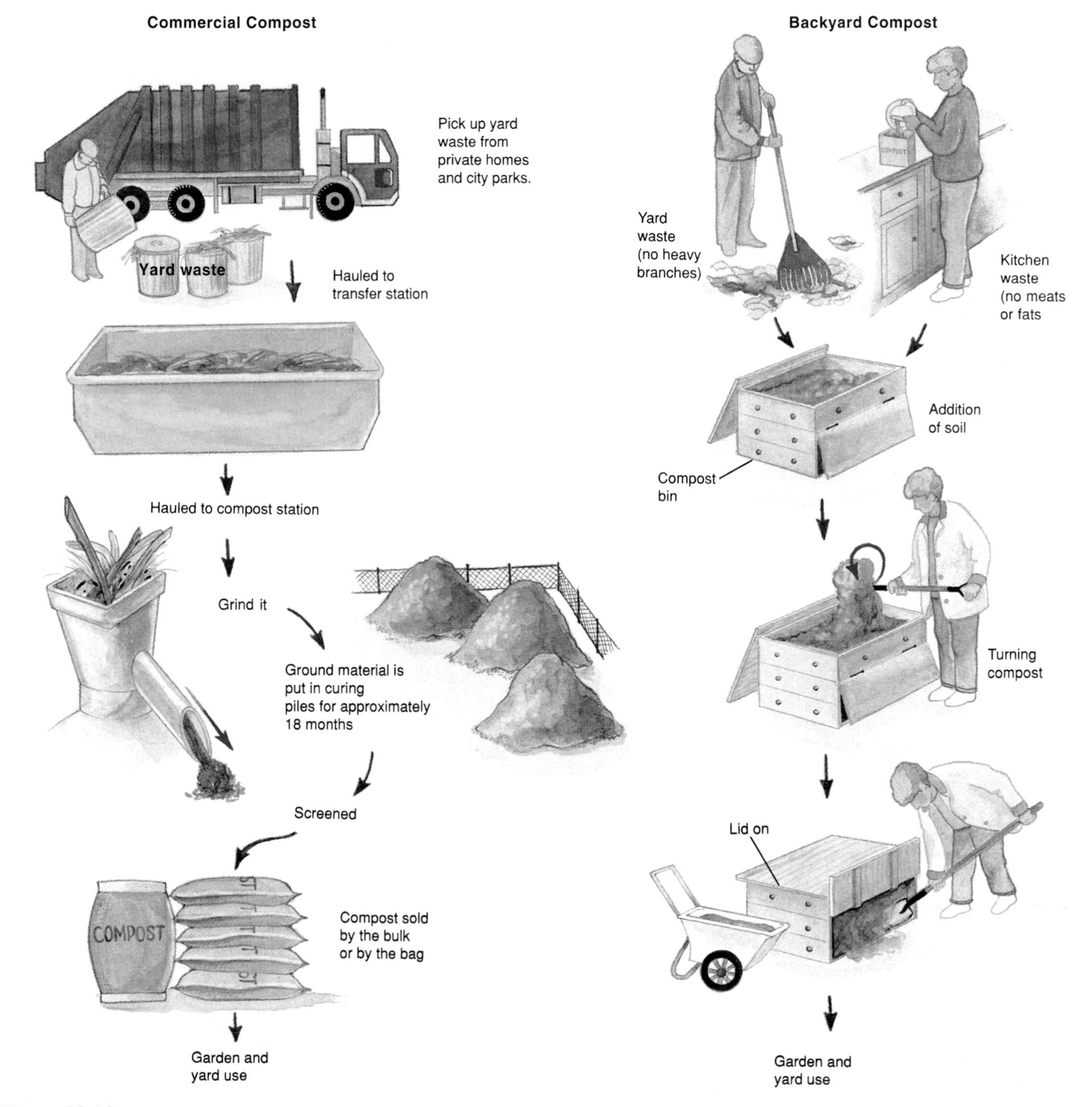

Figure 33.10
Industrial and backyard composting.

and perhaps water are added to it. In a few days, the inside of the pile heats up. At 100°F to 150°F, pathogens are killed but other organisms that are thermophilic are not affected. When the pile cools down, it is aerated by physically stirring it up and turning it. This stirring adds oxygen and allows the bacteria to resume growing and breaking down the organic materials. If a compost pile is turned frequently and other conditions are good for aerobic digestion, the composting can be completed in as little as six weeks.

Composting on a large scale offers cities a way to reduce the amount of garbage sent to their landfills. In Seattle, yard wastes are collected separately from the main garbage. These wastes are then composted and used in various ways, including for improving soils in city parks. Special machinery is used to compost on a large scale, and the composting can be accomplished in a very short time (figure 33.10).

REVIEW QUESTIONS

1. What is meant by the term *sanitary landfill?*
2. What are some alternatives to sanitary landfills?
3. List the steps that must take place in successful composting of organic material.
4. What is the process of mineralization or stabilization?
5. Why might the drainage from a septic tank contain pathogens?

Summary

I. Water Pollution by Microorganisms and Chemicals

II. Water Testing and Treatment

III. Microbiology of Waste Treatment
   A. Microorganisms degrade the components of sewage to inorganic compounds.
   B. Microorganisms reduce the biochemical oxygen demand of sewage.
   C. Methods for waste treatment include primary, secondary, and tertiary treatment.
   D. Digester treatment anaerobically stabilizes sewage.
   E. Pathogens may survive during sewage treatment.
   F. Sewage can also be treated on a small scale.
      1. Lagooning
      2. Trickling filters
      3. Septic systems
      4. Artificial wetlands
   G. Use of treated waste residues is still a problem.

IV. Microbiology of Solid Waste Treatment
   A. Use of landfills has been a low-cost method of disposing of solid wastes.
   B. Backyard and commercial composting is an attractive alternative to landfill.

Chapter Review Questions

1. Name at least one protozoan parasite that commonly causes waterborne disease.
2. Discuss the possible sources of several chemical contaminants found in water supplies.
3. Why do water-testing procedures look for coliform organisms rather than other pathogens?
4. Does chlorination of water supplies guarantee safe drinking water?
5. What is meant by BOD?
6. Explain why primary, secondary, and tertiary treatments of waste are each important in processing sewage.
7. How effective are the different stages of sewage treatment in removing pathogens from the materials?
8. What is the origin of the methane gas produced during sewage treatment?
9. In what situation might lagooning be a good choice for sewage treatment?
10. What are some advantages of using artificial wetlands for dealing with small amounts of sewage?

Critical Thinking Questions

1. It has been said that "the solution to pollution is dilution." Discuss the reasons this is not a valid statement.
2. Water supplies are vulnerable to deliberate contamination by saboteurs. How might enemy agents use microorganisms or their products to poison water supplies and what measures can be taken to prevent or counteract such sabotage?
3. Suppose you own 150 acres of oceanfront property that you want to develop into a community of vacation homes, retaining as much as possible of the natural setting. Safe and effective sewage treatment must be a part of the plan. What are the advantages and disadvantages of each of the following for such a development?
   a. Individual septic tanks for each home
   b. Trickling filter
   c. Artificial wetlands

Further Reading

Edmondson, W. T. 1991. *The uses of ecology: Lake Washington and beyond.* University of Washington Press, Seattle.

McFetters, G. A., (ed.) 1990. *Drinking water microbiology.* Springer-Verlag, New York.

34 Microbiology of Food and Beverages

KEY CONCEPTS

1. A variety of organisms can use food as a growth medium. The end products they produce can be desirable (fermented foods), undesirable (spoilage), or harmful (food poisoning).
2. Factors such as available moisture, pH, storage temperature, and the intrinsic properties of the food product can influence the type of microorganisms that grow and predominate in a food product.
3. A variety of foods such as bread, cheese, and alcoholic beverages result from the metabolic activities of microorganisms.
4. Food spoilage can be eliminated or retarded by altering the conditions under which microorganisms grow.

A variety of foods such as cheeses, breads, and wine are produced with the activities of microorganisms.

PREVIEW LINK

Microbes in Motion

The following books in the *Microbes in Motion* CD-ROM, covering some of the disease-causing microorganisms in foods, may serve as a useful supplement to your reading:

Gram-Negative Organisms: Bacilli. *Gram-Positive Organisms:* Bacilli.

A Glimpse of History

Pasteurization of milk and milk products is so common today that it is hard to fathom that only during this century have federal, state, and local laws mandated it for all milk products. Although most large cities have required milk to be pasteurized since 1900, as late as 1930 most milk sold in rural areas was not routinely pasteurized. As a result, human diseases such as brucellosis and tuberculosis were fairly common.

Alice Catherine Evans, the first woman president of the Society of American Bacteriology (now the American Society for Microbiology), helped establish the connection between unpasteurized milk and brucellosis in humans. A graduate of both Cornell University and the University of Wisconsin, Evans worked for the United States Department of Agriculture, seeking out the sources of microbial contamination of dairy products.

In 1917, Evans reported that cases of human brucellosis were related to the finding of *Brucella abortus* in cows' milk. Her conclusion that *B. abortus* could be transmitted from cows to humans through milk conflicted with the prevailing view of a number of prominent scientists, including Robert Koch. In 1900, Koch had declared that bovine tuberculosis and brucellosis could not be transmitted to humans. As a result, at least 30 years elapsed before many scientists and dairy workers would accept the increasing evidence that diseases could be transmitted from cows to humans through the milk supply. In the 1920s, dairy herds were inspected and vaccinated for tuberculosis. Herds that passed inspection and were vaccinated were called *certified herds*, and their milk could be sold commercially without pasteurization. In the late 1930s, after a number of the children of dairy workers had died of brucellosis even though they had drunk only certified milk, the problem of milkborne disease was finally acknowledged. Today, milk is routinely pasteurized, and only very small amounts of unpasteurized milk are sold in the United States.

Fruits, vegetables, dairy products, and meats all contain a variety of microorganisms on their surfaces. This is not surprising since microbes are ubiquitous and are especially plentiful in the soil and around animals. In fact, at least some microorganisms are likely to occur on most food unless it has been sterilized. Food can be viewed as a type of ecosystem in which the microorganisms metabolize the nutrients in the food to make end products such as acids, alcohols, and gas as discussed in chapter 5.

Microorganisms growing on food can be perceived as desirable if they result in intentional, pleasant tastes or textures. Foods that have been altered due to the carefully controlled growth of microorganisms are called **fermented.** For example, food manufacturers purposely encourage the growth of microorganisms to produce fermented foods such as sour cream and blue cheese. The same biochemical changes, if perceived as unpleasant, are called **spoilage.** Sour milk and moldy bread are examples of foods that have been spoiled as a result of microbial growth. The same biochemical processes that cause the desirable acidic flavor of the sour cream and the blue veining of blue cheese are undesirable in milk and bread. Growth of pathogens such as *Staphylococcus aureus* in a food product generally do not result in perceptible changes in quality of the food, but their growth can result in foodborne illness.

Factors Influencing the Growth of Organisms in Foods

Microbial growth must be controlled in order to encourage desired fermentations or to discourage growth of spoilage organisms and pathogens. Understanding the factors discussed in chapter 4 that influence the growth of microorganisms in foods is an essential step in that control. Factors such as the availability of water and nutrients, the pH, and storage temperature together determine which microorganisms can grow in a particular food product and the rate at which they can grow. For example, bacteria will usually predominate in moist, pH neutral, nutrient-rich food products such as fresh meat because they are able to multiply there so quickly. Other microorganisms such as yeasts and molds can also grow under these conditions, but the bacteria grow faster and rapidly overcome the yeasts and molds. There is even competition among different types of bacteria. On moist, pH neutral foods, members of the genus *Pseudomonas* are among the fastest-growing bacteria. On slightly acidic food products, however, the growth of *Pseudomonas* is inhibited, and the slower-growing lactic acid bacteria can compete. When the growth of most common bacteria is inhibited, such as by dry or very acidic conditions, the otherwise slower-growing fungi will predominate.

The Availability of Water Plays a Role in Determining Which Microorganisms Can Grow

The growth of microorganisms in a food product is influenced by the amount of water available in the food. Food products vary dramatically in terms of how much water is accessible in them. For example, fresh meats and milk have ample water to support the growth of most microorganisms, whereas breads, nuts, and dried foods have relatively little water. Some sugar-rich foods such as jams and jellies are seemingly moist, but they have little available water because the water is chemically interacting with the sugar. Likewise, highly salted foods have little available moisture. The term **water activity (a_w)** is used to designate the amount of water available in foods. The a_w is the ratio of the vapor pressure of a food product to that of pure water. Thus, by definition, pure water has an a_w of 1.0. Most fresh foods have an a_w above 0.98 (table 34.1).

Most bacteria require an a_w above 0.90 for growth, which explains why fresh foods spoil more quickly than dried, sugary, or salted foods (table 34.2). Bacteria grow quickly on fresh, moisture-rich food products and rapidly cause undesirable changes in flavor and texture. Fungi can

Table 34.1 Approximate a_w of Some Foods

Food	Typical a_w*
Fresh meat	0.99
Processed cheese	0.95
Ham	0.91
Maple syrup	0.90
Jam	0.85
Cake	0.70

*The a_w will vary depending on preparation and storage.

Table 34.2 Approximate Minimum a_w for Growth of Some Foodborne Microorganisms

Organism	Minimum a_w
Most foodborne bacteria	0.90
Pseudomonas	0.97
Escherichia coli	0.94
Staphylococcus aureus	0.86
Most yeasts	0.88
Most molds	0.80

Source: Jay, J. 1992. *Modern food microbiology*, 4th ed. Van Nostrand Reinhold, New York.

Table 34.3 Approximate pH of Common Foods

Food	Approximate pH
Milk	6.4
Chicken	6.3
Spinach	5.6
Carrots	5.0
Cherries	3.8
Lemons	2.2

Source: U.S. Food and Drug Administration. 1992. Foodborne pathogenic microorganisms and natural toxins. U.S. FDA: Center for Food Safety and Applied Nutrition.

Table 34.4 Minimum pH for Growth of Some Foodborne Microorganisms

Organism	Minimum pH
Pseudomonas	5.5
Staphylococcus	4.5
Lactic acid bacteria	3.5
Yeasts	1.5
Molds	0

Source: Jay, J. 1992. *Modern food microbiology*, 4th ed. Van Nostrand Reinhold, New York.

grow more readily than bacteria at a lower a_w which explains why forgotten bread, cheese, jam, and dried foods often become moldy. *Staphylococcus*, a genus that is adapted to grow on the dry, salty surfaces of human skin, can grow at a lower a_w than most common spoilage bacteria. It normally does not compete well with other bacteria, but on salty products such as ham and other cured meats, *Staphylococcus* can multiply with little competition. This explains why improperly handled ham is a common vehicle for *S. aureus* food poisoning.

Low pH Inhibits the Growth of Many Bacteria

The pH of a food product is also important in determining which organism can survive and thrive on it (table 34.3). Many species of bacteria, including most potential pathogens, are inhibited by acidic conditions (table 34.4). An important exception is the group of bacteria known as the lactic acid bacteria, used in the production of fermented foods such as yogurt and sauerkraut. This group of bacteria, including members of the genera *Streptococcus*, *Lactobacillus*, and *Leuconostoc*, produces lactic acid as a result of fermentative metabolism. Although they are useful in food production, their growth and accompanying acid production is also an important cause of spoilage of milk and other foods. Fungi can grow at a low pH that inhibits most spoilage bacteria, which is why foods such as fruit are often spoiled by fungi rather than bacteria (figure 34.1).

The pH may also determine whether toxins are produced by contaminating microorganisms. For example, *Clostridium botulinum*, the causative agent of botulism, does not grow or produce toxin below pH 4.5, so it is not considered a danger in highly acidic foods. Thus, the canning process for acidic fruits and pickles is less stringent than that for foods with a higher pH. Newer varieties of tomatoes, however, are less acidic than previous varieties, which means additional acid must be added if they are to be safely canned using the less stringent conditions.

Low Storage Temperature Inhibits the Growth of Spoilage Microbes

The growth of microorganisms is also affected by the temperature. At low temperatures above freezing, many enzymatic reactions are either very slow or nonexistent. The result is that some microorganisms are unable to grow

(a)

(b)

(c)

Figure 34.1
(*a*) Apples infected by the fungus *Venturia inaequalis,* cause of apple scab. (*b*) Orange affected by a *Penicillium* mold. (*c*) Ear of corn and a lemon found after several weeks in the author's refrigerator. There is a mixture of molds and bacteria, as evidenced by the many colors, including *Rhizopus* or *Aspergillus* (the black fungus), *Penicillium* (the blue-green fungus), and *Serratia* (the red bacteria).

at refrigeration temperatures and some still grow but more slowly. Those that do grow are most likely psychrophiles such as members of the genus *Pseudomonas.*

Intrinsic Factors Such as Available Nutrients, Protective Coverings, and Naturally Occurring Antimicrobial Substances Influence Microbial Growth

The nutrients present in a food as well as other intrinsic factors will determine the kinds of organisms that will grow in or on it. For example, if a certain food is deficient in a given vitamin, any organism requiring that vitamin cannot grow on that food. However, other bacteria capable of synthesizing the vitamins can grow if other conditions are favorable. Members of the genus *Pseudomonas* often spoil foods because they can synthesize essential nutrients and grow under a variety of conditions, including refrigerator temperatures.

Rinds, shells, and other coverings aid in protecting some foods from the invasion of spoilage organisms. Even so, sooner or later a microorganism that can utilize those materials will break down the defenses and cause spoilage.

Some foods contain antimicrobial substances that may help prevent spoilage (table 34.5). For example, egg white is rich in lysozyme. Thus, even if lysozyme-susceptible bacteria breach the protective shell of an egg, they are destroyed by lysozyme before they can cause spoilage. Other examples of naturally occurring antimicrobials are benzoic acid, which is found in cranberries, and an antibacterial peroxidase system, similar to that found in phagocytes, which is in fresh milk.

• **Psychrophiles, pp. 91–2**

Table 34.5 Naturally Occurring Antimicrobial Substances in Foods

Food	Antimicrobial Substance	Mode of Action
Eggs	Lysozyme	Lyses cell walls
	Conalbumin	Chelates iron
	Riboflavin	Chelates iron
Fruit	Wax on skin	Retards entry
Horseradish	Allyl iothiocyanate	Inhibits bacteria growth best at 37°C
Garlic	Allicin	Inhibits growth of bacteria and fungi
Onion	Lacrimatory factor	Inhibits growth of bacteria and fungi
Cranberries	Benzoic acid	Inhibits growth of fungi

REVIEW QUESTIONS

1. What type of microorganisms might be expected to grow in orange juice?
2. What is the product of fermentation of the sugars in fruit by yeast?
3. What type of organisms might be expected to grow on fresh meat stored in the refrigerator?
4. Which is more important to refrigerate: homemade stew or bread? Why?

Microorganisms and the Production of Foods and Beverages

The activities of microbes that produce food products have been exploited for thousands of years. The Sumerians and Babylonians produced beer as early as 6000 B.C., and it is known that the Egyptians were using yeast to make bread by 4000 B.C. However, only in the last century have the organisms involved been studied and characterized. As a result, new processes have been developed, while older processes, such as beer and bread making, have been improved to make them commercially more productive.

Ancient methods of preparing foods and beverages using microorganisms or their by-products have often been improved by scientific research. The study of the microorganisms involved has resulted in changes that have increased both the quality of food products and the efficiency of production. Nevertheless, some of these ancient processes, including the production of some fine cheeses and wines, are not fully understood and still remain somewhat of an art.

Some Milk Products Are Made with Lactic Acid Bacteria

Milk is an excellent growth medium for many bacteria because it is high in nutrients. It is normally sterile as it is secreted from the cow's udder, but during the handling of the milk, from the farm to the dairy, it is rapidly contaminated with a variety of organisms. Various species of lactic acid bacteria are commonly found on a cow's udder and are especially well suited to growth in milk. These bacteria ferment lactose, the predominant sugar in milk, to lactic acid, thereby reducing the pH of the milk and souring it. The low pH inhibits the growth of many undesirable bacteria, whereas most lactic acid bacteria are acid tolerant and continue growing. Lactic acid bacteria such as members of the genera *Lactobacillus* and *Streptococcus* are used in the commercial production of fermented milk products. The characteristic textures and flavors of these different products depend on the strains of bacteria used in their productions as well as the temperature and other conditions.

Yogurt

To produce yogurt, milk must first be pasteurized in order to eliminate undesirable spoilage organisms. Then it is inoculated with a mixed culture of *Streptococcus thermophilus* and *Lactobacillus bulgaricus* and incubated at 40°C. These lactic acid bacteria grow well at the relatively high temperature and ferment lactose to produce lactic acid as well as other flavorful end products such as acetaldehyde. The lactic acid lowers the pH, coagulating the milk proteins and changing the texture of the milk. Conditions are carefully controlled so that only the organisms that produce desired end products grow. Each yogurt manufacturer has its own strains of these bacteria and guards them carefully. More than 0.5 million pounds of yogurt are sold each year in the United States alone.

• **Lactic acid fermentation, pp. 119–20**

Cheese

Cheese making is a very old process and truly an art. It is believed that cheese making originated in Asia over 8,000 years ago. The Romans improved the production of cheeses during their occupation of Europe from 60 B.C. to A.D. 300. Even with today's modern production methods, the manufacture of fine cheeses has remained essentially the same, requiring experience, timing, and patience. Most common cheeses are made with cow's milk, but sheep's milk or goat's milk are used for some cheeses to give characteristic flavor differences.

The same initial steps are used to make most cheese (figure 34.2). First, milk is inoculated with a **starter culture** containing lactic acid bacteria. It is then allowed to incubate until fermentation of the milk sugars produces a critical level of lactic acid. At that time, the enzyme renin,[1] which coagulates the milk proteins, is added. The coagulated proteins, or the **curd,** is separated from the liquid portion, called the **whey.** The curd can be used to make cottage cheese or it can be subjected to further microbial processes, known as *ripening* or *curing,* that result in characteristic texture and flavor changes.

Brie and Camembert are ripened by adding a white fungus, usually *Penicillium caseicolum* (also known as *P. candidum*), to the surface of a wheel of cheese. The mycelia of the fungus produce enzymes that alter the texture and flavor of the cheese as they gradually work their way into the cheese wheel. Limburger cheese is made in a similar manner but with the bacterium *Brevibacterium linens.* Roquefort, Gorgonzola, Stilton, and Danabe cheeses are ripened by adding the fungus *Penicillium.* Growth of the fungus along cracks in the cheese gives these cheeses the distinctive bluish-green veins. Swiss cheeses are ripened with *Propionibacterium* spp. These bacteria ferment organic compounds to propionic acid and CO_2. The propionic acid inhibits spoilage organisms such as fungi and imparts the typical nutty flavor of Swiss cheese. The CO_2 gas propionibacteria produce causes the characteristic holes in the cheese.

In addition to varying in the ripening process, cheeses differ in their water content. Cheeses are classified as very hard, hard, semisoft, and soft, according to their percentage of water. Examples of several types of cheeses and the microorganisms used in their production are listed in table 34.6.

Salting, low oxygen levels, acidity, and presence of microbial metabolites such as propionic acid inhibit the growth of pathogens. In the United States, cheeses made

[1]Renin—Enzyme that in the past was isolated from the stomach of calves but today is made using genetically engineered organisms

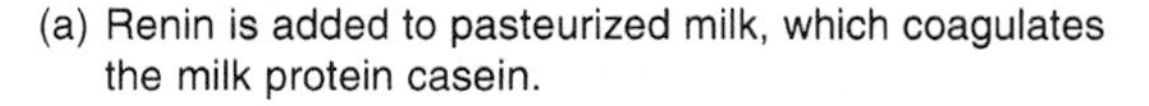
(a) Renin is added to pasteurized milk, which coagulates the milk protein casein.

(b) The milk curdles into solid curd, which is then cut into slabs. Liquid whey is drained away from the slabs of curd.

(c) The whey continues to drain from the curd until a certain determined dryness is attained. The curd is then compressed into blocks for extended aging.

(d) The blocks of cheese are then placed on shelves in the aging room. The longer the ripening period, the more acidic (sharper) the cheese.

Figure 34.2
Basic steps in cheese making.

from unpasteurized milk must be ripened for 60 days in order to kill any pathogenic bacteria that might be present. Regardless of the type of cheese being produced, both the fermentation and ripening processes must be carefully monitored to guard against undesired flavors, textures, and appearances that result from the activities of contaminating microorganisms.

Pickled Vegetables and Olives Are Made with Lactic Acid Bacteria

Pickling of foods originated as a way to preserve vegetables such as cucumbers and cabbage. Today, pickled products such as sauerkraut (cabbage), pickles (cucumbers), and olives are valued for their flavor. Table 34.7 lists some of the products that are made using microorganisms.

Table 34.6 Some Cheeses and the Microorganisms Used in Their Production

Type of Cheese	Cheese Name	Microorganisms*
Very hard, ripened	Parmesan	*Lactobacillus bulgaricus*
	Romano	*Streptococcus lactis* *Streptococcus cremoris* *Streptococcus thermophilus*
Hard, ripened	Cheddar	*Streptococcus lactis*
	Colby	*Streptococcus cremoris*
	Edam	*Streptococcus durans*
	Gouda	*Streptococcus thermophilus*
	Gruyere	*Lactobacillus helveticus*
	Swiss	*Propionibacterium shermanii*
Semisoft, ripened	Gorgonzola	*Streptococcus lactis*
	Monterey	*Streptococcus cremoris*
	Roquefort	*Penicillium roqueforti*
Soft, ripened	Brie	*Streptococcus lactis*
	Camembert	*Streptococcus cremoris*
	Limburger	*Penicillium camemberti* *Penicillium candidum* *Brevibacterium linens*
Soft, unripened	Cottage	*Streptococcus lactis* *Leuconostoc citrovorum*

*Some or all the bacteria listed in each group are used to make each type of cheese.

One of the most well-studied natural fermentations is the production of sauerkraut. The cabbage is first shredded and layered with salt (figure 34.3). The layers are firmly packed and pressed down to provide an anaerobic environment. The salt draws water and nutrients from the cabbage cells, creating a salty brine that inhibits the growth of many bacteria but permits the growth of the naturally occurring lactic acid bacteria. Controlled conditions are important to ensure that the desired lactic acid bacteria grow. For example, too much salt inhibits the lactic acid bacteria, while not enough salt permits the growth of other undesired organisms that can spoil the product. Under the controlled conditions, a natural succession of the lactic acid bacteria grow. These bacteria, *Leuconostoc mesenteroides, Lactobacillus plantarum* and *Lactobacillus brevis*, produce lactic acid, which lowers the pH, further inhibiting undesired bacteria. The lactic acid and other end products of the fermentation give sauerkraut its characteristic tangy taste. When the desired flavor has been attained, usually after two to four weeks at room temperature, the sauerkraut is often canned. Similar processes also relying on the lactic acid bacteria are used to make pickles and olives.

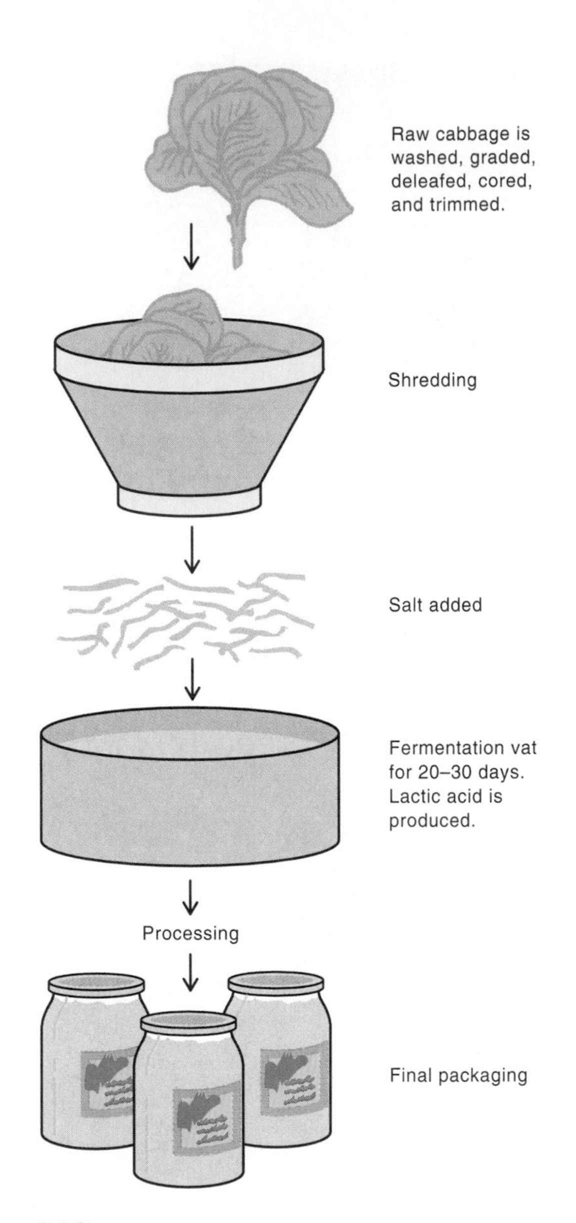

Figure 34.3
Steps in sauerkraut production.

Table 34.7 Some Fermented Foods and Their Ingredients

Product	Starting Ingredient
Kimchee	Cabbage (and other vegetables)
Miso	Soybeans
Olives	Green olives
Pickles	Cucumbers
Poi	Taro roots
Sauerkraut	Cabbage
Tempeh	Soybeans
Soy sauce	Soybeans

Soy Sauce Is Traditionally Made by Fermentation

Soy sauce, a product of the fermentation of soybeans by the fungus *Aspergillus oryzae,* is made by mixing salt, soybeans, and wheat. The mixture, called *koji,* is then allowed to stand for three days, during which time carbohydrates and proteins in the soybeans are broken down to produce fermentable sugars, peptides, and amino acids. After this first step, the mixture is put into a large container with an 18% NaCl solution and the bacterium *Pediococcus soyae* and the yeasts *Saccharomyces rouxii* and a *Torulopsis* sp. This mixture is allowed to ferment for 8 to 12 months, after which the liquid is drained off and sold as soy sauce. The solid materials that remain are used as animal feed.

Yeasts Are Important in the Production of Many Breads

The commercial production of bread in large quantities depends heavily on an associated industry, the commercial manufacture of baker's yeast. Modern techniques are used to produce a high-quality yeast inoculum that gives the reliable performance demanded by large-volume baking.

Leavened bread[1] is made from a mixture of flour, sugar, milk or water, yeast, and sometimes butter or oil. The yeast, usually a strain of *Saccharomyces cerevisiae,* ferments the sugars to ethanol and CO_2 via the alcoholic fermentation pathway. The CO_2 causes the bread to rise and produces the honeycombed texture characteristic of yeast breads (figure 34.4). The ethanol that is formed evaporates from the dough during baking.

Sourdough bread is made with a combination of yeast identified as *S. exiguus,* which ferments sugar to alcohol and CO_2, and a lactic acid bacterium, *Lactobacillus sanfranciscans,* which ferments sugar to lactic and acetic acids, giving the bread its sour flavor.

Alcoholic Beverages Utilize Fermentation by Yeast to Produce Alcohol

It is not surprising that early humans learned to make alcoholic drinks because the yeasts needed to transform fruit juice into alcoholic beverages are found naturally on the skin of many fruits. As many as 10 million yeast cells may be found on the dull waxy film or "bloom" that covers a single grape.

If the juice from any of a variety of fruits such as grapes, pears, plums, or apples is allowed to sit at room temperature for several days, bubbles will usually form on the surface of the juice as the sugars in the fruits are fermented. If the juice is allowed to bubble for several weeks, tasting the mixture will usually reveal that alcohol has been produced. Production of wine, beer, and other alcoholic beverages is really no more than a variation on this process, although it is usually carried out under strictly controlled conditions. The kind of fruit, pH, amount of sugar, and the strain of yeast are very carefully controlled in the commercial production of various alcoholic beverages. Some of the alcoholic beverages produced through fermentation by *Saccharomyces* of various starting materials are listed in table 34.8.

Wine

Commercially, wine is made by crushing carefully selected grapes in a machine that removes the stems and collects the resulting solids and juices, called **must** (figure 34.5). For red wine, the entire must of red grapes is put into the fermentation vat. The red color and rich flavors of red wines are derived from components of the grape skin and seeds. For the production of white wines, only the juice of either red or white grapes is fermented. Rose wines obtain their light pink color from the entire crushed red grape fermented for about one day, after which the juice is removed and fermented alone.

In the fermentation vat, sulfur dioxide (SO_2) is often added to the must to inhibit growth of the natural microbial population of the grape. Lactic acid bacteria and acetic acid bacteria, the most prevalent causes of wine spoilage, are usually controlled by the SO_2 or other inhibitory chemicals that are frequently added. The cultures of specifically selected strains of *S. cerevisiae* are then added. These strains differ from the natural strains in that they are usually more resistant to the antimicrobial action of SO_2.

Table 34.8 Some Alcoholic Beverages Produced by *Saccharomyces*

Beverage	Starting Material	Procedure
Beer	Germinated grain (malt)	Natural fermentation
Table wine	Fruit juice	Natural fermentation
Sake	Rice	Amylase from mold (*Aspergillus oryzae*) converts starch to sugar, which is naturally fermented
Fortified wine (sherry, port)	Fruit juice	Natural fermentation and addition of brandy to increase alcohol content to about 20%
Distilled spirits	Varies	Natural fermentation followed by distillation, alcohol 40%–50%

[1]Leavened bread—bread that is made with an agent such as yeast or baking powder and that causes an expansion of the dough with CO_2

• **Alcoholic fermentation, p. 120**

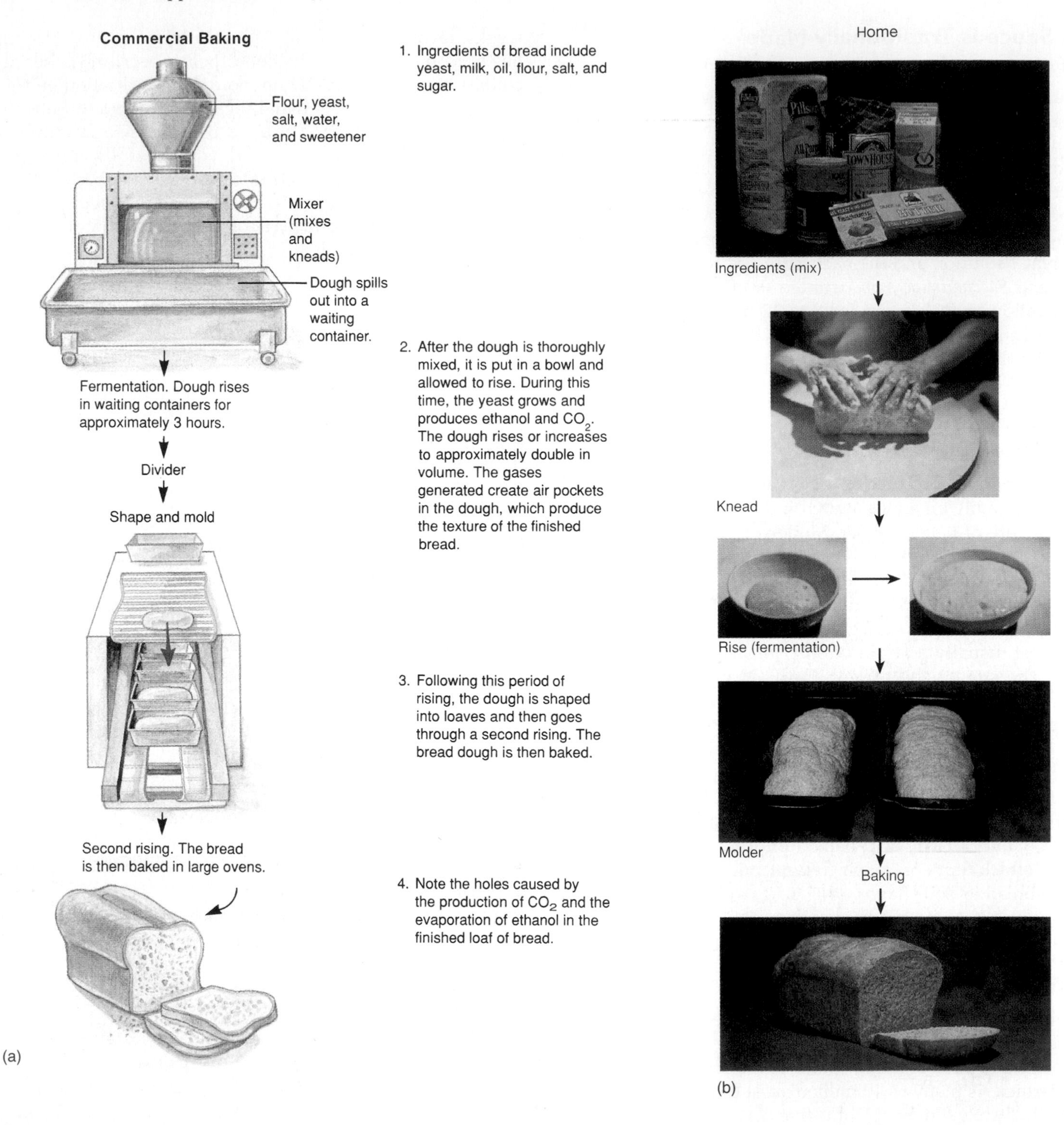

Figure 34.4
Bread production.

Fermentation proceeds at a carefully controlled temperature, which varies with the type of wine, for a period ranging from a few days to several weeks. During this time, most of the sugar is fermented to ethyl alcohol and CO_2, generally resulting in a final alcohol content of less than 14%. At the proper time during fermentations of dry red wines, the juice is separated from the pulp and skins by a wine press, and fermentation is allowed to continue until all of the sugar is fermented. The particulate debris settles, and the wine is cleared by filtration prior to aging.

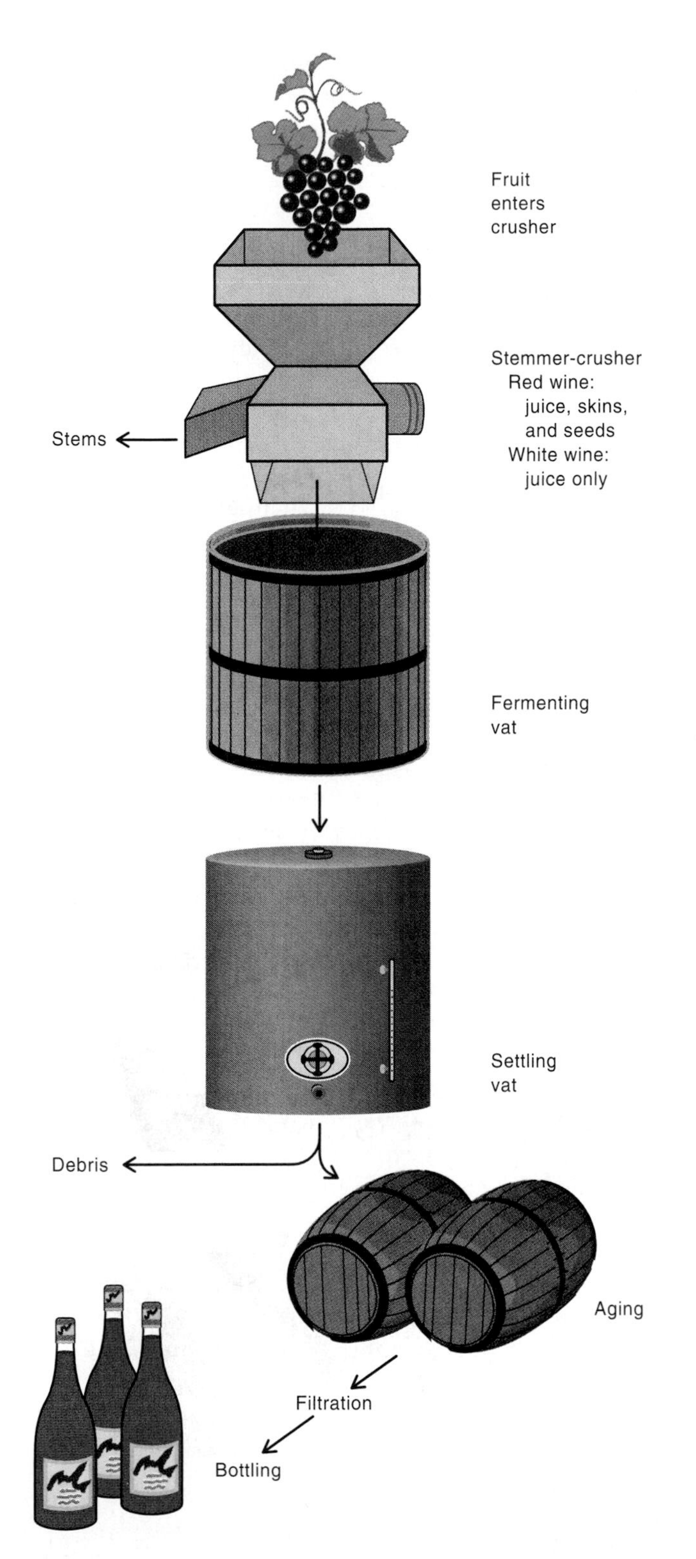

Figure 34.5
Commercial production of wine.

An amazingly complex variety of microbial transformations occurs during the fermentation and aging of wines. Enzymes of yeast and other microorganisms catalyze the conversion of sugars, organic acids, amino acids, pigments, and many other components of fruit into numerous different substances including alcohol. Only a few of these reactions are completely understood.

The fermentation of glucose to ethyl alcohol has been the most extensively studied chemical reaction in wine making. Every mole of glucose or fructose that is fermented to alcohol yields 2 moles of ethyl alcohol and CO_2. This release of CO_2 causes a bubbling or "working" in the wine vat. All of the gas is usually released before the wine is bottled, resulting in a "still" (noncarbonated) wine. Other processes are used to prepare carbonated wines such as champagne.

The Japanese wine, sake, depends on several microbial fermentation reactions. First, cooked rice is inoculated with the fungus *Aspergillus oryzae.* The fungus produces the enzyme amylase which degrades the rice starch to sugar. Then a strain of *Saccharomyces* that converts the sugar to alcohol and CO_2 is added. Lactic acid bacteria add to the flavor by producing lactic acid and other fermentation end products.

Beer

Beer is made by the fermentation of the sugars of grains, traditionally barley, by selected species of *Saccharomyces.* The carbohydrates of grains exist largely in the form of starch that must be first converted to simple sugars before they can serve as a substrate for yeast. This conversion of starch to sugars, called **malting,** is carried out by enzymes that are active on the grain soon after the grains germinate. The germinated grain, or **malt,** is soaked in warm water to yield an extract called malt **wort,** which contains sugars and other nutrients needed for the growth of the yeast. Before the yeast is added, the wort is boiled, in part to destroy the enzymes of the grain and kill most of the microorganisms that may be present and in part to concentrate the wort and improve the flavor. The flowers of the vinelike hop plant are added primarily for their bitterness, but they also contain antibacterial substances that inhibit the growth of unwanted bacteria. Special strains of brewer's yeasts (as opposed to baker's yeasts) are used for the fermentation process, which lasts about a week. This results in an alcohol content ranging from 3.4% to 6%. The beer is then allowed to settle and is filtered to remove microorganisms and other particulate materials before aging, carbonating, and bottling, canning, or other packaging. The process is shown in figure 34.6.

Two different types of *Saccharomyces* strains are in common use. One type often used to make lager beers is called *bottom yeast* because it sinks to the bottom of the fermentation vat. The bottom yeast strains are known as *S. carlsbergensis.* They ferment best at temperatures between 6°C and 12°C and usually take 8 to 14 days to complete fermentation. In contrast, top-fermenting yeasts, such as *S. cerevisiae,* are used to make ales. Top-fermenting yeasts

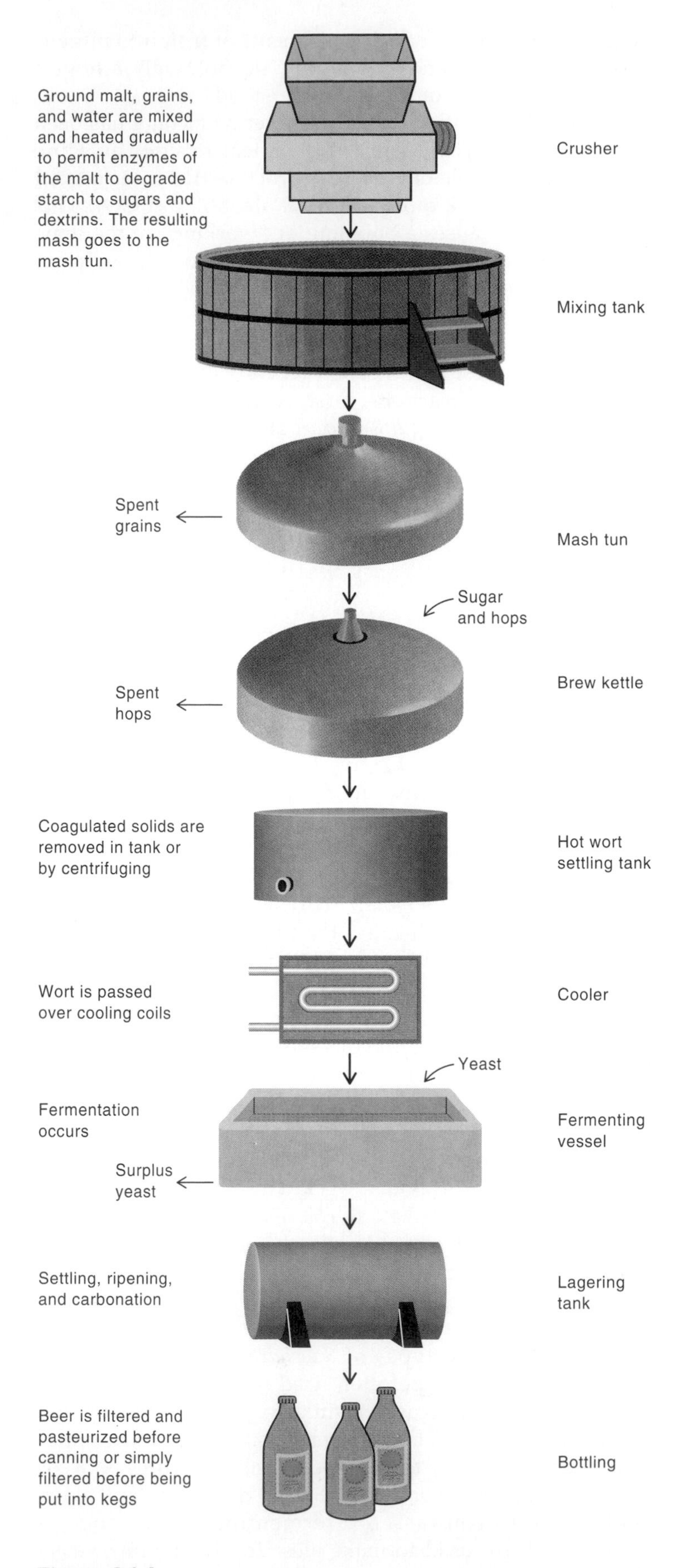

Figure 34.6
Commercial production of beer.

are distributed throughout the wort but are carried to the top of the vat by the CO_2. They also ferment at a higher temperature (14°C to 23°C) and over a shorter period (five to seven days). Most American beers are lager beers produced by the bottom yeasts. Porter and stout beers are made from top-fermenting yeasts.

Distilled Spirits

Distilled spirits such as scotch, whiskey, and gin are manufactured by a process that is initially similar to the production of beer, except the wort is not boiled. Consequently, degradation of starch by the enzymes in the wort continues during the fermentation. When the fermentation is complete, the alcohol is separated by distillation.

Different types of spirits are made in different ways. For example, rum is made by fermenting sugar cane or molasses. Malt Scotch whiskey is the product of the fermentation of barley that is then aged for several years in oak sherry casks. The wood and the residual sherry contribute both flavor and color to the whiskey as it ages. Lactic acid bacteria are used to produce lactic acid in grain mash for making sour-mash whiskey. *S. cerevisiae* subsequently ferments the sour mash to form alcohol. The distilled spirit tequila is traditionally made from the fermentation of juices from the agave plant using the bacterium *Zymomonas mobilis*. This bacterium ferments sugars to ethanol and CO_2 via a pathway similar to the yeast alcoholic fermentation pathway. Table 34.9 lists some common distilled spirits and their starting materials.

Vinegar

Vinegar, which is an aqueous solution of at least 4% acetic acid, is the product of the oxidation of alcohol by the acetic acid bacteria *Acetobacter* and *Gluonobacter*. Acetic acid bacteria are strictly aerobic, gram-negative rods, characterized by their ability to carry out a number of oxidations. They can tolerate high concentrations of acid as they oxidize alcohol to acetic acid (acetification). The transformation of alcohol to acetic acid may be slow or rapid. The best quality vinegars are obtained through a

Table 34.9 Some Distilled Spirits and Their Starting Ingredients

Type of Distilled Spirit	Starting Material
Brandy	Fruit wines
Bourbon	Corn
Gin	Potato starch and juniper berries
Rum	Molasses
Scotch	Barley
Tequila	Agave
Vodka	Potato starch

process of slow acetification, which involves placing oak barrels half full of wine on their sides. The barrels have holes drilled in their upper parts to provide the aerobic environment necessary for the growth of *Acetobacter*. A "starter" vinegar containing *Acetobacter* is added to the wine in the barrels, and the bacteria develop as a film on the surface of the liquid. Approximately one to three months are required for the maximum oxidation of ethyl alcohol to acetic acid by this method.

Rapid acetification in a vinegar generator is widely used to produce most vinegars. In this procedure, alcohol is sprayed onto wood shavings or other materials that provide a large surface area. Acetic acid bacteria grow on the shavings, oxidizing the alcohol to acetic acid as it trickles through. In principle, the vinegar generator operates much as the trickling filter does during degradation of wastes by providing a large surface area for aerobic metabolism.

Food Spoilage

Food spoilage is often due to the growth of microorganisms on or in a food product, resulting in undesirable changes. The types of microbes that predominate on a particular food product and that eventually cause spoilage are influenced by the characteristics of the food product and its storage conditions. For example, milk stored at room temperature is likely to support the growth of lactic acid bacteria that produce acids that eventually impart a sour flavor to the milk. *Pseudomonas* sp. readily grow on refrigerated meat, eventually creating a disagreeable odor and a slimy film of microbial mass.

Food microbiologists have studied extensively the numbers of various kinds of bacteria and fungi in food samples. As would be expected, these counts vary considerably with the kind of food and its previous handling. For example, many foods such as ground beef, fresh shucked oysters, and tofu often exceed 1 million organisms per gram, whereas pasteurized milk has less than 10,000 organisms per milliliter.

The microorganisms that spoil foods are usually not pathogenic for people, which is not surprising when their growth requirements are considered. Most human pathogens grow best at temperatures near 37°C, whereas most foods are usually stored at temperatures well below the normal body temperature. Similarly, the nutrients available in fruits, vegetables, and other foods are often not suitable for the optimum growth of human pathogens. In addition, the pH of foods is often more acid or alkaline than the pH of body fluids. As a result, the nonpathogens can easily outgrow the pathogens when competing for the same nutrients.

Bacteria Commonly Spoil Fresh Foods

A number of genera of gram-negative bacteria are important in food spoilage (table 34.10). *Pseudomonas* spp., for example, can metabolize a wide variety of compounds, and

• **Trickling filter, pp. 753–54**

Table 34.10 Common Examples of Microbial Species that Contribute to Food Spoilage

Food	Organism	Type of Spoilage
Bread	*Aspergillus niger*	Bread mold
	Neurospora sitophila	
	Mucor sp.	
	Rhizopus nigricans	
Milk (raw)	*Streptococcus lactis*	Sour milk
	Streptococcus cremoris	
	Lactobacillus bulgaricus	
Milk (pasteurized)	*Lactobacillus thermophilus*	Sour milk
Chicken	*Achromobacter* sp.	Sliminess
	Pseudomonas sp.	
	Alcaligenes sp.	
	Flavobacterium sp.	
Fish	*Achromobacter* sp.	Characteristic "fishy" odor
	Flavobacterium sp.	
	Micrococcus sp.	
	Pseudomonas sp.	
	Serratia sp.	

Source: Miller, B. M., and Litsky, W. 1976. *Industrial microbiology*, 268, 271. McGraw-Hill, New York.

so it is not surprising that they are able to grow on and spoil many different kinds of foods, including meats and vegetables. Psychrophilic pseudomonads can multiply at refrigeration temperatures and are notorious for spoiling meats as well as other foods. Members of the genus *Erwinia* can produce enzymes that degrade pectin, so they are a common cause of soft rot of fruits and vegetables. Members of the genus *Acetobacter* can transform ethyl alcohol to acetic acid, the principal acid of vinegar. Although this property is very beneficial to commercial producers of vinegar, it is a great problem to wine producers. Milk products are sometimes spoiled by gram-negative rods of the genus *Alcaligenes* that cause "ropiness" of milk and make cottage cheese slimy.

Gram-positive bacteria also frequently spoil foods. Members of the lactic acid bacteria *Streptococcus*, *Leuconostoc*, and *Lactobacillus* all produce lactic acid. Anyone who has unsuspectingly consumed sour milk knows this can be undesirable. The gram-positive rods *Bacillus* and *Clostridium* are particularly troublesome because their heat-resistant endospores survive cooking. *B. coagulans* and *B. stearothermophilus* cause flat-sour spoilage of some canned foods.

Fungi Commonly Spoil Breads, Fruits, and Dried Foods

Many foods are spoiled by fungi such as *Rhizopus* sp., which commonly grows on bread (figure 34.7). Species of *Aspergillus* infect peanuts and other grains, producing aflatoxin, a carcinogen monitored carefully by the Food and Drug Administration. Species of fungi are known to grow readily in acidic as well as in high-salt and high-sugar environments. Thus, fruits are more likely to be spoiled by fungi than by bacteria.

Figure 34.7
Rhizopus sp., the common bread mold.

Foodborne Illness

Food production is carefully regulated in the United States. Federal, state, and local agencies cooperate in inspections to help enforce protective laws. The Federal Food, Drug, and Cosmetic Acts set standards for foods that are shipped between states or territories, and they require accurate labeling of the foods. Other federal laws require that meats and poultry produced for interstate shipment be inspected. State and local agencies enforce many of the laws that protect food consumers, but the United States depends heavily on food manufacturers to act as their own watchdogs.

Generally, the microbiological quality of most foods sold in the United States is very good compared with that of food in other parts of the world, but consumers must employ sound preserving, preparation, and cooking techniques to avoid hazards from both home-prepared and purchased food products. In spite of strict controls on the quality of the food supply, the United States is not free of foodborne illness. Millions of cases are estimated to occur each year from foods prepared either commercially, at home, or in institutions such as hospitals and schools. Table 34.11 lists some bacteria that cause foodborne illness in the United States.

Foodborne Intoxication Occurs when a Microorganism Growing in a Food Produces a Toxin

Foodborne intoxication is an illness that results from the consumption of a toxin produced by a microorganism growing in a food product. When such a food product is ingested, the toxins are what cause illness, not the living organism. The symptoms of the illness, which often appear within a few hours of the food being ingested, reflect the type of toxin ingested. *Staphylococcus aureus* and *Clostridium botulinum* are two examples of organisms that cause foodborne intoxication.

Staphylococcus aureus Intoxication

Many strains of *Staphylococcus aureus* produce an enterotoxin that, when ingested, causes nausea and vomiting. Although *S. aureus* does not compete well with most spoilage organisms, it thrives on moist, rich foods in which other organisms have been killed or their growth has been inhibited. For example, *S. aureus* can grow with little competition on unrefrigerated salty products such as ham (a_w 0.91). Creamy pastries and salads stored at room temperature also offer ideal conditions for the growth of this pathogen because the cooking of the ingredients kills most other competing organisms. The source of the organism is

• **Enterotoxin, p. 417**
• ***Staphylococcus aureus* food poisoning, p. 582**

Table 34.11 A Summary of Common Foodborne Illnesses

Type of Illness/Organism	Symptoms	Foods Commonly Implicated
Foodborne intoxication		
Clostridium botulinum	Weakness; double vision; progressive inability to speak, swallow, and breathe	Low-acid canned foods such as vegetables and meats
Staphylococcus aureus	Nausea, vomiting, abdominal cramping	Meats (particularly ham), creamy salads, cream-filled pastries, sandwich fillings
Foodborne infection		
Bacillus cereus	Two disease types: nausea and vomiting or watery diarrhea and abdominal cramps	Meats, milk, shellfish, rice products
Campylobacter spp.	Diarrhea, fever, abdominal pain, nausea, headache	Poultry, raw milk
Clostridium perfringens	Intense abdominal cramps, watery diarrhea	Meats, meat products, gravy
Escherichia coli 0157:H7	Severe abdominal pain, watery diarrhea, bloody diarrhea; sometimes progresses to hemolytic uremic syndrome	Ground beef
Listeria monocytogenes	Influenzalike symptoms, fever; may progress to septicemia, meningitis	Raw milk, cheese, meats, vegetables
Salmonella spp.	Nausea, vomiting, abdominal cramps, diarrhea, fever	Poultry, eggs, milk, meat
Shigella spp.	Abdominal cramps; diarrhea with blood, pus, or mucus; fever; vomiting	Creamy salads, raw vegetables, dairy products
Vibrio parahaemolyticus	Diarrhea, abdominal cramps, nausea, vomiting, headache, fever, chills	Fish and shellfish
Vibrio vulnificus	Diarrhea, nausea; immunosuppressed people may progress to septicemia	Oysters, clams, crabs
Yersinia enterocolitica	Abdominal pain, fever, diarrhea, vomiting; may mimic appendicitis	Meats, oysters, fish, raw milk

usually a human carrier who has not followed adequate sanitation procedures such as hand washing before preparing the food. If the organism is inoculated into a food that can support its growth, and the food is left at room temperature for several hours, the *S. aureus* can grow and produce the enterotoxin. Unlike most exotoxins, the *S. aureus* enterotoxin is heat stable so cooking of the food will not destroy it.

Botulism

Botulism is caused by ingestion of a neurotoxin produced by the anaerobic, spore-forming, gram-positive rod *Clostridium botulinum.* Canning processes are specifically designed to destroy the endospores of this organism, but errors in the process or postprocessing contamination can allow the germination of the endospores and growth of the vegetative cells that produce toxin. Such errors are extremely rare in commercially canned foods, and most cases of botulism are due to processing errors in home-canned foods. Unfortunately, growth of *C. botulinum* and production of the deadly toxin may not result in any noticeable changes in the taste or appearance of the food. The best way to prevent botulinum when consuming home-canned foods is to boil the product for 10 minutes immediately after opening the can. The toxin is heat labile (sensitive), so the heat treatment will destroy any toxin that may have been produced. Cans of food that are bulging, damaged, or otherwise suspicious should be discarded.

- **Carriers, p. 430**
- **Botulism, pp. 583–85**

Foodborne Infection Occurs when Living Organisms in a Food Product Are Ingested and then Colonize the Intestine

Unlike foodborne intoxication, foodborne infection requires the consumption of living organisms. The symptoms of the illness, which usually do not appear for at least one day after ingestion of the contaminated food, usually include diarrhea but vary according to the type of organism ingested. Thorough cooking of food immediately before consumption will kill the organisms, thereby preventing foodborne infection. *Salmonella,* and *Campylobacter* spp. and *Escherichia coli* 0157:H7 are examples of organisms that cause foodborne infection.

Salmonella and *Campylobacter* Sp.

Salmonella and *Campylobacter* are two different genera of pathogens that are commonly isolated from poultry products such as chicken, turkey, and eggs. Inadequate cooking of these products can result in foodborne infection. Cross-contamination of other foods can result in the pathogens being transferred to those foods. For example, if the same cutting board that is used to cut raw chicken is then immediately used to cut up vegetables for a salad, the salad can become contaminated with *Salmonella* or *Campylobacter* spp.

Escherichia coli O157:H7

The newly emerging strain of *Escherichia coli,* O157:H7, causes bloody diarrhea that sometimes develops into a life-threatening condition called **hemolytic uremic syndrome.** In the past decade, *E. coli* O157:H7 has caused several large food poisoning outbreaks, including an epidemic in Japan in 1996 that sickened over 9,000 people and an outbreak in the United States that spread over four states. The latter epidemic was traced to undercooked contaminated hamburger patties served at a fast-food restaurant. Prevention involves thorough cooking of ground beef, the most commonly implicated source.

Food Preservation

Some methods of food preservation, such as drying and salting, have been known throughout the ages, whereas others have been discovered or developed more recently. The major methods of preserving foods—high temperature treatment, low temperature treatment, addition of antimicrobial chemicals, and irradiation—were discussed in chapter 9.

REVIEW QUESTIONS

1. What group of organisms is used to make bread, wine, and beer?
2. Name one type of food that this group of organisms spoils.
3. What group of organisms is used to make yogurt, cheese, and pickles?
4. Name one type of food that this group of organisms spoils.
5. How is foodborne infection different from foodborne intoxication?

Summary

I. Factors Influencing the Growth of Organisms in Foods
    A. The availability of water plays a role in determining which microorganisms can grow.
    B. Low pH inhibits the growth of many bacteria.
    C. Low storage temperature inhibits the growth of spoilage microbes.
    D. Intrinsic factors such as available nutrients, protective coverings, and naturally occurring antimicrobial substances influence microbial growth.

II. Microorganisms and the Production of Foods and Beverages
    A. Some milk products are made with lactic acid bacteria.
        1. Yogurt
        2. Cheese
    B. Pickled vegetables and olives are made with lactic acid bacteria.
    C. Soy sauce is traditionally made by fermentation.
    D. Yeasts are important in the production of many breads.
    E. Alcoholic beverages utilize fermentation by yeast to produce alcohol.
        1. Wine
        2. Beer
        3. Distilled spirits
        4. Vinegar

III. Food Spoilage
    A. Bacteria commonly spoil fresh foods.
    B. Fungi commonly spoil breads, fruits, and dried foods.

IV. Foodborne Illness
    A. Foodborne intoxication occurs when a microorganism growing in a food produces a toxin.
        1. *Staphylococcus aureus* intoxication
        2. Botulism
    B. Foodborne infection occurs when living organisms in a food product are ingested and then colonize the intestine.
        1. *Salmonella* and *Campylobacter* sp.
        2. *Escherichia coli* O157:H7

V. Food Preservation

Chapter Review Questions

1. The a_w of a food product reflects which of the following?
   a. acidity of the food
   b. presence of antimicrobial constituents such as lysozyme
   c. amount of water available
   d. storage atmosphere
   e. nutrient content
2. What is the minimum pH for growth and toxin production by *Clostridium botulinum?*
   a. 8.5
   b. 7.0
   c. 6.5
   d. 4.5
   e. 2.0
3. What is the principle fermentation end product of *Streptococcus, Lactobacillus,* and *Leuconostoc*?
4. Benzoic acid is an antimicrobial chemical that is naturally found in which of the following?
   a. apples
   b. cranberries
   c. eggs
   d. milk
   e. yogurt
5. What are curds and whey?
6. What causes the bluish green veins to form in blue cheese?
7. What causes the holes to form in Swiss cheese?
8. Name the organism that can survive and grow in improperly canned green beans.
9. Name the pathogen that can grow in relatively salty foods such as ham.
10. Name a food product that is commonly contaminated with *Salmonella* and *Campylobacter* sp.

Critical Thinking Questions

1. Outline some of the general plans for quality control or for microbiological research and development that you might follow, if you owned a large food-processing company.
2. What special precautions should be taken to avoid salmonella contamination during food preparation in one's own kitchen?

Further Reading

Ayres, J. C., Mundt, J. O., and Sandine, W. E. 1980. *Microbiology of foods.* W. H. Freeman, San Francisco.

Block, Eric. 1985. The chemistry of garlic and onions. *Scientific American.*

Board, R. G. 1983. *A modern introduction to food microbiology.* Blackwell Scientific Publications, Oxford.

Center for Food Safety and Applied Nutrition. 1992. Foodborne pathogenic microorganisms and natural toxins (bad bug book). U.S. Food and Drug Administration. http://vm.cfsan.fda.gov/~mow/intro.html.

Doyle, M., Beuchat, L., Montville, T. 1997. *Food microbiology: fundamentals and frontiers.* ASM Press, Washington, D.C.

Felix, Charles W. 1987. *Food protection technology.* Lewis Publishers, Inc. Chelsea, Michigan. Of particular interest is the chapter by Peter S. Elias, *Task force on irradiation processing—wholesomeness studies,* 349–361.

Jay, J. 1992. *Modern food microbiology,* 4th ed. Van Nostrand Reinhold Co., New York.

Kosikowski, Frank V. 1985. Cheese. *Scientific American.*

Mountney, G. J., and Gould, W. A. 1988. *Practical food microbiology and technology.* Van Nostrand Reinhold Company, New York.

Rose, Anthony H. 1981. The microbiological production of food and drink. *Scientific American.*

Appendix I

Microbial Mathematics

Because bacteria are very tiny and can multiply to very large numbers of cells in short time periods, convenient and simple ways are used to designate their numbers without resorting to many zeros before or after the number. In the study of microbiology, it is important to gain an understanding of the metric system, which is used in scientific measurements.

The basic unit of measure is the meter, which is equal to about 39 inches. All other units are fractions of a meter:

1 decimeter is one tenth = 0.1 meter

1 centimeter is one hundredth = 0.01 meter

1 millimeter is one thousandth = 0.001 meter

Because bacteria are much smaller than a millimeter, even smaller units of measure are used. A millionth of a meter is a micrometer = 0.000001 meter, and is abbreviated μm. This is the most frequently used size in microbiology, since bacteria are in this size range. For comparison, a human hair is about 75 μm wide.

Since it is inconvenient to write so many zeros in front of the 1, an easier way of denoting the same number is through the use of superscript, or exponential, numbers (exponents). One hundred dollars can be written 10^2 dollars. The 10 is called the base number and the 2 is the exponent. Conversely, one hundredth of a dollar is 10^{-2} dollars; thus the exponent is negative. The base most commonly used in biology is 10 (this is designated as $\log_{10}$). The above information can be summarized as follows:

1 millimeter = 1 mm = 0.001 meter = 10^{-3} meter

1 micrometer = 1 μm = 0.000001 meter = 10^{-6} meter

1 nanometer = 1 nm = 0.000000001 meter = 10^{-9} meter

The same prefix designations can be used for weights. The basic unit of weight is the gram, abbreviated g. Approximately 450 grams are in a pound.

1 milligram = 1 mg = 0.001 gram = 10^{-3} g

1 microgram = 1 μg = 0.000001 gram = 10^{-6} g

1 nanogram = 1 ng = 0.000000001 gram = 10^{-9} g

1 picogram = 1 pg = 0.000000000001 gram = 10^{-12} g

Note that the number of zeros before the 1 is one less than the exponent. The value of the number is obtained by multiplying the base by itself the number of times indicated by the exponent.

Thus, $10^1 = 10 \times 1 = 10$

$10^2 = 10 \times 10 = 100$

$10^3 = 10 \times 10 \times 10 = 1{,}000$

When the exponent is negative, the base and exponent are divided into 1.

For example, $10^{-2} = \frac{1}{10} \times \frac{1}{10} = \frac{1}{100} = 0.01$

When one multiplies numbers having exponents to the same base, the exponents are added.

For example, $10^3 \times 10^2 = 10^5$ (not 10^6)

When one divides numbers having exponents to the same base, the exponents are subtracted.

For example, $10^5 \div 10^2 = 10^3$

In both cases, only if the bases are the same can the exponents be added or subtracted.

Appendix II

pH

A property of every aqueous solution that is important in all biological systems is the degree of acidity of the solution. This property is measured as pH of the solution (an abbreviation for **p**otential of **h**ydrogen). The acidity of a solution is based on several properties of water. Water itself has a slight tendency to split (ionize) into hydrogen ions, H^+ (a single proton) and OH^- ions. In pure water, only one molecule in 10 million molecules (1 in 10^7) undergoes this breakdown:

$$HOH \rightarrow H^+ + OH^-$$

As this reaction indicates, when water splits into its component parts, the number of H^+ and OH^- ions is equal and, in pure water, the concentration of each is 10^{-7} molar (10^{-7} M). Although these may seem like very small numbers, a small glass of pure water (which weighs about 18 grams) contains 10^{16} H^+ ions and an equal number of OH^- ions, which are extremely large numbers. This is true because a mole[1] of water (which is 18 grams) contains 10^{23} molecules. A mole of any compound contains 10^{23} molecules. Since hydrogen ions play a very important role in the functioning of cells, it is important to understand how their concentration can be controlled.

The concentration of H^+ in water can be increased by adding a compound which contains hydrogen atoms to the water that breaks down to give H^+ but not OH^- ions. HCl is such a compound and ionizes in the following way: $HCl \rightarrow H^+ + Cl^-$. As HCl breaks down, the H+ ions will combine with OH^- to form HOH. Since the product of the concentration of H^+ and OH^- must always be 10^{-14} M ($10^{-7} \times 10^{-7}$ in pure water), each increase in H^+ concentration must be balanced by a corresponding decrease in the OH^- concentration. Therefore, if the concentration of H^+ increases tenfold to 10^{-6} M, then the concentration of OH^- must decrease by a factor of 10 (10^{-8} M). (Recall that exponents are added when numbers are multiplied.) Since the number of 10^{-14} must always hold for the product of H^+ and OH^- concentrations, if the concentration of H+ is known, the concentration of OH^- can be easily calculated. Therefore, it is necessary to specify only the H^+ concentration of any aqueous solution. To simplify the notation, the term pH is used and defined as follows:

$$pH = -\log_{10} H^+ \text{ concentration}$$

If the concentration of H^+ is 10^{-7} M, the pH would be:

$$pH = -(-7) \text{ or } 7$$

The concentration of OH^- must also be 10^{-7} M. Since the concentrations are equal, the + and – charges (of H^+ and OH^-) are equal and the solution is neutral, without any charge. If any substance is added that increases the H^+ concentration, the pH will drop below 7 and the solution will become more acidic. Thus, any substance that increases the concentration of H^+ when it is added to water is termed an **acid.** Any substance that increases the concentration of OH^-, thereby decreasing the concentration of H^+, is termed a **base.**

Most bacteria can live within only a narrow pH range, near neutrality. However, some can live under very acidic conditions and a few under alkaline conditions.

[1]Mole—the amount of a chemical in grams that is equal to the atomic weights of all the atoms in a molecule of that chemical and contains 10^{23} molecules.

Pathways of Glucose Degradation

Glucose

ATP → ADP — Hexokinase

Glucose 6-phosphate

Phosphohexose isomerase

Fructose 6-phosphate

ATP → ADP — Phosphofructokinase

Fructose 1,6-diphosphate

Fructose biphosphate aldolase

Glyceraldehyde 3-phosphate ⇌ Dihydroxy-acetone phosphate

$+P_i$, NAD^+ → $NADH + H^+$ — Glyceraldehyde 3-phosphate dehydrogenase

1,3-diphospho-glyceric acid

ADP → ATP — Phospho-glycerate kinase

3-phosphoglyceric acid

Phospho-glycerate mutase

2-phospho-glyceric acid

Enolase → H_2O

Phosphoenolpyruvic acid

ADP → ATP — Pyruvate kinase

Pyruvic acid

Figure III-1

Embden-Meyerhof Pathway

Note that although we have given the names of substrates and products of the reactions as acids, inside the cell they would occur as salts. For example, 1,3 diphosphoglyceric acid would be 1,3 diphosphoglycerate.

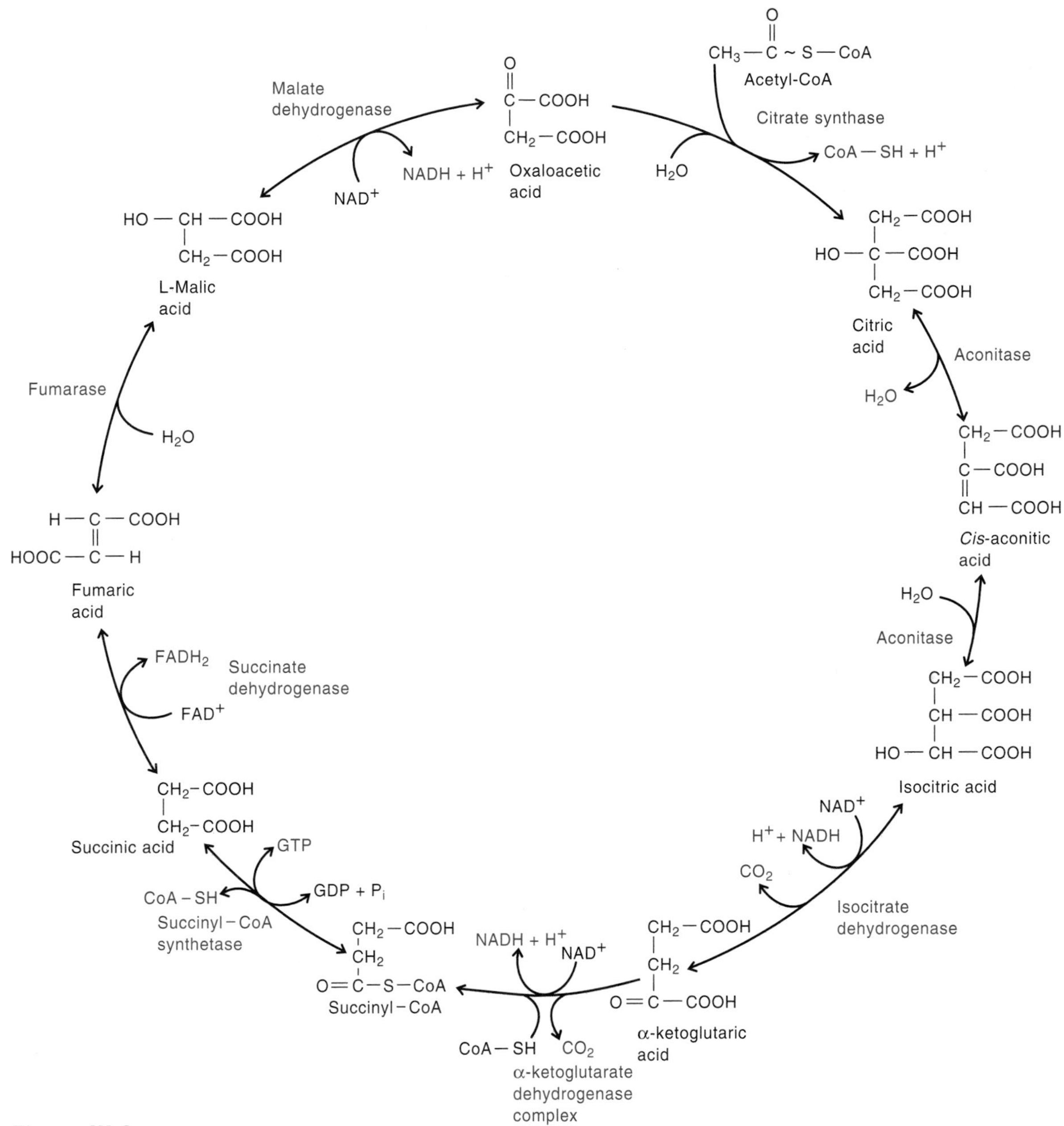

Figure III-2
The tricarboxylic acid cycle (TCA cycle).
Note that although we have given the names of substrates and products of the reactions as acids, inside the cell they would occur as salts. For example, citric acid would be citrate.

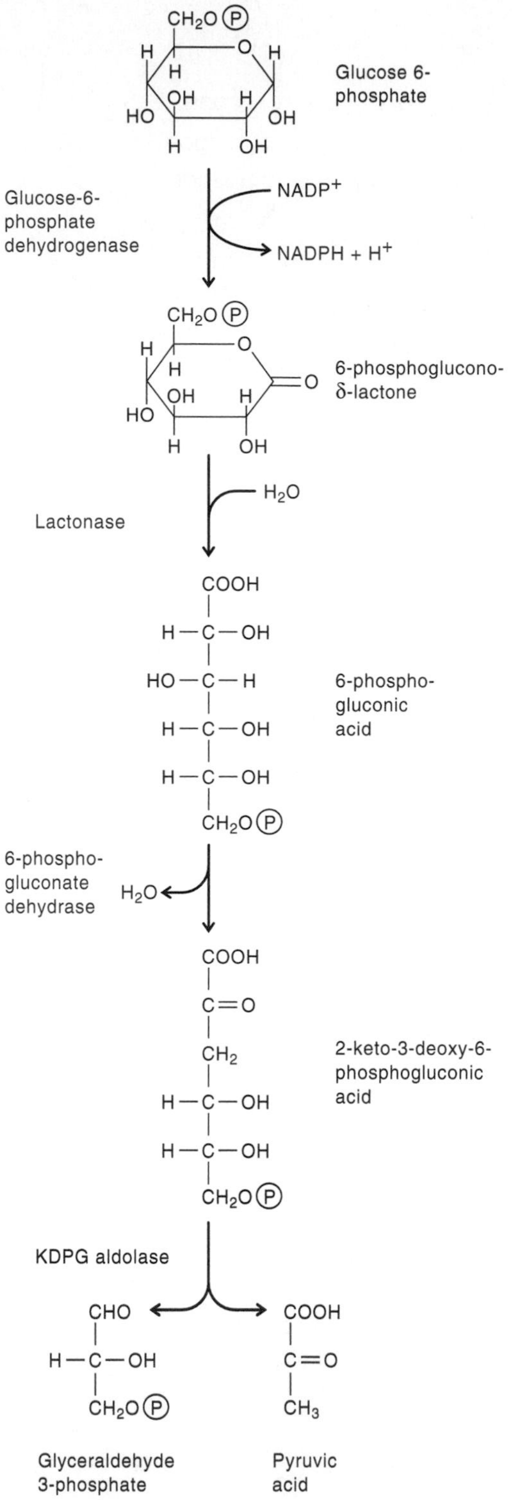

Glucose 6-phosphate

Glucose-6-phosphate dehydrogenase — $NADP^+$ → $NADPH + H^+$

6-phosphoglucono-δ-lactone

Lactonase — H_2O

COOH
H—C—OH
HO—C—H
H—C—OH
H—C—OH
CH_2OⓅ

6-phospho-gluconic acid

6-phospho-gluconate dehydrase — H_2O

COOH
C=O
CH_2
H—C—OH
H—C—OH
CH_2OⓅ

2-keto-3-deoxy-6-phosphogluconic acid

KDPG aldolase

CHO
H—C—OH
CH_2OⓅ

Glyceraldehyde 3-phosphate

COOH
C=O
CH_3

Pyruvic acid

Figure III-3
The Entner-Doudoroff pathway.
Note that although we have given some names of substrates and products of reactions as acids, inside the cell they would occur as salts. For example, pyruvic acid would be pyruvate.

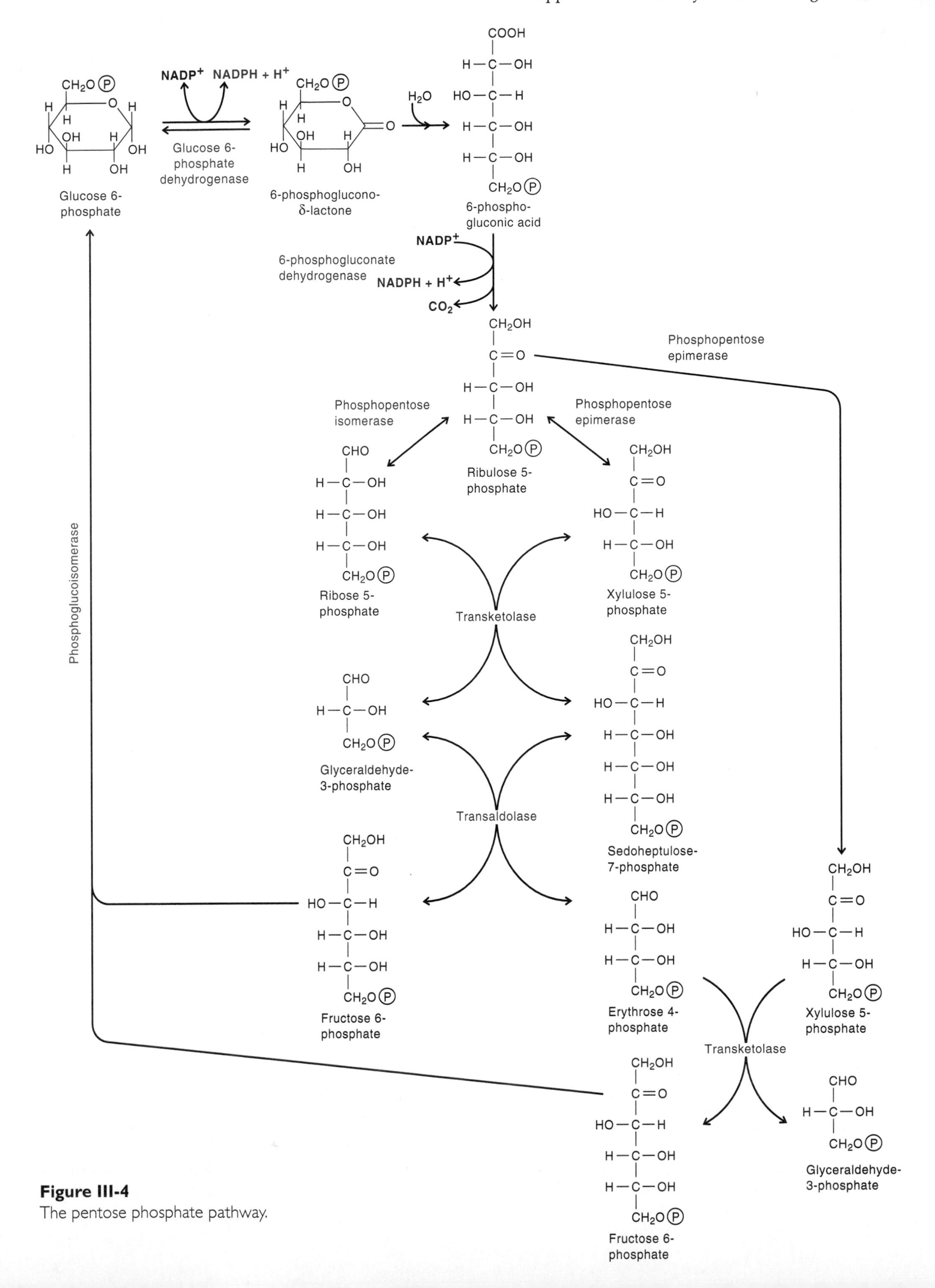

Figure III-4
The pentose phosphate pathway.

Appendix

Additional Bacteria of Medical Importance

Over the last several decades hundreds of new names have appeared among the species of medically important bacteria. This has occurred partly because virulent new species have been discovered. Also, new names have been given to old species as newer methods of examining bacteria have revealed unexpected relationships. Finally, largely because of the AIDS epidemic and the use of immunity-suppressing medications for organ transplants, many species of bacteria formerly considered harmless are now causing serious infectious disease.

The following is a list of some old and new medically important bacteria that were not covered extensively earlier in the text. The list is by no means complete, but it may serve as a helpful reference when unfamiliar names are encountered.

***Acinetobacter* sp.** DNA studies indicate more than a dozen species in this genus. Aerobic, gram-negative rods or coccobacilli in pairs. Oxidase-negative and nonmotile. Commonly found in the environment and on the human skin. Low virulence. Biotype "anitratus" particularly likely to cause nosocomial infections in patients who have undergone endotracheal intubation, bladder or central venous catheterization, or peritoneal dialysis.

***Actinobacillus* sp.** Small, nonmotile, oxidase-positive, gram-negative rods or coccobacilli. One species, *A. actinomycetemcomitans,* is found in the human mouth and is a cause of endocarditis.

***Aeromonas* sp.** Facultatively anaerobic, oxidase-positive, gram-negative rods with polar flagella. Found in brackish and freshwater, and in sewage. Occasionally cause cellulitis, wound infections, acute diarrhea, septicemia, and urinary tract infections.

Agrobacterium radiobacter. Gram-negative rod. Soil organism, some strains of which cause tumors in plants. Other strains are opportunistic human pathogens, causing endocarditis, septicemia, and abdominal infection in persons with impaired host defenses.

Alcaligenes (formerly *Achromobacter*). Aerobic, peritrichously flagellated, oxidase-positive, gram-negative rods, widely distributed in the environment. May be present in bowel flora. Has been found in hospital water supplies. Occasional cause of purulent ear discharges and nosocomial infections.

***"Arizona"* sp.** Commonly used name for organisms encompassing two of the seven subgroups of *Salmonella.* Most strains originate from cold-blooded animals such as snakes and turtles. Some are human pathogens, causing gastroenteritis and septicemia. From 15% to 85% of these strains ferment lactose, unusual for most other salmonellas.

Calymmatobacterium granulomatis. Encapsulated, pleomorphic, gram-negative rods with rounded ends that occur in clusters and singly. Will not grow on most bacteriological media but reported to grow on an egg yolk medium. Responsible for **granuloma inguinale,** a chronic, indolent ulcerogranulomatous disease of skin and mucous membranes usually involving the genitalia. In stained smears of the lesions, the organisms lie intracellularly in phagocytes.

Capnocytophaga canimorsus (dysgonic fermenter-2). A slow growing, gliding, gram-negative rod showing growth enhancement in CO_2. Present in the saliva of normal dogs. Causes septicemia in patients with alcoholism, cancer, and other chronic conditions.

Chromobacterium violaceum. Facultatively anaerobic, motile, oxidase-positive, gram-negative rods that are found naturally in soil and water. Cause chromobacteriosis, a rare systemic infection that occurs mostly in tropical and subtropical climates. Cases have been reported in the southern United States. Most strains produce a violet pigment insoluble in water.

***Citrobacter* sp.** Four species of enterobacteria. Some strains can cause diarrhea, septicemia, or neonatal meningitis.

***Edwardsiella* sp.** Three species of enterobacteria; one, *E. tarda,* can cause septicemia and possibly diarrhea. Present in the intestines of a variety of mammals and reptiles.

Eikenella corrodens. An aerotolerant, oxidase-positive, nonmotile, gram-negative coccobacillus. Requires hemin when grown aerobically. Colonies "pit" the agar surface. Resident microflora of human mucous membranes. Can cause abscesses, endocarditis, meningitis, and septic arthritis.

***Enterobacter* sp.** Genus includes at least eight species of enterobacteria. Can cause septicemia, neonatal meningitis, and bladder infections. Generally more resistant to antibiotics than *Escherichia coli*. Commonly encountered species are *E. cloacae, E. aerogenes,* and *E. agglomerans*.

***Enterococcus* sp.** Catalase-negative, salt-tolerant, gram-positive cocci in chains; some species are motile. *Enterococcus* (formerly *Streptococcus*) *faecalis* is responsible for most human infections. Causes bladder infections, endocarditis, abdominal abscesses, and septicemia. Commonly resistant to many antibiotics, but penicillin and aminoglycosides are synergistic against most strains. Resistance to vancomycin is increasing.

***Flavobacterium meningosepticum*.** Facultatively anaerobic, oxidase-positive, nonmotile, long, thin, gram-negative rods that produce yellow colonies. Utilize glucose and other carbohydrates fermentatively. Found in soil and water. Cause nosocomial infections; particularly, epidemics of meningitis in premature infants.

***Fusobacterium nucleatum*.** Anaerobic, nonmotile, gram-negative rods with a long, slender shape and pointed ends. Normal inhabitants of the mouth and upper respiratory tract. Susceptible to penicillin. Occasional cause of abscesses and blood stream infections. Possible role in **periodontal disease.**

***Moraxella* sp.** Strictly aerobic, nonmotile, oxidase-positive, nonfermenting gram-negative rods. Inhabit mucous membranes of humans and lower animals. Opportunistic pathogens. Six species have been implicated in human disease, including *M. lacunata* and *M. osloensis*. Occasional cause of eye infection, septicemia, pneumonia, and other infections.

***Mycoplasma genitalium*.** Resembles *M. pneumoniae*. Suspected cause of nongonococcal urethritis.

***Plesiomonas* sp.** Facultatively anaerobic, oxidase-positive, gram-negative rods usually with a tuft of polar flagella. Widespread in warm, fresh, or brackish waters. Possible cause of diarrhea. Genetic studies indicate a close relationship to *Proteus* sp.

***Porphyromonas* sp.** Genus includes three species of anaerobic gram-negative rods formerly placed in the genus *Bacteroides* that fail to produce acid from sugars. Their colonies fluoresce under ultraviolet light and produce pigment ranging from buff to black. A probable cause of **periodontal disease.**

***Providencia* sp.** Enterobacteria that resemble *Proteus* but are urease negative. Opportunistic pathogens found in human feces. Cause urinary tract infections and burn wound sepsis. Most human infections are due to *P. stuartii* and *P. rettgeri* (formerly *Proteus rettgeri*).

***Pseudomonas mallei*.** Aerobic, oxidase-positive, nonmotile, glucose-oxidizing, gram-negative rod. Causes **glanders,** a disease of horses, occasionally transmitted to humans. Glanders has been eradicated in the United States and Canada.

***Pseudomonas pseudomallei*.** Aerobic, oxidase-positive, motile, gram-negative rods. Growth accompanied by an odor of decaying flesh. Colonies range in color from orange to cream. Identification is confirmed by serological techniques. Usual habitat is the soil. Cause of **melioidosis,** a disease of humans and animals characterized by pneumonia or bloodstream invasion. Was a problem in American troops stationed in Southeast Asia during the Vietnam War. Rare in the western hemisphere. An AIDS-defining condition in parts of Asia.

***Serratia marcescens*.** Enterobacteria that characteristically produce pink or red pigment at room temperature. Found widely in nature and as an opportunistic pathogen in nosocomial infections. Causes endocarditis, osteomyelitis, septicemia, and wound, urinary tract, and respiratory tract infections. Most strains are resistant to several antibiotics because of the presence of R factors.

***Staphylococcus saprophyticus*.** Facultatively anaerobic, gram-positive, coagulase-negative coccus that occurs in clusters. Unlike other staphylococci, it is characteristically resistant to the antibiotic novobiocin. Found in air, soil, dust, and on body surfaces. Causes urinary tract infections.

***Treponema carateum*.** Spirochete morphologically indistinguishable from *T. pallidum*, but difficult to propagate in laboratory animals. Responsible for **pinta,** a nonvenereal, tropical disease mainly involving the skin.

Treponema pallidum subspecies *pertenue*. A spirochete virtually indistinguishable from *T. pallidum* except by the human disease it causes. Responsible for **yaws,** a nonvenereal syphilislike disease in humans. Found in many tropical regions.

***Ureaplasma urealyticum*.** A mycoplasma that inhabits the normal human urogenital tracts. Is the predominant bacterium in some cases of urethritis and pelvic inflammatory disease, but its pathogenic role is unproven.

***Vibrio parahaemolyticus*.** Facultatively anaerobic, motile, oxidase-positive, curved, halophilic, gram-negative rod, which, unlike *V. cholerae*, fails to ferment sucrose. Cause of seafood-acquired **gastroenteritis.**

***Yersinia enterocolitica*.** Enterobacteria motile at temperatures below 30°C. The organism grows better at 25°C to 30°C than most other enterobacteria. Some strains possess plasmid-mediated virulence factors and cause acute watery diarrhea **(enterocolitis)** in young children, and right-sided abdominal pain mimicking appendicitis in older children and young adults. The pain is due to **mesenteric adenitis** (inflammation of certain lymph nodes in the abdomen), caused when these virulent strains invade beyond the intestinal lining. Prolonged asymptomatic bacteremia can also occur and be responsible for fatal reactions from donated blood. This is because the organisms can multiply to high levels in blood stored in the cold for more than a few weeks.

Appendix

Infectious and Parasitic Diseases

The following sections list some infectious and parasitic diseases and conditions not covered specifically earlier in the text. This is by no means a comprehensive list; it describes only some of the conditions that might be encountered through travel, or in accounts by news media or popular magazines.

Anthrax. Cause: *Bacillus anthracis*. Involves mainly cattle, sheep, horses, and goats. Humans become infected from handling hair, hides, or animal wastes contaminated with *B. anthracis* endospores. If a spore enters a skin abrasion, an inflamed pimple develops that later becomes a blister, which results in a black scab surrounded by swelling. Systemic disease is often fatal. "Woolsorter's disease" is a severe pneumonia caused by inhaling spore-bearing dust. It is often fatal. The virulence of *B. anthracis* depends on capsule and toxin production encoded by genes on two plasmids. Since 1970, groups of U.S. military personnel have been immunized against anthrax because of its possible use as a biological warfare agent.

Aspergillosis. Disease resulting from infection by widespread molds of the genus *Aspergillus*. Inhalation of large numbers of the spores can sometimes cause acute pneumonia. The fungus can also colonize cavities in the lung (due to healed tuberculosis or other conditions) producing a "fungus ball". In patients with immune deficiency from AIDS or other causes, *Aspergillus* species invade all parts of the body and cause death.

Babesiosis. A disease arising from red blood cell destruction by protozoa of the genus *Babesia*, transmitted from domestic and wild animals by ticks. On blood smears, the organisms can be confused with *Plasmodium falciparum*, a cause of severe malaria. Most infections are asymptomatic, but fatalities can occur in people whose spleen has been removed.

Bacillary angiomatosis. A disease whose signs include fever, diarrhea, weight loss, and multiple nodules under the skin. The nodules are the result of inflammation and proliferating small blood vessels caused by infection with *Bartonella henselae* or related bacterium. A complication of AIDS and other conditions that cause impaired immunity.

Bartonellosis (Carrión's disease). An often fatal illness with fever and anemia, caused by *Bartonella bacilliformis*, a small, motile, aerobic, gram-negative coccobacillus. Occurs in Peru, Ecuador, and Colombia. Transmitted by sandflies. Bacteremia may persist for years after recovery from the acute disease.

Blastomycosis. Signs are fever, cough, and weight loss, followed weeks later by progressive ulcerating mucous membranes and skin lesions, abscesses of bone or other tissue. Cause: *Ajellomyces dermatitidis* (*Blastomyces dermatitidis*), a dimorphic fungus. Cases scattered across the southeastern and central United States, as well as Canada, and Central and South America.

Chagas' disease (American trypanosomiasis). Humans, dogs, cats, opossums, and armadillos are reservoirs for the causative agent, *Trypanosoma cruzi*, a flagellated protozoan. Occurs mainly in South and Central America, transmitted by assassin (reduviid) bugs. A leading cause of severe heart disease in the endemic areas.

Clonorchiasis. A chronic disease of the liver caused by the fluke *Clonorchis sinensis*. The worm is acquired by eating inadequately cooked, dried, salted, or pickled freshwater fish containing the encysted intermediate stage of the worm. The disease is common in the Orient and persists many years because of the long life of the worms. Diagnosis is by identifying ova of the worm in the feces. Treatment: praziquantel.

Colorado tick fever. A disease causing sudden severe aching of back and leg muscles, fever, and headache. A rash occurs in some cases. Symptoms generally disappear within a week. Seen mostly in western United States and Canada. Transmitted by ticks. Caused by an orbivirus (reovirus family).

Donovanosis (granuloma inguinale). An ulcerative disease of the skin and lymphatic system in the anal and genital area, endemic in tropical countries. Sexually transmitted. Caused by *Calymmatobacterium granulomatis*, a gram-negative bacterium. The organisms can be seen as clusters of encapsulated coccobacilli (Donovan bodies) in the cytoplasm of macrophages.

Dracunculiasis (guinea worm disease). Caused by *Dracunculus medinensis*, a roundworm up to 80 cm long, occurring mainly in India, Pakistan, and 17 countries of Africa. The worm pushes its uterus through the patient's skin and discharges its larvae into water. The disease is contracted by drinking water containing copepods infested with a larval stage of *D. medinensis*. Approximately 3.5 million cases were reported in 1986. An international effort at global eradication of the disease resulted in a 97% drop to approximately 100,000 cases by the end of 1995.

Ebola Hemorrhagic Fever. Typically, symptoms begin abruptly, with fever, chills, headache, and malaise, and progress to abdominal pain, diarrhea, shock, and hemorrhage. Some infected persons develop a rash 5 to 7 days after the onset of fever. The mortality is 50% to 90%. The disease was first identified in 1976 in the former Zaire and Sudan. Recurrent outbreaks have occurred in parts of Africa since then. Cause: Ebola virus, an RNA virus of the filovirus family. Transmission occurs by close physical contact and exposure to blood, vomitus, and stool. Reservoir of the virus is unknown. Epidemics have arisen from contact with wild chimpanzees that have died of the disease. Incubation period: 2 to 21 days. Strains of the virus vary considerably in virulence. Introductions into the United States have occurred with imported monkeys, and travelers could theoretically bring the virus in during the incubation period. However the virus is not highly contagious, and tests for rapid diagnosis have now been devised. No antiviral treatment is presently available.

Echinococciasis. A disease caused by the intermediate stage of tapeworms of the genus *Echinococcus*. The adult worms live in the intestines of dogs, wolves, and other canines, and the eggs of the worms are excreted in canine feces and contaminate their fur. Human disease is usually acquired from handling dogs that have acquired the worms from eating the flesh of an intermediate host such as sheep, moose, or caribou. Human disease occurs in two forms: a slowly enlarging cyst usually of the liver and an invasive form that mimics cancer.

Ehrlichiosis. A tickborne disease characterized by fever, headaches, muscle aches, low leukocyte count, and low platelet numbers. Severity can range from asymptomatic to life-threatening. Cause: *Ehrlichia* species; tiny, pleomorphic, obligately intracellular gram-negative rods of the rickettsia family that grow preferentially in the cytoplasm of mononuclear or polymorphonuclear cells. Disease caused by *E. chaffeensis* is called **human monocytic ehrlichiosis** and is transmitted by the lone star tick (*Amblyoma americanus*). Another form of the disease is caused by a bacterium closely related to *E. equi*, but has not yet been fully characterized. It is called **human granulocytic ehrlichiosis** and is transmitted by the deer tick (*Ixodes scapularis*) and the dog tick (*Dermacentor variabilis*). Cases of ehrlichiosis have been identified in at least 30 states, mostly in the East, but also in the South and North Central regions, and as far west as California.

Endemic syphilis (Bejel). Chronic progressive disease associated with areas of extreme poverty worldwide. Cause: *Treponema pallidum* subspecies *endemicum*. Mainly a disease of children, manifested mostly by mouth lesions spread by nonsexual direct contact with infected lesions. Congenital infection and central nervous system and cardiovascular involvement generally do not occur.

Enteritis necroticans. An often fatal gangrene of the intestine due to *Clostridium perfringens*. Associated with gorging on food such as pork, and with conditions such as heart failure that impair normal blood circulation to the intestine. The feasting-related disease in New Guinea is known as "pigbel." Infected persons have acute abdominal pain and bloody diarrhea, and can go into shock.

Eosinophilic meningitis. Inflammation of the meninges in which many eosinophils appear in the spinal fluid. Usually associated with severe headache, blurred vision, and shooting pains in the chest, arms, and legs. Most cases are caused by the rat lungworm, *Angiostrongylus cantonesis*, generally acquired by humans in Southeast Asia and Pacific islands when they eat inadequately cooked snails, crabs, or freshwater prawns that may be carrying an infective larval stage of the worm. Treatment: none available.

Epiglottitis. An often fatal infection of the epiglottis by *Haemophilus influenzae*. The onset resembles croup, but X rays of the neck reveal the swollen epiglottis obstructing the airway, which delineates the two conditions. Urgent restoration of the airway is often crucial in saving the patient's life. Now rare in the United States because of vaccination against *H. influenzae* b.

Erysipelas. Localized swelling and inflammation of skin with lymphatic involvement, usually of the face, caused by *Streptococcus pyogenes*.

Erysipeloid. An acute or chronic infection of the skin caused by *Erysipelothrix rhusiopathiae*. The organism enters through minor abrasions from contaminated fish, shellfish, meat, or poultry, causing a spreading, painful red skin eruption, usually on the fingers and hand. *E. rhusiopathiae* causes serious infections in various animals, including swine.

Filariasis. Disease resulting from infection by a number of species of roundworms that live in the lymphatic vessels, deep tissues, or subcutaneous tissues. The female worms discharge tiny larvae (microfilariae) into the blood or tissues, which are then taken up and transferred to a new host by biting arthropods. *Wuchereria bancrofti* and *Brugia malayi* cause **lymphatic filariasis,** causing scarring of the lymphatic vessels, and persistent obstruction and swelling. *Onchocerca volvulus* causes **river blindness** because of microfilarial damage to the eyes. Millions of people in Africa and South and Central America are blind from this disease.

Glanders. Cause: *Pseudomonas mallei*. A serious infectious disease of horses that can be transmitted to humans via broken skin or mucous membranes. Characterized by nodular, necrotic involvement of the nasal mucous membranes, skin, lymphatics, or lungs. Commonly fatal. Isolation of infected animals and humans is necessary. Eliminated in the United States, but sporadic cases still occur in Asia, Africa, and Central and South America.

Granuloma inguinale. See Donovanosis.

Hand-foot-and-mouth disease. A blistery eruption of the mouth, hands, feet or any combination of the three, usually caused by coxsackie A 16 virus.

Herpangina. Signs: sudden onset of fever and extremely sore throat showing pale papules or blisters with surrounding redness. Caused by coxsackie A viruses.

Korean hemorrhagic fever (hemorrhagic fever with renal syndrome). Signs: sudden appearance of high temperature, pain in head, back, and abdomen. After a few days, hemorrhages appear in the conjunctivae, mouth, and skin. Mental dullness and kidney failure develop. Fatalities may reach 30%, but the rate is considerably lower in some epidemics. Cause: hantavirus, an RNA virus of the bunyavirus family, transmitted from house rats by inhalation of dust contaminated by rat feces. Epidemics have occurred in European countries as well as Asian countries.

Lassa fever. Sore throat, fever, inflamed liver and heart, hemorrhages, shock, death in 10% to 50% of cases. Cause: arenavirus contracted from food contaminated with rat urine. Endemic in West Africa. The antiviral medication ribavirin is effective if given early in the illness.

Leishmaniasis. Disease due to flagellated protozoa of the genus *Leishmania*. **Kala azar** (visceral leishmaniasis) is caused by *L. donovani* and is manifested by fever, cough, diarrhea, and enlargement of the spleen. Other species cause ulcerating lesions of the skin and mucous membranes. The disease is transmitted from dogs to humans or from one human to another by sandflies. Leishmaniasis was contracted by a very small percentage of American military personnel involved in the Persian Gulf war of 1990.

Lymphogranuloma venereum. Sexually transmitted disease caused by certain strains of *Chlamydia trachomatis*. The disease begins with a transitory genital blister or ulcer, followed by enlargement and pus formation in the nearby lymph nodes. Late complications include marked swelling of the genitals and scarring of the penis, urethra, or rectum.

Melioidosis. A rare disease caused by *Pseudomonas pseudomallei*. Manifestations range from minor pulmonary infections to an often fatal septicemia with abscesses in many organs. Endemic in Southeast Asia where it commonly complicates AIDS.

Molluscum contagiosum. Chronic, smooth, raised skin lesions, often with a dimpled center, caused by a pox virus. The lesions disappear without treatment, generally within one year. Common in children. Spread by direct contact. Sexually transmitted in young adults.

Mucormycosis. Disease characterized by invasive infection of the lung and sinuses by common molds of decaying organic materials, mostly *Rhizopus* and *Mucor* sp. Infection occurs mainly in diabetics or patients with cancer. Usually rapidly fatal.

Myocarditis. Inflammatory condition involving the heart. May cause death from heart failure or abnormal rhythm of heartbeat. Often an acute illness caused by coxsackie B viruses.

Orf (contagious ecthyma; "sore mouth" in sheep and goats). Signs: painful ulcerating lesions, usually occur on the hands and arms after contact with infected sheep, goats, or deer. Caused by a parapoxvirus of the poxvirus family.

Paragonimiasis. Symptoms are cough and spitting blood because of lung infestation with flukes of the genus *Paragonimus*. The disease can closely mimic tuberculosis of the lung. It is widespread in Africa, Asia, and parts of South America, acquired by eating inadequately cooked freshwater shrimp, crabs, or crayfish containing an intermediate stage of the worm. Treatment: praziquantel.

Pinta. A nonvenereal tropical disease caused by the spirochete *Treponema carateum*, which is morphologically indistinguishable from *T. pallidum*. Months after infection, crops of flat, red skin lesions appear on the hands, feet, and scalp. Serological reactions are identical to syphilis and yaws. Transmission is by direct person-to-person contact, usually during childhood. Occurs in the tropics, particularly in Central and South America.

Pleurodynia (Bornholm's disease). Signs and symptoms: fever and sudden intense lower chest or upper abdominal pain aggravated by coughing and breathing. Most cases caused by coxsackie B viruses. Disappears without treatment within two weeks.

Pseudallescheriasis. Mainly infects immunocompromised hosts. Begins as lung infection that spreads throughout the body, with tissue invasion and abscess formation. Cause: *Pseudallescheria boydii* (*Petriellidium boydii, Seedosporium apiospermum*), a soil fungus.

Psittacosis (ornithosis). Acute pneumonia acquired from parrots and many other kinds of birds. Cause: *Chlamydia psittaci*.

Relapsing fever. An epidemic form is caused by the spirochete *Borrelia recurrentis* and is transmitted from human to human by the body louse *Pediculus humanus*. The endemic form is caused by various *Borrelia* species and is spread from animals to humans by species of *Ornithodorus* ticks. Symptoms: sudden onset of fever that lasts three to four days. The illness subsides, only to recur three to ten days later. This pattern may again be repeated one or more times, with the attacks becoming progressively less severe, until they cease altogether. The organisms in each successive attack differ antigenically from those of prior attacks, suggesting that relapses result from the ascendancy of antigenic variants unaffected by antibodies formed during the previous attack.

Trachoma. Chronic form of conjunctivitis due to certain strains of *Chlamydia trachomatis*. Scarring of the eyelids, ulceration, and scarring of the cornea causes blindness. Transmission is from eye to eye by fingers, towels, and insects. Tends to occur in dusty areas of the world. A leading preventable cause of blindness.

Typhus. Characterized by acute fever, headache, muscle pain, and rash caused by *Rickettsia* sp. **Endemic (murine) typhus** is caused by *R. typhi* transmitted from rats to humans by fleas. **Epidemic (louse-borne) typhus** is caused by *R. prowazekii*, transmitted from human to human by the body louse, *Pediculus humanus*. Recrudescent typhus **(Brill-Zinsser disease)** is a recurrence of louse-borne typhus years after the initial illness. **Scrub typhus** is caused by *R. tsutsugamushi*, transmitted to humans from wild rodents by mites.

Venezuelan equine encephalitis. Mild illness with fever, headache, and muscle pain. Symptoms usually disappear within a week. Recurrent epidemics in South and Central America and the southern United States. Caused by an alphavirus of the togavirus family, transmitted by several genera of mosquitoes, from numerous species of birds and wild and domestic animals. People at highest risk of the disease are workers and soldiers who enter swamps or rain forests.

Vesicular stomatitis. A blistery rash of the mouth and teats seen mainly in cows, horses, pigs, deer, raccoons, and skunks. Humans exposed to infected animals develop fever, muscle and eye pains, headache, blisters in the mouth, and conjunctivitis. Symptoms generally disappear within a week. Caused by a rhabdovirus.

Whipple's disease. A rare disease marked by painful joints, cough, malabsorption, weight loss, and enlarged abdominal lymph nodes. Caused by a bacterium tentatively named *Tropheryma whippelii* based on studies of its ribosomal RNA. It has not been cultivated.

Yaws. A nonvenereal tropical disease caused by the spirochete *Treponema pallidum* serotype *pertenue*, morphologically indistinguishable from *T. pallidum* causing syphilis. The primary lesion, called the mother yaw, appears about four weeks after exposure. It is a painless, red papule surrounded by a zone of erythema, which eventually ulcerates, becomes covered with a dry crust, and heals. Similar lesions occur in successive crops over a period of months to years. Late in the disease, nodules and ulcerations disfigure the nose and face in some cases. Transmission is by person-to-person contact, usually during childhood. Occurs primarily in the tropics.

Appendix

Some Immunizing Agents of Major Importance

Table VI.1 Some Immunizing Agents of Major Importance

Disease	Type of Vaccine	Other Information
Adenovirus	Live attenuated virus of types 4 and 7	This is given only to military recruits.
Anthrax	Inactivated bacteria	Given only to persons who might be exposed, such as laboratory or industrial workers, and military personnel.
Cholera	Inactivated bacteria	Given to persons who may travel in endemic areas. Vaccine is only 50% effective in preventing the disease.
Diphtheria	Toxoid	This vaccine is given routinely to children.*
Haemophilus influenzae type b	Polysaccharide toxoid conjugate	This vaccine is given routinely to children.*
Hepatitis B	Purified antigen from recombinant yeast	This vaccine is given routinely to children.*
Influenza	Inactivated virus	This vaccination is needed yearly as the virus changes its antigens frequently. Vaccination is given primarily to high-risk populations, especially those over age 65.
Mumps	Live attenuated virus	This vaccine is given routinely to children.*
Pertussis	Acellular vaccine; whole-cell pertussis vaccine is an acceptable alternative	This vaccine is given routinely to children.*
Plague	Inactivated bacteria	This vaccine is given to agricultural workers in endemic areas as well as to laboratory workers at high risk.
Pneumococcal infection	Purified multivalent polysaccharide	This vaccine is given once to persons over 65 and to other high-risk groups.
Poliomyelitis	Live attenuated virus (preferred)	This vaccine is given routinely to children.*
	Enhanced inactivated virus	This vaccine is used in unimmunized adults and for the first dose in children.
Rabies	Inactivated virus	Used primarily for persons at high risk, such as veterinarians and other animal handlers.
Rubella (German measles)	Live attenuated virus	This vaccine is given routinely to children.*
Rubeola (measles)	Live attenuated virus	This vaccine is given routinely to children.*
Tetanus	Toxoid	This vaccine is given routinely to children* with a booster every 10 years for adults.
Tuberculosis	Live attenuated tuberculosis bacterium of BCG strain	Widely used in other countries but in the U.S. recommended for high-risk individuals.
Typhoid fever	Inactivated bacteria, live attenuated bacteria	A short-term vaccine that is given to travelers to endemic areas and high-risk laboratory workers.
Varicella	Live attenuated virus	Given to immunosuppressed children and susceptible adults.
Yellow fever	Live attenuated virus	Given to travelers to endemic yellow fever areas; good for 10 years.

*See Table 17.7 for the routine childhood vaccination schedule.

Appendix

Arthropods and Parasitic Worms

A number of disease-causing multicellular organisms are studied using the same microscopic and immunological techniques that are used for microorganisms and viruses.

Most of these medically important multicellular parasites fall into one of three groups: arthropods, roundworms, and flatworms. The arthropods are highly advanced on the evolutionary scale. Their main medical importance is that they are subject to infections, which they transmit to humans. The parasitic worms, on the other hand, are primitive creatures, and only in a few instances do they transmit microbial infections to their host animal. Instead, they cause disease by invading the host's tissues or robbing it of nutrition.

Most multicellular parasites have been well controlled in the industrialized nations, but they still cause death and misery to many millions in the economically underdeveloped areas of the world.

Table VII.1 Some Arthropods that Transmit Infectious Agents

Arthropod	Infectious Agent	Disease and Characteristic Features
Insects:		
Mosquito (*Anopheles* species)	*Plasmodium* species	Malaria: chills, bouts of recurring fever
Mosquito (*Culex* species)	Togavirus	Equine encephalitis: fever, nausea, convulsions, coma
Mosquito (*Aedes aegypti*)	Flavivirus	Yellow fever: fever, vomiting, jaundice, bleeding
Flea (*Xenopsylla cheopis*)	*Yersinia pestis*	Plague: fever, headache, confusion, enlarged lymph nodes, skin hemorrhage
Louse (*Pediculus humanus*)	*Rickettsia prowazekii*	Typhus: fever, hemorrhage, rash, confusion
Arachnids:		
Tick (*Dermacentor* species)	*Rickettsia rickettsii*	Rocky Mountain spotted fever: fever, hemorrhagic rash, confusion
Tick (*Ixodes* species)	*Borrelia burgdorferi*	Lyme disease: fever, rash, joint pain, nervous system impairment
Mite (*Liponyssoides sanguineus*)	*Rickettsia akari*	Rickettsial pox: rash, fever

Arthropods

The arthropods include insects and arachnids. Adult insects have six legs, arthropods have eight. Examples of some important arthropods, the agents they transmit, and the resulting diseases are shown in table VII-1.

Mosquitoes: Accurate Identification Aids Control Measures

Mosquito anatomy is shown schematically in figure VII-1. The mouth parts of the female mosquito consist of sharp stylets that are forced through the host's skin to the subcutaneous capillaries.

One of these needlelike stylets is hollow, and the mosquito's saliva is pumped through it. The saliva increases blood flow and prevents clotting as the victim's blood is sucked into a tube formed by the other mouth parts of the insect. The saliva can cause allergic reactions or transmit infectious diseases.

The mosquito can take in as much as twice its body weight in blood, thus giving it a relatively good chance of picking up infectious agents such as malarial parasites circulating within the host's capillaries. Mosquitoes in an area of arthropod-borne disease can be trapped and identified microscopically, and the blood they have ingested can be tested to see which kinds of animals the different species are feeding on.

Most of the medically important mosquitoes are included in the genera *Aedes, Culex,* and *Anopheles.* Precise identification of species and subspecies of these genera is important because different species of mosquitoes differ greatly in their breeding areas, time of feeding, and choice of host. Identification depends largely on microscopic examination of antennae, wings, claws, mating apparatus, and other features. Correct identification is often essential in designing specific control measures.

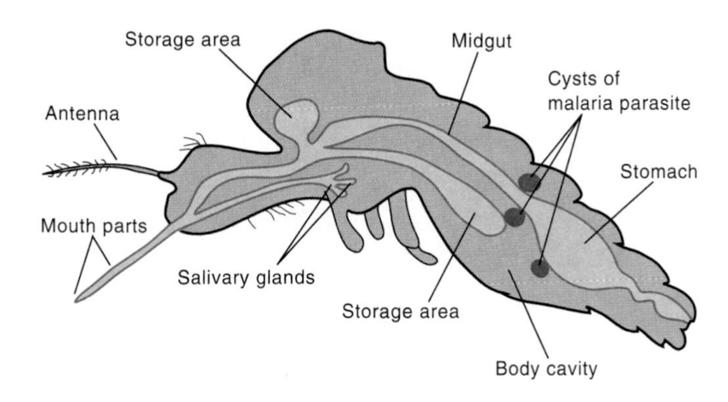

Figure VII-1
Internal anatomy of a mosquito. Note the storage areas that allow ingestion of large amounts of blood, and the salivary glands that discharge pathogens into the host.

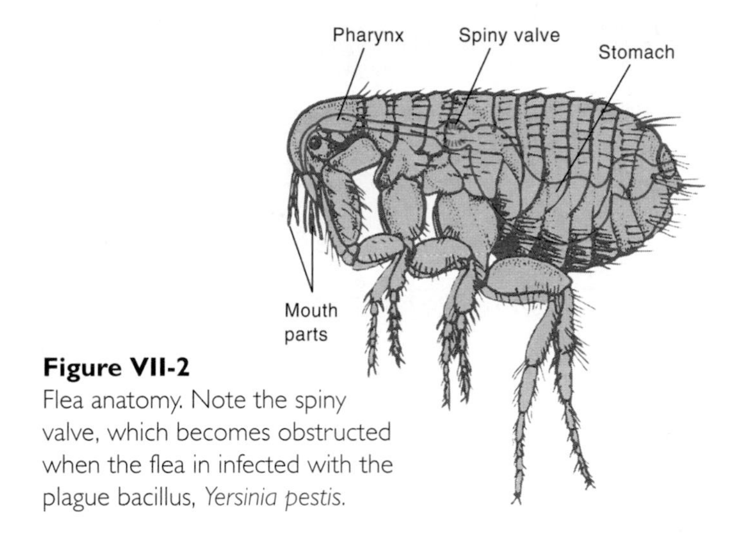

Figure VII-2
Flea anatomy. Note the spiny valve, which becomes obstructed when the flea in infected with the plague bacillus, *Yersinia pestis*.

Fleas: Their Main Importance Is in Transmitting Diseases from Rodents to Humans

Fleas are wingless insects that depend on powerful hind legs to jump quickly from place to place. The anatomy of a flea is diagrammed in figure VII-2. Points of importance in identifying fleas include the spines (combs) about the head and thorax, the muscular pharynx, the long esophagus, and the spiny valve composed of rows of teethlike cells. Fleas are generally more of a nuisance than a health hazard, but they can transmit plague and a rickettsial disease, murine typhus, to humans. Larval fleas have a chewing type of mouth for feeding on organic matter. They ingest eggs of the common dog and cat tapeworm, *Dipylidium caninum*, serving as its intermediate host. Children acquire this tapeworm when they accidentally swallow fleas. Fleas can live in vacant buildings in a dormant stage for many months. When the building becomes inhabited, the fleas quickly mature and hungrily greet the new hosts.

Lice: Transmission from Person to Person Generally Requires Close Physical Contact

Like fleas, lice are small, wingless insects that prey on warm-blooded animals by piercing their skin and sucking blood. However, the legs and claws of lice are adapted for holding onto body surfaces and clothing rather than for jumping. Human lice generally survive only a few days away from their hosts.

Pediculus humanus, the most notorious of the lice, is 1 to 4 mm long, with a characteristically small head and thorax, and a large abdomen (figure VII-3). This louse has a membranelike lip with tiny teeth that anchor it firmly to the skin of the host. Within the floor of the mouth is a piercing apparatus somewhat similar to that of fleas and mosquitoes. *Pediculus humanus* has only one host—humans—but easily spreads from one person to another by direct contact or by contact with personal items, especially in areas of crowding and poor sanitation.

Figure VII-3
Pediculus humanus (body louse) is the vector for *Rickettsia prowazekii*, the cause of typhus.

There are two subspecies, popularly termed head lice and body lice. Body lice transmit **trench fever** and **epidemic typhus,** caused by *Bartonella quintana* and *Rickettsia prowazekii*, respectively, and relapsing fever, caused by *Borrelia recurrentis*. Trench fever occurs episodically among severe alcoholics and the homeless of large American and European cities.

Phthirus pubis (crab louse) shown in figure VII-4, is commonly transmitted among young adults during sexual intercourse. It is not a vector of infectious disease.

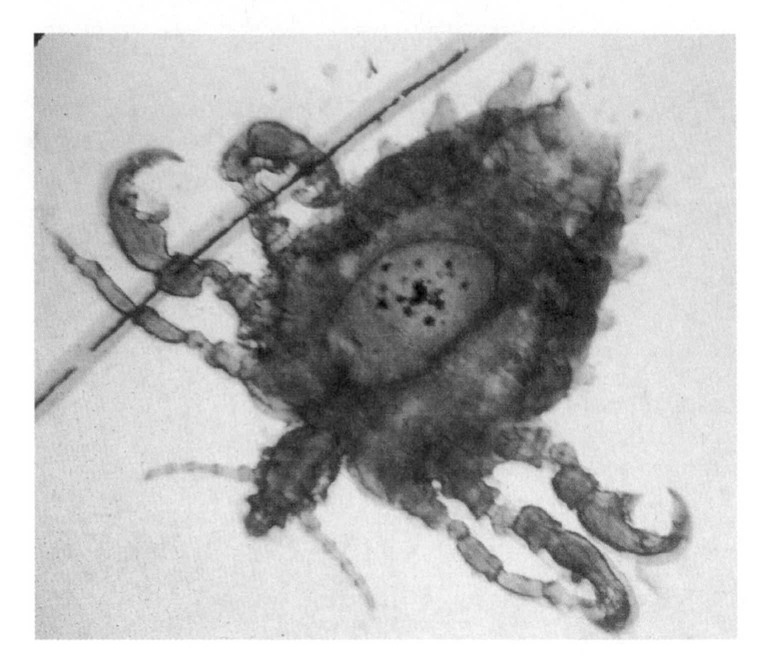

Figure VII-4
Phthirus pubis (crab louse) is commonly sexually transmitted. It is not known to be a vector for infectious agents.

Ticks Have a Wide Host Range

Ticks are **arachnids**. Arachnids differ from insects in their lack of wings and antennae, and their thorax and abdomen are fused together. Although like insects the immature ticks have three pairs of legs, the adults have four pairs. Two tick species, *Dermacentor andersoni* and *Ixodes scapularis*, were discussed in chapter 22 (figures 22.11 and 22.14). The saliva of several genera of ticks can produce a profound paralysis especially in children on whom the tick feeds for several days. Paralyzed humans and animals usually recover rapidly following removal of the tick.

Mites Are Tiny Arachnids

Mites, like ticks, are arachnids. They are generally tiny, fast moving, and live on the outer surfaces of animals and plants. *Demodex folliculorum* and *D. brevis* are elongated microscopic mites that live in the hair follicles or oil-producing glands usually of the face, typically without producing symptoms. Other species of mites cause human disease.

The disease **scabies**, caused by a mite, *Sarcoptes scabiei*, is characterized by an itchy rash most prominent between the fingers, under the breasts, and in the genital area. Scabies is easily transmitted by personal contact, and the disease is commonly acquired during sexual intercourse. The female mites burrow into the outer layers of epidermis (figure VII-5) feeding and laying eggs over a lifetime of about one month. Allergy to the mites is largely responsible for the itchy rash. The diagnosis can only be made by demonstrating the mites since scabies mimics other skin diseases. The mites are best detected by placing a drop of fountain pen ink on the affected skin and wiping off the excess with alcohol. The ink outlines the mites skin tunnels, which are then scraped into a drop of oil with a scalpel. Microscopic examination reveals a mite that is only about 0.33 mm long (figure VII-6). Treatment of scabies is easily accomplished with medication applied to the skin. *Sarcoptes scabiei* is not known to transmit infectious agents.

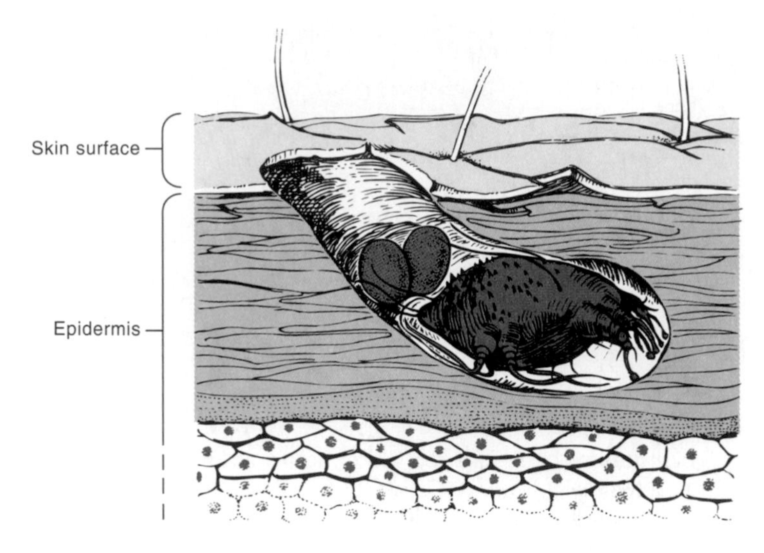

Figure VII-5
Sarcoptes scabiei. Female burrows into outer skin layers to lay her eggs, causing an intensely itchy rash. It is commonly sexually transmitted.

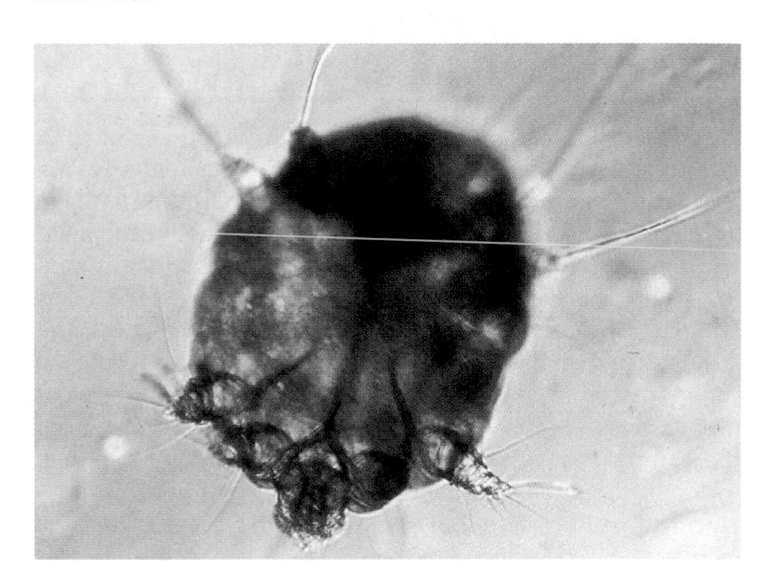

Figure VII-6
Sarcoptes scabiei (scabies mite) is readily identified by its microscopic appearance in scrapings of a scabies burrow. This mite is not known to transmit human infections.

Table VII.2 Nematoda (Roundworms)

Disease and Causative Agent	Entry to Body	Laboratory Diagnosis	Disease Characteristics	Environmental Control	Drug Control
Enterobiasis (pinworm) *Enterobius vermicularis*	Hand to mouth; inhalation	Microscopic examination of transparent tape impression of anal region to identify eggs	Itching of anal region, restlessness, irritability, nervousness, poor sleep	Hand washing, daily change of underclothing and bed sheets	Pyrantel pamoate or mebendazole
Trichuriasis (whipworm) *Trichuris trichiura*	Ingeston of eggs of parasite along with contaminated food or water	Microscopic examination of stool for eggs of parasite	Abdominal pain, bloody stools, diarrhea, and weight loss	Sanitary disposal of human feces; frequent hand washing	Mebendazole
Hookworm disease *Necator americanus* and *Ancylostoma duodenale*	Larvae penetrate bare feet	Microscopic examination of stool for eggs of parasite	Anemia, weakness, fatigue, physical and mental retardation in children	Sanitary disposal of human feces; wearing shoes	Mebendazole
Strongyloidiasis (threadworm) *Strongyloides stercoralis*	Larvae penetrate bare feet	Larvae in fresh stool	Skin rash at site of penetration, cough, abdominal pains, weight loss	Sanitary disposal of human feces; wearing shoes	Thiabendazole
Ascariasis *Ascaris lumbricoides*	Ingestion of eggs of parasite along with contaminated food or water	Microscopic examination of stool for eggs of parasite	Abdominal pain, live worms vomited or passed in stool	Sanitary disposal of human feces	Mebendazole
Trichinosis *Trichinella spiralis*	Eating raw or undercooked meat, usually pork	Biopsy of muscle showing organism; blood tests for antibodies	Fever, swelling of upper eyelids, muscle soreness	Adequate cooking of meat	Thiabendazole in severe cases, corticosteroids are also used
Filariasis *Wuchereria bancrofti; Brugia malayi*	Bite of infected mosquito	Larvae (microfilariae) in blood smear	Fever, swelling of lymph glands, genitals, and extremities	Eliminate breeding places of mosquito	Diethylcarbamazine

Mites of domestic animals and birds can cause an itchy rash in humans, as can mites sometimes present in hay, grain, cheese, or dried fruits. The dust mites that often live in large numbers in bedrooms can sometimes cause asthma when the mites and their excreta are inhaled.

The mites of rodents can transmit rickettsial diseases to humans. **Rickettsial pox,** caused by *Rickettsia akari* transmitted by mouse mites, is a mild disease characterized by fever and rash. Epidemics occur periodically in cities of the eastern United States. Serious rickettsial diseases of other parts of the world such as **scrub typhus** are transmitted by rodent mites.

Roundworms

The roundworms include a large number of species, some of which are human parasites and produce serious disease. Generally, diagnosis of worm infestation depends on microscopic identification of the worms or their ova (eggs), or on blood tests for antibody to the worms. Some examples of human roundworm parasites are discussed below. Table VII-2 summarizes features of these parasites.

Enterobiasis (Pinworm Disease) Occurs Worldwide, Including Cold Climates

Enterobius vermicularis (pinworm) exists worldwide. Infestation develops from ingestion of eggs on dirty fingers and possibly by inhalation of eggs on dust particles. *Enterobius* lives in the large intestine, migrating to the anus where it discharges its eggs usually at night. The resulting inflammation and itching often produce sleeplessness and behavioral disorders in children. Diagnosis is usually made using the transparent tape test (figure VII-7). Figure VII-8 shows the enormous numbers of ova present in a single

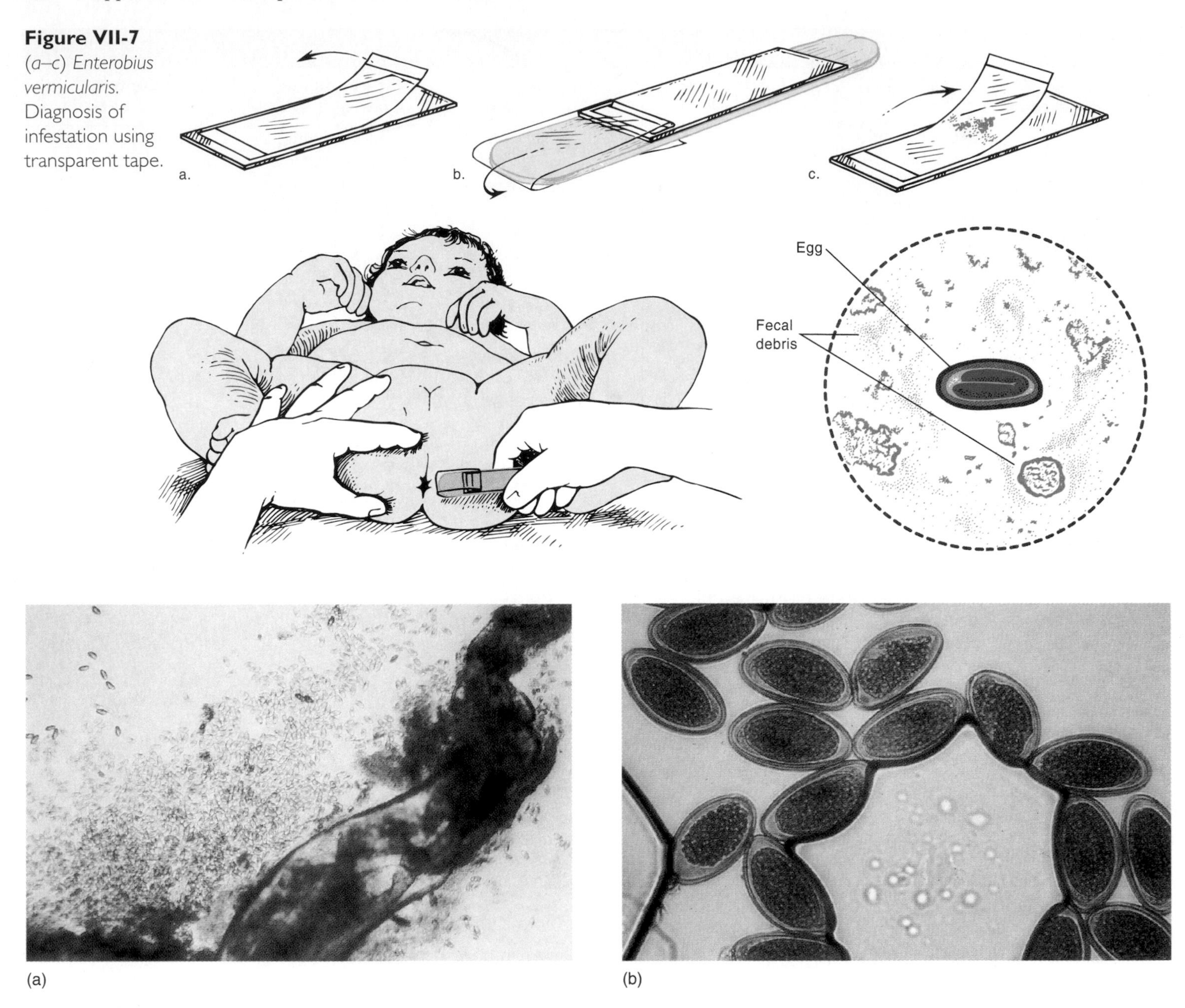

Figure VII-7
(a–c) Enterobius vermicularis. Diagnosis of infestation using transparent tape.

Figure VII-8
Enterobius vermicularis (pinworm). (*a*) Body of adult worm surrounded by thousands of its ova. (*b*) Higher power view showing many of the ova, (×125).

worm. Treatment with oral medication is highly effective, but reinfection is common.

Trichuriasis (Whipworm Disease) Involves Mainly the Beginning of the Large Intestine

Trichuriasis affects an estimated 350 million people in the warm and wet regions of the world. The causative agent, *Trichuris trichiura* causes abdominal pain and diarrhea when many worms are present. The worms are 3 to 5 cm long, with a slender anterior portion and a short, thicker posterior portion, hence the name whipworm. The slender anterior region of the worm is inserted into the intestinal mucosa, with the thicker portion protruding into the bowel lumen. The worms can live 10 years or more, and can be found in people who visited endemic areas years ago. Transmission is by the fecal-oral route. Diagnosis is made by examining the feces for the characteristic ova (figure VII-9).

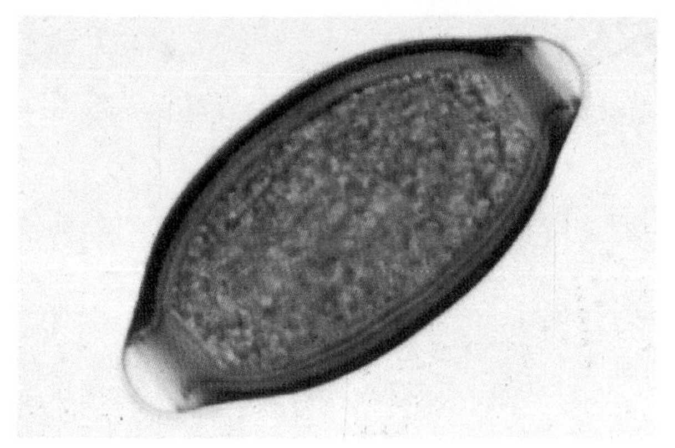

Figure VII-9
Trichuris trichiura ovum. Infestation easily diagnosed because of the characteristic appearance of the ova in fecal suspensions.

Hookworm Disease Is Caused by a Tiny Worm that Penetrates the Skin

Hookworms infest 400 million people and are found in warm, wet areas such as the southern part of the United States, Europe, Southeast Asia, and the Far East. The worm is about 10 mm long and lives in the small intestine of humans, attaching by means of small hooks or plates located about its mouth. It feeds by sucking the blood of the host. More than 1,000 worms may be found in a single person, and the constant loss of blood frequently produces anemia. Children who acquire hookworms may be undernourished already, and the anemia resulting from hookworm disease may cause weakness, fatigue, and physical and mental retardation. The lifespan of the worm is usually one to two years, but may exceed 10 years. Medicine can reduce or eliminate the infestation.

The hookworm life cycle is shown in figure VII-10.

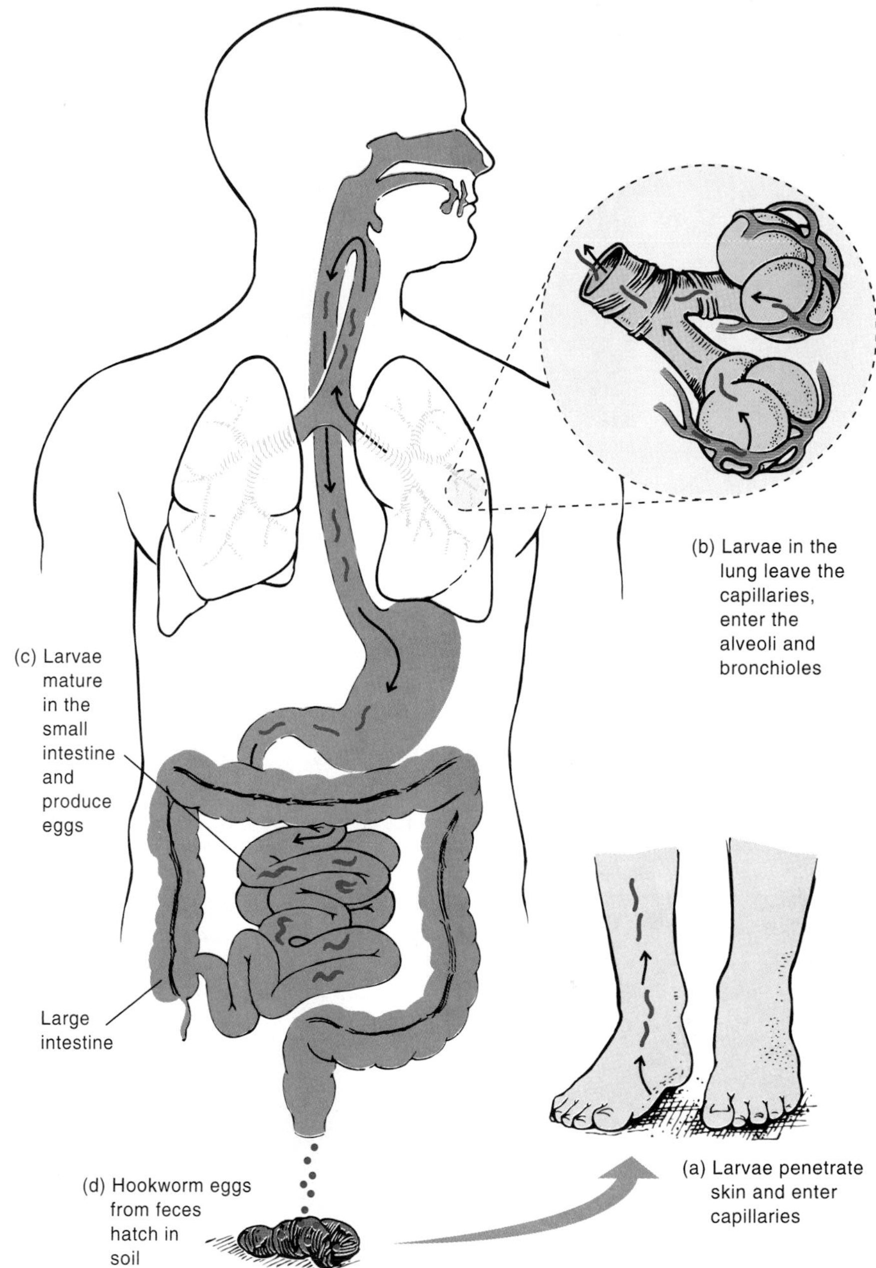

Figure VII-10
(*a–d*) Life cycle of a hookworm. Note that infection begins with penetration of the skin by the worm larvae.

The female worm releases her eggs—perhaps 25 million in her lifetime—and they are discharged with the feces, contaminating the soil in places where toilets are not common. The eggs hatch in the soil, releasing tiny larvae (figure VII-11) that develop in the soil and become infectious. These larvae penetrate the skin, generally of the feet. They then enter the bloodstream and are carried to the heart and lungs. In the lungs, they push through the blood vessel walls into the alveoli and then gain entrance to the gastrointestinal system when lower respiratory tract secretions are coughed up and swallowed. Once in the small intestine, the larvae attach and mature to complete their life cycle.

Diagnosis is made by microscopic examination of fecal material for hookworm ova (figure VII-12). Species identification can be made by allowing the eggs to hatch in a Petri dish and examining the larvae microscopically for characteristic features. *Necator americanus* and *Ancylostoma duodenale* account for most cases of human disease.

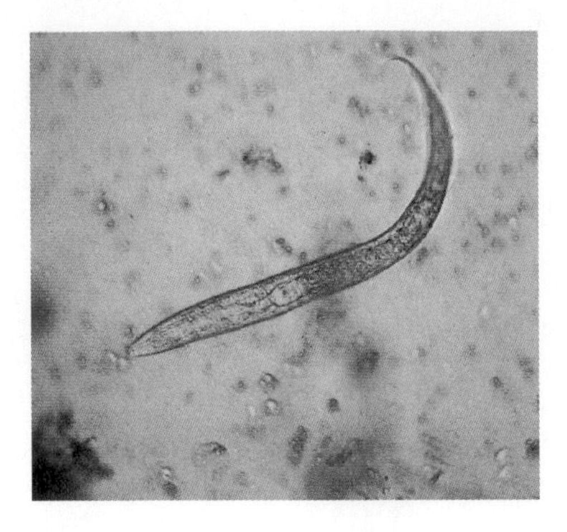

Figure VII-11
Hookworm larva, (×125).

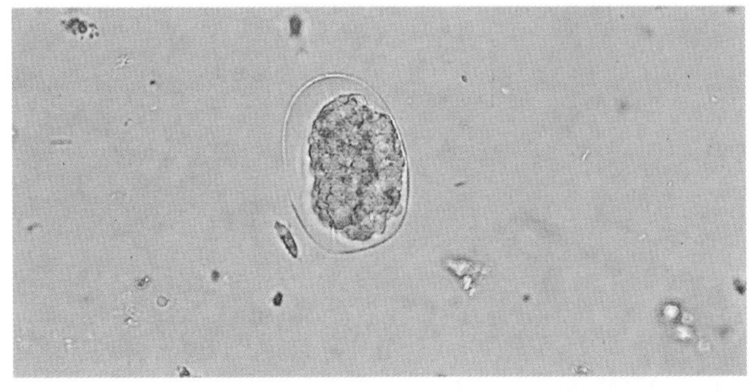

(a)

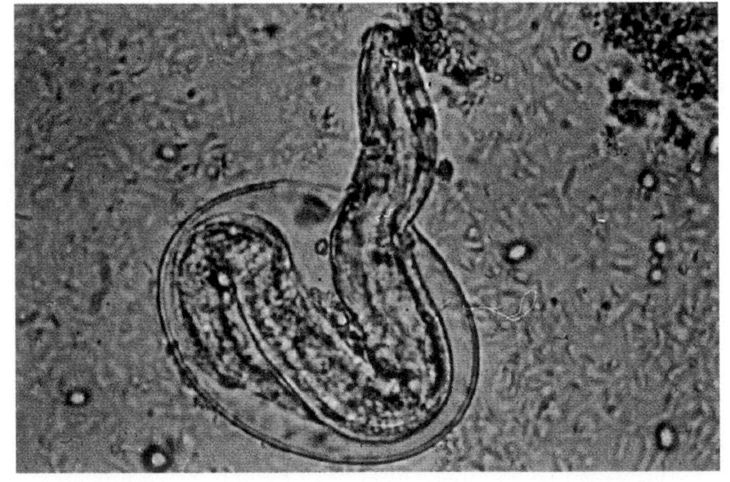

(b)

Figure VII-12
Hookworm ova in feces. (*a*) Early embryo, (×400). (*b*) Larva hatching.

Strongyloidiasis Is a Threat to the Immunocompromised

Strongyloidiasis, caused by *Strongyloides stercoralis*, occurs in a spotty distribution in warm, wet areas of the world, with an estimated 400,000 cases in the southeastern United States and Puerto Rico. Most infections are asymptomatic, but abdominal pain, diarrhea, recurrent rashes on the buttocks and lower back, and recurrent lung problems occur with heavy infestations. The adult worm is only about 2 mm long. Its life cycle is similar to that of hookworms with two important exceptions: (1) the worms can multiply in the soil; and (2) ova produced by intestinal worms hatch before they are discharged in the feces (figure VII-13). Sometimes these intestinal larvae mature to an infectious form, penetrate the intestine or the anal skin, and thereby perpetuate the infection indefinitely. In people who are debilitated, alcoholic, or immunosuppressed (including those with AIDS), massive numbers of the worms arise, penetrating all parts of the body, including the brain. These massive infections are fatal unless treated. Medications used in treatment include thiabendazole, albendazole, and ivermectin.

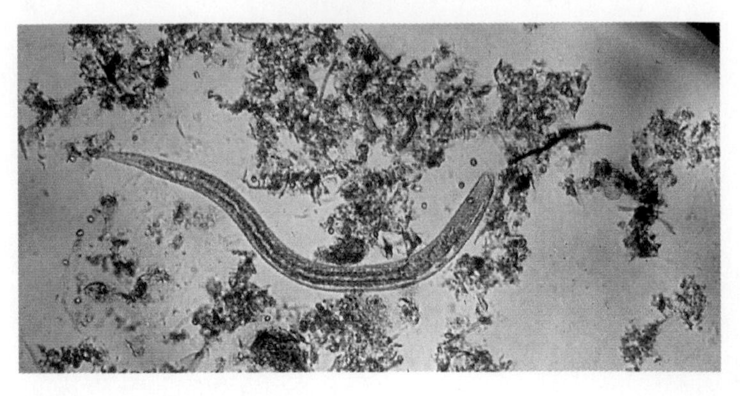

Figure VII-13
Strongyloides stercoralis in feces, (×50). This tiny worm, about 2 mm in length, has a life cycle and distribution similar to hookworms. However, unlike any other intestinal roundworm, *S. stercoralis* ova hatch in the intestine, allowing the worms to multiply in the body of a single host. Multiplication of the worms is often rampant in immunosuppressed patients such as those with AIDS, and can be fatal if not treated.

Ascariasis Is the Most Prevalent of All the Roundworm Diseases

Ascariasis is caused by *Ascaris lumbricoides*, the largest and most prevalent human roundworm. Cases occur worldwide in areas where human feces contaminate the soil. The life cycle of *A. lumbricoides* is shown in figure VII-14. The large female worm (figure VII-15) releases 200,000 eggs per day. Ova from human feces become infectious after a period of maturation in the soil. They remain infectious for years, resistant to drying, cold, and disinfectants and can be spread in airborne dust. A peculiarity of the *A. lumbricoides* life cycle is that when the eggs hatch in the intestine, the larvae migrate through the bloodstream before returning to the small intestine to mature. Although often asymptomatic, ascariasis victims may develop coughing and wheezing

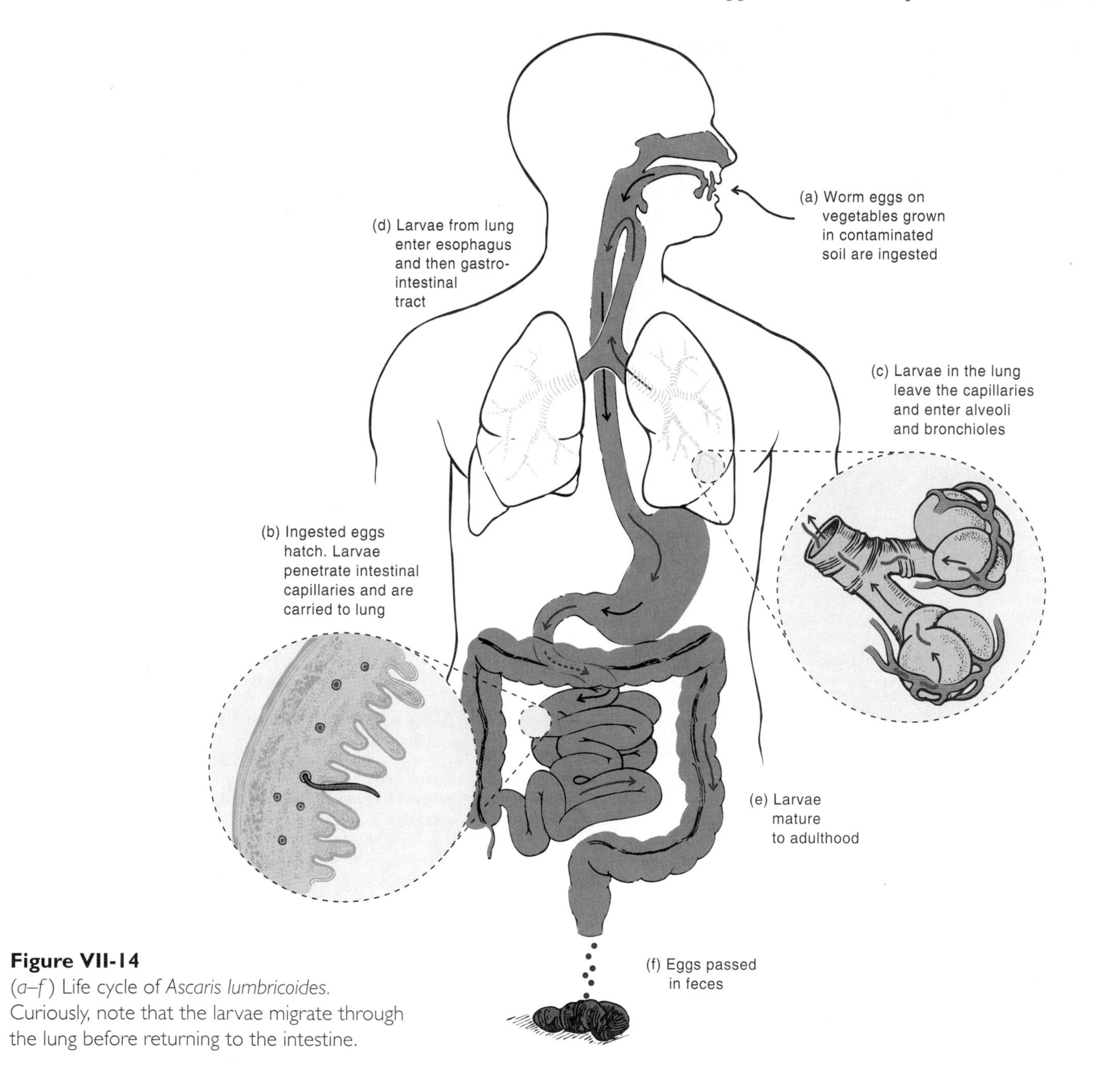

Figure VII-14
(*a–f*) Life cycle of *Ascaris lumbricoides*. Curiously, note that the larvae migrate through the lung before returning to the intestine.

due to larval migration through the lung. Large numbers of worms in the intestine can cause abdominal pain, malabsorption of food, and, in some cases, intestinal obstruction. Diagnosis is made by examining the feces for the characteristic ova (figure VII-16).

Larva Migrans Is Caused by Roundworms that Cannot Mature in Humans

The larvae of some roundworms of animals can penetrate tissues of humans and cause disease even though the worms cannot mature in humans, and they die in the tissues. The movement of the larvae in the tissue is called **larva migrans**. For example, *Toxocara canis*, an intestinal roundworm of dogs that resembles *Ascaris lumbricoides*, can cause blindness. Preschool children are especially susceptible because of their tendency to put rocks and dirt into their mouths. If the material is contaminated with *T. canis* ova from dog feces, the eggs hatch, and the larvae migrate through body tissues and eventually die without maturing. The resulting inflammatory reaction damages body tissues, including the eye. In another type of larva migrans, the larvae of dog and cat hookworms from contaminated soil migrate under the skin before dying (figure VII-17).

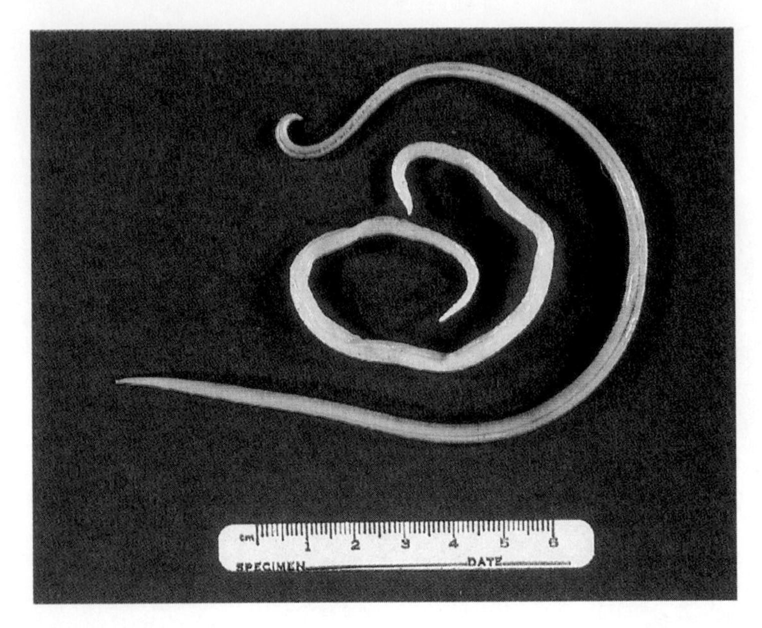

Figure VII-15
Ascaris lumbricoides, the largest of the roundworms infecting the human intestine.

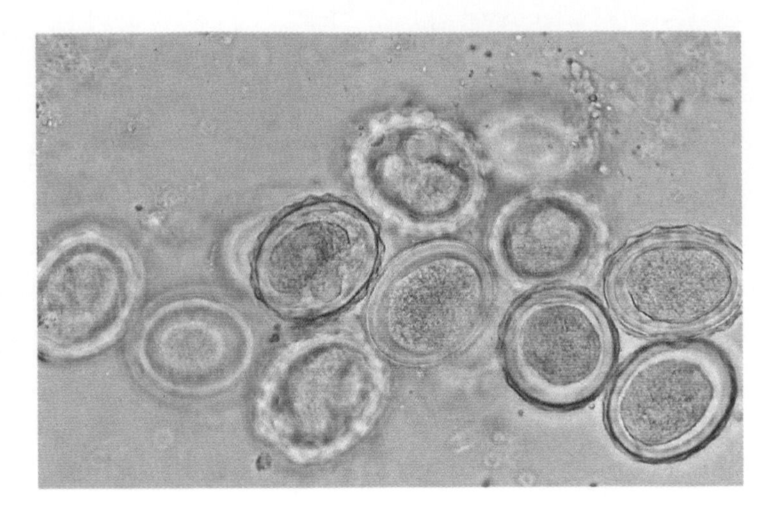

Figure VII-16
Ascaris lumbricoides ovum in feces. The characteristic appearance of the ovum makes diagnosis easy, (×400).

Trichinosis Is Contracted by Eating Inadequately Cooked Meat

Trichinosis is characterized by fever, muscle pain, swelling around the eyes, and sometimes a rash. Occasional cases are fatal because of damage to the heart or brain. The disease occurs worldwide.

The cause of trichinosis is *Trichinella spiralis*, a 1 to 4 mm long roundworm that lives in the small intestine of meat-eating animals, especially rats, pigs, bears, dogs, and humans. Its life cycle is shown in figure VII-18. The female worm discharges her living young into the lymph and blood vessels of the host's intestine without an intervening egg stage, and those larvae are carried to all parts of the body. Most of the larvae are killed by body defenses but some survive in the muscles of the host where they become encased with scar tissue. The worms then stay alive for months or years within the muscle. If the flesh of an infested pig or other carnivorous animal is eaten by humans or other animals, the digestive juices of the new host release the larvae from their cases, permitting them to burrow into the new host's intestinal lining. The larvae then mature and the females begin producing larvae in the new host to complete the life cycle of the worm. Each female adult *Trichinella* may live four months or more and produce 1,500 young. The penetration of these young worms into the tissues of the host's body is responsible for the symptoms of trichinosis. Diagnosis is made by muscle biopsy to identify encysted larvae, and by testing the patient's blood for antibodies to the worms.

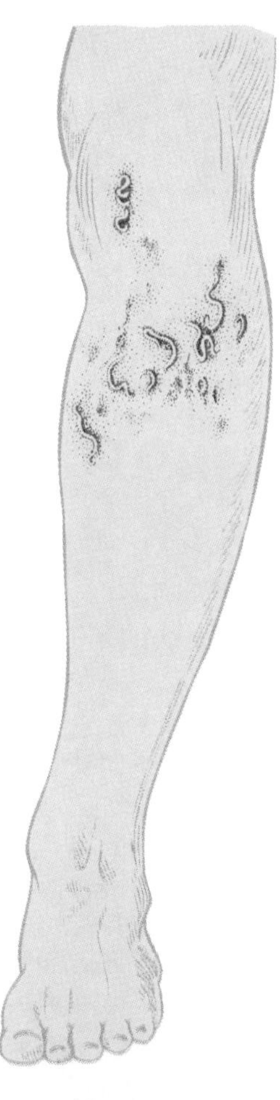

Figure VII-17
Cutaneous larva migrans. This condition is caused by dog and cat hookworms that are able to penetrate the skin of humans but cannot complete their life cycle. They wander under the skin for a time before dying. Contracting these hookworms is one of the occupational hazards of plumbers who have to crawl under buildings where the soil might have been contaminated with cat or dog feces.

Trichinosis is treated with thiabendazole and cortisone-like medicines. Prevention of trichinosis depends on adequate and thorough cooking of meat so that all parts reach at least 150°F. Pork has been the chief source of infections, presumably because pigs were fed uncooked garbage containing meat scraps, a practice now prohibited by law. However, beef ground in a machine previously used for pork can result in cases of trichinosis, and bear meat is a notorious source. Government inspection of meat does not detect *Trichinella* infestation. Larvae in pork are generally killed at 5°F (–15°C) or lower for 21 days if the meat is less than 15 cm thick. However, for reasons unknown, *T. spiralis* in wild game is not reliably killed by freezing.

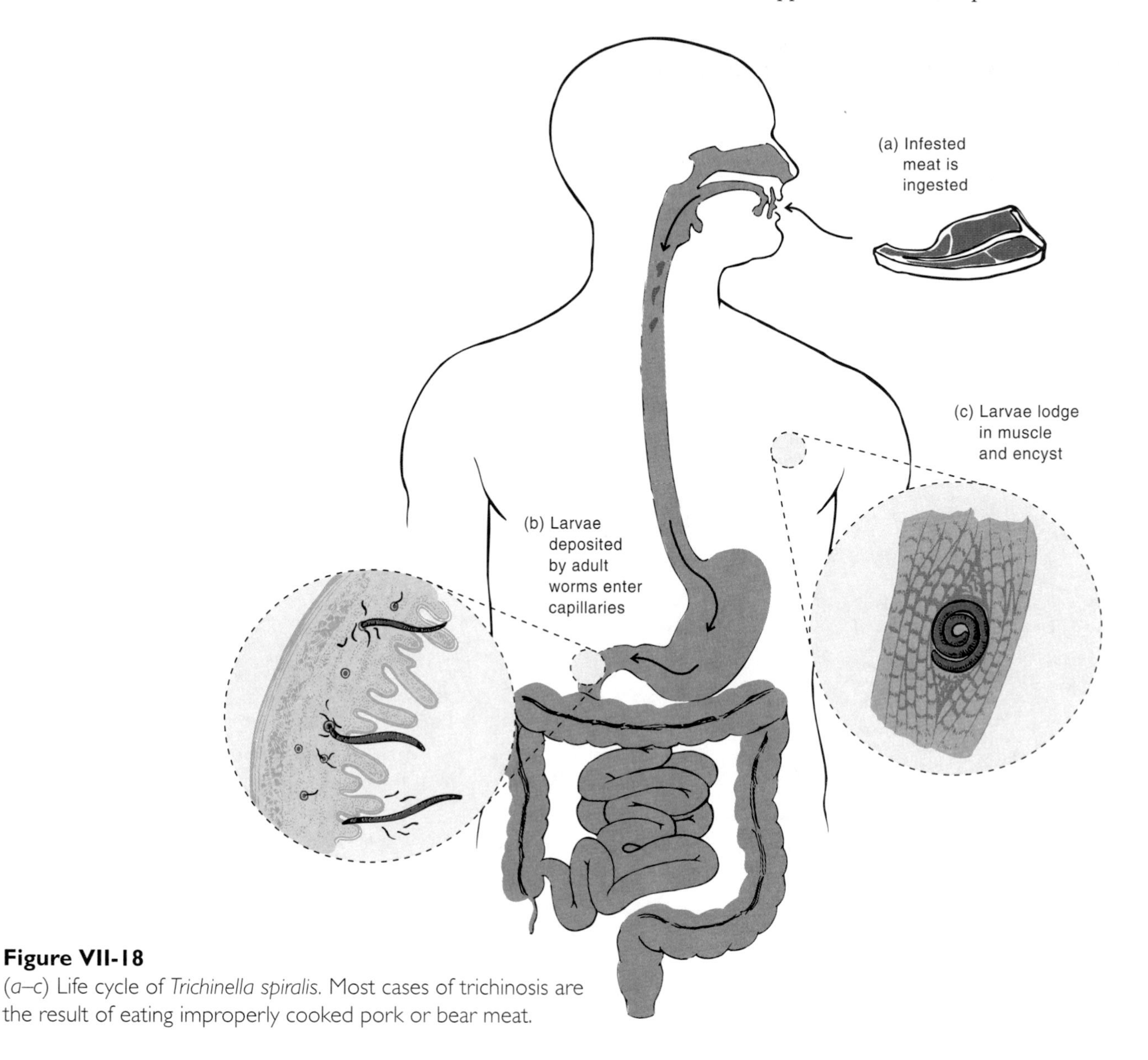

Figure VII-18
(*a–c*) Life cycle of *Trichinella spiralis*. Most cases of trichinosis are the result of eating improperly cooked pork or bear meat.

Flatworms

Flatworms of humans fall into two major groups: **flukes**, which are relatively short, flat, bilaterally symmetrical worms that generally attach by one or more sucking disks, and **tapeworms**, which are usually longer than flukes and are ribbonlike in appearance. Some examples of diseases caused by flatworms are discussed in the following section. Many, but not all, flatworms are hermaphroditic.

Schistosomiasis Is a Serious, Chronic Disease Contracted by Swimming and Wading

Schistosomiasis, endemic in 74 countries of Africa, Asia, the Caribbean and South America, is one of the most common of the diseases caused by parasitic worms. Worldwide the disease is estimated to involve more than 200 million people, 400,000 of whom now live in the United States having emigrated from areas of the world where the disease is endemic.

Schistosomiasis is caused by flukes of the genus *Schistosoma*. Typically, it is a chronic, slowly progressing illness resulting from damage to the liver, with malnutrition, weakness, and accumulation of fluid in the **abdominal** cavity.

Schistosoma mansoni (figure VII-19), a common cause of schistosomiasis, is 10 to 20 mm long and lives in the small veins of the human intestine. Its life cycle is shown in figure VII-20. The slender female worm discharges eggs (figure VII-21) that are carried into the liver by the flow of blood from the intestine, or rupture through the blood vessels and adjacent intestinal lining to enter the feces. The

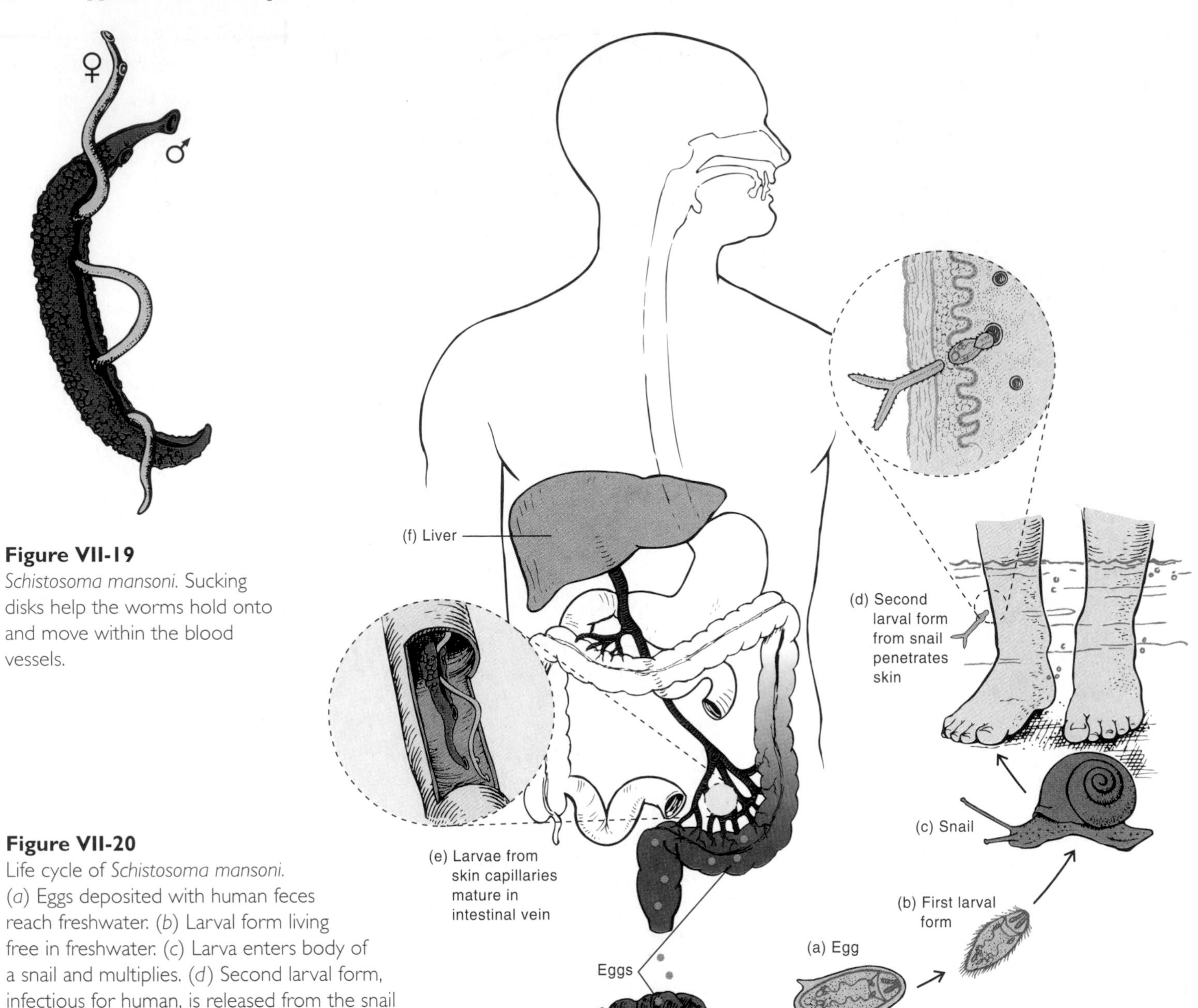

Figure VII-19
Schistosoma mansoni. Sucking disks help the worms hold onto and move within the blood vessels.

Figure VII-20
Life cycle of *Schistosoma mansoni.* (*a*) Eggs deposited with human feces reach freshwater. (*b*) Larval form living free in freshwater. (*c*) Larva enters body of a snail and multiplies. (*d*) Second larval form, infectious for human, is released from the snail and penetrates the human skin. (*e*) Mature adults develop in the blood vessels draining the large intestine, where eggs are deposited. (*f*) While many eggs penetrate the intestine and enter the feces, some are carried by the bloodstream to the liver, where they cause inflammation that, over time, results in extensive scarring.

eggs that lodge in the liver cause inflammation and scarring, and eventual loss of liver function.

Schistosoma enters the body as a cercaria (figure VII-22). Thousands of these infectious cercariae are released into water from the snail intermediate host each day. They burrow through the skin of people who wade in infested waters, leaving their tails at the skin surface. The cercariae then penetrate the blood and lymphatic vessels beneath the skin, reach the venous circulation, squeeze through the lung and intestinal capillaries, and finally lodge in the intestinal veins, where they become mature. The worms can live more than 25 years, continuously producing eggs and causing liver damage.

Measures used in the control of schistosomiasis include treatment of infected individuals with a medicine called praziquantel, sanitary disposal of feces, and chemicals to kill the snail hosts.

The disease **swimmer's itch** is caused by cercariae of schistosomes of birds and other animals. These cercariae penetrate the skin and then die, since they are unable to complete their life cycle in humans. The body's inflammatory response to the cercariae causes an itchy rash.

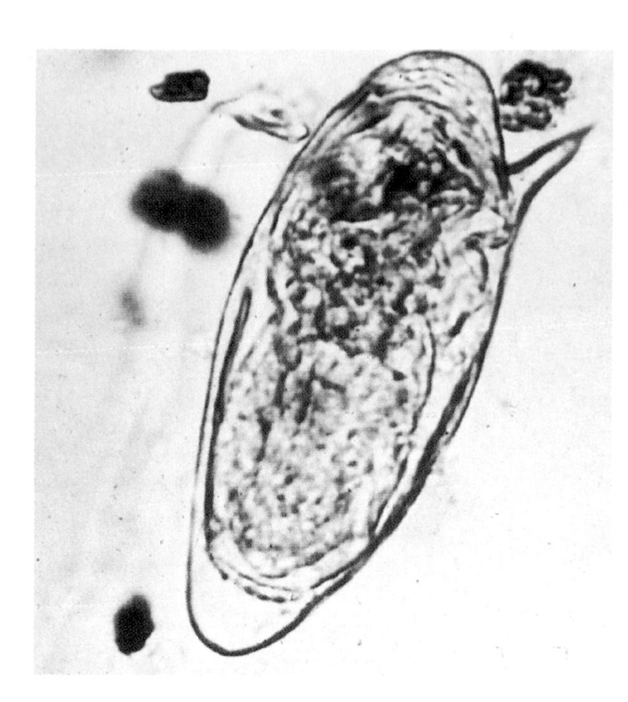

Figure VII-21
Schistosoma mansoni ovum, (×150). Note the lateral spine.

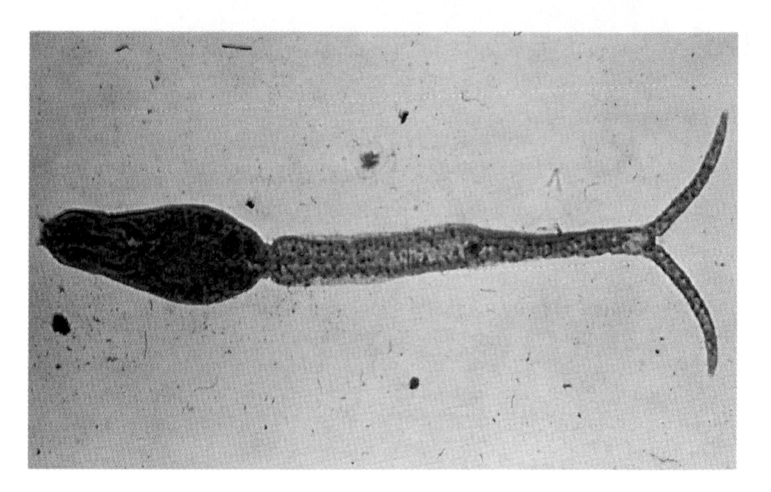

Figure VII-22
Schistosoma mansoni cercaria, the form that penetrates the skin and initiates infestation. Water bird schistosomes may penetrate the skin of swimmers and then die, causing "swimmer's itch."

Tapeworm Disease Is Contracted by Eating Inadequately Cooked Fish or Meat

People become infested with tapeworms mainly by eating inadequately cooked beef, pork, or fish. The adult worms live by absorbing human intestinal juices and usually cause few or no symptoms. However, some large tapeworms absorb enough vitamin pB_{12} to cause anemia, and in unusual circumstances, tapeworm developmental stages can invade the brain and other tissues, causing serious disease.

Figure VII-23 gives the life cycle of the fish tapeworm *Diphyllobothrium latum*. As with many other multicellular parasites, the earlier stages of tapeworm development occur in **intermediate hosts** (for example, the copepod and fish with *D. latum*), whereas adulthood with egg production occur in the **definitive host** (for example, a human). The developing stages of *D. latum* are invasive and grow within the tissues of the intermediate host, while the adult simply attaches to the intestinal lining of the definitive host.

The intermediate developmental forms of tapeworms can infect humans. This usually occurs when tapeworm eggs are accidentally ingested, but infection can also result from applying raw meat to open wounds in the mistaken belief that the meat will aid recovery. If intermediate stage larvae are present in the meat, they penetrate the tissues via the wound.

Diagnosis of intestinal tapeworms can be made by identifying ova in the feces (figure VII-24), or in some cases by microscopic examination of worm segments (proglottids) released into the intestinal contents.

The beef and pork tapeworms, *Taenia saginata* and *T. solium* (figure VII-25) have a simpler life cycle than *D. latum*, since they have only one intermediate host (a cow or pig). Fortunately, *T. solium* is rare in the United States. It is the most dangerous of the three types of tapeworm because on occasion the eggs hatch before being discharged in the feces. When this happens, the emerging larvae invade the various organs of the body, including the brain, where they can mimic a tumor or cause epilepsy.

Control of tapeworms depends on adequate cooking of meat and fish, and on proper disposal of human feces. Effective medicines are available.

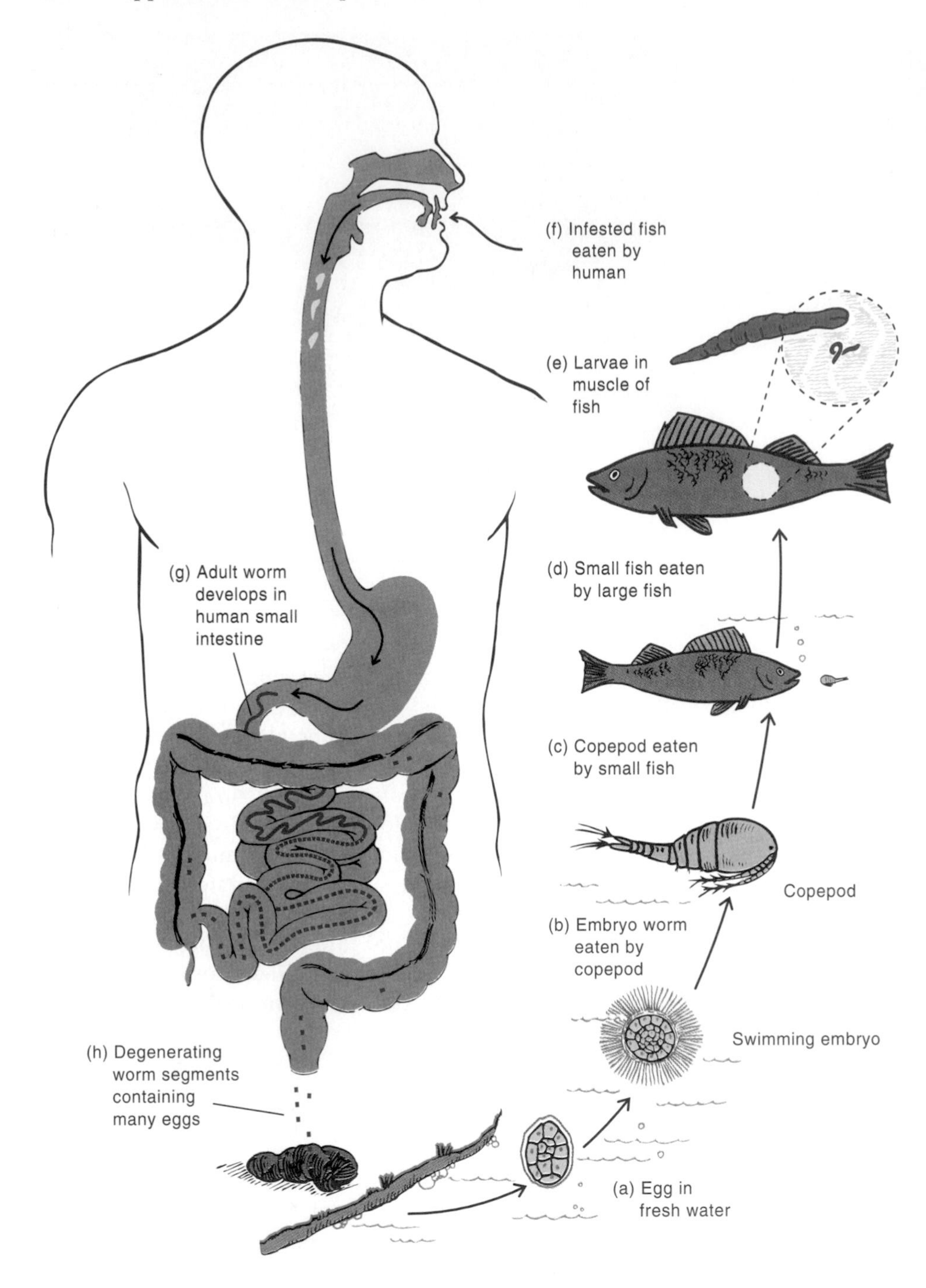

Figure VII-23
(*a–h*) *Diphyllobothrium latum* life cycle. Note that this species requires two or more intermediate hosts (copepod and fishes). Infestation is a risk for those who are fond of eating uncooked fish.

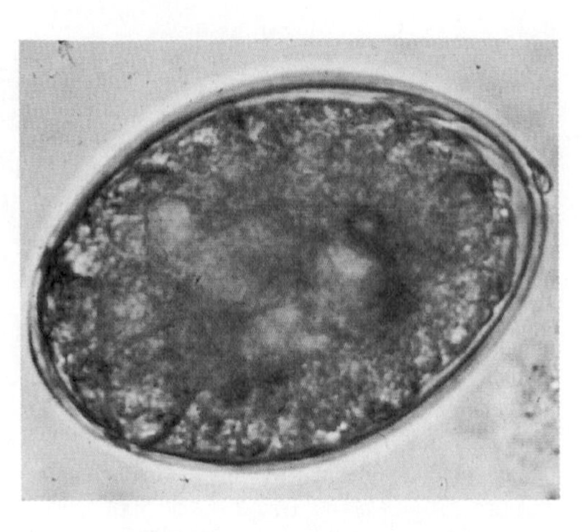

Figure VII-24
Diphyllobothrium latum ovum.

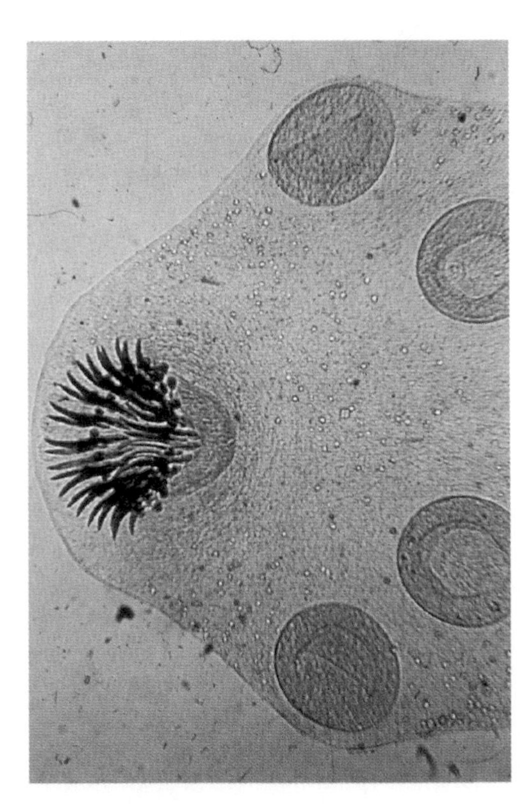

Figure VII-25
Taenia solium scolex, (×25). This tapeworm is acquired from eating inadequately cooked pork. Notice the hooks and suckers on the scolex (head).

Further Reading

"Acute schistosomiasis in U.S. travelers returning from Africa." *Morbidity and Mortality Weekly Report* 39:141 (March 9, 1990).

Murray, P. R. et al., eds. *Manual of clinical microbiology*, 6th edition. American Society for Microbiology, Washington, D.C. 1995. Chapters 107–109 have concise discussions, and clear photos and drawings of arthropods and parasitic worms.

"Cercarial dermatitis outbreak at a state park—Delaware, 1991." *Morbidity and Mortality Weekly Report* 41:225 (April 10, 1992) An outbreak of "swimmer's itch."

Kappus, K. K., et al. "Results of testing for intestinal parasites by state diagnostic laboratories, United States, 1987." In CDC Surveillance Summaries, December 1991, *MMWR 1991:* 40 (No. SS-4):25–47. Laboratories from 44 states reported *A. lumbricoides*, and 39 states reported hookworms. Many cases were in recent immigrants.

Schmidt, G. D., and Roberts, L. S. *Foundations of parasitology*, 4th ed. St. Louis: Times Mirror/Mosby College Publishing, 1989.

"Trichinella spiralis infection—United States, 1990." *Morbidity and Mortality Weekly Report* 40:57 (February 1, 1991).

"Progress Toward Global Eradication of Dracunculiasis." *Morbidity and Mortality Weekly Report* 44(47):875 (December 1, 1995). *Dracunculus medinensis* ("Guinea worm") pokes its uterus out through the skin and discharges millions of larvae into drinking water. In 1986, The World Health Assembly designated dracunculiasis to be the second disease (after smallpox) to be eradicated.

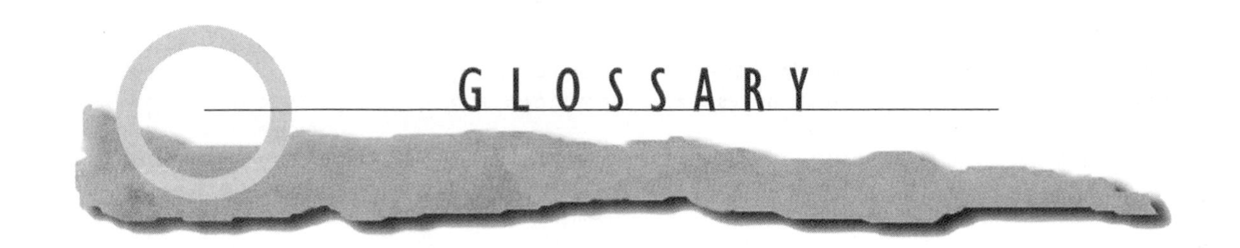

GLOSSARY

A

abscess A localized collection of pus within a tissue.

acid-fast stain Staining technique in which the organism resists decolorization with an acidic solution of alcohol after being stained with a basic dye.

acidophilic Preferring or requiring an acid environment.

actinomycetes Filamentous bacteria, many of which are valuable in the production of antibiotics.

activated macrophages Macrophages that have been stimulated by cytokines to become enlarged and metabolically active, with greatly increased capability to kill and degrade intracellular organisms and materials.

activated sludge method A method of sewage treatment in which wastes are degraded by complex populations of aerobic microorganisms.

activation energy The energy required to elevate molecules to an energy level at which they can react spontaneously.

active site The sequence of amino acids on an enzyme molecule to which the substrate binds, also known as the catalytic site.

active transport The energy-requiring process by which molecules are carried across cell boundaries. Molecules are often concentrated inside cells as a result.

ADCC (antibody-dependent cellular cytotoxicity) Nonspecific killing of target cells by certain cells, such as macrophages, granulocytes and natural killer cells, that contact the target via their Fc receptors binding to Fc of antibodies on the target.

adenosine diphosphate (ADP) The chemical that is formed when the phosphorus atom (P) is split from ATP and energy is released.

adenosine triphosphate (ATP) The principal chemical storage form of energy in cells.

adhesin A carbohydrate-specific binding protein used for adherence.

adjuvant A substance which increases the immune response to antigen.

ADP Abbreviation for adenosine diphosphate.

adsorption The attachment of one substance to the surface of another.

aerosol Material dispersed into the air as a fine mist.

aerotaxis Responding to oxygen.

aerotolerant Able to grow in the presence of O_2 though deriving no benefit from its presence.

aflatoxin Any of several toxins produced by molds such as *Aspergillus flavus*. Some of these are carcinogenic in animals.

agar A polysaccharide obtained from various species of seaweeds used to solidify microbiological media.

agarose A highly purified fraction of agar used in gel electrophoresis.

agglutination The clumping together of cells or particles.

alga (pl. algae) A primitive photosynthetic eukaryotic organism.

allergen An antigen that causes an allergy.

allergy Hypersensitivity, especially of the IgE-mediated type.

allosteric inhibition See *feedback inhibition.*

allosteric site The sequence of amino acids on an allosteric enzyme that binds small molecules and alters the function of the enzyme. It is distinct from the active or catalytic site.

alveolus (pl. alveoli) A small sac-like structure, such as the air-sacs of the lung.

amino acids The subunits of a protein molecule.

amino group $-NH_2$.

ammonification The reactions that result in the release of ammonia (NH_3) from organic nitrogen-containing molecules.

anabolic reactions All biosynthetic reactions in a living organism.

anaerobic jar A glass jar from which oxygen (O_2) can be removed to provide an environment for growing anaerobic microorganisms.

anaerobic organism (anaerobe) An organism that can only grow in the absence of O_2.

anaerobic respiration The metabolic process in which electrons are transferred from the electron transport chain to an inorganic acceptor molecule other than oxygen. The most common acceptor molecules are carbonate, sulfate, and nitrate.

anaerobiosis Living only in the absence of oxygen gas (O_2).

anamnestic response The enhanced immunological response to a second or subsequent dose of antigen.

anaphylatoxins Substances that can directly cause mast cells to release chemical mediators such as histamine from their granules; examples are $C3_a$ and $C5_a$.

anaphylaxis A generalized hypersensitive reaction to an allergen that can cause a profound drop in blood pressure.

anion A negatively charged ion.

anoxygenic A reaction that does not produce oxygen, such as occurs with certain groups of photosynthetic bacteria.

antibiotic A chemical substance produced by certain molds and bacteria that inhibits the growth of or kills other microorganisms.

antibody A protein produced by the body in response to a foreign substance that reacts specifically with that substance.

anticodon The sequence of three nucleotides on tRNA that binds to a complementary sequence on mRNA.

antigen A substance that can cause the production of specific antibodies and combine with those antibodies.

antigen-binding sites The hypervariable regions at the end of the two arms of an antibody molecule that recognize specific antigen; there are two identical antigen-binding sites on each monomer of antibody.

antigen presentation The process in which macrophages ingest and partially degrade antigens which then appear on the macrophage surface in a form that can react with CD-4 T lymphocytes; dendritic cells and B cells can also present antigen to T cells.

antigenic determinant The part of an antigen molecule that combines with the specific antibody.

antigenic drift The slight changes that occur in the antigens of a virus; specific antibodies made to the antigen before the change occurred are now only partially effective.

antigenic shift Major changes that can occur in the antigen of a virus.

antiseptic A substance that inhibits or kills microbes. This term often implies that the substance is sufficiently nontoxic, that it is safe to apply to the skin or other tissues.

apoptosis Programmed cell death, with enzymatic degradation of DNA.
archaebacteria A major group of bacteria that lack cell wall peptidoglycan and differ in many other respects from true bacteria.
aseptic Free of infectious agents; sterile.
asexually Reproduction not preceded by the union of cells or genetic exchange.
ATP The abbreviation for adenosine triphosphate.
attenuated Modified so as to be incapable of causing disease under ordinary circumstances.
autoclave A device employing steam under pressure used for sterilizing materials stable to heat and moisture.
autoimmunity An immune reaction against one's own tissues.
autotroph An organism that can use CO_2 as its main source of carbon.
auxotroph A laboratory-derived mutant of a microorganism that requires an organic growth factor.
avirulent Lacking disease-causing attributes, such as a capsule, for example.
a_w Abbreviation for water activity.
axial filaments Organ of motility found in spirochetes.

B

β-lactamase An enzyme that cleaves the β-lactam ring of penicillin and related antibiotics, destroying their antibacterial activity.
B cells The line of cells of the immune system that secrete antibody.
bacteremia Bacteria circulating in the bloodstream.
bactericidal Able to kill bacteria.
bacteriophage A virus that infects bacteria; often abbreviated "phage."
bacteriostatic Able to inhibit the growth of bacteria.
barophiles Bacteria that can grow under high pressure.
basal body The structure that anchors the flagella to the cell wall and plasma membrane.
base analog A compound that resembles a purine or pyrimidine base closely enough to be incorporated into DNA in place of a natural base.
basophil A kind of granulocyte characterized by large dark-staining cytoplasmic granules. Granulocytes are a subset of white blood cells.
benign tumor A noncancerous tumor.
binary fission An asexual reproduction process in which one cell splits into two independent daughter cells.
binomial system The system of naming each species of organism with two Latin words.
biochemical oxygen demand (BOD) A measure of the amount of biologically degradable organic material in water.
biomass The total weight of all organisms in any particular environment.
bioremediation The process whereby a microorganism is used to remove a pollutant.
biosphere The sum of all the regions of the earth where life exists.
biotechnology The use of microorganisms and other biological systems to aid in manufacturing various goods.
blastospore A fungal spore formed by budding from a parent cell.
BOD Abbreviation for biochemical oxygen demand.
budding An asexual reproductive technique that involves a pushing out of a part of the parent cell that eventually gives rise to a new daughter cell.
buffer A substance in a solution that acts to prevent changes in pH.
burst size The number of newly formed virus or bacteriophage particles released from a single cell.

C

cAMP Abbreviation for cyclic AMP.
cDNA DNA synthesized *in vitro* using an mRNA template and the enzyme reverse transcriptase.
cancer Groups of abnormally growing cells that can spread from their site of origin, also termed malignant tumors.
CAP Abbreviation for cyclic AMP-activating protein.
capsid Protein coat that surrounds the nucleic acid of a virus.
capsomere A subunit of the protein capsid of a virus.
capsule A gelatinous structure that surrounds some bacteria.
carbohydrate Compounds containing principally carbon, hydrogen, and oxygen atoms in a ratio of 1:2:1.
carboxyl group –COOH.
carbuncle A painful infection of the skin and subcutaneous tissues, manifests as a cluster of boils.
cariogenic Aiding the production of decay of a tooth or bone.
carrier A person who is colonized with a potential pathogen, but who is not ill, and can transmit that organism to others.
cascade In biology, a series of reactions which, once started, continues to the final step by each step triggering the next in a special order; activation of complement is an example.
catabolic Metabolic degradation of organic compounds.
catabolite Product of catabolism.
catabolite repression The inhibition of gene activity by an efficiently degraded catabolite.
catalase An enzyme which degrades hydrogen peroxide to oxygen and water.
catalyst A substance that speeds up the rate of a chemical reaction without being altered or depleted in the process.
catalytic site The sequence of amino acids on an enzyme molecule that is concerned with binding the substrate.
cations Positively charged ions.
CD antigens Abbreviation (see cluster of differentiation antigens).
CD-4 lymphocytes T lymphocytes bearing the CD-4 (cluster of differentiation) surface molecules; T helper cells are CD-4.
CD-8 cells T lymphocytes bearing the CD-8 cluster of differentiation surface molecules; T cytotoxic cells and some T suppressor cells are CD-8 cells.
cell-mediated immunity (CMI) Immune responses mediated by T lymphocytes.
cellulase An enzyme that degrades cellulose into glucose subunits.
cellulose A polymer of glucose subunits; principal structural component of plant cell walls.
central dogma of molecular biology The colloquial phrase given to the major theme of molecular biology that the flow of information follows the pattern: DNA to RNA to protein.
cerebrospinal Relating to the brain and spinal cord.
challenge In immunology, to give an antigen to provoke an immunologic response in a subject previously sensitized to the antigen.
chemoattractant A condition or substance that induces positive chemotaxis, meaning that an organism moves toward the substance, light, heat, etc.
chemoautotrophs Organisms that use inorganic chemicals as a source of energy and CO_2 as the major source of carbon.
chemoheterotrophs Organisms that use chemical energy and an organic source of carbon.
chemostat A device used to grow bacteria in the laboratory that allows nutrients to be added and waste products removed continuously.
chemotaxis Movement of an organism in response to chemicals in the environment.

chlamydospore A thick-walled asexual fungal spore formed within or at the end of a hypha.

chlorophylls Light-absorbing green pigments, necessary for photosynthesis.

chloroplast A plastid in eukaryotes that contains chlorophyll and is the site of photosynthesis and starch formation.

chocolate agar An agar medium enriched with blood and then heated under controlled conditions to release hemin and NAD from the blood cells.

chromatin Nucleoprotein structure in eukaryotic cells.

chromatophores A pigmented cell, or pigmented bodies within a cell.

chromosome The array of genes responsible for the determination and transmission of hereditary characteristics.

chronic Of long duration.

cilium (pl. cilia) A short, projecting hairlike organelle of locomotion, similar to a flagellum.

clonal selection The selection and activation of a lymphocyte by interaction of antigen and specific antigen receptor on the lymphocyte surface, causing the lymphocyte to proliferate to form an expanded clone.

clone A group of cells derived from a single cell.

cloning vehicle A replicating piece of DNA into which genes of interest are introduced for the purpose of cloning the DNA.

cluster of differentiation antigens (CD antigens) White blood cell surface molecules identified by specific monoclonal antibodies; used to distinguish subgroups of white blood cells.

CMI Abbreviation (see cell-mediated immunity).

coagulase An enzyme that clots plasma.

codon Three nucleotides that specify a particular amino acid.

coenzyme A small organic molecule that transfers small molecules from one enzyme to another.

coenzyme A A coenzyme that is involved in decarboxylation.

colonization Establishment of a site of reproduction of microbes on a material, animal, or person without necessarily resulting in tissue invasion or damage.

colony A population of bacterial cells arising from a single cell.

colony stimulating factor (CSF) A cytokine produced by certain T cells that cause cells to multiply.

commensalism A relationship between two organisms in which one partner benefits from the association and the other is unaffected.

common source epidemic Outbreak of disease due to contaminated food, water, or other single source of the infectious agent.

communicate To exchange information; cells communicate by means of chemical messages such as cytokines which send signals between cells.

communities All of the living organisms in a given area.

competent A condition in which a bacterial cell is capable of taking up and integrating high molecular weight DNA into its chromosome.

competitive inhibitor A compound that competes with the normal substrate for the active site on an enzyme, thereby inhibiting enzyme activity.

complement A system of serum proteins that act in sequence, producing certain biological effects concerned with inflammation, the immune response, and the lysis of cells.

complement fixation test A serological method in which antibody-antigen reactions are detected by the consumption (fixation) of complement.

complementary structures Two molecules that fit closely together because their composition allows weak bonds to form and hold the molecules together.

complex medium A medium for growing bacteria that has some ingredients that are of unknown chemical composition.

congenital A condition existing from the time of birth.

conjugation A mechanism of gene transfer in bacteria that involves cell-to-cell contact.

conjugative plasmid A plasmid that carries the genes for sex pili and can transfer copies of itself to other bacteria during conjugation.

conjunctiva The membrane that covers the inside of the eyelids and the surface of the eye.

constant region That part of the antibody molecule that does not vary in amino acid sequences among molecules of the same immunoglobulin class.

constitutive enzymes Enzymes that are synthesized at the same rate under all nutritional conditions.

covalent bond A strong chemical bond formed by the sharing of electrons between atoms.

crown gall tumor A tumor on a plant caused by *Agrobacterium* sp.

CSF Abbreviation for colony stimulating factor.

cyclic AMP (cAMP) A molecule important for transcription of certain genes.

cyclic AMP-activating protein (CAP) The protein that binds to cAMP to promote gene transcription.

cyst Dormant resting protozoan cell characterized by a thickened cell wall.

cytochromes Proteins that carry electrons usually as members of electron transport chains.

cytokines A general term for a type of protein released in response to antigen which influences other cells.

cytopathic effect Observable change in a cell *in vitro* produced by viral action such as lysis of the cell.

cytoplasmic membrane The flexible structure immediately surrounding the cytoplasm of all cells.

cytoskeleton The network of microtubules and microfilaments within a cell that help it maintain its shape.

cytotoxic Kills cells.

D

D value Abbreviation for decimal reduction time.

dark repair Enzymes of DNA repair that do not depend on visible light.

dark-field illumination A microscopy technique in which objects appear brightly illuminated against a dark background.

death phase The stage in which the number of viable bacteria in a population decreases at an exponential rate.

decimal reduction time The time required for 90% of the organisms to be killed at a given temperature.

decontamination Destruction of pathogenic microorganisms and viruses and their toxic products.

defective Refers to viruses which do not have all of the genes necessary for virus multiplication and maturation.

delayed hypersensitivity Hypersensitivity caused by cytokines released from sensitized T lymphocytes; reactions occur within 48 to 72 hours after exposure of a sensitized individual to antigen.

dementia A condition involving the brain characterized by loss of the ability to remember and reason as well as a change in personality.

denaturation The disruption of the three-dimensional structure of a protein molecule. The separation of the complementary strands of DNA.

deoxyribonucleic acid (DNA) The macromolecule in the cell that carries the genetic information.

dermis The portion of the skin lying just beneath the surface layers and containing nerves, blood vessels, and connective tissue.

diapedesis The movement of leukocytes from blood vessels into tissues in response to a chemotactic stimulus during inflammation.

diatomaceous earth A sedimentary soil composed largely of the skeletons of diatoms; contains large amounts of silicon.
diauxic growth The two step growth frequency observed when bacteria are growing in media containing two carbon sources.
differential agar medium An agar medium that can distinguish between groups of bacteria according to their different biological properties.
diffusion The movement of molecules from a region of high concentration to a region of low concentration.
dimer A polymer composed of two monomeric structural units.
dipicolinic acid A substance present in large amounts in bacterial endospores.
diploid Having two sets of chromosomes, the normal situation in the nonreproductive cells of higher organisms.
direct microscopic count Counting the number of bacteria using a microscope and counting chamber.
direct selection The technique of selecting mutants by plating organisms on a medium on which the desired mutants will grow but the parent will not.
disaccharide A carbohydrate molecule consisting of two monosaccharide molecules joined together.
disease A process resulting in tissue damage or alteration of function, producing symptoms noticeable by laboratory tests or physical examination.
disinfectant An agent used to kill pathogenic microbes without necessarily sterilizing the material.
disinfection The process of reducing or eliminating pathogenic microorganisms or viruses in or on a material so that they are no longer a hazard.
DNA Abbreviation for deoxyribonucleic acid.
DNA ligase An enzyme that joins short fragments of DNA.
DNA probe A piece of DNA, labeled in some manner, which is used to identify the presence of similar DNA by hybridizing to it.
DNA replication Duplication of DNA.
DNA-mediated transformation Process of gene transfer in which DNA is transferred as a "naked" molecule.
duodenum The first part of the small intestine, extending from the stomach to the jejunum.
dysentery A condition characterized by crampy abdominal pain and bloody diarrhea.

E

eclipse period The time during which viruses exist within the host cell separated into their protein and nucleic acid components.
ecosystem An environment and the organisms that inhabit it.
eczema A condition characterized by a blistery skin rash, with weeping of fluid and formation of crusts, usually due to an allergy.
edema Swelling of tissues caused by accumulation of fluid.
electron transport chain A series of electron carriers that transfer electrons from donors such as NADH to acceptors such as oxygen.
electrophoresis A technique for separating proteins based on the fact that proteins differ in their electrical charges and so will move at different rates in an electrical field.
ELISA Abbreviation for enzyme-linked immunosorbent assay.
embryonated chicken egg Chicken eggs that contain a developing chick.
end product The chemical compound which is the final product in the sequence of a metabolic pathway.
end product repression The inhibition of gene activity by the end product of a biosynthetic pathway.
endemic Constantly present in a population.
endocytosis The process by which cells take up particles by enclosing them in a vesicle pinched off from the cell membrane.
endoplasmic reticulum Internal membrane of a eukaryotic cell to which ribosomes are attached.
endospore A kind of resting cell, characteristic of a limited number of bacterial species, that is highly resistant to heat, radiation, and disinfectants.
endosymbiont An organism that lives within the cell of another organism in a stable, often symbiotic, association.
endotoxin A poisonous substance present in the cell walls of gram-negative bacteria.
enrichment culture A technique used for isolating an organism from a mixed culture by manipulating conditions so as to favor growth of the organism sought while minimizing growth of the other organisms present.
Entner-Doudoroff pathway A pathway that converts glucose to pyruvate and glyceraldehyde-3-phosphate by producing 6-phosphogluconate and then dehydrating it.
enterotoxin A poisonous substance, usually of bacterial origin, that acts on the intestinal lining cells to cause diarrhea and vomiting.
envelope The structure which surrounds the cytoplasm of the cell. It includes the capsule, cell wall, and cytoplasmic membrane.
enzyme A protein catalyst.
enzyme-linked immunosorbent assay (ELISA) A technique used for detecting and quantifying specific antigens or antibodies by using an antibody labeled with an enzyme.
eosinophil One of the granulocyte subsets of leukocytes.
eosinophilic Stainable by the red dye esoin; also a hypersensitivity response characterized by eosinophilic granulocytes.
EPA Abbreviation for Environmental Protection Agency, a federal agency.
epidemic Affecting an unusually large number of individuals within a region or population.
epidemiology The study of factors influencing the frequency and distribution of diseases.
epidermis The outermost skin layers.
episome A DNA molecule that can exist either covalently bonded to the bacterial chromosome or in an extrachromosomal state.
epitope An area of the antigen molecule that stimulates the production of, and combines with, specific antibodies.
ergot A poisonous substance produced by a fungus that causes rye smut.
erythrocytes Red blood cells.
ester linkage A bond found in lipids and nucleic acids, in which –OH is bonded to –COOH with the removal of HOH.
ethidium bromide A colored dye that intercalates between bases of DNA; it is a mutagen and a carcinogen.
eubacteria True bacteria as distinguished from archaebacteria.
eukaryote A complex cell type differing from a prokaryote mainly in having a nuclear membrane.
eutrophic An aquatic system high in nutrients.
eutrophication Nutrient enrichment leading to the overproduction of algae.
exanthema A skin rash.
excision repair A mechanism of DNA repair in which a fragment of single-stranded DNA containing mismatched bases is cut out.
exocytosis The process whereby material is expelled from a cell; the reverse of endocytosis.

exoenzymes Enzymes found outside the cytoplasmic membrane of the cell, in the periplasm or in the extracellular medium.

exotoxin A soluble poisonous protein substance released by a microorganism.

extracellular Outside the cell membranes.

extremophiles Organisms that live under extremes of temperature, barometric pressure, or other environmental conditions.

F

F^- cell Recipient bacterial cell in conjugation.

F^+ cell Donor bacterial cell in conjugation.

F' cell A bacterial cell containing a plasmid that is a combination of the chromosome and the F factor.

F factor Extrachromosomal DNA that codes for sex pilus synthesis.

F plasmid A plasmid found in donor cells of *E. coli* that codes for sex pili biosynthesis. Can be transferred to recipient cells.

facilitated diffusion A type of diffusion of small molecules into cells that does not require energy.

facultative anaerobe An organism that grows best in the presence of oxygen (O_2), but can grow in its absence.

facultatively extracellular An organism that can grow either within or outside another cell.

FAD Abbreviation for flavin adenine dinucleotide.

family A taxonomic group between order and genus.

fastidious Exacting; used to refer to organisms that require growth factors.

fatty acid Long hydrocarbon chains ending in a carboxyl group.

Fc portion of antibody Crystallizable end of the constant region of an immunoglobulin molecule; responsible for binding to Fc receptors on cells, for initiating the classical pathway of complement activation and other biological functions.

Fc receptors Receptors on cells that bind the Fc portion of certain immunoglobulins.

feedback inhibition The inhibition of the first enzyme of a biosynthetic pathway by the end product of the pathway. Also known as allosteric or end product inhibition.

fermentation The metabolic process in which the final electron acceptor is an organic compound.

filament A threadlike structure.

filamentous phage A bacteriophage that has a threadlike appearance.

fimbriae Pili.

fix Strongly attach, as microorganisms are attached to a microscope slide, for example.

flagellum (pl. flagella) A long whiplike organelle of locomotion made up of intertwined molecules of protein.

flavin adenine dinucleotide (FAD) A derivative of the vitamin riboflavin.

follicle A structure resembling a tiny sac or tube, as the skin structure containing a hair; lymphoid follicles are collections of lymphoid cells in a sac-like structure.

fomites Inanimate objects such as books, tools, or towels that can act as transmitters of pathogenic microorganisms or viruses.

foreign Nonself.

fragmentation A form of asexual reproduction in which a filament composed of a string of cells breaks apart, forming multiple reproductive units.

fungus (pl. fungi) An organism of the Kingdom Fungi, a nonchlorophyllic, eukaryotic heterotroph.

fungistatic Able to inhibit the growth of fungi.

furuncle A boil; a localized skin infection that penetrates into the subcutaneous tissue, usually caused by *S. aureus*.

G

GALT Abbreviation (see gut-associated lymphoid tissue).

gametes Haploid cells that fuse with other gametes to form the diploid zygote in sexual reproduction.

gamma globulin A portion of blood serum proteins that is separated by electrophoresis and contains most of the immunoglobulins of the blood.

gamma rays One of three types of rays given off by radioactive substances.

gas chromatography A technique of separating and identifying gaseous components of a substance.

gastroenteritis An acute inflammation of the stomach and intestines; term often applied to the syndrome of nausea, vomiting, diarrhea, and abdominal pain.

GC content The guanine and cytosine content of a DNA molecule.

gel A substance having the consistency of gelatin.

gene A subunit of a chromosome; one gene codes for one polypeptide chain.

gene cloning A procedure by which genes are inserted into a replicon such as a plasmid or bacteriophage which is then introduced into cells where the replicon can replicate.

gene library The sum total of all of the genes of an organism that have been inserted into cloning vectors.

generalized transducing phage A bacteriophage that is capable of transferring any part of the bacterial chromosome from one cell to another. (By contrast, a specialized transducing phage transfers only specific parts of the genome.)

generation time The time required for one cell to divide into two.

genetic code The composition and sequence of all of the sets of three nucleotides that code for the amino acids.

genetic engineering The process of deliberately altering an organism's genetic information by changing its nucleic acid sequences.

genetic recombination The joining together of genes from different organisms.

genome All of the genetic material in a cell.

genotype The entire genetic makeup of a cell. (Contrast with phenotype.)

genus (pl. genera) A category of related organisms, usually containing several species. The first name of an organism in the Binomial System of Classification.

germicide An agent that kills microorganisms and viruses.

germination The sum total of the biochemical and morphological changes that an endospore or other resting cell undergoes before becoming a vegetative cell.

giant cell A very large cell with many nuclei, formed by the fusion of many macrophages during a chronic cell-mediated response; found in granulomas.

global repressor A regulatory protein involved in the control of several different biosynthetic or degradative pathways.

glycogen A polysaccharide composed of glucose molecules.

glycolysis Breakdown of carbohydrates. The initial sequence by which carbohydrates are degraded; often called the Meyerhof Embden Parnas pathway.

glycoproteins Proteins with sugar molecules attached.

Golgi apparatus The structure in eukaryotic cells involved in the secretion of proteins.

Gram stain A staining technique that divides bacteria into one of two groups, Gram-positive or Gram-negative, on the basis of color. Among eubacteria, the staining reaction correlates well with cell wall structure.

granulocytes White blood cells characterized by the presence of prominent granules; basophil granules stain dark with basophilic dyes, eosinophils stain bright red with eosinophilic dyes, and neutrophils do not take up either stain in their granules.

granuloma Found in a chronic cell-mediated response, collections of lymphocytes and stages of macrophages, especially giant cells and activated macrophages; an attempt by the body to wall off and contain persistent organisms and antigens.

gut-associated lymphoid tissue (GALT) Part of the mucosa-associated lymphoid tissue, found in the intestines including Peyer's patches, the appendix and some lymph nodes.

H

halogen Any of a group of chemically related non-metallic elements including fluorine, chlorine, bromine and iodine.

halophile An organism that prefers or requires a high salt (NaCl) medium.

haploid Condition in which each type of chromosome is represented only once.

hapten Substance that can combine with specific antibodies but cannot incite the production of those antibodies unless it is attached to a large carrier molecule.

helper T lymphocytes CD4+ lymphocytes that provide assistance needed for other cells such as B lymphocytes and macrophages to carry out their functions; T helper cells usually act by producing cytokines that in turn act on the other cells.

hemagglutinin A substance that causes red blood cells to clump together.

hematopoietic stem cell Fetal cells that give rise to cells of the blood such as leukocytes, erythrocytes, and platelets.

hemin A brownish-red iron-containing compound crystallized from hemoglobin.

hemorrhage Bleeding.

herd immunity The inability of an infectious disease to spread in a population because of the lack of a critical concentration of nonimmune hosts.

heterofermenters Bacteria that produce various substances such as ethanol, CO_2, lactic, and other acids from glucose. (Contrast with homofermenters.)

heterotroph An organism that obtains energy and its carbon source from organic compounds.

Hfr cells (high frequency of recombination cells) Rare cells in the F^+ population that can transfer their chromosome to an F^- cell.

histamine H substance found in basophil and mast cell granules that upon release can cause dilation and increased permeability of blood vessel walls and other effects; a mediator of inflammation.

histones A type of protein tightly bound to chromosomal DNA in eukaryotic cells.

HLA Abbreviation for human leukocyte antigen.

holoenzyme The total enzyme consisting of all its subunits.

homofermenters Bacteria that convert glucose into lactic acid as the end product of fermentation.

homologous, homology Matching in characteristics; in genetics, one of the pair of genes or chromosomes that carry information for the same trait(s).

homologous recombination Genetic recombination between the same genes as opposed to heterologous recombination.

hook A part of the bacterial flagellum.

host An organism on or in which smaller organisms or viruses live, feed, and reproduce.

host range Refers to the range of cell types that can be infected by a pathogen.

HTST Abbreviation for high temperature short time pasteurization.

human leukocyte antigen (HLA) A group of human cell surface antigens.

humus A dark-colored highly complex organic material in soil that is not readily degraded by microorganisms.

hybridoma A cell made by fusing a lymphocyte, such as an antibody-producing B cell, with a cancer cell.

hydrogen bond A weak attraction between an atom that has a strong attraction for electrons and a hydrogen atom that is covalently bonded to another atom that attracts the electron of the hydrogen atom.

hydrolytic reaction A chemical reaction in which a complex substance such as protein is broken down into its subunits by the addition of water molecules.

hydrophobic bonds Weak bonds formed between molecules as a result of their mutual repulsion of water molecules.

hypersensitivity Also termed allergy. A heightened immune response to antigen.

I

IFN Abbreviation (see interferons).

immune complex A complex of antigen and antibody bound together, often with some complement components included.

immunity The ability of a host to resist foreign invaders.

immunodeficiency The inability to produce a normal immune response to antigen.

immunoelectrophoresis The technique for separating proteins by subjecting the mixture to an electric current followed by diffusion and precipitation in gels using antibodies against the separated proteins.

immunofluorescence A technique used to identify particular antigens microscopically in cells by the binding of a fluorescent antibody to the antigen.

immunoglobulin See *antibody*.

immunological tolerance The inability of the immune system to respond to a given antigen as a result of prior exposure to the antigen.

incidence rate The number of new cases of a disease within a specific time period in a given population.

index case The first identified case of a disease in an epidemic.

indicator medium An agar medium containing a component that is changed in a recognizable way by a particular species of microorganism that helps identify the organism.

indirect selection A technique for isolating mutants by determining which organisms do not grow on a certain medium on which the parents do grow. This often involves replica plating.

induced mutation A mutation which results from the organism being treated with an agent that changes its DNA.

inducer A substance that activates transcription of certain genes.

inducible enzyme An enzyme synthesized only when a substrate on which it can act is present. The synthesis of the enzyme is induced.

induction The process by which a prophage is excised from the host cell DNA; activation of gene transcription.

infection Invasion of tissues by microorganisms or viruses with or without the production of disease.

infectious dose The number of microorganisms or viruses sufficient to establish an infection. Varies from a few to several million, depending on the virulence of the infectious agent and the host defenses.

inflammation A nonspecific response to injury characterized by redness, heat, swelling, and pain in the infected area.

innate immunity Immunity that is not affected by prior contact with infectious agent or other material involved; immunity that is not directly caused by lymphocytes.

inserted repeat A sequence of nucleotides on one strand of DNA which is identical to DNA on another strand when both are read in the same direction i.e., 5′ to 3′. They are associated with transposable elements.

insertion sequence (IS) A piece of DNA about the length of a gene that has the ability to move from one site on a DNA molecule to another. The simplest type of transposable element.

intercalating agents Agents that insert themselves between two nucleotides in opposite strands of a DNA double helix.

interferons A group of antiviral glycoproteins produced by certain cells in response to viral infection.
interleukins A group of proteins produced by macrophages and T cells that regulate growth and differentiation of lymphocytes.
intranuclear inclusion body A structure found within the nucleus of cells infected with certain viruses such as the cytomegalovirus.
intron A part of the eukaryotic chromosome that does not code for a protein. It is removed from mRNA before the mRNA is translated.
ion An atom or group of atoms that carry a positive or negative electrical charge. They gain the charge by giving up or capturing an electron from a nearby atom.
ionic bond Noncovalent bonds resulting from the strong attraction of an atom for electrons of other atoms.
IS Abbreviation for insertion sequence.
isomer A molecule that has the same number and types of atoms as another but differs in its structure.
isotope A form of an element that differs in atomic weight from the form most common in nature. The isotope is usually unstable and decomposes spontaneously.
in vitro In a test tube or other container as opposed to inside a living plant or animal.
in vivo Inside a living plant or animal as opposed to a test tube or other container.

K

keratin A durable protein found in hair, nails, and skin.
kinetic energy The energy of motion.
kingdom The highest category in taxonomic classification.
Koplik's spots Lesions of the oral cavity caused by measles virus that resemble a grain of salt on a red base.
Kupffer cells Macrophages of the liver.
kuru A fatal central nervous system disease thought to be caused by a prion.

L

lactose A disaccharide consisting of one molecule of glucose and one of galactose.
lag phase The stage in the growth of a bacterial culture characterized by extensive macromolecule and ATP synthesis but no increase in the number of viable cells.
langerhans cells Antigen-presenting dendritic cells of the skin, similar in some respects to macrophages.
latent Currently inactive but capable of becoming active; frequently refers to viral infections.
latent infection An infection where the infectious agent is present but not active and remains so until activated at a later time.
leukemia A cancer of the leukocytes (white blood cells).
leukocidins Substances that kill white blood cells.
leukocytes White blood cells.
leukotrienes Substances active in inflammation, leading to chemotaxis and increased vascular permeability; produced by mast cells, basophils and macrophages.
L-forms Bacterial variants that have lost the ability to synthesize the peptidoglycan portion of their cell wall. The defect can be partial or complete and may or may not be reversible.
light repair The process by which bacteria repair UV damage to their DNA; only occurs in the presence of light.
lipid One of a diverse group of organic substances all of which are relatively insoluble in water, but soluble in alcohol, ether, chloroform, or other fat solvents.
lipid A The toxic portion of lipopolysaccharide outer membrane in Gram-negative bacteria.
lipopolysaccharide (LPS) Molecule formed by bonding of lipid to polysaccharide; a part of the outer membrane of Gram-negative bacteria.
lipoprotein The macromolecule formed by bonding lipid to protein.
lithoheterotroph An organism that uses an inorganic compound for energy and an organic compound as a source of carbon.
log phase The stage of growth of a bacterial culture in which the cells are multiplying exponentially.
LTLT The abbreviation for low temperature long time pasteurization.
lymph A clear yellow liquid that flows within lymphatic vessels; it generally contains lymphocytes and may contain globules of fat.
lymphocyte Small, round, or oval white blood cell with a large nucleus and a small amount of cytoplasm; involved in specific immunity.
lymphokines Proteins secreted by lymphocytes that act as intracellular mediators of the immune response.
lysate The remains of cells and virions which are released after lysis of cells.
lysogenic cell Same as a lysogen.
lysogenic conversion The change in the properties of bacteria as a result of their carrying a prophage.
lysogens Bacteria that carry a prophage integrated into their chromosome.
lysosomes Saclike structure in the cytoplasm of a phagocyte which contains enzymes that are capable of destroying microbial invaders.
lysozyme An enzyme that degrades the peptidoglycan layer of bacterial cell wall.
lytic Degradative, leading to destruction.

M

M protein A heat- and acid-resistant protein found in the cell walls of Group A streptococci.
macromolecule A large molecule made up of subunits bonded together in a characteristic fashion.
macrophage A large mononuclear phagocyte, derived from the blood monocyte, which can engulf and destroy many microorganisms and other extraneous materials, can function as an antigen-presenting cell, and can carry out ADCC (antibody-dependent cellular cytotoxicity).
magnetotaxis Movement by bacterial cells containing magnetite crystals in response to a magnetic field.
major histocompatibility complex (MHC) A cluster of genes coding for key cell surface proteins important in cell-to-cell recognition processes.
malarious Referring to the presence of malaria in an area.
malignant tumor Abnormal growth of cells no longer under normal control, which has the potential to spread to other parts of the body.
MALT Abbreviation (see mucosal-associated lymphoid tissue).
mast cell A cell type abundant in connective tissues; it resembles the blood basophil, with prominent basophilic granules and Fc surface receptors for IgE.
MBC Abbreviation for minimal bacterial concentration.
median The middle value of a series of values arranged in ascending or descending order.
mediator A chemical that causes a particular tissue or cell type to react; examples include substances released by nerve cells that cause muscle to contract, and substances such as histamine released from mast cells due to antigen-antibody reactions.
medium (pl. media) Any material used for growing organisms.
meiosis The process in eukaryotic cells by which the chromosome number is reduced from diploid (2N) to haploid (1N).
membrane filter Thin filter generally made of a synthetic polymer having a carefully controlled pore size, which is used to remove microorganisms, and in some cases viruses, from liquids and gases.

membrane transport proteins Proteins in the cytoplasmic membrane which are involved in the active transport of compounds into cells.
memory (anamnestic) response The secondary response (to a second or subsequent dose of antigen) which produces a faster, larger, and longer lasting response than does a primary response to the antigen.
meninges The membranes covering the brain and spinal cord.
messenger In immunology, a substance released by one type of cell that causes another type of cell to react in a specific way.
mesophiles Bacteria which grow most rapidly at temperatures between 20° and 50° C.
messenger RNA (mRNA) Single-stranded RNA synthesized during transcription from DNA that binds to ribosomes and directs the synthesis of protein.
metabolism The sum total of all the chemical reactions in a cell.
metabolite Any product of metabolism.
metachromatic granules Polyphosphate granules found in the cytoplasm of some bacteria that appear as different colors when stained with a basic dye.
metachromatic staining Situation in which binding of a dye to substances within a cell causes a color change in the dye.
MHC Abbreviation for major histocompatibility complex.
MIC Abbreviation for minimal inhibitory concentration.
microaerophilic organisms Organisms that require for growth lower than normal concentrations of oxygen.
microtubules Hollow protein fibers present in the cytoplasm of eukaryotic cells and in their flagella and cilia, important in cell shape, movement, and division.
mineralization The conversion of an element from an organic to an inorganic form.
minimal bactericidal concentration (MBC) Lowest concentration of an antibiotic or other substance killing a microorganism.
mitochondria Highly complex structures about the size of bacteria found in eukaryotic cells; site of energy generation.
mitogen A substance that induces mitosis; a substace that causes a proliferation of cells.
mitosis A process of chromosome duplication in eukaryotic cells.
MMWR An abbreviation for *Morbidity and Mortality Weekly Report*, published by the Centers for Disease Control (CDC).
modification enzyme An enzyme that alters the purine or pyrimidine bases of a DNA molecule, thereby preventing digestion by restriction endonucleases.
mole The amount of a chemical in grams that is equal to the atomic weights of all the atoms in a molecule of that chemical and contains 10^{23} molecules.
molecular weight The relative weight of an atom or molecule based on a scale in which the H atom is assigned the weight of 1.0.
molecule The smallest part of a compound that retains all the properties of the compound.
monoclonal antibodies Antibodies with a single specificity produced *in vitro* by lymphocytes that have been fused with a type of malignant cell called a myeloma cell.
monocyte Immature macrophage produced in the bone marrow and found in blood.
monolayer A single layer of cells, as for example on the container surface of a tissue cell culture.
monomer The repeating subunit of a polymer.
mononuclear phagocyte A defense cell of the body that has an unsegmented nucleus; one of the "professional phagocytes."
mononuclear phagocyte system A system made up of monocytes of the blood and various types of tissue macrophages, found in all parts of the body; formerly known as the reticuloendothelial system.
monosaccharide A sugar. A simple carbohydrate generally having the formula $C_nH_{2n}O_n$, where n can vary in number from three to eight. The most common are five and six.
morbidity rate The number of cases of a specific disease per given population at risk.
morphology The form or shape of a particular organism or structure.
mortality rate The fraction of people who die from a given disease.
mucociliary escalator The moving layer of mucus and cilia lining the respiratory tract that traps bacteria and other particles and moves them into the throat.
mucosal-associated lymphoid tissue (MALT) Lymphoid tissue present in the mucosa of the respiratory, gastrointestinal and genitourinary tracts.
mucus The slimy material that covers the moist surfaces of the body.
mucous membrane A mucus-covered thin layer of tissue that covers a body surface or lines a body cavity.
multiple cloning site A small sequence of DNA that contains several unique restriction enzyme recognition sites into which foreign DNA can be cloned.
murmur A short sound heard over the heart or blood vessels, recurring with the heartbeat, that differs from sounds arising from normal heart valves and heart muscle contraction. A murmur may or may not indicate pathology, depending on its characteristics.
mutagen Any agent that increases the frequency at which DNA is altered (mutated).
mutation A modification in the base sequence of DNA in a gene resulting in an alteration in the protein coded by the gene.
mutualism An association in which both partners benefit.
mycelium (pl. mycelia) A tangled, matlike mass of fungal hyphae.
mycorrhiza A fungus that grows in a symbiotic relationship with certain plant roots.
mycosis (pl. mycoses) A disease caused by a fungus.

N

NAD Abbreviation for nicotinamide adenine dinucleotide, a derivative of the vitamin niacin which is important in the transfer of hydrogen atoms in metabolic processes.
naked capsid virus A virus that has no envelope.
narrow host range plasmid A plasmid that only replicates in one or a few closely related species of bacteria.
necrotic Dead; used when referring to dead cells or tissues that are in contact with living cells as necrotic tissue in wounds.
neurotransmitter Any of a group of substances released from the terminations of nerve cells when they are stimulated, that cross to the adjacent cell and cause it to be excited or inhibited.
neutrophil A type of granulocytic leukocyte in which the granules do not stain intensely.
nitrification The conversion of NH_3 to nitrate (NO_3^-).
nitrogen fixation The conversion of nitrogen gas to ammonia.
NK cell A large granular, non-T, non-B lymphocyte which can recognize and destroy certain cells it senses as foreign in a nonspecific fashion.
nodule A swelling on the root of a leguminous plant caused by Rhizobium; the site of nitrogen fixation.

nomenclature In biology refers to the system by which organisms are named.
non-conjugative plasmid A plasmid which lacks some of the genetic information required for its transfer to other bacteria by conjugation.
normal microbial flora That group of microorganisms that colonizes the body surfaces but does not usually cause disease. Also known as normal flora.
nosocomial infection An infection acquired during hospitalization.
nuclear membrane The membrane that separates the nucleus from the cytoplasm of the cell; only found in eukaryotic cells.
nucleic acid hybridization A technique in which single strands of DNA are mixed together and allowed to associate. The degree to which the single strands associate indicates how similar the strands are.
nucleic acids Ribonucleic acid (RNA) and deoxyribonucleic acid (DNA).
nucleocapsid Viral nucleic acid and its protein coat.
nucleotides The basic subunits of ribonucleic or deoxyribonucleic acid consisting of a purine or pyrimidine covalently bonded to ribose or deoxyribose, which is covalently bound to a phosphate molecule.
nucleus The membrane-bound organelle in a eukaryotic cell that contains chromosomes and nucleolus.

O

obligate aerobes Those organisms that have an absolute requirement for oxygen.
obligate anaerobes Those organisms that can only be grown in the absence of oxygen (O_2). Some members of this group are killed by traces of oxygen gas.
obligately intracellular parasites Organisms that grow only inside living cells.
oligotrophic environment An environment that is deficient in nutrients.
oligotrophs Organisms that can grow in a nutrient deficient environment.
oncogenes A gene whose activity is involved in turning a normal cell into a cancer cell.
operator region The region of DNA that combines with an active repressor molecule.
operon A group of linked genes which are controlled as a single unit.
opportunist An organism that causes disease only in hosts with impaired defense mechanisms, for example, from wounds, alcoholism, AIDS, etc. Also called opportunistic pathogens.
opsonins Substances, generally antibody or complement, that aid phagocytosis.
opsonization Enhanced phagocytosis, usually caused by coating of the particle to be ingested with either antibody or complement components.
order Taxonomic classification between class and family.
organ A structure composed of different tissues coordinated to perform a specific function.
organelle A structure within a cell that performs a specific function.
organic matter Material that contains carbon atoms bonded to other carbon atoms.
osmosis The movement of water across a selectively permeable membrane from a dilute solution to a more concentrated solution.
osmotic pressure The pressure exerted by water on the cytoplasmic membrane as a result of a difference in the concentration of molecules on each side of the membrane.
oxidation The removal of an electron.
oxidative phosphorylation The generation of energy in the form of ATP that results from the passage of electrons through the electron transport chain to a final electron acceptor.
oxygenic photosynthesis Photosynthetic reaction that synthesizes carbohydrate from CO_2 and H_2O with the release of oxygen.

P

PABA Abbreviation for para-aminobenzoic acid, a growth factor for certain bacteria; a precursor of the vitamin folic acid.
palindromic sequence Same as inverted repeat.
pandemic A worldwide epidemic.
parasite An organism that lives on another organism (the host) and gains at the expense of the host.
passive diffusion A process in which molecules flow freely into and out of a cell so that the concentration of any particular molecule is the same on the inside as it is on the outside of the cell.
pasteurization The process of heating food or other substances under controlled conditions of time and temperature to kill pathogens and reduce the total number of microorganisms without damaging the substance.
pathogen An organism that can cause disease.
pathogenesis The process by which disease develops.
pathogenic Disease causing.
pathogenicity The ability to cause disease.
penicillin An antibiotic that affects the synthesis of the peptidoglycan portion of bacterial cell walls.
penicillinase An enzyme that destroys penicillin.
pentamer A polymer composed of five monomeric structural units.
peptide or P site The first site on the ribosome to which the tRNA binds to begin translation.
peptide Any of a group of low molecular weight compounds characterized by yielding two or more amino acids upon hydrolysis.
peptidoglycan layer The rigid backbone of the bacterial cell wall, composed of repeating subunits of N-acetylmuramic acid and N-acetylglucosamine and a number of amino acids.
perforin Molecule produced by T cytotoxic cells and NK cells; functions in killing target cells by forming a pore through the target cell membrane.
periplasm, periplasmic gel The very narrow gel that lies between the cell wall and the cytoplasmic membrane in Gram-negative bacteria.
permeases A system of enzymes concerned with the transport of nutrients into the cell.
Petri dish A two-part dish of glass or plastic often used to contain medium solidified with agar, on which bacteria are grown.
Peyer's patches Collections of lymphoid cells in the gastrointestinal tract, mostly in the small intestine; part of the gut-associated lymphoid tissue (GALT) and mucosal-associated lymphoid tissue (MALT).
pH A scale of 0 to 14 that expresses the acidity or alkalinity of a solution.
phage Abbreviation for bacteriophage.
phage induced Proteins coded by phage DNA.
phagocytes Cells that protect the host by ingesting and destroying foreign particles such as microorganisms and viruses.
phagocytosis, phagocytize Cellular ingestion of particulate materials within membrane-bound vesicles. The particles are engulfed by pseudopods of the cell.
phagolysosome The vesicle that results from the fusion of a phagosome with a lysosome.
phagosome A phagocytic vesicle.
phase variation A reversible change in antigenicity of a structure, such as the flagella of *Salmonella* species.
phenotype The sum total of the observable properties of an organism resulting from the interaction of the environment with the genotype of the organism.

phospholipid A lipid that has a phosphate molecule as part of its structure. One part of the molecule is soluble in aqueous solvents; the other part in nonaqueous solvents.

photoautotrophs Organisms that use light as the energy source and CO_2 as the major carbon source.

photoheterotrophs Organisms that use light as the energy source and organic compounds as the carbon source.

photooxidation A chemical reaction occurring as a result of absorption of light energy in the presence of oxygen.

photosynthesis The sum total of the reactions that certain bacteria and other organisms use to utilize light energy and synthesize cell components.

phototaxis Responding to light.

phototrophs Organisms that use light as their energy source.

phytoplankton The floating and swimming algae and photosynthetic prokaryotic organisms of lakes and oceans.

pili Hairlike appendages on many Gram-negative bacteria. These function in conjugation and for attachment.

pinocytosis Cellular ingestion of liquids, a process similar to phagocytosis.

plankton Primarily microscopic organisms floating freely in most waters.

plaque A clear area in a monolayer of cells. Viral plaques are created by lysis of infected cells within the clear area; in dentistry, a collection of bacteria that adhere to a tooth surface.

plasma The fluid portion of blood, exclusive of cells.

plasma cell The end cell of the B cell series, fully differentiated to produce and secrete large amounts of antibody.

plasma membrane The semipermeable membrane that surrounds the cytoplasm in a cell.

plasmid Small extrachromosomal circular DNA molecules that replicate independently of the chromosome. They often code for antibiotic resistance.

pleomorphic Refers to bacteria that occur in various shapes.

PMN Abbreviation for polymorphonuclear neutrophils.

polar, polarity The degree of affinity that an atom has for electrons. This results in positive or negative charges on atoms in a molecule.

polymer Large molecules formed by the joining together of repeating small molecules (subunits).

polymorphic Having different distinct forms at various stages of the life cycle.

polymorphonuclear neutrophil (PMN) Phagocytic cells that together with the macrophages are known as "professional phagocytes." The nuclei of these cells are segmented and composed of several lobes.

polypeptide A chain of amino acids joined together by peptide bonds; also known as a protein.

polysaccharide Long chains of monosaccharide subunits.

porins Proteins in the outer membrane of Gram-negative bacteria that form channels for small molecules to pass.

pour-plate method A method of inoculating an agar medium with bacteria while the agar is liquid and then pouring it into a Petri dish where the agar hardens. The colonies grow both on the surface and within the medium.

precipitation reaction The reaction of an antibody with a soluble antigen to form an insoluble substance.

precipitin The antibody responsible for causing the precipitation of an antigen.

prevalence rate The total number of cases of a disease in a given population.

primer An RNA molecule that initiates the synthesis of DNA.

prion An infectious protein that has no nucleic acid.

prokaryote, prokaryotic A cell that is characterized by the lack of a nuclear membrane, and the absence of mitochondria and other membrane-bound organelles.

promoter The region on DNA at the start of a gene to which RNA polymerase binds to begin transcription.

propagated epidemic Outbreak of contagious disease in which the infectious agent is transmitted to others, resulting in steadily increasing numbers of people becoming ill.

prophage The latent form of temperate phage that has been inserted into the host's DNA.

prosthecae Extensions of the cytoplasm and cell wall of some bacteria, which increase the surface area in contact with the external environment.

protein A macromolecule containing one or more polypeptide chains.

Protein A A protein produced by *Staphylococcus aureus* which inhibits phagocytosis of the organism by binding to the Fc portion of antibodies.

protoplast A bacterium or plant cell that has no cell wall.

prototroph An organism that has no organic growth requirements other than a source of carbon and energy.

protozoa A group of motile, generally non-photosynthetic, single-celled eukaryotic organisms.

provirus The latent form of a virus in which the viral DNA is incorporated into the chromosome of the host.

pseudopodia Extensions of cells such as occurs in ameba that aid in locomotion.

psychrophile A microorganism that grows best between –5° and 20°C.

pure culture A culture that contains only a single strain of an organism.

purine A component of RNA and DNA. There are two major purines—adenine and guanine.

pus Thick, opaque, often yellowish fluid that forms at the site of an infected wound.

pyogenic Pus producing.

pyrimidine A component of RNA and DNA. There are three major pyrimidines—thymidine, cytosine, and uracil.

Q

quantal assay A method for determining the concentration of a virus.

quaternary ammonium compounds A type of chemical that has four organic groups attached to a central nitrogen atom; used as a disinfectant.

R

receptors Attachment sites on a cell surface.

recognition In immunology, the interaction between antigen and specific antigen receptors on B and T lymphocytes.

recombinant DNA molecules DNA molecules that were joined together *in vitro*.

recombinants Cells containing DNA molecules that are derived from two different sources.

reduction The process of adding electrons and hydrogen atoms to a molecule.

regulatory gene A gene which functions in the control of the rate of synthesis of other gene products.

replica plating A technique for the simultaneous transfer from one medium to other media of organisms representing a large number of separate colonies.

replication fork The Y-shaped structure where DNA separates and new strands of DNA are synthesized.

replicon A piece of DNA that is capable of replicating. It contains an origin of replication.

repressible pathway A pathway in which the enzymes are not synthesized when the end product of the pathway is present.

repression The inhibition of gene function.

repressor A protein that binds to the operator site and prevents transcription.
reservoir Source of a disease-producing organism.
resistance factor (R factor) A plasmid that carries genetic information for resistance to one or more chemotherapeutic agents.
resistance genes (R genes) The genes on the R factor that code for resistance to chemotherapeutic agents.
resistance transfer factor (RTF) A set of genes associated with R factors that are involved in the transfer of R factors from one bacterium to another.
resolving power In microscopy, the minimum distance that can exist between two points such that the points are observed as separate.
respiration The sum total of metabolic steps in the degradation of foodstuffs when the electron acceptor is an inorganic compound.
restriction endonuclease An enzyme that recognizes and cuts DNA at specific purine and pyrimidine sites.
restriction enzyme Same as restriction endonuclease.
retroviruses A group of viruses which carry their genetic information as single-stranded RNA. They have the enzyme reverse transcriptase, which forms a copy of DNA that is then integrated into the host cell chromosome.
reverse transcriptase An enzyme that synthesizes double-stranded DNA complementary to an RNA template.
reversion The process by which a second mutation corrects a defect caused by an earlier mutation.
ribonucleic acid (RNA) The macromolecules in a cell that play a role in converting the information coded by the DNA into amino acid sequences in protein.
ribose A 5-carbon sugar found in RNA.
ribosomes Cellular "workbenches" on which proteins are synthesized.
RNA Abbreviation for ribonucleic acid.
mRNA messenger RNA
rRNA ribosomal RNA
tRNA transfer RNA
RNA polymerase An enzyme that catalyzes the synthesis of RNA from ribonucleoside triphosphates, using DNA as a template. It is composed of a number of subunits. Its action represents the first step in protein synthesis.
RTF Abbreviation for resistance transfer factor.

S

salinity The amount of salt in a solution.
SALT Skin-associated lymphoid tissues.
sanitize The process of substantially reducing the microbial populations on objects to acceptably safe public health levels.
scrapie The common name for a neurological disease of sheep thought to be caused by a prion.
sebum The oily secretion of the sebaceous glands of the skin.
selectable marker A gene that encodes a selectable phenotype such as antibiotic resistance.
selective medium A culture medium that favors the growth of a specific microorganism.
selective toxicity The ability of a chemotherapeutic agent to kill a pathogen without harming the host.
self In immunology, relating to the antigenic components of an individual's own tissues.
self-assembly The spontaneous formation of a complex structure from its component molecules without the aid of enzymes.
semiconservative replication The type of replication of DNA in which new daughter strands complementary to each of the original strands are synthesized in each round of replication.
semipermeable A material which allows the passage of some but not other molecules. This usually depends on the size of the molecules.
septum A wall.
serotonin A basic amine, stored in mast cell granules, that upon release can cause contraction of some smooth muscles and increased vascular permeability; a mediator of inflammation.
serum The fluid portion of blood that remains after blood clots.
sex pilus A thin protein appendage required for attachment of one bacterium to another during conjugation in bacteria.
shingles (herpes zoster) A condition resulting from the reactivation of the varicella-zoster virus.
signature sequence A nucleotide sequence in 16S ribosomal RNA which is different in eukaryotes, eubacteria, and archaebacteria, but the same within the members of each group.
SOS repair A complex, inducible repair process that is used to repair damaged DNA.
southern hybridization A technique devised by Dr. E. Southern for measuring the degree of similarity in DNA by measuring the ability of single strands to reassociate after being dissociated.
species In the binomial nomenclature scheme, the second name given to an organism.
specific In immunology, relating to the unique affinity of an antibody or immune lymphocyte for its corresponding antigen.
specific immunity The immune response that depends on the recognition and elimination of foreign antigens by specialized lymphocytes.
spike A protein on the surface of a virus that is important in attachment to the host cell.
splicing In genetics, rejoining the ends of a genetic strand following removal or introns.
spontaneous generation A discredited theory that organisms can arise from nonliving matter.
spontaneous mutation A mutation that occurs naturally without the addition of mutagenic agents.
spore A specialized reproductive cell.
sporogenesis In bacteria, a complex, highly-ordered sequence of morphological changes during which a bacterial vegetative cell produces a specialized cell greatly resistant to environmental adversity. Eukaryotes can also go through a process of sporogenesis.
static Not moving.
stationary phase The stage of growth of a culture in which the number of viable cells remains constant.
sterilization Rendering an object or substance free of all viable microorganisms or viruses.
steroid A type of lipid which has a specific four-membered ring structure.
sterol A type of lipid that has a characteristic structure. Cholesterol is a common sterol.
strain A population of organisms that has characteristics differing from others within the species.
streak plating A way of spreading out bacteria on the surface of a Petri dish filled with agar medium that results in formation of isolated colonies.
stroke A sudden severe attack; often used to indicate sudden paralysis resulting from brain injury due to a plugged blood vessel.
subcutaneous Below the skin.
substrate The substance on which an enzyme acts to form the product; surface on which an organism will grow.
substrate level phosphorylation The transfer of the high-energy phosphate from a phosphorylated compound to ADP to form ATP.
sucrose A disaccharide consisting of a molecule of glucose bonded to fructose, i.e., common table sugar.
superantigens Substances that react with a part of the T cell receptor not concerned with antigen recognition, thereby stimulating activity nonspecifically in many T cells and leading to illness and sometimes fatal shock.

superoxide (O_2^-) An especially toxic form of oxygen.

superoxide dismutase The enzyme responsible for degrading the superoxide anion.

symbiont An organism that lives in a state of symbiosis.

symbiosis The living together of two dissimilar organisms or symbionts.

synthetic medium A medium in which every component is known in terms of its chemical composition and quantity.

T

taxa The groups into which organisms are classified.

taxonomy The study of biological classification.

T cells A type of lymphocyte formed in the bone marrow that matures in the thymus. T cells are involved in the immune response.

T cell receptors Surface molecules on T lymphocytes, containing hypervariable regions, that recognize antigens specifically; non-immunoglobulin antigen-recognition molecules on T cells.

T cytotoxic cells CD8+ lymphocytes that recognize antigen on the surface of target cells and kill the target cells.

T dependent antigen An antigen that requires help from helper T cells in order to induce an antibody response.

TDT Abbreviation for thermal death time.

temperate phage A bacteriophage that can either become integrated into the host cell DNA as a prophage or replicate outside the host chromosome leading to cell lysis.

template The molecule that serves as the mold for the synthesis of another molecule.

T helper cells CD4+ lymphocytes that produce cytokines, influencing and helping the specific immune response.

therapeutic index The ratio of minimum toxic dose to minimum effective dose of a medication.

thermal death time The shortest period of time needed to kill all the organisms in a microbial population at a specific temperature and under defined conditions.

thermophile An organism that can grow at temperatures above 55°C.

T independent antigen An antigen that can induce an antibody response without the help of T cells.

tissue A group of similar cells that associate into a structural subunit and perform a specific function.

tissue culture A culture of plant or animal cells that grow in an enriched medium outside the plant or animal.

titer A measure of the concentration of a substance in solution; for example, the amount of a specific antibody in serum, usually measured as the highest dilution of serum that will test positive for antibody.

toxic Poisonous.

toxin A poisonous chemical substance.

toxoid A modified form of a toxin which is no longer toxic but is able to stimulate the production of antibodies that will neutralize the toxin.

trachea A tube composed of cartilage and membranous material that branches to become the bronchi and conveys air from the throat to the lungs.

transamination The reaction in which an ammonium ion is transferred from one amino acid to an α-keto acid, resulting in the synthesis of an amino acid from the α-keto acid.

transcription The process of transferring genetic information coded in DNA into messenger RNA (mRNA).

transient microbial flora Microbial flora that does not reside permanently in or on a particular site.

translation The process by which genetic information in the messenger RNA directs the order of amino acids in protein.

transplacental Crossing the placenta, as when an infectious agent enters the circulation of a fetus from the mother's bloodstream.

transposable element A gene that moves from one DNA molecule to another within the same cell or from one site on a DNA molecule to another site on the same molecule. Also called a transposon.

transposition The movement of a piece of DNA from one site in a molecule to another site in the same cell.

trauma A wound or injury, physical or mental.

trophozoite The vegetative feeding form of some protozoa.

T suppressor cells T lymphocytes that specifically suppress immune responses, often CD8+ T lymphocytes.

tubercle A granuloma formed in tuberculosis.

tumor A growth of tissue resulting from abnormal cell growth.

turbidity The cloudiness resulting from a suspension of materials such as bacteria in a liquid.

tyndallization Repeated cycles of heating and incubation to kill spore-forming bacteria.

U

UHT Abbreviation for ultra high temperature pasteurization.

ultra high temperature pasteurization (UHT) A method of rapid pasteurization.

ultraviolet light (UV) Electromagnetic radiation with wavelengths between 175 and 350 nm. The light is invisible.

urticaria Hives; an allergic skin reaction characterized by the formation of itchy red swellings.

UV Abbreviation for ultraviolet light.

V

vaccination The use of a preparation of living or dead microorganisms, viruses, or their components to immunize an individual.

vaccine A preparation of living or dead microorganisms or viruses or their components used to immunize a person or animal against a particular disease.

vaginitis Inflammation of the vagina.

vasopharynx The part of the throat that lies above the roof of the mouth.

vector An organism such as a mosquito that transmits disease. In genetic engineering, a circular self-replicating DNA molecule such as a plasmid or virus into which additional genes are introduced *in vitro*. These new genes are carried into the host cell with the vector and replicate with the vector.

vegetative cell The multiplying or feeding form of a cell.

vehicle An inanimate carrier of an infectious agent from one host to another.

vesicle A small membrane-bound body inside a cell. In medicine, a blister.

virion A viral particle.

viroid A piece of RNA that does not have a protein coat but does replicate within living cells.

virulence The relative ability of a pathogen to overcome body defenses and cause disease. Also, properties of microorganisms that assist pathogenicity.

virus A nonliving, submicroscopic infectious agent.

viscosity Relative stickiness of a fluid.

vitamin One of a group of organic compounds found in small quantities in natural foodstuffs that are necessary for the growth and reproduction of an organism. Vitamins are usually converted into coenzymes.

water activity (a_w) A quantitative measure of the water available.

wavelength The distance between adjacent peaks of two light waves.

wide host range plasmid A plasmid that can infect and multiply in a large number of different bacteria.

wild type The form of an organism that is isolated from nature.

Y

yeast A unicellular fungus.

Z

zoonosis A disease of animals that can be transmitted to humans.

zooplankton The floating and swimming small animals and protozoa found in marine environments, usually found in association with the phytoplankton.

Z value The increase in temperature required to reduce the decimal reduction time to one-tenth of its initial value.

zygote The diploid cell formed by the sexual fusion of two haploid cells.

PRONUNCIATION KEY

FOR BACTERIAL, FUNGAL, PROTOZOAN, AND VIRAL NAMES

A

Acetobacter (a-see′-toe-back-ter)
Achromobacter (a-krome′-oh-back-ter)
Acinetobacter calcoaceticus (a-sin-et′-oh-back-ter kal-koh-ah-see′-ti-kus)
Actinomyces israelii (ak-tin-oh-my′-seez iz-ray′-lee-ee)
Actinomycetes (ak-tin-oh-my′-seats)
Adenovirus (ad′-eh-no-vi-rus)
Agrobacterium tumefaciens (ag-rho-bak-teer′-ee-um too-meh-faysh′-ee-enz)
Alcaligenes (al-ka-li′-jen-ease)
Amoeba (ah-mee′-bah)
Arbovirus (are′-bow-vi-rus)
Archezoa (ar-key-zoe′-ah)
Aspergillus niger (ass-per-jill′-us nye′-jer)
Aspergillus oryzae (ass-per-jill′-us or-eye′-zee)
Azolla (aye-zol′-lah)
Azotobacter (ay-zoh′-toe-back-ter)

B

Bacillus anthracis (bah-sill′-us an-thra′-siss)
Bacillus cereus (bah-sill′-us seer′-ee-us)
Bacillus coagulans (bah-sill′-us coh-ag′-you-lans)
Bacillus fastidiosus (bah-sill′-us fas-tid-ee-oh′-sus)
Bacillus stearothermophilus (bah-sill′-us steer-oh-ther-maw′-fill-us)
Bacillus subtilis (bah-sill′-us sut′-ill-us)
Bacillus thuringiensis (bah-sill′-us thur′-in-jee-en-sis)
Bacteroides (back′-ter-oid′-eez)
Baculovirus (back′-you-low-vir-us)
Beggiatoa (beg-gee-ah-toe′-ah)
Beijerinckia (by-yer-ink′-ee-ah)
Bordetella pertussis (bor-deh-tell′-ah per-tuss′-iss)
Borrelia burgdorferi (bor-real′-ee-ah berg-dor′-fir-ee)
Bradyrhizobium (bray-dee-rye-zoe′-bee-um)
Branhamella (bran-ham-el′-lah)
Brucella abortus (bru-sell′-ah ah-bore′-tus)

C

Campylobacter jejuni (kam′-peh-low-back-ter je-june′-ee)
Candida albicans (kan′-did-ah al′-bi-kanz)
Caulobacter (caw′-loh-back-ter)
Ceratocystis ulmi (see′-rah-toe-sis-tis ul′-mee)
Chlamydia trachomatis (klah-mid′-ee-ah trah-ko-ma′-tiss)
Claviceps purpurea (kla′-vi-seps purr-purr′-ee-ah)
Clostridium acetobutylicum (kloss-trid′-ee-um a-seat-tow-bu-til′-i-kum)
Clostridium botulinum (kloss-trid′-ee-um bot-you-line′-um)
Clostridium difficile (kloss-trid′-ee-um dif′-fi-seal)
Clostridium perfringens (kloss-trid′-ee-um per-frin′-gens)
Clostridium tetani (kloss-trid′-ee-um tet′-an-ee)
Coccidioides immitis (cock-sid-ee-oid′-eez im′-mi-tiss)
Coronavirus (kor-oh′-nah-vi-rus)
Corynebacterium diphtheriae (koh-ryne′-nee-bak-teer-ee-um dif-theer′-ee-ee)
Coxsackievirus (cock-sack-ee′-vi-rus)
Cryptococcus neoformans (krip-toe-cock′-us knee-oh-for′-manz)
Cytophaga (sigh-taw′-fa-ga)

D

Desulfovibrio (dee-sul-foh-vib′-ree-oh)

E

Eikenella corrodens (eye-keh-nell′-ah kor-roh′-denz)
Entamoeba histolytica (en-ta-mee′-bah his-toh-lit′-ik-ah)
Enterobacter aerogenes (en′-ter-oh-back-ter)
Enterovirus (en′-ter-oh-vi-rus)
Epidermophyton (eh-pee-der′-moh-fy-ton)
Escherichia coli (esh-er-ee′-she-ah koh′-lee)

F

Filobasidiella neoformans (fee-loh-bah-si-dee-ell′-ah knee-oh-for′-manz)
Flavivirus (flay′-vih-vi-rus)
Flavobacterium (flay-vo-back-teer′-ee-um)
Francisella tularensis (fran-siss-sell′-ah tu-lah-ren′-siss)
Frankia (frank′-ee-ah)
Fusobacterium (fu′-zoh-back-teer-ee-um)

G

Gallionella (gal-ee-oh-nell′-ah)
Gardnerella vaginalis (gard-nee-rel′-lah va-jin-al′-is)
Giardia intestinalis (jee-are′-dee-ah in-test′-tin-al-is)
Giardia lamblia (jee-are′-dee-ah lamb′-lee-ah)
Gluconobacter (glue-kon-oh-back′-ter)
Gonyaulax (gon-ee-ow′-lax)
Gymnodinium breve (jim-no′-dee-um brev′-eh)

H

Haemophilus influenzae (hee-moff′-ill-us in-flew-en′-zee)
Helicobacter pylori (he′-lih-koh-back-ter pie-lore′-ee)
Hepadnavirus (hep-ad′-nah-vi-rus)
Hepatitis virus (hep-ah-ti′-tis vi-rus)
Herpes simplex (her′-peas sim′-plex)
Herpes zoster (her′-peas zoh′-ster)
Histoplasma capsulatum (his-toh-plaz′-mah cap-su-lah′-tum)
Hyphomicrobium (high-foh-my-krow′-bee-um)

I

Influenza virus (in-flew-en′-za vi-rus)

K

Klebsiella pneumoniae (kleb-see-ell′-ah new-moan′-ee-ee)

L

Lactobacillus brevis (lack-toe-ba-sil′-lus breh′-vis)
Lactobacillus bulgaricus (lack-toe-ba-sil′-lus bull-gair′-i-kus)
Lactobacillus casei (lack-toe-ba-sil′-us kay′-see-ee)
Lactobacillus helveticus (lack-toe-ba-sil′-us hell-veh′-ti-kus)
Lactobacillus plantarum (lack-toe-ba-sil′-us plan-tar′-um)
Lactobacillus sanfranciscans (lack-toe-ba-sil′-us san-fran-sis′-kanz)

Lactobacillus thermophilus (lack-toe-ba-sil′-us ther-mo′-fil-us)
Legionella pneumophila (lee-jon-ell′-ah new-moh-fill′-ah)
Leptospira interrogans (lep-toe-spire′-ah in-ter-roh′-ganz)
Leuconostoc citrovorum (lew-kow-nos′-tok sit-ro-vor′-um)
Listeria monocytogenes (lis-tear′-ee-ah mon′-oh-sigh-to-jen′-eze)

M

Malassezia (mal-as-seez′-e-ah)
Methanobacterium (me-than′-oh-bak-teer-ee-um)
Methanococcus (me-than-oh-ko′-kus)
Microsporum (my-kroh-spore′-um)
Mobiluncus (moh-bi-lun′-kus)
Moraxella catarrhalis (more-ax-ell′-ah kah-tah-rah-liss)
Moraxella lacunata (more-ax-ell′-ah lak-u-nah′-tah)
Mucor (mu′-kor)
Mycobacterium leprae (my-koh-bak-teer′-ee-um lep-ree)
Mycobacterium tuberculosis (my-koh-bak-teer′-ee-um too-ber-kew-loh′-siss)
Mycoplasma pneumoniae (my-koh-plaz′-mah new-moan′-ee-ee)

N

Neisseria gonorrhoeae (nye-seer′-ee-ah gahn-oh-ree′-ee)
Neisseria meningitidis (nye-seer′-ee-ah men-in-jit′-id-iss)
Neurospora sitophila (new-rah′-spor-ah sit-oh-phil′-ah)

O

Orthomyxovirus (or-thoe-mix′-oh-vi-rus)
Oscillatoria (os-sil-la-tor′-ee-ah)

P

Papillomavirus (pap-il-oh′-ma-vi-rus)
Parainfluenza virus (par-ah-in-flew-en′-zah vi-rus)
Paramecium (pair′-ah-mee′-see-um)
Paramyxovirus (par-ah-mix′-oh-vi-rus)
Parvovirus (par′-vo-vi-rus)
Pasteurella multocida (pass-ture-ell′-ah mul-toe-sid′-ah)
Pasteurella piscicida (pass-ture-ell′ah pis-si-si′-dah)
Pediococcus soyae (ped-ih-ko′-kus soy′-ee)
Penicillium camemberti (pen-eh-sill′-ee-um cam-em-bare′-tee)
Penicillium caseicolum (pen-eh-sill′-ee-um cay-see-ih-kol′-um)
Penicillium roqueforti (pen-eh-sill′-ee-um roke′-for-tee)
Peptostreptococcus (pep′-to-strep-to-ko-kus)
Phytophythora infestans (fy′-toe-fy-thor-ah in-fes′-tanz)
Picornavirus (pi-kor′-na-vi-rus)
Plasmodium falciparum (plaz-moh′-dee-um fall-sip′-air-um)
Plasmodium malariae (plaz-moh′-dee-um ma-lair′-ee-ee)
Plasmodium ovale (plaz-moh′-dee-um oh-vah′-lee)
Plasmodium vivax (plaz-moh′-dee-um vye-′vax)
Pneumocystis carinii (new-mo-sis′-tis car′-nee-ee)
Poliovirus (poe′-lee-oh-vi-rus)
Polyoma virus (po-lee-oh′-mah vi-rus)
Propionibacterium acnes (proh-pee-ah-nee-bak-teer′-ee-um ak′-neez)
Propionibacterium shermanii (proh-pee-ah-nee-bak-teer′-ee-um sher-man′-ee-ee)
Proteus mirabilis (proh′-tee-us mee-rab′-il-us)
Pseudomonas aeruginosa (sue-dough-moan′-ass aye-rue′-gin-oh-sa)

R

Rabies virus (ray′-bees vi-rus)
Retrovirus (re′-trow-vi-rus)
Rhabdovirus (rab′-doh-vi-rus)
Rhinovirus (rye′-no-vi-rus)
Rhizobium (rye-zoh′-bee-um)
Rhizopus nigricans (rise′-oh-pus nye′-gri-kanz)
Rhizopus stolon (rise′-oh-pus stoh′-lon)
Rhodococcus (roh-doh-koh′-kus)
Rickettsia rickettsii (rik-kett′-see-ah rik-kett′-see-ee)
Rotavirus (row′-tah-vi-rus)
Rothia (roth′-ee-ah)
Rubella virus (rue-bell′-ah vi-rus)
Rubeola virus (rue-bee-oh′-la vi-rus)

S

Saccharomyces carlsbergensis (sack-ah-row-my′-sees karls-berg-en′-en-siss)
Saccharomyces cerevisiae (sack-ah-row-my′-sees sara-vis′-ee-ee)
Saccharomyces rouxii (sack-ah-roh-my′-sees roos′-ee-ee)
Salmonella choleraesuis (sall-moh-nell′-ah koh′-er-ay-sues)
Salmonella enteritidis (sall-moh-nell′-ah en-ter-it′-id-iss)
Salmonella typhi (sall-moh-nell′-ah tye′-fee)
Salmonella typhimurium (sall-moh-nell′-ah tye-fi-mur′-ee-um)
Serratia marcescens (ser-ray′-sha mar-sess-sens)
Shigella dysenteriae (shig-ell′-ah diss-en-tair′-ee-ee)
Spirillum minus (spy-rill′-um my′-nus)
Sporothrix schenckii (spore′-oh-thrix shenk-ee-ee)
Staphylococcus aureus (staff-ill-oh-kok′-us aw′-ree-us)
Staphylococcus epidermidis (staff-ill-oh-kok′-us epi-der′-mid-iss)
Streptobacillus moniliformis (strep-tow-bah-sill′-us mon-ill-i-form′-is)
Streptococcus agalactiae (strep-toe-kock′-us a-ga-lac′-tee-ee)
Streptococcus cremoris (strep-toe-kock′-us kre-more′-iss)
Streptococcus durans (strep-toe-kock′-us dur′-anz)
Streptococcus faecalis (strep-toe-kock′-us faye′-ka-liss)
Streptococcus lactis (strep-toe-kock′-us lak′-tiss)
Streptococcus mutans (strep-toe-kock′-us mew′-tanz)
Streptococcus pneumoniae (strep-toe-kock′-us new-moan′-ee-ee)
Streptococcus pyogenes (strep-toe-kock′-us pie-ah-gen-ease)
Streptococcus salivarius (strep-toe-kock′-us sal-ih-vair′-ee-us)
Streptococcus sanguis (strep-toe-kock′-us san′-gwis)
Streptococcus thermophilis (strep-toe-kock′-us ther-moh′-fill-us)
Streptomyces griseus (strep-toe-my′-seez gree′-see-us)

T

Thiobacillus (thigh-oh-bah-sill′-us)
Torulopsis (tore-you-lop′-siss)
Treponema pallidum (tre-poh-nee′-mah pal′-ih-dum)
Trichomonas vaginalis (trick-oh-moan′-as vag-in-al′-iss)
Trichophyton (trick-oh-phye′-ton)
Trypanosoma brucei (tri-pan′-oh-soh-mah bru′-see-ee)

V

Varicella-zoster virus (var-ih-sell′-ah zoh′-ster vi-rus)
Veillonella (veye′-yon-ell-ah)
Vibrio anguillarum (vib′-ree-oh an-gwil-air′-um)
Vibrio cholerae (vib′-ree-oh kahl′-er-ee)

Y

Yersinia enterocolitica (yer-sin′-ee-ah en-ter-koh-lih′-tih-kah)
Yersinia pestis (yer-sin′-ee-ah pess′-tiss)

Photographs

Part Openers, Brief Contents, Table of Contents

1: © Tom E. Adams/Peter Arnold; **2:** © George Lower/Photo Researchers; **3:** © Dennis Kunkel/Phototake; **4:** © Manfred Kage/Peter Arnold

Chapter 1

Opener: © Stevie Grand/SPL/Photo Researchers, Inc.; **1.1:** © Science Vu/Visuals Unlimited; **1.2:** Courtesy of J.P. Dalmasso; **1.3:** © Bruce Iverson; **1.5A:** Courtesy of A. Progulske and S. Holt; **1.6A:** Courtesy of Dr. Thomas Fritsche; **BX1.31:** Courtesy of Esther R. Angert and Norman R. Pace, Indiana University; **1.7:** © E. White/Visuals Unlimited; **1.8:** Courtesy of Professor John Fuerst; **1.9A:** Courtesy of L. Santo; L. Santo, H. Hohl, and H. Frank, *J. Bacteriol.* 99:824, 1969; **1.9B:** © T.J. Beveridge/Visuals Unlimited; **1.10A:** © Alex Rakosy/Custom Medical Stock Photo; **1.10B:** © Cabisco/Visuals Unlimited; **1.10C:** © V. Burmeister/Visuals Unlimited; **1.11A:** © A.M. Siegelman/Visuals Unlimited; **1.11B:** © C. Shih & R. Kessel/Visuals Unlimited; **1.12A:** © Kwang Jeon/Visuals Unlimited; **1.12B:** © R. Kessel & C. Shih/Visuals Unlimited; **1.13A:** © K.G. Murti/Visuals Unlimited; **1.13B:** © J.L. Carson/Custom Medical Stock Photo; **1.13C:** © K.G. Murti/Visuals Unlimited; **1.14:** Dr. Diemer/U.S. Dept. of Agriculture

Chapter 2

Opener: © Scott Camazine/Photo Researchers, Inc.; **BX2.11A, B:** Courtesy of David M. Prescott

Chapter 3

Opener: © T. Beveridge/BPS/Tony Stone Images; **3.1:** Photograph courtesy of Leica, Inc., Deerfield, IL; **3.2A, B:** Courtesy of W. A. Jensen; **3.3A:** © E. Chan/Visuals Unlimited; **3.3B:** © J.F. Gennaro and L.R. Grillone/Photo Researchers, Inc.; **3.3C:** Courtesy of Dr. Thomas R. Fritsche, M.D., Ph.D., Clinical Microbiology Division, University of Washington, Seattle; **3.3D:** © Michael Abbey/Photo Researchers, Inc.; **3.3E:** Evans Roberts; **3.3F:** © R. Kessel & C. Shih/Visuals Unlimited; **3.3G:** © Ralph A. Slepecky/Visuals Unlimited; **3.3H:** © H. Aldrich/Visuals Unlimited; **3.3I:** © David M. Phillips/Visuals Unlimited; **3.3J:** © Science Vu/Visuals Unlimited; **3.5A, B:** © Le Beau/Custom Medical Stock Photo; **3.6A:** Courtesy of Mary Bicknell, University of Washington, Seattle; **3.7B:** © Leon J. LeBeau/Biological Photo Service; **3.8:** © John D. Cunningham/Visuals Unlimited; **3.9:** Courtesy of Dr. John Sherris, Univ. of Washington, Seattle; **3.10A, B, D:** © David M. Phillips/Visuals Unlimited; **3.10C:** © Cabisco/Visuals Unlimited; **3.11A:** From Walther Stoeckenius *Walsby's Square Bacterium: Fine Structures of an Orthogonal Procaryote*; **3.11B:** Courtesy of J.T. Staley; **3.13A:** Courtesy of J. Moore, *J. Protozool.* 17:672, 1970, Society of Protozoologists; **3.14A:** © S. C. Holt/Biological Photo Service; **3.15A:** © George J. Wilder/Visuals Unlimited; **3.15B, C:** Courtesy of J.P. Dalmasso; **3.16A:** © Chris Bjiornberg/Photo Researchers, Inc.; **3.16B:** Courtesy of K.J. Cheng and J.W. Costerton; **3.16D:** Courtesy of A. Progulske and S.C. Holt, *Journal of Bacteriology* 143:1003–1018, 1980; **3.17:** Courtesy of D. Birdsell; **3.19C:** © S.C. Holt/Biological Photo Service; **3.21C:** © John J. Cardamone, Jr./Biological Photo Service; **3.23:** Courtesy of E.S. Boatman; **3.24A:** From C. Rensen, F. Valois, and S. Watson, *Journal of Bacteriology* 94:422–433; **3.25:** Courtesy of S. Watson; photograph taken by J. Watebury, *Journal of Bacteriology* 95:1910–1920. American Society for Microbiology; **3.32A:** © Fred Hossler/Visuals Unlimited; **3.32B:** © Science VU/Visuals Unlimited; **3.32C:** © E. Chan/Visuals Unlimited; **3.36:** © D. Balkwill and D. Maratea/Visuals Unlimited; **3.37:** Courtesy of C. Brinton, Jr.; **3.38:** Courtesy Harley W. Moon, USDA; **3.39A:** © CNRI/SPL/Photo Researchers, Inc.; **3.39B:** © G. Chapman/S. Zane/Visuals Unlimited; **3.40A:** Courtesy of R. Kavenoff; R. Kavenoff and O. Ryder, *Chromosoma* 55:13–25, 1976; **3.41A:** Courtesy of Dr. Jack Griffith; **3.42A:** Courtesy of James Lake, from Scientific American 245: 94–97, 1981; **3.43:** Courtesy of E.S. Boatman; **3.44A:** Courtesy of J.F.M. Hoeniger, J. Bacteriol. 96:1835; **3.44B:** © A.M. Siegelman/Visuals Unlimited; **3.45B1:** Courtesy of L. Santo; L. Santo, H. Hohl, and H. Frank, *J. Bacteriol.* 99:824, 1969; **3.46A:** © H.S. Pankratz, T.C. Beaman, P. Gerhardt/Biological Photo Service; **3.47A:** Courtesy J. Moore, *Protozool.* 17:672, 1970, Society of Protozoologists; **3.48A:** © D. Phillips/Visuals Unlimited; **3.49A:** © M. Powell/Visuals Unlimited; **3.50A:** © M. Schliwa/Visuals Unlimited; **3.51B:** © Dr. Gopal Murtis/Photo Researchers, Inc.; **3.53B:** Courtesy of D. Portnoy; **3.54A:** Courtesy of A.L. Olins; A.L. Olins and D.E. Olins, *Science* 183:330, 1974. © 1974 by the AAAS; **3.55:** © Biophoto Associates/Science Source/Photo Researchers, Inc.; **3.56:** © Garry T. Cole, Ph.D., Dept. of Botany, Univ. of Texas/Biological Photo Service

Chapter 4

Opener: © A. Brooks/Tony Stone Images; **4.12B:** © Jack M. Bostrack/Visuals Unlimited; **4.14:** Courtesy of Dr. Edward Bottone; **4.15:** © Fred Hossler/Visuals Unlimited; **4.16:** © Michael Gabridge/Visuals Unlimited; **4.19:** Courtesy of Center for Biofilm Engineering, MSU-Bozeman, Paul Stoodley et al.

Chapter 5

Opener: © Superstock

Chapter 6

Opener: © D. Struhters/Tony Stone Images

Chapter 7

Opener: © Dr. Gopal Murti/SPL/Custom Medical Stock Photo; **BX7.11:** © Matt Meadows/Peter Arnold, Inc.; **7.24:** Courtesy of Dr. Jack Griffith; **7.25:** Courtesy of C. Brinton, Jr. and J. Carnahan

Chapter 8

Opener: © R. Tully/Tony Stone Images; **8.3:** Courtesy of C.L. Hanney; Reproduced by Permission of the National Research Council of Canada. From *Candadian Journal of Microbiology* 1:694, 1955; **8.4:** Courtesy of Milt Gordon; **8.11:** Reprinted with permission from Edvotek, Inc.

Chapter 9

Opener: © B. Ayers/Tony Stone Images; **9.3A, B:** Evans Roberts; **9.6A, B:** © Pall/Visuals Unlimited; **9.7A, B, 9.8:** Evans Roberts

Chapter 10

Opener: © Meckes/Ottawa/Photo Researchers; **10.1A, B:** Courtesy of J. William Schopf; from J. William Schopf, *Science* 260:640–646; **10.8A:** © George Wilder/Visuals Unlimited; **10.8B, C:** Courtesy of Dr. Thomas R. Fritsche, M.D., Ph.D., Clinical Microbiology Division, University of Washington, Seattle; **10.9A:** © Harry Pulschen/Custom Medical Stock Photo; **10.9B:** © E. Koneman/Visuals Unlimited; **10.10A:** © John Cunningham/Visuals Unlimited; **10.10B:** © Richard L. Moore/Biological Photo Service; **10.11A, B:** © Ray Otero/Visuals Unlimited

Chapter 11

Opener: © F. Widdell/Visuals Unlimited; **11.1B1:** From R.K. Nauman, S.C. Holt, and C.D. Cox, *J. Bacteriol*. 98:264, 1969. American Society for Microbiology; **11.2:** Courtesy of B.L. Williams; **11.3:** Courtesy of J.T. Staley and J.P. Dalmasso; **11.4:** © Alfred Pasieka/Peter Arnold, Inc.; **11.6:** Evans Roberts; **11.7A, C, D:** From H.L. Saddoff, E. Berke, and B. Loperfida, J. Bacteriol.105:184, 1971. American Society for Microbiology; **11.9A, B:** Courtesy of J.T. Staley and J.P. Dalmasso; **11.10:** Courtesy of Dr. Willy Burgdorfer, Scientist Emeritus, Rocky Mountain Laborabories, NIAID, Hamilton, Montana; **11.11A, B:** © Alfred Pasieka/Peter Arnold, Inc.; **11.12:** Courtesy of R.R. Friis, *J. Bacteriol* 110:706, 1972. American Society for Microbiology; **11.13, 11.14, 11.15A:** Courtesy of J.T. Staley and J.P. Dalmasso; **11.15B:** Courtesy of Dr. James Staley; **11.17A, B, C:** Courtesy of Mary F. Lampe; **11.18:** Courtesy of J.T. Staley and J.P. Dalmasso; **11.19:** Courtesy of H. Douglas; E. Duchow and H. Douglas, *J. Bacteriol*. 58:409, 1949. American Society for Microbiology; **11.20:** © T.E. Adams/Visuals Unlimited; **11.21, 11.23:** Courtesy of J.T. Staley and J.P. Dalmasso; **11.24:** © Richard L. Moore/Biological Photo Service; **11.25A:** © Bill Beatty/Visuals Unlimited; **11.25B, C:** Courtesy of J.M. Shivley; **11.26A:** © Fred E. Hossler/Visuals Unlimited; **11.26B:** © Richard L. Moore/Biological Photo Service; **11.26C:** © Alex Rakosy/Custom Medical Stock Photo; **11.26D:** © Fred Hossler/Visuals Unlimited; **11.27A:** © Erick Grave/Photo Researchers, Inc.; **11.27B:** © Soad Tabaqchali/Visuals Unlimited; **11.28:** © John Walsh/Photo Researchers, Inc.; **11.30:** From J.C. Esign and R.S. Wolfe, *J. Bacteriol*. 87: 924, 1964. American Society for Microbiology; **11.31:** © John D. Cunningham/Visuals Unlimited; **11.34A, B, C:** Evans Roberts; **11.34D:** © Howard Berg/Visuals Unlimited; **11.35A:** Courtesy of M.P. Lechevalier and H.A. Lechevalier.; **11.35B, C:** From M.W. Rancourt and H.A. Lechevalier, *Can. J. Microbiol*.10:311–316, 1964. National Research Council of Canada; **11.36:** Courtesy of E. Boatman; **11.37:** © Michael Gabridge/Visuals Unlimited; **11.38:** © David M. Phillips/Visuals Unlimited; **11.39A:** © F. Widdel/Visuals Unlimited; **11.39B:** © Ralph Robinson/Visuals Unlimited; **11.40:** © Leon J. LeBeau/Biological Photo Service; **11.1A1, B:** From R.K. Nauman, S.C. Holt, and C.D. Cox, *J. Bacteriol*. 98:264, 1969. American Society for Microbiology

Chapter 12

Opener: © P. Johnson/BPS/Photo Researchers, Inc.; **12.1A, C, D; 12.2A, B; 12.3B2; 12.5A, B, C:** Martha Nester; **12.9A:** © E. Chan/Visuals Unlimited; **12.9B:** © Inga Spence/Tom Stack & Associates; **12.9C:** © Bob Pool/Tom Stack & Associates; **12.14A:** Martha Nester; **12.14B:** © Bruce Iverson; **12.16B, C, D:** Martha Nester; **12.20A, B, C:** Courtesy of Fritz Schoenknecht; **BX12.4B1:** Courtesy of Mary F. Lampe; **12.1B:** © Cabisco/Visuals Unlimited; **12.15Z:** © T.E. Adams/Visuals Unlimited

Chapter 13

Opener: © Meckes/Ottawa/Photo Researchers, Inc.; **13.3A:** © National Institute of Health/Custom Medical Stock Photo; **13.3B:** © Omikron/Science Source/Photo Researchers, Inc.; **13.3C:** © J.L. Carson/Custom Medical Stock Photo; **13.10:** © E. Chan/Visuals Unlimited; **13.13:** Courtesy of Dr. Jack Griffith; **13.18A:** © Prof. L. Caro/SPL/Photo Researchers, Inc.; **13.18B:** Courtesy of Dr. Rafael Martinez, University of California, Los Angeles; **13.21:** Courtesy of B.B. Wentworth; Wentworth and L. French, *Proc. Soc. Exp. Biol. Med.*, 131:590, 1969; **13.22A, B, C:** Courtesy of M. Cooney; **13.23:** Courtesy of F. Eiserling; **13.24B:** Courtesy of M. Cooney; **13.20B:** © Michael Gabridge/Custom Medical Stock Photo

Chapter 14

Opener: © Scott Camazine/Photo Researchers, Inc.; **14.1:** © K.G. Murti/Visuals Unlimited; **14.4B:** Courtesy of Dr. Jack Griffith; **14.12A:** © Dean A. Glawe/Biological Photo Service; **14.12B:** © Bruce Iverson; **14.12C:** © Dean A. Glawe/Biological Photo Service; **14.13:** © Science VU/Visuals Unlimited; **14.14:** From W. Archer and M.J. Fraser, Dept. of Biological Sciences, The University of Notre Dame

Chapter 15

Opener: © David M. Phillips/Photo Researchers, Inc.

Chapter 16

Opener: Boehringer Ingelheim International; **16.3:** Courtesy of N. M. Green; **16.12A:** Courtesy of Dr. James W. Vardiman, University of Chicago; **16.12B:** © NIBSC/SPL/Photo Researchers, Inc.; **16.12C:** © D. Fawcett/Visuals Unlimited

Chapter 17

Opener: © Kevin Noran/Tony Stone Images; **17.6A:** Photo courtesy of M. Tam; **17.8A, B:** © Ed Reschke/Peter Arnold, Inc.; **17.9:** © Terry C. Hazen/Visuals Unlimited

Chapter 18

Opener: © Ralph Eagle/Photo Researchers, Inc.

Chapter 19

Opener: © SPL/Photo Researchers, Inc.; **19.7:** © A.M. Siegelman/Visuals Unlimited

Chapter 20

Opener: © Gianni Tortoli/Photo Researchers, Inc.; **20.3:** © Kent Wood/Photo Researchers, Inc.; **20.4A, B, C:** © Cameramann International; **20.9C, 20.10:** Evans Roberts; **20.12:** Courtesy of Ortwin Schmidt; **20.14:** © McGraw-Hill Higher Education/Bob Coyle, photographer

Chapter 21

Opener: © SPL/Photo Researchers, Inc.

Chapter 22

Opener: © SPL/Photo Researchers, Inc.; **22.3A, B:** © Carroll H. Weiss/Camera M.D.; **22.3C:** © Mary K. Brown/The Stock Shop, MediChrome; **BX22.11A:** Evans Roberts; **BX22.11B:** University of Washington, Seattle, WA; **22.5:** Graves Medical Audiovisual Library; **22.6:** © SPL/Photo Researchers, Inc.; **22.7:** Evans Roberts; **22.8:** © Ken Greer/Visuals Unlimited; **22.9:** Rocky Mountain Laboratory, Hamilton, Montana, Centers for Disease Control; **22.11:** Evans Roberts; **22.12:** © C.C. Duncan/Medical Images, Inc.; **22.13:** © Charles W. Stratton/Visuals Unlimited; **22.14:** © C.C. Duncan/Medical Images, Inc.; **22.16:** © John D. Cunningham/Visuals Unlimited; **22.17:** © SIU/Visuals Unlimited; **22.18:** © Charles W. Stratton/Visuals Unlimited; **22.19:** © Lowell Georgia/Photo Researchers, Inc.; **22.20:** © Dr. P. Marazzi/SPL/Photo Researchers, Inc.; **22.22:** © Carroll Weiss/Camera M.D.; **22.24:** © CNRI/SPL/Photo Researchers, Inc.; **22.25A:** © Dr. P. Marazzi/SPL/Photo Researchers, Inc.; **22.25B:** Evans Roberts; **22.26A:** © Dr. P. Marazzi/SPL/Photo Researchers, Inc.; **22.26B:** © A.M. Siegelman/Visuals Unlimited

Chapter 23

Opener: © Omikron/Photo Researchers, Inc.; **23.5, 23.6, 23.12:** Evans Roberts; **23.13:** Courtesy of M.F. Lampe; **23.16:** © A.B. Dowsett/SPL/Photo Researchers, Inc.; **23.3, 23.4:** Evans Roberts

Chapter 24

Opener: © Science Source/Photo Researchers, Inc.; **24.2:** Evans Roberts; **BX24.11A:** Courtesy of J.T. Staley and J.P. Dalmasso; **24.11B:** Courtesy J.P. Dalmasso; **24.11C:** Evans Roberts; **24.3A, B:** © St. Francis Hospital/ Cameramann Int'; **24.4A:** Evans Roberts; **24.4B:** © Larry Jensen/Visuals Unlimited; **24.6A:** Courtesy of J.B. Baseman; from P.C. Hu, A.M. Collier, and J.B. Baseman, *J. Exp. Med* 145:1328, 1977. Rockerfeller University Press; **24.5:** © Michael Gabridge; **24.7A:** Courtesy Dr. Kenneth Bromberg; **24.7B:** © BioPhoto Assoc/Photo Researchers Inc.; **24.12:** CDC; **BX24.51:** © BioPhoto Assoc./Photo Researchers, Inc.; **BX24.52:** © Alex Rakosy/Custom Medical Stock Photo; **24.13:** © John D. Cunningham/ Visuals Unlimited; **24.14:** © Science-VU/ Visuals Unlimited; **24.20:** © E. Chan/Visuals Unlimited; **24.21:** CDC; **24.22, 24.23:** Evans Roberts

Chapter 25

Opener: © F. Hossler/Visuals Unlimited; **25.3:** Courtesy of W. Fischlschweiger and D.O. Birdsell; **25.4:** Courtesy of K.A. Holbrook; **25.6A, B, 25.7A:** Courtesy of R.C. Page; from S. Schluger, R.A. Yaodel's and R.C. Page, Peridontal Disease, 1977. Lea & Febiger, Philadelphia; **25.7B:** Evans Roberts; **25.9:** © Frederick C. Skvara, M.D.; **25.11:** © Science VU/Visual Unlimited; **25.14:** Courtesy of Mary F. Lampe

Chapter 26

Opener: © Michael Abbey/Photo Researchers, Inc.; **26.3:** © Veronika Burmeister/Visuals Unlimited; **26.4:** Evans Roberts; **26.5A:** © David M. Rollins/Visuals Unlimited; **26.5B, 26.8:** Evans Roberts; **26.14:** © Moredun Animal Health/SPL/Photo Researchers, Inc.

Chapter 27

Opener: © Science Vu/Visuals Unlimited; **27.4:** Evans Roberts; **27.7B:** © Biophoto Assoc./Photo Researchers, Inc.; **27.8:** Evans Roberts; **27.9:** Courtesy of S. Falkow; **27.11:** Courtesy of Morris D. Cooper, Ph.D., Professor of Medical Microbiology, Southern Illinois University School of Medicine, Springfield, IL; **27.12:** Graves Medical Audiovisual Library; **BX27.41A, B:** Courtesy Dr. Sandra Larsen, Division of Sexually Transmitted Diseases Laboratory Research, National Center for Infectious Disease, Centers for Disease Control and Prevention; **BX27.43:** © CDC/Biological Photo Service; **27.13:** © L. Winograd/ Biological Photo Service; **27.14:** Courtesy of F. Schoenknecht and P. Perine; **27.15:** Graves Medical Audiovisual Library; **27.16A:** © Biophoto Assoc./Photo Researchers, Inc.; **27.16B:** From P.R. Murray, *Medical Microbiology*, 2/E, © Mosby-Year Book, St. Louis, 1993; **27.17:** © Mary Stallone/ Medical Images, Inc.; **27.18:** © John Durham/ SPL/Photo Researchers, Inc.; **BX27.61:** CDC; **BX27.6.2:** Evans Roberts ; **27.19A:** Courtesy of Sam Eng

Chapter 28

Opener: © Manfred Kage/Peter Arnold; **28.3:** © Kenneth Greer/Visuals Unlimited; **BX28.11:** © Mary K. Brown/The Stock Shop, MediChrome; **28.5:** © Biophoto Assoc./ Photo Researchers, Inc.; **28.6:** © John Durham/SPL/Photo Researchers, Inc.; **28.7:** © Paolo Koch/Photo Researchers, Inc.; **28.8:** © John D. Cunningham/Visuals Unlimited; **BX28.21:** © Joe McDonald/Visuals Unlimited; **28.9:** Graves Medical Audiovisual Slide Bank; **28.10:** Evans Roberts; **28.12A:** © Tobey Sanford; **28.12B:** © Tektoff-BM/CNRI/SPL/ Photo Researchers, Inc.; **28.13:** Evans Roberts; **28.14:** © Mary K. Brown/The Stock Shop, MediChrome; **28.15:** © Cabisco/ Visuals Unlimited

Chapter 29

Opener: © Alfred Pasieka/Peter Arnold; **29.3:** © Michael English/Medical Images, Inc.; **29.4A:** © Leonard Morse, M.D./ Medical Images, Inc.; **29.4B:** © Carroll H. Weiss/Camera M.D.; **29.5:** © A.M. Siegelman/Visuals Unlimited; **29.6:** Graves Medical Audiovisual Library; **29.7A:** © A.M. Siegelman/Visuals Unlimited; **29.7B:** Evans Roberts; **29.9:** Graves Medical Audiovisual Library; **BX29.11:** © Leon J. LeBeau/ Biological Photo Service; **BX29.12:** Courtesy of M.F. Lampe; **29.10:** Graves Medical Audiovisual Library; **29.11:** Courtesy of Dr. John Slack; **29.12:** © Leonard Morse, M.D./Medical Images, Inc.; **BX29.3.1, Box29.3.2:** Evans Roberts

Chapter 30

Opener: © G. Musil/Visuals Unlimited; **30.2:** Courtesy of J. Roberts & C. Thelon; **30.3:** Centers for Disease Control, Atlanta; **30.5:** © A.M. Siegelman/Visuals Unlimited; **BX30.11:** © Fred Hossler/Visuals Unlimited

Chapter 31

Opener: © Manfred Kage/Peter Arnold, Inc. **31.11:** © Cecil H. Fox/Photo Researchers, Inc.; **31.12A:** © E. Kaneman/Visuals Unlimited; **31.12C:** Courtesy of Dr. Thomas R. Fritsche, M.D., Ph.D., Clinical Microbiology Division, University of Washington, Seattle; **31.13:** © Paula Inhat/Custom Medical Stock Photo; **31.10:** © St. Mary's Hospital Medical School/SPL/Photo Researchers, Inc.

Chapter 32

Opener: © Grant Heilman Photography; **32.3:** Courtesy J.P. Dalmasso; **32.11:** © G. Wilder/Visuals Unlimited; **32.12:** © J. Cunningham/Visuals Unlimited

Chapter 33

Opener: © Earl Roberge/Photo Researchers

Chapter 34

Opener: © Paul Webster/Tony Stone Images; **34.1A, B, C, 34.4B, 34.7:** Martha Nester; **3.45B2, B3, B4, B5:** Courtesy of L. Santo; L. Santo, H. Hohl, and H. Frank *J. Bacteriol* 99:824, 1969

Appendicies

VII.3, 4, 6, 8A, B, 9, 11: Courtesy of S. Eng and F. Schoenknecht; **VII.12:** Courtesy Dr. Elias Bengtsson, Karlinksa Institute, Sollentuna, Sweden; **VII.13, 15:** Courtesy of S. Eng and F. Schoenknecht; **VII.16:** © Bruce Iverson/Visuals Unlimited; **VII.21, 22, 24, 25:** Courtesy of S. Eng and F. Schoenknecht

Illustrations

Chapter 10

Figure 10.2: From J Mandelstam and K. McQuillen, *Biochemistry of Bacterial Growth*, Copyright © 1968 Blackwell Scientific Publications, Limited. Reprinted by permission.

Chapter 13

Figure 13.20: From Kenneth J. Ryan and C. George Ray, "Laboratory Diagnosis of Infectious Diseases" in *Medical Microbiology: An Introduction to Infectious Diseases*, John C. Sherris (ed.). Copyright © 1990 Appleton & Lange, Norwalk, CT. Reprinted by permission.

Chapter 17

Figure 17.7: From Lansing M. Prescott, et al., *Microbiology*, 3d ed. Copyright © 1996 The McGraw-Hill Companies, Inc. All Rights Reserved. Reprinted by permission.

Chapter 25

Figure 25.5: Reprinted with permission from R.M. Stephan, National Institute of Dental Research, *Chemical & Engineering News*, 58(8):30–42, February 25, 1980. Copyright © 1980 American Chemical Society.

INDEX

Page numbers followed by *t* and *f* indicate tables and figures, respectively.

A

C

D

F

G

H

I

J

K

L

M

N

O

P

Q

R

S

T

U

V

W

X

Y

Z

MEANINGS OF PREFIXES AND SUFFIXES (CONTINUED)

Prefix or Suffix	Meaning	Example
-osis	disease of	coccidioidomycosis (disease caused by *Coccidioides*)
pan-	all	pandemic (widespread epidemic)
para-	beside	parasite (an organism that feeds in and at expense of the host)
patho-	disease	pathogenic (producing disease)
peri-	around	peritrichous (having flagella on all sides)
-phag-	eat	phagocyte (a cell that ingests other cell substances)
-phil	like, having affinity for	eosinophilic (staining with the dye eosin)
-phot-	light	photosynthesis
-phyll	leaf	chlorophyll (green leaf pigment)
pleo-	more	pleomorphic (occurring in more than one form)
poly-	many	polymorphonuclear (having a many-shaped nucleus)
post	after	postnatal (after birth)
pyo-	pus	pyogenic (producing pus)
-sta-	stop	bacteriostatic (inhibiting bacterial multiplication)
sym-, syn-	together	symbiosis (life together)
thermo-	heat	thermophilic (liking heat)
trans-	through, across	transfusion
-trich-	hair	monotrichous (having a single flagellum)
-troph	nourishment	autotroph (self-nourishing)
tox-	poison	toxin
zym-	ferment	enxyme

LOWERCASE GREEK LETTERS

α	alpha	ι	iota	ρ	rho
β	beta	κ	kappa	σ	sigma
γ	gamma	λ	lambda	τ	tau
δ	delta	μ	mu	υ	upsilon
ε	epsilon	ν	nu	ϕ	phi
ζ	zeta	ξ	xi	χ	chi
η	eta	ο	omicron	ψ	psi
θ	theta	π	pi	ω	omega

SOME DISEASES AND THEIR CAUSATIVE AGENTS*

Disease	Cause	Description
AIDS	V: human immunodeficiency (HIV)	enveloped, RNA
botulism	B: *Clostridium botulinum*	anaerobic gram-positive, spore-former
chickenpox	V: varicella-zoster	enveloped, DNA
cholera	B: *Vibrio cholerae*	gram-negative curved rod
coccidioidomycosis	F: *Coccidioides immitis*	in tissue, spherules; mold phase, barrel-shaped spores

**A = arthropod (insect, tick, or mite); B = bacterium; F = fungus; P = protozoan; W = worm; V = virus*